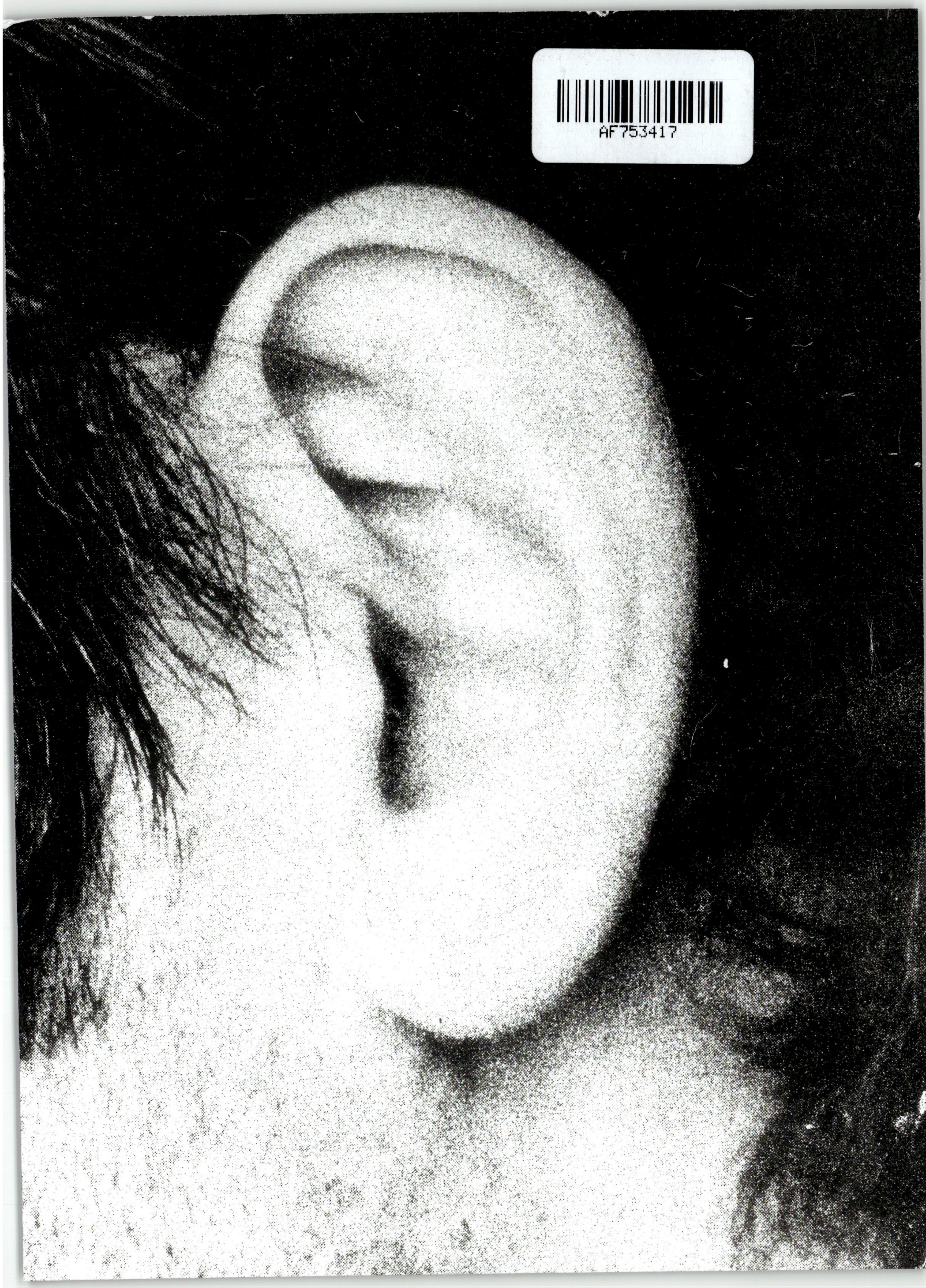

AF753417

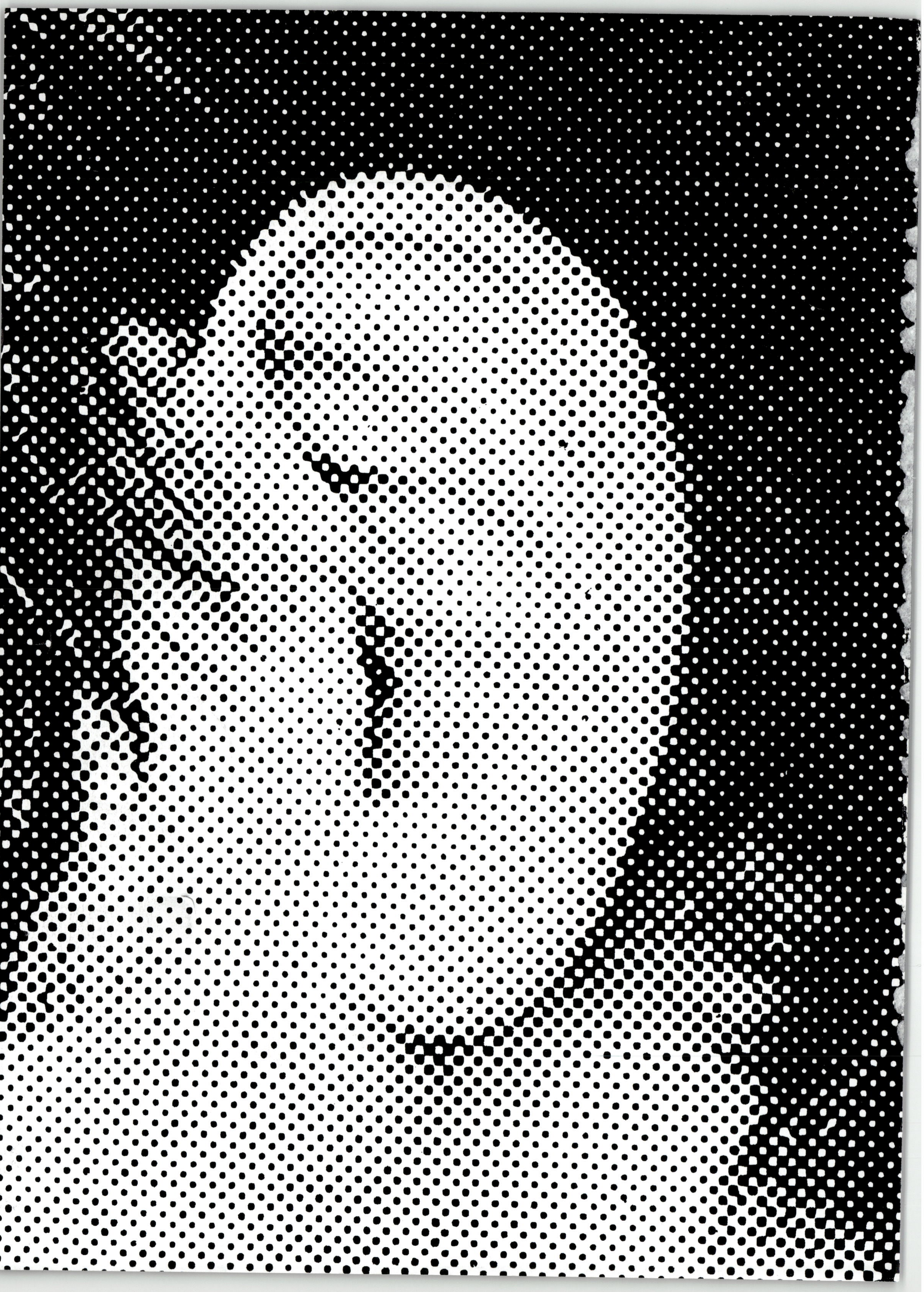

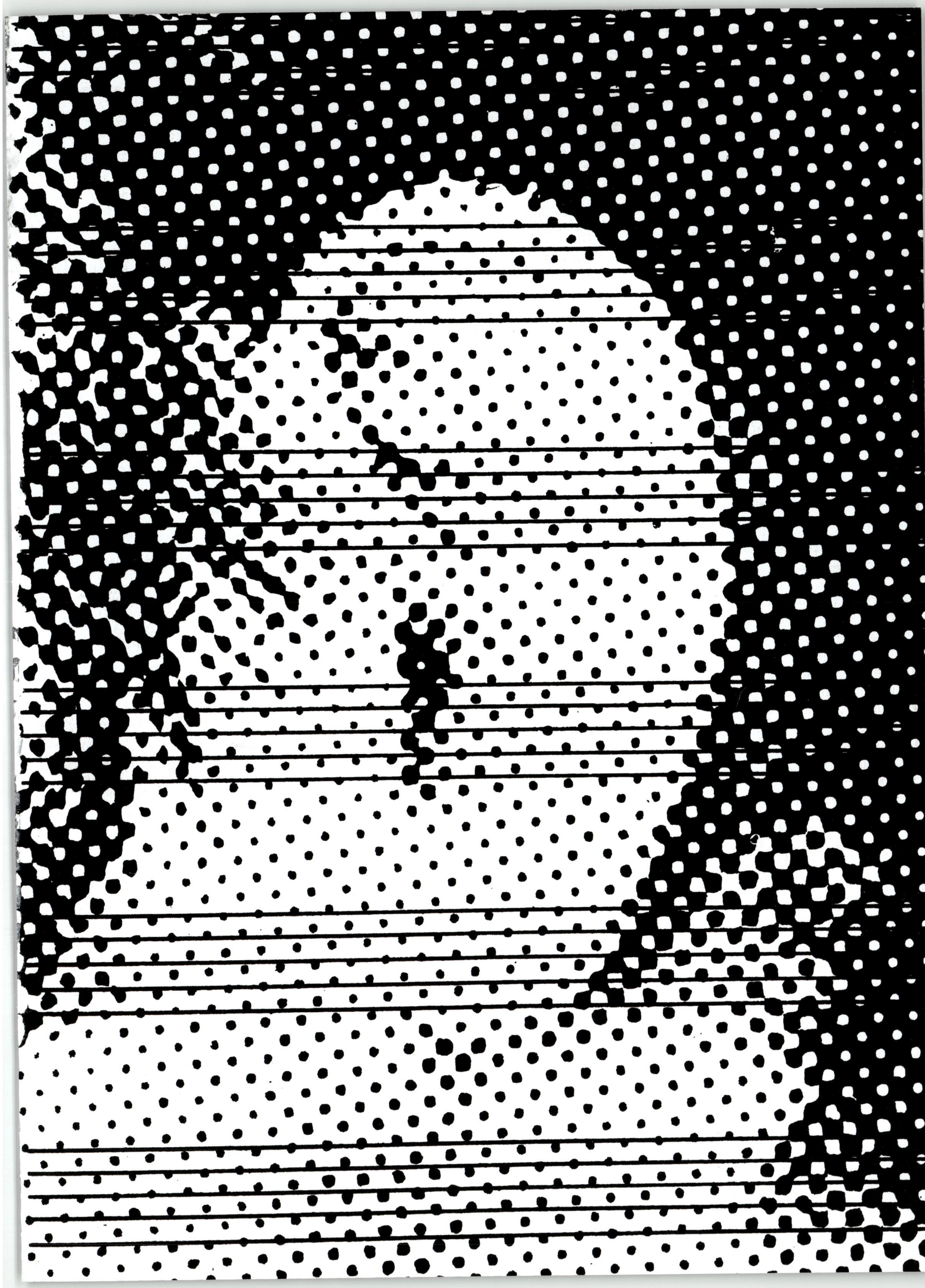

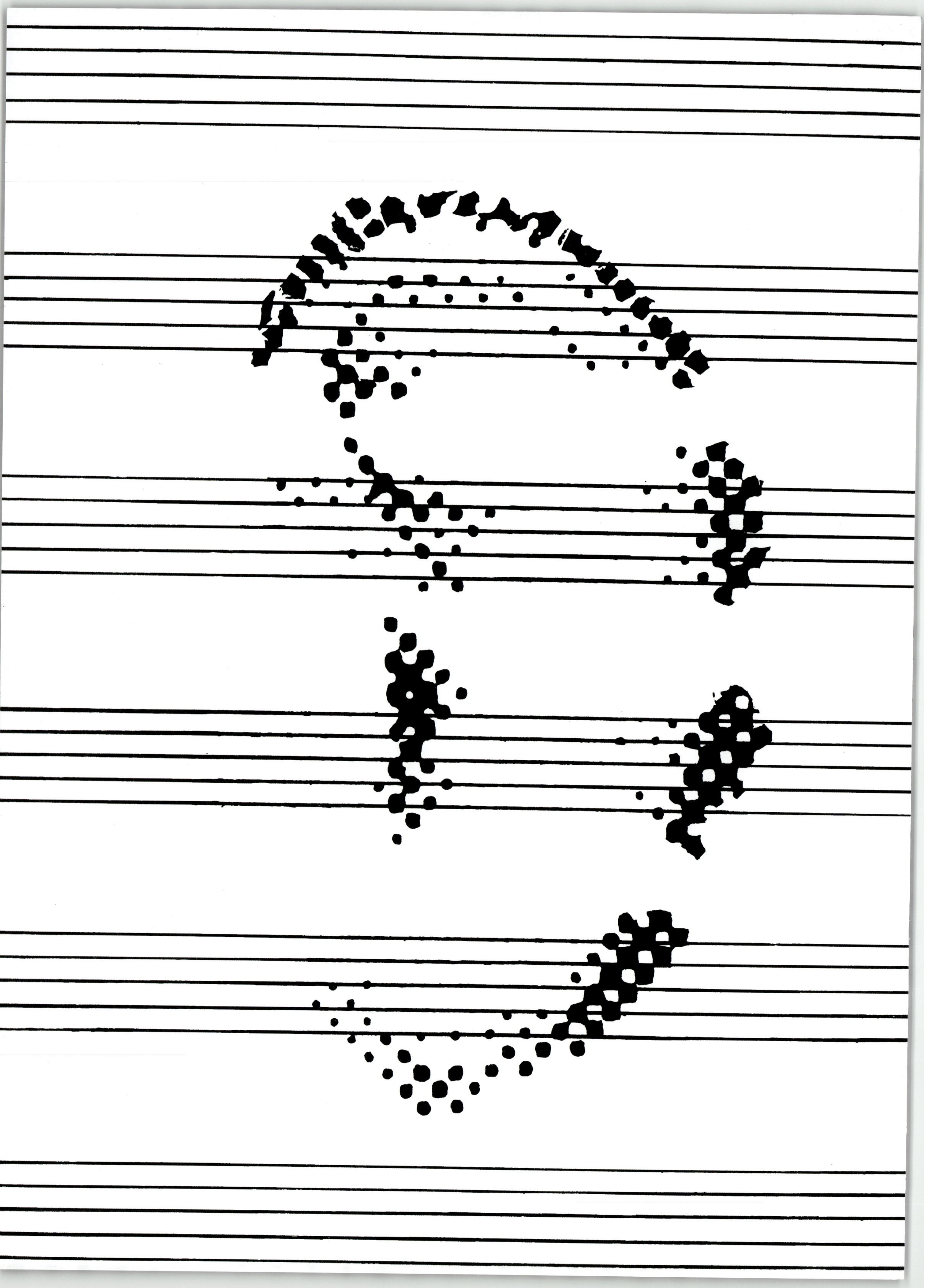

SOUND ART

SOUND ART

SOUND AS A MEDIUM OF ART

Edited by Peter Weibel

Published by
ZKM | Center for Art and Media Karlsruhe, Germany
The MIT Press, Cambridge, MA / London, England

SOUND AS A MEDIUM OF ART

MEDIA TECHNOLOGY IN THEORY AND PRACTICE

SOUND ART IN CONTEXT

HISTORICAL CARTOGRAPHY OF SOUND ART

APPENDIX

Preface

After the hugely successful 2012 exhibition *Sound Art – Sound as a Medium of Art* and the interest it generated, the ZKM | Center for Art and Media Karlsruhe decided to produce an accompanying publication with the same title.

The exhibition presented a major retrospective of fifty years of artists' engagement with sound, and the publication complements this by mapping the artistic and scholarly field of sound in the arts, predominantly focusing on the twentieth century. Since the beginning of the new millennium, several histories of sound art have been written, which have succeeded in establishing sound art as an artistic genre in its own right.[1] As this genre gained momentum in the institutionalized art scene and in academic scholarship, its historical canon became established. By now the story of sound art, from the Futurists, the Dadaists, the Fluxus movement, and Conceptual Art to sound sculptures and sound environments, is an established fact. While this created many possibilities for sound art in the past, which Alan Licht acknowledged when he said "calling oneself a "sound artist" lends a certain legitimacy",[2] it has imposed constraints on the artistic and institutional practices of the present. Young artists, who use loudspeakers without a screen, are inevitably classified as sound artists, and older artists working with a screen and an intriguing installation of loudspeakers, are widely perceived as video artists. This classification has a history, but it should not have a future: it limits practices to particular tools, such as the loudspeaker, or subjects them to certain standards of reception, such as the gallery space. Consequently, this publication endeavors to move beyond the established disciplinary divides to open our eyes and ears to the full potential of sound as a medium of art in the twentieth and the twenty-first century.

Naturally, this ambitious project starts with the art historical canon of sound art to facilitate a broader perspective on sound as a medium of art. In this spirit, Peter Weibel provides an elaborate overview of the history of sound art as media art, in which he considers the work of visual artists, composers, musicians, and architects alike. It goes without saying that Germano Celant's groundbreaking 1973 survey of the vinyl *Record as Artwork*, which he presented at the Royal College of Art Gallery in London and Achille Bonito Oliva's 2012 exhibition *forte piano – the shapes of sound*, which consisted exclusively of pure sound works that were installed inside and outside the Parco della Musica Auditorium in Rome are mentioned as outstanding contemporary exhibitions in the field of sound art.

Our idea is not just to map sound art as a global phenomenon by documenting "soundscapes" on all five continents, but also to show the "deep time" of sound art. Therefore, Siegfried Zielinski undertakes an archeological journey to excavate early sound experiments that took place long before the past millennium

Brandon LaBelle's contemplation of site-specific sound practices refers both to a work's location, and also to the sense of a distinct place that it can evoke in a globalized world. For the sounds that have lost their place in this world, Morten Søndergaard proposes exscription strategies from the archives and curatorial strategies for the gallery, to which Linnea Semmerling and Irene Noy respond with current and historical interrogations of gallery-specific sound practices. The contributions of Seth Cluett and Tony Myatt all focus on the loudspeaker, and the ways in which it has been used by artists to fill, structure, shape, and even build spaces. Seth Cluett surveys the use of the loudspeaker in art productions, ranging from paintings and sculptures to immersive environments. Then Tony Myatt picks up the thread and discusses the trend towards multi-channel speaker systems for spatial music compositions, and introduces sound pavilions as spaces that are specifically built for sound by architects, composers, and artists. Alexandra Supper's contribution scrutinizes sonification in science and art.

To illustrate the multiple histories of sound art, theoreticians and curators from all over the world have contributed "sound art" histories of the regions or countries of their interest. This framing is not intended to reinforce national borders or define artists by their nationality, but to encourage an open approach to the creative practices with sound that have long been – and to a certain extent still are – influenced significantly by local artistic communities. The result is a selection of countries that have hosted an incredibly wide array of practices, ranging from media and sound art in contemporary Australia (Caleb Kelly), from Russian sound poetry during the Soviet Era (Dmitry Bulatov) to site-specific installations in the USA during the 1960s and 1970s (Christoph Cox), from improvised music in England (David Toop) to phonography in China (Dajuin Yao), from Fluxus shows in post-war Germany (Bernd Schulz and Carsten Seiffarth) to the recent audiovisual practices in Japan (Ryo Ikeshiro and Atau Tanaka), from Turkish noise bands (Başak Şenova) to Canadian radio art (Christof Migone) and to Brazilian sound art (Giuliano Obici and Julia Gerlach).

1
See: Douglas Kahn, *Noise Water Meat: A History of Sound in the Arts* (Cambridge, MA: MIT Press, 1999); Brandon LaBelle, *Background Noise: Perspectives on Sound Art* (New York: Continuum, 2006); Alan Licht, *Sound Art. Beyond music, between categories* (New York: Rizzoli International Publications, 2007).

2
Licht, *Sound Art*, 13.

Acknowledgments

The exhibition and this book were only made possible by the enthusiasm and passion of the ZKM team. I would therefore like to thank my co-curator and project coordinator Julia Gerlach for her excellent work on the exhibition, and Till Kniola, who continued the work of project coordination after Julia Gerlach left the ZKM to work in Berlin for the Berliner Künstlerprogramm – Artists-in-Berlin Program. I would also like to thank the curators of the different sections of the exhibition: Morten Søndergaard (Unheard Avant-gardes in Scandinavia), Ursula Block (Broken Music), Georg Weckwerth and Peter Szely (TONSPUR), Hartmut Jörg and Claudia Gehrig (Het Apollohuis), Micah Silver (Maryanne Amacher room), Daniel Teige (Iannes Xenakis room), as well as Björn Gottstein, Rudolf Frisius, Frank Halbig, Luca Vitiello, Herbert Kapfer and Katarina Agathos, and Björn Gottstein (audio stations) for their great work. Furthermore, I am deeply grateful to the entire team of the ZKM | Karlsruhe that contributed to the development of the exhibition.

Sincere thanks to Jens Lutz, Miriam Stürner, Linnea Semmerling, and the entire Publications Department of the ZKM | Karlsruhe for their brilliant editorial work on the volume. They helped me immensely with the incredible task of collecting all the visual material for this book about sound. Furthermore, I want to thank Anett Holzheid, my excellent scientific consultant.

I would also like to thank The MIT Press, in particular our long-time editor Roger Conover, for choosing to include the present volume in their marvelous publications program and for being so patient while waiting for the book to be finalized.

Peter Weibel, June 2019

SOUND AS A MEDIUM OF ART

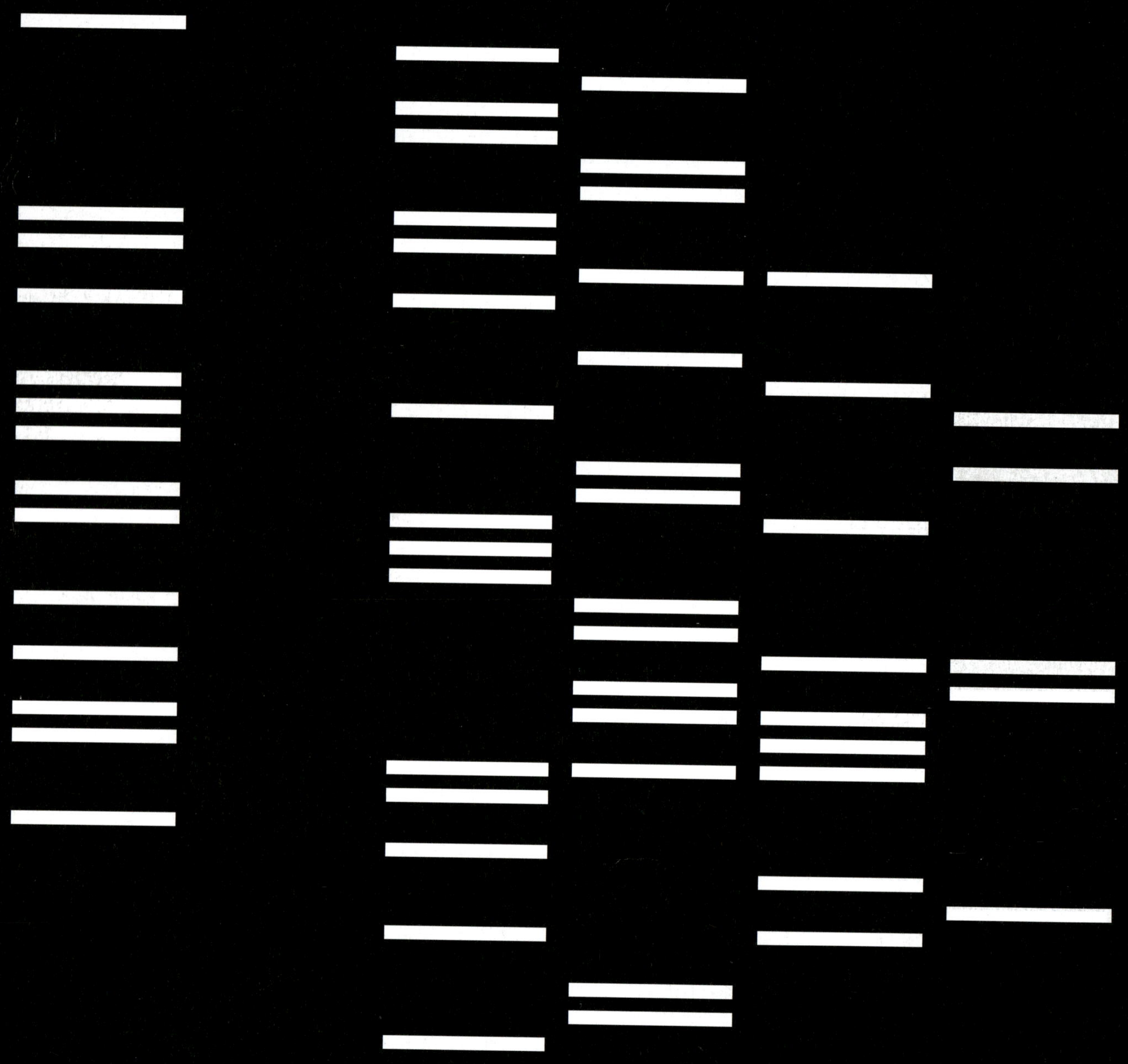

The Liberated Sound between Silence, Noise, and Music

The twentieth century began with a hitherto unknown sound: noise – outcast, suppressed, illegitimate sound. The year 1913 was not only the beginning of abstract painting, with Kazimir Malevich and his black square drawing, or of found-object art, with Marcel Duchamp and his *Roue de bicyclette*, the first readymade. It was also the year Luigi Russolo and his *intonarumori* [noise intonators] proclaimed a new art: sound art. His manifesto *L'arte dei rumori* (*The Art of Noise*) (FIG. 1), dated March 11, 1913, stands at the inception of sound art, musique concrète, bruitism, custom-made instruments and sound generators, loudspeaker sound, sampling, industrial noise, machine music, synthesizers, techno, sound installations, sculptures and environments.

Noise was a product of the industrial revolution, the machine age. With new machines, one could also produce new sounds. The birth of sound art is thus the result of the industrial revolution: machine sound, apparatus-supported sounds as an effect of the industrial machine revolution.

The industrial revolution was a technological revolution of the machines, completely changing the relationship of humans to their environment. The Italian Futurists were among the first artists to take note of these changes and enthusiastically infer the aesthetic consequences for painting, sculpture, and music. Russolo was not a trained musician, but rather a Futurist painter, yet he understood what the machine age meant for music. (FIG. 2) He wrote: "In antiquity, life was nothing but silence. Noise was not really born before the nineteenth century, with the advent of machinery. …

… This revolution of music is paralleled by the increasing proliferation of machinery sharing in human labor. In the pounding atmosphere of great cities as well as in the formerly silent countryside, machines create today such a large number of varied noises that pure sound, with its littleness and its monotony, now fails to arouse any emotion. …

… We must break at all costs from this restrictive circle of pure sounds and conquer the infinite variety of noise-sounds.

… Now we are fed up with [Beethoven and Wagner]. This is why we get infinitely more pleasure imagining combinations of the sounds of trolleys, autos, and other vehicles, and loud crowds, than listening once more, for instance, to the heroic or pastoral symphonies. …

Although the characteristic of noise is to brutally bring us back to life, the art of noises must not be limited to a mere imitative reproduction. …

Here are the six categories of noises for the futurist orchestra that we intend soon to realize mechanically:

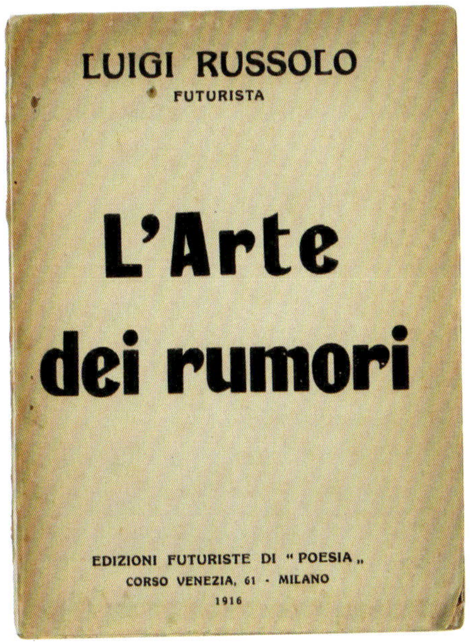

FIG. 1 Luigi Russolo, *L'Arte dei rumori*, 1913 (Milan: Edizioni di Poesia, 1916). Cover.

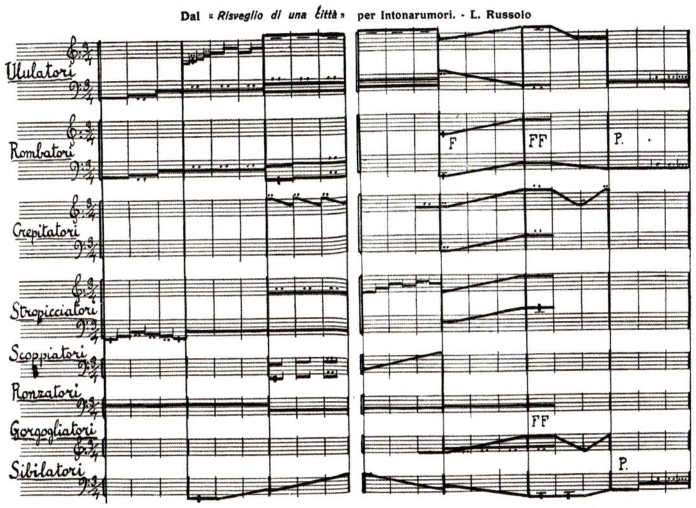

FIG. 2 Luigi Russolo, *Risveglio di una città* [The awakening of a city], 1913. Score for *intonarumori*.

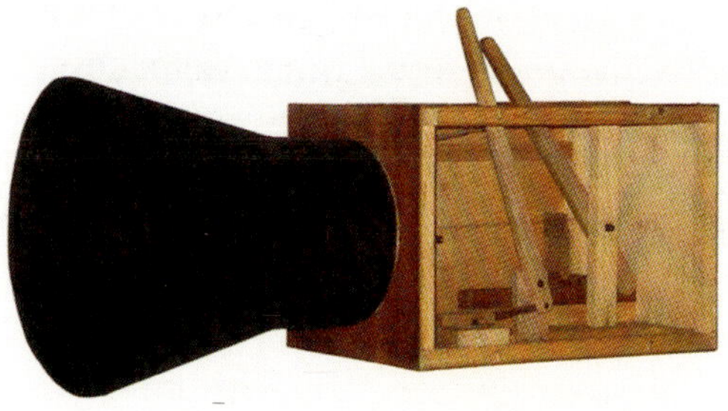

FIG. 3 Luigi Russolo, *Crepitatore*, 1913. Reconstruction 1977 by Pietro Verardo and Aldo Abate. Wood and mechanical elements, 40 × 51 × 40 cm.

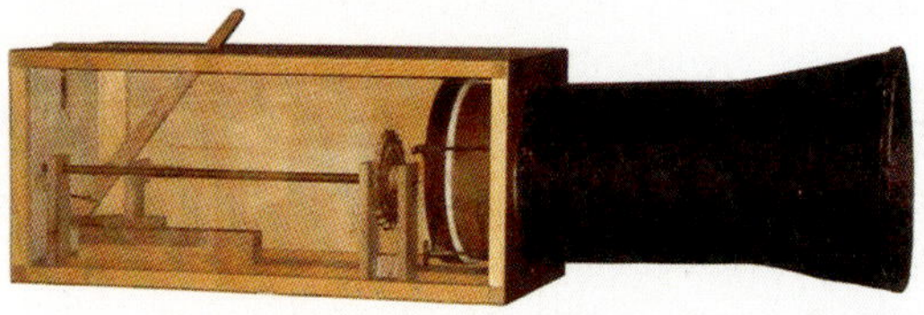

FIG. 4 Luigi Russolo, *Gracidatore*, 1913. Reconstruction 1977 by Pietro Verardo and Aldo Abate. Wood and mechanical elements, 40 × 91 × 40 cm.

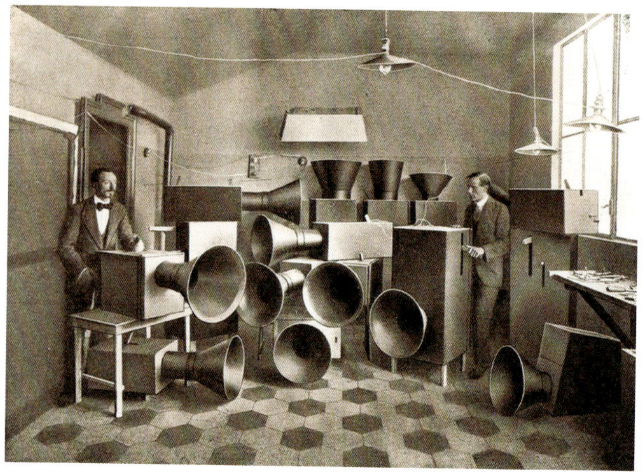

FIG. 5 Luigi Russolo (left) and Ugo Piatti (right) in Russolo's laboratory in Milan, 1914.

1. roars, claps, noises of falling water, driving noises, bellows
2. whistles, snores, snorts
3. whispers, mutterings, rustlings, grumbles, grunts, gurgles
4. shrill sounds, cracks, buzzings, jingles, shuffles
5. percussive noises using metal, wood, skin, stone, baked earth, etc.
6. animal and human voices: shouts, moans, screams, laughter, rattlings, sobs. …
Conclusions:
1 – We must enlarge and enrich more and more the domain of musical sounds."[3]

For Russolo, the limitations of conventional instruments had left music at a dead end. He therefore developed many different instruments for generating noise: the *intonarumori*, a collection of various boxes fitted with amplifying horns and containing membranes specially treated to produce different noises. (FIGS. 3, 4) The photograph of Russolo among his gigantic loudspeakers has become famous. (FIG. 5) Ever since, noise has had a face, sound an image. The age of towering speakers, Stratocasters, guitar armies, and gigantic walls of amplifiers (walls of sound) reared its gorgonian head out of the musical twilight. With these new noise producers, Russolo sought to penetrate into the "infinite variety of noise-sounds."[4] Such "musical noise" exists in the same sonic space to which noises, sounds, and tones all belong.

The Futurist Francesco Balilla Pratella had laid the foundations for a futuristic technological music in his manifesto of March 29, 1911: *La musica futurista. Manifesto tecnico*. Consequently, Russolo's manifesto was indirectly dedicated to Pratella. Even earlier, in 1907, Ferruccio Busoni had published his *Entwurf einer neuen Ästhetik der Tonkunst (Sketch of a New Esthetic of Music)*, which contained ideas for new scales and sixth-tone systems, as well as the first intimations of the possibilities of electrically generated sounds.[5] The publication of a revised edition in 1916 triggered fierce controversy. Hans Pfitzner responded from the conservative side in 1917 with his pamphlet *Futuristengefahr* [The Futurist

3 Luigi Russolo, *The Art of Noise (Futurist Manifesto, 1913)*, trans. Robert Filliou (New York: Something Else, 1967), 3–11.

4 Ibid., 6.

5 Originally published as *Entwurf einer neuen Ästhetik der Tonkunst* (Trieste: C. Schmidl, 1907).

6 Edgard Varèse, "Rhythm, Form and Content," in *Contemporary Composers on Contemporary Music*, ed. Elliott Schwartz and Barney Childs (New York: Holt, Rinehart and Winston, 1967), 201–4, here 201.

danger]. But once the noise virus had been smuggled into the hallowed halls of sound by the Futurists, it could no longer be banished. Noise emancipated itself and became just as legitimate a compositional material as sound, tone, and silence.

Hardly any other twentieth-century musician did as much for the liberation of sound and the conquest of new musical spaces as Edgard Varèse. Like Busoni and Russolo, Varèse attributed a sizable share of the development of new music and the new sonic space to machines and the mechanization of music. He himself provided some key phrases in a lecture at Princeton University in 1959: "My fight for the liberation of sound and for my right to make music with any sound and all sounds has sometimes been construed as a desire to disparage and even to discard the great music of the past."[6] The emancipation of sound was nonetheless to be seen not as a disparagement of music, but rather as a longing for "our musical alphabet [to] be enriched," as he had told the New York *Morning Telegraph* in 1916. He still wanted to make music, just a music that would be able to "keep up with thought"[7] and social change. Consequently, he could not limit himself to old tonal qualities and instruments; he wanted to make music "with any sound and all sounds." His continuous search for new sounds and instruments was an expression of his commitment to the "liberation of sound," a phrase borrowed for the title of a compilation of excerpts from his lectures published in 1966.[8]

No one, then, understood the essence of electronic music better than Varèse – that is, as a way of freeing the composer from indirectness: "Our new liberating medium – the electronic – is not meant to replace the old musical instruments, which composers, including myself, will continue to use. Electronics is an additive, not a destructive, factor in the art and science of music."[9] Instead of working through a score and a performer, with electronics one could work directly with sound, which is why in 1965 Varèse, rather than misinterpreting electronic music as a canned music of dead tones, expressed the only correct view: "To me, working with electronic music is composing with living sounds, paradoxical though that may appear."[10]

Starting at a very early date, Varèse had musical ideas that were difficult or impossible to express with existing instruments. Inspired by the writings of the Polish philosopher and mathematician Józef Maria Hoene-Wroński (music as "the corporealization of the intelligence that is in sound"[11]), and by Hermann von Helmholtz's *On the Sensations of Tone as a Physiological Basis for the Theory of Music* (originally published in German in 1863), Varèse's thinking "even then began turning around the idea of liberating music from the tempered system, from the limitations of musical instruments."[12] Another important influence on his ideas was his friendship, beginning in 1908, with Busoni, who championed new sounds and sonic devices, believing that "the flowering of the art of music is thwarted by our instruments, the flowering of the composer by the study of scores."[13] Busoni, too, strove for "unimpeded technique, [for] tonality without delimitations," because "music was born free, and to win freedom is its destiny."[14] His aversion to the limitations of scores and conventional musical instruments, which doomed "every composer's attempt to fly free,"[15] and his insistence on music's freedom made up important parts of Varèse's "liberation of sound."

This "liberation of sound" from music led to the most radical and consequential upheaval, to the fundamental axiom of music in the twentieth century – that is, the definition of music as "*son organisé*": organized sound, organized tone, organized noise. In the process, Varèse also banished the specter of noise that hangs over all music – for "after all, what is music but organized noises?"[16] By liberating sound, Varèse incorporated noise, the enemy of music, into music itself. If any kind of sound can be the raw material of

7
Edgard Varèse, interview, *Morning Telegraph* (New York), March 16, 1916, reprinted in Louise Varèse, *A Looking-Glass Diary*, vol. 1, *1883–1928* (New York: Norton, 1972), 123.

8
Edgard Varèse and Chou Wen-chung, "The Liberation of Sound," *Perspectives of New Music* 5, no. 1 (Autumn–Winter 1966): 11–19. Varèse's search for new sounds and instruments also led him to open up the field of electronics to music very early on – as early as 1922 – and to seek out opportunities for collaboration with electrical engineers: with René Bertrand, inventor of one of the first electronic instruments, the Dynaphone; and with Léon Theremin, who gave a presentation of his first electronic instrument in 1920 and went on to build a number of devices, including the theremin, which converts gestures made in a magnetic field into tones.

9
Varèse, "Rhythm, Form and Content," 202.

10
Gunther Schuller and Edgard Varèse, "Conversation with Varèse," *Perspectives of New Music* 3, no. 2 (Spring–Summer 1965): 32–37, here 36.

11
Józef Maria Hoene-Wroński, quoted in Edgard Varèse, "The Electronic Medium," in Schwartz and Childs, *Contemporary Composers on Contemporary Music*, 207–8, here 207. Originally presented as a lecture at Yale University in 1962.

12
Edgard Varèse, "Spatial Music," in Schwartz and Childs, *Contemporary Composers on Contemporary Music*, 204–7, here 204. Originally presented as a lecture at Sarah Lawrence College in 1959.

13
Ferruccio Busoni, *Entwurf einer neuen Ästhetik der Tonkunst*, 2nd ed. (Leipzig: Insel, 1916), 34. Translated from the German.

14
Ibid., 35, 8. Translated from the German.

15
Ibid., 34. Translated from the German.

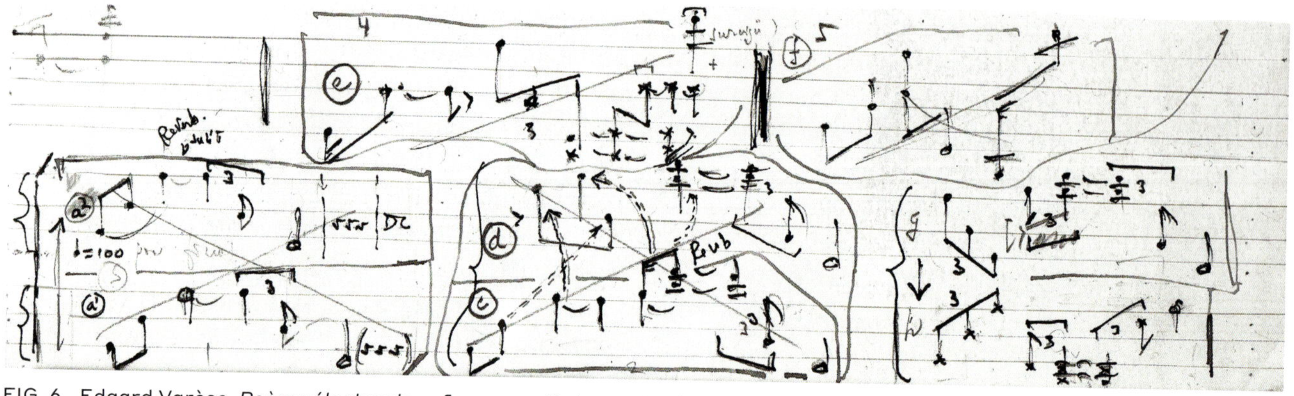

FIG. 6 Edgard Varèse, *Poème électronique* for magnetic tape, 1957/1958. Sketch.

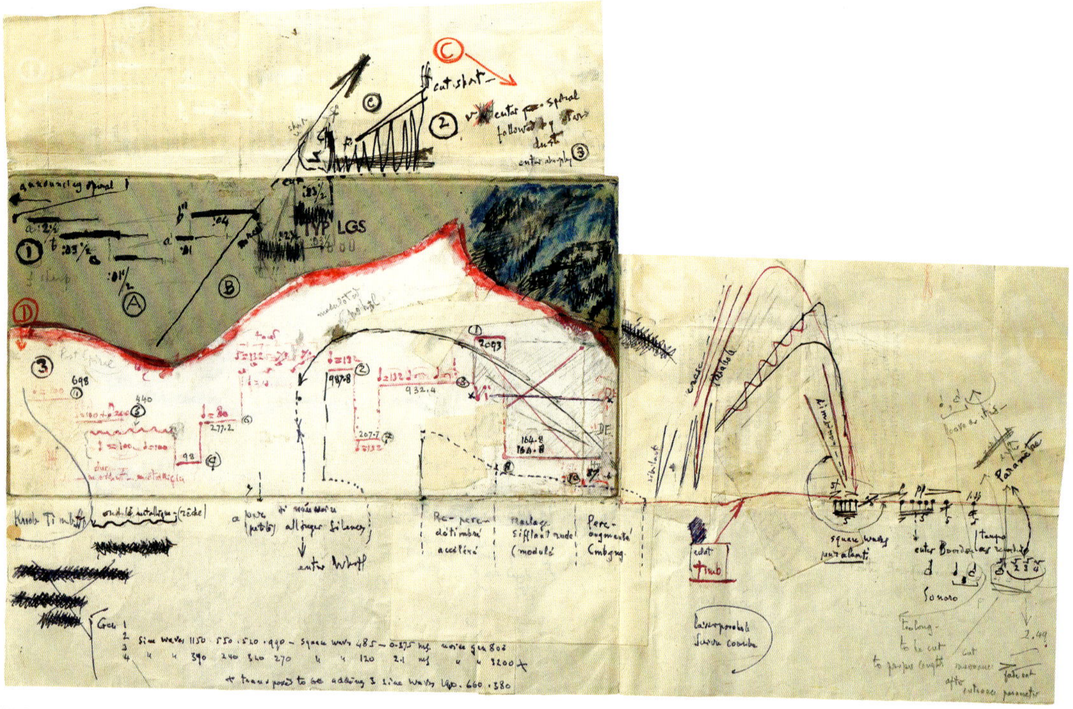

FIG. 7 Edgard Varèse, *Poème électronique* for magnetic tape, 1957/1958. Graphical collage.

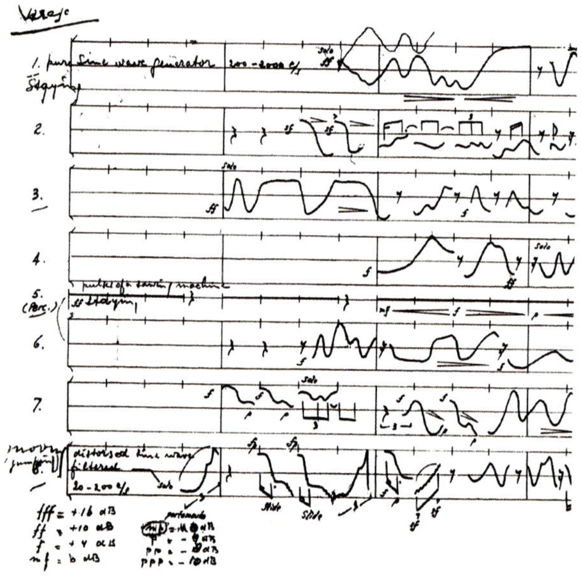

FIG. 8 Edgard Varèse, *Poème électronique*, 1958. Score.

16
Varèse, "The Electronic Medium," 207.

17
Ibid.

18
Paul Schwers, review, *Allgemeine Musikzeitung* (Berlin), March 18, 1932, and Heinrich Strobel, review, *Börsen-Courier* (Berlin), March 1, 1932; both reprinted in Nicolas Slonimsky, *Lexicon of Musical Invective: Critical Assaults on Composers since Beethoven's Time* (New York: Coleman-Ross, 1965), 215–16. Translated from the German.

19
Varèse, "The Electronic Medium," 207.

20
Schuller and Varèse, "Conversation with Varèse," 36–37.

21
Varèse, "The Electronic Medium," 207.

22
Schuller and Varèse, "Conversation with Varèse," 37.

23
Varèse, "Spatial Music," 204.

24
Varèse, "The Electronic Medium," 207.

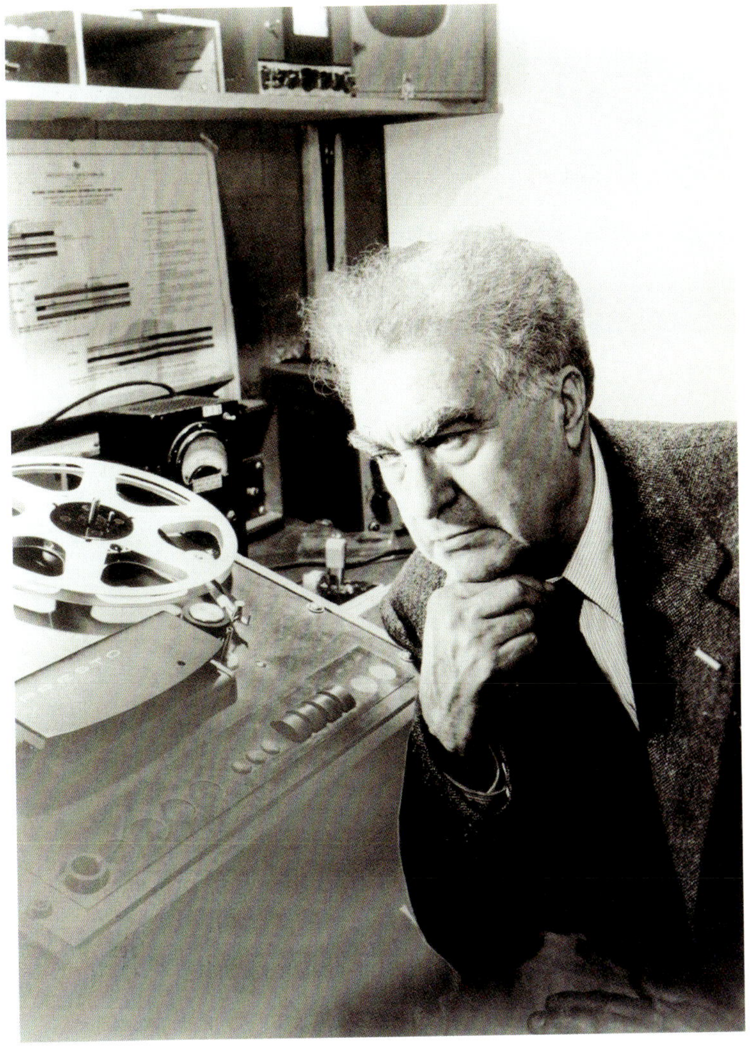

FIG. 9 Edgard Varèse listening to his *Poème électronique*, ca. 1955.

music, then so can noise, which of course is sound as well. And in that case, any sound-producing object is the equal of musical instruments. Given this objective equivalence of all sounds and sound frequencies, noise becomes a purely subjective experience – the experience of "any sound one doesn't like."[17] At the European premiere of Varèse's *Arcana* in 1932, this new music of aural emancipation was repeatedly described as "senseless noise" or "senseless aural filth." This music had "nothing to do with music," it was said.[18] The constant labeling of Varèse's music as noise only shows how new his music was, how his new forms of musical organization and his new instruments attacked the foundations of bourgeois culture. "Indeed, to stubbornly conditioned ears, anything new in music has always been called noise."[19]

"Organized sound – is that still music?" was the inevitable question whenever Varèse's name was mentioned. Today it is apparent that organized sound and liberated sound were an expansion not only of the tonal space, but also of the mental space in which the new music of the twentieth century, from musique concrète to electronic music, could arise. The liberation of sound alone ensured music's freedom. But Varèse also literally expanded music into space; he understood music as movement in space and called it "spatial music." As a result, his work also guaranteed the connection between "free sounds" and "open spaces." The freedom of the art of music does not end, therefore, with new sounds and new instruments; it continues into space: "I think of musical space as open rather than bounded."[20] If sound is the sole musical material, then sound waves themselves – together with space, as the medium in which they propagate – become the stuff of composition for the "worker in rhythms, frequencies, and intensities."[21] As sound waves in space, "after all, sound *is* only an atmospheric disturbance!"[22]

The spatial unfolding of organized sound – through enormous loudspeakers, for example, as with *Poème electronique* (FIGS. 6–9) at the Philips Pavilion at the 1958 Brussels World's Fair – was one of Varèse's earliest conceptions of spatial music: specifically, "as bodies of intelligent sounds moving freely in space."[23] Thus it is clear that free, open space forms a structural part of the freedom of sound. Varèse was one of the first who sought to invent a new sonic alphabet and open up a new aural cosmos for music, including through the use of new technological sound machines. Having defined music as "organized sound,"[24] he wanted to use noise to literally conquer new spaces, to make music that would not just rise up from orchestra pits, but penetrate the listener from every point in the space.

Over the last sixty years or so, there have been repeated attempts to build sound pavilions, beginning with the Philips Pavilion in 1958. The architects were Le Corbusier and Iannis Xenakis, and the composer was Varèse, the great opener of the universe of sound. Four hundred twenty-five loudspeakers were arrayed inside the pavilion: Twenty-five bass speakers were placed at the bases of the walls, while the others, mounted above them, reproduced the high frequencies in a presentation of *Poème electronique*, a 1958 composition by Varèse. *Poème electronique* is a kind of musical assemblage: It combined recordings of

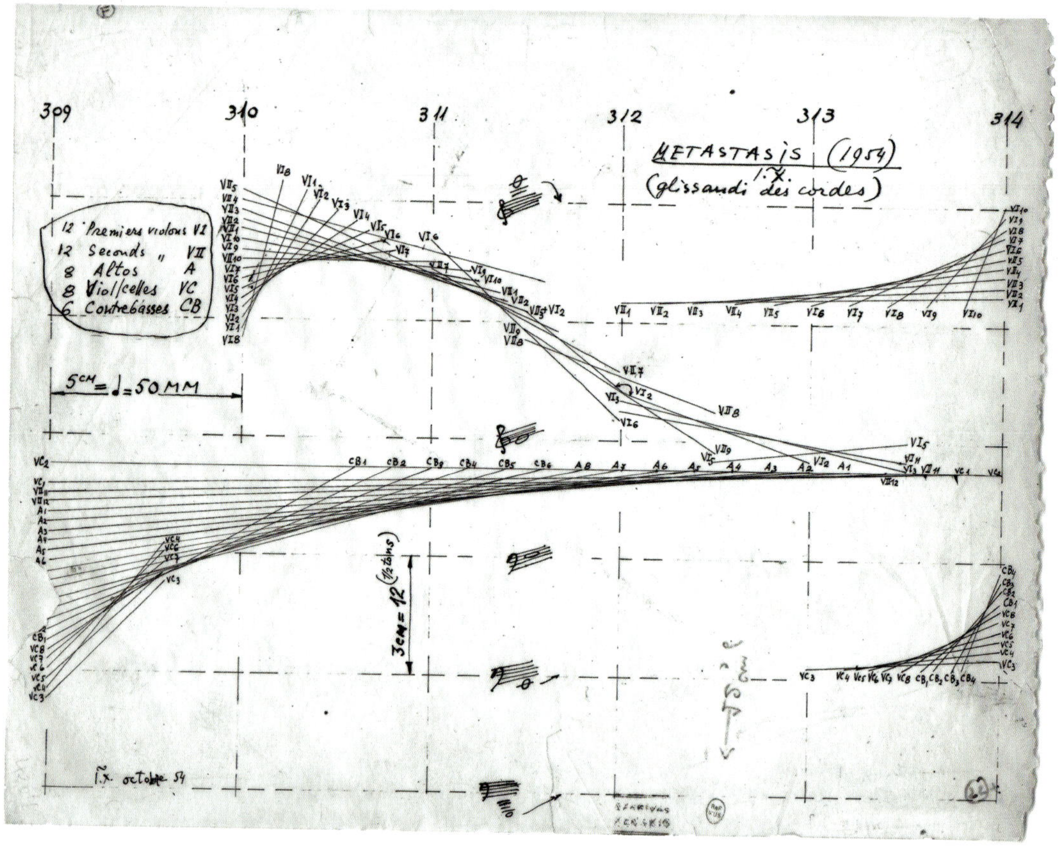

FIG. 10 Iannis Xenakis, *Metastasis*, 1954. Score.

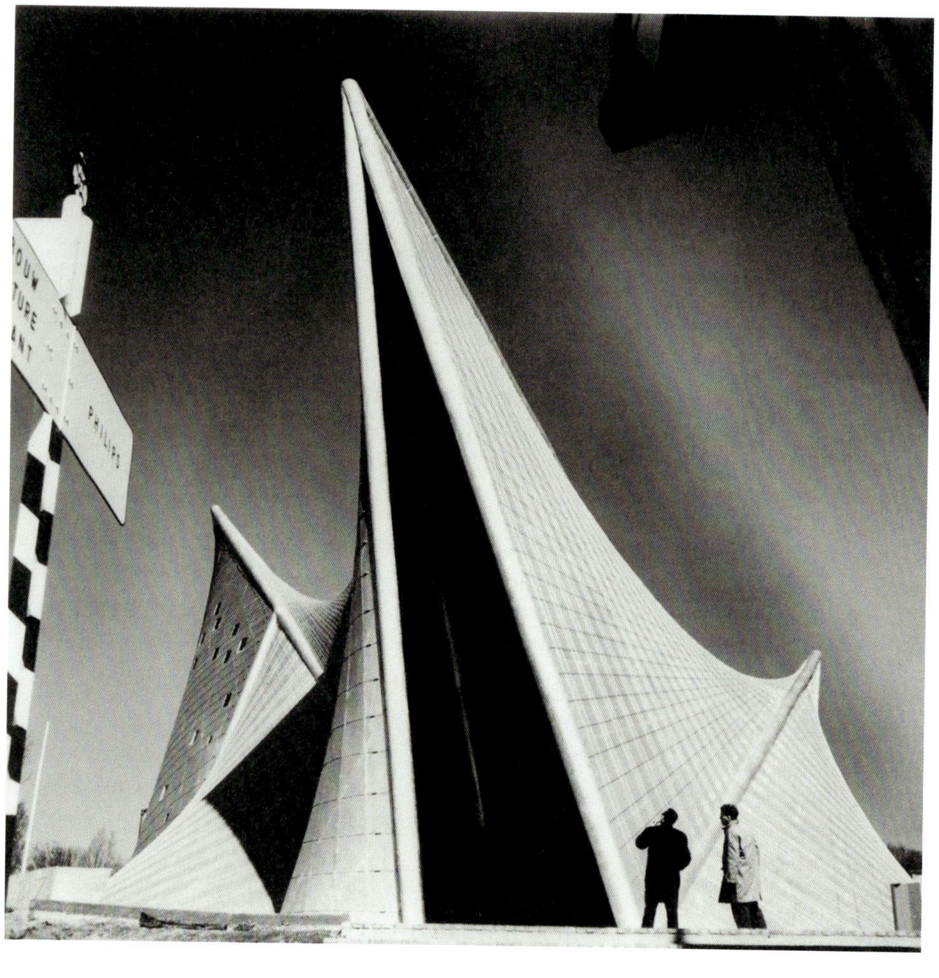

FIG. 11 Le Corbusier and Iannis Xenakis, Philips Pavilion, 1958. Exterior view.

FIG. 12 Iannis Xenakis, Philips Pavilion, 1958. Sketch.

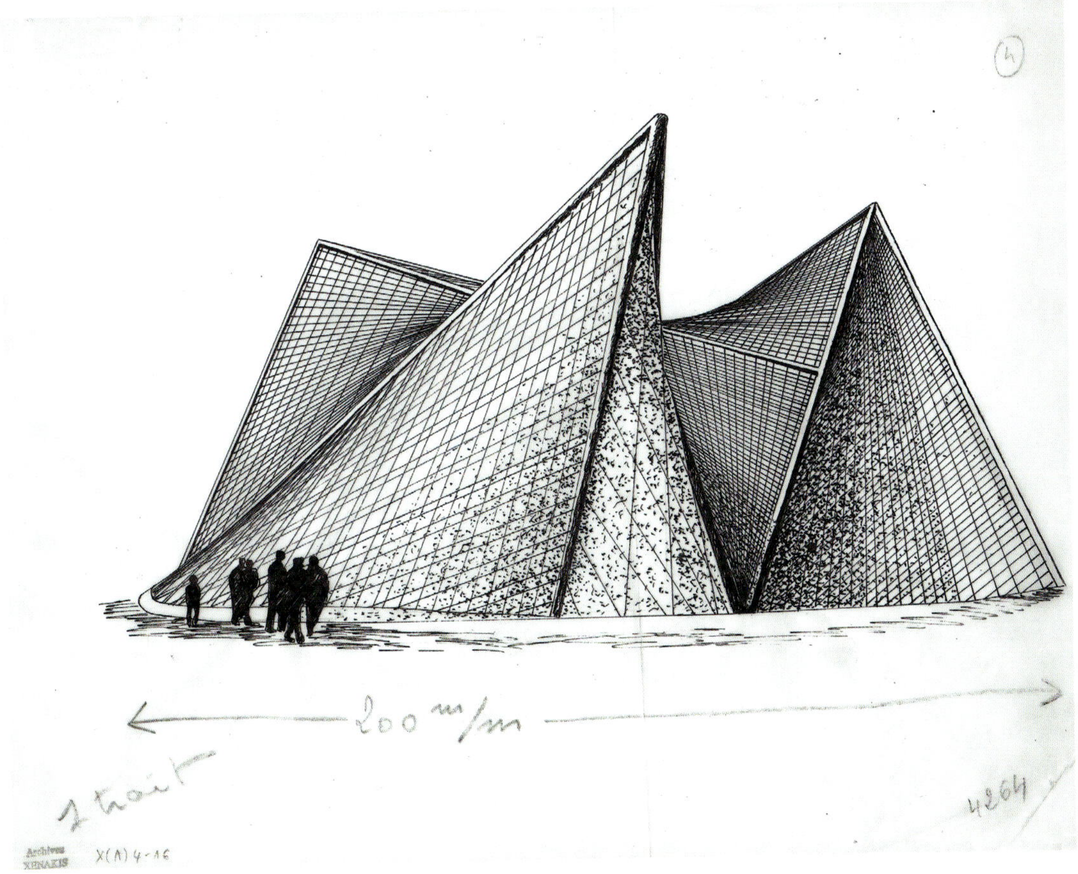

monks' chants and airplane noises with electronically generated sounds recorded by Varèse at the Philips sound studio in Eindhoven. The scores show that the composition was drawn in part from pieces of older compositions, such as *Etude pour Espace* (1947) and *Deserts* (1950–54), and that it integrated a hitherto unpublished jazz piece – the first time Varèse had worked with jazz elements. Parts of the composition were also based on parabolic sound structures.

An eight-minute multimedia show by Le Corbusier ran every half-hour from 10 a.m. to 6 p.m.: a collage of various images projected onto the walls (which were also parabolically curved) in the form of a slide show. It illustrated human evolution from the cave paintings in Lascaux to Le Corbusier's own works, but was also not short on impressions of the delusive achievements of modernity, such as nuclear weapons (a mushroom cloud). The Brussels pavilion was thus the first media museum, featuring a multimedia show of exceptionally high artistic quality, a sound-and-image experience in a custom-built sound dome. While the symbol of the fair, the Atomium, famously still stands today, the sound pavilion has unfortunately been torn down.

If we take a closer look at the origin and form of this sound pavilion, we notice that the visual language of the architecture is quite similar to the graphic notation of the musical avant-garde of the 1950s: We see that the visual notation of Xenakis's composition *Metastasis* (1953–54) (FIG. 10) has been transformed over a series of steps into the sketch of an architectural construction. The characteristic parabolic forms Xenakis borrowed from the score of *Metastasis* also define the architectural form of the pavilion. (FIG. 11) They represent the shortest connection across a diagonal, a principle that Xenakis used for both music and architecture. The musical diagram meant to illustrate a sound structure became, increasingly, a drawing intended to depict a spatial structure, the visual notation serving as a template for the architectural notation. Xenakis's architectural sketches are visual notations (FIG. 12) that have been transformed from the two-dimensionality of a score into the three-dimensionality of an architectural drawing in space. Le Corbusier, too, spoke of architecture as "visual acoustics."[25]

25 Le Corbusier, *My Work*, trans. James Palmes (London: Architectural Press, 1960), 166.

Sound Sculptures, Sound Installations, Sound Environments

Varèse's spatialization of music and legitimation of all kinds of sounds as music highlighted sounds freely moving in space as mobile aural sculptures; he thus originated sound sculpture. His "liberated sound" forged the link to aural sculptures, acoustical sculptures. The unfolding of organized sound in space became the fixed organization of spatial sound: sound sculpture. If any kind of sound can be (the material of) music, and therefore every sound-emitting object is a musical instrument, then it is only logical to build specific sound-emitting instruments, sonic and aural sculptures that bear as little resemblance to traditional musical instruments as noise does to classical music. Thus the liberation of sound – the search for new sounds, sound spaces, and sound-generating machines – freed not only sound, but also instruments and machines, and it made space musically accessible, reimagining music not as a temporal art, as in composition, but as a spatial art. Hence the existence, as the last step (for now) in the liberation of music, of liberated sculptures (spatial structures) that produce "organized sound," sound sculptures – that is, free sounds in open spaces. But these sound sculptures are no longer mere instruments, sound machines that themselves generate sounds; they are independent of, or emancipated from, scores, performers, and even composers. These sculptures draw their sonic raw material from their environments (from water, wind, the street, the atmosphere); they are thus environment-specific. The environmental sounds are played back in mechanically amplified or otherwise modified form. These sound sculptures are autonomous. They play freely in space, independent of humans (whether as producers or listeners). They are not audio media; at most, they are audio mediators. They are not isolated; they live off their environments. Not only do they open up space, they transmit spaces.

Thus Varèse did not limit himself to creating compositions which, owing to their instrumentation (pieces of metal, fur, and wood; streams of air; hammers), are pure and consummate sound pieces such as *Ionisation* (1929–1931). He also spent years working on *Espace* (FIG. 13) (ca. 1932–1947), and he designed structures for literally projecting sound into space. Loudspeaker arrays would make sounds independent of their point of origin, he hoped, enabling them to wander freely in space.[26]

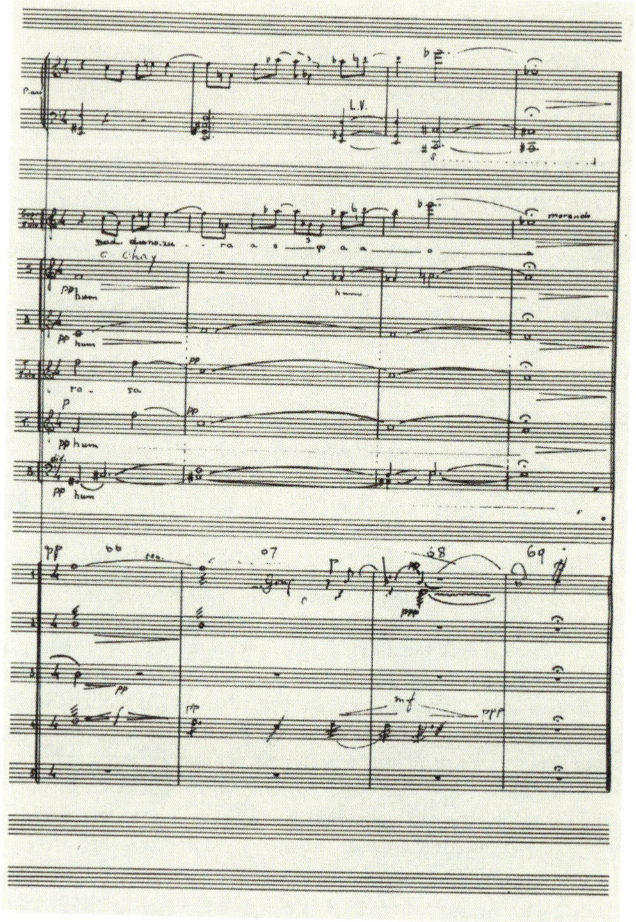

FIG. 13 Edgard Varèse, Étude pour *Espace*, 1947. Score of the conclusion of the fragment with the soprano solo which Varèse then included into *Poème électronique*.

Even in 1917, Varèse did not want to be lumped together with the Futurists. Instead of aestheticizing the technogenous noises of the modern world the way they did, he wanted to counter those noises with his music, to drown them out with it. Thus he himself spoke much less of liberating noises than he did of exploring a new universe of sound. It is true that he signed his name to the 1921 Dada manifesto *Dada soulève tout* [Dada stirs everything up], contributed to Francis Picabia's magazine *391*, held Erik Satie in high regard, and was described by Louis Aragon as "the only musician of the Dada era";[27] nonetheless, he did not feel that he was a Dadaist. The physicality of his sound brought him closer to the physicality of the theater of Antonin Artaud, for example, with whom he had a very close friendship.

One of the very first sound sculptures was therefore also a spatial sculpture: Atsuko Tanaka's *Work (Bell)* from 1955. It was an electromechanical musical work that used noise (the sound of bells). It was also a very early interactive artwork that required the physical participation of the audience.

Work (Bell) was created for the first exhibition of the avant-garde group Gutai in Tokyo. It consists of a series of twenty electric bells that are installed at regular intervals on the floor along the walls of the exhibition space. When a button is pressed the bells begin to ring, one after the other. The piercing sounds at first wander away from the starting point, then return to it, thereby acoustically measuring out the space. Tanaka herself spoke of *Work (Bell)* as a landscape painting that "draws" space with sounds. (FIG. 14)

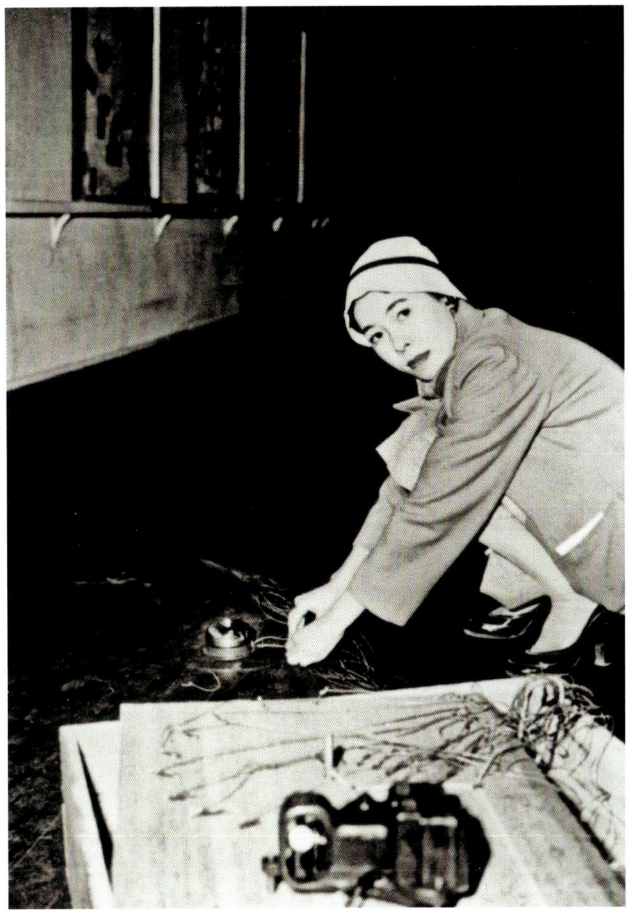

FIG. 14 Atsuko Tanaka installing *Work (Bell)* at the Kyoto Municipal Museum of Art in November 1955.

Obviously, sound and noise, like music, are not confined to the closed spaces of concert halls or galleries. In the course of the second half of the twentieth century, sound art expanded into the city and the landscape – for example, in the Linzer Klangwolke, the annual "sound cloud" of Ars Electronica in Linz, Austria. Numerous installations and environments use liberated sound for events in open spaces such as gardens, rivers, piazzas, woods, seas, fields, and so on, in interaction with natural forces such as wind, water. Ultimately, sound art migrated from physical material space into virtual immaterial space of the Internet.

Liz Phillips's interactive sound sculpture *Graphite Ground* (1988) evokes a Japanese stone garden. The original installation at the Capp Street Project, San Francisco 1988 looked as follows: In all four corners of the room were large pieces of rock containing copper ore. Wooden walkways crossed the room, and slabs of pink stone from Arizona were laid out as paths. The walls and ceiling were painted a soft shade of platinum gray, and a luminous cloud of fluffy wool (the type used for sound insulation) covered the floor. Four large loudspeakers were positioned in the corners of the room, and a computer and synthesizer stood in the middle of the room behind partition walls. Inside this quadraphonic soundscape, sonic events took place. Visitors could move around it and approach every object from various directions, and each movement of the visitors caused the soundscape to change. The sound moved in circles and spread out like a gentle mist. Sonic forms built up and then died away, according to the timing and the positions of the people within the installation space. The electric fields of capacitors were used to electronically sense the distances of people to objects. In her installations, Liz Phillips seeks to use sound as in nature and give physical events a voice. In *Graphite Ground*, sound is both signal and music. (FIG. 15)

Peter Cusack's *favouritesounds.org* (2012) is an interactive sound-mapping website based on Google Maps. It is set up to explore the connections between sounds in the environment and their geography and aims to contribute to the discussion about our fast-changing soundscape. The project asks the seemingly simple question

26
Cf. Varèse's statement in "The *Red Symphony* Is Due to Explode," *New York Times* (March 4, 1937), 25. Anne Jostkleigrewe points out, however, that loudspeakers do not appear in Varèse's written notes; see her essay "Reaching for the Stars: From *The One-All-Alone* to *Espace*," in *Edgard Varèse: Composer, Sound Sculptor, Visionary*, ed. Felix Meyer and Heidy Zimmermann (Woodbridge: Boydell, 2006), 211–19, here 215.

27
"Le seul musicien de l'époque Dada." Louis Aragon, "Sept morts, sept depuis quarte jours…," *Les Lettres françaises*, no. 1105 (November 11, 1965), 11.

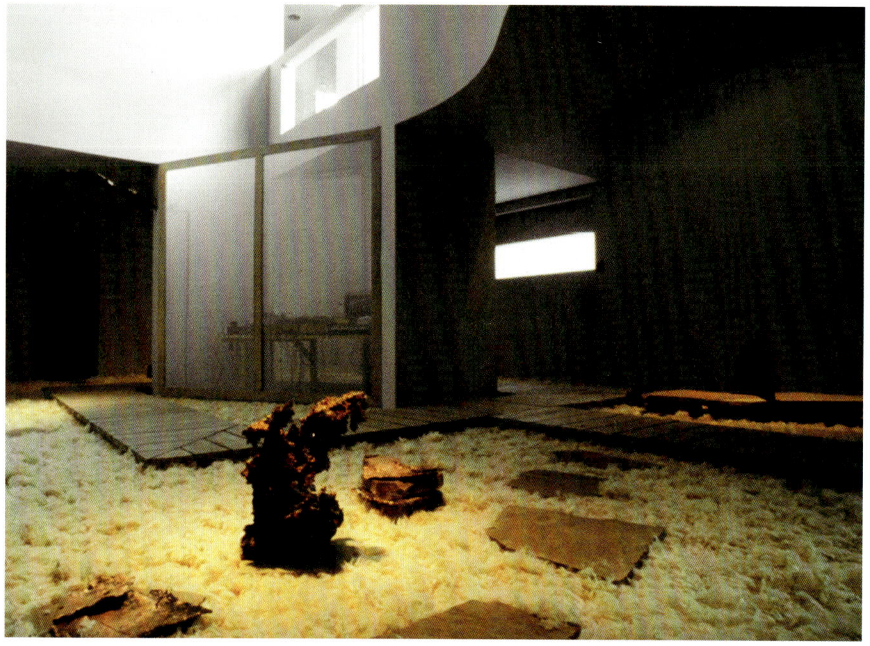

FIG. 15 Liz Phillips, *Graphite Ground*, 1988. Interactive sound installation. Capp Street Project, San Francisco.

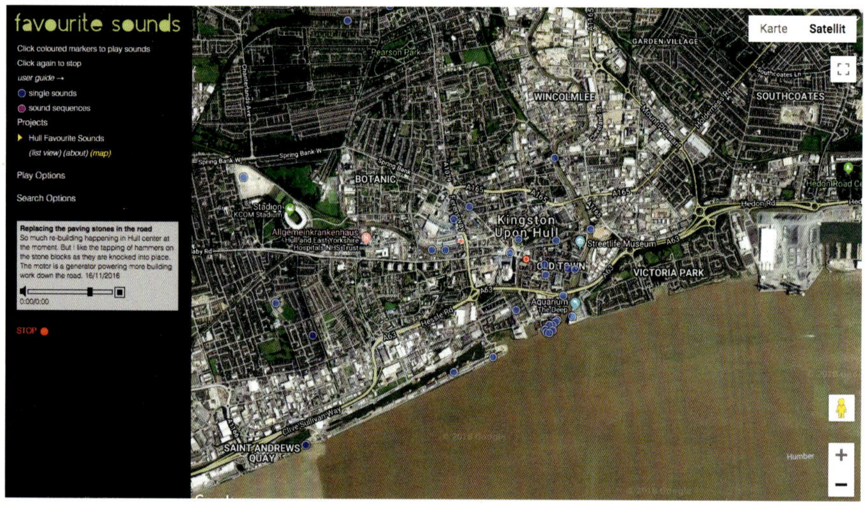

FIG. 16 Peter Cusack, *favourite-sounds.org*, 2012. Sound-mapping website, screenshot.

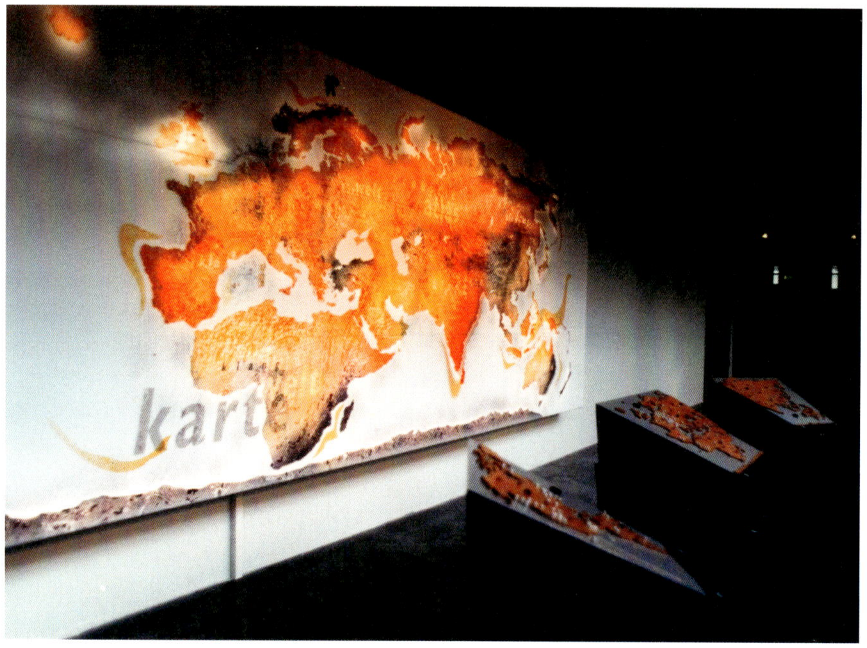

FIG. 17 Thomas Gerwin, *Klang Welt Karte*, 1997. Interactive sound installation. Installation view ZKM | Karlsruhe.

28
"Statement," Bill Fontana's website, accessed March 24, 2019, http://resoundings.org/Pages/Artists_Statement.html.

29
Frank Gertich, „Klangskulpturen," in *Handbuch der Musik im 20. Jahrhundert*, vol. 12, *Klangkunst*, ed. Helga de la Motte-Haber (Laaber: Laaber-Verlag, 1999), 135–88, here 149. Translated from the German.

"What is your favorite sound of London (or any other place) and why?" The replies have been fascinating, both revealing the city to the ear and highlighting the significant role that everyday sounds play in our lives. Many have been recorded and uploaded to this site. Since its beginning as a radio program for London's Resonance 104.4fm, a growing number of towns and cities, including Beijing, Prague, Berlin, Birmingham, Manchester, Southend-on-Sea – have enthusiastically supported the project. (FIG. 16)

Thomas Gerwin's *KlangWeltKarte* (1997/1998) contains 201 different sound files, which are sonic portraits of 156 locations, 37 sonic portraits of regions, 7 sonic portraits of continents, and a "sound of the world." The 156 local soundscapes, ranging in lenght from 0:06 to 1:05 minutes, have been musically "distilled" from acoustic ambient impressions of some of the original venues. Some are little audio plays, and some are conglomerations of essential and significant sounds. Besides the character of the sound file as a specific sonic figure, the pieces are always intended to convey the sounds of a specific place or landscape in a recognizable form. The location where an audio piece was recorded is shown on a large map of the world when it is played. (FIG. 17)

Musicians and sound artists felt very early on that concert halls were too small for their experiments. The spatialization of music called for bigger spaces, wider spaces, free spaces like buildings, cities, and even landscapes. Max Neuhaus, Bill Fontana, and Maryanne Amacher in the USA and Walter Haupt and others in Germany migrated with their music into urban zones.

Walter Haupt created a project in urban space called *Musik für eine Landschaft* [Music for a landscape] in 1973, and in 1978 he realized *Klangwolke über Münchens Innenstadt* [Sound cloud above Munich] a first sound cloud in the city of Munich, which involved a performance of Carls Orff's orchestra piece *Entrada* (1928).

In 1967, Maryanne Amacher broadcasted *City-Links #1* at WBFO FM, the college radio station in Buffalo, New York. She connected eight places in Buffalo with the WBFO studio and mixed the sound sources live for twenty-eight hours. Max Neuhaus's *Drive-In Music* was also realized in Buffalo, in 1967–1968.

Bill Fontana creates works that use recorded, nonlocal sounds, which he calls "sound sculptures." He describes their mode of action as follows: "My sound sculptures use the human and/or natural environment as a living source of musical information. I am assuming that at any given moment there will be something meaningful to hear and that music, in the sense of coherent sound patterns, is a process that is going on constantly. My methodology has been to create networks of simultaneous listening points that relay real-time acoustic data to a common listening zone (sculpture site). Since 1976 I have called these works sound sculptures."[28]

Fontana created his *Sound Recycling Sculpture* (1983) in various locations inside and outside a building and in the surrounding landscape (see this volume, p. 51). With this sound sculpture he explored the compositional aspects of ambient sound. In an even earlier piece, *Telephone Access* (1968), words were converted into sound.

Fontana includes the surroundings in his sculptures and defines a "listening zone" for visitors/listeners. One could also refer to these works as "acoustic sculptures," a term that the visual artist Michael Brewster has used for his pieces since the 1970s: "My works have taken the form of sculpture and drawing installations made by sound when restricted to confined space. … My use of sound as material … began with the realization that the sense of hearing was the most congruent mode for sculptural perception. … I made Acoustic Sculptures by physically matching an architectonic volume with sound waves of a size and power that will boost the room's cavity into resonance. The sculpture that results is a field of palpable sound volumes, of different sizes, densities, and rates of excitement."[29]

R. Murray Schafer summarized these tendencies toward the inclusion of environmental and urban sounds, the sounds of the urban zones and landscapes, in his books *The New Soundscape* (1968) and *The Tuning of the World* (1977). In 2011, the publication *Soundscape i kunsten/Soundscapes in the Arts* edited by Jøran Rudi, was published in Oslo.

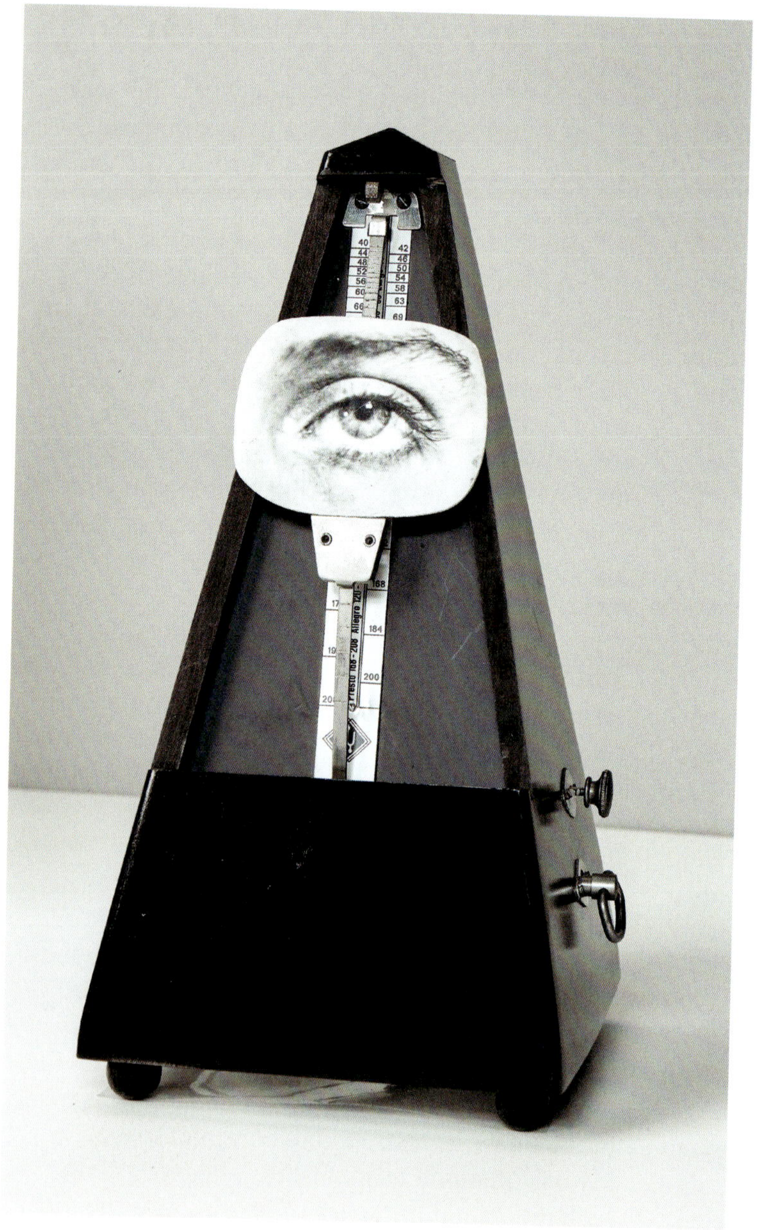

Man Ray, *Indestructible Object (or, Object to be destroyed)*, 1923, replica of 1965 (from: Edition MAT). Metronome with cutout photograph of eye on pendulum, 25 ×11 × 11 cm.

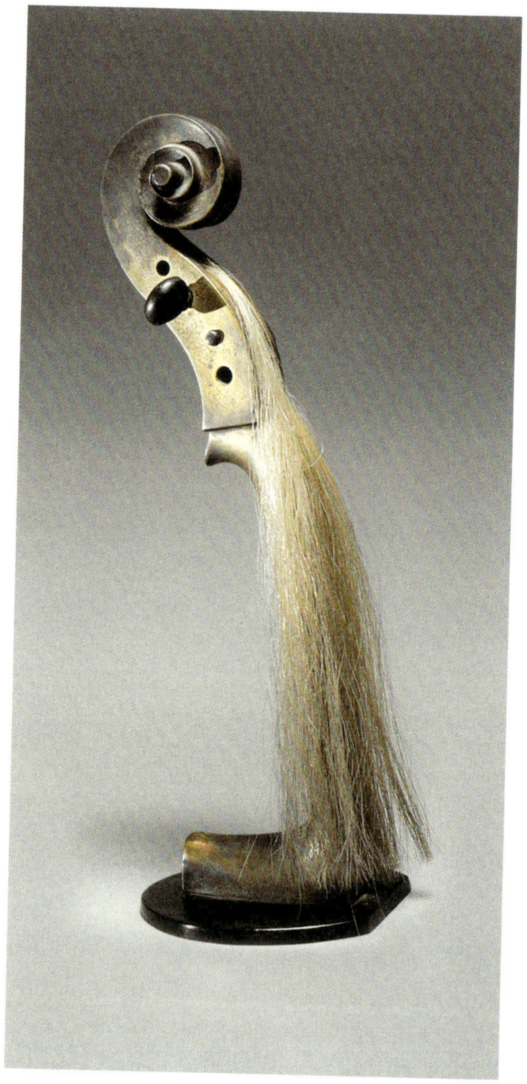

Man Ray, *Emak Bakia*, 1926, replica of 1970. Fingerboard of a cello and horse hair on wooden base, 51 × 19.7 × 26 cm.

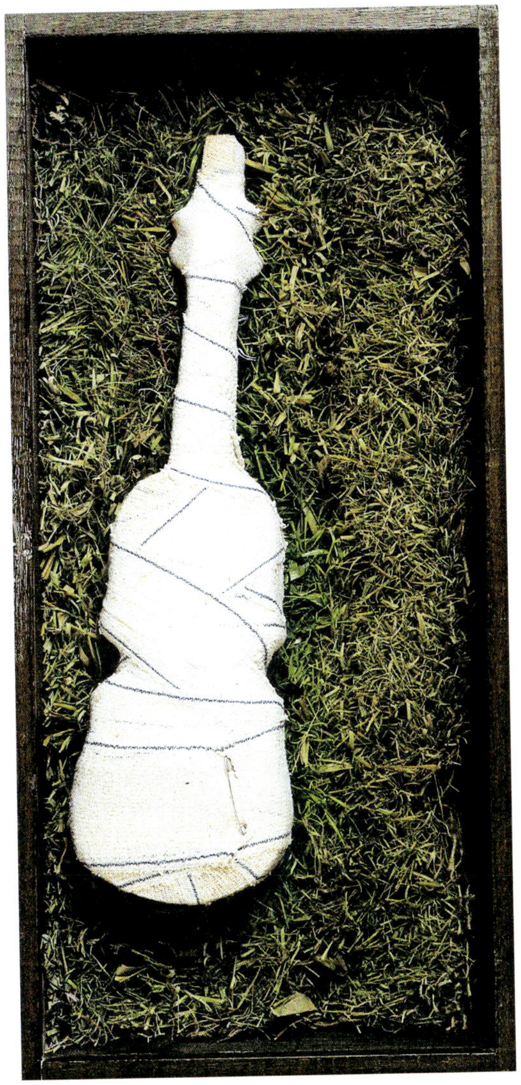

Maurice Henry, *Hommage à Paganini*, 1936, replica of 1968.
Wrapped violin in grass box,
52 × 24.5 cm.

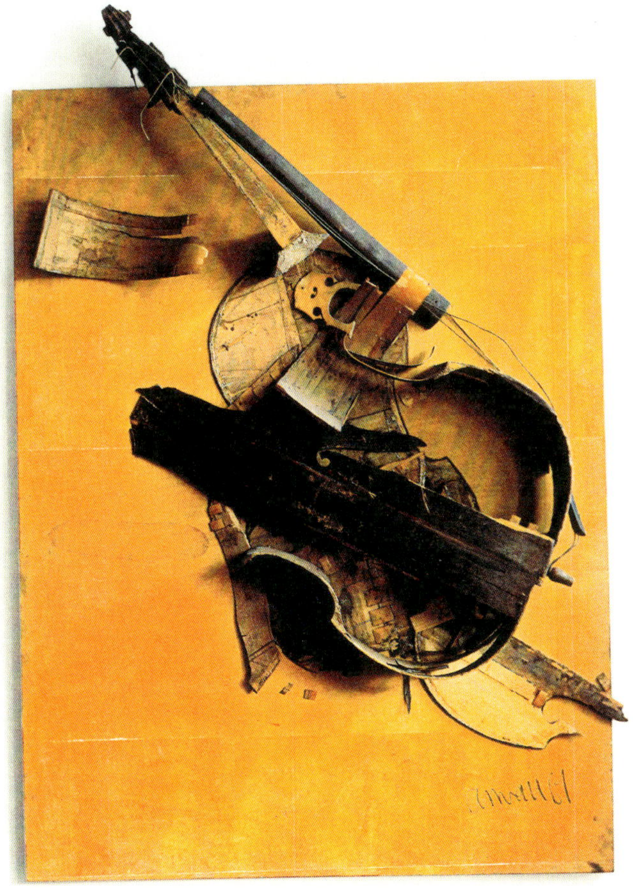

Arman, *NBC Rage*, 1961.
Broken bass fiddle on a wood panel,
78.7 × 61.8 × 9.7 cm.

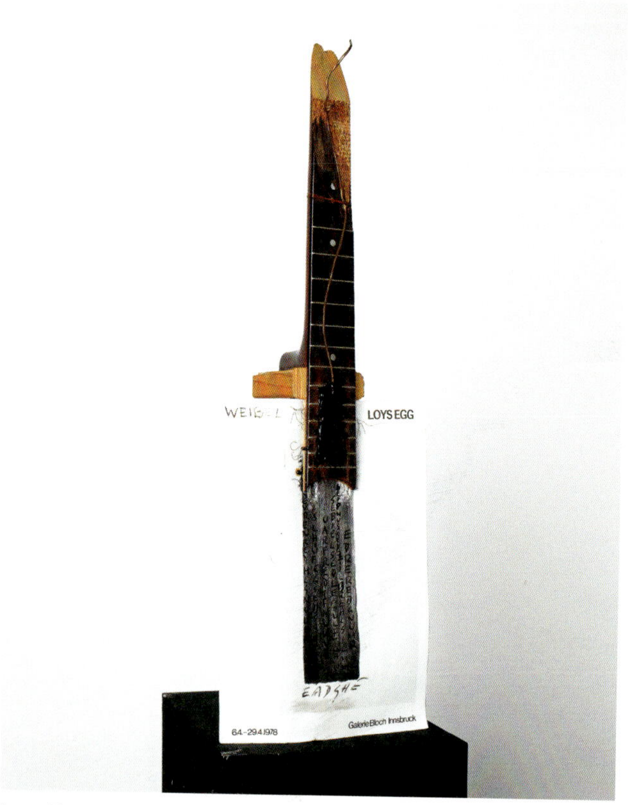

Loys Egg, Peter Weibel, *Die Buchgitarre* [The book guitar], 1978. Front view.

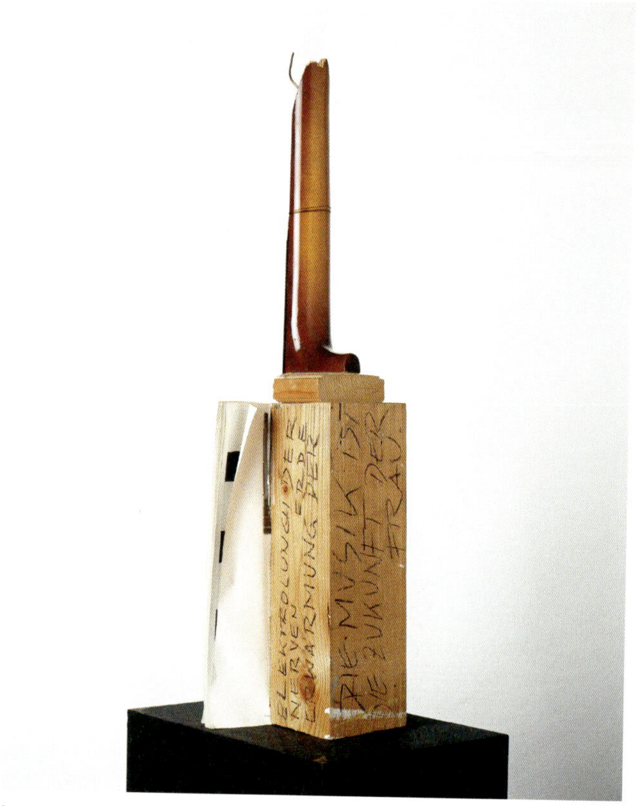

Loys Egg, Peter Weibel, *Die Buchgitarre* [The book guitar], 1978. Back view.

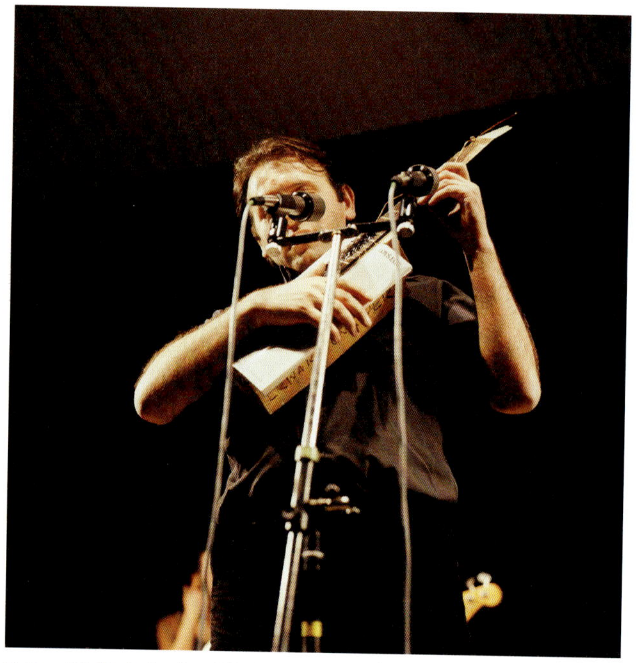

Peter Weibel playing *Die Buchgitarre* [The book guitar], which he invented together with Loys Egg. Performance of Hotel Morphila Orchester at Lenbachhaus, Munich, 1979.

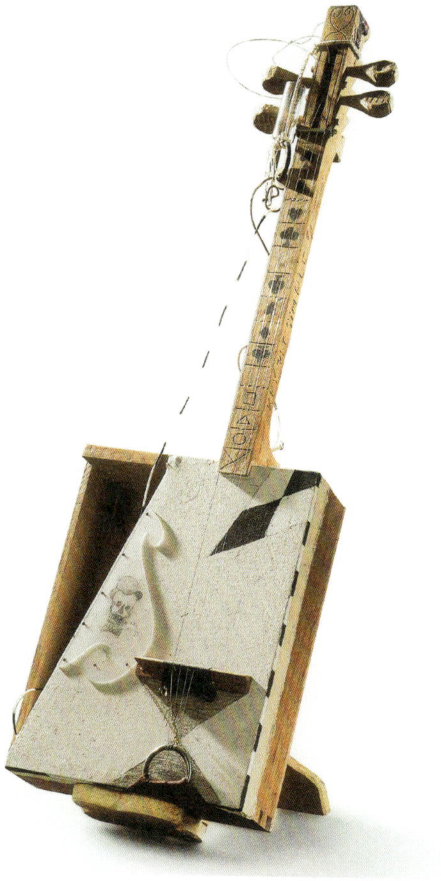

William T. Wiley, *Banjo for J.B.*, 1985. Front view.

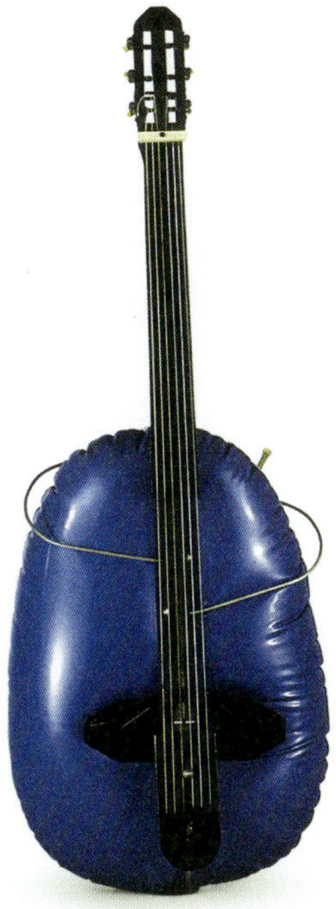

Bernard & François Baschet,
Folding and inflatable Guitar, 1952.

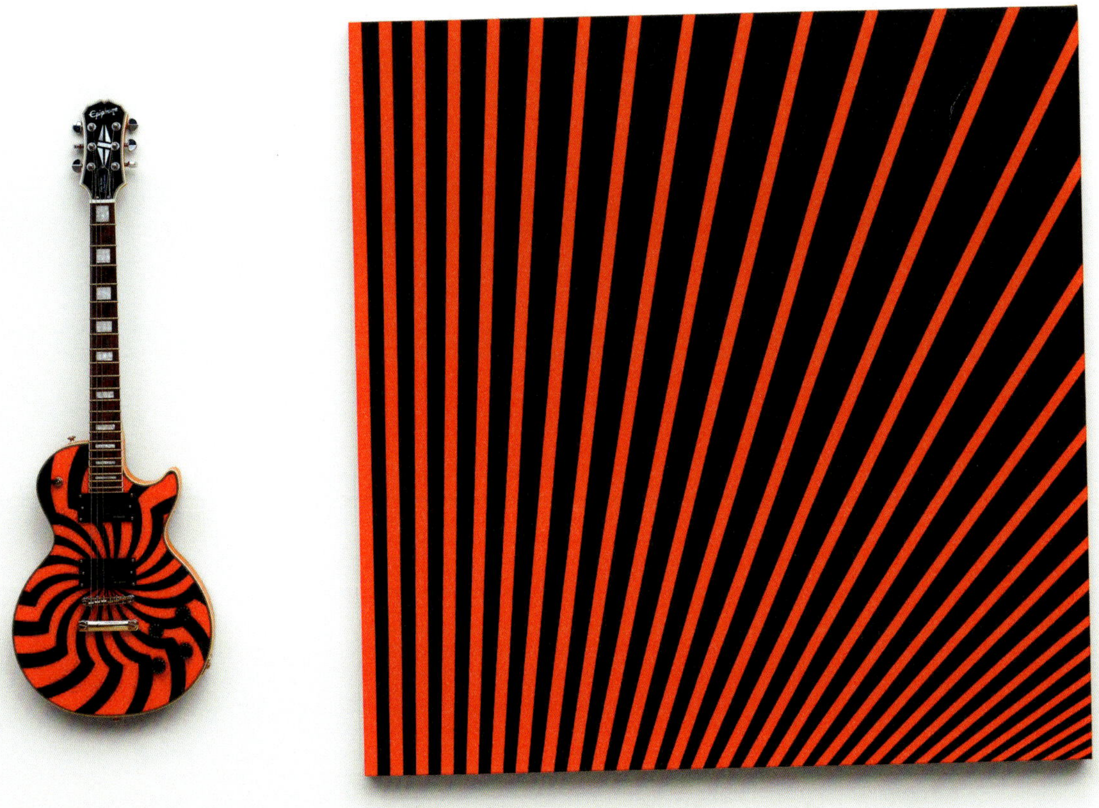

John Armleder, *Zakk Wylde II*, 2008.
Acrylic on canvas, electric guitar, painting: 120 × 120 cm, guitar 102 × 35 cm.

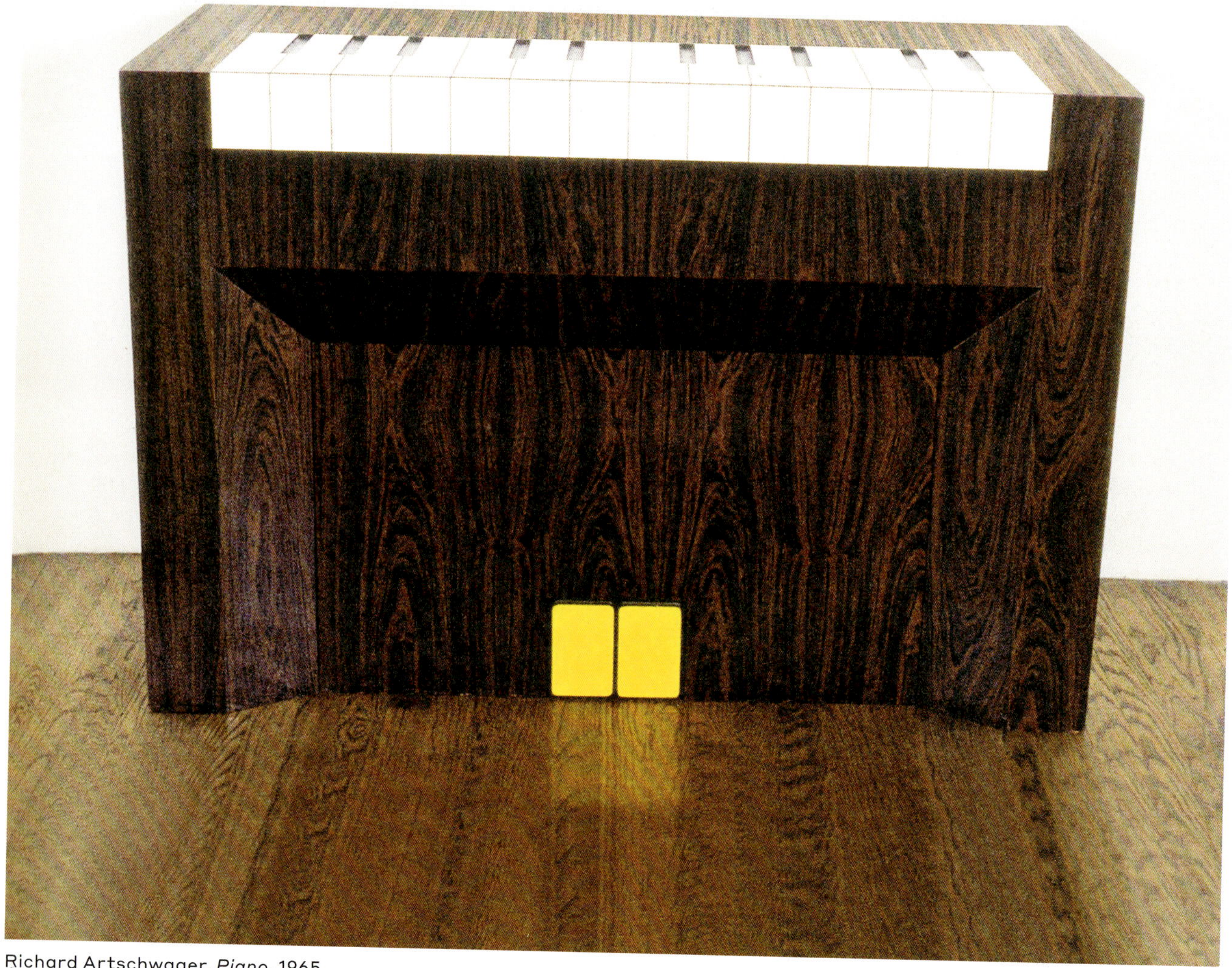

Richard Artschwager, *Piano*, 1965.
Sculpture, laminate on wood, 81 × 122 × 48 cm.

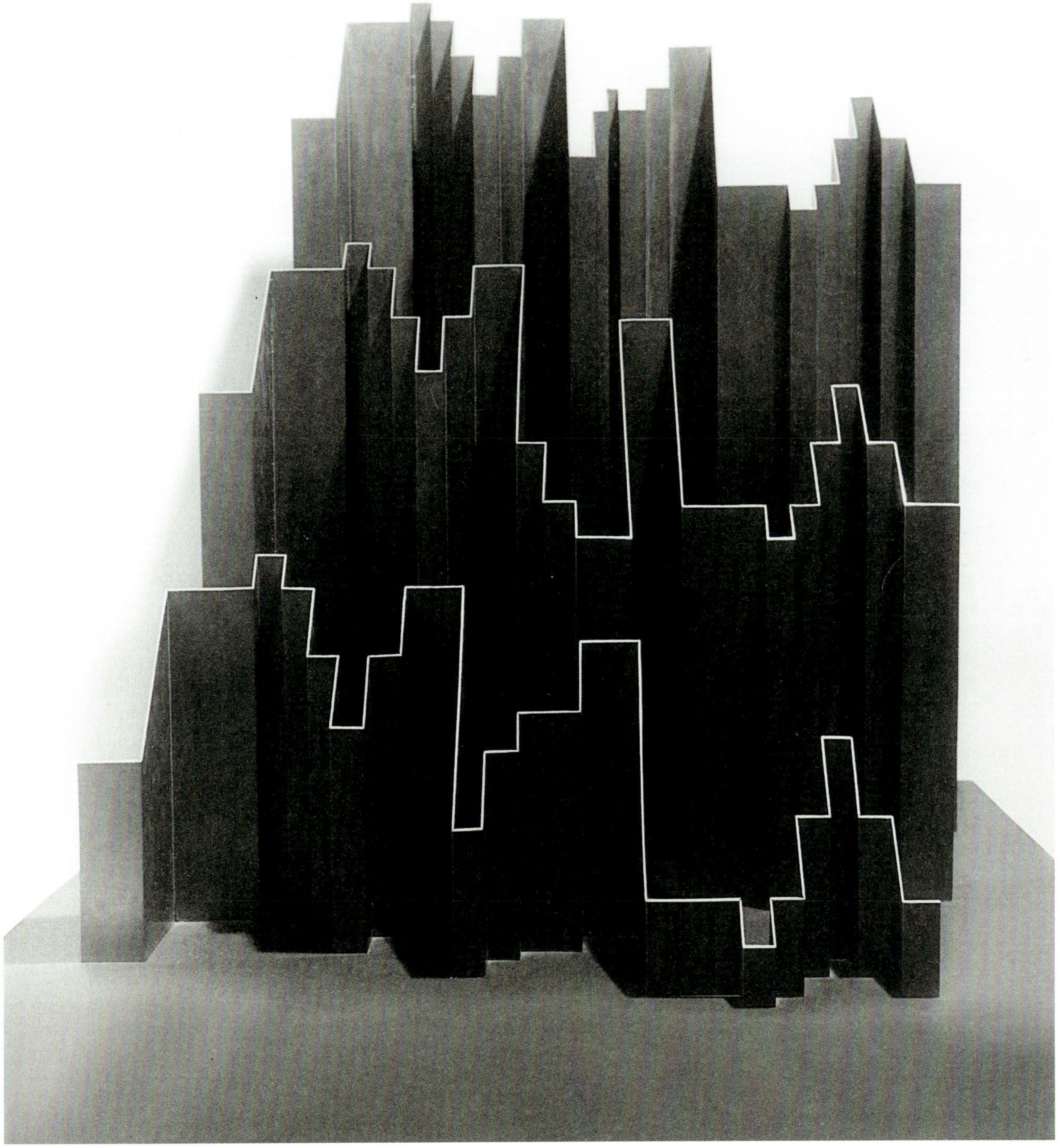

Henrik Neugeboren (Henri Nouveau), *Sculptural Representation of Bars 52-55 of E-Flat Minor Fugue by J.S. Bach. Proposal for a Bach Monument*, 1928, replica from 1966.
Stainless steel on painted blackboard, 74 × 82 × 86.5 cm.

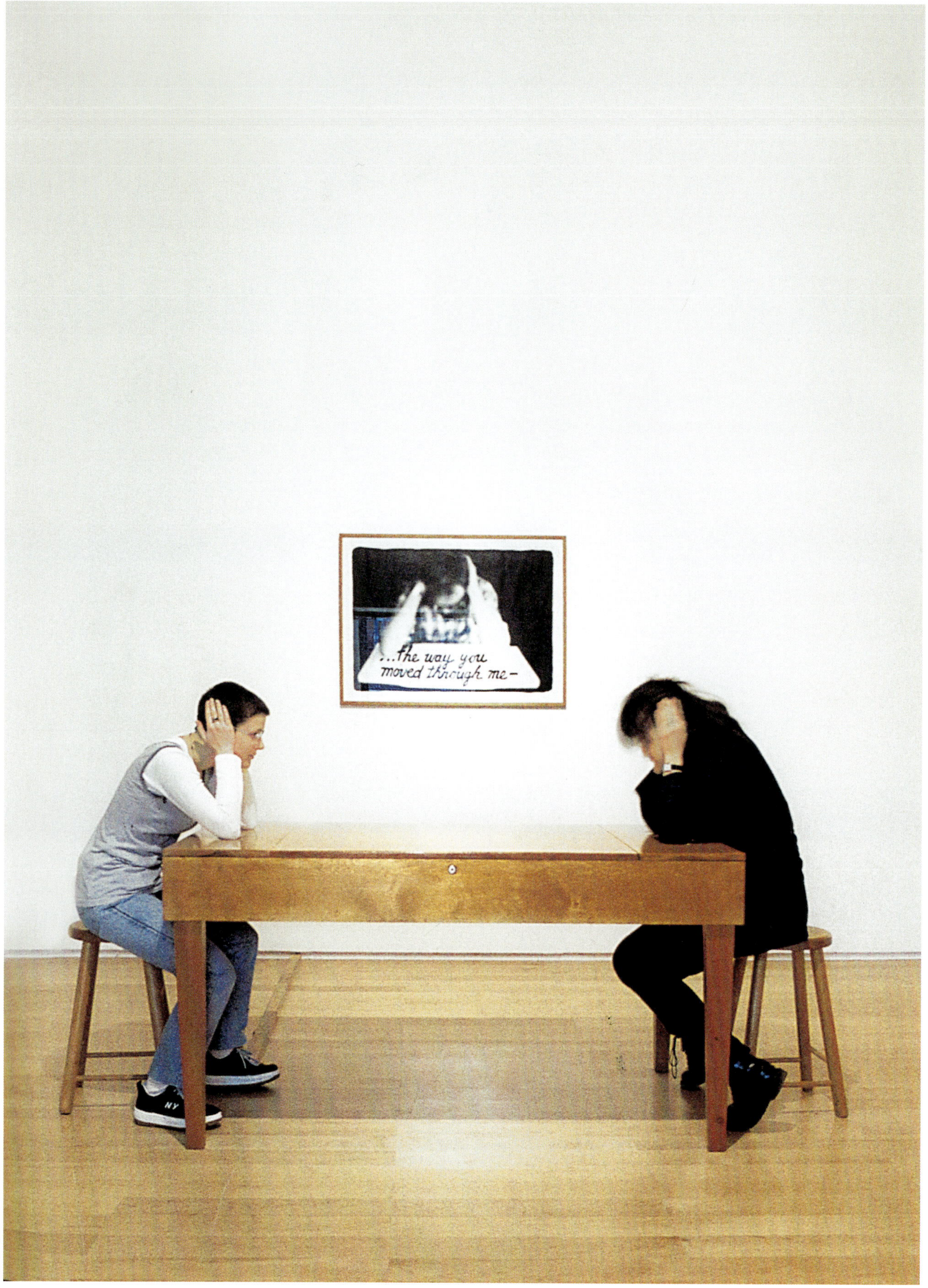

Laurie Anderson, *Handphone Table*, 1978.

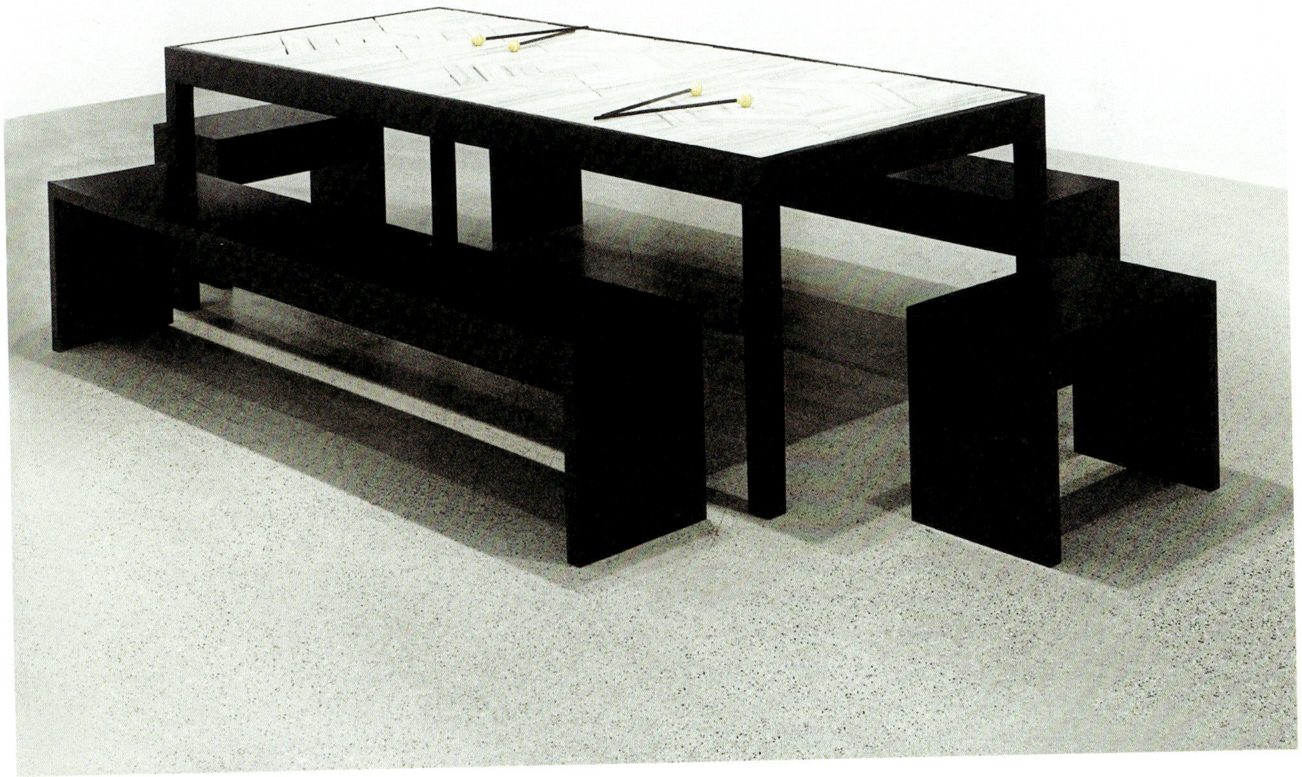

Doug Aitken, *Marble Sonic Table*, 2011.
Sound instrument. Onyx alabaster, wooden table, mallets, benches.

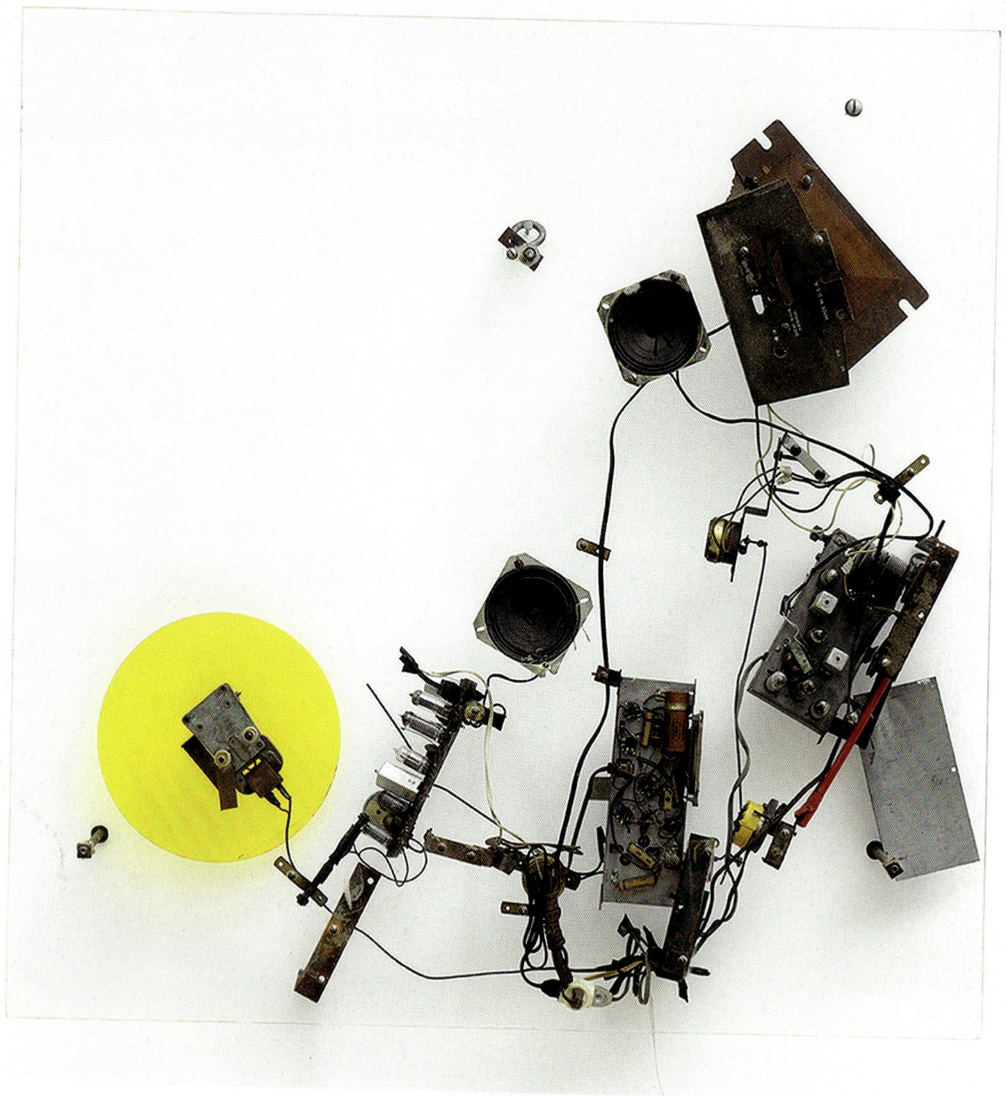

Jean Tinguely, *Radio WNYR 10*, 1962.
Radio sculpture. Plexiglass sheets, metal fixtures, radio, electric motor, 106 × 106 × 18 cm.

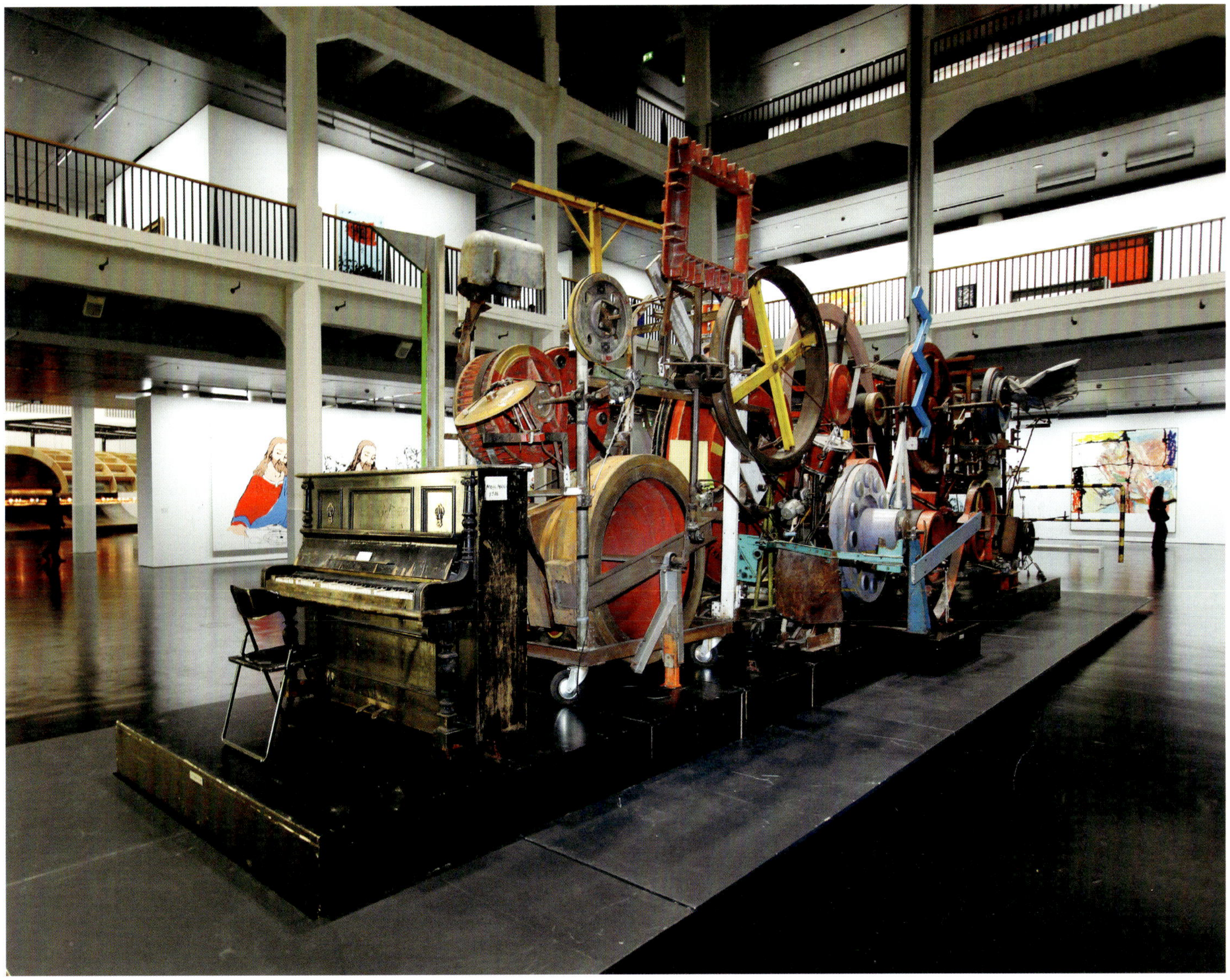

Jean Tinguely, *Grosse Méta-Maxi-Maxi-Utopia Méta-Harmonie*, 1987.
Scrap iron, wooden wheels, aluminium steps, curtain, carousel horse, pottery vase by Ursi Luginbühl, plastic flowers, garden gnome, electric motor, 730 × 1700 × 700 cm.
Installation view *just what is it …*, ZKM | Karlsruhe, 2009/2010.

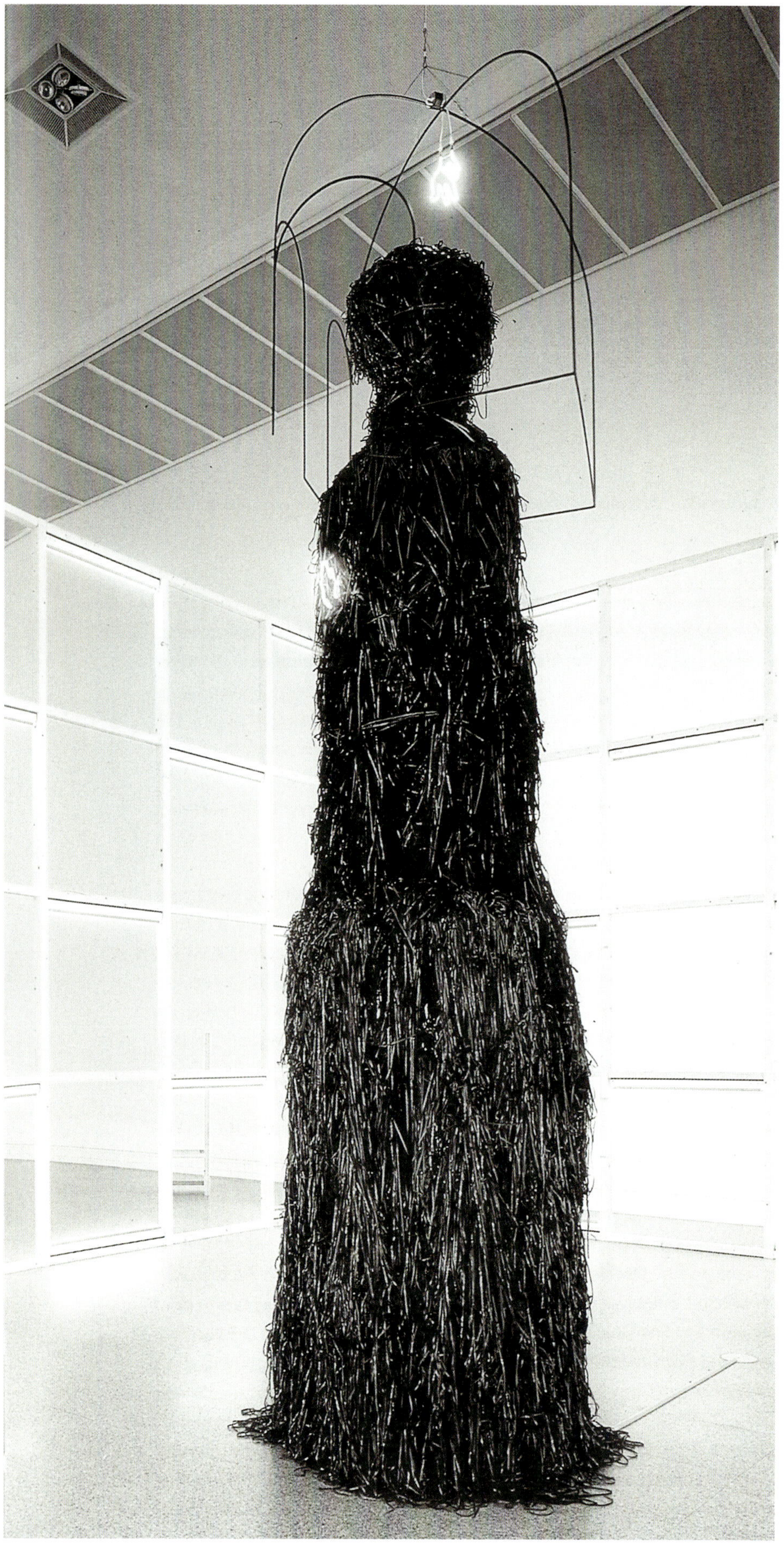

Sarkis, *Geistesblitz*, 1995. Installation view, Kunst- und Ausstellungshalle der Bundesrepublik Deutschland, Bonn.

In 2005–2006, Henri Chopin realized his first *Graphpoemachines* in Verona, Italy. Chopin describes his other three-dimensional pieces as follows: "My sculptures, which are not sculptures, are a neo-alphabet created using second hand electro-acoustic devices or those which have been overtaken by technology which evolves with the passage of time. But old radio sets just like old magnetic tape recorders even if they are tired for me are not to be thrown away."

His curious anthropomorphic totems – sort of self-portraits of the artist in his last years – were made with old instruments to which Chopin had also remained faithful, in both his sound works and public performances.

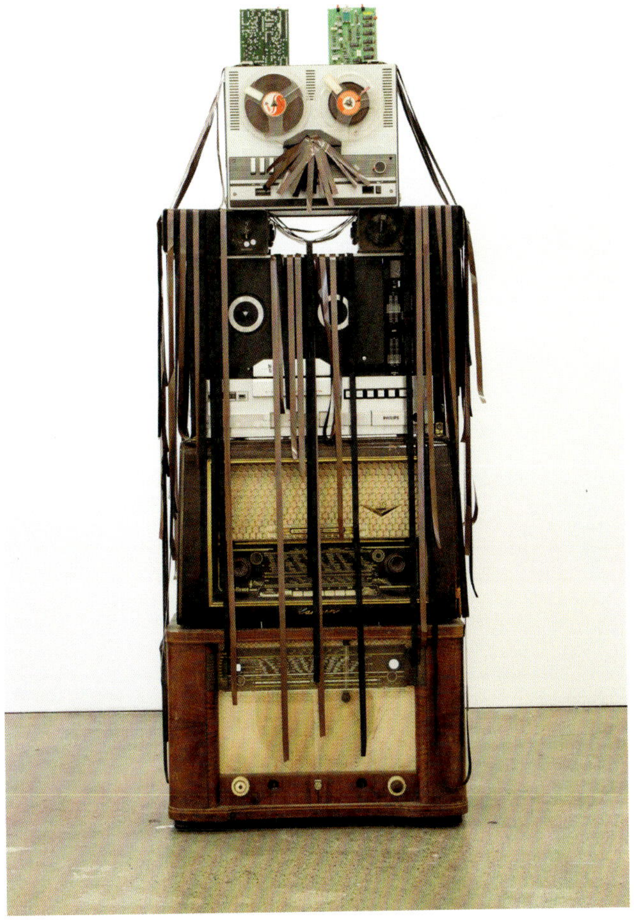

Henri Chopin, *Graphpoemachine #7*, 2006.
Mixed Media, 169 × 67 × 28 cm.
Courtesy Archivio Conz Berlin.

Installation view of the exhibition *Henri Chopin Graphpoemachines*, Supportico Lopez, Berlin 2017.
Courtesy Archivio Conz Berlin.

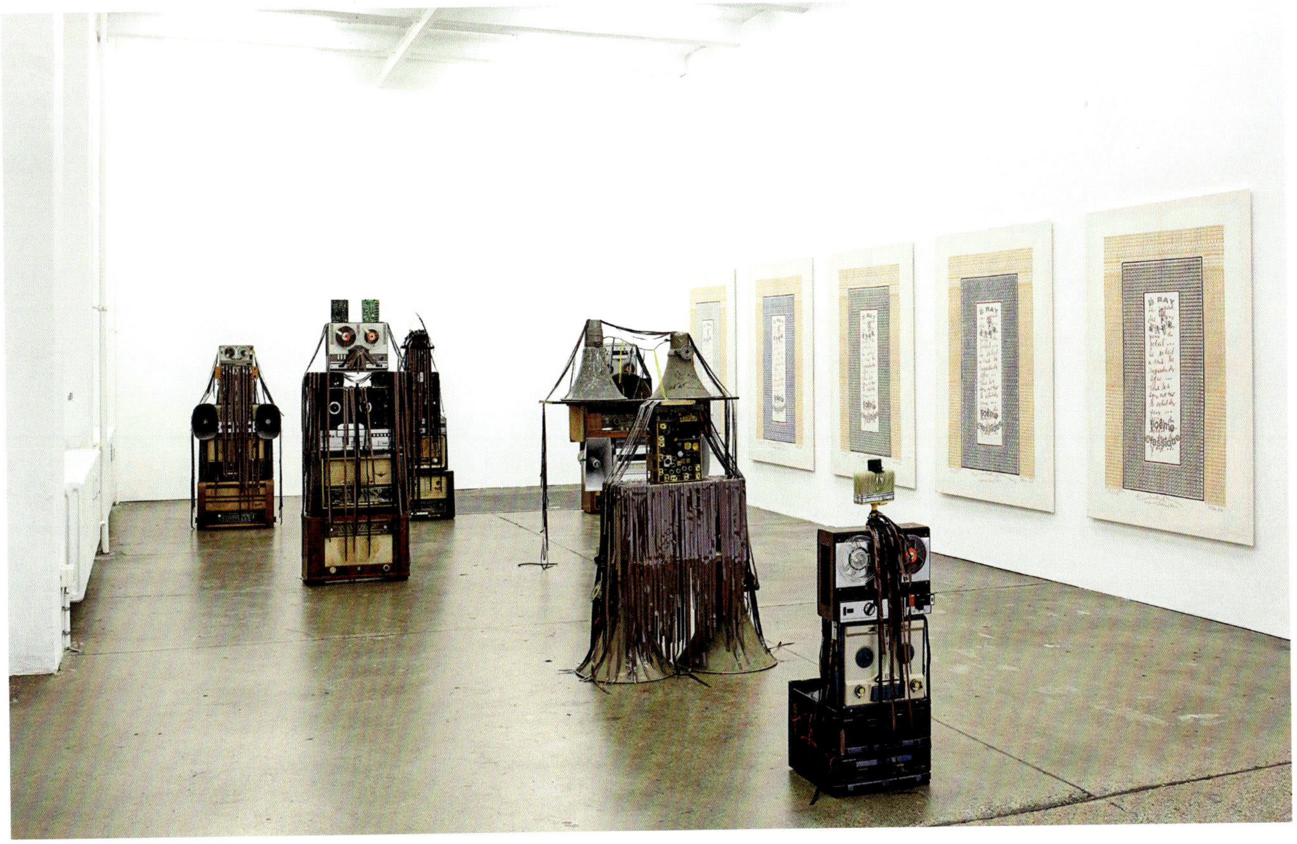

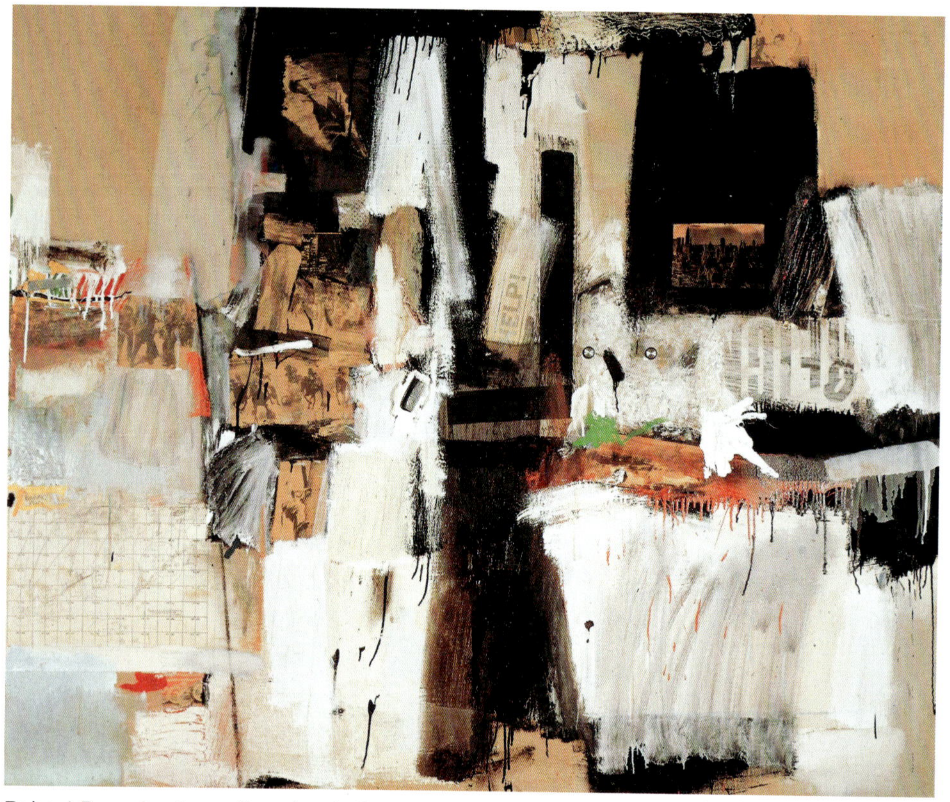

Robert Rauschenberg, *Broadcast*, 1959.
Oil, graphite, paper, fabric, newspaper, printed paper, printed reproductions, and plastic comb on canvas with three concealed radios, 154.9 × 190.5 × 12.7 cm.

Rauschenberg's *Broadcast* is part of his series *Combine*, which dates from 1954–1964. The title alludes to Rauschenberg's method of combining painting and sculpture, thus abolishing the dividing lines between different artistic traditions and giving meaning to otherwise ordinary objects. The hybrid work *Broadcast* features sound, which Rauschenberg introduces using three radios wired to the painting.

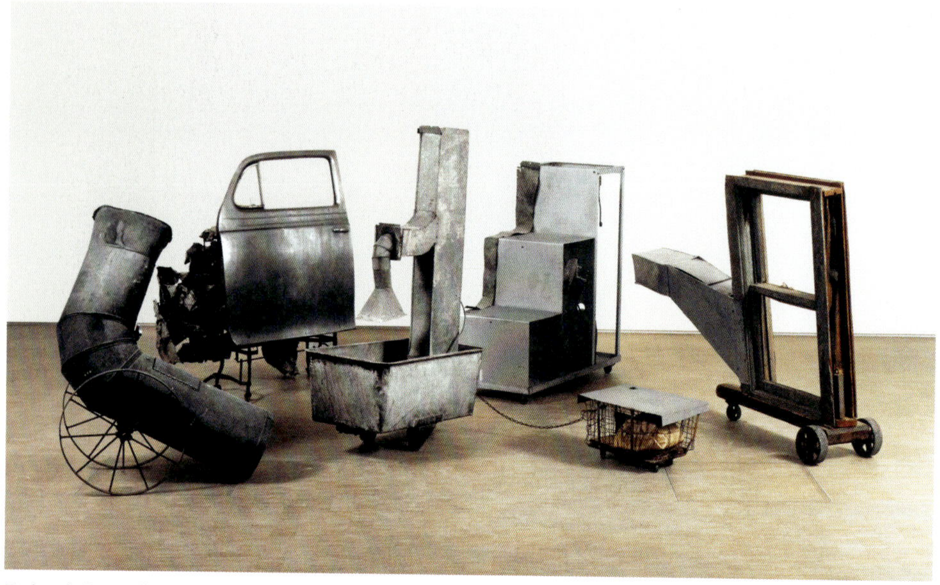

Robert Rauschenberg, *Oracle*, 1962–1965.
Five-part found-metal assemblage with five concealed radios: ventilation duct; automobile door on typewriter table, with crushed metal; ventilation duct in washtub and water, with wire basket; constructed staircase control unit housing batteries and electronic components; and wooden window frame with ventilation duct, dimensions variable.

Oracle is a work that came out of a collaboration with Billy Klüver. To imagine the sound environment, imagine the sound of splashing water. Now add the crackle of a local radio station and finally picture lights flashing in semi-darkness.

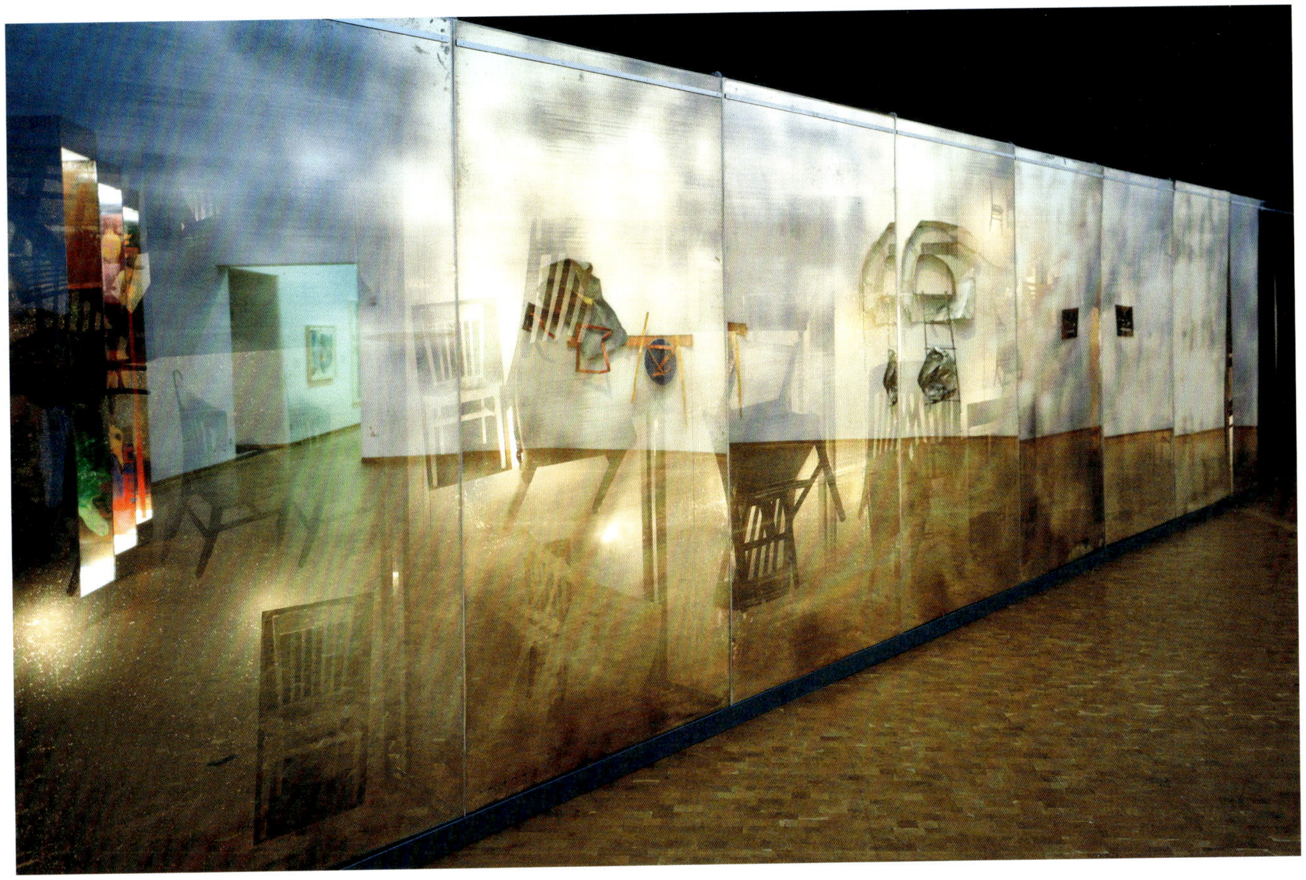

Robert Rauschenberg, *Soundings*, 1968.
Mirrored Plexiglas and silkscreen ink on Plexiglas with concealed electric lights and electronic components, 243.8 × 1097.3 × 137.2 cm.

Soundings is a voice responsive object. When entering the room, the visitor sees his or her own reflection in the silvered panels. The work is electronically activated. What the viewer sees is dependent on his or her particular voice and the sounds he or she makes.

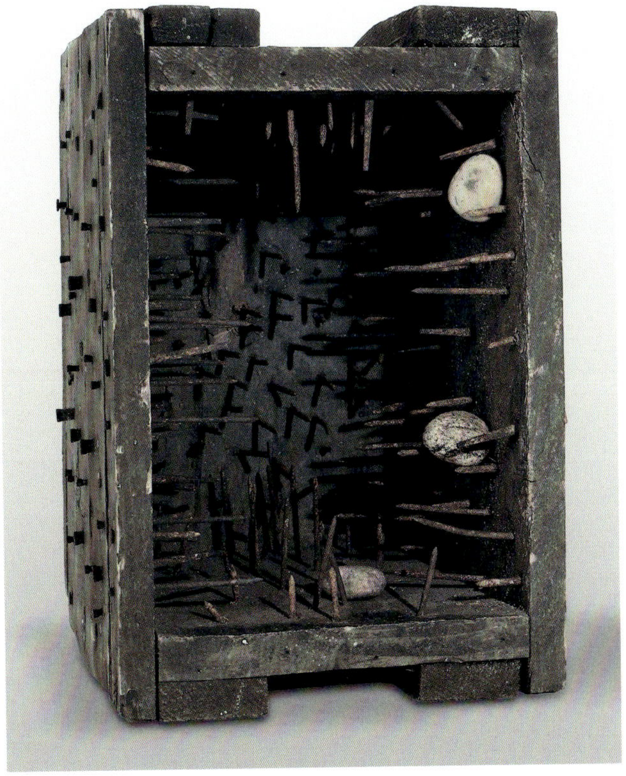

Robert Rauschenberg, *Music Box (Elemental Sculpture)*, ca. 1955.
Wood crate with traces of metallic paint, nails, three unattached stones, and feathers, 27.9 × 19.1 × 23.5 cm.
Collection of Jasper Johns.

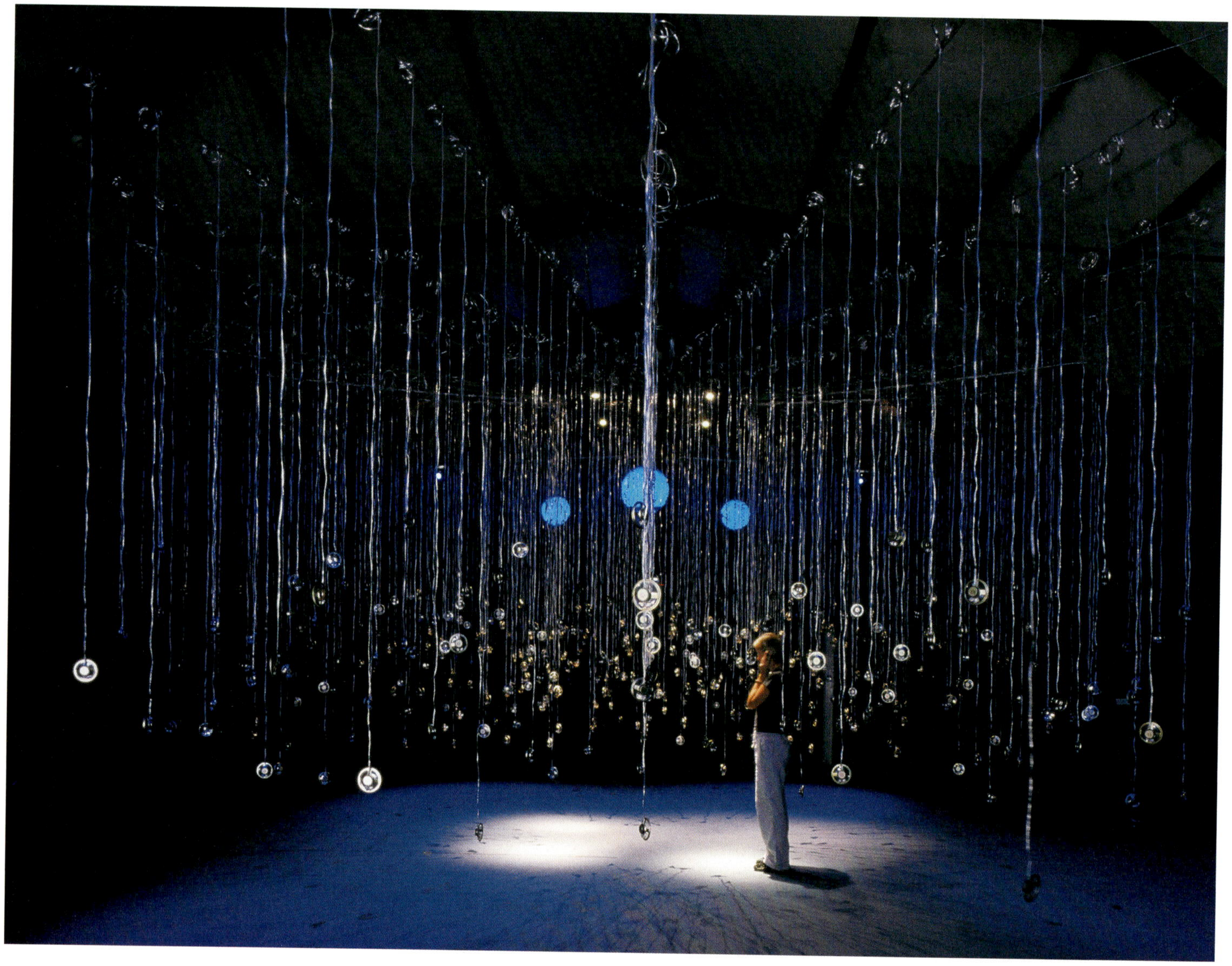

Susan Hiller, *Witness*, 2000.
Audio sculpture. 350 loudspeakers, 10 CD players, amplifiers, wiring, lights.

For many years, Susan Hiller has been collecting stories of unexplained sightings by people from all around the world. Each speaker in the installation *Witness* transmits a voice telling such a story. Whether imagined, real, or hallucinatory, the testimonies demonstrate an essential need for belief in the possibility of a further dimension, in something beyond the rational.

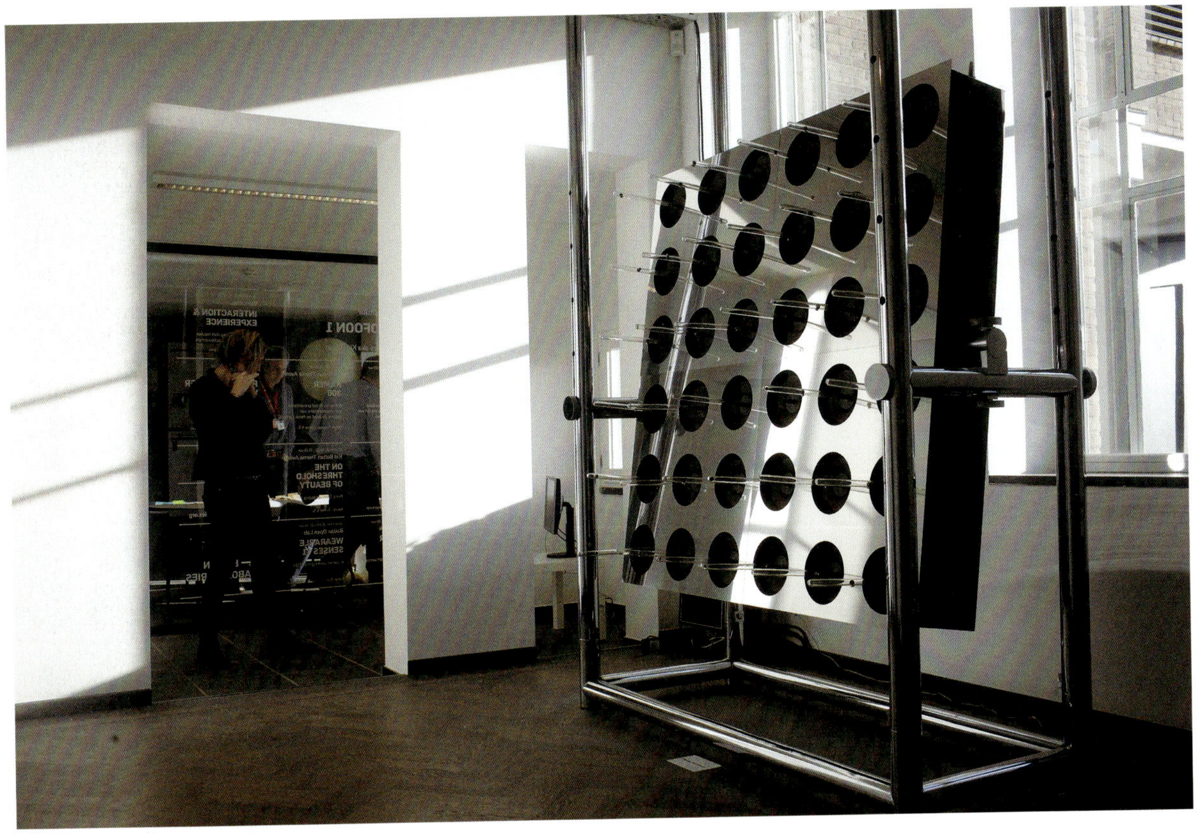

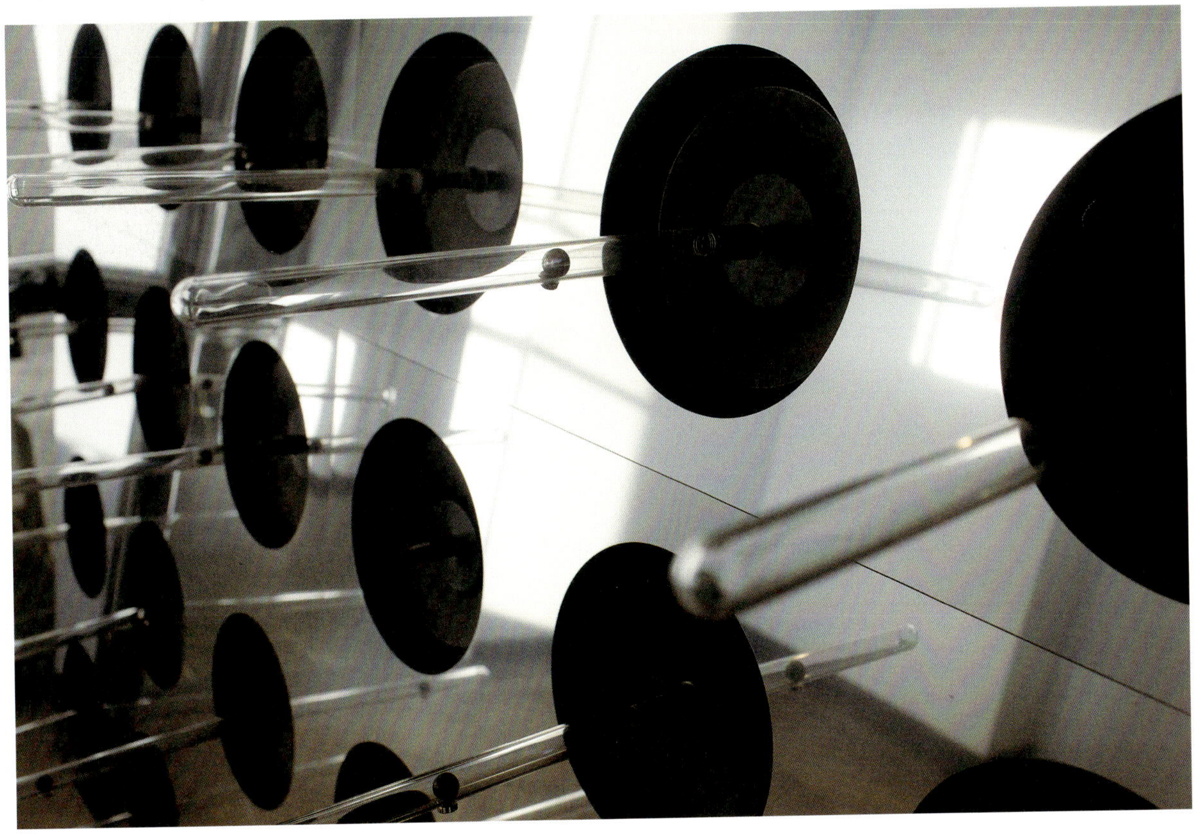

Dick Raaijmakers, *Ideophone 1*, 1970.
Sound sculpture made from loudspeakers.

Ideophone 1 consists of 36 speakers in a metal case that resembles a chalkboard or a tabletop. A glass tube with a steel ball inside is attached to each speaker. The balls inside the tubes bounce whenever the speakers emit sounds creating striking sound effects.

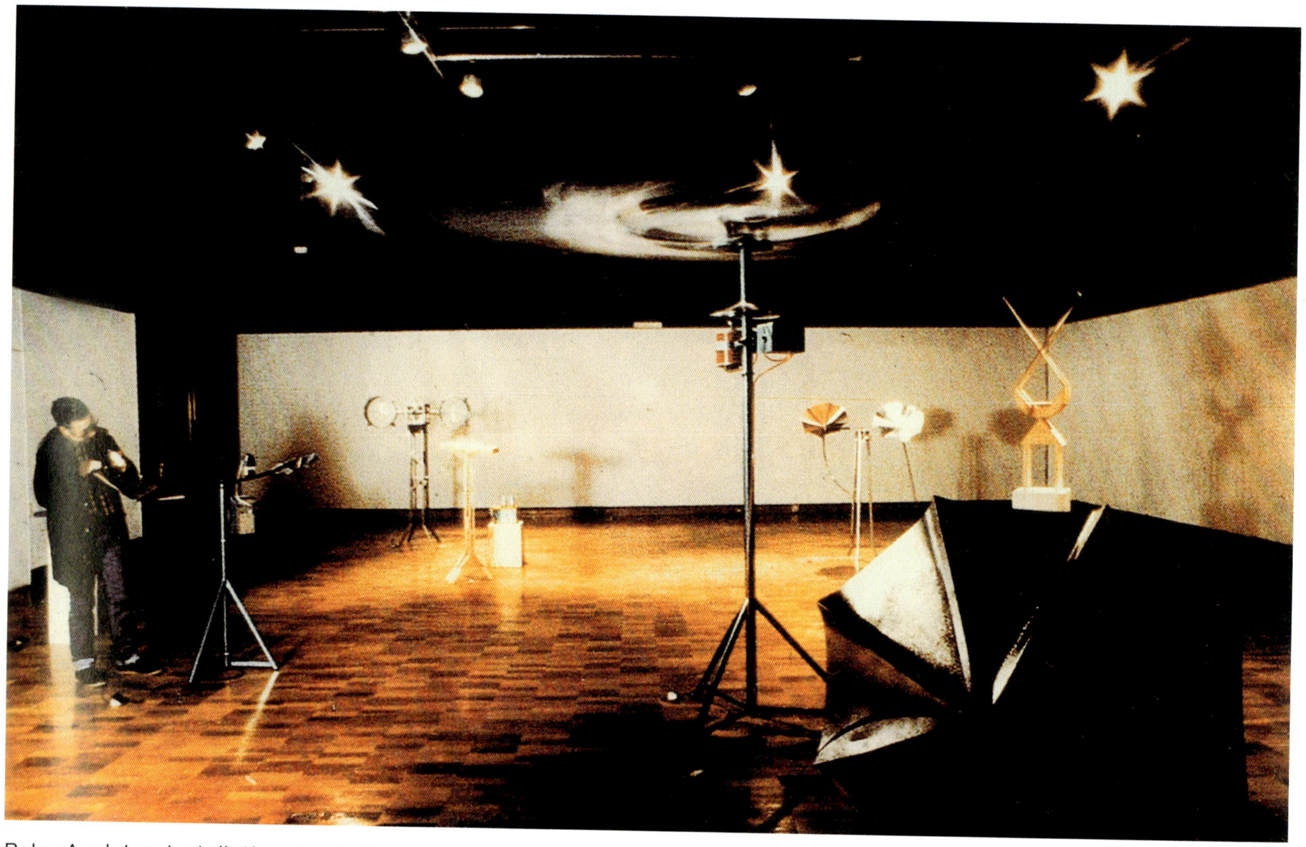

Peter Appleton, Installation of a *Self-regulating Sound Sculpture*, Sheffield, 1985.

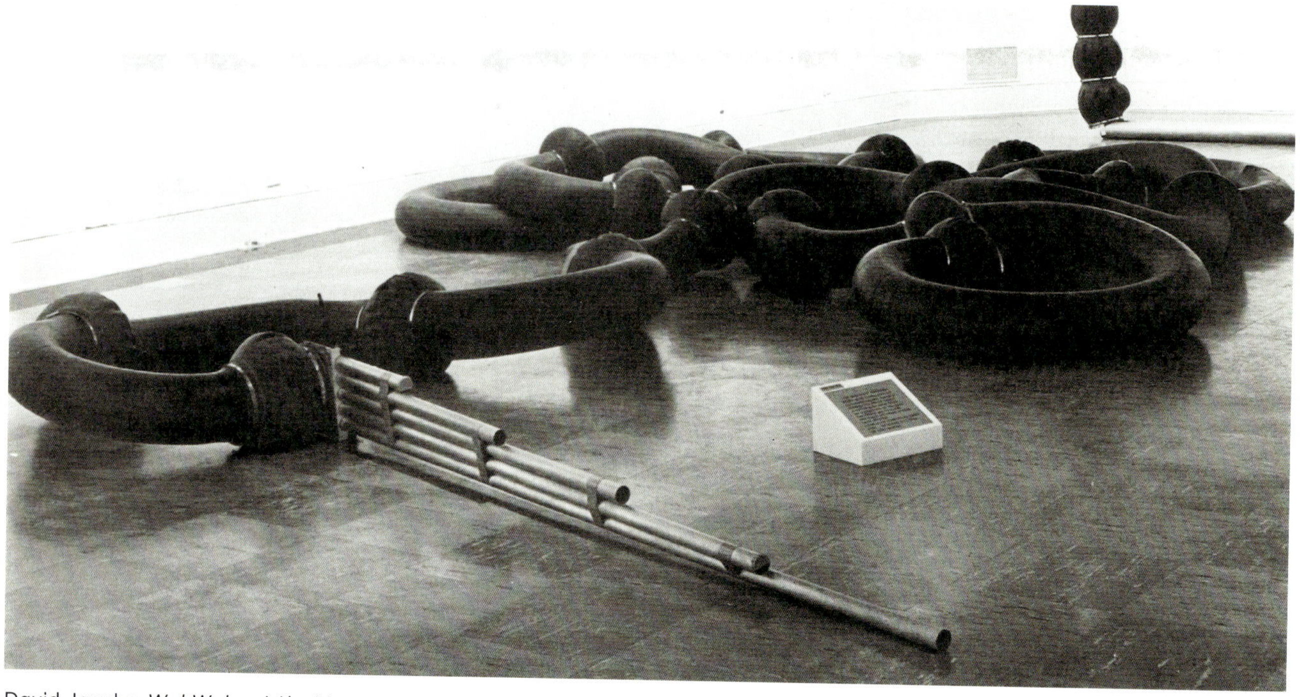

David Jacobs, *WahWahs* at the Vancouver Art Gallery, February 1972.

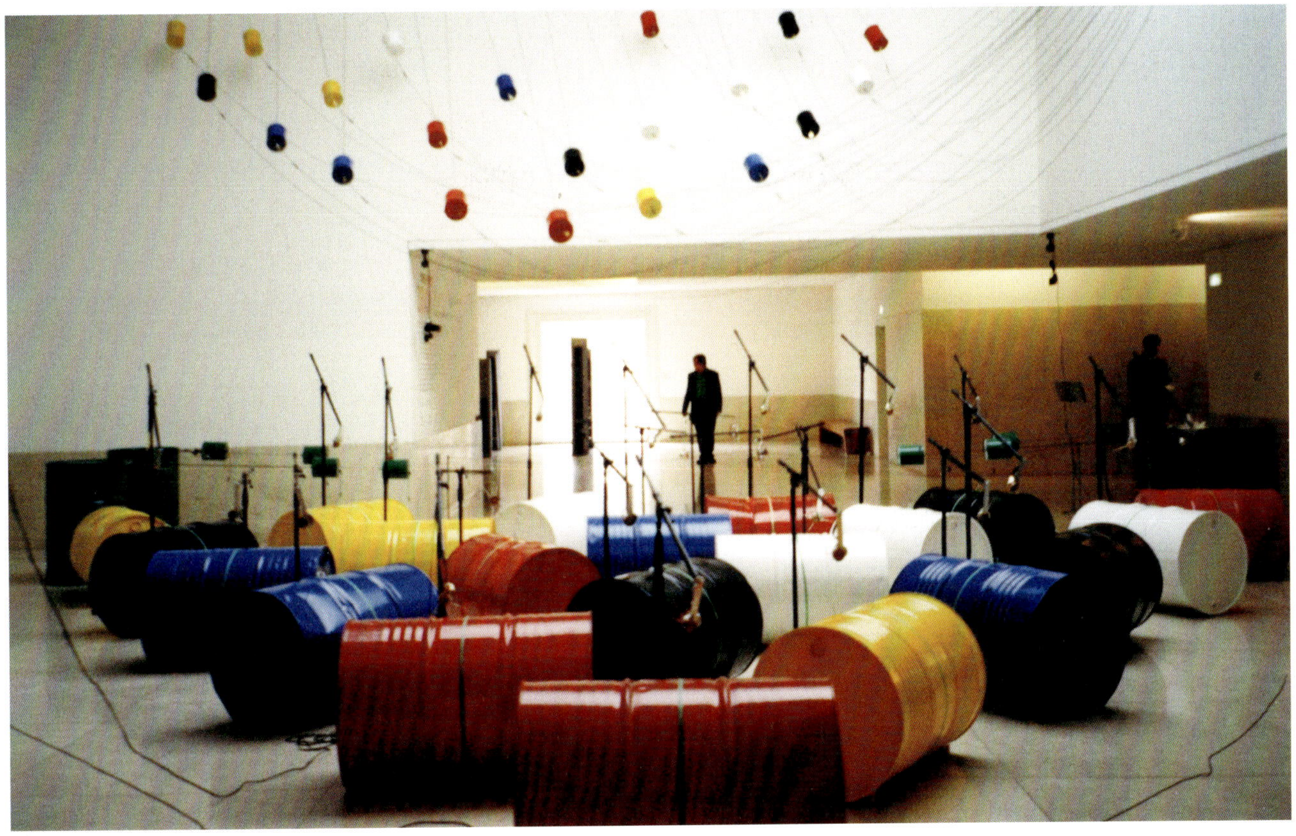

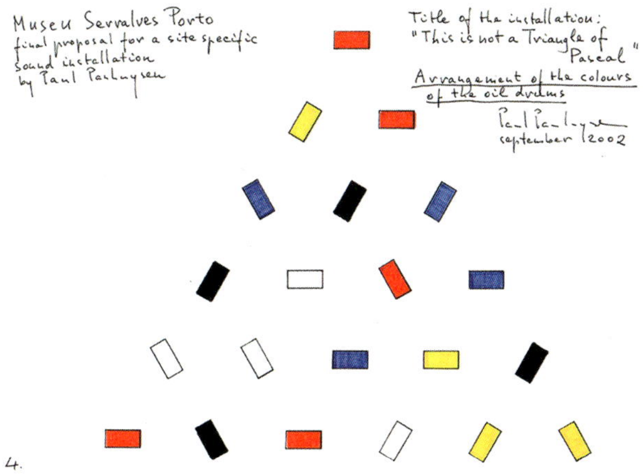

Paul Panhuysen, *This Is Not a Triangle of Pascal*, 2002.
Installation view Auditório Serralves, Museu Serralves, Porto (PT) (top) and drawing (bottom).

Twenty one oil barrels, sprayed in five different colours are laying on their sides on the floor in the shape of an equilateral triangle. The position and direction of the barrels on the floor is deduced from a modulated Fibonacci series, counted by three; its numbers are counted to and fro on the horizontal rows of the grid. From every barrel a thin steel string, going up into the air is stretched to a small resonating can hoovering in the air, and from there to the frontal wall of the vide. The colours of the resonating cans correspond with the colours of the barrels to which they are connected. The lengths of the strings between barrels and cans are all over the same; the lengths of the strings between the cans and the wall of the vide vary. All the cans carry a piezo contact microphone. Close to each barrel on the floor is a microphone stand placed, with a slow electro motor. On each motor is a pick which plucks the string about 2 times a minute. The speed of all the motors varies a little bit. This makes the sound of the installation change all the time. (Paul Panhuysen)

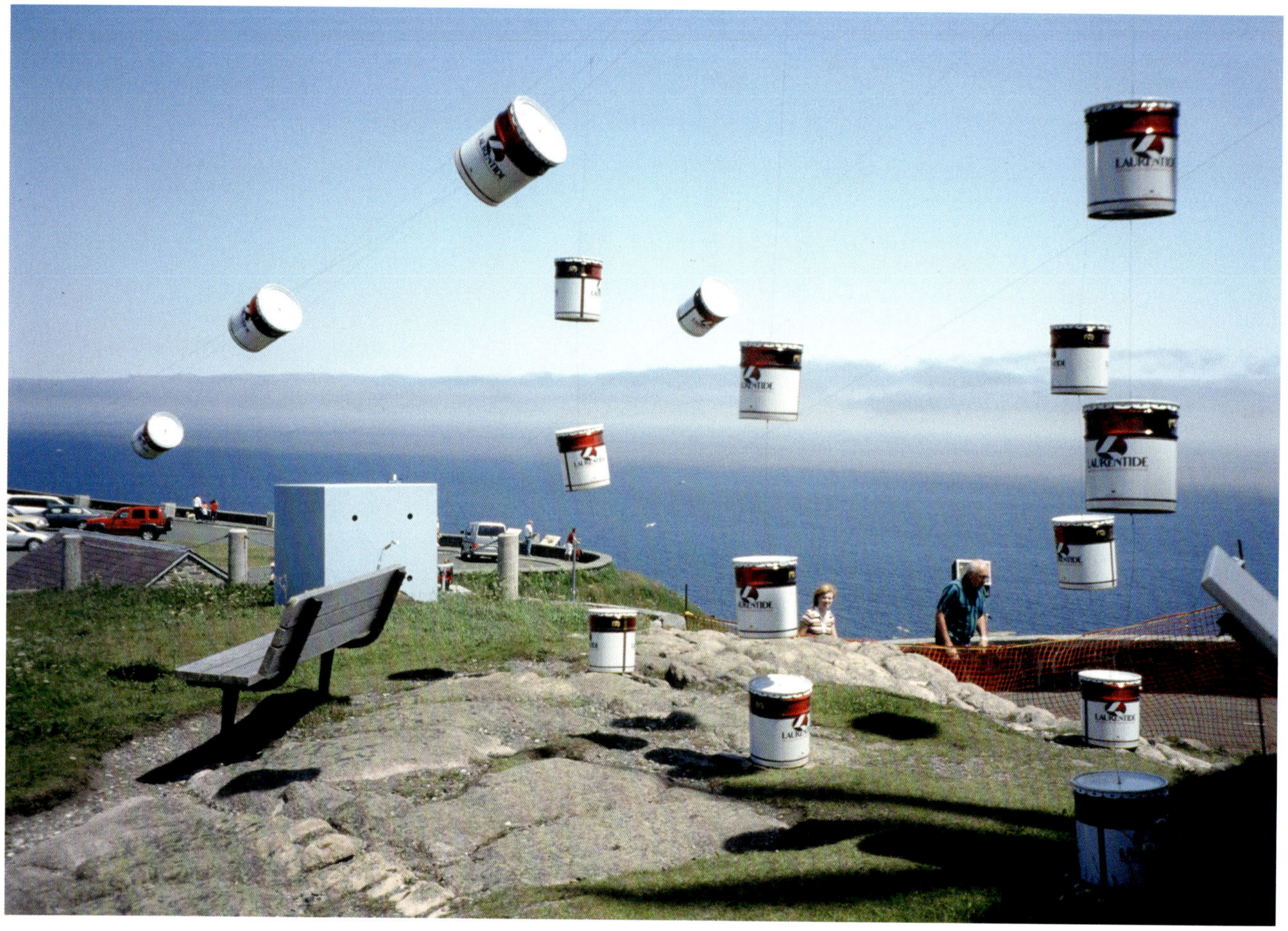

Paul Panhuysen, *Tunes Blown by the Wind*, 2004.
Outdoor sound installation at Cabot Tower on Signal Hill, St. John's, Newfoundland (CA) (top) and drawing (right page)

Four long steel strings are tuned by length, according to the Pythagorean ratio (6:8:9:12) and are stretched between the mast of Cabot Tower on Signal Hill, going to four wooden cubes which are the resonators of the installation at the end of the look out place.
These light blue painted cubes are partly filled with sand or gravel to keep the boxes in place.
On each of the four strings is connected a bell work, a wind chime, a tin on the string and three resonating tins in a vertical string to the ground with each two marbles rolling inside.
The four chimes and the eight tins with the marbles make the installation sounding as the wind is blowing to move them. (Paul Panhuysen)

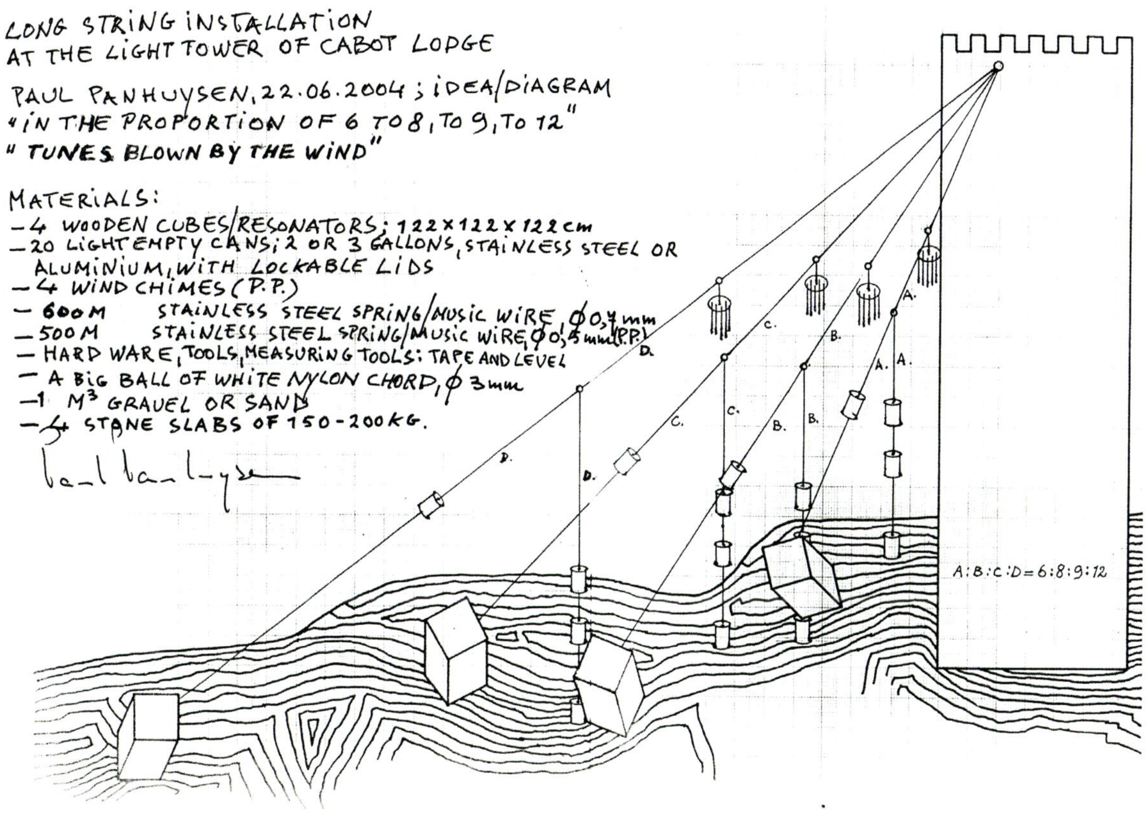
LONG STRING INSTALLATION
AT THE LIGHT TOWER OF CABOT LODGE
PAUL PANHUYSEN, 22.06.2004; IDEA/DIAGRAM
"IN THE PROPORTION OF 6 TO 8, TO 9, TO 12"
"TUNES BLOWN BY THE WIND"
MATERIALS:
– 4 WOODEN CUBES/RESONATORS; 122 × 122 × 122 cm
– 20 LIGHT EMPTY CANS; 2 OR 3 GALLONS, STAINLESS STEEL OR ALUMINIUM, WITH LOCKABLE LIDS
– 4 WIND CHIMES (P.P.)
– 600 M STAINLESS STEEL SPRING/MUSIC WIRE, Ø 0,7 mm
– 500 M STAINLESS STEEL SPRING/MUSIC WIRE, Ø 0,5 mm (P.P.)
– HARD WARE, TOOLS, MEASURING TOOLS: TAPE AND LEVEL
– A BIG BALL OF WHITE NYLON CHORD, Ø 3 mm
– 1 M³ GRAVEL OR SAND
– 4 STONE SLABS OF 150–200 KG.
A:B:C:D=6:8:9:12

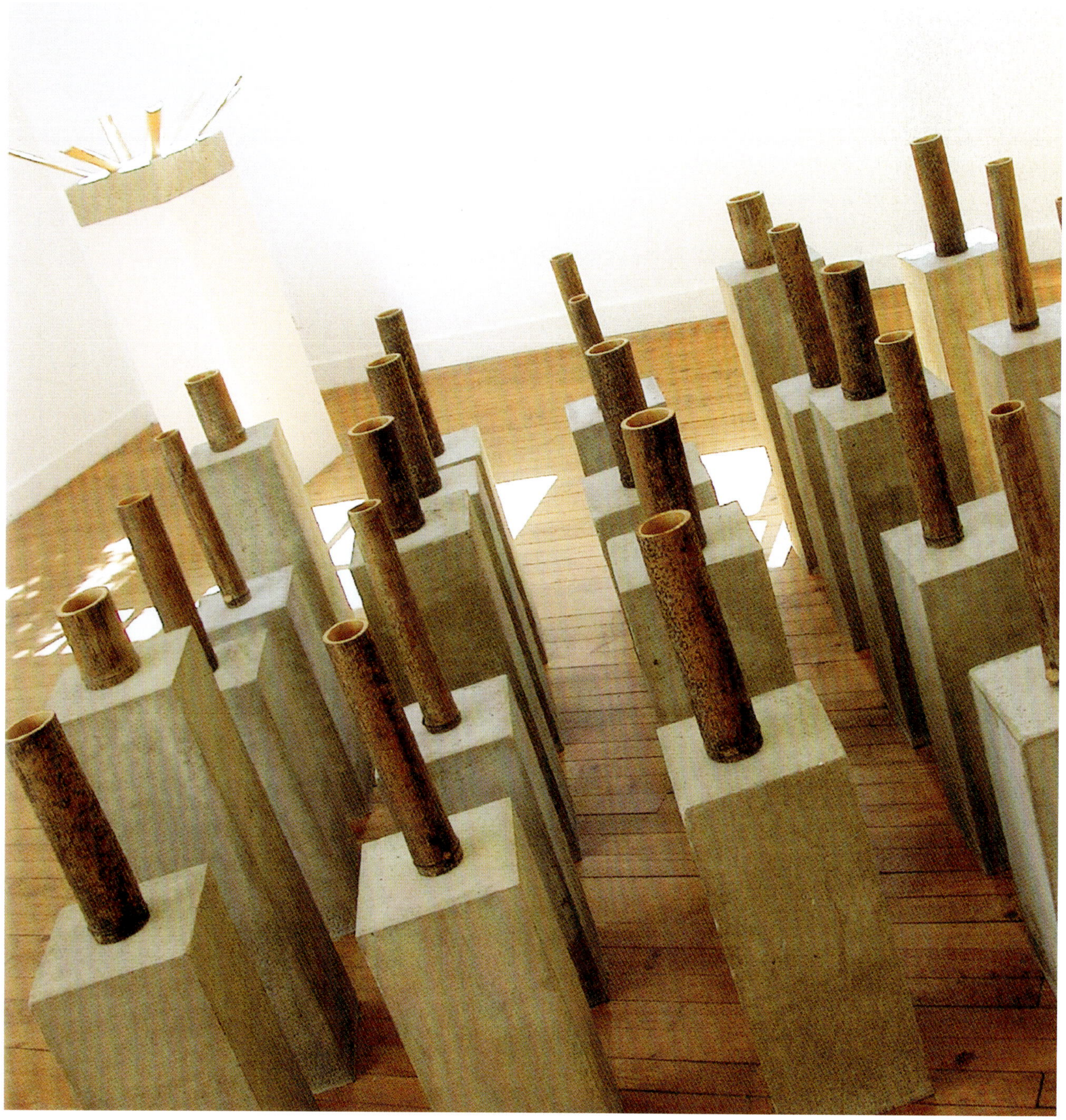

Akio Suzuki, *From One Bamboo*, 2004.
From an exhibition of his work at Zadkine Museum, Paris.
Suzuki used this sculpture as a musical instrument for
the exhibition's opening.

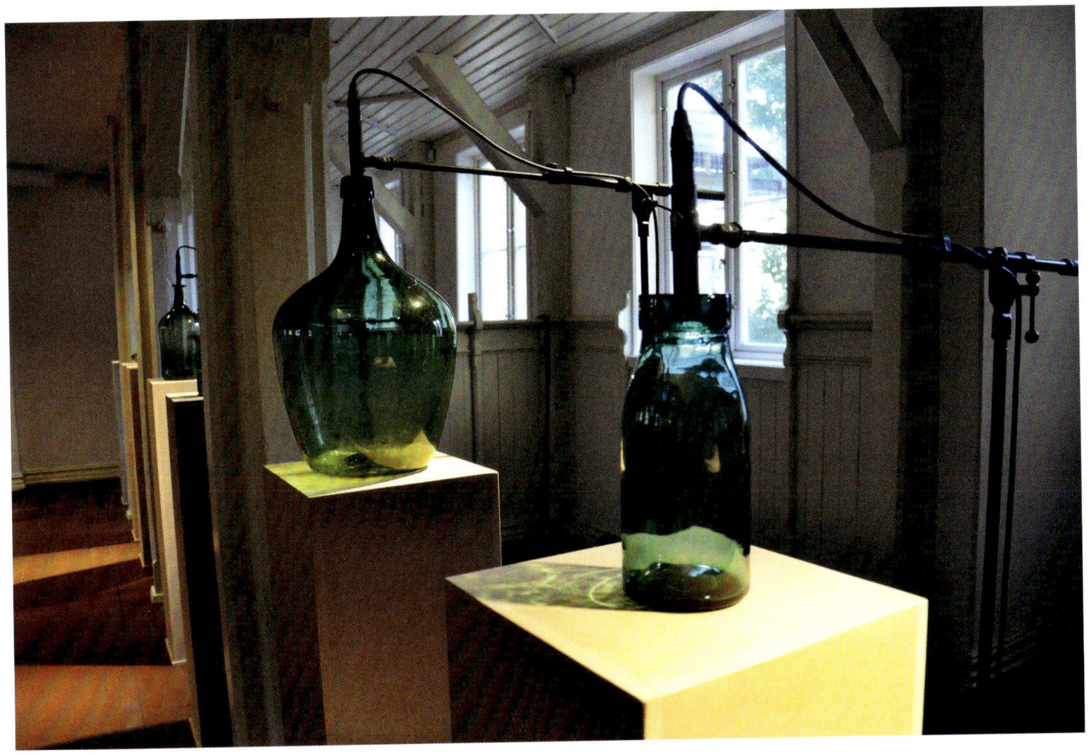

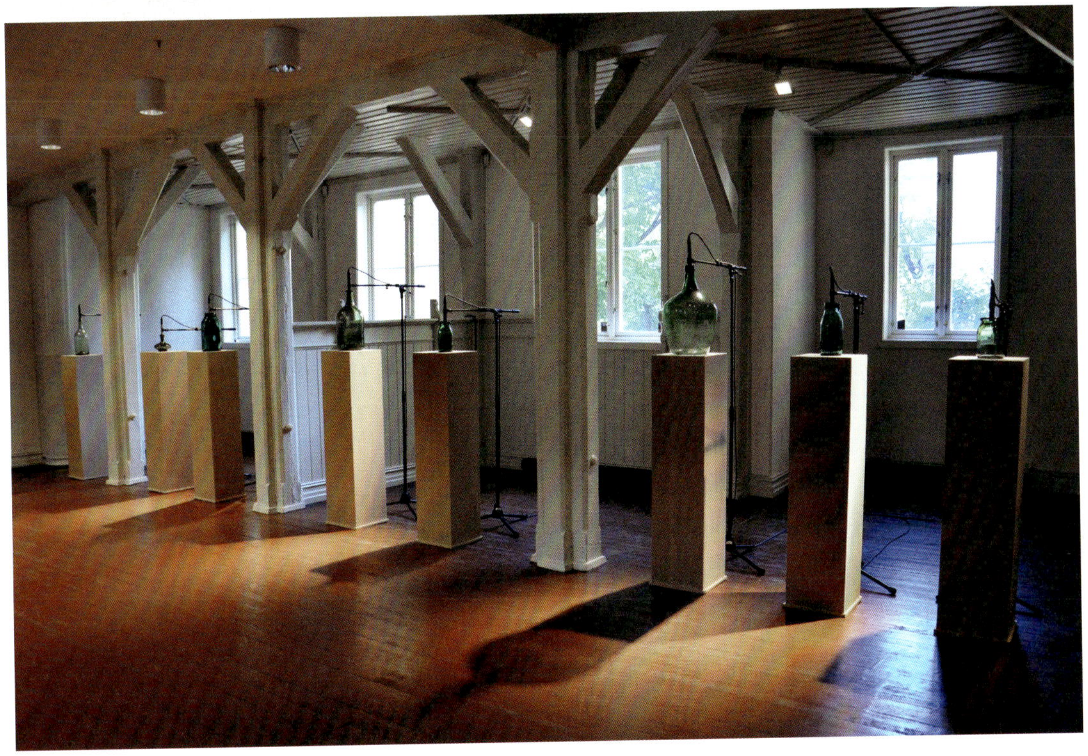

Alvin Lucier, *Empty Vessels*, 1997.
Installation view Atelier Nord, Oslo, 2013.

Empty Vessels belongs to a series of works in which Alvin Lucier studies the resonance characteristics of the smallest interior spaces. He places eight glass water containers and vases on pedestals along the space. Microphones are placed in the openings of the empty vessels, and each is rooted through limiters to speakers. The amplifier in each microphone-loudspeaker system is selected in such a way that each strand of feedback is marked by the resonant characteristics of the respective vessel. This creates a controlled feedback field of resonant tones and their interferences in the gallery space. When entering the field, the visitor disrupts the delicate balance of the system, creating new feedbacks with unexpected frequencies. Even a small head movement is enough to hear a multifaceted, fluctuating spectrum of pitches.

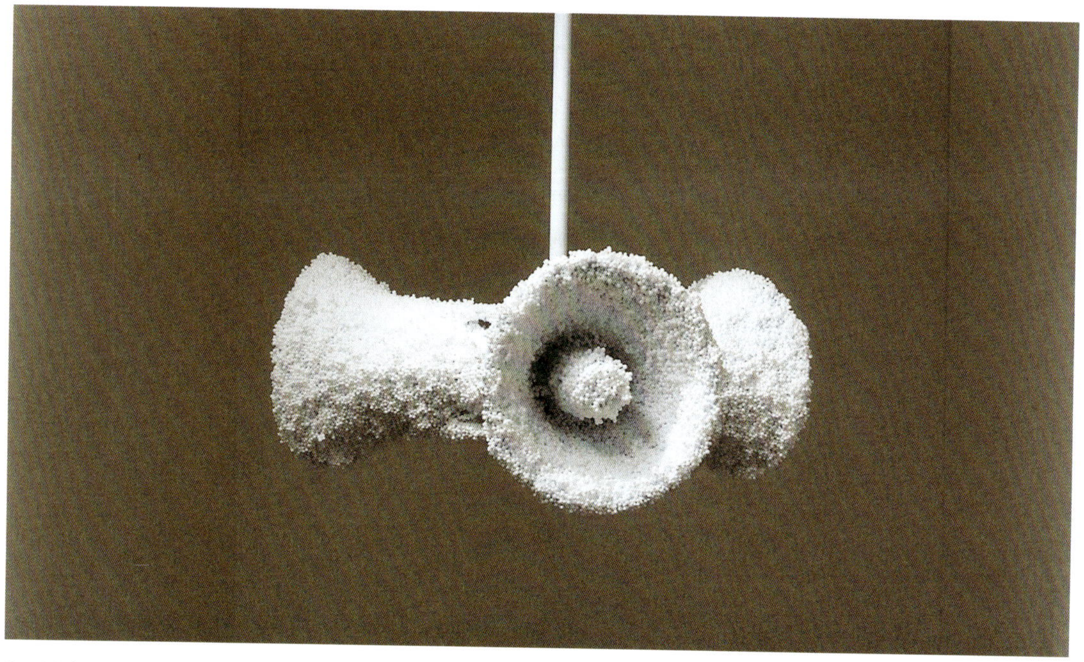

Su-Mei Tse, *Silent Party*, 2010.
Sculpture, 3 megaphones, steel, electric, polystyrene balls, height 151 cm, diameter 80 cm.

Su-Mei Tse replaces the well-known turning mirrored disco ball by three muffled megaphones. The material of the white megaphones seems to have completely absorbed the sound leaving the party silent. The megaphones also refer to the artist's interest in white noise, the audible crackling silence from a vinyl LP disc.

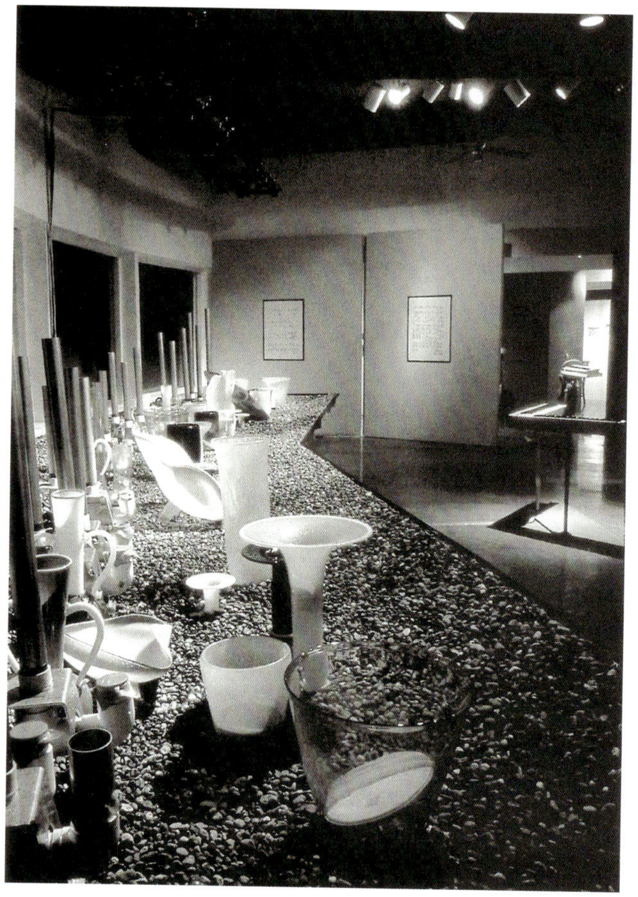

Trimpin, *Liquid Percussion*, 1991.
Installation view Port Angeles Fine Arts Center.

Liquid Percussion is a water instrument created with juice dispensers that have been adjusted to precisely release a single droplet of water into various hand blown glass vessels.

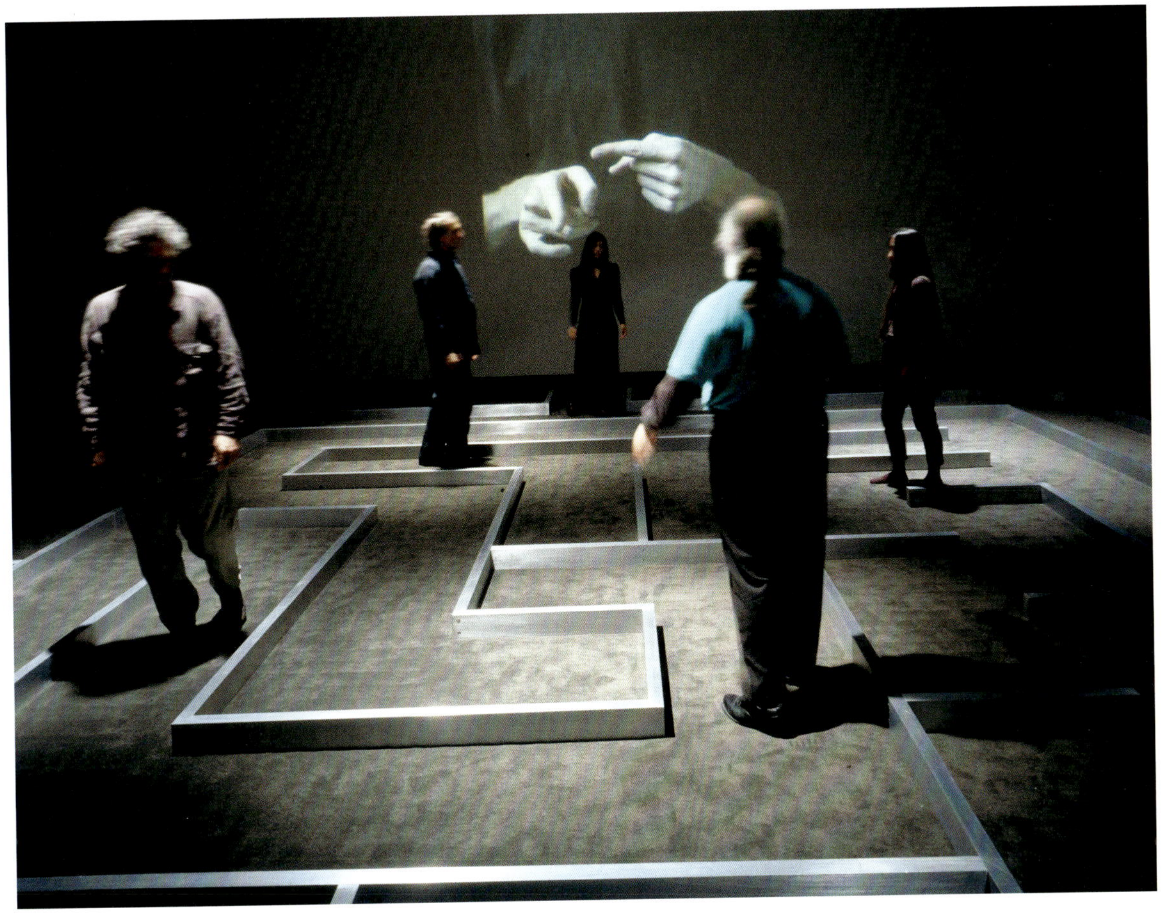

FAX 40 21 7878 #2

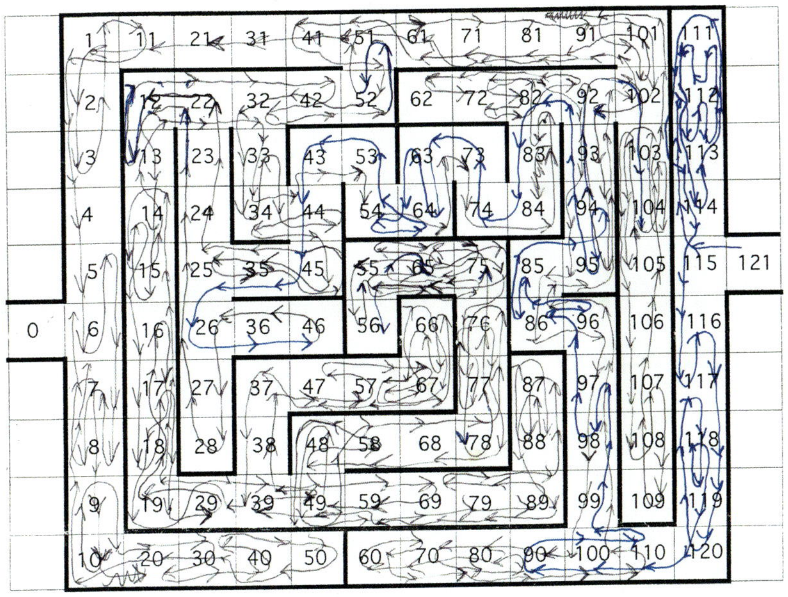

Gary Hill, *Withershins*, 1995.
Interactive sound installation with video projections.

Withershins is an interactive work based on a highly self-reflexive text about left and right handedness and bicameralism in relation to movement, symmetry, and asymmetry. As one moves within the labyrinth spoken phrases are triggered every 76 cm along the way. Two people can be tracked at any given time, simultaneously generating two texts that suggest the possibility of interchange between the participants.

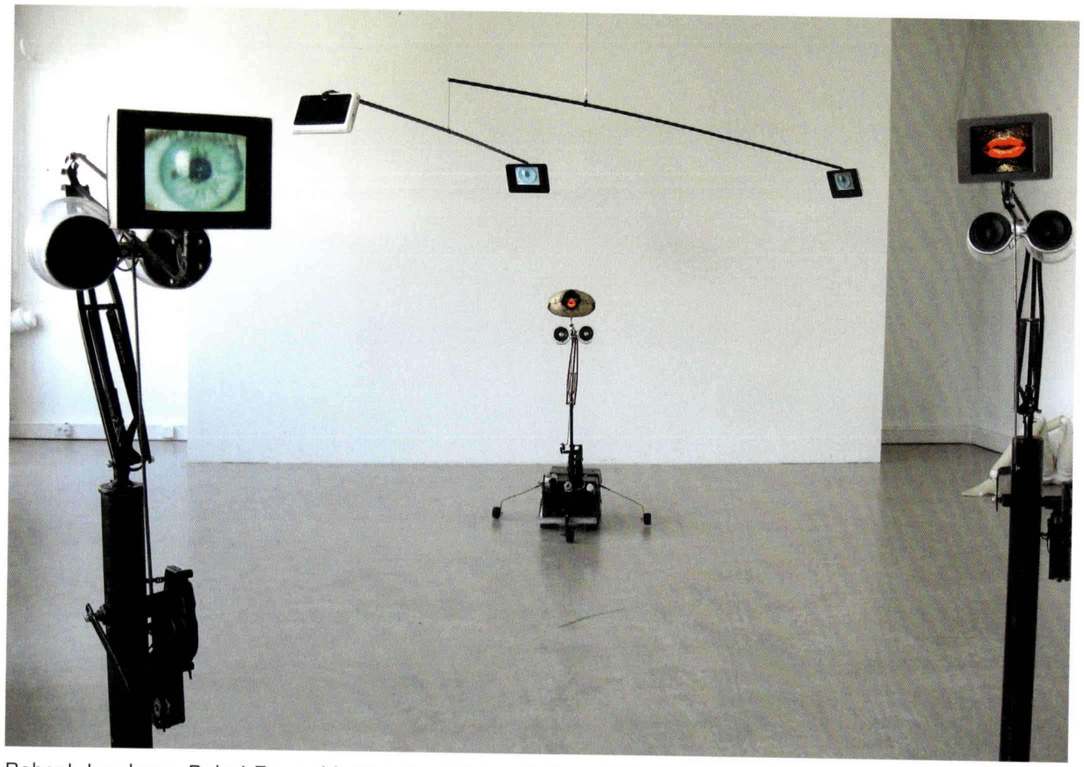

Robert Jacobsen, *Robot Ensemble No I-Funk Robot, It-She-He Robots & Mobile Video Tryptichon Play*, 2008.
6-channel video and sound piece about All & Nothing.
Installation view Wolfs Place Wuppertal, 2008.

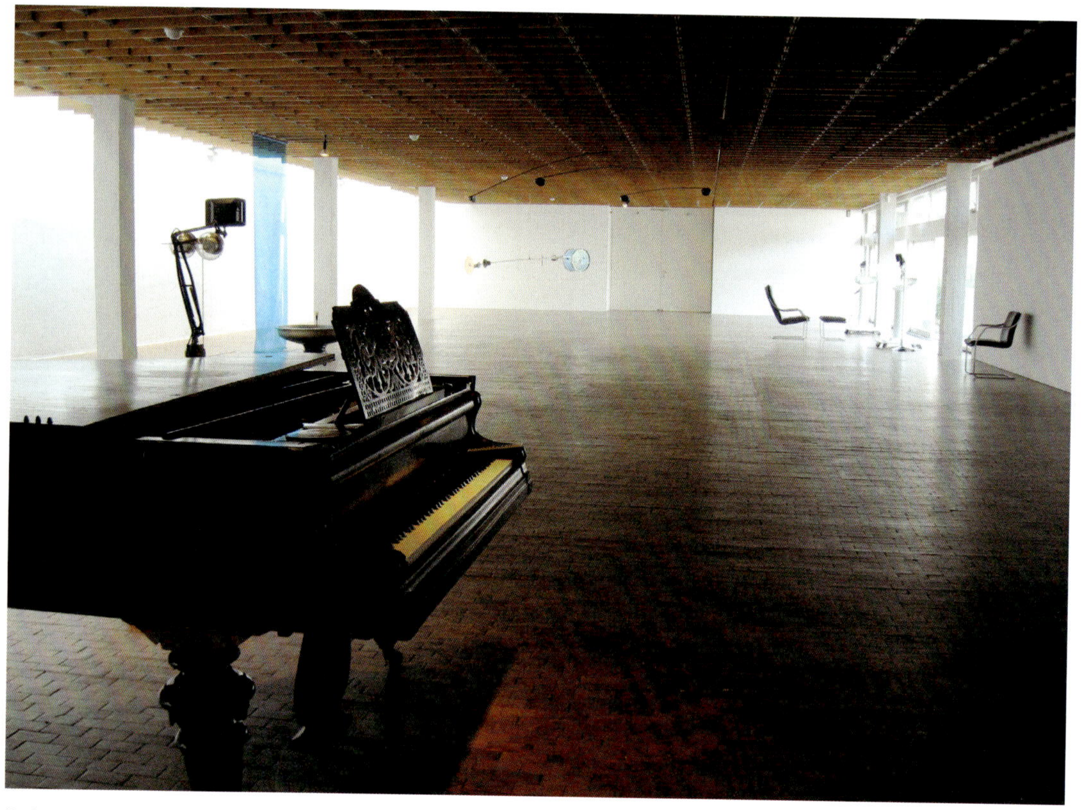

Robert Jacobsen, *Welcome – Spatial composition for a self-playing ensemble.*
From the series *Skulpturales Theater No. IV* [Sculptural Theater No. IV], 2006.
Installation view Sonambiente Festival, Akademie der Künste Berlin, 2006.

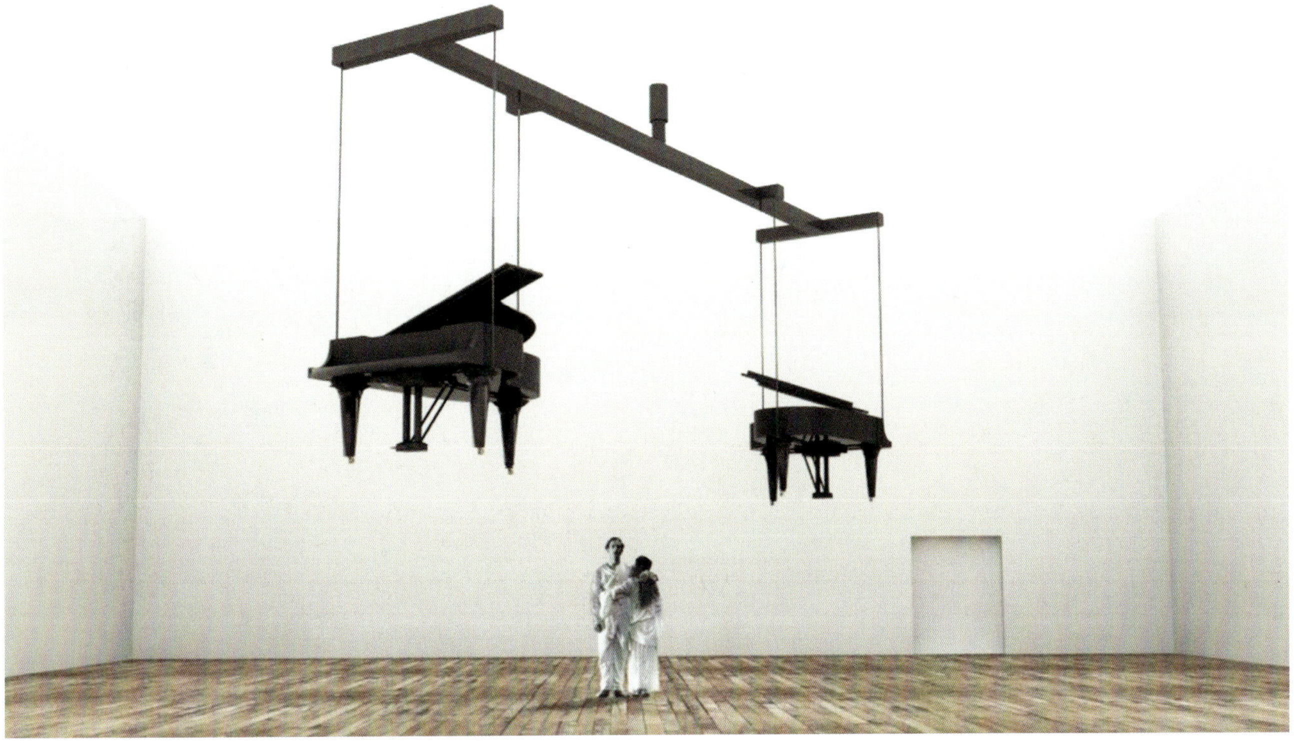

Robert Jacobsen, *Grand Slam Mobile*, forthcoming 2020.
Sound installation.

This intallation is a mobile with two prepared concert grand pianos. The wing lid of the pianos opens very slowly (within 10 minutes) and then strikes suddenly. An enormous sound is created, a kind of sound cloud that floods the room and at the same time appears as an element of surprise. In addition, every three minutes, played in unison, two dissonant intervals sound from the outer layers of the keyboard.

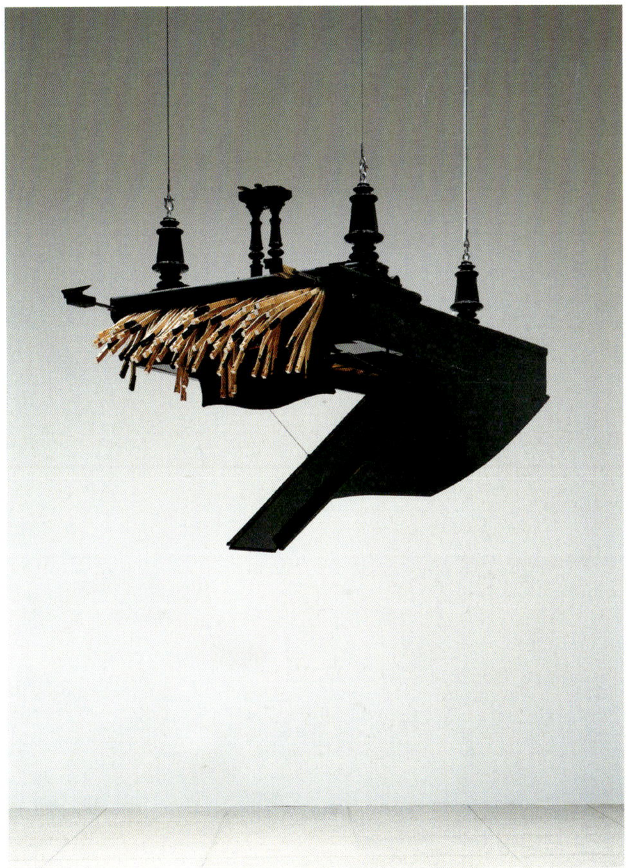

Rebecca Horn, *Concert of Anarchy*, 1990.
Sound sculpture. Grand piano, hydraulic rams, and compressor.

Concert for Anarchy is one of a series of mechanized sculptures Horn began making in the late 1970s. A grand piano is suspended upside down from the ceiling by heavy wires attached to its legs. A mechanism within the piano is timed to go off every two to three minutes, thrusting the keys out of the keyboard in a cacophonous shudder. At the same time, the piano's lid falls open to reveal the instrument's harp-like interior, the strings reverberating at random. This unexpected, violent act is followed between one and two minutes later by a retraction as the lid closes and the keys slide back into place, tunelessly creaking as they go.

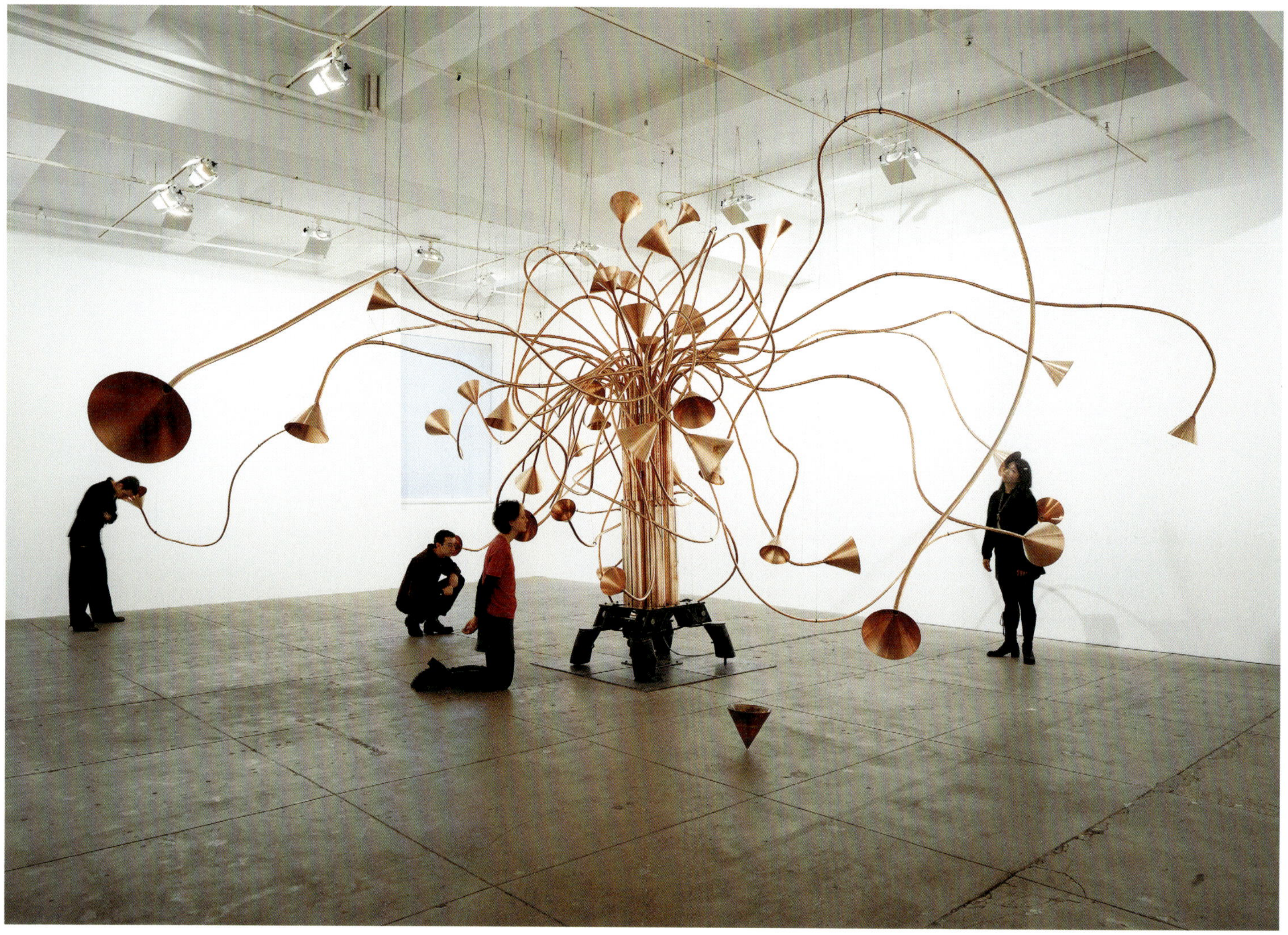

Rebecca Horn, *Der Schildkrötenseufzerbaum / The Turtle Sighing Tree*, 1994. Copper pipes, copper funnels, glass funnels, sound equipment, stone, metal construction, electronics, motor, dimensions variable.

First *Der Schildkrötenseufzerbaum / The Turtle Sighing Tree* shakes its copper branches, which flow into funnels. Then the babble of words rises. People tell of their troubles and sorrows. When you turn to a funnel, you hear a single voice mourning.

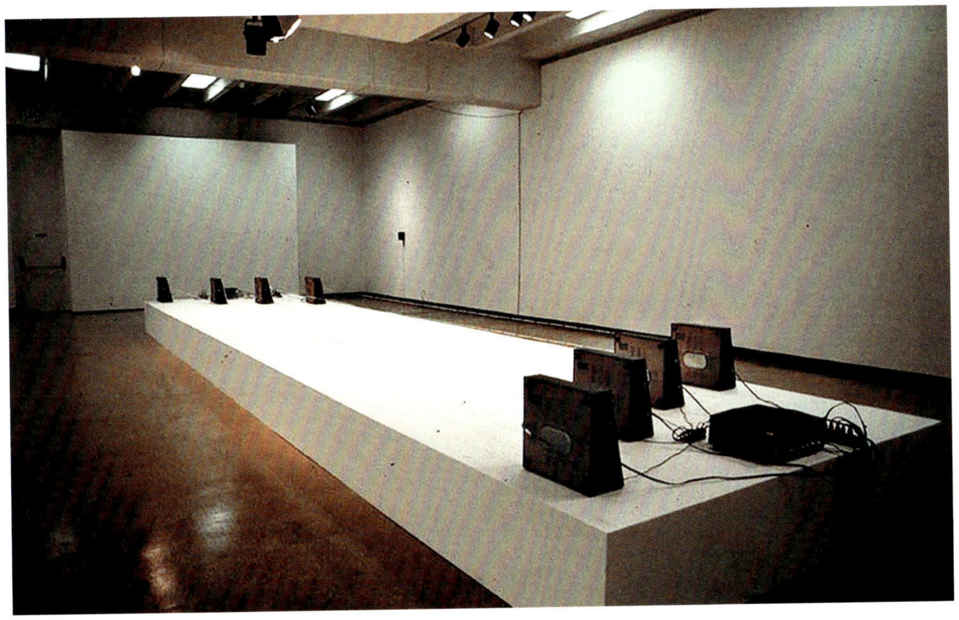

Bill Fontana, *Sound Recycling Sculpture*, 1978–1983. Installation view.

SOUND RECYCLING SCULPTURE

Simultaneously:

Once every 104 seconds a 12 second recording of sounds that happened 52 seconds ago is replayed.

Once every 92 seconds an 8 second recording of sounds that happened 46 seconds ago is replayed.

Once every 84 seconds a 6 second recording of sounds that happened 42 seconds ago is replayed.

Once every 78 seconds a 4 second recording of sounds that happened 39 seconds ago is replayed.

A sculpture that embodies a space by contemplating the sounds in it. Brief random samples of sounds happening in a given place are replayed at long time delays of changing synchronicity. This is accomplished by means of 4 specially constructed tape loops that are mostly silence(leader tape). The ratios of leader tape to recording tape in the loops are: 8.67 to 1, 11.5 to 1, 14 to 1 and 19.5 to 1. Each loop is played between 2 reel to reel tape recorders. One recorder in each set is connected to a sensitive microphone and is always recording, the other recorder in each set is connected to a loudspeaker system and is always replaying The Sound Recycling Sculpture is both visual and aural. As a visual sculpture it is 4 very long black and white tape loops in a spatial configuration. As a sound sculpture it is a relationship to ambient sounds in in a given environment. It is not necesary to have the loops in the environment being recycled for sound. The loops could be indoors within a gallery space and the microphones and loudspeakers placed in a nearby lobby, hallway, stairwell or courtyard.

1980
© Bill Fontana

Bill Fontana, *Sound Recycling Sculpture*, 1978–1983.

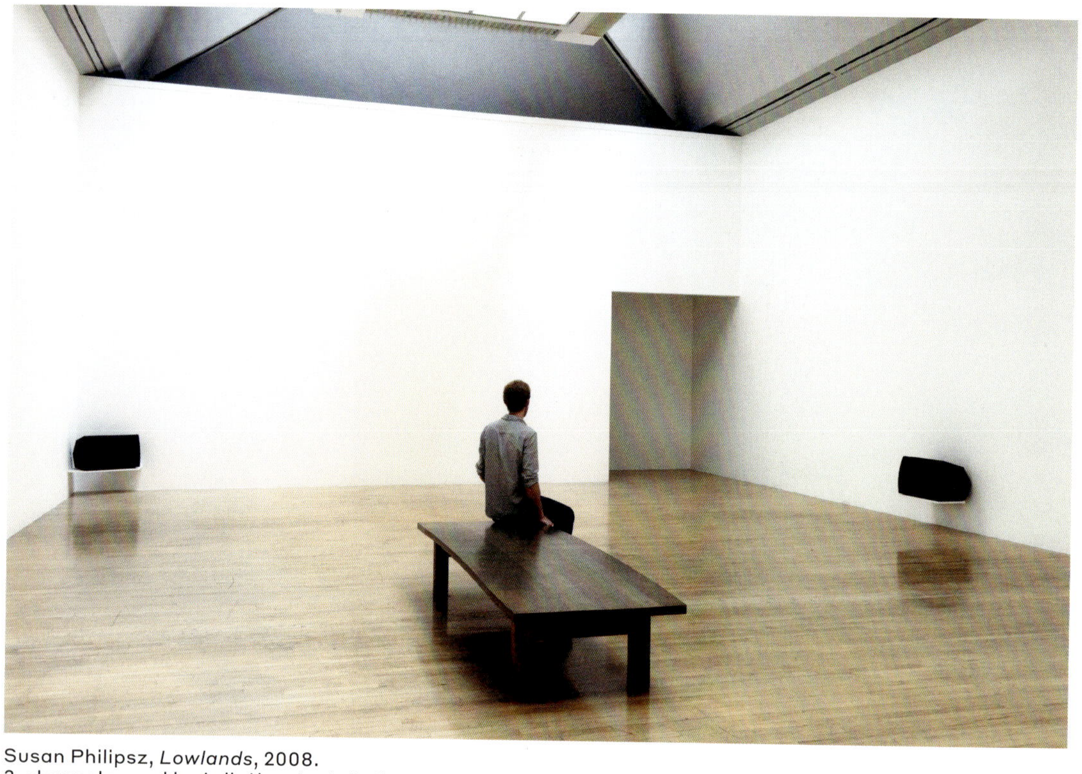

Susan Philipsz, *Lowlands*, 2008.
3-channel sound installation. Installation view, Tate Britain, London, 2010.

For her sound installation *Lowlands*, Susan Philipsz recorded three versions of a sixteenth century Scottish ballad using only her own untrained singing voice. It was originally installed beneath three bridges in Glasgow. In 2010, the piece was awarded the Turner Prize – the first sound installation to win the prestigious art prize.

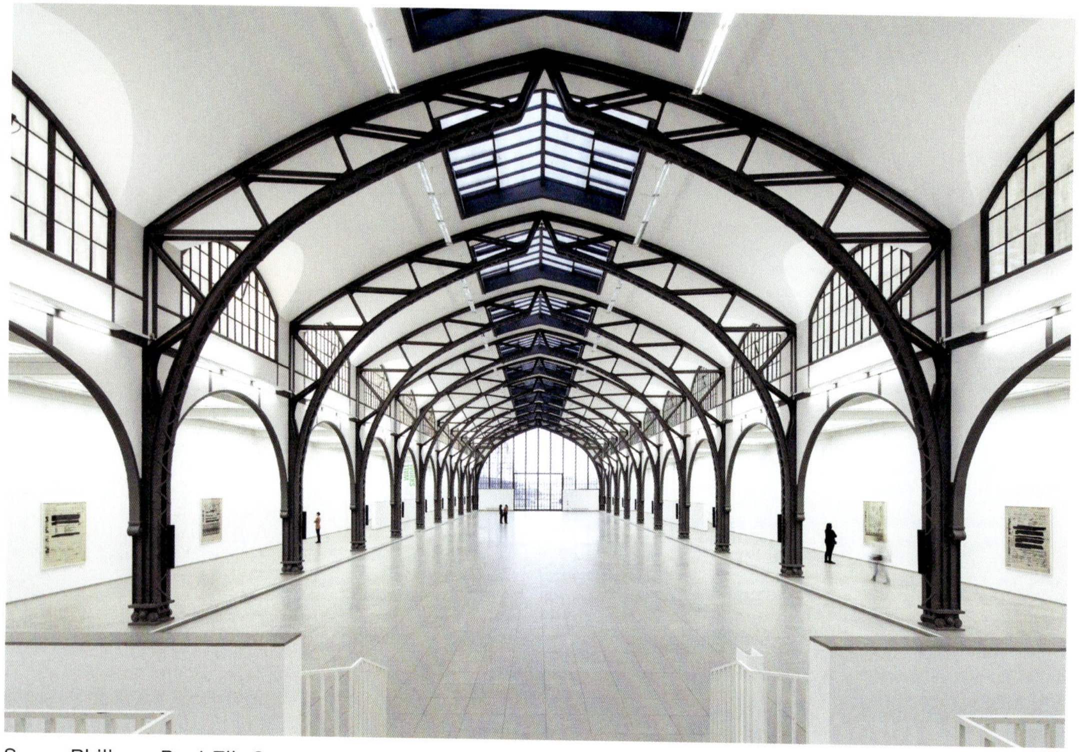

Susan Philipsz, *Part File Score*, 2014. Installation view Hamburger Bahnhof.

Part File Score is Susan Philipsz's artistic approach to the life and work of composer Hanns Eisler (1898–1962). A pupil of Arnold Schoenberg, Eisler emigrated to the United States in the 1930s, but left in 1948 due to his pro-Communist beliefs. In the sound installation, Philipsz isolated single notes from three of Eisler's compositions and each note is played on its own speaker. Wall projections of Eisler's FBI records superimposed over pages of his musical scores accompany the sound piece.

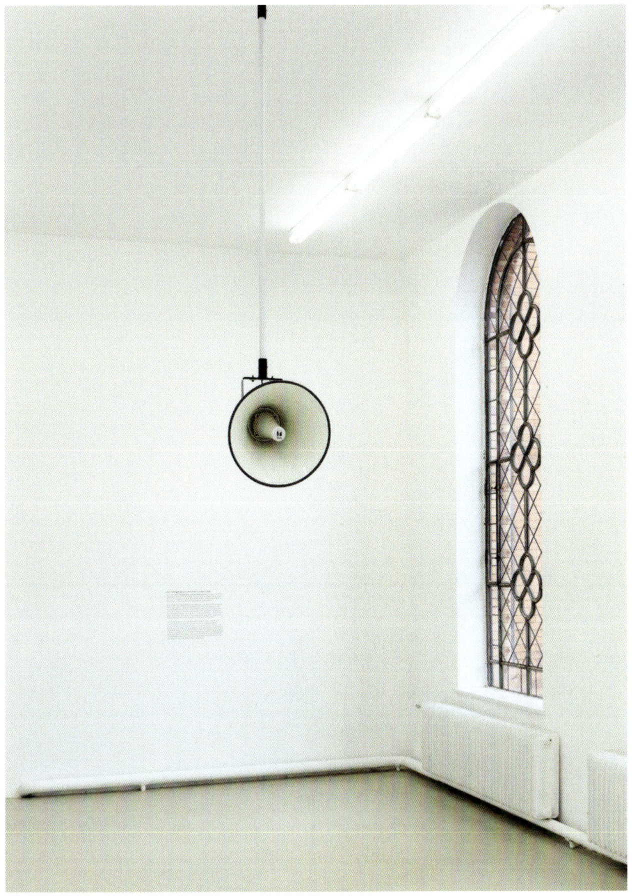

Susan Philipszs, *War Damaged Musical Instruments (shofar)*, 2016.
Installation view Kunstverein Hannover.

Susan Philipsz's series *War Damaged Musical Instruments* features the photograph of a damaged shofar, an ancient Jewish instrument made out of a ram's horn used for religious purposes. The speakers play the sounds that Philipsz produced with the crushed shofar, which belonged to a Jewish family from Hanover that had to leave it behind when they fled from persecution in the 1930s.

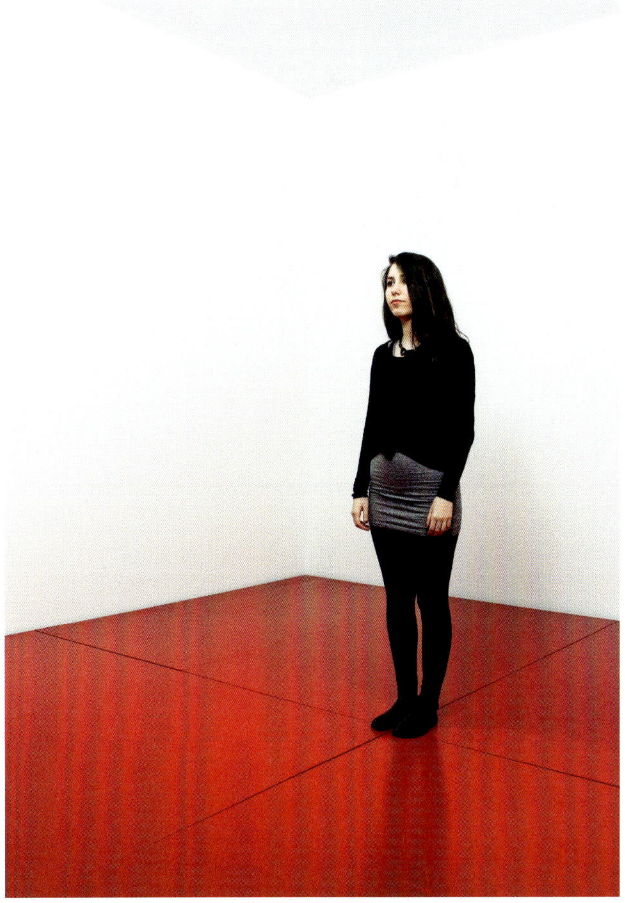

Hanna Hartman, *Acoustic Catacombs*, 2010.
Sound installation.
Floor painted red, loudspeakers, multi-channel audio.

The installation *Acoustic Catacombs* was created in 2010 for a specific room in the Musée Réattu in Arles, France, and was reconstructed in 2012 for the ZKM. Hartman conveys a musical idea of what could happen in the catacombs. Visitors can follow these acoustic visions with their ears and body while standing or lying on the red floor.

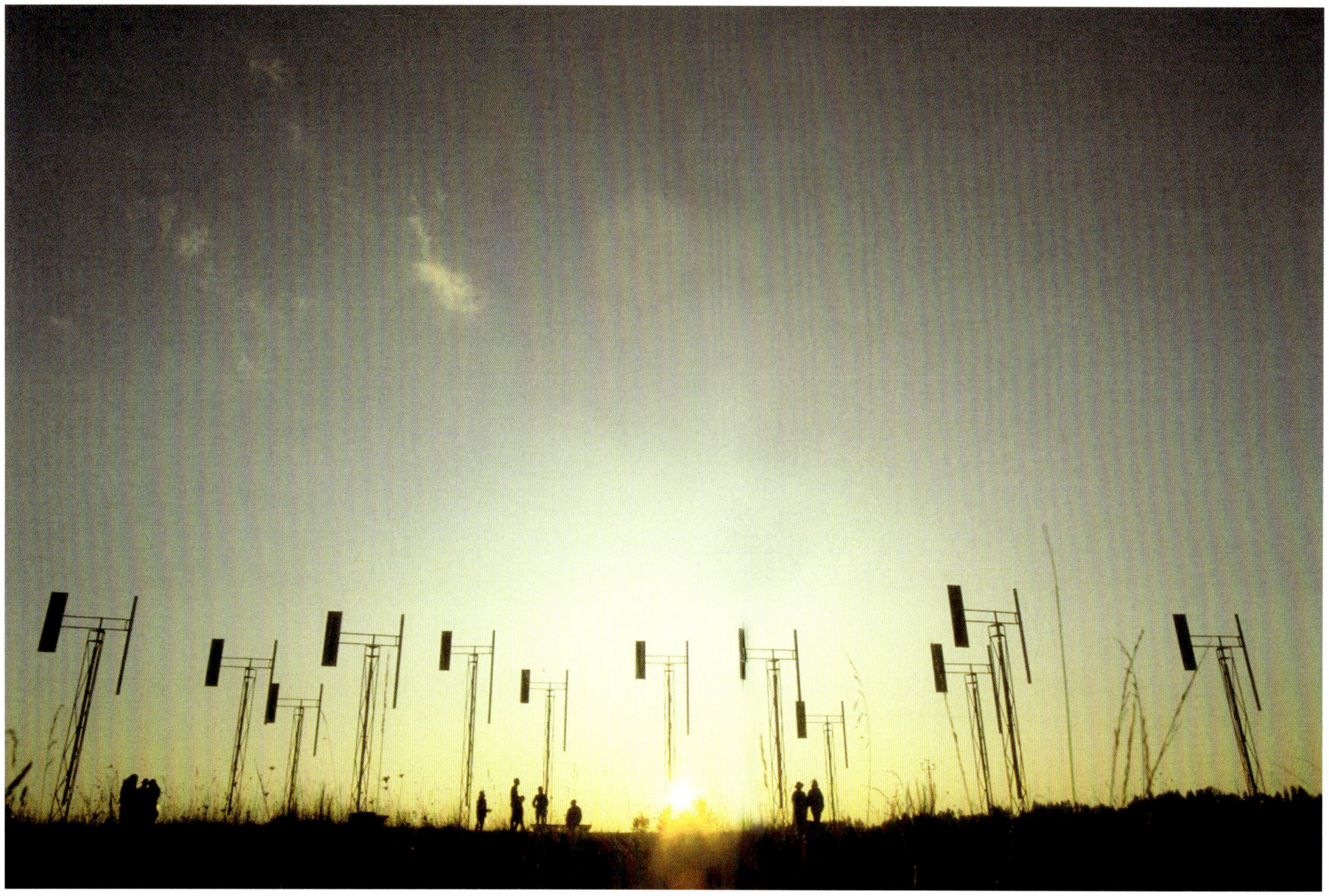

Doug Hollis, *A Sound Garden*, ca. 1993.
Installation in Magnuson Park, Seattle, Washington.

Located on a hill overlooking Lake Washington in Seattle, the giant pipe-like structures murmur, whistle, and howl when the wind blows through them.

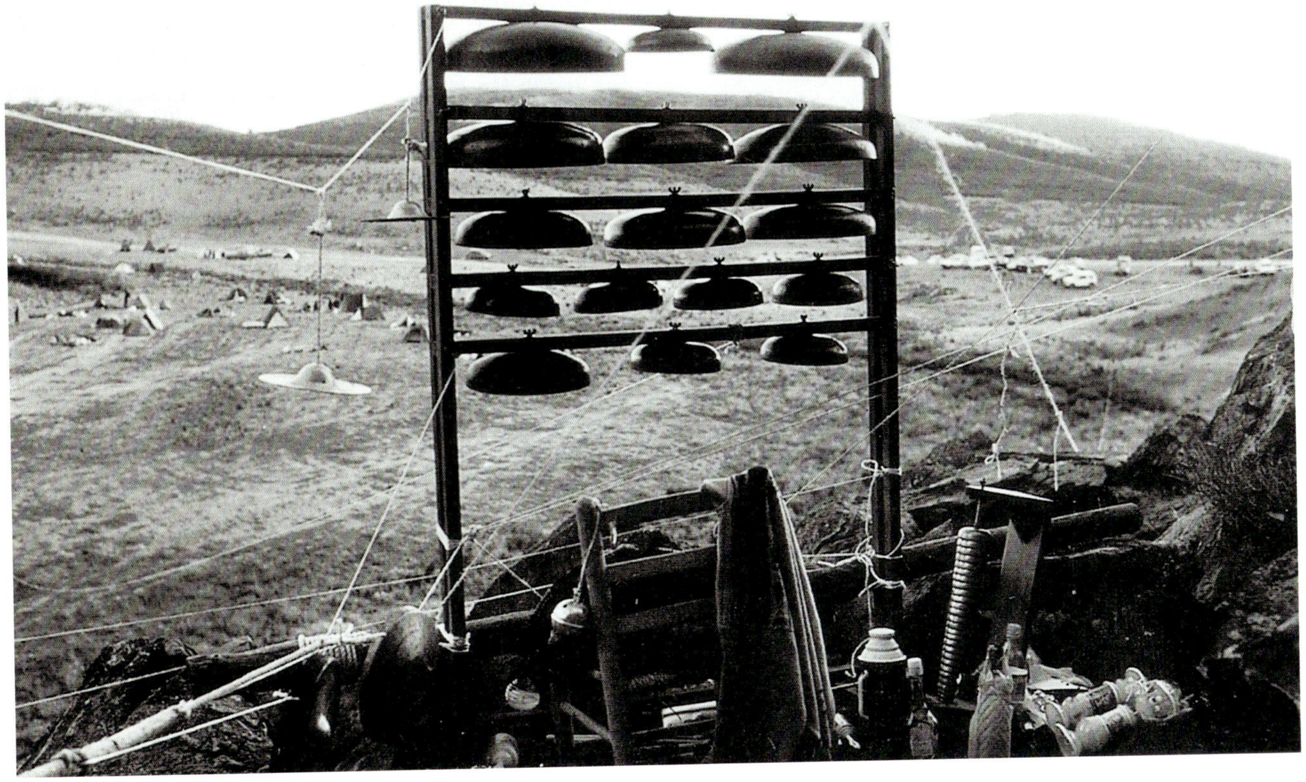

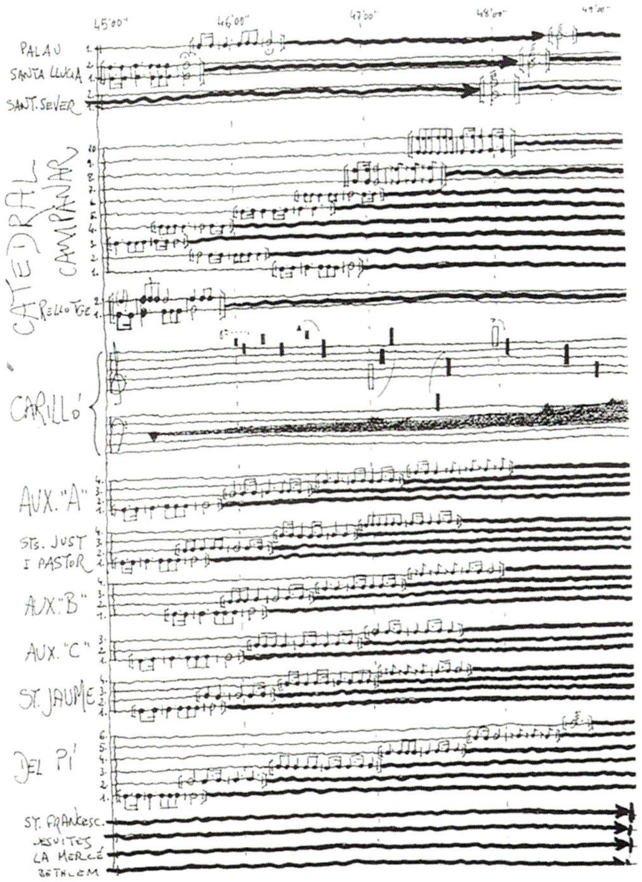

Llorenç Barber, *De Sol A Sol*, 1995, EL Arreciado.

Llorenç Barber is very interested in one of the most ancient sounds of humankind: the sound of bells. For *De Sol à Sol* [From sun to sun] he installed "flying" bells on iron cables strung from one mountain to the other in El Arreciado, Spain. The clappers of the "flying" bells hung down, and when they swung to and fro, a Doppler effect was created with the rocks of the mountain acting as a front wall. Additionally, Barber played his portable bells mounted on a frame.

Sound Towers, Sound Pavilions, Sound Domes

Ever since Russolo's manifesto, loudspeakers have been a source of fascination for musicians and architects. In the 1920s, Herbert Bayer conceived a Bauhaus pavilion with loudspeakers, and in Russia, Gustav Klutsis designed a loudspeaker car for the railroad. Radio speakers laid the foundations for the success of pop music: Without transistor radios, without mobile speakers, the triumph of rock and roll in the 1950s would never have happened. Without the near-mythical Marshall amplifier (1962), there would have been no guitar heroes and no arena rock. Without the electric amplifier, guitars would only be audible in small concert halls or clubs. Only with amplification did the guitar become the central instrument of the rock era. Stacks of amplifiers formed walls of sound.[30] New-music composers proved to be equally fascinated: Karlheinz Stockhausen's 1955 concerto *Gesang der Jünglinge* [Song of the youths] premiered in 1956, consisted of just five groups of speakers distributed around the Westdeutscher Rundfunk (WDR) concert hall, from which the sounds emanated.(FIG. 18)

Mobile loudspeakers also made possible the first attempts at new sonic spaces. Before rooms and pavilions, sound towers were constructed. For the Feria de América, an industrial exposition held in Mendoza, Argentina under the dictator Juan Perón from January to April 1954, the architects César Jannello and Gerardo Clusellas built a fifty-meter-high tower featuring both light and sound: the *Torre Alegórica* [Allegorical tower].(FIG. 19)

Built of steel scaffolding, the tower took up the logo of the exposition with its ten illuminated, pyramid-shaped modules, with both tower and logo displaying variations of two pyramids reflected point to point. While the pentahedrons mounted on the tower's frame could be lit up in red and white (the exposition's colors), the logo was reduced to the essentials: two opposed equilateral triangles. It was designed by Tomás Maldonado, who would later become the dean of the Hochschule für Gestaltung in Ulm (1964–1966).(FIG. 20) From 1951 to 1954, Maldonado was

30
The phrase "wall of sound" derives in part from the music producer Phil Spector, who was known for the dense aural textures of the songs he produced at Hollywood's Gold Star Studios beginning in 1962: Essentially, Spector used studio technology itself as an instrument. Later, the Grateful Dead became famous for their own "Wall of Sound." The sound system used by the band on their 1974 tour took its name from the wall of loudspeakers that dominated the back of the stage and produced a correspondingly thunderous sound.

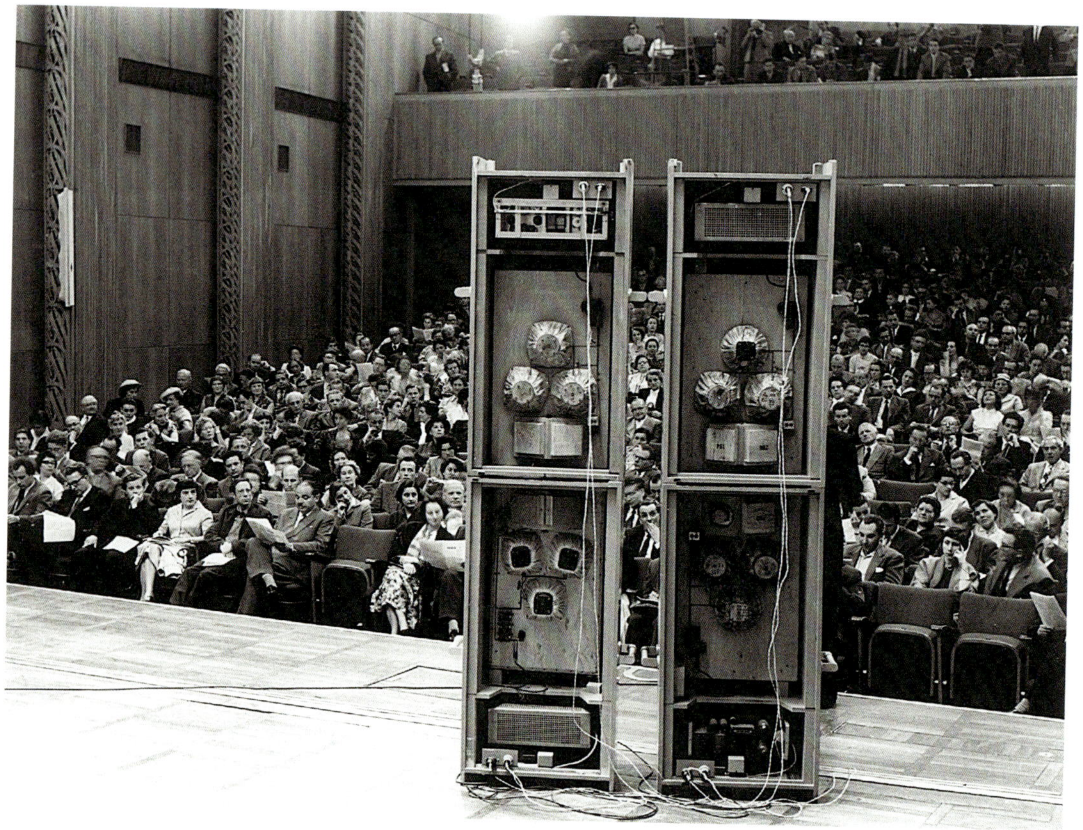

FIG. 18 Karlheinz Stockhausen, *Gesang der Jünglinge* [Song of the Youths], 1955–1956. This photograph was taken at the premiere of the piece on May 30, 1956 in Cologne.

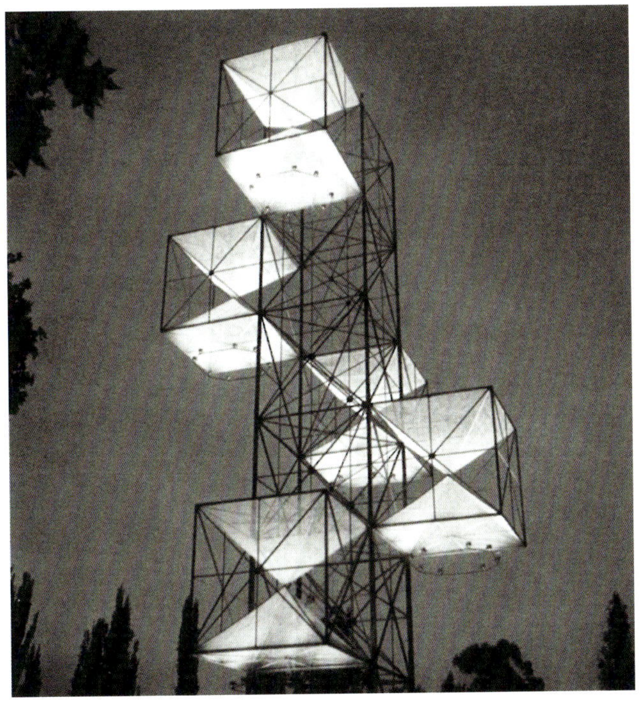

FIG. 19 César Jannello and Gerardo Clusellas, *Torre Alegórica*, 1953/1954.

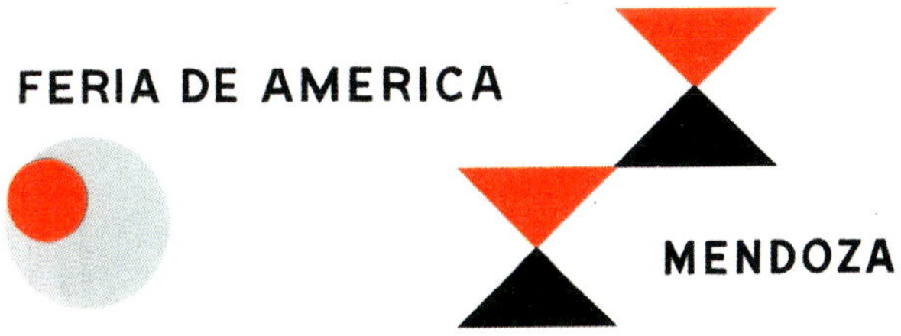

FIG. 20 Tomás Maldonado, Logo design for Feria de América, 1954.

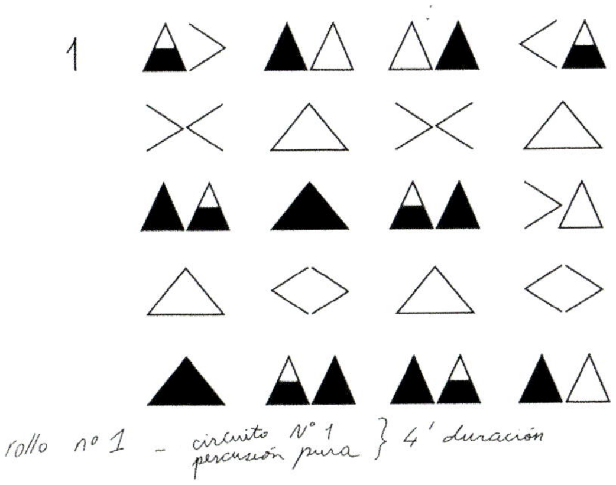

FIG. 21 Mauricio Kagel, Detail of the lighting plan for the five double tetrahedron of the *Torre Alegórica*, 1953/1954.

principal editor of the first four of the nine issues of *nueva visión*, a journal which also counted the composer Mauricio Kagel among its editors. Kagel was the creator of *Música para una torre* (1953–54), a forty-five-minute show of light and music centering on the allegorical tower that was presented for four hours every night starting at dusk.(FIG. 21) The sound was emitted from a total of twenty-four loudspeaker towers arranged around the central tower at a distance of a hundred meters.

A similar gigantomania was displayed by a fifty-meter "spatiodynamic tower": Nicolas Schöffer's *Tour spatiodynamique et cybernétique* (1954–55), designed for an international building exposition at the park in Saint-Cloud, Paris, with music composed by Pierre Henry.(FIG. 22) The homeostatic control of the movable visual and aural elements, which responded to changes in light, temperature, and sound, was the creation of the electrical engineer Jacques Bureau. Schöffer had been creating his "spatiodynamic sculptures" since 1948. These structures generally consisted of a metal frame with loudspeakers mounted on it, and some were kinetic and had light elements like the *Tour spatiodynamique et cybernétique*.

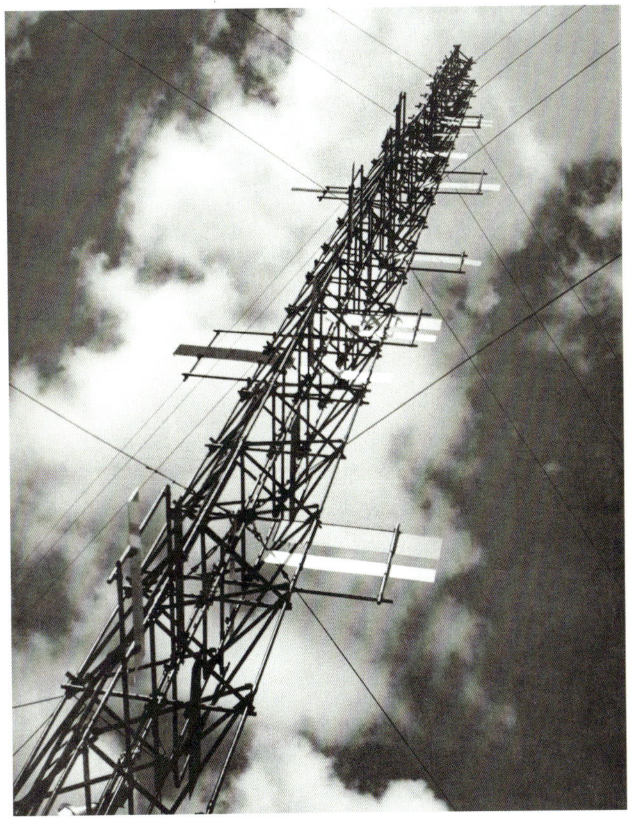

FIG. 22 Nicolas Schöffer, *Tour spatiodynamique et cybernétique*, 1954/1955.

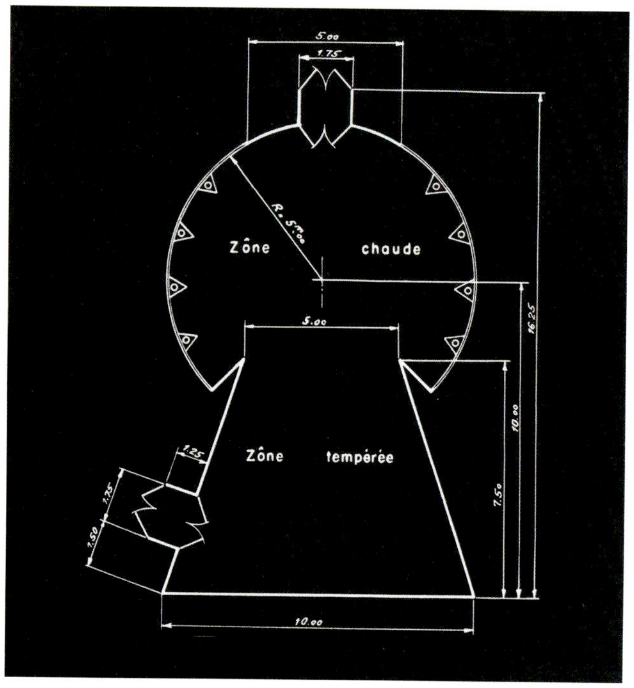

FIG. 23 Nicolas Schöffer, *Trou de serrure*, 1957. Sketch.

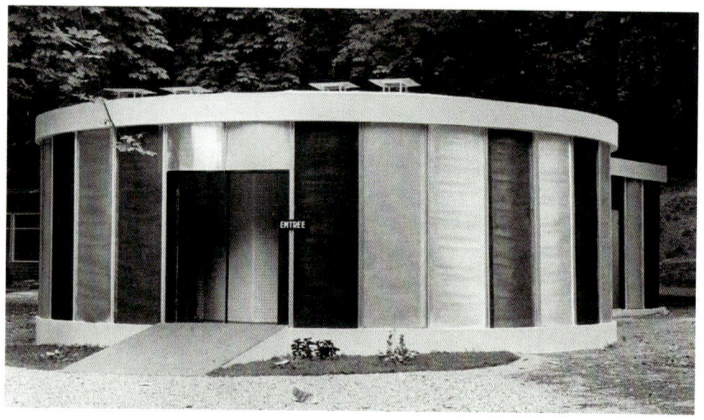

FIG. 24 Nicolas Schöffer, *Trou de serrure*, 1957.

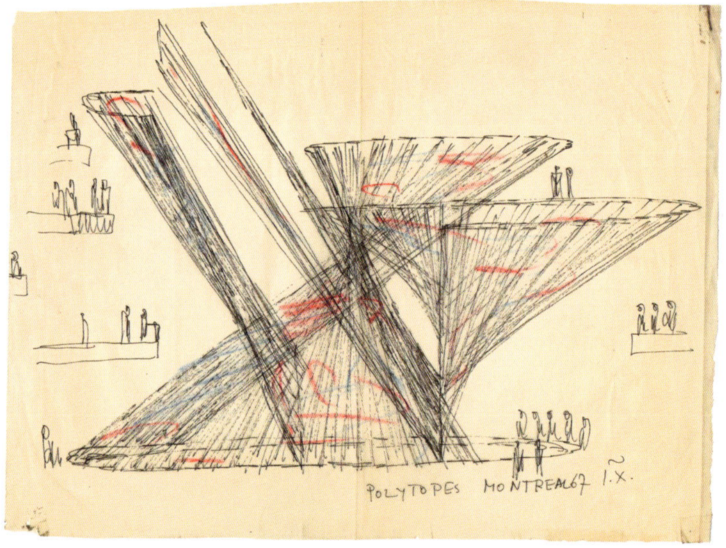

FIG. 25 Iannis Xenakis, *Polytope de Montréal*, 1967. Sketch.

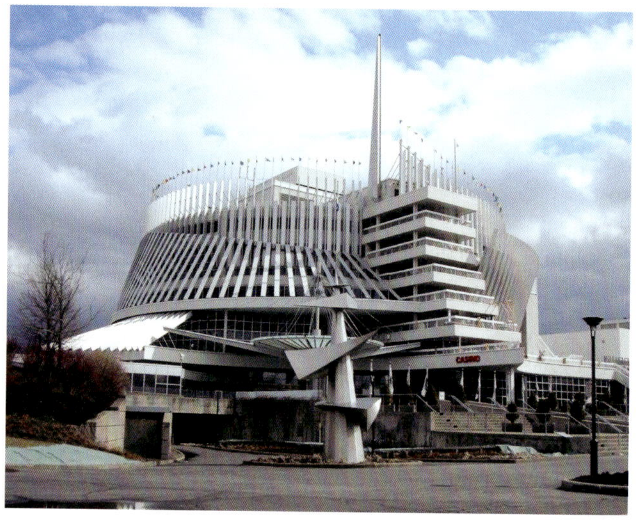

FIG. 26 Iannis Xenakis, *Polytope de Montréal*, 1967.

In Schöffer's own words: "The addition of tone and sound constitutes a further means of extending one's control over space."[31] In 1957, Schöffer exhibited a small sound pavilion in the same park, with sounds by Henry and spatial and aural furnishings provided by the Saint-Gobain and Philips corporations. It was called *Trou de serrure (à cloisons invisibles)*: the house with invisible partitions. Built of glass and aluminum, it had a keyhole-shaped floor plan and consisted of two zones: The round sound zone was red, warm, and loud, while the trapezoidal zone was blue, cool, and quiet.(FIGS. 23, 24)

Xenakis grasped the concept of the sound pavilion and its necessity, and consequently he followed the the Philips Pavilion with a number of other specialized sound pavilions, such as the *Polytope de Montréal*, an installation in the French pavilion at the 1967 World's Fair that featured a light show and a Xenakis composition.(FIGS. 25, 26) His *Diatope*, built in 1978 in front of the Centre Georges Pompidou, Paris, has since been torn down. (FIGS. 27–29)

The UPIC project (Unité Polyagogique Informatique de CEMAMu), which he founded at the Centre d'Etudes de Mathématique et Automatique Musicales circa 1977, was a pioneering attempt to bring musical graphics into the digital age – for example, with styli that produced graphic elements directly on an electronic tablet, which were transformed into tones by the computer in real time and could also be viewed on a screen. This approach pointed the right way forward, for in the future we will use the topological methods of mathematics to create new music.

31
Nicolas Schöffer, "Die drei Entwicklungsstufen der dynamischen Skulptur," in *Europa. Das Jahrhundert der Avantgarde in Mittel- und Osteuropa*, ed. Ryszard Stanislawski and Christoph Brockhaus, vol. 3, *Dokumente* (Bonn: Kunst- und Ausstellungshalle der Bundesrepublik Deutschland, 1994), 112–13, here 112. Translated from the German.

32
Hermann Scherchen, "Was müsste ein ideales Tonstudio leisten?" in Iannis Xenakis, *Kontrolle und Zufall*, ed. Angela Lammert and Hubertus von Amelunxen (Berlin: Akademie der Künste, 2011), 34–35. Originally published in *Werke und Briefe*, ed. Joachim Lucchesi, vol. 1, *Schriften 1* (Berlin: Peter Lang, 1991), 147–48; undated typescript, Hermann-Scherchen-Archiv, 1348, Akademie der Künste, Berlin. Translated from the German.

33
Xenakis to Scherchen, September 9, 1961, in *Xenakis*, by Nouritza Matossian, (London: Kahn & Averill, 1986), 157.

The conductor Hermann Scherchen, who moved to the town of Gravesano in the Swiss canton of Ticino in the early 1950s, opened an experimental studio there in 1954, where he explored the question of what an ideal sound studio would need to provide.(FIG. 30) His answer was as follows: "In 1954, when I discussed my intention to build such a studio with leading practitioners such as Pierre Schaeffer (RTF Paris), everyone said, 'You should just be content to build a classic 'ideal' studio.' To which I always had to reply: 'That's exactly what I want to avoid, because every classic ideal studio is an individual entity that inherently deforms the sound. …
1. The studio should accommodate the sound but not color it (i.e., not deform it with its echo); 2. the usual acoustical defects of a space – the distortion of the sound by standing waves, etc. – should be eliminated to the extent possible; 3. the intrusion of noises from the surrounding environment should be equivalent to zero.[32]

The Gravesano studio would become a venue for annual conferences and a concert hall for young composers such as Xenakis, whose controversial essay "La crise de la musique serielle" was published in 1955 in the first issue of Scherchen's *Gravesaner Blätter*, a quarterly journal for "liminal problems in music, electronics, and acoustics." In 1961, Scherchen commissioned Xenakis to design a concert space that would further develop the ideas found in his Philips Pavilion (FIGS. 31, 32), and which ultimately anticipated the utopian designs of his *Ville cosmique* (1964) and *Cité de la Musique* (1984) (FIG. 33). "It is made of concrete in parabolic hyperboloids like the Philips Pavilion but more composed and interesting. I don't think that such a new and harmonious form exists anywhere. I am happy. I hope it will please you."[33]

Another example is the spherical auditorium designed by Stockhausen, Boris Blacher, and Fritz Bornemann and built as the West German pavilion for the 1970 World's Fair in

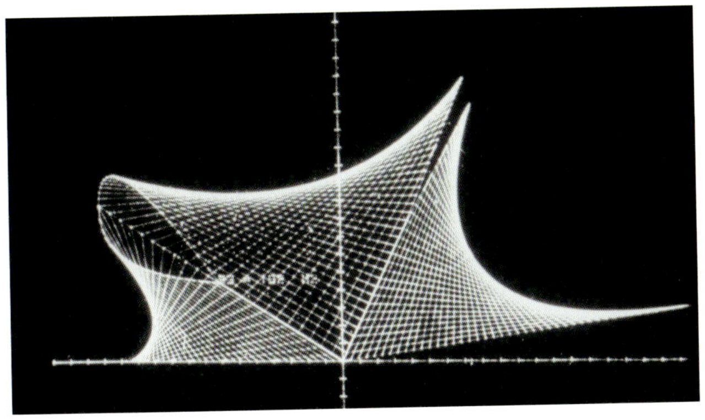

FIG. 27 Iannis Xenakis, *Diatope*, ca. 1978. Digital modell.

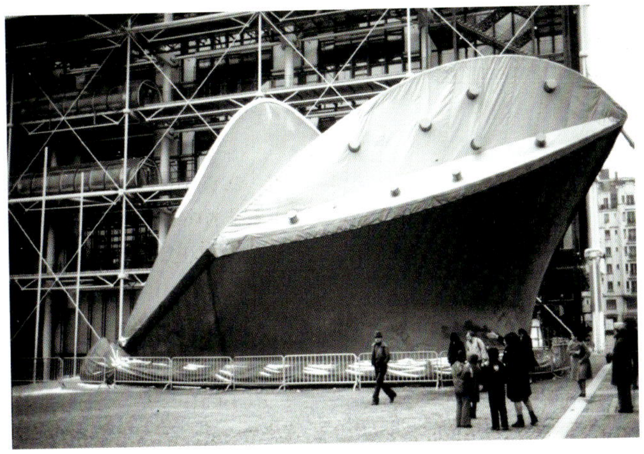

FIG. 28 Iannis Xenakis, *Diatope*, 1978. Exterior view.

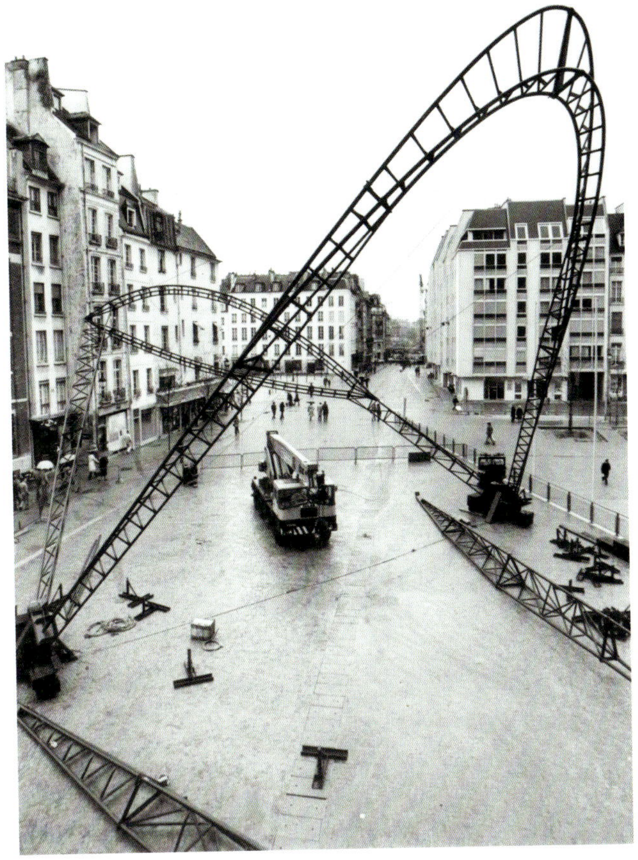

FIG. 29 Iannis Xenakis, *Diatope*, 1978. Construction of the work in front of the Centre Pompidou, Paris.

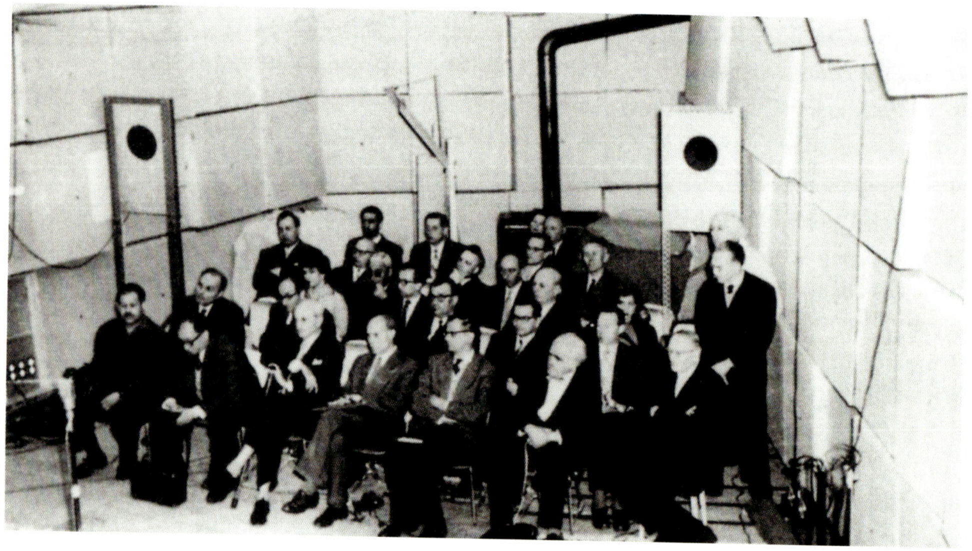

FIG. 30 At the experimental studio of Hermann Scherchen, Gravesano, during a conference in August 1962.

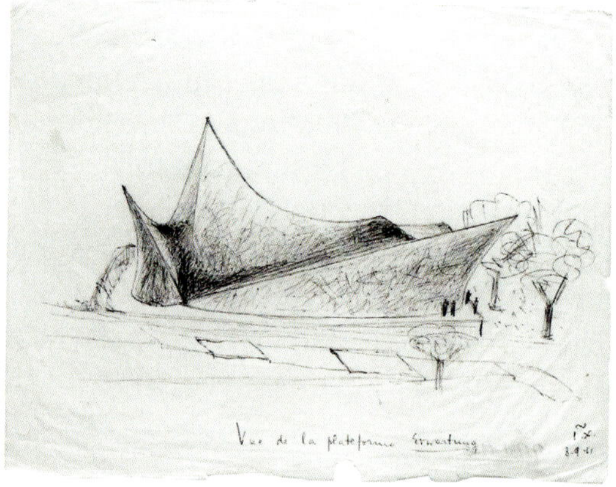

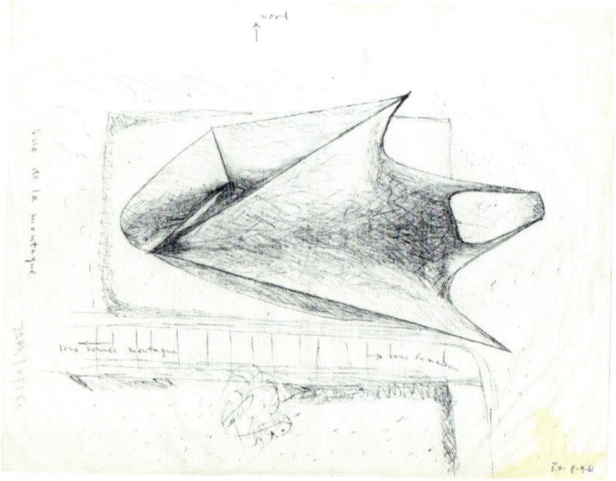

FIGS. 31, 32 Iannis Xenakis, Sketches for Hermann Scherchen's experimental studio in Gravesano, Switzerland, 1961.

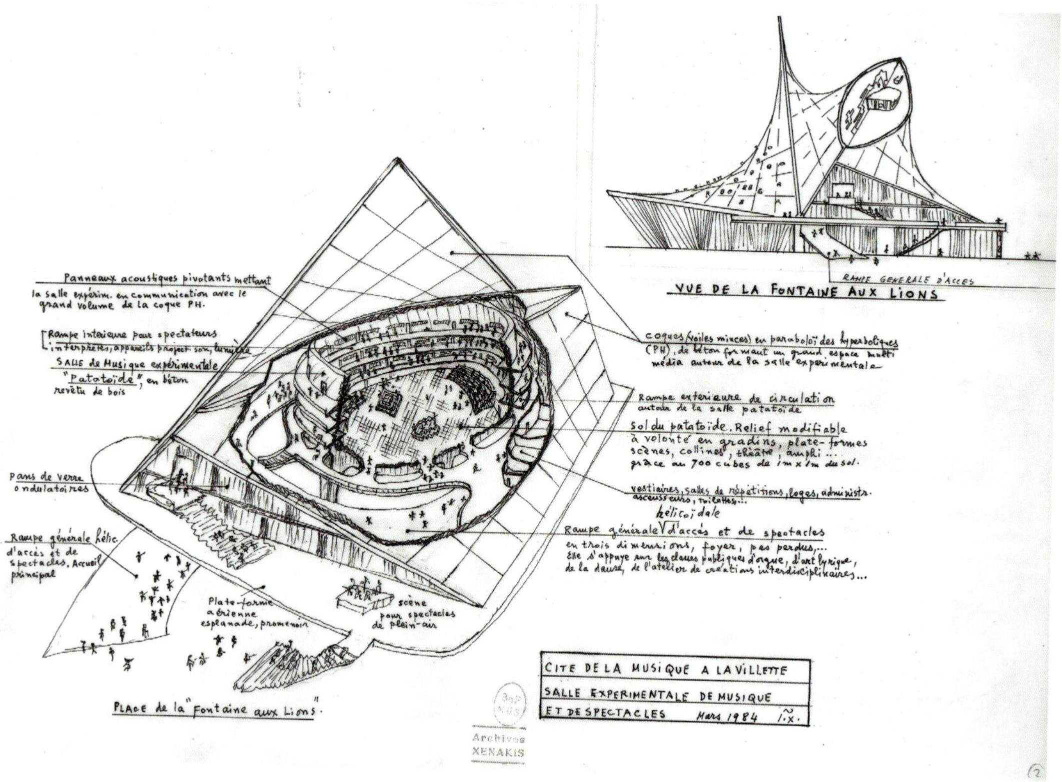

FIG. 33 Iannis Xenakis, Design for Cité de la Musique for La Villette, Paris, 1980s. The project was never realized.

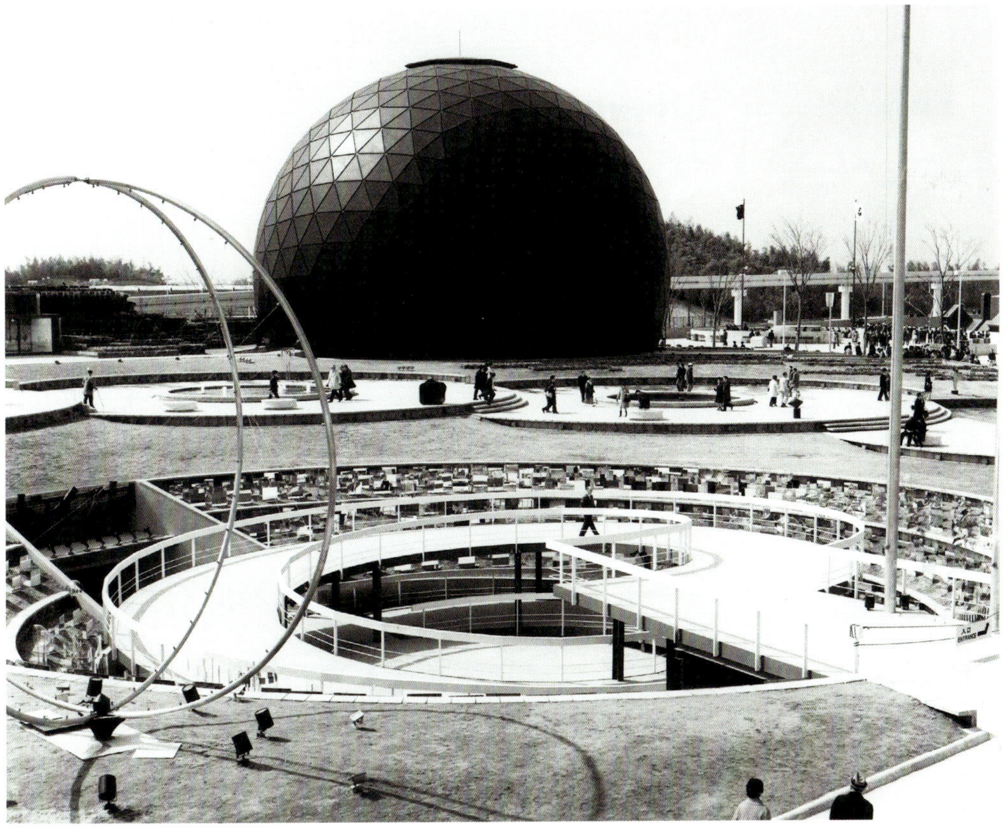

FIG. 34 Karlheinz Stockhausen, Boris Blacher, and Fritz Bornemann (architect), Spherical Concert Hall, West German Pavilion at Expo '70 in Osaka, Japan. Exterior view.

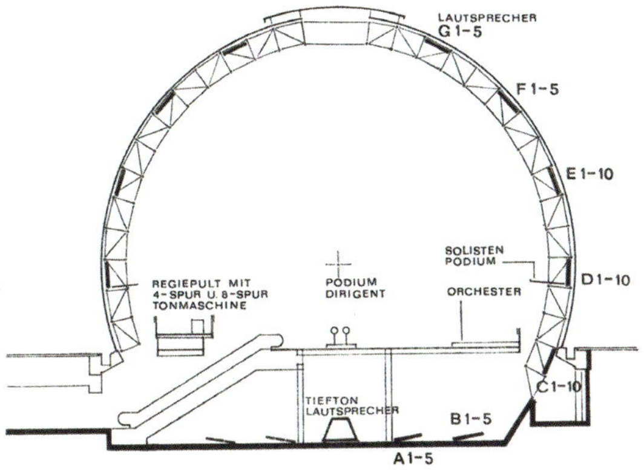

FIG. 35 Karlheinz Stockhausen, Boris Blacher, and Fritz Bornemann, Drawing for the West German Pavilion for the Expo '70 in Osaka, Japan. It was the world's first spherical concert hall.

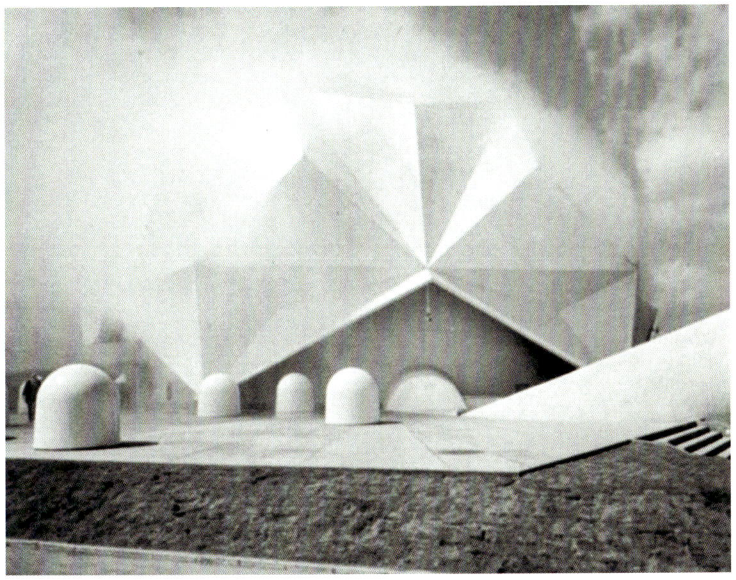

FIG. 36 E.A.T., Pepsi Pavilion at Expo '70 in Osaka, Japan. Exterior view. Architect: John Pearce. Project coordinators: Robert Breer, Billy Klüver. Fog sculpture: Fujiko Nakaya. White sculptures: *Floats* by Robert Breer.

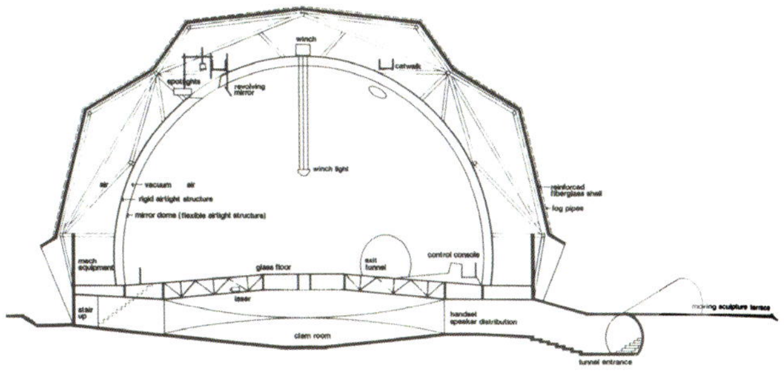

FIG. 37 E.A.T., Drawing for the Pepsi Pavilion for Expo '70 in Osaka, Japan.

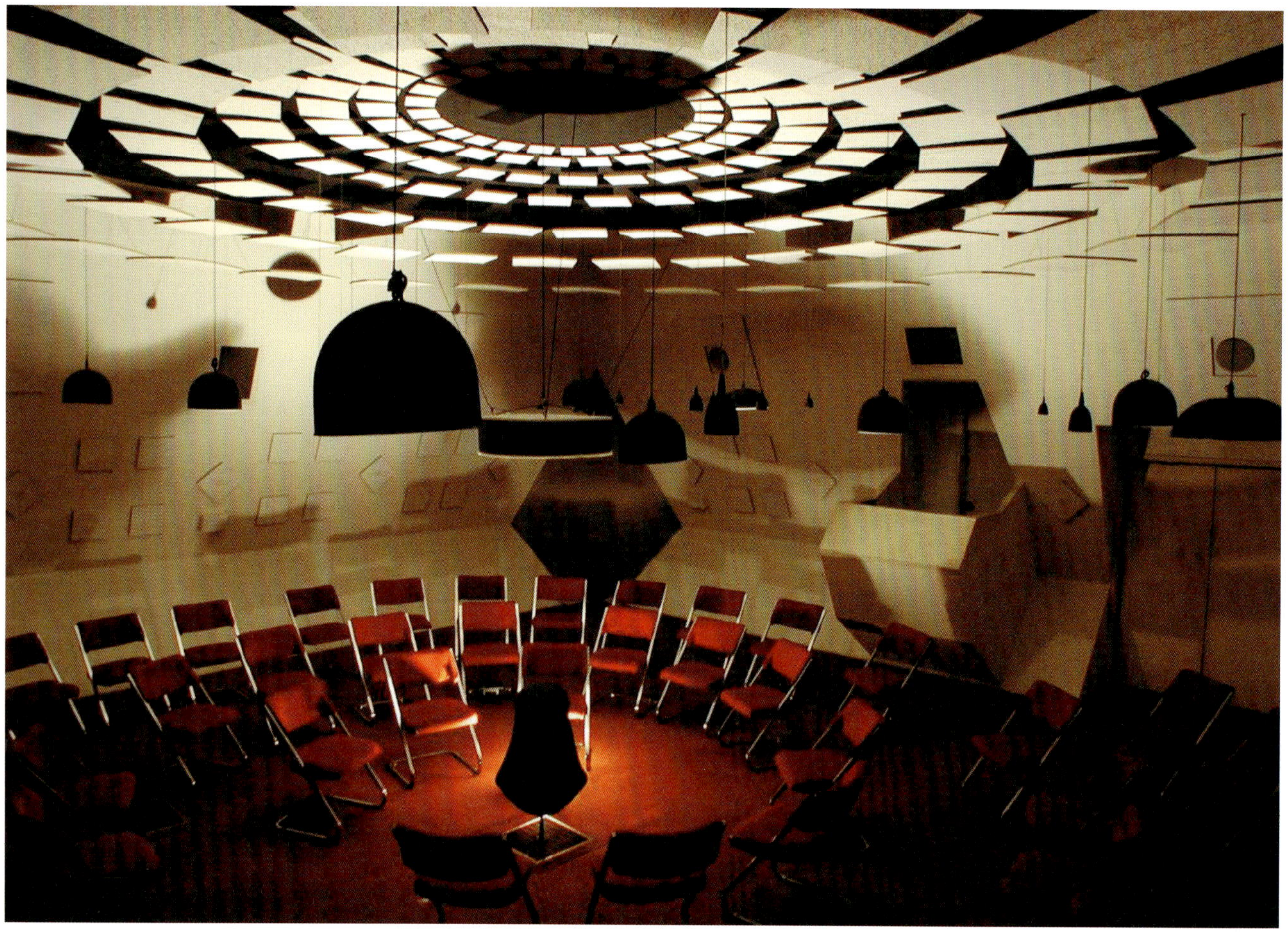

FIG. 38 Stan Shaff and Doug McEachern, Audium, San Francisco, 1967/1975. Interior view.

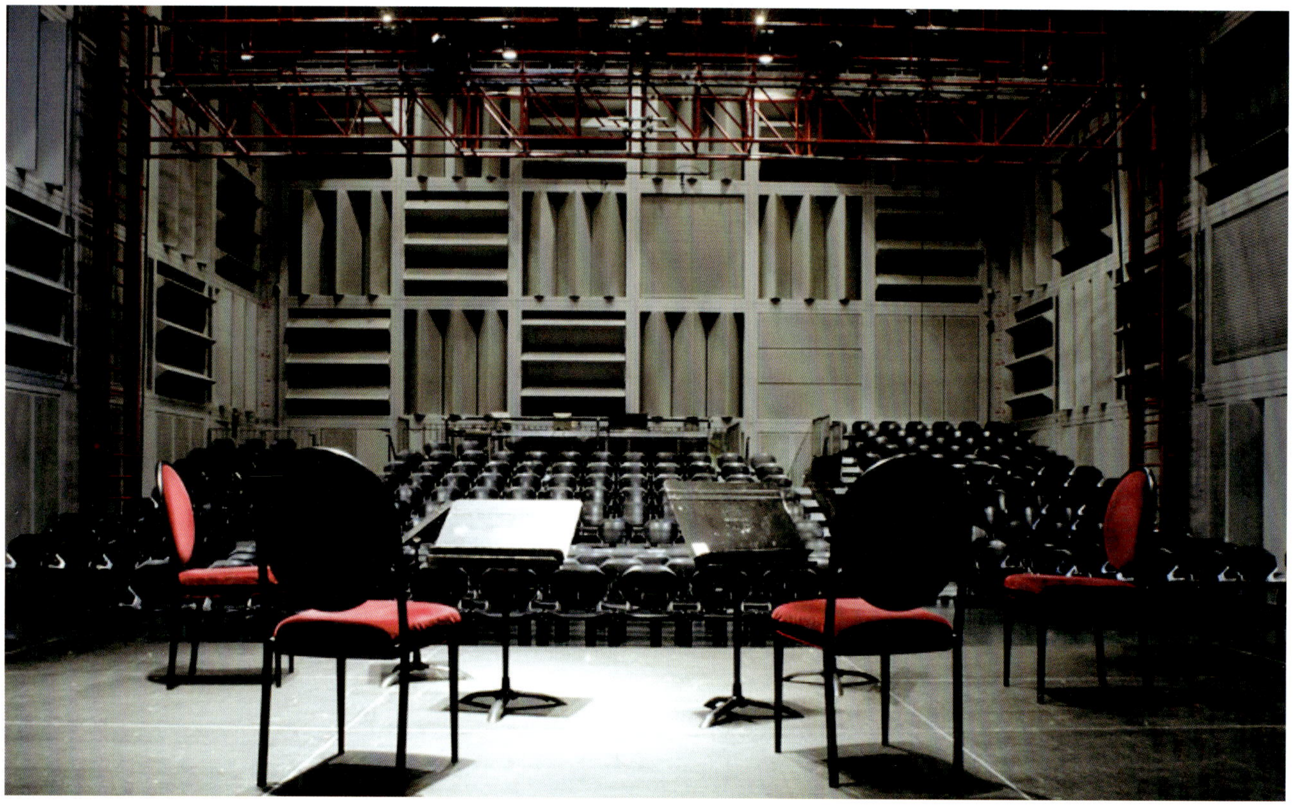

FIG. 39 V. M. A. Peutz, Espace de projection at IRCAM, Centre Pompidou, Paris, 1978. Interior view.

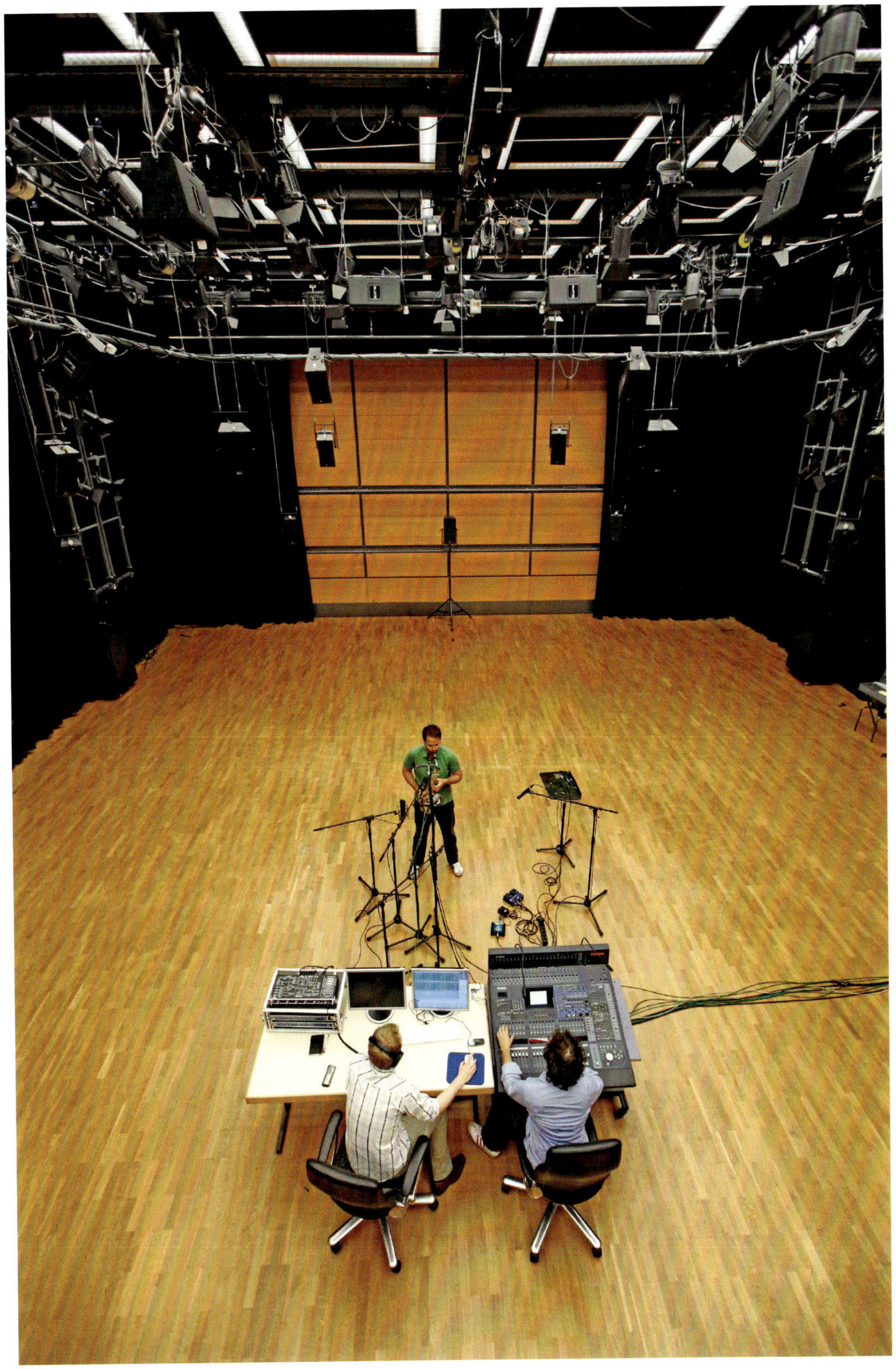
FIG. 40 ZKM Sound Dome, 2003–2006. Interior view.

Osaka. Looking at the working drawings, we can see that that the attempt was being made in 1970 to distribute sound in space, using the mechanical resources of that era (primarily tape players).(FIGS. 34, 35)

A defining characteristic of the 1960s, then, was the musical avant-garde's quest for spatial sound. The experiments Bernhard Leitner conducted in New York after moving there in 1968 were another key component of this movement.[34] More and more, music was changing from a time-based art form to a space-based art, a spatial art. This transformation also fostered collaborations between musicians and visual artists based on new technologies. The 1970 World's Fair was also the site of the famous Pepsi Pavilion designed by the group E.A.T. (Experiments in Art and Technology), which included Billy Klüver, Robert Rauschenberg, Fred Waldhauer, Robert Whitman, Forrest Myers, David Tudor, and Robert Breer – in other words, both artists and engineers.(FIGS. 36, 37) Sound pavilions were an attempt to advance sound art – sound as a medium of art – using the methods of visual art. "Klüver's ambition was to create a laboratory environment, encouraging 'live programming' that offered opportunity for experimentation, rather than resort to fixed or 'dead programming,' as he called it, typical of most exposition pavilions." Visitors were invited "to individually and collectively participate in the experience rather than view the work as a fixed narrative of pre-programmed events."[35]

The music industry fought the evolution of this sonic art, blocking the path to the future. Only a part of the movement has been able to survive in museums as sound art. None of the pavilions mentioned above is still standing today.

San Francisco's Audium, on the other hand – built in 1967 by Stan Shaff and Doug McEachern – does still exist, although it was relocated in 1975.(FIG. 38) When music is played in it, the light is extinguished, so that one listens to the sound alone in darkness. The *Espace de projection* at the Centre Pompidou's Institut de recherche et coordination acoustique/musique (IRCAM), designed by V. M. A. Peutz and opened in 1978, still exists as well.(FIG. 39) The space that has been realized there is completely soundproof: What one hears truly is the pure sound and nothing else. For the ZKM Sound Dome in Karlsruhe

34 See Bernhard Leitner, *Sound:Space* (Ostfildern: Cantz, 1998).

35 Randall Packer, "The Pepsi Pavilion: Laboratory for Social Experimentation," in *Future Cinema: The Cinematic Imaginary after Film*, ed. Jeffrey Shaw and Peter Weibel (Cambridge, MA: MIT Press, 2003), 144–49, here 146.

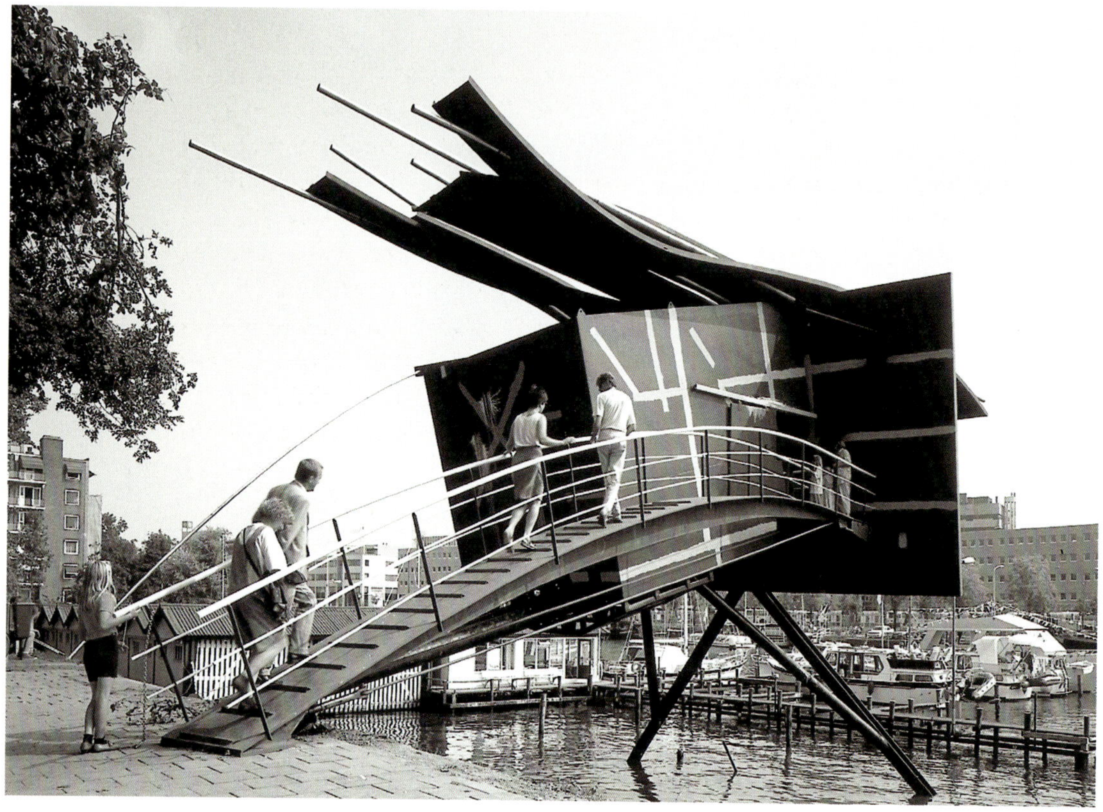

FIG. 41 Coop Himmelb(l)au, *Video Clip Folly*, 1990, Groningen, Netherlands. Exterior view of the video pavilion.

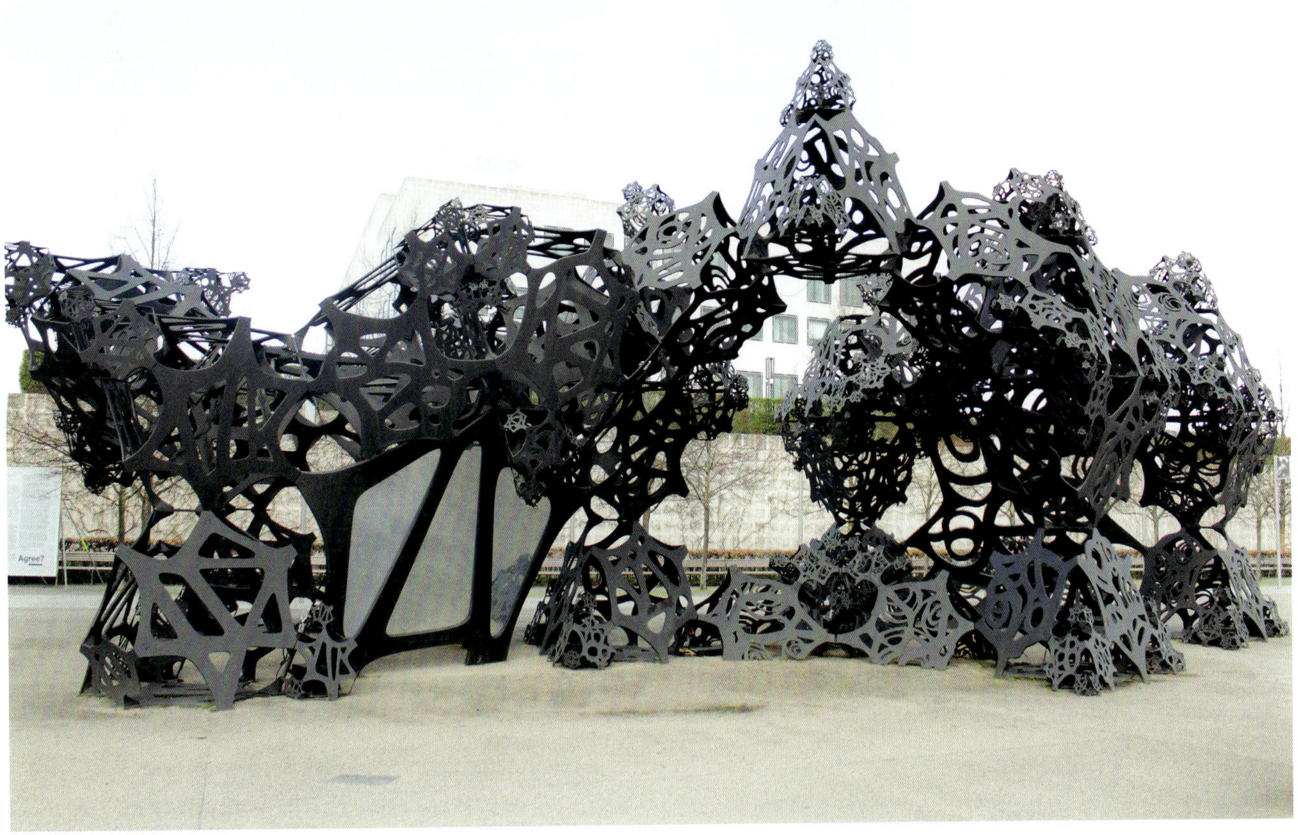

FIG. 42 Matthew Ritchie with Aranda\Lasch and Arup AGU, *The Morning Line*, 2008. Installation view Forecourt of ZKM | Karlsruhe, 2013.

(2005–2006) (FIG. 40), the ZKM | Center for Art and Media Karlsruhe began developing software quite early on under the direction of Ludger Brümmer, with assistance from Götz Dipper, Joachim Goßmann, Chandrasekhar Ramakrishnan, and Bernhard Sturm. Today this software, known as Zirkonium, is used all over the world, even in Hollywood. The rise and spread of music videos in the 1980s and 1990s inspired new hopes for the development of sound pavilions: In 1990, a number of pavilions designed by prominent architects such as Coop Himmelb(l)au were built in the Dutch city of Groningen for an exhibition of music videos. (FIG. 41) In the twenty-first century, however, it seems as though the ZKM | Karlsruhe may be alone in its long-term dedication to the topos of spatial sound.

Hence the great importance of the ZKM's 2013 acquisition of the sound pavilion *The Morning Line* (2008) as a permanent loan from the Thyssen-Bornemisza Art Contemporary Foundation (TBA21) at the instigation of Francesca von Habsburg. (FIG. 42) The accession highlighted the fact that the ZKM, unlike the rest of the world's museums, attaches importance not just to the mediums of image, object, and so on, but also to the medium of sound. The ZKM is the only museum in Germany that is home to a laboratory for sound research.

The Morning Line, created by Matthew Ritchie with Aranda\Lasch, Daniel Bosia, and Arup AGU, will likely remain the only sound dome to be preserved, since it now stands in front of the ZKM with support from the city of Karlsruhe and the state of Baden-Württemberg. It has been shown at the third Bienal Internacional de Arte Contemporáneo de Sevilla in 2008 (directed by Peter Weibel), in Istanbul during the city's tenure as the European Capital of Culture in 2010, and at the Festival for Spatial Sound and Advanced Music Composition in Vienna in 2011–12. For the installation *TBA21 Sound Space*, the English sound artist Tony Myatt took on the responsibility of programming, particularly the programming of the sound and acoustical structure. Thirty compositions have been created specially for the pavilion by artists including Christian Fennesz, Terre Thaemlitz, Lee Ranaldo, Cevdet Erek, Carl Michael von Hausswolff, Jana Winderen, Florian Hecker, and Carsten Nicolai.

Sonic Art

The need to represent industrial, electromechanical society in the musical instruments themselves – that is, to generate a music appropriate to that society, which also necessitated the inclusion of industrial noise and the emancipation of noises in general – brought about a transformation of musical instruments, from hand tools made of wood and brass to magnetic tapes, electronic synthesizers, and, ultimately, the computer.

With the development of the synthesizer around 1960, it became possible to generate tones electronically by means of sound synthesis. The world of sound evolved from the world of natural reality to a purely technological, artificial world, opening up a totally new aural space. With electronic circuits, it also became possible to control the movement of sound in space. The result was spatial music, ambient music. The computer enabled not just new forms of sound, but also new forms of composition. Temporal and sonic deformations and distortions, multichannel memories and repetitions, modulations and the mathematics of stochastic and aleatory processes revealed the field of acousmatic music, which had already been opened up by La Monte Young with his new and unfamiliar frequency relationships. The synthesizer and computer ushered in the age of technosound.

Beginning with tape machines, and even more with electronic music, the role of score, performer, and composer became problematized. With spliced-together tapes (on which some of the tones and sounds were not composed, but rather gathered from the environment or sourced from other sound producers), which, once the "composition" process was finished, were played back by machines – sometimes including live transmissions from multiple radios (frequency shifts) – neither a traditional score nor a performer was necessary. Even the composer, in the classical sense of the author as creator of all his or her material, had disappeared. Of course, working directly with all kinds of sounds, without a score or performer(s), was what had made electronic music so fascinating to Varèse as a way of "composing with living sounds." The technological possibilities achieved for music in the age of repetition – that is, its infinite repeatability and ubiquitous, mobile performability – have resulted in a total temporal and spatial expansion of music's sonic space, now that all sounds have become virtually accessible, performable, and producible everywhere and always, in all places and at all times.

One can now take advantage of the freedom attained through the technological transformation of the musical instrument, the possibilities liberated by the technological and conceptual transformation of music into "organized sound": One can yield to it and speed it on its way, or one can recapitulate it, by representing it in the work and in the compositional technique itself. There are thus two tendencies: The first is represented by Russolo, Satie, Varèse, Arthur Honegger (*Pacific 231*, premiered in 1924), George Antheil (*Ballet mecanique*, 1926), John Cage, and others. Exponents of the second include Pierre Boulez, Stockhausen, Luciano Berio, Henri Pousseur, Xenakis, Terry Riley, and Steve Reich. Naturally, there are musicians who move between the two directions (Young, Kagel, Dieter Schnebel, Henry Cowell, Otto M. Zykan), and composers have occasionally switched sides as well.

The goal of the Boulez-Stockhausen school, one might say, is to use electronic instruments themselves to thematize, through music, the structures produced by the technological society. Serial composition, gradual shifts, monotony, repetition, minimalization (of differences) – how strongly these techniques of musical composition resemble the commercial production techniques found in factories, the rhythms of the world of work (assembly line, serial production, flexible specialization). Through these compositional techniques that echo technological instruments and the technological world, the musician regains control of the material. However, this "ground control" imposed on the attempts of liberated sound to fly free is in fact a vestige of classical composition. The creator of all sounds is transformed into the master of all sounding materials. Yet this compositional position fails to take account of the possibilities opened up by music's technological transformation, the new dimensions of freedom.

The vanishing of classical musical parameters is wonderfully expressed in Earle Brown's *December 1952*. For this composition, Brown created a score that specified neither the notes nor their durations, nor the instruments to be played, thus leaving the length of the piece undefined as well. Not only did such nonspecific, undetermined compositions (indeterminacy being one new dimension of freedom) by the Cagean gang of four – Morton Feldman, Brown, Christian Wolff, and David Tudor – lay the foundations for the brilliant eruptions of Fluxus music (from Young and Riley to Yoko Ono and Joe Jones) that brought with them the emancipation of performer, instruments, and listener; it also initiated the fusion of visual art and music that constituted the basis for sound art. Brown, for example, wrote a piece for percussion quartet in which one of Alexander Calder's mobiles serves as both instrument and conductor.(FIG. 43)

If, in the age of repetition, sound becomes ubiquitous and infinite – if it is therefore true that "any sounds may occur in any combination and in any continuity"[36] – then either this infinite combinatorics can be subjected to rigorous controls, or one can let it happen and heighten it to aleatorism. Cage's chance operations are an attempt to provide compositional freedom in addition to the freedom attained by sound. This sets him apart from his predecessors. The freedom of sound, which opens the door to noise, leads to the freedom of composition, which opens the door to chance. The logical result of new noises, new instruments, and new compositional techniques is a desire to give up one's traditional control over the material of composition, sound itself. This would be the final step in the emancipation of sound, if sound could follow its own rules – if sounds were "allowed to be themselves"[37] – and we simply gave sound activities our attention. "One may give up the desire to control sound, clear his mind of music, and set about discovering means to let sounds be themselves rather than vehicles for man-made theories or expressions of human sentiments."[38] We are still a long way from this ideal, and further still from the ideal of truly liberated sounds in an open space. All too many sounds are still regarded as unmusical.

Sound sculptures and sound installations represent an important step toward the emancipation and liberation of sound. In this sonic art form, many of the goals and stages of the evolution of twentieth-century Western music converge:

1. New instruments, custom-made devices, for which the distinctions between musical and nonmusical instruments, and between musical and nonmusical sounds, no longer apply.
2. Incorporation of the sounds, tones, noise, and silence of the environment. All audible phenomena are accepted as musical material.
3. Disappearance of performer and composer.
4. Investigation of new compositional techniques.
5. Autonomy of sounds.

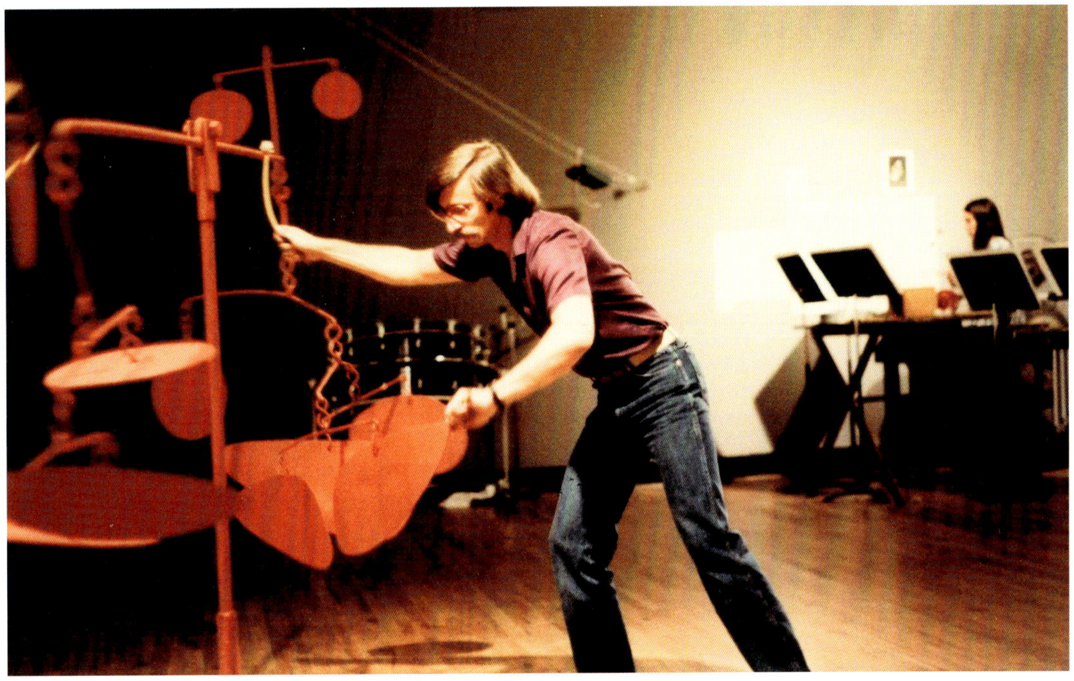

FIG. 43 Earle Brown, *Calder Piece*, 1981. Percussionist Barry Jekowsky perfomring Brown's *Calder Piece* at the Aspen Music Festival 1981. The photo was shot by Earle Brown himself.

36 John Cage, "Experimental Music," in *Silence*, 7–12, here 8. Originally presented as a lecture in Chicago in 1957.

37 Ibid., 10.

38 Ibid.

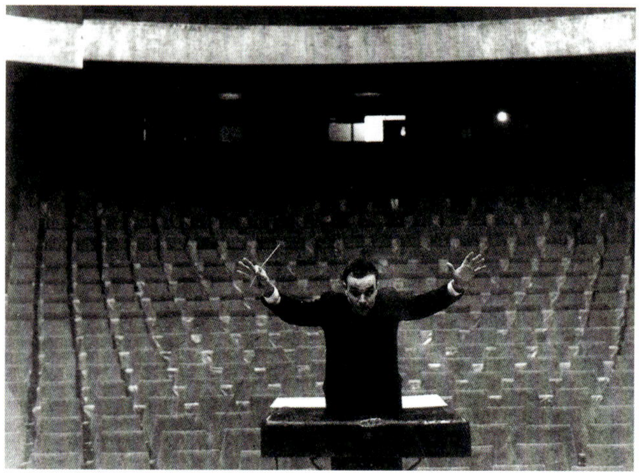

FIG. 44 Yves Klein, *Monotone-Silence Symphony*, 1959.

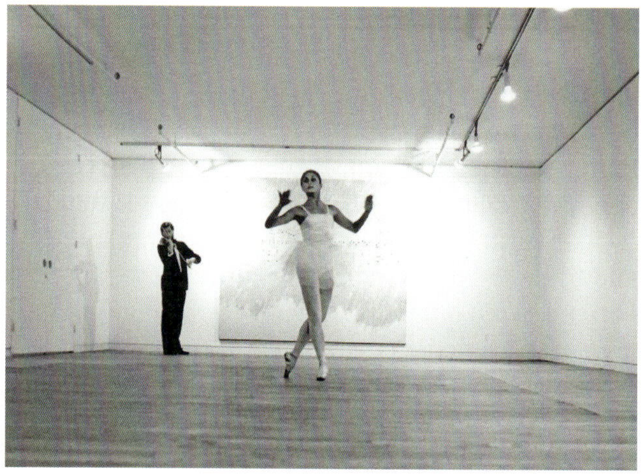

FIG. 46 Jannis Kounellis, *Da inventare sul posto*, 1971.

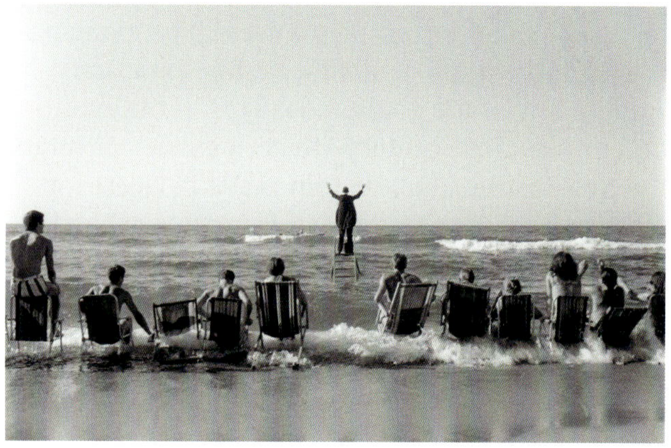

FIG. 45 Tadeusz Kantor, *Panoramiczny Happening Morski* [Panoramic Sea Happening], 1967.

6. Spatialization and objectivization. The trajectory of the liberation of sound runs from the tape machine to sound sculpture, from the audio object to audio space.
7. Emancipation of the listener (following Anton Webern's emancipation of the break and Cage's emancipation of silence). The recipient becomes a participant.

From Futurism to Dada, from Fluxus to happenings, the aesthetic avant-gardes of the twentieth century built on the idea of free sound, of liberated noise, in many different projects and with many newly created instruments. Thus sound art and noise art became part of the visual arts. In the 1950s and 1960s, the representatives of musique concrète and the artists of the happening/Fluxus movement (from Ono to Young) expanded the performative aspect of music to such an extent that the composition could be replaced by chance, music by silence (Yves Klein's *Monotone-Silence Symphony*, 1959 (FIG. 44), Cage's *Silence*, 1961), the orchestra by the ocean (Tadeusz Kantor, 1967 (FIG. 45)), and the musician by a horse or some hay.

La Monte Young composed a monotone piece as well: *Composition 1960 #7*. The score consists of the chords B3 and F#4 and the instruction "to be held for a long time." Young is probably the inventor of the concept of extended duration. Since 1962, he performs with his Theater of Eternal Music different sine waves as drones for seemingly endless hours, and he has conceived pieces meant to last ninety nine-years. His concept was adopted by Andy Warhol in films like *Sleep* (1963) and *Empire* (1964), and by Douglas Huebler with the duration pieces he began in 1969.

In the 1970s and 1980s, even pop music was influenced by industrial noise (Sonic Youth, Art of Noise, Throbbing Gristle, Depeche Mode, etc.), as was punk. Paintings became scores for musicians and dancers – for example, in the work of Jannis Kounellis (FIG. 46) in the wake of Cage, Rauschenberg, Trisha Brown, and Merce Cunningham. Thanks to the fusion of pop and art, more and more visual artists discovered the LP record itself as a medium for visual practices, from Milan Knížák or Jack Goldstein (FIG. 47) to Christian Marclay (FIG. 48), who won the Golden Lion at the 2011 Venice Biennale. Even record covers, not just records themselves, became acceptable as art. Alongside telegrams, books, videos, etc., records became artworks as well.[39] The radio play, too, was conquered as a medium by visual artists in the 1960s.[40] New-media and conceptual

39 Cf. Germano Celant, *Record as Artwork 1959–1973* (London: Royal College of Art, 1973).

40 Cf. Klaus Schöning, *Ars Acustica – Ars Intermedia. Klaus Schöning im Gespräch mit Komponisten, Autoren und Klangkünstlern*, Edition Neue Zeitschrift für Musik (Mainz: Schott Music, 2011). DVD-ROM.

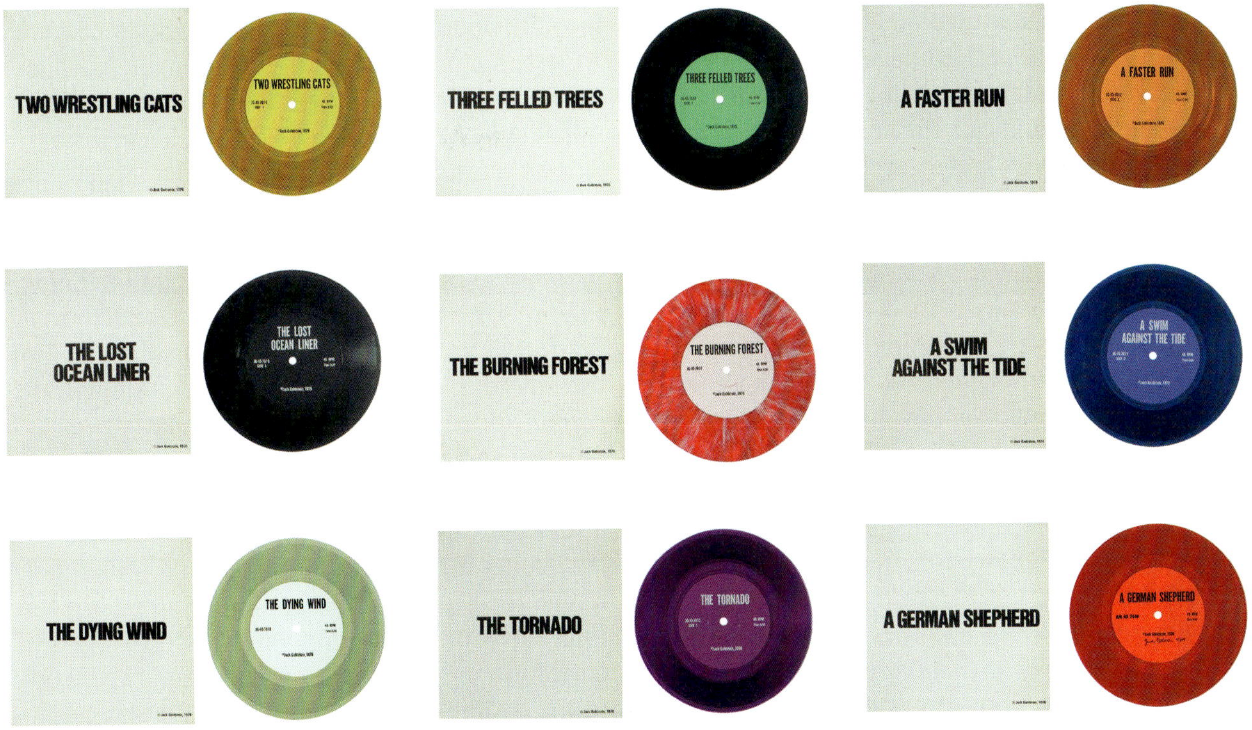

FIG. 47 Jack Goldstein, *A Suite of Nine 7-inch Records*, 1976. Sculpture.

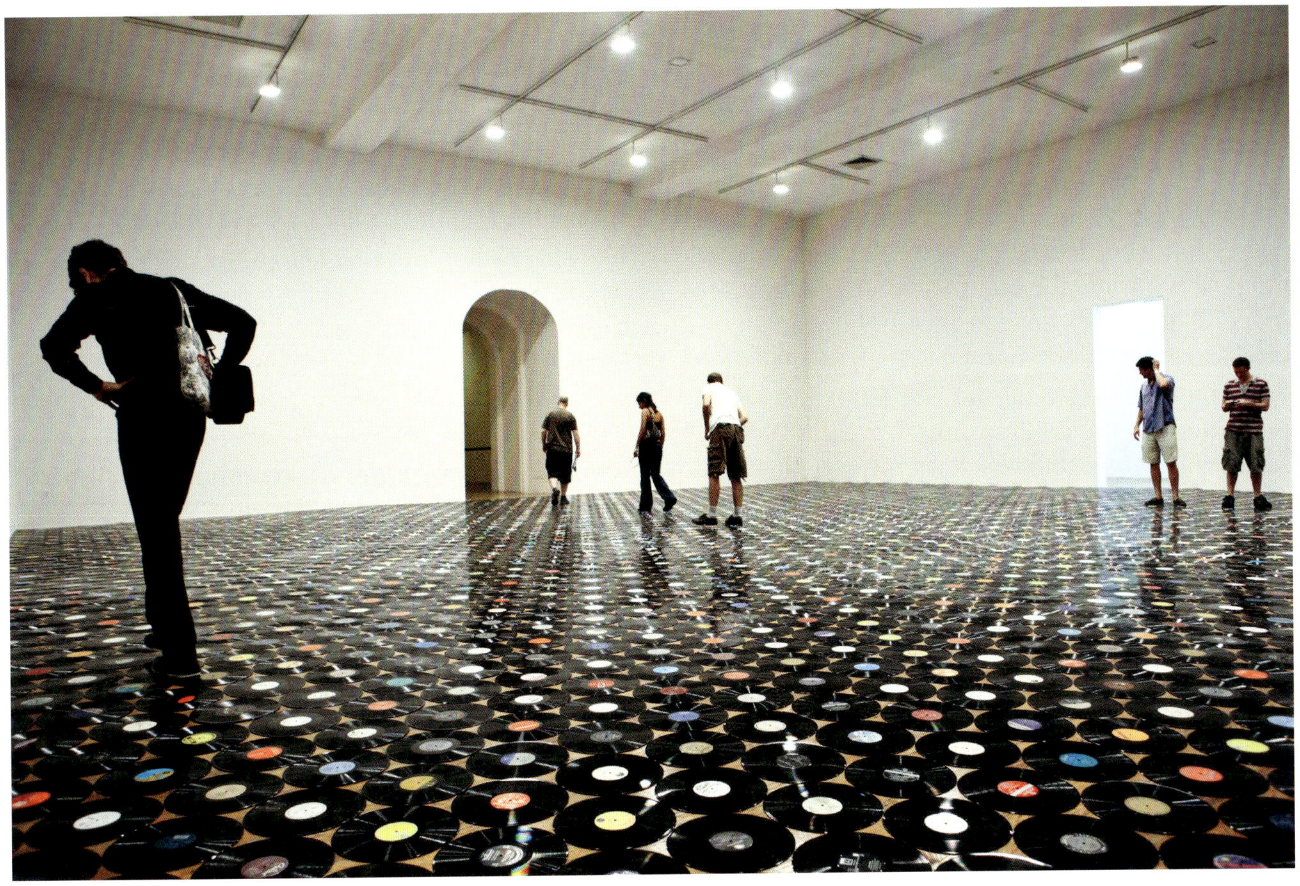

FIG. 48 Christian Marclay, *2822 Records*, 1987–2009. Installation view MoMA (PS1), New York, 2009.

artists were particularly instrumental in advancing sound art, and in the mid-1960s they placed it on a new footing. Sound art became a distinct art form within the visual arts. The development of the world of sound was furthered not by composers and musicians, but by architects and visual artists. The noise that invaded the world of music did not destroy music, but rather expanded the cosmos of sound. Laurie Anderson's performances in the 1980s are in many respects an extension of her sound sculpture and sound spaces into the theatrical. From her homemade tape-bow violin, which had tape in place of bow hairs and a tape head mounted on the body of the instrument, to her 1983 audio room, she created remarkable sound objects and sculptures, such as *Stereo Decoy* (1977) at Artpark in Lewiston, New York. In *Doormat Love Song* (1978), the tape head was on the underside of a door and the tape was placed along the door's arc, so that palindromes were heard whenever the door was opened or closed. In *Handphone Table* (1978), the music could only be heard by placing one's elbows on a table and covering one's ears with one's hands.

As sound sculpture's very name indicates, it is based on a fusion of visual art and music, a tradition of the twentieth century, first in painting and now predominantly in sculpture. Here I would like to recall that Arnold Schoenberg painted, that the Austrian composer Josef Matthias Hauer created a tone-color wheel, much as the American artist Morgan Russell sought to create formal correspondences between the twelve tones of the musical scale and the color wheel in his "synchromies." The American experimental musician Carl Ruggle painted, and the French painter Henri Rousseau composed, as did Duchamp and Kurt Schwitters. There is "tape music" by Jean Dubuffet, Asger Jörn, and Nicolas Schöffer, the creator of cybernetic artworks.[41] Gerhard Rühm, originally a musician, but also a poet and visual artist, composed a monotonic symphony in 1952, like Yves Klein did in 1959. Yves Klein also released a grammophone record in 1959 called *Musique du Vide*.

Of particular present-day interest is the British pop artist Tom Phillips (who was also a teacher of Brian Eno), whose graphic reworking of William Hurrell Mallock's 1892 novel *A Human Document* transformed it into a score called *A Humument: A Treated Victorian Novel*, originally published in 1970, parts of which were reshaped into the opera *IRMA*. The visual artists Larry Rivers, Larry Poons, Walter De Maria, Michael Snow, and Jim Dine have also become known as musicians, as have various filmmakers (from Tony Conrad to Woody Allen), video artists, and computer artists.

Today, the fusion of visual art and music in the field of sound sculpture is striking. This fusion is the result of the aforementioned transformation of the musical instrument itself. The more the boundary between musical and nonmusical sounds faded, the more the boundary between musical and nonmusical instruments faded as well, so that in the end an instrument could look like a sculpture and a sculpture could become a sound-producing instrument. Sculpture and sound merge in the open space of free sound, just as score and image (painting, drawing, photograph) did decades earlier on the two-dimensional plane. The work of the Brothers Baschet, of Jean Tinguely and Takis are paradigmatic of this shift. Around 1952 the brothers Bernard and François Baschet began to create sculptures with metal or glass rods that could produce sounds. In 1957 they also began to give concerts with these instruments, and later they installed sound sculptures in public spaces that could be played or were self-playing (see this volume, pp. 204–207). Also in the 1950s, prompted by Calder's mobiles, Tinguely created a series of moving sculptures that produced sound, as well as mechanical sound-reliefs (*Relief sonore*, 1955, for example), which he called métamatics, and in 1978 he began creating monumental music triptychs. In works such as *3 Totems - Espace Musical* (1981), Takis has worked with metal tubes and spheres, light, gravity, electromagnetism, and sound, and he has been working with "musical sculptures" since 1965. His sculptures are just as compelling as sound generators and audio installations as they are when viewed as visual objects. In the 1970s, he began to expand his sculptures, creating environments which he refers to as "musical spaces."

Around the same time, the minimalist sculptor Robert Morris created his *Box with the Sound of Its Own Making* (1961) (FIG. 50) and Robert Rauschenberg created *Music Box (Elemental Sculpture)* (1953)

41
Nicolas Schöffer, *Spatiodynamisme. Luminodynamisme. Chronodynamisme. Cybernetique. Recherches de 1948–1970* (Paris: Denise René, 1970).

42
Lasting sounds coming from different points and forming a sound sculpture that lasts. Sanouillet, *Marcel Duchamp*, 47.

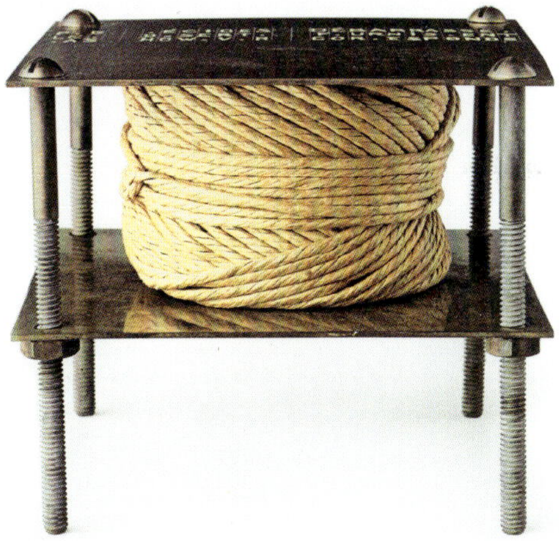

FIG. 49 Marcel Duchamp, *With Hidden Noise*, 1916. Sculpture.

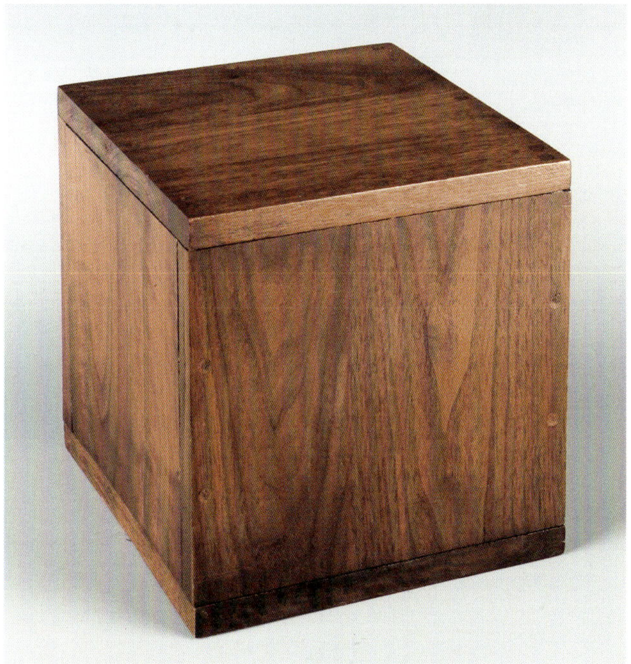

FIG. 50 Robert Morris, *Box with the Sound of Its Own Making*, 1961. Sculpture.

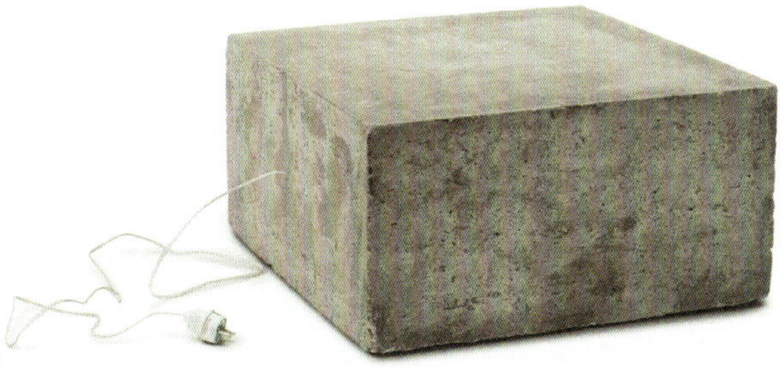

FIG. 51 Bruce Nauman, *Concrete Taperecorder Piece*, 1968. Mixed media sculpture.

after having met Duchamp at the Stable Gallery, New York, the same year.(See p. 37) Both works are an echo of Duchamp's semi-readymade *With Hidden Noise* (1916) (FIG. 49) in more respects than the merely numerical one of dates. Duchamp was a pioneer of sound art as well. His interest in music is demonstrated not only by *With Hidden Noise*, but also by the piece *Erratum musical* (1913). In his *Green Box* from 1934 we find notes on "sculptures musicales" [sound sculptures], which Duchamp describes as "Sons durant et partent de différents points et formant une sculpture sonore qui dure."[42] Duchamp's object reveals the fate of sound in the age of representational music - that is, to remain hidden. Suppressed sound is also the subject of works by Bruce Nauman (FIG. 51) and Douglas Henderson.

After the end of representational art around 1900, when color, line, plane, and image had been liberated, could sound and noise also be liberated. Film music played a decisive role in this development, because, when films showed street scenes or factory scenes, the noises of cars and machines had to accompany the images. Thus, the audience slowly got used to noise. As an example, I would like to cite the contribution of the composer Edmund Meisel to Sergei Eisenstein's film *Battleship Potemkin* (1925), especially the final episode when the rebellious *Potemkin* is allowed to pass safely through the Tsarist fleet. Eisenstein specifically asked the composer to create "music of the machines." (FIG. 52)

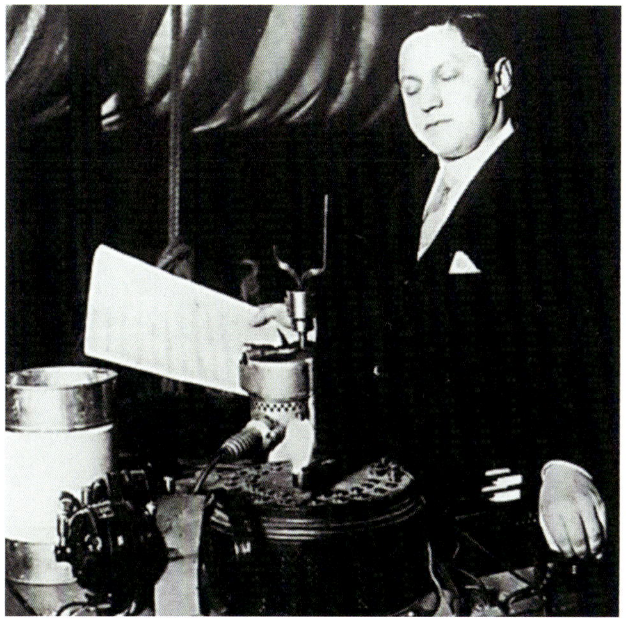

FIG. 52 Edmund Meisel with his noise machine, 1926.

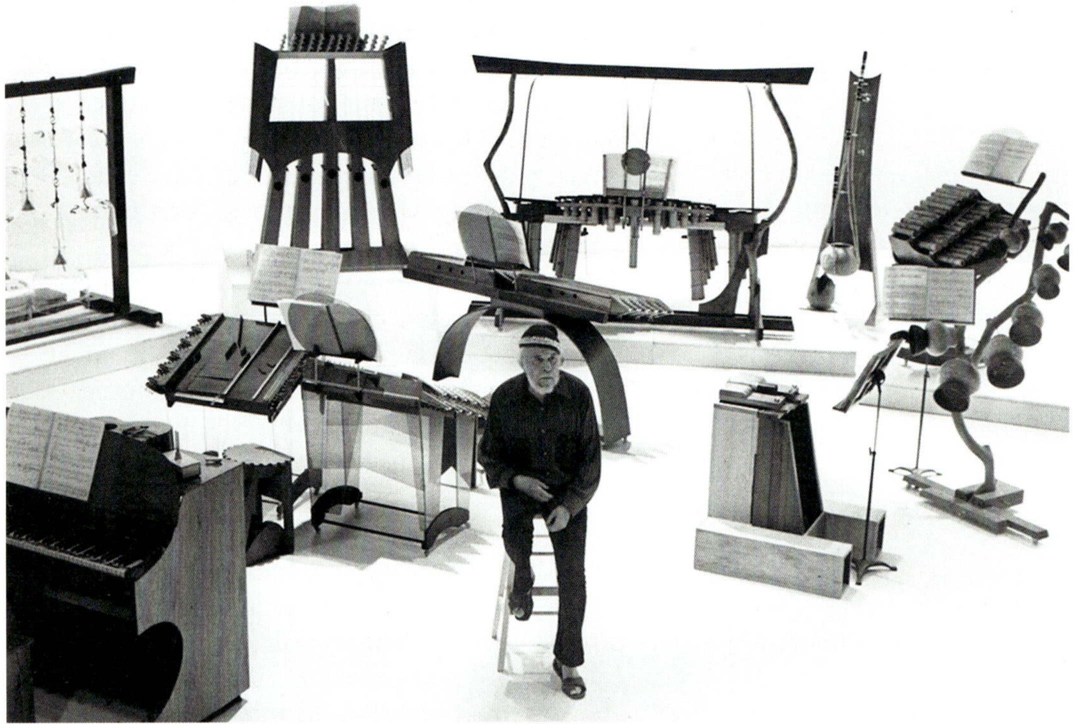

FIG. 53 Harry Partch at the set of the movie *The Dreamer that Remains* (1972) with his self-made instruments.

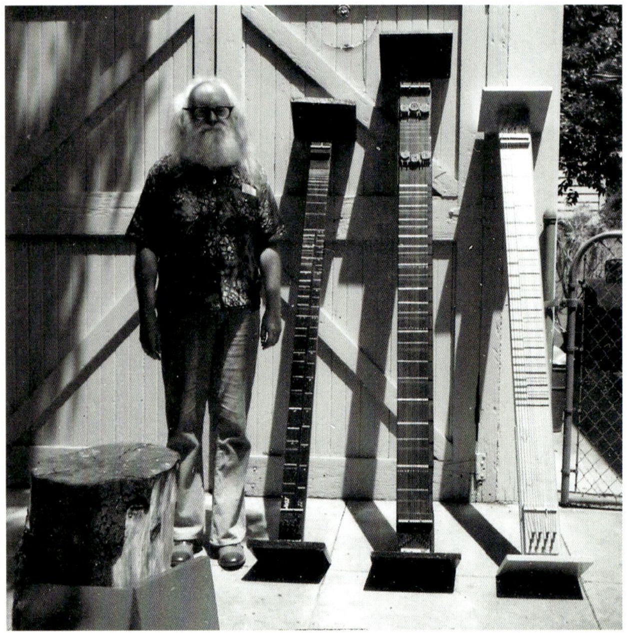

FIG. 54 Ivor Darreg and his mictrotonal instruments, 1970s.

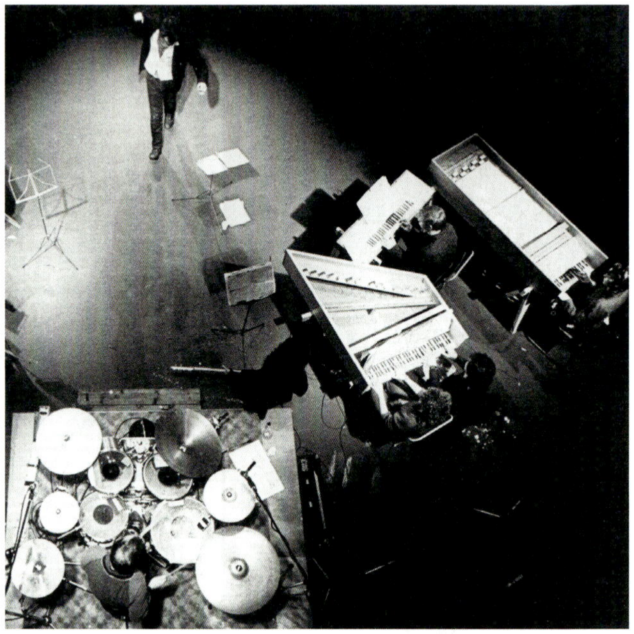

FIG. 55 Glenn Branca and his band performing with their self-constructed instruments at Ars Electronica 1984.

43
Cf. John Grayson, ed., *Sound Sculpture: A Collection of Essays by Artists Surveying the Techniques, Applications, and Future Directions of Sound Sculpture* (Vancouver: A.R.C. Publications, 1975).

44
Bernhard Leitner, *Sound : Space*, 23.

45
Bernhard Leitner, *Kopfräume/ Headscapes*, ed. Peter Weibel (Ostfildern: Hatje Cantz, 2003). Compact Disc.

Audacious builders of instruments, such as the musicians Harry Partch (FIG. 53) and Ivor Darreg (maker of devices for "xenharmonic music") (FIG. 54), or Glenn Branca (FIG. 55) and Ed Romney (FIG. 56), who did so in order to give voice to their new aural visions with the help of appropriate instruments, pointed the way for artists to make sound-generating sculptures: not only Morris, but also Bernard and François Baschet, Harry Bertoia, Bruce Conner, Stephan von Huene, and Bernhard Leitner.[43] (FIGS. 57, 58) Looking back at his early experiments with ambient sound movements between 1969 and 1975, Leitner said that "sound itself was to be understood as building material, as an architectural, sculptural, form-producing material – like stone, plaster, wood."[44] In 2003 he released the audio CD *Kopfräume/ Headscapes*.[45] The sixteen pieces on the CD were created and designed for the interior space of the mind and can only be experienced using headphones. The head functions as a sound space, a place where the space is shaped by the movements of sound lines and sound matter. The artists of minimalist sculpture were also musicians of minimal music (such as Morris and De Maria).

Henning Christiansen is known as the composer who worked on many of Joseph Beuys' performances. The impressive quality of Beuys' actions of was not only indebted to the suggestive visual language of Beuys, but also to the psychic depth of Christiansen's contributions. In addition to his collaboration with Beuys, Christiansen created his own extensive musical oeuvre that opened up new horizons for Fluxus music. (See FIG. 97, p. 126)

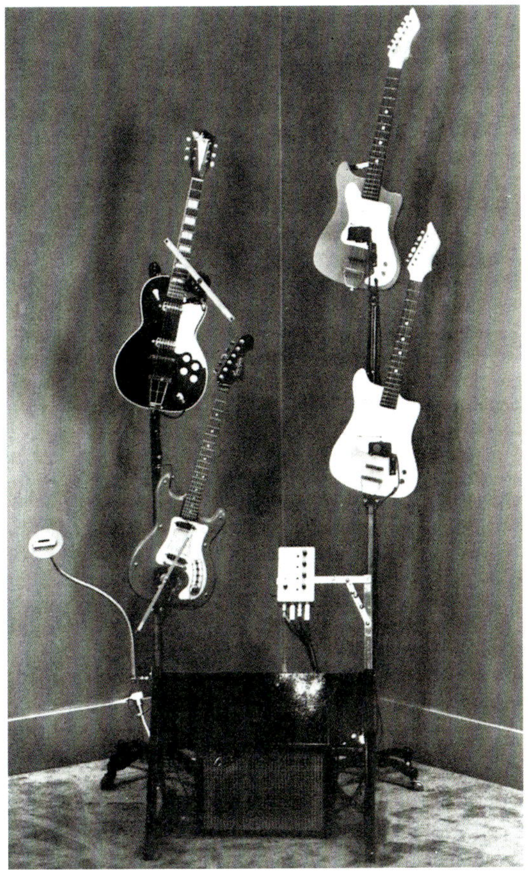

FIG. 56 Ed Romney, *Guitar Tree*, ca. 1980. Sound sculpture.

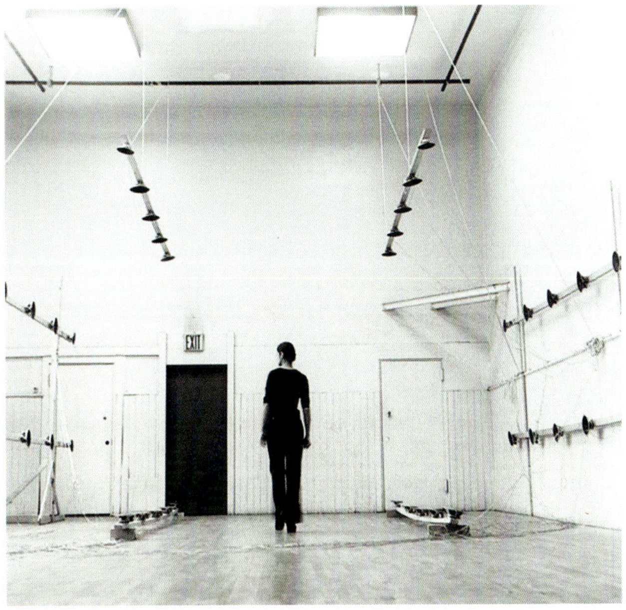

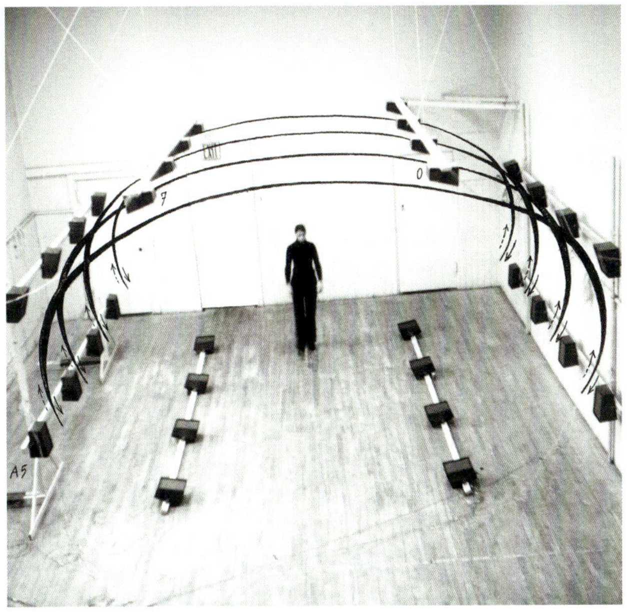

FIGS. 57, 58 Bernhard Leitner, Studies for *Sound Tube*, 1970s.

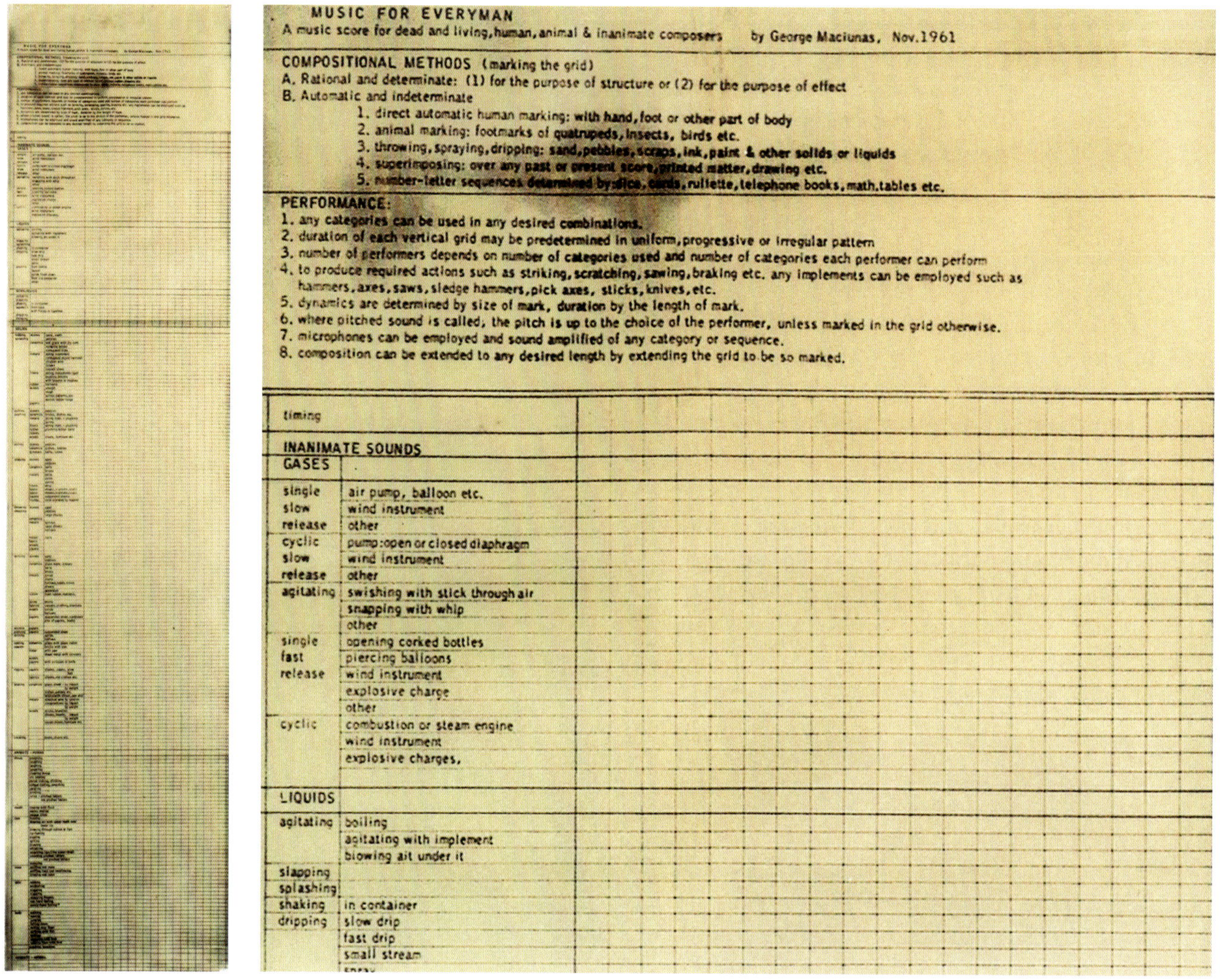

George Maciunas, *Music for Everyman*, 1961.
Score.

The image shows the whole score on the left, a detail on the right. To perform the piece, Paul Panhuysen founded the Maciunas Ensemble together with Jan van Riet and Remko Scha in 1968.

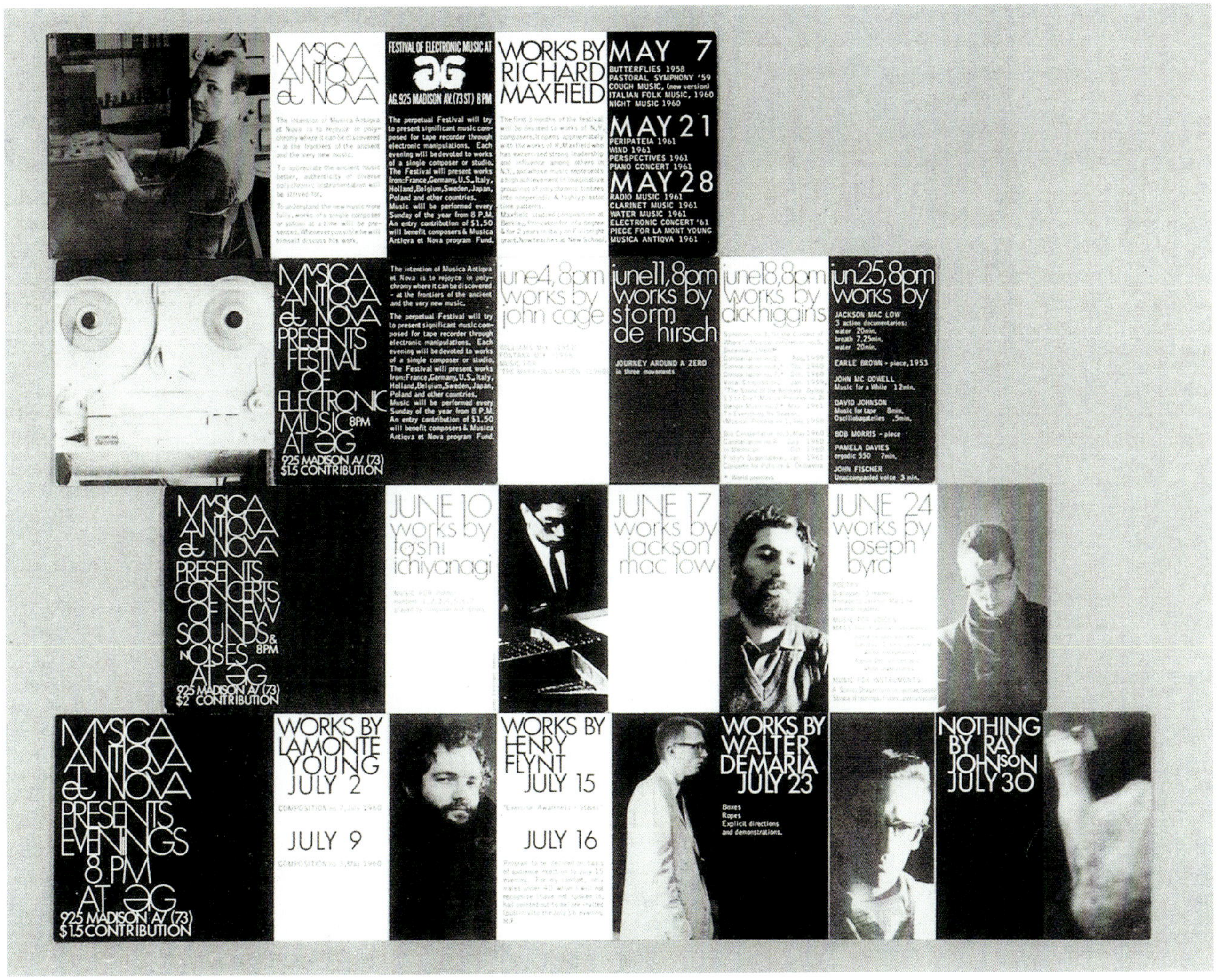

Announcement cards for *Musica Antiqua et Nova*, AG Gallery, New York, July 1961.
Archiv Sohm, Staatsgalerie Stuttgart.

With *Musica Antiqua et Nova*, the two AG gallery owners George Maciunas and Almas Salcius introduced musicians and artists back in 1961, who were then represented in *An Anthology of Change Operations* published by La Monte Young and Jackson Mac Low in 1961/1963.

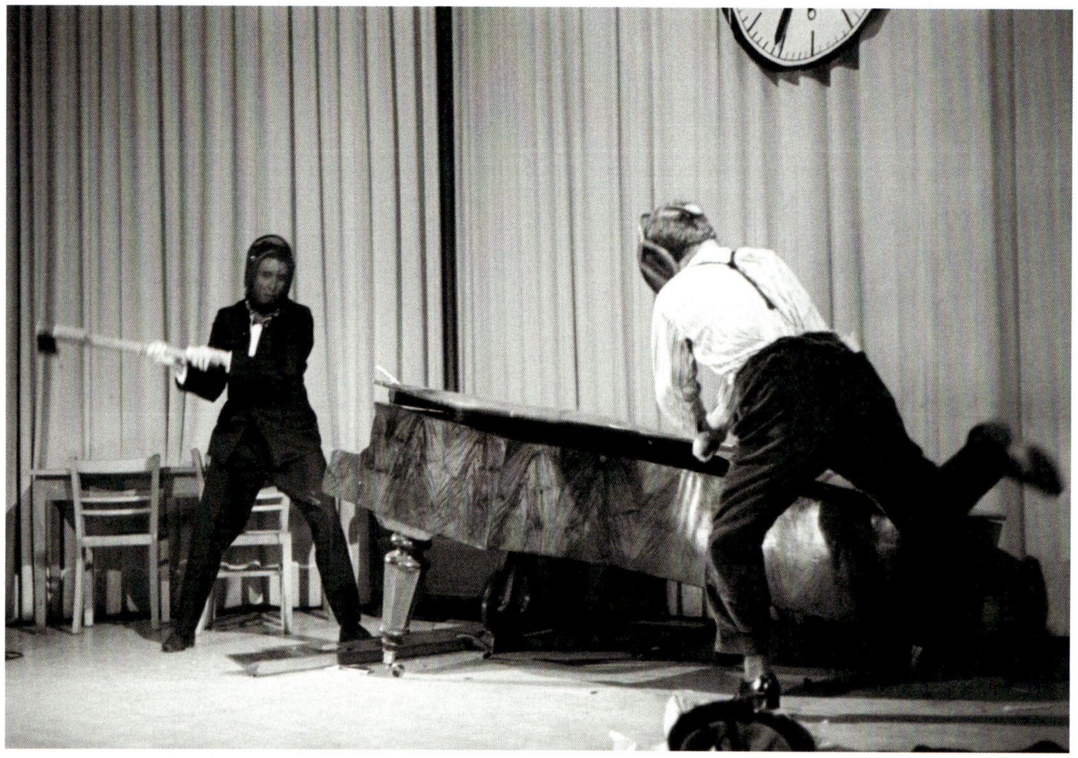

Vienna Group, 2. literarische cabaret [Second Literary Cabaret], Vienna, April 15, 1959. Friedrich Achleitner and Gerhard Rühm smashing a piano on stage.

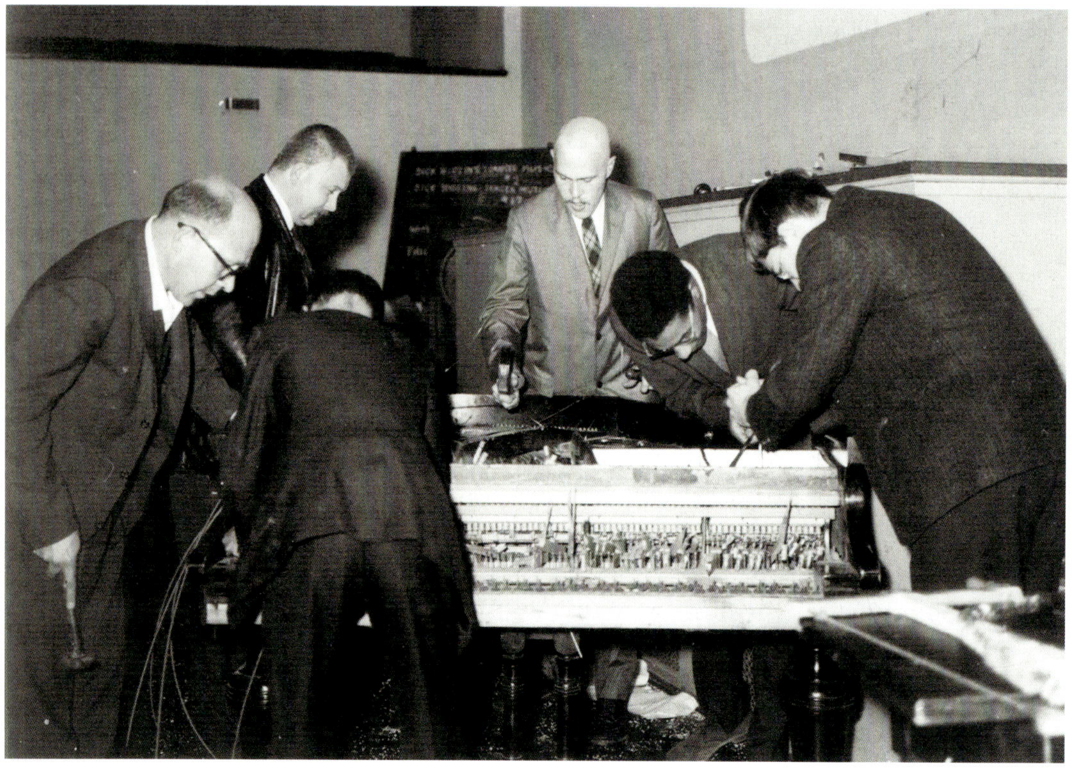

Emmett Williams, Wolf Vostell, Nam June Paik, Dick Higgins, Ben Patterson, and George Maciunas interpret *Piano Activities* by Philip Corner at the Fluxus Festival, Museum Wiesbaden, Germany, September 1962.

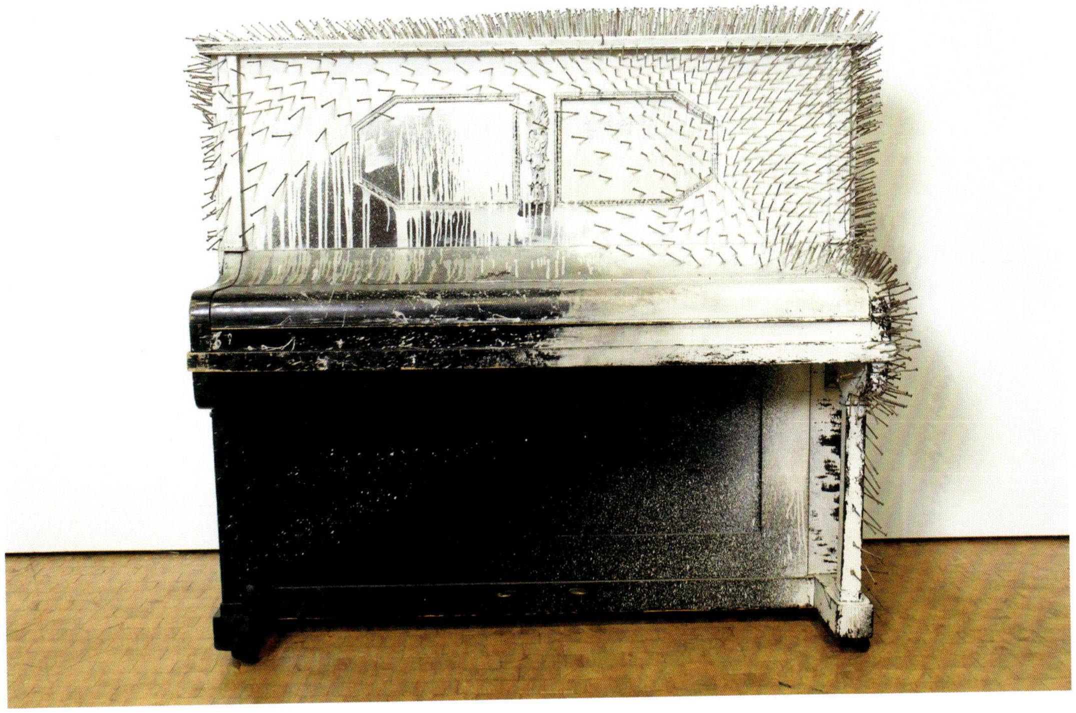

Günther Uecker, *Piano*, 1964. Nails, white paint on black piano, 200 × 150 × 90 cm. Würth Collection.

The nail-studded piano was Günther Uecker's first monumental object: "It became clear to me that people need cult rituals and objects. They washed their cars on Saturday, put TV consoles in their living rooms. And suddenly I wanted my purist structures translated on this objectified cult and consumer world." To this end, Uecker chose the piano, "European culture's greatest good." At the opening of his solo exhibition in Pianohaus Kohl in Gelsenkirchen, Germany (1964), Heinz Mack gave the opening speech, during which Uecker began to hammer nails into a used piano, which he then painted white. The police were summoned but they didn't intervene; the audience was silent. Since Nam June Paik's *Klavier integral* (1963) and Günther Uecker's *Piano*, the piano as the fetish and incarnation of academic art and culture has repeatedly been the object of actions by artists, including Joseph Beuys, Wolf Vostell, and Rebecca Horn.

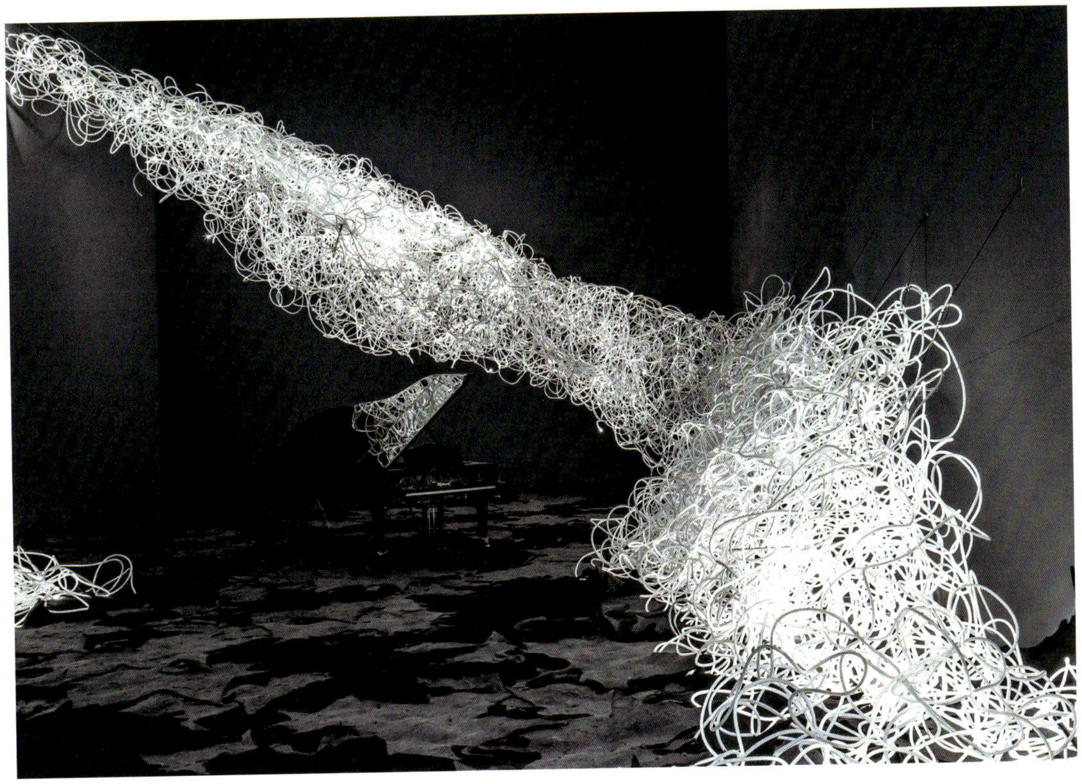

rosalie, *KLINGER – Walkable Landscapes of Melancholy*, 2013. Interactive, kinetic light-sound sculpture, ca. 8 × 9.7 × 19.7 m. Polymer optical fibers, HQI projectors, DMX programmed, cable structure, pneumatic actuators controlled in random movement matrices, pneumatic units, valve station, MIDI grand piano, guitar amplifier, pressure sensors, computer programming. Floor sculpture: leaf forms, multiple layers, black and anthracite. Museum der bildenden Künste Leipzig.

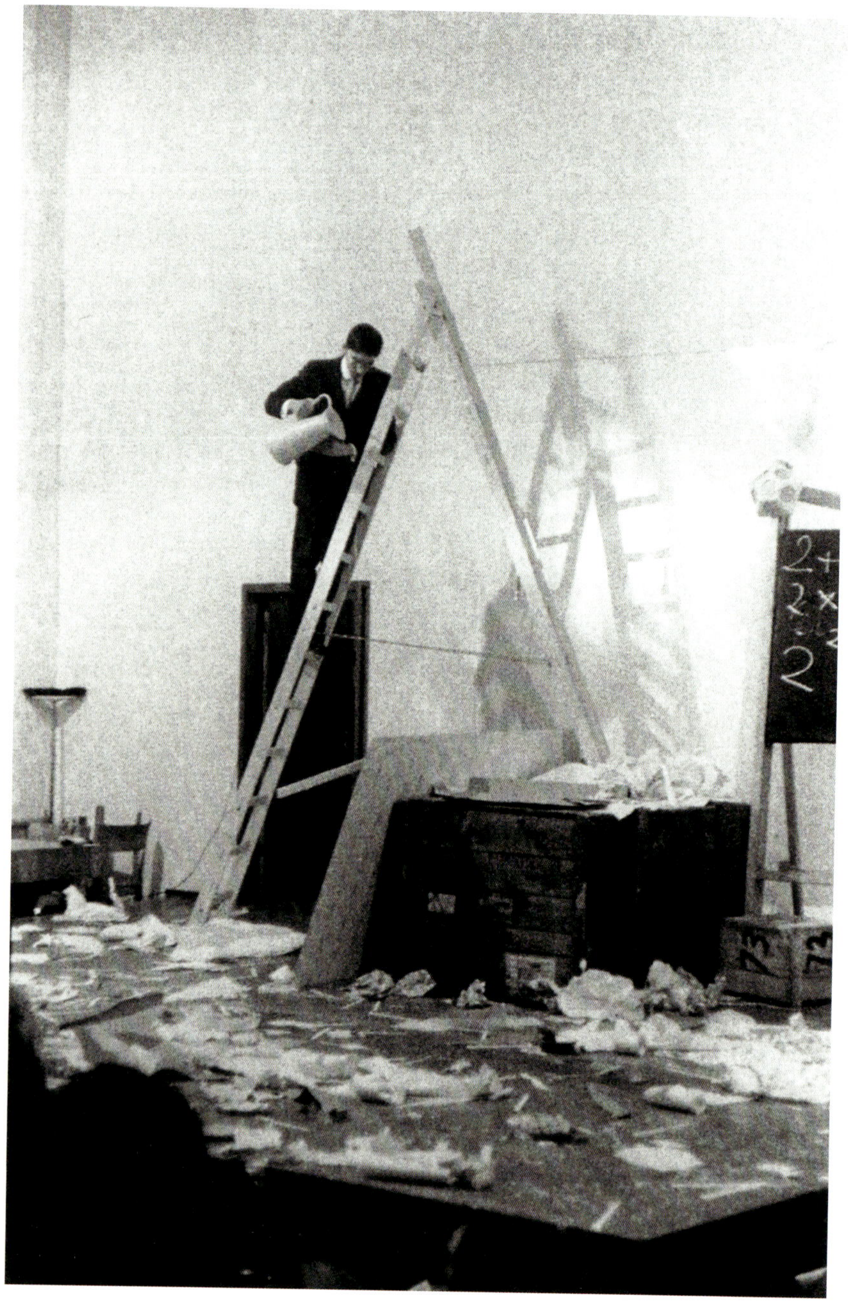

George Maciunas performing George Brecht's *Drip Music* (1959–1962) in the context of *Festum Fluxorum FLUXUS...* at the Düsseldorf Art Academy, February 1963.

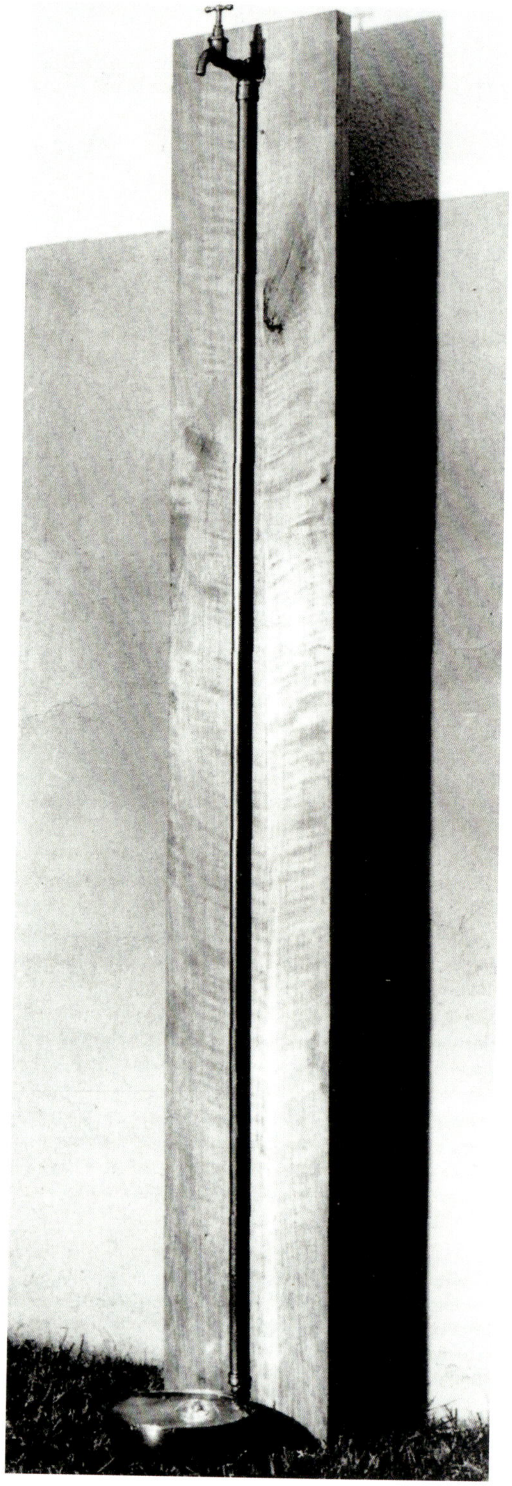

George Brecht, *Drip Music – Fountain*, 1963/1970.
Fluxus sculpture.
Oak plank, vertical, with water tap, bronze pipe and brass drip tray, height: 2 m.

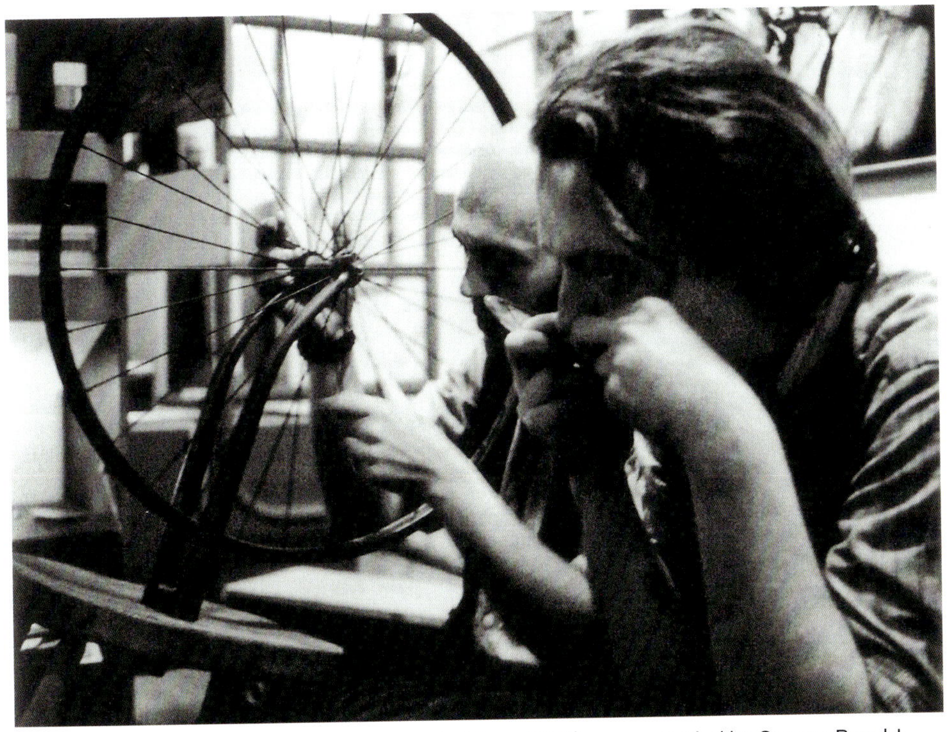

Jim Dine plays on Marcel Duchamp's *Bicycle Wheel* (1913), accompanied by George Brecht on the harmonica, in Richard Hamilton's studio, London, 1968.

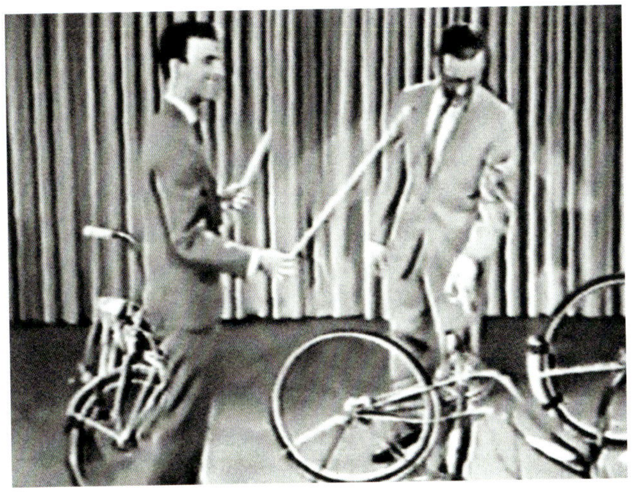

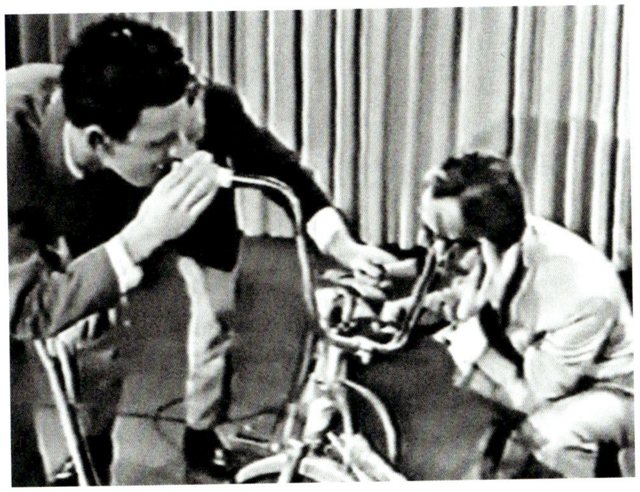

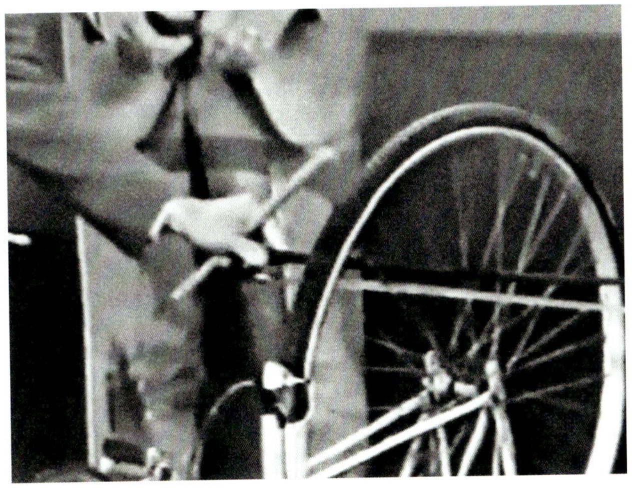

Frank Zappa, *Concerto for Two Bicycles*, 1963.

Only 22 years old and not yet famous, Zappa appeared on The Steve Allen Show and made music with some drumsticks, a bass bow, and two garden-variety bicycles – and nothing more.

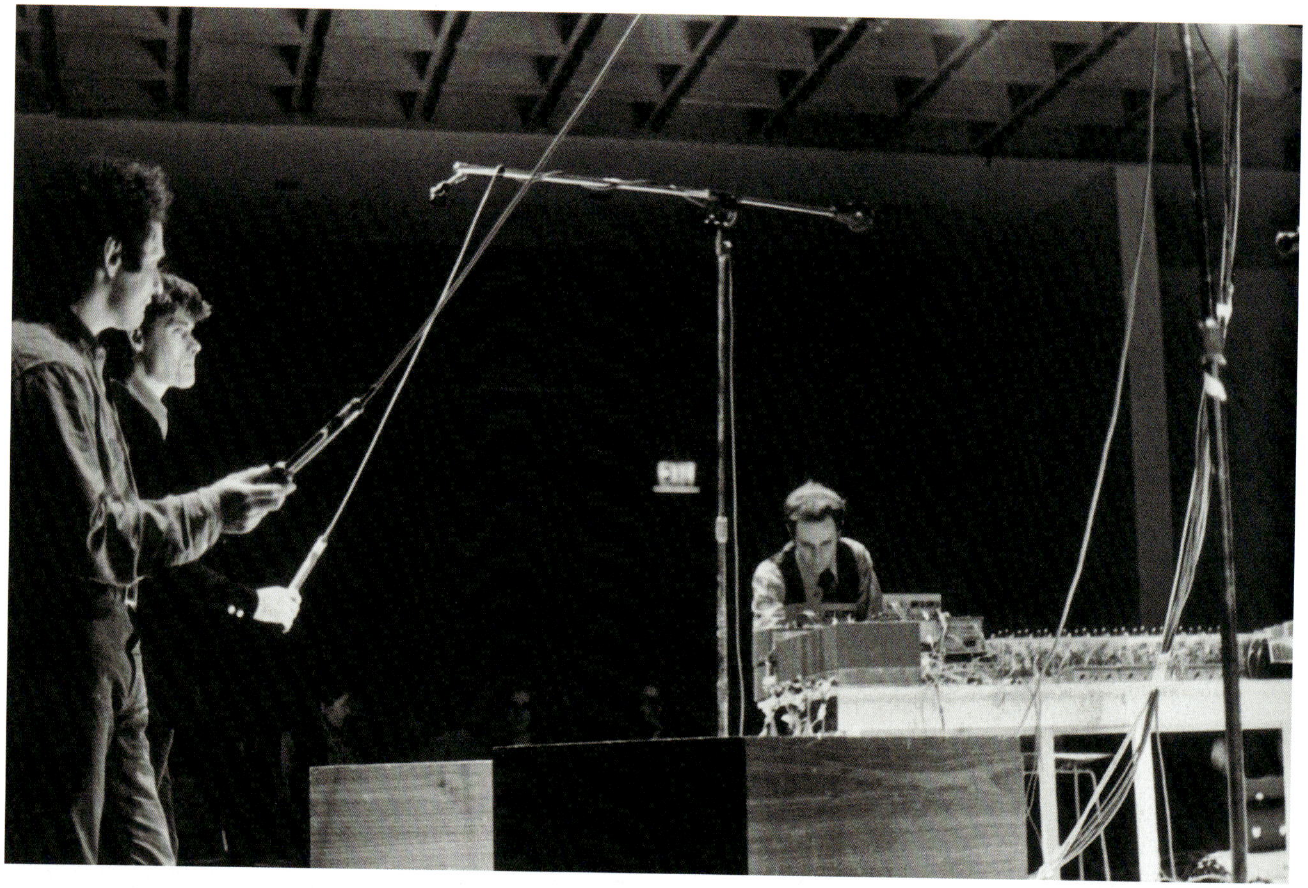

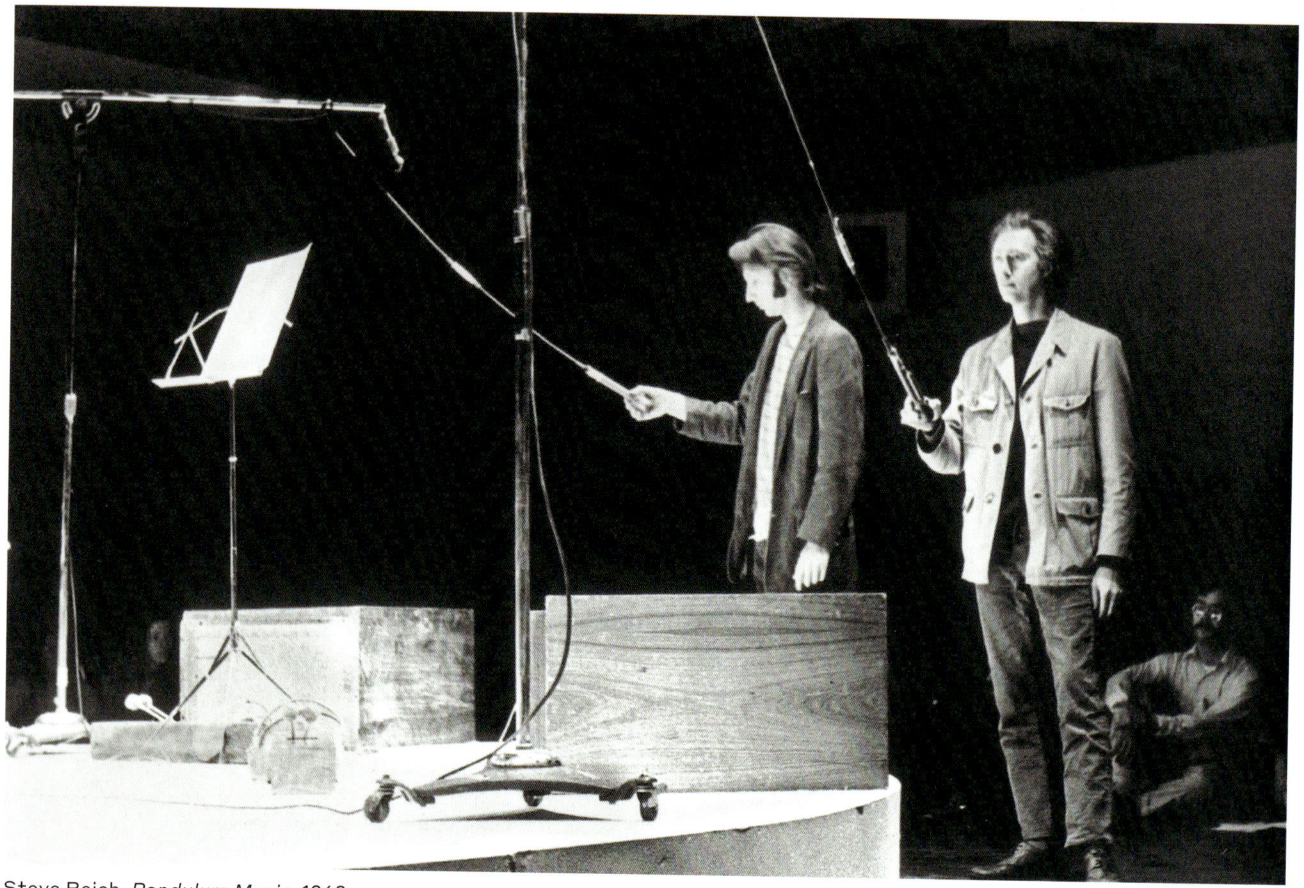

Steve Reich, *Pendulum Music*, 1968.
Performance at the Whitney Museum of American Art, May 27, 1969.
Performers: Richard Serra, James Tenney, Steve Reich, Bruce Nauman, and Michael Snow.

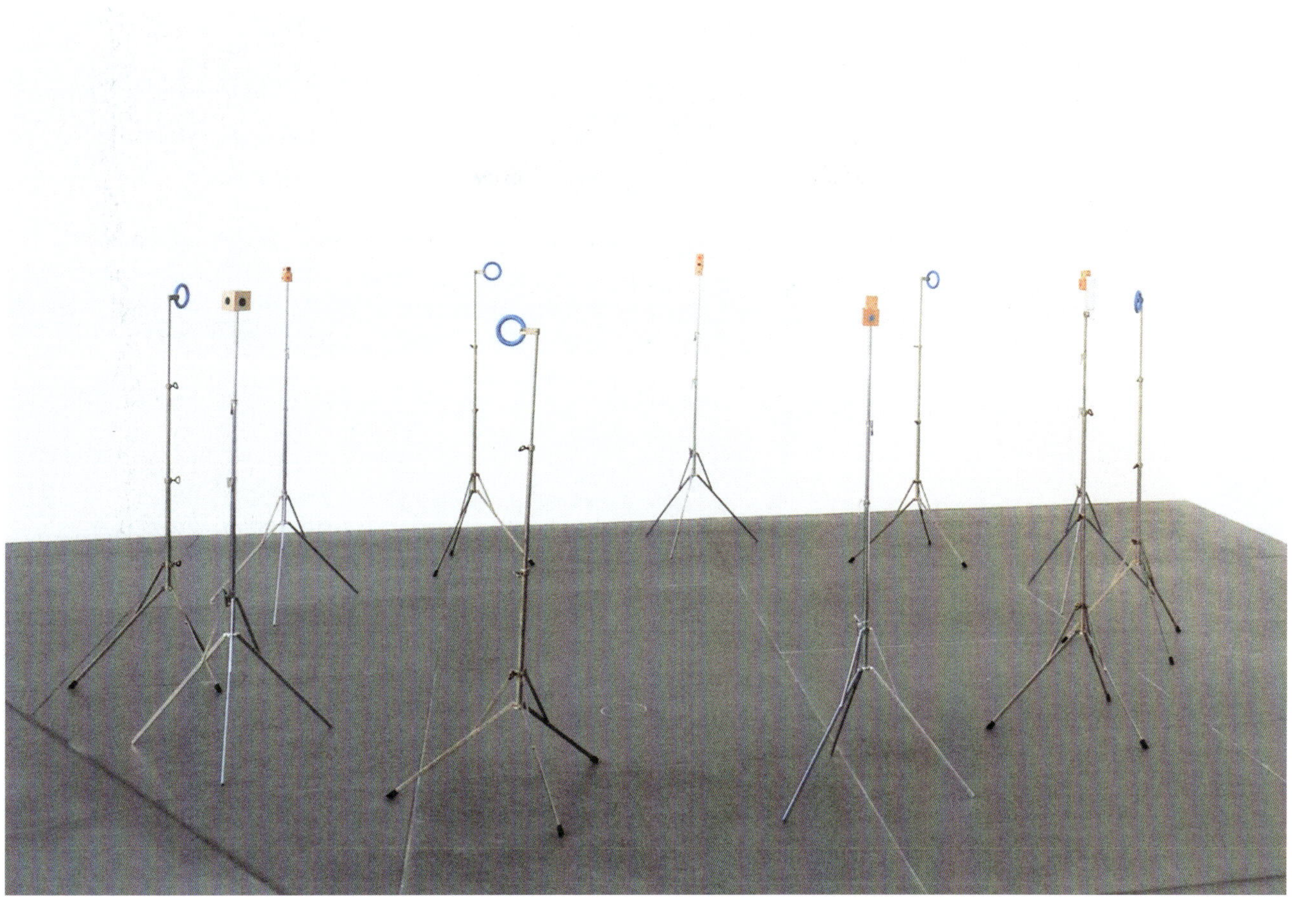

Robert Filliou, *Musique télépathique No. 21*, 1984.
11 music stands, placed in a circle (300 cm diameter),
5 wooden rings, 5 wooden cubes, 1 polaroid, dimensions variable.

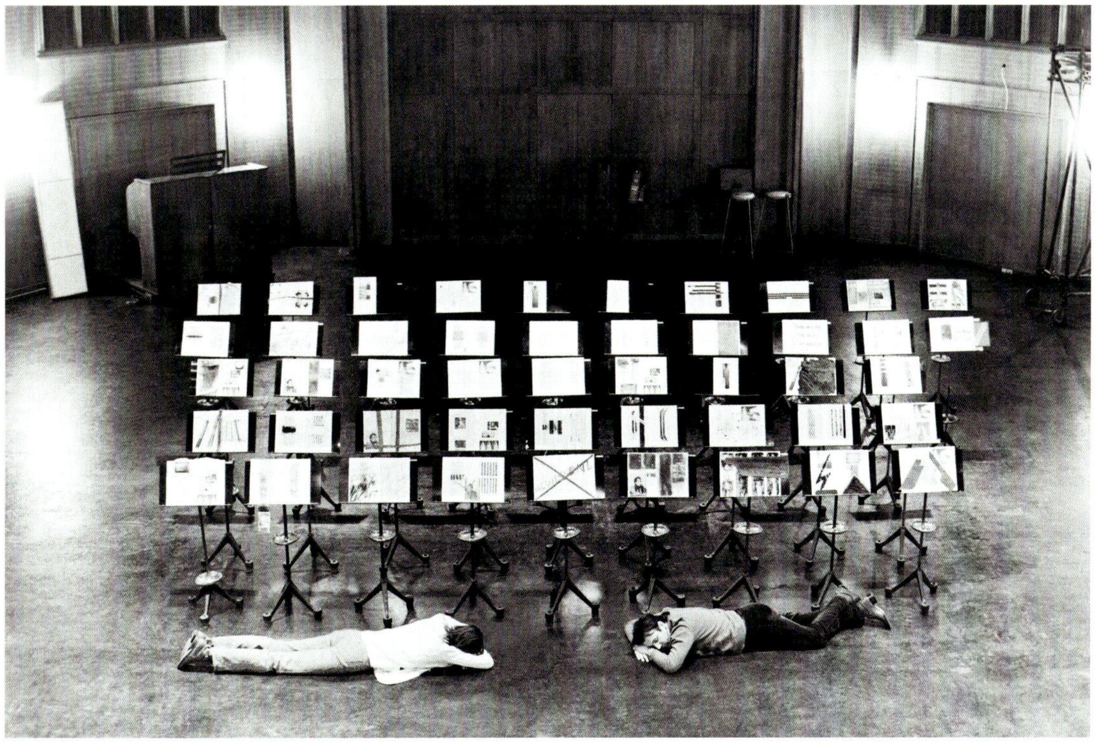

Loys Egg, Peter Weibel, *Bibliotheca Insomnia*, 1979.

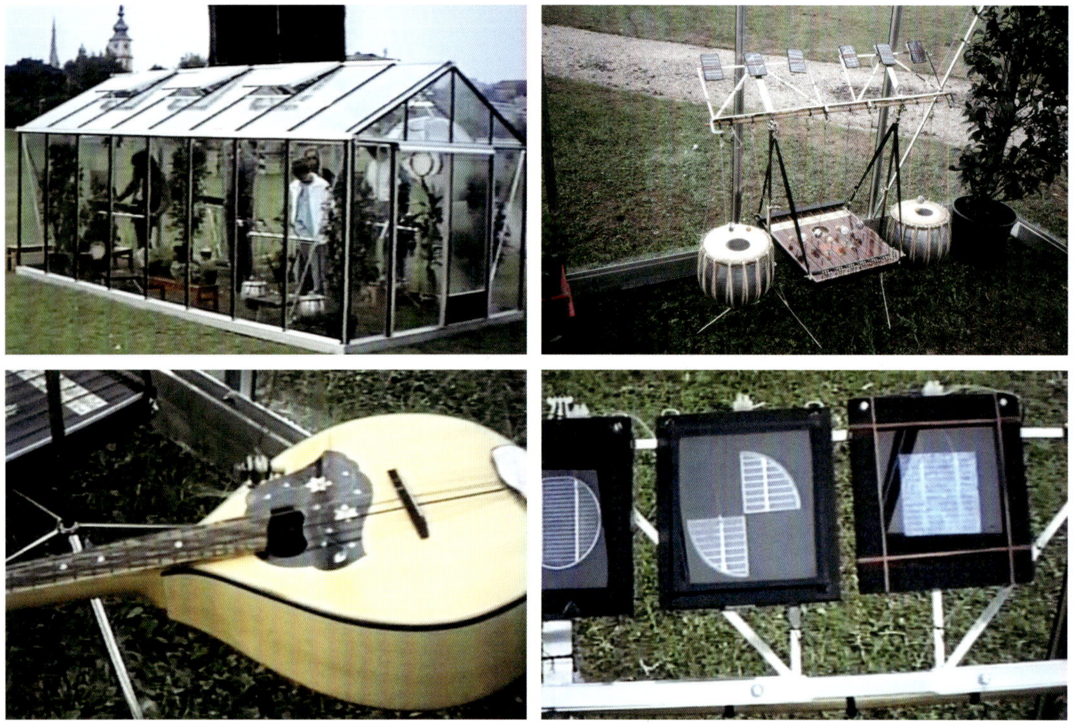

Joe Jones, *Solar Music Hot House*, 1988.
Sound installation at Ars Electronica 1988, Linz, Austria.

For *Solar Music Hot House*, solar-powered music machines and music stands equipped with solar cells were placed inside a hot house amongs plants and flowers with pathways and benches so that people could walk through or sit down in this closed environment. The solar cells on the music stands looked like sheet music to be played, but in this case the solar cells did the playing and the sun supplied the musical score. As the sun rose the instruments started to play. As the sun set, the instruments became silent until the next morning when the sun rose again.

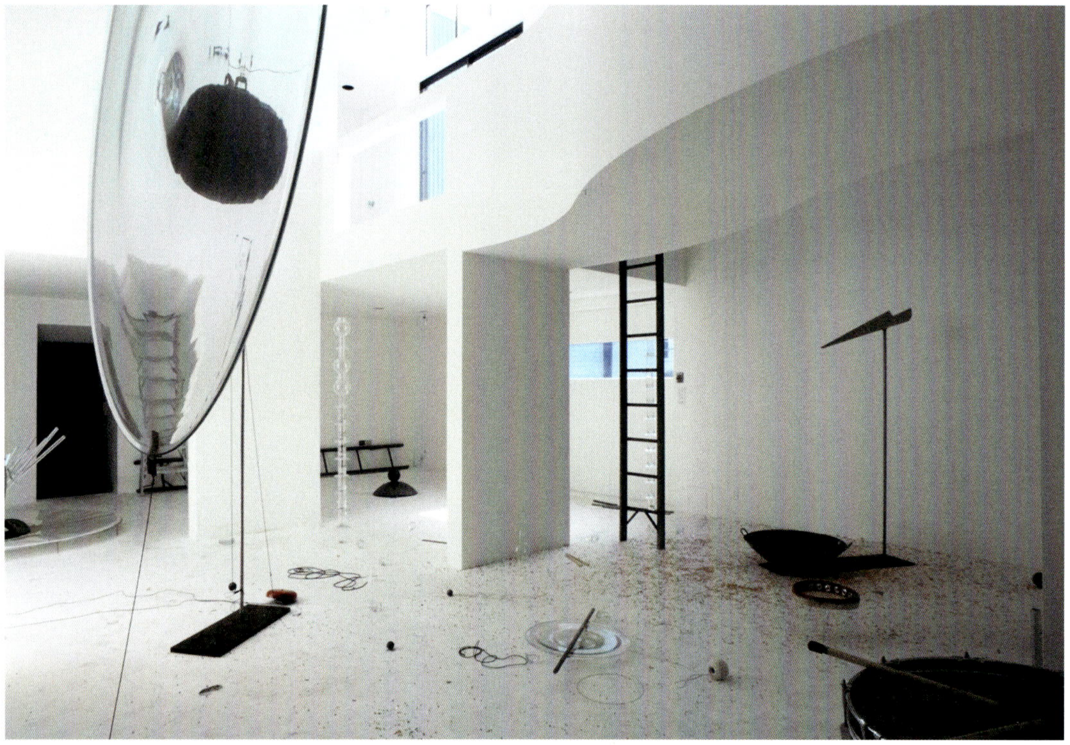

Terry Fox, *Instruments to be played by the Movement of the Earth*, 1987.
Capp Street Project, San Francisco.

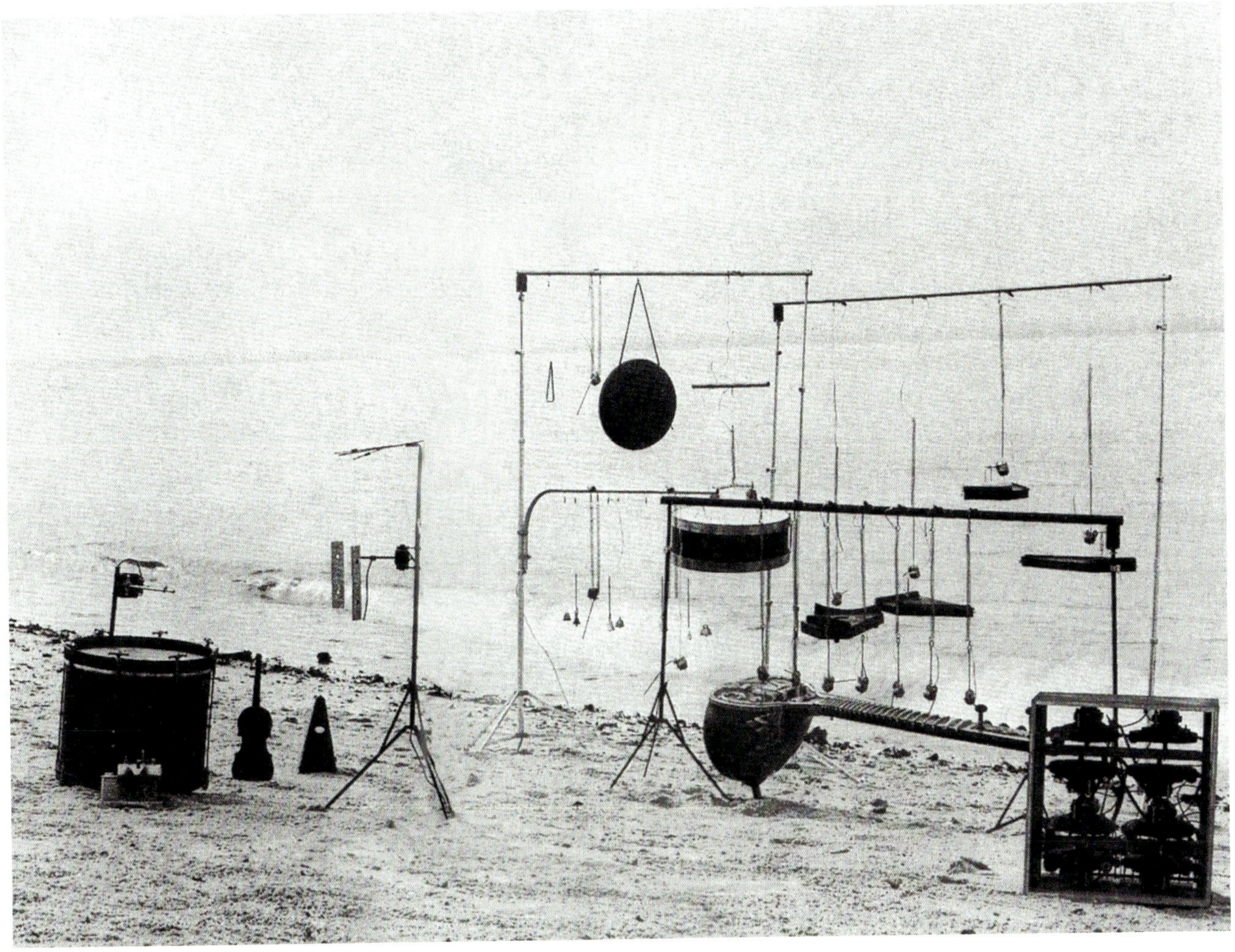

Joe Jones, *Mechanical Fluxorchestra*, ca. 1964.
This photo was used in a collage by George Maciunas on Yoko Ono's *Fly* LP.

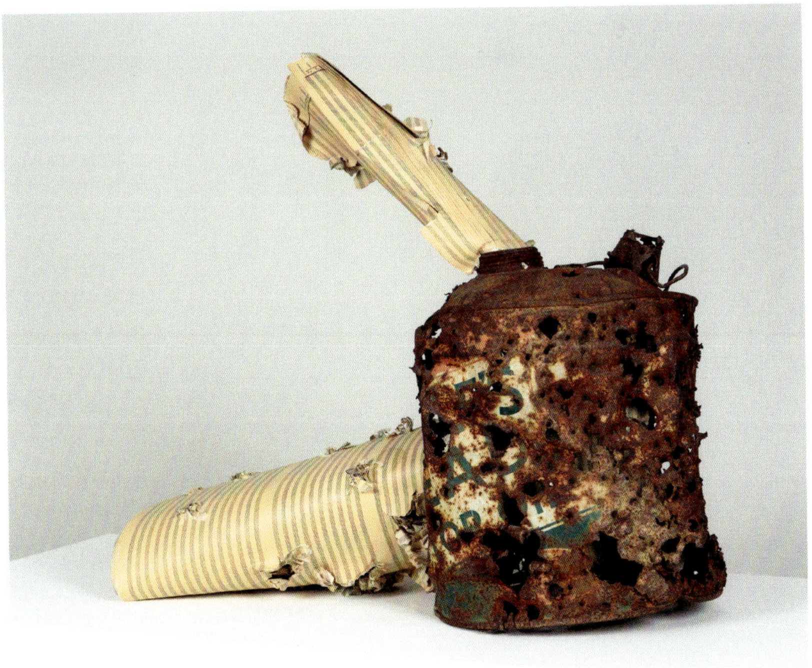

Dick Higgins, *Symphony Dispenser*, 1968.
Tin can, music paper, 62 × 65 × 30 cm.

New Sonic Zones

Ever since Varèse, the reality has been that everything that makes sound belongs to the art of sound. Whether the hissing of hot lead on contact with cold water, the pounding of wooden sticks on rusty metal shutters, or the sounds of a prepared piano, all noises, sounds, and tones are part of the new aural cosmos. Any sound-emitting object can become a sound instrument. Sound is being freed from the prison of music. All kinds of sounds are artistically acceptable.

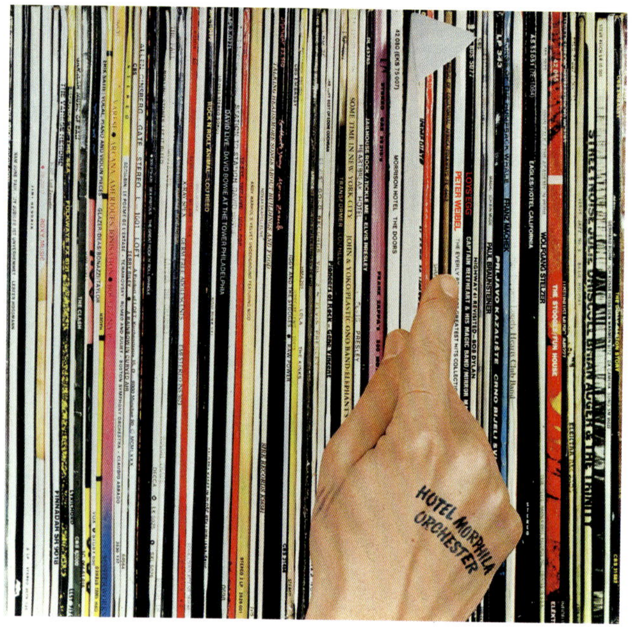

FIG. 59 Hotel Morphila Orchester, *Schwarze Energie* [Black Energy], 1982. Cover.

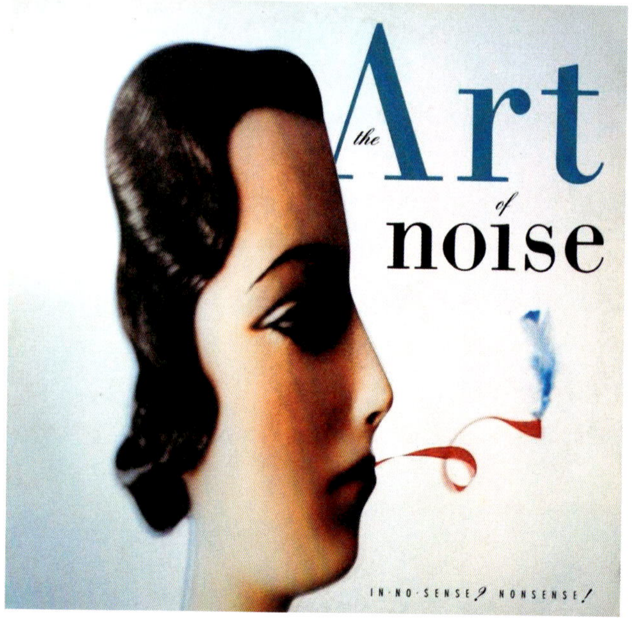

FIG. 60 The Art of Noise, *In No Sense?, Nonsense!*, 1988. Cover.

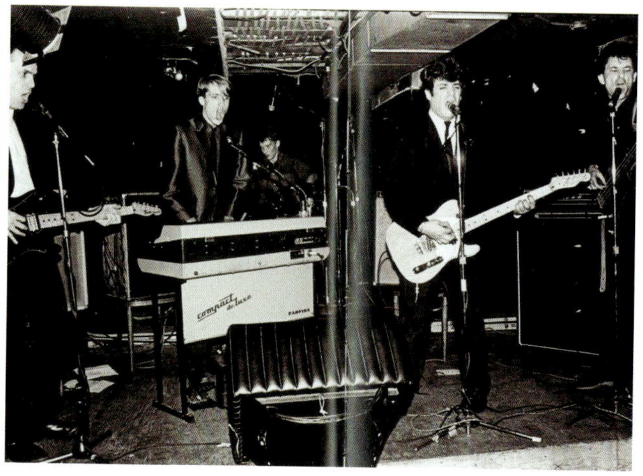

FIG. 61 The band Menthol Wars at the art nightclub Tier 3, New York City, 1980. From left to right: Richard Prince, Joe Hannan, David Linton, Robert Longo, and Jeffrey Glenn.

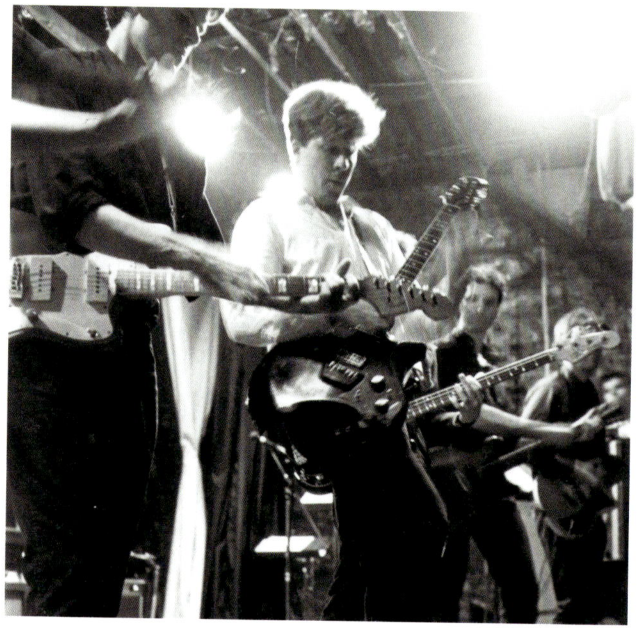

FIG. 62 Rhys Chatham with Catherine Ceresole, Ben Neill, Robert Poss, and Tim Holmes performing Chatham's piece *Die Donnergötter* [The thundergods]. The photograph was most likely shot at the club 8BC in New York City, 1985.

The universe of sound is now coextensive, one might say, with the actual universe. This can also be seen in a curious paradox of physics, a field in which we are accustomed to speaking of the Big Bang. Of course the Big Bang cannot have been a bang, since at the time there were no sound waves. The Big Bang is just a metaphor, an image. If we wanted to stick with a single metaphor, we would have to stop saying "In the beginning was the word"; instead it should be " In the beginning was the bang." The famous harmony of the spheres, therefore, was actually an orgy of noises. And today's music has stepped right into this ocean that is the sonic universe.

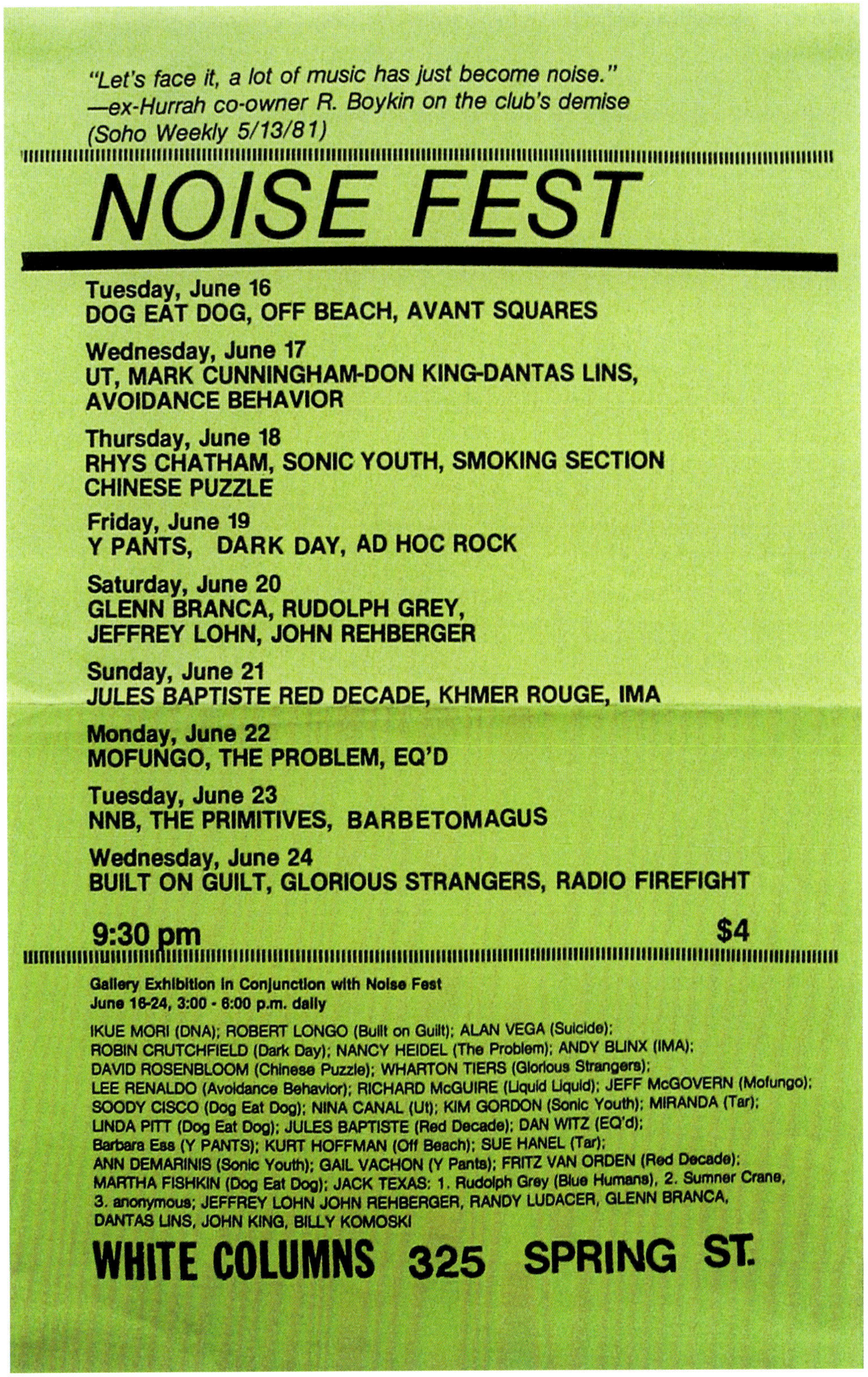

FIG. 63 Poster for the music festival Noise Fest at White Columns, New York City, 1981.

As noted above, the artistic emancipation of noise was realized by the Futurists. Perhaps the most historically significant effect of Russolo's enthronement of noise has been on pop music, from heavy metal to the avant-garde noise music of John Zorn or Arto Lindsay, the Austrian art rock band Hotel Morphila Orchester (FIG.59), the synth-pop group Art of Noise (FIG. 60), Christian Marclay, and the symphonic amplified anthems of Glenn Branca and Rhys Chatham (FIG. 62). Music's struggle against noise is especially apparent in its struggle against rock, new wave, and punk. (FIGS. 61, 63)

To identify the true meaning of the liberation of sound and the emancipation of noise, one must subscribe to Jacques Attali's thesis, presented in his 1977 book *Noise: The Political Economy of Music*, that "the world … is not legible, but audible."[46] In keeping with Plato's dictum that those who move music move the state,[47] Attali hears music as a mimesis of the social order: Music, as dominion over noise, explains the structures of power. Social history is described not just by the philosophy of music, but also by its code.

Attali distinguishes three major periods in the Western evolution of music that explain the use of music by power. The first phase is ritual, in which music is a fight against noise. Music gradually channels noise, socializes it, and shows by its ordered nature that society is possible. In the second phase, the phase of representation, music becomes a spectacle, and society's presentation of itself becomes a concert. The social order is consolidated by the harmony of the notes and their mathematization. The domesticated tone is supposed to instill belief in the harmony of the world, the orderliness of exchange, the legitimacy of money. As a result of money, however, the edifice slowly decays. Stars emerge, masterpieces, a repertoire. Bourgeois society places itself on display. The orchestra, with its director, soloist, and ensembles, becomes a political metaphor for the bourgeois authoritarian society. As a result, from William Godwin (the author of *An Enquiry concerning Political Justice, and its Influence on Morals and Happiness*, 1793, and *Things as They Are; or, The Adventures of Caleb Williams*, 1794), to Federico Fellini (*Orchestra Rehearsal*, 1979), the orchestra has served as metaphor and playing field for experimenting with anarchist insurrection, for the utopia of ungoverned existence, for smashing the bourgeoisie.

Harmony is shattered by the logic of political economy. The logic of political economy accelerates not only the process of the economy itself, turning it into a negotiation with time, but also the process of composing with time – that is, music. Wars and economic crises are reflected in the musical path that leads from Wagner to Schoenberg, from Webern to Cage. Pathos, dissonance, and aleatorism are steps in the decomposition of harmony, in which society's ruptures, its transformation, can be heard.

After ritual and representation, according to Attali, the third era is repetition. In the technological society of the postindustrial age, technological inventions such as radio, television, records, and the tape recorder overturned musical habits, relocating music-listening from the concert hall to the home. And naturally, representational music became, to an even greater extent, a site of mere social representation, a spectacle without music, where the stars, the maestros, are the real event. As a result of the shift from live concert music to the infinite technological reproducibility of canned music, music – once a festive occasion – has now become an everyday phenomenon. From the elevator to the kitchen, there is almost no place anymore where music is not present. As a result of its technological reproducibility, music can be reproduced anywhere; it is at home everywhere (not just in the concert hall), thoroughly inscribed on our surroundings, and ultimately inseparable from our everyday lives.

Of course this is reflected in the work of music itself: from the repetitive structure of pop hits and minimal music to Cage's everyday music, which opened up the sonic space of music to the banal noises of daily life. The endless repeatability of canned music is also a prerequisite for the treatment of music as a consumer good. In Muzak,[48] the characteristics of consumer music

46
Jacques Attali, *Noise: The Political Economy of Music*, trans. Brian Massumi (Minneapolis: University of Minnesota Press, 1985), 3.

47
"For never are the ways of music moved without the greatest political laws being moved." Plato, *The Republic*, trans. Allan Bloom, 2nd ed. (New York: Basic Books, 1991), 102.

48
Especially in the English-speaking world, *Muzak* signifies a kind of popular instrumental background music often heard in department stores and elevators. The word comes from Muzak Holdings, an American company that was known for producing "easy listening" music.

49
Daniel Boorstin, *The Image: A Guide to Pseudo-Events in America* (New York: Harper & Row, 1964), 176.

50
John Cage, "Edgard Varèse," in *Silence* (Middletown, CT: Wesleyan University Press, 1961), 83–84, here 84. Originally published in 1958 in the Swedish magazine *Nutida Musik*, in an issue devoted to Varèse.

51
Kurt Schwitters, "Merz," trans. Ralph Manheim, in *Modern Artists on Art*, ed. Robert L. Herbert, 2nd ed. (Mineola, NY: Dover, 2000), 68–75, here 70.

are simply present in greater concentration than elsewhere; or, as the Muzak executive Donald O'Neill said, "We don't sell music; we sell programming."[49] This programming varies depending on the situation: a restaurant (strings at lunchtime, for example), an office (livelier rhythms in the afternoon, when employees are feeling a bit tired), or a factory. Muzak is that superficial music in which the purpose of representational music becomes audible: that is, to efface the unpleasant noise of work. The music of elevators, department stores, and the radio, that lulling din, is the obverse of the frightening din of reality. It is intended, like the entire industry of mass-media entertainment, to engender that agony of consciousness in which one loses power over oneself and thus succumbs that much more easily to the enticements of consumption. To a great extent, the function of representational music is to agonize consciousness in order to drown out reality. Classical music, such as the music of the women's orchestra that played for the Nazi beasts at Auschwitz, displays music's silence in the face of the real. It muffles and suppresses sinister reality, providing furniture for a vacuum of consciousness in which even the finest orchestration will not render audible the most brutal barbarity.

During the Nazi regime, the educative and ennobling quality that had been ascribed to representational music was revealed once and for all as stupidity or hypocrisy. Art music obscures reality just as much as music made for consumption does. Thus the most varied forms of bruitism – Einstürzende Neubauten, Throbbing Gristle, etc. – are efforts to give voice to reality, to shake consciousness awake. The noise of the street shall be heard in the concert hall as well. In its rebellion against representational music, noise is also a rebellion against bourgeois culture and the bourgeois social order and it is an incursion of money- and capital-ridden reality into the concert hall (society) that has been cleansed of all traces of work, exploitation, and the real. Early forms of rock and roll, new wave, and noise music, with their noisiness and accelerated speeds, were a few such invasions, screams of the real. Music that really wants to be music, intelligible music, something more than audible music, has to open up, open itself up to noise, create an open space through which all forms of sound – whether noise, sound, or tone – and all instruments – whether musical or unmusical – can move.

Classical music is silent about reality. Noise – racket, clamor – is thus often the legitimate sound of the real. Liberated sound (longed for by artists from Busoni to Russolo, from Varèse to Cage) gives reality freedom and a space where it can have a voice. When Cage opened the windows and doors of the concert hall to let in the noise from the street, he consummated in a gesture the liberation and opening of music initiated by Russolo and Varèse, the termination of representational music. Because classical music is silent about reality, Cage had to silence music in order to make the noises, the sounds, the voices, the din of reality itself audible. On August 29, 1952, when he sat silently in front of a piano for 4:33 minutes, it was an indictment of music's silence (about reality). The audience was forced to hear the noise of the real (instead of music). What Cage wrote about Varèse also applied to himself: "He established the present nature of music. This nature … arises from an acceptance of all audible phenomena as material proper to music. While others were still discriminating 'musical' tones from noises, Varèse moved into the field of sound itself. … That he fathered forth noise … makes him more relative to present musical necessity than even the Viennese masters."[50]

In truth, they were both more important than Schoenberg's circle to the liberation of music and its continuing development, because they moved into the open field of tone and noise, where all audible phenomena were accepted as music. The slogan "All sonic events are music," or rather, "Everything that makes sound is music," corresponds to Kurt Schwitters's definition of Merz, which made use of "any material whatsoever" and saw "freedom from all fetters" in the merging of all factors into the total work of art.[51] The totalizing claim that "Everything [audible] is music" anticipated the pronouncements of the 1960s: "Everything is poetry" (Peter Weibel), "Everything is sculpture" (Joseph Beuys), "Everything is architecture" (Hans Hollein).

In the sixties, everything (from concepts to earth) became artistic material, and also sonic material. Everything that produced sound served as a musical instrument, any sound could be classified and interpreted as music, including speech and language, conversations and interviews.

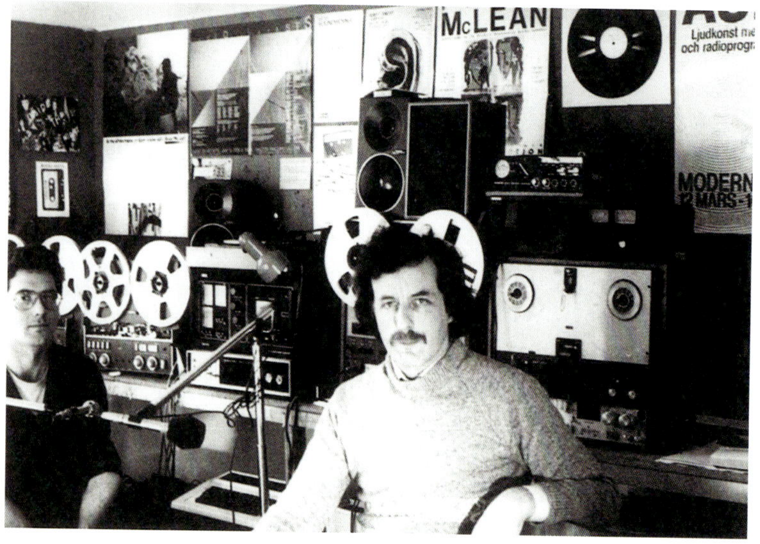

FIG. 64 Michael Archer and William Furlong (right) at the *Audio Arts studio* at the end of the 1970s.

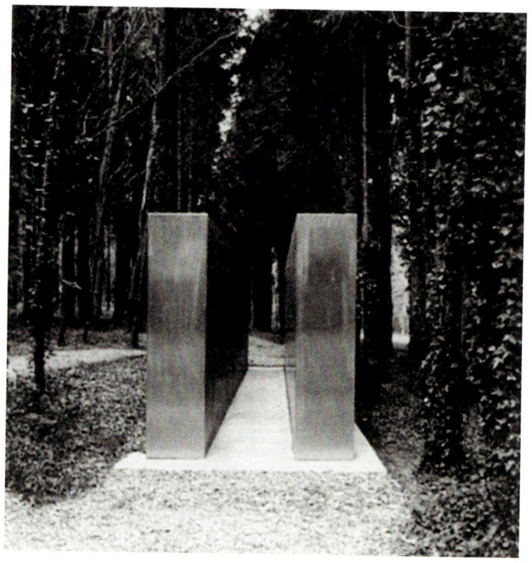

FIG. 65 William Furlong, *Walls of Sound*, 1997. Sound sculpture in urban space.

In the 1970s, Moscow Conceptualists held conversations about art, documented them, and declared them artworks. The Collective Actions group, which was organized by Andrei Monastyrski, and at different times included artists as Ilya Kabakov, Erik Bulatov, and Dmitri Aleksandrovich Prigov, had an agenda for their activities: “living in art.” Therefore even conversation with spoken words could become sonic material and art.

This is why in 1973 the British artist William Furlong established together with Barry Barker an innovative sound magazine on cassettes: *Audio Arts*. For the magazine, which was published four times a year, artists, curators, and scientists were interviewed, or sound performances were recorded. Amongst many others these were Joseph Beuys, Nam June Paik, James Joyce, Noam Chomsky, Andy Warhol, Anish Kapoor, Gilbert & George, Yoko Ono, R. Buckminster Fuller, Hermann Nitsch, Mario Merz, and Gerhard Richter. Furlong recounts: “I started *Audio Arts* ... when the audiocassette was just having an impact. ... people being able to do their own recording ..., make copies and pass them on to their friends. Other people like myself started thinking that the process of recording was now in the hands of everybody, that the audiocassette was a very democratic medium.”[52] (FIG. 64)

In 1982, Furlong commissioned works with a length of five minutes from 45 international artists under the title *Live to Air*, among others Julia Heyward, Bruce McLean, Dan Graham, Arleen Schloss, Peter Gordon, Hank Bull, Dieter Roth, Helen Chadwick, Les Levine, Ulay/Abramović, and Steve Willats. These were published on three cassettes with a twelve-page catalogue in a flat cardboard box. The occasion for this project was a presentation of *Audio Arts* at the Tate Gallery in London. On the front page of the catalogue Furlong writes: “The medium of magnetic tape, and the processes and technologies associated with sound recording, now represent a significant and growing area of art practice that has, up until now, been largely overlooked. This work has no form other than playback from tape ... Furthermore, the processes of working with sound provide artists with similar manipulative, structural and ideological freedoms and possibilities to those normally associated with the traditional media, such as painting, sculpture and collage. In many respects, this audio-technological ‘space’ is parallel to the physical space of a gallery, yet extends it through the potential of widespread dissemination inherent in the multiple production of cassettes and through broadcasting.”[53] A collection of his writings entitled *Audio Arts* has been published in German and English (Leipzig: Reclam Verlag, 1992).

As an established sound artist, Furlong uses voice and sound as materials. This arose from his intimate association with the voice while editing and recording interviews for *Audio Arts*. Many of his sculptures can be interpreted as architectural homes for sound to be pushed and pulled around a space, much like paint is pushed around a canvas. (FIG. 65)

New Sound Instruments

The "expansion of the arts" proclaimed in the 1960s,[54] in which art became the universal quantifier, not only produced new media, materials, and methods, but also gradually replaced representation with reality. This process, however, had already begun in 1913. Up to 1900, the program of classical painting had been committed to mimesis - that is, to reproducing reality as perfectly as possible, using the tools of painterly representation, point, line, plane, and color - a program laid out by Leonardo da Vinci in 1490 in his famous *Treatise on Painting*. With Malevich and the rest began the reductive program of modernism, which would sever the bond between representation and reality. Point, line, plane, and color would no longer serve the imitation and representation of reality; from now on they were only to represent themselves. Think of the title of Wassily Kandinsky's 1926 book, *Point and Line to Plane*. What we call abstraction - the banishment of the object from the picture - is none other than the self-representation of the basic elements of painterly representation. Point, line, plane, and color no longer represent themselves; they constitute reality. The depiction of reality by means of the elements of painting is replaced by the self-depiction of those elements.

Though objects of this sort were banished from the picture, they found their way back into art as real objects. What Duchamp was exhibiting with his readymades, as the name makes clear, were not objects made by the artist, but rather manufactured products, usually industrially produced. The readymades, such as his bottle rack (*Porte-bouteilles*, 1914) (FIG. 66), stand for the self-representation of the object as object. In the course of the twentieth century, almost everything that was representation was replaced by reality: painted landscapes by land art, painted animals by real animals, painted portraits by real bodies (body art, performance art), painted interiors by art environments, painted waterfalls by real waterfalls, painted objects by installations, etc.

At the same time, however, alongside the self-representation of the elements of representation and the self-representation of objects, something new emerged: technological media. The new visual medium of photography had already been introduced by 1840, while film, the technological medium of moving images, emerged around 1900. These were followed by technological media for the storage and distribution of sound: phonograph, microphone, telephone, radio, tape recorder, loudspeaker, amplifier, etc. Artists were particularly fascinated by amplification technology. Abstraction confronted easel painting with a crisis[55] that led to a departure from pictures: The easel picture split into easel (visual object) and picture. The image continued to wander into technological host mediums such as video, while in painting, all that remained of the easel picture was the easel (Francesco Lo Savio, 1960). The result was that the new visual media took the place of the old classical visual media such as painting. Something similar took place in music: As a consequence of the technological revolution, new technological musical instruments stepped into the position of the old musical instruments.

With his *Musique d'ameublement* (1917–1924) (FIG. 67), Satie staked out this new field of sound with a call for a kind of music that would function in much the same way as furniture - that is,

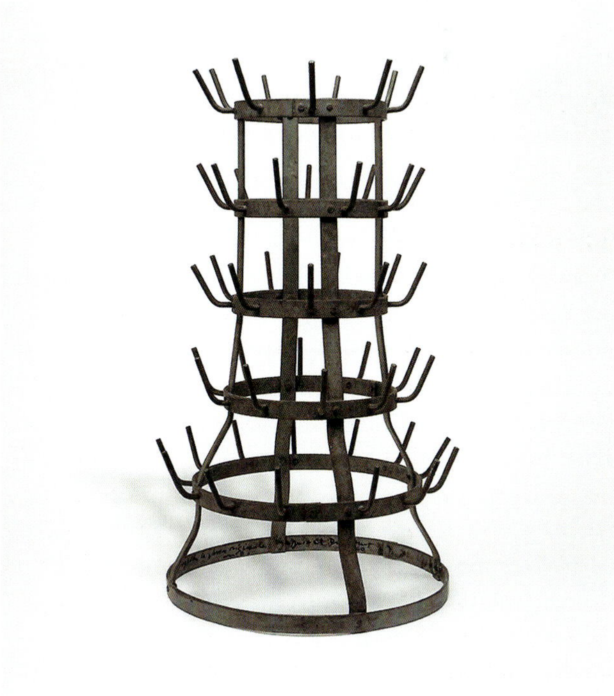

FIG. 66 Marcel Duchamp, *Porte-bouteilles / Bottle Rack*, 1914/1959. Galvanized iron, 59.1 × 36.8 cm.

52
William Furlong, quoted in Heidi Grundmann, "William Furlong: Audio Arts," in *See This Sound: Promises in Sound and Vision*, ed. Cosima Rainer et al. (Cologne: König, 2009) 168–79, here 175.

53
William Furlong, "Live to Air. Artists' Sound Works," in: Inlay for *Audio Arts* 5, nos. 3&4 (1982): 1.

54
Cf. Jürgen Claus, *Expansion der Kunst. Beiträge zu Theorie und Praxis öffentlicher Kunst* (Reinbek: Rowohlt, 1970).

55
Cf. Clement Greenberg, "The Crisis of the Easel Picture," in *Art and Culture: Critical Essays* (Boston: Beacon, 1965), 154–57. Originally published in *Partisan Review*, April 1948.

like noise, it would be part of the environment. This new field of music as a field of sound, noise, tone, and silence was expanded by Cage. When he demanded, in "Experimental Music" (1957), that "the doors of the music" be opened "to the sounds that happen to be in the environment,"[56] resulting in a "total sound-space,"[57] it was an echo of Satie's voice, which lives on in Cage's writings almost as clearly as Varèse's – for example, in Cage's credo "The Future of Music" (1937), the opening text in the anthology *Silence*: "I believe that the use of noise to make music will continue and increase until we reach a music produced through the aid of electrical instruments which will make available for musical purposes any and all sounds that can be heard. Photoelectric, film, and mechanical mediums for the synthetic production of music will be explored. Whereas, in the past, the point of disagreement has been between dissonance and consonance, it will be, in the immediate future, between noise and so-called musical sounds."[58]

Russolo, Satie, and Varèse laid the foundations for a free sound and an open space for music. Two steps seem to have been critical in this process: the quest for new instruments and the incorporation of noise into music. With the collapse of the boundary between musical and nonmusical tones, allowing all tones, noises, and sounds to be organized into music in the "total sound-space," music opened up to the new noises of a new mechanical and industrial reality. No longer did music seek to dominate noise; instead, noise became a part of music. And of course, to obtain new mechanical-industrial sounds, it was also necessary to build new devices.

Mechanical-industrial society could not be adequately conveyed with the handheld wooden instruments that were suited to historical representational music. Busoni, Varèse, Cage, and others set out to find new sound instruments, noise instruments, electromechanical sound generators. Originally intended as channels of distribution and reception, such technological inventions of industrial society as radio, records, record players, reel-to-reel recorders, and cassette recorders (the shift from concert hall to private home) were converted into means of production. Attali's description of the age of repetition in music applies here – meaning that a response to technological devices can be heard in the musical work itself. The new sound instruments operated electromagnetically or electronically. Following the era of experimentation with magnetic-tape devices (Varèse, Pierre Schaeffer, Henry, Cage, Brown, etc.), and beginning with Xenakis, Stockhausen and others, interest today has centered on using electronic instruments to produce sounds synthetically – in other words, to make precisely the sort of music that is only possible because these devices exist. With the digital sampler, the ideal of this school – to "make available for musical purposes any and all sounds that can be heard"[59] – has been attained.

Ever since Russolo, artists and composers have worked together to expand the sonic alphabet, the alphabet of sounds. On the whole, they have sought to make the world of sounds coextensive with the world of objects. Everything that makes a sound could serve as a musical instrument; therefore, any object capable of producing sound became an instrument. No longer satisfied to write compositions for the piano, the violin, etc., they wanted instead to use new instruments to create new tones. They wanted, ultimately, to go beyond the tones that could be produced with classical instruments or everyday objects.

In the 1950s, composers began expanding the sound spectrum to include magnetic, electrical, electromagnetic, and electronic sound sources, which is where computers came in. The computer is not just a calculating machine, as the word *computer* would suggest, but also, and equally, a text machine, an image machine, a sound machine – a universal medium.

For centuries, classical musical instruments such as the violin, flute, trumpet, and piano have given us a wonderful but circumscribed universe of sounds. Using the ingenious "software" of compositional technique and such limited mechanical "hardware" as the piano – a machine of wood, ivory, wire – composers have created an almost otherworldly cosmos of sound. All over the world, academies have been established to educate sufficient numbers of students who would be capable, as performers and composers, of reproducing this cosmos. Buildings have been built all over the world as well – concert halls and opera houses – in which the physical characteristics required for the optimal presentation of

56 Cage, "Experimental Music," 8

57 Ibid., 9.

58 John Cage, "The Future of Music: Credo," in *Silence*, 3–6, here 3–4. Originally presented as a lecture in Seattle in 1937.

59 Ibid., 4.

this music would be provided. A vast arsenal of music venues, orchestra musicians, students, composers, instrumentalists, singers, and critics kept this musical universe constantly alive. Around 1900, however, with the emergence of technological distribution and production media, a rift appeared.

The composers who used these new technological sound instruments in their compositions to produce new sounds were completely barred from the orchestras and concert halls. One might almost conclude that the actual music of the twentieth century – music that was composed in the twentieth century, in a way that was only possible in the twentieth century – could not and did not take place in the concert halls. One might say that what has been played, in the majestic and majority-pleasing concert halls of the twentieth and twenty-first centuries, has been primarily the music of the preceding centuries. New music, then, is still everything that is not played in concert halls because the instruments and musicians that would be suited to it are not to be found there. Above all, sound art is everything that has been shut out of the concert hall.

This is why the composers of new music dreamed of new sound pavilions built specifically for their music. Composers who worked exclusively with the new technological sound instruments wound up in exile: in the museum – for only there happened exhibitions of what was called "sound art," a further evolution of new music that gradually differentiated itself and became autonomous.

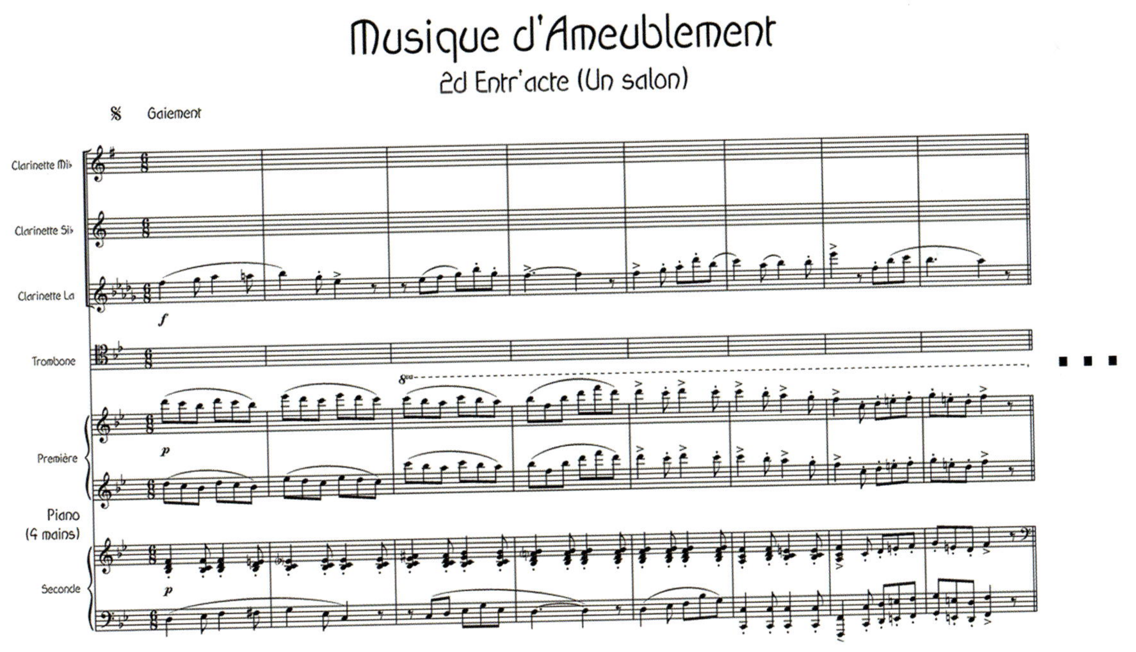

FIG. 67 Erik Satie, *Musique d'ameublement* [Furniture music], 1917. Score.

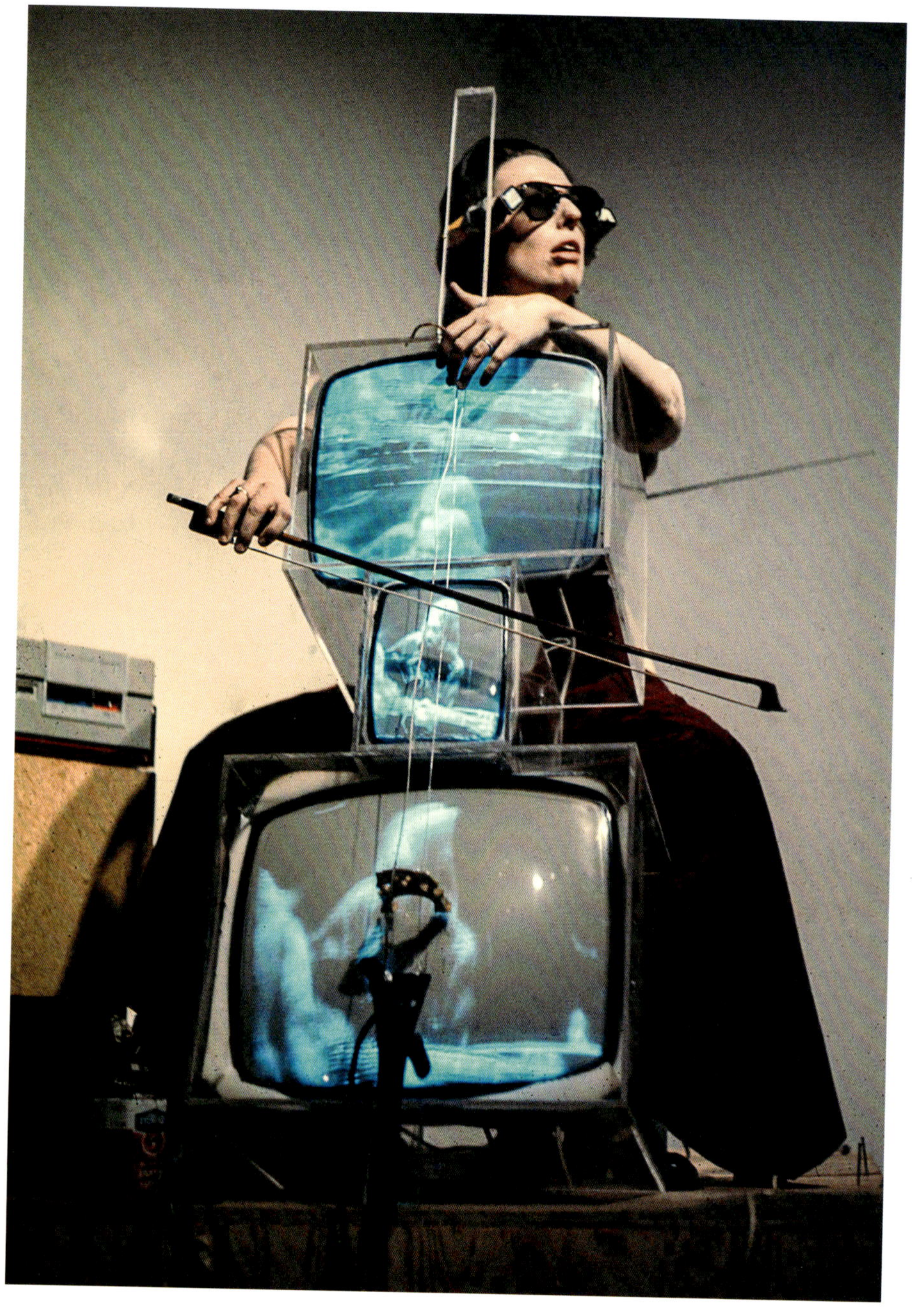

Charlotte Moorman performs Nam June Paik's *TV Cello* wearing Paik's *TV Glasses*, New York, 1971.

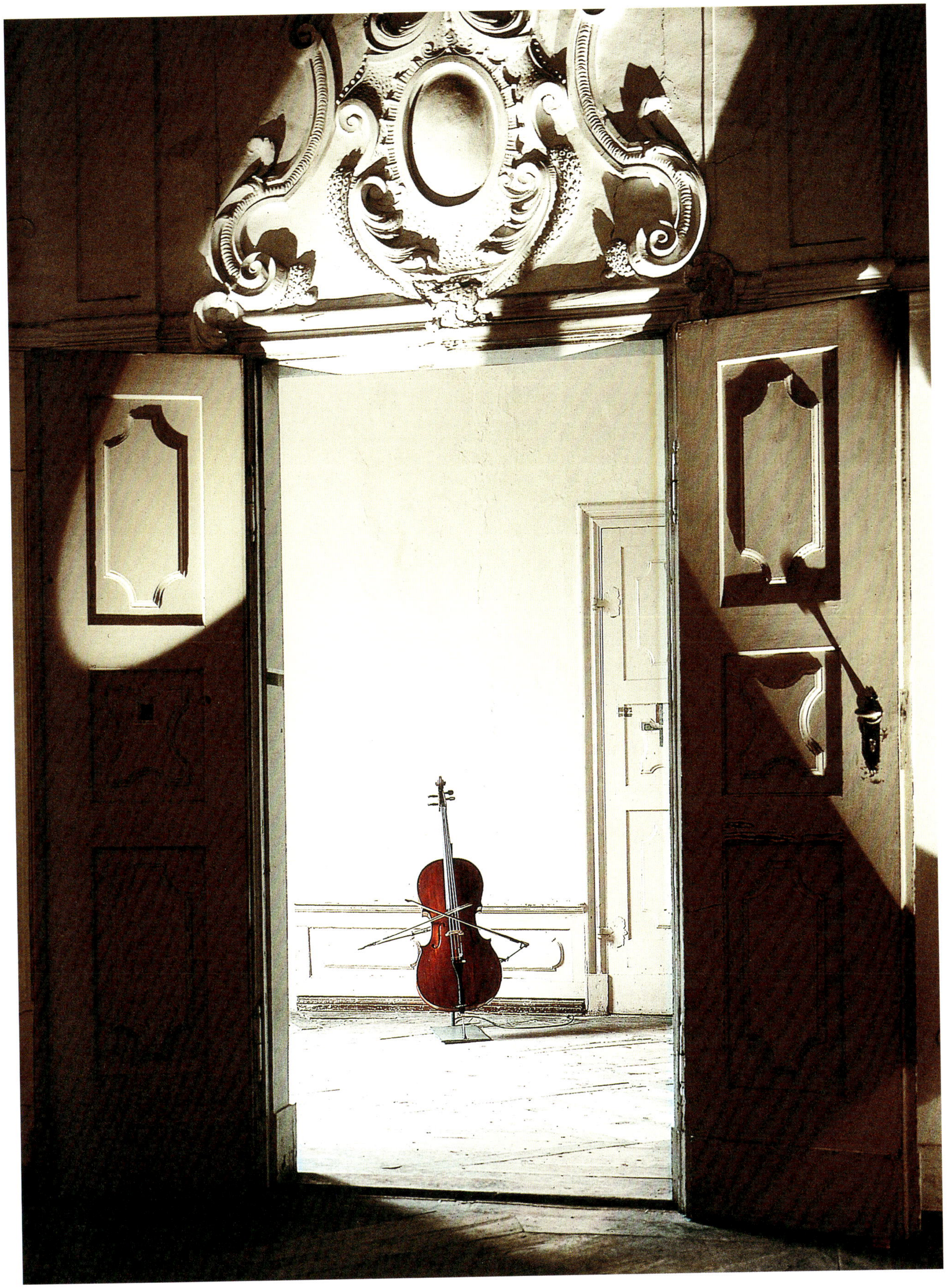

Rebecca Horn, *Concert for the Path of the Bees*, 1999.
Mechanically activated instruments (cello and violin).

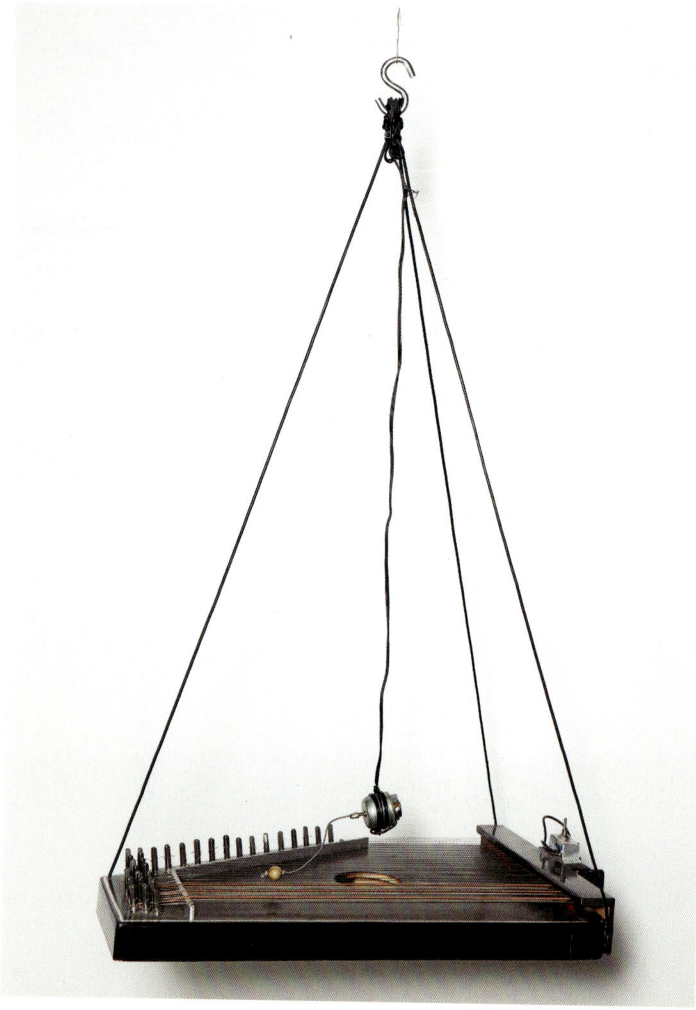

Joe Jones, *Mechanical Surmandal*, 1978.
Surmandal (Indian harp) with mechanical beater.
Created in Asolo, Italy. 6 × 49 × 28 cm.

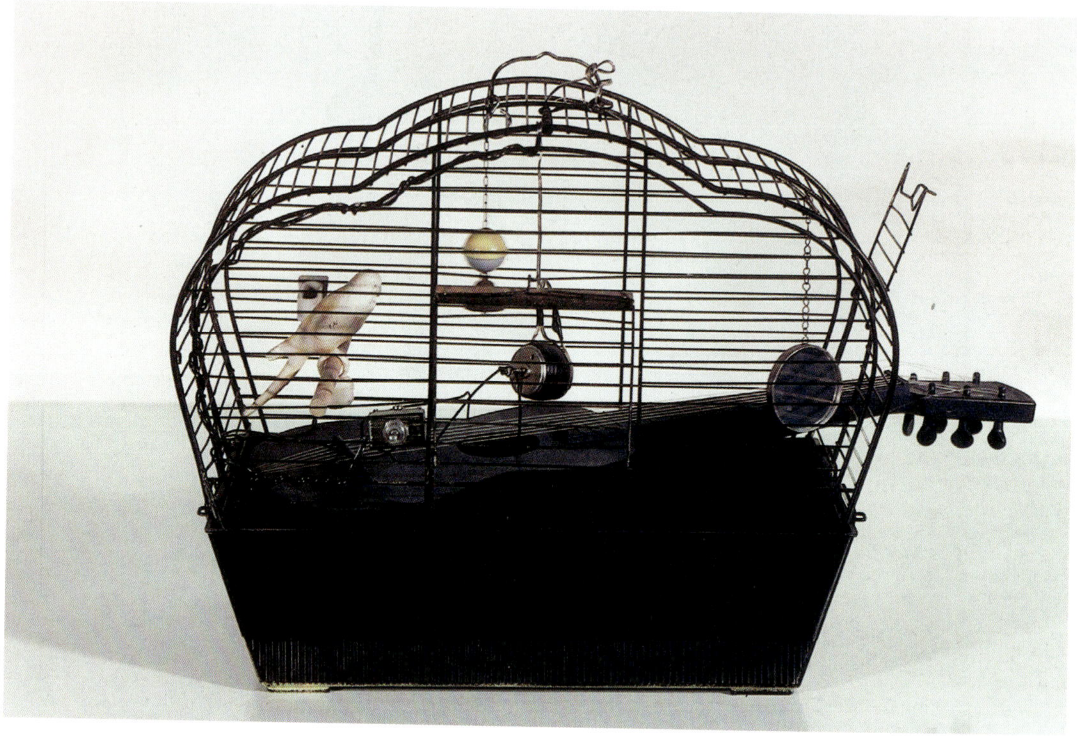

Joe Jones, *Bird Cage*, 1964.

Waltraut Cooper,
Scherzo, 1988.
Installation for piano, neon, and electronics in the tradition of Scribin's color organ and Cage's prepared piano.
Music by Anestis Logothetis (Vienna), Nicola Sani, Luca Spagnoletti (SIM, Rome).
Installation view Ars Electronica Linz, 1988.

To each of the 84 keys of a piano, neon tubes in certain colors and – instead of the strings – stored sound events have been allocated. Maximum duration of these sound events is 8.5 seconds, if the key is held shorter, only part of the sound event is sounded (also a rest).

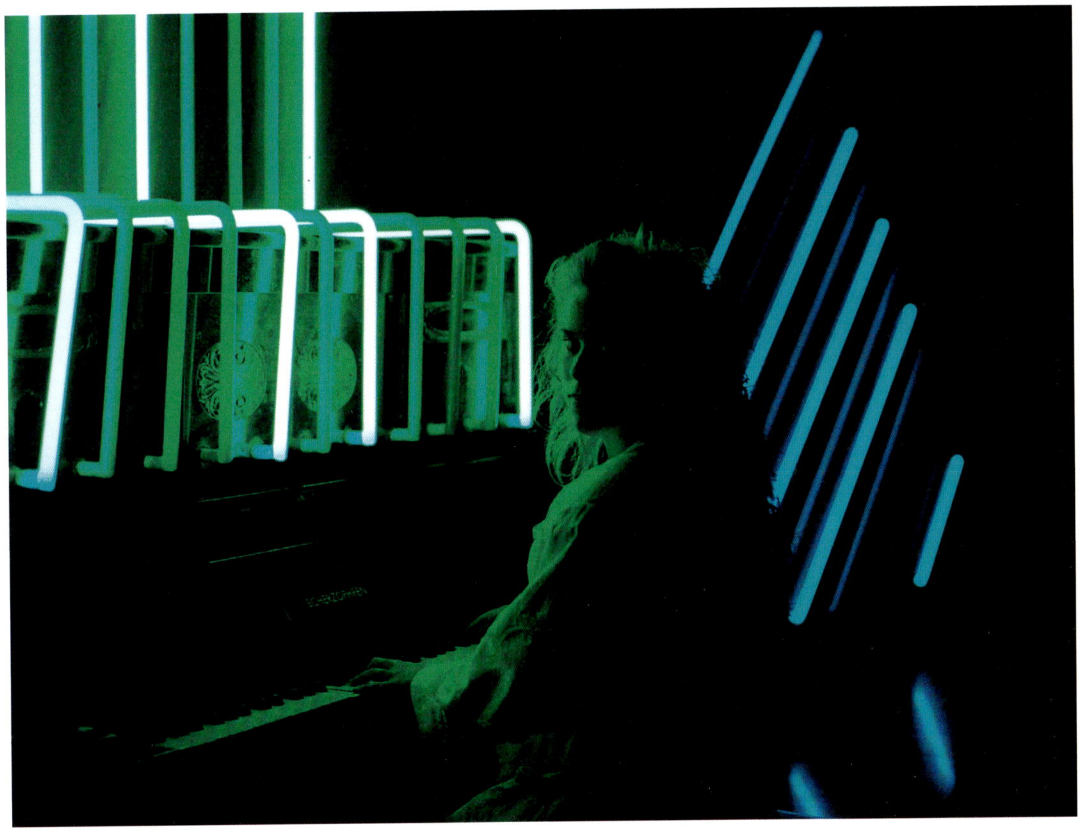

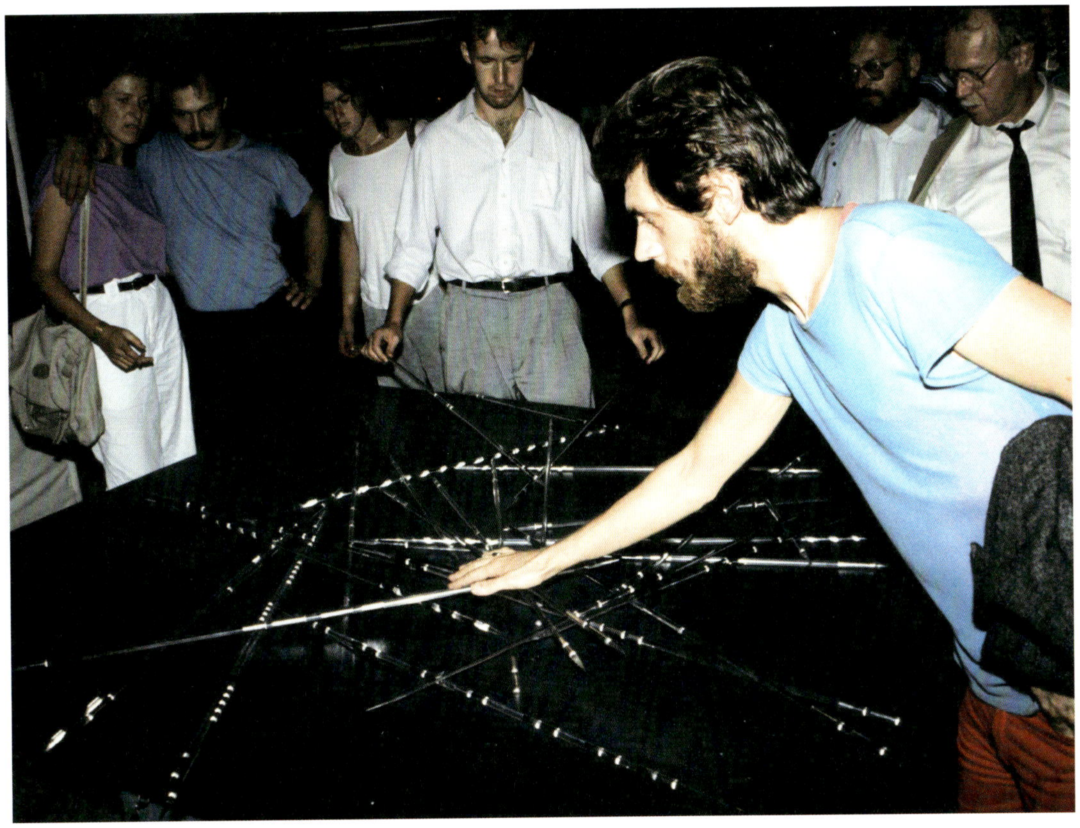

Waltraut Cooper,
Klangmikado [Musical jackstraws], 1987.
Interactive sound instrument.
Music by Gerhard E. Winkler.
Installation view Bronx Museum, New York, 1991.

Each play with the oversized jackstraws creates a new world of sound via electronics and computer and creates a new composition.

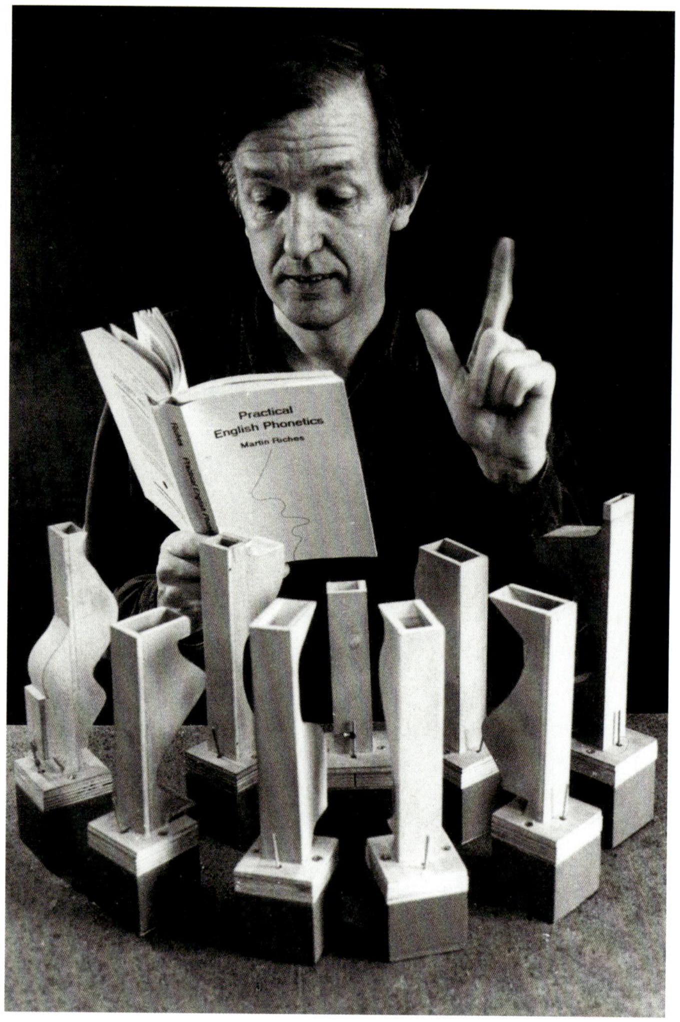

Martin Riches teaching the machine to speak, 1989.

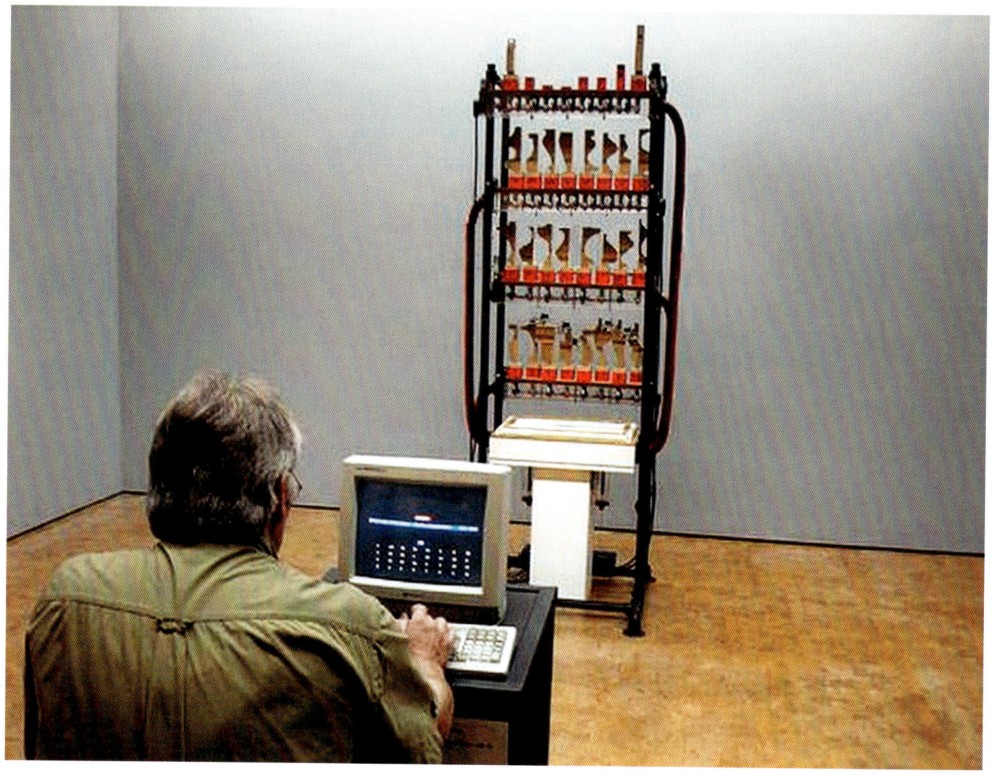

Martin Riches, *Talking Machine*, 1989–1991.
32 pipes and air valves, wind chests, magazine bellows, blower, computer. 230cm. Installation view ZKM | Karlsruhe, 2007.

The white box at the bottom is a sound-proof casing enclosing the blower. The flat white box above that is a magazine bellows which evens out the air pressure no matter how many pipes are being played. The air goes up the red hoses to the four wind chests which carry the pipes. The wind chests are transparent so the movement of the electro-magnetic valves can be seen and this is further reinforced by LEDs attached to each valve. The black cables carry the signals from the computer with the user interface to the valves.

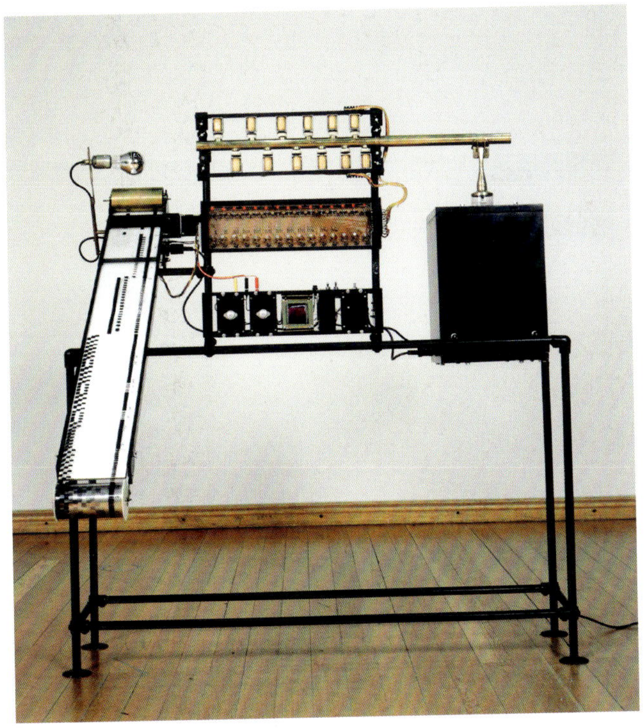

Martin Riches, *The Flute Playing Machine*, 1979–1982.

On the right is the blower encased in its soundproof black box. On top of it is a valve that admits the air into the mouth piece of the flute. The horizontal brass tube is an especially made alto flute with a range of g to g'. It is fitted with 12 electromagnetic keys corresponding to the fingers of a human player. The diagonal slide on the left guides the music roll over a row of 15 photocells. Their signals are amplified by circuitry in the box below the flute and operate the keys and the air valve. The box below that contains the power supplies and the speed controller for the blower motor.

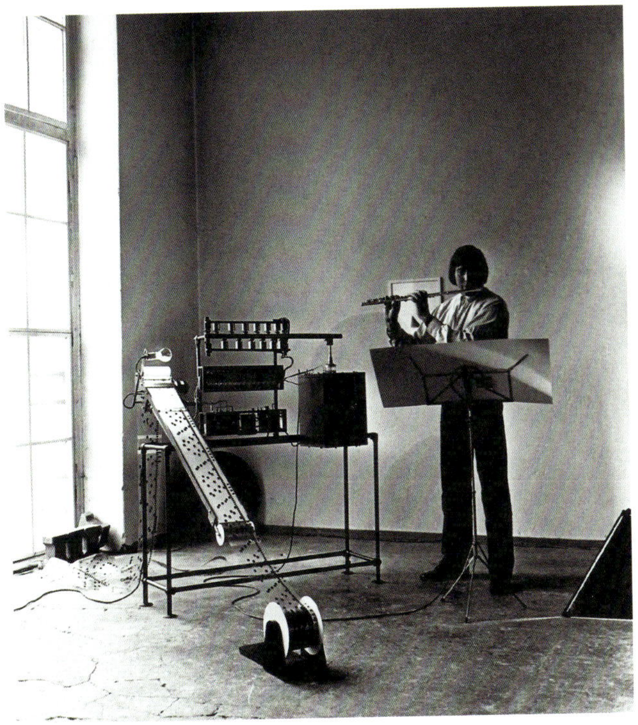

The flute player Eberhard Blum playing a duet with the machine.

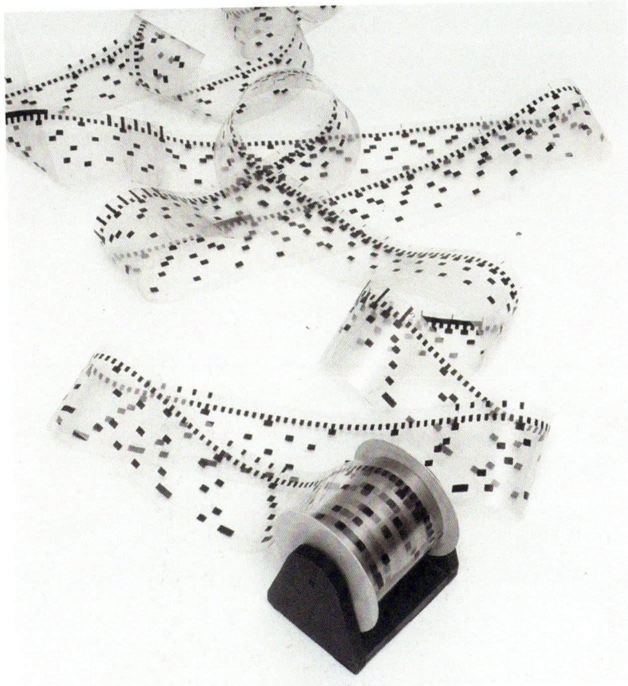

The 27 meter long music roll with the machine part for *All Change* (1982) by Schaun Tozer, a duet for flute and *The Flute Playing Machine*.

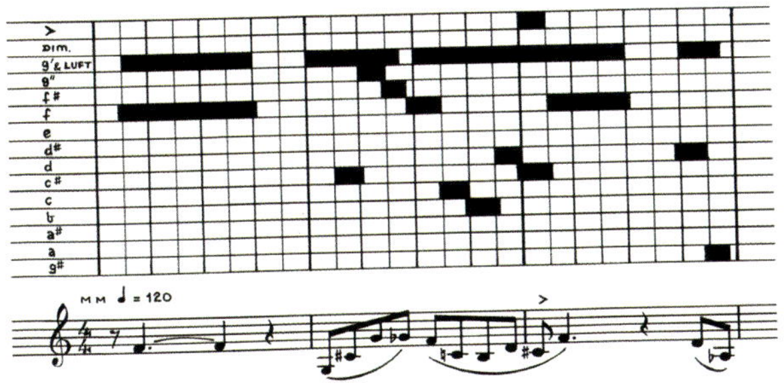

Martin Riches, *Score for The Flute playing Machine*.

The lowest 12 tracks operate the keys of the flute, the next one up opens and closes the air supply and the two upper tracks are spare. In the score above one of these two spare tracks is used to put a stress on a note. When the machine is playing a duet these two spare tracks are used to send cues to the live player – two LEDs on the music stand. The example is taken from *Afternoon Loop* by Hartmut Westphal (1980).

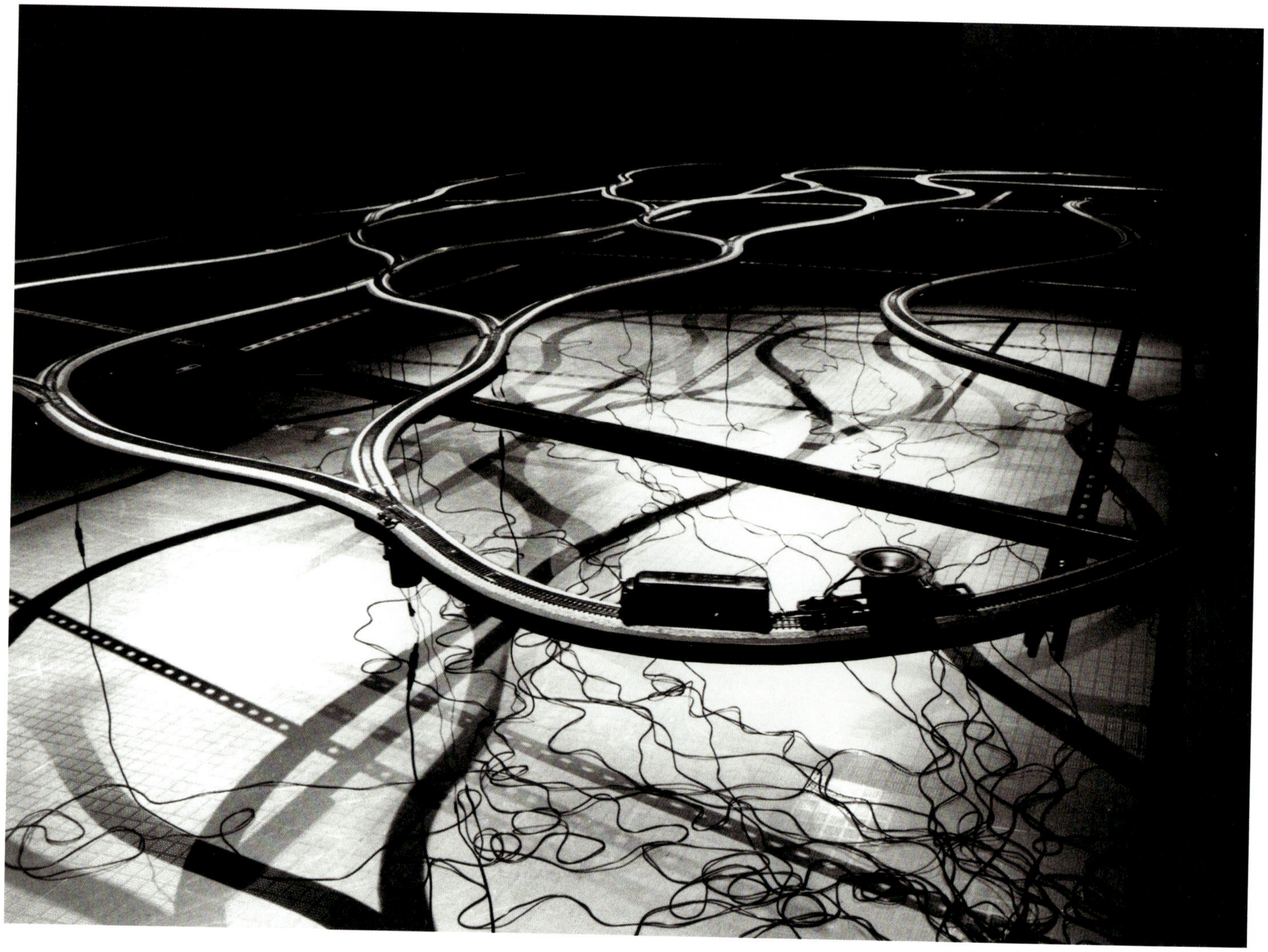

Ed Osborn, *Parabolica*, 1996.
Model train parts, wood, steel, loudspeaker, custom electronics, sound, dimensions variable (approximately 5 × 7 meters). Installation view at the Yerba Buena Center for the Arts, San Francisco, CA, March, 1996.

Playing themes of free will and self-determination against notions of perpetual containment and loss of control, *Parabolica* articulates an individual presence cycling endlessly in a larger unknown. In it, a model train car with a speaker attached to it broadcasts a variety of sounds as it circulates around a track.

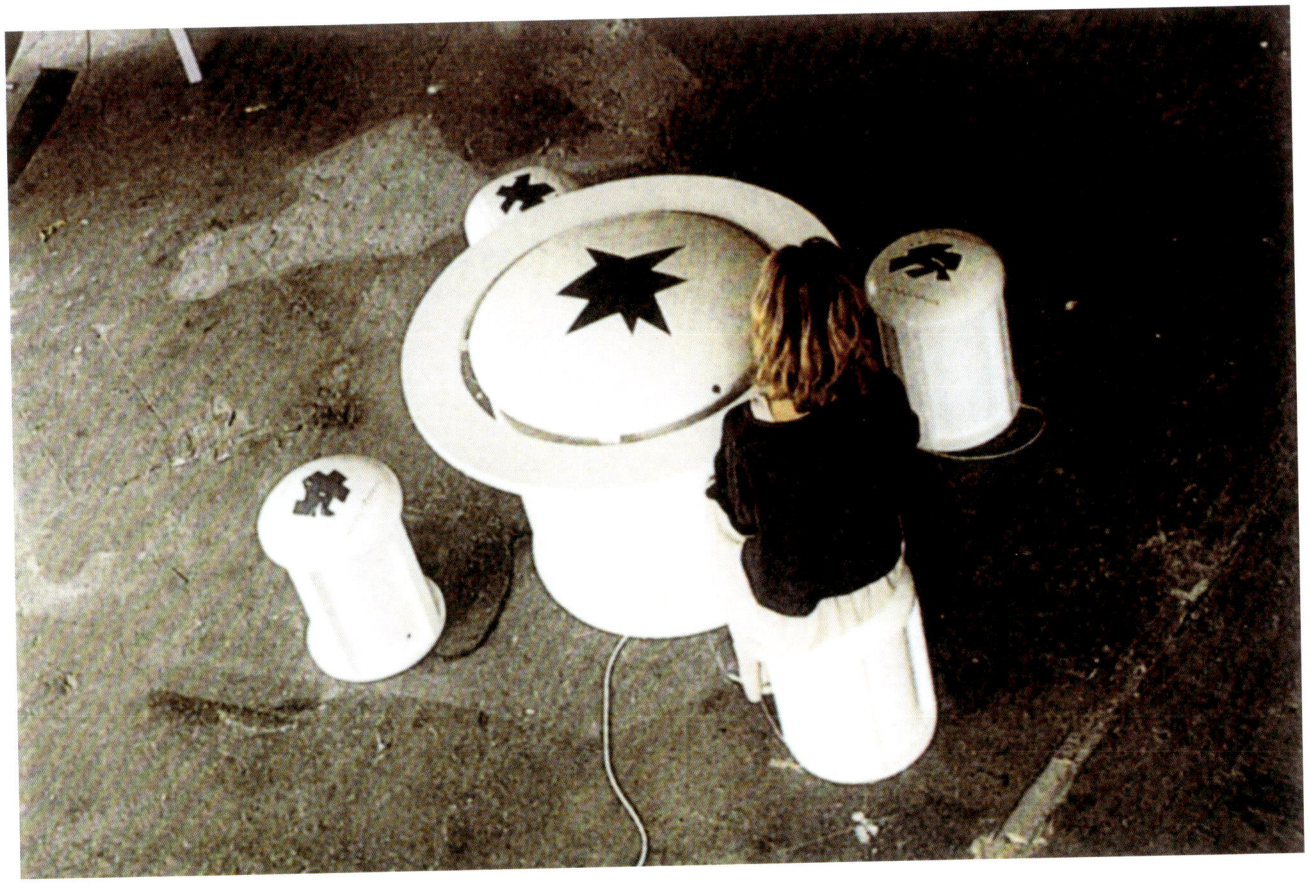

Franz Xaver, *RadioARTives Konferenzmöbel*, 1987.
Acoustic conversion of the buttock heat of the visitors.

For his sound installation *RadoARTives Konferenzmöbel* [RadioARTive conference furniture], Franz Xaver uses picture tubes that he has converted into sculptures. Acoustic signals can be generated by the seats using the body heat of the bottom, on the one hand, and by a cassette deck, on the other. These signals are formed directly into an image with the speed, direction, and black-and-white content of the electron beam. The table establishes the coupling of the four stools acoustically and visually.

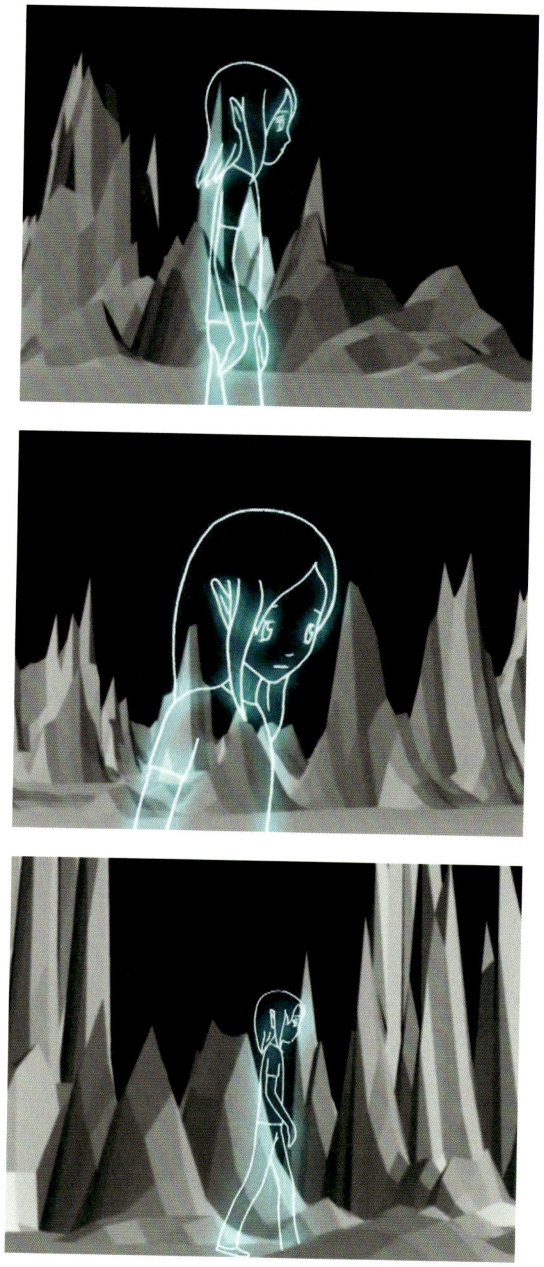

Pierre Huyghe, *One Million Kingdoms*, 2001.
Animated film, color, sound, 7 min., screenshots.

The tension between fact and fiction is at play in *One Million Kingdoms*, a work conceived as part of the collaborative project *No Ghost Just a Shell*, in which a manga character named Annlee is inserted into multiple artistic contexts. In Huyghe's animation, this adolescent girl wanders through a shifting lunar topography and, speaking in a digitally synthesized form of astronaut Neil Armstrong's voice, delivers a narration blending the actual transmissions from the Apollo 11 mission with excerpts from Jules Verne's 1864 novel *Journey to the Center of the Earth*. Annlee appears as a translucent outline, an empty cipher for creative interpretation. Yet at the same time, she is literally the author of her own environment: the mutating features of the landscape through which she walks are generated by the inflections of her own voice.

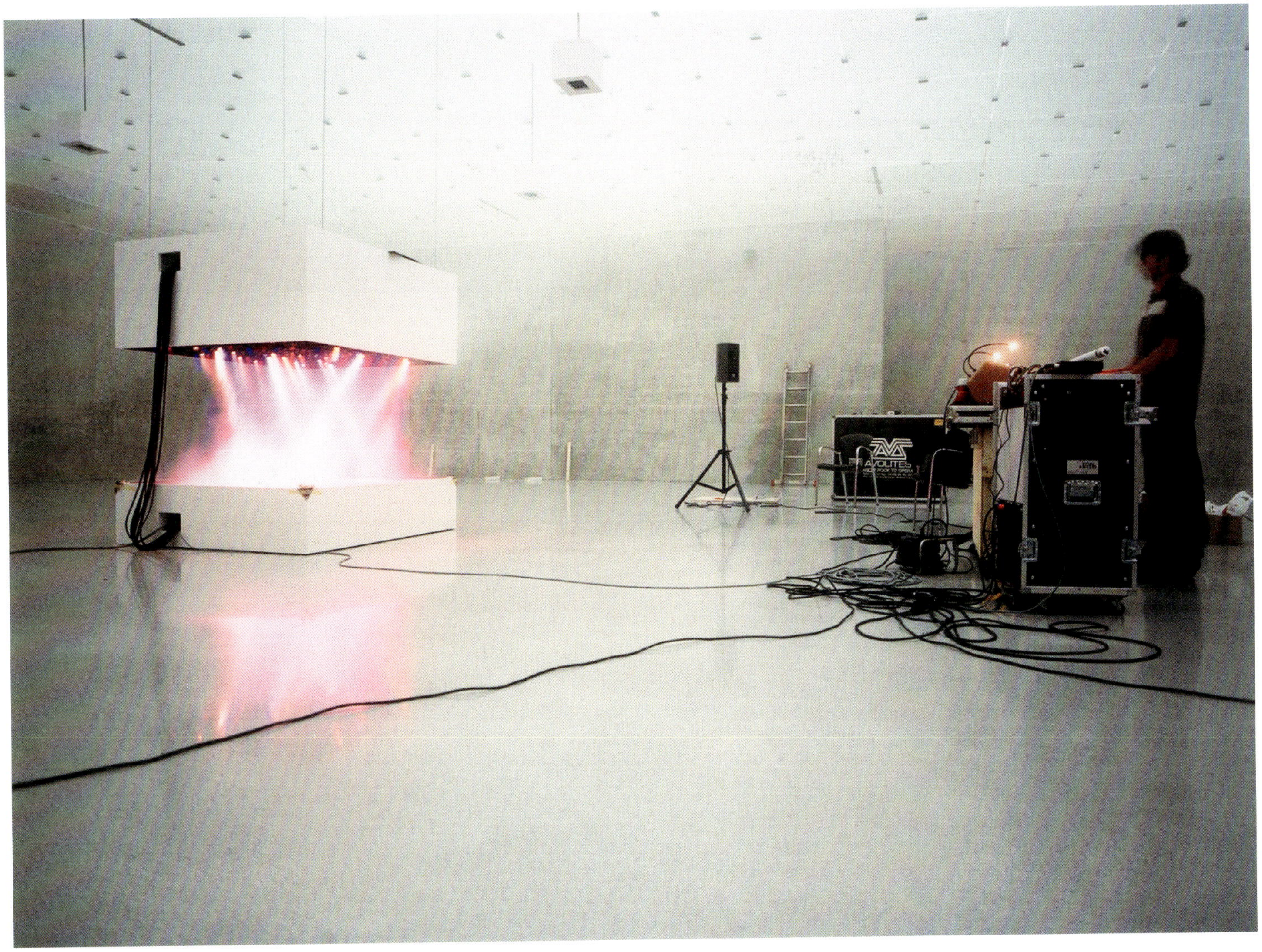

Pierre Huyghe, *L'Expédition scintillante, Acte 2 "Untitled" (Light Box)*, 2002.
Smoke and light system with sound. Installation view Kunsthaus, Bregenz, Austria.

The sounds of Erik Satie reorchestrated by Claude Debussy filter through the space as pulsing lights and smoke emanate from the stagelike sculpture. The ambiance evokes a psychedelic concert. Huyghe recalls Rock and Roll being dubbed the new religion. The artist gives form to the memory of this type of collective experience while conjuring the strange connections between the realm of the familiar and that of the unknown. (Susan Cross)

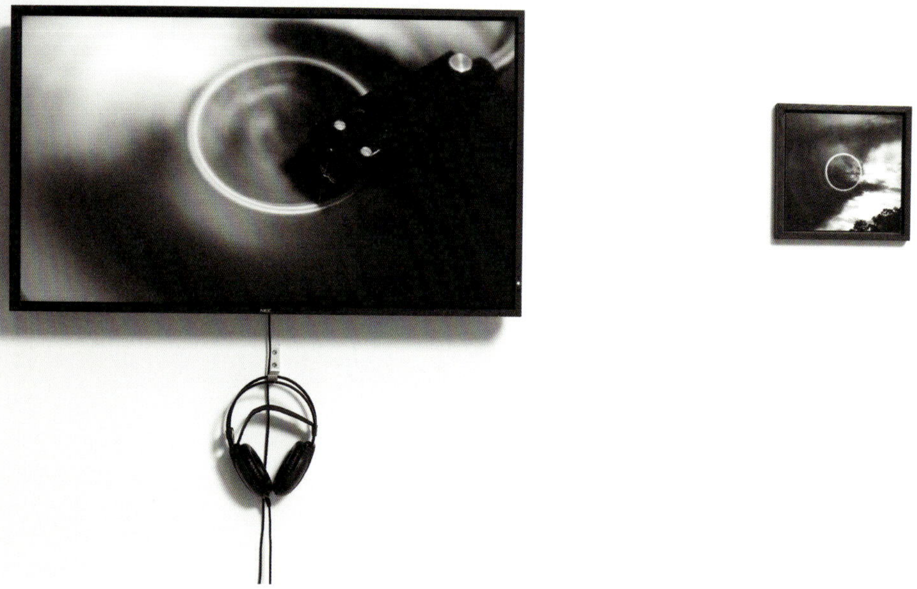

Caroline Heider, *A perfect drawing. (Ooh, a phenomenon!)*, 2011.
Video, paper, HD-document, AAC stereo, 5 min. (loop), Werkstadt Kunst KG, Graz.

This video shows an installation that consists of a modified record player that automatically draws a circle as its stylus scratches the surface of a printed piece of paper. In order to call into question the process of artistic creation, Caroline Heider does not use her hands or other parts of her body for her "perfect drawing" but rather has the record player form a circle on a print she has prepared. The pin is attached to a stylus that plays back the sound of the drawing.

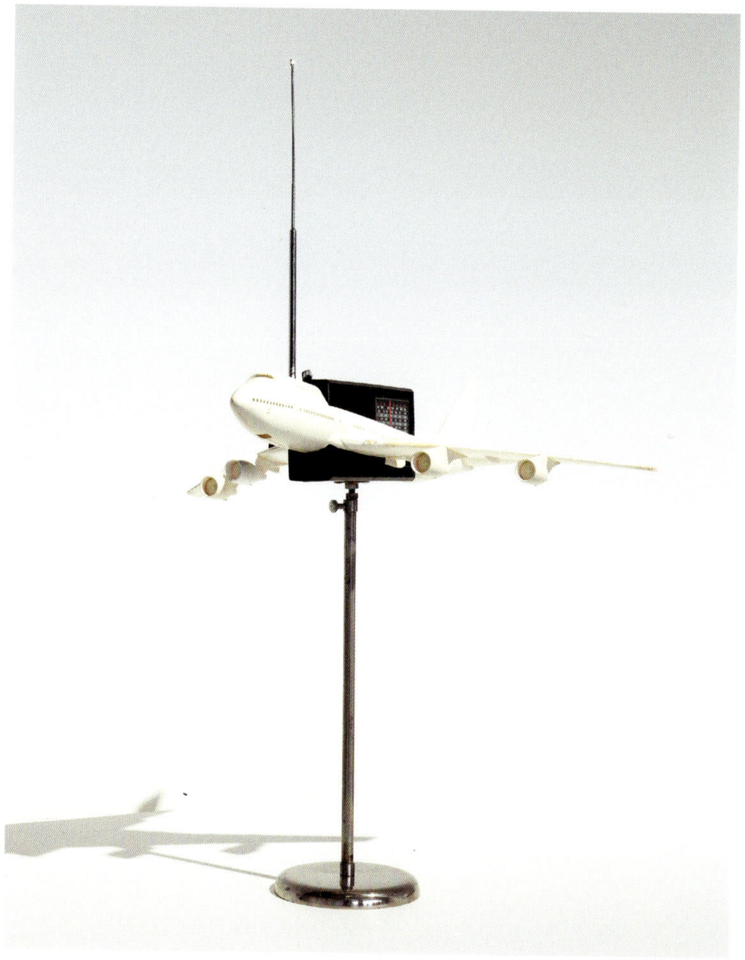

Joachim Baur, *RADIOFLUG*, 1995.
Model aircraft, shortwave receiver, MADE IN HONGKONG / steirischer herbst / baur / klammer / 1995–1997, 45 × 50 × 100 cm.

In the object *Radioflug* [Radio flight] by Joachim Baur we see two forms of communication and transport: on the one hand the classical and material form of transporting goods and people by planes, trains, cars and ships, on the other hand the current and immaterial form of communication via telephone, television and radio.

For the first time in the history of human tools, the message has separated from the messenger due to the inventions and discoveries of various forms of tele-technology and telematic communication, from telegraph and telephone to radio and television. In the past, it took a technical carrier/transfer medium or a messenger, such as man, animal or ship, to distribute a message – apart from light signals and flag language, technological advances made it possible for signs to travel alone in the form of signals. The electromagnetic waves became the invisible carrier medium of messages without messengers.

Joachim Baur's *Radioflug* unites the airplane carried by air waves and the radio based on electromagnetic waves. The noise of radio and airplane reaches the human ear through sound waves.

Uwe Bressnik, *Bressnik's 1210*, 2001.
Wood, stone, berries, sponges, roots, brushwood, moss, 200 × 70 × 75 cm.

At first glance, a deceptively realistic model of the DJ-Set per se: two Technics SL-1210 Turntables and a mixer; but slightly enlarged and made of wood, twigs, stones, roots, berries, sponges and stuff found in the meadows and wood at my grandparent's house. The thing functions in it's own way, with the power of imagination rather than electricity. Global local, high tech low tech, high fidelity low fidelity: Who cares? (Uwe Bressnik)

Uwe Bressnik, *Live Record® live im ausstellungsraum.at*, 2009. Hardboard, acrylic varnish, tempera, silver pencil, diameter 70"/177 cm.

I have been publishing my own records since 1996. The tone grooves are drawn freehand in one continuous line using a pencil or silver pen. For the plates I use oil or tempera, and I paint the tone grooves with clear varnish. The song titles play with the shift of the visual content into the musical context or vice versa. I do original songs like *The Sound of Painting*, as well as new arrangements of known numbers like *Anything Goes* by Cole Porter. (Uwe Bressnik)

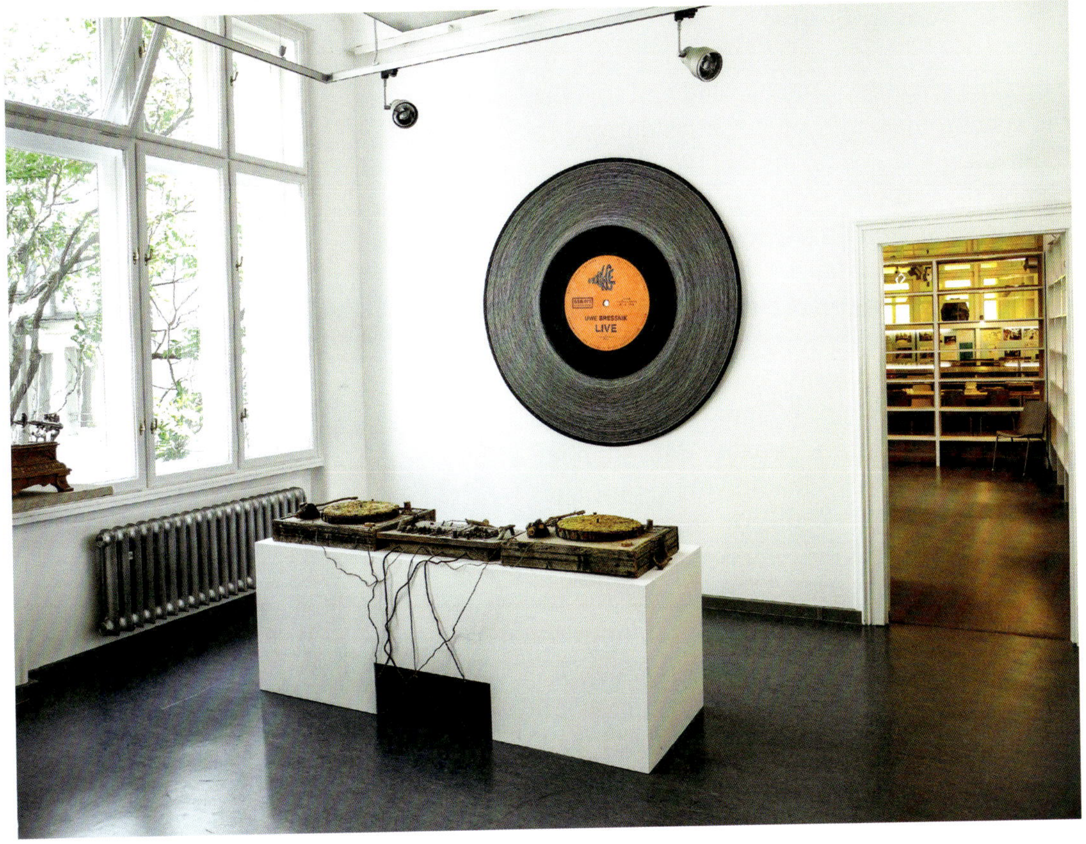

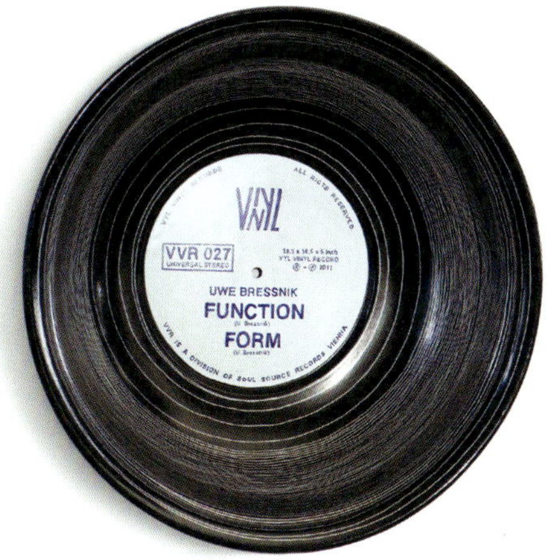

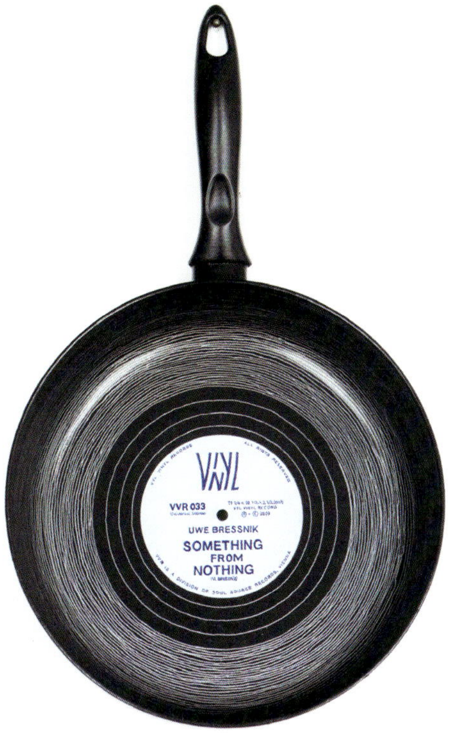

Uwe Bressnik, *Form Follows Function*, 2011.
Wash bowl (metal, enamel), acrylic lacquer and paint, silver pencil, color pencil, 7 × 47 × 12 cm.

Vyl Vinyl Records – since 2003
vyl [fül] = lokal Austrian idiom for 'viel' Vinyl [winül] = lots of vinyl

The acoustic material record outgrows itself and melts with objects from every day life: pots, pans, tee cans, etc. ... The added individual form of the object as a standardized record enhances the game with the (song)titles to another level. (Uwe Bressnik)

Uwe Bressnik, *Something from Nothing*, 2017.
Pan (steel, plastic), acrylic lacquer and paint, permanent and silver pencil, coloured pencil, 47 × 8.5 cm.

The synesthesias of the eye and ear, together with the synchronies and synchromies derived from them circa 1900, were among the driving forces behind the evolution of art in the twentieth century, and also contributed to the rise of an independent sound art. Concordances, correspondences, parallelisms between eye and ear, between painting and music, have been among the constants of experimental physiology, psychology, physics, and art for hundreds of years. The synchronization in the brain of the sensory data supplied by the sense organs – leading to ecstatic states of consciousness and cosmic or religious experiences, which had been experienced for centuries under the influence of specific plants and intoxicants or meditative, ascetic, and religious rituals, but were impossible, as internal, subjective experiences, to control and direct completely – would now, as external events, be controlled and directed through the interplay of all the arts hammering on all the sense organs at once.

From the Middle Ages, and especially the Renaissance, to the multimedia environments of the 1960s and 1970s and the world's discos, the newest technologies were always employed in pursuit of this goal. One thinks, for example, of Louis-Bertrand Castel's famous *musique oculaire* and *clavecin oculaire* (ocular harpsichord), described in his 1735 essay "Nouvelles expériences d'optique et d'acoustique." In 1893, Bainbridge Bishop published *A Souvenir of the Color Organ, with Some Suggestions in Regard to the Soul of the Rainbow and the Harmony of Light*. In 1895, the English painter Alexander Wallace Rimington saw color pianos, organs, and harpsichords as the foundations of *A New Art: Colour Music*. And in 1900, Louis Favre further established the genre with *La Musique des couleurs et les musiques de l'avenir*.

The artistic quest for sound and color harmonies in the nineteenth century rested primarily on the scientific experiments of the eighteenth. In addition to Castel's findings and insights, one might also point to the German physicists Johann Wilhelm Ritter[60] and Ernst Florens Friedrich Chladni.(FIG. 68) Both were considered Romantics; thus their research highlights at least one more aesthetic source for sound art: the Romantic ideal of the fusion of the arts.

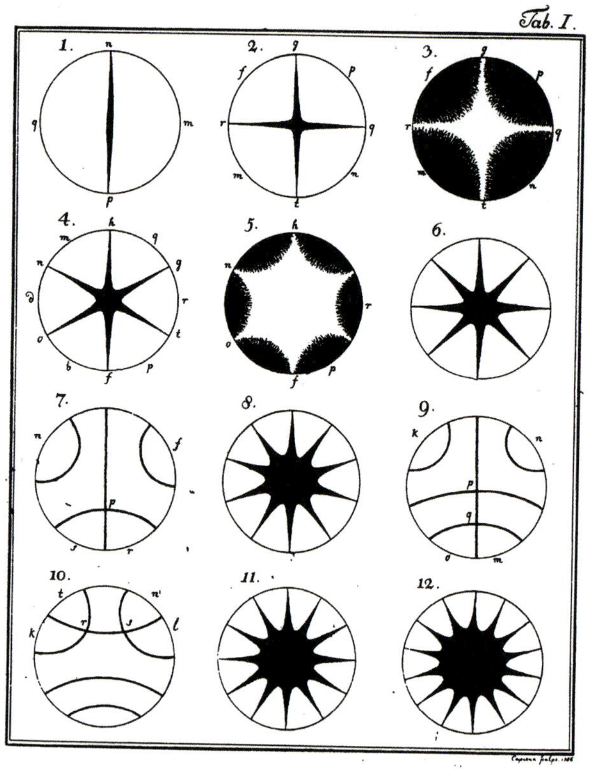

FIG. 68 Ernst Florens Friedrich Chladni, Examples of his socalled "Chladni figures" from his book *Entdeckungen über die Theorie des Klanges* [Discoveries in the Theory of Sound], 1787.

At the beginning of the twentieth century, a series of musicians and visual artists developed color-music instruments upon this historical foundation. A color organ for a universal color music that would supposedly unite the world's peoples was patented in 1908 by an Australian, Alexander Burnett Hector, and demonstrated in Sydney in 1912. The art of color music was propelled forward by both painters and musicians, as well as partnerships between the two. Rimington published *Colour Music: The Art of Mobile Colour* in 1912 and presented a color organ that projected a myriad of colors onto translucent curtains; it was supposed to be used for a performance of Alexander Scriabin's symphony *Prometheus: The Poem of Fire* in 1915. In 1913, the Futurist painter Enrico Prampolini published a manifesto entitled *La cromofonia. Il colore dei suoni* [Chromophony: The color of sounds]. Starting in 1919, Thomas Wilfred, Claude Bragdon (also known for his theories about the fourth dimension, laid out in his 1913 book *A Primer of Higher Space*), and others worked on the idea of a device for projecting

60 See for example Johann Wilhelm Ritter, *Physik als Kunst. Ein Versuch, die Tendenz der Physik aus ihrer Geschichte zu deuten* (Munich: Lindauer, 1806).

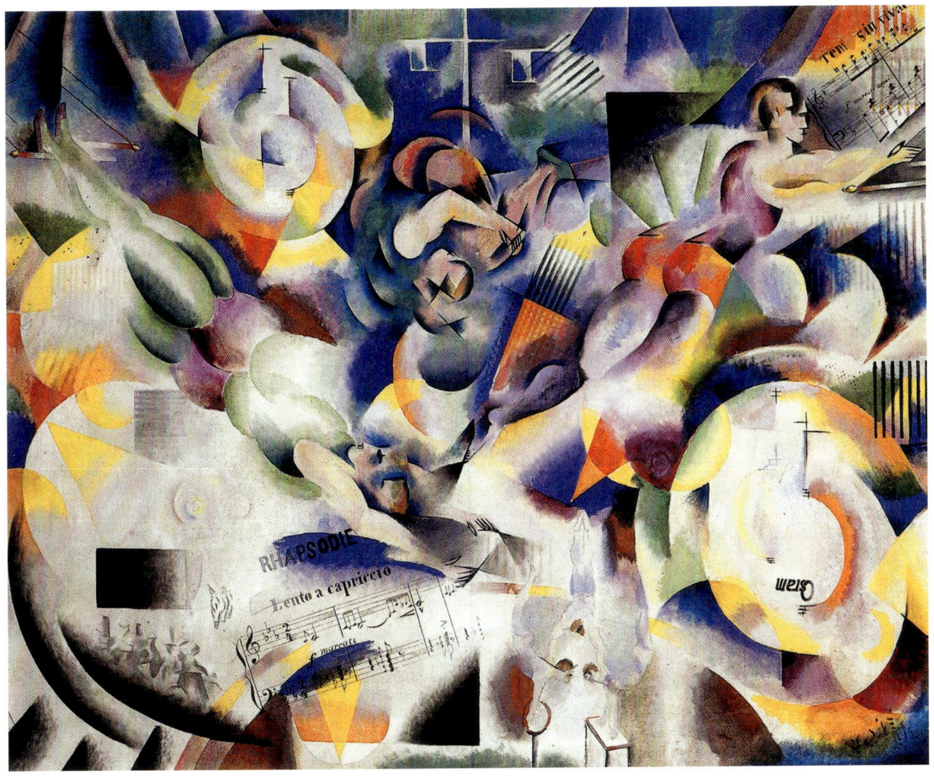

FIG. 69 Vladimir Baranoff-Rossiné, *Capriccio musicale (Circus)*, 1913. Oil and pencil on canvas, 130.4 × 163.1 cm.

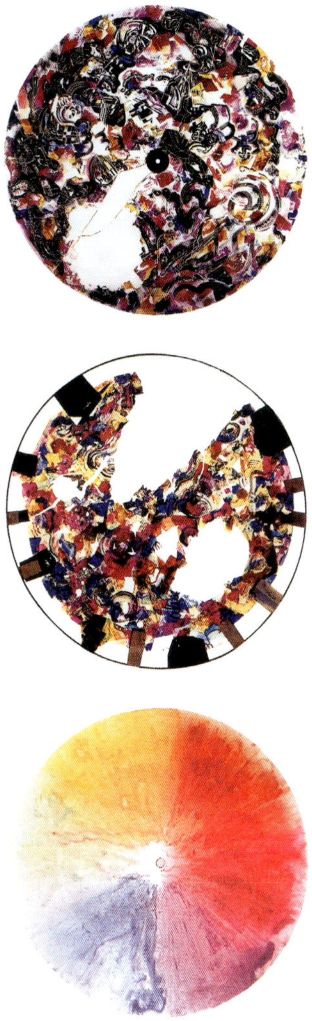

FIG. 70 Vladimir Baranoff-Rossiné, *Disques optophoniques*, ca. 1920–1924. Painting on glass, diameter 40–48 cm.

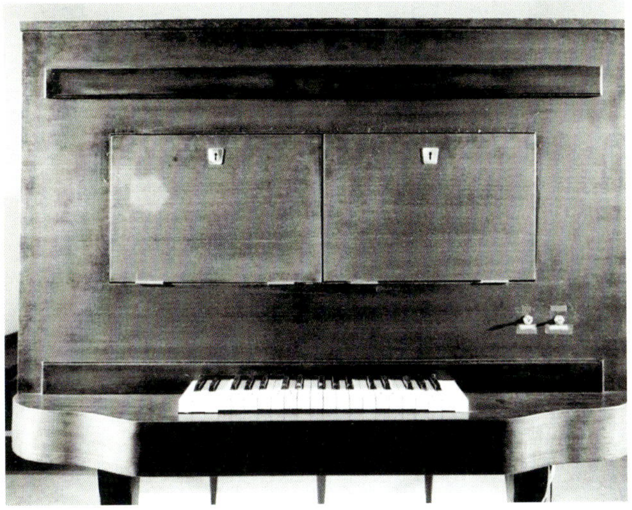

FIG. 71 Vladimir Baranoff-Rossiné, *Optophonic Piano*, 1920–1923. Reconstruction of 1971.

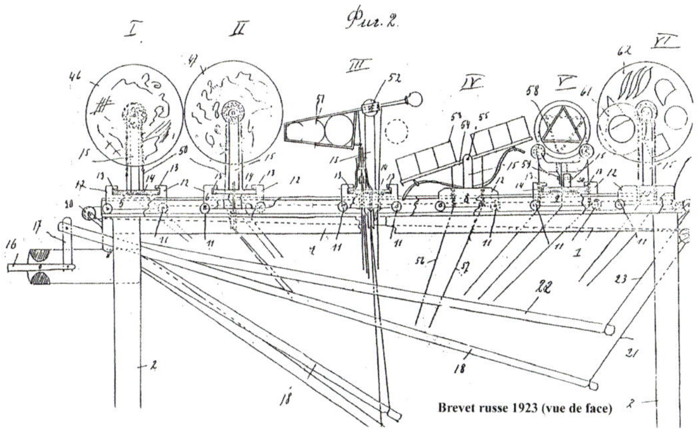

FIG. 72 Vladimir Baranoff-Rossiné, *Optophonic Piano*, 1920–1923. Construction drawing.

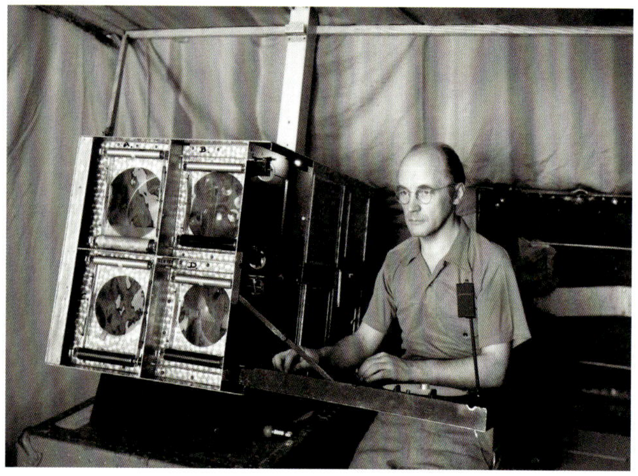

FIG. 73 Thomas Wilfred with his *Model G Recital Clavilux.* Wilfred designed it in 1936, this photograph was taken in the 1950s.

colored light, the Clavilux.(FIG. 73) Favre published *La musique des couleurs et le cinéma* in 1927.

The Russian painter Vladimir Baranoff-Rossiné used pieces of painted glass for projections with his optophonic piano, built in 1923.(FIGS. 69–72) The fashion designer Paul Poiret exhibited a light organ in Paris in 1925. Around the same time, the Hungarian composer Alexander Laszlo built his sonchromatoscope, a machine that projected colored light effects in time with a musical score. Adrian B. Klein's 1926 book *Colour-Music: The Art of Light* lent further support to the synesthetic dreams of color music and optophony. In Graz, the musician Anatol Vietinghoff-Scheel experimented with harmonies of color and music in the late 1920s; his chromatophone projected light from floodlights onto translucent curtains, along with color films set to music by composers from Claude Debussy to Scriabin. Charles Blanc-Gatti, who was a member of the *musicalistes* (a group of artists who called for the musicalization of art), published a book titled *Sons et couleurs* in 1934 and exhibited his ideas for a chromophonic orchestra in *Cyclorama*, a 1937 slide show. For Blanc-Gatti, chromophony meant the unification of sound and light vibrations by a principle that created a mathematical equivalence between the two.

The dreams of chromophony and optophony, of color pianos and musical painting, were also carried forward by the artistic movements of Dada and the Bauhaus. Raoul Hausmann, in particular, is known for his invention of the optophone, patented in 1936, which used selenium cells to electronically translate beams of light into sound waves, and sound waves into light. His theoretical involvement with the idea of the color piano dated to 1921, and in essays from 1922 (in the Viennese journal *MA* and elsewhere) he investigated ways of synchronizing perceptions of light and sound.(FIGS. 74–76) "We demand electric, scientific painting!!! The waves of sound and light and electricity differ only in their length and frequency," he wrote in his 1921 manifesto "Présentismus" [Presentism].[61] In another manifesto from 1933, "Die überzüchteten Künste" [The overbred arts], he wrote: "My dear musicians, my dear painters: You will see through your ears and hear with your eyes...! The electric spectrophone will obliterate your ideas of sound, color, and form."[62]

The optophone, or spectrophone, as Hausmann calls the color piano here, was to be operated – much like a computer – by a keyboard comprising about a hundred keys, which corresponded to a hundred fields, each with a different chrome gelatin relief, whose spectral-line shifts were beamed by a fluorescent lamp into a converging prism. The resulting play of colors was projected onto a screen, while photocells converted the variations in luminance into electronic pulses that produced audio effects via speakers. The device was used for the performance of optical-phonetic compositions. Together with Viking Eggeling, Hausmann published the "Zweite präsentistische Deklaration" ("Second Presentist Declaration") in 1922: "Language, dance, and music were the supreme achievements of intuitive spatiotemporal functionality. Now it is for optics, haptics, etc., to take over, moving along new paths."[63]

Ludwig Hirschfeld-Mack's 1925 "color sonatina" *Farbensonatine (Ultramarin-Grün)*, which was never performed during its composer's lifetime, can be read as a prototypical score for the color music of the Bauhaus. With regard to color, Hirschfeld-Mack's reference was Ewald Hering's theory of color opponency and his proposed primary colors red, yellow, blue, and green. Individual light sources were switched on

61
Raoul Hausmann, "Présentismus. Gegen den Puffkeismus der teutschen Seele," in *Texte bis 1933*, ed. Michael Erlhoff, vol. 2, *Sieg Triumph Tabak mit Bohnen* (Munich: text + kritik, 1982), 24–30, here 27. Originally published in *De Stijl* 4, no. 9 (September 1921): 136–43. Translated from the German.

62
Raoul Hausmann, "Die überzüchteten Künste," in Hausmann, *Sieg Triumph Tabak mit Bohnen*, 133–44, here 144. Translated from the German.

63
Viking Eggeling und Raoul Hausmann, "Second Presentist Declaration – Addressed to the International Constructivists," trans. David Britt, in Benson and Forgács, *Between Worlds*, 476–77, here 477.

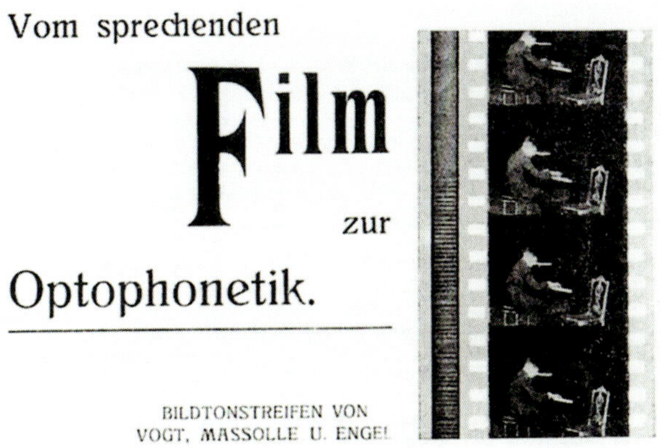

FIG. 74 Raoul Hausmann, “Vom sprechenden Film zur Optophonetik” [From Sound Film to Optophonetics]. The article was first published in: G – *Material für elementare Gestaltung*, 1923.

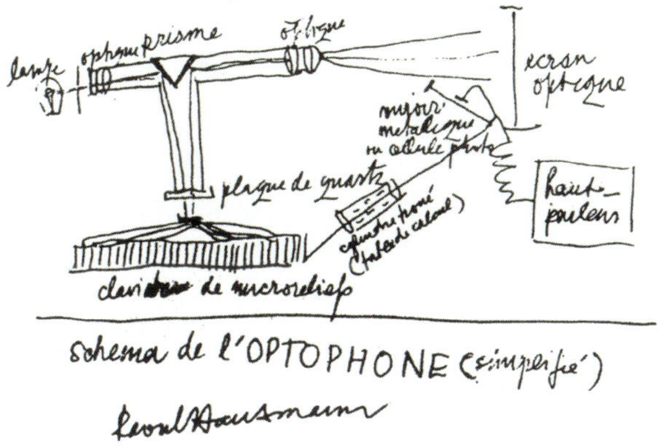

FIG. 75 Sketch of Raoul Hausmann’s Optophone of 1919. Drawn by Hausmann in the 1930s.

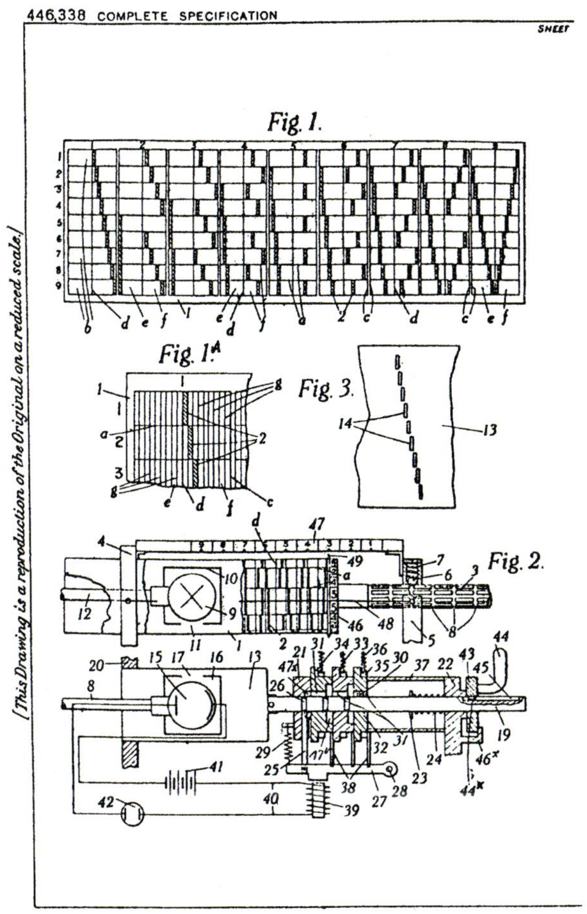

FIG. 76 Raoul Hausmann, “Das Optophon” [The Optophone]. Page from the patent specification 446338, 1926.

simultaneously or sequentially in time with a rhythm. Their form and brightness were varied with the aid of gobos and a fading technique. Rhythmically switching the light sources on and off produced superimposed and juxtaposed planes of light that faded in and out. The resulting temporal structure is the musical element of the *Farbenlicht-Spiele* [color-light plays] Hirschfeld-Mack began creating in 1922. His 1923–1924 piece *Kreuzspiel* [Crossplay], also known as *Lichtfuge* [Light fugue], displayed strong affinities with film and with Laszlo’s color-light music, whose success contributed greatly to the popularity of the term *synesthesia* circa 1920.

The optophonic machines developed into optophonic extravaganzas, a synthesis of the optical and the kinetic, a synesthesia on the stage, as found in Kurt Schmidt’s *Mechanisches Ballett* (1923), Andreas Weininger’s *Revue-Mechanische-Bühne* (1926), Lothar Schreyer’s *Kreuzigung* (1920), and the book *Die Bühne im Bauhaus* (1925), edited by Walter Gropius and László Moholy-Nagy. Painters such as Johannes Itten owed a debt to synesthetic endeavors as well. In *The Art of Color* (1961), Itten writes: “The primeval essence of color is a phantasmagorical resonance, light become music.”[64] The social psychologist Hans Lorenz Stoltenberg also became interested in color and sound analogies early on, and experimented with projections of colored strips of film as early as 1911. He was influenced by Wilhelm Ostwald and his research into the effects of color.[65] The imprint of historical color music is apparent in Stoltenberg’s 1920 book *Reine Farbkunst in Raum und Zeit und ihr Verhältnis zur Tonkunst* [Pure color art in space and time, and its relationship to music].

64
Johannes Itten, *The Art of Color: The Subjective Experience and Objective Rationale of Color*, trans. Ernst van Haagen (New York: Reinhold, 1961), 13.

65
See Wilhelm Ostwald, *Die Farbenfibel* (Leipzig: Unesma, 1917); and also Rolf Sachsse, *Wilhelm Ostwald. Farbsysteme. Das Gehirn der Welt*, ed. Peter Weibel (Ostfildern: Hatje Cantz, 2004).

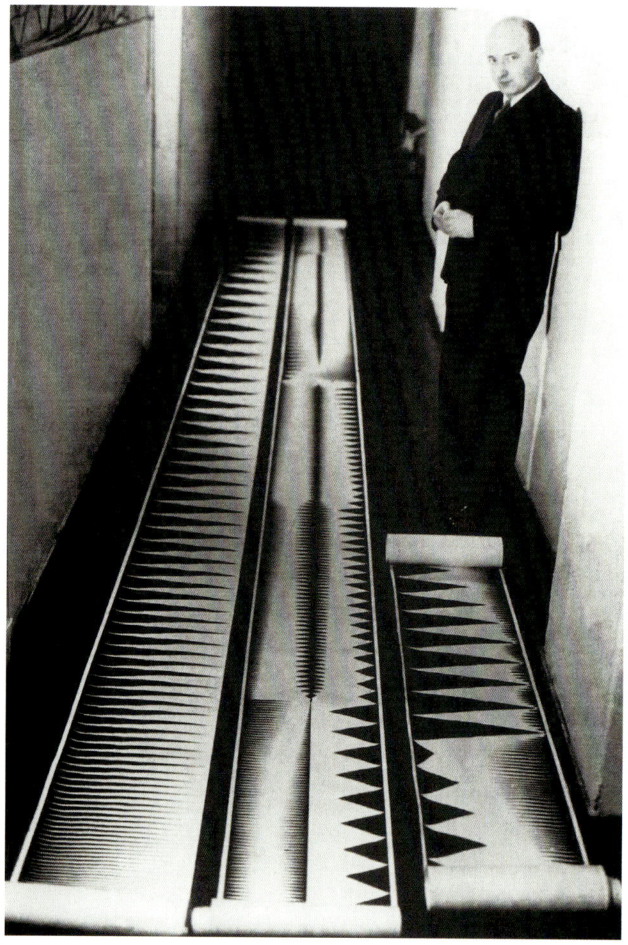

FIG. 77 Oskar Fischinger with sound strips from his ornament sound experiments. These gigantic sound strips were just for publicity, though. The were not used to create sound films.

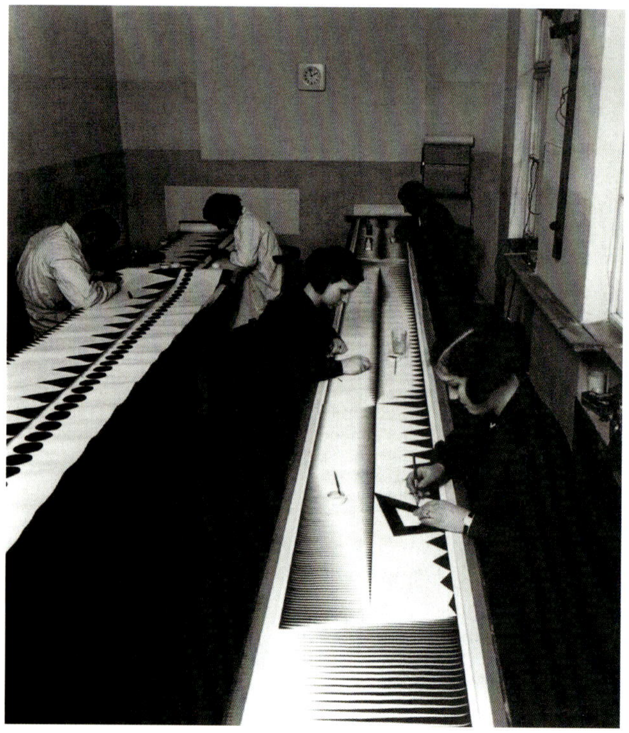

FIG. 78 Working on Oskar Fischinger's sound strips.

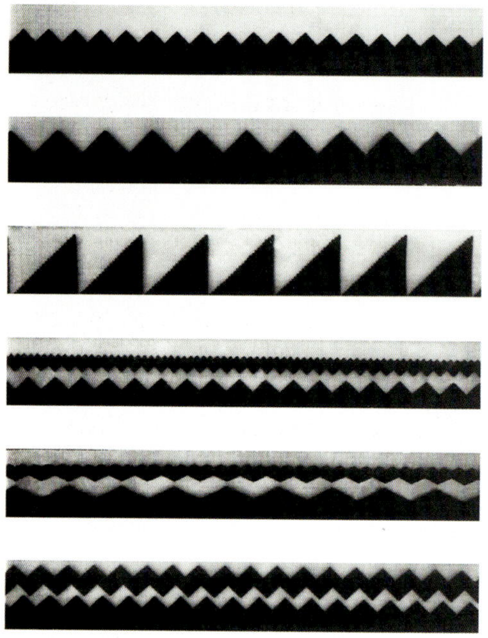

FIG. 79 Examples from Oskar Fischinger's ornament sound experiments.

By 1930, visual research into synesthesia was, for the moment, over; the emphasis in the synesthetic relationship shifted from painting to music, image to sound. And another shift of emphasis occurred as well: Now the search was for synthetic rather than synesthetic approaches. Sound and image were produced synthetically. New musical instruments that produced synthetic sound appeared, followed by picture-machines that generated synthetic images.

For Oskar Fischinger, the desire for a clearer scientific and technological definition of the interrelationships between sound and image was strengthened by the advent of new instruments such as Friedrich Trautwein's *Trautonium* (1930) and Maurice Martenot's *Ondes Martenot* (1928), as well as visits to conferences on color and sound in the 1920s. In 1930, Fischinger realized that there was no fundamental difference between the abstract visual ornaments he used in his films and the patterns on the optical sound track that produced sound. It must be recalled that in that era, optical sound was used rather than magnetic recording, so that the visual and audio information were on the same strip, with no additional magnetic tape involved. Fischinger could thus ask: Which geometric shapes would create both sounds and images; what kind of music could be created by the visual pattern of the sound track? (FIGS. 77–79)

By studying existing sound tracks, Fischinger learned which patterns produced which sounds, and he photographed the drawn ornaments onto the film's soundtrack area. When played through a projector, they were translated into a kind of music. In 1932 he wrote about his experiments in a press release that garnered an great deal of attention and was widely distributed under the titles "Tönende Ornamente" and "Klingende Ornamente" [Sounding ornaments]. In the July 28, 1932 edition of the *Deutsche Allgemeine Zeitung*, he wrote: "Between ornament and music persist direct connections, which means that ornaments are music. If you look at a strip of film from my experiments with synthetic sound, you will see along one edge a thin stripe of jagged ornamental patterns. These ornaments are drawn music - they are sound."[66] Instead of music being modeled on painting, and painting on music, now music was being drawn directly; and by filming the drawn sound track frame by frame, the drawing could be transformed directly into the music and sound of a film.

However, another film artist working in Munich had a very similar idea. In 1929, the animator Rudolf Pfenninger had developed what he called "Tönende Handschrift" [Sounding handwriting], which he presented to the public in 1932.(FIGS. 80, 81) He drew tones on a strip of paper, filmed them individually and directly with a movie camera, and incorporated them into the optical sound track, making him one of the first to produce "synthetic sound." Like the Canadian animator Norman McLaren after him, Pfenninger drew sound directly onto the film. Similar experiments with synthetic sound had already been conducted in Russia by Arseny Avraamov, Mikhail Tsekhanovsky, Evgeny Scholpo, Nikolai Voinov, and Boris Yankovsky.

These different attempts to combine light and sound, painting and music, color and tone, took place on a manual-mechanical level in the first half of the twentieth century. In the second half, the paradigmatic setting of such synesthetic dreams changed - they took place synthetically, on the electronic level. The development of electronic devices for the production of music and pictures, the development of audiovisual synthesizers capable of synthetically generating not just sounds, but also images, resulted in a completely new possibility: the controlled transmission of any generated sound or image, in its every modulation, through space. In 1920 Léon Theremin presented the etherphone, later named the theremin after its inventor: an electronic instrument, built on radio-frequency oscillators, that was controlled by hand movements near two antennas.(FIG. 82) In 1932 Theremin worked with a pioneer of the electronic image, Mary Ellen Bute, to display

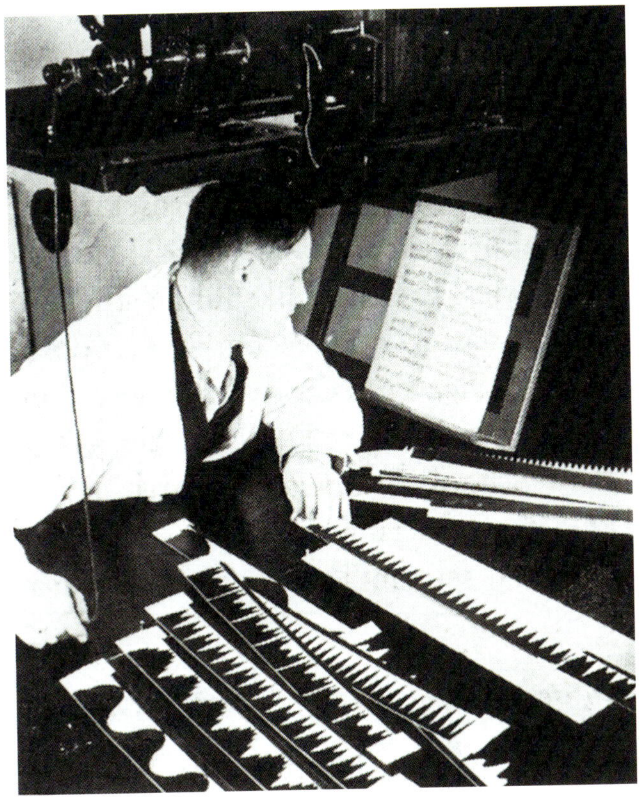

FIG. 80 Rudolf Pfenninger in his laboratory with hand-drawn sound strips.

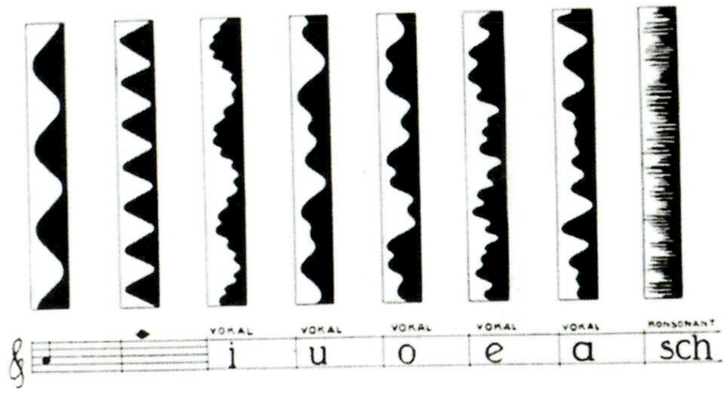

FIG. 81 Rudolf Pfenninger, *L'écriture sonore* [Sounding handwriting]. Advertising for Studio EMELKA, 1932.

66
Oskar Fischinger, "Sounding Ornaments," in *Optical Poetry: The Life and Work of Oskar Fischinger*, by William Moritz (Bloomington: Indiana University Press, 2004), 179–81, here 179.

67
"Bute made a series of Visual Music films which she called 'Seeing Sound' ... *Rhythm in Light*, 1934; *Synchromy No. 2*, 1935; *Dada*, 1936; *Parabola*, 1937; *Escape*, 1937; *Spook Sport* (animated by Norman McLaren), 1939; *Tarantella*, 1940; *Polka Graph*, 1947; *Color Rhapsodie*, 1948; *Imagination*, 1948; *New Sensations in Sound*, 1949 (RCA Commercial); *Pastorale*, 1950; *Abstronic*, 1952; and *Mood Contrasts*, 1953." "CVM's Bute Research Pages: About the Films / Retrospective Program / Upcoming & Recent Screenings," Center for Visual Music, accessed June 20, 2018, www.centerforvisualmusic.org/Bute.htm.

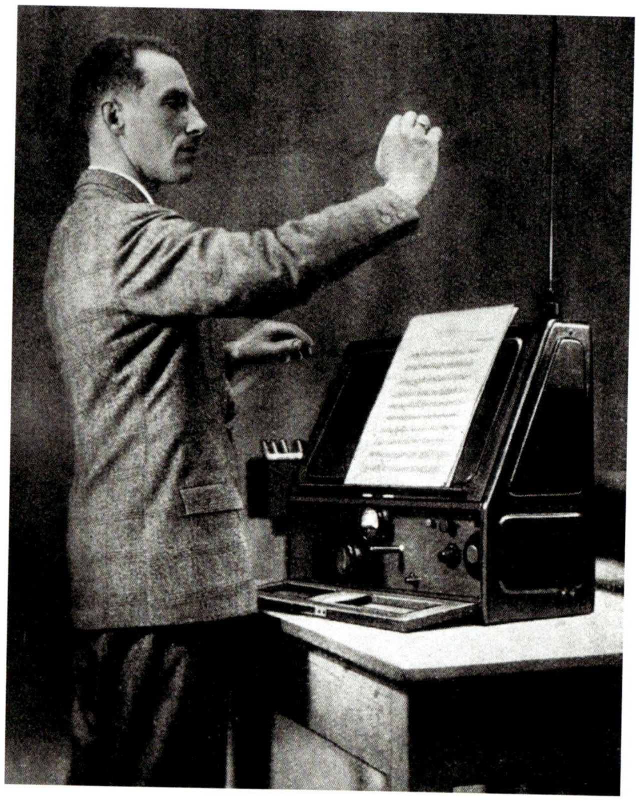

FIG. 82 Léon Theremin demonstrating his instrument, 1927.

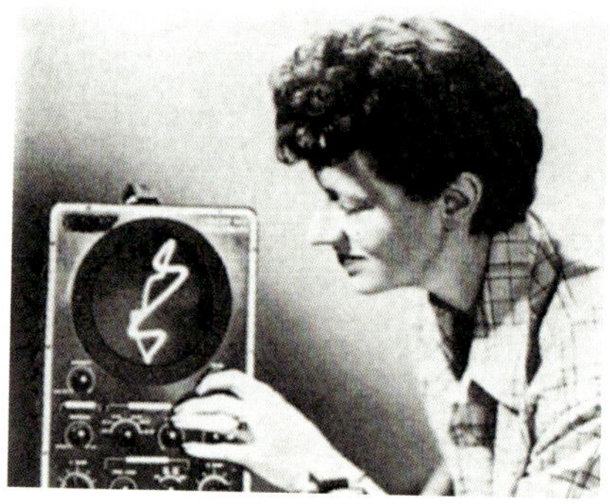

FIG. 83 Mary Ellen Bute with her oscilloscope, ca. 1954.

mathematical formulas synchronized with music in a cathode-ray tube (the same technology as in tube televisions). (FIGS. 83, 84) Bute's aesthetic credo was "seeing sound," or "visual music,"[67] which would later become a slogan for the music-video industry. (FIG. 85) After World War II, following the classic period of handmade synthetic moving images and sounds, the era of the mechanically aided, mechanically generated synthesis of image and sound began. In the late 1960s, video brought the first attempts involving electronic images, culminating in the computer-aided and computer-generated images of today.

Nam June Paik, who in 1963 exhibited prepared TV sets modeled on Cage's prepared pianos, had started out as a musician and was especially inspired by the Studio für Elektronische Musik in Cologne. He also appeared alongside Hans G Helms and others in the 1961 premiere of Stockhausen's *Originale*. (FIG. 97) The shift from the synesthesia of music and painting to the synesthetics or synchrony of music and (electronic) media was evident in Paik's famous exhibition *Exposition of Music – Electronic Television* in 1963. The title alone illustrated, with commendable clarity, the classical relationship between image and music, but now in the electronic medium rather than the medium of painting. Also, the Center for Advanced Visual Studies (CAVS) at MIT, Boston, and MIT's Hayden Gallery created and presented in the 1970s pioneering sound works such as *Sounding* (1972) by Paul Earls in 1972, *Music Box* (1973) by Luis Frangella (water drips through boxes) in 1973, the sonic environments, sonic ambiences by Maryanne Amacher *Everything in Air* and *City-Links* (1967–1981) in 1975, or *Sound Sculpture* (1968) by Joe Moss in 1973.

An excellent overview of subsequent developments can be found in a number of exhibitions and their catalogs: *Sound Sculpture*, Vancouver Art Gallery (Vancouver, 1973), Inge Becker's *Sehen um zu Hören. Objekte und Konzerte zur visuellen Musik der 60er Jahre* (Düsseldorf, 1975), *Sound*, Los Angeles Institute of Contemporary Art (Los Angeles, 1979) and Institute for Art and Urban Resources – MoMA PS1 (New York, 1979), René Block's *Für Augen und Ohren. Von der Spieluhr zum akustischen Environment* (Berlin, 1980), Karin von Maur's *Vom Klang der Bilder. Die Musik in der Kunst des 20. Jahrhunderts* (Stuttgart, 1985), Stephen Vitiello's *I Am Sitting in a Room: Sound Works by American Artists 1950–2000* (New York, 1999, as part of the exhibition *The American Century 1950–2000*), Peter Weibel's *Ars Electronica: Der Freie Klang* (Linz, 1987), Adam Lowe and Simon Schaffer's *N01se: Universal Language, Pattern Recognition, Data Synaesthetics* (Cambridge, 2000), David Toop's *Sonic Boom* (London, 2000), Lawrence Rinder and Debra Singer's *BitStreams* (New York, 2001), Christine Van Assche's *Sonic Process: A New Geography of Sounds* (Barcelona, 2002), *Frequenzen [Hz]: Audiovisuelle Räume*, Schirn Kunsthalle Frankfurt (Frankfurt am Main, 2002), Sophie Duplaix and Marcella Lista's *Sons & Lumières* (Paris, 2004), and Brigitte Felderer's *Phonorama* (ZKM | Karlsruhe, 2005).

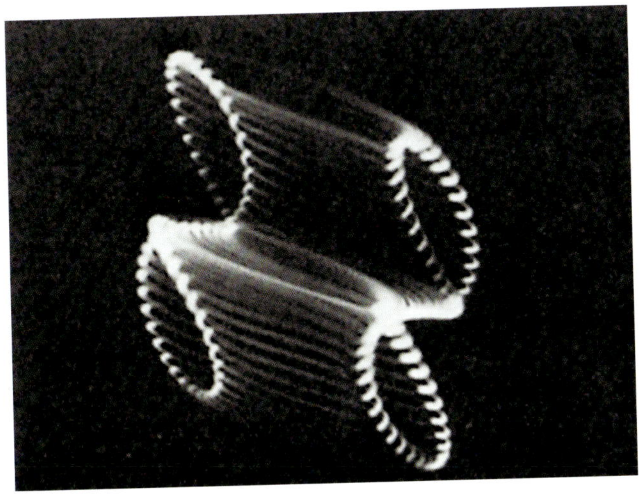

FIG. 84 Mary Ellen Bute, *Abstronic*, 1952. 35 mm film, color, sound, 7 min., music: *Hoe Down* by Aaron Copeland.

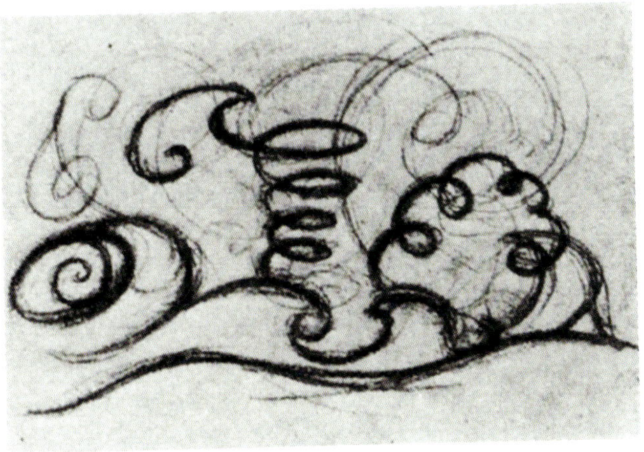

FIG. 84a Unknown artist, Intuitive transcript of a waltz by Frédéric Chopin, ca. 1920. Pencil drawing.
There is an interesting similarity between the oscilloscope drawing by Bute and the pencil drawings.

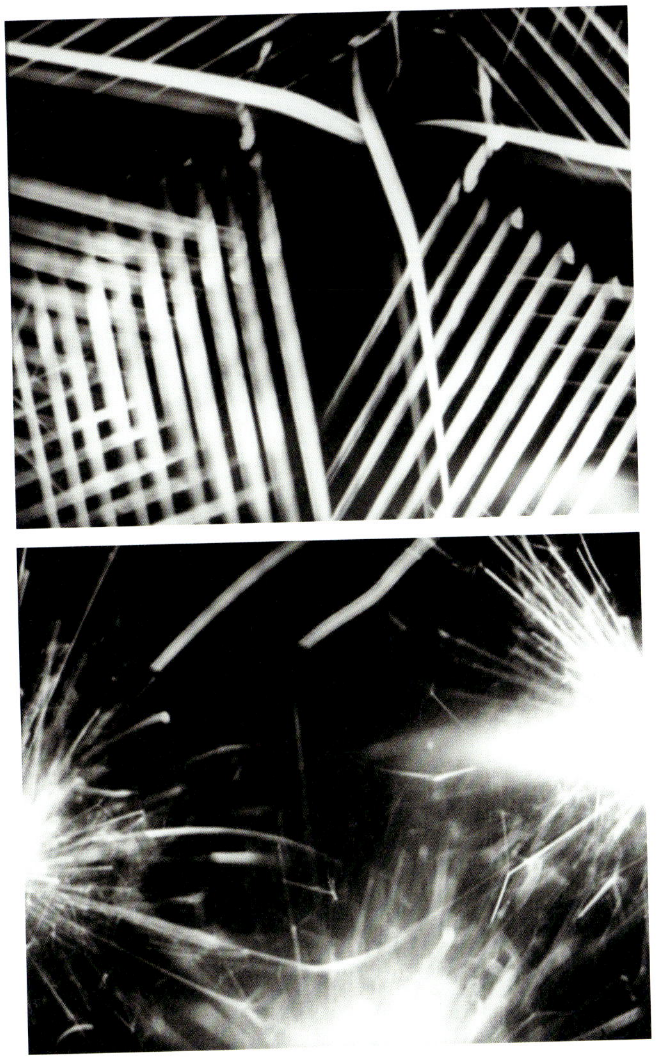

FIG. 85 Mary Ellen Bute, *Rhythm in Light*, 1934. Film stills.

Mary Ellen Bute's first film was the result of the trained painter's many years of searching for a medium for the "kinetic visual art form" that she envisaged. She arranged hand-drawn pictures together with photographic images of moving three-dimensional objects to the *Peer Gynt Suite* by Edvard Grieg.

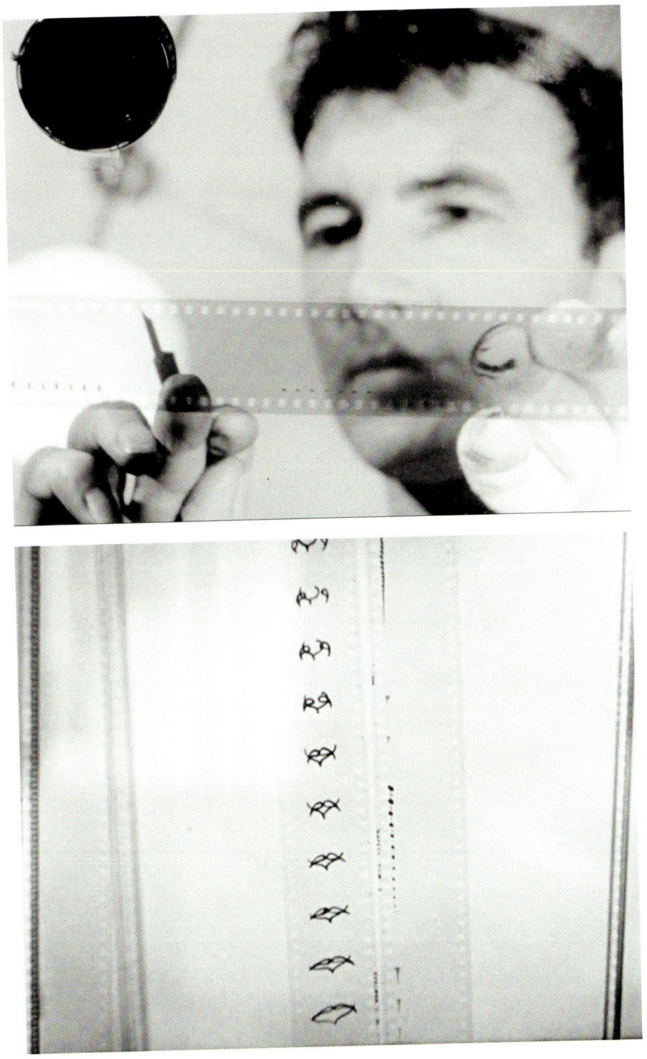

Norman McLaren, *Pen Point Percussion*, 1951. Film stills.

Norman McLaren began experimenting with hand-drawn sound in the late 1930s. One of his first films using this technique was Pen Point Percussion, where he explains his "hand-drawn sound" method. Optical sound makes it possible to transform visual information into sounds.

Graphic Notation – Musical Graphics

The turn to new music prompted by new sound and noise generators brought with it a crisis in classical notation. The musical notation of the score may be understood both as documentation and as the composer's instructions for the vocalists and instrumentalists. The Benedictine monk and music theorist Guido of Arezzo is regarded as the inventor of modern musical notation. The notation system he invented, based on four lines, is described in his major work, the *Micrologus de disciplina artis musicae*, which was written around the year 1025. Before Guido, musical notation involved symbols known as *neumes*, which were usually written above the words of a text to be sung but provided no information on a tone's exact duration or pitch; the actual melody was passed on orally.[68] Symbolic notation using a system of lines enabled the development of a concept of the work that reached its first peak with the historicism of the nineteenth century and the great editions of the collected works of Johann Sebastian Bach, Wolfgang Amadeus Mozart, and Ludwig van Beethoven.

With the arrival of new instruments that produced new tones, classical musical notation was no longer sufficient. Thus the development of graphic notation and musical graphics began in the 1930s and above all in the 1950s. Musical notation became diagrammatic drawings on the two-dimensional surface of the paper: graphic notation. With the help of new technologies such as oscillographs and computers, this evolved into the graphic user interface, the touchscreen, a point of contact between human and machine. Running one's fingers over this surface produced music: The instructions of notation were simultaneously their execution. The user was composer and performer rolled into one. Unfortunately, the significance of graphic notation, which was indicative and symptomatic of the shift from music as a temporal art to music as a spatial art, together with the attendant search for spatial sound, went unremarked by the music industry. As the term graphic notation makes clear, we are dealing with a spatial form of notation. Temporal notation became topological notation. Musicians first sought spatial representations of sound, then techniques and methods of spatial composition.

Attempts to improve, simplify, expand, and complete classical notation got under way in the first half of the twentieth century, as evidenced by books such as Alois Hába's *Neue Harmonielehre des diatonischen, chromatischen, Viertel-, Drittel-, Sechstel- und Zwölftel-Tonsystems* [New theory of harmony in the diatonic, chromatic, quarter-tone, third-tone, sixth-tone, and twelfth-tone systems, 1927], and Oskar Rainer's *Musikalische Graphik* [Musical Graphics, 1925] (FIG. 86).

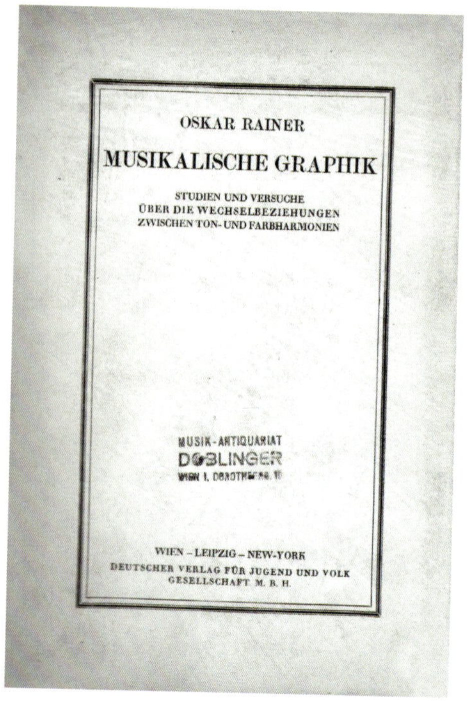

FIG. 86 Oskar Rainer, *Musikalische Graphik* [Musical Graphics] (Vienna: Jugend & Volk, 1925). Cover of the first edition.

In the twentieth century, traditional notation evolved, by way of new notation systems, into the picture-writing of musical graphics, and ultimately into action notation, or verbal instructions. Not only did Schoenberg use new graphic notation, he also constructed slide rules and wheel charts cut out of paper to aid the composition process. His dodecaphonic system was highly formalized to the extent that it was mechanized. To this end, he constructed his mechanical slide rules and wheel charts, which helped him calculate or compose series of notes. These "paper machines" are examples of early algorithmic processes or the kind of mechanical calculating and composing machines that anticipated the digital computer. (See p. 113)

68
Guido of Arezzo is also the namesake of GUIDO Music Notation, a digital format for the representation of music scores.

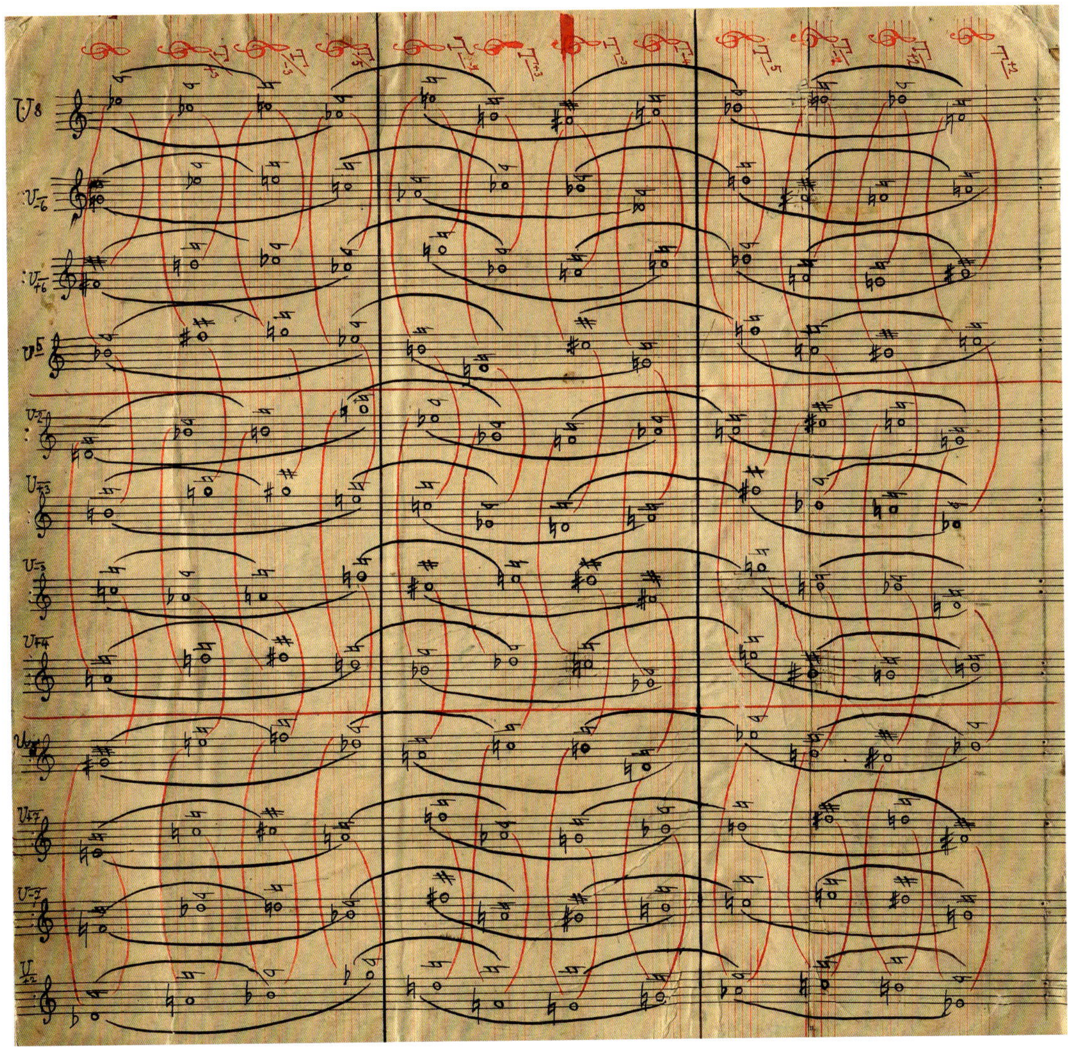

Arnold Schoenberg, *Suite op. 29*, 1925/1926. Bidirectional table, paper on cardboard.

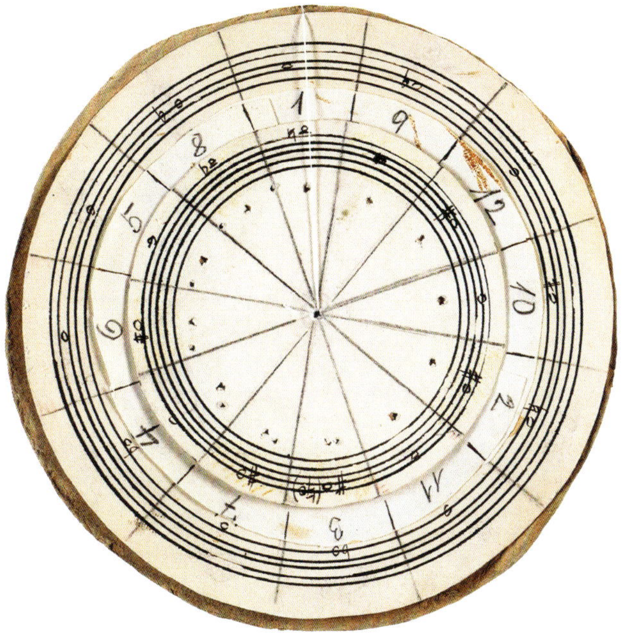

Arnold Schoenberg, wheel chart for *Wind Quintet*, op. 26, 1923/1924.

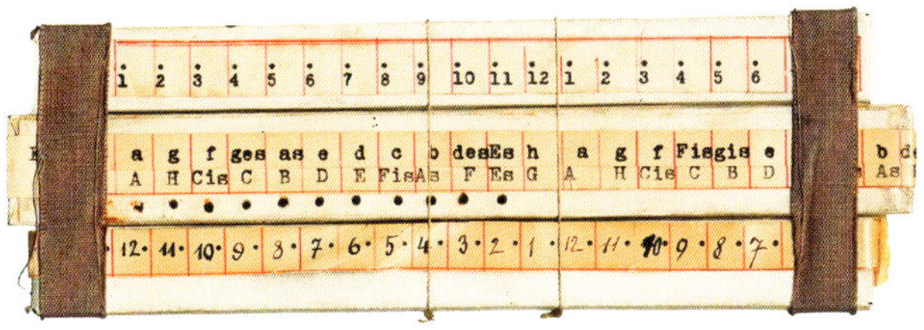

Arnold Schoenberg, slide rule for *Wind Quintet*, op. 26, 1923/1924. Paper and cardboard, 6 × 19 cm.

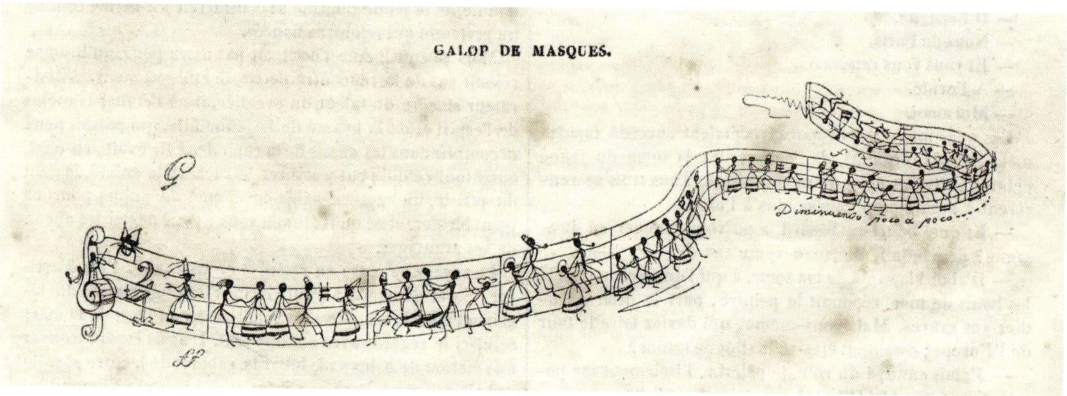

Grandville, *Valse. Galop de Masques. Marche militarie et orientale*, 1840.
Feather lithography. Published in *Le Magasin pittoresque* 31, Paris 1840.

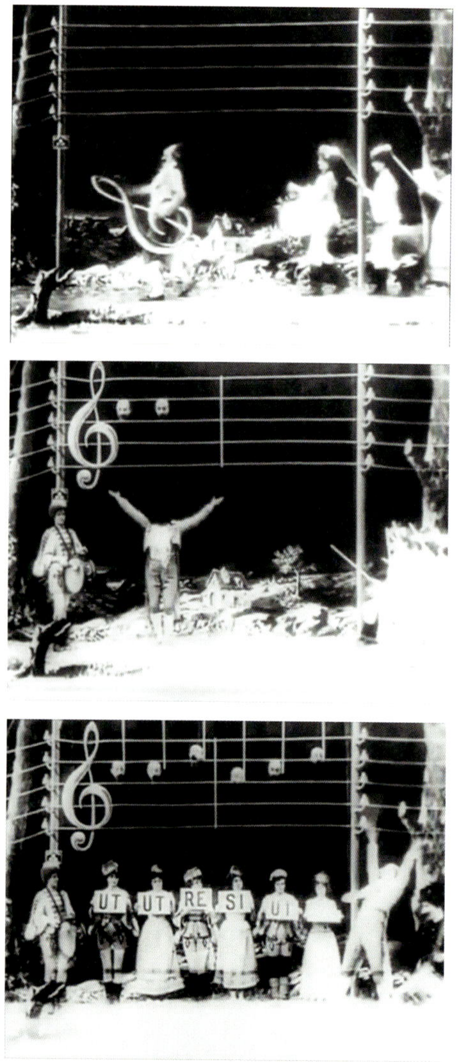

Georges Méliès, *The Melomaniac* (*Le Mélomane*), 1903. Film stills.

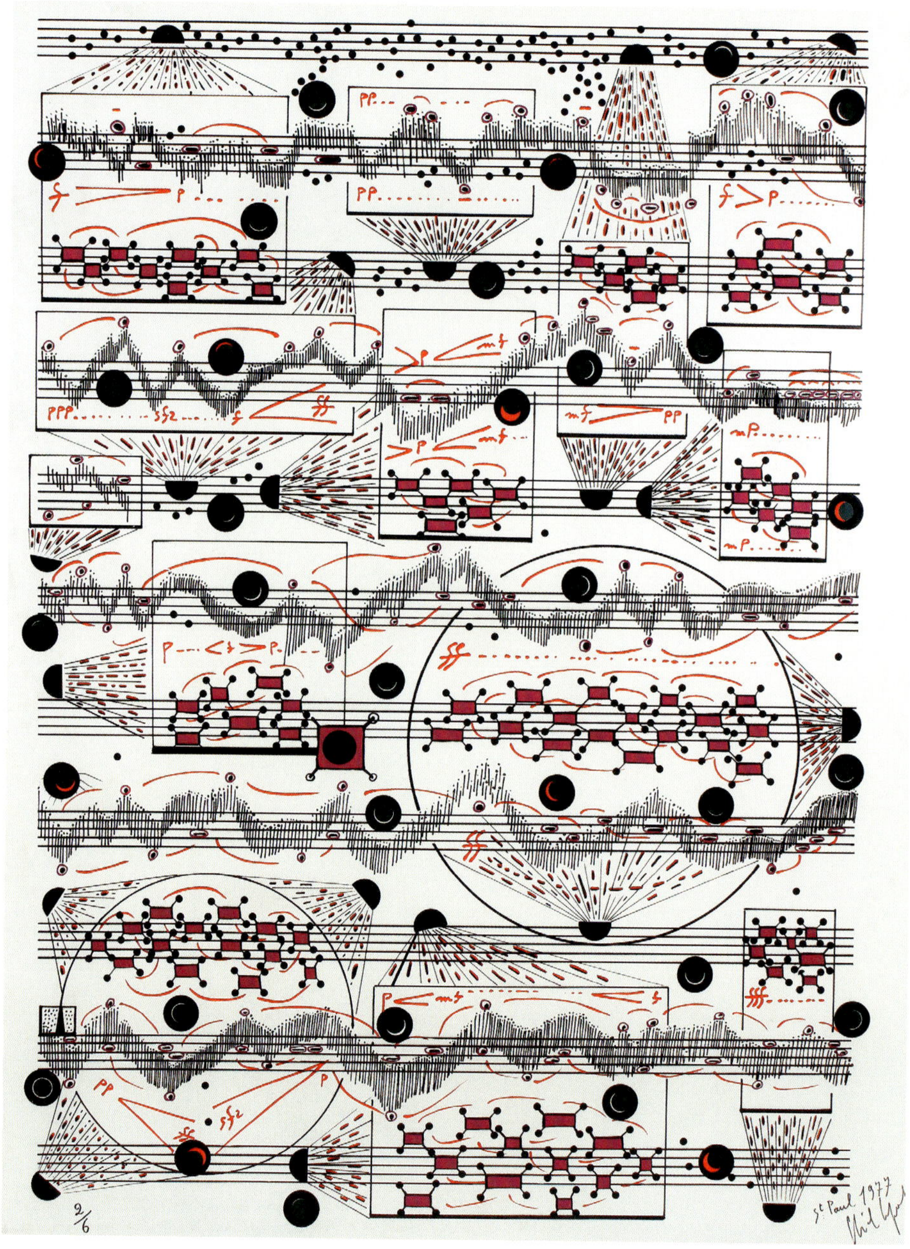

Michel Magne, *Partition de musique visuelle*, 1977.

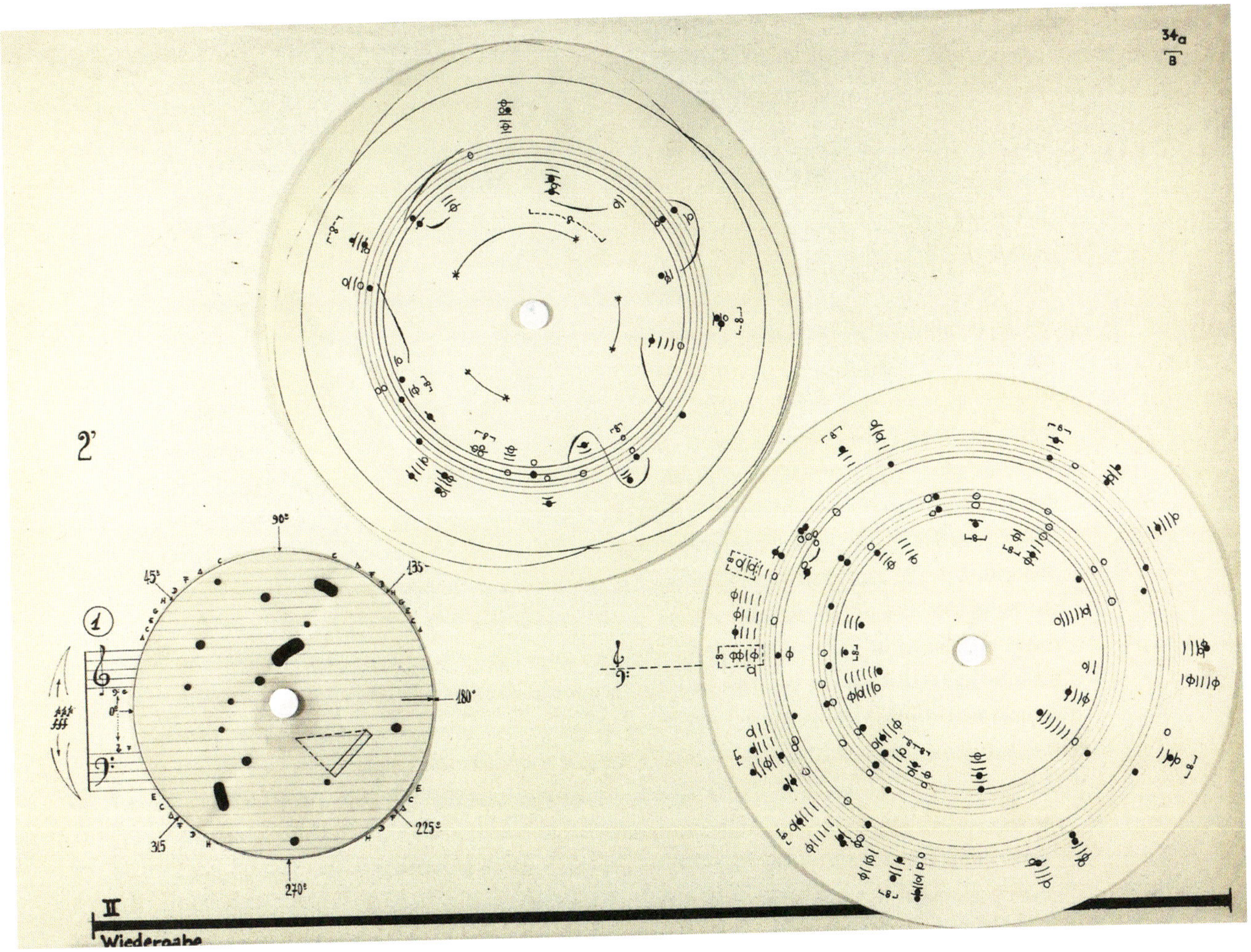

FIG. 87 Mauricio Kagel, *Transición II*, 1958–1959. Score.

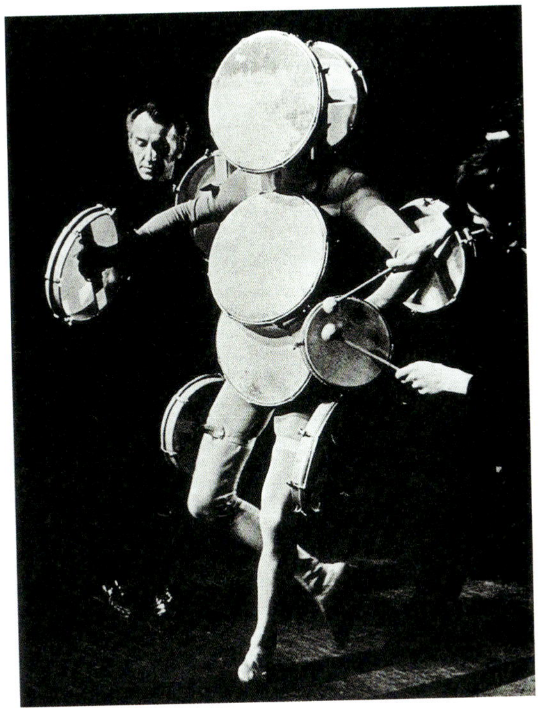

FIG. 88 Mauricio Kagel, *Spielplan: Jungfräulichkeit*. Drum man: 6 to 8 hand drums attached to the players body.

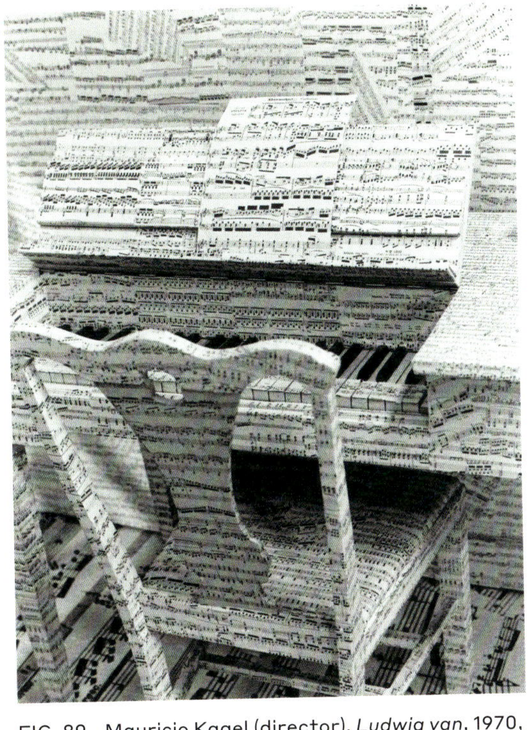

FIG. 89 Mauricio Kagel (director), *Ludwig van*, 1970, 90 min. Commissioned by Westdeutscher Rundfunk. Photograph of the music room in Kagel's documentary about Ludwig van Beethoven.

The new compositional techniques and principles led through dodecaphony and serial and aleatory processes to complicated notations, pictographs, and symbolic instructions, which left performers increasingly uncertain. This led in turn to an "approximate" notation that was open to indeterminacy (as with Cage). The aim was to involve the performer in the act of composition, along with spontaneous occurrences, chance, and noises from the environment. Composers turned to nondeterministic realms intended to free the performer from the rigid framework of a metrical time signature. Random events and spontaneous actions were declared to be constructive compositional principles. The verbal score – verbal instructions, action notation – "does not indicate the resultant sound but the action leading to it."[69] This gradually led to musical graphics as a form of notation, which completely abandoned notation based on traditional interval theory.

Three books published in the mid-1960s took stock of the revolution in new graphic musical notation. In 1965, the ninth volume in the series Darmstädter Beiträge zur Neuen Musik provided a summary of what had been presented and discussed at 1974's nineteenth annual International Summer Courses for New Music, with composers (György Ligeti, Roman Haubenstock-Ramati, Mauricio Kagel, Earle Brown (FIGS. 87 – 89)) as well as instrumentalists (Siegfried Palm, Aloys Kontarsky, Christoph Caskel) voicing their opinions on the topic of musical graphics. In the essay "Neue Notation – Kommunikationsmittel oder Selbstzweck?" [New notation: A means of communication or an end in itself?], Ligeti defines the differences between classical musical notation and musical graphics very clearly: "Fundamentally, then, 'notation' and 'musical graphics' are (to make a more precise distinction) two different realms. …

'Notation' is not a 'representation' of musical events, nor is it a 'depiction' of the movements (actions) that lead to the creation of music (although a subset of notation can relate to such actions); rather, it is a system of signs and at the same time a system of relationships between these signs that brings music into being in consequence of its correspondence to musical relationships. The visual sign system of notation corresponds to a system of auditory events; it denotes musical relationships.

FIG. 90 Earle Brown, *December 1952*, 1952. Musical graphics. Brown had conducted from this "score" at the Darmstadt Summer Courses for Music 1964.

A 'musical graphic,' on the other hand, is not a sign system. It does not denote musical relationships. It can, however, be a representation (depiction) of the events that lead to the creation of music; it can also suggest musical ideas and implementations through association.

There is an interrelationship between music and the visual configurations associated with it. A 'musical graphic' gives rise to an entirely different kind of music than a 'notation' does."[70]

Ligeti describes Brown's *December 1952* (FIG. 90), from *Folio* (1952–53), as particularly illuminating, and elaborates: "Thus the relationship between notation and musical graphics is analogous to that between Arabic writing and the arabesque. If an arabesque is derived from Arabic writing, is it Arabic writing? Yes and no: It possesses the characteristic forms of Arabic writing, yet in consequence of the stylization of the characters and the dissolution of the semantic substrate, it has become an image. A musical graphic can be converted into music, but it need not be; it is also valid as an image. Notation, on the other hand, is not valid in itself, but only in relation to the music it denotes."[71]

Kagel closely investigates the inversion, brought about by new notation, of the classical sequence "composition – notation –

interpretation": "'It took a long evolution for musical notation to travel from indistinctness, vagueness, and ambiguity to precision and an ever more indelible clarity,' wrote Willy Tappolet in 1949 in his book *La notation musicale*. As we see today, it has taken a short evolution for musical notation to find its way back to the indistinct, the vague, and the ambiguous. …

The transformation to which musical notation is today subjected has at the same time blown up the traditional structure

AUDITORY IDEA
(composition)
↓
DOCUMENTATION
(notation)
↓
AUDITORY PROCESS
(performance)

Notation no longer functions as a mediating element, but can now occupy the first and last positions in the configuration. The reform efforts of the past fifteen years have had such a radical effect on our thought processes in the area of notation that the sequence of auditory idea, documentation, and auditory process has by now been put through every possible combination."[72]

And Earle Brown gives the most detailed statement, in "The Notation and Performance of New Music": "The 'decorative' value of a score is in itself a pleasure but I am more concerned with the possibilities of a notational system that will produce an aural world which defies traditional notation and analysis and creates a performance 'reality' which has not existed before."[73]

He continues: "The early development of musical notation proceeded, of course, in the direction of more and more discrete control of all the elements and did not achieve its 'standard' appearance until after 1600 and its standardization of performance practice (the function of the conductor as we know it) until approximately 1800. Obviously we do not negate the music prior to these modern acquisitions merely because they do not conform to present practice. They have their own unique and very beautiful nature and expressive quality. Varèse has said that just because there are other ways of getting there, you do not kill the horse."[74]

Brown gave a great deal of thought to the idea of direct contact between composer and sounds. This led to an engagement with notation and its execution that can perhaps best be seen and heard in *Folio* – and is further developed in *Available Forms I* (1961) (FIG. 91) and *II* (1962) – in which the conductor seems to paint "with a palette of … composed sound events."[75] Brown mentions Charles Ives as one of the first composers whose conceptual approach and powers of musical imagination declared war on classical notation. Ives's music made a genuine effort, he says, "to disengage infinite sound from finite graphics."[76] Brown then quotes Henry Cowell on Ives: "Usually one feels that Ives hopes to induce the performer not to be too bound by any one way of organizing strong and weak beats, playing the passages now one way, now another. Ives' whole approach to his complex rhythms should be

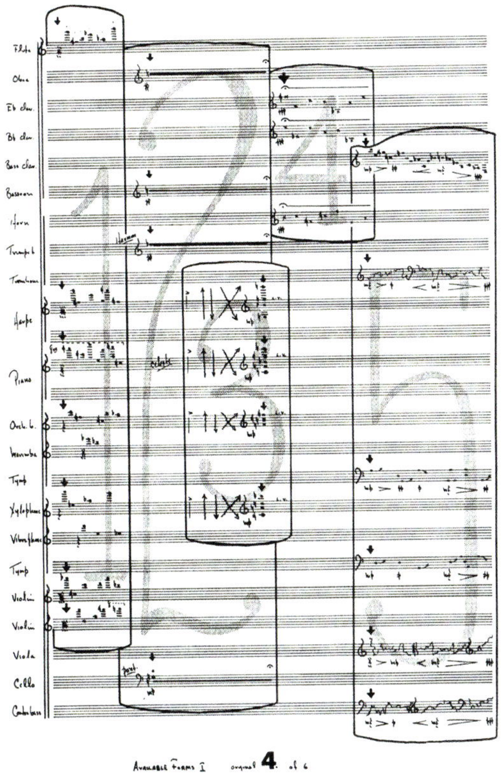

FIG. 91 Earle Brown, *Available Forms I*, 1961. Score for chamber music ensembles.

69 Erhard Karkoschka, *Notation in New Music: A Critical Guide to Interpretation and Realisation*, trans. Ruth Koenig (New York: Praeger, 1972), 3.

70 György Ligeti, "Neue Notation – Kommunikationsmittel oder Selbstzweck?" in *Notation Neuer Musik*, ed. Ernst Thomas, Darmstädter Beiträge zur Neuen Musik 9 (Mainz: Schott, 1965), 35–50, here 36–37. Translated from the German.

71 Ibid., 41. Translated from the German.

72 Mauricio Kagel, "Komposition – Notation – Interpretation," in Thomas, *Notation Neuer Musik*, 55–63, here 55, 59. Translated from the German.

73 Earle Brown, "The Notation and Performance of New Music," *Musical Quarterly* 72, no. 2 (1986), 180–201, here 181. Originally published in German as "Notation und Ausführung Neuer Musik," in Thomas, *Notation Neuer Musik*, 64–86. Brown also says in a footnote: "I should add that I am speaking of new notations whose primary application is toward extending the musical sound possibilities rather than those moving toward verbal or graphic descriptions of more 'theatrical' activities."

74 Ibid., 182.

75 Ibid., 186.

76 Ibid.

understood as an attempt to persuade players away from the strait-jacket of regular beats, with which complete exactness is impossible anyhow. … In fact, Ives has often expressed regret at having to write out a piece at all, since its rhythm will then be hopelessly crystallized."[77]

But Ives and Cowell were not the only ones to mark out and follow new paths. They and Leo Ornstein "devised notation for 'tone clusters' when the standard notation (primarily developed for triadic vertical structures) made the clusters visually contrary to the desired simultaneity. William Russell wrote much percussion music in the 1930s and 1940s and devised a notation for playing on the strings of a piano with a dining fork, and for all of the eighty-eight notes of the piano to be struck simultaneously; Henry Cowell devised notations for playing directly on the strings of the piano, as in *Banshee*. These are, however, primarily adaptations of the standard notation to new sounds and instrumental resources, which while not insignificant, do not basically affect the nature of the musical discourse or performance.

There was an attempt by Cowell to revise the notational situation in terms of rhythm (published as a theory in *New Musical Resources* in 1930, but first used in his music in about 1917). Nicolas Slonimsky said: 'Before he was twenty … [Cowell] delved for the first time into the problems of notation. This latter, not having been taught him as an established religion, he examined without fear – and dissented from its inadequacies. The duple system of rhythmical designations, giving adequate representation of only halves, quarters, eighths, etc., was the particular beam in the eye which Henry Cowell endeavored to extricate. He proposed a special notation for other fractions, so as to avoid the annoying and unscientific methods of setting down triplets, quintuplets, etc.'"[78]

In Brown's description, the approach involved giving the noteheads different shapes, each designating a different division of the beat. He calls Cowell's book *New Musical Resources* (1930) a source of "very reasonable suggestion[s]."[79]

So Ives, Cowell, and the others had already attempted, in the first half of the twentieth century, to free music from the straitjacket of interval theory and the vise of conventional notation. And Russell had apparently invented the prepared piano (à la Cage).

Brown engaged with the notation problem in 1952 in order to address "the problems of 'mobility' and immediacy which had been of great interest to me since the influence of Calder and [Jackson] Pollock in approximately 1948. For me, the mobility (or mutability) of the work had to be activated during the performance of the work (as in a mobile of Calder), and expressed spontaneously and intensely by the performer, as in the immediacy of contact between Pollock and his canvas and materials. These two elements – mobility of the sound elements within the work, and the graphic provocation of an intense collaboration throughout the composer-notation-performance process – were for me the most fascinating new possibilities for 'sound objects' as they had been for sculpture and painting. The necessity for new means of graphic representation is obvious. The compositions, *Folio* (1952–53) and *Twenty-Five Pages* (1953), were my first attempts to cross-relate these two influences and to realize the essence of these involvements in sound."[80]

Brown concludes by listing the reasons for his turn toward graphic "notation systems and performance processes":

"1. Belief that the complexity and subtlety of the desired sound results had passed the point at which standard notation could practically and reasonably express and describe the desired result.
2. The above belief led to a relaxation of finite notational controls and to the conscious inclusion of ambiguity in 'generalized' notations with which the performer and the performance process could collaborate.
3. The search for inherent or 'process' mobility in the work. The work as an endlessly transforming and generating 'organism,' conceptually unified in its delivery (the influence of the work of Calder).
4. The above necessitates a search for the 'conditional' performance state of spontaneous involvement, responsible to the composed materials and to the poetic conception of the work; 'work' in this case being the activity of producing as well as the acquisition of a finite result (the influence of the work of Pollock).
5. The fundamental motivation for all of the above: to produce a 'multi-ordinal' communicative

77 Ibid.

78 Ibid., 188–89.

79 Ibid., 189.

80 Ibid., 192.

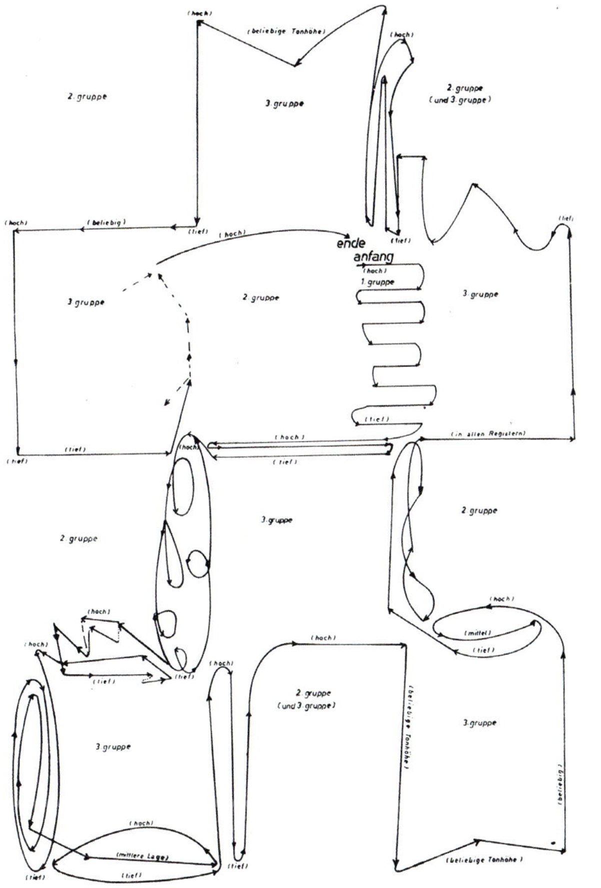

FIG. 92a Anestis Logothetis, *Odyssee*, 1963. Notation. It shows the „path" of the composition. The lines are painted in red color on a tansparent paper. It shows the realization directives fort he three groups of players.

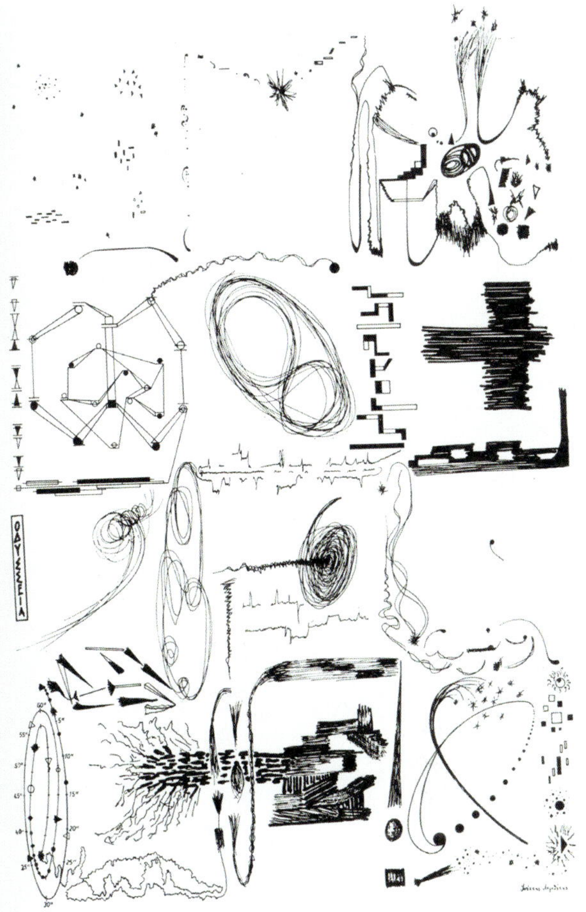

FIG. 92b Anestis Logothetis, *Odyssee*, 1963. Notation. It shows the possibilities of interpretation for each instrument. For example, deep shading means forte, light shading means piano. Rectangles mean clusters (black and white keys), curved lines and shapes mean glissandi.

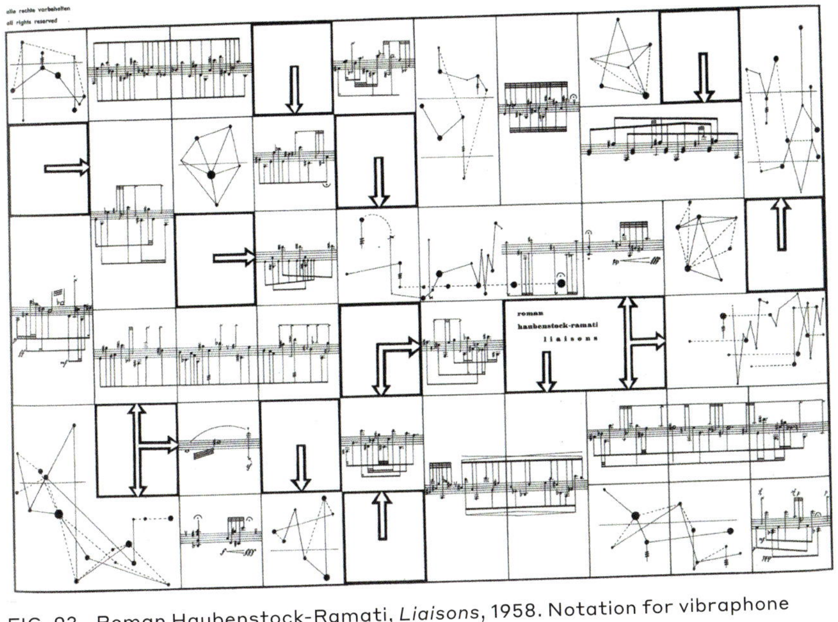

FIG. 93 Roman Haubenstock-Ramati, *Liaisons*, 1958. Notation for vibraphone or vibraphone with marimbaphone.

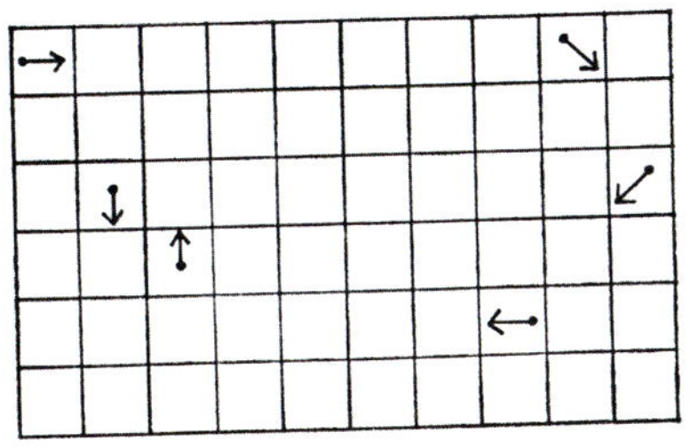

FIG. 94 Roman Haubenstock-Ramati, *Jeux*, 1960. Notation.

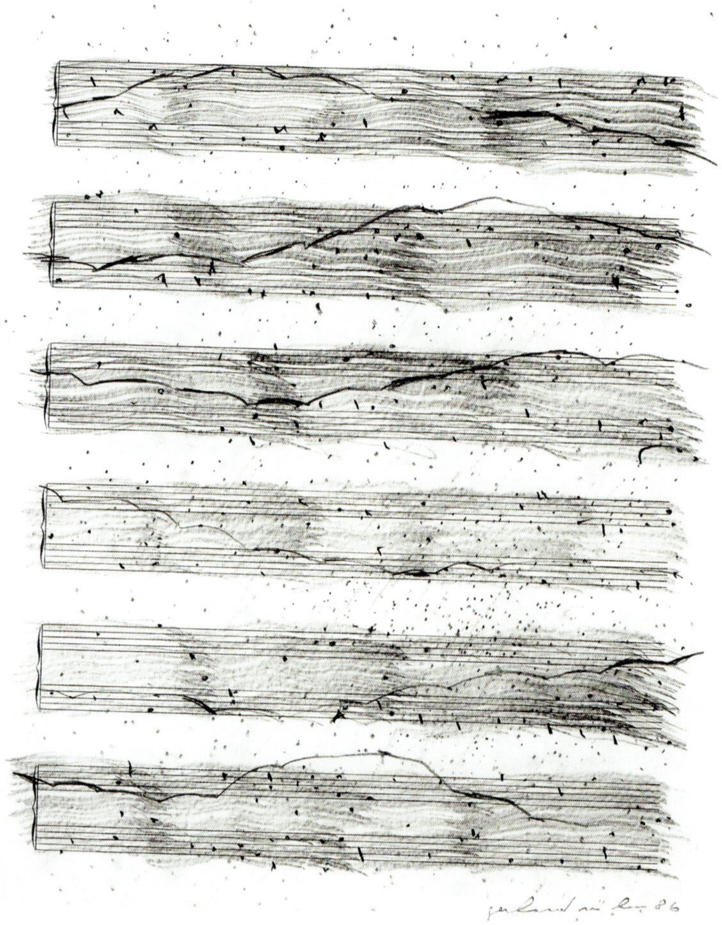

FIG. 95 Gerhard Rühm, *Lesemusik* [Music to read], 1986.

activity between the composer, the work, and the performer, and a similarly 'open' potential of experience for the listener.

The complex word, 'multi-ordinal' seems to me to contain the basic character of communication and meaning to which much of art is addressing itself today. (Joyce speaks of it relative to concepts of Vico.) The effects of this concept are overt in the work of Joyce, Mallarme, Stein, Duchamp, Ernst, Calder, Pollock (to mention only those who were my primary contacts with it), and now in music.

I will presume to say that this is where we are now and where art is most deeply concerned today in its attempt to 'express our time.' For me, this poetic necessity is responsible for the re-viewing of notational performance practice."[81]

The three books that heralded the graphic-notation revolution in music also included *Das Schriftbild der neuen Musik (Notation in New Music)*, by the composer Erhard Karkoschka, which was published in 1966 and contains the author's enlightening explanations and commentaries; and *Notations*, a collection of scores by 269 living composers, which was compiled by John Cage with Alison Knowles and published by Something Else Press in 1969. Rather than providing direct explanations or commentaries, the accompanying texts were, like the typography, the result of chance operations.

In the 1950s, the phrase "musical graphics" became a catchall for any radical experiment in new music. The term originated, however, with Oskar Rainer's book *Musikalische Graphik*. Its focus was a very different one: Instead of the graphic notation of music, the aim was to convert music into painting, with drawings and pictures supposedly expressing the painterly equivalent of tones and noises. Rainer had spent several years conducting studies and experiments with Austrian teenagers on the interrelationships between tonal and color harmonies before ultimately publishing his book on his findings and the results of directly converting music into

81 Ibid., 199.

82 Earle Brown, prefatory note to *Folio (1952/1953) and 4 Systems (1954)* (New York: Associated Music Publishers, 1961), 2.

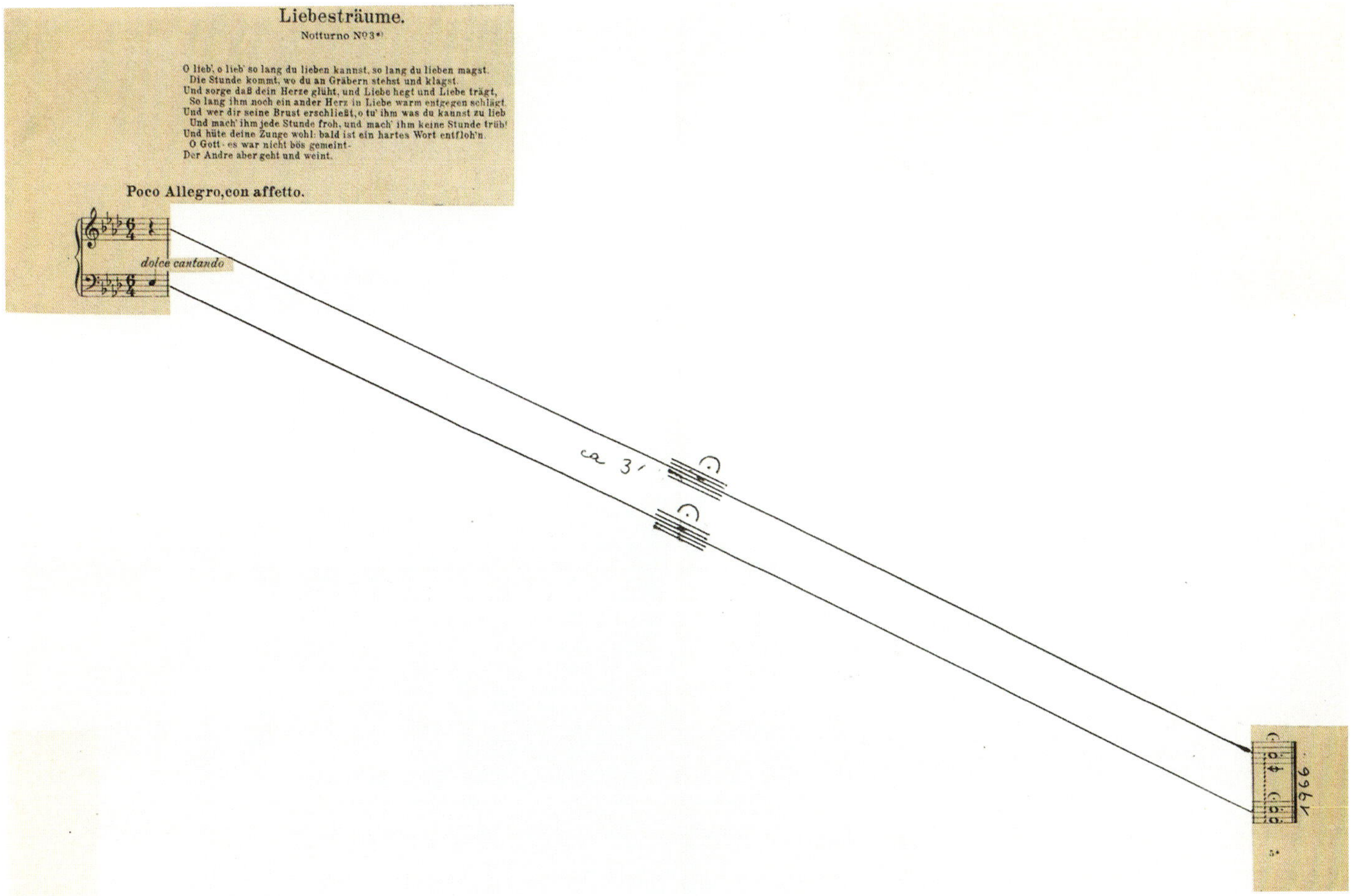

FIG. 96 Gerhard Rühm, *Liebesträume nach Franz Liszt* [Dreams of love à la Franz Liszt], 1966. Score, 31 × 46 cm.

painting. Through the principles of equivalence, analogy, correspondence, and synesthesia, he developed a unique aesthetic that attempted to convert music directly into painting. The process could operate on a subjective (expressive) or objective (quasi-numerical) basis. Rainer thus gave a name to an entire movement, even if it appears today in different forms and under other names (e.g., "visual music") in the visual scores of avant-garde musicians (Anestis Logothetis, Haubenstock-Ramati, etc.) and in the graphic expressions of a musical sensibility (Gerhard Rühm).(FIGS. 92–96)

The decisive factor, however, is the development of a concept of notation that has evolved in post-1950s contemporary music from a symbolic transcription of pitch and duration, dynamics and tonality in a line-based system (the modal notation of the twelfth-century Notre Dame school) to a set of direct playing instructions, as seen in the compositions of Brown, Logothetis, Helmut Lachenmann, Dieter Schnebel, and others. Notation changed from an extremely precise, clearly defined specification to an action for free performers, to the indeterminacy of chance in Cage's work. The score became an arena of action, of the performer.

With the graphically open-ended notation of the score for *December 1952*, which crossed over into the territory of visual art, Brown limited himself to defining the broad parameters of a music that allows for associative and subjective choices of interpretation: "The composition may be performed in any direction from any point in the defined space for any length of time. … It is primarily intended that performances be made directly from this graphic 'implication' (one for each performer) and that no further preliminary defining of the events, other than an agreement as to total performance time, take place."[82]

In the action notation of the musical avant-garde circa 1960, what is defined in the score is not pitches and durations of notes, but rather the exact method of producing the notes, including the place on the instrument where the notes are produced. This was accompanied by the

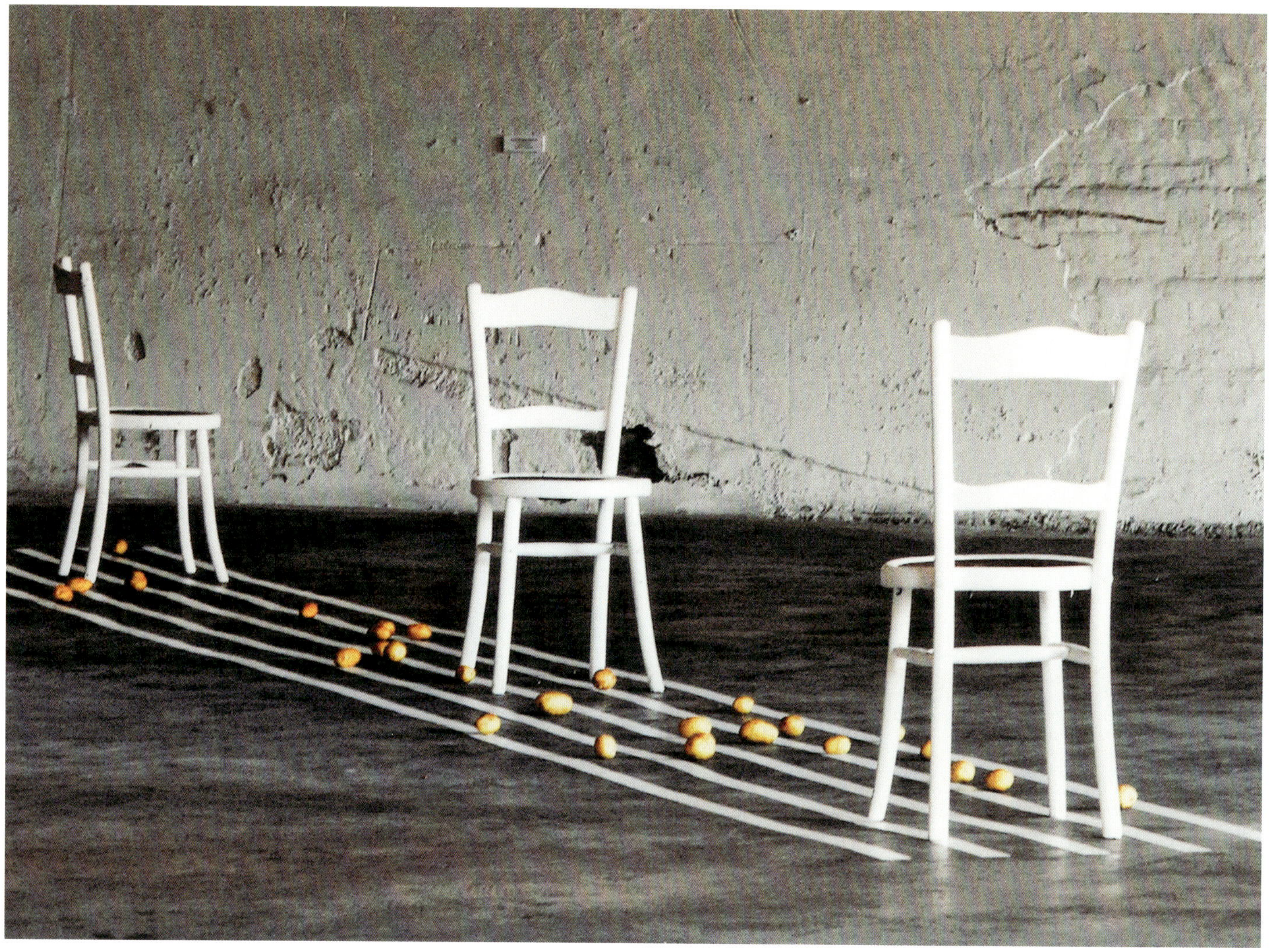

FIG. 97 Henning Christiansen, *Kartoffel-Opera* [Portato opera], 1969/2008. 3 chairs, adhesive tape, potatoes, dimensions variable.

emancipation of noise vis-à-vis the Western concept of the beautiful sound. Helmut Lachenmann, for example, views his noise-speckled music in part as an expression of the human labor that goes into music. The bowing of a violin naturally produces not only a beautiful tone full of vibrato, but also the scraping and rubbing of the bow hairs on the string to be set vibrating. Lachenmann, whose sociopolitical motivations date back to the 1960s, uses a "bridge clef" (which indicates where the sounds are to be produced but not the intervals between them) to make this part of the noise the focus of his work. In this way, he and others created a new spectrum of expression that has flowed into the current vocabulary of all (orchestra) instruments. Thus the evolution of notation can also be read as a pathway from the latency of the work in the score to the immediacy of action notation, to an open set of instructions.

Interestingly, Umberto Eco's 1962 book *Opera aperta* (*The Open Work*), an epoch-making text for art, drew its inspiration not from visual art, but from literature and new music. The concept of the transformation of the work of art, from a seemingly closed semantic and interpretive space into an open work with many possible interpretations, arose out of Eco's contemplation of problems presented by James Joyce's *Finnegans Wake* and musical works by Luciano Berio and Henri Pousseur. Eco mentions Berio's *Sequence for Solo Flute*, a piece in which the performer, although presented with "a text which predetermines the sequence and intensity of the sounds to be played," is "free to choose how long to hold a note."[83]

Eco then moves on to *Scambi* (Exchanges), a 1958 piece by Pousseur, who wrote: "*Scambi* is not so much a musical composition as a *field of possibilities*, an explicit invitation to

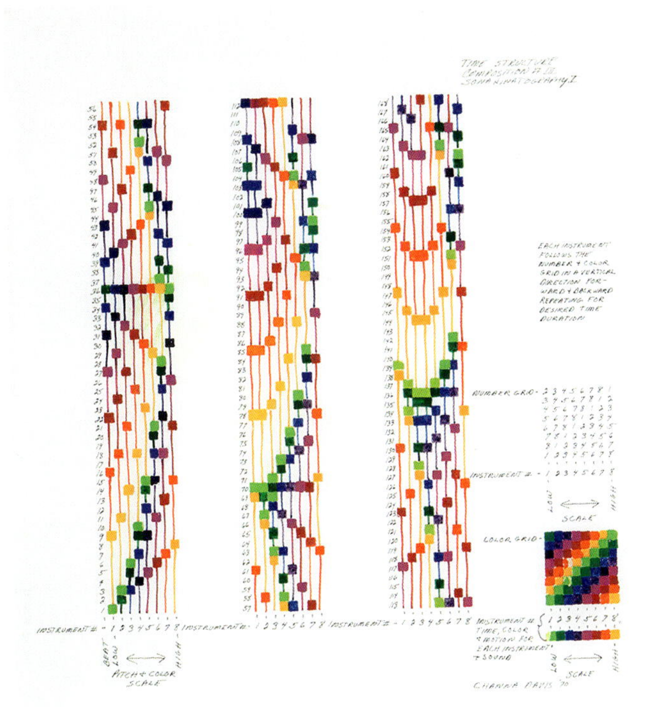

FIG. 98 Channa Horwitz, *Time structure composition III Sonakinatography I*, 1970. Casein on graph paper, 21.5 × 20.9 cm.

exercise choice. … Since the phenomena are no longer tied to one another by a term-to-term determination, it is up to the listener to place himself deliberately in the midst of an inexhaustible network of relationships and to choose for himself, so to speak, his own modes of approach, his reference points and his scale (well aware, however, that his choice is conditioned by the object he is looking at)." The multipolar world of new music elicits from the performer what Pousseur calls "acts of conscious freedom."[84]

Starting around 1960, the leading avant-garde exponents of these new freedoms were Fluxus and action art, which always had a sociopolitically referential aspect, and which confronted the individual with freedom of choice, in the Sartrean sense, in an attempt to emancipate and enlighten. The decisive factor was the evolution of the concept of notation from pure documentation to action notation, i.e., instructions.

Also in the 1960s, Channa Horwitz began developing a minimalistic graphic notation system, based on the principle of deductive logic, which she named "sonakinatography," and which enabled her to visualize time, rhythm, space, and movement in countless variations. She saw this system as a set of precise instructions: "In order to experience freedom, I have to limit my choices."[85] The resulting compositions, which may be read as finely drawn diagrams, musical scores, or choreographic notations, formed the basis for multimedia performances. Horwitz regarded sonakinatography as a visual philosophy and a language comprehensible by all the arts. The notation was based on the numbers one through eight, which derived from the grid of Horwitz's drawing paper. Eight horizontal units were each given a number corresponding to a duration to be counted off, and also a color. Horwitz called these units "energies." She later used an eight-by-eight grid in which the vertical boxes indicated the spatial extent of these energies. In the diagrams, repetition, inversion, and variation of the pattern are used to describe and depict temporal structures, note durations, and movements.(FIG. 98)

Whether "graphic notation" or "musical graphics," "notational image" or "visual music," all these names point to a fundamental transformation of music in the mid-twentieth century in response to the advent of technological devices – to the inception of "spatial music." As the term "graphic notation" already suggests, notation changes from a written image to a sequence of graphic symbols and thus to a drawing, rendering interval theory null and void. Graphic notation, then, showed the way from a time-based compositional technique to a space-based art, to spatial sound.

The graphic notation of the 1950s and 1960s was already a topological rather than temporal method of composition. The transition from classical notation to graphics – musical graphics or graphic music – and ultimately to action notation – that is, to verbal instructions – was the transition, in terms of both reception and composition, to space-based music, to sound art.

83
Umberto Eco, *The Open Work*, trans. Anna Cancogni (London: Hutchinson Radius, 1989), 1.

84
Henri Pousseur, "La nuova sensibilità musicale," *Incontri musicali*, no. 2 (May 1958): 3–37; quoted in Eco, *The Open Work*, 1, 10–11, 4. The parenthetical phrase in the second quoted passage is missing from the English edition of *The Open Work* and has been translated here from the Italian: Umberto Eco, *Opera aperta* (Milan: Bompiani, 1962), 44.

85
"Georg Schöllhammer & Johannes Porsch. Artist Talk with Channa Horwitz, 03/18/2012, ZKM | Center for Art and Media Karlsruhe," in *Moments. Eine Geschichte der Performance in 10 Akten*, ed. Sigrid Gareis, Georg Schöllhammer, and Peter Weibel (Cologne: Walther König, 2013), 426–28, here 427.

From Graphic Notation to Graphic User Interface

Musical notation, properly understood, does not merely indicate the key of a composition but is itself the key to understanding contemporary artistic practice. A classical score is already a set of directions that tells the musician exactly what to do with his or her instrument at specific moments. With a score, instead of a worker's second-by-second movements (with a machine) being photographically recorded and described, a worker's second-by-second movements (with an instrument) are graphically defined and prescribed. A score is a set of instructions – work instructions, performance instructions, instructions for use – for the performer, the interpreter. On one side is the subject, the executing musician; on the other is the object, the machine, the musical instrument. Between them stands the score, in which the transformations of the parts of the subject's body (hands, mouth, etc.) are described, transcribed, and prescribed as phase changes in the interaction with the object (violin, etc.). Musical notation is therefore an interface between the subject and the world of objects, created to realize a previously imagined or intended sound in the here and now. The classical score is an interface between human and instrument (machine); as such, it is interaction design. But it is also an algorithm, a set of directions, which consists of a finite number of rules and clearly defined instructions, and which precisely and completely describes the solution to a problem, or rather, a composition. The question, posed by Schoenberg, of how a composer moves from one note to the next,[86] is answered by the algorithm of the score.

An algorithm is a sequence of instructions, formulated according to the rules of a language, that enables a computer to execute a task. In the same way, a score is a set of instructions, formulated according to rules, for the execution of a task – for example, at the piano, a simple mechanical device. In that respect, the score is an early higher-order machine language, an algorithm.

In modern art, we can distinguish between two implementations of algorithms: the intuitive (e.g., Fluxus) and the strict (e.g., computer art). There are convergences between the two. The Fluxus artist George Brecht produced a work called *The Universal Machine* (FIG. 99), a clear

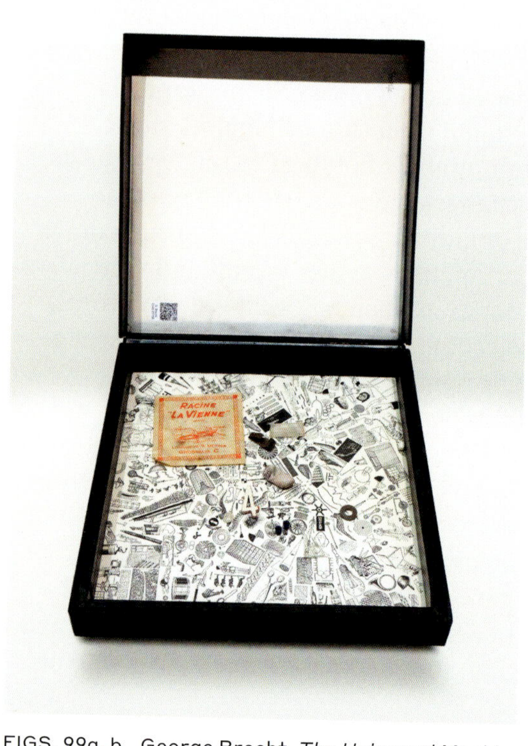

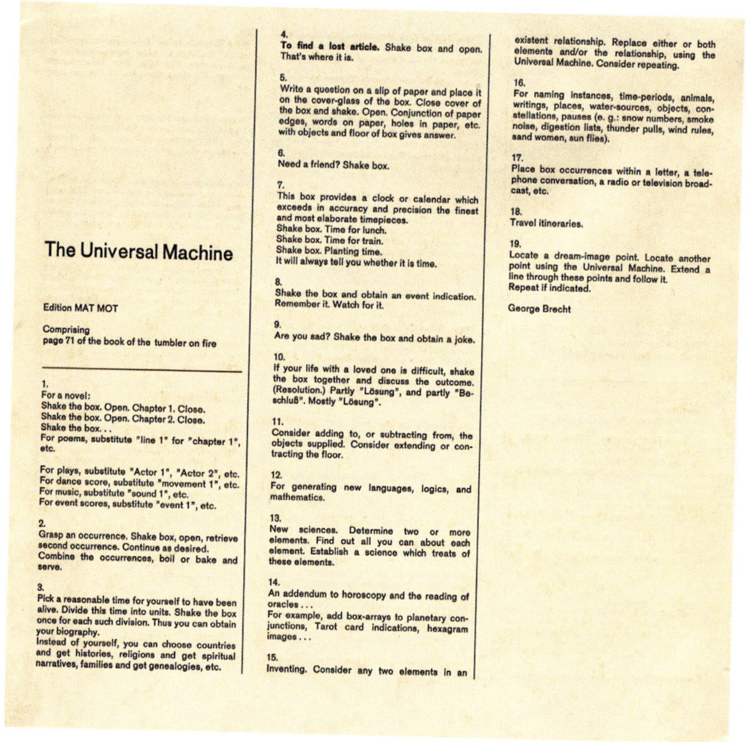

The Universal Machine

Edition MAT MOT

Comprising
page 71 of the book of the tumbler on fire

1.
For a novel:
Shake the box. Open. Chapter 1. Close.
Shake the box. Open. Chapter 2. Close.
Shake the box...
For poems, substitute "line 1" for "chapter 1", etc.

For plays, substitute "Actor 1", "Actor 2", etc.
For dance score, substitute "movement 1", etc.
For music, substitute "sound 1", etc.
For event scores, substitute "event 1", etc.

2.
Grasp an occurrence. Shake box, open, retrieve second occurrence. Continue as desired.
Combine the occurrences, boil or bake and serve.

3.
Pick a reasonable time for yourself to have been alive. Divide this time into units. Shake the box once for each such division. Thus you can obtain your biography.
Instead of yourself, you can choose countries and get histories, religions and get spiritual narratives, families and get genealogies, etc.

4.
To find a lost article. Shake box and open. That's where it is.

5.
Write a question on a slip of paper and place it on the cover-glass of the box. Close cover of the box and shake. Open. Conjunction of paper edges, words on paper, holes in paper, etc. with objects and floor of box gives answer.

6.
Need a friend? Shake box.

7.
This box provides a clock or calendar which exceeds in accuracy and precision the finest and most elaborate timepieces.
Shake box. Time for lunch.
Shake box. Time for train.
Shake box. Planting time.
It will always tell you whether it is time.

8.
Shake the box and obtain an event indication. Remember it. Watch for it.

9.
Are you sad? Shake the box and obtain a joke.

10.
If your life with a loved one is difficult, shake the box together and discuss the outcome. (Resolution.) Partly "Lösung", and partly "Beschluß". Mostly "Lösung".

11.
Consider adding to, or subtracting from, the objects supplied. Consider extending or contracting the floor.

12.
For generating new languages, logics, and mathematics.

13.
New sciences. Determine two or more elements. Find out all you can about each element. Establish a science which treats of these elements.

14.
An addendum to horoscopy and the reading of oracles...
For example, add box-arrays to planetary conjunctions, Tarot card indications, hexagram images...

15.
Inventing. Consider any two elements in an existent relationship. Replace either or both elements and/or the relationship, using the Universal Machine. Consider repeating.

16.
For naming instances, time-periods, animals, writings, places, water-sources, objects, constellations, pauses (e. g.: snow numbers, smoke noise, digestion lists, thunder pulls, wind rules, sand women, sun flies).

17.
Place box occurrences within a letter, a telephone conversation, a radio or television broadcast, etc.

18.
Travel itineraries.

19.
Locate a dream-image point. Locate another point using the Universal Machine. Extend a line through these points and follow it.
Repeat if indicated.

George Brecht

FIGS. 99a, b George Brecht, *The Universal Machine*, 1965.
Named after the invention of Alan Turing , who called the computer a universal machine, Brecht's machine is designed to shaken, so that the items randomly obscure the images. The box contains a text indicating how an interpretation of the objects in relation to the images can answer different questions for the user.

allusion to Alan Turing's 1936 characterization of the computer as a "universal machine."[87] Another Fluxus artist, Dick Higgins, published a pamphlet titled *Computers for the Arts* in 1970, which included a machine score for computer music by James Tenney for which Higgins had written the lyrics.[88] A 1962 text by Umberto Eco, written for the exhibition *Arte programmata*, dealt with the interplay of chance and programming,[89] and the Italian architect Leonardo Mosso was referring to an "architettura programmata" as early as 1969.[90]

Graphic notation introduced ideas into music before they were technologically feasible. A sheet of music is not yet in physical contact with the performer; it is merely a set of readable instructions.

Joseph Schillinger's theories on the mathematical basis of the arts[91] were taken up in 1963 by Robert Baker and Lejaren Hiller to create MUSICOMP: a computer program, based on standard programming languages such as FORTRAN, that was designed to generate musical scores. One of these was their *Computer Cantata* (1963);[92] later compositions bore the telling titles *Algorithms I* (1968) and *Algorithms II* (1972). But the use of computer programs to create sounds proved to be quite laborious, prompting a search for better forms of human-computer interaction. This led Max Mathews and Lawrence Rosler to a graphic-input computer named Graphic 1 in 1965.[93] (FIGS. 100–102) The Graphic 1 console was a computer system with a light pen for real-time graphic input, a keyboard for alphanumeric input, a card reader for binary input, and a cathode-ray tube for graphic output. It also included a DEC PDP-5 computer, a DEC 340 oscilloscope, and the appropriate

86
Cf. Arnold Schoenberg, "Composition with Twelve Tones (1)," in *Style and Idea: Selected Writings*, ed. Leonard Stein, trans. Leo Black (Berkeley: University of California Press, 2010), 214–45.

87
A. M. Turing, "On Computable Numbers, with an Application to the *Entscheidungsproblem*," *Proceedings of the London Mathematical Society*, 2nd ser., 42 (1937): 230–65, here 241.

88
Dick Higgins, *Computers for the Arts* (Somerville, MA: Abyss, 1970), 6–10.

89
Umberto Eco, untitled text in *Arte programmata. Arte cinetica, opere moltiplicate, opera aperta*, exhibition brochure (Milan: Galleria Vittorio Emanuele, 1962).

90
Peter Weibel, "Intelligente Wesen in einem intelligenten Universum," in *Enzyklopädie der Medien*, vol. 1, *Architektur und Medien* (Berlin: Hatje Cantz, 2015), 83–100; see in particular 85–86.

91
Joseph Schillinger, *The Mathematical Basis of the Arts* (New York: Philosophical Library, 1948).

92
See Lejaren A. Hiller and Robert A. Baker, "*Computer Cantata*: A Study in Compositional Method," *Perspectives of New Music* 3, no. 1 (Autumn–Winter 1964): 62–90, as well as Lejaren Arthur Hiller, *Informationstheorie und Computermusik. Zwei Vorträge, gehalten auf den "Internationalen Ferienkursen für Neue Musik" Darmstadt, 1963*, trans. Peter Jansen, Darmstädter Beiträge zur Neuen Musik 8 (Mainz: Schott, 1964).

93
Max V. Mathews and Lawrence Rosler, "Graphical Language for the Scores of Computer-Generated Sounds," in *Music by Computers*, ed. Heinz von Foerster and James W. Beauchamp (New York: John Wiley and Sons, 1969), 84–114, here 107.

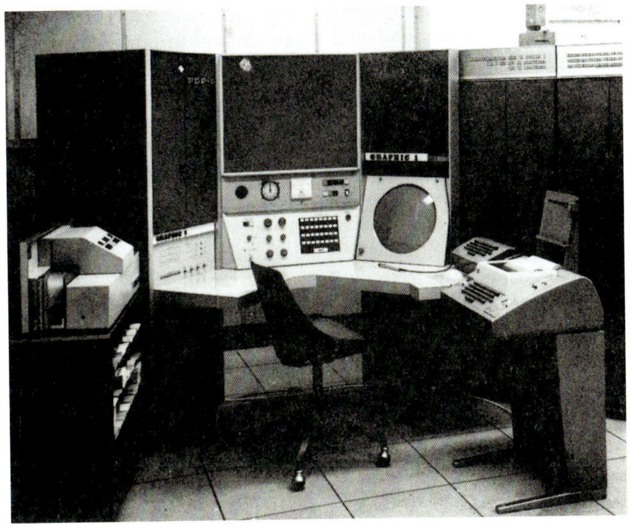

FIG. 100 Graphic 1 terminal, 1965.
Graphic 1 was developed in 1965 by William Ninke (together with Carl Christensen and Henry S. McDonald) at Bell Labs. Lawrence Rosler and Max Mathews used the terminal to develop interactive graphical music.

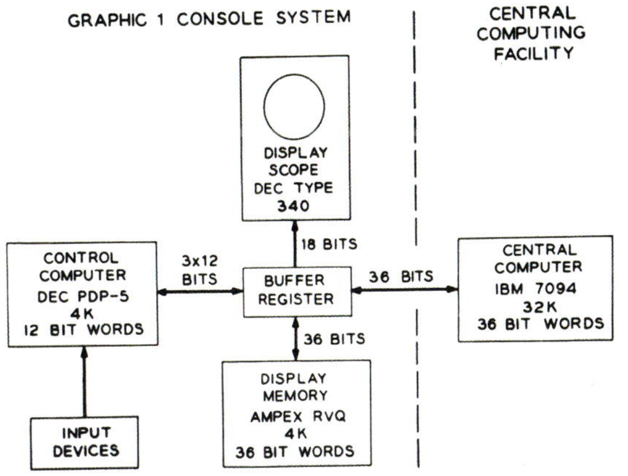

FIG. 101 Graphic 1, 1965. Sketch of the console system.

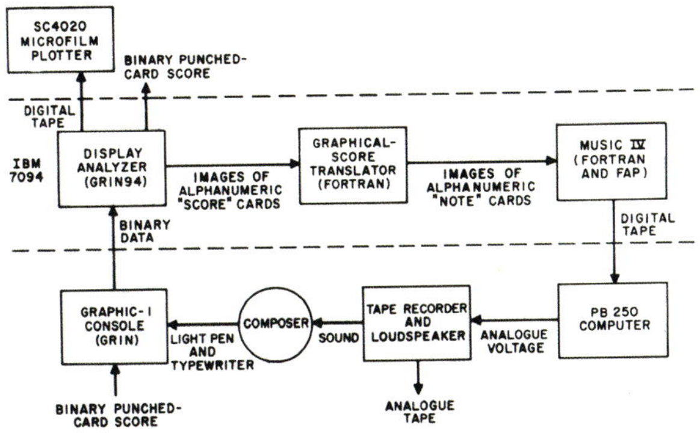

FIG. 102 Max Mathews, Joan Miller, *MUSIC IV* for IBM 7094, 1963. Synthesis software for computer music to be used on the Graphic 1.

hardware interfaces. Tone sequences were represented by graphs, and compositions were produced algorithmically.

The Graphic 1 enabled a person to enter images and symbols directly into the computer's memory by drawing them. The drawings could also be retrieved, deleted, duplicated, and modified. The Graphic 1 was originally intended as a tool for designers, but a musical composition can also be understood as a design problem: Not only can computer sounds be designed, they can also be produced with computers, and algorithms can be useful for creating parts of the music with the aid of the computer. A graphic-notation language was invented so that the score of a piece could be specified as a set of graphs. The heart

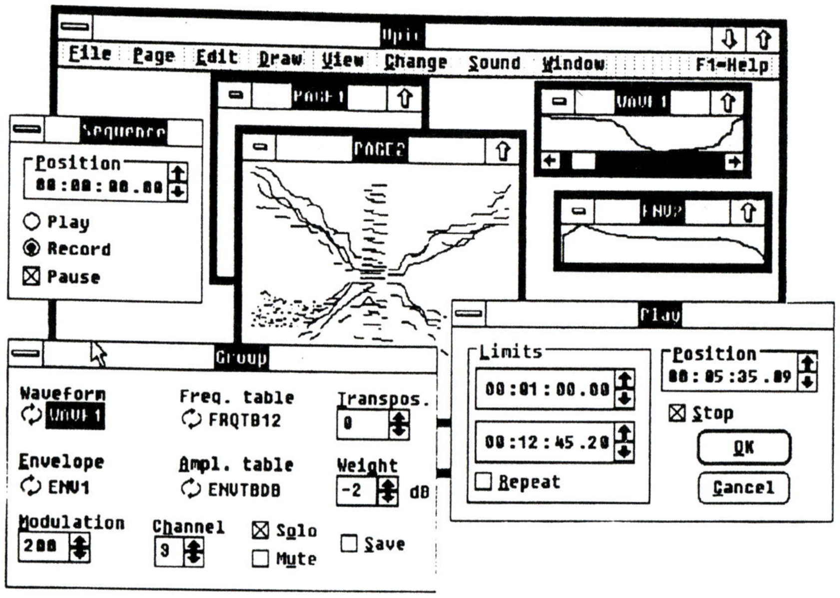

FIG. 103 Iannis Xenakis, Sample Screen of UPIC.

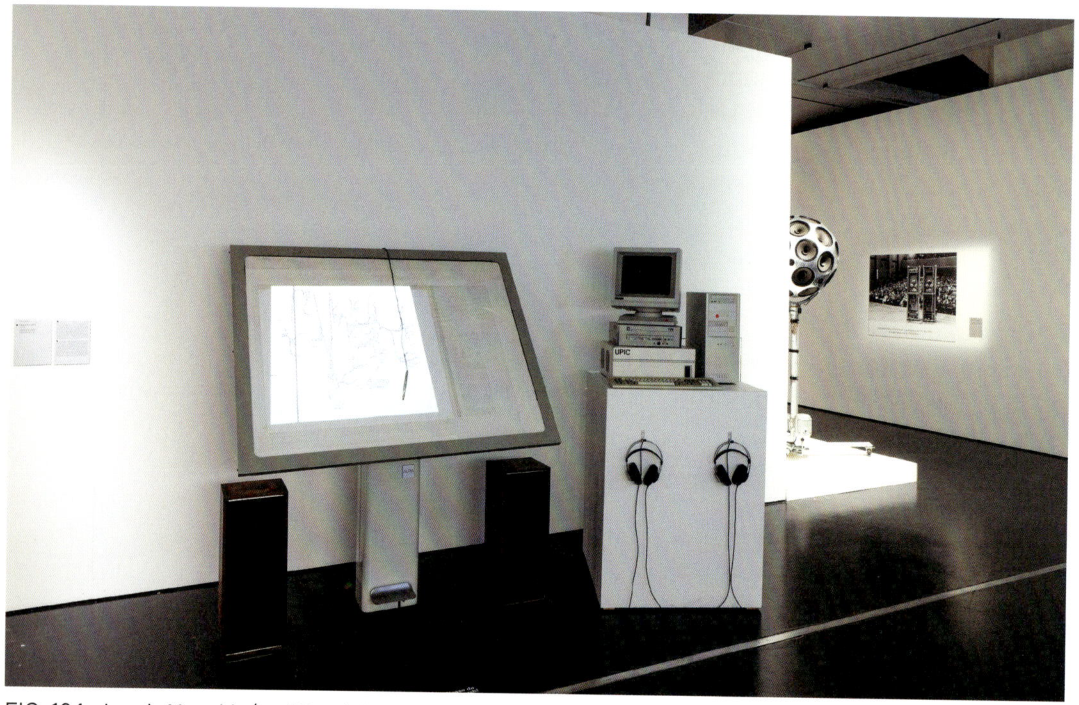

FIG. 104 Iannis Xenakis (and Patrick Saint-Jean), *UPIC – Unité Polyagogique Informatique du Centre d'Etudes de Mathématique et Automatique Musicales*, 1977. Installation view *Art in Motion*, ZKM | Karlsruhe, 2018/2019.

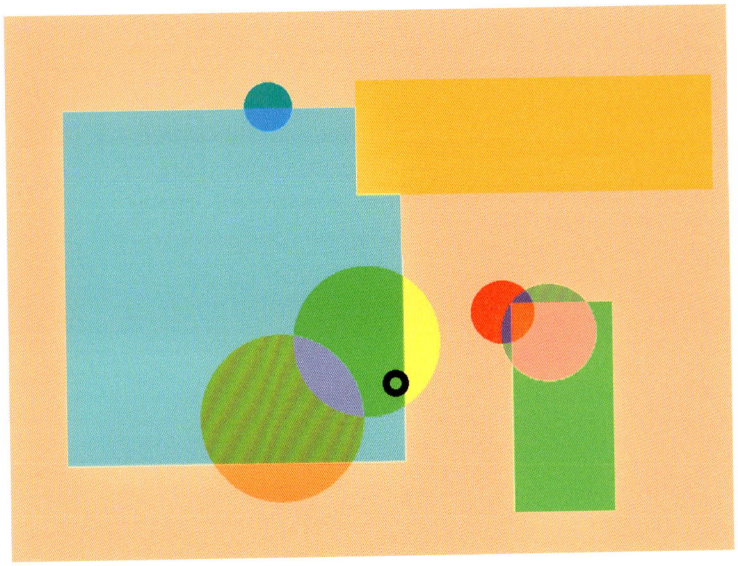

FIG. 105 Masaki Fujihata, Kiyoshi Furukawa, and Wolfgang Münch, *Small Fish*, 1999. Interactive sound installation.

of the system was the program *MUSIC IV* for the IBM 7094 computer.

Mathews and Rosler developed their "graphical language for the scores of computer-generated sounds" based on a computer language called GRIN. The GRIN program they used to create graphic scores divided the console display into two sections: an area for "light buttons" and messages to the composer, and a grid on which the music functions were displayed. "Conventional scores are an insufficient and inconvenient way of describing sound sequences to computers. A procedure is described for drawing scores as graphical functions of time by using a light pen on a cathode ray tube attached to a small computer. The information is transmitted digitally to a larger computer, which synthesizes the sound and reproduces it immediately with a loudspeaker. … The graphical programs provide great flexibility for drawing, copying, erasing, and altering functions. Thus it is easy to develop a sound sequence by a succession of trials. Microfilm and punched-card versions of the score are automatically provided. In addition to being compositional tools, the graphical scores are effective representations of the sound to a listener. In many ways they are easier to follow than conventional scores."[94] In the 1970s, Xenakis's UPIC project captured what was historically inevitable – that is, the transformation of graphic notation into a computer-assisted graphic user interface. Like Graphic 1, the UPIC system aided interactive composition, using an AT 386 microcomputer and a mouse-controlled graphic interface, real-time drawings generated and stored music.(FIGS. 103, 104)

Once the necessary technology was available – specifically, a touchscreen – tones could be generated simply by touching the score. The performer did not have to translate the notes he or she had read into fingerings on musical instruments, but could cause the notes to sound by pressing a finger to them on the graphic user interface of the touchscreen. Consequently, a favorite educational installation at music museums involves a score that is placed behind a pane of glass and also marked on the surface of the glass. Visitors can activate this haptic score with the touch of a finger, which transmits the information to a computer, so that touching the notes produces the corresponding sounds and melodies.

Thus the promise of the revolution in graphic notation was realized only later by the technological revolution of the graphic user interface, the touchscreen. Today, the use of the interface as a graphic score takes place on countless laptops, tablets, and smartphones.[95] The profusion of music applications in the digital realm is the product of graphic notation since the 1950s. A pioneering role was played here by the interactive CD-ROM *Small Fish*, produced by Masaki Fujihata, Kiyoshi Furukawa, and Wolfgang Münch at the ZKM | Karlsruhe in 1999. *Small Fish* includes fifteen different programs for activating and modifying audiovisual compositions.(FIG. 105) Typically, shapes move across the screen, producing tones when they collide or reach the screen's edge. In some of the programs, the shapes can also be moved with the mouse to affect the analog playback of the synthetically generated sounds. The music is created on the graphically formatted score.

The paradigm of notation plays a fundamental and central role in contemporary art, from sculpture as an open practice[96] to performance, but especially in the evolution of new-media art. Additionally, the millions of people who use the global internet as a platform for communication and creativity each day depend on search algorithms, operating instructions, programs, protocols –

94 Ibid., 84.

95 At the 2013 AppArtAwards ceremony at the ZKM | Karlsruhe, Matthias Krebs, an app musician and the founder of DigiEnsemble Berlin, gave a lecture entitled "App Musik: Kunstwerke, Sound Toys und App Instrumente" about the potential uses of smart devices for producing music.

96 Cf. Peter Weibel, "Art as Open Practice," trans. Julia Scheffer, in *Offene Handlungsfelder / Open Practices / Prassi aperta*, ed. Peter Weibel, exhibition catalog, Austrian Pavilion, 45th Venice Biennale (Cologne: DuMont, 1999), 22–35.

digital scores and notations, as it were, which they can or must perform within rule-based frameworks. One might call the twentieth century the century of notation, and it would not be an exaggeration to speak of notation as a universal in twentieth-century art.[97]

The triumph of graphic notation lasted about a decade; sadly, it came to an end for two reasons: First, because music theory failed to recognize the spatial turn, instead regarding graphic notation as a contemporary quirk of the musical avant-garde; and second, because the musicians themselves lacked the mathematical knowledge to develop graphic notation into graph theory. Not until the development of the graphic user interface in conjunction with the computer did graphic notation really begin to take root, as then composers could produce tones directly, without scores or performers.

The word *graph* was introduced in 1878 by the mathematician James Joseph Sylvester in relation to the graphic notations of chemical structures.[98] The first uses of graphs were structural formulas in chemistry.[99] The first textbook on graph theory was published in 1936.[100] Graph theory defined many fundamental concepts: graph, neighborhood, path, cycle, circuit, etc. One important application of graph theory is to find the shortest route between two or more places. These places are defined as nodes and the connections between them as edges. Thus a graph may be generalized as a set of nodes and edges. An edge is a set of exactly two nodes. If the set of nodes is finite, then we speak of finite graphs. If the edges are defined as ordered pairs of nodes rather than a set, then we speak of directed graphs.

Since the solutions to problems in graph theory (e.g., the Seven Bridges of Königsberg problem, the Chinese postman problem, the Hamiltonian circuit problem, and the traveling salesman problem) are often based on algorithms, graph theory plays an extremely important role in complexity theory. In graph theory, musicians might have found a new home for graphic notation. It would have been necessary to integrate graph theory into music theory; today we would then have a music theory befitting the computer. In the musical performances of graphic schemata, the nodes would be notes, the edges vectors, etc.

Music, Graphic Notation, and Graph Theory

There is a branch of mathematics that has taken up the inheritance of geometry: topology. One speaks of topological structures; naturally, these go far beyond the conceptual space of geometry. The famous mathematician Leonhard Euler founded topology in 1736 with the Seven Bridges of Königsberg problem. (FIG. 106)

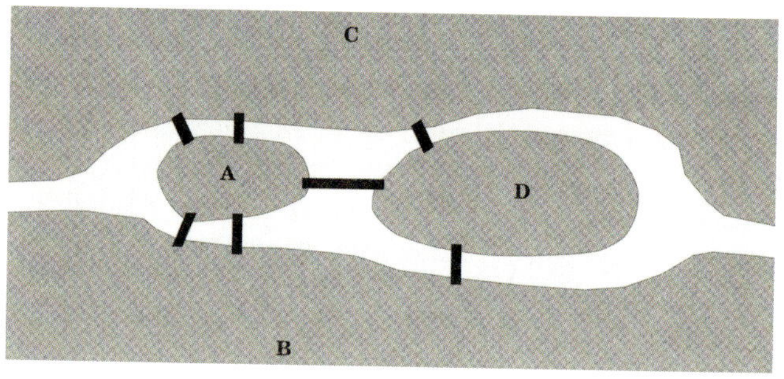

FIG. 106

The banks of the River Pregel, which flowed through the city of Königsberg, were connected to Island A by four bridges and to Island B by two bridges; one bridge connected the islands to each other. The question was whether it was possible to find a circular route, starting from any given location, by which one would cross each bridge just once. Euler proved that there is no such route, since each riverbank or island has an odd number of bridges leading to it. In its abstract formulation, Euler's question is: Can a path be found that starts from any given location (A, B, C, D), crosses each of the seven bridges exactly once, and then returns to its starting point? The four positions constitute the four nodes of a graph, which we differentiate by means of the numbers 1, 2, 3, and 4. We connect two nodes whenever the positions are connected by a bridge. This results in the graph shown in Fig. 107.

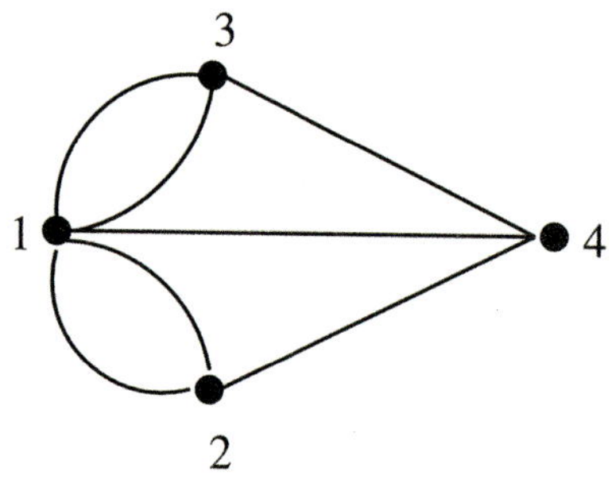

FIG. 107

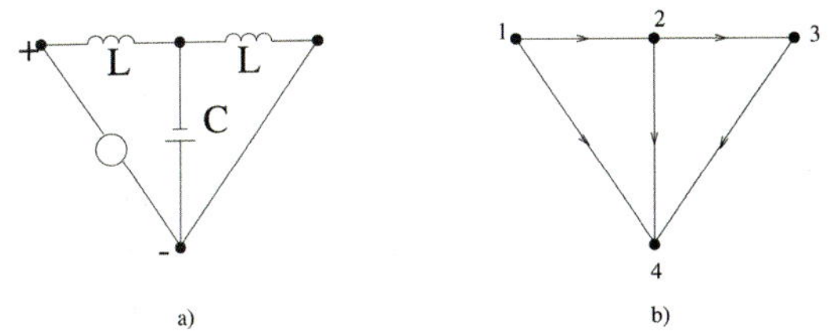

FIG. 108

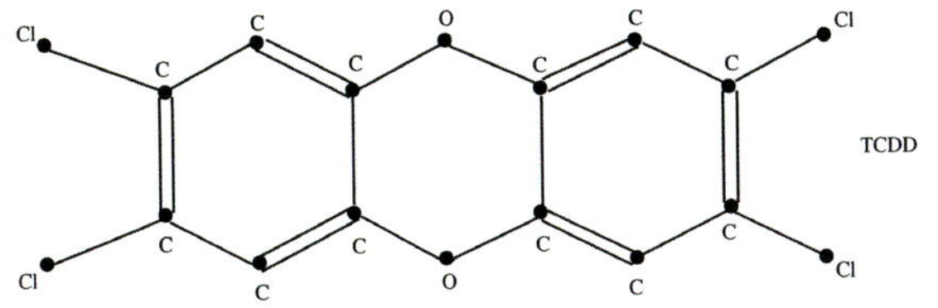

FIG. 109

Different arrangements of the edges produce different graphs. A graph is described as connected if there is at least one path connecting each pair of nodes, as in the graph above. Thus there can be open and closed walks. Let us assume that the edges of a graph model a street system; if we imagine a postman, we can be sure that he would prefer to walk each section of street just once. An extended version of this is the traveling salesman problem: A traveling salesman who has thirty-five cities to visit would obviously rather avoid traveling the same route more than once. He would therefore try to figure out how to visit the thirty-five cities without traveling the same route twice. This topological problem is part of graph theory: Is there a cycle that uses each edge exactly once?

The similarity between the injunction against repeating a note in the twelve-tone series and the avoidance of repeated paths in topology must be noted. The question of how I get from one tone to the next need not be solved with harmonic theory; I can also solve it with graph theory.

Electrical networks can also be correlated with graphs. Graphs can be used to determine a network's capacity. The nodes that appear here are the power sources, the ground points, and the junctions at which the power can be split. Fig. 108 shows an electrical network (left) together with its directed graph (right).

The electricity in the network's individual branches follows Kirchhoff's circuit laws, which can be expressed as linear equations. In 1847, Gustav Kirchhoff showed that the total current in an electrical network can be determined once the current in the branches of a spanning tree is known.[101] *Tree* is a term from graph theory. Trees appear as data structures in the optimization of communication networks, computer-language grammars, etc. We know that electrical networks – that is, graphs – can be physically realized. Thus it follows that computers, through which electricity flows, should compose music not according to old analog rules, but rather according to Kirchhoff's laws and other rules and methods from graph theory.

Today, the traveling salesman problem applies not just to traveling salesmen, or the security staff of an industrial complex, but also to the manipulation of workpieces by robot arms and the positioning of holes drilled in a circuit board (computer chip). Graphs play a role in chemistry as well. Molecules are characterized by their structural formulas, which show how many and what kinds of atoms they are made of. H_2O consists of two hydrogen atoms and one oxygen atom. The possible bonds can be shown as an undirected graph known as a structure graph. Fig. 109 shows the structure graph of the dioxin TCDD. Such graphs display symmetries, automorphisms, and permutations.

Because graph theory seeks solutions for spatial neighborhoods, the application of graph theory to electronic networks presents itself as a way of generating music. Graph theory, then, is a more up-to-date mathematical method for answering the question of how to get from one note to the next.

97
Cf. Hubertus von Amelunxen, Dieter Appelt, and Peter Weibel, eds., *Notation. Kalkül und Form in den Künsten* (Berlin: Akademie der Künste, 2008).

98
J. J. Sylvester, "Chemistry and Algebra," *Nature* 17 (1878): 284.

99
Arthur Cayley, "On the Mathematical Theory of Isomers," *Philosophical Magazine*, 4th ser., 47, no. 314 (1874): 444–46. Cayley is also known for coining the term *tree* for mathematical graphs; cf. Arthur Cayley, "On the Theory of the Analytical Forms Called Trees," *Philosophical Magazine*, 4th ser., 13, no. 85 (1857): 172–76.

100
Dénes Kőnig, *Theorie der endlichen und unendlichen Graphen. Kombinatorische Topologie der Streckenkomplexe* (Leipzig: Akademische Verlagsgesellschaft, 1936).

101
Gustav R. Kirchhoff, "Über die Auflösung von Gleichungen, auf welche man bei der Untersuchung der linearen Verteilung galvanischer Ströme geführt wird," *Annalen der Physik und Chemie* 148, no. 72 (1847), 497–508.

Music and Algebra

There are also other mathematical models that could be used as spatial composition techniques – Clifford algebra, for example. Clifford algebra originated from a number of sources, one of which was Hermann Grassmann's *Lineale Ausdehnungslehre* [theory of linear extension] of 1844. The aim of Grassmann's algebra was to generalize the concept of the vector to higher dimensions in order to describe directed planes, volumes, etc. It is thus a kind of multidimensional algebra. Another source was quaternion algebra, which was discovered by William Rowan Hamilton in 1843. Quaternions are a four-dimensional generalization of complex numbers and are well-suited to the description of rotations in three and four dimensions.

In 1878, William Kingdon Clifford invented the Clifford algebra that bears his name, which he himself called geometric algebra.[102] With it, Clifford hoped to combine Grassmann's algebra and Hamilton's quaternions in a common mathematical framework. Some number systems in mathematics include complex units, the complex numbers, quaternions, and octonions. In these systems, one can define one, three, or seven fixed elements, respectively, which expand the field of numbers into a real vector space that can be multidimensional. Geometric algebra is thus based on a vector space that has meter. A geometric vector is a directed line that possesses both a magnitude and a direction.

A vector, then, is a directed element of a one-dimensional space. In geometry, however, we need not only vectors (lines), but also other objects such as planes and volumes. These can be regarded as a generalization of the vector concept – that is, as multivectors. Thus geometric algebra is a Clifford algebra over a vector space with a quadratic form. The vector space is real two-dimensional space, while the algebra is real four-dimensional.

Clifford algebra is the initial object in a category such that there is a morphism from it to every other object in the category. This gives rise to algebra morphisms. Clifford algebra extends a vector space that goes far beyond the natural numbers. One might thus conclude that the rules that have been written for numbers and for operations with numbers that exist only on a two-dimensional plane would no longer be valid when presented in multidimensional spaces.

The promise of the computer is the possibility of operating in multidimensional space, creating new constellations, connections, paths, and bridges between the notes, letters, and numbers. The computer-aided application of Clifford algebra to music would be the first exemplar of this hope, the beginning of a civilizational revolution: the transition from two-dimensional notation systems to multidimensional systems of notation and composition.

Xenakis never mentioned Clifford algebra but intuitively moved toward it. His musical vision did not operate in a two-dimensional vector space like Haubenstock-Ramati's,[103] but rather in a vector space with a quadratic form. He might have worked with quaternions that described notations as rotations of words in three and four dimensions. The proponents of graphic notation, from Haubenstock-Ramati to Brown, sensed that they were moving toward a new way of composing – that is, the shift from time-based to space-based methods of musical composition. In the anthology *Form in der Neuen Musik*, to which Haubenstock-Ramati, Kagel, and Ligeti also contributed texts, Brown describes the influence of Alexander Calder and Jackson Pollock on his work (see this volume, p. 118), as well as that of Karl Gerstner. He concedes that his compositions do not conform to the general idea of how composers work, and says that he has no objection to being called "a designer of programmes."[104]

Before it was actually possible to compose with computers, the avant-garde composers intuitively saw their methods of composition as programming. Brown makes explicit reference to Gerstner's book *Designing Programmes*:

102 See two books by William Kingdon Clifford: *Mathematical Fragments, Being Facsimiles of His Unfinished Papers Relating to the Theory of Graphs* (London: Macmillan, 1881); and *Mathematical Papers* (London: Macmillan, 1882).

103 Haubenstock-Ramati defined musical form as "a temporal-spatial conception of musical activity" in his essay "Form in der neuen Musik," in *Form in der Neuen Musik*, ed. Ernst Thomas, Darmstädter Beiträge zur Neuen Musik 10 (Mainz: Schott, 1966), 37–39, here 38.

104 Earle Brown, "Form in New Music," in Thomas, *Form in der Neuen Musik*, 57–69, here 63.

105 Ibid., 63.

106 Ibid., 59.

107 Ibid.

108 Iannis Xenakis, *Formalized Music: Thought and Mathematics in Composition*, trans. Christopher Butchers, G. W. Hopkins, John Challifour, and Amber Challifour (Bloomington: Indiana University Press, 1971), 144. Originally published in French as *Musiques formelles: Nouveaux principes formels de composition musicale*, La revue musicale 253–54 (Paris: Richard-Masse, 1963).

109 Ibid., 161–70.

110 Ibid., 169.

111 Ibid., 169–70.

"'Designing a Programme' is similar to the procedures I have described above and somewhat unlike the traditional concept of 'composing' in that (as a negative attitude would say) the final step of definitive arrangement is left out. ... This is a process of *inclusion* and expansion of the concept of 'a work of art' rather than one of deterministic contraction and exclusivity. One does not diminish the amount of 'meaningful' control within a work but seeks to create the work as an entity, a quasi-organism, and to 'programme' a life for it within which it comes to find its shape."[105]

Here, Brown is already envisioning the future of music as the programming of lifelike "quasi-organisms" – the generation of music by cellular automata like Conway's *Game of Life*, perhaps. But what is most remarkable is how rarely Brown uses the word *time* and how often he uses the word *space* in discussing his revolutionary compositions from 1952: "One of the first things that I ever wrote about *Form* is the following, from notebooks, October and November, 1952, under the word 'Synergy': 'to have elements exist in space ... space as an infinitude of directions from an infinitude of points in space ... to work (compositionally and in performance) to right, left, back, forward, up, down, and all points between ... the score (being) a picture of this space at one instant, which must always be considered as unreal and/or transitory.'"[106]

Brown also describes his notation as a vector field: "In the notebook the above is connected to a space diagram of the 'reading' principle of *December 1952*, and the graphics of that work seem to me to activate the above intention. ... The piece *November 1952* ... has as instructions: 'To be performed in any direction from any point in the defined space for any length of time. Tempo: as fast as possible to as slow as possible ... inclusive. Attacks may be interpreted as completely separated by infinite space, collectively in blocks of any shape, or defined exactly within that space. Lines and spaces may be thought of as tracks moving in either direction (horizontally at different and variable speeds) and clef signs may be considered as floating (vertically over the defined space). ... The defined space may be thought of as real or illusory, as a whole or in parts. Either space (vertical or horizontal) may expand, contract, or remain as it seems to be here. Vertical space will vary according to the performer's view of the floating clefs.'"[107] In Xenakis's book *Formalized Music* (1963) it says: "With the aid of electronic computers the composer becomes a sort of pilot: he presses the buttons, introduces coordinates, and supervises the controls of a cosmic vessel sailing in the space of sound, across sonic constellations and galaxies that he could formerly glimpse only as a distant dream."[108]

Searching for nonsubjective, mathematic, scientific composition methods, Xenakis references modern probability theory, quantum mechanics, and Dennis Gabor, the inventor of holography. Like Jean Barraqué (1928-1973), the French master of serial music, Xenakis outlines in the chapter "Symbolic Music," a logical, algebraic form of musical composition, devoting one section to the concept of vector space.[109] "If our musical space has two dimensions, e.g., pitch-time, pitch-intensity, pressure-time, etc., it is interesting to introduce complex variables," he writes.[110] And: "The four forms of a melodic line (or of a twelve-tone row) can be represented by the following complex mappings: ... original form ... inversion ... retrogradation ... inverted retrogradation. ... Other transformations, as yet unknown, even to present-day musicians, could be envisaged. ... Furthermore, for a musical space of more than two dimensions we can introduce hypercomplex systems such as the system of quaternions."[111]

Here we can already see the beginnings of an algebraic theory of music capable of realizing sonic events independent of the temporal axis. Xenakis also mentions algebraic number systems and Hamilton's quaternions. It took more than one step, then, to arrive at my suggested use of Clifford algebra to produce new sonic events and extend new networks across the chasm between two tones. In the piece *Nomos Alpha* (1965), Xenakis used transformations of cubes according to various graphs.(FIG. 110)

The word *network* should not be interpreted in a purely metaphorical sense, either. "We could train a network to generate the next note in sequence by supplying it with some number of previous notes for context," writes the composer and mathematician Gareth Loy. "This would require a feedback arrangement in the network design so that previous outputs could influence subsequent choices. The windowing and context methods could be combined so that the feedback units provide context for whole musical phrases."[112] A process of this sort was described by Peter Todd in 1989.[113]

The neural model of consciousness, which shows that millions of neurons are networked and operate in parallel in the brain, has given rise to the science of artificial neural networks. Because these neurons work in parallel, we speak of "parallel distributed processing" and "connectionist models of cognition." Thanks to the theory of artificial neural networks, an exceptionally large number of models for linking tones have become possible. This approach opens up a pathway for the use of artificial intelligence in computer-assisted algorithmic composition.

As the twenty-first century progresses, music theory will return to its mathematical roots, and acousmatic music will become part of an advanced mathematics, developing spatial models of music in particular, from topology to graph theory. Spatialization of music thus has nothing to do with the sound domes in which listeners have a spatial listening experience provided by dozens of sound sources. Spatialization of music means defining tones as spatial neighborhoods, and using spatial neighborhoods to illuminate the question of how to get from one tone to the next.

Music notation can be compared to the graphic life forms in Conway's *Game of Life*. Notes and staves are two-dimensional beings. If we now translate the idea of Clifford algebra to music, the question arises: What would a notation made up of multidimensional life forms look like? Notes, that is, that connect in space rather than on a plane. The realization of such ideas is only feasible with the aid of the computing capacity of machines with artificial intelligence – of computers. Schoenberg's question about moving from one tone to the next was always conceived as a step on a plane, never as a bridge across a chasm. The future of music lies in spatial composition.

To that end, let us conduct an experiment: Before us lies a sheet of paper divided into four sections, which are labeled (moving clockwise) with the numbers 1, 2, 3, and 4, one in each section. Each of these numbers represents a note. If we flip the paper over on its horizontal axis, the sequence is 4, 3, 2, 1. We thus have a different

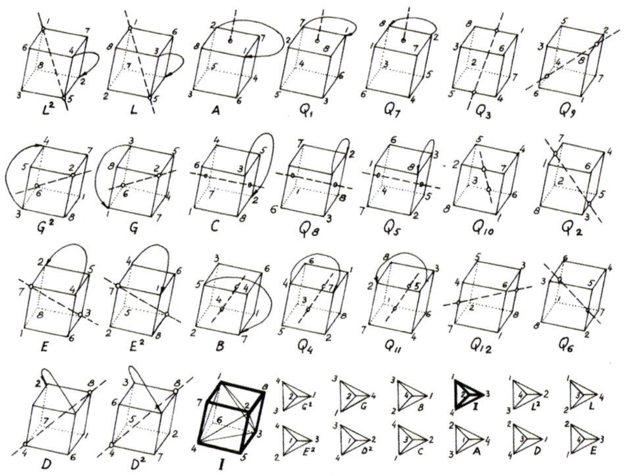

FIG. 110 Iannis Xenakis, Schema of a hexahedron and octahedron group, in *Formalized Music* (French original 1963, Bloomington and London: Indiana University Press, 1971), 220.

sequence of notes just because the numbers indicating the notes are positioned differently in space. If we now flip the paper on its vertical axis, the resulting number sequence is 3, 4, 1, 2. If we flip it on the horizontal axis again, we get the number/tone sequence 2, 1, 4, 3. If we rotate the paper ninety degrees counterclockwise, the tone sequence we get is 1, 4, 3, 2. Flipping it vertically gives us 4, 1, 2, 3, and so on.

We see that the positioning of the numbers, or rather notes, remains the same on the paper, meaning that it is automorphic. What changes is the position of the paper in space. These changes produce constantly changing tone sequences. Spatial parameters give rise to temporal parameters – such as the tone sequence. The Schoenbergian question of how we get from one tone to the next is thus solved spatially, not temporally. A closer look reveals that this procedure and its result is not dissimilar to the four modes of dodecaphony: basic tone row, retrograde, vertical inversion, and horizontal inversion. We see that dodecaphony is essentially an anticipation of spatial composition methods: transformations or automorphisms in space and the use of spatial parameters to shape the progression of the tone rows.

112
Gareth Loy, *Musimathics: The Mathematical Foundations of Music* (Cambridge, MA: MIT Press, 2006), 1:387.

113
Peter M. Todd, "A Connectionist Approach to Algorithmic Composition," *Computer Music Journal* 13, no. 4 (Winter 1989): 27–43. See also Peter Todd and Gareth Loy, eds., *Music and Connectionism* (Cambridge, MA: MIT Press, 1991); and Peter M. Todd and Gregory M. Werner, "Frankensteinian Methods for Evolutionary Music Composition," in *Musical Networks: Parallel Distributed Perception and Performance*, ed. Niall Griffith and Peter M. Todd (Cambridge, MA: MIT Press, 1999).

114
Cluster was also the name of a German band founded in 1969 by Dieter Moebius, Hans-Joachim Roedelius, and Conrad Schnitzler (who left the group in 1971). They had a fruitful collaboration with the British musician Brian Eno.

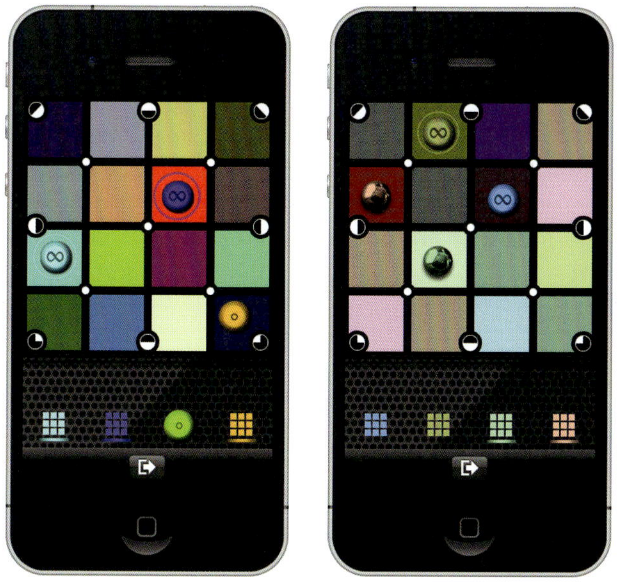

FIGS. 111, 112 Peter Weibel and Jens Barth, *Music Board*, 2011. App, screenshots.

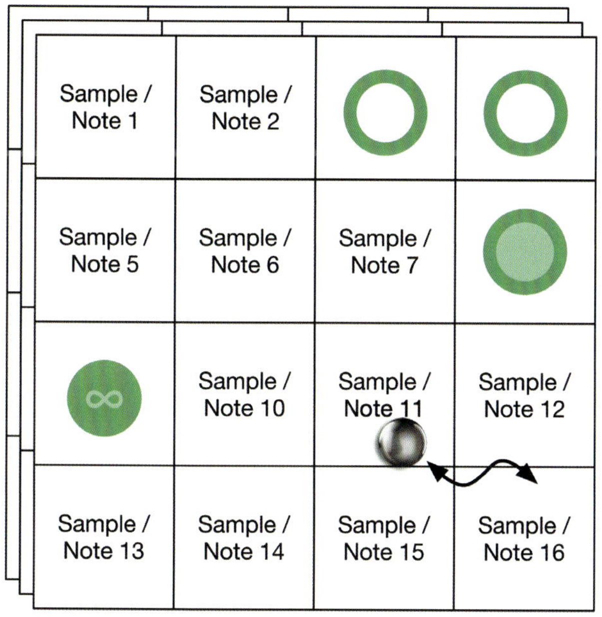

FIG. 113 Peter Weibel and Jens Barth, *Music Board*, 2011. Schematic drawing of the structure of the game board with the different layers for each instrument as well as the individual sound triggering objects: a) a sample object triggers a single sample, b) a sequencer object triggers several samples in a chain one after the other by means of an internal pulse generator, c) a melody object changes its position by tilting the device and collides with other objects.

We are thus approaching a music like that proposed by Morton Feldman, who accused music of subjugating itself to time by means of the metronome. Rhythm, beat, meter, etc. are expressions of music's submission under the dictatorship of time; they make music time's slave. This insight can even be heard in pop songs – one thinks of the Grace Jones song "Slave to the Rhythm" from 1985, or "Hit Me with Your Rhythm Stick," by

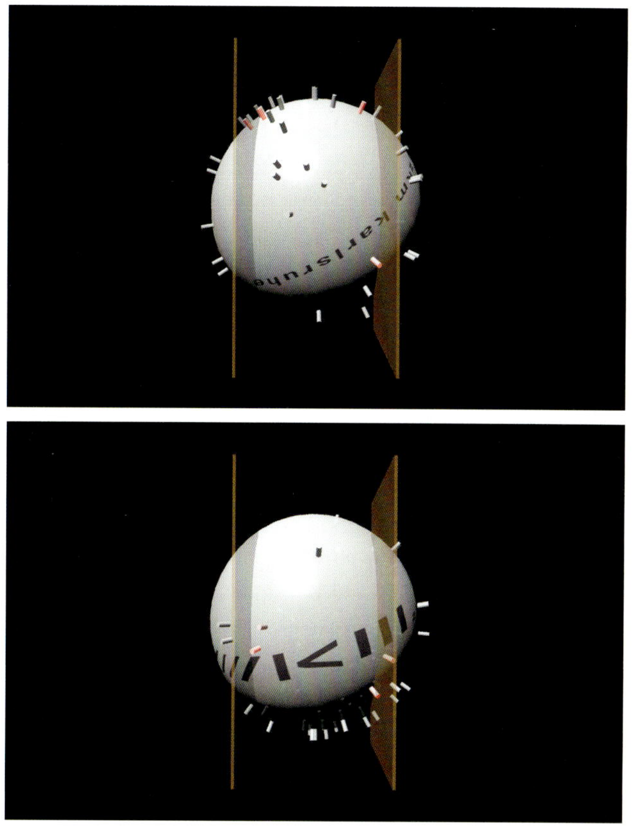

FIGS. 114, 115 Peter Weibel and Chikashi Miyama, *Sound Writer*, 2016. App, screenshots. Sound Writer is a 3-D virtual music box, which can be played with easy finger gestures. Simply rotate the sphere in the middle of the screen by dragging it with your finger. When the pins on the surface of the sphere hit the "tone comb," sound will be played. You can also change the sound color and the layout of the pins by two finger gestures.

Ian Dury and the Blockheads, from 1978. Feldman wanted to create a music that would subjugate time and put an end to its dictatorship. Music was thought of as the enemy of time – essentially, the enemy of interval theory. Computers and the newly proposed topological, geometrical, mathematical composition methods tend to facilitate a breaking up of interval theory by replacing temporal parameters with spatial ones. Perhaps a visual metaphor can explain this timeless music: If you pour white milk or black ink into an aquarium full of water, it creates a cloud, a natural proliferation of organic forms. Rather than point following point, note following note, a continuous field expands, devoid of meter and interval. *Cluster* was the first term chosen, albeit intuitively, to label and investigate this new form of spatial music.[114]

Building on these ideas, I created two apps, *Music Board* (2011, with Jens Barth) (FIGS. 111–113) and *Sound Writer* (2016, with Chikashi Miyama) (FIGS. 114, 115), in which tone sequence and music are determined or generated by the position of the phone in space. Both apps are available on the App Store. When someone plays piano, it makes no difference whether the instrument is sitting on the floor, hanging from the ceiling, or mounted on the wall. The music does not change in any way. The score is played as usual, and the music remains independent of the piano's position in space. With the two apps, on the other hand, the spatial position of the instrument – that is, the smartphone – produces the score or causes the tone sequence to change.

The apps employ the mobile phone as a universal sound instrument whose position in space generates the music. After all, a mobile phone is, among other things, a kind of compass that indicates the device's position in the electromagnetic field. Changing the phone's position by means of hand and arm movements creates the "notation" and the music. The graphic user interface is no longer touched, but becomes a sheet of music that produces music by itself based on its position in space. Rather than the movements of a hand on an instrument or a two-dimensional user interface, it is the movements of the instrument in space that generate the score and the sound. There is no longer any difference between score/notation and sound event. Notation and sound occur in real time. Notation, instrument, and sound event become one.

Sonification

Synesthetic research on the relationship between visual and sonic data produced many results in science and art, from Helmholtz to music video. Finally, art and science converged into the scientific visualization of sound. "Seeing sound" is still the aim, but now in programmed fashion. What is visible should be made audible, but the sound should come from things that are invisible: hidden objects, or objects, too small to be perceptible for the naked eye, such as molecules.

A genuinely new cosmos of sounds, of new microscopic or even nanoscale instruments appear on the horizon. Complex calculations and algorithms make it possible for subatomic or subterranean worlds to resonate. Unseen objects create hitherto unheard sounds. In the words of René Descartes: not only *res cogitans* speak, but now also the *res extensa* speak or make sounds. The formula that everything that creates sound is a musical instrument and every sound is music is now also valid for the hitherto *in*visible zone of nuclear particles.

Thierry Delatour, for example accumulated in 1993 fourteen years of laboratory research experience in the study of molecular vibrations, and earned his spurs in digital synthesis and sound sequencing. He asked himself if it might be possible to translate the microscopic atomic vibrations of molecules into audible sound vibrations. And if so, what kind of sounds might be obtained? Would it be possible to create molecular musical works that were both artistic and useful for scientists?

Delatour presented his works as the result of research of the rigorous and exhaustive translation of scientific data called vibrational spectra into sounds and music. His approach is both artistic (sound design and musical composition) and scientific (obtaining additional information from scientific experiments by means of the unusual auditory display of data). It is also multisensory in that it uses the aural channel in addition to the traditional visual one. In 2015, ZKM | Karlsruhe produced a sound installation together with Delatour called *Molecular Songs*. The installation translates molecules into sounds. Using experimental and/or precalculated data from the vibration spectrum, this work presents an expanded method for the aural translation of molecular vibrations, making audible the inaudible oscillations that naturally occur in molecules. This method leads to molecular tones, molecular scales and even molecular pieces of music. An interactive interface allows visitors to discover different chemical substance families and related vibration spectra and to hear the corresponding molecular tones and pieces of music. (FIGS. 116, 117)

The work shows that vibrational spectra are not harmonious. Accordingly, the perception of a molecular tone is also an unusual one, revealing the complexity and variability of the smallest units that make up our environment. It offers an extraordinary and innovative tool for accessing scientific content and at the same time a musical experience that goes beyond the classical rules of melody, harmony, and the perception of physical phenomena. Another work discovering cellular sounds is Anne Niemetz's and Andrew

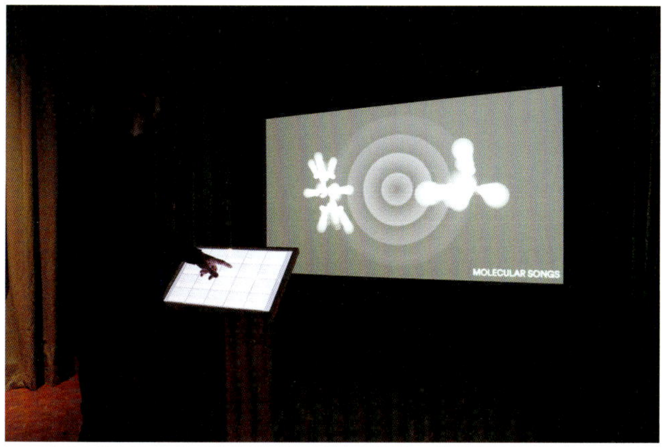

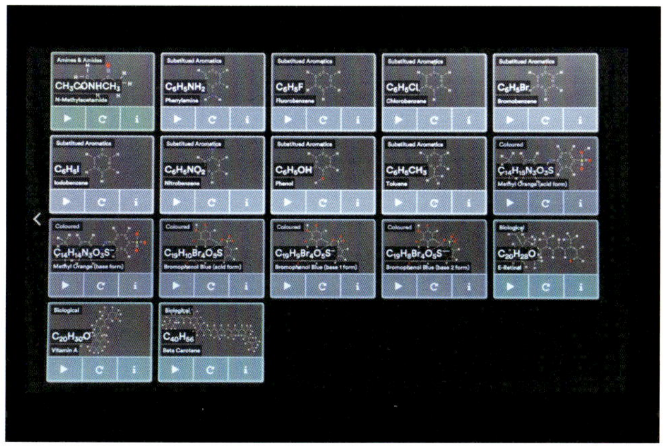

FIGS. 116, 117 Thierry Delatour, *Molecular Songs*. Installation view ZKM | Karlsruhe, 2015.

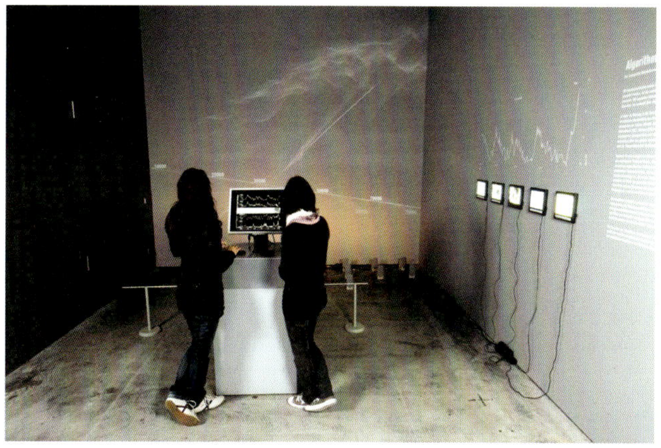
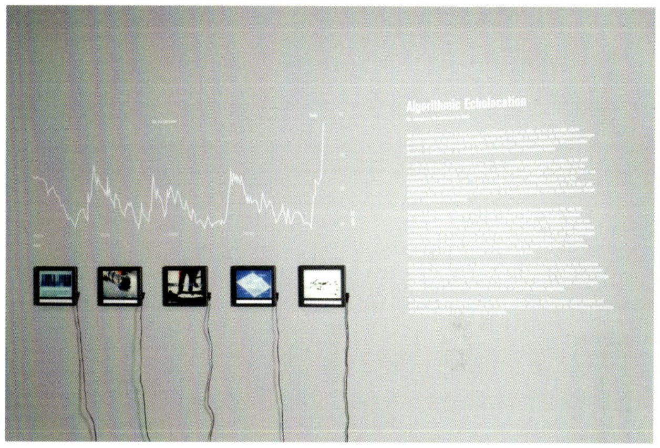

FIGS. 119, 120 Ramon Guardans, Götz Dipper, Matthias Gommel, Adolf Mathias, Martin Schüttler, Renata Sas, *Algorithmic Echolocation*, 2003. Interactive installation. Installation view ZKM | Karlsruhe.

Pelling's composition and sound installation *The dark side of the cell* (2004).

The Japanese artist Tomoko Sauvage's sound installation *In Curved Water* (2015) grew out of her practice as an improvising musician playing amplified bowls tuned by water level – which themselves are inspired by *jal tarang*, a traditional Indian percussion instrument. The title of her work, as well as its association with Indian classical music, harkens back to the oeuvre of American minimalist composer Terry Riley, known for his 1969 album *A Rainbow in Curved Air*. In the gallery's main room, blocks of ice wrapped in twine hang from the ceiling, refracting light from isolated spotlights in the dark; they slowly drip water into bowls, some of which are connected to hydrophones. The resonance of the bowls transforms the pitter-patter of the melting ice into an unpredictable pulse reminiscent of wind chimes. (FIG. 118)

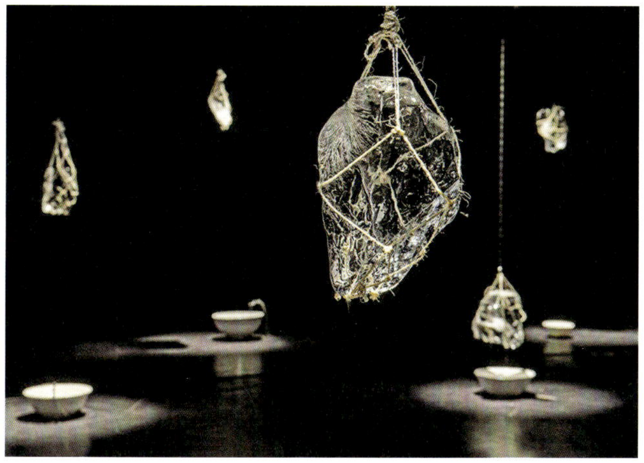

FIG. 118 Tomoko Sauvage, *In Curved Water*, 2015. Sound installation.

Another example is the installation *Algorithmic Echolocation*, which was coproduced by ZKM | Karlsruhe, Soundplots, and MediaLab-Madrid in 2003 (FIGS. 119, 120). This interactive installation deals with the visualization and sonification of complex data, using climate data from the Antarctic as raw material. The users can experience the temporal course, the changes and the mutual interaction of the measured data of these five parameters audio-visually: the data series are subjected to spectral analysis using

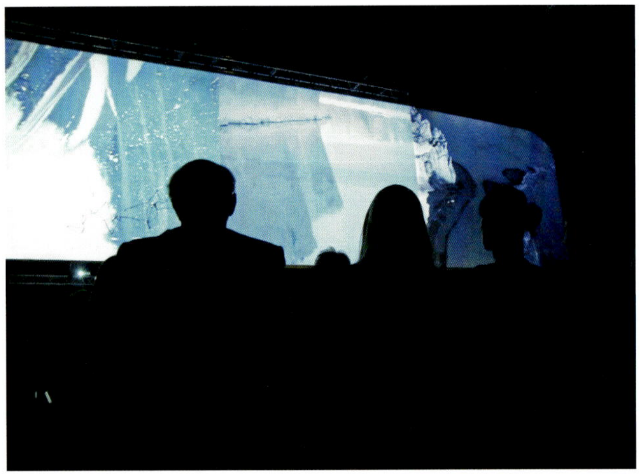

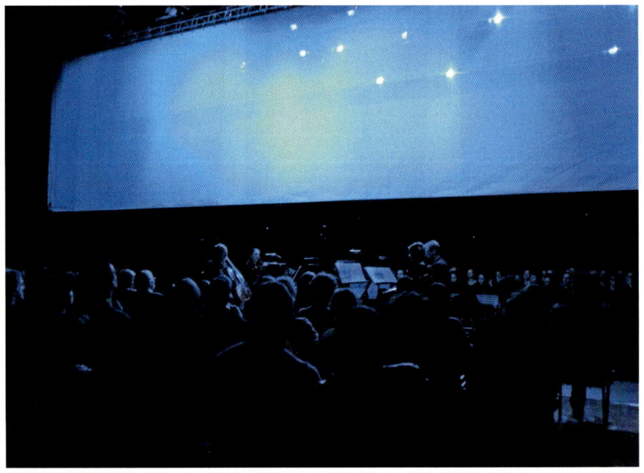

FIGS. 121, 122 Frank Halbig in collaboration with Helios String Quartett, *ANTARKTIKA. A Climatic Time Travel*, 2006. Concert, live electronics, and video. Premiere at ZKM | Karlsruhe on December 10, 2006.

a special algorithm. The resulting dynamic values are displayed in front of a three-dimensional frequency spectrum. At the same time, they are fed into a sound environment that is also generated in real time.

Sonification is also a key word for Frank Halbig's concert for strings, live electronics, and video *ANTARKTIKA – A Scientific Time Travel* (2006) with was realized with the Helios StreichQuartett and premiered at ZKM | Karlsruhe in December 2006. (FIGS. 121, 122) Deep ice cores from the Antarctic allow detailed statements about the history of climatic changes on earth to be made. The data has been extracted by EPICA, the European Project for Ice Coring in Antarctica, and covers 742 millennia of climatic history. On the basis of this data, Halbig sonified the changes found in this data, reflecting the climatic development of our planet in a composition. From the same data, he also generated graphics which were shown together with a video projection composed of images and video footage from the AWI archive (Alfred Wegener Institute for Polar and Marine Research in Bremerhaven).

Another installation that processes scientific data is *SOL* by Florian Grond, Frank Halbig, Jesper Munk Jensen, and Thorbjørn Lausten, which represents data from the sun both visually and sonically. The data show sunspots, the total energy output of the sun, the solar wind, and the sun's solar magnetic field. (FIGS. 123, 124)

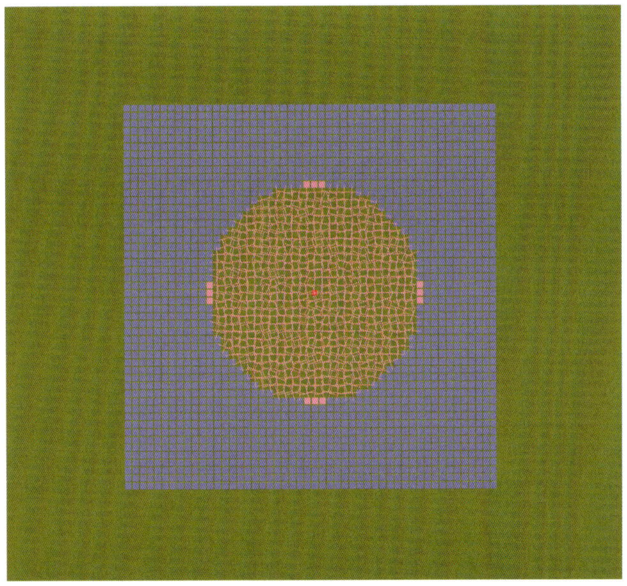

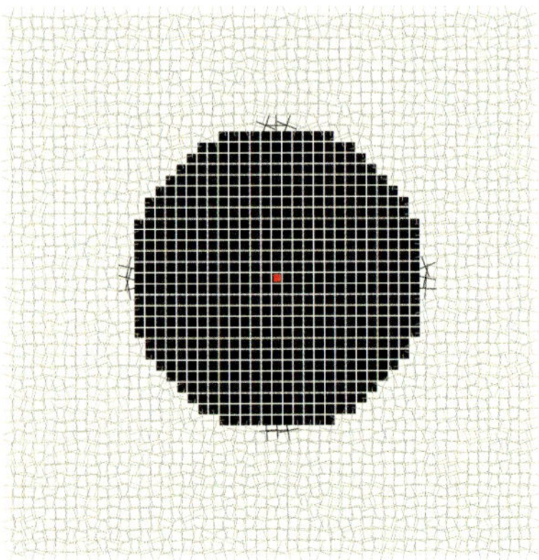

FIGS. 123, 124 Florian Grond, Frank Halbig, Jesper Munk Jensen, and Thorbjørn Lausten, *SOL*, 2004. Visual and acoustic room installation representing data from the Sun.

115 See Klaus Schöning, *Ars Acustica – Ars Intermedia.*

116 Karel Teige, "Excerpts from *Poetism Manifesto*," trans. Gerald Turner, in *Between Worlds: A Sourcebook of Central European Avant-Gardes, 1910–1930*, ed. Timothy O. Benson and Éva Forgács (Cambridge, MA: MIT Press, 2002) 593–601, here 600, 599.

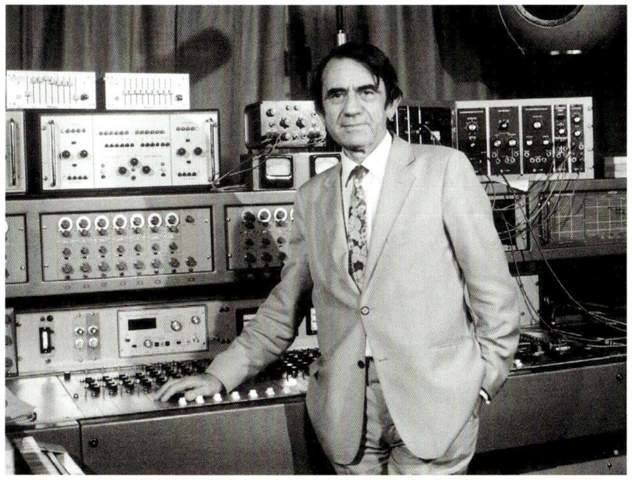

FIG. 125 Pierre Schaeffer at the RTF electronic studio, Paris, 1951.

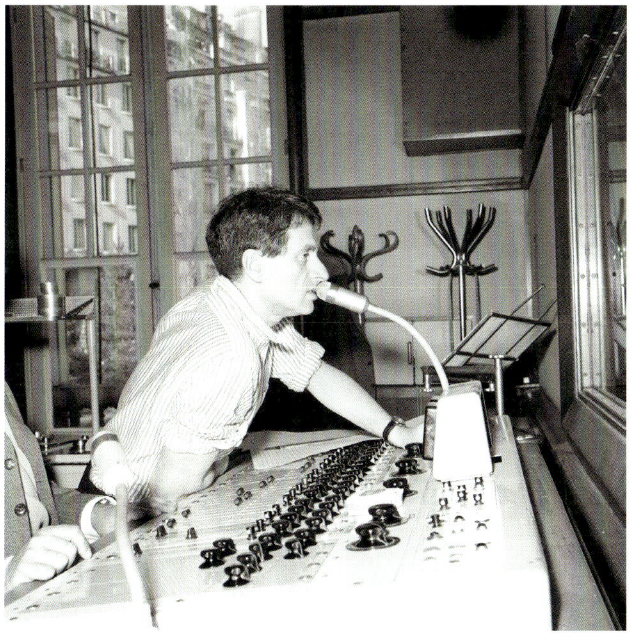

FIG. 126 Iannis Xenakis at the RTF electronic studio, Paris, 1967.

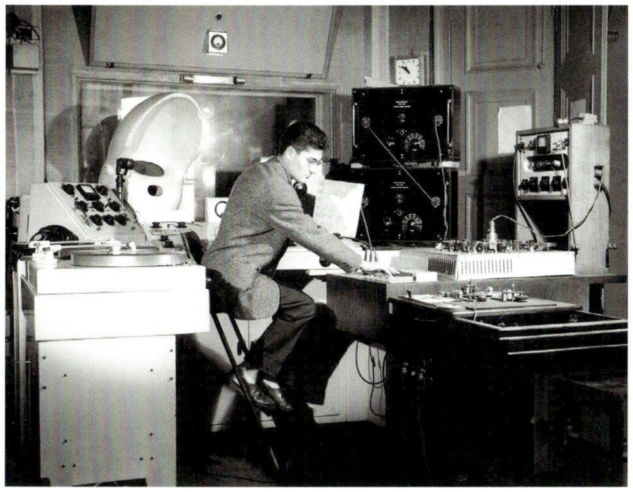

FIG. 127 Pierre Henry at the RTF electronic studio, Paris, 1951.

Electronic Music and Radio Art

There is ample proof to support the thesis that the emergence of electronic music in Europe is inextricably bound up with public broadcasting. For there was a time, from the 1940s to the 1960s, when the public broadcasters were not bastions of triviality and trash, but rather forums for the production and distribution of avant-garde art: new music, new literature, and above all sound art and radio art, the *ars acustica*.[115] This grand and heroic era witnessed the birth of electronic music from the spirit of radio art: Radio art and electronic music are close relatives.

In analogy to the term *visual art*, it would be correct to speak of auditory or aural art, or more specifically, radiophonic aural art. Radio is made up of signals in the audible spectrum, whose physical boundaries are determined by the channel capacity of the human ear. The sound frequencies that can be deciphered by human ears as literature or music, word or sound, can be found between 16 Hz and 20 kHz.

In 1928, the Czech artist Karel Teige envisioned a "poetry for hearing" that would encompass noise music, jazz, and radiotelephony: "In the same manner that poems can be created by film out of light and movement, *radiogenic poetry* can be created as a new art of sound and noise, divorced equally from literature and recitation and from music. … Just like pure kinography and photogenic poetry, radiophonic and radiogenic poems must make use of elementary media (in the former case light and movement, in the latter case sound and noise) and break free of literary and theatrical constraints. Radiogenic poetry [is] a composition of sound and noise, recorded in reality but woven into a poetic synthesis."[116]

After the war, new technological devices greatly increased the potential for condensing particles of reality into a musical synthesis. The composers of the avant-garde began to seek new expressive possibilities in the field between music and literature, sound and language. The electronic, magnetic, and electromechanical devices that offered composers and poets these new possibilities were much too expensive for individuals to purchase when they first came onto the market in the 1950s. Only the public broadcasting companies could afford them. Buoyed by postwar cultural policies which, in countries such as Germany, Italy, France, England, and Sweden, bore the stamp of antifascism and social democracy, the state-owned broadcasters granted themselves a mission to educate, and with it well-equipped studios in which they could support research and experimentation. Only as a result of these technological and social conditions could electronic music evolve in state-owned studios, the products of which mutated into electronic pop music in the age of mass media. Without this electronic-music avant-garde, explored in state-owned radio facilities, our contemporary sonic cosmos, from sound art to pop music, would not exist. The Elektronmusikstudion (EMS) in Stockholm and its predecessors, the Groupe de recherches musicales (GRM) in Paris and the Studio für Elektronische Musik in Cologne, all share responsibility for the worldwide ascendancy of electronic music.[117]

In 1942, Pierre Schaeffer founded Studio d'Essai at Radiodiffusion Française in Paris. After the war it became the Club d'Essai, which sought out new radiophonic forms. The club presented examples of its musique concrète to the public at large in 1948 with *Étude aux chemins de fer* and *Concert de bruits*. In 1951, the Groupe de recherches de musique concrète (which would become the GRM) was founded at Radiodiffusion-Télévision Française (RTF) in Paris. Not only were its director (Schaeffer) and members (Xenakis, Henry, and Luc Ferrari, who was codirector from 1962–63) important progenitors of new music, their work under the heading of *bruit* (noise) has even influenced techno music.(FIGS. 125–127) Thus the birth of musique concrète took place not in the concert hall, but on the radio, not in a private studio, but in a public radio facility. The path was set.

In 1974, François Bayle designed and premiered the *Acousmonium* (FIG. 128), which is an orchestra of loudspeakers arranged in front of, around and among the concert audience. Originally it was used by the GRM at RTF, but it is still in use today for the performance of acousmatic works, or by artists performing mixed musical forms, improvised music, and multimedia – for example, for a series of concerts at the Institute of Contemporary Arts (ICA) in London in 2006.

The world's first studio for electronic music was established in 1951 at Westdeutscher Rundfunk (WDR) in Cologne by the composer Herbert Eimert, the recording engineer Robert Beyer, and the Bonn-based phonetician and information theorist Werner Meyer-Eppler. Karlheinz Stockhausen took over as director from 1963 to 1980. (FIG. 129) The linking of music and language that was characteristic of the early

117 Cf. Sanne Krogh Groth, *Politics and Aesthetics in Electronic Music: A Study of EMS – Elektronmusikstudion, Stockholm, 1964–1979* (Heidelberg: Kehrer, 2014).

118 See also Umberto Eco, "Ai tempi dello Studio," in *Luciano Berio: Nuove prospettive / New Perspectives*, ed. Angela Ida De Benedictis (Florence: Olschki, 2012), 3–16.

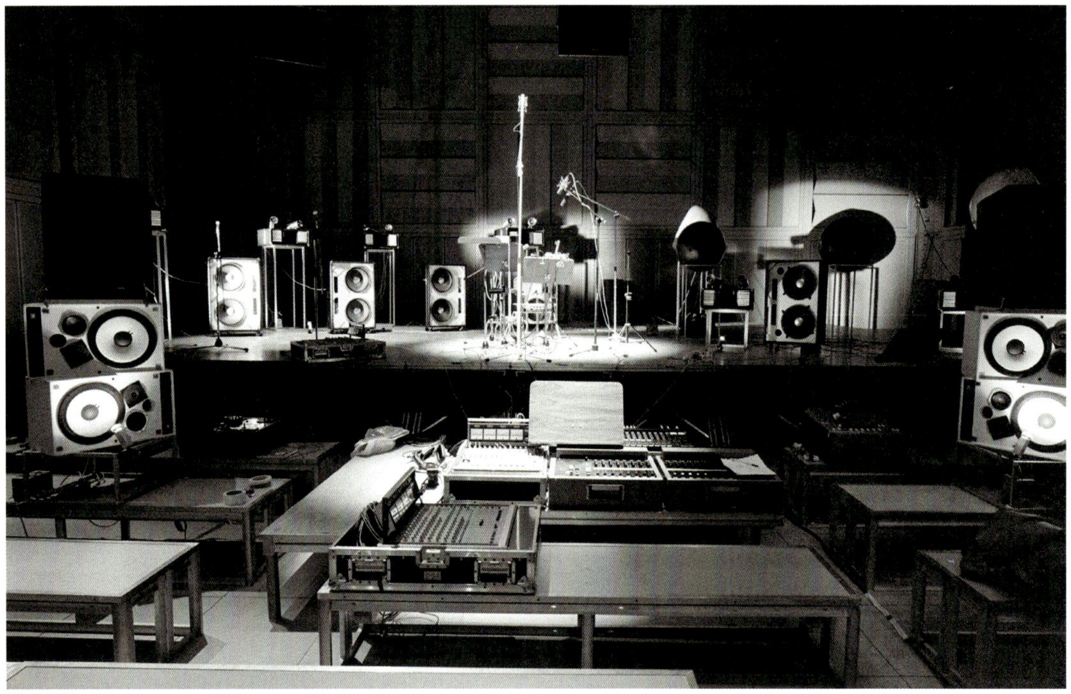

FIG. 128 Acousmonium, ca. 1982.

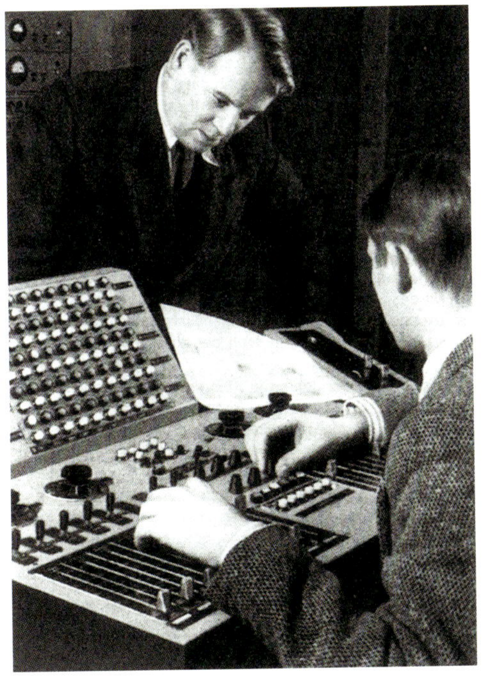

FIG. 129 Herbert Eimert (left) and Karlheinz Stockhausen at the director's table in the WDR studio for Electronic Music, Cologne.

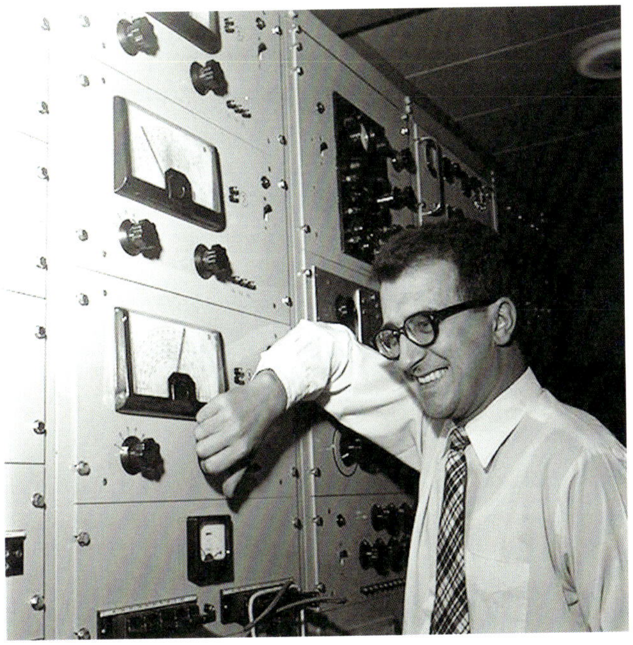

FIG. 130 Luciano Berio at the Studio di Fonologia Musicale, Milan, 1950s.

stages of electronic music can be seen in the title of a 1949 book by Meyer-Eppler: *Elektronische Klangerzeugung. Elektronische Musik und synthetische Sprache* [Electronic sound generation: Electronic music and synthetic language]. In 1956 a concert with the revealing title *Unerhörte Musik* [Unheard-of music] was held at the WDR's broadcasting studio in Cologne. It featured seven musicians, among them Stockhausen, who contributed a concerto with no musicians, only loudspeakers: *Gesang der Jünglinge*. Shocking! From 1955 to 1959, Luciano Berio was the director of the Studio di Fonologia Musicale in Milan, a center for electronic music that he and Bruno Maderna had established at Radiotelevisione Italiana (RAI), where he provided a platform for new music on the radio.(FIG. 130) Berio, Maderna, and the sound engineer Marino Zuccheri developed their own compositions at the studio but also offered production opportunities to many other musicians. Like Schoenberg, Stockhausen, Boulez, and others, Berio exemplified a type of twentieth-century composer characterized by his theoretical awareness of his own activity. Unlike the visual-art world, the music world accepted the unity of artist and theorist, conductor and composer. Berio worked intensively with the possibilities for the aural synthesis of language and music offered by tape recordings and oscillators. A new kind of relationship between word and tone, poetry and music – in other words, between the science of phonology and the science of sound – was his goal. He sought out musical elements in language. Through electroacoustics, in particular, he was able to develop a new form of vocal music that took the human voice (e.g., that of his partner, Cathy Berberian) as its starting point.

His work with literary texts, from Joyce to Edoardo Sanguineti, evoked the polyphony latent in every word, the scope of the semantic space created by phonetic variations. His most famous piece is *Thema (Omaggio a Joyce)* (1958–59). This composition was based on a radio feature titled *Omaggio a Joyce. Documenti sulla qualità onomatopeica del linguaggio poetico* [Homage to Joyce: Documents on the onomatopoeic quality of poetic language], which was created with Umberto Eco for RAI but only broadcast in Belgium. In *Thema* the secret life, so to speak, of the vowels and consonants in Joyce's work, specifically in the eleventh chapter of his 1922 novel *Ulysses*, was amplified in a number of ways via electroacoustic transformations of the voice. Eco, who was a friend of Berio's and conceptually influenced by his music, as well as that of Stockhausen and Pousseur, published his famous book on the open work three years later, in 1962.[118] The field of possibilities implicitly offered by the work was to be made explicit by the freedom of the performer, listener, reader, and recipient.

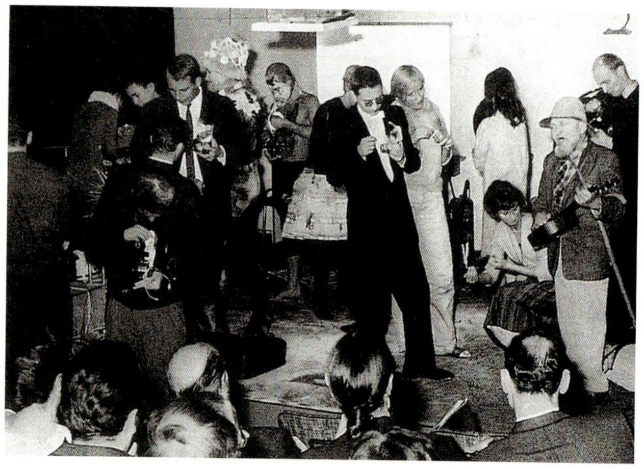

FIG. 131 Performance of Karlheinz Stockhausen's *Originale*, Theater am Dom, Cologne, 1961. Director: Carlheinz Caspari, set: Mary Bauermeister, performers: M. Bauermeister, D. Tudor, C. Caskel, K. Stockhausen, H. G Helms, N. J. Paik, E. Sommer and many others.

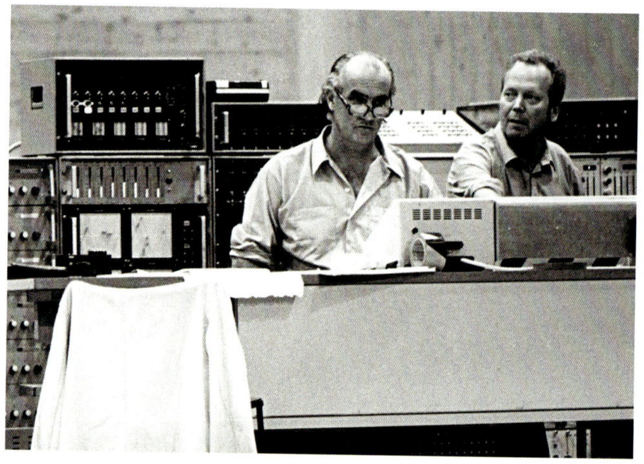

FIG. 132 Luigi Nono (left) and Hans-Peter Haller, director of the Experimentalstudio (1972–1989), 1980s

In 1959 in Germany, Hans G Helms was preparing a work in the tradition of Joyce's *Finnegans Wake*, a similarly polyphonic web of musical and linguistic components.(FIG. 131) In *Fa:m' Ahniesgwow*,[119] purely phonic connections between words alternate with semantic ones, and words from different languages blend together to create an aural ambience that can conjure semantic suggestions or simply be regarded as musical in function. The piece was created at the Studio für Elektronische Musik under Meyer-Eppler's guidance. Together with the composer Gottfried Michael Koenig, Helms conducted sonic analyses and phonetic experiments to supplement his linguistic studies. The order of the signifiers could be freely and individually interpreted. Unsurprisingly, Helms also wrote a text for the Italian composer Bruno Maderna.[120]

In Germany, more new ground was being broken in the field of radio and electronic music: The new-music festival Pro Musica Nova was founded by Hans Otte at Radio Bremen in 1958, and in 1968 Klaus Schöning established the Studio Akustische Kunst at the WDR. Following Heinrich Strobel's creation (also in 1968) of a foundation to facilitate encounters between composers and new technology, the Experimentalstudio was founded at Südwestrundfunk in Freiburg. There, major pioneering works of new music were produced with Luigi Nono,(FIG. 132) Boulez, and Stockhausen. Today, the Experimentalstudio is still one of the world's leading studios and laboratories for electronic and live-electronic music, while the radio play and media art department at Bayerischer Rundfunk, until recently under the directorship of Herbert Kapfer, is a key platform for musical and intermedia experiments and radio art.

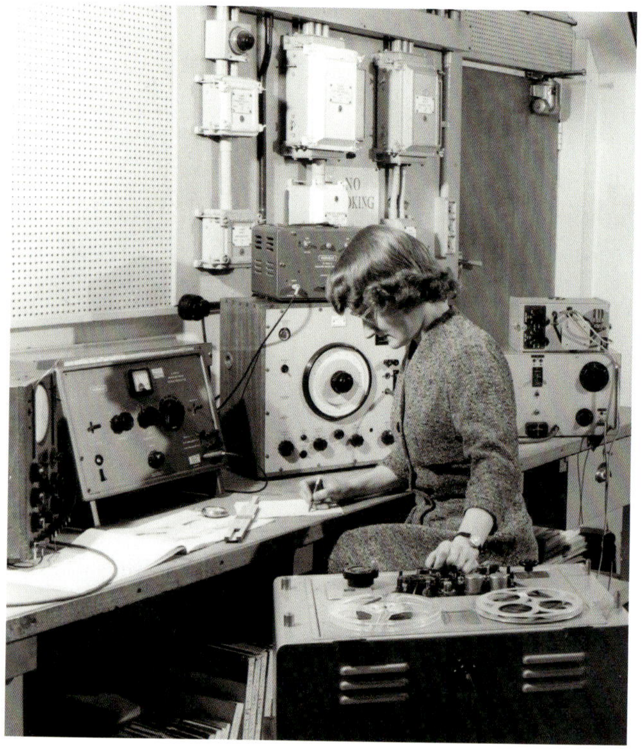

FIG. 133 Daphne Oram at the BBC Radiophonic Workshop, Maida Vale near London, 1958.

Machine-based experiments with music and language began appearing on the radio in the late 1950s in England as well. Having worked with electronic music and musique concrète, in 1958 Daphne Oram founded the BBC Radiophonic Workshop with Desmond Briscoe after a residency at the RTF studios in Paris. (FIG. 133) She left the following year, however, feeling that the Workshop was moving in a different direction

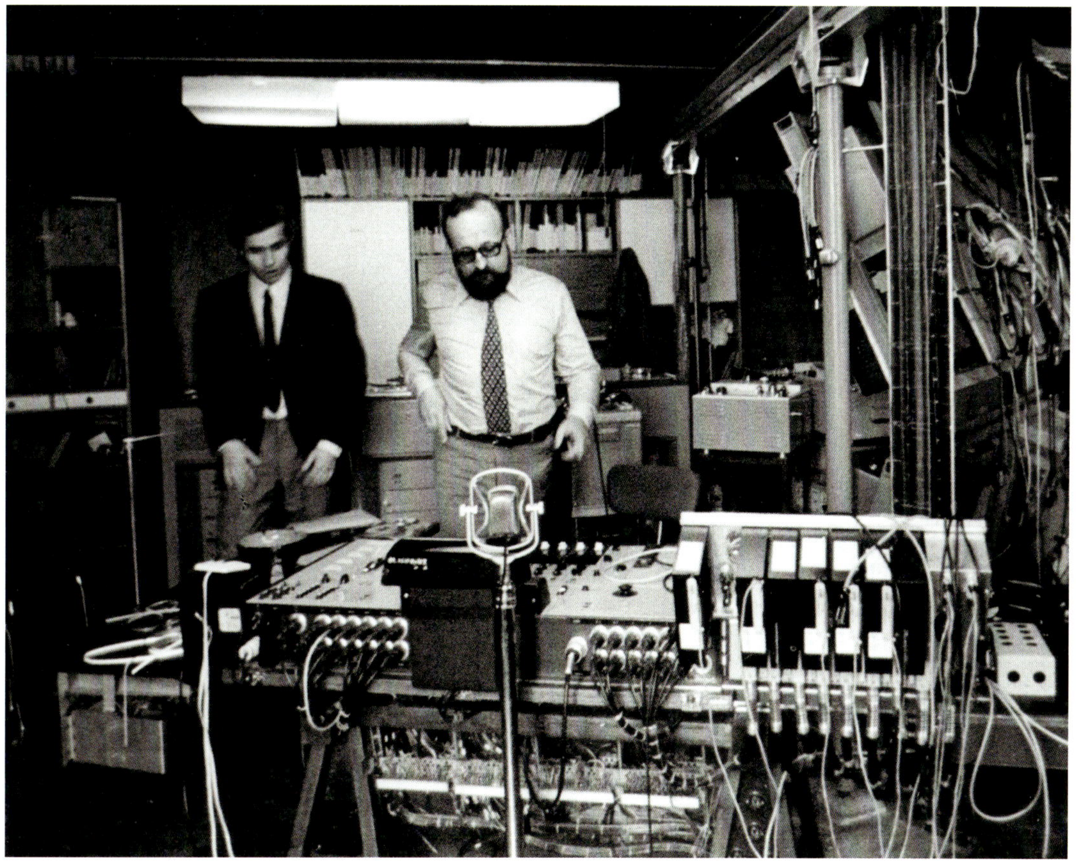

Fig. 134 Eugeniusz Rudnik and Krzysztof Penderecki at the Polish Radio Experimental Studio (PRES), 1960s.

than she had hoped. The same year, she invented Oramics, a process that used drawings on strips of 35-mm film, transferred to photoelectric cells, to generate sound. Her book *An Individual Note: Of Music, Sound and Electronics* was published in 1972, and her archive is housed at the University of London's Goldsmiths' College.

The Polish Radio Experimental Studio (PRES) was founded in autumn 1957 by Józef Patkowski in Warsaw and was the first of its kind in the Eastern Bloc. An early "laboratory" for the production of electronic and electroacoustic music, the PRES appeared to materialize the Soviet claims that science and technology would restore socialist progress in the aftermath of Stalinism. But very little of the studio's output can be described as propaganda: Composers and engineers working there produced extremely original and often challenging electronic music. But PRES was not devoted exclusively to producing electroacoustic pieces; it was also used to create musical "illustrations" for movies, radio, and television. In fact, creating incidental music, as well as what today we would call sound designs for various media, was officially one of the studio's main functions. This meant that its electroacoustic experiments profoundly influenced not only the development of artmusic around Europe, but also popular culture and imagination throughout the Eastern Bloc.

PRES was an exclusive space for creative autonomy. Along with the Warsaw Autumn International Festival of Contemporary Music, it was the only platform that enabled artistic exchange on such a large scale, inviting guests from the United States, Western Europe, and Scandinavian countries. The composers who worked there were, among others, Wlodzimierz Kotonski, Krzysztof Penderecki, Krzysztof Knittel, Hugh Davies, and Lejaren Hiller.(FIG. 134) But the studio also became a laboratory for visual artists and designers, including Oskar and Zofia Hansen, Krzysztof Wodiczko, Kazimierz Urbánski, and Józef Robakowski.[121]

119
Hans G Helms, *Fa:m' Ahniesgwow* (Cologne: DuMont-Schauberg, 1959–1960), book, record, and synchronization map; see also Peter Weibel, ed., *Hans G Helms. Vokale Strukturen* (Cologne: Walther König, 2019), book and two CDs, including original materials and recordings.

120
Maderna used Helms's text as the basis of the piece *Dimensioni II / Invenzione su una voce* (1960), which itself was later incorporated into Maderna's *Hyperion* (1964).

121
See *Through the Soundproof Curtain. The Polish Radio Experimental Studio*, ZKM | Karlsruhe, 2018, exh. broschure; and ZKM | Karlsruhe, Adam Mickiewicz Institute Muzeum Sztuki in Łódź, and David Crowley, ed., *Ultra Sounds: The Sonic Art of Polish Radio Experimental Studio* (Heidelberg: Kehrer, 2019).

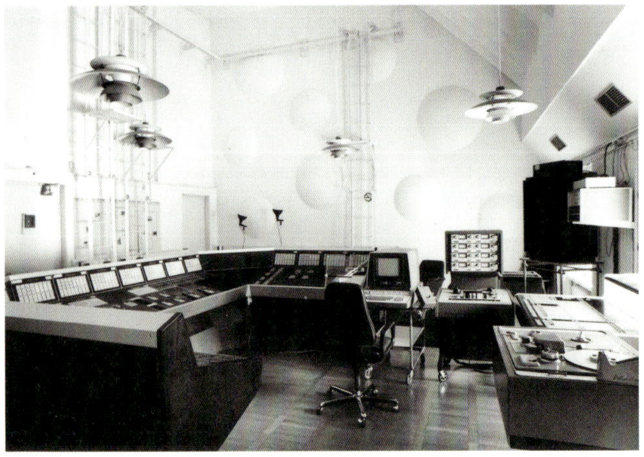

FIG. 135 Elektronmusikstudion (EMS), Stockholm, large studio (Room A).

Electronic-music studios spread throughout Europe in the early 1960s. In 1963, the Institut vor Psychoakoestiek en Elektronische Muziek was founded by the Belgian state broadcaster BRT in collaboration with Ghent University. The movement around the electronic synthesis of sound and language gained critical momentum in Scandinavia as well, and an experimental-music studio was set up by the Norwegian broadcaster NRK in Oslo in 1961.

An especially prominent role was played by the EMS in Stockholm, which grew out of an organization called Fylkingen. Fylkingen was originally founded in 1933 to organize chamber-music concerts, both of new work by contemporary composers and of the traditional repertoire, at various performance venues in Stockholm. The first electroacoustic concert in Sweden was jointly presented by Fylkingen and Sveriges Radio (SR) in 1952. Fylkingen also put on other concerts throughout the 1950s, at Stockholm's Moderna Museet and elsewhere. Moving into the 1960s, the organization's mission grew more radical: happenings, theater, and sound-text compositions were now part of the repertoire as well. The concert programming of this era featured then-new works by Cage, Xenakis, Stockhausen, Boulez, Morton Subotnick, Terry Riley, and La Monte Young, alongside pieces by noted Swedish composers and writers such as Åke Hodell, Lars-Gunnar Bodin, Öyvind Fahlström, and Sten Hanson. Support was also extended to new-media artists and those pursuing interdisciplinary and intermedia work, such as Nam June Paik, Merce Cunningham, Folke Rabe, and Catherine Christer Hennix.

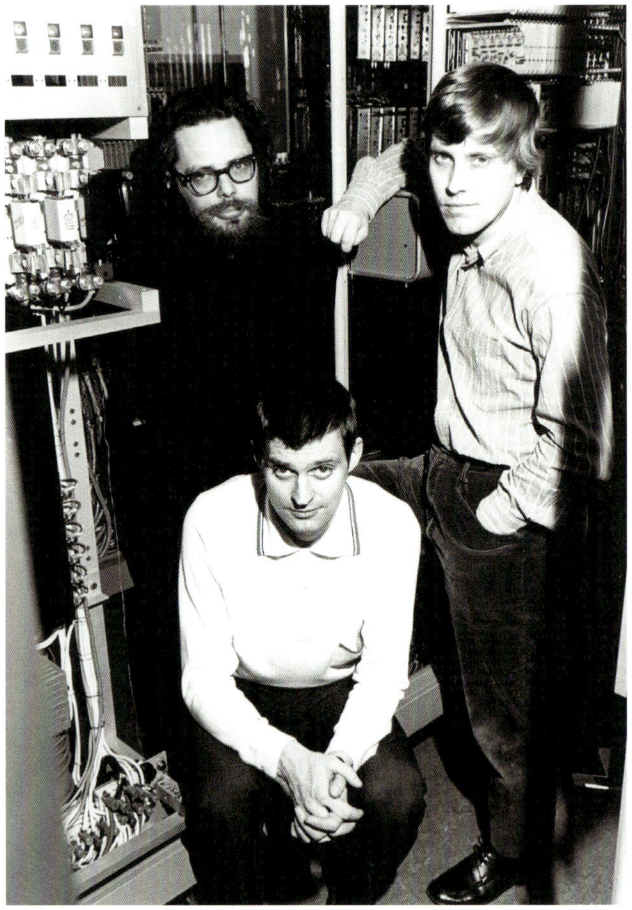

FIG. 136 Sten Hanson, Bengt Emil Johnson, and Lars Gunnar Bodin at the EMS (from left to right).

In keeping with this tradition, the EMS was founded in 1964 by Karl-Birger Blomdahl, the musical director of the SR.(FIGS. 135, 136) Its first director was the Norwegian composer Knut Wiggen, who invited Koenig, Xenakis, Pousseur, and other electronic-music pioneers to Stockholm. The studio cut its ties to the SR in 1969 and has been associated with the Statens musikverk (state performing-arts agency) since 2011. As one of the most modern studios in the world, it became a center for the study and production of text-sound compositions, and for the electronic creation of sound and music, especially after 1969, when it became able to provide a computer for use in its productions. The development of electronic music and its birth from the spirit of radio art, the new synthesis of music and language and the creation of new artistic worlds of sound, can all be read in the history of the EMS. In its evolution, a pivotal chapter in the history of European music and the sound art of the present day comes sharply into focus.[122]

The situation of state-run electronic studios such as the EMS changed significantly in

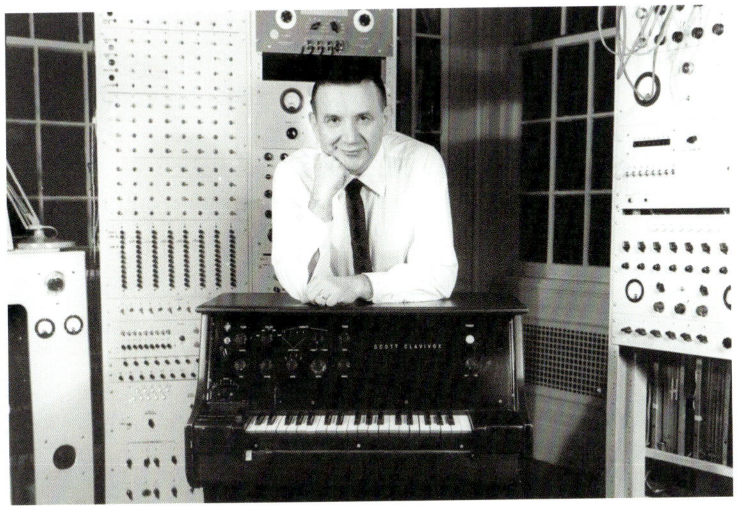

FIG. 137 Raymond Scott with his Clavivox.

the 1970s as a result of the invention of the synthesizer. Under the guidance of the composer Milton Babbitt, a variety of electronic musical instruments were already in use in the 1950s, such as RCA's Mark I and Mark II music synthesizers, the latter built in 1959 at the Columbia-Princeton Electronic Music Center – the same year Daphne Oram created her own synthesizer, as mentioned above. The triumphal advance of the synthesizer began in 1964 with the Moog synthesizer, which could be operated in real time and controlled via a normal keyboard. It was followed by many other synthesizers capable of generating tones electronically through sound synthesis, from Tom Oberheim to Yamaha.

Another of the pioneers experimenting with electronic music and electronic instruments was Raymond Scott in the United States. Scott, born in 1908, was a composer, pianist, sound engineer, and electronic music pioneer. Less well-known are Scott's achievements in inventing electronic musical instruments, which he had already begun in the 1940s. He was the founder of Manhattan Research, which became one of the most advanced studios for the creation of electronic music. Many of his inventions anticipated modules that would later become parts of synthesizers. For example, he invented an early trigger delay device, variable waveshape generators, preset programming devices etc. In the late 1950s, Scott also developed instruments, such as the Clavivox (FIG. 137), a keyboard synthesizer, an automatic composition machine called Electronium or an electronic sequencer he named Karloff. Robert Moog, developer of the Moog Synthesizer, met Scott in the 1950s, designed circuits for him in the 1960s, and acknowledged him as an important influence.[123]

The devices conquered pop music (Pink Floyd, Kraftwerk, Jean-Michel Jarre, Tangerine Dream, etc.). They became more affordable in the 1970s, enabling less successful musicians to use them as well. The modules consisted of signal generators, noise generators, voltage-controlled oscillators, modulators, etc., all connected by a circuit board. But not until these synthesizers became polyphonic – that is, able to produce more than one tone at a time – and capable of storing settings in memory did they gain widespread popularity, let alone mass appeal. With the Fairlight CMI, which was also a sampler, and the Yamaha DX7 in 1983, digital synthesizers displaced the analog models. The Roland D50, introduced in 1987, was particularly popular.

With the advent of the first personal computers in the 1980s, composers were suddenly handed tools and machines that not only matched but, in some cases, surpassed the performance of the old mainframes. Everything that had once been available through the public broadcasters' studios and university laboratories could now be produced by the composers themselves in their home studios – with the result that the equipment from the electronic studios and radio broadcasters wound up in the museum, and the broadcasters closed their studios. By producing and presenting acousmatic music and sound art, the forward-thinking museums took over the state broadcasters' role in promoting the development of one of the most important art forms of the twentieth century.

A good example of this is IRCAM (Institut de Recherche et Coordination Acoustique/Musique – Institute for Research and Coordination in Acoustics/Music), which is located in the Centre Pompidou in Paris. The institute was founded in 1977, the same year that the Centre Pompidou opened, and Pierre Boulez was its founding director.(FIG. 138)

The impetus behind IRCAM's creation was Pierre Boulez's critique of the existing state of affairs: "Radio stations have conducted

122 Sanne Krogh Groth's book (see note 92 above) has the merit of presenting, for the first time, a detailed and informative overview of the evolution of electronic music and other data-based art in the context of the state institutions and cultural policies of postwar Europe.

123 Tom Rhea, "Electronic Perspectives: Raymond Scott's Clavivox and Electronium," *Contemporary Keyboard*, February 1981, 72.

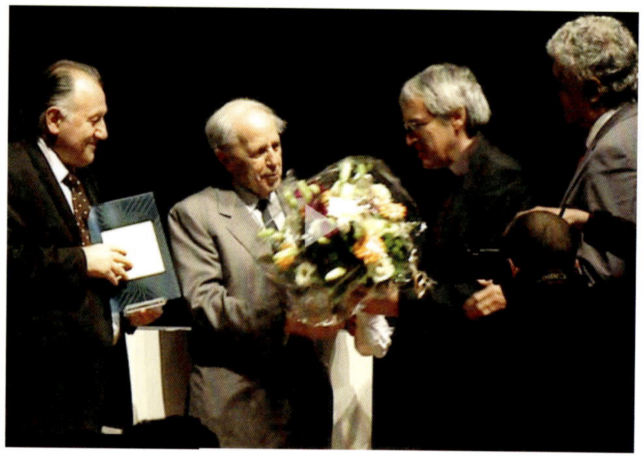

FIG. 138 Peter Weibel, Pierre Boulez et al. at ZKM | Karlsruhe, 2011. Pierre Boulez receives the Giga Hertz Award of ZKM.

prestige operations so as to have electronic music studios wherever possible. But once television budgets began to be so voracious these little studios became a quite secondary concern. In other studios you are dependent on either a university, a telephone company or a general electronics organisation, that engages in research for reasons of prestige but will not channel all its energies into it: research is an 'alibi,' a hobby or pastime. So it seemed to me necessary not to have to rely on institutions which have other aims in mind, other administrative principles and other commitments, even towards society. What is needed is an institution devoted exclusively to research: an institution that can act as a link between other organisations ... and that will concentrate all its resources on a precise aim."[124]

Besides electronic music, experimental poets and visual artists also created new sonic forms in radio that would grow into a distinct art form.[125] As an artistic medium, radio is an experimental system, a medium not only of production, but also of distribution. A legendary example of radio art dates to the year 1938: *The War of the Worlds*, based on the H. G. Wells novel of the same name and produced by Orson Welles and his repertory company, the Mercury Theatre. I will not discuss the content of this radio play further here, as it is sufficiently well known; instead, I would like to call attention to the fact established by the play's legendary success: namely, that radio was becoming not just an artistic medium for literature, but also a genuine medium of sound, of noise, and of music. If we can say "Everything that has wings flies," then we can just as well say "Everything that can be transported on electromagnetic waves is material for sound art, or rather radio art."

In his 1886–1888 experiments with sparks, conducted in Karlsruhe, Heinrich Hertz proved the existence of electromagnetic waves. These Karlsruhe sparks [*Funken*] can still be found in the German word for radio [*Rundfunk*] to this day. Through them, the technological foundations of radio – from the English word *radiotelegraphy*, from the Latin *radius* ["ray"] – were laid.

The story of Welles's *War of the Worlds* makes one thing clear: When we listen to the radio, we are not hearing the radio medium. This is why Marshal McLuhan wrote in *Understanding Media*: "Radio is provided with its cloak of invisibility, like any other medium."[126] It is a bit like a fish in an aquarium: It is unaware that it is swimming in water, because water is its natural environment; for the fish, it is invisible. Only when the fish is removed from the aquarium through human intervention and brought into contact with the air does it suddenly realize that it is in a new medium. We humans live in an aquarium of sound waves. That is one important message of sound art.

When did the phrase "radio art" first appear? As is typical, it was neither a radio technician nor a radio administrator who invented the term; rather, it first appeared in a 1925 essay by the composer Kurt Weill: "Möglichkeiten absoluter Radiokunst" Possibilities for Absolute Radio Art.[127] In this article, Weill essentially introduced the term while also pointing out that radio art is not just about texts, but also about the tones and rhythms of music, sounds from other spheres, the calls of human and animal voices, the sounds of nature, the rushing of the wind, the noises

124
Pierre Boulez, "Research Based on Technology: From *Poésie pour pouvoir* to the IRCAM," in *Conversations with Célestin Deliège* (London: Eulenburg Books, 1976), 108–19, here 111.

125
Cf. Schöning, *Ars Acustica – Ars Intermedia.*

126
Marshall McLuhan, *Understanding Media: The Extensions of Man* (New York: McGraw-Hill, 1964), 302.

127
Kurt Weill, "Möglichkeiten absoluter Radiokunst," in *Gesammelte Schriften* (Berlin: Henschel, 1990), 191–96. Originally published in *Der deutsche Rundfunk* 3, no. 26 (June 28, 1925), 1625–28.

128
M. H., "Der Sender ist verrückt. Ein Versuch am 24. Oktober auf Welle 467," *Die Besprechung*, supplement, *Radio-Umschau*, no. 6 (1924): 44. Translated from the German.

129
"To artistic cinema we oppose kinopravda and kino-eye; to artistic radio broadcasts we oppose radiopravda and radio-ear." Dziga Vertov, "*Kinopravda* and *Radioprav-da*," in *Kino-Eye: The Writings of Dziga Vertov*, trans. Kevin O'Brien (Berkeley: University of California Press, 1984), 52–56, here 56.

130
Cf. Thomas Tode, ed., *Dziga Vertov. Die Vertov-Sammlung im Österreichischen Filmmuseum* (Vienna: Österreichisches Filmmuseum, 2006), 106.

131
John Willett, ed., *Brecht on Theatre* (New York: Hill and Wang, 1964), 51–52.

132
Walter Benjamin, "Lichtenberg. A Cross-Section," in *Radio Benjamin*, ed. Lecia Rosenthal (London/New York: Verso, 2014), 336–59.

133
Rudolf Arnheim, *Radio: An Art of Sound* (London: Faber & Faber, 1936), 14.

of water and trees. And, beyond those, a multitude of new sounds never heard before, which would be artificially produced by microphones and tape recorders – that is, by the recording equipment itself – through the artificial modification of aural images. Weill saw early on that radio art is a technological art, the purpose of which is not only to imitate and reproduce natural sounds, but also to produce new, artificial, heretofore unheard sounds.

An early, truly interesting experiment in radio art was conducted by Hans Flesch in Germany in 1924, fourteen years before Welles's *War of the Worlds*. In *Zauberei auf dem Sender* [Magic at the radio station], Flesch actually played directly with radio waves. According to other sources, the author was Ernst Schoen, the assistant of Flesch. The point was that he was playing with the technological realities of the radio medium, which seemed to be spinning out of the host's control. The result was radio-wave stew: a mix of sounds, snatches of music, and isolated words. In the end, order was restored by the 1867 Johann Strauss waltz *The Blue Danube* – by literal waves, in other words. Superimposing electromagnetic waves with musical water waves, Flesch created a mélange of numbers, music, barking dogs, and the voice of the host. "The station's gone crazy!" wrote one reviewer of the show.[128] And in fact, "crazy radio" was Flesch's slogan.

In his famous essay "From Kino-Eye to Radio-Eye" (1929), Dziga Vertov expanded cinematic art to encompass radio art. However, he made the mistake of describing radio plays in the terms of his theory of the cinematographic eye. Vertov wanted to broaden this idea – of a mechanical eye, a camera eye, which, as an extension of the biological eye, would make possible a new way of perceiving the world – to include radio as an extension of the biological ear. He had earlier spoken, correctly, of the "radio-ear"[129] in a kind of analogy to cinema, which he was already imagining (anticipating sound film) as a combination of film and radio.[130]

In "The Radio as an Apparatus of Communication" (1932), Bertolt Brecht criticized the one-sided use of the medium radio. Instead of distributing from one central point to individuals, it should seek to become a communicative medium through which individuals are able to transmit as well as receive. The problem is that radio can only be used to listen to. As such, it lacks the ability to impart important lessons and simply reformulates traditional principles. Brecht's employment of his educational principles in *Der Flug der Lindberghs* [The Flight of Lindbergh], which premiered in 1929, and later was renamed by Brecht to *Ozeanflug* [The Flight across the Ocean], is exactly the opposite. The audience was allowed to participate – and participation is what was missing for Brecht in the medium radio.[131]

Another way of educating the listeners was Walter Benjamin's radio play *Lichtenberg. A Cross-Section* from 1933.[132] In this radio play, inhabitants of the moon observe the Göttingen based scientist Georg Christoph Lichtenberg (1742–1799) on Earth. The piece was related to Benjamin's deeper interests in Lichtenberg, who in 1777 used a device called an eidophorus to store and copy, positive and negative electrical currents as figures in a light-colored powder against a black background. In the graphic patterns created by tones on resonating plates of the electro-acoustician Ernst Florenz Friedrich Chladni (1756–1827, FIG. 68, P. 104), the electrical figures, controlled by sound frequencies, began to assume a form that Benjamin took an interest in.

Rudolf Arnheim's 1936 text *Radio: An Art of Sound* presents the radio as a highly influencial medium able to leap across national borders and provide the intimacy of sound to spaces of isolation and silence. For Arnheim the longevity of radio and TV depends on the permanence of sound, its pervasive role in the communication of ideas. He writes: "This is the great miracle of wireless. The omnipresence of what people are singing or saying anywhere, the overleaping of frontiers, the conquest of spatial isolation, the importation of culture on the waves of the ether, the same fare for all, sound in silence."[133]

But just as new music was cast out and excluded from the concert halls and opera houses, the state-run broadcasters stopped producing radiophonic and televisual art in the 1960s and 1970s, shutting out radio art. In the same way, artistic forms of film were pushed out of cinema's theaters. For all three, then, the museum was the only place that remained for their existence in exile.

Conclusion

Today, sound sculpture is a distinct musical direction concerned with a contemporary problem, for in the age of repetition there are only two options: first, to decode music – that is, to apply sampling to music's codes, combining and crossfading musical codes rather than tones – and second, to actually realize free sound in an open space at the sacrifice of conventional musical parameters.

Besides the sound sculptures and musical metaphors of Robert Wilhite, Richard Waters, Jim Pomeroy (FIG. 139), Tom Jenkins, Terry Fox, and Laurie Anderson, there are also the sound installations of Christina Kubisch, Alvin Lucier, Ron Kuivila, Doug Hollis, Bruce Fier (FIG. 140), Bob Bates (FIG. 141), David Askevold, and Liz Phillips, as well as the explorers of sound spaces and space sounds (the relationships between sound and space, sound and environment), such as Max Neuhaus, Bill Fontana, Peter Downsbrough, Bernhard Leitner and many others. From sound generator as sculpture (Aeolian harp) to electronically produced ambient sound to environmental installations, this is architectural or sculptural music that realizes the standards attained by the experimental music of the twentieth century: The sounds move through space, and the noise of the environment is admitted into the musical work. There is no score, no composer in the proper sense, for the sounds are almost exclusively the product of extramusical instruments, of our day-to-day industrial or natural environment.

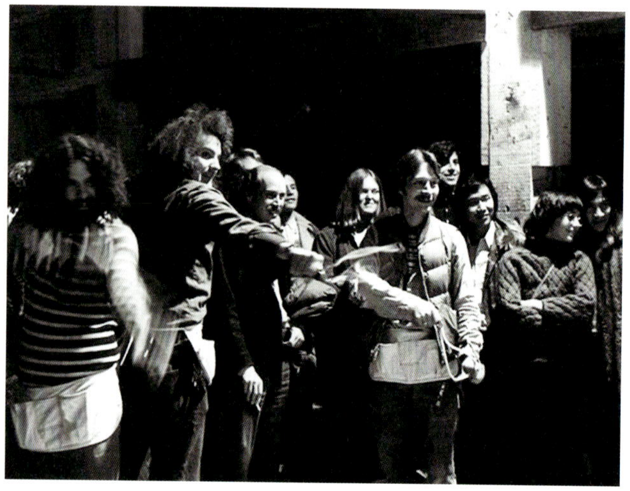

FIG. 139 Jim Pomeroy, *Composition in D*, November 4, 1974. A cantata for string trio and twelve music boxes. Participants: Howard Fried, Jim Pomeroy, Paul Kos. Performance at *South of the Slot*, 63 Bluxome Street, San Francisco.

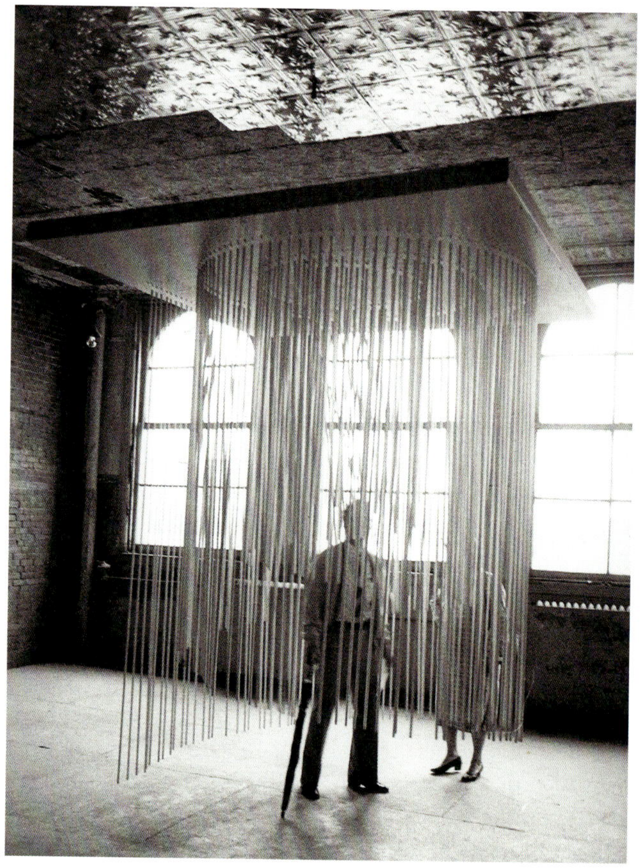

FIG. 140 Installation view of visitors interacting with a work by Bruce Fier. Installation view *Sound* at PS 1, New York City, 1979.

They are not even produced by the device itself, but at most transmitted, amplified, modified. The listener is freed to become a participant, a precipitant; his or her actions are a stimulant to the music.

Thus electronics has turned the sonic cosmos, firstly, into a live experience, which is why we speak of live electronics. Secondly, electronics has opened up the sonic cosmos to the listener: Every listener is a potential composer in a position to work directly with sounds – without a score. Music's freedom, therefore, liberates not only sound, but also performer and listener.

As an interface between art and music, sound sculptures stimulate the ear and the eye simultaneously. Sounds originate from space, move in space, articulate space, sensitize us to space. Spaces give rise to sounds, generate sounds, transform sounds. Sounds are created by the industrial environment (bicycles, traffic lights, cars, factories), by the natural environment (light, water, wind, heat), by media (TV images, telephone, radio), by the recipient, by simple tools or highly complex electronic devices. Sound sculptures are in fact

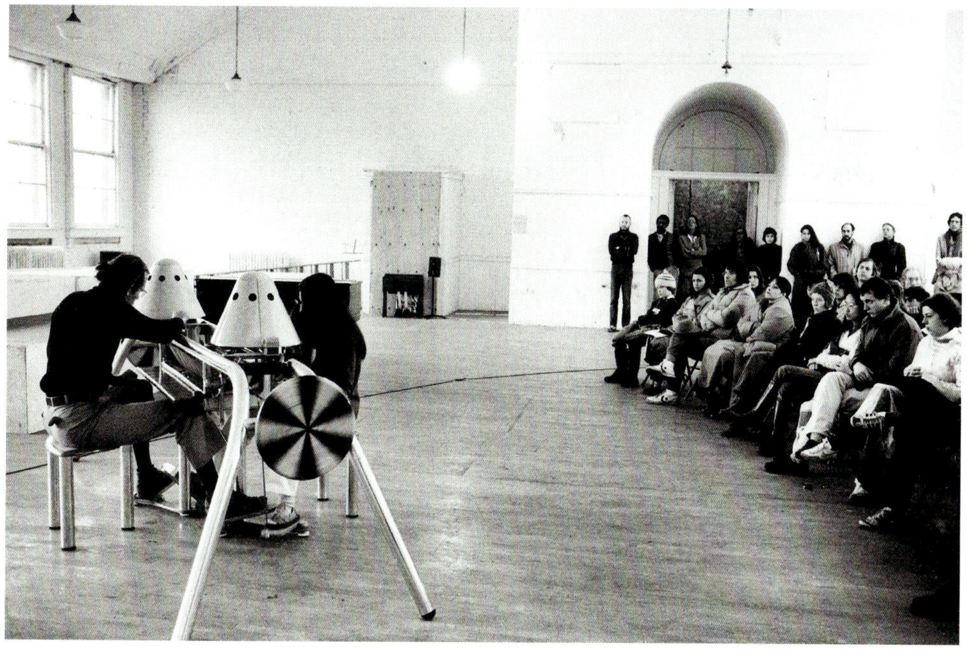

FIG. 141 Bob Bates, *Fuser*, 1978. Installation view *Sound* at PS1, New York City, 1979.

traps, set up to capture sounds and put environmental noises on display. In acoustically configured or tuned spaces, one can actually work with sounds as atmospheric disturbances à la Varèse, sounds in their pure wave form (without any instrument). The important thing with a contemporary sound sculpture is that the sound is generated not by the object itself, as with a musical object, but rather outside the sculpture. The sculpture merely gathers and joins the sound, which originated in some other place: defamiliarizes it, amplifies it, renders it audible in the first place, leaves a free space, untouched by us, for the sound's activity.

The music of sound sculptures is not directly created by humans themselves; it has an apparative existence independent of them. Through their independence and freedom from humans, sounds become more open, freer, purer. For the first time, we truly enter the space of sound. Emancipated from traditional musical instruments, these sound architectures, string installations, and sonic activities advance the legacy of the historical avant-garde, freeing sound to claim its autonomy. In this music devoid of traditional musical parameters and instruments, there is, for the first time, no longer any difference between sound, noise, tone; what sounds, what resounds, is real music. True music is music without music; sound objects and sound sculptures are heading that way. As are the other facets of free sound, such as mobile sound actions, sound events staged in urban space; sound installations and environments; body sounds, where the corporeality of sound, its spatiality, is projected onto the human body (resonating bodies that actually emerge from the body); spatial sound emitted by mobile ensembles and sound generators; or sound images, where the fusion of visual art and music on the level, of the (digital) technological transformation of both image and sound, leads to new forms of interaction. Sounds previously unheard and instruments previously unseen constitute the new sonic art of the twentieth and twenty-first centuries: sound as a medium of art.

Today we are confronted with a new universe of sound. Any object can create sound and be connected by sensors to a computer that stores all kinds of sound samples and compositional techniques. Thus the sound of a wine glass or the opening of a door can be stretched into a symphony. With the aid of sensors and other technology, any object can create the sound of any object. The digital sound universe can create all the sounds of all instruments and objects from all instruments and objects. Not only can any object emit sound, any space can emit sound. We now have a universal aural art at our disposal. Sounds are everywhere, and the ambient is everywhere sonic. Finally, in the course of the twentieth century, sound art has become a universal medium.

László Moholy-Nagy

New Form in Music. Potentialities of the Phonograph

Editor's note:

László Moholy-Nagy (1895–1946) was a Hungarian painter, photographer, typographer, filmmaker, and stage designer. Moholy-Nagy was a teacher at the Bauhaus from 1923 to 1928 and during this time made many lasting contributions as an artist, writer, and teacher.

New Form in Music: Potentialities of the Phonograph exemplifies Moholy-Nagy's experimental approach to modern technology. Having earlier advocated the use of photography to produce abstract light compositions, Moholy-Nagy suggests here a similar approach towards the phonograph.

In the text, he states, that the phonograph should not simply be used as a tool of reproduction but that it be used as a means of musical production. This proposal predates, by several decades, artists' experiments with phonographs (for example John Cage) or even turntable experiments of artists like Christian Marclay.

New Form in Music. Potentialities of the Phonograph

Among present-day musical experiments, an important role is played by researches conducted with amplifiers which open up new paths in the production of acoustic phenomena. The aims of the Italian Bruitists, in constructing new instruments with new sound-formations, have been substantially fulfilled by experiments with the amplification tube as a specific instrument which permits the production of all sorts of acoustic phenomena. However, this alone does not exhaust the potentialities that might be expected as regards the transformation of music. I refer to the excellent paper by P. Mondrian, 'New form in music and the Italian Bruitists' (De Stijl), where the basic principles of innovation in creation with sound are analysed.

Mondrian says among other things, 'Music cannot develop through enrichment in terms of sounds or through refinement, but through the abolition of the duality of the individual and the universal, the natural and the spiritual; in other words, the achievement of human equilibrium is the aim of all creation.' And he goes on: 'Noises in nature result from simultaneous and continual fusion. By having partly destroyed this fusion and continuum, the music of the past has derived from this noise certain sounds which it has arranged in a certain harmony. In order to achieve a more universal mode of creation, the new music will have to attempt a new order of sounds and non-sounds (certain noises). The main point is to deliver ourselves through creation from the "natural", from the "animal", the characteristics of which are fusion and repetition. If the fusion and hereby the predominance of the individual is to be avoided, instruments will have to form the sort of sounds in which both wavelength and frequency must remain as even as possible. Therefore instruments must be constructed in such a way that every after-oscillation can immediately be interrupted. This kind of creation is inconceivable without a different technique and different instruments.'

If they are to be realized at all *externally* by technical inventions, these postulates will actually be met through employing the amplification tube.

My ambitions in the same field of experimental transformation in music are of another kind, though closely connected with the thinking of Mondrian. In what follows I shall pass over the motives for new sound-creation and shall present just one suggested means for its possible realization with the help of a new means of expression.

I have already suggested that the phonograph be transformed from an instrument of reproduction into one of production; this will cause the sound phenomenon itself to be created on the record, which carried no prior acoustic message, by the incision of groove-script lines as required.

Since my description of this process served elsewhere as an example to illustrate another idea, I was very brief in specifying the potentialities, without presenting detailed arguments, for the transformation of our musical conceptions along these lines. In speculative terms, the following is clear:

1 By establishing a groove-script alphabet an overall instrument is created which supersedes all instruments used so far.

2 Graphic symbols will permit the establishing of a new graphic and mechanical scale,* that is, the creation of a new mechanical harmony, whereby the individual graphic symbols will be examined and their relations formulated within a rule. (We may allude here to an idea that sounds rather utopian as yet; namely, the transposing of graphic designs into music on the basis of strict regularities of relationships.)

3 The composer would be able to create his composition for immediate reproduction on the disc itself, thus he will not be dependent on the absolute knowledge of the interpretative artist. So far, the latter was in most cases able to smuggle his own spiritual experience into the composition written in note form. The new potentialities afforded by the phonograph will re-establish the amateurish musical education of our day on a more wholesome basis. Instead of the numerous 'reproductive talents', who have actually nothing to do with *real* sound-creation (in either an active or a passive sense), the people will be educated to the *real* reception or creation of music.

4 The introduction of this system in musical performances will also facilitate to a significant degree independence from large orchestral enterprises, and the large-scale distribution of original creations by means of a simple instrument.

(The efficiency of the phonograph has been substantially improved lately by certain technical innovations. Among others, there are two important inventions in this field. One is electrical operation, the other a newly invented diaphragm which ensures almost completely friction-free reproduction of recorded compositions. I think that if we regard these as a necessary condition, then we shall have technically perfect apparatuses within the shortest time.)

I consider that the following practical experiments with the phonograph in the realm of musical composition should be initiated:

1 Since the grooves on the mechanically produced record are microscopic in size, we shall first have to devise a method for reducing by technological means down to the normal size of a present-day record any large-scale groove-script record that can conveniently be worked on by hand. It would be desirable to make a photograph of a present-day (reproductive) record and to make a photo-cliché or photo-engraving of the photograph by a zincographical or galvanoplastical process. Should such a record prove to be just more or less playable, the basis for subsequent work along these lines will be established.

2 Study of the graphic symbols of the most different (simultaneous and isolated) acoustical phenomena. Use of projectors. Film. (Specialist works on physics already include detailed descriptions thereof.)

3 Examination of mechanical, metallic and mineral sounds. From these, attempts to devise – for the time being, in a graphic way – a special language. Special attention to be paid to symbols created by different tonalities.

4 Graphic production of the largest contrasting relations. (Before beginning experiments on the wax plate, it is suggested that one trace with a needle the graphic wave lines of music on a [reproductive] phonograph disc; these lines will become well known to the experimenter who will acquire therefrom a sense for graphic representation.)

5 Finally, there are improvisations on the wax plate to be considered, the phonetic results of which are theoretically unforeseeable, but which may permit us to expect significant incentives since the instrument is rather unknown to us.

'Neue Gestaltung in der Musik. Möglichkeiten des Grammophons.'
Der Sturm, *Berlin. July 1923, No. 14.*

* Our present scale is approximately one thousand years old, and it is not absolutely necessary to be bound by its inadequacies today.

First published in German as „Neue Gestaltung in der Musik. Möglichkeiten des Grammophons," *Der Sturm*, no. 14 (July 1923): 102–06. First published in English in Krisztina Passuth, ed., *László Moholy-Nagy* (London: Thames and Hudson, 1985) 291–92. Translated from the German by Mátyás Esterházy. Reprinted with kind permission of the Moholy-Nagy Foundation and Thames and Hudson.

Germano Celant

Record as Artwork, 1959–73

Editor's note:

Germano Celant's significant 1973 show at the Royal College of Art in London, *Record as Artwork. 1959–73*, featured his personal collection of records. In 1977 and 1978, the exhibition was extended and shown as *The Record as Artwork. From Futurism to Conceptual Art* in the USA (Fort Worth, Philadelphia, Chicago) and Canada (Montreal), with great success. It included over 150 records and album covers recorded and frequently designed by artists including Karel Appel, Jim Dine, Marcel Duchamp, Richard Huelsenbeck, Yves Klein, Michael Snow, Andy Warhol, Robert Whitman, and many others. Celant's exhibition was the first to view audio recordings and record covers as art to be exhibited in a museum.

In 1988, the exhibition *Broken Music: Artists' Recordworks* at daadgalerie, Germany, had a similar approach. Curated by Ursula Block and Michael Glasmeier the exhibition presented works of visual artists created with and for the medium of the record: records, record covers, record objects, record installations. The exhibition *Sound Art. Sound as a Medium of Art* at ZKM | Karlsruhe, dedicated a whole area to Ursula Block's Broken Music Archive.

In the meanwhile, the subject has been explored by various other exhibtions, for example *Vinyl*, 2005, at MACBA Barcelona and Neues Museum Weserburg, Bremen, or *The Record*, 2012, at the Miami Art Museum.

This text was first published in: Roselee Goldberg (ed.), *RECORD AS ARTWORK*, exhib. cat. (London: Royal College of Art, November 1973). Originally translated from the Italian by The Linguists' Club.

At the end of last century, the role of the art lover underwent a change from active, in the sense of directly commissioning works of art, to passive in the sense of simply collecting them. This, together with the advent of mass media, communications (photography, cinema, press, telegraph, television, gramophone), has contributed to the breakdown of the traditional psychophysical relationship between artist and public.

Whereas before patrons influenced artwork – its topics and trends – in the making, now the art collector exercises his options after the event. From being the outcome of an exchange between artist and public, artwork turned into a more introspective and private affair. The work of art, now it had become a commercial commodity, lost much of its physical impact, and artistic thinking hinged not on outside reality but on the artist's inner consciousness.

The search for new meanings, once the immediate need for visual appeal had lost its edge, led to a move away from the sheer material/physical aspect of the finished product. This polarisation contributed to the loosening of the link between artist and public and, in the present century, has enabled the artist to draw on the full range of media resources available, both technical and technological. With their numerical strength and a-physical quality, these resources have lent themselves to the artist's dematerializing and conceptual process.

It is clear that abstract meaning and mass appeal have had their compensations, after some fifty years. For art has again made contact though superficially with the patron-collector. This turn-about has been achieved by the industrial processing of artistic output. Editions whether of prints or videotapes, books or records, make it feasible for art to reach a bigger audience than before. Technology and information media are part and parcel of this process, which speeds beyond national horizons and the power structures of accepted art institutions like museums and art galleries. The instruments at hand for artists at the present time are supranational and supratribal. The response of these institutions, the museums and art galleries, makes the point. In order to survive the information explosion, they have had to start producing, publishing, and distributing videotapes, records, books, recordings, and films, with a view to preserving their sociocultural standing.

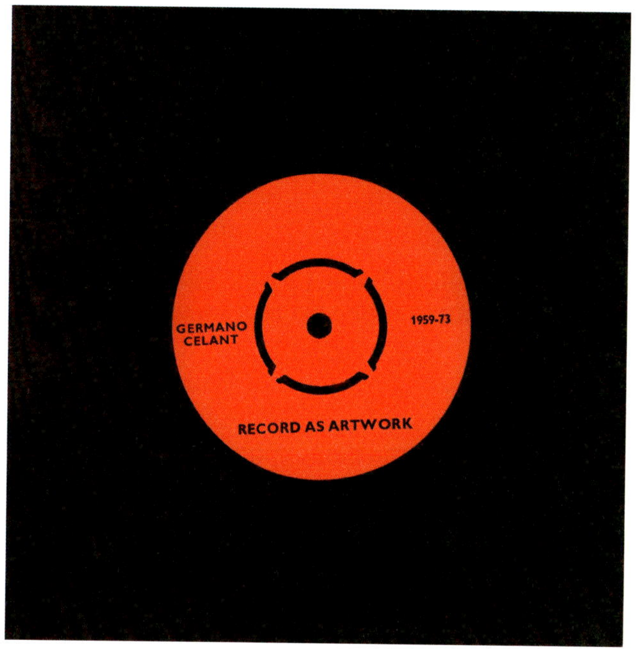

Germano Celant, *Record As Artwork. 1959–73*, 1973. Exhibition broschure, cover.

Naturally enough, when art begins to bypass the gallery's enclosed halls and "tribal" dealings in the world's cultural art market, the use of the media is bound to erode the status of the great national treasure houses of art. Because of this landmark development, institutions have had to liaise at international level and open their doors to artists from other countries. The art world may be a society which is still tribal, but at least it is an open society and functions in many different directions.

Indeed, the media explosion has involved the capitalist superstructures in the realm of art; the giants of television, recording, and publishing have discovered a new market, the art market, which will absorb educational and prestige products. There has also been an enrichment of art itself through reaching more people and devising better means to get the message across. But the art product, its physical presence dwarfed and its presentation standardised has necessarily been subject to some loss of features traditionally associated with its physical and psychological identity. Its presence is more of a commonplace, at the same time it enjoys a high rate of appeal. These factors make art loose charisma; some of its epoch-making power goes by the board in the ranks of consumer goods. Nevertheless, art is given the chance of reaching the public directly in an educative role.

The treasured "mystical" object of former generations is replaced by an anonymous instrument of learning. There is a progression here; for study groups, turning to the periodical and printed book as instruments of communication, begin to operate in the educational rather than art sphere. Among the communications media of videotape, book, telegram, film and photograph, records also contribute to make art more widely known. They do this by converting visual art to pure sound, which can be transmitted anywhere and issued in repeat copies by the thousand. The addition of sound further explodes the old visual approach, conquering new territory for art in the process. A record may therefore be said to treat art data by involution and extension, focusing on, it as an expansion of action and written word. This amounts to a form of audio-scripted page, which is "read" by means of the instrument which has replaced the pen with the needle, namely the gramophone.

The record thus becomes a kind of page, capable of giving coverage to the written word, read or spoken. The technical-aural extension of the written or read word belongs within the framework of abstraction and demands of the artist an inner subjective response to the end product, which explains its usage by nonanalytical and conceptual artists. Sound also means that conceptual artists can get away from the nonprogressive set forms of writing and win back for their message those qualities which the printed page took from live speech.

Recovery of the aural alternative took place at the end of last century, to put the movement in its historical perspective, when poet-artists like Morgenstern and Scheerbart produced a series of phonic poems which attempted to restore to the word its sound content. This reattribution of acoustic quality to writing led to further developments, the history of which is signposted by the phonic poems of Khlebnikov, Ball, Tzara, Hausmann, Schwitters, Marinetti and Depero. Manifestos, such as Marinetti and Masnata's *La Radia* (1933) supported the use of acoustic solutions in art. Immensification of space, the scene, no longer visible or containable, becomes universal and cosmic. It receives, amplifies and transfigures the vibrations emitted by the state of mind dramas, living and dead, which have sound but no words. It receives, amplifies, and transfigures the vibrations emitted by matter ...

In the 1950s, the idea of a word first uttered and then written found poets and artists capable of transferring that word from public speech to recording and disc. Of fundamental importance in this respect was the part played by the review *Cinquième saison*. A phonic disc of Mimmo Rotella was produced in 1950, and this was followed from 1958 onwards by a series of phonetic poem discs. The review's founder-director, Henri Chopin, is one of the phonetic poets who have taken in the threads of former avant-garde concepts, adapting them to present-day communication needs. In 1962, Chopin produced a disc called *Vibrespace*, in this, he used a tape recorder to collate his own breathing, which is distorted and varied by multi-track programming. The idea of mechanising word-sound took immediate hold and poet-artists such as Ferdinand

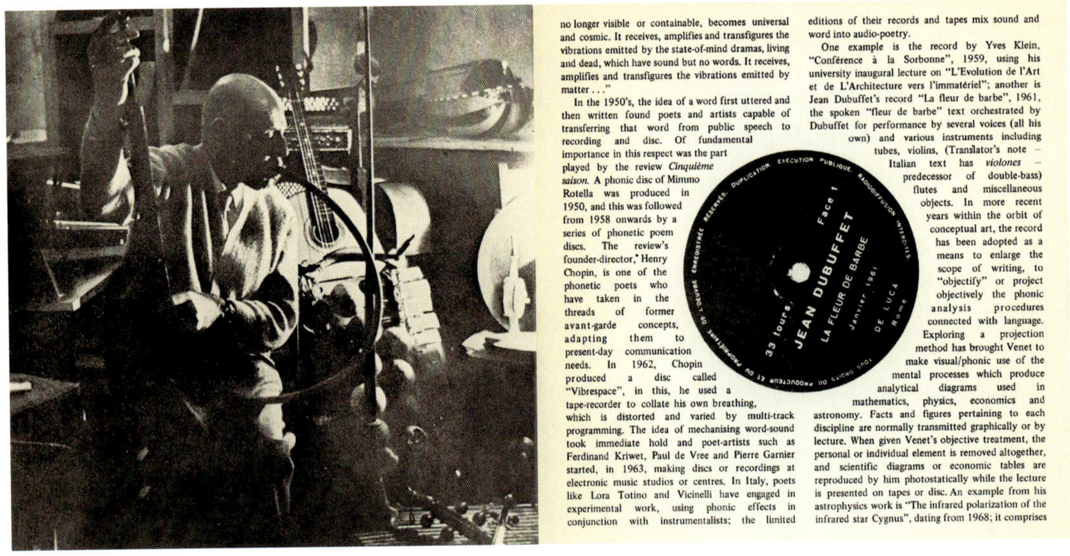
no longer visible or containable, becomes universal and cosmic. It receives, amplifies and transfigures the vibrations emitted by the state-of-mind dramas, living and dead, which have sound but no words. It receives, amplifies and transfigures the vibrations emitted by matter . . ."

In the 1950's, the idea of a word first uttered and then written found poets and artists capable of transferring that word from public speech to recording and disc. Of fundamental importance in this respect was the part played by the review *Cinquième saison*. A phonic disc of Mimmo Rotella was produced in 1950, and this was followed from 1958 onwards by a series of phonetic poem discs. The review's founder-director,* Henry Chopin, is one of the phonetic poets who have taken in the threads of former avant-garde concepts, adapting them to present-day communication needs. In 1962, Chopin produced a disc called "Vibrespace", in this, he used a tape-recorder to collate his own breathing, which is distorted and varied by multi-track programming. The idea of mechanising word-sound took immediate hold and poet-artists such as Ferdinand Kriwet, Paul de Vree and Pierre Garnier started, in 1963, making discs or recordings at electronic music studios or centres. In Italy, poets like Lora Totino and Vicinelli have engaged in experimental work, using phonic effects in conjunction with instrumentalists; the limited editions of their records and tapes mix sound and word into audio-poetry.

One example is the record by Yves Klein, "Conférence à la Sorbonne", 1959, using his university inaugural lecture on "L'Evolution de l'Art et de L'Architecture vers l'immatériel"; another is Jean Dubuffet's record "La fleur de barbe", 1961, the spoken "fleur de barbe" text orchestrated by Dubuffet for performance by several voices (all his own) and various instruments including tubes, violins, (Translator's note – Italian text has *violones* – predecessor of double-bass) flutes and miscellaneous objects. In more recent years within the orbit of conceptual art, the record has been adopted as a means to enlarge the scope of writing, to "objectify" or project objectively the phonic analysis procedures connected with language. Exploring a projection method has brought Venet to make visual/phonic use of the mental processes which produce analytical diagrams used in mathematics, physics, economics and astronomy. Facts and figures pertaining to each discipline are normally transmitted graphically or by lecture. When given Venet's objective treatment, the personal or individual element is removed altogether, and scientific diagrams or economic tables are reproduced by him photostatically while the lecture is presented on tapes or disc. An example from his astrophysics work is "The infrared polarization of the infrared star Cygnus", dating from 1968; it comprises

Germano Celant, *Record As Artwork, 1959–73*, 1973. Exhibition broschure, pp. 6–7.

1 Translator's note: Italian text has *violones*, predecessor of double bass.

Kriwet, Paul de Vree and Pierre Garnier started, in 1963, making discs or recordings at electronic music studios or centers. In Italy, poets like Lora Totino and Vicinelli have engaged in experimental work, using phonic effects in conjunction with instrumentalists; the limited editions of their records and tapes mix sound and word into audio-poetry.

One example is the record by Yves Klein, *Conférence à la Sorbonne*, 1959, using his university inaugural lecture on "L'Evolution de l'Art et de L'Architecture vers l'immatériel"; another is Jean Dubuffet's record *La fleur de barbe*, 1961, the spoken "fleur de barbe" text orchestrated by Dubuffet for performance by several voices (all his own) and various instruments including tubes, violins[1], flutes and miscellaneous objects. In more recent years within the orbit of conceptual art, the record has been adopted as a means to enlarge the scope of writing, to "objectify" or project objectively the phonic analysis procedures connected with language. Exploring a projection method has brought Venet to make visual/phonic use of the mental processes which produce analytical diagrams used in mathematics, physics, economics, and astronomy. Facts and figures pertaining to each discipline are normally transmitted graphically or by lecture. When given Venet's objective treatment, the personal or individual element is removed altogether, and scientific diagrams or economic tables are reproduced by him photostatically while the lecture is presented on tapes or disc. An example from his astrophysics work is "The infrared polarization of the infrared star Cygnus," dating from 1968; it comprises a recording of astronomical observations of the star Cygnus made by the astrophysics laboratory at the university of Arizona.

More personal use of the medium has been made by Lawrence Weiner, with his *A()C Sharpened Carried Done Again Perhaps ... AC()*, of 1972; in this, the language texture of vowels, verbs and adverbs is interwoven with the sound of a flute. The disc expands his investigation of verbal assonance, between words of different languages or of the same language. His basic tools of written word and book are augmented by the record; more recently still he has been experimenting with photographic stills and film. Permutational analysis of the phonetic components of a noun – vowels and consonants – has been carried out by Chia. The word "pietra" is transformed into its syllabic permutations: i.e. trap, etrapi, trapie, rapiet, apietr, by monitoring the variant at two-minute intervals. The aim is to begin to consider spoken language as an objective entity, which can be moulded and built up into a different object, by varying its physical components.

The result of R. Jacobson's investigations into the sound structure of syllabic articulations has been applied by Notargiacomo, in his phoneme or phoneme/word group variations. His *Variazioni del fonema /R/* records the voice of an inter-preter, pronouncing a syllabic /p/ in different languages, Italian, English, French, and German. The study of speech analysis is continued in *ffermazioni*, a record of variants attributable to the emotive stance of the speaker; the same statement is repeated fifteen times by the same person in different tones. Lastly, poetic analysis of the phrase "I like Ike" equivalent "/ay/ /layk/ /ayk/" produces a different poetic meaning according to the scansion of stress and rhythm on a certain pattern.

Records are complementary to the audioscripted dimension, offering the researcher the possibility of reproducing with the utmost fidelity a sound or noise which is realistic. They thus extend the physical/tactile level of experience, absent from documentation and the visual record of actions, performances, and products for the eyes alone. The feeling that performers and events produced and selected share the hearer's living space marks a step forward, in the direction of artwork's audio-tactile integration. For records "touch" almost everyone directly; ensemble of noises and aural messages passes between utterer and listener, capable of recreating in part the psychophysical experience of the event. The art product may be compressed into a record, but there is a gain of sensory value for and to the ear in the process. Use of a record to enlarge a performance has been attempted by Mazzotta with a 33-rpm reproduction of the aural section of the Joseph Beuys presentation at the Kunsthalle in Bern on the occasion of the *When Attitudes Become Form* exhibition. The disc is called *Ja Ja Ja Nee Nee Nee* and contains the chant, in various tones and styles, which accompanied the distribution of butter in the corners of the hill or arrangements of pieces of felt and objects yeilding/unyielding, organic/inorganic that go to make up a Beuys' composition. Isolating a component of the performance does not in fact reduce the scale of achievement, but it does make it appeal to the ear as well as to the eye. For this reason, record and tape are used systematically by Beuys.

Another disc made in conjunction with Terry Fox, presents the last twenty minutes of an hour-long happening at the Kunstakademie in Düsseldorf, November 1970. On both sides of the 45-rpm disc, a series of different sounds is recorded. The first sound comes from Terry Fox, striking together two iron tubes of different lengths. The tube-sound is directed at a window frame of four glass panels. The percussive beat is picked up by the echoing glass but at some points the sound goes dead. When the sound is emitted no further and the acoustic barrier is reached, the glass breaks. At the back of the frame is a lighted candle. When all four glass panels are broken, then the wooden frame is broken and Fox tries to snuff out the candle by striking the tube in order to secure displacement of the air. The smaller tube is put on the floor, to find where the sound goes dead round a lamp. At this juncture, Beuys starts eating a fruit and spitting the pips, singly, into a silver bowl. The disc ends with these two sounds intersecting until the fruit is consumed.

In the same way, Sarkis uses a disc to make a physical record of a series of actions. In *Une lettre à déchiffer* at 33-rpm stereo, there is reproduced the noise-sound made in writing a letter to Turkey which takes thirteen minutes. In *Le chien aboyant courait derrière X* a barking dog is recorded in pursuit of a man who is running away; Sarkis is both the barking dog and the running man.

The work of De Filippi turns back to primitive cultures, from the Oscans to the Celts, and strives continually to re-live the customs and myths of these ancient peoples, in both physical and cultural terms. The Narcissus legend, for example, has been subjected to various tests including sound recording converted to disc. The myth is illustrated by recording the echo, at Passo di Bocca Tribaria on 5 January 1973, produced by De Filippi yelling every three minutes, four times, the tautological word "Eco."

Antonio Dias and Robin Mackenzie use records with another linguistic variant method, in order to document the sound of aural entities, structured according to a conceptual scansion which is arranged beforehand. The album of Dias, *The space between*, was published in 180 copies and records a space continuum arrangement of two sounds, an alarm clock and breathing. In the first instance, the recording is divided every three seconds by three seconds blank or silence on a prearranged schedule, in order to emphasise the pattern of filled and empty space, that is of sound and silence. In the second instance, attention focuses on a breathing continuum which, by its nature, consists of immission and emission, again a positive and a negative, a sound and a silence.

Mackenzie has used records as a compliment to visual work, done in the wilds of Canada. The photo-record, used also on the sleeve of the disc, goes with *A sound work to accompany*; the disc, that is to say, documents the acoustic content of the work by recording, at the same time as the photographs were taken, the wind over the bare ground.

With the aim of providing audiovisual documentation to his "perspective corrections" Dibbets, in 1969, made a record entitled *Afsluitdijk*; it registers the sound produced during a five kilometer journey along the straight causeway across Usselmeer Bay, which links Den Oever and Makkum in the Netherlands, at a uniform speed of 100 kmph. Dibbets' record accompanies a visual art series made on the same causeway and shown at his exhibition in Krefeld. Boltanski's record, like all his work, strikes a more intimate and personal note. It seeks to capture the sound of the nursery songs from his childhood days, including *Fais Dodo Cilas mon petit frère* and *Il était un petit navire*.

Ruthenbeck's record called *Dachskulptur* offers a physical sound experience and records, together with photographic documentation, the technological cries of factory hooters in Düsseldorf. The whistle blasts strike the air in signal form and create an objective sound dimension which is highly effective in aural terms. It serves as a commentary to his sculpture, with objects of rubber and burnished steel.

Records are also useful in giving an aural account of artistic theories or of works which result from these theories. As an example of the first category, mention may be made of a Billy Kluver record, 1963, entitled *Giant size $1.57 each*; it contains live interviews with the leading exponents

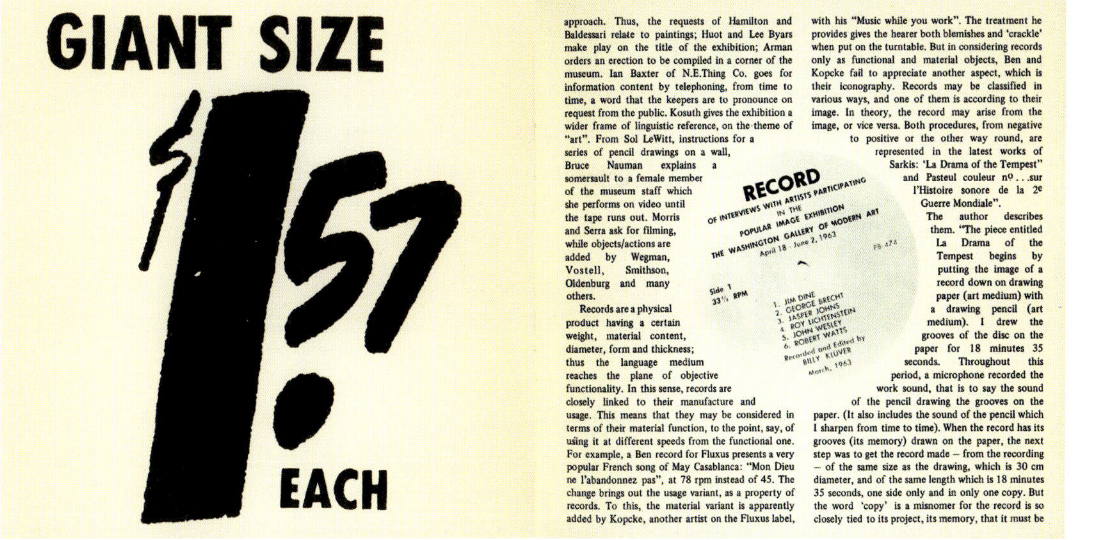

GIANT SIZE $1.57 EACH

approach. Thus, the requests of Hamilton and Baldessari relate to paintings; Huot and Lee Byars make play on the title of the exhibition; Arman orders an erection to be compiled in a corner of the museum. Ian Baxter of N.E.Thing Co. goes for information content by telephoning, from time to time, a word that the keepers are to pronounce on request from the public. Kosuth gives the exhibition a wider frame of linguistic reference, on the theme of "art". From Sol LeWitt, instructions for a series of pencil drawings on a wall, Bruce Nauman explains a somersault to a female member of the museum staff which she performs on video until the tape runs out. Morris and Serra ask for filming, while objects/actions are added by Wegman, Vostell, Smithson, Oldenburg and many others.

Records are a physical product having a certain weight, material content, diameter, form and thickness; thus the language medium reaches the plane of objective functionality. In this sense, records are closely linked to their manufacture and usage. This means that they may be considered in terms of their material function, to the point, say, of using it at different speeds from the functional one. For example, a Ben record for Fluxus presents a very popular French song of May Casablanca: "Mon Dieu ne l'abandonnez pas", at 78 rpm instead of 45. The change brings out the usage variant, as a property of records. To this, the material variant is apparently added by Kopcke, another artist on the Fluxus label, with his "Music while you work". The treatment he provides gives the hearer both blemishes and 'crackle' when put on the turntable. But in considering records only as functional and material objects, Ben and Kopcke fail to appreciate another aspect, which is their iconography. Records may be classified in various ways, and one of them is according to their image. In theory, the record may arise from the image, or vice versa. Both procedures, from negative to positive or the other way round, are represented in the latest works of Sarkis: 'La Drama of the Tempest" and Pasteul couleur nº . . .sur l'Histoire sonore de la 2e Guerre Mondiale".

The author describes them. "The piece entitled La Drama of the Tempest begins by putting the image of a record down on drawing paper (art medium) with a drawing pencil (art medium). I drew the grooves of the disc on the paper for 18 minutes 35 seconds. Throughout this period, a microphone recorded the work sound, that is to say the sound of the pencil drawing the grooves on the paper. (It also includes the sound of the pencil which I sharpen from time to time). When the record has its grooves (its memory) drawn on the paper, the next step was to get the record made – from the recording – of the same size as the drawing, which is 30 cm diameter, and of the same length which is 18 minutes 35 seconds, one side only and in only one copy. But the word 'copy' is a misnomer for the record is so closely tied to its project, its memory, that it must be

Germano Celant, *Record As Artwork*, 1959–73, 1973. Exhibition broschure, pp. 22–23.

of the American pop art scene, in their studios or in galleries, from Johns to Rauschenberg, from Lichtenstein to Rosenquist, from Wesselmann to Oldenburg. An exciting example of the second category is the recording of works dictated by telephone, made on the occasion of the *Art by Telephone* exhibition at Chicago Museum of Contemporary Art in 1969. The designer of the exhibition, Jan van der Marck, had in mind the precedent set by Moholy-Nagy who, in 1922, dictated by telephone a work in hot metal. Van der Marck collected together the descriptions given over the telephone of single works which were to be made in Chicago. Given the common medium of communication – with the artist more absent than present – linguistic resemblances were expected; however, a great deal of variety was forthcoming in the results, because of a nonuniform linguistic approach. Thus, the requests of Hamilton and Baldessari relate to paintings; Huot and Lee Byars make play on the title of the exhibition; Arman orders an erection to be compiled in a corner of the museum. Ian Baxter of N.E.Thing Co. goes for information content by telephoning, from time to time, a word that the keepers are to pronounce on request from the public. Kosuth gives the exhibition a wider frame of linguistic reference, on the theme of "art." From Sol LeWitt, instructions for a series of pencil drawings on a wall, Bruce Nauman explains a somersault to a female member of the museum staff, which she performs on video until the tape runs out. Morris and Serra ask for filming, while objects/actions are added by Wegman, Vostell, Smithson, Oldenburg, and many others.

Records are a physical product having a certain weight, material content, diameter, form, and thickness; thus the language medium reaches the plane of objective functionality. In this sense, records are closely linked to their manufacture and usage. This means that they may be considered in terms of their material function, to the point, say, of using it at different speeds from the functional one. For example, a Ben record for Fluxus presents a very popular French song of May Casablanca: *Mon Dieu ne l'abandonnez pas*, at 78 rpm instead of 45. The change brings out the usage variant, as a property of records. To this, the material variant is apparently added by Köpcke, another artist on the Fluxus label, with his *Music while you work*. The treatment he provides gives the hearer both blemishes and "crackle" when put on the turntable. But in considering records only as functional and material objects, Ben and Köpcke fail to appreciate another aspect, which is their iconography. Records may be classified in various ways, and one of them is according to their image. In theory, the record may arise from the image, or *vice versa*. Both procedures, from negative to positive or the other way round, are represented in the latest works of Sarkis: *La Drama of the Tempest* and *Pasteul couleur* n° ... *sur l'Histoire sonore de la Deuxième Guerre Mondiale*. Sarkis describes them:

> The piece entitled *La Drama of the Tempest* begins by putting the image of a record down on drawing paper (art medium) with 10 a drawing pencil (art medium). I drew the grooves of the

disc on the paper paper for 18 minutes 35 seconds. Throughout this period, a microphone recorded the work sound, that is to say the sound of the pencil drawing the grooves on the paper. (It also includes the sound of the pencil which I sharpen from time to time). When the record has its grooves (its memory) drawn on the paper, the next step was to get the record made - from the recording - of the same size as the drawing, which is 30 cm diameter, and of the same length which is 18 minutes 35 seconds, one side only and in only one copy. But the word 'copy' is a misnomer for the record is so closely tied to its project, its memory, that it must be unique. *Pasteul couleur* n° ... *sur l'Histoire sonore de la Deuxième Guerre Mondiale* comprises 3 discs. I took 3 discs which were on the market entitled *History of the 2nd World War*. Then I took one of the discs, and I drew on this disc with a camouflage colour crayon. I recorded the noises from the crayon rubbing the grooves of the disc. I scribbled over the 2 discs with two different camouflage coloured crayons. With the sounds recorded from these efforts, I had 3 discs cut to 17 cm at 45 rpm in one copy only. On the record a voice is heard, saying: Pasteul couleur n° ... sur l'Histoire sonore de la Deuxième Guerre Mondiale, side 1 ... and the noise ...

Lastly, there is a class of record in conventional use, produced as musical documentation. Examples of the kind are the "concerts" of Tinguely, Agam, Bruynèl, Engels, Schnitzer, Radigue and Sacchi. Agam's *Transformes Musicales* consists of a sound-combination series from different musical instruments (flutes and percussion). Structural treatment, like that in Agam's visual works, is given to sound, first by elements, then by melodic lines and then by entry combinations which are the same and different; the various forms (1, 2, 3, 4 & 5) produce a series of complex compositions, with other instruments making an appearance such as violin[2], clarinet, piano, trombone. The resulting musical "surface" is highly articulated, with reflex motifs and tone color suggesting an audiochromatic pattern. Tinguely's *Sound of Sculpture* is less complex in structure and loser in musical terms; it records his sound-concert of fluid sculpture. The combined objects performance of Tinguely may be compared with Bruynèl's performance; his record *Signs* was made in collaboration with Van den Errenbeent and uses "wind instruments."

In chronological order, there are also the "concerts" - Schnitzer electronic and Engels vocal - with sounds from instruments of the artists' own invention. The *Concerto per un quadro di Adami* by Martin and Adami is different again, illustrating in sound and picture *University of Leipzig* by Valerio Adami. In tune with what is painted, this record presents the talk and sounds, in many moods, exchanged between Martin and Adami in front of the actual picture. The visual ambiguity of the picture is matched by ambiguous sound treatment; phrases, like the images are "recognizable," only by details and by timbres and words, which are often repeated as if to mark the stages of discovery in the meaning of the work's visual and aural illustration.

In conclusion, a few words about some more specifically musical records, by Éliane Radigue and Franca Sacchi. The Radigue record has various speeds between 16 and 78, so that according to duration the ear can explore various avenues through the music she makes. The piece can be taken as a single unit (a=b) at diverse speeds; or else as a summation (a+b) of diverse units at diverse speeds. This is a musical borrowing from mathematical logic, a procedure not shared in the Sacchi record. For she plays two piano pieces with dimensional freedom which takes its time from her own emotional responses. There is preparation and relaxation; Sacchi settles to a rhythm sound of her own. It projects her psychosomatic personality. She brings to the music an archetypal dimension of fundamental importance.

2 Translator's note: Italian text has *violones*, predecessor of double bass.

Record as Artwork, 1959–73

Filippo T. Marinetti, *La Battaglia di Adrianopoli*, 1926. La voce del Padrone, Milan.

Henri Nouveau, *Quatrième trio pour piano, violon et violoncelle,* 1926–1927. Paris, ISO copy, 33RPM LP.

Kurt Schwitters, *An Anna Blume / Die Sonate in Urlauten*, 1958. Lords Gallery, London, 100 copies, 33RPM LP.

Yves Klein, *Conférence à la Sorbonne*, 1959. CNAC, Paris, 33RPM LP.

Jean Dubuffet, *Esperienze musicali*, 1961. 6 records, Edizioni del Cavallino, Venice, 33RPM.
La Fleur de Barbe, 1961. Intersonor, Paris, 33RPM LP.
La Fleur de Barbe, 1961. De Luca, Rome, 33RPM LP.
6 records, Edizioni Galleria del Cavallino, Venice.

Jean Dubuffet, Asger Jorn, *Musique Phénoménale*, 1961. 4 records, Edizioni del Cavallino, Venice, 33RPM.

Yaacov Agam, *Trans-Formes Musicales*, 1962. Edition du Griffon, Neuchâtel, 45RPM.

Ben Vautier, *Mon Dieu ne l'abandonnez pas*, 1962. 45RPM (played at 78 RPM).

Karel Appel, *Musique Barbare*, 1963. The World's Window Barn, Holland, 33RPM LP.
Giant size $1.57 each (Dine, Brecht, Johns, Lichtenstein, Wesley, Watts, Wesselman, Warhol, Oldenburg, Rosenquist, Rauschenberg), 1963. Edited by Billy Klüver, in "Popular Image Exhibition," Washington Gallery of Modern Art, April/June 1963, 33RPM LP.

Jean Tinguely, *Sound of Sculpture*, 1963. Minami Gallery, Tokyo, 1000 copies, 45RPM.

Mimmo Rotella, *Poèmes Phonétiques* (1943–1963), 1965. OU 23–24, Paris, 33RPM LP.

François Dufrêne, *Crirythmes*, 1977. OU 23–24, Paris, 33RPM LP.

Raoul Hausmann, *Poèmes Phonétiques*, (1918–1946), 1968. OU 25–26, Paris, 48 copies, 45RPM.

Marcel Duchamp, *Some texts from "A L'infinitif"*, (1912–1920), 1967. *Aspen* 5, 33RPM.

Richard Huelsenbeck, Four poems from *"Phantastische Gebete"*, (1916), 1967. *Aspen* 5, 33RPM.

Andy Warhol, *Wedding of Lucy*, 1967. Velvet Underground, New York, 33RPM.

Bernar Venet, *The Infrared Polarization of the Infrared Star in Cygnus*, 1968. Letter Edge in Black Press Inc., New York, 33RPM.

Robert Whitman, *Sounds for 4 Cinema Pieces*, 1968. Museum of Contemporary Art, Chicago, 45RPM.

Bruce Nauman, *Untitled*, 1969. Rosa Esman pub, New York, 33RPM.

Art by Telephone (Arman, Armajani, Lee Byars, Cumming, Dallegret, Dibbets, Giorno, Grosvenor, Haacke, Hamilton, Huot, Higgins, Hompson, Jacquet, Kienholz, Kosuth, Levine, Lewitt, Morris, Nauman, Oldenburg, Serra, Oppenheim, Smithson, Uecker, Van Der Beek, Venet, Viner, Vostell, Wegman, Wiley), 1969. Museum of Contemporary Art, Chicago, 33RPM LP.

Jan Dibbets, *Afsluitdijk*, 1969. Museum Haus Lange, Krefeld, 33RPM.

Jean Dupuy, *Untitled*, 1969. New York, 33RPM.

Ton Bruynèl, *Signs*, 1969. Stedelijk Museum, Amsterdam, 33RPM LP.

Joseph Beuys, *Ja Ja Ja Ja Ja, Nee Nee Nee Nee Nee*, 1969. Mazzotta Editore, Milan, 100 copies, 33RPM LP.

Terry Fox, Joseph Beuys, *Isolation Unit*, 1970. Kunstakademie, Düsseldorf, 45RPM.

Richard Artschwager, *Untitled*, 1971. Hedendaagse Kunst. Sonsbeek '71, 45RPM.

Arthur Köpcke, *Music While You Work*, 1958–64. Edition René Block, Berlin, 30 copies, 33RPM.

Ton Bruynèl, *Sound Block*, 1971. Stedelijk Museum, Amsterdam, 33RPM.

Dick Raaijmakers, *Ideofoon II*, 1971. Stedelijk Museum, Amsterdam, 33RPM.

Raymond Hains, *Disque Bleu pour Saffa*, 1970. Galleria Blu, Milan, 250 copies, 45RPM.

Peter Struycken, *Image and Sound programme*, 1970, Stedelijk Museum, Amsterdam, 33RPM.

Sarkis, *Une lettre à déchiffrer*, 1971. Düsseldorf, 45RPM.
Le chien aboyant courait derrière "X", 1971. Düsseldorf, 45RPM.

Marina Abramović, *Suund Environment – Sea*, 1972. Belgrade, 45RPM.

Reiner Ruthenbeck, *Dachskulpturen*, 1972. Städtisches Museum Mönchengladbach, 45RPM.

Jim Dine, Rory McEven, *Songs, Poems, Prints* 1969. Museum of Modern Art, New York, 33RPM LP.

Conrad Schnitzler, *Electric Meditation*, 1970. Edition René Block, Berlin, 33RPM LP.

Pieter Engels, *Fabulous Oldest Hits*, 1972, Engels Third Institute, Amsterdam, 45RPM.

Antonio Dias, *The Space Between*, 1971. New York, Milan, 33RPM LP.

Robin MacKenzie, *A Sound Work to Accompany*, 1972. Carmen Lamanna Gallery, Toronto, 33RPM LP.

Valerio Adami, Henry Martin, *Concerto per un quadro di Adami*, 1972. Studio Marconi, Milan, 33RPM LP.

Christian Boltanski, *Reconstruction de chansons qui ont été chantées à Christian Boltanski,* (1944–1946), 1972. Paris, 45RPM.

Lawrence Weiner, *A () C Sharpened Carried Done Again Perhaps ... A C ()*, 1972. Yvon Lambert Gallery, Paris, 33RPM.

Éliane Radigue, *Σ=a=b=a+b*, 1969–72. Yvon Lambert Gallery, Paris, 16–33–45–78RPM.

Franca Sacchi, *Ho sempre desiderato avere un cane, un gatto ed un cavallo – Ora ho un gatto e un cavallo, mi manca soltanto il cane*, 1973. Toselli editore, Milan, 33RPM LP.

F. De Filippi, *Narciso*, 1973. Rome, 33RPM.

Sandro Chia, *Pietra*, 1973. Rome, 33RPM.

Sarkis, *La Drama of the Tempest*, 1974. Paris.
Pasteul Couleur n° . . . sur l'Histoire Sonore de la 2e Guerre Mondiale, 1973. Paris, 3 records.

Giacomo Notargiacomo, *Variazioni del fonema /r/*, 1973. Rome, 33RPM.
/ay/ /ayk/ /layk/, 1973. Rome, 33RPM.
Affermazioni, 1973. Rome, 33RPM.

Lawrence Weiner, *Having Been Done At / Having Been Done To*, 1973. Sperone-Fischer, Rome, 33RPM.

Jean Dubuffet, *Musical Experiences*, 1973. Atlantic Recording, New York, 33RPM.

Ian Murray, The Top Song, 1973. 250 copies, 33RPM LP.

Angelo Bozolla, *Ticking Clock*, tape, 1972.

Achille Bonito Oliva

Archisculpture of Sound

Editor's note:

The exhibition *forte piano: the shapes of sound,* curated by Achille Bonito Oliva presented the original research of international artists from different generations, who have focused, and continue to focus on the dematerialization of the work of art, attaining unprecedented sound-forms.

The exhibition took place at Parco della Musica Auditorium in Rome and was divided into two phases (Part One: May 2–July 10, 2012, Part Two: July 19–October 31, 2012), each presenting forty works spread across the interior and exterior of the Parco della Musica Auditorium. By this means, the exhibition's highpoint was the acoustic quest for an invisible work, which in turn will intercept the possibly unexpected attention of the viewer.

The presented sound works introduced a new relationship with the spectator that went beyond the inevitable frontality of things. Through the vaporizing of all visual forms the artists were reversing Paul Klee's claim that "Art makes the invisible visible." The outcome of this operation was the development of a new kind of contemplation open to multisensory experiences and the integration within the architecture of places.

This text was first published in Italian in the brochure of the exhibition *forte piano: the shapes of sound*. Translated from the Italian by Robert Booth.

The long-lasting character of the artistic form implies a sense of construction that does not want to be cancelled out by the next work, nor metaphorically duplicate the outward reality. Art is the ambivalent movement that plays on presence and absence, contact and separation, heroism and detachment, attitudes that are not conflicting but complimentary, that affirm the diverse rapport with reality.

"There is no better way to escape the world than through art, and one cannot relate better than through art" (Goethe). This attitude designates the position of artists who work on the concept of total art, the philosophical condition that initially seems to demand distance from things but then adopts them to build a formal order that designates the new reality even if through its contamination. The chosen formal order does not play on pure accumulation, on the presentation of an accidental assemblage that metaphorically duplicates the quantitative dissemination of the exterior world. The work assumes the phenomenological cadence contradicted by a metaphysical bodily order that sustains it. Not the sense of accumulation but that of necessary interpenetration, despite occupation by unforeseen relations.

The stylish unpredictability of the work is underlined by the consistence of materials, sometimes by the desired impracticality of the approaches, and also by the elaboration of links that are pushed to the limits of insertion and structure.

The object is not a highlighted pure mass but the element of a composition that challenges the opaque fullness in favor of an outcome where everything bends to a new formal order. Naturally, this order does not give the single elements a sublimation of specific depth, but plays on converting the individual meanings into a unique and transfigured significance. From here the connotation of involuntary metaphysics, the power of abstraction of these works undoubtedly aimed at the foundation of a new *order of sound beauty*.

The order of beauty that presides over the archisculpture of sound originates from the condensation of the work in its temporality, remains of a cultural attitude that assumes all the responsibility of a typically European tradition that intertwines with its Oriental counterpart.

In this sense the post-modern memory loses its hedonistic trait of simple quotations, to take on the complexity of a historical memory not known for any levity and desire to get rid of itself and its past. Thus the work of art becomes the deterrent of objective resistance that the artist perpetrates with regard to his own present, in favor of his own present, insomuch as capable of producing continuity in a scenario that appears used to the interchangeability of subject and object. The subject, in this case the artist, is he who has the awareness to establish a method through the phenomenology of the said work, an attitude of this mental state capable of safeguarding his own specificity also by passing through the object's opacity.

An object of complexity resides within the works of sound artists made prominent by the evident consistency of the various elements and by the tangible style order that does not just keep them together but weaves them together in an enduring composition that evaporates in space.

Contamination is not simply the pure coincidence in a combinatorial game, but rather the effect of a need of complexities capable of avoiding every simplification. The work is not supported by a simple structure but by a process of abstraction that is not preventive but rather contextual to the physical obviousness of the work.

Since the object cannot be eliminated, to become the effect of a guise, it is more inclined towards a fantastic use that on one side recognizes the tautological presence while on the other it challenges the apparent passivity of use.

From this comes the occasional monumental ironic use, its transfer to a different scale, its inverted use between closed and opened, its decontextualized positioning. There are no privileged objects, as regards recurrences or affectivity, but rather states of necessity that determine the use according to the results to be reached.

The form tends rather to combine the elements in order to push for a double process of awareness, a specific regarding its work and the other more general one regarding its relations with the world.

The new relationship is not established through the simple difference produced by the unknown approach, but rather by the foundation of a formal method rendered even more evident by the affinity of the materials with the outside world. This affinity avoids the easy surprise and consequent amazement, becoming an ulterior challenge for the artist who leaves and re-enters the game at the same time with the things through their adoption that better allows the verification of the new formal order.

Art does not mean putting the world in order, but it does suggest methods of aggregation capable of developing processes of awareness, both internal and external, and interior and exterior. The work is not the fruit of rigidity and paralysis that dry all sedimentations, but rather the practice of appropriation beyond the principal of possession, the basis of a method capable of also adding method to the artist's life.

This does not mean to regressively force the artist within the logic of poetry, within the work's metalinguistic coherence, but rather establish the lightness of being within the evident consistency of the work and through it.

This work is not achieved through accumulation, object even if formalized beside other daily objects. The metaphysical order that without rigidity governs the work assumes the task of establishing a boundary between two universes that live together but at the same time are governed differently.

The method of sound artists does not exalt space in the pure evocation of time but accepts the hindrance of the object in its dense physicality, evaluating its approach to other physicalities certainly not similar for use and destination.

Since archisculpture, moving within a panorama less structured and homogeneous than goods, does not reconfirm the *use value* of the daily object, its being or having been of possible usage, rather the value of the fantastic use that the artist may obtain through the procedure of contamination.

The artist is he who adopts a sort of radioactive process to contaminate and to adopt the daily subject, already per se poisoned by its historical affinity. Another more transfigured story has assailed the daily object, does not redeem like the Duchampian gesture, being as it is impossible to reproduce culturally. Now the object conserves its viscosity and often its decorum, but is bent and adopted for a transversal use that only the artist is capable of doing.

But sound artists do not think of redeeming history, but rather trigger formal processes capable of producing in the present resistant attitudes rendered evident and conspicuous by the work's constructive objectivity. The work assumes the construction of the machine, its internal operation made up of single elements governed by the idea of complexity and unity at the same time.

To run complexity through the construction of a linguistic method now becomes the assumption: not to pathetically duplicate the complexity of the technological universe, but rather the relationship between the individual and society, between the single and the normative apparatus.

A behavioral response through the production of an open and enduring form.

forte piano: the shapes of sound

Part One. May 2–July 10, 2012 at Parco della Musica Auditorium Rome

Participating Artists

Vito Acconci, *Halley II Research Station: First Impression and the Beginning of a Conceptual Approach,* Antarctica, 2004, 6:28 min.

Massimo Bartolini, *Studio Table Moving*, 2012, 3:45 min.

Riccardo Benassi, *Quel che serve ad un pavimento per trasformarsi in pista da ballo (noi),* 2011, 5 min of sound every 5 min of silence (loop).

John Cage, Nam June Paik, *Cage in cage*, 1989.

Maria Thereza Alvez, Jimmie Durham, *Grunewald*, 2010, 13:35 min.

Jan Fabre, *To Bring an End to Royal Judgment and the Judgment of God,* 2009. Live-recorded reading related to an original performance by Jan Fabre at Zerynthia, Rome on October 29, 2009, 31:58 min. From: Jan Fabre, *TO BRING AN END TO ROYAL JUDGEMENT AND THE JUDGEMENT OF GOD (1978-79).* Translation: Gregory Ball.

William Furlong, *Uhms & Aghs*, 1989, 3:07 min.

Al Hansen, Jerry Mulligan, *Headline Poems,* 1994, 13 min.

Douglas Henderson, *On the Road Home,* 2007, 15:53 min.

Jacob Kirkegaard, *Hears of the Other,* 2010, 12 min.

La Monte Young, *The Well-Tuned Piano*, 1987, 30:17 min.

H. H. Lim, *Daily World Music 1.03.2012*, 2012, 9:54 min.

Arrigo Lora Totino, *Fanfara*, 1977, 1:31 min.

Alva Noto, *kern.rotar*, 1998, 5:47 min.

Luca Maria Patella, *L'arrivée du Train à la Gare de l'Artculture (De Mente)*, 1969–2012.

Cesare Pietroiusti, *Bellezza*, 2009. Extract from *Pensiero unico*, 2003, 1:47 min.

Roberto Pugliese, *Anamorphosis*, 2011, 2:37 min.

Steve Roden, *Proximities*, 2010. Recorded in various spaces of the Chinati Foundation, 2010. All sounds generated and organized by Steve Roden, 42:20 min.

Staalplaat Soundsystem, *Yokomono Pro*, 2009, 10:49 min.

Martin Daske, *Unwritten*, 2010. 8 acousmatic miniatures, 40 sec. each.

Michael J. Schumacher, *The Episodes*, 21:26 min.

Other participating artists:

Roberto Benigni, Vettor Pisani, Martux_M, Steve Piccolo, Donatella Spaziani, Peter Schuyff, Walter Marchetti, Gino De Dominicis, Nanni Balestrini, Alberto Tadiello, Emilio Fantin, Marzia Migliora, Liliana Moro, Marc Kostabi and Ornette Coleman, Albert Pinya, Maurizio Nannucci, Cesare Viel, Donatella Landi, canecapavolto

forte piano: the shapes of sound

Part Two. July 19–October 31, 2012 at Parco della Musica Auditorium Rome

Participating Artists

Serge Baghdassarians, Boris Baltschun, *Audioguide*, 2009–2010, 44:38 min.

Gianfranco Baruchello, *Comandi militari e istruzioni per armi inesistenti*, 1965. Tape recording (original) transferred to audio CD (2012), 4:15 min.

Joseph Beuys, Nam June Paik, *Coyote III*, 1984, 58:35 min.

Bianco – Valente, *8106Khz*, 2012, 1:16 min.

Marcel Broodthaers, *Interview with a Cat*, 1970. This is an interview recorded at the Musée d'Art Moderne Département des Aigles, 12 Burgplatz, Düsseldorf, 4:55 min.

Giuseppe Chiari, *Gesti sul piano*, 1965, 20:25 min.

Henri Chopin, *Sauvagerie*, 1987, 10:31 min.

Philip Corner, *Delicate Computations (gamelan)*, 1982, 47:01 min.

Alvin Curran, *Maritime* Rites, 2012, 4:33 min.

Lara Favaretto, *Doing*, 1998. Recorded live sound, 54 min.

Sub-Exhibition *Revolutions Per Minute (The Art Record). A record album of twenty-one original works*, 1982. Concept by Jeff Gordon, Ronald Feldman.

- Jud Fine, *Polynesian/Polyhedron*, 3:17 min.
- Eleanor Antin, *Antinova Remembers*, 4:17 min.
- Terry Fox, *Internal Sound*, 4:26 min.
- Margaret Harrison, *First Lines*, 2:50 min.
- Les Levine, *Would Say No To Some Help*, 4:15 min.
- Hannah Wilke, *Stand Up*, 3:27 min.
- Douglas Davis, *How To Make Love To A Sound*, 4:30 min.
- Komas & Melamid, *Russian Language Lesson*, 3:25 min.
- Helen and Newton Harrison, *A Memoriam to John Isaacs*, 2:34 min.
- Vincenzo Agnetti, *Pieces of Sound*, 4:36 min.
- Chris Burden, *The Atomic Alphabet*, 35 sec.
- Piotr Kowalski & William Burroughs, *You Only Call The Old Doctor Once*, 4:50 min.
- Ida Applebroog, Really, *Is That A Fact*, 3:12 min.
- Edwin Schlossberg, *Vibrations/Methapors*, 5:15 min.
- SITE, *Comments On SITE*, 2:41 min.
- R. Buckminster Fuller, *Critical Path*, 2:57 min.
- Thomas Shannon, *Smashing Beauty*, 4:01 min.
- Conrad Atkinson, *The Louis XIV Deterrent*, 5:36 min.
- David Smyth, *Typewriter in D*, 3:39 min.
- Todd Siler, *Think Twice*, 3:21 min.
- Joseph Beuys, *Excerpt From Cooper Union Dialogue*, 4:15 min.

Francesco Fonassi, *32 rpm (falèna),* 2012, 60 min.

Milan Knížák, *Multisong*, 1992, 8 min.

Christina Kubisch, *Cloud,* 2012, 17:55 min.

Brandon LaBelle, *Preparations for common recognition #1,* 2012, 34:58 min.

Brandon LaBelle, *Preparations for common recognition #2,* 2012, 34:21 min.

Via Lewandowsky, *Intergalactica*, 2011–2012, 8 min.

Robin Minard, *Music for Quiet Spaces*, 1984, 15:46 min.

Phill Niblock, *Hurdy Hurry. Jim O'Rourke, Hurdy Gurdy,* 1999, 15:30 min.

Yoko Ono, *Birdsong (Hiroshima version),* 2011, 5 min.

Benjamin Patterson, *Paper Piece*, 1960, 17:05 min.

Michelangelo Pistoletto, *Chi sei tu?,* 1976, 22 min.

Mimmo Rotella, *America*, 1959, *Suddenly Last Summer*, 1960, *Saint Germain des Pres*, 1964. Sound poems, from the *Epistaltic Manifesto*.

Samon Takahashi, *Sillons Impurs*, 2007, 9:19 min (versione mono, 2012).

Mario Verandi, *Noise Networks,* 2010, 16:30 min.

Stephen Vitiello, *Owls, Drips and Frogs (for disquiet)*, 2012, 3:31 min.

Other participating artists:

Karl Holmqvist, Betta Benassi, Lucia Ronchetti, Nico Vascellari, Mario Schifano, Luca Vitone, Stefano Trevisi, ZimmerFrei, Agostino Di Scipio, Daniele Lombardi, Mario Airò and Enrico Serotti, Alberto Garutti, Giuliano Lombardo, Roberto Paci Dalò, Boris Hegenbart, Alfredo Pirri

The Exhibition *Sound Art* at ZKM | Karlsruhe

Peter Weibel

ZKM | Center for Art and Media Karlsruhe
March 17, 2012–January 6, 2013

The exhibition *Sound Art. Sound as a Medium of Art* presented for the first time in an encyclopedic intention the development of sound art in the twentieth and twenty-first centuries at the ZKM | Karlsruhe and in public. From Futurism to Fluxus, through to Twitter sonifications, the ZKM charts the history of sound art during the twentieth century. However, the focus of the exhibition was placed on contemporary practices: with works from ninety artists the visitor gained insights into the unique sound cosmos of contemporary art. The sound world visualized its own exhibition architecture, and the exhibition visitors became the generators of sounds.

Visual experience dominates in numerous exhibitions. *Sound Art. Sound as a Medium of Art* emphasized auditory experience and transformed the visual experience. The visitors were thus provided with the opportunity to become acquainted with an entirely new sound cosmos, which neither radio, film, nor the music industry has been able to establish to such an extent.

The exhibition made new sound perceptions not only experienceable in the museum: passers-by could encounter sounds in the three installations located in the forecourt of the ZKM and five installations in public areas around the city of Karlsruhe. In addition, a selected concert program with outstanding projects enriched the exhibition: La Monte Young, Iannis Xenakis, John Cage, and Ryoji Ikeda are representative of the program's broad spectrum. The exhibition's wealth of sounds has also been facilitated by the richness of the archives made accessible to the Karlsruhe public for the first time to this extent. Included are the "Unheard avant-garde" from Scandinavia, the Broken Music Archive from Berlin and curated audiopoints from European archive inventories. After an insight into the exhibition, the artists of the exhibition will be presented in alphabetical order.

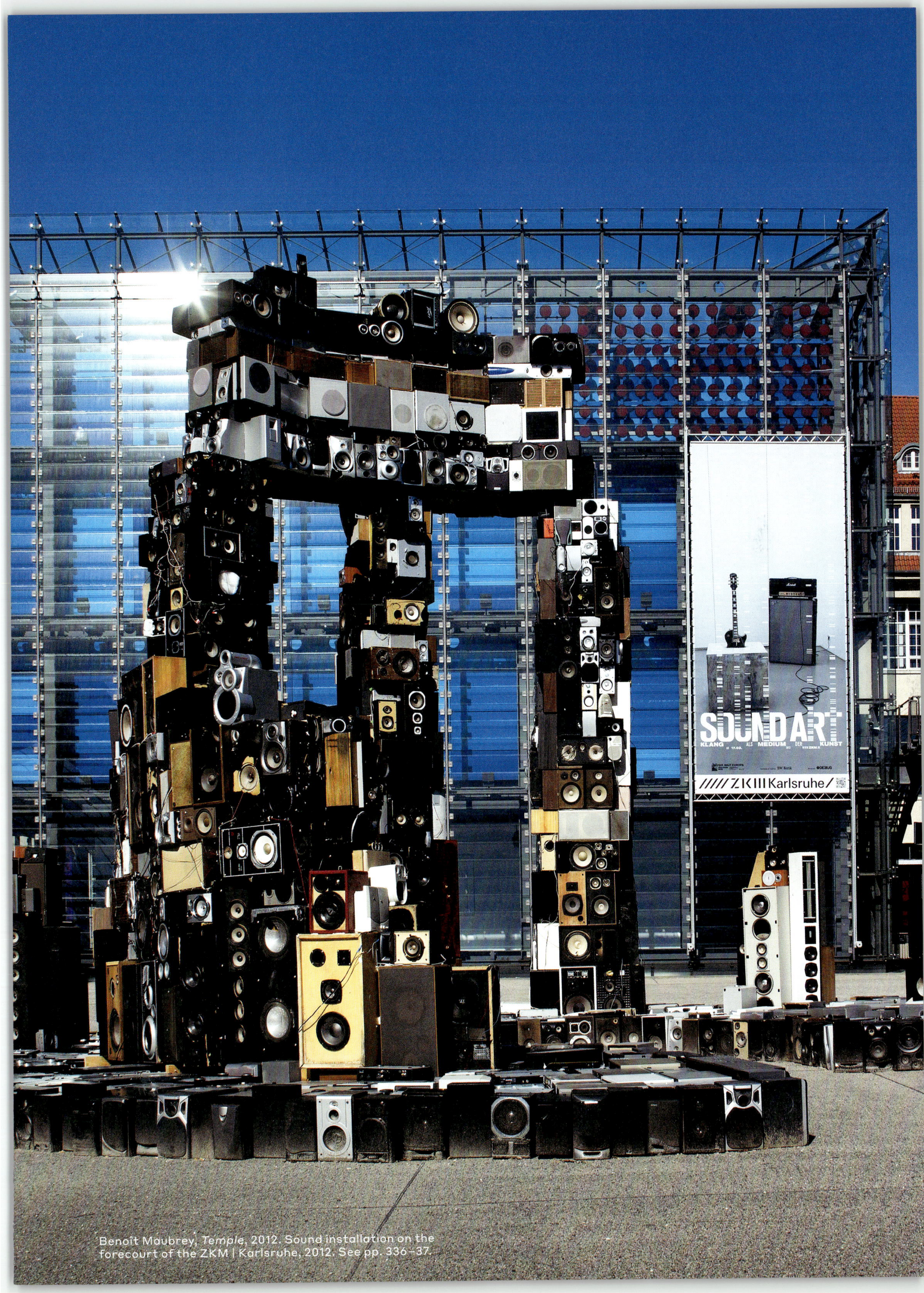

Benoît Maubrey, *Temple*, 2012. Sound installation on the forecourt of the ZKM | Karlsruhe, 2012. See pp. 336–37.

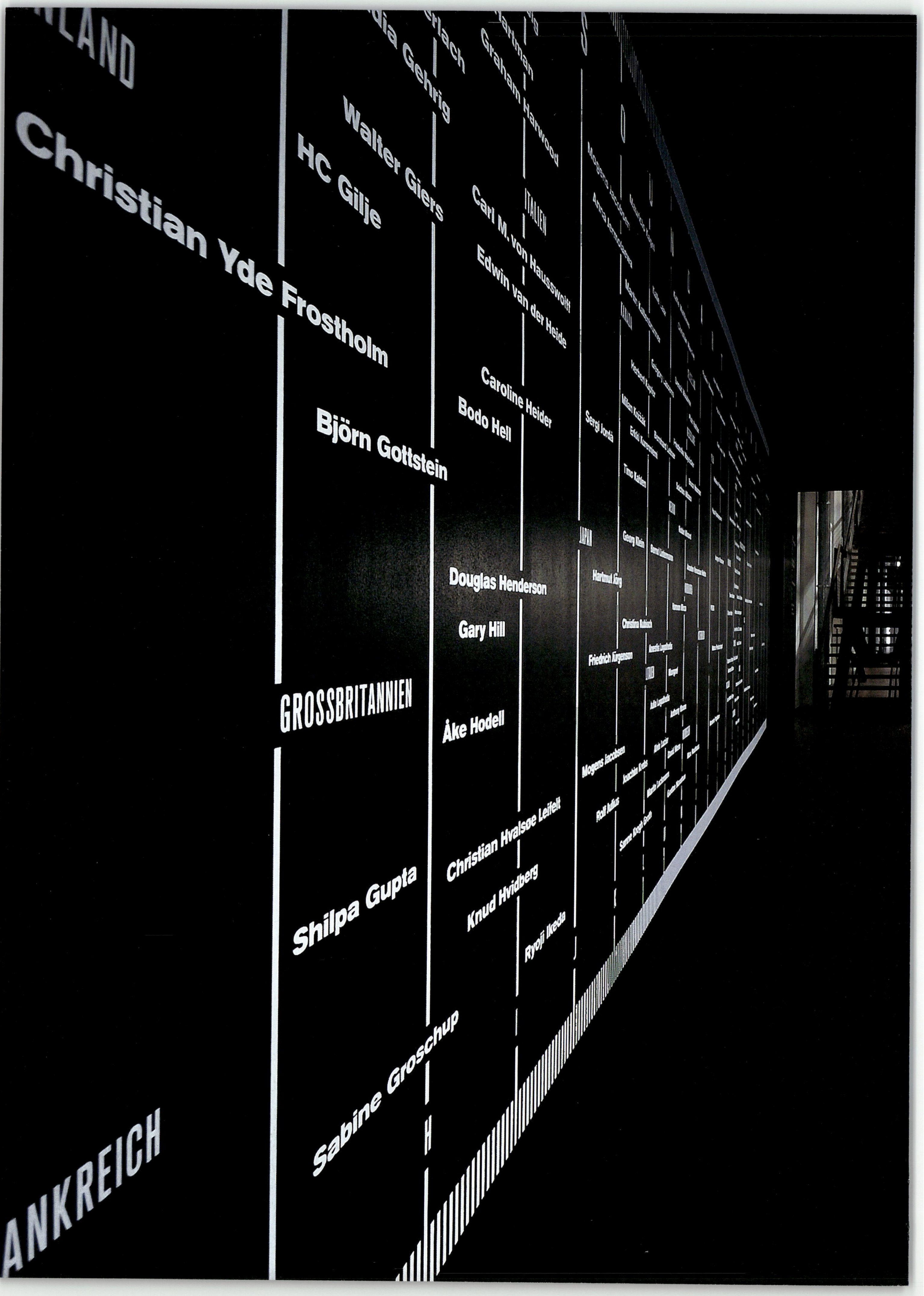
Christian Yde Frostholm
HC Gilje
Walter Giers
Carl M. von Hausswolff
Edwin van der Heide
ITALIEN
Caroline Heider
Bodo Hell
Björn Gottstein
Douglas Henderson
Gary Hill
GROSSBRITANNIEN
Åke Hodell
Christian Hvalsøe Leifelt
Knud Hvidberg
Shilpa Gupta
Ryoji Ikeda
Sabine Groschup
ANKREICH
JAPAN
Sergi Jordà
Hartmut Jörg
Friedrich Jürgenson
Mogens Jacobsen

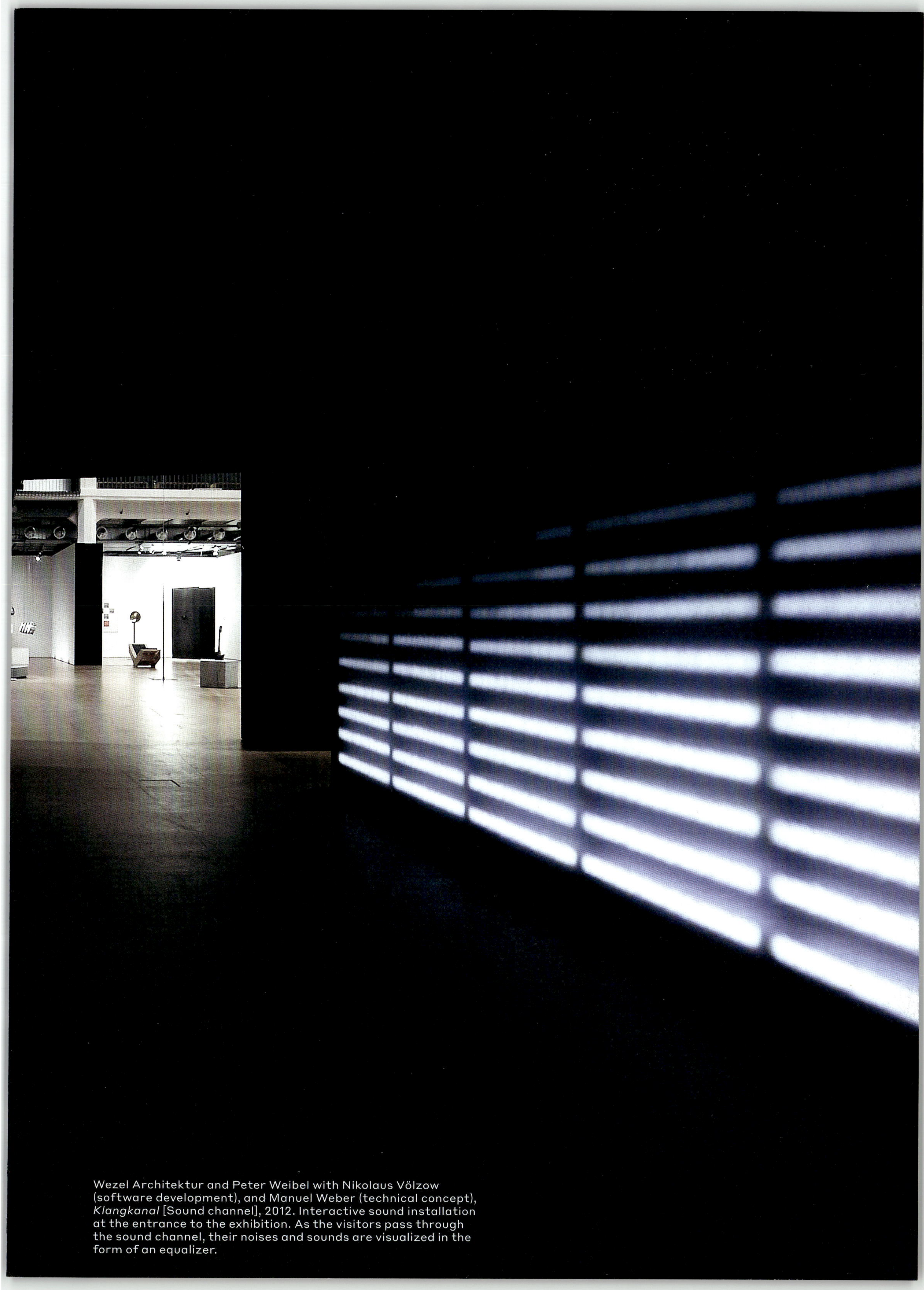

Wezel Architektur and Peter Weibel with Nikolaus Völzow (software development), and Manuel Weber (technical concept), *Klangkanal* [Sound channel], 2012. Interactive sound installation at the entrance to the exhibition. As the visitors pass through the sound channel, their noises and sounds are visualized in the form of an equalizer.

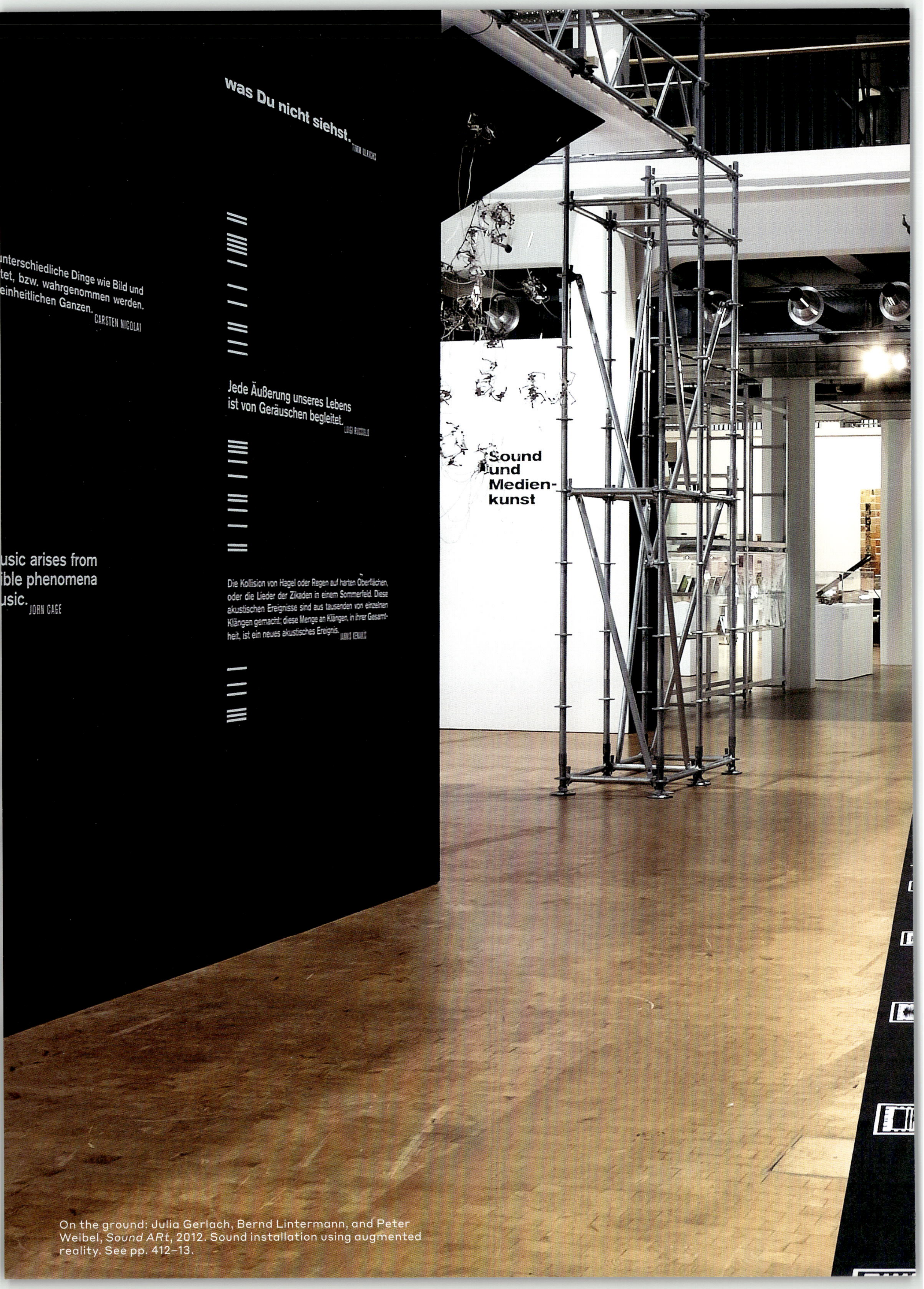

On the ground: Julia Gerlach, Bernd Lintermann, and Peter Weibel, *Sound ARt*, 2012. Sound installation using augmented reality. See pp. 412–13.

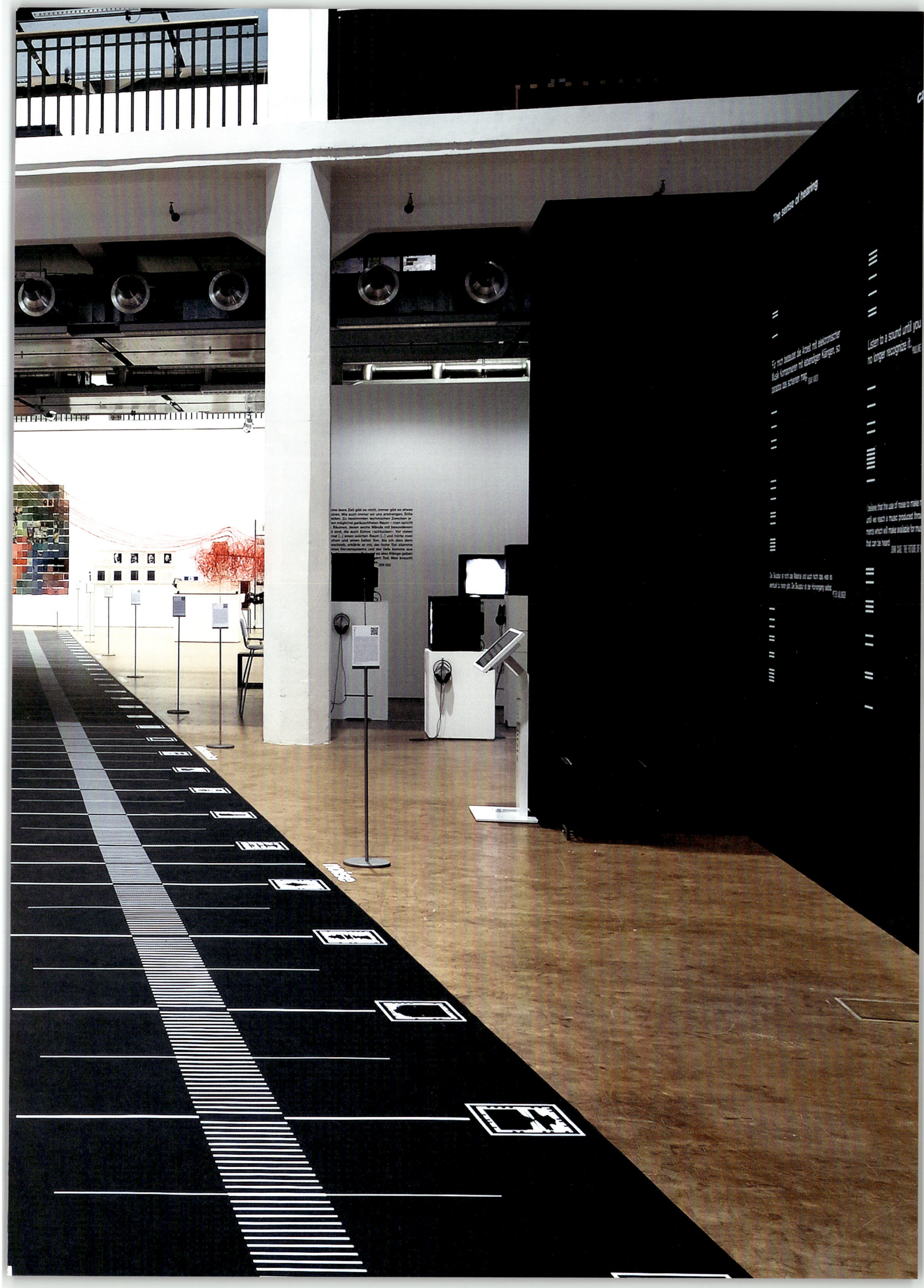
The sense of hearing
Listen to a sound until you no longer recognize it.

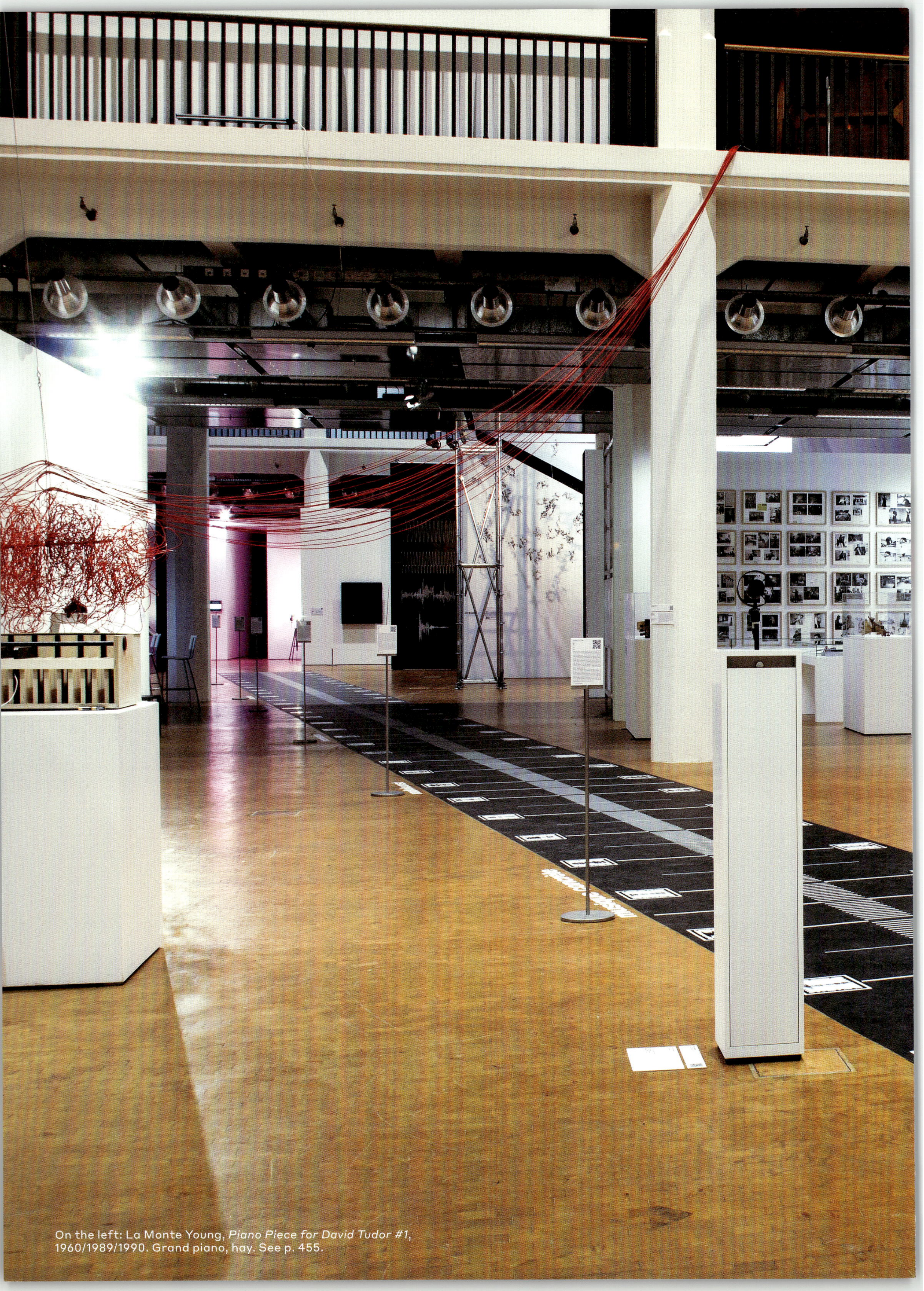

On the left: La Monte Young, *Piano Piece for David Tudor #1*, 1960/1989/1990. Grand piano, hay. See p. 455.

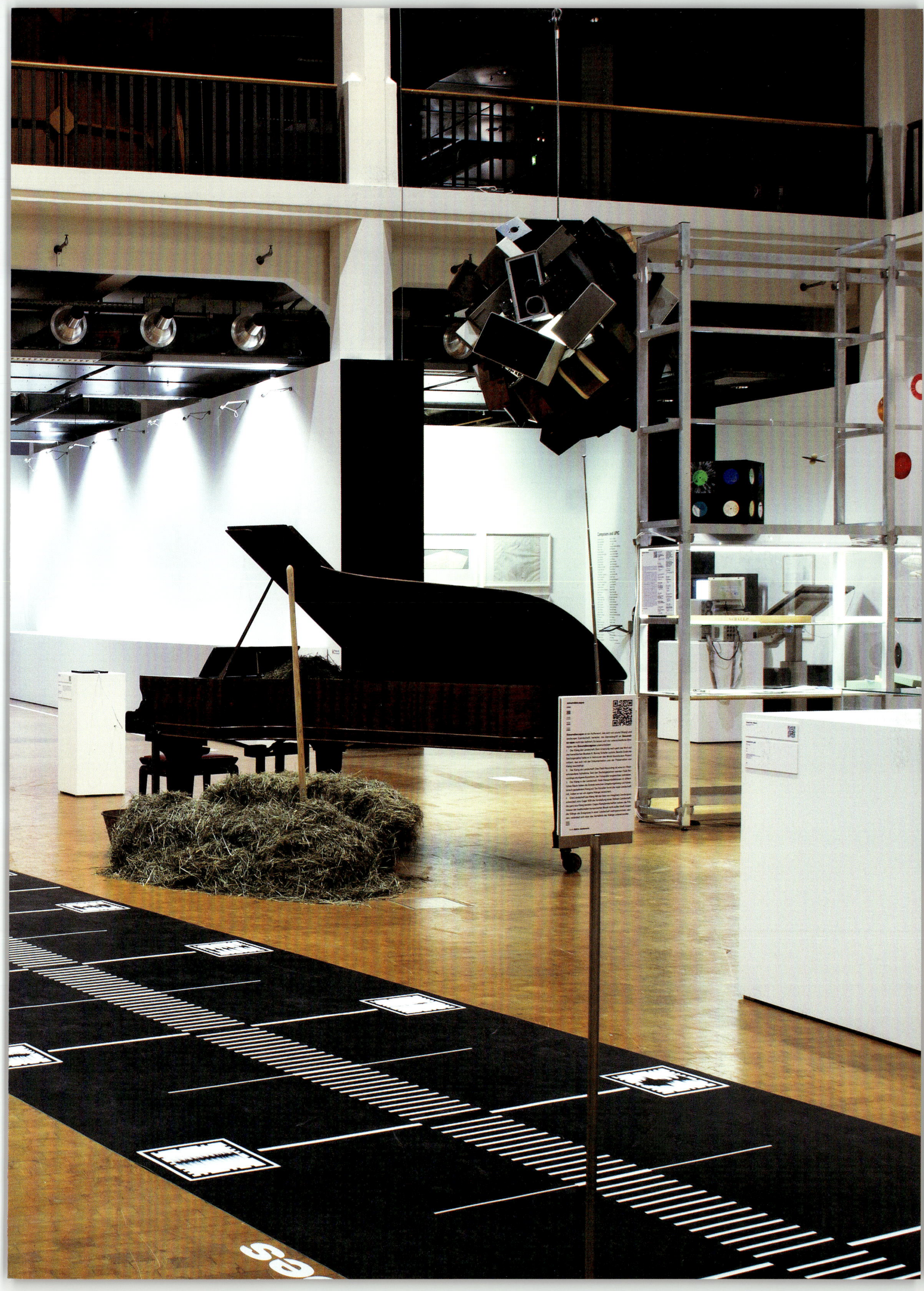

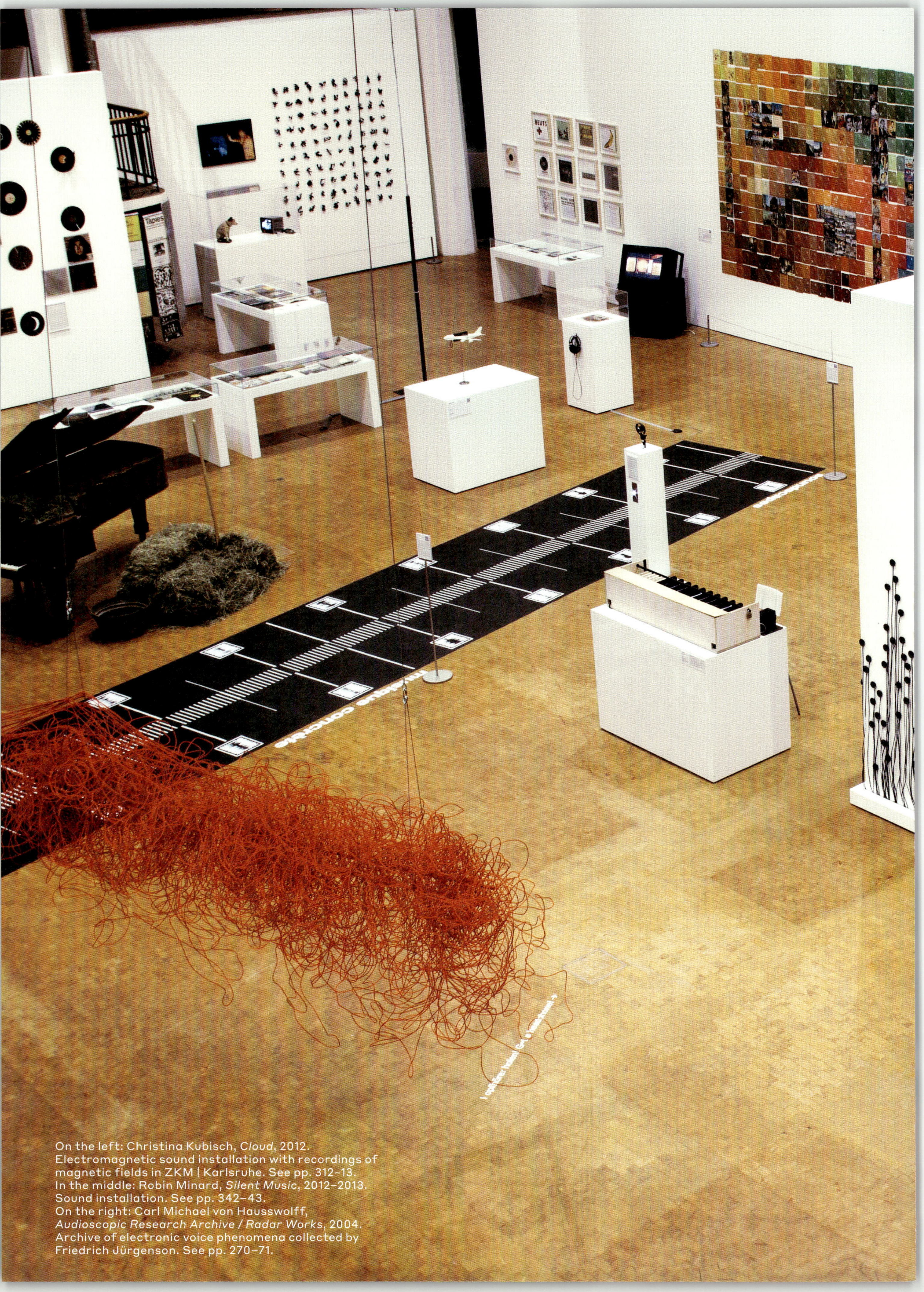

On the left: Christina Kubisch, *Cloud*, 2012.
Electromagnetic sound installation with recordings of magnetic fields in ZKM | Karlsruhe. See pp. 312–13.
In the middle: Robin Minard, *Silent Music*, 2012–2013.
Sound installation. See pp. 342–43.
On the right: Carl Michael von Hausswolff, *Audioscopic Research Archive / Radar Works*, 2004.
Archive of electronic voice phenomena collected by Friedrich Jürgenson. See pp. 270–71.

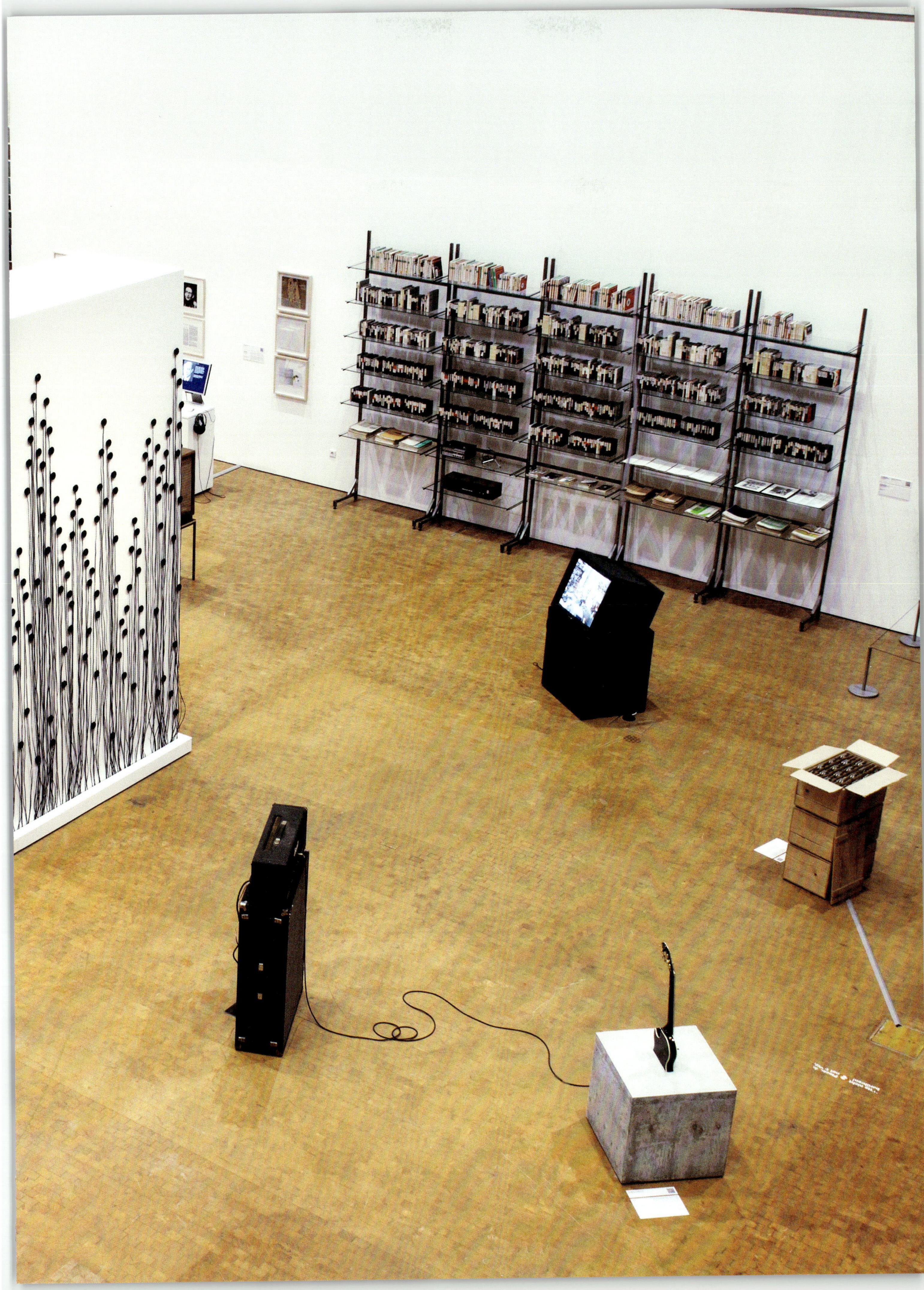

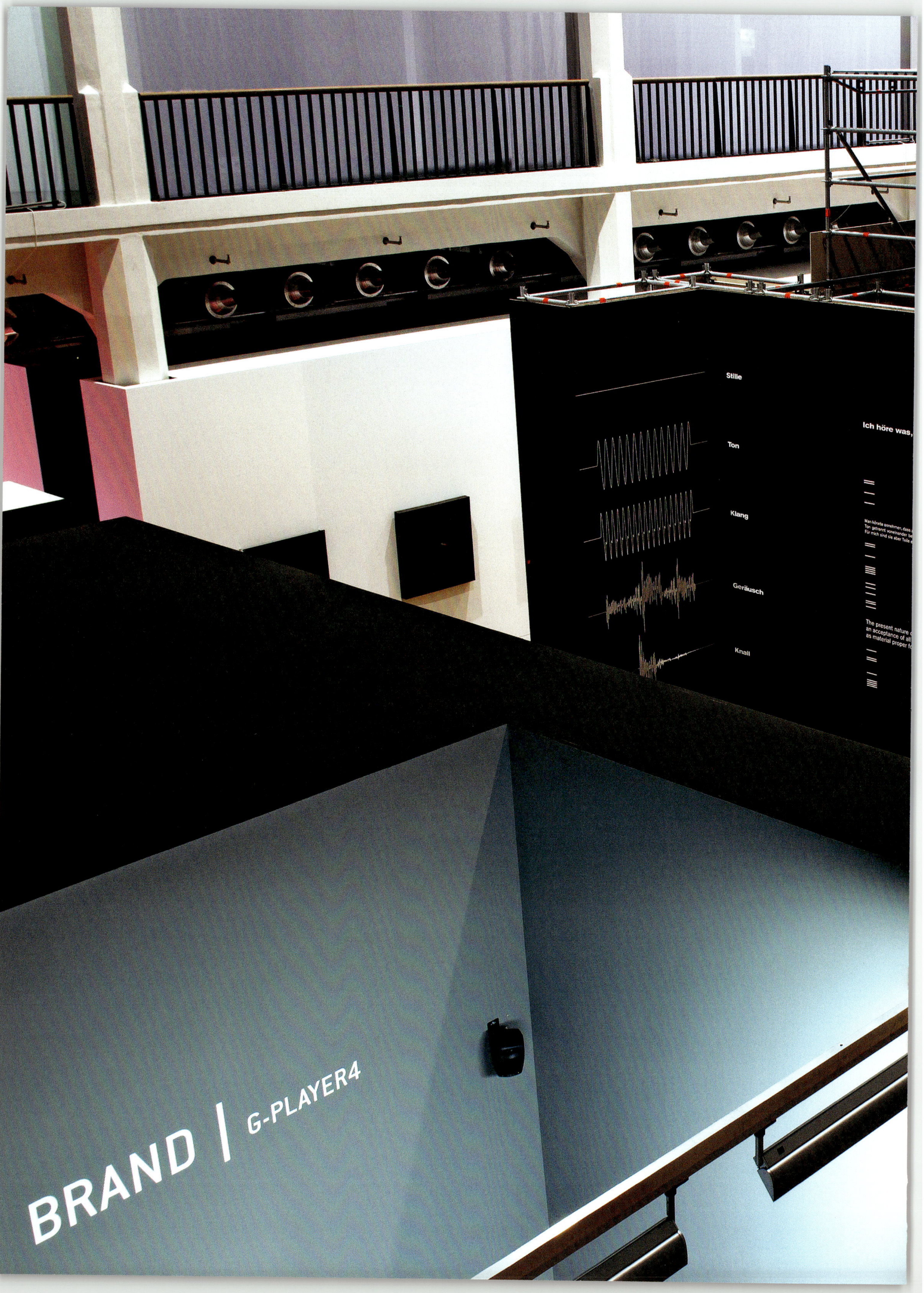
Stille
Ton
Klang
Geräusch
Knall
Ich höre was,
BRAND | G-PLAYER4

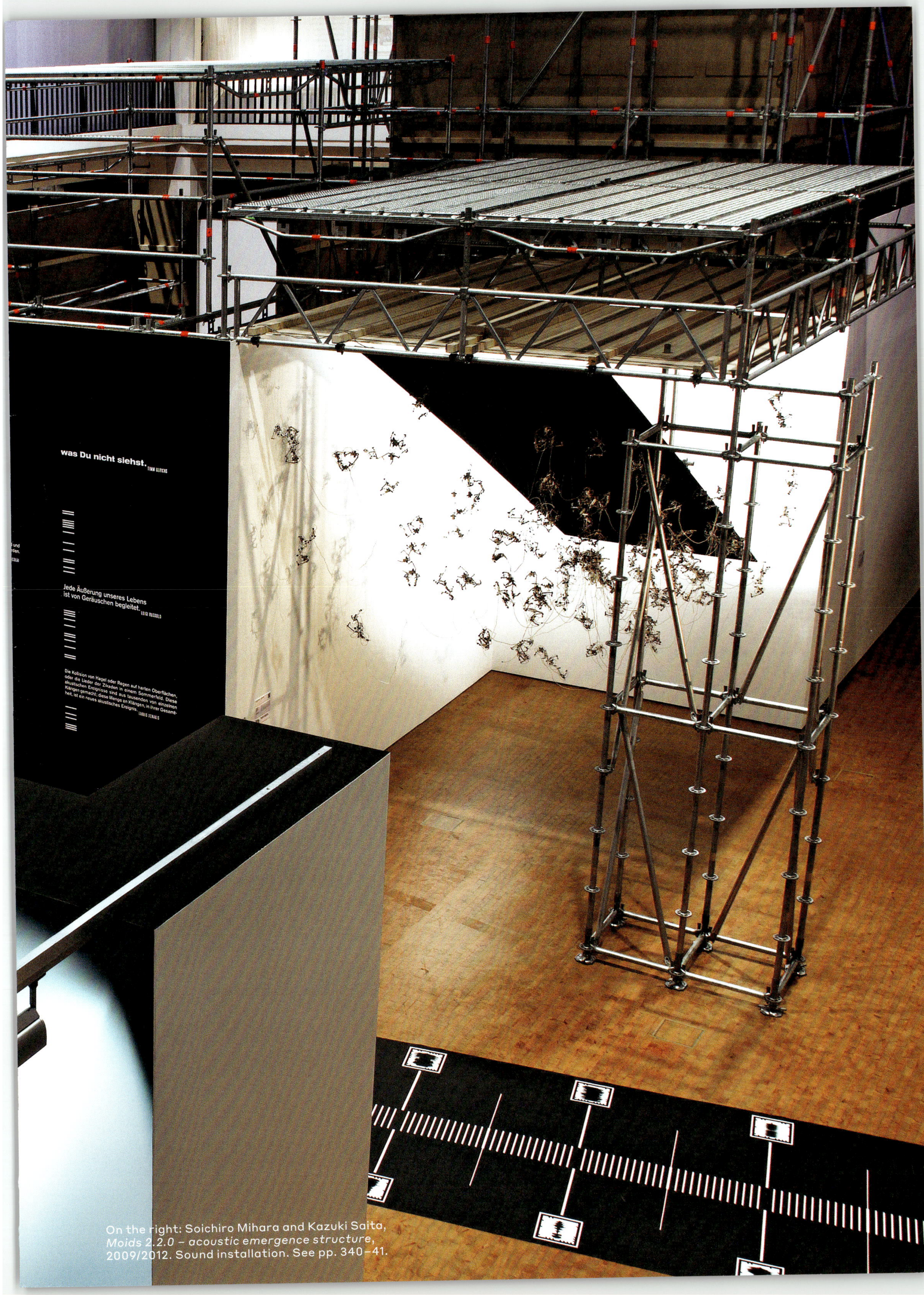

On the right: Soichiro Mihara and Kazuki Saita, *Moids 2.2.0 – acoustic emergence structure*, 2009/2012. Sound installation. See pp. 340–41.

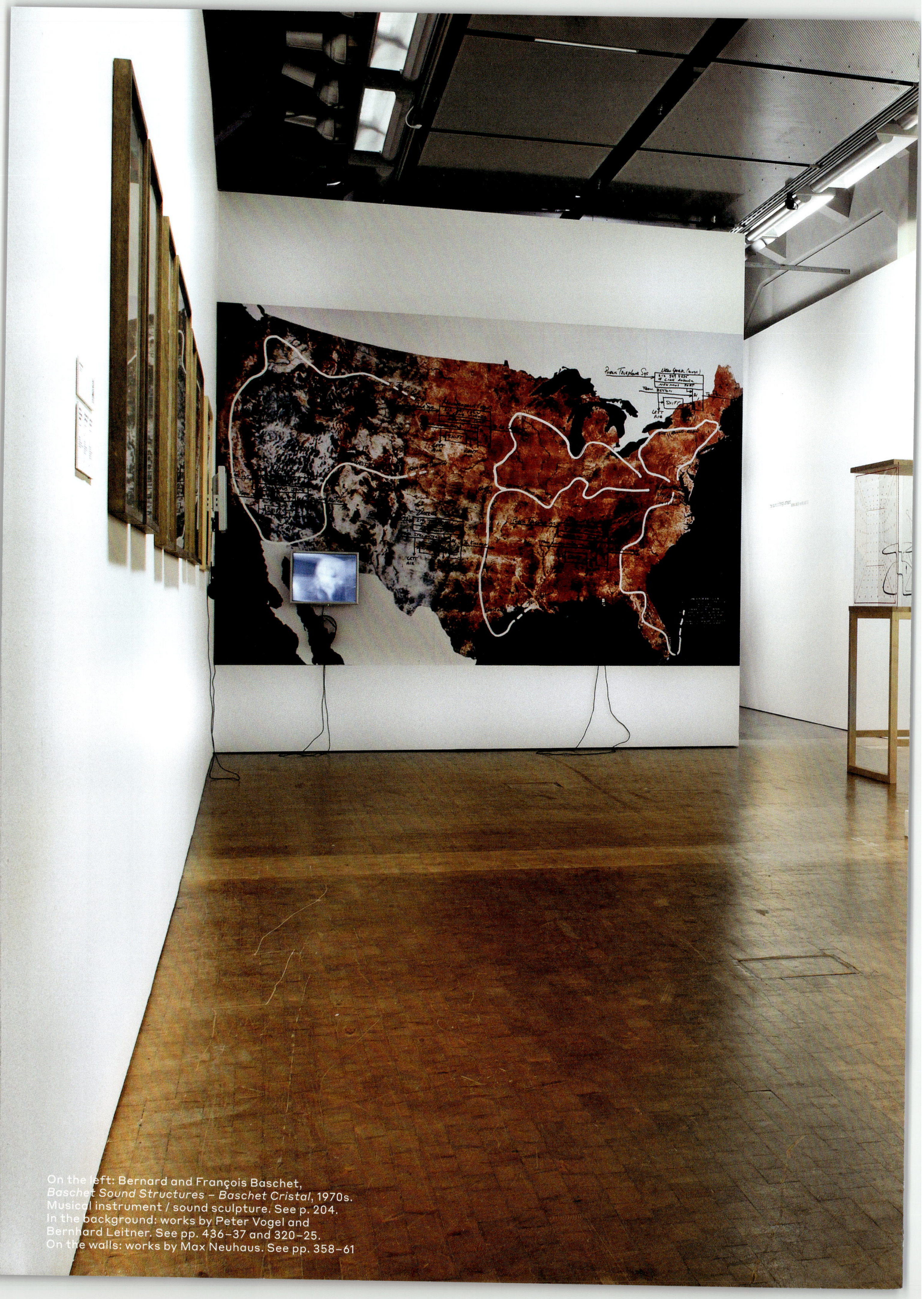

On the left: Bernard and François Baschet, *Baschet Sound Structures – Baschet Cristal*, 1970s. Musical instrument / sound sculpture. See p. 204.
In the background: works by Peter Vogel and Bernhard Leitner. See pp. 436–37 and 320–25.
On the walls: works by Max Neuhaus. See pp. 358–61

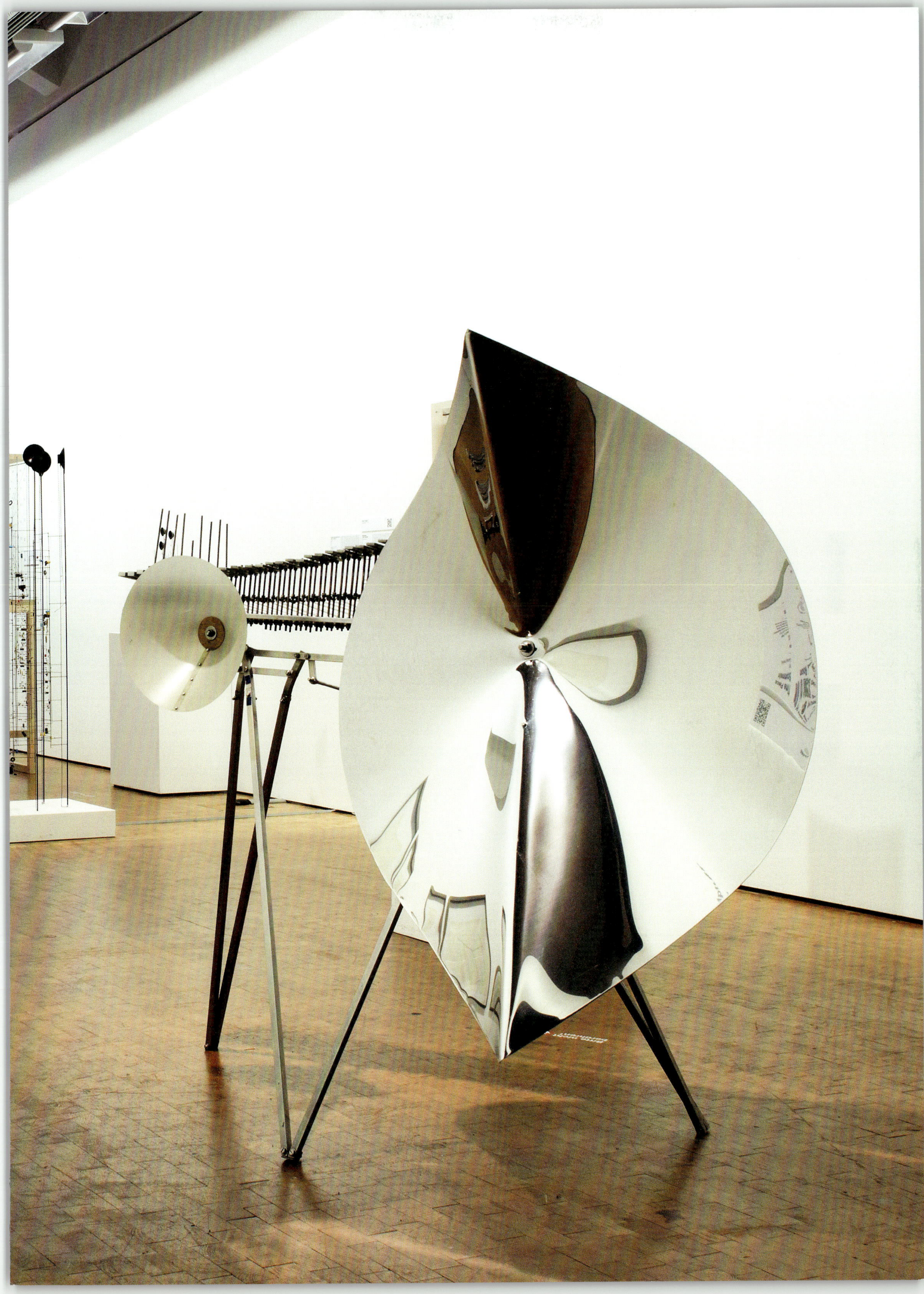

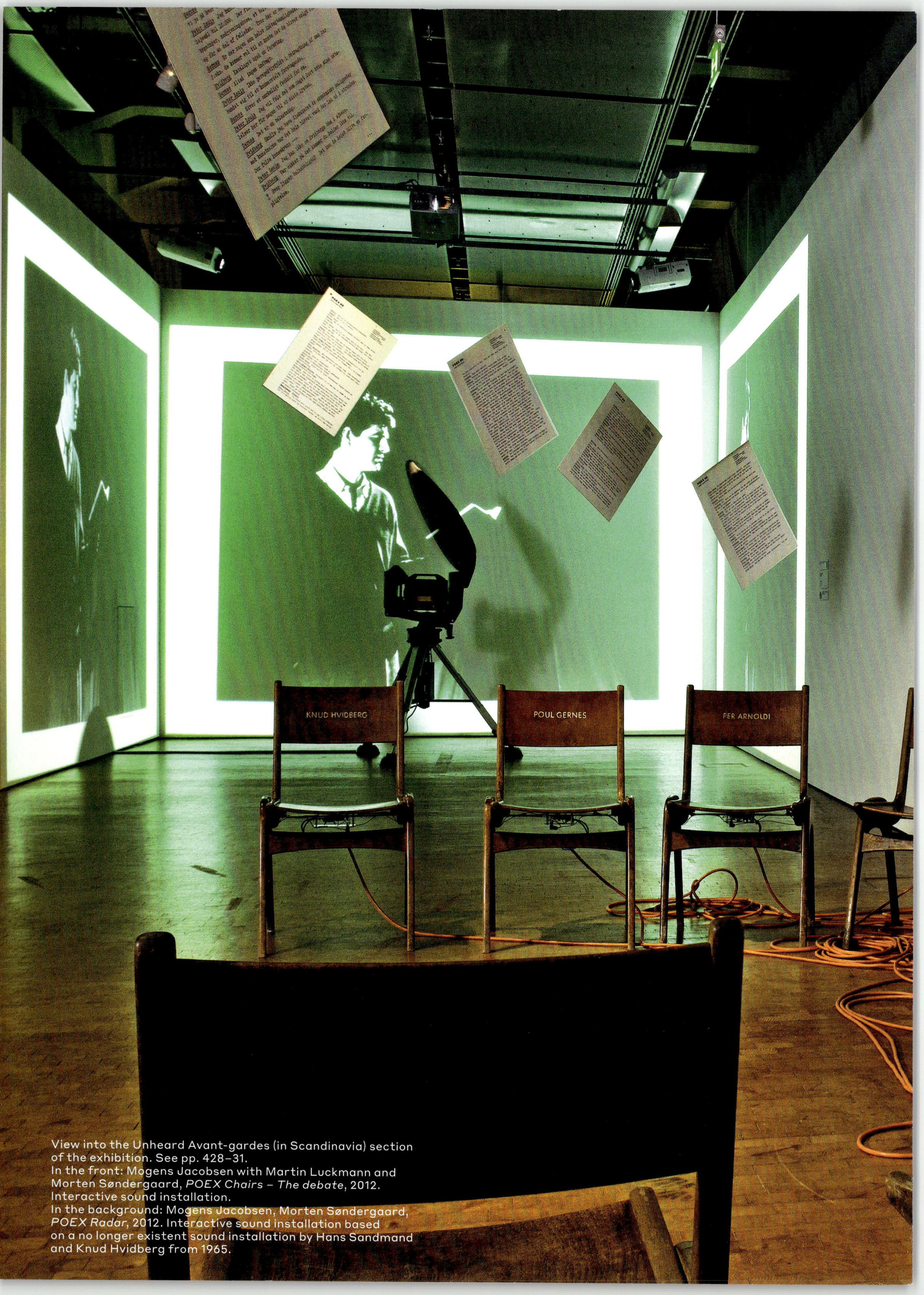

View into the Unheard Avant-gardes (in Scandinavia) section of the exhibition. See pp. 428–31.
In the front: Mogens Jacobsen with Martin Luckmann and Morten Søndergaard, *POEX Chairs – The debate*, 2012. Interactive sound installation.
In the background: Mogens Jacobsen, Morten Søndergaard, *POEX Radar*, 2012. Interactive sound installation based on a no longer existent sound installation by Hans Sandmand and Knud Hvidberg from 1965.

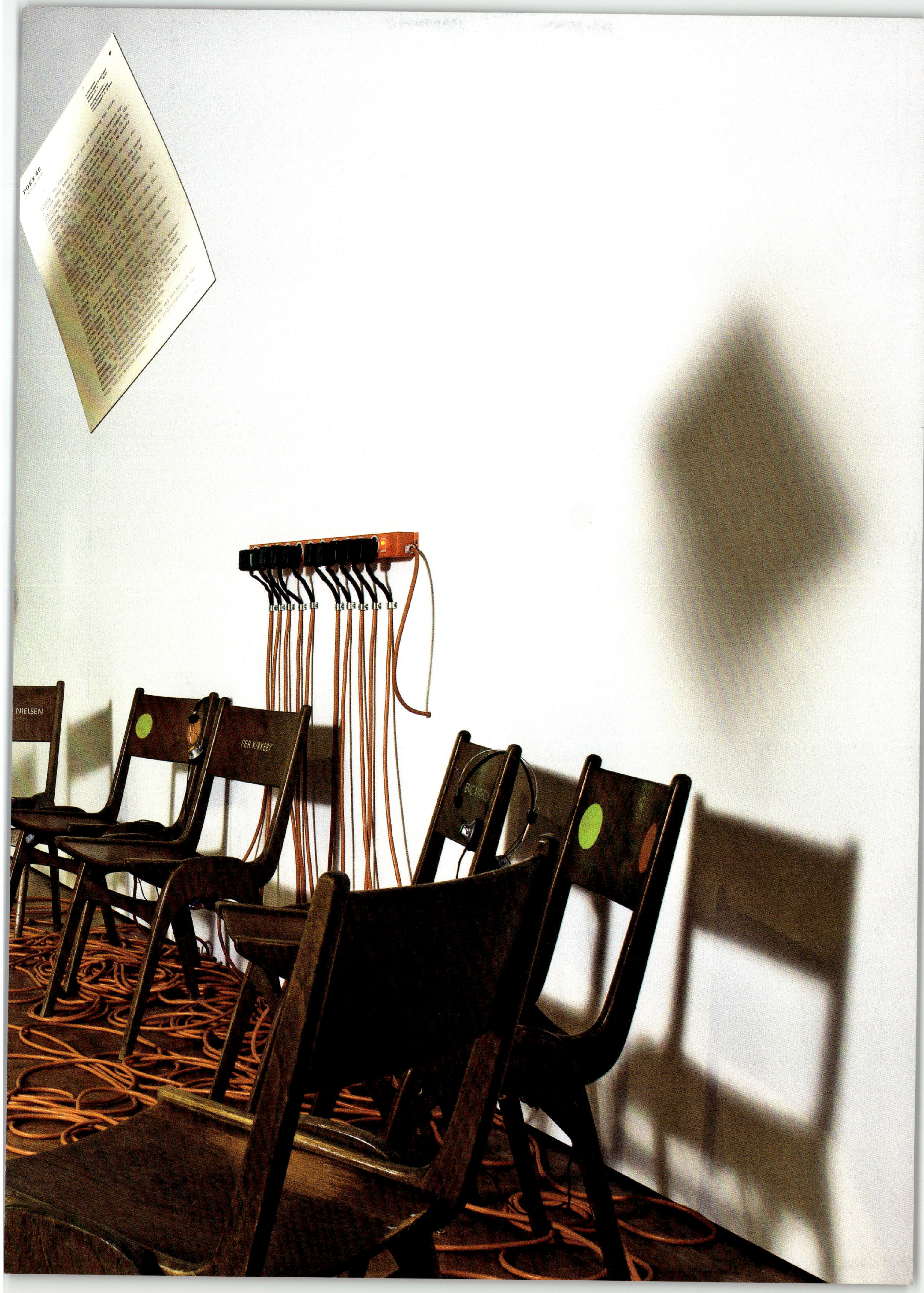
POEX 65
NIELSEN
PER KIRKEBY

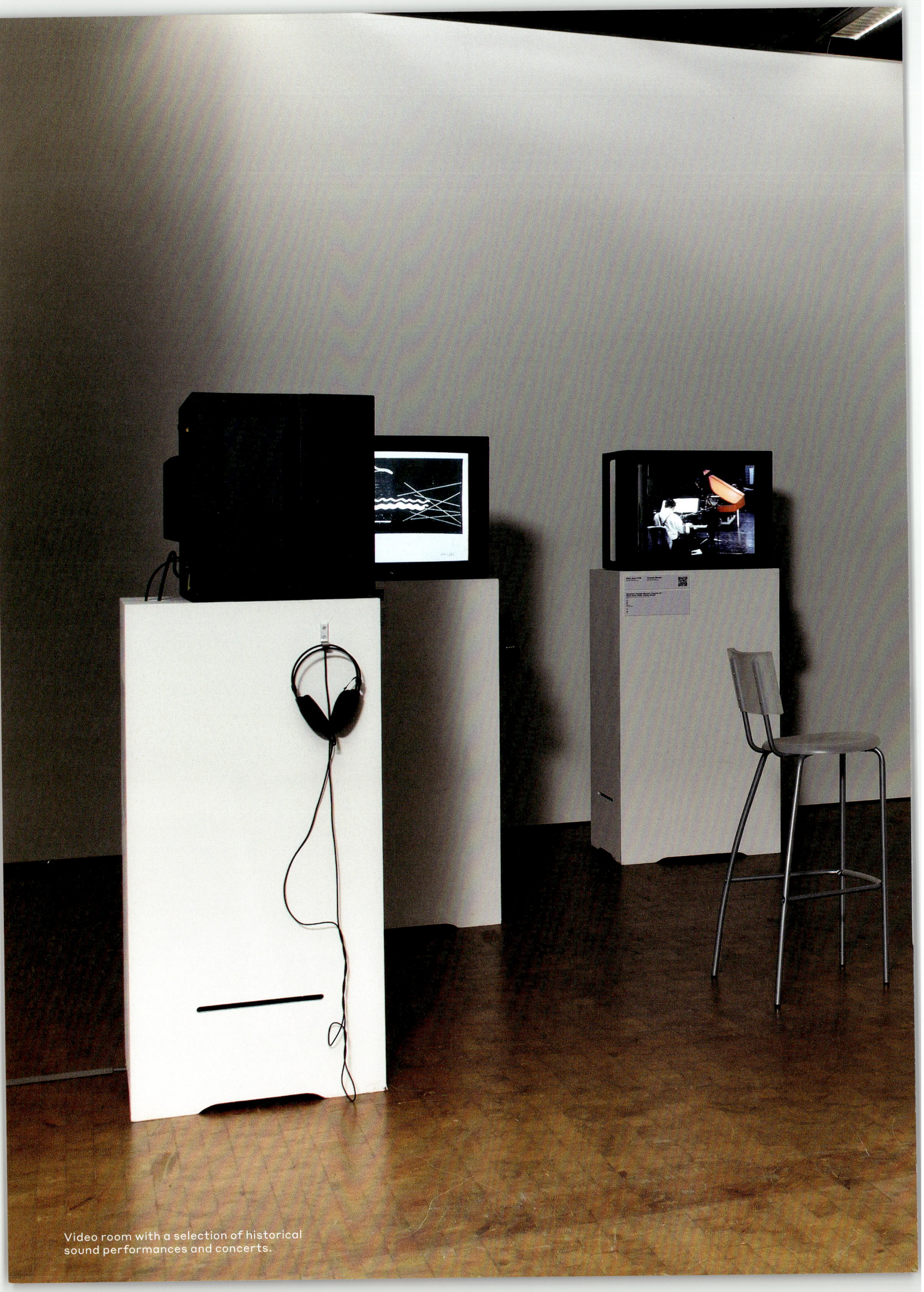

Video room with a selection of historical sound performances and concerts.

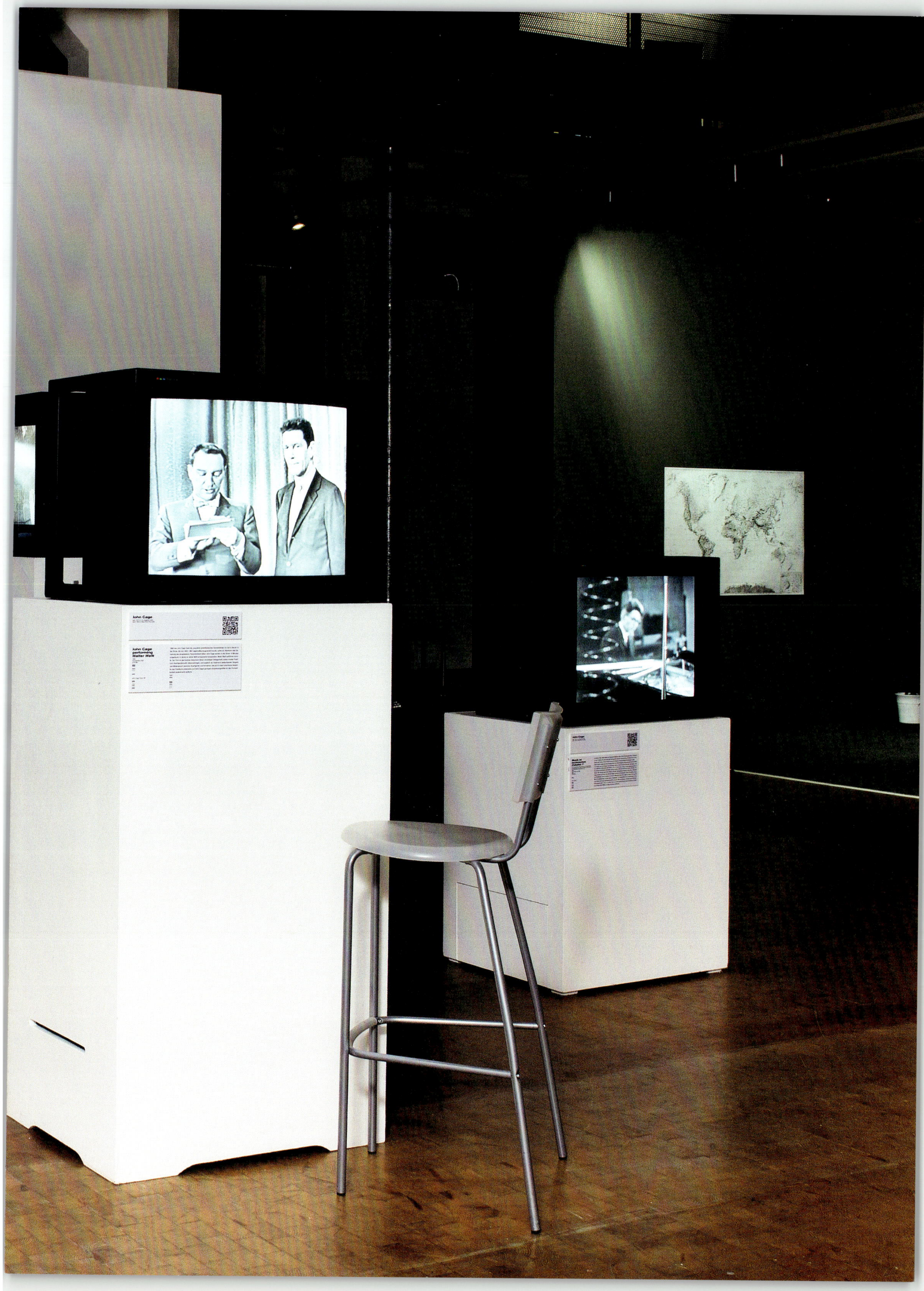
John Cage
John Cage performing Walter Walk
John Cage

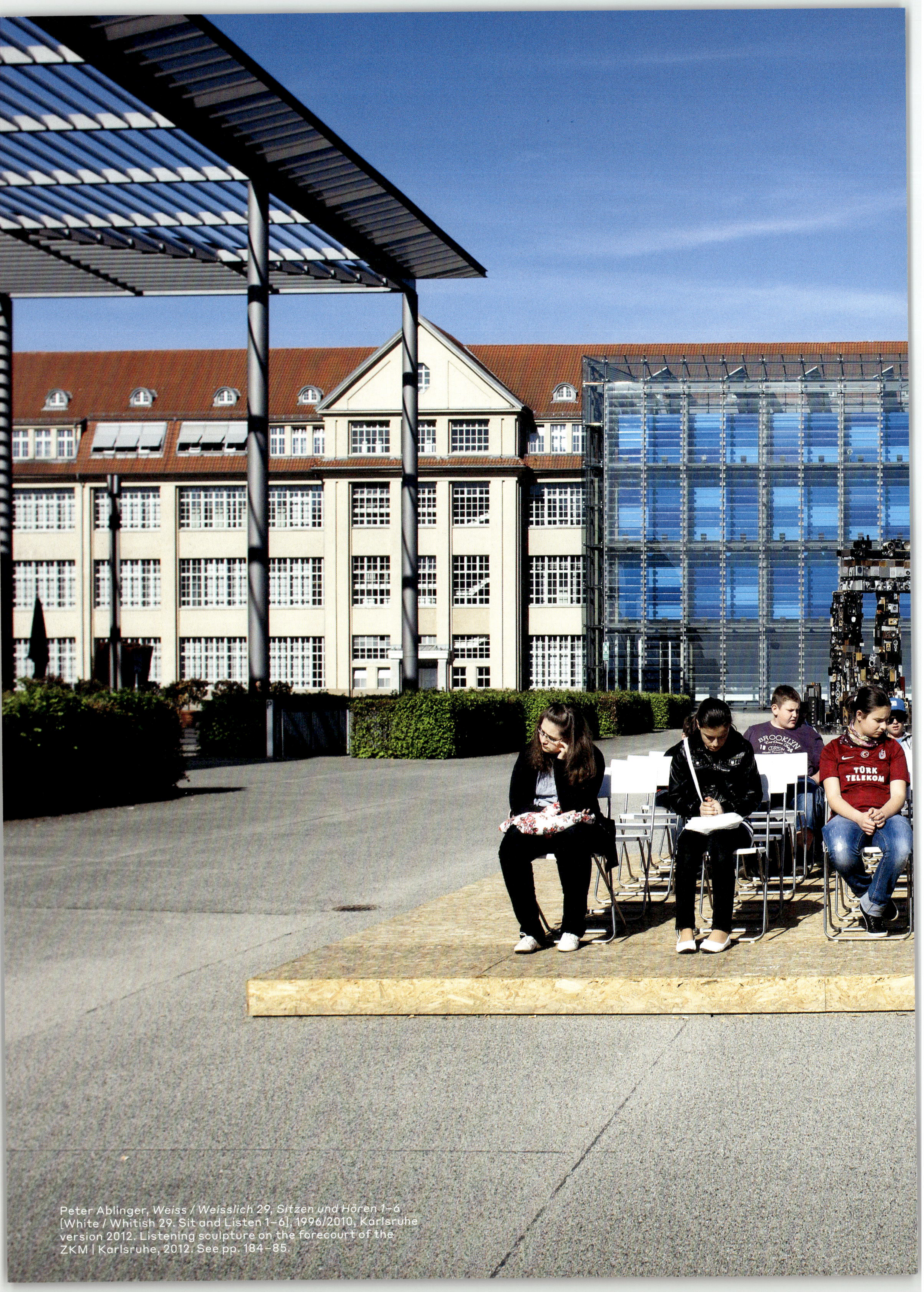

Peter Ablinger, *Weiss / Weisslich 29, Sitzen und Hören 1–6* [White / Whitish 29. Sit and Listen 1–6], 1996/2010, Karlsruhe version 2012. Listening sculpture on the forecourt of the ZKM | Karlsruhe, 2012. See pp. 184–85.

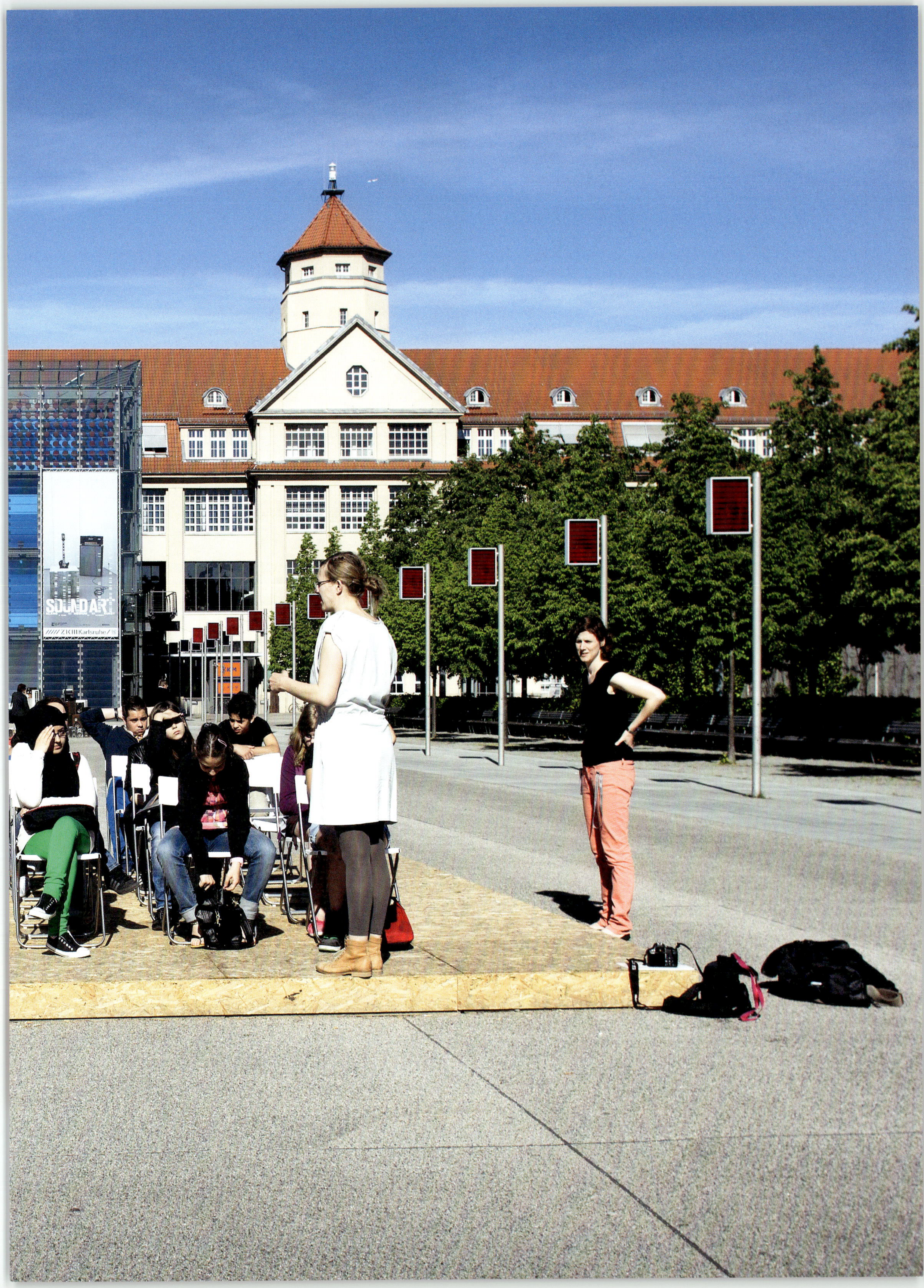
SOUNDART
ZKM Karlsruhe

Peter Ablinger

*1959 in Schwanenstadt (AT), lives in Berlin (DE)

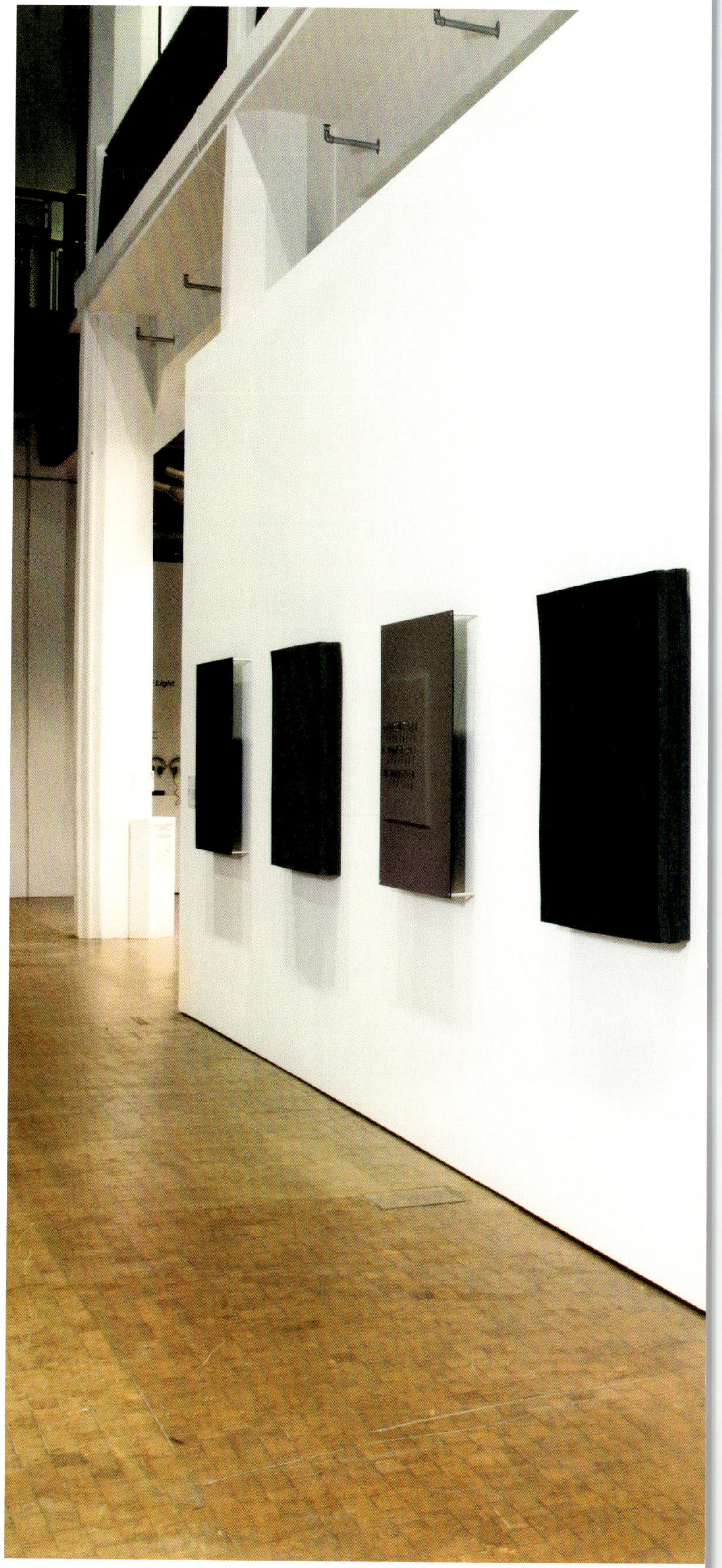

Weiss / Weisslich 3e, Gallery Piece (Absorption/ Reflection), 1997/2004 (first realization 2012)
Plexiglas, insulation, 3 × 3 m
Production: ZKM | Karlsruhe

Squares of sound-absorbing and sound-reflecting materials adorn the wall of the exhibition space like a frieze. Aesthetically, they evoke the impression of a post-minimalist visual work. But in fact, the concept behind the work could be described as acoustic intervention. When people walk by the frieze, the acoustics of the room are almost imperceptibly altered.

Peter Ablinger's *Weiss / Weisslich* [white / whitish] cycle, begun in 1980, focuses on the perception of minimal differences. It encompasses a series of heterogeneous objects and works that sometimes consist only of a title (*walk, and listen to the wind or hold your hand behind your ear/take it away*), sometimes are based on notations, but may also involve instrumentalists and/or the audience.

www.ablinger.mur.at

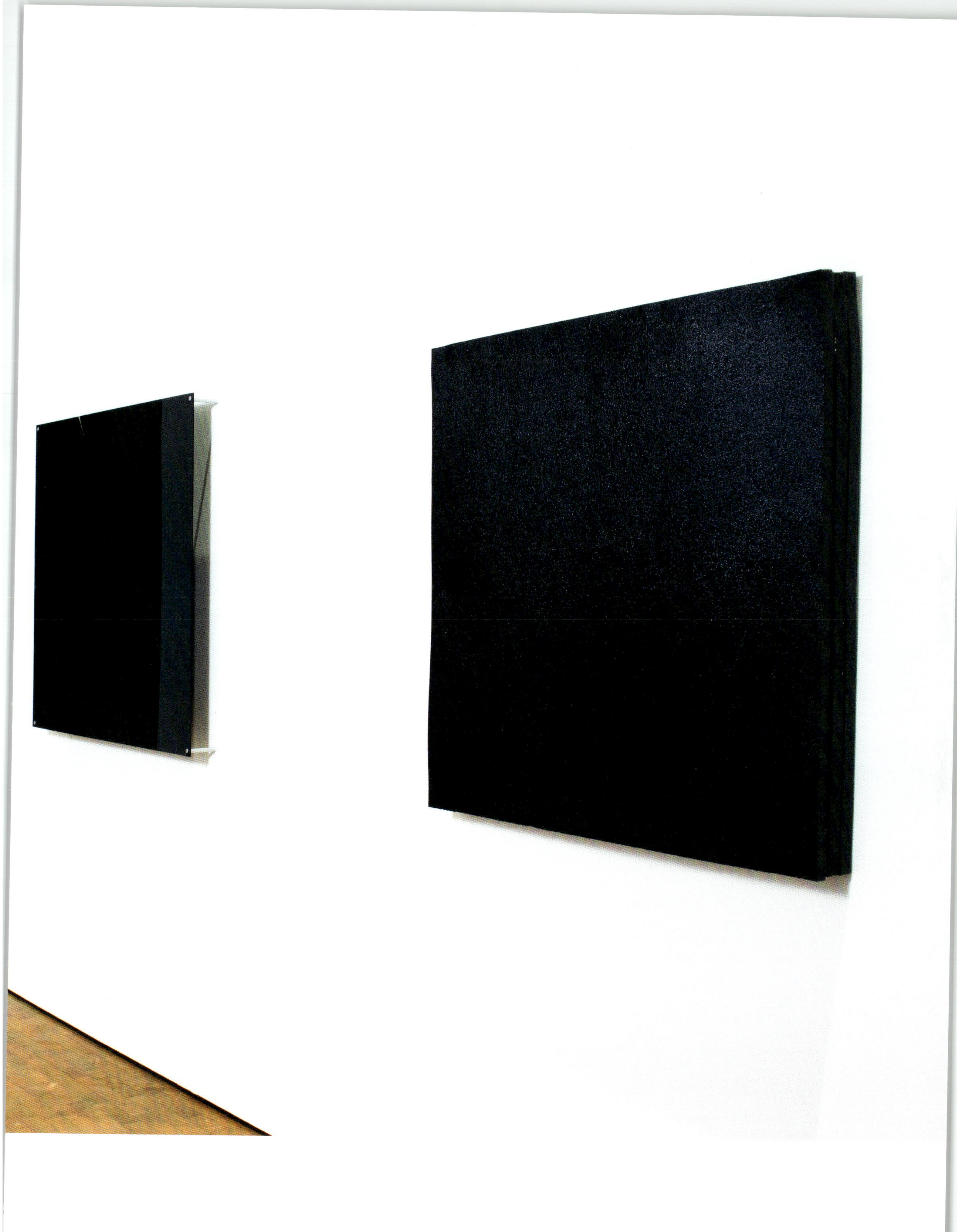

Peter Ablinger

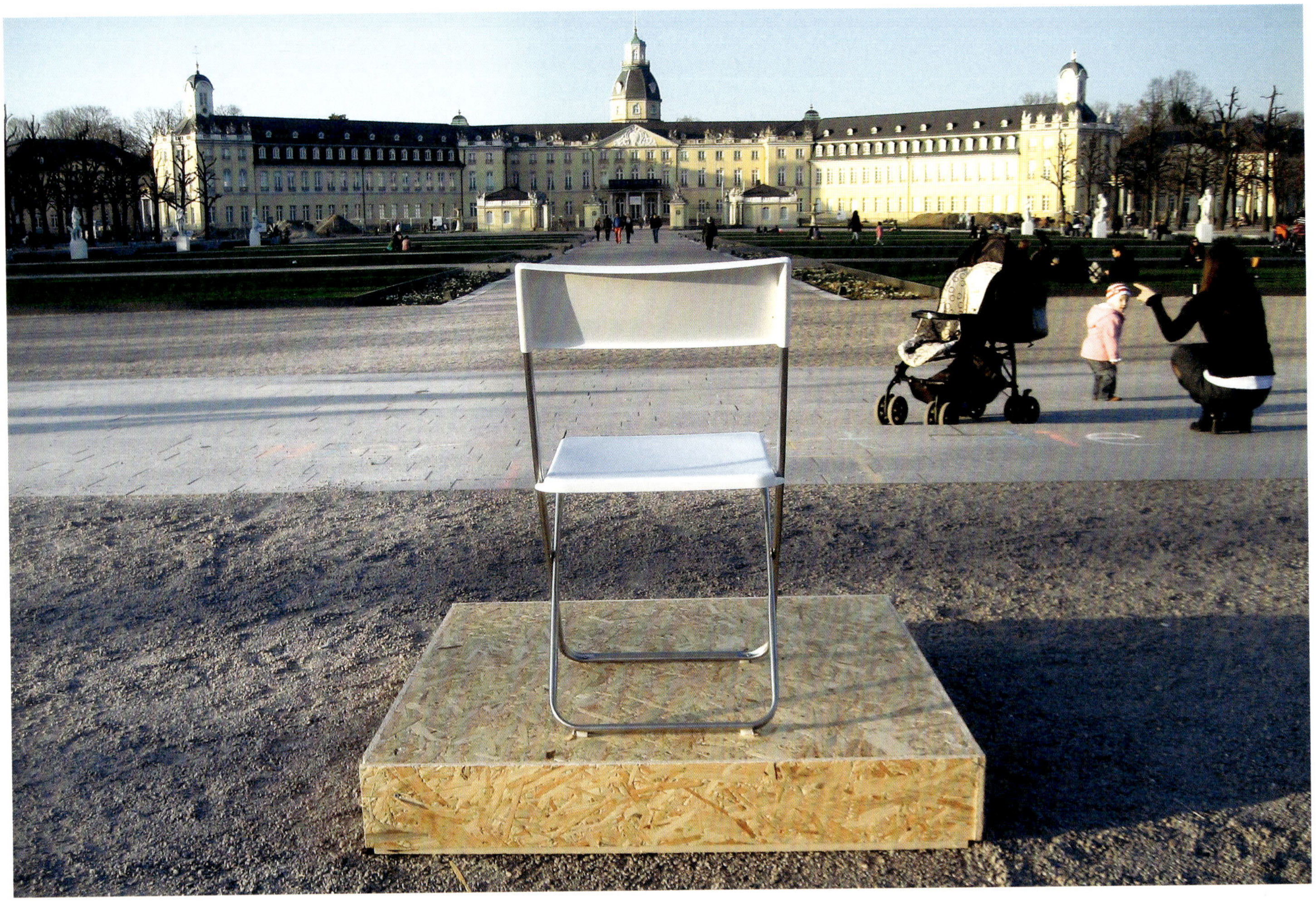

Weiss / Weisslich 29, Sitzen und Hören 1–6, 1996/2010, Karlsruhe version 2012
Listening sculpture in six parts at various locations in Karlsruhe, 6 platforms (6 × 6 m, 5 × 5 m, 4 × 4 m, 3 × 3 m, 2 × 2 m, and 1 × 1 m), 91 chairs
Production: ZKM | Karlsruhe

Sitzen und Hören 1–6 [Sit and Listen 1–6] consists of chairs set up in rows. They form six auditoriums arranged in six different formats and are set up at six different public places in Karlsruhe.

Take a seat and the focus of your perception changes. No longer is it the view of the chairs that counts, but instead what can be perceived from sitting in the chairs. Using or not using the chairs actualizes different facets of the sound sculpture: when the rows of chairs are used, the place becomes a proposal to listen and the situation a concert. Unused, the rows of chairs can be read as a serial sculpture whose subject is the classical Western art of higher listening.

At the center of *Sitzen und Hören 1–6* is the awareness: “The sculpture is not the material and also not that which might be there to hear. The sculpture is the listening process itself.” (Peter Ablinger)

The locations of the six auditoriums were Schlosspark infront of the Karlsruhe Palace, an underground garage, Sonntagsplatz (playground), Schlossplatz infront of the restaurant Multi Kulti, Karlsruhe Hauptbahnhof, and the ZKM Forecourt.

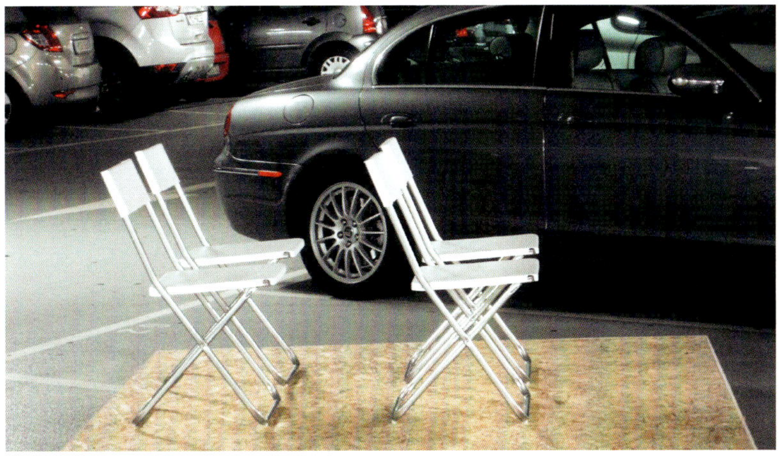

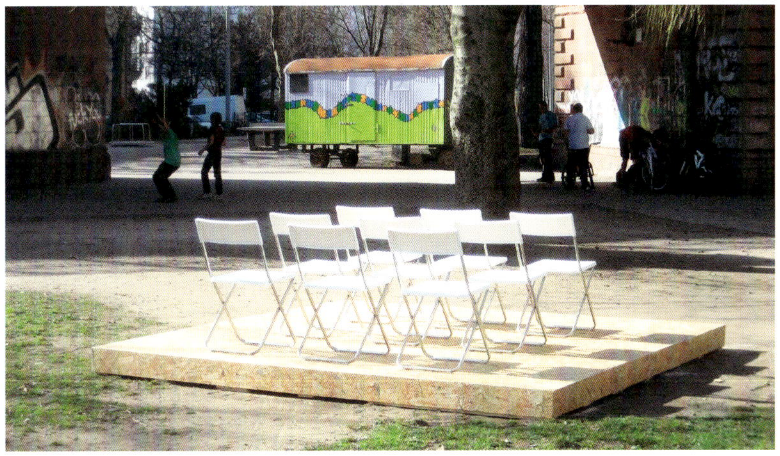

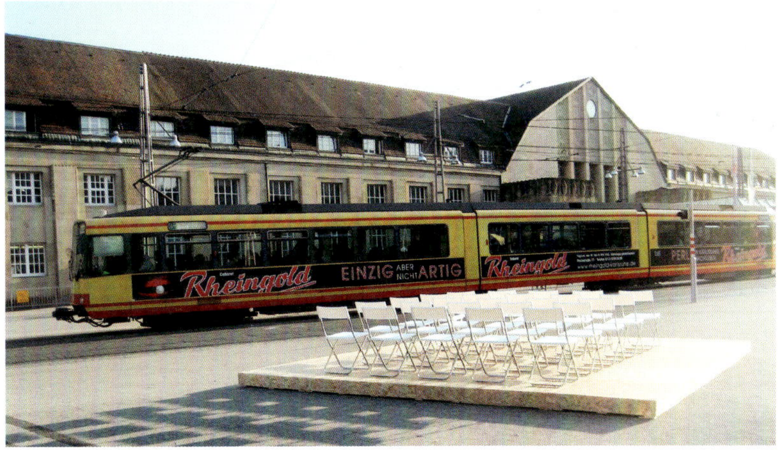
Rheingold
EINZIG ABER NICHT ARTIG
Rheingold

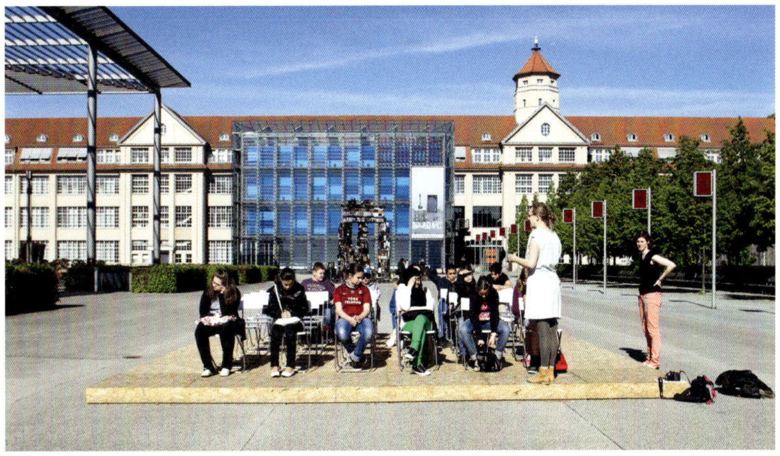

Tyler Adams

*1980 in Los Angeles, CA (US), lives in Los Angeles

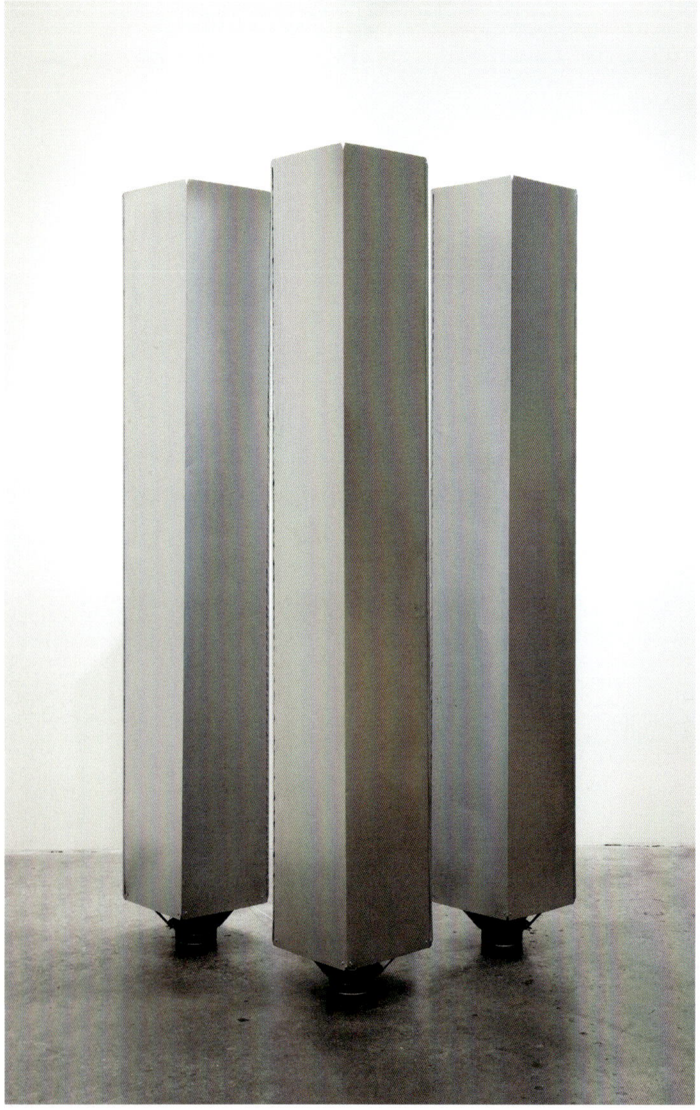

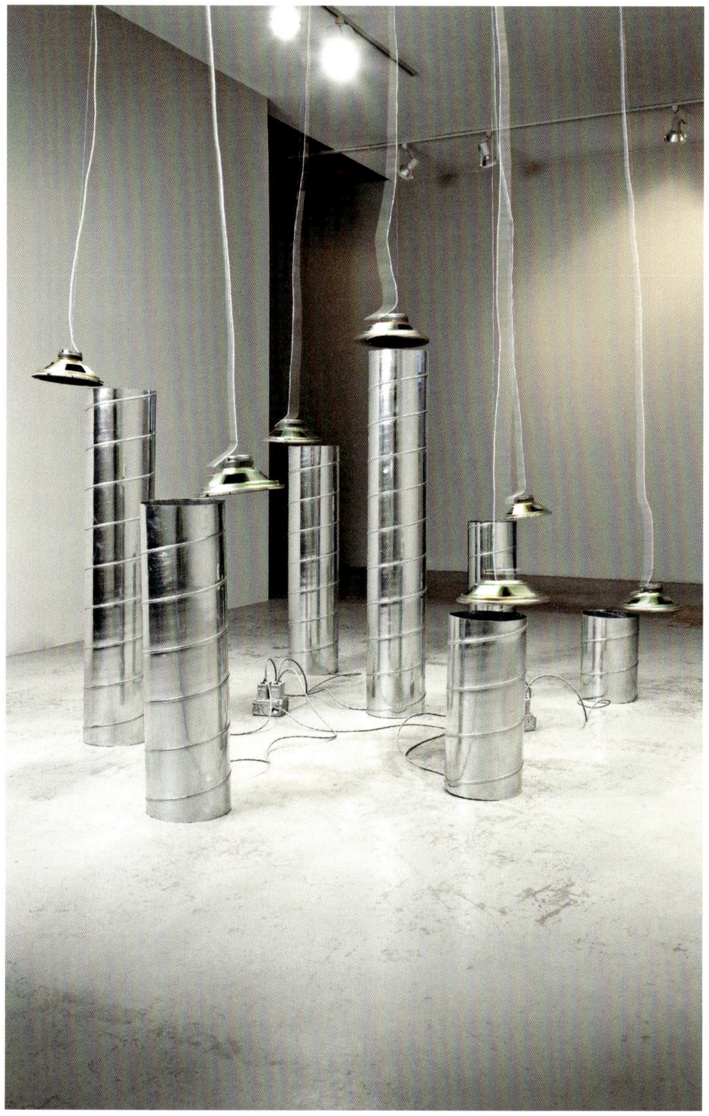

Rumble, 2013
Speakers, galvanized steel sheet metal, amplifier
Tyler Adams

A large speaker is fixed to the bottom of each rectangular duct, generating very low frequencies at the limit of human hearing. The sound generated causes the metal walls of the sculpture to flex and flutter, producing a soft rumbling sound that evokes the flow of air through ductwork. The speakers are tuned to slightly different frequencies, creating subtle acoustic interference patterns between the metal objects.

www.t-adams.com

Interstitial, 2011
Speakers, wire, round duct, fans, amplifiers
Tyler Adams

Speakers suspended by wire each produce a steady unique tone which is tuned in relation to the length of a cylindrical duct over which it hangs. Each duct contains a small fan that blows air upward and sets the speakers in motion. The tone produced by the speaker is not audible until it is directly above the face of the duct, where it is then amplified by the resonance of this hollow tube. The result is an evolving soundscape of overlapping tones fading in and out.

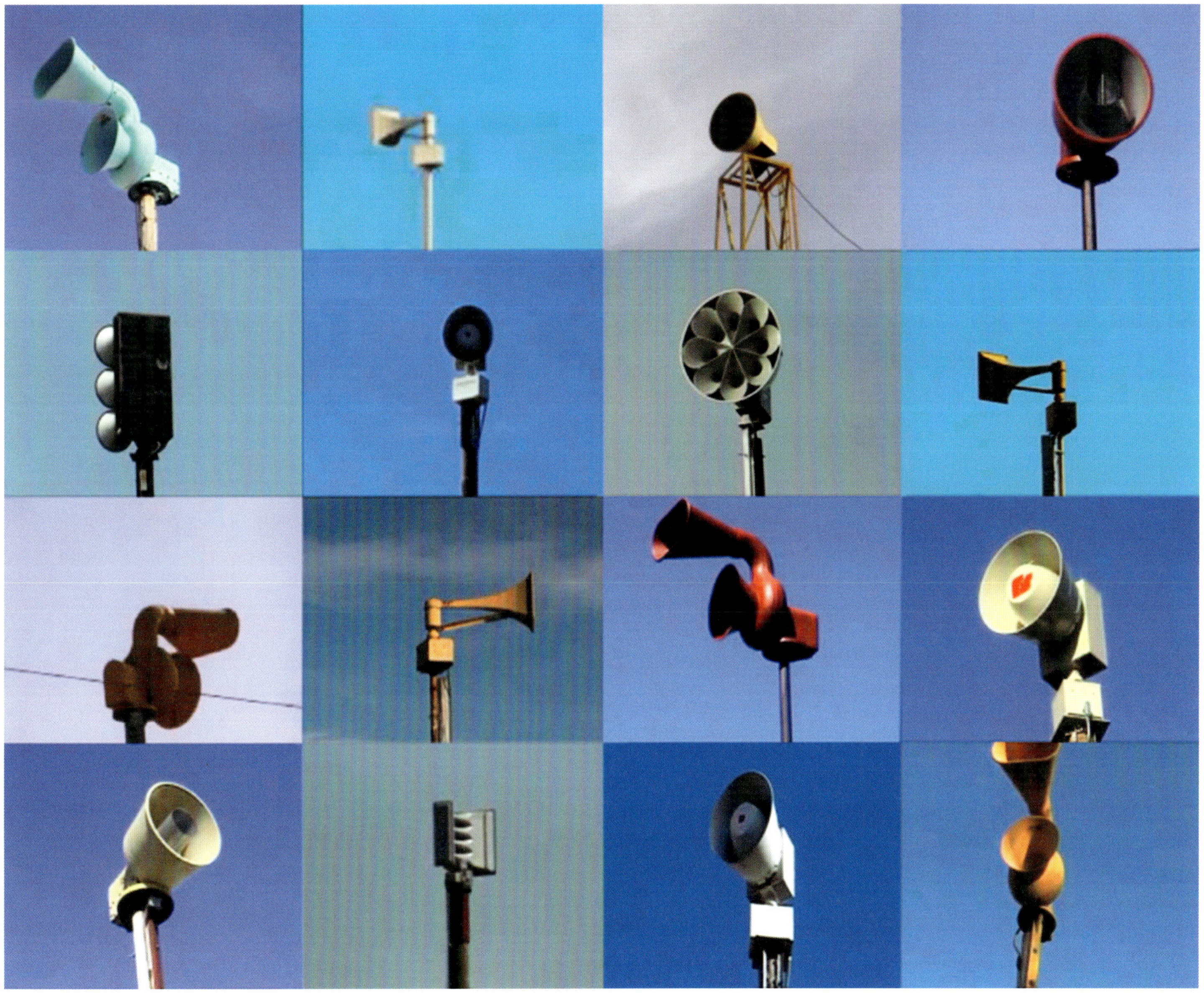

Sirens, 2012
Video, DVD-R, 15 min.
Tyler Adams

Adams collages films of civil defense sirens. The pictures of the automatons rotating against a blue sky reveal nothing about the reason for the alarm. Each siren goes through its own motions and has its individual sound. The video collage documents the array of models and at the same time creates the impression of a choreographed concert.

Tyler Adams's sirens have important antecedents in the futuristic *Intonarumori*, whose siren-like machine sounds were first presented in concert in 1914 in Milan by the Italian painter and composer Luigi Russolo.

Tyler Adams

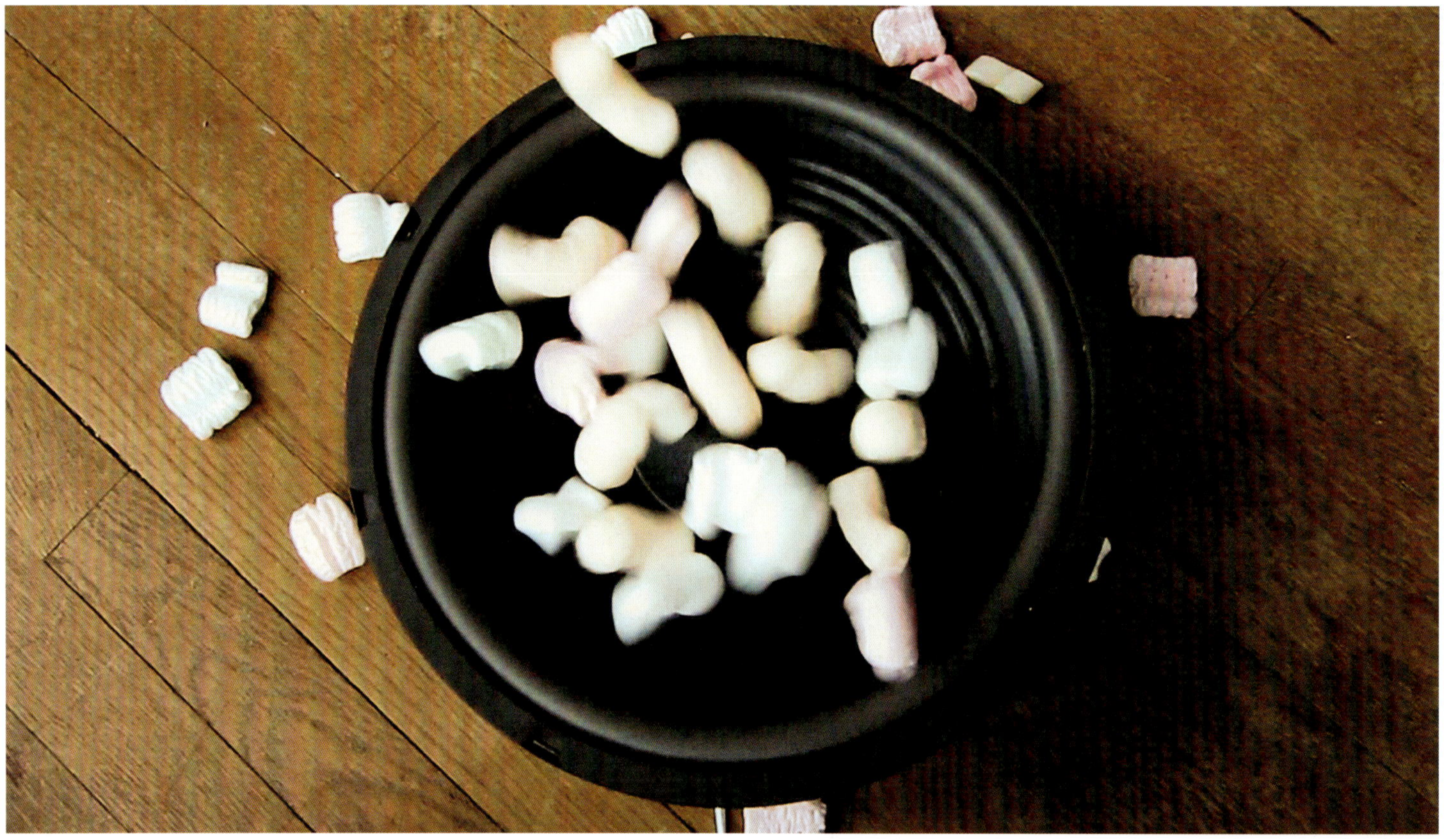

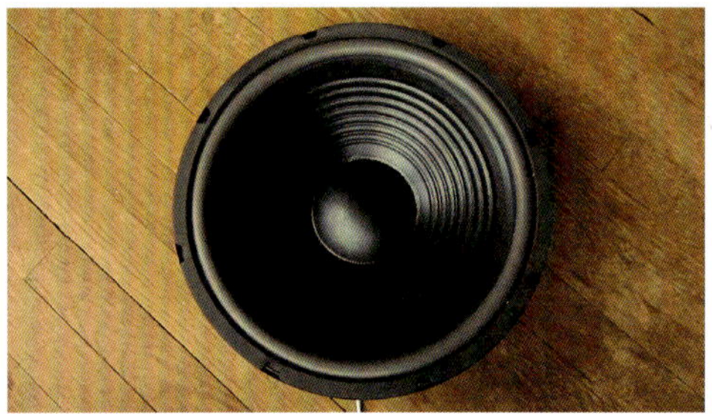

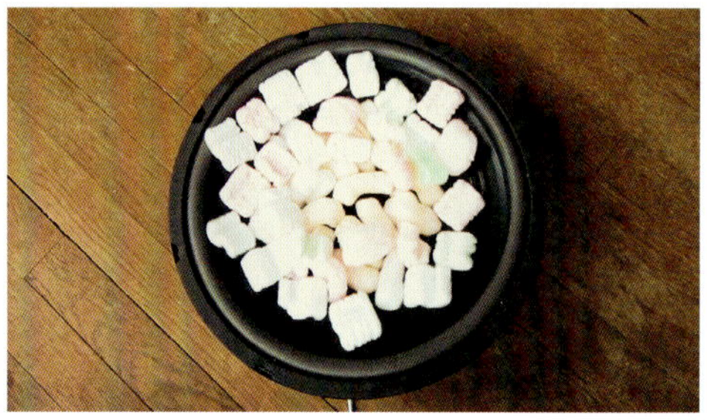

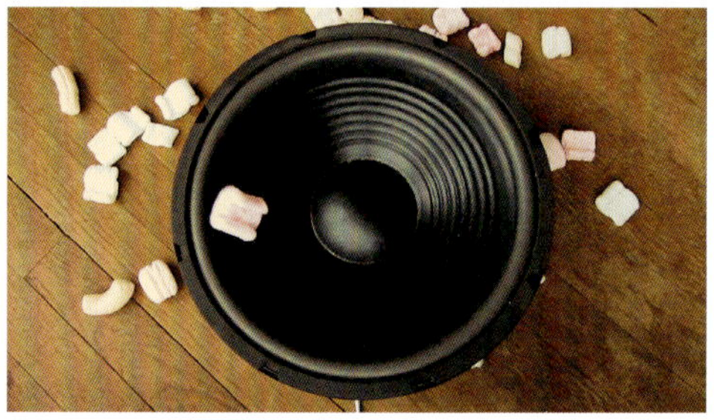

Addition/Subtraction, 2010
Single-channel video, 5:46 min.
Tyler Adams

In this video a single speaker produces a very low frequency that is barely audible. Styrofoam peanuts are individually added to the cone of the speaker. The sounds produced are due to the Styrofoam pieces tapping against the vibrating speaker cone, as well as the pieces bumping against one another. Once the speaker cone is full of Styrofoam, the amplitude is increased and the Styrofoam peanuts gradually begin to escape from the speaker cone.

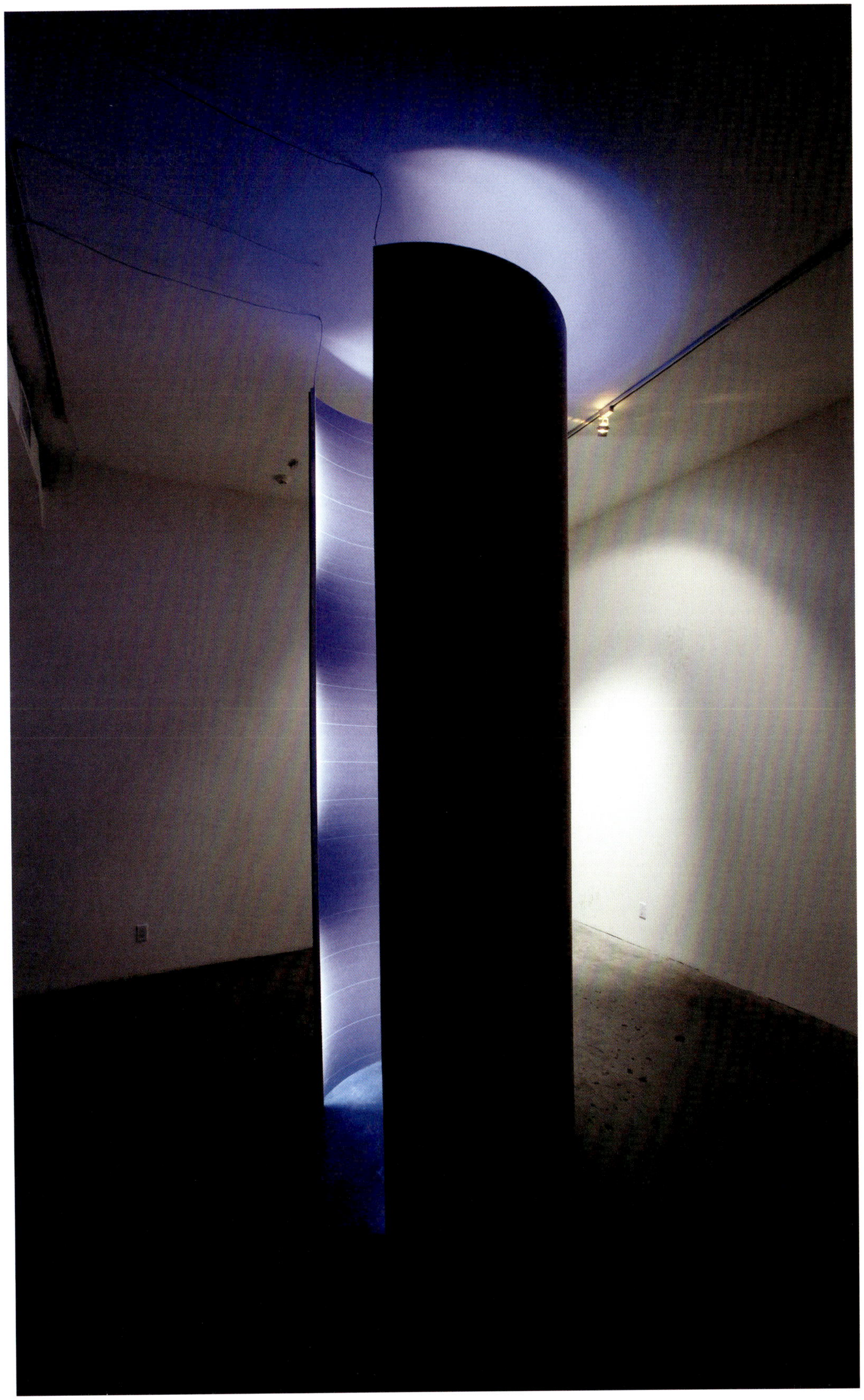

Descending, 2013
Sonotube, aluminum, speaker, amplifier, LED lights
Private collection

The interior of a cardboard pouring form for concrete serves as an immersive portal where visual and aural perceptual effects are employed to evoke a visceral sense of falling. The cylindrical form creates an internal resonance that is not readily perceived until one steps inside. Upward chasing lights and a downward falling shepard-tone-esque soundscape form the experience of perpetual descent.

Maryanne Amacher

*1938 in Kane, PA (US), †2009 in Kingston, NY (US)

Lagrange / Room for Maryanne Amacher, 2006 – unfinished
Paper, video, sound
Sketches and material related to the unfinished site-specific space composition *Lagrange* for EMPAC, Troy, New York
Research, concept: Micah Silver
Maryanne Amacher Archive

In the fall of 2006 Maryanne Amacher began working on what would be her final project. It was commissioned by a new media art center called EMPAC. I initiated the project with excitement about the confluence of a singular, visionary artist and a facility whose aural capabilities seemed to be built uniquely for Amacher's work. The project was a special situation for her as she would be able to have ongoing residencies and production over several years – the opportunity for a long-term research-based project she had been interested in for decades.

For the years leading up to her death in 2009 the piece evolved conceptually through many phases. Primarily: the establishment of a "listening team" for researching the creative potentialities of microgravity (parabolic flight, underwater simulation, space stations), a series of recording and observation trips to underground observatories around the world, the creation of a virtual acoustic system for modeling the aural perception of other species in order to bring the listener into alternate listening bodies.

It proved impossible to bring form to the more sprawling research projects and it was decided to work with stereoscopic video and her Mini-Sound Series presentational mode as an anchor. Stereoscopy was an area she had been interested in since the 1970s when she was at MIT and recently it had linked with her research into the interconnects between microgravity perception, the vestibular system – with fine control over the stereo illusion made more feasible in recent years, one can produce very strong unusual senses of balance and the sense that images pass through a room, not only within it.

The work's dramatic structure would illuminate the research themes. We began recording material for new sound characters based on hearing freight trains in rural New York State during particular weather conditions. In parallel she would develop an elaborate spatialization scheme involving over forty speakers at a time, most of them several levels of the building away from the performance space, acoustically coupled to the listener's air through precisely ajar doorways, vents, orchestra pit apertures opened just so, speakers high above the grid in storage rooms.

At the time of her death she had worked on the piece for nearly two years, enigmatically, methodically – she spent six weeks of residency in the theater working at full scale. The majority of this time was spent solely dedicated to the development of speaker placement and sound shapes – arguably the primary materiality of her work since the early 1980s. In parallel, OpenEndedGroup, the visual team she had invited to join the project, developed stereoscopic material. In the end, the collaboration would not work and Amacher continued on her own, as did OpenEndedGroup (producing *Upending* in 2010).

Those of us who were lucky enough to hear Amacher's work-in-progress, experienced sound images that arrived, seemingly from nowhere – or slowly passed through the space as if glacially emerging from the center of the Earth and passing into space. The train recordings were abandoned as soon as one was recorded that seemed to capture what we had been searching for.

The materials presented at ZKM Karlsruhe were a window into Amacher's thinking at the end of her career – a constellation of sparks around the absence of this final work. The loose papers and animations are from folders she kept with collected research, many emailed to collaborators in the project to articulate how she was positioning her thinking about sound in space and the potential of stereoscopy. Other materials are from physical folders located in her archive. They were presented as a research cluster, a mix of the tangible manifestations of the project and the oblique.

The title, *Lagrange,* refers to Lagrangian Points: locations in space where gravitational forces and the orbital motion of a body counteract each other.
(Micah Silver)

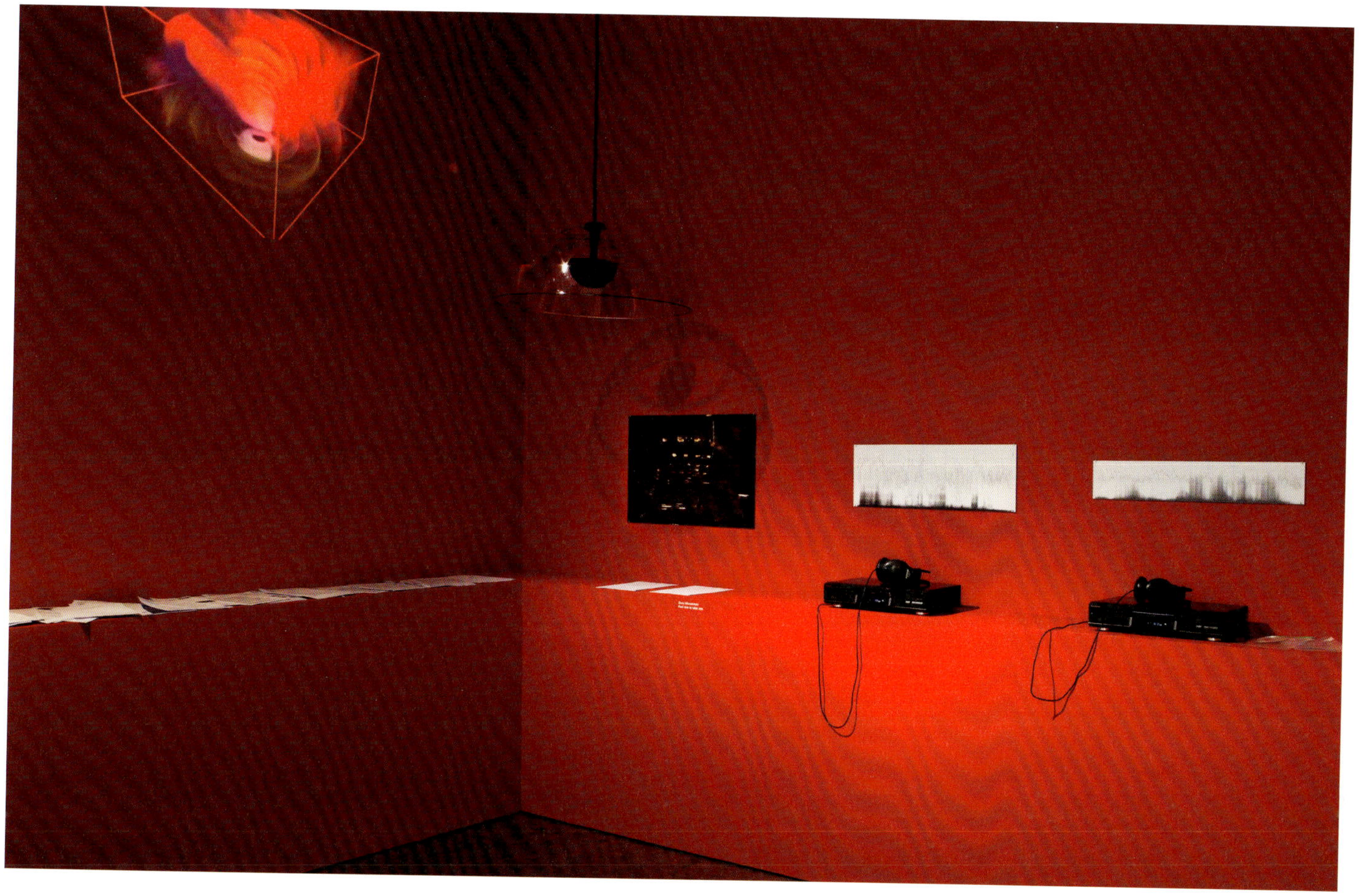

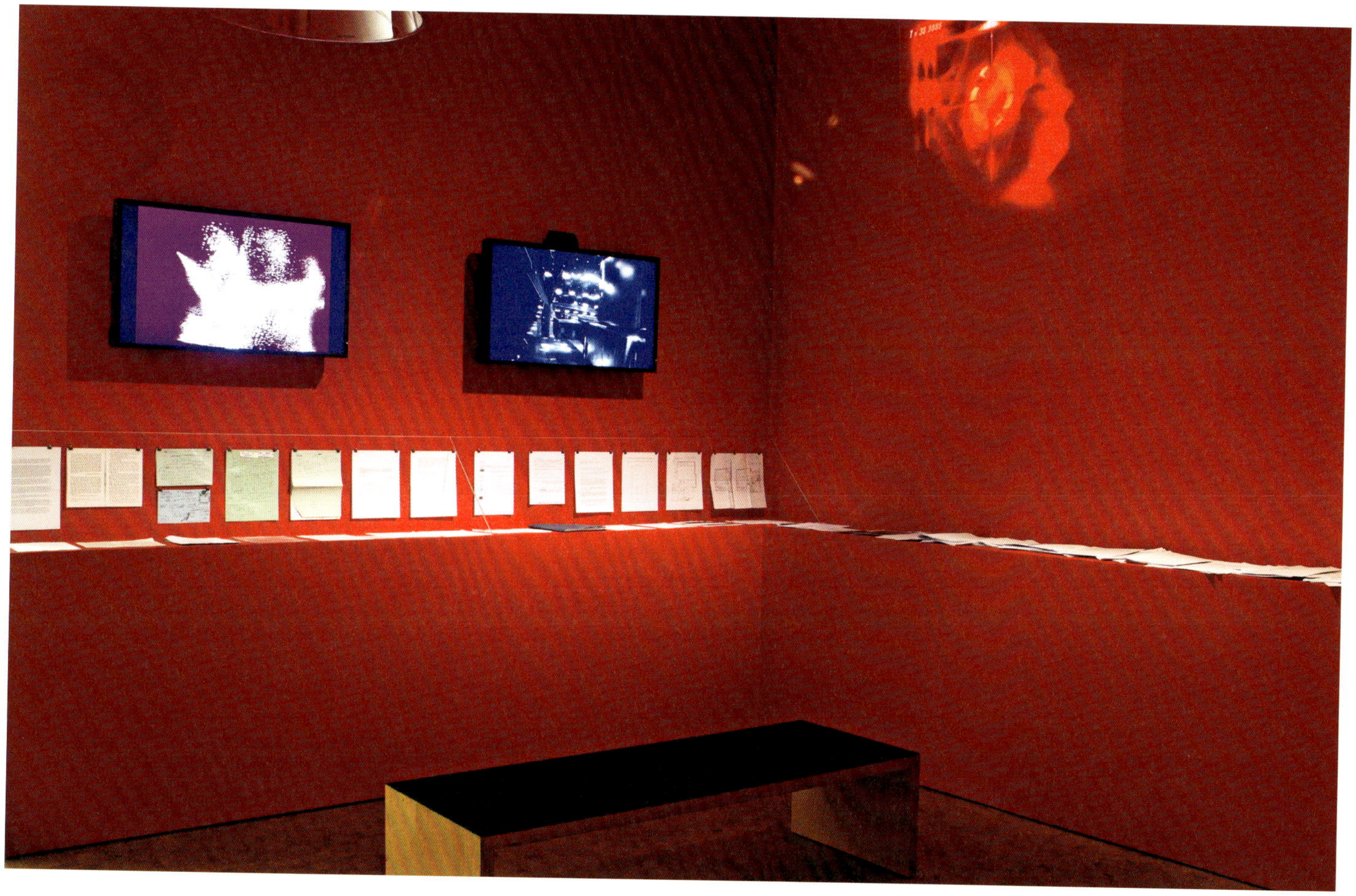

Cory Arcangel

*1978 in Buffalo, NY (US), lives in Brooklyn, NY (US)

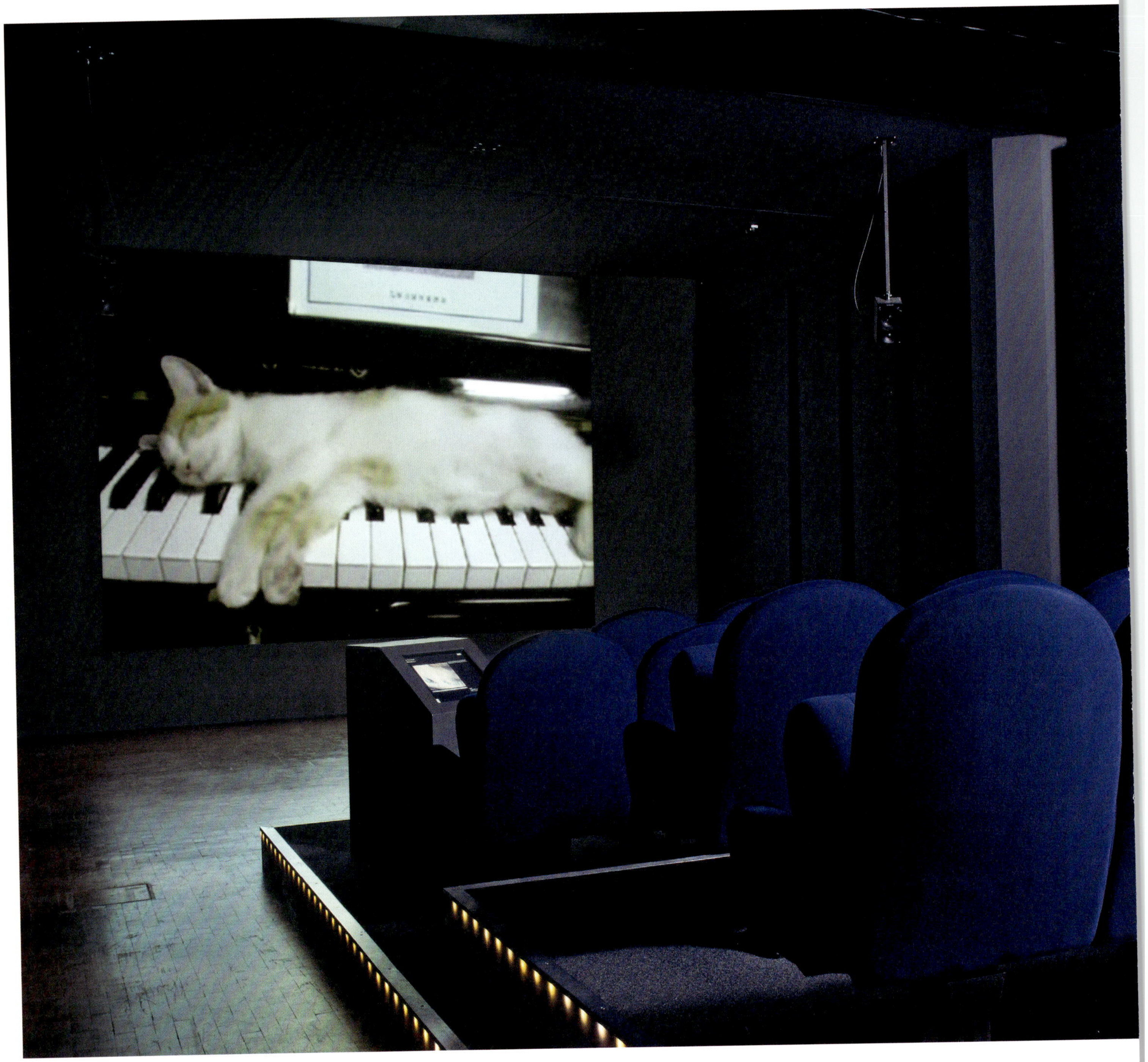

Drei Klavierstücke op. 11, 2009
Single-channel video, 15:58 min.
Courtesy of the artist and
Team Gallery, New York

www.coryarcangel.com

Cory Arcangel reconstructs Arnold Schoenberg's *Drei Klavierstücke* [Three Piano Pieces] (1909) with the help of videos of piano-playing cats. Using the free *Comparisonics* software, which identifies similar sounds within an audio file, he compared the cat soundtracks note for note with Glenn Gould's recording of the *Drei Klavierstücke*. The best correspondences were selected for the "score" of the single-channel video.

The artist, programmer, web designer, and performer uses the Internet as the source of material for many of his digital works – in this case, 170 amateur videos from the video sharing website YouTube. The World Wide Web is also the destination of this copy & paste art. The *Drei Klavierstücke op. 11* can be found on YouTube along with Arcangel's *Paganini's 5th Caprice* (2011) composed using video samples of electric guitars.

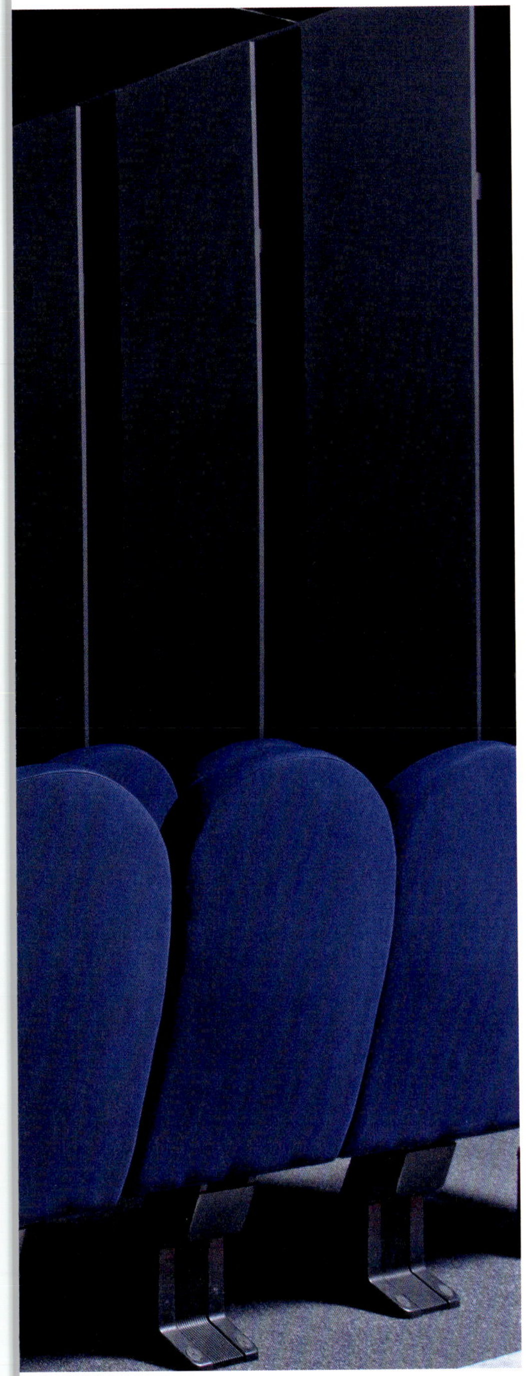

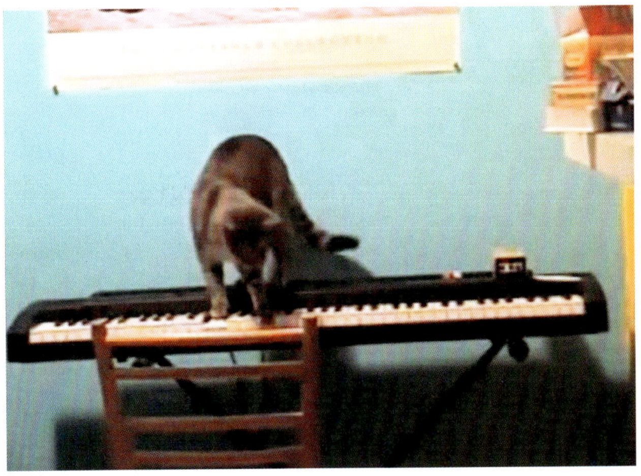

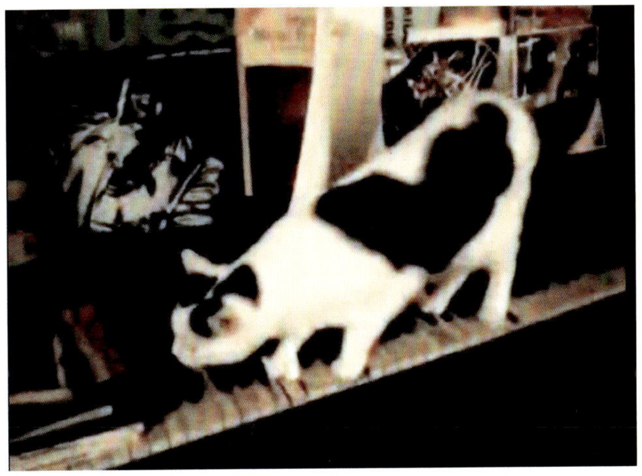

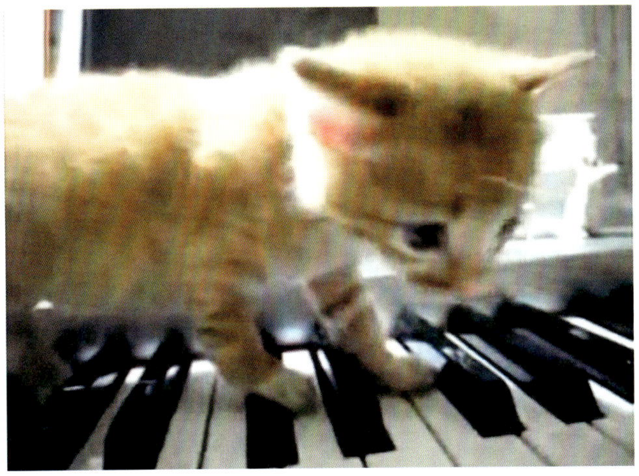

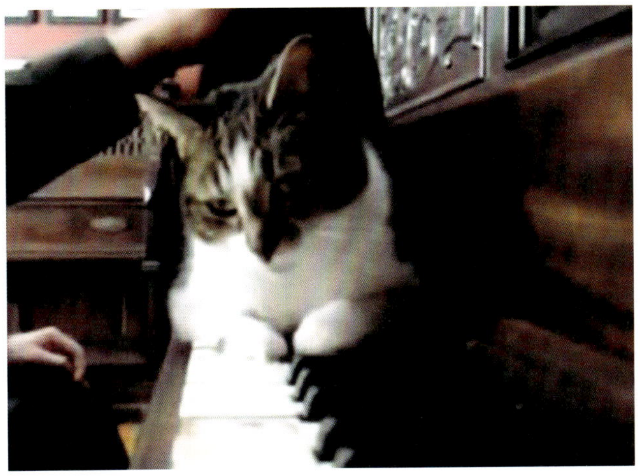

Cory Arcangel

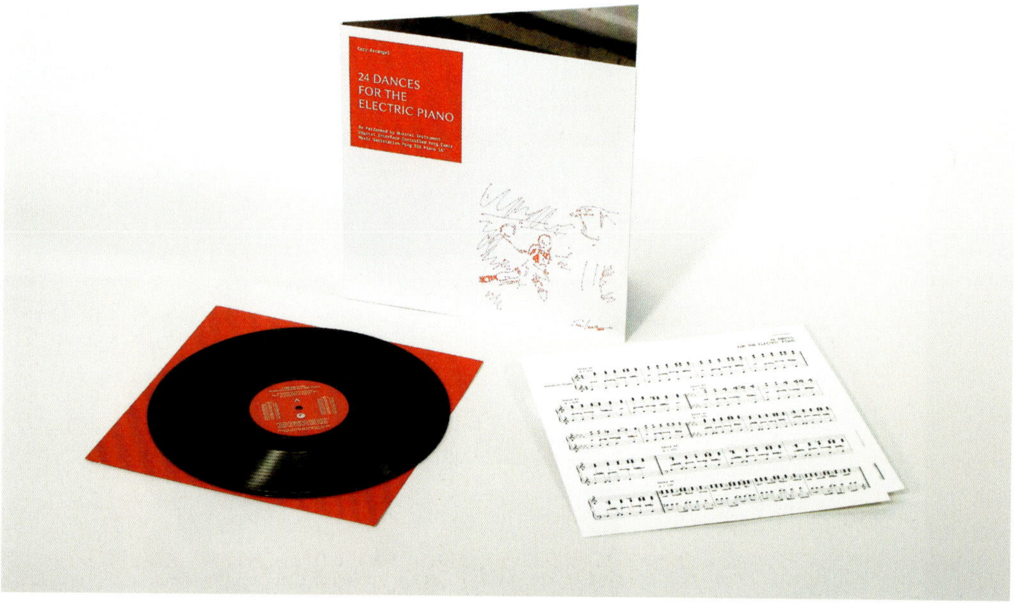

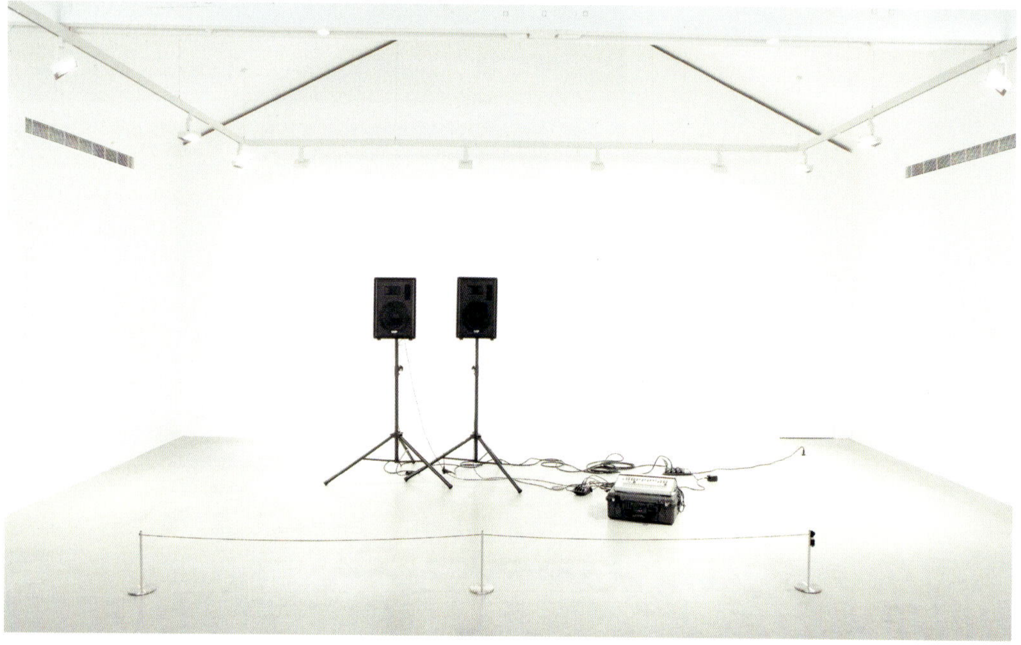

top:
24 Dances For The Electric Piano (Record), 2014
12" LP vinyl disk, composition insert, red pencil plotter print
Courtesy of the artist

24 Dances For The Electric Piano contains compositions Arcangel wrote for the Korg M-1 synthesizer. Inspired by 1990s club music, the compositions take their form from trance and house music riffs, yet played solo on the Korg, lack the percussive element that lends the style's dance club association. This nakedness conveys the compositions more as studies than as mimicry. When performed, Arcangel has a professional pianist play each movement. The pianist decides the duration of each composition and then moves on to the next. Again, here, issues of authorship are at play as Arcangel leaves a large aspect of the presentation out of his own hands.

bottom:
PSK, 2014
Roland TR-909 Rhythm Composer & drum pattern, public address sound system (variable), cables, dimensions variable
Installation view Galerie Thaddeus Ropac, Salzburg, 2015
Courtesy of the artist

For this installation, Arcangel has programmed a rhythm pattern into a Roland TR-909 drum machine. It is the rhythm of the song *P.S.K. What Does it Mean?* (1985) by Philadelphia rapper Schooly D. This rhythm pattern is one of the most heavily sampled loops throughout rap and hardcore music of the late 1980s and 1990s. For Arcangel, the work is a protest against forgetting. His purpose is to save one of the most important technologies (the original drum machine) from becoming obsolete and forgotten.

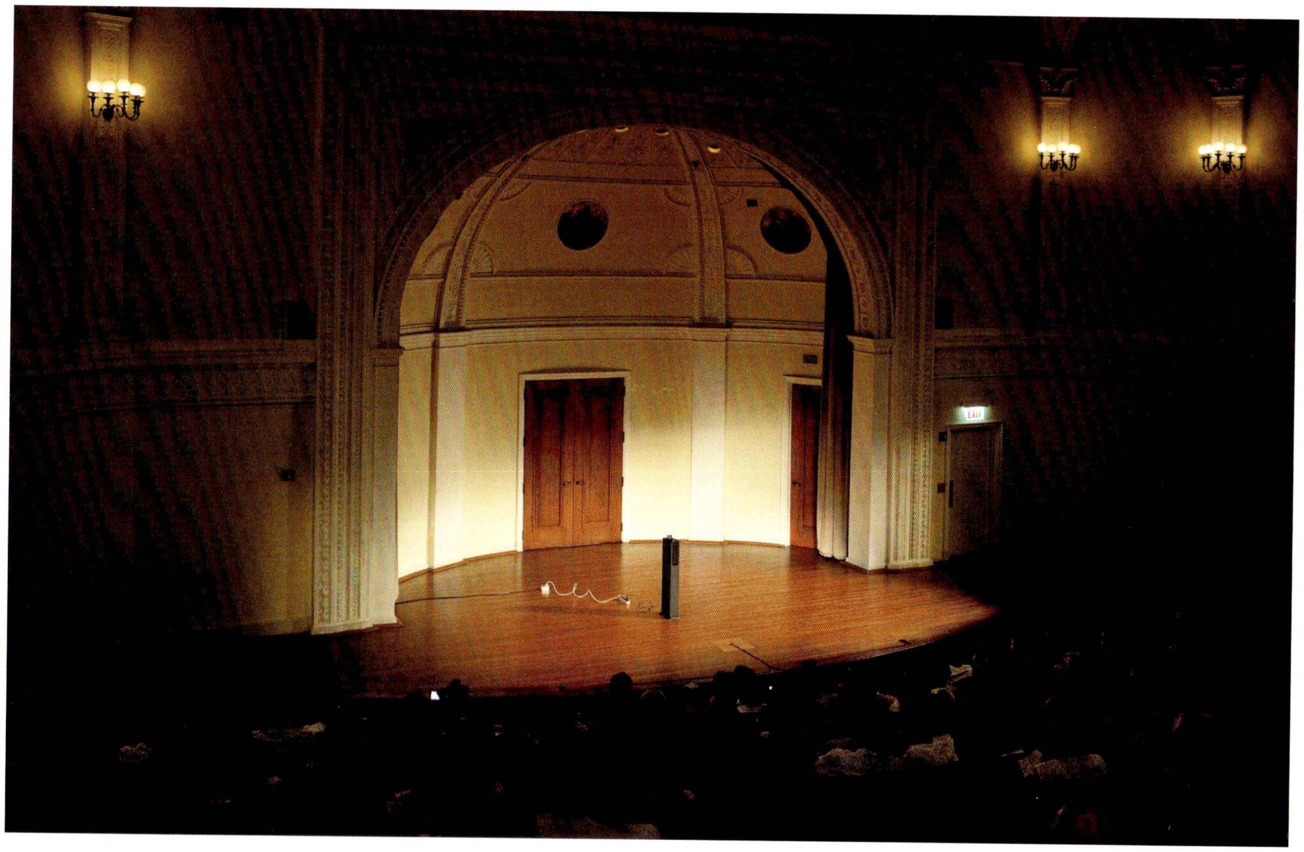

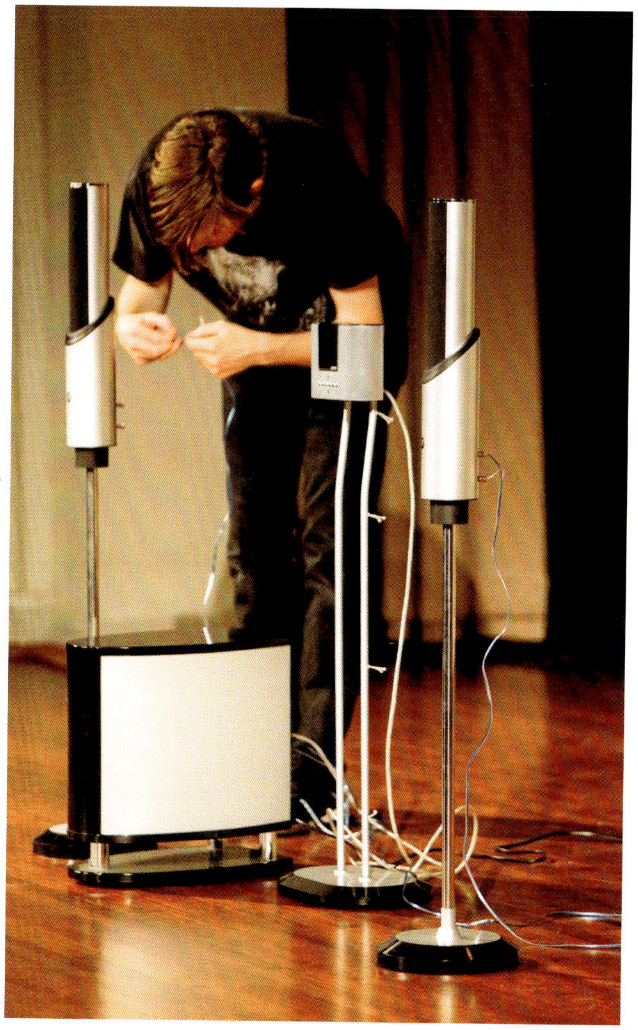

Music for Stereos, 2010–
Performance for home stereos and CDs
Courtesy of the artist

We are taking pop music here, so by style I am referring to several aspects that combined produce what is "of pop." Specifically, melody, harmony, production, and something that we will call the 5th member – a kind of unknown variable which is attitude / image / ??? We are talking some sort of secret ingredient. "If you got it, flaunt it," Warhol would say. So Lil Wayne, Kelly Clarkson, Taylor Swift, and lets tack on a few more for good measure – Miley Cyrus & Avril Lavigne, have all recently released songs that have had similar melodies, harmonies, production, and "5th members." [...] What is going on here? This is music as fashion. Ready to wear, and one size fits all.

My audio performance is an argument for the possible boundaries of this style; its inception, forward leaps, back tracks, tangents, and major players over the last fifty years of contemporary pop music. (Cory Arcangel)

Serge Baghdassarians
Boris Baltschun

*1972 in Fürth (DE), lives in Berlin (DE)
*1974 in Bremen (DE), lives in Berlin (DE)

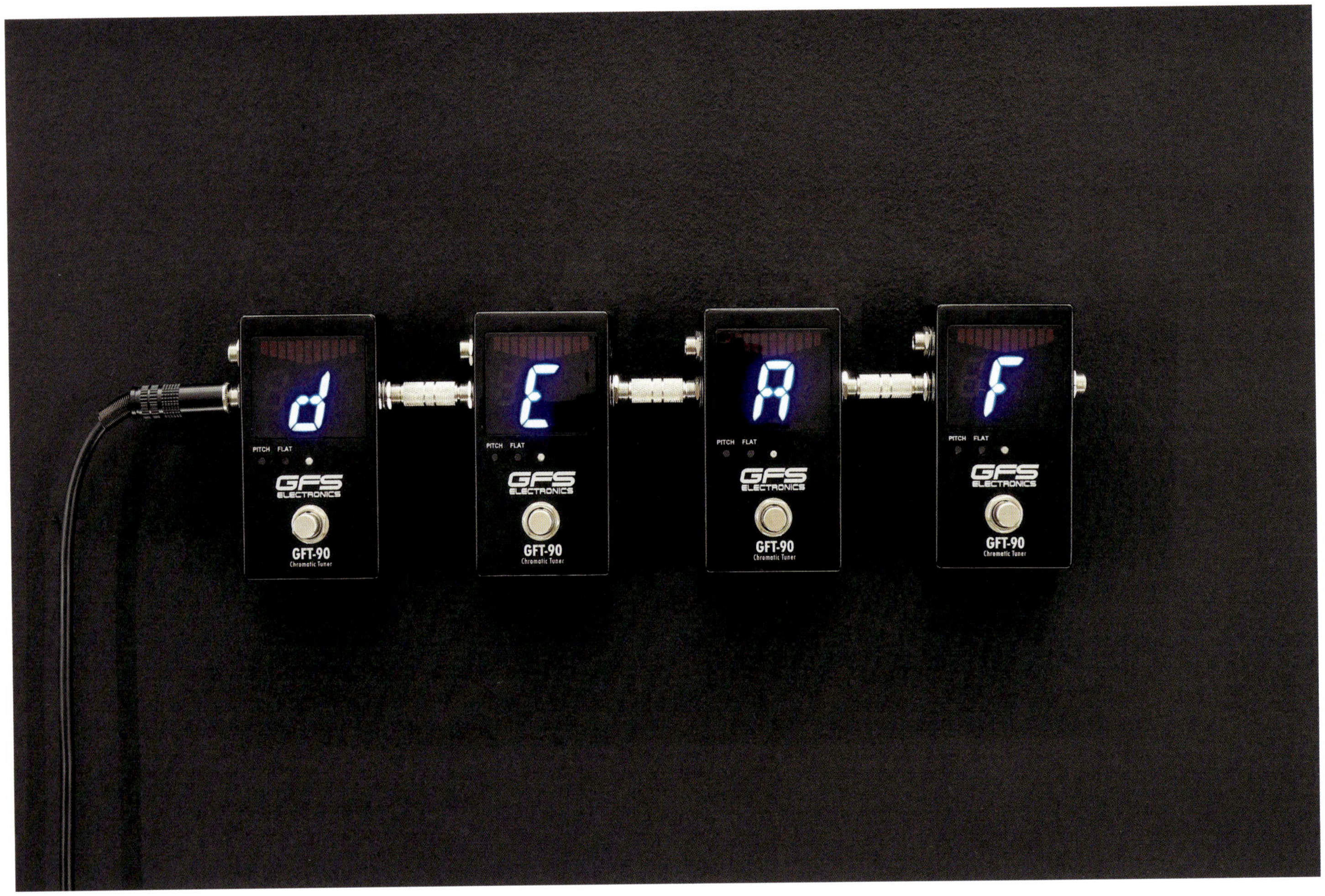

dEAF, 2011
Installation
Modified pitch tuners,
35 × 11 × 4 cm, endless
Studio Baghdassarians / Baltschun

dEAF is part of *Auditorium*, a group of works consisting of modified guitar pedal tuners. These tuners are modified to "tune themselves" without the need for any audio input. The LED displays become screens for the projection of words and sentences and transfer the durational aspect of sound into the visual domain.

www.ohnetitel.org

right side:
Abwicklung, 2011
Installation
Tape recorder, BASF and AGFA reel, amplifier, speakers, wood, book, 54 × 46 (136) × 21 cm, 96 min.
Studio Baghdassarians / Baltschun

Abwicklung in German equally stands for unreeling and liquidation. A tape machine plays back a piece consisting of sine tones slowly descending from A, G, F, and A to B, A, S, and F.

"In 1991 BASF took over the magnetic tape activities of its old competitor Agfa [...], but the transaction did nothing to improve BASF's position. On the contrary, the Agfa label largely collapsed on the market, triggering another round of personnel cutbacks [...]. Persistent overcapacity and losses left no choice but to close the sites in Bedford, Gien, and Ettenheim. Three years later BASF shareholders were still hearing the same message: 'The course of restructuring and streamlining was resolutely continued.' [BASF annual report for 1995, 26] This refrain sounded until the very end, 1996, when the magnetic tape division of BASF, steeped as it was in tradition, ceased to exist." (Werner Abelhauser et al., eds., *German Industry and Global Enterprise. BASF: The History of a Company* (Cambridge: Cambridge University Press, 2004), 588.)

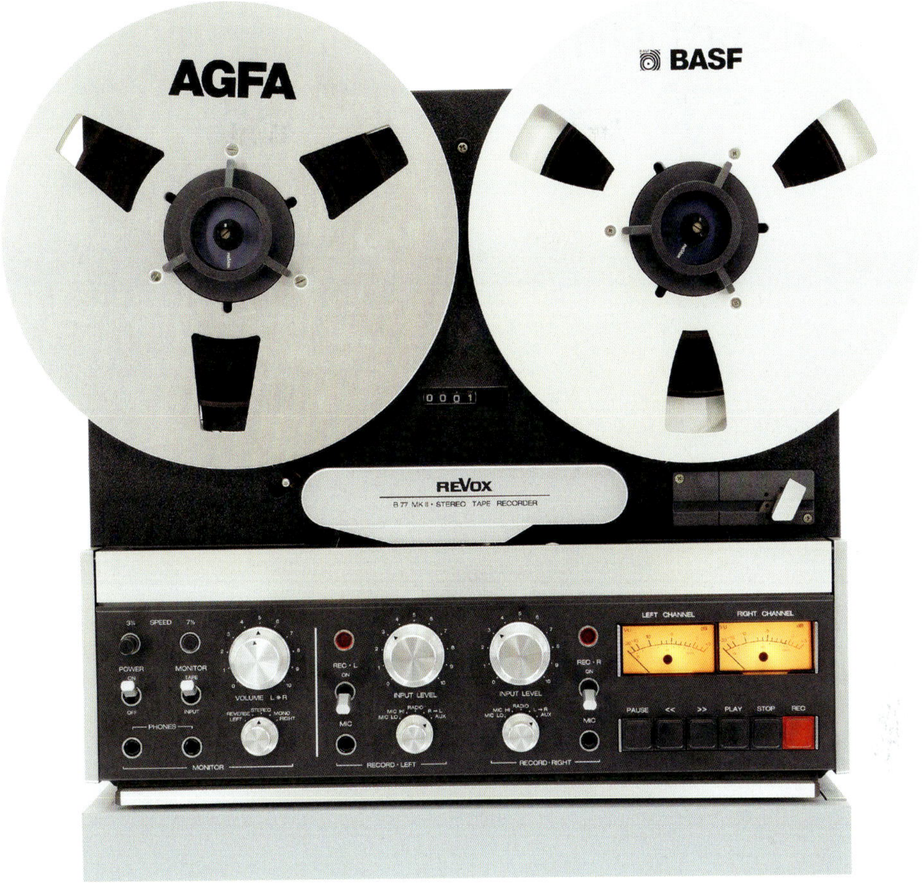
AGFA
BASF
REVOX
B 77 MK II • STEREO TAPE RECORDER

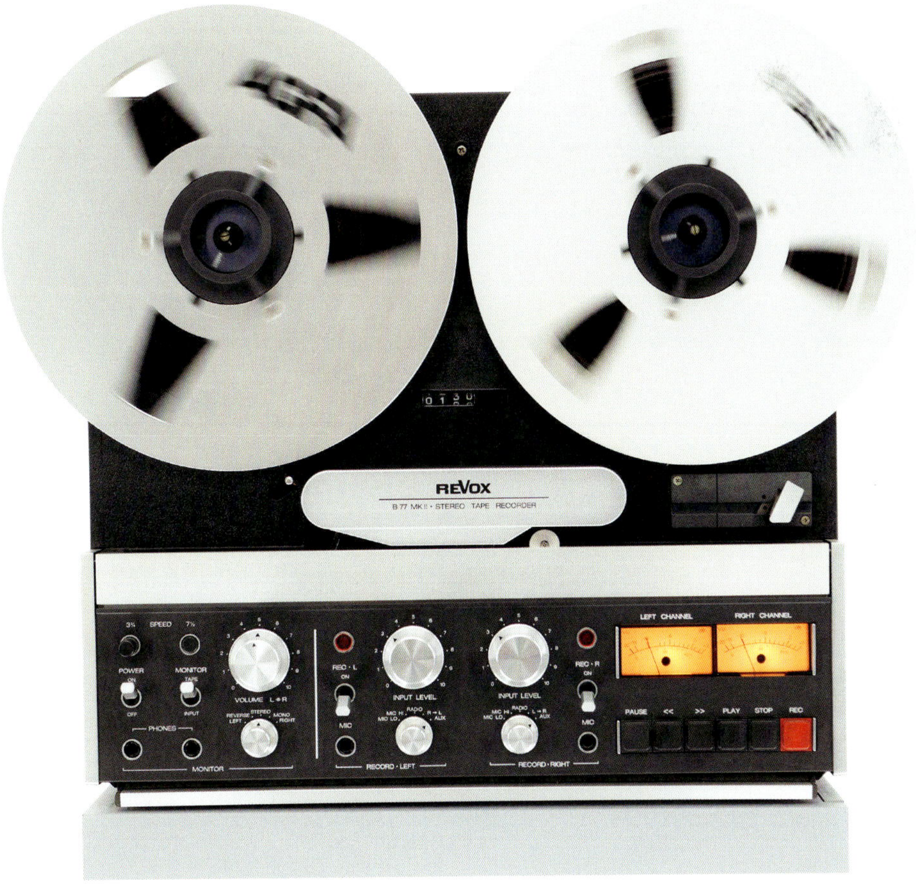
REVOX
B 77 MK II • STEREO TAPE RECORDER

Serge Baghdassarians
Boris Baltschun

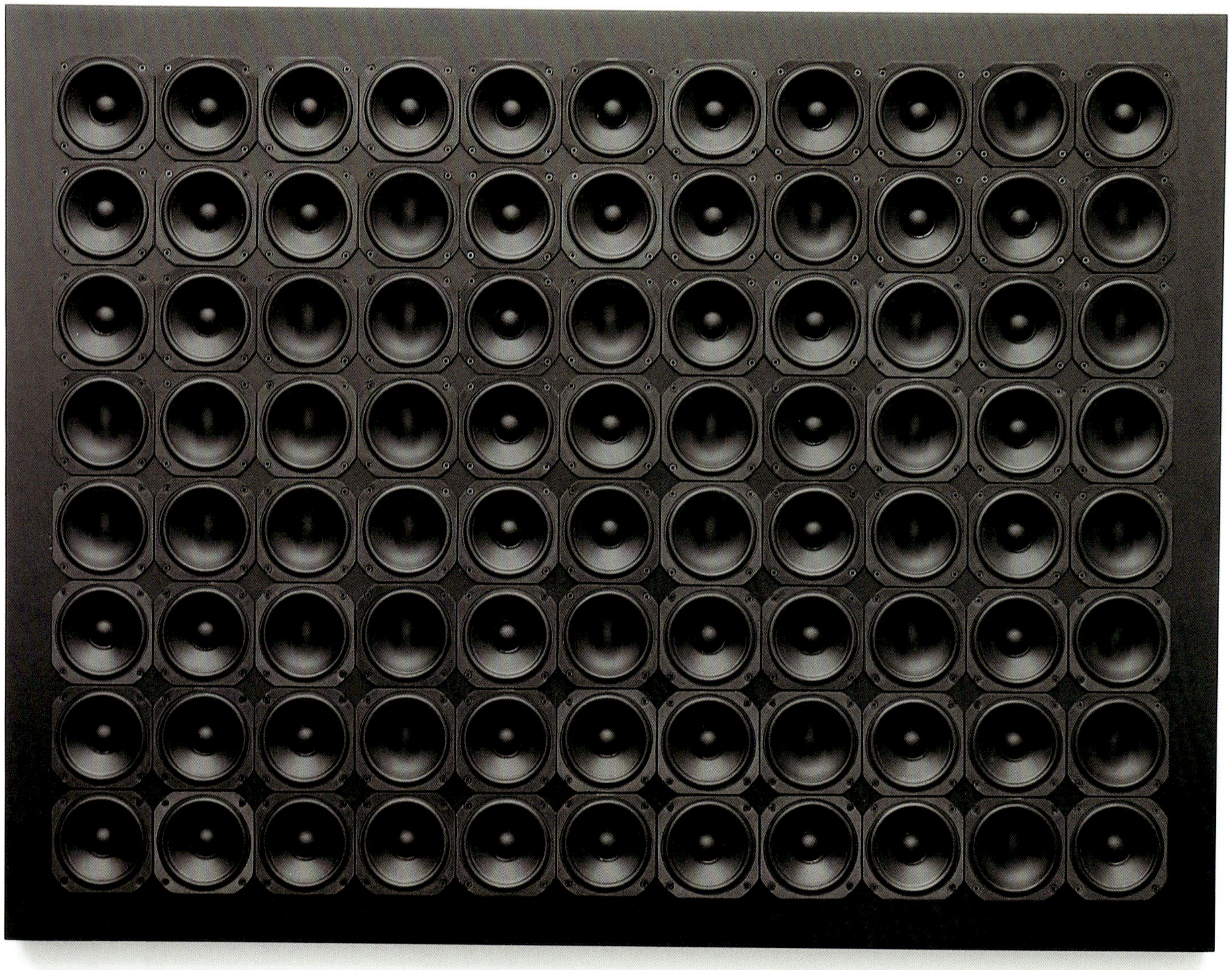

Volume, 2010
Installation
Speakers, amplifier, audio player,
wood, 110 × 76 × 5 cm, endless
Michela Brunetti

The pictogram for *Volume* gets visually animated with an audio frequency below the human hearing range. With maximum amplification the respective speakers silently but visibly vibrate. The animated pictogram consists of a loudspeaker and three sound waves.

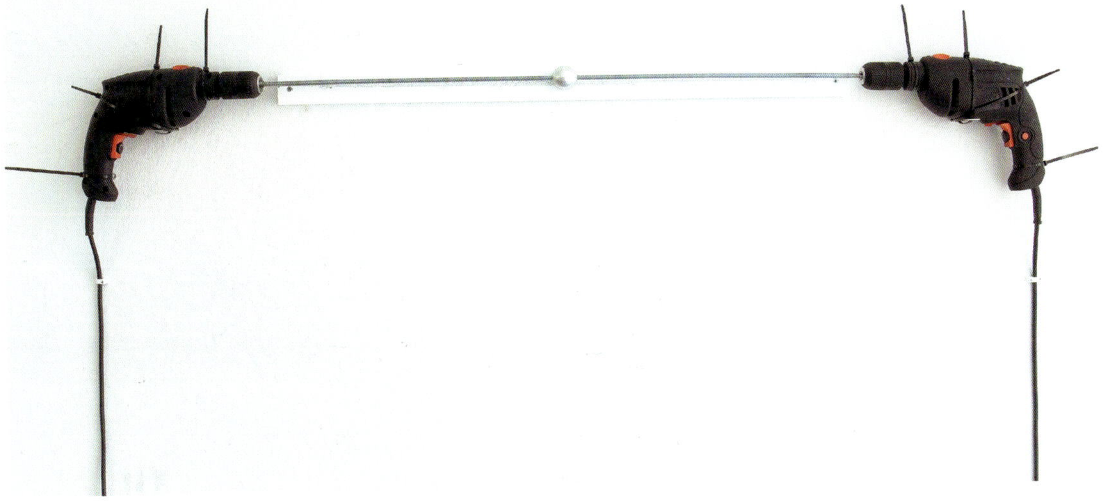

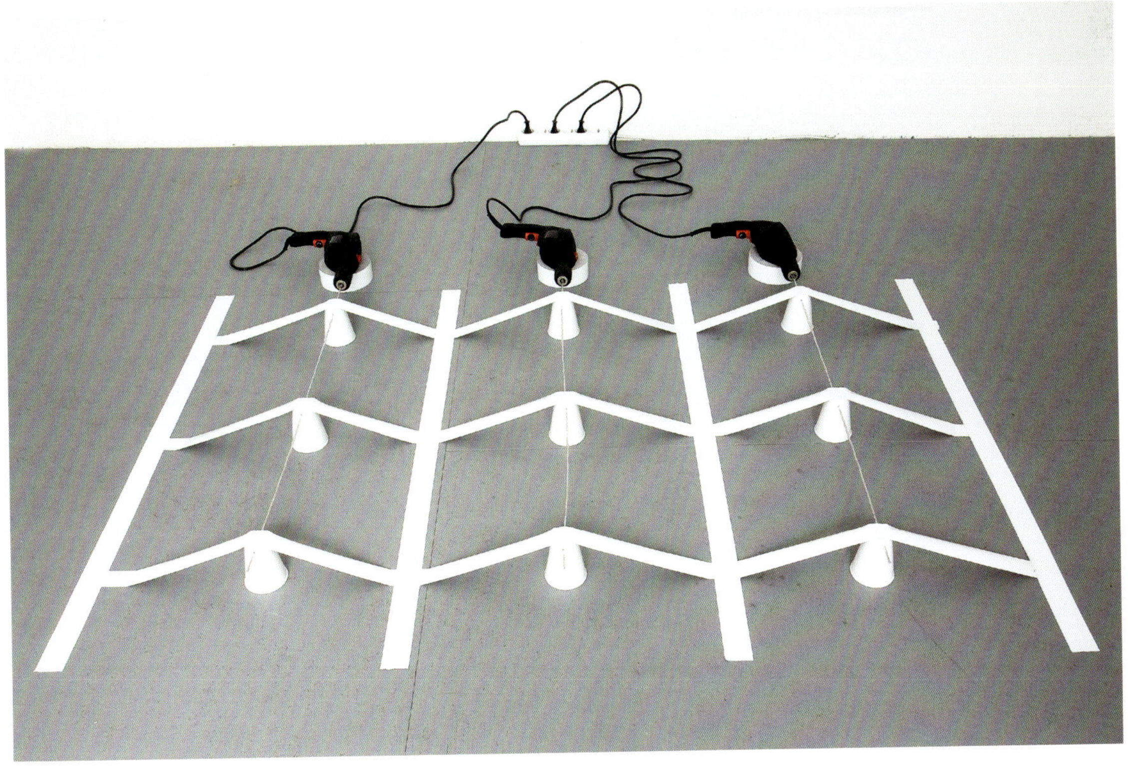

top:
Duell, 2010
Installation
Drilling machines, threaded rod, aluminum sphere, timer, 160 × 30 × 15 cm, endless
Studio Baghdassarians / Baltschun

In the work *Duell* [Duel], two drilling machines run at full speed alternately. Despite the force the aluminum sphere mounted on a threaded rod moves back and forth in slow motion.

bottom:
Ohne Titel, 2010
Installation
Drilling machine, aluminum rod, PVC tube, Styrofoam cup, adhesive tape, 172 × 172 × 20 cm, endless
Studio Baghdassarians / Baltschun

Three drilling machines run at their lowest speed slowly turning aluminum rods coated in PVC tubes. These sticks drill through Styrofoam cups and produce irregular crackling sounds through friction. *Ohne Titel* [Untitled] is part of the work series *Maßnahmenpaket* [Package of Measures] 2009–2010.

Serge Baghdassarians
Boris Baltschun

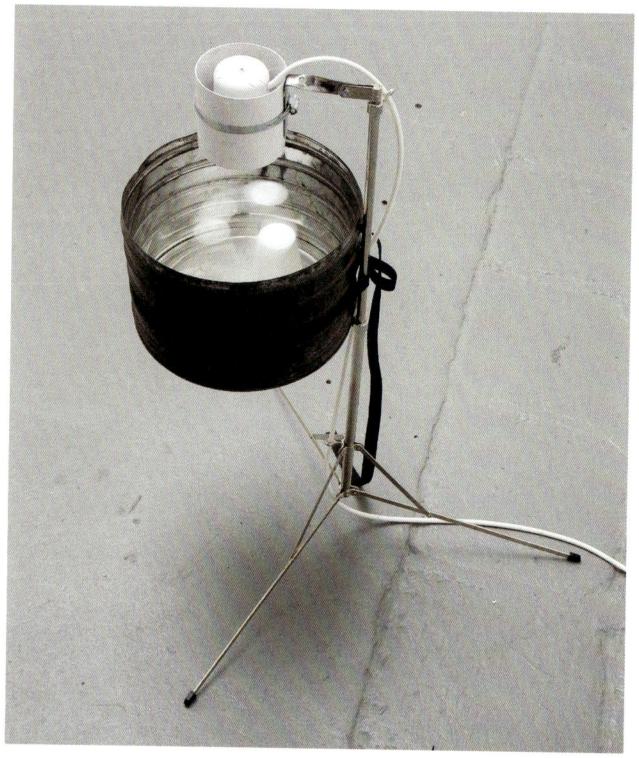

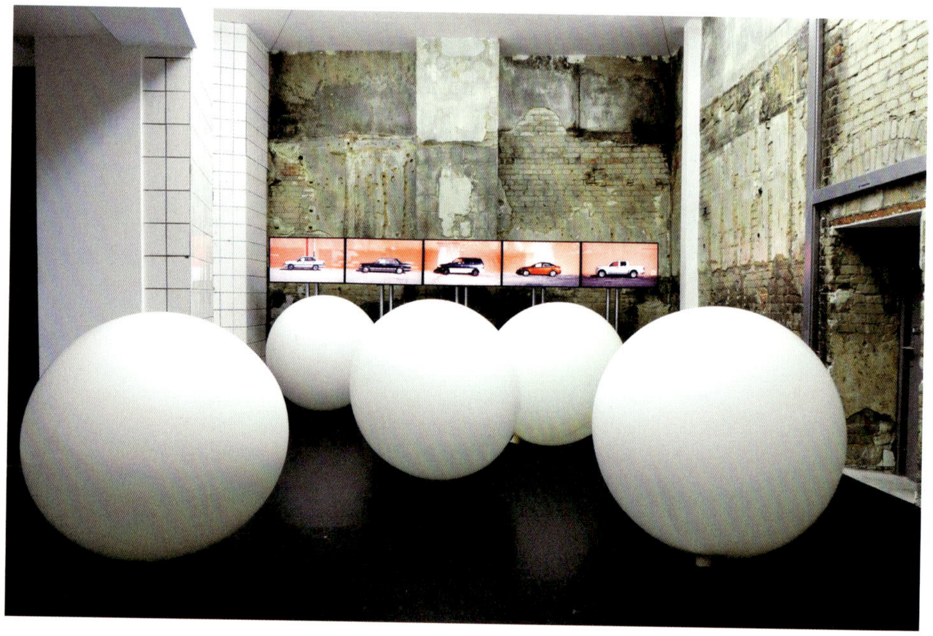

top:
Dreher, 2010
Installation
Tube ventilator, Styrofoam cups, film reel box, lashing strap, music stand, 75 × 40 × 40 cm, endless
Studio Baghdassarians / Baltschun

A ventilator is blowing air into a film reel box animating two Styrofoam cups to move around in various manners. A dance including still standing and changes in direction.

Dreher [Turner] is part of the work series *Maßnahmenpaket* [Package of Measures].

bottom:
Echo Park, 2013/2015
Installation
5 HD videos, 5 balloons modified with animal voices, dimensions and duration variable
Studio Baghdassarians / Baltschun

Echo Park combines a filmed action in urban space with a physical sound object. Parked automobiles transform into reference tone generators while large balloons act as variables. The balloons appear both on screen and – as actors who left the screen for the real world – in the exhibition space.

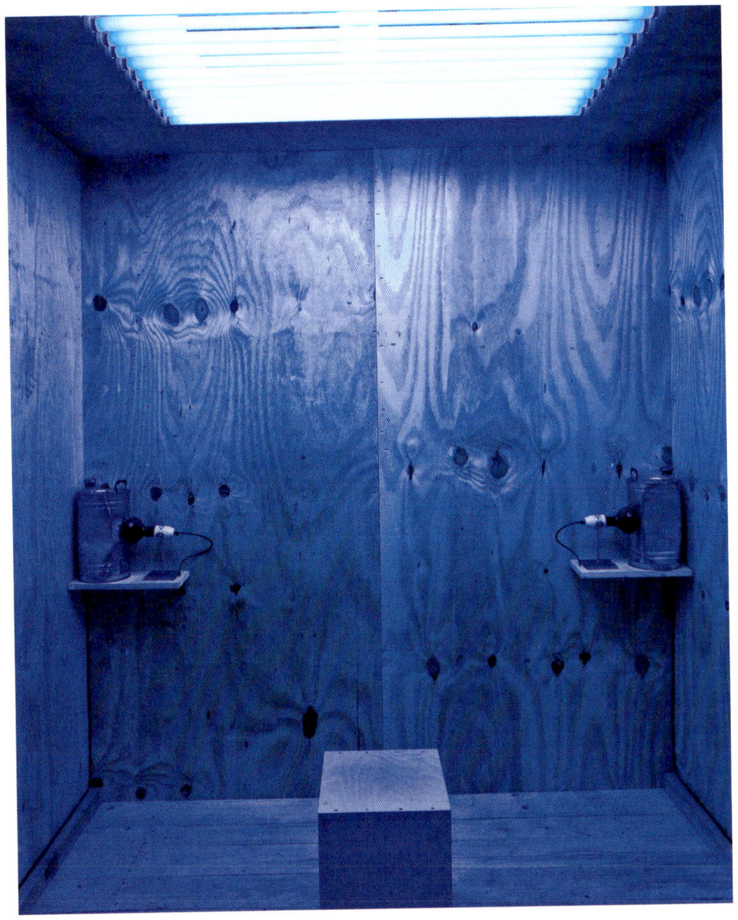

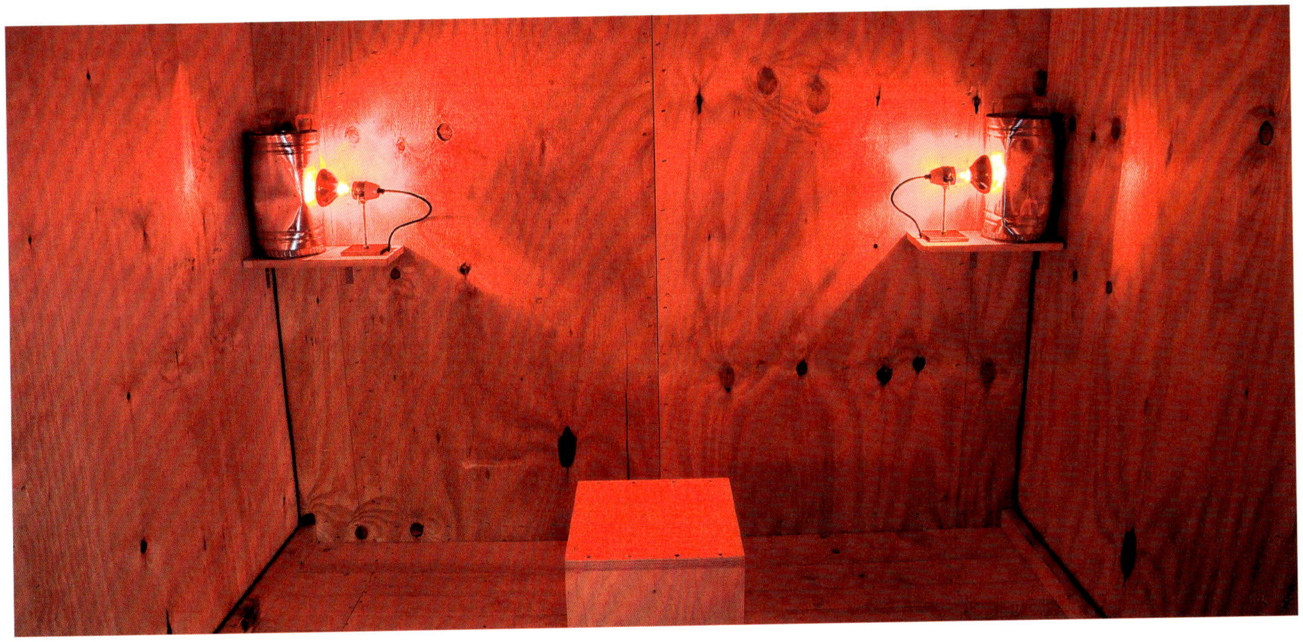

Squaring (Sauna), 2010
Installation
Cans, infrared lamps, blue fluorescent lamps, wood, timer,
20 min., 260 × 220 × 220 cm
Studio Baghdassarians / Baltschun

Four large tin cans are placed in four corners of a wooden container. The inside air temperature of these cans is changed through alternating heating and cooling phases. The temperature change causes the cans to produce a succession of percussive sounds.

Squaring (Sauna) is part of the work series *Maßnahmenpaket* [Package of Measures].

Jens Barth

*1980 in Halle an der Saale (DE), lives in Stuttgart (DE)

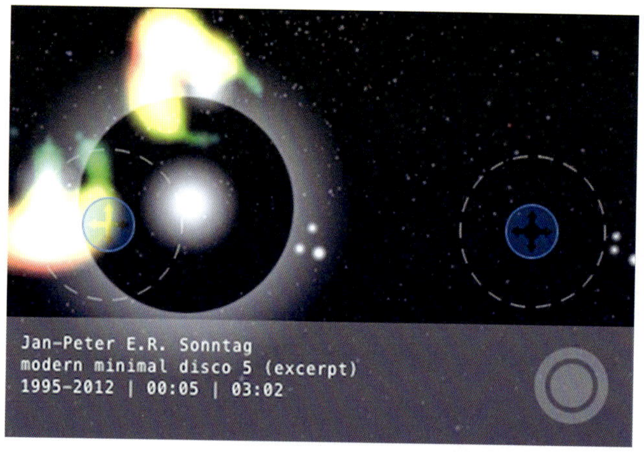

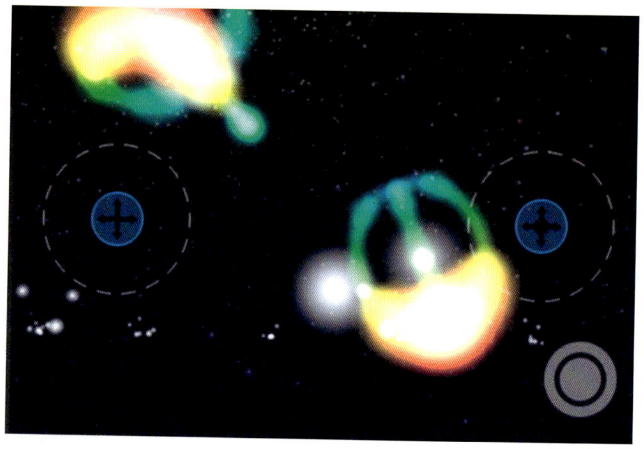

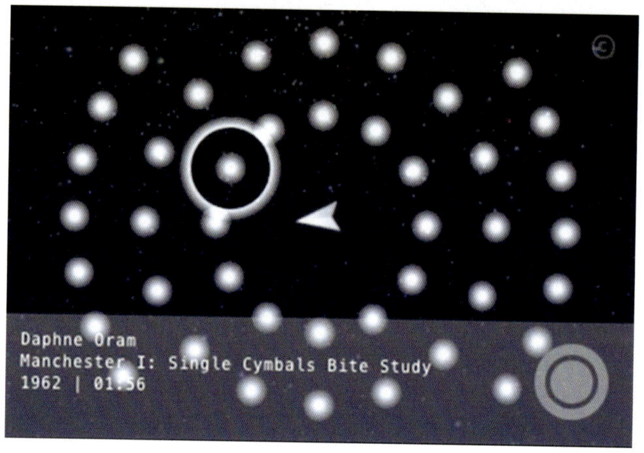

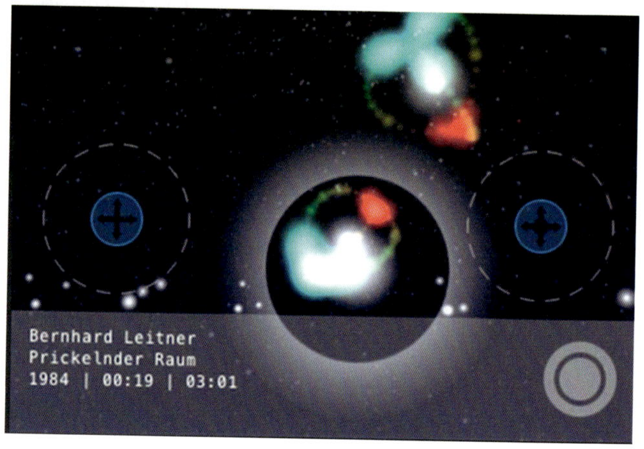

Sonor Field, 2012
App with 40 different artworks and compositions
Audio mastering: Anton Kossjanenko
Commissioned and produced by ZKM | Karlsruhe

www.jensbarth.com

Sonor Field is a 3-D simulation of the exhibition *Sound Art. Sound as a Medium of Art*. Users can travel through the virtual spherical exhibition space and experience more than forty different artworks and compositions by renowned international sound artists. Each sound and its inherent channels can be arranged in space, combined and listened to in 3-D. In this way, users can have a virtually real and individual experience of the exhibited artworks.

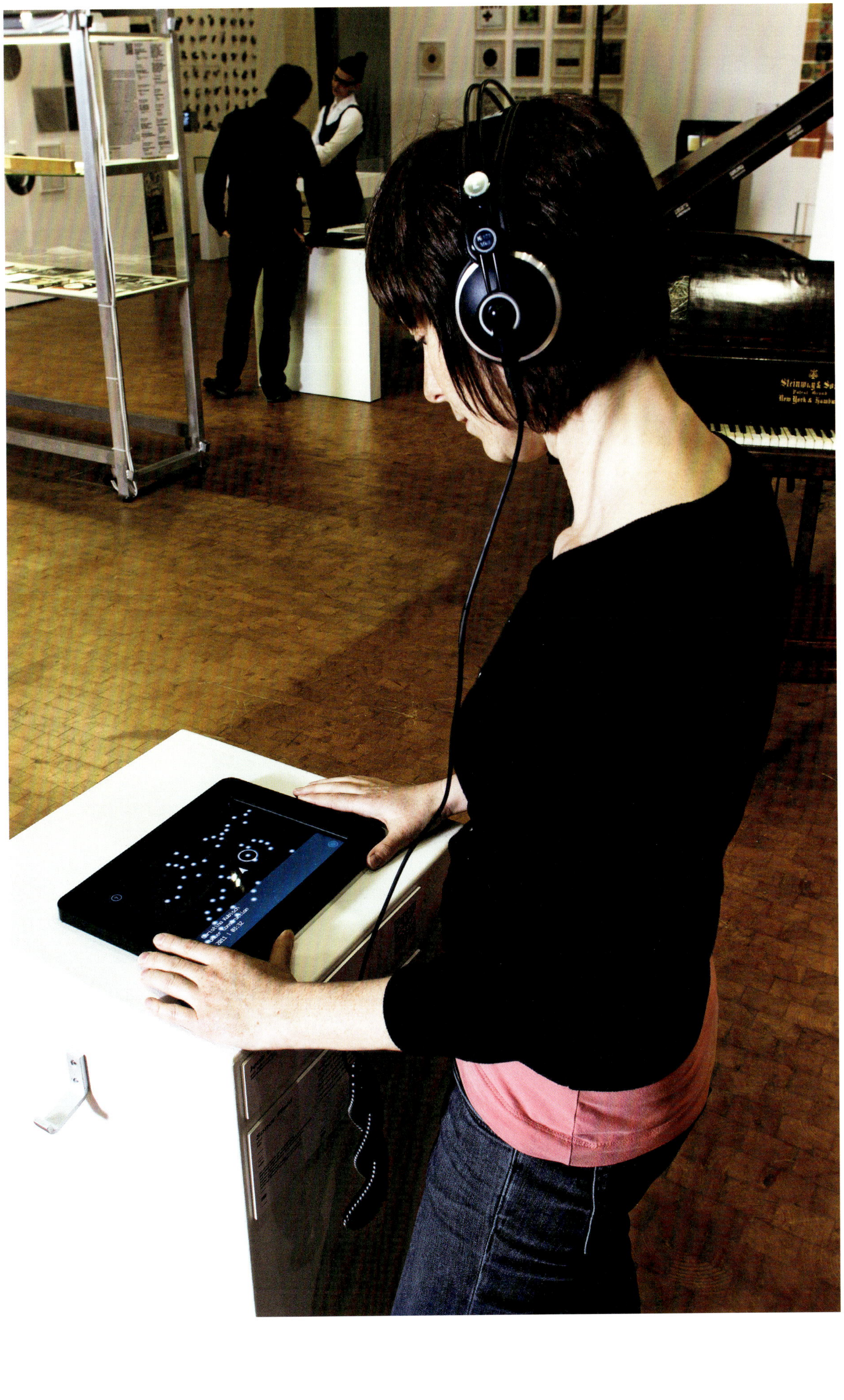

Bernard Baschet
François Baschet

*1917 in Paris (FR), †2015
*1920 in Paris (FR), †2014

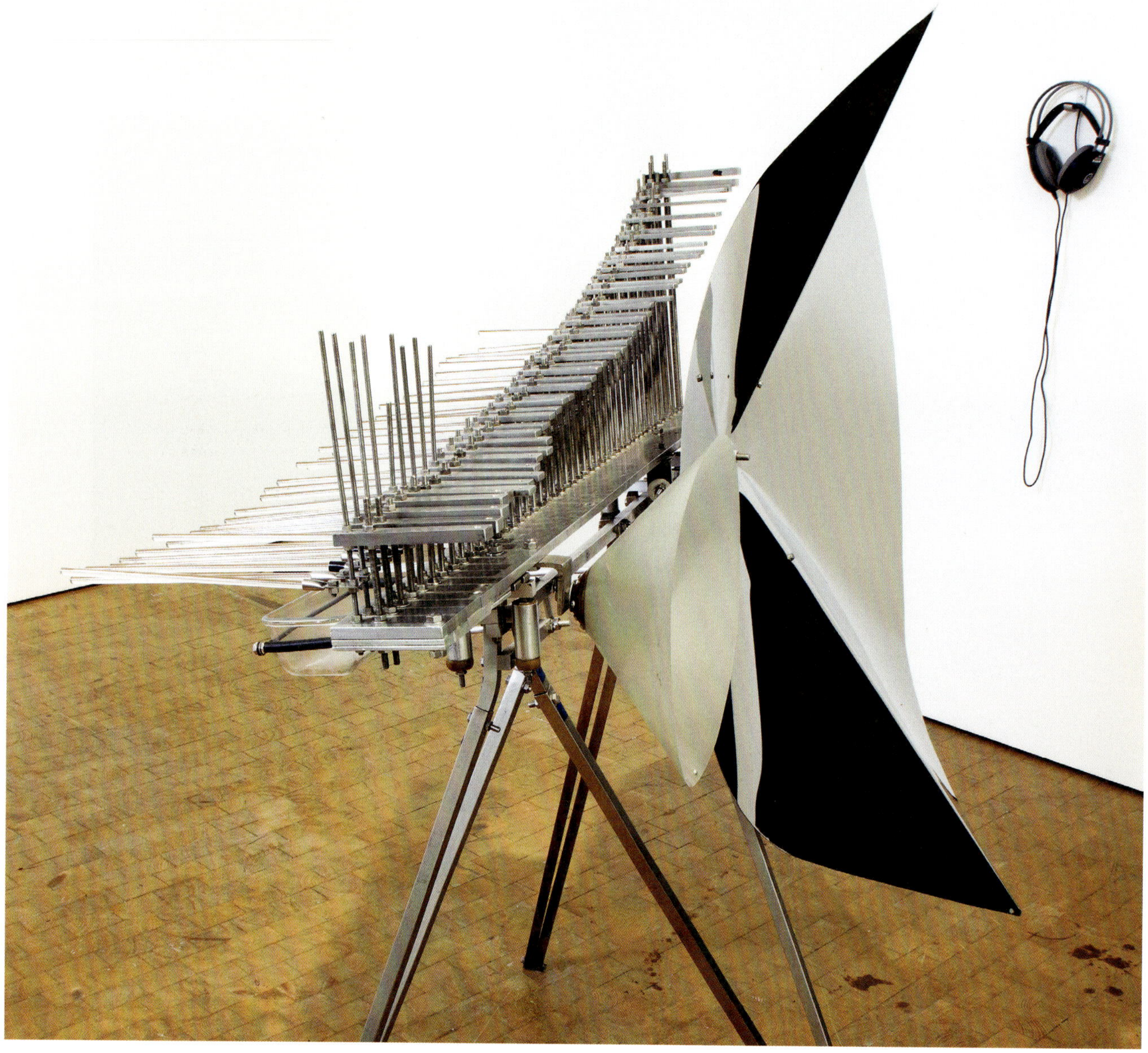

Baschet Sound Structures – Baschet Cristal, 1970s
Musical instrument / sound sculpture
Metal, glass, fiberglass,
140 × 100 × 80 cm, 60 kg
Production: technicians from Baschet's workshop

www.baschet.org
www.francois.baschet.free.fr

The *Baschet Cristal* is a chromatically tuned acoustic musical instrument made of glass and metal. The brothers Bernard and François Baschet created the first *Cristal* in 1952. Since then, some fifty instruments in different registers were produced. As with the glass harmonica, the sound sculpture is played by rubbing its glass rods with moistened fingers. The vibrations are transmitted to metal rods and from there to the amplifier. Fiberglass cones and teardrop-shaped metal forms act as amplifiers.

The brothers' collaboration began with the development of the *Cristal*. At the time, the engineer Bernard and the sculptor François Baschet were studying historic treatises on acoustics and sketches of instrument designs. In addition to *Cristal*, François and Bernard Baschet have also designed other, purely acoustic *Sound Structures* such as the *Multitone Keyboard* and the educational *Baschet Instrumentarium*.

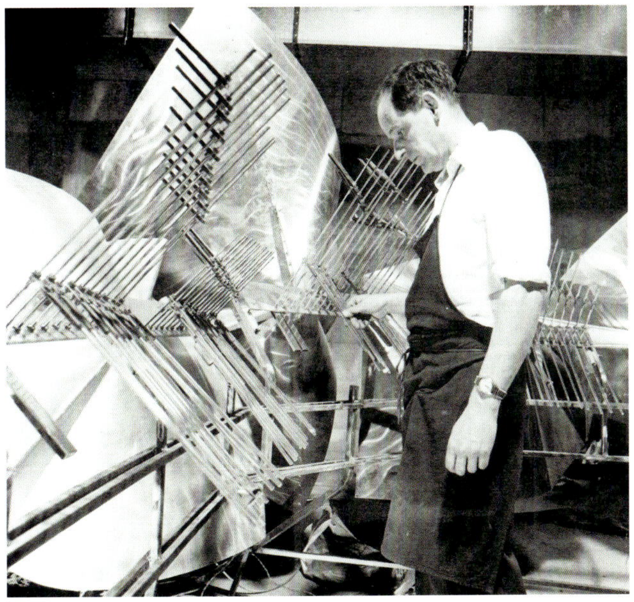

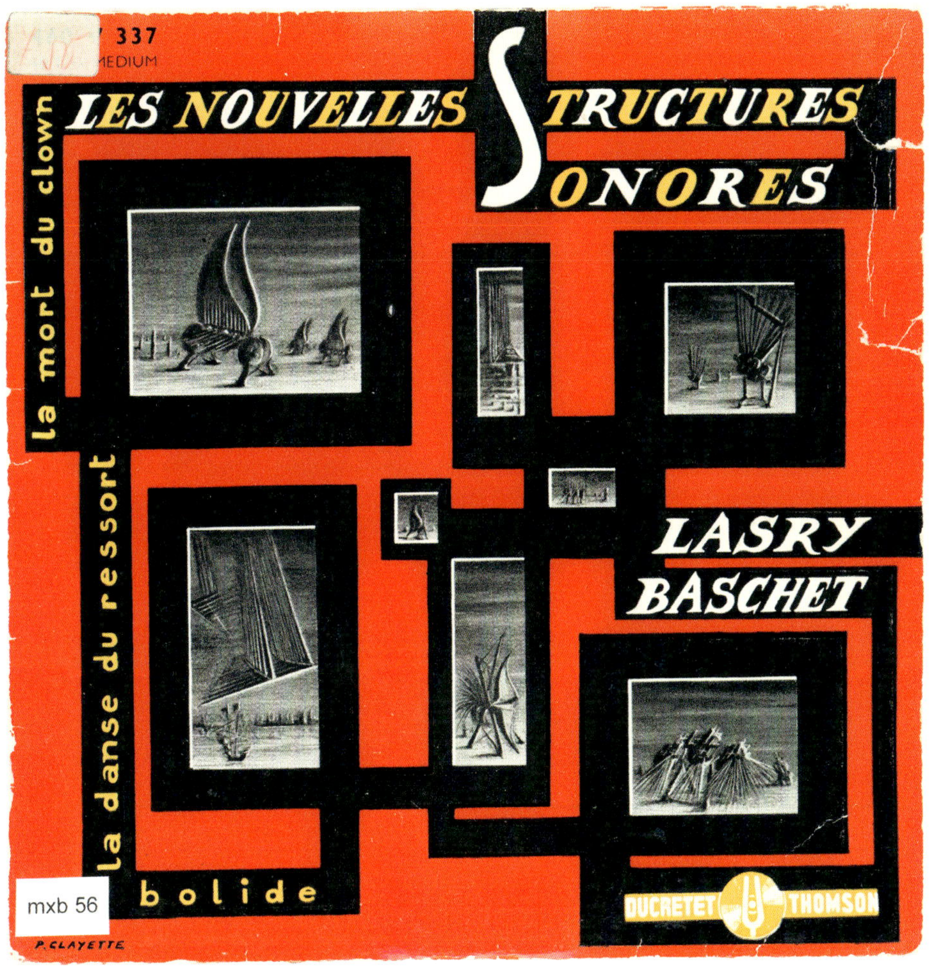

top:
François Baschet in front of the *Baschet Cristal* in the 1970s.

bottom:
François and Bernard Baschet, Jacques, Yvonne, and Teddy Lasry, *Les Nouvelles Structures Sonores. Lasry-Baschet*, 1967, 7'' EP, Cacophonic. The record contains three tracks played with the *Baschet Cristal* and percussions.

Bernard Baschet
François Baschet

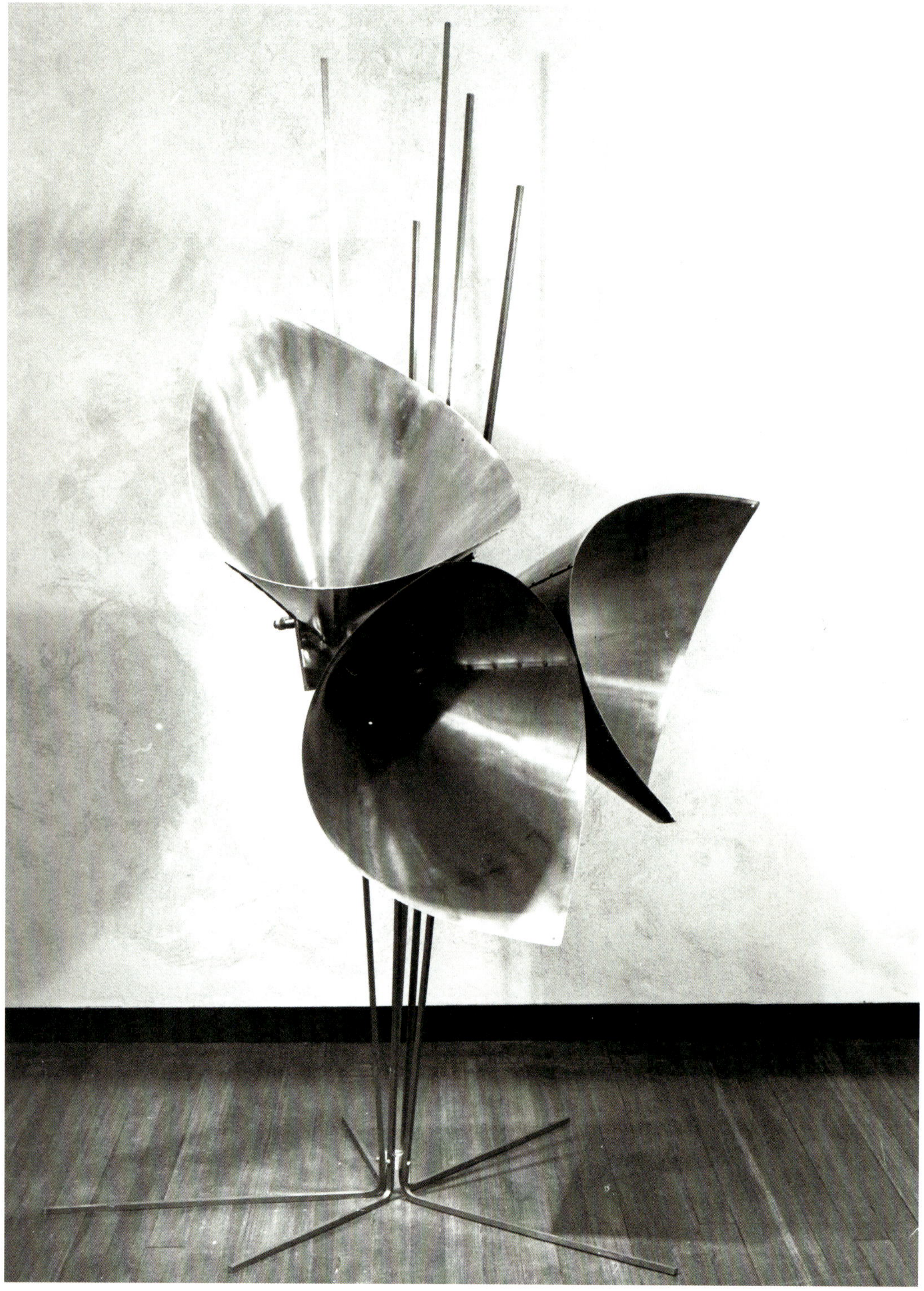

Triple Flower, 1956
Percussion with three cones and
three washers, with short rods
Steel and aluminum, 214.6 cm
Yale University Art Gallery

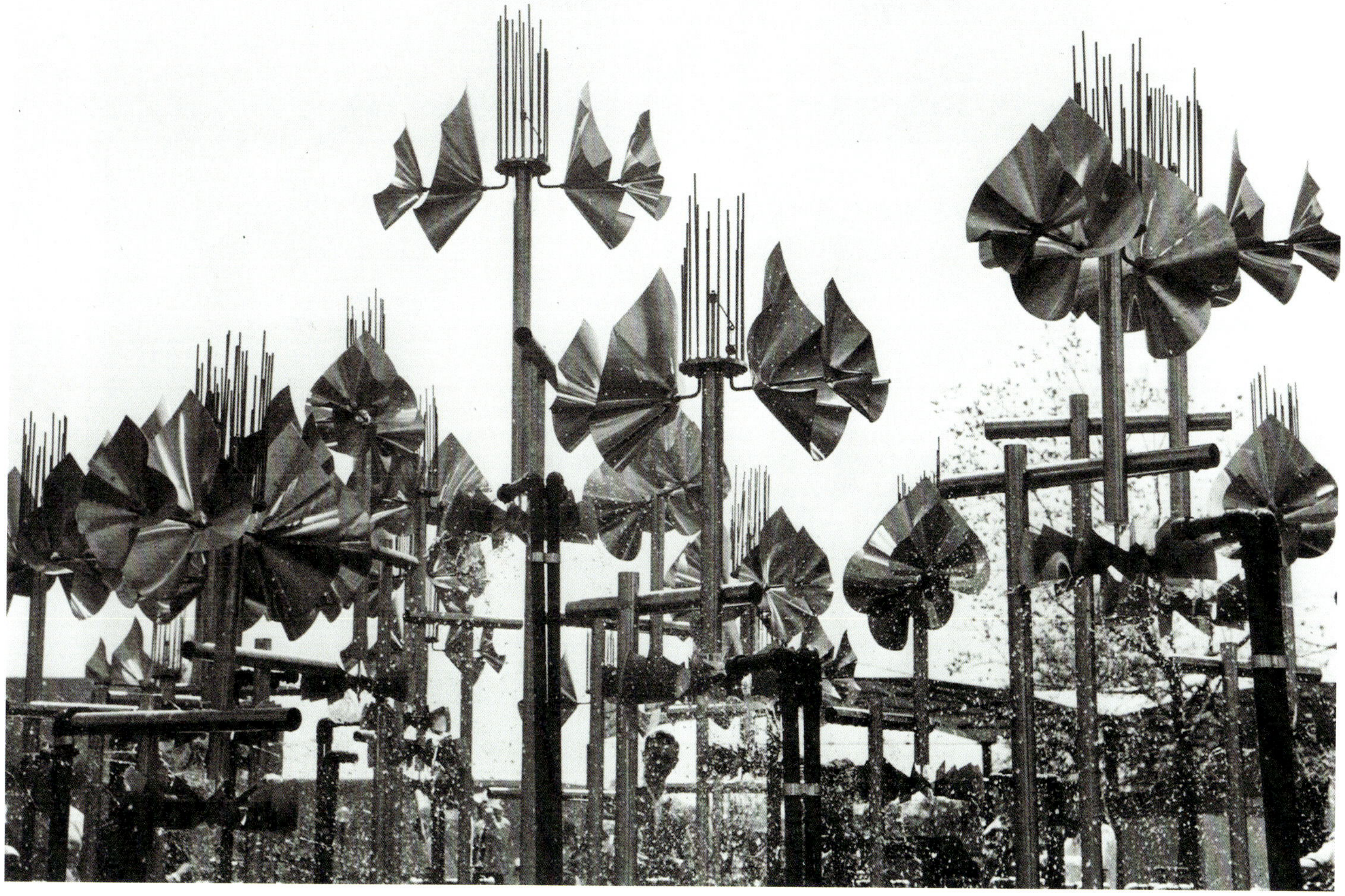

HemisFair Musical Fountain, 1968
Sound sculpture
56 intervalves which operate
28 musical mills

The sound sculpture was made for
HemisFair '68, the official 1968
World's Fair in San Antonio, Texas.

Harry Bertoia

*1915 in San Lorenzo (IT), †1978 in Barto, PA (US)

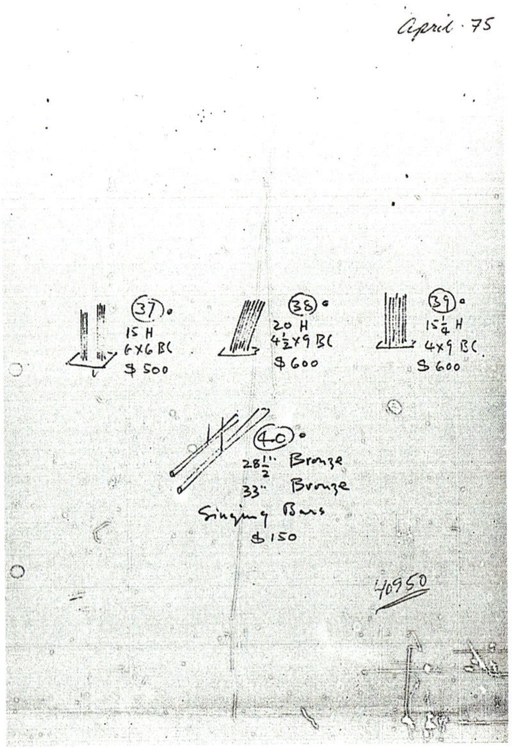

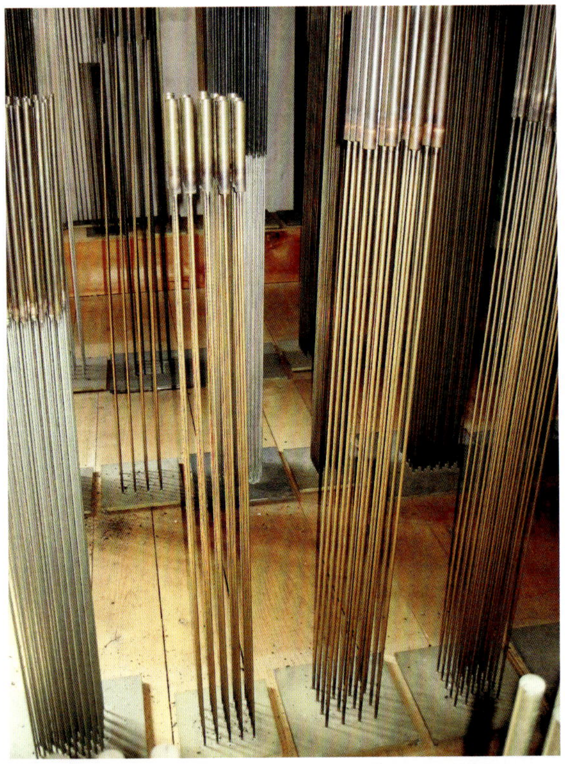

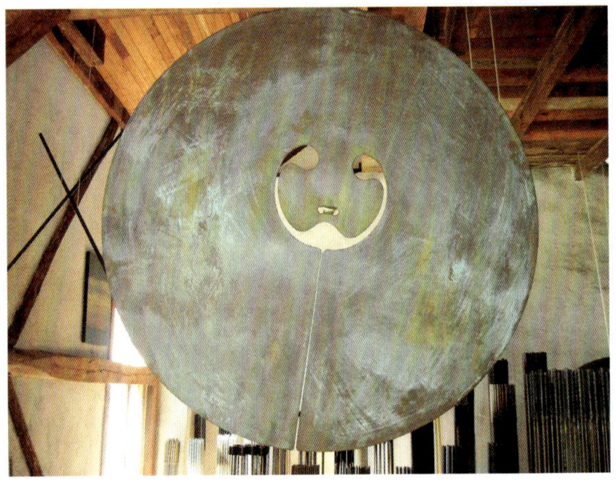

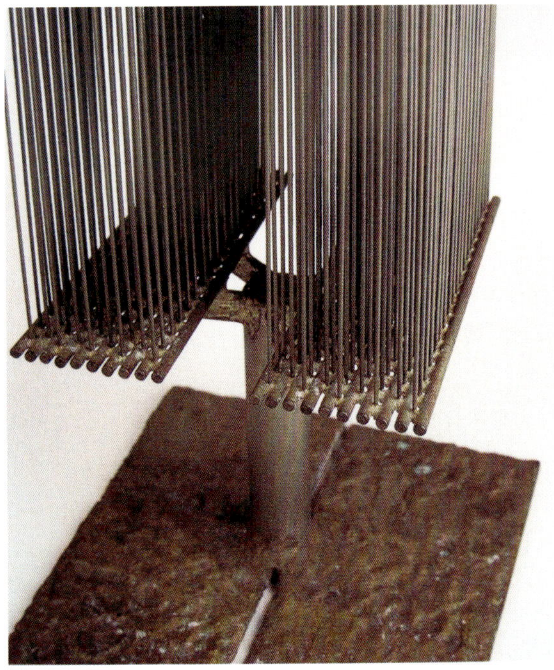

Sonambient, early 1970s
Sound sculpture
Brass, steel, 96 × 30.5 × 11.5 cm
M. u. H. Berlinicke, Wildeshausen

Sonambient, the Sounding Sculpture of Harry Bertoia

Harry Bertoia, my father, always wished there was an instrument that did not require years of training to learn. Having grown up in a musical family, Harry did not play but loved music. He wished for a sound piece on which anyone of any age could be an instant Maestro! Turned out he had to invent it. Working on a wire sculpture in his shop one day, Harry, while bending a wire, bent it too far and accidentally forced a snap and a lovely pinging sound. This seemingly miniscule event in 1960 was the first step to his tonal or sounding sculptures. Harry isolated the moment,

> "[...] while doing this, I heard a sound. I must have heard sounds from wires many times before [...] And the moment when one sort of grasps the new element, immediately the question to mind came. Well, if one wire produces such a sound, what would two rods produce or what would ten or a hundred? And this placed the whole thing on a little more systematic way of investigation so that there was a period of experimentation along that line. Loose ends were coming together, two or three wires of various different alloys and various dimensions and so forth, you know. And this introduced me to the potential that was there and actually evolving sounds that probably were not quite heard before, at least not in that way."

In 1969, Harry remodeled his eighteenth century hay barn into a concert hall of sorts, where he placed about a hundred sounding sculptures. Sonambient, a term he coined to refer to the environment created by the tonal sculptures, lived in the barn and he gave amazing other worldly concerts there, many of which were recorded. He explained that someday he hoped the collection would be moved into a museum hall, intact as a total instrument. (Celia Bertoia)

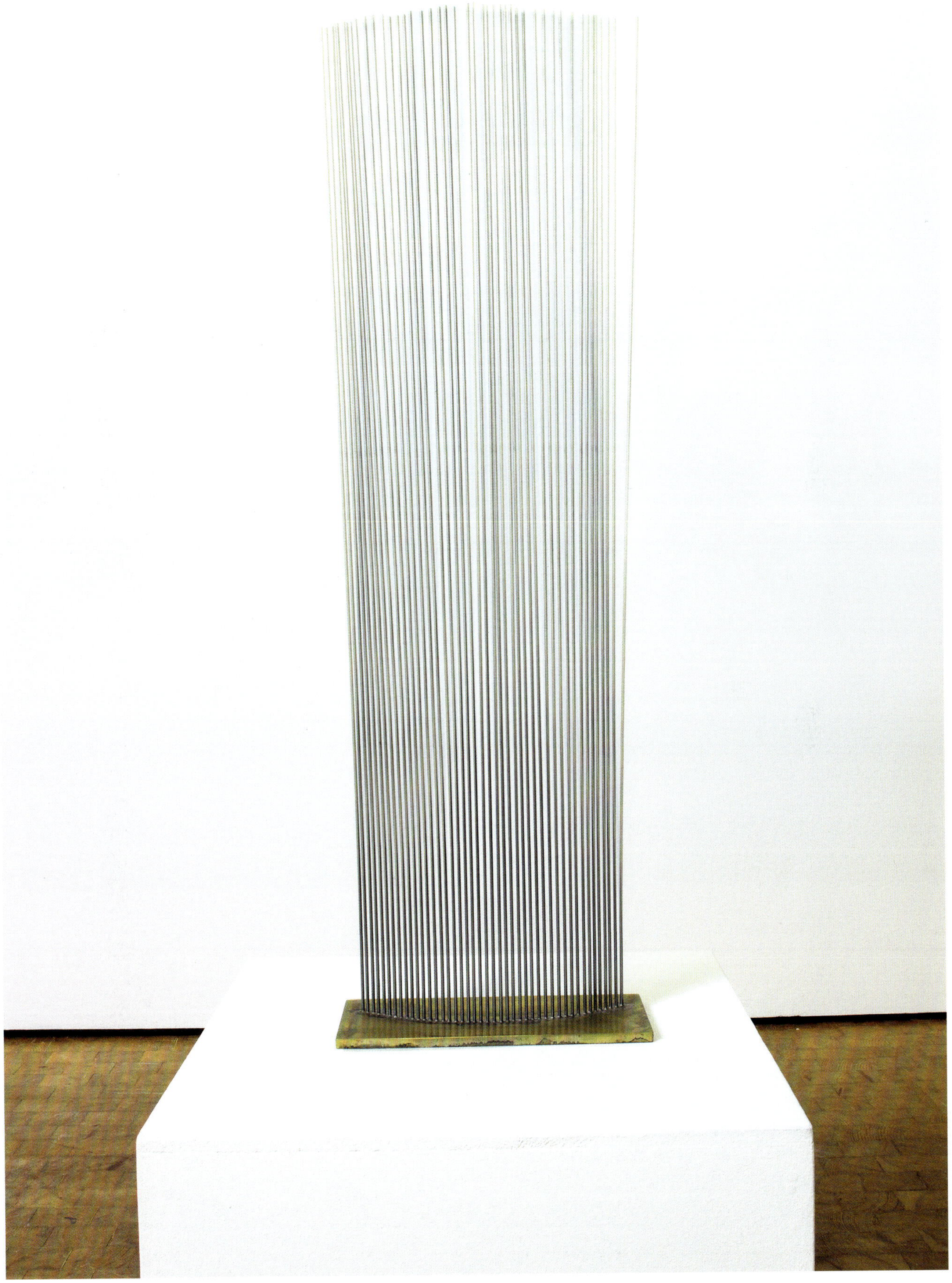

Harry Bertoia

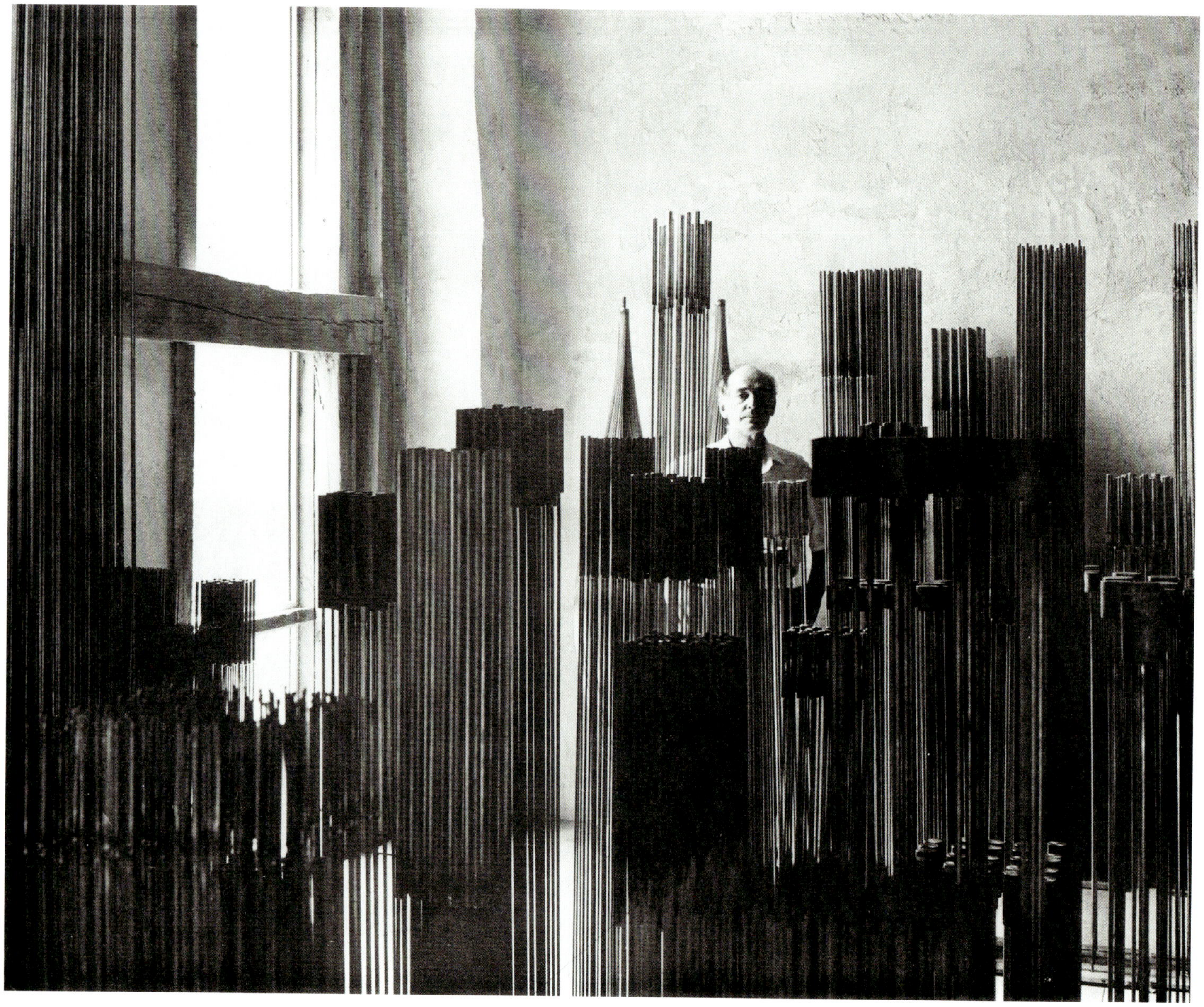

Harry Bertoia standing behind his sounding sculptures,
installed in the barn of his Pennsylvania residence.

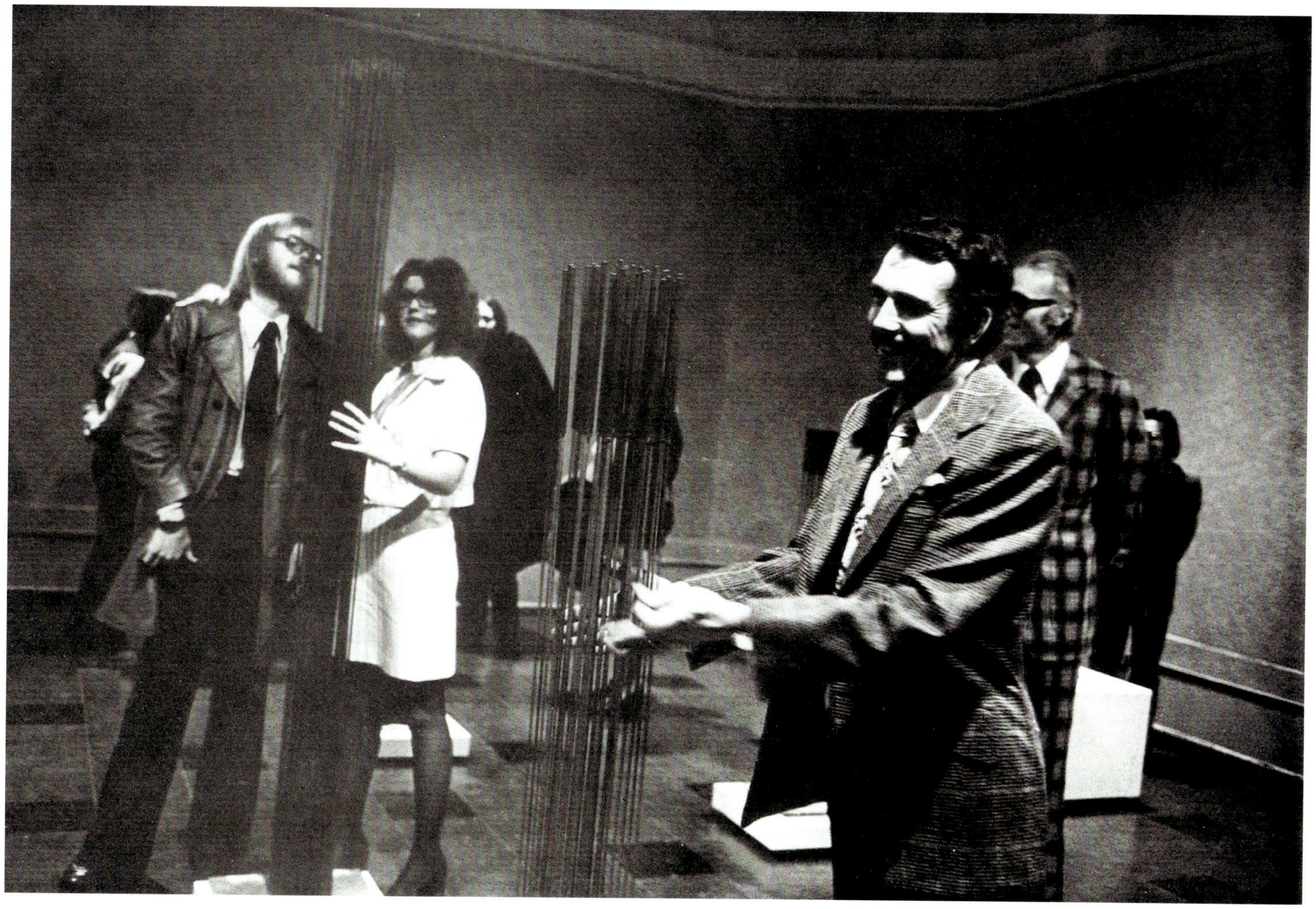

Harry Bertoia's sculptures being played by the public at the Vancouver Art Gallery, 1973.

Joseph Beuys

*1921 in Krefeld, Germany (DE), †1986 in Düsseldorf (DE)

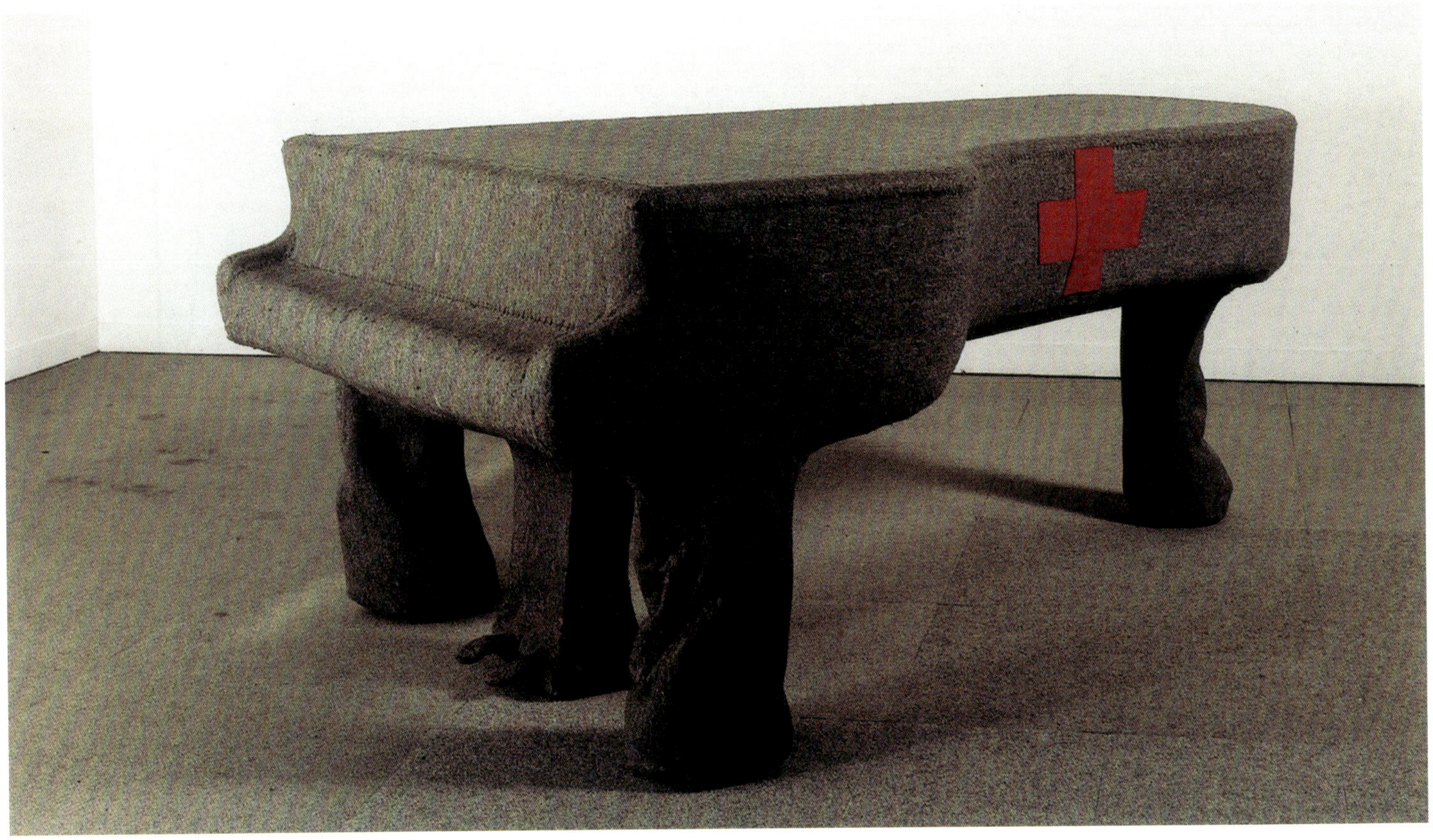

Infiltration homogen für Konzertflügel, 1966
Fluxus sculpture
Grand piano, grey felt, cross in red fabric
Museé national d'Art moderne – Centre Georges Pompidou, Paris

This work originated during Joseph Beuys's 1966 action *Infiltration homogen für Konzertflügel, der größte Komponist der Gegenwart ist das Contergankind* [Infiltration homogeneous for grand piano, the greatest composer of the present is the thalidomide child], performed by Beuys during a Fluxus festival at the Düsseldorf Academy of Art. Beuys silenced a grand piano by wrapping it with felt. With his action, Beuys wanted to highlight the pharmaceutical scandal surrounding thalidomide, a drug that caused serious birth defects in children whose mother took it during pregnancy. The work seeks to indicate the danger that occurs when we remain silent.

right side:
Klavier Oxygen, 1985
Fluxus sculpture
Grand piano, oxygen cylinder, hose, blackboard

This work originated from the *Simultankonzert an drei Klavieren* [Simultaneous concert on three pianos] which was performed on November 29, 1985 at the University of Fine Arts Hamburg by Nam June Paik and Henning Christiansen. Beuys could not attend the action, he was already very ill then. He announced his instructions to the two artists via telephone. Among other things, he gave the instruction to let oxygen escape from an oxygen bottle with a small hose. One would hear the appearance of the oxygen in the room by the noise of the escaping air.

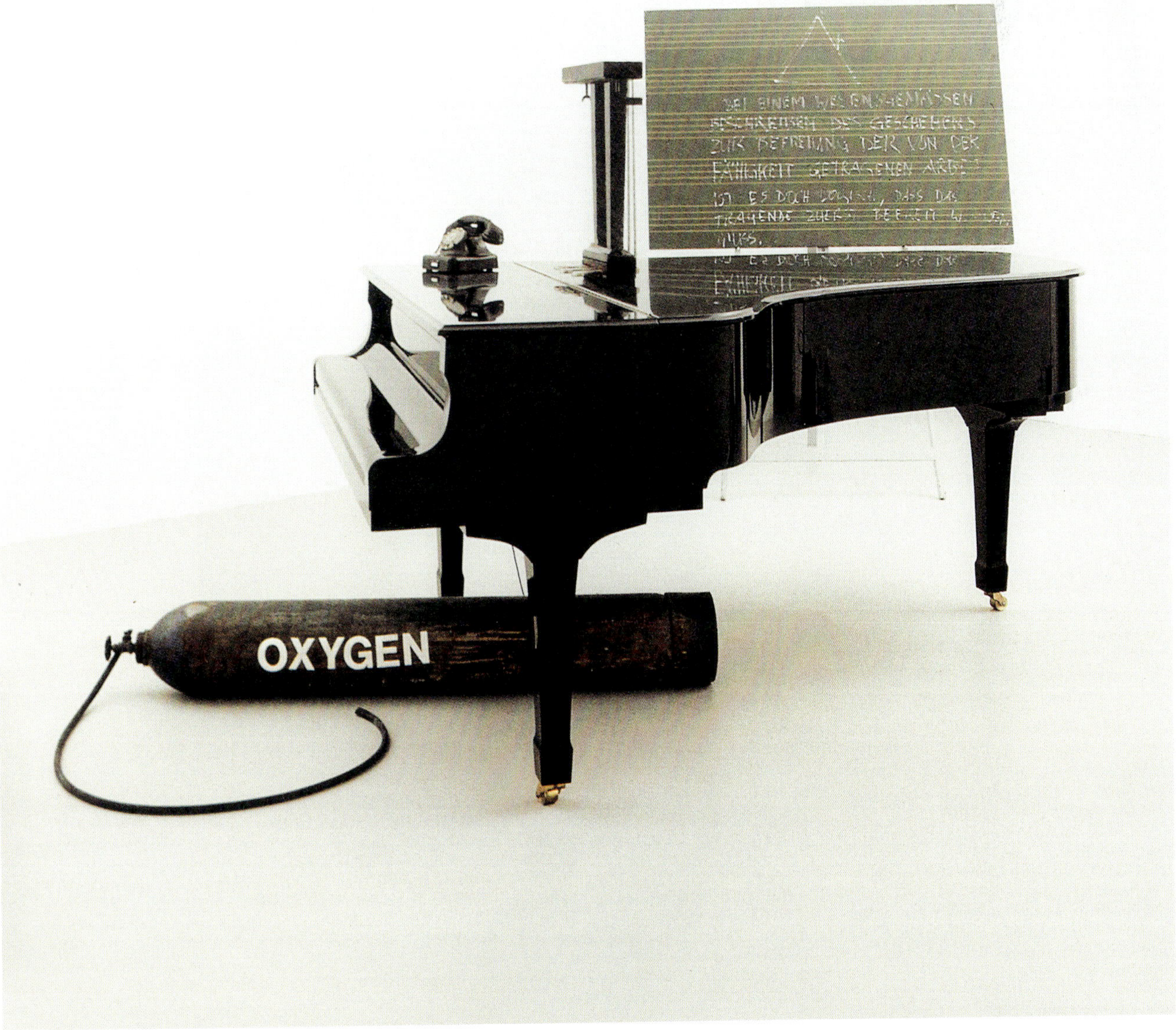
OXYGEN

Jens Brand

*1968 in Dortmund (DE), lives in Berlin (DE)

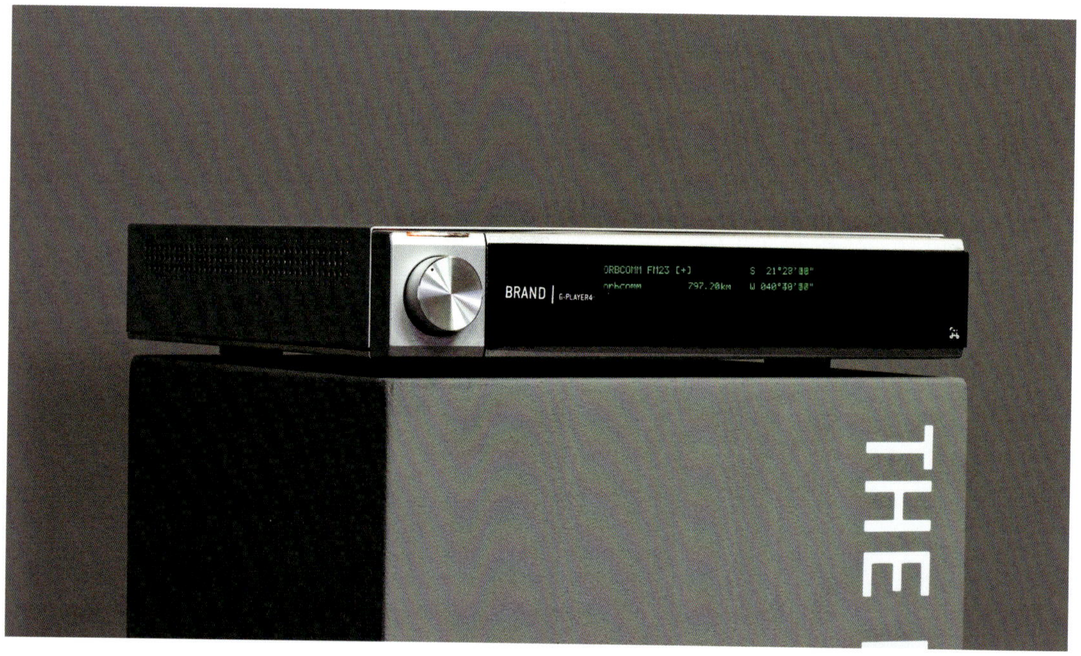

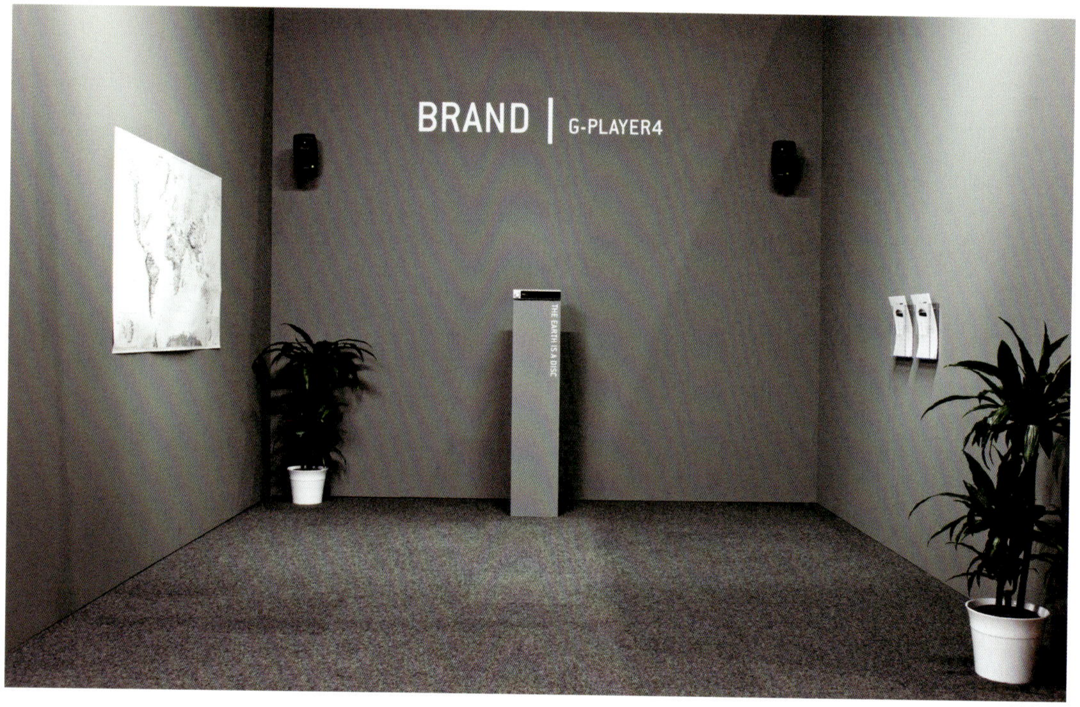

G-Player 4. The Earth is a disc, 2004
Mixed-media-installation
Technical design: Sukandar Kamin Kartadinata
Corporate design: Christa Marek
Production: "per->SON, Raumklang-Klangraum-Raumklang" (special thanks to Peter Kiefer and Georg Dietzler)
Jens Brand

www.jensbrand.com

G-Player plays the Earth like a record. It functions like a regular record or CD player. It recognizes the position of more than 1000 satellites and, by means of a virtual 3-D model of the planet, makes it possible to listen in real time to an imaginary track traced along the surface of the earth. The *G-Player 4* can be integrated into any hi-fi system and is fully functional everywhere. In 2006, the fictional company BRAND expanded its product array to include the user-friendly *G-POD*.

The project concluded in 2007 with the establishment of g-turns.com, an online store. Under the company name BRAND, the *G-Player* also is presented in an installation that simulates a trade fair booth. This staging includes, as an option, a friendly salesperson able to converse with the clientele about the meaning, aim, and beauty of the products on display.

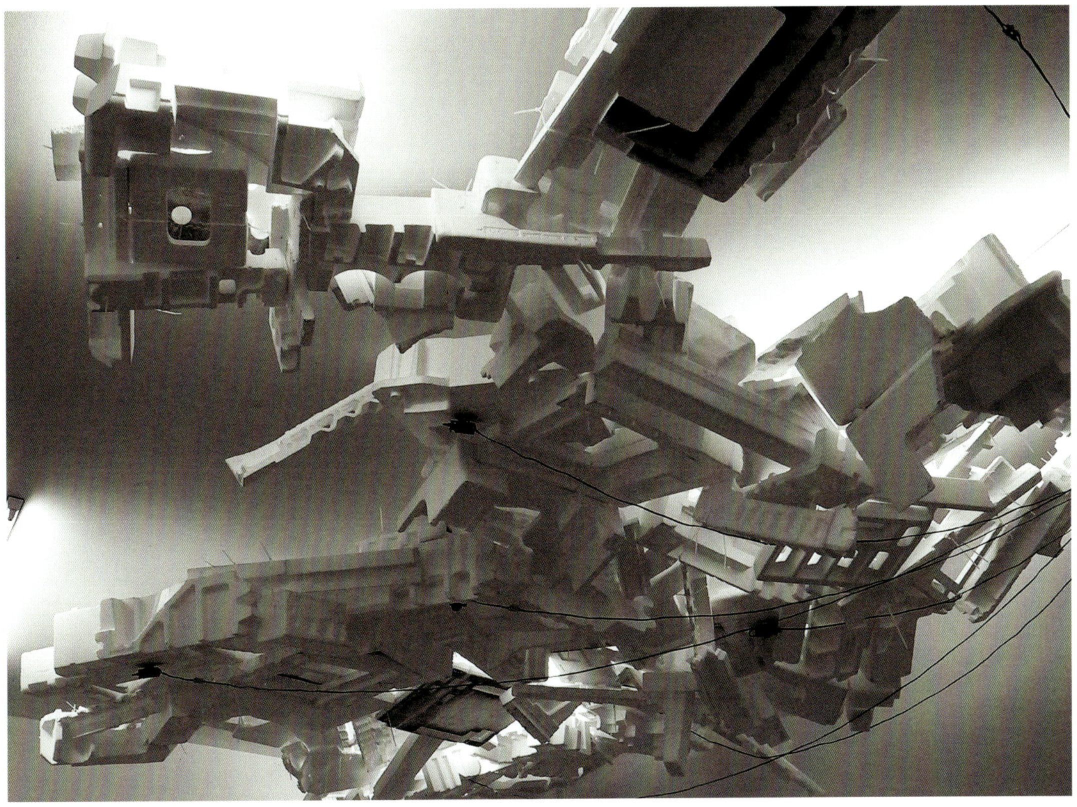

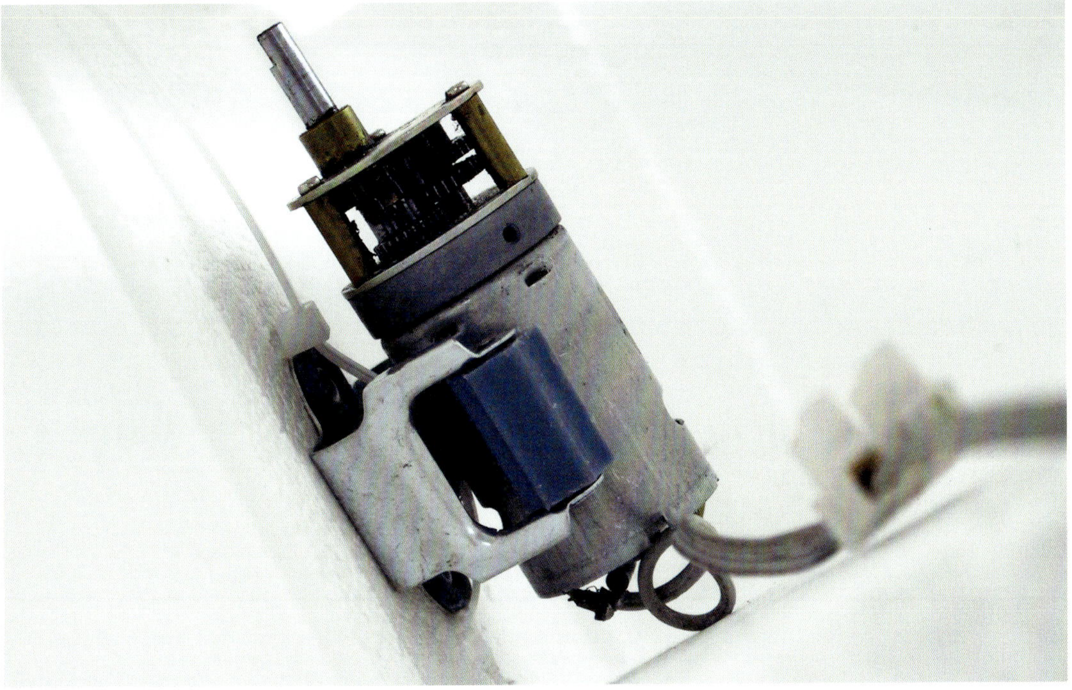

Motors & Styrofoam, 2000/2001
Concert/sound installation, ca. 30 min.
Interface and software:
Sukandar Kartadinata, Jens Brand

Eight motors are mounted to a chaotically assembled, more or less horizontally expanding plane of constructed and found Styrofoam pieces. The sound installation is hanging from the ceiling of an exhibition or concert space above the heads of the seated or standing members of the audience. The Styrofoam serves as a natural amplifier and resonator for the whirring sound of the motors. It supports the development of combining tones resulting from the tight clusters deriving from the speed of the motors and distributes them unequally across the space.

The movements of the motors are computer operated by means of a site-specific sequencer track. The instrument is constructed anew from the pieces of Styrofoam available at each venue. The instrument's dimensions are dependent on the length of time it takes to build it and the amount of Styrofoam collected. The Styrofoam is recycled after the exhibition or concert.

Jens Brand

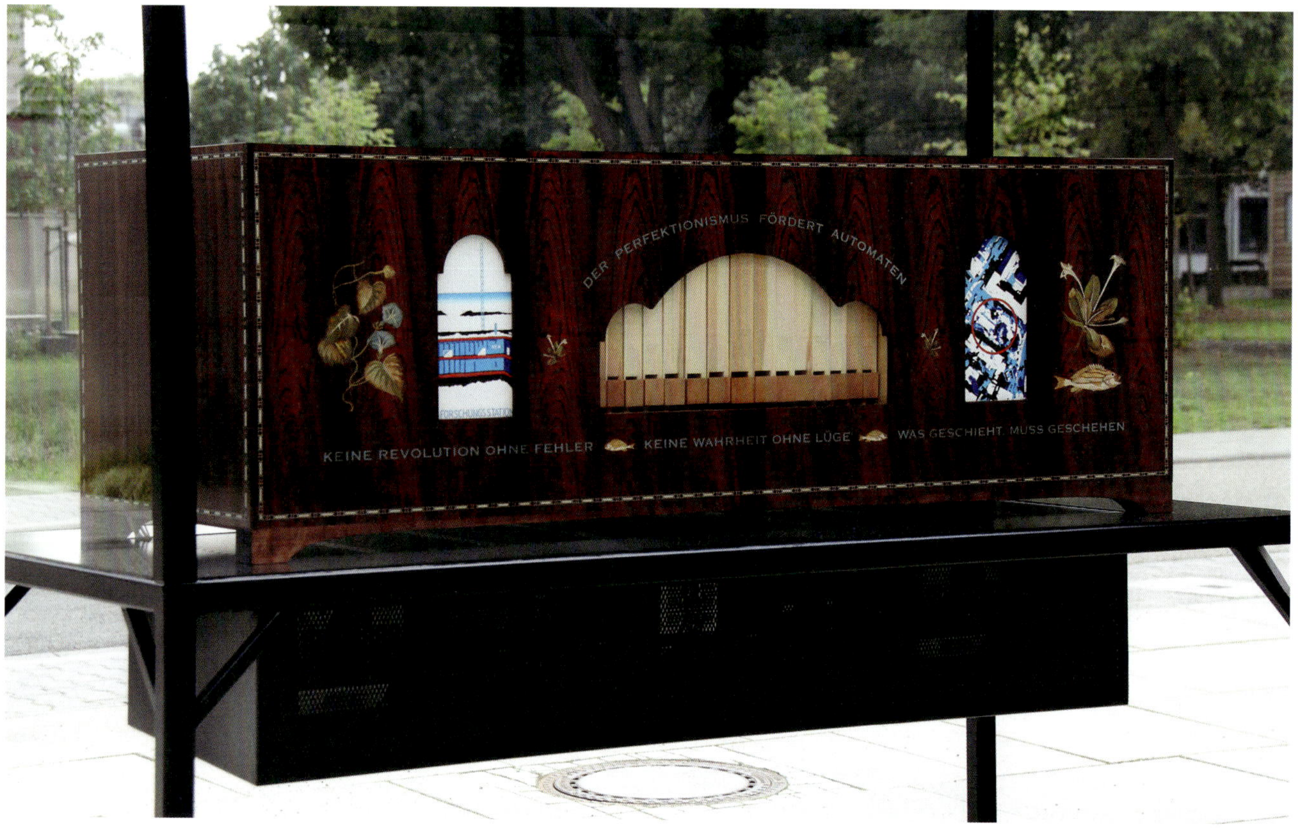

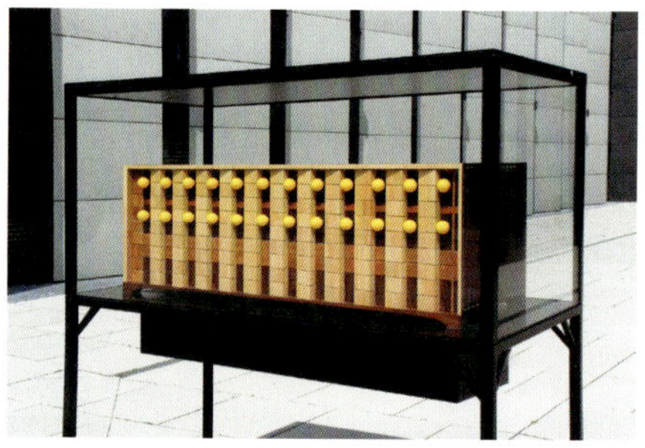

Two-Tone: Rehearsers' Music, 2011
Permanent installation in public space, located at Jakob-Welder-Weg 28, Mainz, Germany
Soft- and hardware construction: Sukandar Kartadinata
Organ builder: Orgelbau Stüber Berlin, Jens Brand

Two-Tone: Rehearsers' Music is a permanent installation in public space, situated at the open courtyard of the School of Music of the Johannes Gutenberg University, Mainz, Germany. The central element of *Two-Tone* is a computer controlled acoustic music machine, similar to a large barrel organ. It comments and quotes the environment of the School of Music.

Directed to the courtyard one can observe 24 rehearsal rooms. These rooms are individually equipped with a network system that gathers the general pitch of each single room, inevitably produced by the practicing students. The information is evaluated and the two dominant pitches and their locations are extracted. This information is being transferred to the organ.

Facing the street and confining the courtyard of the school, a large display cabinet presents a large barrel organ. The organ is equipped with 24 wooden pipes with computer controlled valves. From the perspective of the street the instrument appears to be a scaled down model of the schools rehearsal wing, facing the courtyard, sprinkling it with all sorts of sounds during the day. The windows of the chambers are represented by 24 yellow light bulbs. The other side of the instrument bears two paintings and hand carved veneers that refer to the disappearance of sites that had been dedicated to the German scientist, writer, and revolutionary Georg Forster. Additionally inlayed aphorisms connect truth, mistake, and revolution.

As the title implies, the organ only plays two tones at a time, changing them every 30 to 40 seconds. The tones are depriving from the most dominant frequency in the rehearsing rooms. As the rooms are quite small and not equipped with air conditions, their windows are usually open, filling the courtyard with the cacophonic sound of many different kinds of music at the same time. The drone performed by the organ rearranges the unintended music of the students by trying to add a harmonic basis.

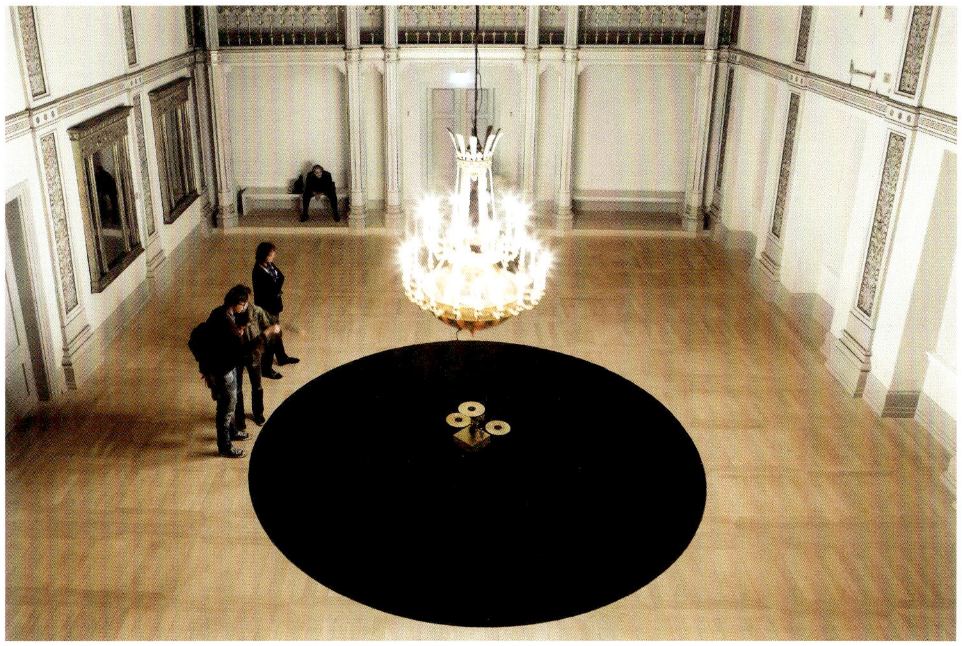

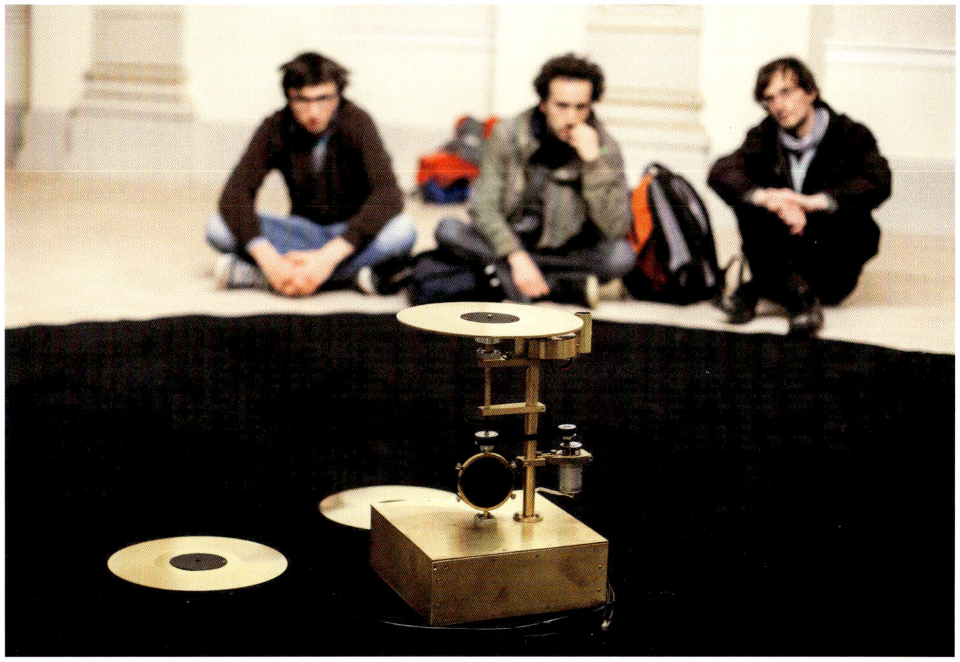

Small World Machine, Donaueschingen, 2011
Copper turntable that turns once a day, 1 rotary speaker that turns at 33 rpm, swinging chandelier executing a pendulum movement, functioning as a virtual stylus, 3 golden records
Turntable construction: Ulrich Panzer
Soft- and hardware construction: Sukandar Kartadinata and Christian Dietz
Records: Gene Autry and Jimmy Long, *That Silver Haired Daddy of Mine* (low resolution); *Golden Voyager Record* (medium resolution); a sonotopic map of the world with Donaueschingen as the center/as the record's hole (high resolution), Jens Brand
Commissioned by the SWR for the Musiktage in Donaueschingen 2011, curated by Armin Köhler. Special thanks to Simone Jung and the Museum Biedermann, as well as to Georg Riedmann and the Bauhof of Donaueschingen for their support.

The chandelier of the Museum Biedermann's hall of mirrors in Donaueschingen, Germany, becomes part of a machine that scratches through different models of the world in different resolutions. Manually changed from day to day, the length of the pendulum is alternating from high to low.

The data handled by the machine materializes in the form of three vinyl records. Like most of the other engines, this machine can be looked at as a tool and metaphor for sorting things out and lining them up. It is also an homage to the work *The Edison Effect* (1993) by Paul DeMarinis.

Broken Music

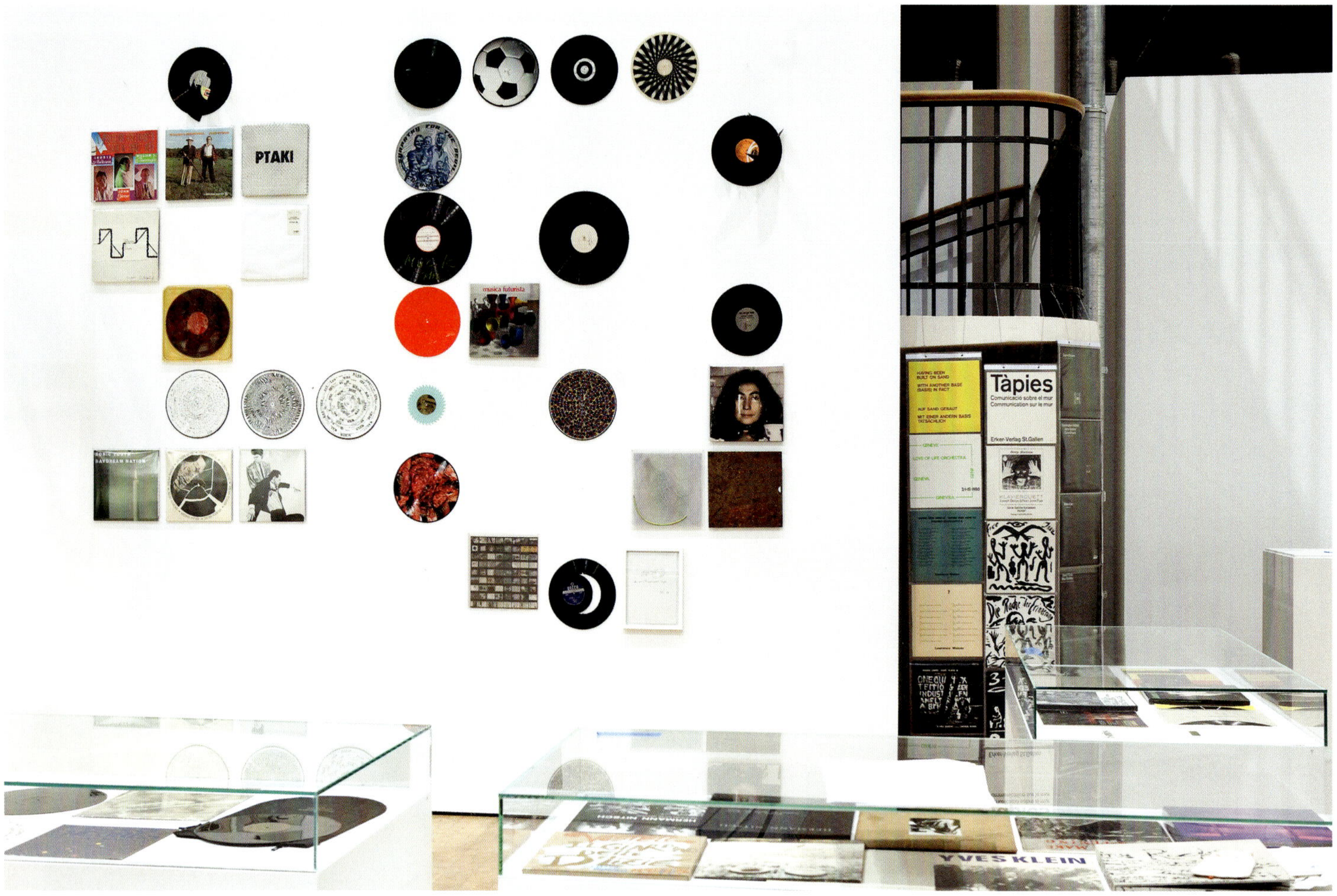

The English expression *to sculpt a sound* is a graphic description of how a record works: it puts a sound into a sculptural form. Conversely, it offers the opportunity of using the formed, "frozen" sound as material for sculpture.

The exhibition area Broken Music showed Ursula Block's collection of works by visual artists that were created with and for the medium of the record: records, record sleeves, record objects, record installations. The artists exhibited view the record not as a mere medium for the storage and reproduction of musical content, but are interested rather in its acoustic and visual presence. In addition to these works, the exhibits included a video installation by Nam June Paik, *TV-Dog*, which the artist made in 1993 especially for Ursula Block. It consists of a sculpture of Ursula Block's dog sitting before a TV and so is reminiscent of Paik's famous work *TV-Buddha* of 1974.

The concept of Broken Music goes back to the Czech artist Milan Knižák, who in the early 1960s began systematically to destroy records by, for example, painting, burning, scratching, drilling, smashing and then gluing them together again; he called the resultant collages *broken music*. Jean Dubuffet also sought new paths out of the smooth and the perfect and towards radically new sounds in his processing of records, while visionary artists such as Lászlo Moholoy-Nagy, Marcel Duchamp, John Cage, Nam June Paik, and Lawrence Weiner likewise occupy key positions in the creative treatment of records. At the conceptual level, Broken Music stands for the break with conventional listening habits, for the striving after the greatest possible diversity of sound.

The theme of the destruction of traditional ideas as the beginning of something new may be extracted as one of the artistic leitmotifs of Knižák's works. Though, of course, it is something of a simplification, it was still strong enough to summarize the entire exhibition.

A common characteristic of all the exhibits in this area was that they displayed a complex discussion between the acoustic and the visual. Ursula Block's presentation of the Broken Music archive sought to highlight these positions and work out as far as possible their historical roots.

Curator: Ursula Bock

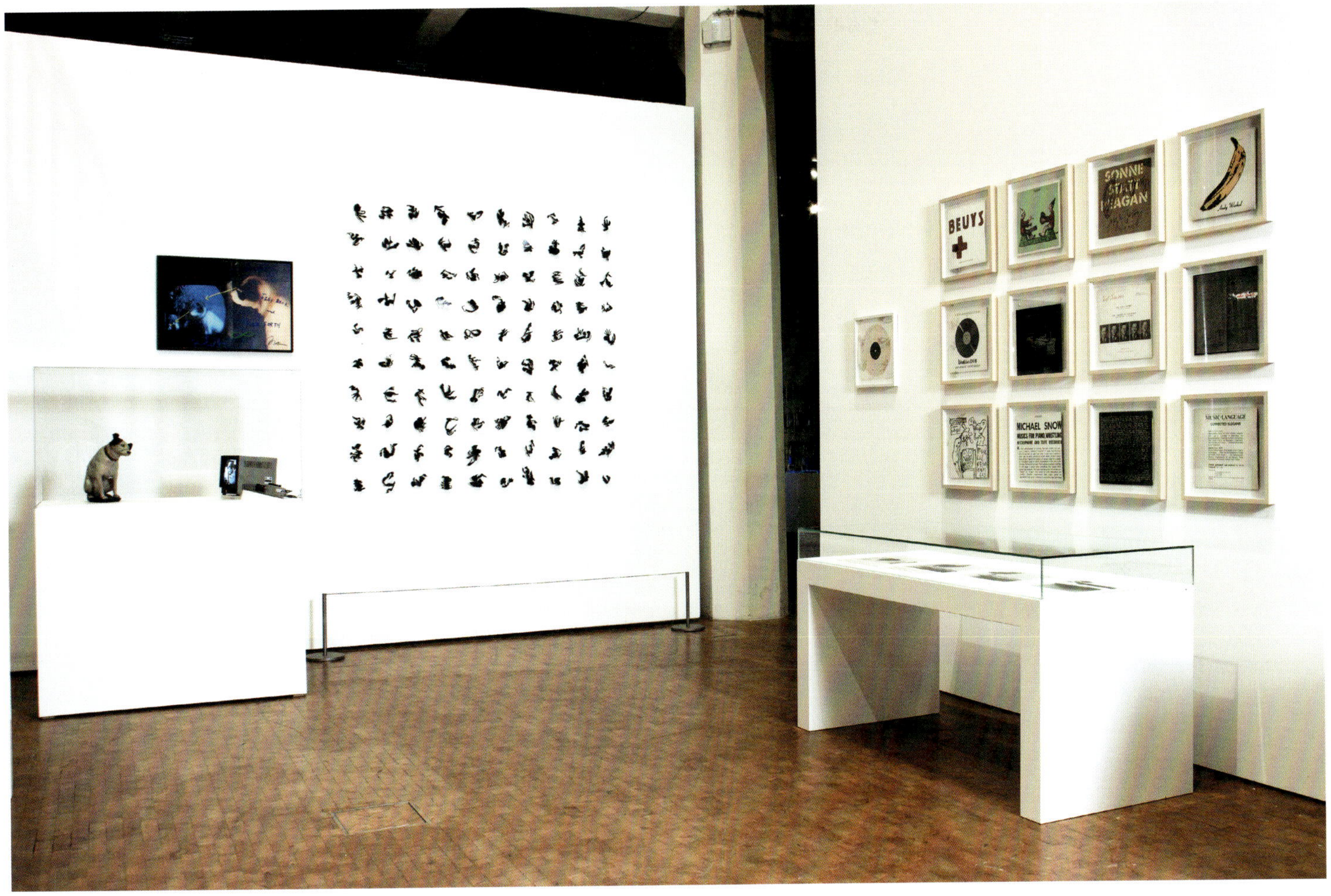

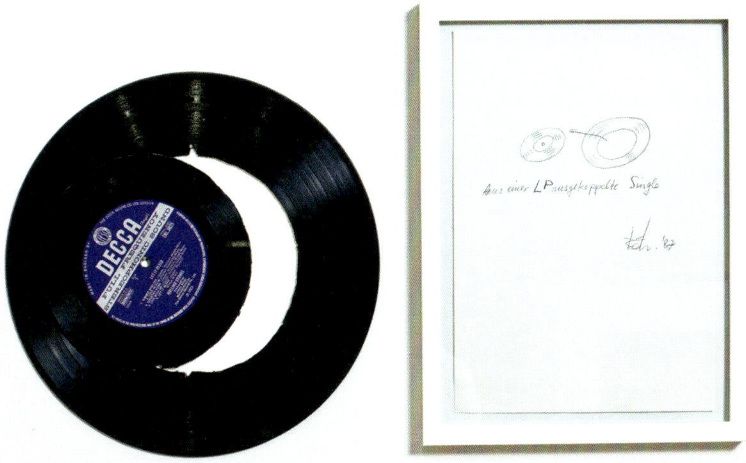

Claus Böhmler, *Aus einer LP ausgekoppelte Single* [A Single Decoupled from an LP], 1987.
Vinyl record out of which a single was cut, drawing.

Broken Music

Selection of Works

Claus Böhmler
*1939 in Heilbronn (DE), †2013 in Hamburg (DE)

Aus einer LP ausgekoppelte Single
[A Single Decoupled from an LP]
1987
Sculpture
Vinyl record (LP) out of which a single was cut; drawing
Ursula Block

KP Brehmer
*1938 in Berlin (DE), †1997 in Hamburg (DE)

Komposition für Tim Wilson II
[Composition for Tim Wilson II]
1986
Drawing
Indian ink and watercolor on graph paper
33 × 33 cm
Ursula Block

Henning Christiansen
*1932 in Copenhagen (DK), †2008

Jouez my pipe
1986
Sculpture
Vinyl record, pipe, color
Ursula Block

Mauricio Kagel
*1931 in Buenos Aires (BR), †2008 in Cologne (DE)

Der Umweg zur höheren SubFidelität
[The Detour to Higher SubFidelity]
4 photo prints on cardboard (photographed by Candida Höfer)
Ursula Block

Milan Knížák
*1940 in Pilsen (CZ), lives in Prague (CZ)

Broken Music
Multhipla Records Nr. 5
1979
LP
diameter 30 cm, 33 rpm

Würfel-Objekt aus Schallplatten Cube
[Cube Object Made from Records]
40 × 40 × 40 cm

Christian Marclay
*1955 in San Rafael, California (US), lives in New York City (US)

Record without a Cover
1985
2 pieces, black and white
diameter 30 cm, 33 rpm
Recycled Records New York

12" Black Vinyl Grooveless Record with Golden Label in Black Suede Poche with Gold Lettering
1987
Edition (50 pieces), signed and numbered
Ecart Editions, Geneva/New York

Piotr Nathan
*1956 in Gdansk (PL), lives in Berlin (DE)

Snowflakes
1987
Mural
100 cuttings from records
Dimensions variable
Ursula Block

Nam June Paik
*1932 in Seoul (KR), †2006 in Miami Beach (US)

TV-Dog
1993
Sculpture and graphics
Camera, monitor
Graphics: 58 × 83.5 cm
Ursula Block

Reiner Ruthenbeck
*1937 in Velbert (DE), †2016 in Ratingen (DE)

Dachskulpturen
[Roof Sculptures]
1972
Edition (issue of 400)
Cardboard (20 × 16.5 cm), record (diameter 17 cm), leporello (120 × 17 cm), 3 double-sided printed cardboard boxes (19.5 × 15.8 cm)
Ursula Block

Tomas Schmit
*1943 in Thier near Wipperfürth (DE), †2008 in Berlin (DE)

tomas schmit's SCHALLPLATTE
[Tomas Schmit's RECORD]
1970
Edition (issue of 10)
Wooden slate, felt pen, hectographed label
99 × 5 × 2 cm
Ursula Block

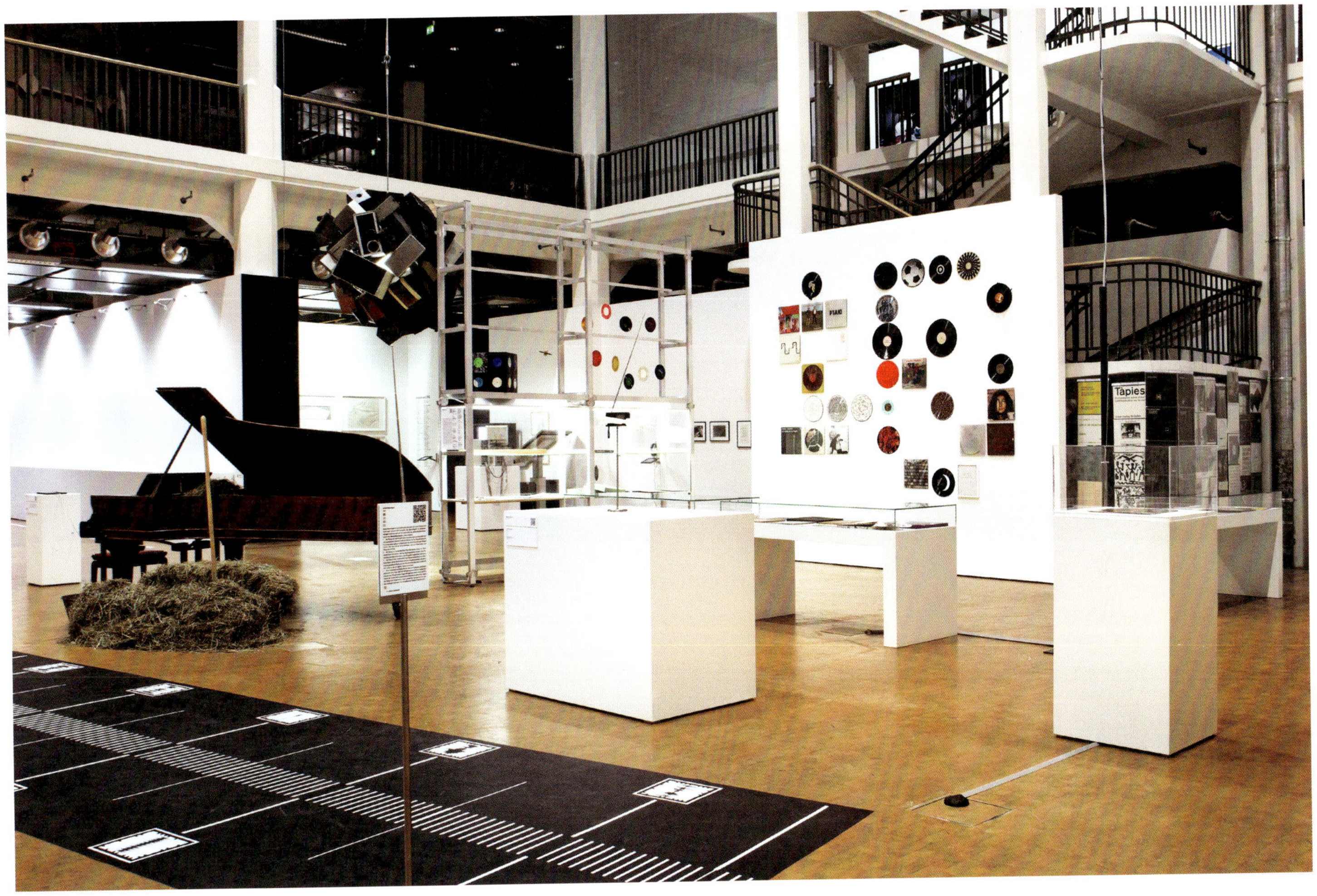

Ludger Brümmer
Götz Dipper

*1958 in Werne (DE), lives in Karlsruhe (DE)
*1966 in Stuttgart (DE), lives in Karlsruhe (DE)

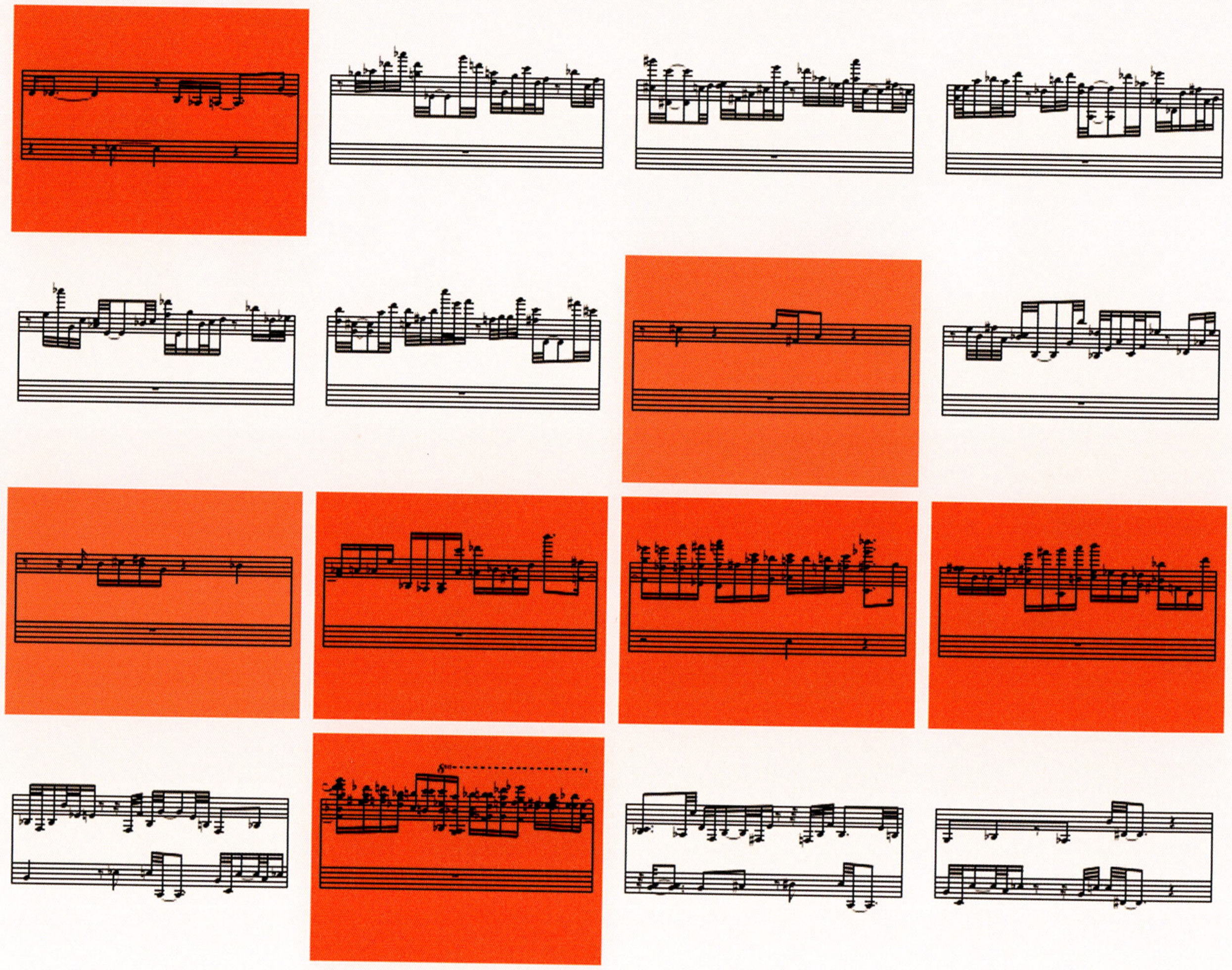

Étude interactive, 2011
Interactive sound installation
Console with touchscreen, loudspeaker/headphone, computer, 70 × 70 × 160 cm
Console construction: Manfred Schmieder
Production: ZKM | Karlsruhe

Étude interactive [Interactive etude] is an installation that constantly restructures itself with the help of chance operations. Ludger Brümmer (composition) and Götz Dipper (concept and visual design) have provided sixteen versions for each of the sixteen measures in the work. The sixteen versions are divided into four groups, each designated by a different color. Each group corresponds to a musical idea: yellow = repetition, blue = scales, turquoise = individual events (single tones and chords), red = polyphony. The musical structures have all been designed using algorithms and share a scheme of pitches that the four groups manifest at different intensities in various constellations.

Étude interactive can also be performed in concert. The concert premiere with the pianist Maki Namekawa took place in the ZKM | Karlsruhe in November 2011. The audience was able to influence the chance mechanism by holding up colored cards to vote on each of the next sixteen measures of the piece.

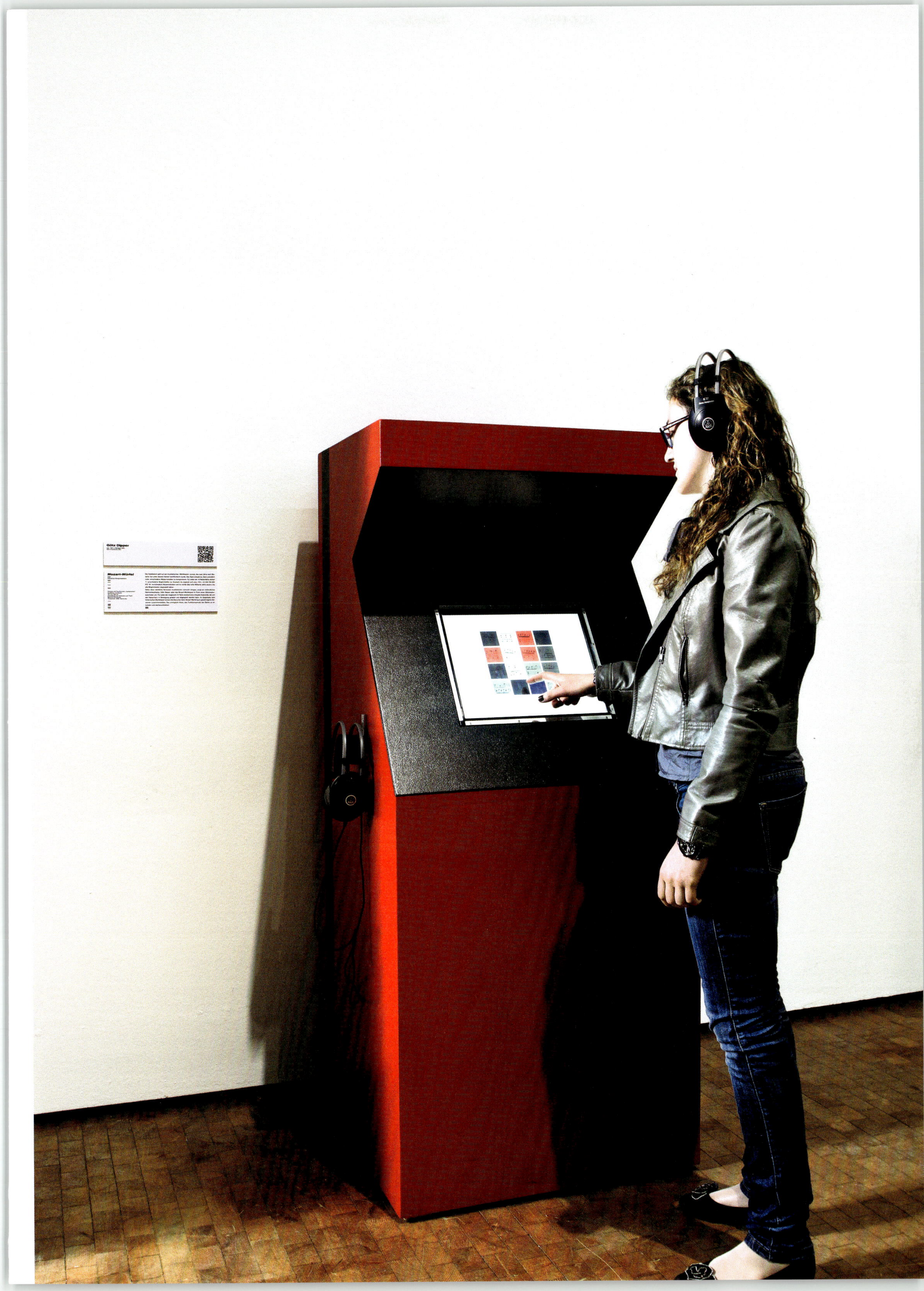
Götz Dipper
Mozart-Würfel

Ludger Brümmer
Götz Dipper

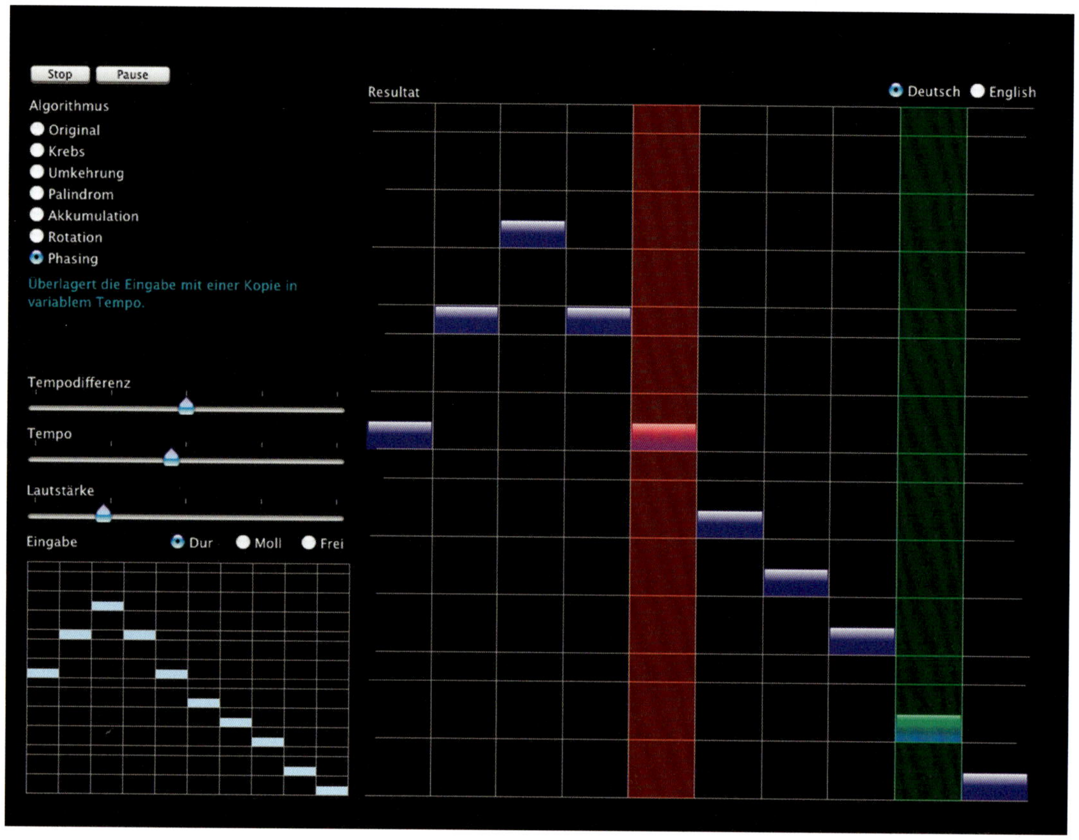

Pattern Machine, 2004
Interactive sound installation
Computer projection,
2 loudspeakers, pedestal with computer mouse, dimensions variable
Production: ZKM | Karlsruhe

Although there is an affinity between the interactive installations *Pattern Machine* and *Random Machine* in terms of space and design, the approaches used have fundamentally different principles of composition. Users of *Pattern Machine* can utilize traditional principles of composition via a graphic user interface whereas in interaction with *Random Machine* twentieth-century methods of composition are apparent.

Pattern Machine engages with the construction and variation of basic musical patterns: such "patterns," also slightly modified, are frequently the foundations of many pieces of music. Via a graphic user interface that controls the data on a monitor visitors can enter a ten-tone melody and create derivatives from this original pattern. For this, pivotal methods of composition from various epochs are available, such as "cancrizans," "retrograde," "accumulation," and "phasing;" instructions for the application of these are provided.

The viewer becomes a composer, a designer of melodies. *Pattern Machine* demonstrates this concept of repeating and varying small segments of music – and Minimal Music composers like Steve Reich were not the first to use this. Johann Sebastian Bach, for example, used rhythmic and melodic patterns in his *Preludes*, and both the accompaniment figures of classical works as well as the arabesques of Claude Debussy and Maurice Ravel are based on the pattern technique.

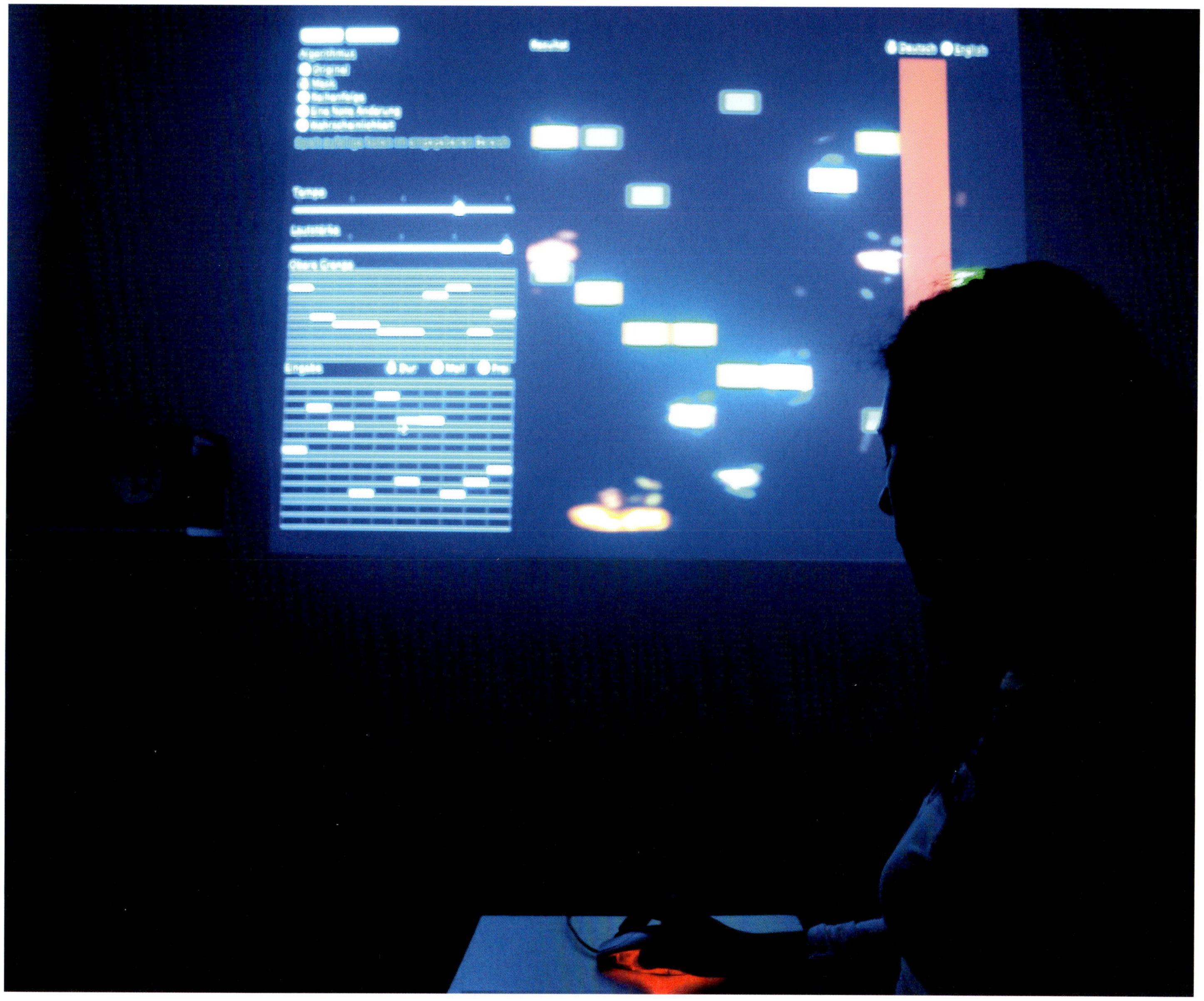

Random Machine, 2004
Interactive sound installation
Computer projection,
2 loudspeakers, pedestal with
computer mouse, dimensions variable
Production: ZKM | Karlsruhe

In the installation *Random Machine* chance is made visible as a creative process with results that are easily controlled and constantly changing. Already Wolfgang Amadeus Mozart understood how to utilize this process in his *Musikalisches Würfelspiel* (KV 294 d) – allegedly with the help of dice – to construct a music piece from two-bar fragments. Chance generates a continuously changing yet uniformly distributed series of values. All users of *Random Machine* have one thing in common: to produce randomness is very easy; the artistic challenge, however, is its application because, to the perception, chance only becomes interesting in connection with clear-cut rules. The installation developed by the ZKM | Institute of Music and Acoustics presents various types of randomness and its applications. *Random Machine* also turns viewers into composers and shows the endless possibilities to use chance creatively – both for discovering new musical ideas and for how to vary them. (Ludger Brümmer)

John Cage

*1912 in Los Angeles, CA (US), †1992 in New York City, NY (US)

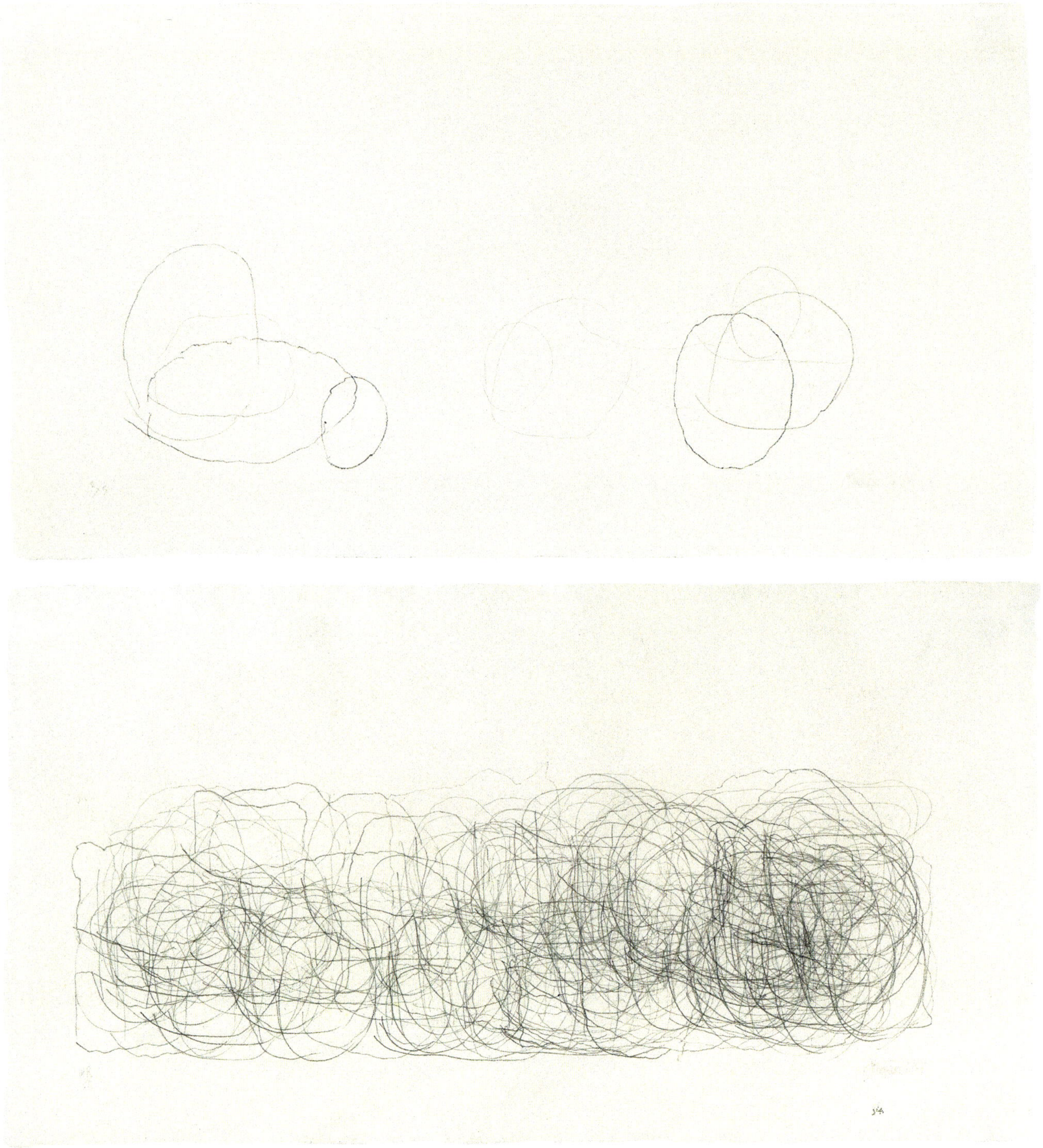

Where R=Ryoanji, 1987 and 1990
Sketches
ZKM Collection

www.johncage.org

Named after the famous zen rock garden in Kyoto, Japan, this piece is an excerpt from a series of two-page drawings Cage made by tracing the spaces around fifteen stones he chose to represent the stones in the Ryōan-ji garden. The resulting drawings were then interpreted by an orchestra of twenty musicians as graphic notations for a music performance, with each two-page set of drawings comprising a song to be performed by soloists (representing the rocks in the garden) and percussionists (representing the sand between the rocks). The curves between the rocks were to be interpreted as glissandi by soloists, and their overlapping lines necessitated the use of pre-taped performances in order to fully interpret the piece through virtual duets or trios. These soloist parts were accompanied by a percussion part consisting of only two unspecified sounds, one made with wood and the other with metal. (John Kannenberg)

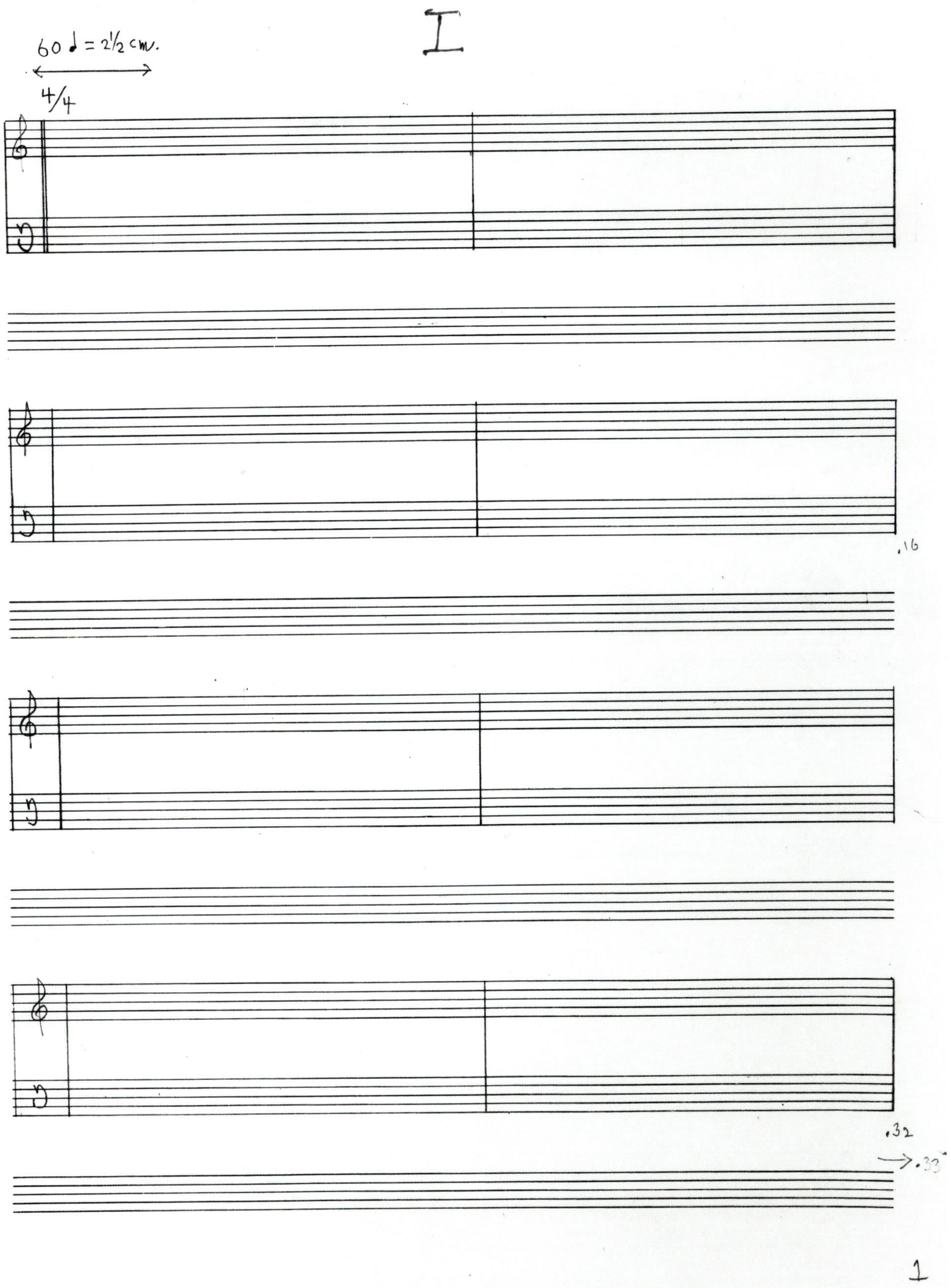

John Cage, *4'33"*, 1952
Reconstruction by David Tudor,
page 1, ca. 1989

John Cage

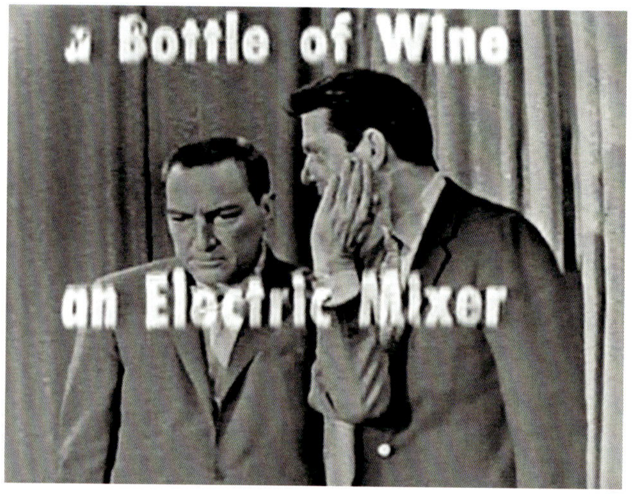

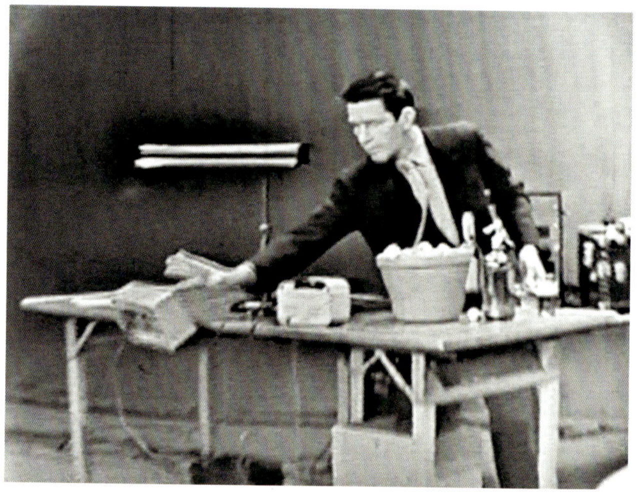

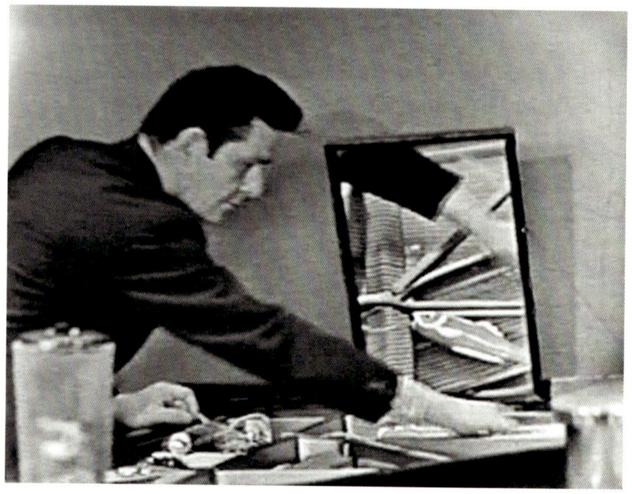

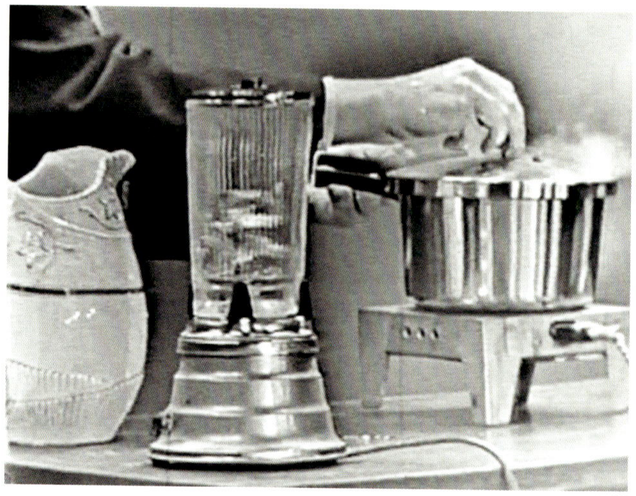

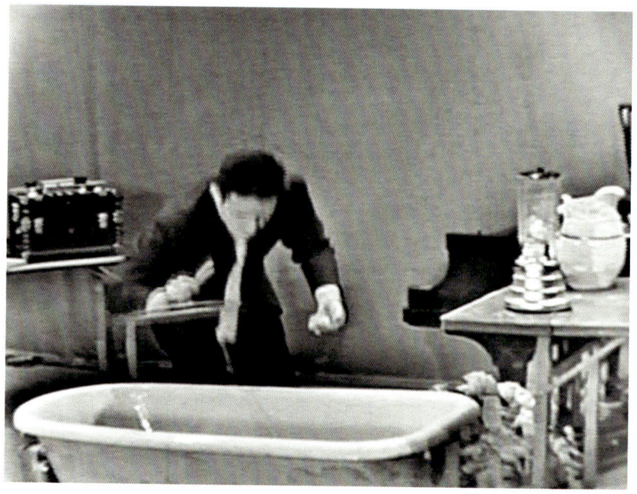

John Cage Performing Water Walk, 1960
Performance in a television show, transferred on DVD
9:22 min.
John Cage Trust, NY

In 1960, John Cage was a guest on the popular American television show *I've Got a Secret.* On that show, which was broadcast regularly from 1952 to 1967, a team was supposed to guess the secret of the personalities invited. John Cage was given ten minutes on the show to perform his composition *Water Walk* of 1959. This film is a curious document of this unique opportunity to introduce a broad audience to avant-garde music as well as historical important evidence of the contradiction between avant-garde and television, expressed by the audience's laughter, on the one hand, and John Cage's limited willingness to adapt to the context of television, on the other.

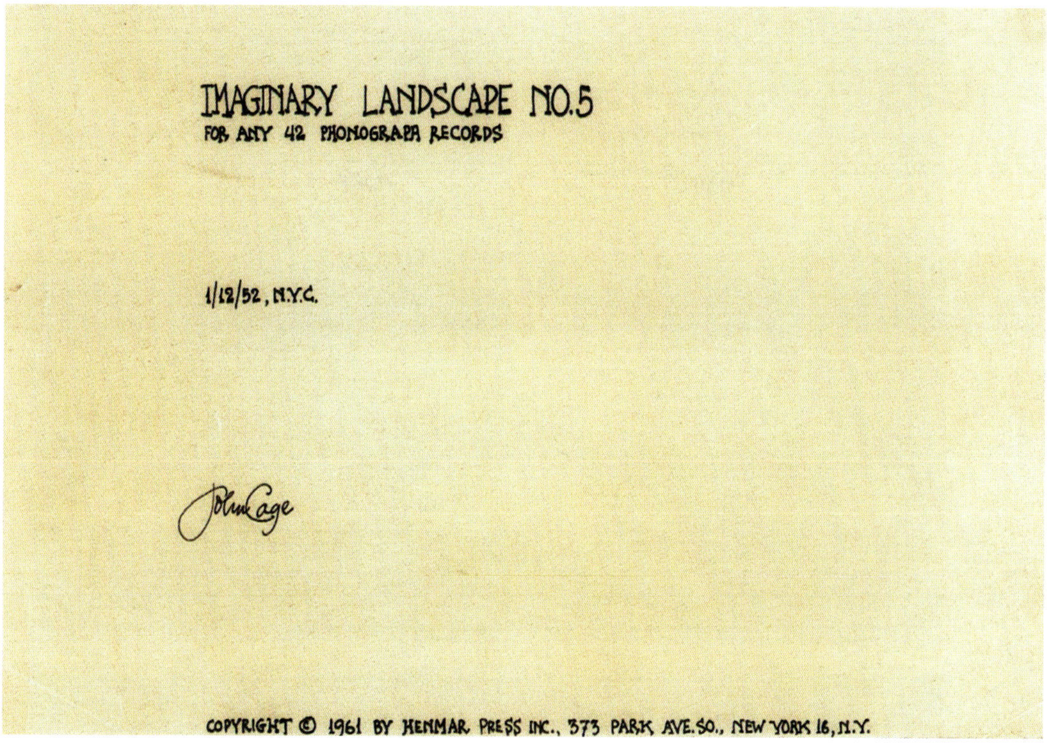

IMAGINARY LANDSCAPE NO.5
FOR ANY 42 PHONOGRAPH RECORDS

1/12/52, N.Y.C.

John Cage

COPYRIGHT © 1961 BY HENMAR PRESS INC., 373 PARK AVE.SO., NEW YORK 16, N.Y.

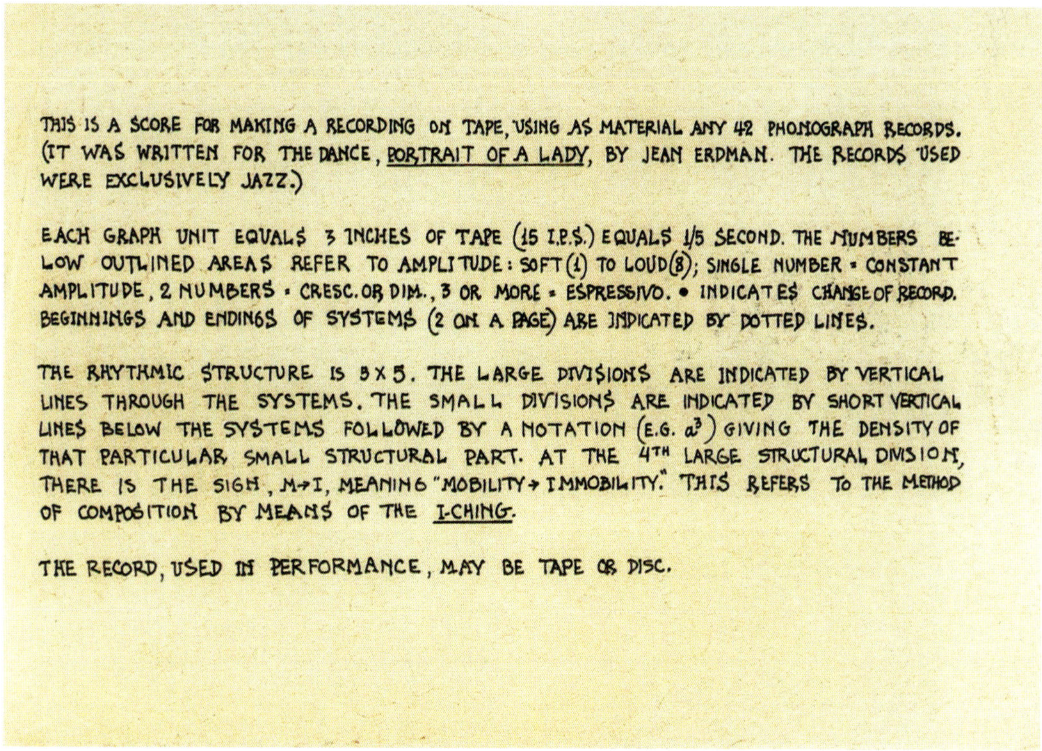

THIS IS A SCORE FOR MAKING A RECORDING ON TAPE, USING AS MATERIAL ANY 42 PHONOGRAPH RECORDS. (IT WAS WRITTEN FOR THE DANCE, PORTRAIT OF A LADY, BY JEAN ERDMAN. THE RECORDS USED WERE EXCLUSIVELY JAZZ.)

EACH GRAPH UNIT EQUALS 3 INCHES OF TAPE (15 I.P.S.) EQUALS 1/5 SECOND. THE NUMBERS BELOW OUTLINED AREAS REFER TO AMPLITUDE: SOFT (1) TO LOUD (8); SINGLE NUMBER = CONSTANT AMPLITUDE, 2 NUMBERS = CRESC. OR DIM., 3 OR MORE = ESPRESSIVO. • INDICATES CHANGE OF RECORD. BEGINNINGS AND ENDINGS OF SYSTEMS (2 ON A PAGE) ARE INDICATED BY DOTTED LINES.

THE RHYTHMIC STRUCTURE IS 5 X 5. THE LARGE DIVISIONS ARE INDICATED BY VERTICAL LINES THROUGH THE SYSTEMS. THE SMALL DIVISIONS ARE INDICATED BY SHORT VERTICAL LINES BELOW THE SYSTEMS FOLLOWED BY A NOTATION (E.G. a^3) GIVING THE DENSITY OF THAT PARTICULAR SMALL STRUCTURAL PART. AT THE 4TH LARGE STRUCTURAL DIVISION, THERE IS THE SIGN, M+I, MEANING "MOBILITY + IMMOBILITY." THIS REFERS TO THE METHOD OF COMPOSITION BY MEANS OF THE I-CHING.

THE RECORD, USED IN PERFORMANCE, MAY BE TAPE OR DISC.

Imaginary Landscape No. 5, for any 42 phonograph records, 1952
Cover and explanatory text of the score
8 sheets, mixed media on paper,
each 17.8 × 25.4 cm

Janet Cardiff
George Bures Miller

*1957 in Brussels (BE), lives in Grindrod (CA) and Berlin (DE)
*1960 in Vegreville (CA), lives in Grindrod (CA) and Berlin (DE)

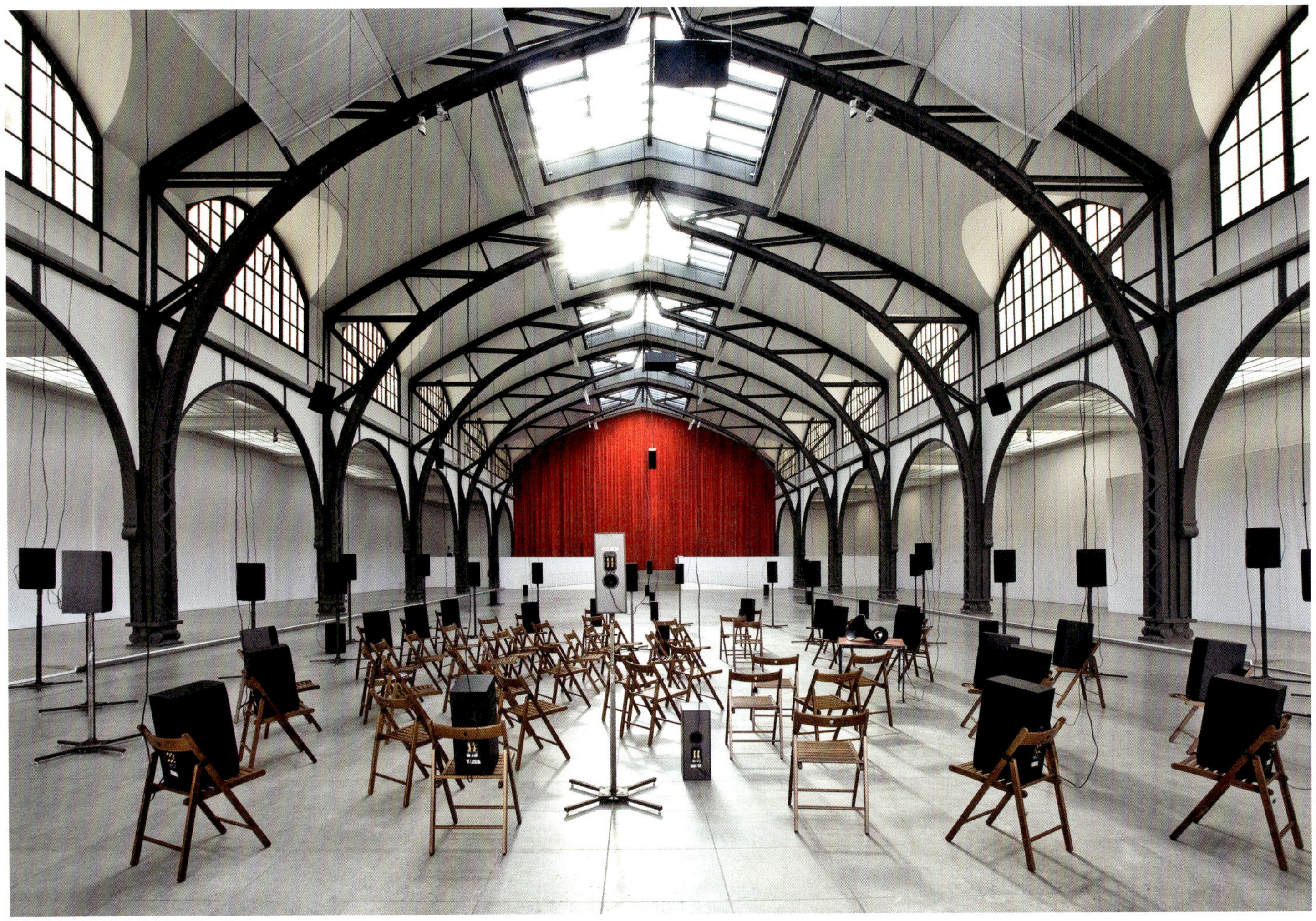

The Murder of Crows, 2008
98-channel audio installation including speakers, table, and chairs, dimensions variable
Edition of 3 and 1 artist's proof

www.cardiffmiller.com

right page, top:
I can't remember (World turning), 2010
Sound sculpture
Telephone, iPod, 2:55 min.
Courtesy of the artists and Luhring Augustine, New York

right page, bottom:
In a convent, 2010
Sound sculpture
Telephone, iPod, 5:12 min.
Courtesy of the artists and Luhring Augustine, New York

In contrast to their frequently monumental and extensive installations, Janet Cardiff and George Bures Miller turn their attention in this work to the intimacy of speaking on the telephone. *In a convent* and *I can't remember (World turning)* consist of vintage telephones connected to iPods. When you pick up the telephone receiver, you hear Janet Cardiff's voice recounting a dream. The installation is part of a series in which Cardiff and Miller focus on the telephone as a medium for storytelling while simultaneously investigating the narrative potential of dreams.

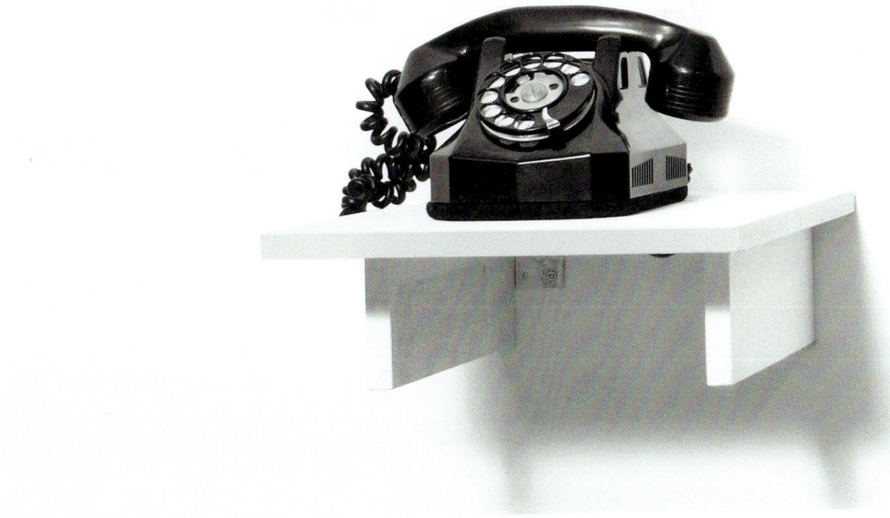

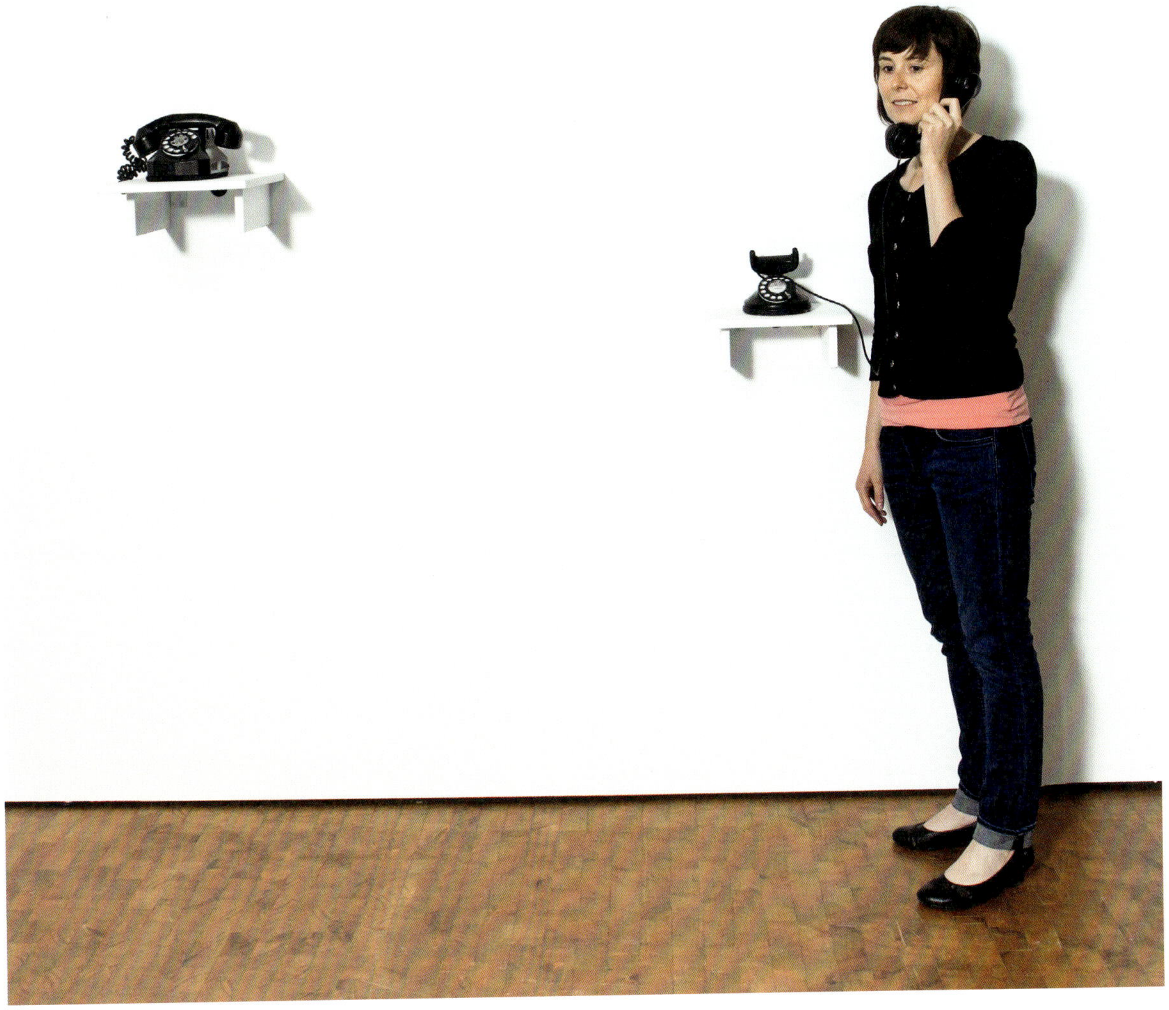

Janet Cardiff
George Bures Miller

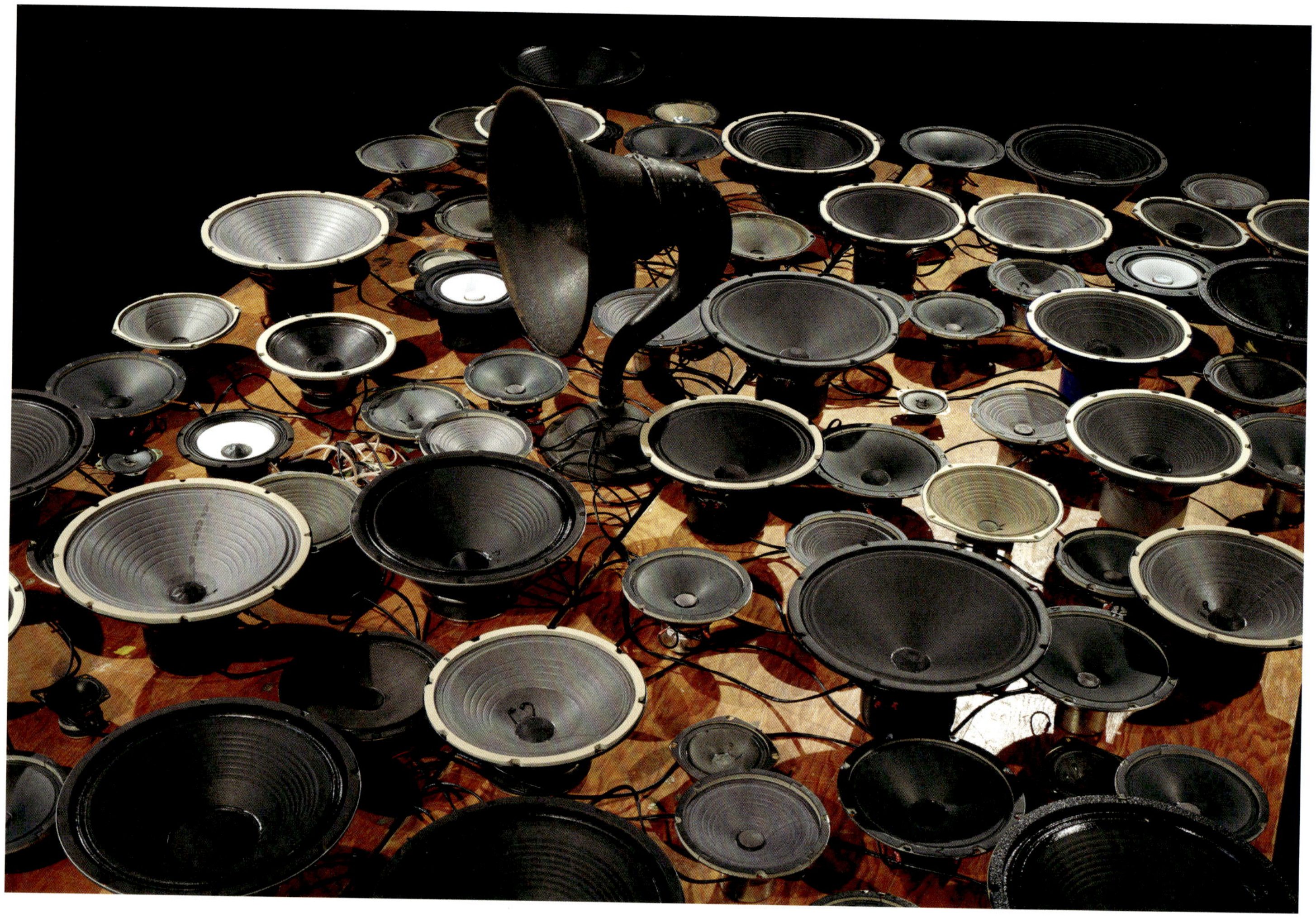

Experiment in F# Minor, 2013
72-channel audio installation including speakers, photosensors, and wooden worktables, 244 × 182 × 76.2 cm
Edition of 3 and 1 artist's proof

On a large table sits a collection of bare speakers of all shapes and sizes. Light sensors are inlaid into the edge of the table and as the viewers move around the room, their shadows cause the various sound and instrumental tracks to fade up and overlap, mingle and fade down. Numerous viewers in the room create a cacophony of musical compositions that vary according to where the audience walks or how many people are in the room. When the space is empty, the table fades to silence.

right page:
The Cabinet of Curiousness, 2010
20-channel audio installation including antique wooden card catalogue and speakers, 132.08 × 44.5 × 68.6 cm
Edition of 3 and 1 artist's proof

The Cabinet of Curiousness is an antique wooden card catalogue with twenty drawers. Functioning as an interactive piece, the opening of each drawer activates a voice or piece of music from within the cabinet. The audience, assuming the role of a DJ, may experience the clarity of sound from one drawer or a cacophony of sounds from numerous drawers opened simultaneously as the cabinet is played like an instrument. A contrast emerges between the obsolete system of cataloguing single pieces of data and our current tendency to inundate ourselves with excessive information. An investigation of knowledge, time, and our relationship to objects and music.

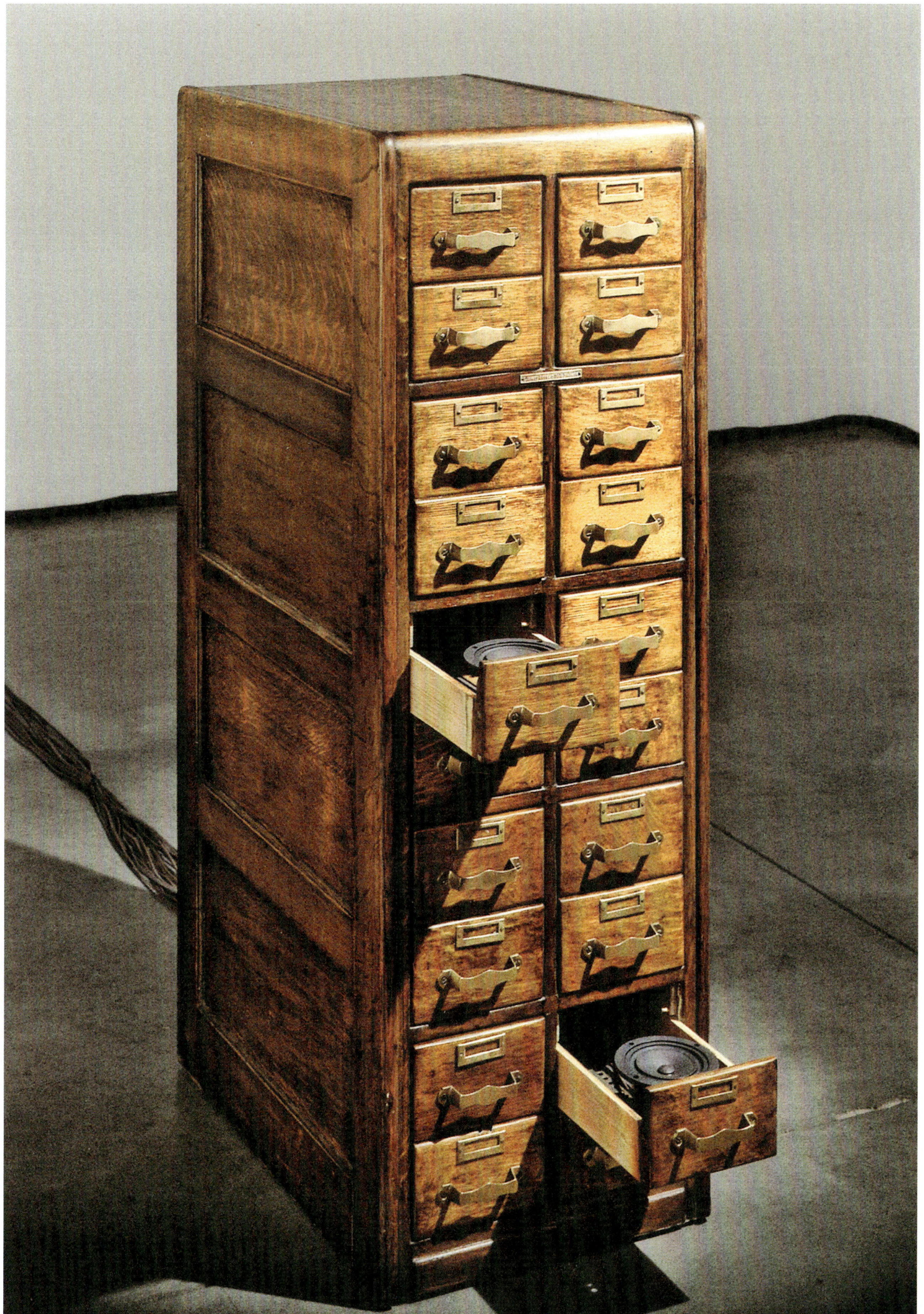

Janet Cardiff
George Bures Miller

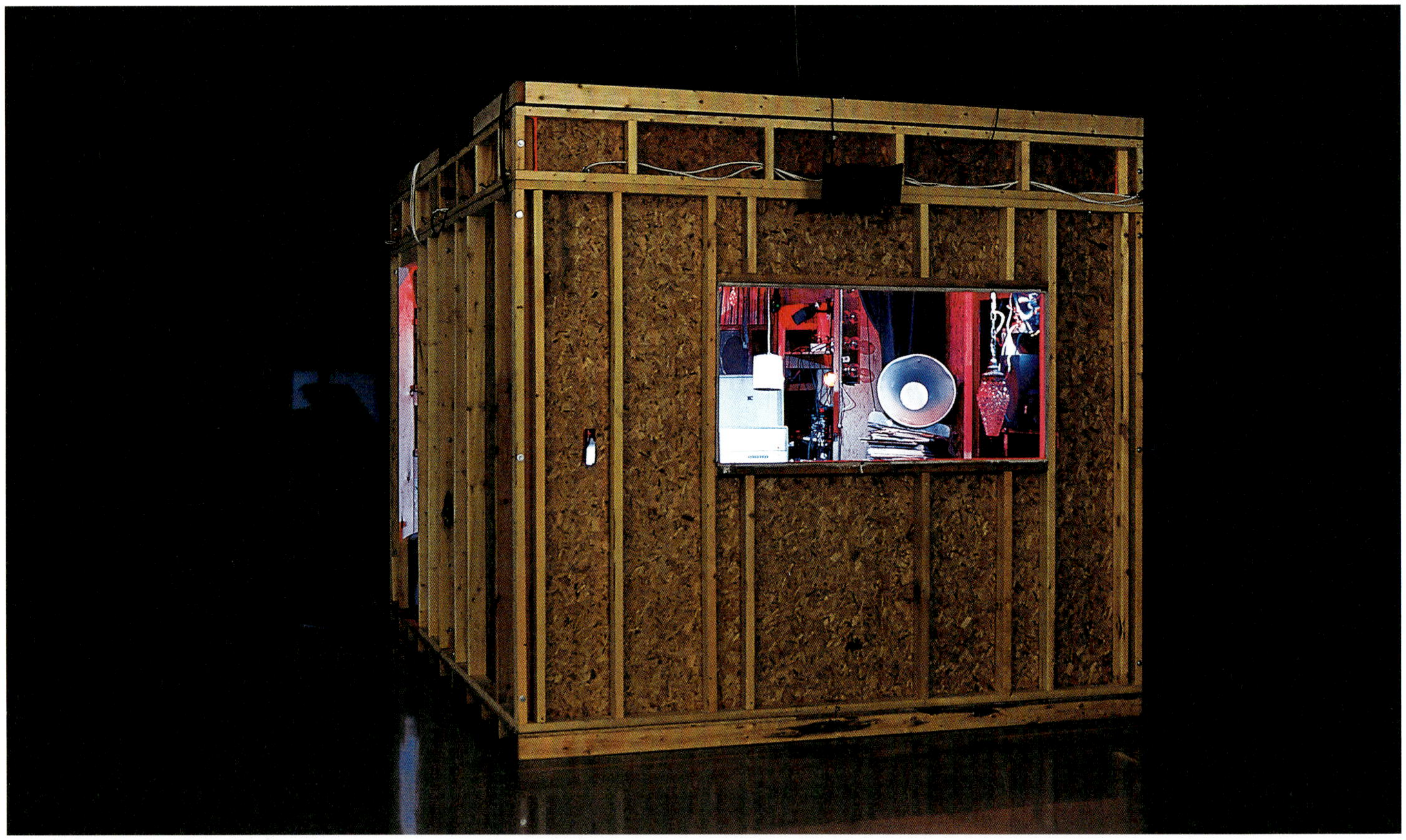

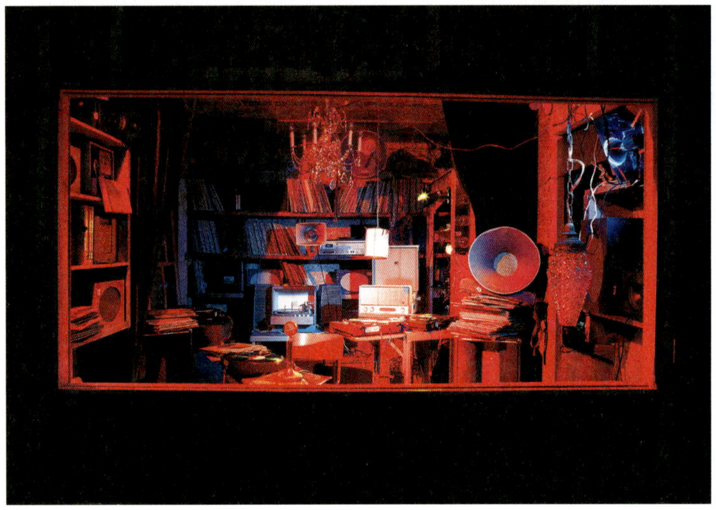

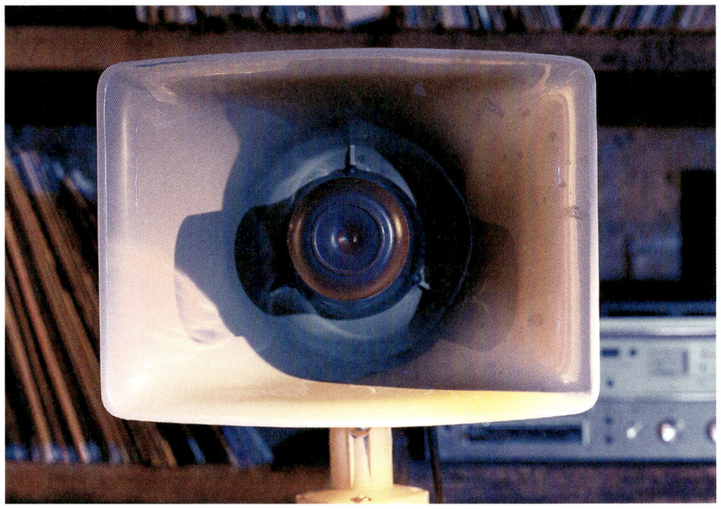

Opera for a Small Room, 2005
Mixed media installation with sound and synchronized lighting, 20 min. (loop), 2.6 × 3 × 4.5 m, edition of 1
Construction: Kyle Miller, Roadkill
Crow Song: Composition voice: George Bures Miller
Guitar, drums, mixing: Titus Maderlachner
Bass guitar, organ, orchestration: Tilman Ritter

There are twenty-four antique loudspeakers out of which come songs, sounds, arias, and occasional pop tunes. There are almost two thousand records stacked around the room and eight record players, which turn on and off robotically syncing with the soundtrack. The sound of someone moving and sorting albums is heard. The audience cannot enter the room. To see and hear his world, they have to look through windows, holes in the walls, and cracks in the doorways and watch his shadow move around the room.

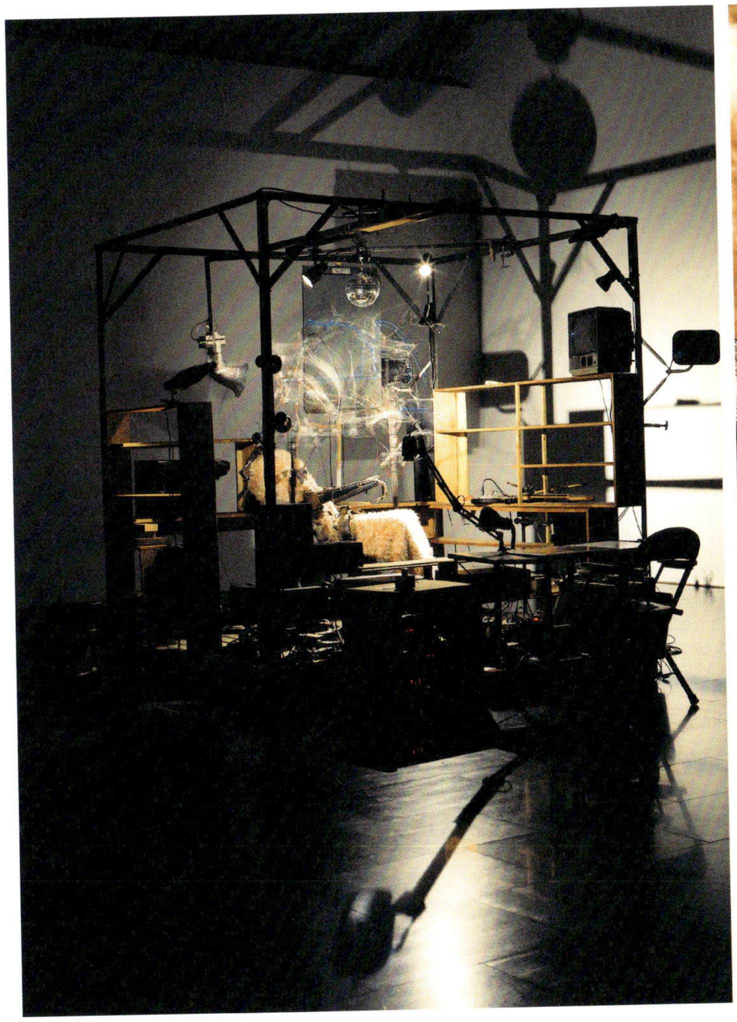

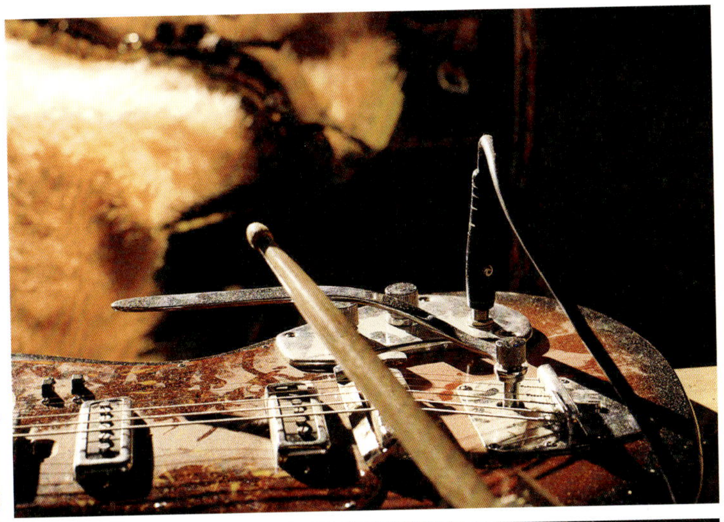

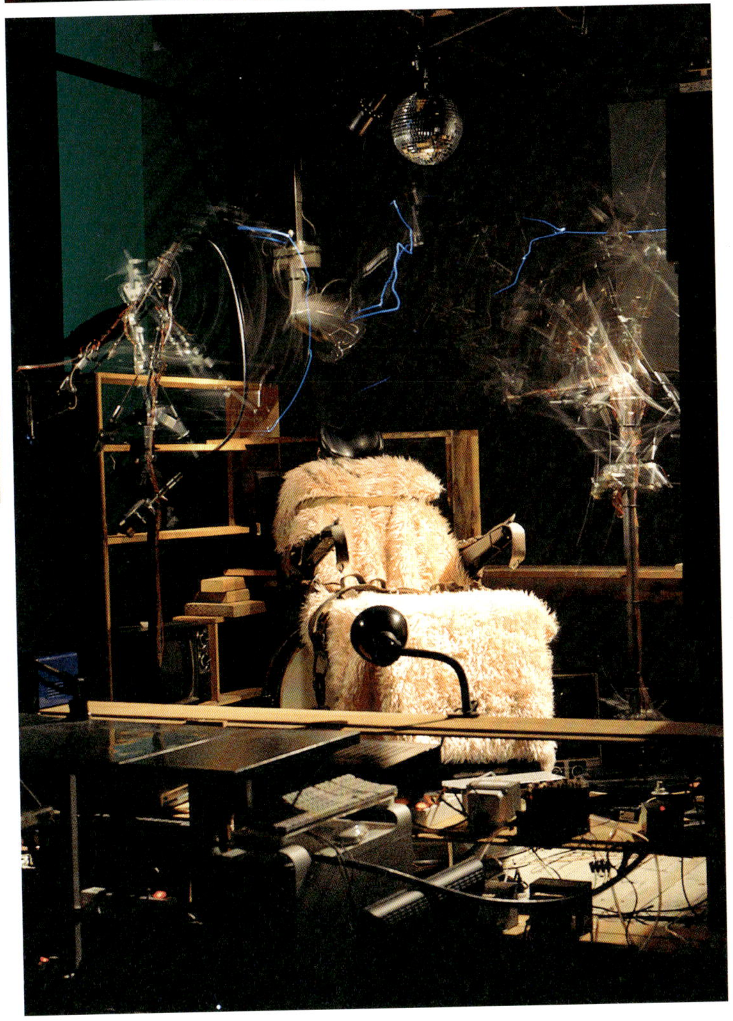

The Killing Machine, 2007
Mixed media installation with sound and lighting, 5 min. (loop), 3 × 4 × 2.5 m
Edition 1 of 2 + 1 artist's proof
Robot art design: Carlo Crovato
Music: *Heartstrings* by Frieda Abtan
Percussion assistance: Titus Maderlechner

Partly inspired by Franz Kafka's *In the Penal Colony* and partly by the American system of capital punishment as well as the current political situation, the piece is an ironic approach to killing and torture machines. A moving megaphone speaker encircles an electric dental chair. The chair is covered in pink fun fur with leather straps and spikes. In the installation are two robotic arms that hover and move – sometimes like a ballet, and sometimes attacking the invisible prisoner in the chair with pneumonic pistons. A disco ball turns above the mechanism reflecting an array of colored lights while a guitar hit by a robotic wand wails and a wall of old TVs turns on and off creating an eerie glow.

Chicks on Speed

Chicks on Speed are a multi-national art and music trio, who joined forces in Munich in 1997, when Alex Murray-Leslie and Melissa Logan met at the Munich Academy of Fine Arts.

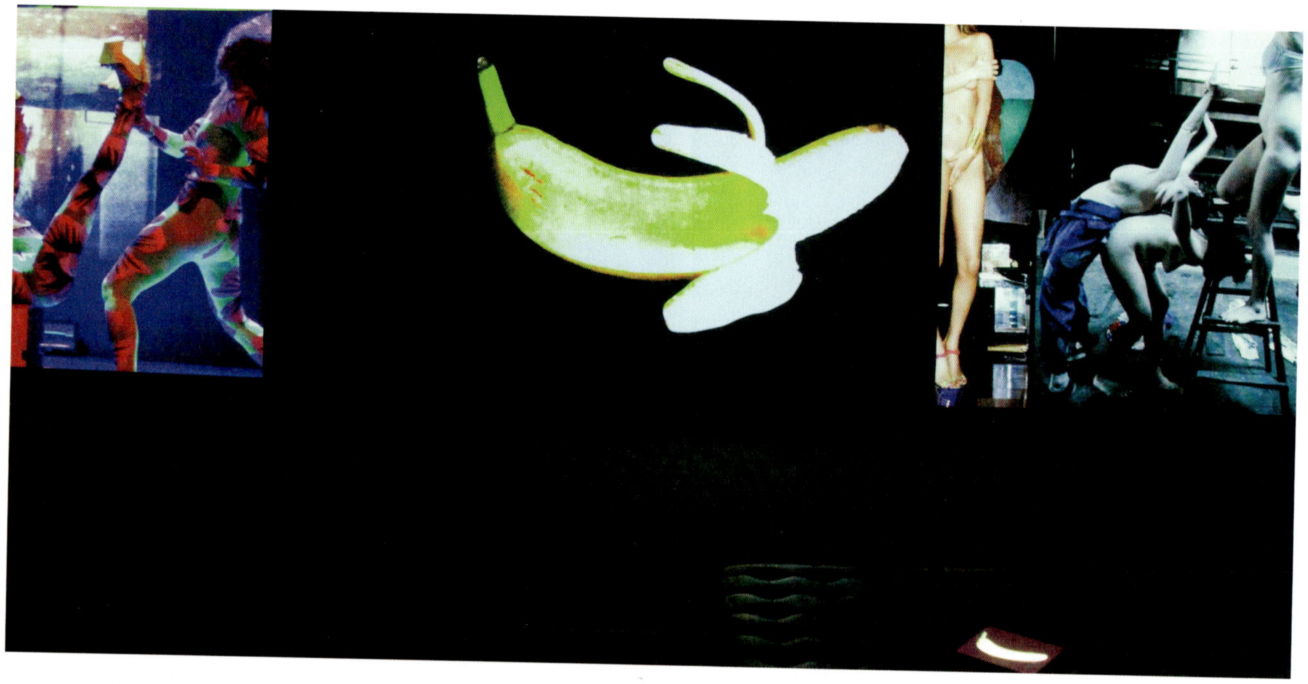

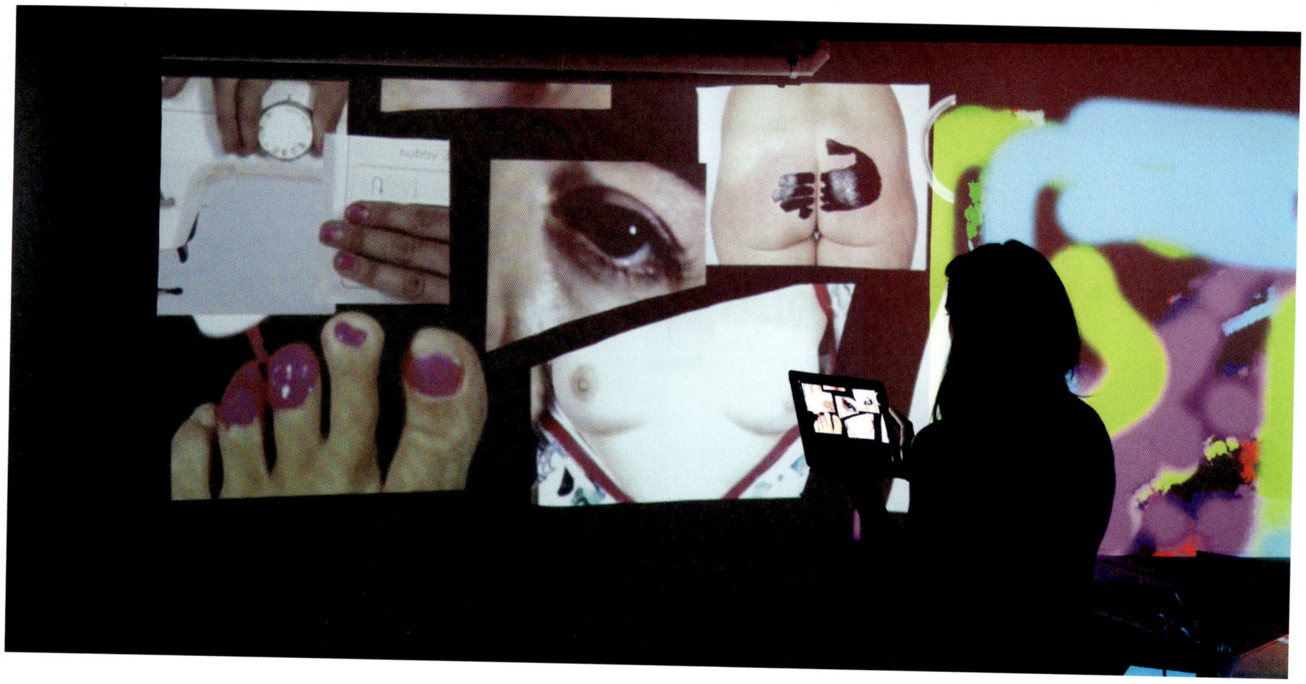

SCREAM app collection, 2012–2013
Face tracking video installation
Commissioned by ZKM | Karlsruhe
Development: Jens Barth
6 interactive iOs apps:
– Turntables, scratch and mix
– Scissors, video collage and sampler
– Spray Can, graffiti and sounds
– Andy Banana, vocoder
– High-Heel Guitar, instrument
– Text, Vodka ...

www.chicksonspeed.com

SCREAM adds a new element into the Chicks on Speed Gesamtkunstwerk: Six interactive apps developed in residency at the ZKM | Karlsruhe that enable audience members to live, compose, and mix audio visual scenes projected into the gallery space. Unleashed through live performances, *SCREAM* features a giant multifaceted sculptural object upon which Chicks on Speed project a visually chaotic collage. This sculptural structure functions as a stage, a canvas and a playable musical instrument both activated through Chicks on Speed's extraordinary array of sounds, rhythms, words and images and "performed" by visitors with iPads.

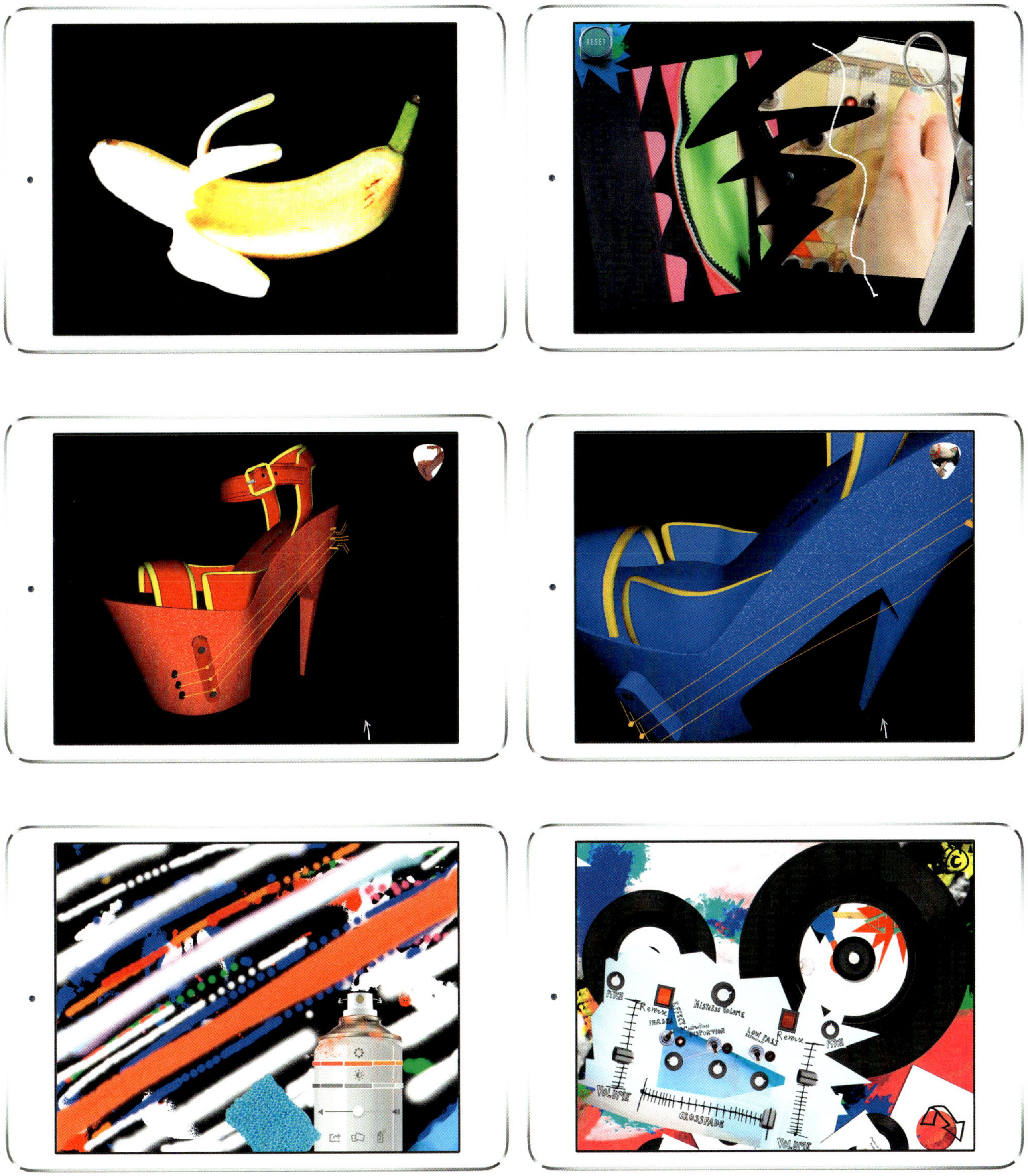
RESET
PITCH
Reverse
PHASER
DISTORTION
MISTRESS VOLUME
LOW PASS
RESONANT
Reverse
PITCH
VOLUME
CROSSFADE
VOLUME

Stephen Cornford

*1979 in London (GB), lives in Bristol (GB)

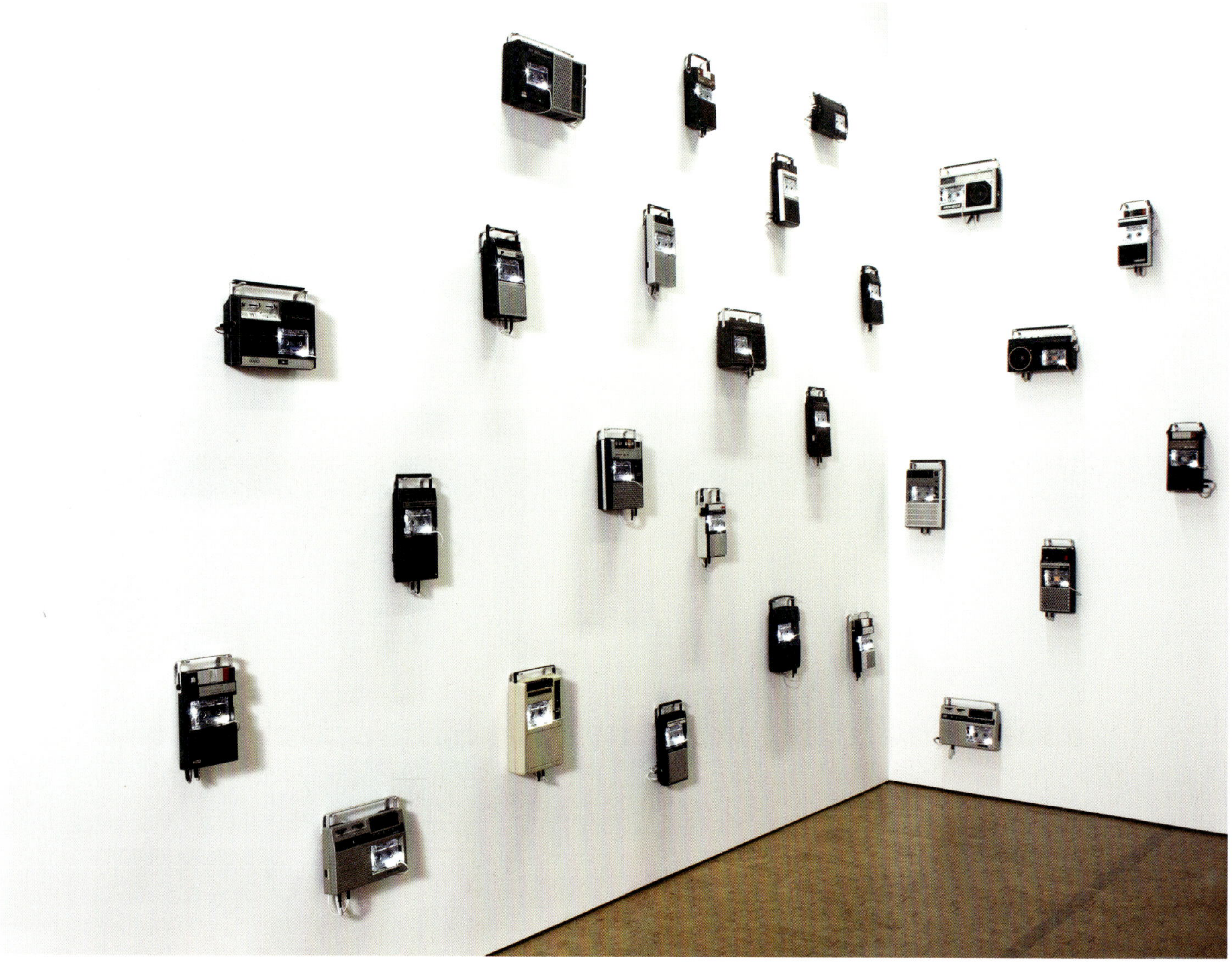

Binatone Galaxy, 2011
Sound sculpture
Portable cassette recorders with self-amplifying cassettes, dimensions variable
Courtesy of the artist

www.stephencornford.net

Stephen Cornford works with used portable audiocassette players. But there is no tape in the cassettes he plays in the devices. Instead, contact microphones amplify the sounds of the recorders themselves. The sound of *Binatone Galaxy* derives solely from the sound of the motors and mechanisms. Movement sensors that control whether or not the devices temporarily pause are also built into the cassettes. In his works, Stephen Cornford uses for the most part analog audio technology and acoustic musical instruments. In *Binatone Galaxy* he resurrects the superannuated medium of a generation without summoning up the musical messages it once conveyed. The installation takes its name from a certain model of British cassette recorder.

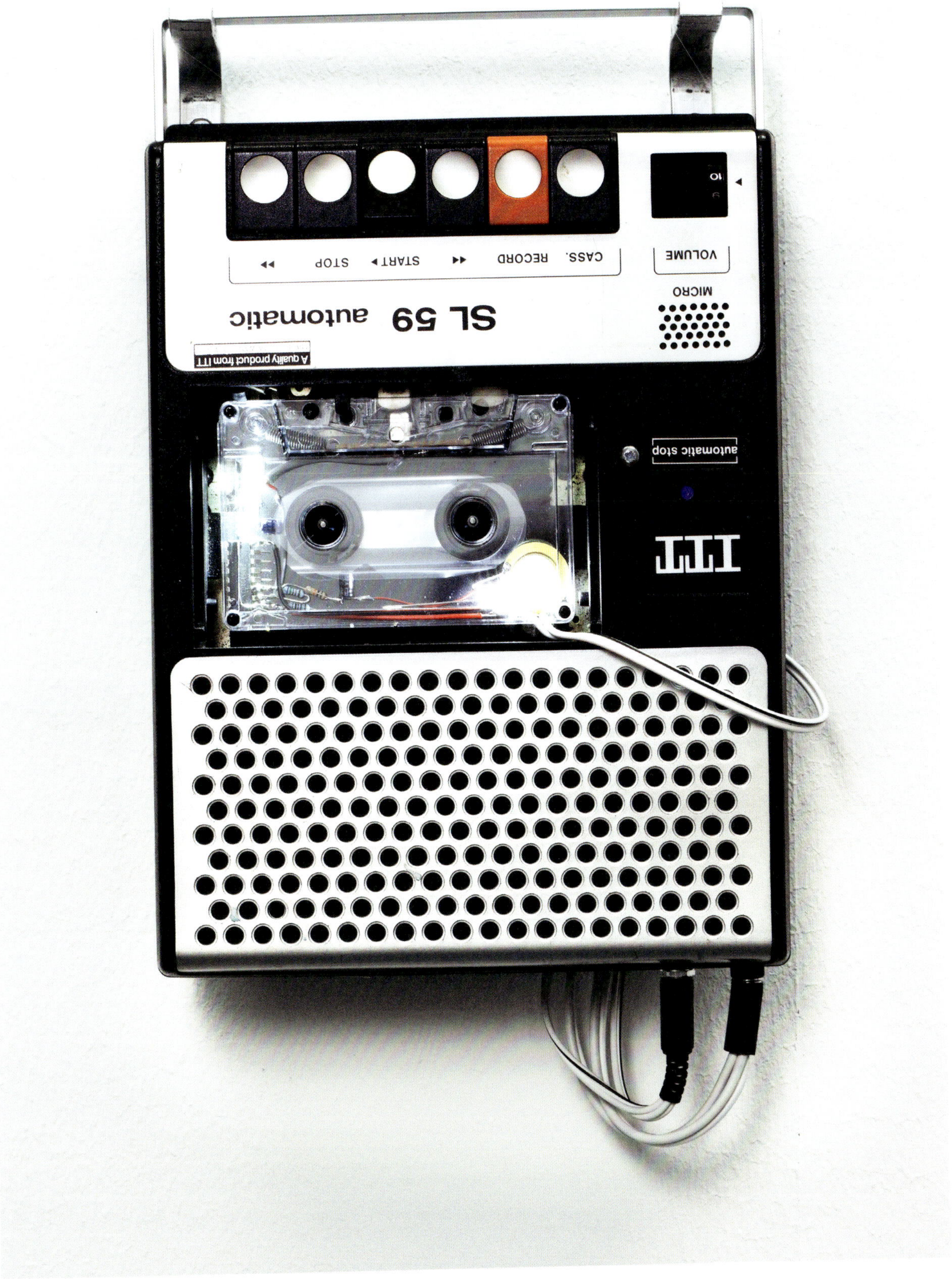
CASS.
RECORD
START
STOP
VOLUME
MICRO
SL 59 automatic
A quality product from ITT
automatic stop
ITT

Stephen Cornford

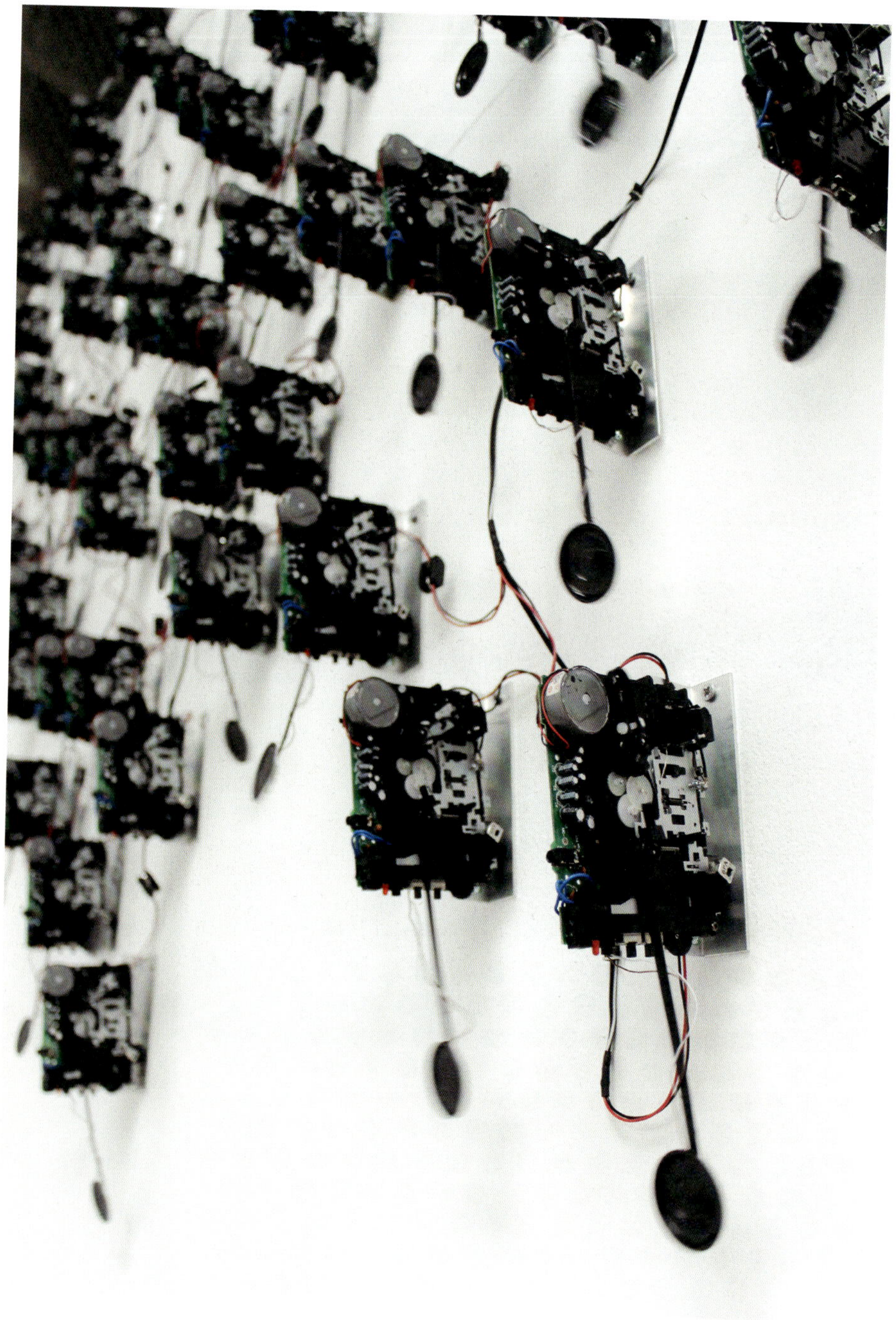

Migration, 2014
Kinetic sound installation
Modified cassette recorders,
dimensions variable
Courtesy of the artist

Migration is an installation for numerous factory-second cassette recorders. By subtly transforming their mechanics and audio electronics, their simple motion, voltage, and operative noise are combined into a visual and aural display reminiscent of the massed migrations of birds or insects. Obliquely referencing the appropriation of the term "migration" as a data process, the work reflects on the global movement of consumer electronics as they pass from production to pollution in the ever accelerating life cycles between generations of technology.

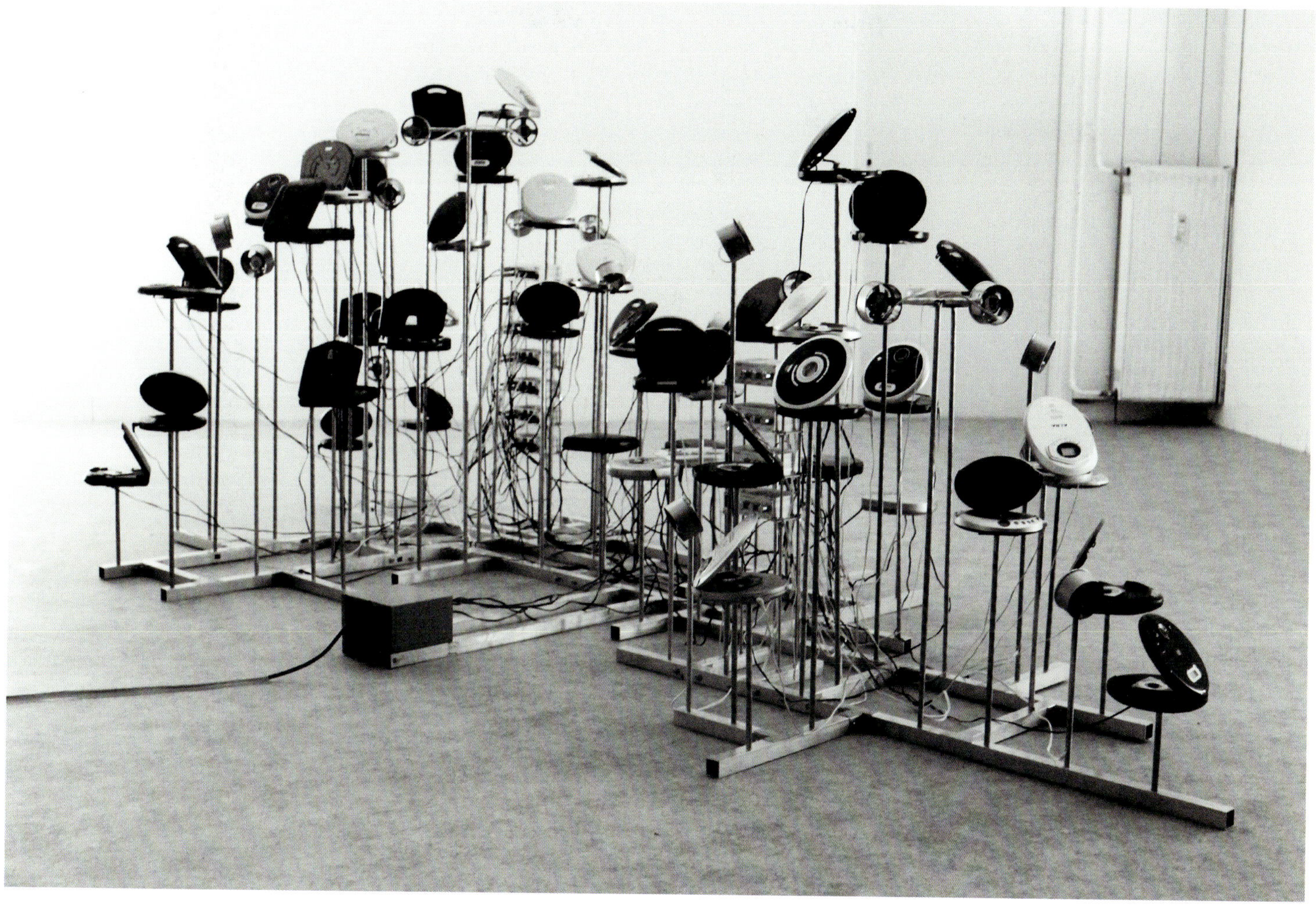

Archipelago, 2013
Kinetic sound sculpture
Modified portable CD players, amplifiers, aluminum, 350 × 160 × 90 cm
Courtesy of the artist

Archipelago is a sculpture for more than 50 portable CD players. The work inverts their normal role as the source of audio, turning them instead into receptors of sound: ears. In doing so, the work asks how these machines might behave were they able to hear. Electromagnetic recordings of the machines are played back into their motors, causing them to twitch, tremble, and spin unpredictably. These mechanized spasms are then amplified through an array of small speakers encircling them. The abandoned portable stereos of the past, a technology designed to enable solitude and isolation in even the most crowded of spaces, are transformed into a network of nervous electronic eavesdroppers.

Stephen Cornford

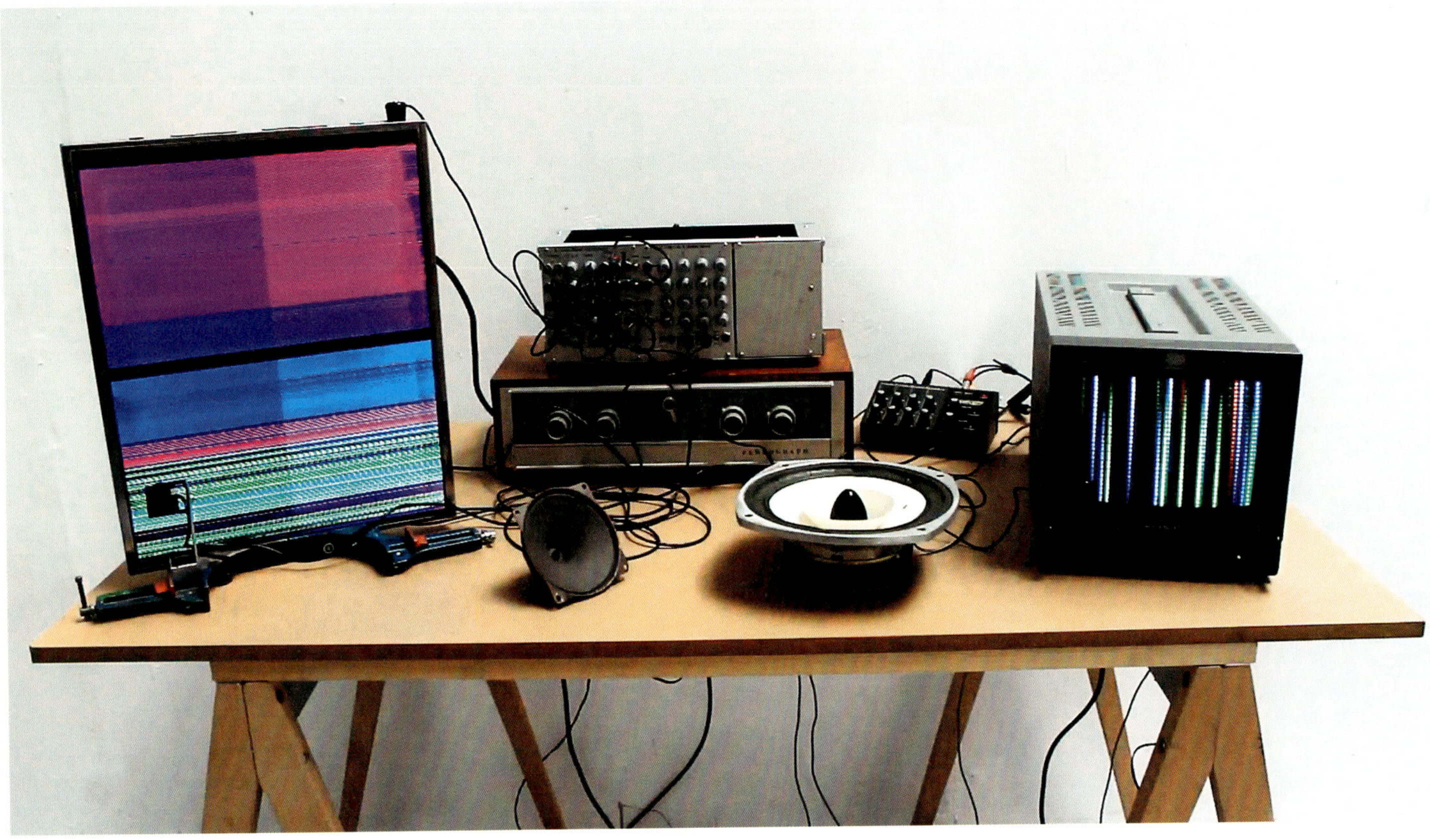

Methodology for a Synaesthetic Screen, 2016
Audiovisual sculpture
Modified LCD screen, CRT monitor, synthesizer modules, amplifier, loudspeakers, 120 × 70 × 80 cm
Courtesy of the artist

Methodology for a Synaesthetic Screen is an audiovisual work exploring the sensory indeterminacy of electrical signals. Two audio signals from a synthesizer module are sent to the input pins of an LCD screen, transducing audio frequencies into chromatic frequencies, producing apparently chaotic patterns of colored stripes. A macro camera captures the raster grid grain structure of this image, which is then displayed on a small CRT monitor. The electromagnetic emissions of both screens are amplified live, allowing us to hear the image of the sound.

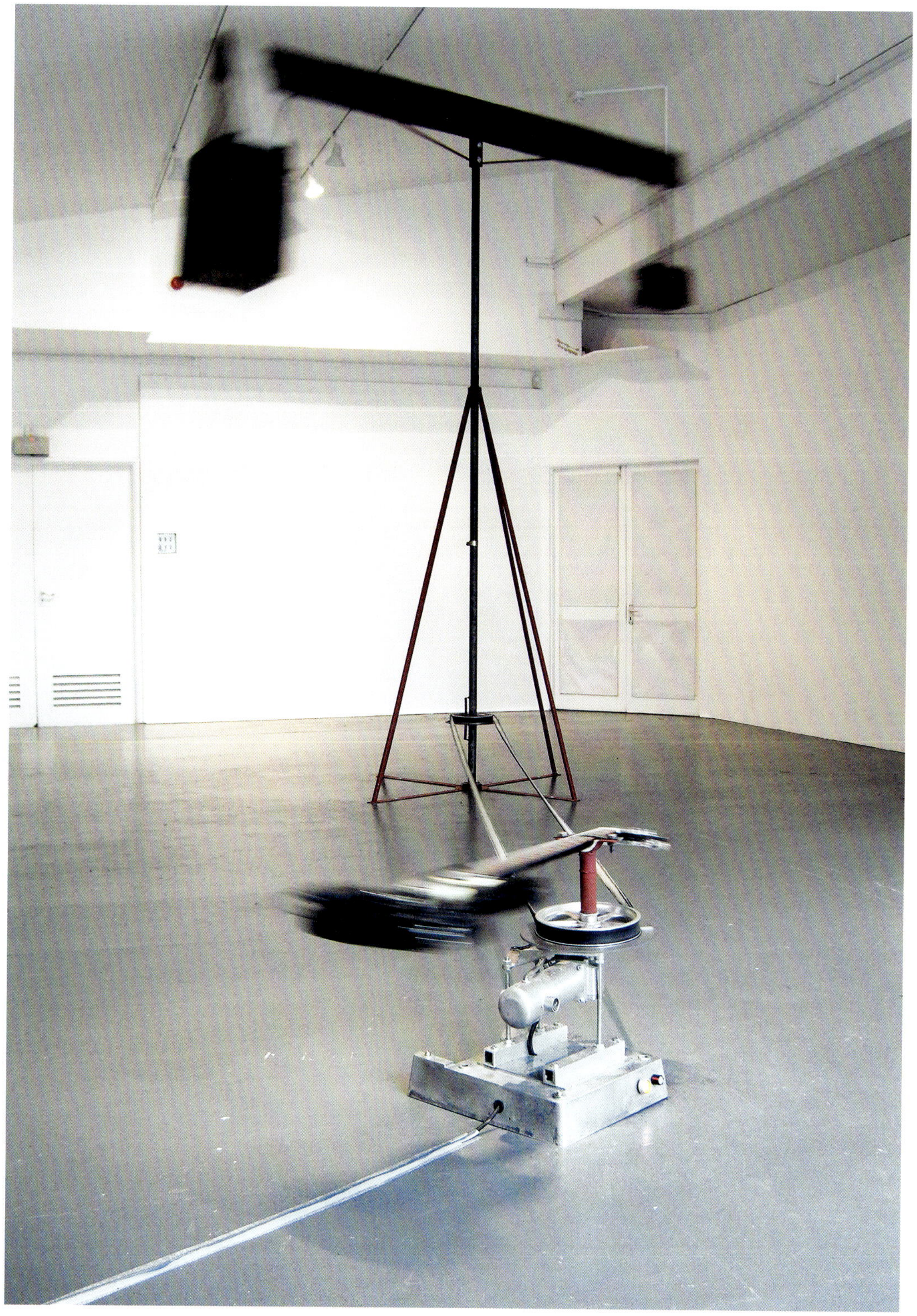

Air Guitar, 2008
Kinetic sound sculpture
Electric guitar, motor, amplifier, battery, wood, metal, dimensions variable
Courtesy of the artist

A single wheelchair motor slowly revolves both guitar and amplifier, creating a droning aeolian loop. The motion that produces the sound is one and the same motion which disperses it around the room. The guitar is simply placed within a kinetic system and allowed to resonate. The combination of doppler and leslie effects achieved by the movement of the speaker and the room's acoustics cause subtle shifts within the eternal chord.

Chris Cunningham

*1970 in Reading (GB), lives in London (GB)

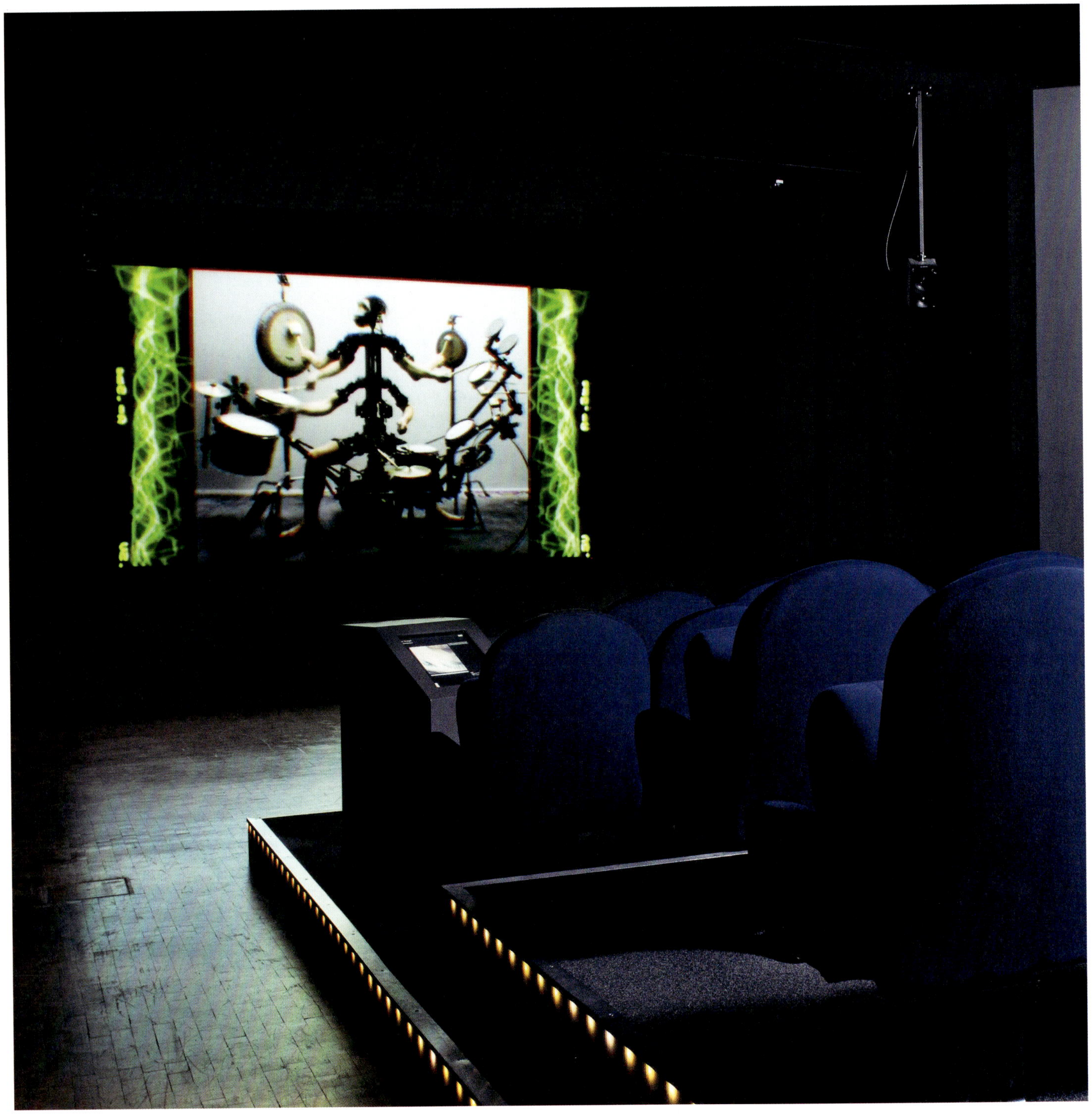

Monkey Drummer, 2001
Video, 2:26 min.
Music: Mt Saint Michel+Saint Michaels Mount (Aphex Twin, album *Drukqs*, 2001)

www.chriscunninghamstudio.com

The historic model for the *Monkey Drummer* is the nineteenth century wind-up toy monkey. The British filmmaker, video artist, director of music videos, musician, and photographer Chris Cunningham wanted to find a modern equivalent for that drumming automaton: "A specimen from the twenty-first century that drums to something as crazy as a really fast track from Aphex Twin." Cunningham filmed the drummer of the Icelandic band The Sugarcubes for the *Monkey Drummer*'s torso and substituted a mechanical torso in postproduction. In this kinetic video study, the monkey's head apparently controls all nine extremities.

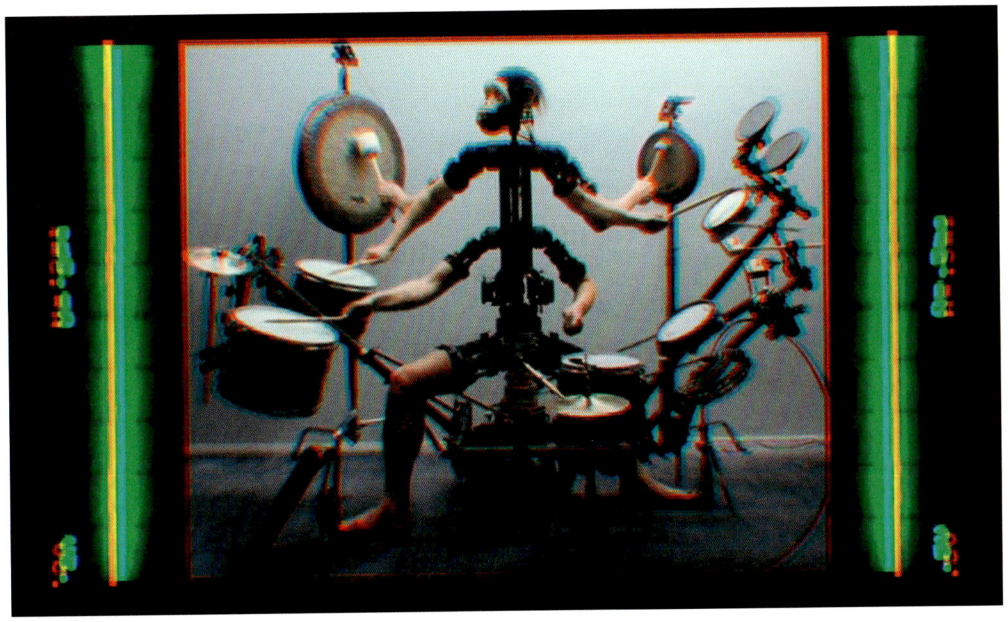

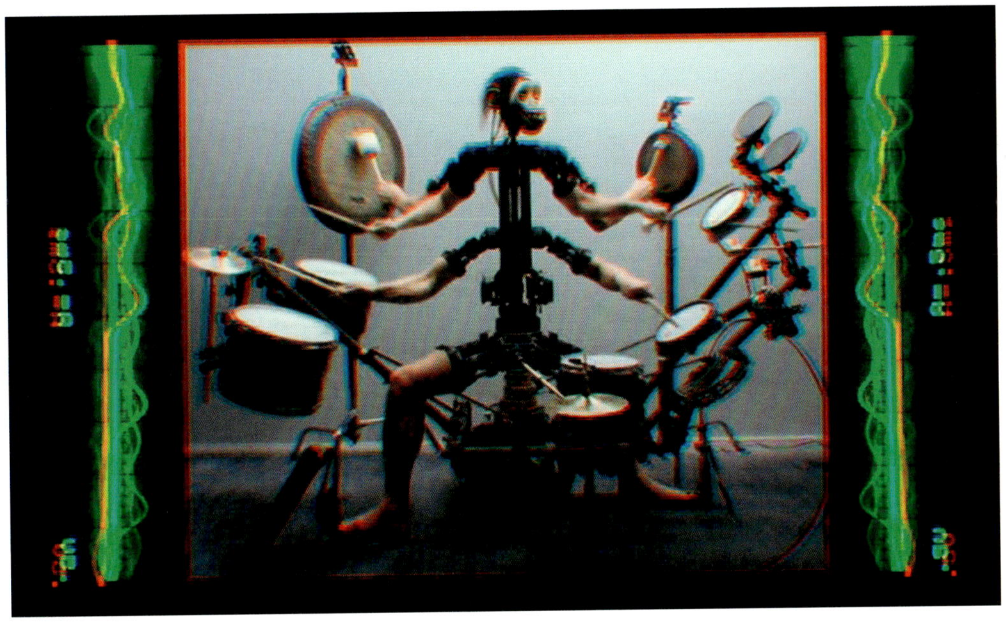

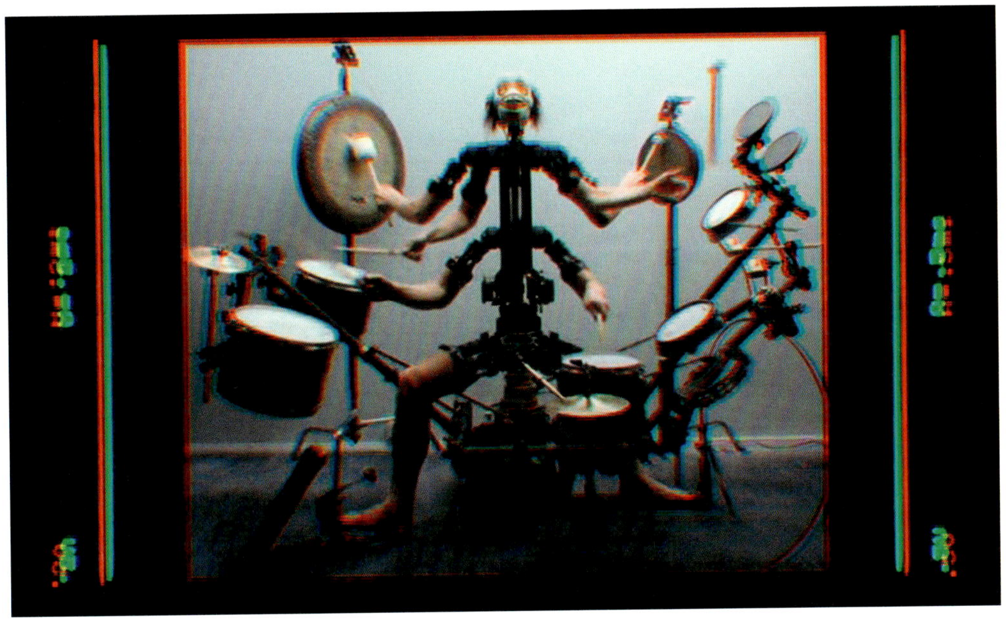

Paul DeMarinis

*1948 in Cleveland, OH (US), lives in Palo Alto, CA (US)

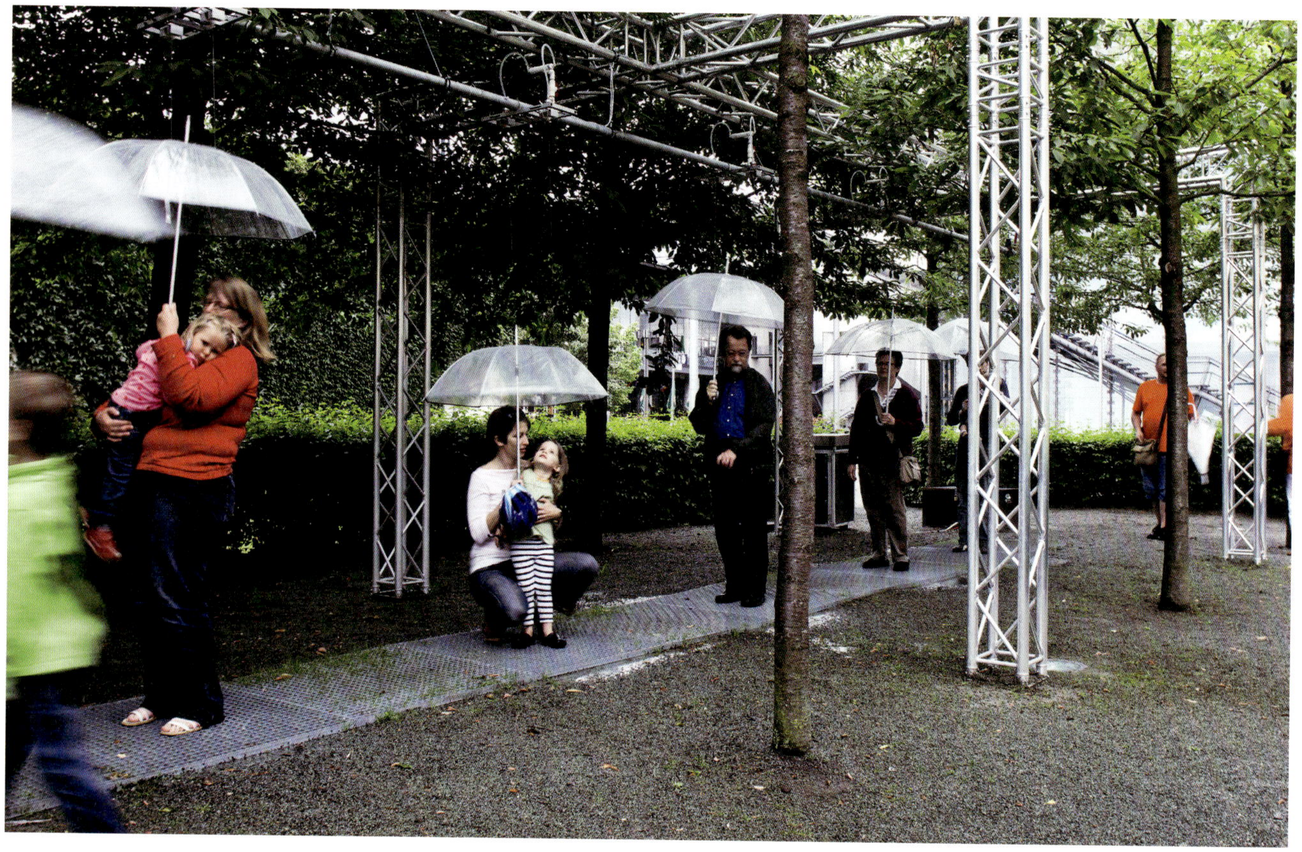

RainDance, 1998
Sound installation in front of the ZKM building, 2012
Water, electricity, sound
Paul DeMarinis

www.pauldemarinis.org

In 1837 the physicist Félix Savart observed that sound vibrations can affect the visual appearance of a jet of water. Subsequent studies determined that the patterns of fluctuations caused by the sound actually reproduce certain aspects of the sound if they fall on a drum.

RainDance builds on this phenomenon of physics to create an interactive and literally immersive sound environment where people can explore "musical" streams of water with umbrellas.

Water is passed through specially designed "modulation nozzles" that impose the vibrations of audio frequencies onto the stream of droplets. For example, 440 vibrations per second results in a stream of 440 water droplets emitted from the nozzle per second. When these droplets fall on a resonant surface such as an umbrella, the tone of A above middle C is reproduced.

In this way various familiar melodies can be reproduced, like Wolfgang Amadeus Mozart's *Eine kleine Nachtmusik (Serenade No. 13)*, Johann Strauss' *The Blue Danube* (Waltz), an Irish folk song, or Gene Kelley's *Singin' in the Rain*. With different streams, multiple-part harmonies or mixtures of disparate materials can be generated.

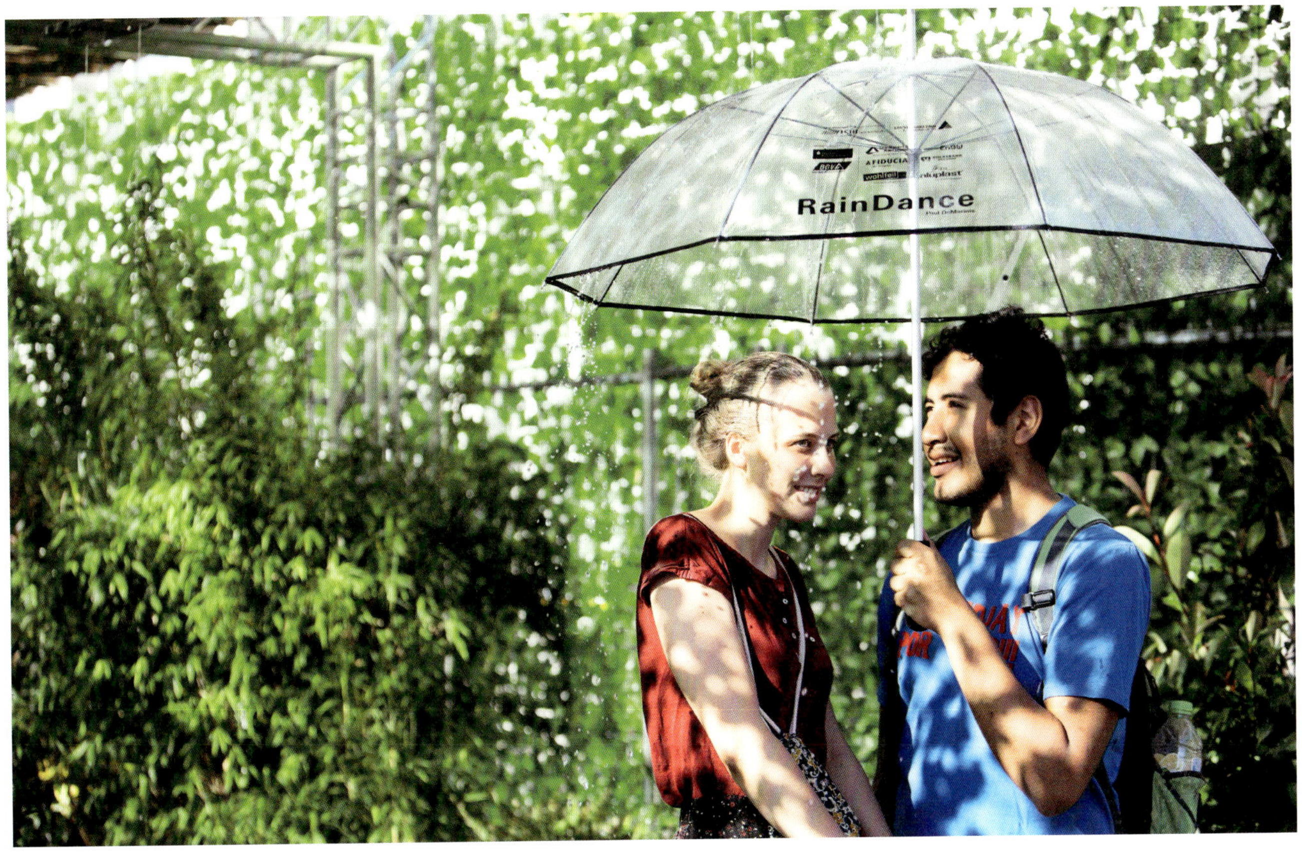
RainDance

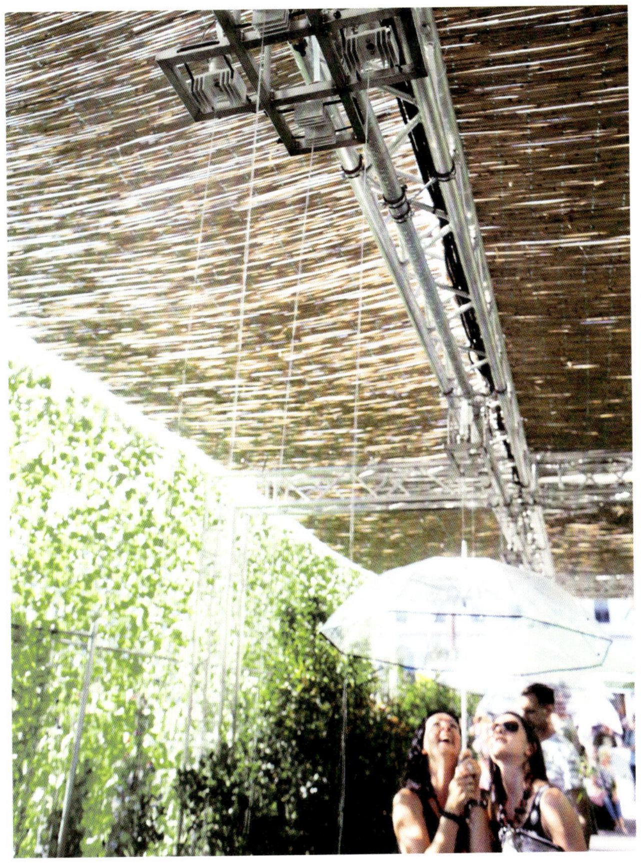

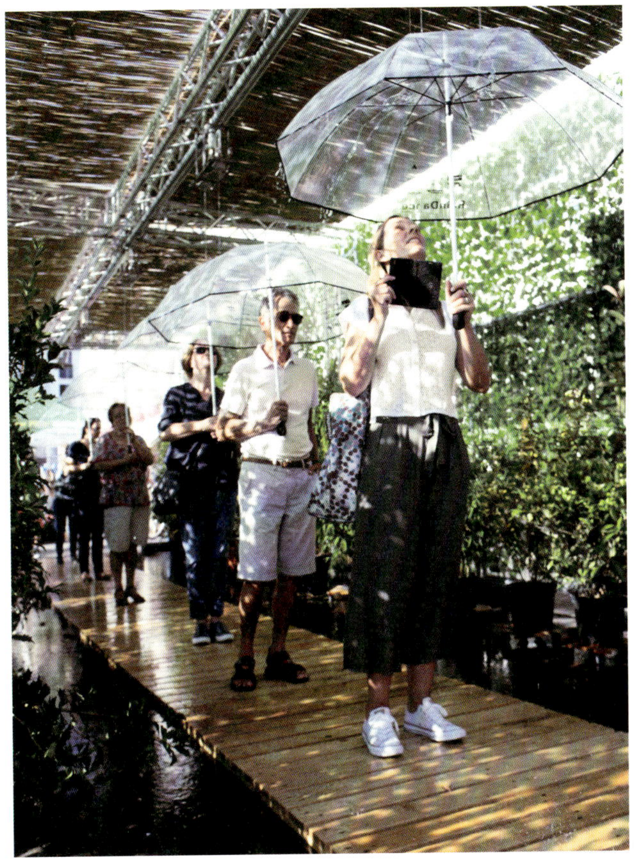

Paul DeMarinis

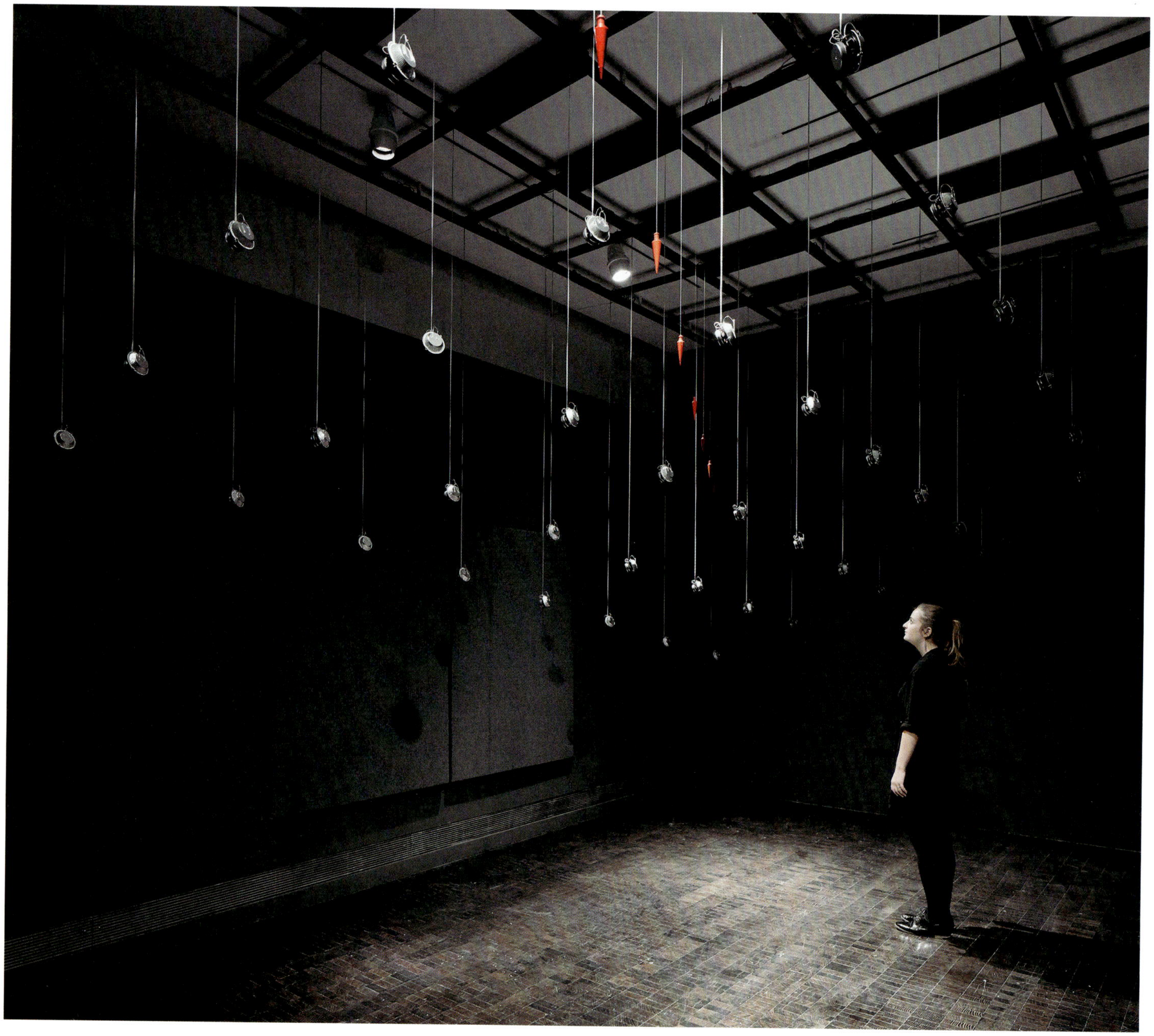

Tympanic Alley, 2015
Sound installation
60 loudspeakers, electronics, dimensions variable
Paul DeMarinis

Sixty small loudspeakers, hanging from the ceiling, are used as motor-oscillators – each speaker is connected in a simple circuit so that when it is activated, it causes the activating current to break, relaxes, and starts over again. The tapper hits on a small metal pie tin that produces a characteristic noise. The net result is a measured but slightly chaotic tapping sound that masses together with the many other speakers to produce the sound and spatial presence of a barrage of rain on a tin roof. A level line of red plumb bobs marks a line of silent orientation in the space.

right page:
Pneuma, 2010
Sound installation
Loudspeakers, electronics, boxes, radio-horns, shelves, disk drives, sound, dimensions variable
Paul DeMarinis

In *Pneuma*, breath and body recordings of sleeping people are played back on loudspeakers. Two signals, both the sound of the sleepers breathing and the measurement of the expansion of their chest as they inhaled and exhaled, are reproduced and superimposed on playback – the paper cones of the loud-speakers slowly rise and fall while simultaneously playing back the sounds of breathing.

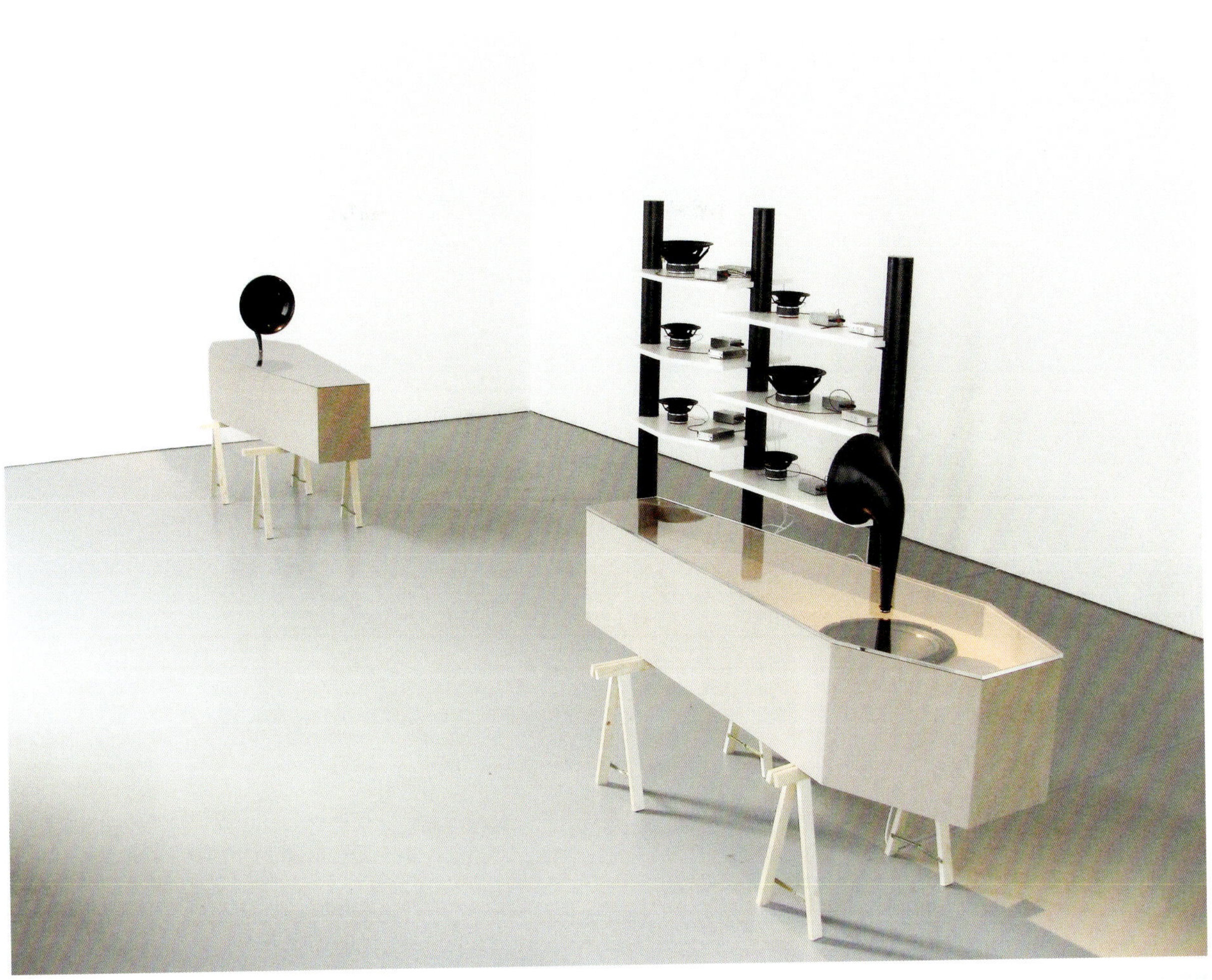

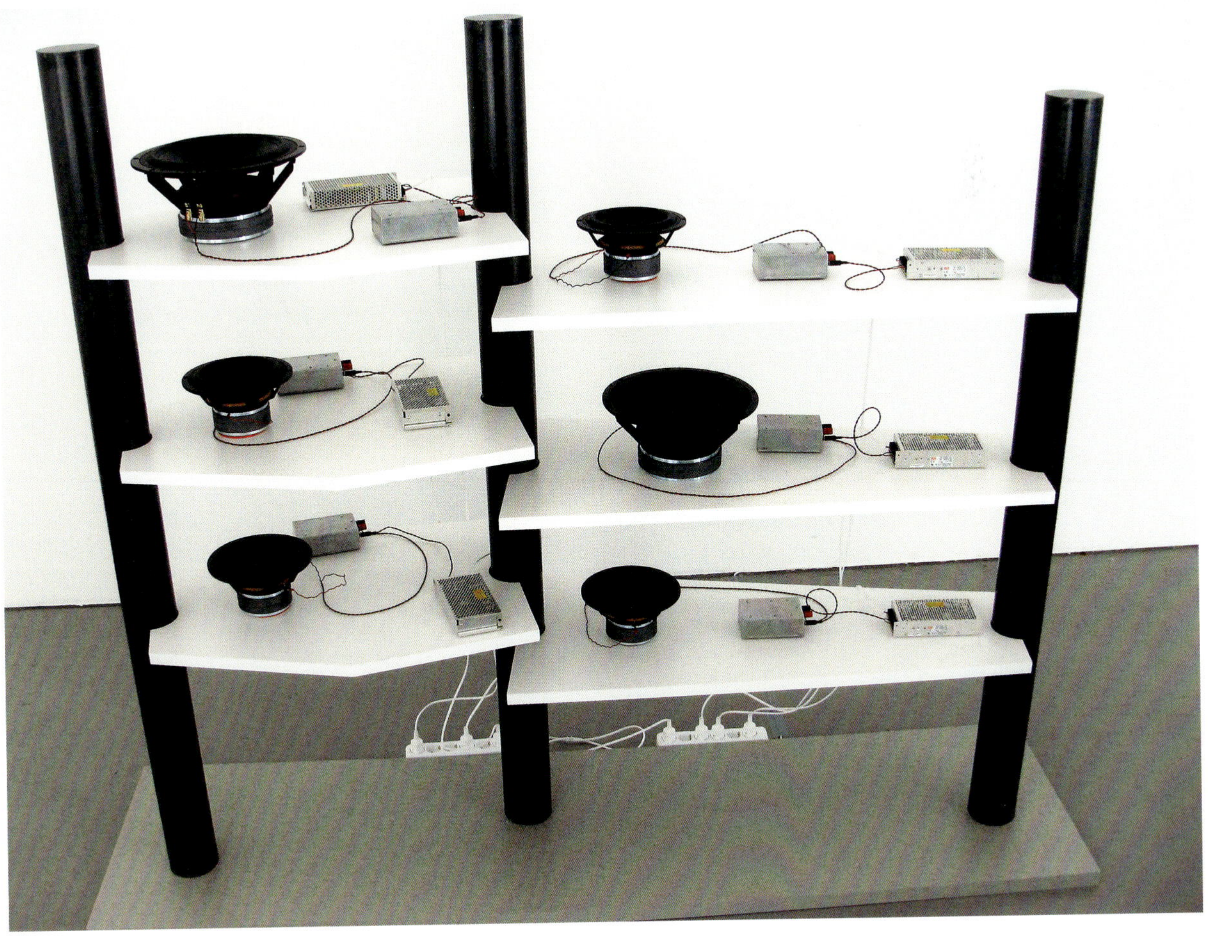

Matthias Deumlich

*1962 in Berlin (DE), lives in Berlin

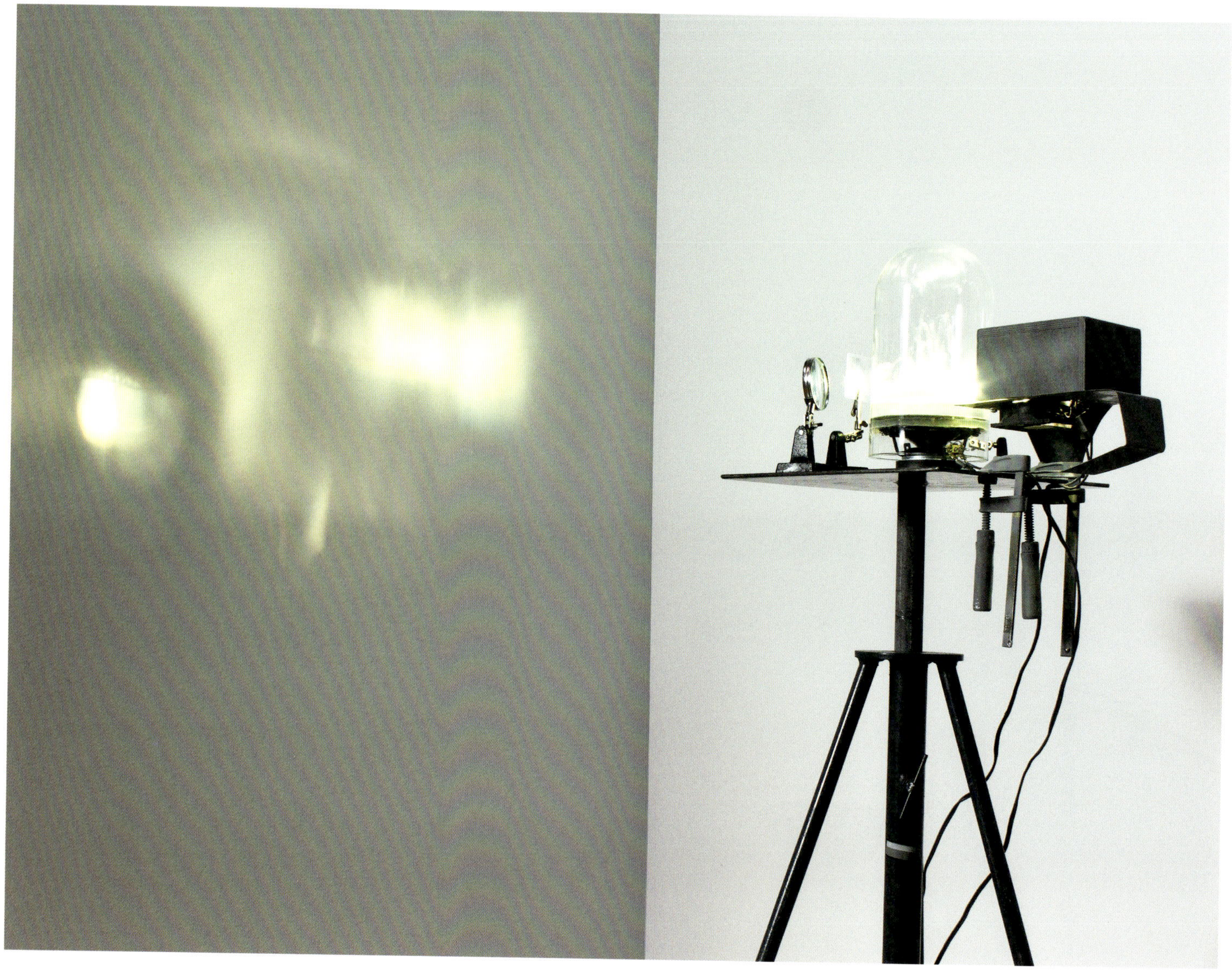

Die spröde Flut, 1999
Light-sound installation
Sound bell: glass with rubber membrane and loudspeaker, metal stand, light projector, lens, transformer, sound control unit
Matthias Deumlich

The sound of a match being struck, slowed down by fifty percent, is the main sound in the installation *Die spröde Flut* [The brittle tide]. Matthias Deumlich uses the movement of the rubber loudspeaker membrane to transmit the wavelike noise to water. Now and then, the sound's structure makes the water in the glass bell spray like a fountain. It reacts sensitively to every change in the slowed-down sound of the match.

The sound generates the motion; motion modulates the sound. In a kind of two-dimensional long-distance sonic effect, the spraying water and the trails of the droplets are projected on the wall by the concentrated light.

www.matthias-deumlich.de

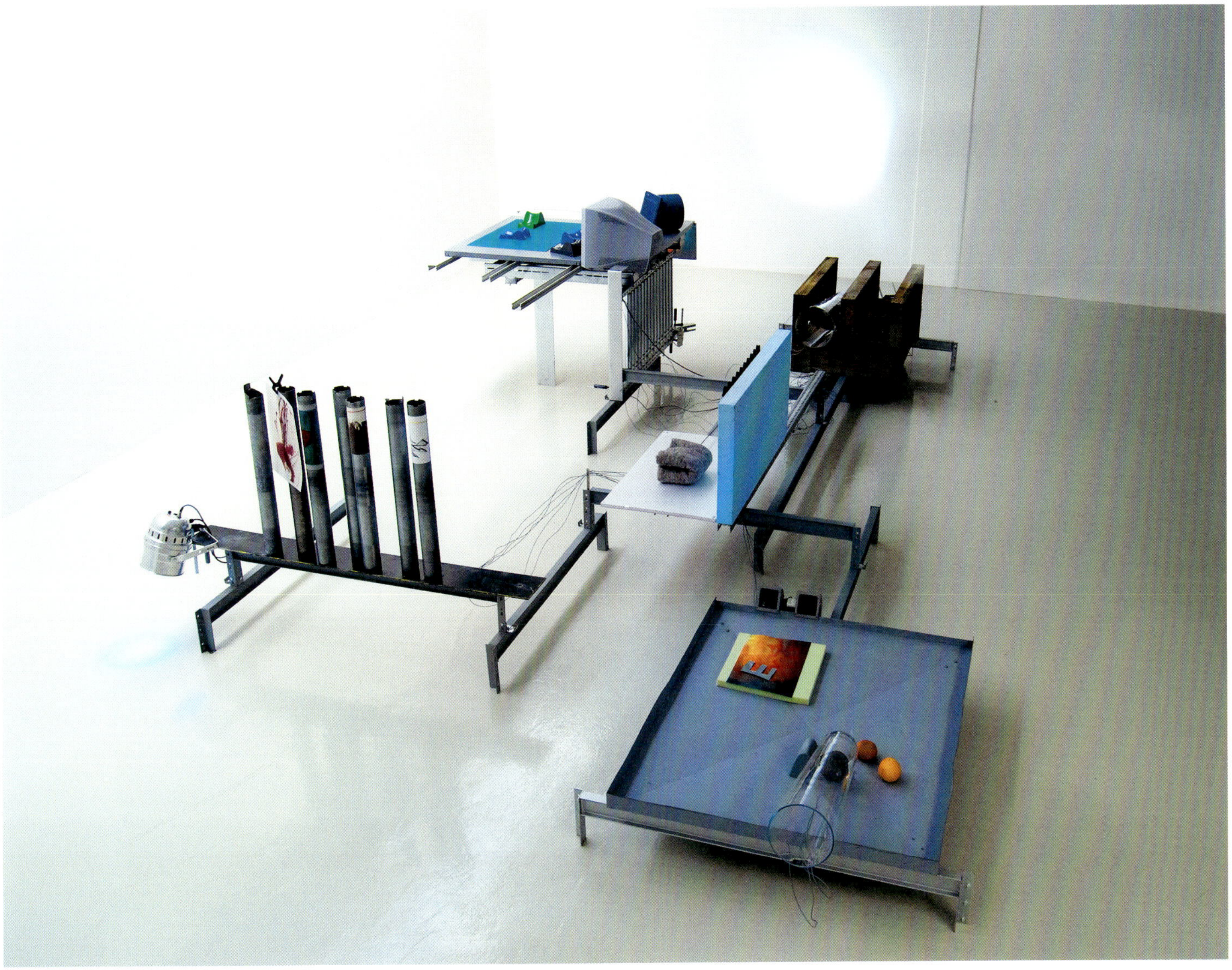

Bergmaschine, 2007
Light, sound, and video installation
5 modules on 9 heavy duty shelf girders (T-profile), various control cables, 6.00 × 4.20 × 1.54 m
Matthias Deumlich

Impressed by the sheer massiveness of a 360-degree Alpine panorama, Matthias Deumlich created the light, sound and video installation *Bergmaschine* [Mountain machine] together with the light and sound object *Neonregal* [Neon shelf] during a four-month working scholarship in the Ticino mountains in Switzerland in 2007.

Owing to the dimensions and the power of the location, the installation has rather the character of a model – its dimensions are more tied to the surface. The highest point is 154 cm tall. At different heights, five different modules are connected to each other on an heavy duty iron construction made of nine former shelf girders.

Bergmaschine focuses on the permanent return to the mountain. What does a mountain "really" look like, regardless of the parts absorbing and reflecting light that reach us, and how does it "sound" in interaction with the approaching noises of humans? The singing sound of the steel cable of the Locarno funicular railway, running over various pulleys, is the raw material of the sound structure that is heard.

Götz Dipper

*1966 in Stuttgart (DE), lives in Karlsruhe (DE)

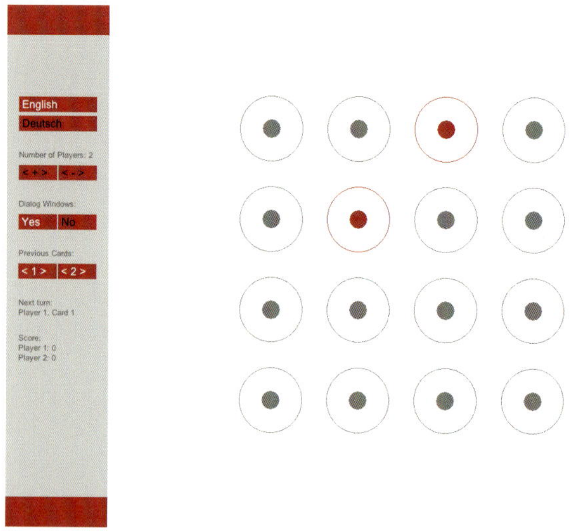

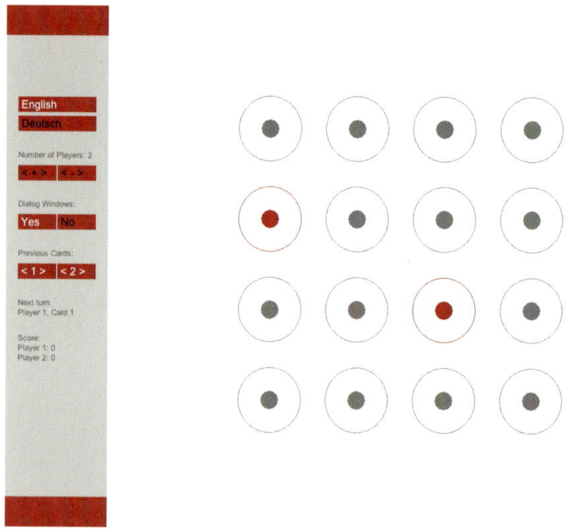

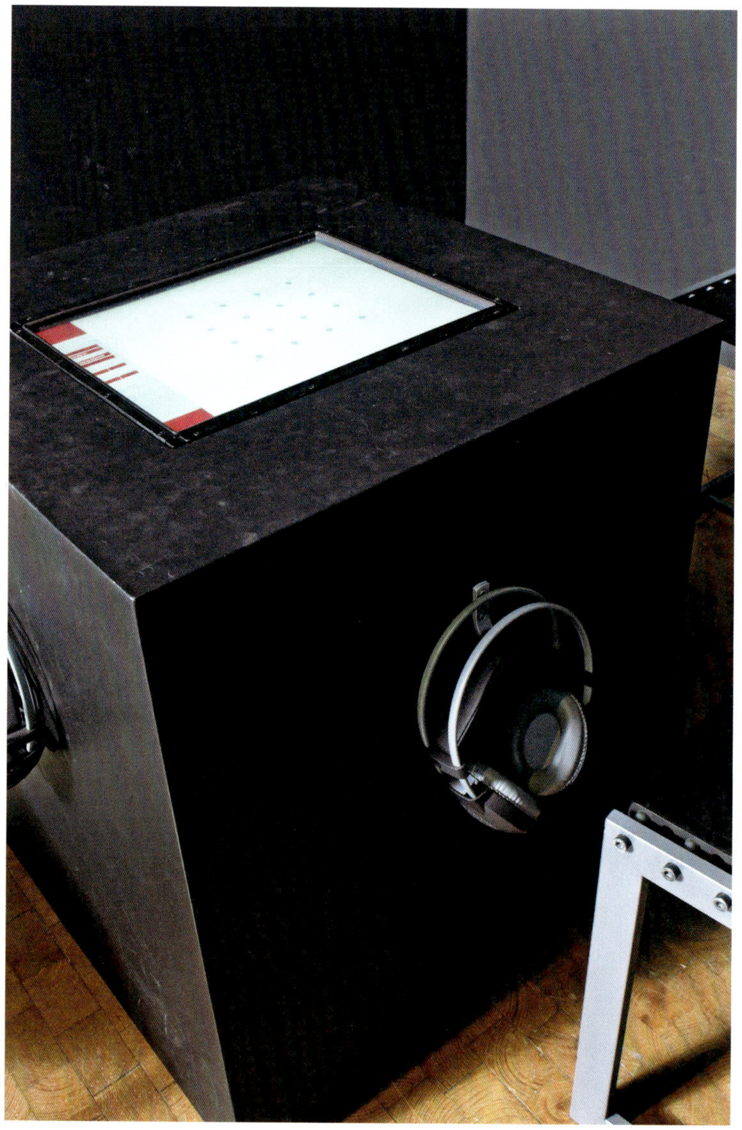

Hör-Memo, 2010/2012
Interactive sound installation,
table with touchscreen, 4 stools,
4 loudspeakers/headphones,
computer, 2 × 2 m
Table made by Manfred Schmieder
Production: ZKM | Karlsruhe

Hör-Memo [Listening memory game] is an acoustic memory game. Visitors can choose between two different versions: *Ton-Memo* [Note memory game], for example, uses individual piano notes as its game material. Visitors are invited to identify notes having the same pitch. The *Klang-Memo* [Sound memory game] version, on the other hand, uses "concrete sounds" and noises. These are primarily everyday sounds like bells ringing, dogs barking, footsteps, or a door slamming.

The contrast between the pitch-oriented sounds of *Ton-Memo* and the noise-like sounds of *Klang-Memo* refers to the emancipation of noise, a central feature of twentieth century Western music history. One of the most important prerequisites for the emergence of sound art occurred when music opened up to other sounds.

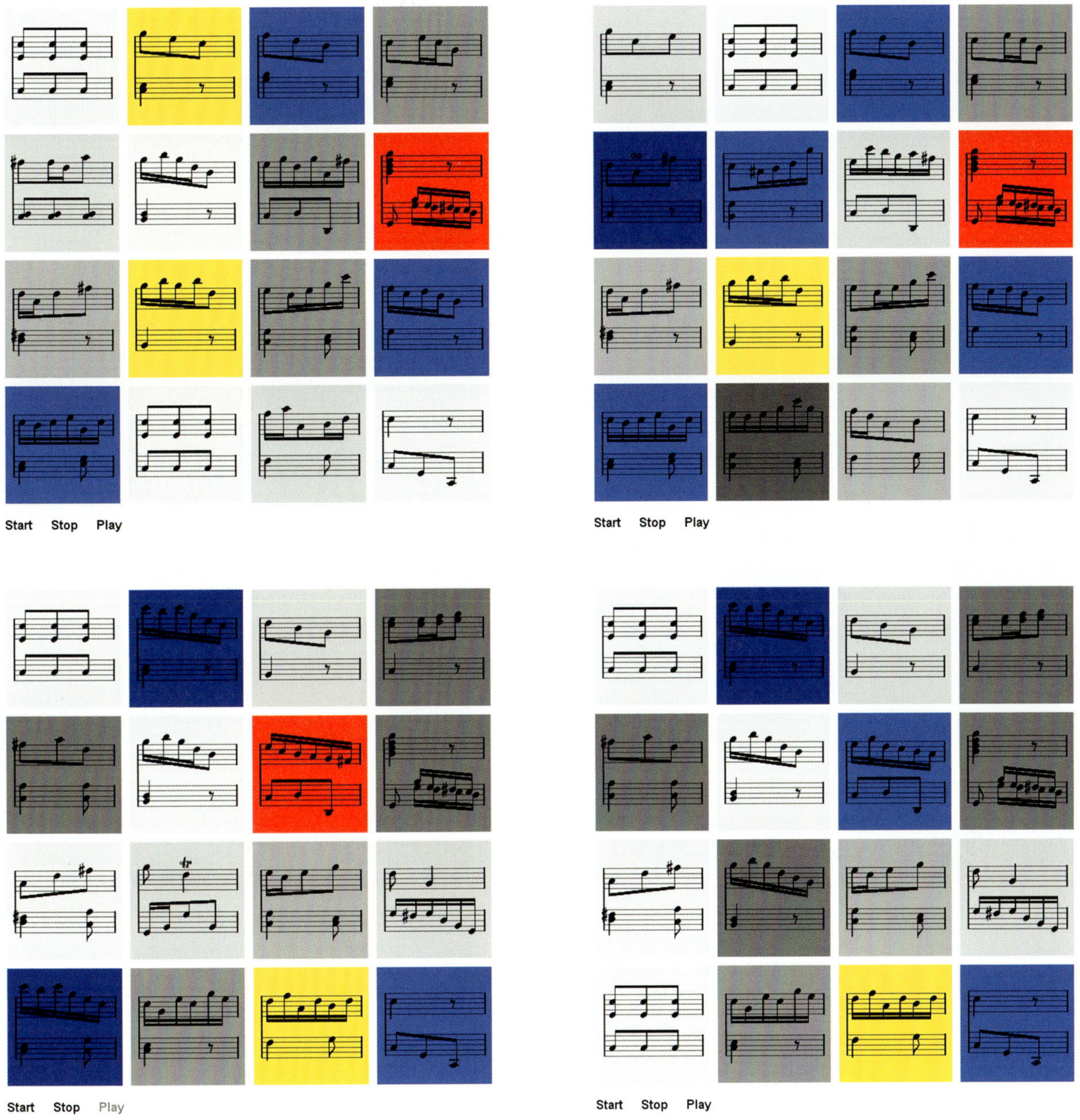

Mozart-Würfel, 2007
Interactive sound installation, console with touchscreen, loudspeakers/headphones, computer, 70 × 70 × 160 cm
Console: Manfred Schmieder
Production: ZKM | Karlsruhe

The installation *Mozart-Würfel* [Mozart dice] traces its lineage to a musical dice game that was produced and popularized under Wolfgang Amadeus Mozart's name two years after his death. Playing the game, one can compose an (almost) infinite number of different waltz melodies. For each of the sixteen measures of the waltz there are eleven different possibilities to choose from. So 1116 = 45 949 729 863 572 161 different waltz melodies are theoretically possible, and it would take more than a billion years to play them all.

A predetermined harmonic schema ensures that all the versions make musical "sense." Götz Dipper has recreated the Mozart dice game as a modern slot machine. For each of the sixteen measures there is a virtual reel of notes that visitors can set in motion and play. Unlike the historical dice game, with the installation *Mozart-Würfel* visitors can also deliberately compose their own versions. This allows them to explore and understand the operative principle behind the game.

Max Eastley

*1944 in Torquay (GB), lives in London (GB)

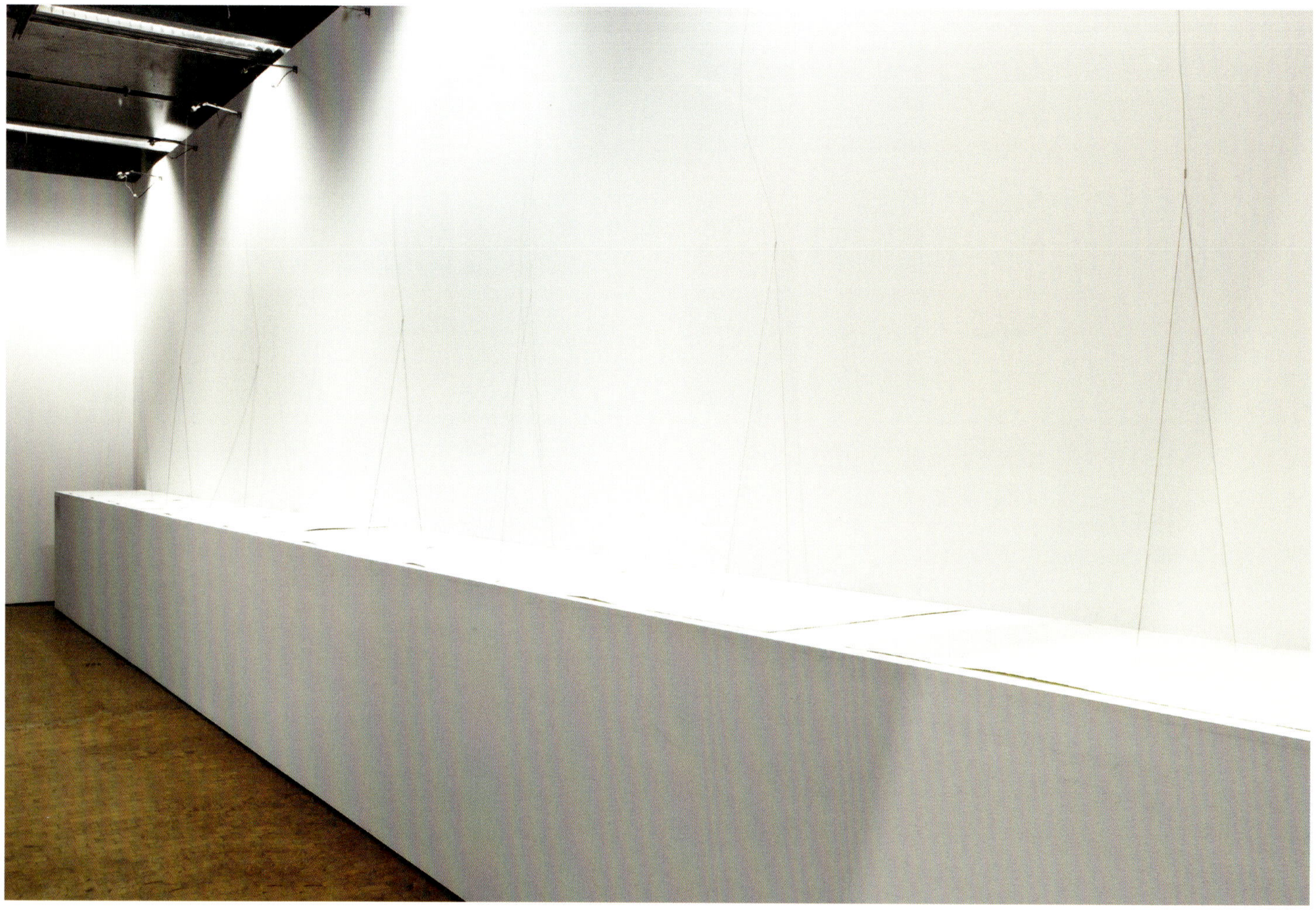

A Procession of Ghosts, 2000
Sound sculpture
Steel cable, steel spring, engines, electronics, paper

www.maxeastley.co.uk

Since the late 1960s artist and musician Max Eastley has explored the relationships between music, art, nature, and chance. With a unique body of work, incorporating kinetic sculptures, music, and environmental forces such as wind and water, Eastley has continually investigated the interplay between these seemingly disparate elements.

"The steel wire figures in the installation were set up in my studio one night and I went to sleep in the next room and left them to run during the night. Some time later I gradually woke from a dream of a spirit medium at a Séance using a Planchette, a small device with wheels and a pencil moved by the medium's hand over paper to write messages from spirits. The movements of the mediums hand and the silent pencil moving over the paper was slowly replaced by the sound of the steel moving over the paper in the next room. Silence is a state of no consciousness, if we perceive, we are alive, however minute the event. Perhaps this is why I strive to find meaning especially through sound and movement, that operate on the threshold of perception, an elusive dream like world which for me is the most vital and closest to the state of life itself." (Max Eastley)

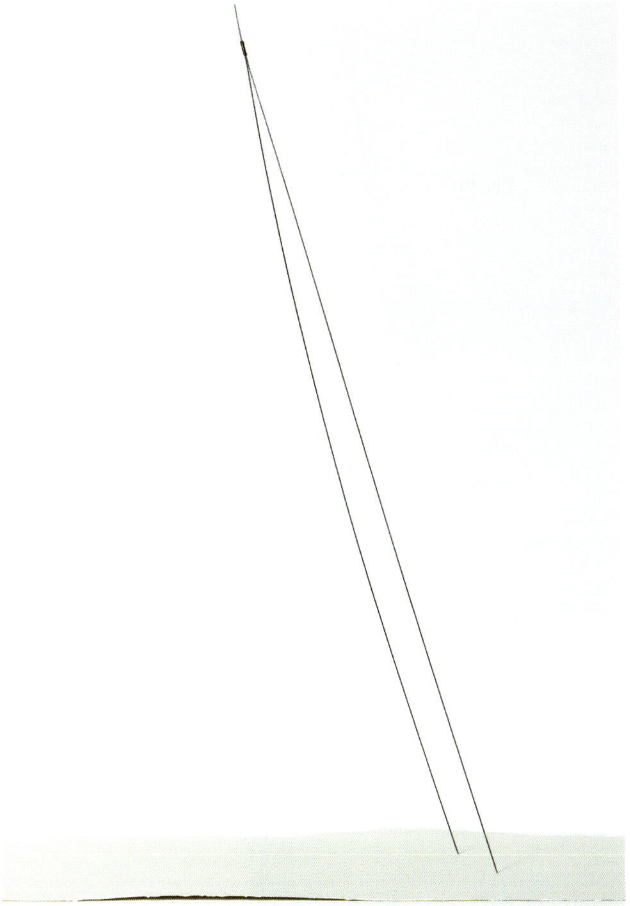
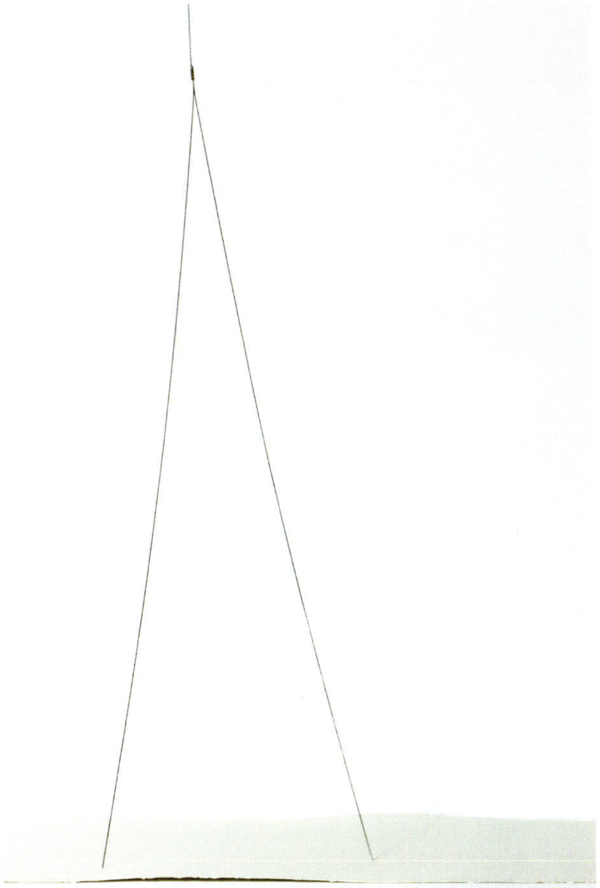
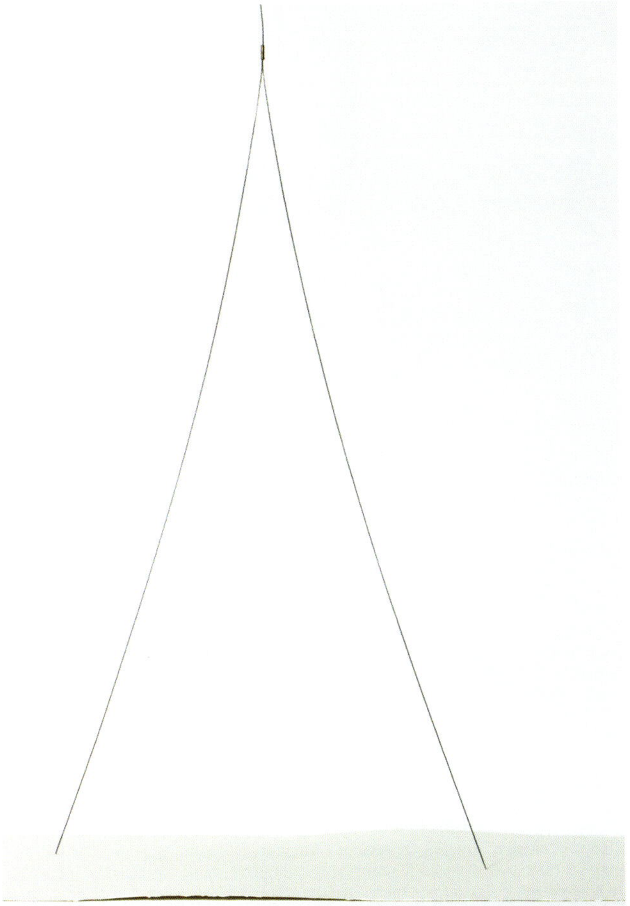

Ulrich Eller

*1953 in Leverkusen (DE), lives in Norderheistedt (DE)

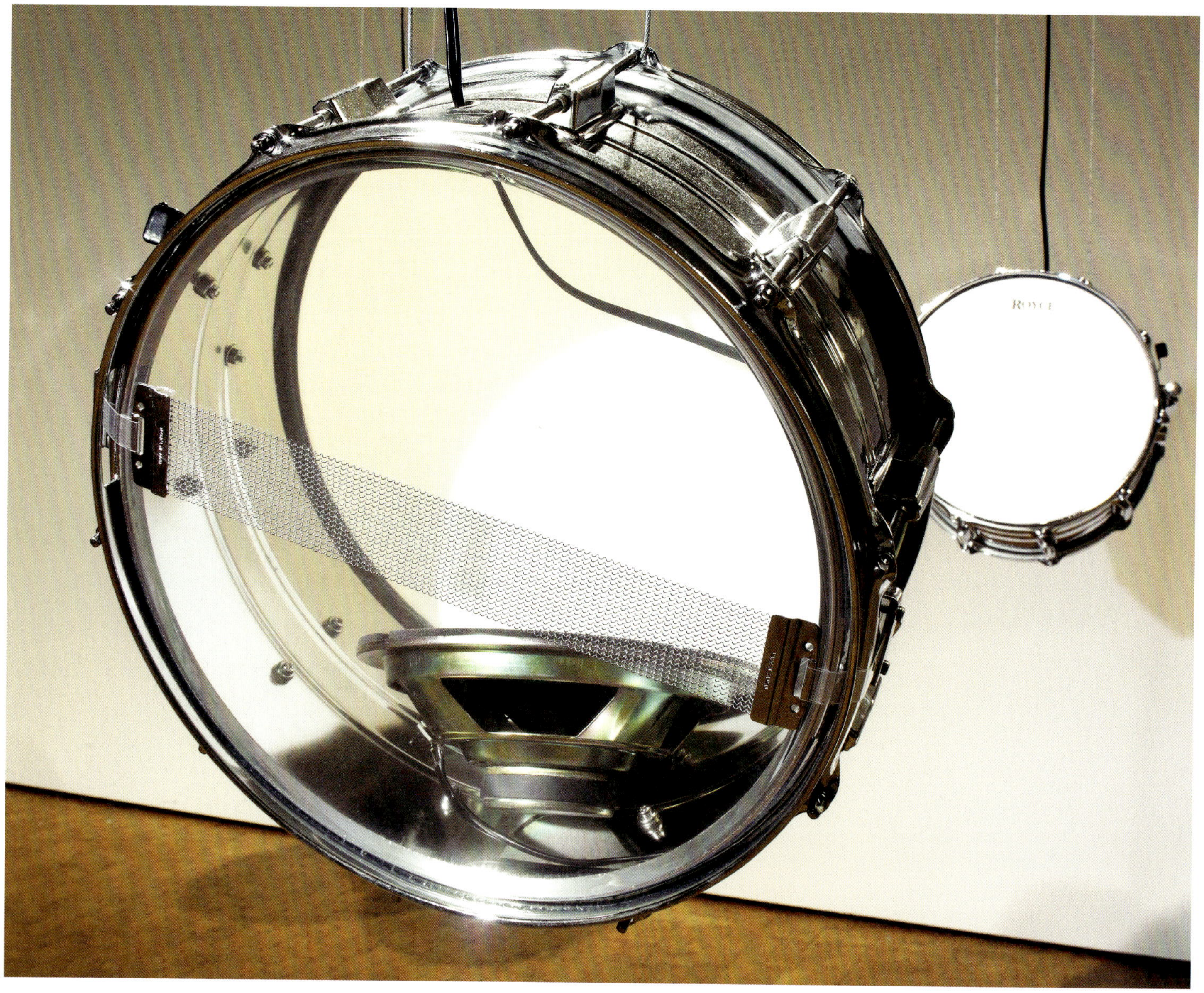

Trommelorchester + Partitur, 2008
Sound sculpture
8 snare drums, 8 loudspeakers, amplifier, matrix, processor, cable, steel cable, pedestal with book object, 29 drawings, each 50 × 50 cm, dimensions variable
Software: Manfred Fox
Production: ZKM | Karlsruhe
Ulrich Eller

www.ulricheller.de

The *Trommelorchester + Partitur* [Drum orchestra + score] has its source in the act of drawing. Ulrich Eller makes his own drawing audible by laying a sheet of paper on the skin of the snare drum. The drum shell amplifies the movements of a hard graphite pencil. This shifts the attention of the person drawing. Not only does the eye monitor the process, the ear also follows the pencil's motions. Listening gains significance in that the activity of drawing follows what is heard, which guides the emerging lines and patterns. The audio recordings created by drawing on the drum skin are the sound sculpture's basic acoustic material. The sounds of drawing are played back via built-in loudspeakers in the eight snare drums. The drum shells react in turn to the feedback. Ulrich Eller views his works as "a compositional game, not in the sense of pitches, chords, and tempi, but as a dialogue between material, form, and sound."

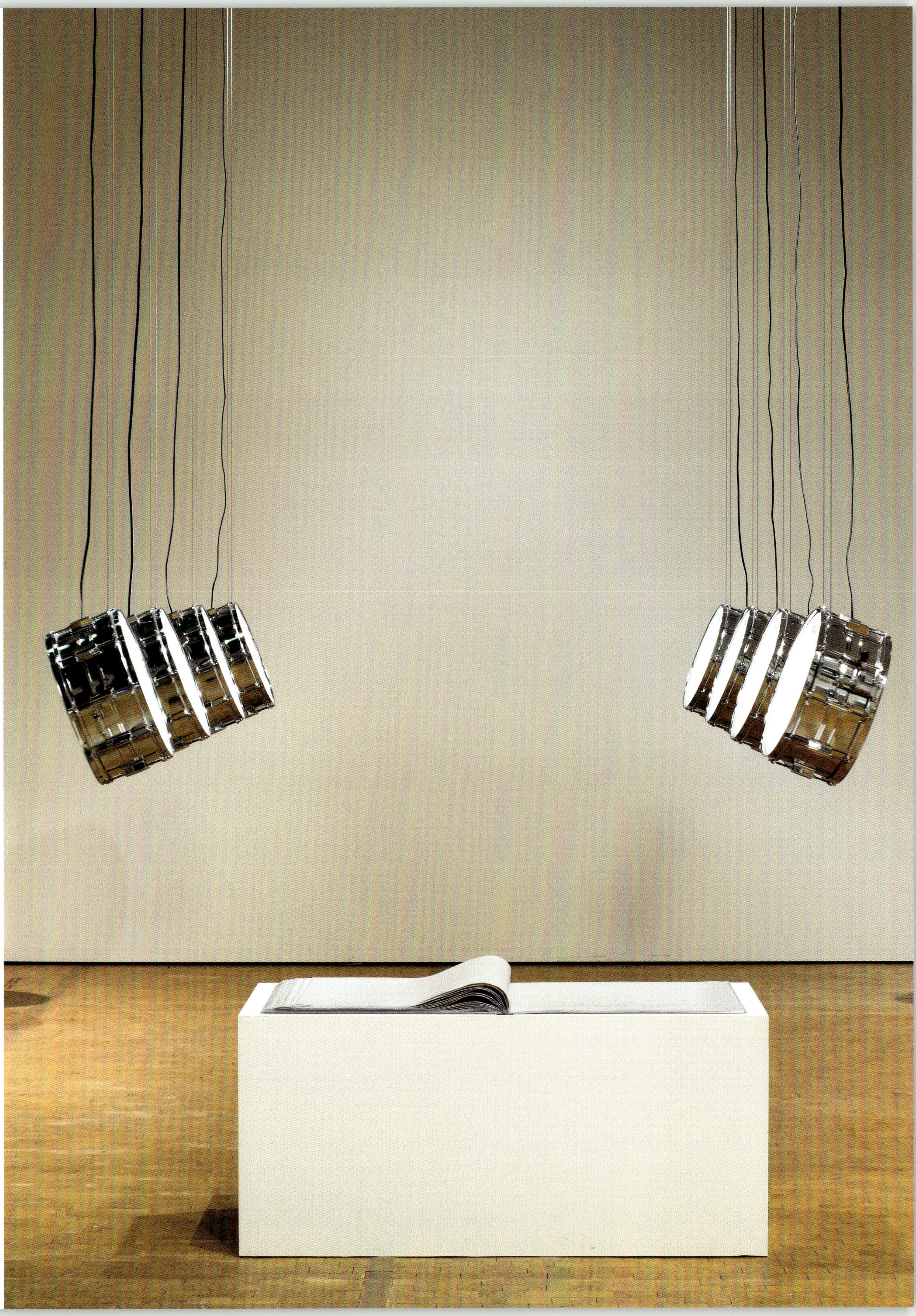

Walter Giers

*1937 in Mannweiler (DE), †2016 in Schwäbisch-Gmünd (DE)

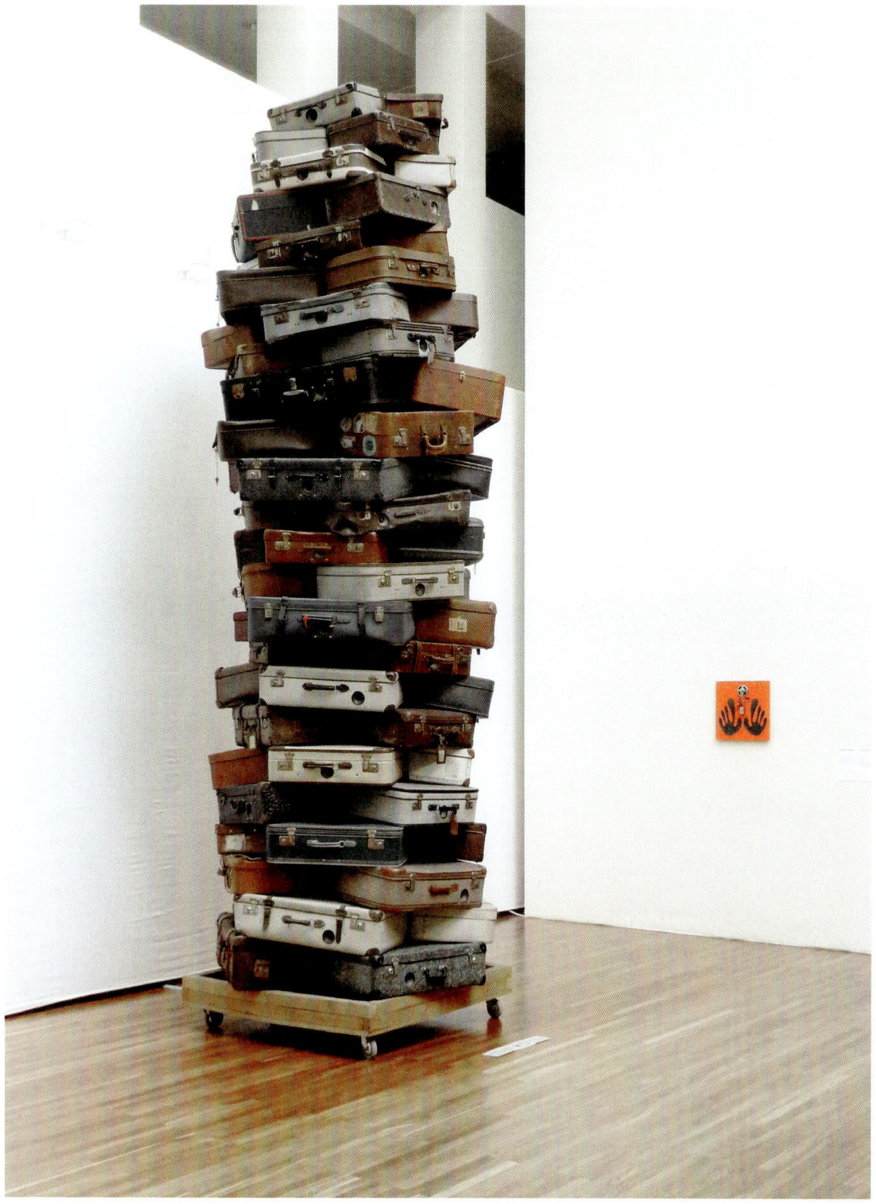

Babel II, 1998
Sound installation
Suitcases, circuits, loudspeakers, 6–8 m high
On permanent loan to the ZKM Collection

Multilingualism as a cultural asset or as a confusing barrier between nations?

Babel II is a tower of much-travelled suitcases, and from each of them issues the sound of a different language. Together, the acoustic sounds from the suitcases result in a relaxing murmur with sporadic dominance of one or other language. With the help of the imagination the listener can interpret different content in this noise, because humans have a faculty whereby they recognise patterns in apparent randomness. The title *Babel II* is "version numbered," the original Tower of Babel being number one – Giers' typical sense of humor. (Victor Giers)

www.waltergiers.de

right page:
Art-Statement-Art, 1993
Sound sculpture
50 loudspeakers, audio recordings, dimensions variable, ca. 200 × 170 cm
ZKM Collection

The artist, engineer, and jazz musician Walter Giers transformed the visual and acoustic elements of technological building blocks into sound and light artworks situated between music, visual art, and technology. Random generators control many of the works. Each evolves its own unforeseen dynamics through the constantly new variations in its complex inner life.

For *Art-Statement-Art* Walter Giers assembled fifty old loudspeakers and arranged them as a sculpture. Voices emanate from every box. For the recordings, Giers had asked a number of people the question: "What is art?" The answers sound simultaneously from all fifty loudspeakers – a portrayal of the heterophone opinions.

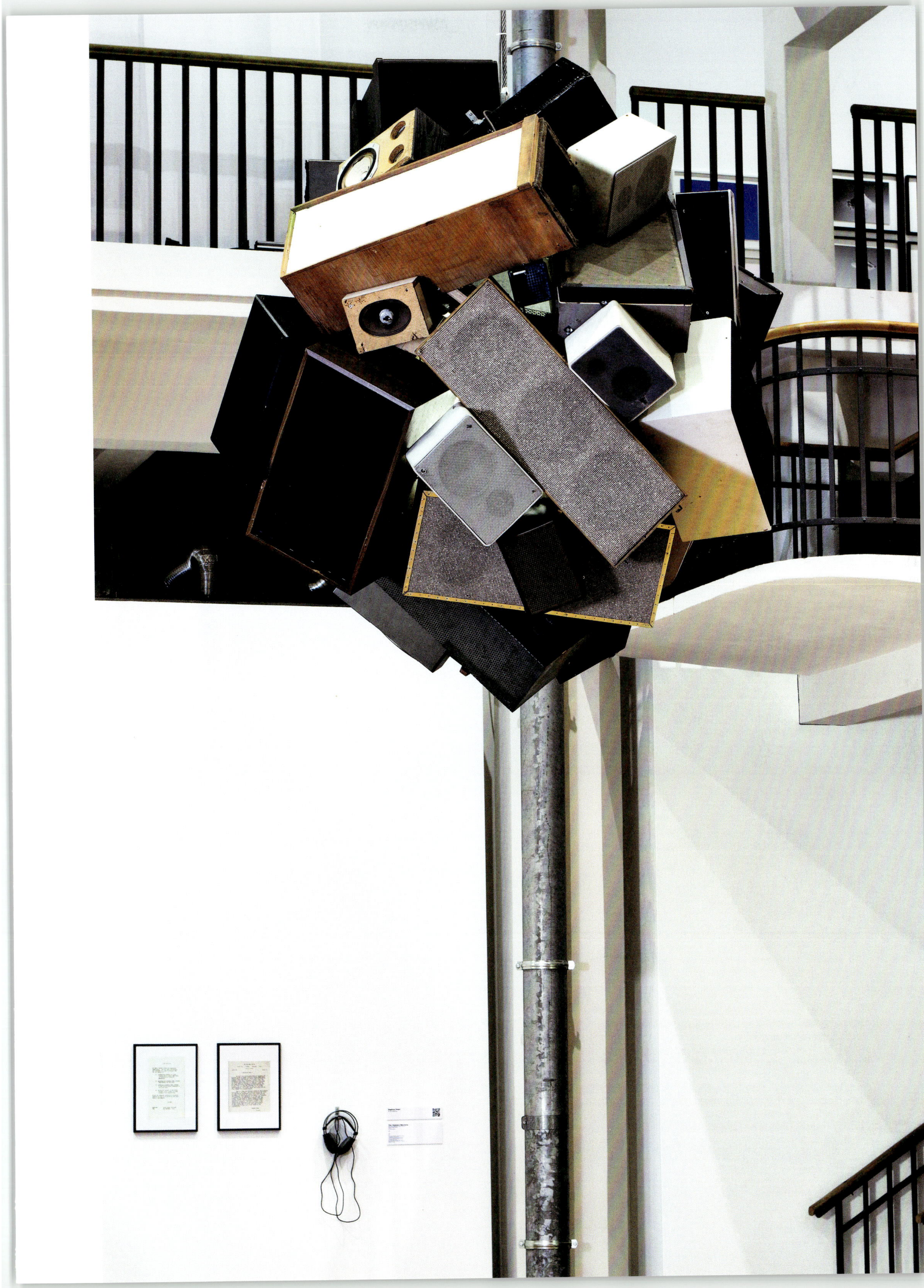

Walter Giers

*1937 in Mannweiler (DE), †2016 in Schwäbisch-Gmünd (DE)

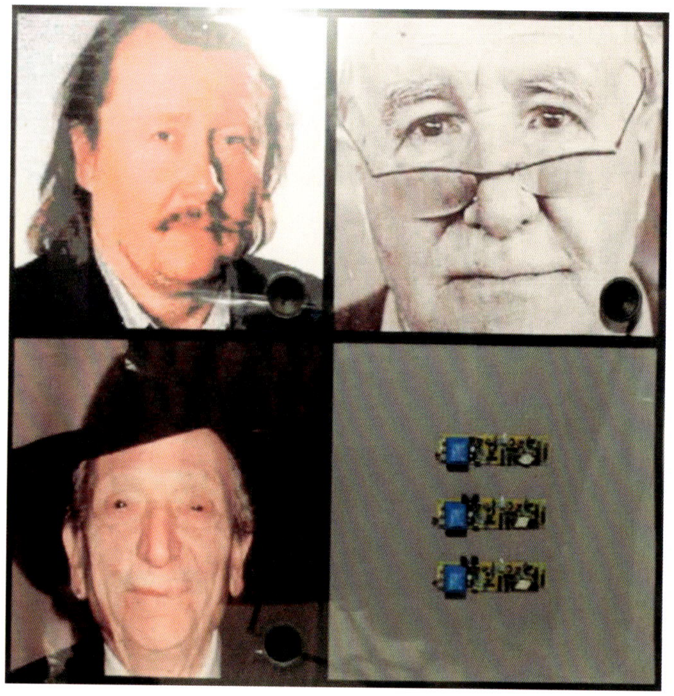

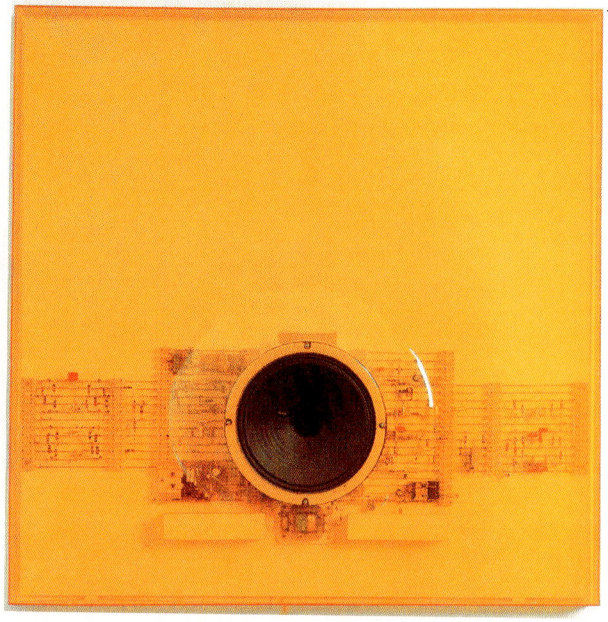

Neue Simultanlesung (Philosophenrunde) /
New simultaneous lecture (panel of philosophers), 2011
Sound installation
Acrylic, circuits, loudspeakers, photographs, 75 × 75 × 6 cm
Private collection

Peter Sloderdijk (*1947, philosopher and cultural theorist), Max Bense (1910–1990; philosopher, writer, and publicist), and Kurt Weidemann (1922–2011; typographer, graphic designer, and author) "philosophize" via random generator.

Impertinent, 1976
Interactive sound installation
Acrylic, circuits, loudspeakers, 75 × 75 × 6 cm
Private collection

In Walter Gier's work *Impertinent* [Impertinent], sound structures unpleasant to the ear are manipulated by random generators and by a glaring light source.
The object builds up panic.

If the viewer turns the object off, it switches over to an internal source of energy. Even if the viewer pulls the plug out of the electric socket, the object will only cease its impertinence when the random generator decides it will do so. Here again the background to this work is the issue concerning the autonomy of the machine. (Victor Giers)

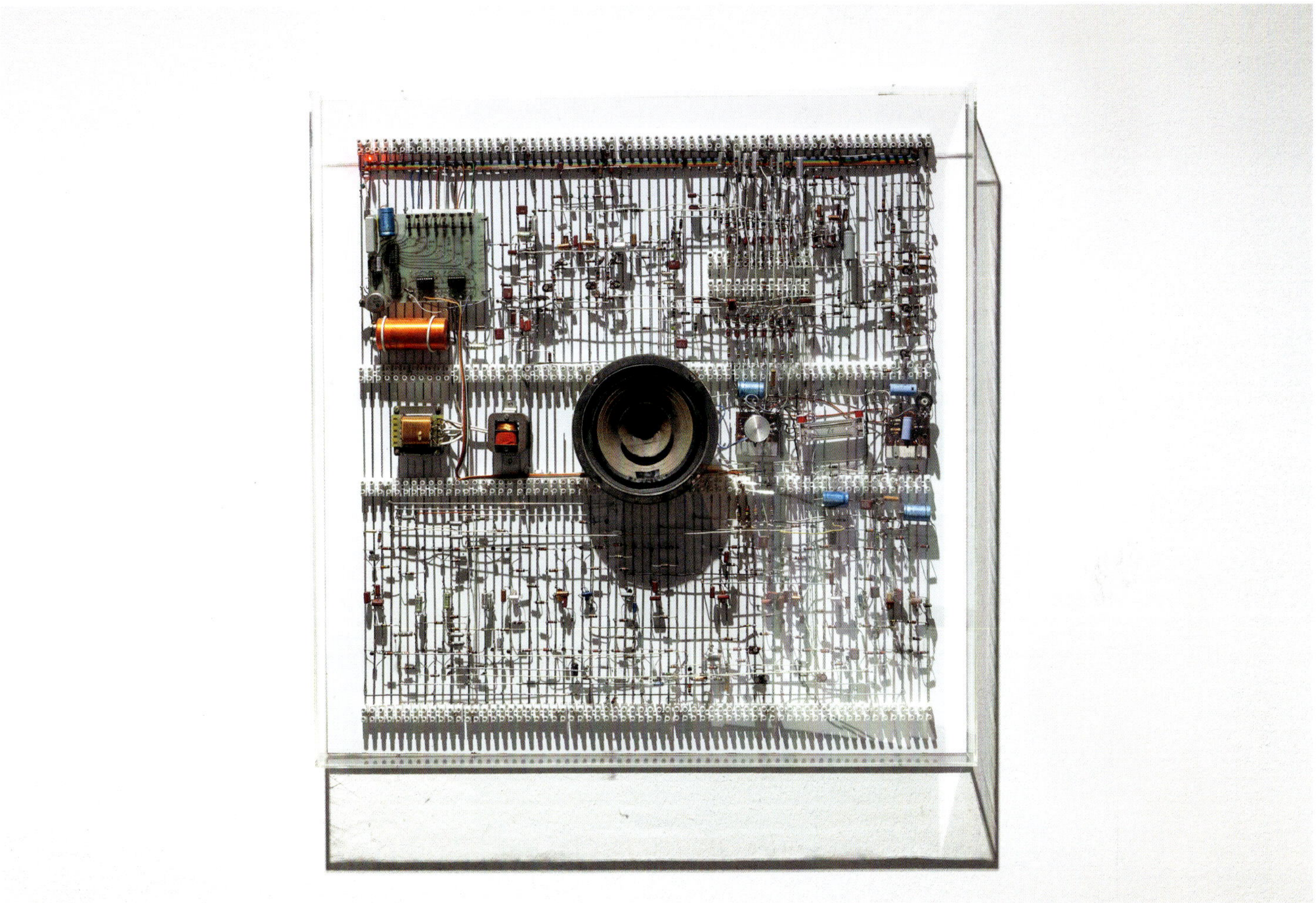

Erotischer Zyklus, 1992
Interactive sound installation
Acrylic, circuits, LEDs, loudspeakers, 75 cm × 75 cm × 6 cm
Private collection

Erotischer Zyklus [Erotic cycle] is Walter Gier's first so-called "concert picture." It depicts the communication between man and woman from the perspective of a bachelor.

Sabine Groschup

*1959 in Innsbruck (AT), lives in Vienna (AT) and Berlin (DE)

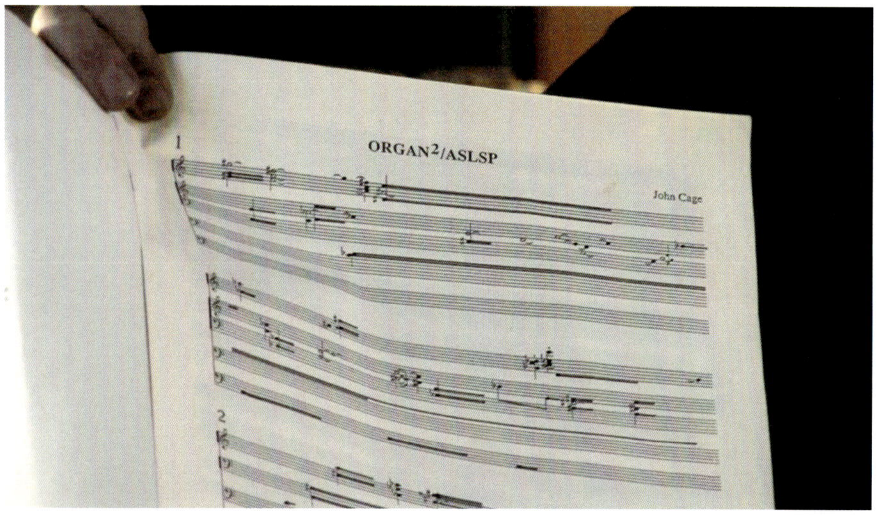

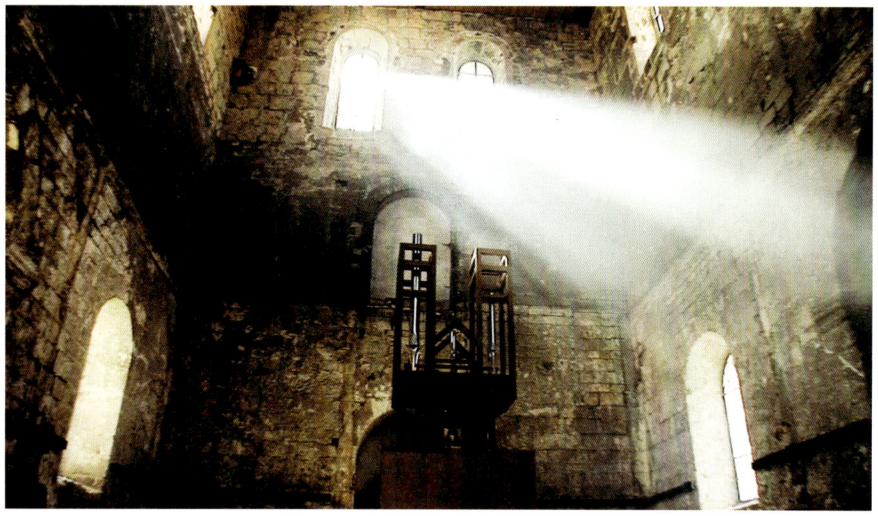

(JC{639}), 2006/2012–
A film in 89 aleatory variations on John Cage's *ORGAN2/ASLSP* at the St. Burchardi Church in Halberstadt (DE), 29:14 min.
Courtesy of the artist, the John Cage Trust, New York, the John Cage Organ Project Halberstadt & Edition Peters, Leipzig–London–New York

Sabine Groschup's experimental film *(JC{639})* on John Cage's *ORGAN2/ASLSP* at the St. Burchardi Church in Halberstadt, Germany, documents in striking images (director of photography Jerzy Palacz; black-and-white photography Barbara Klemm) and superb sound design (Eric Spitzer-Marlyn) the 639-year performance (2001–2640) of Cage's composition for organ that is attracting worldwide attention. The Austrian artist, filmmaker, and author uses a technique quite possibly unique in the history of the film medium, one that relates directly to John Cage: she allows chance to determine the course of her film about the musical work that is to be played "as slow as possible" (*ASLSP = As SLow aS Possible*).

The DVD with the first variation of the film (*chance operations #1* by Jozef Cseres) includes a reader with texts by Christoph Bossert, Franziska Bruckner, Sabine Groschup, Wulf Herzogenrath, Ray Kass, Laura Kuhn, Rainer O. Neugebauer, Dieter Schnebel, Hannah Schwegler, Georg Weckwerth. Photos Barbara Klemm.

www.sabinegroschup.at
www.jc639.sabinegroschup.at

"*(JC{639})* is an experimental film dedicated to John Cage on the occasion of his 100th birthday. *(JC{639})* is 29:14 min. long, precisely the same length as the original performance of John Cage's *ORGAN2/ASLSP* by Gerd Zacher in Metz in 1989. *(JC{639})* commemorates John Cage both by way of the critical enquiry into his composition *ORGAN²/ASLSP* by the Halberstädter Cage Project, and through an approach to John Cage's access to chance. In the case of *(JC{639})* I found it important not to pursue a conventional path, but to step into unknown territory, and to attempt something new under John Cage's influence. My film *(JC{639})* is defined by chance, or by the absence of intentionality, through a random selection of stills from single film scenes. The selection is to be made by persons for whom John Cage's work and his influence are of personal significance. The single scenes may be divided into four varying forms of expression each of which exhibit fixed structures within the units. Each scene may be considered as a note. When the notes thus randomly string together, then the composition of *(JC{639})* emerges."
(Sabine Groschup)

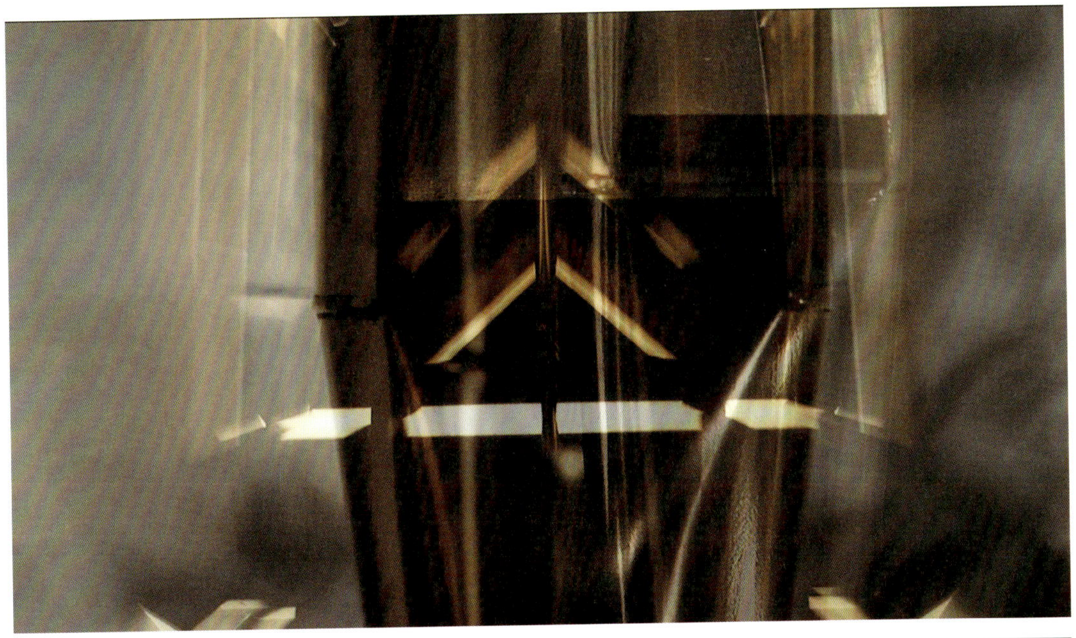

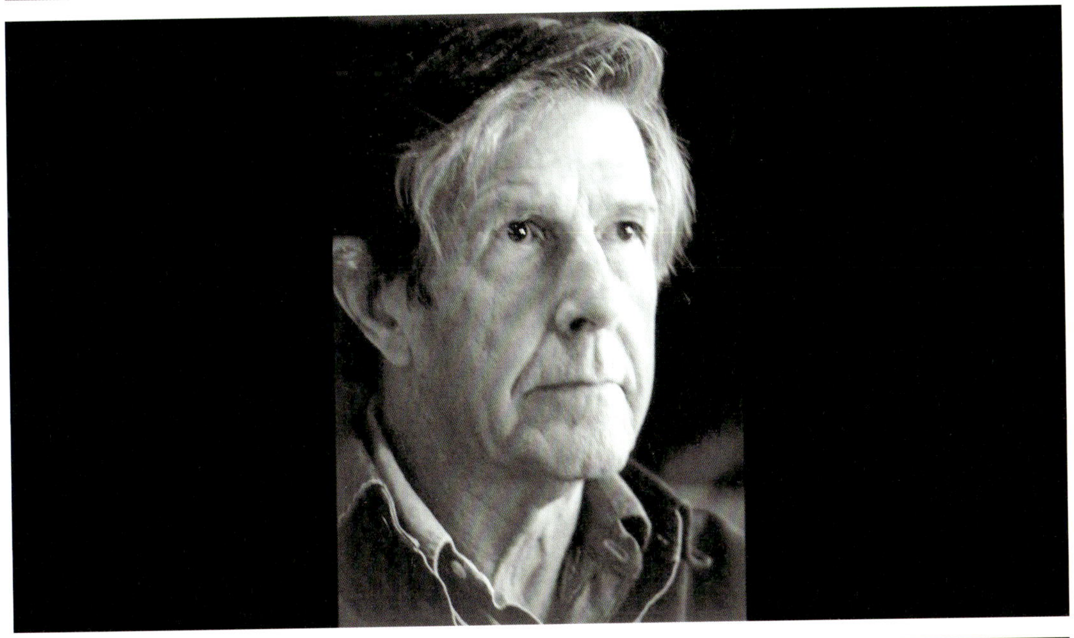

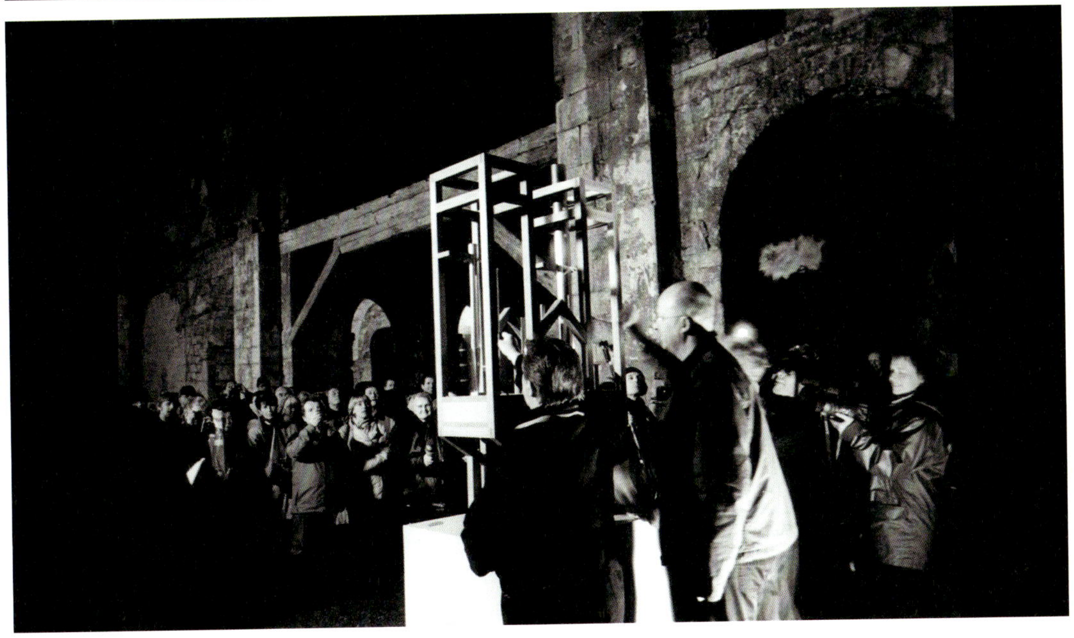

Sabine Groschup

Prototype Easter Memories

sound without air
touches the shoulder and goes on
wind, or

idiophone ancient and
old wed with golden chocolates

soundmakers from China
Easter-rabbit bells chiming
tinkle
the tones

spoons on the pot rim
prototype
red ribbon
Lindt-gold at ease in beige cherrywood

Sabine Groschup, 2013
(Translation: AK Lerner)

Prototyp Ostererinnerung, 2013
10-part sound installation
Little bells by Lindt mounted
on cherrywood cooking spoons,
dimensions variable
Courtesy of the artist

"A traditional Chinese musical instrument gave me the idea for the design of *Prototyp Ostererinnerung* [Prototype Easter Memories]. The ancient flat wooden object decorated with painted flowers is part of the collection of the Technisches Museum Wien (Vienna). This approximately 20 cm long strip of wood tapering to a handle at the bottom end has two circular holes in its upper part in which small bells are fixed.

In our latitudes, the chocolate "Gold Bunny" by Lindt, with its golden bell on a red ribbon collar, has become a tradition. In TV advertisements the rabbit becomes "alive" and rings the soundmaker around its neck.

Year after year these little bells are left over after savoring the chocolate rabbit. I went on a search for a suitable piece of wood for the tiny bells and found the perfect hole for them in a mundane/ordinary object: a kitchen utensil, the wooden mixing spoon." (Sabine Groschup)

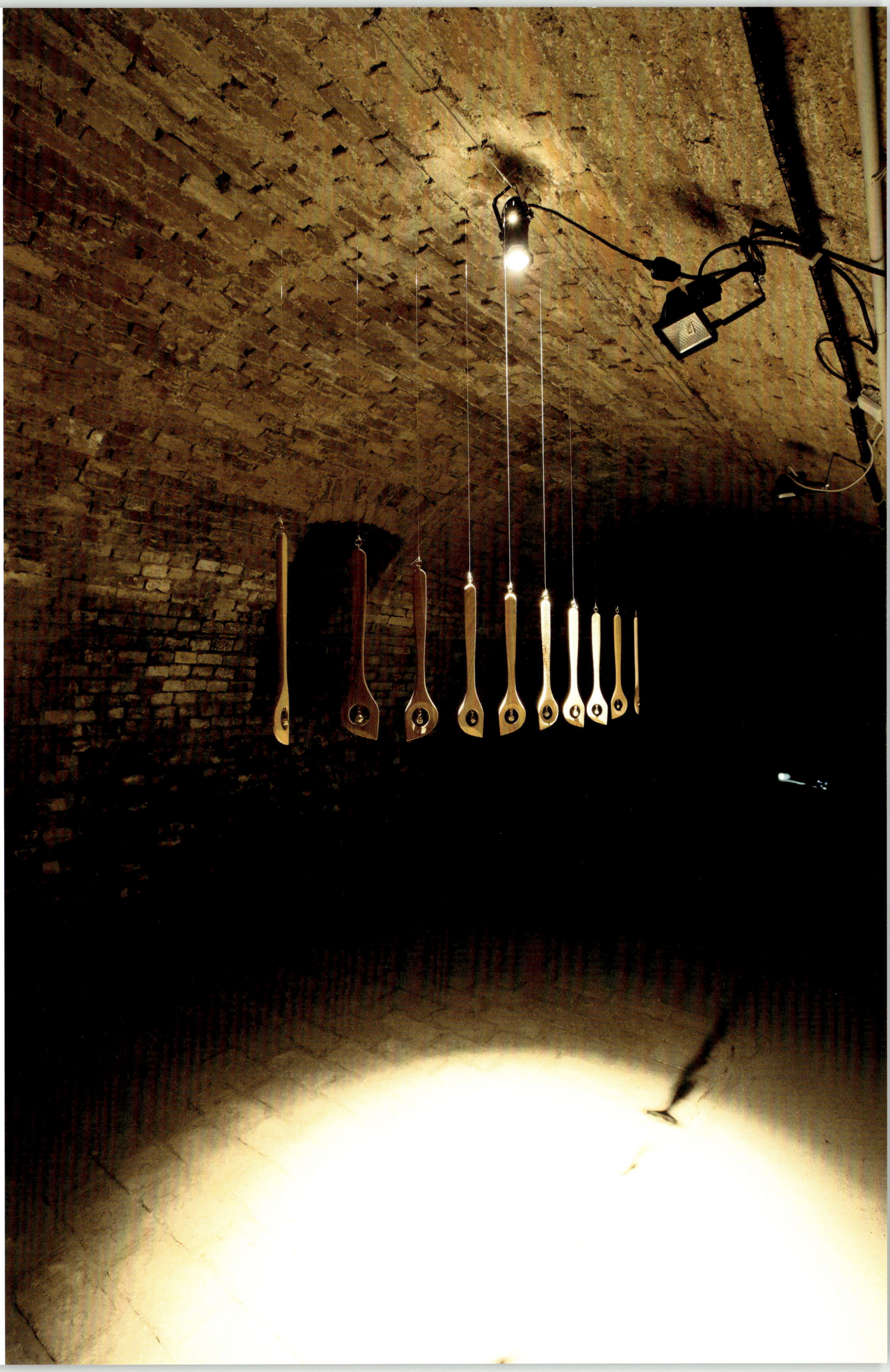

Milan Grygar

*1926 in Zvolen (SK), lives and works in Prague (CZ)

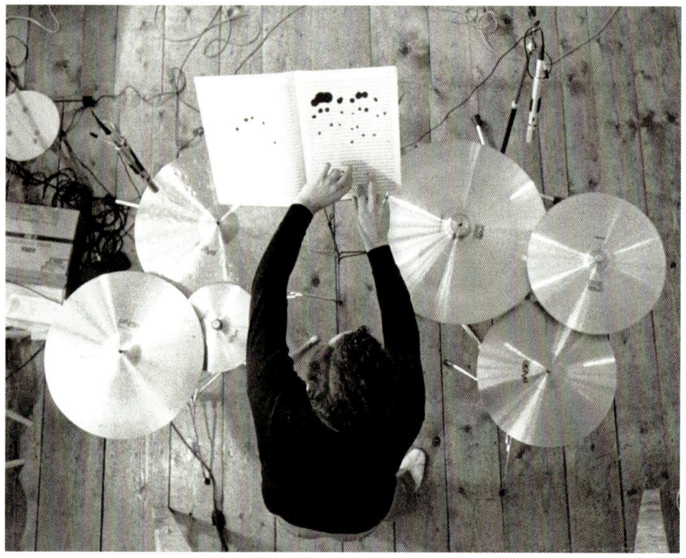

The photograph shows Grygar realizing the work at his studio, 1981

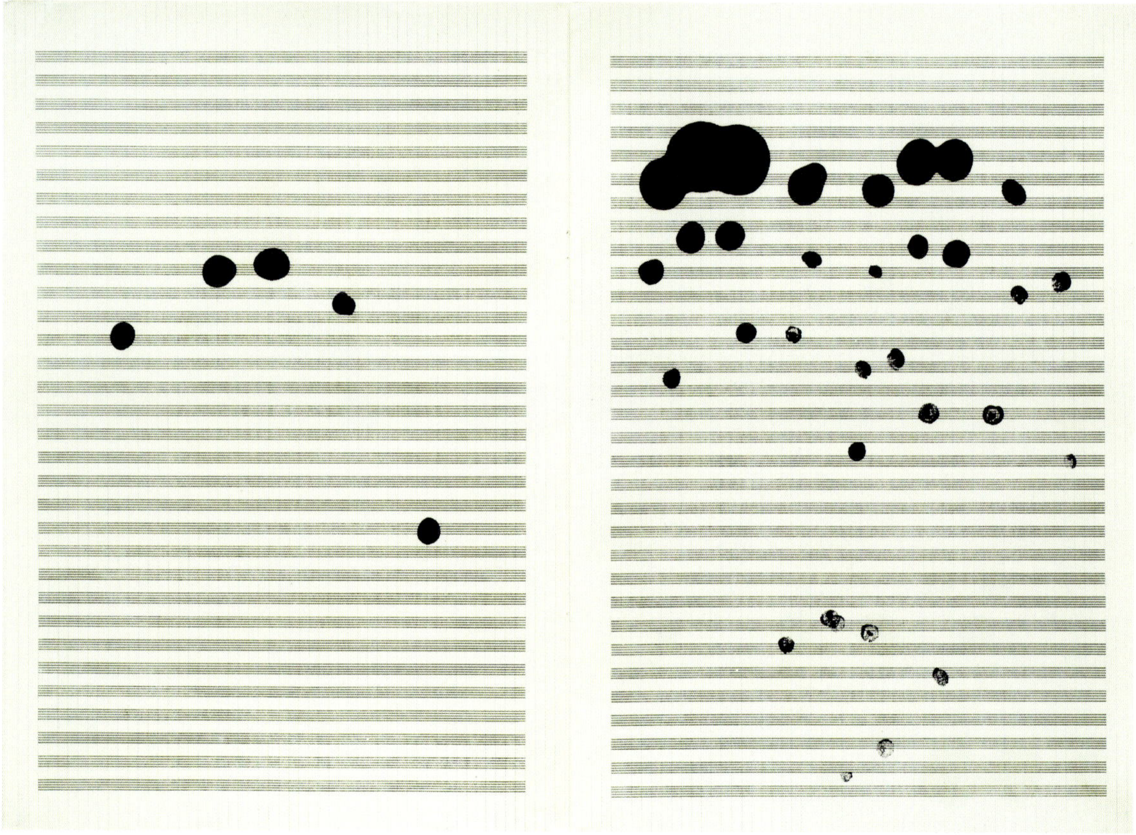

Finger score, 1972
Color for street signs, music paper, 20 pages, 33 × 25.5 cm

Both are signs; one is an acoustic sign, the other is visual, whereby the visual always carries acoustic elements within it as, vice versa, the acoustic always contains visual elements. It occurred to the painter Milan Grygar in the early 1960s that a line also conveys the idea of a longer sustained note and a dot conjures up a one-off short note. From then onward Grygar investigated the relations between visual and acoustic signs in his drawings, paintings, films, and performances. In 1972, for example, he used black traffic sign paint to make a formation of dots on music paper, which a drummer could then translate into sound with his/her fingers (hence the title *Finger score*) through touching cymbals. Thus images become sounds – one cannot imagine a closer linkage of sound and image. (Noemi Smolik)

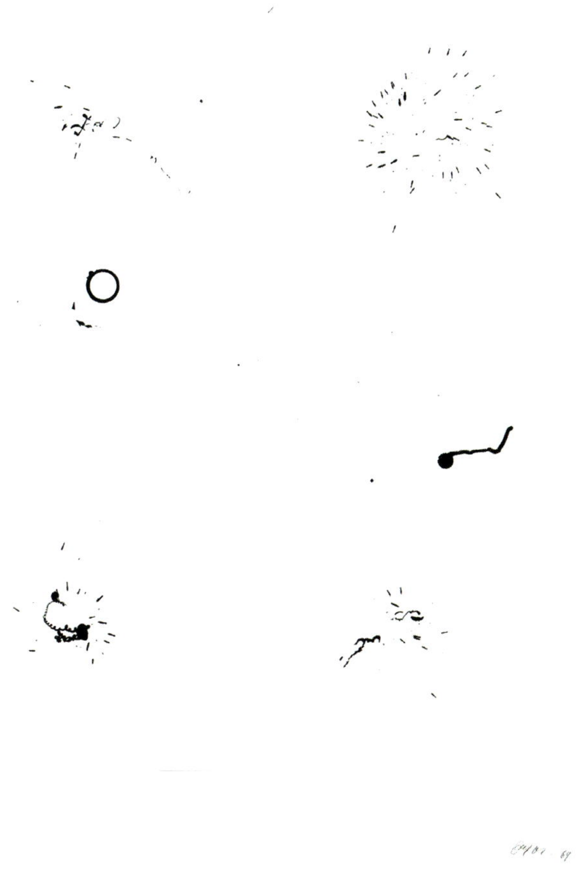

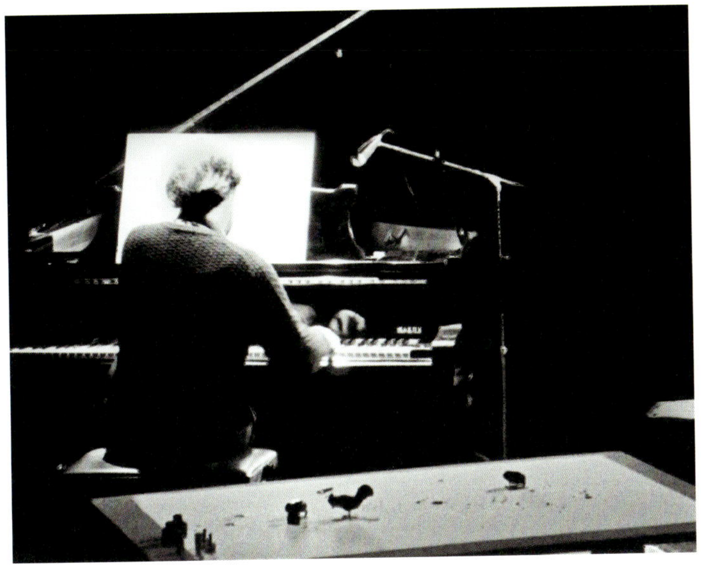

The photograph shows Grygar realizing the work at the OFF-OFF festival in gent, 1986

Acoustic drawing, 1970
Ink, small gear-wheel, mechanical toys, paper, 103 × 73 cm

At the beginning of the 1960s the painter Milan Grygar became aware of the fact that drawing with a pen or a pencil produces sounds. These sounds, which accompany the creation of a drawing, he then recorded and presented together with the exhibited visual artwork. He named these combinations of acoustic and visual signs “acoustic drawings.” Later, Grygar began to experiment with drawing using a variety of objects and the sounds they made, such as a comb, a burnt matchstick, a cogwheel, and a hopping mechanical toy chicken made of metal. Experiments such as these he also performed before audiences, for example, at the OFF-OFF Festival in Ghent in 1986. Grygar also worked with a pianist who used a complex system to translate his signs on paper into notes on the piano. Here again it was about the reciprocity of image and sound. (Noemi Smolik)

Shilpa Gupta

*1976 in Mumbai (IN), lives in Mumbai

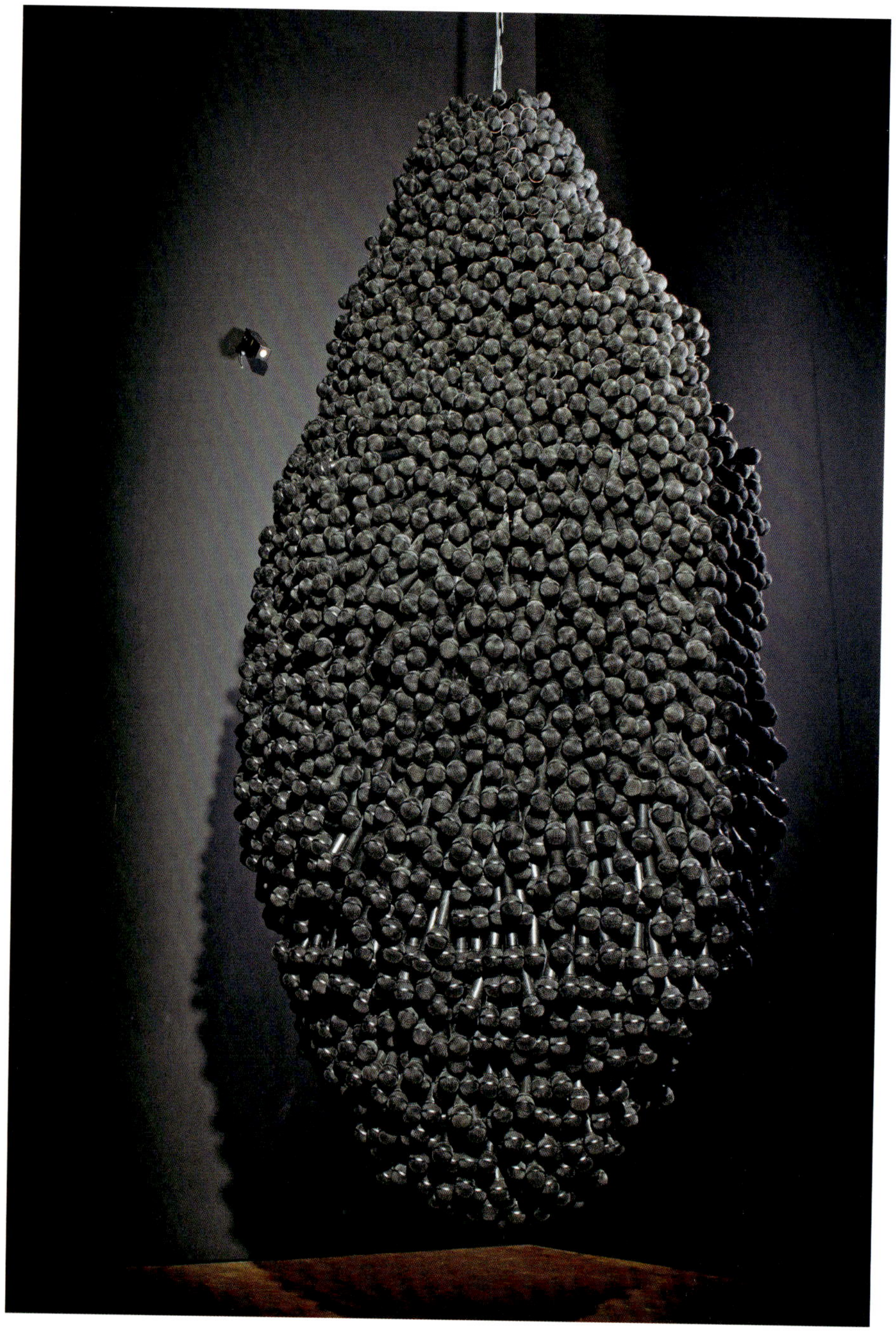

I keep falling at you, 2010
Sound sculpture
Microphones, loudspeaker,
multi-channel audio, 3:12 min.,
380 × 130 × 200 cm
Courtesy of the Tiroche DeLeon
Collection and Art Vantage PCC Ltd.

www.shilpagupta.com

A bundle of more than one thousand microphones hangs from the ceiling like a bunch of ripe grapes. Shilpa Gupta has rewired the microphones to create loudspeakers that play a multi-channel sound and speech composition. The text picks up on the visual impression of heaviness. *I keep falling at you* deals with opposites: with endless falling, with burial, being buried, and disappearing.

I keep falling at you
But I keep falling at you (chorus, repeat)
Your garden is growing on me
I will take it away with me
To a land which you can mark no more
Where distances don't grow anymore
I keep falling at you
But I keep falling at you (chorus, repeat)

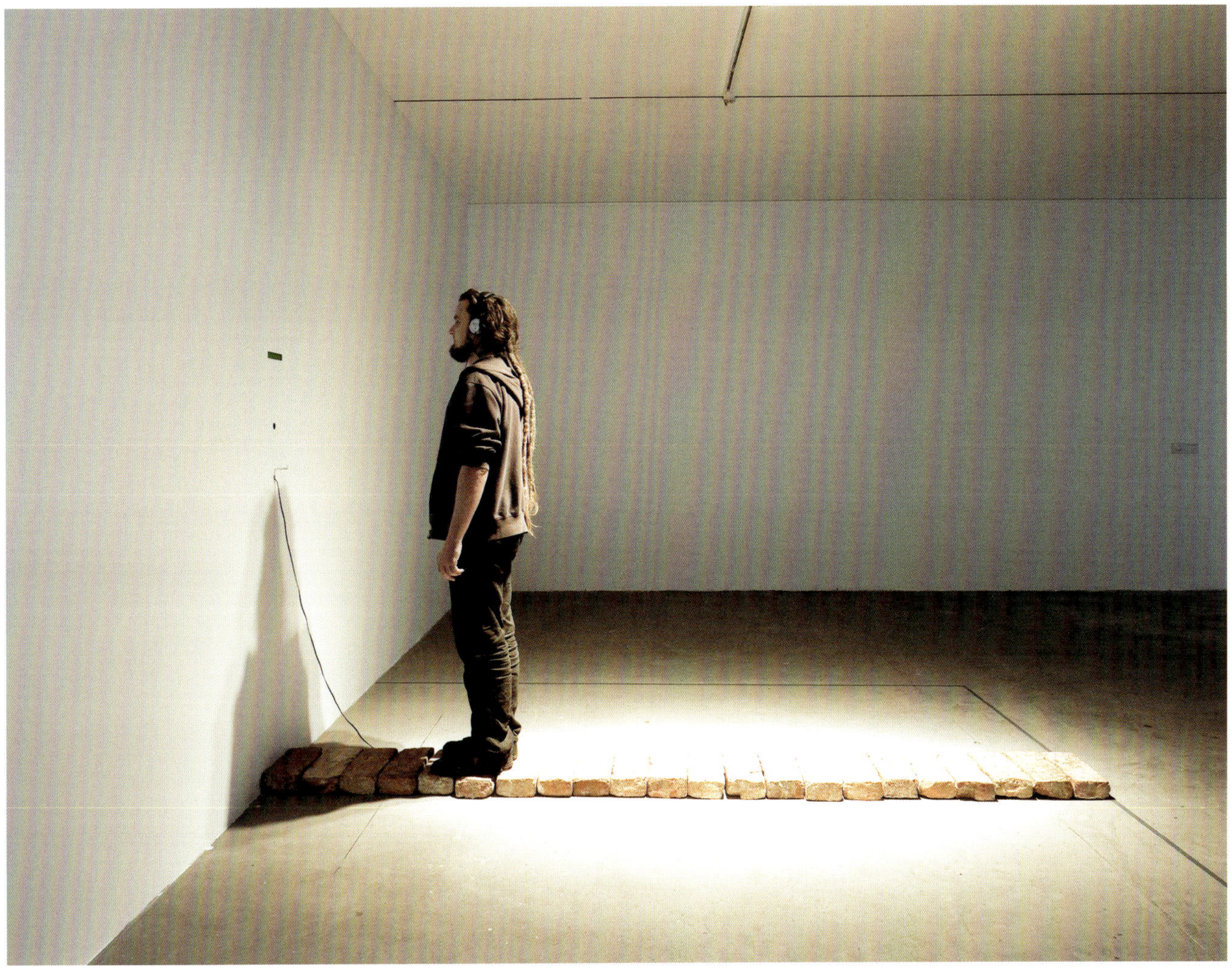

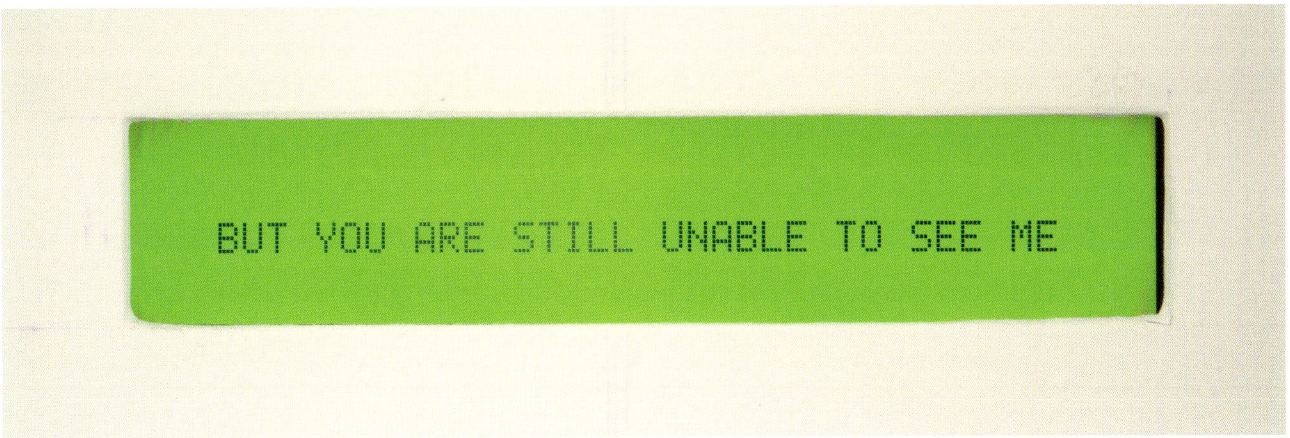

Speaking Wall, 2009–2010
Interactive sensor-based
sound installation
LCD screen, bricks, headphone,
8 min. (interaction loop),
300 × 300 × 300 cm

The viewer steps up onto a row of narrow bricks that dead-ends into the wall, and puts on the headphones. The wall has an embedded distance sensor which is able to detect how far or how close the viewer is from the wall. A mounted LCD panel displays electronic text which can be heard via the headphones.

During the eight minutes conversation/interaction that begins, the wall tells the viewer to sometimes step forward, a little more, or sometimes back. This interaction which transforms the viewer into a performer is complemented by the voice over the head-phones beginning to speak about shifting borders.

Carl Michael von Hausswolff

*1956 in Linköping (SE), lives in Stockholm (SE)

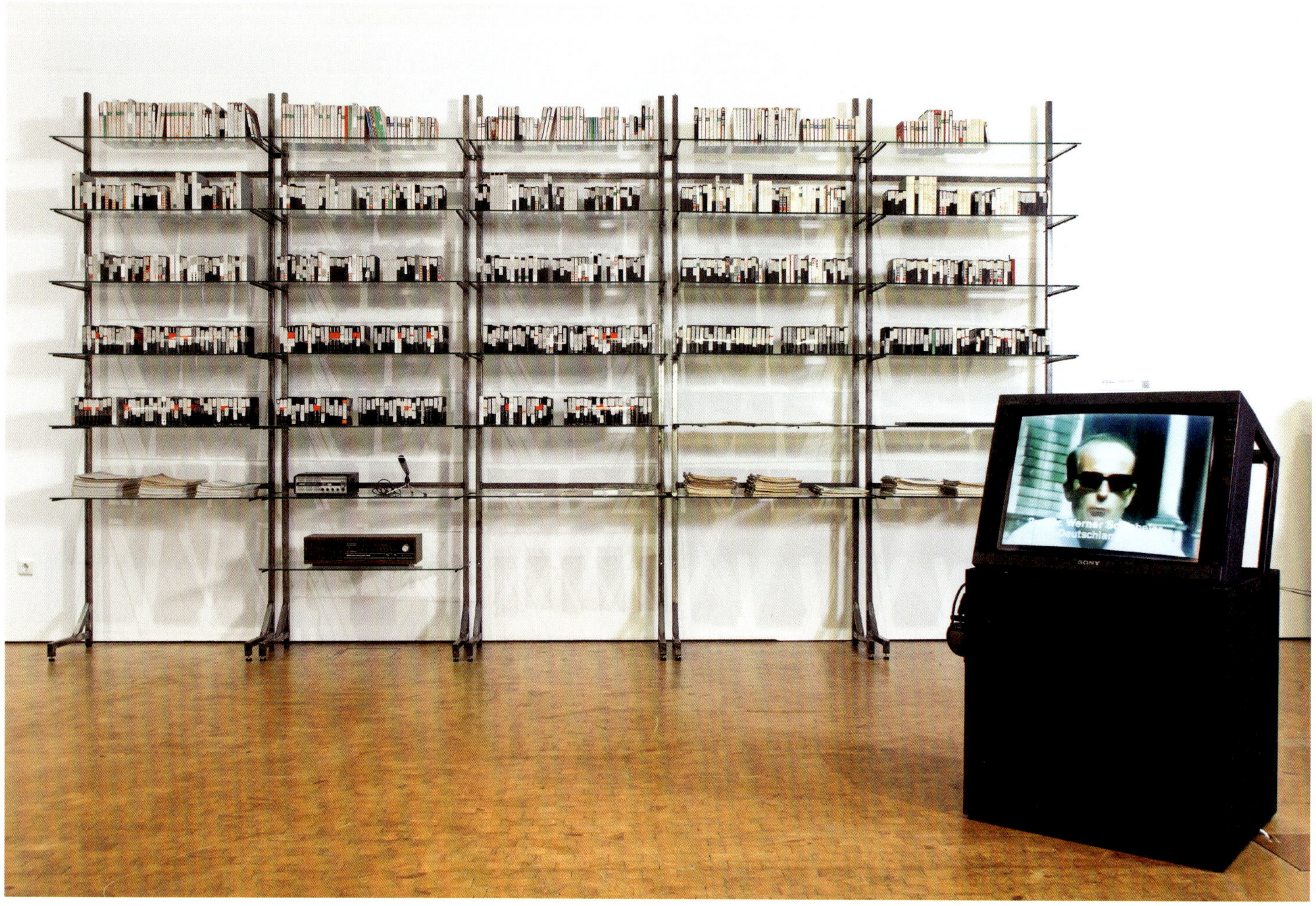

Audioscopic Research Archive / Radar Works, 2004
Archive of electronic voice phenomena
Shelf, glass plates, audio tapes, video, ca. 4 × 2.5 m
ZKM Collection

www.cmvonhausswolff.net

Opera singer, philosopher, and documentary filmmaker Friedrich Jürgenson (1903–1987) was the discoverer of electronic voice phenomenon (EVP). While recording his voice in 1959, he came upon strange artifacts in the tape signal that he eventually recognized as voices speaking in a variety of languages. After making many such recordings, one of the voices suggested to Jürgenson to use radio waves in his research. He left behind a large archive of audio recordings and other research related to trans-communication, the study of the voices of the dead.

The artist Carl Michael von Hausswolff has been working with Jürgenson's archive for several years, using its contents as source material for installations in the series *Radar Works*, an absurd archive exploring his shared quest for contacting intermediate realms. (John Kannenberg)

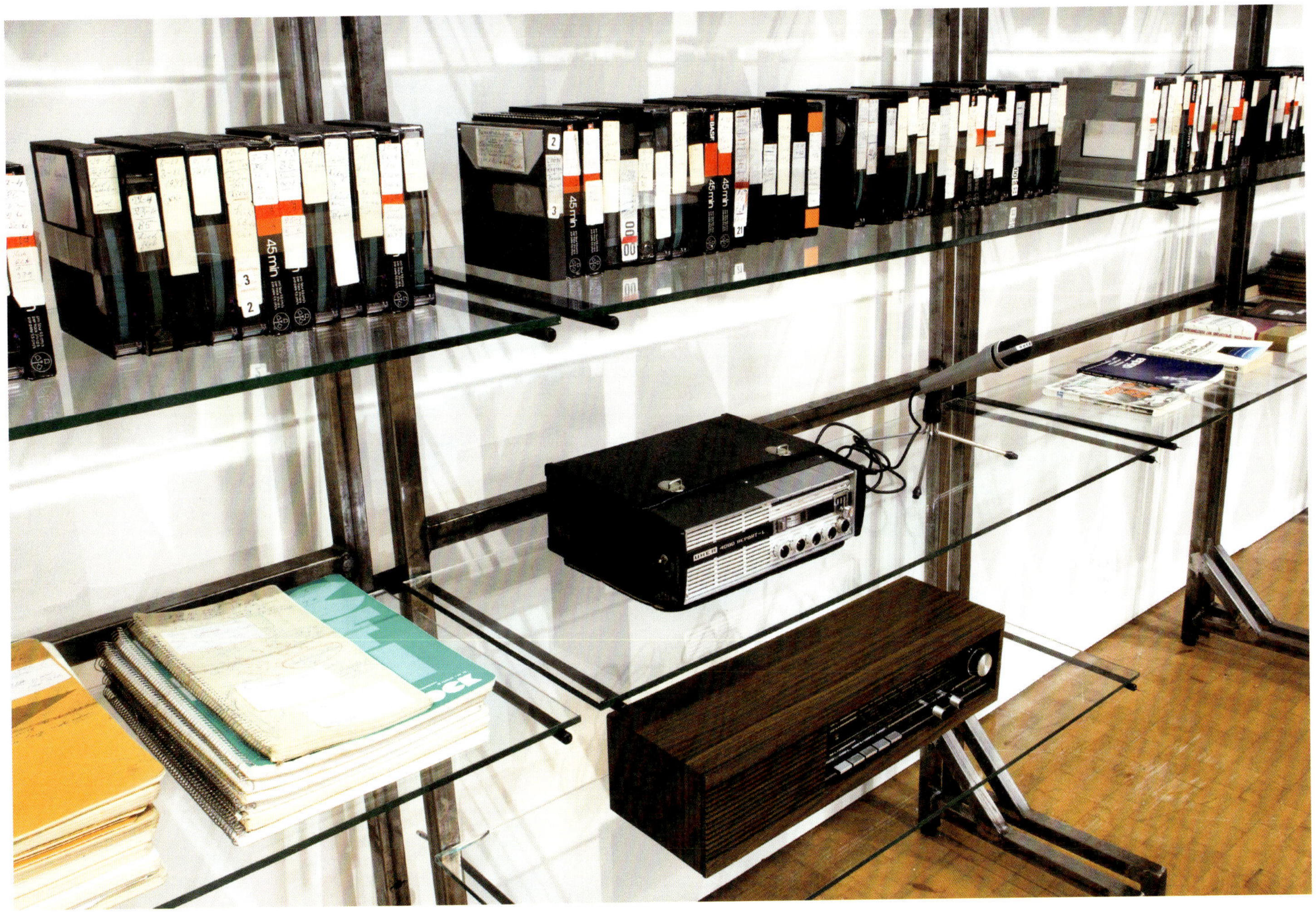

Edwin van der Heide

*1970 in Hilversum (NL), lives in Rotterdam (NL)

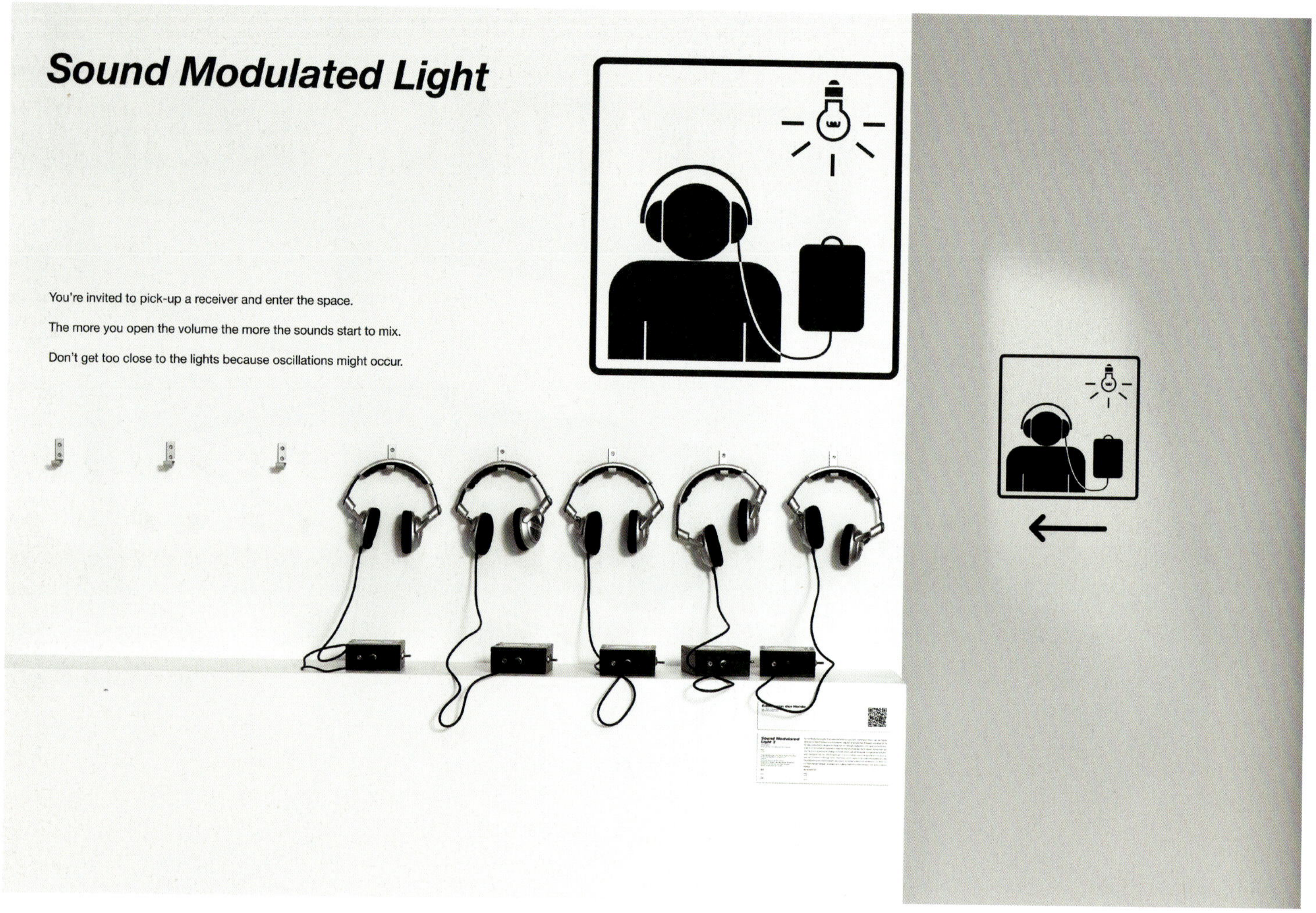

Sound Modulated Light 3, 2004/2007
Exploratory light/sound environment
Portable receivers, computer, audio interface, software, headphones, light bulbs, 5 × 5 m
Concept: Edwin van der Heide
Realization: Edwin van der Heide, Eckehard Güther, Thijs Eerens, Bart Koppe
Studio Edwin van der Heide

www.evdh.net

Sound Modulated Light 3 is an installation made of sound and light. Edwin van der Heide works with the flickering of light bulbs whose typical frequency of about 50 Hz is invisible to the human eye. Sound waves alter the intensity of "sound modulated light." By minimally varying the current, van der Heide composes a dynamic sound-light field through which visitors can navigate with special receivers. The receivers are equipped with light sensors and translate the flickering into sounds. Each light bulb has been assigned its own soundtrack. The soundtracks are composed so they can be combined with each other. Moving the light receiver results in different combinations of the composed sounds.

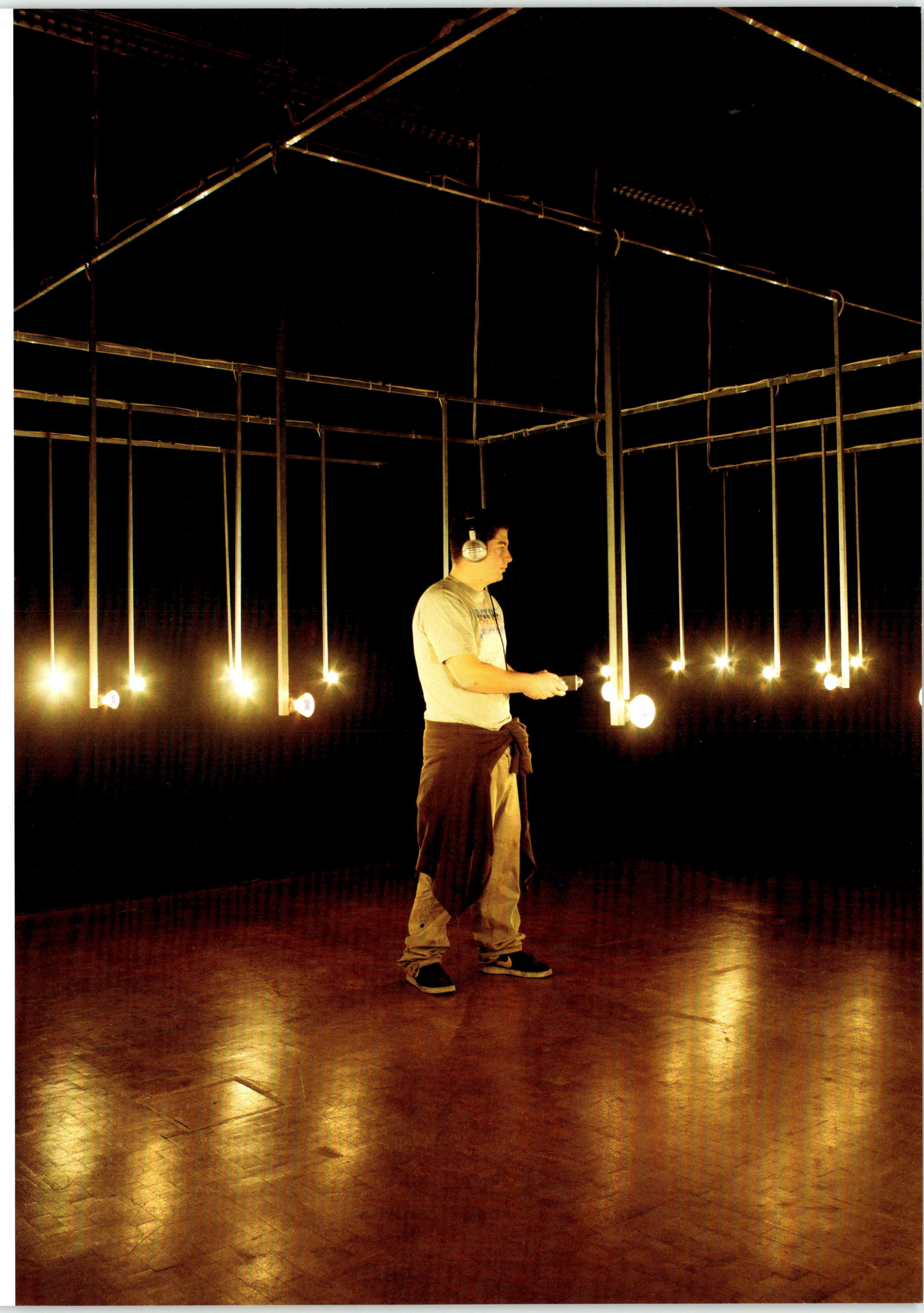

Douglas Henderson

*1960 in Baltimore, MD (US), lives in Berlin (DE) and Brooklyn, NY (US)

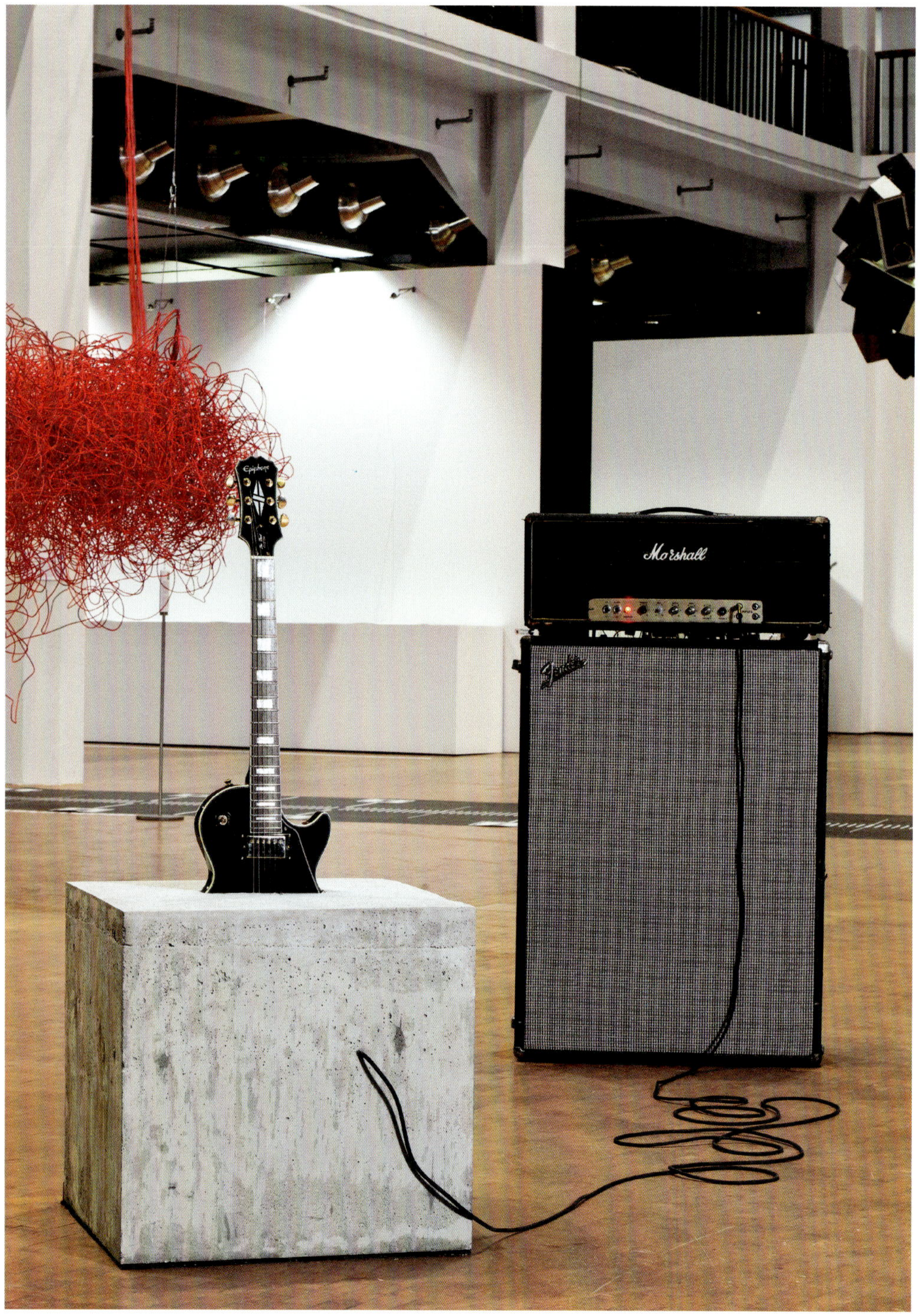

stop., 2007
Sound sculpture
Concrete, wood, electronics, amplifier, 140 × 60 × 60 cm (guitar block) and 130 × 75 × 30 cm (amplifier)
Douglas Henderson

www.douglashenderson.org

A brand-new electric guitar is half buried in a massive block of concrete. It is plugged in, its cable leads to a massive 1970s amplifier stack that is switched on, turned up, humming and crackling with latent energy: a frozen solo attached to the engine of rock history. It is as if an effort to preserve the instrument has condemned it to a kind of half-life. Like an homage to the guitar-destroying rock legends Jimi Hendrix or Pete Townsend, the work evokes a lost past in the style of Pompeii. Outside the museum, this instrument has no future.

Douglas Henderson's works are governed by visual criteria. Trained as a composer, Henderson sees himself first and foremost as a visual artist: "I don't think that I make music, I'm trying to make sound visible."

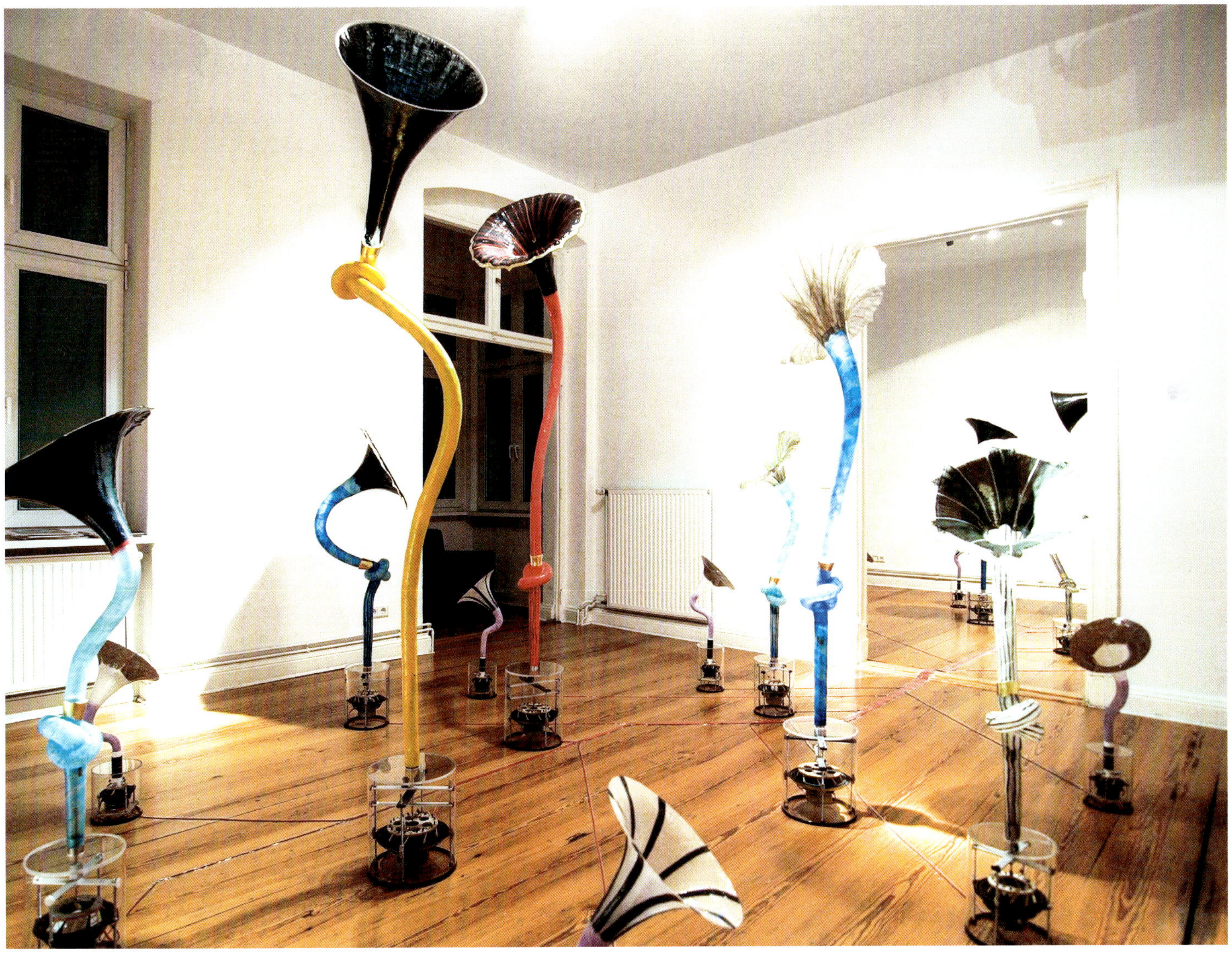

summer of love, 2015
Sound installation
Carbon fiber, Kevlar, fiberglass, flax cloth, 22K gold, acrylic, wood, steel, aluminum, speakers, kinetic, 16-channel audio, 23 min. (loop), dimensions variable
Douglas Henderson

Douglas Henderson's *summer of love* is a large, multi-component sound/sculpture installation consisting of 21 giant kinetic flowers made of Kevlar, carbon fiber, and fiberglass. Ranging up to 250 centimeters in height, each flower sits atop an upward-pointed loudspeaker and a mechanism which allows it to rotate according to the vibrations of the speaker membrane. A 16-channel electroacoustic composition constructed around beat poet Gregory Corso's controversial poem *BOMB* (1958), in which he satirically idolizes nuclear weaponry, broadcasts from the flower blossoms. This in turn drives the whirling choreography of the garden and creates a highly complex sound field, as frequencies, reflections, and Doppler shifts continually re-spatialize the soundtrack.

summer of love plays on the paradoxical fear of, and dependence on, the threat of global annihilation. The obsessive fears of apocalypse, nourished by the media, seem to become the subject of a bizarre fashion show: each generation has its threat. In 1967 it was the atom bomb, while today global warming competes with ISIS on the catwalk of doom. With this new work Henderson embraces these fears as a means of conquering them, cultivating a dancing garden to survive the flood.

Featuring readings by actors Meret Becker, Anna Clementi, Julia Stefani Möller, and Hedda Oledzki, the composition weaves together the readings of the poem, cross-synthesized with archival material and percussion phrases played by Tony Buck.

Douglas Henderson

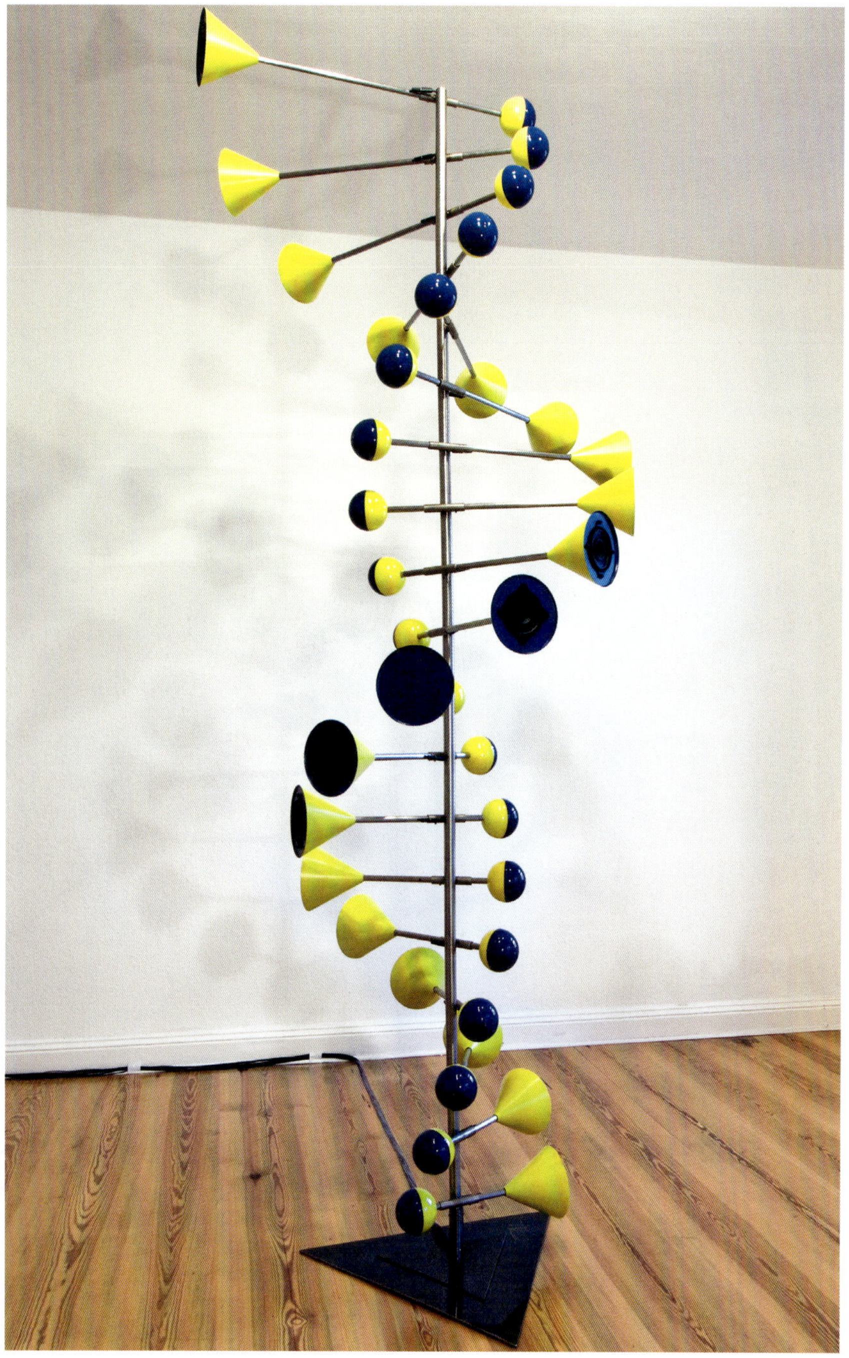

Babel V: Dream Man, 2012
Sound installation
Fiberglass, stainless steel, wood, speakers, 7-channel audio, 16 min. (loop), 110 × 90 × 285 cm
Douglas Henderson

The *Babel* series are an optimistic retelling of the story of the Tower of Babel, positing that the multiplication of languages, instead of dispersing civilization, is a source of serene beauty: the poetry of variation. Here a poem by Russell Edson, *Dream Man*, is the vehicle for the exploration not only of how text sounds and what it means, but foremost a concern not often addressed: where is the text? Using a double layered composition, the text was first read one word at a time, with different people reading up to three disconnected words, in different physical locations around Berlin and in Italy. These were then edited as a stream, so that the text becomes a structure for travel amongst a hundred or so ambiences. The second layer is drawn from recordings of the full text read by 35 different people, of different nationalities and facility in English, cut up word by word, and sprayed in aggregate phrases, to develop an atomized rendering of the text, swirling around this helical vortex. The poem is absorbed, rather than literally understood, in an ex-temporal reading.

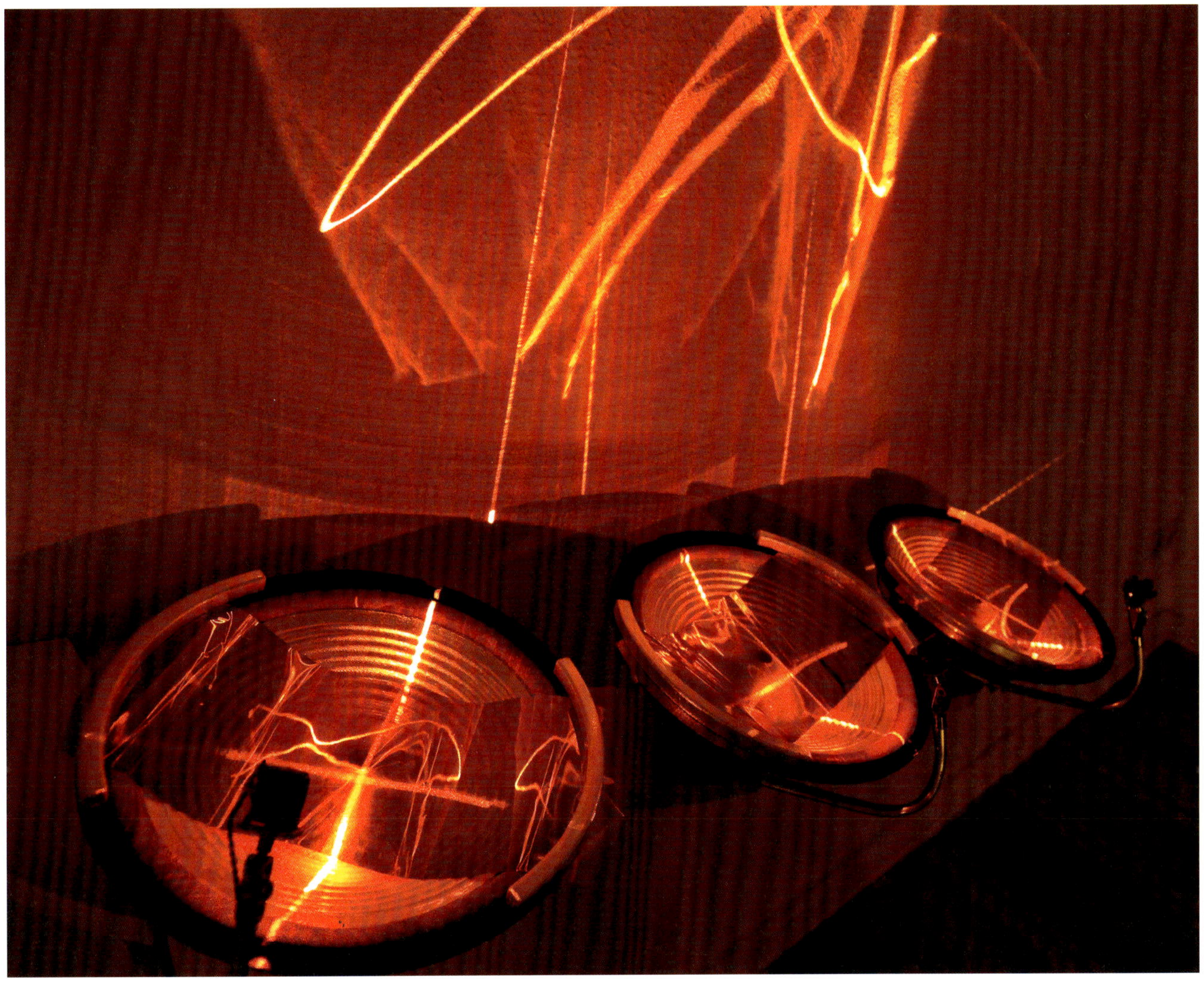

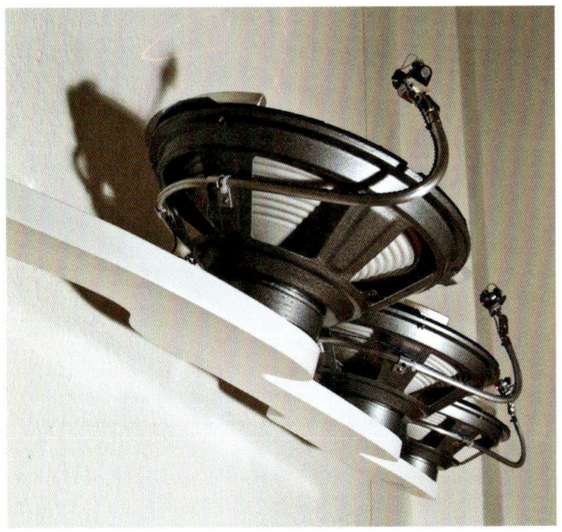

Silver Surfer #2: Exile, 2012
Sound installation
Wood, loudspeakers, sterling silver, Mylar, PVC, aluminum, lasers, 4-channel audio, 8:40 min. (loop), dimensions variable
Douglas Henderson

Silver Surfer #2: Exile is a wall-mounted 4-channel sound and light sculpture incorporating three 12" woofers and a smaller satellite. Three linear lasers are bounced off Mylar film on each speaker, which ripples as low frequencies are played, creating kinetic, ever-changing patterns on the wall. The 4-channel electroacoustic composition uses intricate phase and Doppler shifts to develop a multi-dimensional sound space throughout the room.

Superheroes condense our deep longing for justice and our desire for safety. One of a series of works devoted to the subject, this abstract portrait of the lonely and tireless Silver Surfer expands the comic-book rescuer into a giant, hypnotic amalgam of whirling sound and throbbing light.

Het Apollohuis

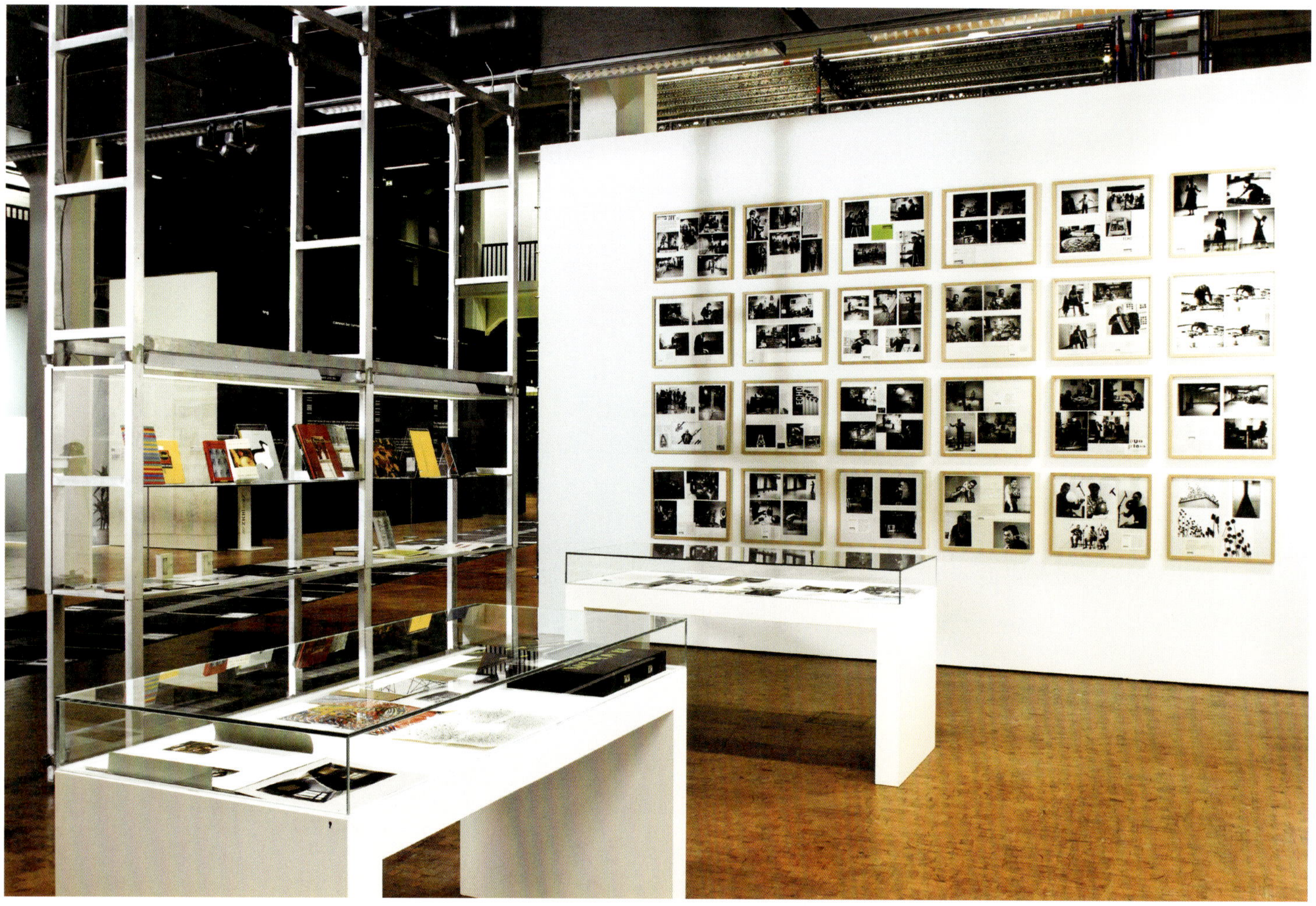

An artist's artist space for artists – this is how one could describe the significance that Het Apollohuis in Eindhoven in the Netherlands had for two decades. Founded by Paul Panhuysen and his wife Hélène, Het Apollohuis [The Apollo House] offered a tremendous spectrum of events, ranging from exhibitions, concerts and performances, lectures and symposia, to big festivals for contemporary art music and bi- and multilateral exchange programs.

Het Apollohuis opened on October 28, 1980 with a concert performance of the *Out of Tune Guitar* by Rhys Chatham, Robert Longo, Joe Hannan, and David Linton [on drums]. The same evening its exhibition program also opened with paintings by Paul Panhuysen and Johan Goedhart. When the building of Het Apollohuis was closed to the public in January 1997, more than 140 exhibitions, 500 concerts, and 100 lectures, discussions, and symposia had taken place in its rooms. Even after the building was closed , the name Het Apollohuis continued to be used in the Netherlands until 2001 for curatorial purposes such as festivals as well as lecture and concert series.

Het Apollohuis published nearly 50 artists' books in small print runs, 25 books for the international book market, and around 20 music editions in the form of cassettes, LPs, and CDs.

Since 2009 the archive of Het Apollohuis, with its numerous audio, textual, and visual documents, has been held at the ZKM | Karlsruhe, where experts are documenting and preserving it and making it accessible.

The exhibition *Sound Art. Sound as a Medium of Art* devoted a space to the Het Apollohuis archive, which provided an overview of the history of this important center for the avant-garde situated between visual art and music.

(Excerpts from: Rolf Sachsse, "Klangkunst@Het Apollohuis," in: *Sound Art@Het Apollohuis*, CD, Edition ZKM milestones, Wergo, 2011.)

Curators: Hartmut Jörg, Claudia Gehrig

In 2011 the Edition ZKM milestones released the CD *Sound Art@Het Apollohuis*. It contains previously unreleased excerpts from recordings of concerts and performances that took place at Het Apollohuis.

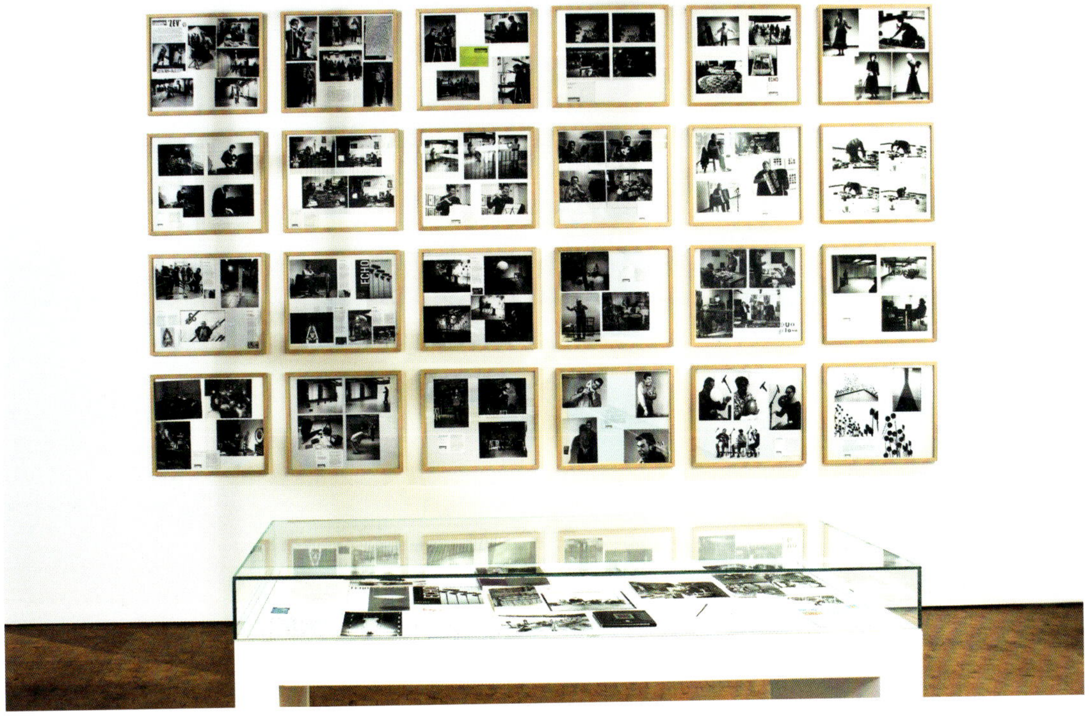

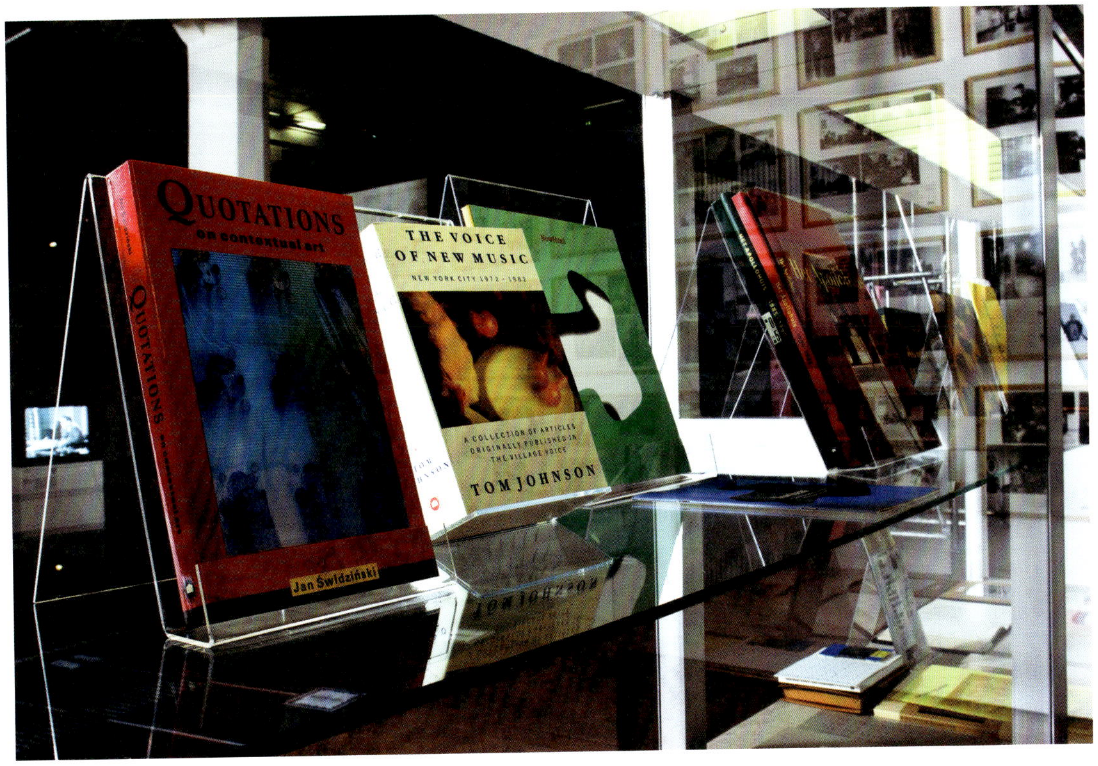

Showcase 1
Black Box Frankfurt
Wooden box with 5 original artworks, folder with 49 sheets, audio cassette, newspaper article from the *Eindhovens Dagblad*, 4 June 1983

Showcase 2
Publications by Het Apollohuis Publishing House: 19 books/brochures, 5 CDs, 4 LPs, 8 MCs, 5 artists' books

Showcase 3
Echo Festival: The Images of Sound I and II
1 CD, 2 program booklets, 2 catalogs, original letters from the participating artists (Max Eastley, Rolf Julius, Christina Kubisch, Richard Lerman, Godfried-Willem Raes, Jon Rose) and VANDALIA [Symposium], photographs

Photo wall
Photographs of concerts at Het Apollohuis from 1980 to 1995, 24 pieces, framed

Gary Hill

*1951 in Santa Monica, CA (US), lives in Seattle, WA (US)

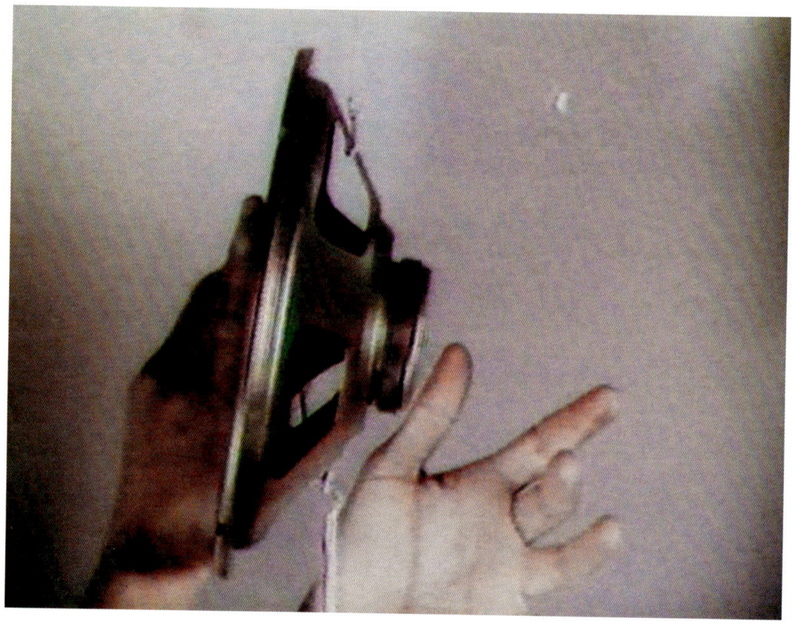

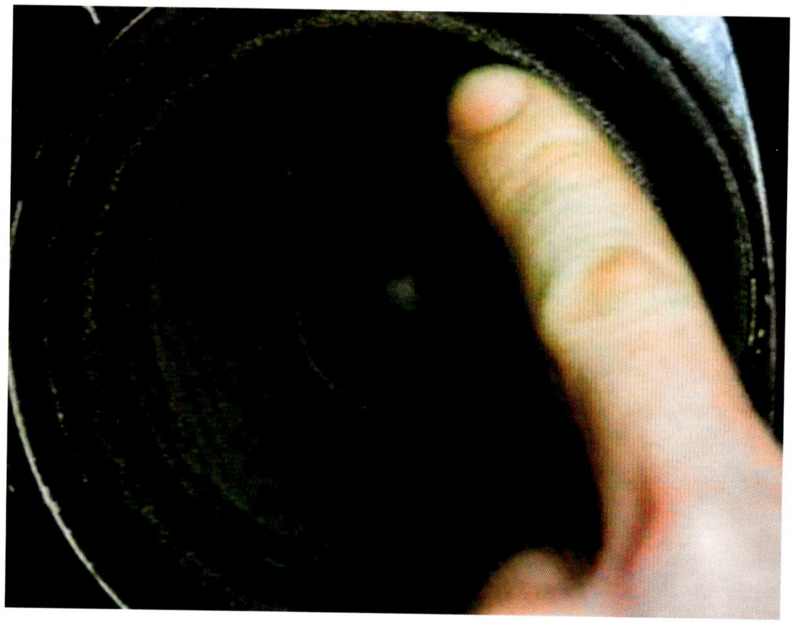

Soundings, 1979
Video, 18:03 min.
Courtesy of the artist and
Donald Young Gallery, Chicago

www.garyhill.com

Concerned with the translation of image into sound and sound into image, Gary Hill's film explores the phenomenology of sound. A close-up view of a speaker that visibly vibrates while playing the sound of his own voice is then physically manipulated by Hill as he attempts to change the sound by drowning, burning, burying, and tearing apart the speaker. These physical manipulations take on a feeling of poeticized scientific experimentation, with the artist's physical actions in dialogue with the words of his voice that emanate from the speaker. (John Kannenberg)

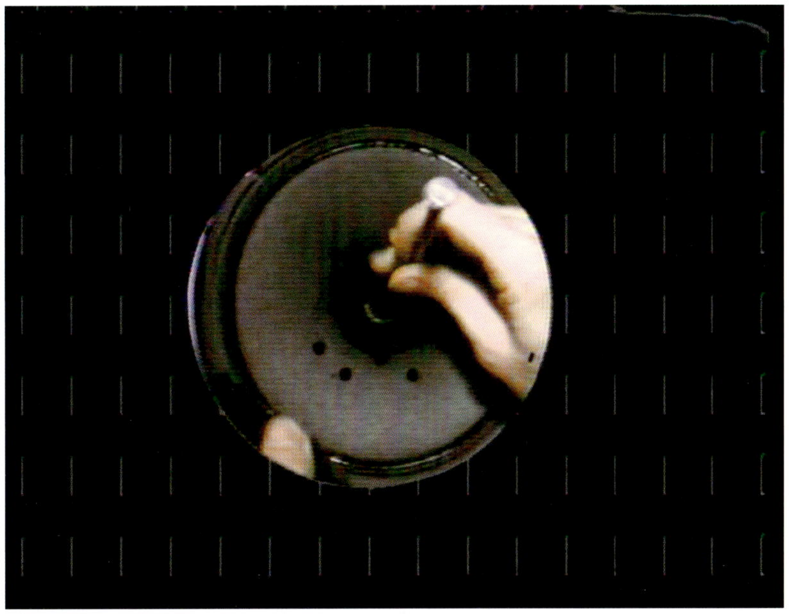

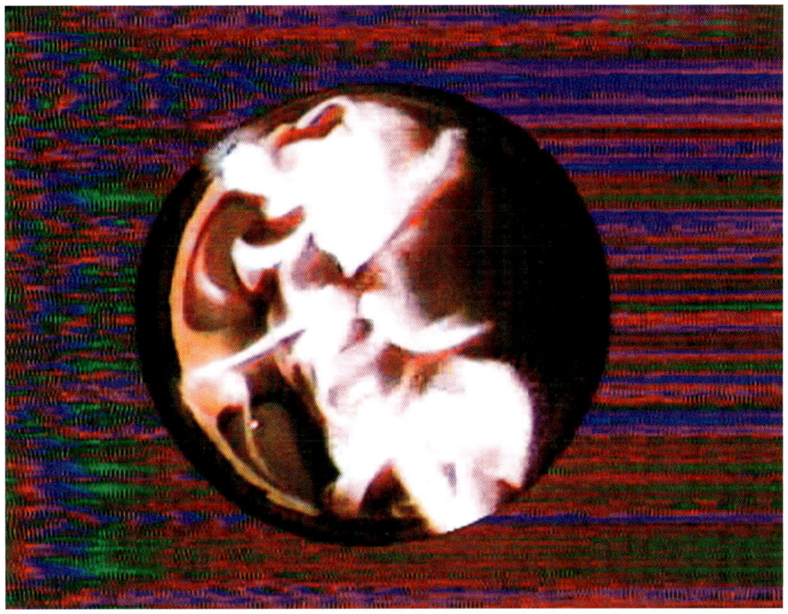

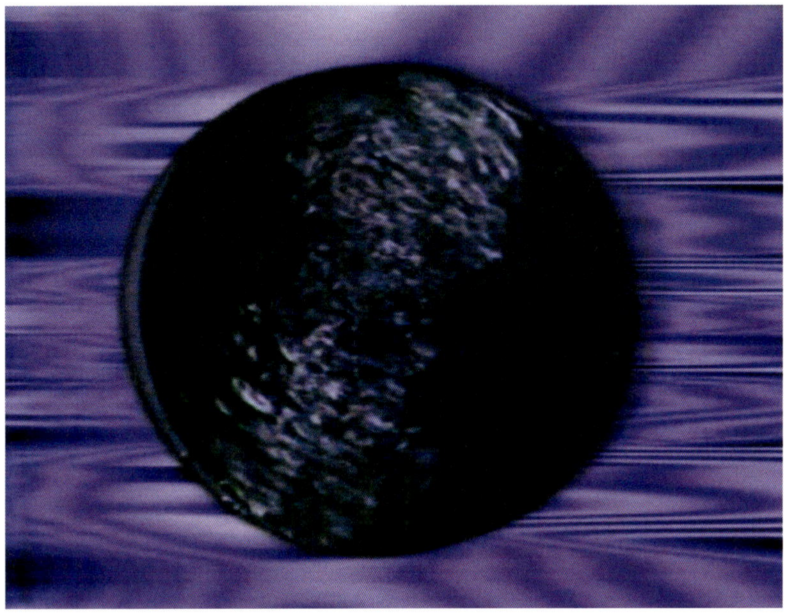

Stephan von Huene

*1932 in Los Angeles, CA (US), †2000 in Hamburg (DE)

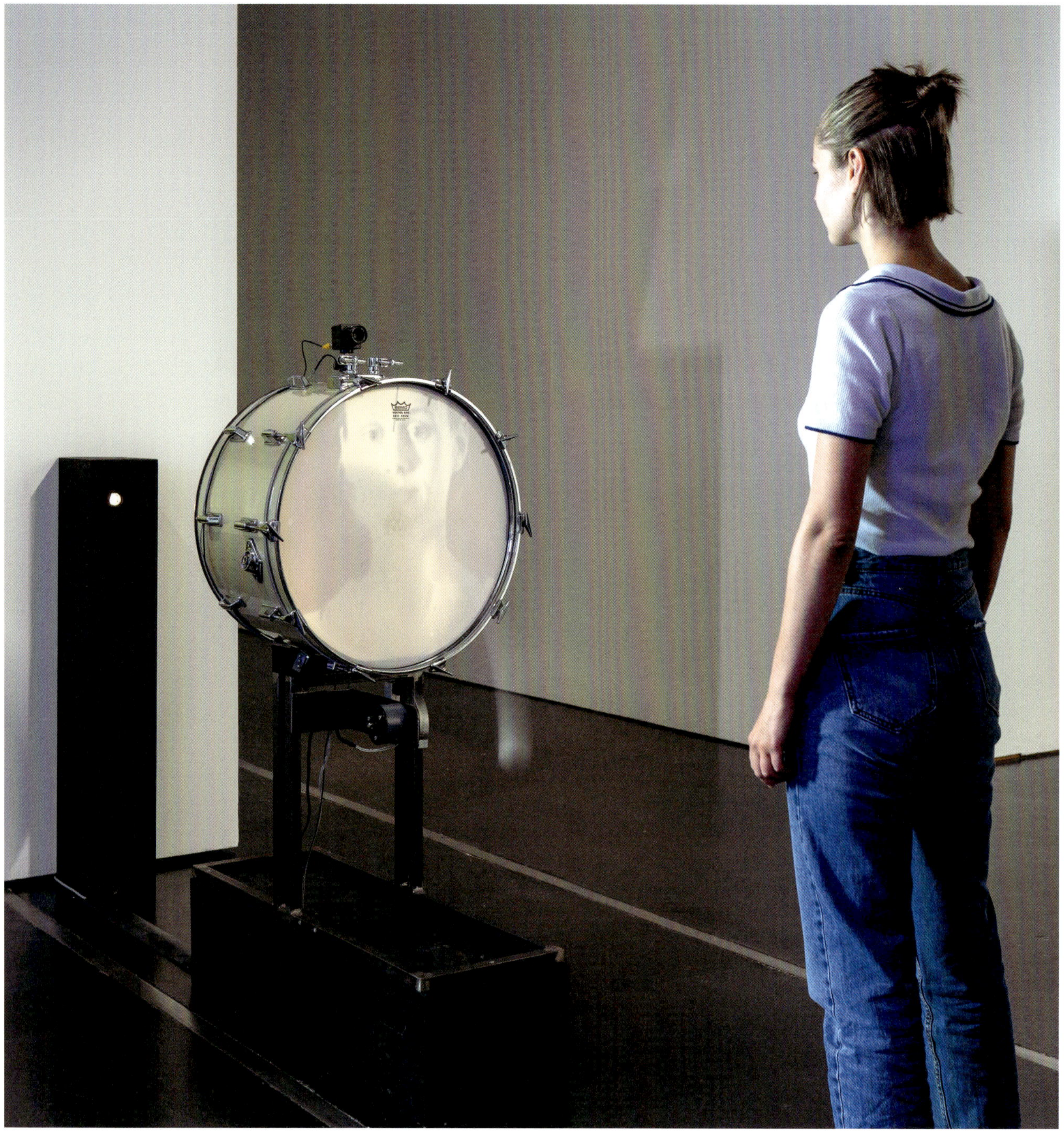

Greetings, 1996
Sound installation
Drum mounted on a wooden box, drumsticks, mechanics, video projection, 154 × 64 × 11 cm
ZKM Collection

www.vonhuene.de/stvh

The relationship between man and machine plays a special, often unpredictable role in Stephan von Huene's works. It is the visitor who brings the "Theater of Machines" to life. Throughout his life, von Huene was interested in art that lies more in the audience's experience than in the art object itself. When visitors approach this sound sculpture, their image is projected from the rear onto the translucent drum skin. If they come even closer to the drum, a powerful strike to the bass drum makes the skin vibrate, along with the projected image. It is as if the sculpture demands viewers' entire attention, so that they grasp that they themselves are protagonists in this mechanical–synaesthetic system. The cybernetic feedback makes viewers the content of what they see.

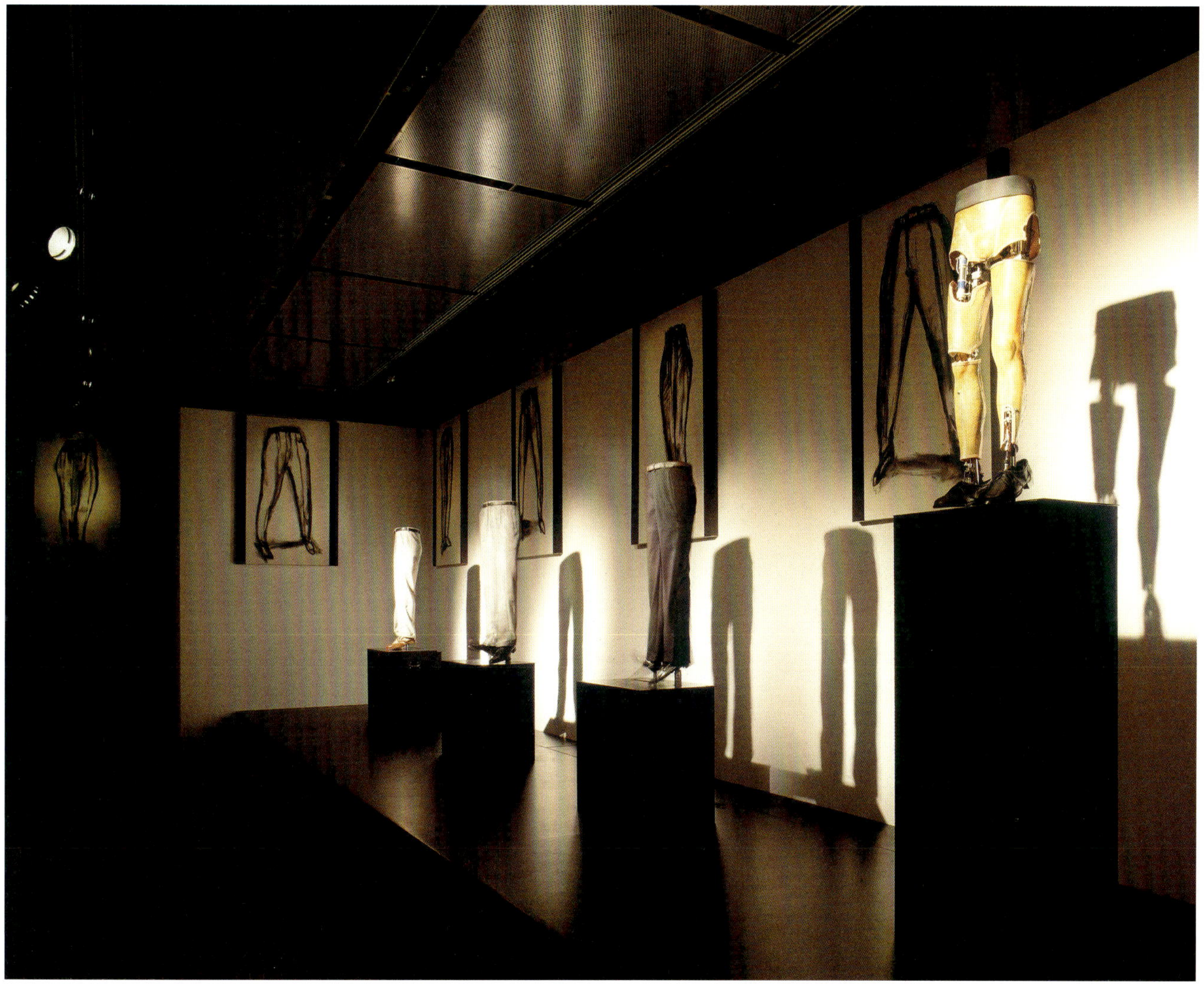

Dancing on Tables (Tisch Tänzer),
1988–1993
Sound installation
4 computer-controlled sculptures with ultrasonic sensor, 14 framed drawings, computer-controlled choreography, 2 compressors, sculptures ca. 200 × 80 × 50 cm each; drawings 170 × 115 × 5 cm each
ZKM | Collection

In *Dancing on Tables* three of the four half-figures wear men's pants and shoes, and move to speeches by American politicians (General Eisenhower, President Lyndon B. Johnson, Jesse Jackson). The fourth, unclothed dancer, standing on its base, moves to arias from *The Pearl Fishers* by Georges Bizet and Georg Friedrich Händel's *Rinaldo*.

Fourteen large-format drawings, which are bordered on both sides with the kind of material used in making men's suits, hang behind the sculptures. When the sculptures are set into motion, this triggers a triple, staggered optical effect, from the sculptures, to the shadows they cast, to the drawings on the wall.

Stephan von Huene

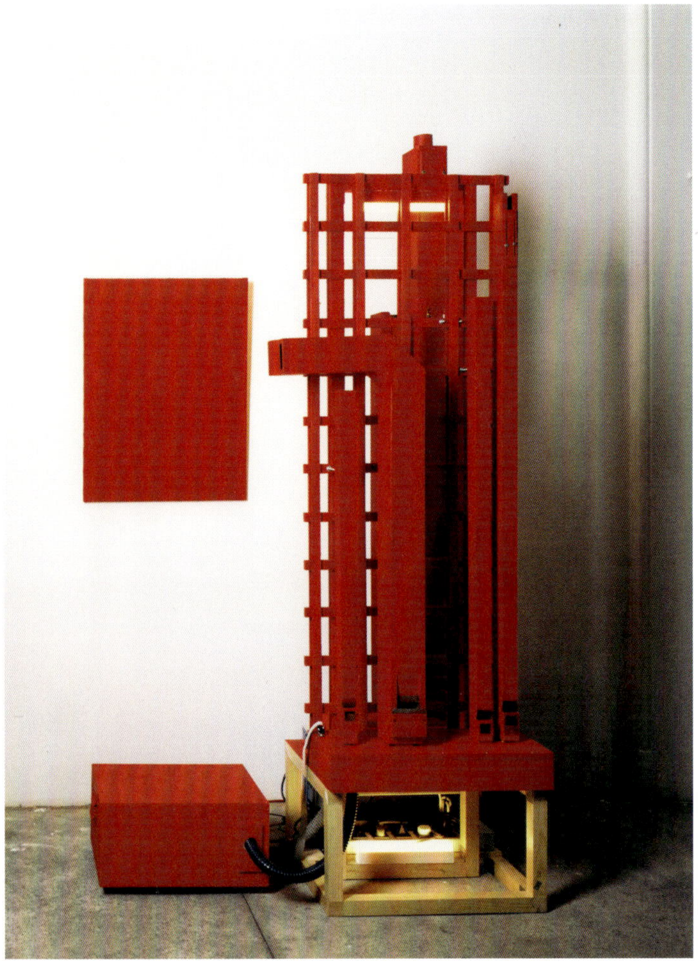

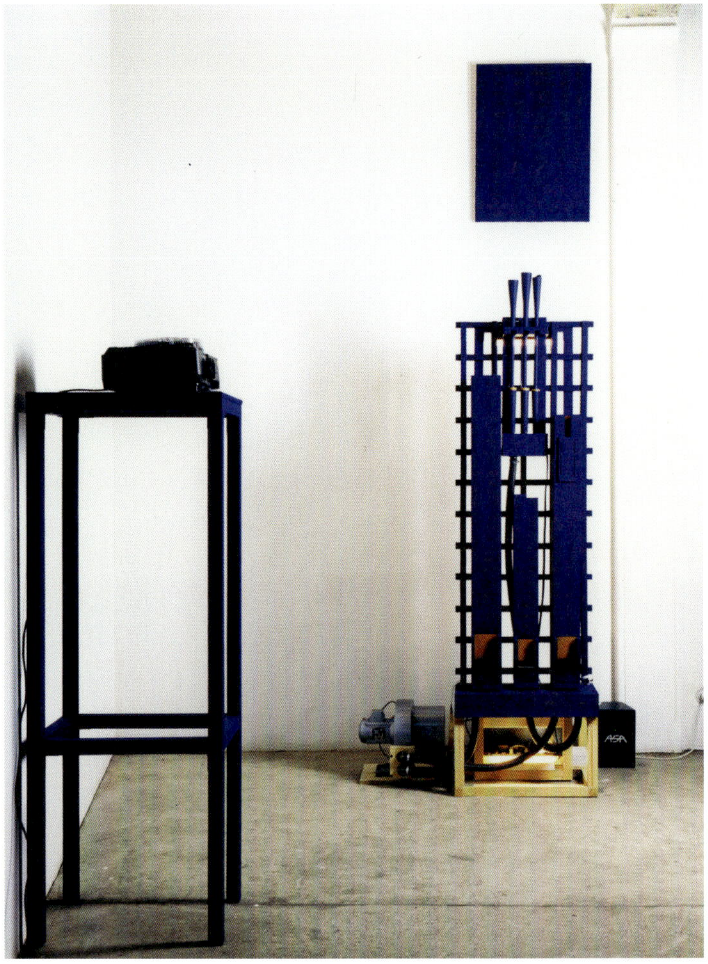

What's Wrong with Art?, 1997
Sound installation
Wood, metal, organ pipes, organ valves, 3 computers, 3 speakers, 1 slide projector, dimensions variable
Voices: Stephan von Huene, Maria Poelchau, Olav Westphalen
ZKM Collection

Catchphrases used by artists and art critics are taken up and mixed together with refrain-style repeated questions about and judgments of art. The enigmatic language of the one is confronted with the doubt and rejection of the other. The organ pipes repeat the rhythm of the sentences in a tristrophic order. For the iconoclasts, there is no doubt: "Art is always wrong."

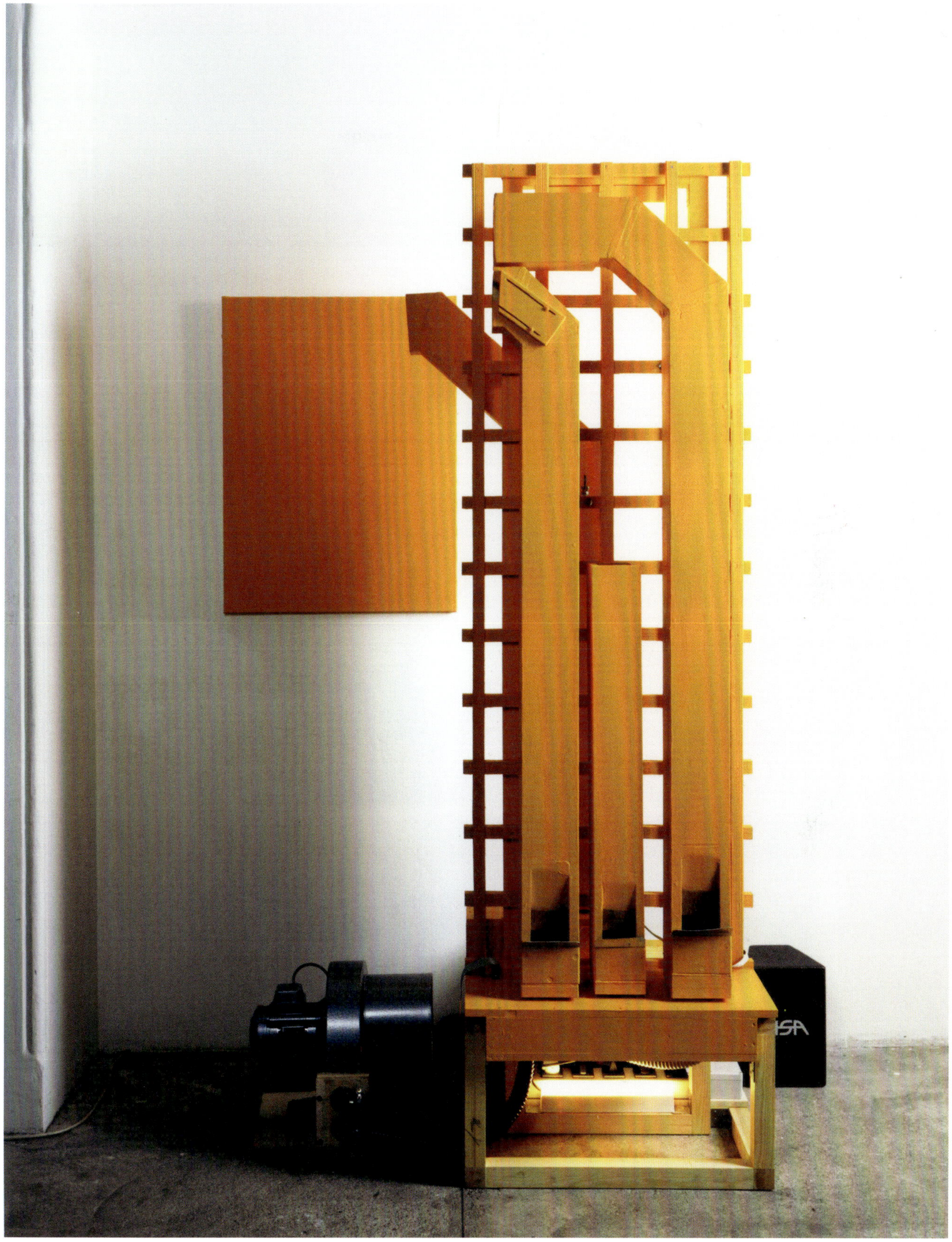

Stephan von Huene

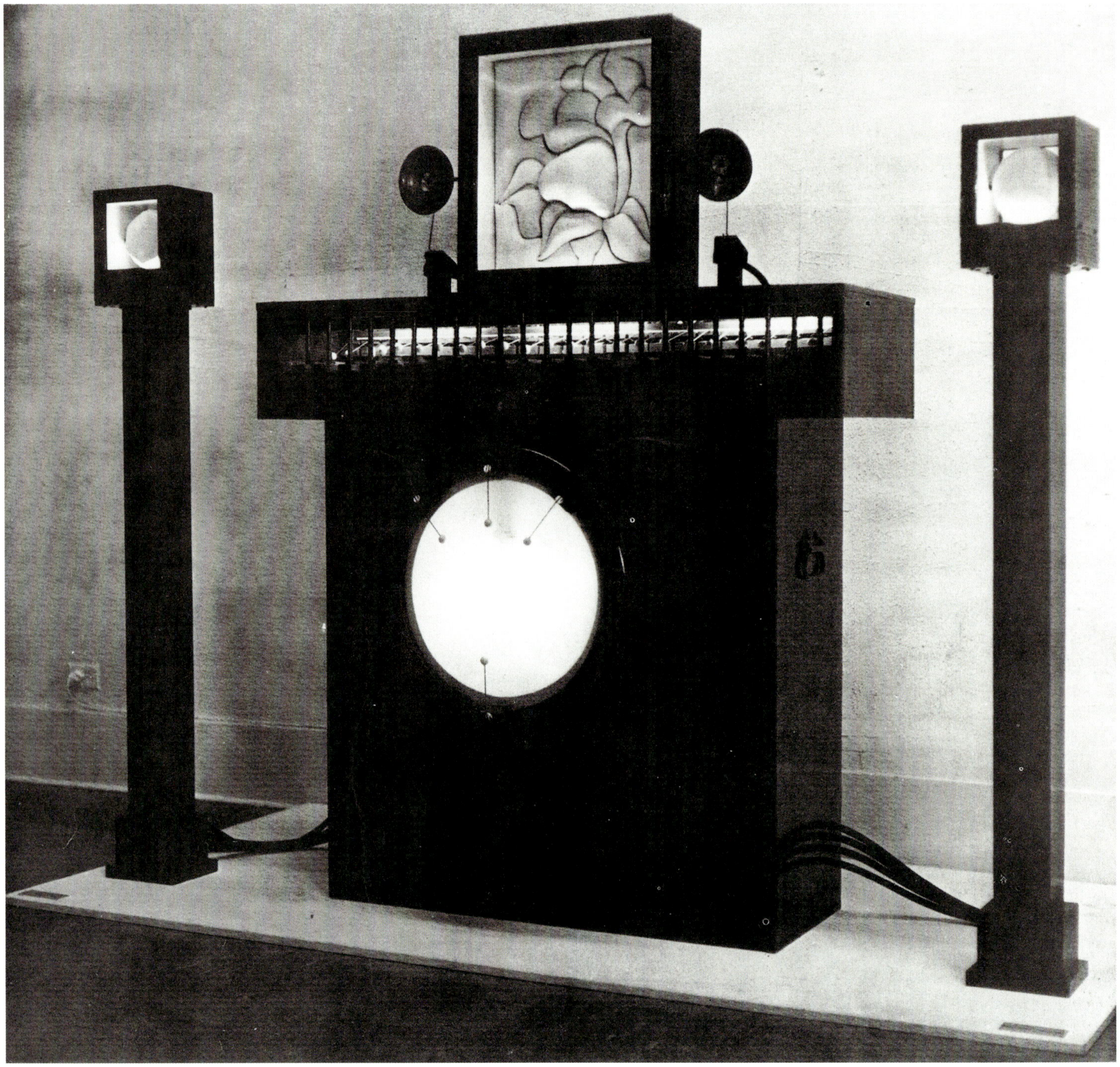

Rosebud Annunciator, 1967
Kinetic sound sculpture
Wood, formed leather, pneumatic parts, pneumatic system,
213 × 243 × 122 cm
Museum Ludwig, Cologne

"[*Rosebud Annunciator*] plays like an automatic piano. [...] The sound patterns come from the music and with the help of photocopies, scissors and adhesive tape create a collage; they are programmed, but not controlled, do not owe themselves to pure chance, but rather to the hope of a happy coincidence. [...] *Rosebud's* optical appearance comes from the architecture of the houses in which I lived in Los Angeles – violins, cellos and other instruments used to be built in connection with the rooms in which they were played and in accordance with the furniture."
(S. v. H. in: *Stephan von Huene. Die Retrospektive*, exhib. cat., Munich, Duisburg, Hamburg: Hatje Cantz, 2002 / 2003), 187. Translated from the German.)

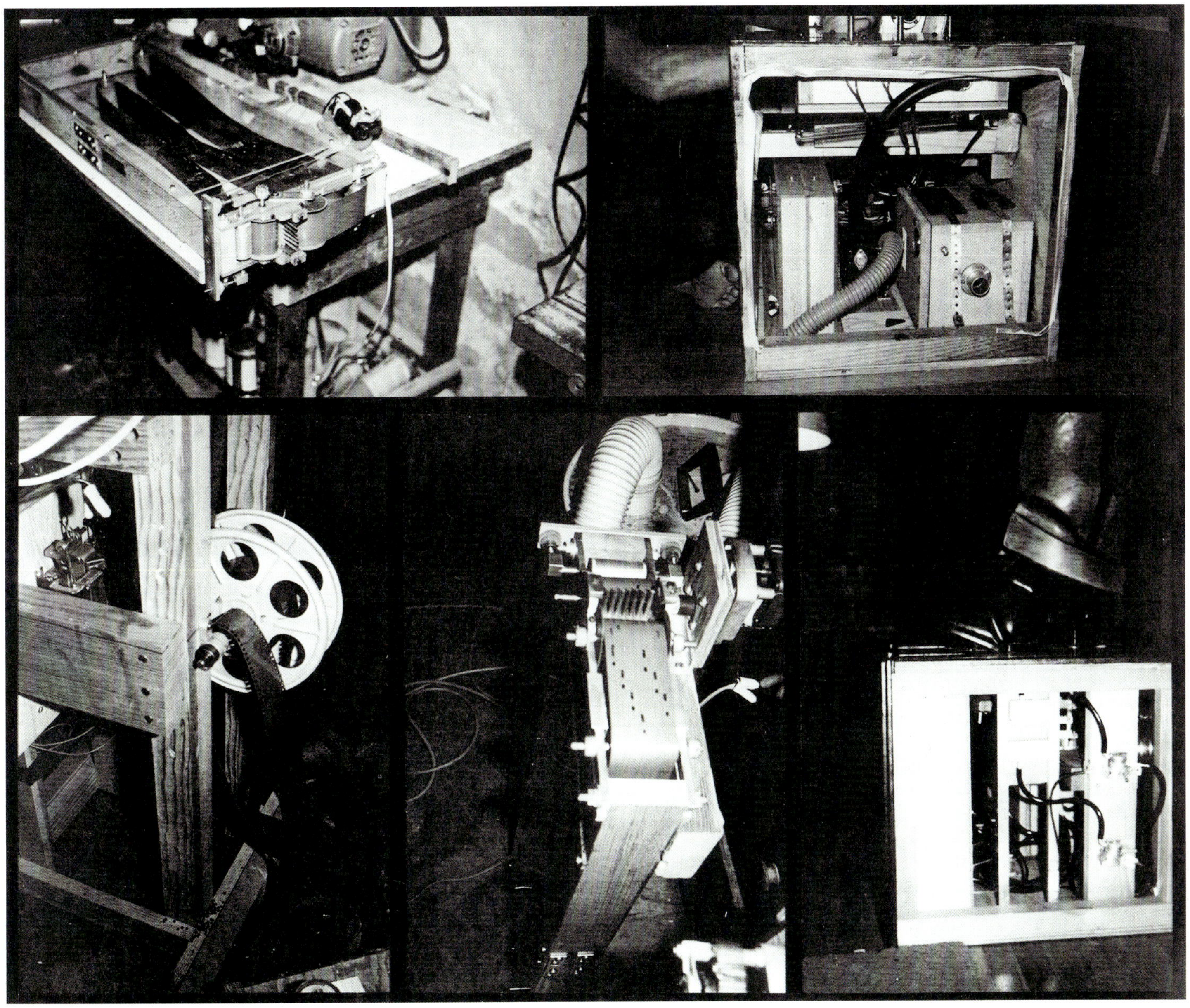

A look inside the machines of Stephan von Huene.
He felt that this was a useful way to depict his way
of working in sound sculpture.

Ryoji Ikeda

*1966 in Gifu (JP), lives in Paris (FR) and Kyoto (JP)

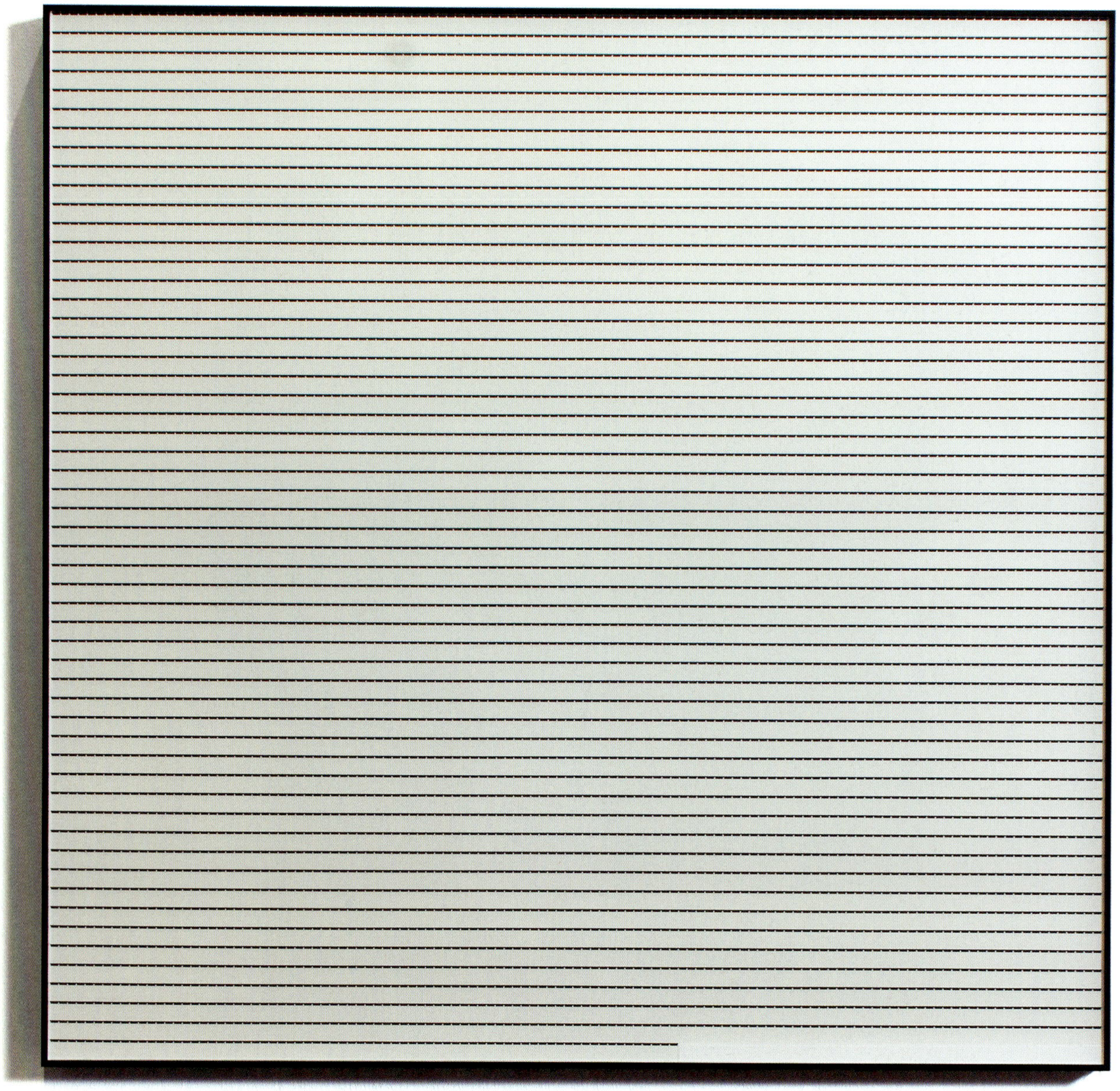

4'33", 2010
Score, framed 16mm film (blank with AATON time code), 88 × 92.5 cm
ZKM | Karlsruhe

www.ryojiikeda.com

The framed work *4'33''* consists of undeveloped 16mm film with AATON time codes. The duration of the strips of film is exactly 4 minutes and 33 seconds and the work is part of the series *time and space*, which the Japanese composer and visual artist Ryoji Ikeda began working on in 1998. He refers here to the eponymous composition by John Cage. Cage's score contains no notes, only the instructions: "I Tacet, II Tacet, III Tacet."

At the premiere of Cage's *4'33''* in Woodstock, NY on August 29, 1952, pianist David Tudor indicated the three movements by closing and opening the piano lid. Before the performance, he tossed a dice to determine the relative length of the three movements at 33 seconds, 2:40 min, and 1:20 min, which add up to 4:33 min.

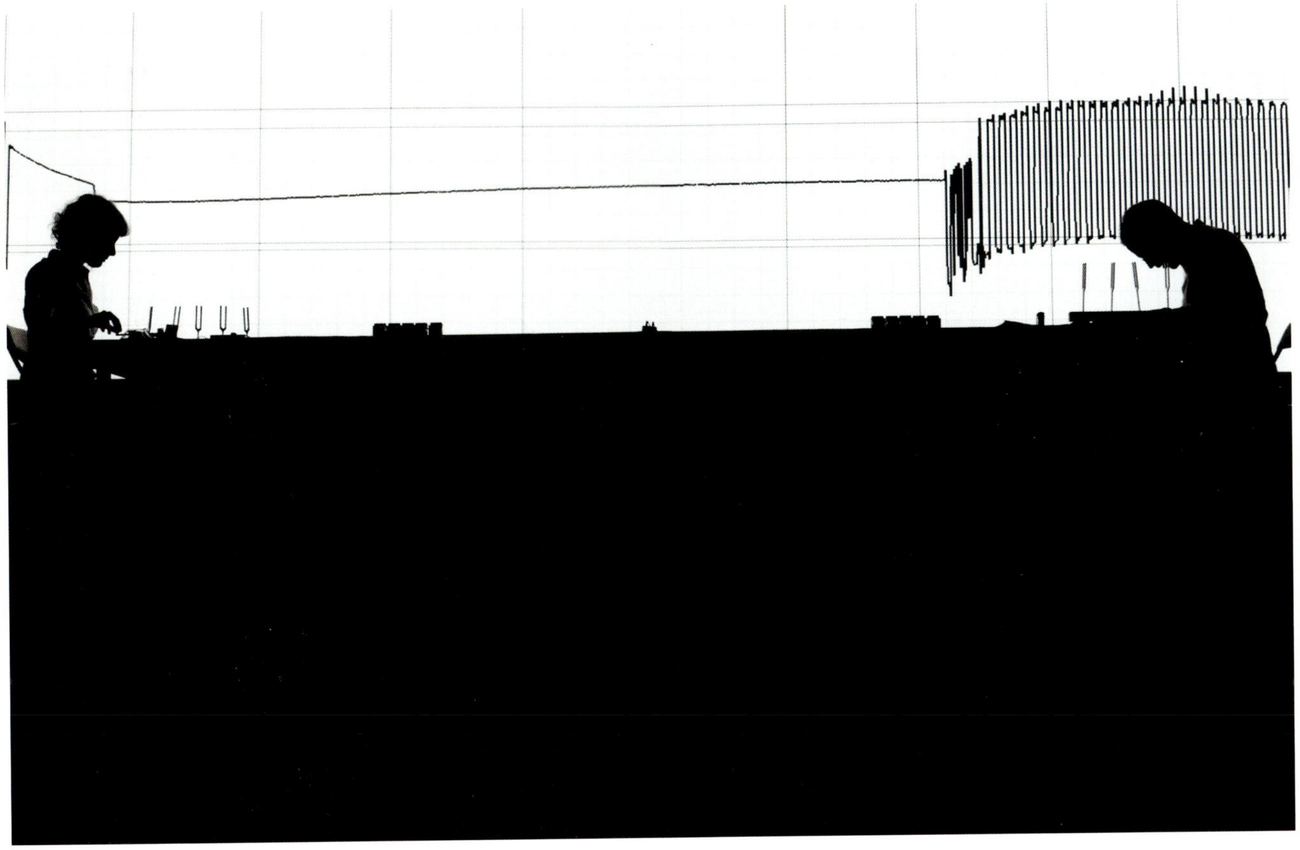

superposition, 2012
Performance

superposition is a project about the way we understand the reality of nature on an atomic scale and is inspired by the mathematical notions of quantum mechanics. Performers will appear in Ikeda's work for the first time, performing as operator/conductor/observer/examiners. All the components on stage will be in a state of superposition; sound, visuals, physical phenomena, mathematical concepts, human behavior, and randomness – these will be constantly orchestrated and de-orchestrated simultaneously in a single performance piece.

Concept, direction, music: Ryoji Ikeda; performers: Stephane Garin, Amélie Grould; programming, graphics, computer system: Tomonaga Tokuyama, Norimichi Hirakawa, Yoshito Onishi; optical devices: Norimichi Hirakawa; stage manager: Simon MacColl; technical manager: Tomonaga Tokuyama; production assistant: Daisuke Sekine; production: Ryoji Ikeda Studio (artistic direction: Emmanuelle de Montgazon; administration: Yuko Higaki), Quaternaire (producer and artist management: Sarah Ford; associate producer: Laurie Uprichard; administration: Kathleen Aleton; coordination and marketing: Joanna Rieussec), Forma (artistic director: David Metcalfe); commissioned by the Festival d'Automne à Paris for the musical part; created and developed at Parc de La Villette (FR), Yamaguchi Center for Arts and Media (JP), ZKM | Center for Art and Media Karlsruhe (DE); coproduction: Festival d'Automne à Paris, Les Spectacles Vivants – Centre Pompidou (FR), Barbican (UK), Concertgebouw Brugge (BE), Festival de Marseille _ danse et arts multiples (FR), Parc de La Villette (FR), Kyoto Experiment (JP), ZKM I Center for Art and Media Karlsruhe (DE), STRP Art and Technology Festival (NL), Stereolux / Festival Scopitone / le lieu unique (FR)
With the support of the DICReAM-CNC (FR)

Ryoji Ikeda

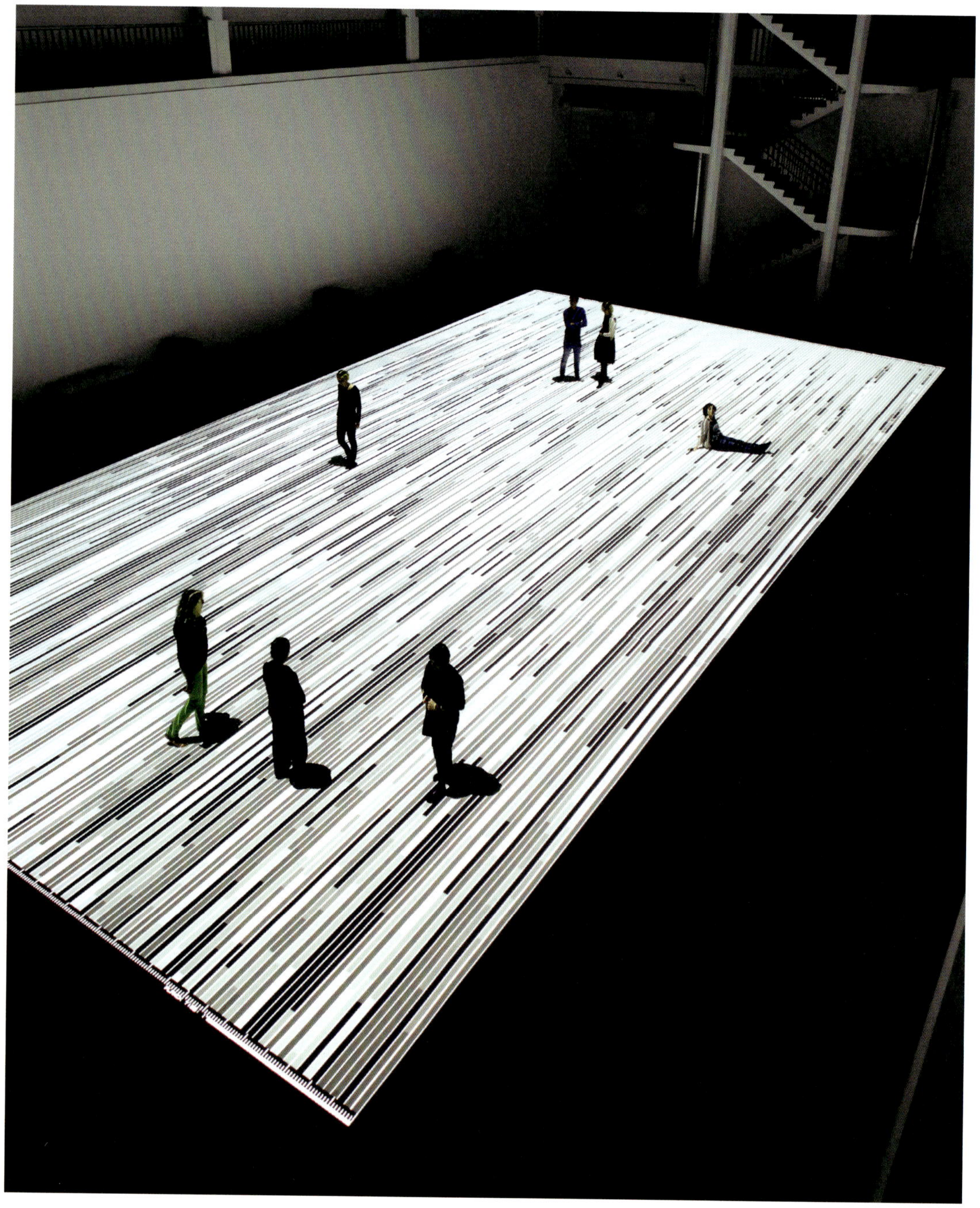

the planck universe [micro] and *[macro]*, 2015
Large-scale floor projection with sound
3 DLP video projectors, computers, loudspeakers
Concept, composition: Ryoji Ikeda
Programming, computer graphics: Norimichi Hirakawa, Tomonaga Tokuyama, Yoshito Onishi, Satoshi Hama
Commissioned and coproduced by the ZKM | Karlsruhe

In *the planck universe [micro]*, visitors can perceive nature transferred from the Planck scale (10^{-35}) to human scale. Scientists use the Planck scale to describe the smallest entities of the universe, so-called strings. The human mind cannot comprehend the extremely small size of those strings, let alone measure or register them empirically in any other way. While scientists agree on the invalidity of all known physical laws at the Planck scale – which implies that space and time cease to exist – they have not yet formed any consistent ideas as to what might take their place. However, they are positive that conditions at the Planck length might be similar to those before the Big Bang, which would imply that the questions about the universe at its smallest scale are quite closely connected with the questions concerning the origins of the universe.

the planck universe is a series of works by Ryoji Ikeda, commissioned and coproduced by ZKM | Karlsruhe specifically for GLOBALE, developed during the artist's residency at CERN, the European Organization for Nuclear Research, Geneva, Switzerland, in 2014/2015 as the winner of the 2014 Prix Ars Electronica Collide@CERN Artists Residency Prize.

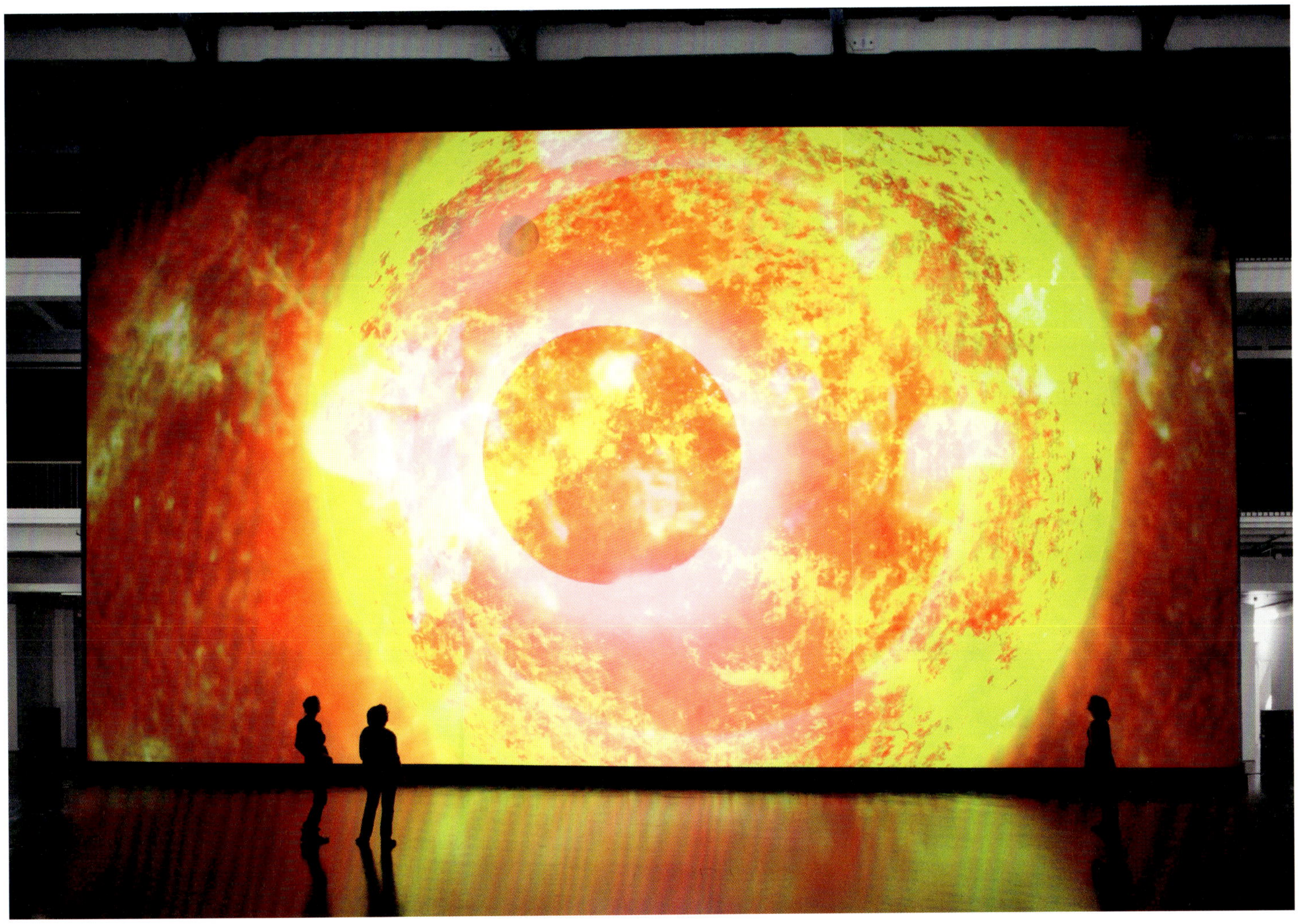

the planck universe [macro] provides visitors with an experience of nature that has been shifted from the human scale to the cosmological scale existing beyond the observable universe (over 1026m). Since the universe may be infinite in size, but is certainly finite in age, we can see out to a distance that is equivalent to the age of the universe, which is currently 13.8 billion light years. Whatever may lie beyond the observable universe is thus too far away for the light emitted since the Big Bang to have had enough time to reach the Earth. However, as the universe is expanding, and the expansion is accelerating, the observable horizon could disappear, so in about a billion years we might not be able to observe anything anymore.

Ryoji Ikeda

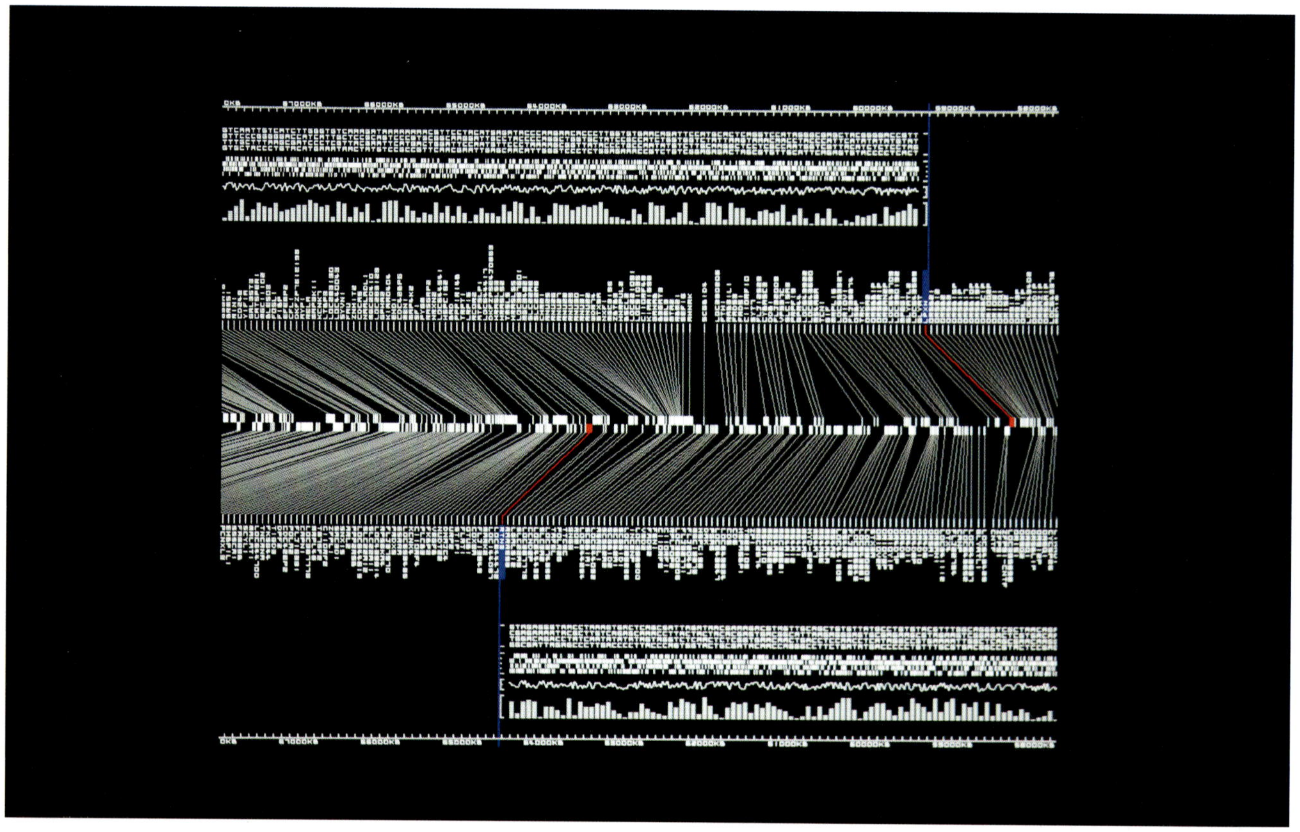

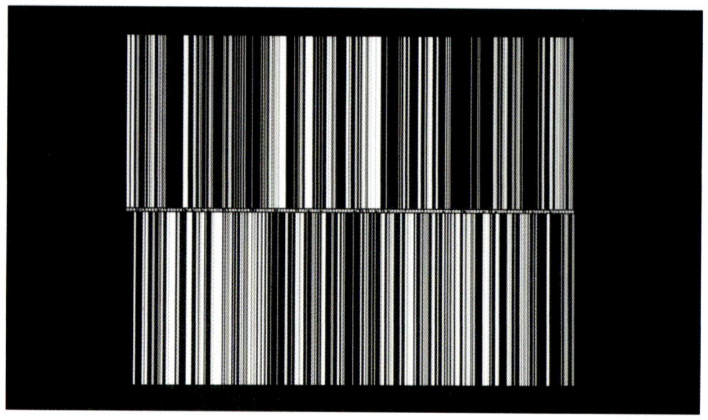

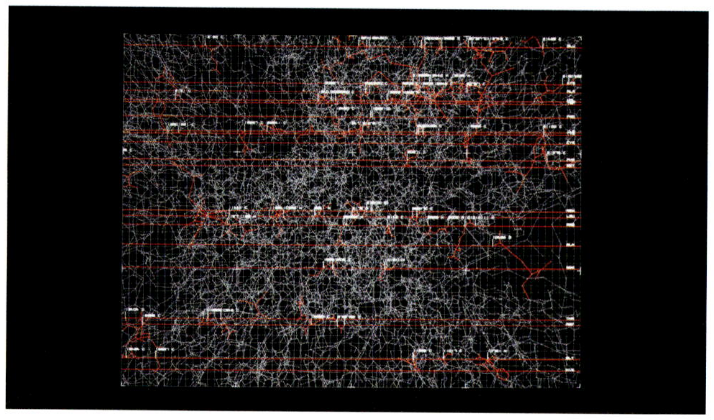

datamatics [prototype-ver.2.0], 2006–2008
Audiovisual concert
Directed by Ryoji Ikeda
Concept, composition: Ryoji Ikeda
Computer graphics, programming: Shohei Matsukawa, Norimichi Hirakawa, Tomonaga Tokuyama
Tour technical director: Kamal Ackarie
Co-commissioned by AV Festival 06, ZeroOne San Jose & ISEA 2006, 2006
Co-produced by Les Spectacles vivants, Centre Pompidou, Yamaguchi Center for Arts and Media (YCAM), 2008
Supported by Recombinant Media Labs
Produced by Forma

Using pure data as a source for sound and visuals, *datamatics* combines abstract and mimetic presentations of matter, time, and space. *datamatics* is the second audiovisual concert in the *datamatics* series. Projecting dynamic computer-generated imagery in pared down black and white with striking color accents, the intense yet minimal graphic renderings of data progress through multiple dimensions. From 2-D sequences of patterns derived from hard drive errors and studies of software code, the imagery transforms into dramatic rotating views of the universe in 3-D, whilst in the final scenes four-dimensional mathematical processing opens up spectacular and seemingly infinite vistas. A powerful and hypnotic soundtrack reflects the imagery through a meticulous layering of sonic components to produce immense and apparently boundless acoustic spaces.

datamatics [prototype-ver.2.0] is the full-length version of this audiovisual concert. With a commissioned second part added, *datamatics [prototype-ver.2.0]* is significantly developed from the earlier version of this piece, which premiered in March 2006. Driven by the primary principles of *datamatics*, but objectively deconstructing its original elements – sound, visuals, and even source codes – this work creates a kind of meta-datamatics. Real-time program computations and data scanning are employed to create an extended new sequence that is a further abstraction of the original work. The technical dynamics of the piece, such as its extremely fast frame rates and variable bit depths, continue to challenge and explore the thresholds of our perceptions.

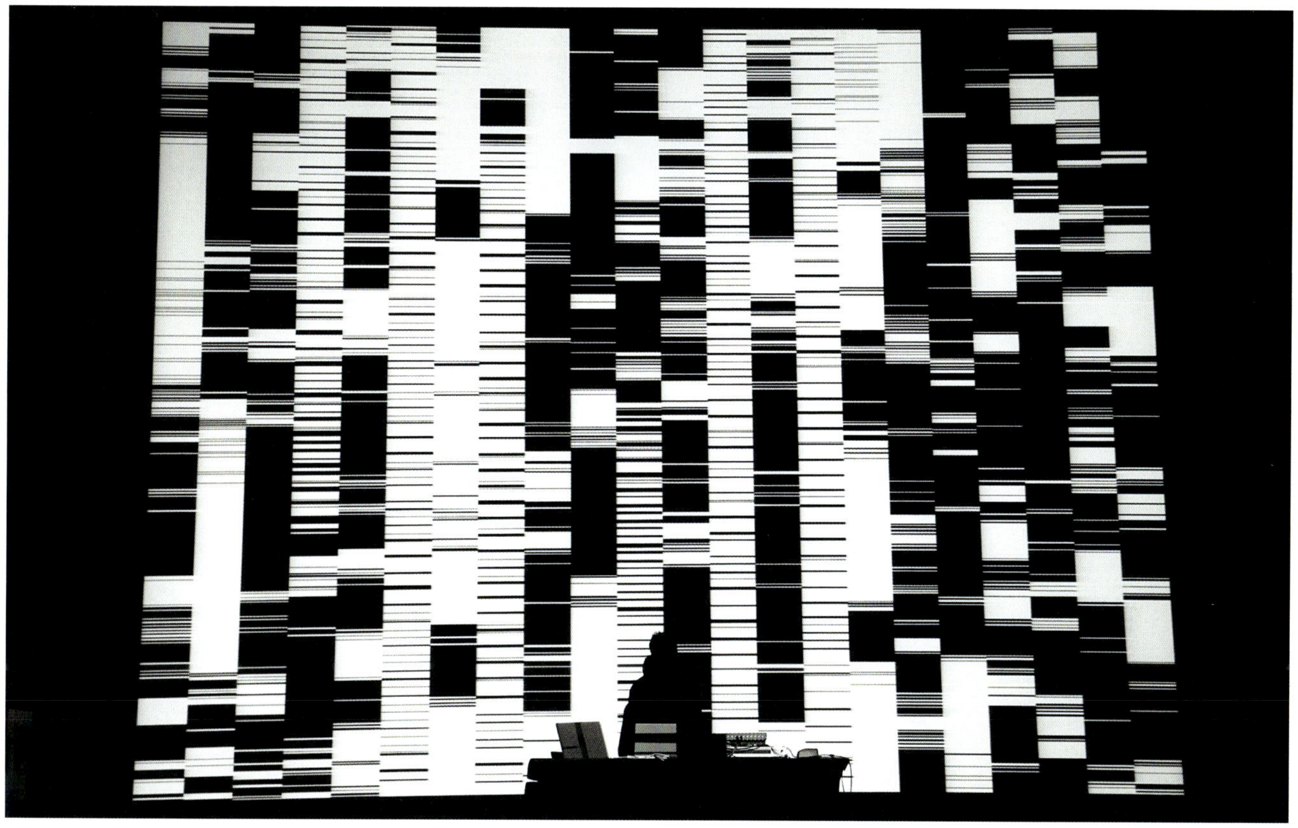

test pattern, 2006
Concert and installation

test pattern [live set], 2008
Audiovisual performance
Concept, composition: Ryoji Ikeda
Computer graphics, programming:
Tomonaga Tokuyama

test pattern [nº5], 2013
Audiovisual installation, 5 DLP projectors, 6 computers, 6 speakers, dimensions variable, ca. 44 × 21 × 25 m
Concept, composition: Ryoji Ikeda
Computer graphics, programming:
Tomonaga Tokuyama

test pattern is an ongoing project started in 2006. It is a creative system that converts any type of data (text, sounds, photos, and movies) into barcode patterns and binary patterns of 0s and 1s. Through its application, the project aims to examine the relationship between critical points of device performance and the threshold of human perception.

As an audiovisual installation series, *test pattern* is comprised of visual patterns converted and generated from sound waveforms in real-time. Each iteration of the installation series responds to the specific context of the space or venue itself.

test pattern [live set] presents intense flickering black and white imagery which floats and convulses in darkness to a stark, powerful, and highly synchronized soundtrack. Through a real-time computer program, *test pattern* converts audio signal patterns into tightly synchronized barcode patterns on screen.

In both concert and installation, the velocity of the moving images is ultra-fast, some hundreds of frames per second, so that the work provides a performance test for the audio and visual devices, as well as a response test for the audience and visitor perceptions.

Mogens Jacobsen

*1959 in Rome (IT), lives in Copenhagen (DK) and Istanbul (TR)

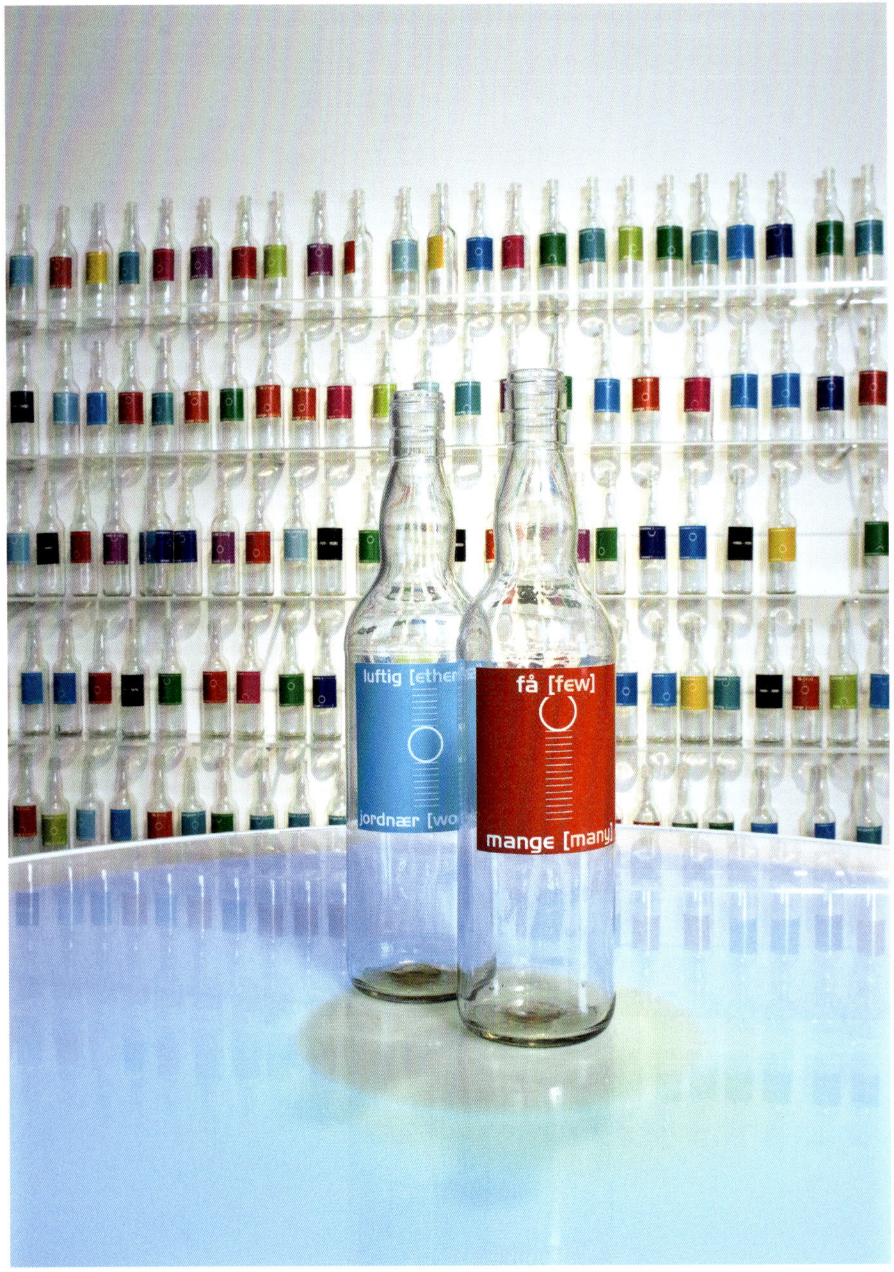

Hørbar / Audiobar, 2007
Interactive sound installation
260 glass bottles, shelves, table, radio frequency identification tags (RFID-tags) and readers, 2 workstations, 2 flat-screens, light, electronics, audio database, amplifier, loudspeaker
Concept: Mogens Jacobsen, Morten Søndergaard
Courtesy of the Museum of Contemporary Art, Roskilde (DK)

www.mogensjacobsen.dk

The *Hørbar / Audiobar* is a sound database that functions on the self-service principle. Colorful labels indicate the contents of the bottles. For example, one bottle contains "mostly noises," another "80% vocals" or "20% rhythms." There are twelve different categories designated by color. Some of the bottles have black labels with white lettering; these contain audio tracks from a specific period, such as "1960–1965."

When the bottles are placed on the illuminated table, the corresponding sound files play. Visitors can put up to four bottles at a time on the table. Each bottle is identified by a tiny radio frequency tag (RFID). The computer reads the tags and selects the corresponding soundtracks. The sounds of the *Hørbar* come from the audio database of the Museum for Contemporary Art in Roskilde, Denmark, whose continually expanding collection of audio art soundtracks goes back to 1890.

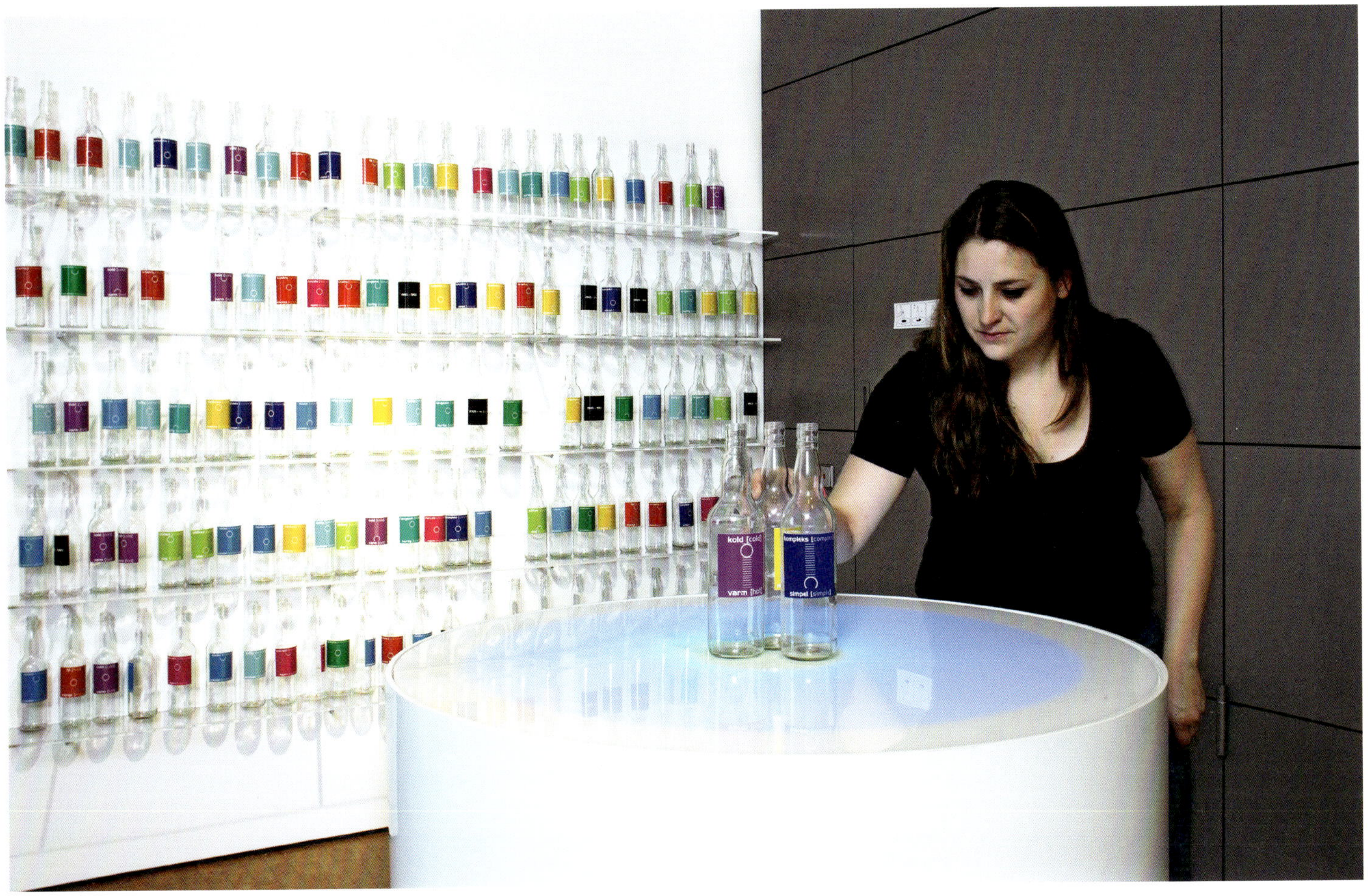
kold [cold]
varm [hot]
simpel

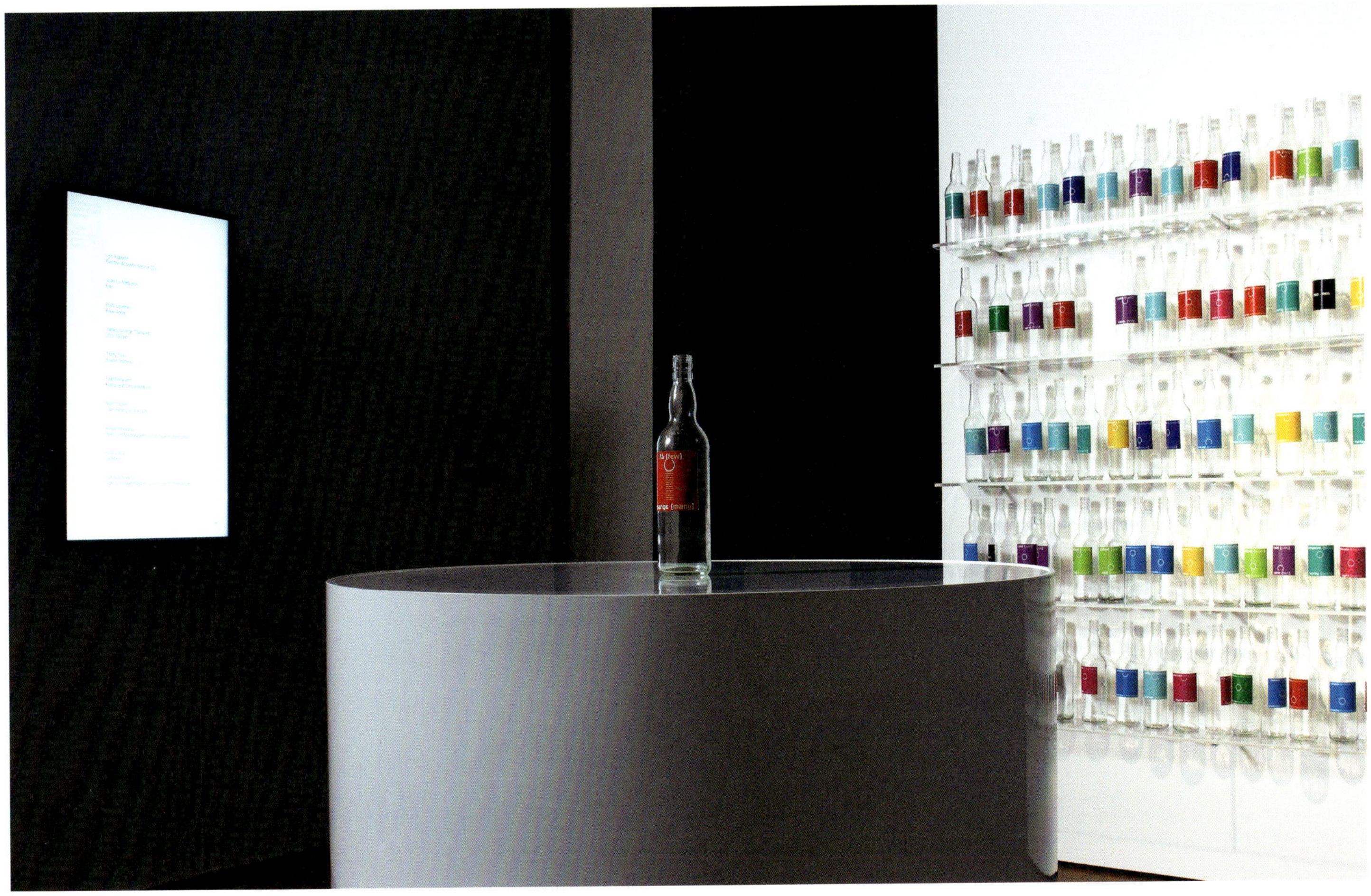

Anna Jermolaewa

*1970 in Saint Petersburg (RU), lives in Vienna (AT)

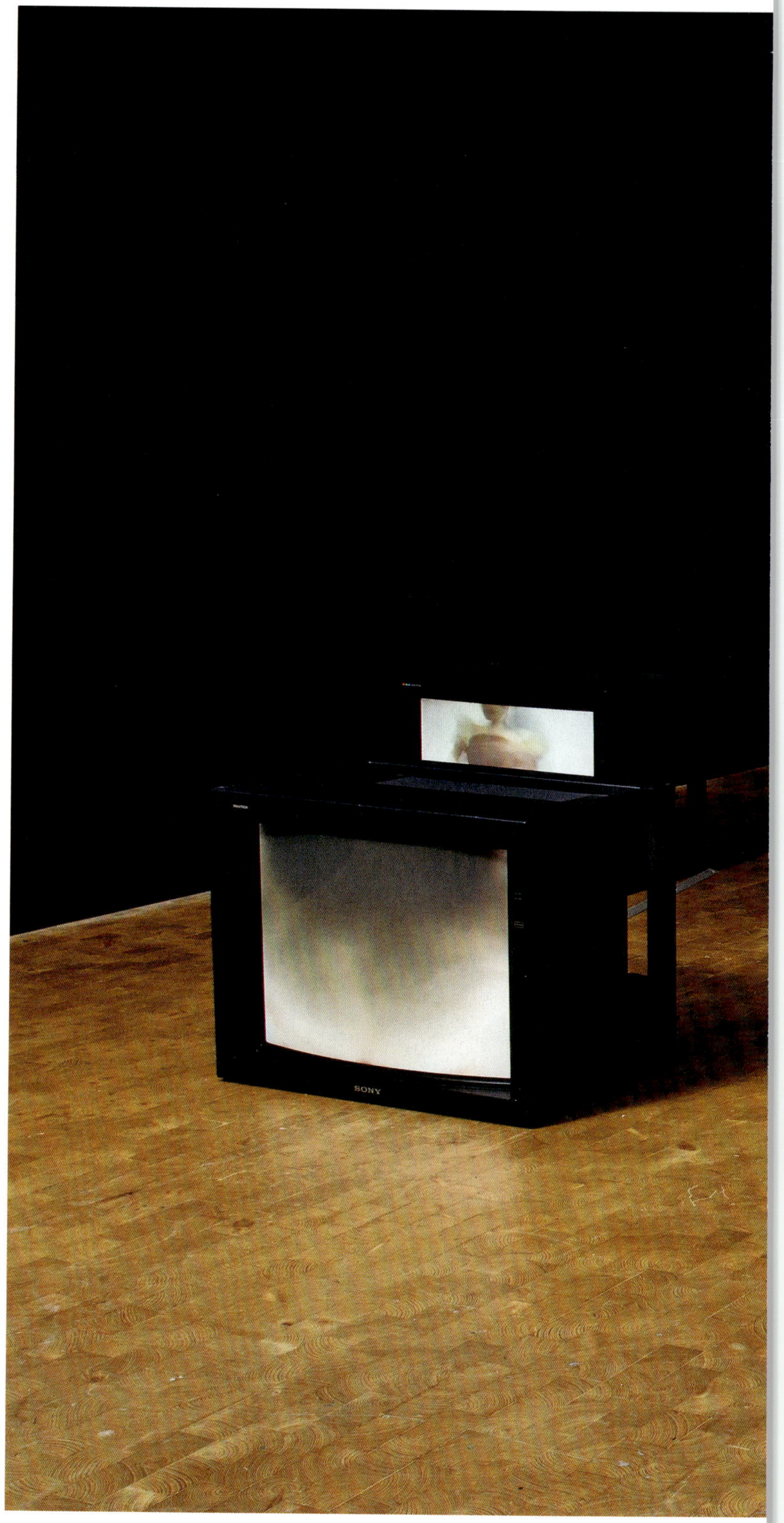

Orchester reloaded, 2000–2007/2012
8-channel video installation
8 monitors, 8 DVD players, 1–4 min.
Anna Jermolaewa and Kerstin Engholm Galerie

Children's wind-up toys go through their paces on the fifteen screens. Anna Jermolaewa substitutes small automatons, cheap industrial products, for the musicians and instruments of the video orchestra. In her other works too, Jermolaewa transmutes childlike miniature worlds of pop-up figures, remote-controlled cars, and stuffed animals into uncanny, even threatening scenarios. The sounds of music boxes and tinkling bells can be heard on the soundtrack. There is nothing in *Orchester reloaded* that evokes the masterpieces of mechanical toy art. Endless loops direct the attention to repetition. As with György Ligeti's *Poème Symphonique* for one hundred metronomes, this mechanical orchestra requires no conductor. What happens on the monitors is not coordinated: the individual loops play in complete isolation. A polyphony of tireless movements and sounds arises – obsessive, disconnected, and at the same time liberated from all restraint.

www.jermolaewa.com

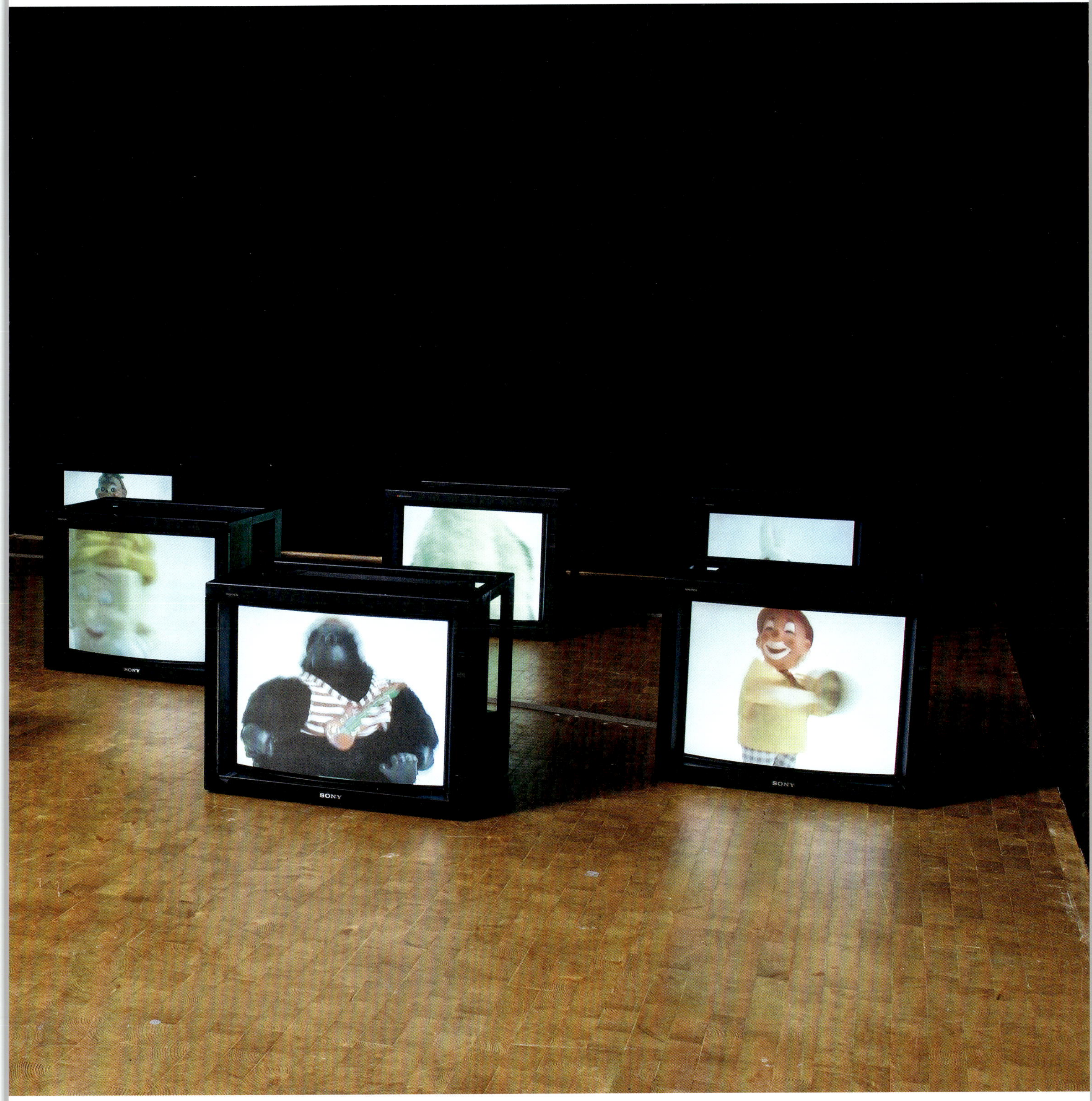
SONY
SONY
SONY
SONY

Sergi Jordà

*1961 in Madrid (ES), lives in Barcelona (ES)

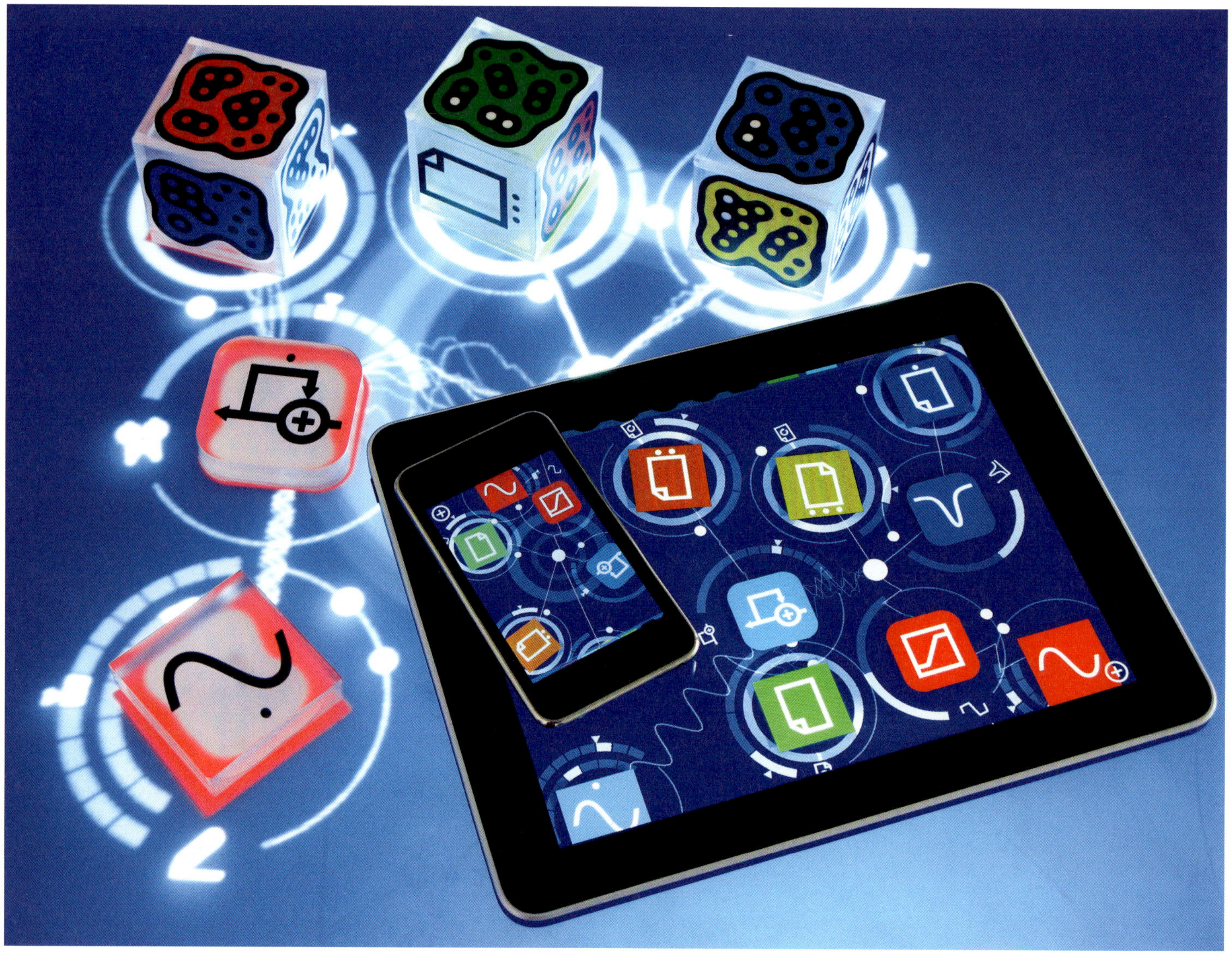

reacTable, 2005
Interactive audiovisual installation / electroacoustic, musical instrument
Interactive table, video, multi-channel audio
ZKM Collection

With Marcos Alonso, Martin Kaltenbrunner, Günter Geiger

www.reactable.com

The *reacTable* is an electronic musical instrument with an interface that transforms sound, rhythm, and music into a visible and palpable experience. By moving, spinning, and creating connections between Plexiglas objects on the table, several persons can play the instrument at the same time. The objects illuminate when played, and their acoustic interaction is visible on the table's surface. The instrument stands in the tradition of classical modular synthesizers but can also be played intuitively by amateurs.

The *reacTable* was created by a team of digital musical instrument builders from the Music Technology Group of Pompeu Fabra University in Barcelona. Several international prizes have been awarded to the multi-disciplinary group of scientists, musicians, and designers, including the Prix Ars Electronica Golden Nica, two D&AD Yellow Pencils and the Prize of the City of Barcelona.

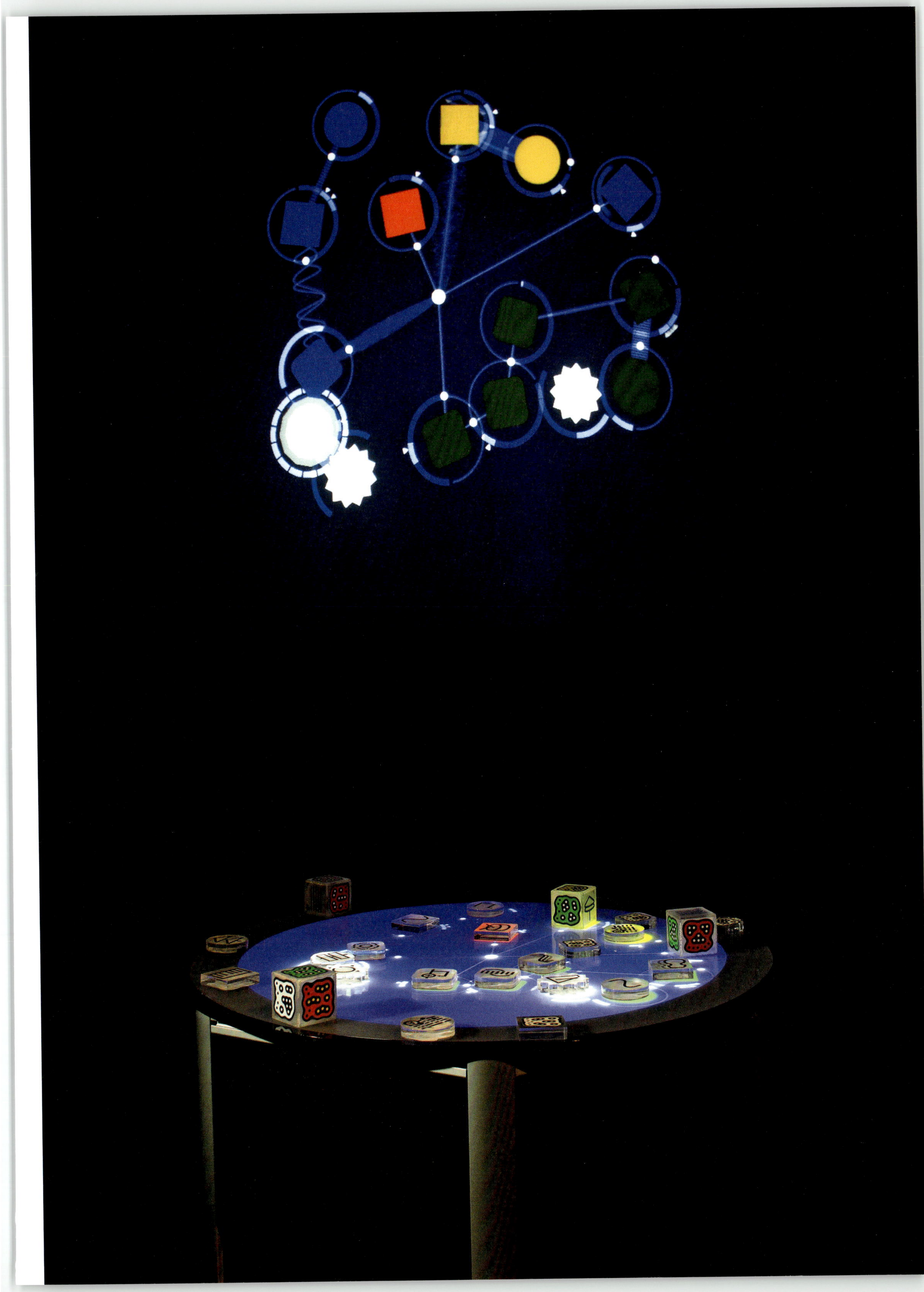

Rolf Julius

*1939 in Wilhelmshaven (DE), †2011 in Berlin (DE)

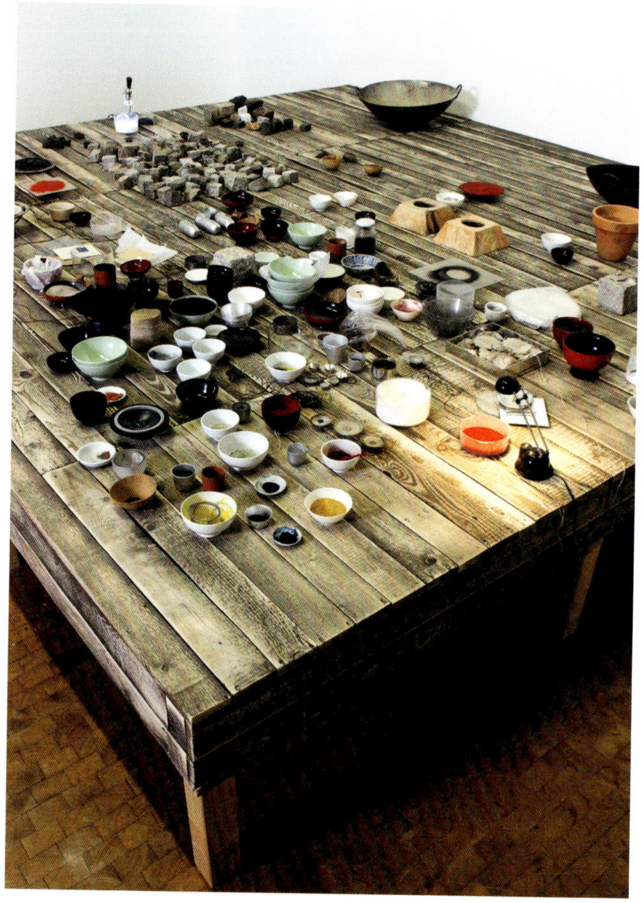

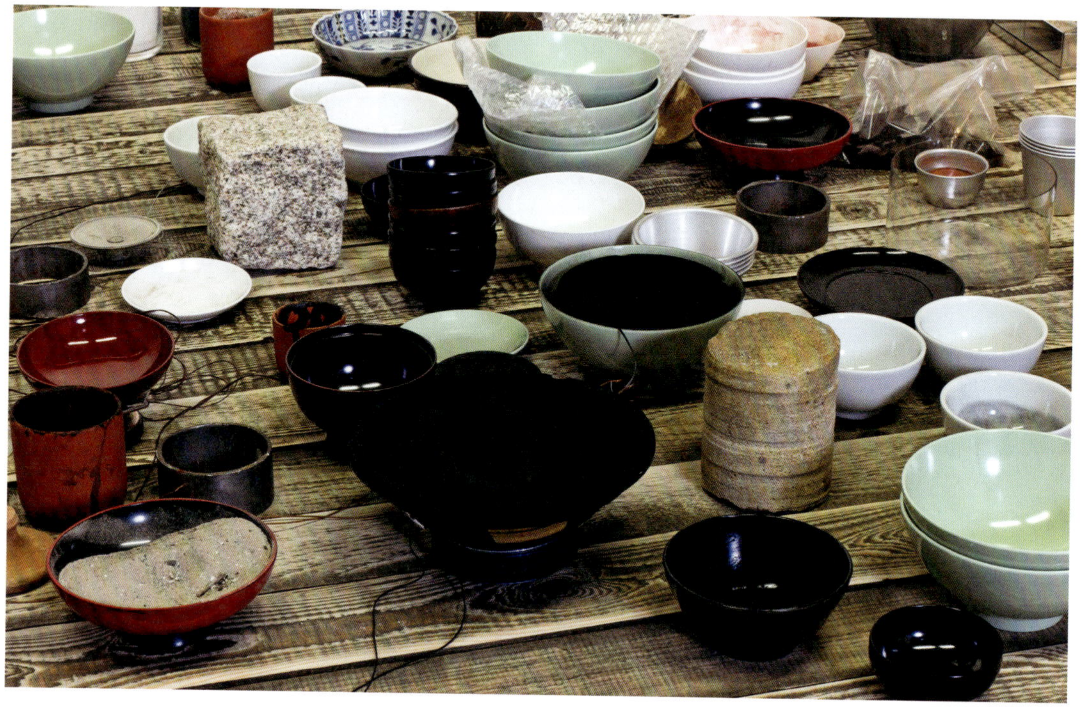

Backstage, 2008
Mixed media installation
Ca. 300 objects from the artist's studio, 60 × 400 × 400 cm
Collection of the Hamburger Bahnhof – Museum für Gegenwart, Berlin

The sound installation, which was created in 2008, can be regarded both as a snapshot of Rolf Julius's workshop and as the quintessence of his oeuvre. Thus some of the objects evoke his early works such as the cobble stones (*Musik in einem Stein* [Music in a stone], 1982; *Stonefield*, 1999) or the panes of glass sprinkled with color pigments (*Musik für einen fast leeren Raum* [Music for an almost empty space], 1998). *Backstage* preserves a transient state; in this place, all the threads of the work accomplished come together and new possibilities appear in outline. *Backstage* is a portrait of an artist without the artist, who lays both his material and spiritual landscape at the feet of the audience.

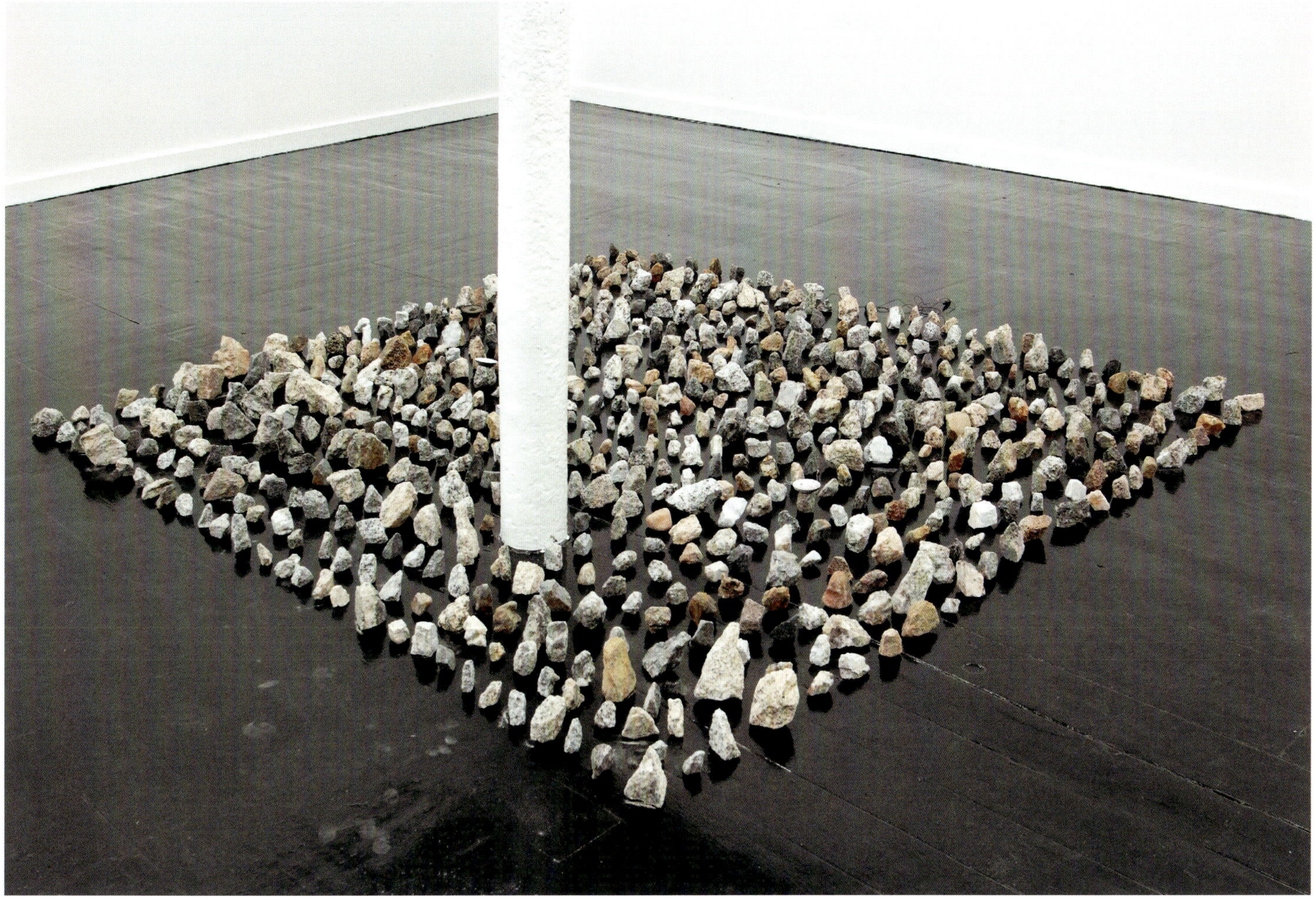

Large Stonegarden (monochrome), 2003
Mixed media installation
Stones, 3 loudspeakers, wire, audio, ca. 165 × 165 cm
Estate of Rolf Julius

Stones from the area around the artist's workshop in Finland form a square stone garden on the floor, upon which small loudspeakers repose. Some of them are lightly colored. Through their soft and abstract sounds (these are insect sounds recorded in Japan) the installation comes to life and starts to move, whereby the sounds cannot be located. It is as though a shimmering layer spreads itself over the stone landscape. Together with the sounds, their acoustic dimensions are also transferred to the stone landscape. When looked at close up, the true proportions shift.

Rolf Julius

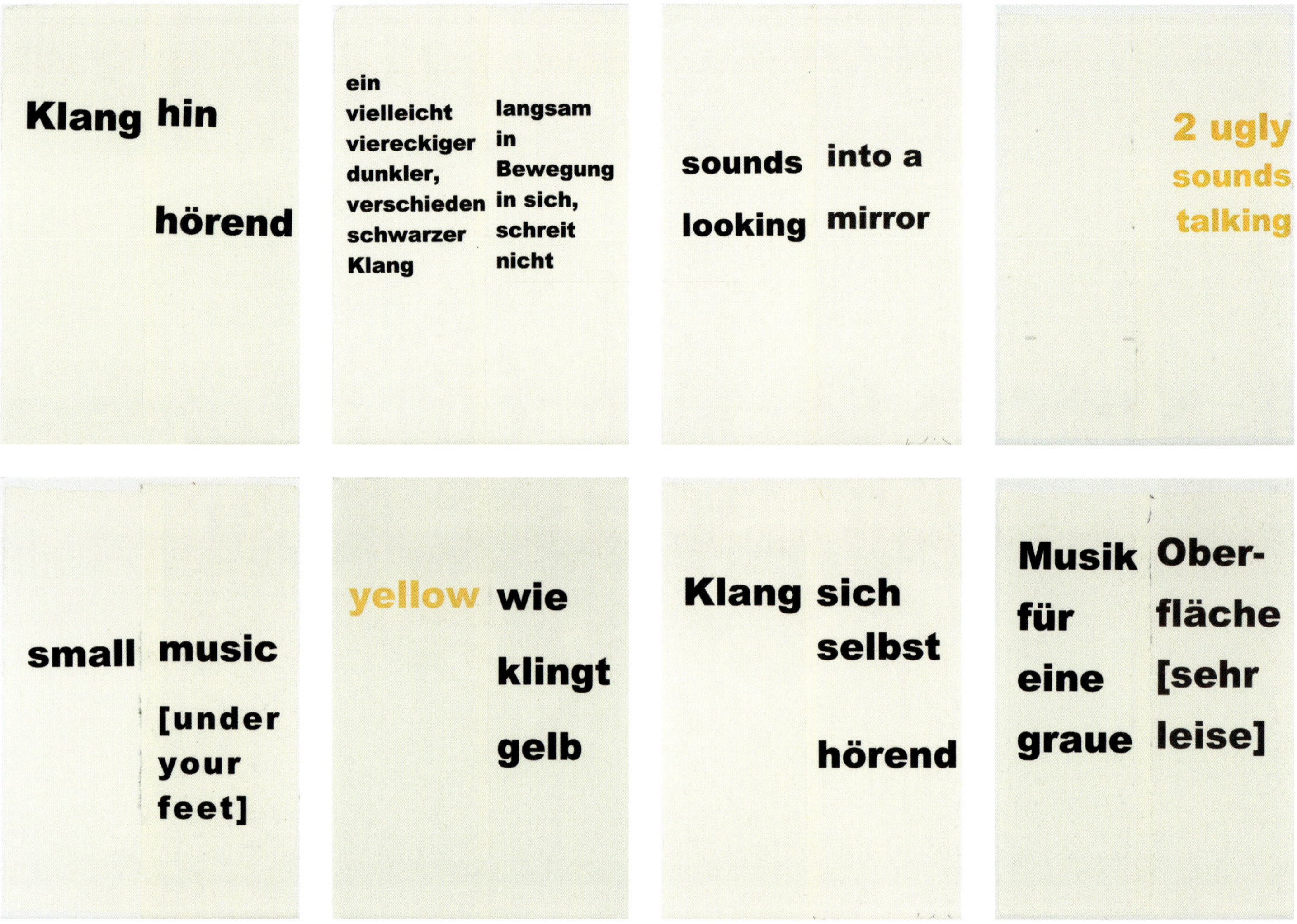

Yellow, wie klingt gelb, 2003
Score
8 inkjet prints on paper
35 × 50 cm each
Estate of Rolf Julius

Whereas at the beginning he produced mainly single framed prints (mostly text prints, Partituren [scores], 1982), in the 1990s and 2000s Rolf Julius changed to working on wall-filling series of inkjet prints, which thematized to a greater or lesser extent the intermediate area between visual art and music. *Yellow, wie klingt gelb* [Yellow, how does it sound] of 2003, consists of eight image texts, whose slightly varying positioning and movement stems from the montage of various pairs of slightly overlapping lengths of paper. The individual text parts visually combine to form a contemplation on the interferences between image and sound.

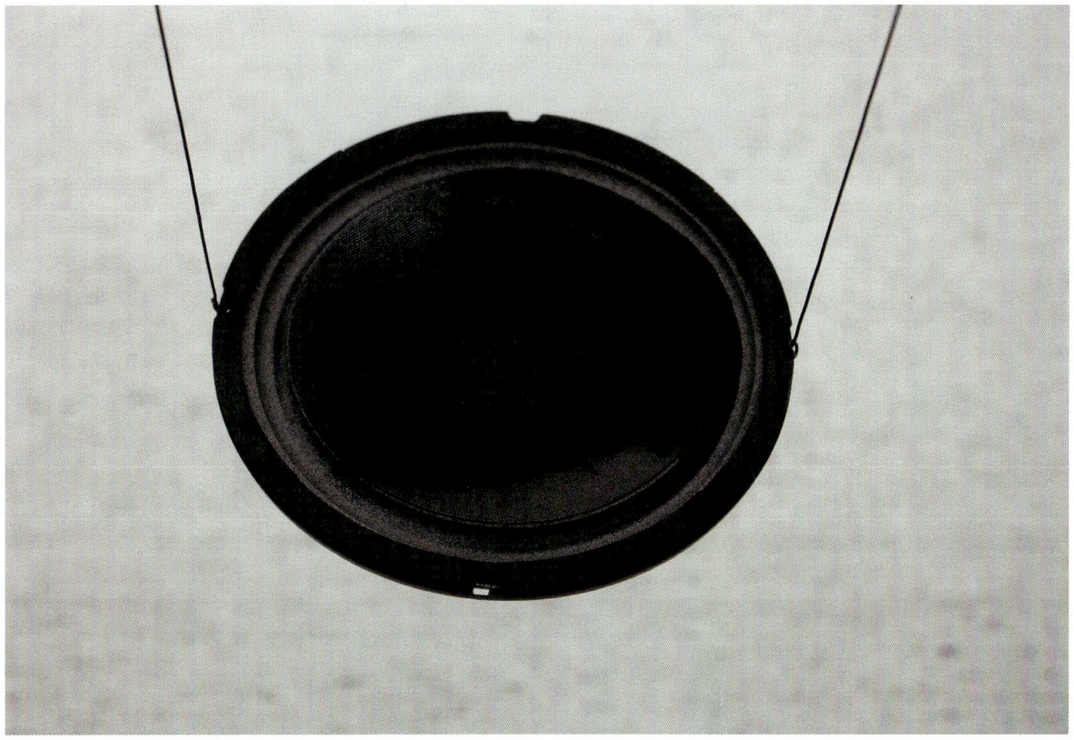

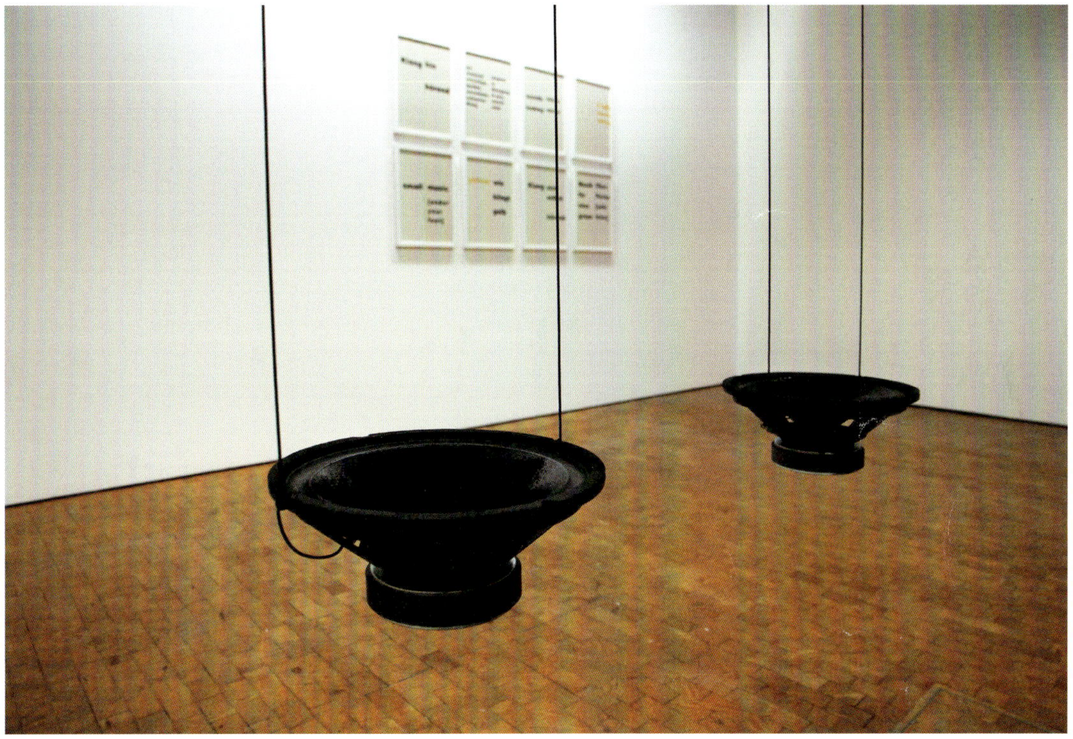

Singing II, 2006/2012
Mixed media installation
Loudspeakers, wire, pigment, audio
Estate of Rolf Julius

The works by Rolf Julius manifest themselves through the ephemeral. They are conceived with total openness. Openness towards their contexts and in regard to their own shape and design. In *Singing II* two loudspeakers hang freely from their cables into the space – through this reduction similar to a drawing. A vibration in the upturned membranes is triggered by the quiet composition that is running in the piece, thus causing the sieved pigments on the membranes to dance softly. The fragile and brooding sounds seem to give them life.

Timo Kahlen

*1966 in Berlin (DE), lives in Berlin

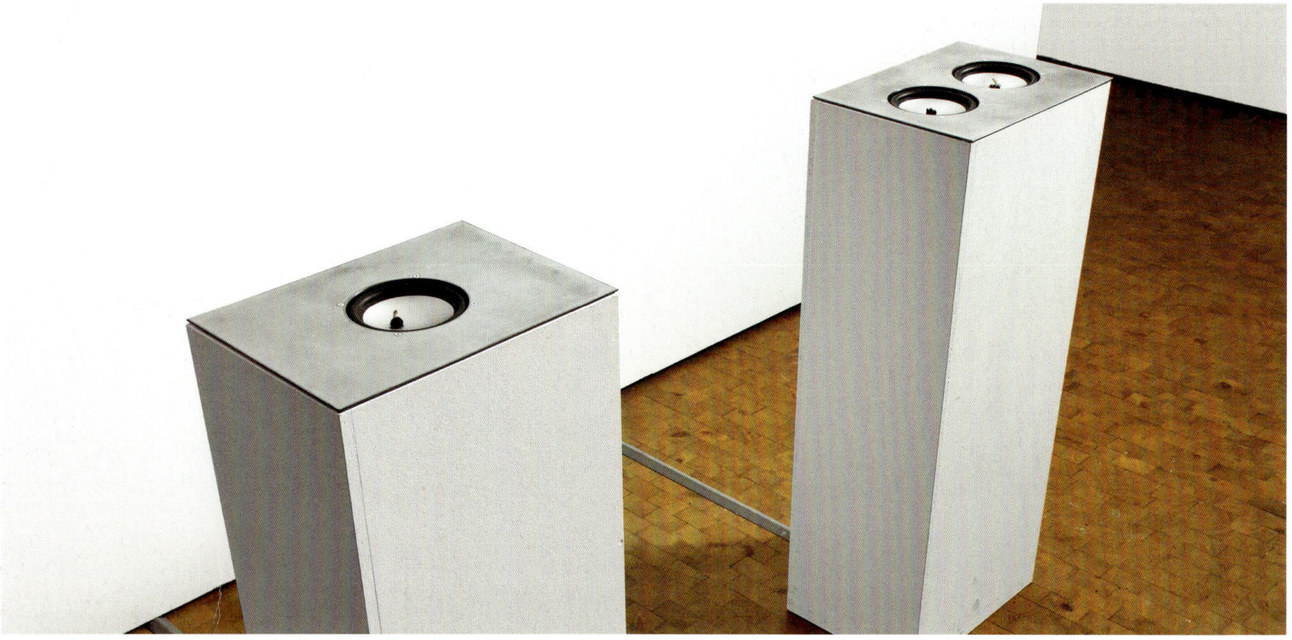

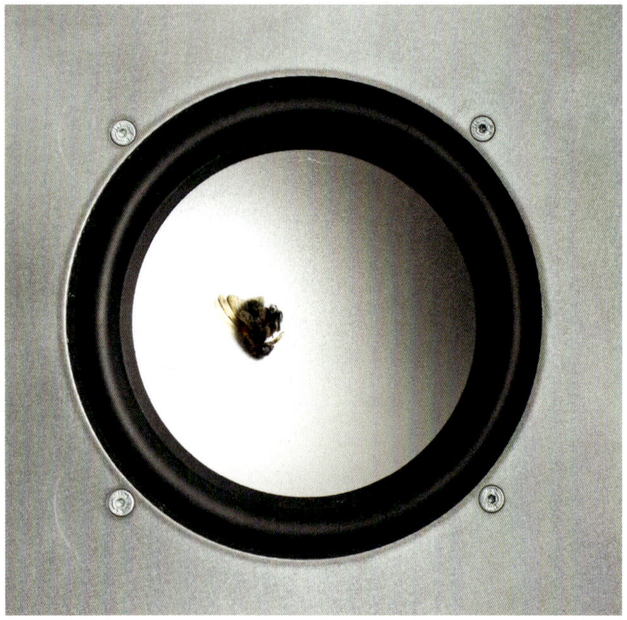

Tanz für Insekten, 2010
Sound sculpture
Pedestals, loudspeakers, insects, sound, ca. 100 × 140 × 31 cm
Collection of the artist, supported by a scholarship of Stiftung Kunstfonds, Germany 2010

Three dead insects (objets trouvés) dance and levitate, shiver and float on the surface of three vibrating loudspeaker diaphragms. A low frequency, nearly inaudible sound makes the membranes vibrate, and with each bounce the fragile chitinous exoskeletons make a light, clicking percussive noise. Caught in an endless series of random and nervous movements, the insects appear to be more alive than dead. Timo Kahlen's kinetic sound sculpture *Tanz für Insekten* [Dance for Insects] seems to shift the thin line between life and death.

www.timo-kahlen.de

right page:
Audio Dust, 2011
Interactive sound work, visual projection with interactive sound surface
ZKM Collection and collection of the Rose Goldsen Archive of New Media Art, Cornell University, US; supported by a scholarship of Stiftung Kunstfonds, Germany 2010
See www.staubrauschen.de/audiodust

Audio Dust is about the beauty of noise. Windjammer, poodle, or deadcat: the jargon of sound engineers has many names for the windshield made of soft and tactile fur that makes it possible to record with the microphone even under extreme wind conditions. Timo Kahlen takes the image of filtering sound as the starting point for his interactive sound work. *Audio Dust* makes audible the incidental and interfering noises that a windshield absorbs. Our attention shifts from the microphone to the "audio dust" in the thicket of fur.

The invisible depositions become audible when the visitor moves the computer mouse across the image. Static and mobile sound fragments form multilayered gritty, creaking, buzzing, and rustling microcosms that are generated – always different – according to the direction, speed, and voluntary pauses of the mouse as it hovers and clicks on the touch-sensitive sound surface.

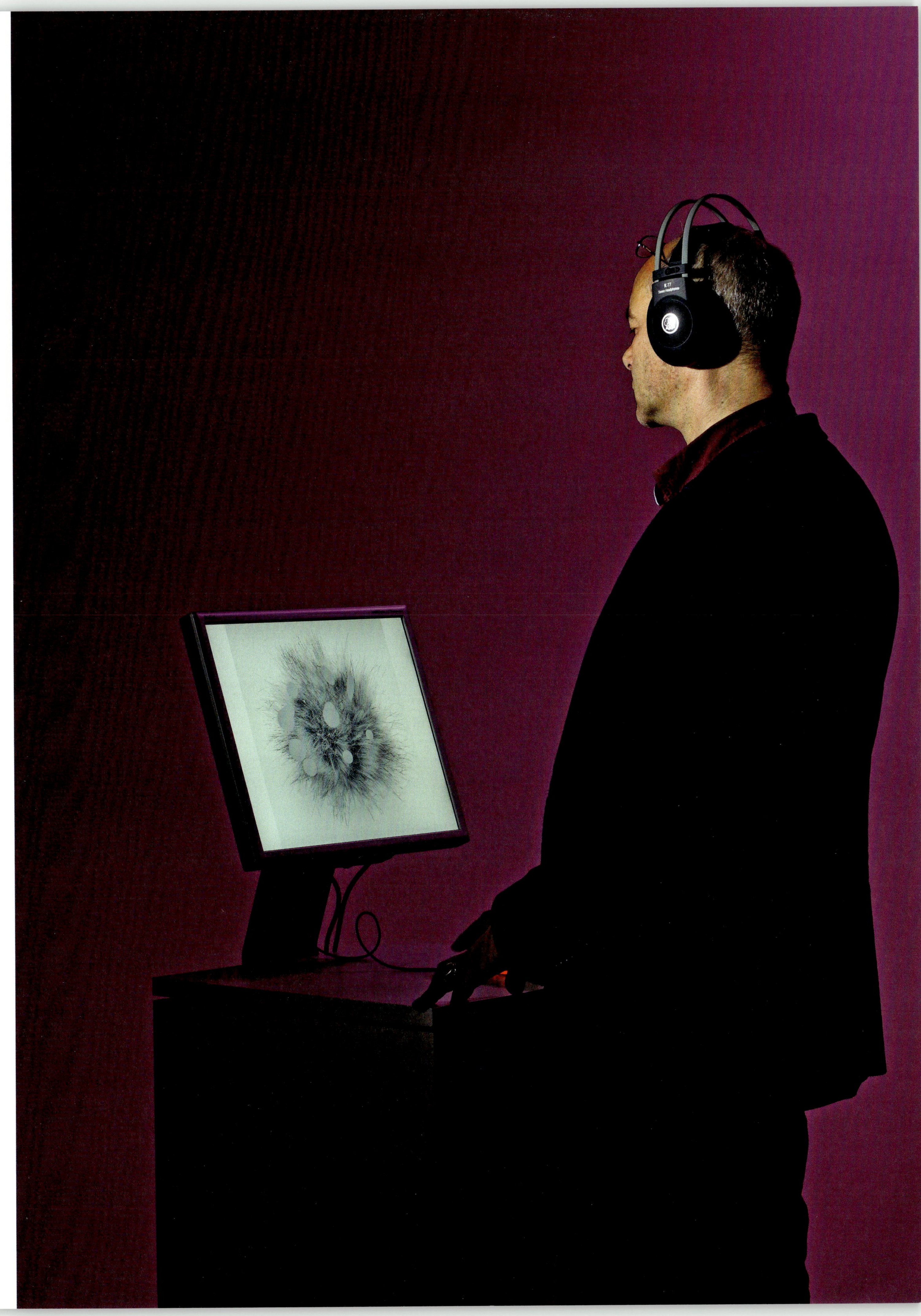

Timo Kahlen

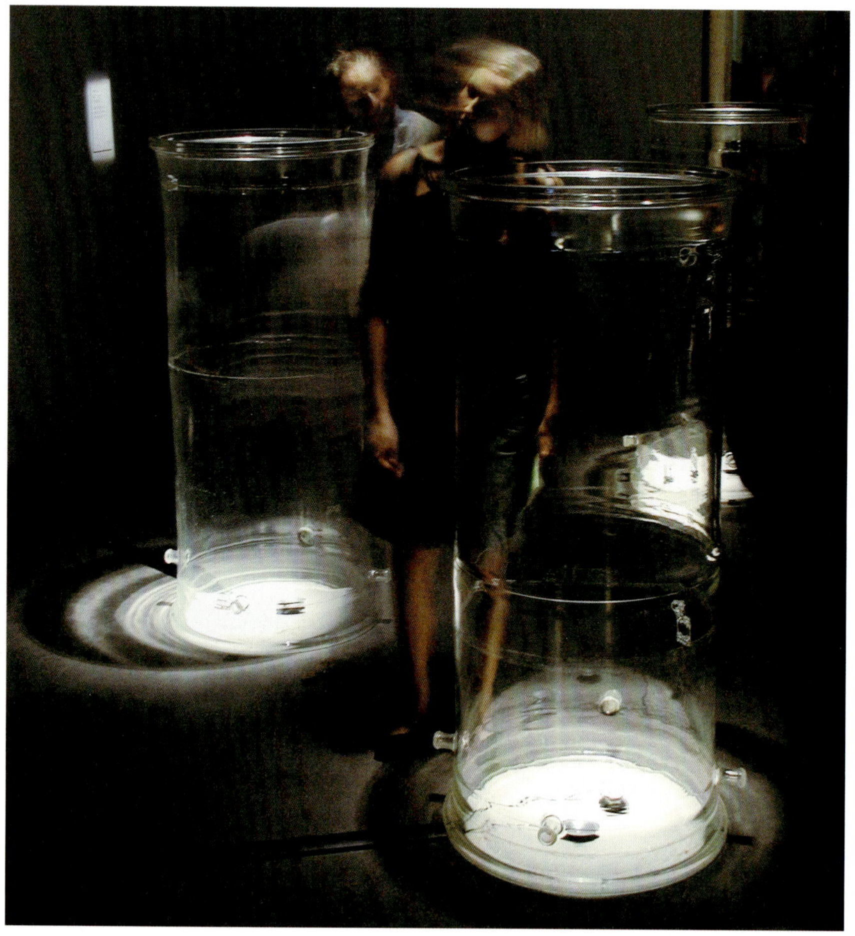

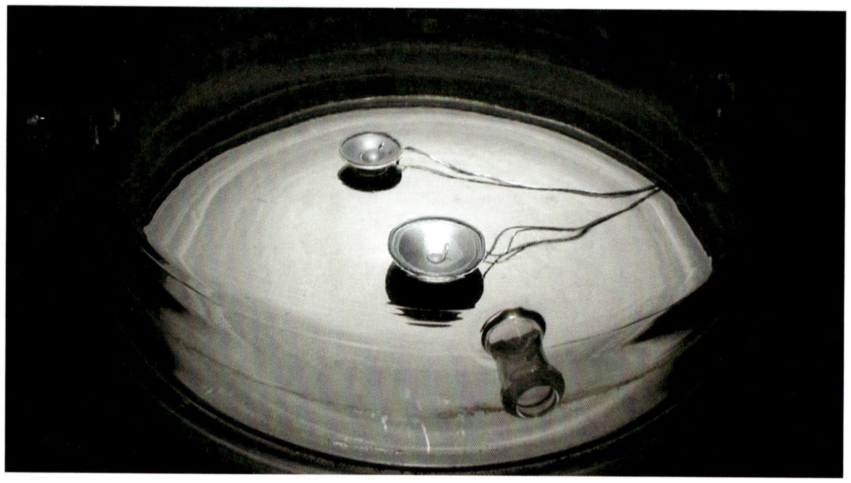

Media Dirt, 2004
Sound installation, glass cylinders (150 × 67 × 67 cm each), loudspeakers, sound of interfering radio waves in between radio stations
Collection of the artist

Several of Timo Kahlen's early sound installations and sound sculptures make use of empty spaces (voids) to visualize sound waves and radiation occupying or filling the apparently "empty," transparent volume of air. *Media Dirt*, the most iconic installation of this type, consisting of three glass cylinders filled with the dirty, raw, hissing, gurgling, scratching, whirring, singing sound of interfering radio waves, was conceived for the *Wireless Experience* exhibition at ISEA 2004, Kiasma Museum of Contemporary Art in Helsinki, Finland, before its presentation at the Ruine der Künste Berlin in 2005.

Three glass cylinders, several loudspeakers, and the sound of interfering radio waves: "temporary, immaterial noise and beauty found in between radio stations [...] creating a complex, artificial, purely technological, yet seemingly 'natural' (sometimes insect or bird-like) soundscape. In this way, the glass cylinders become the laboratory for a re-construction, a new abstract form of nature." (Werner Ennokeit)

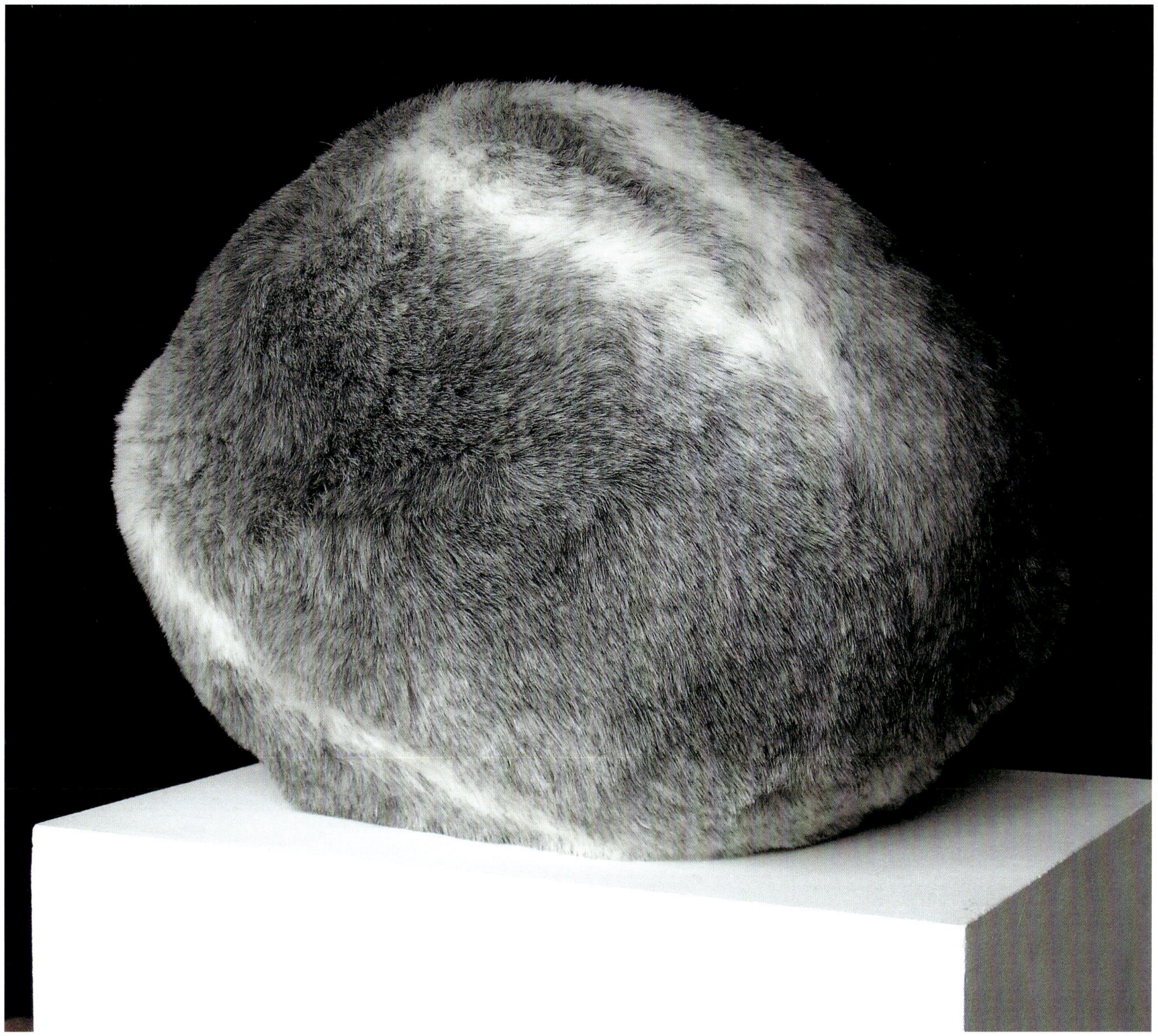

Eins (One), 2005
From a series of softly purring, vibrating, kinetic sound sculptures
Artificial fur, feather stuffing, enclosed loudspeakers, pulsing sound and perceptible vibration, ca. 30 × 45 × 40 cm
Collection of the artist

"A soft, oval fur object, in different shades of grey and with a dark streak along its ridge, is placed on a white pedestal. Careful! It's a trap for our senses, taking our emotions hostage and muffling all rational thought. Approaching the object, you can hear the soft, pulsing purr, feel the slight vibration and warmth with your hands and will begin to stroke the object. You just have to. We feel, what our senses know. And yet: It is not the warmth of a living body, rather the heat created by the friction of the mechanical parts enclosed; not genuine fur, rather a synthetic imitation of its properties; not skin and bones that can be felt beneath the soft surface; not a curled-in, comfortly purring cat. Rather a thing, a kinetic sculpture [...] radically reduced to the basic shape of an ovoid, that turns out to be a precise technical machine: made of synthetic fur, of low frequency and infra-sound pulsation, of two subwoofers stitched into a soft filling, of various cables and a sound source. 'In his installations, Timo Kahlen defines the relationship between nature and its representation as a process of assimilation. Imitating nature thus creates a new form of being, a new form of existence that is both distinct from and similar to nature' (Martin Schönfeldt, Galerie Pankow 2001)."

(Excerpt from: Werner Ennokeit, "Noise & Beauty," in *Timo Kahlen: Noise & Beauty. 25 Years of Media Art*, ed. Werner Ennokeit and Leon Zwart (Berlin: Edition Ruine der Künste Berlin, 2010), 13–15, here 14.)

Georg Klein

*1964 in Öhringen (DE), lives in Berlin (DE)

*Der gelbe Klang*² *– Hommage an Kandinsky*, 2012 (premiere)
Sound-light installation in public space
2 emergency exits in an underground parking garage in Schlosspark Karlsruhe, 2 × 6-channel audio with 3-channel lighting, sound and light controls, 2 audio loops, each 17 min.
Production: ZKM | Karlsruhe

www.georgklein.de

Anchoring this installation of sound and light are two concrete staircases that seem to lead nowhere. From the emergency exits in an underground parking garage emanate two "yellow sounds" that merge in the middle of the space.

An acoustic inventory of the location served as Georg Klein's source material. From the site-specific sounds he created two abstract 6-channel compositions, which are accompanied by two 3-channel light projections. Georg Klein's works are often characterized by a single color that, together with a foundation sound, defines the prevailing atmospheric mood of the installation space.

Here, Klein evokes the visions of Wassily Kandinsky with two "pure yellow sounds." Exactly one hundred years ago in the almanac *Der blaue Reiter*, Kandinsky presented his project *Der gelbe Klang: Eine Bühnenkomposition* [The Yellow Sound. Composition for Stage], in which the "sound of one art [...] supports the identical sound of another art."

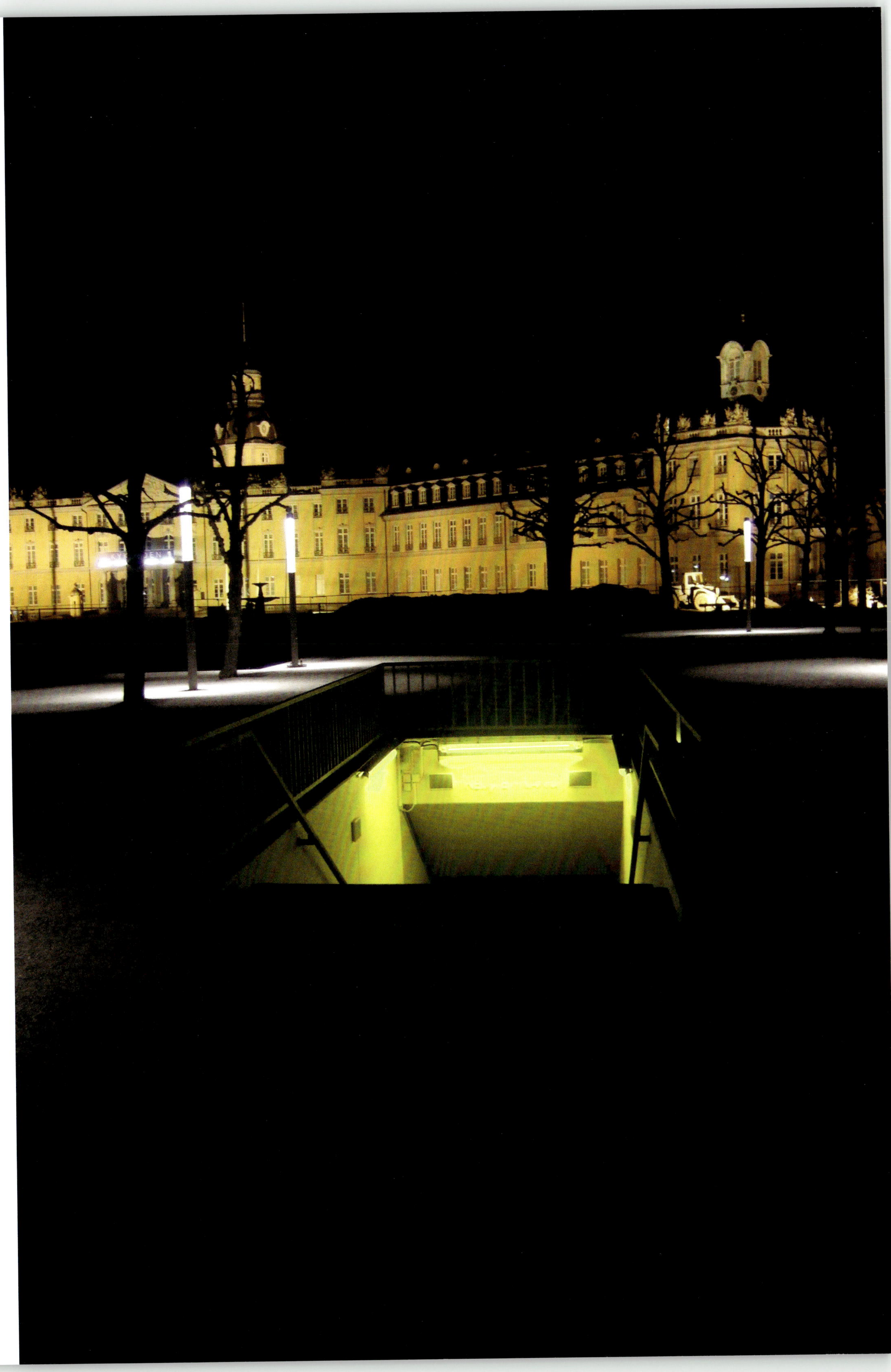

Georg Klein

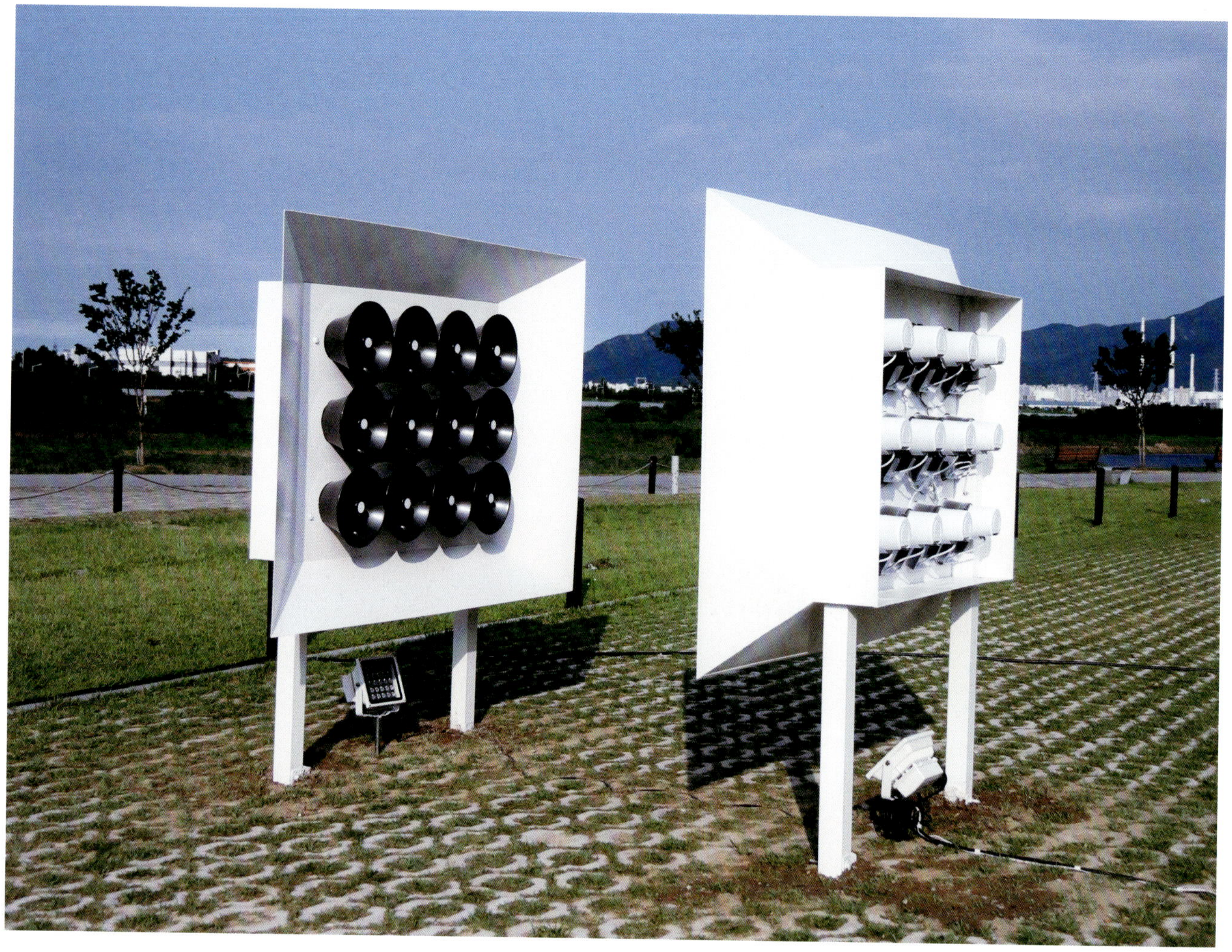

Deep Difference Unit, 2016
Loudspeaker sculpture in 2 parts
24 horn speakers, audio player,
metal frames 9 min. (loop),
5 pairs of electronic sinus waves,
voices of 3 birds, Korean song
Saeya, saeya, parang saeya

The "difference tone" is a psychoacoustic phenomenon of an additional tone that is artificially perceived when two pure tones are sounding at the same time. The *Deep Difference Unit* produces four difference tones specially referencing Korea: four states of a virtual, invisible unity of two different but close wave motions, an ambivalent unity of two different but closely related sides.

The loudspeaker sculpture reminds of the huge loudspeaker units installed by both states at the border between North and South Korea to manipulate the other side with propaganda and music. In *Deep Difference Unit* they are standing extremely close together, producing a melody of difference tones, only audible for visitors in the zone in between. The visitors here can pass through – in real life that's not possible as the border zone is totally blocked. In the real in-between-zone, the Demilitarized Zone (DMZ), nature resurrects to full life. Wild animals feel safe because humankind is not disturbing them. Thus the voices of three birds out of that zone sometimes appear among the deep difference tones by the sculpture: the crane, the magpie and the golden oriole. For birds borders don't exist, and it is an old Korean song of the "Blue Bird" (*Saeya, saeya, parang saeya*) which tells a sad story from the Donghak Rebellion in 1894 when the Korean people fought against their own corrupt rulers and the growing Japanese dominance. The four tones of this melody appear transposed as deep different tone frequencies. In its complexity *Deep Difference Unit* condenses different layers of the Korean history.

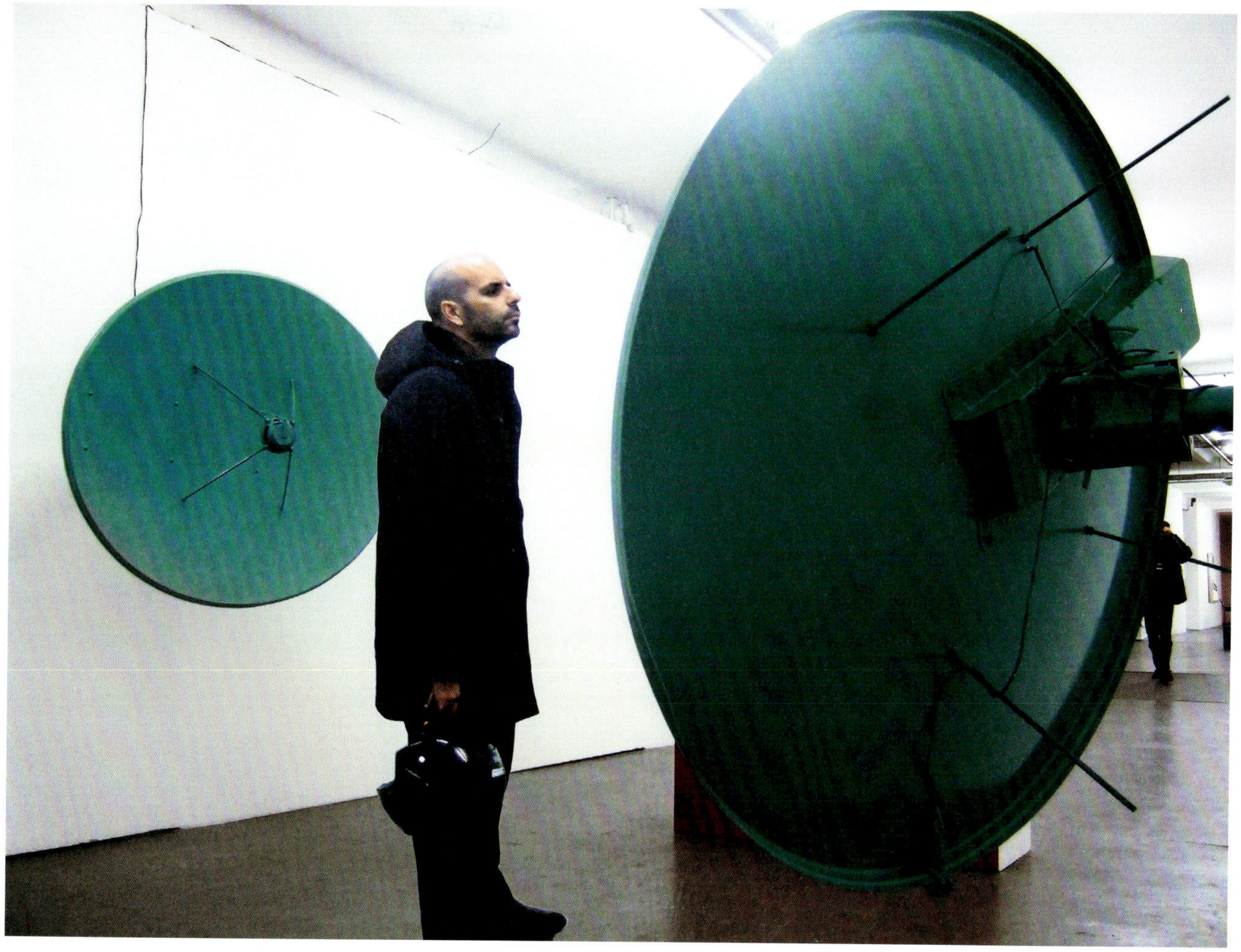

Na Na, 2009
Interactive installation
2 parabolic satellite dishes with loudspeakers, motion detector, sound files, computerised process control, duration variable, 13 min. total (stereo loop), 66 motion-sensitive interaction sounds
Voice: Sun-ju Choi (Korean/German)

The installation consists of two satellite dishes converted into loudspeakers and mounted on opposite walls. The opposition of these green parabolic reflectors, which recall the radar antennas on Cold War surveillance posts, creates a sharply defined beam of stereophonic sound. Thus passing the space in between the two satellites visitors find themselves walking into a sound focus, a point of convergence between front lines.

Musically, the installation operates on a minimum of source material: the German word *ich* [I] and its Korean equivalent *na*, both spoken by a Korean woman living in Germany. These words appear on two auditory levels: as a changing loop (a long, overlaid fragmentary transit between the two words) and as a motion-sensitive sound interaction triggered by people walking between the speakers. The 66 interaction sounds, appearing in random order, are musical fragmentations of the two short words, multiplied, splintered, and abstracted to the point of unrecognizability. Broken down into microscopically minute particles of speech, they simultaneously reassemble themselves into noise and rhythm.

The visual and auditory layout of the installation reflects self-awareness from the perspective of a threatening initial situation: a multiple I, which is aurally and spatially mirrored in a situation of polar tension, in an ambivalent "between." Thus identity is deconstructed in the installation in two ways: in the experience of emigration, and in the experience of political division that once existed in Germany and continues to exist in North and South Korea.

Christina Kubisch

*1948 in Bremen (DE), lives in Hoppegarten (DE)

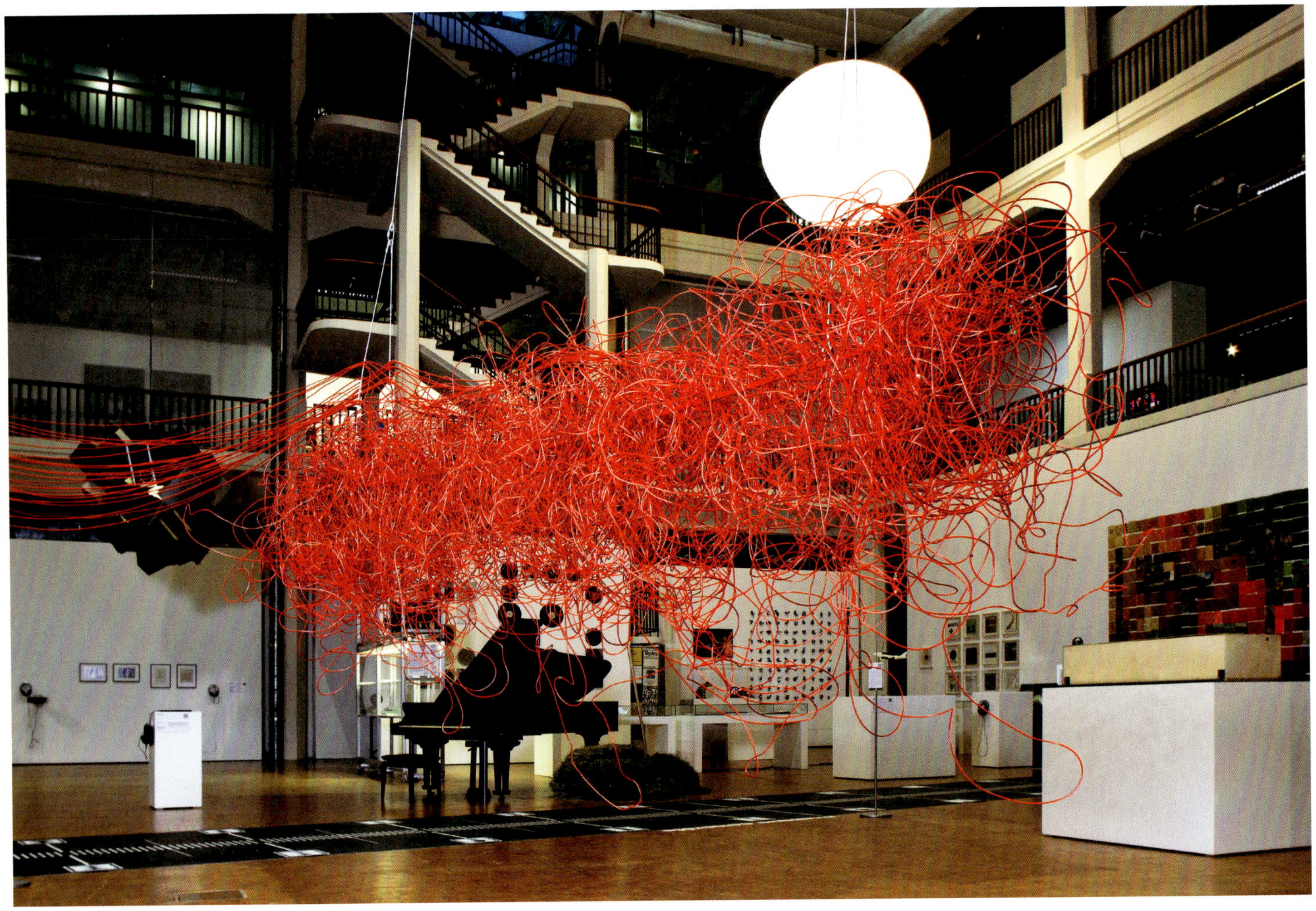

Cloud, 2012
Electromagnetic sound installation with recordings of magnetic fields in ZKM | Karlsruhe
1200 m electric cable, metal rods, induction headphones, 6 amplifiers, 6 DVD players, 12 resistors on metal plates, 240 × 340 × 150 cm
Audio engineering: Eckehard Güther
Production: ZKM | Karlsruhe
Christina Kubisch

www.christinakubisch.de

Christina Kubisch works with special induction headphones that she developed, which help make audible electrical fields that exist around us but cannot be perceived by the naked ear. The title of the electromagnetic sound installation *Cloud* denotes more than just a meteorological phenomenon. Since the mid-2000s, metaphorical data clouds have formed: public clouds, private clouds, hybrid clouds, and community clouds. These terms stand for the evolving worldwide delivery models of information technology.

A preoccupation with the visible and the invisible is one of Christina Kubisch's main themes: "Nothing sounds the way it looks. Nothing looks the way it sounds." By translating magnetic fields into the audible range, the artist enables us to experience sounds that are hidden, casually perceived, or unheard. The intensification of the recorded sound material is essential to the generation of sonic atmospheres.

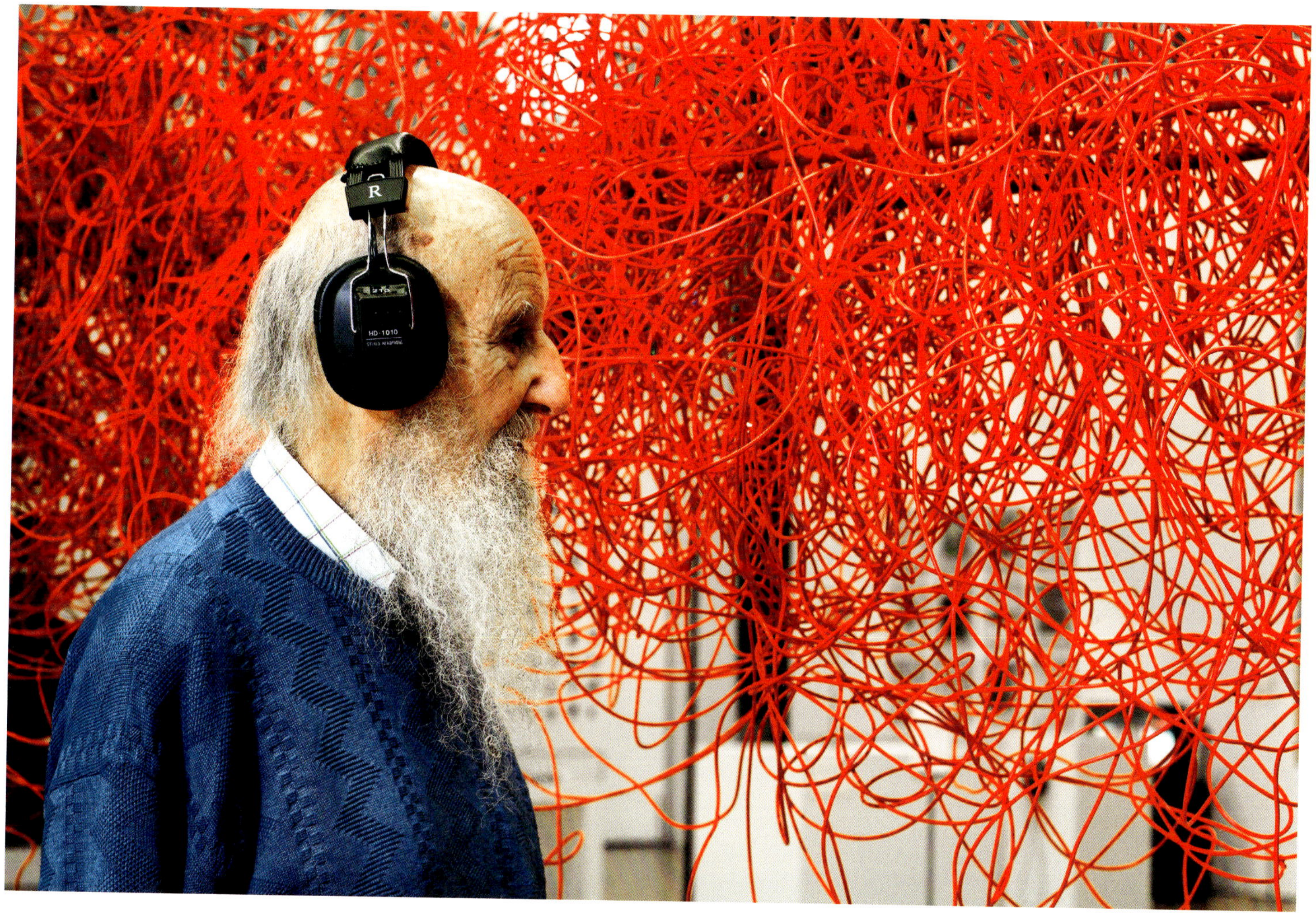

Christina Kubisch

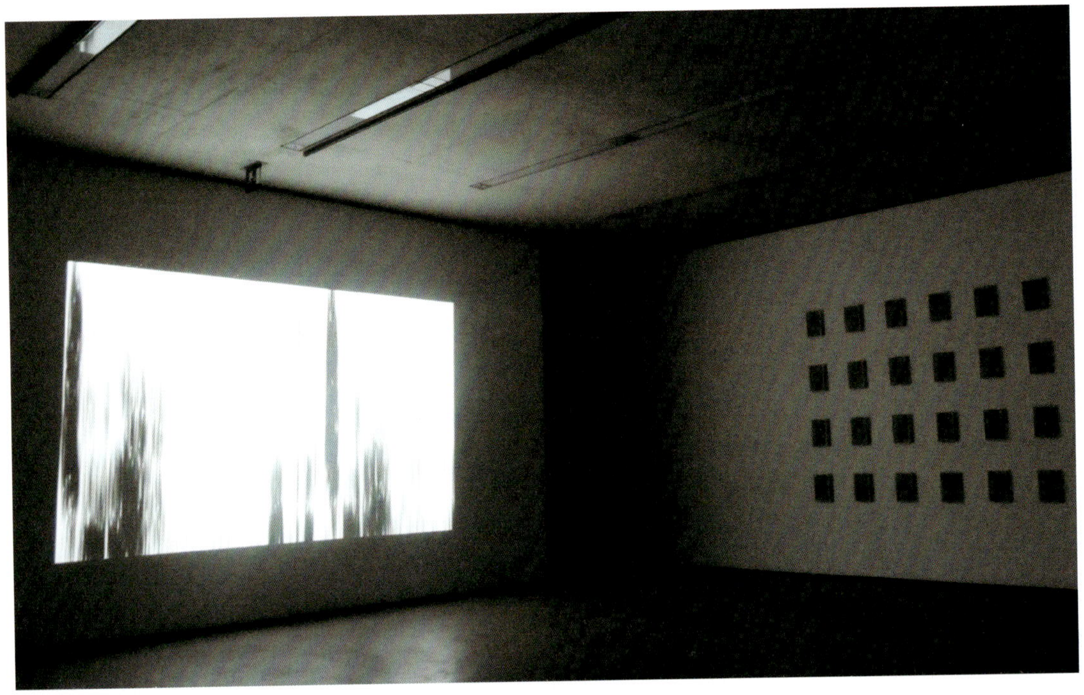

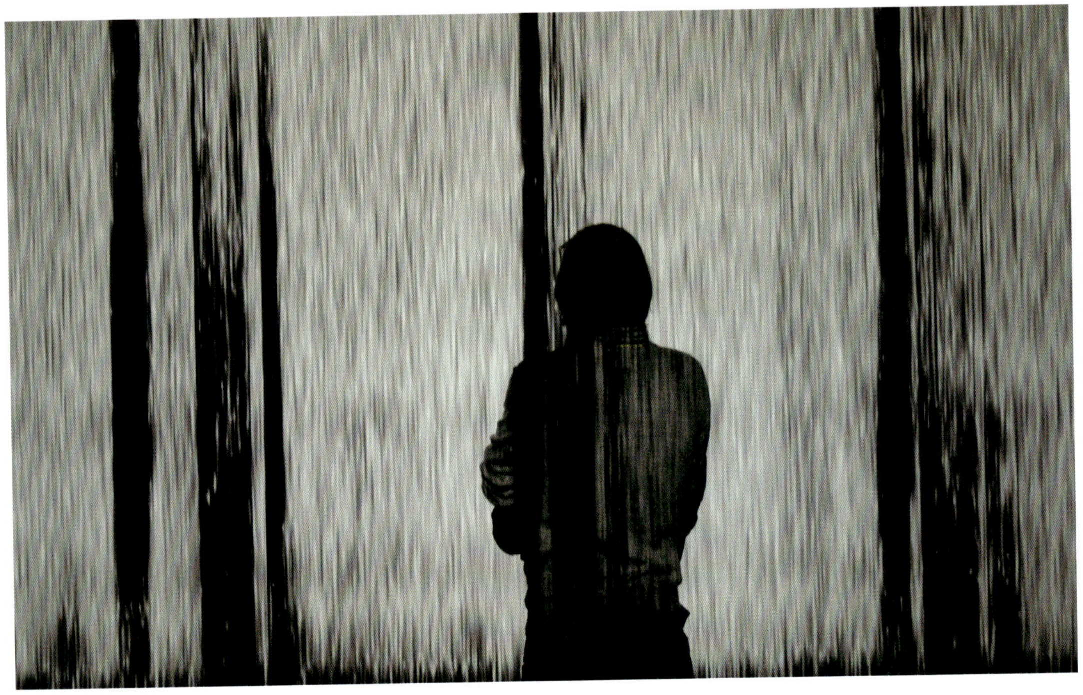

Silent Exercises, 2011–
Multi-channel installation
and film, duration ca. 120 min.

Silent Exercises is an audio-visual installation, whose presentation form is open to adapting to the architecture and acoustic character of each exhibition space. The basic material is the spoken word. For several years now I have asked people to repeat the word "silence" in their native language for a longer time and have recorded it. The meaning and the significance of the term "silence" are changing among the countries, languages and nations. Some of them have no word for silence but describe it as "the place without sound." Some words, mostly of spoken languages like those of African or Australian natives, seem magic and mysterious. Some "silences" sound similar or close to each other. The film *Silent Exercises* is a video of about two hours duration showing the sonagrams of the various recordings in black and white. They slowly fade into each other and form a quiet visual landscape which is projected not onto a screen but onto the architecture itself. Is there ever silence? (Christina Kubisch)

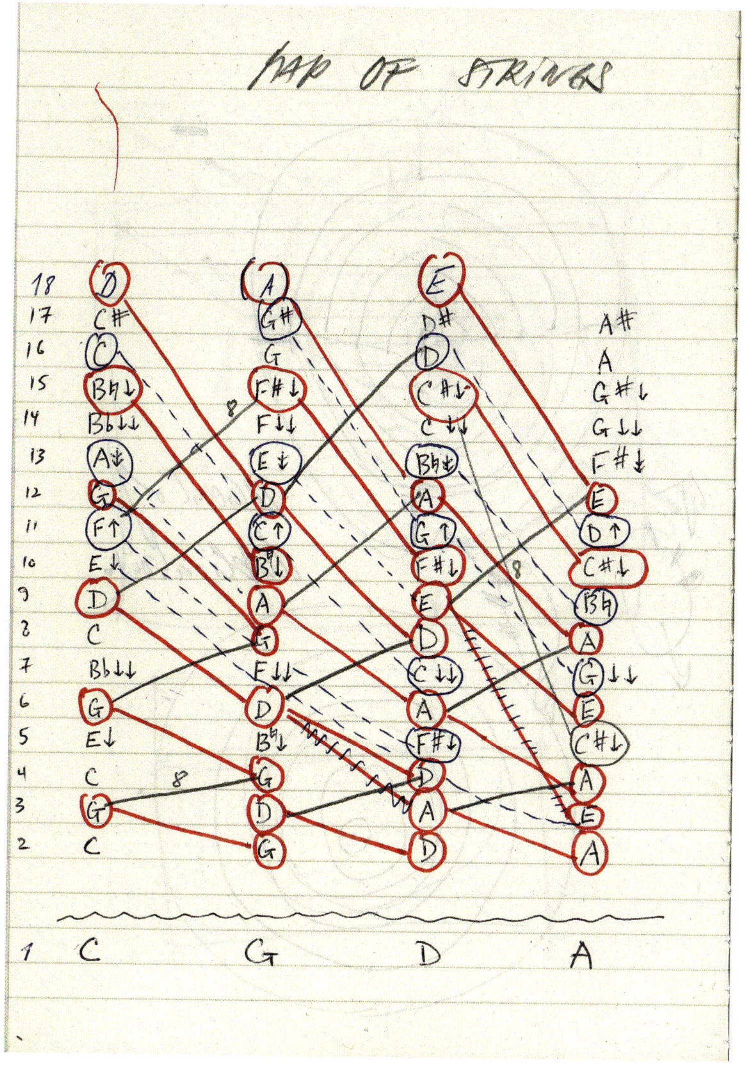

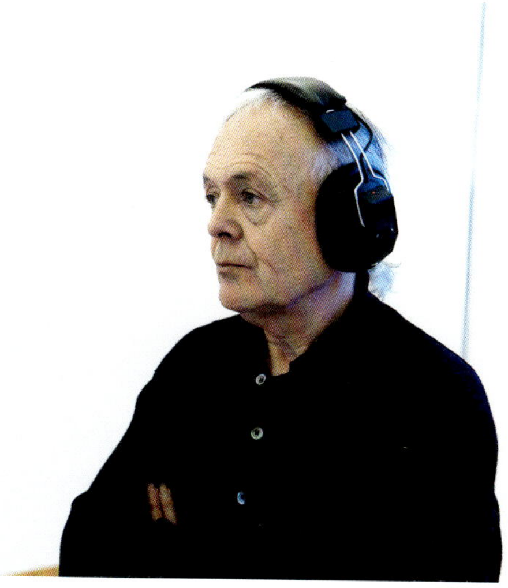

Rolf Julius listening to the sound installation *Klangkörper*.

Klangkörper (Die Zeit des Wartens dehnt sich endlos), 2009
Sound installation
Electric cables, electromagnetic headphones, 10-channel composition with the 24 flagcolet notes of the cello

Hermann Hoernes, former resident of Villa Rot, was forced to end his career as a cellist and conductor during the Nazi era because of his alleged "Jewish appearance." Nevertheless, he continued to play his cello for two hours a day and, together with his wife Feodora, who owned the house, turned Villa Rot into a center for artists who could meet, relax, and exchange ideas there.

Unfortunately, no audio documents of Hoernes playing the cello or conducting have yet been found.

Kubisch's installation *Klangkörper (Die Zeit des Wartens ist vorbei)* [Sound-emitting body (The time of waiting is over)] was set up in the former salon of Villa Rot overlooking the park, and blue electric cables formed a horizontal net above the heads of the visitors. The cables not only formed a visual structure, but were also sound carriers. All visitors were given a wireless electromagnetic headset with which they could move freely about the room.

The sounds were recordings of the flageolet tones of a cello, which were produced as overtones by the partial vibrations of a string. The recordings reproduced both the sound of the string and also the rough bowing of the bow.

Seen from the outside, one saw a silent audience moving about the room and hearing something that remained hidden from the non-participating visitor.

Seen from the outside, one saw a silent audience moving in the room and listening to something that remained hidden from the non-participating visitor.

The movements of the listeners within the sound fields gave rise continually to different overlays of the flageolet tones. They did not have the typical warm, full sound of the cello, but appeared brittle, fleeting, and distant.

Christina Kubisch

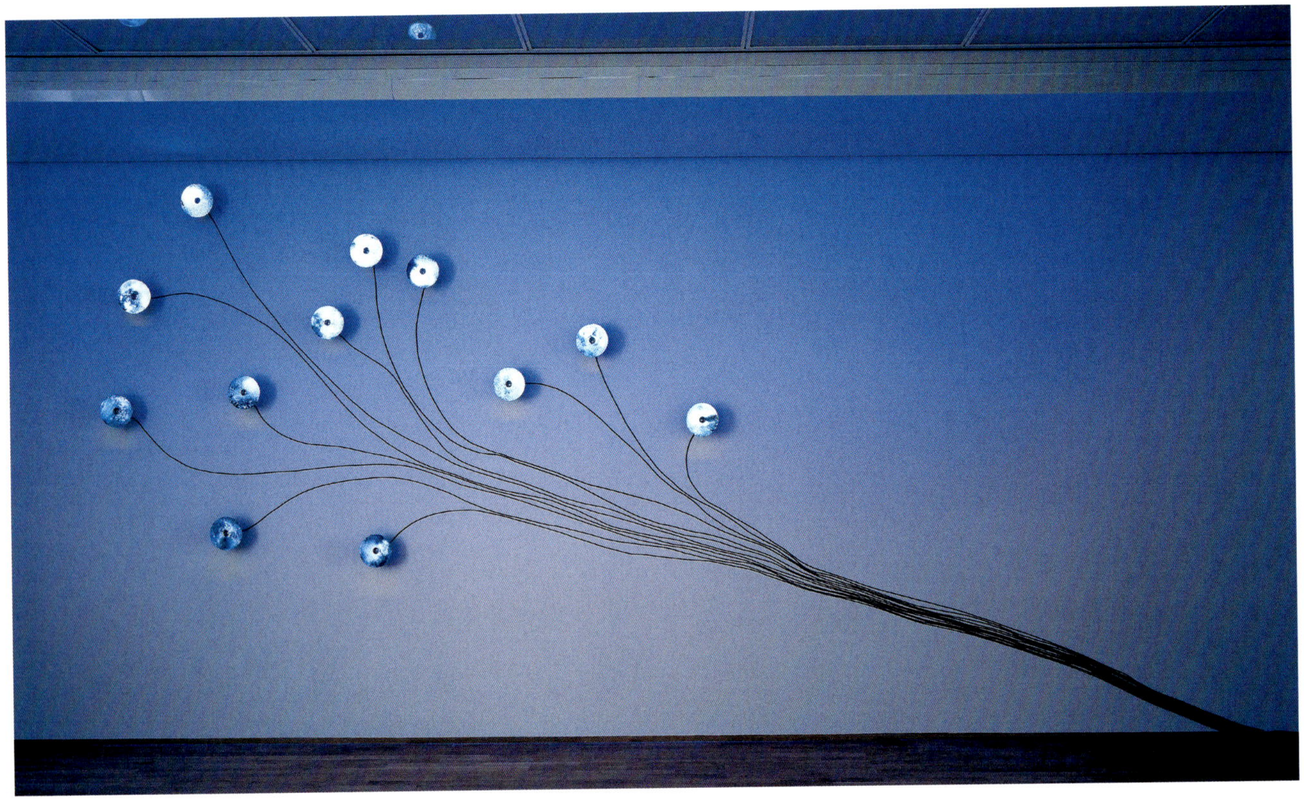

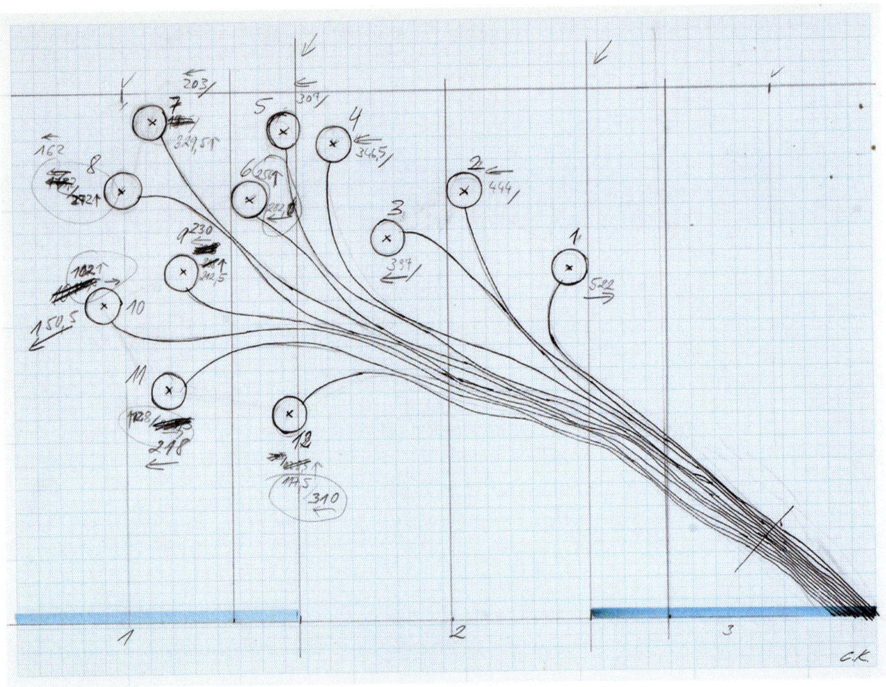

The True and the False I, 1992
Sound sculpture
Grinding disks, pigment, cable,
black light, 12-channel composition
Collection Hamburger Bahnhof,
Museum für Gegenwart

In the 1990s, Christina Kubisch developed solar-controlled techniques that influence sounds and sound sequences depending on how the light falls. *The True and the False I* is one of these works.

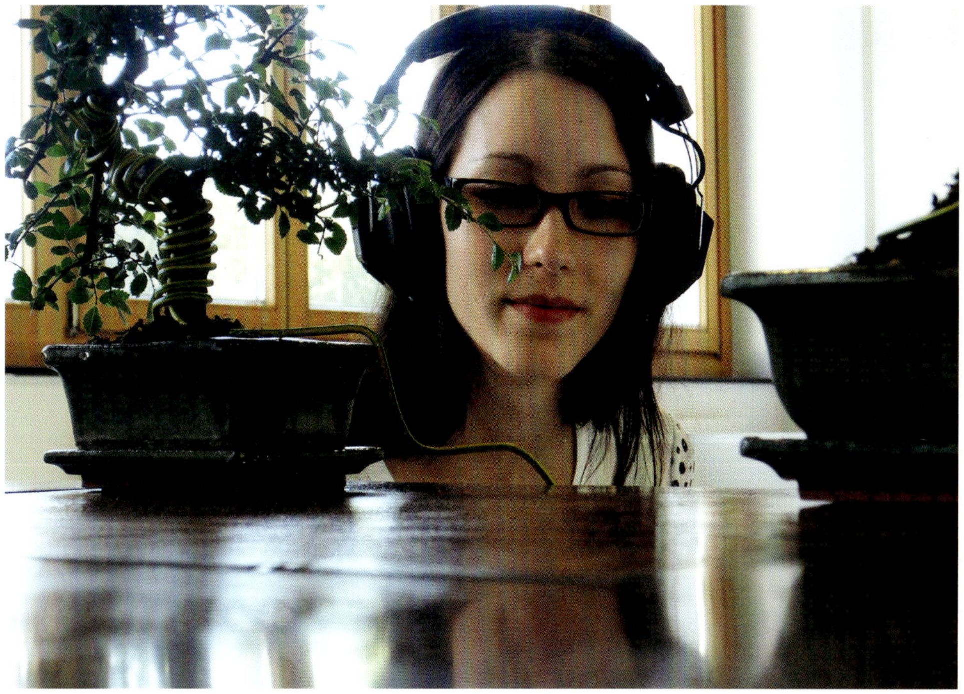

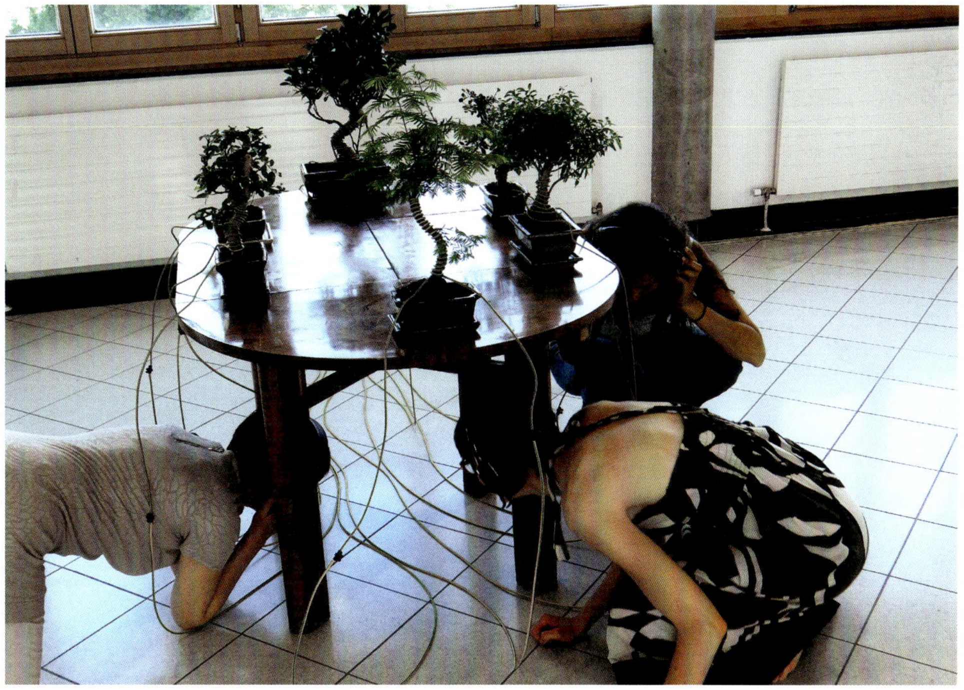

Die Konferenz der Bäume, 1988
Sound installation
Conference table, bonsai trees, ground cable, electromagnetic headphones, 5-channel composition

In the sound installation *Die Konferenz der Bäume* [Conference of the trees], five bonsai trees of different sizes and types are placed at regular intervals on a round conference table. Their small trunks are wrapped with a spiral of yellow-green electrical cable in the same way that young bonsais are usually shaped with thick wire in tree nurseries. The cables lead down to the ground; they hang root-like over the edge of the table and connect to a wire on the floor.

Visitors are provided with an electromagnetic headset with which they can listen to the trees as they walk around the table. Depending on the speed of movement and the proximity to the bonsais, the sounds merge faster or slower. Each bonsai is assigned a "characteristic" sound. One hears a whisper, a noise, etc. All the sounds are processed field recordings.

Kalle Laar

*1955 in Germany, lives in Krailling (DE) and Vienna (AT)

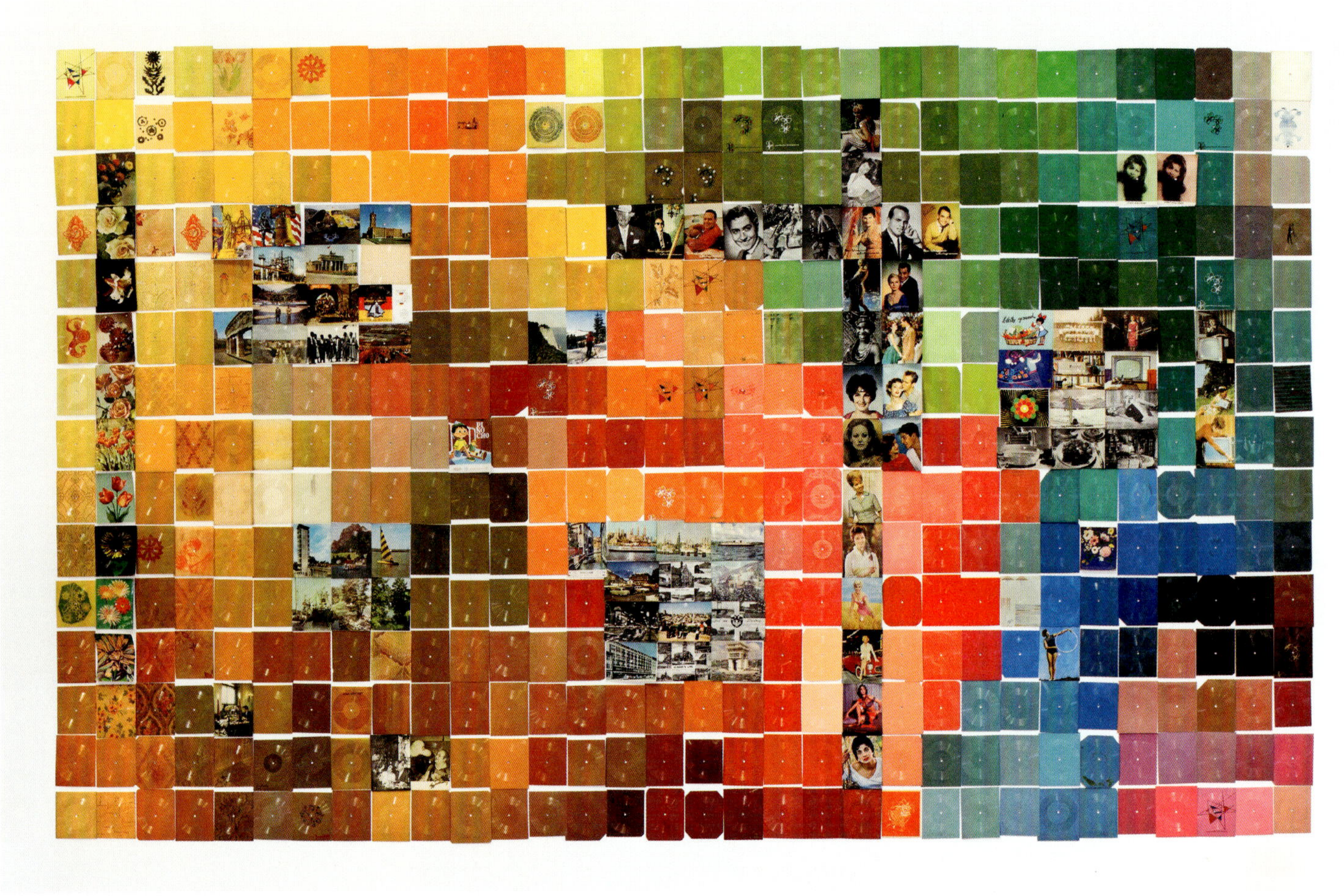

Klangpost 4, 2012
Installation
Sound postcards, video, wireless headphones, 300 × 500 cm, video ca. 120 min.
Production commissioned by ZKM | Karlsruhe
Kalle Laar

www.soundmuseum.com

Klangpost 4 [Soundmail 4] consists of a collage and a video. A tableau composed of more than 400 sound postcards hangs on the wall. The sound artist, radio dramatist, and DJ Kalle Laar collects these playable postcards, now a thing of the past. In this installation he uses just the plastic ones produced in Poland between 1960 and 1980. On the wall, the cards form an ensemble designed with only visual criteria in mind. The video conveys the acoustic messages of the lo-fi audio medium. In the video, each individual postcard is played for 10 to 20 seconds, one after the other, on a record player.

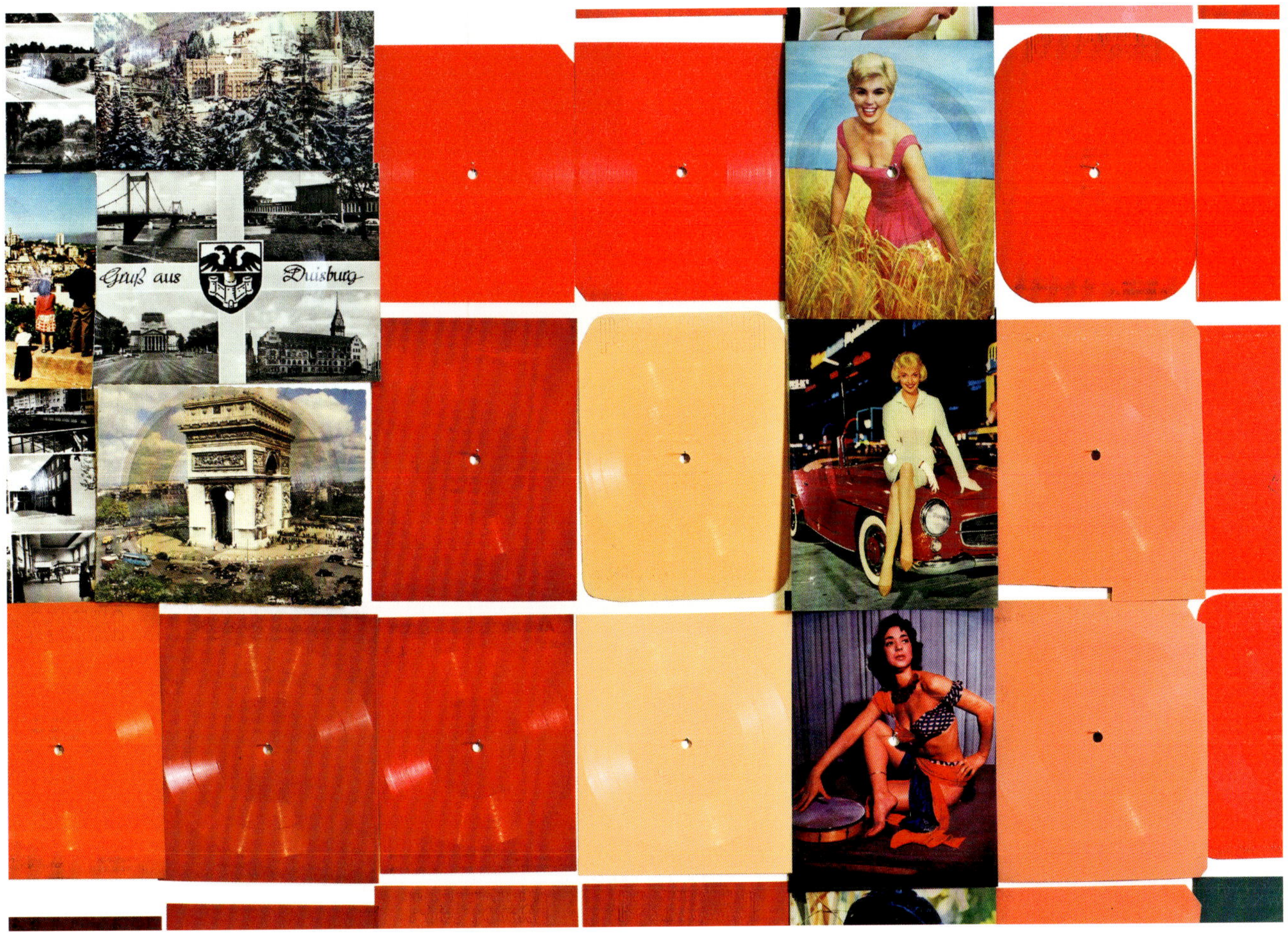
Gruß aus
Duisburg

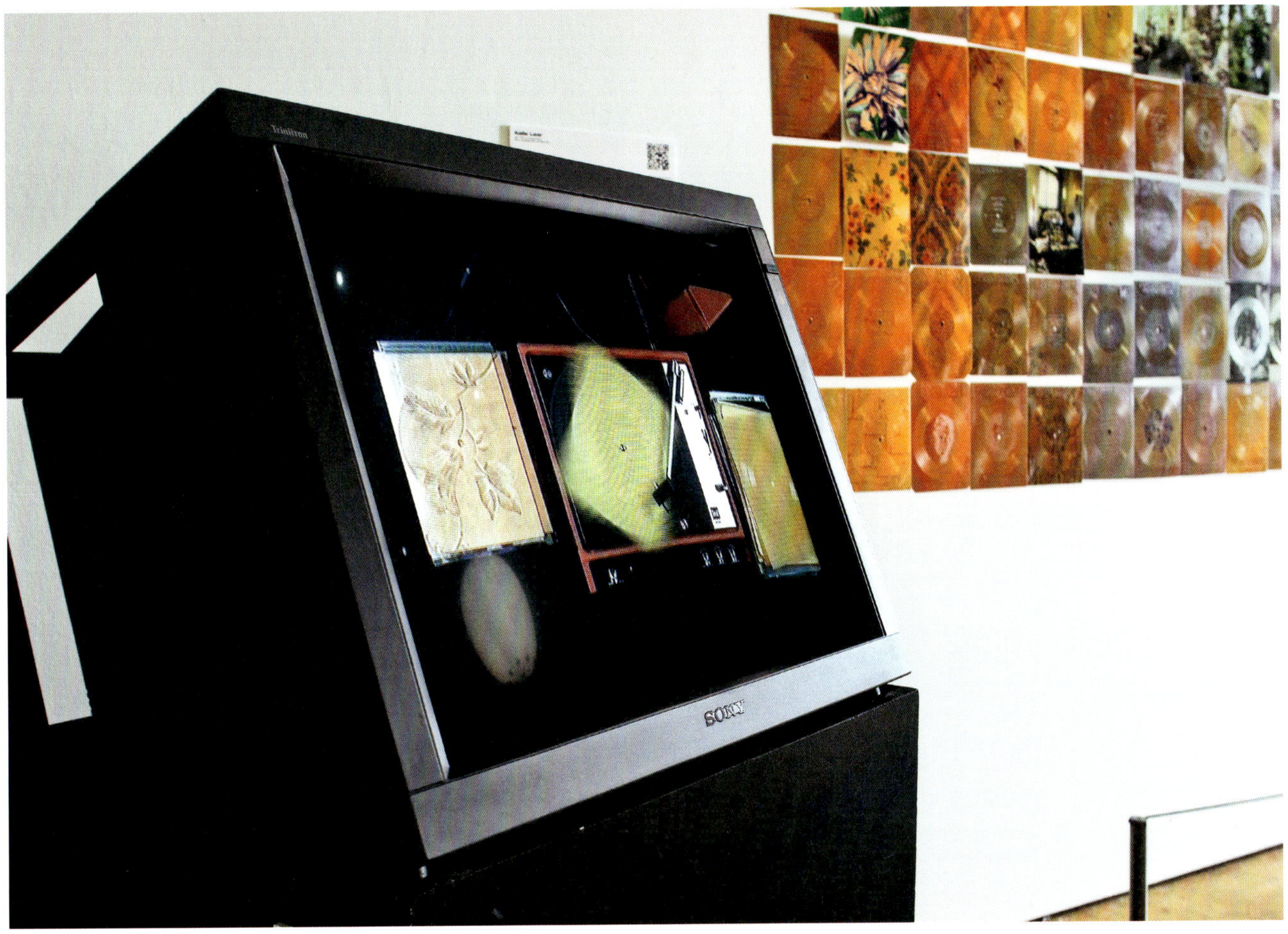
SONY

Bernhard Leitner

*1938 in Feldkirch (AT), lives in Vienna (AT)

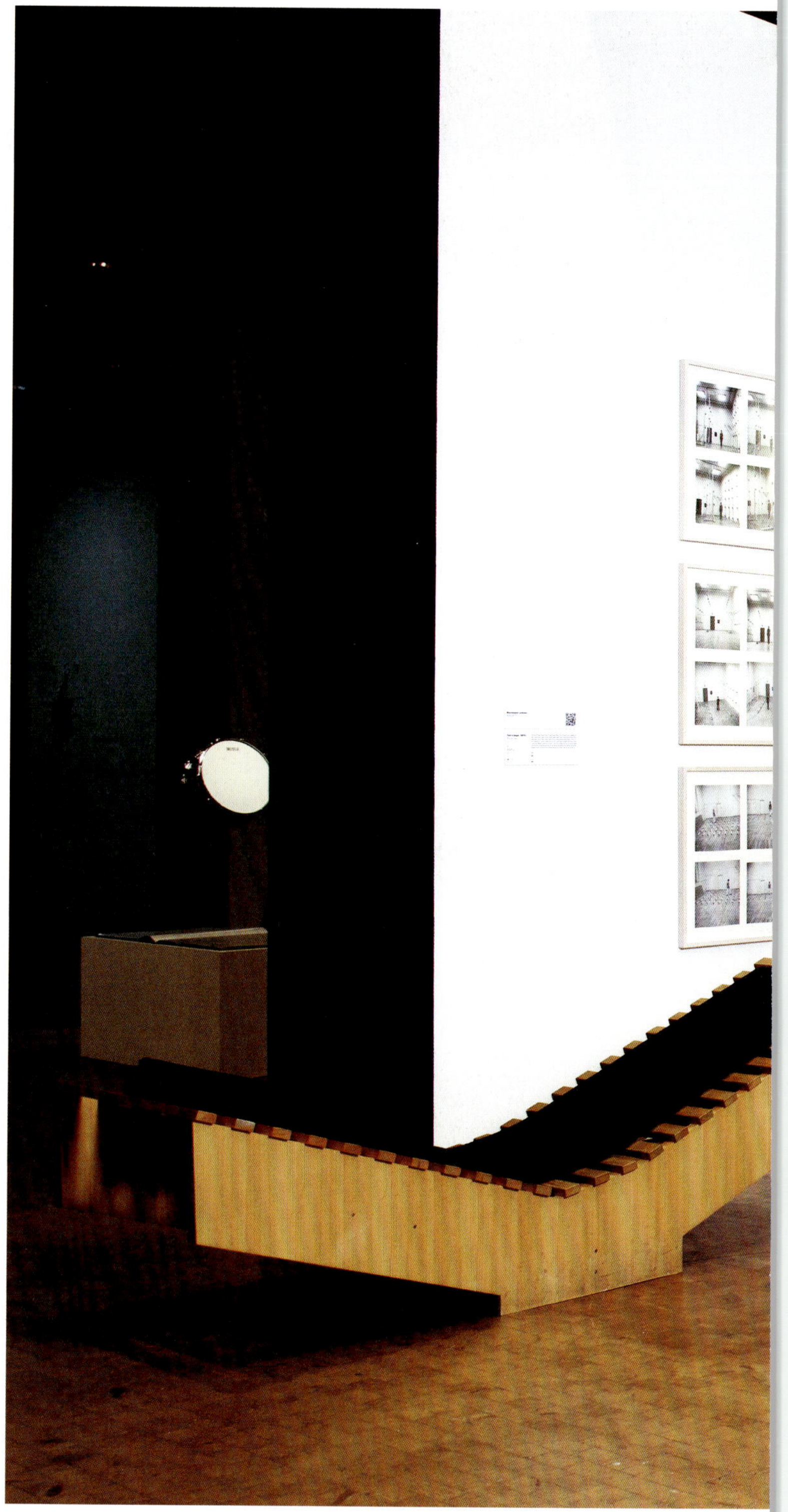

View into Bernhard Leitner's exhibition space at the exhibition *Sound Art. Sound as a Medium of Art*
From left to right:
Ton-Liege 1975 / Sound Chair 1975, 1977
TON-RAUM-UNTERSUCHUNGEN / Sound-Space-Investigations, 1971–1974
Variationen einer Gang-Röhre / Variations of a Sound Tube, 1973
StahlFederWellen / SteelSpringWaves, 2003
Pulsierende Stille / Pulsating Silence, 2007

www.bernhardleitner.at

In New York in the late 1960s, the Austrian architect and sound artist Bernhard Leitner began to design his first sound-space sculptures. Initially these remained sketches and models. It was not yet possible to technically realize sound architecture using multiple audio channels.

Bernhard Leitner's first walk-in sound-spaces, whose visible architecture consists of wooden slats and loudspeakers, were *Sound Lock*, *Sound Gate*, and *Sound Tube*, created in 1971. In these sound-space sculptures, he worked with architectonic, visual, choreographic, and musical parameters. The acoustic potential of the soundspace sculptures lies in a variety of possible sound-spaces and motions, which Leitner sketches in scores for the spaces.

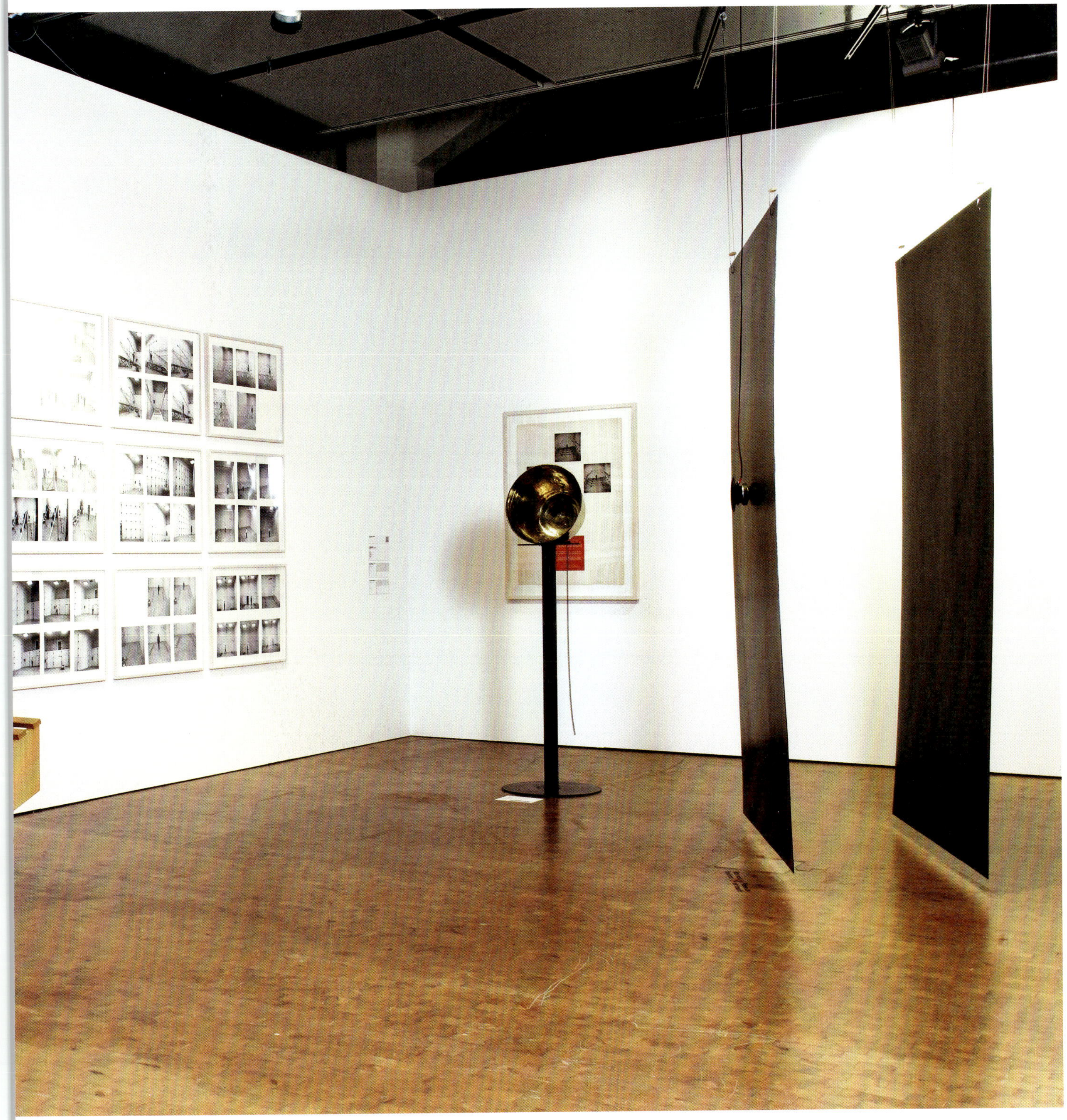

right side:
Pulsierende Stille / Pulsating Silence, 2007
Walk-through sound-space sculpture
Composition 1: tuba multiphonics
Composition 2: sound frequencies
74 Hz and 85 Hz
Bernhard Leitner

The two suspended sheets of metal create a passage-like architecture. A loudspeaker chassis is attached to the outside of each one at a precisely determined point. The loudspeakers transmit a beat frequency to the metal. The beat consists of the interference between two frequencies (74 and 85 Hz) that lie beneath the human auditory threshold.

The vibration of the metal panels reciprocally affects and defines the enclosed area, transforming the narrow corridor into a softly pulsating acoustic space that not only envelops but also penetrates and pervades the body. Walking through the pulsating interior is a deeply acoustic and sensuous experience, while outside the sculpture stillness reigns.

Bernhard Leitner

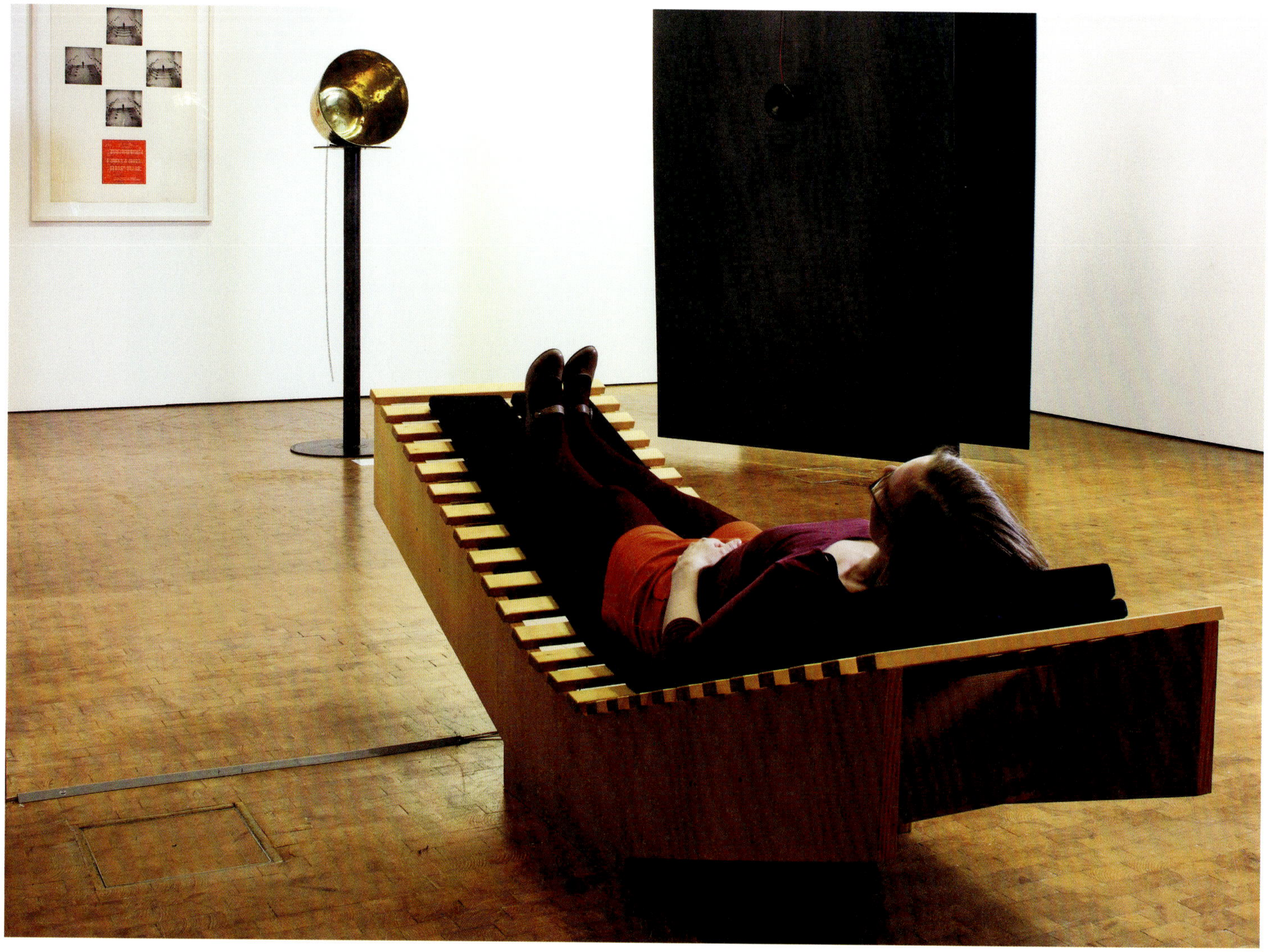

Ton-Liege 1975 / Sound Chair 1975, 1977
Sound-space composition
Object 2006
Bernhard Leitner

On the *Ton-Liege 1975*, the listener's body lies in the axis of sonic motion. Two loudspeakers under the lower back and calves of the reclining body define the soundspace. Processed cello and horn sounds sweep the body, moving back and forth so that the sound shifts into the body. The body itself – with legs and feet propped up – becomes a vibrating form. The uniformly monotone sonic sweeps enfold the body and alter self-perception. Bernhard Leitner here investigates the threefold relationship of sound, space, and body.

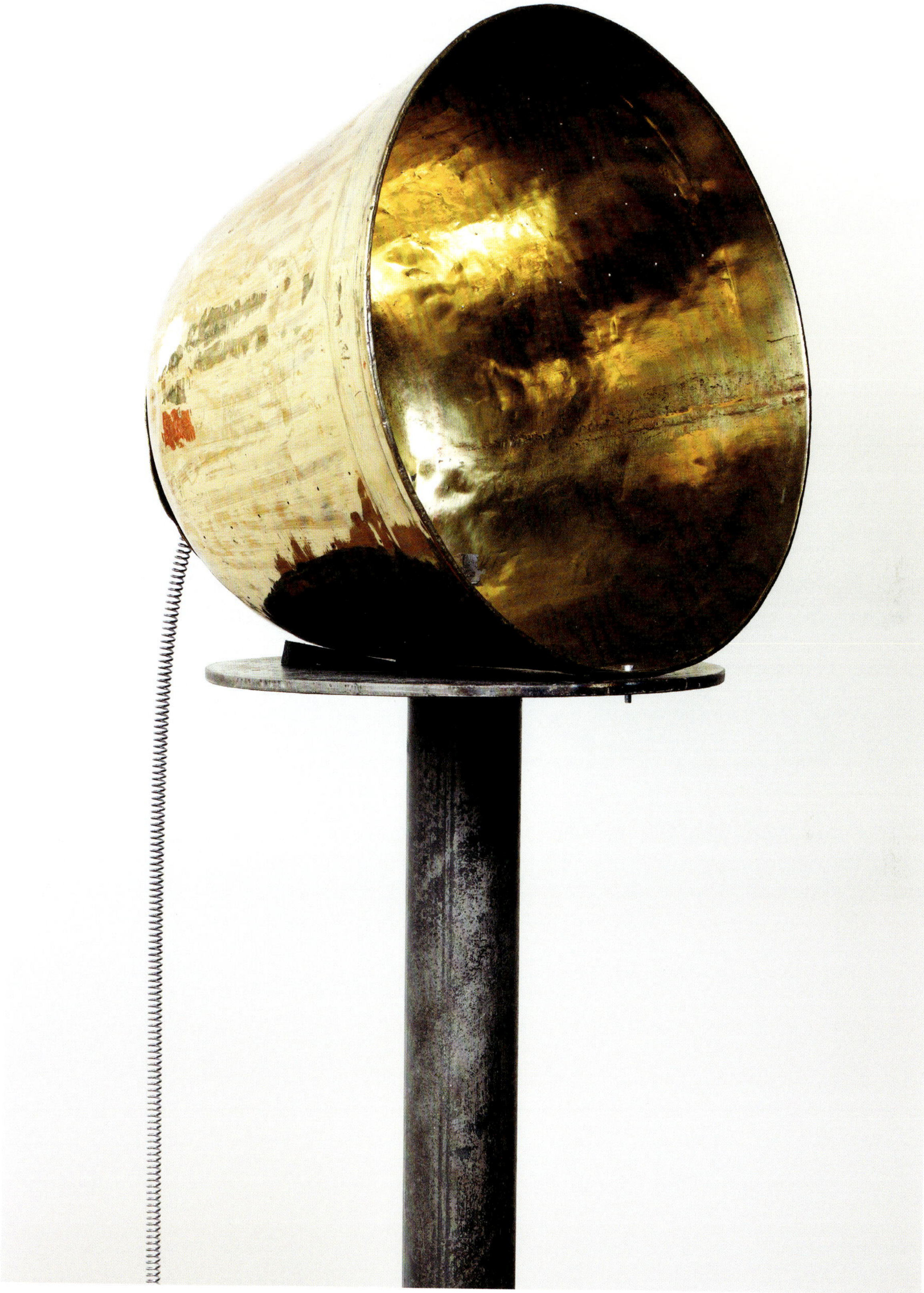

StahlFederWellen / SteelSpringWaves, 2003
Sound-space sculpture
Bernhard Leitner

A plain steel spring hanging down from the back of a brass bowl delivers the sound of this sculpture. If the bottom third or end of the spring is lifted and then allowed to drop, complex mechanical movements comprised of transversal and longitudinal vibrations arise. As with a string instrument, the waves are transferred to a resonator. Here they transform the interior of the brass vessel, which lies horizontally on the steel pedestal like a megaphone, into a powerfully resonating soundspace.

Bernhard Leitner

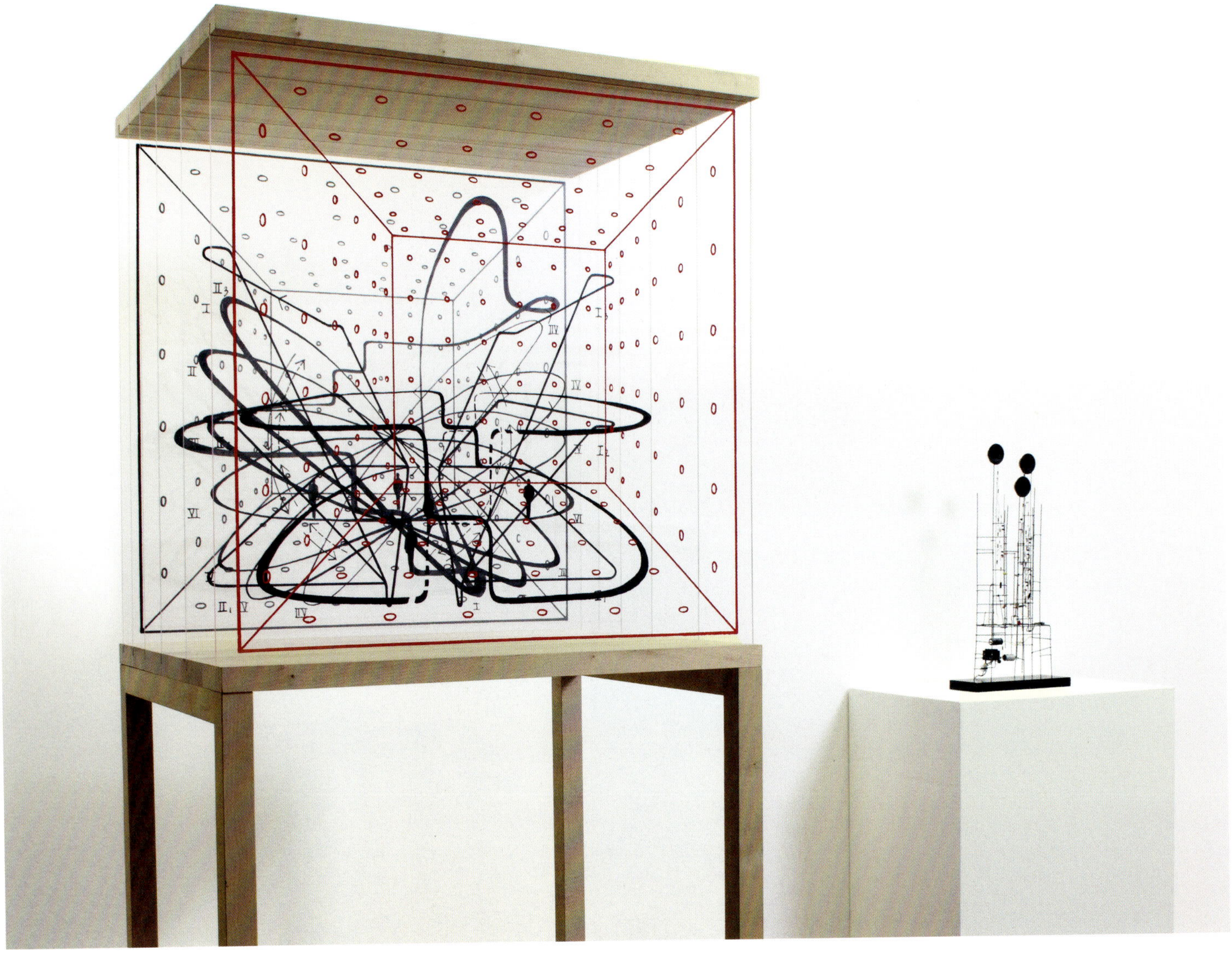

SOUNDCUBE 1970.
Vier Ton-Architekturen, 2008
Notation sculpture
Bernhard Leitner

"Movements of sound as tool to create and to characterize space. The soundcube allows one to achieve this. 64 loudspeakers on each of the sides of the soundcube. The sound is programmed to travel from loudspeaker to loudspeaker," Bernhard Leitner noted this description under a 1969 drawing of the *SOUNDCUBE 1970*, which was designed to heighten people's acoustic and physical experience of sound. But the soundcube is just the medium; the artwork is complete only with its corresponding sound-space program. These scores establish the positions and movements of mostly simple sounds. It's a matter of points, lines, planes – and spaces, too. One of the variables is the audience, which moves freely through the three-dimensional composition in the cube. In the four "sound architectures" printed on Plexiglas, Leitner noted the spatiotemporal routes of the sounds and the possible positions of the listeners.

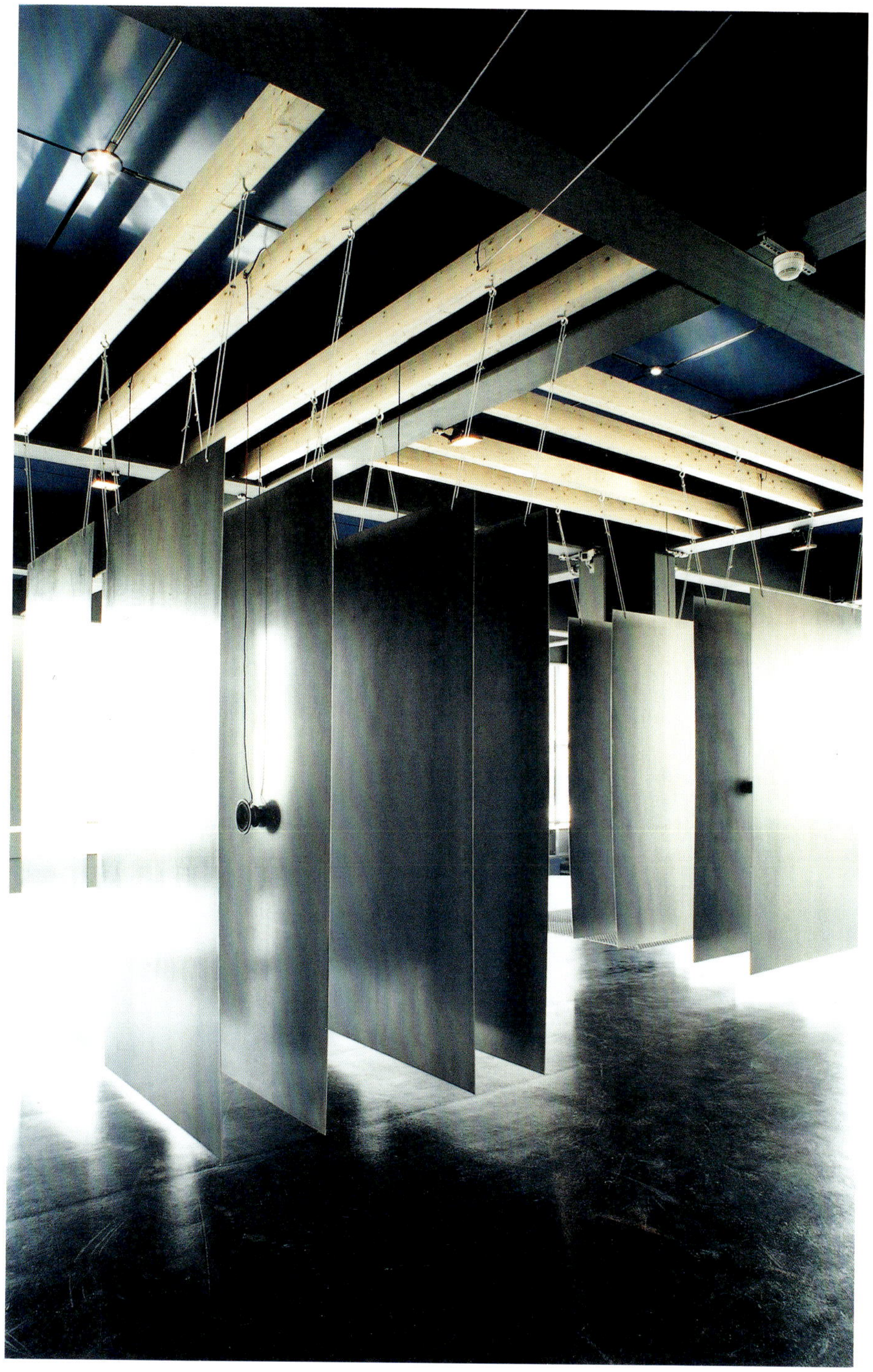

Tuba Architecture, 1999
Sound installation, installation view, ZKM | Karlsruhe, 2002
60 metal sheets, loudspeakers, dimensions variable
Bernhard Leitner

Sixty metal sheets hanging from the ceiling (hot rolled industrial metal sheets, 3 mm thick, 2500 mm long, 1500 and 1250 mm wide) form the walls of a passage-like, acoustic architecture. A tuba multiphonics composition (bass tuba with overtones) sets the ramified architecture into acoustic vibration to the point that all surfaces resonate in unison and parallel in time. Loudspeakers magnetically mounted in the central areas of the metal sheets transmit the sound (building) material to the steel surfaces. Thereby the metal sheets are transformed into vibrating resonance walls dissolving the boundaries of the narrow passage spaces in between the walls by contracting and expanding it. In the pauses of the sound (building) material (lasting up to fifteen seconds) the passages gradually fade out in the time axis of the architecture. In contrast to the static tuba material of a consistent tone, a staccato-like trombone sequence imparts rhythmic pulses to the walls, which then reverberate in waving echoes.

The heavy metal sheets suggest weight, mass, and timelessness to the eye; in their sound they generate spaces that are limited in time, acoustically body-intensive, yet repeatedly tending towards lightness and weightlessness: an architecture that dissolves itself into pianissimo.

Anestis Logothetis
Julia Logothetis

*1921 in Burgas (BG), †1994 in Vienna (AT)
*1945 in Vienna (AT), lives in Vienna and Athens (GR)

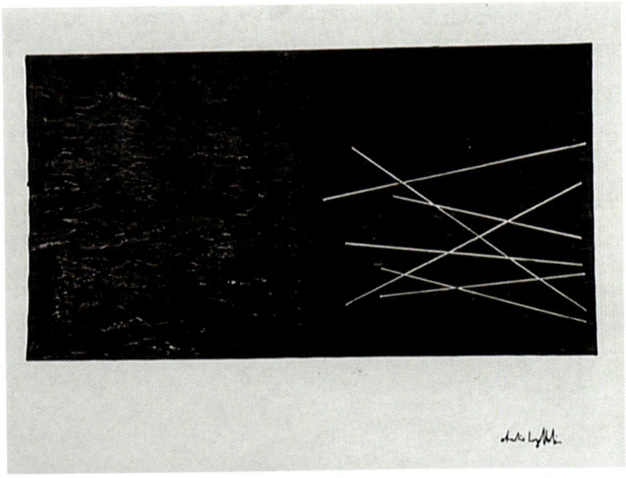

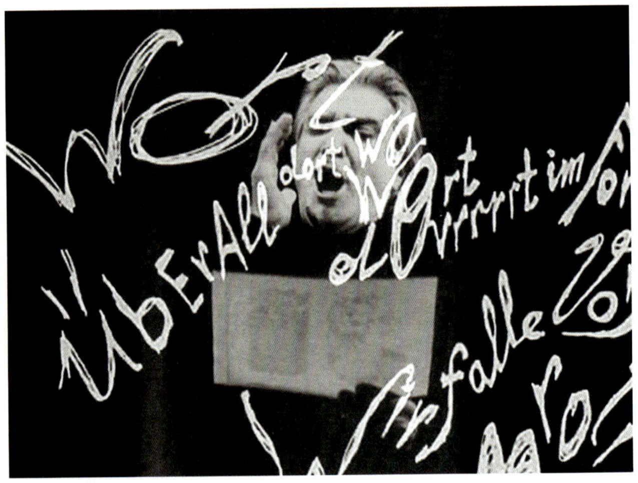

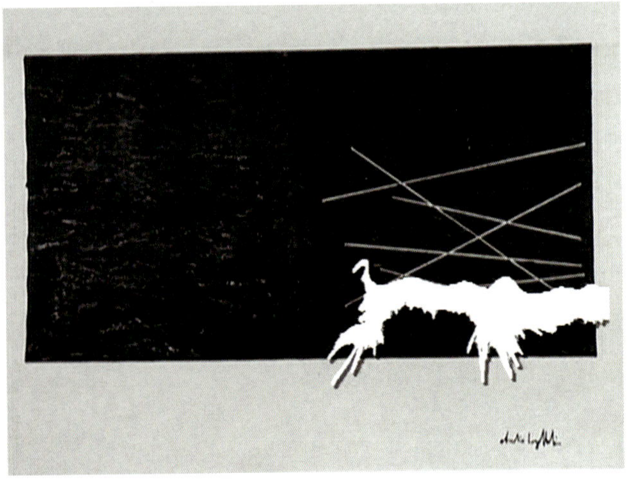

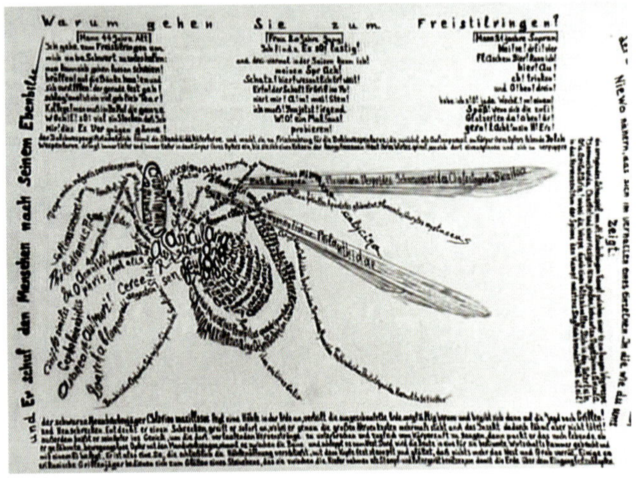

erSCHAUTE KLÄNGE, aus *Chor der Puppen aus Karmadharmadrama*, 1967/2002
Video, 13 min.
Julia Logothetis

www.logothetis.at

The video *erSCHAUTE KLÄNGE* [heLOOKED SOUNDS] is the centerpiece of a video installation by the painter Julia Logothetis about the graphic notation of her father, Anestis Logothetis. In 1959 he designed a notation "whose symbols for notes embody the characteristics of the sound and sonic events. The symbols and ways of reading them, in light of their variously oriented associative affinities, also depict the movement of sounds in space." The composer wanted to create notation that could portray an intended sonic event, including the expressive gestures involved.

Julia Logothetis assumes that the viewer must move in order to do justice to the integration of eye and ear. In *erSCHAUTE KLÄNGE* she animates her father's graphic score to the *Chor der Puppen aus Karmadharmadrama* [Dolls' Chorus from Karmadharmadrama] and synchronizes it with the sound recording.

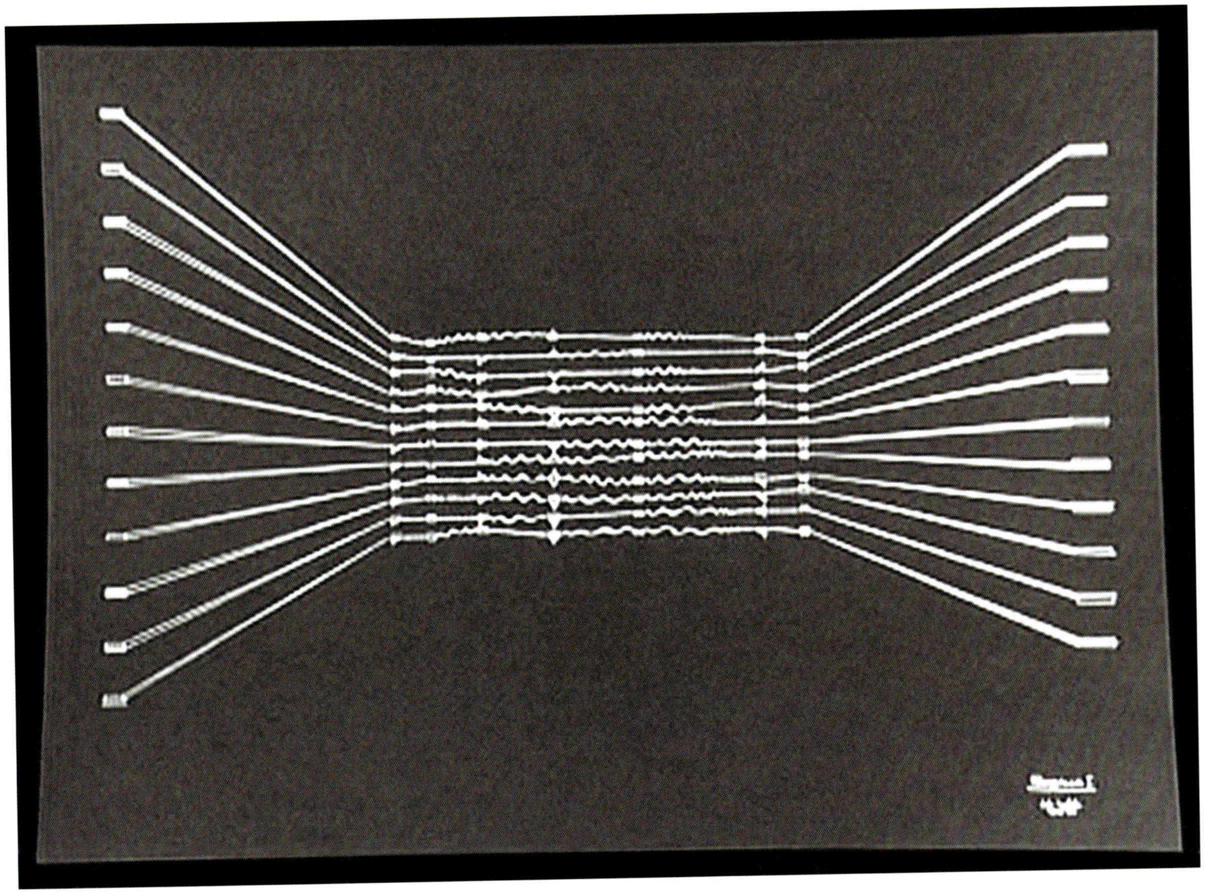

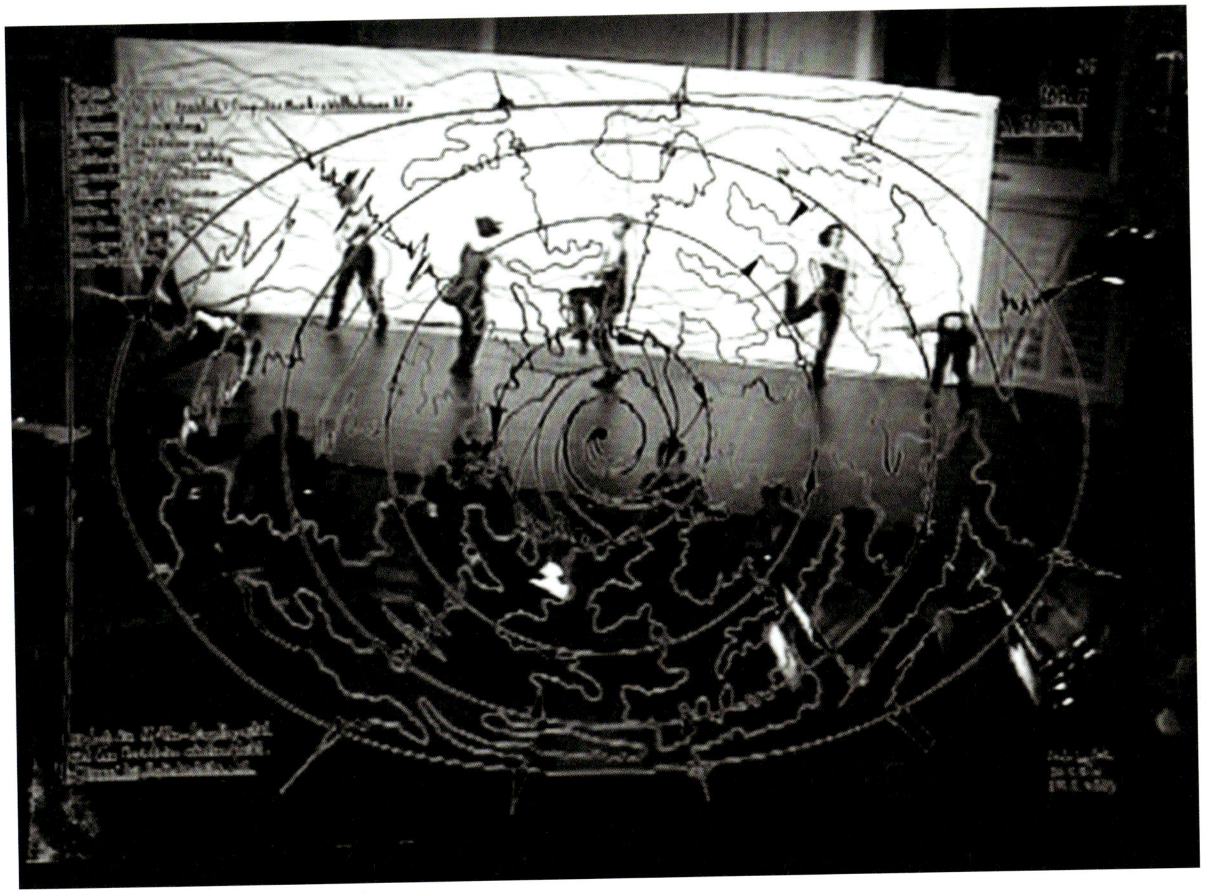

Alvin Lucier

*1931 in Nashua, NH (US), lives in the USA

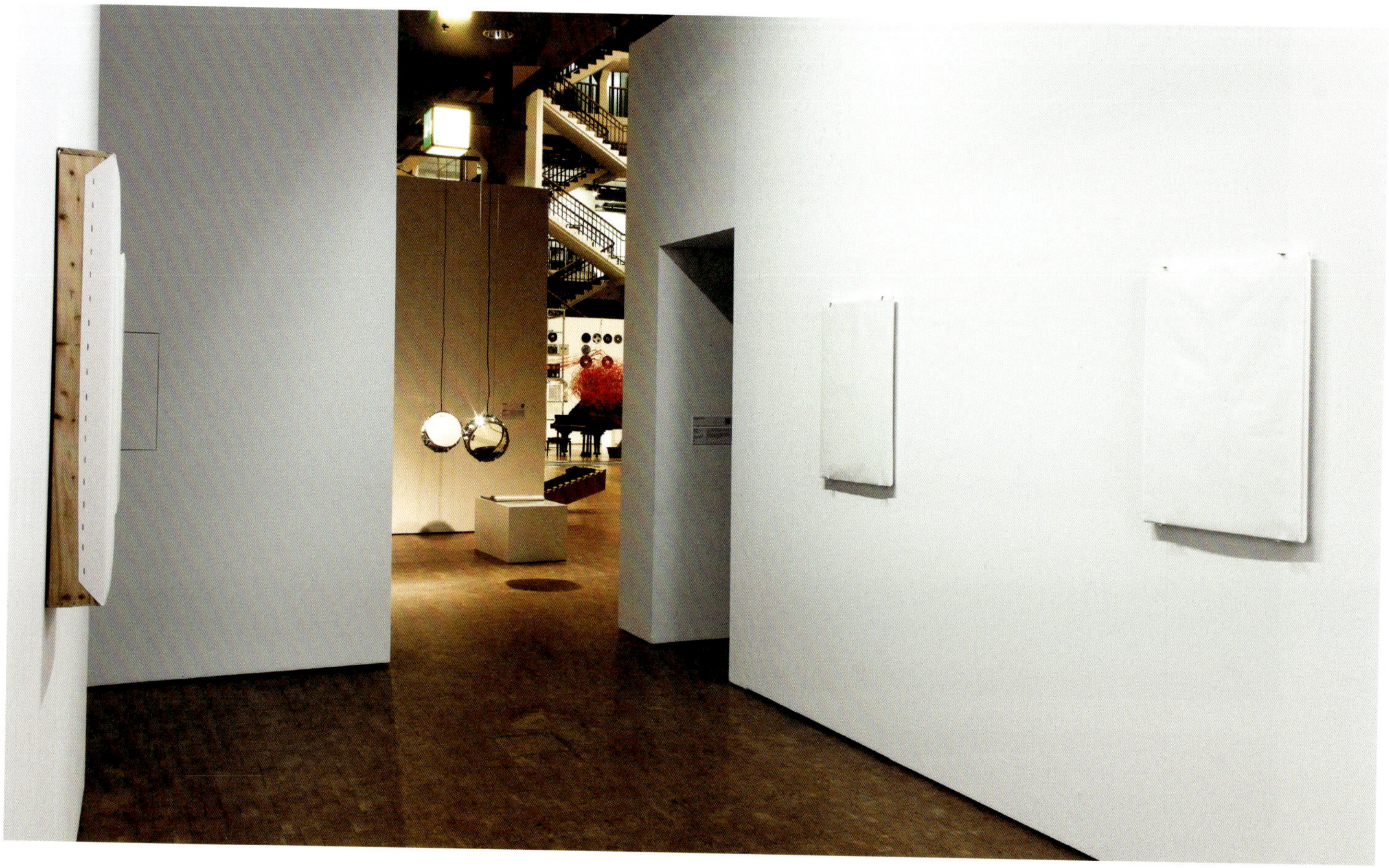

Sound on Paper, 1985
Installation
Wooden and metallic frames, watercolor and rice paper, loud-speakers, computer, 60 min. (loop)
Realization: ENSEMBLE UIACX (University of Indianapolis | Athens Campus eXperimental), Keith Moore, Michael Glover, Delaney Stockli, Christos Vernardos, Ioanna Korakis

www.alucier.web.wesleyan.edu

Different kinds of paper are stretched on two wooden and two metallic frames. Behind each frame, a loudspeaker is installed. Each emits same low frequency sine tone, which makes the paper vibrate audibly and visibly, in turn coloring and projecting the tone. The frequency sent to all the loudspeakers is identical, between 32 and 44 Hz, but each kind of paper reacts differently to the signal.

Only the loudness of the signal sent to the four speakers changes independently; each speaker has its own timetable. So the objects are heard not only as a quartet, but also in every other possible combination: as trios, duos, solos, and at varying intensities. *Sound on Paper* was first realized as a concert piece that accompanied a dance performance, with Alvin Lucier controlling the amplitude of the signals live.

The installation for ZKM | Karlsruhe was realized by ENSEMBLE UIACX from the University of Indianapolis (US) in collaboration with Alvin Lucier.

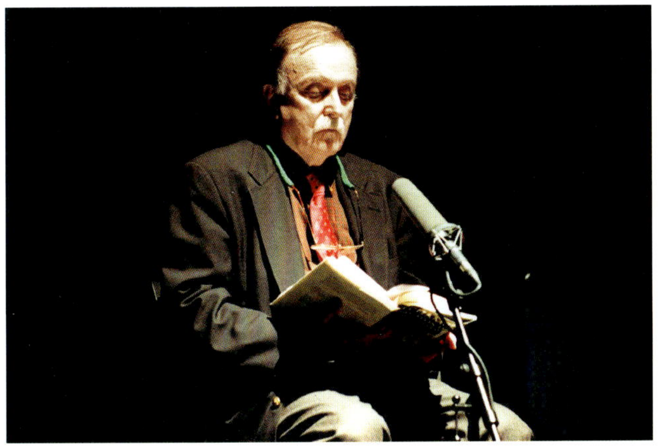

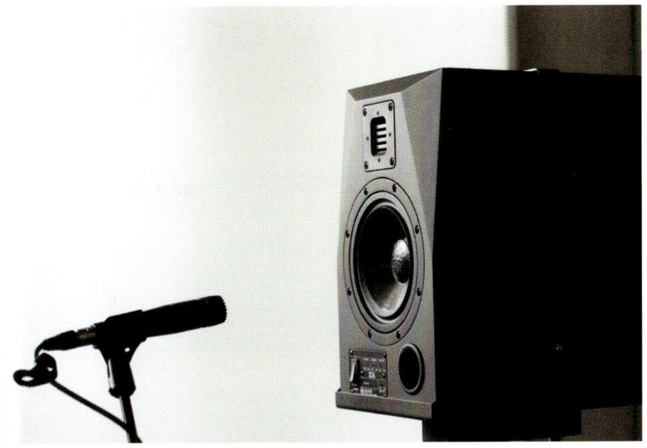

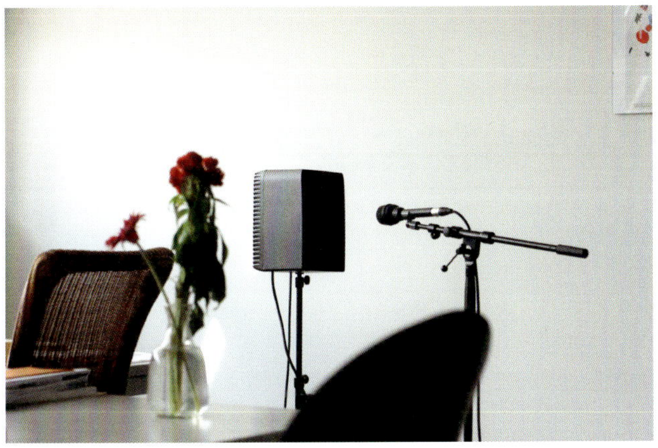

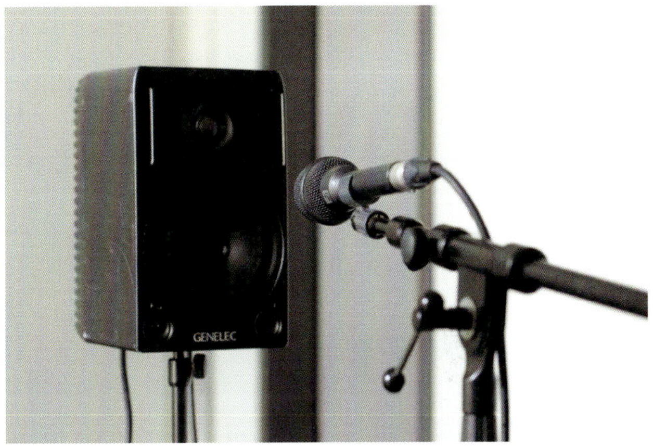

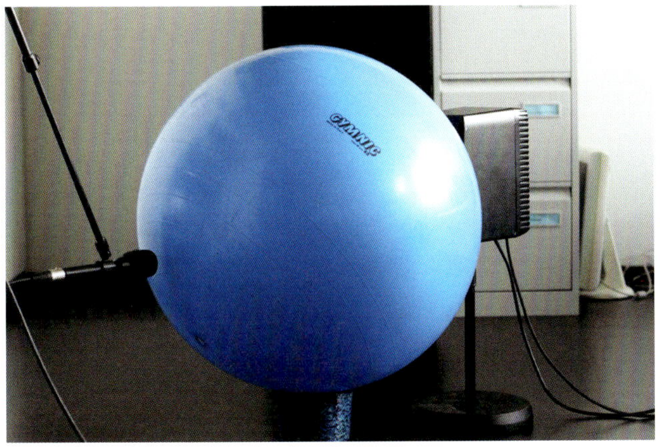

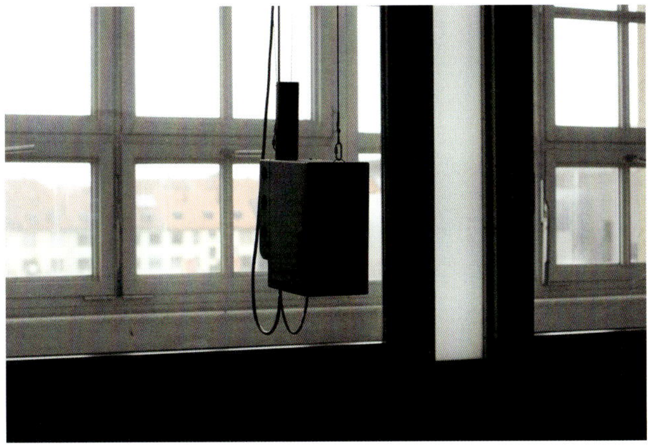

I Am Sitting in a Room, 1969/2000
Composition for fixed media (acousmatic composition)
Concept: Johannes Goebel, Matthias Osterwold
Audio network: Alexander Noelle, Bernhard Sturm
Sound projection: Christian Venghaus, Thomas Sauer

I Am Sitting in a Room was composed and also premiered in 1969. Alvin Lucier recorded a text spoken by him that describes the circumstances of the production: "I am sitting in a room different from the one you are in now. I am recording the sound of my speaking voice and I am going to play it back into the room […]." The recording is immediately recorded again via a microphone as it is played back in the recording room, creating a "second-generation" recording. This recording is then again played back and recorded, and the process is repeated with subsequent recording generations. The spoken words slowly become incomprehensible with each new recording; a sound tape is created that merely maintains the rhythm of the spoken words as sound compositions change. The sounds of the original speaking voice that correspond to the resonances of the room slowly increase and come to the fore, creating the specific color of the sound tape while other language sounds that do not find a "partner" in the room fall silent. The resonances of the room are sounds that are determined by the room dimensions. The characteristics of the room surface determine how strongly sounds assert themselves. The newly conceptualized concert version for the ZKM | Karlsruhe was performed live and integrated eight rooms: Alvin Lucier read his text at the ZKM Media Theater where the audience was present as well. This recording was played back and recurrently recorded in eight rooms of the ZKM that differ according to volume and surface characteristics while the recordings were transmitted back to the media theater via eight loudspeakers. The audience was, therefore, able to track how the rooms articulated at different speeds and in different tones while "performing their concert."

Christian Marclay

*1955 in San Rafael, CA (US), lives in New York City, NY (US)

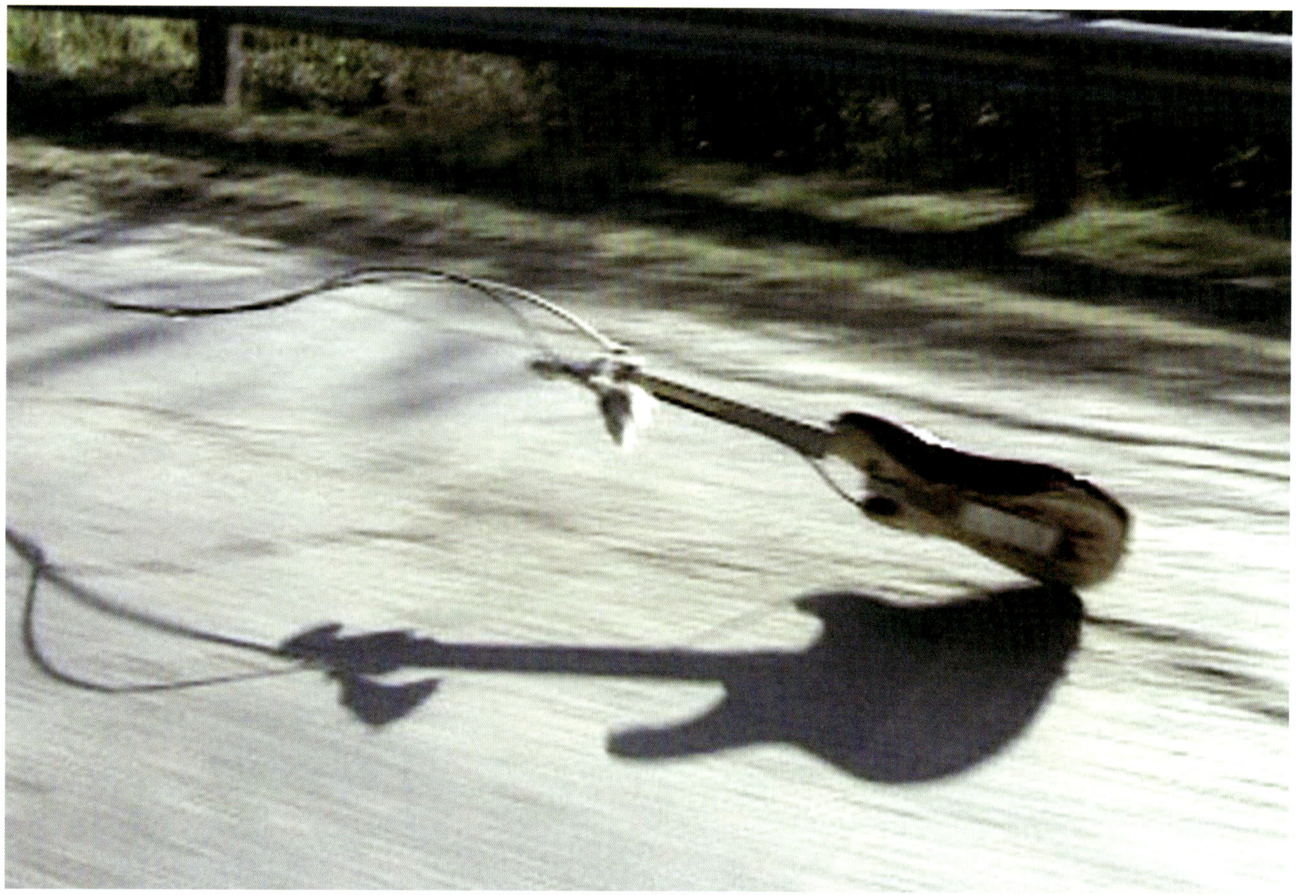

Guitar Drag, 2000
Video installation
Video, sound, 14 min.
Christian Marclay,
courtesy Paula Cooper
Gallery, New York

From the first sounds of duct-tape being ripped, the familiar sound of someone gently slamming the strings of the guitar, to the sounds of an engine starting and then slowly the beginning of a sound that is hollow and at the same time human, and that takes us through the deepest roaring to high pitch screams until the very end when it all slows down and stops, unwillingly it seems. Knowing the story behind the work or not, the sound itself triggers your imagination, leaving your soul no peace. For even if this could be seen, or heard, solely as a beautiful noisy sound piece, there is too much to the idea behind *Guitar Drag*, for it to not leave you restless and thoughtful. *Guitar Drag* has many different layers of references, it alludes to the ritual of smashing guitars in rock concerts, it recalls Fluxus and its many destructions of instruments. It is also like a road movie, with reference to the landscape of Texas, where it was filmed, with references to cowboys and rodeos. It is about violence in general and more specifically about the lynching of James Byrd Jr. who was dragged to his death behind a pickup truck. I want the video to have these multiple layers and trigger people's imagination in contradictory ways. The piece ends up being seductive and repulsive at the same time.
(Christian Marclay)

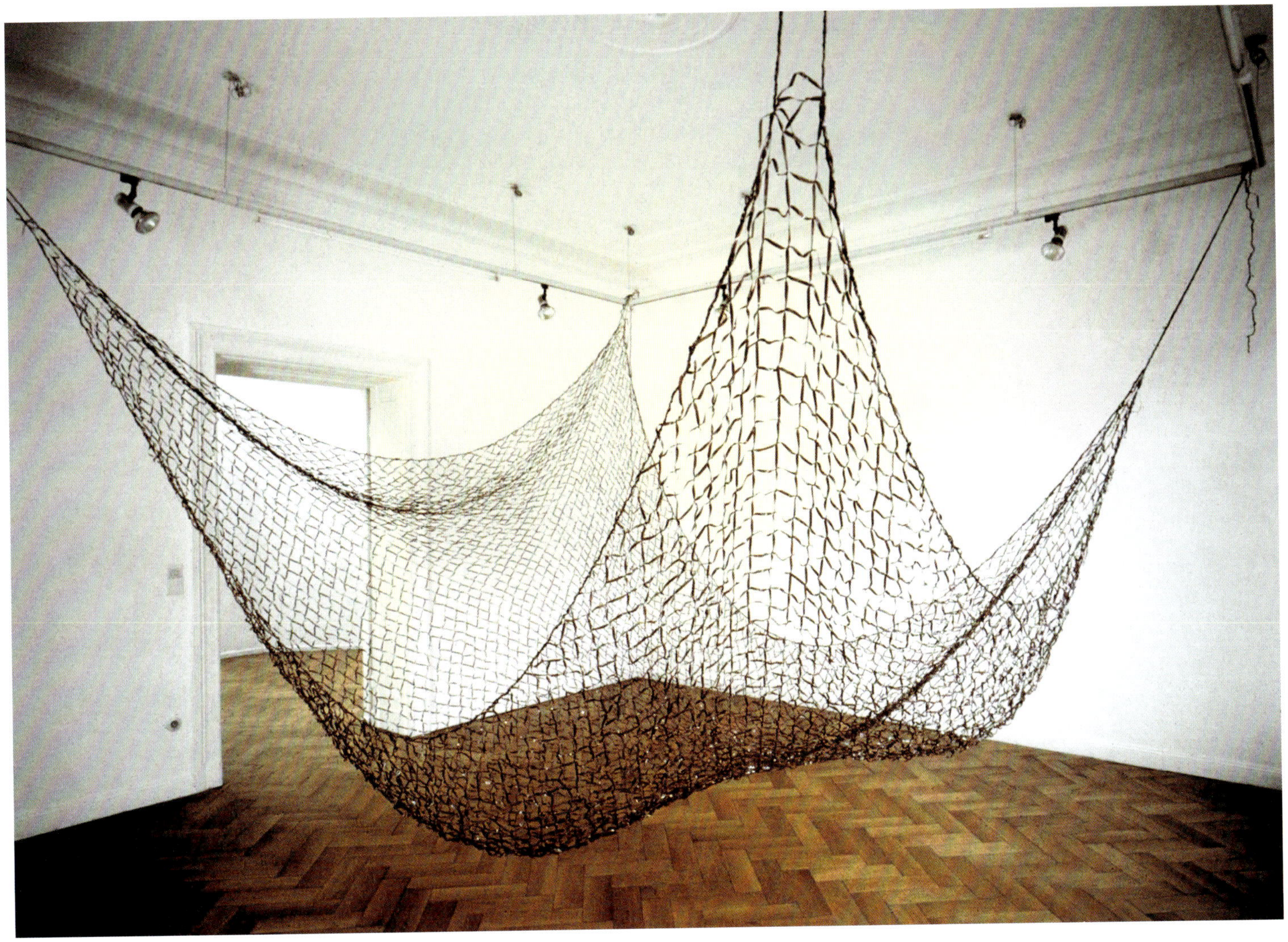

Net, 1991
Sound installation
Magnetic recording tape,
ca. 4.27 × 4.27 m
Christian Marclay,
courtesy Paula Cooper
Gallery, New York

Christian Marclay used magnetic tape on which the sound of rain has been recorded.

Christian Marclay

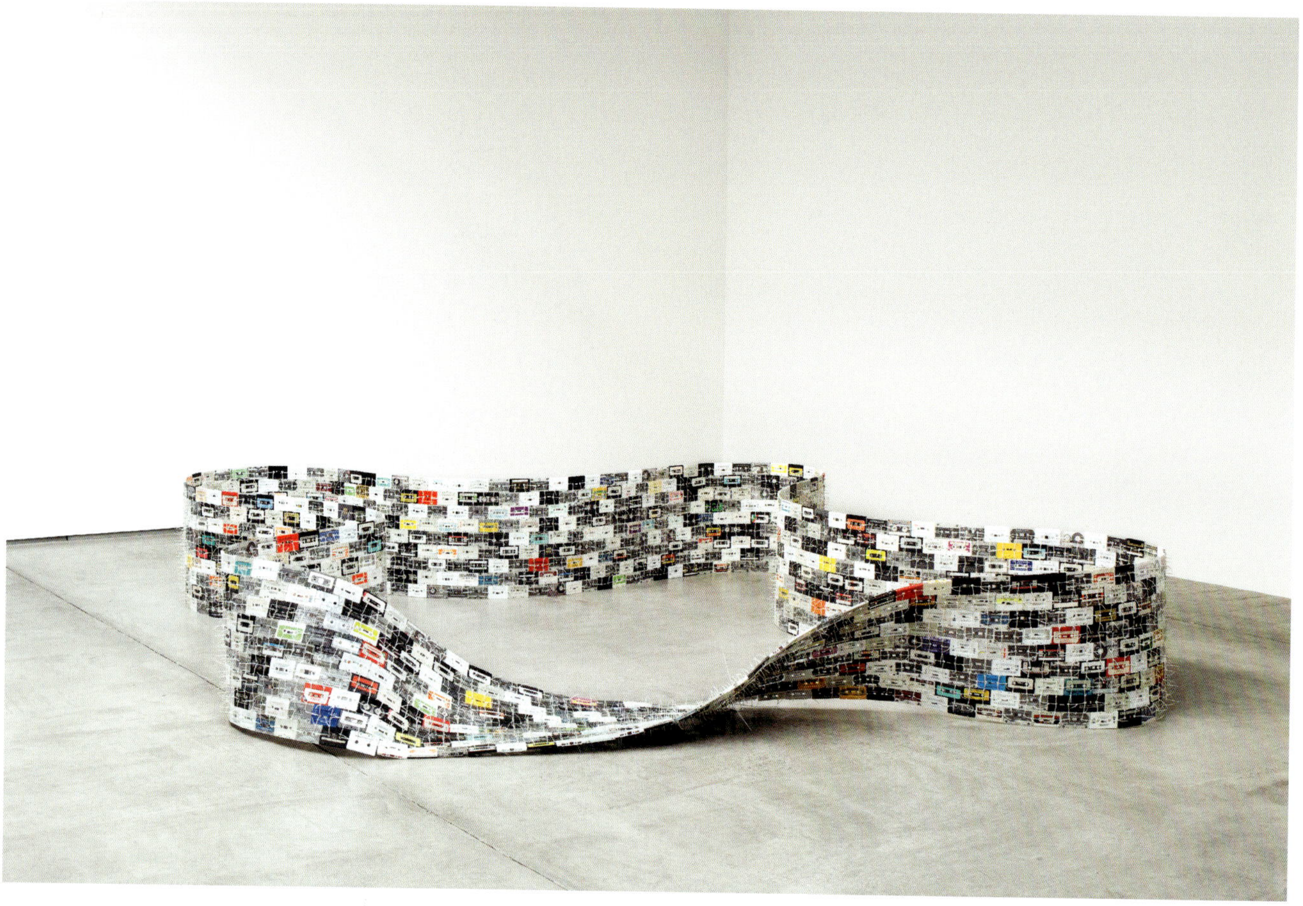

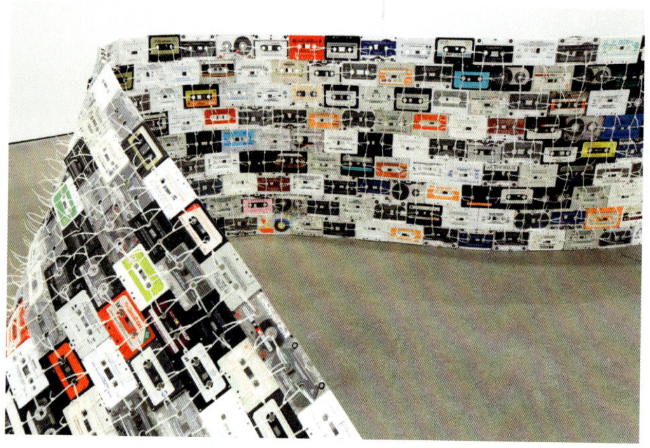

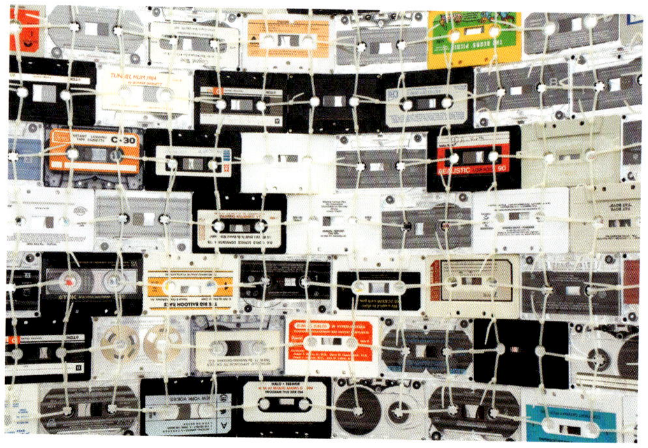

Moebius Loop, 1994
Installation
Cassette tapes, nylon ties, 61 × 213.4 × 609.6 cm
Christian Marclay, courtesy Paula Cooper Gallery, New York

In an iTunes era, *Moebius Loop* has a nostalgic appeal. An eclectic collection of genres – from mix tapes and Crystal Gayle, to Disney's *Little Red Riding Hood*, *The Aristocats*, symphonies and a Hewlett-Packard training tape – was used to complete the work. Its graphic impact is formed by both the shape and the random quilt-like pattern of color created by the selection of cassettes.

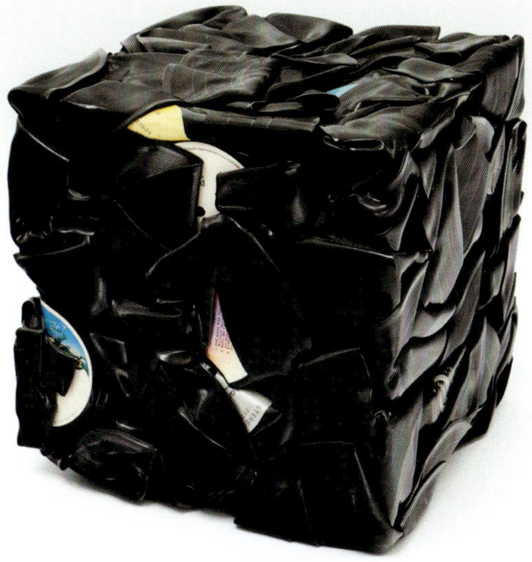

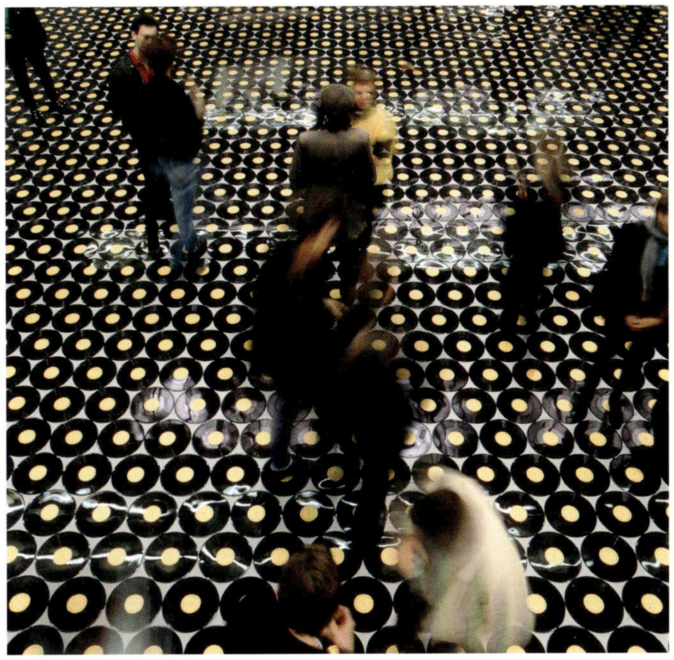

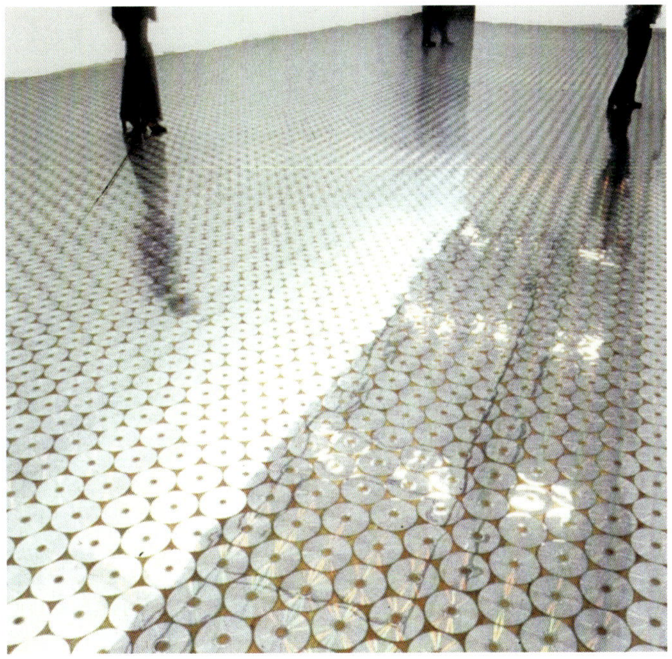

top:
Footsteps, 1989
Installation
Single-sided, 12 inch vinyl records, dimensions variable
Christian Marclay, courtesy Paula Cooper Gallery, New York

bottom:
Echo and Narcissus, 1992–1999
Installation
Compact discs, dimensions variable
Installation view Tokyo Opera City Art Gallery, 1999.

Melted Record Cube, 1989
Object
Melted vinyl records, 30.5 × 30.5 × 30.5 cm
Christian Marclay, courtesy Paula Cooper Gallery, New York

Melted Record Cube is a powerful visual symbol for the artist's discourse on music, art, and the uncharted histories of objects.

In the Summer of 1989, the floor of one of the Shedhalle galleries, Zürich, Switzerland, was covered with copies of a record titled *Footsteps. Footsteps* had been recorded in December 1988 in the deserted hallways of the Clocktower, New York, and in the studios of Harmonic Ranch, New York, where Keiko Uenishi's tap dancing was mixed in. During the six weeks of the installation, people were invited to walk on the records. Willingly or not, they had to step on them to reach the adjacent galleries where other sculptures were exhibited, contributing to the final composition by altering the records' surface with dirt and scratches.

Kaffe Matthews

*1961 in Essex (GB), lives in Berlin (DE)

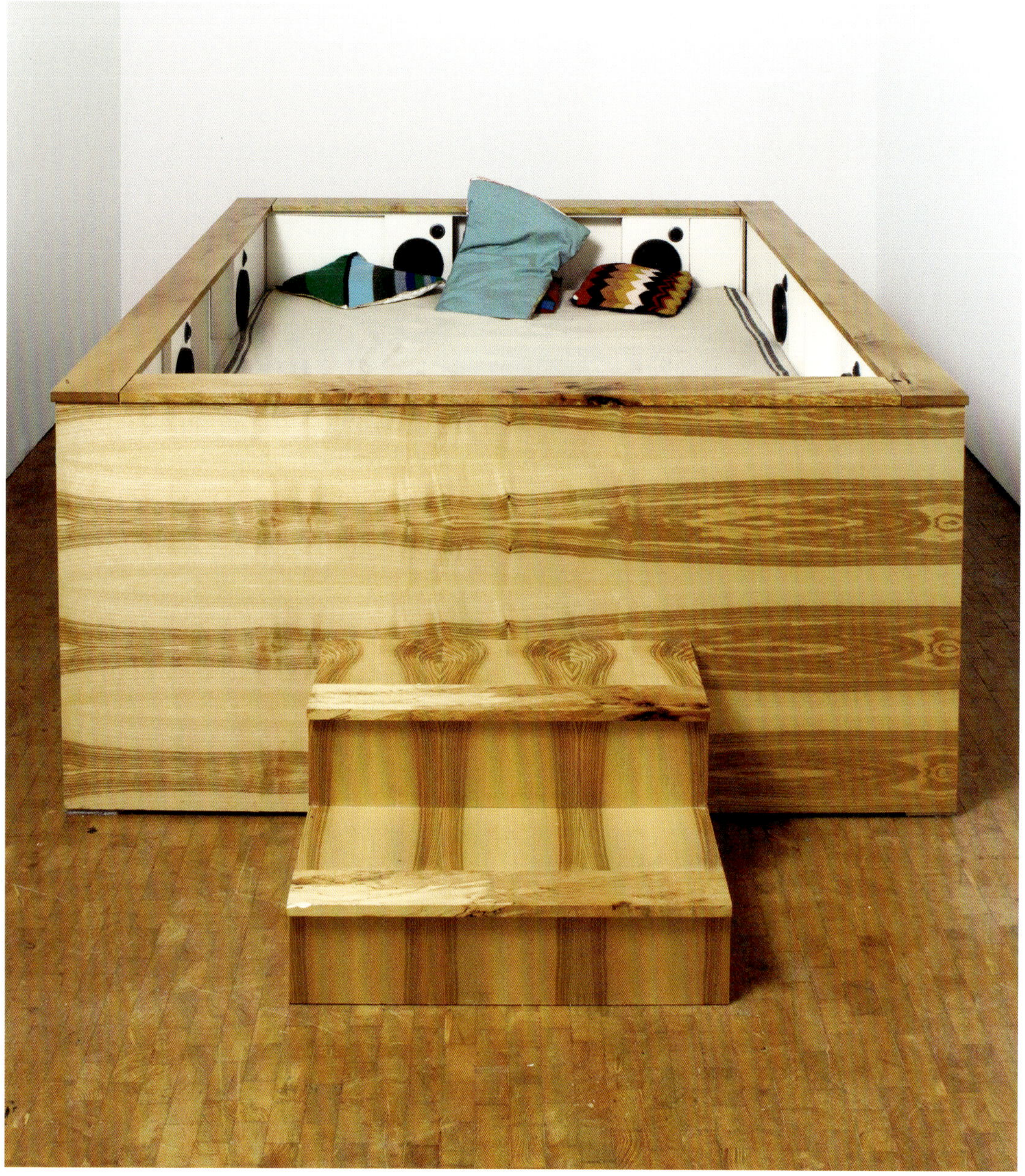

Sonic Bed_Scotland, 2005
Installation with the compositions *Bend* by Kaffe Matthews and *Omnht* by Éliane Radigue
Mattress, cushion, cover, music, 12-channel sound system, Mac mini, webcam
Software interface: David Muth
Music for Bodies

www.kaffematthews.net
www.musicforbodies.net

Sonic Bed_Scotland is an instrument to lie in and feel music. The music moves under, over, and around your body in ever-changing paths of vibration. Here it plays *Bend*, an original work Kaffe Matthews composed for the sonic bed using the sounds of her theremin, which she played and processed while lying in the bed, also moving the theremin around the bed, as well as Éliane Radigue's composition *Omnht*.

Bend can be heard when you stand next to the bed. But visitors who lie down in the instrument will feel it make another sense, becoming an immersive and intimate experience. *Sonic Bed* is an acoustic and social experiment that explores our perception of sound. It was developed as part of the ongoing research project *Music for Bodies*.

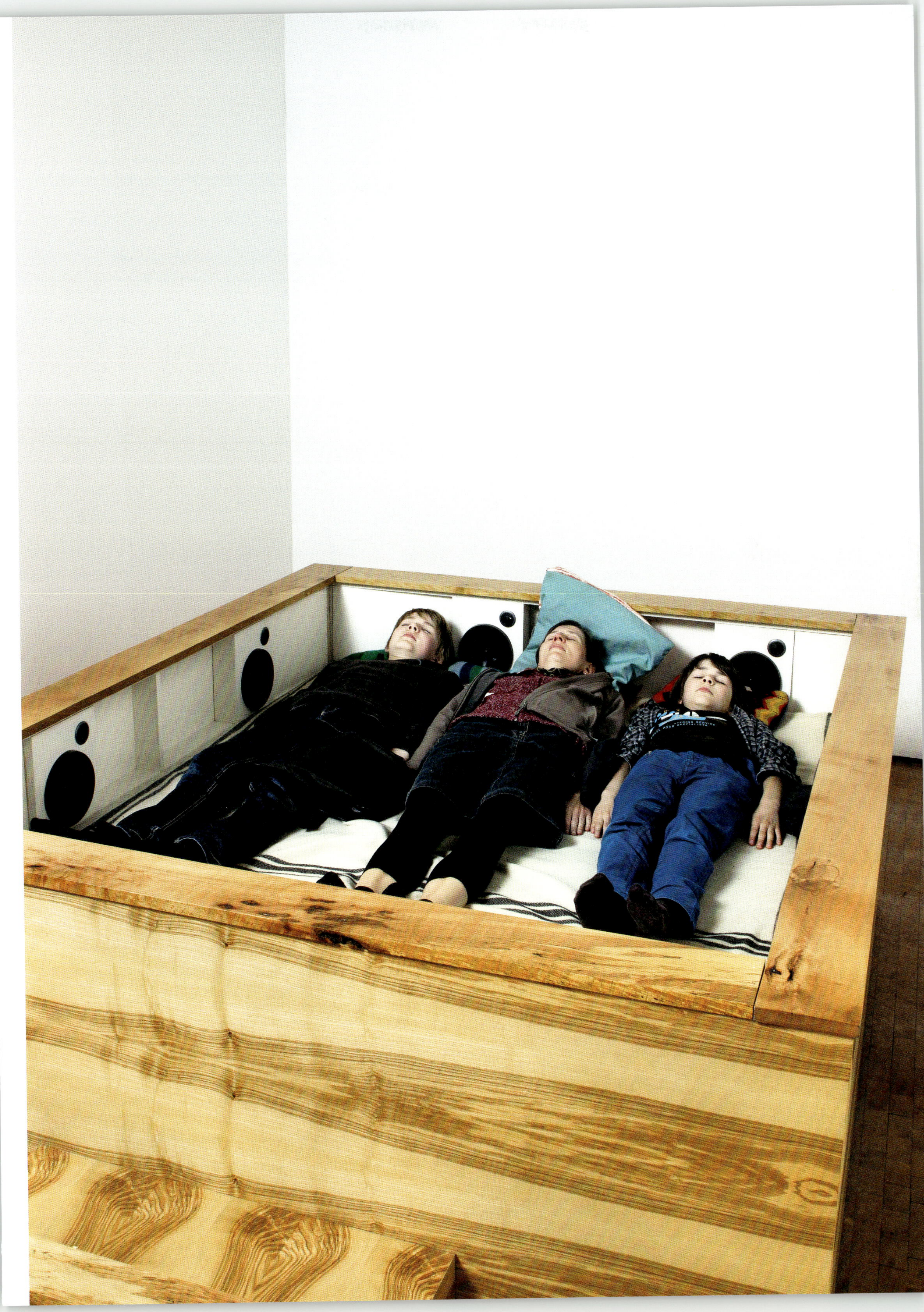

Benoît Maubrey

*1952 in Washington, DC, (US), lives in Berlin (DE)

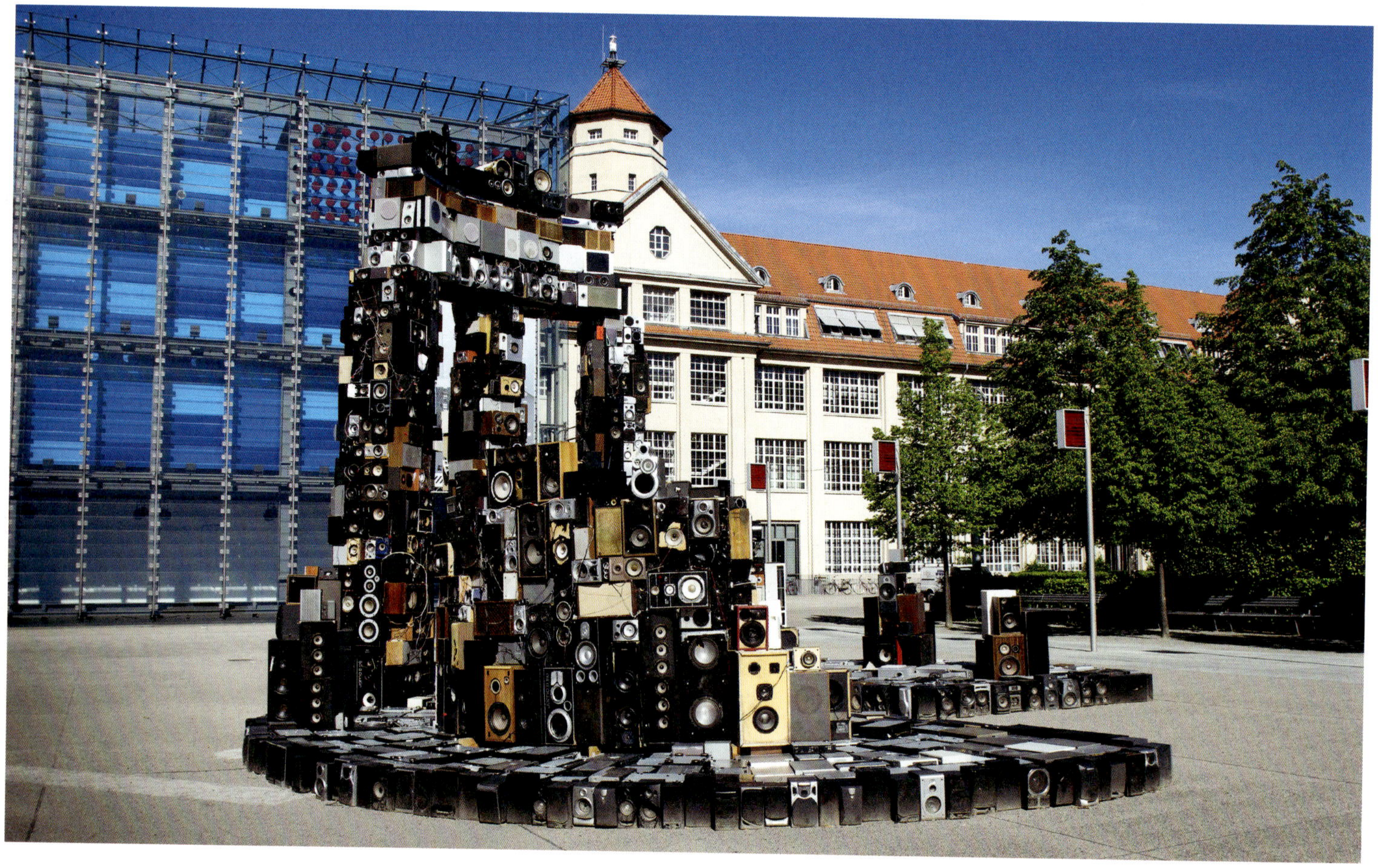

Temple, 2012
Sound installation on the forecourt of the ZKM | Karlsruhe
3000 recycled loudspeakers and assorted electronic parts,
10 recycled amplifiers, 10 recycled radios/tuners, mixing board
Production commissioned by ZKM | Karlsruhe

www.benoitmaubrey.com

With his *Temple*, Benoît Maubrey replicated the Tholos of Delphi, Greece, using electronic junk. Ancient tholoi were circular structures that originally served as sacred spaces for communicating with the gods.

Voices and "white noise" from radio receivers were the soundscape of the "second-hand shrine." Radio receivers conveyed the sounds of the ether to the loudspeakers. In addition, any local, but also absent visitor could call the German telephone number +49-721-8100-1818 and contribute his or her own sound material to *Temple's* flow of communication. The *Temple* broadcasted each input for three minutes.

Maubrey recalls that he was actually looking for a new place for his installation *Speakers Wall* and/or its components, that is why he contacted Julia Gerlach and Peter Weibel who were organizing the *Sound Art* exhibition at ZKM | Karlsruhe at that time. As they were looking for a suitable sculpture to be placed in front of the museum, it was during a brainstorming session with Mr. Weibel that the idea for *Temple* was born.

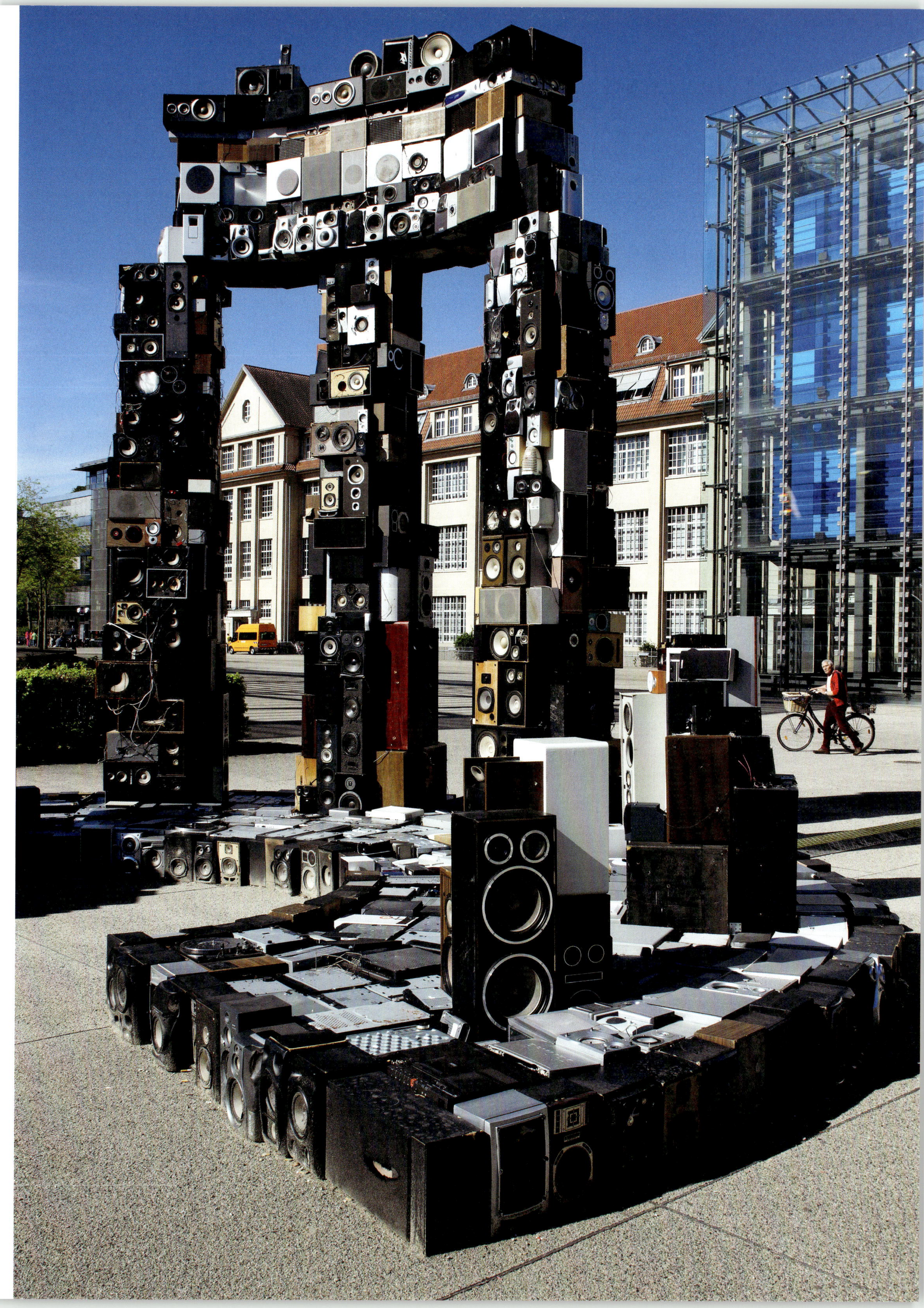

Benoît Maubrey

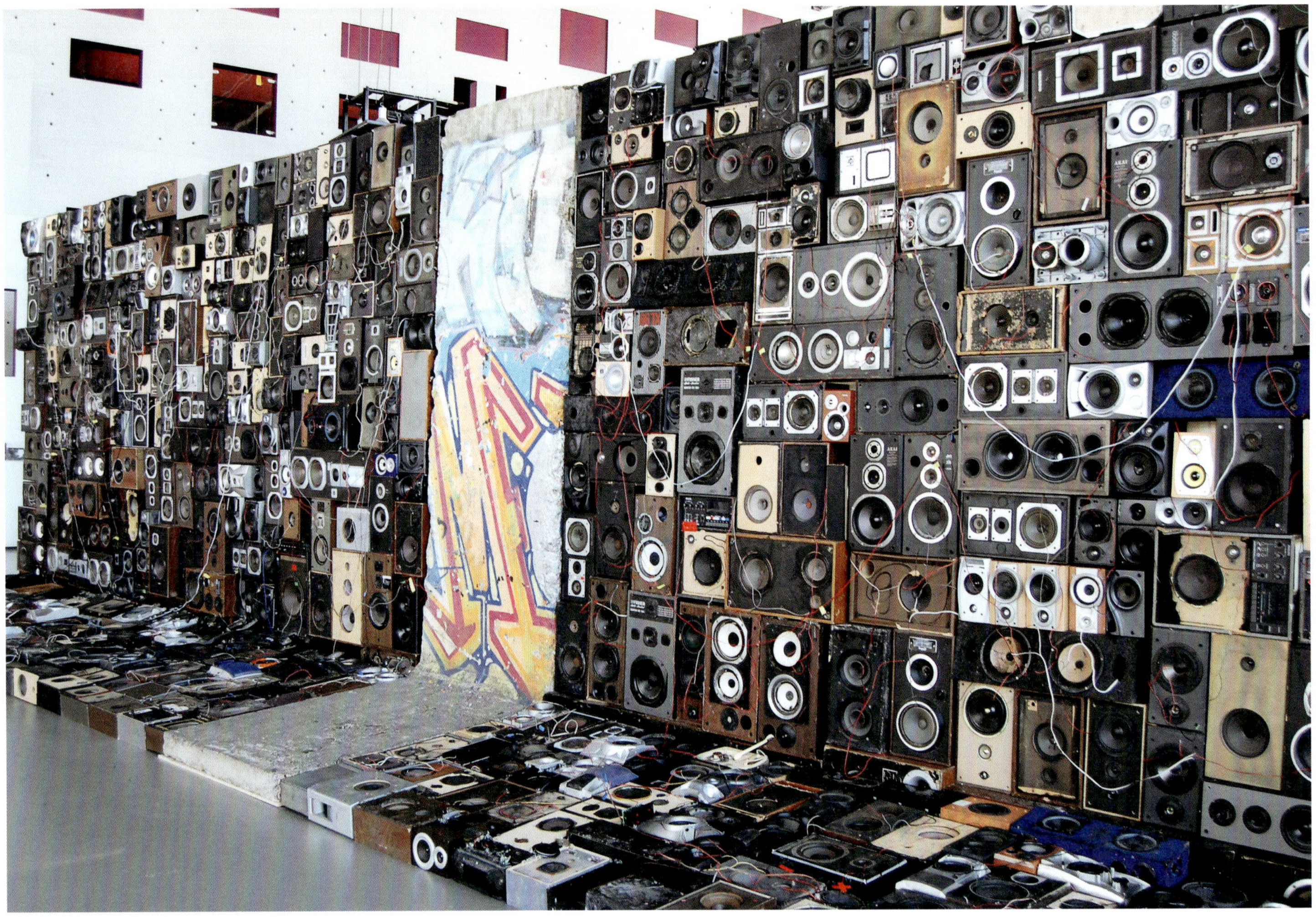

Speakers Wall (le Mur Sonore), 2011
Interactive electroacoustic sculpture
700 recycled loudspeakers, total length: 15 m, height: 3.60 m
Benoît Maubrey

Speakers Wall is an interactive electroacoustic sculpture built with an original segment of the Berlin Wall as the centerpiece. On each side of the centerpiece are 350 recycled loudspeakers mounted on panel boards. The total length is 15 meters, and the height is 3.60 meters. The 700 loudspeakers are soldered in series of eight to ten speakers, which are in turn "hooked up" in parallel circuits, so that in the end each channel has a resistance of four to six ohms. The speakers are connected via a mixing board to 4 recycled radios and 4 amplifiers, which generate eight channels of "white noise" as "background" sound. The low-level white noise changes with the time of day and the local atmospheric conditions. A telephone answering machine allows people to call a special telephone number and leave a message, music, or sound up to three minutes long. The telephone messages are quite audible. (Benoît Maubrey)

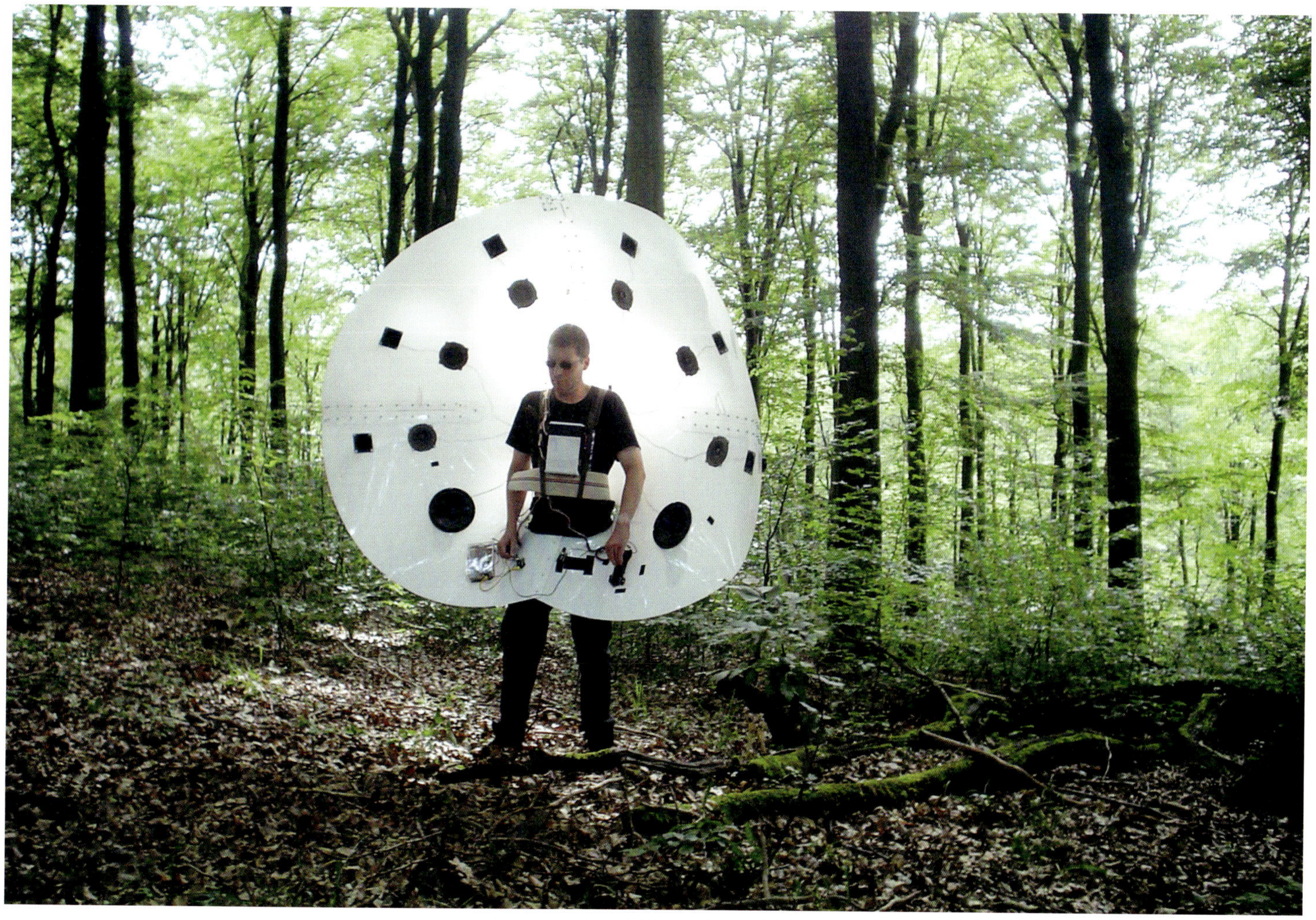

Audio Peacock, 2002
Wearable sound instrument
2 mm thick polycarbonate material, 16 loudspeakers, batteries, car amplifier, microphone, digital recording device ("sampler")
Benoît Maubrey

In 1990, following the fall of the Berlin Wall, I moved from West Berlin to a small village in the state of Brandenburg, Germany. There I made the acquaintance of a neighbor's colorful and vociferous courtyard pet: a peacock. Impressed by the sound and perambulatory qualities of this bird and inspired by the recent success of the *Audio Ballerinas*[1], I accepted an invitation from Lieux Publics (National Center for Creation in Public Spaces) in Marseille, France, to create 6 *Audio Peacock* costumes during a residency program.

The performers wear electroacoustic instruments shaped to resemble a peacock's fan-like plumage. The instruments are highly directional, projecting the sound into a space like an oversized radar dish. In much the same way as the *Audio Ballerinas'* tutus, they are equipped with digital samplers and loop/pitch mechanisms that allow them to record and alter their voices and local sounds. Just as the courtyard peacock "struts his stuff" in front of a pea-hen and impresses with his awesome cry, so does the *Audio Peacock* stalk his architectural prey – using sound as a scalpel that cuts through air and sculpts it, transforming it into a new acoustic space.

These electroacoustic "birds" can also perform as *Video Peacocks*. Their costumes – in this case made of white polycarbonate as opposed to the clear material of the *Ballerinas* – are used as mobile projection screens: whenever their paths intersect the light of a video projector, the costumes metamorphose into multi-colored screens. Images and videos (movies, Internet blogs and closed-circuit camera views) are "beamed" onto them as they play their sounds. Even the sound patterns of their own voices can be simultaneously visualized onto them as they "*squawk*". As a multi-phonic performance, the perambulations of these *CyberBirds* (performance title at Hohenrausch Festival, Rostock, Germany 2008) are choreographed vis-à-vis the emplacement of the projectors in the architectural space. (Benoît Maubrey)

1 The *Audio Ballerinas* were originally created for the Festival Les Arts au Soleil (1990). In this mobile installation project, dancers are equipped with "audio tutus" which permits them to produce sounds via their movement and the space around them using a variety of interactive electronic instruments.

Soichiro Mihara
Kazuki Saita

*1980 in Tokyo (JP), lives in Yamaguchi (JP)
*1981 in Tokyo, lives in Tokyo

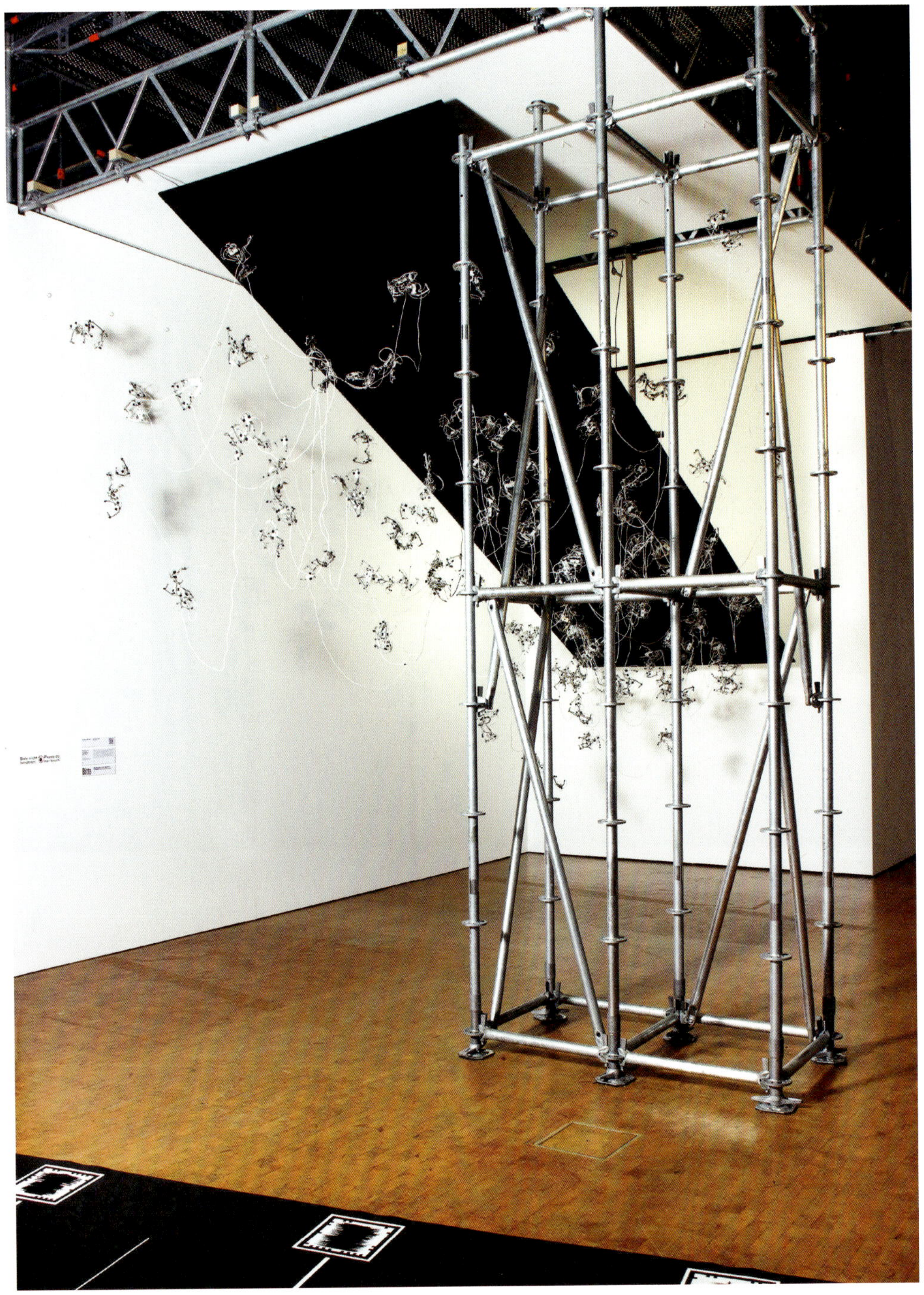

Moids 2.2.0 – acoustic emergence structure, 2009/2012
Sound installation
Microphones, microcontrollers, switches, loudspeaker units, cable
Conceptual adviser: Hiroko Mugibayashi
Soichiro Mihara, Kazuki Saita

Moids 2.2.0 is an interactive system made up of 1024 autonomous circuits. Each circuit consists of a microphone that picks up ambient sound, a microcontroller that analyzes the incoming sound, a switch, and a loudspeaker component. The installation reacts both to the sound of the environment and to the sound it produces itself.

The Japanese artists and computer experts Soichiro Mihara and Kazuki Saita have been working on the development of organic sound systems since 2004, when they attended a lecture on imaginary energy fields by synthesizer pioneer Robert Moog. Their goal is the creation of a decentrally organized alternative nature, which will make the "living echo of technology" audible.

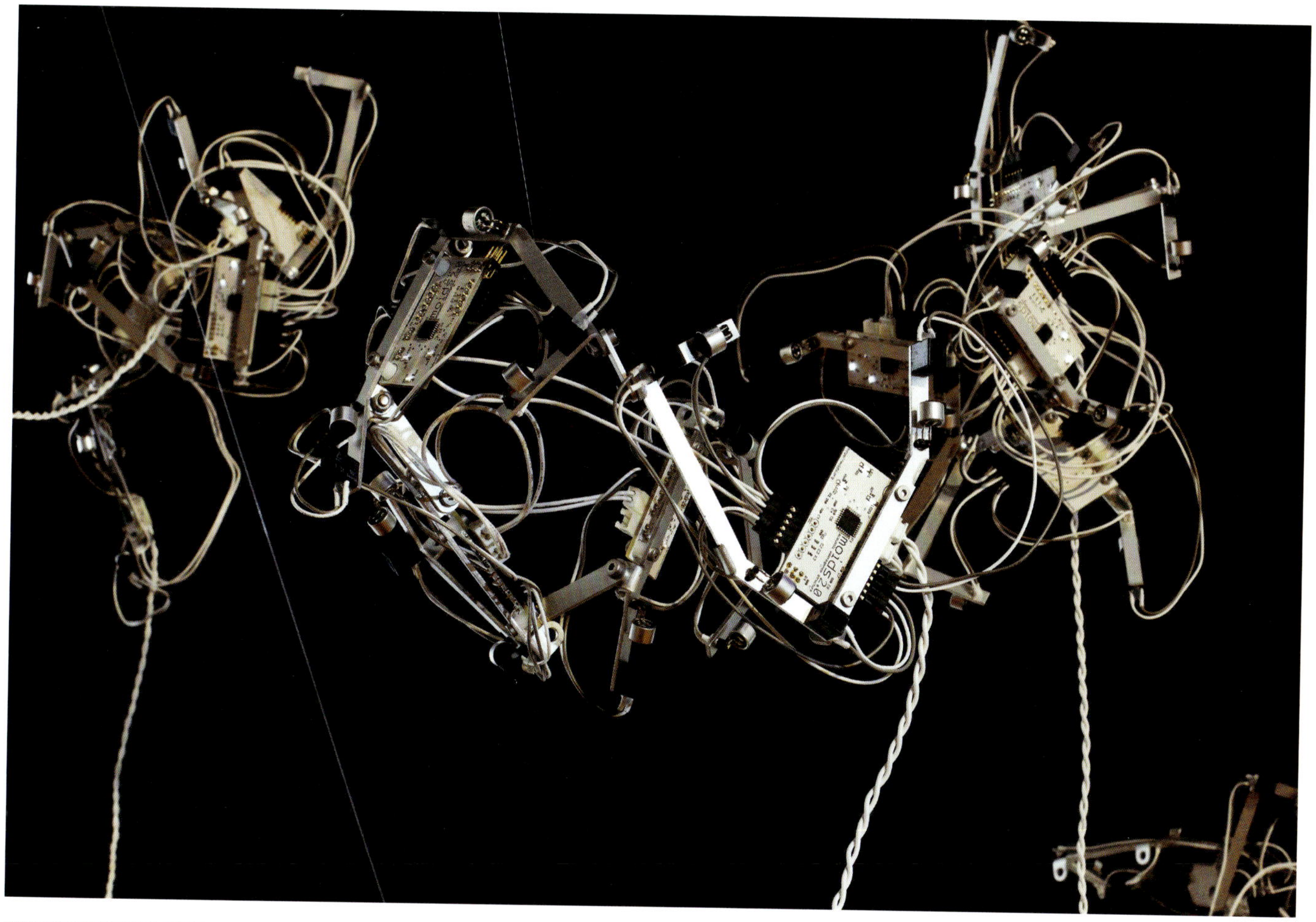

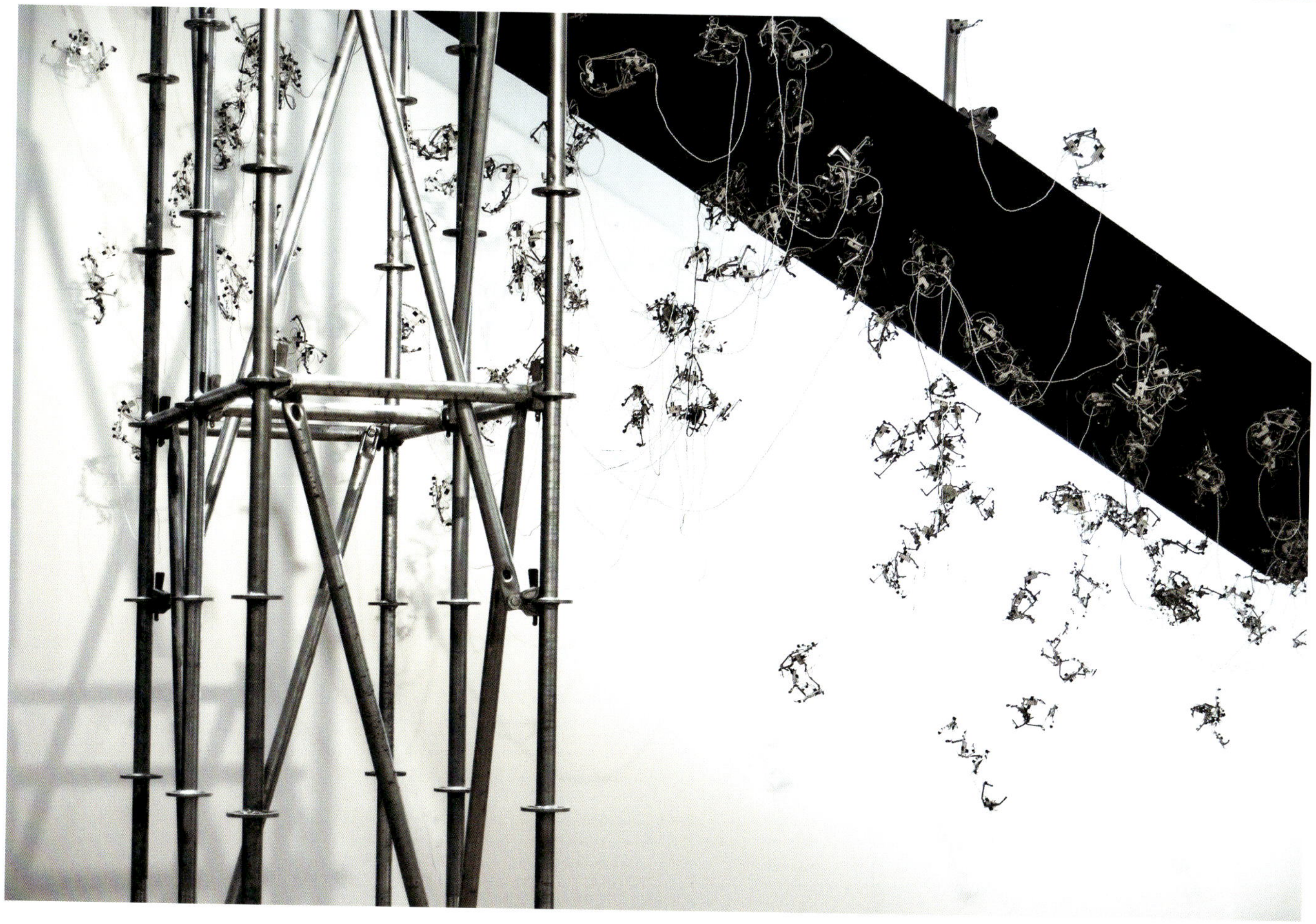

Robin Minard

*1953 in Montreal (CA), lives in Weimar (DE)

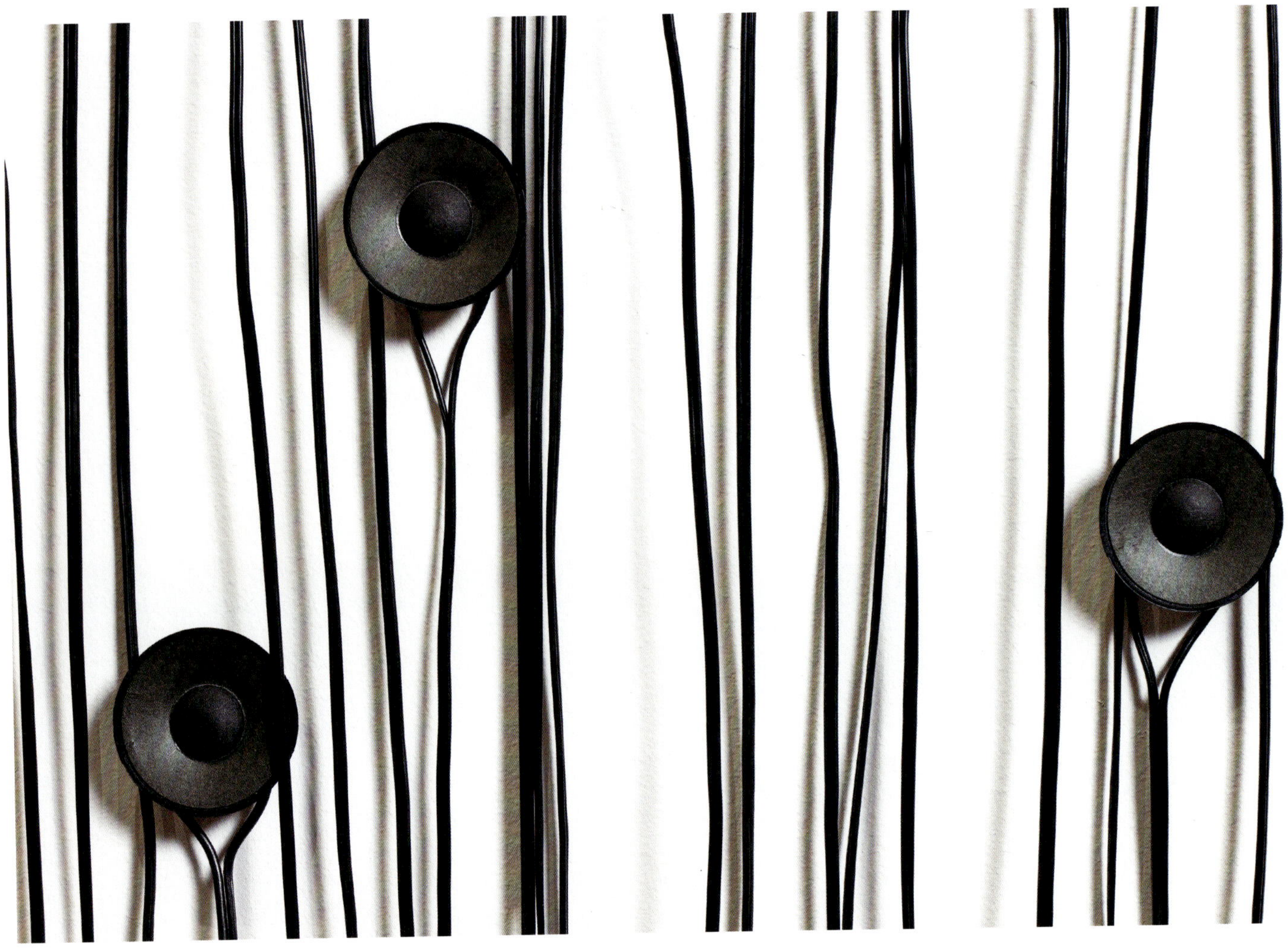

Silent Music, 2012–2013
Sound installation
Piezo loudspeakers,
4-channel audio,
dimensions variable
Robin Minard

www.robinminard.com

The installation *Silent Music* consists of several hundred piezo loudspeakers (ferroelectric loudspeakers). Together with their cables, the tiny sound transformers form plantlike structures. High, soft sounds from natural and synthetic sources emanate from the loudspeakers. The work also connects natural and synthetic elements visually. Although the installation consists solely of loudspeakers and cables, the arrangement of materials suggests life and growth.

Silent Music is a series of works started in 1994, which Robin Minard redesigns expressly for each exhibition venue. He has also created network-like wall and floor installations using the tiny sound transformers, composing the multichannel audio recordings for the many point-like sound sources so that differentiated effects arise in the space.

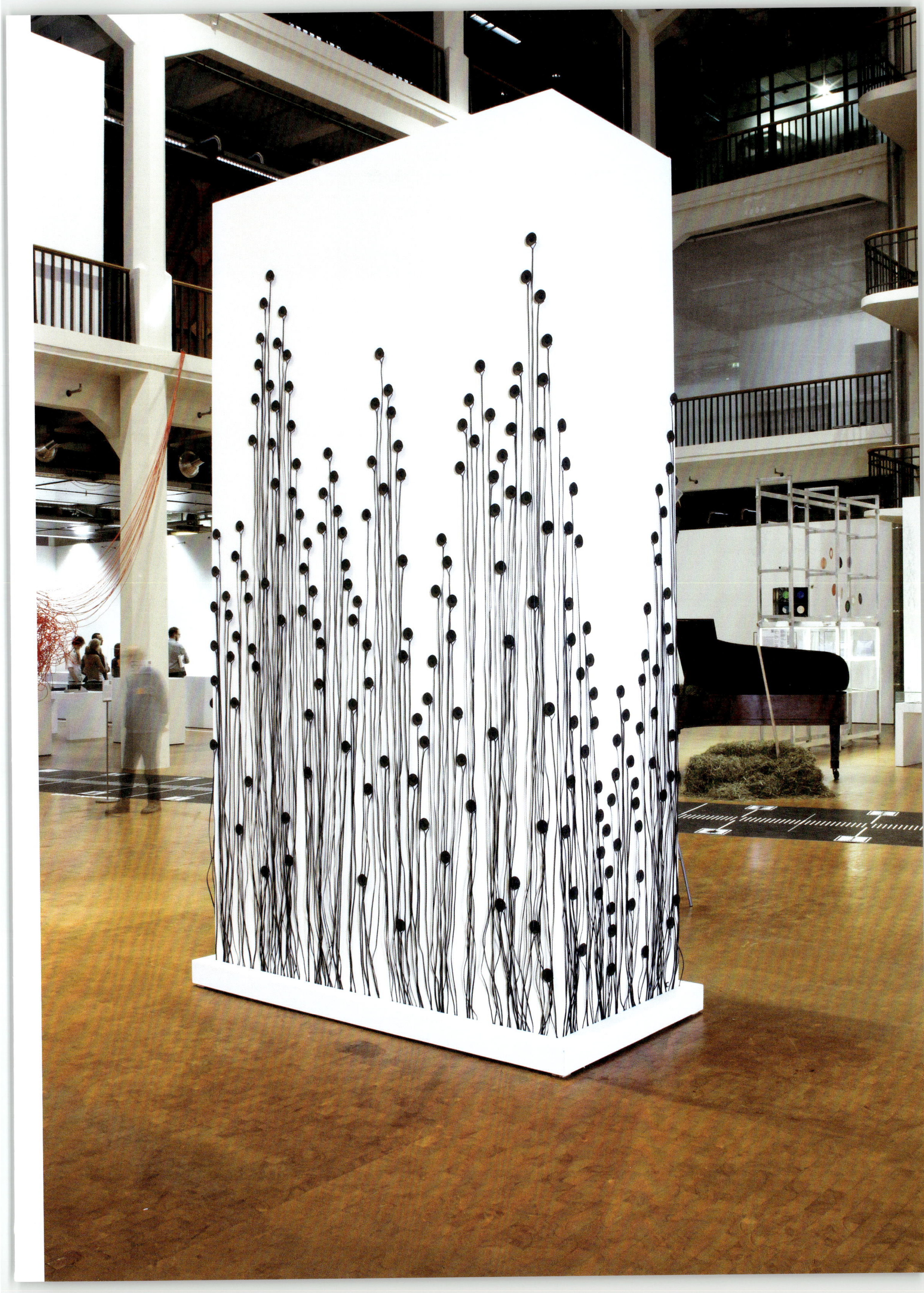

Haroon Mirza

*1977 in London (GB), lives in London and Sheffield (GB)

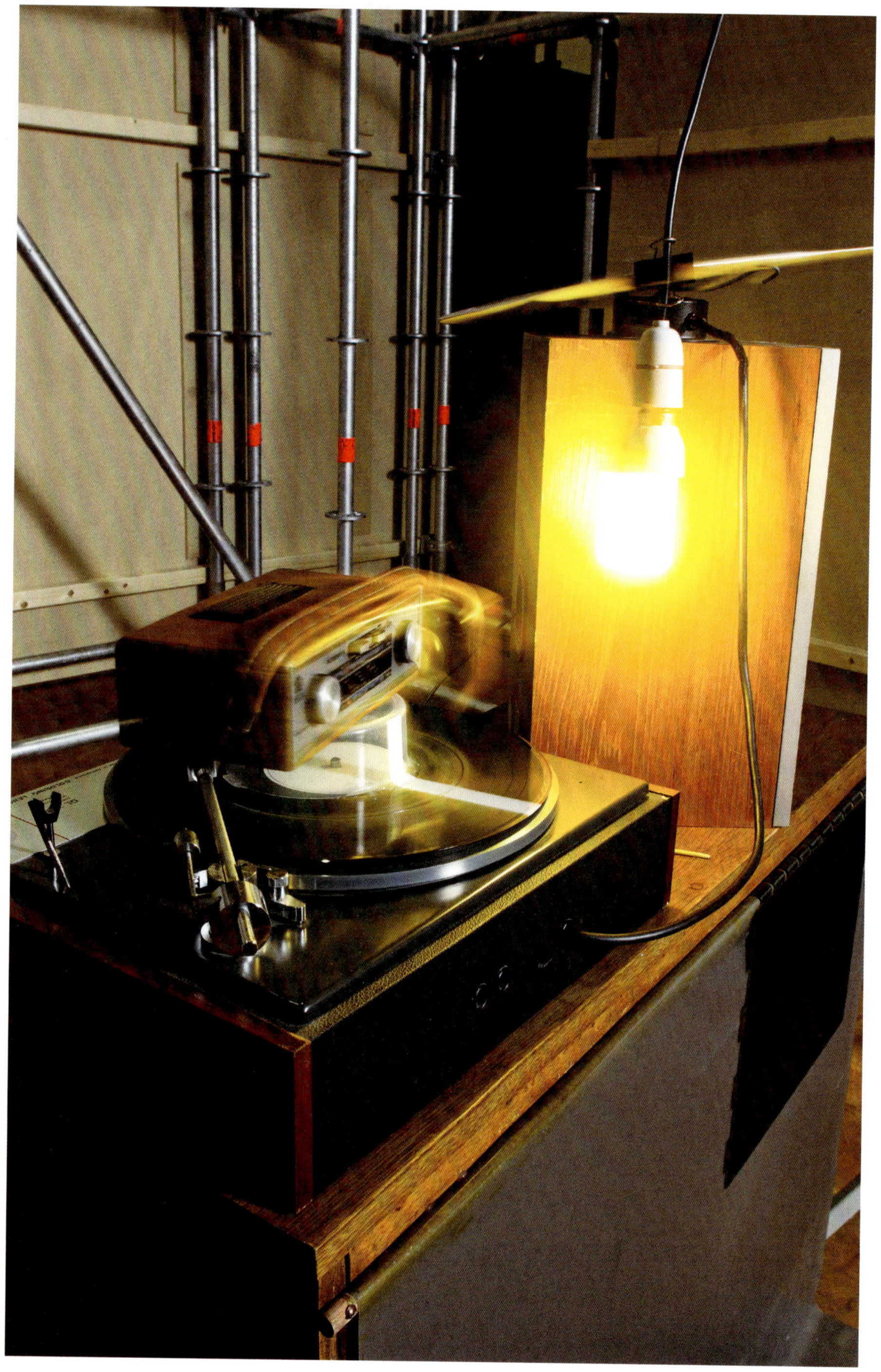

The Last Tape, 2010
Sound installation
Record player, radio, loudspeaker, lamp, cupboard, LCD screen, amplifier, record covers, stroboscope, motor, transparencies, media player, 150 × 110 × 70 cm
Studio coordination: Gabriela Cala-Lesina
Technology: Ben Barwise
Private collection

In *The Last Tape* Haroon Mirza reinterprets Samuel Beckett's one-act play *Krapp's Last Tape* (1958). In Beckett's play, the protagonist, Krapp, looks back at the events of his life, as recorded onto tape each birthday. The film of this staging of the Beckett drama features the actor and musician Richard "Kid" Strange enacting previously unpublished lyrics by Ian Curtis onto magnetic tape. Haroon Mirza draws a parallel between Krapp (who "has nothing to talk to but his dying self") and the frontman of the post-punk band Joy Division. Strange connects the lyrics and the sounds created by the accompanying sculptural elements. A stroboscope blinks intermittently, an allusion to Curtis's epilepsy. Haroon Mirza treats furniture, electronics, and sounds as musical material and makes no distinction between audio and visual composition: in the process of arrangement, each material bears aesthetic significance.

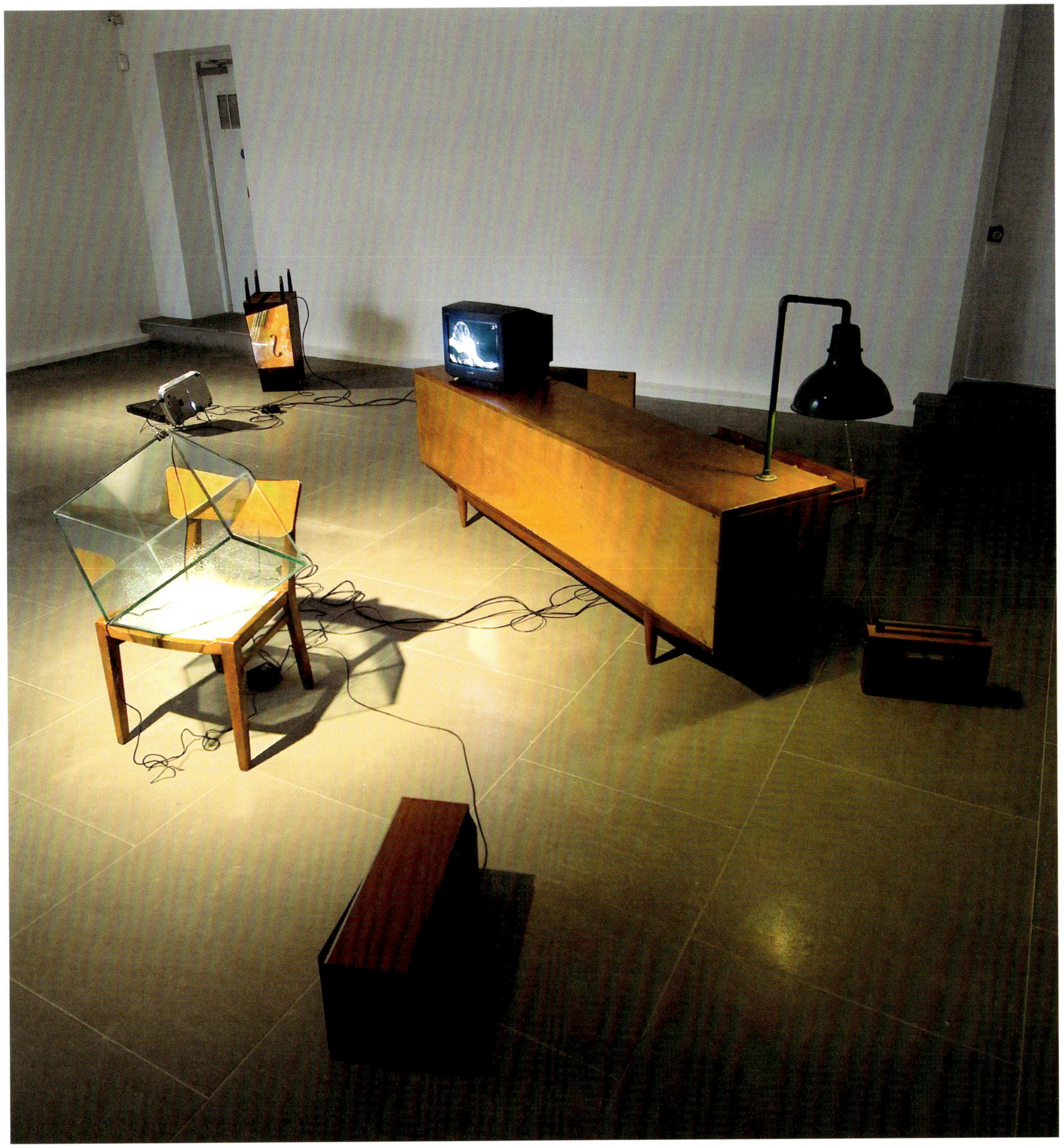

Adhān, 2009
Mixed media installation
Dimensions variable

The Arabic word *adhān* is the Islamic call to prayer, recited five times a day by a muezzin. This work is a mixed media installation which resulted from the same residency in Pakistan as *Taka Tak*, where Mirza was researching the place of music in Pakistani culture.

Operating in their own separate ways, the discrete elements within the installation form part of an interconnected composition or system, both literally (the sonic rhythm is echoed by a light display) and conceptually. Mirza's installation draws on his film of the same title which was screened in the Sheffield Pavilion during the 10th Istanbul Biennial in September 2009.

Gordon Monahan

*1956 in Kingston (CA), lives in Meaford (CA) and Berlin (DE)

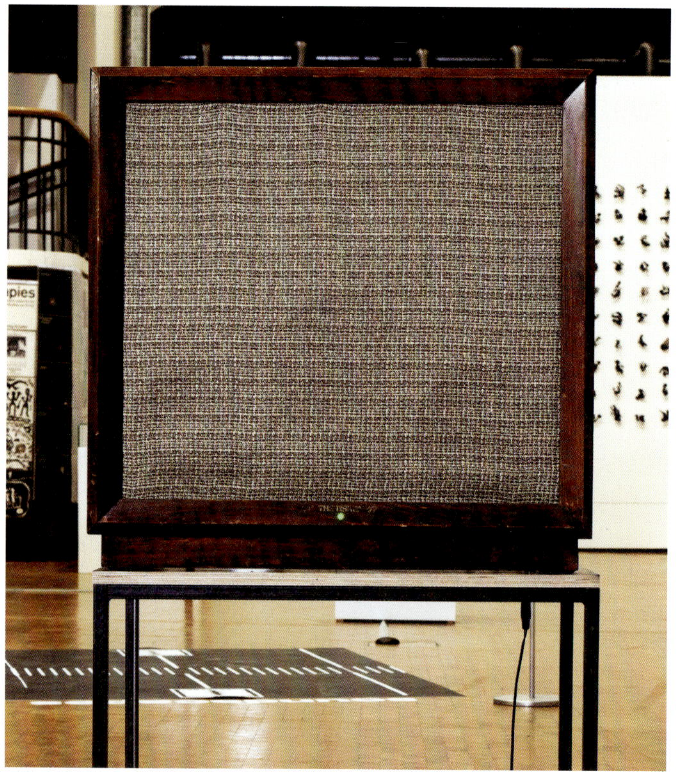

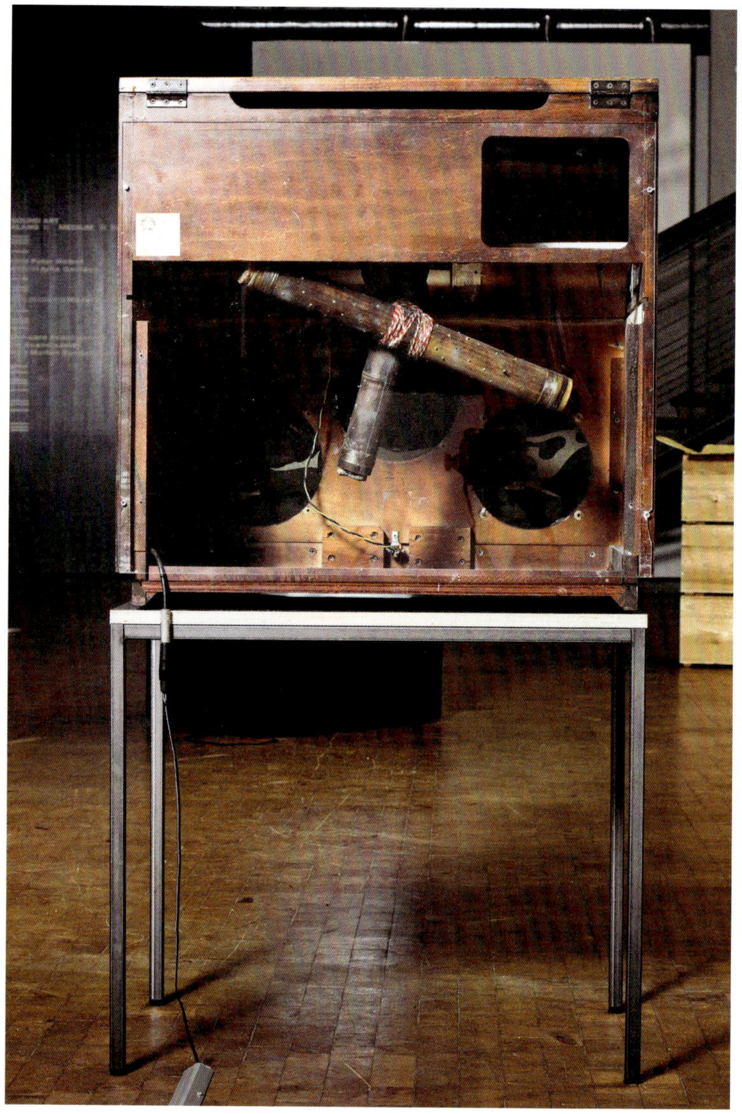

Music From Nowhere, 1989–1990
Sound object
Loudspeaker box, motor, rainsticks, dimensions variable
Gordon Monahan

www.gordonmonahan.com

In the work series *Music From Nowhere*, Gordon Monahan transforms loudspeaker cabinets into autonomously operating music machines. The loudspeaker chassis were gutted and the speakers replaced with devices that mechanically produce sound. All of the built-ins work automatically. The Canadian sound artist has repurposed tiny fountains, mechanical vibrators, and other motor-driven systems to generate sounds and noises. At the same time, the construction takes mechanical variables into account, anticipating that the sounds generated will vary in unpredictable ways. This indeterminacy evokes the illusion that the loudspeaker boxes are broadcasting radio recordings or LPs. But *Music From Nowhere* has no sound source other than the mechanical music machines inside the "fake" loudspeakers.

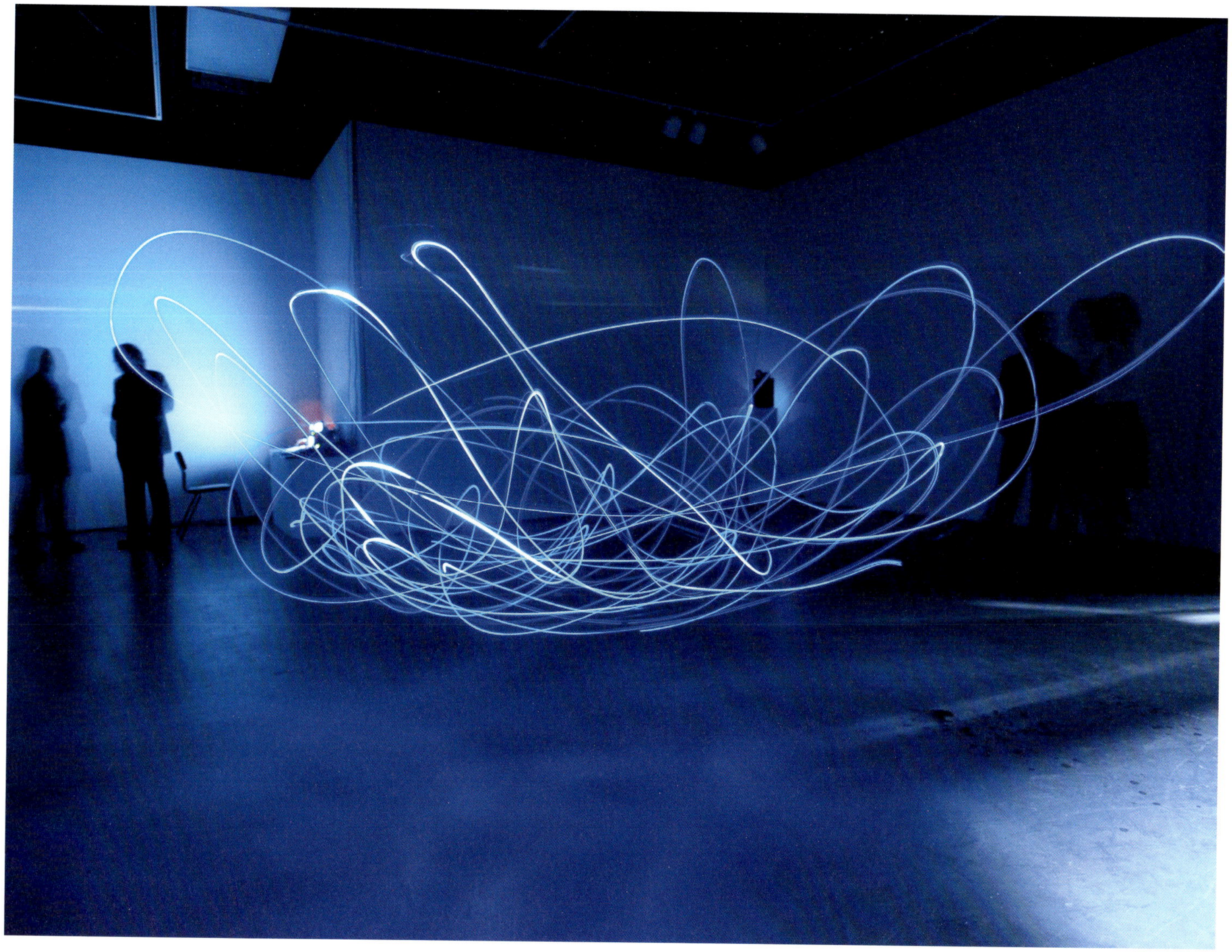

Theremin Pendulum, 2008
Installation
Theremin, motors, computer running Max/MSP, motion sensor, midi interfaces, motor controllers, audio system
Created with the assistance of a research and production grant from the Canada Council for the Arts, and supported by an artist-in-residency with the New Media Sound Lab at the University of Regina, Saskatchewan, Canada
Gordon Monahan

A theremin antenna made of flexible wire is mechanized so that it swings spastically as a chaotic pendulum. This causes the theremin to change pitch according to the swinging motion of the pendulum. The sound produced is multiplied by a delay system so that we hear the sound of eight theremins at the same time. This is mixed automatically by a computer-mixing program (MAX/MSP), and is then broadcast into the gallery. Also attached to the end of the antenna pendulum is an LED light that swings with the antenna, so that when you enter the darkened gallery all you see is a light moving in a seemingly unpredictable pattern that mirrors the frequency wave changes in the theremin. (Gordon Monahan)

Gordon Monahan

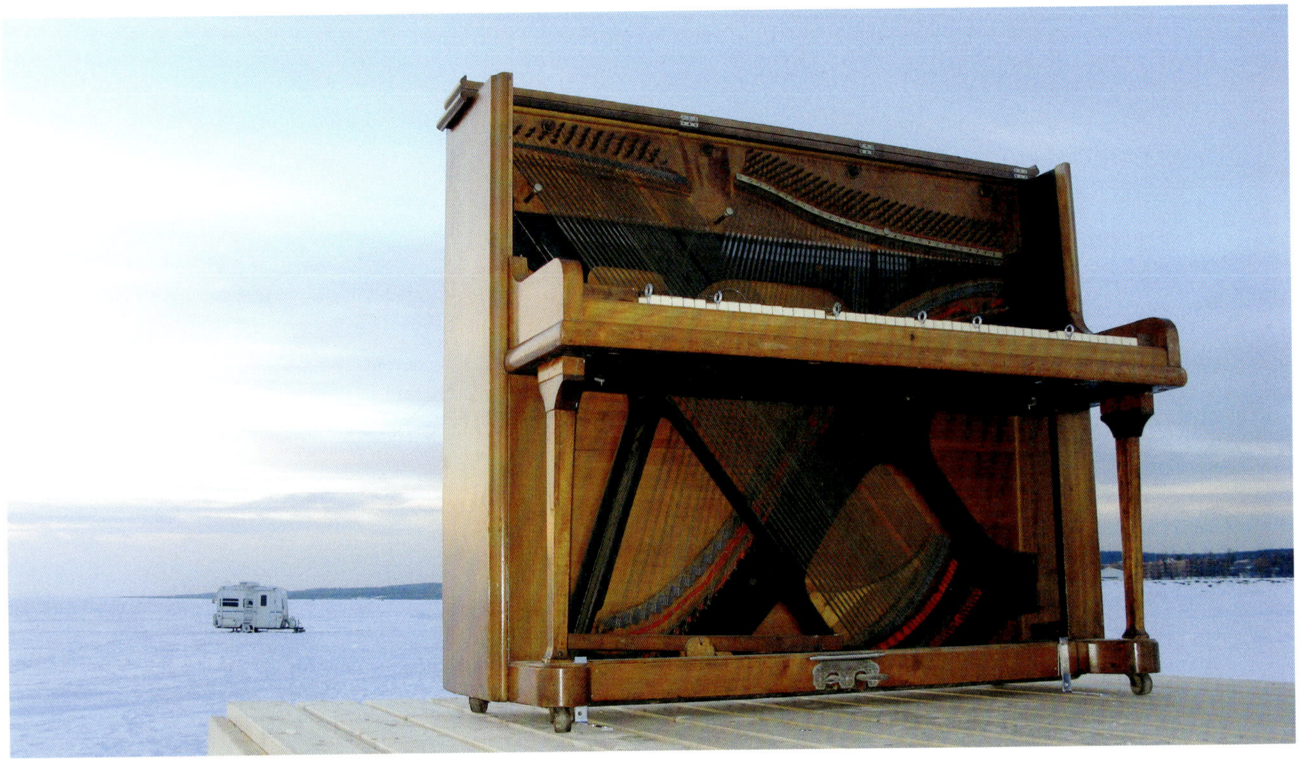

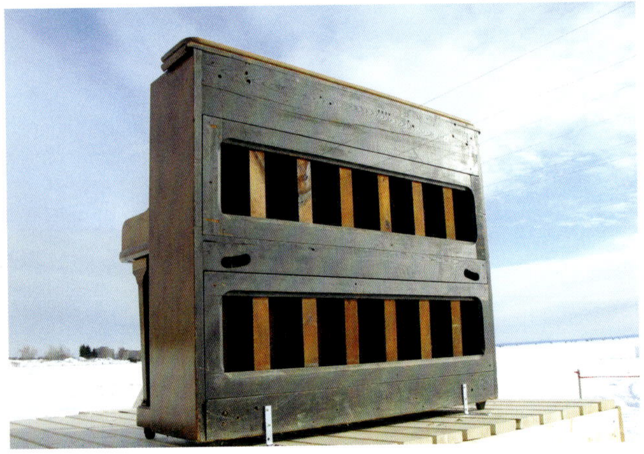

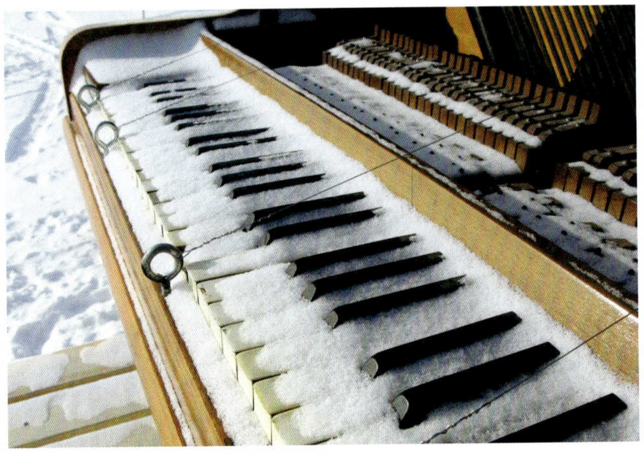

Piano on Frozen Lake Nipissing, 2014
Installation on the frozen ice of Lake Nipissing ca. 100 meters offshore from central downtown North Bay, Ontario, Canada, piano, piano wires, motors, amplifier, mp3 player, hardware fasteners
Presented by the Near North Mobile Media Lab, North Bay, Ontario, with thanks to Dermot Wilson and Holly Cunningham

A piano is transported onto the ice of frozen Lake Nipissing, and 220-foot-long piano wires are stretched across the ice-scape and anchored to a trailer positioned on the ice specifically for this purpose. Inside the trailer are an audio player and amplifier, that send audio recordings to six small motors, each attached to the ends of the six long piano wires. The motors reproduce the audio signals and send those vibrations into the long wires, which transmit them to the soundboard of the distant piano. The audio recordings consist of 26 recomposed works of Frederic Chopin, and eleven original and recomposed works of Henry Cowell.

In addition, aeolian tones are created when wind blows across the long piano wires, and this happens quite often. When a breeze blows mildly across or obliquely to the wires, the result is a blend of prerecorded transmissions and aeolian tones; when the wind blows directly across the wires at low, medium, or strong velocities, the aeolian tones dominate and the prerecorded sounds are (mostly) drowned out. The piece runs 24 hours a day.

Thus, a musical composition of quasi-indeterminate form is realized. A man-made physical structure is constructed to interface with natural forces so that spontaneous musical forms are manifest. These spontaneous manifestations are the result of a physical sculpture reproducing prerecorded audio signals that are unpredictably interacted-upon and dominated by wind-generated overtone vibrations.
(Gordon Monahan)

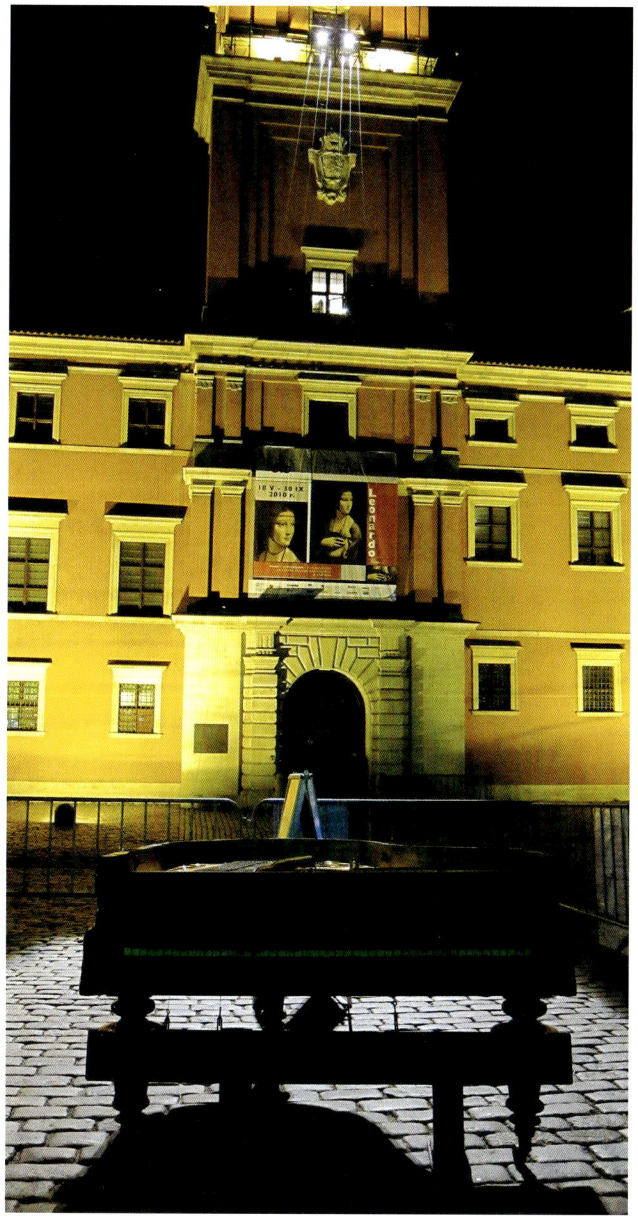

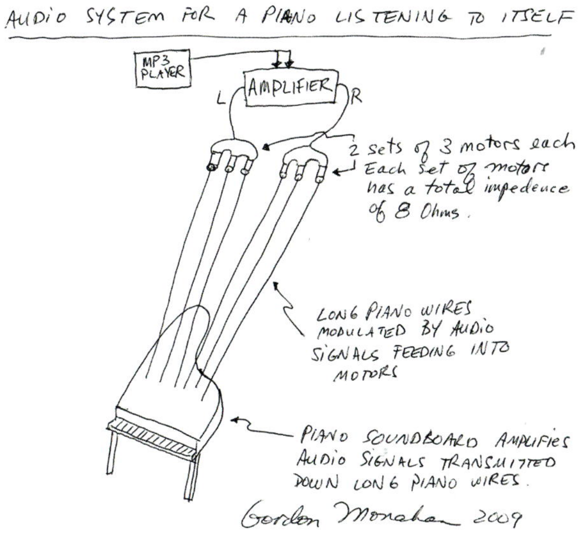

A Piano Listening To Itself – Chopin Chord, 2010
Installation
Piano, piano wire, motors, amplifier, mp3 player, hardware fasteners
Commissioned by Warsaw Autumn
Gordon Monahan

Six long piano wires are suspended from the tower of the Royal Castle in Warsaw, Poland and are connected to a piano positioned in the middle of the square below. At the connecting point high above on the balcony of the tower, audio recordings played into an amplifier are transmitted into the long piano strings using vibrating coils in small motors attached to the wires, which cause the piano strings to vibrate in sympathy with the audio signals. The vibrations in the long piano strings are transmitted down the long piano wires and are amplified by contact to the piano soundboard below. The audio recordings are thus reproduced without any loudspeakers – instead, the motor coils, the long piano wires, and soundboard become substitutes for a loudspeaker system.

Audio recordings transmitted into the piano strings consist of note sequences derived from piano compositions by Frederic Chopin. The note sequences were extracted from midi files of Chopin's piano pieces and "recomposed" using midi software, to form new pieces that were then recorded using a software-based piano. In total, 26 "recompositions" were derived from twenty preludes, two etudes Opus 10 and 25, the *Fantaisie-Impromptu* Opus 66, one Nocturne Opus 27 Nr. 1, one Polonaise Opus 53, and the first movement of the Sonata Opus 35.

A second version was installed at the Centro Nacional de los Artes in Mexico City, for the Festival de Artes Electrónicas y Video Transitio_MX 04 in 2011.
(Gordon Monahan)

Gordon Monahan

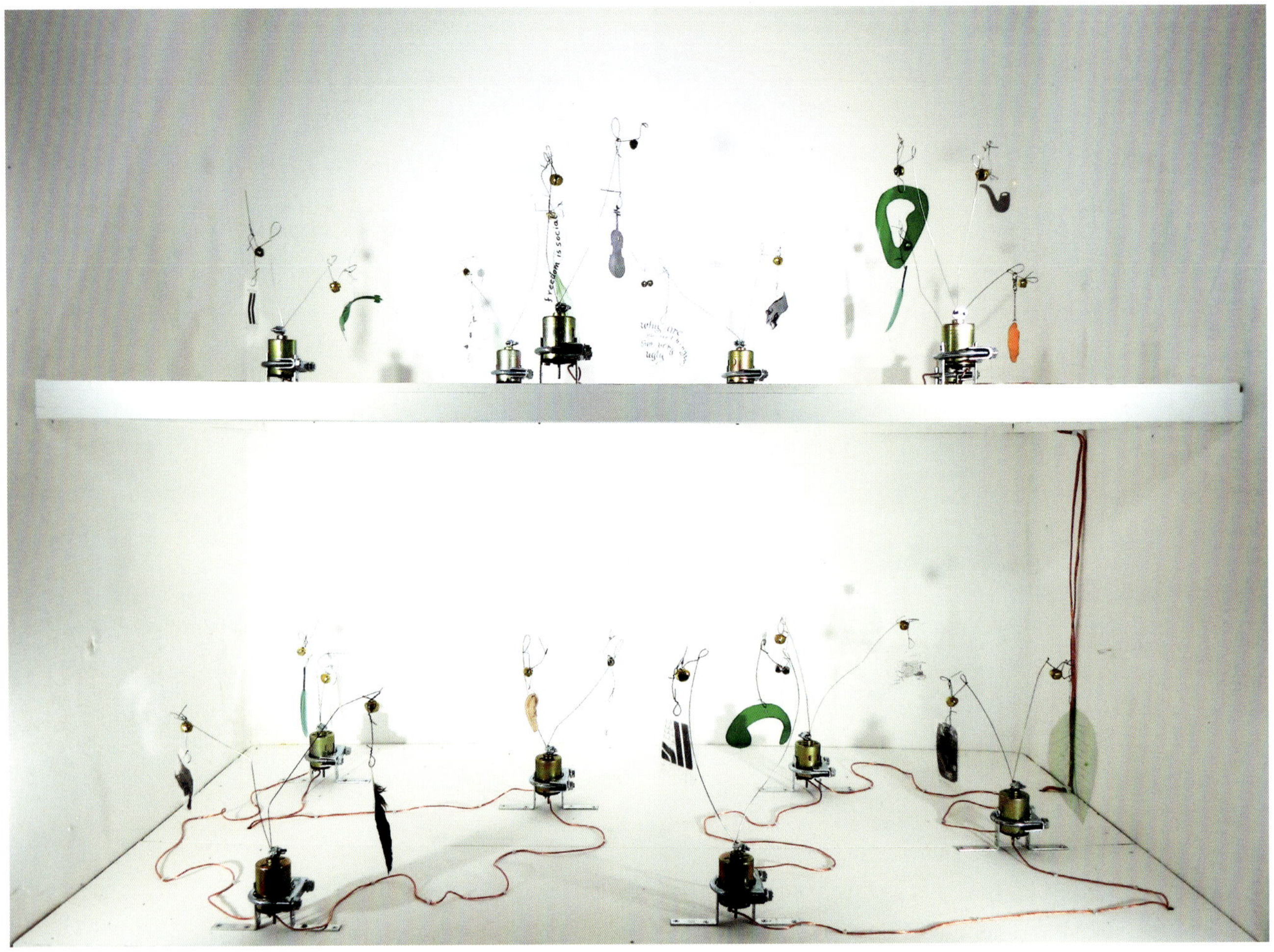

Trembling Antennae for Henning Christiansen, 2013
Installation
Motors, plastic, wire, amplifiers, computer running Max/MSP, audio interface, motion sensor
Commissioned for the group exhibition *The Hammer Without a Master: Henning Christiansen's Archive*, curated by Chiara Giovando for Kunsthal 44, Moen, Denmark
Gordon Monahan

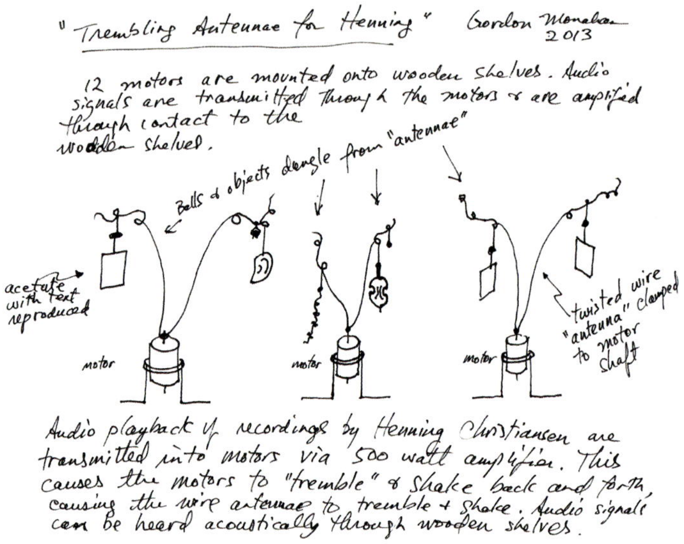

Twelve electric motors are attached to the speaker outputs of audio amplifiers, so that the motors shake and quiver in relation to the audio signals transmitted into them. Attached to the motors are twisted metal wires resembling antennae. Each "antenna" has bell(s) and/or a small object hanging from it. Each object relates to a specific theme or symbol in Henning's work.

The motors are mounted onto wooden shelves, which act as sound diaphragms, so that the audio signals and trembling sounds of the motors are reproduced and amplified into the exhibition space.

The audio recordings transmitted into the motors are excerpts from Henning's solo and collaborative projects released as LPs and CDs, including *Abschiedssymphonie*, *Dust Out of Brain*, *Hulemaned*, *Schafe Statt Geige*, *Schottische Symphonie*, *Stone Song*, *Sulemaned*, *Symphony Natura*, and *Vogel Symphonie*. (Gordon Monahan)

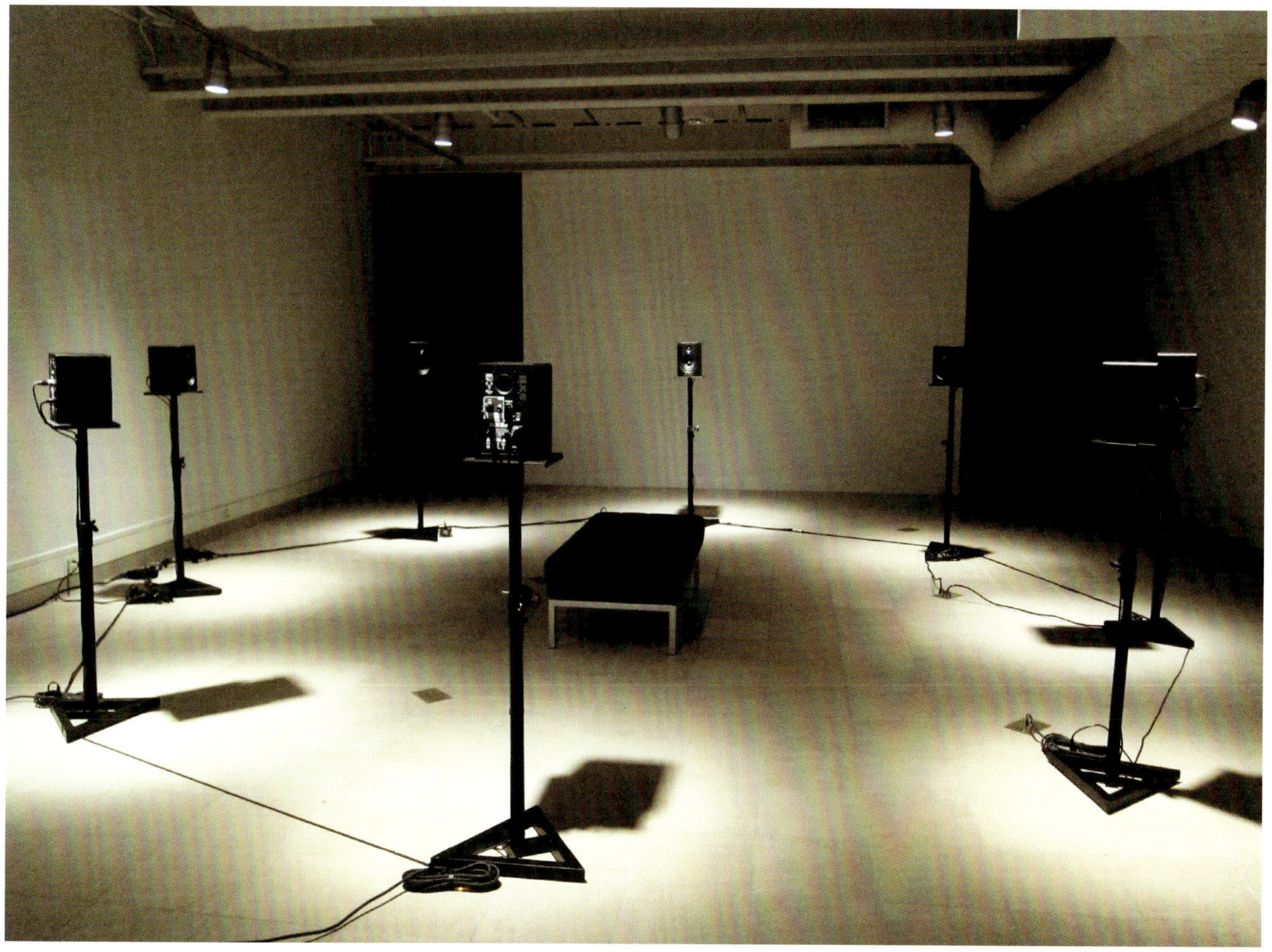

A Very Large Vinyl LP Constructed in Acoustic Space, 2007
Installation
8 loudspeakers, computer running Max/MSP, audio interface
Created with the assistance of a research and production grant from the Canada Council for the Arts, and supported by an artist-in-residency with the New Media Sound Lab at the University of Regina, Saskatchewan, Canada
Gordon Monahan

Eight loudspeakers are configured in an octagonal formation and arranged in a circle in the exhibition space. The octagonal formation refers to the spatial placement as well as the wiring configuration of the eight speakers to the audio distribution interface, such that each consecutive speaker is wired in a step-by-step clockwise pattern from speaker number one through to speaker number eight. When the listener sits in the middle of the circle of speakers and a sound source pans through the eight consecutive speakers, the sound rotates around the listener in a circular pattern.

An audio collage of vinyl LP surface scratches and vintage easy listening records fade in and out, spinning either clockwise or counterclockwise or both directions simultaneously, creating an illusion of a large virtual vinyl LP spinning in acoustic space around the listener. (Gordon Monahan)

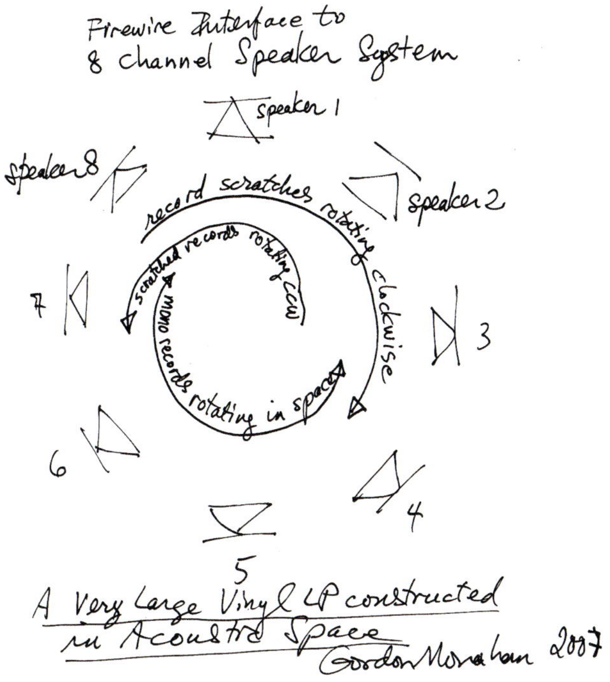

Mongrel

Mongrel, the group around Graham Harwood, Matsuko Yokokoji, Richard Pierre-Davis, and Mervin Jarman, saw itself as an open collective of human beings and machines making socially engaged culture, which sometimes means making art, sometimes software, sometimes setting up workshops or helping other mongrels to set things up.

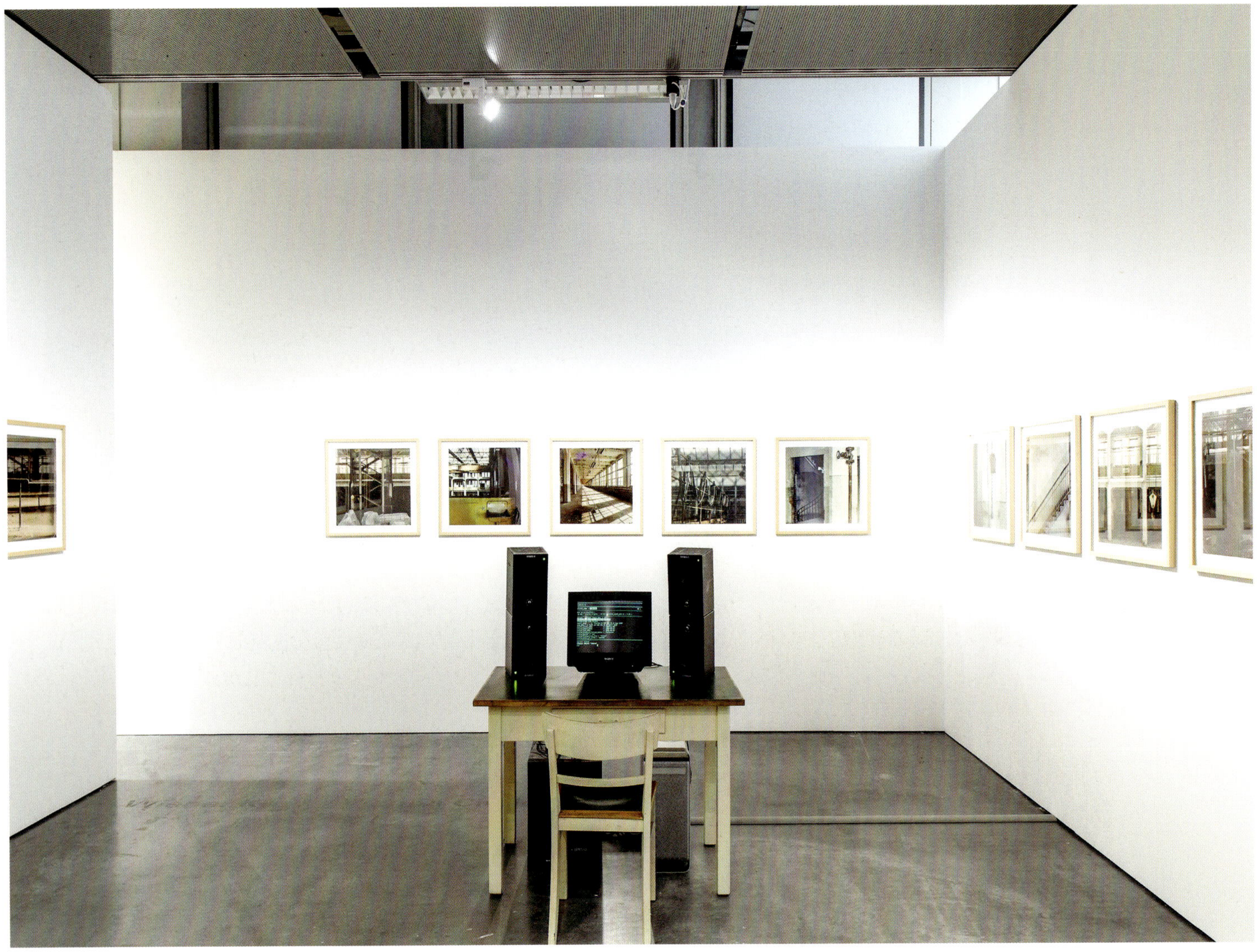

Lung: Slave Labour, 2005
Sound-based installation using anonymous data from the Generallandesarchiv Karlsruhe
Table, chair, computer, monitor, loudspeaker, dimensions variable
ZKM | Karlsruhe

Lung: Slave Labour is an installation in memory of the 4500 forced laborers who in World War II were compelled to work in Hall A of the former German Weapons and Munitions factory, which houses today the ZKM | Karlsruhe.

Graham Harwood and his colleagues in the English artists' collective Mongrel (1997–2008) used self-written software, *Lungs.pl*, to calculate the lung volume of the forced laborers during their time in Karlsruhe. Their collective breathing is audible via the loudspeaker. In concert with the inhaling and exhaling, information about the individuals, gleaned from the Generallandesarchiv Karlsruhe, appears on the screen. The project's aim was to connect the perception of data with social experience by confronting the machine "software" with the machine "human."

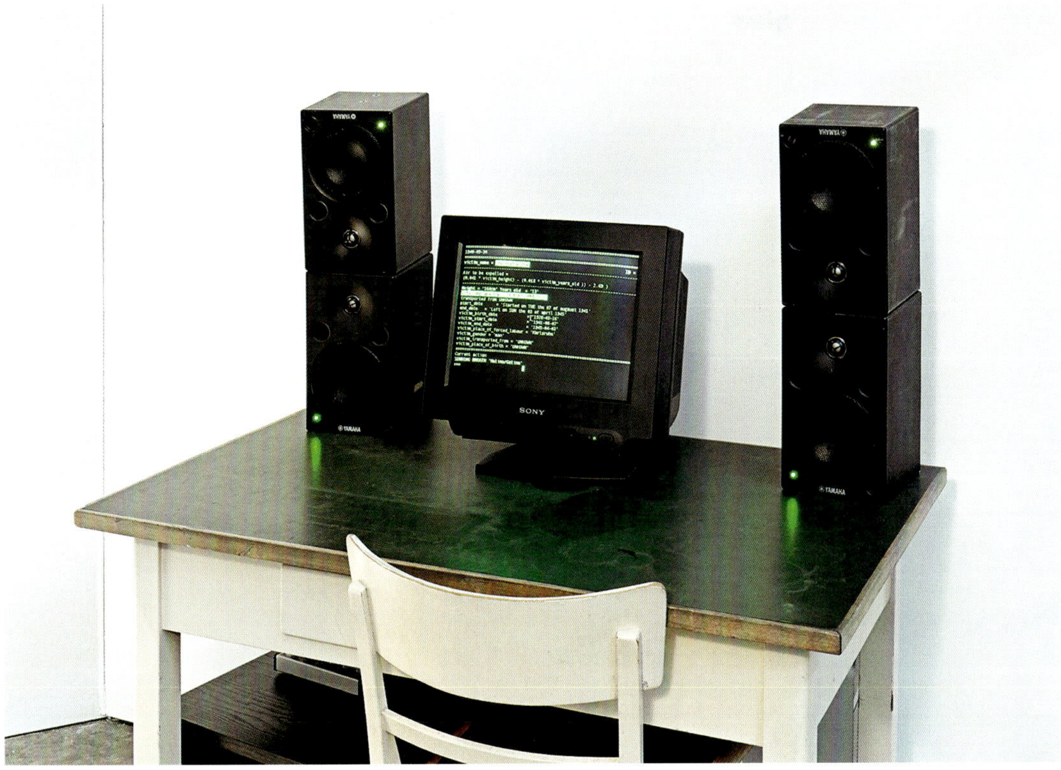

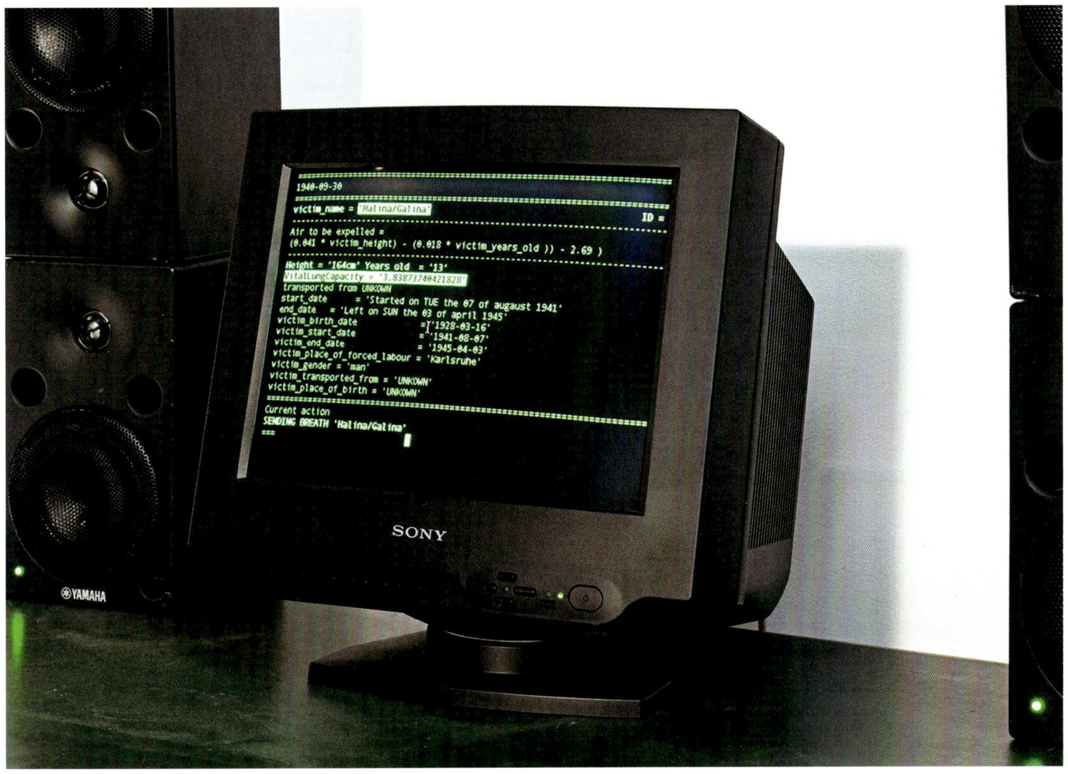
1940-09-30
victim_name = 'Halina/Galina'
ID =
Air to be expelled =
(0.041 * victim_height) - (0.018 * victim_years_old)) - 2.69)
Height = '164cm' Years old = '13'
VitalLungCapacity = '3.83873740421828'
transported from UNKOWN
start_date = 'Started on TUE the 07 of augaust 1941'
end_date = 'Left on SUN the 03 of april 1945'
victim_birth_date = '1928-03-16'
victim_start_date = '1941-08-07'
victim_end_date = '1945-04-03'
victim_place_of_forced_labour = 'Karlsruhe'
victim_gender = 'man'
victim_transported_from = 'UNKOWN'
victim_place_of_birth = 'UNKOWN'
Current action
SENDING BREATH 'Halina/Galina'
SONY
YAMAHA

Bruce Nauman

*1941 in Fort Wayne, IN (US), lives in Galisteo, NM (US)

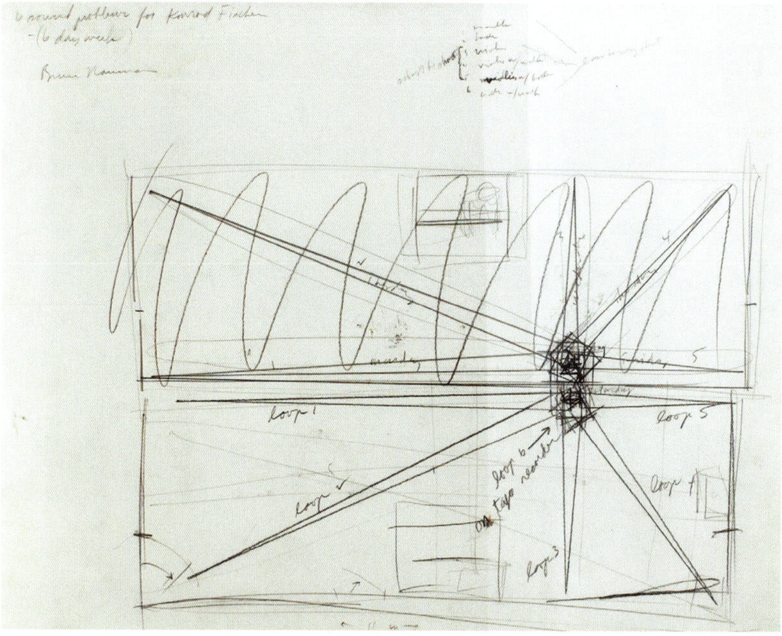

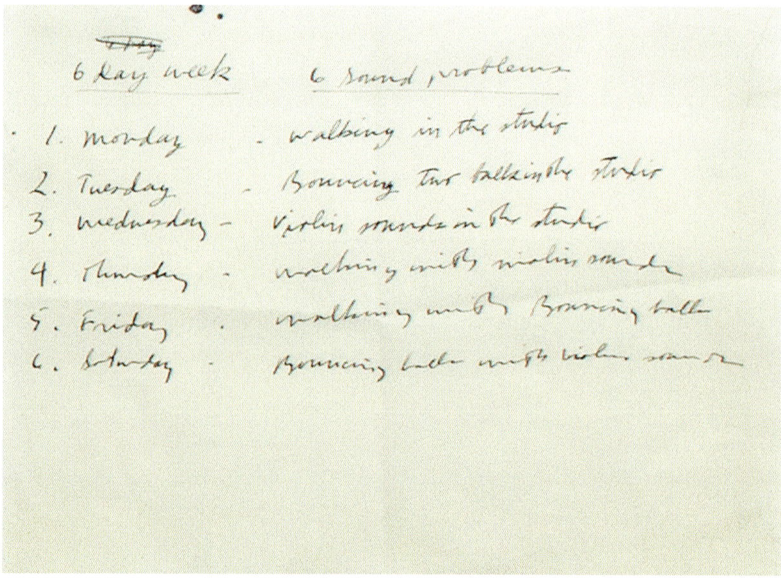

6 Day Week – Six Sound Problems for Konrad Fischer, 1968
Installation
Tape deck with six soundtracks of varying duration, chair with pencil, dimensions variable
Collection Dorothee and Konrad Fischer

In July 1968 Bruce Nauman designed *Six Sound Problems* for the Konrad Fischer Galerie in Düsseldorf, Germany. For four weeks he played a different tape every day the gallery was open. The sound events were related to the exhibition venue and were recorded there: on Mondays the sounds of footsteps, Tuesdays two balls, Wednesdays the sounds of his violin, Thursdays footsteps and balls, Fridays footsteps and sounds of the violin, and Saturdays sounds of the violin and balls.

The tapes are stretched like a ribbon between the tape player and a pencil attached to a chair. The title indicates that the work deals not only with documenting everyday site-specific sounds and their possible permutations. Nauman presents the recordings as a discursive riddle. The sounds labeled as problems beg for solutions, without any concrete questions having been formulated.

Monday – Walking in the gallery
Tuesday – Bouncing two balls in the gallery
Wednesday – Violin sounds in the gallery
Thursday – Walking and bouncing balls
Friday – Walking and violin sound
Saturday – Violin sounds and bouncing balls

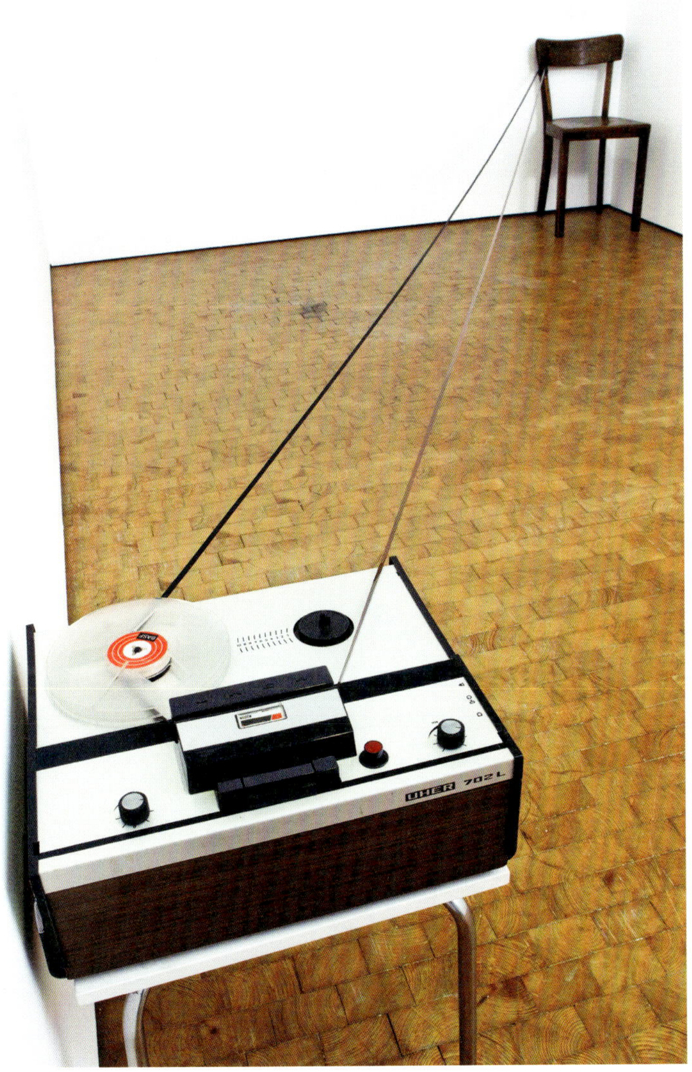
UHER 702 L

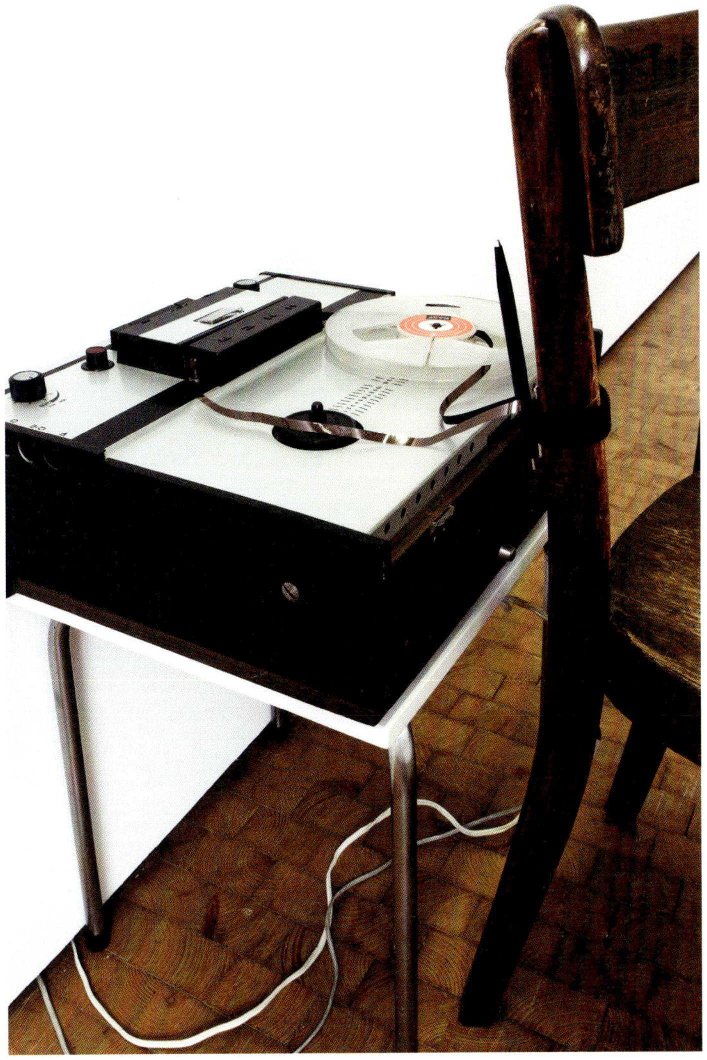

Anselm Venezian Nehls
Tarik Barri

*1981 in Berlin (DE), lives in Berlin
*1979 in Arnhem (NL), lives in Berlin

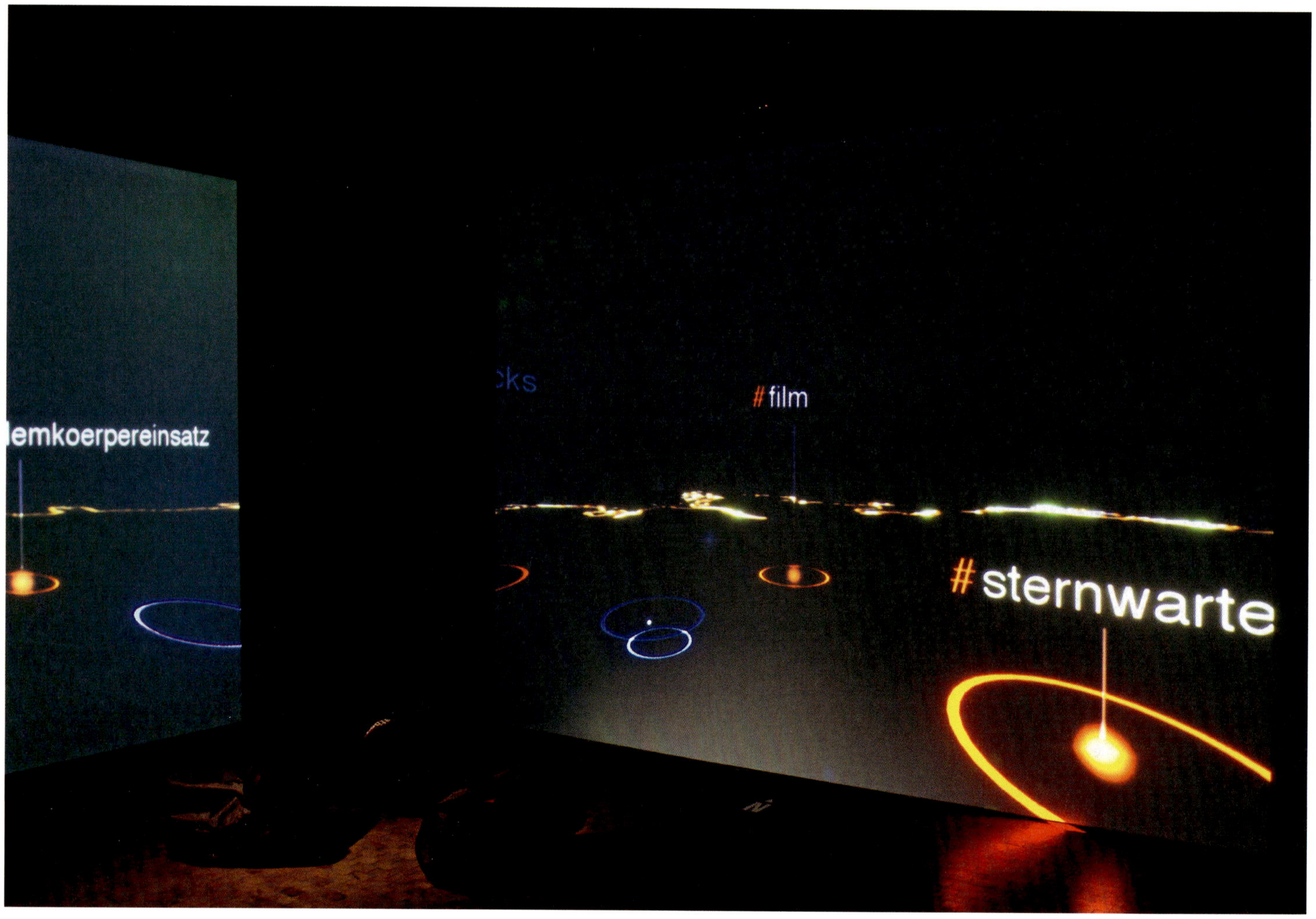

#tweetspace, 2011
Interactive audiovisual surround installation
Computer, loudspeakers, projectors, self-developed software
Additional programming: Florian Eitel
Scientific consultant: Dr. Thomas Hermann
Artistic consultant: Sam Auinger
Dramaturgy: Marcus Gammel
Courtesy: © HEAVYLISTENING
Production: *#tweetspace* is part of the *#tweetscapes* project, a collaboration between Deutschlandradio Kultur, HEAVYLISTENING, CITEC Bielefeld, and MA Sound Studies, Berlin University of the Arts

www.tweetscapes.de
www.venezian.de
www.tarikbarri.nl

#tweetspace offers a new sensory mode of access to the social web. Around the clock, all Tweets in German language are transformed into abstract sounds and images – in real time, right after being sent. This happens according to a set of rules: different sounds are assigned to #hashtags (thematic keywords), @replies (personal messages), Retweets (forwards), and Trending Topics (dominant themes). The computer automatically works out a unique characteristic sound for every possible subject.

In this way, the up-to-the-minute mood at Twitter and the evolution of nationwide themes of discourse become evident: the initial flare of a message, the rapid spread of its echoes and repeats, and the "basic undertone" of a day marked by a certain theme. *#tweetspace* is both an interactive audiovisual composition and a scientific sonification and visualization of data, which should facilitate new hypotheses about the underlying information.

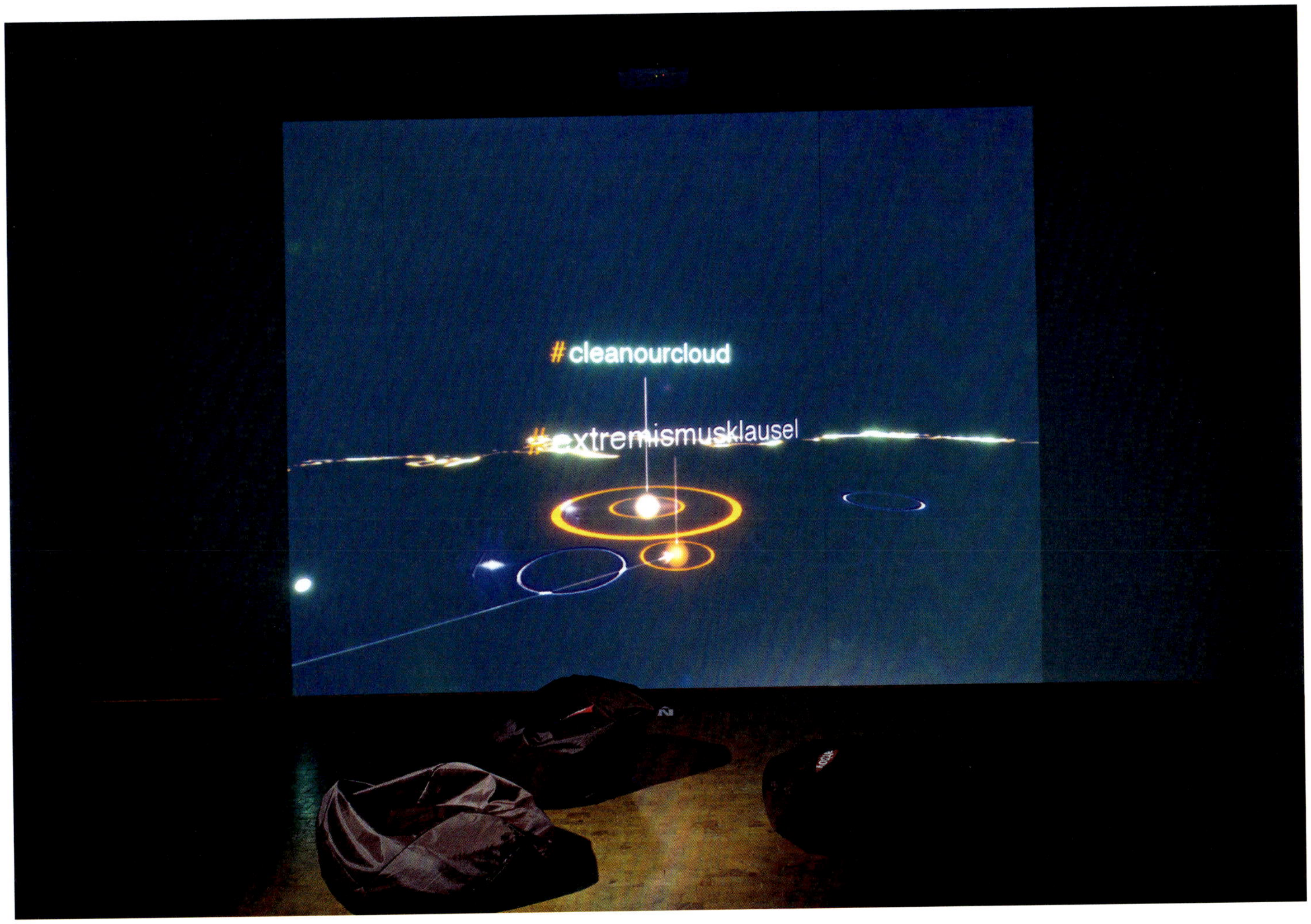
#cleanourcloud
#extremismusklausel

Max Neuhaus

*1939 in Beaumont, TX (US), †2009 in Maratea (IT)

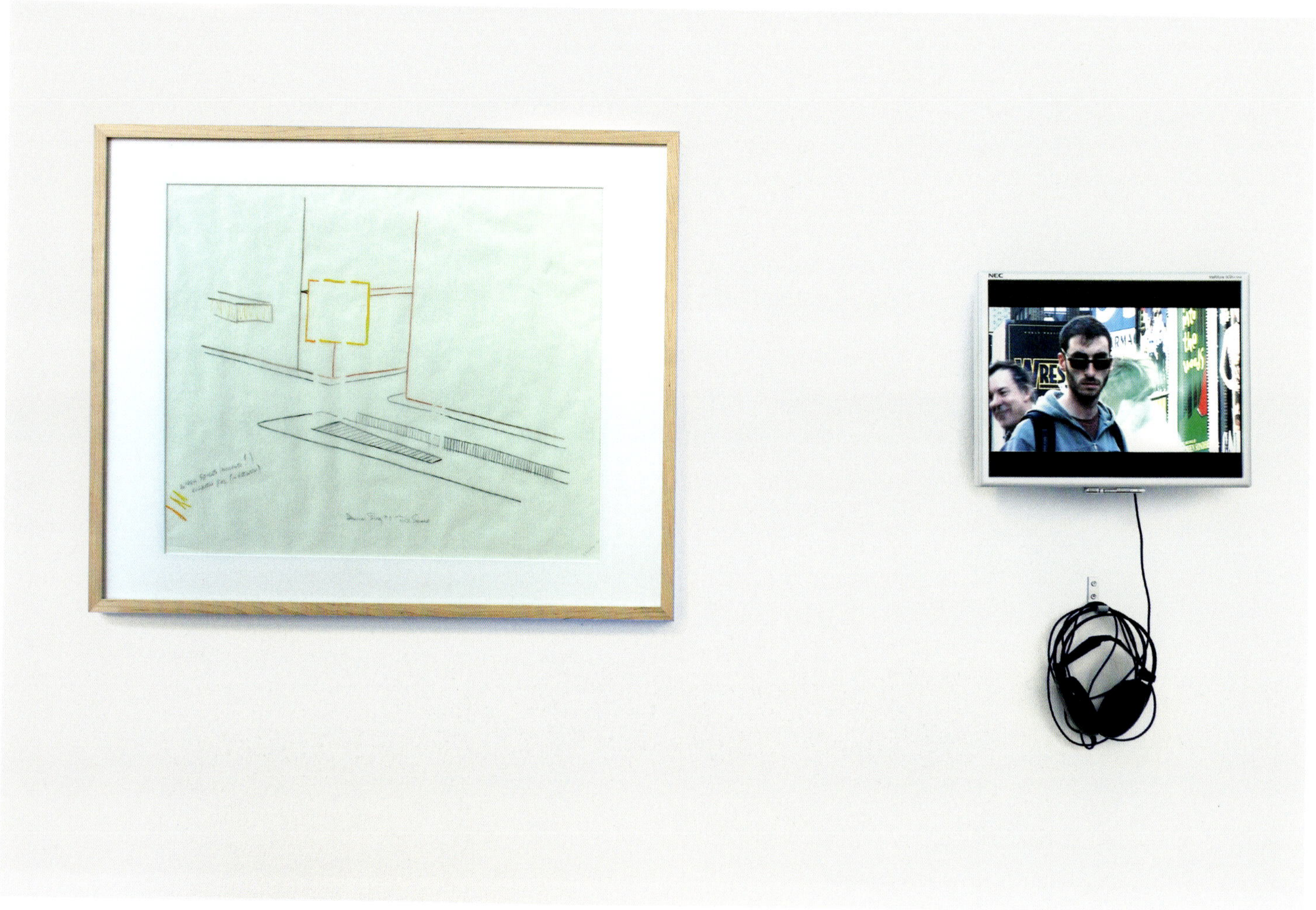

Times Square, Drawing Study 3, 1992
Drawing, colored pencil on paper, 71 × 85 cm
Place of the reference sculpture: pedestrian island between 46th and 45th Street, New York City (1977–1992 and 2002–today)
The Estate of Max Neuhaus

right page:
Times Square, 2002
Film by Rory Logsdail
Production: Firefly Pictures for Rai Sat Art, interviews with Max Neuhaus, Christine Burgin, Christine Burgin Gallery, Lynne Cooke, Dia Center for the Arts, David Andrews, Andrews Audio
DVD, 7:57 min.
© Firefly Pictures 2002, The Estate of Max Neuhaus

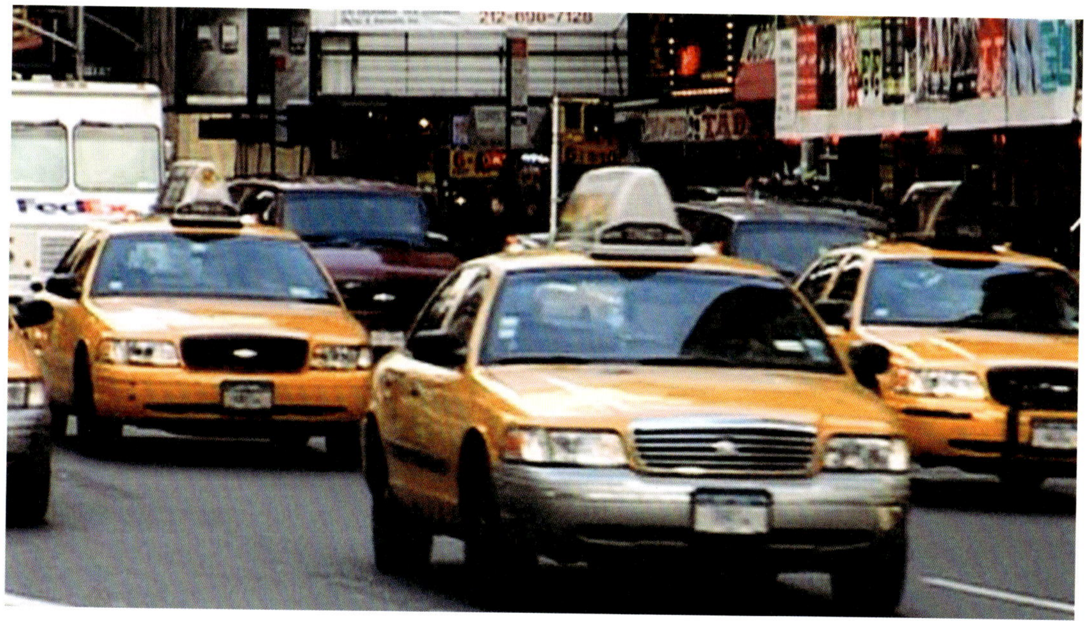

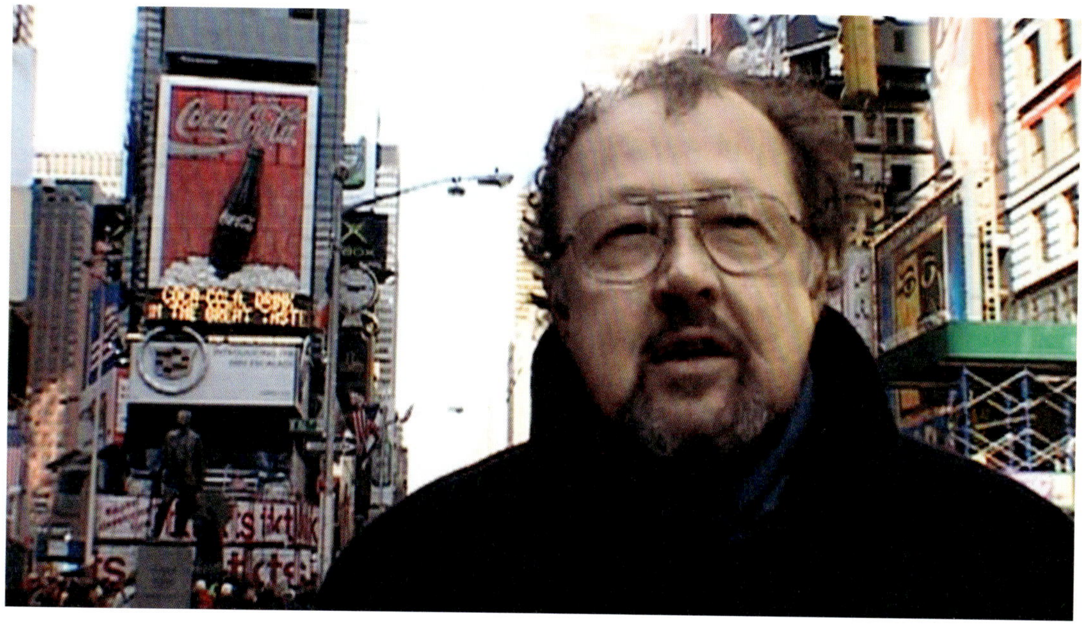
Coca-Cola

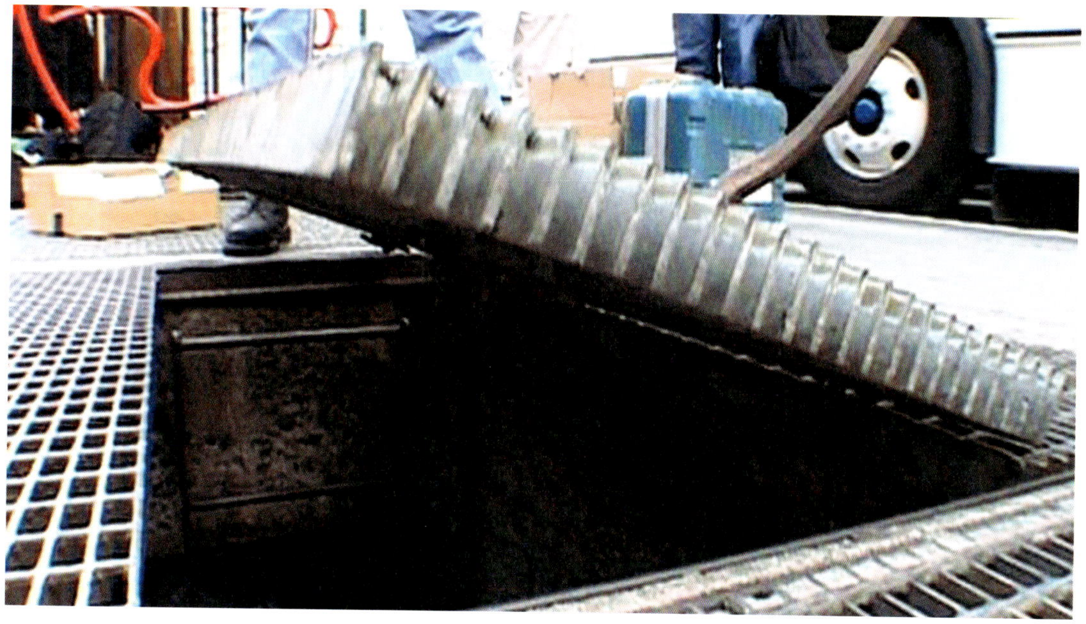

Max Neuhaus

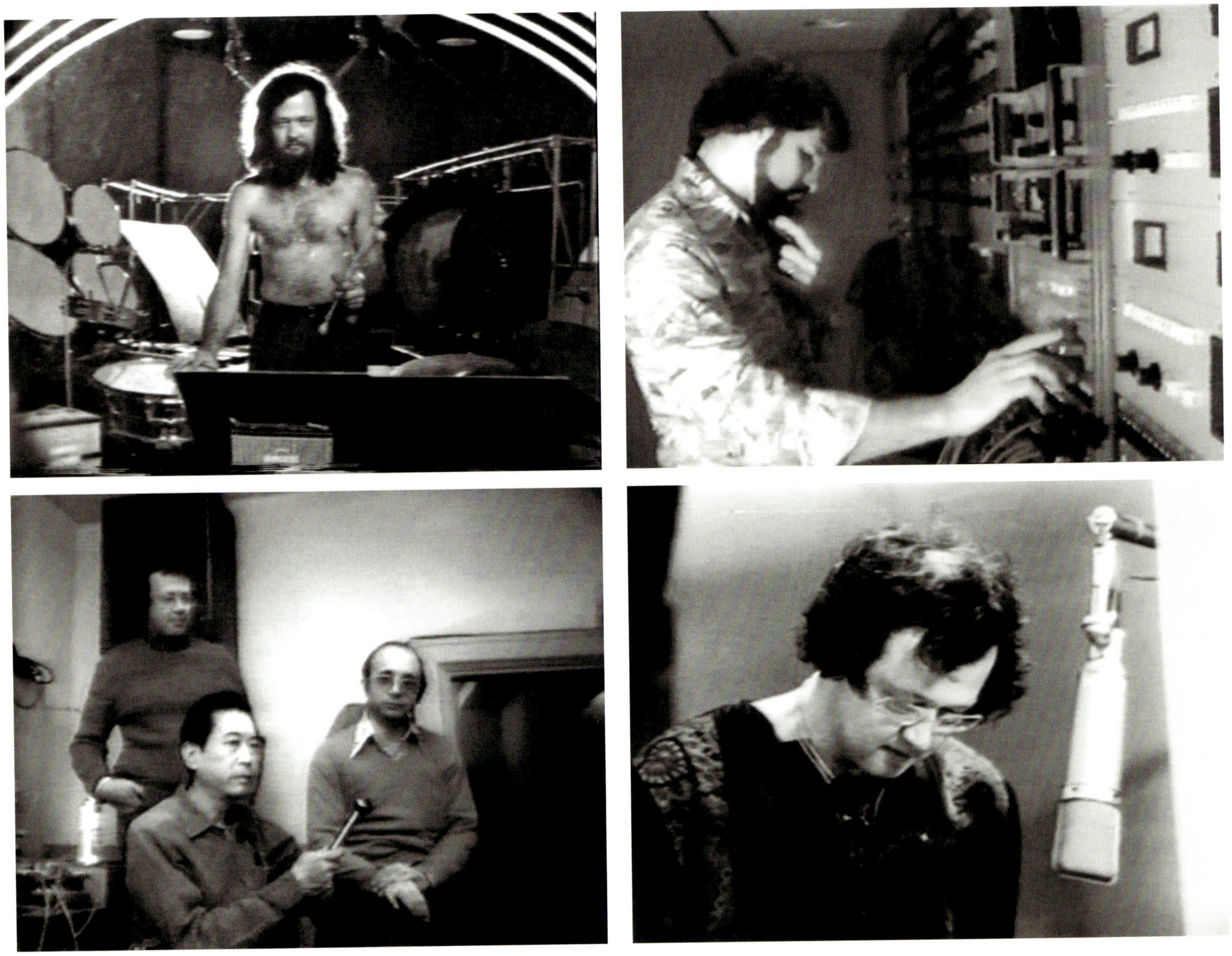

A Look behind the scenes at Radio Net's process, 1977
Documentary, DVD, 25:09 min.
© Broadside Television
The Estate of Max Neuhaus

In the late 1970s, the radio programs of the American public radio network NPR were transmitted over telephone lines connecting all 200 NPR stations throughout the country.

In 1977, sound artist Max Neuhaus took over the NPR radio waves with his work *Radio Net*. Neuhaus invited the listeners to call and produce a particular sound, which then went through a mixer and was looped at different pitches and began to overlap until it gradually ended. The more calls came in, the more complex the sound became.

His ambitious experimental sound installation and live broadcast lasted two hours and consisted of the specific sound generated by callers from all over the country.

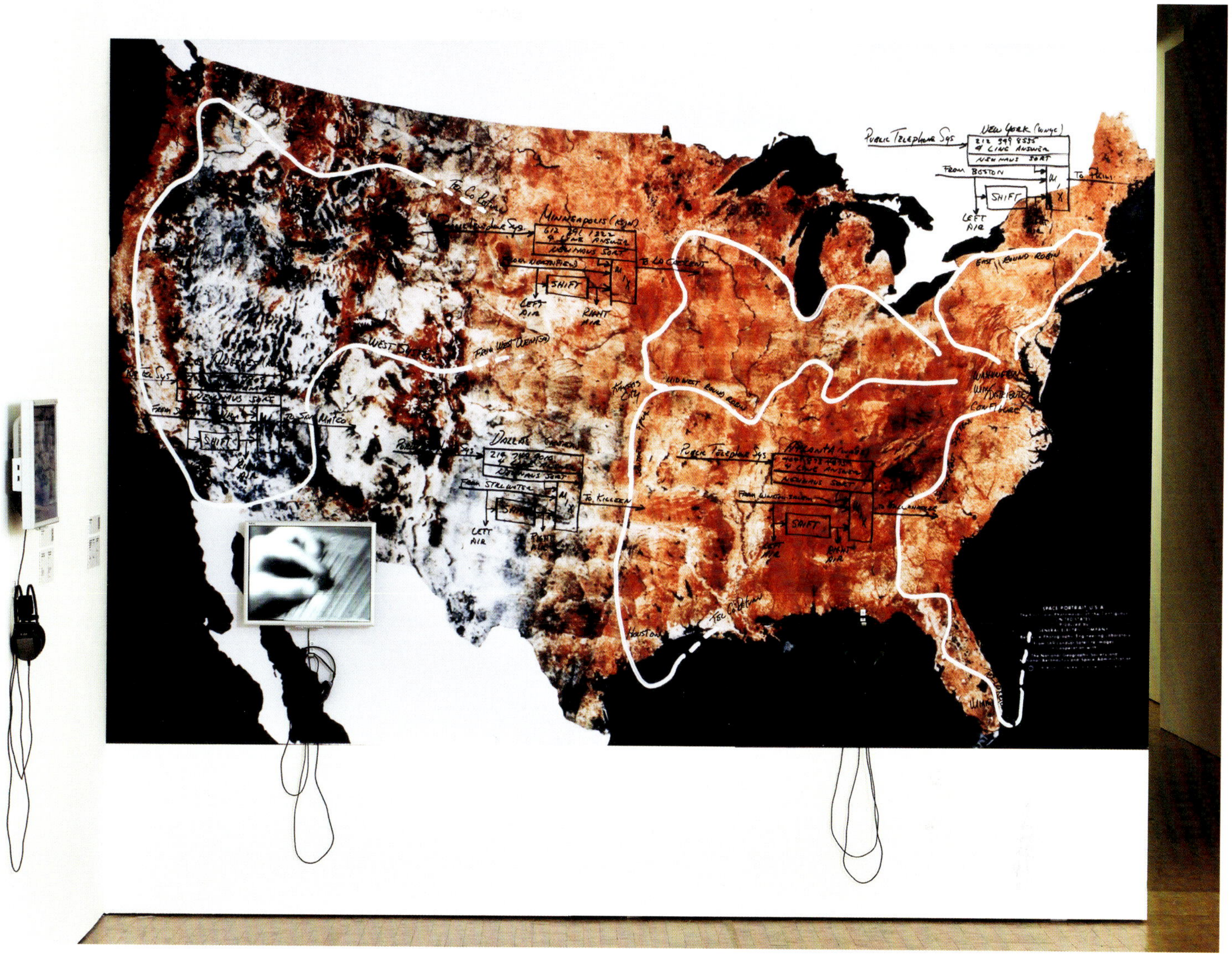

Sound Paths, Radio Net, 1978
Cibachrome print and paint,
video documentary of *Radio Net*,
dimensions variable
The Estate of Max Neuhaus

Carsten Nicolai

*1965 in Karl-Marx-Stadt (GDR), today Chemnitz (DE), lives in Chemnitz and Berlin (DE)

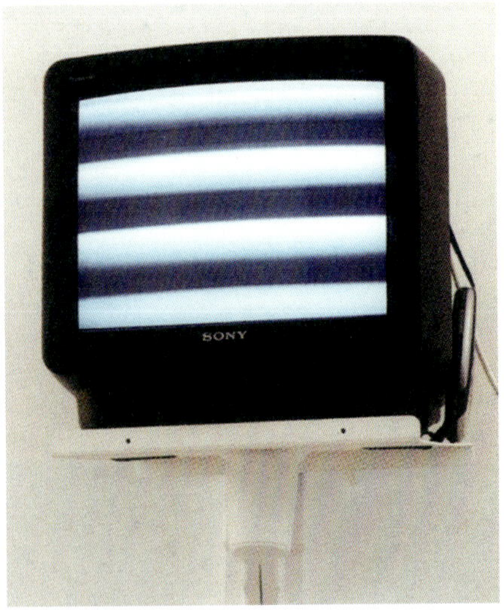

Telefunken, Audiosignal for Television, 2000
CD player, 4 Sony TVs, audio CD *Telefunken*,
dimensions variable

The audio signal of a CD player is connected to the video input of a television. The audio tracks on the CD – impulse frequencies and various test sounds – are translated by the television into an abstract image, interpreting the impulse frequencies as horizontal stripes on the television screen. The respective intervals or the width of the impulse correspond with the density or width of the line of the image on the screen. This installation instantaneously demonstrates the visualization of acoustic material.

www.carstennicolai.de

right page:
aoyama space no. 4, 2009
Sound installation
Anodized aluminum box, electronics,
light, sound, 160 × 100 × 200 cm
Courtesy Galerie EIGEN + ART and The Pace Gallery

aoyama space takes its name from a concave, apparently endless space in a photo studio in the Aoyama district of Tokyo. The works in the series are models of imagined spatial installations. Their small scale makes it possible to realize radical structures with complex features. The model rooms have curved surfaces, illuminated slits and projection surfaces, and are individually lighted.

aoyama space no. 4 is the fourth model in the series. On the rear interior wall of the box, which is painted black, three fine horizontal lines can be discerned. The lighting is controlled by especially composed sounds, ranging from deep bass frequencies to high-frequency clicks, whose frequency modulations and rhythm influence the interplay of light and space. Deep tones create linear one-dimensional slivers; flashes of light give the impression of two-dimensional surfaces; flashes of light escaping the box generate three-dimensional impressions of the space.

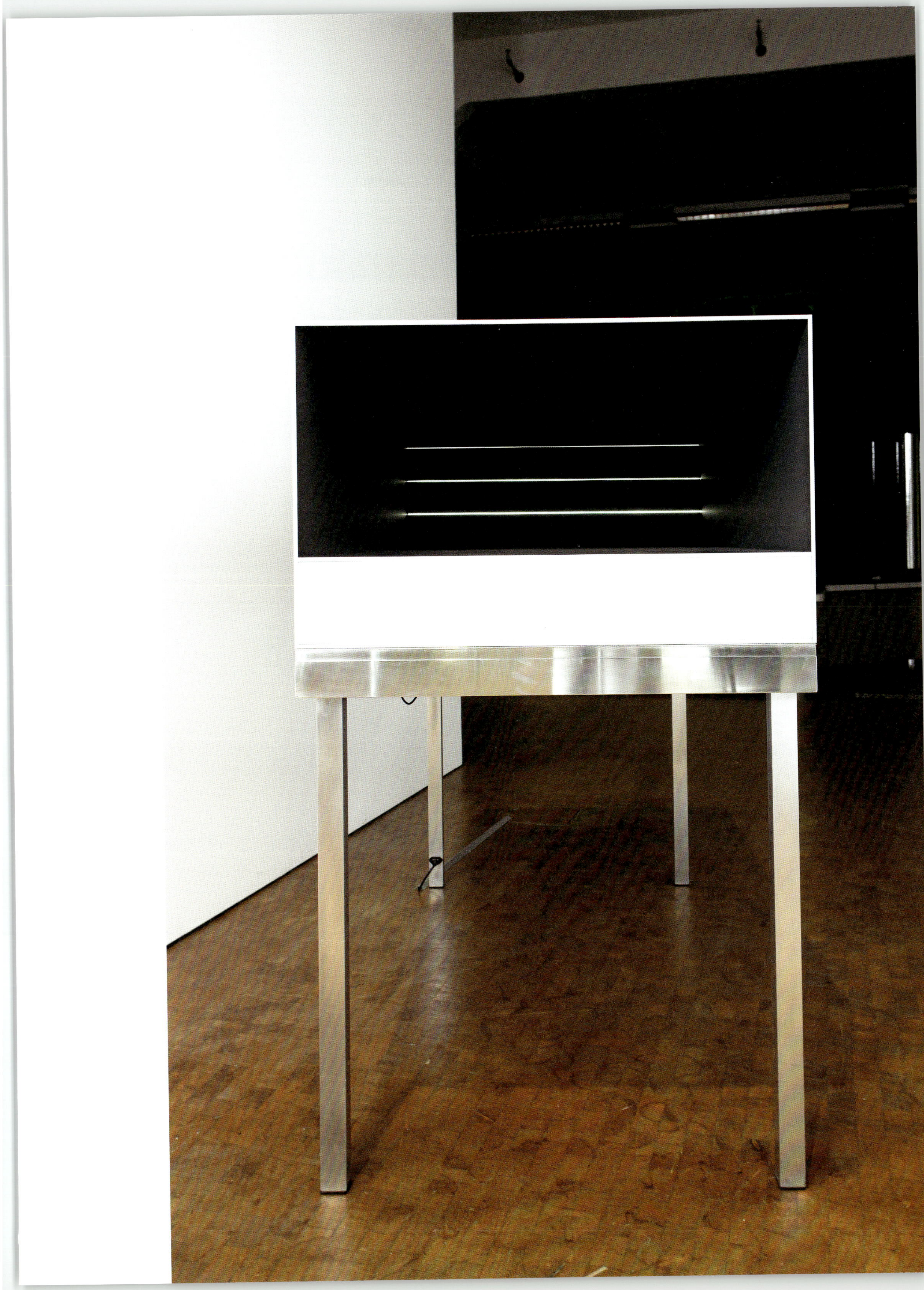

Pauline Oliveros

*1932 in Houston, TX (US), †2016 in New York City, NY (US)

SONIC MEDITATIONS

by Pauline Oliveros

INTRODUCTION I

Sonic Meditations are intended for group work over a long period of time with regular meetings. No special skills are necessary. Any persons who are willing to commit themselves can participate. The ♀ Ensemble to whom these meditations are dedicated has found that non-verbal meetings intensify the results of these meditations and help provide an atmosphere which is conducive to such activity. With continuous work some of the following becomes possible with Sonic Meditations: Heightened states of awareness or expanded consciousness, changes in physiology and psychology from known and unknown tensions to relaxations which gradually become permanent. These changes may represent a tuning of mind and body. The group may develop positive energy which can influence others who are less experienced. Members of the Group may achieve greater awareness and sensitivity to each other. Music is a welcome by-product of this activity.

Copyright © 1974
By SMITH PUBLICATIONS
All Rights Reserved.

Sonic Meditations, 1974
Score, paper
Selection of 4 facsimile pages from the original version
(16 pages), Smith Publications, 1977

www.paulineoliveros.us

Pauline Oliveros's work as a composer centered around her practice known as Deep Listening, a meditative approach to hearing and composing. She was also a keen educator, and held workshops that instructed people in ways to listen more closely to their environment and to use these attention-centering skills in musical composition. The *Sonic Meditations* are excerpts from a series of group exercises created by Oliveros in 1974 that are meant to guide groups of listeners through the following acts: actually making sounds, actively imagining sounds, listening to present sounds, and remembering sounds. Her intention was that students would work at these exercises over long periods of time to achieve heightened states of awareness.

INTRODUCTION II

Pauline Oliveros has abandoned composition/performance practice as it is usually established today for Sonic Explorations which include everyone who wants to participate. She attempts to erase the subject/object or performer/audience relationship by returning to ancient forms which preclude spectators. She is interested in communication among all forms of life, through Sonic Energy. She is especially interested in the healing power of Sonic Energy and its transmission within groups.

All societies admit the power of music or sound. Attempts to control what is heard in the community are universal. For instance, music in the church has always been limited to particular forms and styles in accordance with the decrees of the Church Fathers. Music in the courts has been controlled through the tastes of patrons. Today Muzak is used to increase or stimulate consumption in merchandising establishments.

Sonic Meditations are an attempt to return the control of sound to the individual alone, and within groups especially for humanitarian purposes; specifically healing.

Each Sonic Meditation is a special procedure for the following:

1. Actually making sounds
2. Actively imagining sounds
3. Listening to present sounds
4. Remembering sounds

Because of the special procedures involved, most all of the meditations are available to anyone who wishes to participate regardless, or in spite, of musical training. All that is required is a willing commitment to the given conditions.

Sound making during the meditations is primarily vocal, sometimes hand clapping or other body sounds, sometimes using sound producing objects and instruments.

Sound imagining is encouraged through the use of various questions designed to trigger auditory fantasy. Individuals are then asked to share what was heard inwardly, with members of the group using any means to describe the experience. Conditions given for listening to present sounds are intended to expand awareness of the auditory environment, both within and without of the individual.

Auditory memory is also encouraged by trigger questions with subsequent sharing of these memories in the group. Some of the meditations involve body movement as well. The term meditation is used simply to mean dwelling with or upon an idea, an object, or lack of object without distraction, or divided attention.

Healing can occur in relation to the above activities when 1) individuals feel the common bond with others through a shared experience. 2) when one's inner experience is made manifest and accepted by others. 3) when one is aware of and in tune with one's surroundings. 4) when one's memories, or values, are integrated with the present and understood by others.

In process a kind of music occurs naturally. Its beauty is not through intention, but is intrinsically the effectiveness of its healing power. This may be felt by the group, and the music relates to the people who make it through participation and sharing, as a stream or river whose waters offer refreshment and cleansing to those who find it.

Copyright © 1974
By SMITH PUBLICATIONS
All Rights Reserved.

Pauline Oliveros

–I–

Teach Yourself to Fly

Any number of persons sit in a circle facing the center. Illuminate the space with dim blue light. Begin by simply observing your own breathing. Always be an observer. Gradually allow your breathing to become audible. Then gradually introduce your voice. Allow your vocal cords to vibrate in any mode which occurs naturally. Allow the intensity to increase very slowly. Continue as long as possible naturally, and until all others are quiet, always observing your own breath cycle.
Variation: Translate voice to an instrument.

Copyright © 1974
By SMITH PUBLICATIONS
All Rights Reserved.

—XIII—

Energy Changes
(For Elaine Summers' movement meditation, Energy Changes)

Listen to the environment as a drone. Establish contact mentally with all of the continuous external sounds and include all of your own continuous internal sounds, such as blood pressure, heart beat and nervous system. When you feel prepared, or when you are triggered by a random or intermittent sound from the external or internal environment, make any sound you like in one breath, or a cycle of like sounds. When a sound or a cycle of sounds, is completed re-establish mental connection with the drone, which you first established before *making another sound or cycle of like sounds.*

Daphne Oram

*1925 in Devizes (GB), †2003 in Maidstone (GB)

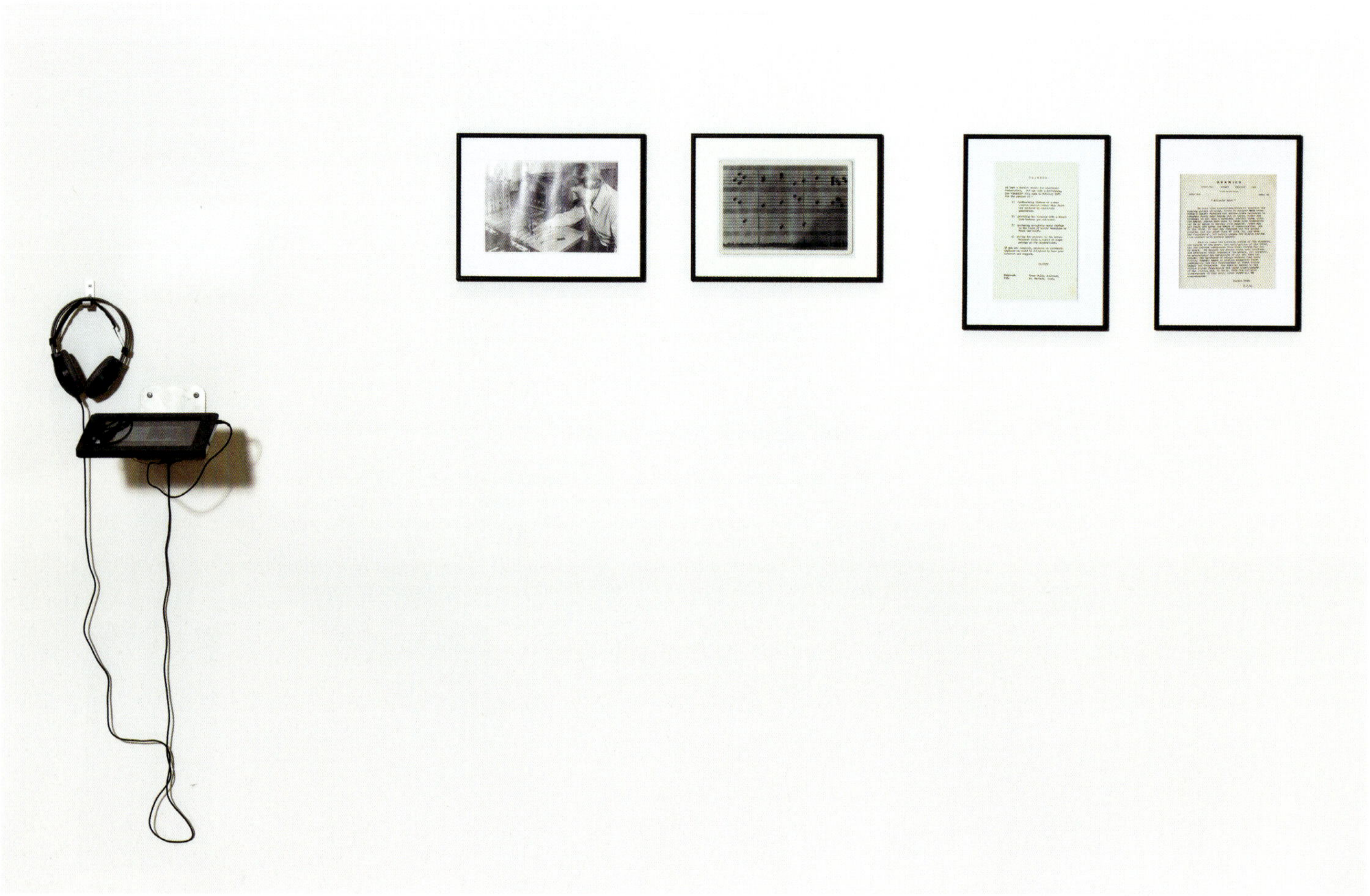

The Oramics Machine. The work of Electronic Music Pioneer, Daphne Blake Oram, 1959/2011
Mixed-media installation
iPad-Oramics-Emulator, sound recording, photographic prints
Programmer app: Parag Mital
Dr. Mick Grierson, Director, The Daphne Oram Collection, Goldsmiths, University of London in association with soundandmusic.org and The Science Museum, GB, all material supplied with permission of the: Daphne Oram Trust, The Daphne Oram Collection, Goldsmiths College

www.daphneoram.org

This iPad app emulates Daphne Oram's groundbreaking drawn sound technique known as *Oramics*. In 1957, Oram developed a system of sound composition based on drawing. Shapes from a symbolic alphabet were drawn by hand onto ten strips of clear 35mm film that were laid on a table and fed through a machine. As the film was passed through a machine containing photocells, the drawn lines acted as masks that modulated the light projected through the film. The machine would then translate the light into synthesized music output onto magnetic tape. In this iPad app, users can draw shapes on simulated 35mm film strips to effect pitch, reverb, and other aspects of the waveforms produced by a virtual Oramics machine. The app translates the sound instantly, seamlessly looping the virtual film through the Oramics translation algorithms. Unlike the original machine, the app allows the user to easily erase or edit the shapes drawn onto the virtual film. (John Kannenberg)

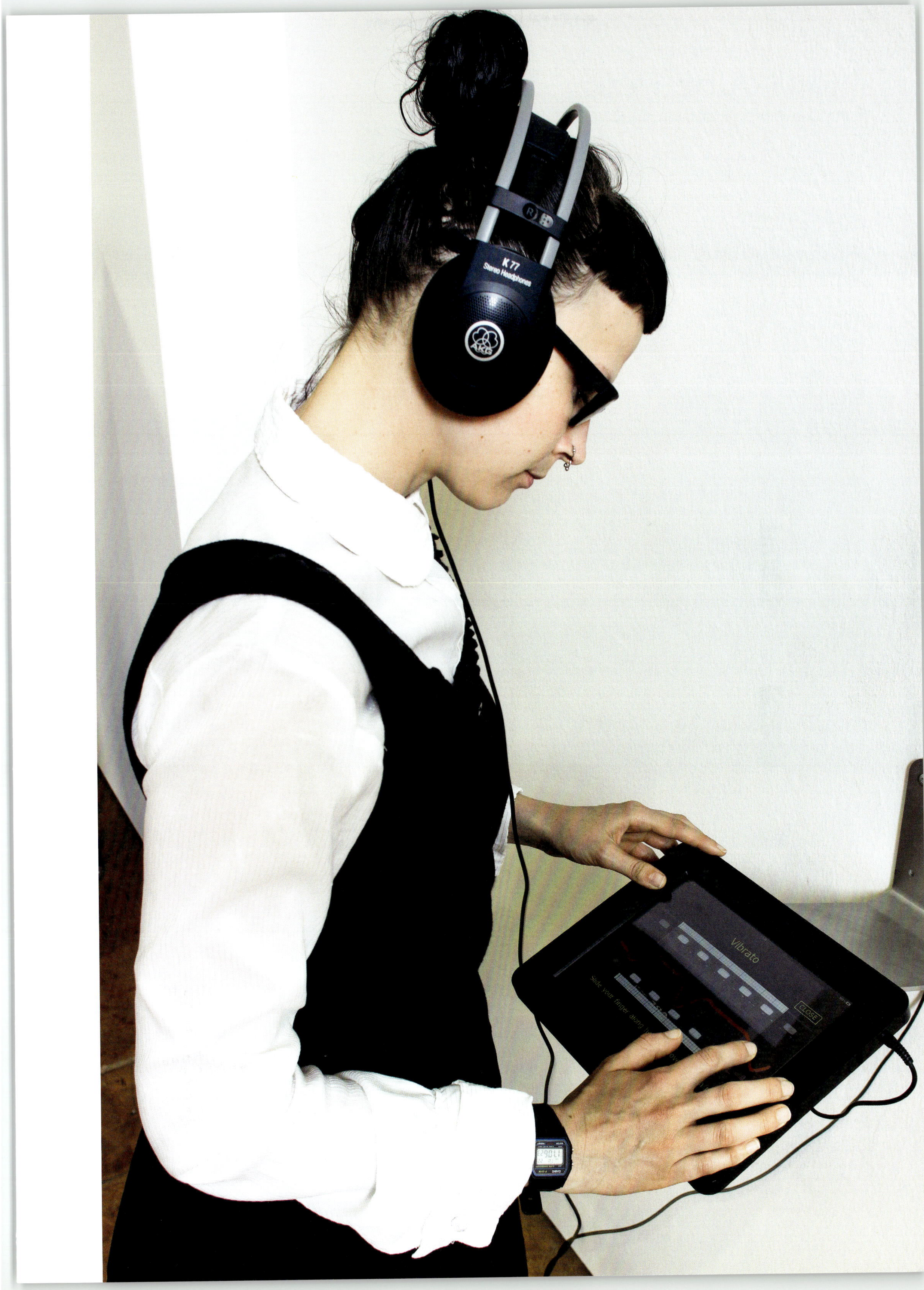
K 77
Stereo Headphones
AKG
Vibrato
CLOSE

Nam June Paik

*1932 in Seoul (KR), †2006 in Miami Beach, FL (US)

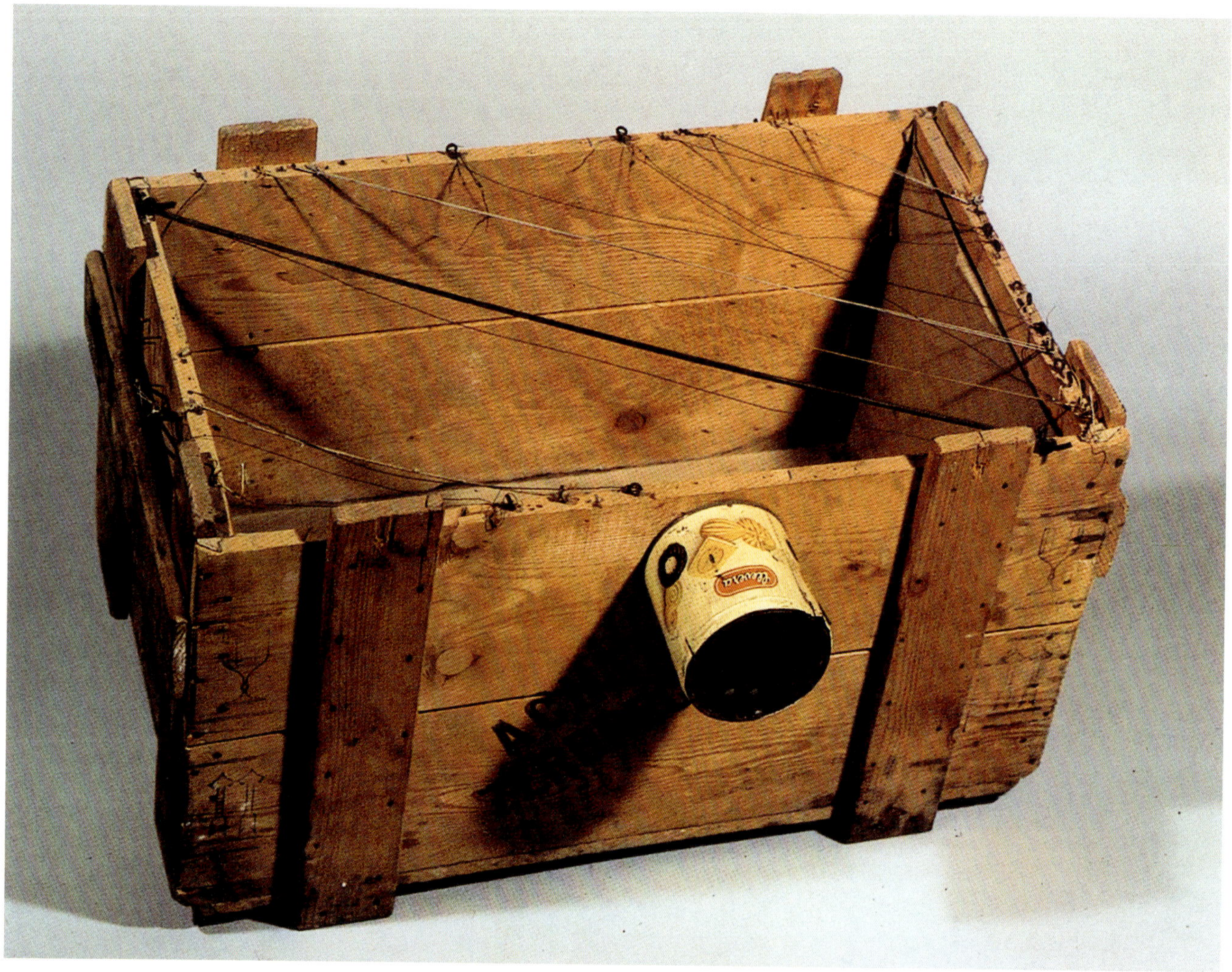

Urmusik, 1961
Wooden box, can, wire, 52 × 92 × 68 cm
Museum Moderner Kunst Stiftung Ludwig Wien

Nam June Paik's *Urmusik* [Primeval Music] consists of a simple, rough wooden box, with a rotatable tin can holding a ball mounted on the outside. Wire strings are stretched across the open top of the box. There are no particular pieces of music specifically composed for this sound box. Instead, viewers are invited to play the instrument as they like by rotating the box or plucking the strings.

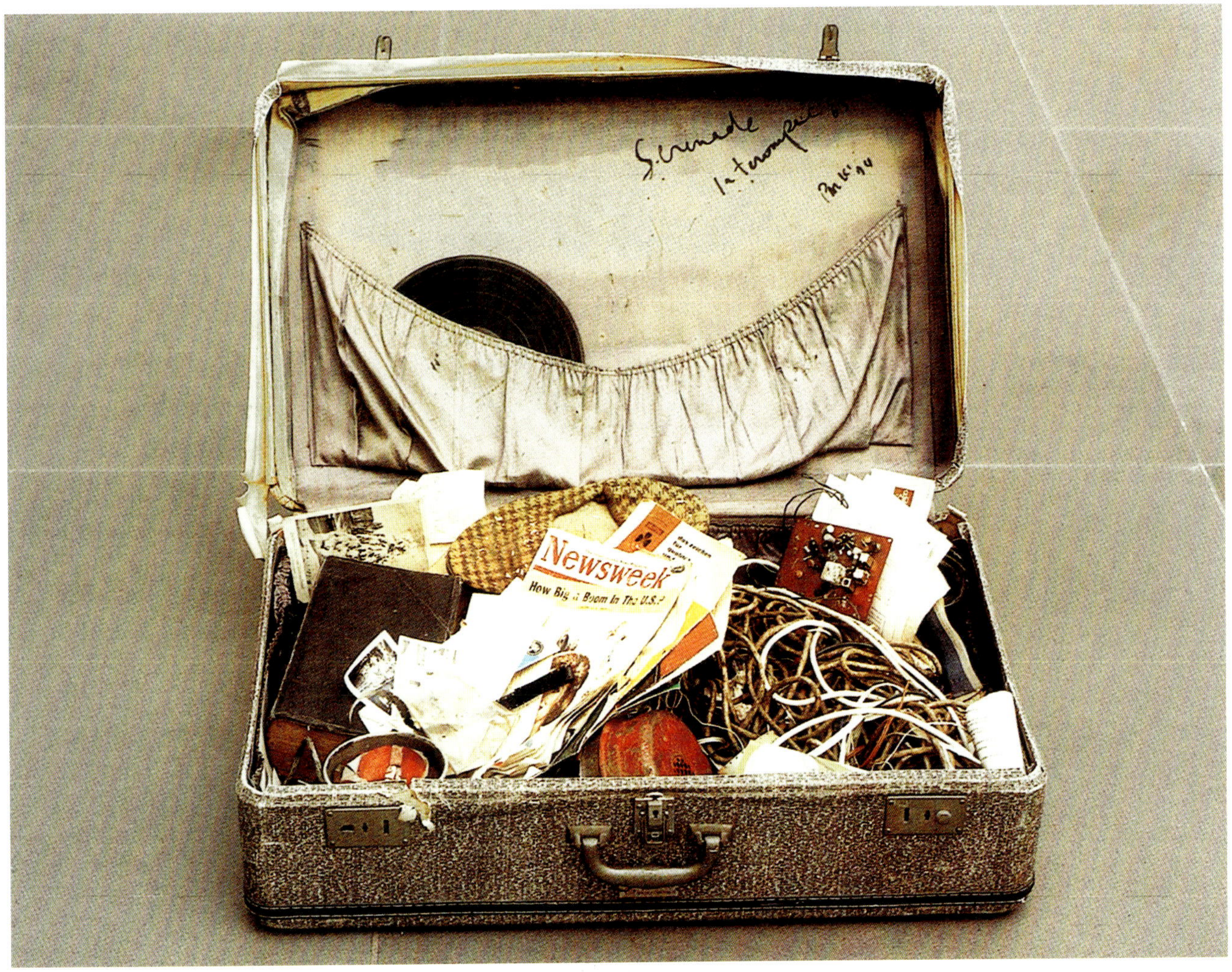

Box for Zen (Serenade), 1963
Suitcase with different sound objects, 35 × 55 × 80 cm

In his 1963 work *Box for Zen (Serenade)*, Nam June Paik shows a suitcase filled with a jumble of sound objects from the artist's early Fluxus performances.

Nam June Paik

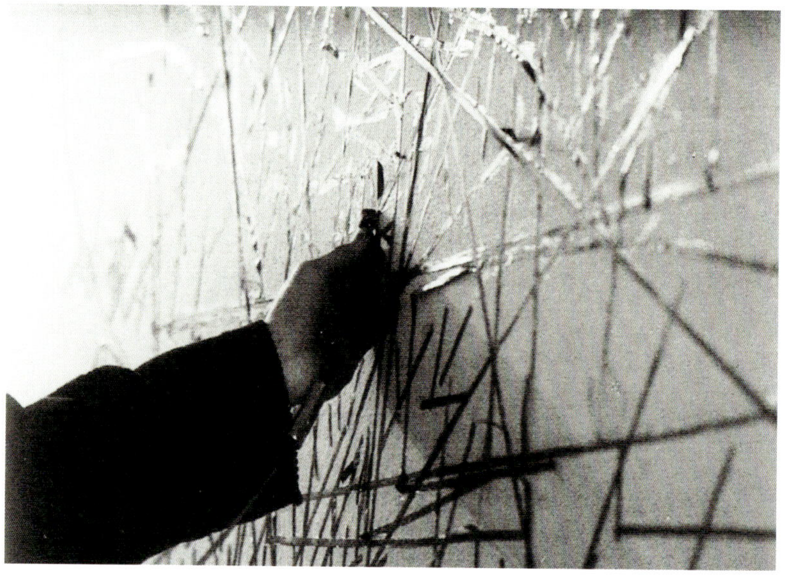

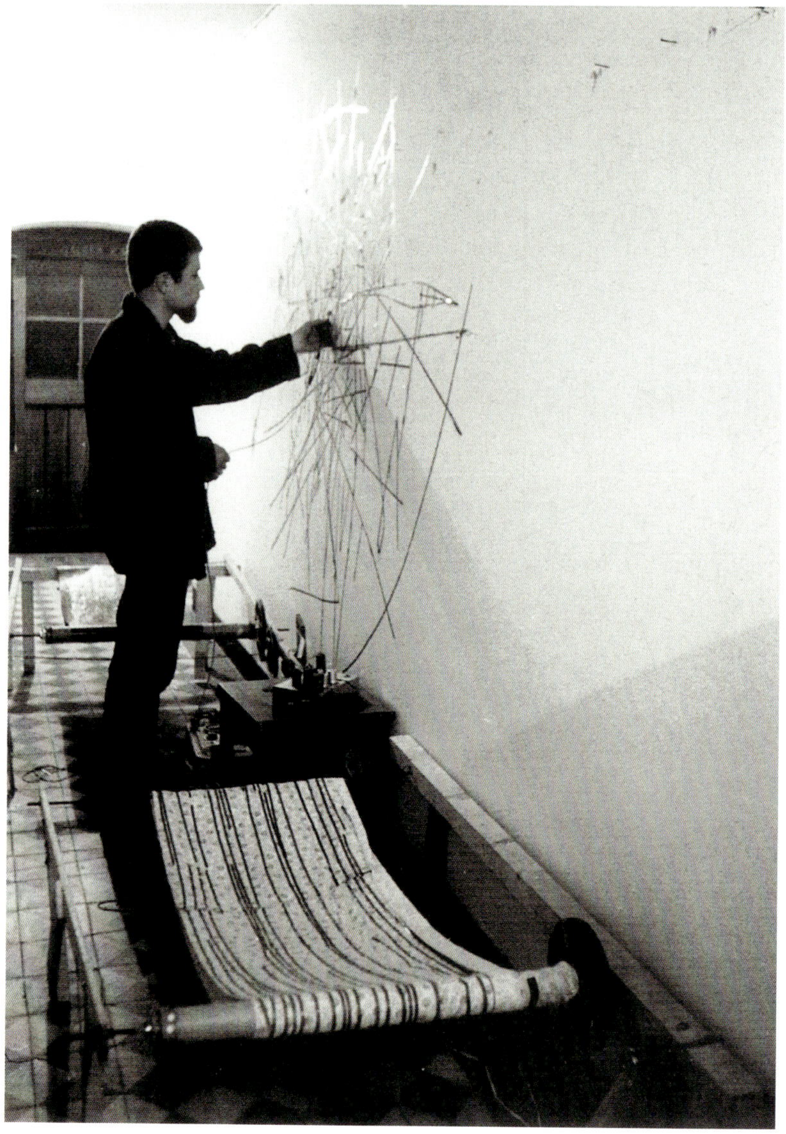

Peter Brötzmann demonstrates *Random Access*, 1963
Strips of audiotape, open-reel audio deck, extended playback head, speakers
Installation view *Exposition of Music-Electronic Television*, Galerie Parnass, Wuppertal, Germany, March 11–20, 1963

In his 1963 installation *Random Access*, Nam June Paik literally deconstructed a tape player, and invited the viewer to play strips of audiotape scattered across a wall using a playback head wired to a pair of speakers. By controlling the speed and position of the playback head, the viewer is able to create new sounds. Through exposing the insides of the tape player, *Random Access* not only offers instant access to the different segments of the tape, but also breaks up a formerly linear medium in favor of the possibility of endless combinations.

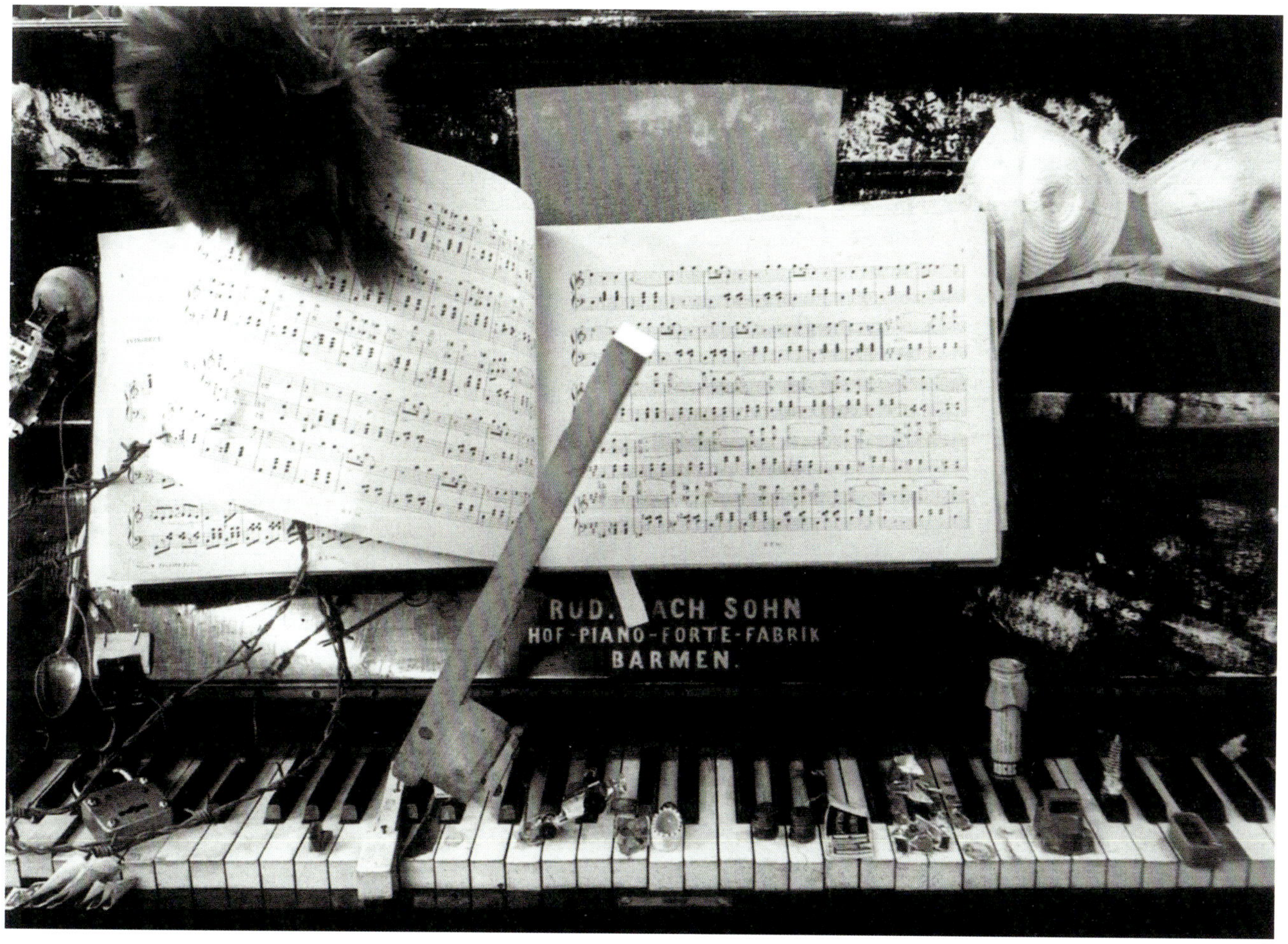

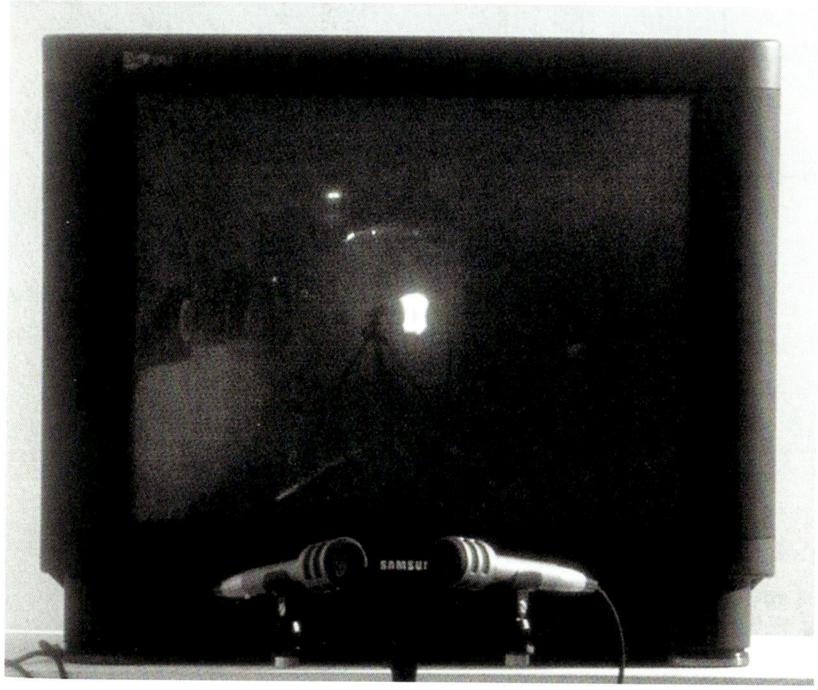

top:
Piano Integral, 1963
Fluxus-Piano prepared with different materials, 136 × 140 × 65 cm
Installation view *Exposition of Music-Electronic Television*, Galerie Parnass, Wuppertal, Germany, March 11–20, 1963

bottom:
TV Experiments (Mixed Microphones), 1963/1969–1995
Monitor, 2 amplifiers, 2 microphones, 2 sound generators, temperature controller

Nam June Paik
Joseph Beuys

*1932 in Seoul (KR), †2006 in Miami Beach, FL (US)
*1921 in Krefeld (DE), †1986 in Düsseldorf (DE)

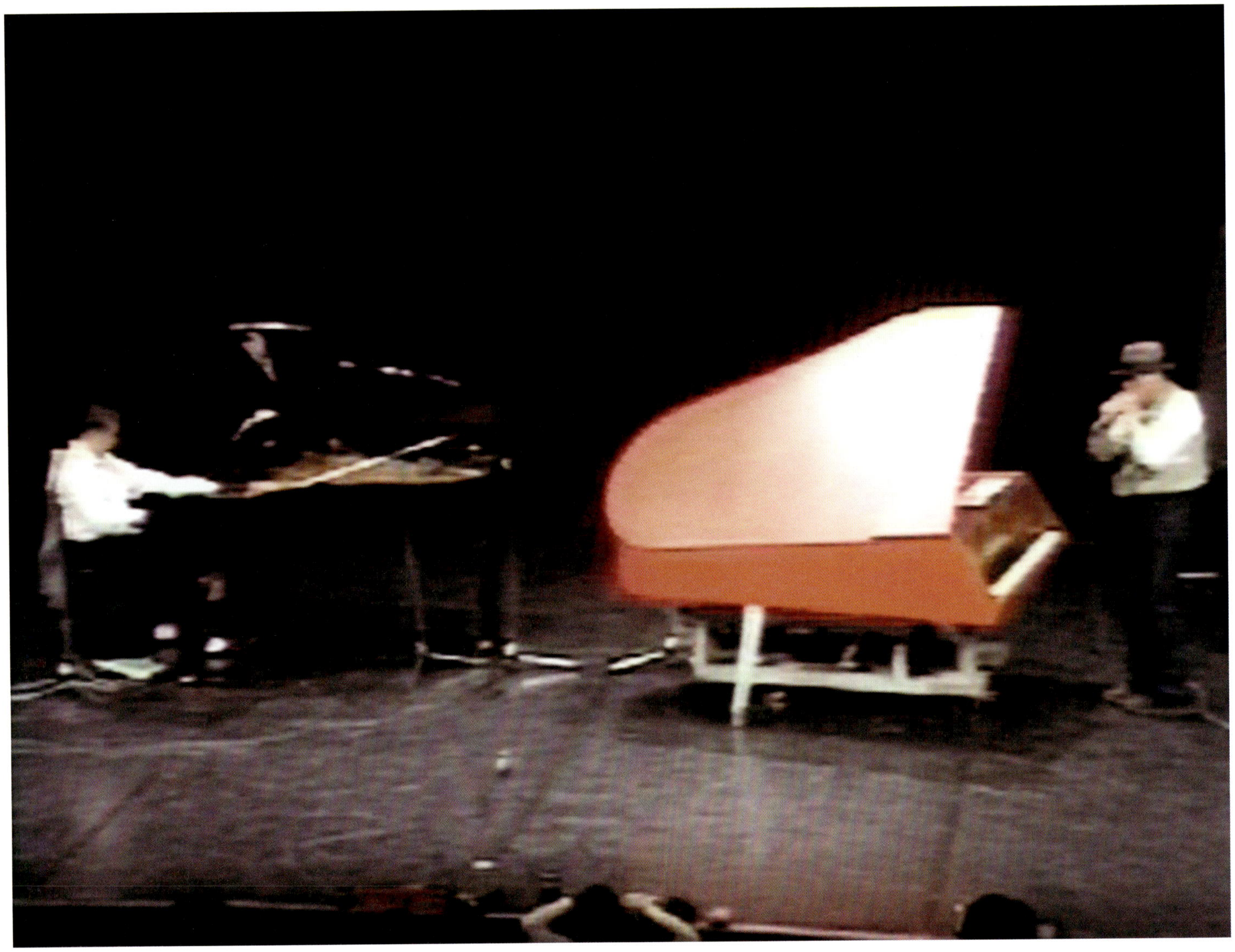

Joseph Beuys: Coyote III /
Nam June Paik: Piano Duett
Seibu Museum, Sôgetsu Hall,
Sezon/ Tokyo, Japan, June 2, 1984
Video, 62:10 min.
Wulf Herzogenrath

A duet between the two artists filmed at Sôgetsu Hall in Tokyo in June 1984: this performance features Nam June Paik improvising on piano while Joseph Beuys conducts a vocal performance that varies from animalistic grunting to sophisticated sound poetry. The only onstage decoration is a blackboard on which abstract chalk marks are written that resemble Morse code; Beuys adds to these markings with textual writing as the concert progresses.

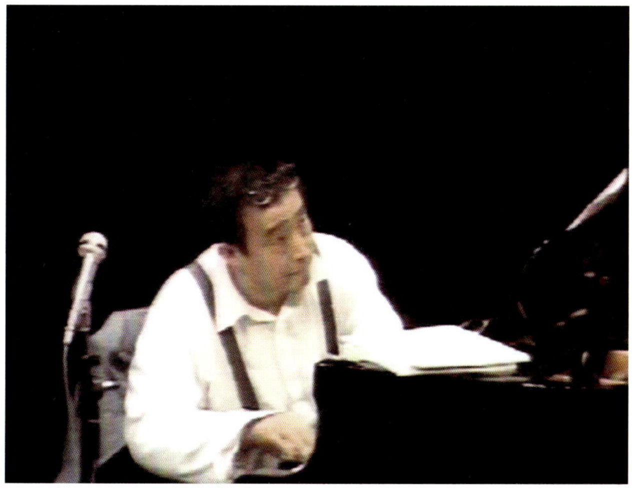

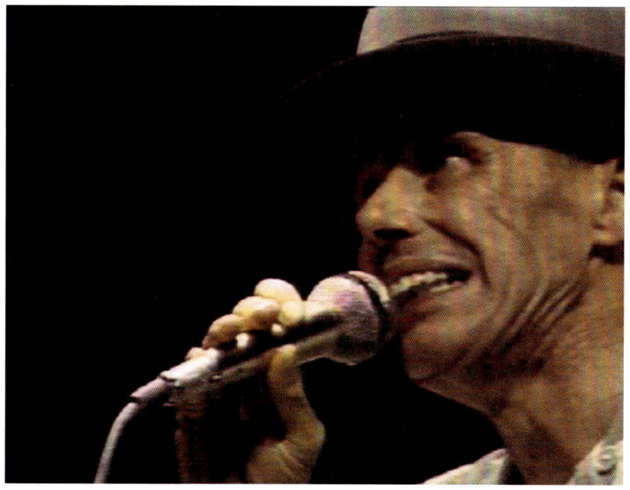

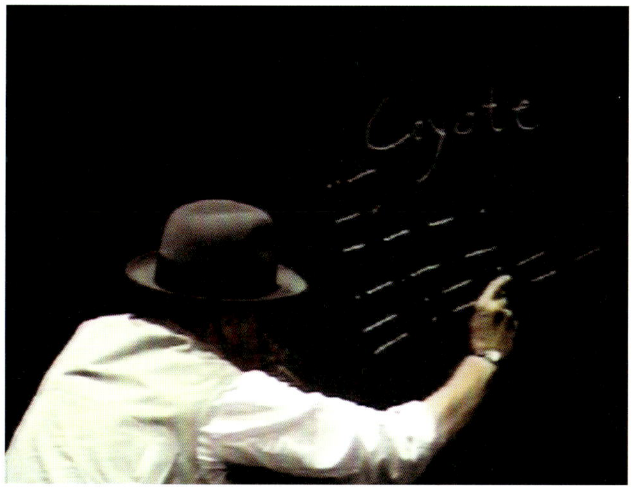
Coyote

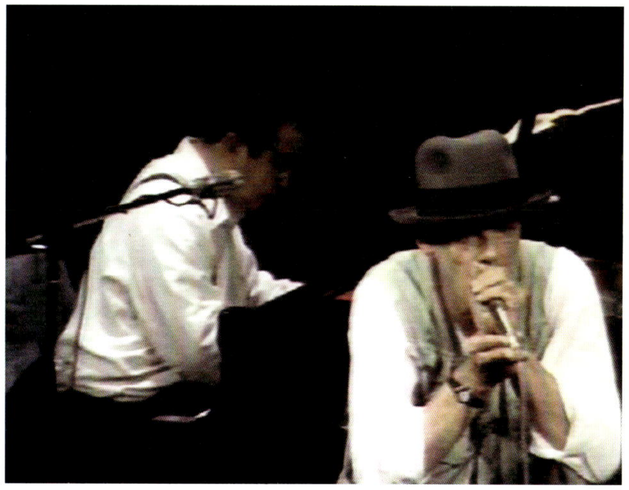

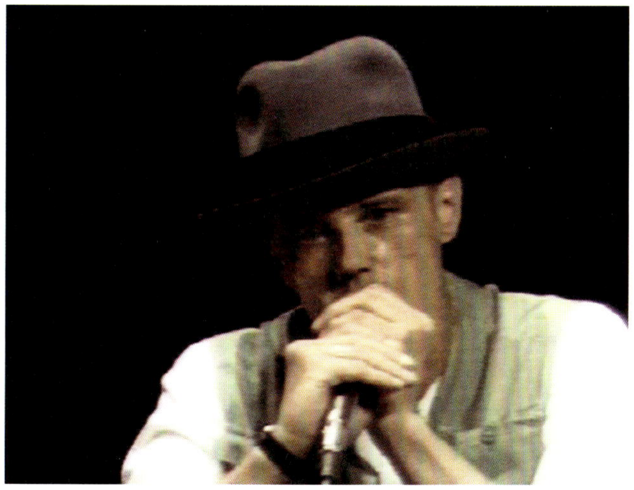

Marco Preitschopf

*1967 in Bretten (DE), lives in Karlsruhe (DE)

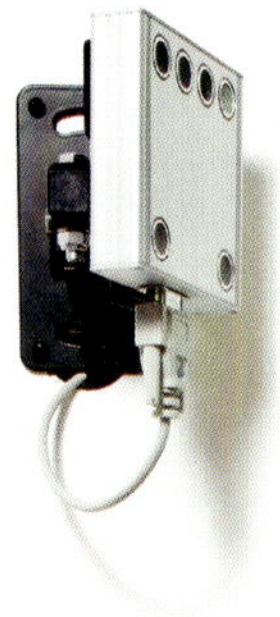

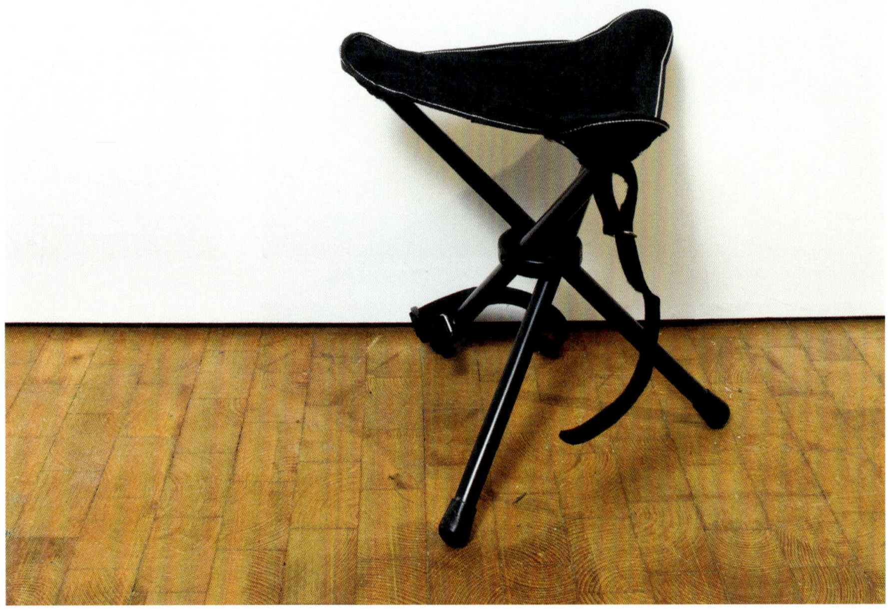

post event of showing
presque vu 03.02.96
illusion (body) I'd like to bury
meeting me half way
a sound of someone who's going away
gone / went / gone 2012, 1996/2012
Interactive mixed-media installation
Camping stool, show case plinth,
158 × 45 × 45 cm, 7:30 min. (loop)
Commissioned by ZKM | Karlsruhe

post event of showing consists of a showcase plinth and a black camping stool. The further the viewer moves away from the empty showcase towards the stool, the clearer the sound issuing from the showcase is heard.

The sound source material is a recording of an exhibition opening that took place on February 3, 1996. The processed recordings were presented as an installation with the title *presque vu 03.02.96 / a sound of someone who's going away*, as part of a follow-up exhibition at the same exhibition venue. The *post event of showing* pursued this original approach of the temporal shifting and mirroring of a recurring event, and interwove the model of art as presentation with the reality model of art.

post event of showing works with distances: the showcase is both receptacle and site of a sound documentation. The stool is the site and distance to the work. The interaction/relationship of the viewer to the work of art and the relationship of the author to the work are recapitulated. (Marco Preitschopf)

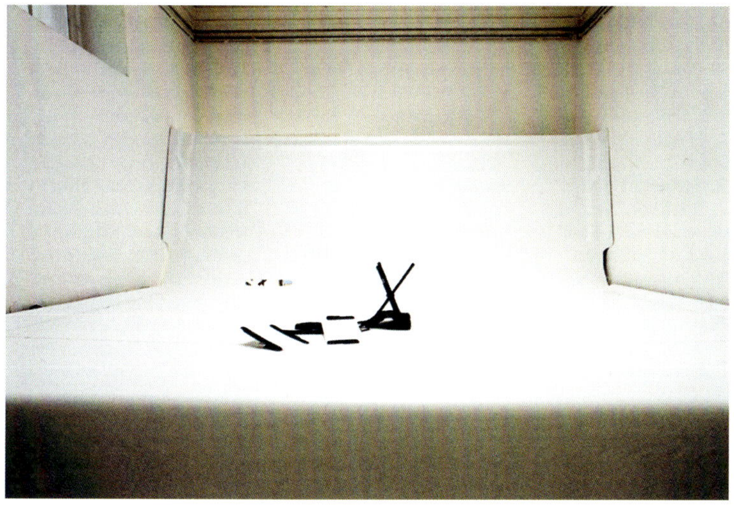

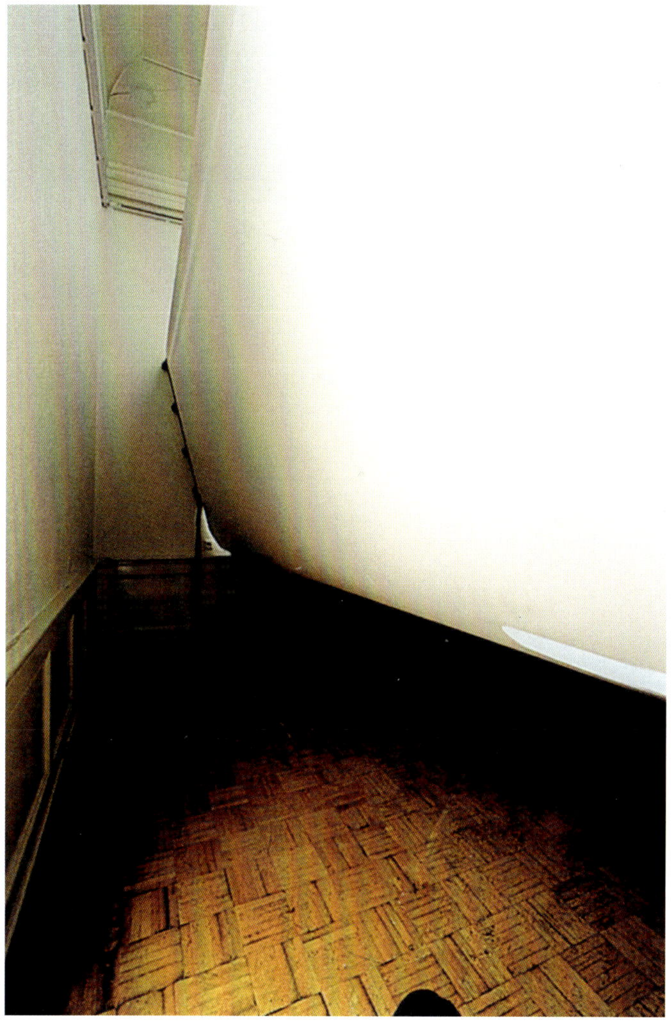

stealing all the light // blow abstracted used hands, 2000
Interactive mixed-media installation
PVC, belts, stool, small components, plastic covers, audio system

A white tarp, called black soul by experts, because it does not let any light through, is mounted into the space like a photographic hollow. The work turns its back on the visitors, leaves them to the curved projection wall to expose themselves. The visitor is (in) the picture. Through a small cut in the trap the reconstruction of a white cube can be seen. Another white tarp comes into view. Like a second floor, an artificial level of abstraction it stretched through the room horizontally. Behind these artificial horizons created by the different traps, a deeper acoustical room of imagination comes into focus. Marco Preitschopf has edited bits of a "square dance" recording repeatedly. One can hear distorted audio fragments of folk music, that measure the room geometrically, giving instructions for the worked out set of rules for the bodies. Voices that command and voiced that break up into breathing, inhaling themselves instead of communication with the outside. Voices remind of bodies and mediate cut up spoken language. (Angelika Stepken)

Roberto Pugliese

*1982 in Naples (IT), lives in Naples

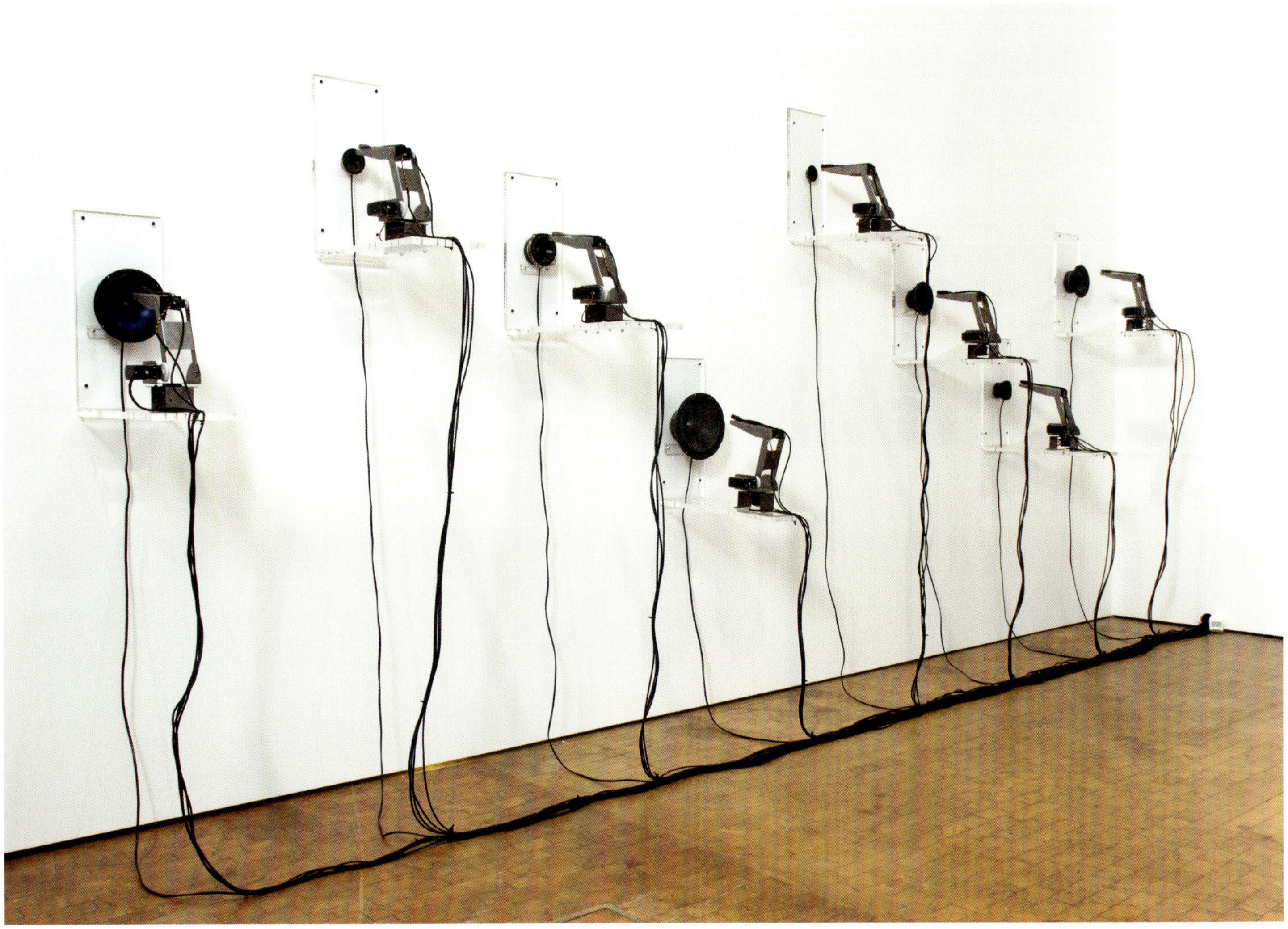

Equilibrium, 2011
Sound installation
Metal, Plexiglas, motors, electronics, loudspeakers, microphones, dimensions variable
Technical assistance:
Walter Pugliese
Courtesy of Galerie
Mario Mazzoli, Berlin

www.robertopugliese.com

In *Equilibrium*, Roberto Pugliese plays with the acoustic phenomenon of feedback. Feedback arises when a microphone is so close to a loudspeaker that the microphone's acoustic signals create a circuit that gets amplified with each new cycle.

In this work, the Neapolitan sound artist and composer has positioned a lever with an attached microphone in front of every loudspeaker. The seemingly natural movements of the computer-controlled levers result from an attempt to keep the feedback sound in balance. Roberto Pugliese programmed special software for this purpose. Since the sound tends to amplify itself, it is impossible to maintain it at a constant volume, so the levers stay in constant motion. The robotic units behave differently due to the different features of the loudspeakers.

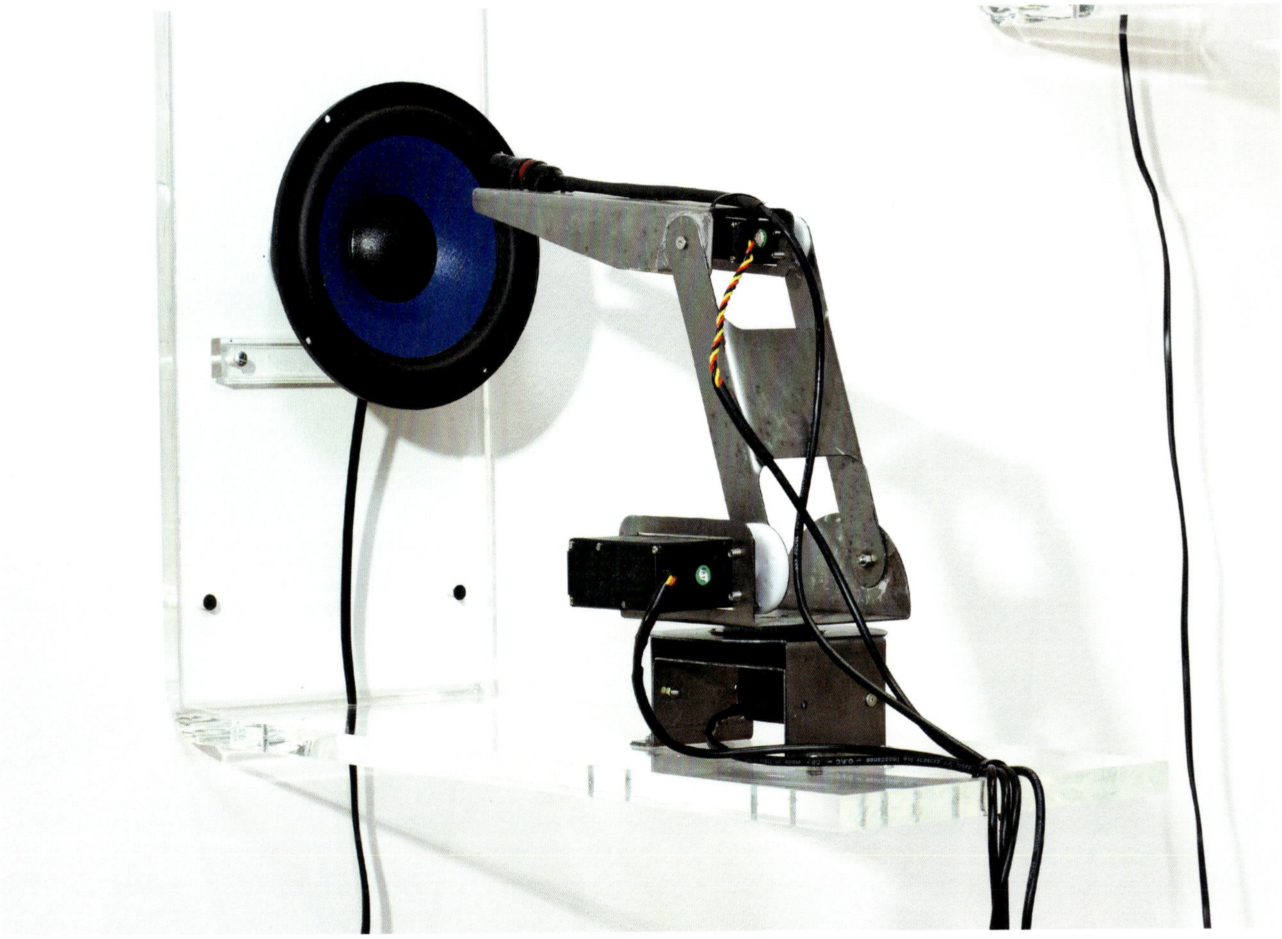

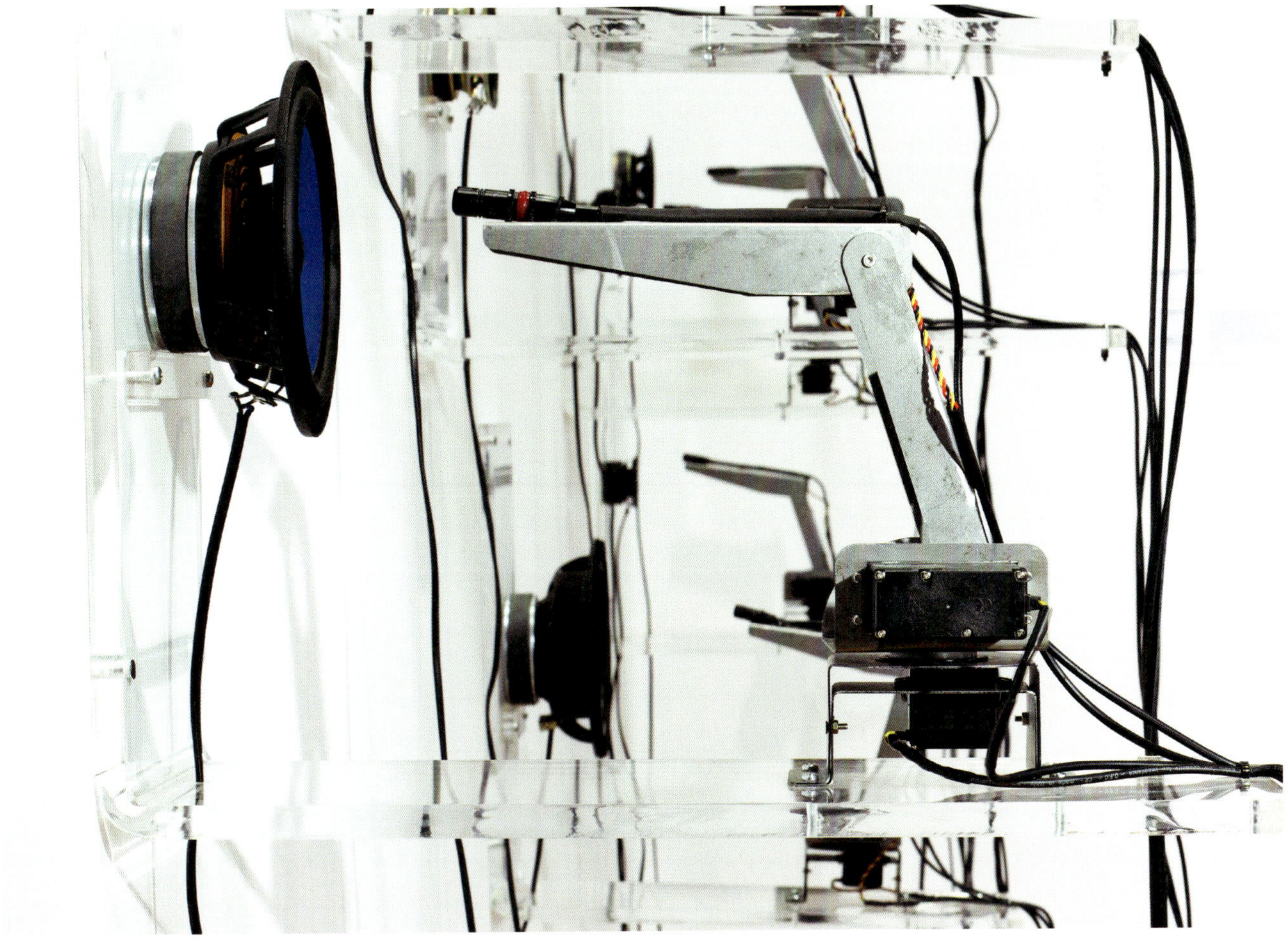

Roberto Pugliese

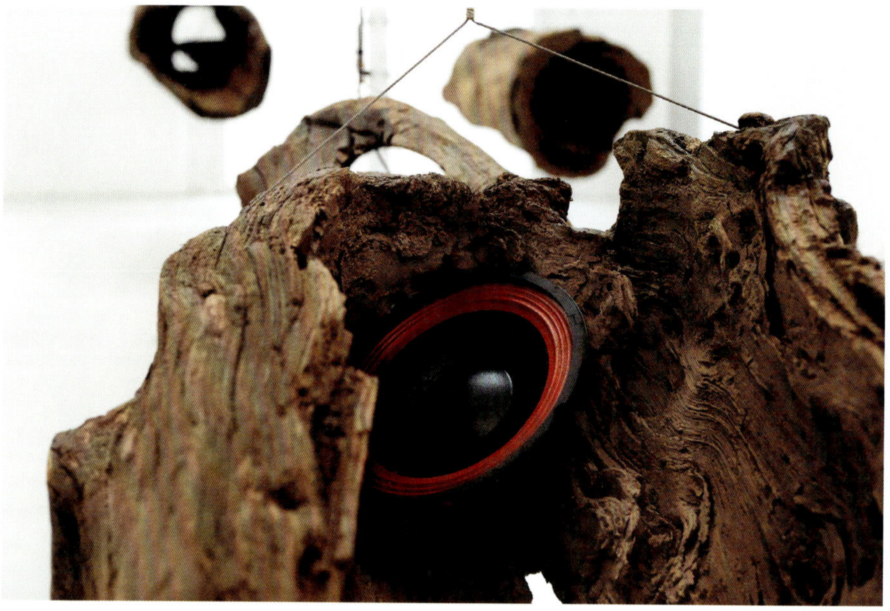

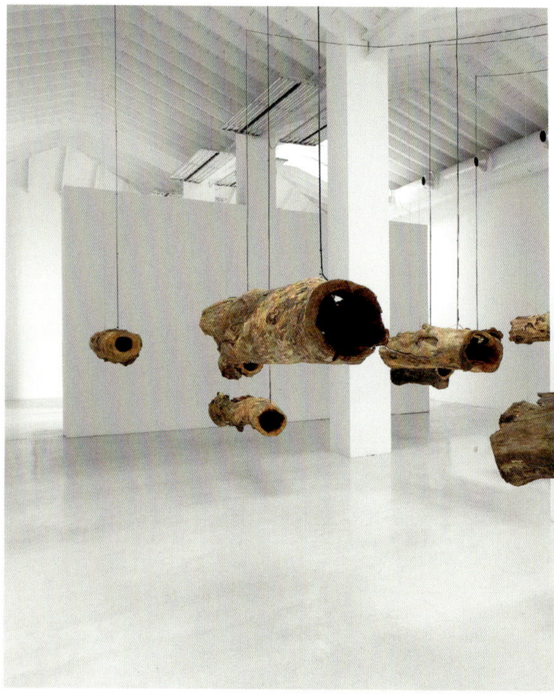

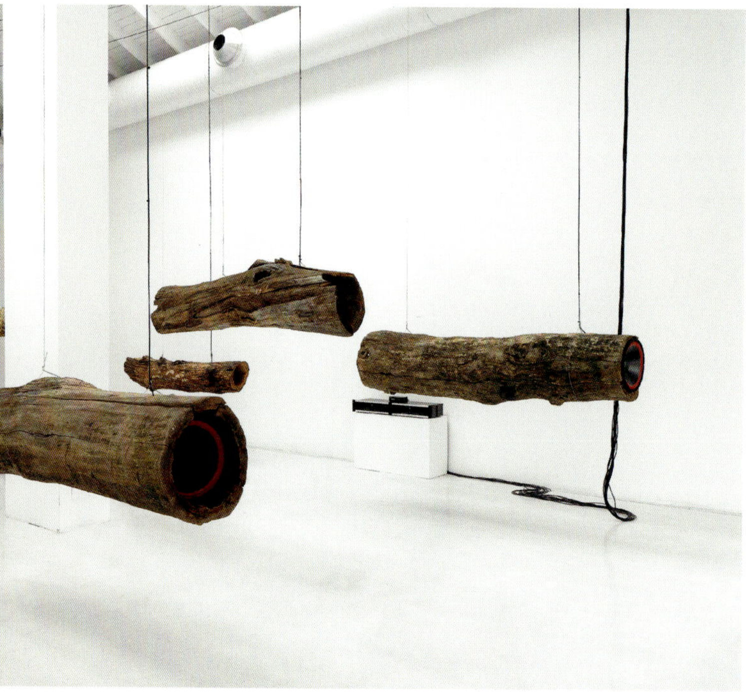

Concerto per natura morta, 2014
Sound installation
Tree trunks, iron, speakers, audio cables, multi-channel audio system, audio composition, dimensions variable
Courtesy of Studio la Città, Verona

Concerto per natura morta [Still life concert] features thirteen chestnut tree trunks suspended from the ceiling in horizontal positions and at different heights from the floor. The trunks are from dead trees, and have been hollowed out. Inside each trunk is a loudspeaker and the differing volumes and shapes of the trunks amplify and modify the sound that comes from the speakers. The sounds reproduced within the installation are both recordings from the places where the trees originally stood and recordings made during the process of hollowing them out. These sounds were then digitally processed and compositionally assembled. In this way the viewer is immersed in a "still life" where the presence of the sound combined with the imposing and impressive visual texture give the work a great emotional and engaging impact.

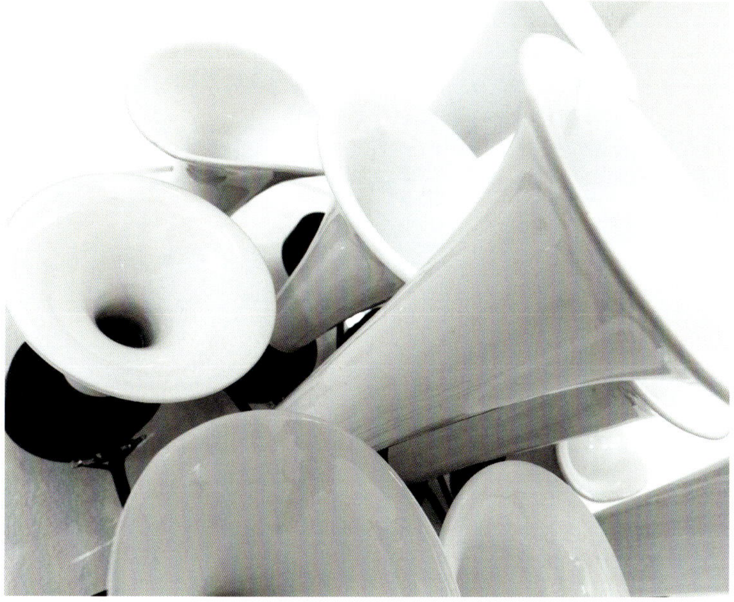

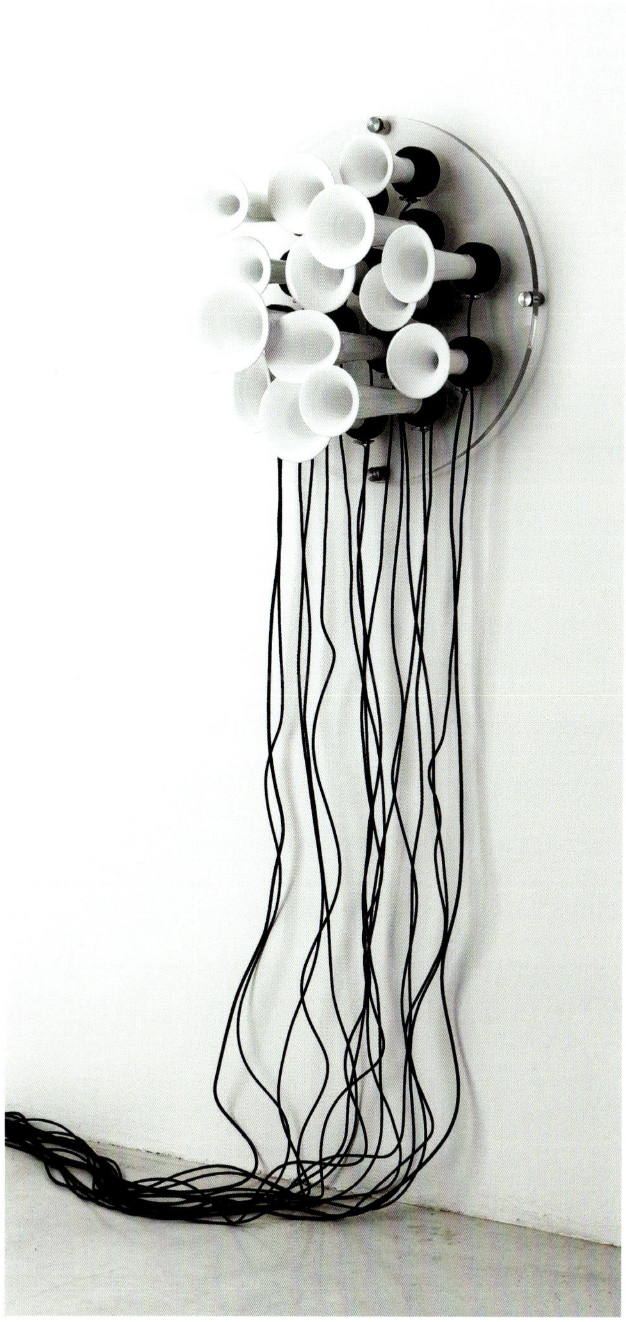

Risonanti pressioni materiche, 2014
Sound installation
Acoustic ceramic horns, iron, Plexiglas, audio composition, audio playback system, 60 × 60 × 35 cm
Courtesy of Studio la Città, Verona

Risonanti pressioni materiche [Resonant material pressures] correlates its sound and its visual appearence. Some iron supports are placed on a circular pane of Plexiglas. Anchored on each of these is a ceramic horn. The 6-channel audio composition played by the horns consists of processed and assembled audio recordings made during the production of the ceramic horns. The idea behind the piece is to create an ad hoc electroacoustic composition for its "playback support," taking into account the characteristics of the acoustic ceramic horns, their number, and their spatial location as the audio derives from the actual manufacturing of the work. The intention is to make the sound composition an integral and essential part of the sculpture and capture in the work itself the acoustic energies developed during its creation.

Kirsten Reese

*1968 in Kiel (DE), lives in Berlin (DE)

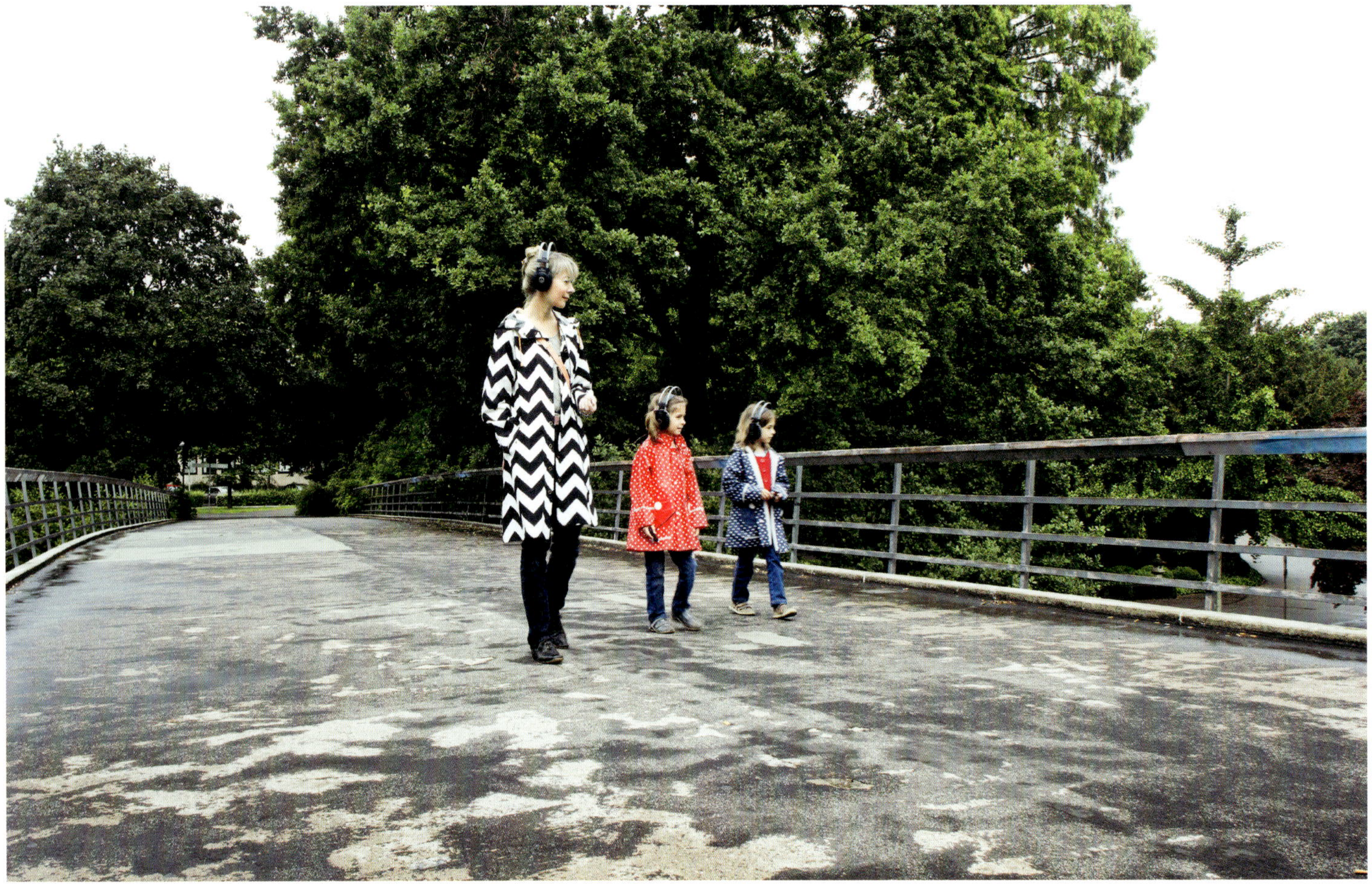

Zoobrücke
Audio passage on Karl-Birkmann-Brücke (Tiergartenweg), Karlsruhe, 2012
Sound installation / audio walk
QR code/smartphone, headphones, loudspeakers
Technical consulting: Manfred Fox
Production commissioned by ZKM | Karlsruhe

www.kirstenreese.de

"… Lauscht ihr Ohr nach allen Seiten, sucht nach wild vertrauten Tönen …" [Their ears listen to all sides, looking for familiar wild sounds.] (from: Joachim Ringelnatz, *Giraffen im Zoo* [Giraffes at the Zoo])

On the footbridge that cuts across the Karlsruhe Zoo, the Berlin composer and sound artist Kirsten Reese had set up an audio trail. Passers-by could access the sound files by using the cameras on their smartphones to take pictures of the QR codes mounted along the way. Alternatively, they could borrow audio players and headphones at the zoo's east entrance.

Sounds accompanied listeners as they walked across the bridge. The composed audio terrain became superimposed on the view of "artificial nature" and the animal enclosures in the zoo. In the soundtrack too, artificially generated material mixed with samples of actual animal and natural sounds. The sound files were created with the help of the Fairlight, one of the first digital musical instruments of the late 1970s and the first digital synthesizer that also had sampling technology. The audio trail ran along the border zone between nature and culture at their most extreme – as wilderness. There it encountered the latest in QR code and smartphone technologies of that time.

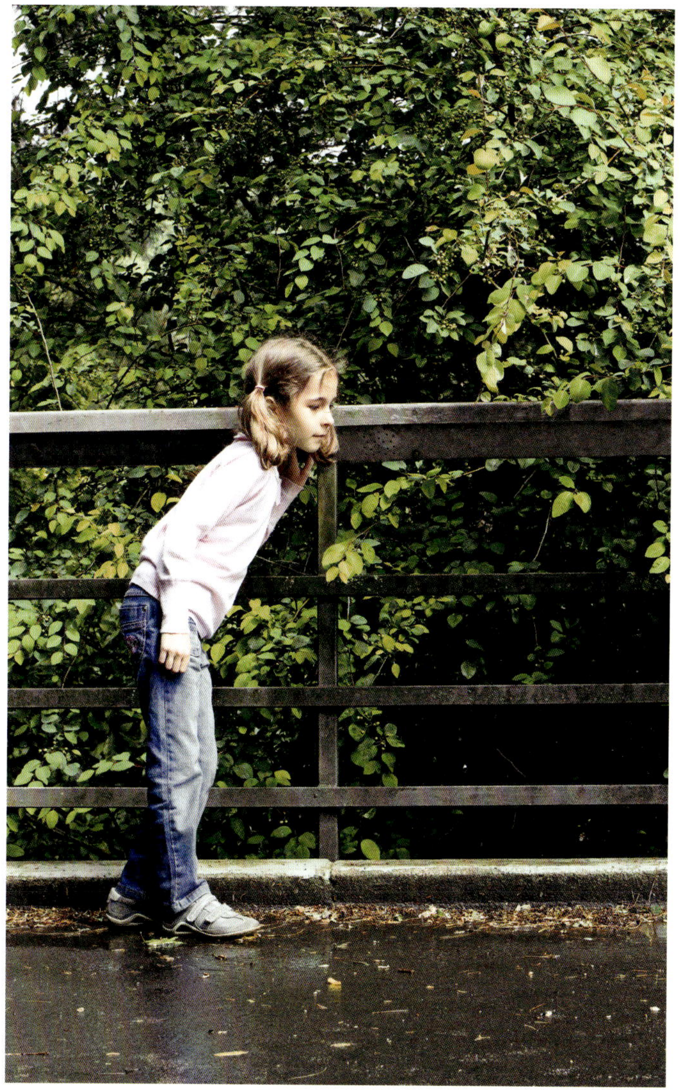

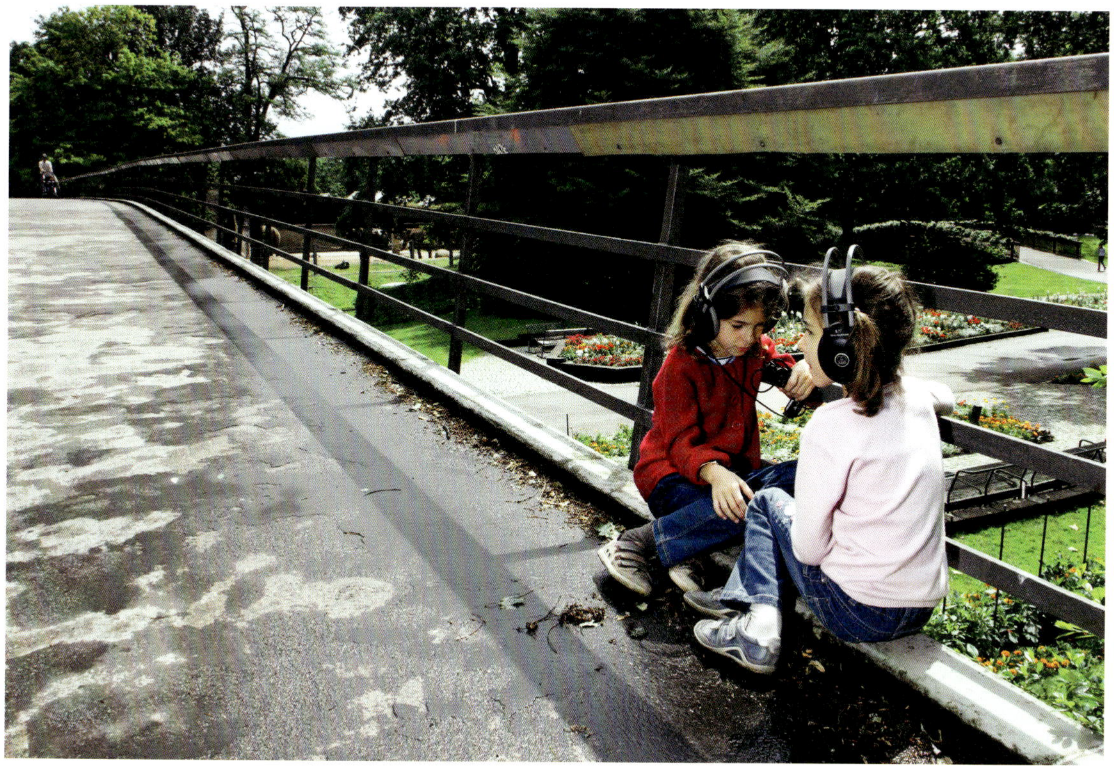

Werner Reiterer

*1964 in Leibnitz (AT), lives in Vienna (AT)

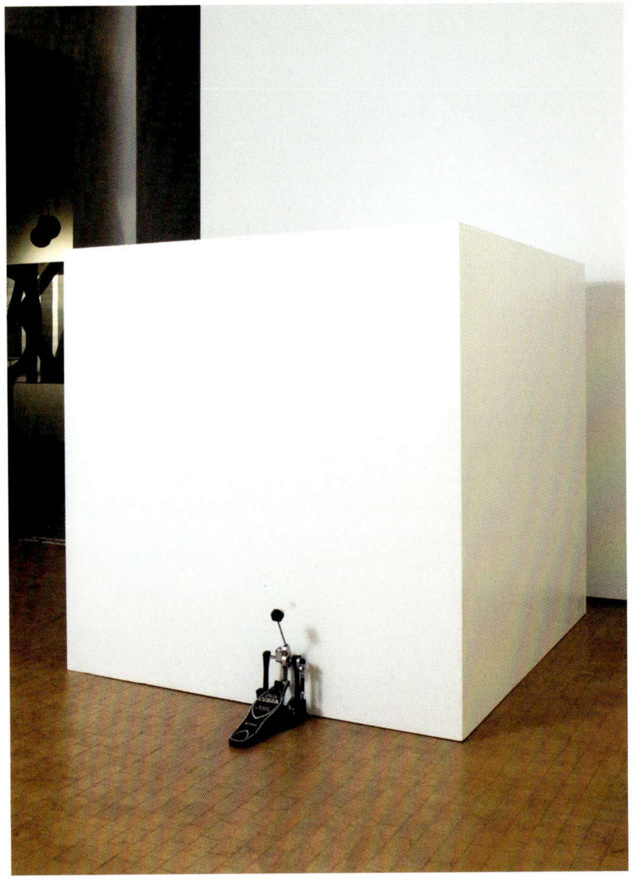

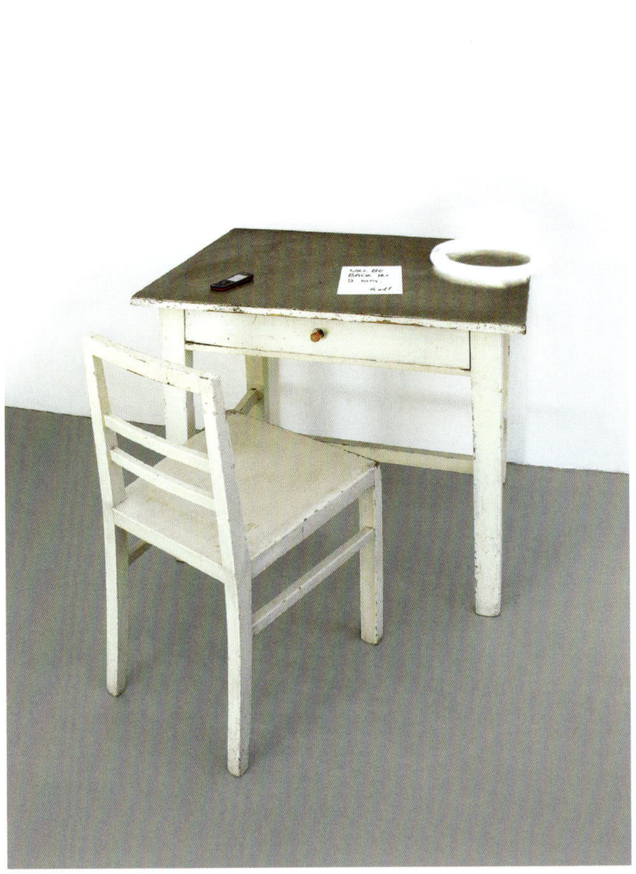

Life counts Death, 2008
Interactive installation
Wood, bass drum pedal machine, loudspeaker, electronics,
150.5 × 150 × 206 cm
Werner Reiterer

Werner Reiterer's sculptures, installations, and interventions confront us with the finiteness of life and often challenge the observer to act. The visible components of *Life counts Death* consist of a wooden cube and a drum pedal. Stepping on the pedal unleashes a powerful peal of bells. Bells marks an arbitrary moment when one person comes into the world and another person departs. Life and death are collapsed into one.

Draft for an Altar, 2009
Sound installation
Table, chair, neon, mobile phone, sensor, loud speaker,
83 × 144 × 136 cm
Courtesy of Galerie Loevenbruck, Paris

In *Draft for an Altar* God has put down his halo and left a note "Will be back in 5 min. God." Even on his mobile phone, God is not available. It rings every five minutes, but God is not there to accept the call.

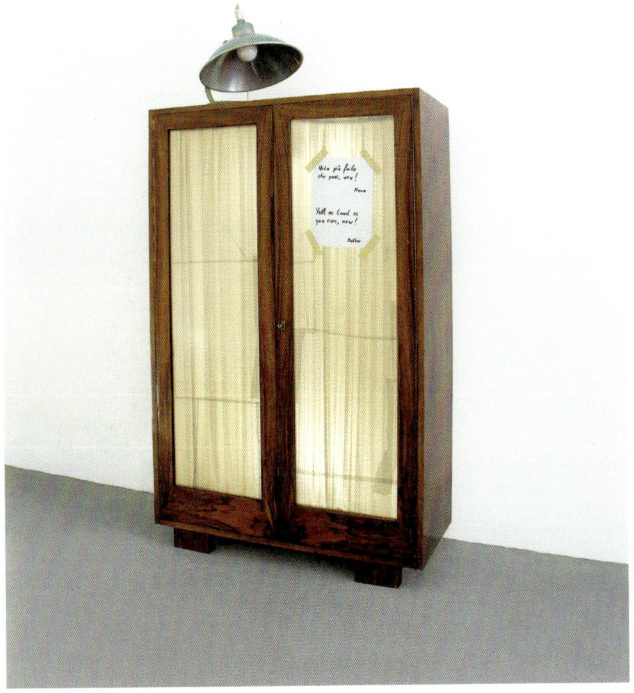

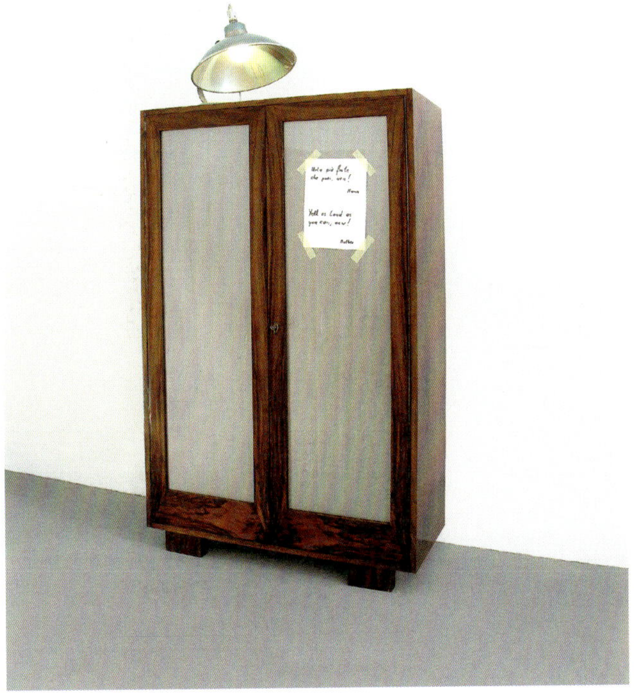

Mothers Advice, 2010
Interactive sound installation
Cupboard, light, electronic, paper, loudspeaker, 208 × 101 × 46 cm

On the cupboard's front there is a note, telling the visitor to "Yell as loud as you can, now!" If the visitor does this, he gets an advice from the mother.

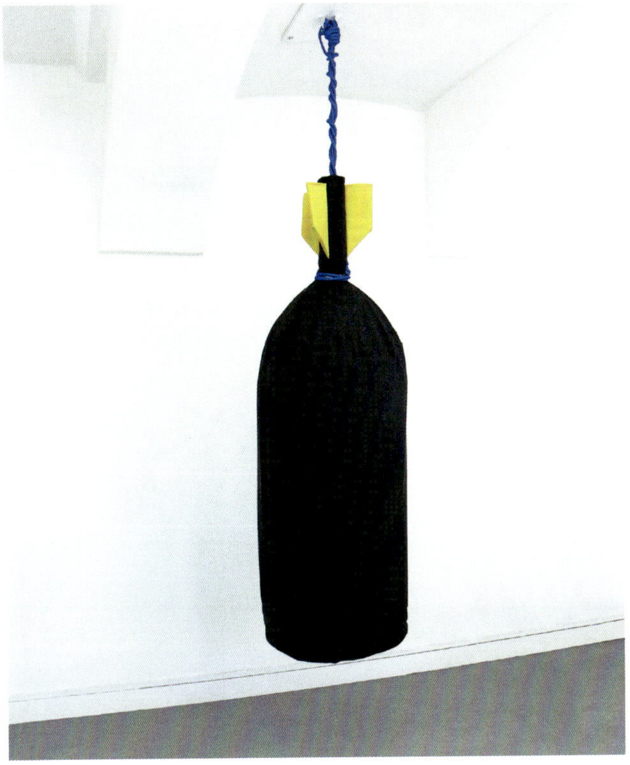

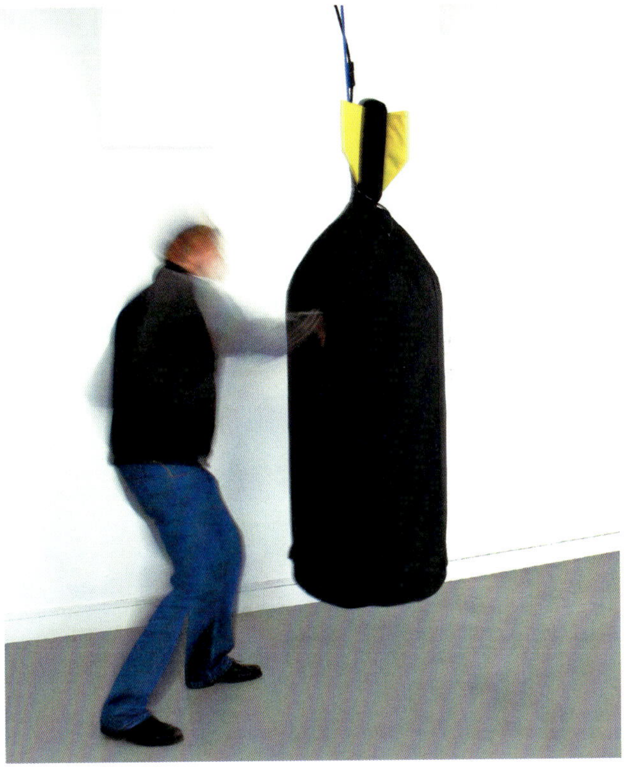

Bang the Bomb!, 2008
Interactive sound installation
Foam, synthetic material, diverse electronics, speakers, rope, 190 × 60 × 60 cm

A punching bag in the form of an aerial bomb hangs from the ceiling. If the visitor strikes hard enough, the sound of a loud explosion will sound.

Steve Roden

*1964 in Los Angeles, CA (US), lives in Pasadena, CA (US)

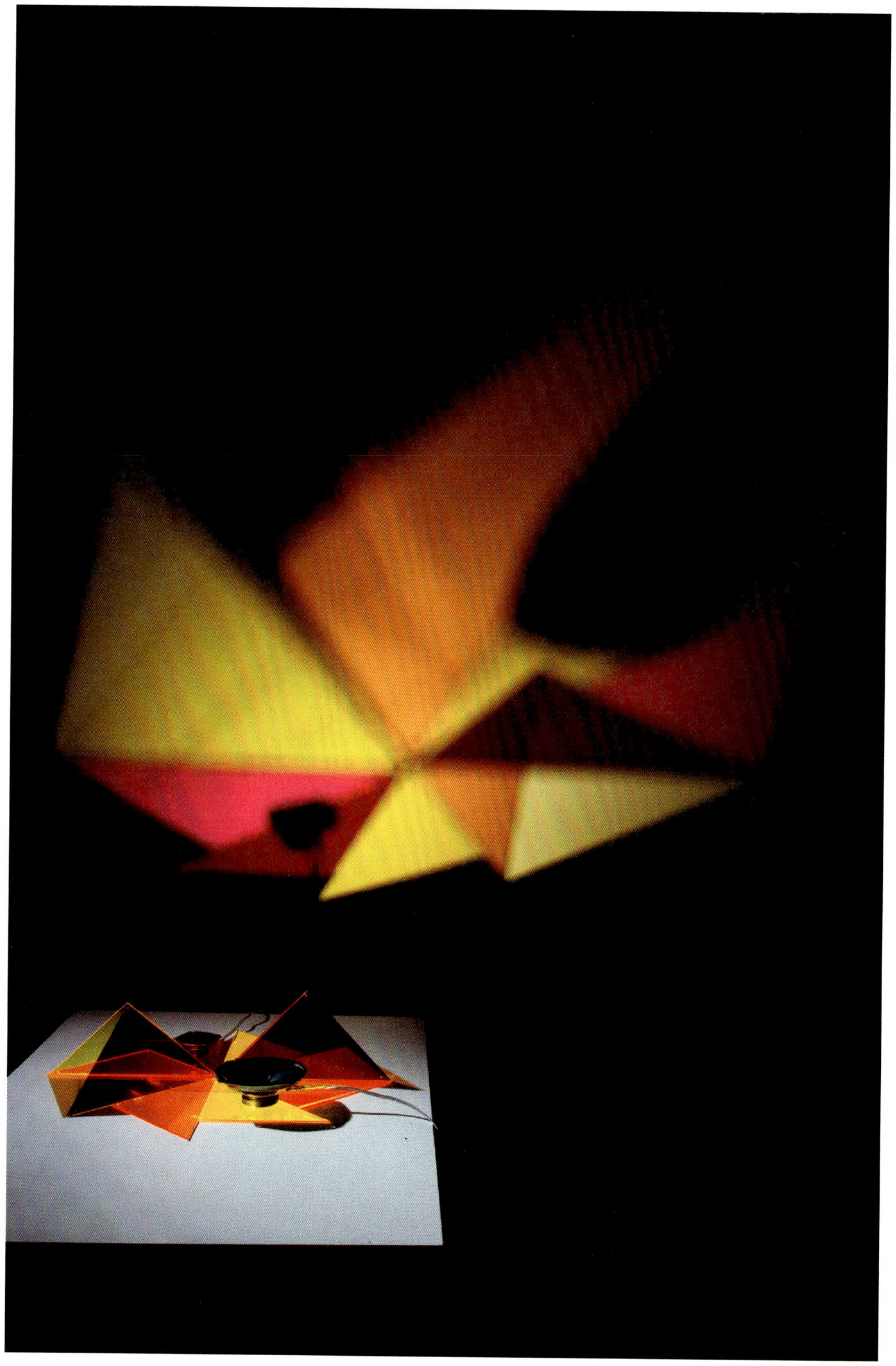

gradual small fire 5, 2012
Sound sculpture
Plexiglas, cable, 2 loudspeakers, stereo soundwork
Courtesy of the artist, CRG Gallery, New York, and Susanne Vielmetter, LA Projects, Los Angeles

www.inbetweennoise.com

Steve Roden's *gradual small fire 5* is one of a group of small sculptures made of Plexiglas and loudspeakers. The first five audiovisual fires were created for Daniel Libeskind's building The Run Run Shaw Creative Media Centre in Hong Kong. Colors, light refractions, and stylized flames evoke real fires. The rigidity of the Plexiglas contrasts with the idea of the burning process. From the soundtrack comes hidden rustling, interspersed with instrumental and everyday sounds that undergo just gradual alterations. Rodan's group of sculptures is a contemporary response to Ed Ruscha's artist's book *Various Small Fires and Milk* (Los Angeles, 1964), an "open experiential space" without texts or explanations.

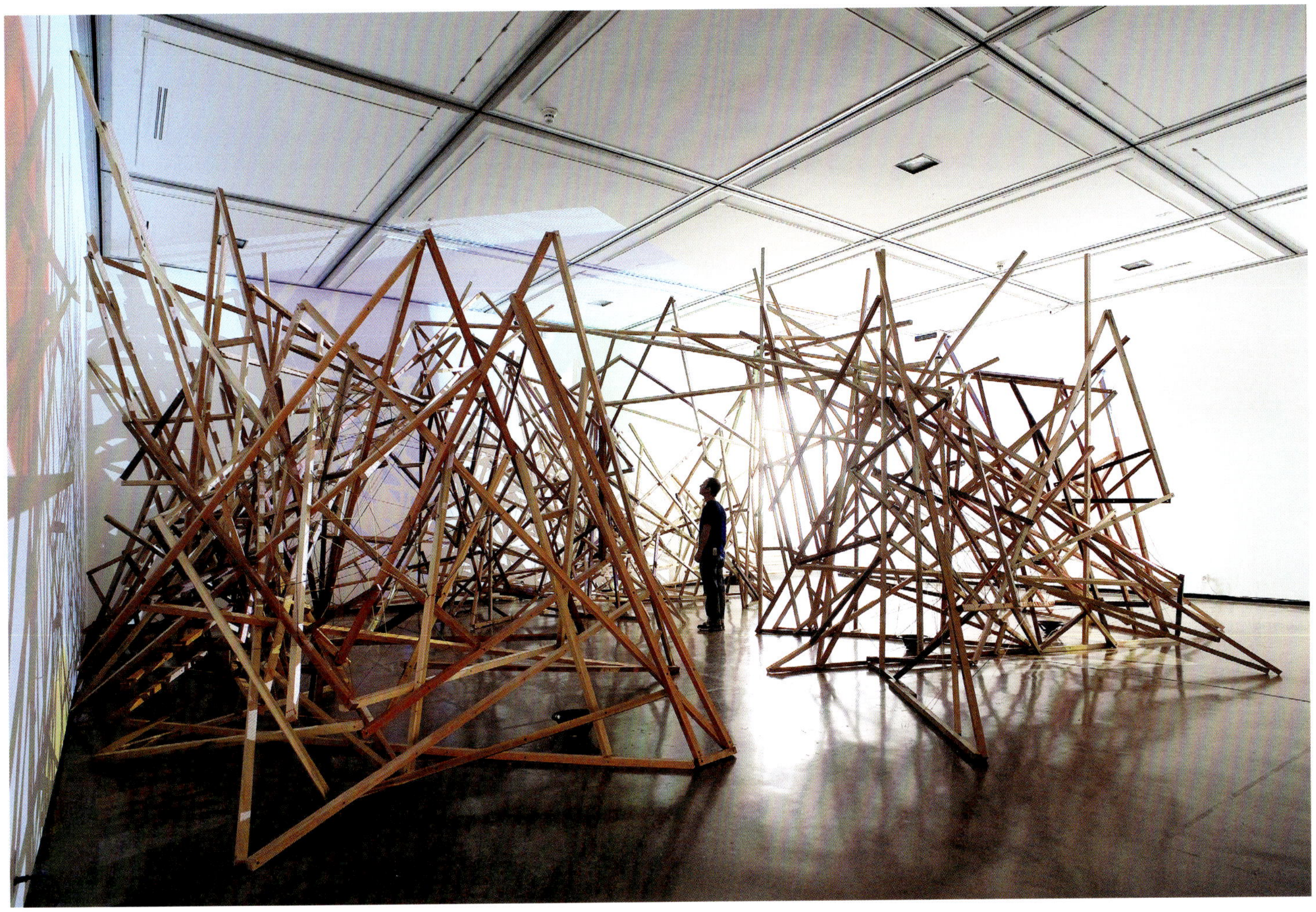

bowrain, 2010
Sculpture/sound/film installation
Wood, wire, twine, sound composition, 6 audio speakers, 3 CD players, 3 stereo amplifiers, ink on 16mm film stock transferred to video, 3 DVD players, 3 video projectors
Commissioned by the Pomona College Museum of Art, Claremont California

The piece was inspired by a small notational drawing by Buckminster Fuller. The sculpture is made of 480 pieces of work, constructed using chance operations. The sound and hand drawn film used the same 480 elements that determined the sculpture to form their compositions.

Tim Otto Roth

*1974 in Oppenau/Black Forest (DE), lives in Oppenau and Cologne (DE)

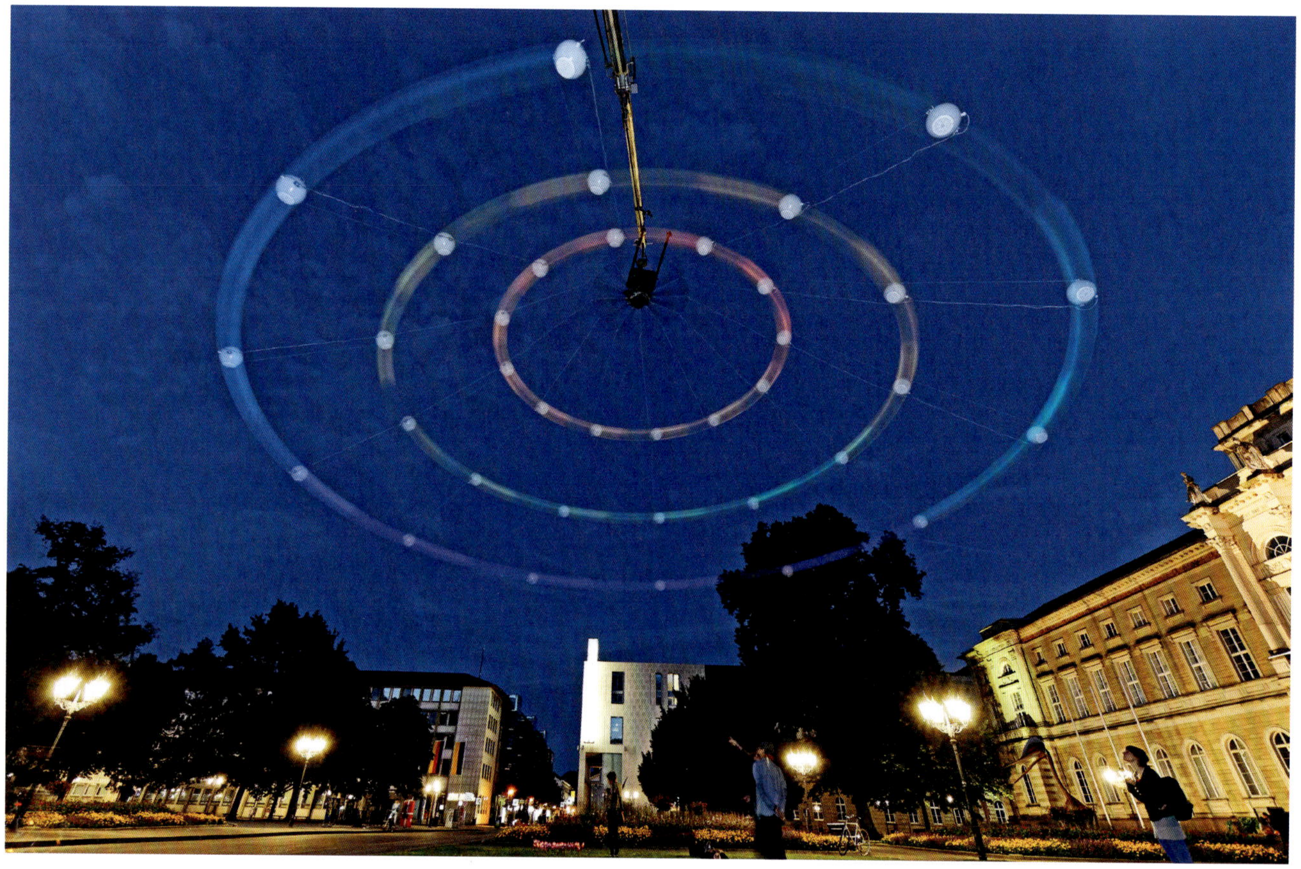

Heaven's Carousel, 2014/2015
Outdoor kinetic sound and light sculpture
36 illuminated loudspeakers, poles, crane, diameter 16 m

Tim Otto Roth's sound and light art is situated at the intersection of art and science. Roth has applied Walter Benjamin's concept of the aura to a sound installation accompanied by light, which follows a calculated algorithm. It was utilized here eponymously because it indicates very well the way he works. Roth takes scientific findings, which are not just a source of inspiration for him, but also an occasion to probe them profoundly. That which is objective and calculable he endows with an independent aesthetic shape.

Roth's works are capable of translating things quite vividly, as demonstrated by his astrophysics related laser installation *From the Distant Past* projected to public facades in Venice, Baltimore and New York. They can be allegories of models, as is the case with his cellular automata related *Pixelsex* works, to which *aura calculata* also belongs. And they can, for example, the light art works, draw attention to how little we know about our perception and constructions of reality.

A sensational work by Roth is the huge light and sound kinetic sculpture *Heaven's Carousel*. Its height is 10 m and the span of its outspread arms is up to 16 m. Mounted on the 12 rotating doubles strings are 36 colored, light-emitting loudspeakers creating sine waves and sound, whose pitch changes with the velocity of the rotations and the movements of the visitors, forming a mix of sound when the strings of the carousel spread out with increasing speed. The faster the rotation during one of the works, the higher the sounds. The sonic events become extremely complex. Through the motion of the loudspeakers playing sounds visitors experience the Doppler Effect.

Roth also uses different music scales, for which he draws on models (including Ivan Wyschnegradsky). Experimenting with a metric scale, he increased the intervals between the tones as they became higher, with which he took into account the ability of the human ear to differentiate, although not with the customary logarithmic scale. Controlled by a computer, the tone pitches can be overlaid in three layers. This results in novel kinds of harmonies on a scale from rough to soft. They differ considerably depending on the underlying scale, upon which the colors of the light also depend. A music of the spheres becomes an allegory of something that is amenable to scientific research, yet inaccessible to naked human perception, namely, the expanding universe. The effect of *Heaven's Carousel* is greatest at night, when it also makes visibly clear that the night sky is not nearly as static as we believe we see it. (Helga de la Motte-Haber)

As part of ZKM | Karlsruhe's exhibition *The City is the Star. Art at the Construction Site* – organized during the Karlsruhe festival summer in 2015 and located in the inner city – Tim Otto Roth's *Heaven's Carousel* could be seen in the evening hours at Friedrichsplatz in Karlsruhe from June 19–September 27, 2015.

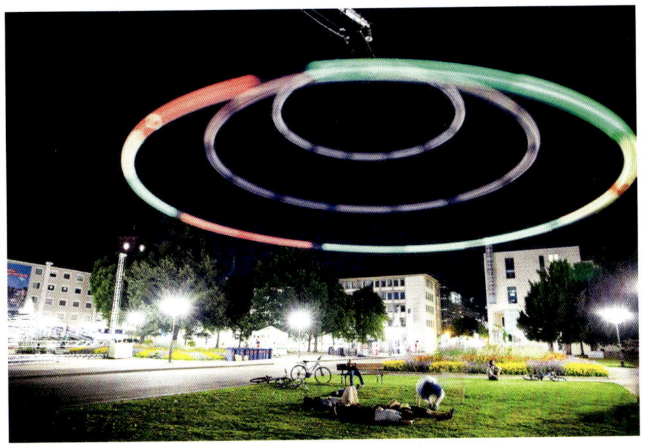
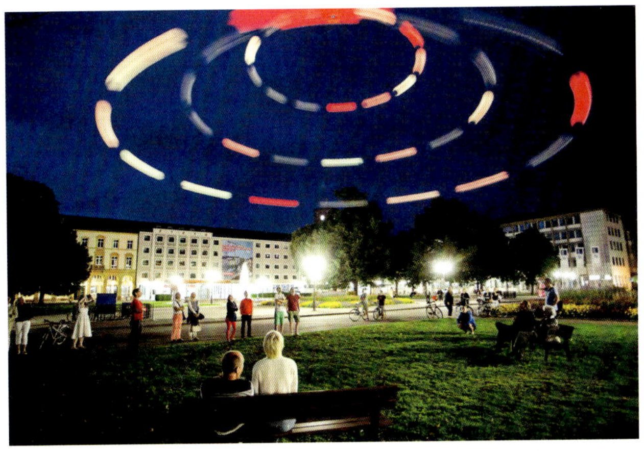
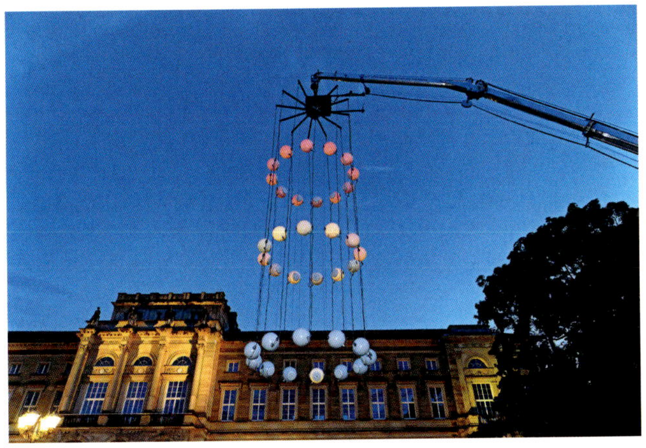
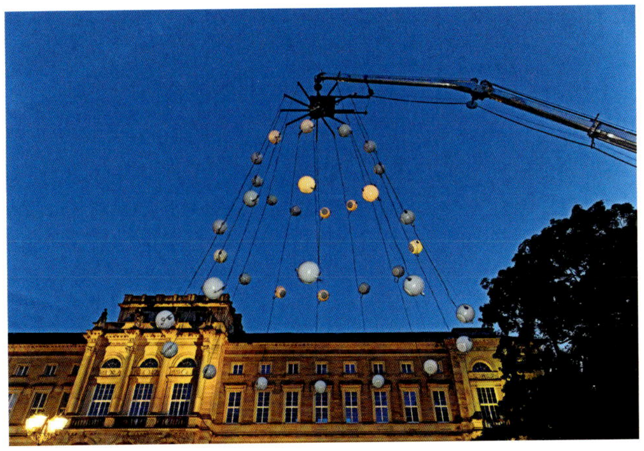
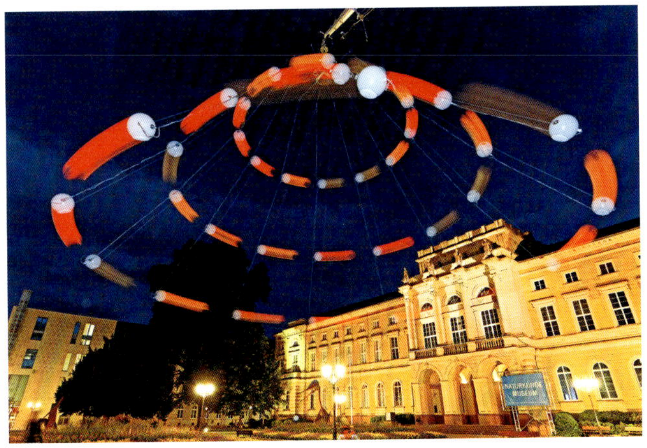
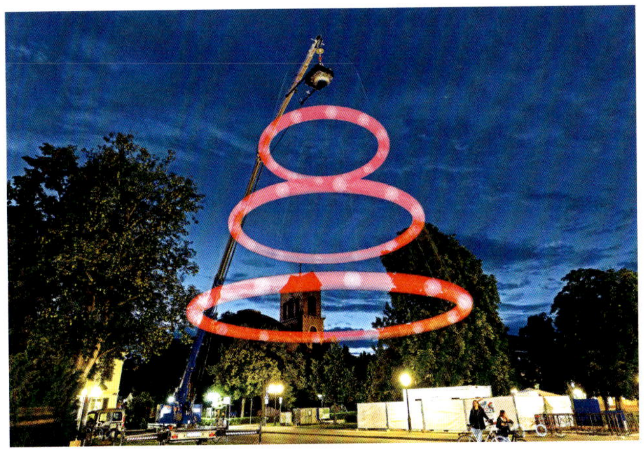

Tim Otto Roth

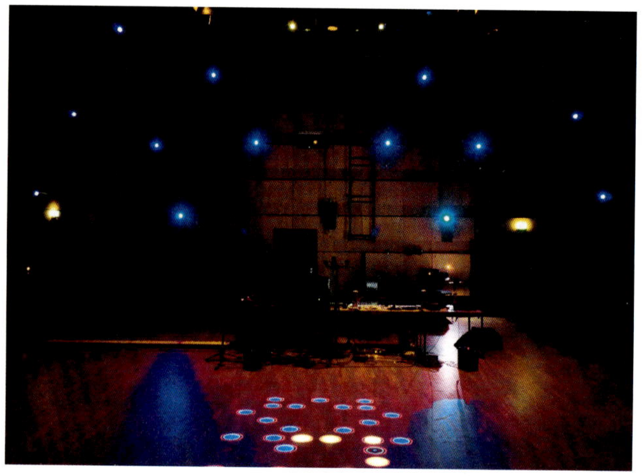

ZKM_Sound Dome with LED illumination and floor projection

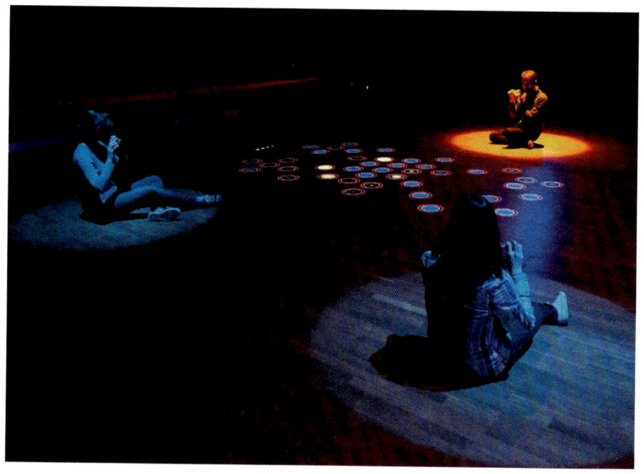

The three piccolo players sit around the floor projection, which displays changing membrane potentials of the audio neurons.

The Sonapticon – A Space as an Acoustic Neuronal Network, 2012
With Benjamin Staude

Sonapticon: Space as an Acoustic Network

The *Sonapticon* connects acoustics and neuroscience conceptually with space, changing the Sound Dome at the ZKM | Center for Art and Media Karlsruhe literally into a musical mindstorm.[1] The core of the system is the so-called audio neuron communicating not with electric currents but with sound waves. An audio neuron registers impulses (*spikes*) from its acoustic environment by a microphone and fires its own impulse by a loudspeaker. The clue of that system is that every neuron gets assigned a sine tone with an individual frequency. This unique translation of neuronal systems' electrical communication into acoustic communication is combined with the deceleration of biological neuronal processes that take place on the timescale of milliseconds. In this way, in the *Sonapticon* neuronal communication mechanisms become both a sensual and a comprehensible experience. At the concert-like premiere in the context of the ZKM's IMATRONIC festival and the symposium Neuroaesthetics[2] in November 2012, three solo piccolo players revealed with a simple combinatoric play[3] the different facets of the *Sonapticon*, whether a microtonal sound wash or a pseudo-repetitive pattern in the inaudible sound spectrum of ultra sound.

The *Sonapticon* at the ZKM Sound Dome
Thanks to the invitation of the Institute of Music and Acoustics at ZKM | Karlsruhe, we – the author and the neuro-mathematician Benjamin Staude – could explore together the *Sonapticon* as guest composers from 2010–2012 in a very sophisticatedenvironment. In collaboration with the sound engineer Holger Stenschke from ZKM, we were changing the ZKM Sound Dome into an acoustic neuronal network. The Sound Dome consists of 43 loudspeakers building an impressive hemisphere of 15 meters in diameter. The loudspeakers are joined up via microphones. Intuitive understanding of the acoustic feedback through sine tones in the Sound Dome space is complemented visually by LED lights mounted on the speakers flashing up in a blue or yellow color when the neuron is firing. Most importantly, a video projection on the floor shows a visualization of all the active neurons with their changing membrane potentials. In the ZKM Sound Dome the audio analysis uses a framework based on MaxMsp, programmed by Holger Stenschke. The adaptation of the neuron model and the composition framework was realized in Python by Benjamin Staude. For the visualization Tim Otto Roth used Gem in combination with Puredata.

A New Instrument
Thus, with the *Sonapticon* we transformed the Sound Dome into a new instrument for which we have developed a special method of composing in and with the Sound Dome space: choosing scales up to the eighth tone range, giving tones a location by assigning frequencies to an audio neuron, setting the connections between the audio neurons, and finally, varying the tempi by playing with the system's update rate. The piece at the concert premiere comprised eight movements starting with a single neuron. Step by step, the number of neurons the three piccolo players interacted with was increased. Mainly, the composition used different tonalities as a Western chromatic, an Indian Shruti, or a microtonal eighth tone scale. In the last movements the frequencies were transposed subsequently by perfect fifths, ending in a finale with inaudible ultra sound.

The models themselves are a playfield for composition. Parameters can be changed globally, but also the characteristics of individual neurons can be varied. Modifying the ratio of neurons and interneurons has a significant effect. Until now, this was studied only rudimentarily, so it is a future task for the *Sonapticon* to explore the phenomenon of synaptic plasticity acoustically also using different neuro-biological models. (Tim Otto Roth)

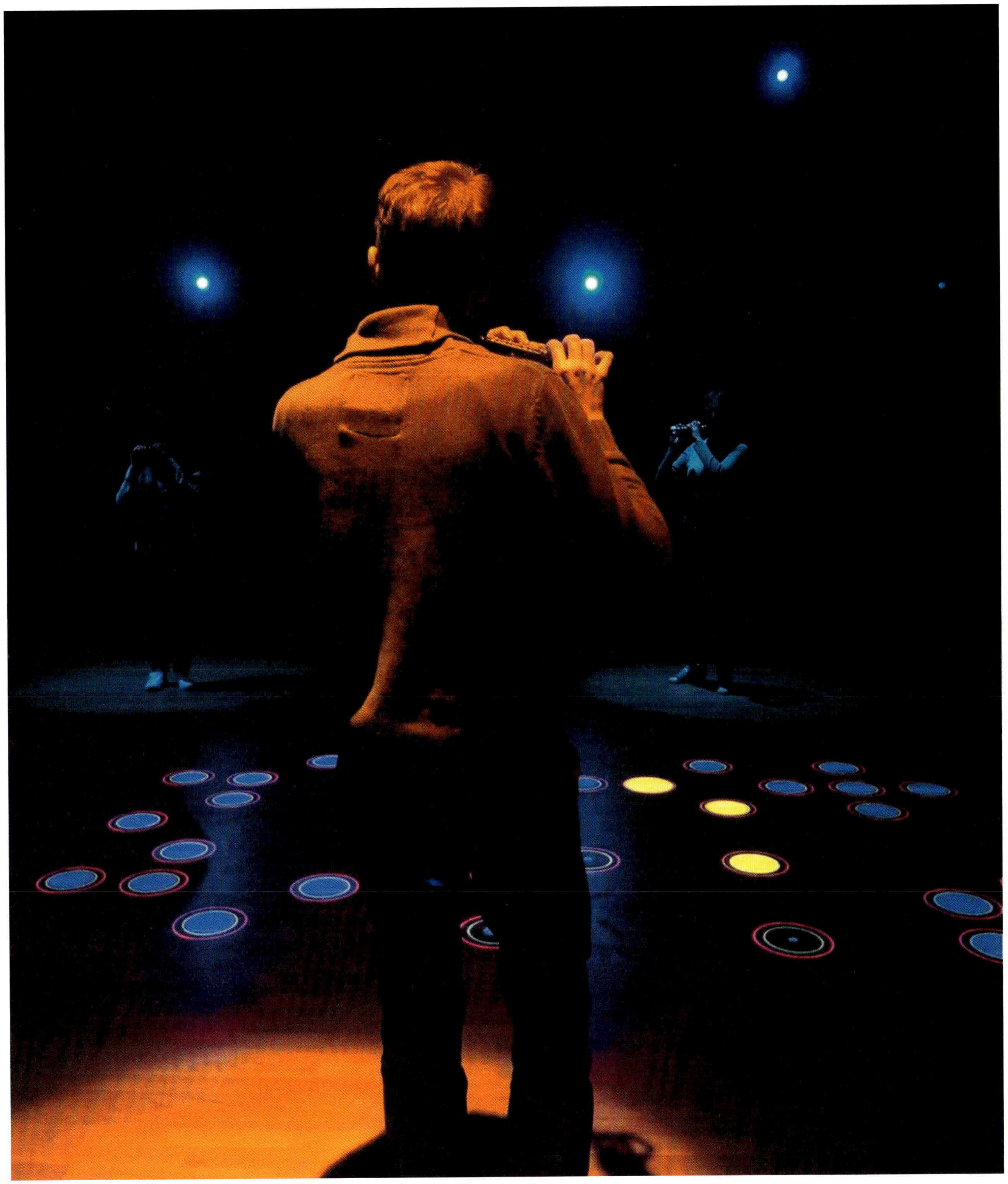

Eric Thanbichler, one of the piccolo players

1 An in-depth documentation of the project, including two video introductions, is provided on the project page: http://www.pixelsex.org/sonapticon.
2 Record of Tim Otto Roth's and Benjamin Staude's talk on the Neuroaesthetics symposium at ZKM | Karlsruhe: https://zkm.de/en/media/video/benjamin-staude-tim-otto-roth-the-sonapticon.
3 Only one tone is played by each player in a specific sequence, so together the three soloists play all possible tone combinations. This time-based combinatory play is inspired by Tom Johnson's counting pieces. See: http://www.editions75.com.

Tim Otto Roth

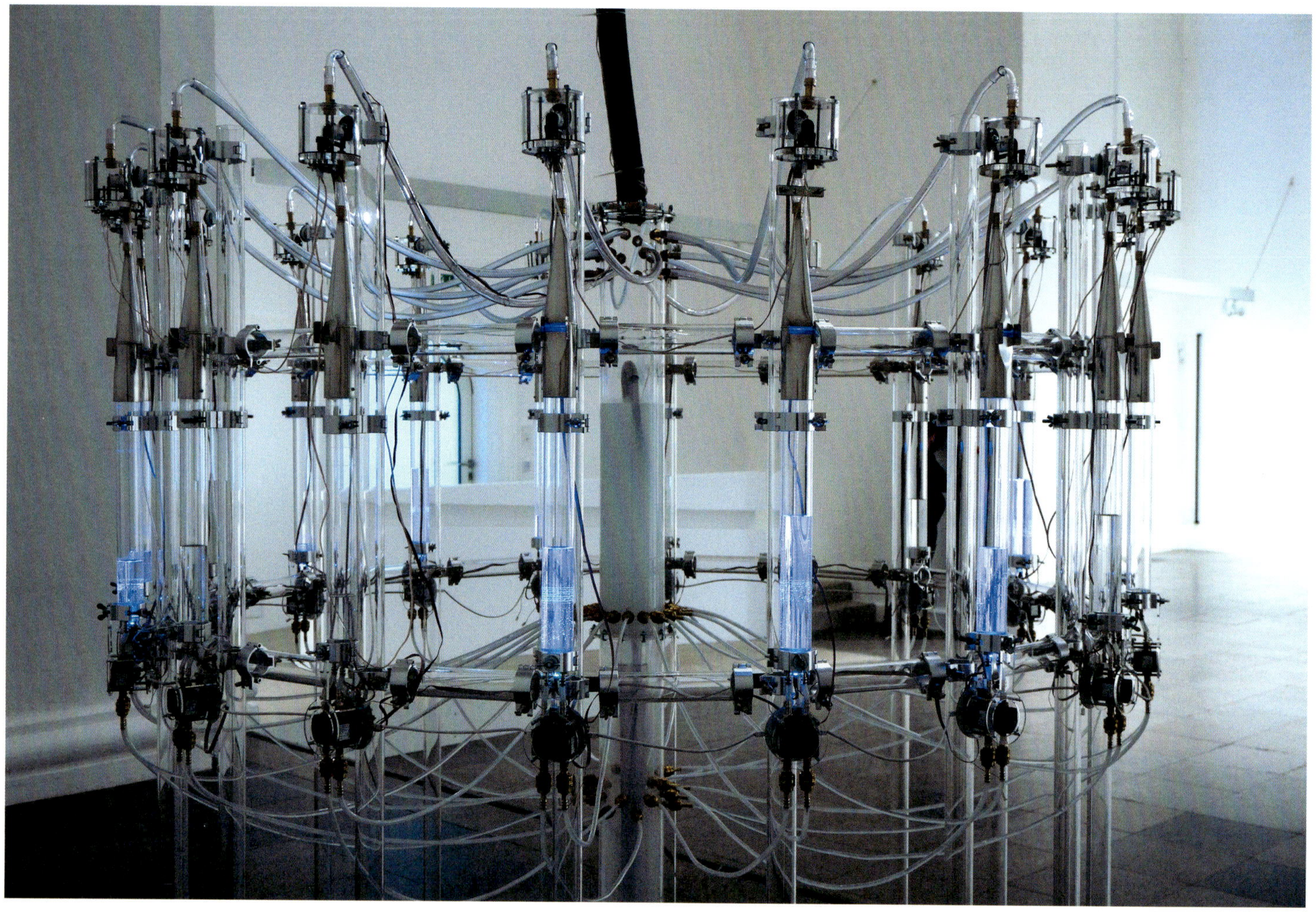

The Sonapticon – A Space as an Acoustic Neuronal Network, 2012
With Benjamin Staude

The installation *aura calculata* is a permanent installation at the high performance computing center of the Technical University in Dresden. One can describe it as a narrow relief of light and sound, which clings unobtrusively to the wall of a room. It references the theory of cellular automata. 39 "cells" of light and sound react to adjacent cells according to four simple rules. These rules were predefined by the artist for the system, which then operates independently. Beginning with the color green, a kaleidoscopic linear figure of colored light runs along the wall. For the visitor, the dynamics of the system that unfold in the room develop especially on the acoustic level. With standard pitch A, which corresponds to the hue of green in the middle of the color spectrum, an increasingly complex acoustic process begins, which transforms into a sound mix with many beats of the interfering sine waves. The sound swells upwards and particularly downwards. If one listens closely, a small melodic phrase is occasionally heard. Where does it come from? Clearly distinct in the set metric pulse, yet unpredictable for human perception, are also the pulsating and changing dynamics. Cellular automata are creative, they demonstrate growth processes, yet in terms of mathematics quite a few aspects seem unpredictable to me, even when one knows the rules of the system. Surprising effects are the result. (Helga de la Motte-Haber)

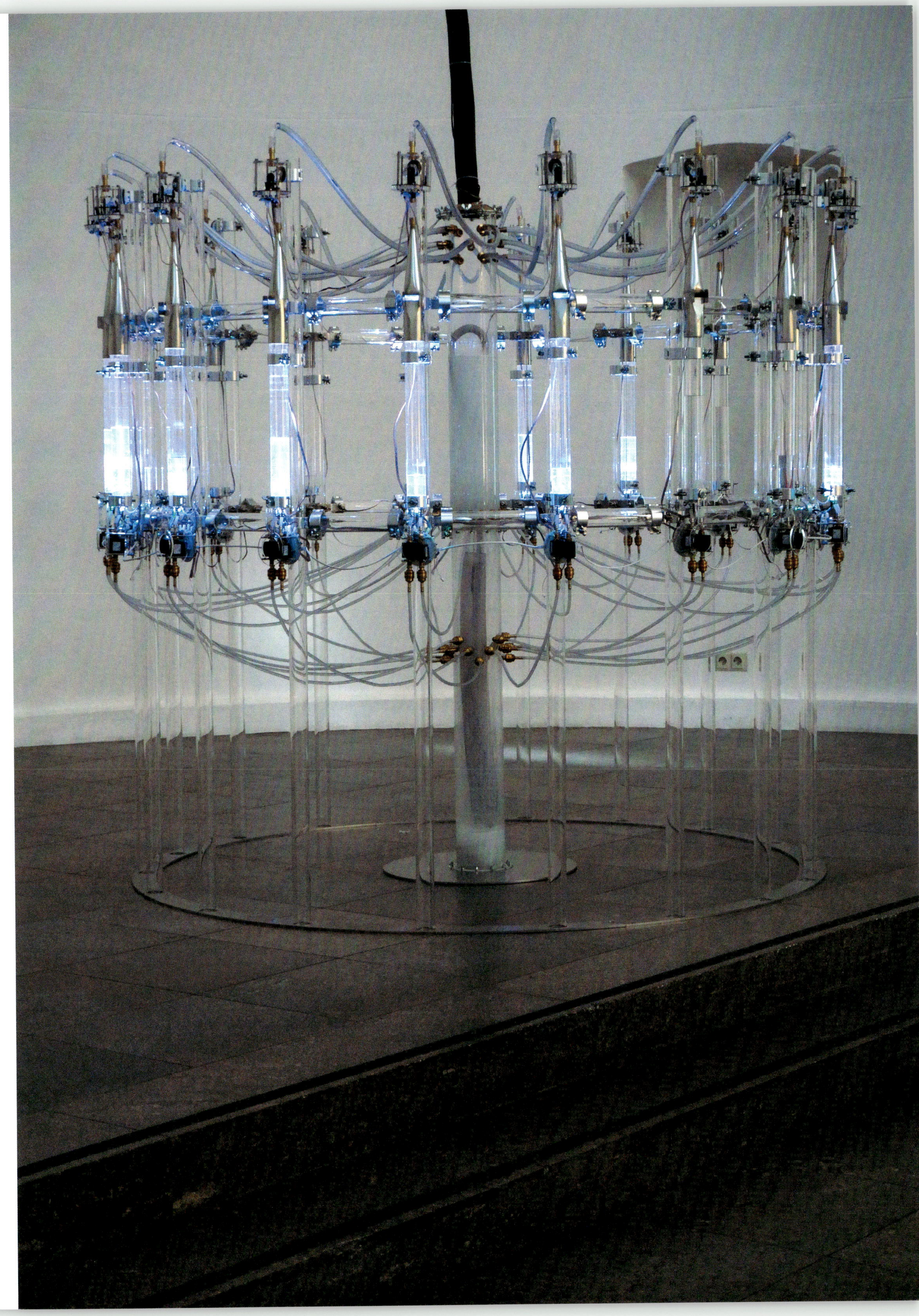

<SA/JO>
Sabine Schäfer
Joachim Krebs

*1957 in Karlsruhe (DE), lives in Karlsruhe
*1952 in Karlsruhe, †2013 in Ettlingen (DE)

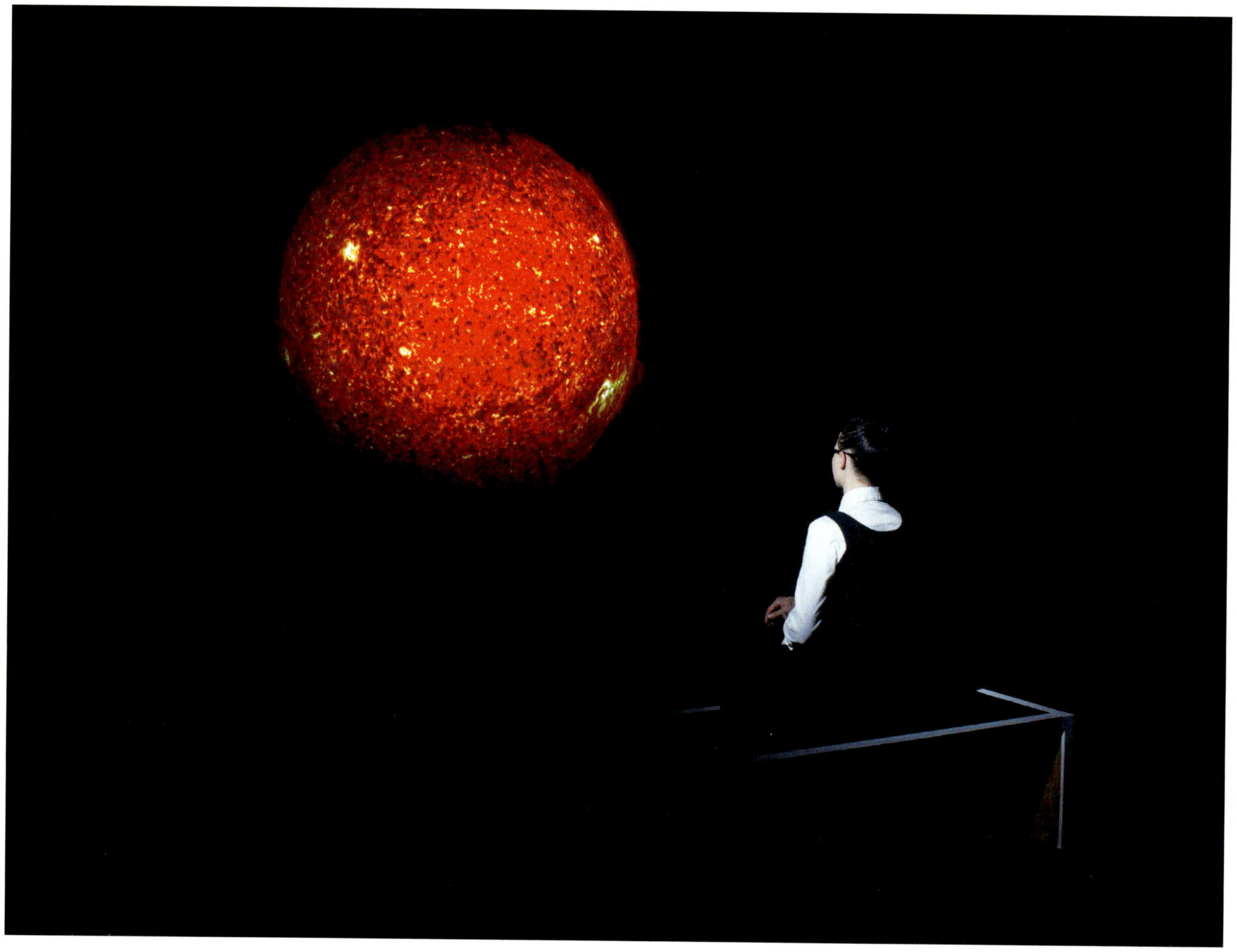

SolarSonical Insects – An enterable microSound colorLight event sphere, 2012
Audio-video installation
1-channel video, 2-channel audio played through 5 loudspeakers, 10:35 min. (loop)
Data conversion of NASA images of the sun: Dr. Davina Markiewicz-Innes, Borut Podlipnik, Max Planck Institute for Solar System Research (MPS) Katlenburg-Lindau
Data transfer: Sebastian Nagel (KIT)
Animation: Moritz Büchner (ZKM | Karlsruhe)

www.sajo-art.de

7.200 NASA satellite pictures of the SUN are accelerated, animated at freely varying speeds of 10 to 50 pictures per second. (Ultrasound-)microphone recordings of INSECTS are acoustically decelerated and transposed into the human audible range by means of procedures specially developed by the artists.

Through reducing-shortening ACCELERATION of the infinitely long durations of the sun, the INVISIBLE becomes VISIBLE and, through enlarging-extending DECELERATION of the infinitely short durations of the insects, the INAUDIBLE becomes AUDIBLE.

Thus, the more TIME manages to transform itself somewhat into a SONIC imagined all-enveloping COLOR-LIGHT-resonance-BODY of "sun," the more intensively the molecular-acoustic vibrations of the insect sounds and noises transmute into a SHINING imagined all-enveloping MICROSOUND-resonance-SPACE. The insects are transformed into continuously oscillating "solar" instruments, played by the LIGHT of the sun, with which they are communicating!

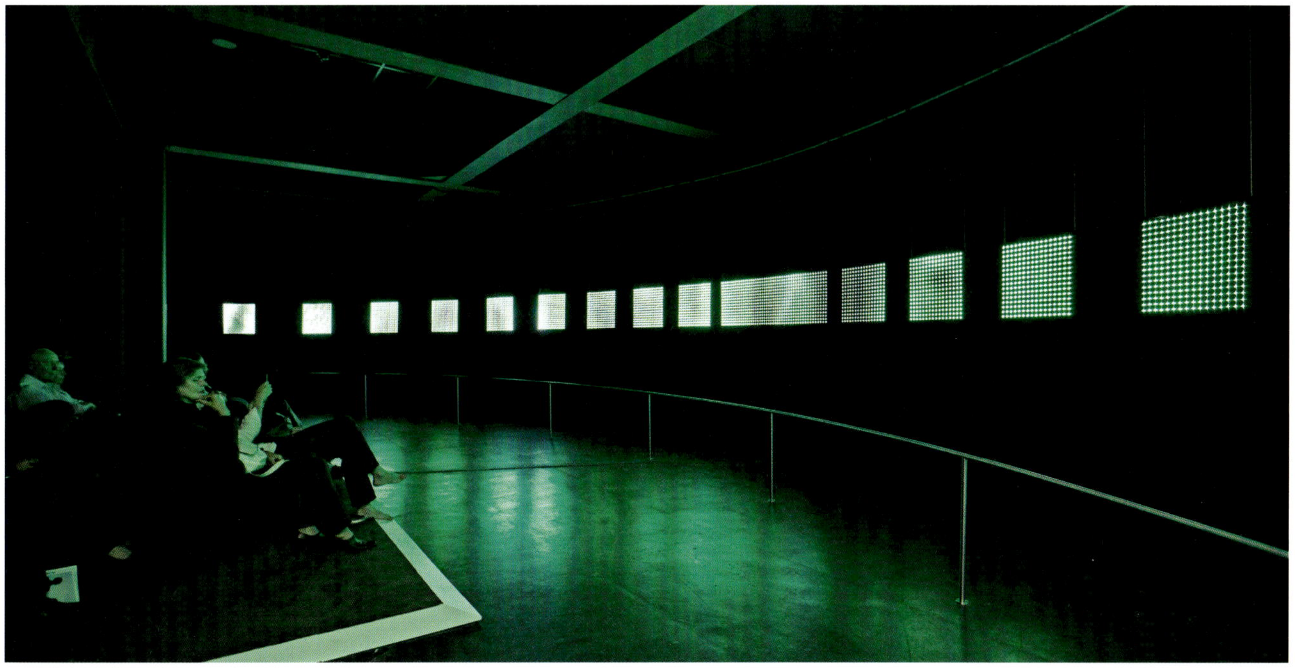

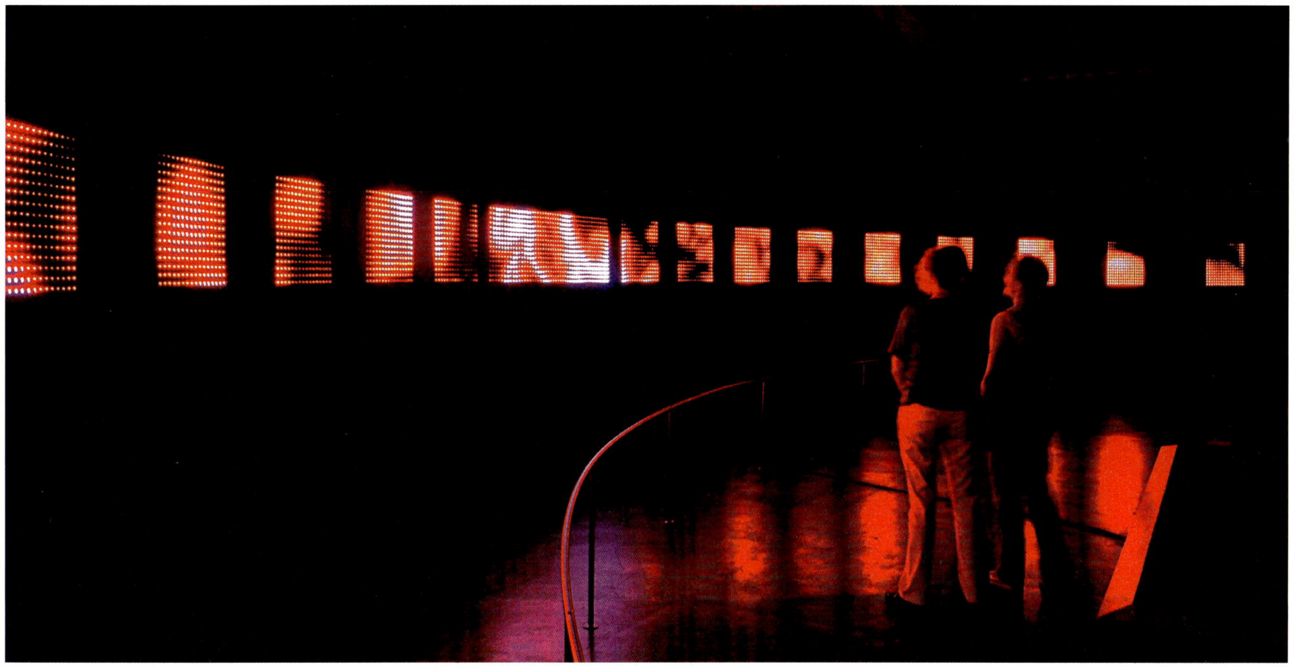

MicroSonical Shining Biospheres No. 1 An enterable microSound colorLight event sphere at the ZKM_Subspace, 2009
Audio-video installation
Loudspeakers with audio interface, LED panels with DVI controller, computer with Windows operating system
Project partner: Karlsruhe Institute of Technology (KIT)
Opening project for the cooperation of the ZKM | Karlsruhe and the KIT started in 2009; sponsored by: Fiducia Karlsruhe, Badisches Staatstheater Karlsruhe, KIT, Joint Research Centre (JRC) of the European Commission
Video production: Angelika Münch
ZKM | Karlsruhe

"Luminous SOUND" as "emerging existence" in space and time – "sounding LIGHT" as "existing emergence" of time in space.

With this motto, media artists <SA/JO> present their "walk-in spatial-sound colored light body" for micro-acoustic animal sounds and three-dimensional LED-colored light, produced for ZKM |Karlsruhe's twentieth anniversary. In creating this sound-light environment, they were inspired by the latest breakthroughs in neuroscientific aesthetics and bionics, among other things. At the center is the normally inaudible and until now largely unknown acoustic micro cosmos of animal noises and natural sounds.

So-called "micro acoustic, colored light biospheres," quasi artificially created imaginary environments for multifarious "audio-light hybrids" arise by means of the acoustic-scientific "sound microscopics" (endo-sono-microscopics) specially developed by Joachim Krebs in the mid 1990s, and "audio slowmotion."

These in-between beings, continually moving between real-unreal, concrete-abstract, and natural-artificial, can exist solely in such artificially generated "sound light habitats," and can only be experienced by museum visitors live, on-site.

The two natural phenomena color and light, on the contrary, are "composed" as a largely abstract, reference-less "event" through monochrome color oscillations and specific light-dark movement choreographies in time and space.

<SA/JO>
Sabine Schäfer
Joachim Krebs

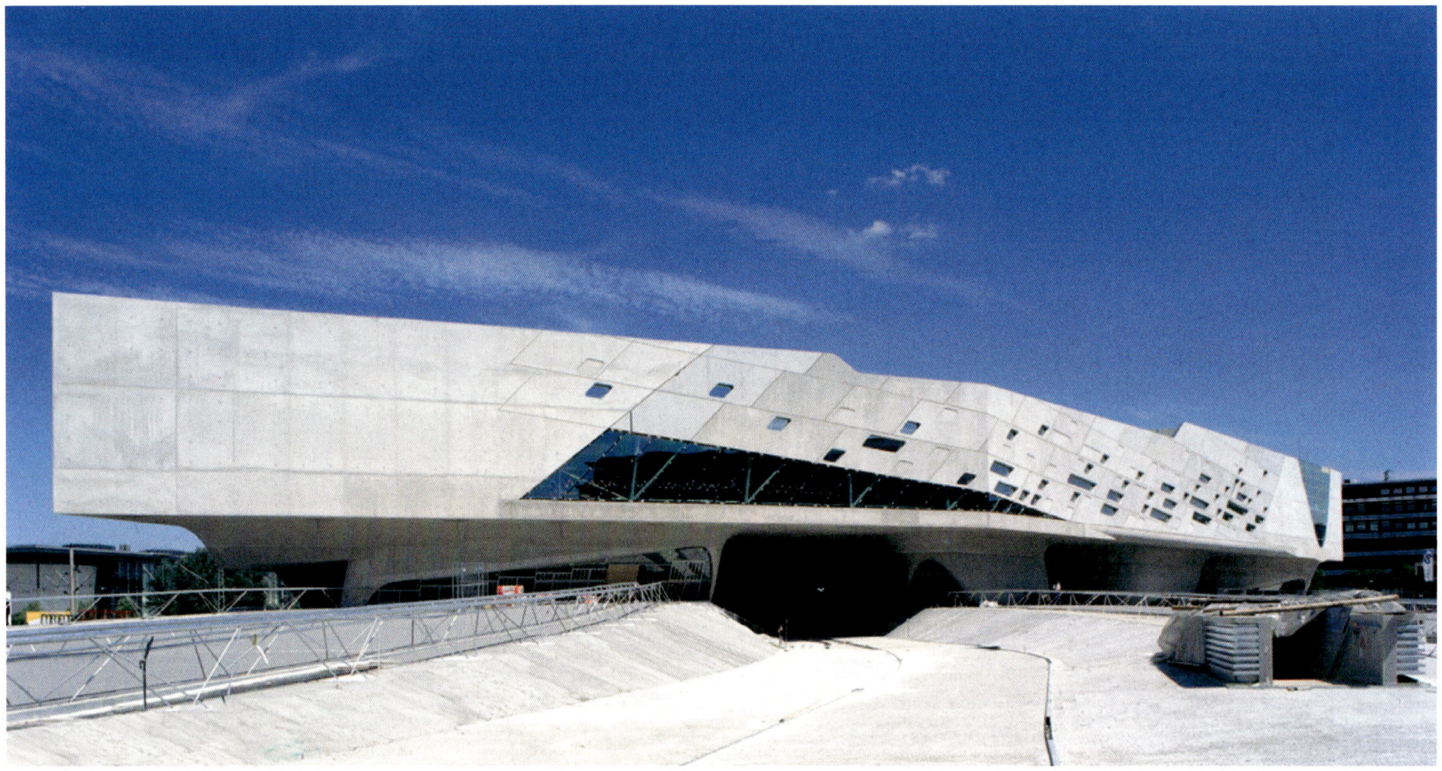

TopoSonic Tunnel. An enterable Space-soundBody, 2005
Permanent installation with luminous glass seat, designed for the architecture of the Science Centre phaeno Wolfsburg (DE)
Audio composition: 16 min. (loop)
Architecture: Zaha Hadid
Curated by Joe Ansel, Ansel Associates Inc. USA
Commissioned by the city of Wolfsburg
Project partner: Ansel Associates Inc. CA, USA
Head of technical construction team: Matthias Ossmann
Collection of the Science Centre phaeno Wolfsburg

Sound – Space – Movement

The space-sound installation *TopoSonic Tunnel* by the media artists <SA/JO> conceived specifically for one of the most extraordinary architectonic zones in the Centre phaeno Wolfsburg, Germany, Zaha Hadid's visionary architectonic landscape functioning as a congenial resonance space, again focuses on the normally inaudible and thus to this day almost completely unknown acoustic micro-dimensions of natural sounds and noises recorded in the animal kingdom, in nature and in everyday human environment.

Even basic physical phenomena such as time, space, and movement are shaped, in this work, in a purely artificial way and thus made experienceable "just by listening" as elementary and, at the same time, universal features.

Thanks to a computer supported digital PA system based on the principle of the bending wave – recently and innovatively developed by Siemens AG/"puren" – it was possible to distribute the space-sound surfaces at will and to integrate them invisibly in the walls, thus creating the impression of being surrounded by multiply moving space-sounds practically coming out of nowhere.

In the *TopoSonic Tunnel* one may listen to the acoustic micro-dimensions of nature and technic environments as well as of sounds and voices from the animals' world.

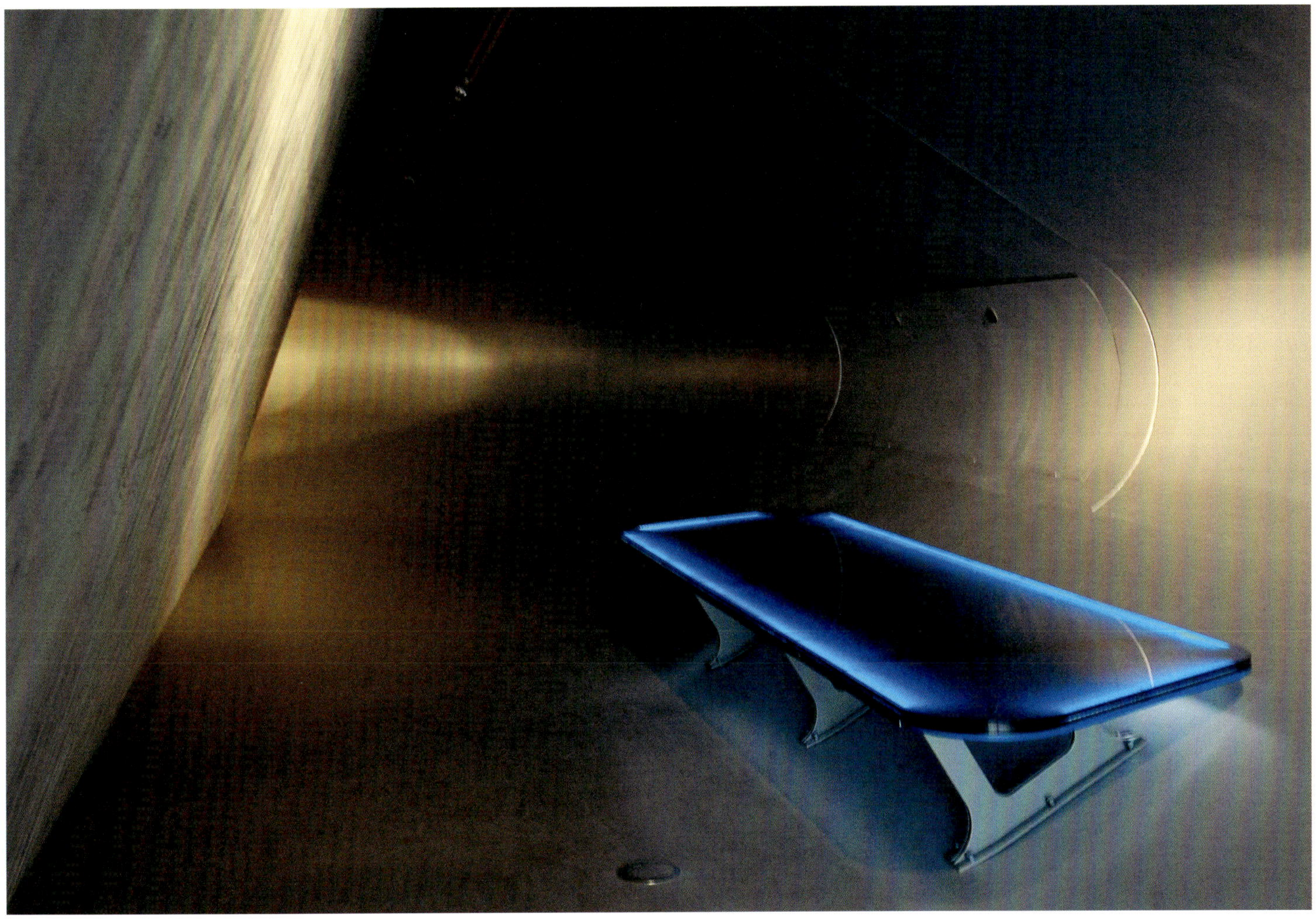

Michael Saup

*1961 in Hechingen (DE), lives in Berlin (DE)

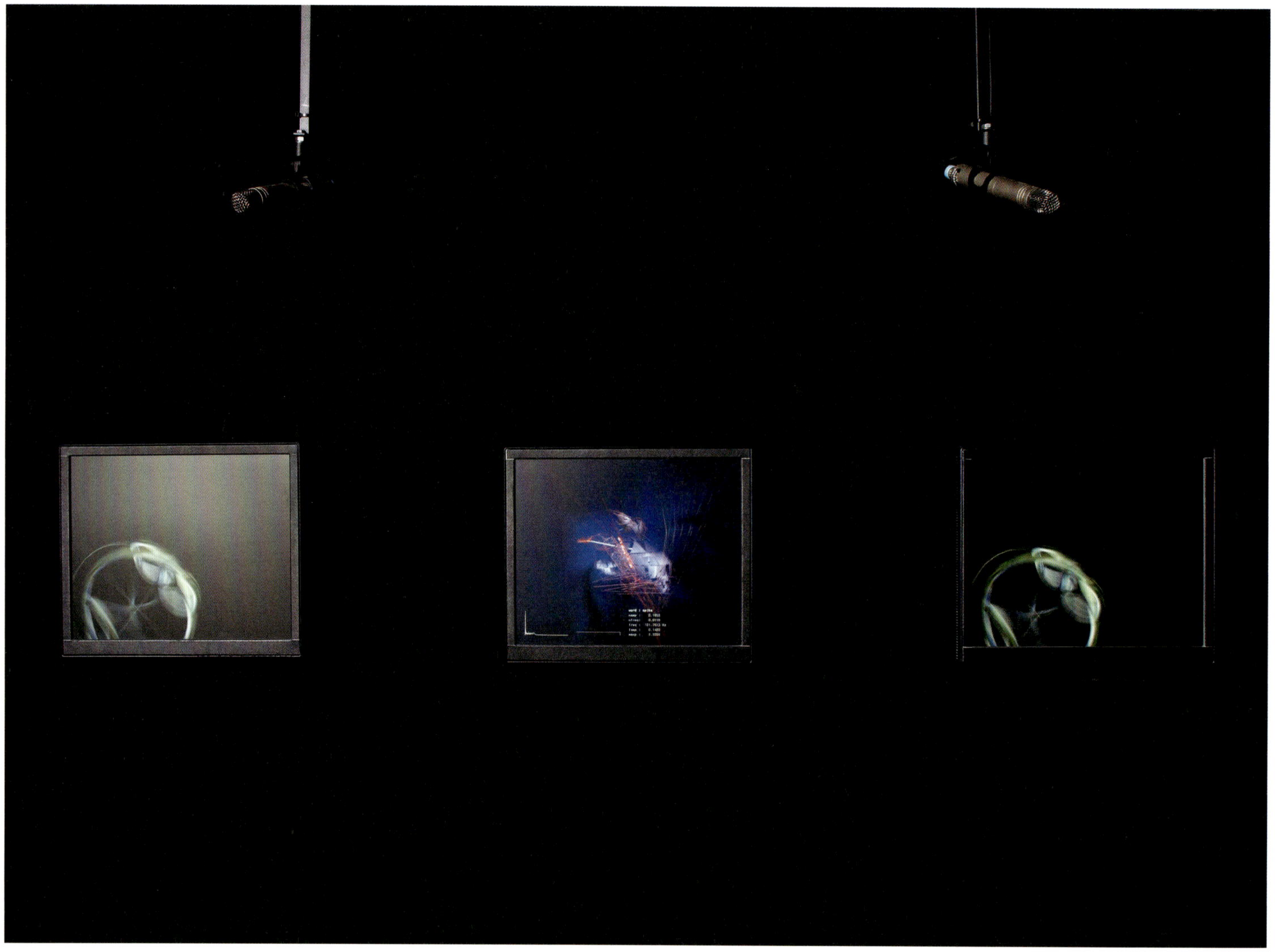

Pulse8, 1992/2004
Interactive computer installation
Computer, 2 monitors, 4-channel audio, gravel, microphones, dimensions variable
Sound: Paul Modler
ZKM | Karlsruhe

www.kunst.1001suns.com

In an alcove in the exhibition area, the floor is covered with gray gravel. Visitors are invited to walk on the field of stones and approach monitors that hang on the opposite wall. When someone treads on the field, a sound arises that transforms the screen images. The pulsing particles, the "supreme particles" of virtual material, transmogrify. Visitors influence the volume and vectors of the visual elements through their footsteps as well as their whispers, speaking, and singing in the vicinity of the screens. A random algorithm and the laws of gravity govern the onscreen motion.

The field of stones symbolizes a radioactive field, the stones can be seen as radioactive waste, the destructive end product of an information society that gorges itself on energy. Human beings similarly transform, living on their own graves only as long as the beating heart generates a pulse. Banished from eternal paradise, and yet in the hands of everlasting godly benevolence.

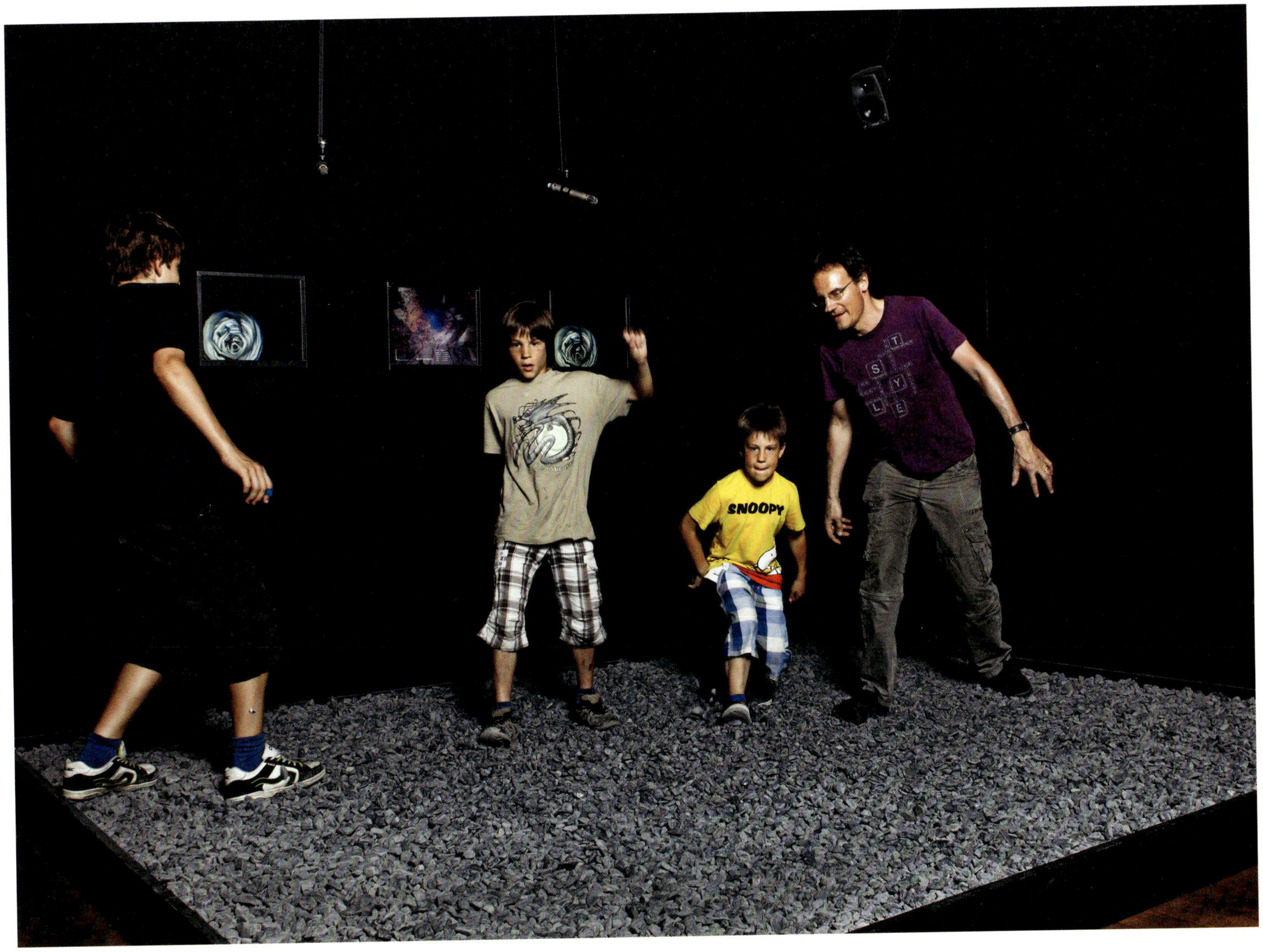
SNOOPY

Dawn Scarfe

*1980 in Darlington (GB), lives in London (GB)

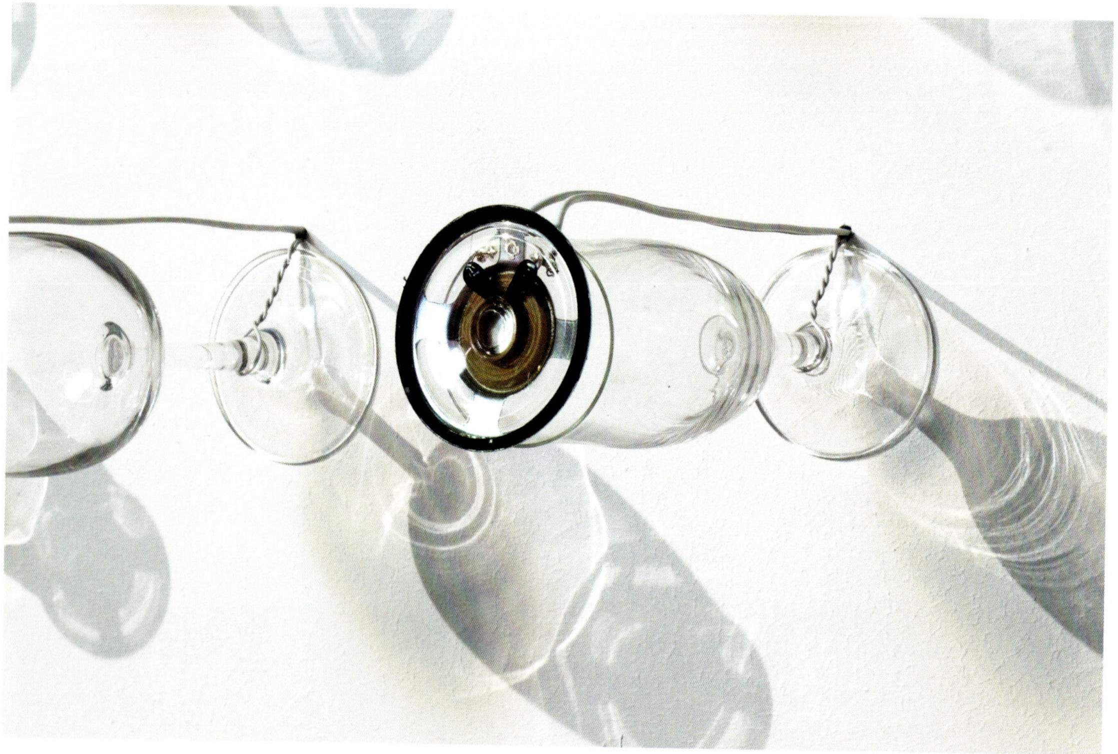

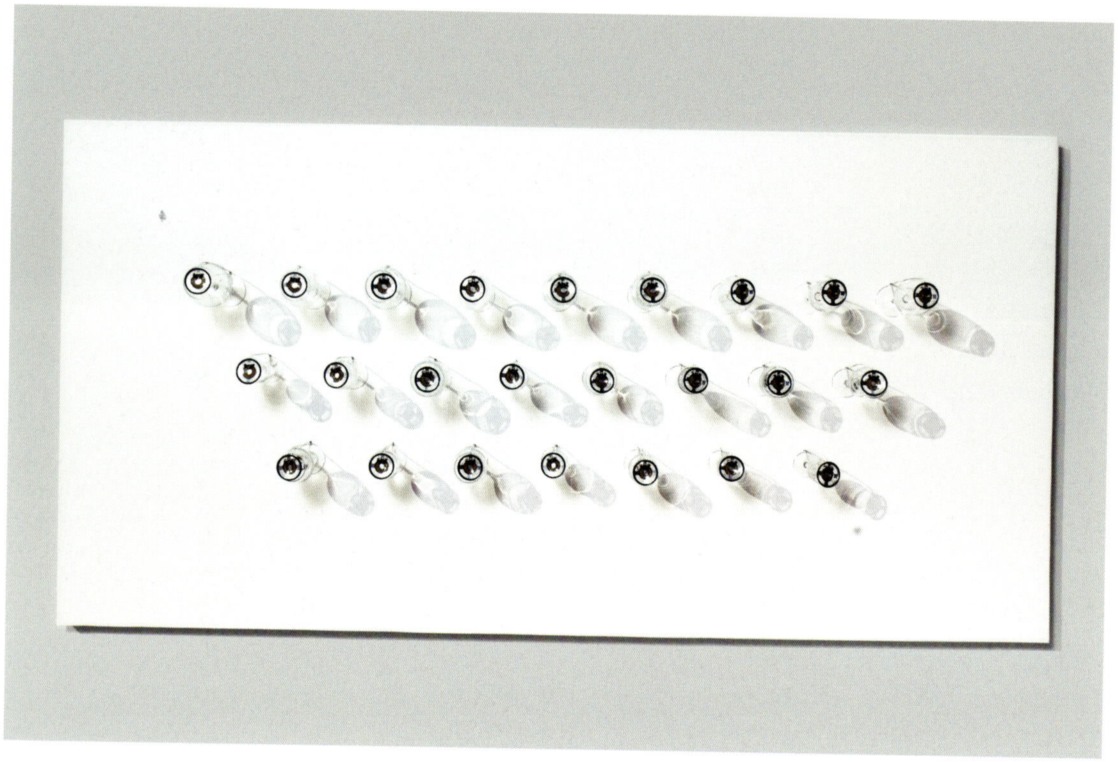

Lenses, 2008/2012
Sound installation
24 wine glasses,
24 loudspeakers,
240 × 120 × 26 cm
Courtesy of the artist

www.dawnscarfe.co.uk

Each of 24 wine glasses has a loudspeaker installed just above its rim. Each glass has a different pitch, which was made to sound for an acoustic recording by rubbing a moist finger around its rim. From these recordings, Dawn Scarfe compiled a soundtrack that follows a pattern of "change ringing," which stipulates that all permutations of tone sequences are played so that no row of tones is repeated. The only exceptions are first and last sequences, in which the tones are arranged in descending order.

In *Lenses* the same soundtrack plays from each loudspeaker. The sounds are amplified and transformed by the glasses.

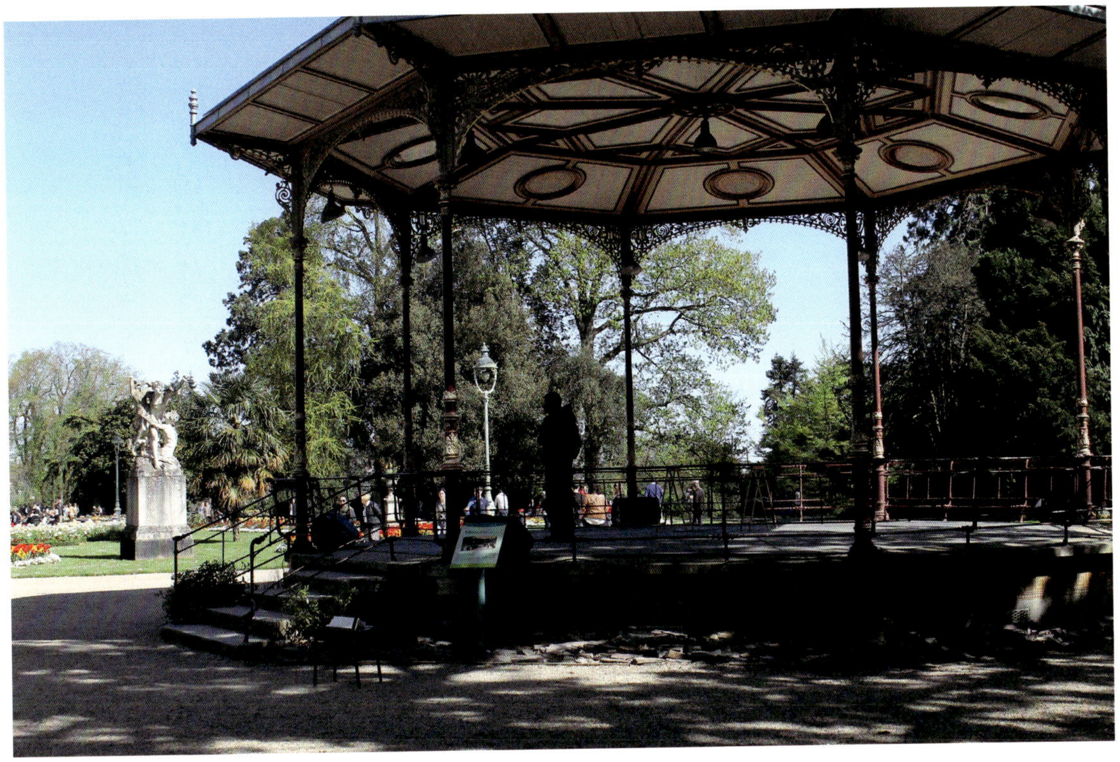

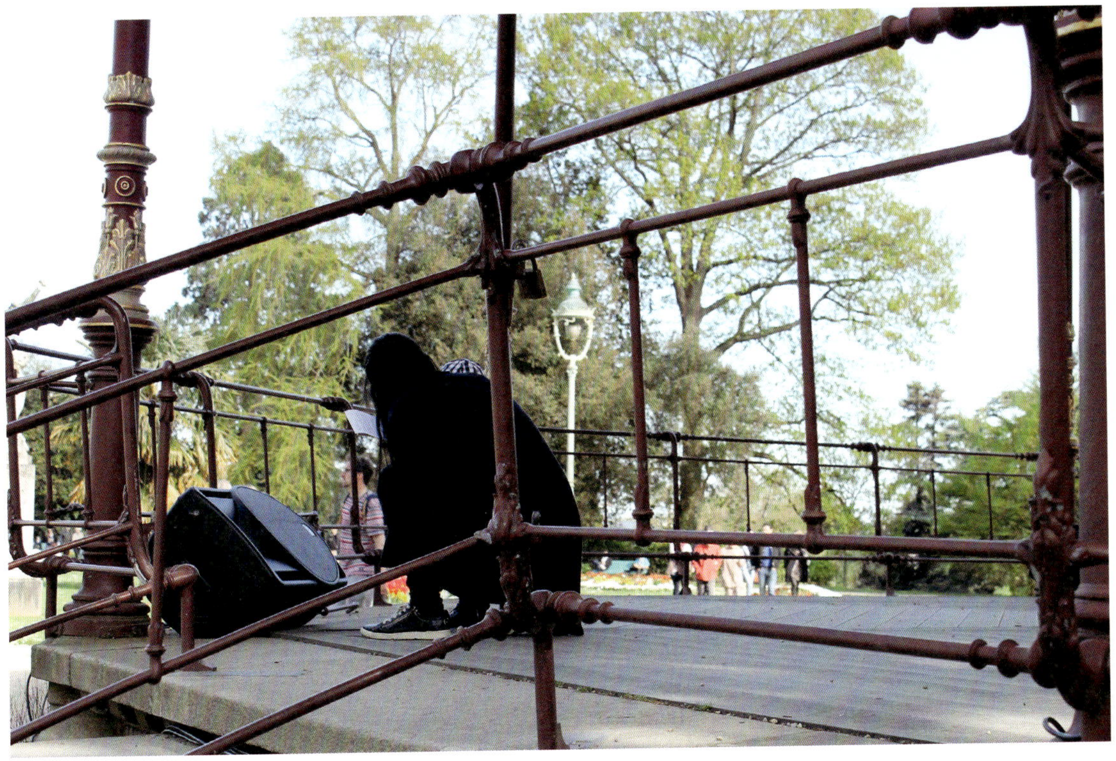

Bee Strings, 2014
Sound installation in public space, arranged for 4 speakers (Ruark Audio), 10 min. (loop), ca. 10 × 10 × 3 m
Funded through a residency with Forestry Commission England and Sound and Music
Courtesy of the artist

Bee Strings explores the tiny, rapid movements of bee wings with layered recordings of bowed violin strings.

Natural Philosopher Robert Hooke proposed tuning violins to the sound of bees in order to grasp the frequency of their wingbeats. Following his logic, if we know the rate at which musical strings vibrate, we can infer that a bee sounding a G# beats it's wings just over 200 times per second.

Scenocosme
Grégory Lasserre
Anaïs met den Ancxt

*1976 in Annecy (FR), lives in St. Etienne (FR)
*1981 in Lyon (FR), lives St. Etienne

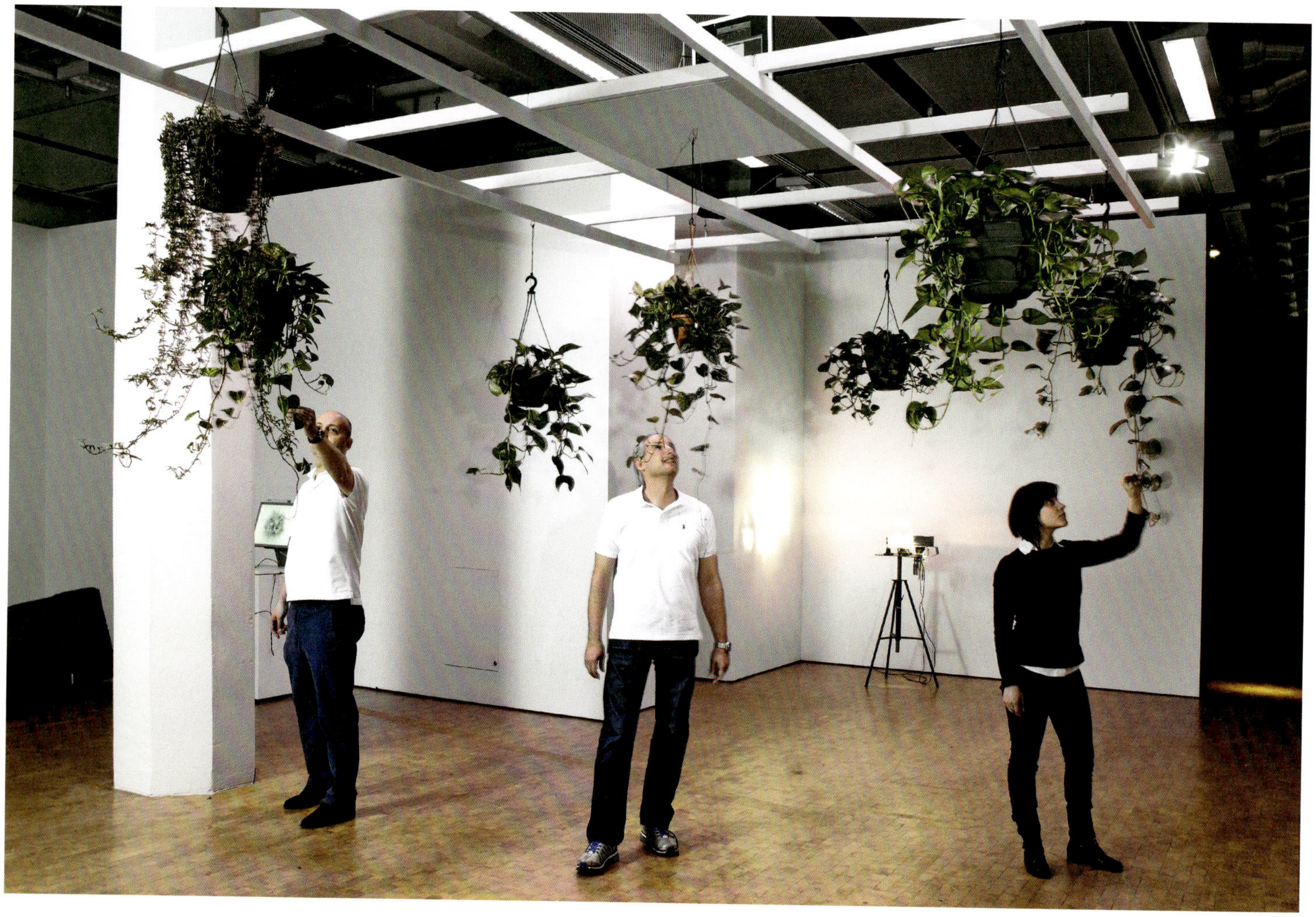

Akousmaflore. Sensitive and interactive musical plants, 2007
Visual-acoustic installation
Plants, electronics, audio system
ZKM | Karlsruhe

www.scenocosme.com

The hanging garden of Scenocosme reacts to the visitor's touch. Every plant responds with its own sound. *Akousmaflore* is a hybrid between plants and digital technology. Grégory Lasserre and Anaïs met den Ancxt use plants' natural sensitivity to the electrostatic aura of human beings. To hear the "Akousmaflores," one has to carefully touch the leaves and stem, hold or stroke them slightly. Sometimes physical proximity alone is enough. Electronic interfaces and computer programs translate the plants' reactions into sound.

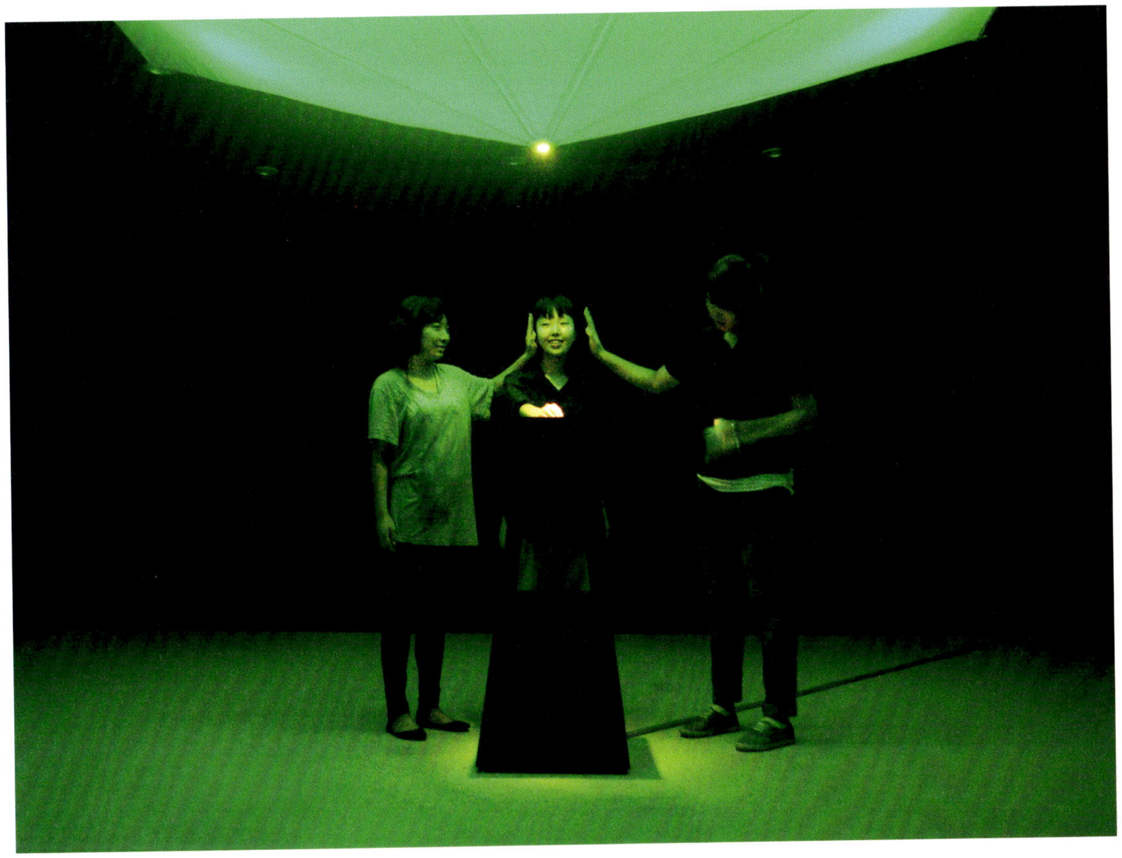

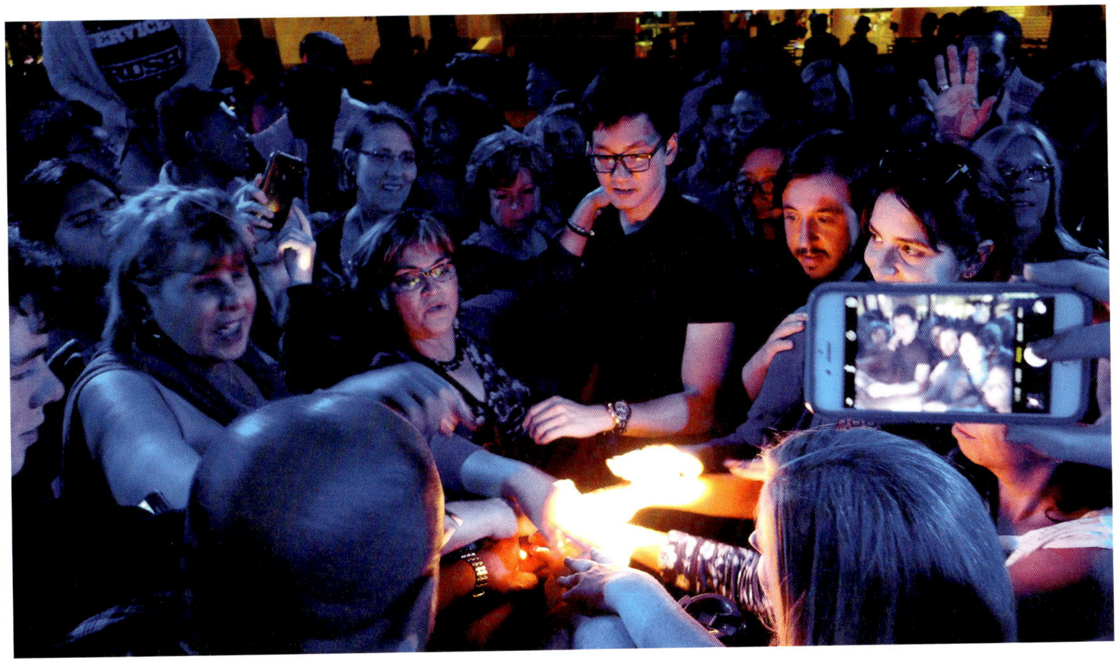

Lights contacts, 2009
Interactive sound
and light installation
Humans, electronics,
audio system, lights
ZKM | Karlsruhe

Lights contacts is a sound and light installation interacting and engaging with the physical presence of bodies and skin of the audience. Its social dimension questions the perception of the other participants. The installation is composed of a pedestal and a shiny ball on the top interacting with the body and the skin of two people and more. Once a person puts his/her hand on the shiny ball, his/her body gets sensitive and reactive to other living bodies. To cause a reaction, the first person has to invite another person to touch his/her skin. Only then each touch generates variable sounds and lights, which evolve according to the proximity of contacts and spectators. The moment nobody interacts, the sounds and lights switch off. It induces a variety of stagings among the spectators' bodies who are transformed into real sonorous human instruments.
In this intimate artwork, the electrostatic energy of human bodies becomes perceptible through music and light.

Scenocosme
Grégory Lasserre
Anaïs met den Ancxt

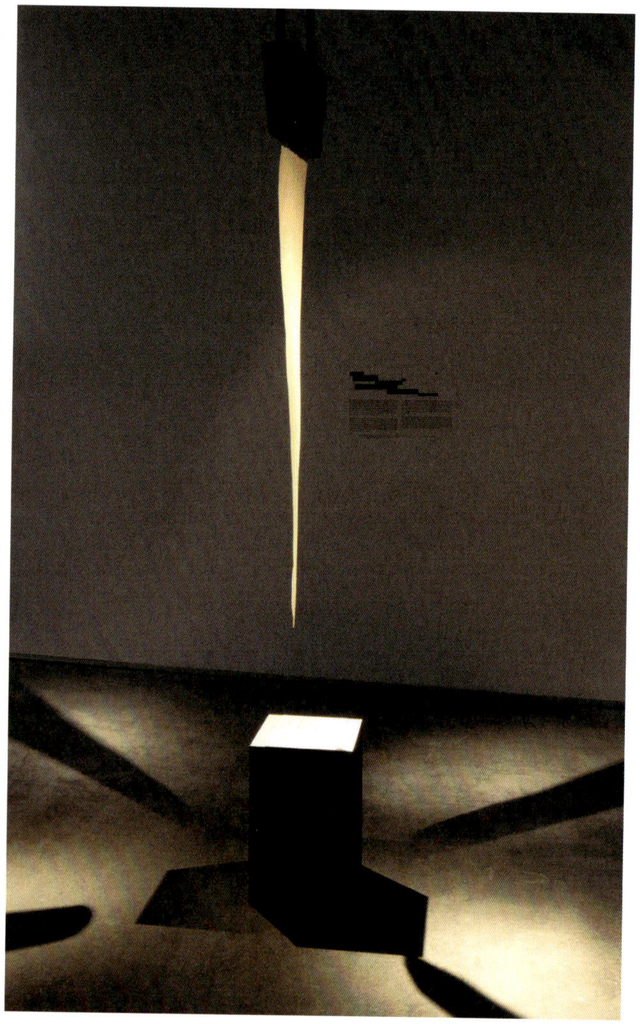

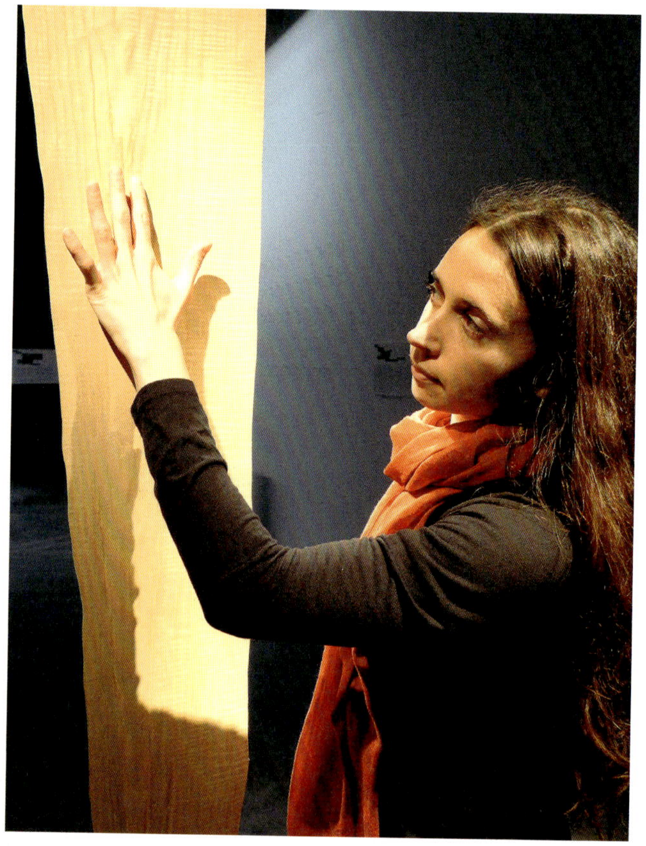

Matières sensibles, 2013
Sonorous interactive marquetry
Wood, electronics, audio system,
awarded with the Human Interface
Award 2015 of the Science and
Art Festival Phaenomenale
at the Phaeno Science Center,
Wolfsburg (DE)

Matières sensibles [Sensitive matters] is a sculpture made of a very thin ash wood veneer sheet. This sheet has distinct sonorous touch zones that follow the natural veins of the wood. The human body's electrostatic energy acts as the trigger for the sculpture. An invisible design concept allows the artists to define a musical score, spread over different areas of the wood. This process of bio-hacking, invented by the artists, is called "interactive marquetry."

The wood sculpture produces sounds when the spectator touches it. Sounds are used to stimulate haptic and gestural behavior. Thus, the design of the sculpture resembles an instrument, using touch to produce various kinds of sound.

Matières sensibles fosters a sensory and intimate relationship between the wood and the body of the viewer, able to remember and reproduce the specific sound induced by the touch of a particular viewer.

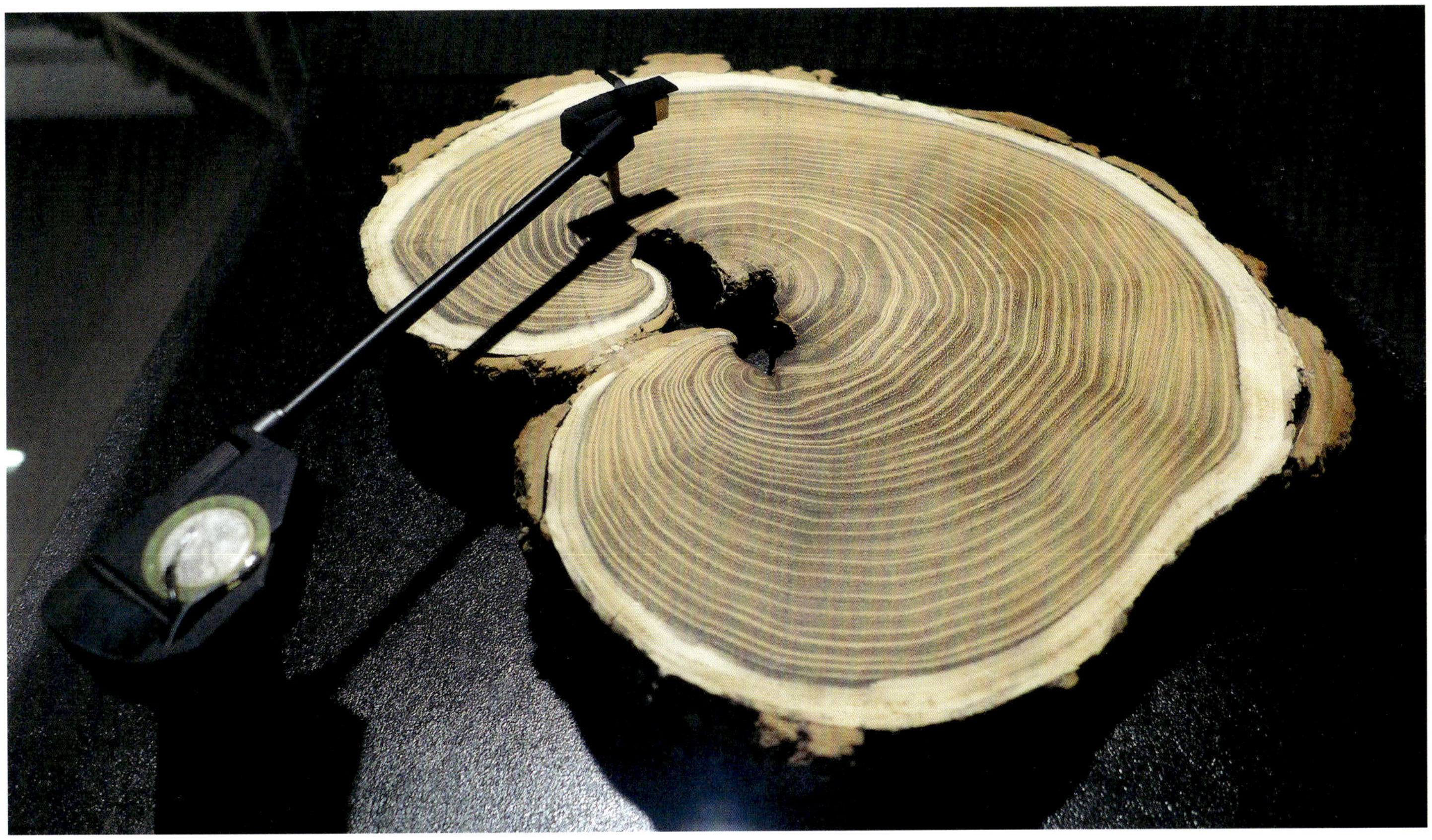

Echoes, 2011
Sound installation
Wood, electronics, audio system,
Grégory Lasserre & Anaïs met
den Ancxt

The reliefs of the times engraved in the rings of this tree are feeling and reading in real time and produce rhythmic sound.

Dieter Schnebel

*1930 in Lahr (DE), †2018 in Berlin (DE)

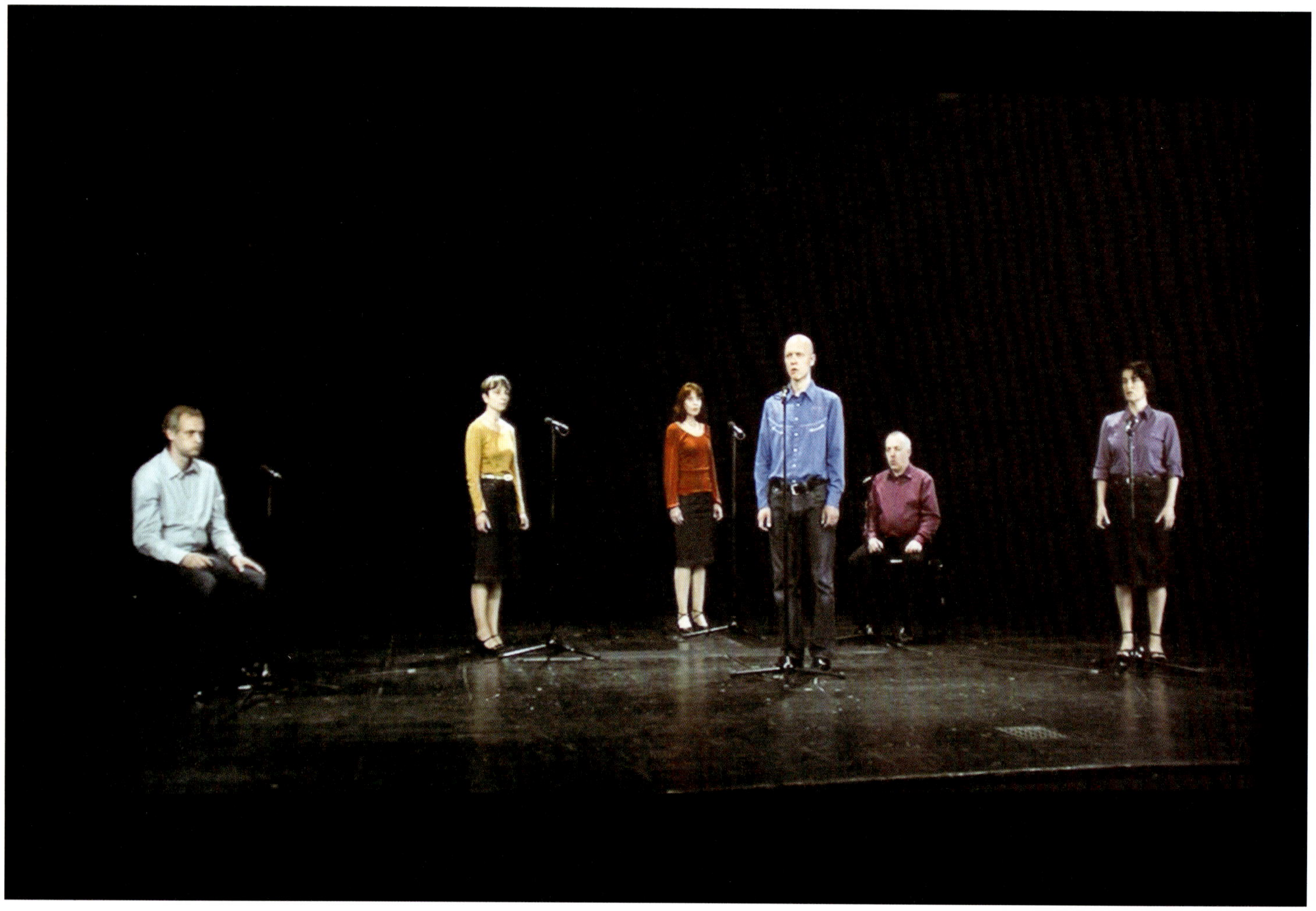

Maulwerke, 1968–1974
For speech organs and reproduction devices, film version 2010, DVD, video
Participants: Die Maulwerker [The Mouth workers] (Michael Hirsch, Ariane Jessulat, Henrik Kairies, Christian Kesten, Katarina Rasinski, Steffi Weismann)
Concept: Susanne Elgeti, Christian Kesten, Julia Gerlach, and Steffi Weismann
Director: Susanne Elgeti
Musical concept: Martin Supper
Die Maulwerker in collaboration with the Academy of Arts Berlin and the Academy of Arts' studio for electroacoustic music, supported by the PRO MUSICA VIVA Maria Strecker-Daelen Foundation and filmArche non-profit organization and film school. The film version was made possible by a grant from Germany's Capital Cultural Fund.
DVD © Schott Music & Media / WERGO. By courtesy of www.wergo.de

With his *Maulwerke* [Mouth works] the composer Dieter Schnebel has written music history. It is one of the most significant works of non-verbal vocal expression in which the articulations are the sole musical material. The score initiates a process of communication and composition that demands an extreme degree of voice control and drafts a new, processual concept of a work. Schnebel's verbal score organizes the process of development in three phases – exercises, productions, communications – through to the final opus for public performance.

To mark the occasion of the composer's 80th birthday in 2010, Die Maulwerker ensemble produced a new version of *Maulwerke – Opus* – the film version of which is presented here. Following a note of Schnebel in the score that technological media could be used, the camera becomes a co-contributor: the images become an autonomous "voice" and shape sequences independently.

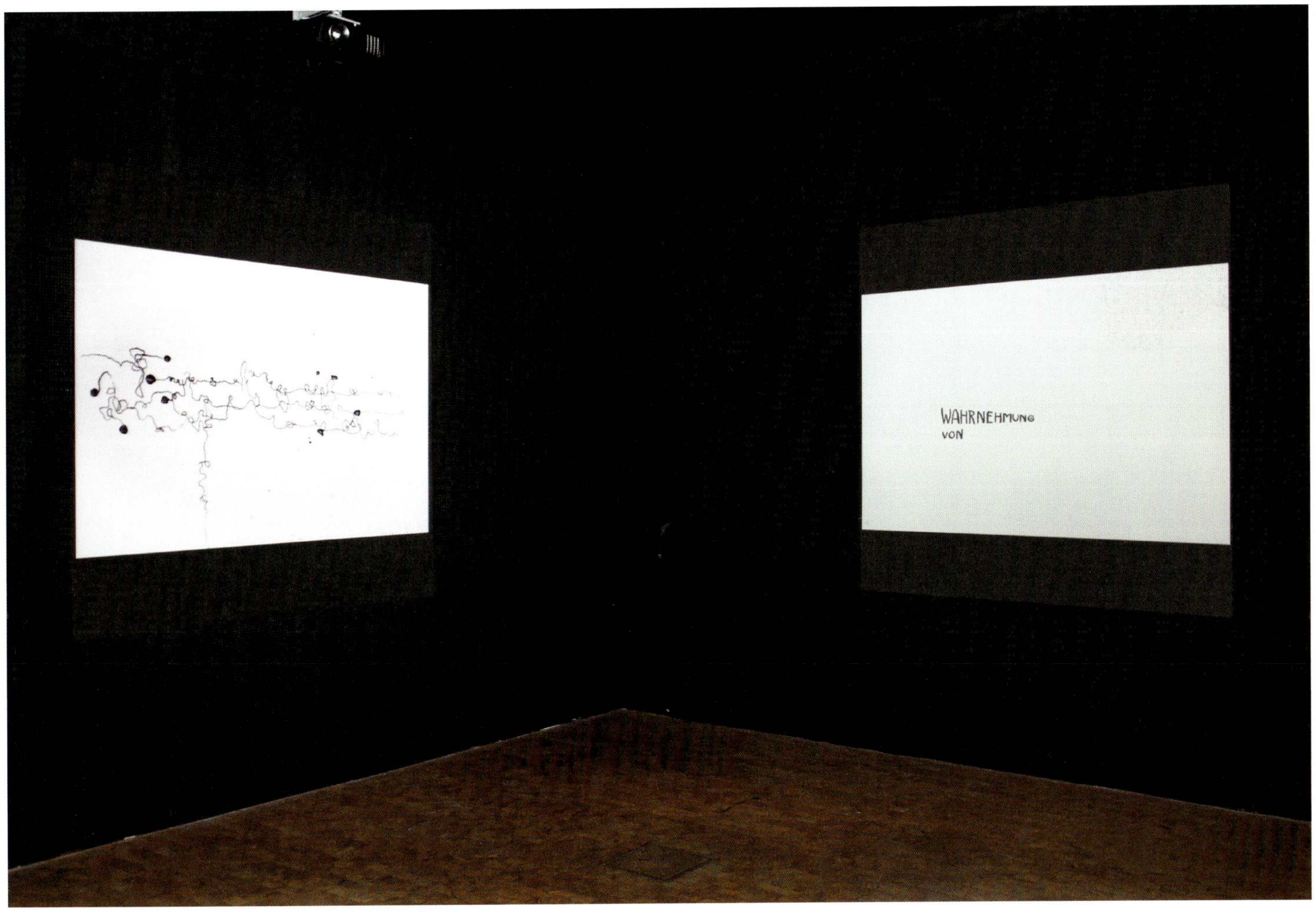

ki – no. Nachtmusik für Projektoren und Hörer, 1963/1967, 2012
Video and audio installation
3-channel video, stereo-audio, screens
Digital video: Andreas Brehmer
Audio: Anton Kassjanenko
Production of the digital version: ZKM | Karlsruhe

ki – no. Nachtmusik für Projektoren und Hörer [ki – no. Serenade for projectors and listeners] is predominantly visual music. Slides are projected onscreen in specific rhythms. The performance material consists of series of slides on different themes: Rezeptionen [receptions] I/II, Imaginationen [imaginations] I – VI, Auswüchse [excesses] I – VIII, Fenster [windows] I – XXI, Umrisse [silhouettes] I/II.

On the one hand, the pictures show texts that lead us toward musical perception of actual or fictive acoustic events. On the other hand, they contain notes for imagining music or for the imaginative production of "impossible" sounds (tones beyond the threshold of hearing, pauses that make sound, etc.). This Noh theater stages the kind of music that arises in the mind of the listener and is interpreted there.

Cornelia Sollfrank

*1960 in Feilershammer (DE), lives in Berlin (DE)

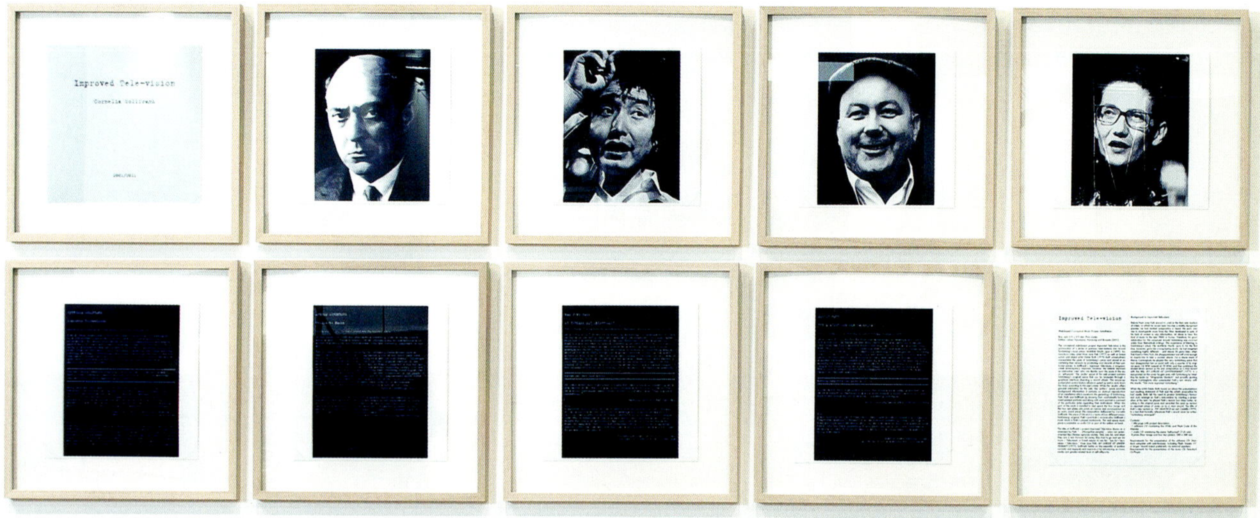

Improved Tele-vision. Make your choice and become avant-garde, 2001
Interactive Internet installation
Website, 2 audio CDs, prints, 3 records, dimensions variable
Cornelia Sollfrank, Broken Music/Ursula Block

www.arturarez.org

Is Arnold Schoenberg's string quartet *Verklärte Nacht* [Transfigured Night] nothing but "Wagnerian rubbish"? That was how video artist Nam June Paik described it in 1977. He "improved" the music by pressing the 78-rpm vinyl record at 16 rpm or ¼ speed. The Swiss action and object artist Dieter Roth responded in 1979 with his own edition, the original tempo restored. Cornelia Sollfrank's installation *Improved Tele-vision* adds a new dimension to the controversy.

The web-based part of the installation consists of a sound studio and a virtual gallery. In the studio, the user can control the music's tempo. The gallery displays the "Schoenberg, Paik, Roth, Sollfrank" geneaology. For the installation in real space, Cornelia Sollfrank composed her own music, titled *Self-paced*. The work's title refers to a statement by Paik, in which he describes his Mongolian forebears' question about what lies beyond the horizon: "To be able to see more, you have to go there – 'tele-vision' in Greek means 'seeing-into-the-distance' = television."

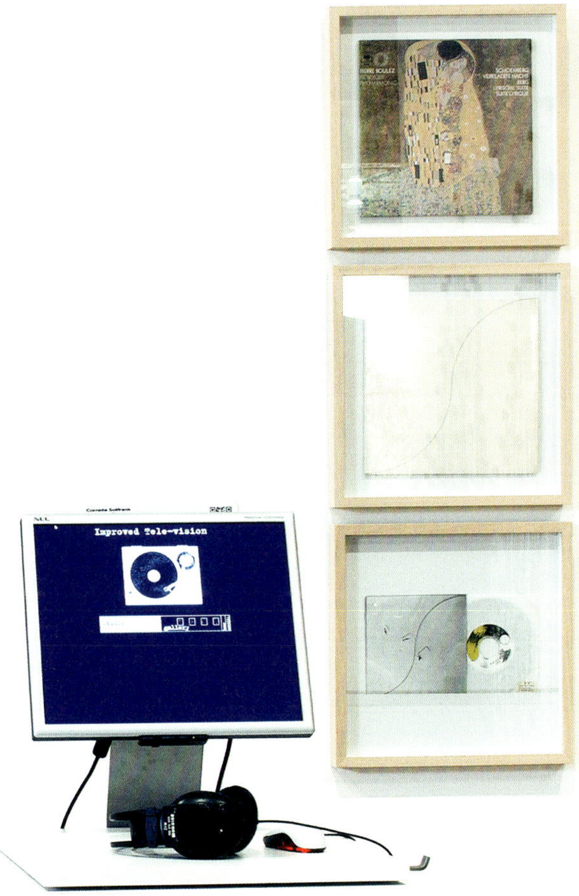
Improved Tele-vision

Jan-Peter E.R. Sonntag

*1965 in Lübeck (DE), lives in Berlin (DE)

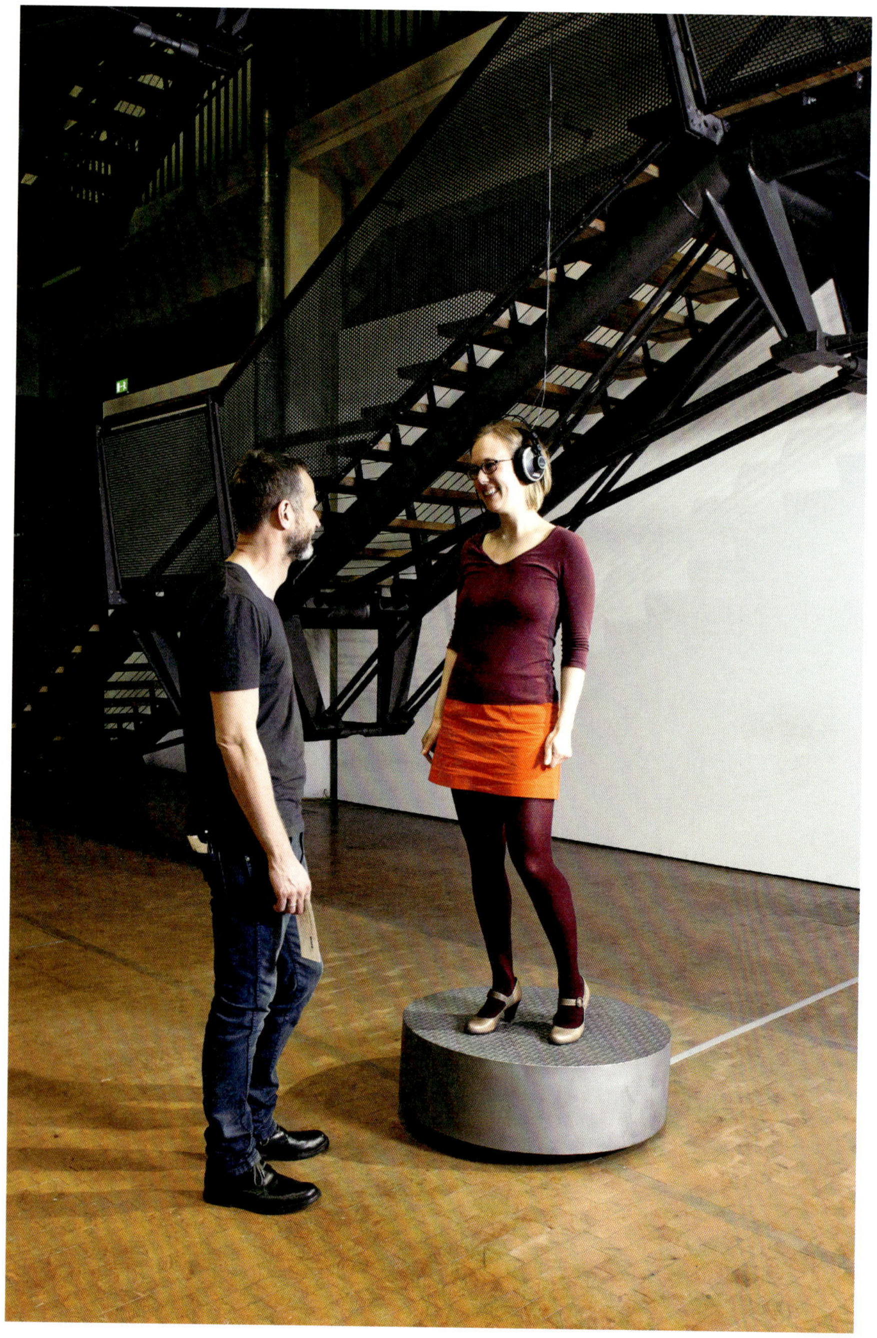

modern minimal disco 5. a psychoacoustic immersive space, 1995–2012
Installation
Infrasonic stainless steel pedestal, headphones, control system, composition spotlight, N-solab
Jan-Peter E. R. Sonntag

With *modern minimal disco* the visitor stands on a steel pedestal in the center of the exhibition space so that she/he is able to put on the headphones. In this exposed, elevated position the viewer at the center of the exhibition submerges acoustically in a paradoxical space of experience – a space that is continually and endlessly speeding up. Here the sub-frequency parts of the accelerating pulse, perceptible only as shock waves, are directly transmitted as vibrations via special headphones and pedestal to the listener. Who – now separated – becomes part of the work and as such is being exhibited while observing. (Jan-Peter E.R. Sonntag)

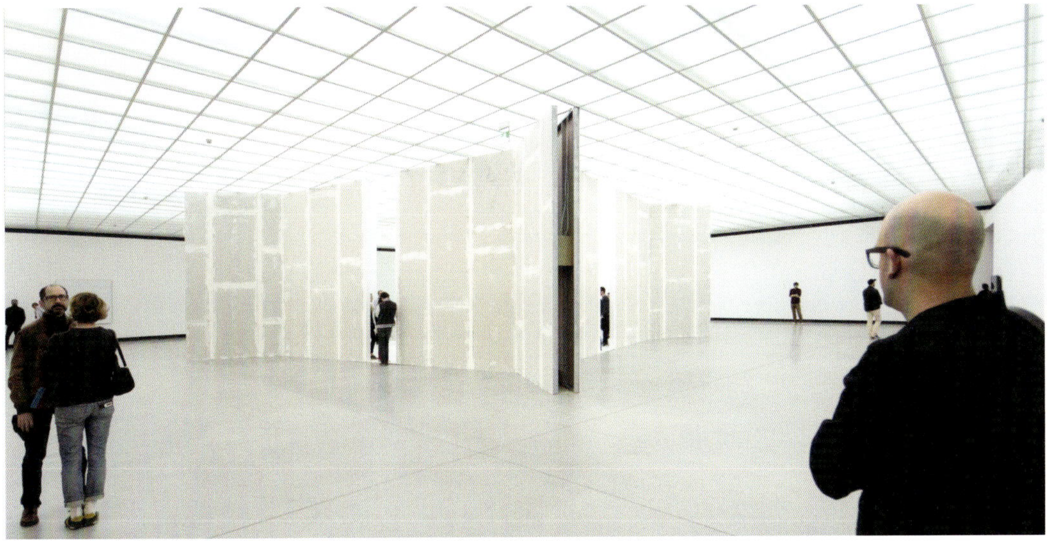

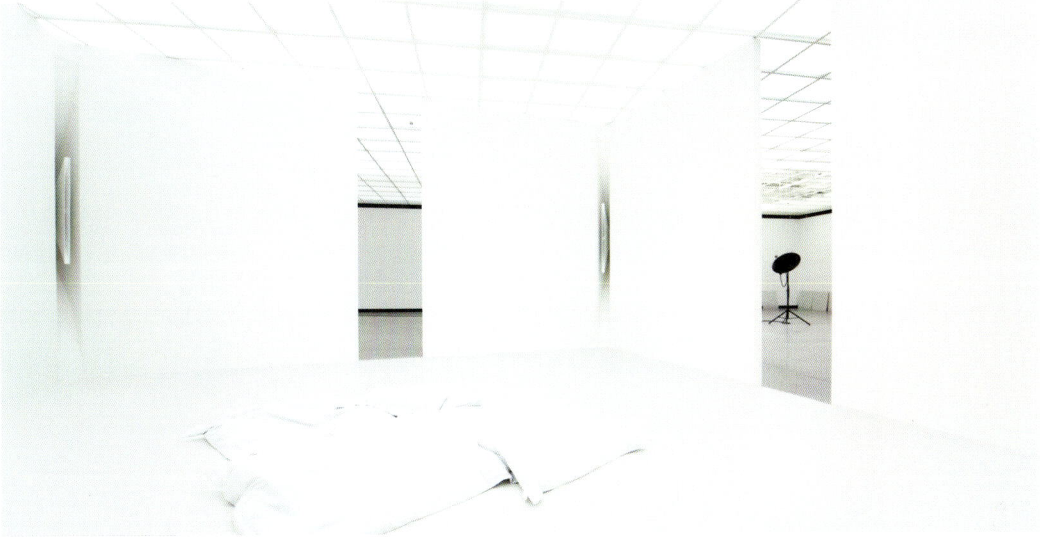

RAUSCHEN, 2015
Spatial-sound-composition/
architecture

The German word *Rauschen* is full of meaning: it describes the sound of leaves rustling in the trees, waves beating against the shore, the whisper of silk, or the patter of rain. Not coincidentally, *Rauschen* and the word *Rausch* – meaning, a state of intoxication – share the same origin. As a verb, *rauschen* not only describes the audible information produced by external movement, it also characterizes an internal frame of mind. In the age of digital mass communications, *Rauschen*, or noise, is not only the supposedly disturbing opposite of signal, as in the "signal-to-noise ratio." Mathematicians pay a great deal of attention to white noise. It contains equal measures of all values – only its sequence cannot be predicted.

As in *modern minimal disco* the sonic space perceptual illusion developed by Sonntag and his team for the first time is also at the center of *RAUSCHEN,* which was created for the Württembergische Kunstverein in Stuttgart in 2015. Together with the programmer Jörg Spix, Sonntag produced a column of endlessly rising noise for the oval, empty space surrounding the eight-story spiral staircase in the Ministry of Science and Culture in Hanover at an IRCAM work station in 1993. Until the early 1990s it was only possible for the artist and composer to work with consumer electronics, so he needed to collaborate with developers and physical-acoustic laboratories. This led to the creation of the N-solab, a team of engineers and developers with whom Sonntag has continued to work on developing and building the circuitry and controls for his installations. Built in the center of the White Cube was a quadrophonic Rausch-room with four interconnected coaxial diffraction horns (developed in Sonntag's N-solab) each with mouths measuring 52 square metersthat turned the open white cube into a sonic interface. Inside, one lost all sense of direction, while all information – images, sounds, structures, and patterns – faded away. *RAUSCHEN* turned the Kunstverein's gallery into a space and sound composition. Over the sonic course of the work, *Rauschen* (noise) from technical components mixed with soundscapes of *Rauschen* from forests and oceans, and Tristan's nearly sclerotic passage from Wagner's *Treibhaus*, culminating in an endlessly rising, digitally produced spiraling wind that ultimately drenched the space in white noise. Ever since Helmholtz, the experiential space of music has existed somewhere in between the simple oscillation, the simple tone, which can be mathematically calculated as a sine or cosine function, and white noise, or *Rauschen*. Contrary to Rudolf Arnheim's statement that sound art is a disembodied art for Sonntag, as he states in the concept for his radio opera *RUNDFUNK ÄTERNA* comissioned by documenta 14, sound art is a body-addressing spatial art.

SoundARt – AR installation by ZKM

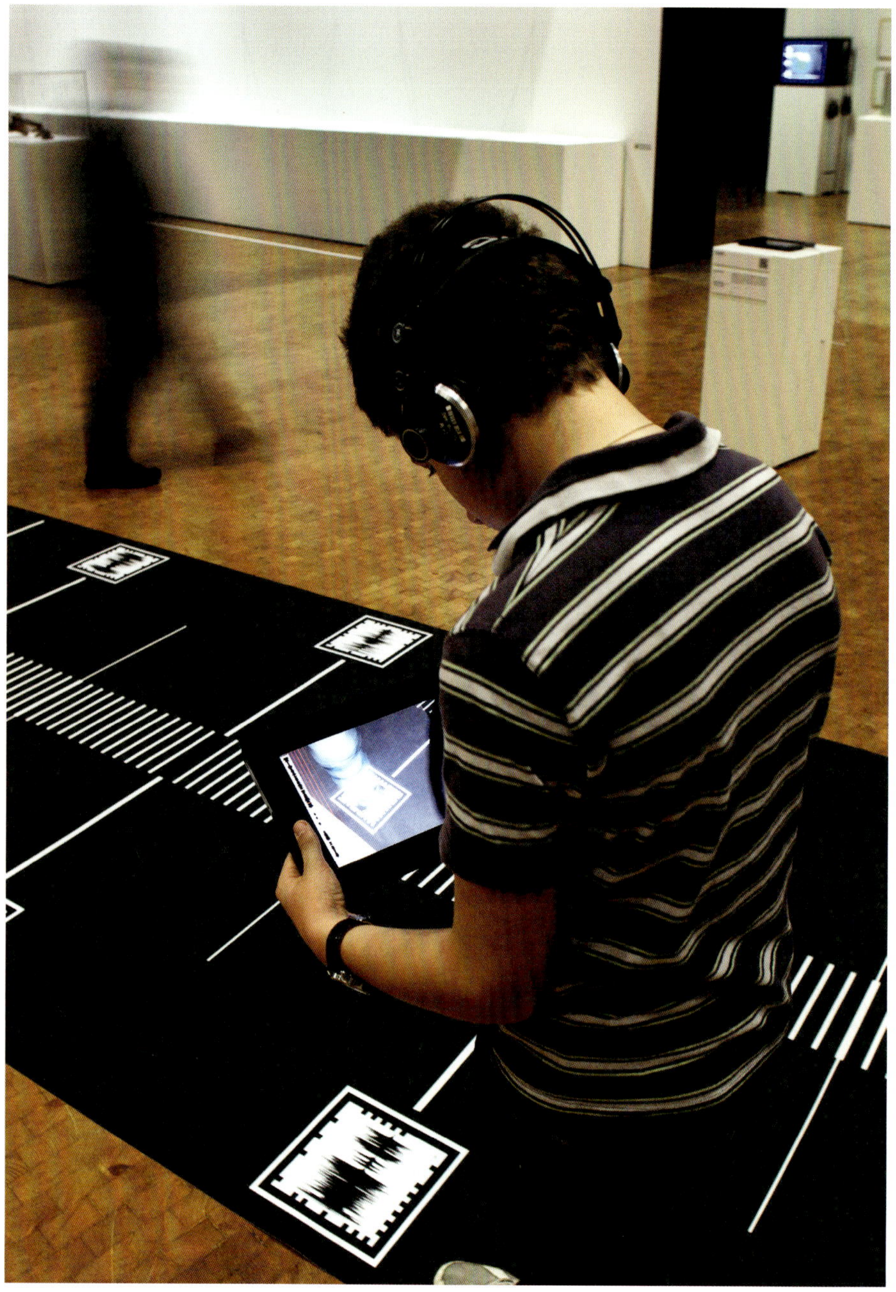

SoundARt, 2012
7 listening stations in Augmented Reality
iPad, code, headphones
Concept: Julia Gerlach, Bernd Lintermann, Peter Weibel
Software: Bernd Lintermann (software development for O.R.pheus – www.orpheus2012.com)
Technical coordination: Manfred Hauffen
Curators: soundscape (Björn Gottstein), musique concrète (Rudolf Frisius), radio art (Frank Halbig), radioartemobile (Luca Vitiello), intermedium (Herbert Kapfer and Katarina Agathos), electronica (Björn Gottstein), noise (Björn Gottstein)

SoundARt goes far beyond installations and sculpture. At seven stations acoustic works in the areas of soundscape, musique concrète, radio art, radioartemobile, intermedium, electronica, and noise are staged in unique ways and made available to visitors. The themes were curated by specialists. A black beam, which connects two atria of the ZKM | Karlsruhe and which reminds one of the display band on a radio, carries the titles of the seven thematic areas (with introductory texts). On the floor are QR codes: each work has its own QR code as well as an image of the work as an oscillation graph. Using an iPad, which can be loaned at the exhibition, the works can be experienced by visitors: they are rendered audible and also visualized in a 3D simulation as an Augmented Reality experience as soon as the iPad camera captures the code. The iPad is the pick-up of the radio station, which samples the frequency. Information about the works can be retrieved with the iPad. Soon after its display at Karlsruhe, *SoundARt* was presented at Fórum Eugénio de Almeida, Évora, Portugal.

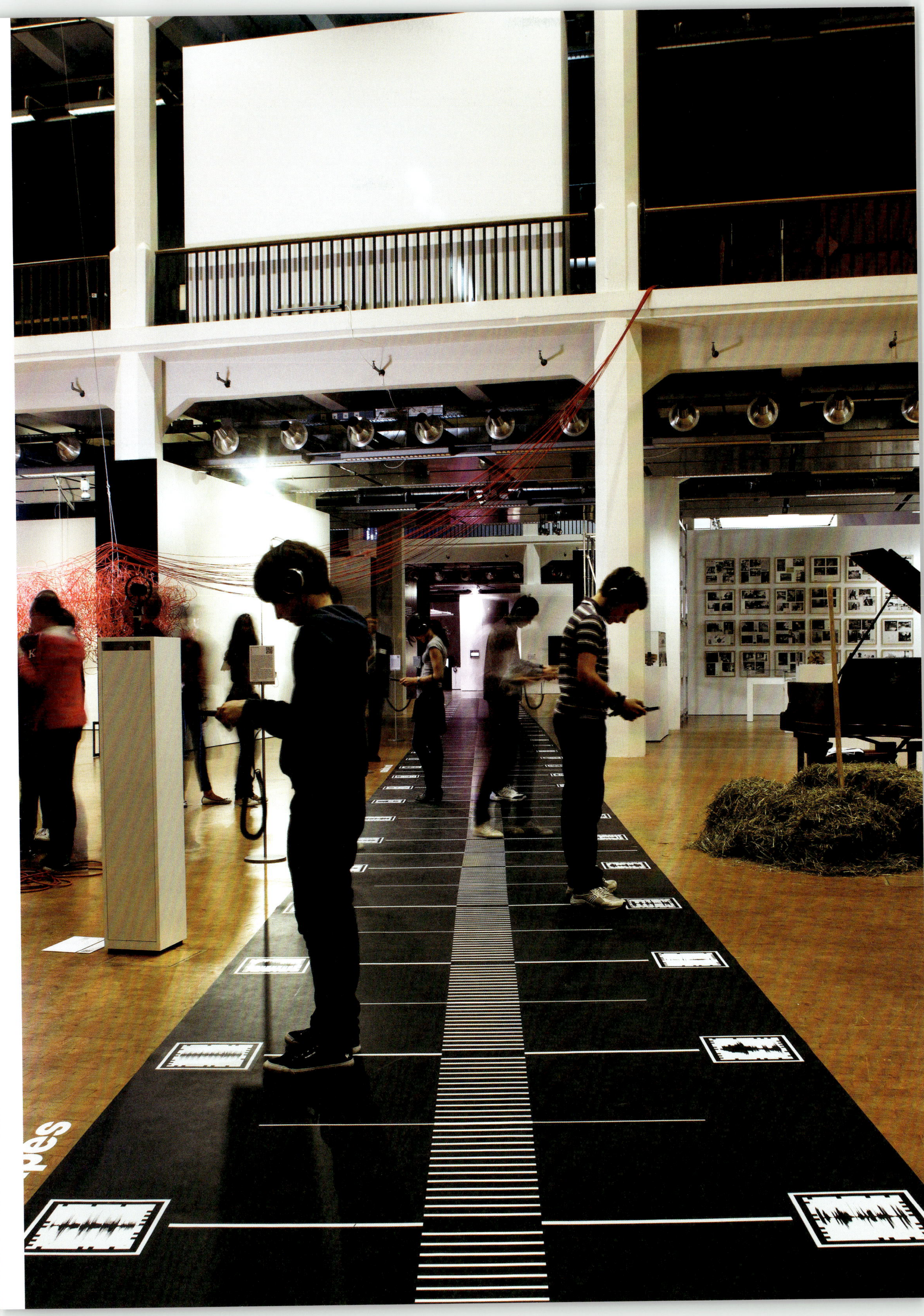

Joulia Strauss

*1974 in Saint Petersburg (RU), lives in Berlin (DE)

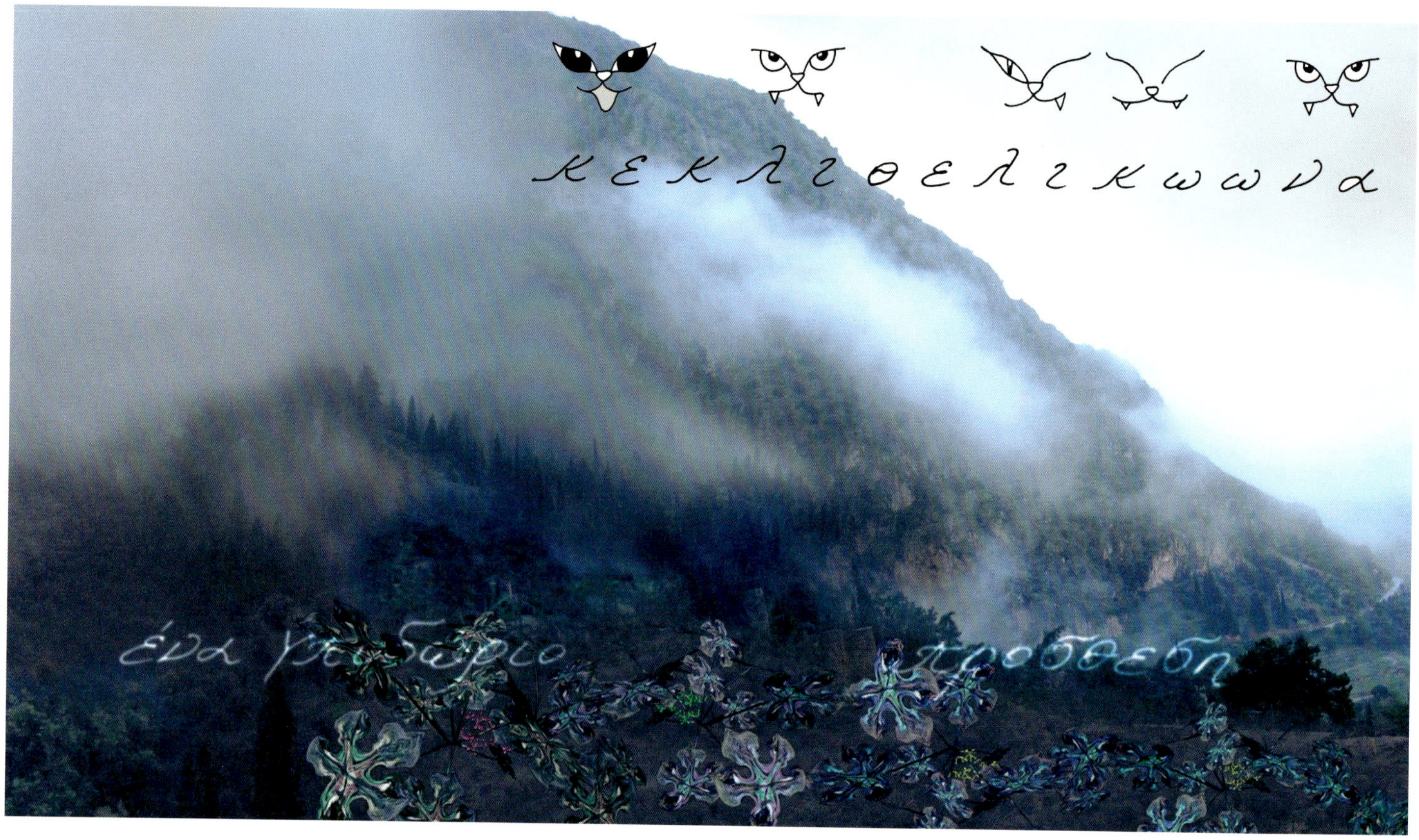

First Delphic Hymn to Apollon, 2009
Video installation
HD video, sound system, 5:19 min.
Video compositing: Moritz Mattern, recording: Andreas Köhler, filming: Hilary Koob-Sassen, voice, lyre: Joulia Strauss Eye-square; HAECCCC, 0+865, and the artist

www.joulia-strauss.net

In the *First Delphic Hymn to Apollon*, video and performance artist Joulia Strauss stages herself with lyre in a fantasy landscape. The recordings were made in the vicinity of Delphi. A series of fantastical animals moves on lines that evoke notation systems. Their names allude to ancient Greek modes. The "Dorian humming cats," "depressed Lydian minor dogs," "cathartic Hypolydian division turtles," and "integral swans" appear. The video is part of a work series of wire sculptures and oil paintings (*Second Delphic Hymn to Apollon*, 2010) that sounds variations on the theme of "animal mathematical operations" at the intersection of archaeology and art.

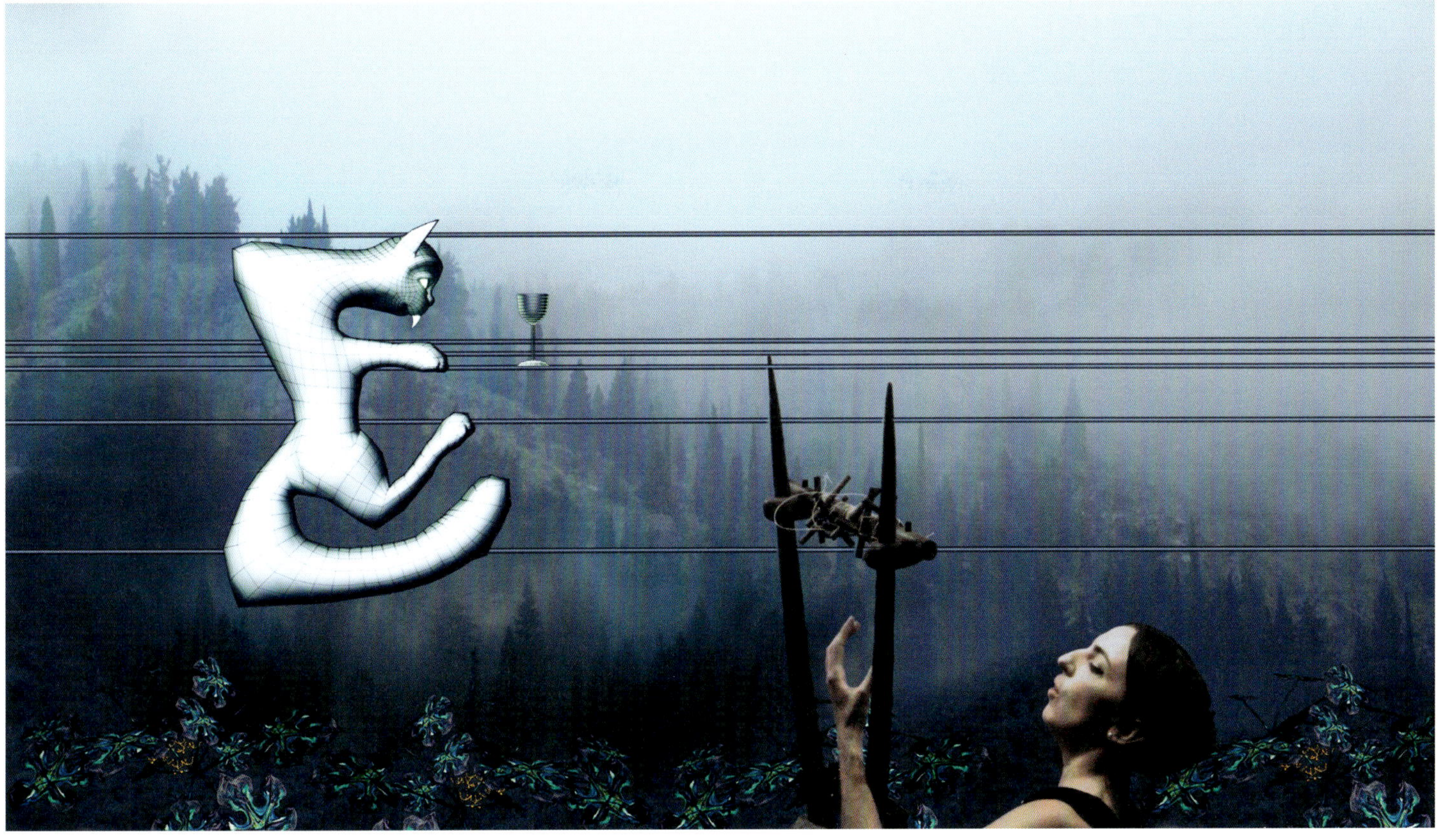

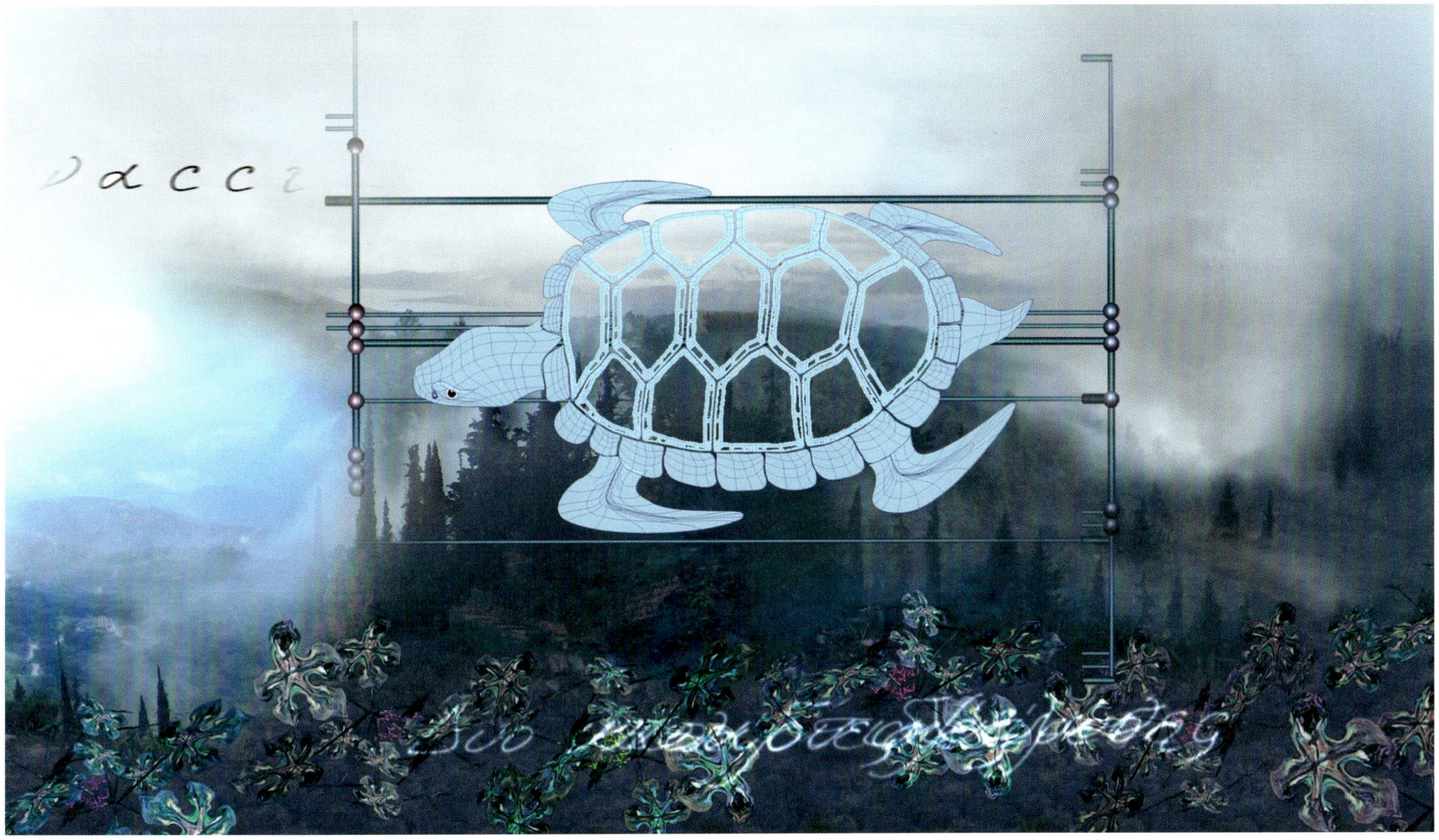

Akio Suzuki

*1941 Pyongyang (KP), lives in Japan (JP)

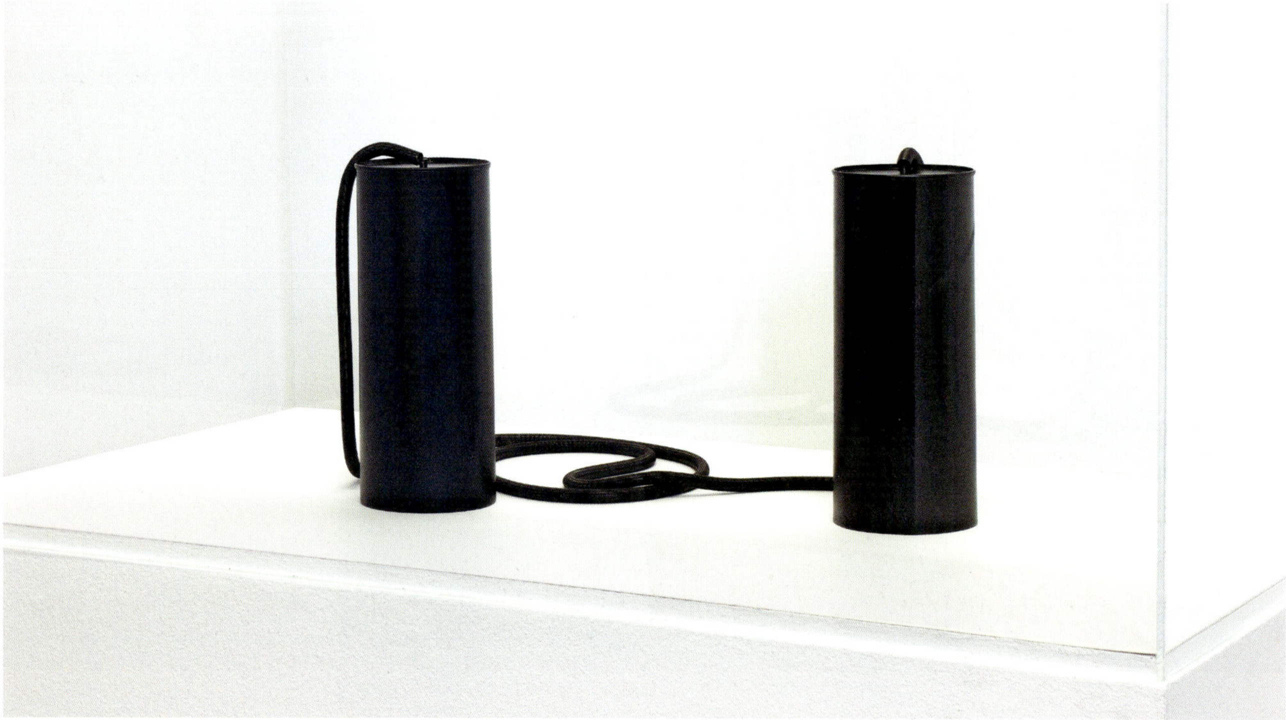

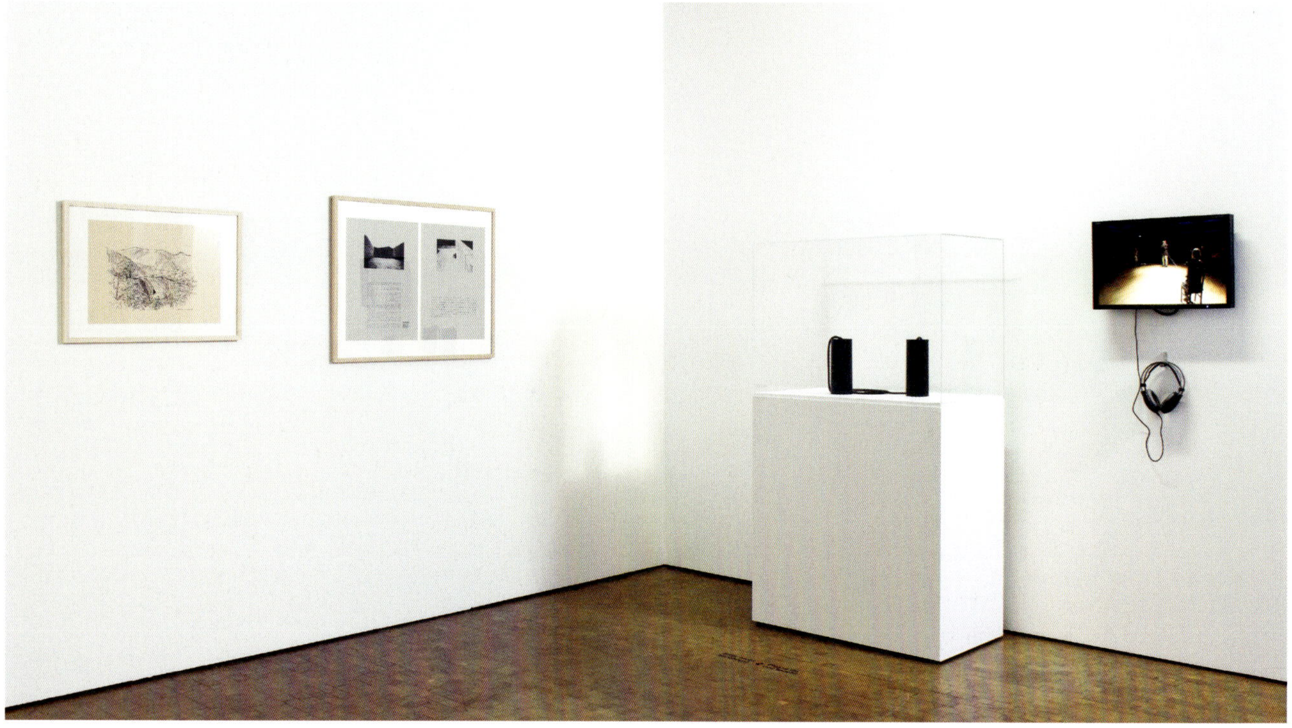

ANALAPOS-A, 2011
Musical instrument
Coil spring, iron cylinders
Akio Suzuki

www.akiosuzuki.com

Since the 1960s, Akio Suzuki has been investigating the acoustic quality of selected locations and creating corresponding topographies. His intensive involvement with the phenomenon of pulse and echo led him to develop his own instruments in the 1970s. One of these is the spiral echo instrument *ANALAPOS-A*. It consists of a coil spring and two iron cylinders that function as resonating chambers, and is played with the voice or by hand. The name ANALAPOS combines the terms "analog" and "postmodern" and stands for analog listening in the digital age.

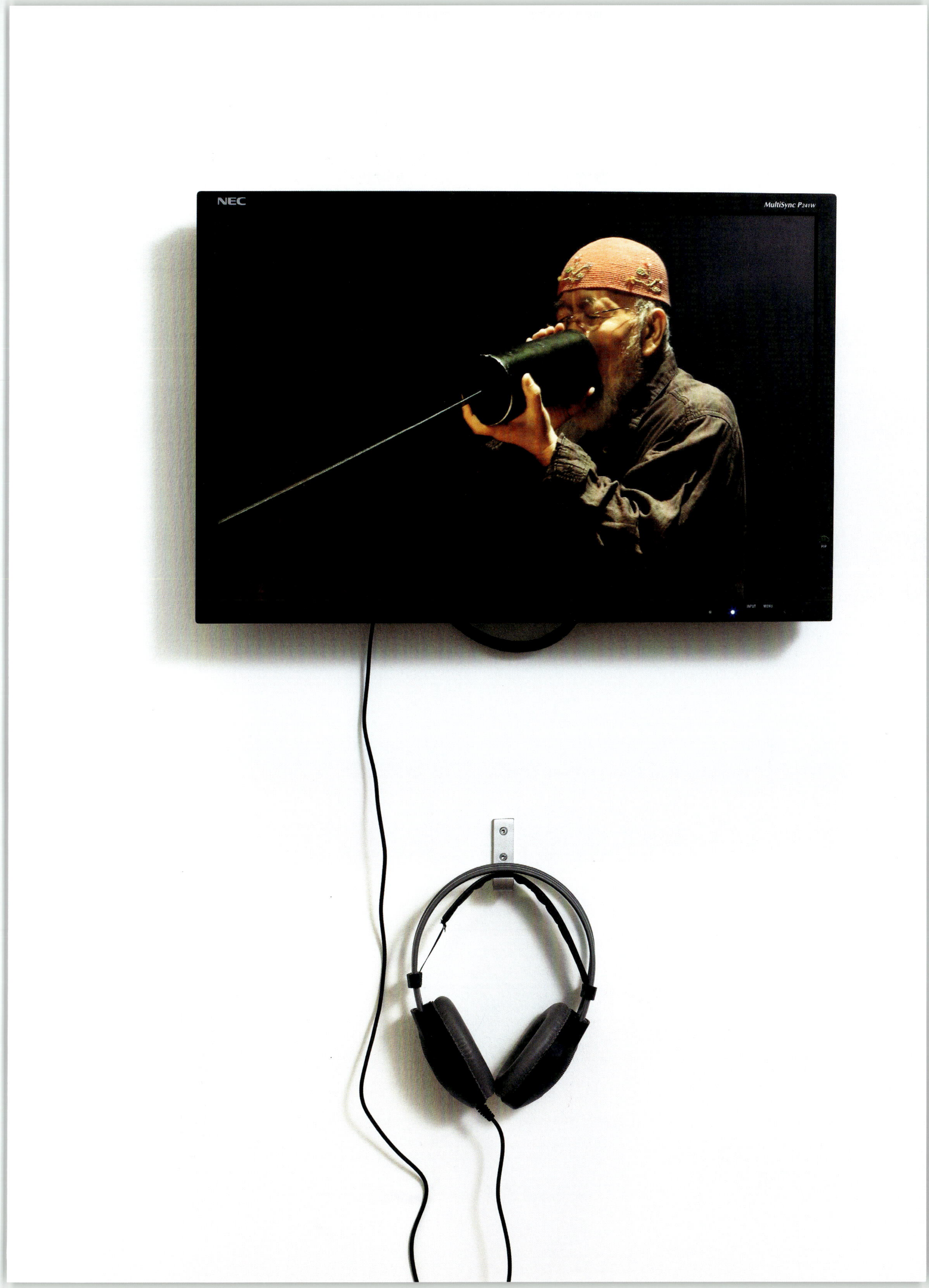
NEC
MultiSync P241W

Akio Suzuki

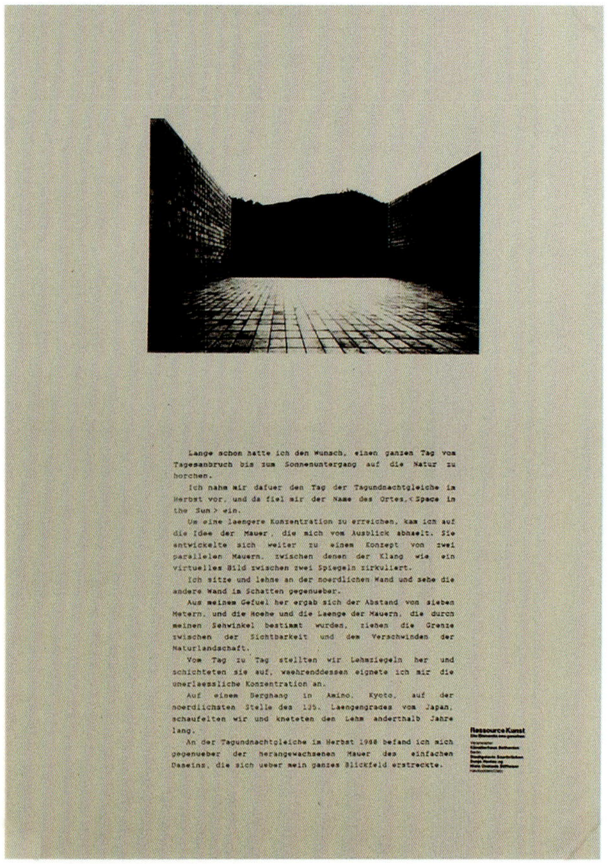

Lange schon hatte ich den Wunsch, einen ganzen Tag vom Tagesanbruch bis zum Sonnenuntergang auf die Natur zu horchen.

Ich nahm mir dafuer den Tag der Tagundnachtgleiche im Herbst vor, und da fiel mir der Name des Ortes, < Space in the Sun > ein.

Um eine laengere Konzentration zu erreichen, kam ich auf die Idee der Mauer, die mich vom Ausblick abhaelt. Sie entwickelte sich weiter zu einem Konzept von zwei parallelen Mauern, zwischen denen der Klang wie ein virtuelles Bild zwischen zwei Spiegeln zirkuliert.

Ich sitze und lehne an der noerdlichen Wand und sehe die andere Wand im Schatten gegenueber.

Aus meinem Gefuel her ergab sich der Abstand von sieben Metern, und die Hoehe und die Laenge der Mauern, die durch meinen Sehwinkel bestimmt wurden, ziehen die Grenze zwischen der Sichtbarkeit und dem Verschwinden der Naturlandschaft.

Vom Tag zu Tag stellten wir Lehmziegeln her und schichteten sie auf, waehrenddessen eignete ich mir die unerlaessliche Konzentration an.

Auf einem Berghang in Amino, Kyoto, auf der noerdlichsten Stelle des 135. Laengengrades vom Japan, schaufelten wir und kneteten den Lehm anderthalb Jahre lang.

An der Tagundnachtgleiche im Herbst 1988 befand ich mich gegenueber der herangewachsenen Mauer des einfachen Daseins, die sich ueber mein ganzes Blickfeld erstreckte.

Ressource Kunst

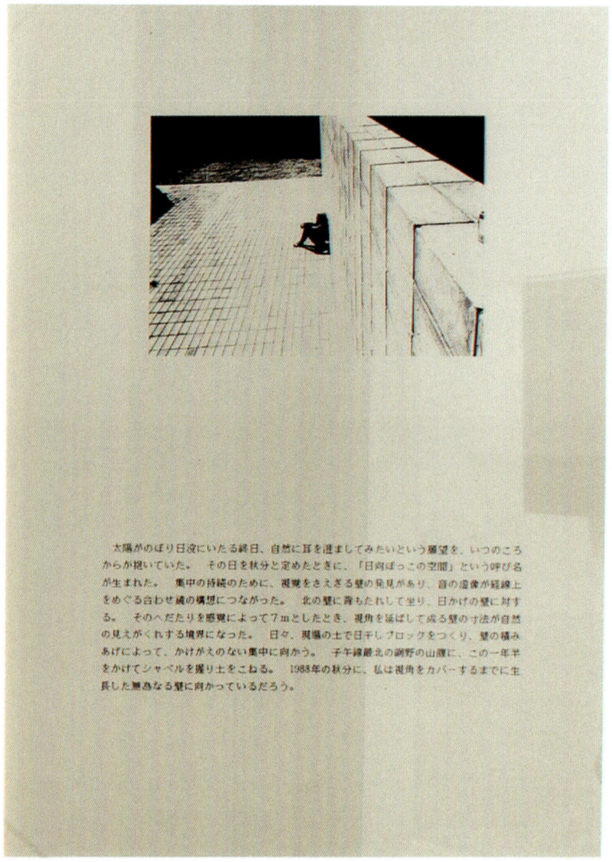

太陽がのぼり日没にいたる終日、自然に耳を澄ましてみたいという願望を、いつのころからか抱いていた。 その日を秋分と定めたときに、「日向ぼっこの空間」という呼び名が生まれた。 集中の持続のために、視覚をさえぎる壁の発見があり、音の虚像が経線上をめぐる合わせ鏡の構想につながった。 北の壁に背もたれして坐り、日かげの壁に対する。 そのへだたりを感覚によって7mとしたとき、視角を延ばして成る壁の寸法が自然の見えがくれする境界になった。 日々、現場の土で日干しブロックをつくり、壁の積みあげによって、かけがえのない集中に向かう。 子午線最北の網野の山腹に、この一年半をかけてシャベルを握り土をこねる。 1988年の秋分に、私は視角をカバーするまでに生長した無為なる壁に向かっているだろう。

Space in the Sun, 1988–2017
Environmental sculpture
and listening point on Mount
Takatenvama (JP)
Two photographic reproductions
with text, sketch
Akio Suzuki

Built in 1988 on a hilltop in Kyotango city in the Tango region outside of Kyoto, Japan, *Space in the Sun* comprised a brick floor and two brick walls and created a space in which to listen to the area it inhabited. It took eighteen months to complete the structure, with Akio Suzuki working alongside his then-wife Junko Wada and others, and consisted of 10,000 handmade earthen bricks. Suzuki had envisioned that the work would deteriorate naturally over time, but in 2017 the work had to be demolished for safety reasons.

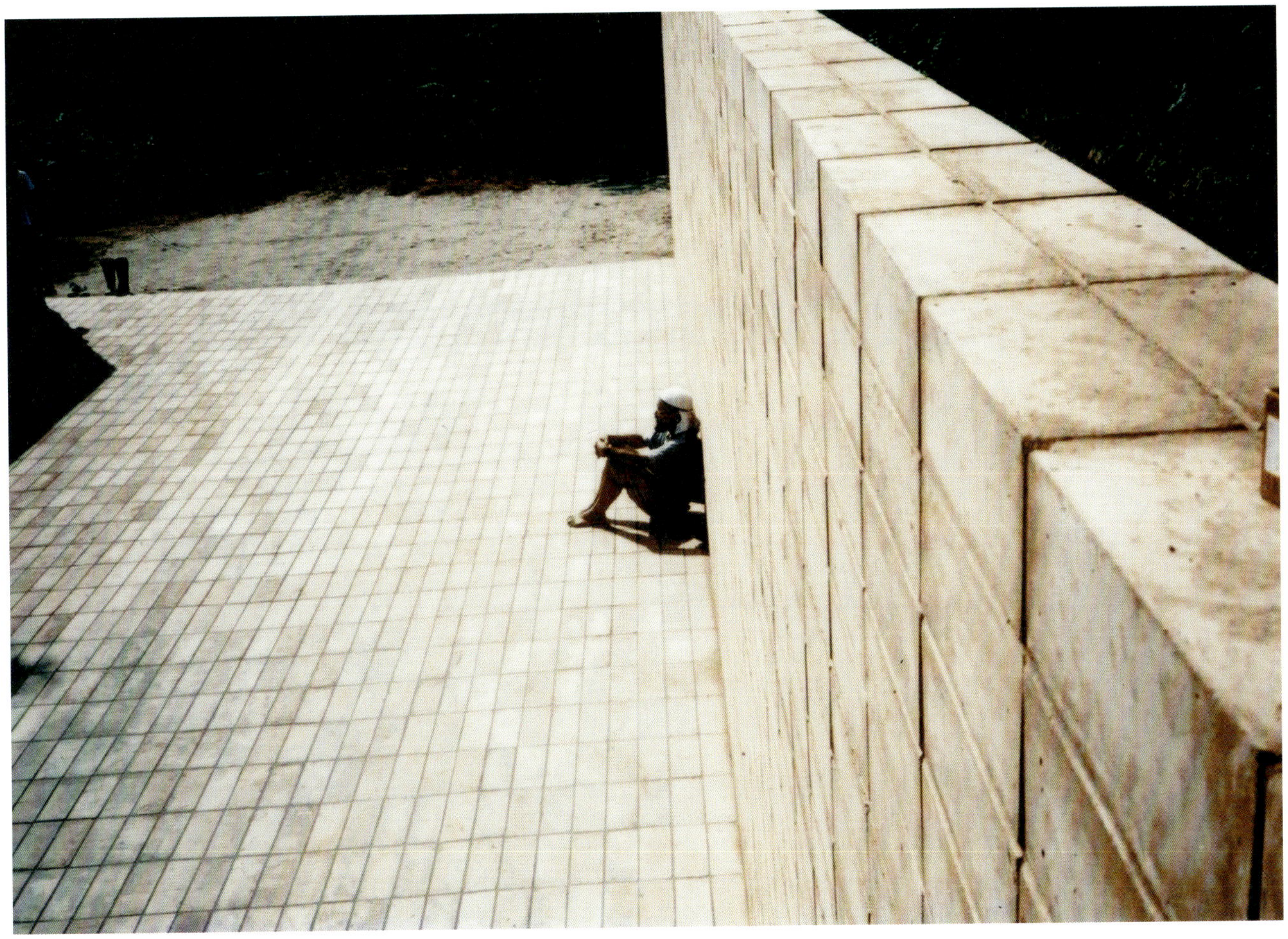

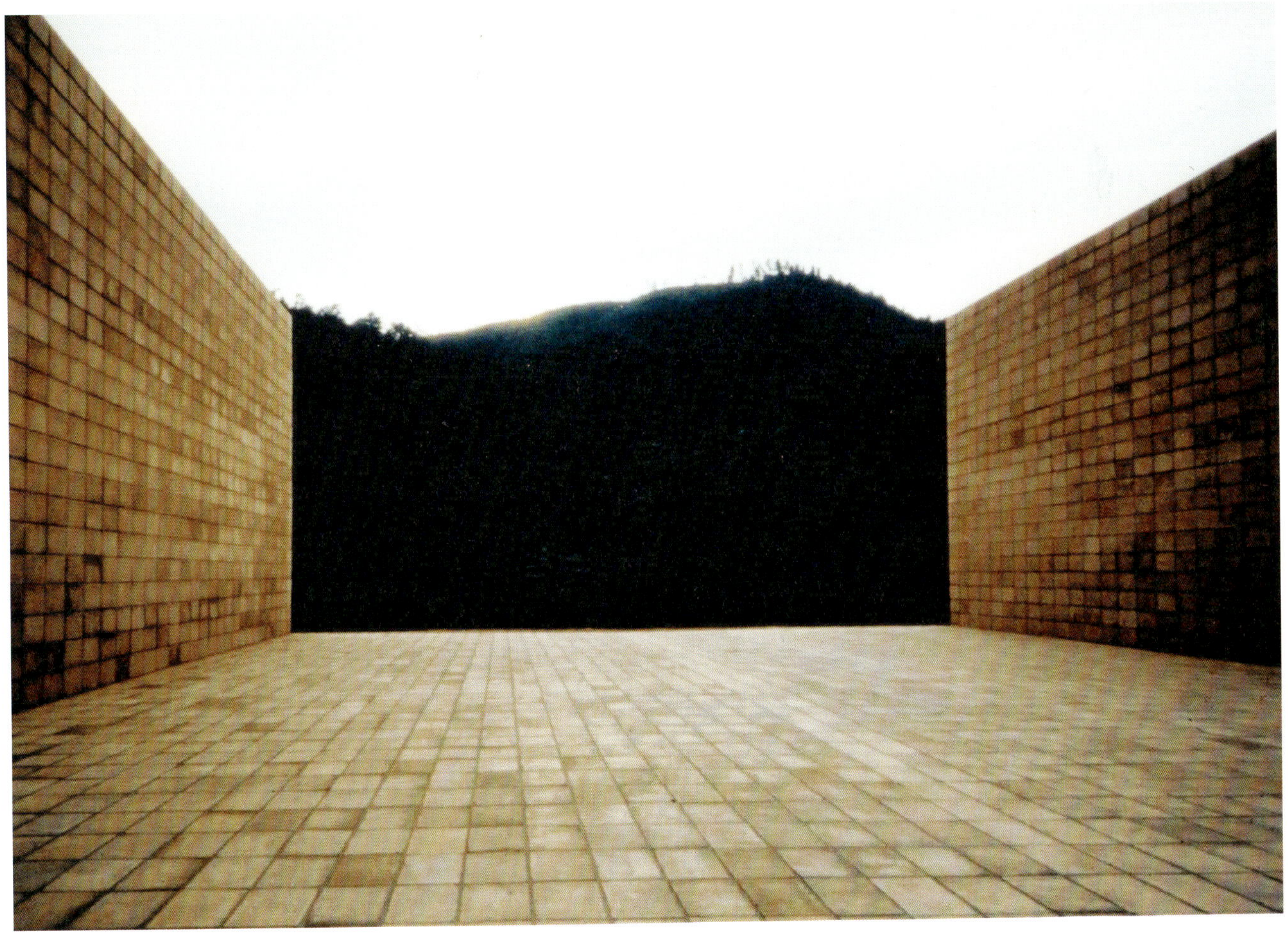

Takis

*1925 in Athens (GR), lives in Paris (FR), London (GB), and Athens

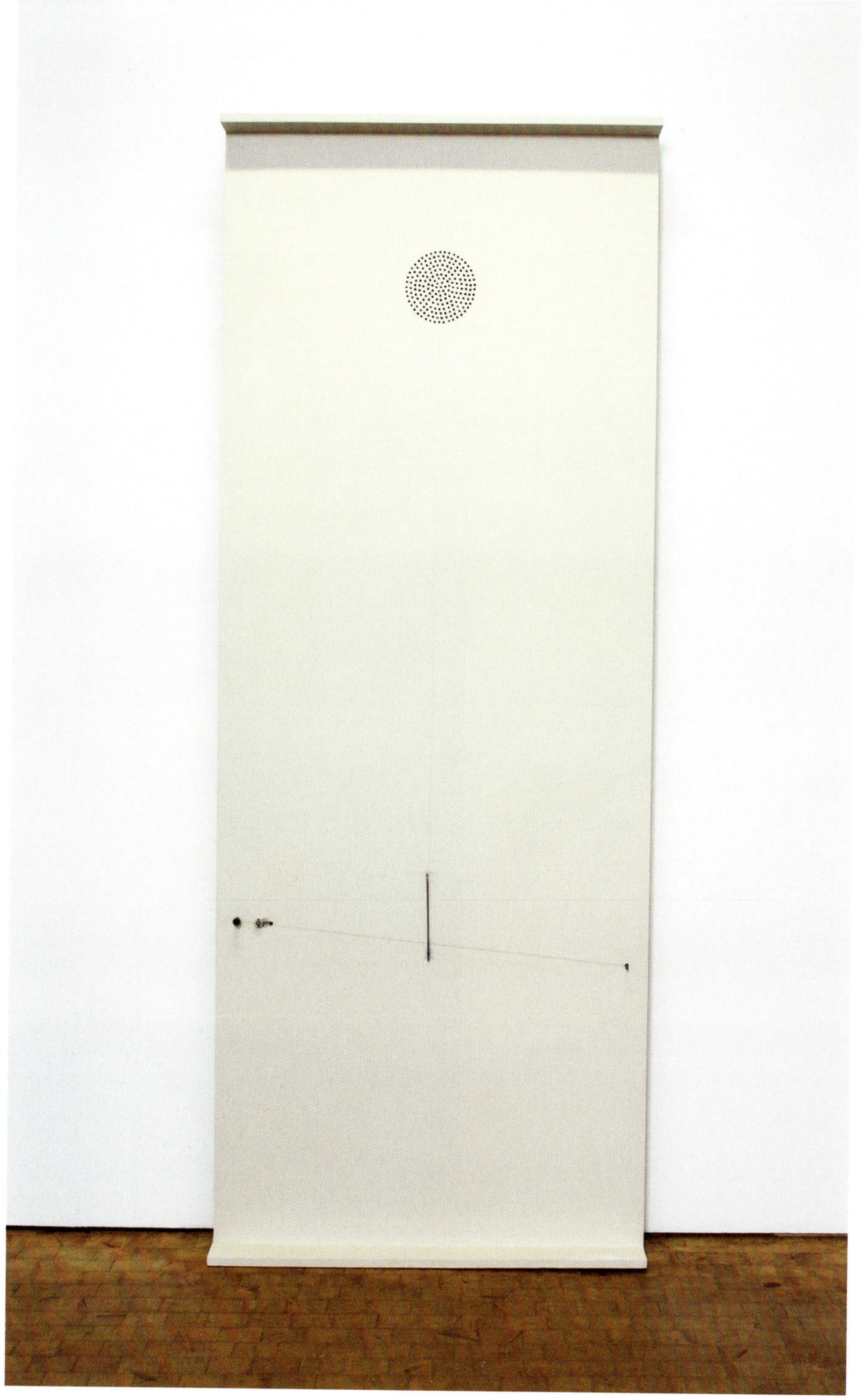

Musical Hannover, 1974
Sound sculpture, wood, string, and sticks, dimensions variable
ZKM | Karlsruhe

www.takisfoundation.org

The Greek artist Takis first began exhibiting musical sculptures such as *Musical Hannover* in 1974. Concerned with the repetition of sound, these sculptures use magnetic waves caused by electricity to activate the plucking of strings or other sound-making materials. Takis's sound sculptures seek to implement a natural means of generating sound, separated from the arbitrary decision-making process of a lone artist. By using magnetism, Takis presents sound as a discoverable object, a series of events outside of his control. (John Kannenberg)

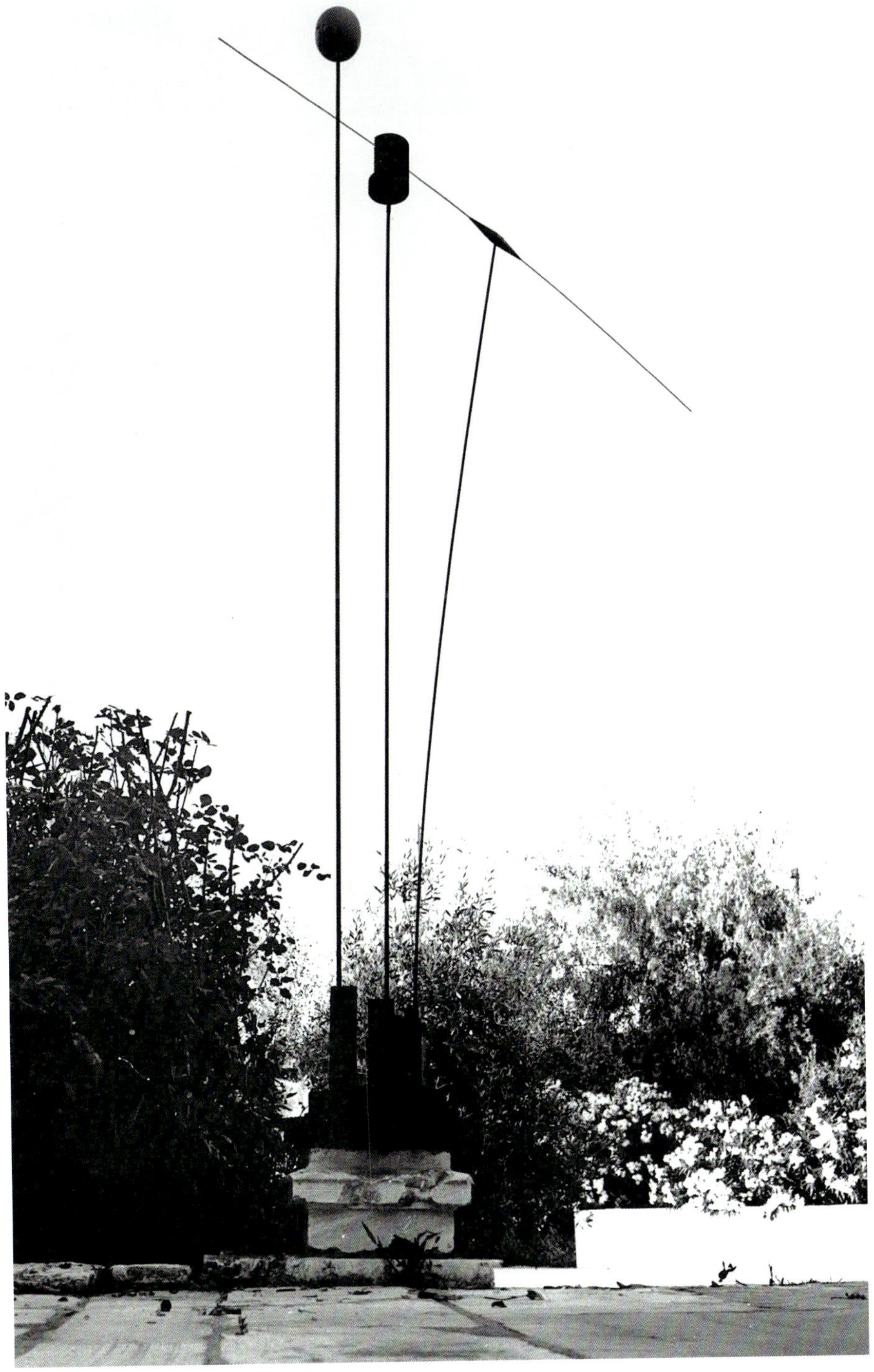

Takis, *Signals*, 1955–
Sculpture
Metal, found objects,
dimensions variable
Takis Foundation

In 1955 influenced by the invention of the radar and the technological landscape of the station of Calais, Takis constructed his first *Signal* sculpture. Takis' *Signals* are his most renowned trademark. They are an ongoing series of tall rods springing from a base on the ground and tapering towards a finial at the top. Generally speaking, these finials are of particular symbolic significance and consist of either sculptures or found objects. The sculptures consist of various elements, whose form and shape evoke primordial symbols. In response to their interpretation, Takis said: „I am responsible only for manufacturing the symbols. Their meaning depends for interpretation on each and every one of us."

TBA21 Sound Space

Thyssen-Bornemisza Art Contemporary Foundation, founded in 2002 by Francesca Thyssen-Bornemisza

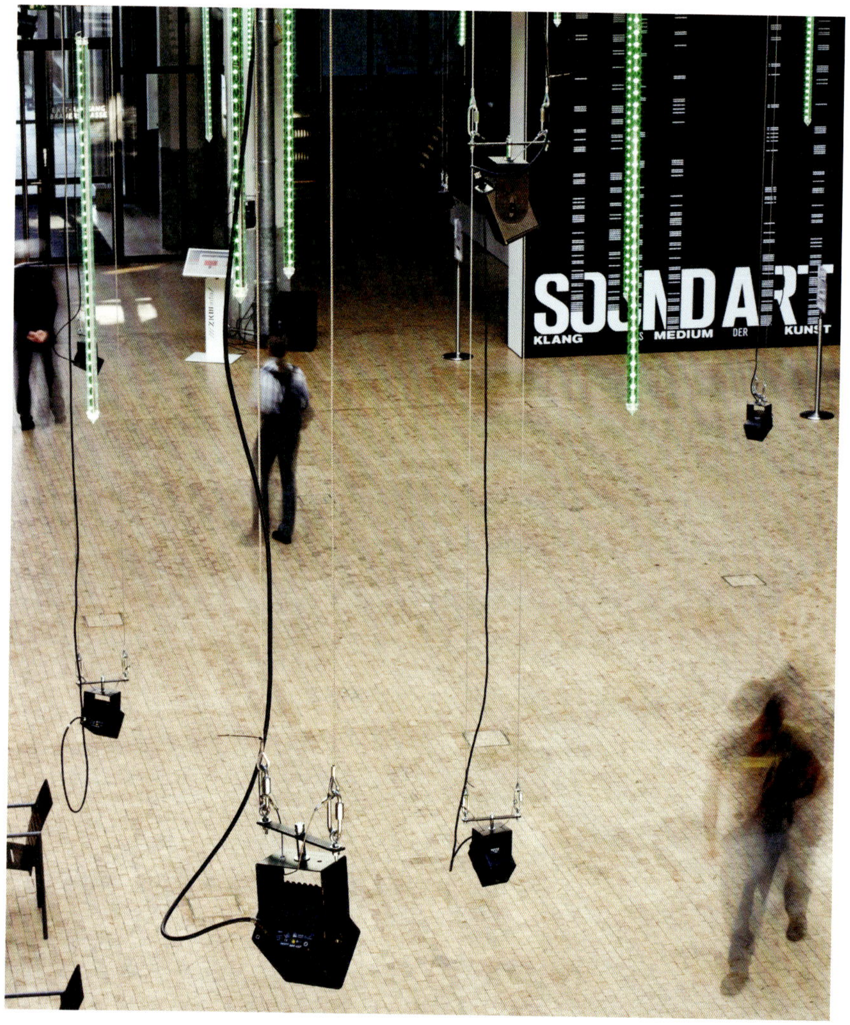

TBA21 Sound Space, 2012–2013
Installation with interactive and multispatial sound system
Technical development/programming: Tony Myatt
Curated by Florian Hecker, Bryce Dessner, Russell Haswell, Franz Pomassl
Composers: Alexei Borisov, Batuhan Bozkurt, Bryce Dessner in collaboration with David Sheppard, Evan Ziporyn, Cevdet Erek, Mark Fell, Roc Jiménez de Cisneros, Christian Fennesz, Ghostigital, Bruce Gilbert, Tommi Grönlund, Petteri Nisunen, Carl Michael von Hausswolff, Florian Hecker, Erdem Helvacıoğlu, Jónsi, Alex, Carsten Nicolai, Mehmet Can Özer, Zsolt Olejnik, Finnbogi Pétursson, Franz Pomassl, Lee Ranaldo, Terre Thaemlitz, Yasunao Tone, Chris Watson, Thom Willems, Jana Winderen, Zavoloka, and Peter Zinovieff
Commissioned by Thyssen-Bornemisza Art Contemporary

www.tba21.org

The *TBA21 Sound Space* is a project commissioned by Thyssen-Bornemisza Art Contemporary to bring to a larger audience the unique series of compositions that were produced for *The Morning Line*. Conceived as an interdisciplinary platform structure by artist Matthew Ritchie and architects Aranda\Lasch, where artists, architects, engineers, physicists, and musicians would each contribute their own specialized information to create a new form, *The Morning Line* uses a unique interactive multispatial sound system saturated with over fifty speakers conceived by Tony Myatt and a team from the Music Research Centre, University of York. It is installed at the ZKM Forecourt.

The novel spatialization software suggested a new form of music written especially for *The Morning Line* which has now been extended to be reproduced in the *TBA21 Sound Spaces*. The first *Sound Space* was developed in the context of Peter Weibel's landmark exhibition *Sound Art. Sound As a Medium of Art* exhibition at ZKM | Karlsruhe. To celebrate the evolution of the medium, TBA21 has put on loan its sound archive of nearly thirty experimental sound compositions. These were commissioned over the last four years from an international group of composers whose work reaches beyond the boundaries and the programming of traditional concert halls. The existing archive of music and soundscapes has been reconfigured through a specially written software by Tony Myatt and his team, to recreate the multispatial audio installation. The system has multiple surround sound fields, some adjacent to each other and some nested inside each other in layers. This provides an experiential three-dimensional sound reproduction system within a larger scale spatial sound environment. (TBA21)

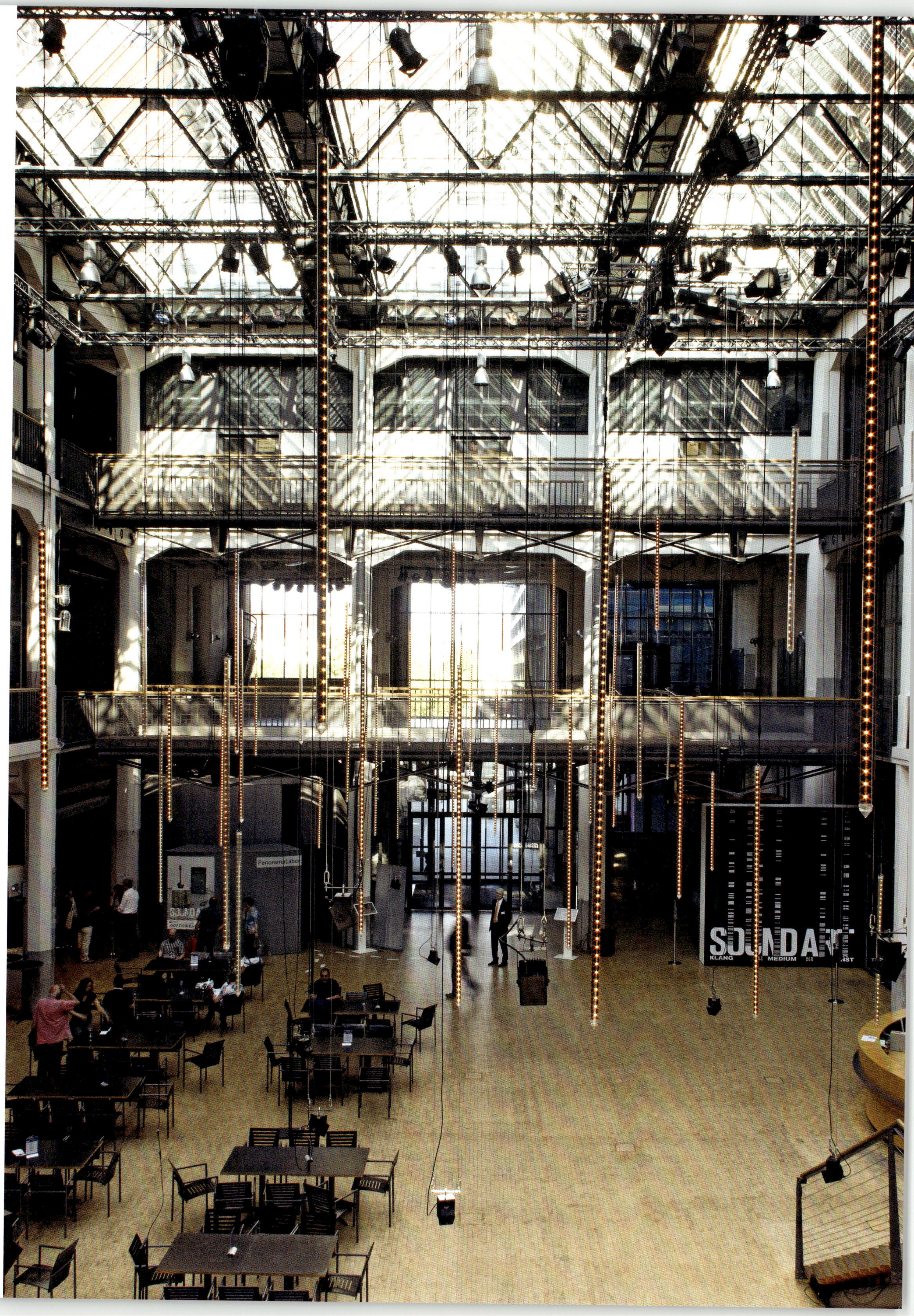
KLANG
MEDIUM
DER

TONSPUR @ ZKM

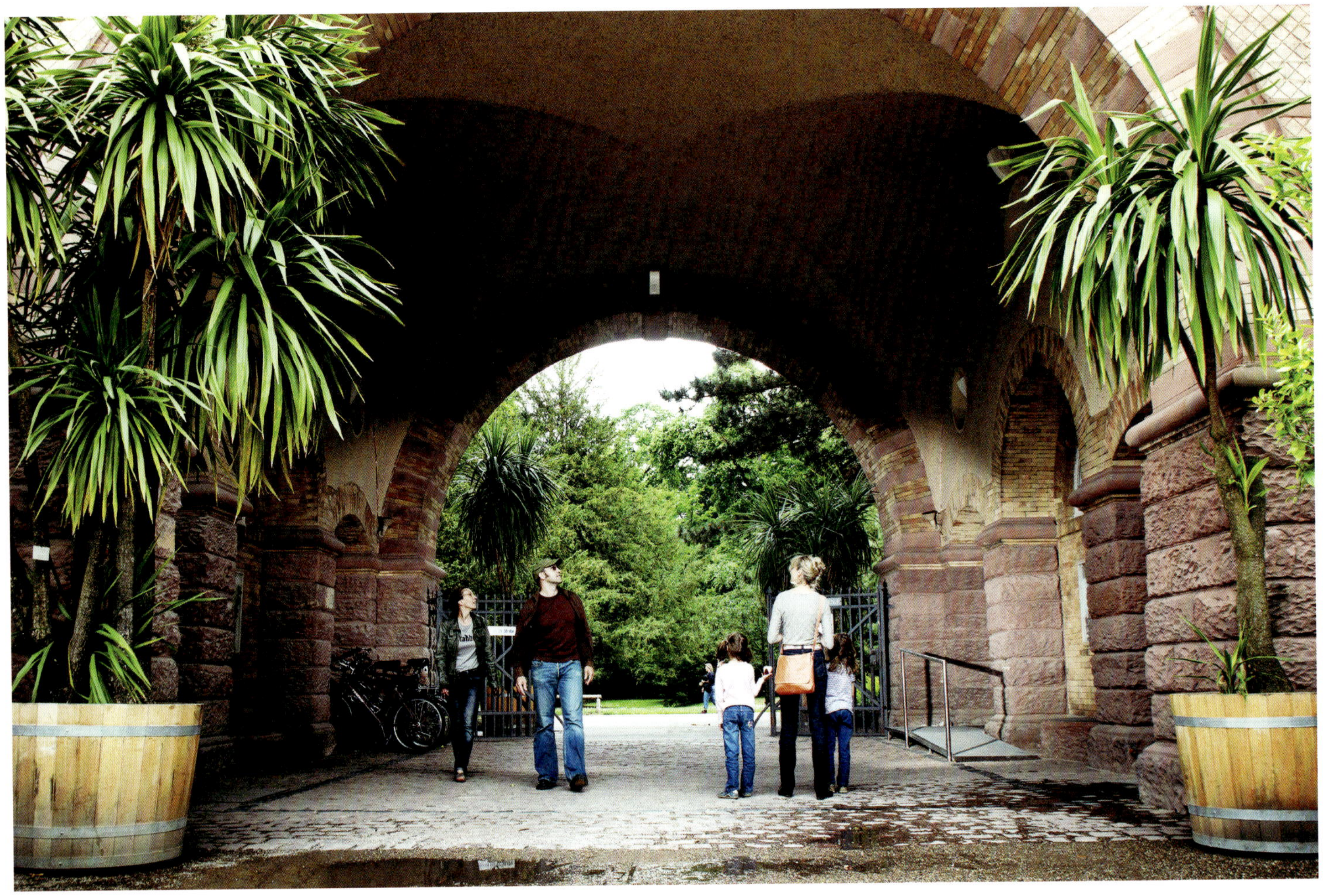

TONSPUR @ ZKM
Works from the TONSPUR_ collection
Installation view Karlsruhe Botanical Gardens, 2012
Sound works
8-channel digital audio (computer controlled)
Various lengths
Sound direction: Peter Szely
Artistic director: Georg Weckwerth

www.tonspur.at

People automatically pass through the TONSPUR_passage in the MuseumsQuartier in Vienna and the Schloßplatz in the center of Berlin as they make their way across, or tour, both cities. In the TONSPUR series, international artists working in all genres have developed and realized computer controlled sound works for these transitory venues. Their multi-channel compositions go well beyond the standard stereo image, creating fascinating sound architectures and walk-in sound spaces.

TONSPUR extends contemporary art's diverse and interdisciplinary examination of sound as a plastic and malleable material. The project brings together artists from various disciplines via sound, while guiding the audience towards an extended notion of reception and of art. In this spirit TONSPUR for a public space[1] engages with everyone who encounters the city and art with open eyes and ears.

For the exhibition *Sound Art. Sound as a Medium of Art*, TONSPUR established a temporary venue in the centrally located Torbogen Building of the Botanical Gardens in Karlsruhe for the presentation of five sound works selected from the collection of over fifty that have been created since the start of the project in 2003.

The sound pieces by Gerhard Rühm, Friederike Mayröcker + Bodo Hell, Peter Weibel, David Moss, and Gary Hill, which interweave language with other sounds, could be experienced free of charge by all from Wednesday to Sunday from 11 a.m. to 6 p.m.

Curators: Georg Weckwerth and Peter Szely

1 At the 2010 Prix Ars Electronica, TONSPUR for a public space received an award in the category of Digital Musics & Sound Art.

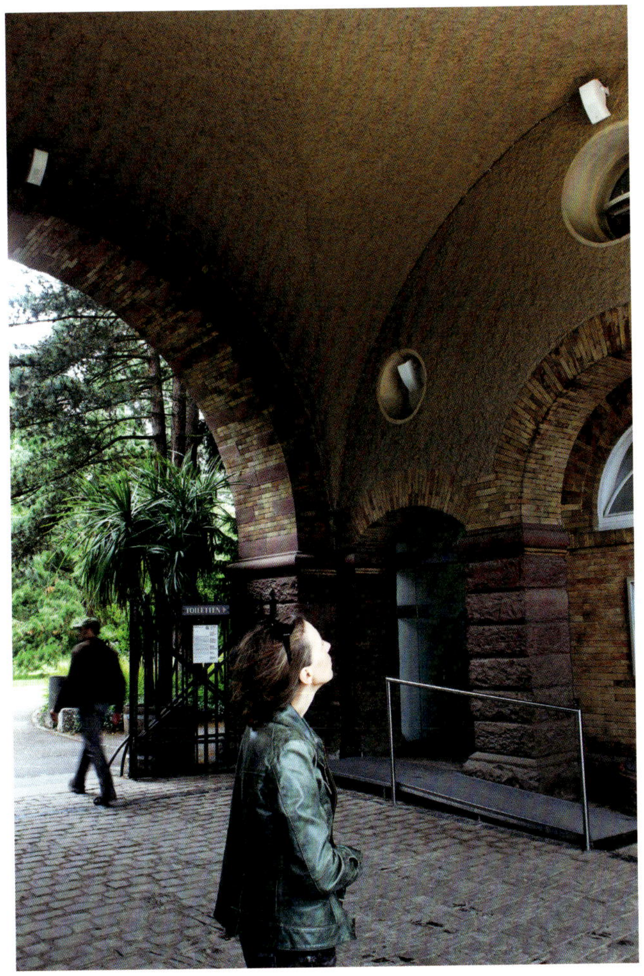

TONSPUR 46: David Moss, *23 ways to remember silence, but only 1 way to break it*, 2011
Presentation period: March 17–April 15, 2012

TONSPUR 42: Peter Weibel, *Das Leben im 20. Jahrhundert: 225 Millionen Morde. Ein Oratorium* [Life in the Twentieth Century: 225 Million Homocides. An Oratory], 2011
Presentation period: April 16–May 13, 2012

TONSPUR 35: Friederike Mayröcker + Bodo Hell, *vom Umarmen des Komponisten auf dem offenen Soffa* [on the composer's embrace on the ophen sopha], 2010
Presentation period: May 14–June 10, 2012

TONSPUR 48: Gary Hill, *Child's Play*, 2011
Presentation period: June 11–July 8, 2012

TONSPUR 25: Gerhard Rühm, *Paradiesische Passage* [Paradisiacal Passage], 2008
Presentation period: July 9–August 5, 2012

Tonspur (lit. soundtrack): A classical German term from visual and acoustic media has furnished the name for globally unique projects with changing sound works for public space based in Vienna since 2003 and in Berlin since 2009.

Timm Ulrichs

*1940 in Berlin (DE), lives in Hannover, Münster, and Berlin (DE)

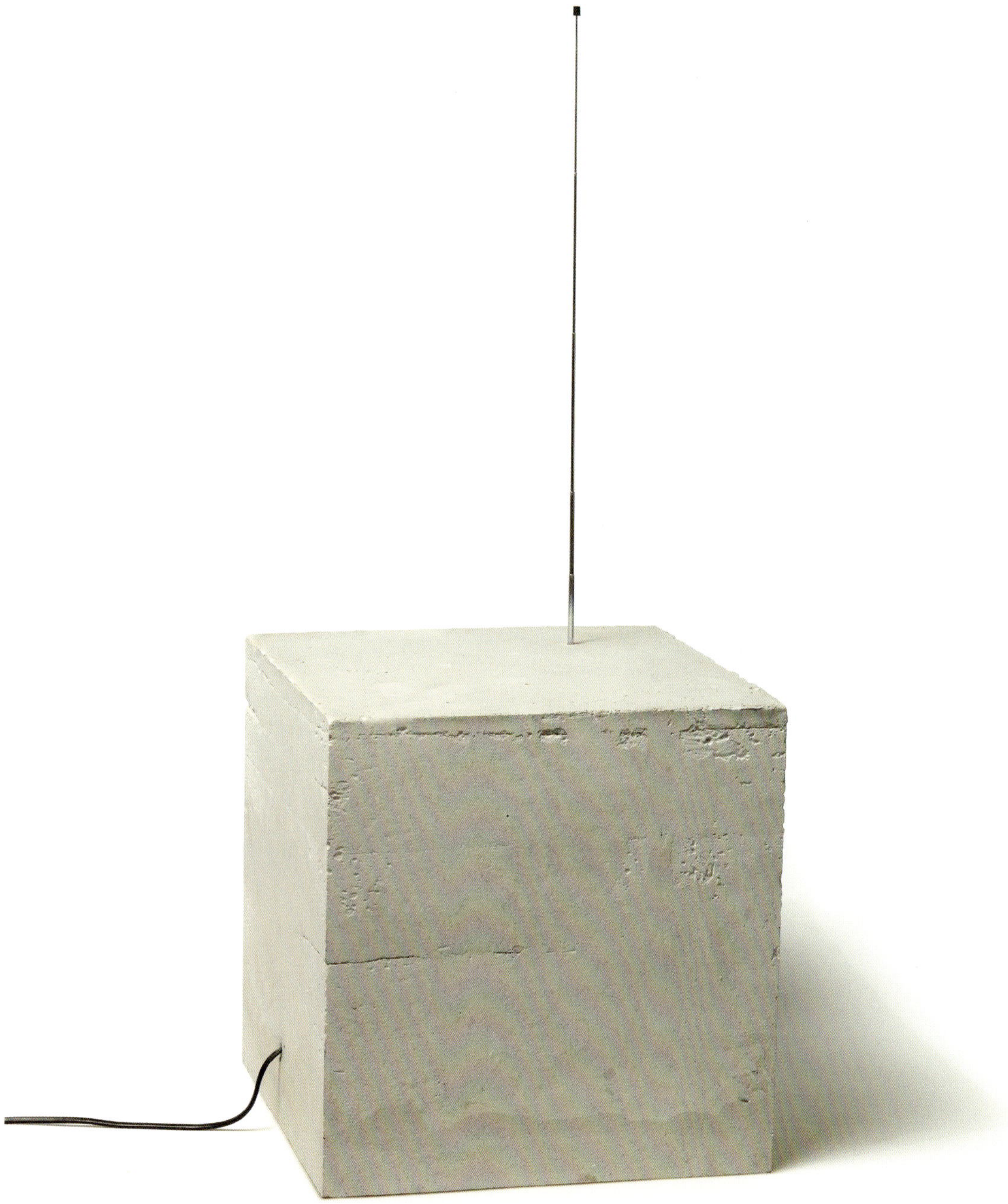

Radio, 1977/1990
Sculpture
Plaster block, FM radio
ZKM Collection

In this work by "Total Artist" Timm Ulrichs a FM radio has been sunk in a block of plaster. The radio is fully functional. On moving very close to the object, the visitor can hear the crackling sound of the radio inside. *Radio* encloses sound, entraps it, and then emits it again. One gets the impression that the plaster cube had swallowed the sounds and is now broadcasting them again.

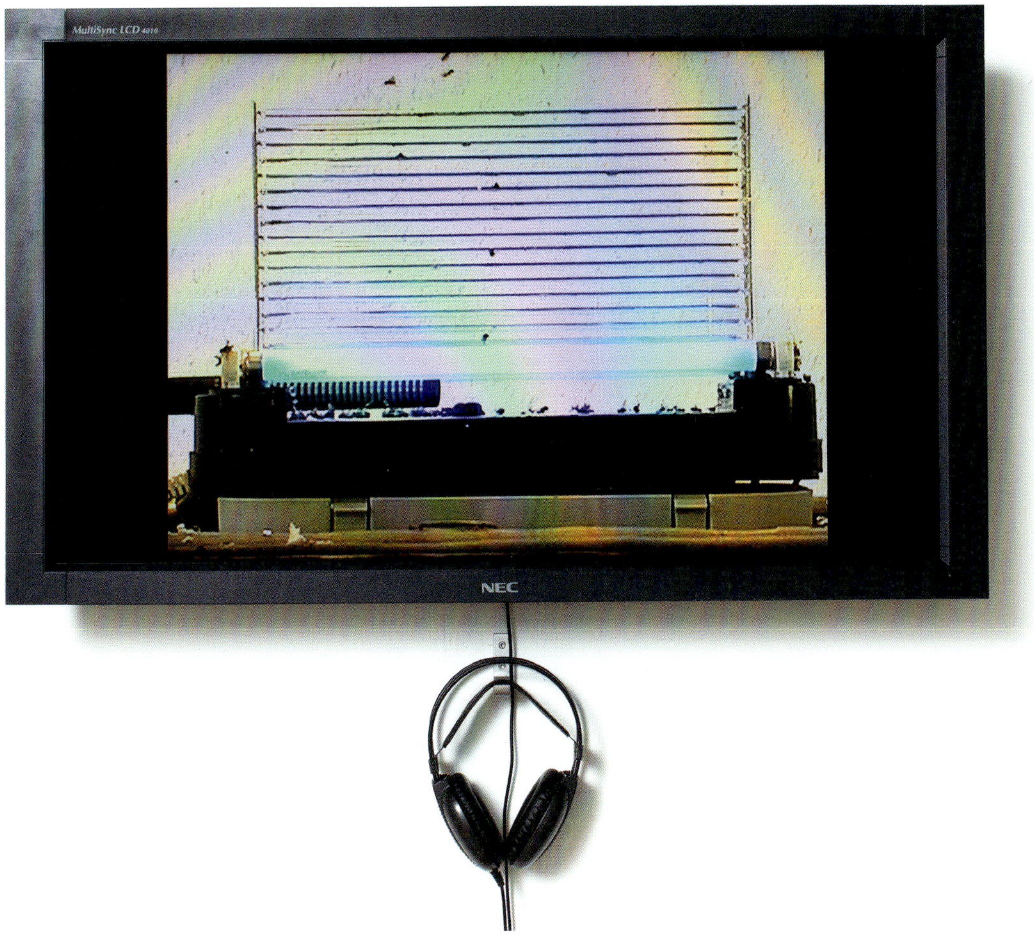

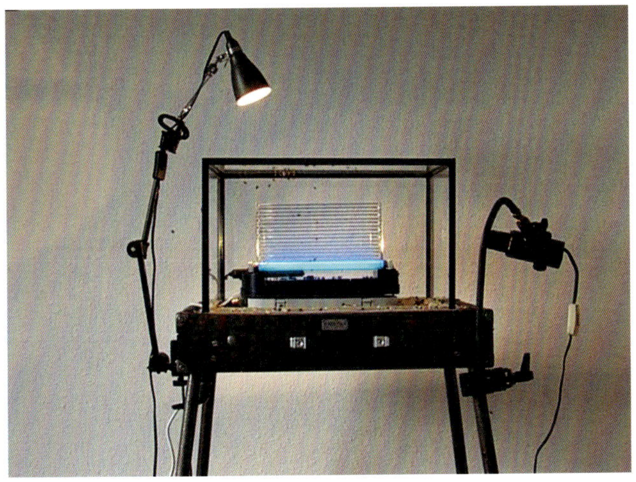

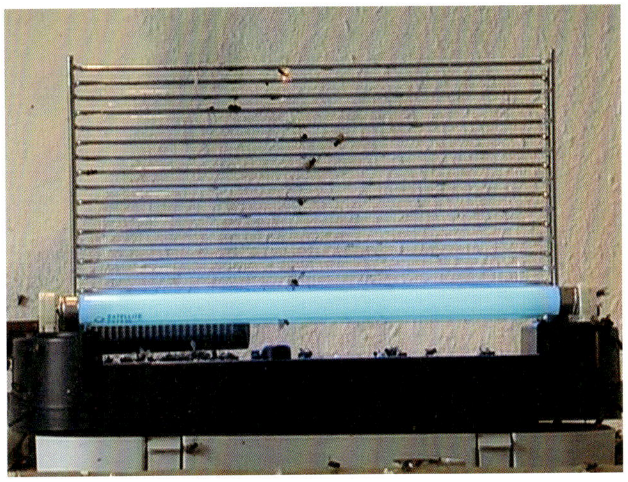

Requiem. Tier-Film III, 1989/2006
Filmed installation, sound,
image ratio 4:3, 5:35 min. (loop)
Timm Ulrichs

Unheard Avant-Gardes (in Scandinavia)

The exhibition area *Unheard Avant-Gardes (in Scandinavia)* presents the research project of the same name, which wants to rescue believed-to-be lost archives of Scandinavian media art and bring them back into the current discussion. But *Unheard Avant-Gardes* is more than "merely" an archive project, for in addition to its function as an historical archive, it also examines the basic principles and importance of experimental avant-garde "media art." Thus, the project can be seen as an attempt to re-arrange (the academic ideas about) the relations between technology, media, and art and to formulate a theory of this re-alignment. The various platforms of the exhibition present the project's hitherto obtained findings.

Unheard Avant-Gardes (in Scandinavia) consists of 3 + 1 platforms. Each platform represents a major "node" of the Scandinavian experimental scene where technologies, media, and artistic positions are mixed and re-mixed into new sound art. The invited artists designed the individual platforms as independent works and new artistic "spaces" so as to give the unheard avant-gardes a voice in history.

Curator and project coordination: Morten Søndergaard
Artists and designers of the platforms: Mogens Jacobsen, Martin Luckmann, Christian Leifelt, Hans Sydow, Ane Mette Ruge, HC Gilje, Christina Ann Sydow
Curatorial assistants: Anna Holm, Karen Mette Fog Pedersen, Emil Alenius Boserup
Production coordination: Sebastian Frese Bülow (AAU cph), Henrik Juul Jensen (DR)
Programming: Martin Luckmann, Alex Usbergo, Brian Groth
Research director / concept: Morten Søndergaard
Production of *EMS documentation space*: Sanne Krogh Groth and Mats Lindström (EMS)
Project partner: DR – Danish Broadcast Company, AAU cph – Aalborg University Copenhagen, LARM – sound archive infrastructure in Denmark, ZKM | Karlsruhe

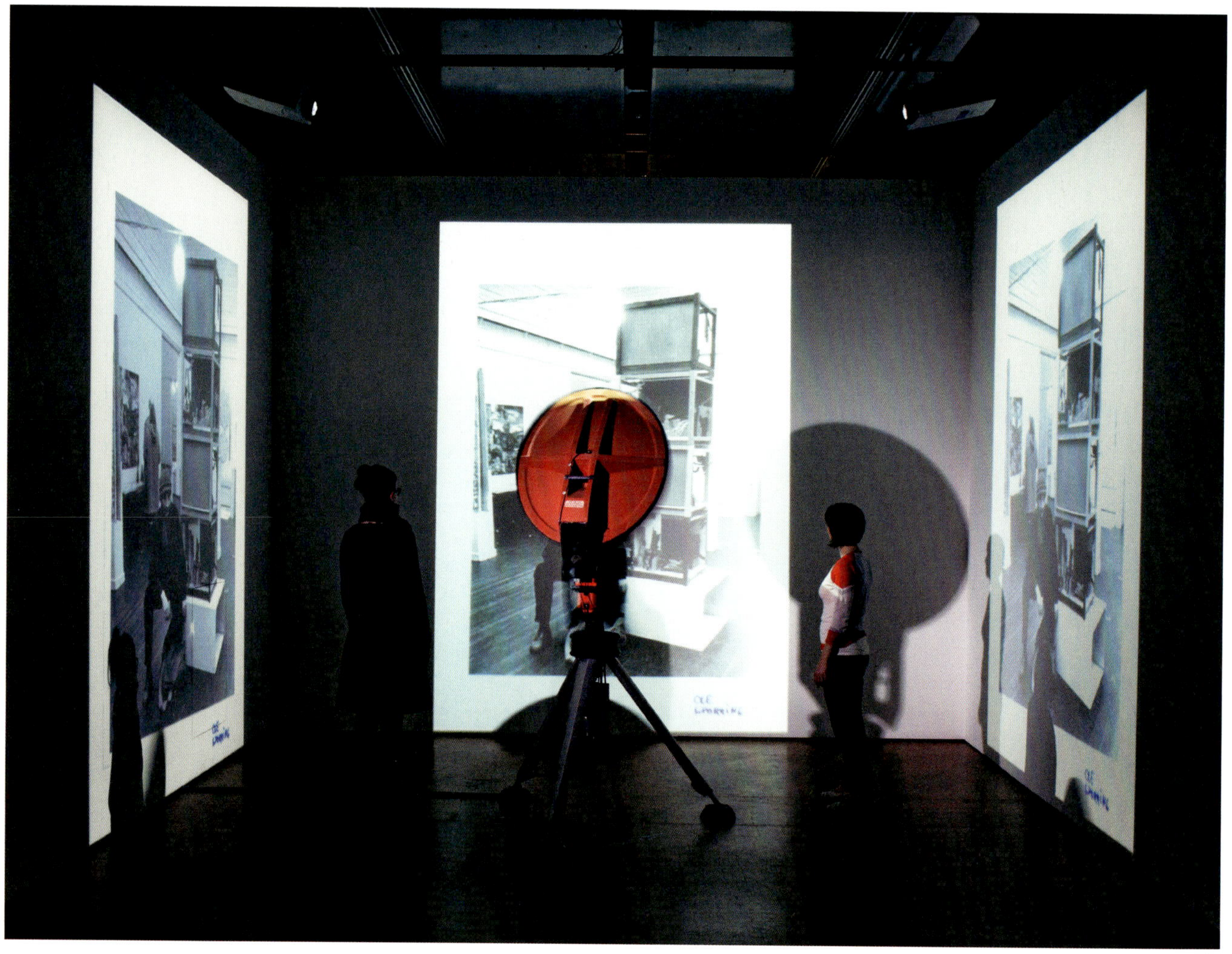

Mogens Jacobsen, Morten Søndergaard, *POEX Radar (homage to Hans Sandmand and Knud Hvidberg)*, 2012. Installation view ZKM | Karlsruhe, 2012.

Platform 1
EMS (Elektronmusikstudion, Stockholm)

The focus of the Platform 1 is the Elektronmusikstudion (EMS) in Stockholm, Sweden. Formerly known as Electro-acoustic Music in Sweden, this institution was founded in 1964 and serves as a national center for electronic music and sound art.

Hans Sydow
born 1968, lives in Odder (DK)
Christina Ann Sydow
born 1967, lives in Odder (DK)
Hansons Head. An audiovisual remix of Sten Hanson
2012
Audiovisual installation
3 large PVC banners, 1 amplifier, 8 speakers, speaker cable, 1 8-channel playback device
Hansons Head is sponsored by AV-huset (Denmark) and RESONANCE.DK
Thanks to Sten Hanson, EMS (Stockholm), Mats Lindström, Sanne Krogh Groth, Leif Elggren

EMS (Elektronmusikstudion) documentation space
Produced by Sanne Krogh Groth and Mats Lindström / EMS
2012

Platform 2
Finnish Electronic Music Studio & Errki Kurenniemi

Platform 2 is marked by the unconventional electronic aesthetics of the Finnish pioneer of electronic music Errki Kurenniemi and his music studio. In 1962 at the Institute for Musicology at the University of Helsinki (Finland), Kurenniemi set up an electronic studio with the intention of developing there an automated system of composition. In the 1960s he built his own synthesizers and in the 1970s his own invented musical instruments of the DIMI series.

Christian Hvalsøe Leifelt
born 1970 in Copenhagen (DK), lives in Copenhagen
Erkki Kurenniemi
born 1941 in Hämeenlinna (FI)
Erkki vi Erkki
2012
Interactive sound installation
3 projectors, media archive (video, audio, visual and textual documentation of the works of Erkki Kurenniemi), programmed musical instrument with optical input (remake of Erkki Kurenniemi's *Dimi-O* from 1971), video camera, iPad control, 2 speakers

Platform 3
POEX65

The third and largest platform throws light on the interdisciplinary event POEX65, which took place in December 1965 in Copenhagen, Denmark. The platform includes the *POEX Radar*, the *POEX Chairs* and a huge archive of the unheard avant-garde.

POEX En_acts:

Christian Yde Frostholm (CYF) on Åke Hodell
born 1963 in Copenhagen (DK)
I Gevär (Present Arms)
1965/2007
Animated sound poem after a sound poem by Åke Hodell (1923–2000)

Eric Andersen
born 1943 in Antwerp (E), lives in Copenhagen (DK)
OPUS 65
2012 (1965)
Digitalized version of original 16-mm film and sound installation

HC Gilje on Else Marie Pade
born 1969, lives in Oslo (NO)
7 cirkler
1958/2012
Light installation and sound composition
Metal rings, LEDs, microcontrollers, computer, speakers

Ane Mette Ruge on Jørgen Plaetner
born 1955, lives in Copenhagen (DK)
PLAeTnEr
2012
Sound installation based on a live recording of the POEX65 event
Projection on old parabolic antenna, mono and stereo audio
Original composition: Jørgen Plaetner, 1965
Sound engineer: Emil Alenius Boserup

Mogens Jacobsen
born 1959 in Rome (IT), lives in Copenhagen (DK)
Morten Søndergaard
born 1966 in Copenhagen (DK), lives in Copenhagen (DK)
Look down! (Kig ned!)
2012 (1965)
Homage to Hans Sandmand and Knud Hvidberg
Interactive sound installation based on a no longer extant sound installation by Hans Sandmand and Knud Hvidberg (1927–1986) from 1965
Tires, mirrors, sensors, light, computer, MP3 player, voice recording

POEX Re_Act:

Concept: Martin Luckmann, Mogens Jacobsen, Morten Søndergaard

Mogens Jacobsen
with Martin Luckmann and Morten Søndergaard
POEX Chairs – The debate
2012
Interactive voice interface
Modified chairs, sensors, computer, MP3 player, voice recording
Contents, sound, and research: Morten Søndergaard
Concept: Mogens Jacobsen, Martin Luckmann, Morten Søndergaard
Mix: Emil Alenius Boserup
Sound production: Emil Alenius AEOEAA

Martin Luckmann
born 1982 in Aabenraa (DK), lives in Copenhagen (DK)
with Morten Søndergaard
POEX Tape
2012
Interactive sound archive interface
Infrared LEDs, Wii controllers, projectors, database, Arduino platform, plexiglass, openFrameworks
Contents and research: Morten Søndergaard
Concept: Martin Luckmann and Morten Søndergaard
The tape material comes from private archives and boxes belonging to now deceased participants in the POEX65 event.

Mogens Jacobsen
Morten Søndergaard
POEX Radar (homage to Hans Sandmand & Knud Hvidberg)
2012
Interactive sound installation based on a no longer extant sound installation by Hans Sandmand und Knud Hvidberg (1927–1986) from 1965
Modified radar, motor, sensor, computer, MP3 player, voice recording, digitalized photographs
Contents, sound, research: Morten Søndergaard
Concept: Mogens Jacobsen, Martin Luckmann, Morten Søndergaard

Unheard Avant-Gardes (in Scandinavia)

MAP – Media Art platform

The Media Art Platform (MAP) occupies a special place in the exhibition area. It is based on the provisional media art archive compiled by Mogens Jacobsen and Morten Søndergaard between 2005 and 2009 for the Museum for Contemporary in Roskilde, Denmark, and pursues basic ideas and experimental intentions similar to those of the Unheard Avant-Gardes project.

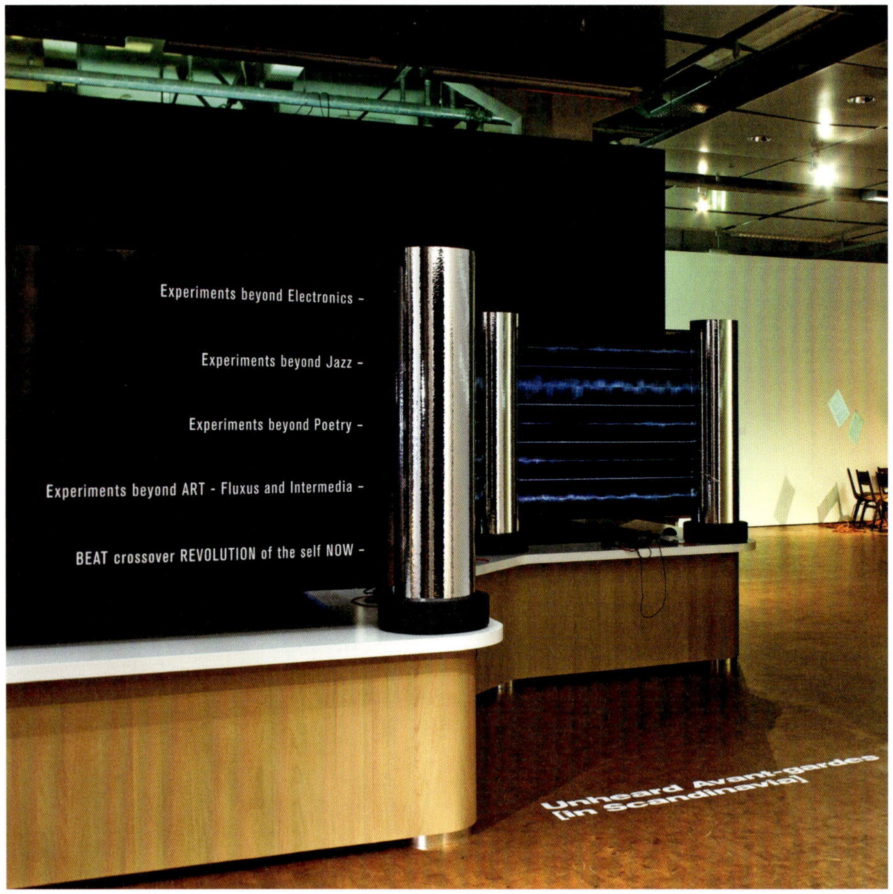

Martin Luckmann with Morten Søndergaard, *POEX Tape*, 2012.
Installation view ZKM | Karlsruhe, 2012.

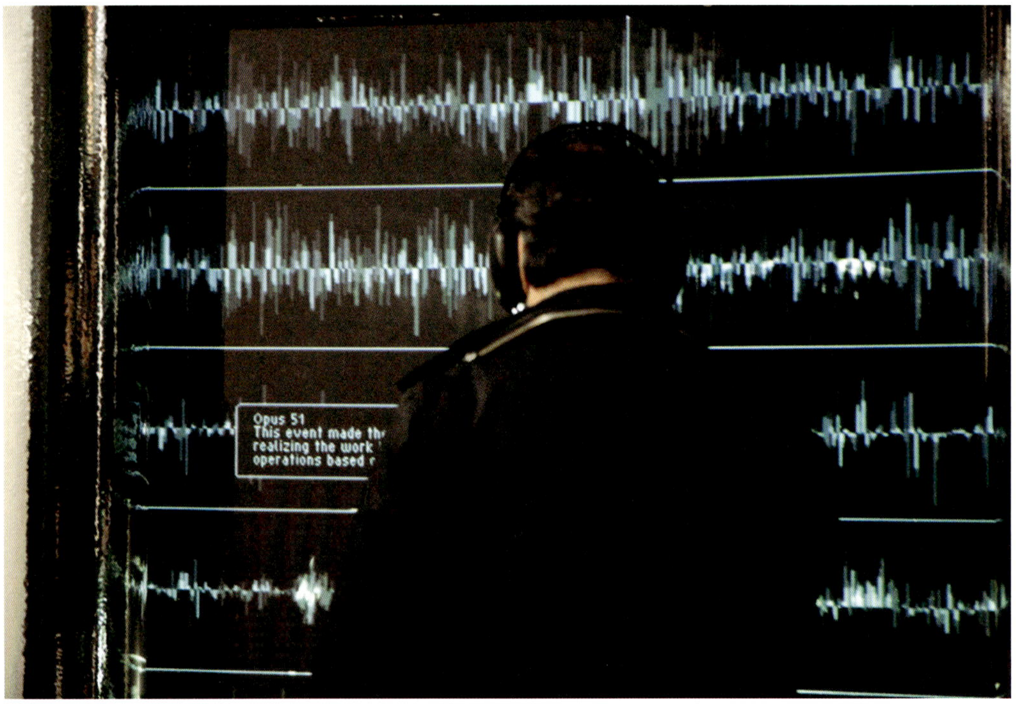

Martin Luckmann with Morten Søndergaard, *POEX Tape*, 2012. Installation view ZKM | Karlsruhe, 2012.

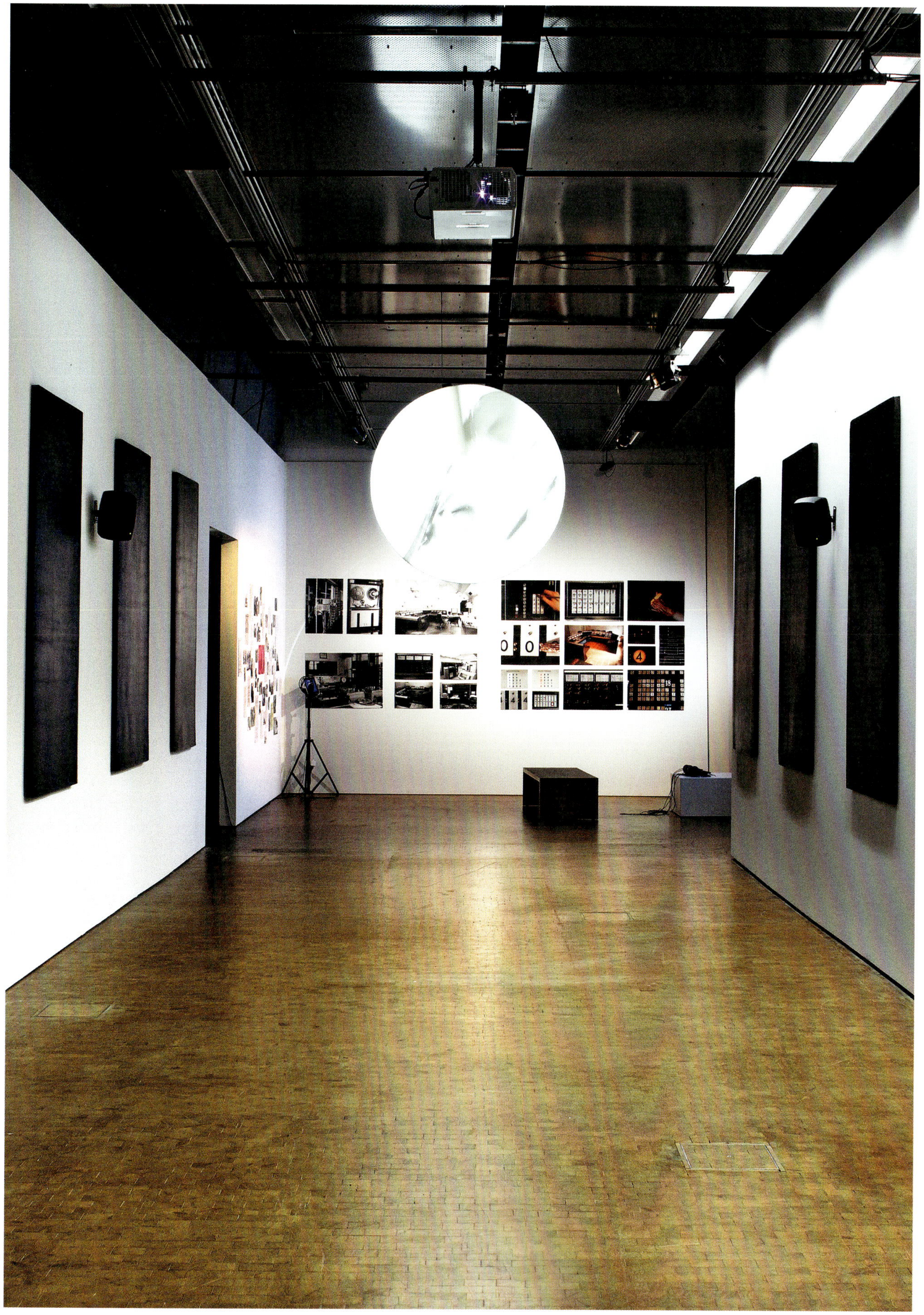

Ane Mette Ruge, *PLAeTnEr*, 2012. Installation view ZKM | Karlsruhe, 2012.

Julijonas Urbonas

*1981 in Klaipeda (LT), lives in Vilnius (LT) and London (GB)

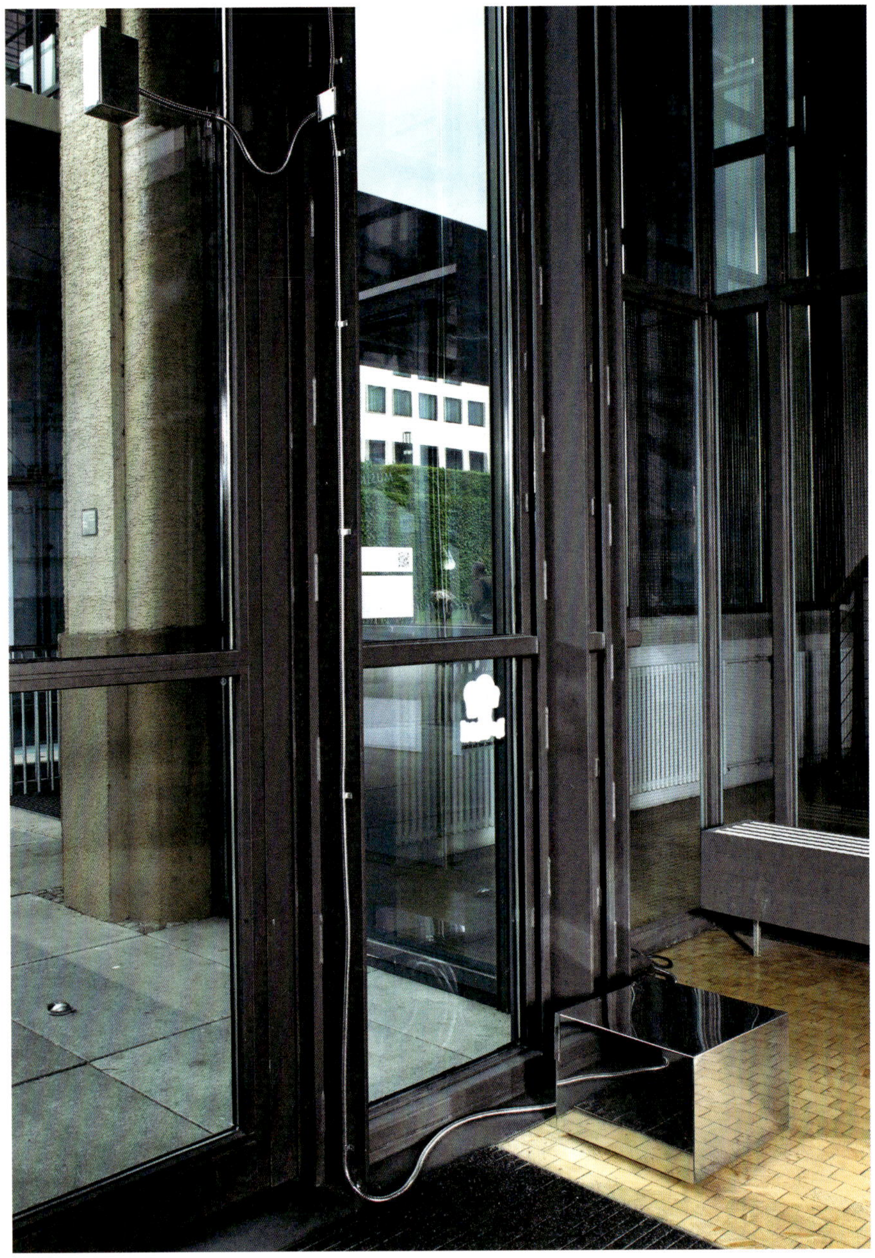

Sounding Door, 2009
Interactive sound installation
Terfenol-D-based sound transducer, subwoofer, audio amplifier, capacitive rotary encoder, custom electronics and software, 50 min.
Programming: Julijonas Urbonas, Sebastian Lexer
Compositions: Andrea Vigani, Augustas Ribokas, Franziska Windisch, Frauke Schmidt, Julijonas Urbonas, Vincent Wikström
Commissioned by "Vilnius – European Capital of Culture 2009"
Julijonas Urbonas

www.julijonasurbonas.lt

Sounding Door is an interactive sound installation that transforms any door into an idiosyncratic musical instrument and any place into a stage. The movements of the door, its position, speed and degree of acceleration, determine the sound it produces. The Lithuanian designer, artist, writer, and engineer Julijonas Urbonas uses custom-designed electronics and software in the installation.

In a time when designers are busy perfecting the sounds of cornflakes and hair dryers, the sound of architecture is still all too seldom discussed, maintains Urbonas. For *Sounding Door* he asked composers, musicians, sound designers, and sound artists the question: "Why do we choose a door for our home based on its visual characteristics alone and not, for instance, on the particular sound of its hinges?" More than one hundred compositions were submitted in reply. Urbonas has chosen five pieces for this exhibition; each plays for ten minutes every hour.

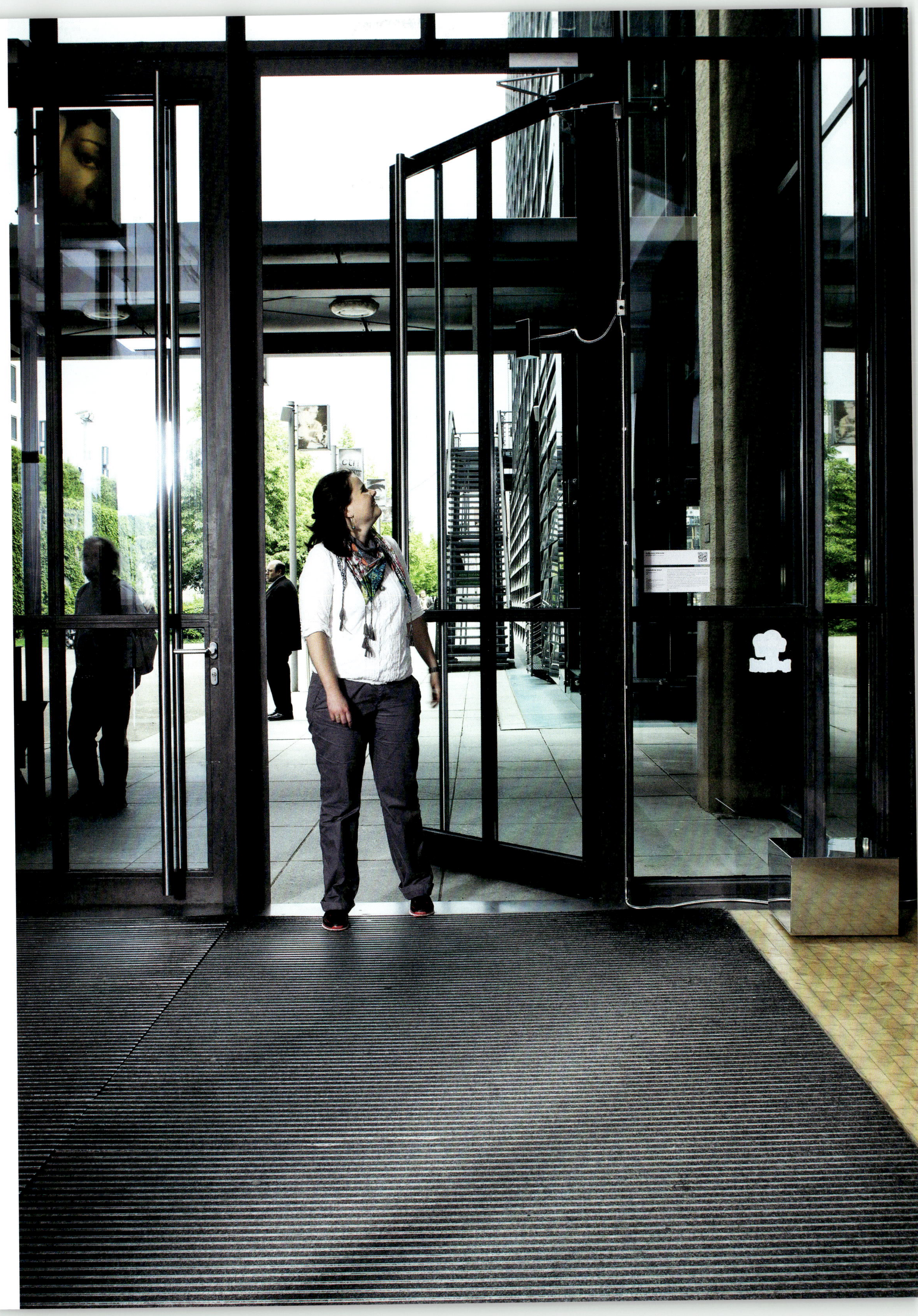

Peter Vogel

*1937 in Freiburg im Breisgau (DE), †2017 in Freiburg im Breisgau

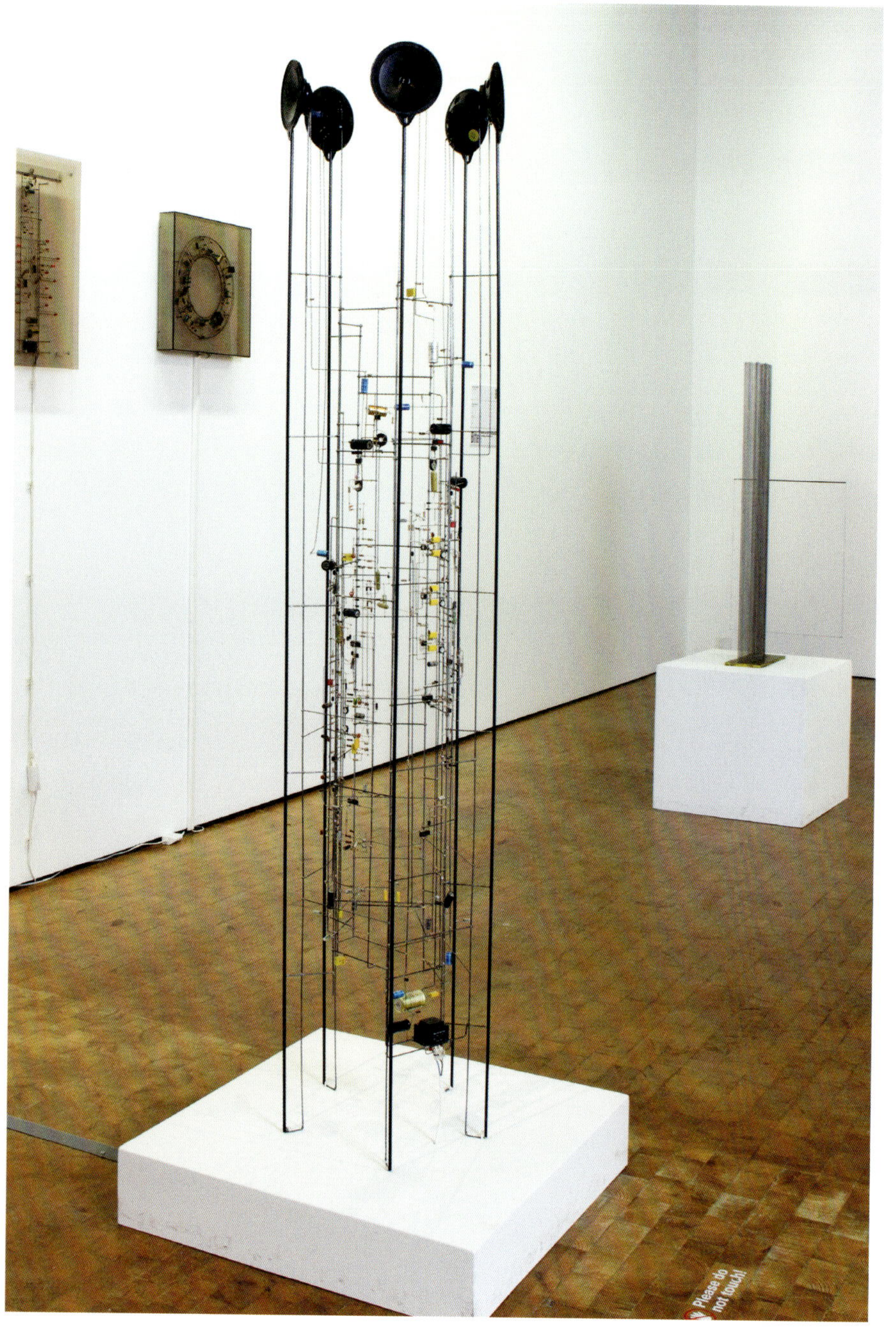

right page:
Zwitschermaschine „Auf und ab“, 1988
Sound sculpture
Electric circuit, mechanics, loudspeakers, ca. 61 × 10 × 25 cm
ZKM | Karlsruhe

Paul Klee’s famous painting of the same name was the stimulus for a whole group of works titled *Zwitschermaschine „Auf und ab“* [Twittering machine “Up and down”]. Here, loudspeakers replace the birds; they move and imitate chirping birds when a crank is turned. Peter Vogel sees Klee’s painting as “a parable for the scientization of nature and an allegory of the mechanization of desire.” Visitors must turn the crank to trigger the twittering. The wire framework may be touched; the work’s mechanical construction is easily grasped. But the electronic mimicry of the birds’ warning calls remains mystifying. The artist draws an analogy to our use of technology: “We get our hands all these apparently harmless little machines and with them set things in motion that we can no longer understand and that insidiously intrude into our lives. We are confronting a system that is developing its own independent dynamics and slipping out of our control.”

Römischer Turm, 1989
Sound sculpture
Electric circuit, light sensors, loudspeakers
ZKM | Karlsruhe

The parapets of the *Römischer Turm* [Roman tower] consist of round loudspeakers. Several sensors inside the transparent tower register the shadows of observers. The timing of the shadows’ appearance triggers certain sound events. The laws of transformation governing the conversion of movement structures into sound structures are not immediately evident. Yet they convey the feeling that the sculpture’s behavior is based on constantly fresh variations of movements. The title is an allusion to little-known structures in ancient Roman music.

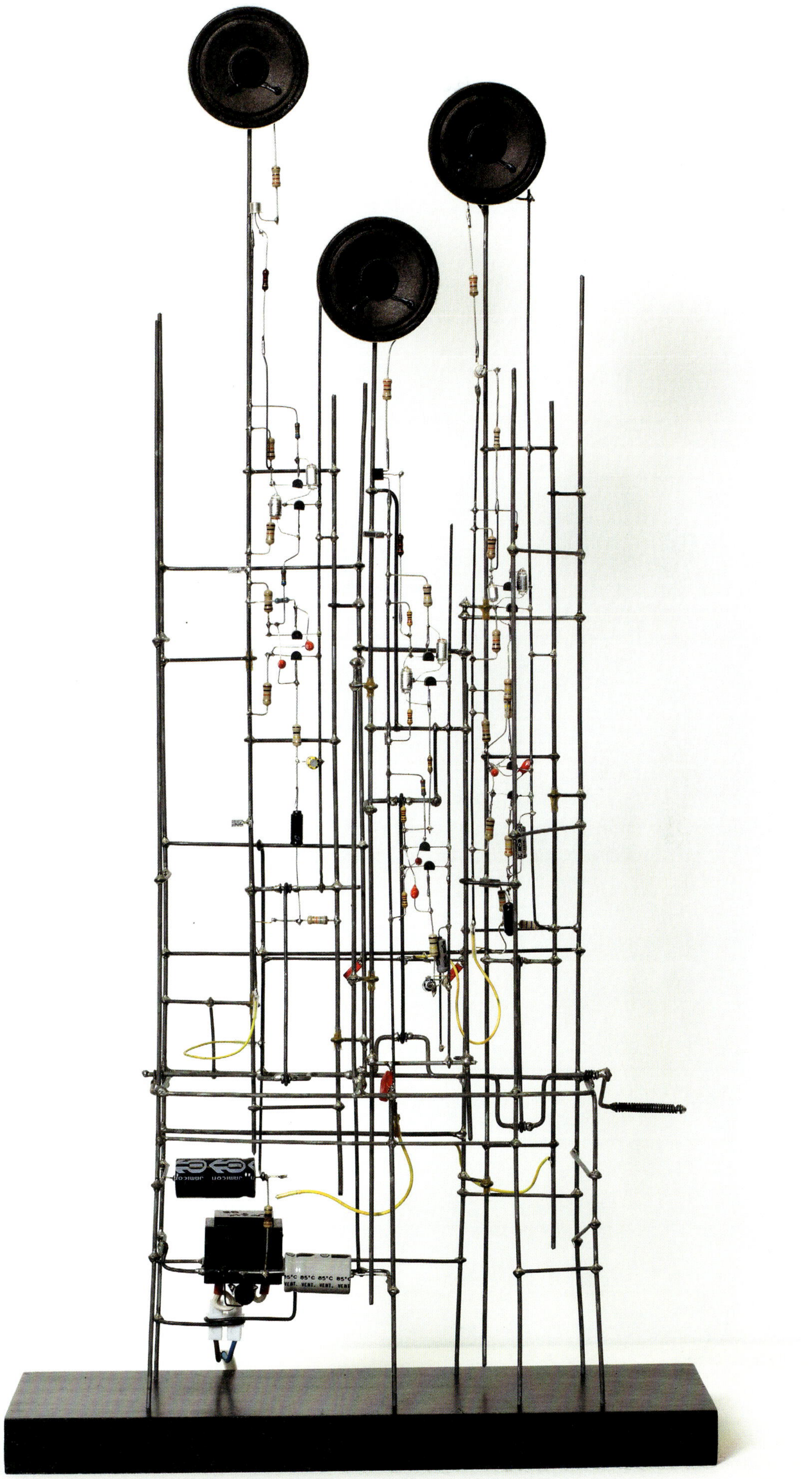

Peter Vogel

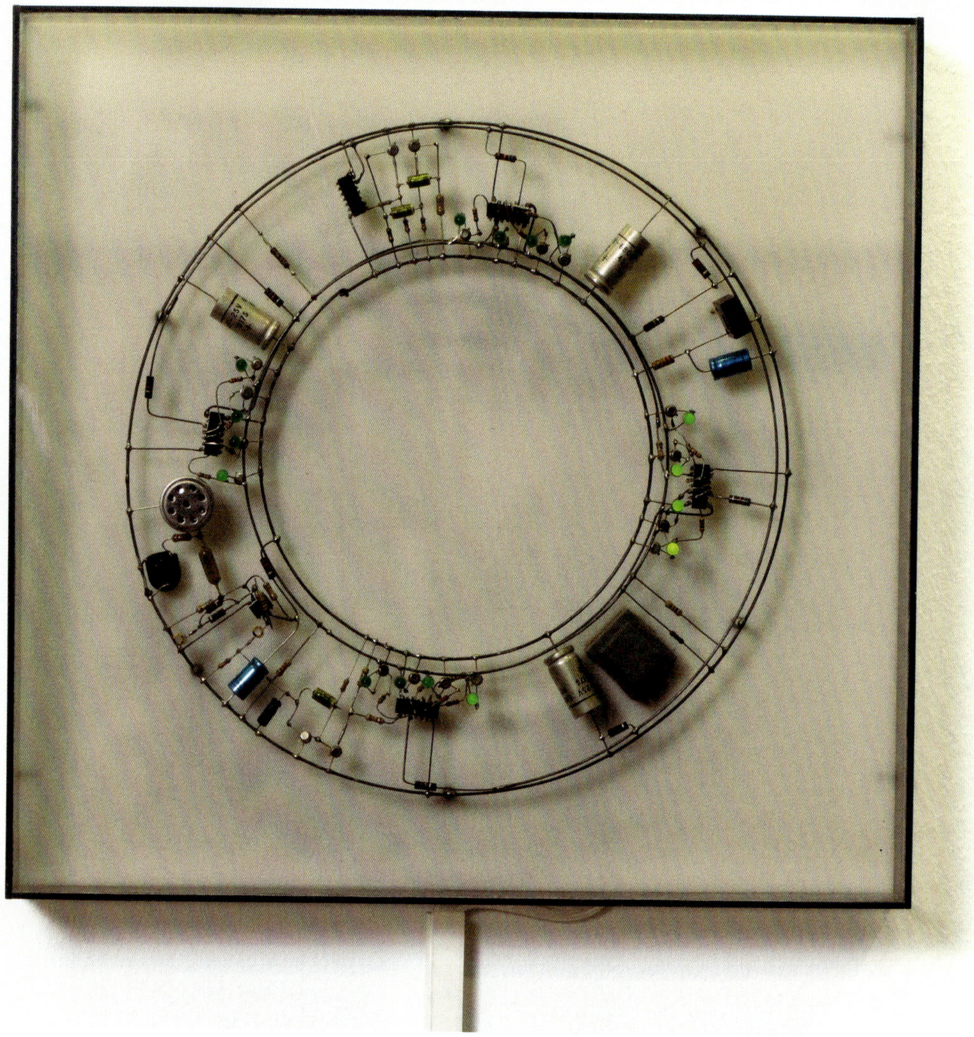

Grüne Lichtpunkte, 1984
Sound sculpture
Electric circuit, microphone, LEDs,
Plexiglas, 40.6 × 39.9 × 6.5 cm
ZKM | Karlsruhe

After having studied physics, the artist and performer Peter Vogel was initially involved with brain research. He developed medical devices but also, early on, composed electronic music and designed instruments. Vogel's sonorous, luminous, and as a rule interactive objects tempt people to play. In his first light objects, illuminated light bulbs functioned only as a signal; later on, the visual aspect gained more significance. Specific formal structures determined the arrangement of the bulbs. In the sound sculpture *Grüne Lichtpunkte* [Green light points], the timing of the light signals reacts directly to acoustic stimuli. With every stimulus, a stable condition is transformed, the lights alternate, or the constellation of bulbs changes. Specific acoustic stimuli are assigned to specific luminous reactions. But since the triggering sounds vary in volume, phases get skipped, and so the originally deterministic sequence effectively becomes an aleatory (random) one.

right page:
Sensible rote Punkte, 1989
Interactive sound sculpture
Electric circuit, microphone, LEDs, ca. 70 × 40 cm
ZKM | Karlsruhe

The LEDs react to sound; they flicker like an amplitude graph. When the noise stops, the last two activated LEDs continue to shine.

Peter Vogel's sound sculptures are transparent. They permit a glimpse into the way they are constructed. *Sensible rote Punkte* [Sensitive red dots] also follows the aesthetics of a circuit diagram. Yet the electronic programming is opaque, offering scope for speculation and interpretations. What reactions does the visitor's behavior trigger? To which sounds and noises are the red dots of light reacting? In his interactive sculptures, Peter Vogel investigates social modes of behavior and puts the emphasis on the interaction between human being and machine.

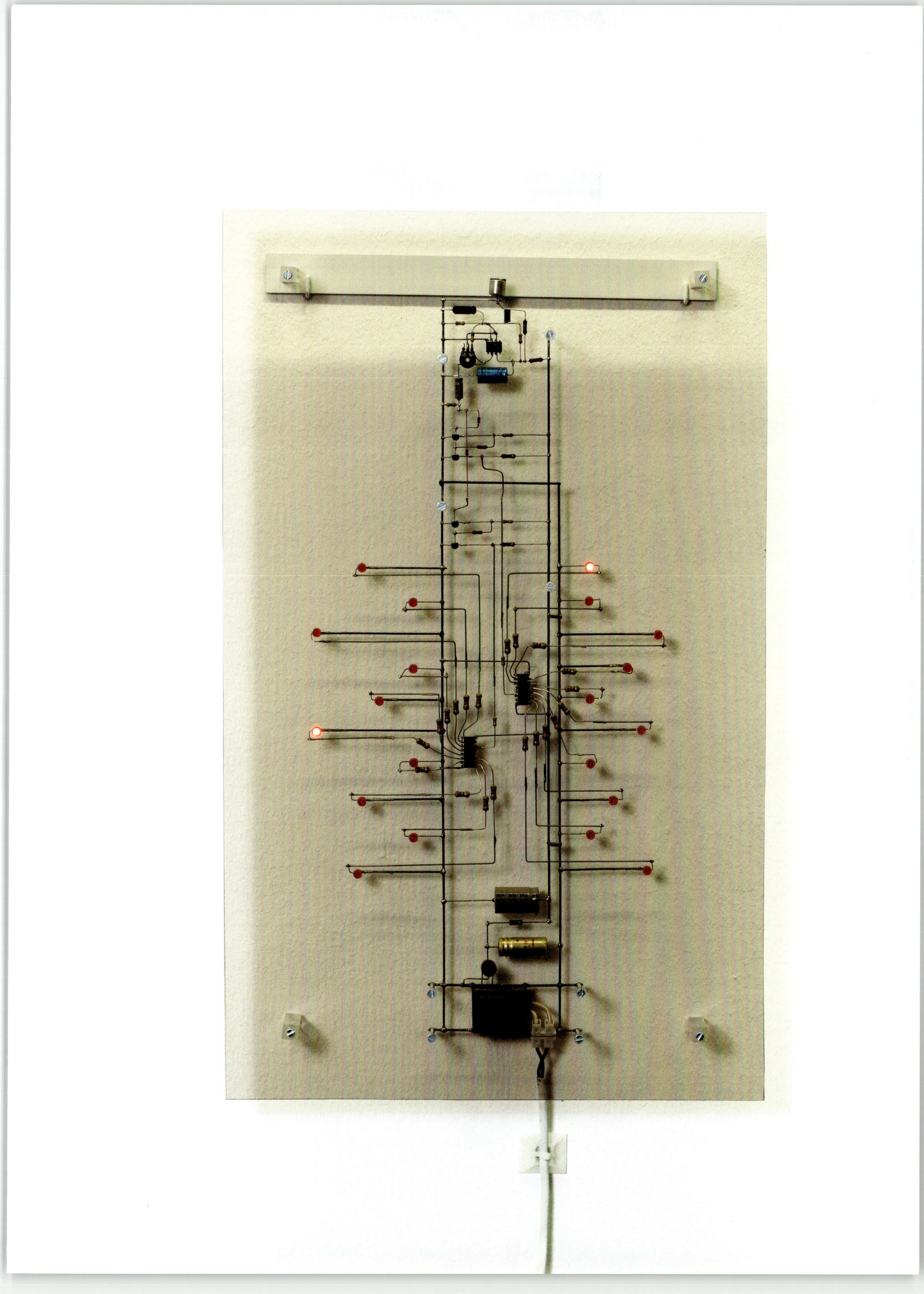

Bram Vreven

*1973 in Ghent (BE), lives in Den Haag (NL)

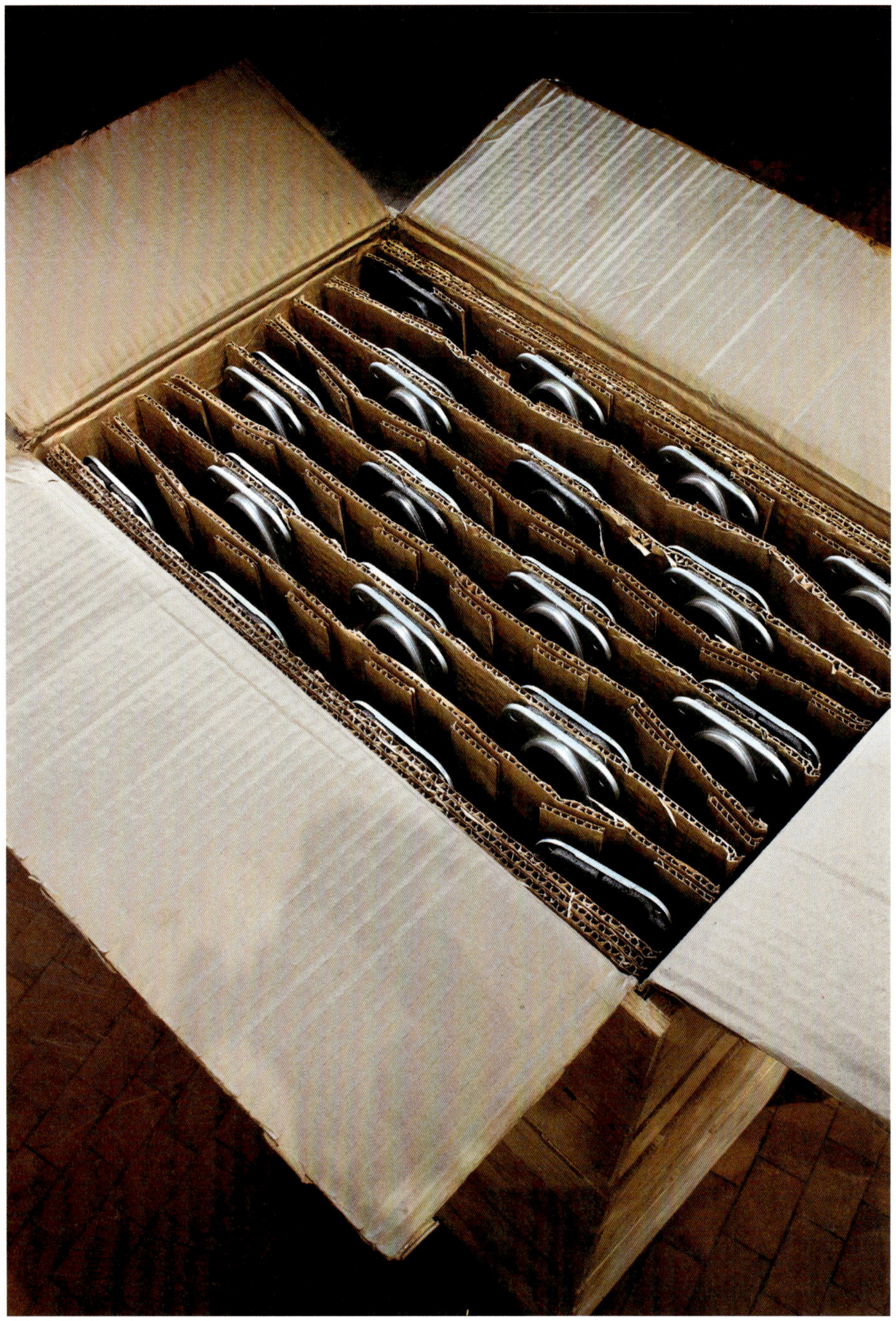

untitled, 1999
Sound installation
Wood, cardboard, loudspeaker, electronics, 90 × 55 × 35 cm
Electronics: Jo Scherpenisse
Bram Vreven

www.bramvreven.com

The three coarse wooden crates contain small loudspeakers. Several layers of cardboard protect the loudspeakers during transport. The crates are unlabeled, so where the randomly dumped delivery is going is an open question – to a production hall, or a storeroom?

untitled is one of the first sound installations by the former jazz musician Bram Vreven. The stack of crates gives no visual indication that all 28 loudspeakers are not only functional but also active. To contrast with the muffled cardboard acoustics, Vreven has chosen vivid, clearly contoured sound material. The noises are reminiscent of the sounds and movements of tiny insects. A lively animalistic miniature soundtrack emanates from the loudspeaker chassis and "escapes" from the crates. Here a theme that characterizes almost all of Vreven's works is already evident (though as yet immaterial): motion.

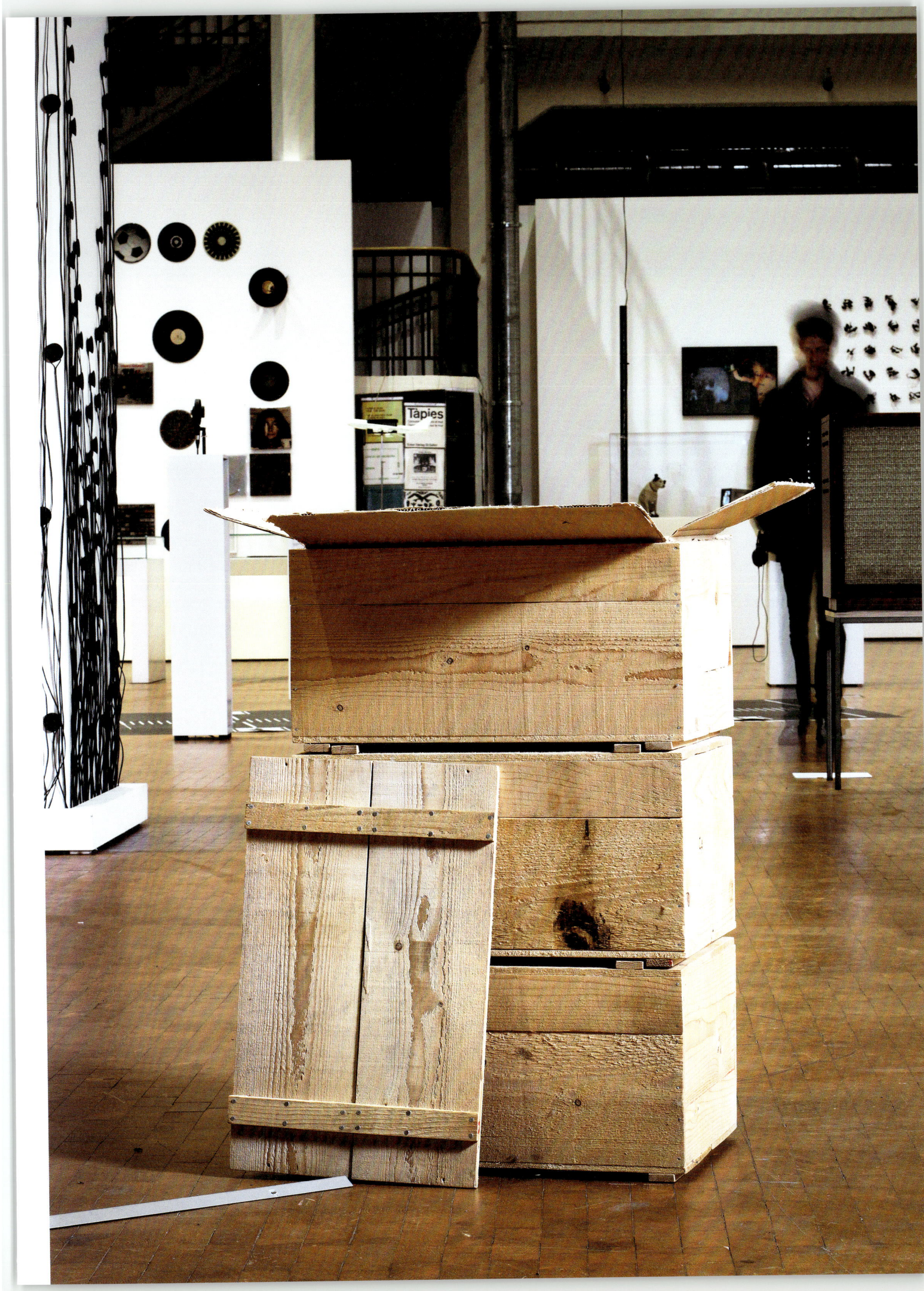
Tàpies

Marijke van Warmerdam

*1959 in Nieuwer Amstel (NL), lives in Amsterdam (NL)

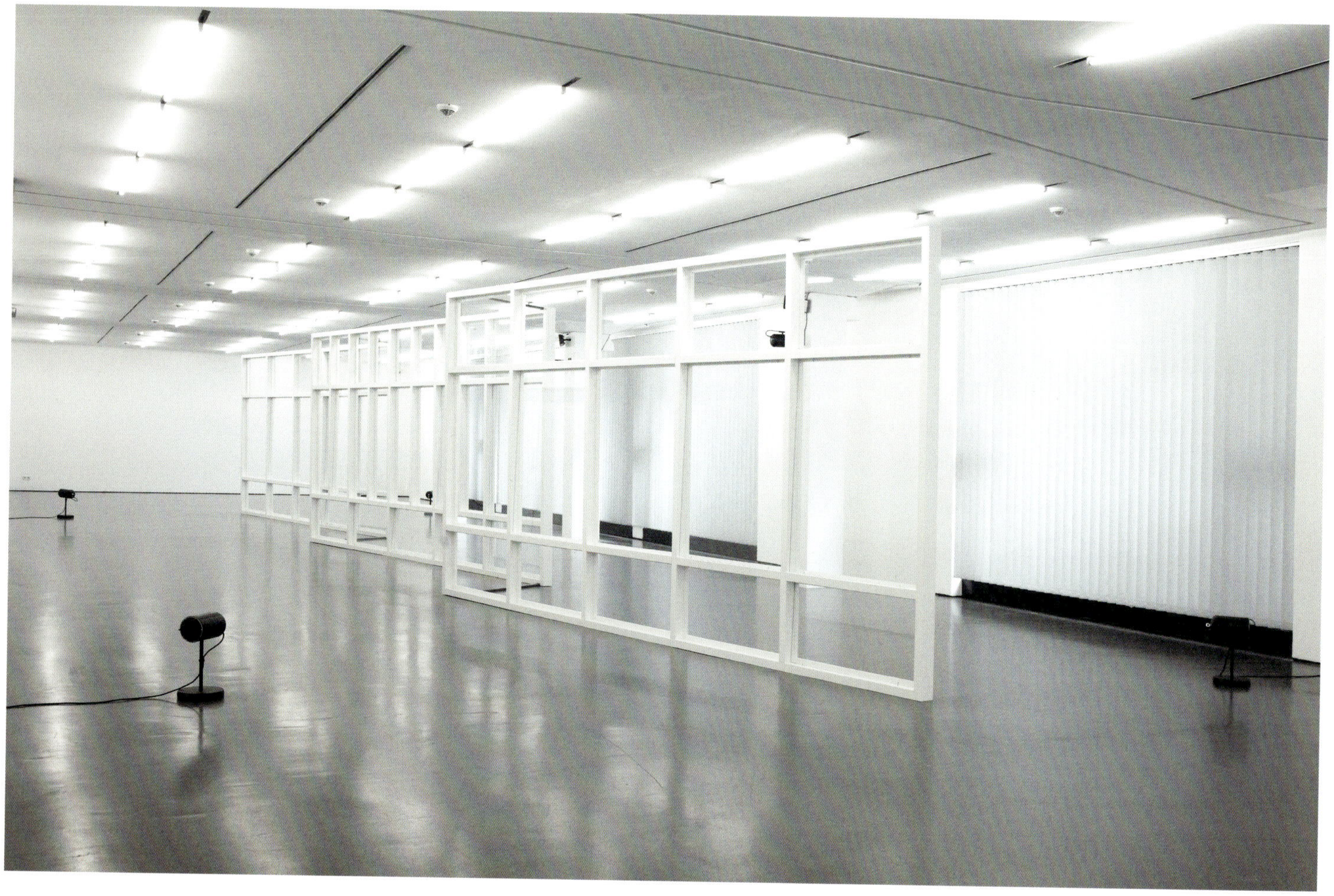

New balls, please!, 2014
Multi-channel sound installation
Wood, glass, metal, sounding equipment, 1900 × 280 × 135 cm, 8:30 min.
Installation view Kunsthalle Düsseldorf, 2014

Marijke van Warmerdams minimalistic sound sculpture *New Balls, please!* was exclusively created for the exhibition *Nahbei in der Ferne* [Nearby in the distance] by the Kunsthalle Düsseldorf. The work features a zigzag shape and vibrating windows, which interact with its environment, mainly the long window front of the Kunsthalle.

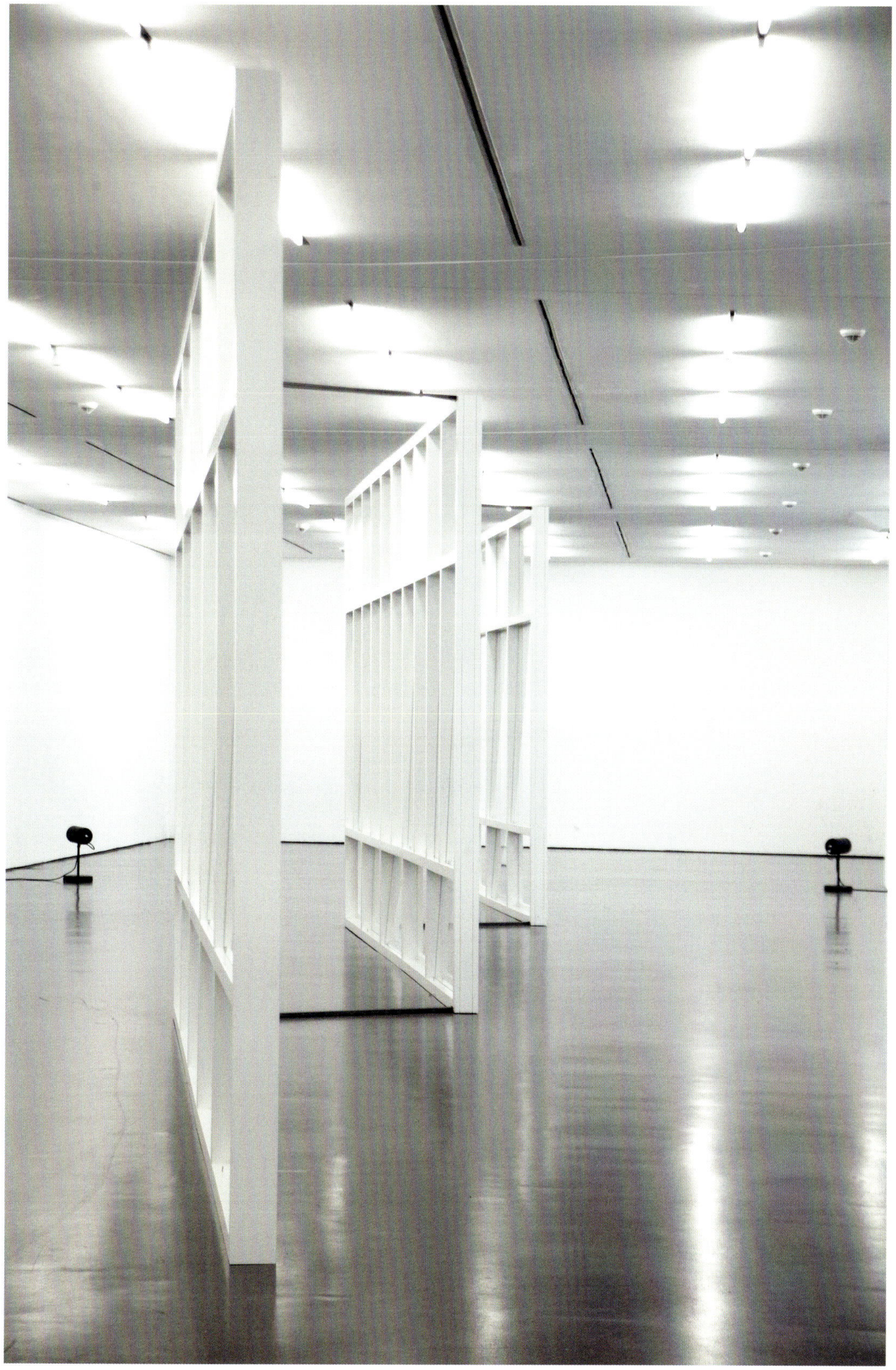

Peter Weibel

*1944 in Odessa (UA), lives in Karlsruhe (DE)

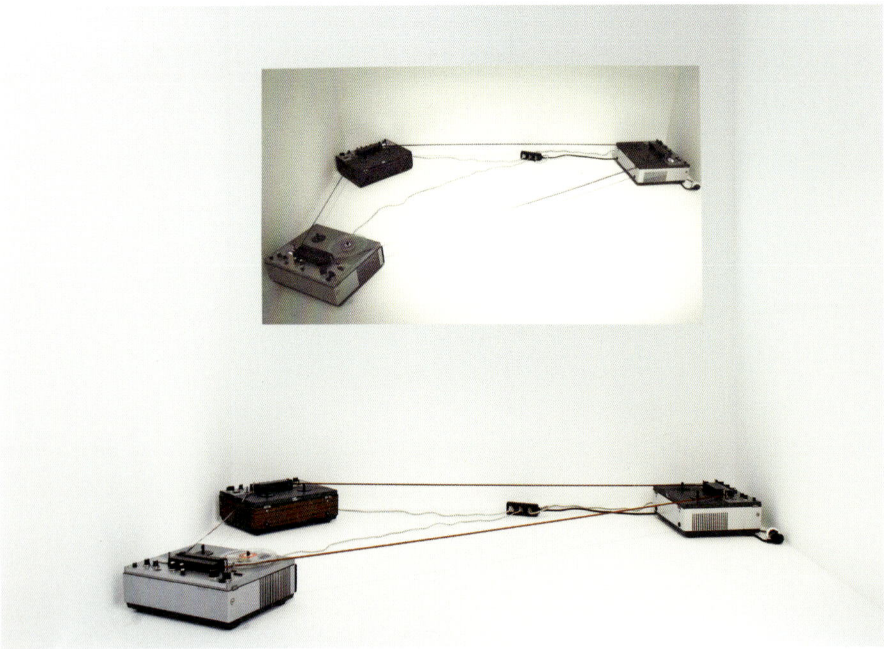

ichmasse – Masseich, 1977,
reissue 2012
Sound sculpture
3 tape players, tapes, loop and
projection

An enormous tape loop stretches across three tape recorders in Peter Weibel's *ichmasse – Masseich* [I mass – mass I]. On the loop, a recording of the spoken word "*Ich*" ["I", or "me"] travels from one tape recorder to the next. When the word arrives at a magnetophon, the "*Ich*" sound is heard. The machine says "*Ich*" (me). Each machine asserts its own identity but uses the same word (*Ich*) [me] as the others. A paradox! By proclaiming their individuality, their uniqueness, and their singularity, but by using the same word, all individuals form a homogeneous mass. The real installation is not operating, only in the projected image, the installation works.

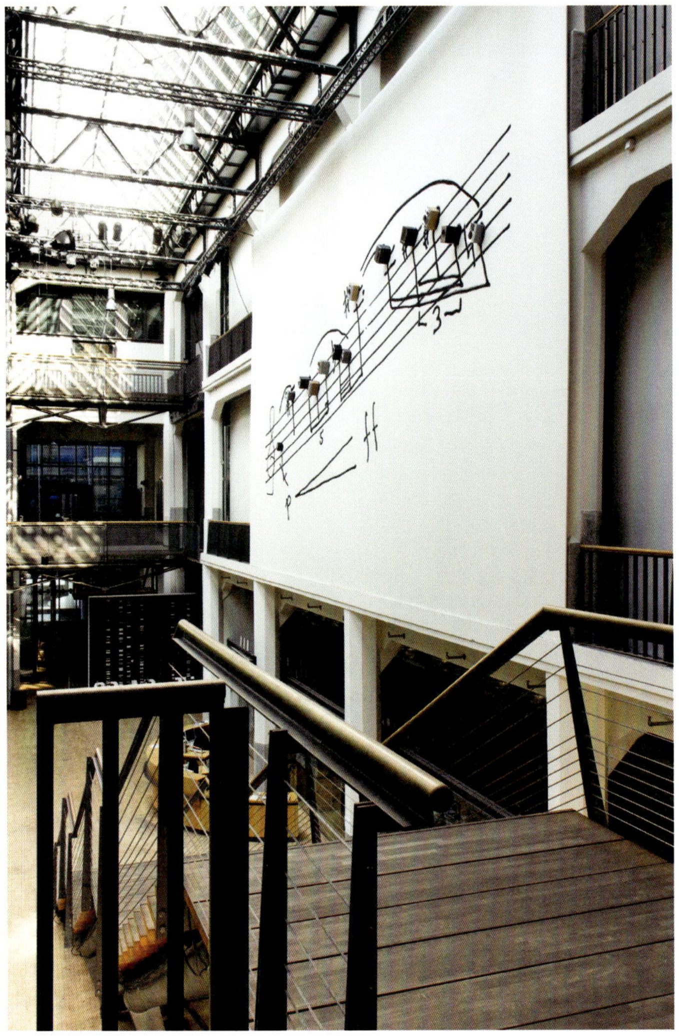

Lautsprecher als Noten, 2012
Sound installation at ZKM | Karlsruhe, 2012
Canvas, loudspeakers, drawing, 8 × 13.24 m
Composition: Ludger Brümmer

In *Lautsprecher als Noten* [Loudspeakers as notes] the symbols of the notes in musical notation are at the same time sounding objects, the loudspeakers. In fact the note is an element of a written code of sound information, and the loudspeaker stands for technical encoding of sound information. The score is encoded instruction, and the loudspeaker executes the code. In this work instructions and execution are identical. The historical and the current technical notation of sound combine the written note-loudspeakers.

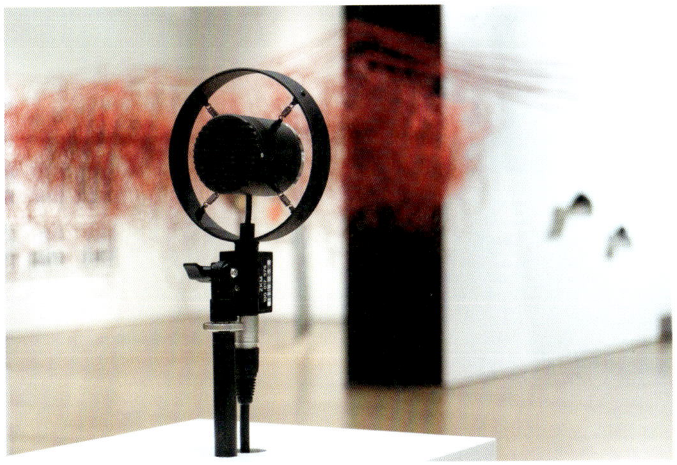

oh cet echo, 2012
Interactive sound installation
Microphone, 2 loudspeakers, computer
Software development: Götz Dipper
Artistic-scientific assistance: Barbara Nerness

The sound installation *oh cet echo* enables visitors to speak into a microphone, while their voice is simultaneously simulated by a musical instrument (piano, marimba, vibraphone, or flute, each using short or longer tones). Visitors can change from one instrument to the next using a push-button, and are also provided with a short recording of previous visitors (with a time delay of one second). Over time, the words of the visitors create a joint musical composition.

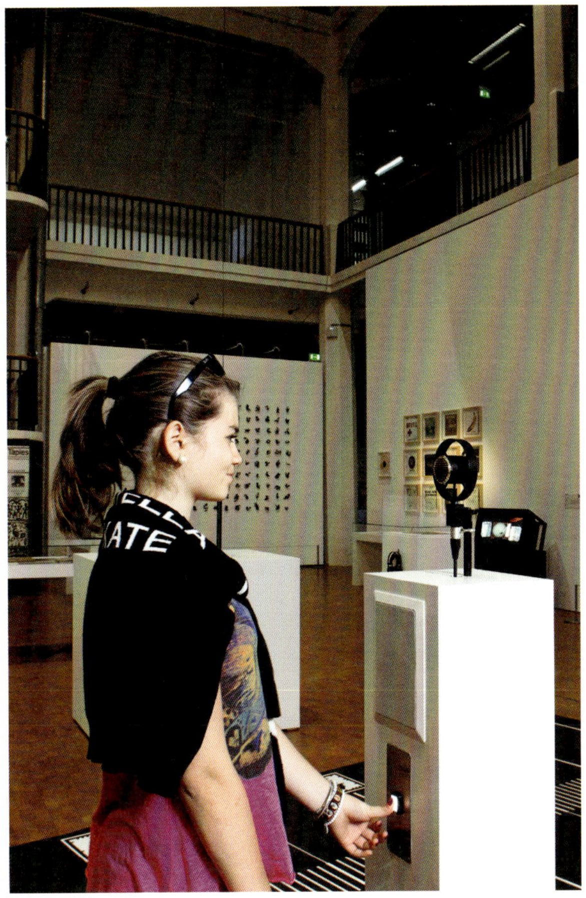

Data Music, 2016
Interactive sound installation
Software development: Nikolaus Völzow
Sound and interface design: Derek Hauffen

The installation consists of a music stand with a photo of a guitar on it. The six guitar strings on the photo are real and connected to a computer in which several musical compositions are stored, for example, famous rock riffs. Visitors are invited to pluck the guitar strings, which are both the "notation" and the "instrument", thus becoming both the "composer" and the "performer" of the music. As the music is stored physically as data in the computer, that is, preprogrammed, we could say that the player performs music via an interface (music stand with strings), which is programmed by an algorithm.

In addition, the performance of the player is recorded by a Kinect camera and displayed on a screen on the wall. As the images of the performance are pure data, the configuration of the visual data is influenced by the configuration of the acoustic data. Sound and image are synchronized with each other. Thus, whenever a player is in front of the music stand, they see their mirror image displayed on the screen. The moment the user starts performing with the music stand, their image distorts into a flow of visualized acoustic data.

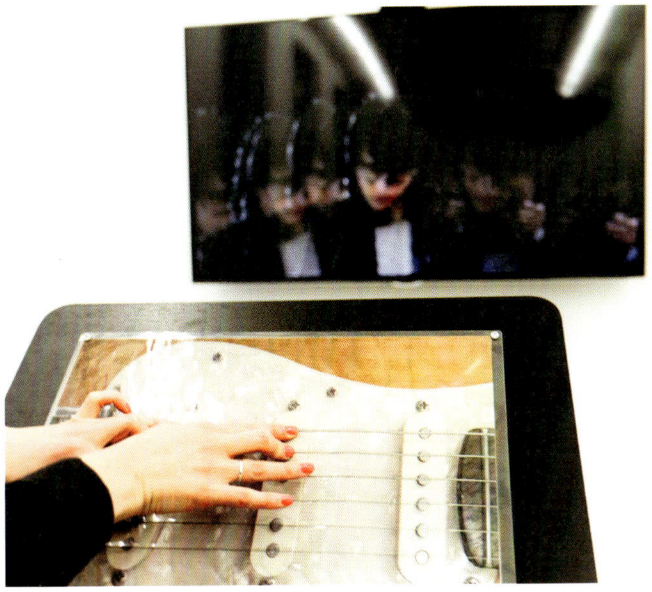

Peter Weibel

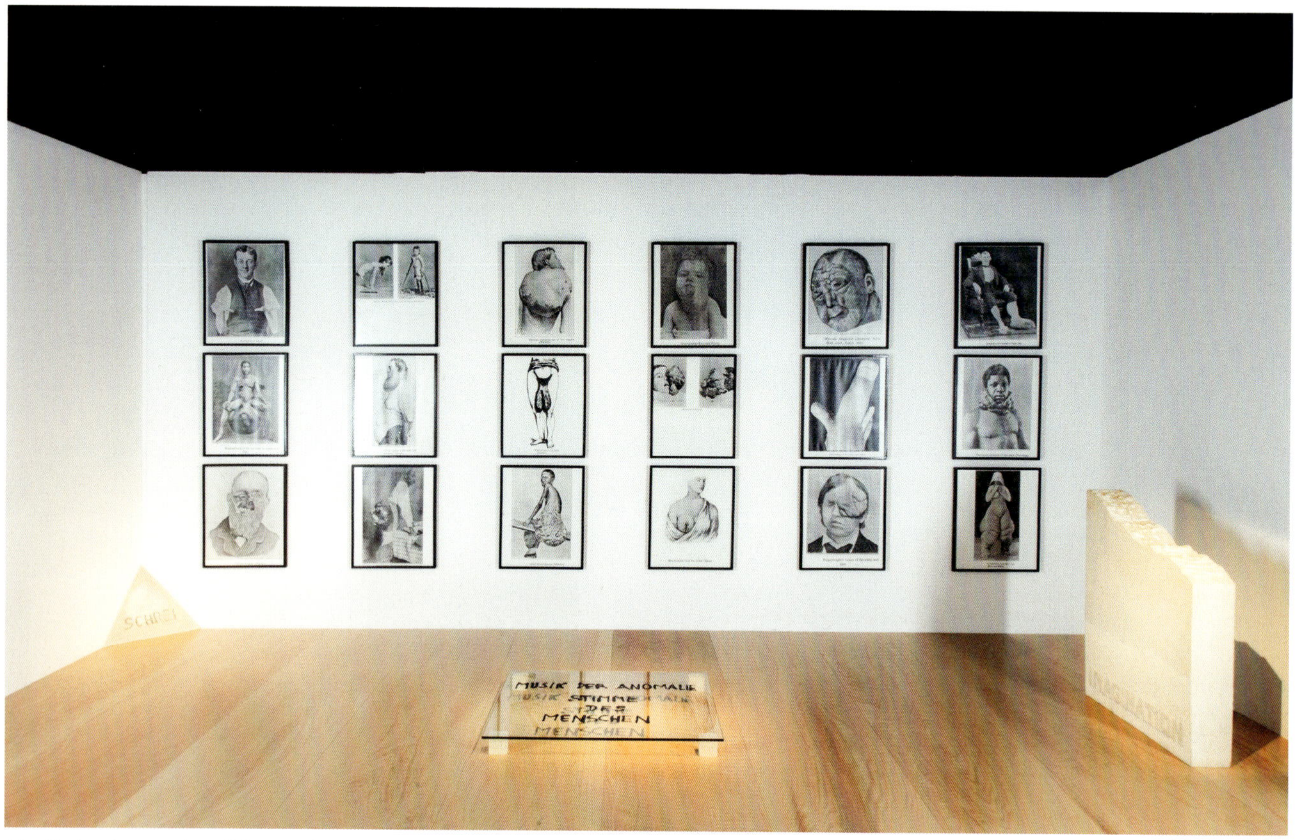

Musik der Anomalie – Stimme des Menschen, 1982
Exhibition view, 21er Haus, Vienna, 2014.

Musik der Anomalie – Stimme des Menschen [Music of the Anomaly – Voice of Man] shows an installation with portraits of physically disfigured people, with a sandstone sculpture, a concrete corner sculpture with the German word “SCHREI” (scream) and a glass plate with the words “Musik der Anomalie – Stimme des Menschen” in asphalt.

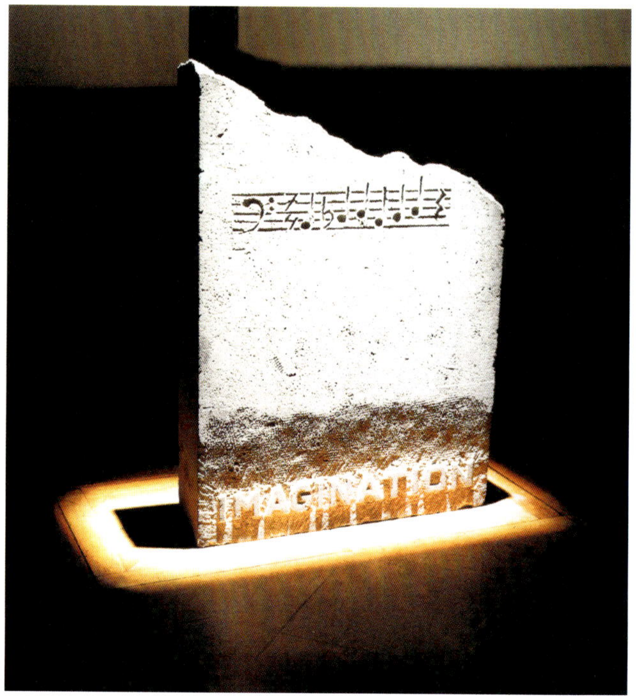

Part of the installation is a sanstone sculpture entitled *Imagination* from 1974.

The first notes of Edgard Varèse’s composition *Arcana* (1927) are carved into the sandstone. The score of *Arcana* bears a quotation from Paracelsus as an epigraph: “There is one star, higher than all the others. This is the star of the apocalypse. The second is the Ascendant. The third is that of the elements which are four in number. There are therefore six firmly established stars. In addition, there is another star, the Imagination, which produces a new star and a new sky.” Varèse himself says: “It is the imagination that gives form to dreams.”

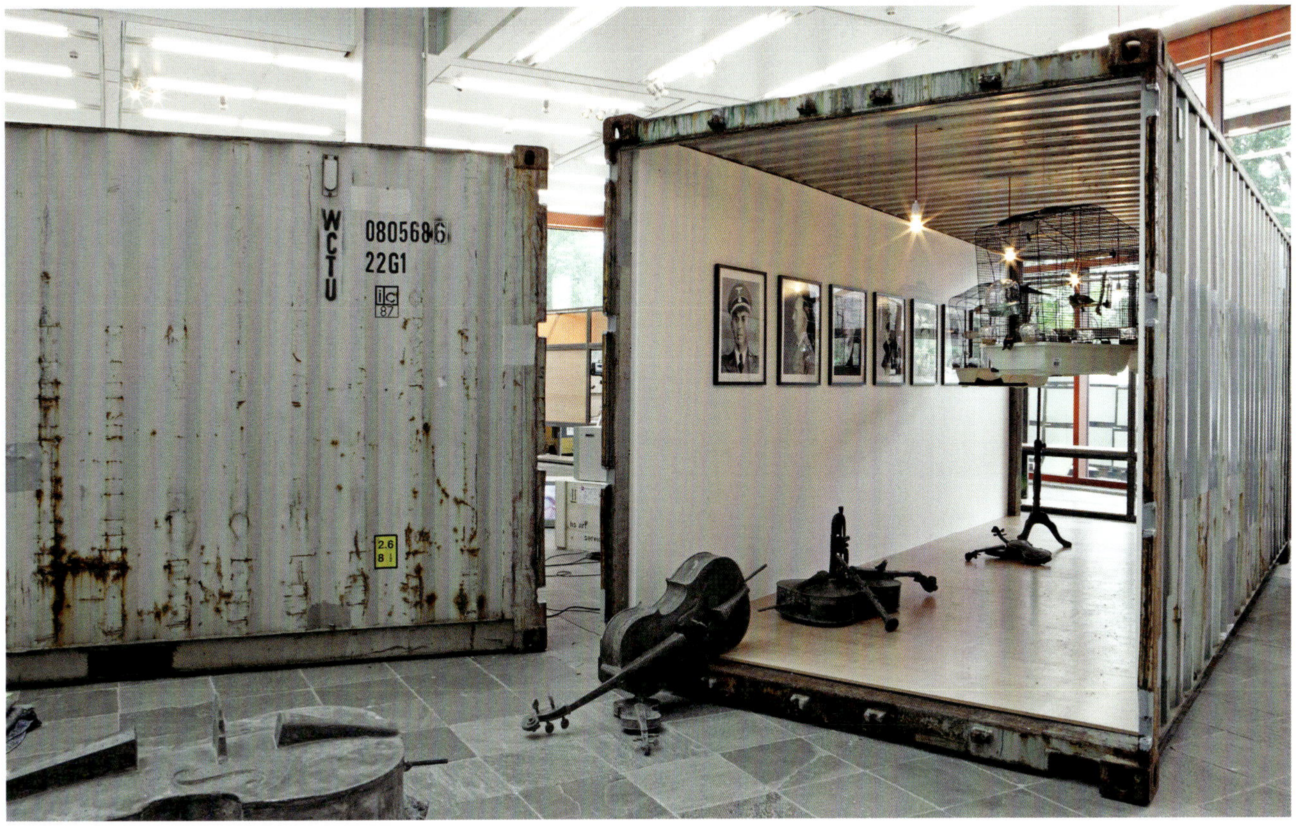

Musikausstellung [Music exhibition], 1975
Installation view, 21er Haus, Vienna, 2014.

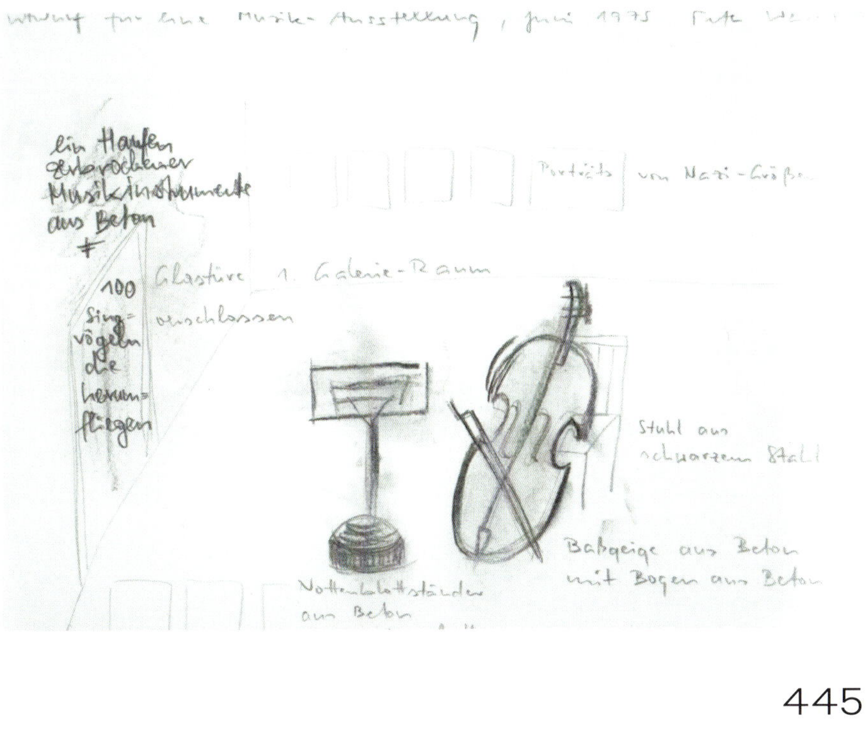

Draft for a music exhibition, 1975
Pencil and crayon on cardboad, 22 × 30 cm, 7 photographs of prominent Nazis, 7 photographs of victims, 100 birds (behind a wall of glas), music stand and violin made of concrete, a heap of broken musical instruments made of concrete, steel chair.

Peter Weibel

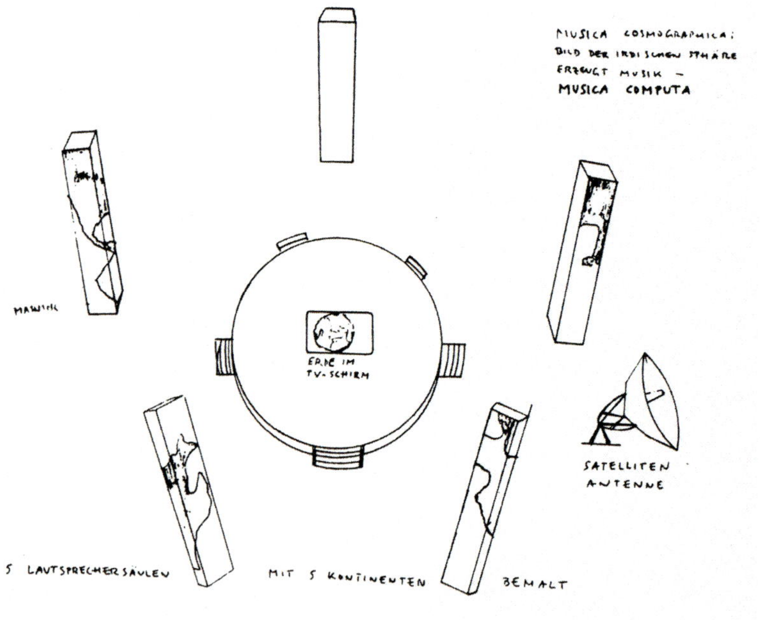

Sketch for *Spektralmusik. Musica Cosmographica [Orbitale Skulptur IV]*, 1987.

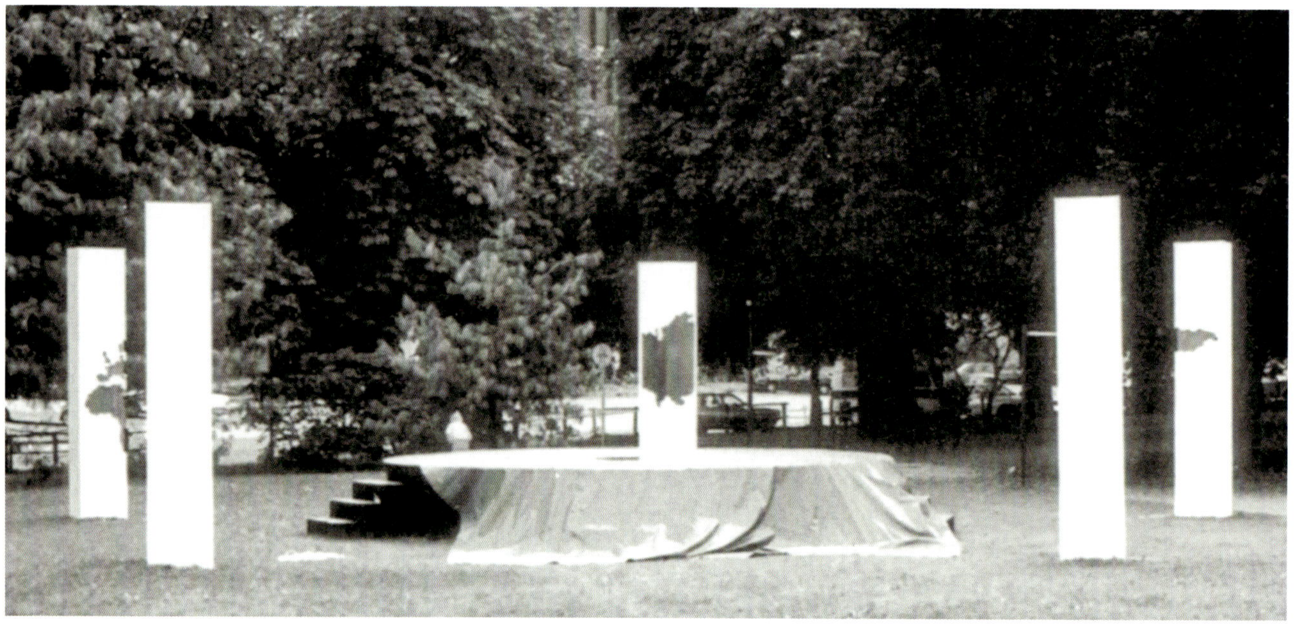

Installation view in front of the Brucknerhaus, Linz. Ars Electronica 1987.

Spektralmusik. Musica Cosmographica [Orbitale Skulptur IV], 1987
Sound installation
Monitor, computer, loudspeakers, satellite
A stage in the form of a circle with a monitor at the center showing live satellite images of the Earth (delivered by the Meteosat weather satellite). 5 columns with loudspeakers with drawings of the 5 continents surround the stage
Technical assistance: Franz Xaver

The sound installation not only draws its material directly from the natural environment, but also from satellites and other telemedia. The images of the atmosphere and the cosmos are converted into sounds, generated and calculated with the aid of a computer. Environmentally specific computer music is created. People enter the stage and look down to the monitor. The monitor shows the earth seen from the orbit. The visitor looks at the earth in a double way: he or she stands with his or her feet on the earth, but his eyes look at the earth via satellite.

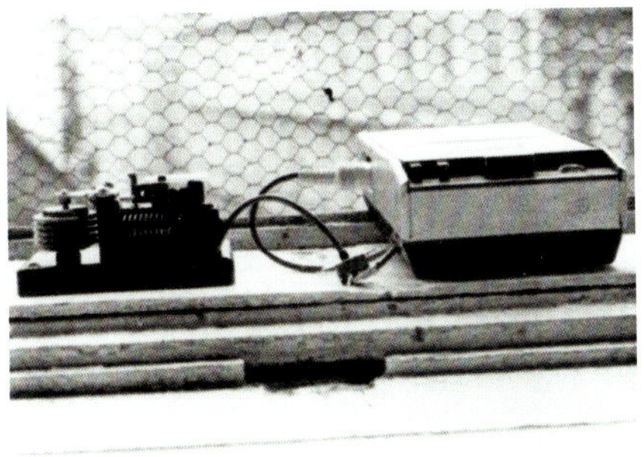

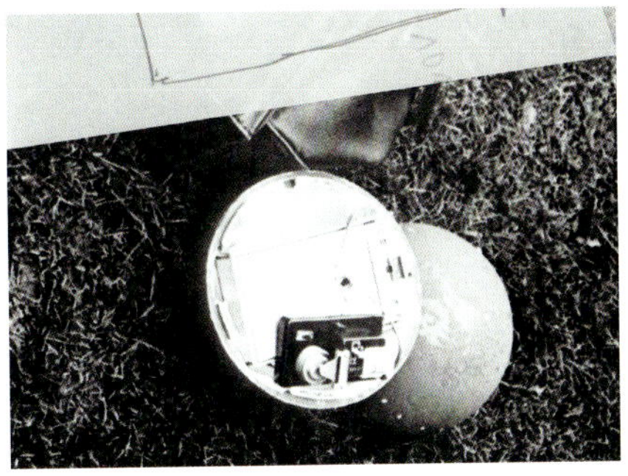

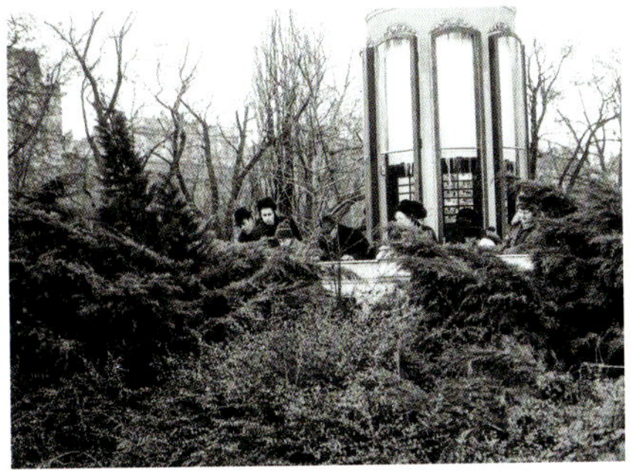

projekt

für eine landschafts-plastik .
viele dieser 'stöhnenden steine' entlang der donau-ufer von wien
nach budapest legen. so erhält man des nachts ein stöhnendes
donau-ufer.

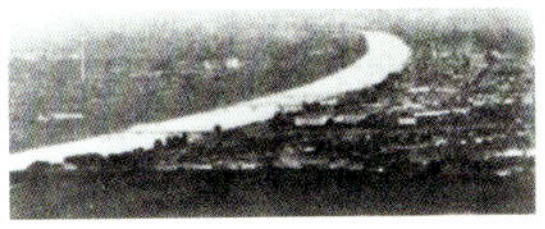

Detail

Stöhnender Stein. Nicht-humanes Gedicht [Moaning stone. Non-human poem], 1969
Sound sculpture

In an artificial stone, I concealed a battery-powered tape recorder with an endless loop of the moaning of a sick person. The tape recorder is connected to an electronic thermal circuit and the expansion or shrinking of gas in a container. The circuit reacts automatically to temperature changes. If the temperature falls, the phosphorus gas cools down: the gas container shrinks. This causes a coil spring – below a certain level at night – to switch on the tape recorder. You hear the stone moan. When the temperature rises again in the morning, the gas and the container expand and the tape recorder is switched off by the coil spring.

I hid the moaning stone in the Vienna Stadtpark. Visitors heard a moan and called the police. I was arrested for "creating a public nuisance." Later, I installed the stone in different places, for example, at the river Danube. I wanted the whole Danube between Vienna and Budapest to moan.

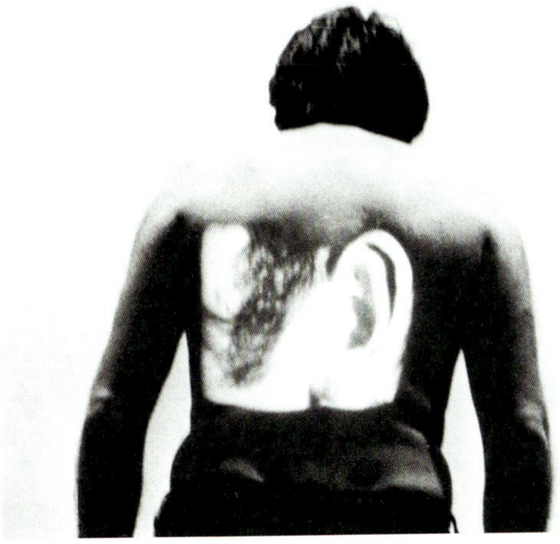

Hör zu [h(ear)], 1968
Expanded cinema, 8 mm, 3 min. each

The audience sees an ear projected on the author's body, texts by the author are read live aloud. The receptive organ is reproduced, the productive organ is live. People listen with their ears, and their eyes see a listening ear: body sound. The act of hearing has been visualized. Variations have been a projected microphone for the live sound of the audience or a projected turning record or a mouth singing.

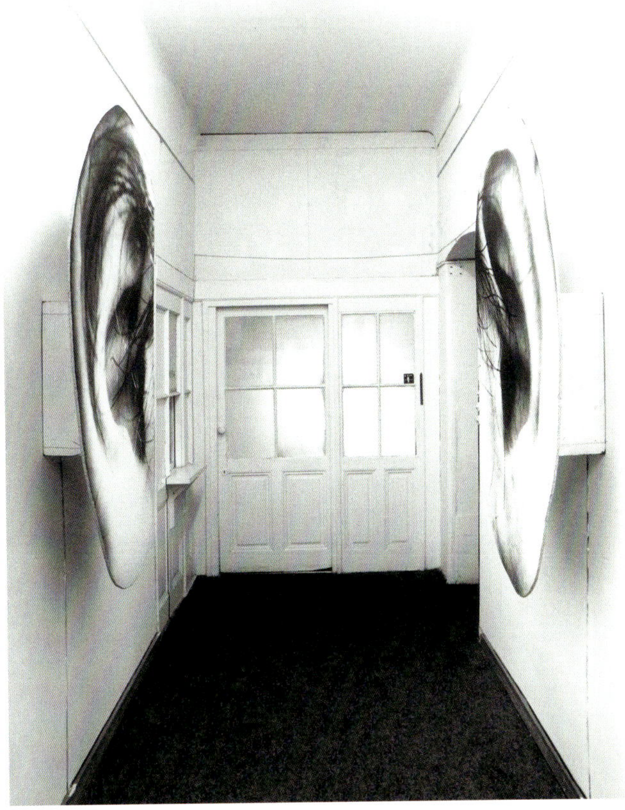

Peter Weibel, *Gehörgang*, 1979
Sound installation.
Installation view Audio Scene '79, Modern Art Galerie

The visitor who stands between the two ears of the installation *Gehörgang* [Auditory passage] hears in his ears how a text (from an endless loop) sounds from the photo ears. The ears, normally receptors, are paradoxically loudspeakers.

John Wynne
Tim Wainwright

*1957 in Zweibrücken (DE), lives in London (GB)
*1954 in Brisbane (AU), †2018 in London

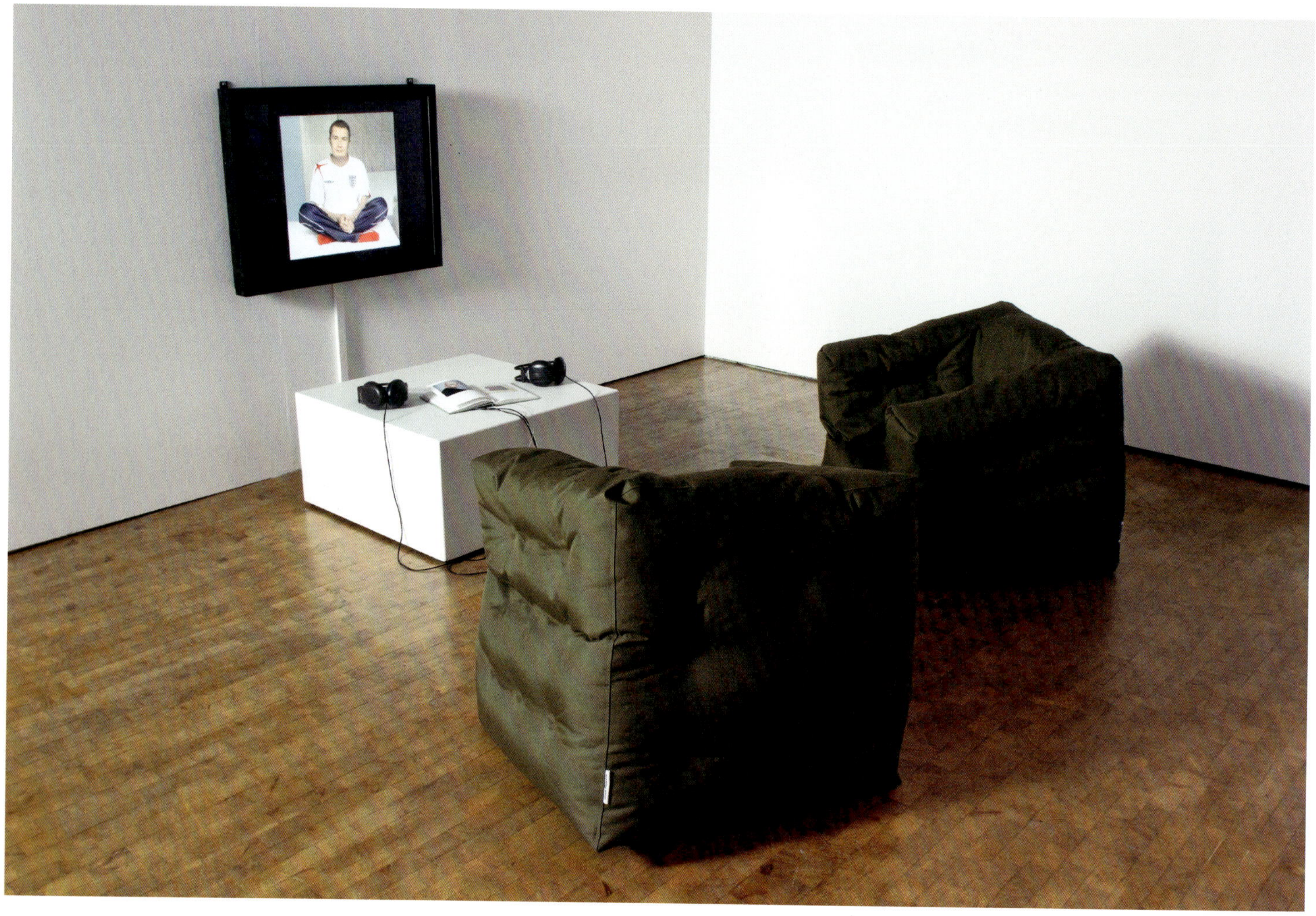

Transplant, 2008
Video with stereo sound,
36:11 min., book
John Wynne and
Tim Wainwright

www.sensitivebrigade.com

The Canadian sound artist John Wynne and the British photographer Tim Wainwright spent a year as artists-in-residence at Harefield Hospital, a world-renowned center for heart and lung transplantation near London. The sound recordings and photographs of patients, medical devices, and hospital rooms made during that time constitute the material for a 24-channel installation, a book, a video, and a half-hour BBC Radio commision. The original audio recordings were somewhat edited; often the editing relates to the patients' comments or observations.

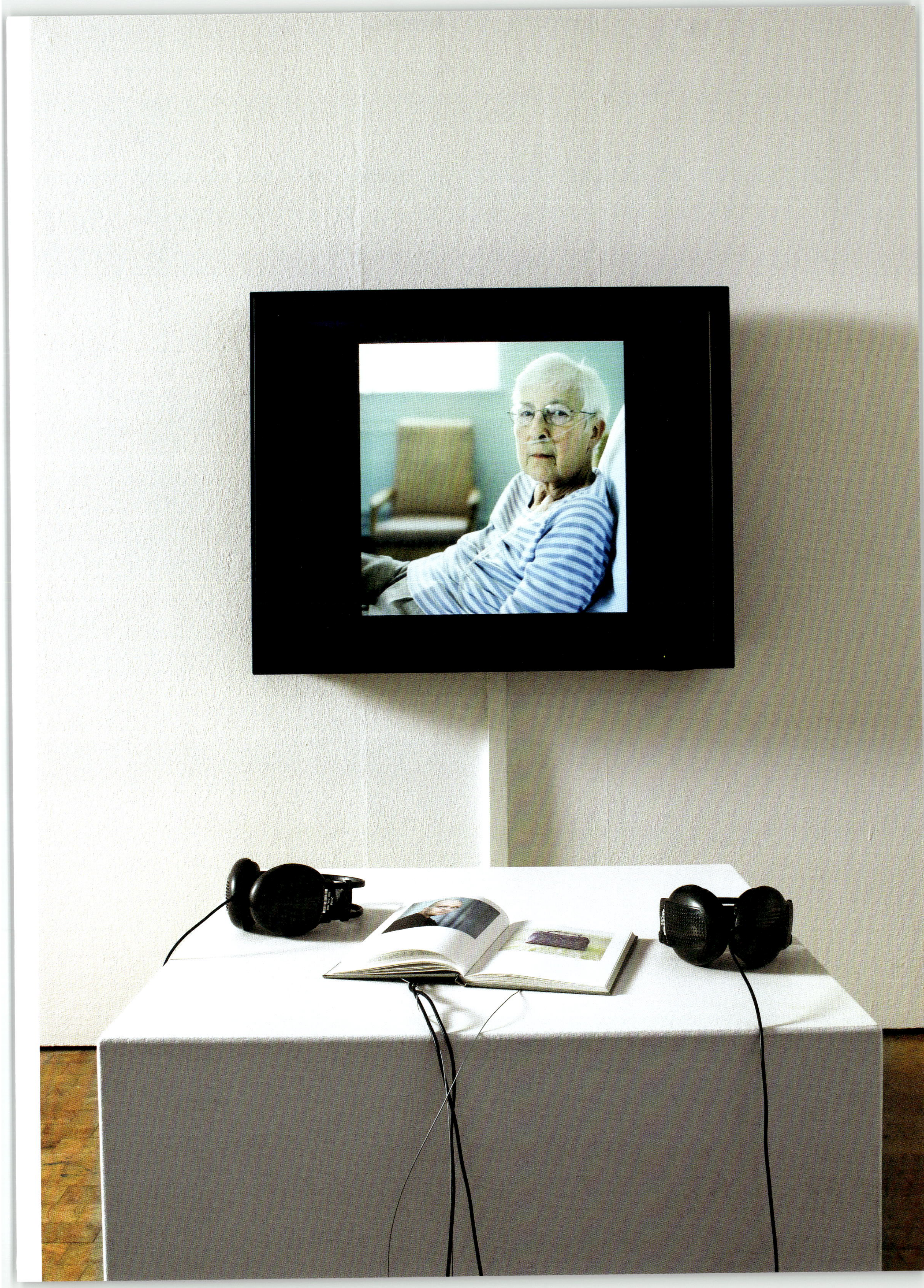

Iannis Xenakis

*1922 in Brăila (RO), †2001 in Paris (FR)

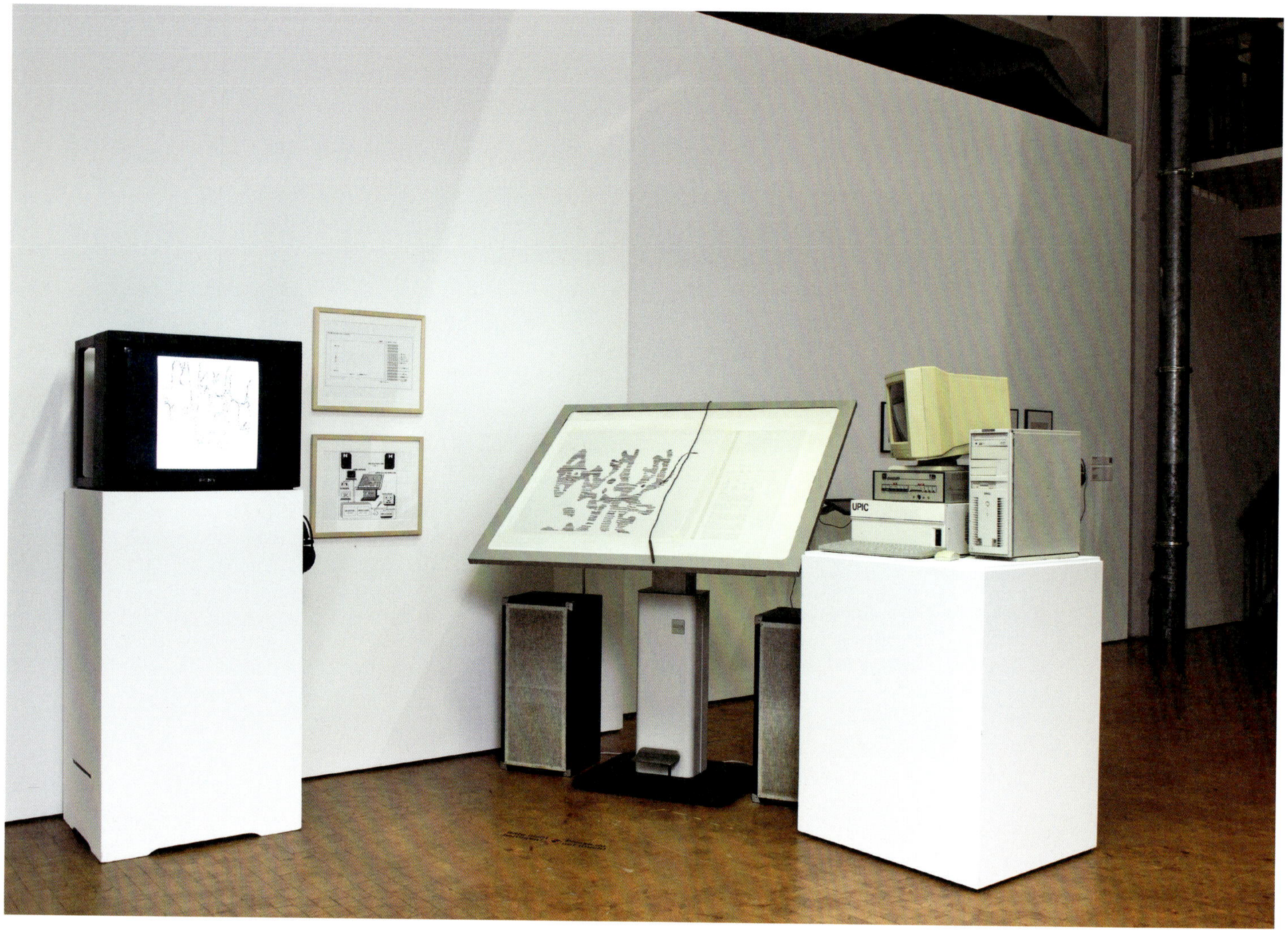

UPIC Space, 2012
Xenakis room at the exhibition
Sound Art at ZKM | Karlsruhe
Sketches by Iannis Xenakis
for *Mycenae Alpha*, UPIC device,
film, compositions, timeline,
dimensions variable
Research and concept: Daniel Teige
Research and texts: Cyrille Deyhaye,
Rudophe Bourotte
Centre Iannis Xenakis, Rouen

www.iannis-xenakis.org

From the 1950s onward, Iannis Xenakis pursued the idea of developing a machine that would enable him to liberate himself from the yoke of traditional musical notation.

"In retrospect I would say that drawings flowed easily from my hand back then: my drawings represent musical symbols. But that was not the way to liberate my thinking. I was convinced that music could be written in a different way as well. I began to imagine sound phenomena by drawing: a spiral, interlocking planes ..." (Iannis Xenakis, 1979)

For Xenakis, agogics embraces the totality of notational parameters that serve musical expression. Starting from that premise, Xenakis wanted to create possibilities for combining structures, sounds, and dynamics in a system of signs. This machine was intended to simplify the research for his new methods of composition. But only in 1977, at the Centre d'Études de Mathématique et Automatique Musicales (CEMAMu), a research center (founded in 1966 by Xenakis, the prototype for a machine was developed that would bring together drawing, sound synthesis, and music. It was named the UPIC (Unitée Polyagogique Informatique du CEMAMu).

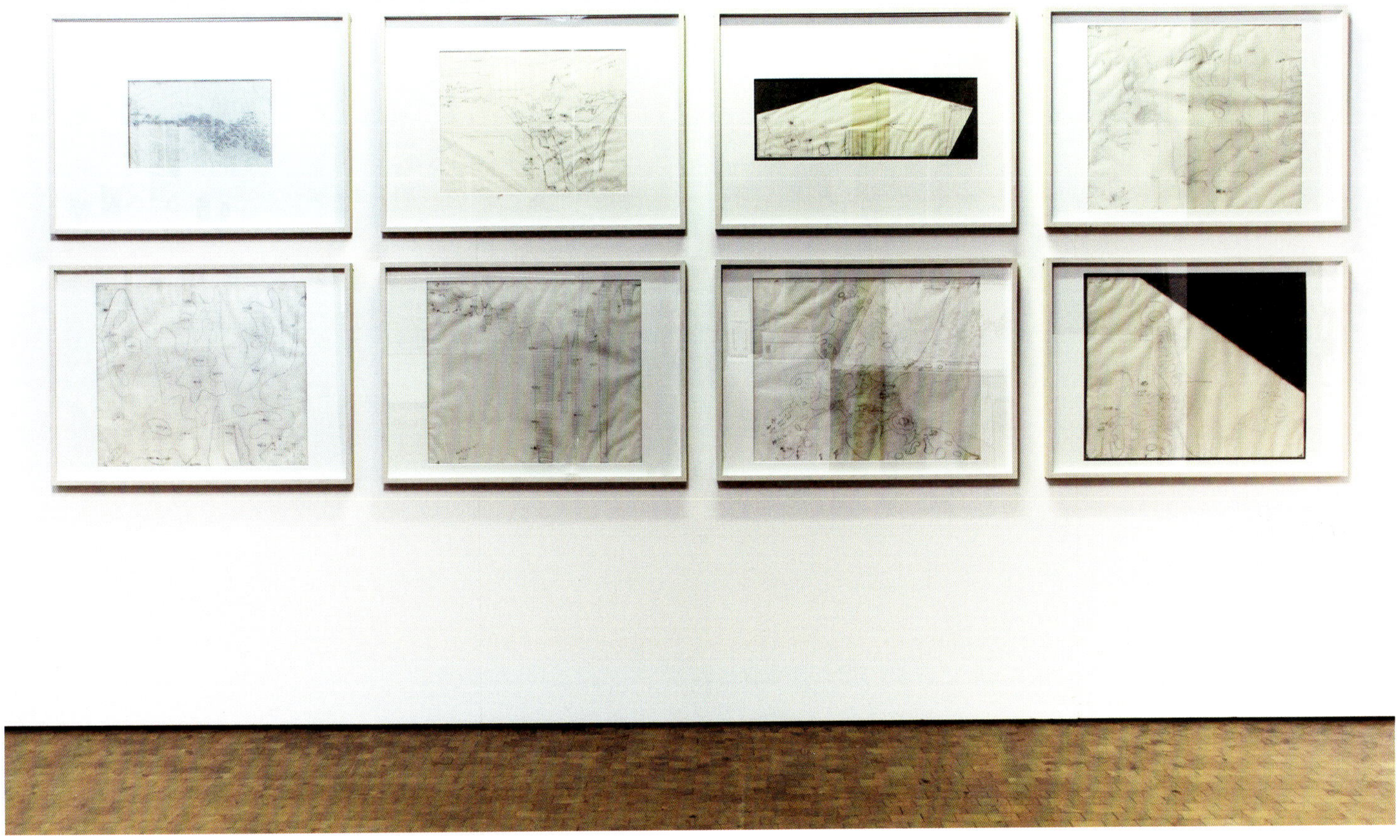

Mycenae Alpha, 1978
8 sketches from the 1970s, paper
Centre Iannis Xenakis, Rouen

Mycenae Alpha was the first work by Iannis Xenakis to be composed entirely with the UPIC. The composition is one part of *Polytope de Mycènes*, which was premiered September 2–5, 1978, in the ruins of the fortress of Mycenae. The synthetic sounds of *Mycenae Alpha* were generated with the aid of the UPIC A prototype developed at the CEMAMu. This polytope, which Xenakis called "a celebration of light, movement, and music," was intended to be an amalgamation of the experiences and achievements of his previous multimedia productions with light, sound, space, and architecture (*Polytope de Montréal*, 1967; *Polytope de Cluny*, 1971 and 1972; *Polytope de Persepolis*, 1971; and *Diatope* 1978).

The music lasts nine minutes and forty-five seconds and consists of eleven pages drawn with the UPIC, one of which is repeated. The exhibited drawings are sketches by Xenakis and are shown here for the first time.

La Monte Young
Marian Zazeela

*1935 in Bern, ID (US), lives in New York City, NY (US)
*1940 in New York City, lives in New York City

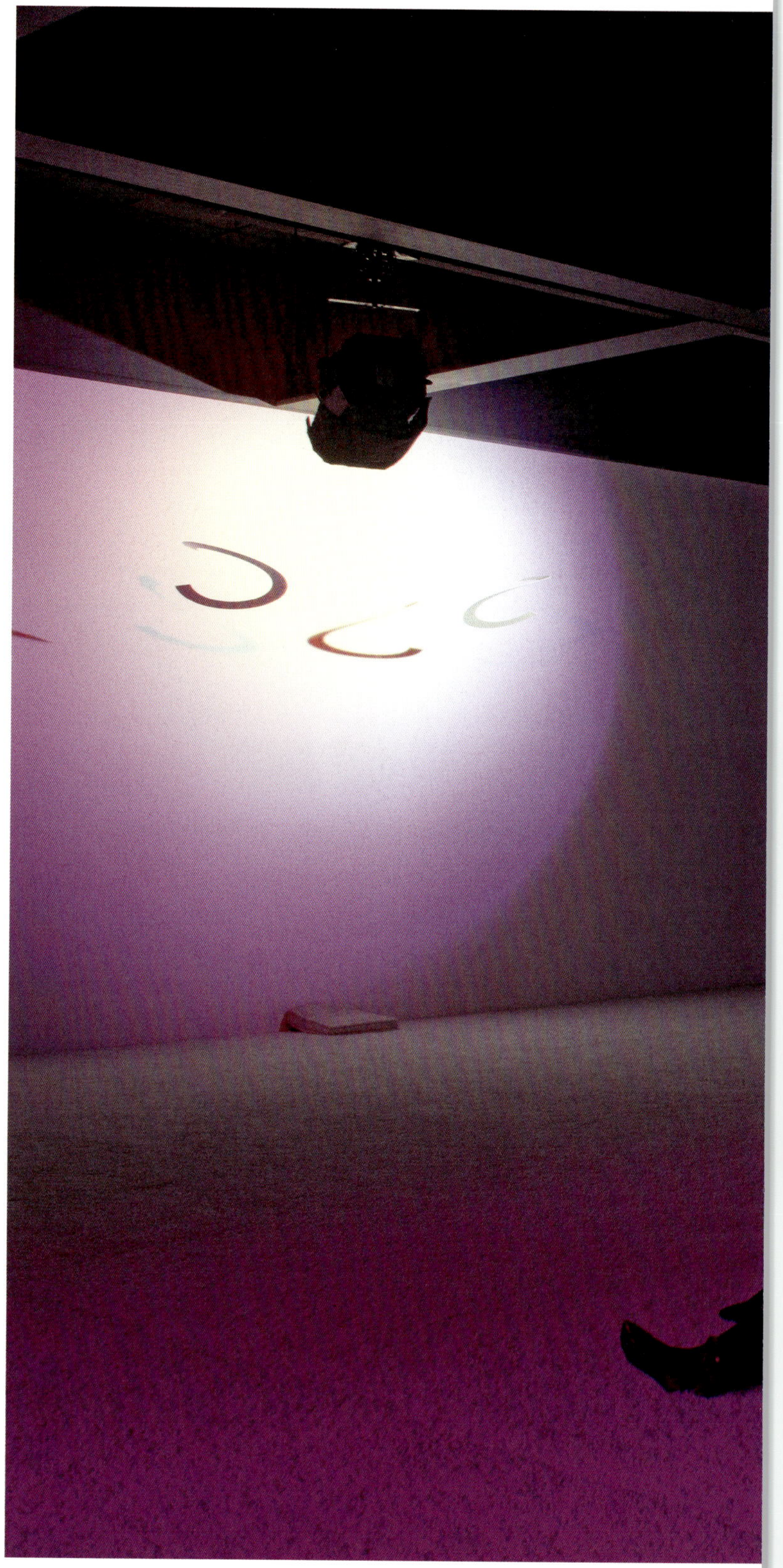

Dream House, 1989
Sound and light environment
Wall sculpture, neon, speaker, light, plastic film
Lighting consultants: Jim Conti, Uli Schaegger
Technical coordination at ZKM | Karlsruhe: Jürgen Galli
Collection du Musée d'art contemporain de Lyon, Regenbogenstadl Polling, La Monte Young, Marian Zazeela

Dream House, a mythical work, is a sound and light environment created by the composer La Monte Young and his wife the artist Marian Zazeela.

As early as 1963, the two conceived their first light and sound installation, which they called *Dream House*. Marian Zazeela devised a system of evolutive and colored lights that she placed on mobiles. La Monte Young used various sine wave oscillators, oscilloscopes, amplifiers, and loudspeakers to create continuous frequency environments. The suspended mobiles react to the music played, which consists of held notes that can be extended indefinitely, in an extremely minimal way. The first installation of *Dream House* at an arts venue took place in July 1969 at Galerie Friedrich in Munich, and many more followed.

Visitors entering the light and music flooded environment are invited to immerse themselves in the sound and color. Whether standing, sitting, or lying, motionless or moving in their own rhythm within the space, every visitor can enjoy the sound modulations caused by their own movements, however small these may be. This unique experience stimulates introspection, meditation, and dreaming. The viewer/listener sees the music as well as hearing the color, in pure synaesthetic perception. Time seems to slow down and the relationship to space and duration changes completely between the real and the imaginary.

www.melafoundation.org

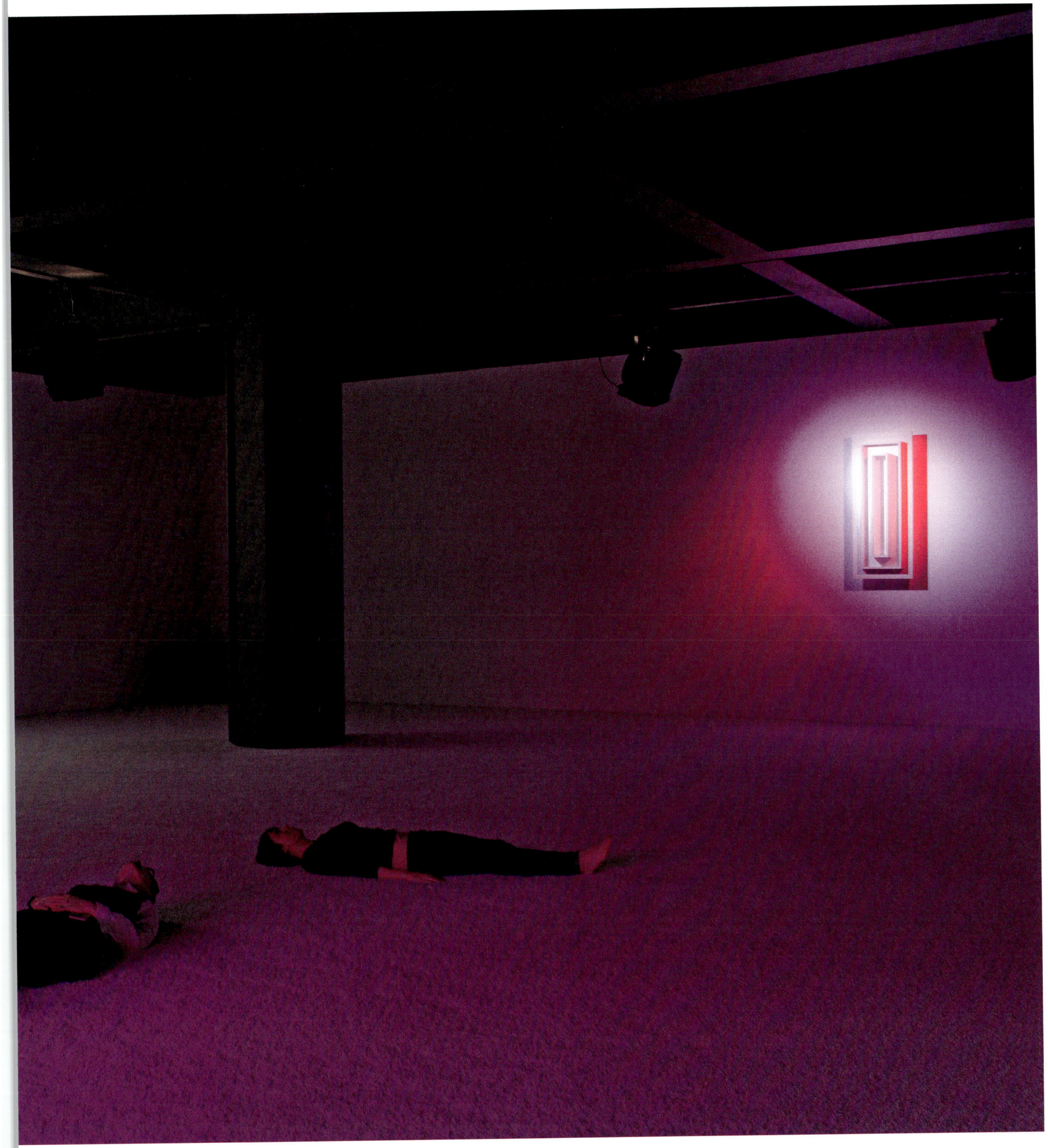

La Monte Young
Marian Zazeela

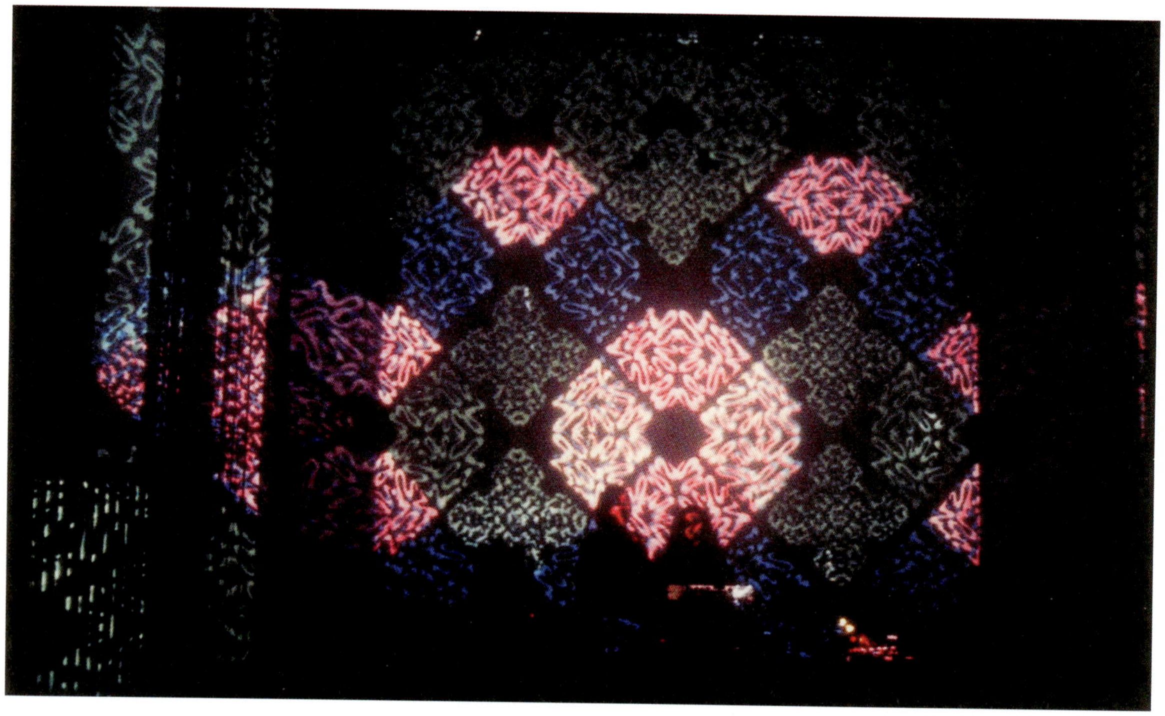

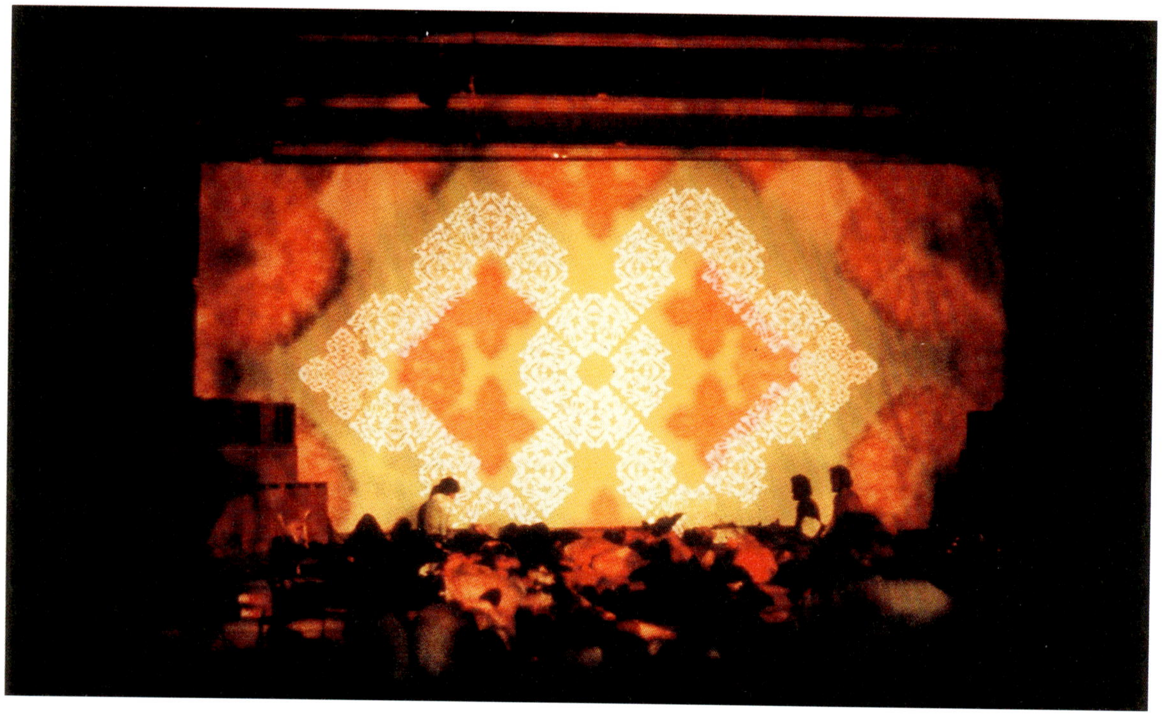

top:
Dream House "Map of 49's Dream The Two Systems of Eleven Sets of Galactic Intervals Ornamental Lightyears Tracery," 1969
Installation view Gallery A 37 90 89, Anvers, Belgium.

bottom:
Ornamental Lightyears Tracery, 1970
Sound and light concert by Marian Zazeela at Moderna Museet, Stockholm, Sweden
Multi-projections of superimposed slides with recordings of La Monte Young.

La Monte Young

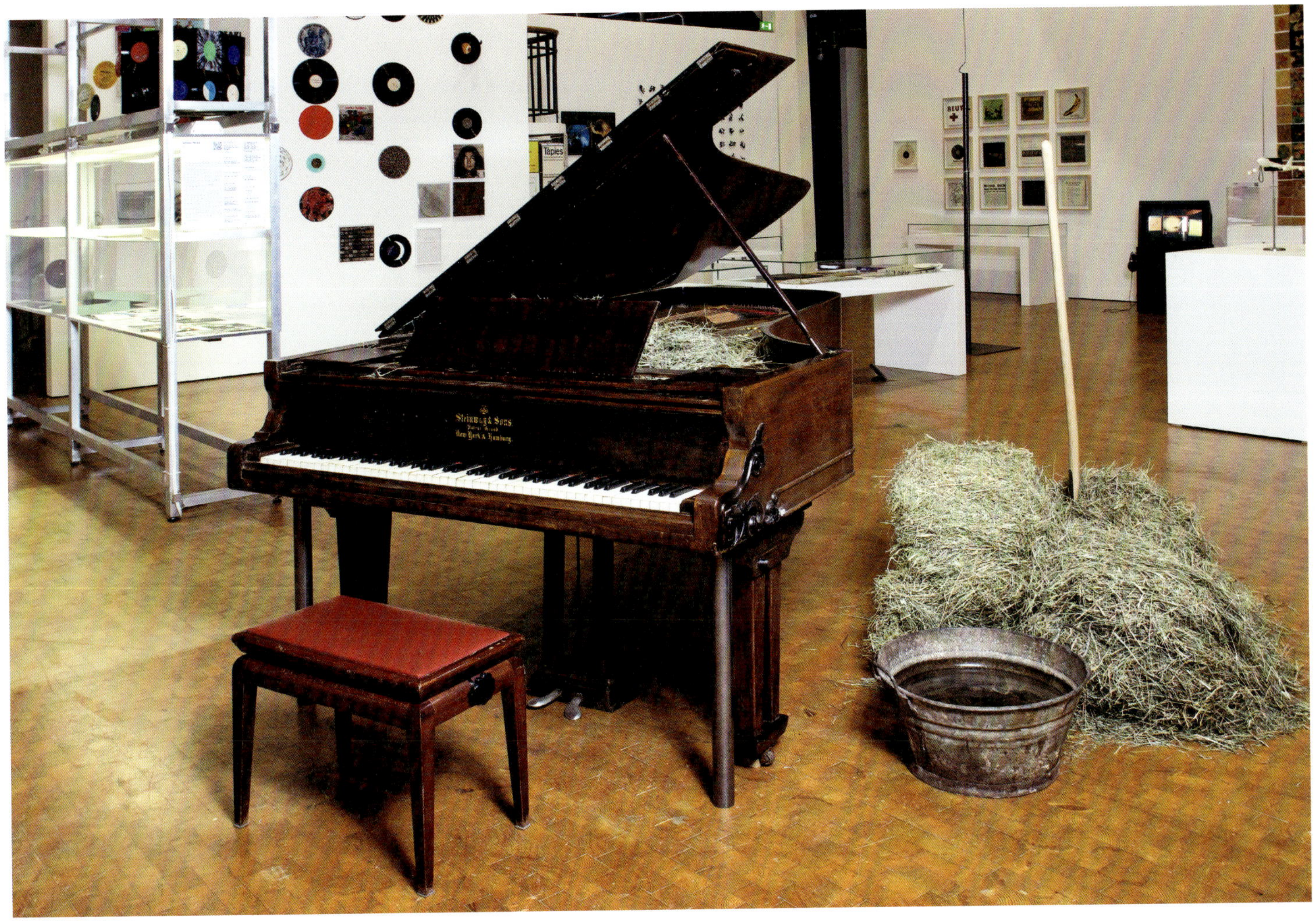

Piano Piece for David Tudor #1,
1960/1989/1990
Object
Grand piano, hay
FONDAZIONE MUDIMA, Milan

In this highly theatrical text composition, the performer – in this case, celebrated experimental pianist David Tudor – is asked to "[b]ring a bale of hay and a bucket of water onto the stage for the piano to eat and drink. The performer may then feed the piano or leave it to eat by itself. If the performer, the piece is over after the piano has been fed. If the latter, it is over after the piano eats or decides not to." Depending on the choices of the performer, a performance of the piece can move from something comedic into something highly conceptual, with the audience waiting for a sign from the piano that it is finished eating. (John Kannenberg)

MEDIA TECHNOLOGY IN THEORY AND PRACTICE

Siegfried Zielinski

Lüology, Techno-Souls, Artificial Paradises:[1] Fragments of an An-Archaeology of Sound Arts

Introduction

Sound art relates to phenomena, events, processes, and installations that produce acoustic material for intellectual and sensual enjoyment with the aid of or through constructed artifacts. Genealogies of sound art that conforms to this definition, as τέχνη *(téchne)* in the broadest sense of the Greek term, including craftsmanship, are frequently written in such a way that they have virtually nothing to do with the complex genesis of artificial worlds of sound. Analogous to the visual arts, in which the mathematization of the image and the geometrization of the gaze began during the second, that is, European Renaissance (with occasional references to the mathematically unsophisticated Greek and Hellenistic tradition[2]), the mechanization and automatization of the arts of sound commenced in the High Middle Ages, or early modern period, respectively (with occasional references to the Pythagorean, that is, Greek and Hellenistic tradition). Curtis Roads, for example, in his reconstruction of the automation of music, which he refers to as "informed music," naturally includes some of the Greek concepts dating from the second century BCE, but then he jumps immediately to the thirteenth-century mechanical carillons from the Netherlands.[3]

Translated from the German
by Gloria Custance.

In his fascinating conceptual discussion of composing experimental music, Christoph von Blumröder does stress the fundamental link between composition and scientific research. However, for the author its history only starts to get interesting when the word *experimentum* appears in Latin medieval treatises in Europe, and only becomes truly relevant for practice and theory with Francis Bacon's operational definition of the word "experiment" as "sought-after experience" in *Novum Organum*,[4] published in 1620. And the legendary 1980 exhibition of sound art objects curated by René Block and organized by the Berlin Academy of the Arts in collaboration with the DAAD artists' program and the Berlin Musical Instruments Museum of the State Institute for Music Research (SIMPK) prefaced its fascinating displays with a genealogical diagram which featured only Western dates and names.(FIG. 1)

The introduction to the catalogue did at least carry an illustration of an automaton for telling the time, which also produced sounds, attributed to the twelfth-century Kurdish engineer al-Jazarī.[5] The automaton pictured had in fact been designed and built three hundred years before that, in the early ninth century, at the Baghdad court of the ruling dynasty; it was subsequently updated in al-Jazarī's mechanical encyclopedia. In 806 the magnificent horologium of Caliph Hārūn al-Rashīd was loaded on board a ship at the port of Sidon (now in Lebanon) that narrowly escaped being attacked by a Byzantine fleet while sailing from the eastern Mediterranean to the Adriatic; the gift clock was offloaded at Venice, and then taken via Treviso over the Alps to its destination, Aachen, the court of Charlemagne, where it arrived in 807. Four elephants were sent along with the clock, but unfortunately only one survived

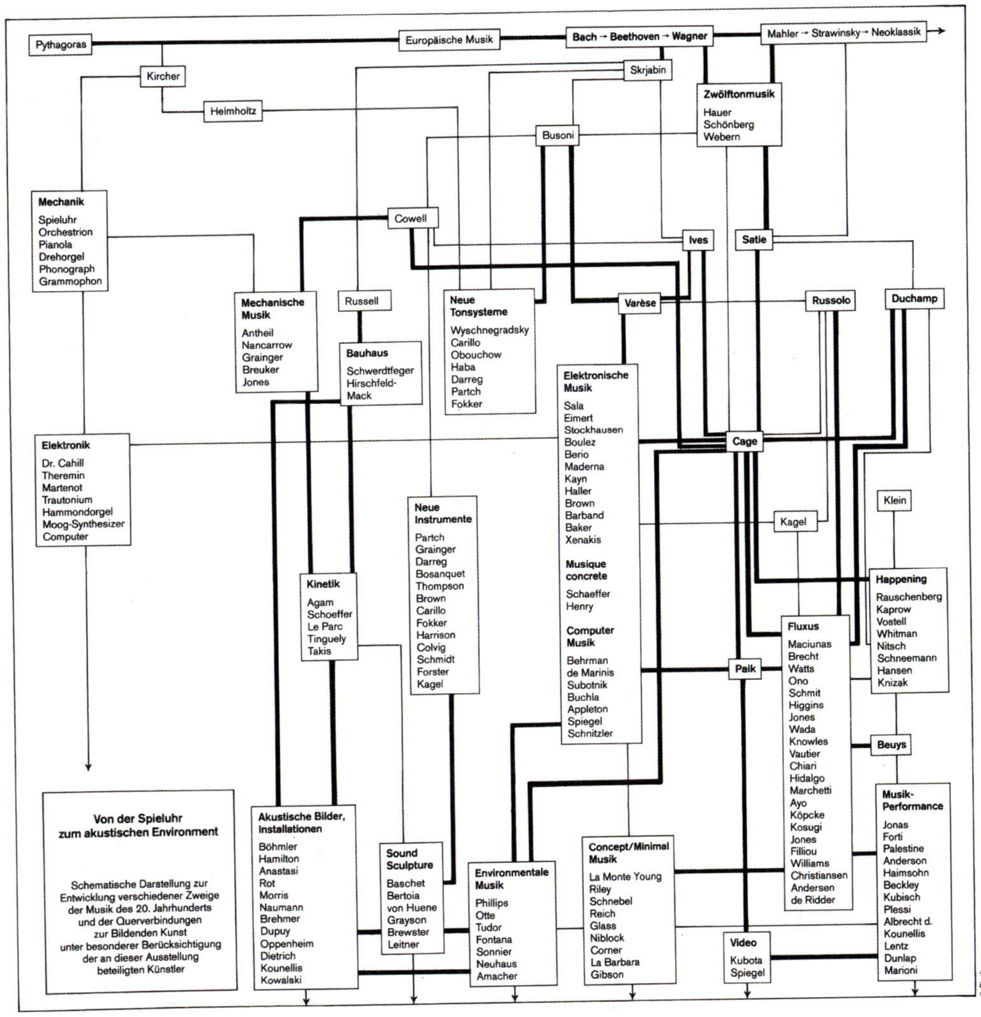

FIG. 1 Development scheme of the different branches of music in the twentieth century and interconnections to visual arts. Opening diagram from *Für Augen und Ohren – Von der Spieluhr zum akustischen Environment. Objekte Installationen Performances*, exhib. cat. (Berlin: Akademie der Künste, 1980), 6.

1
The term is a reference to Charles Baudelaire's famous text *Artificial Paradises* of 1860, which can also be applied to heavenly machines that produce sounds. Cf. Charles Baudelaire, *Les Paradis artificiels: Opium et Haschisch* (Paris: Libraires – Éditeurs, 1860).

2
On the much older Egyptian theory of harmonious proportions and the corresponding tools, see the works by Albert Presas i Puig, especially *Numbers, Proportions, Harmonies, and Practical Geometry in Ancient Art*, Preprint 262 (Berlin: Max Planck Institute for the History of Science, 2004).

3
See Curtis Roads, "Research in Music and Artificial Intelligence," *Computing Surveys* 17, no. 2 (1985): 163–90.

4
Christoph von Blumröder, "Experiment – Experimentelle Musik," *Feedback Papers* 29 (July 1982): 1–24, here 1, 2. Translated here from the German.

5
See René Block et al., *Für Augen und Ohren: Von der Spieluhr zum akustischen Environment. Objekte Installationen Performances*, exhib. cat. Musikinstrumenten-Museum des Instituts für Musikforschung Preussischer Kulturbesitz (Berlin: Akademie der Künste, 1980), 6, 7.

the journey, whose name was Abul Abas. Legend has it that his tusks were later used in Aachen to make the keys of keyboard instruments. Baghdad did not really expect any gifts in return from the Northern barbarians that would be anywhere near as magnificent and accurate as the Caliph's present.[6] And it was naturally a water clock that Hārūn al-Rashīd sent to Aachen and not a solar-powered one, because they assumed in Baghdad that the sun would not shine enough in the northwest for the horologium to work properly.

The following fragments are connected with three genealogically oriented projects which intersect variously with the sound arts: (1) With the concept of "expanded animation," I seek to develop a concept of animation for the present-days of the past and future, which is entirely focused on movement, ensoulment via technology, and thus on the time-based dimension of this cultural technique. In this effort, sound- and image-producing machines play a preeminent role. (2) The first renaissance of the knowledge of European classical antiquity did not take place in Italy and Europe, but in Persia and the Arab region. The translation of Ancient Greek manuscripts and further development of the ideas they contained by Muslim scholars, particularly in the half-millennium between the eighth and thirteenth centuries – also referred to as the Golden Age of Islamic culture – gave us access to important worlds of knowledge about technology and the arts. Study of the deep time relations between arts, (3) sciences, and technologies, which we call *variantology* and have been working on for some time now, has uncovered various significant challenges in the Near and Far East. In a few years' time I want to conclude this research with an international symposium on the winds and the necessity of learning to read them again. Ideally, it will take place in Shanghai or Hong Kong.

This highly diverse research is held together by an idea which is also important for sound art. With the invention of *modernity*, European culture declared itself the center of the universe. What was considered old-fashioned, primitive, passé, or innovative, complex, progressive in the arts, sciences, and technology was defined from the perspective of the self-appointed avant-gardes operating between Rome, Paris, London, and, intermittently, St. Petersburg. Modernity and the modern became paradigmatic postulates in the battle between cultures, religions, and nations, which at the beginning of the third millennium is revealed once again to be primarily a battle for economic and ideological hegemony.

From the perspective of the deep time relations between arts, sciences, and technologies, it is above all a certain attitude toward the world that I understand as *modern*. The hallmark of this mindset is an experimental relationship to everything that surrounds the individual, and in the midst of which he or she lives; it is not a relationship characterized by testing, appropriating, or even exploiting. The world is understood as something that can be changed, in the same way that, from the perspective of others, any individual is capable of changing or being changed. In an ideal scenario both the individual and the surrounding environment will be changed, to their mutual advantage. Modern attitudes of this description are found in various parts of the world in very different epochs. They are not confined to European countries in the period from the fifteenth to the eighteenth century. And in the vanishing point of the diverse inquiries, a space of possibilities opens for which we potentially do not need the hegemonial concept of modernity any longer.

6 See Dzevad Karahasan, "Hārūn al-Rashīd schenkt Karl dem Großen einen Elefanten," *Trajekte – Zeitschrift des Zentrums für Literatur- und Kulturforschung Berlin*, no. 19 (2009): 36–40. On the clock see particularly Ulrich Alertz, "The Horologium of Hārūn al-Rashīd Presented to Charlemagne: An Attempt to Identify and Reconstruct the Clock Using the Instructions Given by al-Jazarī," in *Variantology 4, On Deep Time Relations of Arts, Sciences and Technologies in the Arabic-Islamic World and Beyond*, eds. Siegfried Zielinski and Eckhard Fürlus (Cologne: Walther König, 2010), 19–42.

Lüology

The school of the Pythagoreans was a hard one and had more prohibitions than wishes or desires. It is said that after rising in the morning, members were required to smooth the sheets of their beds so that all traces of their bodies from the time they were sleeping and dreaming – and thus time that was not subject to discipline or mathematization – were no longer visible. For the Pythagoreans music expressed cosmic and divine harmony, which human earthlings must attempt to translate into harmonious proportions perceptible to the senses. Symbolic values do not produce sounds. The medium of sound was understood as numbers, *numerus sonorus*, as it was later termed in the Baroque theory of harmony: the musical tones of numbers, or rather numbers that have been made to produce sounds. The same numerical ratios that determine the distances of the planets to the central fire of the Earth, and the organs of the body to the *anima mundi*, the soul, also applied to the discrete notes of the diatonic scale and vice versa. The freemason Robert Fludd, philosopher and physician, as such responsible for the health of the body, charged Theodor De Bry in Oppenheim near Frankfurt with the production of impressive engravings of this genesis of the macrocosm and microcosm.(FIGS. 2, 3) Neo-Platonism, which resonates in this unity of the great and the small, we shall encounter again in the course of this text in connection with automata in the Muslim tradition.

Until the early modern period, music as a discipline was subordinate to mathematics. The faculty for music theory and practice was disdainfully referred to as *scientia subordinata*. The vibrations of tones and the intervals were subordinated to the geometry of ratios and arithmetic. The Atomists, such as the Greek natural philosopher Democritus, who described vibration as the only real form of movement, were largely ignored for a very long time by the discipline of music theory. Amongst others, Marin Mersenne took up their ideas again in the seventeenth century; he understood frequency not as measures of length, but as oscillations, that is, as a temporal phenomenon.

The history of Greek-Hellenistic hegemony in founding European arts of sound is well-known and has been described by many commentators, from Adorno and Xenakis ("we are all Pythagoreans more or less") to contemporary media theorists (Friedrich Kittler). A part of this history – in the tradition of the Pythagoreans and the preeminence of the *mathematikoi* – is the school of Aristoxenus. A pupil of Aristotle, he wrote *Elements of Harmony* in the fourth century BCE, the first great musical treatise. Aristoxenus rejects the idea that numbers are the determining element in harmony. Only in the second instance can they formulate the harmony produced by experience, by the hearing and musical intuition: "using calculations to create intervals that no voice and no instrument could produce and that the ear could not identify"[7]

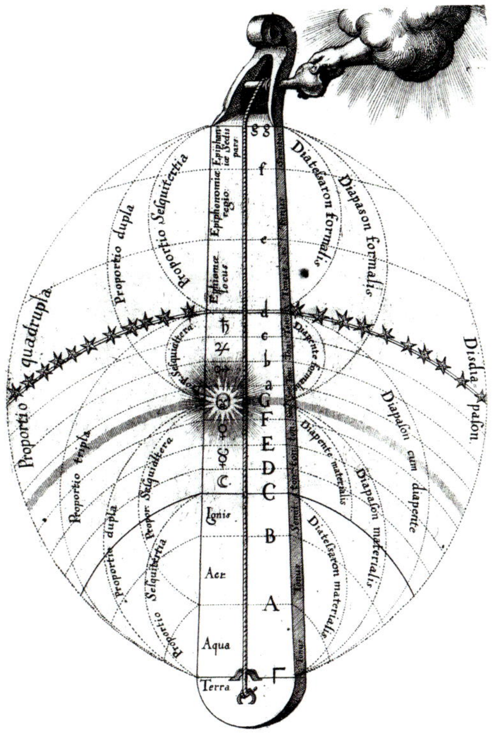

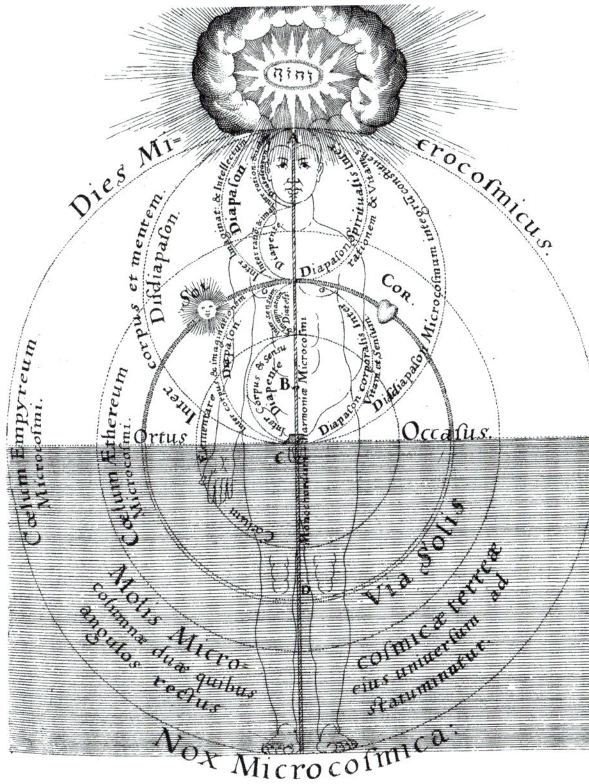

FIGS. 2, 3 The great cosmos (above) as Monochord, carefully tuned by the hand of God, the small cosmos on Earth with the same proportions/intervals – a completely harmoniously tempered relationship. Illustrations from Robert Fludd, *Utriusque Cosmi Maioris scilicet et Minoris, Metap(h)ysica, Physica atque Technica Historia*; Tom. 1, Tract. 1 (Oppenheim: de Bry, 1617), 90 (left), and Tom. 2, Tract. 1 (Oppenheim: de Bry, 1629 [i.e. 1619]), 275 (right). Sächsische Landesbibliothek – Staats- und Universitätsbibliothek Dresden, Sig. 1.B.3237-1 and 1.B.3237-2.

makes no sense at all. Archaeologies of science and technology offer other topics and stories. But for these we must leave the sedate regions around the *mare internum*, the Mediterranean, and cross the ocean or the great deserts of Asia to China.

In his magisterial study of science and civilization in China, Joseph Needham emphasizes that in Ancient China the cultural technique of tuning was primarily oriented on physical sensations and phenomena of movement – activity.[8] Analogous to taste and color, sound is understood as an active experience. In Ancient China various high tones and their ratios to each other were investigated, for example, using the sounds produced by bubbling cooking pots. According to how deep and wide the pots were, and how much water was brought to the boil in them, they produced different sounds. The steam, the energy that was released during boiling and that pushed up the lid with a hissing noise, was termed onomatopoetically by the Chinese as *Qi*, which is undoubtedly one of the most important categories in Chinese natural philosophy. It is safe to say that *Qi* was and is extremely complex.[9] *Qi* stands for a variety of very heterogeneous concepts, such as breathing, temperament, the principle of life, as well as for offering a meal. Applied to music, according to Needham a particular form of pneumatic action is articulated in *Qi*, which later played a very important role in early music automata which utilized air pressure.[10]

Singing requires pneumatics-based actions. Raising the voice, the specific way a note is intoned, is the most active part of using the voice, and expresses most clearly the uniqueness and thus the *kairos* nature of the voice. "Kairos" means the auspicious moment, the time we use, or rather how we use the time, how we decide to use it. Articulating a vowel, for example, pulls apart the vocal folds in the larynx, and generates a short explosive sound. This is clearly to be avoided at all costs when using a sensitive microphone. To articulate a vowel softly, the vocal folds are only pushed apart to the extent necessary to make the desired sound and then they bounce back. In this case, the vocal sound is preceded by a very soft sound, which the Czech experimental physiologist Jan Evangelista Purkinjě discovered in his Breslau (Wroclaw) laboratory in 1836, and which he called a "lekki chuch" [gentle whisper].[11]

With regard to the steaming cooking pots, an analogy occurs to me from media archaeology which concerns an archaic instrument featured in the Bible. When Jacob flees after betraying his twin brother and his blind father, he lies down at night to rest with his head on a stone and sees a ladder to heaven with angels ascending – like the steam from a cooking pot. This image of Jacob's ladder obviously can also be interpreted as an acoustic scale. In the parable about a possible ascent to the higher spheres, Jacob did not see the ladder in his dream, he heard it (or perhaps both). Led Zeppelin memorialized this idea impressively in rock music with their "Stairway to Heaven."

Xu Fei, eminent physicist, archaeologist, and musicologist of the deep time of Chinese knowledge cultures, views the development of standard pitch instruments for tuning as an important indication of the success and degree of humans combining arts, science, and technology. In his essay "Exploration and Achievements of the Acoustics of *Yuelü*,"[12] he refers particularly to the brilliant studies on musical tuning methods by the mathematician and acoustics researcher Chen Cheng-Yih, who in his book *Early Chinese Work in Natural Science* (1996) retains the lovely Chinese word *Lüxue* (律学), which literally means the study of *Lü* 律 [pitches], that is the study of the ritual tone system, and which is often translated as "temperament." As Chen considers this an inadequate translation, he suggests better not to translate the term and

7
See the excellent essay on "Harmonics" by Annie Bélis in Jacques Brunschwig and Geoffrey E. R. Lloyd, ed., *The Greek Pursuit of Knowledge* (Cambridge, MA: Harvard University Press, 2003), 161–76, here 170.

8
Remarkably, J. W. Ritter, the Silesian physico-chemist, used this idea and the same term when he developed his operational anthropology of the arts around 1800. Human activity for Ritter is the essential reference for the four stages within which the relationship between art and science develops. Sound as art within time is for him the ultimate perfection of activity. See my short essay: "Im Zustand der Schwingung kann es keine Ruhe geben: Ein kurzes Portrait des Physikochemikers Johann Wilhelm Ritter (1776–1810). Peter Weibel zum Geburtstag," in *05-03-04: Liebesgrüsse aus Odessa. Für/For/à Peter Weibel*, ed. Ecke Bonk, Peter Gente, and Margit Rosen (Berlin: Merve, 2004), 8–12.

9
On *Qi* see, for example, Chen (Joseph) Cheng-Yih, "Cultural Diversities: Complementarity in Opposites," in *Variantology 3: On Deep Time Relations of Arts, Sciences and Technologies in China and Elsewhere*, ed. Siegfried Zielinski and Eckhard Fürlus in cooperation with Nadine Minkwitz (Cologne: Walther König, 2008), 153–88; Dagmar Schäfer, "*Ganying*–Resonance in Seventeenth-century China: The Examples of Wang Fuzhi 王夫之 (1609–1696) and Song Yingxing 宋應星 (1589–ca. 1666)," in *Variantology 3*, 225–54.

10
See Joseph Needham, *Science and Civilisation in China. Volume 4: Physics and Physical Technology, Part I: Physics* (Cambridge: Cambridge University Press, 1962), 132–33. For a comparison of the Greek and Chinese traditions see also: Paul Feigelfeld, "Chinese Whispers. Übertragung und Übersetzung: Die stille Post des Wissens zwischen China und Europa 1500–1700" (MA thesis, Humboldt University Berlin, Faculty of Humanities and Social Sciences, 2008), esp. ch. 4, 80–89.

11
On Purkinjě's experimental work see my *Deep Time of the Media – Toward an Archaeology of Hearing and Seeing by Technical Means* (Cambridge, MA: The MIT Press, 2006), 196–203.

12
See Xu Fei, "Exploration and Achievements of the Acoustics of *Yuelü*, the Theory of the Tone System in Ancient China," in *Variantology 3* (see note 9), 257–94.

13
Chen Cheng-Yih, *Early Chinese Work in Natural Science: A Re-examination of the Physics of Motion, Acoustics, Astronomy and Scientific Thoughts* (Hong Kong: Hong Kong University Press, 1996), xxxii. Chen Cheng-Yih's most important study on this thematic complex is *Two-Tone Set-Bells of Marquis Yí* (1988), published by the Húbei Provincial Museum Wuhàn and the University of California.

14
Ibid., 260. During a workshop held in 2005 at the Academy of Media Arts Cologne, Xu Fei played a replica of a Gudi flute. The flute is held vertically and slightly tilted and produces a range of tones from powerful to muted. That it is possible to play music on an eight-thousand-year-old instrument which we recognize as modern, was for me one of the most impressive experiences of researching the deep time of media archaeology.

15
Ibid., 261.

16
Ibid., 258.

17
See Siegfried Zielinski, "Expanded Animation: A Short Genealogy in Text and Images," in *Pervasive Animation*, ed. Suzanne Buchan (London: Routledge, 2013), 25–51.

instead speak of "Lüology" to refer to the ancient Chinese methods of tuning as an important "branch of early science in Chinese civilization."[13]

From surviving ancient documents and recent archaeological discoveries, Xu Fei and his team have determined that the Chinese began to manufacture standardized wind-pipe music instruments around 6000 BCE – eight thousand years ago. The large number of artifacts evidence the existence of established musical practices and methods, which must have been shared by many people. The earliest yet found, the *Gudi* (骨笛) flutes carved out of animal bone and excavated in large numbers from the Jiahu Neolithic site were clearly very popular instruments.(FIG. 4) Xu Fei has studied their technical and acoustic qualities and found that when they are played while held vertically at an angle, the results are comparable to today's musical instruments. "Most of the 6- and 7-hole Gudis can perform two true octaves when played with the old technique of vertical and inclined end-blowing. This result […] affirms that the Jiahu bone flutes actually served as functioning musical instruments. Furthermore, […] even at that time, humans had obviously already mastered suitable technologies for making musical instruments."[14] Not later than the end of the Yin dynasty and the beginning of the Zhou dynasty, the ancient Chinese invented pitch pipes (*lüguan*, 律管) "to regulate the pitches of bell chimes. During the Spring and Autumn Period (722–476 BCE) and the Warring States Period (475–221 BCE), the pitch-pipe had already become a commonly used tuning instrument. The widespread utilization of pipe-tuning instruments in ancient times met its own social demand."[15]

The above sentence is succinct, but on closer inspection there is a lot behind it

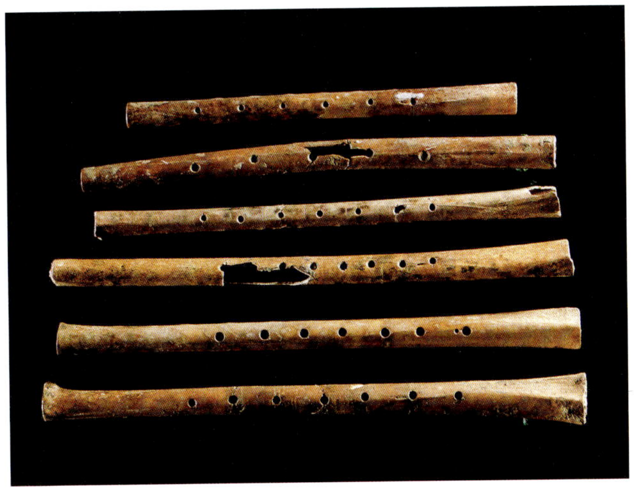

FIG. 4 The six Gudi bone flutes (M511:4) excavated from the Neolithic site Jiahu, China.

philosophically. The thinking of scientist Xu Fei, grounded in *dialectical* and *historical materialism*, makes reference here to the connection between social practice and the cultural technique of tuning, which still echoes today in our expression "in tune with." The Pythagorean School also understood its practice as political and ethical. Precise tuning of instruments becomes necessary in situations where several instruments are played together and must harmonize. For individual players to play together without descending into cacophony or chaos, they must agree on and set a common pitch. "Ritualised official ceremonies with music and dance […] were a national institution […]."[16] This was the main reason why flute-like instruments were used for standardization. And it remained the most important reason why research on instruments and study of the tone system never stopped in Ancient China; indeed, was actively fostered by the rulers of the various dynasties.

Techno-Souls

The genealogy of automata that produce sounds I shall discuss in the context of a concept of expanded animation.[17] *Anima* in Latin means in a physiological sense a breath of air, the breath of the body, the breath of God, the *pneuma* that flows through the subconscious of a man and directly motivates, that is, moves, his intentions and actions. The soul, *anima*, also has female connotations in Latin. Its male counterpart is *animus*. *Animus* is energy, the current of air that activates the vital spirit of the man and penetrates the female principle completely. As in Chinese natural philosophy, this does not refer primarily to biological gender differences and their potential compatibility. *Yin* and *Yang* represent opposite principles that are actually perfectly compatible in the way they relate to each other. In animation they work together cooperatively. In a wider sense *anima*, as well as *animus*, stands for the principle of life. This dimension of meaning also resonates in the concept of animality. On the one hand, the body and everything to do with it is quite alien to the *anima* and *animus*, but at the same time they ensure the vitality of the body and are a constituent part of its corporeality. The Latin verb *animare* not only means to animate, to ensoul; in Baroque music theory it still meant the act of composing.

No kinetic device can function without motive force, without energy. "A machine may be defined as *a system of interruptions* or breaks (*coupures*),"[18] according to Deleuze and Guattari in *Anti-Oedipus*. "Every machine, in the first place, is related to a continual material flow (*hylè*) that it cuts into."[19] In order that an artifact moves, there must be energy and also organic or technical power transmission that carries the energy to the artifact. Such technical installations take over the position formerly occupied by the soul, or spirit, as the motive force in the human body. The earliest material forms of such artificial souls were water and air, frequently combinations of hydraulics and pneumatics. Rotating elements of automata were driven directly by falling water, or vessels produced air pressure by filling with water and then emptying again which drove the mechanical movements of the artifact. In the traditions of Hellenistic Alexandria, Byzantium, Persia, and Mesopotamia, a vast number of these technical solutions were developed and over the centuries – up to the first renaissance in the Arab-Islamic countries from the ninth to the thirteenth century – became increasingly elaborate and sophisticated.[(FIG. 5), 20]

Beginning at the latest in the fifteenth century, explosive substances and fire were used in Europe to set immobile matter in motion. Pyrotechnics, in turn, is very close to the form of energy that has become the soul of all modern machines and automata; namely, electricity and electronics.

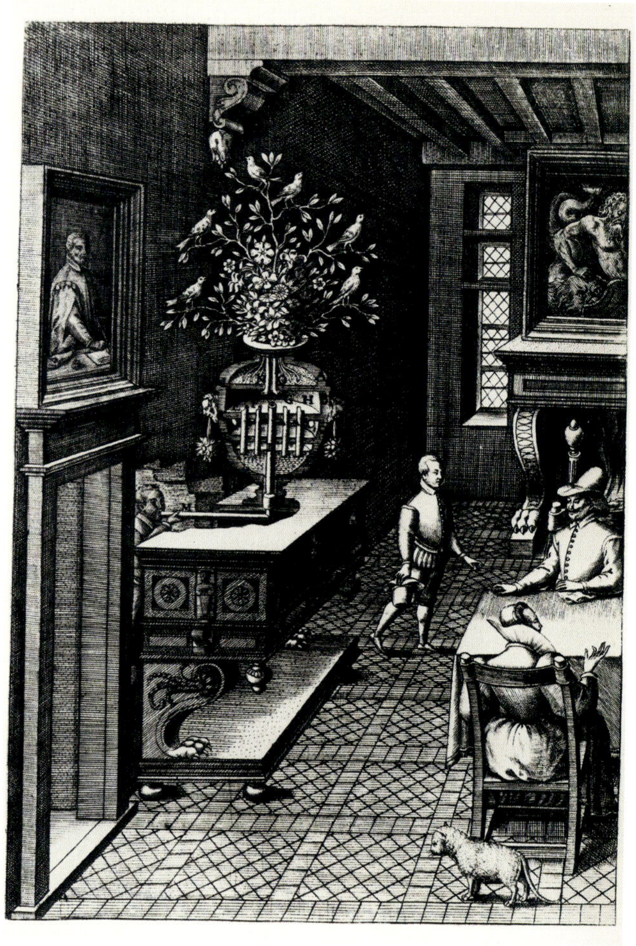

FIG. 5 Many magnificent and ornate sound installations are described in the *Tales from the Arabian Nights* (see page 466), automata that seven hundred years later will grace the Renaissance interiors of the aristocracy, albeit in forms that are far more modest. Here the metal birds are supplied with the pneuma not by water power, but by a servant. The engraving is from Capitano Agostino Ramelli's *Le Diverse et Artificiose Machine del Capitano Agostino Ramelli* (1588). Annette Beyer, *Faszinierende Welt der Automaten* (Munich: Callwey, 1983), 32.

To begin with, in connection with the concept of magnetism, electricity was celebrated as the mover of everything and was even the object of religious veneration. In the tradition of physicotheology, the *theologi electrici* developed in Germany and Switzerland, as well as in England – a movement that viewed electrical phenomena as representing the presence of God in the world.[21]

John Freke (1688–1756), for example, was a theologically well-versed English surgeon who worked at St. Bartholomew's Hospital in Smithfield, London. Freke specialized

18
Gilles Deleuze and Felix Guattari, *Anti-Oedipus: Capitalism and Schizophrenia*, trans. Mark Seem, Helen Lane, and Robert Hurley (London: Continuum, 2004), 38.

19
Ibid., 38–39.

20
On the European tradition, see the chapter "Die Anfänge der Hydromechanik," in *Geschichte der mechanischen Prinzipien und ihrer wichtigsten Anwendungen*, ed. István Szabó (Basel, Stuttgart: Birkhäuser, 1977), 143–56, but which unfortunately leaves out the Persian and Arab contexts entirely.

in diseases of the eye, invented many medical instruments, and caused a stir with an early natural philosophical-theological essay on electricity, "A Treatise on Electricity," which he sent to the Royal Society in 1746.[22]

For Freke, electricity is the noblest phenomenon in the sublunary world of the earthlings, "the First Principle in Nature," and is clearly responsible for all movement. Freke places special importance on stating that the "electrical fire" does not originate in the "apparatuses" or in any of their physical components. Instead, electricity resides in the air, which he calls "paebulum vitae," nourishment or food of life. Thus, for Freke, electricity is inextricably bound up with all things living, and it permeates the animal as well as the vegetable and the mineral. It makes the blood red ("rubefying the blood"), and is therefore "flamma vitalis," the very flame or fire of life. Analogous to Robert Fludd's designation for the sun, Freke refers to electricity as "anima mundi," the soul or mover of the world.

In the technical section of his treatise, Freke describes exactly what happens when an electrified body meets a non-electrified body; when the electricity jumps over from one body to the other and the electrical fire discharges acoustically with a crackling sound and optically with a spark: electrical audiovision in its most archaic form.

One of the most striking and mysterious formulations for a theology of electricity is found in Freke's preface. With electricity one becomes directly acquainted with the "*Officer* of God Almighty." This term not only connotes the meaning of executing a function, but also includes the notion of a visual indicator: electricity is a display of the actions of God, a medium of His ensoulments.[23]

Divine Automata

If we project the idea of one almighty god onto our semantic field, what we get is the prime mover. "Deus est semper movens immobilis" [God: the ever-moving immobility] is the nineteenth answer given by the twenty-four philosophers to the question "What Is God?" in the famous *Book of 24 Philosophers*.[24] Expressed in technological terms, God is the eternally moving – and thus mover of all things – perfect meta-automaton.

At the same time what moves is what motivates (from *motus*, movement) and what creates everything. (For Marcel Duchamp, movement was the most important marker of modernity.[25]) Out of water and earth, God created the creatures who brazenly call themselves *Homo sapiens* because He gave them a soul and gave them dominion over all other living organisms on Earth. This created the primal split which cannot be overcome in reality, only in the imaginary where cinema and other media reside. The ponderous and lethargic physical material is moved by the meta-physical soul, becomes flesh, has desires and hunger, struggles, suffers, loves, and dies. The soul is inside humans. Thus it appears to be automotive, a mover of the self, and this touches upon a highly complex theological question. It concerns the issue of free will and the way in which free will fits in (or not) with the concept of the omnipotence of the mover of all things.[26] In technological parlance, we might refer to humans as God's automata.

Over the last two and a half thousand years this automaton of God's has consistently sought to praise God's creation – that is, first and foremost itself – by building presentations of the world-machine which produced harmonious sounds. This applies equally to Christians and Muslims. All of the automata, for example, designed or actually built by the Kurdish engineer Ibn al-Razzāz al-Jazarī (1136–ca. 1206) from the Al-Jazira region[27] were created to glorify the prophet Mohammed and to honor al-Jazarī's powerful patrons who gave him commissions and paid him royally for his work. Al-Jazarī designed

21
See my essay "Theologici electrici: Einige Passagen," in *Theologie und Politik: Walter Benjamin und ein Paradigma der Moderne*, ed. Bernd Witte and Mauro Ponzi (Berlin: ESV, 2005), 254–68; and with reference to the work of a contemporary sound artist, "God is Electric, My Soul Is Electric, Nature Is Electric / For Paul DeMarinis," exhib. cat. DAAD Gallery (Berlin: DAAD, 2010).

22
See John Freke, "II. On Electricity," in *A Treatise on the Nature and Property of Fire: In Three Essays* (London: W. Innys, and J. Richardson, 1752).

23
John Freke, *A Treatise on Electricity, An Essay to show the Cause of Electricity and why some things are Non-Electricable. In which is also Considered Its Influence in the Blasts on Human Bodies, in the Blights on Trees, in the Damps in Mines; and as it may affect the Sensitive Plant*, 2nd ed. (London: W. Innys, 1746), 3–59.

24
Liber XXIV philosophorum [Book of 24 Philosophers], is an anonymous medieval Latin text first cited in the twelfth century, of which until today no printed English translation exists. The translation cited here is provided online by The Matheson Trust For the Study of Comparative Religion at http://themathesontrust.org/papers/metaphysics/XXIV-A4.pdf.

25
See, for example, Calvin Tomkins, *Marcel Duchamp: The Afternoon Interviews* (Cologne: Walther König, 2013), 67–93.

26
Henning Schmidgen describes the way in which physiopsychology has attempted to measure this problem for over 150 years in his book *Hirn & Zeit: Die Geschichte eines Experiments 1800–1950* (Berlin: Matthes & Seitz, 2014).

27
See Alertz, "The Horologium of Hārūn al-Rashīd" (see note 6).

and described his devices as though they were mechanical objectifications of the cosmic spirit: Allah's automata.

Such installations and devices are familiar from *Tales from the Arabian Nights*, which reached the Arabian Peninsula from Persia in the fifth century and was translated into Arabic in the eighth century. In "The Adventure of Djaudar the Fisherman from Kahirah and His Meeting with Sultan Baibars," there is a particularly impressive description:

> In the middle of the hall was a fountain with four golden lions and four peacocks adorned with pearls and precious stones above them, and opposite the lions were four brass statues that each held a trumpet to its mouth. Next to the statues were four Greek female slaves with tambourines covered in the skin of gazelles and four female Franconians with lutes. The girls looked so fresh that one thought they were alive and could talk [...] the brass statues bowed and the slave girls began to play on their instruments. I remained seated and listened to them, until Heifa said to me, "Even if you sit on that seat for a thousand years, the girls won't stop playing; they do not tire because they are not alive and they move by the power of magic."[28]

In addition to cultic functions and their entertainment value, many of the automata fulfilled practical purposes, which competed with the deity's claim to omnipotence. Superficially, a great many of the hydraulic and pneumatic automata served the purpose of getting the guests at orgies or revels drunk as quickly as possible. However, as large mechanical apparatuses they also intervened in the given natural environment and changed it greatly to the benefit of its inhabitants by turning extremely dry and barren swathes of land into fertile green landscapes. This entailed transporting water from deep underground to the surface. In this way pulleys, pumps, wheelworks, and other instruments produced instances of animation in the direct sense of the word: ensouling as bringing to life. The origins of these instruments can be traced far back in time to the cultures of Ancient Egypt, Alexandria in the Hellenistic period, and Byzantium, where the goal was also to eliminate hard physical work, particularly for female slaves, as illustrated by a poem from the first century BCE, which celebrates the water wheel:

> Let your hands now rest, you milling girls, and sleep
> Long; may your slumbers be undisturbed by the morning cockcrow.
> Ceres has decreed that from now on the nymphs will do your work,
> They leap and rush over the rolling wheel
> With many spokes which turns on its axis
> And drives four heavy, grinding millstones.
> Now we enjoy again the golden days of yore
> And eat the goddess's fruit without burdensome toil.[29]

The masterpieces of these automata were the clocks; they could be read and they were also audible from a considerable distance. They were horological apparatuses, which one can also designate as archaic visual installations, complex structures in which the passage of time was inscribed so that it was visible, but in which time also resounded. The man-made artifacts counted the hours (or smaller temporal units), and thus intoned the structure of time within which people acted – the function of the timepieces was comparable to that of the muezzin who calls the hours of prayers from a minaret at regular intervals and in this way endows the urban or rural environment with a sacral structure. Etymologically, "minaret" derives from the Arabic word for lighthouse, *manāra*, which actually does not match this practice. For the minaret is first and foremost an acoustic installation, comparable to the bell towers of Christian churches.

In a long chapter subdivided into ten sections with densely interwoven texts and illustrations, al-Jazarī dismantles once again the legendary eighth-century clock of Caliph Hārūn al-Rashīd. He describes its components so precisely that with the requisite skills it is possible to build the timepiece and reconstruct it today.[30] This is a further notable aspect of a genealogy of technology-based arts: in the beginning (avant-garde), the engineers opened up their machines and even wrote manuals so that their customers could construct replicas. It was only with industrialization that the apparatuses increasingly mutated into black boxes, which is what the vast majority of them are today. When the first and second avant-gardes of the twentieth century again opened them and exposed their mechanical hearts to demonstrate that the machines themselves were producing reality, it was actually a throwback to the early protagonists of technology-based sound art. It is difficult to imagine futures without going back to the sources.

What becomes apparent is the extent to which the Arab engineers and geometers availed themselves of the knowledge of various Ancient Greek authors, for example, when al-Jazarī meticulously describes a semi-circular disk on which the twelve hours of daylight are represented as signs of the Zodiac. In the third century BCE, Archimedes of Syracuse (ca. 287–212 BCE) and Ctesibius of Alexandria (ca. 285–222 BCE) were already working on mechanical solutions for regular timekeeping based on the principle of the clepsydra. And like his pupil Philo of Byzantium, Ctesibius worked on the construction of water-powered clocks and organs.[31] Both genealogies, of the horological devices and of musical instruments, are very closely connected as time-based arts of sound in the archaeology of music.

In his writing, the engineer al-Jazarī devotes a great deal of attention to perfecting the hydraulic drive of the water clock (*binkām*) so that the timepiece operates without interruption, and he also describes various mechanical and pneumatic principles for producing air pressure to activate models of birds, flutes, and trumpets. The hydraulic mechanism is based on the clepsydra ("water thief"). These simple water clocks with consistently spaced markings were developed in Ancient Egypt and dated by Arab historians of technology back to 1500 BCE. Analogous to the hourglass, or sand clock, except that in the "water thieves" the water flows from one vessel into another, clepsydrae were constructed to measure predetermined time periods, for example, time limits for speakers during judicial proceedings or political debates. Via China, Hellenistic Alexandria, Byzantium, and Iran, this specialist knowledge was refined and then spread to the Arabian Peninsula and from there to Andalusia in Europe. It is reported that in the eleventh century there were two handsome water clocks on public squares in Toledo.

The most impressive water-powered horologium, also regarding the sound, was built at the end of the eleventh century by the Chinese engineer Su Song in the palace gardens at Kaifeng, approximately at the same time as Ken Shuo's experiments with the *camera obscura*. Su Song's installation, which could be heard from far away, was ten meters high, driven by a waterwheel with a diameter of four meters, and used an escapement mechanism, the key invention which made the construction of all-mechanical clockwork in Europe in the High Middle Ages

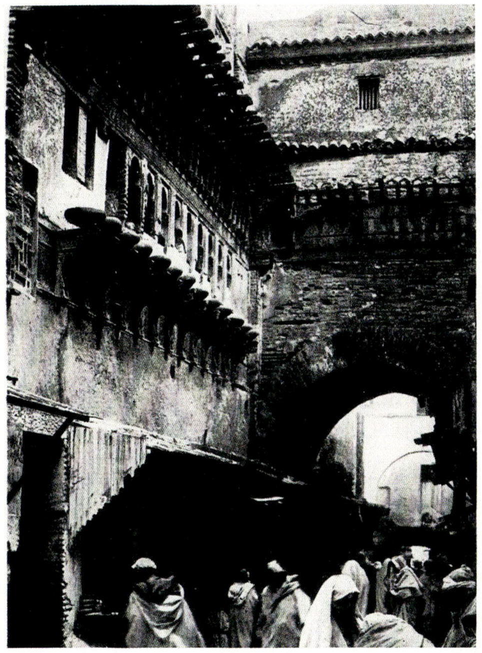

Fig. 35 Vestiges de l'horloge de Abou Inane, à Fez (Maroc). *Vestiges of the Abou Inane clock, at Fez, Morocco.*

FIG. 6 The Dar al-Magana clockhouse (1357) in the Medina district of Fès, commissioned by Sultan Abu Inan Faris. A hydraulically powered mechanism opened one window per hour and released an iron ball which fell into a brass vessel below and made a loud clanging sound like the ringing of a bell. Opposite the clock tower is a famous madrasa, the Islamic college of Bou Inania, whose diurnal rhythms the sound installations also structured. Alfred Chapuis and Edmond Droz, *Les automates: Figures artificielles d'hommes et d'animaux. Histoire et technique* (Neuchâtel: Édition du Griffon, 1949), 40.

28
Quoted from Gustav Weil's version of 1838: *Tausend und eine Nacht*, vol. 4, chapter 63, http://gutenberg.spiegel.de/buch/tausend-und-eine-nacht-vierter-band-3447/63. Translated here from the German.

29
Poem quoted from Conrad Matschoss, *Die Entwicklung der Dampfmaschine. Eine Geschichte der ortsfesten Dampfmaschinen und der Lokomobile, der Schiffsmaschine und Lokomotive*, vol. 1 (Berlin: Julius Springer, 1908), 13–14, fn. 1, www.digitalis.uni-koeln.de/Matschossd/matschossd_index.html. Translated here from the German.

30
See especially Alertz, "The Horologium of Hārūn al-Rashīd."

31
Several Arabic manuscripts make reference to Archimedes as the constructor of a large chiming water clock; classical scholars and archaeologists, however, doubt his authorship; see A. G. Drachmann, *Ktesibios, Philon and Heron: A Study in Ancient Pneumatics* (Copenhagen: Enjar Munksgaard, 1948), particularly 16–40. On the genealogy of the ancient water organ, see particularly the outstanding collection of source material in Michael Markovits, *Die Orgel im Altertum* (Leiden: Brill, 2003), 401–14.

possible (as well as the mechanism of the film camera later).[32] Su Song's chiming timekeeper is reported to have featured kinetic mannequins with cymbals, trumpets, and gongs. In the case of the famous weight-powered water clock in Fèz, Morocco, completed in 1357, it is fortunately still possible to experience the imposing physical presence of such public horological devices. The clock's engineer was Abou al-Hassan Ibn Ali Ahmed Tlemsani. Its twelve windows can still be seen in a photo of 1913 in the legendary book *Les automates* by Alfred Chapuis and Edmond Droz.(FIG. 6) The chimes from the Dar al-Magana clockhouse synchronized the sacred and profane time periods of everyday life for Moroccans, dating back to the time when Fèz was their capital city.

In the chapter which describes the functions of the device – designed as twice the height of a man and therefore an impressive size – al-Jazarī focuses on the production and performance of sounds. The sounds were intoned by a group of five female musicians: two trumpeters, two players of different drums, and a cymbalist. At certain hours of the day and night, the figurines were set in motion via a mechanism involving two model falcons; the falcon is an important symbol in Islamic culture denoting virtuality and virility. From the wide open beaks of the birds, small brass balls drop into beakers positioned below them, which sets in motion a system of hoists, connections, and pulleys that keeps the musicians animated for a while before they finally come to rest and then wait for their next mechanical appearance. At the beginning of the night, all of the twelve windows above the musicians' heads are closed and in darkness; at the end of the night's twelve hours they are all open and brightly illuminated. This is a powerful image for Allah's automaton: the divine light, which causes day to break; the Almighty who is all-knowing. And with the words "God is all-knowing," al-Jazarī ends this subchapter of his manual.[33]

At the turn of the first millennium, automata that produced sounds, which al-Jazarī developed for feasts and orgies, were particularly popular at the courts of the rich and powerful in Mesopotamia, Baghdad, and Basra. Made of valuable metals like silver or copper, these artifacts played out what the big water clocks achieved on a larger scale. Mechanical ducks or other birds squawk, or mechanical musicians play, as long as enough wine remains in the vessels to exert hydraulic pressure on the movable mechanisms. In al-Jazarī's descriptions, an overt social and gender-specific component appears time and again, which is familiar from classical Greece, the *Tales from the Arabian Nights*, and even more drastically in the compendium *101 Nacht*.[34] These technical devices were operated by female slaves (*jāriya*), and the musicians as well are referred to as slave girls.(FIGS. 7a–d)

The geographic and political hot spot of the knowledge culture of the Golden Age of Arab-Islamic sciences was the Bayt al-hikma, the "House of Wisdom," founded in Baghdad in the early ninth century. The initiator of this research and teaching facility, Caliph Al-Ma'mūn, who ruled for twenty years (813–833), not only commissioned many translations of classical Greek texts in natural philosophy, he encouraged young scholars who were eager to learn to think independently and pursue an experimental approach to the world. This made the House of Wisdom an early modern laboratory.

Among those who profited from this institution were the three brothers Muhammad, Ahmad, and al-Hasan, the sons of Mūsā Ibn Shākir, who as a mini-cooperative combined a whole universe of scientific qualifications in their brotherhood: from mathematics and geometry to astronomy, natural philosophy, and medicine, as well as music and engineering. They have gone down in the history of science and technology as the Bānū Mūsā. Their achievements

32
See Ahmad Y. al-Hassan and Donald R. Hill, *Islamic Technology: An Illustrated History* (Paris: UNESCO/Cambridge University Press, 1986), 55–56.

33
The Institute of the History of Arabic Sciences at the University of Aleppo in Syria has conducted a major research project to reconstruct this valuable knowledge in which al-Jazarī's manuscript, completed probably in early 1206, plays a central role. Ahmad Y. al Hassan has re-edited it under the title *A Compendium on the Theory and Practice of the Mechanical Arts* (Aleppo: University of Aleppo, Institute for the History of Arabic Science, 1979). Recently I had the good fortune to receive a copy via a middleman in Andalusia. For German-speaking scholars studying al-Jazarī, the early works of Eilhard Wiedemann from the beginning of the twentieth century are invaluable, and for English-speaking scholars the meticulous translations and commentaries by the English engineer and historian of science and technology Donald R. Hill, particularly in Ibn al-Razzāz al-Jazarī, *The Book of Knowledge of Ingenious Mechanical Devices (Kitāb fī ma'rifat al-iyal al-handasiyya)*, trans. and ed. by Donald R. Hill (Dordrecht: Reidel, 1974), here 19. In 2002 the Turkish-born orientalist Fuat Sezgin presented a completely reconstructed version of the manuscript, of which only a few copies are extant. Sezgin's reconstruction integrates single pages held in various libraries all over the world into the master manuscript held in the Topkapı Sarayı Library in Istanbul. I am very grateful to Fuat Sezgin for this exquisite gift.

34
101 Nacht [101 Nights], translated from the Arabic into German by Claudia Ott from an Andalusian manuscript in the Aga Khan Museum, dating from 1234 (Zurich: Manesse, 2012).

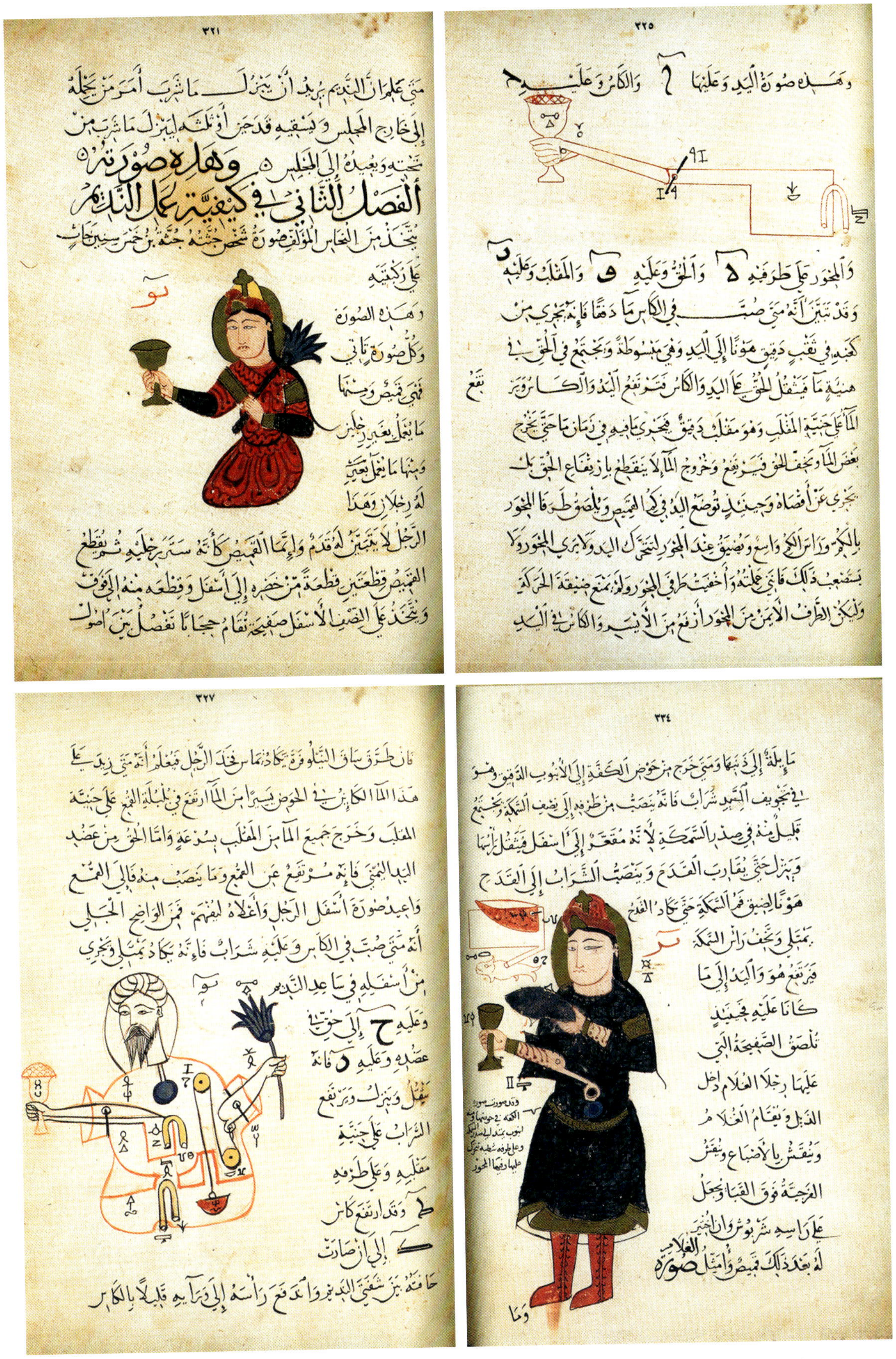

FIGS. 7a–d Four facsimile pages with various drinking automata from al-Jazarī's *Compendium on the Theory and Practice of the Mechanical Arts* (1206). Fuat Sezgin, ed., *Compendium on the Theory and Practice of the Mechanical Arts* (Frankfurt am Main: Institute for the History of Arabic-Islamic Science, Goethe University Frankfurt, 2002), 321, 325, 327, and 334.

include rescuing books V–VII of Apollonius of Perga's treatise on conic sections and saving them for posterity. They edited the translation of this classical Greek work on geometry and mathematics, which did not survive in the original.[35] Particularly prominent in their work are traces of the great Alexandrian automaton constructor Heron, whose texts they carefully translated from the Greek into Arabic. (Some of Heron's treatises on mechanics have only come down to us in Arabic translations, which were then translated directly into German.)[36]

Ahmad is considered to be the greatest engineer of the three brothers. He is thought to be the main author of their *Kitāb al-Hiyal* [Book of Ingenious Devices] from the mid-ninth century.[37] The book is a compendium filled entirely with sketches and exact instructions for building over one hundred mechanical models (*shakl*). These include a variety of artifacts, mechanical components, kinetic sculptures, and automata, the latter in the direct sense of devices that move of their own accord: scooping and drinking devices operated hydraulically and mechanically, pneumatically activated animals that make noises, oil lamps that not only refill automatically, but also have draught shields that self-adjust to keep the flame protected and eternally lit – in short, items that functioned as a special kind of *perpetuum mobile*.

That things can be perpetually in motion, without interruption, was evidently very important to the Bānū Mūsā in regard to music, for one of their masterpieces exhibits this characteristic. This particular automaton is not included in the three surviving copies of the *Book of Ingenious Devices.* Three hundred and fifty years later, al-Jazarī must have had access to the manuscript describing the Bānū Mūsā's automaton because in his *Compendium on the Theory and Practice of the Mechanical Arts*, he makes explicit reference to the three brothers in the introduction to the chapter in which he describes mechanical devices with fountains and perpetually playing flautists.[38]

"The Instrument Which Plays by Itself"[39] is the title of a manuscript that Henry George Farmer describes in his famous work *The Organ of the Ancients: From Eastern Sources* as a "solitary exemplar"[40] that is now considered lost. The only known copy was held for many years in the library of the Three Moons monastery, today a college, of the Greek Orthodox Church in Beirut. With the aid of the Lebanese science historian George Saliba, who teaches at Columbia University in New York and is a member of our *Variantology* network, we learned that it was stolen from the college in the nineteenth century and is now possibly in the library of the Greek Orthodox patriarch in Damascus or was taken to Deir El-Balamand in the El-Koura region in northern Lebanon. At the present time it is difficult to visit either of these places.[41] However, around 1900 a photographic negative must have been made of the manuscript which is now kept in the Bibliothèque Orientale of Saint Joseph University in Beirut. Louis Cheikho, the founder of the journal of Arabic studies *al-Mashriq* [the East], described the instrument of the Bānū Mūsā in an issue published in 1906 (vol. 9). George Saliba arranged for us to get access to the negative. We were able to establish that the manuscript in the negative is identical to the text in *al-Mashriq* of 1906, so it is highly likely that we now possess a photographic copy of the master text.[42]

The "instrument which plays by itself" from mid-ninth-century Baghdad, is well worth the research effort of an investigation that was almost criminological. For the manuscript not only contains a description of an artificial, perpetually playing flautist, but also an early programmable universal music machine. The title evidences that the Bānū Mūsā were quite aware of the significance of

35
Apollonius, *Conics, Books V to VII. The Arabic Translation of the Lost Greek Original in the Version of the Bānū Mūsā*, 2 vols., ed. and trans. Gerald J. Toomer (New York: Springer, 1990). This text machine also begins with the invocation of Allah: "In the name of God, the merciful, the forgiving. I have no success, except through God." (preface of the Bānū Mūsā, vol. II, 620).

36
For example: *Herons von Alexandria Mechanik und Katoptrik*, ed. and trans. L. Nix and W. Schmidt (Leipzig: Teubner, 1900).

37
See the new edition of *Kitāb al-Hiyal* by Ahmad Y. al-Hassan (Aleppo: University of Aleppo, Institute for the History of Arabic Science, 1981). He translates the title as *The Book of Ingenious Devices*.

38
See "Category IV: On fountains which change their shapes (*tabaddala*) at known intervals, and on perpetual flutes," in al-Jazarī, *The Book of Knowledge of Ingenious Mechanical Devices* (see note 33), 157ff.

39
Henry George Farmer, *The Organ of the Ancients: From Eastern Sources (Hebrew, Syriac and Arabic)* (London: William Reeves, 1931), 88.

40
Ibid., 87.

41
Pers. comm. George Saliba, 12/05/2014.

42
The first translation into German was by the science historian Eilhard Wiedemann using the text published in *al-Mashriq*. "Über Musikautomaten bei den Arabern," was published in the Festschrift *Centenario della nascita di Michele Amari*, vol. 2 (Palermo, 1910), 164–85, and also distributed as an offprint (cf. the reprint in vol. 1 of the *Gesammelte(n) Schriften zur arabisch-islamischen Wissenschaftsgeschichte*, revised by Dorothea Girke for the *Veröffentlichungen des Instituts für Geschichte der Arabisch-Islamischen Wissenschaftsgeschichte*, ed. Fuat Sezgin (Frankfurt am Main, 1984). For our exhibition project *Allah's Automata* (2015/2016) at the ZKM | Karlsruhe, Imad Samir has recently produced a new translation into the German. Compared with the earlier translation, Samir's is complete and more exact in many details. It was published in *Allah's Automata: Artifacts of the Arab-Islamic Renaissance (800–1200)*, eds. Siegfried Zielinski and Peter Weibel (Ostfildern: Hatje Cantz, 2015), 69–86.

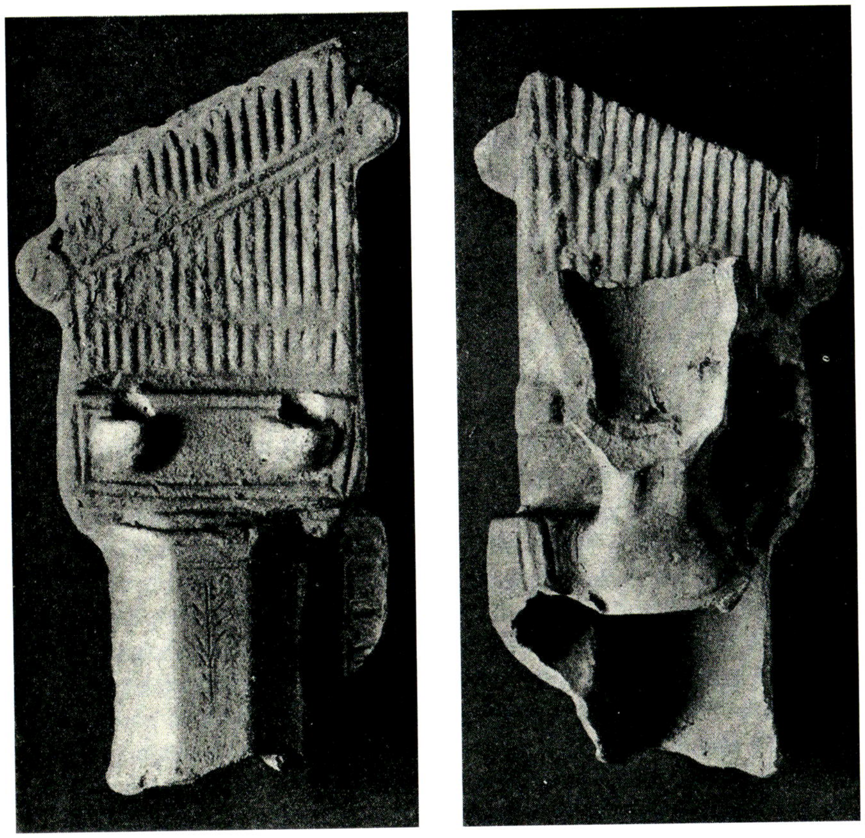

FIGS. 8a, b Terracotta model of a water organ modeled on Ctesibius' invention, National Museum of Denmark in Copenhagen. A. G. Drachmann, *Ktesibios, Philon and Heron* (Copenhagen: Munksgaard, 1948), 11.

their technology and that it could be generalized. Clearly, they wanted their invention to be understood independently of any specific form in which it was realized, such as the flute player. They describe a complete music automaton, which can vary the rhythm of the music and has the potential to be primed with different melodies. At the very beginning of the text, the Bānū Mūsā state their intention: "We wish to explain how an instrument […] is made which plays by itself continuously in whatever melody […] we wish, sometimes in a slow rhythm […] and sometimes in a quick rhythm […], and also that we may change from melody to melody when we so desire."[43]

Supplied with air pressure via hydraulic systems, animated song birds or flute players have their place in ancient Chinese literature just as in sources from the traditions of Hellenistic Alexandria and Byzantium. Both Ctesibius and Philo of Byzantium described such apparatuses in connection with their chiming clocks and organs.(FIGS. 8a, b) In the first century, the great experimenter Heron of Alexandria elaborated, embellished, and evolved these inventions into a vast phylum of theatrical machines, sometimes at the cost of their technical functions.[44] The technologically most advanced solutions – with regard to design of the driving mechanism and system – are attributed in the Greek tradition to Apollonius.[45] He had already developed a hydraulic-pneumatic mechanism of such complexity that his anthropomorphic synthetic figure could play the flute endlessly – provided it was supplied with a constant flow of water. As a result of his circular design for an air chamber system, whereby a second water container filled up while the first was emptying and pressed out air for the flautist, the automaton had a constant inflow of energy in the most literal sense. The Arab concept of sound was physical (one meets with more references to Aristoxenus in the literature than to Pythagoras): "The musical tone is a simple sound that during a perceptible period of time remains in the body where it is produced,"[46] as Wiedemann quotes the universal natural philosopher al-Fārābī (ca. 872–950).(FIGS. 9, 10)

43
Henry George Farmer, *The Organ of the Ancients: From Eastern Sources (Hebrew, Syriac and Arabic)* (London: William Reeves, 1931), here 88.

44
Drachmann, *Ktesibios, Philon and Heron* (see note 31), 114–19.

45
In the ancient literature this flute player is incorrectly attributed to a text by Archimedes, which is actually by Apollonius. On the issue of its provenance see also E. Wiedemann and F. Hauser, "Byzantinische und arabische akustische Instrumente," in *Archiv für die Geschichte der Naturwissenschaften und der Technik* (Leipzig: F.C.W. Vogel, 1918), 140–66.

46
Quoted from Eilhard Wiedemann, "Ueber Musikautomaten bei den Arabern," in *Centenaro della nascita di Michele Amari*, vol. 2 (Palermo: Stabilimento Tipografico Virzì, 1910), 164–85, here 169, note 1. Translated here from the German.

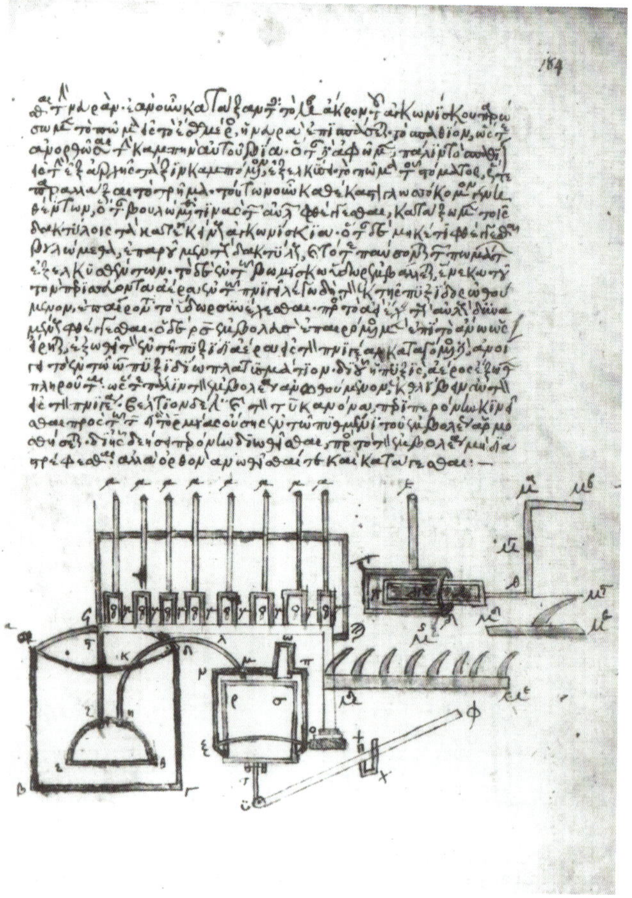

FIG. 9 Heron's adaptation of the water organ of Ctesibius from the first century CE (Alexandria) from a thirteenth-century copy of his book *Pneumatica* in the Biblioteca Nazionale Marciana, Venice. Michael Markovits, *Die Orgel im Altertum* (Leiden: Brill, 2003), 751.

The mechanical heart of the automaton is a hydraulically driven cylinder that rotates on its own axis, a *barbach*, as it is called in the Arabic manuscript, a cylinder carved out of wood. It has roughly the same length as the instrument to be played, and is positioned parallel to the instrument. In the Bānū Mūsā's description the instrument is a *sornay*, a wind instrument that is played with a double reed, similar to the European oboe. The sornay, too, reached the Arab region via Persia. On the surface of the cylinder are bands made of wood or metal, which can carry small protruding pins of different lengths.(FIG. 11) (In Giovanni Battista della Porta's terminology around 1600, these pins then become *digits*, fingers.)

Prographein means to prescribe. The way the pins are arranged on the cylinder formulate the musical prescriptions or instructions, the *program* of the instrument: the pins and bands are notation translated into hard material. Depending on how the pins are positioned on the bands and how fast the cylinder revolves, a mechanical translation opens or closes the valve over one of the sornay's holes, a single organ pipe, or moves another sound-producing element. If the diameter of the cylinder is increased, it is possible to put several melody fragments on one band, which enables changing from one melody mode to another without a break. The brothers also gave

47
Quoted from the new German translation by Imad Samir of the Bānū Mūsā manuscript (see note 42), 82. Translated here from the German.

48
Ibid. Translated here from the German.

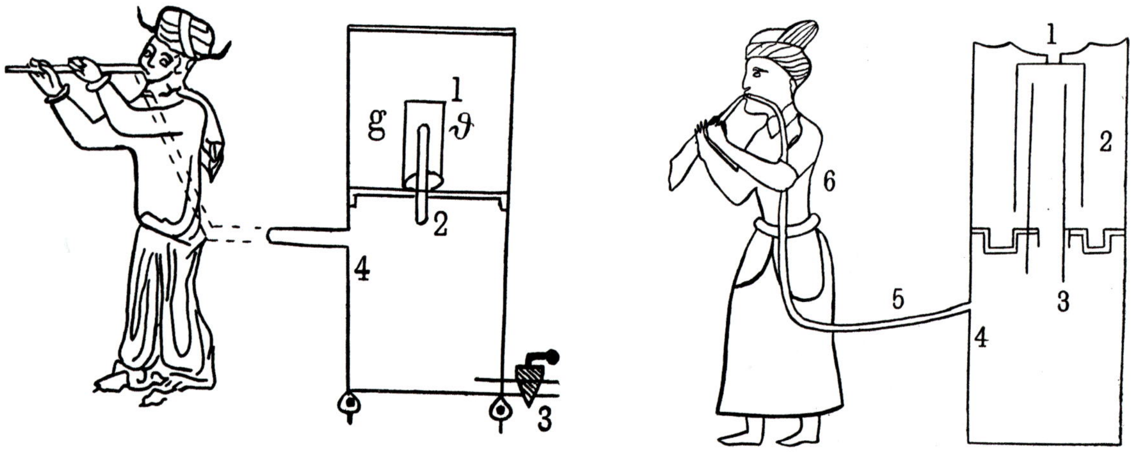

FIG. 10 Depiction of a hydraulically animated flautist based on the treatise by Apollonius. Figures from Eilhard Wiedemann's text on musical automata in *Sitzungsberichte der physikalisch-medizinischen Sozietät zu Erlangen*, no. 46 (1915), 17–26; translated into English by Baruch Gottlieb in *Allah's Automata: Artifacts of the Arab-Islamic Renaissance (800–1200)*, ed. Siegfried Zielinski and Peter Weibel, exhib. cat. (Ostfildern: Hatje Cantz, 2015), 49–55.

thought to the speed of rotation and its effects on the progression of the sound performance.

The Bānū Mūsā attach particular importance to the possibility of controlling any instrument with their automaton and this underlines the universality of their aims. Using the same method that they describe in detail for the sornay, it is "sometimes proper that we should make an image which plays (*lit.* beats) on the lute (*'ūd*), or on an instrument of strings [...]."[47] It sounds as though the Bānū Mūsā want to create an entire band of mechanical musicians: "Then each of the two images conforms to the other, for the 'organ' (zamr) conforms to the string [instrument], and the string [instrument] conforms to the 'organ.' And it is also possible that we should make figures of images which dance and follow this 'organ' and these strings. And the contrivance in all this is like the contrivance of the 'organ,' so that every note of the strings corresponds with every note of the 'organ' to the end of the piece of music (nauba)."(FIGS. 12a, 12b), [48]

Thus there existed around 850 CE the concept of a master machine for automatically controlling mechanical bodies that generate sounds and movements. The description is couched in awkward language, but from an engineering point of view it is entirely comprehensible. The hardware of the Bānū Mūsā is virtually identical to the programmable cylinders with pins which were used five hundred years later in the heavy European glockenspiels of the late Middle Ages,

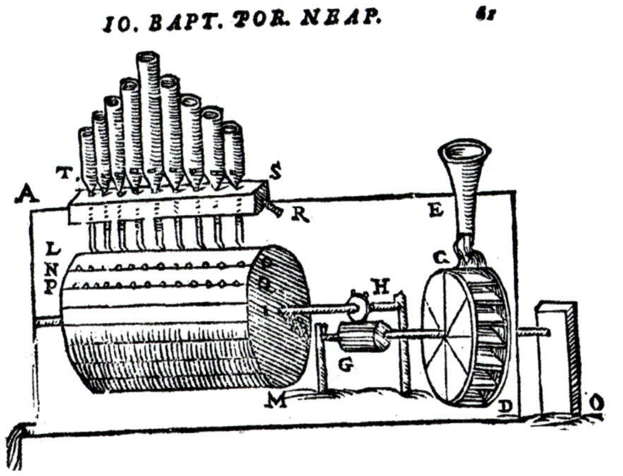

FIG. 11 The movements of the fingers, the discrete digital units (the pins on the cylinder), the mechanical transmission to the organ pipes, and the driving mechanism of a water wheel: depiction of the ancient hydraulic organ in the tradition of Ctesibius and Heron of Alexandria by the Neapolitan Giovanni Battista della Porta in his *Pneumaticorum libri tres*. Very similar depictions are found later in the works of Robert Fludd, Athanasius Kircher, and others. Giovanni Battista della Porta, *Pneumaticorum libri tres* (Naples: Iacobus Carlinus, 1601), 61.

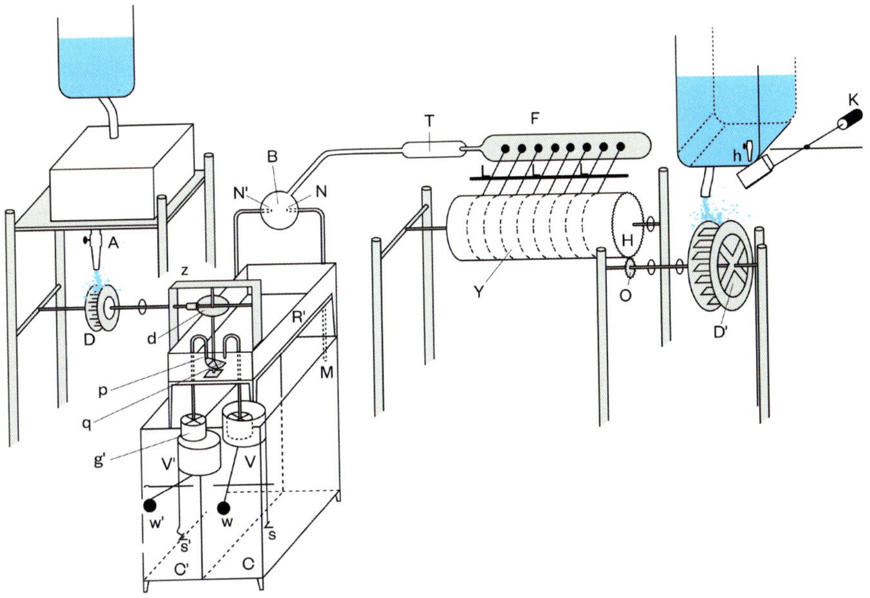

FIG. 12a The basic construction of the programmable, universal music automaton of the Bānū Mūsā; left: the hydraulic-pneumatic ensemble for the permanent supply of energy; center: the apparatus of the air pump; right: the programmable mechanical heart, the rotating cylinder with discrete pins and bridges, which in this example controls the valves of organ pipes. The depiction is taken from Zielinski and Weibel, *Allah's Automata*, 88. Visualization by Clemens Jahn.

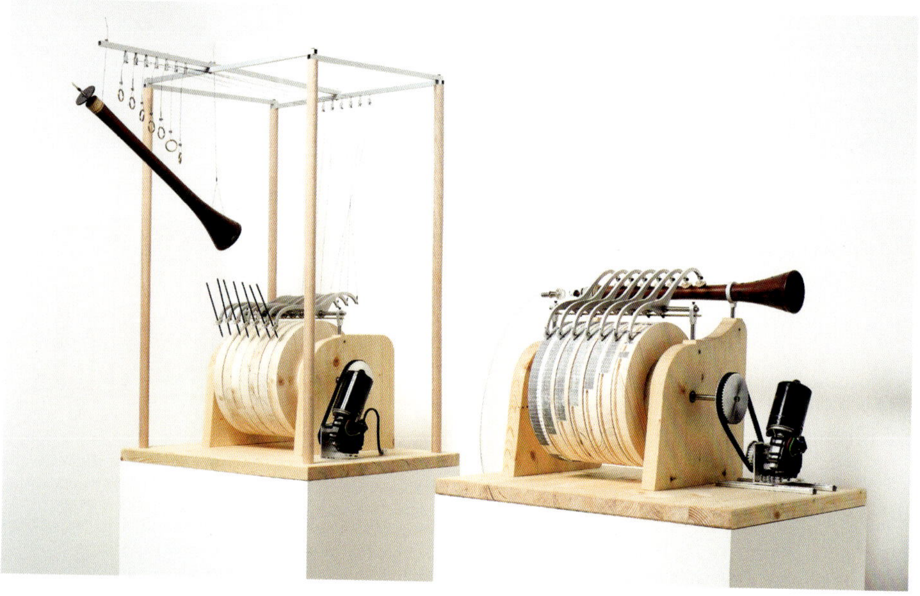

FIG. 12b Reconstruction of the Bānū Mūsā's music automaton according to their description. Exhibition view *Allah's Automata*, ZKM | Karlsruhe 2015.

and again even later in the mechanical organs of the Renaissance, as well as for writing automata and automatic music instruments in the Age of Enlightenment. Giovanni Battista della Porta, Robert Fludd, Athanasius Kircher, Jacques de Vaucanson, and many others not only drew on the work of the ancient Alexandrians and Byzantians, but also systematically exploited the knowledge of the Muslims.(FIG. 13) One of the only three extant

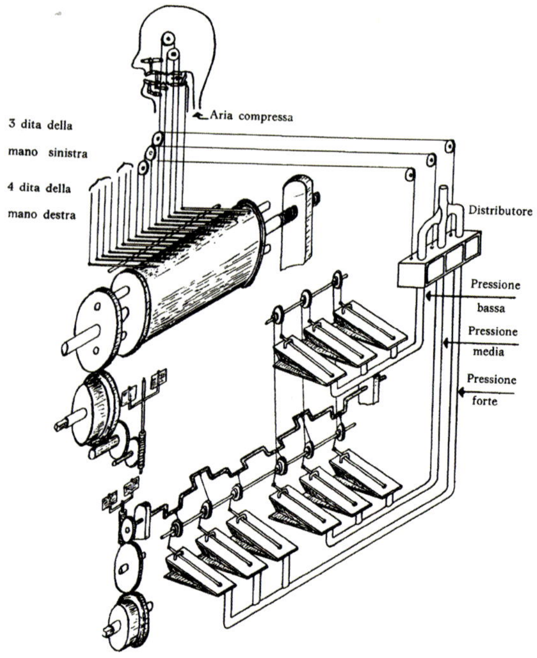

FIG. 13 Nine hundred years after the invention of the rotating cylinder with pins, Jacques de Vaucanson (1709–1782) built an android, whose command administration functioned in a similar way. The figure shows a reconstruction by Edmond Droz, the legendary conservator of the automata museum in Neuchâtel (Switzerland). *Guida introduttiva agli strumenti musicali meccanici, Collezione Marino Marini* (Ravenna: Tipografia Musiani, n.d.), 25.

manuscripts of *Kitāb al-Hiyal* [Book of Ingenious Devices] is held in the Oriental Department of the Biblioteca Apostolica in the Vatican.

With Kircher we can see how a further step was taken in the technological genesis of automata. In *Musurgia Universalis*, his main work on the theory and practice of the arts of sound of 1650, one finds hydraulically powered automata that use cylinders with pins. A two-dimensional projection of the three-dimensional information on the cylinders(FIG. 14) gives rise to the technical solution of the punch card, which was used to control the music instruments of the Age of Enlightenment, as well as nineteenth-century looms.(FIG. 15) The step forward to abstraction, to the program, can be discerned more clearly from a projection onto a flat surface than with the three-dimensional revolving cylinder of the three brothers from Baghdad circa 850 CE. They constructed the mechanical heart which powered nearly all the music automata of (early) modern times in Europe.

Universalization and automatic control of mechanical ensembles, including sound, are features that we define as of modern origin. Our brief excursion into the deep time of the automata of Islamic provenance has attempted to show that the European variant is not the starting point of modernity, but instead must be viewed as a combination and concentration of knowledge and cultural techniques that were in circulation many centuries before. They were widespread, moreover, in regions that were excluded from modernity after the establishment of European hegemony.

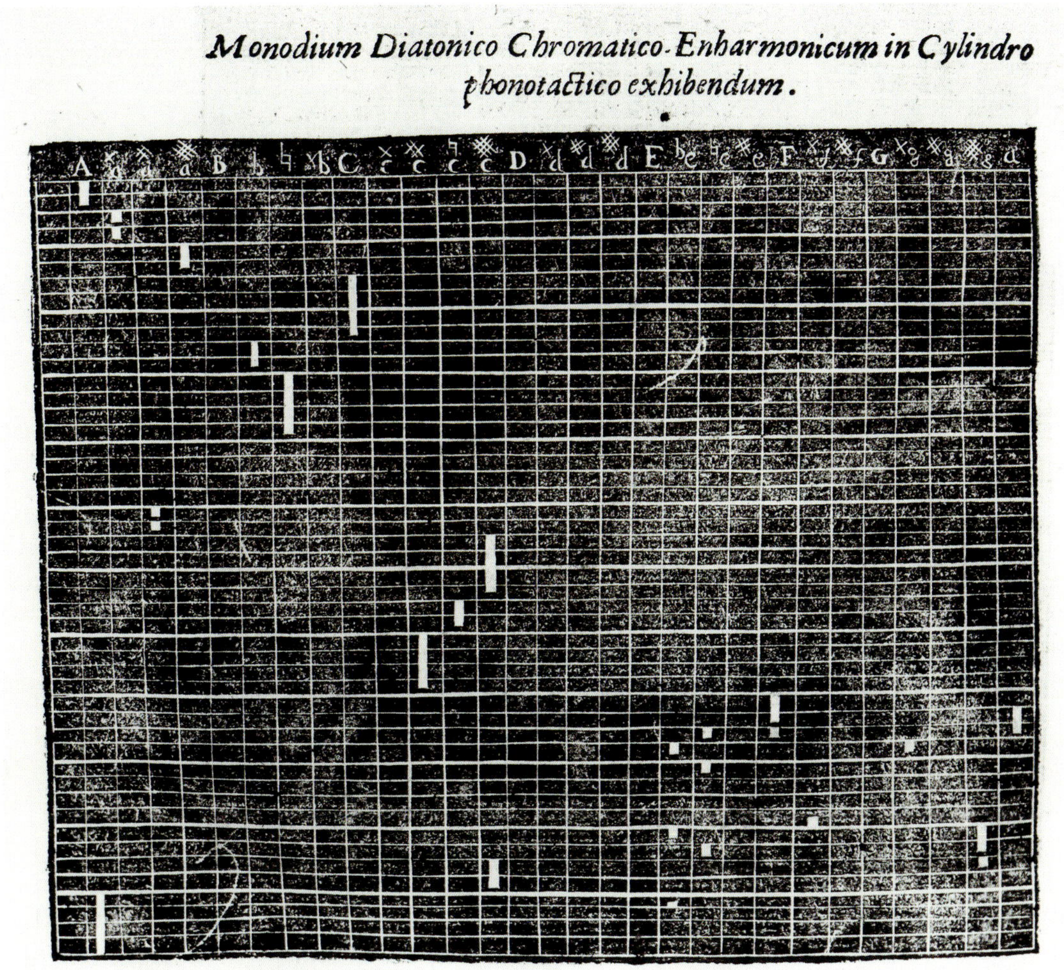

FIG. 14 In this section of his *Musurgia universalis* Athanasius Kircher demonstrates how, by projecting a phonotactic cylinder onto a two-dimensional surface, the outline of a program emerges that we will encounter 200 years later in the punch cards used to control machines. Athanasius Kircher, *Musurgia universalis* (Rome: 1650), 520.

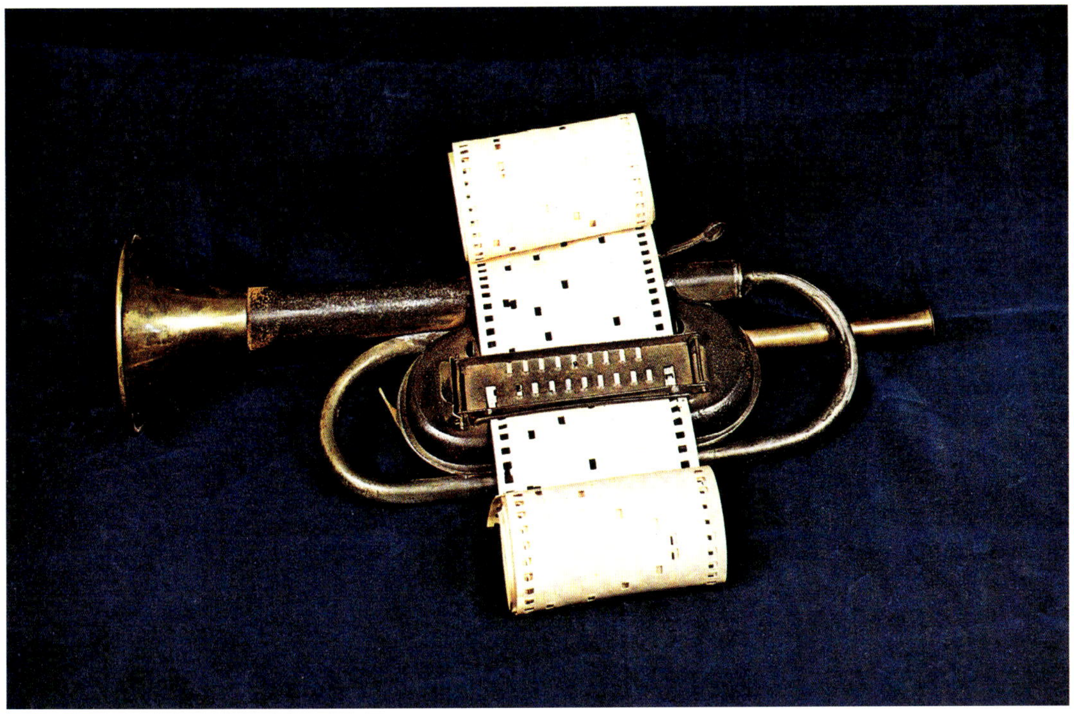

FIG. 15 A punch card-controlled trumpet from the eighteenth century. *Guida introduttiva agli strumenti musicali meccanici. Collezione Marino Marini* (Ravenna: Tipografia Musiani, n.d.).

Seth Cluett

The Loudspeaker in Art Practice: A Preliminary Taxonomy

The loudspeaker is emblematic of and synonymous with sound as a creative medium in the second half of the twentieth century. From Robert Rauschenberg's combine painting *Broadcast* (1959), Jean Tinguely's *Radio Sculptures* and *Radio Drawings* (1960–1972), and Dick Raaijmakers *Ideophones* (1970–1973) to the immersive environment of Char Davies's *Osmose* (1995) and [The User]'s hybrid sound work *Silophone* (2001), the loudspeaker, functioning as a form of acoustic display, projects the audible content of artworks, whether abstract or conceptual, representationally mimetic or expressively poetic. The critical potential of the loudspeaker's contribution to aesthetic practices, however, has all too frequently been subsumed under discussion of the ontological status of the sound it carries, amplifies, displaces, reproduces, translates, and transforms. By focusing on the loudspeaker as a component inextricably embedded in works mobilizing sound, it is possible to productively address the irreducibly mixed, multimodal condition of contemporary art practices.

The history of both media technologies and audible art practices can be seen orbiting around the fixed point of the loudspeaker. Since Johann Philipp Reis's first experiments with electromagnetic transduction in 1860, Alexander Graham Bell's patent for the telephone in 1876, and Ernst von Siemens's patent for the moving coil transducer in 1877, the loudspeaker has undergone only incremental change for technical improvements to yield greater clarity or increased power. Unlike nearly all other forms of media-technological innovation, the fundamental mechanism of the loudspeaker has not changed substantially over its history: the technology produced for Bell's telephone is still in use in much of the most advanced sound-making devices of the twenty-first century. But even though the loudspeaker has been present in nearly all audio technologies that produce sound electrically or digitally since the third quarter of the nineteenth century, the loudspeaker does not want to be seen. Examining the role of the loudspeaker in art practice affords a critical position from which to interrogate the immunity granted to a ubiquitous technology that has had a profound effect on the way society both listens to and amplifies itself.

After framing the status of the loudspeaker as technical apparatus, this essay reads through selected sound works and reveals a preliminary taxonomy of affordances and deployments of the loudspeaker for artistic practice. From the analysis of these works, three areas of focus emerge that center on the manner in which the loudspeaker is used in practice. The first examines the creative disposition of the loudspeaker in the form of a bare cone, pointing up an attempt by artists to pare down sound technology in the service of embedded audio within both objects and the environment. The second looks at works that overtly use or draw on the cultural references associated with commercial speaker enclosures to critique modes of listening and communication. The last section addresses the exploration of headphones not only for the creation of individualized listening experiences but also to function as the subject of works examining interior psychological and perceptual phenomena.

With an ear toward the effects produced by the loudspeaker in the interpretation of these works, its role as a component in both artistic and cultural practice is rendered audible. Attention to the operation of the loudspeaker in the forms of the bare cone, the commercial enclosure, and the headphone begins to develop a mechanism to address a range of issues in artistic production, from reproductive veracity, audiovisual parity, and geographic proximity to acoustic abuses of power and the narrative intimacy of amplified speech.

For all the breadth and range of aesthetic possibility suggested by the loudspeaker, it is not a neutral object. From a technical standpoint, the frequency response of most loudspeakers – the amplitude at which it reproduces each frequency – is rarely if ever an accurate reproduction of the source. This alteration of tone color in amplified sound production redoubles what film theorist Rick Altman describes, discussing audio reproduction in film, as the spatial signature of audio recordings generally: the acoustic fingerprint left by a recorded object or location is carried forward by the mechanism of reproduction.[1] Never an accurate mimetic double, the loudspeaker itself codes the transmission of the sound being reproduced, sometimes suggesting other reproduction mechanisms of like kind, but altering, however subtly, the character of the sound.

In consumer products, the loudspeaker is almost always hidden from view. Imitating the screen that hid Pythagoras from his pupils, the tweed Tolex cloth of a guitar amplifier, the plastic earpiece of the telephone, the screen of the cinema, and the housing of the headphone each hide the loudspeaker from the listener, concealing the source of the sound.[2] The conflation of sound and technology produced by the loudspeaker hidden in these devices obscures the sound producing mechanism and reinforces the normative media-specific relationship between the viewer/subject/audience and the information delivered. The audience of artworks employing consumer audio technologies therefore engages the content reproduced by the loudspeaker in a form that is both coded by the media-specific conditions of its use and inseparable from its multi-modal context. The loudspeaker is ubiquitous in the early history of video art, for example, enabling the television to function not only as a display device

1 See Rick Altman, *Sound Theory, Sound Practice* (New York: Routledge, 1992), 24.

2 Known as "acousmatics," the pupils of Pythagoras only heard their master from behind a screen, purportedly to ensure that they follow the content of his lecture unsullied by his visage. This term is mobilized by the French composer/theorist Pierre Schaeffer in his 1966 *Traité des Objets Musicaux* to indicate a sound presented in the absence of its source.

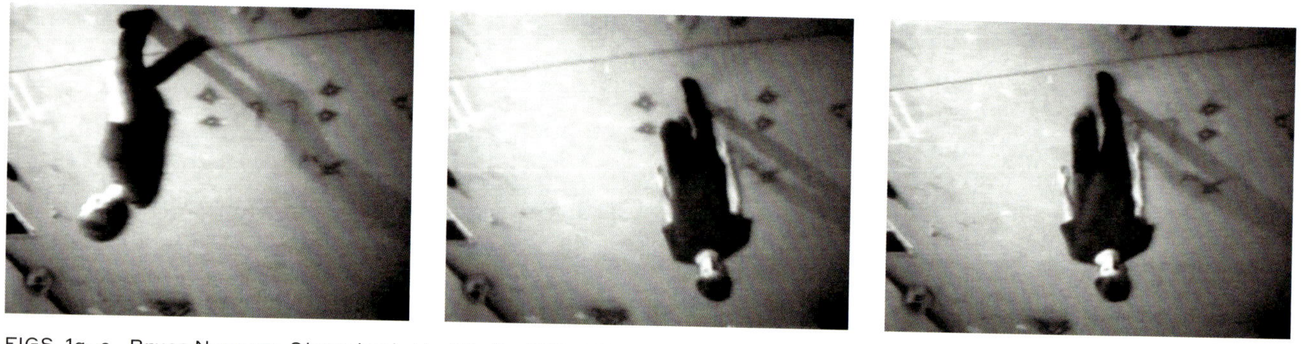

FIGS. 1a–c Bruce Nauman, *Stamping in the Studio*, 1968. Video, black-and-white, sound, 62 min.

for video but for audio as well. Works like Bruce Nauman's *Stamping in the Studio* (1968), Carolee Schneemann's *War Mop* (1983), and Gary Hill's *Mediations (Excerpt from a Remake of Soundings)* (1979/1986) connect action to audible trace effectively demonstrating Samuel Weber's point that "television – despite its name – involves the transmission of sight and sound."[3] As an embedded component masquerading as the source, the loudspeaker presents sound as a material support, a sound that is reciprocally coded by the apparatus in which it is housed.(FIGS. 1a–3f)

As a device enabling the amplification of sound, the loudspeaker allows sound to travel across substantial local distance, relaying explicit information via signals and sirens, implicit instruction by way of the voice, or musical content operating as background or foreground. Before the first large-scale public address loudspeaker systems were used to amplify the 1919 Liberty Loan Rally in New York City, the role of conveying acoustic information across geographic space was limited to bells, air sirens, and amphitheaters.[4] In the twentieth century, the loudspeaker takes on many of the roles once held by bell towers in public squares, churches, temples, and mosques, knitting together geographically proximate communities. Bells, and by extension sirens and other mass-signaling devices, "prescribed an auditory space that corresponded to a particular notion of territoriality."[5] Public loudspeaker-based works in artistic practice share some of the same aural effects of this use of bells within the landscape, such as Sam Auinger and Bruce Odland's public sound works (1987 to present), many of Walter Fähndrich's *MUSIK FÜR RÄUME (MUSIC FOR SPACES)* (1980–), and Max Neuhaus's *Time Pieces* (1989–2007). Each of these artists mobilize loudspeakers to spread sound over distance, enabling a source to reach exponentially greater numbers of listeners, highlighting communal experience while recontextualizing the regularly occurring sounds of the environment around the work.(FIGS. 4, 5)

As the component that renders transmission systems such as telephony, radio, and the Internet audible, the loudspeaker creates the condition of possibility for the spatial displacement of sound over non-local distance as well, and further, when transmitting recorded audio, it grants access to temporally displaced traces of captured sound. In this guise, the loudspeaker allows the sounds of different locations in cities – or between several cities – to be mixed together as in Maryanne Amacher's decades-long *City-Links* series (1967–1981). Similarly, Keith Sonnier's architectural dislocation of telephonically connected gallery

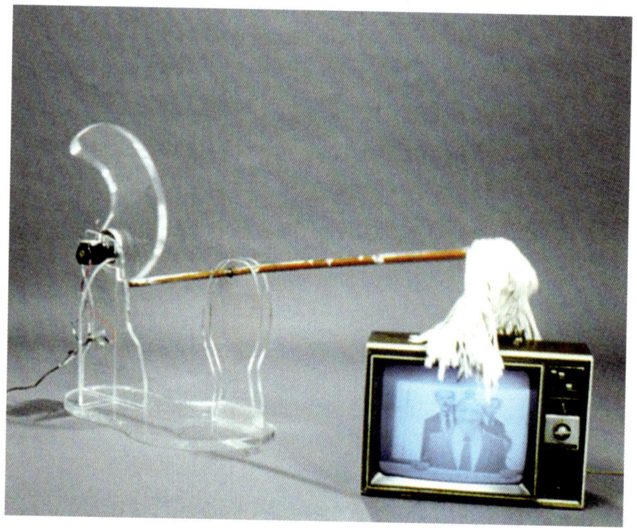

FIG. 2 Carolee Schneemann, *War Mop*, 1983. Kinetic sculpture: Mop, Plexiglas, motor, custom hardware, and monitor. On monitor: *Souvenir of Lebanon*, 1983. Video, color, sound, 6 min.

3
Samuel Weber, *Mass Mediauras: Form, Technics, Media* (Stanford: Stanford University Press, 1996), 109.

4
See Seth Cluett, "Acoustic Projection and the Politics of Sound," *Tacet Revue* 1, no. 03 (2014), 83–119.

5
Alain Corbin, *Village Bells: Sound and Meaning in the 19th-Century French Countryside*, trans. Martin Thom (New York: Columbia University Press, 1998), 95. Originally published as *Les cloches de la terre* (Paris: Albin Michel, 1994).

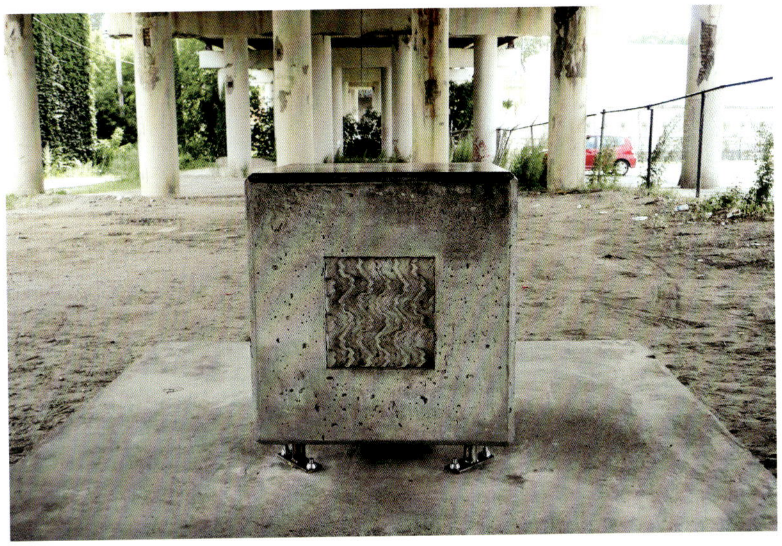

FIG. 4 Sam Auinger and Bruce Odland, *Harmonic Bridge*, 1998. Permanent sound installation, MASS MOCA, North Adams, MA. The cement "Cube" loudspeaker plays back a hemispheric field of harmonic resonance generated in a tuning tube on top of highway bridge just above.

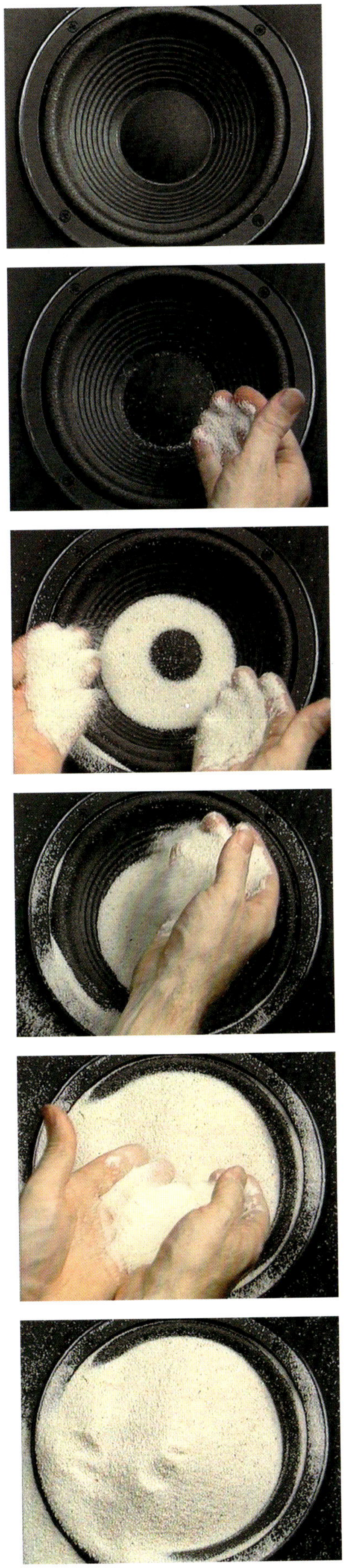

FIGS. 3a–f Gary Hill, *Mediations (Excerpt from a Remake of Soundings)*, 1979/1986. U-matic, NTSC, color, sound, 4 min.

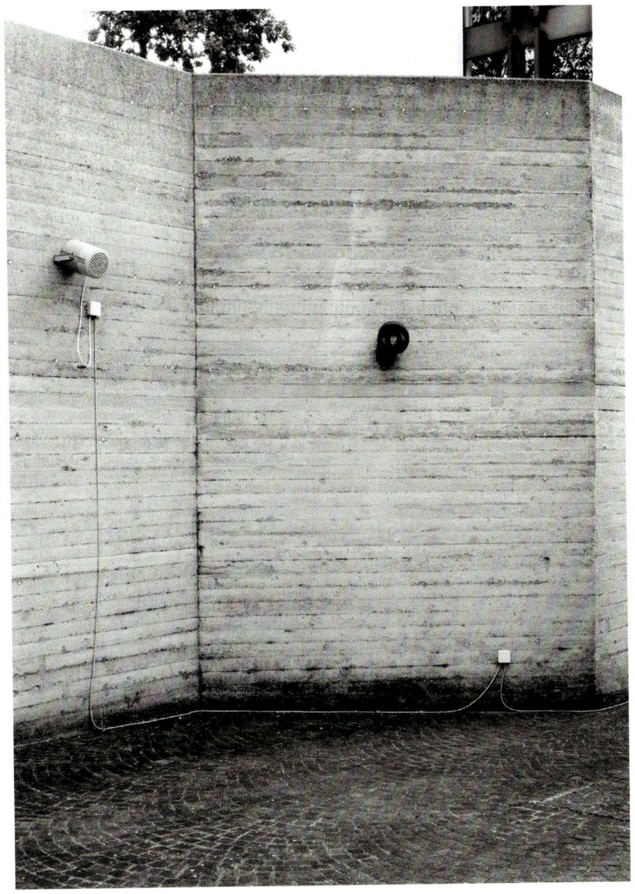

FIG. 5 Walther Fähndrich, *MUSIK FÜR RÄUME (MUSIC FOR SPACES)*, Sprengel Museum Hannover, 1995.

spaces in *Air to Air* (1975) and Bill Fontana's unidirectional re-presentation of live sounds from the Brooklyn Bridge transplanted to One World Trade Center in his *Brooklyn Bridge* (1983) demonstrate a common fascination with the uncanny effect of the dislocation of sound from source through technological means.(FIGS. 6a, b)

Deployed as an aesthetic component, the loudspeaker has functioned alternately as a presentation space, frame, substitute, surrogate, and prosthesis. This polyvalent potential makes it productive as a critical device because, through it, the manner in which sound operates within works of art can be critiqued, considered, and assessed with greater clarity. By attending to the cultural productions of the twentieth and twenty-first century from the point-of-audition of the effects produced by the loudspeaker, many of the factors that limit the clarity of arguments regarding sound as an artistic medium come clearly into focus.

Bare Cones

The bare loudspeaker cone is one of the most curious artifacts of the artistic use of sound. The bare cone is commonly employed in situations in which the loudspeaker is intended to function architecturally or blend seamlessly into the built environment, an object, or a landscape. When the loudspeaker is divorced from the coded consumer trappings of media technology (the cabinet, screen, or brand markings of a commercial enclosure), a direct cultural referent for the technology is substantially though not completely diminished. However, even the outward appearance of the cone materials themselves – paper, Mylar, plastic, Kevlar – still suggest a technological context for their use; the quality, cost, and origin of the bare cone is often coded in the materials employed by the artist. The bare cone becomes a voice excised from its media-specific body, and regardless of content the absence of an acoustically tuned enclosure dramatically alters the sound while situating it more completely in a new spatial condition.

Ulrich Eller's installation *Geräuschelager* (2000) employs 213 bare cones lined up one after the other on industrial metal shelving in a warehouse. Making use of myriad brands,

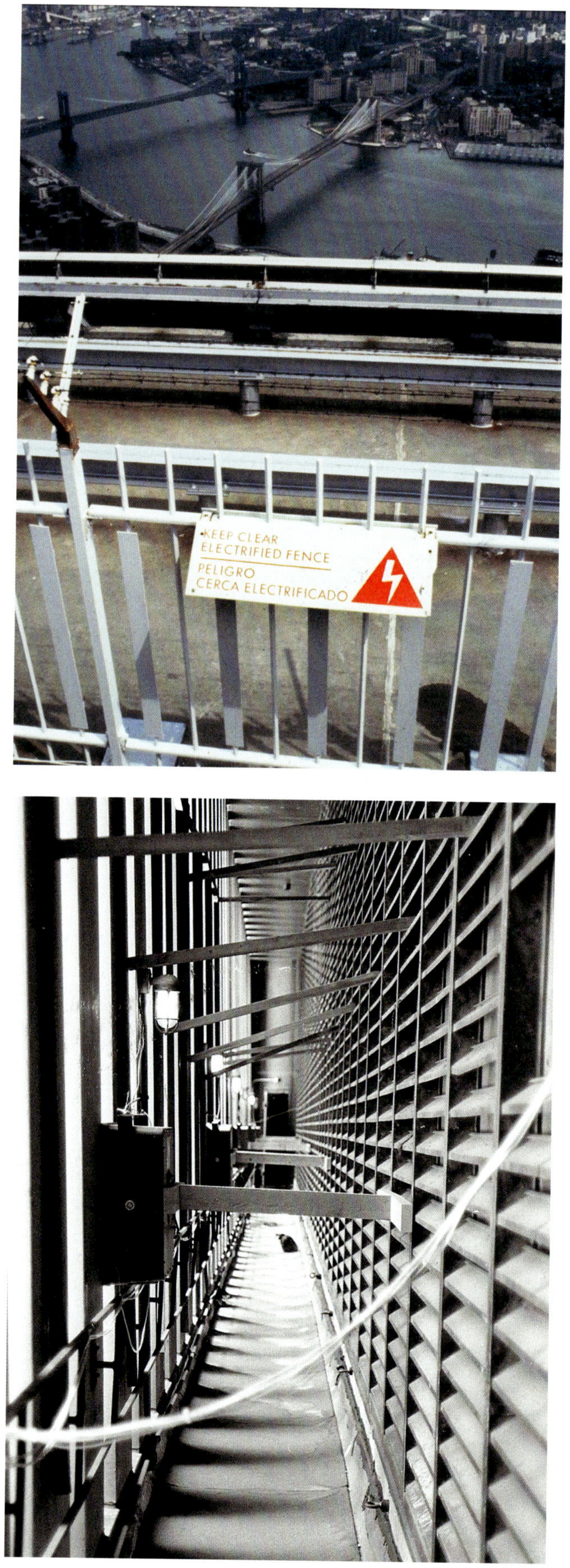

FIGS. 6a, b Bill Fontana, *Brooklyn Bridge*, 1983. Sound sculpture.

6
René Hirner in *Qin Yufen* exhib. cat. Kunstverein Göttingen et al. (Tübingen: Ernst Wasmuth Verlag, 2001), 106.

shapes, and sizes of historic speaker cones extricated from their housing, the varied materials and technological origins of the loudspeakers transmit a single, noise-based sound source. Though the sound source sent to each is identical, each loudspeaker imparts its audio-technical signature on the sound it (re)produces. As the listener moves through the work, each speaker colors the volume and timbre of playback, and the individual character normally eschewed by works employing multiples becomes a subtle revelation for the listener. Movement through the work reveals change and difference, as the sound of the whole gives way to the local awareness of each singular sounding object. Highlighting the technological multiple as an archive of sonic difference, Eller's clinical display of bare speaker cones points up an index of the loudspeaker as a discarded marker of technological consumption and overproduction.(FIG. 7)

In *Microtonal Wall* (2011), Tristan Perich displays fifteen hundred identical miniature speaker cones in a regular grid, fitted perfectly into a laser-cut aluminum housing mounted to the wall. Perich inverts the conceptual strategy

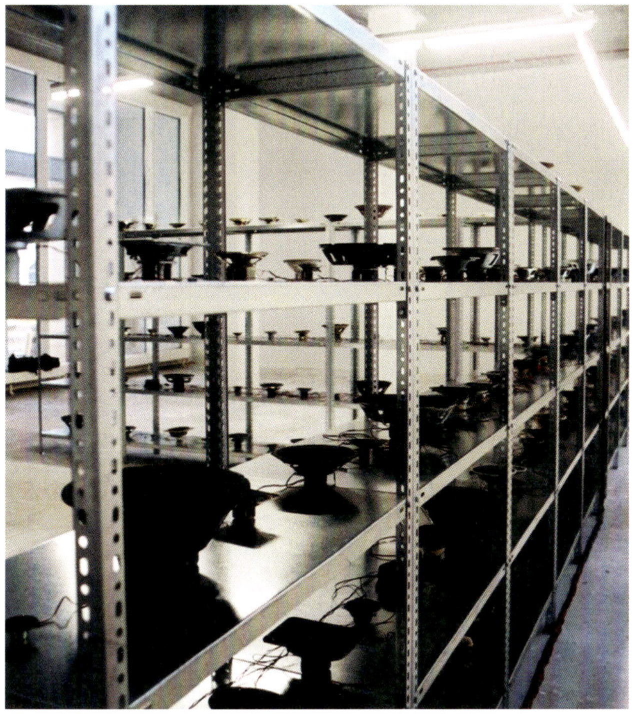

FIG. 7 Ulrich Eller, *Geräuschelager*, 2000. Klangkunstforum, Park Kolonnaden Potsdamer Platz, Berlin, 2000. 16 galvanized shelves positioned in T-shape, neon ceiling lights in identical form and dimensions, 213 loudspeakers of different type, 24-channel composition, cables, feed technology.

of Eller by sending different, single sine waves through each of the identical loudspeakers. The resulting gestalt, heard from afar, is perceived as white noise, while from close-up the atomized components of the whole are revealed as singular elements within a field of sameness. Where the installation condition of horizontal display in the Eller encourages the listener to observe a single source within and across time and technology, Perich emphasizes the individual experience of the listener as they become bodily aware of change performed through their movement.(FIGS. 8a, b) Like Pe Lang and Zimoun's *Untitled Sound Objects* (2008) and Richard Garet's *30 Cycles of Flux* (2013),(FIGS. 9a–10) Perich's use of loudspeaker cones creates a modern display of the multiple that causes tension where locally individuated and globally unified perceptions oscillate in and out of focus.

In the 1992 site-specific work *Partitur* by Qin Yufen, dozens of three-inch black cones playing the sound of cloth rustling are draped seemingly haphazardly over common metal clothes dryers. Beyond a mere delivery device, the bare cone here functions as a trace of, or surrogate for an implied material, providing the audible presence of a visually absent source. In contrast to the varied formal conditions of the works of Eller and Perich, Qin Yufen's use of sound is often carefully constructed within theatrical installations that make use of everyday household items. The bare cone in *Partitur* is one of many iterations in Qin Yufen's output that examine similar objects (sheets of fabric, items of clothing, laundry drying racks) mobilized to work with and against the conditions of the exhibition space. *Partitur* uses the folding aluminum laundry drying racks seen throughout many works in her oeuvre: but where previous iterations see the drying racks draped with clothing, paper, or stacked architecturally, in this work, loudspeakers play the sounds of hands working with fabric. As also in her work *Home is Here, Home is There*, for example, the loudspeaker thus becomes a placeholder for a physical action critiquing the "irksome parts of housework"[6] evoked as a memory rather than an explicit material condition.(FIG. 11)

Eller, Perich, and Qin Yufen's works deploy the bare loudspeaker to open the interpretive potential of sound in installed environments, lessening the effect of an explicit relationship to a

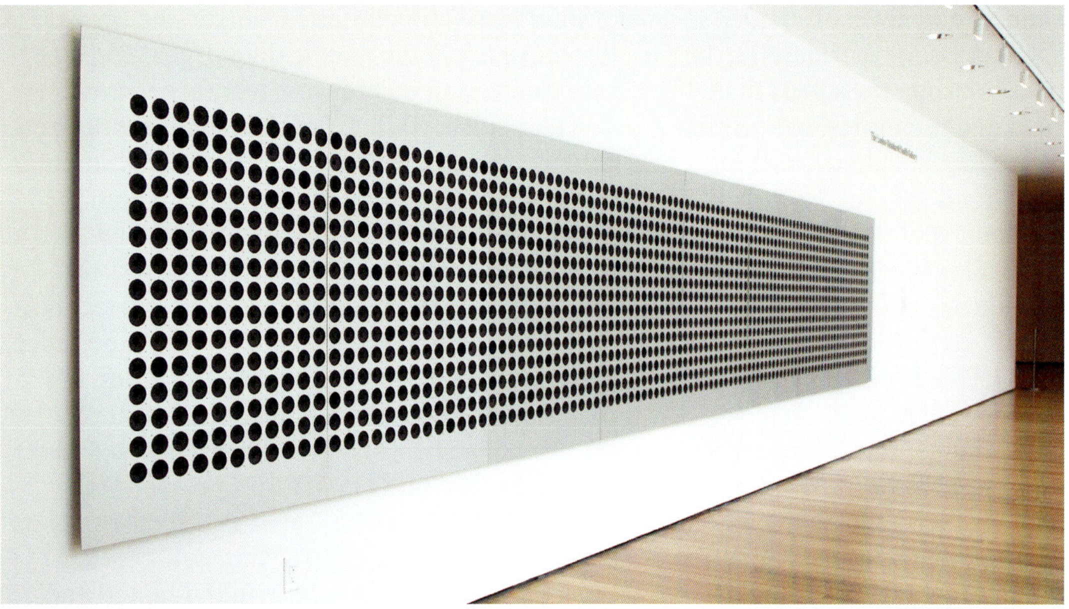

FIG. 8a Tristan Perich, *Microtonal Wall*, 2011. 1,500 1-bit speakers, microprocessors, aluminum.

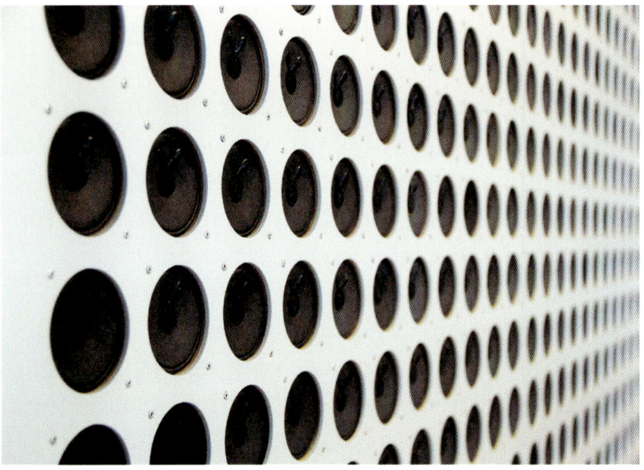

FIG. 8b Tristan Perich, *Microtonal Wall*, 2011. Detail.

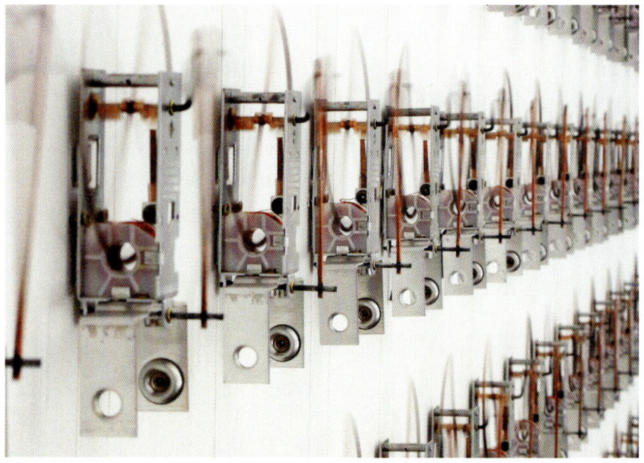

FIG. 9a Pe Lang + Zimoun, *Untitled Sound Objects*, 2008. Detail.

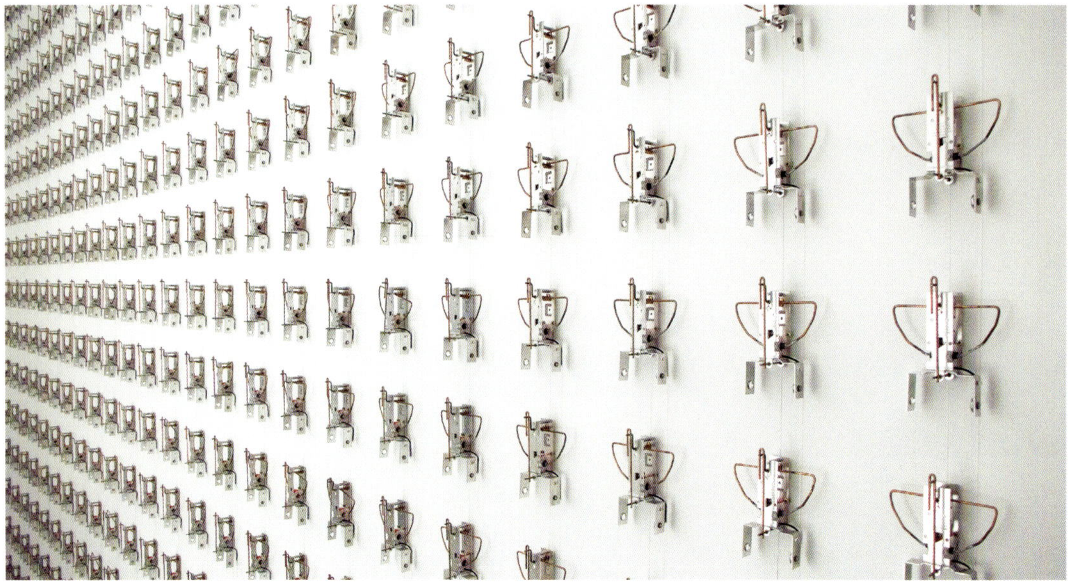

FIG. 9b Pe Lang + Zimoun, *Untitled Sound Objects*, 2008. 450 pendulum motors on wall, 570 × 290 cm. Installation view Bon Accueil Rennes.

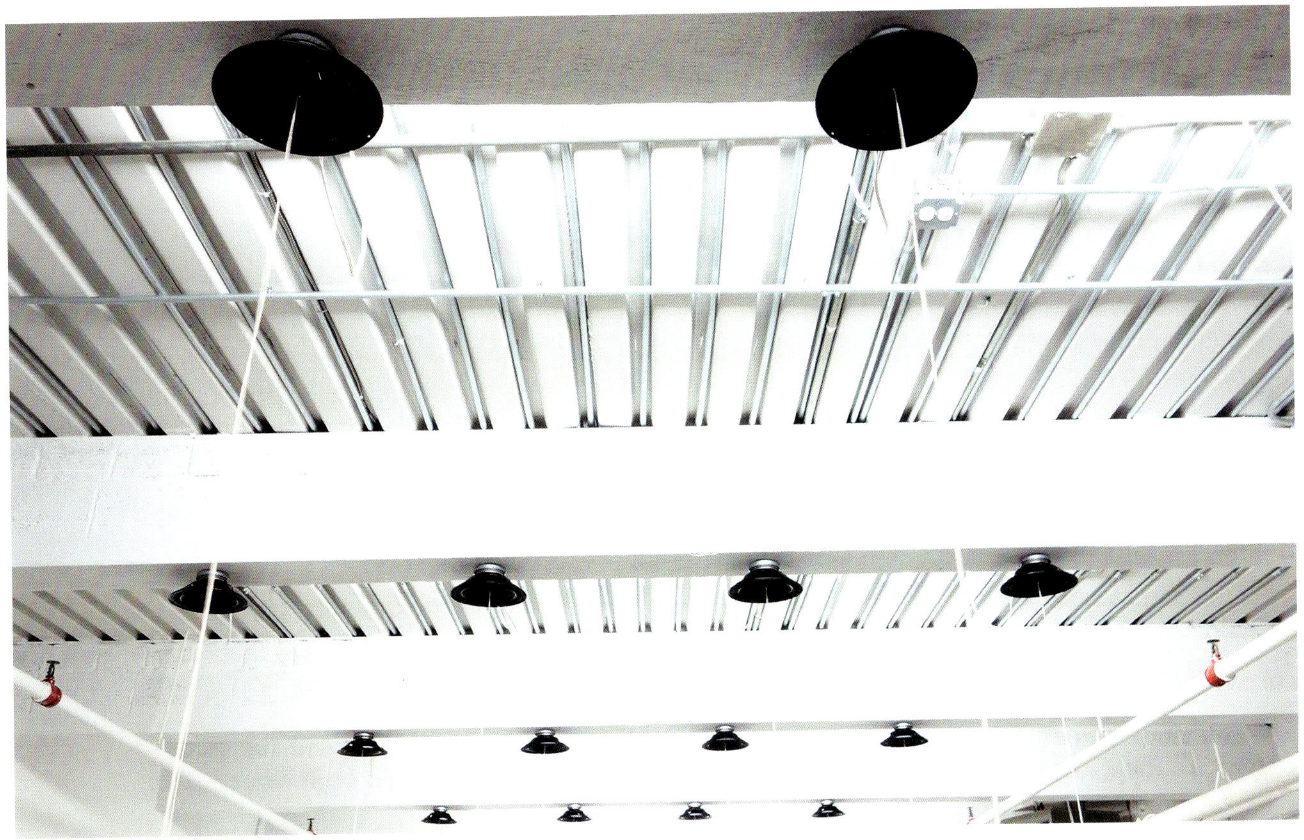

FIG. 10 Richard Garet, *30 Cycles of Flux*, 2013. Sound installation. Installation detail from *Extraneous to the Message* at Julian Navarro Projects, NY, 2013.

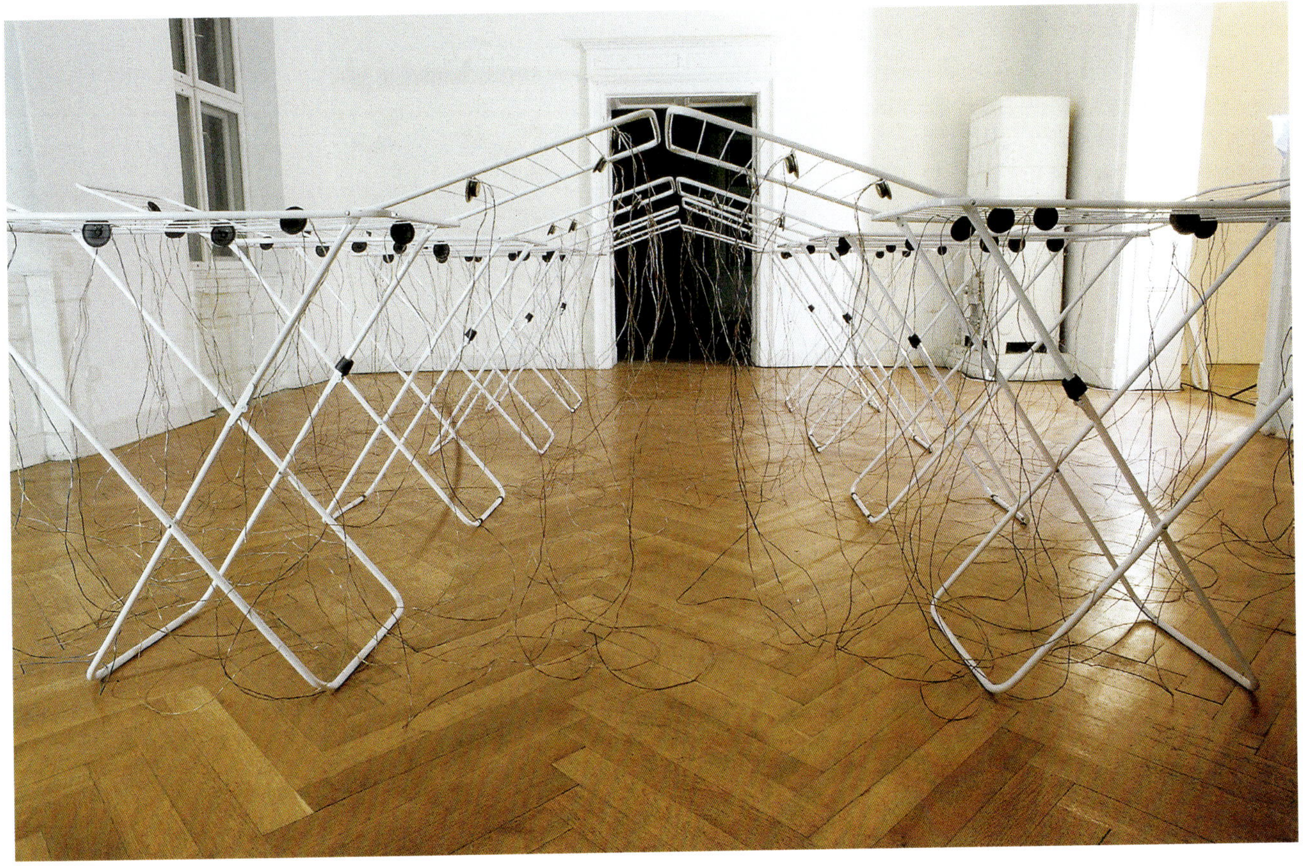

FIG. 11 Qin Yufen, *Partitur*, 1992. Sound installation (laundry rack, loudspeaker, cable, Walkman, car amplifier, audio tape, stereo) ca. 120 × 440 × 500 cm.

media device, and allowing the audible to operate more as phenomenon than product. The degree of occlusion, the physical placement, and the choice of material become crucial components in the way a piece is read by the public experiencing the work. The outward spatial condition of bare cones becomes a sculptural decision whose effects are compounded by the relationship between the content of the sound produced and the type and manner of their distribution. By shedding the enclosure and reducing the footprint needed to project sound, the bare cone efficiently embeds sonic phenomena and semiotic content into both architectural spaces and soundless objects.

Enclosures

The consumer loudspeaker enclosure carries with it the legibility of the media of which it is a part or from which it was repurposed. Enclosed loudspeakers overtly cue a focused listening space in the surround or ambisonic multichannel work of artists like Jacob Kirkegaard, Jana Winderen, and Stephen Vitiello, or provide material for sculpted structures built from the post-consumer detritus of disused hi-fi components, boom-boxes, and radios in the work of Benoît Maubrey, Maia Urstad, and Alexis O'Hara. Functioning symbiotically within haptic and image media technologies, enclosed loudspeakers project the soundtracks of video art and film, complete the immersive space of virtual reality, or enable the communication potential of the telephone. Further, the ubiquity of loudspeaker enclosures in public space and the proliferation of sound systems for use in both homes and entertainment venues provide an accessible vocabulary of experience to exploit toward the expressive or critical intent of the work.

During the same period as his now-iconic sound works *Forming Sounds* (1971)(FIGS. 12a–c) and *Echo* (1973), Dennis Oppenheim produced a large-scale kinetic installation entitled *Table Piece* (1975). In this work, two marionette figures (without strings) of the kind seen so often in Oppenheim's work, one in black, one in white, sit at opposite ends of a sixty-foot long table; black at one end and white at the other, the table traces the gradient between the colors of the marionettes. The mouths of the marionettes move as the speaker inside each repeats the words "white, light, black, dark" into a microphone. The words are broken up and randomized by phoneme and projected over four loudspeakers placed at regular intervals on the underside of the table. The resulting spatialized performance of language gathers meaning as the individual phonemes congeal into words. *Table Piece*, like much of Oppenheim's body of work, deals with issues of the body and power that view the site of production as a ritual performance space. By exaggerating the table dimensions, using the visual dichotomy of black versus white, and diffusing the sound over the physical distance between the marionettes, Oppenheim elicits the impression of tension and misunderstanding. The deliberate placement of consumer audio devices emphasizes the spatial expanse of the table, critiquing the transmission-loss produced during the reconciliation of verbal exchanges by people holding polarized views.(FIG. 13)

Camille Norment's installation *Dead Room* (2000) creates an oversized loudspeaker enclosure that may be entered by the listener.(FIG. 14) By inverting the position of the speaker cabinet and the listener, the room is encapsulated by acoustic treatments and lined with colored vinyl-wrapped foam. Legible alternately as a stylized car audio subwoofer and the padded cell of solitary confinement, this work extends from

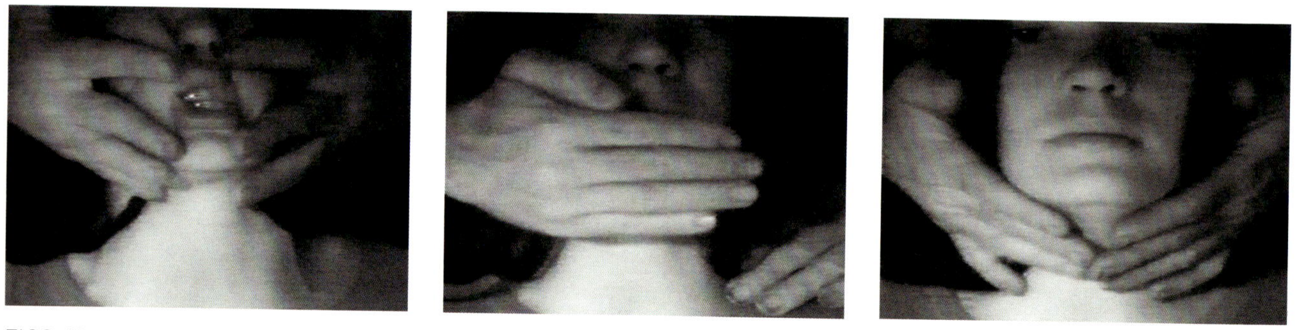

FIGS. 12a–c Dennis Oppenheim, *Forming Sounds (Phyllis and Dennis Oppenheim)*, 1971. Video, Betacam SP, black-and-white, mono, 7:11 min. Video stills.

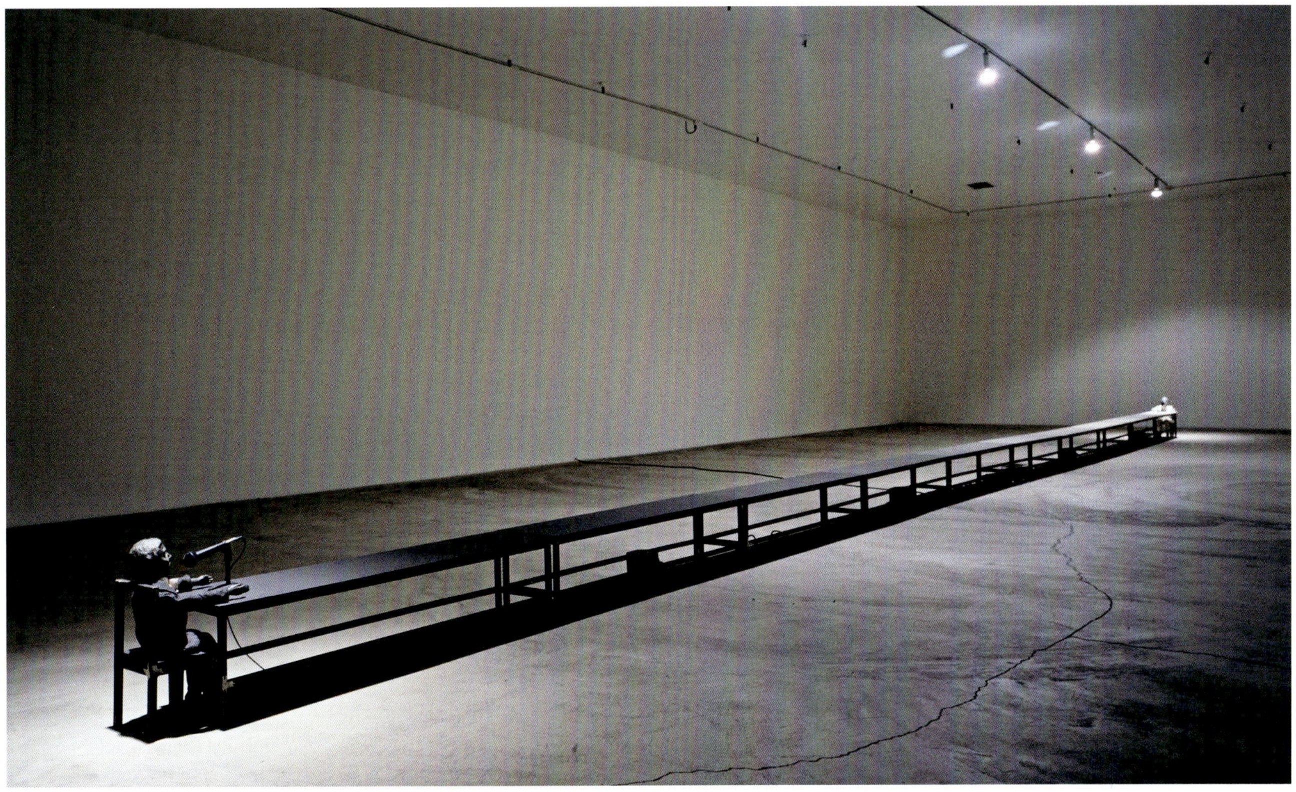

FIG. 13 Dennis Oppenheim, *Table Piece*, 1975. Figures at each end equipped with a speaker system located inside the head, plus an exterior control device that activates the lower jaw in perfect lip sync to external voice control input. Both units are attached to a stereo system which allows the figures to carry on a dialogue. Steel, wood, and formica tables and chairs; 2 wood, felt and cloth figures with mechanical jaws which activate an exterior sound system consisting of four speakers, 2 microphones with stands, an electronic control box and mixer for audio tape. 120 × 180 × 60 cm, approximate dimensions. Collection, Cartier Foundation, Paris.

Norment's long-term engagement with the social politics of designed objects to address topics of technically-mediated experiences, identity, and the body. When the listener enters *Dead Room*, an air pressure increase is created by sub-audible tones emanating from subwoofers enclosed in the space, creating a feeling of suffocation and anxiety. In the absence of any audible sound, the work makes the speaker enclosure itself a habitable space that can be entered and explored. The acoustically dense air caused by the tones inside the enclosure additionally cancels frequencies in the voice of anyone who tries to speak within it. This denial of voice points to the history of noise abatement and silencing in African American communities and the public use of sound as a delivery system for oppression by highlighting the use and abuse of auditory technology in the built environment.

In his 2009 solo exhibition at the Abbaye de Maubuisson, Dominique Petitgand uses hi-fi loudspeakers distributed throughout the grounds to create a series of audible characters for the viewer/listener to encounter while traversing the space of a medieval monastery. The works in the exhibition point up the solemnity of the site by presenting sounds that refocus the spatial experience of a structure intended for silence. In *Les ballons* (2006/2009), the sound of ricocheting rubber balls is used to superimpose notions of play over a space of contemplation;

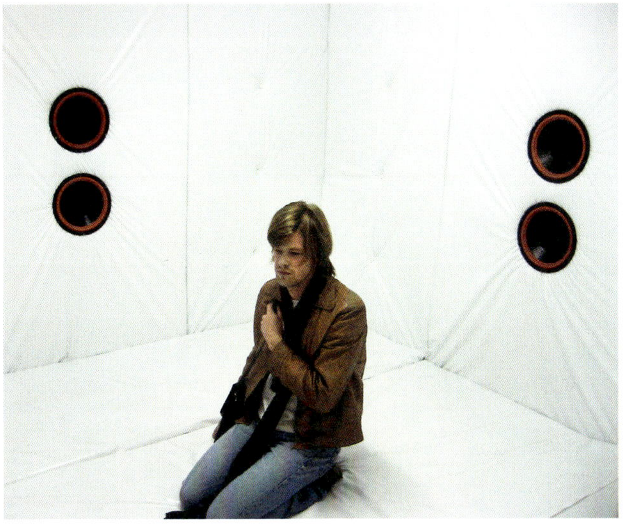

FIG. 14 Camille Norment, *Dead Room*, 2000. Architectural sound installation, 360 × 360 × 240 cm.

in *Je parle* (2009), children's voices break the silent focus of the space, interjecting broken phrases: "jamais [...] on ne se regardait pas" ["never [...] we were not looking at each other"].(FIGS. 15, 16) In contrast to the visually charged works of Norment and Oppenheim, Petitgand eliminates any outwardly visible referent, using the loudspeaker to create aural spaces that unfold into an awareness of mental imagery for the listener. Mounted on plinths as well as gallery walls and ceilings, the overtly placed consumer loudspeakers in Petitgand's work explicitly overlay a conceptual space on the extant architectural condition to examine culturally situated listening while ensuring that the work is heard as having been added to the space. On par with the multichannel video

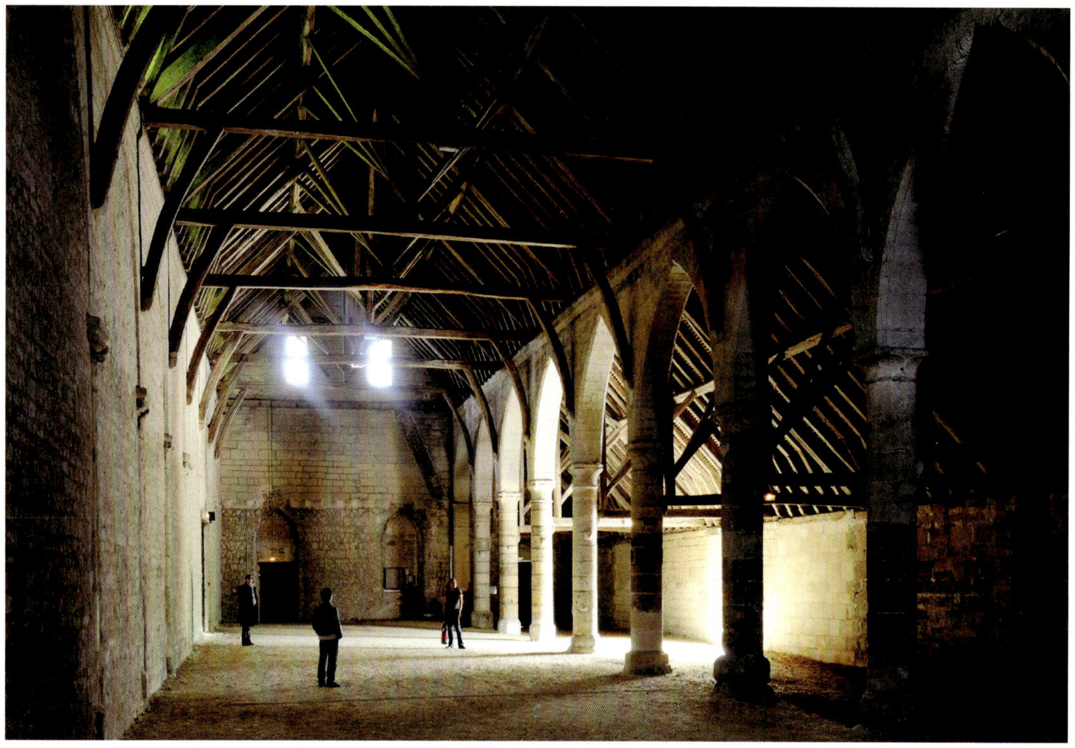

FIG. 15 Dominique Petitgand, *Les ballons*, 2006/2009. Sound installation with 4 speakers.

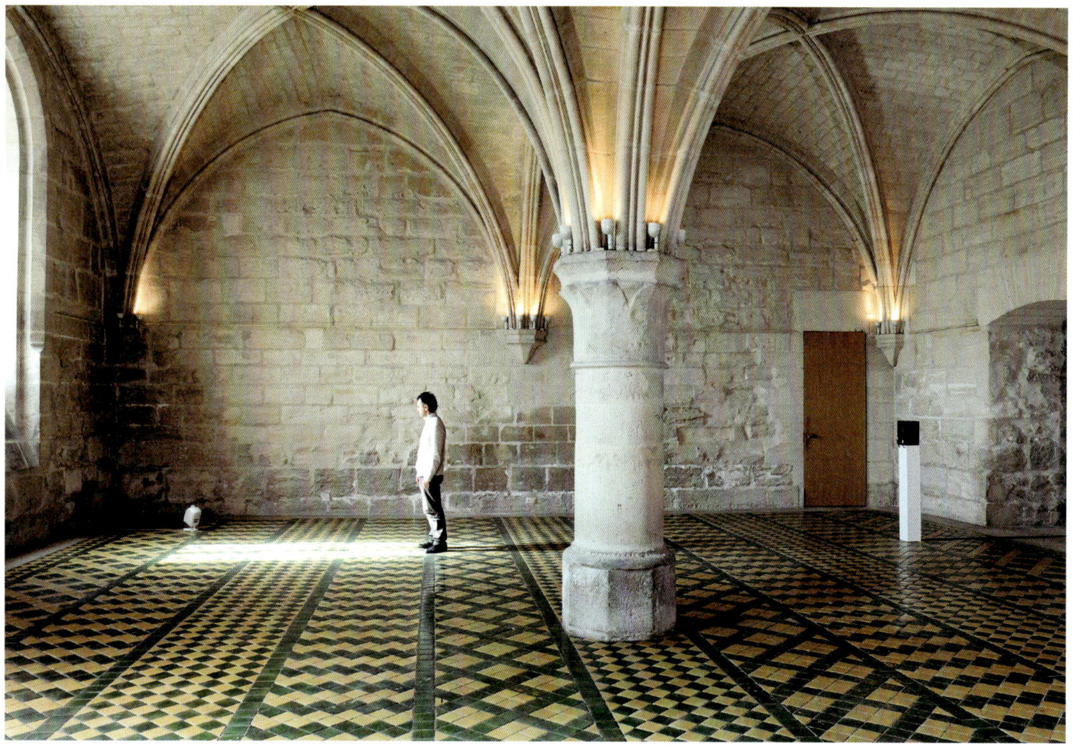

FIG. 16 Dominique Petitgand, *Je Parle*, 2009. Sound installation with 4 speakers.

works of Douglas Gordon, Christian Marclay, and Mary Lucier, Petitgand creates auditory environments using loudspeaker enclosures to deliver material at once arrestingly literal and poetically obfuscating.

Within the outward visual trappings of branded enclosures or the explicitly stylized use of loudspeakers in purpose-built structures, the boxed loudspeaker brings with it the legibility and implied use of media and architectural spaces familiar to viewers. All loudspeakers can operate in art practice as a means of presenting sound in parallel to other content, and by extension, the sound they produce within the works is done so both with and against other sensory modalities. Consumer loudspeaker enclosures, however, compound the dynamic semiotic relationship between fixed or moving images and the always already time-based content of sound, further complicating the multimodal experience of the work by referencing their use outside of artistic practice.

Headphones

At once a generalized curatorial space solving the problem of sound isolation, a ubiquitous consumer space enabling personal soundtracking, and a professional listening environment in the service of fidelity, the headphone has long been a component in the production of sound works, a subject within multimedia practices, and the display apparatus for linear audio works as an autonomous artistic medium. Not simply a convenience for isolated presentation and distribution, the headphone, as a culturally prevalent technology, codes and is coded by its everyday use.

As a common curatorial conceit, the headphone has been explored as an exhibition space with increasing frequency since the turn of the millennium. The two most notable examples of this practice are the *Volume: Bed of Sound* exhibition mounted by MoMA PS1 in 2000, where numerous headphones are strewn about a large comfortable surface presenting a survey of artists' sound works, and more recently in the 2008 exhibition *21:100:100* at the Melbourne International Arts Festival, where headphones descend from the ceiling of the exhibition space, resting on placards providing written descriptions of sound works by one hundred contemporary artists.

The use of headphones in this form, however, begins with the genre of artists' cassettes in the late 1970s in exhibitions such as *Sound* (1979) at the Los Angeles Institute of Contemporary Art and *A Sound Selection* (1978) at Artists Space in New York.[7] Headphone listening stations, whether for purpose-built works, playback of music, or used in sound works related to the exhibition theme, have become standard in sound exhibitions since the widespread consumer adoption of stereo headphones in the mid-1960s.[8]

In 1974, Richard Serra and Nancy Holt created *Boomerang* for broadcast on a local television station in Amarillo, Texas. A classic of early video art, *Boomerang* is a single-channel video with monaural audio consisting of a close-up of Holt wearing headphones against a solid background.(FIG. 17) As Holt speaks, her voice is echoed back to her through the headphones she is wearing; her spoken voice on camera and the sound of her vocal echo returned to her are mixed and audible for the viewer. The headphone, employed here as a conceptual conceit, exposes the operation of an interior listening that mirrors thought in its opposition to and practical exclusion of exterior stimuli. By hearing herself at a delay through headphones, Holt reflexively confronts the improvisation of her externalized speech in the same

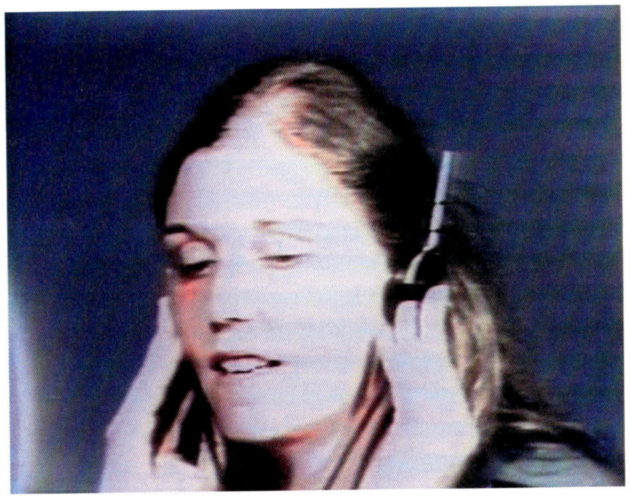

FIG. 17 Richard Serra, Nancy Holt, *Boomerang*, 1974. Video, color, sound, 10 min.

7
For an in-depth discussion of these early exhibitions, see Seth Cluett, "Ephemeral, Immersive, Invasive: Sound as Curatorial Theme 1966-2012," in Nina Levent and Alvaro Pascual-Leone, *The Multisensory Museum: Cross-Disciplinary Perspectives on Touch, Sound, Smell, Memory, and Space* (Lanham: Rowman & Littlefield, 2014), 114–115.

8
While available since the turn of the century and used throughout both World Wars, the introduction of the first stereo headphone by the American company Koss in 1958 is a milestone in consumer audio consumption. Since the increased competition from Philips, Onkyo, and Sennheiser in the 1960s, the headphone has remained a fixture.

form in which she is generating that speech. This chaotic mental juxtaposition of the present moment with a reflected echo of the recent past derails the thought process of content generation and destabilizes her vocalizations. And while traditionally conceived notions of feedback from cybernetics suggest that the return processing of information enables adaptation and continual refinement, the normative feedback loop (hearing oneself speak) is countered in *Boomerang* by the delay initiated through the headphones. The headphones in use here isolate the speaker from the expected feedback loop exposed to the listener through the layering of the source and its modification.

Bernhard Leitner's project *Kopfräume/Headscapes* (2003) "is a three-dimensional sculptural work exhibited on audio CD."[9] Literally translated as "Headspaces" or "Headrooms" and described by the artist as "Headscapes," this work is presented with the explanation that these pieces "do not represent exterior space, but were conceived and created specifically for the interior of the head" and that it is an "audio CD for earphones only." Like much of Leitner's monumental, architectural, minimal works, *Kopfräume* presents a series of isolated sound objects subjected to panning manipulation to imply movement, as well as reverberation, equalization, and amplitude cues to suggest distance. More than just an artistic application of the 3-D stereo test record fads of the 1960s, Leitner aims to construct architectural geometry in the mind by taking advantage of the headphone's ability to control the spatial perception of sound. By eliminating the dependency on real physical space, the headphone makes possible the condition of isometric head placement required to ensure repeatable experience of intangible yet perceivable shapes made with sound.(FIG. 18)

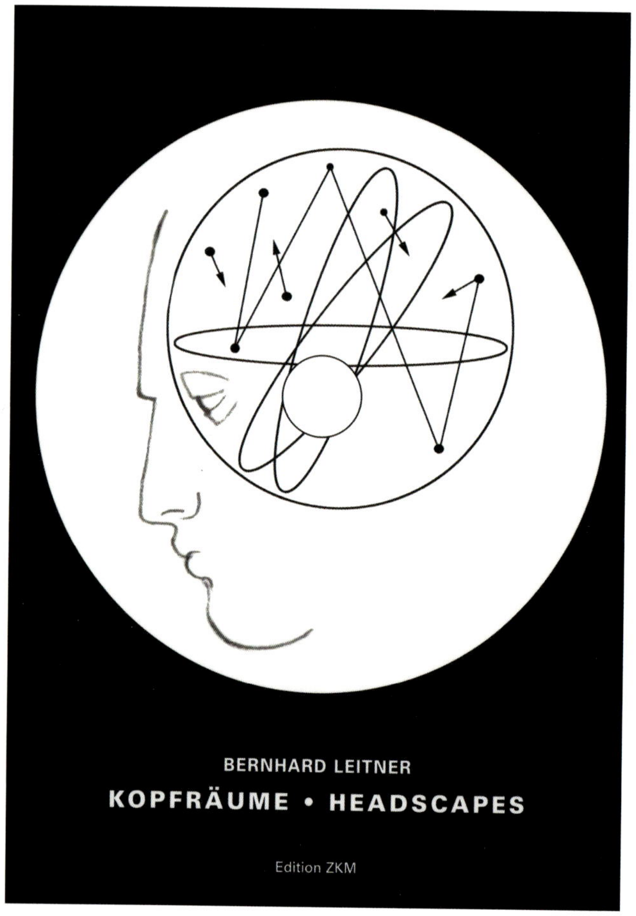

FIG. 18 Bernhard Leitner, *Kopfräume/Headscapes*, 2003.

In *Reasons for Knocking at an Empty House* (1983), Bill Viola incorporates headphones as an integrated component of a larger, multimodal scene. In this work, a CRT television monitor sits atop a white plinth opposite a roughly constructed object that vaguely resembles an electric chair. Two loudspeakers are placed in the far corners of the room and a pair of headphones is mounted to the back of the chair so that, when sitting, the gallery-goers are able to sit at eye-level with the video screen and easily place the headphones on their ears. The video image is a medium close-up at eye level of a very tired, almost delusional Bill Viola.(FIGS. 19a–c) As Viola sits, drinks water, and breathes, the sound of these actions, close-mic'd through a lavalier microphone, are playing in the headphones above a bed of whispering, thoughtlike voices. When a silhouetted figure occasionally strikes Viola from behind, the sound comes through the speakers placed in the installation space. In this work, interior sounds that are conceptually mimetic of the non-linearity of thought are paired with the close-mic clarity and unreal detail of body sounds played directly to the ear. In bypassing the air and entering the body directly via the headphones, the juxtaposition of these sounds with the loudspeakers' jarring exterior rendition of Viola being struck turns the room into an exploded representation of Viola's own body – the raw sounds of his pragmatic interiority and the invasive, inescapable violence of the exterior world.

In the form of headphones, the loudspeaker serves as a technology that can become

9
Bernhard Leitner, *Kopfräume / Headscapes* (Ostfildern: Hatje Cantz, 2003).

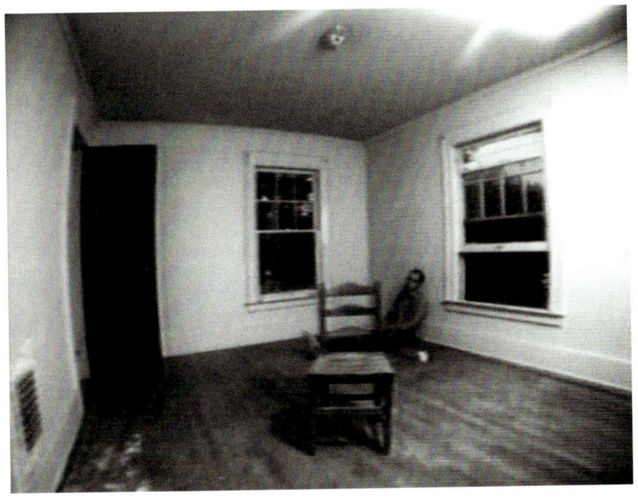

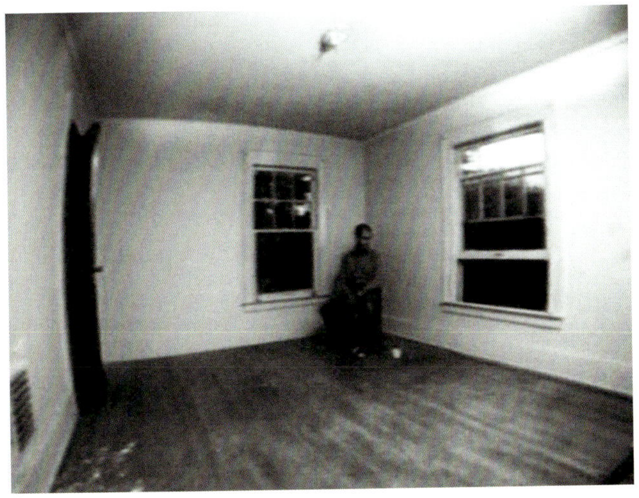

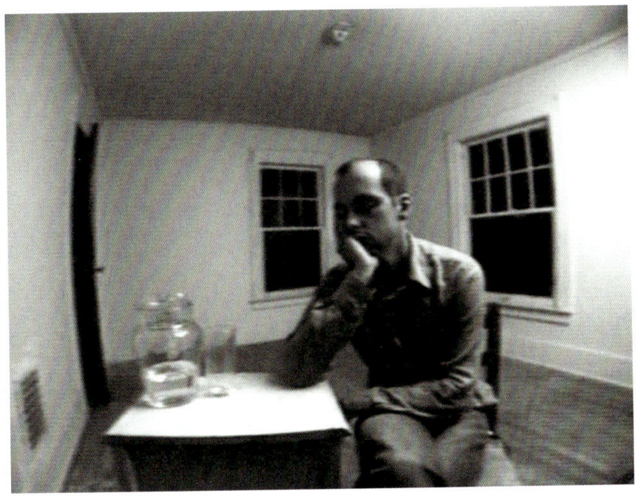

FIGS. 19a–c Bill Viola, *Reasons for Knocking at an Empty House*, 1983. Video, black-and-white, stereo, 19:12 min.

instrumentalized through action in either the physical space of sound-walks or on-screen in video art. Functioning as an isolated listening space, headphones allow for the controlled fabrication of aural space, enabling both the abstracted experience of structure and the narrative experience of place and story. Further, the direct, private connection to the ear and the corded tether turns the headphone into a device emblematic of and critical toward power relationships, at once reflecting the intimate trust of close listening and the potential danger of sonic abuse of power. The ubiquitous use of the headphone as a display device for curatorial convenience has obscured the rich conceptual territory explored by artists and composers. In current use, the headphone provides both solitude and soundtrack for commuters as well as mobility and monitoring for freelancers. And while the recent ubiquity of headphone listening may place these portable listening systems in the critical territory of devices aimed at distracting the masses, they also create a productive venue for artists willing to engage its myriad affordances.

Conclusion

Functioning, in its various uses, as a material support, a cultural object, a static technology, and an invisible source, the loudspeaker is only one lens through which to understand the history of sound in the visual arts and music. It would also be productive to consider this history in the context of some of the other through-lines that move across the past fifty years of sound practice: long strings (Ellen Fullman, Alvin Lucier, Gordon Monahan, Paul Panhuysen, Johan Goedhart), the voice (Janet Cardiff and George Bures Miller, Christof Migone, Cathy Lane, Martin Riches), or kinetic movement (Ed Osborn, Trimpin, Edwin van der Heide, Terry Berlier). Understanding the loudspeaker's operation as a component – and by extension sound-as-component – reveals a technology whose effective capacity is derived from a productively multivalent relationship to the sound source(s). By both tacitly facilitating the auditory characteristics embodied by media in everyday life and actively representing the content, material, and intent of the artist, the loudspeaker has productively extended the audible reach and expressive potential of the creative work of the last fifty years.

Alexandra Supper

Singing Data: Sonification and the Relation between Science and Art

During an event related to the *David Bowie is* exhibition in April 2013, the musicians Alexis Kirke and Martyn Ware took the stage with an eccentric set of compositions at the Victoria and Albert Museum in London. Their pieces, which they referred to as a "career sonification" of David Bowie, translated statistical analyses of Bowie's work over the years into a series of sometimes quite dissonant-sounding electronic pieces. One piece used an analysis of the emotional content of Bowie's lyrics to drive a "hyperspeed piano" tune, while another mapped his album sales onto pitch – resulting in both painfully high tones and low rumbling sounds, to mark the huge fluctuation in commercial success over the years.(FIG. 1), 1

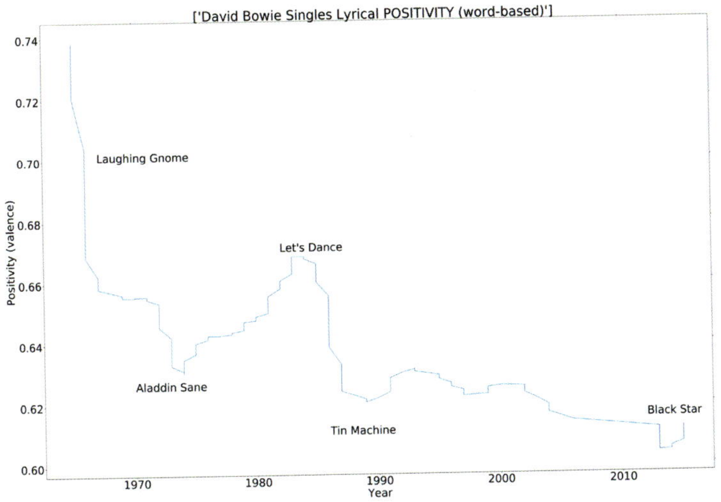

FIG. 1 Alexis Kirke and Martyn Ware, *Career Sonification of David Bowie*, 2011. Chart of David Bowie Singles Lyrical POSITIVITY (word based).

The career sonification of Bowie's work may seem like a somewhat unusual example of sonification, the auditory representation of data. After all, the idea of sonification is that data – which in and of themselves usually have nothing to do with sound – are translated into an audible signal, for the purposes of facilitating communication or interpretation.[2] Some prominent examples include the sonification of scientific data from domains such as neurology or seismology.[3] The career sonification, on the other hand, is based on data that already deal with music. Nonetheless, the piece follows the principles of sonification: rather than playing back a recording of one of Bowie's songs, it creates new sounds that are driven by parameters in a dataset. Also, there is nothing unusual about the fact that elements of scientific data display and music are brought together – on the contrary, the blurring of boundaries between art and science is quite typical for sonification. In this essay, I want to trace the relationship between science and art in sonification.

Singing Stars and Banging Brains: Musical Precedents for Sonification

The idea of making data audible has a long musical tradition. Some of the canonical early examples include John Cage's *Atlas Eclipticalis*, composed in 1961/1962, which transposed a map of the stars onto a graphical score; Alvin Lucier's *Music for Solo Performer*, first performed in 1965, which amplified brainwaves and then used the resulting pressure waves to trigger percussion instruments placed around the stage; and Charles Dodge's *Earth's Magnetic Field*, released in 1970, which used electromagnetic data to control computer-generated sound.[4] Compositions like these are not intended to enable scientific insights or communicate facts, but rather to open up a compositional approach. They can be understood as part of a shift in the course of the twentieth century from conceptions of music that stress its individuality, subjectivity, and expression toward a more materialist and objective one, emphasizing artistic restraint and preaching "the resolute elimination of the artist's ego or personality from the artistic product."[5] Sonification, as a way of ostensibly handing over some compositional decisions to scientific data rather than an individual artist, fits this tendency perfectly.

However, this is not to say that compositional decisions are indeed delegated entirely to data. Lucier's *Music for Solo Performer* is a case in point: although Lucier has a reputation for using his compositions as a means of exploring natural and scientific phenomena,[6] the piece is by no means directly driven by brainwaves. A series of aesthetic decisions and technological interventions shape the musical output – not in the least because an assistant decides which loudspeakers, and thus which adjacent percussion instruments, to activate during any given part of the performance. As Volker Straebel and Wilm Thoben have recently pointed out, Lucier has sometimes downplayed these decisions and boosted the illusion of a direct translation of brainwaves into music to create theatrical effects – an interpretation that has been eagerly picked up by the interdisciplinary community dedicated

1
A brief explanation and sound examples of the two sonifications discussed above can be found on Alexis Kirke's website: www.alexiskirke.com/projects/sonification-of-david-bowies-career.

2
The following texts provide an overview and introduction to the field of sonification research: Gregory Kramer et al., "Sonification Report: Status of the Field and Research Agenda. Report prepared for the National Science Foundation by members of the International Community for Auditory Display" (1997), www.icad.org/websiteV2.0/References/nsf.html; Thomas Hermann, Andy Hunt, and John G. Neuhoff, eds., *The Sonification Handbook* (Berlin: Logos Verlag, 2011). For a critical discussion of the efforts to define sonification in relation to science and art, see Alexandra Supper, "The Search for the 'Killer Application': Drawing the Boundaries around the Sonification of Scientific Data," in *The Oxford Handbook of Sound Studies*, ed. Trevor Pinch and Karin Bijsterveld (New York: Oxford University Press, 2012), 249–70.

3
See Florian Dombois, "Using Audification in Planetary Seismology," in *Proceedings of the 7th International Conference on Auditory Display*, July 29–August 1, 2001, Espoo, Finland, http://hdl.handle.net/1853/50660; Gerold Baier, Thomas Hermann, and Ulrich Stephani, "Event-based Sonification of EEG Rhythms in Real Time," *Clinical Neurophysiology* 118, no. 6 (2007), 1377–86.

to sonification, declaring Lucier's piece and similar compositions as artistic forerunners of sonification.[7]

Near, but Out: Demarcating Sonification from Music

The musical examples discussed in the previous section have predated the coining of the term "sonification," and have emerged independently of the dedicated sonification research community. This community has developed over the last three decades and has found its institutional embodiment in the International Community for Auditory Display (ICAD), founded in 1992. It convenes at annual conferences, in which an interdisciplinary group of researchers – ranging from psychologists to physicists, and from computer scientists to composers – come together to present their latest research about the use of non-speech sound to display data and convey information.

Much of the work of the community concerns the systematic technical development and evaluation of techniques for transforming data into sound, especially ones that can be used by scientific specialists to complement existing, usually visual, data analysis techniques. Many scientists outside of this research community, however, react with scepticism to the idea of listening to data: the notion that only seeing can generate objective insights, while hearing creates subjective impressions, is widespread. Consequently, an important activity of the sonification community consists in "lobbying for the ear": seeking to establish sonification as a scientific and objective method by emphasizing the capabilities of the human hearing system in systematically processing information.[8] In order to understand their strategies for doing so, the way in which they draw the boundaries between science and art is revealing.

From the very beginnings of this community, composers and musicians have been prominently involved in sonification research; and to this day, a broad consensus exists that the involvement of artists and composers is beneficial for the success of sonification: not just because they can help to create public interest, but also because their skills related to technical implementation, aesthetic design, and listening practices can make important contributions to sonification research.[9] At the same time, there is widespread agreement (even among artists and composers) within the community that sonification should not be equated with music, and there have been some pleas to limit the definition of the term "sonification" to cases where it is used systematically as a scientific method, effectively excluding artistic approaches.[10]

The research community dedicated to sonification thus engages in a delicate balancing act to position itself in relation to artistic approaches. They seek contact with sound artists and musicians, but are careful to present sonification as something other than music or sound art. Quite characteristically for this delicate balancing act, the sonification researchers Gerold Baier and Thomas Hermann have welcomed their audience at a presentation in the context of a festival for contemporary music by expressing their gratitude for the audience's interest "even though this has nothing to do with music."[11] The fact that they regard their work on the sonification of brainwaves as having nothing to do with music was apparently no reason for them to turn down an invitation to present their work at a music festival; quite on the contrary, they very enthusiastically took this chance to introduce an artistically informed audience to the principles of sonification while also cautioning them that it is not the same as music.

Similarly, many of the annual sonification conferences organized by the ICAD

4
See Douglas Kahn, "Alvin Lucier, Edmond Dewand und *Music for Solo Performer*," in *Klangmaschinen zwischen Experiment und Medientechnik*, ed. Daniel Gethmann (Bielefeld: transcript Verlag, 2010), 211–29; Andi Schoon and Florian Dombois, "Sonification in Music," in *Proceedings of the 15th International Conference on Auditory Display*, May 18–22, 2009, Copenhagen, http://hdl.handle.net/1853/51415; Volker Straebel, "The Sonification Metaphor in Instrumental Music and Sonification's Romantic Implications," in *Proceedings of the 16th International Conference on Auditory Display*, June 9–15, 2010, Washington, DC, http://hdl.handle.net/1853/50066; Alexandra Supper, "Sublime Frequencies: The Construction of Sublime Listening Experiences in the Sonification of Scientific Data," *Social Studies of Science* 44, no. 1 (2014), 34–58.

5
Richard Taruskin, *The Oxford History of Western Music. Volume 5: The Late Twentieth Century* (New York: Oxford University Press, 2005), 55. Also see Ian Willcock, "Composing Without Composers? Creation, Control, and Individuality in Computer-based Algorithmic Composition," in *Electronics in New Music*, ed. Claus-Steffen Mahnkopf, Frank Cox, and Wolfram Schurig (Hofheim: Wolke Verlag, 2006), 221–35.

6
Indeed, in a personal interview I conducted with Lucier (Vienna, November 1, 2008), he mentioned that his desire to explore such phenomena through his work was a point of disagreement between him and John Cage, who accused him of being too invested in cause-and-effect relationships rather than sound on its own terms.

7
See Volker Straebel and Wilm Thoben, "Alvin Lucier's *Music for Solo Performer:* Experimental Music Beyond Sonification," *Organised Sound* 19, no. 1 (2014), 17–29.

8
See Supper, "The Search for the 'Killer Application'" (see note 2, above); Alexandra Supper, "Lobbying for the Ear: The Public Fascination with and Academic Legitimacy of the Sonification of Scientific Data" (PhD diss., Maastricht University, 2012).

9
Ibid.

community feature an evening program with concerts, often with sonification-based music. The inclusion of these concerts stimulate cross-pollination between artistic and scientific applications of sonification; but they also convey the message that such artistic applications are best suited to the evening program rather than the scientific sessions during the day. Art and music are thus kept near, but out.[12]

Sonification in Sound Art: Stellar Epistemologies

While the sonification community has been attempting to keep artistic and musical sonification projects at arm's length (though no further than that), sonification has nonetheless been sprouting outside of this community, not least because it has become a popular ingredient of science popularization initiatives. In those cases, it is used not so much as a data analysis technique for scientific specialists, but rather as a means of communicating (sometimes simplified) messages to lay audiences. During the 2014 Rosetta mission, for instance, the European Space Agency published a press release with an mp3 file promising the listeners to hear the sounds of a "singing comet." The accompanying sound file, created by German musician and sound designer Manuel Senfft, succeeded in creating quite a stir on social media and contributed to the public excitement about the Rosetta mission in the days leading up to the first successful landing of a space probe module on a comet.[13]

It may seem easy to dismiss sonification as something of a gimmick to grab public interest for scientific research among laypeople; however, that is not my intention in this essay. Rather, in the following two sections, I want to discuss a few selected examples where sonification has been used by artists

10
See Thomas Hermann, "Taxonomy and Definitions for Sonification and Auditory Display," in *Proceedings of the 14th International Conference on Auditory Display*, June 24–27, 2008, Paris, France, http://hdl.handle.net/1853/49960.

11
For this event, part of a series on "Music and the Brain" at Vienna's 2008 *Wien Modern* festival, the presentation of Hermann and Baier's sonifications was paired with a performance by Alvin Lucier of his *Music for Solo Performer*. Lucier also acted as a test subject for a live demonstration of the scientific sonification by Baier and Hermann.

12
This phrasing is borrowed from Thomas Gieryn, who has remarked that the boundary work of science involves keeping politics "near but out." See Thomas F. Gieryn, "Boundaries of Science," in *Handbook of Science and Technology Studies*, ed. Sheila Jasanoff et al. (Thousand Oaks, CA: SAGE Publications, 1995), 393–443.

13
See http://blogs.esa.int/rosetta/2014/11/11/the-singing-comet.

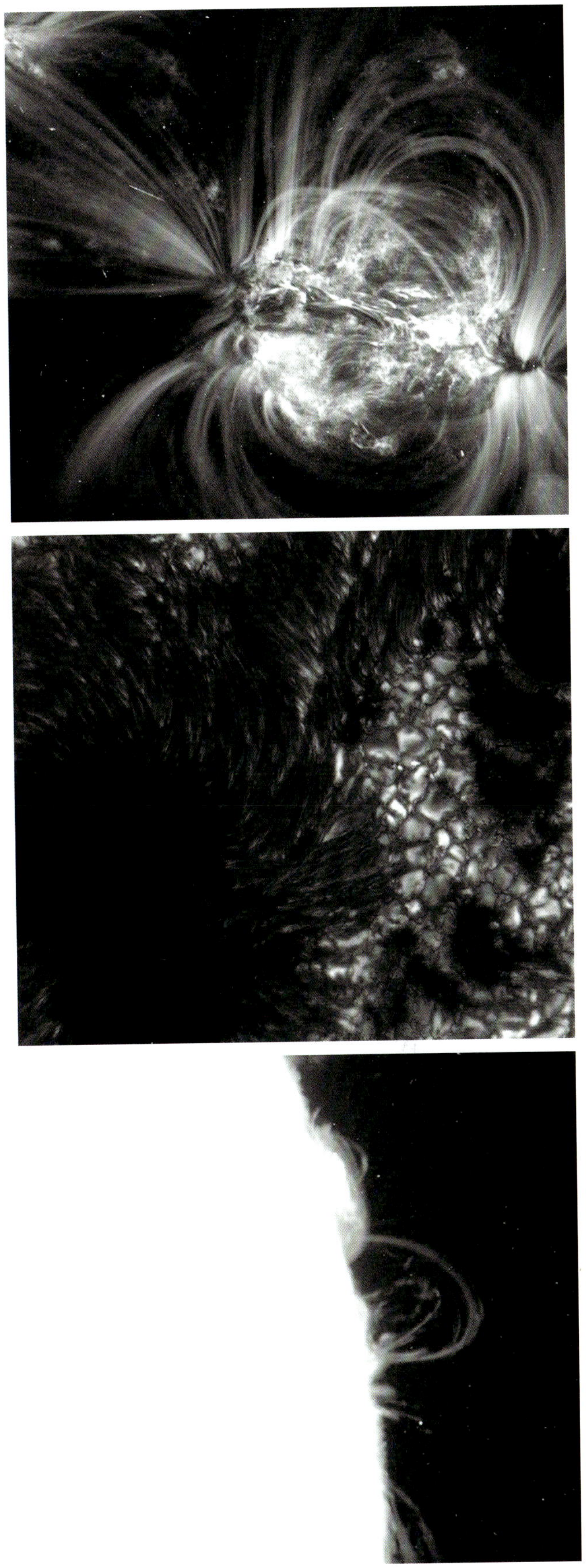

FIGS. 2a–c Semiconductor, *Brilliant Noise*, 2006. Filmic installation, various lengths, SD / HD / single-channel and multichannel versions. A Semiconductor work by Ruth Jarman and Joe Gerhardt.

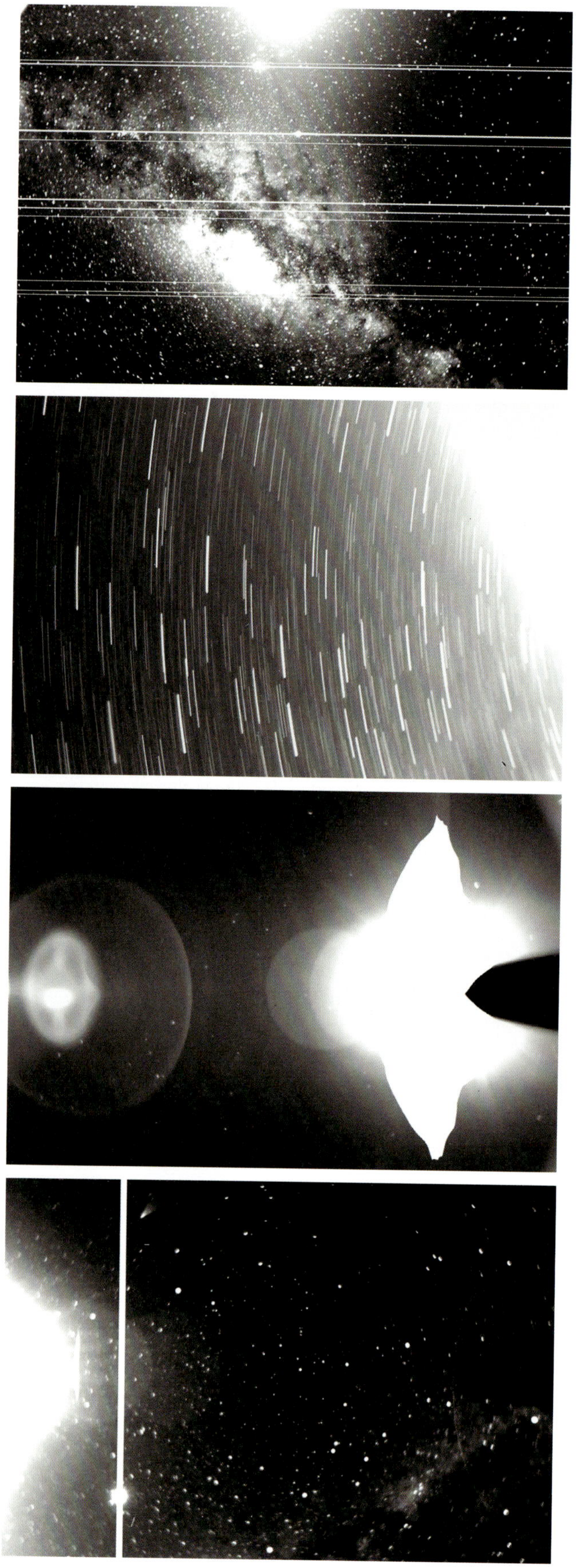

FIGS. 3a–d Semiconductor, *Black Rain*, 2009. Filmic installation, 3 min. / 17 min. loop, single-channel and installation. A Semiconductor work by Ruth Jarman and Joe Gerhardt.

and musicians to do more than that. Unlike the musical examples mentioned earlier, these artistic sonification projects have been inspired and influenced by scientific terminology and approaches for translating data into sound; yet, as I want to show, they also open new possibilities for thinking about scientific phenomena. With a few examples of sonifications of astronomical data, I show how they can open up a space for epistemological reflection. Subsequently, I discuss how sonifications of geophysical data have been used to raise environmental awareness.

The British artist duo Semiconductor have used the sonification (and visualization) of solar astronomical data as the basis of their multimedia installations *Brilliant Noise* and *Black Rain*.(FIGS. 2a–c, 3a–d)[14] As the artists explain in a video interview for the Creators Project, they are interested in exploring "how man experiences the nature of the physical world around him and how the tools of science come into that."[15] An important aspect of this exploration is that both of the aforementioned pieces work with raw data from satellite transmissions, rather than cleaned-up and processed datasets. Aesthetically, the grainy texture of the resulting sounds and images provides a stark counterpoint to the frequently quite polished and glossy representations of outer space in the popular scientific imagination. As art historian Inge Hinterwaldner points out, "Semiconductor embraces what is normally filtered out":[16] the noises and distortions that are usually painstakingly removed from scientific datasets become an essential part of the work of art. By leaving in and drawing attention to artefacts and glitches, the artists also remind us "of the presence of the human observer who endeavours to extend our perceptions and knowledge through technological innovation,"[17] and thus of the many human interventions that make scientific representation possible in the first place.

The piece *Bonner Durchmusterung* [Bonn Patternization] by electronic composer Marcus Schmickler (in collaboration with composer/sonification researcher Alberto de Campo, visual artist Carsten Goertz, and astronomer Michael Geffert) similarly uses sonification as a space for epistemological reflection.(FIG. 4) Unlike in the films of Semiconductor, the sounds of *Bonner Durchmusterung* are not driven by its images. If anything, the piece reverses the usual hierarchy

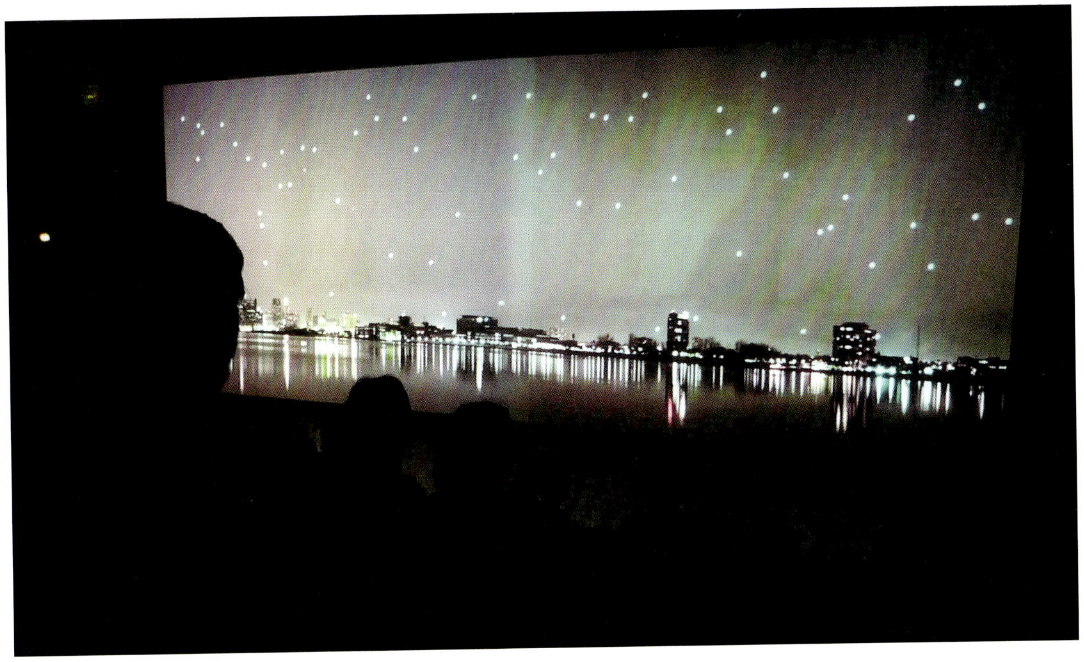

FIG. 4 Marcus Schmickler et al., *Bonner Durchmusterung*, 2009. Composition based on sonifications of astronomical models and data with projections, 10.2 channel audio, projections, 33 min. Commissioned by IYA 2009 (International Year of Astronomy) and Deutscher Musikrat.

between sound and vision: considering the prominent role of visualization in the scientific enterprise, one might expect that the visual projection aims for a more objective representation of scientific facts, while the sonification serves as an illustrative soundtrack. In *Bonner Durchmusterung*, however, much more artistic license was taken with the visualization than with the sonification. While the sounds are quite straightforwardly driven by scientific datasets and sound fairly abstract, the visualization "treats the subjects more freely or associatively,"[18] evoking concrete, urban spaces.

The program notes for the piece not only briefly explain the ten astronomical phenomena that have been sonified (including solar eruptions and gamma ray bursts), but also promise "an epistemological exchange between music and astronomy." Starting out from the statement that even "raw data are deeply dependent on a theoretical model of the world on which the measuring procedure is based," Schmickler and his collaborators reflect on the role of mathematical models and visualizations in astronomical research, and call into question "the relationship between data and reality of the observed objects."[19] How, for instance, does one come from a complex series of numbers to an understanding of the objects or even to a consistent phenomenology of the cosmos? The essay concludes by encouraging artists and scientists to work together on a mutually enriching "poetic interpretation, depiction and contextualization of scientific knowledge." Indeed, in my interview with the composer, he stressed that he was interested in sonification as a means of triggering interdisciplinary reflexivity.[20]

Sonification in Sound Art: Sonic Ecologies

Not only astrophysical data have inspired sound artists and musicians in their work, however. In this section, I want to discuss several examples of artistic sonifications based on data from another domain, that of geophysics. With their *Aftershock* sound installation,(FIGS. 5a–c) which debuted at the Norwegian art gallery ROM in 2011, the sound artist Natasha Barrett and geoscientist Karen Mair promise listeners the ability to "experience the intricacies of the inherently 3-D process of rock deformation from the inside." *Aftershock* is in fact not a single sound installation, but rather a set of four interrelated works, which are placed in spatially separate but adjacent zones of the exhibition

14 See http://semiconductorfilms.com/art/brilliant-noise/ and http://semiconductorfilms.com/art/black-rain.

15 See video file on http://thecreatorsproject.vice.com/blog/meet-semiconductor-the-mad-scientists-of-the-visual-art-realm.

16 Inge Hinterwaldner, "Semiconductor's Landscapes as Sound-Sculptured Time-Based Visualizations," *Technoetic Arts: A Journal of Speculative Research* 12, no. 1 (2014), 15–38, here 30.

17 See http://semiconductorfilms.com/art/black-rain.

18 Interview by the author with Marcus Schmickler, Cologne, April 15, 2010.

19 See http://piethopraxis.org/projects/bonner-durchmusterung for the quoted program notes and some sound examples.

20 Interview with Marcus Schmickler.

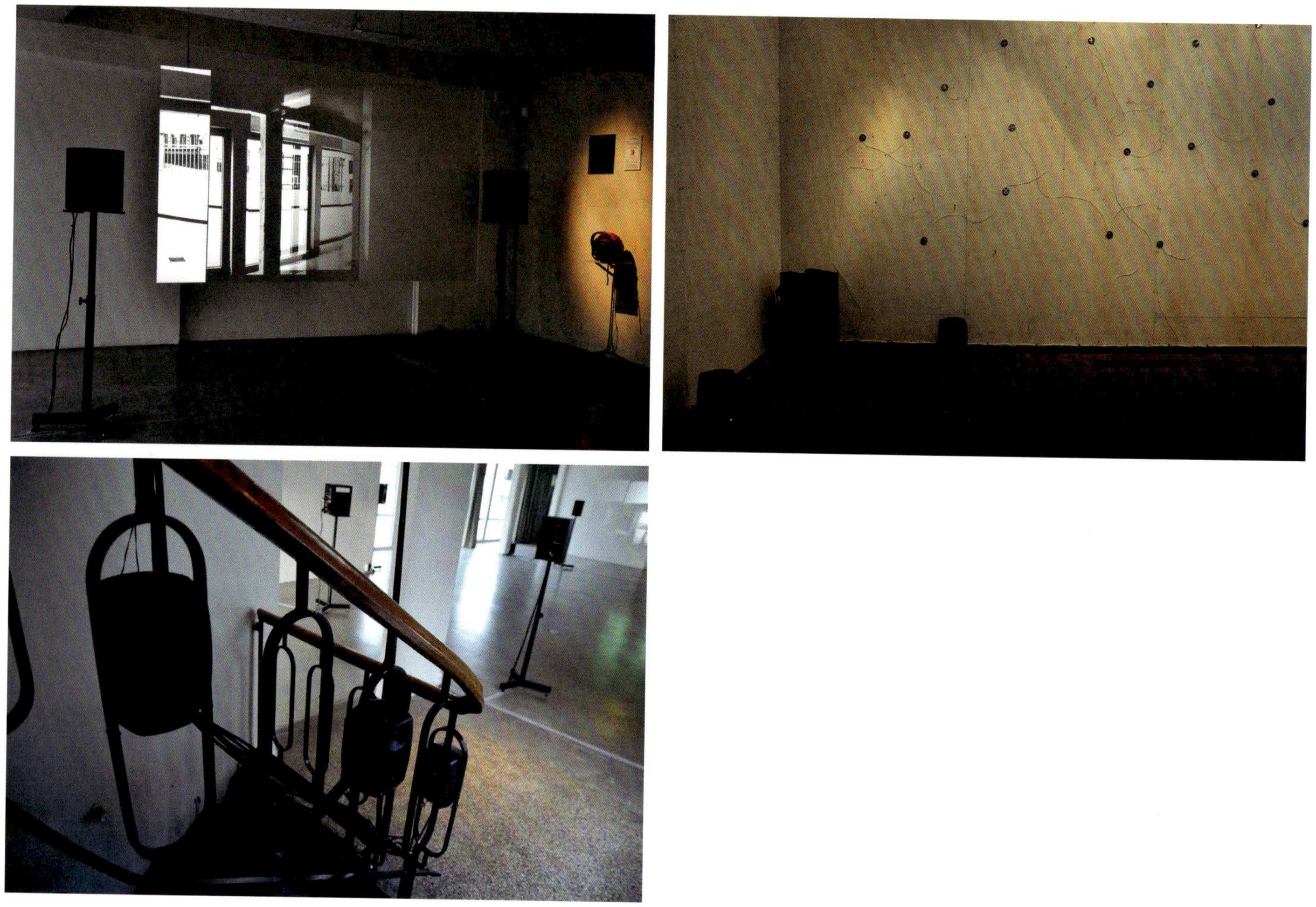

FIGS. 5a–c Natasha Barrett and Karen Mair, *Aftershock*, 2011. Interactive sound art installation.

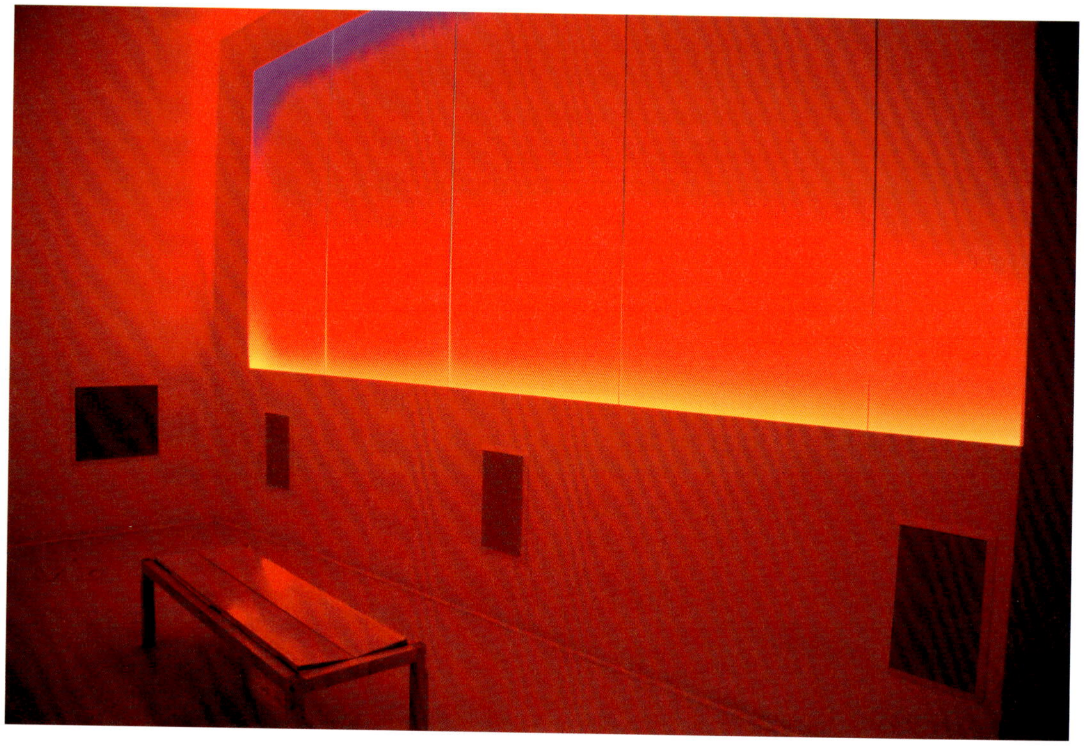

FIG. 6 John Luther Adams, *The Place Where You Go to Listen*, 2004–2006. A continuous sound and light environment at the University of Alaska, Fairbanks Museum of the North.

space, thus allowing the sound of the four installations to intermingle. The installation is meant to encourage an active exploration of geophysical phenomena; in fact, the installation remains silent unless an "exploring listener" is present and triggers the sounds.[21]

Aftershock is not the only recent example of a sound installation based on geophysical data. The sound-and-light installation *The Place Where You Go to Listen* by composer John Luther Adams, installed at the University of Alaska's Museum of the North in Fairbanks, is even more ambitious in its translation of geoscientific data into sound.(FIG. 6) Rather than sonifying a single dataset, Adams has opted to tackle several at once, combining seismological, geomagnetic, and meteorological datasets – all of which are translated into sound in real time. The result is a complex sonic tapestry based on a number of scientific parameters (from the current position of the sun and moon to seismic and geomagnetic activity), described as follows by music critic Kyle Gann:

"You walk in, separate yourself from the world directly outside, sit on the bench, and slip into the red-and-violet, or blue-and-yellow, moods of the five glass panels in front of you. A continual hum greets you, and after a moment you begin to sort out the strands of the complex tapestry that the hum turns out to be. There are sustained chords, an intermittent rattle of deep bells overhead, and an irregular boom of extremely low frequencies that you have to focus on to remain aware of."[22]

In a book companion to the installation, Adams emphasizes that the piece transforms "inaudible forces of nature into audible sound" and that these forces therefore drive some of the decisions traditionally made by the composer; yet he also insists that it is "a work of art" rather than a "scientific demonstration of natural phenomena." What's more, Adams points out, the piece does not aspire to simulate the natural world, but rather "is a heightened form of experience itself." Unlike *Aftershock*, the piece continuously emits sound (and light) whether a listener is present or not, but nonetheless, the listener's experience of a sense of place stands central according to the composer: "The essence of this work is the sounding of natural forces interacting with the consciousness of the listener."[23]

In a recent article about the piece, musicologist Tyler Kinnear suggests that Adams's sounding of the inaudible voices of nature contains a "gesture of environmental advocacy."[24] Adams, argues Kinnear, "is not interested in representing an environment; rather, he seeks to transform it."[25] Indeed, Adams deliberately tried to avoid "conventional modes of musical expression and picturesque evocations of nature,"[26] and thus used sine tones and filtered noise, rather than sounds that mimic natural processes or musical instruments. Rather than offering a simulation of natural phenomena, Adams tries to evoke an experience of space and the environment which is fundamentally driven by ecological concerns about "overpopulation, overconsumption, pollution, deforestation and widespread extinction."[27] Climate change and environmental destruction are recurrent themes of the composer's diary written during the making of the installation.[28]

John Luther Adams is not the only composer or sound artist who has drawn upon sonification to express environmental and ecological concerns: such issues form a red thread running through the works of sound artist Andrea Polli – in a recent journal article, Polli muses about the potential of sonification to raise environmental awareness.[29] For instance, her 2009 album *Sonic Antarctica*(FIG. 7) combines meteorological sonifications, field recordings, and snippets of interviews with climate researchers who talk not only about their scientific findings, but also about their environmental motivations; another project, called *Heat and the Heartbeat of the City*,(FIG. 8) makes concerns about climate change audible by translating predicted temperature changes in New York's Central Park into sound.[30] Both of these projects engage in a dialogue with the scientists studying the phenomena that

21
See Natasha Barrett and Karen Mair, "Aftershock: A science-art collaboration through sonification," *Organised Sound* 19, no. 1 (2014): 4–16. Also see www.natashabarrett.org/crush3.html for more information and a short video.

22
Kyle Gann, "Fairbanks: A Long Ride in a Slow Machine," *NewMusicBox* (2006), www.newmusicbox.org/articles/fairbanks-a-long-ride-in-a-slow-machine.

23
John Luther Adams, *The Place Where You Go to Listen: In Search of an Ecology of Music* (Middletown, CT: Wesleyan University Press, 2009), 5, 8.

24
Tyler Kinnear, "Voicing Nature in John Luther Adams's *The Place Where You Go to Listen*," *Organised Sound* 17, no. 3 (2012), 230–39, here 237.

25
Ibid., 231.

26
Adams, *The Place Where You Go to Listen*, 139. Also see David Rothenberg, "Go There to Listen: How Music Based on Nature Might Not Need Natural Sounds," in *The Farthest Place: The Music of John Luther Adams*, ed. Bernd Herzogenrath (Boston: Northeastern University Press, 2012), 107–15.

27
Adams, *The Place Where You Go to Listen*, 9.

28
Ibid.

29
Andrea Polli, "Soundscape, Sonification, and Sound Activism," *AI & Society* 27, no. 2 (2012), 257–68.

30
See www.andreapolli.com/centralpark.

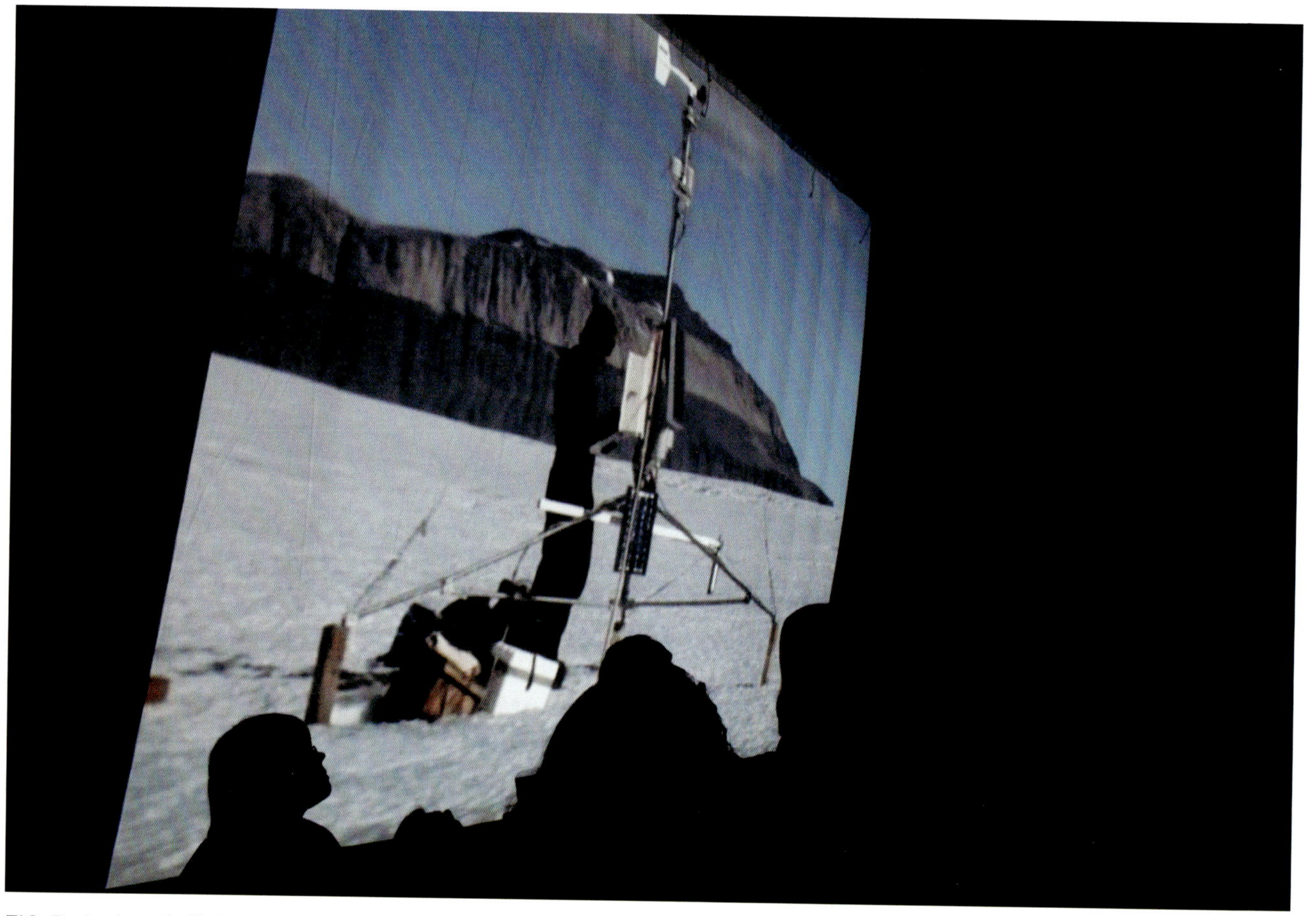

FIG. 7 Andrea Polli, *Sonic Antarctica*, 2007/2008. Project including a radio broadcast, live performance as well as a sound and visual installation.

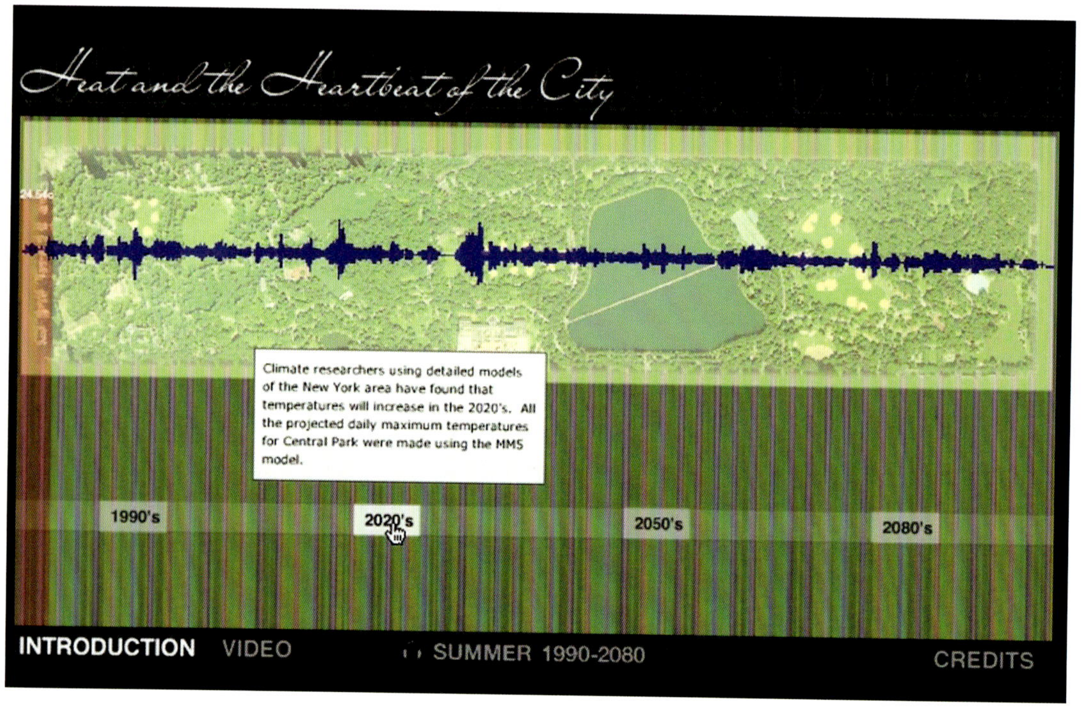

FIG. 8 Andrea Polli in collaboration with Cynthia Rosenzweig, David Rind, and Richard Goldberg, *Heat and the Heartbeat of the City*, 2004. Web-based interactive work including sonifications illustrating projected climate changes focusing on the heart of New York City. Screenshots.

Polli translates into sound, and in both cases, the voices of those scientists have become part of the works themselves. If Polli uses her work to raise environmental awareness, then, that does not just mean that she draws attention to those issues; her work also becomes a place for interdisciplinary dialogue between artists and scientists on issues related to the environment.

Conclusions

Contrary to the musical examples in the first section of this essay, all of the artistic projects described in the previous two sections were to some extent informed by scientific terminology and approaches for sonifying data – and indeed, in several cases, the artistic projects have fed back into scientific research. For instance, the *Aftershock* collaboration, which started out as an artistic collaboration between sound artist Barrett and geophysicist Mair, has led to a scientific poster presentation at the 2014 general assembly of the European Geosciences Union,[31] while sound artist Andrea Polli has presented one of her projects to an audience of sonification specialists at the 2004 ICAD conference.[32]

Clearly, then, sonification is a domain that opens up possibilities for exchange and interactions between science and art. These exchanges and interactions can take a myriad of forms. In an article about the emerging practice of art-science (though not primarily in relation to sound), Georgina Born and Andrew Barry have identified three different "logics" according to which art-science initiatives can operate: the logic of accountability, in which art helps to make science accountable and accessible to a broad public; the logic of innovation, in which interdisciplinary research fuels industrial innovation and economic growth; and the logic of ontology, directed at altering "ways of thinking about the nature of art and science" and "transforming the relations between artists and scientists and their objects and publics."[33] The latter, in particular, is often neglected in discussions about the value of interdisciplinary and art-science research.

The distinction between different logics of interdisciplinarity can illuminate how the relationship between science and art is negotiated within the scientific community dedicated to sonification research. As I have discussed, this community values the contributions of musicians and composers for increasing the public appeal and accessibility (logic of accountability), as well as the technical and aesthetic design of sonification (logic of innovation). But at the same time, the boundary between scientific and artistic uses of sonification, while interpreted flexibly, is not altogether dissolved: the community is engaged in a careful balancing act between science and art. This balancing act is constitutive of the (inter)disciplinary identity of the sonification community, and fuels many of the debates going on within the community about what sonification is and should be. In other words, encounters with sound art and music help the sonification community to rethink the nature of their work and its relation with its objects and publics – the logic of ontology is at play within the sonification community, too.

However, the relationship between science and art is not one-sided: the case of sonification shows that not only science can benefit from the engagement with artistic practices, but that the reverse is also true. Indeed, the three different logics of interdisciplinarity also run in the opposite direction: for artists – especially those working in traditions that lack mainstream exposure – the engagement with scientific topics can also help to attract new audiences (logic of accountability), while encounters with scientists working on sonification can lead to new tools that can also be used in artistic contexts (logic of innovation). Perhaps most intriguingly, however, the engagement of composers and sound artists with scientific data can also take the form of a logic of ontology. Indeed, as many of the examples in this essay have shown, their ability (or at least intention) to stimulate epistemological and ontological reflection, and to second-guess taken-for-granted scientific assumptions and conventions, is a particularly fascinating aspect of many artistic sonification projects.

31 See http://meetingorganizer.copernicus.org/EGU2014/EGU2014-4489.pdf. Originally, Barrett and Mair had also proposed a special session dedicated to sonification, but the session was eventually withdrawn (see http://meetingorganizer.copernicus.org/EGU2014/session/15515).

32 Andrea Polli, "Atmospherics/Weather Works: A Multi-Channel Storm Sonification Project," in *Proceedings of the 10th International Conference on Auditory Display*, July 6–9, 2004, Sydney, www.icad.org/websiteV2.0/Conferences/ICAD2004/papers/polli.pdf.

33 Georgina Born and Andrew Barry, "Art-Science: From Public Understanding to Public Experiment," *Journal of Cultural Economy* 3, no. 1 (2010), 103–19, here p. 105.

SOUND ART IN CONTEXT

Linnea Semmerling

Shhh...! In Search of the White Cube's Soundtrack

"Next time you enter a gallery or museum, close your eyes."
Christof Migone[1]

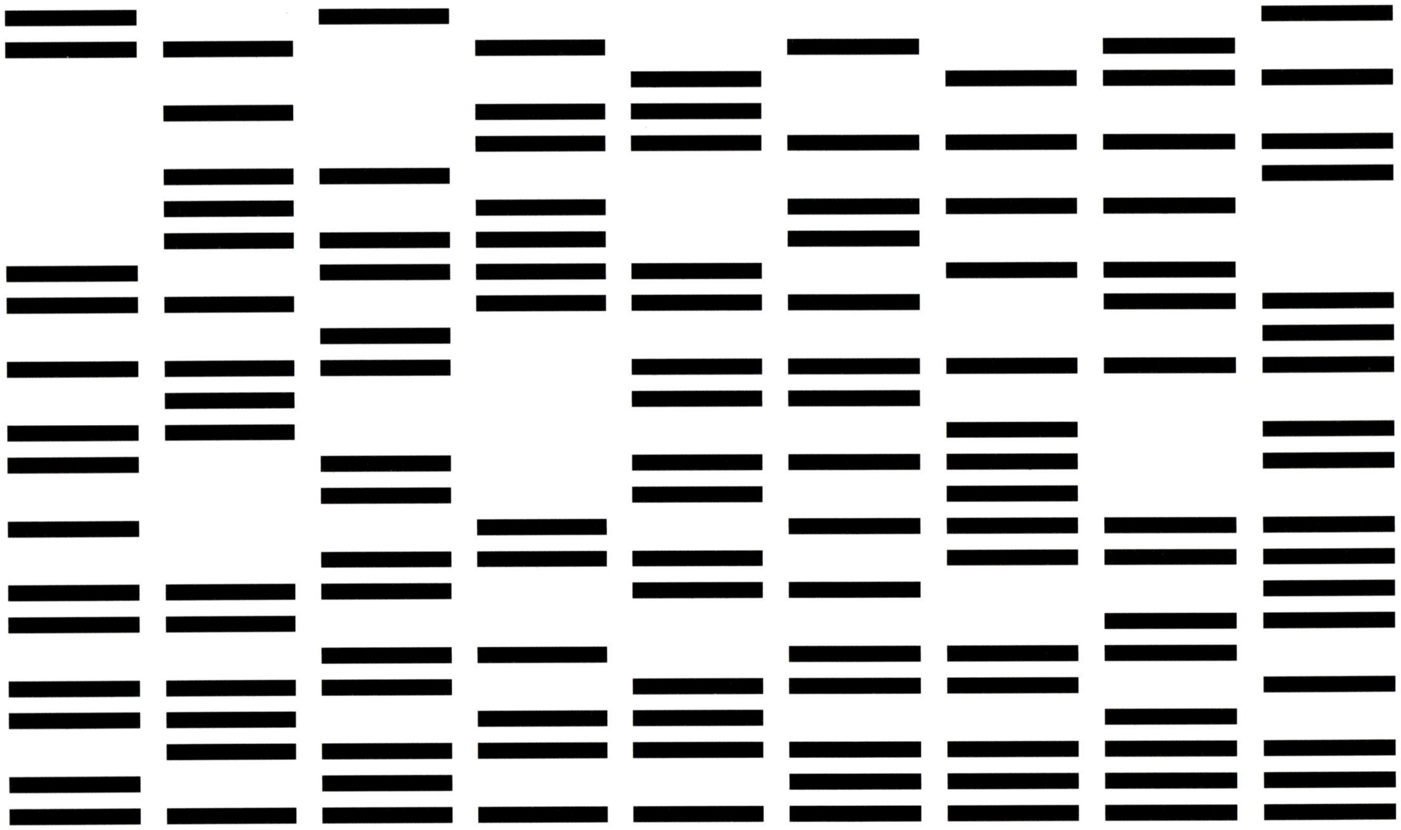

"Videos have soundtracks. Galleries don't. In the museum, sound, music, even speech are invasive."[2] This is how Sean Cubitt introduces an essay on the spatiality of sound in video installations in 1996. He continues by pointing out some of the practical challenges in exhibition situations, where "more than one soundspace in a gallery not only wrecks the fictive silence of contemplation; they appear to interfere with each others' autonomy. No pure work, but mélange, spillage, mess."[3]

About twenty years later, the black box pervades contemporary gallery spaces in various sizes and shapes, and appears to have become a widely accepted solution to at least some of the problems that Cubitt describes. The carpet minimizes sound reflections and dampens visitor noises, while the dark walls and thick entrance curtain provide sound isolation, allow the projections to shine bright, and prevent visitors from distracting one another visually. When it comes to art with sound, but without projections, such an aesthetic and social standard of presentation does not seem to be found just yet.[4] Accordingly, Christof Migone points out that "exhibiting sound art poses challenges to the white cube, sound epitomizes leakage, sound confirms the porosity of space. Sound art's presence in museums is increasingly prevalent but remains problematic for spaces conceived for viewing instead of listening. Wide open, reverberative galleries are not generally conducive to focused listening. Even prior to an intentional sound entering the equation, every space has its own soundtrack, its room tone."[5] As a prologue to an in-depth investigation of exhibition strategies for sounding artworks in the gallery spaces of museums and other contemporary art venues, this chapter sets out to examine the ways in which contemporary artists working with sound have appropriated, challenged, and experimented with the soundtrack of the gallery space. Their grappling with the space as an institutional facilitator and oppressor that challenges and at the same time limits experiences, through habitualized rituals or social experiments, sets the stage for a curatorial research endeavor.[6]

The present analysis will be guided by Dorothea von Hantelmann's notion of performativity. She describes how "this notion brings into perspective […] the contingent and difficult to grasp realm of impact and effects that art brings forth both situationally, i.e. in a given spatial and discursive context, and relationally, e.g. in relation to a viewer or public."[7] She stresses that every artwork has a reality-producing dimension to it, which is not always consciously shaped or dealt with as a level of meaning production in its own right. Exposing the ways in which an artwork either depends on or defies certain conventions of presentation can expose this performativity. This chapter will focus on the ways in which artworks' sounds either perpetuate or challenge the conventions of display at contemporary gallery spaces. An analysis of these repetitions and disruptions should not just give us an understanding of the auditory situation that the artworks produce and the manner in which they situate their viewers as listeners, it should also allow for an assessment of the values, conventions, ideologies, and meanings associated with the soundtrack of the contemporary art museum gallery.[8]

A Soundtrack of Silence

An investigation of the contemporary gallery space's soundtrack has to go back to its early days as a "white cube" or "white flexible art container" in the middle of the past century, as these conventions of display continue to set the standard for exhibitions up until today. Brian O'Doherty coined the notion of the "white cube" in a series of *Artforum* essays between 1976 and 1986, later published as the canonical *Inside the White Cube*. He describes a gallery space where "the outside

1 Christof Migone, *Sonic Somatic: Performances of the Unsound Body* (Berlin: Errant Bodies Press, 2012), 183.

2 Sean Cubitt, "Video Installation and the Neo-Classical Soundtrack," *Essays in Sound: Diffractions* 3 (1996), 8–20, here 8.

3 Ibid.

4 The black box is certainly not just a curatorial measurement. Ralf Beil describes how artists have long developed the isolated back spaces "beyond pure screening rooms for films and videos. When executed accurately, the black box offers them a maximum of control over the context, the conditions of presentation, and thus also the perception of their own work." Ralf Beil, *Black Box: Der Schwarzraum in der Kunst* (Ostfildern: Hatje Cantz, 2001), 9, 21. Translated from the German by the author.

5 Migone, *Sonic Somatic*, 165–66.

6 Parallel to the development of this text, the author began a NWO-funded PhD project *Listening on Display: Exhibiting Sounding Artworks 1960–Now* with Prof. Dr. Karin Bijsterveld and Dr. Peter Peters at Maastricht University. The project will investigate how sounding artworks have come to challenge curatorial strategies at museums and other contemporary art venues in Germany and the US since the 1960s, and how research into the listening experiences of artists, curators, and visitors may inform new curatorial strategies.

7 Dorothea von Hantelmann, *How to Do Things with Art*, trans. J. Gaines and M. Turnbull (Zurich, Dijon: JRP | Ringier, Les presses du réel, 2010), 18.

8 Ibid. Whereas von Hantelmann strives to understand the visual arts' impact on society at large through the museum exhibition, the present article is interested primarily in the museum exhibition itself.

world must not come in, so windows are usually sealed off. Walls are painted white. The ceiling becomes the source of light. The wooden floor is polished so that you click along clinically, or carpeted so that you pad soundlessly."[9] Charlotte Klonk has disproved some of O'Doherty's assumptions about the white-walled art space in her historical survey of art gallery interiors. She dismisses the idea of an enclosed cube as a fiction serving the rhetoric of institutional critique, and hence replaces it with the concept of the "white flexible art container" opening out to its urban environment with its adjustable white walls, its flowing enfilade architecture, and its facilitation of natural light.[10] Whereas the neutrality of the color white purveyed a sense of seclusion to O'Doherty, it was actually meant to achieve permeability according to Klonk's research. None of the two authors, however, devotes much consideration to the sounds of the white gallery spaces, they both just vaguely imply a certain quietude.[11] In fact, a lot of curatorial writing about the white cube appears to act on the assumption of such quietude, or even silence. Lorenzo Benedetti, for example, describes how "modern exhibition space is defined by rigorously designed lines, a predominance of white colors and a somewhat candid neutrality. In its pure state, it can be interpreted as a space of silence."[12] In a less metaphorical manner, Germano Celant speaks of the white cube as a visual machine, where "any hint of noise or ruffling of any of the senses has been banished."[13] Contemporary gallery spaces thus seem to be understood as silent, or at least quiet, mostly in relation to the visual, either by association, when whiteness, brightness, and emptiness suggest silence, or by subordination to a form of monosensory contemplation, when focused looking requires quietude.

Both silences have long come to be challenged by sounding artworks. Not only Migone, also Cubitt recognizes how "sound installations tend to make you more aware of the absence of silence in the soundless areas of the museum."[14] However, Cubitt would not refer to this absence of silence as a soundtrack. His literal soundtrack comprises the sounds of a video, with dialogue, music, and sound effects that are purposefully composed and carefully orchestrated to accompany moving images. A viewer can only adjust the volume of this soundtrack or helplessly follow it shifting out of synch with the image track.[15] On the contrary, Migone's figurative soundtrack emphasizes the sonic characteristics of any particular gallery space by itself. This soundtrack would thus be randomly produced by a combination of the gallery space's acoustic properties and the sounds of its equipment, the sounds of the artworks on display in adjacent galleries, and the sounds that the visitors produce in the space either intentionally or by accident. By elevating these accidental sounds to the status of a soundtrack, Migone conceptualizes an important development that also shows through in his and other artists' creative usage of museum sounds as part of artworks that question their own conditions of display: the increasing appreciation, and consequent investigative exploration of the art museum as a space for listening, which appears to be in constant tension with the traditional institution's supposed silence or rather "soundlessness."

Ticking Thermo-Hygrographs

Quite recently, the Swiss artist Thomas Moor has demonstrated the gallery's "soundfulness" by presenting a switched-on thermo-hygrograph as part of his drawing performance *We spotted the ocean* (2014). The conservational device, which is usually hidden in a corner of the exhibition space, is here presented on a high pedestal in the center of the gallery, conveying the aura of a sculpture.(FIG. 1) On the plain white gallery walls behind it, framed thermo-hygrograph papers document other exhibition spaces' temperature and humidity levels. As the

9
Brian O'Doherty, *Inside the White Cube: The Ideology of the Gallery Space* (Berkeley: University of California Press, 1999), 15.

10
See Charlotte Klonk, *Spaces of Experience: Art Gallery Interiors from 1800 to 2000* (New Haven: Yale University Press, 2009), 137–49, 218.

11
See O'Doherty, *Inside the White Cube*, 15, 49, and Klonk, *Spaces of Experience*, 207; see also a conversation between Niklas Maak, Charlotte Klonk, and Thomas Demand, www.tate.org.uk/context-comment/articles/white-cube-and-beyond.

12
Lorenzo Benedetti, "The Shape of Silence," in *MARTa schweigt: Garde le silence, le silence te gardera; Die Kunst der Stille von Duchamp bis heute; Das Mysterium der Etrusker*, exhib. cat. (Herford: MARTa, 2007), 90–93, here 91.

13
Germano Celant, "A Visual Machine: art installation and its modern archetypes," in *Thinking about Exhibitions*, ed. Reesa Greenberg, Bruce W. Ferguson, and Sandy Narine (London: Routledge, 2005), 371–86, here 382; originally published in *documenta 7*, ed. Rudi Fuchs, vol. II (Kassel: Paul Dierichs, 1982), XIII–XVIII.

14
Cubitt, "Video Installation and the Neo-Classical Soundtrack," 15.

15
Chion has refuted the notion of the soundtrack for the equality it suggests between sound and image. He is convinced that the sound of a film does not constitute an entity in its own right, but can only be understood in relation to the images. However, this does not keep Chion from drawing on the soundtrack as an experiential construct or sensory experiment, which is similar to how I am trying to evoke it here. See Michel Chion, *Audio-Vision. Sound on Screen* (New York: Columbia University Press, 1990/1994), 40 and also Michel Chion, *The Voice in the Cinema* (New York: Columbia University Press, 1982/1999), 3–4.

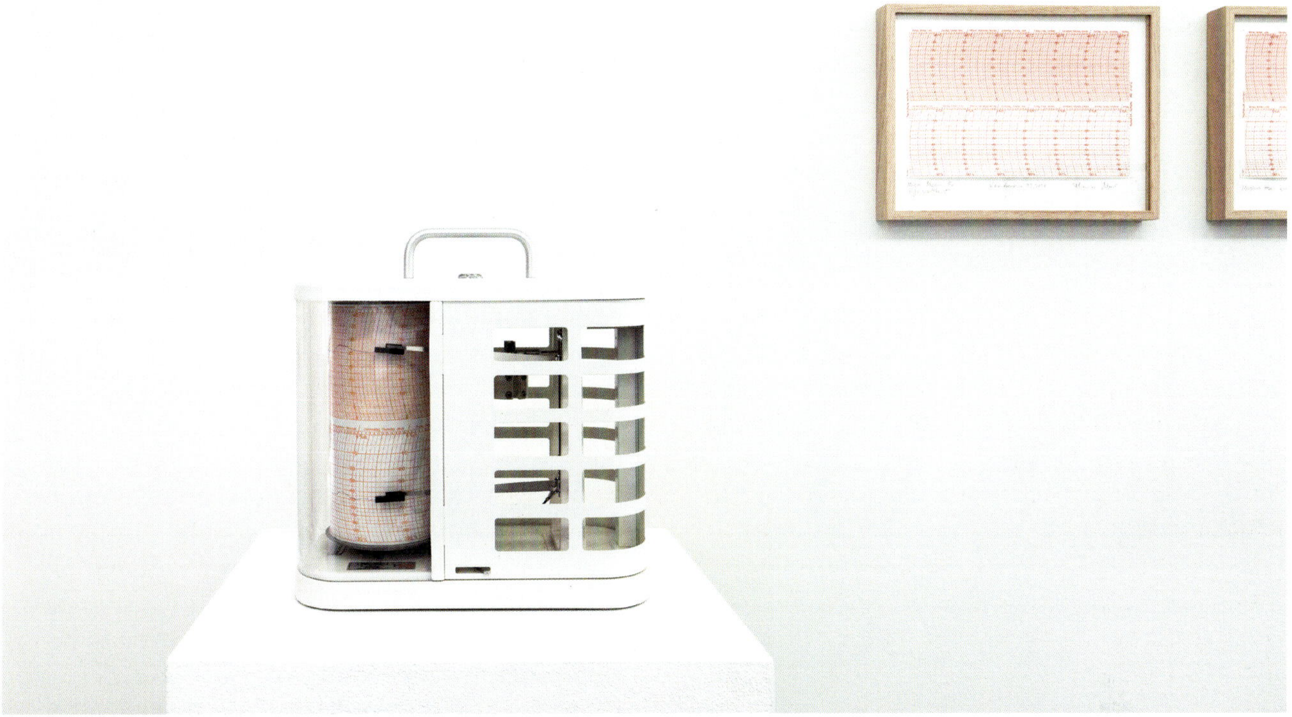

FIG. 1 Thomas Moor, *We spotted the ocean*, 2014. Installation, thermo-hygrograph and thermo-hygro-graph paper (left), drawings, thermo-hygrograph pen on thermo-hygrograph paper, framed (right). Installation view Nationale Suisse Art Prize, LISTE, Basel, 2014.

thermo-hygrograph is switched on to also document the temperature and humidity levels of the current exhibition space, it makes its familiar ticking sounds. Consequently, the work does not just pun on the well-known white-cube-effect of mistaking the fire extinguisher for a work of art;[16] it elaborates this effect by introducing an auditory dimension to it. Inside the white, reverberant gallery space, it may not just be the fire extinguisher that turns out to look like art, the thermo-hygrograph may also start to sound like art. By elevating a conservational device to the status of an artwork through its form of display, Moor thus caricatures visitors' tendency to see and also hear within the institutional confines of the museum. In addition to this performative dimension, a certain reflexivity about museum conditions may be conveyed by the thermo-hygrograph's actual documentation of the temperature and humidity levels in the space. As the little machine ticks along, visitors can follow the graph emerging on the sheet of paper and compare it to the graphs from the other exhibition spaces on the wall. The question arises whether the conservational device as artwork documents its own exhibition conditions or the viewing conditions of the visitors.

By presenting a working museum device as a readymade art object, Moor's work achieves an effect that is reminiscent of the way O'Doherty describes the American conceptual artist William Anastasi's 1965 project at the DWAN gallery in Los Angeles, for which he hung silk-screened paintings with literal renditions of the gallery walls on those very same walls in a slightly smaller scale. O'Doherty describes how "when the paintings came down, the wall became a kind of readymade mural."[17] Moor's work is likely to have a similar effect, the auditory dimension of which is particularly revealing. The thermo-hygrograph is not just a familiar sight that tends to be overlooked, it also creates a common gallery sound that is usually unheeded. By inviting visitors to hear the sounds of the thermo-hygrograph, the work may achieve a lasting awareness not just of the ticking sounds, but of the soundtrack that museum equipment in general provides. Just as

16
Svetlana Alpers describes this as "the museum effect – turning all objects into works of art." Svetlana Alpers, "The Museum as a Way of Seeing," in *Exhibiting Cultures*, ed. Ivan Karp and Steven D. Lavine (Washington: Smithsonian Institution Press, 1991), 25–32, here 26.

17
O'Doherty, *Inside the White Cube*, 34. It is an interesting coincidence that, in his next show at the same gallery, Anastasi installed a radiator in the space that was equipped with a loudspeaker presenting its recorded and amplified sounds.

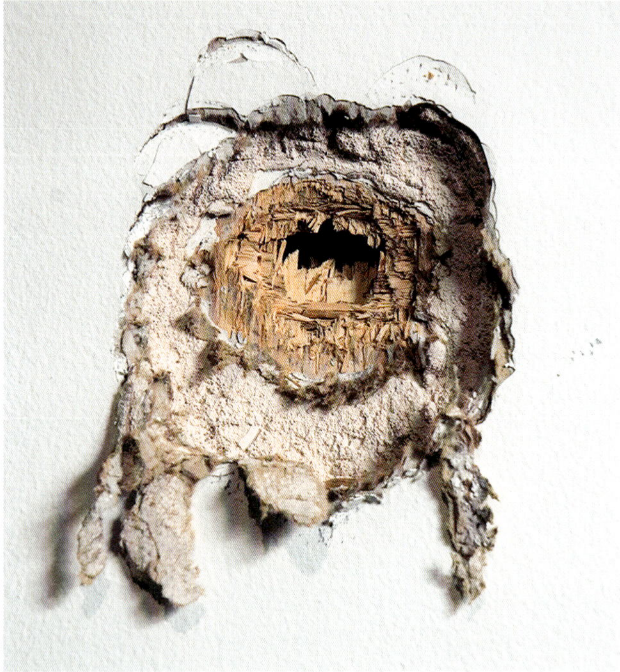

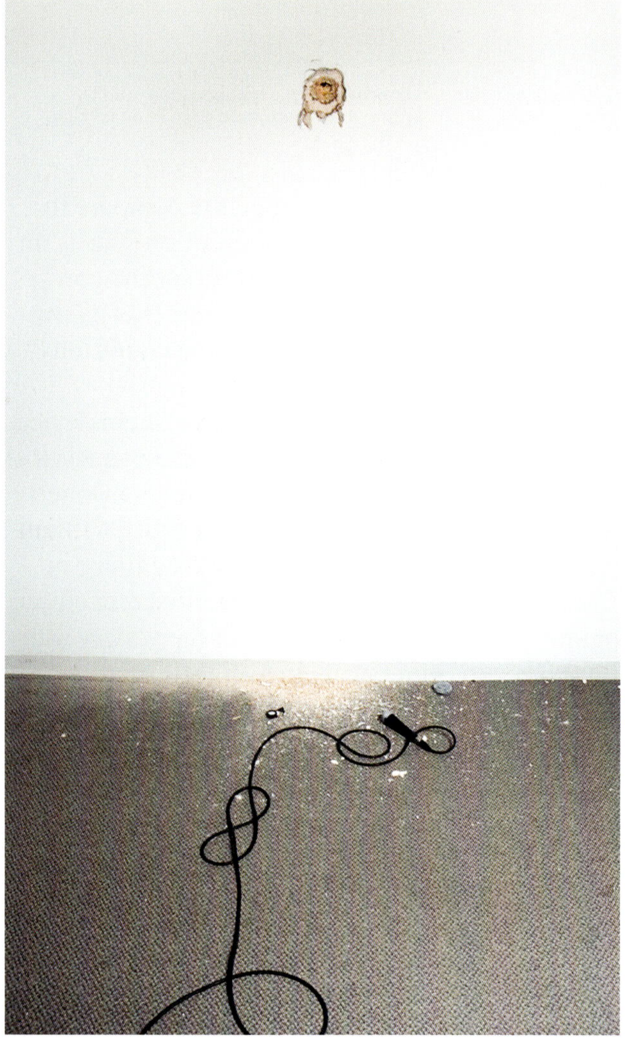

FIGS. 2a, b Christof Migone, *Microhole*, 2006. Installation with a recording of the hitting action. Installation view Galerie de l'UQAM, Montreal, 2006.

the visitors of the DWAN gallery may never have been able to see the gallery walls as a plain white cube again, the viewers and listeners of Moor's work may never be able to hear the white cube's silence again. Instead, they will be haunted by the ticking of the thermo-hygrograph, the humming of the air-conditioning, or the drone of the projectors' ventilation. They are the kinds of sounds that have received little attention until sounding artworks entered the museum galleries and intentionally or unintentionally amplified them.[18]

Porous Walls

Another prevalent issue in sound art exhibitions is sound isolation, which the modular display units that are commonly used as gallery walls simply cannot provide. Our ability to hear the sounds of artworks through these walls has proven to be a considerable problem, in particular with group shows featuring several artworks that sound.[19] Christof Migone has already been quoted describing how "sound epitomizes leakage, sound confirms the porosity of space."[20] He demonstrates this with his *Microhole* (2006) installation. During the mounting of an exhibition, the artist hits a wall with a microphone "until the wall yields under the pressure of the repeated action.(FIG. 2a) In the exhibit, the recorded action is played back through the hole in the wall."[21] The damaged microphone is left lying on the ground in front of the wall.(FIG. 2b) While the work has been conceived as a sound piece that could be executed with any wall, its site-specific realizations in gallery spaces have charged the ensuing installations with a peculiar performativity.

With his *Box with the Sound of Its Own Making* (1961), one of the earliest and most canonic sound sculptures, Robert Morris famously exposed the art-making process by presenting a wooden box with a loudspeaker inside that played the sounds of the box's construction, which is three and a half hours of measuring, drawing, sawing, and hammering. By presenting a hole in the exhibition wall "with the sound of its own making," Migone similarly exposes his exhibition-making process. As the sound recording from Morris's studio broke open the box's sculptural entity to convey a tedious building process, the sound recording and the gaping hole resulting from Migone's engagement with the gallery wall lay bare the challenges of installing a sound work in a standardized white cube space. The hammering microphone performs a peculiar double-task as it hits the wall and simultaneously registers the sound of itself hitting the wall. The hitting starts out quite regularly at about 90 beats per minute, but becomes slower and more irregular with ever-longer breaks for the artist to recuperate his strength during the course of the action, which appears to demand considerable physical effort. The microphone hits still convey information about the space's acoustics and the material properties of the wall, while slowly but steadily demolishing it. The work's creation from the wall thus goes hand in hand with its destruction of the wall. The microphone becomes a sledgehammer with which Migone attacks the gallery space, indicting it for its visual bias and convicting it by exposing its poor construction to both eyes and ears.

Whispering Visitors

With his *Conversation Piece* (2010) performance, Brandon LaBelle invites visitors to directly engage with the gallery walls that Migone attacks, and that Moor hangs his framed papers on. He describes how "the performance is imagined as a narrative in the making: the audience enters and is invited to listen to the empty room – is there anything there? Something starts to appear, to gather in the corners, to fill the walls. The audience whispers to each other: no, there is nothing... and they start to move around, to write on the walls, trying to capture this silence, to express it, to fill the space with writing, markings."[22] With nothing to see in the gallery space except for handwriting on the walls, the visitors start to listen to the ambient sounds of the space, but also to their fellow visitors' walking, whispering, and writing noises.(FIGS. 3a, b) The performative dimension of this experiment is particularly interesting to understand in relation to William Anastasi's collaboration with John Cage on October 31, 1977, as part of the performance project *You Are*. Cage sat in the Clocktower Gallery in New York with his back turned to the entrance to describe the sounds of the visitors that came to Anastasi's exhibition. Anastasi summarized the performance, explaining that "a narrator describes the audience for ninety minutes; a court stenographer takes down the narration; a typist types up a longhand version of this from the shorthand notes; a 'page stapler' fastens each page when ready to a wall at eye level."[23] In Anastasi's performance, the walls are thus still the place where the work is hung, this work consisting of the audience's sounds in the gallery space mediated by the composer or, on other evenings, the audience's appearance and behavior described by visual artist Les Levine and writer Carl Kielblock. As Cage transforms the sounds into words that describe them, he notices that "the situation im in is some-what impossible, because in order to hear i must be quiet. but in order to narrate which im obliged to do, i have to speak and when i speak my attention isnot on what im hearing. so that my mind is divided between listening and saying."[24] Consequently, his descriptions are interrupted by silences, which he starts to refer to as "a solo for the typewriter" and other things.[25] In the development from the sounds as heard to the sounds as read, it is the spoken word that stands central in its paradoxical conflict with listening as well as in its problematic phonetic transcription.[26]

LaBelle's *Conversation Piece* overcomes Anastasi's strict progression of hearing, speaking, writing, typing, stapling, and reading, which opposes hearing on one end to reading on the other end, through the visitors' own bodily involvement in writing and thus enables a quiet yet dynamic conversation to unfold in the gallery space. The work's title references eighteenth-century informal group portraits in which the

18 An example of such intentional amplification is Jeppe Hein's *Invisible Cube* (2004). The artist has installed movement-sensitive surveillance cameras in an empty white cube gallery that do not allow visitors to stand further than one meter away from the exhibition walls or an alarm sounds. Rather than keeping a safe distance from the artworks on the walls, visitors have to be aware not to move too far away from them in order to remain outside of the invisible cube that can be experienced only through that alarm. However, effective steps toward the abatement of museum equipment noises have also been taken, for example, with Manon de Boer's *Two Times 4'33''* (2008), for which she instructs exhibition organizers to build a projector case into the wall of the black box in order to prevent the projector's ventilation from reaching the ears of her audience listening to a video performance of John Cage's *4'33''*.

19 See also Cubitt, "Video Installation and the Neo-Classical Soundtrack," 14.

20 Migone, *Sonic Somatic*, 165.

21 Christof Migone, "Microhole," www.christofmigone.com/html/projects_gallery/Microhole.html.

23 Brandon LaBelle, "Conversation Piece," www.brandonlabelle.net/conversation_piece.html.

23 William Anastasi, *The Cage Dialogues: A Memoir*, ed. Aaron Levy, Contemporary Artist Series 6 (Philadelphia: Slought Books, 2011), 4.

24 Ibid., 7.

25 Ibid., 88.

26 Also Alison Knowles' 1983 performance piece *Scribe* has to be mentioned here. During an evening event that took place at Emily Harvey Gallery on February 14, she "sat blindfolded at the table and recorded with three brown markers and a graphite pencil the sounds in the room and the voices of people nearby." See http://www.emilyharveyfoundation.org/ehg_1.html

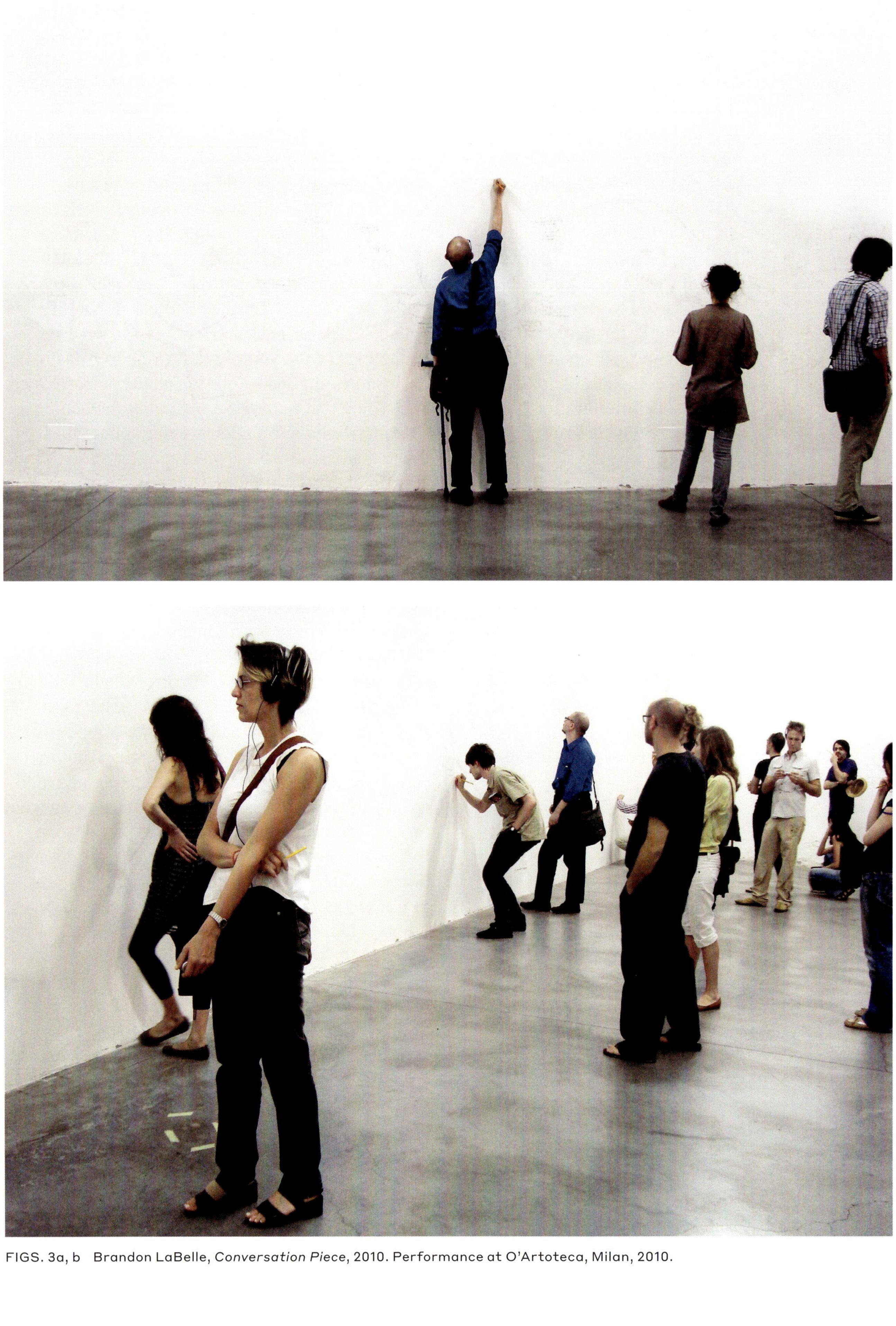

FIGS. 3a, b Brandon LaBelle, *Conversation Piece*, 2010. Performance at O'Artoteca, Milan, 2010.

27
www.brandonlabelle.net/conversation_piece.html.

28
Caleb Kelly, "Can someone please turn the work up? The Exhibition of Sound in Contemporary Australian Art," *Art Monthly Australia* 225 (2009), 7–10, here 8.

portrayed are engaged in casual conversations. With the white gallery walls as his medium, LaBelle redirects the focus of the visitor group from a centralized conversation around a work of art to a decentralized conversation across the gallery walls. These walls that visitors are not even allowed to touch in a conventional gallery situation are here neither background, as in *We spotted the ocean*, nor subject, as in *Microhole*, but a medium for visitors' experiences of museum sounds. As the conversation unfolds on those walls, the gallery space is increasingly co-shaped by a listening audience that is not only leaving a momentary audible impression on each other, but also lasting visual traces in the gallery. What evolves is a gallery situation, which LaBelle also describes as a "choreography," in which all visitors relate to one another situationally through their listening, and lastingly through their evolving wall conversations. An additional level of complexity is achieved by recordings of people dancing to music on headphones that are also provided on headphones. LaBelle describes this as "another space to enter, a soundtrack to one's own moving, a secondary choreography to this one."[27] The sounds of the bodies that are actually moving through the gallery space may thus be replaced by the sounds of dancing bodies at will, allowing visitors to step out of the museum ritual and open up to different kinds of movements.

Silence Fiction

Caleb Kelly has described how "the gallery space has been cleared of all unnecessary elements and painted clinically white, but sound is not so easily cleared away."[28] Moor's *We spotted the ocean*, Migone's *Microhole*, and LaBelle's *Conversation Piece* do not approach the white cube as a sacral space of silence, but they acknowledge the ritualistic traditions of the clutter-free spaces with their white walls and bright skylights to devise exaggerations, provocations, challenges, and experiments for specific auditory aspects of these kinds of galleries. They achieve their situational and relational impact on the basis of fundamentally different artistic strategies: by presenting a noisy conservational device as a readymade art object, Moor points to the gallery's functional equipment noises, its ticking thermo-hygrographs and humming air-conditioning systems; by hitting the gallery walls with a microphone, Migone exposes the architectural space's sonic characteristics, its porous walls, and its peculiar reverb; and with an experimental instruction piece for the space's visitors, LaBelle draws attention to the social venue with its hushed conversations and ritualistically choreographed bodies. All these sounds and sound-determinants are there prior to any artwork making or evoking a sound. By elevating them to the status of a soundtrack to which any artwork with or without sound will eventually have to relate, the gallery space becomes a space for listening. However, the performative gestures of the works discussed in this chapter do not only achieve a heightened awareness for the white cube's soundtrack, they transform the space by finding new ways of ascribing meaning to its objects, reforming the power relationships implied in its architectures, and opening its social rituals out to dynamic models of communication. The museum visitors become complicit in the development of this new kind of space not just as listeners to a work, but as co-creators of an auditory realm, which is that of the soundful gallery.

Irene Noy

Touching Sound Art: Curatorial Practices in West Germany[1]

The title of one of the most renowned exhibitions associated with sound art, namely *Für Augen und Ohren* [For Eyes and Ears], held in Berlin in 1980, highlighted the synthesis and central role of sight and hearing. The title, however, disregarded the involvement of at least one other sense at which a great number of works aimed – the sense of touch. This was also the case with two less well-publicized exhibitions, which preceded the event in Berlin: *Sehen und Hören: Design und Kommunikation* [Seeing and Hearing: Design and Communication], held in Cologne in 1974, and *Sehen um zu hören: Objekte und Konzerte zur visuellen Musik der 60er Jahre* [Seeing in Order to Hear: Objects and Concerts of Visual Music of the 60s], held in Düsseldorf in 1975. While the former was oriented toward industrial and technological innovation, the latter highlighted experimentation in musical composition. All three exhibitions exemplified a growing institutional interest in transforming their exhibition spaces into interactive hubs.

By taking these exhibitions as case studies, this essay considers the crucial role of the sense of touch, coupled with the importance of the conjunction between the senses of sight and hearing. Closely examining some of the curatorial aspects of these exhibitions sheds light on the links between art institutions, their integration of experimental music, technological innovation, and the manner in which these elements contributed to the expansion of the experiences offered in the "white cube" exhibition environment.

I.

The exhibition *Für Augen und Ohren* took place in Berlin in 1980 and is referred to as one of the main events that gave rise to what became known as sound art in the second half of the twentieth century.[2] This exhibition was a large-scale international event with seventy-four participating artists and musicians. Its great scope aimed to show the widespread popularity of art aimed at eyes and ears and was supposed to present a long-lasting trajectory connecting visual arts and music. The exhibition included works by deceased artists such as Luigi Russolo and Man Ray, as well as works by living artists such as Christina Kubisch and Bernhard Leitner. Although the earliest works were from the beginning of the twentieth century, the exhibition catalogue included references as early as the fourteenth century, such as the automatically operated bells in an astronomical clock in Strasbourg from 1354.[3] This early reference shows that the emphasis was laid particularly on the ability to mechanize and automate the connection between seeing and hearing. The fascination with automation laid the foundations for a visual and sonorous event free of human interference in its operation.

At the exhibition in Berlin, mechanization, technology, and media were at the core of its curation. The curators, the musicologist Nele Hertling and the gallerist René Block, stated clearly that the works for the exhibition were chosen on the premise that they incorporated aural and visual elements simultaneously alongside an emphasis on technological progress.[4] A statement that opens their introductory essay in the exhibition catalogue exemplifies this point: "The perception of our time is mainly determined by communication media. Wherever we are, we are surrounded by invisible waves, which can be received everywhere and anywhere and can be made sonorous and vivid."[5]

This assertion acknowledged the pervasiveness of technology and communication media and the categories used to divide the exhibited works are a testimony to that. These included, for instance, a category which incorporated technological developments in music reproduction and electronic instruments. Another category was sound objects *(Klangobjekte)* – where composers and visual artists used mechanical and electronic techniques. Consequently, what stood at the forefront of this exhibition was an extensive interest in innovative technology and its involvement in and mediation of the connection between the senses of sight and hearing.

There was, however, at least one other sense inquisitively employed by the visitors and that is the sense of touch. The visitors to the exhibition were allowed to touch and play with the exhibited works, which meant that the limited experiences encouraged so far in the white cube context were significantly expanded. The debate on the role of the white cube took off, particularly since the release of the influential text on the subject, *Inside the White Cube: The Ideology of the Gallery Space* by the artist Brian O'Doherty.[6] Published just a few years before the exhibition in Berlin, O'Doherty's book discussed the mediated environment of the modernist gallery, which is in fact integrated into the artworks' content.[7] The central principal of the modernist exhibition space of the white cube came to be understood

1 I elaborate further on these exhibitions and situate them within the context of sound art's formation in Germany in my book entitled *Emergency Noises: Sound Art and Gender* (Oxford: Peter Lang, 2017).

2 The full title of the exhibition was *Für Augen und Ohren: Von der Spieluhr zum akustischen Environment.* Ojekte Installationen Performances (For Eyes and Ears: From a Music Box to an Acoustic Environment. Ojecta Installations Performances). Folkmar Hein, one of the organizers of the Electronic Music Studio at the Technical University of Berlin, sees this exhibition as belonging to the beginning of sound art. See Folkmar Hein, "Klangkunst im elektronischen Studio" (Berlin: TU, 2010), www2.ak.tu-berlin.de/Geschichte/themen/TU-Klangkunst.html. Yet the exact date of sound art's origin is still debated. Unlike Hein, the musicologist and theorist Helga de la Motte-Haber claims that a specific date for the establishment of this field would be misleading. See Helga de la Motte-Haber, *Klangkunst: Tönende Objekte und Klingende Räume*, Handbuch der Musik im 20. Jahrhundert 12 (Laaber: Laaber-Verlag, 1999), 9.

3 See Helmut Danninger, "Daten zu Musik, Elektronik, Synästhesie, Environment und Performance," in René Block et al., *Für Augen und Ohren: Von der Spieluhr zum akustischen Environment*, exhib. cat. (Berlin: Akademie der Künste, 1980), 284–94, here 284.

4 See Nele Hertling and René Block, "Einführung," in Block, *Für Augen und Ohren*, 7–8.

5 Ibid., 7. Translated here from the German by the author.

6 See Brian O'Doherty, *Inside the White Cube: The Ideology of the Gallery Space*, expanded edition (Berkeley: University of California Press, 1999). This was first published in 1976 and 1986 as a series of essays in *Artforum* magazine.

7 See ibid., 15.

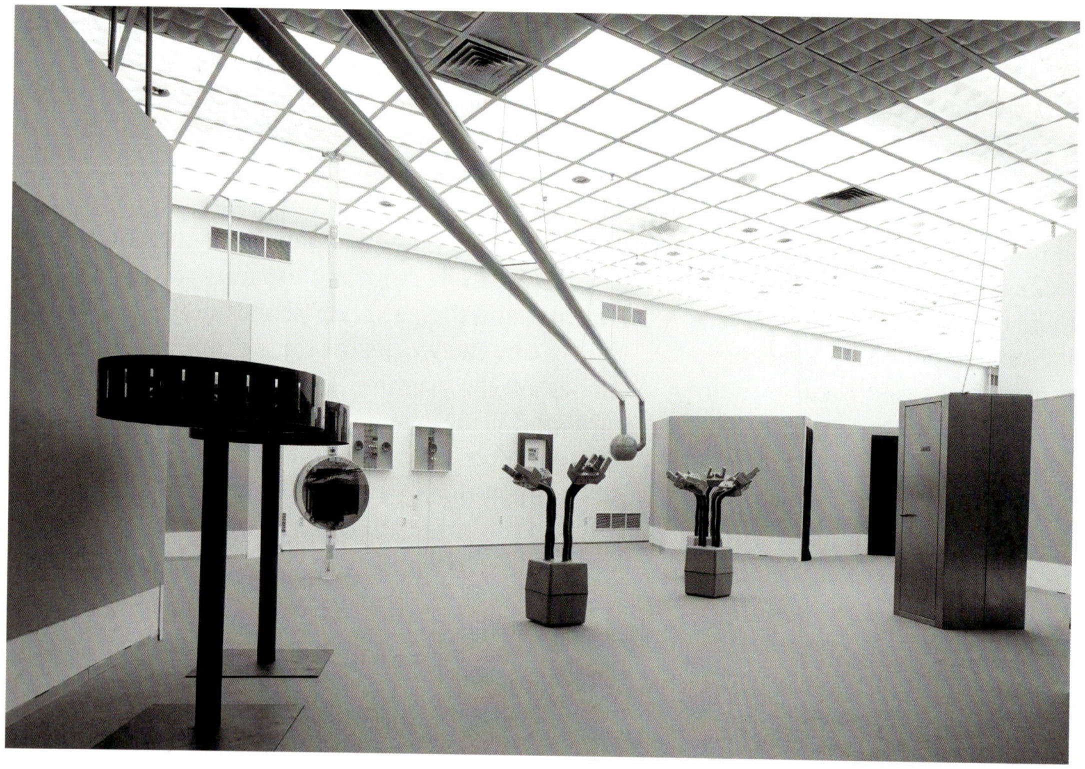

FIG. 1 View of the exhibition *Sehen und Hören*, Josef Haubrich-Kunsthalle, Cologne 1974.

as a neutral space, where works are displayed in a setting that allegedly allows optimal concentration on their perception. Sealed windows, artificial illumination, and walls painted white create a "clean" and "artificial" environment in which the artworks are not subject to time and space and are kept disconnected from the world outside of it.[8] Such a space is thus dedicated to the Kantian and Greenbergian understanding of an optimized concentration on the sense of sight. This is why exhibitions that attempted to include other senses, such as those of hearing and touch, destabilized the traditional institutional policy of display.

This exhibition in Berlin was not the first one concerned with the expansion of sensorial experience. It was preceded by a number of events in the 1970s, albeit on a smaller scale. The three exhibitions mentioned here emphasized the correlation between the senses of sight, hearing, and touch and yet the discourse around those events as well as the remaining documentary material is predominantly visually driven. This is particularly interesting given the fact that these exhibitions were supposed to move art institutions away from their preoccupation with the sense of sight. I argue that despite a general openness toward experimentation and innovation in terms of the kind of experiences such institutions were offering, their outcomes were in fact ambiguous.

II.

In seeking precedents to *Für Augen und Ohren*, one could think of one of the largest exhibitions dedicated solely to the latest technological innovations. The title of the exhibition is fascinating in this context; *Sehen und Hören: Design und Kommunikation* alluded directly to the importance of technology to the senses of sight and hearing. Incorporating six hundred objects from sixteen different countries, this major international exhibition was organized at the art exhibition space of the Kunstgewerbemuseum in Cologne in 1974.(FIG. 1), [9] Building on the long-lasting tradition in Germany of the connection between arts and crafts, an exhibition of the newest innovations in technology seemed to be appropriately situated in an institution dedicated to this tradition.

Although facilitated at an art exhibition space, *Sehen und Hören* was directed at consumers and the objects exhibited in it were referred to as products. The curator of the exhibition, the industrial designer Peter Frank, divided the exhibited items into three broad categories, which included: "How we expand communication" (products such as film cameras and radios), "How we orientate in time and space" (digital watches and clocks), and "How we process information" (calculators

8
Ibid.

9
Today, this institution is titled Museum für Angewandte Kunst Köln [Museum of Applied Arts Cologne].

10
See Guy Debord, *The Society of the Spectacle*, rev. ed. (Detroit: Black & Red, 1977, repr. 1983), e.g. *§5, 6* (unpaginated).

11
See ibid., e.g. §10.

12
See ibid., e.g. §17.

and computers). Some of these newly designed products were placed in darkened rooms and were supposed to be used mainly for listening. Other spaces were dedicated to watching videos on multiple screens. In most spaces, the objects were exhibited in sterile-looking rooms and, apart from the colors of the walls, which had darkly painted panels, the rooms did resemble a typical white cube setting. In some cases, there was a high density of works, or indeed products, which were positioned on shelves, as if in a shop. In other cases, the spacious positioning of the items and their size could have resembled a visual arts exhibition.

Despite this sterile, perhaps cold site, visitors to this design and communication exhibition were allowed to touch and test all the items on display.(FIG. 2) The inclusion of the sense of touch allowed visitors to familiarize themselves with the products, which they would later probably be exposed to in shops. Such a commercially driven situation meant that although *Sehen und Hören* emphasized the senses of sight and hearing, the sense of touch offered a corporeal closeness to the displayed items, which stood in contrast to the bodily distance habitually practiced in the white cube.

Such an expansion beyond the sense of sight may have signaled also a reaction to the critique uttered by Guy Debord only a decade earlier. Debord reacted to the social role and impact of the "spectacle," a term originating from the Latin *spectaculum* (public show) and *spectare* (to look). By reviving this concept in the 1960s, Debord famously created a Marxist account of contemporary society, which is oriented toward materialism and commodity. This social state is governed by an overemphasis on the sense of sight, which is mediated by visual images and results in a state of spectacle. Spectacle, Debord argued in his *La Société du Spectacle* (1967), is not to be understood as a mere product of mass media but as a worldview that has become ubiquitous.[10] In this condition, spectacle functions as a totalizing ideological frame that reduces experience to appearance.[11] Such mediated representation and appearance in society is a spectacle that replaces life, once lived directly through tangible experience.[12] Tactility is then essential to experience since its mediation

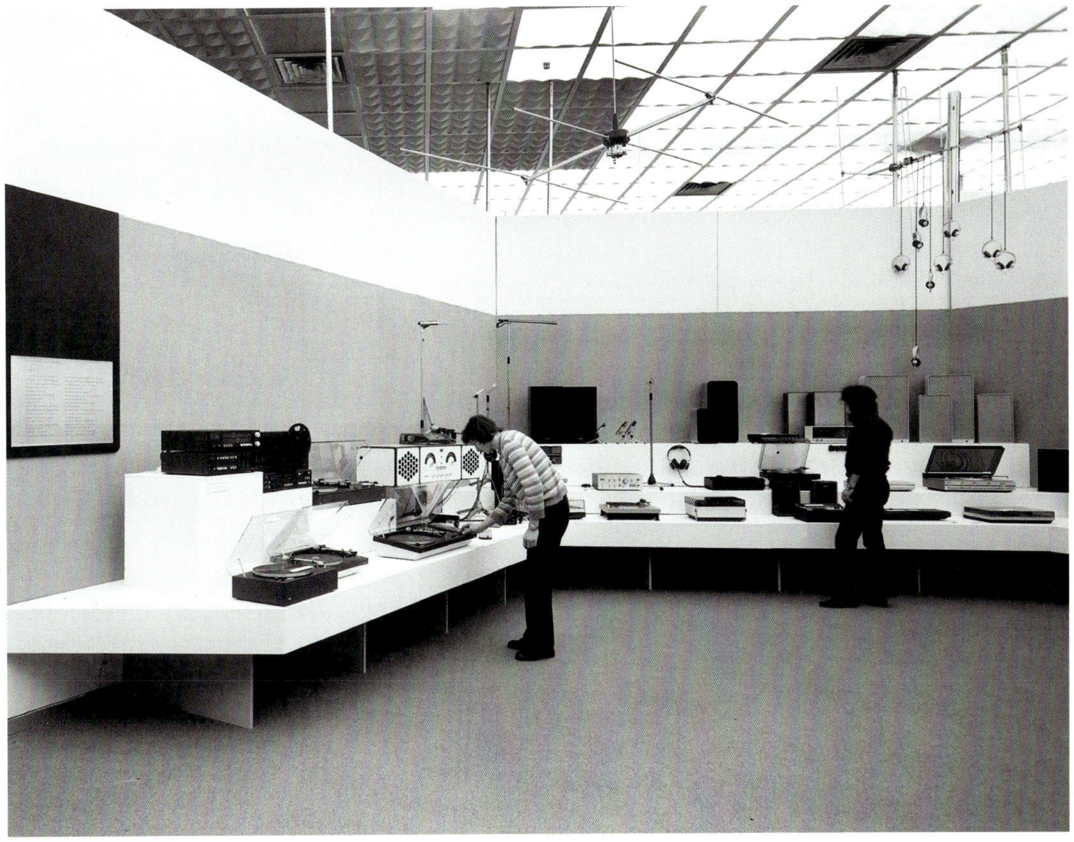

FIG. 2 View of the exhibition *Sehen und Hören*, Josef Haubrich-Kunsthalle, Cologne 1974.

via sight alone is reductive – an experience that alienates spectators in favor of viewed objects. If sight alone was alienating, then the inclusion of the sense of touch, as employed in *Sehen und Hören*, was supposed to intensify the connection between the exhibited objects and the visitors. Such an understanding has been prevalent at least since the Enlightenment. While discussing the different qualities of the senses and differentiating between inner and outer senses, Immanuel Kant suggested that touch is considered to allow an "immediate external perception of an object."[13] In *Sehen und Hören* this was precisely what the curator achieved with the integration of the sense of touch and at the same time, these encounters were expanded beyond "external perception" via the senses of seeing and hearing. Steven Connor ruminates on the various connections between the senses and advocates touch as an accompaniment to the sense of hearing. Whereas the relation between seeing and hearing is that of immediacy, that is, sound is immediate and sight confirms or allows understanding of what is heard, the relation between hearing and touch is that of mimesis.[14] In other words, touch "imitates and performs" what we hear rather than translates and defines it, as, according to Connor, sound touches us.[15] Contact generates sound and thus sound is dependent upon touch. What we hear then, according to Connor, is the "event" which involves the object, not the object itself.[16] This was not necessarily the case in the technology-driven context of *Sehen und Hören*, where objects were programmed to have particular sounds. Sound depends upon what comes into contact with an object yet this was not the aim of the exhibition. The purpose was not to experiment with the actual tactility of, for example, a synthesizer by tapping on it but to listen to the various sounds such an object can generate when activated.

The inclusion of the sense of touch in *Sehen und Hören* turned the exhibition and the products in it into an event where visitors were the performers who had to follow particular instructions. It was hoped that touching the items on display would allow familiarization with those new objects. The architecture of the space closely resembled that of a white cube and what was now offered to the visitors was mediated representation and appearance that included at least three senses.

III.

A further deployment and yet subversion of the white cube principles took place in an additional precedent to the Berlin exhibition. Organized only a year after the exhibition in Cologne, *Sehen um zu hören: Objekte und Konzerte zur visuellen Musik der 60er Jahre* took place in the neighboring city Düsseldorf in 1975. Facilitated in an institution habitually dedicated to the visual arts, namely the art exhibition space of the Städtische Kunsthalle, *Sehen um zu hören* included objects and installations, accompanied by performances and film evenings. Although the event in Düsseldorf was on a much smaller scale than *Für Augen und Ohren* in Berlin, its reputation was supported by the inclusion of some of the most prominent practitioners of that period. That is, artists and musicians such as John Cage, Joe Jones, and Stephan von Huene were included in the exhibition in 1975 and would go on to exhibit in 1980.

In Düsseldorf, their works were exhibited in a white cube setting and again, similarly to the design and communication exhibition in Cologne, artists and curators encouraged visitors to interact with the works. This was the case, for instance, in John Cage's *33 1/3* from 1969 composed of twelve records simultaneously playing different music.(FIG. 3) Visitors were able to take the role of a DJ and choose the records they wanted to hear. The exhibition was also turned into an event every evening when some of the works from the exhibition were incorporated into the concert and performance program such as those by John Cage,

13 Immanuel Kant, *Anthropology from a Pragmatic Point of View*, trans. Victor Lyle Dowdell (Carbondale: Southern Illinois University Press, 1978), 41.

14 See Steven Connor, "Edison's Teeth: Touching Hearing," paper written for the conference *Hearing Culture*, organized by the Wenner-Gren Foundation and scheduled to take place in Morelia, Mexico, October 4–12, 2001. See www.stevenconnor.com/edsteeth.

15 Ibid.

16 Ibid. Douglas Kahn, for example, also associates the creation of sound with an event, that is, something needs to happen for sound to be produced. Douglas Kahn, *Noise, Water, Meat: A History of Sound in the Arts* (Cambridge: The MIT Press, 1999), 158. Yet researchers are still arguing whether sound could be a property of an object or whether it is both event and object, whether it is a wave or a phenomenon of perception. See Mark Grimshaw and Tom Garner, *Sonic Virtuality: Sound as Emergent Perception* (New York: Oxford University Press, 2015), 21.

17 See Inge Baecker's introduction to *Sehen um zu hören: Objekte & Konzerte zur visuellen Musik der 60er Jahre*, exhib. cat. (Düsseldorf: Städtische Kunsthalle, 1975), 9–11, here 10. Although Inge Baecker stated that the visitors were invited to look at the objects, the press reviews of that period reveal that the audiences were also encouraged to touch and experiment with the instruments. For example, see Helga Meister, "Variationen in der Tonne geben," in *Westdeutsche Zeitung*, October 18, 1975.

18 See references to both terms in Helga Meister's aforementioned article or in Emil Fischer, "Kagels Geräusche-Show, 'Acustica' – Klangdemonstrationen in der Kunsthalle," in *Düsseldorf Nachrichten*, October 25, 1975.

19 See Jürgen Harten, 'Vorwort,' *Sehen um zu hören*, 3.

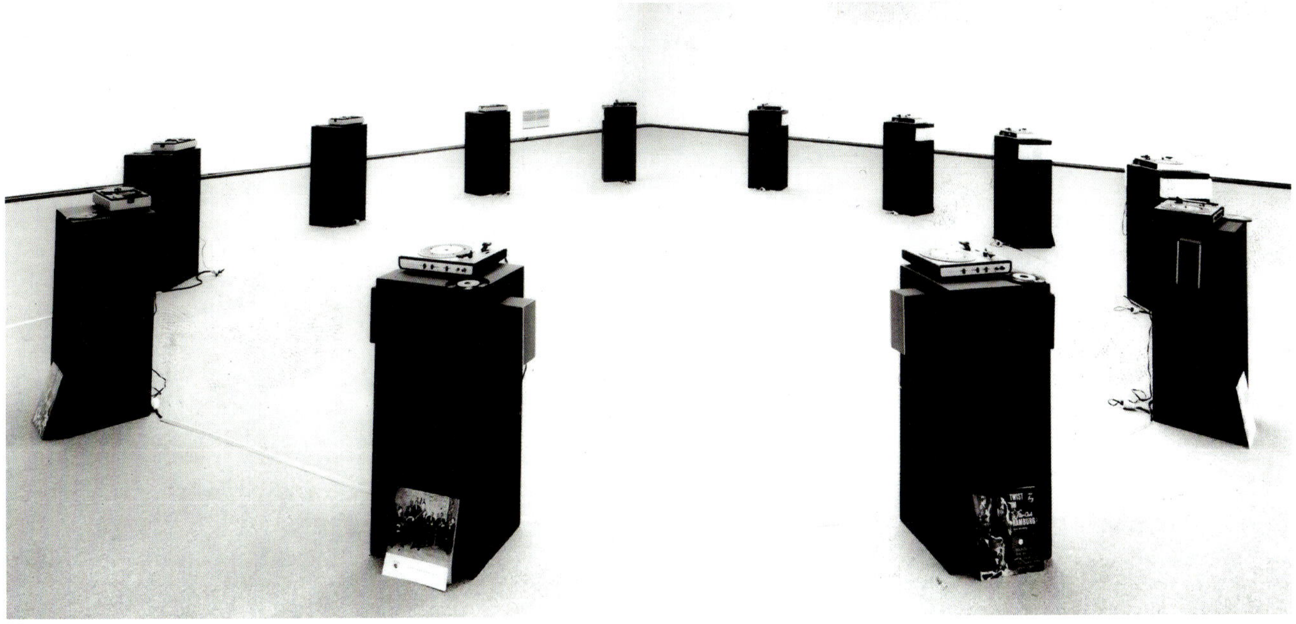

FIG. 3 John Cage, *33 1/3*, 1969. Twelve turntables, twelve stereo amplifiers. Installation view *Sehen um zu hören: Objekte & Konzerte zur visuellen Musik der 60er Jahre*, Städtische Kunsthalle Düsseldorf, 1975.

Charlotte Moorman, Nam June Paik, and Mauricio Kagel.(FIG. 4), [17] The fact that it was possible for everyone to touch and test the exhibited objects enabled the inclusion of senses beyond the sense of sight. This intensified corporeal involvement with the artworks meant that there was a process of demystification of the technology used in some works and of art as a whole.

Some of the terms utilized to describe the exhibited works exemplify the sought-after corporeal relation. *Klangkörper* (sound bodies) were transformed musical instruments, for instance, as in the case of the altered pianos by Cage or newly constructed instruments by Kagel. *Klangobjekte* (sound objects) was another term in circulation.[18] Both *Klangobjekte* and *Klangkörper* refer to the potential materiality of sound. However, there is a clear difference between the term "object" – associated with plastic, inorganic substances – and the term "body," which elicits notions of organic and living substances. If we reflect on the distinction between both of these notions, then *Klangkörper* may suggest an interconnectedness between the body of the musical instrument, the space in which it is situated, and the body of the person using that object. This is particularly interesting in cases where *Klangkörper* refers to an automated object, for example, Stephan von Huene's *Tap Dancer* (1967). In such instances, sound becomes an event for which the body of the performer is no longer needed for its operation. The term *Körper* (body) is perhaps a reminder of that human body, which once was involved in the object's operation, but now refers to that of a machine. The cue for an involvement of a human body disappears altogether in *Klangobjekte*, which alludes only to sounding "objects" that one can enjoy via seeing and hearing alone. While the sense of hearing became a definite component of the experiences offered in the white cube, the sense of touch was offered as an addition providing an element of choice. The mediation and involvement of all three senses was presented as a welcome option and a shift in curating approach.

Adding the senses of hearing and touch into the curatorial strategy indicated not only a change in how the exhibited artworks were experienced, but also in how visitors were expected to perceive the institution. To explain why works containing sound entered the art gallery, the director of the Städtische Kunsthalle, art historian Jürgen Harten, tried to forge links between the exhibited works and assemblage art. In his foreword to the exhibition catalogue, he constructs a parallel between this form of visual art and the exhibited works in terms of sound equipment, noise machines, and altered usage of traditional instruments.[19] The idea was to contextualize sound artworks within already known traditions and simultaneously make a statement about the open approach being taken by the institution.

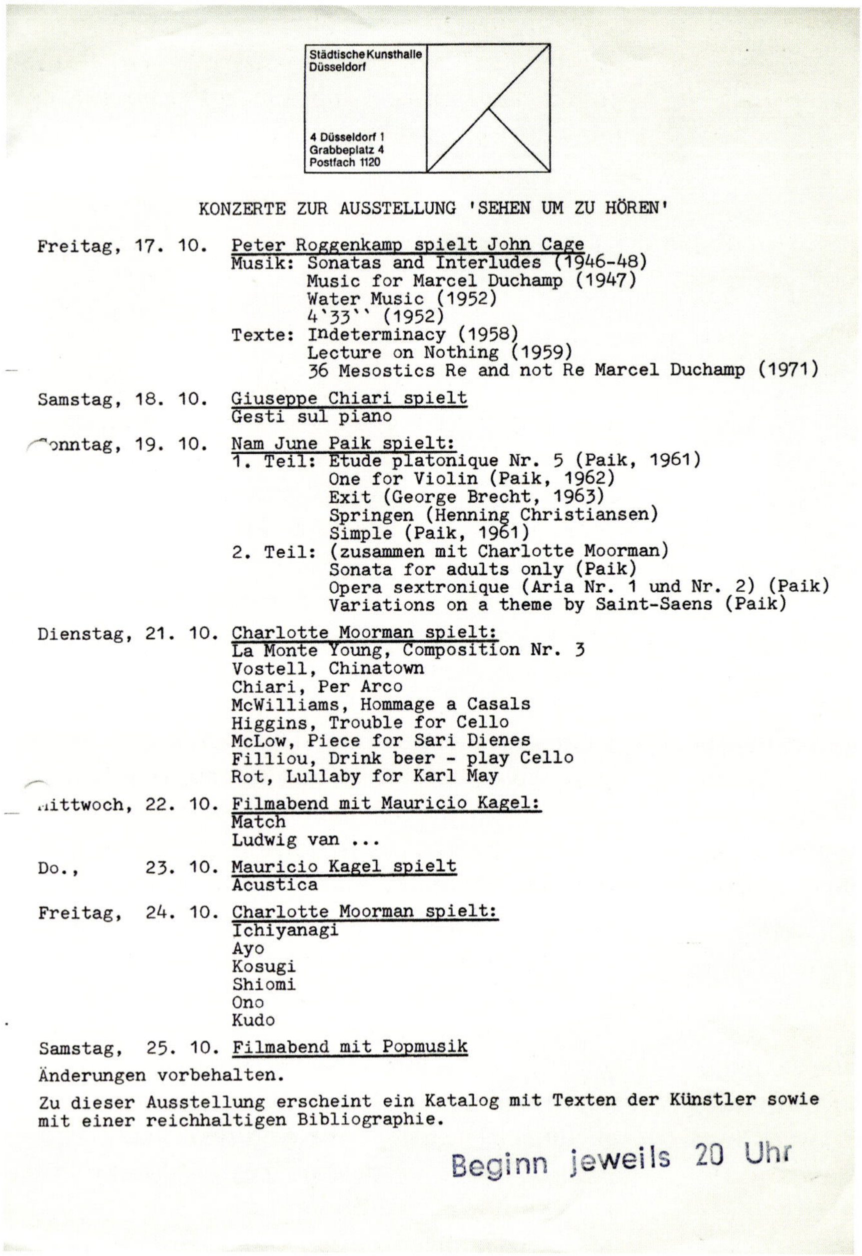

Städtische Kunsthalle
Düsseldorf

4 Düsseldorf 1
Grabbeplatz 4
Postfach 1120

KONZERTE ZUR AUSSTELLUNG 'SEHEN UM ZU HÖREN'

Freitag, 17. 10. Peter Roggenkamp spielt John Cage
Musik: Sonatas and Interludes (1946-48)
Music for Marcel Duchamp (1947)
Water Music (1952)
4'33'' (1952)
Texte: Indeterminacy (1958)
Lecture on Nothing (1959)
36 Mesostics Re and not Re Marcel Duchamp (1971)

Samstag, 18. 10. Giuseppe Chiari spielt
Gesti sul piano

Sonntag, 19. 10. Nam June Paik spielt:
1. Teil: Etude platonique Nr. 5 (Paik, 1961)
One for Violin (Paik, 1962)
Exit (George Brecht, 1963)
Springen (Henning Christiansen)
Simple (Paik, 1961)
2. Teil: (zusammen mit Charlotte Moorman)
Sonata for adults only (Paik)
Opera sextronique (Aria Nr. 1 und Nr. 2) (Paik)
Variations on a theme by Saint-Saens (Paik)

Dienstag, 21. 10. Charlotte Moorman spielt:
La Monte Young, Composition Nr. 3
Vostell, Chinatown
Chiari, Per Arco
McWilliams, Hommage a Casals
Higgins, Trouble for Cello
McLow, Piece for Sari Dienes
Filliou, Drink beer - play Cello
Rot, Lullaby for Karl May

Mittwoch, 22. 10. Filmabend mit Mauricio Kagel:
Match
Ludwig van ...

Do., 23. 10. Mauricio Kagel spielt
Acustica

Freitag, 24. 10. Charlotte Moorman spielt:
Ichiyanagi
Ayo
Kosugi
Shiomi
Ono
Kudo

Samstag, 25. 10. Filmabend mit Popmusik

Änderungen vorbehalten.

Zu dieser Ausstellung erscheint ein Katalog mit Texten der Künstler sowie mit einer reichhaltigen Bibliographie.

Beginn jeweils 20 Uhr

FIG. 4 Accompanying program to the exhibition *Sehen um zu hören*, Düsseldorf, 1975.

Even if nothing were to change about the architectural structure of the white cube, there could at least be a reconsideration of how such spaces and the works exhibited in them are experienced. In light of such tendencies, the art historian Jonathan Harris states that in order to break free from the critique associated with the white cube, art galleries sought to establish themselves as open and welcoming cultural spaces.[20] Institutions that were seen to be isolated from cultural developments aimed to update their image and promote themselves as educational and interactive spaces. Such new aspirations, together with the institution's architectural features, as O'Doherty went on to suggest, became part of the content of the exhibited works, which were enhanced with the label of "art," and later, "sound art."

In this instance, an institution dedicated primarily to the visual arts took the novel step of welcoming musicians and composers for an exhibition of what would come to be known as "sound art."[21] This meant that the leading concept of *Sehen um zu hören* was not considered as a dialogue between visual artists and musicians but as an event curated primarily from music's perspective. The curator of the exhibition, Inge Baecker, confirmed this by pointing out that all the works in *Sehen um zu hören* were reflected through a "Cagian spectrum."[22] The ability to connect the source of what was heard to a direct visual object, at a time when computer-generated sounds added a sense of confusion regarding experimental composition, served as a form of translation and definition for the aural. Music expanded its disciplinary parameters by entering into a visual arts institution, and at the same time, this institution was able to broaden its horizons beyond the visual.

The sense of touch acted as a form of subordinate mediator between the two senses. Although tactility, traditionally categorized among the "lower" and "feminine" senses, was included in the exhibition, its role was that of a bystander to the two principal senses of seeing and hearing.[23] Yet the disciplinary advances proposed in this exhibition were ambivalent with regard to music's hegemonic inheritance. Such an approach is evident, for instance, when examining the aesthetics of the exhibition catalogue's cover.(FIG. 5) Featuring Beethoven's portrait styled in a Pop art manner, the design of the cover can be interpreted as

FIG. 5 Cover of the exhibition catalogue *Sehen um zu hören: Objekte & Konzerte zur visuellen Musik der 60er Jahre* (Düsseldorf: Städtische Kunsthalle, 1975). Designed by Inge Baecker.

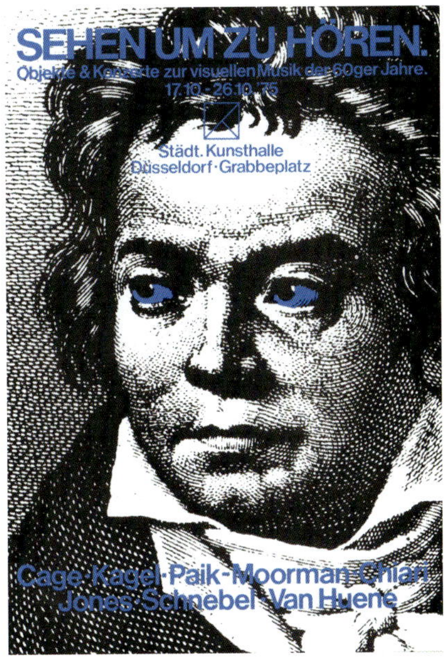

a biased statement in terms of politics and gender. This is because it reaffirms a particular canon and its implications in relation to the male-dominated metanarrative of musical history. Baecker argues that Beethoven was chosen due to his composition *Wellington's Victory* from 1813 – a piece which is considered to be one of the first historical examples of visual music.[24] The choice of Beethoven's portrait restated the influential German legacy in music. The embrace of Pop art elements signaled the North American addition to or, indeed, the continuation of this transnational cultural trajectory. This is particularly interesting considering the fact that the catalogue is one of the main reliquaries of the exhibition and is, of course, exclusively visual.

IV.

Technological development made the mediation of our senses ever more pervasive. In addition, mechanization of objects meant that the human body was no longer required for their operation, although the ability to touch drew visitors' attention to the corporeality of their bodies and those of artworks. Yet this sense was neither integrated in any of the exhibitions' titles nor meaningfully discussed by the curators and reviewers. Instead, the union of sight and hearing stood at the forefront of the exhibitions' promotional strategy, although it did not mean an interdisciplinary collaboration between music and visual arts. What this union expressed was a subversion of the developing notion of the white cube, which occurred parallel to the formation of the emerging field of sound art. Such curatorial decisions implied that instead of perceiving the exhibited artworks as aesthetic objects, in the Greenbergian signification, artworks in white cube settings could be explored with an intensifying corporeal intervention. When senses other than sight were more consciously involved, the function of such a space started to shift toward a more interactive concept. In light of Steven Connor's emphasis on the embodiment of sound as an "event" of objects, touching objects became an accompanying event to hearing and seeing them.

Für Augen und Ohren in Berlin constituted one of the major moments in the formation of sound art in Germany and can be considered as a kind of hybridization between the innovation in musical composition sketched out in Düsseldorf and the developments in the newest technology outlined in Cologne. The curatorial decisions made in these three exhibitions responded to an expanded form of spectacle – in addition to the sense of sight, aural and tangible experiences were now also incorporated into the white cube. These were, however, still conducted in mediated and isolated environments. Guy Debord's work from the 1960s served as an example of the intensified reexamination of the symptoms of spectacle that occurred with ever greater intensity in the years following his book's publication. In thinking of Debord's critique of the spectacle, by adding the sense of touch to the conjunction between seeing and hearing, the curatorial practice developed and simultaneously subverted the modernist spectacle of the untouchable commodity.

20
See Jonathan Harris, "Introduction: Elements Towards a Historical Sociology of Contemporary Art," in *Art, Money, Parties: New Institutions in the Political Economy of Contemporary Art*, ed. Jonathan Harris (Liverpool: Liverpool University Press, 2004), 11–38, here 21.

21
Seven out of the eight participants were chiefly educated in music. The composers whose work was played and exhibited included John Cage, Giuseppe Chiari, Stephan von Huene, Joe Jones, Mauricio Kagel, Charlotte Moorman, Dieter Schnebel, and Nam June Paik. Stephan von Huene was the only practitioner whose primary education was in visual arts. The catalogue also includes an interview with John Lennon. This meant that in addition to edgy experimental composition the parameters of this exhibition were also opened towards mainstream pop music.

22
Baecker, *Sehen um zu hören*, 11.

23
See Constance Classen, *The Colour of Angels: Cosmology, gender and the aesthetic imagination* (London: Routledge, 1998), 65–67.

24
Baecker, *Sehen um zu hören*, 9.

Brandon LaBelle

Sonic Site-Specificities

Something stirs – a movement, a voice, *commotion*. Active, and yet steady. It comes right into my listening to construct, through its rather ephemeral presence, a spatial envelope, a place to be. A place within a place. Already there is more than a single direction, more than one temporality; something close by, suddenly, quick and piercing, and also in the distance, floating – is it overhead, or behind, I can't be sure. It sustains itself across this moment, folding into and through all that passes by. Within such multiplicity, though, I find a place; I would say, I take up residence within this sonorous cacophony, this *occurring*. In short, I listen, and in so doing I figure myself amidst the plethora of audible emanations.

Is not listening then a process of inhabitation? To nestle, to dwell, to be absorbed, captured, and to occupy (or be occupied). Even in the midst of a flood of agitating sounds, one may surrender to build for oneself a sense of belonging, or to search for escape, an alternative listening.

Listening and belonging; sonority and the processes by which place is found, made, and from which relationships are nurtured. I'm interested to position these topics and processes in relation to sound art, and especially sonic expressions that lead us deeper into public environments. It is my concern to follow sound art as it engages with questions of site-specificity, querying how audition may enable a *dwelling in the world*. By extension, this will lead to understandings of sound art as a range of practices that search for a means to live side by side with the foreign, the strange, and the unknown.

Site / Siteless

Historically I may turn to the seminal work by Alvin Lucier, *I am sitting in a room* (1969), as a way to open up these topics, and specifically to identify forms of site-specificity articulated by sound art. *I am sitting in a room* is based on the composer recording himself speaking a set of instructions, and then playing it back into the same room while recording, again and again, for a duration of 32 cycles.[1] In doing so, what comes forward is a slow dissolution of his articulated speech, as the room's acoustical resonances become more pronounced over time, until by the end we hear a mysteriously undulating drone remains. Lucier's speech is altered through this acoustical operation, where the interior of a room captures his speech, *overtakes* it. In this regard, *I am sitting in a room* stages a phenomenological fact, as well as a primary relation between oneself and the world, revealing how the materiality that inevitably surrounds us performs an intensely conditioning effect. In short, acoustical experiences immerse us and lend to processes of orientation (and disorientation) that are constantly at play.

The sonic undertaking central to *I am sitting in a room* can be heard to reveal sound – even that of a spoken utterance – as inherently site-specific, while also, as Steven Connor is keen to suggest, being *untethered* from any ground.[2] Lucier's speech is both shaped by the room in which he speaks while floating freely, leaving his physical body behind as nothing but audible traces. Such might be one of the primary lessons of sonic experience: as listeners we are continuously located within a dynamic exchange between sitedness and sitelessness, between phenomenological fact and sonic experience, and between singularity and its diffusion, its becoming. Sound immerses us within such an unsteady oscillation, connecting us to the immediate while integrating more phantasmic, ineffable, and temporal relations to what we perceive. I'm interested in considering how sound art draws upon this relation, accentuating the drive of a site-specific practice while also locating this within a greater project in which immateriality and the unfamiliar participate. In doing so, sound art often searches for the foreign and the strange explicitly found within the materials and architectures around us; it may emphasize the facticity of things, while also relating this to the energies of the invisible, or the foreign matters of more distant or hidden locations. This may be found already in the discussions and works occurring in the postwar period of the 1950s, in which the material emphasis found in the compositions of John Cage, for example, are contrasted within the more de-contextualized, non-referential compositions of *musique concrète* defined by Pierre Schaeffer, Pierre Henry, and others working at the electroacoustic studios of the Groupe de Recherches Musicales (GRM) in Paris. The relation between material and immateriality, phenomenological fact and sonic experience, and further, between the familiar and the unfamiliar, is thus articulated as a central historical thread around which sound art may be heard to take shape.

I would emphasize to consider this duality less as a strict line and more as a network of possibilities, to ultimately suggest a view onto sound art that, rather than withstand the confrontations and discrepancies between fact and fiction, matter and dreams, *conducts* them.

The works of Maryanne Amacher can be appreciated as further expressing these tendencies, where sound functions to both emphasize the specifics of

1 See Alvin Lucier, *I am sitting in a room*, CD (New York: Lovely Music, 1980).

2 See Steven Connor, *The Matter of Air: Science and Art of the Ethereal* (London: Reaktion Books, 2010).

places while opening onto a sphere of an affective labor: a dreaming-listening leading to what Salomé Voegelin refers to as the "possible impossible" to which sound art may lend an important platform.[3] From the late 1960s through to her passing in 2009, Amacher consistently sought to extend what a sound art could be by creating elaborate, telepresent, and environmentally engaged works. In particular, her *City-Links* series was based on live amplifications of sounds culled from specific spaces and transmitted to other locations; often incorporated into live performance or radio broadcasts, the works intensified a relation to material reality while uprooting and diffusing such materiality toward a plane of "musicality." As the artist describes: "*City-Links* uses electronic means to connect spaces distant from each other. Together in time. Like music. Here there is no boundary to the outside. No wall to the outside. To receiving. But the wall might be with us. Inside."[4] Collapsing or transcending boundaries, Amacher, in linking as well as diffusing the fixity of places, accentuates sound as a medium for another type of site-specificity: "My hope is that the split which now exists between these two worlds - that of musical language and of environmental sound - one day will be closed."[5]

Domestic Practices

It is my view that sound art is driven by a general interest and engagement with experiences of *intensity,* of finding place by confronting alien energies, and deepening our listening sense precisely where it is most difficult. Sound art brings us into intimate contact with such intensities, bridging often-sharp boundaries - between not only musical language and environmental sound, but equally between myself and another, between this human figure and other species, between the seen and the unseen - to produce dramatic experiences of embodied and affective sounding. In this regard, sound art creates the conditions for relating to what is often foreign or distant, conditions that I might speculatively highlight as *domestic* and in support of forms of care and attachment.

Site-specificity, in this regard, is a methodology for giving attention to both the physical environment as well as to the social and cultural constructions that often shape such environments and our experience of them; in addition, though, within sound art, site and sitedness operate to question the facticity of what is there before us. Subsequently, a sonic site-specificity is one that moves beyond the surface appearance of things, to listen *into* the architectures, the social experiences, the ecologies, and the often quiet or non-human soundings found therein. It further aims to register, through an experiential or affective intensity, what is often only described within the discourses and designs that attempt to articulate what a place is or has been.

The field recording work of the Norwegian artist Jana Winderen, for example, deepens a view onto the rather hidden regions of oceans and rivers, capturing the sounds of underwater creatures as well as the complex murmurings of all that lurks below.(FIG. 1) In this gesture, I might suggest we discover an essential sonic activity found within sound art, that of relating to the unknown that lives beyond our experiences or habitats. A kind of primary fieldwork is embedded in sound art - an itinerant curiosity - leading artists to extend our listening sense toward (other)worldly understanding, affective intensity, and animate attunement. In Winderen's recordings, which she presents through audio releases, sound installations, and performances, we hear the movements and forces of alterity itself; these differences are brought *up against* the body through experiences of concentrated listening. Domestic spaces and working environments, performance venues and gallery situations become contexts for such listening, allowing for the distant to become intensely proximate, to *touch* us.

Yet, there is something more taking place here: while we may listen intently to the recordings of distant places, experiencing situations of worldly intensity, we also find a more quiet and considered relation by doing so - we *dwell* within the sonic curvatures and temporalities, aligning

3
See Salomé Voegelin, *Sonic Possible Worlds: Hearing the Continuum of Sound* (New York: Bloomsbury, 2014).

4
Maryanne Amacher, *City-Links* pamphlet (New York: Ludlow 38, 2010), 8, http://ludlow38.org/files/mabooklet.pdf.

5
Ibid. According to the pamphlet, the citation first appeared in *Chicago Style – Chicago Today*, May 10, 1974, 37.

6
Jana Winderen, in *Manual for the construction of a sound as a device to elaborate social connection*, ed. Brandon LaBelle (Berlin: Errant Bodies Press, 2010), 92.

7
See website of the artist: http://maaheli.ee/main/archives/5000.

where listening becomes a process of *making* space for the fragile ecologies of worldly life.

I'm interested in these processes as indicative of a greater trend within sound art, where work is focused on relating, however concretely or imaginatively, to what is *out there*: whether underwater creatures, derelict buildings, foreign conditions, or marginal communities, sound art often tunes us to what we cannot see, and what is mostly charged by unknown or unknowable phenomena – as well as by the urgency and desire to construct a space for deeper attention. A position of concern and care, and even, I would say, love.

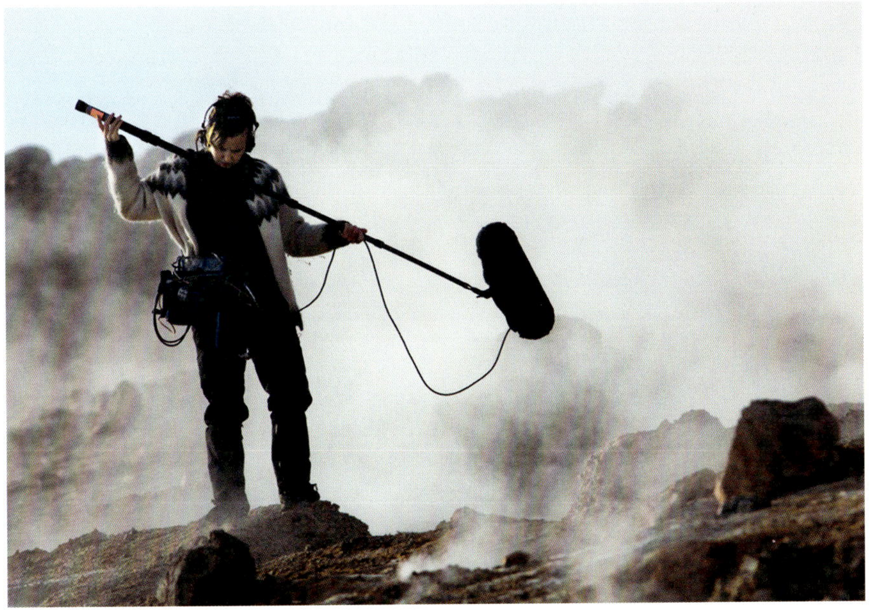

FIG. 1 Jana Winderen, recording above water with a sound field microphone, Iceland, 2014.

ourselves with the animate presences found in places far from our own. The intensities of worldly presence are literally *housed* within a situation of care and consideration, of attunement and concentration – a composed listening by which we attend to this unfamiliar world.

> Little is still known of the effect man-made noises like seismic testing, motor noise from ships or industrial noises has on the ocean. The only time I can record quiet sounds like the sound of sea urchins filtrating the water, saithe grunting as they are hunting smaller fish, or just listening to the current as it moves, as I believe fish are doing to orientate themselves, is at four o'clock in the morning, before fishing boats, ferries or other boats enter the environment and totally overpower the sounds in the ocean.[6]

Winderen, in this sense, undertakes her fieldwork with great dedication, driven by a feeling of joy and discovery, as well as compassion and concern, where environmental issues and strategies of sonic composition are interwoven, sending our listening toward heightened sensuousness as well as states of affective labor: sound art becomes a labor aimed at discovering the extremes of the phenomenal, where the unnameable hovers and

Subsequently, sound art forces us to rehear the familiar through methods that I might characterize as "domestic." Here I emphasize the domestic not solely as related to private or familial life, to what is close by or factual, but as a position by which to nurture deeper relations. Sound art, while often seeking out the intensities of the open and the stark publicness of certain environments, does so through gestures of care and concern, creating experiences that surprise us with their intimacy.

John Grzinich's work, which is equally based on acts of fieldwork, gives further expression to this sonic site-specificity. Often focused on collective improvisations located in marginal environments, such as military bunkers or derelict industrial sites, Grzinich seeks to activate a sonic interaction between people and places. For instance, the artist has been making recordings of interacting with a disused water tower near his home in Mooste, Estonia, for a number of years.(FIG. 2) This large, metal structure provides an opportunity for numerous interventions, which the artist calls "portraits of a sounding object."[7] Bowing and playing the object, with microphones placed in or on the structure, Grzinich captures the sonic textures and atmospheric soundings of this found object. In doing so, these portraits exemplify a sonic site-specificity in which materials and structures of the world function as

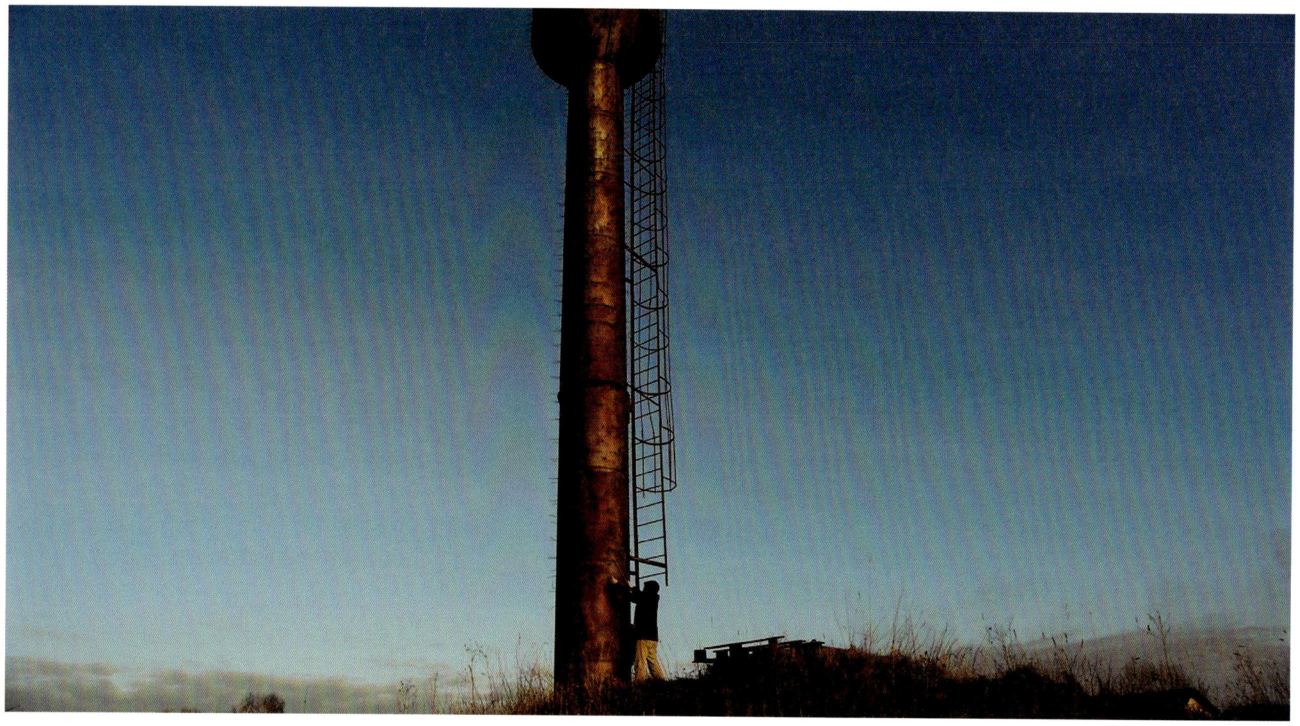

FIG. 2 John Grzinich, *Portrait of a sounding object II*, 2014, video documentation of an interaction with an unused water tower in Mooste, Estonia. Video still.

instrumental opportunities. While we may see the water tower as a form of industrial detritus, Grzinich moves closer, to touch its rusty surface, to scrape and to drum, bowing its cylindrical shape into tonal reverberance. These haptic labors suggest a form of attention that transforms the leftover and the ruinous; through such gestures Grzinich makes home within abandoned environments, and I would say, his sonic approach articulates a domestic mindfulness, especially for the disused and the forgotten.

What I'm arriving at is the expression of a relationship, a rather charged sonic dialogue that moves back and forth between the domestic to that of the worldly, from singularity to publicity, and from the desire to create to its material manifestations that often spark a range of negotiations, confrontations, and surprising commingling. In this regard, I'm curious to ask: What types of community might such sounding arts demand or inspire? Is there a new auditory consciousness attached to the global changes today, one that might interweave the intimacies of the familiar and immediate with the intensities of (other)worldliness? And by extension, that can support forms of solidarity that are equally about mindfulness and imagination? As well as practices of care that are publicly minded?

Stranger Relations

The works of Jana Winderen and John Grzinich return us to the critical dialogue passing historically between John Cage and Pierre Schaeffer, and how the duality between material reality and otherworldly sonic experience would come to define a conceptual territory for sound art. With its fuller emergence starting in the 1960s and 1970s, such a dialogue becomes generative for a rich, diverse culture of sonic thinking and sounded practices. Winderen and Grzinich are exemplary of a contemporary approach, where acts of sounding and listening equally engage with what is to be found in local or distant environments, drawing this into a sphere of attention. Yet, what comes forward additionally is how such attention is bound up with a relation to strangeness, intensity, rupture, and profound agitation. In short, the domestic, sonic imaginary is shot through with figures and forces of strangerhood.

As I trace in my book *Background Noise,* the work of acoustic ecology and soundscape composition is key to appreciating this dynamic within sound art. Based primarily on field recording, and the collecting of sonic information from specific sites, much of acoustic ecology research leads to composed sound works that bring the

8
See Budhaditya Chattopadhyay, "Auditory Situations: Notes from Nowhere," *Journal of Sonic Studies* 4, no. 1 (May 2013): n. p., http://journal.sonicstudies.org/vol04/nr01/a06.

9
Heimo Lattner, in *A voice that once was in one's mouth* (Berlin: Errant Bodies Press, 2014), 71.

foreign, the distant, and the alien into structures of musicality; a musical listening surrounds soundscape compositions, as expressed in the works of Hildegard Westerkamp for example, allowing us to hear unfamiliar (or rehear familiar) territories, as well as the sonic imaginaries that, as Budhaditya Chattopadhyay is keen to articulate, envelope our listening experiences.[8] In other words, soundscape compositions are important journeys into the interaction between the familiar and the unfamiliar, putting each into a compositional structure and giving narrative to a certain relational intensity: that what may exist beyond our immediate environment is often deeply informative for how we might live in this world. Phenomenological fact and sonic transformation, the immediate and the imagined, these are no longer separate territories but emerge as an intensity of relations, *auditory composites* that broaden our listening toward what we hear as well as what we don't hear. Soundscape compositions lead us into a more holistic as well as "migratory" listening, where the physicality of the here and now is always already in dialogue with the more ineffable, imaginary properties of sonic experience. In short, acoustic ecology reveals the particulars of what listening is always already doing: confronting the differences to which I am always related.

I Love To You, a project by the artist group e-Xplo, additionally captures this relation by way of the singing of others. Developed as part of the Sharjah Biennial in 2007, the project aimed to engage with the migrant workers living in the United Arab Emirates. e-Xplo was concerned to intervene and reflect upon the conditions of their displacement, and general precarity, searching for ways not only to highlight such precarity, but to equally turn it toward gestures of agency, and of newfound expressiveness. The group set out to record workers singing songs from their respective homelands; workers from parts of Africa, as well as throughout the Middle East, from Egypt to Tunisia to Saudi Arabia, were recorded singing remembered songs. In this gesture, singing became equally a process of *returning home*.(FIG. 3)

For the exhibition, e-Xplo shared the recordings with local shops and cafés, asking them to play the work over installed speakers; throughout the duration of the Biennial, the voices of migrant workers could be heard, puncturing the life of the city, and the general soundscape, with their singing, their dialects and accents – voices equally suffused with an emotional energy that relocated home onto a place of strangeness, a foreign territory. As Heimo Lattner, a member of the group, states: "These voices in this particular place, under those circumstances, conveyed what cannot be said or addressed publicly."[9] Place and displacement, home and itinerancy, are woven into a complex sonic fabric to capture our listening in a sphere of concern; a tender map leading us deeper into the politics of belonging and global labor, migrant laws and capital – all subsumed by a single voice reminding of the essential landscapes of the heart.

e-Xplo's project captures an interaction between strangers, as well as articulating the position of strangerhood not as a border or barrier, but as a force of difference that can equally produce relations of

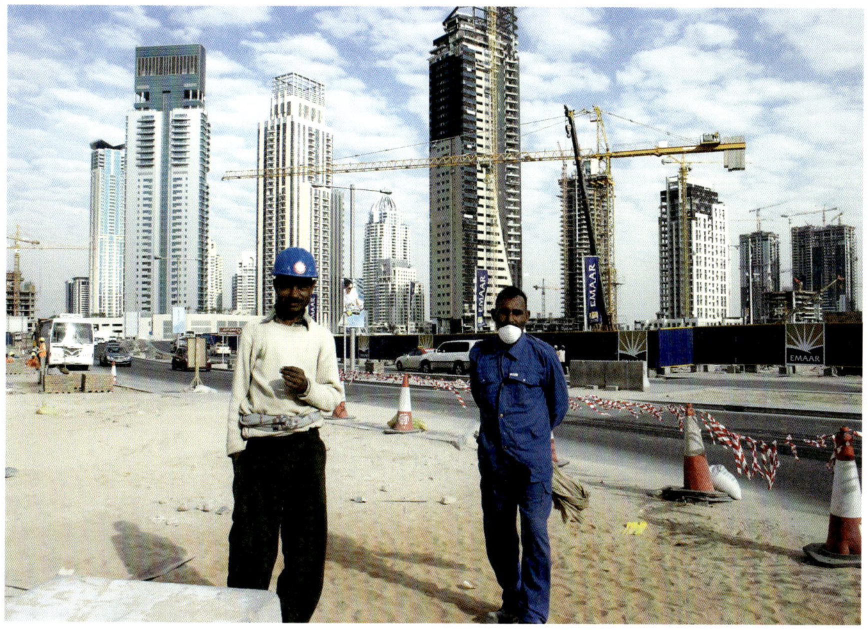

FIG. 3 e-Xplo (Rene Gabri, Heimo Lattner, Erin McGonigle) with Ayreen Anastas, *I Love To You. Workers Voices from the U.A.E.*, 2007. Sound installation, map, CD, and booklet. Installation view Sharjah Biennial 8, 2007. Commissioned and produced by Sharjah Art Foundation, courtesy of the artists.

closeness and affective labor – the singing voice may bring forward a critical entry into the immediate by explicitly introducing the foreign.

Michael Warner, in his work on publics and counterpublics, highlights how the figure of the stranger participates within public life by appearing on the horizon: the stranger is someone to the side of an identifiable public group, and that nonetheless is addressed in public discourse. Warner terms this "stranger sociability."[10] That I may speak, addressing a group or crowd, for instance, and by which public assembly may be created consisting of people I may know and others that I may not – in this moment, I am always speaking a little beyond what I know; I attempt to address, and to share with everyone, and therefore my words are delivered with the stranger in mind. The stranger, in this sense, is an important figure, for he or she is a catalyst for new forms of speech and identification. In short, I *reach* for the stranger, and in turn I must listen for all that may speak back.

The stranger thus appears while also remaining nameless, intervening within the space of appearances as a peripheral figure nevertheless central to public engagement, as well as a figure to which speech and listening are directed. In the work of artist Sarah Vanhee, the stranger additionally operates. Her performance work *I Screamed and I Screamed and I Screamed*, developed in 2013 working at the prison in Mechelen, Belgium, seeks to consider the hidden figure of the prisoner. Through a series of interactions, Vanhee set out to record the screams of prisoners, with the overall ambition to create a choir of screaming voices. Her project, in this regard, searches for a way to break in, to craft a fissure through which the prisoner may give voice. As with e-Xplo's work, Vanhee focuses on vocalization as a means for critically navigating the politics of marginal or subordinated groups and related systems of abuse.

After a series of attempts to make her recordings, a process that led her through a number of challenging and thwarted interactions, Vanhee finally presented a performance outside the prison walls.(FIG. 4) Standing in front of the prison, the artist screamed to a gathered crowd consisting of prison staff, police, and an art audience:

> There is nothing more personal than a voice. Screaming is something very intimate. To open your mouth and let off sounds from a hole in the middle of your face. But you do it, you speak, or you scream, primarily, in order to be heard. And the idea of being heard, of possessing a voice nearly coincides with that

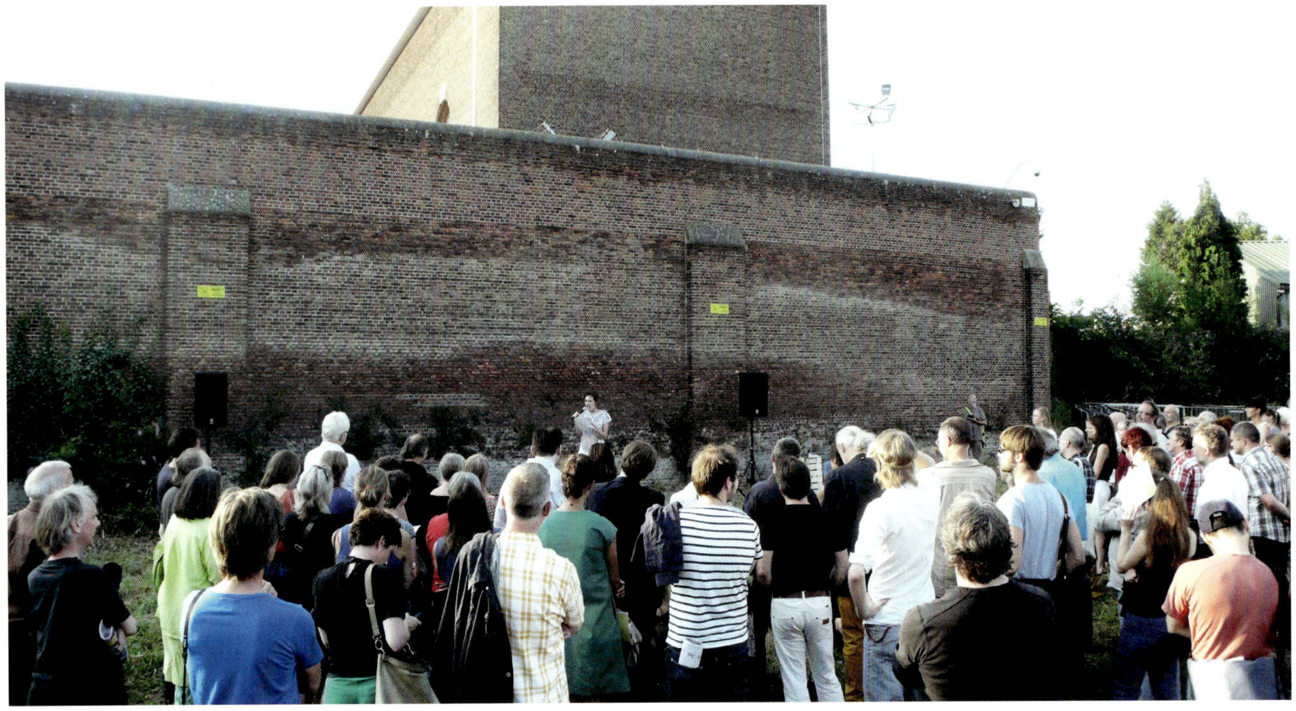

FIG. 4 Sarah Vanhee, *I Screamed and I Screamed and I Screamed*, 2013. Performance outside the wall of the jailhouse in Mechelen, Belgium.

of human and civil rights. Having a voice is much the same as having a vote.[11]

Vanhee searches for a route over the wall; she struggles for a means to nurture conversation precisely with the strangers inside, shifting interior and exterior through an act of concern and compassion. Hers is a voice longing for others, and one that tries to unsettle the lines of separation.

Audible Dwelling

I've been considering sound art as a particular mode of site-specific creativity, one that focuses upon questions of the immediate and the factual while relating this equally to experiences of the foreign and figures of strangeness. As such, sound art can be understood to approach the border, the social impasse, and the foreign territory not so much as a barrier, but rather as a surface of vibration, an animating conductor that supports passage between interior and exterior, domesticity and worldliness, dreams and work – and that further drives a negotiation with the powers of exclusion. Subsequently, we are led toward a deepening of a sonic imaginary from which I might begin to ask: Is it possible to evoke an acoustic space that assembles the nurturing of a domestic life with the strangeness of what it connects to, always already just outside the door? Can we hear in sonic experiences a celebration of how privacy is more public than we often realize? And that constructs, through its radical animations, a common good according to the drives of care and concern?

As the artist Toshiya Tsunoda has demonstrated, walls of buildings are often replete with vibrations, signals that traverse the surfaces and depths of architectures and that we might hear as embedded animations imparting additional messages and sensations onto our experiences of being somewhere.[12] Tapping into such vibrational surges, amplifying their hidden energies, Tsunoda challenges spatial understandings of separation, boundary, and volume, allowing us to pass beyond the visual construct toward the affective intensities of place, and further toward all that may lurk just on the other side. To hear what is seemingly foreign and yet absolutely so close.

As in the works of Amacher, Tsunoda articulates a sonic site-specificity in which architecture and the built environment become partners in an expansive event of listening. This expansiveness, however, may also lead to deeper questions about the qualities of public space, and the tensions found in public assembly. I'd like to follow this through by considering a work by the artist team Learning Site. Their *Audible Dwelling* project is a mobile dwelling that travels to different locations.[13] It consists of two structures – which act explicitly as a stereophonic doubling – each of which is divided into two rooms or sections. The first section functions as a small bedroom or office, as well as integrating recording equipment; the second room can work as a living room, as well as a sealed speaker enclosure operating to transmit sounds from the first room to the exterior of the dwelling. The *Audible Dwelling* is thus an architectural unit that, as in Tsunoda's work, conducts a vibrational or sonic signal, in this case by specifically functioning as a transmission line loudspeaker.[14]

For its presentation in Columbus, Ohio, in 2009–2010, Learning Site responded to particular communities related to the site (in this case, the parking lot of the Columbus College of Art and Design) and its connection to immigrant histories.(FIGS. 5, 6) *Audible Dwelling*, in short, operated as a "speaking machine" supporting local expressions and speech acts aimed at the public sphere. Interestingly, *Audible Dwelling* equally locates such public concerns alongside practices of dwelling, occupation, and assembly; their project articulates how such speech acts are connected to the sphere of the domestic, giving expression to a striking mutuality between public life, and its related politics, and experiences of dwelling. This is furthered in their decision to construct the interior of *Audible Dwelling* in reference to modernist architecture, in particular the work of Eileen Gray and the De Stijl movement. Gray's *De Stijl table* was specifically chosen as the main reference, and used as both a functional object within the dwelling and as part of the transmission line, supporting

10 See Michael Warner, *Publics and Counterpublics* (Cambridge, MA: The MIT Press/Zone Books, 2005).

11 From the script for the performance by Sarah Vanhee found on the artist's website, www.sarahvanhee.com.

12 For more on Tsunoda's architectural work, see his installation *Monitor Unit for Solid Vibration* presented at ICC, Tokyo, 2000, www.ntticc.or.jp/Archive/2000/Sound_Art/Works/monitor.html.

13 See Learning Site, *Audible Dwelling* (Chicago: Half Letter Press, 2011).

14 Transmission line loudspeakers operate by expanding the cabinet of the speaker, usually into a series of absorbent pathways, through which to send the signal, eliminating sonic anomalies and resulting in greater control of the sound.

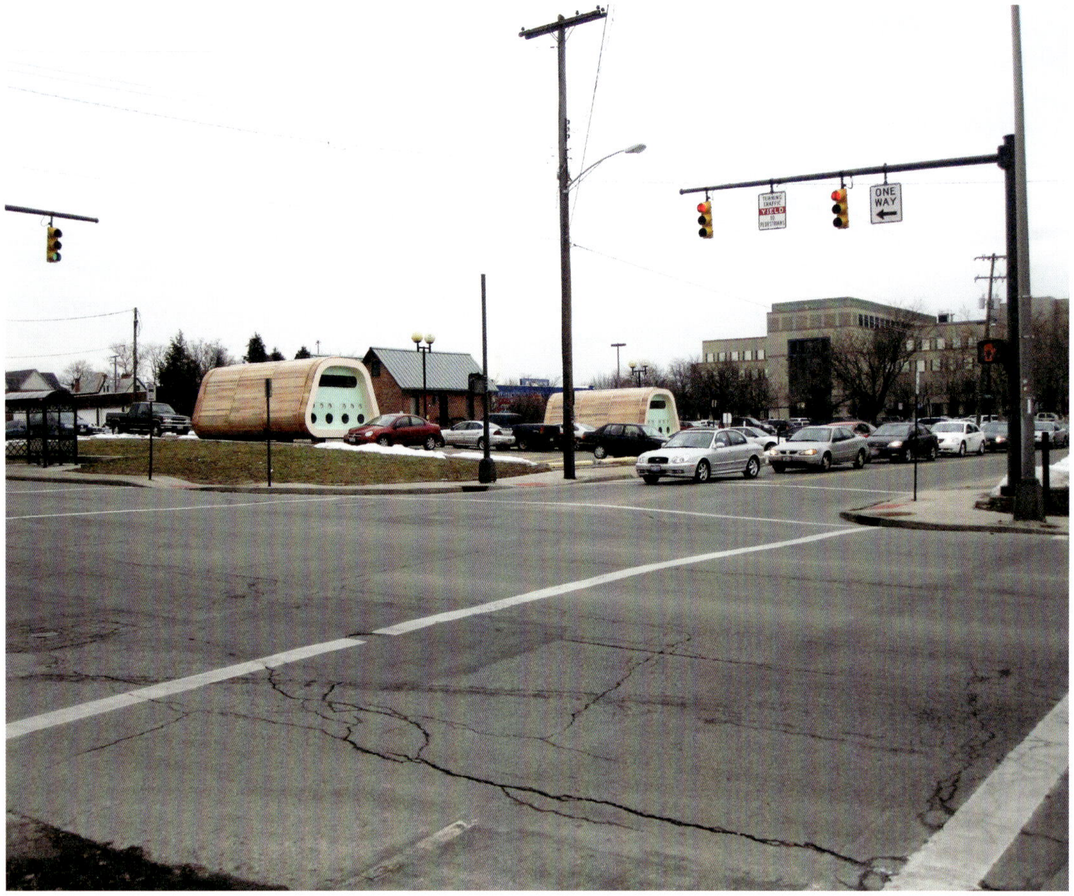

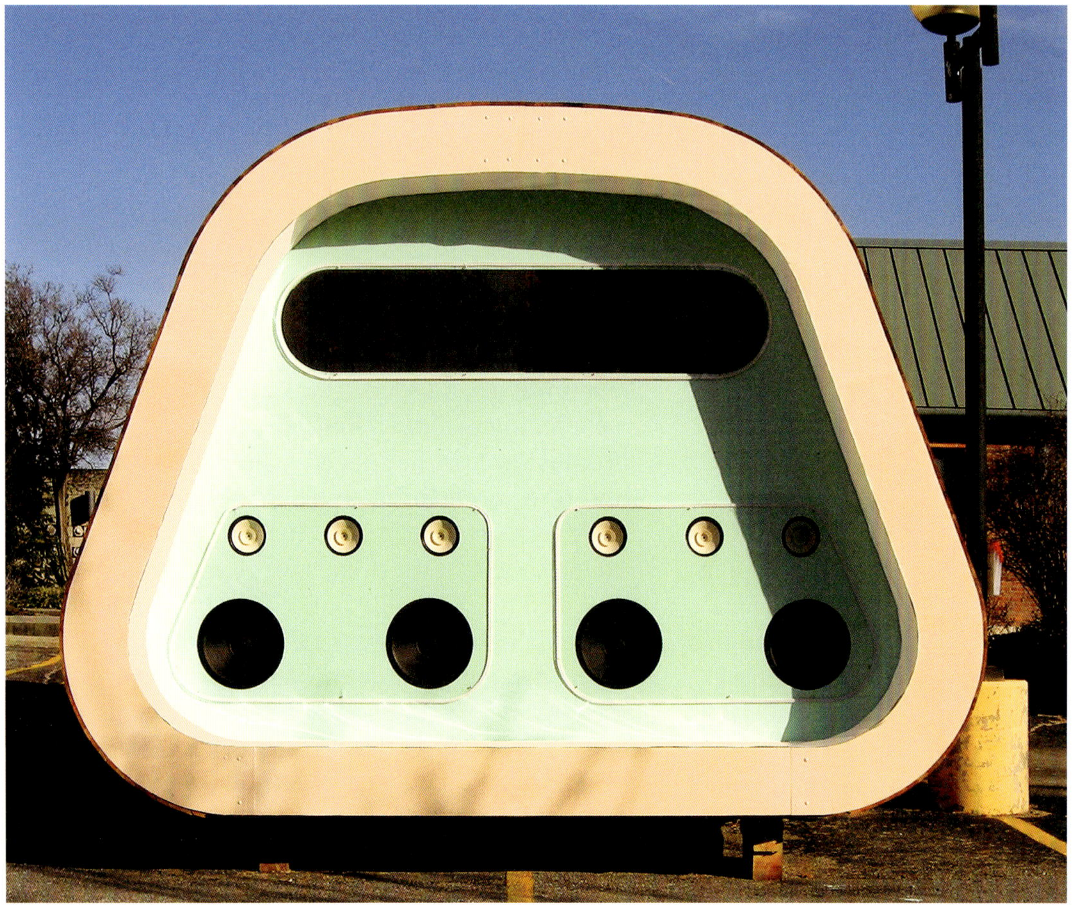

FIGS. 5, 6 Learning Site (Rikke Luther and Cecilia Wendt), *Audible Dwelling 0.1*, 2009/2010, Columbus, OH. Text created in exchange with Jaime Stapleton, sound performance by Cassandra Troyan, audio file available online at: www.learningsite.info/Audible-Dwelling-web.mp3.

the movement of the sound signal toward its amplification. In this regard, participants come to occupy an architecture of sounded activity, living and working, recording and amplifying within the body of a sonic enclosure that is part-public demonstration and part-living room, and where furniture actively participates in the making of a public broadcast, interweaving the space of the home with worldly concerns. Gray's table literally animates the sonic actions – we might even detect some of its materiality in the voices that sound forth from *Audible Dwelling*, announcing social concerns, reflections on the city, and visions of future societies by way of spaces of living.

From Specificity to Solidarity

Following the affective labors posed by sound art, I'm led to an understanding of sonic site-specificity as a mode of sounding and listening that may allow us to approach the intensities of global life today, specifically forcing recognition as to the possible associations by which the familiar and the unfamiliar, neighbors and strangers may come together. Sound art, in other words, explicitly tunes us not only to all that resides here and there, but sets the conditions for greater solidarity. Vibrating walls and unseen creatures, songs and screams that produce publics, or counterpublics, and the sonic drive that leads us to lend a caring ear – such phenomena and gestures point toward a central thread within the sonic imagination: the desire to discover new openings for collectivity, sharing, and togetherness, especially with what may appear through the movements of a mysterious or unforgettable sound. This takes on greater urgency if we consider the uprisings, conflicts, and general unrest that have swept over the globe in recent years in the wake of financial, democratic, and public crises. The level of charged debate, political battling, social movement, and the narratives that thread themselves around issues of public life and that increasingly appear through social projects, new formations of public life, commoning, self-organization, and alternative pedagogies, are indications of just how pressing questions of togetherness are. The intensities of the desire for more direct modes of governance might be the flipside, within a European context at least, to the dismantling of state welfare and the "crises" that face a great many today: that in that gap we are forced into what I would characterize as an intensely creative *and* domestic approach toward worldliness. Caring for resources and each other, extending compassion beyond the immediate and the visible, radically shifting the dichotomy between neighbors and strangers, interiority and exteriority – new openings for collectivity, sharing, and togetherness. Within such public mindfulness, listening, sharing, and finding place take on pronounced urgency, and the platforms instigated by sonic site-specificities may act to nurture imaginative solidarities.

Tony Myatt

Sound Pavilions

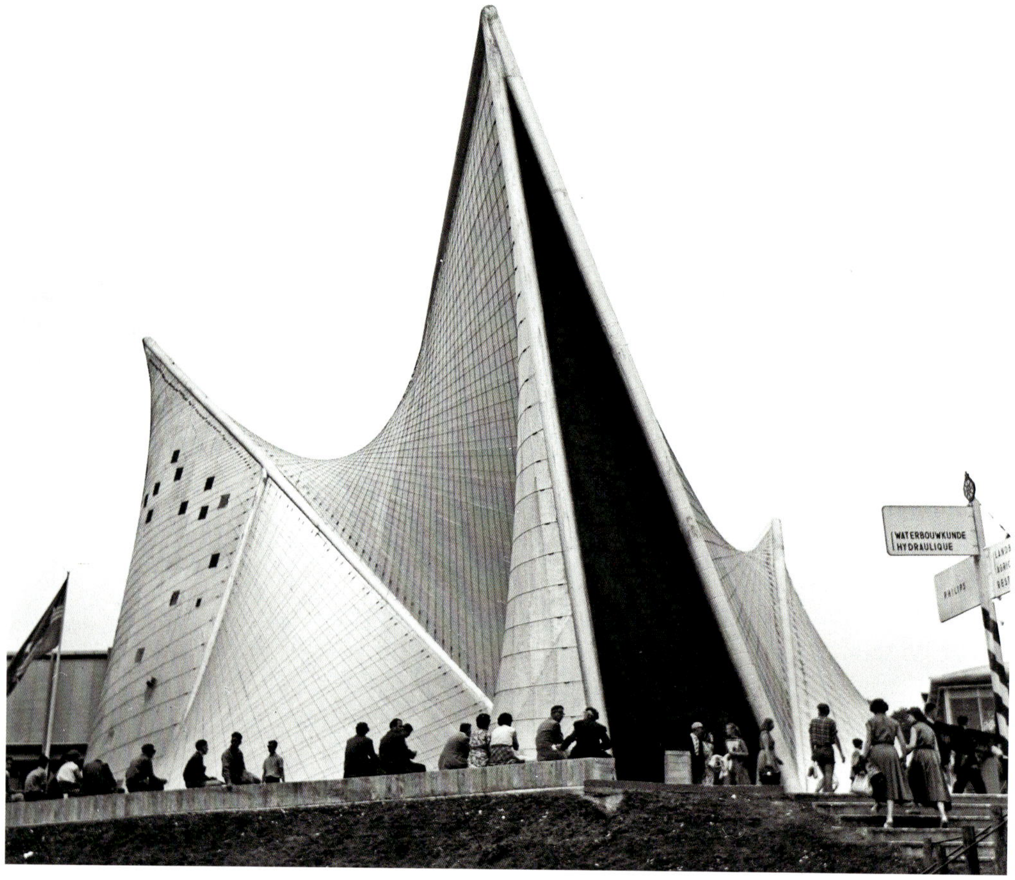

FIG. 1 Iannis Xenakis and Le Corbusier, *Philips Pavilion*, Brussels, 1958.

The hyperbolic parabolas of Iannis Xenakis's Philips Pavilion(FIG. 1) for Expo 58, the Brussels World's Fair, have resonated through the canon of postwar architecture. The structure has become an icon of the modern and heralded as one of the most significant moments in the history of twentieth-century electronic music for its use of spatialized sound.

Architectural pavilions are regularly exhibited throughout the world. They range from student experiments such as the London-based Architectural Association School of Architecture's Summer Pavilions series to site-specific interventions in the manner of Benoît Le Thierry d'Ennequin and Yves Pagès's entrance pavilion for the Palace of Versailles in France. London's Serpentine Gallery has commissioned temporary pavilions from some of the world's most prominent architects and artists for their annual series since 2000. This has included work by Zaha Hadid, Daniel Libeskind, Olafur Eliasson, Frank Gehry, Herzog & de Meuron, and Smiljan Radić. Pavilions as a vehicle to explore architectural form, architectural sites, and new construction materials, or simply to express the joy of architectural virtuosity, have become rooted in contemporary architectural practice.

Curiously, sound and spatialized sound have become prominent infiltrators in this architectural laboratory, following the example of the Philips Pavilion, which was designed principally by Xenakis on behalf of Le Corbusier's studio. Writings and discussions of Xenakis's work for the Brussels construction have established the Philips Pavilion with an almost mythical status. Even today, the pavilion exerts a fascination, despite our inability to experience the work or understand our own response to it following its demolition in 1959. We cannot but be intrigued by the pavilion's sonic and multimedia experience; by its sound projections of Edgard Varèse's *Poème Électronique* and Xenakis's *Interlude Sonore* (subsequently titled *Concret PH*), which could be heard as the audience entered and exited the building; and most particularly, by the technologies used to create its postwar technical wonder.

But the work has left a thread of an idea that has remained with us: the potential to combine and potentially unify architectural, sonic, and visual experiences. As technologies and practices in the field have developed since 1958, we now look toward a future where greater and more complex control of sound may begin to rival the architectural substance of the pavilion, perhaps bringing a shift toward the realization of Varèse, Karlheinz Stockhausen, and many other composers' visions of true spatial music – an architecture of sound alone.

There are also experiential dimensions within current spatial audio practices, and through the use of spatial audio technologies, questions about the nature of our spatial perception and our engagement with space and sound art are beginning to emerge. Audio that can reproduce sophisticated, polyphonic, and perceptually "believable" sound environments has the potential to introduce convincing representations of remote sound environments, to generate superpositions of space, or to create powerfully convincing sound-field reproductions that might lead our perception beyond architectural delineations of space. This perceptual space, apprehended through sound alone, can juxtapose inside- and outside-spaces, allow listeners to move through a work of spatial music as they wish, to explore the audio constituents of a work for themselves in the same way that they might explore a physical architectural construction.

A number of architects and sound artists have sought to integrate spatial constructions of sound, light, and structural form following the model of the Philips Pavilion, often striving for a holistic observer experience. As one might expect within the early twenty-first century's zeitgeist of conceptual, post-structural, and environmentally focused artistic themes, the unification of architectural and audio space has taken many directions away from the goals of Varèse, Xenakis, and Le Corbusier, but their approach still permeates the discipline.

The concept of a sound pavilion moves beyond theaters, the cinema, or notionally "immersive" virtual environments. It has become an experiential space where artists often strive to dissolve the dividing lines between disciplines; it blurs distinctions between what might constitute a work of architecture, sound art, spatial music, a designed environment, a performance, a sound installation, or simply a framing or re-presentation of our experience of the world per se through sound.

The following text highlights works that have taken forward the concept of the sound pavilion. It discusses current practices along with the impact and potential of technological developments for sound projection and architecture. The text discusses links between architectural space and contemporary sound art, audio evolutions from the sound pavilion's modernist starting points toward contemporary focuses on materiality and experiential work in audio, and themes that have emerged from more than fifty years of sound pavilion practice.

Multisensory Experience

Le Corbusier's guiding metaphor for the Philips Pavilion was a cow's stomach,[1] ingesting, digesting, then excreting its World Fair audience. While this may not have been every viewer's understanding of their experience, the work was clearly engaging, frightening for some, perhaps bewildering, but its lasting impact was the connection of architectural space, light, and spatialized electronic sound.

Details about the methods and technologies used in the Brussels pavilion have emerged since its first exhibition. Writings by the pavilion's original creators and detailed research, such as Marc Treib's *Space Calculated in Seconds*,[2] the work of Kees Tazelaar[3] and Vit Zouhar et al.,[4] have revealed details that allow us a more definitive understanding of the pavilion. We can also glimpse the work through Fabio Turcheschi's extraordinary virtual simulation of the structure and its audiovisual content.[5]

What we are still required to imagine, of course, is our true encounter with this work, the more than three-hundred loudspeaker sound reproduction that reached high inside the structure, the impact of the pavilion's acoustics, or the juxtaposition of the physical architectural spaces, the sonic architecture, and the visual presentations.

The Philips Pavilion showed a film with a visual montage by Le Corbusier and music by Varèse that evoked the evolution of modern humans and technology, images of science and the brave new world reaching forward from the postwar era in an archetypal futurist dream. The pavilion was a technological showcase, an experiential encounter with a carefully crafted and controlled environment. As Le Corbusier stated:

> I will not make a facade for Philips, but an electronic poem. Everything will happen inside: sound, light, color, rhythm. Perhaps, a scaffolding will be the pavilion's only exterior aspect.[6]

Varèse's interest in the notion of spatial music[7] was evident in his plans and drawings of the work's spatial presentation concepts, but this didn't translate to a direct engagement with the precise, architectural, spatial presentation of *Poème Électronique* in the pavilion. Xenakis generated the scheme of sound motions and trajectories(FIGS. 2a, b) for the pavilion's sound projection, which was used to distribute three discrete tracks of audio. Varèse's work was reproduced simultaneously with the films and images projected in the pavilion.

Le Corbusier's vision of an organic synthesis of media and forms occurred in the audience encounter with these elements. The juxtaposition of materials was unified in part by their simultaneous presentation, but also by a structuralist notion that the conception and implementation of common processes, underlying rules for the creation of the work's constituents – for example Xenakis's mathematical curves, which defined both the architectural structure of the pavilion and sound movement curves, or later the patterns of flashbulbs in *Le Diatope* – might lead to an inherent synthesis of disparate elements; but the "poem" Le Corbusier described was essentially formed by a juxtaposition of discrete sound, visual, and architectural components.

The Philips Pavilion audience was, in practice, ingested periodically; audience members observed the spectacle, digested the images projected on the walls, listened to the most contemporaneous sonic expression of abstracted sirens, voices, and instrumental sounds moving through the space, before leaving the performance to be excreted back into Expo 58. Looking back, they could marvel at the technological possibilities of the modern age: a multimedia cinema presentation on a grand and spectacular scale.

1
See Joseph L. Clarke, "Xenakis and the Philips Pavilion," *The Journal of Architecture*, vol. 17, no. 2 (2012), 213–29.

2
Marc Treib, *Space Calculated in Seconds: The Philips Pavilion, Le Corbusier, Edgard Varèse* (Princeton: Princeton University Press, 1996).

3
Kees Tazelaar, *On the Threshold of Beauty: Philips and the Origins of Electronic Music in the Netherlands 1925–1965* (Rotterdam: V2_Publishing, 2013).

4
Vit Zouhar et al., "Hearing Varèse's Poème Électronique Inside a Virtual Philips Pavilion," in *Proceedings of ICAD 05-Eleventh Meeting of the International Conference on Auditory Display, Limerick, Ireland* (2005), under http://iem.kug.ac.at/fileadmin/media/iem/altdaten/projekte/publications/paper/poeme_electronique/poeme.pdf.

5
Fabio Turcheschi, "Padiglione Philips e Poème Électronique, una ricostruzione virtuale," animated visual reconstruction, 2003, www.youtube.com/watch?v=YVJgOtZT-b8.

6
Treib, *Space Calculated in Seconds*, 9.

7
J. M. Strawn, "The Intégrales of Edgard Varèse: Space, Mass, Element, and Form," *Perspectives of New Music*, vol. 17, no. 1 (Autumn–Winter 1978), 141.

8
"Musique: Sons et lumières. Une interview d'Yannis Xenakis par Maurice Fleuret," *Le Nouvel Observateur*, October 16, 1972, 75–76.

9
Ibid.

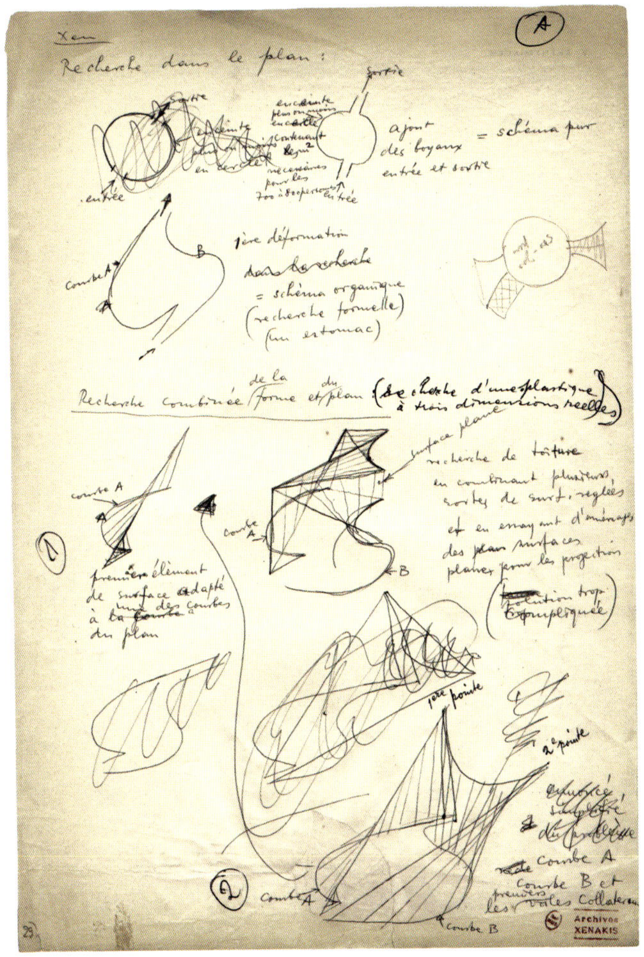

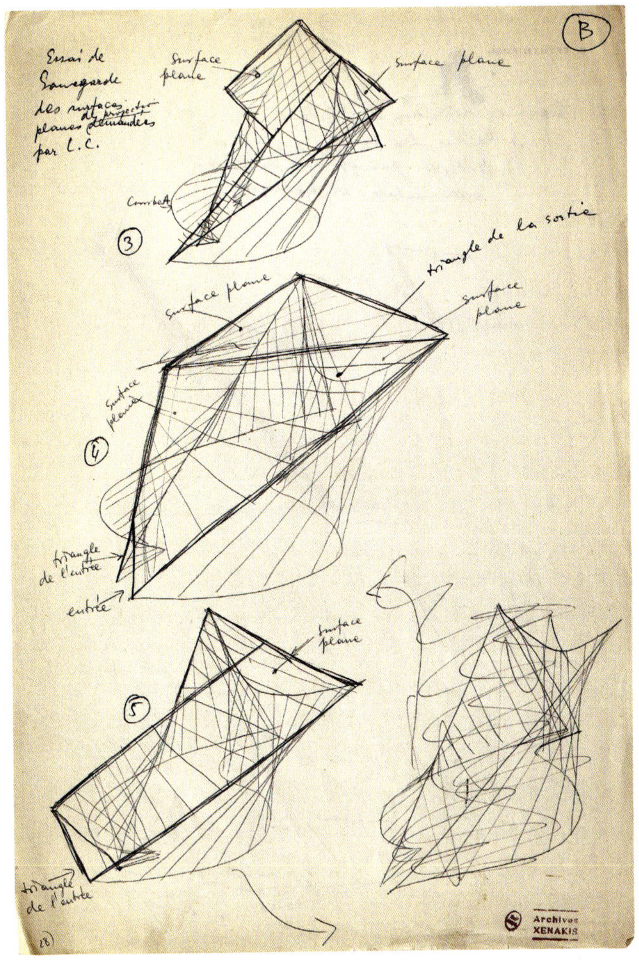

FIGS. 2a, b Iannis Xenakis, Sketches for the *Philips Pavilion*, Brussels, 1958.

Polytope Evolutions and Performance Spaces

From 1967 and his *Polytope de Montréal* to the 1978 presentation of his audio composition *La Légende d'Eer* in *Le Diatope* in Paris, Xenakis continued his exploration of design, light, and sound projection. Polytopes are geometric forms bounded by lines, planes, or hyperplanes, typical of the geometric elements used to construct the "curves" of the Brussels pavilion. In an interview with Maurice Fleuret about the *Polytope de Cluny,* published in *Le Nouvel Observateur* in October 1972,[8] Xenakis embraced the literal Greek origins of the term to encapsulate the multiparametric nature of the works:

> From "poly," many, and "topos," place[…] For indeed there are many different things in one place. Not only light and sound but the thought structures, the selected systems, the technical means employed.[9]

A complex series of works resulted, variously featuring lasers, hundreds of flashbulbs with sophisticated computer and electronic control systems, architectural structures, and sound. Not all of the polytope series combined all media: *Persépolis* (1971) and *Polytope de Mycènes* (1978) were outdoor works with theater, fireworks, and projections, *Polytope de Montréal* a light- and sculpture-based work with audio, and *Polytope de Cluny* (1972 and 1973) an installation within an existing architectural space.

Sound, light, and architecture were combined within *Le Diatope*'s curved geometry that complemented the extraordinarily powerful audio work *La Légende d'Eer*. *La Légende d'Eer* was projected over eleven loudspeakers alongside computer-controlled flashbulbs and complex patterns of lasers directed by mirrors. The density and complexity of this work, composed in sound, light, and space, shares parametric control between many of its materials. This parametric abstraction facilitated a composition that spanned all media – creating a multisensory experience through composition rather than through a juxtaposition of elements. The application of similar parametric control for aspects of *Le Diatope*'s architectural design exposes some of the compositional devices used to control the audio and visual elements of the piece.

The connections between *La Légende d'Eer* and *Le Diatope* are articulated with a temporal displacement between a viewer's apprehension of the structure, as the audience enters and exits the pavilion, and the period when *La Légende d'Eer* is presented in the darkened interior space. The architecture becomes a silent extension of certain compositional processes. Xenakis's arrangements of lights, lasers, and loudspeakers during the performance of the work develop their own spatial arena.(FIG. 3a) The performance work might, potentially, be configured within any large enclosing space, such that the experience of *Le Diatope* occurs in phases: firstly an architectural experience without sound and light, then the primarily sound-and-light experience of the *Légende d'Eer* performance, followed by a reengagement with the architecture alone at the conclusion of the performance.

In a sense the collection of polytope works are hard to describe collectively as sound pavilions, yet they reveal one of the central questions: the distinctions between sound installations, sound sculptures, audio performances, and sound pavilions. In Xenakis's work, these questions are perhaps clarified by the unifying methods used to generate material for different media; arcs or trajectories equally applied to define the movement of a sound's pitch, a pattern of flashbulbs, or the surface of an architectural pavilion. But in *Le Diatope* the primary purpose of the structure is to deliver a light-and-sound work to an audience; an architecture designed to support a performance, a concert venue of a very special kind that also articulated some of the methods central to the generation of Xenakis's audiovisual work.(FIG. 3b)

In Xenakis's vision, the audio and spatial juxtaposition, the lights, flashbulbs, or lasers, fused in the space to create a unified performance experience; but a performance with beginning and end, designed for an audience to attend and then leave as they would a concert presentation.

Xenakis's sonic elements used multichannel sound to distribute audio throughout the

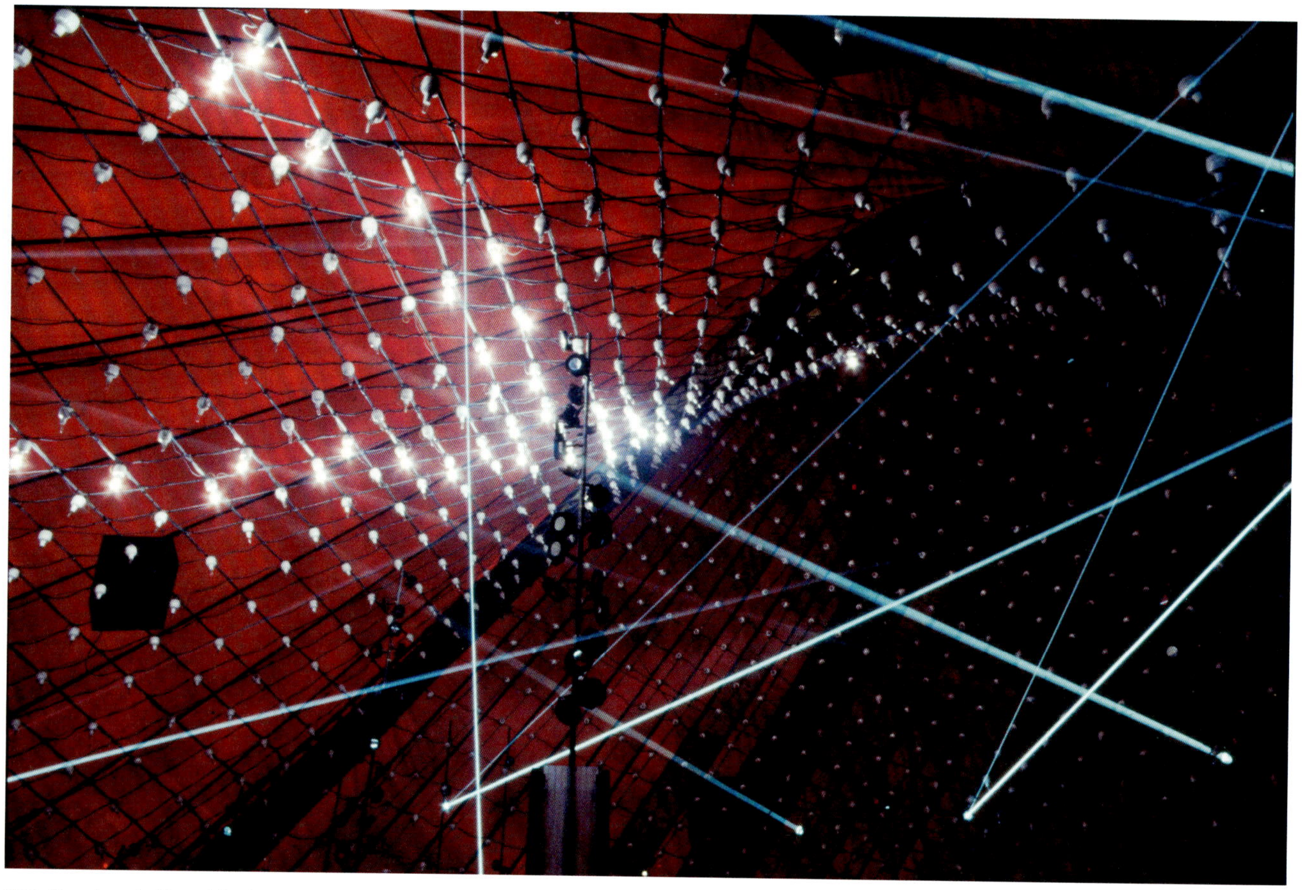

FIG. 3a Iannis Xenakis, *Le Diatope*, 1978–1979. Interior view, flashlights and blue lasers.

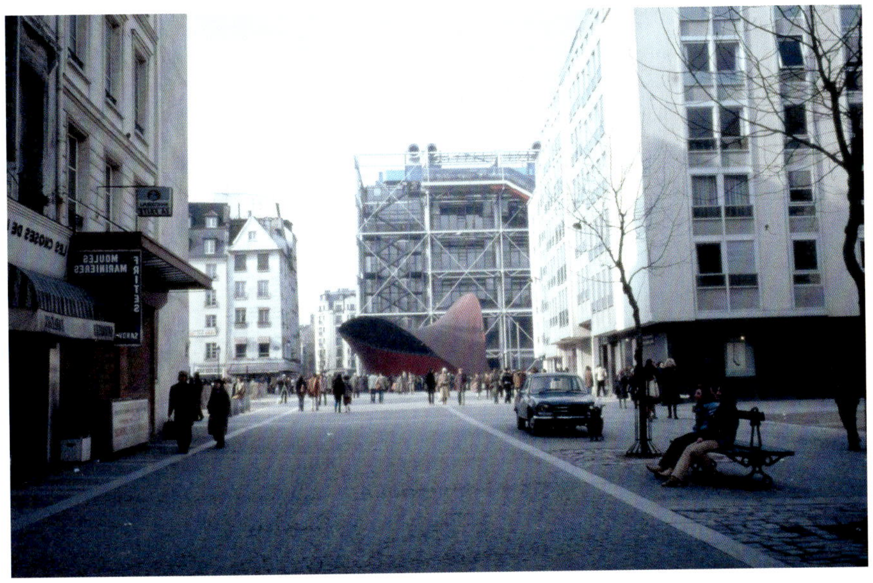

FIG. 3b Iannis Xenakis, *Le Diatope*, 1978–1979. Exterior view, Centre Pompidou, Paris.

performance areas, seven channels distributed over twelve loudspeakers in *Polytope de Cluny*, eight channels for the outdoor *Persépolis*, seven tracks over eleven loudspeakers for *La Légende d'Eer*, twelve tracks in his Steel Pavilion composition for Expo '70 in Osaka, Japan.[10] For each work, Xenakis created spatial audio plans that would allow him to identify where the soundtracks would be reproduced and how the juxtaposition of the discrete audio tracks might interact in the space.

The audio tracks of *La Légende d'Eer* create powerful spatial sound, supported by the density and complexity of their electronic audio materials. However, the loudspeaker reproduction of the seven tracks over eleven loudspeakers represents a significant quantization of the spatial audio field that surrounds its audience. The audio reproduction is also mediated by its loudspeaker presentation (loudspeakers afford information about their nature as discrete sound sources that radiate sound waves, primarily in one direction from the front of an enclosure). In perceptual terms, this low spatial audio resolution via loudspeakers confines the sound element of the work to its own, discrete, architectural configuration of loudspeakers – inside and to a large extent separate from its enclosing architecture.

It is not clear if Stockhausen had an opportunity to experience the *pupitre de relief* – an early experiment in concert-hall sound projection developed by the Groupe de Recherche de Musique Concrète (GRMC) – during his time in Paris in 1952.[11] In later works, however, he sought a purer, modernist approach to spatial audio presentation. From his early *Gesang der Jünglinge* (1955–1956) in five audio channels, through *Kontakte* (1958–1960) for four audio channels, Stockhausen pursued "spatial music" and called for new kinds of concert halls to reproduce it:

> New halls for listening must be built to meet with demands of spatial music. My idea would be to have a spherical chamber, fitted all round with loudspeakers. In the middle of this spherical chamber, a platform, transparent to both light and sound, would be hung for the listeners. They could hear music, composed for such adapted halls, coming from above, from below and from all directions.[12]

Stockhausen's call was answered through another world's fair, Expo '70, where he collaborated with the architect Fritz Bornemann to realize the German Pavilion in Osaka, Japan.(FIG. 4a)

Under Stockhausen's guidance, Bornemann designed a near-spherical auditorium *(Kugelauditorium)* with loudspeakers distributed over its surface in seven rings at various elevations.(FIG. 4b) The auditorium's audio reproduction system could place electronic sounds in loudspeakers around, above, and below an audience seated on an elevated platform. Electronic circuits were used to generate rotations of sound (from one loudspeaker to the next) around the loudspeaker rings and in spirals over the inside surfaces of the sphere.

10
See Iannis Xenakis, *Formalized Music – Thought and Mathematics in Composition* (Stuyvesant: Pendragon Press, 1992), 379.

11
Circumstantially, it seems likely that he was aware of GRMC's approach to the spatial presentation of *musique concrète*, as the *pupitre de relief* was used in 1952 to perform the *Timbres-Durées* by Olivier Messaien, with whom he studied composition at the time. Stockhausen also visited Pierre Schaeffer's GRMC studio in March 1952 and subsequently studied *musique concrète* to complete his own *Etude (musique concrète)* in December of that year. Cf. Simon Emmerson, *Living Electronic Music* (Aldershot, UK: Ashgate, 2007), 155.

12
Karlheinz Stockhausen, "Music in Space," *Die Reihe* 5 (1959; Eng. ed.: Bryn Mawr: Theodore Presser Co., 1961), 67–82, here 69.

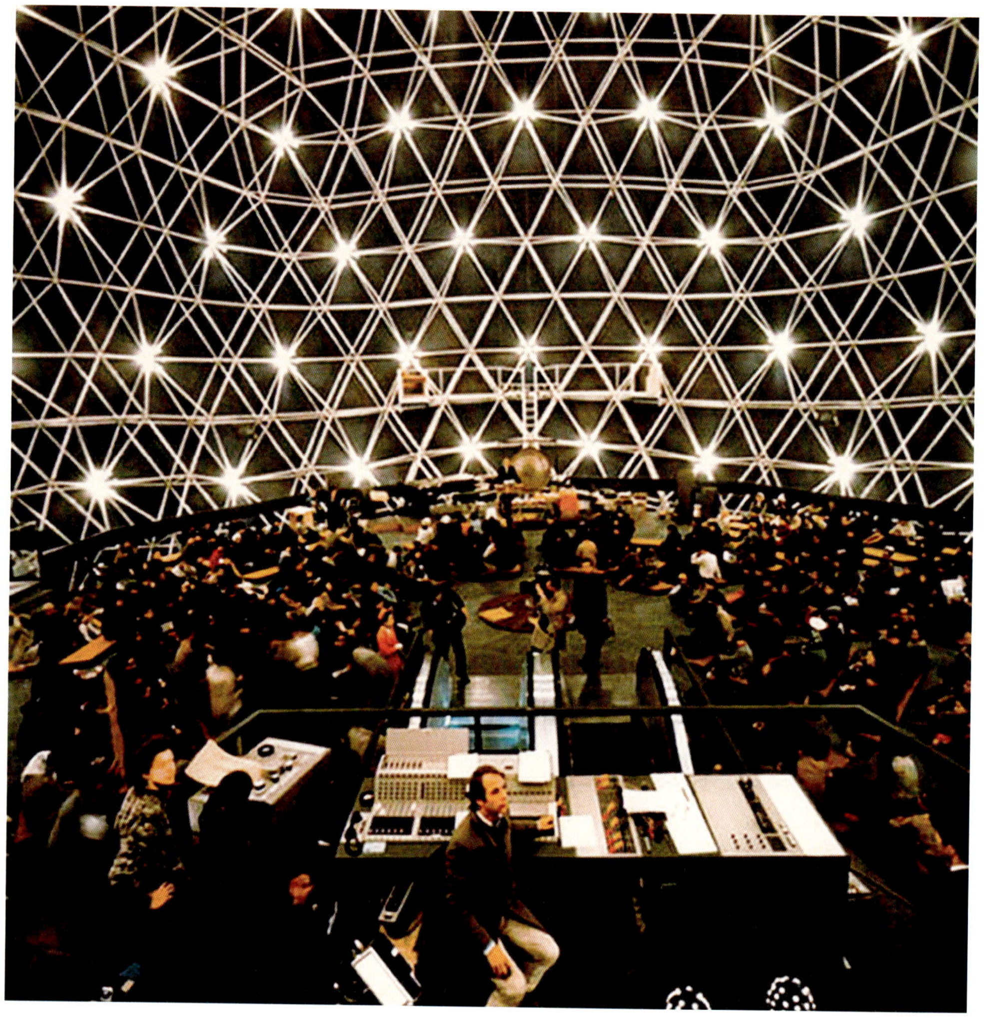

FIGS. 4a, b Fritz Bornemann and Karlheinz Stockhausen, German Pavilion, Osaka, 1970. Interior view (a, top) and elevation view (b, bottom).

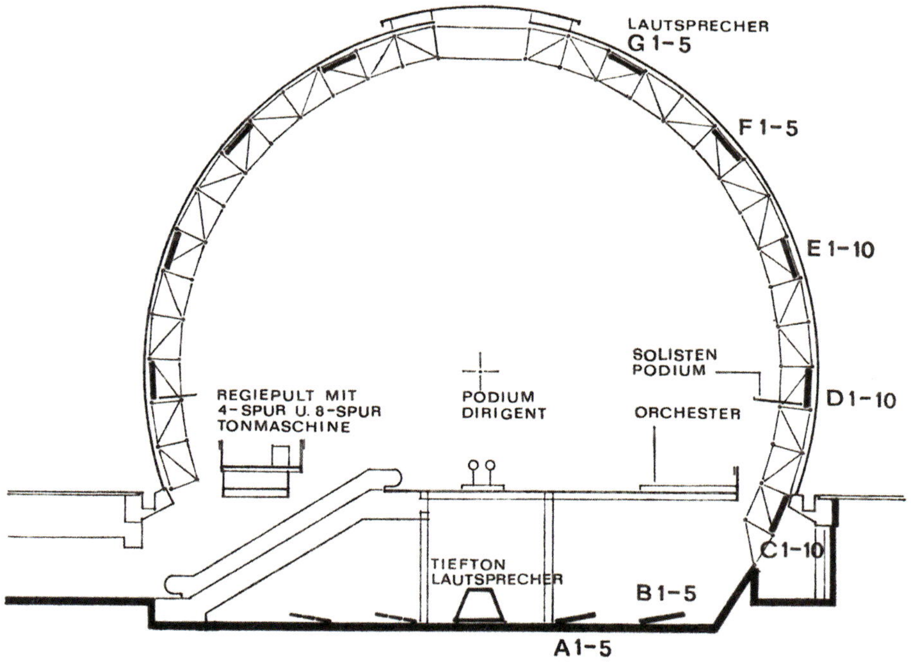

13
Seth Kim-Cohen, "The Hole Truth," *Artforum*, November 2009, 99–100, here 99.

The compositional focus of the pavilions of Xenakis (after Brussels) and of Stockhausen established them primarily as performance spaces: unique or dedicated spaces for the presentation of sound and multimedia works. In both these works, audio processes informed architectural design, which in turn illuminated and amplified aspects of the music they presented.

Listening Posts

Away from the concert experience, pavilions as controlled and unique environments have facilitated listener encounters with rare and intriguing sounds, an approach adopted by Doug Aitken with *Sonic Pavilion*, Marianne Decoster-Taivalkoski and Hanna Haaslahti with *Sonic Seascape Terrace*, Luke Jerram with *Aeolus*, and Katarzyna Krakowiak with the Polish Pavilion for the 2012 Venice Biennale.

These works establish site-specific listening environments that allow access to remote or obscure sound phenomena. The relationship between sound and architectural space in these works remains associated with the idea of holistic perception, but focused on a particular collection of sounds unique to their geographic location and presented within a complementary architectural environment that is crafted to support the audio encounter.

Doug Aitken's *Sonic Pavilion* (2009) is located at the Centro de Arte Contemporânea Inhotim, Brazil, set on a hill within Bernardo Paz's private sculpture park, surrounded by dense forest. The pavilion provides an audition environment for sounds transduced by microphones and accelerometers buried one mile deep below the pavilion. The microphones and accelerometers are employed to detect vibrations caused by the rotation of the Earth and the movement of tectonic plates. The signals, collected deep underground, are "transposed" for human audition and then reproduced by the pavilion's amplifiers over eight loudspeakers.

> Everything about *Sonic Pavilion* – the emptiness of the interior, the austere geometry, the foggy visual frame – is designed to induce the impression of unprecedented access to secret sensory experience.[13]

This kind of "secret sensory experience" – achieved through the re-presentation of elements of the natural environment that are normally inaccessible – can be found in other site-specific sound pavilions and in a number of contemporary audio works developed for presentation in pavilions. Audio materials that project actions or events of the natural world, particularly those which lie beyond our everyday experience, offer a privileged, often intimate, experience to the listener. Sound materials of this type are often used to articulate messages and concepts beyond the audio domain; themes of conservation or environmental awareness are regularly the subject of discourse in relation to these works.

Decoster-Taivalkoski and Haaslahti's *Sonic Seascape Terraces* and Jerram's *Aeolus* both integrate natural phenomena through the behavior of natural elements – water and wind.

Sonic Seascape Terraces was exhibited during the *Curated Expedition to the Baltic Sea* as part of the European Capital of Culture program in 2011.(FIG. 5) One of the two pavilions created for

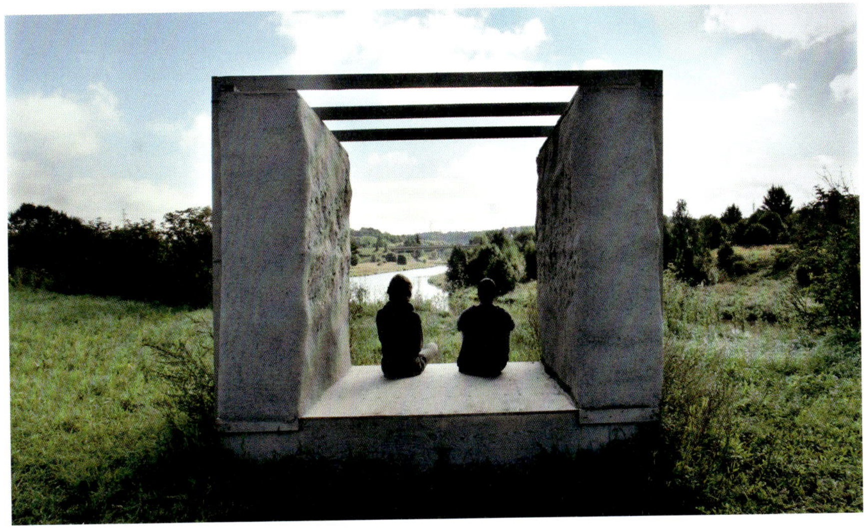

FIG. 5 Marianne Decoster-Taivalkoski, Hanna Haaslahti, and Alejandro Montes de Oca, *Sonic Seascape Terraces*, 2011. Site-specific installation in Turku, Finland (May 31–September 1, 2011), which explores connections between the seascape and sounds emanating beneath its surface.

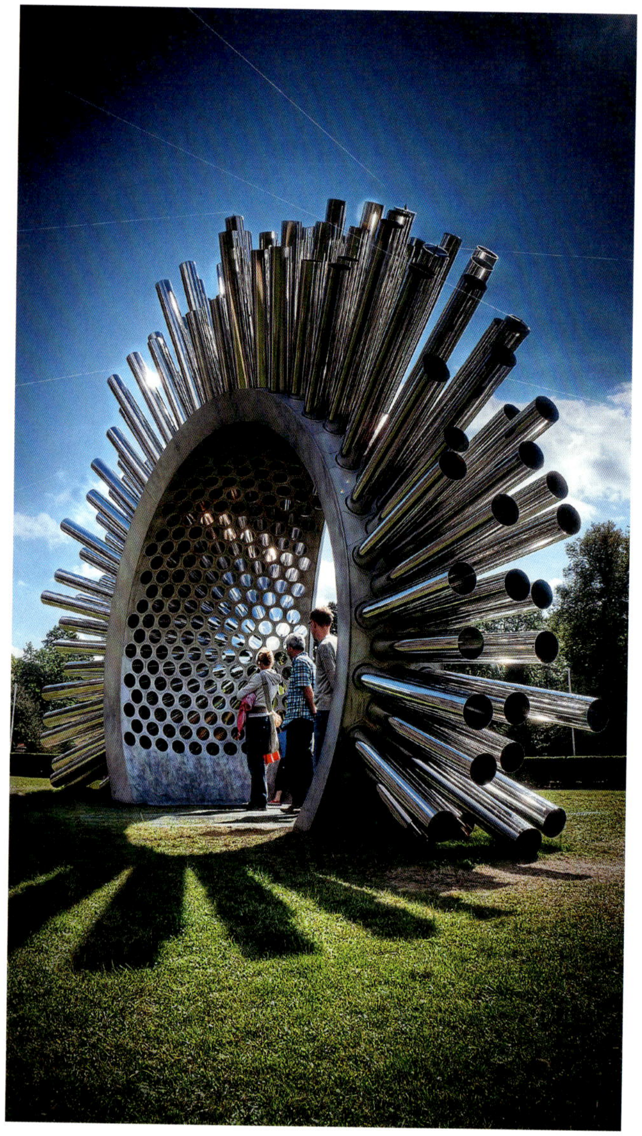

FIG. 6 Luke Jerram, *Aeolus*, 2011. Acoustic wind pavilion in the shape of a giant Aeolian harp.

this work was situated in a remote field overlooking a bend of the Aura River in Turku, Finland. Its modified straw-bale construction provided a listening station that framed a view across the surrounding countryside and a bend in the river where underwater hydrophones had been placed. The sound captured by the hydrophones was relayed, processed, and reproduced for listeners through loudspeakers built into the pavilion's walls. The work was conceived in relation to theories of acoustic ecology, after R. Murray Schafer.[14] It presented an insight into a distant audio world by transporting sounds from the river and its inhabitants to the site of audition; this process made it possible to represent these sounds in the wider context of the landscape that supported the ecosystem.

Aeolus places its focus on its listeners, who are thus positioned at the "sweet spot" for sounds generated by the structure.(FIG. 6) Here, Jerram "transposes" natural events into sound using a series of resonant volumes that form the structure of the pavilion. These volumes amplify taut strings set in motion by the prevailing wind. The pavilion articulates the force, direction, variability, and power of the wind through polished steel tubes that also reflect and reinterpret the visual environment surrounding the work (and act as light-tubes that reflect light inwards). *Aeolus* spatializes sound in an arc over the listener; it auralizes air movements, channels light, and reframes its environment.

R. Murray Schafer argued that sounds might be categorized according to those that resulted from human activity and those of the natural world.[15] Many interpret this as a suggestion that humans might be seen to exist outside the natural world, as a Platonic vision, in contrast to an ecological perspective that might see people and their actions as fundamentally connected and integrally linked to the world they inhabit. Murray Schafer's position does, of course, support environmental awareness and a motivation to highlight the potential negative impact of human-made sound on other species or environments (and on humans). This is poignant in particular where human impacts are negligent or ignore the significance of sound for communication between, or propagation of, other species.

Sound pavilions have become sites to articulate these messages, but as sites where humans gather they also provide an opportunity to look toward human activity.

Thilo Frank's *EKKO* pavilion of 2012 harvests and replays sounds made by its visitors.(FIGS. 7a, b) These are distributed throughout the pavilion – delayed in time such that visitors might hear themselves – and added to a collection of sounds made by previous visitors. All captured sound materials are archived as an audio-record of all those who have visited the structure.

Similarly, *Making the walls quake, as if they were dilating with the secret knowledge of great powers*, the Polish Pavilion developed by

14
R. Murray Schafer, *The Soundscape: Our Sonic Environment and the Tuning of the World* (Rochester, VT: Destiny Books, 1994).

15
Ibid.

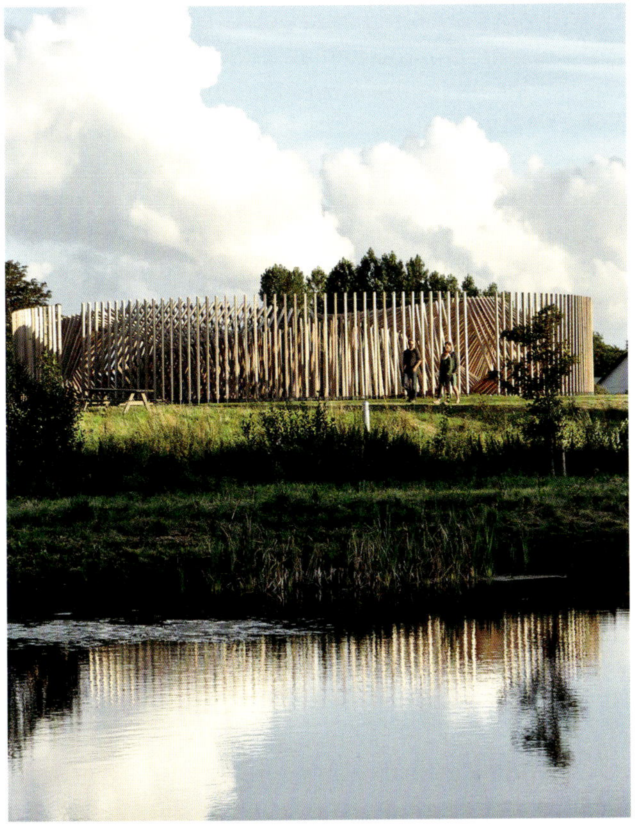

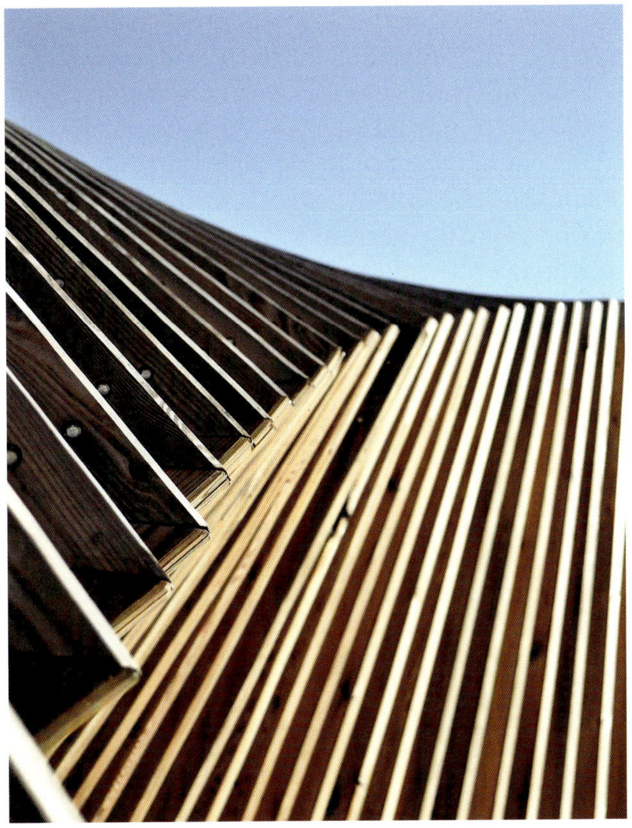

FIGS. 7a, b Thilo Frank, *EKKO*, 2012. Permanent Public Installation, Ø 20 m H 3 m, modified wood, stainless steel wire, zinc coated steel, aluminum, concrete, microphones, amplified computer controlled sound-system. Installation view Hjallerup, Denmark 2012.

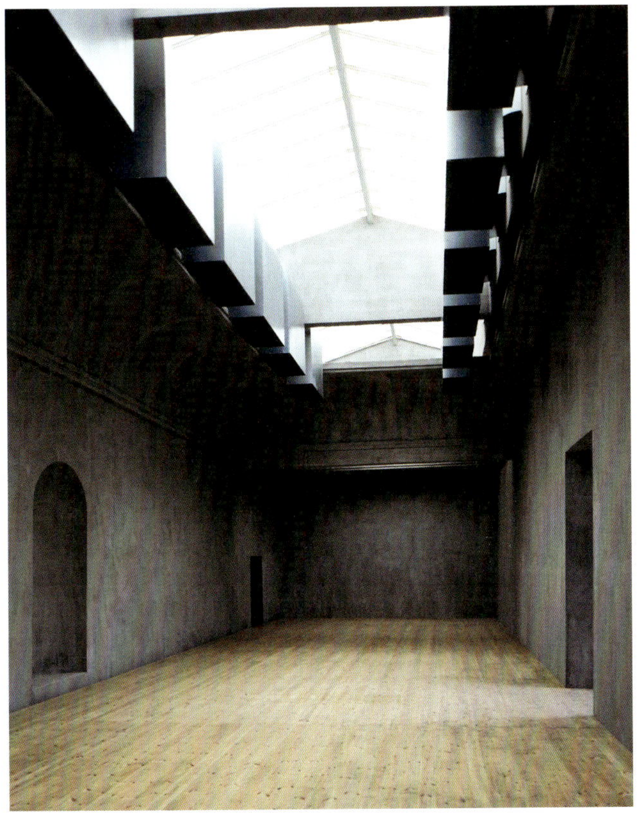

FIG. 8 Katarzyna Krakowiak, *Making the walls quake, as if they were dilating with the secret knowledge of great powers*, 2012. Polish Pavilion, 13th Biennale of Architecture, Venice.

Katarzyna Krakowiak for the 13th Venice Architectural Biennale in 2012 was conceived as a site to reveal sounds made by exhibition visitors in the other national pavilions and in the Giardini gardens.(FIG. 8) Background sounds, those often "tuned-out" (in Murray Schafer's terminology) by people who occupy noisy environments, were brought to the fore in this work. The piece established a listening space created by a near neutral, empty room; a site to "tune-in" to the sounds made by other people within the building that were transferred into the space and modified by the building's architecture and ventilation systems.

These "listening post" sound pavilions incorporate natural and human sounds from the vicinity of the installation site in order to reveal what normally goes unseen or unheard. They integrate conceptual, functional, and ideological concerns and extend perception to previously hidden realms: the past in Frank's 2012 pavilion, the stirrings of underwater animal and environmental life in the *Sonic Seascape Terraces*, background human activities in Krakowiak's work, air movement in *Aeolus*, and sounds from the depths of the Earth in Aitkin's *Sonic Pavilion*.

As with earlier pavilions, these works continue the approach of combining architecture and sound, but they also integrate site-specific and conceptual materials: the sound pavilion acts to reveal aspects of the environment or to highlight particular sounds to articulate the concept of the work. The pavilion's architecture, in part, takes on a practical function to present, direct, and facilitate audition.

Accessing Acoustic Phenomena

The installations and sculptural works of Bernhard Leitner carefully position loudspeakers to reproduce electronic sounds and acoustic phenomena. These works are often presented in gallery spaces but many of Leitner's installations, e.g. *Serpentinata 04* (2004) and *Grosse Röhre* (2008), have architectural forms that also establish architectural audio spaces. Loudspeakers are arranged in combination with sound absorbing materials or sound reflectors to place sound in a spatial context and to reveal a variety of acoustic phenomena.

TON-TOR (1990) exhibited at the Karlsplatz, Vienna, for the 175th jubilee year of the Vienna Technical University moves beyond the gallery context to create both material and immaterial architectural forms; a wooden structural arch and a sound arch articulated by twelve loudspeakers. Leitner describes the perception of the work as one in which "listening and seeing merge,"[16] fused and bound together by their structural foundation in the notion of the curve.

This "merger" has parallels with Xenakis's approach to the unification of sonic and visual experience: both audio and visual elements of the work are articulated by a common metaphor. But where does the merger occur in *TON-TOR* in relation to seeing and listening? The curved sound path and the curved architectural geometry inhabit two distinct domains: one is a spatial geometric structure, the other a temporal acoustic phenomenon. Listeners are, however, drawn by its moving audio materials to scan the curved structure visually; the perception of sound movement is then led by the physical geometry of the red circle that surrounds the listener. The merger occurs in an interaction between seeing and listening, the perception of an acoustic phenomenon that draws the listener's gaze around the curved structure.

Edwin van der Heide has been prolific in his exploration of audio and architectural forms. His works, including *Water Pavilion* (1997–2002), *Son-O-House* (2004) and *Pneumatic Sound Field* (2006–2008), unite contemporary technologies, architectural design, and interactive technologies. Many explore acoustic and auditory phenomena and exhibit their acoustic attributes through interactions with architectural forms.

Van der Heide's *Pneumatic Sound Field* in its 2008 incarnation in Beijing featured an open-form architectural pavilion design by Lars Spuybroek as a container for the audio pavilion.(FIG. 9) The audio generation equipment used for this work has been installed in a number of architectural contexts, for example the Tesla-Sonambiente festival in 2006 – a contemporary garden pergola-style incarnation of the work. The work employs a series of pneumatic valves to generate sound, rapidly switching pulses of compressed air. Forty-two pneumatic valves are arranged in a grid above the presentation space and produce rhythms, tones, and spatial audio effects through time-control release of the valve triggers. The expulsion of compressed air generates the sound component of this work, while also producing a haptic element as air pulses toward the listeners.

Acoustic interactions of synthetic sound materials are also created in the audio works by the duo of Mark Fell and Roc Jiménez de Cisneros and by Carl Michael von Hausswolff for *The Morning Line*, a sound-system pavilion produced by the Thyssen-Bornemisza Art Contemporary 21. As with the works already discussed, these artists' audio creations for *The Morning Line* establish sound fields that allow listeners to engage with timbral acoustic phenomena. In both *A very short proof of Forester's rigidity result* by Fell and Jiménez de Cisneros (2008) and *No Rest Even for the Static (Matter II)* by Hausswolff (2010), synthetic tones are distributed over a large multichannel audio system comprising forty-seven channels. Each sound component of the synthesized timbres is spatially distributed to define the sound field of the works. The location of listeners, surrounded by the work or viewing it from some distance, determines the acoustic balance of sound components heard. This develops a constantly shifting mix of tones and timbre as listeners

16
Leitner Studio, "Sound Gate 1990," under www.bernhardleitner.at/works.

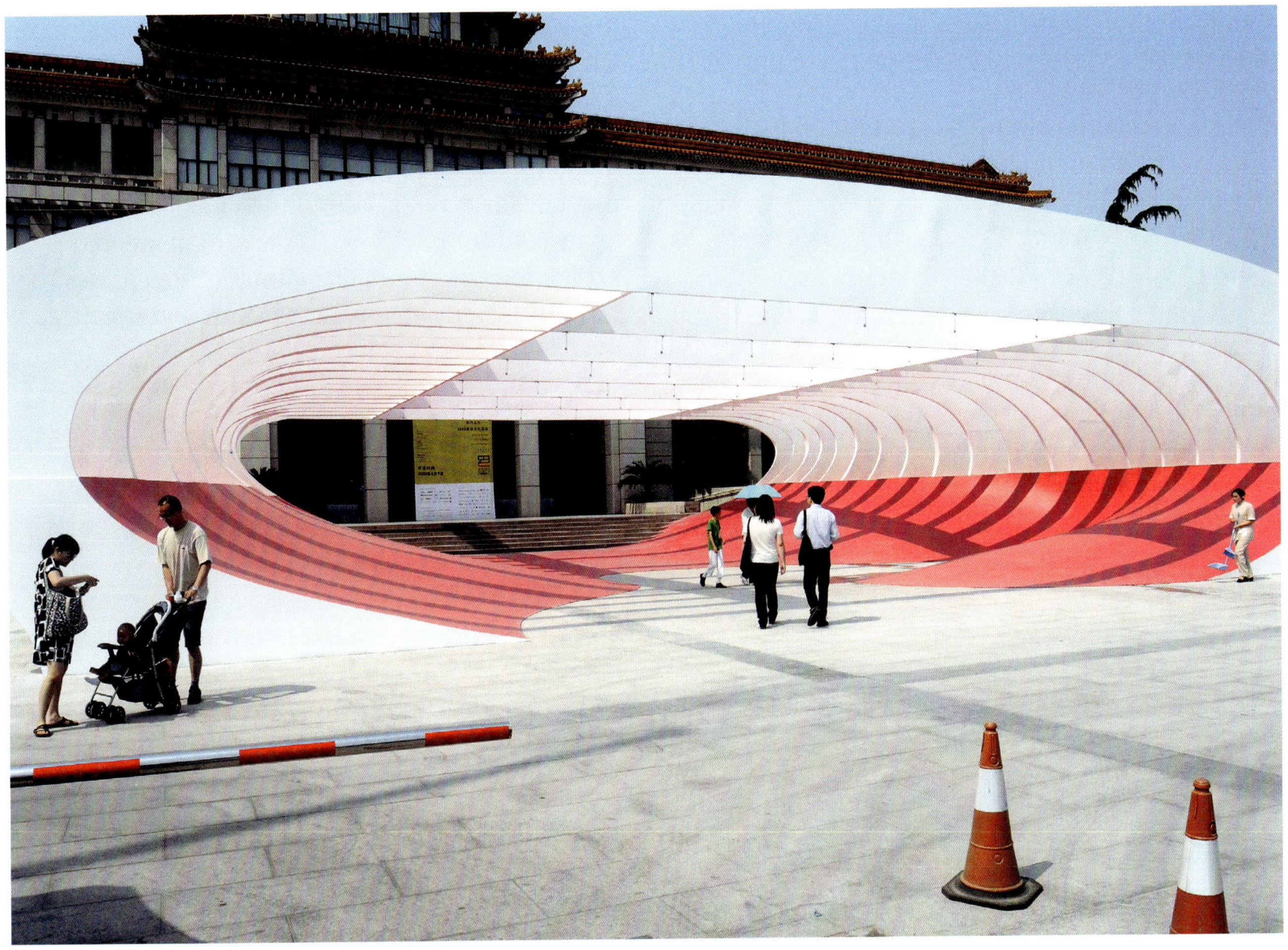

FIG. 9 Edwin van der Heide, *Pneumatic Sound Field*, 2008. The pavilion was specially designed for this artwork by Lars Spuybroek. Installation view *Synthetic Times* exhibition, NAMOC, Beijing, 2008.

move through the sound field, created by both the position of the listeners and the sound localization strategies employed by the artists.

The spatial sound phenomena of the works by van der Heide, Fell/Jiménez de Cisneros, and Hausswolff are, to a large extent, independent of the pavilions that contain them. The pavilions facilitate an arrangement of loudspeakers and actuators, and also establish listening sites and modes of presentation, but the works focus on listeners' encounters with spatial acoustic phenomena. In this sense, the compositions exist independently from the architectural design of their containing pavilions to exhibit their own audio architecture, distinct from the physical architecture of their presentation spaces.

Audio Technologies

Sound integration in many pavilion concepts requires controlled sound reproduction via the loudspeakers. In early pavilions, this was achieved using custom-designed electronic circuits to switch or pan audio to loudspeakers positioned in various locations. Early works were also restricted by the recording and audio-replay technologies of their period and often limited to a small number of discrete audio streams.

Digital audio systems have facilitated a revolution in spatial sound production. Computer systems can now store and reproduce large numbers of discrete audio streams and perform complex audio signal-processing. This capability has reignited interest in spatial audio presentation and long-established spatial audio technologies such as Ambisonics. The latter has developed from a system that rivaled Quadraphonic sound (four-channel sound) during the 1970s to refined high-order implementations that can combine many loudspeakers to reproduce sophisticated surround-sound fields. Wave-field synthesis has also become an established method to project sounds in three-dimensional space, using large numbers of coordinated loudspeakers to produce 3-D acoustic projections.

These technologies, combined with widespread use of computer tools for audio and visual processing, such as the Max/MSP software developed by Cycling '74, are being employed by sound artists and producers in increasingly innovative pavilions.

Loudspeaker technologies for public audio presentations have also evolved away from large public address systems, that might distribute audio energy by pushing high-power sound from one or two locations, to embrace smaller, distributed loudspeaker systems that project sound from many locations. This distributed sound-reproduction approach is also well suited to spatial sound reproduction. Combinations of contemporary loudspeaker reproduction technologies and multichannel computer audio processing methods have facilitated greater access to audio spatialization and new opportunities for architectural sound.

Contemporary Technology-Enabled Pavilions

Complex computer-designed forms are an important aspect of works such as the two parts of the *Water Pavilion* installed on the artificial island of Neeltje Jans, Netherlands, and created by van der Heide and Victor Wentink, with architecture designed by Spuybroek (NOX) and Oosterhuis Associates.(FIG. 10) Both the *Freshwater Pavilion* and the *Saltwater Pavilion* employ high numbers of computer-controlled loudspeakers and lights to generate a holistic, dreamlike environment focused around a theme of water and abstractions of water behavior transposed into the domains of light and sound. Van der Heide's *Son-O-House* (with architects NOX/Lars Spuybroek) also integrated sophisticated computer control, mapping twenty-two motion sensors to influence an algorithmic audio composition projected over twenty loudspeakers; the resulting work presents a sound field that forms an acoustic union of listener presence, movement, and synthetic audio.(FIGS. 11a, b)

My own experience of sound pavilion development was with *The Morning Line (TML)*, the sound-system pavilion produced by the Thyssen-Bornemisza Art Contemporary 21, known as TBA21. *TML* was conceived by artist Matthew Ritchie, whose line drawings were extrapolated by architects Aranda\Lasch into fractal-like tetrahedral tessellations to form a large, open outdoor pavilion. Ritchie implemented projections of computer animations as part of the fixed environment of the work and specified concepts for the project to integrate sound and to react to the presence of people.

Departing from the strategy taken with many pavilion projects, TBA21 commissioned a series of artists to produce sound works for the pavilion. Interestingly, there was a desire to facilitate all the potential approaches to pavilion contexts that have been described above: namely, *TML* sought to act as a site-specific public exhibit for sound works, as a live performance venue, and as a platform for works that explore synthetic audio materials and acoustic phenomena, conceptual works, and works that re-present environmental soundscapes.

TML was exhibited in Seville (2008), Istanbul (2010), Vienna (2011–2012) and at ZKM, Karlsruhe (2013),(FIGS. 12a–d) where it remains as part of the permanent collection of ZKM. Over the period since *TML*'s first incarnation, TBA21 has commissioned over thirty sound artists to create new works for the pavilion.[17] While the pavilion structure has undergone some physical reorientation over this period, revisiting the integration of sound and architecture by so many sound artists has placed an emphasis on the work's sonic components and has supported audio experimentation over an extended period of time.

I led the small team of audio programmers who developed the audio system for *TML*.[18] The system was focused on perceptually informed sound production and reproduction methods. Dr. Peter Lennox and I have published a number of papers around this theme.[19] The implementation of the *TML* audio system aimed to establish an information-rich spatial audio environment that could appeal to human spatial audio perception.

17
TML composers: Peter Zinovieff, Zavoloka, Jana Winderen, Thom Willems, Chris Watson, Carl Michael von Hausswolff, Yasunao Tone, Terre Thaemlitz, Lee Ranaldo, Franz Pomassl, Finnbogi Pétursson, Zsolt Olejnik, Mehmet Çan Özer, Richardas Norvila/Benzo, Carsten Nicolai, Kollektiv/Rauschen, JG Thirlwell, Jónsi Birgisson and Alex Somers, Erdem Helvacıoğlu, Florian Hecker, Tommi Grönlund and Petteri Nisunen, Bruce Gilbert, Ghostigital, Christian Fennesz, Mark Fell and Roc Jiménez de Cisneros, Cevdet Erek, Bryce Dessner, Evan Ziporyn and David Sheppard, Batuhan Bozkurt, Alexei Borisov, Auxpan, Asfast, Ludger Brümmer.

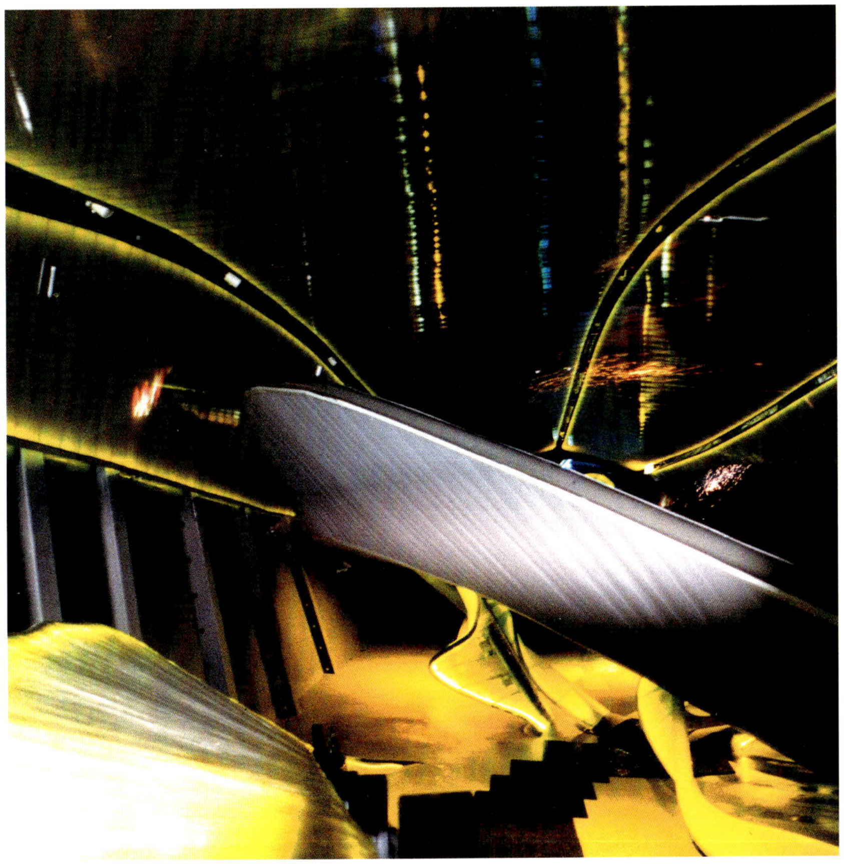

FIG. 10 Edwin van der Heide, Victor Wentink, Lars Spuybroek, and Oosterhuis Associates, *Water Pavilion*, 1997–2002.

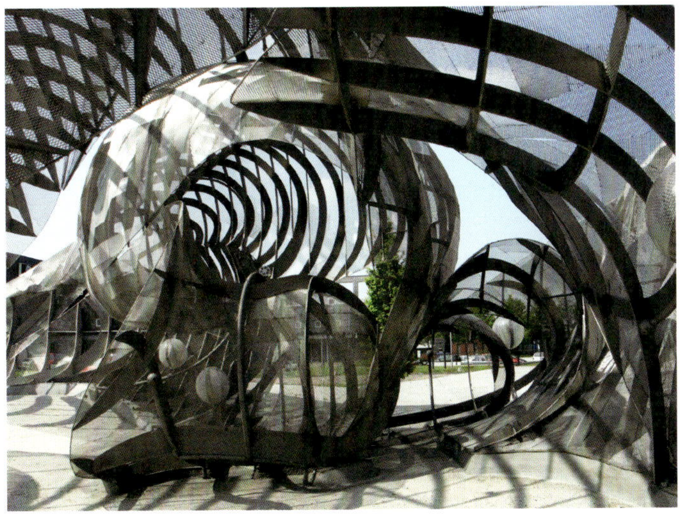

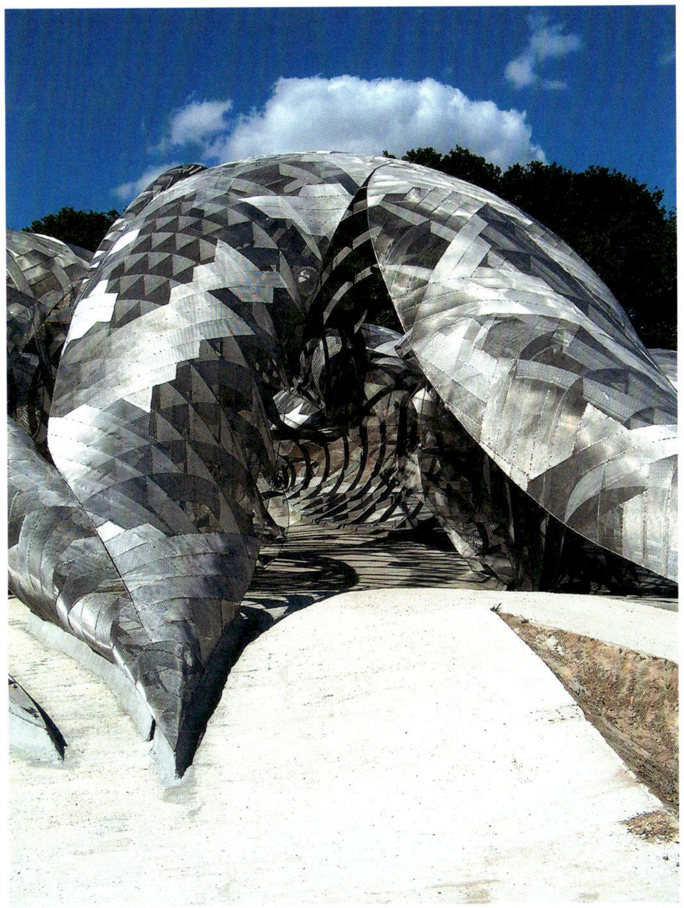

FIGS. 11a, b Edwin van der Heide and Lars Spuybroek/NOX, *Son-O-House*, 2004.

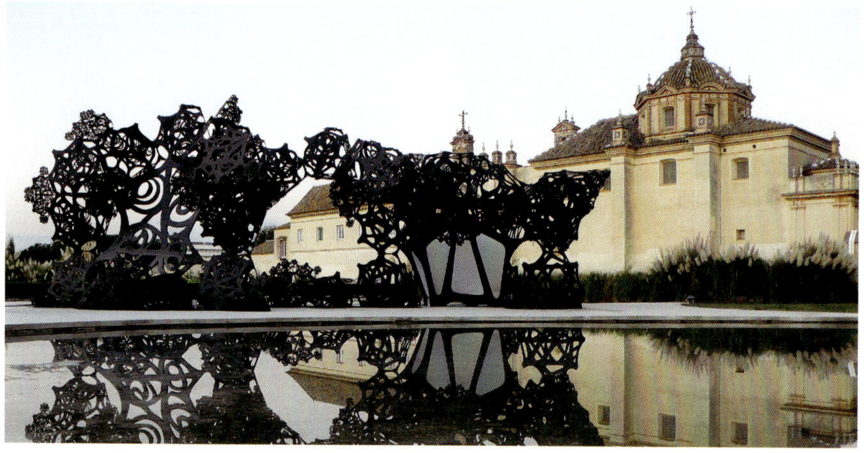

FIGS. 12a–d Matthew Ritchie with Aranda\Lasch and Arup AGU, *The Morning Line*, 2008. Architectural interactive structure immanent musical instrument. Modular construction: Aluminum alloy, black epoxy with aggregate coating. Audio: 47 channels, 41 speakers, 12 subwoofers. Video: 6 video projections. Commissioned by Thyssen-Bornemisza Art Contemporary.

FIG. 12a Matthew Ritchie with Aranda\Lasch and Arup AGU, *The Morning Line*, 2008. Installation view Seville, 2008.

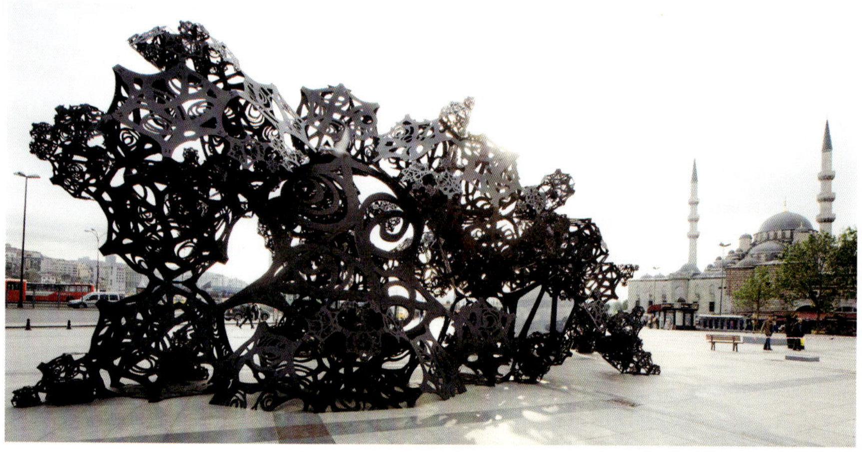

FIG. 12b Matthew Ritchie with Aranda\Lasch and Arup AGU, *The Morning Line*, 2008. Installation view Istanbul, 2010.

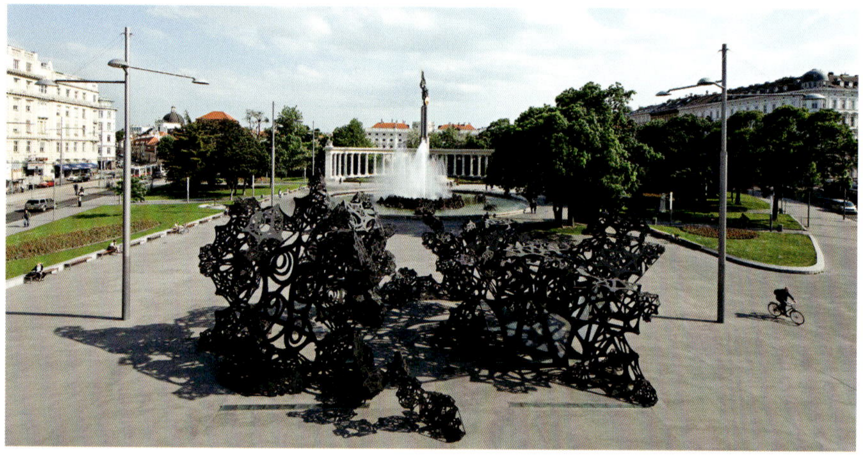

FIG. 12c Matthew Ritchie with Aranda\Lasch and Arup AGU, *The Morning Line*, 2008. Installation view Vienna, Schwarzenbergplatz, 2011.

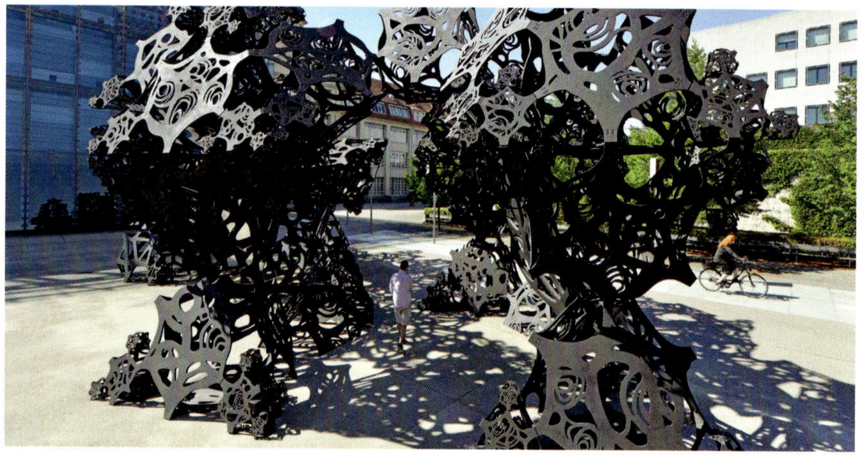

FIG. 12d Matthew Ritchie with Aranda\Lasch and Arup AGU, *The Morning Line*, 2008. Installation view Forecourt of ZKM | Karlsruhe, Karlsruhe, 2013–.

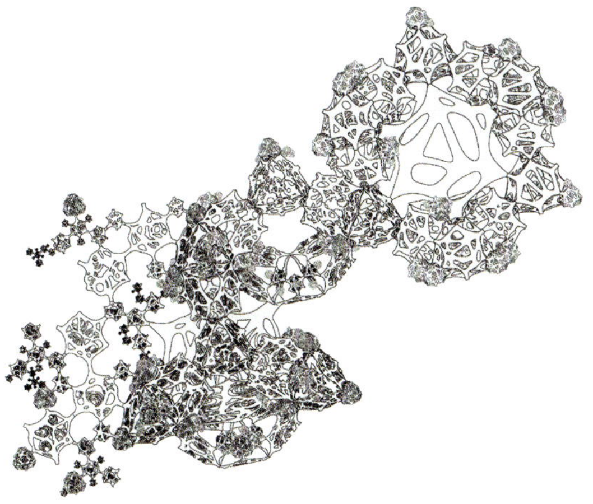

FIG. 13a Plan of *The Morning Line*.

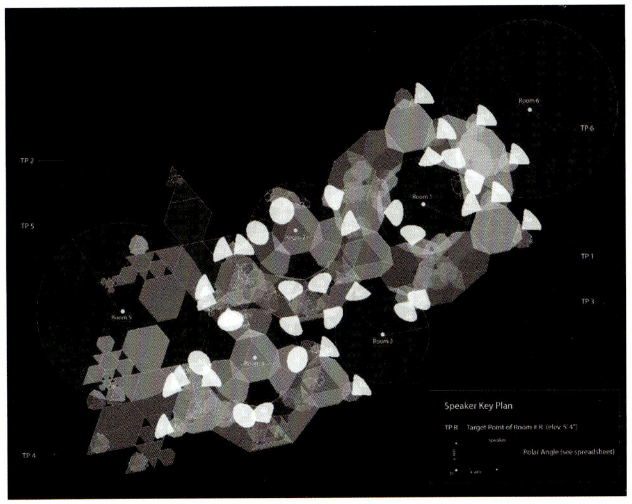

FIG. 13b Speaker key plan of *The Morning Line*.

The technical implementation of the *TML* sound system established a series of full with-height sound fields, distributed throughout the structure. Each sound field area reproduced a coherent local sound environment and formed adjacent or distant sound fields from other listening positions. Audio could be explored by listeners as they moved through the structure. The multispatial approach also required the development of new audio production techniques and various spatial sound encoding methods to place, move, and reproduce sounds throughout the structure.(FIGS. 13a, b)

TML's audio system rendered each sound work from its constituent audio components as they were presented. It used an object-based audio approach that consisted of a series of "spatial audio objects" that combined sounds and information about their relative spatial location or movement trajectory. The system was implemented for the Seville exhibition of *TML* in 2008, along with software interfaces to allow spatial audio objects to be created prior to live rendering or for live performances.

TML's configuration of loudspeakers formed a three-dimensional array of fifty-three loudspeakers. The software could send audio to individual loudspeakers within the array, and employed combinations of sound-field reproduction techniques (including Ambisonics, quad, stereo, etc.), and the application of spatial panning approaches such as Ville Pulkki's Vector Base Amplitude Panning method.[20] This combination of audio technologies and perceptually informed spatial audio production techniques supported the creation of works with a high degree of spatial polyphony.

Chris Watson's *Snæfellsnes* for *TML* was inspired by the extreme black aggregate-textured finish of *TML*. The work recreated a series of sound fields from sounds recorded in the vicinity of the *Snæfellsnes* volcano in Iceland, where black volcanic rock is prevalent. Watson presented each sound environment as a coherent surround sound field reproduction in six areas of the pavilion. Listeners could traverse the surround-sound reproductions. The acoustic transparency of the pavilion's open structure allowed an understanding of the entire audio perspective of the work, of sounds presented beside the listener, some meters away, or in the far distance. Listeners explored a series of sound fields from the sea to the top of the volcano, via the beach, grasslands, boiling mud pools, through bird colonies to the movement of ice recorded deep within the *Snæfellsnes* glacier.

After nightfall, during the playback of a complex synthetic work *Rotating psychoacoustic tuning curves* by Florian Hecker at the Seville exhibition of *TML*, many listeners reported that it was possible to perceive a large, multifaceted sound object. This consisted of certain sounds

18
TML software developers: Tony A. Myatt, Matthew Paradis, Peter Worth, Oliver Larkin, Theo Burt.

19
Peter P. Lennox, Tony A. Myatt, and John M. Vaughan, "From Surround to True 3-D," in *Proceedings of the AES 16th International Conference: Spatial Sound Reproduction, Rovaniemi, Finland* (1999); Peter P. Lennox, Tony A. Myatt, and John M. Vaughan, "3-D Audio as an Information-Environment," in *Proceedings of the AES 19th International Conference on Surround Sound, Schloss Elmau, Germany* (2001); Peter P. Lennox and Tony A. Myatt, "Concepts of Perceptual Significance for Composition and Reproduction of Explorable Surround Sound Fields," in *Proceedings of the 13th International Conference on Auditory Display, Montreal* (2007); Peter P. Lennox and Tony A. Myatt, "Perceptual Cartoonification in Multi-spatial Sound Systems," in *Proceedings of the 17th International Conference on Auditory Display, Budapest, Hungary* (2011).

20
Ville Pulkki, "Virtual Sound Source Positioning Using Vector Base Amplitude Panning," *Journal of the Audio Engineering Society*, vol. 45, no. 6 (June 1997), 456–66.

oscillating in one area of the work, some sweeping across another area, sound weaving through the structure and spiraling around various centers – all simultaneously. In the dark, without clear sight of the pavilion, its audio components appeared to afford information consistent with the perception of a large three-dimensional spatial sound object: an object one could observe from afar, walk up to, step inside, and perceive from any perspective.

TML, *Water Pavilion*, and *Son-O-House* are technology-enabled, multifaceted works that use large computer-controlled loudspeaker arrays to generate multiple simultaneous sound fields. They deliver a significant quantity of spatial audio information, approaching the complexity of the spatial audio information we encounter in our perceptual explorations of the natural world and architecture.

Additionally, in these works where large numbers of loudspeakers are controlled to produce complex audio sound fields, individual loudspeakers can disappear to our ears. Loudspeakers that operate together with many others can produce sound images that appear in free space, outside loudspeaker enclosures (in the same manner as the central, between-speaker image created by Blumlein stereo reproduction). In these circumstances, loudspeakers afford less information about their nature as individual sound sources and, consequently, the perception of many surrounding sounds that appear not to emanate from loudspeakers can also change the context of listeners' spatial audio apprehension.

Technologies of this type facilitate works that can embrace the concept of spatial music and sound fields that can be explored and navigated by listeners.

Moving Forward and Backward in Time

The small selection of works discussed here illustrate approaches to sound pavilion design and implementation that have emerged in the last half-century, from the structuralist applications of mathematical models to control architecture, light, and sound in the Philips Pavilion and Xenakis's *Polytopes*, to contemporary conceptual and experiential works. Sound pavilions now include a diverse and evolving canon of spatial audio and architectural work, to the point where recent developments have seen sophisticated audio technologies potentially reverse the concept of the sound pavilion as a dazzling world-fair architectural exhibit (with sound and image), toward one of architectural sound design and spatial music (with architecture).

How architecture and sound might unite in future sound pavilions is an interesting question, one likely to be determined in part by contexts established for sound pavilions over the last fifty years – pavilions as technological demonstrators, cinematic/multimedia exhibitions, concert halls and performance venues, sites of contemplation and engagement with sound phenomena, interactive public structures and homogenous sound pavilion artworks. Methods of audio reproduction and contemporary architectural design also suggest that the integration of new and emerging technologies will continue to be an incentive for a foundation of new projects and the presentation of new works.

Dynamic Architecture

In some respects, sound is the more dynamic, malleable medium in the sound pavilion partnership, in contrast to the static structures and materials of architecture – "frozen music," as Schelling would have it. Sound is able to embrace the element of time, to change or morph, to move, and to depict found environments. In some ways, this dynamism might be seen to disconnect audio art from architecture, so we might ask how future sound pavilion architecture might intersect with the technical and aesthetic developments in contemporary sound art.

The integration of spatial sound and architecture continues to present opportunities and challenges to sound artists and architects, but now also to materials designers, tangible and embedded computing researchers, structural engineers, and those engaged in new fabrication experiments.

Contemporary technological developments in these areas point toward potential synergies between sound works and the substance and processes associated with architecture. Contemporary technologies have supported developments in architectural design, in materials design, and

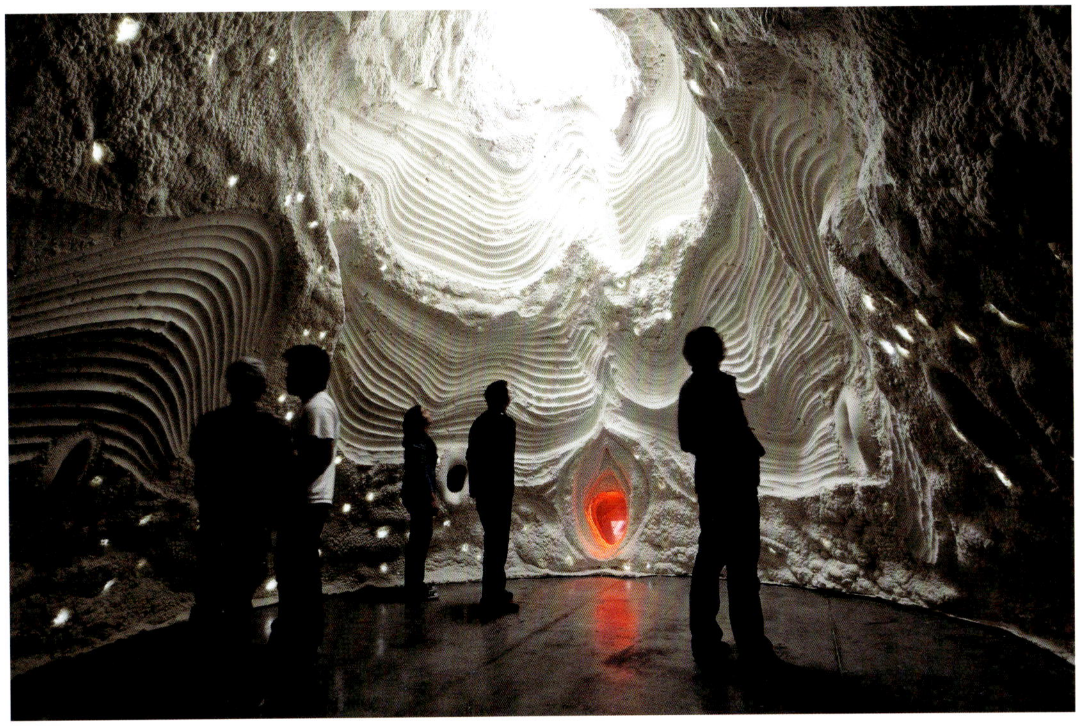

FIG. 14 Patrick Tighe with Ken Ueno (composer), *Memory Temple*, 2012.

in manufacture. Intelligent, “smart” materials have been created; form-finding techniques are now integrated in contemporary architectural practices, generating new process-based or derived architectural forms such as Patrick Tighe’s *Memory Temple*.(FIG. 14) Tensile and impermanent constructions that embrace sound have also become exciting architectural interventions.(FIGS. 15a–c) Embedded computing has introduced active building elements, capable of changing their function or form. Telemedical devices now transmit detailed and real-time personal data that can be interpreted to support interactive encounters between an artwork and an audience. Materials pioneers such as Skylar Tibbits have also demonstrated self-assembly and programmable materials[21] that can independently initiate dynamic structures, capable of assuming preprogrammed forms following an input of energy.

The integration of these new architectural design methodologies and dynamic, contemporary construction methods will undoubtedly provide fertile ground for future sound pavilions. It becomes possible to conceive of dynamic acoustic (and other) modifications of constructed spaces via building automation. One might also envisage the integration of electroacoustic sound reproduction methods to create sounding structures within architectural and materials design; or significant automated interactions between architectural elements choreographed to sound events. High degrees of interaction between a structure, people, and their biomedical data are also possible, perhaps to create pavilions that could, for example, display a form of empathy with their occupants and viewers.

Spatial Music

Three-dimensional spatial music now exists. If it is to be pursued, where might one experience it?

As Stockhausen suggested, the sound pavilion, rather than the concert hall, may continue to be an answer, where unique and specific architectural and acoustic designs can unite to support the presentation of spatial audio artworks, to create bespoke exhibition and audition environments for contemporary sound art.

Auditioning and Perceiving Space

For an ambulant audience there is an experiential sensing and an unfolding apprehension of space as we encounter architecture and move through it. There are parallels between the human experience of spatial music and spatial sound fields and the experience of architecture, particularly where both can be navigated by people.

Recent decades have seen a considerable body of research activity relating to human

21 Skylar Tibbits, “Design to Self-Assembly,” *Architectural Design*, vol. 82, no. 2 (2012), 68–73; Skylar Tibbits and Kenny Cheung, “Programmable Materials for Architectural Assembly and Automation,” *Assembly Automation*, vol. 32, no. 3 (2012), 216–25.

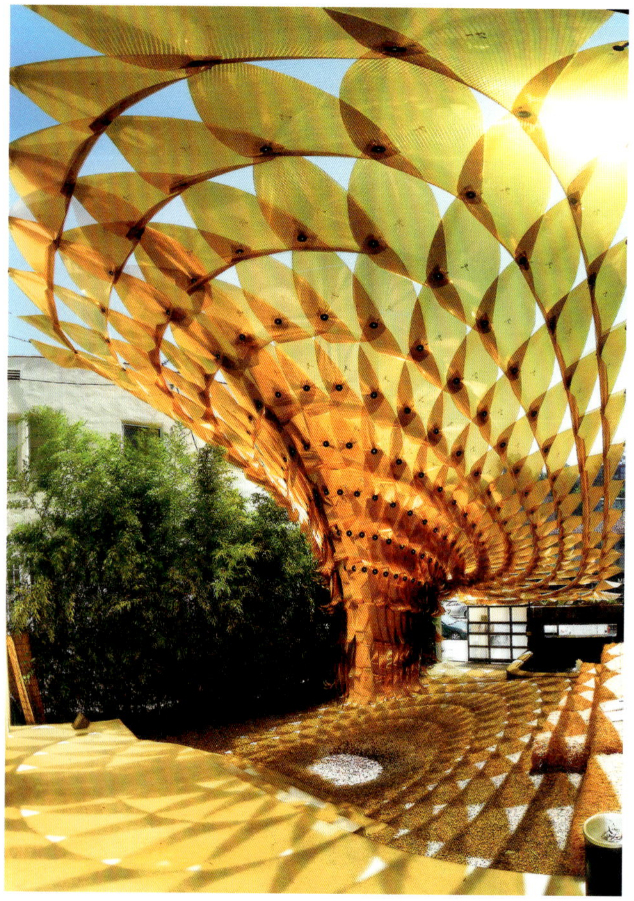

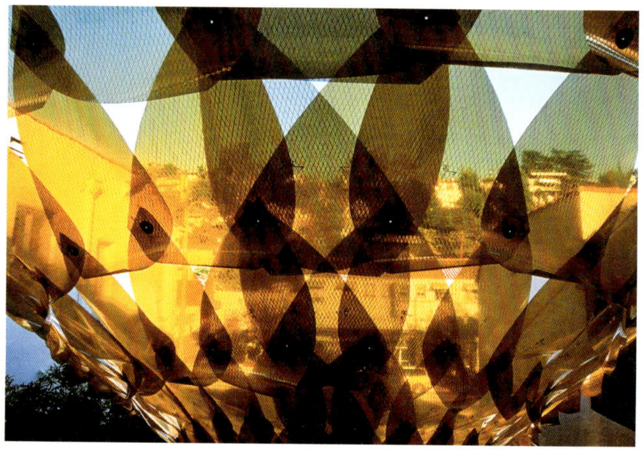

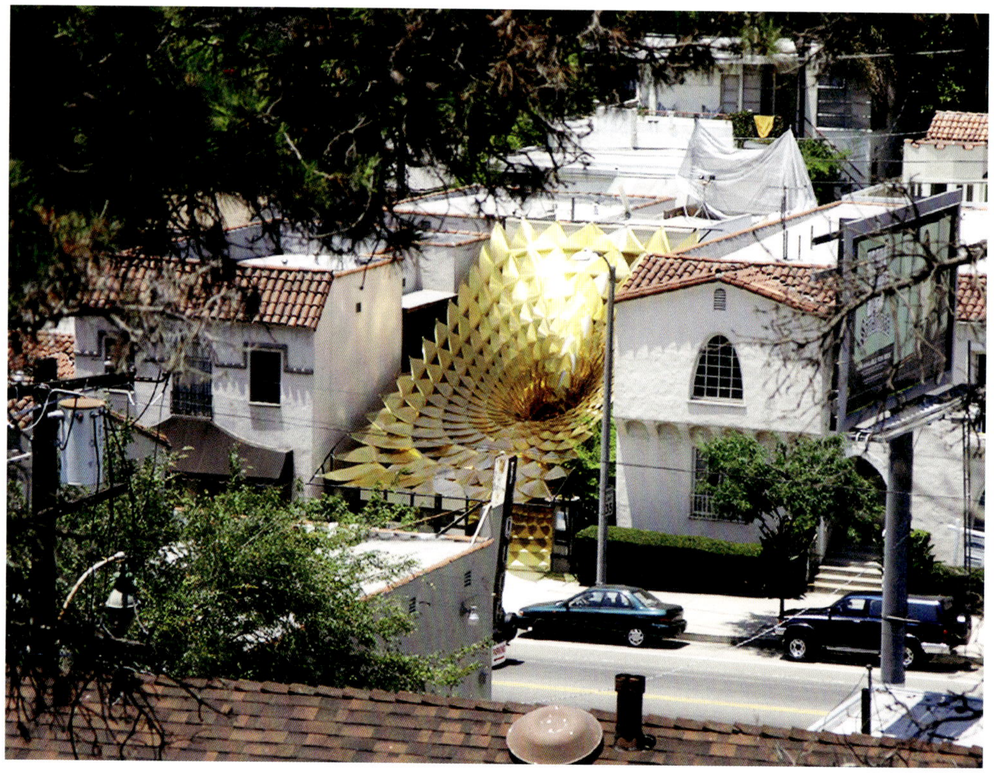

FIGS. 15a–c Benjamin Ball and Gaston Nogues, *Maximilian's Schell*, 2005. Vortex-shaped, temporary outdoor installation in the Los Angeles exhibition space of Materials & Applications, with integrated ambient sound installation by composer James Lumb entitled *Resonant Amplified Vortex Emitter*.

perception and our understanding of how humans might comprehend and interact with spatial environments. Some of these contemporary theories of human spatial perception may shed light on our encounters with spatial artworks, both to inform the development of new work and to quantify our apprehension of sound spaces.

While many audio engineering texts often focus on hearing alone as the primary communicator of spatial sound information – interaural time delays and amplitude differences, head-related transfer functions, etc. – Knill and Richards,[22] among others, promote Bayesian models of multimodal perception, where each sense contributes a votive input to our perceptual understanding. This suggests that spatial music and audio materials presented via methods that can appeal to visual and spatial perception per se may better support the apprehension of spatial audio. The legacy of J. J. Gibson's[23] work in visual perception and the work of Timo Järvilehto,[24] for example, suggest that we actively explore the world and engage with it to uncover information afforded by the objects and environments we observe. Might future sound pavilions incorporate these notions in their presentation of spatial sound materials? The outcomes of R. L. Gregory's[25] experiments in visual perception might also guide pavilion creators. Gregory proposes that we regularly rely on our own cognitive knowledge and information-based models of the world to guide and inform our perception. Within the context of a sound pavilion, which may not conform to our cognitive models of spatial sound behavior, might his thoughts influence how we present spatial music and spatial audio within architectural constructs?

These and other related theories of perception suggest that we stand to gain rich spatial experiences from encounters with artworks that facilitate exploration, that are real artifacts in our world, that afford information about audio space, and that support the reorientation of our perceptual mechanisms while retaining appropriate spatial audio cues. Such models may inform future sound pavilion creation or perhaps help to avoid designs and concepts that might, conversely, inhibit spatial perception.

Collaboration

The significance of collaboration between artists, architects, scientists, and engineers in the realization of sound pavilion projects should not be overlooked. Such interdisciplinary engagements establish pertinent connections between contemporary aesthetic or artistic concerns and contemporary technical and engineering practice. Sound pavilion projects can encompass fields that are so complex and expansive that collaboration becomes essential to ensure the potency of a work in all domains. Successful collaborations that unite engineering, science, technology, art, aesthetics, architecture, music, sound, light, biomedical interactions, and many other contributory fields are inevitably demonstrations of significant human achievement, not dissimilar from those we might see result from large-scale film projects or the opera house.

Sites of Innovation

However, the subsequent and continued interest in the Philips Pavilion, discussions of its motivations, construction, ideas, and concepts for spatial music also demonstrate the lasting impact of collaboration in the construction of sound pavilions. Sound pavilions have been, and remain, valuable sites of experimentation and innovation in contemporary media, technology, and architecture. They continue to be a fertile ground for explorations of environmental design, media presentation, spatial music, sound art, architecture, sound reproduction, and structural innovation; they simultaneously demonstrate state-of-the-art technology and cross-disciplinary unity in a common artistic endeavor. Ideas flow from activities of this type into many domains – ideas that can, as we've seen, resonate for many decades.

22
David C. Knill and Whitman Richards, eds., *Perception as Bayesian Inference* (Cambridge: Cambridge University Press, 1996).

23
James J. Gibson, *The Ecological Approach to Visual Perception* (Boston: Houghton Mifflin, 1979).

24
Timo Järvilehto, "The Theory of the Organism-Environment System: I. Description of the Theory," *Integrative Physiological and Behavioural Science*, vol. 33, no. 4 (October 1998), 321–34.

25
Richard L. Gregory, "Knowledge in Perception and Illusion," *Philosophical Transactions of the Royal Society – Biological Sciences*, vol. 352, no. 1358 (1997), 1121–1128.

Morten Søndergaard

Exscribing the Archive: Curating the Unheard Avant-garde (in Scandinavia)

Introduction

Sound is immaterial and fugitive. It is time-based, and it is nothing without being experienced. In that sense, it is the perfect material for "divergent" or "experimental" art practices. It is often considered "difficult" or even "outside" what is, by habit, thought of as an art experience; and because of the invention of technologies that make it possible to liberate sound from its source, sound even became the first technologically emerging artistic material; as such, sound art could be seen as the first media art practice, historically. Moreover, on a more philosophical level, sound art seems to elude the academic trends and paradigms throughout the twentieth century and operates into or with "ontological gaps" in the (cultural, political, and social) construction of knowledge. For those reasons, among others that I will investigate further in this chapter, sound art practices, as most media art practices, become "unheard" (of) in archives. They do not fit into the art-historical or cultural matrices of their day (and still do not, to a large extent). So, historically, or nonhistorically rather, sound art flies beneath the radar of cultural memory and knowledge like a sub-archive of parallel "sound art histories" (I have no better word for it at this time).

It is from this subarchival and nonhistorical status of sound art that the Unheard Avant-garde project finds its momentum: not only in giving the unheard a voice, but also in addressing some fundamental issues concerning the way new transdisciplinary domains are renegotiated across disciplines and boundaries of competences.

The Unheard Avant-garde research project, initiated in the context of the Danish Research Infrastructure Project named LARM (i.e. "noise"), which ran from 2010 to 2014, focuses on sound and radio culture in archives.[1] The Unheard Avant-garde project is exploring, investigating, and contextualizing the sound art practices and experiments that never made it into the formal and traditional archives – and how, if possible, to "reactivate" them.

The Unheard Avant-garde, therefore, is more than just another archiving project: it is a reinvestigation of the fundamental conditions for writing the histories of sound art (as a media art). As such, it may also be viewed as an attempt to reconfigure (our ideas about) the relations between technology, media, art, and cultural memory.

This was the curatorial point of departure in the making of the Unheard Avant-garde in Scandinavia section at the exhibition *Sound Art – Sound as Medium of Art* at ZKM (March 17, 2012–January 6, 2013). The focus is on sound in every aspect: sound as noise, as document, as material in artistic practices, and as a medium of cultural memory and exploration.

This essay will present, analyze, and discuss the Unheard Avant-garde section from the different perspectives of the competencies involved – and, especially, focus on how interaction-designers, systems-designers, media artists, (post) humanistic researchers, and media art curators collaborated and negotiated in a truly transdisciplinary configuration, and reactivated the unheard avant-gardes of Scandinavia in Karlsruhe.

The Archive in the Age of Digital Media

No archive is perfect. No one, I assume, will contest this claim – especially, in this day and age, where archives are everywhere and "everything is miscellaneous."[2] It is a reasonable claim that from the use of digital media is emerging a new understanding of what archives are, and what they could be. At least, it seems much more logical today to claim that any archive, including the "grand" and professional (scientific) archives of national and international libraries and museums, are imperfect and even selective. People are selective in their choices and tastes, and their archives are a mirror of their biases and behaviors. Personal taxonomies are contextualizing social networks. It is not hard to imagine that the formation and construction of the "grand" archives would display processes of choice and taste similar to those within professional and scientific networks.

There are huge gaps in the construction of the "grand" archives, and the Unheard Avant-garde project focuses on those lacunae and gaps from the particular angle of sound art. It claims that the private and social networks and "small" archives have emerged as authentic fields of "postarchiving": they establish a much-needed counterindicative position to the traditional art-historical positions (which I will not touch further upon in the following). These private collection spaces and networks could be considered precursors (avant-garde) of the social media revolution (however, most members were artists or collectors). It is perhaps not a coincidence that we "needed" social media to be able to imagine how to be documenting and disseminating the ephemeral and always mediated practices of sound art – as well as other innovative and technologically experimental modalities of time-based and nonvisual cultural production. According to Walter Benjamin, the archive as a medium of memory hides more than it reveals and, thus, should not be considered an instrument for *exploring* the past.[3] He continues:

> the "matter itself" is no more than the strata which yield their long-sought secrets only to the most meticulous investigation. That is to say, they yield those images that, severed from all earlier associations, reside, as treasures in the sober rooms of our later insights – like torsos in a collector's gallery.[4]

1 https://larm.sites.ku.dk/about-larm/.

2 See David Weinberger, *Everything Is Miscellaneous: The Power of the New Digital Disorder* (New York: Henry Holt, 2007).

3 See *Walter Benjamin's Archive: Images, Texts, Signs*, ed. Ursula Marx et al. (London: Verso, 2015), 40–42, 7.

4 Ibid., 42.

The archive is a manifestation of a certain kind of "logic" of memory. Traditionally, that logic has primarily been conducted in a scientific vernacular with objects, data, files, indexes, categories, contextualizations, and conceptualizations – the "matrix" of the scientific archive. However, as Benjamin points out, there is another archive situation at hand: one that does not address that which was immediately indexed and categorized, but rather that which constitutes emergent substrata of secrets. They are to be visited and found later, in a different pace. They constitute a hidden matrix beneath the scientific matrix of the traditional archive – one that the medium does not mediate.

The situation of postmedia and a culture of ubiquitous information, which is moving past the second phase of computing,[5] is transforming the traditional scientific matrix of the archive. And with it, the unmediated parts of the archive are transformed as well. They resurface.

Today, it is safe to say that the original scientific matrix of the archive is being challenged by many different developments, and thus finds itself projected onto a number of other matrices. This is accompanied by a different understanding of what memory is in an age of ubiquitous computing – rather than a tool of remembering, it could be described as a process of reformatting in a collective act of "writing elsewhere":

> Memory as a collective act is evident in our daily experiences with computers, which do make their memory seem – against physics – permanent and ubiquitous. If things remain now it is not because digital memory is more robust than paper; the opposite is true. Silicon degrades far more quickly than paper; websites – even seemingly as solid and permanent as geocities – disappear. Clouds are clouds – mercurial mixtures of vapor, liquid, and solid. Despite this, digitization can be considered a form of "saving" because it preserves content through a process of "reformatting": a reading, which is writing elsewhere. All computers read by writing elsewhere, or, to use Nancy's terms, by exscribing: by copying and disseminating. Importantly, exscription has little to nothing to do with meaning, but everything to do with communicating.[6]

In what follows, I will describe and reflect on the processes and practices involved in curating the Unheard Avant-garde in Scandinavia section. I will structure my presentation and discussion around the production and content of the three "platforms" which were, in turn, structuring the section in 2012 – each platform framing an important "hub" of exchange and network in the production of sound art in Scandinavia. Furthermore, I will showcase and analyze how curating involves a practice of exscription as method and a strategy; the act of curating writes the unheard avant-garde elsewhere… The questions that faced me in the summer of 2011, were: Where and how to exscribe and communicate them? How to operate with exscribing as practice in a curatorial strategy? And finally, in the last part of this essay, I will discuss how the exscribing practices, and the curatorial strategy, are enabling a domain-game to be played between the audience and sound art histories.

Exscribing Practices I: Preparing for Dissemination

The complexity of both the material and the questions of how to disseminate it obviously needed to be exposed into a creative process working across many different competencies and disciplines. A transdisciplinary process cannot run on curatorial individuals, but is kept alive and sustained by cross-domain exchange among many (albeit relevant) competences. Consequently, the first methodological step was a step back from the curator role in order to achieve a different, more open and creative, situation: collaborative and collective design-making.[7] However, even though the focus on the Unheard Avant-garde is a set framing from the beginning, and it would be my research supplying the sonic and contextual material, the matter of practicing exscription itself is uncharted territory for all of us – the issues of how to copy and disseminate are shared

5 See Wendy Hui Kyong Chun, *Programmed Visions: Software and Memory* (Cambridge, MA: The MIT Press, 2011).

6 Wendy Hui Kyong Chun, "Ubiquitous Memory: I Do Not Remember, We Do Not Forget," in *Ubiquitous Computing, Complexity, and Culture*, ed. Ulrik Ekman, Jay David Bolter, Lily Díaz, Morten Søndergaard, and Maria Engberg (New York: Routledge, 2016), 161–74, here 169.

7 However, I did choose the members of the group and also had a budget to invite people. Which is different from when a group establishes itself based on shared visions or goals. But I worked to make everyone in the group as equal as possible in the creative and artistic decision processes.

challenges. To follow in the footpath of Wendy Chun (see above), those challenges have everything to do with communicating.

A core team was quickly assembled,[8] key partners secured,[9] and a number of artists/designers invited[10] to take part in the process. When we all met for the first time we had less than five months to the opening date, and for that reason I suggested an overall structure: three platforms, focusing on "Scandinavia."[11]

The first stage was the digital copying and formatting process of the material collected in the research project. To facilitate this process, I commissioned a data-curation tool from the Danish Broadcast Company (DR), which could access digital material and organize it in "areas" of special interest for dissemination purposes. The big relational database/data cloud running DR's digital infrastructure – from production to streaming – is called CHAOS.[12] Since DR was a full LARM partner, all the collected audio material was digitized and copied into the CHAOS infrastructure and put to our disposal as digital copies from there. To be able to work with the material, and since the only interfaces that existed were developed for either production or streaming purposes, we needed to develop our own data interface in order to work with the digital material. And, moreover, the data-interface had to prepare for dissemination, somehow "fitting" the formats and designs for the exhibition, which were still only vague ideas and sketches. The tool had to be capable of organizing the media files in CHAOS visually onto curated "data-platforms," generating metadata and creating the relations between the media files and layers of content. Moreover, it had to be possible for the dataplatforms to function as "backspace" for public interfaces, streaming both media files and metadata from a cloud infrastructure that could be accessed anywhere in the world. Ultimately, the idea was to make an exscription tool that would work in the field, and could be used when "stumbling" on unheard avant-garde material (often in private and hidden "archives" – as further explained below).

Immediately, when I presented this idea to Martin Luckman at the Department of Innovation at DR, he suggested to make an application (aka an "app"), rather than a "tool." Since CHAOS is constructed as a relational cloud structure, and since it should be possible to work on and disseminate existing data by the curators/users (and also because of Martin Luckman's unique competences), the work began on the "CHAOS app."

To fully explain what the CHAOS app would be doing in terms of facilitating the exscription of the unheard avant-garde, I will go further into the methodological implications of exscription as curatorial strategy when dealing with, in effect, material which was deferred.

Exscribing Practices II: Beyond the Deferred Medium

To repeat Walter Benjamin's claim (from above), a huge portion of "cultural memory" is never immediately indexed or categorized. It constitutes a kind of substrata of secrets and hidden matrices, beneath the scientific matrix of the traditional archive – one that the medium (of the archive) does not mediate. This point is repeated by Foucault,[13] and later also by Wolfgang Ernst,[14] with a twist: the discourses are constructing this situation of the archives and consequently rule out other ways of expressing cultural content or materiality. I will return to the ways of working around this situation when I discuss the effects of exscribing practices as curatorial strategy on the construction of (our experience of) sound art histories in the last part of this essay.

The exscription of the Unheard Avant-garde as curatorial strategy takes literally Benjamin's idea of severed images that require later

8 Based partly on people I had collaborated with before and some from new partners and work situations: media artist Mogens Jacobsen (independent), programmer and designer Martin Luckman (Danish Broadcast Company), designer and programmer Christian Leifelt (independent), hardware programmer and visual artist Sebastian Frese Bülow (Aalborg University IT-department).

9 Danish Broadcast Company, EMS in Stockholm (Mats Lindström and Sanne Krogh Groth), Finnish National Archive in Helsiniki (Pertty Rastas), and The Museum of Contemporary Art in Roskilde.

10 Ane Mette Ruge (DK), HC Gilje (NO), Hans Sydow (DK), Eric Andersen (DK), Vagn Steen (DK), and Christian Yde Frostholm (DK).

11 Primarily from Sweden, Finland, and Denmark. It was not possible to acquire enough information about possible unheard avant-gardes in Iceland, Faroe Islands, or Norway in the short time we had available; but we satisfied ourselves with the fact that Icelandic and Norwegian artists were active on both the EMS and the POEX platforms, and that one of the most significant directors of EMS, Knut Wiggen, was in fact from Norway. However, given more time, it would have been a fine and important addition to the Scandinavian section of the 2012 exhibition at the ZKM to have sound art practices from Norway, Iceland, and Faroe Islands included. It calls for continued work on the unheard avant-garde in Scandinavia.

12 CHAOS is a cloud of data rather than one database, and it is codeveloped by a number of media broadcasting companies and cultural heritage institutions.

13 Michel Foucault, *The Archeology of Knowledge* (New York: Pantheon Books, 1972). Originally published as *L'archéologie du savoir* (Paris: Gallimard, 1969).

14 Wolfgang Ernst, *Digital Memory and the Archive*, ed. Jussi Parikka (Minneapolis: University of Minnesota Press, 2012).

insights (data without knowledge attached to it, in the broadest sense). Into this emerging materiality, or to help it emerge, the existence of recorded sounds (from private archives), alongside the possibility of interpreting and disseminating hidden phenomena through sound, addresses a level of postcognition and (again taken literally in the curatorial strategy) the potential of "technogenesis" in the sense of N. Katherine Hayles. According to her,

> [g]rasping the complex ways in which the time scales of human cognition interact with those of intelligent machines requires a theoretical framework in which objects are seen not as static entities that, once created, remain the same throughout time but rather are understood as constantly changing assemblages in which inequalities and inefficiencies in their operations drive them toward breakdown, disruption, innovation, and change.[15]

That technogenesis is framing and motivating a resurfacing of the hidden matrices of cultural knowledge on both a bodily and technological level is the curatorial claim here. The question that then needs to be addressed is: how to operationalize the deferred medium in a cultural production of knowledge that takes into account both recording and technogenesis?

According to Arthur Koestler, most cultural discourses are based on a "verbal logic" that is more or less dogmatized, based on the codes of various disciplines that are organized in what he calls a matrix. Contrary to this somewhat stifling layer of "common sense," the "creative act" is the situation that arises when new matrices emerge and "the codes of disciplined reasoning are suspended."[16] Matrices are operational in all matters of knowledge and communication as, for instance, "matrices of perception" or "matrices of reasoning," etc. Koestler argues that, when these "independent matrices" at some point interact, the result is a collision, fusion, or confrontation. Furthermore, according to Koestler, the collision brings laughter, the fusion new intellectual synthesis, whereas the confrontation results in aesthetic experience.[17] To simplify matters a little (for argument's sake), the situation of the archive may be understood as an extreme dogmatized "creative act." The archive is certainly traditionally based on a "verbal logic" and embedded in codes of various disciplines. Whether this has produced a kind of "common sense" of the archive could easily – maybe too easily – be claimed. However, I am suggesting taking a step back from this dogmatized creative act, and asking about what creative acts emerge in the deferred archive – operating in and with technogenesis? What matrices are in play here? It seems to be of greatest importance to understand how to describe the "old" codes, as they are transforming, and the "new" codes, as they are emerging. Since I do not have pages enough to describe the old, nor all new matrices of the archive in the age of technogenesis, it seems relevant to revisit briefly one such "creative act" of analyzing "new codes," which may be found in the (very) short history of new media criticism.

In his seminal study of the effect of the database on our patterns of representation, *The Language of New Media*, Lev Manovich calls for a new way of understanding the database as medium – or rather, a new way of understanding how we use this new medium in particular as a cognitive reference-tool.[18] The *language* of new media, then, is not to be understood as a new "scientism" or "language of computers"; rather it should be seen as a demarcation of a transformation in the configuration of our cognitive faculties using conceptual metaphors drawn from programming and computer hardware. It is a very human process Manovich describes, despite its apparent mask of technology: it is really about how we may understand memory as a medium with the database as a metaphor. Computers, technology, and culture influence each other and we may very well discover something new about ourselves and the way we navigate in our own media, Manovich argues.

However, *The Language of New Media* is also an account of an archive that was never realized – that of the (of its day) new media: the cinema. Which means that there are no critical texts to build a cultural science of cinema on.

It is perhaps not surprising, then, that much of the more recent work of Lev Manovich is focusing on the archive in various ways – especially on the ways that we may represent our memory drawn from the digital data we use and produce in our everyday lives.

In this way, Manovich is on the same page as Benjamin, in that it is not the new in itself (as new principles for understanding) that is at stake – rather, it is the renewal of the old principles of understanding:

> I, however, had something else in mind: not to retain the new but to renew the old. And to renew the old – in such a way that I myself, the newcomer, would make what was old my own – was the task of the collection that filled my drawer.[19]

Neither Manovich nor Benjamin refer directly to "matrices" in the way I am arguing for it in this paper (although Manovich refers to the Hollywood film *The Matrix* in ways that are close[20]), but the issue of bringing out material – organizing and referencing hidden materiality – as compared to pointing at the hidden processes of ideas involved in that very same process of collecting, I would claim, is exactly that: a way of mapping unseen and invisible matrices (invisible in the sense of Wittgenstein and the language game in which the visible is merely hiding) onto and into each other. New ideas, and processes of ideas, emerge from that dynamics – ones that would not have appeared if the movement had not taken place. Thus, Manovich and Benjamin are both, each in his very different way, addressing the notion of an "archive" which does not exist and therefore is not part of any critical tradition until a connection is "created" – between the new and the old matrices.

What – or who – makes this connection happen, we might ask?

If we look at the methodological work of Manovich in *The Language of New Media*, it should appear that he is very attentive to both the codes and matrices of the archive, as well as the apparatus condition the archive is entering. In fact, he describes five "principles of new media" present in most new media objects, which "should be considered not as absolute laws but rather as general tendencies of a culture undergoing computerization."[21] In other words, he is pointing out matrices – but not only that, he describes ways in which old and new matrices are intersecting with each other. This intersection is often due to artistic practices, which Manovich mentions are the real drivers of the dynamics behind the emergence of this "language" of new media.

What brings about the intersection of old and new codes, and reveals hidden matrices, is the framing of art and technology in a creative process of data and sound representation. This process is not about how things are depicted or visualized, but about how we approach and process the real through remote or deferred representations of it, yet without ever actually "reaching" it. The creative process of data and sound representation brought about by the intersection of old and new matrices is, I would suggest, very near to Andrew Pickering's concept of "agency realism":

> If the Western tradition aimed at representational realism, the works I have in mind aim at what one could call agency realism – not the portrayal of how things look but how things go [– …] an experimental openness to emergence, and adaptation rather than control.[22]

Of course, what Pickering is discussing is ontology in the age of the increasingly "ubiquitous" computational "real," and the issue of the hidden matrices of cultural memory is a matter of making things real, albeit in a different cultural context. Thus, as a possible way of grounding my curatorial strategy, I worked towards mapping this notion of agency realism to the model of independent matrices of perception that Koestler introduces; in particular, I focused on the convergences of intellectual synthesis and aesthetic experience in the recorded (and copied) sound art practices. The stage was now set to develop and produce ways to disseminate the Unheard Avant-garde. And mobilize the CHAOS app into a curatorial strategy of exscribing the unheard avant-garde.

15
N. Katherine Hayles, *How We Think: Digital Media and Contemporary Technogenesis* (Chicago: University of Chicago Press, 2012), 13.

16
Arthur Koestler, *The Act of Creation* (London: Hutchinson, 1964; reprint 1976), 45.

17
Ibid., 176.

18
See Lev Manovich, *The Language of New Media* (Cambridge, MA: The MIT Press, 2001).

19
Walter Benjamin, *Selected Writings*, vol. 3: 1935–1938, ed. Howard Eiland and Michael W. Jennings (Cambridge, MA: The Belknap Press of Harvard University Press, 2002), 403.

20
Manovich, *The Language of New Media*, 330.

21
Ibid., 27.

22
Announcement for Andrew Pickering's keynote lecture "Art, Agency and the Technological Unconscious," at the international interdisciplinary conference *SOFT CONTROL. Art, Science and the Technological Unconscious*, November 16–17, 2012, Maribor, Slovenia, published in the accompanying brochure, p. 46.

The Unheard Avant-gardes (in Scandinavia) – The Platforms[23]

In what remains of this essay, I will be showcasing and analyzing the research methods and methodologies involved in disseminating the experimental practices of Scandinavian sound and media artists in the late 1960s, and discuss them in the context of the activity of rereading and rewriting those sound and media art practices in a different place and time, beyond the medium of the archive.

For the purpose of exscribing the sound art practices, and to present the methodologies and materials of experimental practice and production, an adapted methodology was used, in three steps. 1) Instead of reenactments, which would focus on restaging artworks, I want to reactivate an entire event and context – including its ideas and methodologies. 2) I am dividing the reactivation-strategy into two modalities, Enacts and Reacts, which are focused on either the works and contexts of a single artist (Enacts), or the idea and context of an entire event. 3) I am using "new" media artists to reactivate the unheard media artists – adding a third element into the reactivation strategy (Reacts): that of reworking and reactualizing works, technology, events, and processes.

For the purpose of dissemination, the Unheard Avant-garde project at the Scandinavian section was structured around three platforms:

1) The EMS Platform. This revealed the Elektronmusikstudion in Stockholm and all the amazing productions from more than fifty years that were unarchived until a Danish PhD student, Sanne Krogh Groth, via the Museum of Contemporary Art in Roskilde (and my contact with Mats Lindström in 2003–04), made it her research project to recover them.[24]

2) The Erkki Platform. This was a tribute to Erkki Kurenniemi, the Finnish pioneer, innovator, engineer, artist, and self-archivist and creative user or inventor of most known media and technologies, who was also unheard (of) outside very specialized circles until this exposure and a parallel (but uncoordinated) event at documenta in 2012, curated by Joasia Krysa.[25]

3) The *POEX* Platform. The largest of the three platforms at the Scandinavian section in 2012, this focused on the POEX65 (POetry EXperiment) event, which took place in Copenhagen on December 10–20, 1965. However, it became unheard of and virtually "unwritten," not represented in any archive or art-historical account,[26] until the exhibition, which featured a number of Scandinavian sound and media artists and poets from the original event (but far from all of them). In particular, the platform presented the extremely varied experimental and pioneering performative and transdisciplinary sound art program from POEX65, including experiments beyond electronics, jazz, poetry, art (Fluxus and Intermedia), and "beat" – a pre-pop genre fusing and experimenting with all the above and more. (As is often the case, beat is composed and performed by many of the same musicians, poets, and artists involved in other lines of experiments.)

Finally, yet another platform was presented in 2012 that was not produced for this section specifically, but could be seen as a forerunner of the design processes and methodology going into the Unheard Avant-garde project: the Media Art Platform,[27] represented here by the *Audiobar* by Mogens Jacobsen. As an example of a preliminary media art archive project at the Museum of Contemporary Art in Roskilde (MFSK), Denmark, between 2005 and 2009, it represents a similar concept and interest in experimental practice as the Unheard Avant-garde project. I will be discussing the impact of the Media Art Platform on the 2012 section in the final part of this essay.

23 Special section curated by Morten Søndergaard for the *Sound Art* exhibition at the ZKM | Karlsruhe. Contributions by: Mogens Jacobsen, Martin Luckman, Christian Leifelt, Hans Sydow, Ane Mette Ruge, HC Gilje, Eric Andersen, Christina Sydow, Karsten Vogel, Per Højholt, Vagn Steen, Hans Sandmand, Knud Hvidberg, Sten Hanson, Jørgen Plaetner, Else Marie Pade, Erkki Kurenniemi and many more. EMS docu space by Sanne Krogh Groth (RUC) and Mats Lindström (EMS). Curatorial assistants: Anna Holm, Karen Mette Fog Pedersen, Emil Alenius Boserup. Production coordination: Sebastian Frese Bülow. Programming: Martin Luckman, Alex Usbergo, Brian Groth, Sebastian Frese Bülow, HC Gilje, Mogens Jacobsen. Sound production: Emil Alenius Boserup. Head of research / conceptual design: Morten Søndergaard. Partners: DR (Danish Broadcast Company); Aalborg University of Copenhagen (AAU CPH); LARM (sound archive infrastructure in Denmark); ZKM | Zentrum für Kunst und Medien Karlsruhe; EMS, Stockholm.

24 Sanne Krogh Groth, *Politics and Aesthetics in Electronic Music. A Study of EMS – Elektronmusikstudion Stockholm, 1964–1979*, trans. Juliana Hodkinson and Isabel Thomson (Heidelberg: Kehrer, 2014).

25 *Online Archive: Erkki Kurenniemi (In 2048)*, project initiated by Joasia Krysa, commissioned by KURATOR (Joasia Krysa) and dOCUMENTA (13) in partnership with Central Art Archive of the Finnish National Gallery (Perttu Rastas), and Contemporary Art Museum KIASMA, Helsinki. Supported by Arts Council England.

26 With two important exceptions: In 1991, POEX65 was mentioned in a catalogue for a Knud Hvidberg exhibition; see Marianne Barbusse, "Knud Hvidberg," in *Knud Hvidberg*, exhib. cat. (Herning: Herning Museum of Contemporary Art, 1991), 48. And in 2009, POEX65 was officially, and for the first time, recorded in the "official" history of the Experimental Art School (Eks-skolen) in Denmark, but even then, it only filled half a page in a book of more than 400 pages (Lars Morell, *Broderskabet: Den eksperimenterende Kunstskole 1961–69* (Copenhagen: Thaning & Abel, 2009).

The EMS Platform[28]

Since 1964, EMS (Elektronmusikstudion in Stockholm) has been the main "hub" in Sweden not only for the production of electronic music and sound art, but also for networking between artists and all other kinds of practitioners. Originally part of the National Swedish Broadcast Company, Swedish Radio (SR), it was later reorganized and became an independent and (heavily) state-funded institution in 1969.

From the late 1960s, the EMS studio hosted a state-of-the-art computer studio, all designed in modern Scandinavian design.[29] But it was mainly because of the ability to accommodate an impressive host of international artists and practitioners in sound and computing, and to operationalize it widely also (and perhaps mainly) outside Sweden, that EMS became known as one of the world's most advanced platforms for experimental sound production. The history of EMS is not only a story of ideological pioneers, who wished to put Sweden on the international avant-garde map; it is as much the histories of sound artists getting an opportunity, for the first time, to practice sound art in a transdisciplinary setting. Because of the high-tech framing and the fine opportunities in terms of public funding, EMS gradually developed a certain sound art "genre" reflecting the requirements and possibilities of the state-of-the-art recording facilities. However, this "genre" did not always suit the more conceptual artists, and for that reason, and other reasons beyond the scope of this essay, a number of creative discussions and debates were conducted with EMS as center stage. And maybe those debates are as important, one could argue, for sound art histories, as the actual recordings that took place at the time.

In the EMS Documentation Space, both the works and some of the research, technology, and music executed and produced at EMS are shown in photographs and text documents.(FIGS. 1, 2) You may also listen to a soundtrack of some of the seminal productions from the studio.

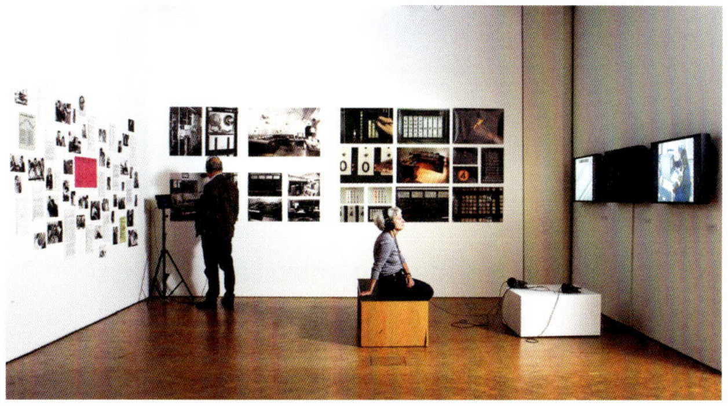

FIG. 1 Sanne Krogh Groth and Mats Lindström, *EMS (Elektronmusikstudion) documentation space*, 2012. Installation views ZKM | Karlsruhe, 2012.

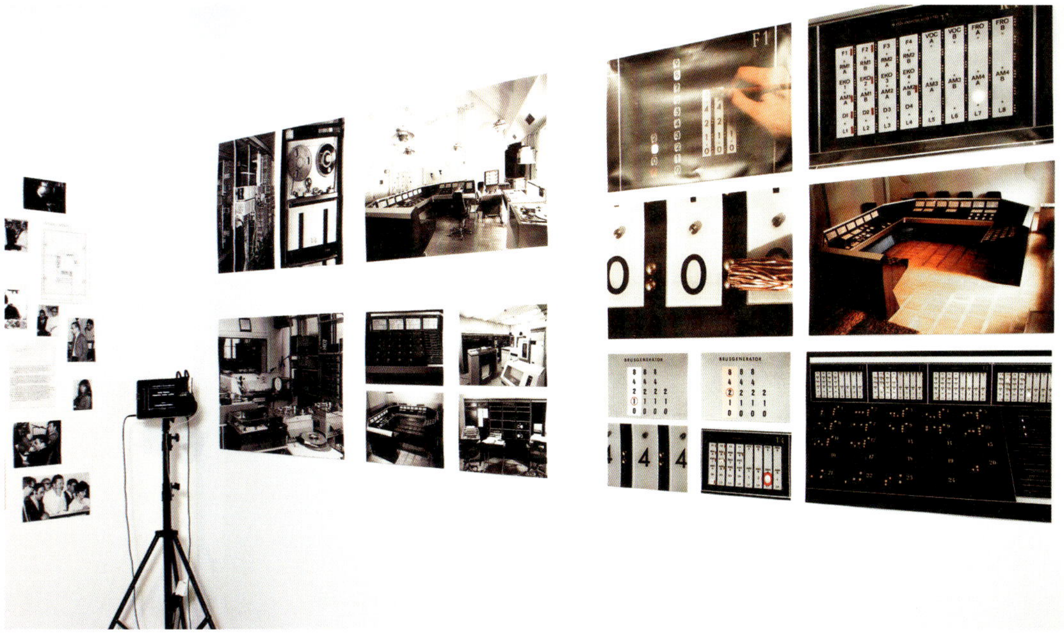

FIG. 2 Sanne Krogh Groth and Mats Lindström, *EMS (Elektronmusikstudion) documentation space*. 2012. Installation views ZKM | Karlsruhe, 2012.

27 Mogens Jacobsen and Morten Søndergaard, eds., *RE_ACTION – The Digital Archive Experience: Renegotiating the Competences of the Archive & the Museum in the 21st Century* (Aalborg: Aalborg Universitetsforlag, 2009).

28 For the EMS Platform, Sanne Krogh Groth and Mats Lindström were invited to select and design a documentation space. All materials are from the archives of EMS and Fylkingen. Exhibited material digitized and printed by EMS / Statens musikverk.

29 "In the early days, EMS was located in an old radio theater studio. The studio was meant to be temporary, but considerable resources were allocated to build a very prestigious and, for its time, high-quality computer music studio, which opened in 1970. The purpose of the studio was both to compose music, and to be used as a tool for research into sound and sound perception. The technical solution chosen for the studio was a so-called hybrid studio, i.e. a computer controlled analog studio. A nine-meter-long console with 2000 numeric buttons was specially designed to serve as an alternative tool to the data programming and as a control of the programming." Excerpt from the explanatory text by Sanne Krogh Groth.

EMS Enact: Sten Hanson / Hans and Christina Ann Sydow, *Hanson's Head*[30]

A number of Enacts were commissioned, as part of the curatorial strategy of exscribing the unheard avant-garde. For the EMS platform, I asked Danish composer Hans Sydow to choose one artist from the EMS studio to work with. The very fast answer to that question was Sten Hanson and his peculiar *How are you* from 1969, a classic text-sound composition, produced at EMS in Stockholm.

> The composer Hans Sydow (*1968) often takes his starting point in the area between poetry and music, and works with language as musical material. He makes a virtue out of mixing traditions – pop and avant-garde, the catchy and the noisy. In 1998, the CD *Glimtvis* ("In Glimpses"), based on Tom Kristensen's poetry, was released. In 1999, the work *Entropy*, in a revised version, won the electro-acoustic music competition for young composers held by the National Arts Foundation. In 1999–2000, Hans Sydow received the National Arts Foundation's two-year enterprise grant. Alongside his composing work, he manages the sound art project *Crossfield* at the Museum of Contemporary Art in Roskilde. See www.resonance.dk.
>
> Sten Hanson (1936–2013) made his first appearances in the early 1960s as an experimental poet and composer. From an early stage he was aware of the importance of tape-recording techniques in the renewal and development of poetry's resources. Text-sound-image, often combined with intensely personal "live" performances, are vital ingredients in Sten Hanson's artistic workmanship, and he is one of the forerunners in the field of multimedia art. His works include electro-acoustic pieces as well as instrumental and vocal compositions.

The result is an audiovisual remix, entitled *Hanson's Head*.

Sten Hanson made different variations on his piece *How are you* as a written text, as graphic art, and as a text-sound composition, all based on the two main statements: "How are you?" and "I'm OK, thank you," that open and end the piece.

In *Hanson's Head*,(FIG. 3) all these variations are rewritten into the content and context of the remix, which consists of a text-sound composition and graphic art. As Sydow writes in the text following the remix:

> As the words are being filtered in the sound composition, they fade into different nuances of black and gray in the graphic art. In that way the graphic art and the sound art create a counterpoint in the remix, and between them new statements appear.[31]

This is a perfect example of exscription. Sydow continues:

> In *Hanson's Head*, the sound composition opens with a spatialization of the original piece by Sten Hanson, diffused to the eight head-shaped loudspeakers hanging in a circle from the ceiling, to make an illusion for the audience of being inside Hanson's head, while fragments of his text and sound move around their own heads.[32]

The composition of Sten Hanson is short but it nevertheless contains a series of sonic variations on its own theme referencing different musical genres (i.e. electronic music, modern

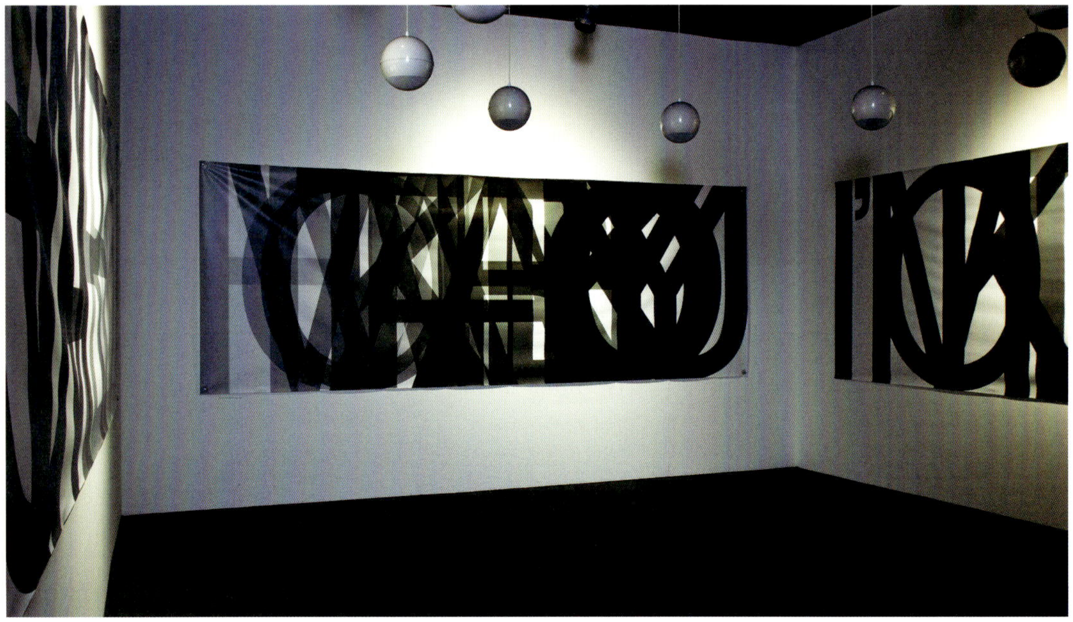

FIG. 3 Hans Sydow, Christina Ann Sydow, *Hanson's Head*, 2012. Audiovisual installation, 3 large PVC banners, 1 amplifier, 8 speakers, speaker cable, 8-channel playback device. Installation view ZKM | Karlsruhe, 2012.

30
Hanson's Head, sketch by Christina Ann Sydow, soundfile by Hans Sydow.

31
Hans Sydow, *Hanson's Head*, artistic statement for the Unheard Avant-garde Project, Copenhagen 2011–2012. Unpublished.

32
Ibid.

33
Christian Leifelt / Erkki Kurenniemi, interactive sound installation, 2012.

jazz, contemporary music, and soundscapes). The remix continues with a sonic retelling, or exscription, based on the musical genres. Ten sound statements from the original composition are now transformed into notes and rhythmical structures, played back with other instruments and sounds to emphasize the artistic grips of the text-sound scene, of which Sten Hanson was a cofounder. Just as Sten Hanson made us aware of the sound within the spoken word, this remix hopefully makes listeners aware of the world of music being exscribed inside Hanson's sounds.

The graphic design in *Hanson's Head* unfolds the sound as a large-scale abstract visual score that creates a timeline across the walls, based on the same font as Sten Hanson used in his graphic art. It, too, is exscribing what is happening on the sonic and textual level in *How are you*.

The Erkki Platform: *Erkki vs Erkki* – Claviature for Recontextualization and New Understandings[33]

Erkki vs Erkki(FIG. 4) is a tribute and an artistic examination of the life and work of the Finnish designer, scientist, futurologist, engineer, and artist Erkki Kurenniemi. The installation brings together extraordinary and very different aspects of his work – from his 1960s electronic compositions, experimental short films and innovative musical instruments, to his later vlog project and theories on mathematical harmonies and artificial intelligence from recent years.

> Erkki Kurenniemi (1941–2017) was a pioneer of electronic art in Finland, a former nuclear scientist who composed computer-based music and designed his own instruments, founding the University of Helsinki Electronic Music Studio in 1962. Kurenniemi was a pioneer of industrial automation at Rosenlew in the 1970s, an automation designer in Nokia's cable division in the early 1980s, and head of exhibition planning at the Finnish science center Heureka from 1987 to 1999. The exploratory search for new species of user interfaces for musical instruments and the semi-automatic generation of music were among Kurenniemi's main goals throughout his career.

The basic elements of the installation are materials from his media archive (with video, sound, photo, and text documentation of Erkki Kurenniemi's work) and a programmed musical instrument with optical input (a partial reconstruction of Erkki Kurenniemi's DIMI-O from 1971). Historically, the original DIMI-O has primarily been used to generate sound and music based on visual input of human gestures (see DIMI-Ballet and DEAL performance). The installation makes it possible for the audience to feed the DIMI-O with the visual documentation of Erkki Kurenniemi's own life – creating an audiovisual interactive dialogue across context, time, and media.

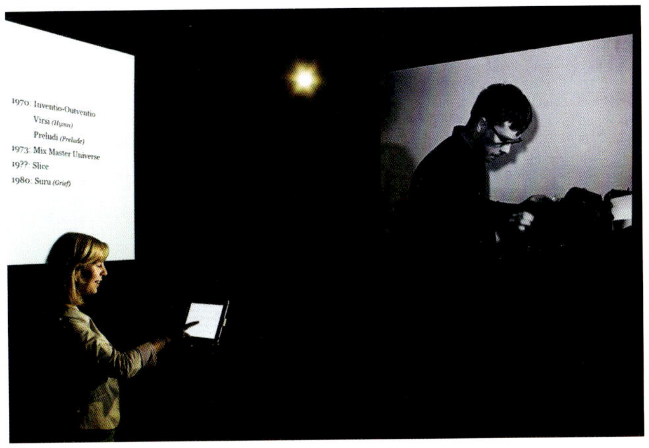

FIG. 4 Christian Leifelt, Erkki Kurenniemi, *Erkki vs Erkki*, 2012. Interactive sound installation, 3 projectors, media archive (video, audio, visual and textual documentation of the works of Erkki Kurenniemi), programmed musical instrument with optical input (remake of Erkki; Kurenniemi's DIMI-O from 1971), video camera, iPad control, 2 speakers. Installation view ZKM | Karlsruhe, 2012.

The installation builds directly on the private media archives of Erkki Kurenniemi, which are multifaceted and extremely versatile. They mostly document Kurenniemi's life and actions, alone and with others that he shared his life with. The production of the archive is not strictly about art processes, but it is not autobiographical either. Moreover, it has been recorded and produced on many different media platforms over more than four decades, involving sound, images, and footage in various different formats. This has allowed Kurenniemi's archives to serve other purposes or have another status than the archives you find in museums or libraries. For the installation, only works by Kurenniemi on YouTube are used, making it a double feedback loop of his archive into the pre-quantum computer and the dissemination of those into social networks, ultimately exscribing his footage and sounds from a semantic data-space into the exhibition space of the Scandinavian section.

Thus, Erkki Kurenniemi's media archives are playing with the role of "authentic" exscription strategies (for example, if they are going to appear in a quantum computer space, then what are the requirements, etc.?), while highlighting and questioning the status of "real" life, as it is lived, while it is lived, with sex, drugs, art, anxieties, car drives, food, and a long list of other meticulously recorded, seemingly insignificant events of life. The media archive of Erkki Kurenniemi is thus not to be seen as a result of documentation processes in any traditional sense; underneath the media archive, or rather ahead of the production of archive material, which is primarily Kurenniemi's own artistic and personal action-space, another and different submedia stage is unfolding. Also here, there is a submedia archive running parallel to (and pointing ahead of) the media archive of Kurenniemi, which conceptually is evolving from the postcybernetic metaphors he lived by, charting out a future space between storing machines and memory technologies. This future space may be understood as directed toward a construction of interfaces of future authenticity. And, even more radically, since the archive is intended as a submedia reservoir for the construction of the future human version of Erkki Kurenniemi, then the real sender is in fact only a real self in a future bio-technological configuration. So, how does Erkki authenticate Kurenniemi in the future? How do we?

We cannot approach Kurenniemi curatorially without placing his practice somewhere in the vicinity of second-order cybernetics and cybernetic arts. A quick survey of selected early cybernetic art represents elements of modern technological phenomena. In the 1960s, cybernetics is absorbed in popular culture and art begins working with systems on different levels, and becoming systems itself. Influenced by second-order cybernetics, artists such as Nam June Paik work from a double-focus on art and Buddhism; Karlheinz Stockhausen works on electronic feedback sound systems; Steina and Woody Vasulka investigate feedback systems in video; these are only a few, selected examples. All of these artists, different as they are, share a systemic quality and method in their artistic practice; and they all experiment within and with the system.

> The bridge between art and cybernetics had to be constructed by creating metaphorical parallels. [T]he application of cybernetics to artistic concerns depended on the desire and ability of artists to draw conceptual correspondences that joined the scientific discipline with contemporary aesthetic discourses. The merging of cybernetics and art must be understood in the context of ongoing aesthetic experiments with duration, movement, and process.[34]

Thus, in this movement of ideas, artists play an important role in making diverse fields feed into each other – thereby actively facilitating those ideas being concretized and historicized into cultural configurations. In this way, the discourses of science and technology coming out of second-order cybernetics facilitate and transform artistic production radically. But what is perhaps less clear in all this, and what leads to a postcybernetic condition, is that parallel to this movement of ideas, a ubiquitous discourse about authenticity emerges. It is not a discourse about solid states and frozen moments, the unique object and true soul, which in the epistemology of second-order cybernetics is deemed inauthentic; it is a discourse of the future of technology and art, where the submedia future-space of the media archive is played out for real. It is this discourse in particular that Erkki Kurenniemi feeds into through the act of playing with (his own) archive, messaging the imagination with metaphors and other figurative representations of a future postcybernetic authentic consciousness.

The human life is not reducible to files or bits, yet in Erkki Kurenniemi's archive it seems we are reduced to authenticate ourselves by a variety of technological stylistics or formats. Even so, he records and records, incessantly, throughout his active career as scientist and artist, on all accessible platforms and in obscure analog or digital formats. The implicit human-computer/human-machine interfaces, which Kurenniemi develops and unfolds incessantly throughout his work with the archive are not focused on creating interfaces to his biography, which seems irrelevant to him; they are focused on how

34
Edward E. Shanken, "From Cyberbetics to Telematics: The Art, Pedagogy, and Theory of Roy Ascott," in Roy Ascott, *Telematic Embrace: Visionary Theories of Art, Technology, and Consciousness*, ed. Edward E. Shanken (Berkeley: University of California Press, 2007), 21.

35
Knud Hvidberg, *Invitation to Participate in POEX 65* [1965], Copenhagen, unpublished letter, archive of the Museum of Contemporary Art, Roskilde. Trans. Morten Søndergaard.

relations may occur and develop between digital and human phenomenological systems; how they, in turn, react to one another based on implicit patterns. His goal is, I would claim, an interface to a future human-technological existence.

This issue is important to stress because it points towards a completely different way of looking at the world as a non-intended, probabilistic set of possible interactions. It is a symptom of a search for future authenticity.

At the Scandinavian section in 2012, the audience could enter into that search themselves, using Kurreniemi's material as playground for exscription practice.

The POEX Platform

The third platform at the Scandinavian section had POEX65 as turning point. POEX65 was a transdisciplinary experiment and event, which took place in Den Frie Udstillingsbygning (The Independent Exhibition Building – or, simply, The Independent) in Copenhagen, December 11–18, 1965. Short for POetry EXperiment, POEX65 was an exhibition event curated and created by Danish artist Knud Hvidberg (1927–1986). It aimed at breaking the boundaries of art genres, the false division of professional and amateur, as well as the autonomy of the "work of art" through the active use of technological and mediated platforms such as Flexowriters, Punch Paper Poetry, and Electronic Visual Music. As such, it was a very important event in Scandinavian media art history, with more than eighty participants from five countries – many of whom became part of the leading class of artists in Scandinavia. Nevertheless, POEX65 somehow was forgotten – and almost erased from the academic memory and public archives.

Knud Wedel Hvidberg (1927–1986) worked from around 1960 in a constructivist idiom inspired by artists from the Danish group Linien II. Especially Gunnar Aagaard Andersen's serial works with variations on the same basic elements resonated with Hvidberg when he worked with optical motion pictures. He made kinetic sculptures, mostly in the form of mobiles, but soon also coupled motors and electric lights to them. In 1961, he exhibited in The Trinity, along with William Soya, an audio and light installation. In 1965, Hvidberg founded and coorganized POEX65, with which he wanted to move away from speculative art production, and towards one that involves an audience and is based on experience. By extension, he later sought inspiration in Eastern mysticism, which he would translate into sound, light, and movement.

POEX65 was not a traditional event or exhibition in any sense of those words, but a platform on which the public was invited to play an active role. In this way, the artist/curator Knud Hvidberg wished to distance himself from "art as an object of economical speculation, from traditional attitudes, from the traditional divisions between various arts and finally from official art exhibitions."[35] The event featured more than eighty professional artists, of whom about twenty-two are present in the group photo.(FIG. 5) Those twenty-two are installation

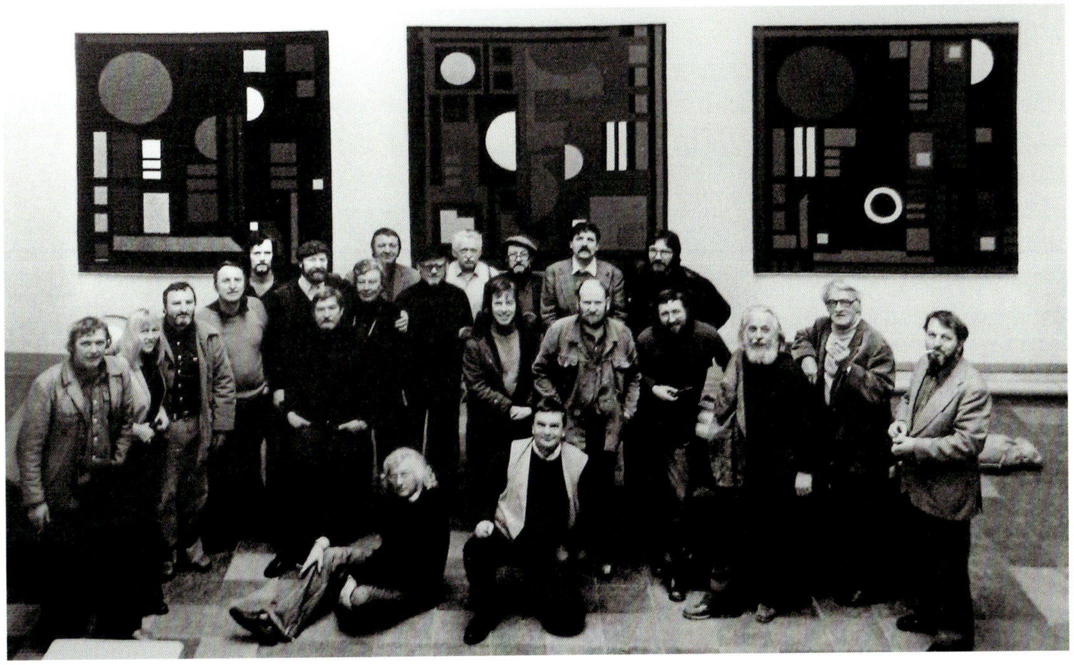

FIG. 5 Participants of POEX65, 1965.

artists of a variety of different orientations, whereas the rest – the number is uncertain and may never be accurate – were involved in the program of sound art performances, crossover events, and electronic, jazz, and beat concerts. The (artistic/curatorial) aim of POEX65, according to Hvidberg, was to make an exhibition-in-progress or, as he explains,

> [the] exhibition as a mental and physical performance ["udstillingen som forestilling"…] where poets, painters, and composers could meet and collaborate so that, each day, a new program would emerge.[36]

It was the intention to break with the established and elitist art circles – and invite non-elitist and non-professional artists to participate in the transformation, as Hvidberg saw it, of the art space itself. POEX65, he continues, would be

> an exhibition of poems. […] Poems as concretistic, as paintings, graphic works that are poems; performing poems, declamations, recitations, singing, electronics, sound, light, movement; performing paintings, theatre, films and happenings that are poems.[37]

Together with Knud Hvidberg, the Danish visual poet Vagn Steen and the Danish composer Karsten Vogel were the core of the POEX65 curatorial group. Since the early 1960s, and together with two other Danish semioticians and theoreticians, Peter Madsen and Per Aage Brandt (the latter was also a jazz pianist), Vogel had run an informal study group of new tendencies and theories of collaborative work formats and actions of art. POEX65, in many ways, fed into the ideas and thoughts coming out of that group – which also participated with an event during the ten days of activities. Vogel, however, had his main role as the experimental jazz musician who inspired the music-paint-light-situated works happening at POEX65. A short clip from the Danish Radio *Jazz News* broadcast in December 1965 was recovered from the personal archive of Kristen Lockenwitz, from which one could hear a "glimpse" of what was going on. From this recording we hear Vogel and Brandt playing in a set-up, where certain "signals" or "motions" by the painters or light-artists, triggers a chaotic outbreak of sounds from a number of acoustic instruments.

Five "stages" were active in POEX65: The ante-room (visual poetry), the large exhibition room (happenings and live music), the center hall (theater), the reading room (experimental poetry), and the small exhibition room (showing a so-called "co-ritus" environment by Jørgen Nash, Jens Jørgen Thorsen, Hardy Strid, and Carl Magnus).[38] There was a "Musicbar" alongside the Flexowriter in the reading room.

POEX65 did eventually manage to bring together a large variety of artists from all kinds of art genres and "ideologies." And they did work together in a number of ways. So far, some 128 "actors" have been identified – artists, engineers, dancers, musicians, critics, active audiences, etc.

Perhaps the most radical part of POEX65 is the "experimental" and "innovative" vision that Knud Hvidberg introduced. His was a vision – borrowed from the natural, social, and political sciences – that asked (today we might say "nudged") artists and nonartists to work together across genres, domains, and competencies.

POEX65 and the Archive

A complete archive of POEX65 does not exist. This is perhaps not surprising. What is more surprising is that *nothing* exists in the archives from that event in the Danish Broadcast Company or at any museum in Denmark when I began looking in 2006–2007. Nothing at all.

A preliminary search in the media archive of Denmark Radio (DR) in 2008 for POEX65 material proved unsuccessful. However, a search in the so-called White Programs (analog, paper programs of all radio and television broadcasts) revealed that a program *The Sunday Walk*, edited by Ella Wang, had been aired Sunday, December 19, 1965. Also, the program *Jazz News* had reported from the experimental jazz-painting concerts at POEX65 on the same day at 10 p.m. However, the tapes corresponding to this entry do not exist anymore; they have most likely been erased. A true unheard avant-garde situation!

Thus, apart from radio listeners and visitors to POEX65, until the opening of the Scandinavian section of ZKM's *Sound Art* exhibition, no one had heard what that experiment sounded like. However, from two boxes donated to the Museum of Contemporary Art in Roskilde, some old tapes appeared that, apart from other things yet to be deciphered, revealed the two lost radio programs and a lot of other unidentified recordings from POEX65.[39] From these sources, it seems fair to establish as a fact that a program of performative events and concerts took place during the ten days that the event lasted.

A very important part of the POEX65 event was the large number of experiments with crossovers between musical art forms, from electronica to beat, to other art forms. This may also explain why the event never became part of official records or archives – and was erased from the DR broadcast archive: it was transdisciplinary before that word even existed.

At this point in the process of the exscription of the unheard avant-garde (via dissemination and communication), it is worth noting that a number of experimental aesthetic paradigms are at play on the platform that result in transformations of artistic practices on two levels, mainly.

1) POEX65 witnesses the emergence of that which has later been termed the noncochlear sonic art – the fundamental break with the phenomenological notion of the moment, and self-presence of things or sounds. Seth Kim-Cohen, in theoretical dialogue with the French philosopher Jean-François Lyotard, writes that "sound-in-itself is just as inconceivable as self-presence [...]. A non-cochlear sonic art moves beyond the territory of the ear, resisting sound-in-itself in much the same way that conceptualism in the visual arts resists Greenberg, opposing not the focus on materiality as the central issue but the very notion of a central issue."[40] At POEX65, visual art moves beyond the territory of the eye and sound art moves beyond the territory of the ear and so on. Hvidberg is clearly favoring and encouraging intermedia art practices, such as Eric Andersen's *Opus*, as well as media art practices like film poetry, "diamusic" consisting of slides and electronic music, and jukebox actions with records of sound art made for the occasion – everything in different levels of dialogue with the paintings and installations on site.

2) POEX65 is one of the first clear indications in Denmark – although it may not have been possible to see this at the time – of a technology and media consciousness taking hold in art practices, beyond the territory of technology, media, and art. POEX65 is all about what Clay Shirky, writing in 2009, terms "collaborative production" and "collective action." In defining these terms, Shirky points out that collaborative production is "where people have to coordinate with one another to get anything done [which] is considerably harder than simple sharing, but the results can be more profound. New tools allow large groups to collaborate, by taking advantage of nonfinancial motivations and by allowing for wildly differing levels of contribution."[41] Collective action, on the other hand, is "even more complex than collaborative production, but here again new tools give life to new forms of action. This in turn challenges existing institutions."[42] It is striking how close Shirky is to a description of what I believe was the actual drive of POEX65 where, intentionally, technologically and media-motivated practices are both sensual and collective – perhaps best exemplified in the experiments of the Flexowriter, and new work categories such as Punch Tape Poetry. POEX65 moves beyond the territory of poetry and technology, beyond the known territory of conceptual categories, and toward new domains. There is a strong sense of collective action in everything that is going on in POEX65. However, elements of collaborative production are also evident, especially in the performances – the live events, dance-paintings, sculpture-concerts, and electronic happenings at POEX65.

It should be noted, however, that it is exactly this somehow strange way of using technologically inspired ideas of production and action that makes it unrecognizable to humanistic research. As interdisciplinary as it may seem, performance studies do tend to focus on an ontology of performance, which Peggy Phelan has described thus: "Performance's only life is in the present. Performance cannot be saved, recorded, documented, or otherwise participate in the circulation of representation *of* representations: once it does so, it becomes something other than

36 Ibid.

37 Ibid.

38 Ibid.

39 Magnetic tapes (12-inch and 15-inch), as well as a 6-inch tape which previously had been used in a telephone answering machine; they are now restored in the DR archives.

40 Seth Kim-Cohen, *In the Blink of an Ear: Toward a Non-Cochlear Sonic Art* (New York: The Continuum International Publishing Group, 2009), 259.

41 Clay Shirky, *Here Comes Everybody: How Change Happens When People Come Together* (London: Penguin, 2009), 109.

42 Ibid., 143.

performance."[43] This ideal of the performance as essentially ephemeral and disappearing, and running contrary to any kind of archiving is strong within the area of performance research, although not unrivalled.[44]

However, it is the performance as "something other" that interests me here: I would argue that what Peggy Phelan also describes here, without intending it perhaps, at least not in the archive context, is exactly the ontological gap of the unheard avant-garde. This idealism of the *Augenblick* of performance, or the *Ohrenblick* ("blink of an ear") as Kim-Cohen would term it,[45] and the phenomenological theory behind it, is also partly guilty of an almost nonarchival approach to performance art strategies. However, this is only a small part of the whole picture – and an ontology of performance is not nearly as rare in archives as one may think. Names like John Cage, Yoko Ono, and Marina Abramović are good examples of the opposite. What is really rare to find in archives, however, are the practices of media art or transdisciplinary practices that work outside the territories of known ontologies, or in-between them, as we find in POEX65. The particular approach of Knud Hvidberg, as well as the focus on collective action in an almost post-digital sense, making the movements beyond traditional territories a question of an institutional and political critique, pushes POEX65 toward hidden matrices and obscure, private reservoirs of memory-fragments – fragments the *POEX* Platform, like the other platforms in the Unheard Avant-garde section, focus on exscribing (by unwriting them in sound and data visualizations) into a myriad of sound art histories.

The *POEX Tape*

The *POEX Tape* is at the center of the exscription practice of the POEX Platform, and moreover the end result of the work on the CHAOS app. The *POEX Tape* allows the users/audience to navigate the entire archive of uncovered sound art from the POEX65 event in Copenhagen.

The dissemination idea was to create a large-scale reel tape recorder customized for material related to the unheard avant-garde of POEX65. The installation consists of a tape onto which is projected audio material relating to the POEX65; the audience is in a position to listen to and search through all the sound art performed or played at POEX65 (which were recorded, at the time or later).

The interface, in effect, is the visual representation of the sound waves of the works in the archive as they are playing. Visitors are able to listen to the works in the archive in real time, but can also navigate it differently – allowing them to hear the sound of fast forward or rewind, or even listen vertically through all the "layers" of the archive, while browsing. In this way, the *POEX Tape* interface becomes an investigative instrument whereby the audience may discover the domains of the unheard avant-gardes in their own way – and have completely different listening experiences. The audience is playing their own domain-game, designing their own processes of listening.

Behind the canvas (of glass), which forms the tape, a series of projectors connected to a database via an open framework platform provides the visual part of *POEX Tape*. To be able to track the position of the listening stations and the "listening stethoscope" on the tape, these projectors are equipped with infrared light and a camera to detect the light, also positioned behind the canvas. The audio is visualized by the waveforms of the sounds and the tape is constantly moving, like in an actual reel-to-reel tape recorder. The listening stations consist of a set of headphones and a "stethoscope." When the listening funnel is placed onto an area of the tape, the sounds of this specific area will reach the audience member through the headphones.

The redesign and reconfiguration of the domains of experience and sound in the *POEX Tape* installation is based on the analysis that sound art is a multiplicity of genres and practices coming together in an experimental game of domains. Moreover, it assumes that future curatorial practices should build on transdisciplinary methodologies. I have named this methodology a domain-game, which conditions the language-games and conditions certain competences in that domain. Thus, the transdisciplinary methodology at play in the responsive interfaces of the *Audiobar* and *POEX Tape*, I would claim, are redesigning the way we listen to and use sound as experience – and, at the same time, interlayering this into a game of making unheard historical material and context present and active for an audience.

Locating the Unheard

With the *POEX Tape* you can be your own researcher – and learn how to search and identify the unheard avant-garde in (or, rather, outside) existing archives. Firstly, we looked for sound material from collaborative, performative, time-based intermedia art, media art, and new media art projects from the 1920s to the present day; we particularly looked in small and private archives and we triangulated what was found here – dates, persons, places, other artists – with the databases (or, in most cases, before 1983, written or printed records) of the Danish Broadcast Company, Statsbiblioteket [The Danish National Library], and the Museum of Contemporary Art in Roskilde – that one could also access via the *Audiobar* in the exhibition at ZKM. This archive encompasses art that is practicing "fusions of words, images, and sounds," often in different variations of performative, fluid, conceptual, and mediated situations.

The archival status is marked by varied and fugitive material, to say the least. One priority was to find adequate framings for the categorization and contextualization of the archive of unheard material, which in the "grand" archives proved inadequate on closer inspection and formulation. This certain affinity with the ear in the archive context further sparked the attempt to formulate a systematics and methodology of the dissemination of the unheard; this, in turn, enhanced the more ambitious attempt to implement a transdisciplinary practice and a sensuous and embodied technology into the core of the construction of the archive, allowing a dynamic flow of domains to establish themselves in a real-time dialogue with an audience. This attempt resulted in the *POEX Tape*.

The sensuous and embodied material of sound, and the modalities of using the ears in new ways – exploring sonic perception and sound as document – are the issues at play in *POEX Tape*.

43
Peggy Phelan, *Unmarked: The Politics of Performance* (London: Routledge, 1993), 146.

44
See Camilla Jalving, "Inventing reality: On truth and lies in the work of Hayley Newman," in *Performative Realism*, ed. Anne Jerslev and Rune Gade (Copenhagen: Museum Tusculanum Press, 2005), 145–80.

45
See Kim-Cohen, *In the Blink of an Ear.*

The Content of the *POEX Tape*

The content of *POEX Tape* are sound art pieces or sound documents from the POEX65 event. They constitute five main layers of experiments with sound – thus the interface is structured into five streams, represented (visually) as sound waves, all present at the event in December 1965: experiments beyond jazz, electronics, art (Fluxus), poetry, and beat. To structure this material and to create metadata, we used the CHAOS app. The CHAOS app is actively used to organize (and maintain) the sound files and metadata in the *POEX Tape* – making it an interface to the first archive of unheard avant-garde material. The metadating and structuring of the sound material is de-facto the visual output of the interface – and the textual pop-up with information about tracks and artists etc. that the interactive listening creates on the screen is also controlled by the app.

POEX Tape is exscribing audio works, interviews, and radio broadcasts from POEX65, as explained above. In the POEX Tape interface the hidden layers of cross-genre experimental practice and artistic agency at POEX65 become visible as five streams of archival sound:

Experiments beyond electronics
Experiments beyond jazz
Experiments beyond poetry
Experiments beyond art (Fluxus and intermedia art)
Beat crossover "Revolution of the Self Now"

In the following, I will describe what constituted each stream in more detail.

Experiments beyond Electronics

Electronic composition and sound was new to the Danish art scene in 1965, but it was not in its infancy. Even though Denmark did not have an institution like EMS in Sweden, composers such as Else Marie Pade and Jørgen Plaetner, who were both inspired by Pierre Schaeffer and *musique concrète*, as well as by Stockhausen (who also performed at POEX65 according to the program but no documentation of that performance has emerged at this time), were experimenting with the possibilities in unique ways. Else Marie Pade, who was the only Danish (and only female) member of the Schaeffer group, live-premiered her seminal work, *Syv Cirkler* [Seven Circles]. Originally produced in the studio of the Danish Radio in 1958 after a visit to a planetarium at the World's Fair in Brussels the same year, it found a new audience and purpose at POEX65 when it was performed live for the first time. Based on Dante and on Pade's experience of being interred in a concentration camp during World War II, the piece takes the listener through these seven circles of sound loops, the composition descends into the pits (or gaps) of the (psychologically and materially) real with an element of cruelty always lurking behind the ephemeral beauty of electronic sounds. The recording used in the *POEX Tape* was on the unlabeled reel tape in Knud Hvidberg's personal archive.

Else Marie Pade (1924–2016) is one of the pioneers of electronic music in Denmark. From the beginning of the 1950s, she, in close cooperation with technicians and assistants on Radio Denmark, produced a substantial amount of concrete and electronic music, partly in the shape of independent works for radio broadcasts, partly in the shape of accompaniments to various radio dramas. So far, the music has existed only on tapes in the archives of the Danish Radio.

Jørgen Plaetner (1930–2002) studied composition under Vagn Holmboe, and attended contemporary music courses at Darmstadt in 1950, where he encountered the experimental practices of Stockhausen. Despite living in Sweden from 1977 until his death in 2002, he became known as one of Denmark's foremost electronic composers.

Also on the reel tape was a live-recording of Jørgen Plaetner performing together with the poet Poul Borum and visual artist Helge Ernst. This piece, simply called *Elektronisk Musik* [Electronic Music][46] in the program (there were many errors in the original program and not all were corrected), was performed live for the first time at POEX65. No other known recording, previous or later, of this particular piece by Jørgen Plaetner is known to exist.

Two other live-electronic pieces by Plaetner, *Continuo* and *8 Pieces*, were premiered at POEX65. They both feature a performative and collaborative element that is emblematic of all POEX65 productions, including spoken words and live visuals – in the latter case by Knud Hvidberg.[47] The live electronics at POEX65 inspired several later activities that sought to get electronic music out of the studio and into other "social" spaces.

Syv Cirkler by Else Marie Pade and *Elektronisk Musik* by Jørgen Plaetner both served as material for two POEX Enacts. The dialogue between contemporary media artists and the sound artists from POEX65 was a key element in the curatorial dissemination strategy – writing and unwriting, mixing and remixing the unheard avant-garde. Therefore I commissioned two contemporary media artists, HC Gilje and Ane Mette Ruge, to enter into a dialogue with these two recordings. The results of this dialogue created two very strong visual and sonic artistic expressions in the Scandinavian section of ZKM's *Sound Art* exhibition.

HC Gilje (*1969) works with installations, live performance, set design, and single-channel video. He graduated from Kunstakademiet in Trondheim in 1999. He is a former member of the video-improv trio 242.pilots and was also the visual motor of the dance company Kreutzerkompani. Apart from various screenings, performances, and collaborations, Gilje's works have been exhibited at Woodstreet Galleries, Pittsburgh, 2012; IMAL, Brussels, 2011; and the Museum of Contemporary Art, Roskilde, 2010. HC Gilje lives in Oslo (NO).

Ane Mette Ruge (*1955) is an artist. Her work is interdisciplinary and confronts and combines ephemeral media like sound, video, photography, light, and text with lasting materials such as ceramics and concrete. She has had several exhibitions, including at Den Frie Udstillingsbygning, Copenhagen, 2011; Skive New Art Centre, Skive, 2009; A-Trans Pavilion, Berlin, 2008, and Charlottenborg, Copenhagen, 2004. Ane Mette Ruge lives in Copenhagen (DK).

Brief Excursion 1: POEX Enact – *Syv Cirkler* by Else Marie Pade / HC Gilje

Syv Cirkler [Seven Circles] is the first piece of electronic music that Else Marie Pade composed in 1958 after a visit to a planetarium at the World's Fair in Brussels the same year. The piece is a composed constellation in sound, based on seven layers of sine-wave-generated layers of sound, building up from one layer to all seven layers then winding down to the first layer again. Each layer or circle is double the speed of the previous one.

Similar to the structure of Else Marie Pade's *Syv Cirkler*, HC Gilje also uses seven circles in his light installation based on Pade's composition.(FIGS. 6, 7) He creates circles of moving light patterns, where each circle of light corresponds to a layer of sound in the composition.

The circles are stacked on top of each other and mounted in the ceiling of the exhibition space. They manifest themselves as an actual source of light, from the inside of the circles, and as reflections on the floor and as shadows cast on the walls, giving a sensation of rotation in the whole space and in the experience of the piece.

In *PLAeTnEr*,(FIG. 8) Ane Mette Ruge remixes Jørgen Plaetner's POEX65 live performance *Elektronisk Musik*. *Elektronisk Musik* consists of a sequence of loops and amorphous

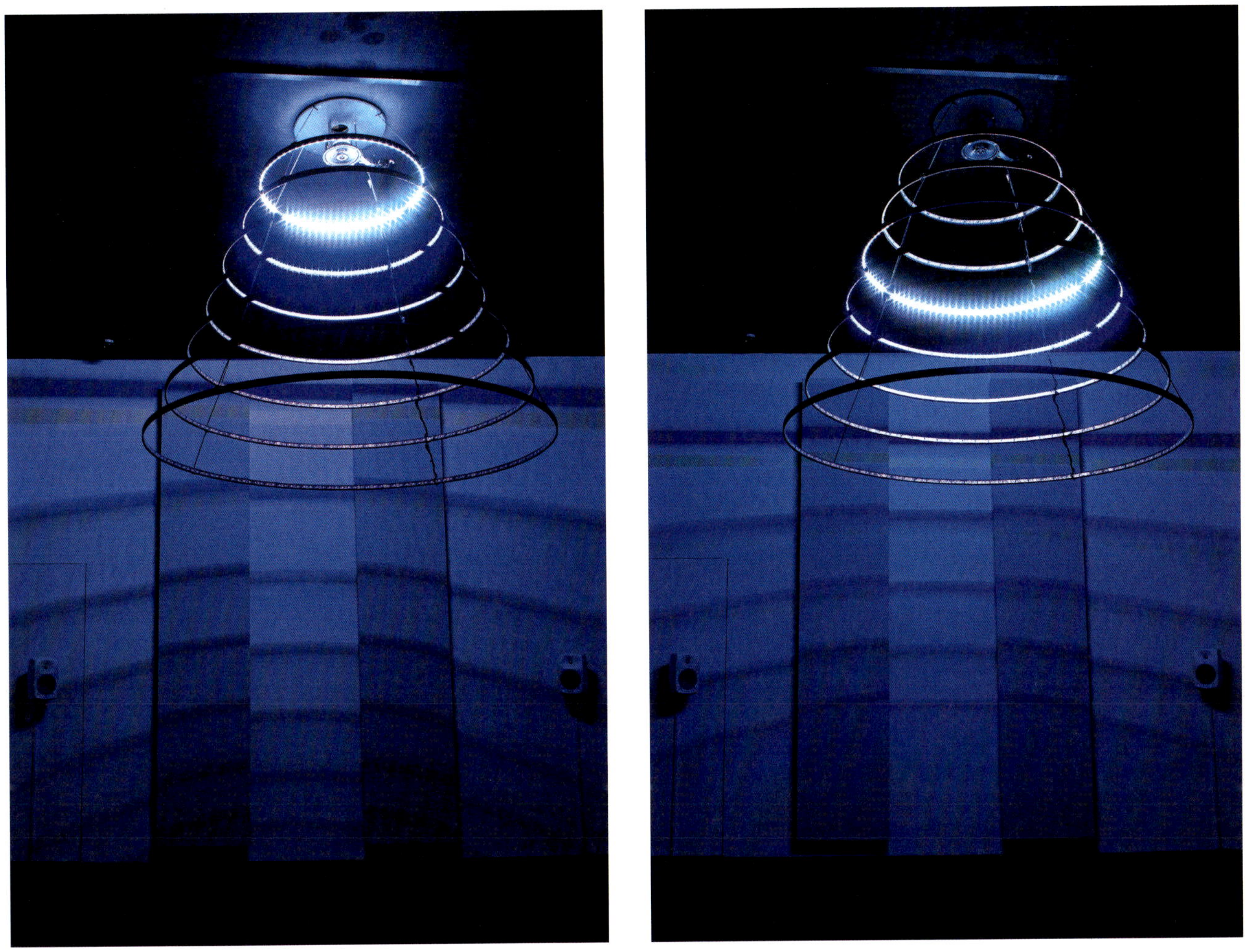

FIGS. 6, 7 HC Gilje on Else Marie Pade, *Syv circler* [Seven Circles], 1958/2012. Light installation and sound composition, metal rings, LEDs, microcontrollers, computer, speakers. Installation views ZKM | Karlsruhe, 2012.

audio-objects. As a counterpoint, the remix adds an animated video sequence with a synchronized soundtrack created by Ruge to the original Plaetner audio piece. Ruge's video and soundtrack add a poetic narrative to Plaetner's composition, while also reinforcing it by establishing moments and sequences of synchronicity between the elements in play. The video projection on the large circular parabolic dish shows a delicate mess of objects in slow progression. The objects are out of mutual scale and devoid of function, and as such beyond general semantic categorization. Being mapped onto the parabolic dish, the animation may be seen as an allusion to the process of exchanging information over a long distance, such as a time span of forty-seven years.

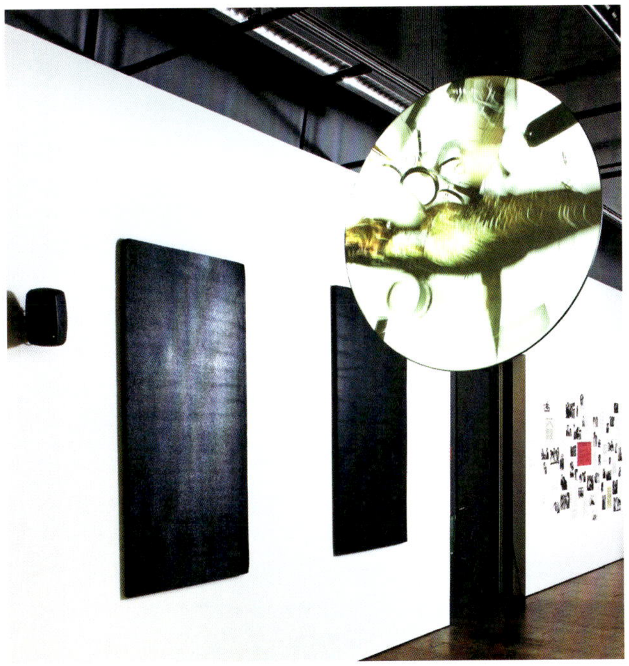

FIG. 8 Ane Mette Ruge on Jørgen Plaetner, *PLAeTnEr*, 2012. Sound installation based on a live recording of the POEX65 event. Projection on old parabolic antenna, mono and stereo audio. Installation view ZKM | Karlsruhe, 2012.

46 *Elektronisk Musik* (1965): Jørgen Plaetner: composer, performer; Poul Borum: poet, performer (voice); Helge Ernst: visual artist, performer (visuals).

47 *Continuo* (1965): Jørgen Plaetner: composer, performer; Bent Irve: performer (spoken words); Børge Sornum: visual artist (live visuals). *8 Pieces* (1965): Jørgen Plaetner: composer, performer; Knud Hvidberg: performer, visuals.

Experiments beyond Jazz

Because of the impact of the beat movement (which was somewhat delayed and stretched out, at least in Denmark), avant-garde and experimental jazz were still important elements on the art scene in 1965, and even though the impact of jazz on art practice would wane considerably in only a few years, its influence on performative and improvisation elements in artistic production cannot be overlooked (or diminished).

Moreover, as it turns out, the upcoming beat and more political rock and fusion experiments in music were very much founded by the same musicians and composers (at least, in Denmark) as those that were active in the avant-garde jazz environment.[48]

At POEX65, the borders of collaboration and exploration between explorative improvisation and art were pushed even further. The jazz-painting event called *Spirals* is a good example of this.[49] From what we can hear in the recording, the Brothers Vogel Quartet and Per Aage Brandt Trio are simultaneously playing a repetitive spiraling piece (like a "canon"), prompting the Danish conceptual artist (and professor at the Art Academy in Copenhagen) Gunnar Aagaard Andersen to cut spirals of paper. This concretist "translation" of sonic material into visual material and form is very typical of POEX65.

Also typical, but on the level of practicing exscription as curatorial method, is a recording from an unlabeled tape found in Knud Hvidberg's personal archive. It most likely originates from *Jazz News*, which was a show on Danish Radio Program 2, airing every Sunday evening in the mid-1960s. This particular show, from December 13, 1965, does not exist in the Danish Radio Archive anymore – and we only have this short bit of sound recording and live reporting from Knud Hvidberg's tape reel.

The pieces *FTSS*[50] and *Sinistra*[51] (playfully referring to acid jazz [and drugs] and deconstructing old-school jazz icons like Frank Sinatra) were performed by LUX Gruppen, composed by visual artist Paul Lindegaard and the Brothers Vogel Quartet. Both pieces included live jazz and visuals. No recording documents the live visuals, so it is only Karsten Vogel's account that may give us a sense of the action and artistic ideas. The recordings in the *POEX Tape* are, according to Karsten Vogel, the only surviving live recordings of the exact same quartet that played at POEX65. The recordings were lost for many years due to a theft from Karsten Vogel's cellar – and only just recovered in 2011 from a tape-recording of a tape-recording of a tape-recording.

As another example for such experiments beyond jazz, the performances by Freddy Fræk could be mentioned. Fræk was a Danish sound artist, satiric musician, and composer who in the 1960s and 1970s performed on television and was quite popular with his skewed and "a little dirty" (the literal meaning of "Fræk" in Danish) portraits of bourgeois Danish (non)sexuality. For the *POEX Tape*, two pieces were recovered from Knud Hvidberg's tape – both are documentations of live performances at POEX65: *Digte for en diller* [Poems for a Penis] and *Dulle Poesi* [Bimbo Poetry]. The skiffle-like style of playing (mandolin) and singing combines a postjazzy feeling with creative and subtly affronting lyrics that could also be seen as an experiment beyond poetry.

48 Anecdotal evidence of this, apart from Karsten Vogel, would be Franz Beckerlee, who performed in The Contemporary Jazz Quartet at POEX65. Then a saxophonist, but originally educated as a painter at the Royal Danish Art Academy, he was to be the iconic lead guitar player in the rock band Gasolin'.

49 *Spirals* (1965): Gunnar Aagaard Andersen: performer, visuals (spirals of paper); Karsten Vogel Quartet: avant-garde jazz; Per Aage Brandt Trio: avant-garde jazz.

50 *FTSS* (1965): Brothers Vogel Quartet, constellation 2: Karsten Vogel: composer and performer, alto saxophone; Peter Ingemann: performer, bass; Henrik Vogel: performer, drums; Martin Dam: performer, trumpet; Paul Lindegaard: performer, textuals. This piece was performed at 4 p.m. on Saturday, December 18, 1965, at POEX65. No recording of this performance has been recovered. However, a live recording with the same line-up from Strandvejen 83A from June 25, 1966, was generously made available by Karsten Vogel.

51 *Sinistra* (1965): Brothers Vogel Quartet, constellation 1: Karsten Vogel: composer, performer, alto saxophone; Per Aage Brandt: performer, piano; Søren Gorm Kristensen: performer, bass; Henrik Vogel: performer, drums. Recorded live at Radiohuset, February 6, 1964.

Freddy Fræk (1935–2016) was a composer, actor, musician, and sound artist. Originally educated as an ironsmith, Fræk was active as a musician and performer in Denmark from early 1960, mainly in television and film. From the early 1980s, Fræk concentrated on sound sculpture. His work in sculpture led him to be included in the Galerie Denise René, Paris, in 1989.

Experiments beyond Poetry

POEX, "Poetry Experiment," the title of the event, can be interpreted in many ways. But in all interpretations, poetry plays an important role, in one way or another. There is no doubt, however, that it is a concrete, visual, and experimental (non-capitalist) poetry, sound poetry, or media poetry, that informs the conceptual and practical setup of POEX65. Poetry is another word for crossing over and for working across domains, genres, and disciplines. And for involving the audience into

52
Look Down! (1965/2012): Created by Hans Sandmand; restaged by Mogens Jacobsen and Morten Søndergaard. Reactive sound installation. Car tires, mirror, sensor, light, built-in-computer, and MP3 player, recorded voice.

53
Radar (1965/2012): Morten Søndergaard: content, sound and research; Mogens Jacobsen, Martin Luckman, and Morten Søndergaard: conceptual design. Reactive sound installation. Modified radar, motor, sensor, built-in-computer, and MP3 players, recorded voice, digitized photos.

the production of art. Together with Danish visual poet Vagn Steen and visual artist Sven Dalsgaard, Knud Hvidberg worked on making the "poetic element" of POEX65 as strong and vibrant as possible. A poetry café and "Punch Card Poetry," as well as the "Record Bar" and a poetry room with "exploded books," were all a testament to the importance of the visuality and mediality of poetry at POEX65.

Hans Sandmand (1910–1983) was originally educated as an electrician and served as a conductor on the Copenhagen Railways in the 1940s and early 1950s. A serious disease forced him to find another path, and he first became the owner of an elecric supply shop, and then as an artist around the late 1950. Together with William Soya, he pioneered interactive sculpture in Denmark. Very little is written about Hans Sandmand's work and life.

William Soya (1927–1994) began his career in the late 1940s as a naturalistic landscape painter, but from the late 1950s he began painting "compositions" and making sculptures, which were clearly influenced by Constructivism. From around that time, he also pioneered the use of sound and light as artistic materials in Denmark. Together with Knud Hvidberg, he attempted to cross borders, in philosophically, politically charged, artistic expressions. Together, they experimented with sound and light, and "radio sculptures." Soya saw technological advances as anecdotal statements about our time and he wanted to demonstrate that the technology, though created for use by the military, was itself neutral. He later also pioneered the use of the computer as an artistic medium in Denmark. From around 1970, he was experimenting with digital expressions, creating black-and-white computer graphics and using the computer screen for showcasing constantly changing, colorful images that illustrated the countless possibilities for creating non-figurative art.

Sven Dalsgaard (1914–1999) experimented with expression and materials and, throughout his long career, drew inspiration from different artistic movements, including Surrealism, Dadaism, and finally Conceptual Art. His conceptual works often had a quirky or humorous aspect.

What gives a brief glimpse of the ironic and potent poetic style of Sven Dalsgaard, who in the 1960s, among other things, was grappling with conceptual issues in a "postreligious" time, is the *Markus Evangeliet* [Gospel of Mark], a tape-recording directly from one of the records in the Record Bar at POEX65. This is the only sound work known to exist by Dalsgaard.

One of the very central instances of experiments beyond poetry at POEX65 is the artist Hans Sandmand. Sandmand is one of the pioneers of interactive art in Denmark (who is largely overlooked). In the 1950s and 1960s he produced a large number of mechanic or kinetic pieces, whereof some are operating with sound. All of them were meant to be touched and used, which is the reason for his unheard status: his works were worn down and broken from many years of use. Also, he sought out public places like libraries or train stations for his works, and not museums. Knud Hvidberg, William Soya, and Hans Sandmand worked together on a number of occasions and made a number of technology-based interventions and projects between 1957 and 1965 – and, even though William Soya did not participate in person at POEX65, his influence is evident.

For POEX65, Sandmand created a number of interactive sound poetic sculptures. For the 2012 ZKM exhibition, we decided to do a remake, or more exactly, to reenact these sculptures using present-day technology: *Poesien er noget du har I dig selv* [Poetry is something you carry within yourself] became *Kig ned!* [Look Down!] – an homage to Knud Hvidberg and Hans Sandmand, and our third *POEX* React.(FIGS. 9, 10) In the *POEX Tape* we hear a live recording from POEX65 documenting the original voice (in Danish) from the original sound installation.

Visitors walk up to a stack of tires with a sign asking them to "Kig ned" [look down]. When they look down, two things happen: a light turns on and the visitors see themselves in a mirror at the bottom of the tires; and a voice says: "Poetry is something you carry within yourself."

The *Look Down!* React[52] is a remake of an interactive sound installation by Knud Hvidberg and Hans Sandmand that was shown at POEX65. We have had the conceptual idea written down (by Sandmand) and a short description by Knud Hvidberg from a radio interview in 1965 where the original (Danish) voice from the installation is also heard in the background. This original voice was included in the *POEX Tape* installation (in the "experiments beyond poetry" stream).

Radar,[53] another piece by Sandmand using a turning radar device, was located in the central hall of the exhibition, right where people entered the space. The radar "saw" visitors walking into the semi-confined space and registered their presence. Then two things happened: a voice repeated in Danish: "jeg venter på den store intuition" [I am waiting for the great intuition] and the

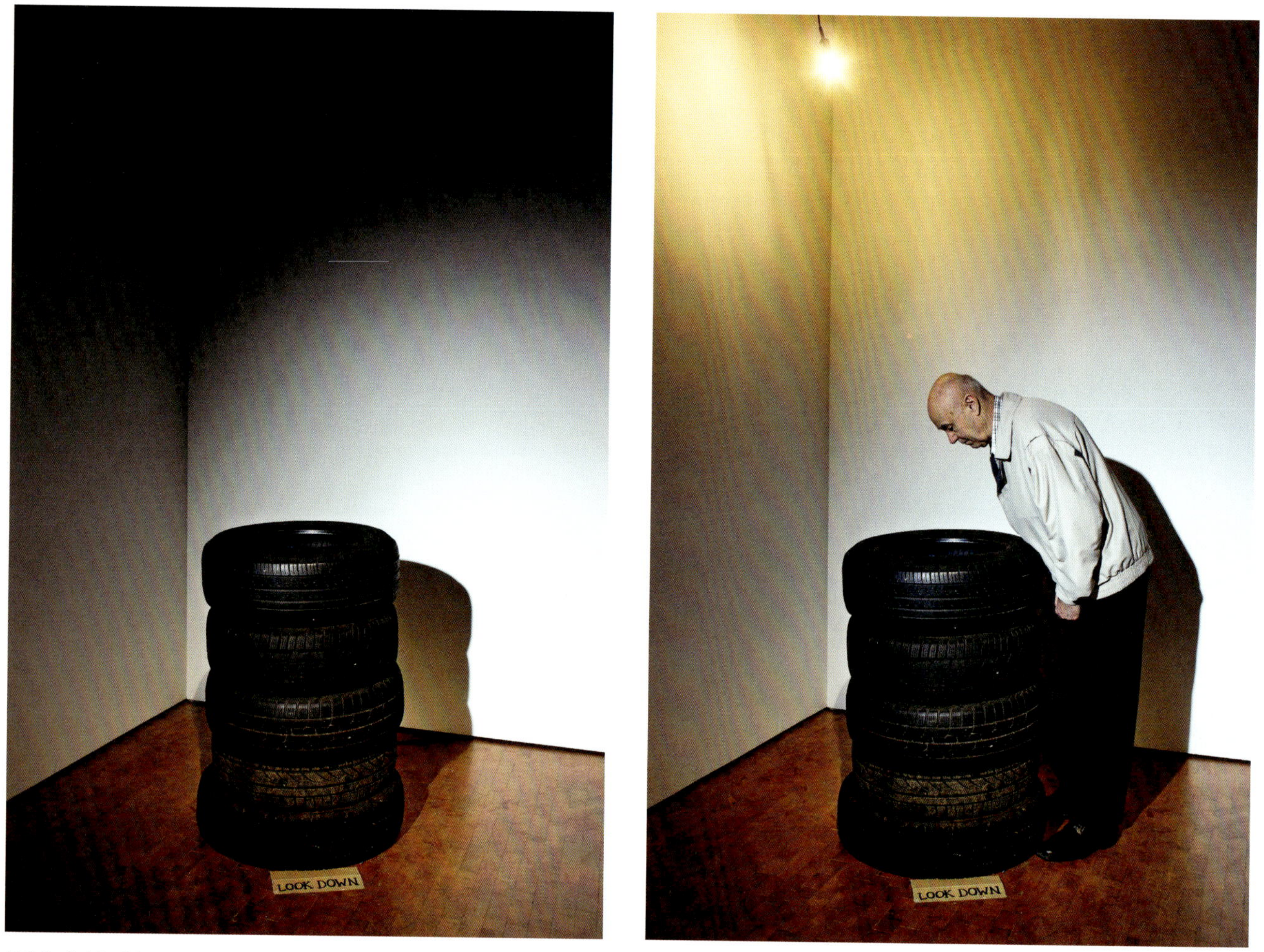

FIGS. 9, 10 Mogens Jacobsen, Morten Søndergaard, *Look down! [Kig ned!]*, 2012. Homage to Hans Sandmand and Knud Hvidberg. Interactive sound installation based on a no longer existent sound installation by Hans Sandmand and Knud Hvidberg from 1965, tires, mirrors, sensors, light, computer, MP3 player, voice recording. Installation views ZKM | Karlsruhe, 2012.

images on the three walls changed. Sandmand's conceptual idea as well as a short description of its realization can be found in a letter, and is also explained by Knud Hvidberg himself in a radio interview from 1965.

For the Unheard Avant-garde section at the ZKM exhibition *Sound Art – Sound as Medium of Art*, the original ideas from these two pieces by Sandmand were reactivated and altered slightly, in a process of writing and unwriting the vision of Knud Hvidberg (involving the non-artists and the audience into art production) into a 2012 context.(FIG. 11) So, in the words of Knud Hvidberg, they became metaphors of POEX65. Thus, in 1965 Hvidberg wished to "pick up" the invisible movements as well as the audience involvement in art production. POEX65 produced art; it did not just "show" it. Therefore, the great intuition became a kind of "landmark" for all the experimental projects "at sea" during the event.

The images are all from the private album of Knud Hvidberg – and they show the opening, the events and concerts, the rehearsing and setting up; the actors, artists, and audiences at POEX65. Together with the letters and the sound tape, and participating artists still alive, this is all that is left of POEX65 – fragments and shadows of experimental sound art practices fifty years ago.

Time and chance would send William Soya, Hans Sandmand, and Knud Hvidberg – despite their visionary and innovative use of technology and sound as media in visual arts at a very early stage (for Denmark as well as internationally) – more or less into obscurity. Lost in translation … from performative and moving media and sound to text and (still) images; from one ontology to another.

As an example of some of the crossover practices that materialized after POEX65 and were directly inspired by it, one could mention

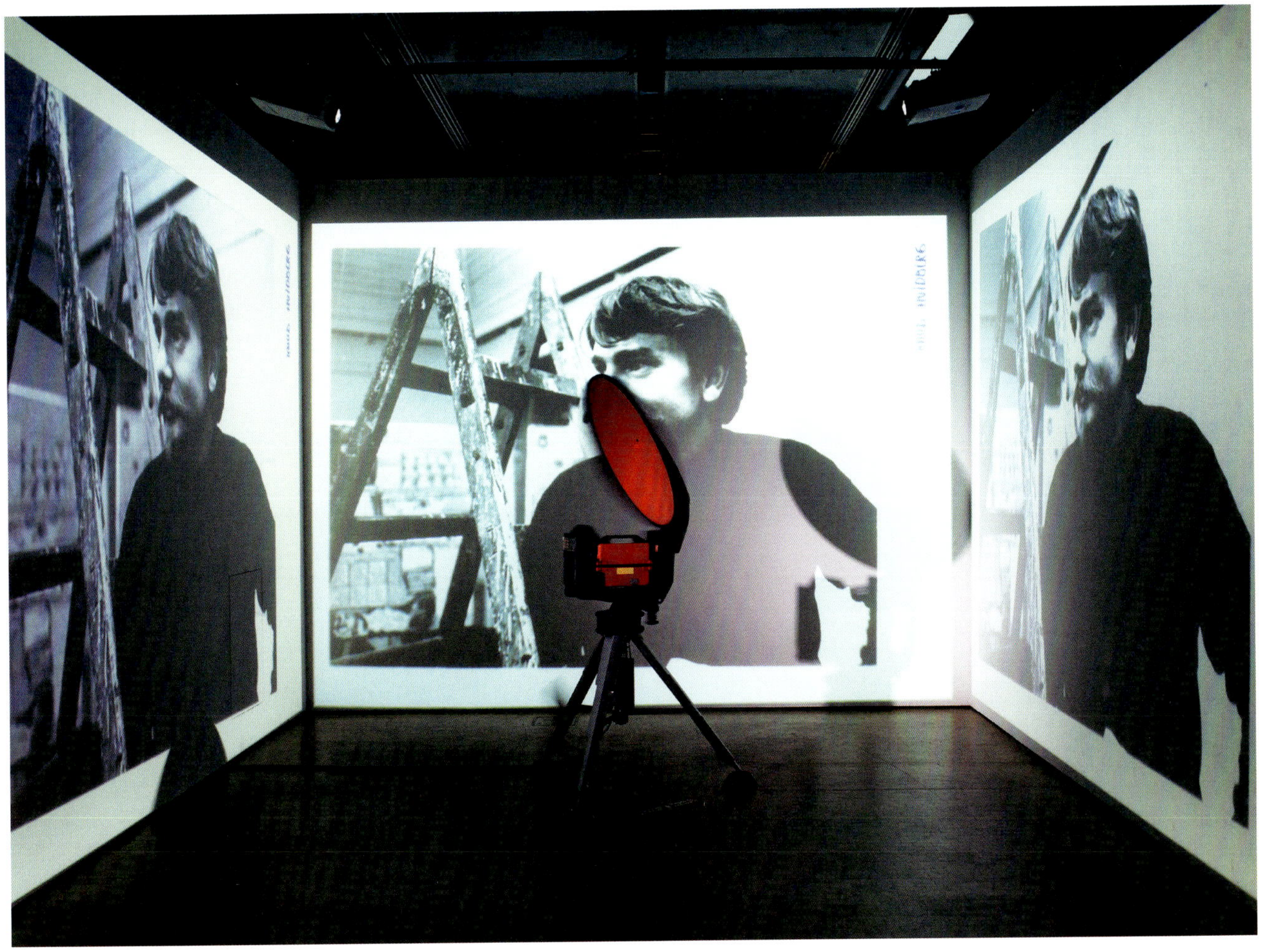

FIG. 11 Mogens Jacobsen, Morten Søndergaard, *POEX Radar*, 2012. Homage to Hans Sandmand and Knud Hvidberg. Interactive sound installation based on a no longer existent sound installation by Hans Sandmand and Knud Hvidberg from 1965, modified radar, motor, sensor, computer, MP3 player, voice recording, digitalized photographs. Installation view ZKM | Karlsruhe, 2012.

Omringning [Surrounding] by the Danish media poet Per Højholt. From 1966, Højholt purposely sought out collaborations with other artistic genres - and ventured to be a "media-aware artist." *Omringning*, first performed in 1969, is a sound document from a live performance in which the artists (Per Højholt, Gunner Møller Pedersen, and Poul Ib Henriksen), using the alphabet and recorded electronic sound in five channels, encircle (surround) the audience, who in turn were placed on chairs in the middle of the space. The piece is about circulation and movement, of ideas, words, materials - and about making the audience participate in a change of mindset.

Also a main source of inspiration for Knud Hvidberg was Åke Hodell. With his Pistol Teatern [Gun Theater] in Stockholm and activities on the very political part of the Swedish avant-garde scene, he was one of the most influential

Per Højholt (1923–2004) was a Danish visual and concrete poet, who from around 1966 began working as a media poet. He experimented with and innovated the radio-montage genre in Denmark. He also made media performances, with strong elements of glitch and noise, inspired by sources as varied as Jimi Hendrix and Åke Hodell. Influenced by Jacques Derrida, his writing style has strong philosophical and self-explorative overtones, with a notable touch of irony and very dry humor. Even if he often wrote in a complex and obscure style, he was constantly looking for more popular formats. His most widely popular experiment was *Gitte monologer* [the "Gitte Monologue" (first as a radio piece, later as a live performance), in which Per Højholt impersonates the suburban (very Danish) woman Gitte in dialogue with her (equally very suburban and Danish) friend Susanne. In the 1990s, Per Højholt's *Praxis* series was partially translated into German by Hans Magnus Enzensberger.

Åke Hodell (1923–2000) was an experimental Swedish composer who began working as an artist in the early 1960s and was still composing work up until his death in 2000. He produced many works in text-sound composition. Text-sound composition is the collage of field and locational recordings, with an emphasis on using the human voice and its many transformations to tell narratives not dissimilar to radio drama or *musique concrète*. His often aggressive, polemic, and somewhat sarcastic work using voice collages could be defined in the realm of electro-acoustic or avant-gardism, but is more closely associated with the ethics and aesthetics of the international conceptual artists known as Fluxus.

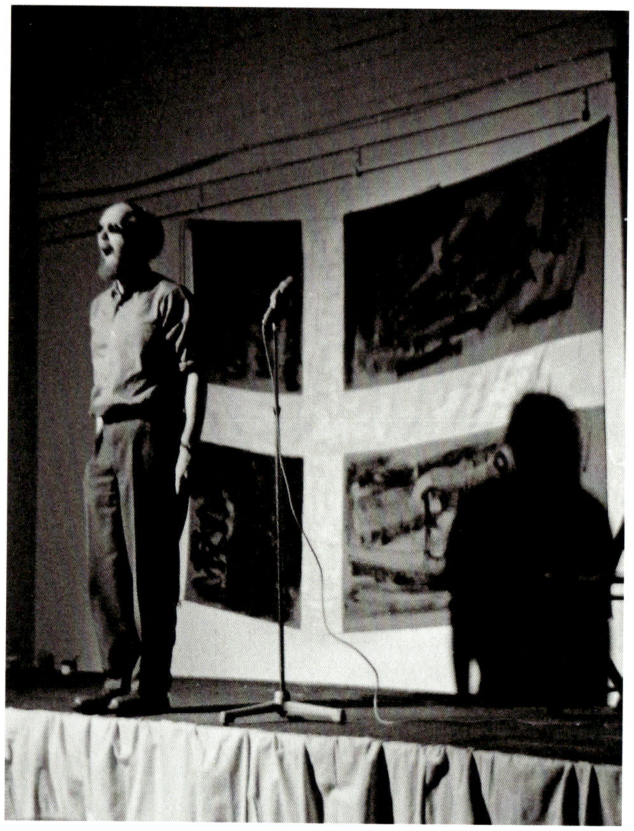
FIG. 12 Åke Hodell, *Large Exhibition Room*, happening.

Scandinavian performative sound artists of his time, and in 1965 at his peak (and influencing the entire experimental scene, including Per Højholt).

Two sound poetry pieces are included in the *POEX Tape*, representing Hodell's impact: *I Gevar* [Present Arms] and *General Bussig* [The Cozy General]. Formally known as part of the "verbal brainwash" that Åke Hodell wanted to impose upon his audience, both pieces are experiments beyond the boundaries of poetry.

Hodell also performed live at POEX65, where he repainted a Danish flag from red and white into the Swedish colors, blue and yellow.(FIG. 12) *I Gevar* was enacted on the *POEX* Platform by Danish visual poet Christian Yde Frostholm. In the enactment, which is an animated text-sound composition, Frostholm makes visual and concrete what is happening in *I Gevar*. The manipulated stretch of the *iiiiii…* in the composition gradually fills the screen, or invades it.

Christian Yde Frostholm, or "CYF" (*1963) is a Danish writer and visual artist who was a member and cocreator of Afsnit P (www.afsnitp.dk) – the online exhibition space based in Copenhagen which existed from 1999 to 2009. Apart from writing and being active as a visual artist, CYF has been an active blogger for many years, in more private as well as professional contexts.

Experiments beyond Art – Fluxus and Intermedia

This track of experiments in the exscription of art itself could also be called the "Eric Andersen track," because at POEX65 he was the main (only) exponent of the art practice (including sound), otherwise known as Fluxus and Intermedia.

In 1962, Andersen cofounded the international network Fluxus and took part in the second concert ever given by Fluxus held during the Festum Fluxorum in the Nikolaj Kirke [Nicholas Church] in Copenhagen. Before then, he was a pioneer of intermedia art. In his *Opus* works from the early 1960s, Andersen explored the open interaction between performer and public, developing open self-transforming works, such as "arte strumentale." Andersen's performances depend very much on the public. This is true of not only his performance works but also his installations, to which the public may be prompted to contribute. Between 1962 to 1966, he and Arthur Køpcke worked closely together.

Eric Andersen (*1940) is one of the artists who (from 1959 on) took an early interest in intermedial art. As a cofounder of Fluxus, he took part in numerous Fluxus festivals and actions. In the late 1960s, he turned to mail art, then in the 1970s was keenly concerned with geographical space. His most eminent works include *Hidden Paintings*, *Crying Spaces*, *Confession Kitchens*, *Lawns that turn towards the Sun*, and *Artificial Stars*.

At POEX65, Eric Andersen participated with several works, performing in the field between and especially beyond music and art. These four pieces, *Composition No 1*, *Composition No 2*, *Composition No 3*, and *Opus 51*, were originally designated to be played in the Record Bar at POEX65, but it is uncertain if the Record Bar ever happened due to technical problems. The records for the bar were made, yet it has not been possible to trace any of them, anywhere. The recording in the *POEX Tape* was found on an unlabeled tape in Knud Hvidberg's personal archive (a cardboard box labeled "miscellaneous").

In *Composition No 3*, Andersen is repeating the sentence "I regret the bad circumstances for recording" using different filters and technical manipulations to alter the voice by recording it on top of previous recordings (and ironically pointing out what was intended but not wanted by "normal" listeners: to distort and explore the recording space). In some ways, the sound recording experiments by Eric Andersen could be seen as a prequel to the much-celebrated piece by Alvin Lucier from 1969, *I'm Sitting in a Room*.

Eric Andersen was also present outside the *POEX Tape*, notably with the invisible film and soundtrack, *Opus 65*, which was commissioned for the Unheard Avant-garde section at the ZKM in 2012. *Opus 65* is an artistic/curatorial exscription of *Opus 51* – a performative event operating within the intermedia methodology and aesthetics at POEX65. Photographs exist showing Andersen sitting in the "central hall," waiting for the audience to arrive. This event made the audience actively collaborate in realizing the work by running through a combination of operations based on rules defined by Andersen. *Opus 65* is both old and new. It is based on a 16 mm film that was made for POEX65 and never used since. This film was made out of leftover clippings from other films by him that were over- or under-exposed; other parts were simple recordings of a row of windows at one of the floors of a hotel in Copenhagen, with the camera scanning back and forth – repetitively. The whole film was edited so equal amounts of light and darkness were presented. It served as backdrop for a live event in which Andersen had asked the audience to bring certain kinds of food, and while he read aloud the timetable of the S-bahn trains in Copenhagen, a series of actions and exchanges (and eating) of the food were carried out.

For the *POEX* Platform at the 2012 exhibition, Eric Andersen created a new soundtrack for the (now digitized and restored) original 16 mm film, and the two elements combined emerged as: *Opus 65*.

Experiments beyond Popular Music: Beat Crossover "Revolution of the Self Now"[54] According to Karsten Vogel, who participated intensively at POEX65, these recordings from the Berliner Philharmonie almost four years later are as close as you can get to the effect and energy present at POEX65. The people who founded the group Burnin Red Ivanhoe a short year later met each other and got the idea and impetus at POEX65. It expresses the energy during the late nights in December 1965. At the time, the Danish Media Council had forbidden the playing of beat music in public spaces (and rock or "barbed wire," as it was called in Denmark), so it is impossible to find any recordings of beat in the period from 1962 until the early 1970s. This makes the experiment that beat was – in social life forms and within musical genres – totally unheard (of) in its early stages. Almost nothing can be found documenting experimental beat and pop music in Denmark from this early period.

54
Recording from live performance at the Berliner Philharmonie (1969): Karsten Vogel: composer, sax, keyboard; Kim Menzer, Steen Claësson, Jess Stæhr, Bo Thrige Andersen: performer; John Tchicai: guest performer.

Burnin Red Ivanhoe is a Danish rock band that was formed by Karsten Vogel in 1967 in Copenhagen, after having gotten the inspiration in 1965 during the POEX65 event. The band was one of the first Scandinavian bands that gained international attention and success. On April 10, 1972, the band was dissolved and Karsten Vogel formed the purely instrumental group Secret Oyster. Burnin Red Ivanhoe has come together again several times since then. The band plays a fusion of free-form jazz, blues, and hard rock and has done both concise melodic songs and instrumental and complex compositions of longer duration.

The crossover experiment between avant-garde jazz and everything else from contemporary sound (music and nonmusic), including rock and the spoken word tradition is unique, as witnessed in *Saxophone Piece*, *Ivanhoe I Brondbyerne*, and *Medardus*.

Exscribing Practices III: Indexing and Framing Knowledge Construction

POEX Tape functions as an index of the *POEX* Platform. It is an investigation and documentation of the methodologies needed in order to disseminate (as part of practicing exscription) the domains of the unheard avant-gardes: What are the categories? How do we describe them? Do they manifest themselves into (new) domains? And what would be best practice for metadating and documenting those domains? Furthermore, these questions also point toward more fundamental questions about the experience of art in a mediated cultural context and environment.

The curatorial methodology operating behind *POEX Tape*, the *POEX* Platform, and the other platforms was conceptually structured around the idea and practice of exscribing, and the notion of reading and writing the hidden matrices of the deferred archive of artistic experiments in other places (such as installations and interfaces). The design of the responsive

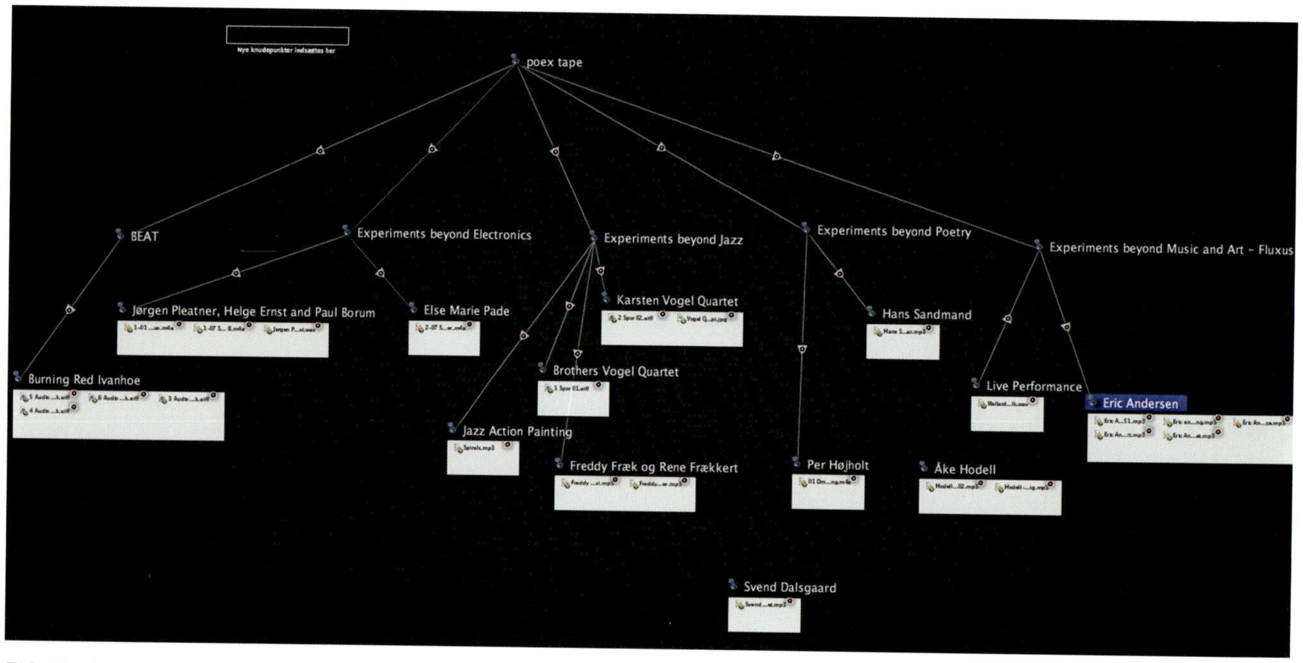

FIG. 13 Screenshot from the POEX App – the structuring of the content in the *POEX Tape* responsive database.

sound interfaces begins in the curatorial assumption that we may gain access to the hidden matrices of the unheard avant-garde by systematically delaying or replaying the sonic or visual material in no particular predefined order. In short, the idea was to find ways for the audience to gain auditory access to the hidden matrices, rather than to a particular "story" about what happened, or a certain aesthetic discourse.(FIG. 13)

In the process of curating, this assumption gradually grows into other areas of knowledge construction, primarily by framing concretely transdisciplinary exchanges between the semantics of reading and the syntax of listening.

In March 1965, eleven artists met at a hotel in Copenhagen to discuss POEX65. They had all a few weeks earlier been invited by Knud Hvidberg to participate in this event where the experiment was to be the center of attention – and there should be no differences between the various art genres. No one genre was favored – the idea that Hvidberg expressed was that the spaces should be free and collectively "owned" for the ten days the event took place. Everyone was asked to cross genres.

The Eks-skolen (Experimental Art School) in Copenhagen called the meeting because its representatives (Peter Louis Jensen in particular) were not convinced by nor pleased with Hvidberg's framing. Peter Louis Jensen from Eks-skolen and Knud Hvidberg met with a number of other prominent Danish artists at the time: the Danish visual artists Poul Gernes, Per Kirkeby, Per Arnoldi, Bjørn Nørgaard, the Fluxists Arthur Køpcke and Eric Andersen, and the poets Søren Krøjer and Hans Jørgen Nielsen.

We have an idea about what was said at the meeting, because Kirsten Lockenwitz, the self-proclaimed "secretary" (and participating visual artist) of POEX65, took detailed minutes from the meeting. The big issues were: artistic freedom, money, artistic collaboration, autonomy of the artwork, crossovers between genres, manifesto/no manifesto, amateur/professional, and involvement of the audience.

An installation entitled *POEX Chairs – The Debate*,[55] which was especially created for the 2012 exhibition, reactivated the dialogue from the meeting and gave the visitors the possibility of participating in the debate that took place forty-five years ago.(FIGS. 14, 15) It seems strangely relevant today, where discussions abound on topics ranging from collaborative artistic strategies and audience involvement, to the media avant-garde as a transdisciplinary practice.

As a member of the audience, one could find a chair with one of the participant's names on it (limited to eight artists) and then sit and listen

55
POEX Chairs – The Debate (2012): Morten Søndergaard and Mogens Jacobsen: content; Morten Søndergaard: research; Mogens Jacobsen, Morten Søndergaard, and Martin Luckmann : conceptual design; Emil Alenius Boserup: mixing. Responsive voice interface: modified chairs, sensors, built-in microprocessors, and MP3 players, recorded voice.

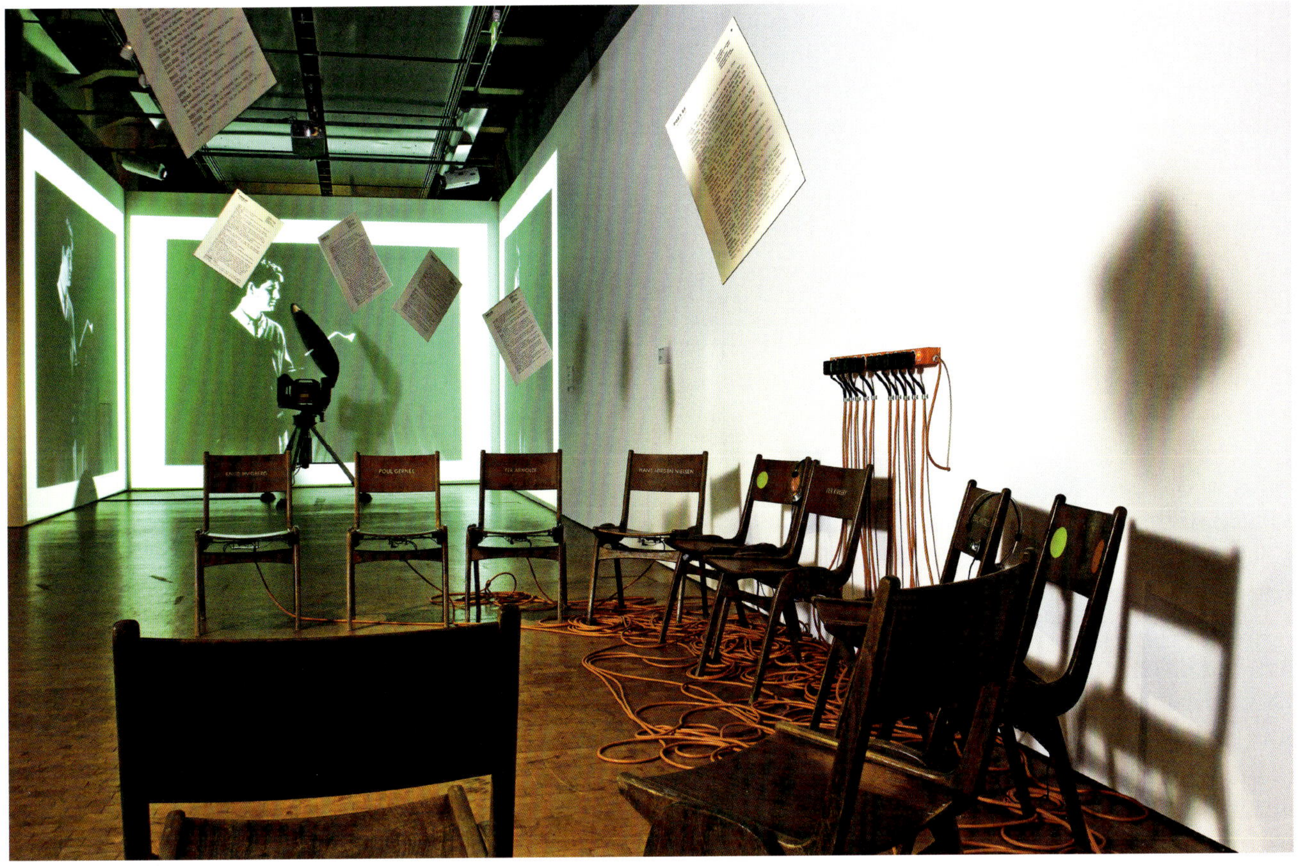

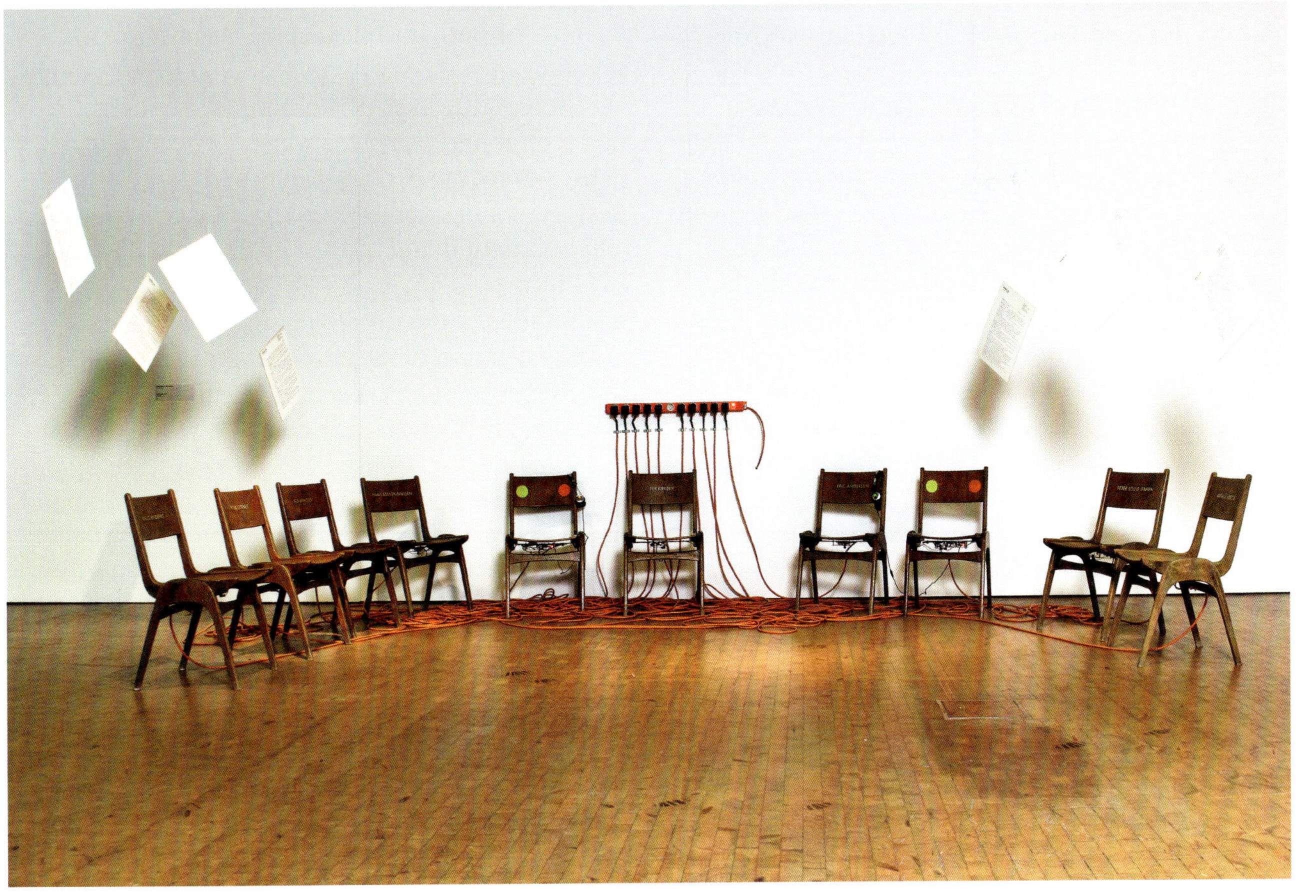

FIGS. 14, 15 Mogens Jacobsen with Martin Luckmann and Morten Søndergaard, *POEX Chairs – The Debate*, 2012. Interactive voice interface, modified chairs, sensors, computer, MP3 player, voice recording. Installation views ZKM | Karlsruhe, 2012.

to that artist's point of view. One could then move the chair next to another artist's chair and listen to a fragment of their discussion. Visitors could continue moving the chair to other chairs and listening to fragments of the debate – until they had made up their own minds or maybe wanted to start their own debate. It was possible to hear the whole dialogue if all configurations were tried out – as presented in the matrix.

Copies of the original typed minutes from the debate (in Danish) – complete with the original POEX65 letterhead – could also be studied.

Back in the day, however, not everyone was convinced of the quality of the project and its intentions.[56] It appears that it created kind of a scandal within the established art circles, most notably witnessed in the argument between Knud Hvidberg and the Eks-skolen, which at the time was establishing itself as the leading platform for conceptual and environmental artistic expression.[57] The main issue, first of all, was the problem of artistic autonomy (but this issue is a bit unclear to us today and perhaps mostly a matter of "clashing personalities"), as documented in the installation *POEX Chairs – The Debate*. Another critical matter was Hvidberg's idea of letting amateurs mix with "professional" artists – excluding no one from the artistic production in advance. Lastly, the very idea of a "poetry of action," which intentionally moved beyond the idea of the "happening" (which, according to Hvidberg, was too "heavy labeled" and "speculative" – meaning that too much was expected from this genre already before you started working, leaving little room for improvisation and creativity) was not Eks-skolen's cup of tea either. Eks-skolen eventually chose not to participate as a group, although a few of its members participated as individual artists.

It was not until very recently that the remaining members of the Eks-skolen, in the words of Lars Morell, recognized the importance of POEX65 – as one of the most important experimental events in Denmark of the 1960s.[58]

Exscribing the Unheard Avant-garde

In the process of curating the Unheard Avant-garde section, the practice of exscription as curatorial strategy enabled the collaborative working group to address the dissemination and communication of a complex set of data, recordings, ideas, and motivations. By writing and unwriting all these circumstances in front of an audience, and with the audience, the recovered recordings and echoes of art practices from 1965 are reactivated. The hidden matrices of the artistic networks, and the deferment of sound art into cultural memory, enabled the team to develop unexpected points of departure for communicating (with) different sensibilities in an audience. This attempt to highlight the unheard avant-garde as a complex network – experimenting with sound and technology in many different domains and directions as a new sensibility directly involving the audience – informs the dissemination strategy and interface "philosophy" exercised in the Scandinavian section.

In considering today's status of the EMS experiments with music (unheard until recovered by several research projects), Erkki Kurenniemi's Innovative Electronic Music Study (unheard until restaged by documenta and the exhibition at the ZKM), and POEX65 (unheard despite, or perhaps because of, its attempt to revolutionize the way artists and audiences interact through the uses of sound, performance, and technology), our response is more than a matter of reconstructing tapes and digging up past techniques and technology by "art historians." The submedial and hidden practices of EMS, Erkki Kurenniemi, and POEX65, and the key instabilities of the ontological gaps in the archives hiding them run much deeper than any technological determinism would explain. Moreover, it would be difficult to "remediate" the impact of EMS or POEX65 in the 1960s. Erkki Kurenniemi was aware of this, and even aimed his self-archiving at the invention of the quantum computer to solve and restage. Also, the use of the Flexowriter and the activity of Punch Tape Poetry at POEX65 is a good example of this.

The Flexowriter was an electric typewriter that could write "punched tape" (or "paper tape") that could be used in a computer. It could also write and punch a new tape that is rolled up on a "cardboard core" (Hasselblad, 1965). Even though the Flexowriter was a common instrument in larger offices from Word War II until the early 1970s, it is largely unknown today.

The idea of using the Flexowriter was an important part of the conceptual framing of POEX65. The idea for POEX65 was to create "a punch tape library with poem and other art expressions."[59] Knud Hvidberg writes in a letter dated June 12, 1965, that

> as for media there will be ongoing hanging of photos of poetry as in Sweden Flexowriter Television Gramophone Music-box with recordings of the participating poets exhibition of books […] theatre a box for performing paintings a hall for happenings and an entire empty hall for performing poems in sound and light.[60]

The writings and discourses that are reread and rewritten on the *POEX* Platform indicate that the Flexowriter was an active part of POEX65. However, the deferment of the mediation of technology and its uses, which could be termed the "Flexowriter condition," is a good example of the key instabilities at play in an archival situation involving material from collaborative and experimental events like POEX65. Moreover, the Flexowriter is literally an exscribing-machine, performing automated readings and writings elsewhere in time and/or space. The Unheard Avant-garde constitutes an "archive under the archive" – a kind of a laboratoria condition, in the sense of Foucault, where archive documents (Foucault calls them discoursive 'statements') are more akin to aesthetic spaces, media art installations, or performance events.[61] Michel Foucault calls this laboratoria condition the "heterotopia" – a condition where stages of time and space are organized differently.[62]

As a curatorial strategy, as I am suggesting in this essay, the laboratoria condition affords exscription. It has one main purpose: to enable the user of the archive to navigate actively and experience the different material across contexts and domains.

Although I hesitate to call the exscription of the unheard avant-garde a truly heterotopian condition – I do not wish to adopt the position of Foucault (or Wolfgang Ernst for that matter) in this way without further considerations and criticism which fall outside the purpose of this essay – I would claim that any reactivation of that which Wolfgang Ernst calls "der submediale Raum," the submedia space of archives, is bound to be hinged upon an awareness of the exscribing of cultural memory.[63] However, my aim in this essay is to demonstrate how this submedia space is not a static condition, but an active one that calls for curatorial agency and audience experience.

Data, Sound, and Experience

The status of the intersections of "data" and "the real world" has been investigated in several artistic visions or visualizations throughout the twentieth century. Man Ray's method in *Rayograph* (1923), for instance, was to throw dirt and salt directly onto strips of film, creating a grainy depiction of tiny real-world objects flickering on top of film recordings making visual parts of naked bodies or moving art objects. In 1923, Man Ray is pointing towards one of the most central premises of representing hidden matrices – that it dissolves the very premise of representation as visualization. Instead, a new matrix of representation aesthetics is needed.

Man Ray, to put it very briefly, stages the transformation of the relation between depicting and visualizing "the real" by using film as an archive for analog data of the world's basic phenomena and agency. Thus, the film in Man Ray's vision and version of it becomes the first deferred medium in the sense that it reduces the constituent elements of what we consider "real" to "agents" in a confrontational creative act where the representational function is "revealed" as an "aesthetics."

This, however, does not involve digital data or sound, but does abstractly refer to the metaphorical status of technology in cultural memory, in that the film is used by Man Ray both as a technology (for "physical" recording) and a deferred medium (dissecting and confronting

56 Notes on discussion between Knud Hvidberg and the Eks-skole [May 1965], Copenhagen, 7 pages, unpublished, collection of the Museum of Contemporary Art in Roskilde.

57 See Lars Morell, *Broderskabet*, 69.

58 Ibid, 70.

59 Letter from Morten Hasselblad to Knud Hvidberg (1965), unpublished, collection of the Museum of Contemporary Art in Roskilde.

60 Letter from Knud Hvidberg to Vagn Steen (June 12, 1965), trans. Morten Søndergaard, unpublished, collection of the Museum of Contemporary Art in Roskilde.

61 See Foucault, *The Archeology of Knowledge*, 79.

62 Foucault, "Different Spaces" in Faubion, J., ed: *Aesthetics, Ethics, and Epistemology* (New York: The New Press, 1982), 175-186, here 179.

63 Wolfgang Ernst, *Das Rumoren der Archive. Ordnung aus Unordnung* (Berlin: Merve 2002), 19-22.

the scientisms involved in the definitions of "reality" in his age).

According to Ulric Neisser the "internal information-processing model of perception" is the most accepted representation of an optical perception process leading to a consciousness about the information given.[64] This model is a strong undercurrent in most theories on computational cognition and explains – up to a certain point – the idea of process as a crucial element in having something on our mind. But it does not explain very well how we may have different things on our minds, for instance; or how it may appear that the same information-processing in different people constructs different ideas about the information entered in the first instance. Another question is how we may operate with information that is not conventionally understood within a cultural semantics or sign system. And what about information about something that is outside the reach of any human perception – like data?

Neisser furthermore sees expectations, exploration, and modification as playing a central part in the shaping of the understanding and metaphorization of human consciousness.[65] Consciousness is processing (and not just a mental one) of several layers of information, possible meanings, and alternative associations that take place almost "automatically." In this sense, perception-of-something is a construction, but this does not mean that just anything may be constructed. The construction process has constraints. Neisser suggests that we are cognitively engaged in a "perceptual cycle" based on patterns – or, as I am suggesting in this essay, in the context of the archive as deferred medium, on hidden matrices.

In the context of this essay, the interesting point to make here is that although scientists and psychologists primarily address these questions, artists and computer engineers work in the latent gray zones of unheard sounds and lost knowledge, highlighting them as an experimental framework for a "new practice" (instigated by media and technology in the twentieth century) of sound and sonification.

Man Ray, alongside a great number of artists experimenting with mechanic and electronic media, is indeed one of the early practitioners of the hidden matrices in or beneath scientific archives – actively investigating the generative relationship of sound, moving images, and text to cultural memory and "consciousness" (in the sense of Neisser). Although scientists and psychologists primarily address these questions to "explain" or "answer" them, artists practice them as creative intersections of hidden matrices. These art practices work across the fields of science, art, and technology to claim the delayed or postponed mediation of the unheard avant-garde in its own right. And from this "takeover" of the unarchived sound art nonhistory emerge several creative confrontations with the scientific definition of the "real," which may serve as momentum in the development of exscription as a curatorial strategy.

Finally, this makes the act of listening to the unheard itself a matter of curatorial concern as well: After all, how do we embody something that was delayed from entering the "perceptual cycle" of lived, shared memory? The whole point of exscribing anything is precisely to make the inscription of something present in an embodied and tangible process. In the final part of this essay, I will go deeper into the process of redesigning ways of embodying and listening, with a focus on how complex sets of data and sounds were exscribed into audience experience, with the last platform in the Unheard Avant-garde section, the Media Art Platform, and the *Audiobar* as example.

Sound Art Curating

With the experience of the Unheard Avant-garde section in mind, it is possible to claim that sound art curating is a "domain-game." Echoing Wittgenstein's notion of the language-game,[66] and coupling it with Peter Weibel's notion of the transdisciplinary domain,[67] I am suggesting that in sound art curating, we might add another level of epistemology to that of the language-game: thus, the domain is the framing of the language-game, or that which conditions the language-game. One might suggest that, at some level, a certain domain makes certain language-games probable, and others improbable. A domain, therefore, not only conditions which

64 Ulric Neisser, *Cognition and Reality* (San Francisco: W. H. Freeman, 1976), 23

65 See ibid, 146–47.

66 A language-game is also defined as "consisting of language and the activities into which it is woven." Ludwig Wittgenstein, *Philosophical Investigations*, ed. P. M. S. Hacker and Joachim Schulte, trans. G. E. M. Anscombe (Oxford: Wiley-Blackwell, 2009), 8e.

language-games are most likely to be "played," but also how they are played. A domain is thus framing certain kinds of competencies. And also setting certain limits to what can be exscribed, and where.

Domains are part of the same conditions as language. There is an ongoing domain-game happening in a transdisciplinary setting. In the domain-game, the competencies of certain domains are being renegotiated. This could be the domain of art institutions, or it could be the domain of sound.

Habitual and conventional domains of knowledge and art are shattered and renegotiated, and in their place new domains are reconfigured.

We are faced with the question of formulating qualitatively new phenomena and inventing concepts and strategies in-between theory and practice. The construction of new domains is based on highly different competencies that are working together, in a dynamic exchange of solutions and differences.

A historic example of this might be the transformative explorations by the Futurists into the "cognitive space" of the modern man, which signaled a new way of operating and producing relational patterns of meaning. It ran counter to the Positivistic paradigm that flourished around the same time, but it also differed in a number of important ways to the historicisms of humanistic academic traditions. Today, the Futurists and their experiments remain an enigma to the more disciplined scientific traditions exactly because, I will argue, they exscribe a new transdisciplinary domain for the first time: the domain of sound art. Moreover, they do so in a manner that includes an element of modern design strategy into their artistic practice: they replicate actual experienced sounds with invented instruments.[68] Thus, from the outset, and even before the advent of electronic media, transdisciplinary art practices within sound art point out an alternative way of producing art that differs completely from the modern (and modernistic) domains of specific genres and art categories. The Futurists were, in short, playing domain-games.

Playing domain-games in the field of curation and sound opens up the possibility of generating new ways of listening, which is evident in the responsive sound interfaces of *Audiobar* – and *POEX Tape* (as we have seen above).

Audiobar[69] (or *Hørbar* in Danish, which has the double meaning of "making things audible" and the concrete reference to the physical bar) is a responsive environment designed by Mogens Jacobsen for exploring a huge collection of sound art. *Audiobar* was part of the Media Art Platform project, which was documented in a publication discussing the domain of digital archiving; however, here it should be discussed particularly within the context of a responsive sound design aimed at a wider audience.

In the *Audiobar*, the audience interacts with the sound art pieces in the collection using a tangible interface: by placing bottles on the bar. The bar consists of a central table and 260 colorfully labeled bottles placed on shelves around the bar. The entire collection was indexed and described according to twelve pairs of emotional parameters. This was done by listening to all the tracks (more than 1,600 of them) and "fitting" them within the system of description. The labels indicate a "game of categories," which on one hand, and not without irony and humor, points away from the heavy, historicized categories habitually used in art museums; on the other hand, it works within an "embodied" relation to the sound art pieces via associations to the act of experiencing sound - and how we may describe those experiences (which is hard to do when operating "outside" the symbolic and iconic representational paradigm, as argued above). Therefore, the curatorial methodology of the Media Art Project and *Audiobar* places the audience inside a new domain where their own associations and

67 In the 2004 *Ars Electronica* catalogue, Peter Weibel describes and analyses the "transdisciplinary ways of working" that are reconfiguring the domains of the contemporary avant-garde. He sees the contextual challenges of "transdisciplinarity" as a condition for the way avant-garde practice is reconfiguring the traditional domains of art. When taking into consideration the transdisciplinary condition as part of a curating methodology, it is engaging on the notion of Peter Weibel, that "a transdisciplinary mapping of competences is taking place in the fields of art." He continues: "the redistribution of competence [... is precisely the progressive practice of the contemporary avant-garde: [... its protagonists attempt to undertake new comparisons and redistributions of competence, and to make the transition from the formal level to the level of practice. Extension of competence is the latest phase in the expansion of the concept of art. Art expands from the object to the practice and its practice expands from its field of work into new domains [...]." Peter Weibel, "New Protagonists and Alliances in 21st Century Art," in *Ars Electronica 2004: Timeshift – The World in Twenty-Five Years*, ed. Gerfried Stocker and Christine Schöpf, exhib. cat. (Ostfildern: Hatje Cantz, 2004), 38–40, here 39, http://90.146.8.18/en/archives/festival_archive/festival_catalogs/festival_artikel.asp?iProjectID=12928.

68 See Luigi Russolo, *The Art of Noises*, trans. Barclay Brown (New York: Pendragon Press, 1986).

69 *Audiobar* (2006). Mogens Jacobsen and Morten Søndergaard: conceptual design. The *Audiobar* was commissioned for the Media Art Platform project, and premiered at the exhibition *Social_Action* at the Museum of Contemporary Art in Roskilde, 2006. It was exhibited at the *YOU_ser: Century of the Consumer* exhibition at ZKM in 2007–2008, and included again, on special request from Peter Weibel, into the Scandinavian section of the *Sound Art* exhibition in 2012.

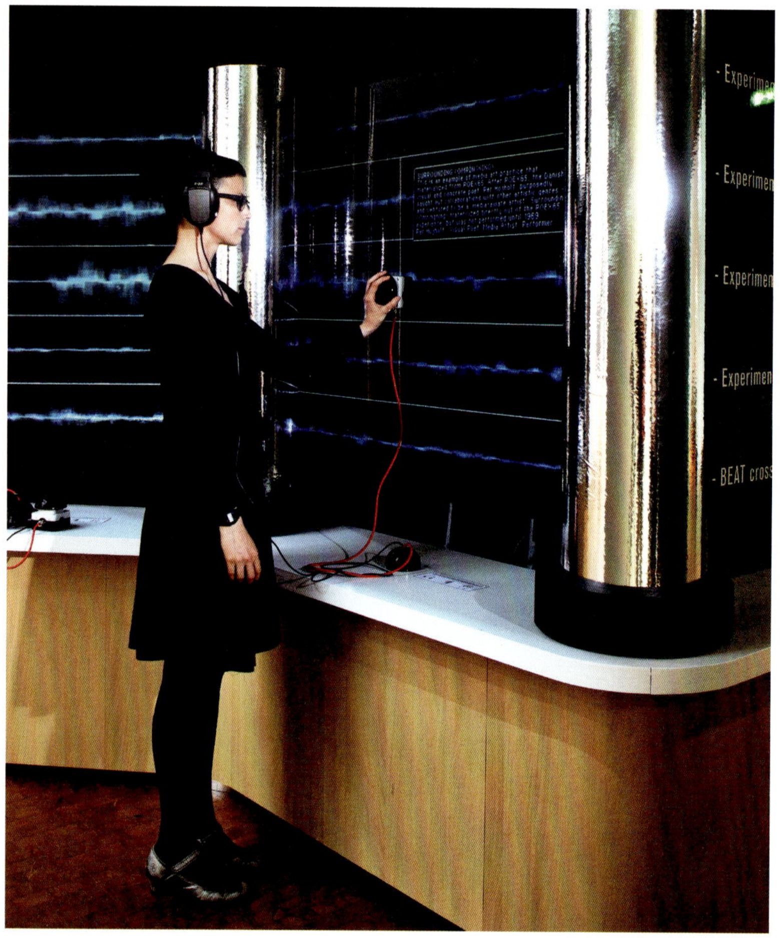

FIGS. 16, 17 Martin Luckmann with Morten Søndergaard, *POEX Tape*, 2012. Interactive sound archive interface, infrared LEDs, Wii controllers, projectors, database, Arduino platform, plexiglass, openFrameworks. The tape material comes from private archives and boxes belonging to now deceased participants in the POEX65 event. Installation views ZKM | Karlsruhe, 2012.

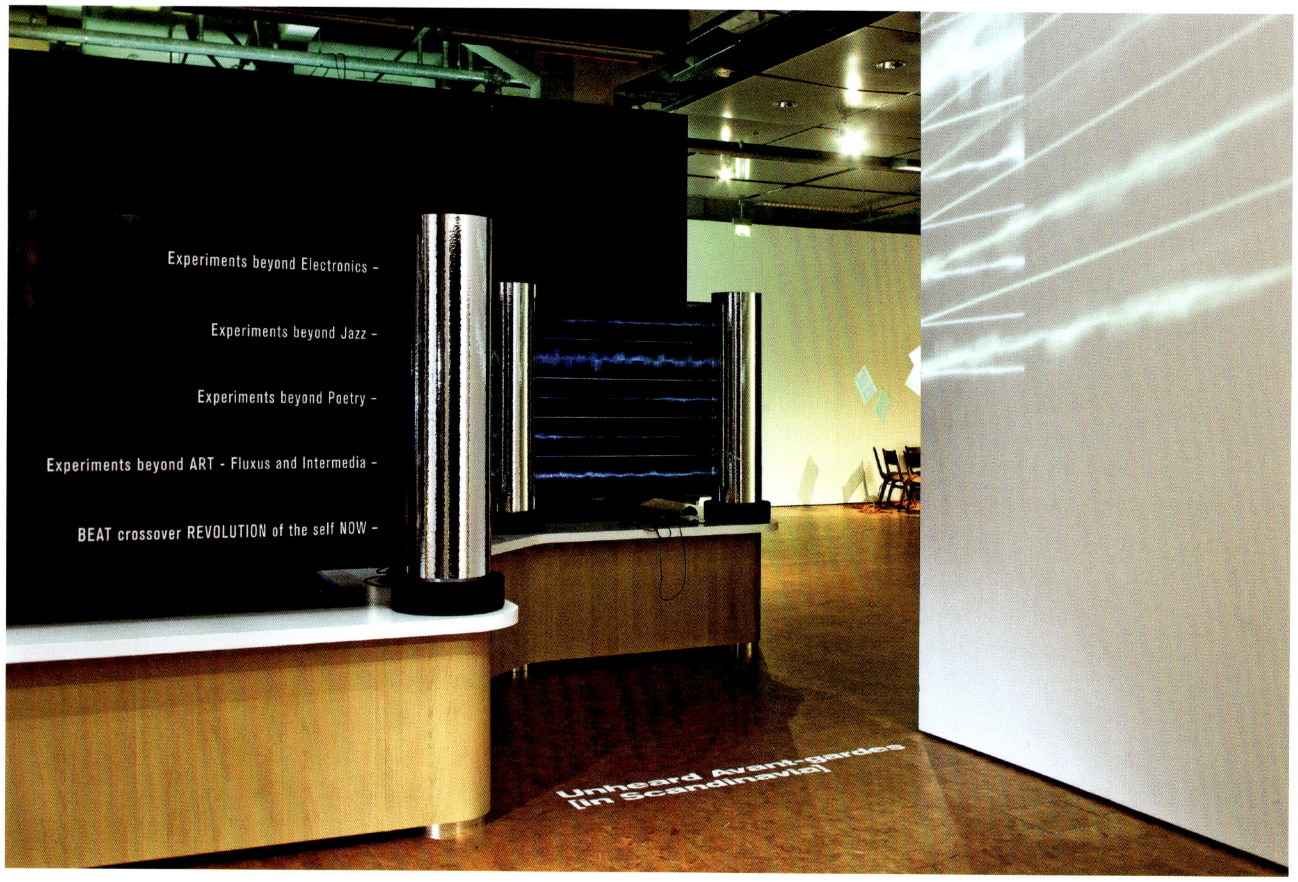

experiences serve as initiators – I use the word can-openers – for witnessing this extraordinary, amazing thing called sound art… and find that they can grapple with a little bit of complexity and have fun at the same time.

The action begins when a visitor places one or more bottles on the table. The light below the semi-transparent table changes color from green to bright blue. After less than a second audio begins to play and the table is lit by white light. If other visitors place additional bottles, the table blinks again, and a track is selected that fits the combination of bottles. If all bottles are removed, the table returns to its silent state with a green illumination.

At first, the bar activates the listening senses of the audience by subjecting them to the "unprepared" listening of the sound art pieces that are played back to them after placing the bottles. Then the audience is drawn into a critical position that may even bring about curiosity, asking questions such as: How and when was this piece made? Why does it sound like that? Who made it? Why is it more noise than silence?

A similar strategy is employed in the *POEX Tape*, which allows the users/audience to navigate the sound streams in the archive of sound art from POEX65; this, in effect, is a new domain for which the audience is an unprepared listener.(FIGS. 16, 17) At the time of the POEX65 exhibition, the main recording technology was the reel-to-reel tape recorder. The ability to (fast) forward and (fast) rewind, and to manipulate the tape, was the main inspiration for the interface of the *POEX Tape*: to give people a sense of what that was like, in sound. We wanted to make an interactive interface for the archive of the unheard avant-garde of POEX65 by playing a domain-game with audiovisual representation.

But, whereas the embodied interaction-process in the *Audiobar* involves using the bottles as instruments in a domain-game of listening, acted out between the choices made by the audience, the *POEX Tape* puts the creative focus more directly on the interplay between technology and listening, and representations of it. Both are playing domain-games that are engaging the audience to listen to the avant-garde in completely new ways. However, whereas the *Audiobar* turns the domain-game into a collective playful situation where listening is conditioned by probability and negotiated content which is not predefined but a result of the choices and opinions of the audience, the *POEX Tape* is an individual listening and sensing situation.

In this essay, I have documented, described, and discussed a methodology for curating the Unheard Avant-garde section at the ZKM exhibition *Sound Art – Sound as a Medium of Art* in 2012, which suggests and mobilizes the exscription of hidden matrices in subarchival or nonarchival material as a curatorial strategy for disseminating and communicating sound art histories. Playing transdisciplinary domain-games and searching for unheard avant-gardes, as well as maturing the posthumanistic methodologies operating with archives in a postcognitive culture of technogenesis, will be a priority for many years ahead.

HISTORICAL CARTOGRAPHY OF SOUND ART

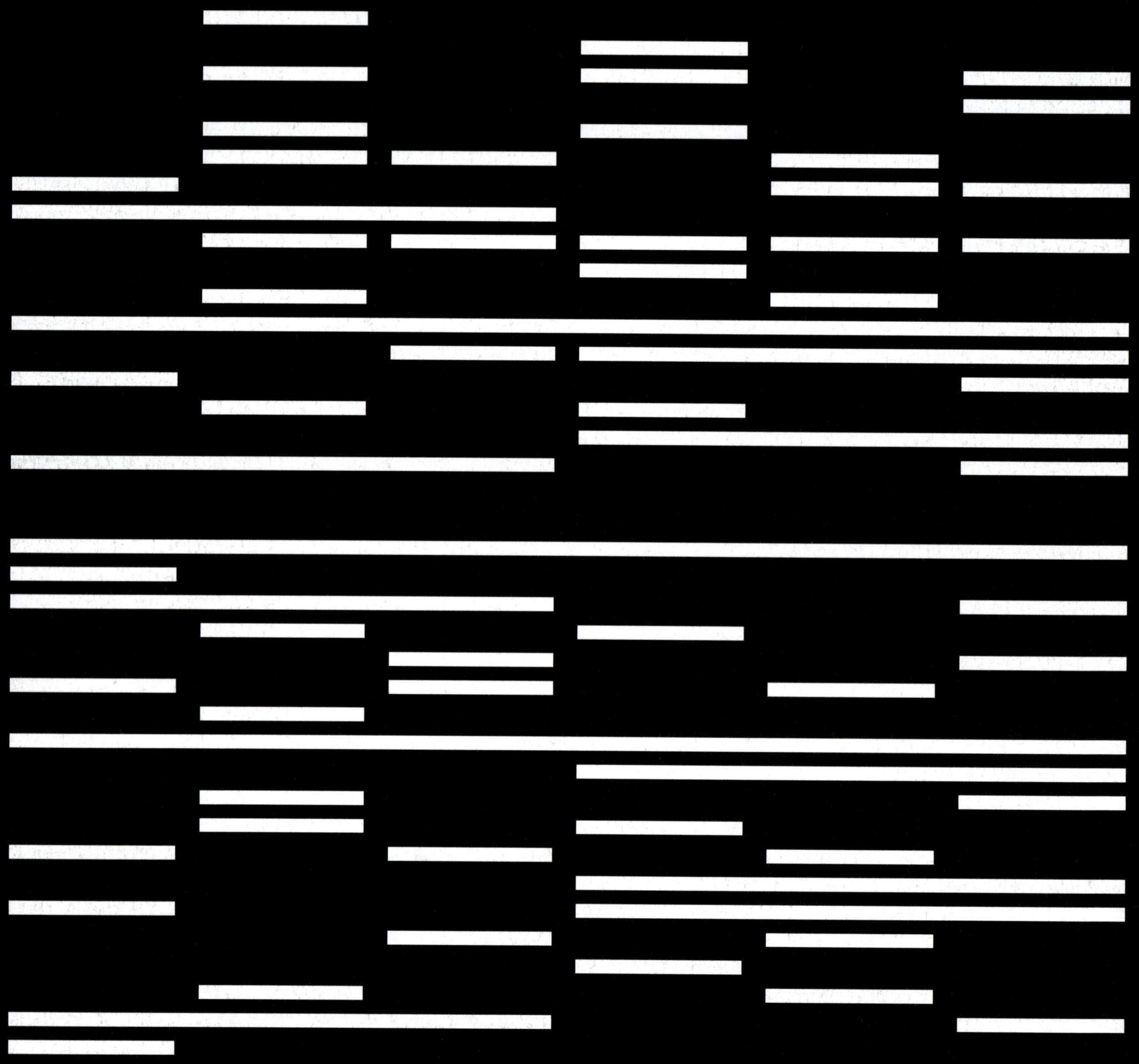

David Toop and Adam Parkinson

Unfinished Business: A Conversation on Sound Art in the United Kingdom

The following is a transcript of a conversation between David Toop and Adam Parkinson, conducted in April 2015 as a public performance at the London Metropolitan University. This format – the public unfolding of a text – was conceived in the context of a "manifesto" written in 2014 for a series called *Offering Rites*, curated by David Toop: "What I'm trying out at this stage of my life is new formats, or new settings maybe, or formats and settings that have been tried before but then been forgotten or pushed aside because established formats have such a powerful hold on our thinking. I'm frustrated by [...] all the familiar routines that frame practice and discourse [...] what I want to do is modest, small scale, quiet, and uncertain – just a slight shift of conditions [...] a big thing but small, a conversation that can be quiet but loud." The aim was a reframing of what are known as sound art practices in the UK. What ensued is partial and fragmentary, but a beginning.

The conversation was edited collaboratively by Adam Parkinson, Atau Tanaka, and David Toop.

Adam Parkinson: Do you find the term "sound art" useful? What did people do before the term came into use in 1983?

David Toop: In the 1970s we called it "sound work," which I prefer and still use. One of the reasons for that was a very 1970s idea of aligning yourself with a worker. We tried to detach ourselves from the art world (which we were detached from whether we liked it or not) and also from the music world. We were also conscious that we were making work that couldn't be accommodated by existing orthodoxies of music or art galleries. But sound art to me is problematic for a number of reasons. One, because it is so closely associated with a particular world and a particular economy – the art world – and there are all sorts of reasons why that's difficult. It puts a strong focus on the creation of some kind of object, even if that be a conceptual, text, or installation object. The idea about sound *work* was that it was always at heart a process. That shift was important. I think there is a relationship to Christopher Small's term "musicking," and it is interesting that a number of us involved in these marginal areas worked with Chris in a group.[1] "Musicking" is about being engaged in some way with the making of music, but not necessarily making music. The process – in the widest sense – is important. That idea of process came in part from the discourse and practice of the early 1970s.

Parkinson: Has sound art emerged as a practice distinct from, say "art" or "music"? Is it even healthy to box it off as one?

Toop: It is a hybrid form, maybe more hybrid than most. That is particularly true in the UK, where there has been more resistance to accepting the term sound art. There is a strong musical element which doesn't go away, so you have musical aesthetics, music practice, orthodoxies, administration, and economy attached to it and then you have fine art practice, painting, and sculpture, and then performance art and conceptual art and so on.

All these different ideas come together in a field which is undefined, and in a sense they struggle with each other all the time. For me the important thing is keeping a discourse open, but for other people in this field, the important thing is to narrow it down. Maybe in a more academic sphere where it's important to define terms before you can begin to really talk about them, the necessity of clarifying and delimiting this term sound art is important, but the effect on the practice can be very negative. For example, on a practical level, if somebody has defined what sound art is and then there is an opportunity to get the benefit of some kind of funding or residency and you fall outside that definition, you're going to suffer, so I think it's important to keep breaking it open all the time.

1 Christopher Small, *Musicking: The Meanings of Performing and Listening* (Hanover, NH: University Press of New England, 1998).

Parkinson: Let's look more broadly at the origins of sound art practices in the UK. How do you see this history as being different to, say, how the history of sound art is seen in America? What unique forces do you see at play in the UK?

Toop: It's a difficult question to answer because you have to dismantle the whole conception of what sound art is now. There was an impulse to go beyond a certain kind of art or music practice, and the historical reasons for that are very complex. I can give you examples, one of them is Gilbert and George, *Singing Sculpture*, which was a work they produced at St Martin's School of Art in 1969. The piece involved playing a gramophone record called *Underneath the Arches*. This was very famous – sung by a musical duo called Flanagan and Allen who were part of the variety scene, very closely associated with the wartime generation, so there was a strong sense of nostalgia about it, not only as a type of singing, but also the nature of the song which had a sentimentality about being down-and-out, so I think it would speak very directly to people who had gone through the 1930s, the Great Depression, the Second World War, and the severe austerity of the postwar period. Gilbert and George painted themselves, to some degree anonymizing themselves, and played the record. I think it was very significant for a number of reasons. One reason was that it became a durational performance. Also the idea of living beings being artworks in a gallery was fairly radical at that time, but it was also the insistence of this record playing over and over again. Its genealogy was very complex, and this is one of the areas where I depart from orthodox histories of sound art because I feel very strongly that comedy has played a big part in the evolution of sound work in this country. There was a high level of comedy in the Gilbert and George piece, and at the same time it was an interesting conceptual piece. They thought of themselves as a sculpture. They were associated with the sculpture department at St Martins at that time, which was a very radical department where sculpture could become sound, music, living beings, or actions.

The comedy element in their piece is traceable back to strange variety acts like Wilson, Keppel, and Betty. This was two thin men dressed in mock ancient Egyptian clothes with a female dancer. They would do the *Sand Dance* where they'd throw sand on the floor and do a mock Egyptian dance, a strange tap dance thing, very surreal, no words, just music and the sound of the sand. There's a history of acts where there is a tremendous amount of physical humor and humor of expression, a lot of silence and strange uses of sound. There's a film made by Laurel and Hardy called *Busy Bodies* – they are driving along in Los Angeles, and Oliver Hardy suggests listening to music. Stan Laurel pulls a string and music starts. At a certain point the music stops, they stop the car, Stan Laurel opens the bonnet, and there's a gramophone in there with the tone arm attached to a string. He puts the needle back on the gramophone, puts the bonnet down, gets back in the car, pulls the string, the music returns and they go off again – that to me is part of a continuum which will lead you to Janek Schaeffer or Christian Marclay.

There is another film which was extremely influential on my generation called *The Running Jumping and Standing Still Film* (1959), by Spike Milligan, Peter Sellers, Bruce Lacey, Mario Fabrizi, and directed by Richard Lester. The Beatles loved it, so Lester went on to direct their films. There was a kind of postwar craziness, either people who had been in the armed forces and, like Spike Milligan, had suffered shell shock, or people who had been called up for national service after the war, and found it ridiculous and extremely damaging. Military discipline seemed to lead to this near madness, and this was on top of very strong feelings that the war itself had been a demented passage in history. The response to that was radio shows like the *Goon Show,* a kind of surrealism involving music and electronic sounds created by the BBC Radiophonic Workshop. *The Running Jumping and Standing Still Film* was basically people doing stupid things in fields: there's one scene where somebody is standing in front of a sawn-off tree trunk and he produces a record from his overcoat, takes a tone arm out, and runs round the tree trunk with the needle in the record.

Parkinson: I think this connection between comedy and sound art is a really interesting one. Although people talk about "the British sense of humor," I'm suspicious of this and think there are too many unfunny British people who think that being able to quote Monty Python or whatever makes them witty. Nonetheless, humor seems connected to improvised music in the UK: through Derek Bailey, there is a very real connection, playing guitar for Morecambe and Wise. I recently read a Clive Bell article about the London Musicians' Collective where he quotes Thurston Moore specifically mentioning the humor in the London improv scene.[2] In the work of Hugh Davies, it feels there was always humor, there's Stewart Lee and Steve Beresford performing together. It seems like comedy and free improv provide spaces where you can explore freedom and fantasy.

Toop: There is a Gavin Bryars piece called *Marvellous Aphorisms Are Scattered Richly Throughout These Pages* – a raincoat which opens up with pockets inside containing various devices for making sound. It's a kind of conceptual piece. You could relate it to Fluxus but if you come from the UK and you are my age you're more likely to relate it to the things that Gavin and I were exposed to as young people, this very bizarre humor which often involved music. To give you another example, there was a famous comedian, Max Wall. He had this character called Professor Wallofski who was terrifying. He wore this crazy wig with hair sticking out, black tights, big boots, and a strange coat that made him look like a bird. It was all about his relationship to the piano, deconstructing the act of playing the piano. There was a lot of business that was about not playing – I think that's interesting in terms of free improvisation – but it was also about physical hindrances to playing. He could do this thing where one

2 See Clive Bell, "History of the LMC," www.variant.org.uk/8texts/Clive_Bell.html.

arm would suddenly seem to get longer. It is very bizarre, partly because it's comedy from a different era, but for me it is interesting in being partly about negation, which is very much a part of musical humor. There's an American comedian called Jack Benny and a lot of his routine was about not playing the violin. There was a complex double bluffing going on, because obviously people had come to see him playing *and* not playing the violin. Experiences of seeing these people were very formative, and they were part of this thing derived from the music hall called variety, which was exactly what it says: if you look at a variety theatre poster you'll see there were about thirty acts, with a lot of specialist acts who would do just a few minutes. There was one man I discovered whose act was to unfurl flags of the world. He would come on with poles and unfurl all these flags all around the stage and then roll them up again. That was his act.

Parkinson: Moving onto another institution that might have shaped a particularly British scene, could you talk a bit about the BBC Radiophonic Workshop and the role that played?

Toop: The BBC Radiophonic Workshop is certainly important. It started around 1958, and one of the most important people was Daphne Oram, who is definitely relevant to any history of sound art in this country. In 1958 Oram worked with a BBC Radio producer called Madeau Stewart, and they did a radio play which used electronic music, *Prometheus Unbound*. Madeau Stewart was a BBC archive producer, and I met her in 1971 because I sent a letter to the BBC saying that I thought they had a responsibility to do something about the fact that many natural sounds and many forms of music were disappearing. Hubris of the young! The ecology movement was building momentum at that time, and I had the feeling that this movement had a relevance for the way I was perceiving the relationships of music, sound work, and environment. I wrote to *Ecology* magazine asking if they would be interested in a piece on musical ecology and music's relationship to the environment, and they wrote back saying they couldn't see any possible connection. Then I got a very interesting reply from Madeau Stewart, saying she felt sympathy for these ideas. What developed from that was me making programs for Radio 3. I was given free access to the BBC sound archives which she had helped set up, with *carte blanche* to make strange radio programs about connectivity and ecologies. Madeau was a fascinating character. I had got interested in bioacoustics, in the strange sounds made for example by seals, and mentioned this to Madeau, who said she had played flute to seals. She was related to the Mitford sisters who had an island in the northwest of Scotland: sometimes she would stay there, and she would go out and play her flute to seals. This would have been in the 1950s. She got a BBC recordist to record her doing this – she would play flute to seals, and the seals would answer. Now there are very strong connections here. One is with the BBC Radiophonic Workshop and Daphne Oram, and those two were quite mystical, they were interested in Atlantis and folklore, and I think there was a powerful connection, certainly from Madeau in her ideas of musical significance in society and these mystical ideas, but I think it is this idea also of making direct connections of music to the environment, in other words playing flute to seals. I see that as a kind of sound art, but like Gilbert and George as being a completely invisible part of the history. You can't help feeling that this new category is created for the convenience of exclusions, whether they be to do with gender or nationalism or whatever, so that's why I'm keen to include Max Wall and Madeau Stewart in the history.

I produced a book in 1974 called *New/Rediscovered Musical Instruments* with work by Hugh Davies, Paul Burwell, Max Eastley, Evan Parker, Paul Lytton, and myself. Madeau Stewart wrote the introduction: "Suddenly we are enjoined to hear and listen to our present environment with greater attention. From the materials about us we are encouraged, all of us are encouraged, to conjure up the latent music. Some of this music would be familiar to people on other distant continents whose musical life we are so successfully destroying. Where isolated tribes use insects as buzzers attached to or trapped in some casually cut length of local wood, so the new instrument makers, who are also composers, casually adopt a clock spring or an ex-army shell case as a sound maker. Incidentally, a society surrounded by disposable or throwaway goods is not new. Sometimes the wind is pressed to perform and often the abundant sources of electricity are channelled to contribute to the creation of a composition" – and there I suppose she's thinking about working with Daphne Oram as much as anything – "and audiences are mobile, free to chat and circulate as in the earliest days of the proms."[3] So, musical instrument making I would say was a very strong part of this movement that became known as sound work.

Parkinson: With the Radiophonic Workshop, variety theatre, and comedy, it seems to be through everyday channels like TV and Saturday night entertainment that this stuff affects people. These are things that aren't positioning themselves as art in any way, and I wonder whether the influence of Fluxus allowed people to start bringing in these influences and framing aesthetics shaped by these things as art?

Toop: That is the conventional art historical view, but for me there is another generative force, and that is a kind of repression. There's a very good book by Louis Niebur about the BBC Radiophonic Workshop. He has gone back into the paperwork from the beginnings, and one of the things that comes through is that the establishment as personified by music critics of the time and the BBC itself did everything it could to suppress the kind of experimental electronic music that was happening at that time in Germany, America, Italy, France, and Japan. It's extraordinary the lengths that were gone to. The BBC even drafted an edict which said that recording engineers should not work with electronic music for anything

3 David Toop, *New/Rediscovered Musical Instruments* (London: Quartz/Mirliton, 1974).

longer than a specific duration, because it would have a bad effect on their nerves.[4]

Parkinson: Might turn out to be true.

Toop: It might turn out to be true! But, true or not, they were determined to implement it. A lot of electronic music happened under the auspices of state radio in other countries and was encouraged, facilities were given and recording studios were developed. But in this country, that was not the case, and what happened instead was the Radiophonic Workshop. The irony was that it created a type of electronic music that was far more popular than anything produced by Karlheinz Stockhausen, Luciano Berio, or Pierre Henry. It was a kind of populist, subversive version of academic electronic music. One of the most famous examples of electronic music is the *Doctor Who* theme by Delia Derbyshire, and it's only relatively recently that she has been given any credit at all. She was a forgotten pioneer, as was Daphne Oram, but what they produced – because it came through the television set in the context of popular entertainment – had a powerful effect. For example, there was a television series in the 1950s called *Quatermass and the Pit*. I watched that as a child, and it was terrifying. There was a strange mix of themes, occultism, people coming from outer space millions of years ago and being abandoned. It's about an excavation in which a spaceship is found with dead creatures inside which cause bizarre effects. Daphne Oram worked on some of the music, and it was electronic music to create the effect of these insectivorous creatures. I remember the scene which terrified me most: one of the people working on the site late at night physically changes into an insect – he runs in this insectivorous way, then falls to the ground, the ground ripples and you hear eerie insect-like electronic music. I saw that as a child and the effect of it was just overwhelming. I've watched it since and still find it very eerie. One of the most important things is that it doesn't fit within the art historical narrative; it's not conceived of as a complete piece like *Telemusik*, you catch a fragment without knowing what you're hearing. So if I were to distill down a thesis here, I would say many of us who were involved in the early evolution of sound art in this country, our formative influences came from these sources, from comedy or television horror, hearing accidental bits of electronic music, and those are the kind of experiences that you really absorb deeply, you can't shift those.

I made a piece in 1971 called *Bionic Rock*, which was just a text. At the time I was using a very extreme guitar distortion pedal called a Rush Pep Box. If I turned it up full it made these clicking sounds and to me those sounded very similar to the sound of bats when you detune them down to human hearing. I saw what I was doing at that time in the context of rock music, not experimental music. I was very interested in bionics – the science of systems based on living things – and so I came up with something called "bi/s/onics," which was the science of sound systems based on living things. It was tongue-in-cheek, but there was a relationship being made between this electronic guitar pedal and the sounds bats produce. I think that was my first bit of sound art and it was a massively important thing for me, that you could make a sound piece that was completely silent, that was just a text. I was not really aware at that time of much conceptual art, but I was aware of Yoko Ono's book *Grapefruit*, a very important work in terms of sound art and quite underrated.[5] About roughly the same time, the Scratch Orchestra with their *Nature Study Notes* were doing similar things. Michael Parsons made a piece called *Table Music* – it was just a text, something like "make whatever sound you can from a table" – and I think that it is a really significant piece, because it's saying anything can be an instrument, any action can be music, and a text can be a composition.

Parkinson: It's changing the way we listen, and changing our relationship with the sounds around us.

Toop: Very much so, yes. Michael was not so well known, sadly he's still not so well known, and I have tremendous respect and admiration for him, but Yoko Ono by then was a massive public figure because of her role in the Beatles, and so there is this strange area where you've got a crossover between the most popular art form of the time, which was pop music, and the least known, which would've been free improvisation and these esoteric areas of sound art. There was a concert in 1969 with John Lennon, Yoko Ono, John Stevens, and John Tchicai in Cambridge. It's funny, because listening back, the most extreme music comes from John Lennon and Yoko Ono; the John Stevens and John Tchicai bits sound more conventional.

Parkinson: It feels like there's almost a *Man in the High Castle* alternative history to be imagined, if AMM's support slots for Pink Floyd had gone a bit better, and there had been a commercial explosion of free improv, something quite different could almost have unfolded.[6]

Toop: I think about that a lot, a sort of Philip K. Dick reimagining of the future. In a way it was poised, you had this moment when there were many people in rock bands interested in this stuff, like Paul McCartney and Syd Barrett. You had Jack Bruce playing free improvisation with Jon Hiseman before joining Cream, so, yes, you could reimagine a future in which the really massive stadium stuff was AMM and the Spontaneous Music Ensemble.

Parkinson: I was wondering if you could talk about some of the specific institutions – we've already mentioned the Radiophonic Workshop and touched on the Scratch Orchestra – but other groups and institutions, what kind of role have they played?

4 See Louis Niebur, *Special Sound: The Creation and Legacy of the BBC Radiophonic Workshop* (Oxford: Oxford University Press, 2010).
5 Yoko Ono, *Grapefruit: A Book of Instructions and Drawings by Yoko Ono* (New York, et al.: Simon & Schuster, 1964/2000).
6 See Philip K. Dick, *The Man in the High Castle* (New York: Mariner Books, 1962/2011).

Toop: I'd like to go back a bit to the evolution of free improvisation in this country, which is complex, but I think very relevant. Again, it was a distancing from music orthodoxy and institutions. Groups like AMM and the Spontaneous Music Ensemble all began in relatively conventional ways. Three of the Spontaneous Music Ensemble began playing in the Royal Air Force bands, going to Germany, hearing visiting Americans, meeting some of the German musicians, beginning to play something like free jazz. AMM, slightly different trajectory. Keith Rowe was an art student, who then played with Mike Westbrook, playing modern jazz, and Eddie Prévost, playing modern jazz, Lou Gare, also an art student, playing modern jazz, later joined by Lawrence Sheaff who was playing modern jazz with Mike Westbrook, and by Cornelius Cardew. Nowadays, we feel we can do what we like – our problem is what to do with that freedom. Back then, there were huge barriers to doing something that we might consider simple, like playing without regular time or moving away from harmonic structures. How to do that, if you've grown up playing those things? It's fascinating analyzing the strategies people used, some of which were quite convoluted, to achieve that. Take an example like John Stevens, a drummer who was the driving force behind the Spontaneous Music Ensemble. In many ways, John was quite a conventional musician in the same way that Derek Bailey was quite a conventional musician. They had values based on African-American music, to do with subtleties of time-playing and so on. For John, I think there were three main elements in shedding this. One was hearing the Albert Ayler Trio and the way they approached time-playing – a weird contradictory thing – but if you listen to the Albert Ayler Trio, you hear something so fast it's almost shimmering, but at the same time it's slow underneath. Then he heard Japanese *gagaku*, and the slowness of *gagaku* fitted into this idea of how you could create a new music. He was listening to Webern – this is a working-class man who's educating himself, who would never normally come across a composer like Anton Webern, but he did.

Parkinson: Do you know how?

Toop: There was an obsession amongst these musicians about Anton Webern. Derek Bailey, Tony Oxley, and Gavin Bryars were listening to Webern in Sheffield. Tony Oxley partly because he had also been in the army, he'd played in the Black Watch band, and studying classical music had taken him onto Schoenberg and then Webern. I think it was to do with this rapid fluctuation of tonal coloring, and the way that time seems infinitely slow and infinitely fast at the same time in many Webern pieces, pieces that seem to go by so quickly because they're so short, but condensed into them and utterly transparent is this extraordinary range of coloration. So there is a triangulation there of ideas which come from very different sources: one is a sixth-century music from Japan which is a hybrid of music from India, China, Korea, and Cambodia, another is twelve-tone music and serialism of extreme compression and brevity, and the third is African-American free jazz which is deeply rooted in spirituals and gospel. All of those different source materials came together to produce a utopian idea of group music which had no director, composer, score, or predetermined course.

Parkinson: Still, it's a real break to go from a lot of stuff that is quite rule-based to the kind of freedom, the agenda, that John Stevens put forward in *Search and Reflect* – to go from rules that require a lot of training to this really open, exciting place for group music.[7]

Toop: There was a massive jump for all of them – it caused tremendous friction, but I think it was a type of de-conditioning process. I met John Stevens in 1971 and took part in his free improvisation workshops at Ealing College. He had almost distilled music-making down to two elements: click and sustain. He had a piece called *click piece* and another piece called *sustain piece*. For *click piece* you played the shortest sound you could on your instrument, and focused on playing the short sound, you didn't listen to what other people were doing, and *sustain piece*, you only played on your out breath, so you played a sustained sound that corresponded to your out breath. I do that piece with my students because I think – particularly if they are working with laptops – it's interesting to relate it back to the body. Those were the two elements that John was working with at that time, and he started running improvisation classes at Ealing College, under the auspices of Christopher Small.

Parkinson: I hadn't realized Christopher Small was involved, that there was that close connection between his concept of *Musicking* and these early communal music experiments in free improvisation.

Toop: John started a club called the Little Theatre Club, just off St Martins Lane, in 1966, where people could come together to experiment and develop this music, because you couldn't do it in commercial clubs. Very few people were going, but there are these brilliant stories about the audiences – Derek Bailey played to one person, the one person being Fred Frith, or he played to two people, the two people were John McLaughlin and Dave Holland, who were working with Miles Davis. Anyway, some of the audience started thinking, "Well, I want to do this too." Chris Turner, a music student at Ealing College, suggested to Christopher Small, the senior music lecturer, that John Stevens run improvisation workshops. The drummer I was working with, Paul Burwell, was doing all sorts of things – grave digger, playing music in strip clubs – he was desperate to get out of it. He had no conventional artistic skills, but he applied to Ealing and he got accepted because it was always a revolutionary place, with people like Gustav Metzger and Roy Ascott doing courses; Gustav Metzger lectured about destruction in art, and that's where Pete Townshend got his ideas about destroying guitars. Anyway, I started going with Paul to these improvisation classes. They were extraordinary because

7 John Stevens, *Search and Reflect: A Music Workshop Handbook* (London: Rockschool, 1985/2007).

they took you down to the foundational elements of working with sound with an instrument. We were still working with conventional instruments, though I was playing electric guitar which was considered a bit *risqué*.

Parkinson: It must have built communities and a focus on community building and group interaction as much as it did the listening practices that emerged out of it.

Toop: It had a knock on effect. Paul and I had started working with a sound poet – Bob Cobbing – who I also include in this story. Bob was a well-known sound poet and very involved with the 1960s avant-garde. He was the manager of Better Books book shop and he was involved in organizing big poetry events, but he was also beginning to use the tape recorder as a way of expanding the possibilities of the voice. There are astonishing recordings of him and Annea Lockwood. I think of her also as being one of the most important people in this early history of UK sound art. She made a sound hat which was just a hat with objects hanging from it, a bit like an Australian cork hat, but her sound hat was included in Henri Chopin's magazine *OU*, and she worked with Bob, she worked with Hugh Davies, so there was a tight nexus of people in this same area [...] a lot of this work was done in collaboration with Jeff Keen, the filmmaker from Brighton, who was also making sound pieces using toys and basic sound-making tools, so I think Jeff Keen is also part of this history. There's that overlap between what is experimental film, what is sound poetry, what is tape music, all of this work falls uncertainly across fields of activity which have grown wild and merged with each other.

Paul and I were working with Bob Cobbing in a trio, but then we formed a sextet with another vocalist, Lyn Conetta, an alto saxophone player, Herman Hauge, and Christopher Small. That sextet – abAna – played together for quite a few years and the trio played on and off right up until Bob's death, so there's another area of experimentation which was very much about the expanded voice and the way the voice can work together with instruments and sound-making devices. Bob worked with several people who were involved in what was thought of then as experimental music – Hugh Davies, for example. Hugh studied music, then became an assistant to Stockhausen, and there he learnt a lot from Jaap Spek, a Dutch technician who taught him about contact microphones. He came back to England and started making extremely eccentric instruments using contact microphones – I'm looking at one here which is a star map, springs, the internal workings of a telephone, bits of wire, a small mixing desk, a 3-D photograph of swans, a violin bow, toothbrush, screwdriver. He used an egg slicer which was held against one of the microphones, and you could pluck that or blow that. He called them *shozygs*, which came from Encyclopaedia Britannica – sho to zyg. The instrument was built inside a gutted encyclopedia, and he said the encyclopedia was gutted to substitute direct experience for learning, which I think encapsulated Hugh's approach.

Parkinson: What did it look like to see him play that, in terms of spectacle, comedy, or virtuosity?

Toop: Hugh deliberately added comic elements and was very explicit about trying to make witty things. He liked doing workshops with children or people who had never come into contact with any kind of music-making before, and he and Max Eastley would make instruments in workshops. To be humorous about what you were doing was a big advantage in that situation. Hugh worked in one of the early improvisation groups, Music Improvisation Company in 1968, with Derek Bailey, Evan Parker, Jamie Muir, and later a singer, Christine Jeffrey. One of the things Derek Bailey said about that group was that he had spent years developing a technique in which he had completely eliminated the sound of an open string, and suddenly along comes Hugh with these strings stretched across boards, so what they are essentially is open strings! They were kind of a direct challenge to these refined personal languages that were building up amongst free improvisers. There's an idea of conflict, which is useful to pursue, and important to me in terms of rethinking these supposedly watertight classifications that we have, like "sound art." They're umbrella terms for people who are very diverse and often opposed in their ideas, theories, and aims. Some of the free improvisation groups were really about conflict resolution: they were people who were diametrically opposed in what they were doing yet were supposed to be working together in a situation where nobody was directing them. There was no third party to produce an external focal point; everything was down to them working this out through the music.

Parkinson: I'm always a bit suspicious of the projection of democratic ideals onto free improv, because in practice it seems more complex.

Toop: It's often a struggle. Hugh spoke about this in Derek Bailey's book *Improvisation*, and he gave a specific instance: there was a performance by Music Improvisation Company.[8] Evan Parker had started a very high repeated thing on the saxophone, and Hugh knew that he was the most suitable person to join in at that point because of the resources he had compared to the other people playing; instead of joining in straight away he waited until he felt that Evan was at the end of being able to do that repeated high thing on the saxophone. And then he joined in, which made Evan continue longer than he actually wanted to. So there's game playing, a type of conflict in which people are working together but there are antagonisms, aggressive aspects, or passive-aggressive aspects to it which I think make the music much more than just an exercise in anodyne democracy.

There was a feeling that something was broken, music had come to lack a social connection and become an empty ritual, and so you looked for pockets of resistance. Some of the political upheavals

8 Derek Bailey, *Improvisation: Its Nature and Practice in Music* (Ashbourne, UK: Moorland Publishing, 1980).

of the early 1970s were about that. For example, Keith Rowe and Cornelius Cardew believed the music that they had been making was a response to that but then they became dissatisfied with it and became Maoists. For a while that seemed to be the answer for them. There was a lot of interest in mysticism, a lot of interest in politics of various kinds, and whatever seemed to be a path to developing more meaningful relationships of music to wider society.

There are two people I wanted to talk about whom I haven't mentioned yet: one is Max Eastley, and the other, Annabel Nicolson. Max started making instruments that were activated by weather. He was a student at Hornsey College of Art and his final-year show in the early 1970s was one of the most important events in developing my thinking about sound art. It was like an alchemist's laboratory of devices that made sound, and they were autonomous. This is a crucial part of this story: how to make sound when no human beings are involved.

Parkinson: An interesting contrast to the absolute involvement of human beings in John Stevens's work!

Toop: Yes, polar opposites who might end up working together! The John Stevens thing was very much person- and community-centered, and then Max, in 1968 or 1969, was thinking about these instruments where you could build something and just leave it, and it would carry on making sound because of the water running through or the wind or whatever. This became part of this movement we thought of as instrument-building which was a kind of sound art practice. It was a way of thinking about site-specific installation using sound. Annabel Nicolson came from another direction: she was an independent filmmaker interested in expanded film. She moved into something way beyond film – using sound and text and light and water.

She made a beautiful book called *Escaping Notice*, and Annabel's story is quite poignant because she has kind of escaped notice, despite her importance. There's a fantastic conversation between her and Max Eastley in this collectively produced magazine called *Musics*: "I've been thinking recently that performances are almost like lectures, focussing thought as a means of sounding out what is most urgent in one's mind [...] and no gaps, everything just immediate, just the directness of talking. Nothing else is needed, just the means you have, like your voice [...] Performance is a struggle and in a sense things are coming from something silent and making a huge leap towards being audible. Something very ancient about it."[9] I think all of these ideas, they're really powerful. Speaking personally, they're the kind of ideas I continue to try to work with, because I think there's a lot of unfinished business with them. Unfinished business: that's what we've got here, isn't it?!

9 Annabel Nicolson and Max Eastley, "Max Eastley talking to Annabel Nicolson," Musics, No. 18, July 1978, 11–12.

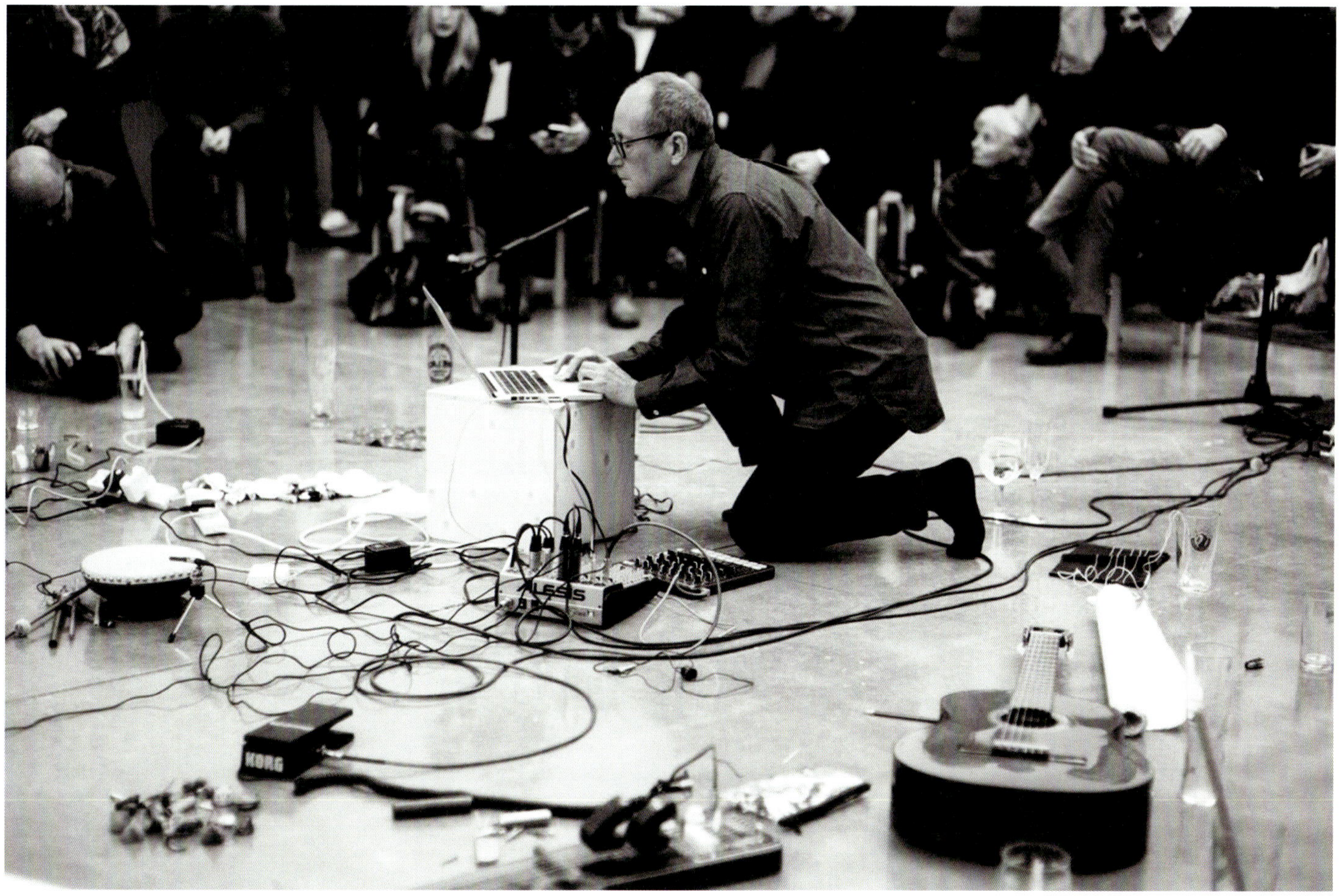

David Toop performing at White Cube, London. January 31, 2015.

Carsten Seiffarth and Bernd Schulz

Sound Art in Germany

Carsten Seiffarth and Bernd Schulz in conversation with Linnea Semmerling, Berlin, April 2015.

Linnea Semmerling: In the art world, the term *Klangkunst* is established well beyond the German-speaking context. What does it mean to you? Is there such a thing as "German sound art"?

Bernd Schulz: There's what you might call a scholastic rage, in the German world, a rage for categorizing everything, putting everything in little boxes and sticking labels on them. In all the years I've been involved with art, I've never really considered that there might be such a thing as German sound art. I think the more art evolves, the more these categories will melt away. I think that today, in the wake of globalization and the "erosion"[1] of materials Adorno talked about – and of media, which we have to add these days – the only thing that matters is whether or not it's good art. Good art benefits our perception. The categories that are operative today are entirely different from the ones that were recognized at the time "German" or "European" sound art was evolving.

Carsten Seiffarth: From a contemporary perspective, I agree with you. But it's quite interesting to see how certain developments in the art world have been shaped by different terminologies. We both know that *Klangkunst* is really a very weak translation of the English term sound art. But apart from that, sound art stands for an extremely broad field of sonic activity beyond concert music, while the

Translated from the German by Patrick Hubenthal.

term *Klangkunst*, in the German-speaking world, is much more narrowly defined, because its key criterion is space. This means that *Klangkunst* is seen more as a form of visual art, whereas sound art is typically closer to experimental music. On top of that, to add to the confusion, we have sonic art, a mainly British genre that's closely linked to electroacoustic music. So it can certainly be said that *Klangkunst* occupies a special position, and also that it's often used in contrast to the more general term sound art, in international contexts, to discuss space-oriented works of art involving the material or medium of sound. The term *Klangkunst*, incidentally, appears quite late in German literature. That has to do, in part, with the fact that some of the artists who had already been working with sound for some time started calling themselves *Klangkünstler*, to differentiate themselves from others. I myself, for example, have been using terms like *Klanginstallation* and *Klangskulptur* more and more in recent years, and in English I talk about sound installation art instead of sound art.

Schulz: I've made a point – specifically with the intention of educating the public – of always talking about "so-called *Klangkunst*" in order to draw attention to the translation problem. The term *sound*, in English, has a completely different set of associations than our German *Klang*. The reasons for that are not only linguistic, but also historical. In my opinion it goes back to Hermann von Helmholtz, who distinguished pure *Klang* from *Geräusch*[2] – and this, interestingly, is what first made *Klang* an object. So *Geräusch* had to be laboriously reclaimed by the artists. With *sound*, the Americans had an entirely different understanding of these concepts from the very beginning. For us, though, *Klang* almost immediately calls up an association with musical tone.

Seiffarth: Right, primarily tone organized into musical forms.

Schulz: I learned a lot about this from Rolf Julius. His source material often consisted of real acoustic sounds, sometimes from instruments, such as the piano. He then broke this material down into microtones and recombined it electronically. As a result, the "noise" parts of the sounds became particularly interesting – for example, the transients produced by piano strings.

Semmerling: Today the whole world associates your work with sound art.

Seiffarth: [...] Well, that pretty much seems to happen by itself. You keep working more and more intensively in a certain subject area, and at some point you're left holding the bag – from now on you're a "sound art curator."

1 Theodor W. Adorno, "Art and the Arts," trans. Rodney Livingstone, in *Can One Live after Auschwitz? A Philosophical Reader*, ed. Rolf Tiedemann (Stanford: Stanford University Press, 2003), 368–88, here 369.
2 See Hermann von Helmholtz, *On the Sensations of Tone as a Physiological Basis for the Theory of Music*, trans. Alexander J. Ellis (London: Longmans, Green, Reader, and Dyer 1875), 11–12. The relationships between the German and English vocabularies for auditory phenomena are complicated; in certain contexts, both *Klang* and *Geräusch* may be translated as *sound*, while in others, *Klang* is closer to *tone* and *Geräusch* to *noise*.

Schulz: You just can't do anything against these processes. [...] I've often wished that those exhibitions in Saarbrücken that are presented as sound art exhibitions today, had just been seen as art exhibitions. [...] In my work – with the exhibitions and the publications – I always just wanted to explore how art sharpens our awareness of so-called reality (FIGS. 1–10) Sound art is just one part of that. [...] But there are limits to this business of sharpening public awareness. They're determined by individuals' social experiences, by their confined milieus, etc. For example, in Saarbrücken we also had a very coherent concert program, with improvised music as well as serious written music. But it wasn't embraced by the public the way we'd hoped. The public remained split: the opera and theater fans appreciated the new classical music, but they didn't show up for the improvised-music events, or for the exhibitions involving sound, where we were trying to chip away at the boundaries of ossified perception. That showed me that the whole thing actually goes much deeper, and that getting comfortable with this new art form is quite a long process.

Seiffarth: These are the labels that get put on things. Of course, to some extent we produced them ourselves. Even if you're always talking about "so-called" sound art, you're obviously still producing the expectation that something is going to happen with sound and art. [...] But really, all along, the point has just been to show good art. So it's completely ridiculous to make sound, as an acoustic phenomenon, your sole criterion. There are important works of sound art where sometimes there's nothing to hear – pieces by Akio Suzuki, Terry Fox, and Joseph Beuys, for instance. [...] Ever since 2010, when the Turner Prize went to a sound installation for the first time, the visual-art world's interest in so-called "loudspeaker art" has been growing. All at once the market started paying attention; it was suddenly clear – like when the flat screen arrived in video art – that there was something to buy and sell there. It's happening more and more now, whether you have four, twenty-four, or a hundred speakers in a room. The visual-art scene finds it exciting if every now and then, in the middle of the exhibition gauntlet, you have a kind of movie for your ears, to activate the mind's eye. That's just how banal the market-friendly segment of today's sound art is, most of the time. It reminds me of the familiar reception model in classical music: just close your eyes and create "beautiful" mental images. But just because it uses a material that's in vogue right now and seemingly easy to generate – sound – that doesn't automatically mean it's good art – and most of it's not space-oriented sound installations, either. But on a very general level, it's really a good thing that sound is finally starting to make real inroads into visual art as a material and a medium. Though in actual fact, of course, it was always there. Just think about documenta 8 (1987). It's just that the art industry still lacks experience and critical expertise in the area of sound.

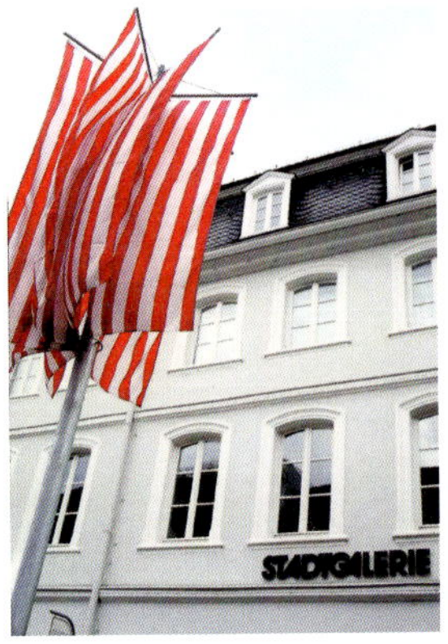

FIG. 1 Ernest W. Uthemann, ed., *Stadtgalerie Saarbrücken 1994–2001: Eine Dokumentation* (Saarbrücken: Stadtgalerie, 2011). Cover.

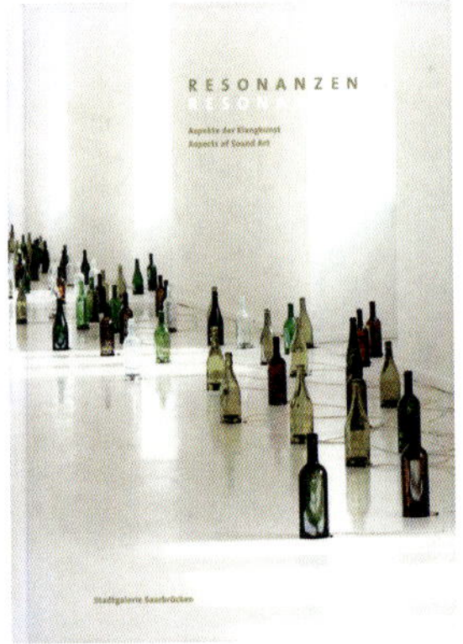

FIG. 2 Bernd Schulz, ed., *Resonanzen/Resonances. Aspekte der Klangkunst/ Aspects of Sound Art*, exhib. cat. Stadtgalerie Saarbrücken (Heidelberg: Kehrer Verlag, 2002). Cover.

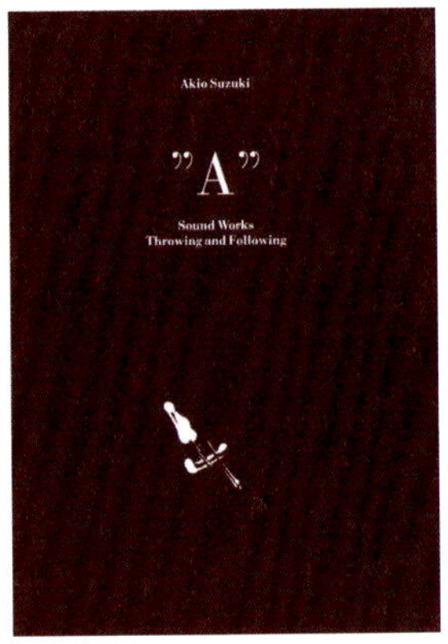

FIG. 3 Bernd Schulz, ed., *Akio Suzuki „A": Sound Works: Throwing and Following* (Saarbrücken: Herziger, 1998). Publication accompanying the exhibition of Akio Suzuki at Stadtgalerie Saarbrücken. Cover.

FIG. 4 Bernd Schulz and Irmgard Heesche, eds., *Christina Kubisch Zwischenräume*, exhib. cat. Stadtgalerie Saarbrücken and HBK Saar (Saarbrücken: Stadtgalerie, 1996). Cover.

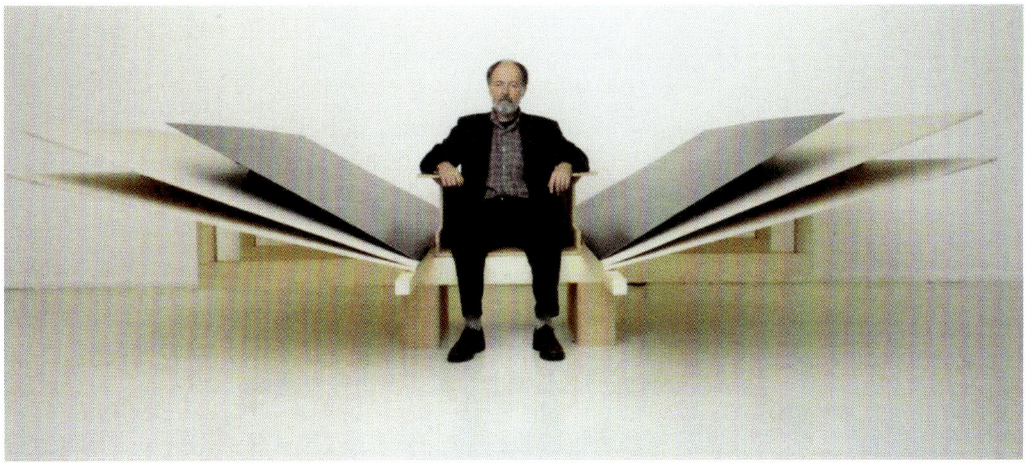

FIG. 6 Bernhard Leitner, *Innen-Weiten* [Inside Expanding], 2002. Installation view Stadtgalerie Saarbrücken with Bernd Schulz sitting on the sound object.

FIG. 5 Bernd Schulz, ed., *Klangräume* [Soundspaces], exhib. cat. featuring Bernard Baschet, Francois Baschet, Gunter Demnig, Ulrich Eller, Stephan von Huene, Rolf Julius, and Christina Kubisch (Saarbrücken: Stadtgalerie, 1988). Cover.

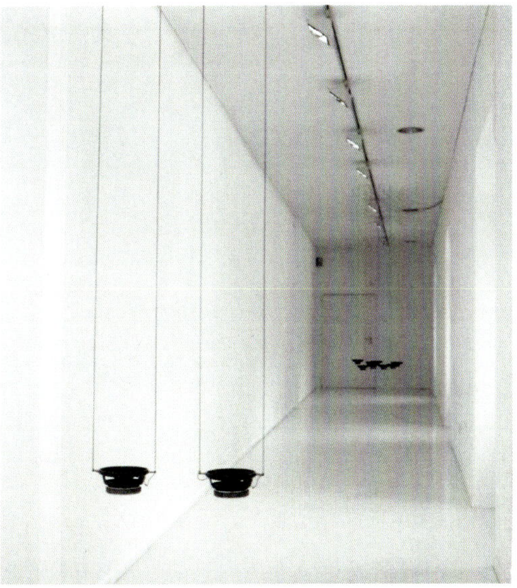

FIG. 7 Rolf Julius, *Sound cooking*, 1994. Installation view Stadtgalerie Saarbrücken, 1994.

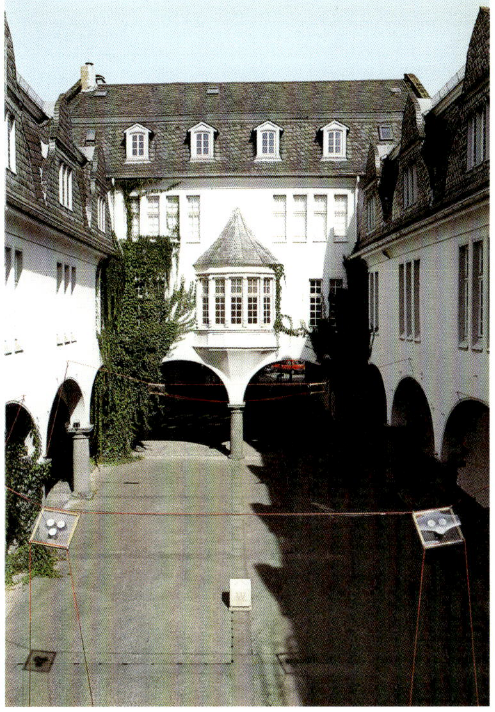

FIG. 8 Andres Bosshard, *Zirruswolken werfen keine Schatten – Klanginsel für vorbeifliegende BallonfahrerInnen* [Cirus Clouds cast no shadows – sound island for passing balloonists], 2002. Installation view Stadtgalerie Saarbrücken, 2002.

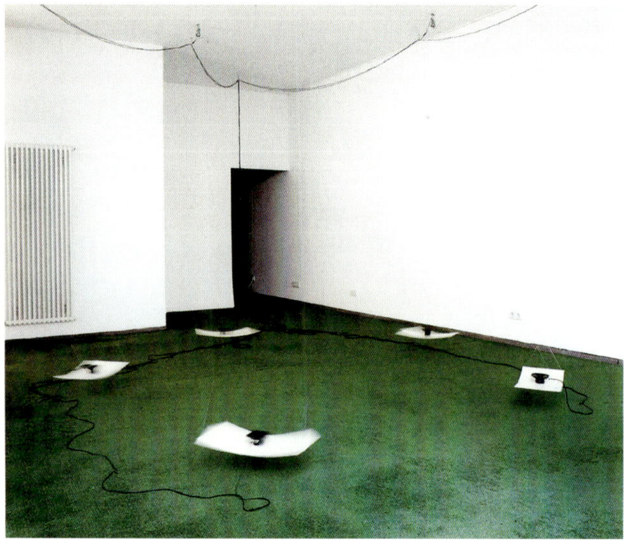

FIG. 9 Ed Osborn, *Swing Set*, 2002. Installation view Stadtgalerie Saarbrücken, 2002.

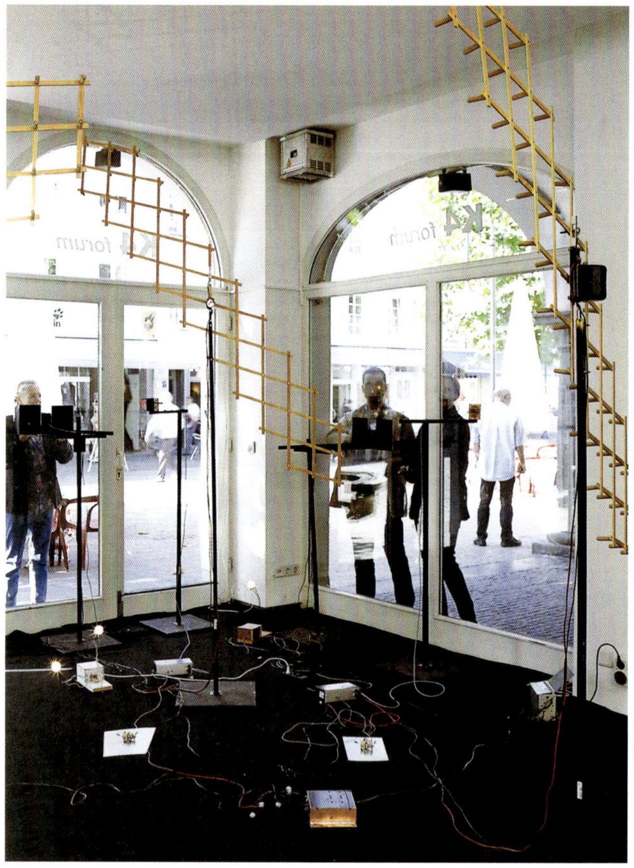

FIG. 10 Erwin Stache, *trans Formation*, 2002. Installation view Stadtgalerie Saarbrücken, 2002.

Semmerling: What happened in the 1970s?

Schulz: One would have to study the influence of the musical avant-garde of the 1950s and 1960s more closely. The influence of information theory, artificial intelligence, structuralism on a new generation of artists. Who, ultimately, is the typical "sound artist"? Personally, I think of Beuys's student Claus Böhmler (FIG. 11) and his mediacritical works. He had a lot of influence as a university lecturer. Artists like Gunter Demnig (FIG. 12) and Otmar Sattel worked with sound. Dieter Roth! But you don't see the "sound art" label on them today. There are lots of overlaps with performance art, where the bodily, the processual, comes clearly into play. One would also have to include artists like Terry Fox, Emmett Williams, Paul Panhuysen, and others who were present in Germany. If we narrow our focus to Germany, then we have, after Rolf Julius, the first generation, such as Christina Kubisch and Ulrich Eller: artists who were already interested in the intersection of visual art and music as students – Kubisch as early as the beginning of the 1970s, and Eller at the end of that decade. Christina Kubisch, flautist; Ulrich Eller, rock musician. In France, influenced by Pierre Schaeffer, the Baschet brothers. For many students, especially in Germany, John Cage was already a role model by then. Then in 1987, Kubisch and Eller were at documenta 8 with Rolf Julius. Julius was originally in visual art.

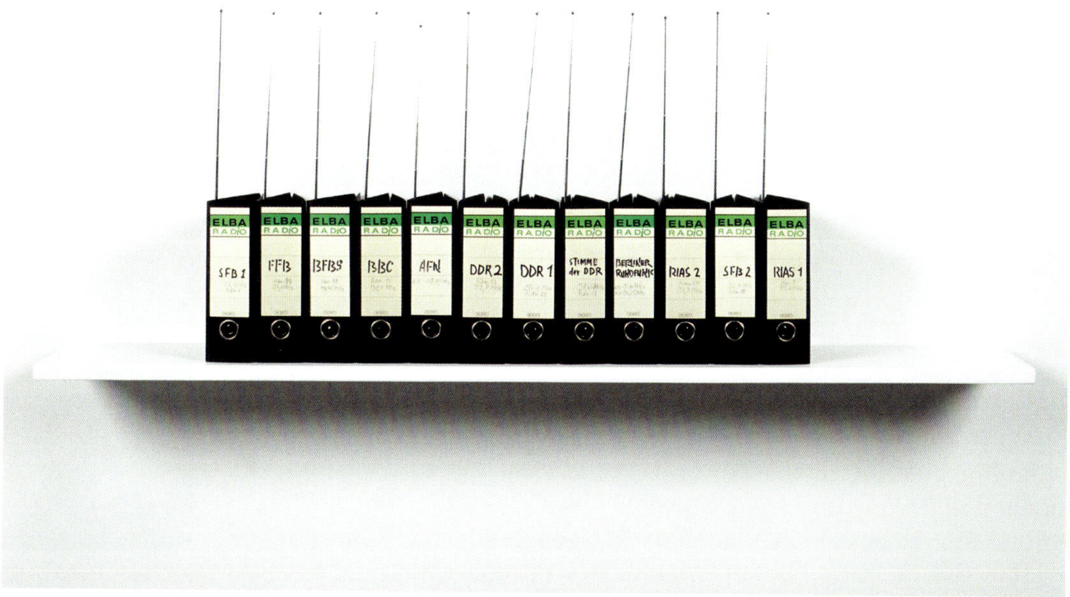

FIG. 11 Claus Böhmler, *Berliner Rundfunkvereinigung* [Berlin broadcasting union], 1984.

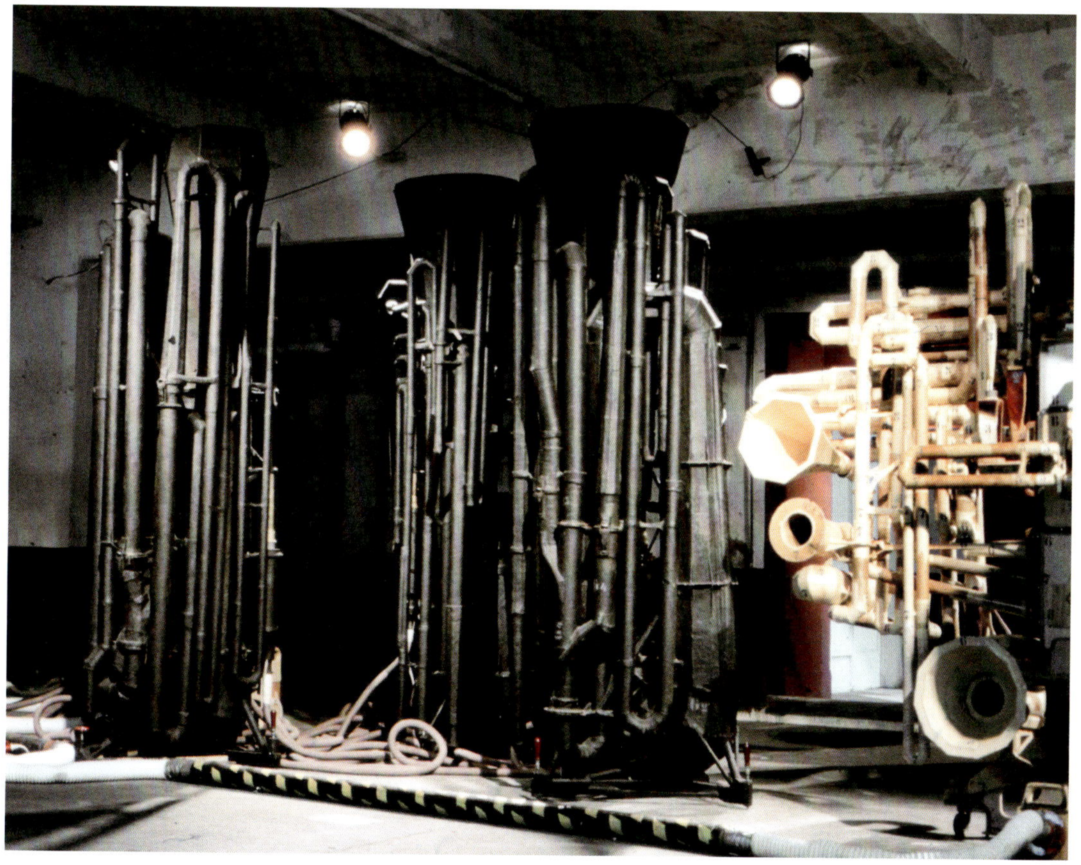

FIG. 12 Gunter Demnig, *Infraschall* [Infrasound], 1991. Installation view *MultiMediale 2*, Karlsruhe, 1991.

Hans Gercke, who was the director of the Heidelberger Kunstverein, brought him to my attention in the early 1980s. He showed me photos and told me about this strange guy who made sounds for trees. I looked at the photos and recognized Arvo Pärt in the audience. He was in Berlin in 1981–1982 as a DAAD scholar. And I said to myself, if Arvo Pärt is listening to this *Concert for a Frozen Lake* with such an intense look on his face, then there must be something to it, I need to look into it. And then, of course, we can't forget Bernhard Leitner, who combined architectural space with sonic space. He was already at documenta 7 (1982) with his *Soundcube* and *Sound Tower*. [...] For the first time, there began to be a focus on an art that seemed to be working with new possibilities of perception. And that also launched the discussion about the term *Klangkunst*. If I remember correctly, many of the players were against the label *Klangkunst*.

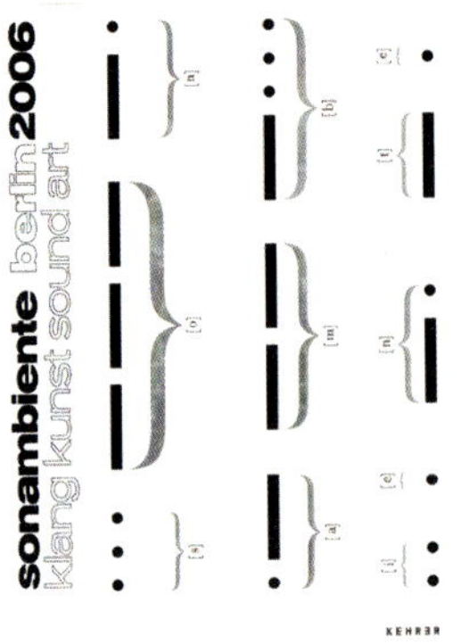

FIG. 13 Helga de la Motte-Haber, Matthias Osterwold, and Georg Weckwerth, eds., *Sonambiente Berlin 2006: Sound Art* (Heidelberg: Kehrer Verlag, 2006). Cover.

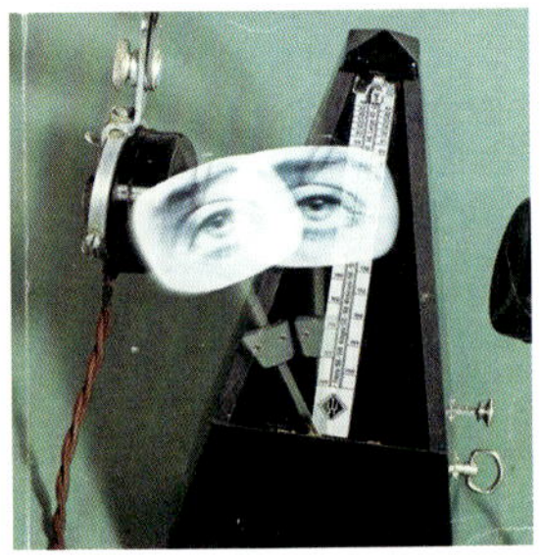

FIG. 14 René Block, ed., *Für Augen und Ohren. Von der Spieluhr zum akustischen Environment* [For Eyes and Ears: From the Music-box to the Acoustic Environment] (Berlin: Academy of Arts, 1980). Cover.

Seiffarth: Up to that point it was still a very small field, with a relatively small number of places where you could actually find this so-called *Klangkunst*. Bernd had already started in the 1980s, and then I came along in the 1990s. And by coincidence, 1996, when we started singuhr – hoergalerie in parochial – we got the idea in 1995, and we opened in 1996 – was also the year Michael J. Schumacher started Studio Five Beekman in New York. That evolved into Diapason Gallery, which closed in 2012. Schumacher himself, though, has a very strong background in musical composition, so it was more of a gallery for electroacoustic environments and installations. That same year the big sound art exhibition *sonambiente* was held at the Akademie der Künste in Berlin – that was basically the bandwagon we jumped on with singuhr. I mean, we'd already had the idea before *sonambiente*, but thanks to the festival we suddenly had the means to produce and mount five exhibitions in our first season, and we appeared in the catalogue (FIG. 13) as a no-name institution too. That was certainly quite important at the beginning, since as an alternative project, coming essentially out of nowhere, we really had nothing. [...] And Berlin, in particular, had been the melting pot for sound art since the 1970s; some people call it the unofficial capital of sound art. The big exhibition *Für Augen und Ohren* [For Eyes and Ears] was held here in 1980 (FIG. 14); the DAAD's Artists-in-Berlin Program was active here, bringing in international sound artists; Rolf Langebartels was here with his art association, Giannozzo; and many important German and international sound artists lived here, and still live here today. And Helga de la Motte-Haber taught systematic musicology at the TU Berlin – she was one of the few academics who was consistently involved with sound art and knew how to get her students interested in it too. And she was always in contact with the artists and the people putting on the shows.

Semmerling: It sounds like Helga de la Motte-Haber contributed a lot to German sound art in the 1990s and the early 2000s.

Seiffarth: Of course, I also have a personal connection to her, because I had the opportunity to study with her, and I'm still always happy to work with her today. It's rare to find someone who's still willing and able to keep exploring new areas at her age, including ideas that really have nothing to do with her own territory. For the latest singuhr catalogue, for instance, she wrote an important essay illuminating the concept of site- and situation-specificity.[3] All in all, I think she's one of the most important people for sound art in Germany, not just in terms of educating a generation for whom sound art is perfectly normal, but also in the way she combines scholarly reflection with the practical suggestions she gives artists and producers for their work.

Schulz: Naturally, she's also been very important for me as a curator. As a musicologist and phenomenologist, she's looked – and listened – quite closely. She's helped us develop a terminology for what's been happening with sound in art.

Semmerling: So far, we've been dealing exclusively with West Germany; was there no such thing as sound art in East Germany?

Seiffarth: Well, you can't really say there was no such thing. I grew up in the East, but back then I didn't know enough about the margins, about what was going on there – in the 1980s I was active as a listener in modern classical music, for one thing. Only later did I try to investigate things a little. For example, there were some quite interesting composers who were open to working outside the realm of classical concert-hall music, such as H. Johannes Wallmann and Robert Linke. In the late 1980s Armin Köhler, who would go on to direct the Donaueschingen Festival, put on an event called *Klanghaus* [Soundhouse] at the Kulturpalast in Dresden. That was a kind of walking-concert project, which at the time was quite innovative, because for a brief moment it allowed him to create an artistically open space. That was really unique under the cultural policies of "real socialism." In visual art, by contrast, they were working a lot with sound/*Klang* in the underground – the Autoperforationsartisten in Dresden, for example. And let's not forget Carlfriedrich Claus and his "speech sheets" (FIG. 15); later, in the 1990s, he did a sound installation too. But I think Erwin Stache (FIGS. 16a, b) was the only sound artist – although he didn't present himself that way officially; at that time he was more of a sound tinkerer – and, oddly enough, we started singuhr in May 1996 with what was essentially the GDR's only sound artist.

Schulz: In my opinion, Stache is one of the few artists in this field who brings some humor to the proceedings. That humor is very helpful for loosening up inhibited or ingrained perceptions. That makes me think of Henri Bergson, who won the Nobel Prize primarily for

3 See Carsten Seiffarth and Markus Steffens, eds., *singuhr – hoergalerie 2007–2014: sound installation art in berlin* (Heidelberg: Kehrer, 2014).

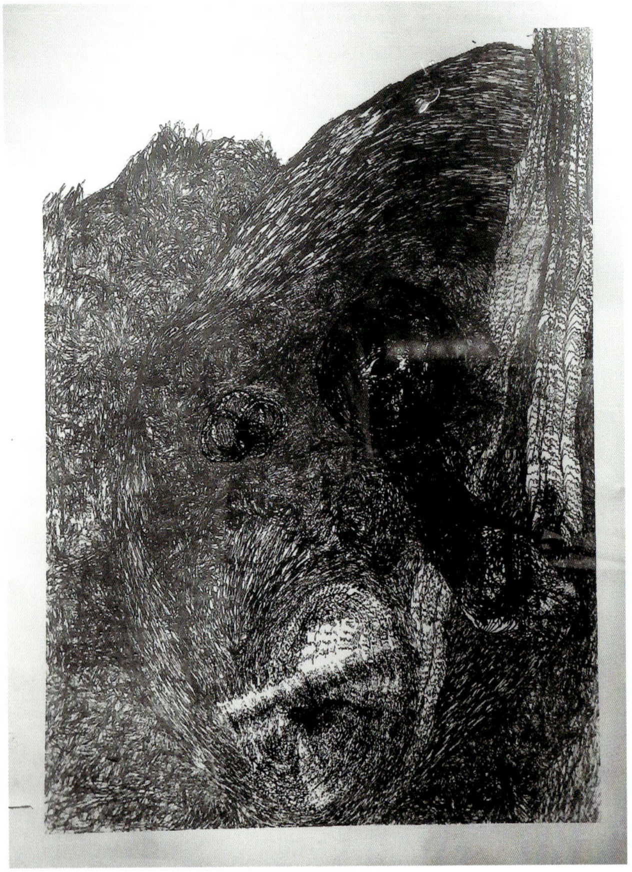

FIG. 15 Carlfriedrich Claus, *Schweigen* [Silence], 1968.
Created on the occasion of the Prague Spring.

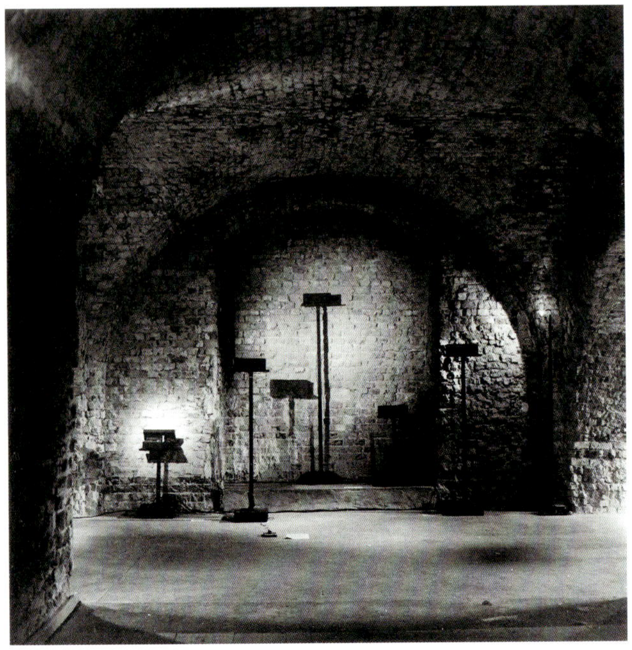

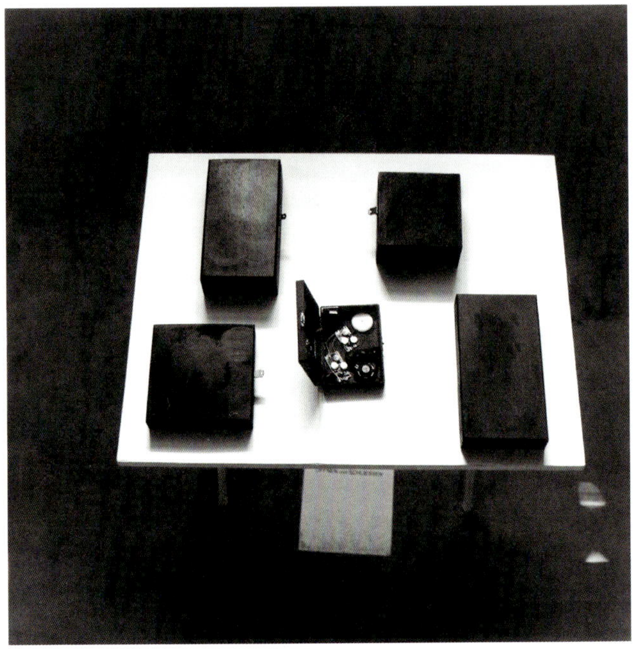

FIGS. 16a, b Erwin Stache, *Klangkästen* [Musical Boxes], 1996. Sounds from 27 black boxes, wooden boxes of different sizes, light sensors, prepared/manipulated „everyday electronics," and sound chips made of toys, greeting cards etc. Installation view singuhr–hoergalerie, 1996.

his book on laughter.[4] That's something that's sunk completely out of view in art – maybe because of the unusual seriousness with which we tend to approach things these days in Germany.

Semmerling: Bernd Schulz, you founded the Stadtgalerie Saarbrücken and were its director from 1985 to 2002. Carsten Seiffarth, you founded singuhr in 1996 in Berlin, first in the Parochialkirche and later in two underground cisterns. How did you position yourselves back then? What were you trying to do, how did you do it, and how have you pursued it over all these years?

Seiffarth: In the early 1990s, when I first heard of sound art – I was still studying with Helga de la Motte-Haber then – festivals were pretty much the only place you could experience that sort of thing. So if you wanted to see and hear sound art, you had to travel – to Hanover, for example, for SoundArt 95, or to Rotterdam for the Dutch Electronic Arts Festival (1994, 1995, 1996, and 1998). [...] Sound sculptures and sound installations were often shown in the context of new-media art as well. The only consistent place for this art in Germany was the Stadtgalerie Saarbrücken – and even though it was really far away from Berlin, and I had no money, at some point I made it to almost every one of Bernd's sound art exhibitions. It was simply the only place where sound art was being regularly and continuously maintained. They were doing research there – just think of all the important monographs – and someone was paying attention to these interesting, unconventional artists. Then, later, music festivals started presenting sound art too – the Donaueschingen Festival, for example, and later the Wittener Tage für neue Kammermusik. [...] The problem with festivals in general is that most of the time they're not producing anything new; they're just buying what already exists. The work obviously gets tremendous attention, quantitatively speaking, but it doesn't work to create a good situation for taking in a work of space-based sound art at a two- or three-day festival, there's just too much going on. When we started the gallery back then [...] it was precisely that kind of antifestival gesture. Festival culture was everywhere, and it was crowding out the culture of permanent alternative spaces. And if you wanted to be involved at all, you essentially had to be perpetually traveling from festival to festival. With singuhr, we wanted the opposite: peace and quiet, a chance to downshift, to slow down – to realize new installations for specific spaces. And of course we also wanted to show things that didn't fit into the festival circuit. As is still the case, the productions got passed around from festival to festival, which meant that more and more you were seeing things you'd already experienced. Unlike Bernd's place in Saarbrücken, at singuhr the idea was simply to establish a space where sound installations and sound sculptures could be experienced over relatively long periods of time, and where all the pieces would be produced specifically for the site. That was the directive. And at that time there existed nothing like it on that scale.

4 See Henri Bergson, *Laughter: An Essay on the Meaning of the Comic*, trans. Cloudesley Brereton and Fred Rothwell (London: Macmillan, 1911). Originally published as *Le Rire. Essai sur la signification du comique* (Paris: Félix Alcan, 1900).

Schulz: [...] Carsten brings up a point that was very important for both of us in our work. This so-called sound art sensitizes us to space – to a large space, in the sense of Murray Schafer's soundscapes (FIGS. 17, 18), in which the real external space and the perceiver's internal emotional space overlap and permeate one another, but also to architectural space: sounds settle around objects and alter them. So for example, art with sound can call attention to the historical significance of public places and reclaim them. That's especially important in Germany, because our public space was largely destroyed in World War II, and also by contemporary architecture after the war. Suddenly there was a new kind of art in public space that made people aware of their environment and its history, in a way that was not only visual but also auditory. Places acquired a new aura. Carsten had the good fortune – or maybe the misfortune – to establish a space in the public consciousness that already had a significant aura of its own. The Parochialkirche in Berlin had not only a special sonic and architectural quality, but also a historical aura, which many artists did wonderful things with.

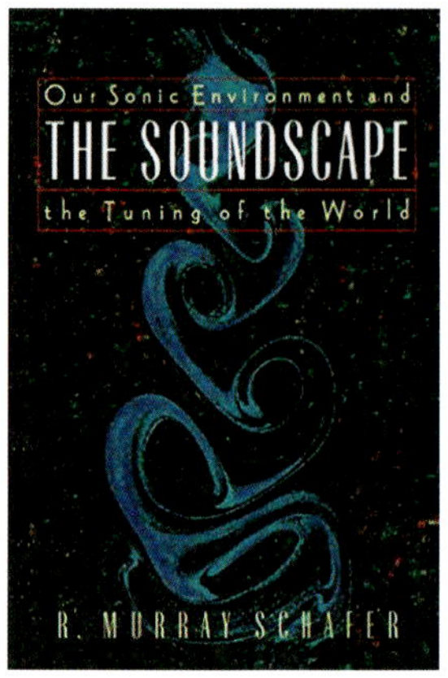

FIG. 17 R. Murray Schafer, *The Soundscape: Our Sonic Environment and the Tuning of the World* (Lochester, VT: Destiny Books, 1977). Cover.

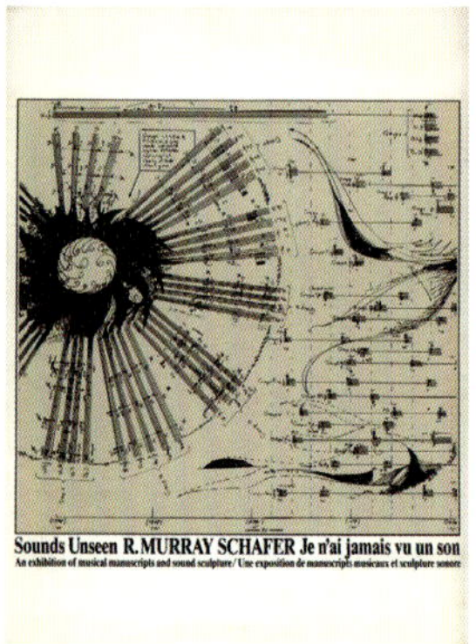

FIG. 18 R. Murray Schafer, *Sounds Unseen. An exhibition of musical manuscripts and sound sculpture.* Cover image of a flexi disk which was produced 1981 on the occasion of R. Murray Schafer's exhibition at Gallery Stratford in Stratford, Ontario (CA).

Semmerling: Bernd Schulz, you've stepped down as director of the Stadtgalerie Saarbrücken. Carsten Seiffarth, what's next for singuhr?

Seiffarth: [...] Every project is like a child. In that sense, you can't just finish projects and do something completely new. In the end we decided not to totally abandon singuhr, after eighteen years, but just to close the gallery, for political and financial reasons. I'm quite pleased to have finally shed the "hoergalerie" label in the process. [...] For one thing, singuhr was always more of a project space – or in Berlin cultural-policy jargon, a "project initiative" – than a gallery. [...] For another, *hearing* was often misconstrued, unfortunately, as musical, purely auditory listening. But of course if you really think about it, hearing is much more than just listening with the ears. It's a kind of full-body perception that integrates the other senses rather than excluding them. So since 2014, singuhr – projekte has been operating without a permanent location, working with various partners to use other spaces and, mainly, to realize sound installations via subsidized, project-oriented artist's residencies. We're also trying to counter the isolation of the hoergalerie, to take the artists and our expertise to other institutions and spaces and have a lasting impact there. That's a completely different way of working – without permanent locations, as we did for the first eighteen years. Most of all, we're trying to create, maintain, and defend spaces for art with sound. [...] That's what Bernd did too: he simply said to the artists, I'll give you spaces, and you'll also have the opportunity – which hardly existed back then – to publish your first catalogue. With singuhr – projekte, we're no longer presenting sound art projects in a sort of isolated special sound art space; they're right where they belong. I think it's like new-media art – we don't need dedicated sound art festivals anymore; what we really need is just to take sound art where it belongs: museums, art galleries, public spaces. And the main priority is not to show sound art, but to show good art. So the projects have to have good curatorial oversight, and the technology needs to be at least as good as what we provided when we were a hoergalerie. Our first project with singuhr – projekte, incidentally, was in summer 2014 with the Estonian sound artist Raul Keller (FIGS. 19a, b). We invited him to Berlin for a project residency and supplied the space, the technicians, and the materials. The residency was in the exhibition space itself, which brings out incredible insights into space and its qualities. When you're working with sound in space, you really need time and quiet on-site. After a four-week residency, the final piece had a month-long public showing. For singuhr – projekte, it's essential that production and presentation are on an equal footing. Of course the public will still be important to us, but my heart is first and foremost with the artists.

Semmerling: What role does the pop music of this Western/European world play for artists who work with sound?

Schulz: The significance of so-called pop music for sound artists in Germany, Europe, and America is a topic Carsten and I don't always agree on. My impression is that things developed differently in England and the U.S. after the war than they did in Germany. Brian Eno, for example – who was actually awarded the ZKM's Giga-Hertz Award – would be unimaginable without John Cage and the Artists' Club in New York. […] There's this mixing over there, this interpenetration of pop music and so-called serious music. Just think of Frank Zappa – even Pierre Boulez bowed down before him.

Seiffarth: And Stockhausen. Two of his students started Can.

Schulz: Yes, and also between jazz and pop music. Speaking of jazz, it's strange that jazz and the African musical tradition seem not to have played any part in Cage's music. There's this interpenetration of popular music and serious music in the U.S. and England, but not in Germany, oddly enough. And I'm not so sure that Kraftwerk can really be viewed as the successful confluence that their celebrated appearance at the Neue Nationalgalerie in Berlin would suggest.

Seiffarth: But that has nothing to do with it, at least not directly.

Schulz: I think it absolutely has to do with our understanding of sound, and with our understanding of music. The artists whose work we're talking about here wanted to get away from the beat that divides time into equal parts. Kraftwerk – and many others in popular electronic music – are carrying that forward. It should not be forgotten that Cage was actually a percussionist. He had a specific feel for complex rhythms. That's different from traditional meter, which quickly turns into a mechanical pulse.

Seiffarth: Well, that's a separate aspect, since America and England have a completely different pop culture than Germany.

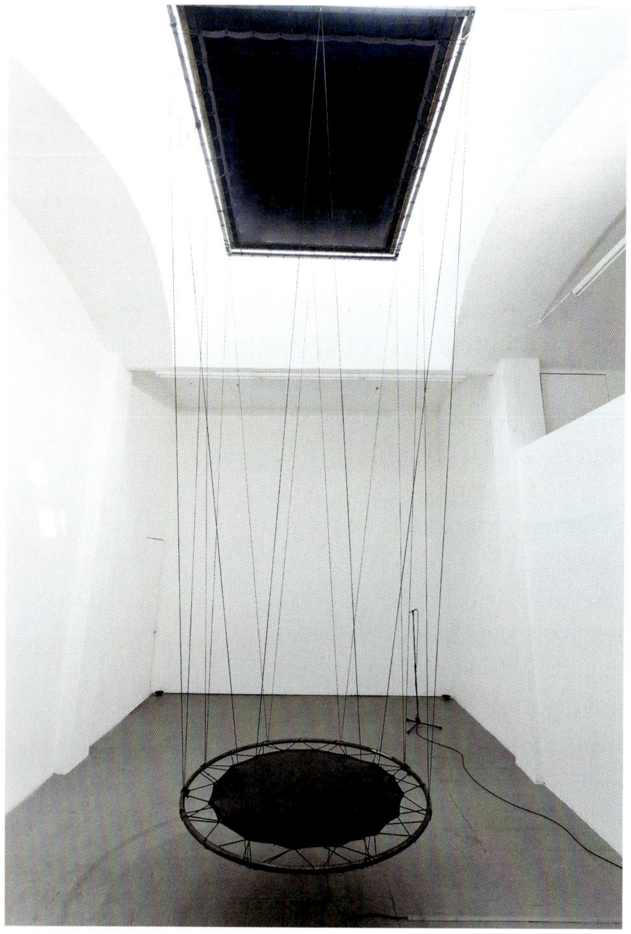

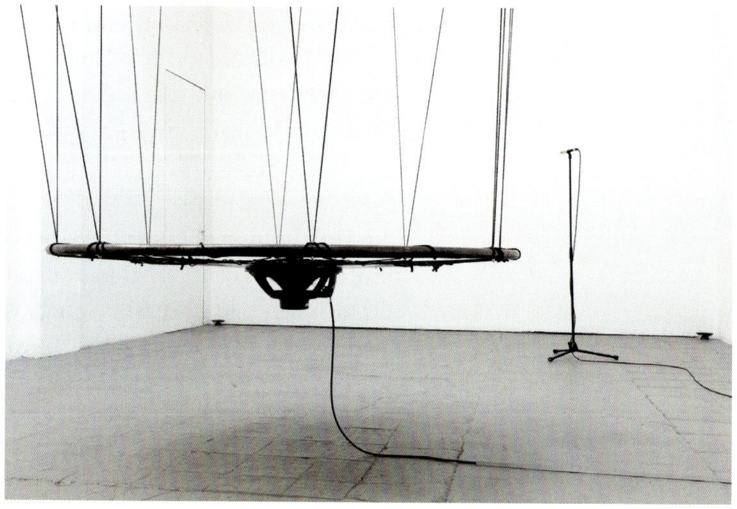

FIGS. 19a, b Raul Keller, *Moondur/Shifter*, 2014. Commissioned by singuhr–projects. Installation view Meinblau project space.

Schulz: True, but it's an important topic that deserves to be examined more closely. It's certainly notable how these two worlds are walled off from one another in Germany.

Seiffarth: I agree, without making any judgment. It's not just Brian Eno, by the way; the members of the Talking Heads were art students too, for instance. And some of them – David Byrne, for one – are making art again, sometimes with sound. That sort of thing never happened in Germany, and definitely not in sound art.

Schulz: Laurie Anderson is another one of these links to the art world.

Seiffarth: [...] In America it always seems to be possible to go into pop culture with experimental or unusual ideas and get big and famous with them; that's hardly ever been true in Germany. Just think of the pioneers of Krautrock, who have been making a comeback over the last few years – who was aware of them back then? At any rate, their music was not part of what you'd call popular culture.

Schulz: Maybe this can be linked to the spirit of the 1950s avant-garde – to Darmstadt in the 1950s, for example. That was a time when people in Germany, especially the younger generation, were just happy to have survived. The "economic miracle" had given them a new freedom. The European avant-garde was heavily focused on rehabilitating the prewar avant-garde that had been expelled and obliterated by the Nazis. The New York avant-garde, which had known the war only from a distance – Cage wasn't in it, you know – had positioned itself against American consumer culture, which was already quite developed by then. That probably strengthened their resolve to break with European tradition. At the same time, the European avant-garde took inspiration from the American scene. The elegance with which Cage critiqued the consumer world, and also the media, can be seen in *Water Walk* (1959). You can find his 1960 performance on the American TV variety show *I've Got a Secret* on the Internet. In America, they laughed at the way Cage made music with household objects arrayed around a bathtub. Everyone knew David Sarnoff, the founder of NBC, who made Arturo Toscanini famous but consistently ignored the avant-garde composers. For him, good European music was all that mattered. […] Almost everyone in America knew that he had said, "I regard radio as a sort of cleansing instrument for the mind, just as the bathtub is for the body."[5] That explains the two central gestures of the performance: on the one hand the bathtub, into which Cage dumped all sorts of things, and on the other, the knocking over of the radios. As a former radio person, I'm also reminded that our radio history in Germany has been totally different from America's. Here, state-controlled content was the only media experience for an entire generation, at a time when hundreds of local stations were free to grow in the U.S. So a curiosity about the possibilities of electronic media developed much sooner than it did here.

Seiffarth: Which brings us back to pop culture. It's important to note that the show Cage appeared on was part of the network's entertainment program, not the Toscanini program. In Germany, by contrast, Kagel's *Papiermusik* [Paper Music] was aired as high culture in the ZDF late-night program at 11:00 p.m. That was the difference.

Schulz: Maybe because it was fairly humorless, so it could pass for serious music? Speaking of humor, or humorlessness: Cage's laughter really must have made the attendees at the Darmstadt International Summer Courses for New Music in 1958 quite grumpy and aggressive. I'm sure he encountered a concentrated "scholastic rage" back then too – even among the representatives of Europe's musical avant-garde. […]

Semmerling: What institutions have shaped the history of German sound art? Did you experience the early exhibitions in Berlin, for instance – the projects by René Block and Giannozzo?

Seiffarth: René Block and Nele Hertling's *Für Augen und Ohren* is definitely one important project, since it was the first attempt to take a historical approach to sound art. Of course I only know that show from the catalogue, since it was held at the Akademie der Künste in West

5 Alex Ross, *The Rest is Noise: Listening to the Twentieth Century* (New York: Picador, 2007), 288.

Berlin in 1980, and I grew up in East Berlin, but it was clearly a milestone. Then came *sonambiente* in 1996, another milestone, now in the 1990s. Berlin was always called the unofficial capital of sound art. [...] By 1988 Rolf Langebartels had put on something like 34 exhibitions here with Galerie Giannozzo, in basement and storefront spaces in Schöneberg. I wasn't able to go there either, nor did I know anything about it at the time, unfortunately; how could I have? [...] Then there was also the DAAD's Artists-in-Berlin Program, which brought in international sound artists, even though that category didn't actually exist then. Most of the artists were invited as part of either the music program or the visual art program, and they were presented at exhibitions or the Inventionen festival. They included Stephan von Huene (1976), Max Neuhaus (1977), Milan Knížák and Joe Jones (1979), Terry Fox (1981), Robin Minard (1990), and Gordon Monahan (1992). So the Artists-in-Berlin Program was an important stimulator of the sound art scene, or whatever you want to call it.

Schulz: That exhibition title, *Für Augen und Ohren*, reminds me of Rolf Julius's piece with the little speakers on the eyes, *Music for the Eyes* (1981). That was just brilliant. But there were also other curators, before Block and Langebartels, who showed work that appealed to more than one sense - for example, Harald Szeemann and his 1969 exhibition *Live in Your Head: When Attitudes Become Form (Works - Concepts - Processes - Situations - Information)*. In that tradition, I see the title *Für Augen und Ohren* more as a metaphor for the area between the eyes and ears. The brain is there too, for instance, and the rest of the body, which we tend to forget. [...]

Seiffarth: Sure, there's the approach via the senses, and there's the historical approach. To trace the origins that way, you certainly don't have to have seen the projects yourself; you just know the documentation. Nam June Paik's *Exposition of Music: Electronic Television* at Galerie Parnass in Wuppertal in 1963 was, I think, one of those exhibitions that were extremely important for what came after.

Schulz: Fritz Schwegler is interesting too. He taught in Düsseldorf for a long time, in Beuys's day, but Harald Szeemann showed his *Effeschiadiana* at documenta 5 in 1972, which was quite early. That was great, that such an idiosyncratic artist got invited to documenta. He was an excellent harmonica player, he sang, wrote absurd texts, painted funny little pictures. And he was a wonderful person, there's no way to describe it. Terry Fox was the same kind of odd outsider. (FIG. 20) These artists never worried too much when people shook their heads in incomprehension; they just kept sensitively making their work. And it's very important that we don't just pigeonhole that as sound art, because artists like Fritz Schwegler would never fit into that way of looking at it. I just think it's great that Harald Szeemann had his antennae out for everything. The viewers were shocked and unsettled, and only ten, maybe twenty years later did people understand and appreciate that exhibition.

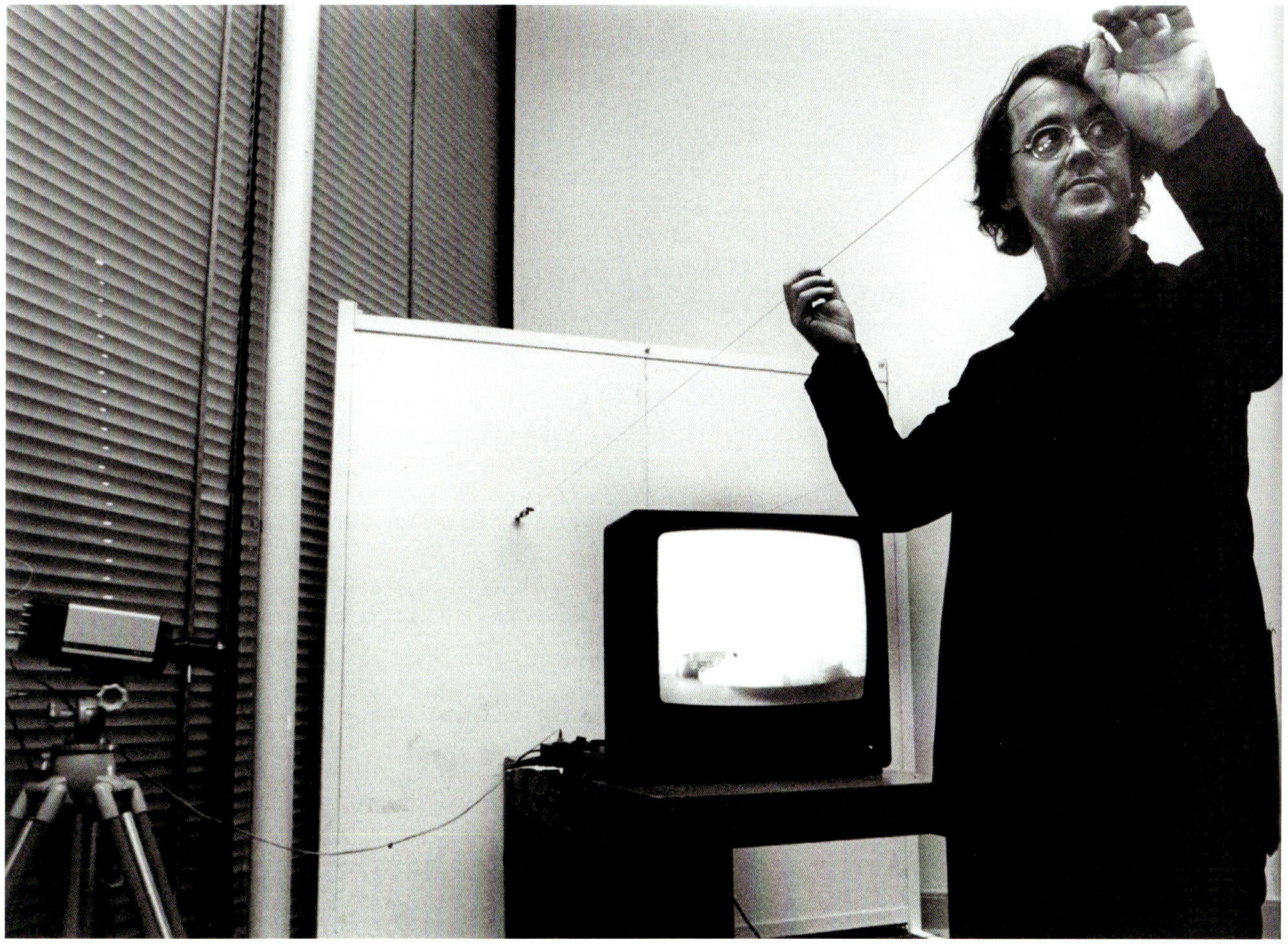

FIG. 20 Terry Fox, *Radiation*, 1979. Installation view Museum Folkwang Essen.

Daniel Muzyczuk

PRES in Search of Multidimensional Sound Spaces

Krzysztof Szlifirski, the then technical director at the Polish Radio Experimental Studio (PRES), began his talk at the UNESCO conference on electronic music in Stockholm in 1970 with a paradoxical statement: "Present-day technology provides a composer with […] further reasons for doubting!"[1] Some of the questions that might be raised by this statement were voiced during the regular internal seminars at the facility. Others were left in the notebooks of Eugeniusz Rudnik, one of the engineers turned composers that worked at PRES.

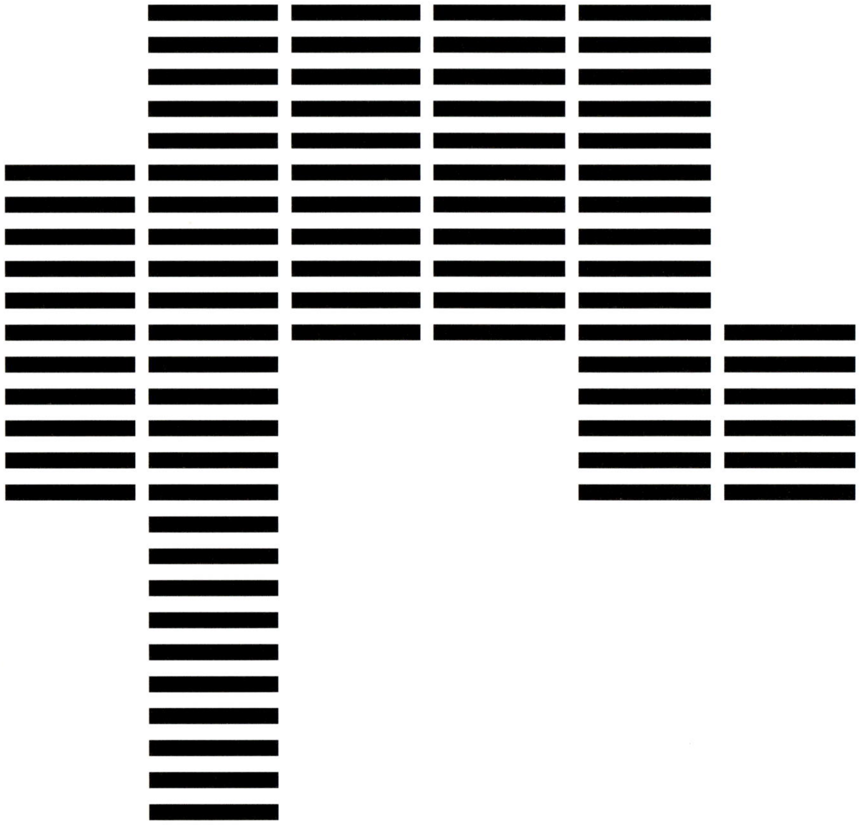

A detailed study of the PRES can be found in: ZKM | Karlsruhe, Adam Mickiewicz Institute, Muzeum Sztuki in Łodź, and David Crowley, eds., *Ultra Sounds: The Sonic Art of Polish Radio Experimental Studio* (Heidelberg: Kehrer, 2019).

In the documentary film *15 Corners of the World* (2014) made by Zuzanna Solakiewicz, she reads fragments of Rudnik's journals aloud. Even if the statement is more about the problems of notation, strangely it brings into the fore the issue of spatiality of sound produced in the studio: "Sound is found and happens in two-dimensional space. Sound can be described using three values that may be visualized in the form of lines mapped along the axis of time, the axis of frequency, and the axis of dynamics. This is how a geometric image of sound is created. A structure of a certain shape. If sound is a three-dimensional object, that means you can play such an object as music. You are viewing a camera shot of a building. It clearly has three axes, right?" Rudnik replies: "Well, one could play a building, why not?"[2] The juxtaposition of these two statements will be used as a signpost for the investigation into sound spaces created at PRES since its origin in 1957. The critical approach of Szlifirski is an example of the spirit of experimentation itself. Rudnik's words are expressing something more specific: a research into the most basic qualities of sound as it is projected into space.

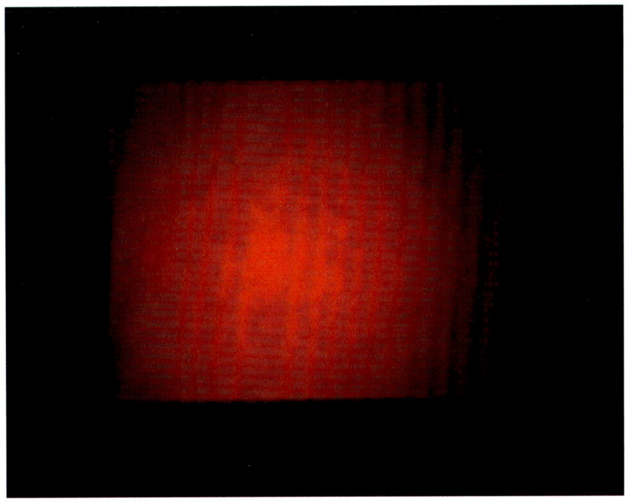

FIG. 1 Józef Robakowski, *Prostokąt dynamiczny* [Dynamic Rectangle], 1971. Film still. The rectangle shown is shaped to music composed by Eugeniusz Rudnik.

The investigations exceeded the limits of disciplines and mediums. There are several reasons for this expansion. Józef Patkowski was not only a musicologist, but also an enthusiast of visual art and new technologies. His concerns extended into the usage of new music in film. Both basic forms of activity – autonomous compositions and soundtracks and effects for radio and film – proved to be fertile ground for introducing constantly new problems. The second function naturally gained more attention from the general public and resulted in some of the most interesting film illustrations in Polish cinema. The otherworldly nature of the sound objects constructed with concrete music methods was ideal for science fiction and horror films. Zofia Lissa, a Polish musicologist and film music theorist made an enumeration of different uses of these sounds in 1962. Among them were "a. science fiction, b. weirdness, c. unusual phenomena, unparalleled situations, d. subjectively unreal situations (visions, hallucinations), e. artificial formations of a para-real world, and f. film visuality of abstract nature."[3] (FIG. 1) Film scores made in PRES not only stressed the oddity of futurological computers, but could also work as a way to deepen the psychological layer of the film.[4] The excessively reverberating spaces suggested a projection of internal states into the environment. These tools – obvious for anyone familiar with the usage of delay (slowing down the sound) – became the basis for an expanded field of inquiry that resulted in some of the most interesting sound installations of the era.

The spatial aspects of sound on tape were at the center of the conceptual framework for at least some of the artists. Bogusław Schaeffer, a composer and playwright interested in this expanded view of music and who worked at PRES since the early 1960s, made some notes which we can treat as a starting point for this investigation: "A method of spatial treatment of material became an important element in electronic music. This method is responsible for 'illustrating' the course of music to a new listener almost metaphorically as a process that is replacing the 'missing' form. It appears that in composers' practice it is possible to use spatiality or its equivalents equally to the sound material. Each sound can be presented spatially or in other words it can be spatially

1
Krzysztof Szlifirski, "New Technology and the Training of Composers in Experimental Music," in *Music and Technology. Stockholm Meeting June 8–12, 1970* organized by UNESCO (Paris: La Revue musicale, 1971), 151–161, here 151.

2
Quoted from *15 Corners of the World: Motion Picture Sound Essay*, CD (Warsaw: Bołt Records, 2015).

3
Zofia Lissa, *Estetyka muzyki filmowej* [The Aesthetics of Film Music] (Krakow: Polskie Wydawnictwo Muzyczne, 1964), 334. Translated here from the Polish.

4
For a closer examination of aspects of the activity of PRES such as graphic scores, soundtracks, and design experiments, see essays by Michał Libera, Agnieszka Pindera, and David Crowley respectively at the *post* feature on PRES: Magdalena Moskalewicz and Daniel Muzyczuk, "Polish Radio Experimental Studio: A Close Look," in *post: Notes on Modern & Contemporary Art Around the Globe*, December 19, 2013, http://post.at.moma.org/themes/14-polish-radio-experimental-studio-a-close-look.

projected – a characteristic of which is that a spatial, stereophonic playback of music is better audible and can be better distinguished than any other minute differences in pitch or intensity."[5] It is interesting also to note that Schaeffer was responsible for some of the most interesting graphic scores in Polish music, as well as the creation of some pieces whose titles and methods of work were directly influenced by visual art: *Little Symphony: Scultura* (1960), *Collage and Form* (1963), or *Assemblage* (1966).

Questions posed by Schaeffer resulted from a simple realization that was summed up in an introduction to electronic music published by another composer who worked at the studio from the very beginning – Włodzimierz Kotoński. He notes: "Since by nature the sounds obtained from electronic devices do not form any acoustic space, then this space has to be produced for them either by the virtual position in stereophonic arrangement or by an artificial reverberation which will allow for a virtual approximation and distancing of the virtual source of sound and creation of simulation of a multidimensional space."[6] The problems related to spatial projection of sound on tape became the area that generated some of the most innovative projects realized at PRES.

An interesting example demonstrating the usage of the studio as a set of tools that can enrich music performance with an expanded variable space comes in the form of Tomasz Sikorski's 1960 composition *Echa II* [Echoes II]. The piece is described as a composition for one to four pianos, percussion, and magnetic tape. The tape part is based on the same sequence as the live part, but it should be recorded in a specific way. The composer imagined two components: "the sounds produced by the performers and their resonances. Two microphones have been used for the first recording, one for the 'normal sounds' and the other, very close (twenty centimeters) to the sound source – for the reverberations. The dynamic envelopes of the reverberations and their scale should be worked out by the sound engineer during the recording."[7] The original tape was prepared by Bohdan Mazurek and it was meant to be projected with four loudspeakers. The intention of the composer seems to be obvious: an interference into the perception of the

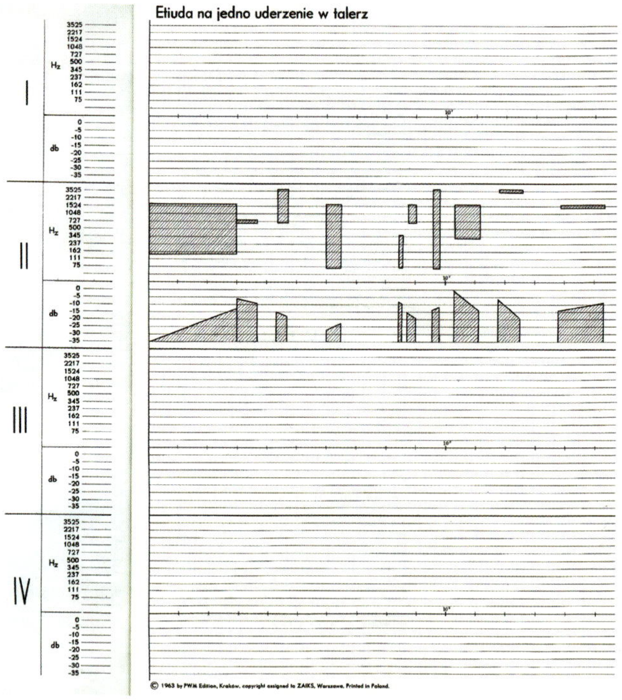

FIG. 2 Włodzimierz Kotoński, *Etudia na jedno uderzenie w talerz* [Study for One Cymbal Stroke], 1959. First published in 1972 by the Polish music publishing house PWM, Kraków, and PRES, Warsaw, in 435 copies. The original publication included a vinyl record.

space where the performance takes place. An expressionistic piece thus gained another emotive level. Reverberation as a basic tool in recorded music offers an effect similar to one of the usages of electronic music and concrete sounds in film listed by Lissa: enhancement of subjectively unreal situations.

Włodzimierz Kotoński, who previously prepared some soundtracks for short films, was responsible for the first autonomous work in 1959: *Etiuda na jedno uderzenie w talerz* [Study for One Cymbal Stroke].(FIG. 2) The short piece did not serve any other purpose like film scores or radio play effects usually do, but was entirely centered on one sound object that was the basis of every event in the composition. The published score looks similar to *Studie II* (1954) by Stockhausen – a piece of music that was completely synthesized. Hence Kotoński was able to prepare a work that satisfied both electronic music enthusiasts (because of the purity of used material) and musique concrète proponents (as the sound object was based on the

5
Bogusław Schaeffer, "Z notatnika" [From the Notebook], in *Muzyka w studio*, Forum Musicum 2, ed. Teresa Chylińska and Bogusław Schaeffer (Kraków: Polskie Wydawnictwo Muzyczne, 1968), 2–11, here 9. Translated here from the Polish.

6
Włodzimierz Kotoński, *Muzyka elektroniczna* [Electronic Music] (Kraków: Polskie Wydawnictwo Muzyczne, 2002), 125. Translated here from the Polish.

7
Tomasz Sikorski, *Echa II* (Kraków: Polskie Wydawnictwo Muzyczne, 1963), 4. Translated here from the Polish. Two recordings of the concert version have been published so far: the composition played by Sikorski himself with the tape part prepared by Bohdan Mazurek, published on *Solitude of Sounds: In Memoriam Tomasz Sikorski* 2 CDs (Warsaw: Bołt Records, 2013); and a revisited version played by Zygmunt Krauze, Szábolcs Esztényi (pianos), and Hubert Zemler (percussion), published on John Tilbury and Zygmunt Krauze, *Grand Tour* CD (Warsaw: Bołt Records, 2016).

8
Each sound can be divided into attack and decay. Attack is the moment when, for example, a string is being hit. Decay happens just after that and means the dying of the original sound. The two phenomena have totally different sound qualities.

9
See Małgorzata Gąsiorowska, *Rozmowy z Włodzimierzem Kotońskim* [Conversations with Włodzimierz Kotoński], (Warsaw: Warszawska Jesień, 2010), 44.

10
Letter from Arne Nordheim to Józef Patkowski, Amsterdam, April 22, 1966. Collection of the Arne Nordheim Centre of Artistic Research at the Norwegian Academy of Music. Quoted from Ola Nordal, "The Sound Sculpture *Ode To Light*: Arne Nordheim's First Project at the Studio Eksperymentalne," http://eog.chopin.nifc.pl/public/files/1468918039a62b1a70a7fecedbf35b96482153c062.pdf.

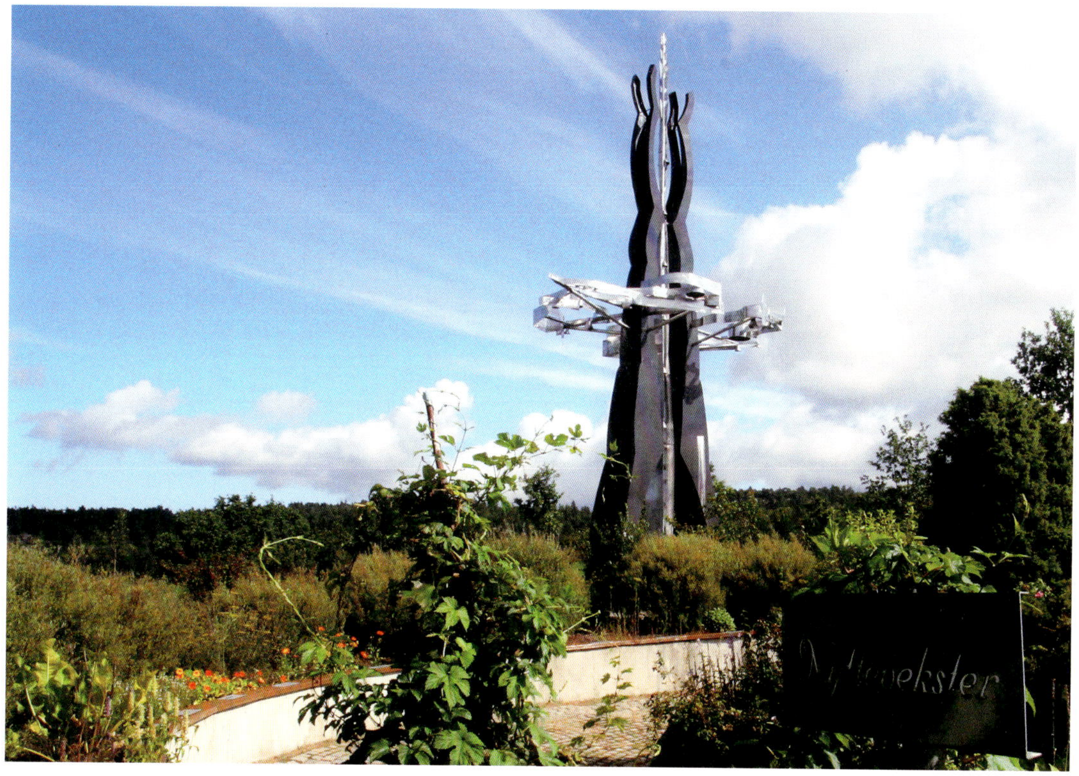

FIG. 3 Arnold Haukeland, *Ode til Lyset* [Ode to Light], 1968. Electroacoustic sound installation in the flower garden at Storedal Kunstsenter.

microphone recording). The creation of Kotoński's second autonomous work is remembered by Eugeniusz Rudnik as one of the most intensive and rewarding learning processes he had experienced. *Mikrostruktury* [Microstructures] from 1963 took the methodology previously used in *Etiuda na jedno uderzenie w talerz* to a completely different level. Kotoński decided to use non-musical sounds: these were made with wood, metal, and glass, but also created by performing actions where these materials would meet, for example cutting through wood with a metal saw). The aim was to use only the attacks (not the decays)[8] and group them so densely that they would form something the composer referred to as "sound-clouds." These were the "microstructures" mentioned in the title – assemblages so dense that individual sounds could not be distinguished.[9] For the composer they were of almost sculptural quality.

It was exactly this diversity of sound-making and recording tools that attracted Norwegian composer Arne Nordheim to PRES when he was looking for methods of work and sounds to be used in a sculptural commission. "From our friend Kotoński have you heard that I am interested in the possibility of working in the Warszawa-Studio. As probably also [Włodzimierz] have told you I am making a kind of music-machine, and the work I eventually will do in the studio is a making of the basic sound-material for this machine. [...] I am especially hunting for pieces of hard metallic sound, and I am bringing with me som[e] basic material as we eventually can use for research and for my understanding of the working prin[ciple] of the studio."[10] Kotoński's recommendation to his friend almost certainly referenced the experience he gained while working on *Mikrostruktury*. At this time PRES was also known for preparing the quadrophonic versions of some pieces.

The commission Nordheim came to Warsaw to record was composed as an accompaniment to Arnold Haukeland's *Ode to Light*, a 19.5-meter-high sculpture created for the Storedal Kunstsenter; located in Skjeberg, Norway, the park was specially geared to the blind and visually impaired.(FIG. 3) Nordheim's composition relies on the positioning of twenty-eight loudspeakers inside and around the sculpture. Built of

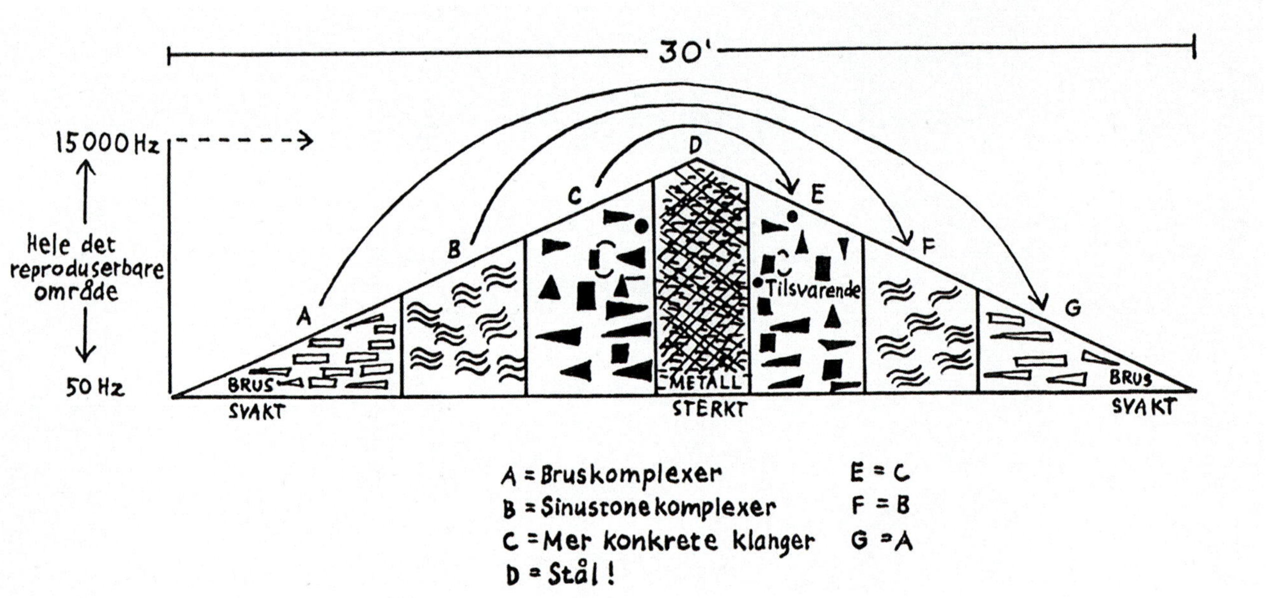

FIG. 4 Arne Nordheim, *Ode to Light*, 1968. Score.

contrasting tones, it reflects the struggle between light and darkness. At the same time, it is an attempt to describe the sculpture's structure through the sounds – it is created for people who most often find themselves in its proximity: the blind.(FIG. 4)

The composer used similar means to contrast sounds with "pleasant" connotations against those with "unpleasant" connotations in his subsequent spatial piece: the Scandinavian pavilion at the 1970 World Expo in Osaka.(FIG. 5) In this work, the aleatoric structure describes human beings living in accordance with their natural environment as well as human beings posing a threat to nature. In a number of ways, the Osaka work was one of the first manifestations of the ecological movement. The sounds recorded with the assistance of Rudnik became part of a multimedia piece that was built of slide projectors, loudspeakers, and tape recorders. Together with the photographs, the sounds present the positive and negative sides of human influence on the environment. The sound installation was created through the chance relationships of tape loops of different lengths. Rudnik was also responsible for designing the setup of sound equipment and preparing the length of loops. Nordheim considered the material so interesting that he went on to prepare three autonomous pieces on its basis: *Lux et Tenebrae*, *Poly Poly*, and *Fem Osaka-Biter*.

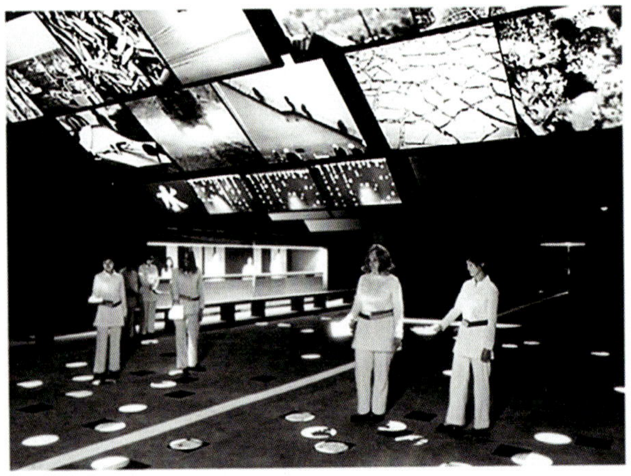

写真提供:日本万国博覧会記念協会 小澤徳太郎

FIG. 5 Scandinavian Pavilion at the Osaka Expo in 1970. Architecture by Bent Ole Severin.

Another frequent visitor at PRES from Norway was Kåre Kolberg, who created two audiovisual works with the help of Bohdan Mazurek. The first one is thematically strangely similar to the theme of the Osaka pavilion. The Norwegian Broadcasting Company commissioned the composer to prepare a soundtrack for *Omgivelser* [Surroundings], a short film directed by Jan Horne in 1970.(FIG. 6) For the work, Kolberg used tapes he brought to Warsaw from home, including the voice of jazz singer Karin Korg. The piece was designed to be an unspoken commentary that would enhance the anxiety caused by the images. The second work created by Kolberg and Mazurek was *Nova* (1972), a composition for a multimedia installation by Anders Kjær installed at the Henie Onstad Kunstsenter.(FIG. 7) The synchronized music and light show again focused on the destruction of the natural environment. The space was filled with paper and tires. The expressive and brutal appearance of the installation was enriched by music characterized by altered metallic sounds as well as one of the hallmarks of PRES – heavily manipulated human voices.

The Norwegians' work, with its critical focus on the environment, was done at the same time as works by Krzysztof Wodiczko, who was starting at another level – the relation between technology and the human body. In the late 1960s, Wodiczko, at this time an industrial designer evolving into a visual artist, became a frequent collaborator of PRES. He was the son of the famous conductor Bohdan Wodiczko, and a friend of pianist and composer Szabolcs Esztényi. In collaboration with the latter, Wodiczko was able to prepare a graphic score for an ensemble playing transistor radios. *Same tranzystory* [Just Transistors] from 1970 is a work that allows the use of

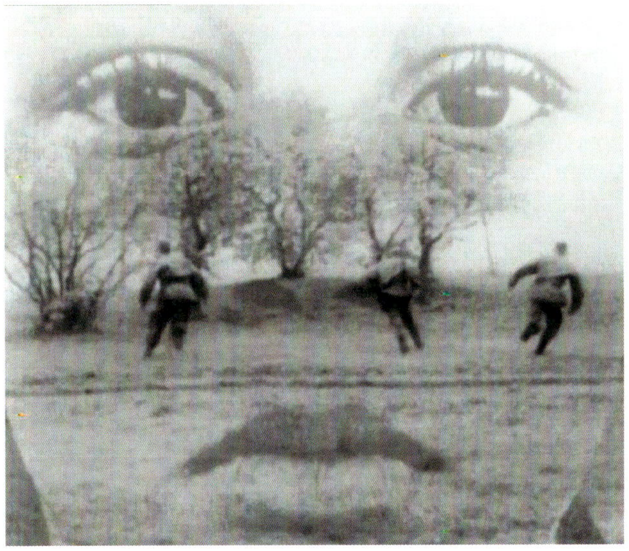

FIG. 6 *Omgivelser* [Surroundings], 1970, short film directed by Jan Horne. Soundtrack by Kåre Kolberg.

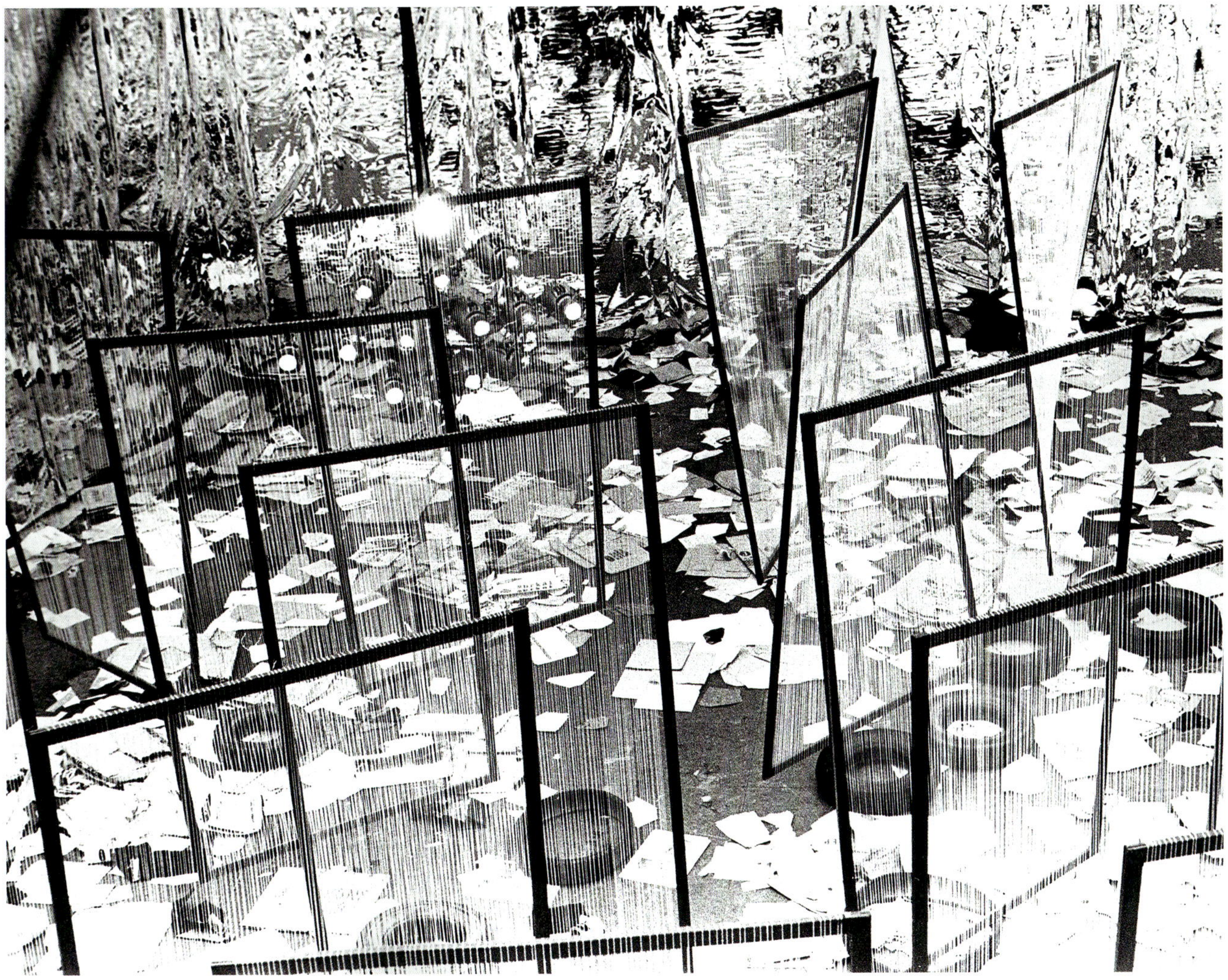

FIG. 7 Kåre Kolberg and Anders Kjær, *Nova*, 1972. Space-filling mixed-media installation, dimensions variable. Installation view Henie Onstad Kunstsenter's Studio Hall, 1972.

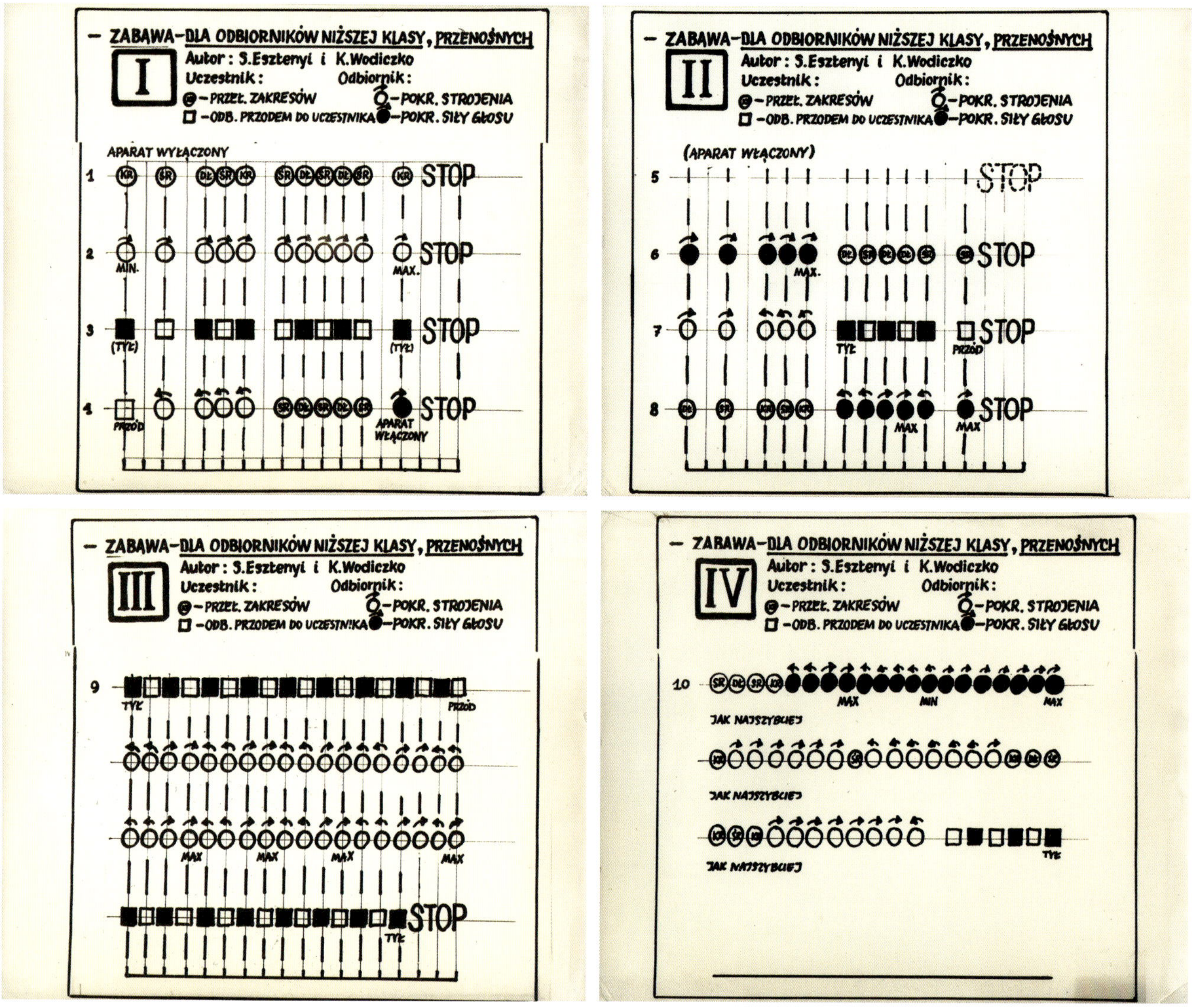

FIGS. 8a–d Szabolcs Esztenyi and Krzysztof Wodiczko, *Same tranzystory* [Just Transistors], 1969. Multimedia installation, technical details, measures. Graphic score.

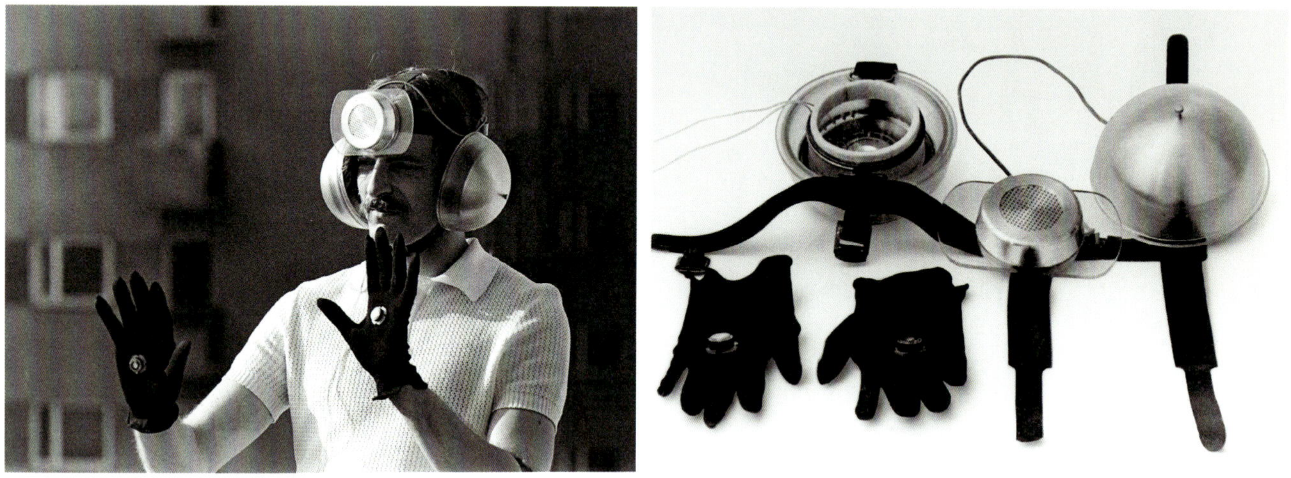

FIGS. 9, 10 Krzysztof Wodiczko, *Instrument osobisty* [Personal Instrument], 1969. Performance and Object.

radio noise to reach a critical moment.(FIGS. 8a-d) The resulting unison piece was strangely similar to the radio jamming sound that everyone who used to search for the Radio Free Europe signal knew too well. For both artists, the act of operating the radio as an instrument was a bodily experience.

Wodiczko's *Instrument osobisty* [Personal Instrument] created in 1969 was a device worn on the head and hands. Responding to the movements of the wearer, the device made it possible for the individual to amplify or diminish the flow of sounds from the environment.(FIGS. 9, 10) The gloves were fitted with microphones and light sensors. Wodiczko's instrument was both a critical comment on the distribution of space and surveillance in the Polish People's Republic and an attempt at creating an autonomous auditory experience.

The third piece realized in the PRES milieu was *Instrument-Laboratorium perkusyjne* [Instrument-Percussion Laboratory] from 1968. It was commissioned by Józef Patkowski, who thought of extending the studio with another room where microphone recordings of a wide array of sound sources could be used to prepare the material for further treatment. This idea obviously was connected to a previous project at PRES conducted by Michael Ranta, who prepared a set of sound objects that were meant to be a resource for use by composers working at the studio.[11] Bogusław Schaeffer, for example, made use of the prerecorded sounds for his "radio opera" *Monodram* (1968). It is also important to note that PRES was closely involved in reflecting on the practice of percussion: Włodzimierz Kotoński, for example, was not only responsible for some of the pieces that made use of percussion sounds as basic material but also published an important study on the role of percussion in the orchestra in 1963: *Instrumenty perkusyjne we współczesnej orkiestrze* [Percussion Instruments in the Contemporary Orchestra]. Wodiczko's *Percussion Laboratory* was the work of a designer reacting to the set of problems that a percussionist faces while performing. It was a metal frame that allowed the performer to adjust the environment and instruments to the specificity of his or her needs as well as the characteristics of the desired sound.(FIGS. 11, 12)

11
The article by Ranta was published along with the description of *Instrument – Percussion Laboratory* in *Res Facta* 5 (1971), a magazine co-edited by Józef Patkowski.

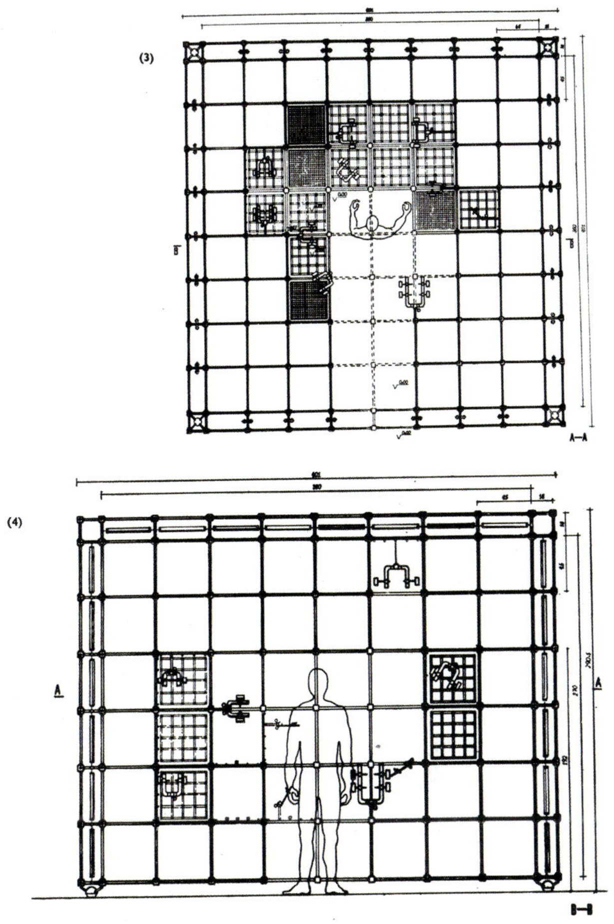

FIG. 11 Krzysztof Wodiczko, *Instrument–Laboratorium perkusyjne* [Instrument-Percussion Laboratory], 1970. Plans for an extension of PRES.

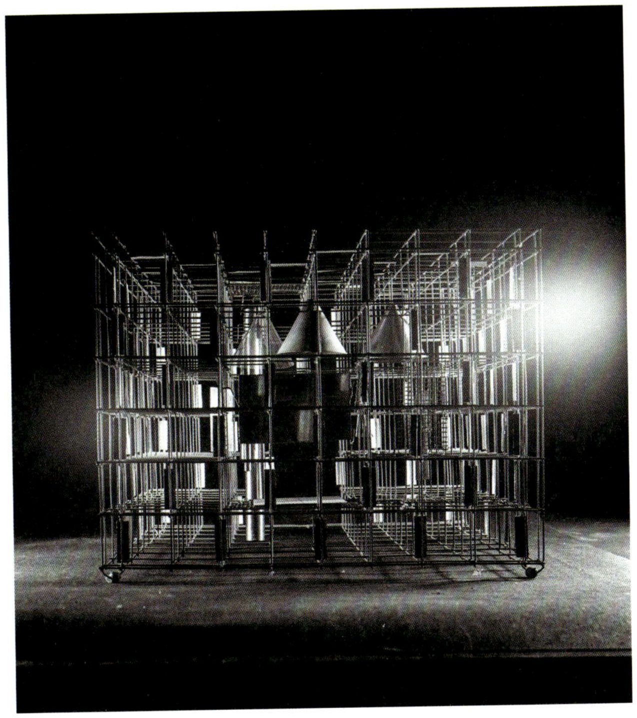

FIG. 12 Krzysztof Wodiczko, *Instrument–Laboratorium perkusyjne* [Instrument-Percussion Laboratory], 1970. Model.

This environment had a predecessor: the design of the room where the studio was located in the building of Polish Radio had been treated similarly itself. A few years before, Patkowski commissioned Oskar Hansen, an architect and teacher at the Academy of Fine Arts in Warsaw to prepare an interior design that would serve the functions of the facility. Hansen understood his task as a designer who respects the environment while offering a maximum of creative freedom.(FIGS. 13, 14) In an interview made at the time, he spoke mainly about the possibilities offered by the design: "My main difficulty was to arrange all the interior elements so as to allow for maximum adaptability to the varying sound recording demands of the future studio staff operating the electronic equipment. As you see, this I have managed to achieve and all the parts are movable, easily reversible. The generators and filters are held in position from the rear by special pincers on a horizontally revolving axle, and the walls and curtains are elastic and can be also placed in desired positions, though for quite different reasons."[12] The relation between the project by Hansen and the unrealized design of Wodiczko seems obvious and they both signify a will to expand PRES into new areas. Hansen's project was meant to facilitate microphone recordings and the spatial effects performed on them. In reality, these recordings were rarely made on site as the room occupied by PRES was rather small and so the functions were mostly limited to equipment-based operations. The failure of Wodiczko's project meant that the studio would not grow into a multitask center for experimental music and would remain a space mainly devoted to music on tape.

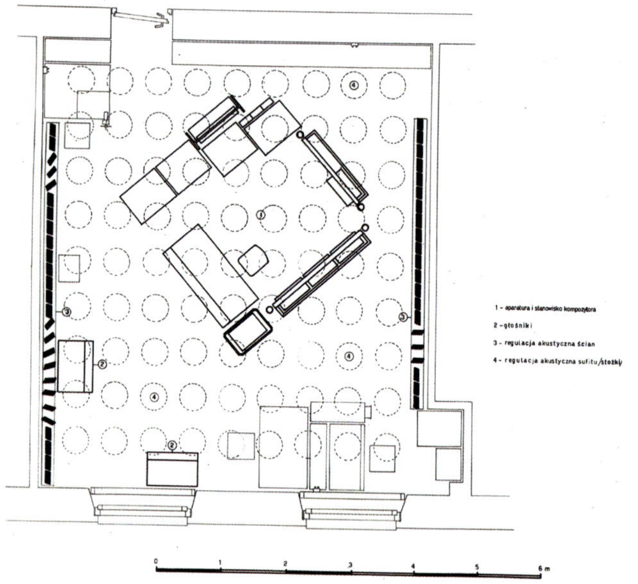

FIG. 13 Oskar Hansen, Zofia Hansen, design for the Polish Radio Experimental Studio, 1962.

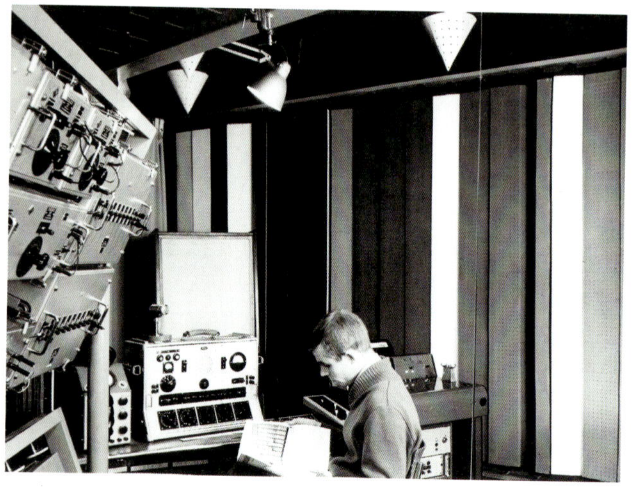

FIG. 14 Polish Radio Experimental Studio.

What should also be mentioned here is that PRES indirectly influenced the first sound installation in Poland. *Spatial-Musical Composition* (1968) was a sound environment prepared by architect Teresa Kelm, composer Zygmunt Krauze, and sculptor Henryk Morel for Galeria Sztuki Współczesna in Warsaw.(FIGS. 15, 16) The audience was given an opportunity to control the sounds and colors of the installation. Moving freely between the spaces, one could build up his or her own version of the composition using the tools delivered by the artists. For the latter, it was a way to overcome the canonical, passive reception inscribed in the concert hall. But there is more that inspired the piece than just the Polish avant-garde of the interwar period. By assigning the audience freedom for their own audiosphere, the artists revealed their debt to magnetic tape processing. The modular character of the composition makes it easy to separate and blend the tracks, and so the liberation of the audience in the installation is based on acknowledging the principles of recording and mixing tools. The artists described their intentions in the booklet as follows: "The piece is composed of several audio layers, each being emitted at the same time by a single speaker

12
Stanisław K. Stopczyk's interview with Oskar Hansen was published in *Projekt*, no. 1 (1965): 8–10. Here quoted from David Crowley and Daniel Muzyczuk, *Sounding the Body Electric: Experiments in Art and Music in Eastern Europe 1957–1984* (Łódź: Muzeum Sztuki, 2012), 139–142.

13
Teresa Kelm and Zygmunt Krauze, *Spatial-Musical Composition* (Warsaw: Galeria Współczesna, 1968). Here quoted from Crowley and Muzyczuk, *Sounding the Body Electric*, 149.

14
Bogusław Schaeffer, "On Audiovisual Music: On Action Music," in *Atelier '72* (1972), n. p.

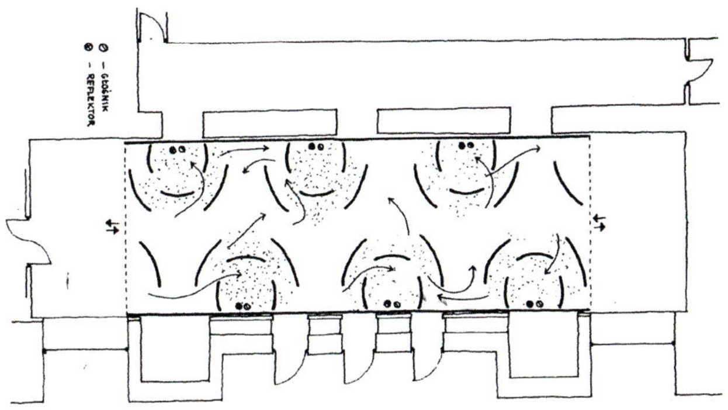

FIG. 15 Teresa Kelm, Zygmunt Krauze, and Henryk Morel, *Spatial-Musical Composition*, 1968. Sound environment prepared for the Galeria Sztuki Współczesna, Warsaw. Scheme.

installed inside a soundproof booth. Moving from booth to booth, the listener integrates the separate tracks into a single whole. Unlike in the concert hall, the piece does not unfold in time in front of a passive listener, but rather the listener can shape the piece depending on the path they take and the time they spend in the designed space."[13] Even if he had never worked in the studio, Krauze embraced the logic of music on tape as well as the aleatoric aspect of the unequal lengths of loops and set out to create a machine where each listener created the composition just by walking.

Bogusław Schaeffer noted a few observations useful for this investigation in an article published in the catalogue of *Atelier '72* – an exhibition of Polish art curated by Richard Demarco in Edinburgh and organized in cooperation with Galeria Foksal and Muzeum Sztuki in Lódź. On the basis of the aforementioned examples, with special consideration for *Spatial-Musical Composition*, he argued that "the listener/spectator is given partial information in audiovisual music, for such is the nature of contact between the composition and receiver in this case. The deficiencies in reception of the whole, as a whole, which, besides, does not exist for anybody in this situation, are made up for by the participation, more direct than ever before, in the reception of music."[14]

The invention of magnetic tape and synthesized sound forced composers to question the seemingly stable parameters of sound spaces and prepare critical instruments and environments that used reverberation more as a dialectical tool than an enrichment of the otherwise conservative constructions. For a few years, PRES was treated as a toolbox for revealing the institutional conventions of sound in space and emancipatory and participatory sonic experiences.

FIG. 16 Teresa Kelm, Zygmunt Krauze, and Henryk Morel, *Spatial-Musical Composition*, 1968. Reconstruction of the installation, Muzeum Sztuki, Łódź, 2013.

Dmitry Bulatov

The Other Side of Sound: Sources and Trends of Sound Art in Russia

Some Boundaries and Nonboundaries

An attempt to identify the evolutionary stages of any phenomenon in art is a thankless undertaking, and not only due to obvious "blank spots," but also because of the impossibility of determining with any precision which events may be considered points of departure, and which not. For this reason, a number of the reviews dedicated to the sources of contemporary sound art practices in Russia give attention to the division that occurred at the beginning of the twentieth century in the so-called standard literature between the visual and phonetic contents.[1] If the works of visual poetry, in which, as part of a single compositional element, texts, inscriptions, and words appearing side-by-side with visual representation signaled the arrival of transgenre fields combining literature and the visual arts, then the art of sound poets, utilizing among their methods various phonetic-intonational elements, offers a point of contact between literature and music.

Translated from the Russian by Alexander Cigale.

Other researchers[2] note the remarkable intensification of the struggle which unfolded in those years in the field of musical art in the aftermath of the intention of the new generation of composers to widen the very definition of music, incorporating into this concept all of sound that is in one way or another organized and structured. At times, views of this sort are connected with an analysis of unusual performative methods of the so-called "noise" orchestras[3] – a ubiquitous attribute of the theatrical performances of the 1920s, activities of the theatrical workshops or folkloric studios,[4] in which the new methods of the apprehending and composition of sound were developed. Or, with the innovative discoveries in the sphere of engineering in the invention of musical instruments, which lay the foundation for electronic music.[5]

But no matter how much the stories of the experts, the dates, and the causative relationships which hypothetically guided the development of sound art in Russia diverge, this cultural phenomenon possesses unifying features. The majority of historians and theoreticians of experimental acoustics converge in their opinion that the origins of this particular art field, which subsequently received the appellation of "sound art," are largely circumscribed in Russia by the brief, striking, and uniquely tragic period of the decades of the 1910s to 1930s. It was precisely during this period – a time of innovation in all spheres of artistic creativity and unprecedented in dimension and scope – that a new perception of the role of acoustic means as a formative element, organically comprising the artistic work, emerged. The characteristic feature of this phenomenon may be termed its conceptual foundation on the basis of tendencies toward *épatage*, arising against the logocentric and logo-directed brain: provocatively radical gestures, intended toward the destruction as much of language as of the limits of language itself. It was to a great degree due to the subsequent emancipation from the established linguistic norms in poetry, music, theater, and cinematography that the autonomy of sound art was established. And it was the same strategy that accompanied the erasure of genre and typological distinctions, the desire to broaden the territory of acoustic art as a product of its moving beyond the limits of the customary functional, separate domains of art. The appearance of experimentalist authors in various fields, the direction of whose work becomes interdisciplinary acoustic practice, seems entirely understandable under these circumstances.

At the Juncture of Word and Sound

In our time, there is every reason to pair the achievements of Russian linguists with the experiments of the Russian Futurists in the decomposition and rearrangement of the word in the development of the language of *zaum* and subsequent phonetic experiments. At the end of the nineteenth century, the leading linguist J. A. Baudouin de Courtenay laid the foundation for this new direction in science – phonology and the incipient conception of "phonemes" as the simplest units of language. He proposed that "phonemes are not like separate notes, but are like chords composed of several elements."[6] It is customary in Russia to connect the achievements of phonological theory with the name of Baudouin's student, Lev Shcherba, who already in 1912 published his book *Russkie glassine v kachestvennom I kolichestvennom otnoshenii* [Russian Vowels in Their Qualitative and Quantitative Relationship]. It was precisely his lectures that the leading representative of the Russian avant-garde, the young poet Velimir Khlebnikov, managed to attend while briefly a student in the department of philology at St. Petersburg University. The poets who had by that time become associated around the press Hylaea – Alexei Kruchenykh, Velimir Khlebnikov, Elena Guro, Vasily Kamensky, Vladimir Mayakovsky –

1
Dmitry Bulatov, ed., *Homo sonorus: An International anthology of sound poetry* [in Russian and English, with boxed set of 4 CDs] (Kaliningrad: National Center for Contemporary Arts, 2001); Dmitry Bulatov, ed., *Tochka zreniia: vizual'naia poéziia, 90-e gody* [A Point of View. Visual Poetry: the 90s] (Kaliningrad: Simplicij, 1998); and Sergei Birjukov, *Zevgma: Russkaja poezija ot man'erizma do postmodernizma* [Zeugma: Russian Poetry from Mannerism to Postmodernism] (Moscow: Nauka, 1994).

2
M. Katunyan, ed., *Sto let russkogo avangarda* [100 Years of Russian Avantgarde, incl. 1 CD] (Moscow: Nauchno-izdatelskij centr "Moskovskaya konservatoria," 2013).

3
S. Rumyantsev, "Samodejatel'nie orkestry I instrumental'nye ansambli" [Amateur Orchestras and Instrumental Ensembles], in *Samodejatel'noje khudozestvennoje tvorchestvo v SSSR: Ocherki istorii* [Amateur Art in the USSR: Historical Sketches 1917–1932] (St. Petersburg: Bulanin, 2000), 293–313.

4
K. Listov, "Ekscentricheskij shumovoj orkestr 'Sinej bluzy'" [The Eccentric Noise Orchestra 'Blue Blouse'], *Sinjaya bluza* 40, 16 (1926).

5
Andrey Smirnov, *Sound in Z: Experiments in Sound and Electronic Music in Early 20th-Century Russia* (London: Walther König, 2013); and Dmitri Zakharine and Nils Meise, eds., *Electrified Voices: Medial, Socio-Historical and Cultural Aspects of Voice Transfer*, vol. 1 (Göttingen: V&R Unipress, 2013).

6
Jan Baudouin de Courtenay, *A Baudouin de Courtenay Anthology: The Origins of Structural Linguistics*, trans. and ed. by Edward Stankiewicz (Bloomington: University of Indiana Press, 1972), 272.

were already actively inventing not only lexical hybrids, but also new words, which they qualify as "sound painting." Even earlier, in 1908–1909, Velimir Khlebnikov produced clear examples of the possibility of working with a single word. His poems "Zakljatie smekhom" [Incantation by Laughter] (published in the collection *Studija impressionistov* [Impressionist's Studio] in 1910) and "Opus 13" (published in the collection *Poshchechina obshchestvennomu vkusu* [A Slap in the Face of Public Taste] in 1912) represent absolutely unique work in the utilization of neologisms, in which, in the words of the poet himself, "the sound element lives a self-sufficient life."[7]

In 1913, at David Burliuk's suggestion, Alexei Kruchenykh published his famous poem "Dyr bul shchyl," "written for the purpose of emphasizing the phonetic aspect of the Russian tongue."[8] As a result, this and the other poetic liberties of the Futurists brought them into conflict with the scientist-linguists (bringing upon themselves the open criticism of Baudouin de Courtenay). Polemics of this kind were important contributions in the convergence of theory and practice. For example, it was out of one such clash of opinions between Kruchenykh and the literary critic Andrei Shemshurin that the well-known theory of "shift" that Kruchenykh later called "sound shift" emerged. The abstract phonetic experiments occupied an important place not only in the poems of Khlebnikov and Kruchenykh, but also in the work of Olga Rozanova, Vasilisk Gnedov, Alexei Chicherin, and Igor Terentiev. According to the records of contemporaries, Ilya Zdanevich recorded the sound experiments on a phonograph,[9] Alexander Tufanov sang his own "phonetic music" (his performance clearly demonstrated the doubling of vowels by doubling the performers and thus utilizing two voices), and Vasily Kamensky even presented his "soundings" to the accompaniment of a ballerina.[10] However, despite the orientation of the Futurist poets toward verse as performance, it would not be valid to say that the vocal content took precedence in their work. The central achievement of Russian Futurist poetry was still zaum – the transrational poetic language, in which the phonetic content was but one of the elements of rejection of linguistic norm.

It was precisely as a counterweight to zaum poetry that in 1916, Kazimir Malevich

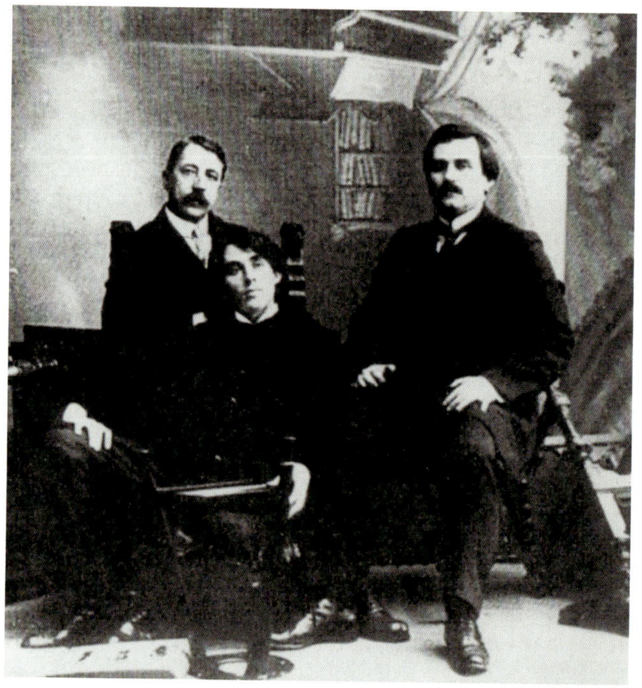

FIG. 1 Karl Bulla, *Alogical Portrait of the Futurists Mikhail Matyushin, Alexei Kruchenykh and Kazimir Malevich*, St. Petersburg, December 1913.

undertook the attempt to create a theory of sound poetry. He outlined the dominant points of this theory in his letters to Mikhail Matyushin and Kruchenykh.(FIG. 1) Malevich wrote: "The new poets must definitively take the side of sound (not music)," and, he continued, "arriving at the idea of sound, we obtained note-letters, expressing sound masses. Possibly, in the composition of these sound masses (former words), the new path will be found."[11] Most likely, these formulations were inspired by conversations with another famous Russian linguist, Roman Jakobson, who parted with Khlebnikov and Kruchenykh, stating: "I was not in agreement when after 'The Word as Such' there followed 'The Letter as Such'; for me it was 'the sound as such.'"[12] If Malevich wrote of sound masses and sound, then his followers in the GINKhUK (State Institute of Artistic Culture) he had led discovered a passion for noise. In the phonological section of GINKhUK, the associates of the Futurist poet Igor Terentiev proposed the categorization of sounds according to "the nature of the material and the form of the instrument producing the sound."[13] The musicologist and pianist Mikhail Druskin, in his pamphlet "Zvuchaschee Veschestvo" [The Sounding Matter], summarizes the situation, closely unifying the principles of sound poetry with the noise experiments of the 1920s: "Each and every human speech is answered by furniture, the walls of the room," and, furthermore, "the throat is a sort of projection system for all kinds of noises."[14] Undoubtedly, as an active participant of "left-wing" art actions, Druskin was well-familiar with the various conceptions of noise music, which were being developed in those years in compositional circles and in the theatrical studios of Proletkult, MastFor (Nikolai Foregger's workshop), and the Blue Blouse movement.

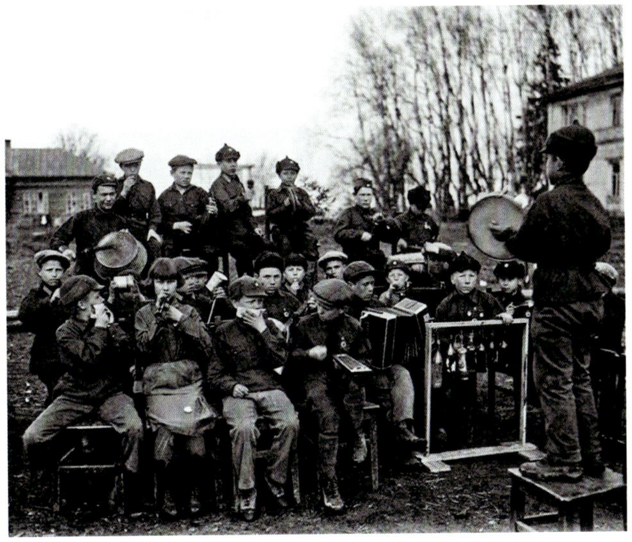

FIG. 2 Rehearsal of the Noise Band of Shelychensk orphanage in Vologda, 1927.

The Noise Idea

The idea of noise orchestras, which the Italian Futurists were first to propagate, found a fertile soil in Soviet Russia. Proletkult and LEF, as well as other artist groupings of revolutionary character, variously celebrated the injection into life of "new dissonances, engendered by the noises of the city, the rattle of machines, the roar of whistles and sirens."(FIG. 2), [15] In 1922, in the inner circle of one such collective – the theatrical workshop of Sergei Eisenstein – a particularly extravagant project of "proletarian noise orchestras" developed: the "Pere Tru" noise orchestra (the touring company of the Moscow Proletkult). The leading role in creating the orchestra was played by the Futurist poet Boris Kushner. The ideologist of the noise movement, Boris Yurtsev, expressed the essence of the project in his article "The Orchestra of Things":

"The instruments making up the orchestra are manufactured by all of the world's factories. Instruments are essentially everything that may produce some sound. At the present time, a selection is ongoing of instrumentsthings and materials. Made of bottles, pots, bowls, glass, nails, needles, churns, wood, metal."[16] In Yurtsev's conception, the plan was to create a noise orchestra in every field of industry (metallurgy, lumber, etc.) which would use the materials specific to that industry. The metallurgist orchestra, for example, would make noise "on the various brands of steel, copper, sheet and formed iron, on cast iron in various stages of production,"[17] and the carpenter's orchestra, on the various species and types of trees and lumber. Afterward, through the Worker's Unions, all the industrial orchestras would have been united into a single noise orchestra of the VTsSPS (All-Union Central Council of Trade Unions). It was intended that the music for it would be written by the composer Grigory Lobachev, whose "March of the Metallurgists" (1928) was preserved in the archives of Gosizdat and was subsequently reconstructed.[18]

A special place among the sound art experiments of the 1920s is occupied by Arseny Avraamov's *Symphony of Sirens*, a grandiose project whose purpose was to "musicalize" the cityscape. This noise performance, as it would be called today, astonished with its scale and was performed live twice – in 1922 in Baku, and a year later in Moscow. Avraamov's project consisted in "noising out," as part of the celebrations dedicated to the anniversary of the October Revolution, well-known revolutionary songs, of course first and foremost "The Internationale," "Varshavianka,"

7
Velimir Khlebnikov, "On Contemporary Poetry" (1920) in *The Collected Works of Velimir Khlebnikov*, trans. Paul Schmidt, ed. Charlotte Douglas, vol. 1 (Cambridge, MA: Harvard University Press, 1987), 373.

8
V. Nechaev, "Vspominaja Kruchenykh" [Remembering Kruchenykh], *Minuvshee* [The Bygone]: *A Historical Almanac* 12 (1991), 383. Translated here from the Russian.

9
See Luigi Magarotto, Marzio Marzaduri, and Giovanna Pagani Cesa, *L'avanguardia a Tiflis: Studi, ricerce, cronache, testimonianze, documenti* (Venice, 1982), 306.

10
Serge Segay, "The Correct Way to Slam a Door," in Bulatov, *Homo sonorus*, 192.

11
E. Kovtun, "K. S. Malevich: Pis'ma k M. V. Matyushinu" [K. S. Malevich: Letters to M. V. Matyushin], in *Yezhegodnik rukopis'nogo otdela Pushkinskogo doma na 1974* [Annual of the Manuscript Department of the Pushkin House for 1974] (Leningrad: Nauka, 1976), 191. Translated here from the Russian.

12
Roman Jakobson, *My Futurist Years*, ed. Bengt Jangfeldt, trans. Stephen Rudy (New York: Marsilio, 1997), 24.

13
G. Demosfenova, "Iz materialov Fonologicheskogo otdela GINKhUKa" [From the materials of the Phonological Department of GINKhUK], in: *Terentev album No.1*, ed. S. Kudrjavtsev (Moscow: Hylaea, 1996), 125. Translated here from the Russian.

14
Ibid. 123, 126.

15
Boris Yurtsev, "Orkestr veschej" [The Orchestra of Things], *Zrelischa* 6 (1922), 22. Translated here from the Russian.

16
Boris Yurtsev, "Orkestr veschej" [The Orchestra of Things], *Zrelischa*, no. 6 (1922): 22. Translated here from the Russian.

17
Ibid.

18
The concert was performed at the Fabrika project in Moscow, on October 8, 2009. Video under http://vk.com/video59510249_165867037.

and “The Marseillaise.” For this purpose, all the acoustic possibilities of the city were harnessed – the whistles of factories, docks, and depots, the foghorns and whoops of ships and of trains, church bells, wind instrument orchestras, and even the sounds of military aviation.[19] This musical picture was rounded out by a group of “percussion instruments”: the role of a drumroll was played by machine-gun and rifle firing, and for the base drum – salvos of heavy artillery. During the performance of this opus, the “symphony director” (as Avraamov called himself) was stationed on the roof of a tall building, directing the action with the aid of signal flags.(FIG. 3) The main instrument of the “symphony” was a steam organ (the so-called Magistral) – the sound-producing mechanism of which consisted of fifty locomotive whistles and three sirens. In accordance with the author’s intentions, the whistle “Magistral” could be made stationary (secured on factory furnaces) or mobile – for example, in Baku, it was installed on the torpedo boat “Dostoyni.” In 1923, in the journal *Gorn*, Avraamov published the paper “Symphony of Sirens”[20] in which he presented not only the technical description of the project, but also expounded his vision for this supermusical epic.(FIGS. 4–6)

The grandiose pathos of Avraamov’s production was not universally accepted. One of the critics wrote in the paper *Izvestia*: “instead of music, many heard primitive cacophony, previously unknown to musical culture.”[21] This comparison of the sound-noise performance to primitive chaos and dissonance seems coincidental at first glance only. Avraamov’s *Symphony of Sirens* cannot be viewed in isolation from the central idea of his life – a reconsideration of the equal temperament. Avraamov believed Johann Sebastian Bach, with his well-tempered twelve-tone system, to be “the greatest criminal in the eyes of history, crippling the hearing of millions of people.”[22] Having given a mathematical, physical, and even administrative rationale behind this controversy in musical art (during the 1920s, he applied to the Commissariat of Public Enlightenment to develop a project for the incineration of all of the country’s grand pianos[23]), Avraamov developed his own method for coordinating tones. Utilizing this method, among the pioneering efforts in microtonal music (1914–1916), he was able to transcribe with precision the folk melodies of the Northern Caucasus. Avraamov believed that by means of the “continuous scale” he had introduced, humanity will be able to “unify arbitrarily captured tones into a musical harmony and construct out of them any melodic mode.”[24] Such striving to discover a primitive, irreducible pattern, on which the past was built and the future may be constructed, was one of the main features of the Russian avant-garde. Wassily Kandinsky wrote of the reduction of all figures to various combinations of color and form, Kruchenykh emphasized the meaning of “pre-sound,” and Aleksei Kapitonovich Gastev and Nikolai Bernstein the derivation of the movements of the human body out of a “primal temporhythm.”(FIG. 7) In this context, Avraamov’s desire to return the musical scale to its “prerational” roots was part of the grand idea proposing a new way of hearing the world. From “quantumization” of the acoustic material to its algorithmization and programming, only a step remained. And this step was, as a matter of course, taken.

19
See S. Rumyantsev, *Ars Novij, ili Dela I prikljuchenija bezustal’nogo kazaka Arsenija Avramova* [Ars Nova, or the Deeds and Adventures of the Indefatigable Cossack Arseny Avraamov] (Moscow: Deka-BC, 2007), 127–28. Translated here from the Russian.

20
Arseny Avraamov, “Sinfonia Gudkov” [Symphony of Sirens], *Gorn*, no. 9 (1923), 109–16.

21
“Simfonija gudkov” [Symphony of Sirens], *Izvestiya VTsIK* (Nov. 9, 1923). Translated here from the Russian.

22
Arseny Avraamov, “Smychkovij polikhord” [The Bowed Polychord], *Muzikalnij sovremennik*, no. 3 (1916), 48. Translated here from the Russian.

23
See Tatiana Sukhanova, *Ljubitel’skoe khudozestvennoe tvorchestvo v Rossii XX veka* [Amateur Art in 20th-Century Russia] (Moscow: Progress-Tradicija, 2010). 7. Translated here from the Russian.

24
Arseny Avraamov, “Nauchnaja organizatsija khudozestven-nogo materiala” [The Scientific Organization of Creative Material], *Sovetskoe iskusstvo*, no. 4 (1928). Translated here from the Russian.

Instruments of the Future

The name of Léon Theremin – Russian engineer, physicist, and musician – is closely associated with novel technologies that turned the world of music upside down. In 1919, Theremin constructed one of the world’s first electroacoustic instruments – the *etherophone*, now known as the *thereminvox* or *theremin*.(FIG. 8) This instrument undoubtedly belongs to a select number of revolutionary inventions. And not only for its unusual sound: the technology it is based on gave rise to entirely novel possibilities for generating sound. Playing the theremin consisted in the player modulating the distance between his or her hands and the instrument’s antennas, which caused

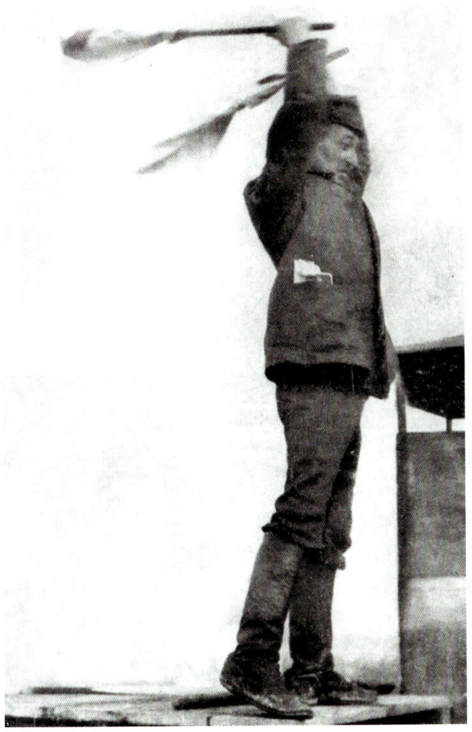

FIG. 3 Arseny Avraamov conducting *Symphony of Sirens* with the aid of signal flags, Moscow, Nov. 7, 1923. Published in René Fülöp-Miller, *Geist und Gesicht des Bolschewismus* (Vienna: Amalthea, 1926), 247 (Tafel 140).

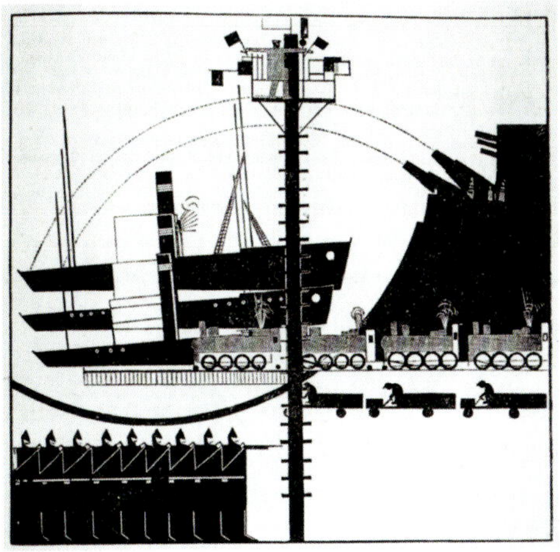

FIG. 4 Illustration of *Symphony of Sirens* in Baku, 1922. Image from Arseny Avraamov, "Symphony of Sirens," *Gorn* 9 (1923), 113.

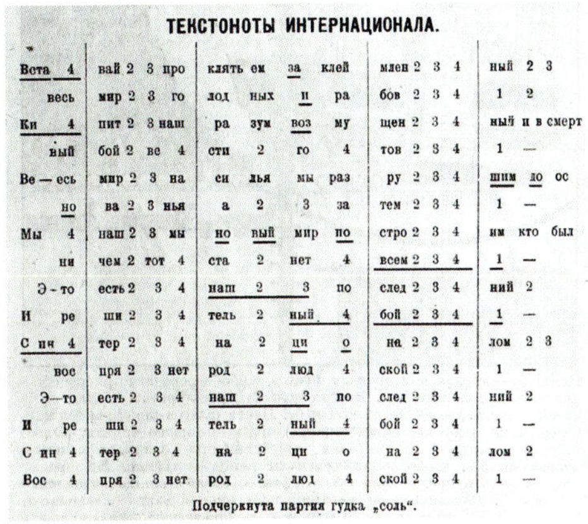

ТЕКСТОНОТЫ ИНТЕРНАЦИОНАЛА.

Вста 4	вай 2 3 про	клять ем за клей	млен 2 3 4	ный 2 3
весь	мир 2 3 го	лод ных и ра	бов 2 3 4	1 2
Ки 4	пит 2 3 наш	ра зум воз му	щен 2 3 4	ный и в смерт
ный	бой 2 ве 4	сти 2 го 4	тов 2 3 4	1 —
Ве — есь	мир 2 3 на	си лья мы раз	ру 2 3 4	шим до ос
но	ва 2 3 нья	а 2 3 за	тем 2 3 4	1 —
Мы 4	наш 2 3 мы	но вый мир по	стро 2 3 4	им кто был
ни	чем 2 тот 4	ста 2 нет 4	всем 2 3 4	1 —
Э - то	есть 2 3 4	наш 2 3 по	след 2 3 4	ний 2
И ре	ши 2 3 4	тель 2 ный 4	бой 2 3 4	1 —
С ин 4	тер 2 3 4	на 2 ци о	на 2 3 4	лом 2 3
вос	пря 2 3 нет	род 2 люд 4	ской 2 3 4	1 —
Э—то	есть 2 3 4	наш 2 3 по	след 2 3 4	ний 2
И ре	ши 2 3 4	тель 2 ный 4	бой 2 3 4	1 —
С ин 4	тер 2 3 4	на 2 ци о	на 2 3 4	лом 2
Вос 4	пря 2 3 нет	род 2 люд 4	ской 2 3 4	1 —

Подчеркнута партия гудка „соль".

FIG. 5 "Textonoti" (text-scores) of the second episode of *Symphony of Sirens*. Published in Arseny Avraamov, "Symphony of Sirens," *Gorn* 9 (1923), 114.

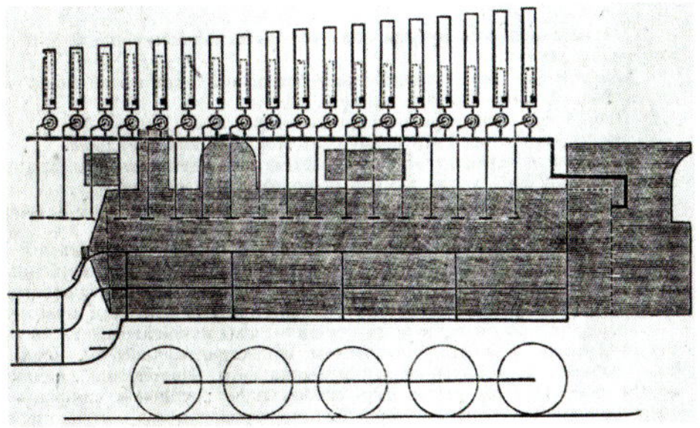

FIG. 6 The Steam Organ "Magistral." Published in Arseny Avraamov, "Symphony of Sirens," *Gorn* 9 (1923), 110.

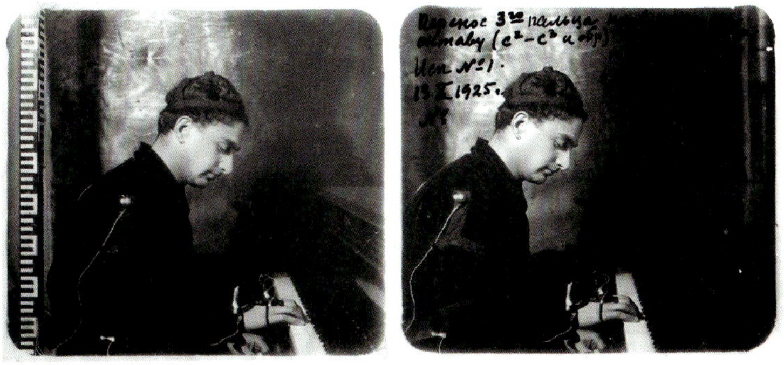

FIG. 7 Physical trajectories of the corporeal joints of a wired pianist, The Central Institute of Labor (CIT) and the State Institute for Musical Science (GIMN), Moscow, 1925.

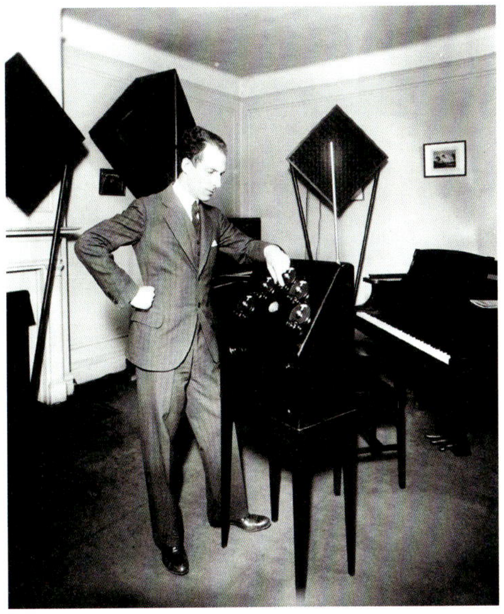

FIG. 8 Léon Theremin, Russian inventor of the theremin, one of the first widely used electronic musical instruments.

a change in the volume contour of the amplitude and, as a consequence, in the frequency of the sound. The vertical straight antenna was responsible for the sound's pitch, and the horizontal, horseshoe-shaped one, for its volume. Through invisible means, the melody arose as though out of thin air. The performer no longer needed strings or keys - only the flights of his fancy and good coordination. The introduction of the theremin also marked the appearance of the first compositions for electronic instruments. These scores were significantly distinct from works written for acoustic instruments. With the harnessing of electricity, the relationship of "performer-instrument" underwent a fundamental transformation. Theremin wrote:

> "The new instruments present a possibility for entirely freeing performers from expending mechanical work to produce sound and thus enable them to direct all their energies into the mastering of the acoustic material, its multitude of parameters. All of the performer's abilities must be expressed precisely in such a cutting of the musical cloth, and not be distracted by the generation of the sounds and the necessity to convey to the instruments their own mechanical energy."[25]

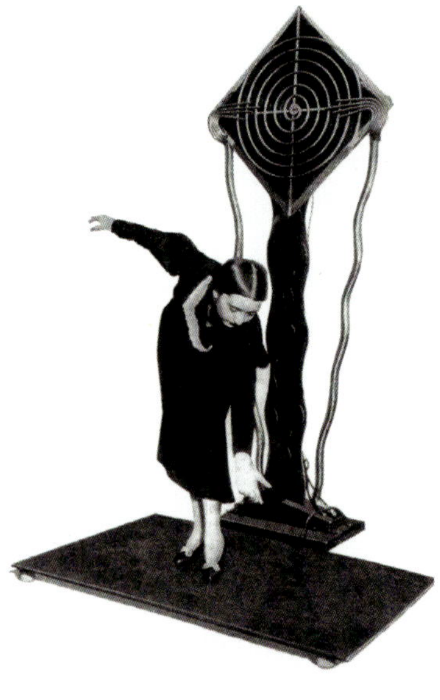

FIG. 9 Clara Rockmore performing on the terpsitone, Carnegie Hall, April 1932. Published in *Popular Science Monthly*, 1932.

This principle of technological mediation of sound became the key staple of many of Theremin's apparatus instruments. In part, this was done by producing a polyphonic clavier machine, the *harmonium*, in which each tunable (micro)tone was reproduced with its own amplifier and speaker. In another version, these tones and timbres could be switched by a look, the movement of the eyes - the photoelement tracked the pupil. Finally, in the complete version of the theremin - the *terpsitone* (1932), musical compositions were produced not with the movement of the hands, but with those of the entire human body - through dance.(FIG. 9) Yet another of Léon Theremin's babies was the world's first rhythm-machine, the *rhythmicon* (1930). It allowed for an unconstrained variation of the sound's duration and ability to manipulate these to produce the richest rhythmic variations. Working on a multitude of projects, Theremin strove to enhance "radiomusic" with light, color, the body's plasticity, tactile sense, smell. It is difficult to imagine the existence of any sort of musical synthesizers, security systems, or motion detectors today without his inventions. In the last years of his life, Léon Theremin was preoccupied with the problem of immortality. According to the reminiscences of Bulat Galeev, Theremin developed a method of "microscopy of time," which made it possible to work with the cells of the human body. The essence of this method consisted in being able, in a timely fashion, to select and culture the body's red blood cells, "which exist in a variety of types in a proportion that changes with a person's age."[26] For culturing these cells, Theremin constructed "a cell incubator" (a sort of bioreactor) and carried out some of the earliest experiments at the Institute of Biophysics in Moscow. By the way, in parallel with these experiments, Theremin invented a means of hearing the sounds of spermatozoa - "for all these beings, you know, under the microscope, carry on a choir with their singing."[27]

25
Léon Theremin, *Fizika I muzikal'noje iskusstvo* [Physics and Musical Art] (Moscow: Znanie, 1966), 31. Translated here from the Russian.

26
Bulat Galeev, *Soviet Faust: Leon Theremin – Pioneer of Electronic Art*, trans. Sally Brown (Sydney: ETT Imprint, 2010); translated here from the Russian 2nd ed. *Sovetskij Faust: Lev Termen – pioneer elektronnogo iskusstva* (Kazan. Gos. Konservatoria, 2010), 134.

27
Ibid., 154.

28
Arseny Avraamov, "Novaja era muzyki" [The New Era of Music], *Sovetskoe iskusstvo*, no. 3 (1925), 67. Translated here from the Russian.

29
N. Izvolov, "Iz istorii risovannogo zvuka v SSSR" [On the History of Drawn Sound in the USSR], *Kinovedcheskie Zapiski*, no. 53 (2001), 293.

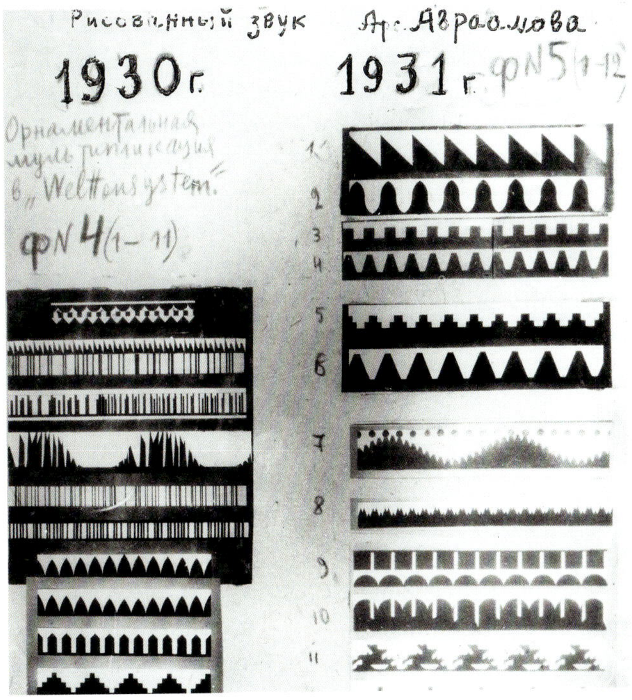

FIG. 10 Ornamental soundtracks, drawn by Arseny Avraamov in Moscow, 1930 and 1931.

The development of sound film at the end of the 1920s added a powerful stimulus to the development of systems of sound recording, generation, and manipulation of sounds. In a matter of only a few years, in the Soviet laboratories working on adding sound to film, more than twenty types and systems of sound recording and instrumentation for producing sound were prototyped. The most widely distributed among them were the systems of Pavel Tager (the *tagephone,* 1926), Alexander Shorin (the *kinap,* 1928), and Vadim Okhotnikov (1929). All of them were based on the principles of optical preparation of phonograms through the modification of the density of the sound wave on the film strip. Yet another of Shorin's systems, under the name of *shorinophone* (1931) was broadly used for recording sound in field reporting by means of mechanical incision of a soundtrack onto a strip with the aid of a needle. Parallel experiments were carried out with a variety of sound recording media – tin sheets, steel wire, and paper. One result of such experiments was the invention in 1929, by Avraamov, Evgeny Sholpo, and Mikhail Tsekhanovsky, of the technique of "drawn" ("graphic" or "synthetic") sound. This technique made it possible, with the aid of artificially created sound-ornament tracks, to generate any sounds and effects. The musician had to literally draw the music. The sources of the sounds thus produced were geometric figures – squares, triangles, and circles, from which not only any visual form, as Kandinsky had asserted, but also an optical phonogram of any form on the film strip could be created.(FIG. 10) It seemed that the music of the future – "where there will be no limit to technical virtuosity, nor to the gradations of available scales, nor impediments to dynamic expression"[28] – was becoming a reality.

Drawn Sound and ANS

The main thrust of the research conducted by Arseny Avraamov and Evgeny Sholpo was the creation of systems of synthesis and transformation of sound without the performer's participation (through the utilization of the achievements of technology only). But if Avraamov was more taken with the possibility of analyzing the sources of sound and representation,[29] then Sholpo's guiding idea was the creation of a "mechanical orchestra" – of a machine capable of automating the process of musical composition. In 1931, working together with the composer Georgy Rimsky-Korsakov, he built the first version of the *variophone* – an optical synthesizer which made possible the production of artificial phonograms by means of automated graphical sound techniques. In appearance, the instrument looked like a film projector with a transmitter that transferred the revolutions of the electromotor to the mechanism which pulled the tape through the acoustic wave pattern. (FIGS. 11, 12) The pitch of the tone thus generated varied according to the speed with which the tape moved through the mechanism. This equipment

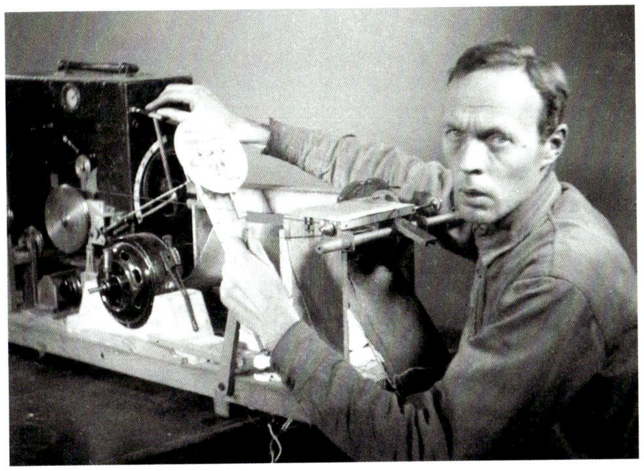

FIG. 11 Evgeny Sholpo working on with the first version of the variophone, Leningrad, 1932.

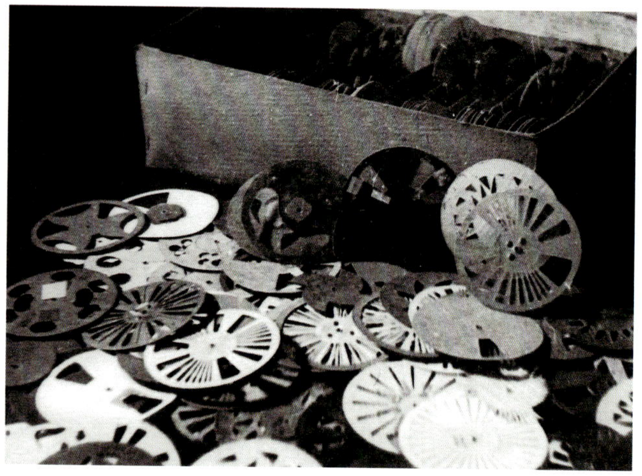

FIG. 12 The variophone optical discs with cut wave shapes.

utilized revolving discs – Sholpo incised on them the form of the waves he required. It would seem that this instrument represents a prototype of electro-optical samplers and the wave-table synthesis of sound. In the paper "Drawn Sound" (1935), dedicated to the developments in the field of audio technologies, the reviewer V. Solev remarked that Sholpo's method was distinguished for the quickness and ease of its manipulation. But these distinctions, in the opinion of the author, had an obverse side: the variophone became an independent musical instrument, possessing its own "technique of performance," the manipulation of which required a great deal of skill and knack.[30]

Yet another variation on the technology of "drawn" sound was developed by the painter-animator Nikolai Voinov. In 1930, he was a member of Avraamov's group Multzvuk and participated in the work on the first drawn-sound tracks. However, Voinov soon left the group and began his own experiments, as a result of which the instrument *nivotone* was created. The nivotone is an apparatus designed to optically read sound profiles cut from paper and translate them into sound via a photo-electrical process. The synthesis of soundtracks was performed by means of cutting profiles of sound waves from paper and using the nivotone for one-frame-at-a-time photographing of the segments of the soundtrack on the animation lathe.[(FIGS. 13, 14), 31] Beginning in 1931, Voinov participated in the creation of animated films with synthetic sound tracks: *Barynya* (1931), *Preliud Rakhmaninova* [Rachmaninov's Prelude] (1932), *Tanets voroni* [Dance of the Crow] (1933), and *Vor* [Thief] (1934). Another piece of equipment in the technology of "drawn" sound – the *vibroexponator* – was created by the artist and acoustician Boris Yankovsky. The vibroexponator produced an animated graphing of sound onto a normal

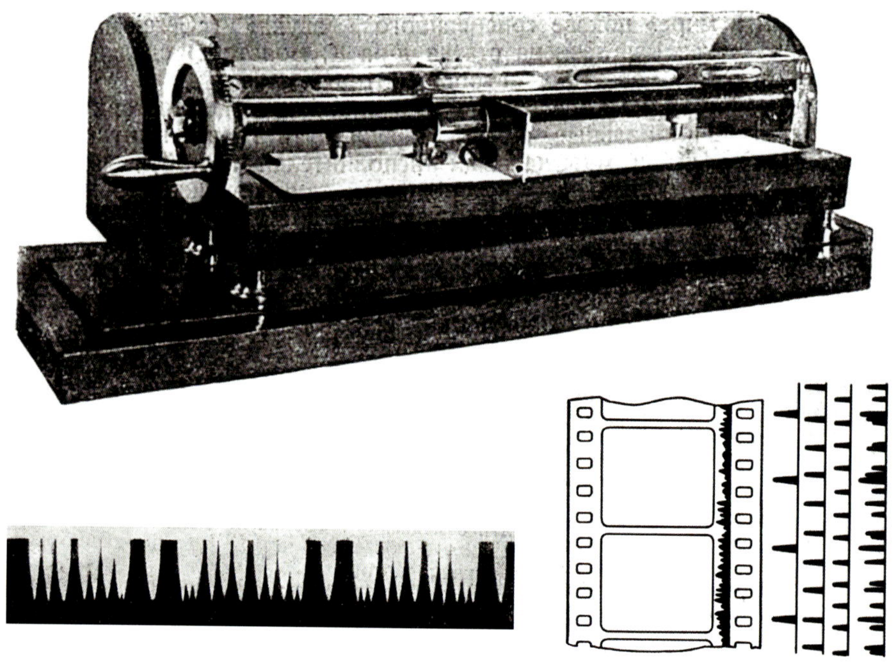

FIG. 13 Nivotone tool and illustration of paper sound method with a final soundtrack, 1931.

30
See V. Solev, "Risovannij zvuk" [Drawn Sound], *Radio-front*, no. 3 (1935).

31
See N. Izvolov, "Iz istorii risovannogo zvuka v SSSR, ili medium bez mediuma" [On the History of Drawn Sound in the USSR, or Medium without Medium] in *Sovietskaia vlast i media* [Soviet Power and Media], ed. Hans Günther and Sabine Hänsgen (St. Petersburg: Akademichesky proekt, 2005), 366.

32
See Boris Yankovsky, "Teorija I praktika 'graficheskogo zvuka'" [The Theory and Practice of 'Drawn Sound'], manuscript fragments, Leningrad, 1939–1940, www.theremin.ru/archive/yankovsky1.htm.

33
See Gleb Anfilov, "Chapter 10: Composer as a Painter," in *Physics and Music* [1962], trans. Boris Kuznetsov (Moscow: Mir Publishers, 1966).

34
Ibid.

35
L. Pereverzev, *Iskusstvo i kybernetika* [Art and Cybernetics] (Moscow: Iskusstvo, 1966), 128.

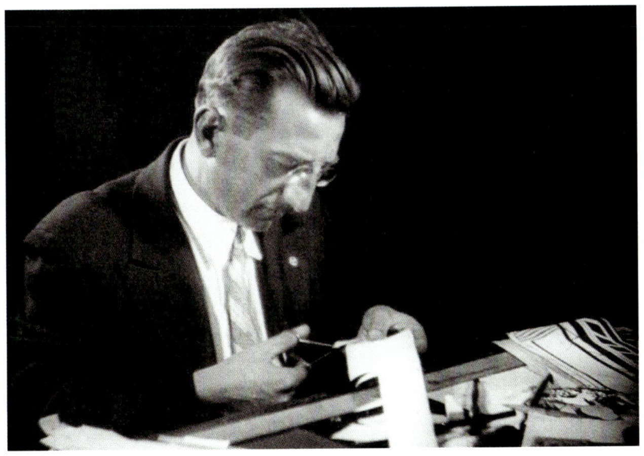

FIG. 14 Nikolai Voinov cutting his paper soundtracks, Moscow, 1931.

35 mm film strip. Yankovsky believed in the possibility of creating a universal library of sound elements, something akin to Dmitri Ivanovich Mendeleev's table of elements. His photographed "spectral templates" – were essentially semiotic units which, through combination, would yield new sound hybrids.[32] Although, by 1936, Yankovsky compiled a collection of one hundred spectroscopic templates, "syntones," he did not succeed in creating a table analogous to that of Mendeleev. His ideas were too far ahead of their time, and their further development required a foundation in technology that was not yet in existence.

In more complicated form, the principle of the synthesis of optical phonograms was used by the inventor Evgeny Murzin. In 1959, he patented the instrument "ANS" (the initials of the composer A. N. Scriabin) – the first fully polyphonic musical synthesizer in the world. Boris Yankovsky, who in 1939–1940 participated in the development of this instrument, tested the first working version of the synthesizer in 1957.[33] In the ANS, the octave was divided into seventy-two intervals, and the process of composition, as in the first efforts of Avraamov and Sholpo, also consisted in a gesture or, more precisely, the drawing of a "sonogram" – a special graphical score representing the dynamic spectrum of a sound, with the possibility for synthesizing it in real time. Murzin's instrument made it possible for the composer to operate with any modes and scales, to imitate any sounds or noises, and even to generate human speech. The second version of the synthesizer (1964)(FIG. 15) contained a set of five optical discs with a drawing inscribed on them, and utilized 720 controlled generators covering a range of ten octaves of sound. Besides the synthesis of sound, Murzin's apparatus also made it possible to record the sounds created onto magnetic discs. Moreover, with the aid of a system of tape recorders, the composer was able to record and replicate entire tracks, adding different effects, such as reverberation. For the first time in history, he succeeded in unifying the processes of composition, recording, and performing of music.[34] In 1967, with the synthesizer ANS at its foundation, the first studio of electronic music in the USSR was established in Moscow. From the 1950s to the 1970s, the instrument was worked on by such composer-innovators as Sofia Gubaidulina, Alfred Schnittke, Edison Denisov, and others. The instrument was used for producing the soundtracks of numerous films, including Andrey Tarkovsky's films *Solaris, Mirror,* and *Stalker*.

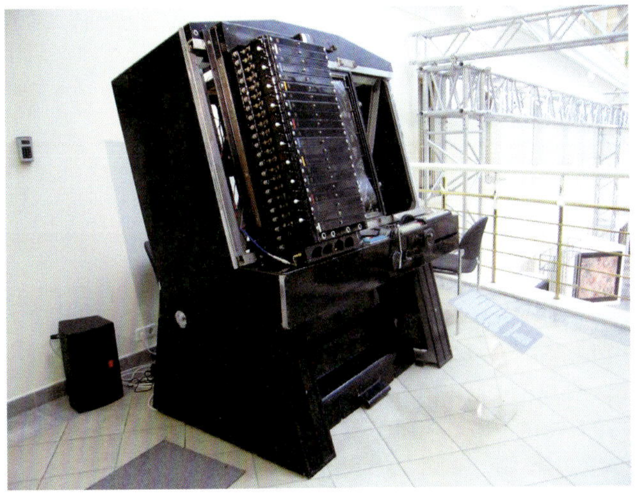

FIG. 15 The second version of the ANS synthesizer by Evgeny Murzin, 1964.

Cybernetics and Synaesthesia

Finally, the decades of the 1950s, 1960s, and 1970s brought with them a "cybernetic boom" of sorts, connected with the ubiquitous use of electronic calculating machines (computers) in the most diverse fields – in biomedicine, mathematics, linguistics, and pedagogy. This popular trend naturally influenced the creative sphere,[35] so that Rudolf Zaripov's experiments in the programming of sound compositions ("automaton-composer") became widely known among the circles of scientists and musicians. In 1959, working on the "Ural-1" computer, Zaripov synthesized monophonic

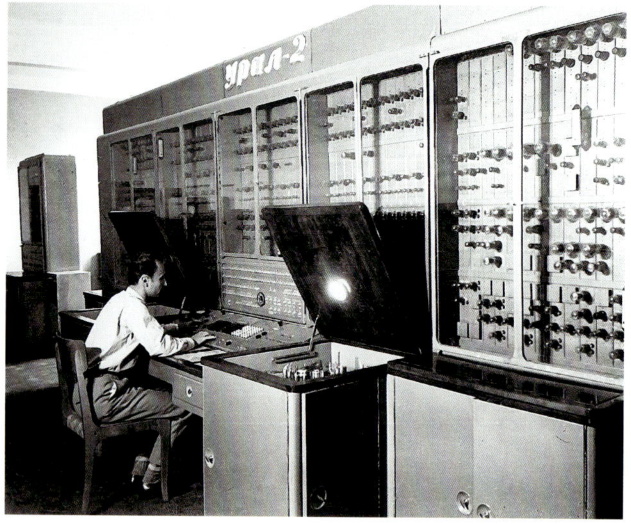

FIG. 16 Rudolf Zaripov's experiments in the programming of sound compositions on the "Ural-2" computer, 1960.

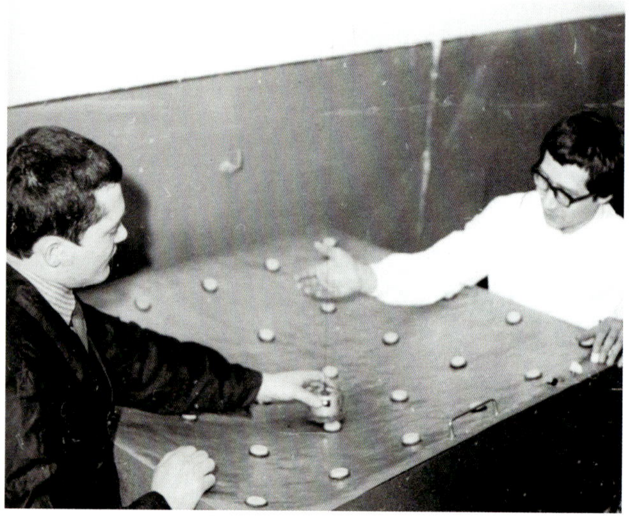

FIG. 17 Bulat Galeev (right) and Rustam Saifullin working with the first version of the spatial sound instrument, 1974.

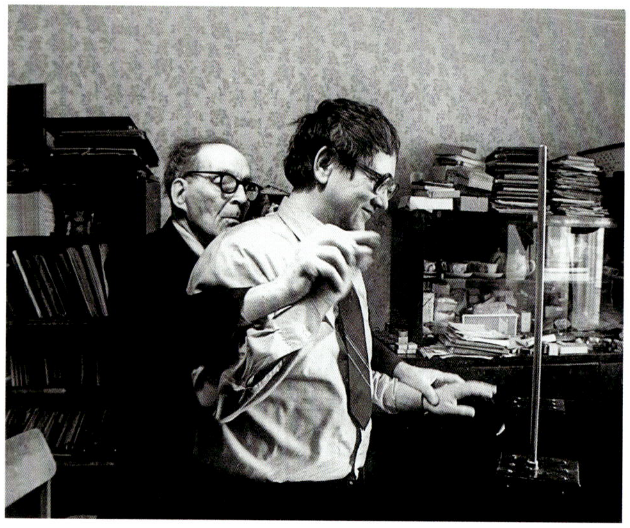

FIG. 18 Léon Theremin teaches Bulat Galeev how to play the theremin, 1986.

musical pieces that became known as the "Ural melodies." The algorithm they were based on was a precisely scripted process controlling the various elements of musical texture – form, rhythm, pitch, etc. Zaripov deduced an entire set of rules for the composition of such pieces, and also carried out the harmonization of melodies as well.(FIG. 16) He summarized his experience in the synthesis of sound compositions on the computer in an article published in 1960 in *Proceedings of the USSR Academy of Sciences*.[36] For the first time, a Soviet science journal published the score of the "automaton-composer" – the embodiment of the idea first developed by Evgeny Sholpo, who in 1917 had already offered a full description of generative musical machines. Subsequently, Zaripov devised a comprehensive research program, with the goal of creating a system of AI capable of reproducing the activities characteristic of the creative processes. His book, *Cybernetics and Music* (1963) was translated and published in Italy, and the publications on automatic compositions created by means of his algorithms circulated around the world. He became a recognized authority in the field of cybernetic music. The article "Modeling Melodic Variations on a Digital Computer" (1983) summarized the results of the newest stage of research dedicated to the creation of models for the process of generating sound and other works of art.[37]

In 1962, in Kazan, where Zaripov had begun his work, Scriabin's symphony *Prometheus* was performed, with the accompaniment of a light show, as it was conceived by its author at the beginning of the twentieth century. From this time on, the idea of the synthesis of sound with the newest technologies became a

36
Rudolf Zaripov, "Ob algoritmicheskom opisanii processa sochinenija muziki" [On an Algorithmic Description of the Process Involved in the Composition of Music], *Doklady Akademiya Nauk SSSR* 132, 6 (1960), 1283–86.

37
Rudolf Zaripov, *Kybernetika y muzyka* [Cybernetics and Music] (Moscow: Znanie, 1963); Rudolf Zaripov, "Modelirovanie melodicheskikh variatsii na tsifrovoi vychislitel'noi mashine" [Modeling Melodic Variations on a Digital Computer], *Problemy Kibernetiki* 40 (1983), 201–54.

38
Bulat Galeev, *Svetomuzika: stanovlenije I suschnost' novogo iskusstva* [Light-Music: The Establishment and Essence of the New Art] (Kazan: Tatknigoizdat, 1976); Roberta Reeder, Claas Cordes, "Light Music in the Soviet Union," in *Ars Electronica 1992: Endo Nano (part 02)* (Linz: Ars Electronica, 1992), 225–40, http://90.146.8.18/en/archives/festival_archive/festival_catalogs/festival_artikel.asp?iProjectID=9553.

39
Bulat Galeev, *Chelovek, iskusstvo, tekhnika (Problemi sinestezii v iskusstve)* [Mankind, Art, Technology (The Issue of Synesthesia in Art)] (Kazan: KGU, 1987).

40
Segay, "The Correct Way to Slam a Door," 197.

41
Rea Nikonova-Tarshis, "Ekologija pauzi" [Ecology of Pause] in *Experimental'naya poezija: Izbrannie stat'i* [Experimental Poetry: Selected Articles], ed. Dmitry Bulatov (Kaliningrad: Simplicij, 1996), 181.

42
A. Khlobystin, fragment from an unpublished book "Russian Schizorevolution," provided by the author. Translated here from the Russian.

guiding principle for a group of artists and scientists united first into an engineering firm and then into a research institute of experimental aesthetics, Prometheus. Bulat Galeev, its leader and the guiding spirit behind all the projects of the collective, and coeditor of the journal *Leonardo* (MIT Press), regarded the creation of works at the junction of science and art as its own independent orientation, in which the imagination of the authors, with the assistance of contemporary engineering possibilities, is given free reign and an unbounded diversity to the manifestations of creative thought.(FIG. 17), [38] In their works, the members of Prometheus produced out of a conception of an expanded system of art, developing that which had already been started by their compatriot pioneers: Scriabin in visual music, Theremin in electronic music,(FIG. 18) Vladimir Tatlin and Gustav Klutsis in kineticism, Nikolai Voinov in "drawn" sound, Zaripov in cybernetic music. Multimedia performances, scores of unique apparatuses and machines, projects in light and sound architecture – in all these fields, the artists from Kazan searched for and found unusual resolutions to their experiments. A separate line of their research was functional sound (controlled audiovisual media) and cosmic psychology (indicators of the "man-machine" system).[39] Bulat Galeev and the Prometheans took part in many Russian and international festivals of contemporary art (Ars Electronica, ISEA, EMAF, and so forth), and the films of the collective from the 1960s to the 1980s are deservedly considered classics of avant-garde cinema.

Transpoetry and Multimedia

The 1970s and 1980s brought with them a second wind for experiments in the field of sound poetry. These efforts were in many respects connected with the reemergence of the tradition of the Russian poetic avant-garde and the praxis of the "noise" orchestras. The poet and literary critic Serge Segay described the acoustic means of the "transpoet" group (Serge Segay, Rea Nikonova, Boris Konstriktor, A. Nik, and V. Earl) in the following way: "As soon as the idea of a sound chronicle became too limited, new instruments were invented: polyethylene bags full of stones, plastic boxes containing pigs' bones, paper flutes and paper harps."[40] All these instruments not only accompanied the voice of the author, they created those very same "sound masses" that Malevich spoke of, that were capable of replacing any, even the most poetic of words. Collaborating with the transpoets, who published their own [samizdat] journal of theory and practice, *Transponance* (1979–1986), were the entire spectrum of independent experimental artists – from A (Alchuk) to U (Unksova). Contributors to the journal included Dmitri Prigov and Ilya Kabakov, Lev Rubinstein and Genrikh Sapgir, Igor Bakhterev and Leonid Aronzon – a polyphony of the living and the departed on a single stage.[41] Sound-poetry jams were organized, in which, beside the transpoets, those who took part included the German sound poet Valeri Scherstjanoi and the St. Petersburg poet Alexander Gornon, who subsequently turned to digital manipulation of the voice to create his "poem-operas." There also existed other groups, not indifferent to sound poetry and the field of sound more generally – the praxis of sound performance was important in the work of Andrey Monastyrsky (for example, *Rechevoj potok o pustom dejstvii* [Stream of speech regarding a meaningless action] and *Tufli* [Shoes], both 1984), elements of cacophonous declamation in the performances of Konstantin Zvezdochetov and the group The Fly Agarics. A significant place in this history also belongs to the group the New Composers, organized in Leningrad in 1983 by Igor Verichev and Valery Alakhov.

In their works, the New Composers utilized the quotidian noise background of the Soviet people produced by radio and television. It was comprised of fragments of radio plays, morning communal calisthenics programs, news flashes, film fragments, etc. Utilizing a variety of techniques of mixing on reel-to-reel tape recorders, the New Composers transformed this "ideologically overloaded gurgling of the collective subconscious"[42] into avant-garde experiments in the style of the "industrial music" of the 1920s. The works of the New Composers were held in high esteem by Western "electronica" musicians, including Brian Eno, who visited them in Leningrad in 1987. At times, the group's projects were joined by friends, giving rise to spontaneous collectives, one of which subsequently grew into the famous Pop-Mekhanika (1984–1996) of Sergey Kuryokhin.[43] Speaking

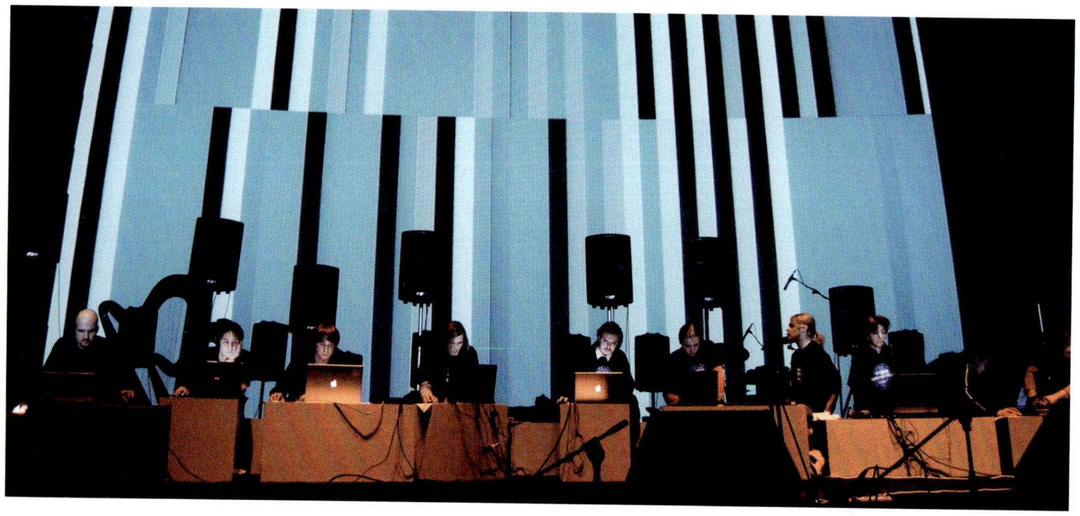

FIG. 19 The Moscow Laptop Cyber Orchestra ("CybOrk," artistic director: Andrey Smirnov), 2006.

of the "noise" branch of the St. Petersburg electronic sound, it is impossible not to recall the techno-shamanic trance of "Bardoseneticcube" (Igor Potsukailo), and on the other hand the experiments with industrial themes of the group ZGA, the permanent participant of which, Nikolay Sudnik, is also one of the founders of the Center of Experimental Music GEZ-21 (since 1999). The interdisciplinary possibilities of science art, harking back to the ideas of Theremin on commanding sound with the aid of bioelectricity, found their expression in the project "Neurointegrum" of the new generation of media artists, Yury Didevich, Anton Makarov, Patrick K.-H., and others. In a series of large-scale performances realized on the New Stage of the Alexandrinsky Theatre, the authors turned their attention to one of the key story lines of culture at the crossroads of two centuries – the interaction of man and machine, and the attempt to establish a dialogue between them.

To imagine these visual and acoustic projects in their complex interactions – from estrangement and autonomy to independent mutual interconnections, the synthesis and osmosis between them – is a task that the SoundArtLab project (Danil Akimov) has been engaged in for a period of many years. The central idea of the shows he organized as part of the Baltic branch of the NCCA festivals "Sound Around Kaliningrad" – is no longer "the art of sound," nor "art = sound," as it was previously, but a dilemma: art or sound? During the past two decades, a key role in the development of the Russian sound-art scene has been played by the Theremin Center (Andrey Smirnov). Established in Moscow in 1992 and named after Léon Theremin, the center unites young artists, musicians, programmers, etc., who share an interest in using the newest electronic technologies in creating art.(FIG. 19) To list only a few of its activities, the Theremin Center has delved into serious archival explorations and shows, practical workshops and lectures by the world's leading specialists, and the development of new technological means.[44] It was precisely under its roof (or under its influence) that a new generation of artists was produced – artists who once again directly embrace the question: What is the point of departure for sonic construction? Modulation of patterns, as during the times of phonetic poetry and "noise" orchestras, new instruments or novel strategies, utilizing sound outside the scope of rules and conventions? And what about the experience and traditions of Russian and international sound art? Artists such as Dmitry Morozov, Sergey Kasich, Evgeniy Cherniy, Alexander Senko, and the community SoundArtist.ru have developed their own answers to these questions.(FIGS. 20–22b) At heart, the answer may be very simple – it is the full spectrum, which offers an infinite number of possibilities for developing the potential of the medium of sound. Through it, the energy of sound-images is liberated, which is, it would seem, the true purpose of the sound artist.

43
Ibid.

44
http://theremin.ru.

Acknowledgments:
The author would like to thank copyright holders who greatly assisted in this publication, including photographers or their representatives, archives, foundations, and collecting institutions: Konstantin Dudakov, Oleg Ivanov, Irina Aktuganova, Andrey Smirnov, Marina Sholpo, Anastasia Maksimova, Sergey Zorin, the Nikolay Nikiforov Collection, the Multimedia Art Museum, the Prometheus Center, the Alexandrinsky Theatre, the Glinka State Central Museum of Musical Culture, and the Center for Electroacoustic Music at Moscow State Conservatory for their help and assistance.

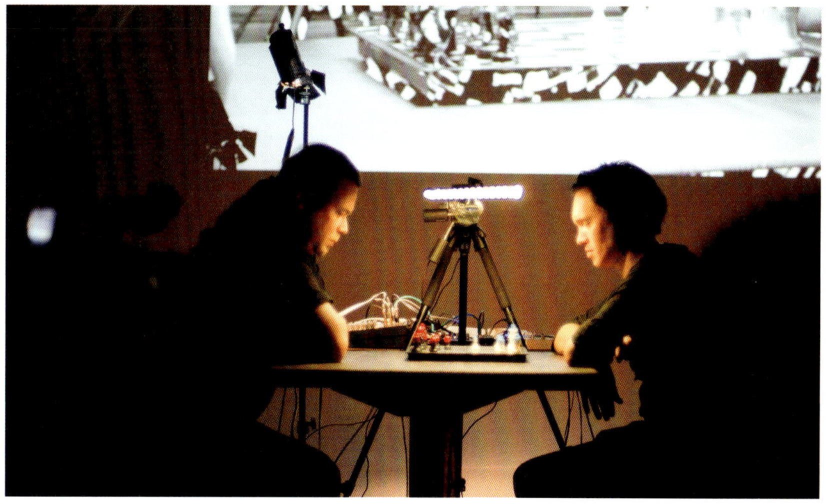

FIG. 20 Patrick K.-H. (aka Anton Iakhontov), Oleg Makarov, *Room Sketch: Chess*, 2011. Multimedia performance, where sonic and visual material is to be structured by chess play in real time.

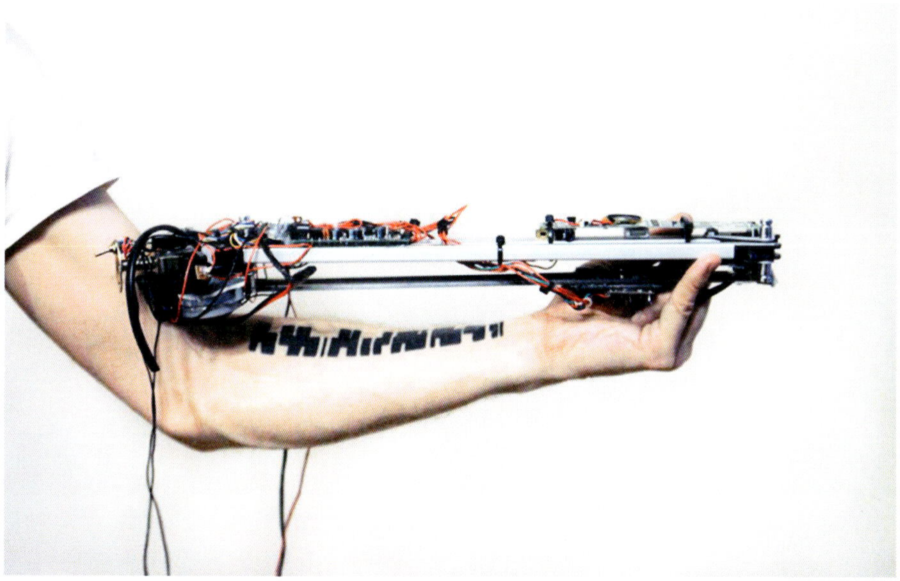

FIG. 21 Dmitry Morozov (aka ::vtol::), *Reading My Body*, 2013. A special instrument that uses tattoo as a music score and combines human body and robotic system into a single entity.

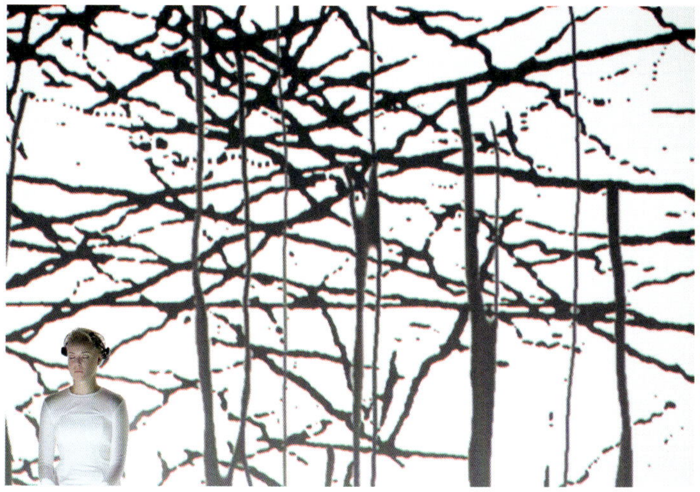

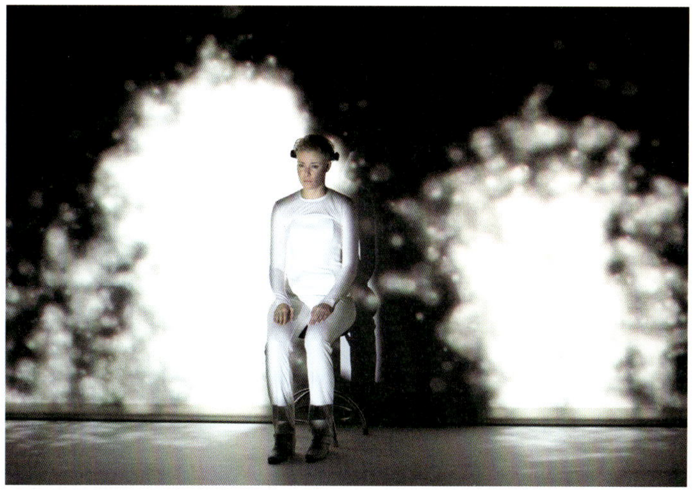

FIGS. 22a, b Yury Didevich at al., *Neurointegrum*, 2014. Series of large-scale biofeedback performances realized on the New Stage of the Alexandrinsky Theatre, St. Petersburg.

Başak Şenova

Sonic Notes and Records from Turkey – A Short Story of Fifteen Years: The Shift from the Stage to the Exhibition Space

In the early 1960s in Turkey, only a few outstanding musicians were experimenting in music.[1] However, until the 1990s, these early productions were extended only as experiments following influences mostly from the West. The 1970s and 1980s saw some individual, if isolated, efforts, but the coup d'état in 1980 brought a serious rupture in the developing art scene. The militaristic government apparatus and pressures of the state-initiated neoliberal policy created social, economic, and cultural changes and unleashed new flows of migration. These developments were followed in the 1990s with intensifying civil war between the Turkish State and the PKK (Kurdistan Workers Party, designated as a terrorist group by the Turkish authorities). The country witnessed unaccounted murders, while unknown numbers of people were held in custody, ratcheting up pressure on the universities and the youth.

The Istanbul scene in the 1990s was mostly dominated by musical experiments, such as the work of ZeN – a groundbreaking band from the city's alternative music world. The group fused the energies from a variety of the modern era's radical inputs, including Dada, punk, and psychedelic together with Sufi transcendentalism, Turkish folk music, and gypsy tunes. Their music also bore the mark of both the historical and contemporary music schools that were embedded in the city's memory. Founded in the 1980s, ZeN continued to evolve as a band, and by the mid-1990s the group transformed into a new ensemble called BaBa ZuLa. In the beginning, BaBa ZuLa's music combined recorded natural sounds with both traditional and modern acoustic instruments. Electronic instruments were also incorporated, for an overall electronic effect. From a basis in improvisation – later developing into fixed musical elements and scored material – the group developed their own distinct method of "defined improvisation" for their performances.

The transformation that marked ZeN's evolution into BaBa ZuLa was echoed in the Istanbul scene, as a number of groups began experimenting with electronic effects and recorded "found sound." The group Fairuz Derin Bulut, founded in Istanbul in 1996, grew out of a solid base in rock, oriental, funk, and arabesque styles. They used keyboard sound and darbuka along with other traditional Turkish percussion instruments. In the early 2000s, the band composed works drawn from the sounds they recorded on the streets and in their studio, albeit with relatively mainstream results. The band is still active and address themselves as "a psychedelic music orchestra."

Considerably more avant-garde is the material produced by Turkish multimedia artist Serhat Köksal.(FIG. 1) Köksal formed his 2/5 BZ project in 1986 and has performed widely on the international circuit, including the Club Transmediale festival in Berlin (2003, 2006), Phonem by Miller (2003) and the Norient Musikfilm Festival in Bern (2011). His live performances, either solo or with a large entourage, typically include video projection and spoken word elements, set against a mesmerizing sonic tapestry of tapes, samplers, saz, darbouka, electronics, drums, and vocals. With a markedly electronic and experimental sound, 2/5 BZ performances blend traditional music, electronica, and sampled elements from Turkish cinema and media. Outside these performances, Köksal's multimedia work has ranged from audiotapes to fanzines to video collages from Turkish films and political propaganda, which he uses to investigate "culturalistic clichés and their effects on the economical and political situation of individuals."[2] In a similar vein, the group Anabala[3] also worked with projected imagery during their performances, though typically with lighter results. Founded at the beginning of the twenty-first century by Ceren Oykut together with Murat Ertel of ZeN, Anabala was a multidisciplinary project concentrating on the sounds and sub-cultures of Istanbul. The duo created multimedia pieces taking the aspects of humor, parody, and surprise as the basic elements of their performances. Integral to their onstage work was the use of projected digital images, which they created and rendered in real time.

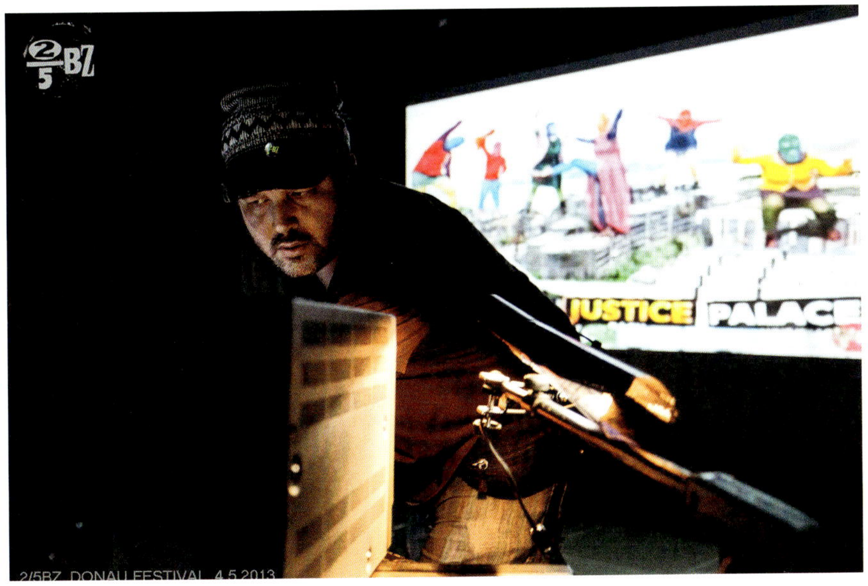

FIG. 1 2/5 BZ aka Serhat Köksal, *NO Touristik NO Exotic Militant Oriental Sounds & Deconstruct Films & Nurkish Sound Art*, 2006. Sound perfomance at Donau Festival Krems, 2013.

1 For example, blending Turkish folk tunes with Western rock elements and instruments; using synthesizers and incorporating psychedelic rock and Anatolian rock features; adapting Western progressive music sounds; and experimenting with amplifiers.

2 "Köksal, Serhat | 2/5 BZ," CTM–Club Transmediale participant list, see http://archive.ctm-festival.de/archive/all-participants-a-z/k/koeksal-serhat-25-bz/photo/1.html.

3 Anabala continued performing both in Turkey and abroad until 2011.

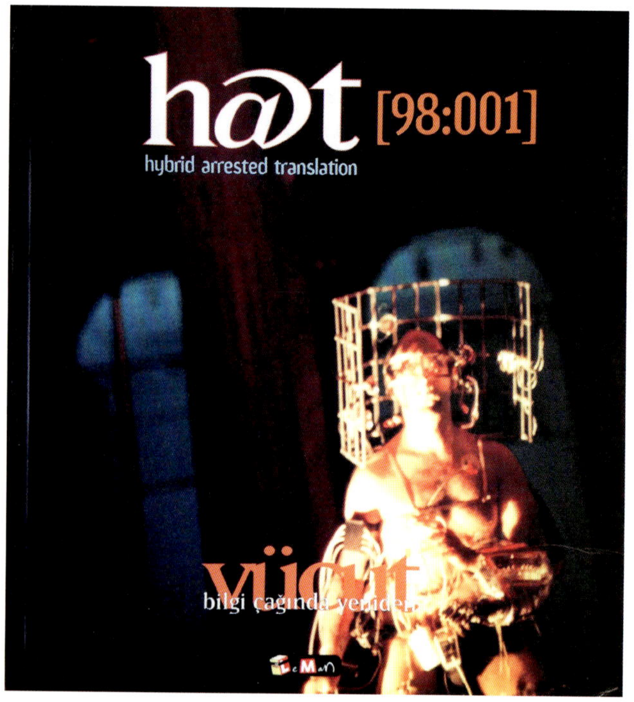

FIG. 2 Fatih Aydoğdu, *HAT (Hybrid Arrested Translation)*, 1998. Magazine cover.

Parallel to these productions, a new magazine was launched focusing specifically on new digital arts and media theory. Published by Vienna-based artist and designer Fatih Aydoğdu, *HAT (Hybrid Arrested Translation)*[4] appeared in 1998.(FIG. 2) Although there was only a single issue of the magazine – devoted to "the body in the information age" – , it can be considered a groundbreaking first attempt to create a theoretical infrastructure for the field of sound art in Turkey.

By the end of the 1990s and the beginning of the twenty-first century, the most important direction was to sustain connectivity for accumulative production in the creative fields of the country. From this line of thought, a core group of designers, engineers, architects, curators, and writers came together in 2002 to found NOMAD.(FIG. 3), [5] The group aimed to produce and experiment with new patterns in the digital art sphere by using the lenses of various other disciplines. NOMAD's production network built connections across territorial borders through projects oriented around digital culture – initially in Europe and the Middle East, then in other areas. The main goal of these projects was to establish a productive communication channel that would enable access to new resources of information. NOMAD has developed numerous local and international projects, including festivals, exhibitions, performances, multimedia events, experimental film screenings, lectures, panels, and publications. In 2001 and 2002, the founders of NOMAD started to develop online projects with the task of connecting ideas, projects, techniques, functionalities, and locations across time and space. Since its inception, NOMAD has sought to establish a recognition for sound art in Turkey. Due to the lack of congruous discourses on sound practices in art, it was essential to frame a common milieu for related production, consumption, apprehension, and reception. This platform had to go beyond a mere contemplative mode of presentation through a series of concerts, or isolated installations. NOMAD managed to realize ctrl_alt_del,[6] the first sound art festival in Turkey, by connecting the nodes of the international art network and linking them to local actors and situations.(FIGS. 4–7) The project was presented in three editions, which took place between 2003 and 2008 in collaboration with the Istanbul Biennial. The presentation included panels, performance series, workshops, exhibitions, site-specific installations, radio programs, screenings, and publications, in addition to CD and DVD releases in 2003, 2005, and 2007–2008.

For the development of sound-based works, two schools have been important in Istanbul: Yildiz Technical University's (YTU) Audio Design Department, headed by Alper Maral, and Istanbul Technical University's (ITU) Center for Advanced Music Research (MIAM) with director Pieter Snapper. At both institutions, faculty and students staged performances and CD releases that played a pivotal part in the sound art scene of Turkey. In addition to Maral and Snapper, participants have included Cem Ömeroğlu, Emre Can Öziş, and Eray Düzgünsoy at YTU, and Reuben de Lautour, Kerem Aksoy, Murat Yakin, Zeynep Bulut, Deniz Arat, and Aykut Şahlanan at ITU.

Another important institution for Turkish sound art is Bilkent University in Ankara. Bilkent founded a school of music and performing arts in 1986, making it the first university in Turkey to offer higher education in these fields. It is still one of the most important

4 The magazine was published in the printing press of Turkish satirical magazine *LeMan*, where Aydoğdu was once a cartoonist.

5 NOMAD was founded in 2002 as an independent formation and registered as an "association" in 2006. See www.nomad-tv.net and www.goethe.de/ins/tr/lp/prj/art/med/bgt/trk/enindex.htm.

6 See http://project-ctrl-alt-del.com.

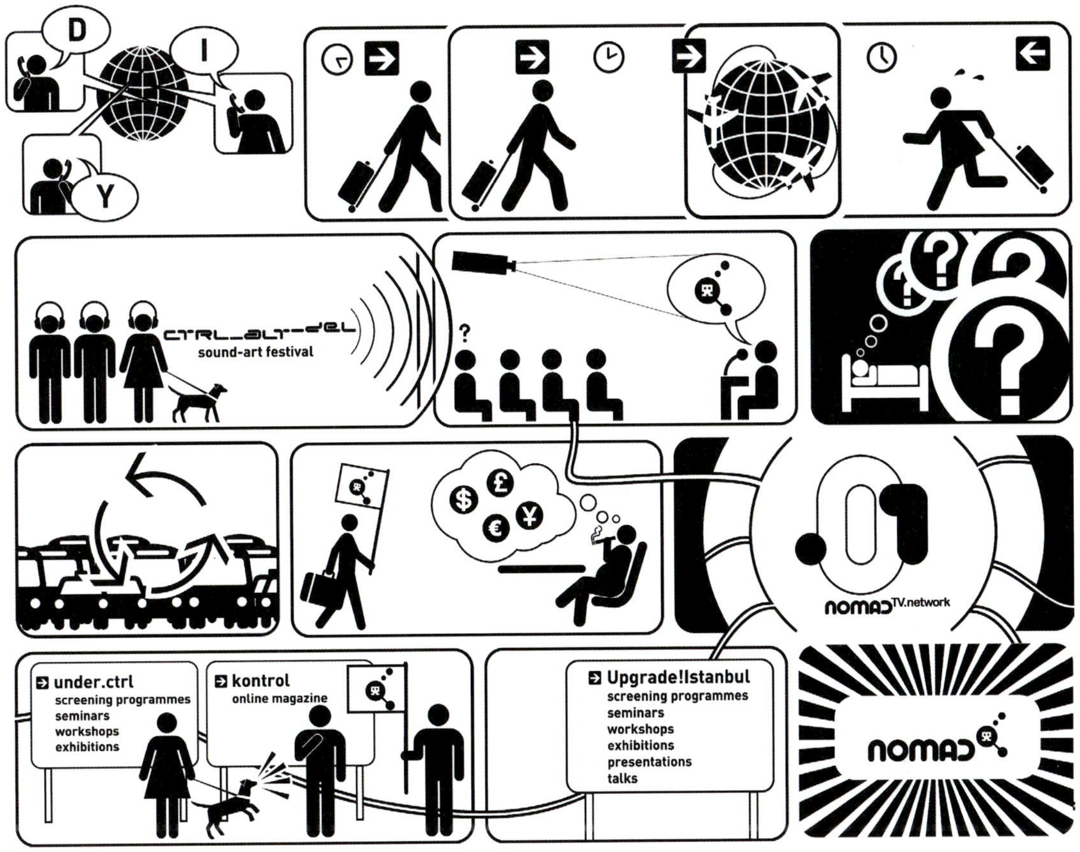

FIG. 3 NOMAD, Istanbul-based association. Installation view showing the different fields and disciplines they work in.

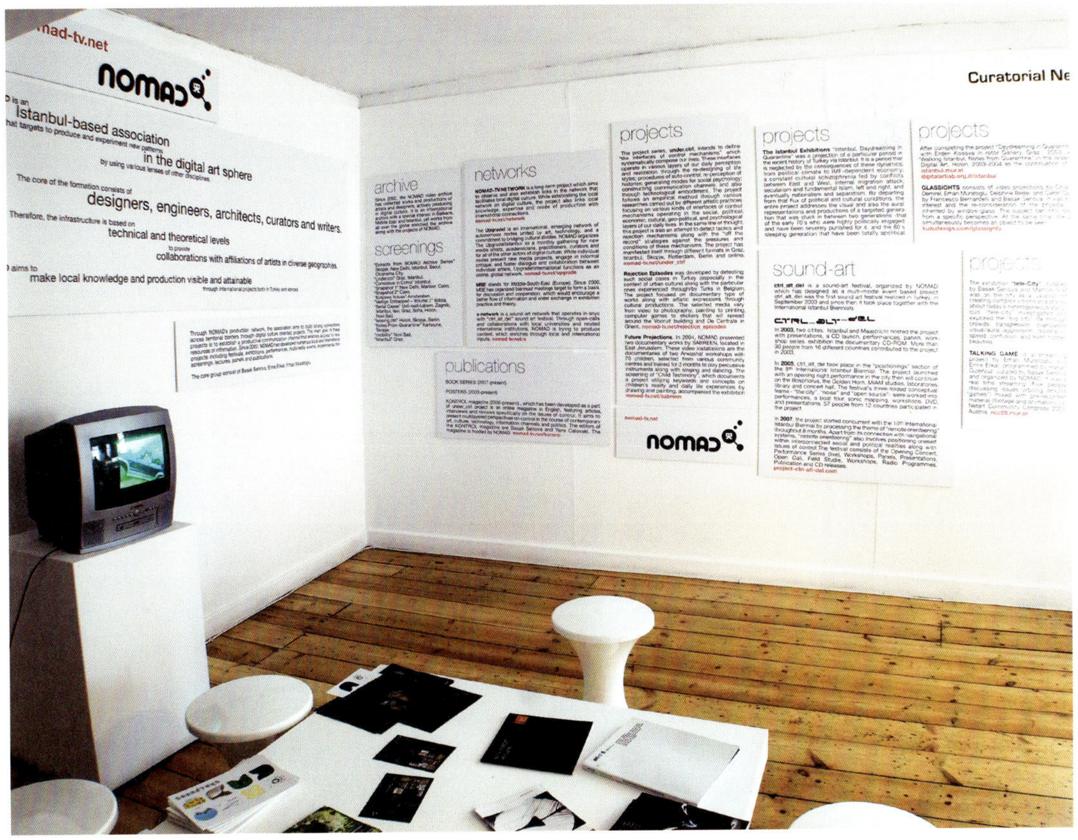

FIG. 4 NOMAD. Iconic poster explaining the ctrl_ alt_del sound art festival.

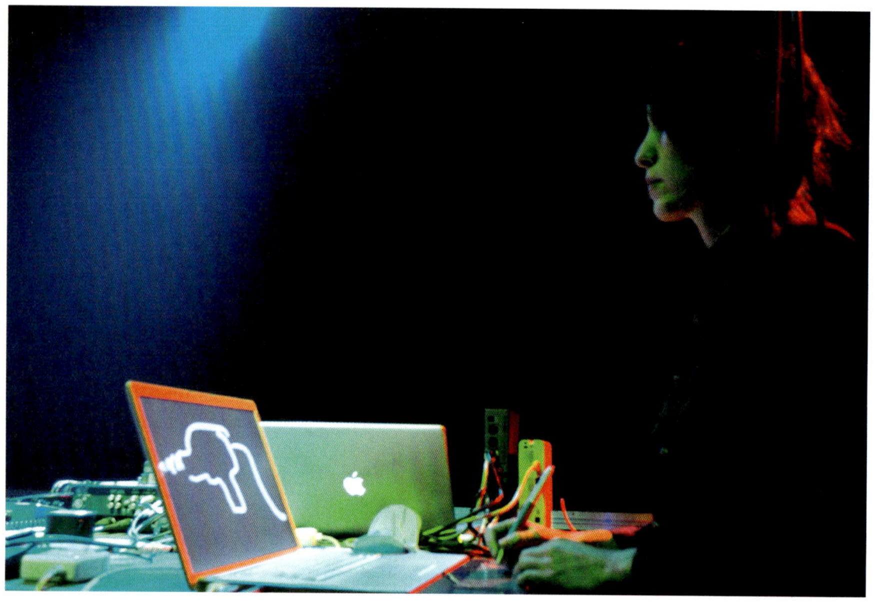

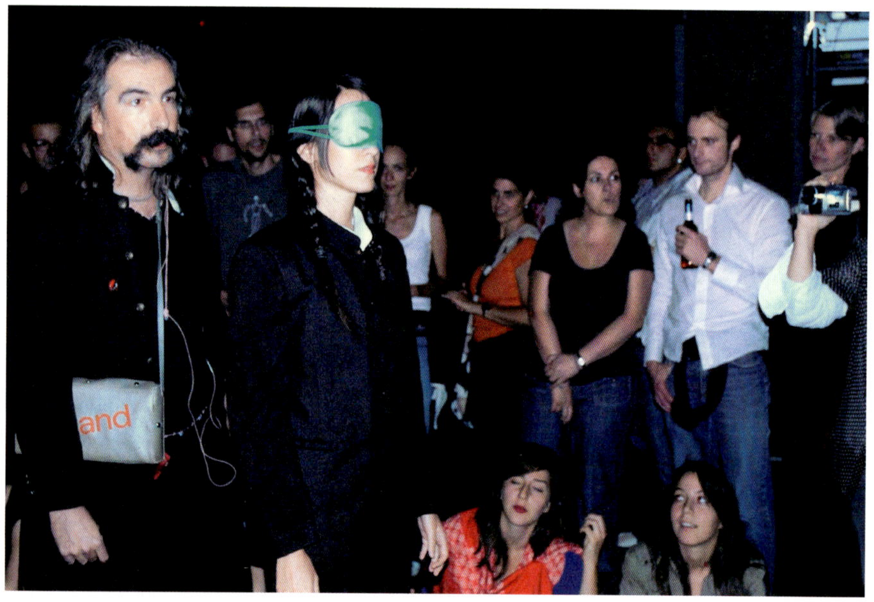

FIGS. 5, 6 Performance of the group Anabala at the ctrl_alt_del sound art festival in Istanbul, 2007.

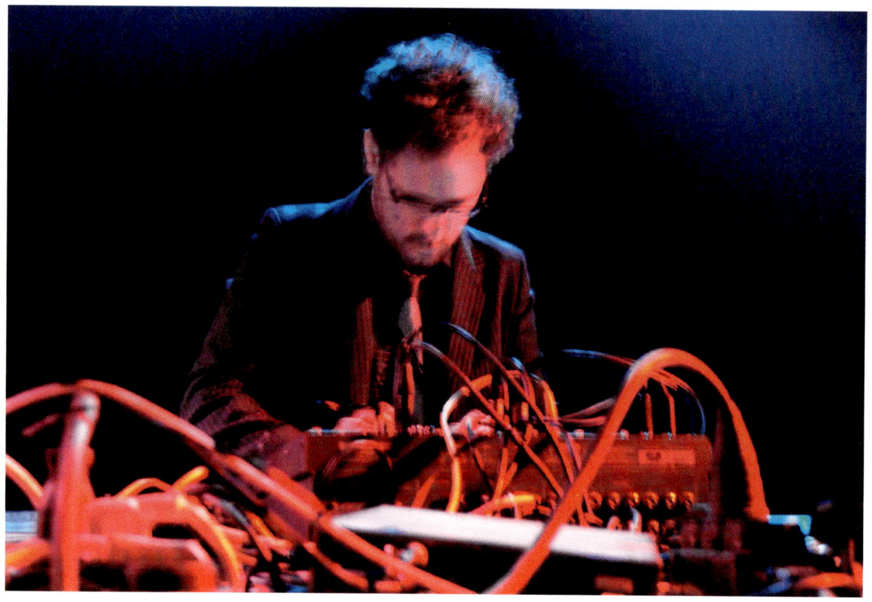

FIG. 7 Performance of Eran Sachs at the ctrl_alt_del sound art festival in Istanbul, 2007.

hubs for classical music in Turkey with its early music training program, music preparatory primary and high schools, various choral and orchestral groups, and the Bilkent International Anatolian Music Festival. Meanwhile, teaching sound design in Bilkent's Communication and Design Department is composer, sound designer, filmmaker, and audio recording engineer Ufuk Önen. Önen is the author of *Audio Recording and Music Technologies*, the first reference book in the Turkish language for the fields of audio recording and music technology, alongside a number of other books and more than forty articles.

Performances and concerts have naturally accompanied the expansion of sound art in Turkey. An electronic music platform, Phonem by Miller, organized annually by KOD Müzik and the Istanbul Foundation for Culture and Arts, was operational between 2003 and 2010. Aside from individual performers such as Koray Tahiroğlu, Erdem Helvacıoğlu, Korhan Erel, and Zafer Aracagök, many young groups and sound artists have started to perform in alternative spaces in Istanbul. Their styles vary from analog to digital, from industrial and noise-based experimental sounds to improvisation. Among these groups and individuals are Tolga Tüzün, Aykut Şahlanan, Attila Kadri Şendil, Serkan Emre Çiftçi, Batur Sönmez, Unit, Analog Suicide, Extreme Parking, Dadaloop, Radiophonik, Audio, Werk, Not at Home, Dead Frequencies, Subjekt, AVST, B/S, and Teletrans. A particular standout in this list is Ankara-based Munir Tireli, aka munimonde, who has done extensive research into Anatolian pop and rock, documented in a number of experimental LPs and recordings.

Two meetings have proven to be key moments of pause for the analysis, investigation, and understanding of Turkey's developing sound art scene. The first was the S-Network Colloquium, organized by NOMAD and Yildiz Technical University's Faculty of Art and Design and Combined Arts Department in 2006. Prior to the colloquium, NOMAD and the YTU team conducted a set of meetings over the course of two months to determine the main goal of the event. The organizers decided to structure the S-Network Colloquium as a public event, discussing the current state of the Turkish sound art scene by establishing a network. The event's participants thus included artists, institutional representatives, university scholars, and members of a variety of organizations and initiatives that were active within the fields of sound.

The second meeting was realized under the coverage of the 2014 project Remote: Relocating Sound.[7] The project and the meeting, hosted by the Goethe-Institute Ankara and Bilkent University, were designed to trace new mental maps through the field of sound art with installations, performances, presentations, and workshops. The meeting was staged as a parallel event to ATMM – Audio Technologies for Music and Media.

Both meetings gathered all the past and current information about sound art production, and they have had an important and lasting effect on the scene. They started discussions on the new trends and the international collaborations and content-based developments in the field. Additionally, the infrastructural and intellectual deficits along with the difficulties and constraints in the field were discussed collectively. Each in its own way, the two meetings framed the developments of sound art in Turkey during the past fifteen years.

Reviewing the content of these meetings and activities, we can observe a shift in both production and reception in the field. Between 2003 and 2008, a vital scene was marked by various NOMAD projects such as loosing.ctrl,[8] City-Sense: Surveillance, Secrecy, Security, Control (FIGS. 8, 9),[9] Upgrade!Istanbul,[10] and the ctrl_alt_del project, which attracted a mass audience, as well as other stage-based activities such as the events Analog Suicide, Noiseistanbul, Transmit, Transmit Records, Spektro Festival, and Signals From Outside, all created by organizer Batur Sönmez, and additional performances that took place in different settings and contexts in Istanbul, such as the Amber – Art and Technology Festival. In the wake of such feverish musical exploration, the scene has started to change direction in favor of a more production-based approach with experiments as they typically occur in the field

7
See www.goethe.de/mmo/priv/11791394-STANDARD.pdf.

8
A joint project, processing actions and reactions of "control mechanisms," produced by NOMAD and the Israeli Digital Art Center, Holon.

9
A project pondering media, control, surveillance, secrecy, and art in urban environments, curated by Nat Muller in collaboration with NOMAD, and held at GarajIstanbul in 2008.

10
Monthly gatherings for new media artists, academics, practitioners, curators, and other actors of digital culture to establish a local archive in this field. The meetings presented key speakers with discussions and related performances and presentations of new projects.

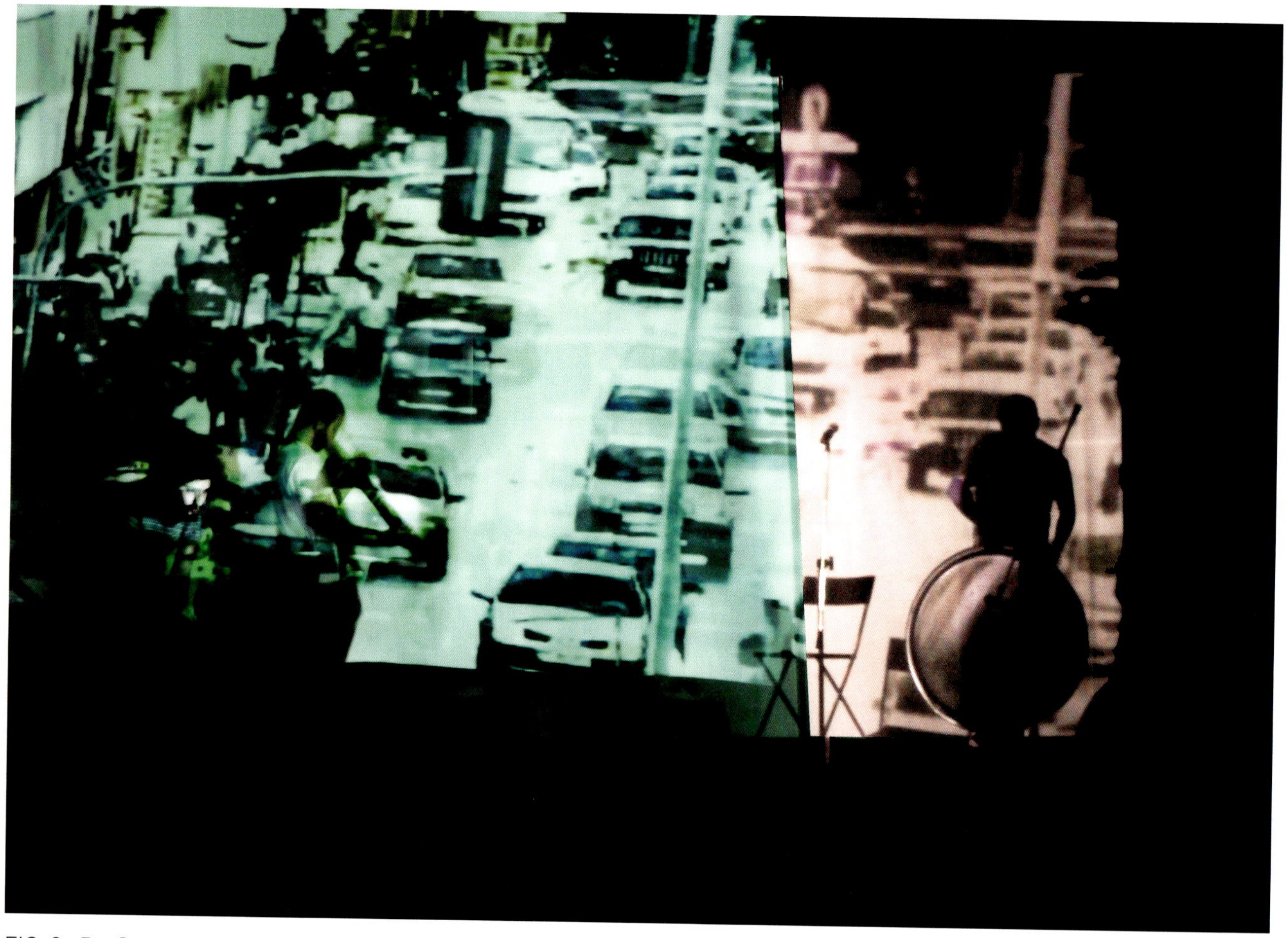

FIG. 8 Perfomance of the group Not at Home at the CitySense festival at Garajistanbul in Istanbul, 2008.

of contemporary art. In the meantime, the majority of the actors of the sound scene either started to work in the realm of electroacoustic music or to merge their works with contemporary art practices. Artists such as Emre Erkal, Cevdet Erek, Erdem Helvacıoğlu, Selçuk Artut, Fatih Aydoğdu, and Candaş Şişman produce contemporary art projects that include elements of sound, sound production, and aural perception.

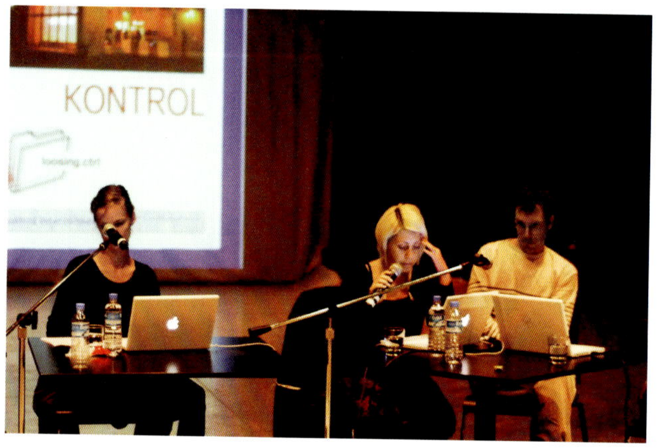

FIG. 9 Talks at CitySense festival at Garajistanbul in Istanbul, 2008.

This shift into the contemporary art world is currently exemplified by projects that are highly diverse in terms of research and production interests. These include such varied offerings as the performances of artist collective Ha Za Vu Zu[11] on new social orders and alternative modes of political engagement, Buğra Karabey's installation-based works on the bio-electric signals of plants,[12] and a series of site-specific sound installations by Candaş Şişman.[13] The group DeCol has addressed the mounting emphasis on electroacoustic music production through a range of sound-oriented works.

Artist and researcher Koray Tahiroğlu works on embodied approaches to music performance and experience, multimodal physicality in sound and interaction, and new interfaces for musical expression. Tahiroğlu's musical composition exploits recently developed affordances in new musical instruments, and he has contributed to the artistic practice of interactive music technologies as well as to their design and implementation.[14]

Another important figure is Gökçe Akçelik (of Replikas), whose artistic research analyzes the relationship between "sound object" and the collective/personal memory by the creation of soundscapes culled from data collected in cities (especially Istanbul).

Erdem Helvacıoğlu,[15] who is a composer, sound designer, guitarist, and producer, has been blending experimental microphone techniques, avant-garde performance ideas, and sophisticated live electronics systems to create albums such as *Resonating Universes*. Helvacıoğlu introduced live sound diffusion and acoustic music techniques to the sound art audience through the performance of his own pieces released on the album *Wounded Breath* and others. He has also contributed to research and documentation in the music field through the release of the album *An Anthology of Turkish Experimental Music*,[16] produced in collaboration with Batur Sönmez, which takes stock of much of Turkey's avant-garde music in the past half century.

Artists working in both the aural and visual field of contemporary art since the beginning of their careers have become more influential and acceptable in contemporary art spaces. As a notable example, Fatih Aydoğdu's installations appear as interventions in the exhibition space. He translates verbal and visual data to sound to be perceived and interpreted within the territory of his works, which are mostly politically oriented conceptual installations. Candaş Şişman, on the other hand, investigates the relationship between the physical and the digital. His works process sound with the physical components of his installations by either creating digital sound compositions or by interpreting acoustic sounds together with digital visuals. Exploring the spatial aspects of sound, Emre Erkal weaves sonic perceptual processes and haptic experiences into the form of architectural installations. His works are realized in contexts that are both historically rich and austerely modern, yet he almost always attempts to rework the roots of our sensory encoding, where inhabitancy of space becomes indistinguishable from auditory interaction.

Sound art in Turkey debuted by means of performances and concerts, but today, the stage has mostly been left behind. Stage-based performances now occur primarily in the context of conferences, festivals, and projects. Sound production is either fusing into the layers of contemporary art exhibitions or preserving itself as a purely aural experience on LPs, CDs, and online music platforms.

11
See http://hazavuzu.blogspot.co.at.

12
The installation was part of the Amber Festival in 2009; see http://amberadmin.webfactional.com/amber/wp-content/uploads/2015/07/amber09.pdf.

13
See www.csismn.com.

14
See http://mlab.taik.fi/~korayt.

15
See www.erdemhelvacioglu.com.

16
See www.subrosa.net/en/catalogue/electronics/anthology-of-turkish-experimental-music-1961-2014.html.

Dajuin Yao

Sound Art in China: Revolutions Per Minute

One morning in late February 2015, in the city of Xi'an, a water spraying truck was heard playing its regular tune loudly while watering the streets. The daily routine appeared as normal as can be for the local residents. Yet anyone with the slightest aural sensitivity not yet destroyed by this sonic bombardment would notice that the truck was actually playing a heavily cut-up remix of three Western Christmas songs: "We Wish You a Merry Christmas," "Jingle Bells," and "Santa Claus Is Coming to Town." To the average Chinese ear, these sounds simply signal Western, pop-music tunes (both concepts desirable); their role as a signifier for a specific season or purpose is not relevant anymore. When the daily sonic environment in China is made up of random noises lacking a coding system, and individuals can create public noise without sensitivity to the surroundings, how can we expect sound art produced there to be an easy cultural translation of "sound art" as practiced and understood elsewhere? How likely is it to fit into Western-style art spaces squarely, neatly, and without any awkwardness?

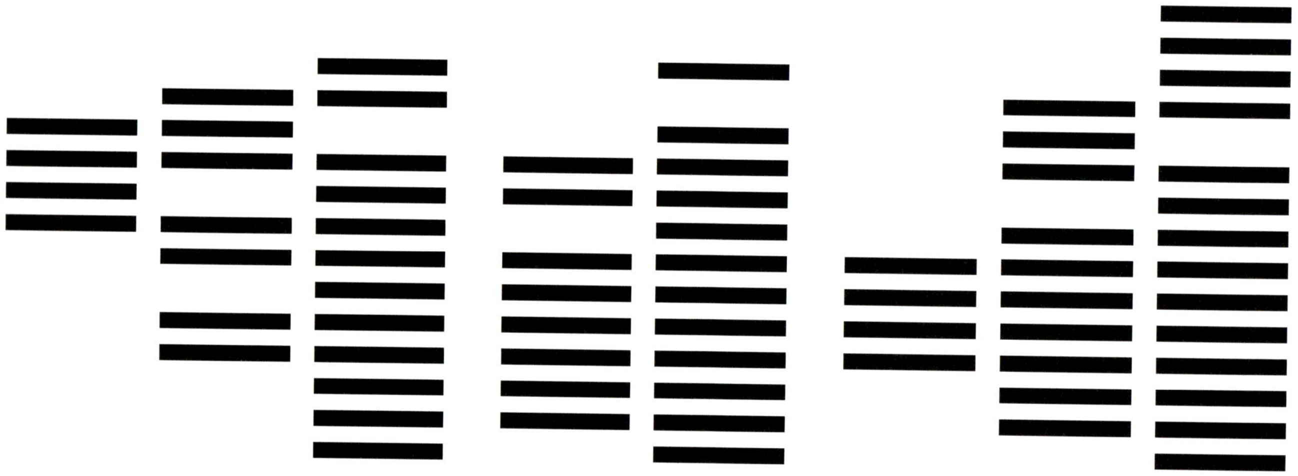

Ancient China: Expanded Sound Art?

To say that China lacks a long sound art tradition is misleading. For sure, "sound art" in its contemporary Western definition did not exist in China before the 1990s, but the culture did produce the most advanced sound theories and auditory aesthetics among world cultures. The ancient Chinese made a connection between music/tuning and the social order that was tighter and more direct than the ancient Greeks or Jacques Attali[1] did, to the point where each official history of the successive Chinese dynasties has a separate chapter on the tuning system, documenting how the court managed to control the tuning of the period. For thousands of years, the successive Chinese dynastic rulers had seen the calendar system and the tuning system as the foundations of their sovereignty. And, the Chinese interest in sound has never been in the sounds themselves or acoustics, but in their correlations, references, and the interplay of phenomena. Sound, music, and tuning have been correlated to cosmology, astronomy, astrology, philosophy, medicine, and so forth.

The concepts of sound of ancient Taoist philosophers such as Lao Zi and Zhuang Zi resonate well with the twentieth-century ideas of sound of John Cage, the Futurists, and the like. Ancient texts such as *Yue Ji* [The Book of Music], (ca. 10 BCE) and Ji Kang's *Treatise on the Non-Emotive Nature of Sound* (ca. 260 CE) are invaluable texts in aesthetics, and have been the foundation of Chinese musical aesthetics for two thousand years. Not to mention equal temperament, or the well-tempered scale, the invention and publication of which by the Chinese Prince Zhu Zaiyu predated its discovery in Europe.[2]

There are numerous acoustic inventions and examples of intriguing acoustic contraptions throughout Chinese history. One ancient architectural construction worthy of examination from today's sound art point of view is the Echo Wall at the Temple of Heaven in Beijing (built in 1420). It is a circular wall 200 feet in diameter, to which one can whisper or talk and the sound can be heard by someone on the diametrically opposite side of the circle.(FIG. 1) The famous set of over sixty bronze bells made 2,500 years ago excavated from the tomb of Marquis of Zeng(FIG. 2) would be a remarkable piece of sound art if exhibited today in an art gallery, not only because of the visual and sculptural impact, the craftsmanship, the metallurgic techniques involved, but also the science of tuning behind it – the bells are tuned to a twelve-tone scale, not well-tempered, but based on a cycle of fifth theory. And let's not forget the true nature of this piece, that it was designed as a burial object, to be buried underground forever, never to be seen by living beings, not to mention being played at all. It was supposed to be silent, and yet is fully resonant in our minds. Does it really matter that it is not shown in a contemporary art space, and not signed by a sound/conceptual artist? Perhaps our concept of what "sound art" is calls for further expansion.

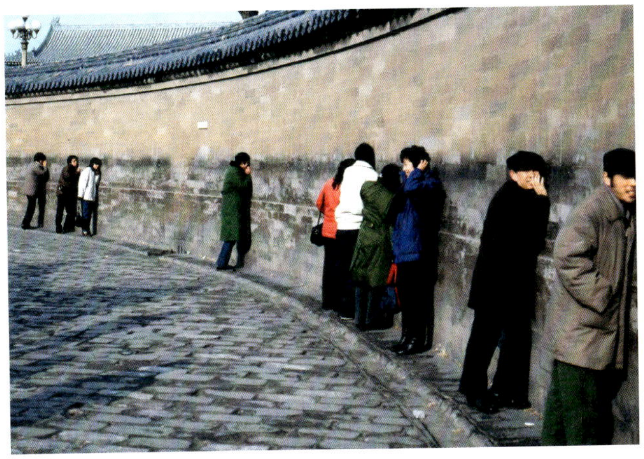

FIG. 1 Echo Wall at the Temple of Heaven in Beijing, 1988.

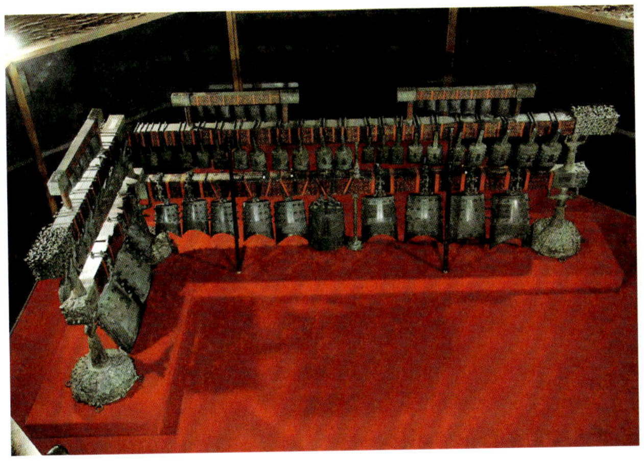

FIG. 2 The Bianzhong of Marquis Yi of Zeng.

1 See Jacques Attali, *Noise: The Political Economy of Music*, Theory and History of Literature, 16 (Manchester: Manchester University Press, 1985); originally published in French in 1977.

2 See Joseph Needham, *Science and Civilisation in China*, vol. IV: *Physics and Physical Technology* (Cambridge: Cambridge University Press, 1971).

Modern Context: Between Spaces, Hidden Traditions

While sound art in other countries either belongs to or follows the international movements in experimental sound, sound art practice in China must be seen from a different vantage point, for China was not in sync with the major countries in aural experience for the entire second-half of the twentieth century. Experimental sound practices blossomed at the beginning of the new millennium, thanks to the rise of the personal computer and the Internet – the former lowered the threshold for making sound experiments, the latter broke the conceptual and aural isolation.

The individual sonic experiments by young artists who were not even aware of one another led to two culminating landmark events in 2003. *China: The Sonic Avant-Garde*, produced by and released on Post-Concrete in the United States, was the first comprehensive compilation CD introducing Chinese experimental sound to the world.[3] Later in the same year, Post-Concrete produced *Sounding Beijing*, the first international sound and media art festival in China, a historic event that gathered together Western and Chinese artists, and an audience from all over the country under one roof.[4] The two events declared the existence of an experimental sound art scene in China.

Starting in 2013, the *Sound Art China: Revolutions Per Minute* series of survey exhibitions, curated by myself, began touring worldwide, including New York, Shanghai, and Hong Kong. By now, some thirteen years after the historic 2003 events, we can find Chinese artists working in a wide range of sound genres: extreme noise, *musique concrète*, electroacoustic, field recording, mixer feedback, circuit-bending, interactive controller, live audio-visual, glitch, minimal electro, gallery-based sound art, free improvisation, etc. And the subjects and issues involved include: space make-up through sound, urban sound culture, schizophonia, sound and oral narrative, Buddhahood and sonic space, Cold War politics and the Art of Noise, non-spatial sounds and surveillance structure, social media and collective sound production, etc. It has indeed been a compressed development of sound art in a short period of time, hence the title *Revolutions Per Minute*.

In the early years, it was contemporary visual artists, with their curatorial connections and established visibility in the art circle, who first made attempts at "sound art" that fit the narrow definition of the term currently used in the West today, namely concepts and sound installations that can be shown in a gallery, described by text, and so on. However, sound in such works was more or less a crude byproduct and the works were really side adventures for these artists. They never amassed enough energy to create a body of work or their own aesthetics of sound. In 2001, well-known contemporary artist Qiu Zhijie gave a solo sound art exhibition in Japan, titled *Sound of Sound*.[5] Curiously, all three works in that exhibition were silent. Nevertheless, they do refer to the sound in Chinese tradition in various ways through writing and language.

It was not until experimental sound practitioners entered this territory that the works started to show both the materiality of sound and the tangibility of form fused together in a coherent manner. The *Buddha Machine* (2005) by electronica duo FM3 (Christiaan Virant and Zhang Jian) is a rare success.(FIG. 3) It uses cheap,

3
China: The Sonic Avant-Garde, compiled and edited by Dajuin Yao, Post-Concrete, POST 005, 2003. 2 CDs.

4
Sounding Beijing: International Electronic Music Festival, The Loft New Media Art Space, Beijing, November 1–4, 2003.

5
Sound of Sound, Gen Gallery, Koshigaya, Japan, 2001.

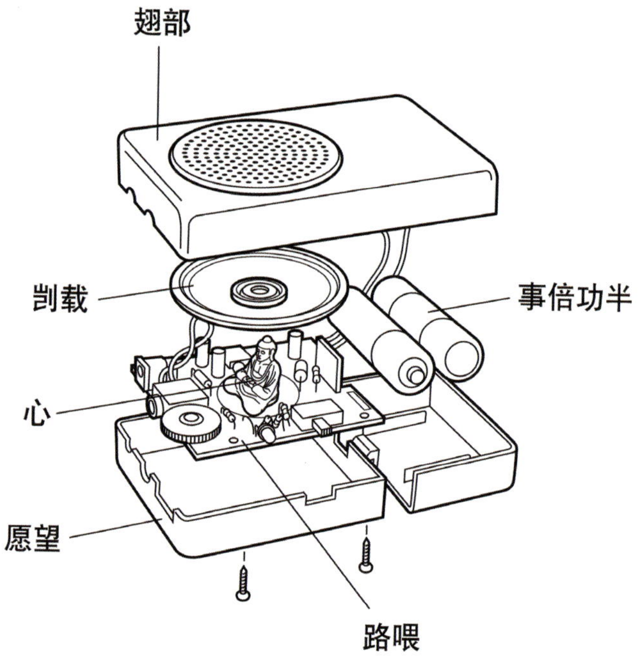

FIG. 3 FM3, *The Buddha Machine*, small plastic box that plays music composed by Christiaan Virant and Zhang Jian, first released in 2005.

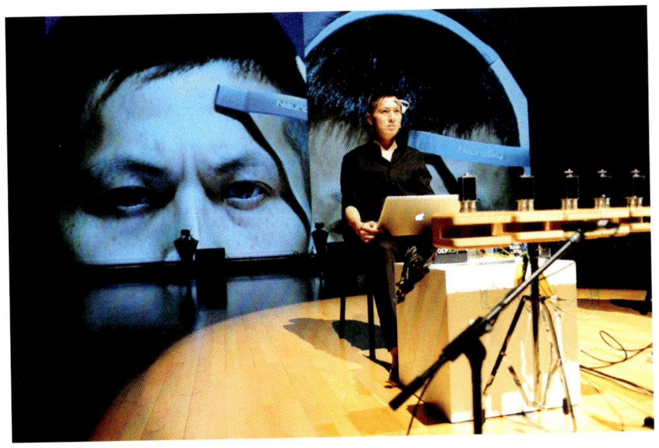
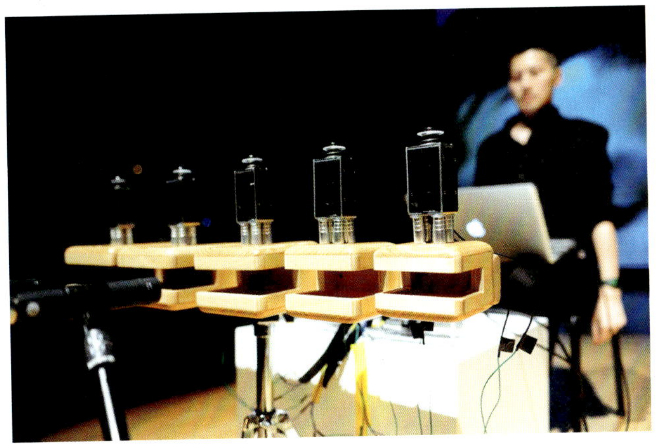

FIGS. 4a, b Samson Young, *I am thinking in a room, different from the one you are hearing in now (homage to Alvin Lucier)*, 2011. Installation and performance for two performers with brainwave sensors, custom electronics, snare drum, tom-tom, woodblock, duration 15 min. Installation views Microwave International Media Art Festival, Hong Kong Design Institute, 2011.

mass-produced small Buddhist prayer machines, replacing the chanting with FM3's own electronic ambient clips. The success results not only from its concept, but also their worldwide marketing techniques. The *Buddha Machine* can likewise be seen as a tongue-in-cheek look at mass production culture and the superficial religiousness rampant in China now.

Samson Young's *I am thinking in a room, different from the one you are hearing in now* (2011) is a performance/installation piece that allows the performers' brain waves to control a sound-making device.(FIGS. 4a, b) Unlike other such experiments in interactive technique, this work shows its depth through its knowledge of and linking to the tradition of sound art, as an homage to pioneer Alvin Lucier and his milestone pieces in brain waves and room acoustics. It is significant to note that this work comes from Hong Kong, where, as in Taiwan, Western sound art history is better known, and works from these two regions fit more snugly into the so-called "sound art" category.

Phonography artist Sun Wei's *Song for 100 Children* (2006) is an installation that may also be presented as a performance. The work features two Chinese printing machines, which are fed the text of a Sichuan folk rhyme. The sound of printing is picked up, amplified, and then modulated by computer. The resulting output is a spectacular, psychedelic sound mass despite its traditional source input.

Yan Jun's *How To Eat Sunflower Seeds* (2011–2013) is a quadraphonic installation playing back the sound of four persons cracking sunflower seeds, a popular Chinese snack and pastime, as if in a conversation.(FIG. 5) This relaxed and unambitious work is typical of Yan Jun's style of making art out of everyday life. His other work *Still Life, Wild Life* (2013) sets up a telescope in front of a window, through which the spectator can watch a video of someone setting off a firecracker

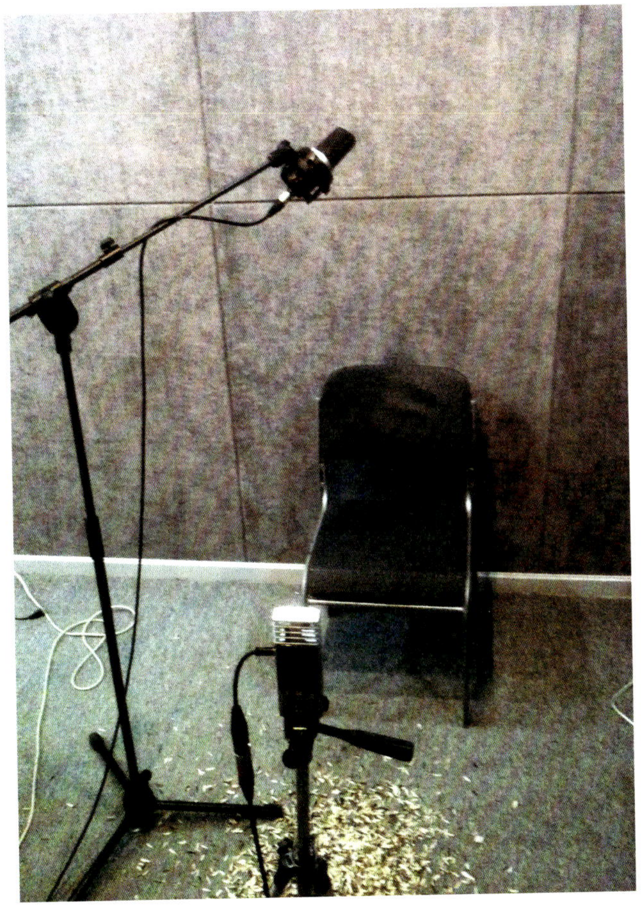

FIG. 5 Yan Jun, *How To Eat Sunflower Seeds*, 2011–2013, photo of the studio recording in 2011. Concept and re-composition: Yan Jun; based on a text score by Natalia Kamia. First exhibited as a site-specific quadrophonic installation for Per Hüttner and Elias Arnér's 2011 project *(In)Visible Dialogues* in Stockholm. Recording musicians: Fan Shijie, He Xiaoyu, Song Yuzhe, and Yan Jun; recording engineer: Fan Shijie (Nice Tune Studio, Beijing); assistant: Zhang Yan.

in different ways on a monitor outside, and at the same time hear the sound through headphones.

Sound artist Weng Wei's *Twice-Cooked Pork* (2012) is a unique combination of an interactive sound device and video projection. The title, the name of a famous Chinese dish from Sichuan, refers to a cross-modulation between ancient calligraphy – video projected onto the controller – and the sound of ancient Chinese phonetics, which the audience can control in real time through moving sliders. The technique behind the work involves Max/MSP/Jitter, Arduino, Processing, as well as custom hardware design.

References to traditional or contemporary Chinese culture can be seen in most of the works mentioned above, even Yan Jun's seemingly casual cracking of seeds and firecracker ignitions are inescapably Chinese.

One work that integrates ancient culture successfully is Jiang Zhuyun's *Sound Medicine Chest* (2007). Eighty-one sound samples separated into categories of machine, body, fire, wind, and water are stored in an old Chinese medicine chest, the kind used in old-time pharmacies. By opening a combination of different drawers, the audience can hear different sounds played back from a medicine boiler pot, as if taking different prescription medicine. The healing power of sound, traditional Chinese medicine, phonography, digital signal processing, and interactive installations are all subtly and smoothly united in this work.

My own *Geophone Nanking* (2005) is a rare case of referencing the mysterious and fascinating ancient Chinese auditory heritage. The sound installation hinges on the analogy of the ancient geophone, a military surveillance device the Chinese used in battles two thousand years ago: a large pottery urn which was buried underground and a scout who was placed inside to listen for the attacking enemy's tunnel digging. The physical principle behind this device is known in the West as "Helmholtz resonance." In the gallery, *Geophone Nanking* is set up as a large black box in which the audience can sit in total darkness and listen to a surround-sound collage of field recordings made in Nanking. This site-specific work is an homage to the war-ridden city of Nanking both in sound and in concept.

Wenhua Shi's *Futurist LoudSpeakers: Roars, Thunderings, Explosions, Hissing Roars, Bangs, Booms* (2013) is not only a centennial tribute to the Futurist Luigi Russolo's *Intonarumori*, but also a sarcastic look at Cold War politics with the large loudspeakers playing the US Armed Forces sound effects from 1967 and Chinese Mandarin language familiarization LPs from 1960.(FIG. 6)

Phonography: Weirder Than Fiction

The bizarre reality of life in contemporary China is simply too powerful to ignore; same with its sounds. The sounds from real life in China are often much more intriguing than any noise-generating gadgets, software, or installations in a gallery; hence the popularity of phonography in this country. Yet, in direct opposition to the denial of the humanistic or the conceptual dichotomy between man and nature (evil versus good) deeply embedded in most Western phonography practice since R. Murray Schafer's Acoustic Ecology movement – with the Gordon Hempton–type of paranoia over machine sound being the extreme example –, the Chinese phonographers celebrate the domain of human affairs, the non-utopian, the this-worldly, the social, the interpersonal, the personal, the bodily, the vocal, the vehicles, and all.

Since 1997, China Sound Unit, an open-ended guerrilla phonography group founded by myself, has focused on non-invasive observation, surveillance, documentation, and re-contextualization of the Chinese sonic phenomena. From the very beginning, China Sound Unit has insisted upon a humanistic direction, recording interesting sounds related to humans and the society, and has shown no interest in the natural environment or the animal world. The archival recordings made by China Sound Unit have been used in installations, concerts, and also have been released simply as untreated sound archives. Most Chinese phonographers share this humanistic view in attitude, including those who are not members of the China Sound Unit. An example is the super active PlayBack Unit (Zhong Minjie and Li Zhiying), which operates around the city of

6 http://sjp.blogbus.com.

7 Words sharing the same sound are differentiable only through pitch variations.

8 John Oswald, "Plunderphonics, or Audio Piracy as a Compositional Prerogative," in *Musicworks*, 34 (1986), 5–8; see also Christof Migone's contribution in this catalogue.

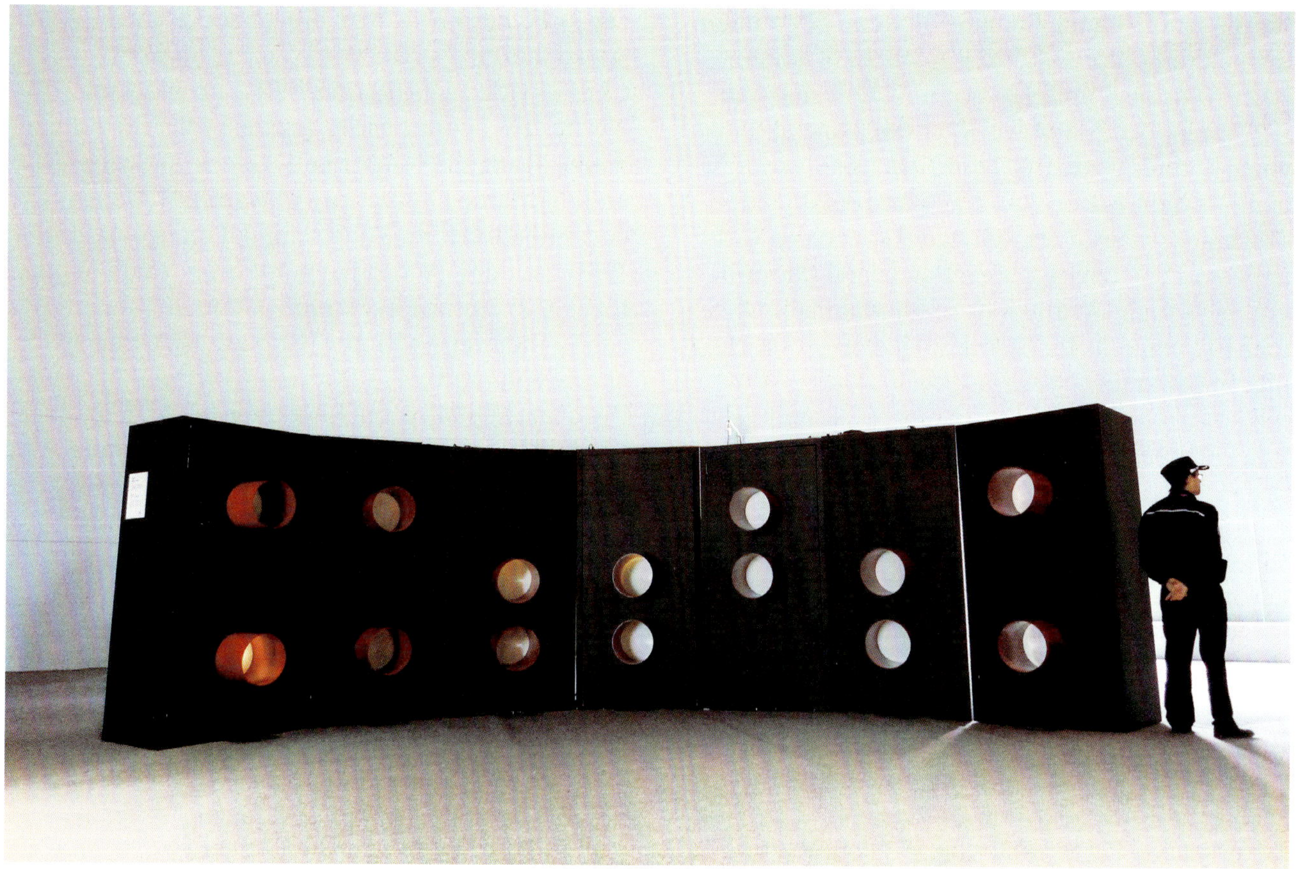

FIG. 6 Wenhua Shi, *Futurist LoudSpeakers: Roars, Thunderings, Explosions, Hissing Roars, Bangs, Booms*, 2013. Sound installation, wood, 14 transducers, 26 metal funnels or painted PVC pipes, Arduino, 7 infrared sensors, Processing + Max, 600 audio samples, 7 speakers, 100 × 200 × 70 cm each. Installation view *West Bund 2013: A Biennale of Architecture and Contemporary Art*, October 2013, Shanghai.

Guangzhou and documents the acoustic environments there extensively, once even posing as customers in a massage parlor wearing concealed recording devices. Besides releasing CDs and playing live, PlayBack Unit, from 2007 to 2009, posted new recordings together with pictures of the sites recorded on their website.[6]

Voice, Language, and Confucianism?

This strong interest in human-related sonic phenomena might be considered a defining trait of Chinese phonography, or Chinese sound art in general. In a way, this tendency can be seen as Confucian, because of the emphasis on the world of human affairs and human relationships. Together with this humanism comes the fascination with the human voice, and language. From a practical point of view, when listeners are facing highly homogenized Chinese urban developments, it is often the local dialects that function as Schafer's soundmarks in the sonic environment. Moreover, Chinese culture is above all a culture of text, writing, and literature. Language has always been the core and soul of Chinese culture. Acoustically speaking, for artists using language as sound samples for signal processing, the Chinese language family, being tonal[7] and with numerous dialects and regional variations, offers a tremendously rich resource for all types of experimentation in digital signal processing and modulation.

Early works exploring the sonic possibilities of the Chinese language include my own *Dream Reverberations* (1997) and *Cinnabar Red Drizzle* (1999), in which algorithmic poetry, granular synthesis, and other digital signal processing methods are combined to reveal an unknown acoustic microcosm hidden inside the everyday language, and Hu Mage's *Firewood, Rice, Oil, Salt, Soy, Vinegar & Sugar* (2002), in which Chinese opera rhythms, granular synthesis, and pop song "plunderphonics"[8] are all mixed up with humor.

Works featuring the human voice can be found in two major compilations. *Oral Works* (Kwanyin, 2010), a project initiated by Hong Qile and Yan Jun, is a collection of works showcasing extended voice techniques. *Talk Talk Talk*,[9] on the other hand, is a daring project and massive collection of Chinese speech art, unaltered speech recordings made by over a dozen artists and phonographers found in all kinds of unimaginable, real-life contexts that are amazing, even shocking, to listen to.

The Apolitical Unconscious

Contrary to most Westerners' expectations and insistence, politics is not a major concern and is rarely found in sound art in China. This is due not so much to government censorship or the resulting self-censorship as a shared sense of futility and apathy in the post-1989 era. Furthermore, to read the thriving Chinese harsh noise scene as a form of critical protest would prove to be too naive. On the other hand, genuinely and explicitly political sound works are found more often in Hong Kong and Taiwan.

Jasper Fung's installation work *Encode* (2014) employs a fire alarm bell which plays percussive patterns in Morse code for government-censored keywords on Chinese social media.(FIG. 7) Samson Young's 2012 work *Liquid Border* (with the Chinese title *Violence Border*) uses contact microphones to record barrier fences in the forbidden zones along the border between Hong Kong and mainland China. Edwin Lo's 2011 *Mourn* is dedicated to Hong Kong's annual mourning of the 1989 incident. Audio from TV broadcasts of the gatherings are excerpted and manipulated to subtly intervene with the listener's time and space at low volume.

The most outspoken political sound artwork comes from Taiwan: Hwang Dawang and Zhang Yousheng's *Minkoku Hyakunen* [100 Years of the Republic of China] is a political satire on the current Nationalist regime, which was then celebrating its centennial, using invasive phonography, where the artists comment and even "act" on the various scenes being recorded. This political use of creative phonography earned the artists an Honorable Mention in Prix Ars Electronica, 2012.

Toward Noise 2.0: From Cellphones to Social Media

One direction of sound practice in China worthy of special attention is based on the platform of cellphones, the Internet, post-Internet smart-

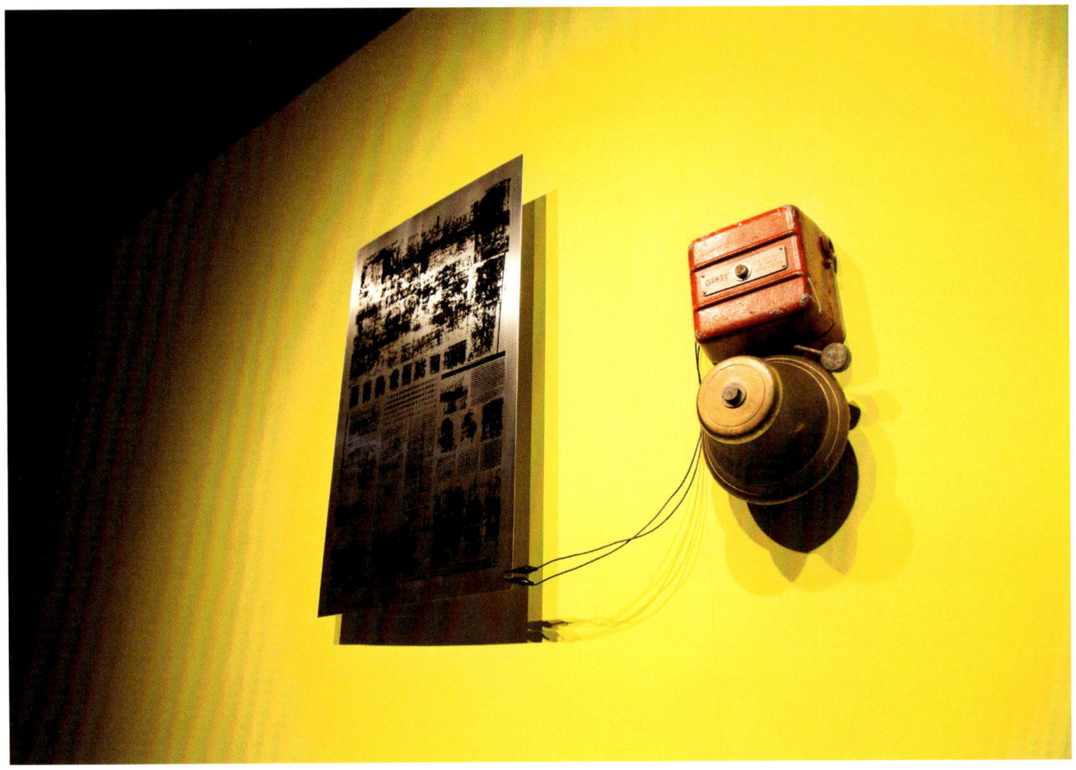

FIG. 7 Jasper Fung, *Encode*, 2014. Mixed media installation, silk print on stainless steel plate (58 × 38 cm), Gents' of Leicester fire alarm bell, electronics. Installation view *Revolution Per Minute – Sound Art China*, Run Run Shaw Creative Media Centre, Hong Kong, 2014.

9
Talk Talk Talk: Chinese Speech Art, Post-Concrete, POST 010, 2015. 2 CDs.

10
Such as *Beijing Sound Unit – Archive.XX1*, recorded 1998–1999, Post-Concrete, 2008.

phones, and social media. Beyond the surface of merely trendy and geeky experiments, a highly self-reflexive look at the medium, the platform, and, often, the element of the social itself, runs through these works.

As early as 2001, Fu Yu (half of the pioneer media art duo 8GG), created *Internet Soundscape*, a work documenting the sound of the Internet via the Netscape Web browser. Zafka (Zhang Anding) took the idea of the virtual soundscape one step further with his 2007 limited edition vPods, knock-off iPods with sound tracks preinstalled, and the 2008 CD release *i•MIRROR*, which contains a soundscape created from sound snippets recorded on the virtual world platform Second Life. In the mid-2000s, Xu Cheng made extensive recordings of online chatroom conversations and other net broadcasts, covering a wide range of topics, from underground Christian sermons to S/M bondage master classes.

These works dig into and open up an acoustic world inside the Internet, of life on the Net, which is usually considered only as a carrier of sound files and sound information exchange. Personal computers with Internet connection played a crucial role in the beginning phase of sound art in China, from retrieving information and sound works, to the production of sound, networking among artists, dissemination of works and event info, etc. As we move away from the PC and the Web in this post-Internet era, new mediums are offering new potentials.

In 1997, China Sound Unit began paying attention to the increasingly important sonic medium of the cellphone and started surveillance recordings in various cities in China. This project has since accumulated an archive of sounds of the society. As the society transforms at an unprecedented pace, the archive has become an increasingly important resource for sociological, historical, phonological, or psychological research. Whether this collection of sound objects can be called sound art is irrelevant; it certainly can be if exhibited in public or released on CD.[10]

In 2005, a few years before smartphones and iPhones became the standard in China, Jiang Zhuyun performed a unique cellphone work entitled *Start with Hello*, at the opening of *Sound Imaginations*, a China Academy of Art sound art exhibition featuring sound installations and performances by more than twenty artists, in which he had two cellphones crossed, with the earpiece of one phone facing the mouthpiece of the other, and then dialed two different numbers on the two phones. And so the spontaneous, unpredictable, yet totally passive, human interaction and dialogue ensued.

And as the iPhone landed in China, works making use of the new device, which, among other things, is a convenient recorder, surfaced all over the map. Edwin Lo made a CD (2011) of monophonic iPhone phonography recordings in Hong Kong. The piece *iPhone Improvisation* (2012) by Vavabond (Wei Wei) is a performance in which she experiments with attaching various transducers to the body of an iPhone and then performs various tasks on the iPhone body as well as on the iOS system, thus picking up a stream of unpredictable noises.

In 2012, Wang Changcun wrote and released the first Chinese sound art app for the iPhone, titled *Cicadas*, in which the user can design and control his own cicada sound by changing the parameters on a series of sliders. The Soul Lemon band and I performed in Shanghai in 2010 on iPhones and iPads only. Different from other iPhone orchestras, Soul Lemon improvised with only non-musical apps, such as utility and game apps, presented smartphone sounds "as is," and even handed over their setup to the audience, who took over the show halfway into the concert.

Social Media Sound as Sound Art

Early examples of social participation in sound can be found in *Harbin Sound Map* by Zhang Liming (aka Hitlike), a 2008 Web project in which the users can freely post and share their own recordings according to geographical location. A similar construction and idea were used by Zhang Liming and Wang Changcun in their 2009 collaboration *China Sound Map*. With the advent of social media in China, we see a new frontier on the horizon. Using the new Chinese social media app PaPa, anyone can make a live sound recording and share it. Though perhaps not an original concept, PaPa has fostered a huge user-base, high traffic, and a level of creative use of sound manipulation unseen elsewhere. With tens of millions of active users, PaPa is the most fascinating window to the sonic China. The cultural significance of these audio social media platforms cannot be emphasized enough. Firstly, they are carrying the largest collective production, exchange, and consumption of "sound objects"[11] in history, especially when the PaPa app offers the user various modulation effects used in *musique concrète* such as pitch-shifting, mixing, etc. Secondly, the audio social media have internalized the sound objects. They have almost rendered pointless many earlier formats of social participation in sound, for example, the soundmaps mentioned earlier. PaPa offers much fuller participation, and neither the production or consumption is geographically or visually oriented.

Furthermore, beyond the idea of UGC (user-generated content), these audio platforms have, in a way, realized what Jacques Attali envisioned long ago as the stage of composing, where everyone creates their own music, or noise. This is several steps beyond the art of phonography – giving up the artist-oriented concept of sound art altogether. If "noise is power," then this is power in the hands of every mobile user; every sentient being is a Buddha. Social media sounds have liberated creative sound-making from the gallery/concert/recordings ecosystem of contemporary art, and even challenge the very idea of art. It is also the largest collective archiving of our sonic life. It is *Noise 2.0*.

The Mirror Stage and the Aural Awakening

In late 2014, China's most popular social media platform had shifted from Weibo, a Twitter knock-off with photo and video built-in, to Weixin, a knock-off of an audio messaging platform, TalkBox. This rise of the most pervasive, number one Chinese social media platform, the audio-based Weixin (with half a billion users), together with PaPa, is signaling a social perception change from the visual to the auditory. As every mobile user in China is discovering the technique and joy of recording and repeated playback of one another's sound objects, people's way of listening and even of talking have changed. Many are hearing their own voices for the first time. Through repeated recording and playback, some people are changing their way of pronunciation, acquiring the self-consciousness of a broadcaster. People are gaining an aural consciousness. These audio platforms are indeed triggering an auditory discovery of the sonic self. It is alarmingly similar to Lacan's theory of the "mirror stage."[12] Without our realizing it, these auditory social media are teaching the society how to listen, and even behave, which will prove to be historically and socially significant in the long run.

Future Sound of China

Today's China is one of the noisiest countries in the world. Noises run wild, especially noise in public space, such as the nonstop Muzak at subway stations, railway trains, individuals conducting private cellphone conversations at high volume, etc. Ironically, this different sonic ideology is also what imbues the society with inexplicable acoustic energy.

Contemporary sound art in China is still young; too young to predict its future. But one thing is certain – the sounds currently running uncontrolled in the society will be gradually disciplined as the country moves rapidly toward its goal of "civilizing," while what is known as "art sound" will come to flourish within its art world ecology, as sound art.

It may be enlightening to see these recent developments of sound art in China as an extension of one of the oldest and most important heritages in world auditory culture, where ancient Chinese made a direct link between sound and the wider realms of cosmology, astrology, philosophy, politics, and ethics. In other words, "Chinese sound art" is only a tiny, albeit significant, part of Chinese auditory culture, the recognition of which may lead us to more fruitful views of the current condition, as well as of the future of sound art. What might be most worthy of our

11
As defined by Pierre Schaeffer in his *Treatise on Musical Objects* (1966).

12
See Jacques Lacan, "Le Stade du miroir comme formateur de la fonction du Je: telle qu'elle nous est révélée dans l'expérience psychanalytique," *Revue française de psychanalyse* 13 (1949), 449–55.

expectations would be a different kind of sound practice, mode of listening, or acoustic energy, one that has little or even nothing to do with installation, artist, or even art, one that is out of sync with the global sound art agenda. That energy is continuing its search for viable outlets and platforms. What we can do now is to lend our ears.

Appendix: Sonic Hotspots

A network of organizations, labels, and festivals are the driving forces behind much sound art activity in China. Although these hotspots are spread out all over the country, as well as in Hong Kong and Taiwan, some of the more active ones include:

- Noise Asia: Organization and label founded in 1997 by Dickson Dee in Hong Kong, promoting independent music concerts and workshops
- Post-Concrete Records: Label founded in 1997 by Dajuin Yao in California, promoting Chinese sound art in a global context
- Media art festivals: Sounding Beijing 2003, Sounding Taipei 2004, Shanghai eArts 2008 Opening Concerts, SONART series (all curated by Dajuin Yao)
- 2pi Festival (2003–2007): Highly regarded experimental sound festival organized by Li Jianhong
- Sub Jam, Kwanyin: Organization and label founded by Yan Jun, promoting experimental music and free improvisation since 2004; largest volume release of Chinese experimental music
- Waterland Kwanyin (2005–2010), Mini MIDI Festival (2005–2010), and Miji Concerts series: Experimental music performance series curated by Yan Jun
- Sound Art China: Ongoing touring survey exhibition of Chinese sound art, curated by Dajuin Yao in 2013, showing in New York, Shanghai, Hong Kong, etc.
- China Academy of Art: Includes an active sound art program, and promotes sound art through festivals and exhibitions
- NoiShanghai: Performances organized by Cao Junjun, aka Junky, of Torturing Nurse, a key proponent of harsh noise
- Soundpocket (Hong Kong): Non-profit sound art organization
- School of Creative Media, City University of Hong Kong: Features an active sound art program
- Kandala Records (Taiwan): Label and organization founded by Chang Yousheng
- Department of New Media Art, Taipei National University of the Arts (Taiwan): Features an active sound art program

For academic electroacoustic and computer music, activities are centered around the Central Conservatory of Music in Beijing and the Shanghai Conservatory of Music. The former's Center for Electronic Music of China (directed by Zhang Xiaofu) hosts an annual festival of electroacoustic music, MusicAcoustica Beijing, and has also hosted ICMC (International Computer Music Conference) and workshops by GRM. The Music Technology Department (directed by Chen Qiangbin) at the Shanghai Conservatory of Music has recently started an annual festival, Electronic Music Week, which features academic electroacoustic music as well as interactive sound installations. These two conservatories have frequent exchange programs with INA-GRM, EMS, and other academic music organizations. Professor of musicology, Xiao Mei, of the Shanghai Conservatory of Music, also leads a research program focusing on the social and documentary aspect of soundscape in China.

Ryo Ikeshiro and Atau Tanaka

Sound in Japan: Silence, Noise, Material, and Media

Introduction

Sound holds a special place in Japan. The country's distinctive aural signatures are closely bound to its history, nature, and culture. It is hard to imagine the sound of cicadas without recalling the unbearable heat and humidity of summer. The sound of trickling water punctuated by the occasional bamboo hitting a rock from *shishi-odoshi* in Japanese gardens is integral to the contemplative calm with which it is associated, as are the chanting of sutra by monks and the bell chimes in temples. In contrast, its urban megalopolises are filled with the incessant din of advertising, alerts, and public transport announcements, shop attendants continually greeting customers, and talking machines – the effects perhaps of lenient noise regulation and rampant consumerism, but also possibly a curious counterpoint to an aesthetic of tranquility. These soundscapes can be attributed in part to the influence of Buddhism, and to historical events – World War II and the highly accelerated economic development that followed. They have had a profound influence on the polar opposites of serenity – associated with nature and tradition – and chaotic urban din. They do not pose a contradiction, but reflect two views into the same thing. Noise is not considered an "other" to silence. Similarly the relationship of sound and sound art to other art forms, including music, is fluid.

Wabi-sabi is a Buddhist concept that the philosopher Muneyoshi Yanagi describes as the "beauty of irregularity." In contrast to Western ideals, characterized by symmetry, equal temperament, purity of timbre, and four-square phrases, in Japan asymmetry, imperfections, and incompleteness are considered to be attractive and articulations of beauty.[1] The notion of *wabi-sabi* may be seen as the aesthetic grounding of the Japanese reinterpretation of Buddhism, reflecting plainness and simplicity, aged and solitary states.

In *Noh* theater, long temporal development, a concentrated and measured sense of pulse, and additive rhythms are far removed from the metric repetitive structures of Western music. In *Gagaku* court music, melodic lines shared in loose synchrony across instruments yield rich heterophonies distinct from Western functional harmony. The frequency spectra of Japanese musical instruments, such as the *shakuhachi*, are highly irregular in comparison to their Western counterparts. Like the typical vocal style used in *Noh*, which hovers over the phase transition point between normal voice and falsetto, the playing styles of the *shakuhachi*, *sho*, and *koto* often deliberately explore unstable regions at the limits of chaos.

The pronunciation of syllables in Japanese vocal music also differs. Whereas consonants receive greater emphasis in most Western song, in Japanese, vowels follow almost all consonants and are more prominent, creating contrasting senses of phrasing and rhythm. The frequent use of onomatopoeia in Japanese reflects a direct relationship to sound where ideophones form a sort of "sound symbolism." A particular type of onomatopoeia, *gitaigo*, depict phenomena which produce no sound. Silence is described as *shīn*, demonstrating how the absence of sound can be symbolized by an actual sound.

The fact that the Japanese landmass is a collection of islands has played a significant role in shaping its cultural identity. In addition to obvious limits in trade and exchange, it has experienced alternating cycles of opening and closing to external influence. After over two hundred years of isolation under the Tokugawa regime's policy of *sakoku*, it was forced with the arrival of Admiral Perry's Black Ship to open to the West in the mid-nineteenth century, bringing about the Meiji Reformation. The rise of nationalism in the early twentieth century leading to the Pacific War and World War II again isolated Japan, with the postwar acceptance of defeat resulting in the avid modernization that has led to today's cosmopolitan society. Despite the open flows of information that the Internet provides, Japan remains an island distinct in its ways and customs – no matter its assimilation of and influential position within global culture. Embracing the West but holding it apart, it continues the incorporation of opposites and the indifference to contradiction that is fundamentally Buddhist.

1
See Muneyoshi (Sōetsu) Yanagi, *Cha To Bi* [Beauty and Tea] (Tokyo: Kodansha, 2000).

Sound art remains a relatively new term in Japan. Art trends there often favor the creation of obsessively detailed lower-level labeling while eschewing precise higher-level definitions – a myriad of new species belonging to vague genera. In audio art forms, works may be described as sound sculpture, *sound objet*, self-made instrument, sound installation, or soundscape, but broader categories such as sound art and its distinction from music can be said to be considered secondary. This fluidity and lack of broad and straightforward boundaries reflects a situation where sound and sound art have existed without the need for strict categorization. Here, we follow a similar course as we discuss movements and artists without recourse to questions of what may or may not constitute sound art. Our approach includes elements of the historical and theoretical, seen through the lens of practice. Context is a driving force providing the emerging themes we identify. This chapter will be of interest to readers both inside and outside Japan curious about sound practices in relation to other art forms and its connections to the global sound art scene.

Historical Movements

Jikken Kōbō

Jikken Kōbō (or Experimental Workshop) (1951–1957) was the first mixed-media group in Japan and an early example of continued collaborative practice between different art disciplines. In addition to their leader Shūzō Takiguchi being pivotal in introducing Dada and Surrealism to Japan through connections to André Breton and Marcel Duchamp, the group organized the Japanese premier of notable European contemporary classical works by the likes of Arnold Schoenberg and Olivier Messiaen. There were fourteen members including visual artists, composers, critics, choreographers, and poets, many of whom would later pursue successful individual careers such as the composers Tōru Takemitsu and Joji Yuasa. Collaborative spirit was perhaps most present in their works such as *Another World* (*Mishiranu sekai no hanashi*) (1953) made for automatic slide projector or auto-slide produced by the company that would eventually become Sony. The device, which consisted of a slide projector synchronized with a tape recorder, provided the perfect medium for combining abstract photography with *musique concrète*.[2]

Jikken Kōbō also collaborated with Toshio Matsumoto for *Ginrin* (1955), the first experimental colour film made in Japan.(FIG. 1) Successful combinations of different media through collaborations continued after Jikken Kōbō group activities ceased. Examples include subversive and experimental films using contemporary music such as *Funeral Parade of Roses* (1969) and *Eros + Massacre* (1969) with soundtracks by Yuasa and Toshi Ichiyanagi respectively.

Gutai

The Gutai Art Association (1954–1972) is arguably the most significant artist collective of postwar Japan. While Jikken Kōbō was closer to multimedia – as a combination of traditional media – Gutai was decidedly more intermedia – in consideration of the interstices of media they explored. Their outdoor group events may be seen as precursors to the happenings of Allan Kaprow and Fluxus. Furthermore, the range of practices from the fifty-nine members in Gutai's eighteen-year history was diverse and also included sound, installation, interactivity, and environmental art. It both prefigured and then later developed in tandem with a rise in left-leaning politics and non-hierarchical modes of organization associated with the protests against the renewal of the Treaty of Mutual Cooperation and Security between the United States and Japan (*Anpō*) in 1960 and the rise of the student movements in the ensuing decades.[3]

Atsuko Tanaka's *Work (Bell)* (1955) was Gutai's first sound art piece.(FIG. 2) A sign which

2
See Mélanie Mermod, "On the Autoslides – Interview with Jōji Yuasa," in *Jikken kōbō: Atelier experimental/ Experimental Workshop*, exhib. cat., ed. Mélanie Mermod (Paris: Bétonsalon, 2011), 10.

3
See Ming Tiampo, "Gutai Chain: The Collective Spirit of Individualism," *positions* 21, no. 2 (2013), 383–415.

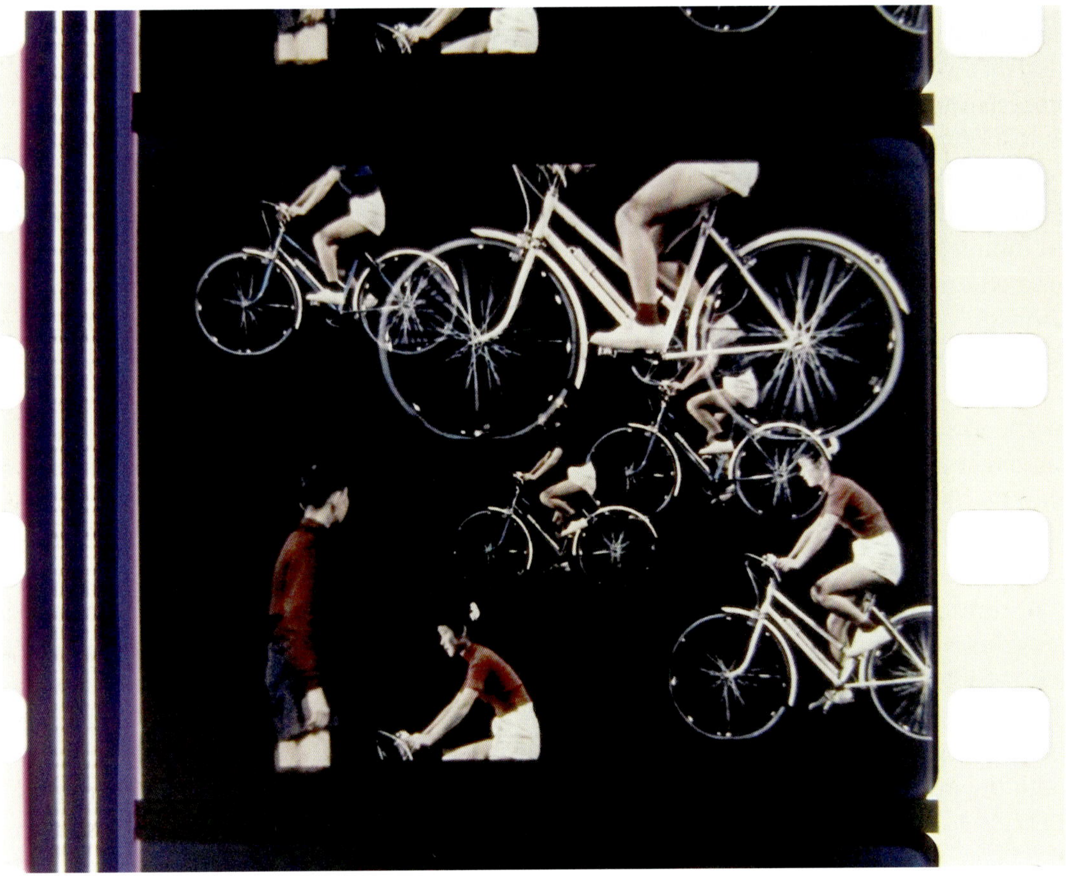

FIG. 1 Toshio Matsumoto, *Ginrin*, 1955. Film, colour, 12 min.

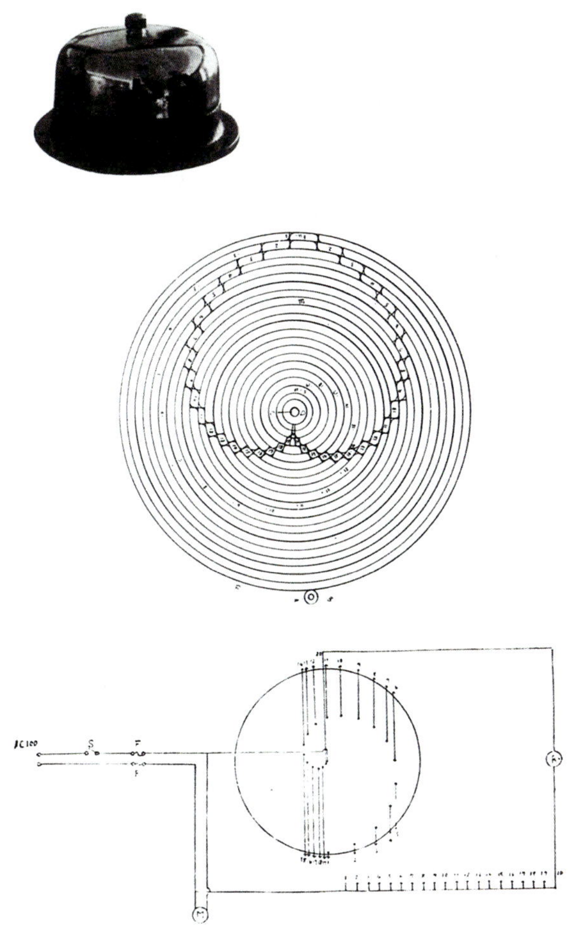

FIG. 2 Atsuko Tanaka, Drawing for *Work Bell*, 1955. Published in *Gutai journal*, July 1956, p. 24.

read "Please push this button" invited visitors to trigger a sequence of loud and intrusive electric bells. The playful use of high-pitched sounds usually associated with warning signals and the incorporation of interactivity blurred the separation between art and the everyday. Sonically, it defined the architecture of the space, being described by Shōzō Shimamoto as "perhaps the first ever invisible work in the history of art."[4] It is certainly a contender for the earliest work of sound art in the world.

In 1957, Shimamoto began recording everyday sounds and experimenting with tape recorders, producing "concrete" (or *gutai*) music. However, he eschewed the heavily cut-and-spliced nature of *musique concrète* which he considered to be constrained to the framework of classical music and composition. Instead, he recorded his material in a natural way and left the recordings unstructured. The results recall more the aesthetics of field recordings and presage long-durational installation sound works.[5]

NHK Studios

Unlike many Western European countries, Japan lacks a centralized government infrastructure with subsidies for art and culture. Artist groups often develop outside of institutional contexts such as with Jikken Kōbō. Broadcast media in Japan and Europe were similar, however, in that a state-run network dominated the sector. The first major period of electronic music production in Japan was realized with the establishment in 1954 of the electronic music studio at Nippon Hoso Kyokai (NHK), national radio and television, similar to the WDR studio in Cologne and GRM in Paris. The studio at NHK became the nucleus for experiments in Japanese electronic music, which included Toshiro Mayuzumi, Makato Moroi, Toshi Ichiyanagi, and Joji Yuasa.

Group Ongaku

Group Ongaku was a loose improvisation ensemble started by Shuko Mizuno and Takehisa Kosugi and soon joined by Yasunao Tone and Mieko (Chieko) Shiomi in 1960. They rejected dodecaphony and the official lineage of classical music, taking inspiration from artists such as Jackson Pollock. For them, improvisation and the incorporation of non-Western instruments and tape technology became a form of anti-art, analogous to the Surrealists' concept of automatic writing. As evident in their work entitled *Automatism No. 1* (1960), they consciously portrayed themselves as neo-Dada. Shiomi notes how she came to recognize how actions such as throwing keys into the air became inseparable from the sounds produced within a performance context.[6] The incorporation of such everyday activities into the realm of music was a reflection of the politics of the day.

Tone's work *Tēpu Rekōdā* (*Tape Recorder*) (1962) for the *Yomiuri Independent* exhibition was what would now be described as a sound sculpture. It consisted of a large white cloth bag containing a reel-to-reel tape machine playing

4 Shōzō Shimamoto, "Can a Piece of Cloth Be a Work of Art?" [Ichi-mai no nunokire demo geijutsu sakuhin ka?], trans. Reiko Tomii, in *Electrifying Art: Atsuko Tanaka, 1954–1968,* exhib. cat., ed. Ming Tiampo (Vancouver and New York: Morris and Helen Belkin Art Gallery, University of British Columbia; Grey Art Gallery, New York University, 2004), 107–09. Originally published in *Gutai* 4 (1956), 31.

5 See "Shimamoto Shozō," *Gutai.com*, n.d., http://articide.com.pagesperso-orange.fr/gutai/fr/ss.htm.

6 See David T. Doris, "Zen Vaudeville: A Medi(t)ation in the Margins of Fluxus," in *The Fluxus Reader*, ed. Ken Friedman (London: Academy Editions, 1998), 91–135.

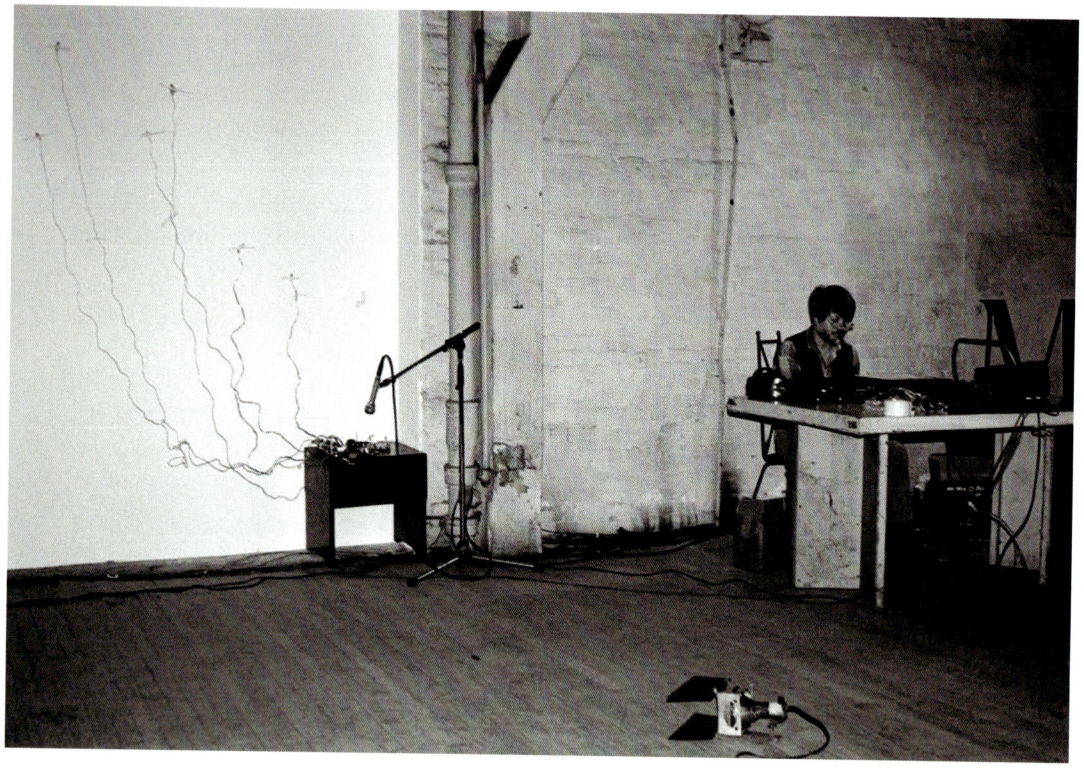

FIG. 3 Yasunao Tone performing "Music for 2 CD Players," 1986. Experimental Intermedia Foundation, New York, 1986.

back a loop of intermittent noises. Tone and Kosugi were also closely associated with Hi-Red Center and participated in many of their political public interventions. Armed with portable tape players and radios, Tone and Kosugi participated in the collective's now-infamous *Yamanote Line Event* (1962), an improvised happening on the circular train line running through Tokyo.[7] Tone's intermedia works include *TV is a chewing gum for eyes* (1969) which involved live performance, televisions, and moving microphones as a critique of mass media.[8] There has been a recent revival in interest in and recognition of Tone's work including the Golden Nica prize for digital music at *Ars Electronica* in 2002. Notable later works include *Music for 2 CD Players* (1986)(FIG. 3) and *Solo for Wounded CD* (1997) created from compact discs with scratches and adhesive tape. Errors in reading the data produce skipping and glitches, exploring the limits of the medium.

Fluxus

Fluxus is the international art or Dada-inspired anti-art movement that began in the 1960s with principal poles of activity in New York and Europe. Despite being comprised of a large and fluid ensemble of artists from around the world with its activities taking place in the West, a strong Japanese flavor was evident in Fluxus due in part to the self-conscious incorporation of the influence of Zen Buddhism as well as the significant contributions made to the movement by Japanese artists in Japan and abroad. Many Japanese Fluxus artists had attended lectures by John Cage introducing the possibilities of silence, noise, and everything else outside music. His ideas were the initial springboard for Fluxus and the most significant influence for the group. Cage himself was a student of Daisetsu Teitarō Suzuki, the key writer responsible for introducing Zen Buddhism to the West. Cage visited Japan for the first time in 1962 at Ichiyanagi's behest, and after several years of correspondence with Kuniharu Akiyama/Jikken Kōbō in the 1950s. Although Cage's ideas had been known through those familiar with the composer and his works, his visit had a huge impact in Japan and caused what many describe as "Cage Shock."

Similarities can be observed between numerous Fluxus "instruction" pieces and the *kōan*, a pedagogical tool in

7
See William Marotti, "Sounding the Everyday: The Music Group and Yasunao Tone's Early Work," in *Yasunao Tone: Noise Media Language*, ed. Brandon LaBelle (Los Angeles: Errant Bodies, 2007), 13–33.

8
See Dasha Dekleva, "In Parallel," in *Yasunao Tone*, ed. LaBelle (see note 7), 39–61.

9
See Doris, "Zen Vaudeville."

10
Yayoi Uno-Everett, "Scream Against the Sky: Japanese Avant-Garde Music in the Sixties," in *Sound Commitments: Avant-Garde Music and the Sixties*, ed. Robert Addlington (Oxford: Oxford University Press, 2009).

11
See Midori Yoshimoto, "Expo '70 and Japanese Art: Dissonant Voices – An Introduction and Commentary," *Review of Japanese Culture and Society* 23 (2011), 1–12.

Zen in the form of a sutra. Both unfold through demonstration rather than discourse to aid experience and understanding.[9] "What is the sound of one hand clapping?" is a well-known example of a *kōan*. Without being audible, it contains the idea of sound, prefiguring a common theme explored in the field of sound.

Yōko Ono was involved with Fluxus from its early stages along with La Monte Young. Ono is an example of an artist who consciously embraced Japanese characteristics. But as noted, these were not necessarily opposed to the general spirit of Fluxus. Perhaps as an expatriate, her sense of Japanese identity was heightened as is sometimes the case, particularly as an Asian female within the white male-dominated context of the New York art world. Although originally unappreciated in her native country, Ono was able to capture the zeitgeist of those turbulent times. Her instruction "scream against the sky" from her *Voice Piece for Soprano* (1961) became the title of the first survey of postwar Japanese art, curated by Alexandra Munroe in 1994 and held in the Guggenheim Museum SoHo and Yokohama Museum of Art. As partner of Beatles singer John Lennon, she is also an example of the cross-fertilization of pop culture and the avant-garde.

The basement of the Sōgetsu Art Center (1958–1971) was significant in providing a platform for avant-garde artists including those associated with Jikken Kōbō, the NHK electronic studios, and Group Ongaku. Fluxus-related events included Ichiyanagi's 1961 concert, which featured elements of performance, loudspeakers, improvisation, and the use of graphic scores and IBM mainframe computer punch cards exhibited in the lobby, as well as Nam June Paik's 1964 performance using a hammer and saw to destroy the inside of an upright piano.[10]

Expo '70

Following 1958 in Brussels, where Varèse premiered the *Poème Electronique* at the Philips Pavilion designed by Le Corbusier and the composer/architect Iannis Xenakis, Expo '70 Osaka was a seminal moment for Japan on the world stage. International artists such as Karlheinz Stockhausen produced a multichannel concert in a geodesic dome.

Gutai embraced the opportunity to participate in the generously funded and well-attended World Fair, or *Banpaku*.[11] Their exhibits included *Work* (1970) by Senkichirō Nasaka and Michio Yoshihara which references both the two earlier Gutai sound pieces.(FIG. 4) It consisted of 150 meters of aluminum piping which traversed the entire group exhibition in the Midori Pavilion through which "concrete" music was played back. Sonically

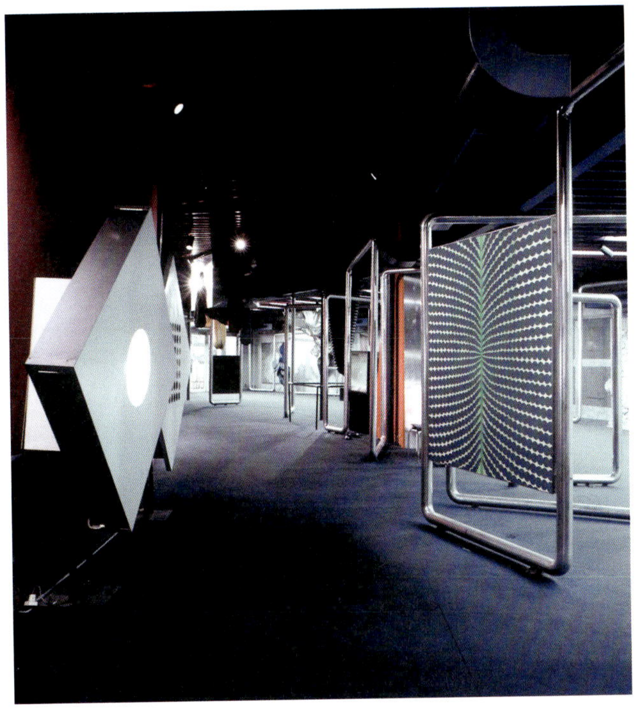

FIG. 4 Senkichirō Nasaka and Michio Yoshihara, *Work*, 1970. Stainless steel pipe, recorded sound, ca. 150 meters long. Installation view, Gutai Group Exhibition, Midori Pavilion, Expo '70, Osaka, March 15–September 13, 1970.

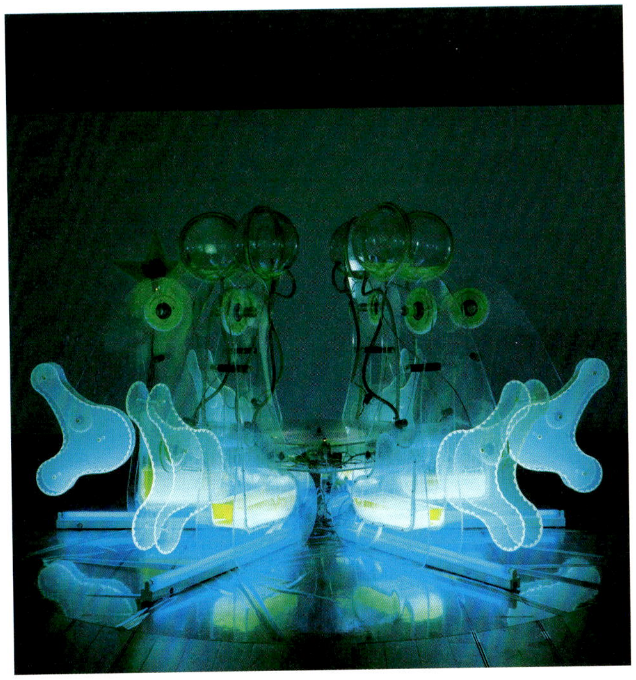

FIG. 5 Minoru Yoshida, *Bisexual Flower*, 1970. Plexiglas, motors, electrical circuitry, tubes, bath salts, water, sound.

and visually, it connected the disparate practices of the artists exhibited; a few of the other artworks were displayed along the pipe. Furthermore, it articulated the space, creating an environment. Also featured was Minoru Yoshida's *Bisexual Flower* (1969), a self-pollinating psychedelic plant made of Plexiglas, a new material at the time.(FIG. 5) With its motor-powered moving parts and gurgling colored water alongside a self-constructed analog synthesizer, it is an ambiguous reflection on technology and progress in the form of a kinetic/sound sculpture.[12]

Sonic Materialism

Perhaps as a reaction to the World Expo and the technological progress and economic development it celebrated, works questioning technology began to emerge already in the 1970s. These were characterized by a predilection for creating sounds using simple objects and materials and exploring their natural resonance. Pure sonic characteristics and their relationship to the environment without interference can be described as sonic materialism. Terms such as "sound sculpture," "sound installation," and "sound objet" were used as descriptions for these works. Distinct from Pierre Schaeffer's *objet sonore*, the Japanese "sound objet" – a combination of the English word "sound" and the French word "objet" – refers to readymades which make sound, recalling Marcel Duchamp's *With Hidden Noise* (1916), a ball of twine sandwiched between brass plates that makes a noise when shaken, sonically exposing an unknown object inside.

Akio Suzuki is a key exponent of this approach and continues to perform today, appearing alongside younger Japanese artists like Aki Onda and internationally with artists like David Toop. Active since the 1960s, he began constructing objects/instruments such as the *analapos* (1970–), essentially a metal version of a "string telephone," which normally consists of a piece of string attached to the bottom of two paper cups.(see p. 418) It was first exhibited as an installation at Minami Gallery, a groundbreaking exhibition space for contemporary art in Nihonbashi, Tokyo. He was encouraged by the gallery owner Kusuo Shimizu to perform with the work. Takemitsu then asked him to perform with the analapos on his soundtrack to Nagisa Oshima's 1978 film *Empire of Passion*. This led to Suzuki's international debut, via Takemitsu's introduction, at the Paris Festival d'Automne in 1978, where he participated alongside Kosugi in an exhibit entitled *MA: Espace-Temps du Japon* organized by the architect Arata Isozaki.

Suzuki also creates "silent" installations, consisting of no more than an indication of a place to stand or sit. Works such as *Otodate* (1996), *Flower* (1997), and *Pyramid* (1999) explore the notion of listening through becoming perceptive to environmental sounds. The first of such works was *Space in the Sun* (1988), where, in preparation, he spent a year and a half constructing two adobe brick walls, three meters high and seventeen meters wide and facing each other.(see p. 420) On the day of the autumn equinox, he would sit on one wall, facing the other and just listen for a whole day.[13] The goal to become one with nature and experience sound without any semantic mediation is reminiscent of Schaeffer's notion of reduced listening.

Keijiro Sato, a former member of Jikken Kōbō, is another artist whose works explore sonic materialism. He began creating what he described as "electric objets" from the late 1960s onwards. With the dictum "all things shine" – his translation of Cage's famous phrase "everything is expressive" – he constructed intimate kinetic sculptures in which small spheres and magnets oscillate and vibrate as they move along straight or wave-shaped axes.[14]

The *Sound Garden* exhibitions (1987–1994) featured an early use of the term "sound art" in Japan by Hiroshi Yoshimura.[15] Participating artists included Tatsuji Ushijima who produced "useless machinery" pieces such as the *For the Remembrance* works (1987–) where small rocks are rotated in contact with other material, producing inscriptions and sounds. These sound exhibitions often featured performances where the exhibits were also used as instruments. The convergence between the fine arts and music

12
See Ming Tiampo, *Gutai: Decentering Modernism* (Chicago and London: University of Chicago Press, 2011).

13
See Yutaka Tanimoto, "Interview with Akio Suzuki" [in Japanese], website of Phoenix Hall, Osaka (2007), http://phoenixhall.jp/interview/2007/03/16/631/.

14
Tama Art University Museum, "Monohikaru Ten" [Everything is Expressive – Vibratile Objets of Keijiro Sato], exhib. (2012) [in Japanese], see www.tamabi.ac.jp/museum/exhibition/121101.htm.

15
See Katsushi Nakagawa and Tomotaro Kaneko, "The Development of Sound Art in Japan – *Sound Garden* (1987–94) and the Classification of Sound Art by Yoshimura Hiroshi" [in Japanese], *Cross Sections* 4 (2011), 56–62.

was productive for artists such as Suzuki, Akinori Matsumoto, Kenichi Kanazawa, Christophe Charles, the collective WAY and Yukio Fujimoto. Fujimoto would go on to organize *Audio Picnic at the Museum* (1997–2006), an annual one-day happening-like exhibition in Nishinomiya. It featured his works, including "sound objets" made from music boxes that draw attention to our sonic environment through listening.

What we have described as sonic materialism continued into the 1990s and to the present day. Seiji Morimoto began incorporating buzzers amongst other objects as a source of vibration and sound in his performances and installations. Tetsuya Umeda and Rie Nakajima explore the use of found objects and low-tech kinetic sculptures in both performance and installation. Miki Yui employs a similar approach with the addition of multichannel audio.

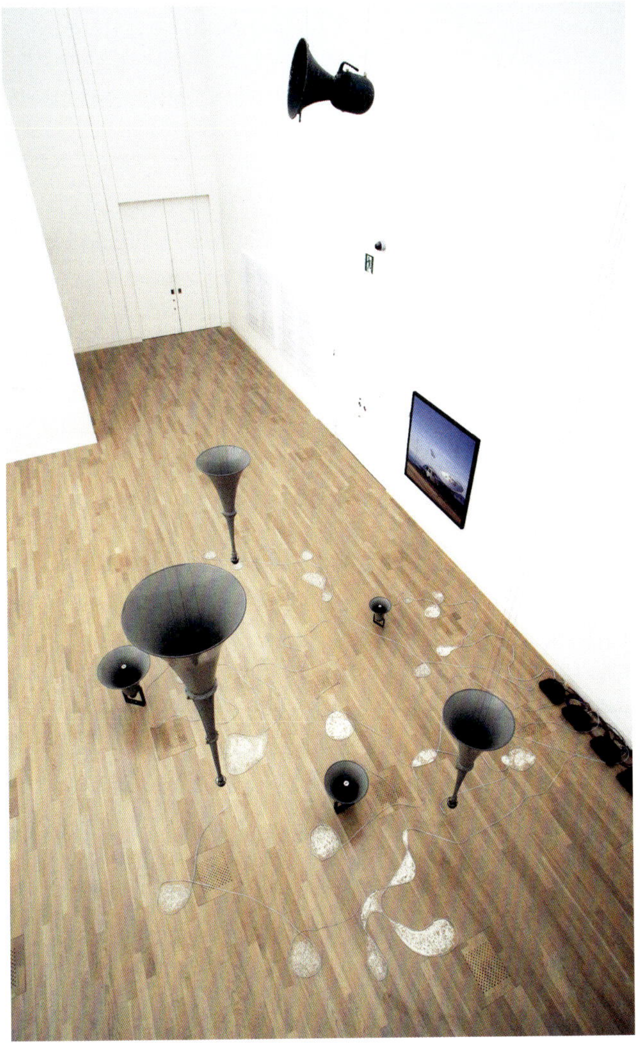

FIG. 6 Hitoshi Nomura, *COWRA*, 2009. Installation view at the exhibition *Nomura Hitoshi: Perceptions-Changes in Time and Field*, The National Art Center, Tokyo, 2009.

Mono-ha

Mono-ha focused on invisible processes and natural phenomena. Translated as the "school of things," it was an art movement from the late 1960s to the early 1970s which used natural material such as stones, trees, paper, paraffin, or industrial material such as paper and steel plates, both by themselves or in combination in almost unaltered states. It sought to re-evaluate relations and interdependencies between ourselves, objects, and the surrounding environment.

Hitoshi Nomura's early multimedia works explored the natural process of decay, as seen in *Tardiology: Tokyo* (1969): made from giant cardboard boxes, it grew to over seven meters tall before reaching the point of collapse. In other works, Nomura's processes were less visible and translated more directly to sound – for example, in the piece *COWRA (Cosmic Waves & Radiation)* (1987–1992) which receives electromagnetic waves from the sun and the galaxy and converts them to audio for playback over speakers.(FIG. 6)

Kazumichi Fujiwara's *Resonance Orientation [Onkyō Hyōtei]* was a project begun in 1968 whose incarnations included a giant outdoor sound object in Shibuya, Tokyo, in 1974, reminiscent of the large-scale installations by Mono-ha artists. He went on to open Otonokinoko, described as the world's first store and website to sell sound.

Processes/Context

A gradual distancing from the privileging of reduced listening and sound-in-itself toward a greater emphasis on processes and context may also be observed. Several themes emerge, an example being the exploration of invisible phenomena. WrK (1994–2006), the collective comprised of Minoru Sato, Toshiya Tsunoda, and Jio Shimizu, explored changes in the relation between time-based phenomena and our perception and behavior. Sato, as the group's leader, investigated phenomena such as electromagnetic fields, vibrations, and wind – and our perception of them – by translating them into sound and light in works such as *perturbation field* (1995)(FIG. 7) and *Surface* (1995). Tsunoda's work with Luke Fowler was part of *Soundings* (2013), the sound art

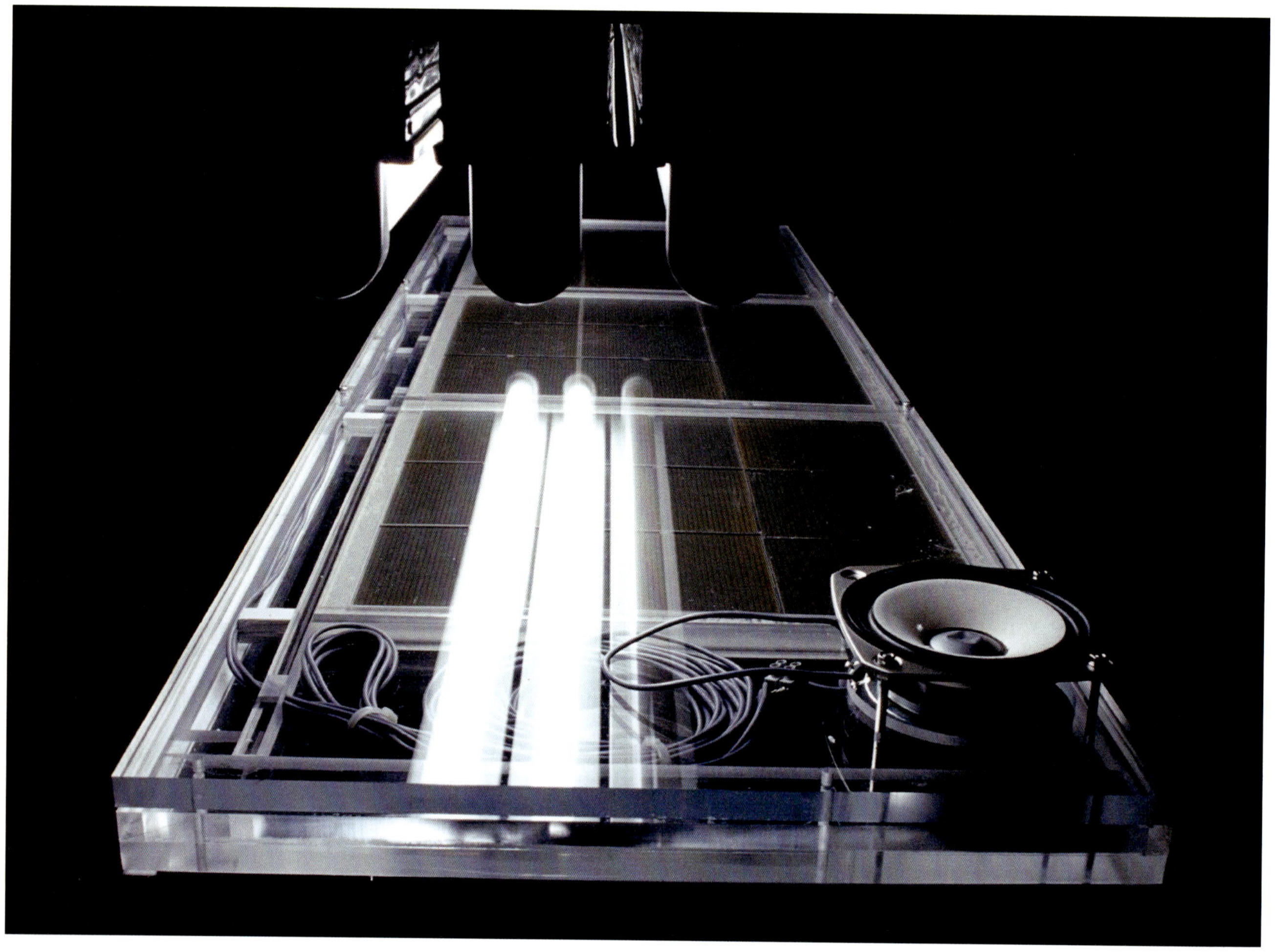

FIG. 7 Minoru Sato, *irregularity / homogeneity: emerging from the perturbation field*, 2007. Solar cells, fluorescent lamps (including "half-broken" irregularly flickering lamps), loudspeaker unit, acrylic case, graphical text of silkscreen with electro-luminescence sheets, text on cutting panel of stainless steel. Installation view WAKO University, Tokyo, 2007.

exhibition at the Museum of Modern Art in New York. One of his most striking releases is *Low Frequency Observed at Maguchi Bay* (2007). It was recorded using contact mics placed on piers at a fishing bay which picked up low frequency sounds from the water current. Each track is followed by a second, edited version in which all sound above 20 Hz has been filtered out, effectively removing anything audible to the human ear. Although the track seems silent, the speakers still move, due to the low frequency sounds. The work provides a physical representation of the low frequencies directly related to the structure and history of the location: the construction of piers and breakwaters to calm the waves which was necessary for its transformation into a fishing bay.(FIG. 8), 16

Artists exploring similar phenomena and media include Keitetsu Murai, whose work *georama* (1997–1999) generated sound out of signals drawn from a graphite probe in contact with ultra-thin aluminum foil and a teaspoon holding physiological saline solution, and Kanta Horio, whose *particle* (2008) playfully makes clicking sounds with broken paperclips invisibly jittering under the control of an electromagnet.(FIG. 9) Examples of more visible processes which take precedence over purely aural characteristics include racquet sports in Atsushi Nishijima's *Mondrian Ping Pong* (1991), whose sounds were determined through participants playing table tennis. Ecosystems which include both visible and invisible processes often involving visitors are created in installations by Soichiro Mihara – for example *moids* (2004–), a collaboration with Hiroko Mugibayashi and Kazuki Saita(see p. 340) – and Yuko Mohri – for example *I/O* (2011–2013).

Another factor that provides the context for many of the works is Japan's consumer society, as seen in Muneteru Ujino's practice. Works such as the *Love Arm* series feature refurbished electrical appliances in arrangements which

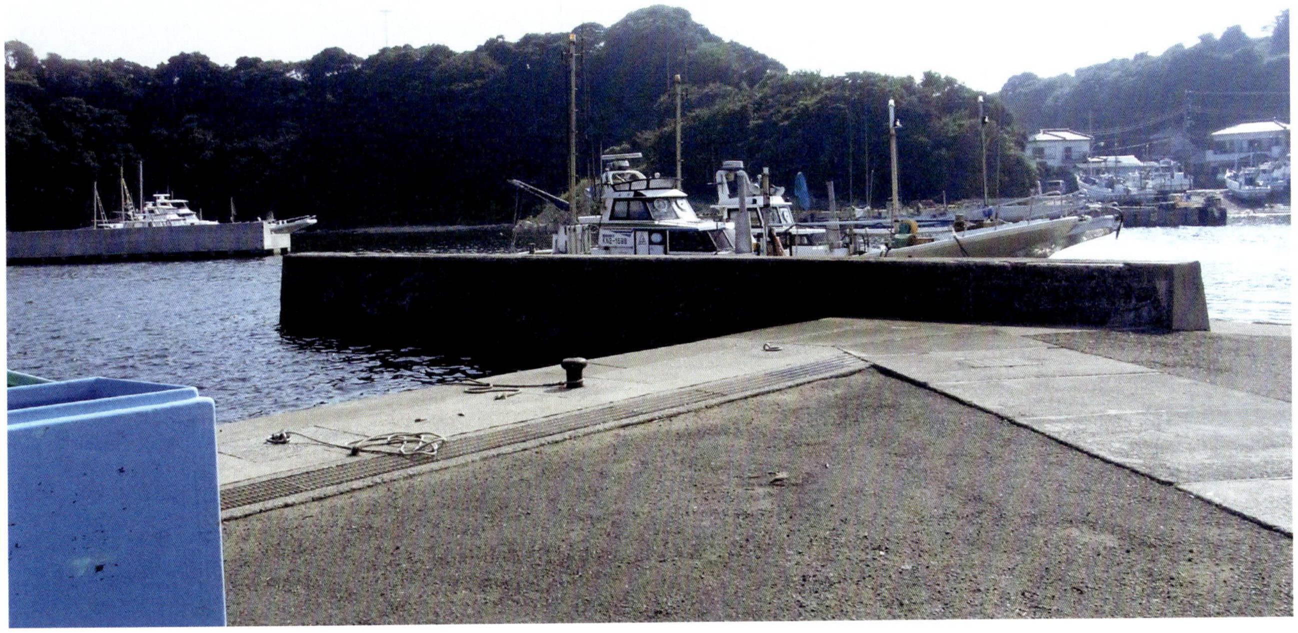

FIG. 8 Toshiya Tsunoda, *Low Frequency Observed at Maguchi Bay*, 2007. The (smaller) breakwater at Maguchi Bay from which the low frequency vibrations were recorded.

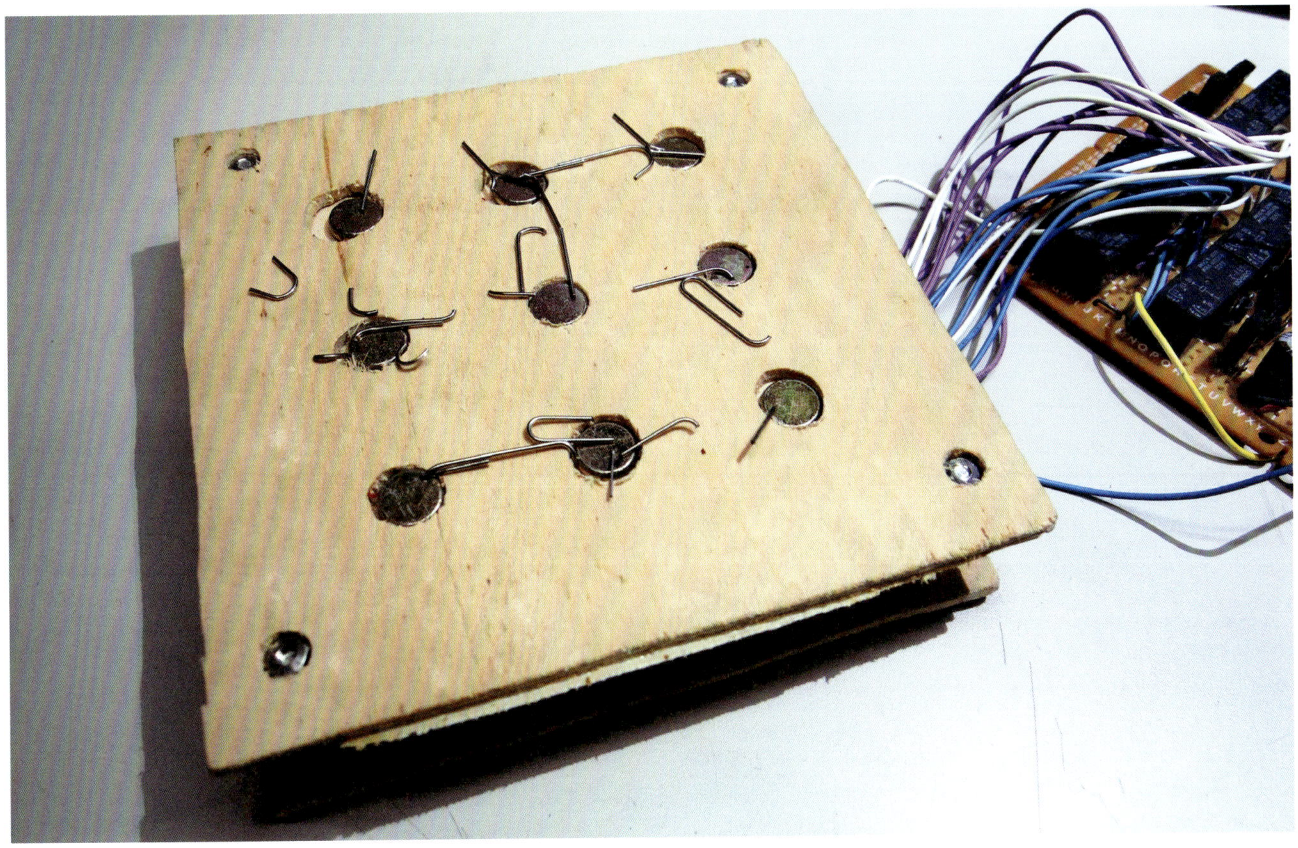

FIG. 9 Kanta Horio, *particle*, 2003. Nine electromagnets, segments of paper clips.

he describes as being akin to Japanese bento and ikebana in installations and performances.(FIG. 10), [17] On a related theme, two contrasting examples in the use of old media may be found in Lyota Yagi's *Vinyl* (2005–2008), a record made of ice,(FIG. 11) and Ei Wada, who uses reel-to-reel tape machines and CRT monitors.

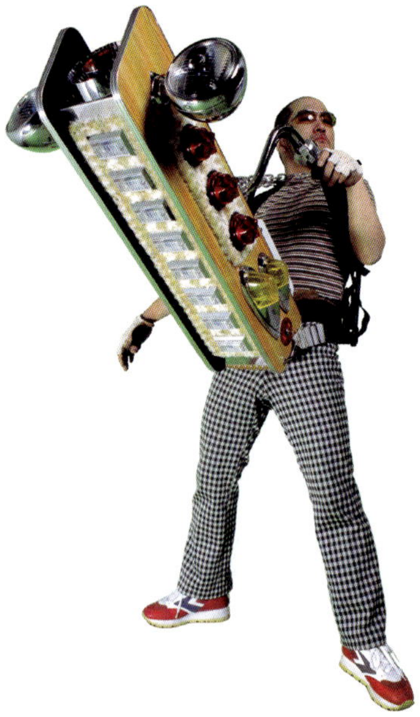

FIG. 10 Muneteru Ujino, *Love Arm 2PRO*, 2001. Mixed media.

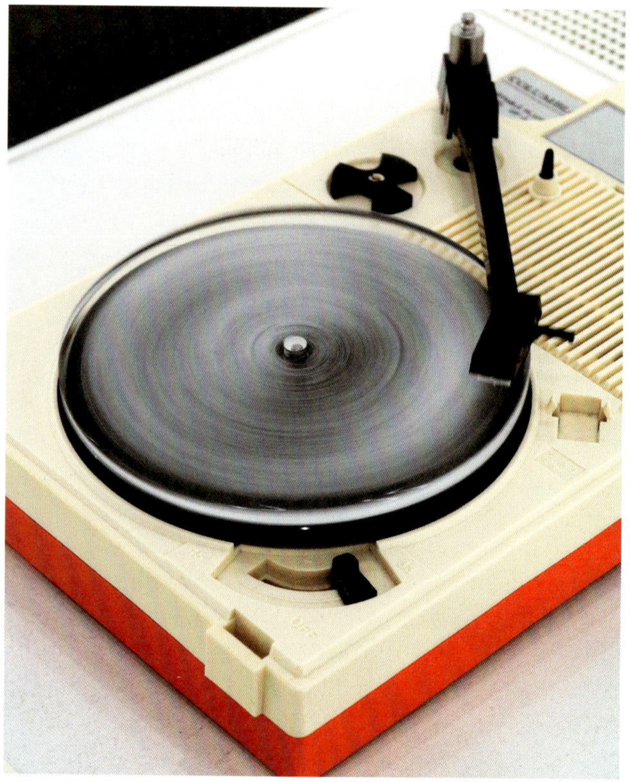

FIG. 11 Lyota Yagi, *Vinyl*, 2005–2008. Ice record (silicon, purified water), record player, freezer.

Media Art

With the development and increasing availability of computer technology, its use in the arts also became more widespread. Artists within the field of media art creating sound works include Masahiro Miwa, Masaki Fujihata, and Kiyoshi Furukawa.[18] Miwa's work *Le Tombeau de Freddie* with Nobuyasu Sakonda as the Formant Brothers, received Honorary Mention at *Ars Electronica* (2009) and re-synthesizes the voice of Freddie Mercury to sing the *Internationale* in Japanese.(FIG. 12)

Digital technology brought with it possibilities to manipulate sound as data and perform in laptop ensembles such as French expatriate Philippe Chatelain's Tokyo-based *Laptop Orchestra* (2002). Ikue Mori also switched to using a laptop after first using drum machines in improvised music, connecting Tokyo to the downtown New York scene in the 1980s at the Knitting Factory and John Zorn's The Stone. Akitsugu Maebayashi's *Sonic Interface* (1999) proposes a sound walk where participants wear headphones and a laptop in a backpack that increasingly delays incoming sound, creating a disconnect between felt and heard reality. Atau Tanaka's *Global String* (Ars Electronica Honorary Mention, 2002) creates a physical and networked monochord spanning two cities where geographical distance and Internet packet transmission delays create a virtual resonating chamber for the string.(FIG. 13), [19]

The musician and composer Ryuichi Sakamoto, originally from the band Yellow Magic Orchestra, defies genre boundaries. His collaborations, such as with Toshio Iwai in the audiovisual networked *Music Plays Images X Images Play Music* (Ars Electronica Golden Nica, 1997), places Iwai's playful interactive imagery as an extension to Sakamoto's piano. A design aesthetic and crossover appeal takes hold in works by Maywa Denki (Ars Electronica Award of Distinction, 2003), Kouichi Okamoto (Kyouei

16
See Toshiya Tsunoda, "On Portrayals (a reflection)," in *Santa: Sound and word review* 10 (2008), 3–4.

17
See Régine Debatty, "Interview with Ujino Muneteru" (2007), http://we-make-money-not-art.com/archives/2007/10/intro-image-1-h.php.

18
See Minoru Hatanaka and Tomotaro Kaneko, "Oto: Kikukoto, Mirukoto, Chikaku surukoto--On=Gaku=Āto no Genzaikei" [Sound: Listening, Seeing, Perception--Mu=sic=the Current Situation of Art] (2013), http://artscape.jp/study/artwordtalk/10094695_18145.html.

19
Atau Tanaka and Kasper T. Toeplitz, "Prix Ars Electronica 2002, Interactive Art, Honorary Mention: Global String" (Linz: Ars Electronica, 2002), http://90.146.8.18/en/archives/prix_archive/prix_projekt.asp?iProjectID=11733.

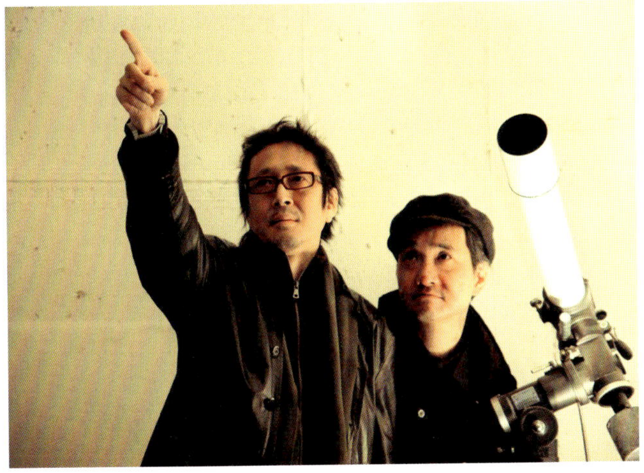

FIG. 12 Masahiro Miwa, Nobuyasu Sakonda (Formant Brothers), *Le Tombeau de Freddie*, 2009.

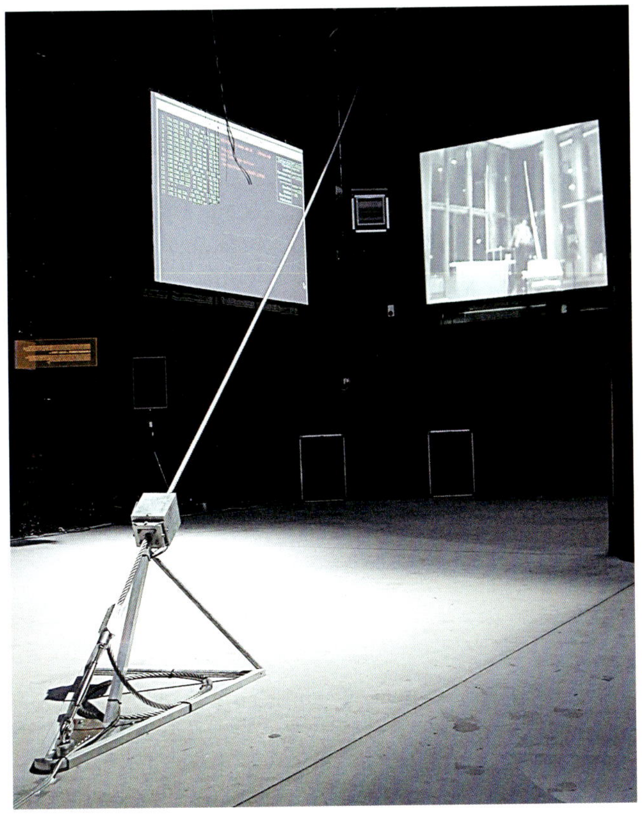

FIG. 13 Atau Tanaka, *Global String*, 1998–present. Multi-site network music installation, network-connected string, vibration sensors. Installation view, Dutch Electronic Art Festival, Rotterdam, 2000.

Design), Daito Manabe (Rhizomatiks), Ken Furudate, Kazuhiro Jo, Daisuke Ishida, and Mizuki Noguchi (SINE WAVE ORCHESTRA), Keiichiro Shibuya and evala. Yuri Suzuki has performed with Maywa Denki, and has gone on to produce model trains on vinyl record tracks at the Sonar Festival in Barcelona (2009), and networks of acoustic tubes at the Biennale Internationale Design St. Etienne (2015).

Audiovisual

Currently in Japan, sound art generally implies works incorporating audio displayed in gallery settings. Ryoji Ikeda worked in the 1990s with the theater collective Dumb Type and developed minimal and sparse aesthetics, incorporating extremes of register and frequency – piercing treble and rumbling bass – and timbre – pure sine tones and harsh noise. His recent *datamatics* series takes data as the theme and material for audiovisuals, objects, and new media pieces. Ikeda was the artist-in-residence at the CERN particle accelerator in Switzerland in 2014. Other artists exploring the combination of audio and visual media in a related manner include Ryoichi Kurokawa, Takashi Makino, and Ryo Ikeshiro.

Noise

Noise music in the 1980s was the antithesis of the electro-pop music of the time. Originally a reaction to all that was formal in music and all that was technical, "Japanoise" had its roots in a misuse of technology for its sonic ends. Masami Akita of Merzbow, for example, created networks of feedback through interconnected guitar effects, while Tetsuo Furudate performed renditions of Shakespeare's plays, smashing his sampler keyboard. Guitarist K. K. Null removed the guitar from his setup to create "nullsonics," while Pain Jerk used electronics in an anti-technological stance.

Social rebellion may be the most obvious motivation for this genre, but this stereotype is a far-too-reductive catchall to interpret the complexity of this movement. The Incapacitants are a duo of real-life salarymen who arrive at gigs in suits and transform into noise warriors onstage. Noise music goes beyond pure physical rebellion to take on a social and environmental dimension – a social noise. The very fact that noise can be considered as sonic material recalls the Japanese relationship to silence and nature. While the West tends to appropriate elements of nature (citing birdsong as an inspiration to create melodies unrelated to birds), the Japanese instinct observes nature *in situ* (the sound of insects defines the sense of environment of each season). Traditional Japanese music, *gagaku*, and music of *Noh* theater create a sense of space across a span of time.

Noise music is no exception. Masami Akita notes how the physical experience of noise is primordial.[20]

Onkyō

Onkyō, meaning sound or resonance, typically in the context of acoustics, describes a style of improvisation practiced at the monthly Meeting at Off Site (2000–2005) concerts. Run by the artists Atsuhiro and Yukari Ito, Off Site in the Yoyogi district of Tokyo epitomized the convergence of art and music through its dual function as a gallery and a performance venue. Although onkyō belongs to the broader context of improvised music, its demand for concentrated listening of sparse textures, often at almost imperceptible volumes, shares affinities with sound art. This is in some ways the byproduct of pragmatic issues of modernization and the densely populated urban landscape of Tokyo – the quiet playing volumes were due to the venue's thin walls and consideration for its neighbors rather than being a direct reaction against the excesses of noise music. The site-specific limits imposed often resulted in the intermingling of noises from the environment, while the musicians consciously embraced the concept of *ma* through the use of long pauses and silence in contrast to more frenetic Western free improvisation.[21] This scene shaped the development of Toshimaru Nakamura's "no-input mixing board" feedback and Sachiko M's "empty" sampler playing sine-tones. Other notable participants included Tetuzi Akiyama, Taku Sugimoto (both of whom also ran the series), Yoshihide Otomo, and Taku Unami.

In Closing/ Post-Fukushima

What we have sought to describe is one perspective on the plurality of practices dealing with sound in Japan. Cultural exchange with the West continues to be bi-directional, and a range of approaches occurs across various genre boundaries. Themes include Zen, nature, silence/ noise, and technology, which may be attributed to factors that have also shaped the country in general: Buddhist traditions, World War II, and subsequent modernization.

These themes continue to resonate today, as evident in arguably the most significant event in Japan's recent history, the Great Tōhoku Earthquake and Tsunami on March 11, 2011, and the ensuing nuclear disaster and public outcry. The catastrophe, its death toll and massive devastation, the subsequent nuclear meltdown in Fukushima necessitating the evacuation of hundreds of thousands of residents, and the slow response of government and the utility company, has triggered large-scale public demonstrations. This in many ways recalls the 1960 *Anpō* protests following the signature of the Treaty of Mutual Cooperation and Security between the United States and Japan that was the backdrop to the Gutai movement. The recent jolt to the national psyche has similarly been seized upon by the artistic community as a call to action, including direct involvement in the organization of protests such as Amateur Revolt. The most notable artistic project in response to, and in support of, survivors of the disaster has been Project Fukushima!, a series of free music festivals co-organized by journeyman guitarist and turntablist Yoshihide Otomo, who himself had grown up in Fukushima. These acts continue to raise awareness of the ongoing crisis and signal a return to political engagement among artists, sound and other, in Japan.

20 See Masami Akita, *Noizu Wō: Noizu Myūjikku to sono Tenkai* [Noise War: Noise Music and its Development] (Tokyo: Seikyusha, 1992).

21 See Clive Bell, "Off Site" (2003), under www.clivebell.co.uk/offsite.

Acknowledgments:
The authors gratefully acknowledge the assistance of Minoru Hatanaka, Christophe Charles, Kazuhiro Jo, Carl Stone, Masayoshi Kosugi, Michiko Oki, and Inigo Wilkins.

Selected Exhibitions/Events

1949–1963
Yomiuri Independent (annual exhibition), Tokyo Metropolitan Art Museum

1962–1970
Gutai Pinacotheca (museum), Osaka

1962
John Cage and David Tudor's visit to Japan including concert at Sōgetsu Arton (October 24)

1969
Intermedia Arts Festival, Nikkei Hall, Tokyo (January 18–20)

Cross Talk/Intermedia, 2nd Yoyogi National Gymnasium, Tokyo (February 2–5)

1970
Expo '70 (*Banpaku*), Osaka (March 14–September 13)

1978
MA: Espace-Temps du Japon, Festival d'Automne à Paris, Musée du Louvre, Paris (October 11–December 11)

1984
R. Murray Schafer's visit to Japan

1987–1994
Sound Garden series of six exhibitions, Striped House Art Museum (now Striped House Gallery), Tokyo

1989
Moments Sonores (Oto no aru Bijutsu) [Fine Art with Sound], Tochigi Prefectural Museum of Fine Arts

1994–1995
Japanese Art After 1945: *Scream Against the Sky* (traveling exhibition),
Yokohama Museum of Art (February 5–March 30, 1994);
Guggenheim Museum SoHo, New York (September 14, 1994–January 8, 1995);
San Francisco Museum of Modern Art in association with the Center for the Arts at Yerba Buena Gardens, San Francisco (May 31–August 27, 1995)

1997–2006
Audio Picnic at the Museum (annual exhibition), Ontani Memorial Art Museum, Nishinomiya

2000–2005
Meeting at Off Site (concert series), Tokyo

2000
Sound Art – Sound as Media, NTT InterCommunication Center, Tokyo (January 28–March 12)

2001
JU-JIKAN, Phyllis Wattis Theater, San Francisco Museum of Modern Art (September 21–29)

2011
Jikken Kōbō, Bétonsalon – Centre d'art et de recherche, Paris (September 8–October 29)

Simple Interactions: Sound Art from Japan, Museum of Contemporary Art, Roskilde, Denmark (September 24–December 18)

2013
Tokyo 1955–1970: A New Avant-Garde, Museum of Modern Art, New York (November 18, 2012–February 25, 2013)

Gutai: Splendid Playground, Guggenheim Museum, New York (February 16–May 8)

Caleb Kelly

A Brief Introduction to the Sound Arts in Australia

As I write this, the Museum of Contemporary Art in Sydney is preparing for a major retrospective of Australian media artists David Haines and Joyce Hinterding.[1] The pair work both as individuals and as a collaborative team, producing media art that engages the senses – especially the visual, aural, and olfactory – within speculative and often otherworldly contexts. Together they have imagined a house emptied of its contents by a freak weather event (*The Blinds and the Shutters*, 2001), produced a computer game set in a dream space without strategic targets or weapons (*The Outlands*, 2011), placed an anechoic chamber within a huge resonant space (*Telepathy*, 2008),(FIGS. 1a, b) and made audible the analog TV signal at the point of its discontinuation (*Purple Rain*, 2004). These installations represent a body of work that draws on their equally diverse individual practices.

Solo, Haines has performed electronic music internationally as d.haines, juxtaposed a churning primordial mass with coded messages being sent via the switching on and off of hundreds of house lights (*The Seventeenth Century*, 2002), and imbued lumps of coal with the scent of damp earth (*Slow Fast Mountains (earth aroma)*, 2014).(FIG. 2) Hinterding has wound twenty to thirty kilometers of copper wire around columns to produce an antenna that draws power from the space of its installation (*aeriology*, 1995),(FIG. 3) played recorded and live VLF signals through handmade electrostatic speakers (*Electrical Storms*, 1992), and used graphite to draw antenna shaped like mathematical algorithms transmitting the sound of earth hum (*Aura*, 2009). This deep engagement with the senses runs in tandem with other Australian practices that explore sound, demonstrating how contemporary art itself is already and always was engaged in the exploration of the senses. Theirs is not an art in which sound (or vision, or smell) is imagined decoupled from the other senses, nor from the cultural, social, and imaginative spheres.[2]

Recently, I have been asked a number of times why there is such depth within the sound arts in Australia, and perhaps the reason most often recounted is distance. While we now have technologies that are said to close distances, no Skype session will change the fact that many of us are six hours flight from Indonesia (the nearest Asian country) and some twenty hours flight from Europe.[3] This distance creates both a zone of contemplation and a space for prolonged and focused research and practice. However, despite this isolation, Australia looks out at its neighbors and farther afield to America and Europe, creating an extended and multicultural array of arts practices.[4]

In 1996, Douglas Kahn wrote about some of the factors behind the Australian sound arts scene, conceptualizing a context in which mimicry was central and focusing on the figure of the lyrebird (a bird renowned for its ability to mimic sounds from its environment).[5] I would argue that actually the Australian bowerbird is a more apt example. The bowerbird collects objects to make its nest attractive to a mate. Australian arts practitioners are outwardly facing, aware of international practices, and able to collect concepts and most importantly processes for the creation of work. Hence, artists such as Hinterding and Haines are able to draw from a variety of sources, not just art or music, to produce generative conceptual frameworks from which to produce installations and media art.

Australian artists have a longstanding tradition of engaging with sound in a variety of

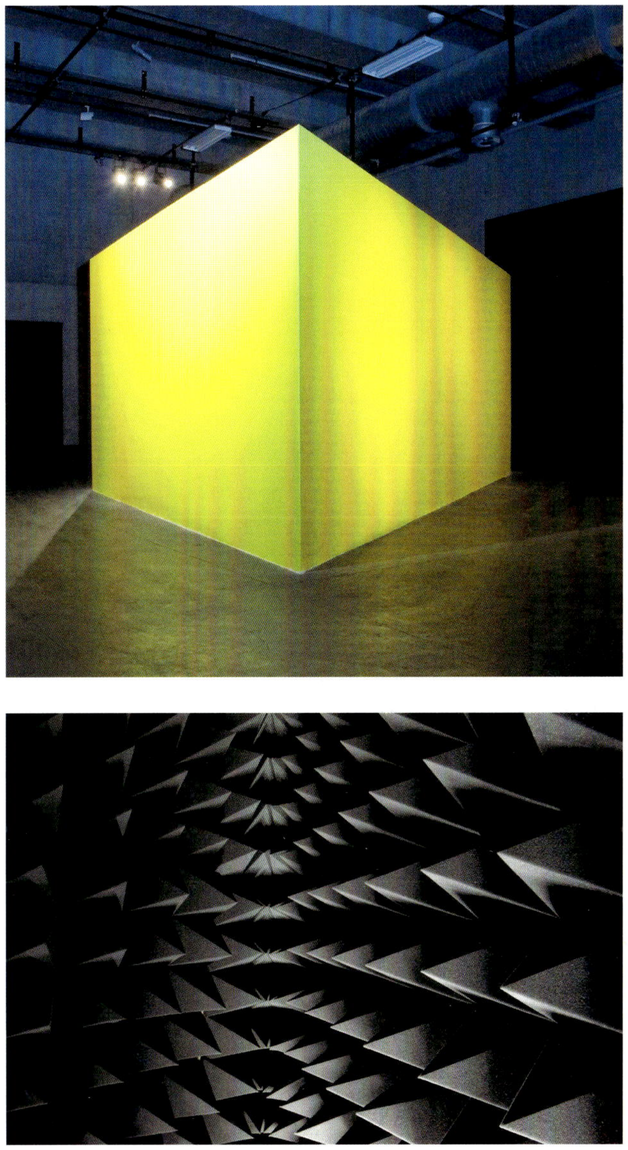

FIGS. 1a, b David Haines and Joyce Hinterding, *Telepathy*, 2008. Mixed media installation, timber, custom anechoic acoustic tiles, acoustic blanket, plasterboard, 5 × 5 × 7 × 3.6 m. Installation views of exterior and interior, Carriageworks Bay 19, Sydney.

1 *Energies: Haines & Hinterding*, Museum of Contemporary Art, Sydney, 2015.

2 For details on David Haines and Joyce Hinterding's work, see www.haineshinterding.net.

3 New Zealand is closer to Australia but it shares many of the same cultural and social situations as Australia in terms of distance.

4 There have been a small number of texts written specifically about the sound arts in Australia. Alongside the catalogue essays for exhibitions discussed below, two texts should be singled out: Douglas Kahn, "The Lyre's Island: Some Australian Music, Sound Art and Design," *Leonardo Music Journal* 6 (1996): 89–93, and Gail Priest, "Preface: Starting Somewhere," in *Experimental Music: Audio Explorations in Australia* (Sydney: UNSW Press, 2009), 1–8.

5 See Kahn, "The Lyre's Island."

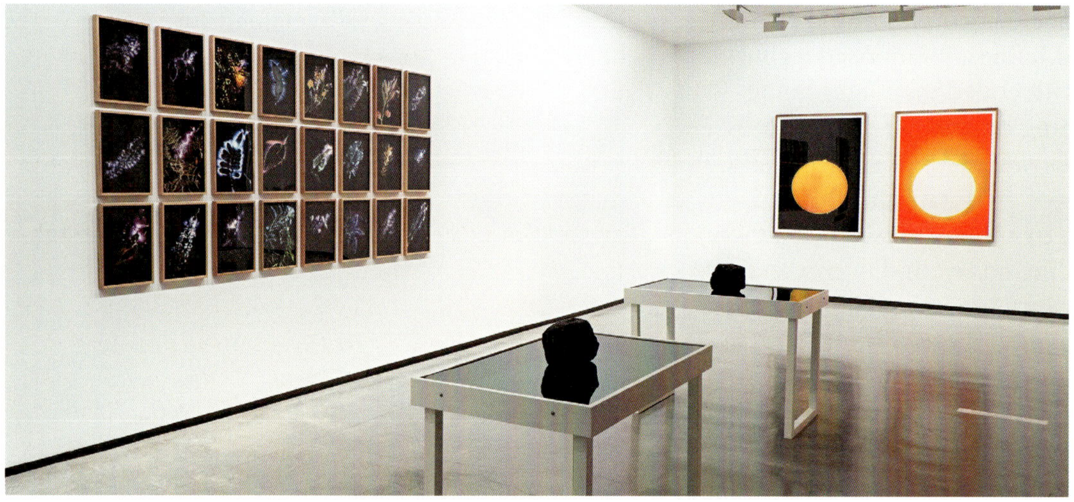

FIG. 2 David Haines, *Slow Fast Mountains (earth aroma)*, 2014. Mixed media installation, timber, perspex, coal samples, aroma chemistry, sound system. Installation view *Energies: Haines & Hinterding*, Museum of Contemporary Art Australia, Sydney, 2015.

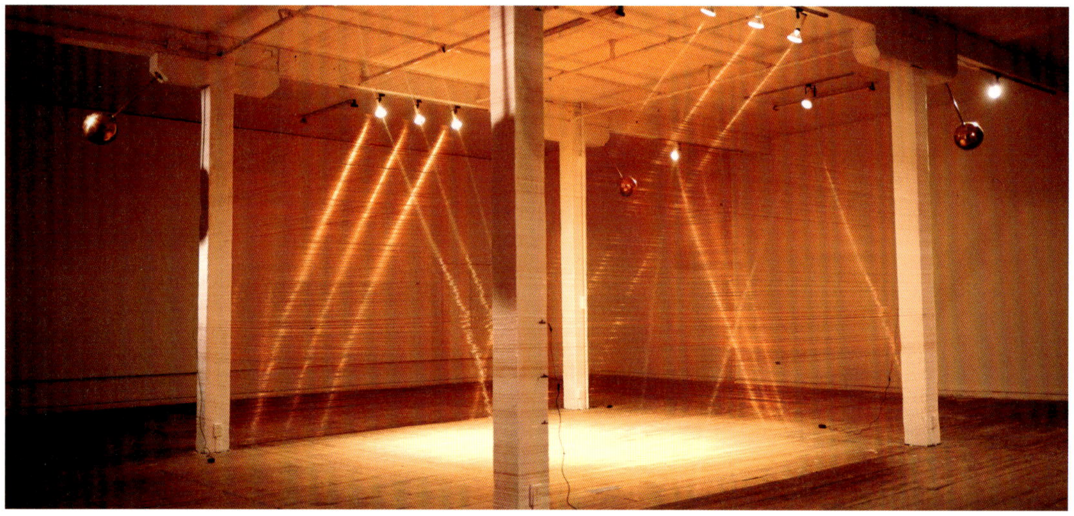

FIG. 3 Joyce Hinterding, *aeriology*, 1995. Large-scale detuned custom built antenna, 18 kilometers copper magnet wire, speakers, spun copper balls, 4.6 × 4.6 × 3.3 m. Installation view Artspace Auckland New Zealand, 1995.

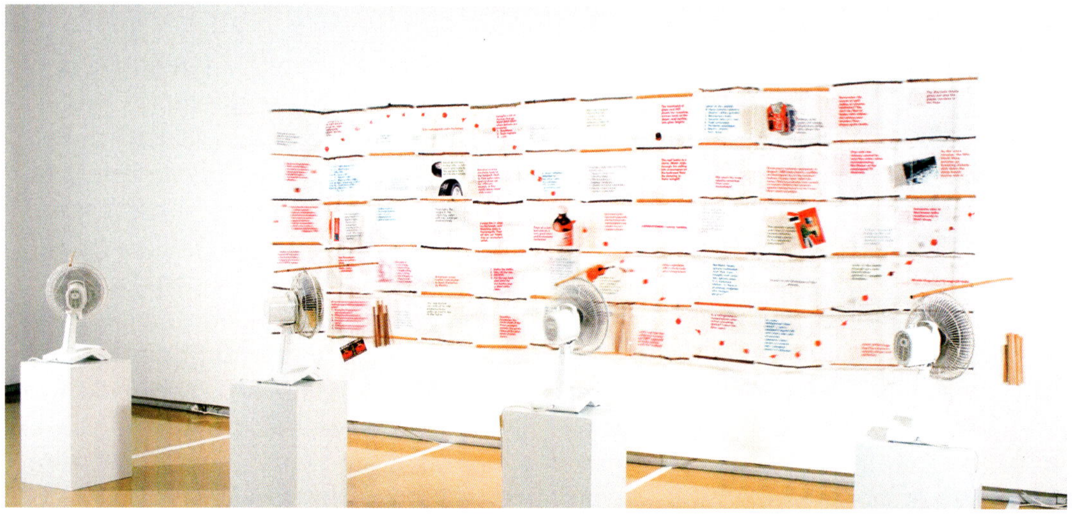

FIG. 4 Ernie Althoff, *See Breezes*, 1995. Installation view *Sound in Space: Adventures in Australian Sound Art*, Museum of Contemporary Art Australia, Sydney, 1995.

arts practices. I use the term *sound arts* in this chapter as a way to broaden the discussion from simply sound art, allowing the field to open up beyond a practice and discourse often limited to a predominant focus on sound itself. Sound has always been a part of the arts and will continue to be so for as long as we have them. John Cage in the early 1950s argued explicitly for the impossibility of silence, creating a piece of very noisy silence in *4'33''* (1952) in the process. If silence is impossible then it stands to reason that art is not, and is never, silent. The argument put forward here is that if we can come to grips with the impossibility of silence and understand that all of the arts come with sound, then we can comprehend the arts and their engagement with our senses in a broader, and I would say, more human manner.

As attested to by the many recent sound-focused publications – including this one – there has been a shift in attention to sound. It is also the case that this "new" interest in sound is anything but new. Australia's history confronts the recent glut of ever-repeating "firsts," such as *Soundings* (2013) at MoMA (a non-first first if ever there was one). In Australia, we can begin by looking to the 1980s, when the proliferation of publications, festivals, and exhibitions were difficult to ignore. A raised awareness of the sound arts in the period was presented in publications such as *New Music Articles* (1982–1992) and a dedicated issue of *Art & Text* (1989); festivals such as the similarly titled *Soundworks: A Festival of Experimental Sound* in Perth (1985) and *Soundworks Festival* (part of the Biennale of Sydney in 1986); and the long-running ABC National radio program *The Listening Room* (1988–2003).[6] Subsequently, interest in sound grew rapidly in the 1990s, in part due to a very committed and productive new media art scene in the major Australian cities.[7]

Crucial to this development was the support of the Australia Council for the Arts, the major government body responsible for funding arts nationally. The Australia Council had established a Hybrid Arts Committee in 1994, and in 1996 established a New Media Arts Board charged with supporting hybrid and interdisciplinary practices with a deep commitment to fund artists engaging new technologies – at the time, this meant predominantly emerging digital practices and their use within art, music, and performance.

Alongside new media art's development in the 1990s was further commitment in the form of contemporary art and experimental music, as well as swiftly developing academic and critical engagement with the field. The seminal sound-focused exhibition *Sound in Space: Adventures in Australian Sound Art* (1995), at the Museum of Contemporary Art in Sydney, made clear that sound was a common and accepted medium and an area of conceptual focus within contemporary art.[8] *Sound in Space* included a set of artists working with sound as a phenomenon, a carrier of information, a material, and a conceptual tool. The artists included were almost evenly distributed between male and female practitioners, a continuing and extremely visible element of sound practice in Australia not always reflected in sound scenes in the Northern Hemisphere.

While *Sound in Space* was a stand-out exhibition for an emphasis on sound's role within contemporary art in the 1990s, it was not alone. Festivals are a mainstay of sound practices, and in Australia this is no different. SoundCulture, for example, began in Sydney in 1991, providing a platform for sound investigations across a number of countries in the Pacific Rim. The Melbourne-based festival experimenta '94 included the event *Earwitness: Excursions in Sound* that sought to "re-examine the role of the aural in the production and reception of ideas."[9]

Australian sound arts practice has a long-term focus on materiality, with much of the sound arts situated within discourses embedded in the material. *Sound in Space* and *Earwitness* exhibited works on the cusp of a major shift to digital production and those artists' works were resolutely materials-based. Many of them were handmade,

6
This chapter is not an attempt to survey the field; all examples should be understood to be exemplars and a sampling of the many things happening during the period.

7
Media art is an art that is entangled with audio. Australian artists, such as Stelarc, produced large-scale performances in which sound was a dominant element in the spectacle, and many others working within a digital studio produced rich audio soundtracks for new media installations.

8
Curator Rebecca Coyle states in her catalogue essay that, "in Australia, sound art practice reaches into fields of multi-media [...], sculpture, experimental and new music, performance, sound poetry, radiophony, and sound design." Throughout the essay she makes clear that sound as a medium has a history and place within the contemporary arts. Rebecca Coyle, "Sound in Space: A Curatorial Perspective," in *Sound in Space: Adventures in Sound*, exhib. cat. Museum of Contemporary Art, Sydney (1995), www.sounddesign.unimelb.edu.au/web/biogs/P000451b.htm.

9
As quoted from the passage on *Earwitness* in the event brochure of experimenta '94 available online at www.accaonline.org.au/sites/default/files/1994_Earwitness%20Excursions%20in%20sound%20associated%20Experimenta%20events%20brochure.pdf. *Earwitness* included many of the artists shown in *Sound in Space*. Venues were also at the center of the sound arts. In Sydney, Performance Space remains a major supporter of experimental performance and music practices dating back to the 1980s, alongside Artspace. In Melbourne and Brisbane, similar commitments to the sound arts and media art occurred.

including those by Ernie Althoff, who assembled various mismatched sounding objects that mechanically interacted;(FIG. 4) Sherre DeLys and Joan Grounds installed women's shoes from the 1920s into the botanical gardens, placing inside them small speakers playing the sounds of people mimicking birdcalls, among other things; Nigel Helyer produced a large wooden vertical horn speaker; Hinterding was in both exhibitions, exhibiting a hand-drawn circuit in the former and electrostatic speakers playing VLF radio audio in the latter.

Experimental music festivals and events staged in institutional and underground venues became emblematic of the scene in the 1990s. The most influential of these was What is Music? (1994–2005).[10] The festival was conceived and directed by the musicians Oren Ambarchi and Robbie Avenaim (both from Sydney), who brought together an eclectic ensemble of musicians and performers, merging experimental and outsider music scenes. A key objective of the festival was to bring highly regarded international musicians to the country and join them in collaborations and on the bills with local and often young or unknown musicians. This produced a number of very strong international networks and numerous musicians have continued to play a role within the experimental music scenes internationally and in Australia (not least Oren Ambarchi himself).[11]

Ambarchi and Avenaim both take an eclectic and outward facing approach to their improvisations. Ambarchi is known for his base heavy explorations, while Avenaim in recent years has developed mechanized drumming tools

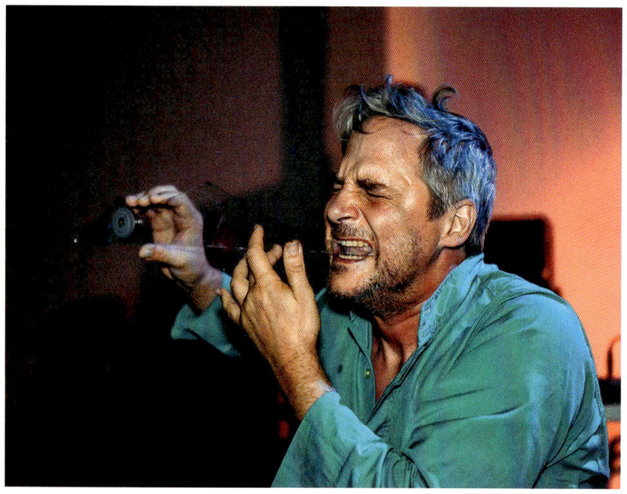

FIG. 5 Lucas Abela, performance at VIVO Media Arts Centre, Vancouver, 2015.

that augment his own physical approach. During this period, noise musician Lucas Abela developed a series of noise instruments, many of which had very physical consequences. In performance, Abela in recent years has focused on an instrument created from a sheet of glass.(FIG. 5) This glass has a contact microphone attached, and he blows, rubs, scrapes, scratches, cuts, and slashes his mouth with the glass during performance.

Noise is also a central theme in contemporary art. Marco Fusinato has engaged in a politically oriented noise practice over the last twenty years – his artwork is an example of the crossover between music and art. Many of these works intervene in the expected nature of sound within an art gallery context. *Spectral Arrows*, for example, is a performance work in which Fusinato performs his brand of heavily amplified guitar noise within an art gallery for the duration of its opening hours. *Aetheric Plexus (Broken X)* (2013) is a sensorially intense installation, for which 36,000 watts of light and 105 decibels of white noise were inflicted on his unsuspecting audience.(FIG. 6) These works set out to examine the gallery as a venue for entertainment and the desire in the audience to have an art experience.

Sound practices such as Fusinato's that slide between art and music are prevalent in recent process-driven works in Australia. Artists such as Peter Blamey, Vicky Browne, Pia van Gelder, Ross Manning, and Kusum Normoyle produce art and live performances via a heightened awareness of materials.(FIGS. 7–11) This includes recycling, scavenging, feedback, re-use of older technologies, everyday playback technology, and media ecology. These emerging approaches to media have a strong sound component that is bound to the media itself.

10
While What is Music? continues as a banner used by Robbie Avenaim, 2005 was the last year Oren Ambarchi co-directed the festival and at this time the festival changed its format.

11
The success of What is Music? in turn prompted other producers to start their own festivals and events, including Liquid Architecture (Melbourne, 1999–ongoing) and the NOW Now festival (Sydney, 2001–ongoing). There have also been numerous regular events that started in the late 1990s and many that began in the early years of the 2000s, including the Make It Up Club, Fabrique, Audiopollen, If You Like Improvised Music We Like You, and impermanent.audio among others.

12
Douglas Kahn has returned to Sydney, where he continues to academically lead the discussion on the history and theory of sound in the arts.

13
Douglas Kahn, *Noise, Water, Meat: A History of Sound in the Arts* (Cambridge, MA: The MIT Press, 1999).

14
See the Sound and Materials Research Group website, under www.soundandmaterials.com.

15
Transmission, curated by Carrie Miller and Matthew Hindson, Campbelltown Art Centre, Sydney, 2012; *Sonic Spheres*, curated by Victoria Lynn, TarraWarra Museum of Art, 2012; *Sound Full: Sound in Contemporary Australian and New Zealand Art*, curated by Caleb Kelly and Aaron Kreisler, Dunedin Public Art Gallery, New Zealand, 2012.

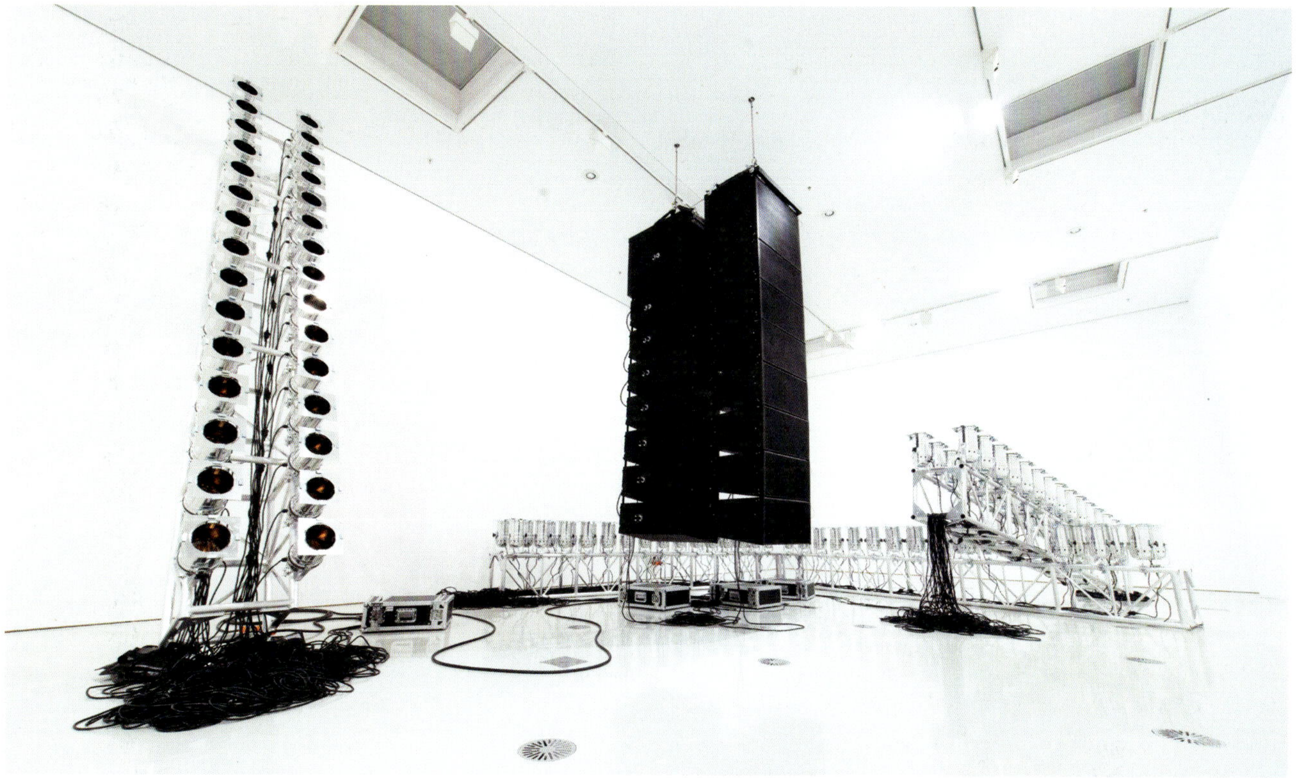

FIG. 6 Marco Fusinato, *Aetheric Plexus (Broken X)*, 2013. Installation, 36,000 watts white light, 105 db white noise, 40 × 500 cm box truss, Par can 56 lights, couplers, Lanbox LCM, DMX controller, dimmer racks, DMX mp3 player, line array PA, sensor, extension leads, shot bags, dimensions variable. Installation view *Melbourne Now*, National Gallery of Victoria, Melbourne, 2013/2014. Collection of the National Gallery of Victoria, Melbourne.

Finally, Australia was (and is) host to an academically rigorous discussion of sound within the arts. Douglas Kahn was living in Australia in the 1990s before leaving for the United States in 2001.[12] Kahn wrote his influential *Noise, Water, Meat* while in Australia – a book that in many ways initiated the current academic interest in the sound arts.[13] In addition to Kahn, Frances Dyson, Martin Harrison, Norie Neumark, and John Potts were also writing about the sound arts. This academic influence has marked the practices of sound in Australia and continues to influence younger generations. Kahn supervised my own PhD and we are now both supervising a swiftly growing group of students in practice-based PhDs in the sound arts. These artists approach their practices through materials, ecologies, and cultural/social contexts: in doing so, they use technologies and media in an eclectic manner, employing home-made electronics, energies, and a vast array of sound-filled processes.[14]

In Australia the emergence of sound within the arts gained prominence during a period of ongoing discussion as to what it meant to be Australian and what Australian culture was (if we even could be said to have one). Despite this debate, which was often centered on British invasion, colonization, and immigration, few words even to this day have been given to a meaningful discussion of Indigenous Aboriginal culture and its engagement with sound and the sound arts. This is an ongoing problem in Australia and one still in need of attention.

Perhaps due to the long-term commitment to the sound arts, Australia has bypassed the trend of "sound art" and jumped straight into a much more fertile territory where sound is examined within the purview of the so-called visual arts and experimental music. Three exhibitions mounted in 2012 profiled sound in Australian contemporary art: *Sonic Spheres*, *Transmission*, and *Sound Full*.[15] The artists and musicians represented in these exhibitions would not be described in Australia as "sound artists," as we are comfortable in describing them simply as artists or musicians. Sound has certainly become a very present area of practice in recent years, yet it remains deeply embedded in wider discourses within the arts and maintains a critical edge within emerging contemporary practices.

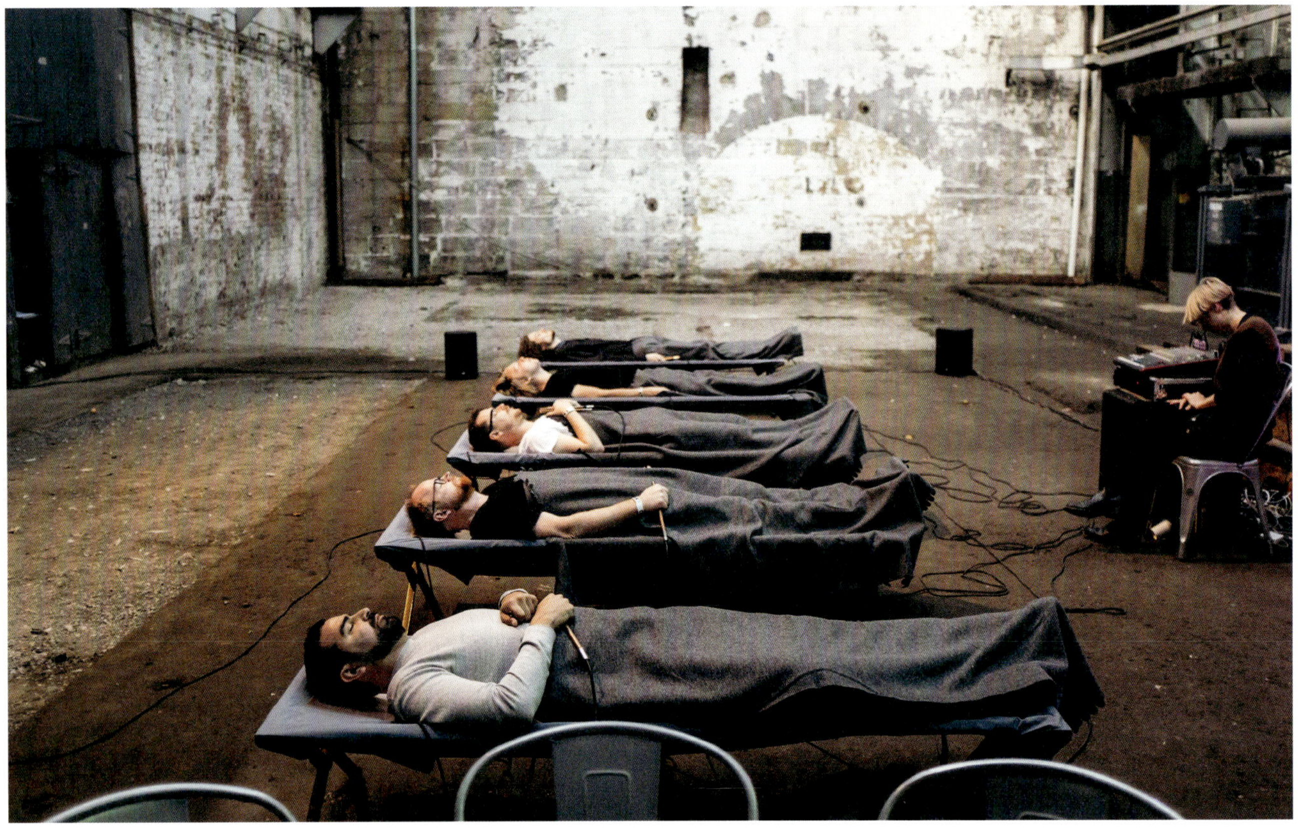

FIG. 7 Pia van Gelder, *Relaxation Circuit*, 2015. Interactive participatory installation, electronics, copper, textiles, video. Installation view Underbelly Arts Festival, Cockatoo Island, Sydney, 2015.

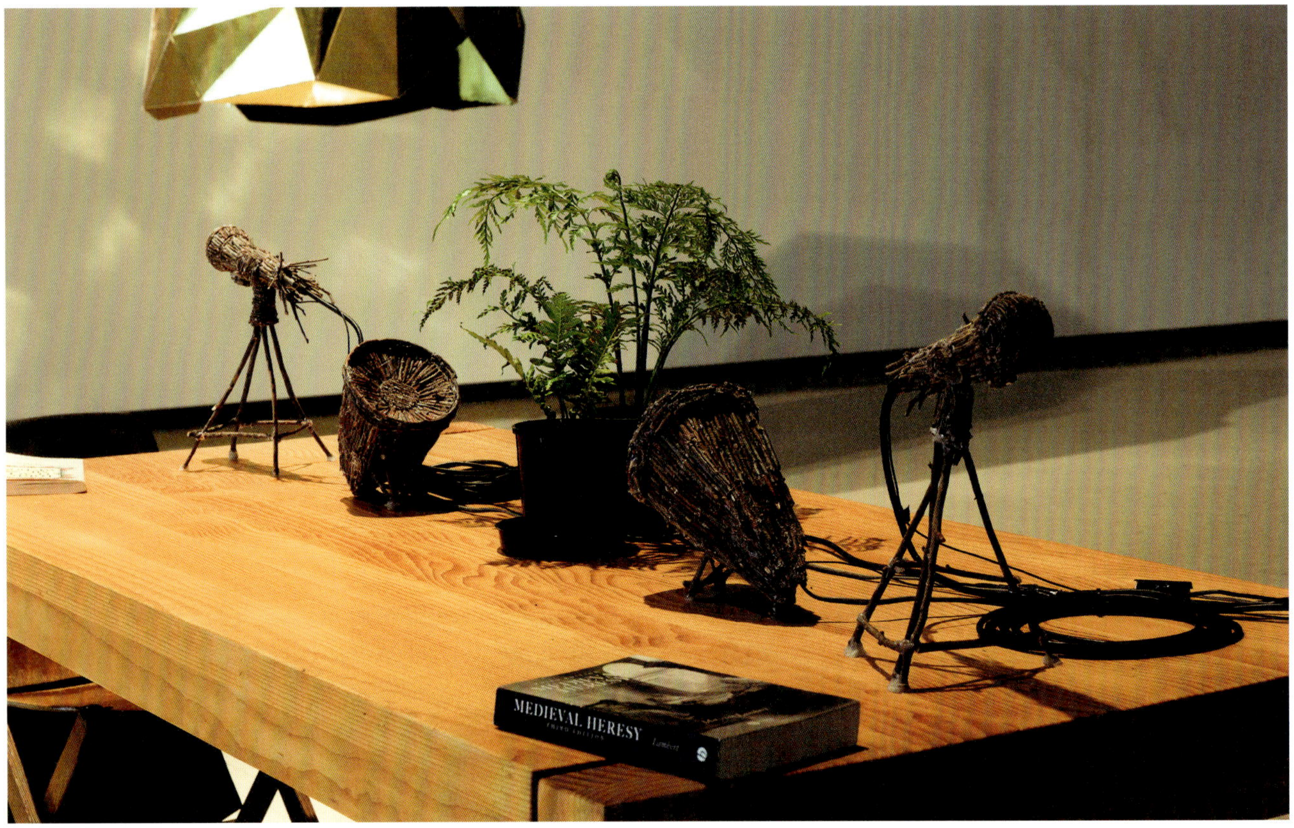

FIG. 8 Vicky Browne, *The Sound of Plants and Music*, 2012, mixed-media installation. Installation view *Sound Full: Sound in Contemporary Australian and New Zealand Art*, City Gallery Wellington, 2013.

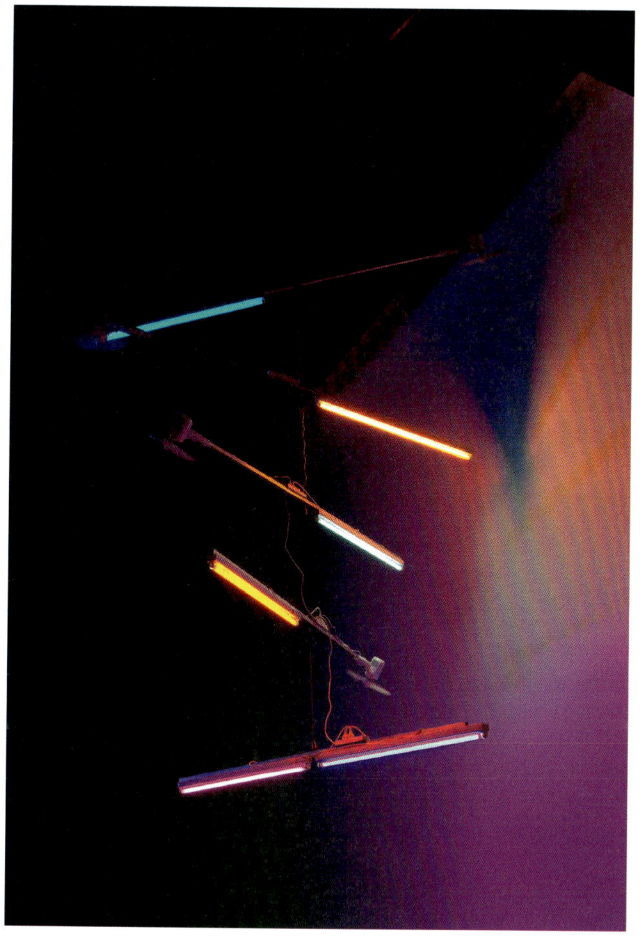

FIG. 9 Ross Manning, *Spectra II*, 2012. Oscillating fans, fluorescent lights, theater gels, rope, aluminum, cables. Originally commissioned by the Australian Center for Contemporary Art (ACCA) for NEW12.

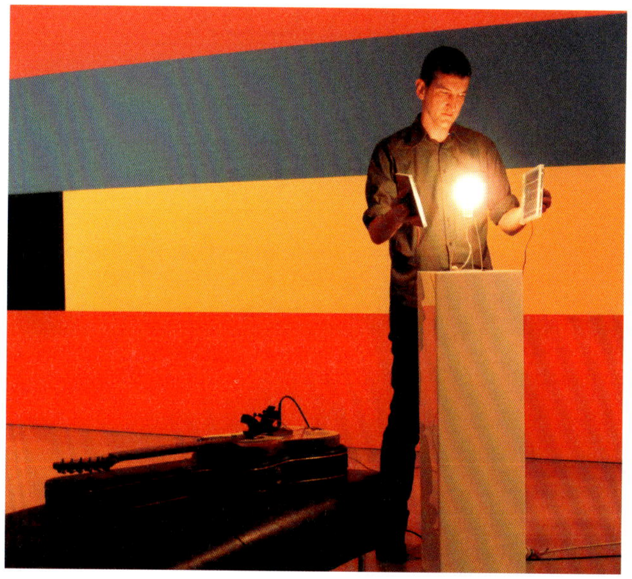

FIG. 10 Peter Blamey, *Double Partial Eclipse*, live performance project. Performance in the course of *Rapture / Rupture*, a part of the Sonic Social project, ArtBar, Museum of Contemporary Art, Sydney, 2014.

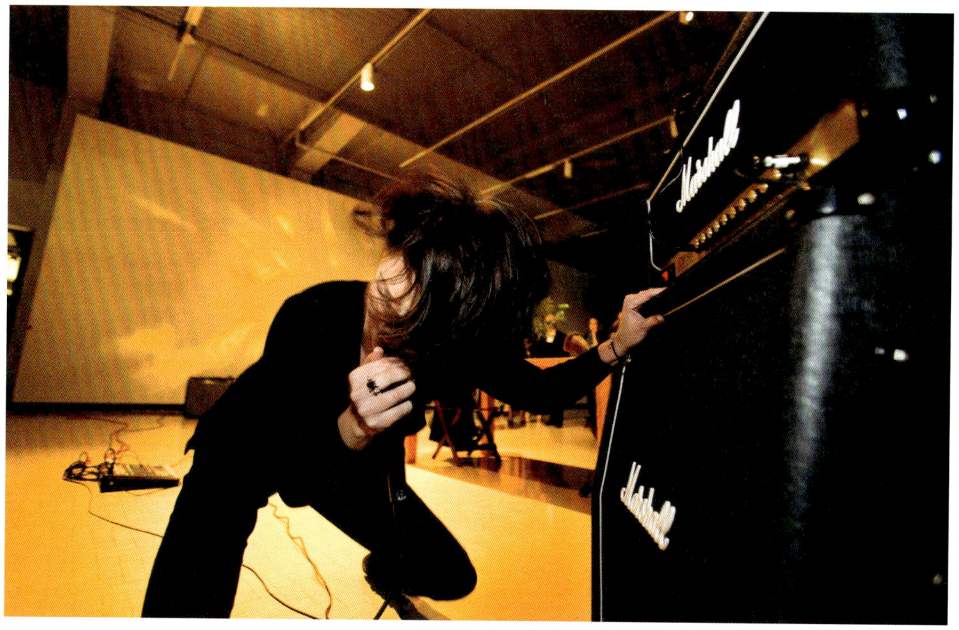

FIG. 11 Kusum Normoyle, performance at *Sound Full: Sound in Contemporary Australian and New Zealand Art*, City Gallery Wellington, 2013.

Christoph Cox

Sound Art in America: Cage and Beyond

"Sound Art"?

In American art discourse, "sound art" is a thorny label. Practitioners shy away from it, preferring to call themselves simply "artists" or "composers." Critics often reject the phrase as well in favor of more generic descriptions such as "sound in the arts" or "the sonic arts." Curators, too, are reluctant to put those two words side by side in exhibition titles.[1] The reasons for this wariness seem to be twofold: (1) that "sound art" might just be a passing fad with which soon no one will want to have been associated, and (2) that the description is too narrow to capture work that invariably involves more than sound. In a statement written as wall text for the 2000 exhibition *Volume: Bed of Sound* at MoMA PS1 in New York City, the American pioneer of sound installation Max Neuhaus leveled both these arguments against the label. After declaring the flurry of sound-themed exhibitions to be "an art fad," he dismissed the phrase "sound art" as a category mistake, the equivalent of grouping everything from steel sculpture to steel guitar music under the title "steel art."[2]

Yet, for all that, "sound art" keeps sneaking back into the discourse - and for good reason. While no more adequate to its content than the terms "video art" or "performance" are to describe the wild variety of work that falls under those labels, "sound art" helpfully marks the fact that, in the past two decades or so, sound has indeed become more prominent in venues of contemporary art in the US and around the world, and that this sonic artwork tends to be markedly different from musical performance and from other art forms (video and film, for example) in which sound most often plays a merely supportive role. "Sound art" is as good a term as any to describe works in any artistic medium or modality (installation, sculpture, drawing, film, video, recorded sound, etc.) that draw particular attention to the sonic and consider it aesthetically. Surely this category overlaps with "music," and no firm division need be made between them. But, as I have argued elsewhere, "sound art" often marks an ontological distinction between work that *employs* the sonic and work that *examines its conditions of possibility*.[3]

Several Origins

On this (and just about any other) definition, sound art has multiple origins, and Americans figure prominently in its history. Prior to the contributions of composers and artists, the deaf polymath Thomas Edison laid the groundwork for sound art with his invention of the phonograph, which severed sound from its present performance and allowed it to be *installed*, played back independently of the live event, repeated in the absence of the performer and even the listener.(FIG. 1) Moreover, as Friedrich Kittler later observed, the phonograph expanded the aesthetic appreciation of sound beyond music and speech, for it registered audible vibrations indiscriminately, heedless of whether they were emitted by a musical instrument, a human voice, wind through the trees, or a passing locomotive.[4] Music was thus subsumed within the broader field of sound or noise and was no longer the only sonic art.

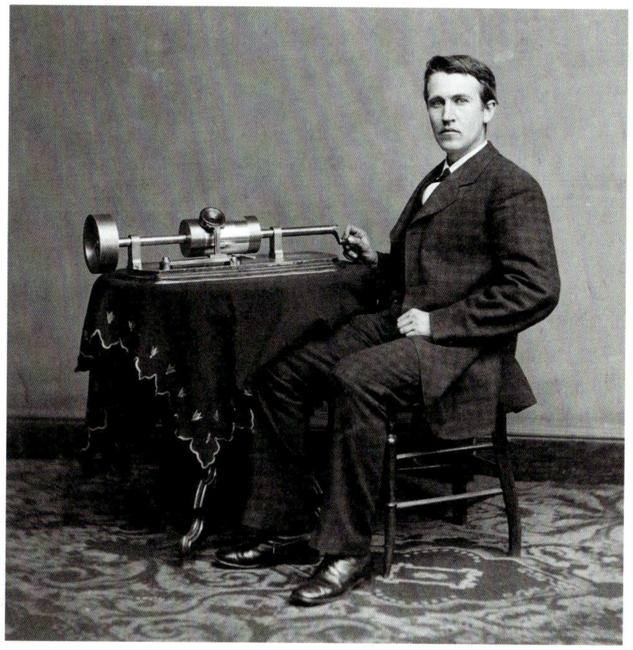

FIG. 1 Thomas Alva Edison with his second phonograph, photographed by Mathew Brady in Washington, April 1878. Part of the Brady-Handy Photograph Collection (Library of Congress).

Luigi Russolo, Edgard Varèse, and Pierre Schaeffer explored this newly discovered domain of noise. But all three were content to *make music* with noise, to capture the sounds of the world and use them to musical effect. A more profound contribution was made by John Cage. It is customary to think of Cage as the composer of silence. This is true, of course, but misleading. Instead of exploring musical silence or making silence musical, Cage's most famous piece, the so-called "silent piece" *4'33"* (1952), serves as a window or door through which music opens out to what Cage called "the entire field of sound."[5] Indeed Cage is not so much the thinker and composer of *silence* as the thinker and composer of *noise*, considered as the entire sonic field of which music is but a tiny part. Cage repeatedly reminded us that "there's no such thing as silence"[6] and that "wherever we are, what we hear is mostly noise."[7] In short, for Cage, *silence* = *noise*, and not just any noise, but the hubbub of the whole audible world, the impersonal and anonymous sonic flux that precedes and exceeds us: "Until I die there will be sounds," he remarked. "And they will continue following my death."[8]

1 See, for example, the conversation between Stephen Vitiello and Marina Rosenfeld in *NewMusicBox* (March 1, 2004), www.newmusicbox.org/assets/59/interview_vitiello.pdf, and Douglas Kahn, "The Arts of Sound Art and Music," www.douglaskahn.com/writings/douglas_kahn-sound_art.pdf.

2 See Max Neuhaus, "Sound Art?," in *Volume: Bed of Sound* (New York: P.S. 1 Contemporary Art Center, 2000), www.max-neuhaus.info/soundworks/soundart/SoundArt.htm.

3 See Christoph Cox, "Sound Art and the Sonic Unconscious," *Organised Sound* 14, no. 1 (2009): 19–26. I develop this idea (and many of the ideas in this essay) in my forthcoming book *Sonic Flux: Sound, Art, and Metaphysics*.

4 See Friedrich A. Kittler, *Gramophone, Film, Typewriter*, trans. Geoffrey Winthrop-Young and Michael Wutz (Stanford: Stanford University Press, 1999), 23.

5 John Cage, "The Future of Music: Credo," in *Silence: Lectures and Writings by John Cage* (Middletown, CT: Wesleyan University Press, 1986), 3–6, here 4.

6 Richard Kostelanetz, *Conversing with Cage*, 2nd ed. (New York: Routledge, 2003), 70.

7 Cage, "The Future of Music: Credo," 3.

Reflecting on *4'33"* in 1974, Cage told an interviewer: "I have felt and hoped to have led other people to feel that the sounds of their environment constitute a music which is more interesting than the music which they would hear if they went into a concert hall."[9] Cage invited us to leave the concert hall and attend to the sounds of the environment. Yet he did not relinquish "music," hoping that others would accept his expansion of the term to encompass everything that can be heard. It was another artist, Max Neuhaus, who took the decisive step outside of music toward what we know as "sound art." A musical prodigy specializing in avant-garde composition for percussion, Neuhaus had performed pieces by Russolo, Varèse, Cage, and other composers eager to incorporate everyday sounds into their work. Yet, by the mid-1960s, Neuhaus began to worry that this strategy was insufficient. "Few [concert goers] were able to carry the experience over into an appreciation of these sounds in their daily lives," he remarked. "I became interested in going a step further. Why limit listening to the concert hall? Instead of bringing these sounds into the hall, why not simply take the *audience* outside – a demonstration in situ?"[10] Neuhaus intended this exit from the concert hall to be quite literal. In 1966, he initiated a project called *LISTEN* in which he would invite audience members to meet at a concert venue, stamp their hands with those six letters, and then silently lead them outside on a walk through power plants, highway underpasses, and city neighborhoods.

Neuhaus' own final exit from the concert hall came two years later. After recording an LP of his percussion repertoire, he left the world of music and performance for good, turning instead to what he called "sound installations," continuous fields of sound – generally complex drones – that shaped and colored their chosen sites. "In music the sound is the work," he noted, while "in what I do the sound is the means of making the work, the means of transforming space into place."[11] This shift of interest from temporally bounded works toward site-specific works that defined a place, he thought, connected his work more fully with sculpture and the visual arts than music. "In terms of classification," he told an interviewer, "I'd move the installations into the purview of the visual arts even though they have no visual component, because the visual arts, in the plastic sense, have dealt with space. Sculptors define and transform spaces. I create, transform, and change spaces by adding sound. That spatial concept is one which music doesn't include; music is supposed to be completely transportable."[12]

Neuhaus began installing unmarked sound pieces in stairwells, subway stations, swimming pools, and elevators, filling them with lush drones, phased clicks, or other sounds that were at once unobtrusive and subtly transformative. In 1973, he happened upon a subway vent on a pedestrian island in New York's Times Square and was struck by a desire to use the cavernous space as the resonant chamber for a sound work. Four years of arduous negotiation with the Metropolitan Transit Authority and Con Edison

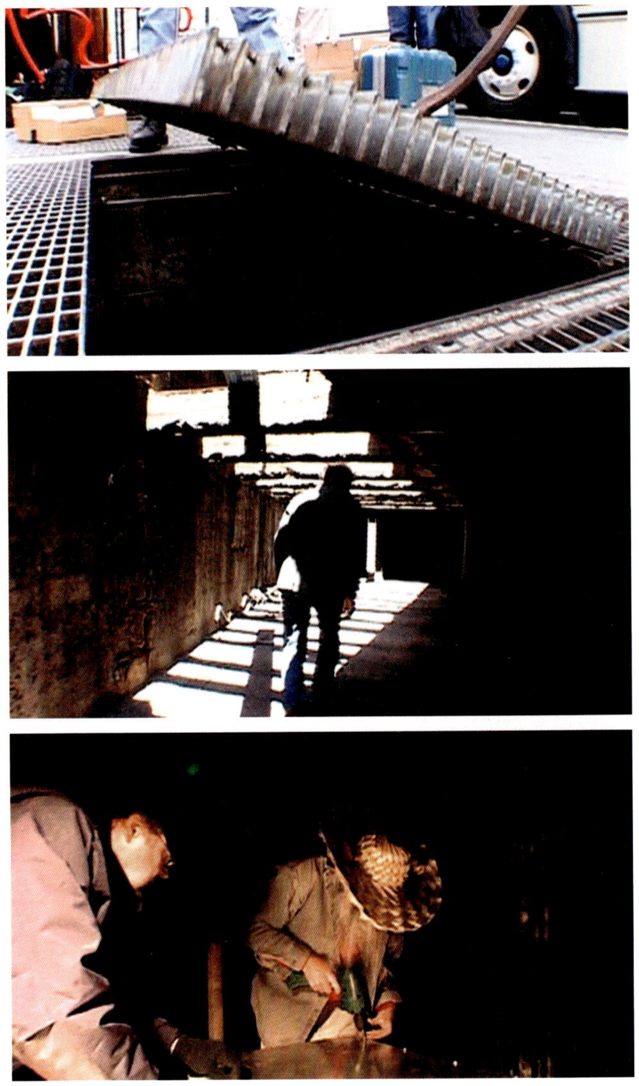

FIGS. 2a–c Max Neuhaus, *Times Square*, 1977/2002. Permanent sound installation. Video stills from Rory Logsdail's short film *Max Neuhaus – Times Square*, production: Firefly Pictures for Rai Sat Art, 2002, 7:75 min.

8
John Cage, "Experimental Music," in *Silence: Lectures and Writings by John Cage* (Middletown, CT: Wesleyan University Press, 1986), 7–12, here 8.

9
Kostelanetz, *Conversing with Cage*, 70.

10
Max Neuhaus, "Listen," in *Sound by Artists*, ed. Dan Lander and Micah Lexier (Toronto: Art Metropole, 1990), 63–67, here 63.

11
Max Neuhaus, "Conversation with Ulrich Loock," in *Max Neuhaus: Sound Works*, vol. 1, *Inscription* (Ostfildern: Cantz, 1994), 122–135, here 130.

12
Max Neuhaus, "Interview with William Duckworth," in *Max Neuhaus: Sound Works*, 42–49, here 42.

13
Walter Cianciusi, "Max Neuhaus: Pioneer of Invisible and (Almost) Inaudible Sound Installations," in *International Computer Music Conference Proceedings* (2013), 248–253, here 249.

ensued until Neuhaus finally received permission to climb down into the vent shaft and install a loudspeaker and some homemade electronic sound generators that he jerry-rigged to the city's lighting grid.(FIGS. 2a–c) Neuhaus built the sound by ear, listening carefully to the sonic environment, layering frequencies and timbres the way a painter layers color, and shaping mass like a sculptor working with invisible material. As in all of Neuhaus's installations, the sound was to be, he liked to say, "almost plausible" in the context and yet also a bit out of place, a slight dislocation of the aural topography. The result was a dense drone that, as Neuhaus described it, resembled "the after ring of large bells,"[13] a sound that summoned the restless clamor of its environs and bathed it in a consistent aural hue.

Launched in September of 1977, the piece defined an aural field that remained in place twenty-four hours a day for fifteen years before Neuhaus dismantled it. In 2002, the Dia Art Foundation relaunched *Times Square* as a permanent installation that is now one of New York City's great works of public art.

Drones and the Sonic Flux

Neuhaus coined the phrase "sound installation"; but he was not alone in developing ongoing, site-specific sound environments during the 1960s and 1970s. As early as 1962, the minimalist composer and improviser La Monte Young began to envision what he called a Dream House, a space that "will allow music which, after a year, ten years, a hundred years or more of constant sound, would not only be a real living organism with a life and tradition all its own but one with a capacity to propel itself by its own momentum."[14] Four years later, Young and his wife, the vocalist and light artist Marian Zazeela, experimented with sine wave oscillators to create a private Dream

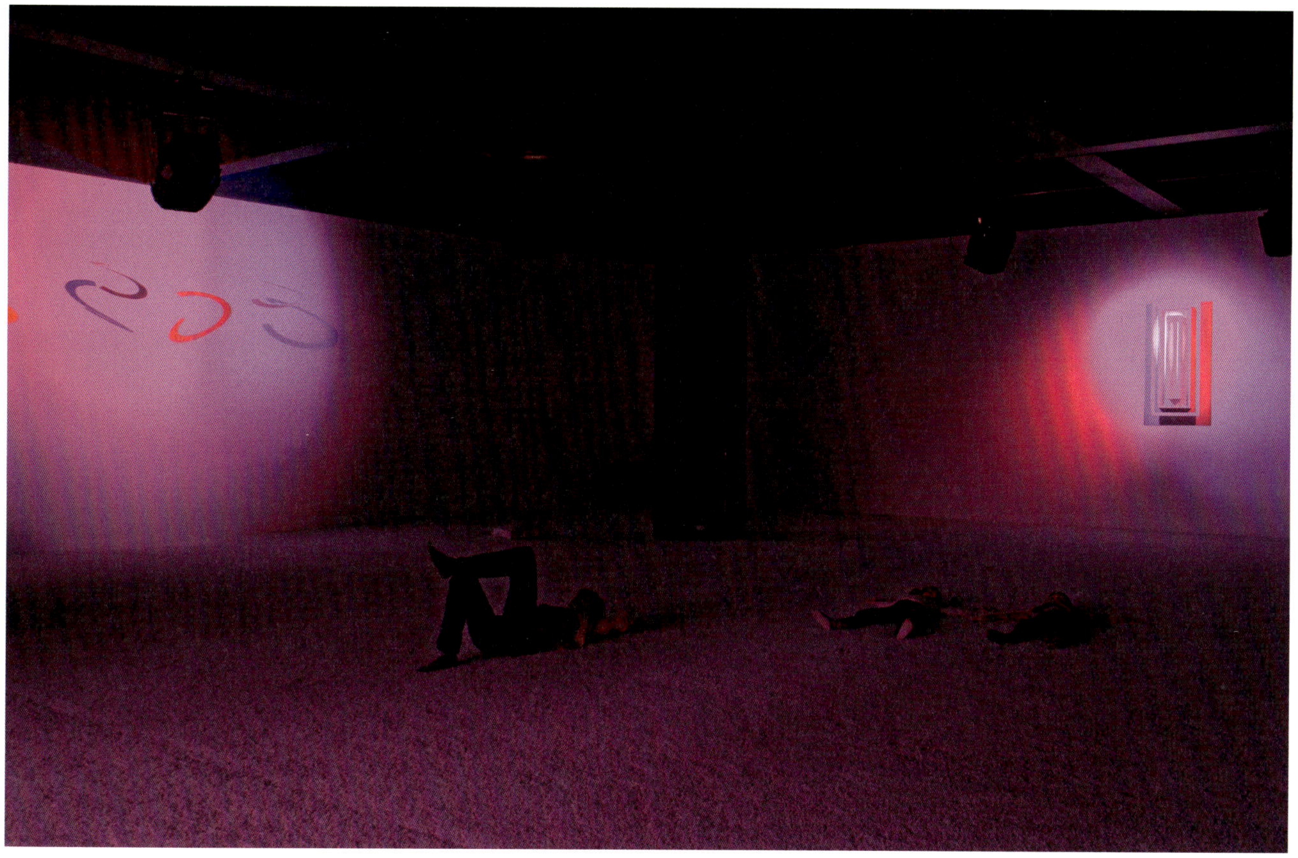

FIG. 3 La Monte Young, Marian Zazeela, *Dream House*, 1989. Sound and light environment, wall sculpture, neon, speaker, light, plastic film. Installation view ZKM | Karlsruhe, 2012.

House in their Tribeca loft. The project was first presented in public for two weeks in July 1969 at Galerie Heiner Friedrich in Munich, where Young's oscillators generated a field of sine tones fluctuating around 50 Hz ("the underlying drone of the city and all AC-powered equipment"[15]) and Zazeela projected pure light frequencies at metal mobiles to give the impression of "self-luminous colored bodies freely suspended in an atmosphere of continuously moving calligraphic strokes."[16] A decade later, Friedrich's Dia Art Foundation funded a "permanent" *Dream House* in the former New York Mercantile Exchange building in lower Manhattan. The installation ran for six years before disputes within the foundation led to its dismantling and reinstallation elsewhere, finally in a space on Church Street, where it remains today. Temporary versions of the sound and light environment have been presented in museums and art spaces around the world.(FIG. 3)

While Neuhaus's installations unobtrusively mark and color public space, Young and Zazeela's are overpowering presentations of sound in interior spaces that allow them precise control over frequencies and their psychoacoustic effects. Nonetheless, these early sound installations are kindred in their use of complex drones, a defining feature of early sound installation in America. Cage's *4'33"* set out to draw attention to the sonic flux of the world, the ceaseless din that surrounds us at every moment. In a grand sense, this background noise is surely a drone – the continuous and simultaneous sounding of all that is audible. Like Cage, Young attempted to tap into this perpetual flux, drawing our attention to an experience of time that vastly exceeds not only the temporality of musical composition but also the span of human life. Both Cage and Young affirmed the drone in all its manifestations, whether natural or technological, human or nonhuman. Cage was inspired equally by the droning sounds of "a truck at fifty miles per hour," "static between the stations," and "rain,"[17] La Monte Young by "the sound of the wind going through the chinks of the log cabin"[18] where he was raised, "the sounds of insects; the sounds of telephone poles and motors; sounds produced by steam escaping from such as [his] mother's tea-kettle or train whistles; and resonation from the natural characteristics of particular geographic areas such as valleys, lakes, and plains."[19]

Sculpting Sound

A no less cosmic but more scientific or phenomenological approach to the drone was pursued by other American sound artists in the 1970s. Working without any direct connection to the New York art and music scenes, the southern California artist Michael Brewster combined Young's interest in controlled environments with Neuhaus's desire to articulate space through sound. Unlike Cage, Neuhaus, and Young, all of whom came from music, Brewster's background was in the visual arts and, specifically, sculpture. "I think like an artist and I behave like an artist, not like a musician,"[20] he told an interviewer. "I work with sound because of its sculptural potential – it is a physical material to me."[21] "It has size and dimensions, viscosities, textures, even excitements,"[22] he continued.

> At lower frequencies, its wavelengths are of human scale. Its densities are such that we can walk through them, inhabiting their interiors, considering their particulars. I love the elegant physics of sound's wave-form behavior. I think this whole cosmic show is driven by what physics studies: the very stuff of life. So I feel close to the essential motors of our world when I'm working with the quasi-physical oscillations of sounds, blending and coaxing them into place. I had hopes that the resulting fields of sound space would be a way to expand and reinvigorate the sculptural experience.[23]

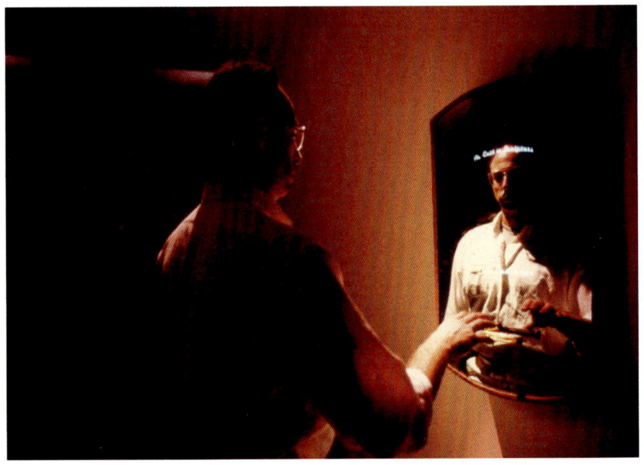

FIG. 4 Michael Brewster, *An Exit to Sculpture*, 1985. Acoustic sculpture with neon, audio, and narration, dimensions variable. Installation view Temporary Contemporary building, Museum of Contemporary Art, Los Angeles, 1985/1986, the artist activating the installation.

In 1970, Brewster began producing what he called "acoustic sculptures"(FIG. 4) consisting of sine tones generated by oscillators and broadcast into bare rectangular gallery spaces via loudspeakers. Brewster tuned his equipment to generate standing waves, invisible but nonetheless audible and palpable forms that created a sonic geography of peaks and valleys to be traversed by visitors. From a musical perspective, the work was utterly simple and unimpressive. Yet, for Brewster, these sine tones were simply tools for sculpting space and shaping forms. And, at this, he felt, they were superior to hammers and chisels. "I realized that hearing was more congruent with the goals and ambitions of sculpture than was seeing,"[24] Brewster noted. "We see through the flat while we hear fully in the round, hence we can hear more sculpturally than we see."[25]

A sculpture made of metal, stone, or wood remains an object set apart from us in the world, an entity to be surveyed with the eye as we circle around it. But sound, Brewster thought, enables audiences to inhabit the *interior* of the form. Just as Young often spoke of "living inside the sound," Brewster maintained that his acoustic sculptures collapse the distinctions between subject and object, here and there, now and then.[26] "No longer is the object of desire over there. We are in it, a part of it, here, now. I've always wanted to bridge the distance, to be closer to the core of my own experience, to overcome that awayness, the apartness that consciousness generates, to have my desire become one with its object. If I'm providing anything at all, I hope it is a full-on, all-around experience, from toes to head. Acoustic sound wraps itself elaborately all around us."[27]

Sonic Architecture

This physical exploration of sounds in space was also a driving force for Alvin Lucier, whose work since the late 1960s has constantly shuttled between musical composition and sound installation. Lucier's sound installations are inspired less by sculpture than by architecture – by the way that sounds can map the built environment, which in turn acts as a sonic filter and amplifier. The early work *Vespers* (1968), for example, paid homage to the common bat *(Vespertilionidae)*, equipping blindfolded performers with hand-held echolocation devices that they used to maneuver through a darkened space toward a central point.(FIGS. 5a–c)

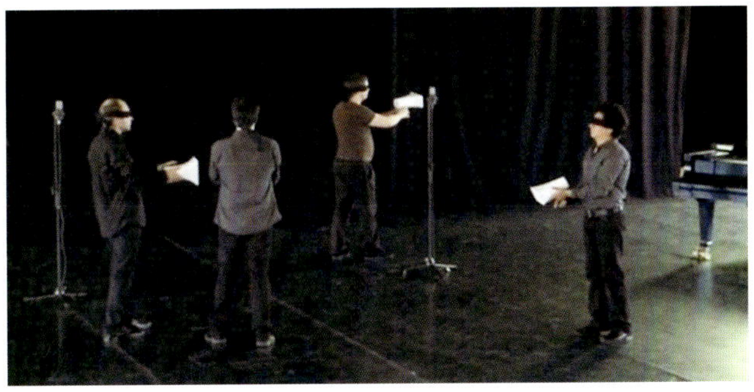

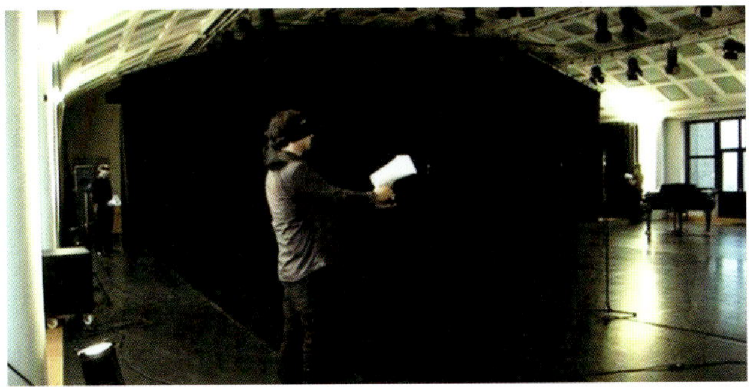

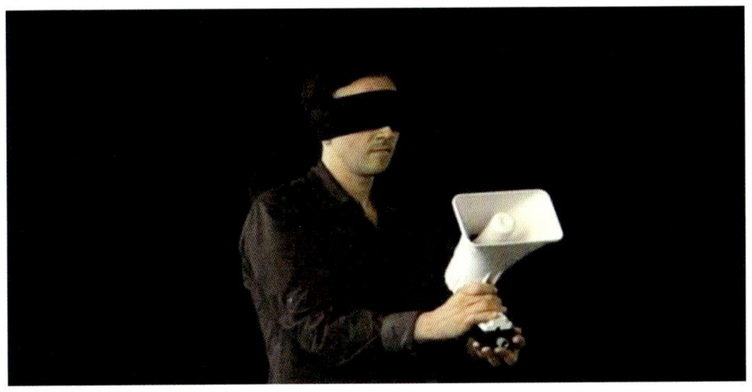

FIGS. 5a–c Alvin Lucier, *Vespers*, 1968. Acoustic orientation by means of echolocation, for players with hand-held echolocation devices. Performed by Pepe Garcia, Joey Marijs, Juan Martinez, and Niels Meliefste of Slagwerk Den Haag, Festival Dag in de Branding, The Hague, 2010, video stills. Video available online at: vimeo.com/89093088.

14
La Monte Young and Marian Zazeela, "Dream House," in Young and Zazeela, *Selected Writings* (Munich: Heiner Friedrich, 1969), reissued by ubuclassics in 2004, 10–16, here 16, www.ubu.com/historical/young/young_selected.pdf. Biographical details about Young are drawn from Jeremy Grimshaw, *Draw a Straight Line and Follow It: The Music and Mysticism of La Monte Young* (Oxford: Oxford University Press, 2011).

15
Young and Zazeela, "Dream House," 12.

16
Ibid., 13.

17
Cage, "Future of Music: Credo," 3.

18
Richard Kostelanetz, "Conversation with La Monte Young," in Young and Zazeela, *Selected Writings* (2004), 17–63, here 20.

19
Ibid., 33.

20
Michael Brewster, "An Interview with Brandon LaBelle," in Michael Brewster, *See Hear Now: A Sonic Drawing and Five Acoustic Sculptures* (Los Angeles: Los Angeles Contemporary Exhibitions, 2002), 26–30, here 28.

21
Ibid.

22
Ibid., 26.

23
Ibid.

24
Ibid.

25
Ibid.

26
See Kostelanetz, "Conversation with La Monte Young," 35, and La Monte Young, "Lecture 1960," in Young and Zazeela, *Selected Writings* (2004), 66–76, here 74–75. On "living inside the sound," see also Young's erstwhile collaborator Tony Conrad, "LYssophobia: On *Four Violins*," in *Audio Culture: Readings in Modern Music*, ed. Christoph Cox and Daniel Warner (New York: Continuum, 2004), 313–18.

27
Brewster, "Interview with Brandon LaBelle," 27.

"A performance of *Vespers* gives you an acoustic signature of the room," remarked Lucier, "as if one were taking a slow sound photograph over a long period of time. You hear what the room sounds like. That was mysterious to me and wonderful. It really turned me on."[28]

Lucier's scientific inquiry into echolocation transformed his conceptions of music and sound. "I began thinking of sounds in terms of short and long wavelengths, not as high and low pitches or notes written in time from left to right on a page,"[29] he recalled. "Thinking of sounds as measurable wavelengths [...] has changed my whole idea of music from a metaphor to a fact and, in a real way, has connected me with architecture."[30] Lucier pursued these architectural investigations in several other key works, notably his most famous piece *I Am Sitting in a Room* (1970). A wonderfully self-reflexive variant on the "text scores" that, via Fluxus and conceptual art, became prominent in the 1960s, *I Am Sitting in a Room* consists simply of a four-sentence description of the piece read into a tape recorder and played back into a room repeatedly until the resonant frequencies characteristic of the space pile up on one another and overwhelm the sounds of speech.(FIG. 6) In the 1980 recording of the piece, Lucier's characteristic voice (and evident stutter) becomes more distant and indistinct with each iteration of the text, eventually dissolving into a wavering metallic drone.

The drone formed the basis of another of Lucier's installations, *Music on a Long Thin Wire* (1977).(FIG. 7) In its debut version, Lucier clamped each end of a ninety-foot wire to two wooden tables under the dome of the US Custom House in New York City and set the wire into vibration using an oscillator and an electromagnet, amplifying the movement of the wire with contact microphones. Lucier explained:

> By carefully tuning the oscillator, the wire could be left to sound by itself. Fatigue, air currents, heating and cooling, even human proximity could cause the wire to undergo enormous changes. In a dance studio in Kyoto, for example, visitors' footsteps on the Marley floor caused extremely slight shifts in the positions of the tables to which the wire was clamped, causing spectacular changes in the sound of the wire. Shin Nakagawa, who arranged my visit there, slept overnight under the wire and reported that even with no movement in the room it would mysteriously erupt into triadic harmonies.[31]

Lucier, Brewster, and Young were concerned with the ways that sound filled spaces and the ways that enclosures shaped and modified sounds. The American composer and sound artist Maryanne Amacher pressed beyond the architectural volume to investigate the very

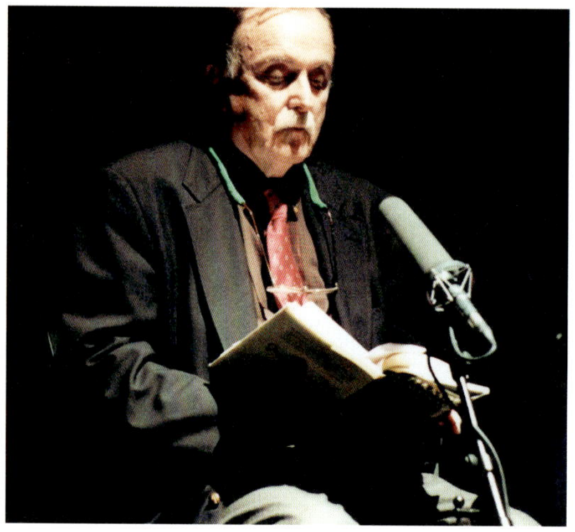

FIG. 6 Alvin Lucier, *I Am Sitting in a Room*, 1970. Audio installation for voice and electromagnetic tape.

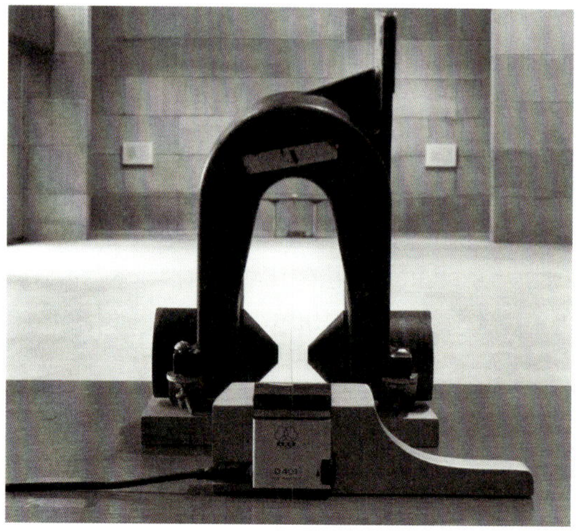

FIG. 7 Alvin Lucier, *Music on a Long Thin Wire*, 1977. Sound installation for audio oscillator and electronic monochord. Installation view, Wesleyan University, 1988.

28
Alvin Lucier, *Music 109: Notes on Experimental Music* (Middletown, CT: Wesleyan University Press, 2012), 87.

29
Ibid., 85.

30
Alvin Lucier, "Every room has its own melody," in Alvin Lucier, *Reflections: Interviews, Scores, Writings* (Cologne: MusikTexte, 1995), 94–103, here 98.

31
Alvin Lucier, liner notes to *Music on a Long Thin Wire*, Lovely LCD 10011, 1992.

32
See Maryanne Amacher, "Music for Sound-Joined Rooms," http://maryanneamacher.org/memorial/Maryanne_Amacher/Amacher_Archive_Project/Entries/2009/10/24_music_for_sound_joined_rooms.html, and http://maryanneamacher.org/memorial/Maryanne_Amacher/Amacher_Archive_Project/Entries/2009/10/23_Writing_on__Living_sound%2C_patent_pending.html.

33
See Amacher, "Composing Perceptual Geographies," http://maryanneamacher.org/memorial/Maryanne_Amacher/Amacher_Archive_Project/Entries/2009/10/23_Composing_perceptual_geographies.html, and Kabir Carter and Alan Licht, "Sound/Klang," *Parkett* 89 (2011), 6–20.

34
See Douglas Kahn, "Christian Marclay's Early Years: An Interview," *Leonardo Music Journal* 13 (2003), 17–21.

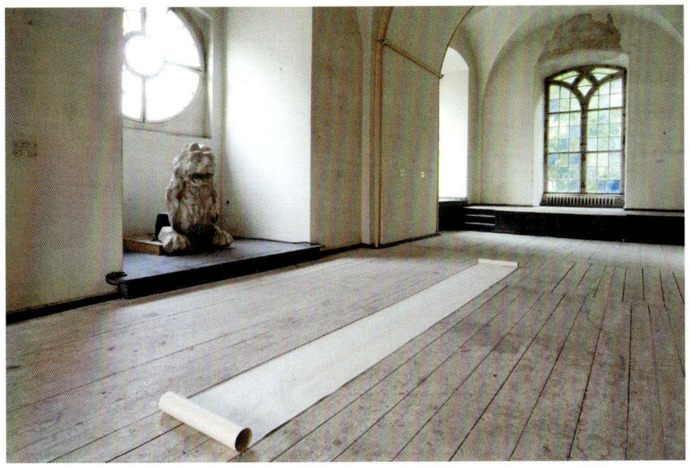

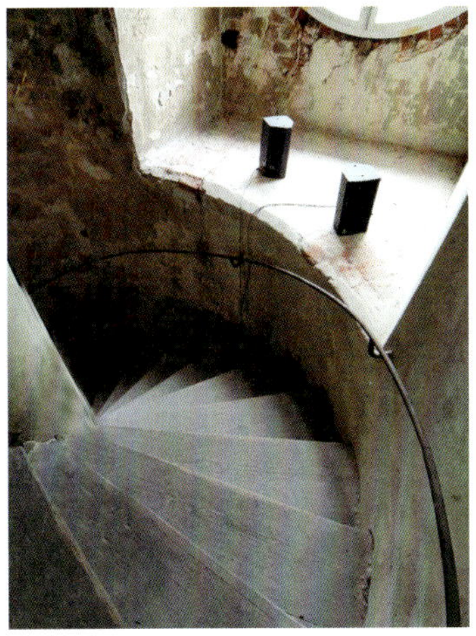

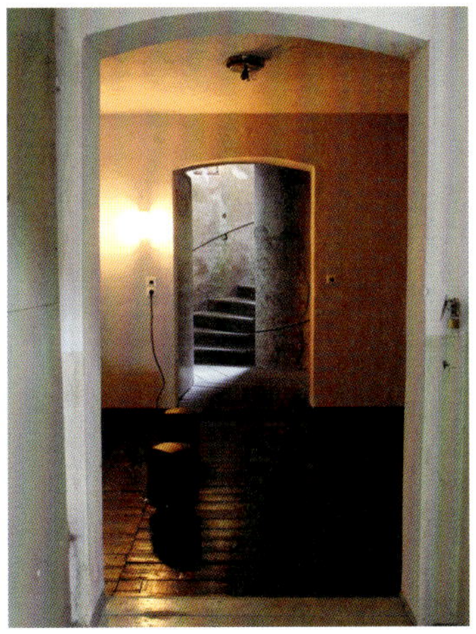

FIGS. 8a–c Maryanne Amacher, *Gravity* from the *Music for Sound Joined Rooms* series, 2006. Site-specific installation, created for the tower room of the parochial church in Berlin. Installation view singuhr – hoergalerie in parochial berlin (a), 2006. Detail view of the northern spiral staircase in front of the bell room (b) and detail view of the former vestry between the northern and the southern spiral staircase (c).

infrastructure of buildings and spaces. In 1980, she launched the project *Music for Sound-Joined Rooms*,(FIGS. 8a–c) which called for the installation of loudspeakers and transducers throughout a building to generate "structure-borne" sound, designed to travel not so much through the air but through the wood, metal, stone, and plaster surfaces of an entire house, gallery, or museum. After receiving a commission, Amacher would spend weeks at the site investigating its material features and acoustic potentials. Likening the process to choreography, theater, and cinema, she aimed to create spaces of aural and tactile intrigue, carefully placing and sequencing vibrations to suggest distance or to produce sonic close-ups, generate sonic illusions, entice visitors into neighboring rooms, lead them through pockets of intense pressure, or deposit them into spaces of ethereal calm.[32]

Between Sound and Image

Amacher's fascination with immersive sound environments and sonic-tactile experience drew her to club culture and to the idea that the dance club might be considered a sort of sound installation.[33] The early work of Christian Marclay also crossed this divide between the gallery, the club, and the concert hall. Marclay attended art school where, inspired by Cage, Marcel Duchamp, Joseph Beuys, and punk rock, he began to perform using cheap turntables and records found on the street or purchased at thrift stores.[34] Marclay dealt with this detritus of pop culture in all manner of ways – cutting the records apart and gluing them back together in different configurations; hammering nails into the record surface so that the tone arm was buffeted back and forth, etc. – devising a practice that connected with the scratching and sampling that hip hop DJs were developing at the time.

In the 1980s, Marclay began to consider the culture of recording more broadly, exhibiting photographic collages, sculptures, and installations composed of records, album covers, magnetic tape, and found images of singers or instrumentalists. Working between music and visual art, Marclay was fascinated with the connections and disjunctions between sound and image. *Chorus* and *Chorus II* (both 1988), for example, present found photographs of mouths open in song, the gap at the center of each image drawing attention to a fundamental lack, the incapacity of the image to supply its sonic content.(FIG. 9) Later projects investigated the indeterminacy of translation between sound and image and, in particular, the generative capacities of the musical score, which, even in its most conventional form, requires the performer to render a set of visual symbols as sounds. For *Graffiti Composition* (1996), Marclay posted blank sheets of musical paper on kiosks and walls throughout Berlin, inviting passersby to mark on them. Marclay photographed the results and printed them on cards to be interpreted by musicians as prompts for improvisation.(FIGS. 10a, b)

Such translations between sound and image are central to the work of Steve Roden, an audiovisual artist from Los Angeles whose work is

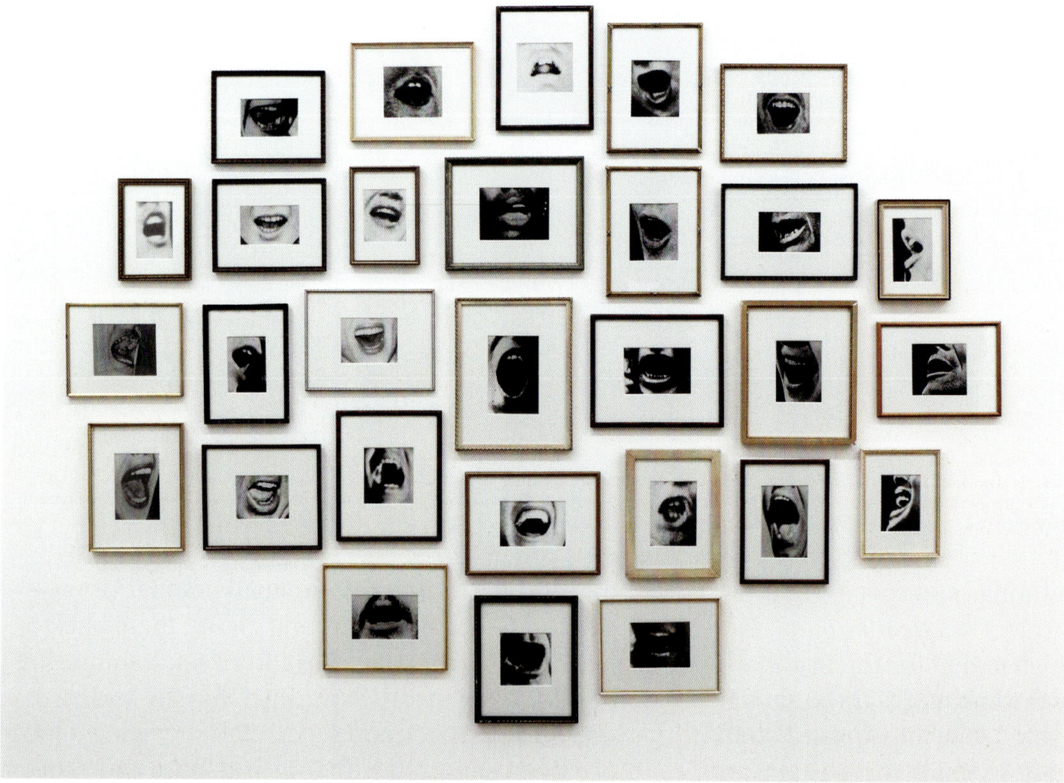

FIG. 9 Christian Marclay, *Chorus II*, 1988. Installation, 29 black-and-white photographs, framed. Installation view *Honk If You Love Silence*, fifth episode of the *Rolywholyover* exhibition cycle, MAMCO, Musée d'art moderne et contemporain, Geneva, June 25–September 21, 2008.

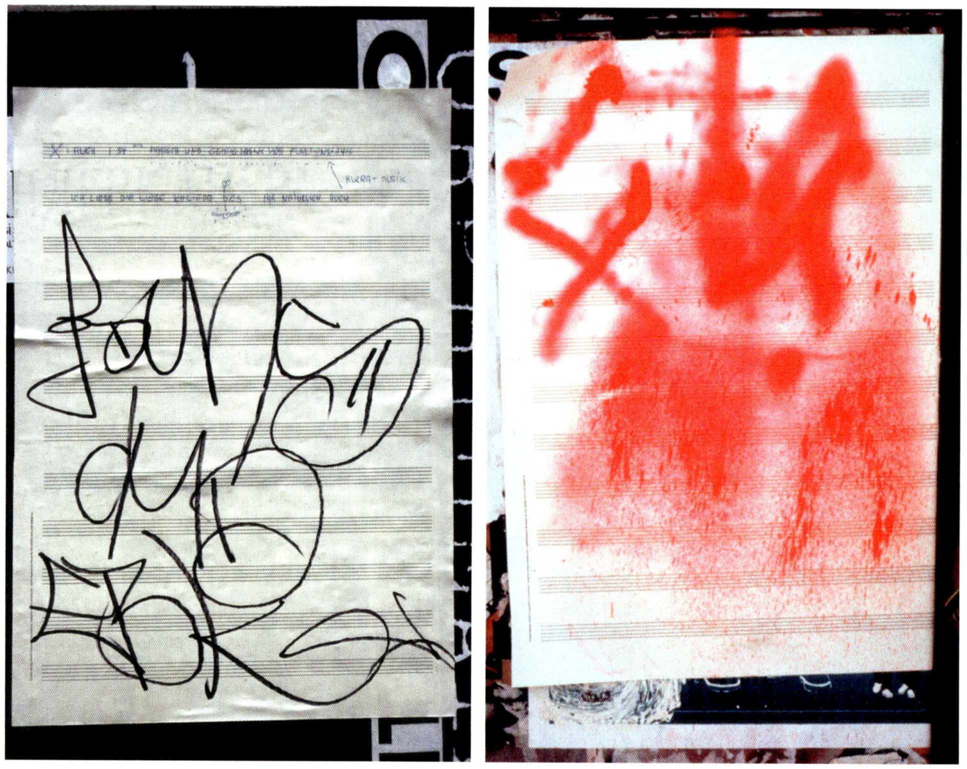

FIGS. 10a, b Christian Marclay, *Graffiti Composition*, 2002. Portfolio of 150 images, Indigo prints on Cougar stock, 35.56 × 24.13 × 7.62 cm, published by Paula Cooper Gallery, limited edition of 25, plus 5 performance copies and 5 artist's proofs.

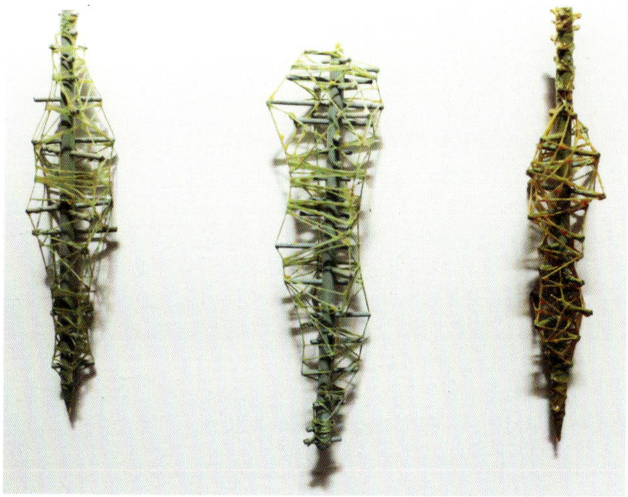

FIG. 11 Steve Roden, *letter forms*, 2002. Installation, 26 objects, wood, embroidery thread, polyurethane, enamel paint, ca. 28 × 10 cm each. Installation view *Treble*, SculptureCenter, New York, 2004.

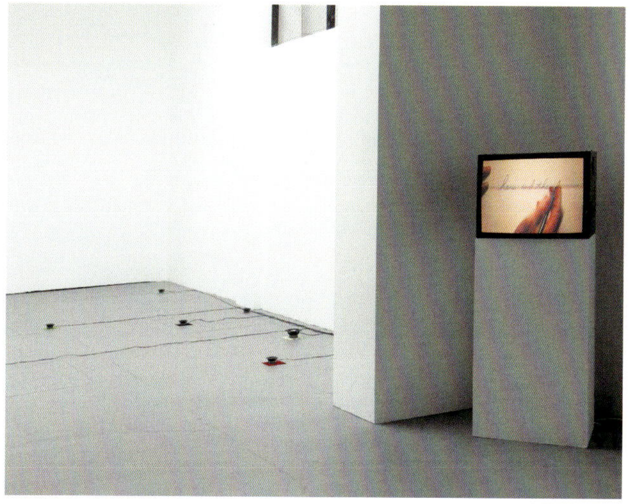

FIG. 12 Steve Roden, *ragpicking (berlin)*, 2012. Mutimedia installation, on the floor: *symbol/cymbal*, 2012, 8-channel sound installation with a single cymbal as sound source, 20 min., looped; on the monitor: *a lexicon of walter benjamin's silences*, 2012, video, 22:19 min., silent. Installation view singuhr – hoergalerie berlin, partial view, MEINBLAU project space, Berlin.

rooted in sound. In the late 1990s, Roden began to devise idiosyncratic translation schemes in order to spark new ideas and chance conjunctions. His *letter forms* (2002), for example, are a set of wood and thread sculptures that model the sound waves of the artist's voice speaking or singing each letter of the alphabet,(FIG. 11) the vowels painted in accordance with letter-color associations laid out in Arthur Rimbaud's poem "Alchemy of the Verb." The more recent installations *rag picker (los angeles & new york)* (2013) and *ragpicking (berlin)* (2012)(FIG. 12) derive from visual elements discovered in the notebooks of German cultural critic Walter Benjamin, whose color-coded symbols Roden transformed into graphic notation that guided a percussive/electronic score and determined color combinations in a set of accompanying drawings.

Sonic Politics

For much of its history, American sound art – like sound art in general – has either concentrated on the physical, psychoacoustic, sculptural, and architectural capacities of sound or considered the tensions between sound and image. Yet, particularly recently, in the context of American military aggression in Iraq and Afghanistan, sound artists have attuned themselves to the political potentials of sound, examining the role of the sonic in what philosopher Jacques Rancière has called "the distribution of the sensible" in social space, and employing sound as a means of altering that distribution.[35] Via videos, sculptural installations, and performances, the artist duo Allora & Calzadilla (Jennifer Allora and Guillermo Calzadilla) probe the relationships between music and war, investigating sound as an instrument of military violence, territorial domination, and the mobilization of human bodies.(FIG. 13) Likewise, in their practice, Angel Nevarez and Valerie Tevere test the populist potentials of radio and the protest song, and examine the uses of voice recognition for biopolitical control.(FIG. 14)

Surely the most sustained investigation of audio politics has been undertaken by the collective Ultra-red. Formed in Los Angeles in 1994, the group initially dedicated itself to AIDS activism before turning to broader questions of housing and immigration. Ultra-red's early recordings – which mixed samples of political speech with experimental electronic dance music – allowed its members entry into music festivals that became platforms for their political interventions. Since 2006, the collective has primarily dedicated itself to the politics of silence and listening in public space. Deeply inspired by John Cage and developing the political potential of *4'33"* for what the group calls "Militant Sound Investigations,"[36]

35
See Jacques Rancière, *The Politics of Aesthetics*, ed. and trans. Gabriel Rockhill (London: Bloomsbury, 2013), esp. part 1: *The Distribution of the Sensible*.

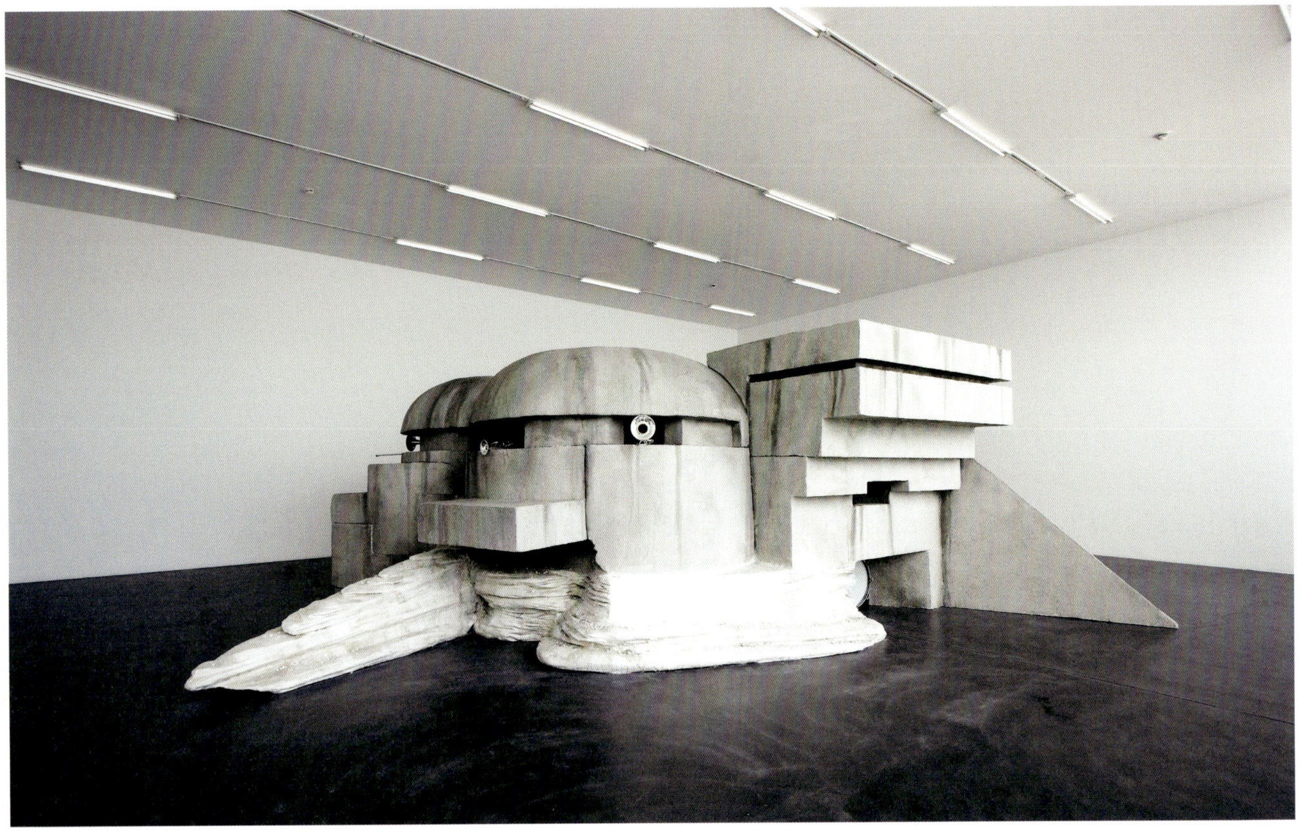

FIG. 13 Allora & Calzadilla, *Clamor*, 2006. Multimedia installation, plaster, foam, pigment, 1 tuba, 1 trumpet, 2 trombones, 1 flute, 1 drum kit, pre-recorded sound, and live musicians. Installation view Kunsthalle Zurich, Zurich.

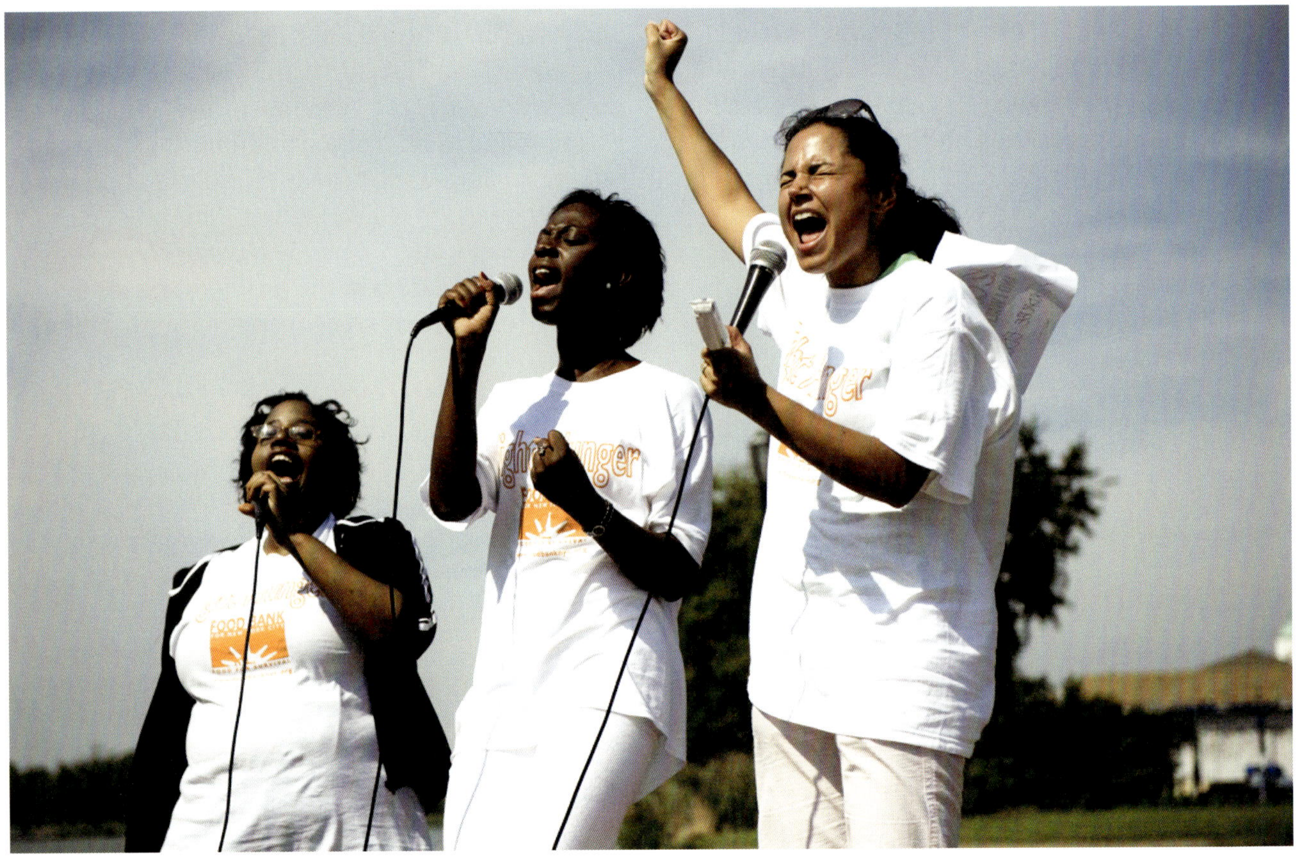

FIG. 14 Angel Nevarez and Valerie Tevere, *Another Protest Song: Karaoke with a Message*, 2008. Public performance with karaoke system, Flushing Meadows / Corona Park, Queens, NY.

36
Ultra-red, "Mission Statement" (2000), www.ultrared.org/mission.html.

37
Ultra-red, *Five Protocols for Organized Listening: With Variations*, 2012, www.ultrared.org/uploads/2012-Five_Protocols.pdf.

38
Ultra-red, "Mission Statement."

39
Dont Rhine, quoted in Mark Fisher, "Public Space is the Place," *The Wire* 295 (September 2008), 28–33, here 29.

40
See Alfred North Whitehead, *Process and Reality* (New York: The Free Press, 1978), 39.

Ultra-red has laid out a set of "protocols for organized listening"[37] aimed at "exploring acoustic space as enunciative of social relations."[38] Via site- and community-specific discussions,(FIG. 15) the group facilitates inquiries around broad questions of sonic politics, such as: "What is the sound of anti-racism?" "What is the sound of the war on the poor?" "What is the sound of freedom?" Participants are invited to undertake sound walks or to produce recordings that they then subject to analysis guided by the deliberately open question "What did you hear?" Ultra-red's aim is to attune communities to the affective and social forces of the soundscape, empowering them to compose it differently. The project richly manifests Rancière's proposal that every social space is aesthetic insofar as it determines what and who can be seen and heard, and, conversely, what is forced to remain silent or invisible. As Ultra-red cofounder Dont Rhine put it: "With Cage the idea was that the composer is not composing sounds but is composing new ways of listening. You are organizing listening, so the sound artist organizes sound as a political strategy. If we're going to take that seriously, then organizers are already involved in aesthetic operations."[39]

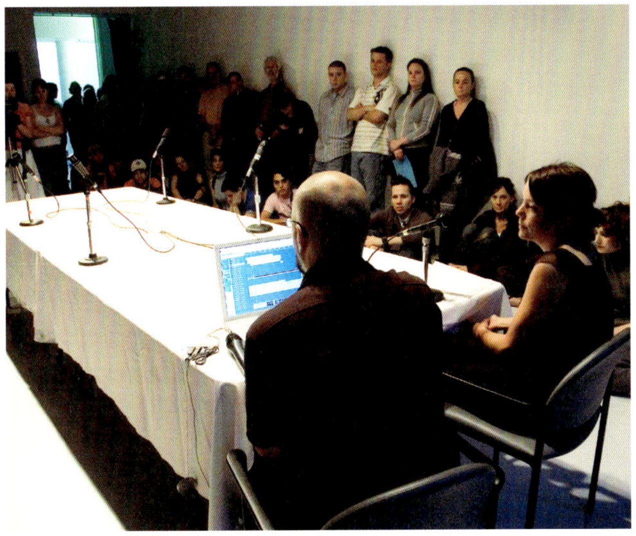

FIG. 15 Ultra-red, SILENT|LISTEN, listening session, Banff Center for the Arts, Banff, Canada, June 23, 2005. Shown seated (left to right), Dont Rhine and Nicole Neve. Further information on the project SILENT|LISTEN is available online at http://ultrared.org/pso8.html.

Ultra-red's practice reveals how tremendously generative Cage's questions and proposals have been for American sound art. While artists such as Neuhaus, Young, Lucier, and Amacher pursued Cage's ontological and epistemological inquiries concerning noise, sound, silence, and listening, Ultra-red, Allora & Calzadilla, Nevarez & Tevere, and others have amplified the politics of these sonic relationships. Alfred North Whitehead famously remarked that the European philosophical tradition consists of a series of footnotes to Plato.[40] Likewise, it can reasonably be said that American sound art has and continues to be a series of footnotes to Cage, its progenitor and the source of its problems and prospects.

Christof Migone

Canadasonic

Sound art was invented in Canada.

According to Alan Licht in *Sound Art: Beyond Music, Between Categories*, the term was "coined by Canadian composer/audio artist Dan Lander in the mid-1980s."[1] That is, of course, a distinct claim from the apocryphal absurdity of my opening provocation. After all, naming does not necessarily coincide with the incipience of the activity being named. While the historical accuracy of Licht's assertion is questionable, it functions as a useful point of departure, for it echoes R. Murray Schafer's statement that "radio existed long before it was invented."[2] In other words, slippage and fluidity are not only recurring characteristics of physical sound, but the related nomenclature operates with similar schizo-ontological and anachronistic confusion. Jonathan Sterne's notion of a field of study which "reflexively *mind[s] sound*"[3] is productive here for it emphasizes intersections without eschewing a critical stance, especially toward trite theoretical presuppositions based on the differences between the physiological functioning of hearing versus that of vision.

1
Alan Licht, *Sound Art: Beyond Music, Between Categories* (New York: Rizzoli, 2007), 11. Rahma Khazam's "A Brief History of Sound Art," *Volume – What You See Is What You Hear* 1 (2010): 6–9 also points to a mid-1980s Lander moment of origin. In Douglas Kahn's "Sound Art, Art, Music" (from 2005 but recently reprinted in *Tacet* 3, [2014]: 328–47), he traces his own disinclination with the term "sound art" in connection with his friendship to Dan Lander in the period leading up to his 2001 book *Noise, Water, Meat* (Cambridge, MA: The MIT Press).

2
R. Murray Schafer, "Radical Radio," *Ear Magazine* (1987), 18–20, here 18.

3
Jonathan Sterne, "Sonic Imaginations," in *The Sound Studies Reader*, ed. Jonathan Sterne (New York: Routledge, 2012), 1–17, here 10. One could posit that I am replicating one of the entries in what Sterne calls *the audiovisual litany* (9) here by suggesting a correlation between the physical characteristics of sound and a descriptive theory. That is partly true, though I have avoided (and have no interest in) making that linkage in relation to a stereotypical critique of visuality.

Categorical delineations tend to ignore nuanced definitions, but artists are apt to be unperturbed by blurs. Correspondingly, Canadasonic embraces a pansonic approach, one which acknowledges and attempts to encompass, in the brief space allotted, the various streams of art forms and practices which incorporate and foreground sound, whether these call themselves sound art or not. Emblematically, Michael Snow speculated he could dub himself a "time-light-sound poet" in 1968.[4] In other words, for the purposes of this text, the inclusion of media and visual arts, *musique actuelle*, electroacoustics, acoustic ecology, communications, expanded cinema, sound studies, sound poetry, radio art, and performance art is not only strategic but also reflective of the intermeshing of a plethora of artistic and research practices. In short, sound art in the expanded field. All that being said, Lander's editorial work on *Sound by Artists* in 1990 and *Radio Rethink* in 1994 did serve as a considerable coalescing and catalyzing force in the early stages of what has become, now twenty-five years later, a multilayered and multivalent scene.[5]

Listen! and Its Ramifications

The focus on listening as a pedagogical imperative with a sociopolitical agenda was advanced by the World Soundscape Project (R. Murray Schafer, Barry Truax, Hildegard Westerkamp, among others) starting in the late 1960s in Vancouver and percolatign to this day.[6] The corollary: a fostered awareness of the role sound plays in the everyday environment – in short, acoustic ecology. A caveat that haunts the acoustic ecology project, at least in its initial didactic (read moralistic) incarnation, is succinctly articulated by Lisa Robertson: "the concept of acoustic ecology [is] a direct expression of late capitalism's division not only of labor but of laborer from consumer."[7]

That being said, the original lesson still informs the compositional stratagems employed by artists, many of whom operate in a variety of territories and approaches at once, such as Darren Copeland (co-founder of New Adventures in Sound Art in Toronto), Andra McCartney, Magali Babin, and Hélène Prévost, to name only a few. The listener was also on Glenn Gould's mind when he bemoaned the unrepeatability of live performance and dreamed of "issuing a kit of variant performances and let the listener assemble his own performance"; he follows with a prescient sentence:

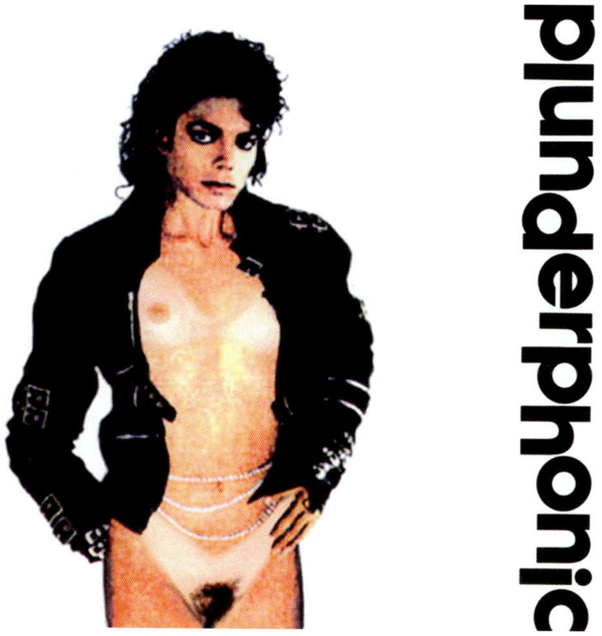

FIG. 1 John Oswald, *Plunderphonics*, 1989. CD jacket. The image is a cut and paste collage made by John Oswald.

"It would draw the audience in the re-creative process."[8] John Oswald's notoriously censored *Plunderphonics*, ongoing since 1985,(FIG. 1) could be examined through this filter – the artist as listener for recombinable structures culled from the entire history of recording. The act of recording presumes a listening that accents the verb form of the latter – i.e. an activated reception, an active listening which injects liveness into the supposed deadness or fixity of recording. Recordings also manifest materially, and thereby become manipulable.[9] With the faux ethnographic work *The Last LP* (1987), Michael Snow takes recording to one endgame of listening in relation to veracity and subterfuge (*What am I listening to?*).(FIG. 2) A fascinating parallel work of deceit occurs in the slyly titled *Sound* (2001) by David Cecchetto wherein he manipulates a live performance by

4
Michael Snow, *The Collected Writings of Michael Snow* (Waterloo, ON: Wilfrid Laurier University Press, 1994), 43.

5
Dan Lander and Micah Lexier, eds., *Sound by Artists* (Toronto and Banff: Art Metropole and Walter Phillips Gallery, 1990). Reissued in a facsimile edition by Blackwood Gallery and Charivari Press, 2013. Dan Lander and Daina Augaitis, eds., *Radio Rethink: Art, Sound and Transmission* (Banff: Walter Phillips Gallery, 1994). A few years later, additional publications solidified the scholarship: Jim Drobnick, ed., *Aural Cultures* (Toronto and Banff: YYZ Books and Walter Phillips Gallery Editions, 2004); and Nicole Gingras, ed., *S:ON: Sound in Contemporary Canadian Art/Le son dans l'art contemporain canadien* (Montreal: Éditions Artexte, 2003). These publications (plus the long-running magazine *Musicworks*) reflect a predilection not only for historicization but also for theorization of the art form – a persisting parallel current. They also denote the willingness of publishers to examine the field in a focused manner, in contradistinction to any of the major Canadian museums, thus far.

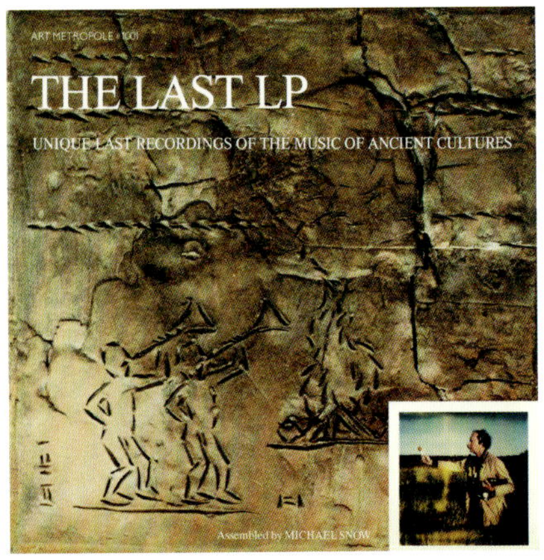

FIG. 2 Michael Snow, *The Last LP*, 1987. LP jacket.

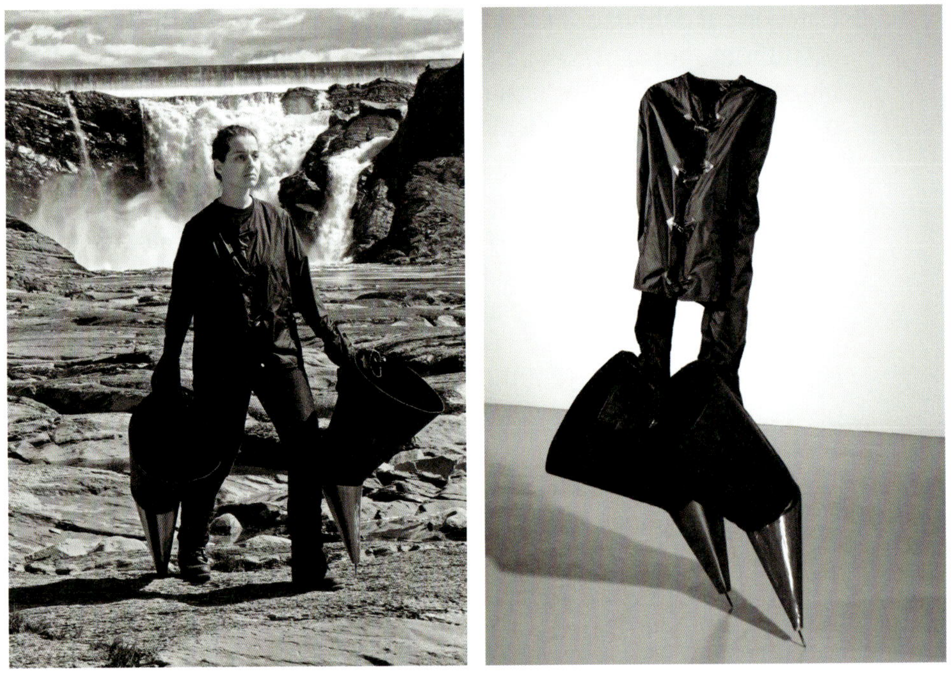

FIGS. 3a, b Nataliya Petkova, *Geophonies*, 2011, sound walk tool. Steel, textile, radio transmitters.

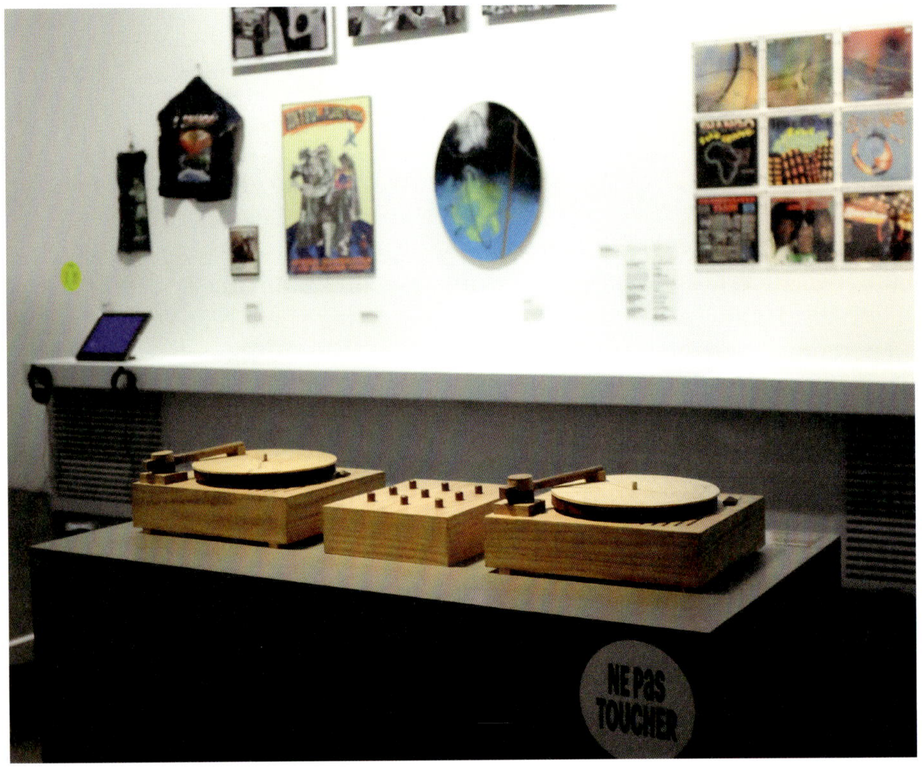

FIG. 4 Jordan Bennett, *Turning Tables*, 2010, sound installation. Installation view Institut du monde arabe, Paris.

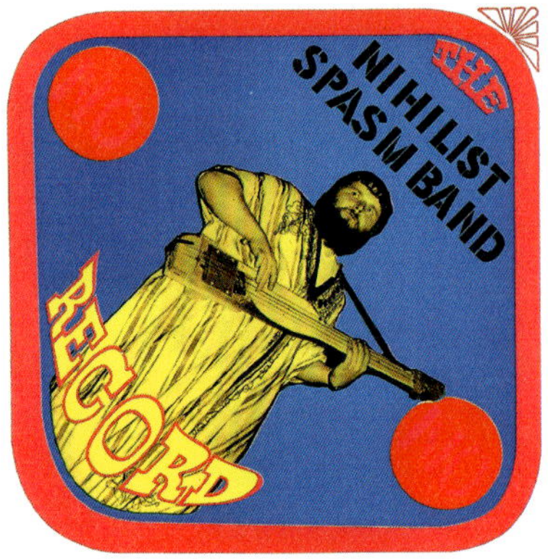

FIG. 5a The Nihilist Spasm Band, *No Record*, 1968. LP jacket.

tampering with the customary visual cues of a piano accompanied by a mic-computer-speaker rig, the latter of which is there just for show, yet is somehow "heard" by the audience. This piece also illustrates Cecchetto's general intent to explore the idea of a "sound that isn't sounded" and "sound [that] is never just sound."[10] These notions will reverberate asonically with those by Raymond Gervais and Jocelyn Robert that will appear at the conclusion of this text. Documenting a non-performance, Brady Cranfield and Jamie Hilder in *Night Shift* (2012) sound out the invisible labor of repeatedly painting gallery walls white on an LP where effectively background becomes foreground – static yet always moving, sans denouement (*Why am I listening to this?*). Nataliya Petkova with *Geophonies* (2011),(FIGS. 3a, b) and Douglas Moffat with *Montréal Phonographe* (2007) turn the stylus into recording devices in order to turn geographies into grooves (*Is the ground I walk on playable?*). Jordan Bennett's *Turning Tables* (2010) displays a functioning DJ setup all made out of walnut, oak, and pine, which might be deemed quaint Canadiana, until one hears Mi'kmaq being learned on one of the tree slices turned playable record: the perseverance of First Nations (*Can one rescue a language from being lost via recording?*).(FIG. 4)

Beyond and Between (Noise Voices)

"« listen » / The known world is a noisy ball."[11] The poetic pithiness of Mystery Tapes (aka John Oswald) encapsulates the relationship of listening to an outside that is often beyond our control. In the case of The Nihilist Spasm Band, as they enter their fiftieth year of activity, the outside din is interiorized.[12] Based in London, Ontario, the band (including visual artists Greg Curnoe, Murray Favro, and John Boyle) have been resolutely saying *yes* to *NO (Canada)* for the past half a century.(FIGS. 5a, b) Their brand of autodidactic noise and idiosyncratic instrument building has found ears worldwide. The enthralling racket is punctuated by Bill Exley's anarchic pronouncements, for example, from *Destroy the Nations*: "Destroy the Nation! Let's start with America! Canada is already dead!"[13] Destroying language,

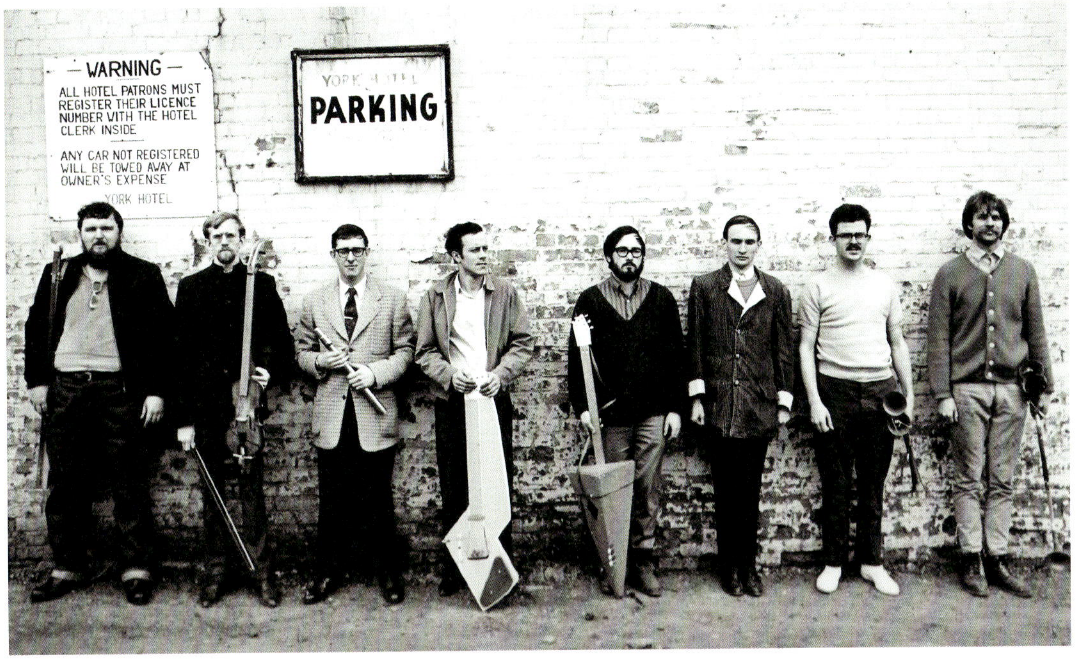

FIG. 5b The Nihilist Spasm Band in 1968, posing against the back wall of the York Hotel, the pub in which they played every Monday night for many years in the 1960s and 1970s. From left: Hugh McIntyre (bass), Art Pratten (Pratt-A-Various), Archie Leitch (slide clarinet), Murray Favro (guitar, drums), John Clement (guitar), Bill Exley (vocals, theremin), John Boyle (kazoo), Greg Curnoe (kazoo, drums).

6
The lasting influence of Schafer cannot be underestimated, for instance the following three recent tomes on sound feature his writing as the intro piece (in first two cases) or in the first section: Mark M. Smith, ed., *Hearing History* (Athens, GA: University of Georgia Press, 2004); Michael Bull and Les Back, eds., *The Auditory Culture Reader* (New York: Berg, 2003); and Christoph Cox and Daniel Warner, eds., *Audio Culture: Readings in Modern Music* (New York: Continuum, 2004).

7
Lisa Robertson, *Nilling* (Toronto: BookThug, 2012), 68.

8
Glenn Gould in Richard Kostelanetz, "Glenn Gould: Bach in the Electronic Age," in *Glenn Gould: By Himself and His Friends*, ed. John McGreevy (Toronto: Macmillan, 1983), 134.

9
There is an abundant list of artists who mine the history of recording for both its material properties and its cultural aspects, including: Martin Tétreault, Dave Dyment, Daniel Olson, Ian Murray, Rodney Graham, Tim Lee, Charles Stankievech, Brian Joseph Davis, Pierre-André Arcand, Érick d'Orion.

10
David Cecchetto, *Humanesis: Sound and Technological Posthumanism* (Minneapolis: University of Minnesota Press, 2013), 126.

11
Mystery Laboratory, "Mystery Tapes," in Lander and Lexier, *Sound by Artists*, 203. Also used as an epigraph by Jim Drobnick, *Aural Cultures*, 9. One would think, however, that the unknowable world would be even noisier, especially since one recurring definition of noise is sound that is uncategorizable.

12
Notable other sound art groups *avant la lettre* include: Canadian Creative Music Collective (or CCMC) which began in the mid-1970s, current members: Paul Dutton, Michael Snow, John Oswald; Artist's Jazz Band (1962–?); L'infonie (1967–1974).

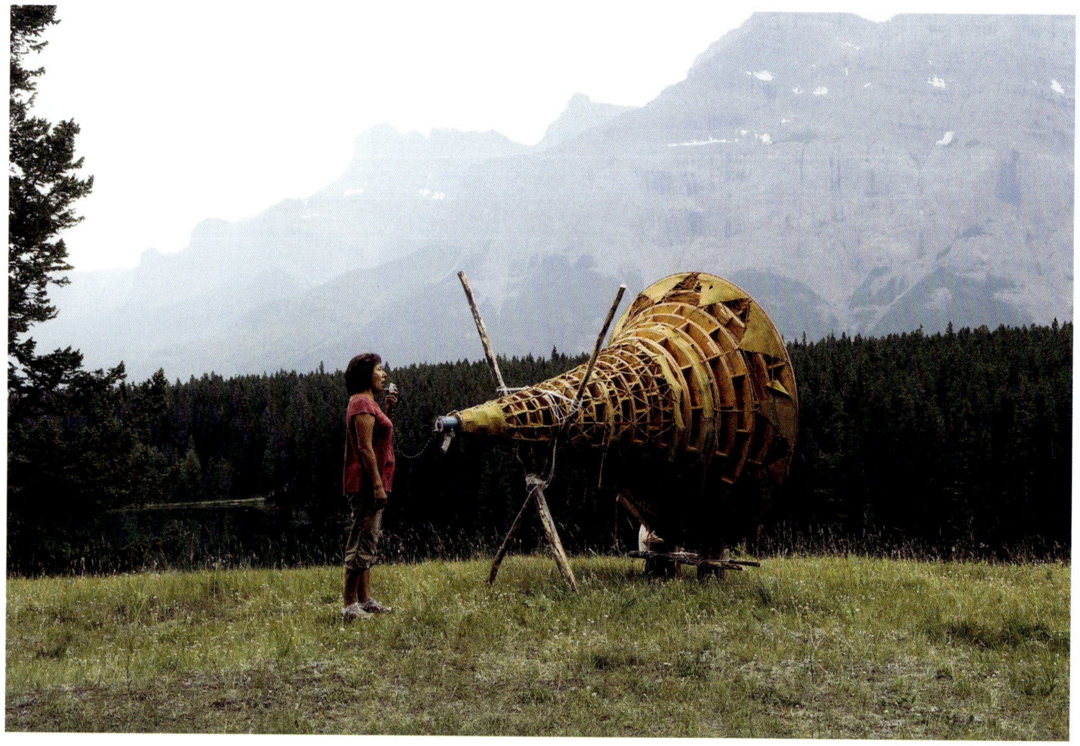

FIG. 6 Rebecca Belmore, *Ayum-ee-aawach Oomama-mowan: Speaking to Their Mother*, 1991. Presented by the Walter Phillips Gallery as part of the exhibition *Bureau de Change*, July 12–September 28, 2008. Banff National Park, Johnsons Lake, 2008.

or at least perverting and subverting it, propels the sonic properties of the voice to the fore. The sound poetry of The Four Horsemen and Claude Gauvreau, for instance, furthered the European avant-garde tradition of exploring and exploiting glossolalic utterances through inventive scripts and powerful live performances.[14] Voice also features prominently and poignantly in Rafael Lozano-Hemmer's *Voz Alta* of 2008 and Rebecca Belmore's *Ayum-ee-aawach Oomama-mowan: Speaking to Their Mother* of 1991.(FIG. 6) Both works stage voice at the center of a collective enterprise performing a spectacle of testimony. Both use oversize megaphones to cast wide and loud. Both impress through their scale even before they are heard. Voice is intimately tied to territory and *self-determination*;[15] in myriad manifestations, it clamors claims through narrative or visceral outflowing, from Tanya Tagaq's melding of Inuit throat singing into hybridized soundscapes to Alexandre St-Onge's animalistic breaching of "xenocommunicative channels [via an] intensification of the molecularities of vocal utterance."[16] In Lozano-Hemmer's *Microphones* (2008), the circuitry of voice/ear, speaking/listening merges through the layering of contributions by gallery visitors within the selfsame iconic mics.(FIG. 7)

In the narrative-centric soundwalks of Janet Cardiff (and sometimes George Bures Miller), a similar *mise en abyme* unfolds, but the mobility injects a prosthetic extension to the experience: "The audio works as a time machine. Janet is beside/behind you talking to you, taking you with her into her universe."(FIG. 8), [17] The time machine is of course the recorder/playback device, what Sterne terms to be "an extension of ephemerality"[18] – which is itself an extension of Marshall McLuhan's foundational concept regarding the immersive role of media and technology.[19] The energy required to enable exponential extensionality is arguably destroying the nation, and beyond. The artistic activities of the nervous membrane, to paraphrase McLuhan, jacked on the power amplification affords, are compromised as

13
The Nihilist Spasm Band, *No Record*, LP, Allied Record, 1967 (reissued on CD, Alchemy Records, 1996).

14
The Four Horsemen (1972–1988) were Rafael Barreto-Rivera, Paul Dutton, Steve McCaffery, and bp Nichol. Ron Mann's 1982 film *Poetry in Motion* captured their vociferousness: www.youtube.com/watch?v=843O0bTVKHQ. Claude Gauvreau's 1970 appearance at *La Nuit de la poésie* (www.youtube.com/watch?v=5g5lXcFlwd0) had a resounding impact on Rober Racine, see his postcript "Étranges et Foudroyants," in Claude Gauvreau, *Étal Mixte et Autres Poèmes 1948–1970* (Montreal: Éditions de l'Hexagone, 1993), 239–43.

15
Phrase used by Jolene Rickard in "Rebecca Belmore: Performing Power," in *Rebecca Belmore: fountain*, ed. Cindy Richmond and Scott Watson, exhib. cat. (Kamloops and Vancouver: Kamloops Art Gallery/The Morris and Helen Belkin Art Gallery, 2005), 69.

16
Marc Couroux on St-Onge's *l'indécidable crowbar cosmogonique* (2010) from "PREEMPTIVE GLOSSARY FOR A TECHNOSONIC CONTROL SOCIETY (with lines of flight)," in *Volumes*, ed. Martin Arnold and Christof Migone (Mississauga: Blackwood Gallery, 2015), 61.

17
Janet Cardiff and George Bures Miller, "[untitled text]" in *Voices Voces Vox*, ed. Barbera van Kooij and Christopher Phillips, exhib. cat. (Rotterdam: Witte de With et al., 1998), 89.

18
Jonathan Sterne, "Lost Recordings," in *Traces*, ed. Nicole Gingras (Montreal: Leonard and Bina Ellen Gallery, 2006), 74.

19
Colette Urban's *Oh My* (2002), Jessica Thompson's *walking machine* (2003), Darsha Hewitt's *Personal Soundtrack Emitters* (2006) also expand from traditional soundwalks; in these works, they bypass recording in favor of mediated liveness – stuck in the here and now.

20
See Marshall McLuhan, *Essential McLuhan*, ed. Eric McLuhan and Peter Zingrone (Concord, Ontario: House of Anansi Press, 1995), 94. "Electricity is in effect an extension of the nervous system as a kind of global membrane."

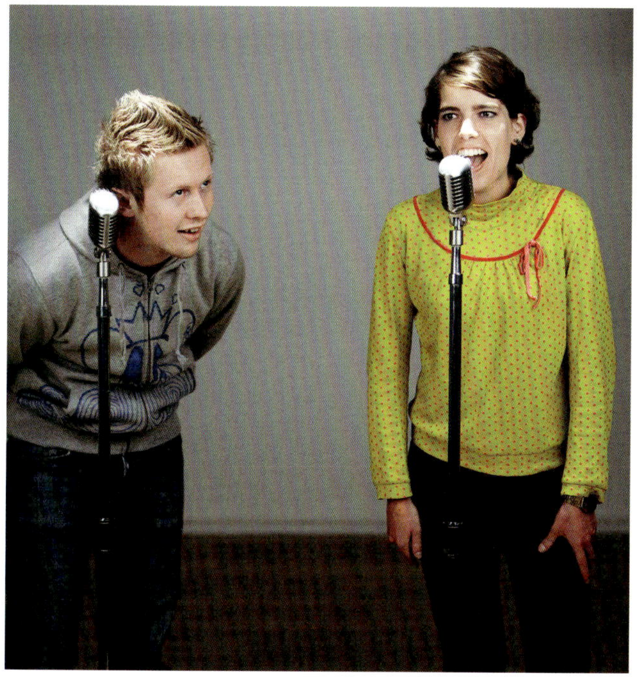

FIG. 7 Rafael Lozano-Hemmer, *Microphones*, 2008. Modified vintage microphones, electronics, computers and loudspeakers. Installation view Manchester Art Gallery, 2010.

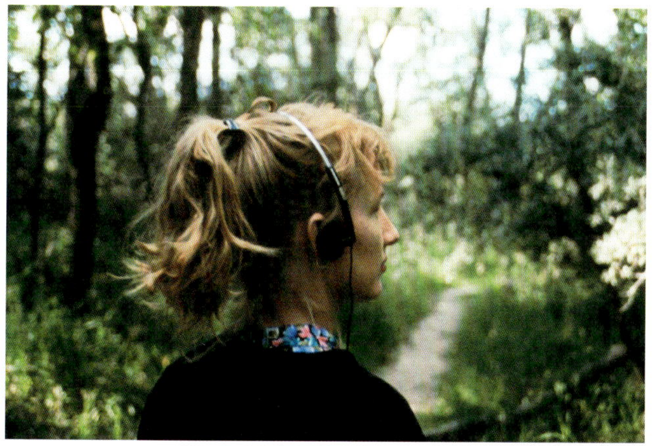

FIG. 8 Janet Cardiff, *Forest Walk*, 1991. Audio walk, 12 min. Banff Centre for the Arts, Canadian Artist in Residence Program.

soon as the power is switched on and the mouth approaches the microphone.[20]

From Coast to Coast to Coast (to Border) (Radiating Kinosonics)

Northrop Frye commented that Canadian culture is "obsessed with movement and transportation,"[21] and while the sheer expansive geography of the country lends itself to such an observation, it must be considered in tandem with the technological advents that have accelerated and amplified such preoccupations. A considerable number of artists make this manifest through work which occupies space via a tangible temporal skew. In other words, they kinetically invest space – be it within the white cube, black box, or in the streets. Jean-Pierre Gauthier's *THORAX* of 2010(FIGS. 9a, 9b) presents a breathing exo-skeletal singing body electric.[22] It all starts with air, and our breathing of it, everyday, all the time. Then it's what you do when you move about, how you interact with objects and with others. Diane Landry's neologism *mouvelle* fits aptly here for it combines (in French) temporality with movement and newness. The repurposing of everyday objects is a recurring method utilized by numerous artists to sound space:[23] Landry with keys, salad spinners, washing machines, kettles, shredders;(FIG. 10) Marla Hlady with teapots, toys, cocktail shakers, rice balls; Béchard/Hudon with paper, glasses, fishing poles, cones;(FIG. 11) Peter Flemming with garbage cans, bricks, buckets; Ken Gregory with fire alarms, kites, bottles. With *My Teeth Are in Slivers* (1994), *In Bocca al lupo* (1991), and *Taking It to the Teeth* (1993), Rita McKeough peeled the familiarity of domesticity to reveal the violent tensions that sometimes pervade the home – "sharp rhythm of breath, in and out, perforates, punctuates space."(FIG. 12), [24] Formally, a lot of these works straddle the interstice between *performed* installations and *installed* performances. A germane sound-art-historical example of this would be the project by Martin Messier and Nicolas Bernier (with Jonathan Villeneuve), *Machine_Variation* (2014),(FIGS. 13a, b) which resembles an updated and perverted *intonarumori* on a gigantic scale – played from the inside, it engulfs the performers who fittingly engage it with FM Einheit-like

21
Northrop Frye, "National Consciousness in Canadian Culture," in idem, *Divisions on a Ground* (Toronto: House of Anansi Press, 1982), 50.

22
A fuller discussion of this work in Christof Migone, *Sonic Somatic: Performances of the Unsound Body* (Berlin: Errant Bodies Press, 2012), 191–95.

23
Of course, sound and the everyday need not be entwined together nor necessarily linked to usage of household objects. Works which merit mentioning in a broader category of sonic spatiality include David Rokeby's pioneering *Very Nervous System* (1986–1990), and various works by Steve Heimbecker, Diana Burgoyne, Adam Basanta, Diane Morin, Robyn Moody, Caroline Gagné, Annie Martin, Thomas Bégin, Manon Labrecque.

24
Paula Levine, "Rita McKeough's *My Teeth Are in Slivers*: A Calling from the Other Side of Trauma," in Lander and Augaitis, *Radio Rethink*, 157. Anne Carson makes a pertinent declaration in her "The Gender of Sound" essay: "Every sound we make is a bit of autobiography" (in *Glass, Irony and God* [New York: New Directions, 1995], 130).

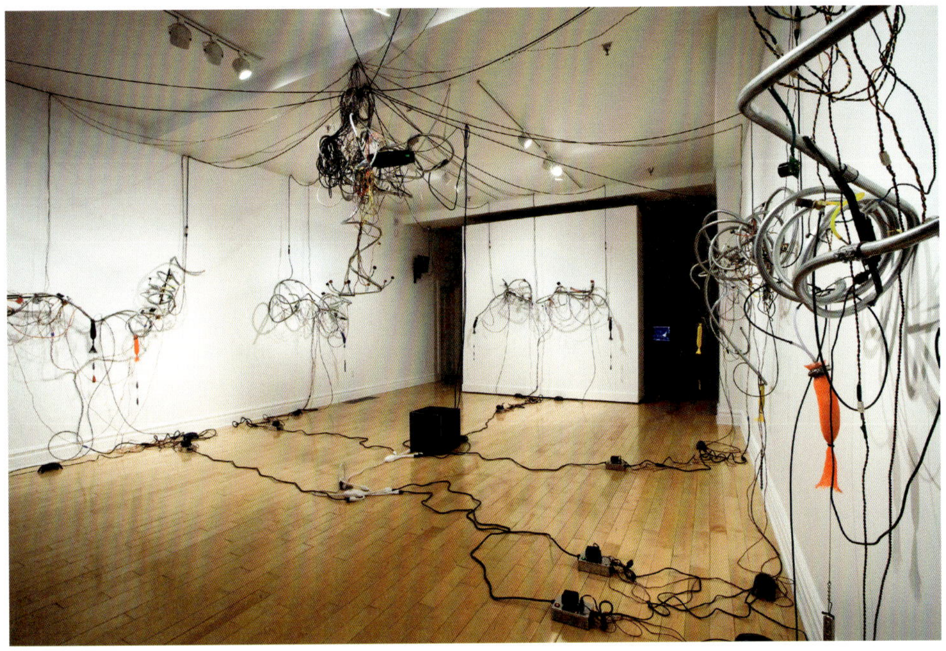

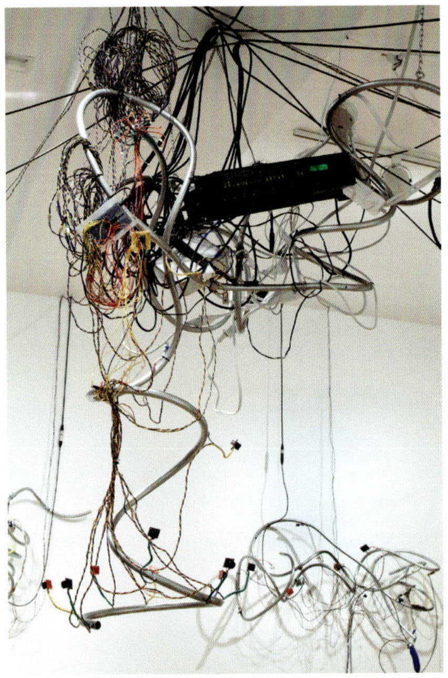

FIG. 9a Jean-Pierre Gauthier, *THORAX*, 2010, interactive sound installation. Installation view Action Art Actuel, Quebec.

FIG. 9b Jean-Pierre Gauthier, *THORAX*, 2010, interactive sound installation. Detail.

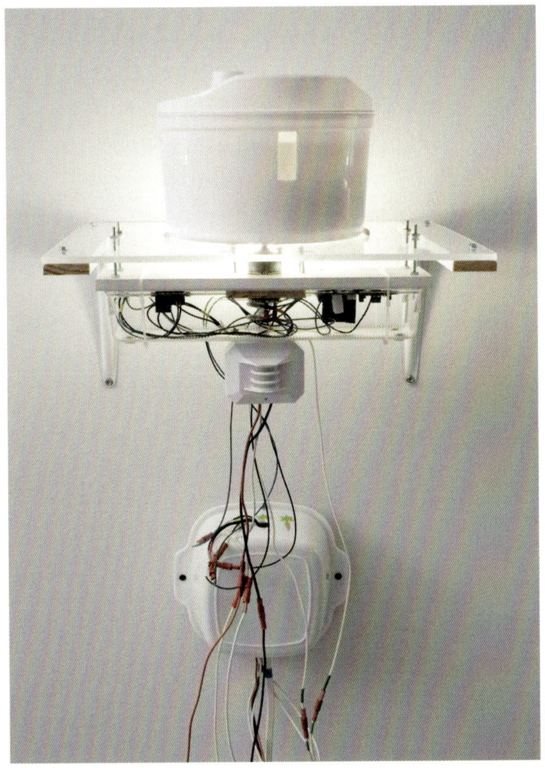

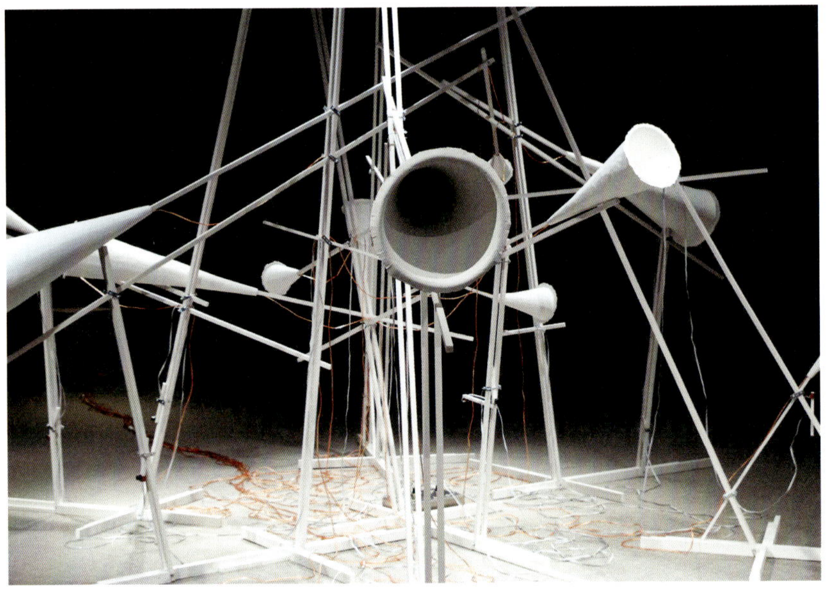

FIG. 11 Béchard/Hudon, *La circulation des fluides I / The Circulation of Fluids I*, 2008–2009, sound installation. Wood, paper, speakers, amplifiers, audio interface, infrared sensors, electronics, cables, computer, 5.18 × 6 × 2.5 m.

FIG. 10 Diane Landry, *Je ne trouve pas ma montre, elle ne s'est pourtant pas envolée / I can't find my watch, yet it hasn't flown away*, 2006, motorized installation. Salad spinners, selected objects, inkjet photographs, lights, timer switches, motion sensors, 6 objects: 90 × 36 × 36 cm (overall), total length: 536 cm.

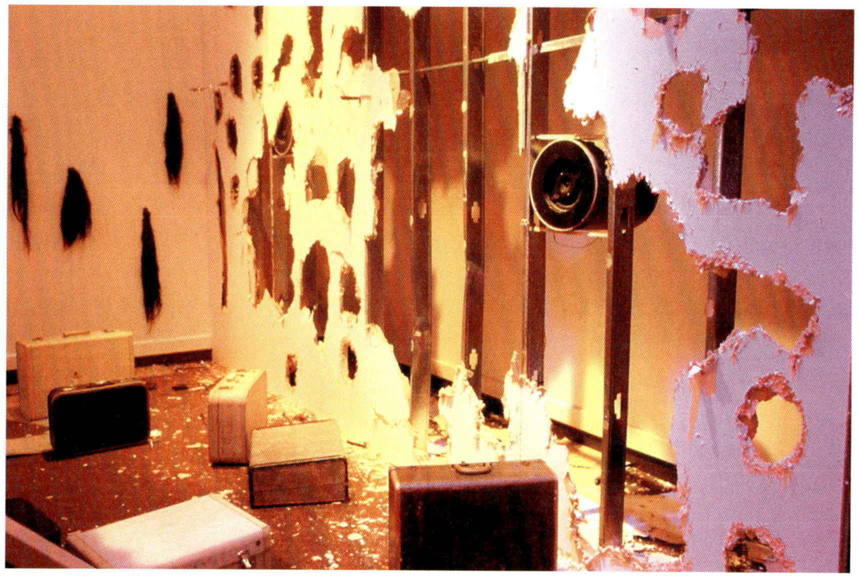

FIG. 12 Rita McKeough, *Taking It to the Teeth*, 1993. Installation/performance at Glenbow Museum, Calgary, January 1993.

FIGS. 13a, b Martin Messier and Nicolas Bernier, *Machine_Variation*, 2014. Sound performance, 30 min.

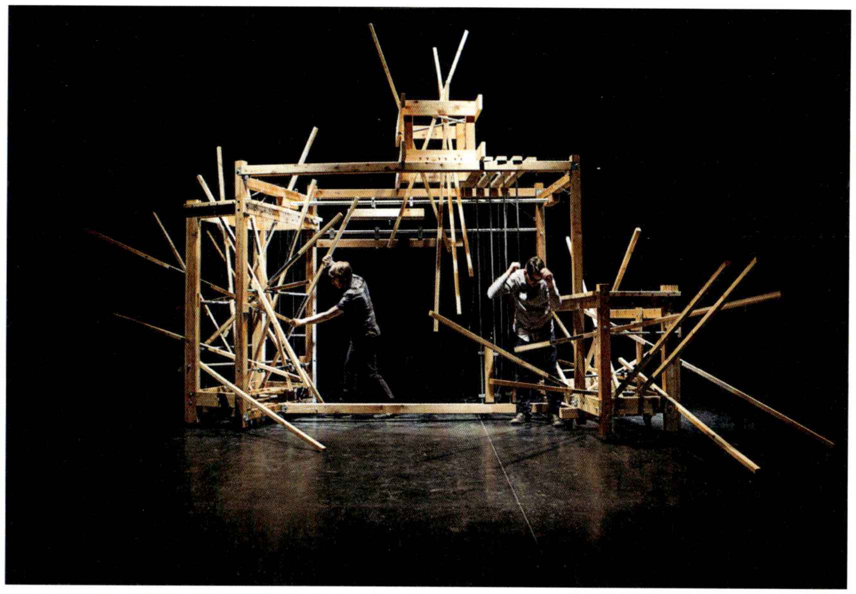

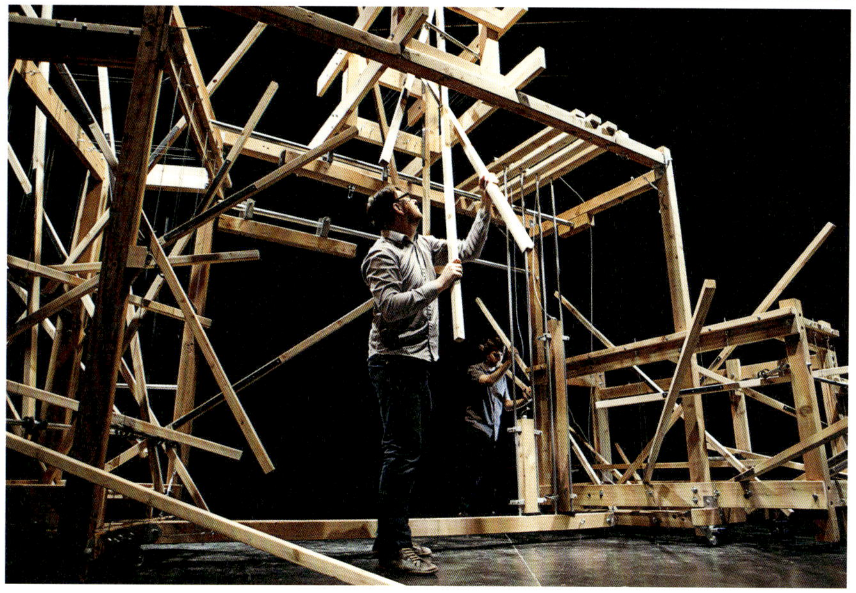

fervor. Sound is part and parcel of this equivocal dance. One could posit that these strategies lie at the heart of the notion of *experiment* as articulated by Gordon Monahan in an interview regarding his seminal 1982 *Speaker Swinging*.(FIG. 14), [25] He advocates for "experiment" *tout court*, not as a qualifier of music. In that spirit and in continuity are the intersections currently explored by Sarah Peebles with entomology (apiaries in particular), Erin Sexton with alchemy and crystallography,(FIG. 15) and Philippe-Aubert Gauthier with acoustics from an engineer's perspective.

Radio involves the invisible kinetics of electromagnetic waves. Radio's currents also activate the sociopolitical realm with the impact of its presence and liveness in localities as well as networks. Hank Bull recalls a pronouncement by General Idea that "artists, like Hermit crabs, can invade the dead forms of mass media."[26] Radio is by no means the sole exhibition space for sound, but it is primed for it. Some may argue that it is its own distinct art form, but the entwinement with sound art is undeniable. However, there are specific tactics beyond mere use of radio as diffusion vehicle that artists such as Anna Friz, Peter Courtemanche, Steve Bates,(FIG. 16) Chantal Dumas, Jean-Pierre Aubé, Kristen Roos, Matt Smith, Emmanuel Madan, André Éric Létourneau, Eleanor King, and Martine Crispo regularly investigate. The Canadian radio art context continues to thrive; it should come as no surprise that it was at the center of Lander's activities from the mid-1980s to the mid-1990s (in addition to being active at Art Metropole – which was founded by the aforementioned General Idea in 1974).

25
Gordon Monahan, "Speaker Swinging," interview conducted by Robert Johnson in 1987, in Lander and Lexier, *Sound by Artists*, 141–45, here 145. "Rather than the genre of experimental music, *Speaker Swinging* deals with the reality of experiment."

26
Hank Bull in Candice Hopkins, "Hank Bull: From the Centre to the Periphery," in *Re-Inventing Radio: Aspects of Radio as Art*, ed. Heidi Grundmann et al. (Frankfurt am Main: Revolver, 2008), 220.

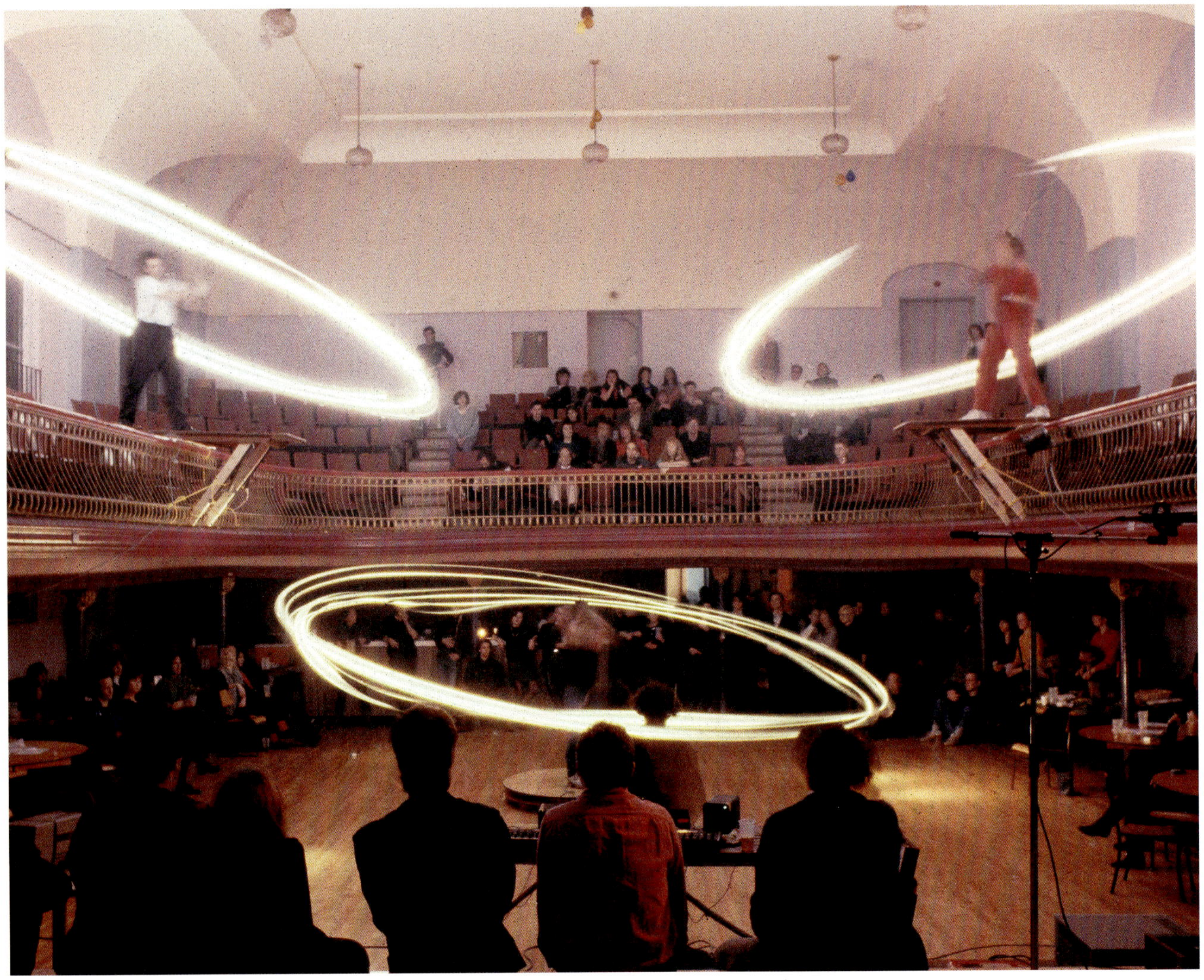

FIG. 14 Performance of Gordon Monahan's *Speaker Swinging* at the Music Gallery, Toronto, 1987.

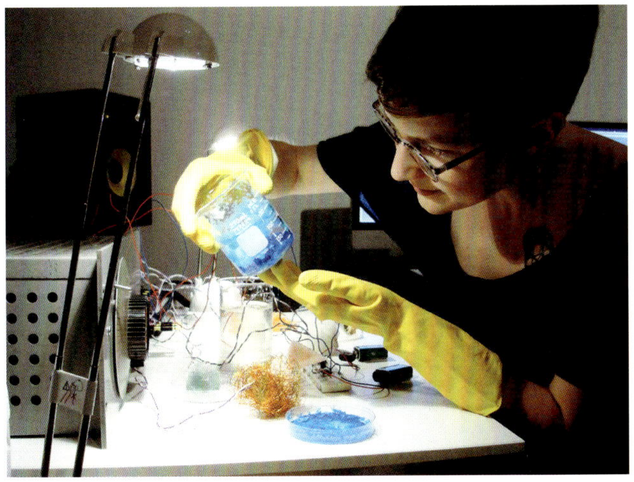

FIG. 15 Erin Sexton, *crystalline domain*, 2014, installation series.

Crisscrossed Ecologies

Experiments with broadcasting imply an institutional enabler - it is a governmentally regulated medium after all. The national broadcaster (CBC/Radio Canada) has played a role here, most notably by producing Gould's *Solitude Trilogy* (*The Idea of North*, *The Latecomers*, and *The Quiet in the Land*) between 1967 and 1977; on the French side, at Radio Canada, Mario Gauthier and Hélène Prévost in their respective programs did considerable work between the 1980s and early 2000s featuring sound and radio art, as well as experimental musics. However, support for experiments in sound have atrophied since and are now negligible. On the other hand, the country-wide network of community radio stations and key partners like the Banff Center, Western Front, and Avatar have been and continue to be fervent in their commitment to the artistic bending of the waves. The examination of the radio context prompts us to pay attention to the structural conditions that provide the means for sound art to thrive. Space is scant here, but at least an enumeration will identify some of the key players and give the reader a sense of the diversity of the scene. Festivals like Send+Receive (Winnipeg), Tone Deaf (Kingston), Sounds Like (Saskatoon), Deep Wireless (Toronto), Sound Bytes (Halifax), Open Ears (Kitchener), Sound Thinking/Open Sound (Vancouver), Mois Multi (Quebec City), FIMAV (Victoriaville), Mutek, Suoni per Il Popolo, Elektra/BIAS, and Sight+Sound (Montreal for the last four) provide much-needed cross-pollination platforms for local, national, and international artists. Artist-run centers and collectives are the backbone of the arts in general in Canada, those that have a particular (but not exclusive) focus on sound include the Music Gallery, Avatar, Western Front, Studio XX, Perte de Signal, Undefine, Centre for Art Tapes, Open Space, and Oboro. A number of curators (artists themselves in many of the following cases) have a predilection for

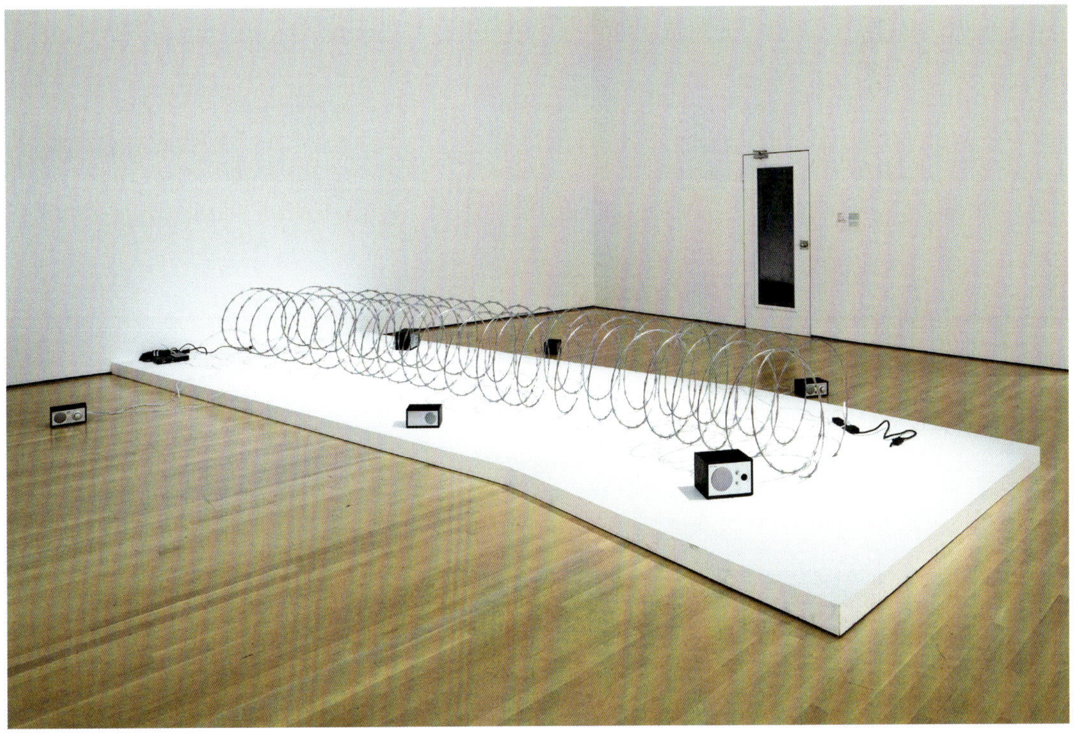

FIG. 16 Steve Bates, *Concertina*, 2011, barbed wire installation.

sound, including Nicole Gingras, Eric Mattson, France Jobin, Jim Drobnick, crys cole, jake moore, and Paul Walde. There is a notable concentration of sound art activities in the province of Quebec (and Montreal especially) that is reflected throughout this text: it is due to comparatively greater funding support by government agencies, but only in part – the reasons for this magnet-effect are manifold (and thus beyond the scope of this text).

Unsound Theories

"It is the listening not the sound that is rendered in space. It is the gaze which plays the music, in silence."[27] The reversals in this quote encapsulate why Raymond Gervais operates as an apt éminence grise to track the evolution of sound art's conceptual or non-cochlear dimension in Canada. For Gervais the incipient moment for sound's conceptual turn can be traced all the way back to 1901 on Signal Hill in St. John's, Newfoundland, when Marconi was waiting to receive the wireless signal of the letter "S" in Morse code. The fortuitous sharing of that letter by the words "sound" and "silence" offers Gervais an endlessly rich set of paradoxes to mine.(FIGS. 17a, b), [28] It is in the referential capacities of images and objects that he stages this interplay: "works that would relate to sound, but without sound."[29] This shift to the sonic potential of images seems a different type of investigation than the sound-centric video work done by Le Révélateur, Skoltz_Kolgen, Tasman Richardson, Nelson Henricks,(FIG. 18) Ian Birse and Laura Cavanaugh, Gary James Joynes, Pascal Grandmaison, and Frédéric Lavoie, among many others.[30] Although, since Gervais precedes all of these generationally, it could be tempting to ascribe a lineage or at least a connection. But the point is more to highlight the breadth of approaches, and to be sure, conceptual strategies populate contemporary practices in countless ways. With Gervais and interlocutors of his like Rober Racine, whose constellation of preoccupations consists of "silence, retreat, solitude, creation, meditation, infinity, night, time and invisible,"[31] the approach is resolutely lowercase.

Lowercase suggests an apposite tone by which to describe the general constitution of sound art in Canada, especially in terms of its international profile. Without question, Canadian sound artists rove the globe constantly and garner deserving praises and successes (some have stayed expatriates like Robin Minard and John Wynne). And internally, sound art in Canada has reached an impressive maturity and diversity that is reflected in the sheer quantity of those named here – or left unnamed for lack of space.[32] But perhaps borders, be they territorial or categorial, miss the mark: they counterproductively delineate where no such containment is needed. Perhaps through the lowercase effacement and erasure lies an uncharted expanse – one that we are fully immersed and invested in already. A plural space that echoes Jocelyn Robert's insight that "in audio art, sound is secondary."[33] The appropriate response, and pithy conclusion, here might therefore be, *in Canadian sound art, Canada is secondary*.

27
Raymond Gervais, *Via Charles Ives: La Symphonie Universe (via La Question sans réponse)* (Orford: Centre d'arts Orford, 2001), 14.

28
Raymond Gervais, "From sound to image" in *Raymond Gervais 3x1*, ed. Nicole Gingras, exhib. cat. (Montreal: Leonard and Bina Ellen Gallery/VOX, 2006), 59.

29
Ibid.

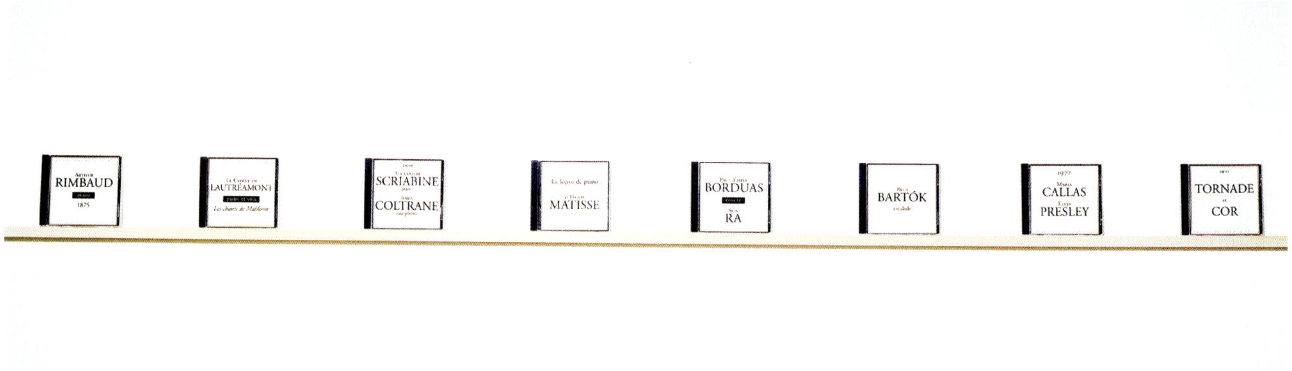

FIG. 17a Raymond Gervais, *The Theatre of Sound*, 1997. CD cases and text, 12.4 × 14.2 × 1 cm (each element). Installation view The Montreal Museum of Fine Arts, gift of the artist.

30
By the same token, one could track a number of sound-centric choreographers, dance companies, film and theater directors (as well as the sound artists who work extensively in those contexts). A mix of these would include Michel F. Côté, Marie Chouinard, Louis Dufort, La La La Human Steps, kondition pluriel, Ame Henderson, The Holy Body Tattoo, Jacob Wren, Lynda Gaudreau, David Kristian, Nancy Tobin, and Alexis O'Hara.

31
Rober Racine, "Everywhere, the Invisible," in Gingras, *Traces*, 57.

32
It bears mentioning that on the academic side, sound art as a distinct entity is comparatively late in coming to the fore, but its pseudo-parasitical presence in various communication, visual art, and music departments cultivates a fertile grounding in interdisciplinarity. That being said, both pedagogy and research (theoretical and practice-based) are active and growing. The work of Sterne at McGill (Mitchell Akiyama and Tim Hecker are notable recent PhD graduates of his), Jocelyn Robert at Laval, Christof Migone at Western, Jessica Thompson at Waterloo, Paul Walde at UVic, Matt Rogalsky at Queen's, Marla Hlady at UTSC, Lukas Pearse at NSCAD, Marcus Boon at York, Duncan MacDonald at Brock are just a few examples of where insertions of sound art into the academy are taking place. The newly minted Sonic Research Institute (Marc Couroux, David Cecchetto, Eldritch Priest) and its related Tuning Speculation conferences also provide a promising sign.

33
Jocelyn Robert in Christof Migone, "The Parachutist and His Piles," in *Jocelyn Robert*, CD-ROM (Québec: La Bande Vidéo, 2002), 8. The utterance has heightened paradoxical force knowing that it comes from one of the founders of Avatar and a prolific sound artist. Gregory Whitehead makes a remarkably similar statement: "Radio happens in sound, but sound is not really what matters about radio." (*Wireless Imagination: Sound, Radio, and the Avant-Garde*, ed. Douglas Kahn and Gregory Whitehead (Cambridge, MA: The MIT Press, 1992), 254).

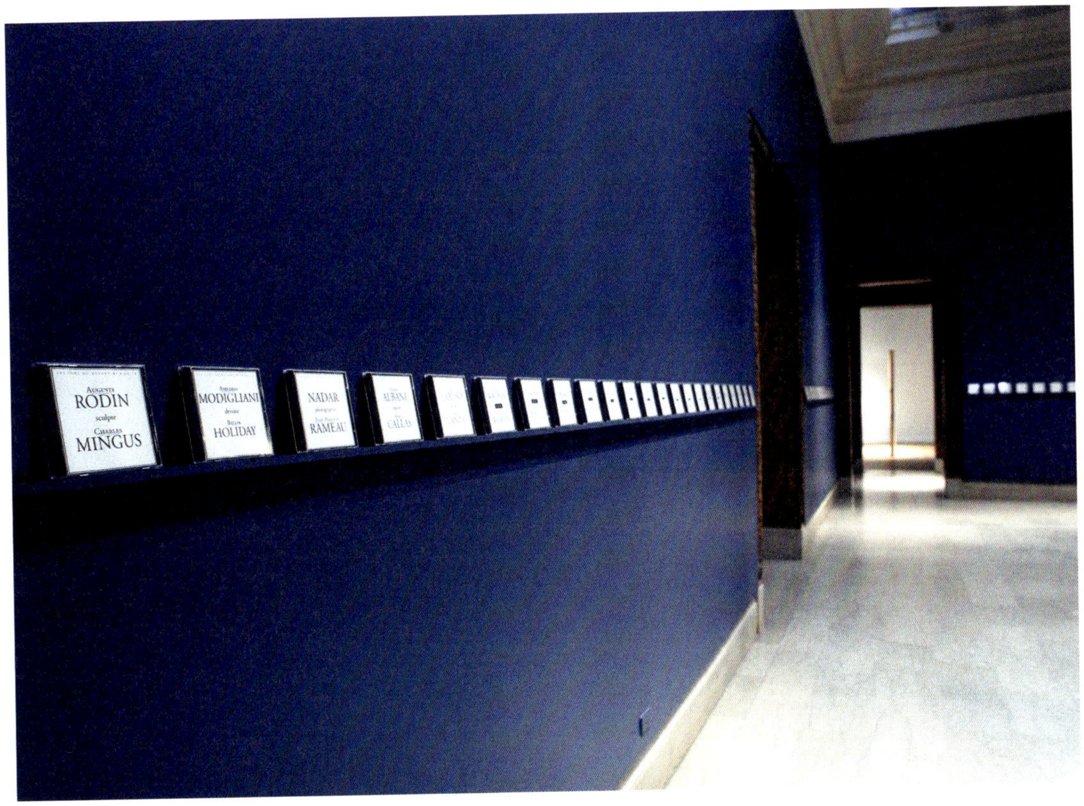

FIG. 17b Raymond Gervais, *The Theatre of Sound*, 1997. CD cases and text, 12.4 × 14.2 × 1 cm (each element). Detail. The Montreal Museum of Fine Arts, gift of the artist.

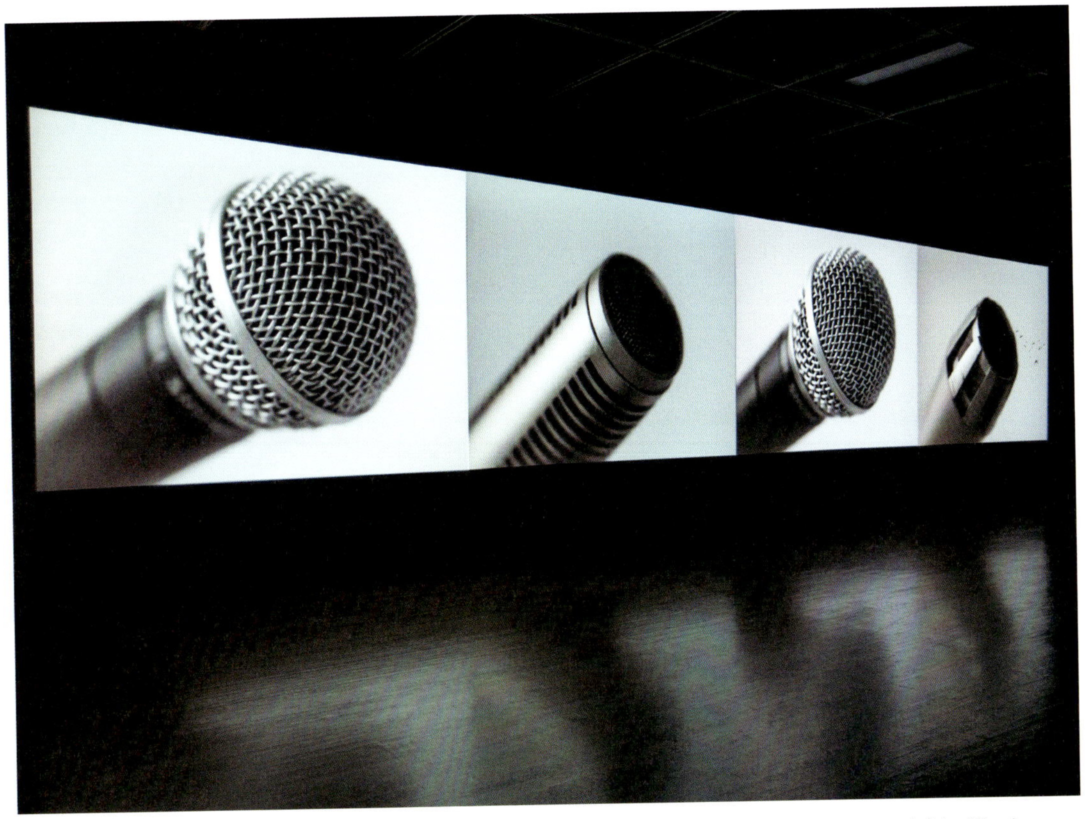

FIG. 18 Nelson Henricks, *Unwriting*, 2010. Four-channel video installation, dimensions variable, 12 min.

Giuliano Obici

Brazilian Sound Art from a Gambiarra Perspective

This paper presents Brazilian sound art from the perspective of the *gambiarra* – a popular Portuguese expression that indicates an improvised and informal way of solving everyday problems without having the appropriate resources. It is divided into two parts. First, it briefly describes the scenario of Brazilian sound art over the last fifteen years. Second, it introduces the notion of *gambiarra*, as a way to think about relevant aspects of the sound art universe, as well as looking at experimental music as a technological mediation.

Translated from the Brazilian by John Laudenberger.

A Brief Contextualization

The production of sound art has been growing in Brazil over the last two decades. This has been perceived, for example, through the art biennials that began to include a greater number of sound works with the use of radiophonic diffusion. *Rádio Visual* at the 7th Mercosul Biennial in Porto Alegre (2009), *Mobile Radio* at the 30th São Paulo Biennial (2012), and *Rádio Documenta 14* (2017), which included a program with Brazilian artists.

In addition to biennials, much of sound art production has been circulating through integrated art festivals: Video-brasil (São Paulo), Multiplicidade (Rio de Janeiro), FILE (São Paulo), FAD (Belo Horizonte), On-Off (São Paulo); contemporary and experimental music festivals: International Music and Sound Art Encounter – EIMAS (Juiz de Fora), Novas Frequências (Rio de Janeiro), Hipersônica (São Paulo), International Festival of Experimental Music (São Paulo), OVO (Porto Alegre); and independent spaces: Audio Rebel and the extinct Plan B in Rio de Janeiro, Ibrasotope, Fita Crepe and Da Haus in São Paulo, to name a few.

Since the beginning of the 2000s, there have been events and exhibitions dedicated to sound art. *Instalasom* (Capacete, Rio de Janeiro, 2003), *Paisagens Plásticas e Sonoras* (SESC Pinheiros, São Paulo, 2005), *Arte e música* (Caixa Cultural, Rio de Janeiro, 2008), *Sound Art* series (Lage Park, Rio de Janeiro, 2010), *Sound Art Occupation* (Castelinho, Rio de Janeiro, 2015), *Sound Cartography* (Espaço do Conhecimento, Belo Horizonte, 2016), *Sound Art* course (Lage Park, Rio de Janeiro, 2017); as well as exhibitions by artists such as *Smetak Imprevisto* (MAM, São Paulo, 2008) and *Guilherme Vaz: A Fraction of the Infinite* (CCB, Rio de Janeiro, 2016) are some. At the same time, several active sound artists have performed in galleries and cultural centers.[1]

The Solar Ferrão public museum in Salvador that houses Walter Smetak's work and the private Inhotim Institute in Minas Gerais, with installations like *Forty part motet* (2001) by Janet Cardiff and Doug Aitken's *Sonic Pavilion* (2009), are two exceptions of spaces that permanently exhibit sound art works.

With regard to the literature on the subject, a number of academic texts are worth mentioning, such as "Return to the Future: Smetak and his Sonorous Plastic" from 2001 that resulted in the book *Walter Smetak: The Alchemist of Sounds* by Marco Scarassatti (2008); Lilían Campesato's "A Metamorphosis of the Muses: Referential and contextual aspects in sound art," *Organised Sound* 14, 1 (2009); Felipe Vaz's "Elementos da arte Sonora" [Elements of Sonorous Art] (2008); Ricardo Cutz Gaudenzi's PhD dissertation "Arte Sonora: Entre e Plasticidade e a Sonoridade" ("Sound Art: Between Plasticity and Sonority," 2008); Vivian Caccuri's book *O que Faço é Música* (*What I Make Is Music*, 2013); my own doctoral dissertation *Gambiarra e experimentalismo sonoro* (Gambiarra and Sound Experimentalism, 2014); the article by Constanza Ipinza Olatte and Aldo Hidalgo Hermosilla, "Cartografías sonoras: Instrumento disciplinar para pensar-experimentar el espacio," *Revista Planeo* 30 (2016); and the compilation edited by Lilian Zaremba, *Rádio Nova: Constelações na Radiofonia Contemporânea* (*New Radio, Contemporary Radio Constellations*, 1999). Likewise, we should also take note of the following books and exhibitions: *Simbologia dos instrumentos* [Symbolism of the Instruments] (Walter Smetak, 2001 [1980], Salvador: Associacão Amigos de W. Smetak); *Guilherme Vaz: uma Fração do Infinito* (*A Fraction of the Infinite*), curated by Franz Manata (2016); *Paisagens Plásticas e Sonoras* (*Visual and Sound Landscapes*), curated by Marco Scarassatti (2005); *Arte e música*, curated by Luiza Duarte and Marisa Flórido (2008); and *Smetak Imprevisto*, authored by Arto Lindsay and Jasmin Pinho (2008). And lastly, we cannot overlook the production of the artists' groups Chelpa Ferro and O Grivo,[2] as well as several valuable online platforms: the *Arte Sonora* podcast,[3] which collects interviews with artists and researchers in a radio format and *Nendú*,[4] which has created a cartography of local production between 2016 and 2017.

Having done this brief contextualization, we will now analyze aspects of sound art from a gambiarra perspective.

1 Chelpa Ferro, O Grivo (Nelson Soares and Marcos Moreira), Paulo Vivacqua, Vivian Caccuri, Paulo Nenflidio, Floriano Romano, Chiara Banfi, Cadu, Lilian Zaremba, Ricardo Siri, to cite a few.

2 Different catalogues can be found at http://www.ogrivo.com.

3 See http://exst.net/artesonora/. Initiated by Franz Manata and Saulo Laudares in 2009.

4 See http://www.nendu.net. Nendú "A possible archive (realized between 2016–2017) of brazilian sound art" is a postdoctoral project by Rui Chaves. It was made possible by a grant from FAPESP (São Paulo Research Foundation) and based at NuSom – Research Centre on Sonology (University of São Paulo).

About Gambiarra

Gambiarra is a popular Brazilian expression that describes an improvised and informal way of solving an everyday problem when the necessary tools or resources are not available. The expression *gambiarra*, as applied to art, broadly describes an improvisational method of working with materials, devices, technology, and/or institutions within the local art scene in Brazil.[5] The term encapsulates productions from visual arts to music,[6] from sound art to media art and activism.[7] It includes diverse practices and trends, such as DIY (do-it-yourself), instrument sound design, circuit bending,[8] hardware hacking,[9] dirty electronics,[10] cracked media,[11] *povera* technology,[12] residualism,[13] and technographic marking.[14] Later in this article, we look at the application of *gambiarra* to sound art, but first, we examine *gambiarra*'s genealogy and its global and local contexts.

Gambiarra's Genealogy

Originally (1881), *gambiarra* referred to string lights, or extension wire with attached lights and the illegal installation of electrical cables.(FIG. 1) [15] Popularly, the word *gambiarra* varyingly means to fit, to fix, to repair, to mend, to adapt, to improvise, or to assemble; it can also refer to a handyman, to patchwork, to tricks, or to DIY. It may describe an unconventional approach to problems involving inventiveness, intelligence, creativity, and the ability to come up with immediate, unexpected, or alternative solutions to problems; an uncommon, unusual practice; vernacular, autochthonous, popular art; the act of taking advantage of a situation; or an irregular, illicit, dishonest, marginal, illegal, or fraudulent custom.[16]

Gambiarra twists industrial design logic, establishing short circuits between a product's form and functionality. In principle, it emerges from an existing design, but, depending on the degree of interference, it can also result in a new design object.

Gambiarra carries distinct local characteristics relevant within Brazilian artistic movements such as the 1920s modernist Anthropophagic movement[17] and Tropicalism in the late 1960s.[18] Cultural portrayals of local culture commonly connected to the *gambiarra* approach include

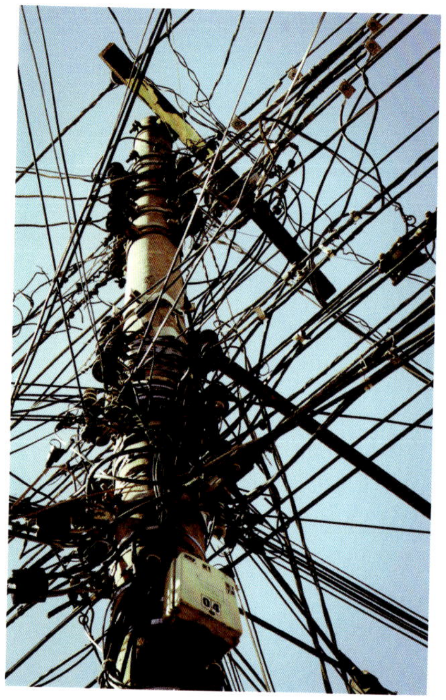

FIG. 1 Electric cables installed illegally in a utility pole also called gambiarra.

jeitinho, *malandragem*, and *carnival*.[19]

In the global context, *gambiarra* is close to several other common expressions[20]: *Revolico* and *rikimbili* in Cuba; *rasquache* in Mexico; *chapuza* or *arreglo temporal* in Uruguay; *solución parche* in Chile; *arreglo hechizo* or *reparación hechiza* in Colombia; *desenrascar* in Portugal; *jugaad* in India, Pakistan, and some African countries; *hacking*, *kludging*, *jury-rigging*, *workarounds*, *do-it-yourself*, and *makeshift* in English; or even *bricolage*, from the logic that "something can always be used for something."[21]

Technology Without Edges

The diverse practices represented by these terms reveal a global culture based on the benefits, but also the limitations, of techno-products as residual effects of product design. This paradoxical gap between the advances and limitations of techno-consumption is described by cyberpunk

5 See Giuliano Obici, *Gambiarra e experimentalismo sonoro* (PhD diss., São Paulo: ECA-USP and Berlin: Technische Universität, 2014).

6 Lisette Lagnado, "O malabarista e a gambiarra," *Trópico: Revista de Cultura e Turismo* 3 (2003).

7 Ricardo Rosas, "The Gambiarra: Considerations on a Recombinatory Technology," *SESC - Caderno Video Brasil* 2 (2005): 36–53.

8 See Reed Ghazala, *Circuit-Bending: Build Your Own Alien Instruments* (Indianapolis, IN: John Wiley & Sons, 2005).

9 Nicolas Collins, *Handmade Electronic Music: The Art of Hardware Hacking* (London: Taylor & Francis, 2006).

10 John Richards, "Getting the Hands Dirty," *Leonardo Music Journal* 18 (2008): 25–31, here 17.

11 Caleb Kelly, *Cracked Media: The Sound of Malfunction* (Cambridge, MA: MIT Press, 2009).

12 Erkki Huhtamo, "Thinkering with Media: On the Art of Paul Demarinis," in *Paul DeMarinis/Buried in Noise* , ed. Ingrid Beirer, Carsten Seiffarth, and Sabine Himmelsbach (Heidelberg: Kehrer, 2010): 33–46.

13 Kim Cascone, "Residualism," in *Sound*, ed. Caleb Kelly, Whitechapel: Documents of Contemporary Art (Cambridge, MA: MIT Press, 2011).

14 Rodolfo Caesar, "O loop como promessa de eternidade," *Anais do XVIII Congresso da ANPPOM*, Salvador (2008): 286–290

15 Antônio Houaiss, *Dicionário eletrônico houaiss da língua portuguesa* (Rio de Janeiro: Ed. Objetiva, 2001).

16 Rodrigo N. Boufleur, *Fundamentos da gambiarra: a improvisação utilitária contemporânea e seu contexto socio-econômico*, unpublished doctoral diss. (São Paulo, Universidade de São Paulo, 2013).

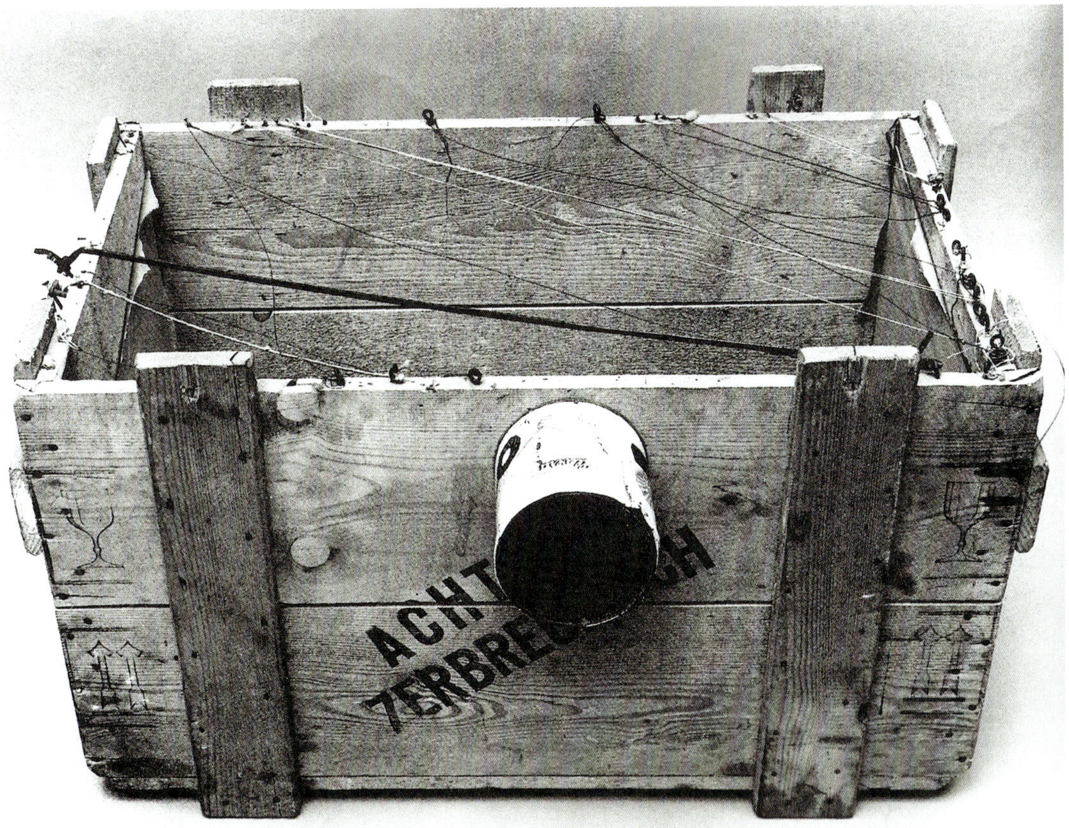

FIG. 2 Nam June Paik, *Urmusik* [Primeval music], 1961. Wooden box, wire, can, and various materials.

writer William Gibson in terms of high technology and precarious life, "high tech and low life";[22] likewise, Nam June Paik regards anti-technological technology as an "electronic humanization"[23], or even a search for the primordial music.(FIG. 2)

Neither low nor high technology, *gambiarra* is closer to being a "technology without edges" that combines archaic techniques with modern machines.[24] Beyond identifying creativity and invention as ways to overcome restrictive conditions, *gambiarra* includes and subverts polarized logic and discourse. It approximates a "geometry of life," as the sound artist Guilherme Vaz suggests, based on "Indian geometry, freehand made, reached a certain category. Within this geometry there is a philosophy, a world thinking. It is a geometry of life. […] which accepts curves. […] in a different type of order, without exclusion. The curve is social and inclusive."[25]

Sound and Gambiarra

Gambiarra in music lies between "playing within the rules" and "playing to subvert the rules" of a musical system. It's similar to hacking. As Richard Stallman, founder of the free software movement, wrote: the "1950s musical piece by John Cage, 4'33" [which has no notes] is more of a hack than a musical composition."[26]

When lack of resources and precariousness are a work's foundational elements, the need for adaptation and repair – as well as the risk of failure, glitches, and cracking – is high. This implies an openness to error and unplanned occurrences – conditions that permeate cracked media, circuit bending, or hardware hacking. Likewise, "dirty electronics" emphasize the exploration of invented instruments and prioritize gesture and social interaction.[27]

17
Based on the practice of cannibalism by Brazilian aborigines, the anthropophagic movement defended a plural culture that aims to digest everything that comes from outside, as proposed by the poet Oswald de Andrade. See Oswald de Andrade, "Manifesto antropofágico: em piratininga ano 374 da deglutição do bispo sardinha," *Revista de Antropofagia* 1, 1 (May 1928).

18
Tropicalismo exposed the contradictions of modernity in Brazil: the optimism generated by the capitalist development and strong urbanization of the 1950s and 1960s (including the capital Brasília); on the other hand, its position of underdevelopment and economic disadvantage in relation to Europe and North America. See Flora Sussekind, "Chorus, Contraries, Masses: The Tropicalist Experience and Brazil in the Late Sixties," in *Tropicália: A Revolution in Brazilian Culture (1967–1972)*, ed. Carlos Basualdo (São Paulo: Cosac Naify, 2005): 31–56, here 31.

19
For more information on these aspects, see Guiliano Obici, "Gambioluthiery: Revisiting the Musical Instrument from a Bricolage Perspective," *Leonardo Music Journal* 27 (December 2017), 87–92, and Rosas, "The Gambiarra: Considerations on a Recombinatory Technology," 36–53.

20
See Obici 2014.

21
Claude Lévi-Strauss, *The Savage Mind* (Chicago: University of Chicago Press, 1966 [1962]).

22
"The overall feel of Gibson's cyberspace works with its low-life ambiance." Quoted from David Ketterer, *Canadian Science Fiction and Fantasy* (Bloomington, IN: Indiana University Press, 1992), 143.

23
"Ich möchte die Elektronik humanistischer, bewusster für die Problematik der darzustellenden Realität und diese sichtbarer machen. Ich liebe die anti-technische Technologie." ["I would like to make the electronics more humanistic, more aware of the problem of the reality to be presented, and to make it more visible. I love the anti-technical technology."] Nam June Paik quoted in Wulf Herzogenrath, *Nam June Paik: Fluxus, Video*(Esslingen: Schreiber, 1983): 116.

"Dirt" and "lack" here refer not to substandard quality but rather to the practice of contradicting technologies, supposedly universalizing character as a way to reveal aspects of the dream that lie hidden in technology. Along similar lines, Paul DeMarinis has proposed rebuilding obsolete and absurd technologies,[28] a proposal that resembles *gambiarra*'s "technology without edges," or the carnivalization of design.

Gambioluthiery

A neologism formed by the prefix gambio- (from *gambiarra*) and luthiery (the construction of musical instruments), gambioluthiery refers to the construction of instruments oriented around the logic of *gambiarra*. It involves activities such as composing, decomposing, inventing, proposing, constructing, collecting, adapting, and appropriating materials, objects, artifacts, devices, instruments, or system setups.[29] As such, gambioluthiery intervenes between the form and the purpose of objects and devices, resulting in a new instrument, performance, intervention, action, music, sound file, installation, or sculpture.

Gambioluthiery works in a peripheral zone, with instruments that are not yet or no longer musical, between the audible and visible, the musical and sound arts, performance and installation, and even between musical project and sound design. By exploiting sound without necessarily drawing from the syntax of traditional musical instruments, its practice does not necessarily exist outside music but rather emerges from a tension within the concept of a musical instrument and its material context.

Reviewing the history of certain musical instruments, one notes the blurred lines that separate an everyday tool from its use in a musical context. Consider what unites and separates a blowpipe from a flute; the stretched rope of a bow and arrow from a monochord, lyre, or berimbau; a ceramic vessel from a percussion instrument; a calabash and *maracas* with seeds from a match box shaker[FIG. 3]; a wooden box as flamenco *cajon* or a grater as *guiro*, reco-reco. What distinguishes the mouth that eats, the mouth that speaks, and the mouth that sings? Gambioluthiery thus reconnects with this broad feature, in which the instrument overlaps with the utilitarian object.

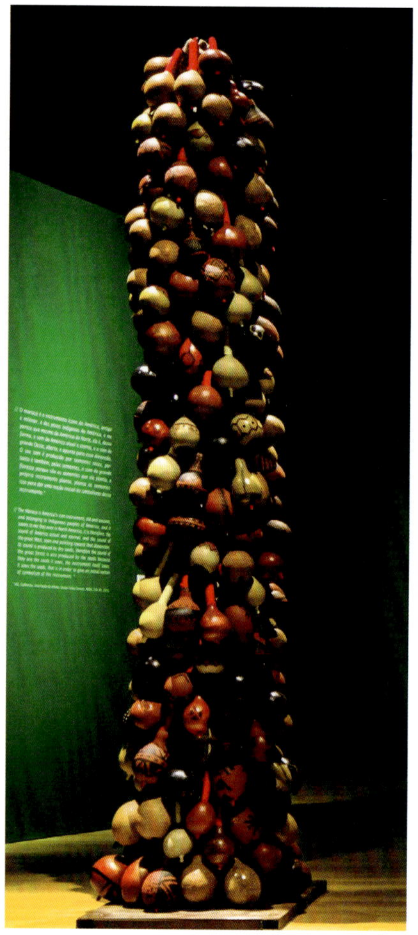

FIG. 3 Totem, 2015. Maracas (indigenous instruments).

Gambioluthiery reinforces connections between sound and its materiality as well as the paradoxical gaps between advantage and limitations that techno-consumption produces globally. While the term refers locally to a particular Brazilian repertoire,[30] - examples of which are described below - it enhances a general tendency of expanding the idea of the musical instrument through sound art.

Repertoire

The term *gambiarra* has always been a part of Brazilian musician and sound artist discourse.

24 As the composer Tato Taborda used to describe his multi-instrument: "All the technology used in *Geralda* has a *gambiarra* way of being. It's rounded. It's technology without edges." Unpublished interview between Tato Taborda and Guiliano Obici (2013). Personal archives.

25 Guilherme Vaz quoted in *Guilherme Vaz: uma fração do infinito*, exhib. cat. (Rio de Janeiro: EXST, 2016): 77–78.

26 Richard Stallman, "On Hacking," blogpost, http://stallman.org/articles/on-hacking.html (2002).

27 John Richards, "Getting the Hands Dirty," *Leonardo Music Journal* 18 (2008): 25–31.

28 Beirer, Seiffarth, and Himmelsbach, *Paul DeMarinis/Buried in Noise*.

29 Despite luthiery's origin referring to stringed instruments, here it describes makers of instruments and/or sonorous artifacts in general.

30 Obici 2014.

31 Franz Manata and Saulo Laudares, "Chelpa Ferro," podcast *Arte Sonora*, available at http://exst.net/artesonora/podcasts/pod-06-chelpa-ferro/2013.

32 Fred Paulino, Gambiólogos: a gambiarra nos tempos do Digital. (Belo Horizonte: Kludgist, 2010). Available at: https://issuu.com/gambiologia/docs/catalogo_gambiologos_web.

33 Guiliano Obici and Alexandre Sperandéo Fenerich, Jardim das gambiarras chinesas: uma prática de montagem musical e bricolagem tecnológica. In: Daniel Quarante and Alexandre S. Fenerich (eds), *10 olhares sobre a música de hoje: seleção de artigos apresentados no eimas* de 2010 a 2013. Juiz de Fora: Garzia Edizioni, 2014.

34 Lagnado, "O malabarista e a gambiarra."

35 Rosas, "The Gambiarra: Considerations on a Recombinatory Technology."

36
For a detailed view on Walter Smetak, see the contribution by Julia Gerlach in this volume.

37
Vivian Caccuri, V. (2011). Ouvindo as artes visuais: sonoridades de Waltercio Caldas, Cildo Meireles, Chelpa Ferro e Hélio Oiticica" (Master diss.: Universidade Federal do Rio de Janeiro, 2011): 58.

Chelpa Ferro describes *gambiarra* instruments as "half-adapted, constructed in a way [... and] the sound is also half-limited, half-raw";[31] Paulo Nenflidio describes *gambiarra* as a process he used in creating his piece *Polvo* (2010) "to modify an original function of these materials"[32]; similarly, En Minus One (n-1) explains that "adapting, hacking, building, programming, improvising, pirating, and appropriating were all part of the process" of their work.[33] In addition, *gambiarra* was present in the local discourse of visual and media art, where it was pointed out that the interest in "establishing relations with a political and aesthetic accent"[34] was a feature of its technological culture.[35]

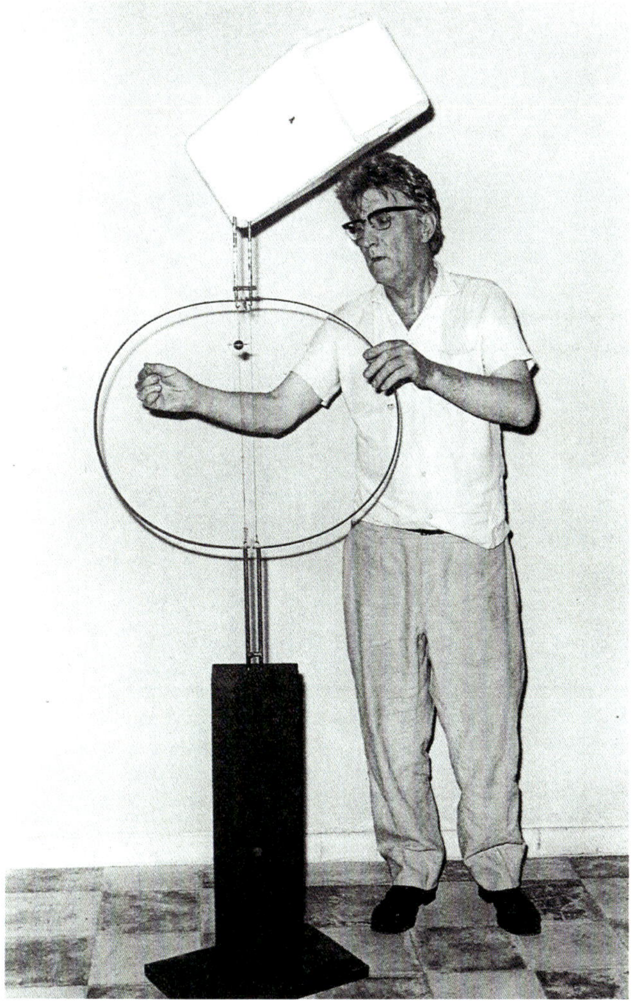

FIG. 4 Walter Smetak, *Disco Voador* [Flying Disk], 1974. Wooden base, metal rim, strings, metal disc, two metal rods, styrofoam box as resonator.

A precursor who traced the route toward gambioluthiery is Walter Smetak. Several of Smetak's instruments can be related to *gambiarra* by his use of available materials: *Piston Cretino* (1976), comprising an aluminum kitchen funnel, a plastic hose, and a piston nozzle; *Disco Voador* (FIG. 4); kinetic instruments like *Treis Sóis* (1971), made with wood, metal, Styrofoam, and PVC pipe; collective instruments such as *Pindorama* (1973), a 2.2-meter-high wind instrument made of gourds, plastic tubes, bamboo, PVC pipe, wood, and metal; and even silent sonorous plastics like *Caosonância* (1972), to name a few.[36]

Inspired by Cage's piece *Imaginary Landscape no. 4* (1951), Paulo Nenflidio (b. 1976) created *Decabráquio Radiofônico* to simultaneously play ten radios.(FIG. 5) As something between instrument and installation, Nenflidio's work is a hybrid ins(trumen)tallation, as the term gambioluthiery suggests. He adapts traditional and nontraditional instruments to the electric context, as in the case of *Berimbau Elétrico* (2003); to mechanisms of movement, as in *Bicicleta Maracatu* (2000); or in the use of existing devices: in electronic gadgets such as razors (solenoids); or in electromechanical mechanisms such as hammers or PVC pipe (*Teclado Sismico*, 2008). Nenflidio's works exist somewhere between "handmade contraptions" and "design products" – a realm where the work of inventor, luthier, and sculptor merge.

Chelpa Ferro's work *Samba* (FIG. 6) uses a table with a sewing machine on one side, a fishing reel on another, and a snare drum in the middle. The reel connects to the sewing machine by a fishing line that is stretched over it. When the

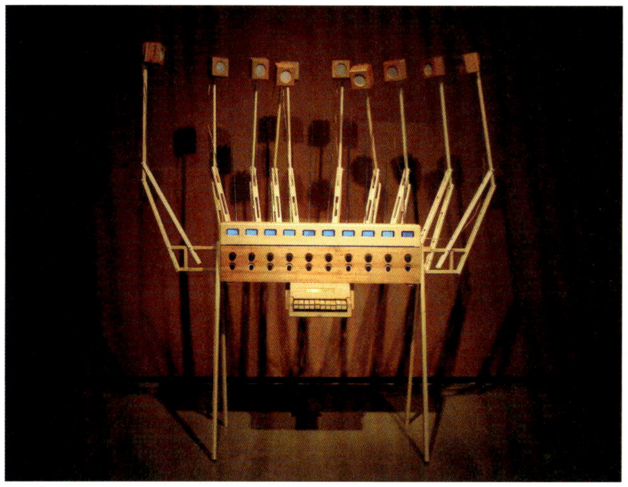

FIG. 5 Paulo Nenflidio, *Decabráquio Radiofônico* [Radio Decabration], 2006. Keyboard, wood, electronic circuit, radios, speakers.

machine is turned on, the line hits the snare drum, creating a rhythm similar to that of samba. Due to the line's malleability, the sewing machine motor produces a regular rhythm with occasional faults, creating a noisy oscillation that adds to the noise from the sewing action and spool gears which is then picked up by microphones and placed in each source (drum, reel, machine). The swing is not very convincing, because the source is a machine: "Samba is treacherous: a machine that wants to have swing (*ginga*), a kind of *gambiarra* that wants to be precise."[37] Some features of this work are related to gambioluthiery: assemblage, bricolage, trickery, sonorous visual installation, and the unstable procedural rhythm that juxtaposed mechanisms generate.

Geralda(FIG. 7) is a mix of multi-instrument and electroacoustic orchestra created by Tato Taborda (b. 1960), essentially a one-man-band instrument; the player activates more than seventy sound sources using hands, elbows, knees, head, and feet. The instrument was designed between 1992 and 1993 and continued to evolve over the following decade as Taborda added sound layers and both incorporated and abandoned materials, instruments, and devices. It emerged as an exclusive project based on instruments and acoustic objects, which were later microphoned (1988–1989) and, in 2001, incorporated live electronics. Over its trajectory, *Geralda* accumulated three layers of sound (acoustic, electric, and digital), suggesting a possible taxonomy for gambioluthiery. Several other works point to gambioluthiery. *Babel* (2001)(FIG. 8) by Cildo Meireles, which collects hundreds of radios in unison on different stations; *Music for paper sheet* and *Crude* (1973)(FIG. 9) by Guiherme Vaz, which uses minimal elements as part of the sound experience; *Bass and Guitar* (2012)(FIG. 10) by Chiara Banfi, which turns walls into a body of resonance with its disembodied instruments; the audiovisual meta-instruments by En Minus One and myself, such as the installational instruments from O Grivo(FIG. 11), to name a few. Proposals such as *Silent Walk* (2013) by Vivian Caccuri(FIG. 12) and *Sonorous Driftings* (2011) by Marco Scarassatti will point to other perspectives as we relate it to "listening as instrument."

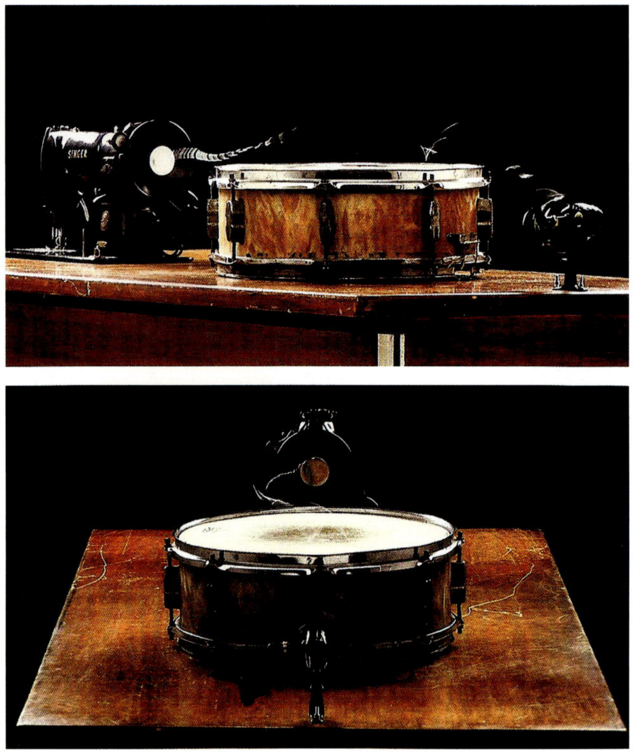

FIG. 6 Chelpa Ferro, *Samba*, 2009. Table, sewing machine, fishing reel and line, snare drum.

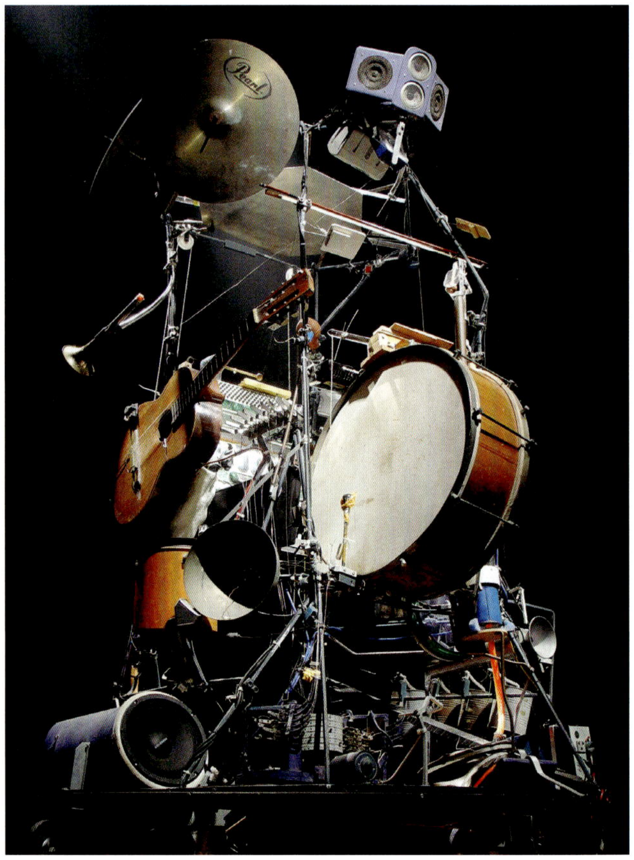

FIG. 7 Tato Taborda, *Geralda*, 1988–2001. Multiple sound sources.

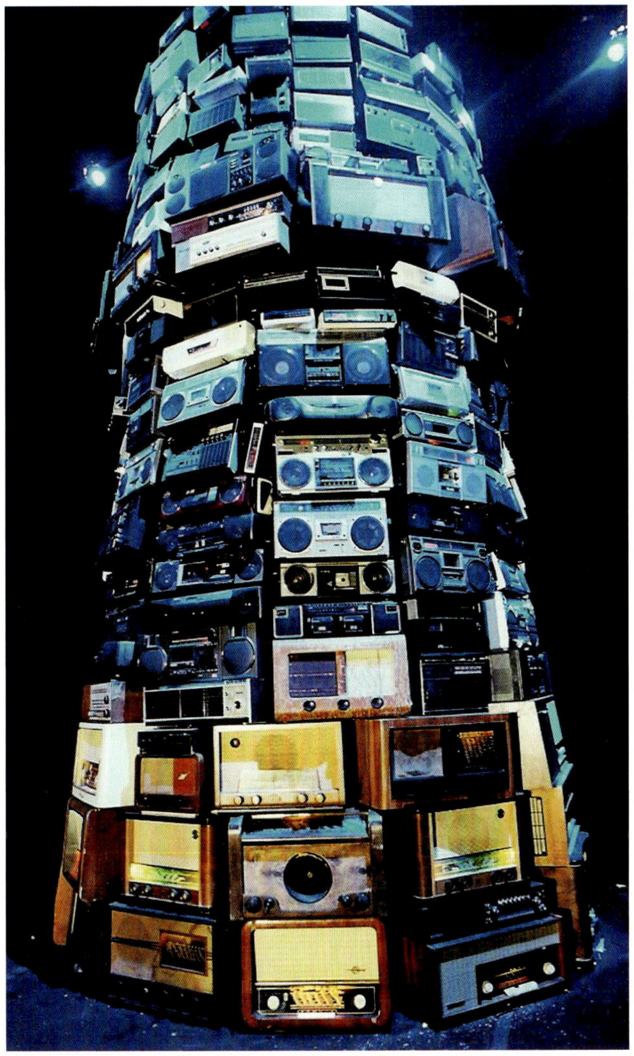

FIG. 8 Cildo Meireles, *Babel*, 2001. Radios, lighting and sound, overall display, dimensions variable.

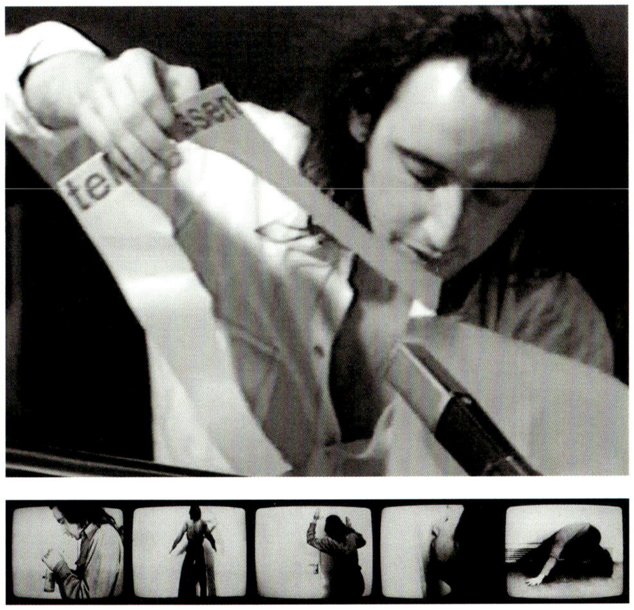

FIG. 9 Guilherme Vaz, *Music for Paper Sheet and Crude*, 1973

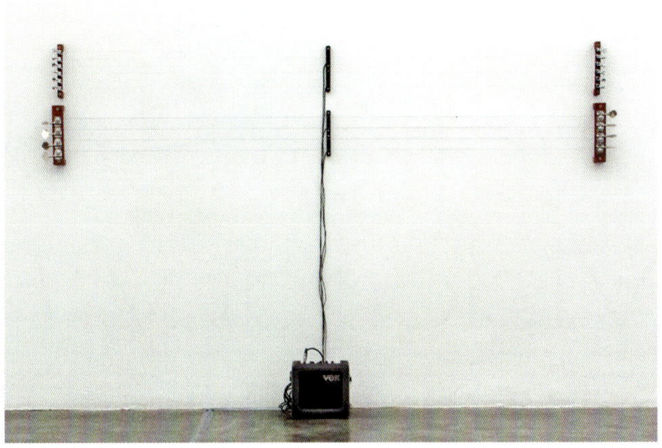

FIG. 10 Chiara Banfi, *Baixo e Guitarra* [Bass and Guitar], from the series *Desenho Sonora* [Sound Design], 2012.

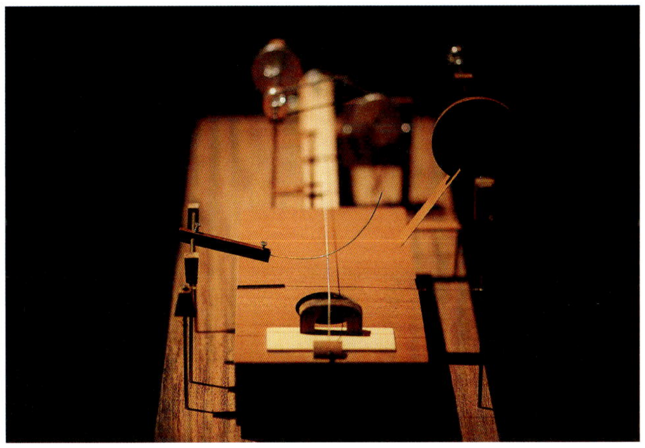

FIG. 11 O Grivo (Nelson Soares / Marcos Moreira), *Quadros Sonoros / String Boxes Machines*, 2009

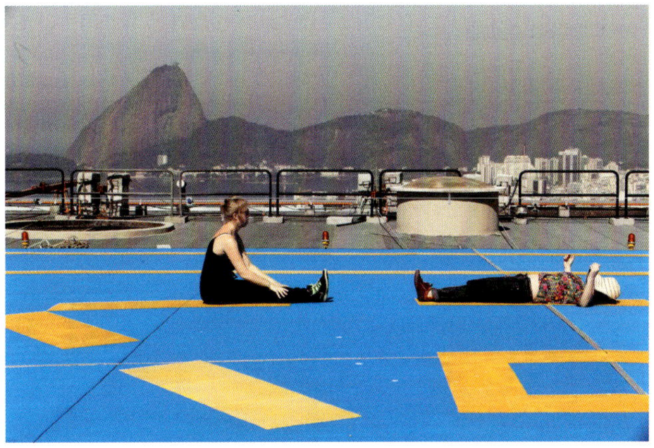

FIG. 12 Vivian Caccuri, *Silent Walk*, 2013

Julia Gerlach

Walter Smetak's Visionary *Plásticas Sonoras* — Influences on Sound Art and Experimental Music in Brazil

Brazil boasts a lively sound art scene whose lines of aesthetic development can be traced back to one artist in particular: Walter Smetak (1913–1984). Swiss-born Smetak, who had studied music, emigrated to Brazil in 1937 where he at first worked as a cellist in various orchestras in Rio de Janeiro and São Paulo. In 1957 he began to teach at the Federal University in Salvador de Bahia (UFBA), where Hans-Joachim Koellreutter, who had also emigrated from Germany, had been head of the music department with its international music seminars since 1954. In the years that followed, the music department developed under Koellreutter's influence into the most active and sophisticated center for contemporary music in Brazil.

Translated from the German by Gloria Custance.

1
In view of the complexity of Smetak's work, the characterization of the artist as a "sound alchemist" is very apt. See Max Nyffeler, "Der Klangalchimist von Bahia: Auf den Spuren von Walter Smetak (1913–1984)," in *Swiss Made: Die Schweiz im Austausch mit der Welt*, ed. Beat Schläpfer, exhib. cat., Musée d'Art et d'Histoire, Geneva (Zurich: Scheidegger & Spiess, 1998), 197–206. See also the first comprehensive publication on Walter Smetak by Marco Scarassatti, *Walter Smetak: O Alquimista dos Sons* (São Paulo: Edições Sesc, 2008).

2
Of a total of twenty-two surviving manuscripts, only two have been published: *Simbologia dos Instrumentos* (1980) and *Retorno ao Futuro* (1982).

3
As early as 1966, his *Plásticas Sonoras* were exhibited at the Biennial of Salvador. Set up on pedestals in a large courtyard, they were occasionally used by musicians for live performances. One year later, the *Plásticas* were on show in the exhibition *Nova Objetividade Brasileira* in Rio de Janeiro. To date, Smetak's sound boards have been part of innumerable solo and group exhibitions, also outside Brazil. The first solo exhibition in Europe, titled *Smetak's Inventions* and curated by Julia Gerlach, ran from March 17 to April 9, 2017, at the daadgalerie, Berlin, with a selection of *Plásticas Sonoras*, drawings, and writings curated by Julia Gerlach.

4
See Helga de la Motte-Haber, ed., *Klangkunst: Tönende Objekte und klingende Räume* (Laaber: Laaber, 1999).

Within the politically and aesthetically charged spirit of optimism and change prevailing in Salvador, Walter Smetak quickly became part of a lively, culturally diverse, and highly productive scene of artists. That Smetak, who described himself as a "decomposer," is today considered one of the fathers of Brazilian sound art is above all due to his visionary perception of sound, his experimental approach to the construction of instruments, and his abandonment of traditional patterns of thought and sound, which all significantly influenced the work of subsequent sound artists and composers in Brazil.[1]

Although Smetak wrote concrete poetry, texts on music theory, stories, and theater plays,[2] it was his *Plásticas Sonoras* that made him famous, a collection of around 150 instruments and sound sculptures, which he built in his studio at the university beginning in 1960.[3] Like Bernard and François Baschet, Jean Tinguely, Nicolas Schöffer, Sarkis, Joe Jones, and Harry Bertoia, Smetak with his sound sculptures, too, sought to extend the body through which sound waves resonate, and to explore in an entirely new way the relationship between temporality, spatiality, and movement which paved the way for the development of sound art as a genre, situated between music and the visual arts.[4]

Taking its cue from the different groups of instruments in an orchestra, the spectrum of Smetak's sound boards ranges from woodwinds, brass, and percussion to strings. It includes instruments that are clearly intended for making music, as well as sculptures that can only generate a single tone and the *Plásticas*, which are completely unplayable; these remain silent and only exist to be looked at and to imagine the sounds they might produce. The starting point of Smetak's sound boards was stringed instruments similar to cellos, such as the *Choris*, which Smetak fitted with sculptural elements on the scroll.

Smetak's principal instrument belongs to this group of instruments: the *Vina*, completed in 1966, is a multi-instrument which can be plucked, bowed, and played percussively.(FIGS. 1a, 1b) In his paper *Simbologia dos Instrumentos*, which was published in 1980, Smetak gives instructions that the instrument should be played turned away from the audience, and the player should wear a cloak

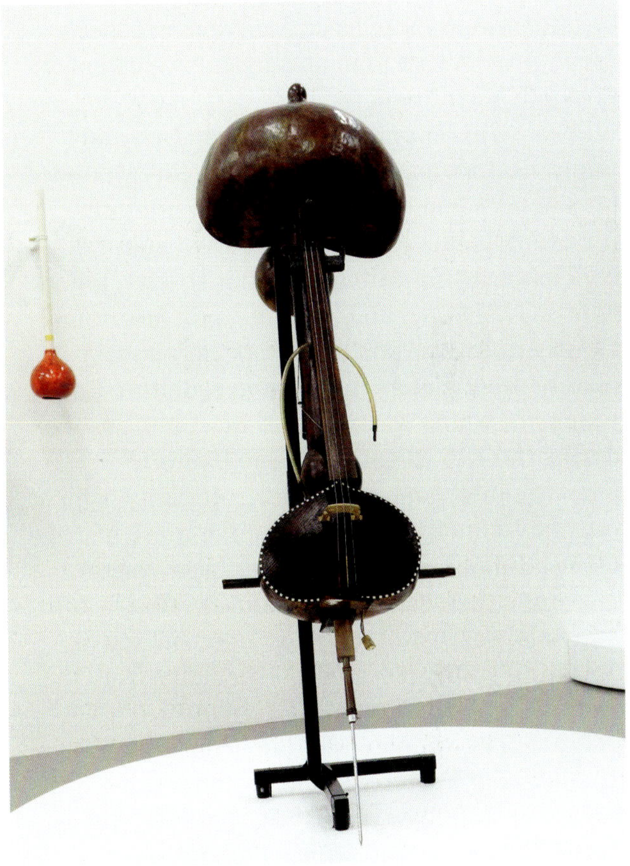

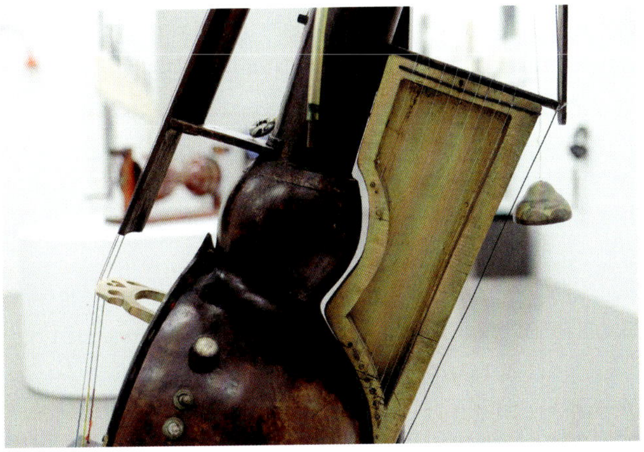

FIGS. 1a, b Walter Smetak, *Vina*, 1966, Sound sculpture. Installation view *Smetak's Inventions*, daadgalerie, 2017. Front view and detail.

and a mask. Although no documentary evidence survives that the *Vina* was actually played in this way evocative of a priest celebrating a rite, his explicit instructions certainly evidence how important the act of playing was to Smetak. He obviously attributed much greater importance to the act of playing the instrument than, for example, to accuracy of pitch, for he gave no indication of the tuning for many of his string instruments, and always preferred improvisation to adhering to the composition. Therefore, the very few of his scores that have survived should be understood more as collective instructions for improvisation than as musical notation for possible performances.

This approach with its random and improvisatory aspects is especially foregrounded by Smetak's wind-activated sound sculptures, for example, his Aeolian microtonal guitars, which are strung with six similar strings, tuned within a halftone, and generate a diffuse sound-cloud within a certain timbre, or his approximately three meters high sound sculpture *Caossonancia*, which was installed in a public space during the XIV. Biennial of São Paulo.(FIG. 2) Smetak's understanding of microtonality, which was the foundation for developing these instruments, did not actually aim for a new system that was organized on a certain division of tone intervals, but instead should be considered as deviating from all systems, especially from the diatonic scale system of Western music.

The simple materials from which Smetak constructed his sound sculptures were found, recycled, or typical for the Bahia region like the bottle gourds, which the artist used in various sizes as resonators for almost all types of instruments. His intuitive, rather playful assembly of highly disparate materials, especially recycled materials (piano strings, iron spirals, iron bars, bamboo, hoses, plastic pipes, rubber gloves, various woods, Styrofoam, small bells, but also guitars, records, audio tapes, and electronic components), is found also, for example, in the kinetic meta-instruments by Jean Tinguely, the mechanical orchestra by Pierre Bastien, or the large-scale sculptures by Benoît Maubrey. It was also subject to the principle of improvisation and intuition and can be described as *gambiarra*[5], which refers to a typical Brazilian creative (survival) strategy.

This term, which comes originally from sociology, also demonstrates that Smetak, with his collage-like assembled sound sculptures, contributed to the preservation of cultural heritage and to the cultural identity of Brazil by assimilating its indigenous culture and including significant elements in his figurative and narrative constellations of material.(FIGS. 3, 4) Among these are, in particular, collective practices of the Brazilian art and performance scene, which appear, for example, in the 1970s performances by painter, sculptor, installation artist, and pioneer of Brazilian interactive art Lygia Clark.

Through these connections to cultural traditions, Walter Smetak's *Plásticas* are in line with an aesthetic that was formulated and practiced in South America in the 1960s by various composers and theorists: Hans-Joachim Koellreutter, mentioned above, in Brazil; Coriún Aharonián and Graciela Paraskevaídis in Montevideo, Uruguay; Cergio Prudencio in La Paz, Bolivia; the inventor of instruments Joaquín Orellana in Guatemala, and many more.[6] They all saw the aesthetic necessity of harmonizing the ideas of the avant-garde with South American culture and local sociopolitical conditions, and of preserving in contemporary culture the memory of the indigenous culture of South America, indigenous languages and music, as well as the natural environment – this was also intended as a strategy serving the development of identity and decolonization.

Walter Smetak's versatile sound boards were used at numerous improvisation sessions in his studio to which, over the years, he invited highly productive exponents of Salvador's contemporary artist scene. These included artists like Caetano Veloso, Gilberto Gil, and Tom Zé, who were part of the Tropicália movement known for its experimental approaches, as well as members of the art scene working in intermedia from the 1966 and 1968 Biennials, and also the Grupo de Compositores da Bahia, the first association of contemporary composers at the UFBA. Like Smetak, these three movements that originated in Brazil were striving for change or a renewal of the existing traditional understanding of sound and also to develop their own aesthetic language.

5
See the contribution by Giuliano Obici in this volume.

6
After studying electroacoustic music in Buenos Aires, Joaquín Orellana invented and built many percussive instruments in Guatemala for his vision of adequate music, which were exhibited at documenta 14 (2017).

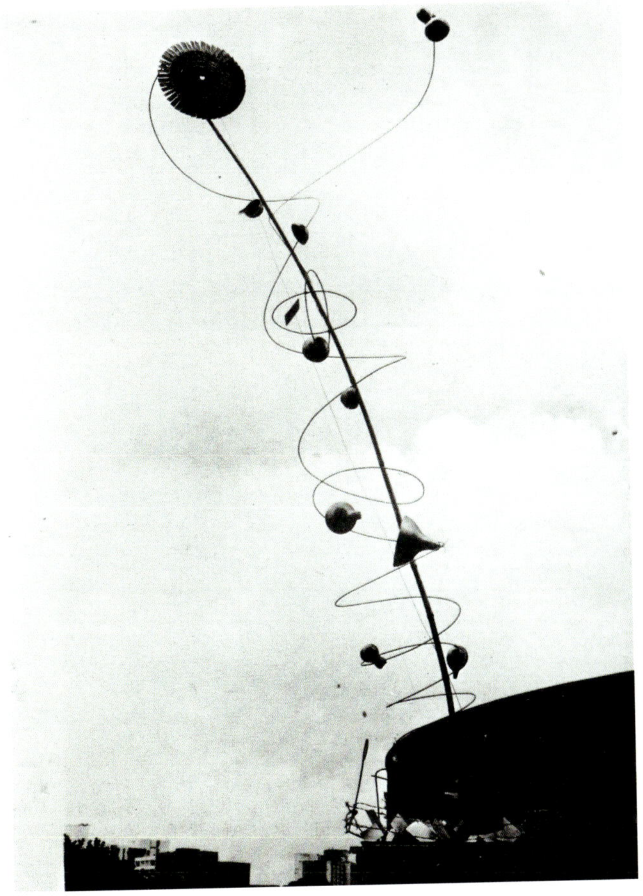

FIG. 2 Walter Smetak, *Caossonância* [Caossonance], 1976 (idea), 1977 (sound sculpure), 1978 (improvisation). Installation in public space at the XIV. Biennale of São Paulo.

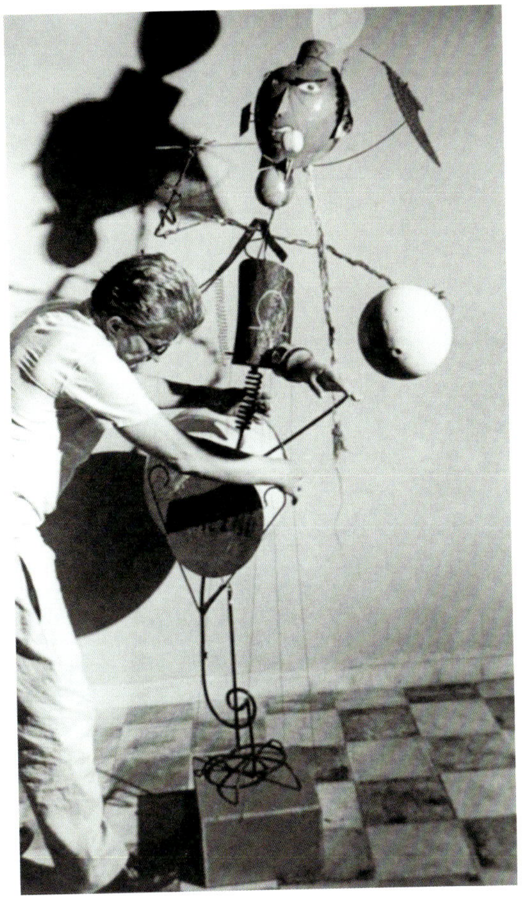

FIG. 3 Walter Smetak, *M 2005*, 1967, sound sculpture. The musical play changes with a theatric action.

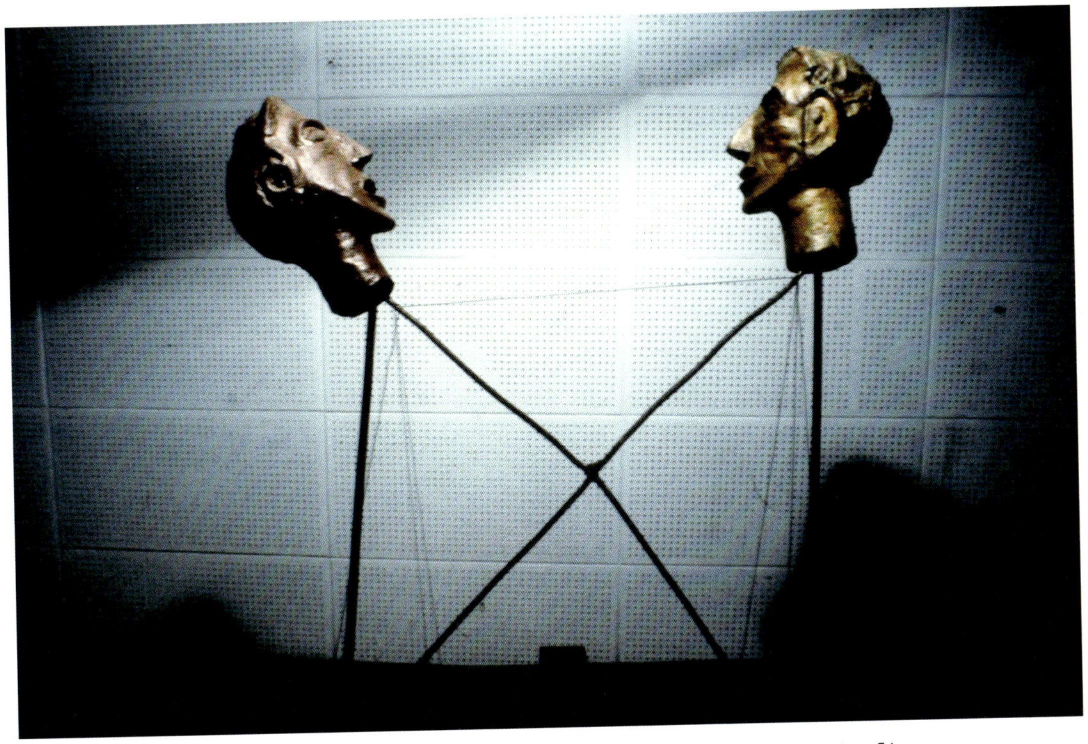

FIG. 4 Walter Smetak, *Colóquio* [Colloquim], 1969, sound sculpture. The movements of two golden-painted heads, facing each other as if they were in a conversation, modulate the pitch of a string stretched between them.

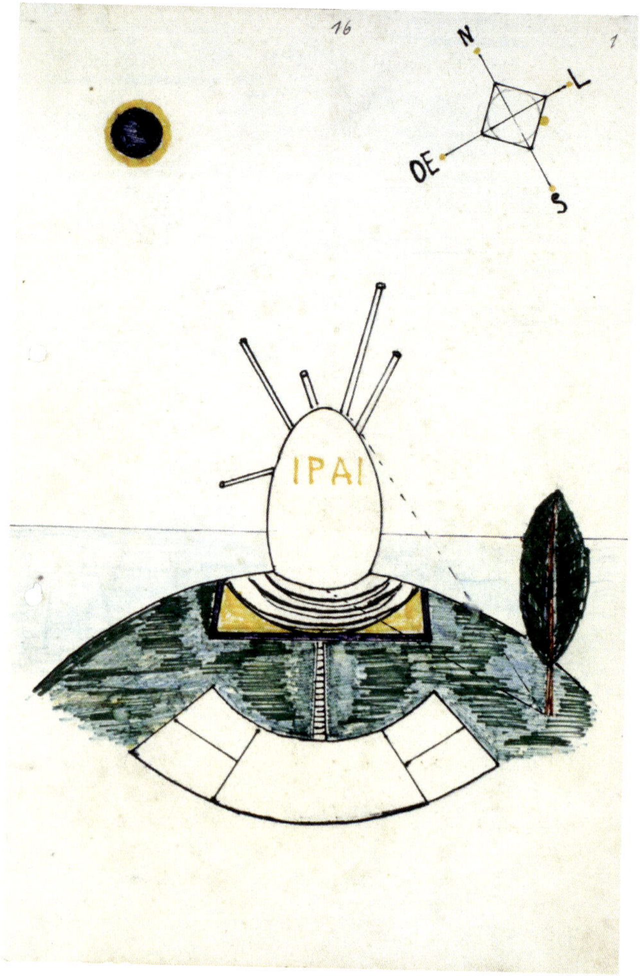

FIG. 5 Walter Smetak, *Projeto do Estudio Ovo* [Egg Studio Project Study], 1970s, sketch.

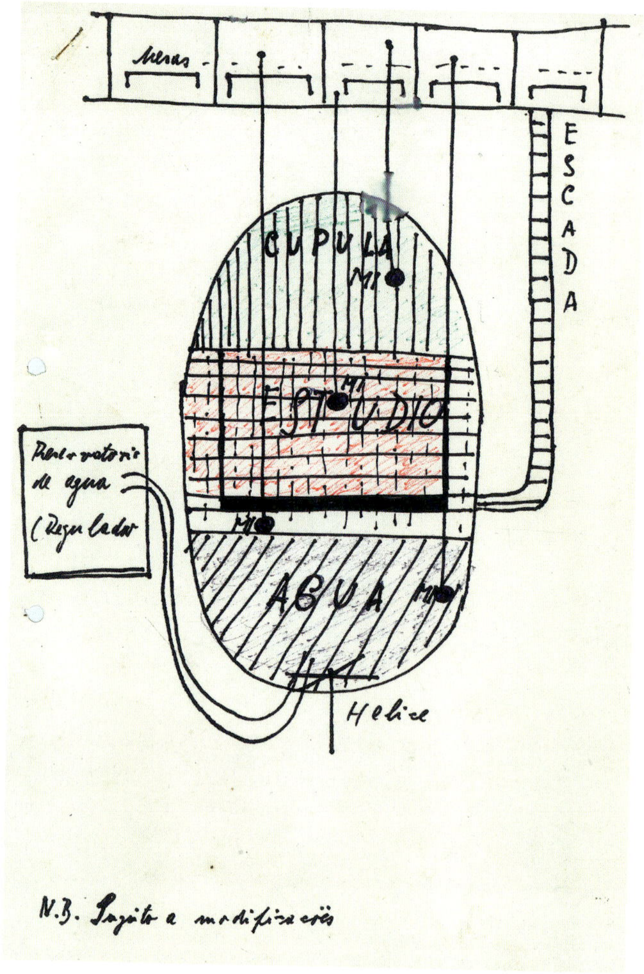

FIG. 6 Walter Smetak, *Projeto do Estudio Ovo*, 1970s, sketch.

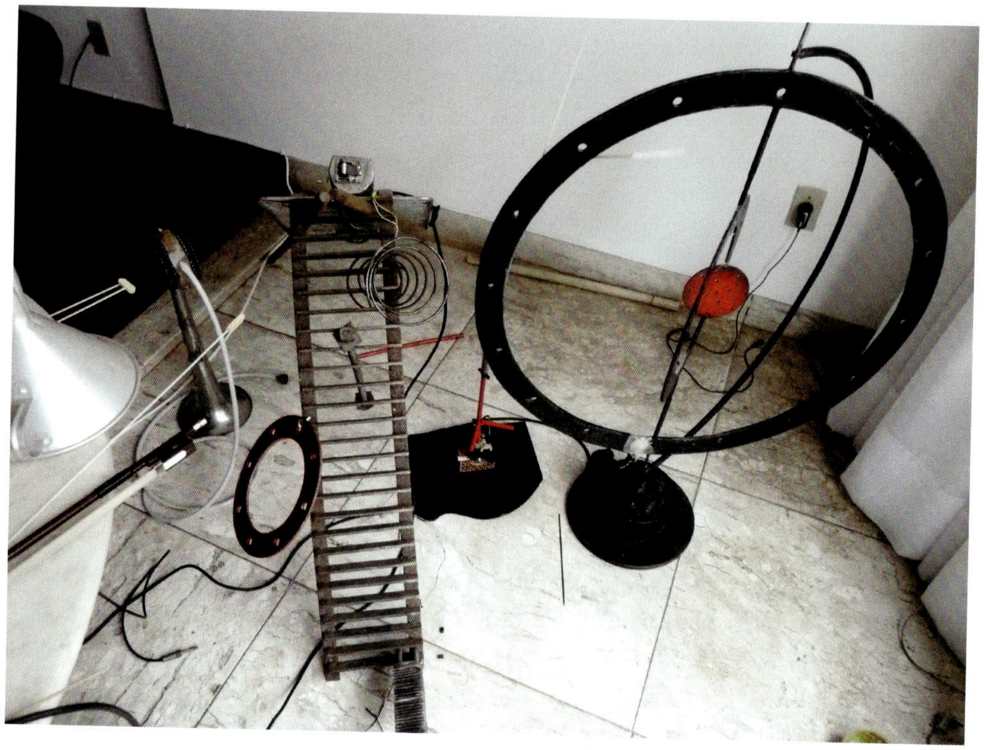

FIG. 7 Marco Scarassatti, *Escultura Sonora* [Sound Sculpture], constellation of instruments.

7
See Franz Manata (ed. and curator), *Guilherme Vaz. Uma fraçao do infinito*," , exhib. cat. Centro Cultural Banco do Brasil (Rio de Janeiro: EXST, 2016), p. 80.

8
Tuzé de Abreu continues to teach at the Universidade Federal da Bahia in Salvador, where he wrote his dissertation on the instruments invented by Smetak. Parts of the dissertation have been translated into English: Tuzé de Abreu: "Smetak and His Instruments," *ART Music Review*, no. 25 (Sep. 2013). Available online: https://www.revista-art.com/smetak-and-his-instruments.

9
See the image of the work on page 706 (FIG. 7) in this volume.

In the 1970s, to give his sound experiments and the collective improvisations a more solid framework, Smetak designed the *Projeto do Estudio Ovo*. This visionary sound research center, an oviform sound and listening studio, was intended to allow both natural sound reproduction as well as sound absorption or the controlled amplification of resonance; it would thus be a place for conducting many types of acoustic experiments. The designs for the studio brought together all of the areas of interest in which Smetak had worked during the preceding years: microtonality, echo, resonance, nature, improvisation, and unpredictability. Although the studio was never built, the texts and sketches that have survived make it quite clear just how much potential Smetak saw in his *Projeto*.(FIGS. 5, 6) And collective improvisation, which he had hoped to provide with an institutional frame, was without doubt the form of artistic creativity that suited him the most. The influence that this had on musicians from very different fields, and thus on the genesis of the sound art scene of Brazil, cannot be overestimated.

From making music together with Walter Smetak, the already mentioned exponents of the Tropicália movement Caetano Veloso, Gilberto Gil, and Tom Zé took over experimental elements and integrated them in their popular music. Tom Zé to this day remains the truest of all to experiment and, to a certain extent, uses instruments he has built himself.

The ritual aspect and the close connection with the origins of Brazil was further developed by the concept artist and composer Guilherme Vaz, who from 1965 to 1967 studied under Walter Smetak. Alongside Smetak himself, Vaz is known as one of the most important figures of Brazilian sound art. His body of work, which was presented for the first time in a major solo exhibition in Rio de Janeiro in 2016, includes conceptual art, drawings, movie soundtracks, video, and sound installations. In the 1970s, Vaz was closely linked to the conceptual art scene in Europe and the United States, and participated in a number of exhibitions there. Already in 1975 Vaz made the radical demand that the dividing line between sight and sound and between noise and sound should be abolished.[7] Consistently pursuing this approach, he lived for twenty years with indigenous groups in the west and north of Brazil, got to know their music and painting, and, on the basis of intensive communication with and knowledge of the people, developed new artistic forms of expression. His compositions are all long, and – similar to many works by Smetak – are not notated; they are based on sketches, processes, and improvisations, which are used to create sound environments where continuous rhythmic displacements suggest a tranquil trance mood, in which a balance between the contemporary and the ritualized seems possible.

The basic principles of spiritual improvisation were also taken up and continued by the flutist Tuzé de Abreu, who played with Smetak for many years. Abreu still lives in Salvador, and to this day appears live on stage as a musician with instruments that are similar to the *Plásticas Sonoras*.[8]

Of particular note is the influence of Smetak on his student Marco Antônio Guimarães, who already in 1978 founded the group UAKTI, a band consisting of four musicians who built their own instruments, developed a rhythmic popular music language, and played live on stage with their program at home and abroad.

Indeed, there are a large number of contemporary sound artists whose work exhibits references to Smetak, for example, Tato Taborda, who is known for his standing multi-instrument *Geralda*[9], which is in the tradition of Smetak, or Marco Scarassatti with his music instruments and Aeolian sound sculptures.(FIG. 7) Edbrass Brasil's interventions in public space are also in the tradition of Smetak as are Gustavo Alfaix's installation made of Aeolian guitars, the expanded techniques of the guitar of Silvia Ocougne, and the approaches to music of the bands Chelpa Ferro and O Grivo. Based on the ideas of Smetak, composer Ricardo Eizirik has even integrated new instruments in an orchestra ensemble.

APPENDIX

Peter Weibel

Sound as a Medium of Art

Sons durant et partant de différents points et formant une sculpture sonore qui dure. [Lasting sounds coming from different points and forming a sound sculpture that lasts.][1]
– Marcel Duchamp

The term *sound art* dates back to William Hellermann's Sound Art Foundation, founded in 1982, which primarily seemed to work with "experimental music" or "new music," although it did organize an early show of sound sculpture and other exhibitable work at the Sculpture Center in 1983.[2]
– Alan Licht

Translated from the German by Luca Branchetti.

1
Marcel Duchamp, note from the *Green Box*, 1934. See Michel Sanouillet, *Marcel Duchamp: Duchamp du Signe. Ecrits* (Paris: Flammarion, 1975), 47.

2
Alan Licht, *Sound Art: Beyond Music, Between Categories* (Rizzoli: New York, 2007), 11.

Authors' Biographies

Dmitry Bulatov is an artist, curator, and art theorist. His research focuses on different aspects of interdisciplinary art media (sci-art, techno-biological arts, etc.), as well as on submediality aesthetics. He has published many articles on contemporary art in Russia and abroad, as well as books and anthologies. His artworks have been presented internationally, including at the Venice Biennale (2001, 2003) and the Ars Electronica Festival (2002). He is a two-time winner of the Russian Innovation Award for contemporary visual arts (2008, 2013) and a Golden Nica nominee of *Prix Ars Electronica* (Austria, 2014) in the category Visionary Pioneers of Media Art. In 2007 one of his artworks was named in the *Top 10 New Organisms of the Year*, selected by *Wired* magazine. Bulatov has curated more than twenty major exhibitions in Russia and abroad; since 1998, he has been the curator at the Baltic Branch of the National Centre for Contemporary Arts and the leading expert at the Innovation Park of the Baltic Federal University in Kaliningrad, Russia.

Seth Cluett is an artist whose work includes installation, concert music, photography, and critical writing. His research investigates auditory memory, architectural acoustics, and the history and documentation of sound in the arts. His work has been presented at venues such as the Palais de Tokyo, the Whitney Museum, MoMA PS1, Apexart, and Eyebeam. He is recipient of awards from Meet the Composer and the Foundation for Contemporary Arts Emergency Grant. Cluett holds a PhD in composition from Princeton University and an MFA in electronic art from Rensselaer Polytechnic Institute in Troy, USA. He has published writings with *Tacet Revue*, *Leonardo Music Journal*, and the *Journal of the Acoustical Society of America*, and Line Imprint and Errant Bodies Press as well as the recording labels Sedimental and Winds Measure Recordings have documented his creative work. He is jointly appointed in the programs in visual arts & technology and music & technology at Stevens Institute of Technology in Hoboken, USA.

Christoph Cox is Professor of Philosophy at Hampshire College in Amherst, USA, and visiting faculty at the Center for Curatorial Studies, Bard College, Annandale-on-Hudson, USA. He is the author of *Sonic Flux: Sound, Art, and Metaphysics* (University of Chicago Press, 2018) and *Nietzsche: Naturalism and Interpretation* (University of California Press, 1999), and co-editor of *Realism Materialism Art* (Sternberg, 2015) and *Audio Culture: Readings in Modern Music* (Continuum/Bloomsbury, 2004/2017). Cox is editor-at-large at *Cabinet* magazine and recipient of an Arts Writers Grant from Creative Capital/Warhol Foundation. His writing has appeared in *October*, *Artforum*, *Journal of the History of Philosophy*, *The Wire*, *Journal of Visual Culture*, *Organised Sound*, *The Review of Metaphysics*, and elsewhere. He has curated exhibitions at the Contemporary Arts Museum Houston, The Kitchen, CONTEXT Art Miami, New Langton Arts, G Fine Art Gallery, and other venues.

Julia Gerlach studied musicology in Berlin and Paris and since then works in the field of contemporary music and sound art. In 2003 she founded KlangQuadrat.büro für Klang- und Medienkunst together with the sound artist Georg Klein. From 2008 to 2012 she taught at the Frankfurt Conservatory of Music and Performing Arts. In 2009, she took up the position of project coordinator at ZKM | Karlsruhe; in 2011/2012 she was co-curator and project director of the exhibition *Sound Art. Sound as Medium of Art*. Since 2012 she is head of music at the DAAD Artists-in-Berlin Program. In this position she founded the annual festival mikromusik and curated over seventy concerts, solo-exhibitions, and discussions. She is especially interested in transcultural projects and initiated discourse on "Sound Art in the Global Context" and the project "Re-inventing Smetak" with Ensemble Modern and Goethe-Institut. She has contributed to numerous juries and artistic boards, is editor of books and catalogues and author of articles in the field of sound art and contemporary music.

Ryo Ikeshiro is a UK-based Japanese artist whose practice deals with sound, media, and computation. He is interested in the artistic potential of computational technology and algorithmic processes as well as the cultural and political dimension of computation and media and their critique within art. His output includes audiovisual performances, interactive installations, generative music and data visualization and sonification. He is featured in the *Electronic Music* volume of the *Cambridge Introductions to Music* series (Cambridge University Press, 2013) and he is a contributor to *Iannis Xenakis. La musique électroacoustique* (LHARM, 2015). He was part of the Asia Culture Center's inaugural exhibition in Gwangju, South Korea, and his Teletext art pages (2015) have been broadcast on German, Austrian, and Swiss national TV. He also works as a lecturer.

Caleb Kelly is an academic, event director, and curator working in the area of the sound arts. In 2009 he published his first book, entitled *Cracked Media: The Sound of Malfunction*, with The MIT Press. He also edited *Sound* (2011), part of the *Documents of Contemporary Art* series published by The MIT Press and Whitechapel Gallery in London. He was co-curator with Aaron Kreisler of *Sound Full: Sound in Contemporary Australian and New Zealand Art* (2012–2014), an exhibition that included fourteen sound-intensive installations and a performance series at two major New Zealand art museums, the Dunedin Public Art Gallery and The City Gallery, Wellington. He was the director of the sound gallery Pelt in Chippendale, Australia and director of the series *impermanent.audio*, which ran for six years. He is a member of the National Institute of Experimental Arts at UNSW Art & Design in Paddington, Australia.

Brandon LaBelle is an artist, writer, and theorist working with sound culture, voice, and questions of agency. He develops and presents artistic projects and performances within a range of international contexts, often working collaboratively and in public. His work has been presented at documenta 14 in Athens (2017), South London Gallery (2016), and Liquid Architecture, Melbourne (2015). He is the author of *Sonic Agency: Sound and Emergent Forms of Resistance* (The MIT Press, 2018), *Lexicon of the Mouth: Poetics and Politics of Voice and the Oral Imaginary* (Bloomsbury Academic, 2014), *Diary of an Imaginary Egyptian* (les presses du réel, 2012), *Acoustic Territories: Sound Culture and Everyday Life* (Continuum, 2010), and *Background Noise: Perspectives on Sound Art* (Continuum / Bloomsbury, 2006/2015). He is Professor in new media at the University of Bergen, Norway.

Christof Migone is an artist, curator, teacher, writer, and assistant professor in the Department of Visual Arts at the University of Western Ontario, Canada. His research delves into language & voice, bodies & performance, intimacy & complicity, sound & silence, rhythmics & kinetics, translation & referentiality, stillness & imperceptibility, structure & improvisation, play & pathos, pedagogy & unlearning, failure & endurance. His writings have appeared in numerous publications and journals, including *Aural Cultures*, *S:ON*, *Musicworks*, *Semiotext(e)*, and *Esse*. He has curated independently since 1990 and was the director/curator of the Blackwood Gallery at the University of Toronto from 2008 to 2013. A monograph, *Sound Voice Perform* (2005), and an exhibition catalog, *Trou* (Galerie de l'UQAM, Montreal, 2006) have been published on his work. His writings on sound art are compiled in *Sonic Somatic: Performances of the Unsound Body* (Errant Bodies Press, 2012). He is a founding member of Avatar.

Daniel Muzyczuk is head of modern art department at the Muzeum Sztuki in Lodz, Poland. His curatorial projects include: *Gone to Croatan* (with Robert Rumas), CoCA Torun, 2009 and HMKV, Dortmund, 2011; *MORE IS MORE* (with Agnieszka Pindera and Joanna Zielińska), CoCA Torun, 2010; *Melancholy of Resistance* (with Agnieszka Pindera), CoCA Torun, 2010; *Sounding the Body Electric: Experiments in Art and Music in Eastern Europe 1957–1984* (with David Crowley), Muzeum Sztuki, Lodz, 2012 and Calvert 22, London, 2013; *Notes from the Underground: Eastern European Alternative Art and Music Scene 1968–1994* (with David Crowley), Muzeum Sztuki, Lodz, 2016 and AdK, Berlin, upcoming, 2018 and *Museum of Rhythm* (with Natasha Ginwala), Muzeum Sztuki, Lodz, 2016. He was co-curator of the Polish Pavilion at the 55th Venice Biennale in 2013 (with Agnieszka Pindera). He is the vice president of AICA Poland.

Tony Myatt is the Head of Music and Media at the University of Surrey, UK. He has pursued research in spatial audio production throughout his career, and developed technologies and methods informed by theories of spatial audio perception. Myatt's sound installation work includes spatial audio for Thyssen-Bornemisza Art Contemporary's *The Morning Line* sound pavilion project, exhibited from 2008 in Seville, Istanbul, Vienna, and now permanently installed at the ZKM | Karlsruhe, and collaborations with Jana Winderen for MoMA and Park Ave Road Tunnel, New York. Myatt regularly works with sound artist Chris Watson, has developed audio software for Yasunao Tone and Russell Haswell, and spatial audio releases that include Haswell's *IN IT (Immersive Live Salvage)* (eMego 115) and *Leyfðu Ljósinu* (Touch TO:90) for Hildur Guðnadóttir. Professor Myatt was a founding editor of Cambridge University Press's *Organised Sound: An International Journal of Music and Technology*.

Irene Noy is the author of *Emergency Noises: Sound Art and Gender* (Peter Lang, 2017). She holds a PhD from The Courtauld Institute of Art, London, where she also completed a post-doctoral fellowship (with generous funding from the Marie-Louise von Motesiczky Charitable Trust). Prior to that, Noy received her education from the University of Bonn, Germany; University of British Columbia, Canada; and University of Edinburgh, UK. Her research explores twentieth-century aural and visual cultures in Britain and Germany. Other publications include "A Sound Walk into Susan Hiller's *Monument* (1980–81)," in: *n.paradoxa* (vol. 37, January 2016); "Conversing 'trina chéile'," in: *Corporeality and Cultures: Bodies in Movement* (Routledge, 2015); "Art That Does Not Make Noise? Mary Bauermeister's Early Work and Exhibition with Karlheinz Stockhausen," in: *Immediations* (vol. 3, no. 2, 2013). More recently, Noy has been experimenting with alternative methods of research dissemination by leading sensory tours that explore the sensorial and meditative perceptions of art in institutions such as the National Gallery and the Courtauld Gallery, UK.

Giuliano Obici is artist and professor of Art Department at Universidade Federal Fluminense in Rio de Janeiro. He holds degrees in music, communication, and psychology; is the author of *States of Listening* (7Lettras, 2008) and *Gambiarra and Sound Experimentalism* (PhD thesis, USP São Paulo and TU Berlin, 2014). His artistic and research projects has been presented in different contexts such festivals, galleries, biennials, and conferences in Vienna, New York, Rio de Janeiro, Berlin, London, Bogotá, Oslo, Kassel, Lisbon, São Paulo among others. Obici received a production award at Giga-Hertz Prize 2014 (ZKM and SWR Experimentalstudio), as well as prizes and grants form Red Bull Residência Artísitica, DAAD, CAPES, and FAPESP. His interests are focused on perception, sound and media art as well as the political, ethical, and aesthetical aspects of these technologies.

Adam Parkinson makes electronic music, and is a researcher in the EAVI group in the Department of Computing, Goldsmiths, University of London. He completed his PhD in music at Newcastle University, UK, and prior to that studied history. He makes music under too many aliases, including glitchy algorithmic techno as Dane Law and skewed electronic pop in the band Emotional, with releases on Quantum Natives, Fat Cat, Milk Records, Upset the Rhythm, and Entr'acte. As a musician and programmer he has worked on recordings, performances, and installations with Kaffe Matthews, Arto Lindsay, Phill Niblock, and Rhodri Davies. He also organizes various concerts, including the long-running EAVI gig series in South East London, which have hosted a variety of artists including Trevor Wishart, People Like Us, AGF, Phil Minton, Jennifer Walshe, and Maria Chavez.

Bernd Schulz (1941–2017) was a German curator, journalist, and art educator. After studying natural sciences at the University of Freiburg im Breisgau, Germany, he worked as a journalist in the fields of culture and science and was the manager of the Department for Culture and Science at Saarland Broadcasting – SR. From 1985 to 2002 he was the director of the Stadtgalerie Saarbrücken, Germany. He published numerous texts on contemporary art, especially focusing on the usage of sound in artworks. Schulz was a honorary professor of art and science at the Academy of Fine Arts in Saarbrücken, and had taught at the National Schools of Fine Arts in Nancy and Paris, France. Since 1978 he was a member of the International Artists Forum (IKG).

Carsten Seiffarth studied orchestral music at the University of Music Franz Liszt Weimar, Germany (trombone) and musicology/sociology at the Technische Universität Berlin. An independent curator and producer since 1991, Seiffarth has worked on numerous sound art exhibitions and contemporary music projects in cooperation with festivals, galleries, museums, theaters, and foundations across Europe, as well as in the US and Mexico. He is the editor of, among others, *singuhr – hoergalerie in parochial* (Kehrer, 2010), *tesla berlin: medien kunst labor* (Alexander, 2010), *Paul DeMarinis: Buried in Noise* (Kehrer, 2010), and *Gordon Monahan: Seeing Sound* (Robert McLaughlin Gallery, 2011). He was one of the artistic directors of Berlin's Inventionen festival (2000–2008) and tesla media art laboratory (2005–2007). In 2011 and 2012, he was artistic director of the project *Sound Exchange. Enclaves of Experimental Music Culture in Central Eastern Europe*. Since 1996, he is curator and director of singuhr – hoergalerie berlin, since 2010 of bonn hoeren, and since 2014 of singuhr – projekte.

Linnea Semmerling is currently a PhD candidate at Maastricht University's Faculty of Arts and Social Sciences, Netherlands, where she is working on a NWO-funded research project *Listening on Display: Exhibiting Sounding Artworks 1960–Now*. She holds a BA in arts and culture from Maastricht University and a research MA in art studies from the University of Amsterdam. From 2013 to 2015, she was assistant curator at the ZKM | Karlsruhe, realizing exhibitions such as *Lynn Hershman Leeson: Civic Radar* and *Ryoji Ikeda: micro | macro* as well as assistant editing this publication. She is also a member of CAS Collective, an Amsterdam-based group of humanities researchers, putting out academic articles and art events.

Başak Şenova is a curator and designer. She studied literature and holds a MFA in Graphic Design and PhD in Art, Design and Architecture from Bilkent University, Turkey. She attended the 7th Curatorial Training Programme of Stichting De Appel, Amsterdam and is a founding member of NOMAD, as well as the organizer of *Upgrade! Istanbul* and *ctrl_alt_del* sound art projects. As an assistant professor, she lectured in various universities in Turkey. She is an editorial correspondent for *ibraaz.org* and Turkish correspondent of *Flash Art International*. She co-curated the *UNCOVERED. Nicosia International Airport* Project (Cyprus) and the 2nd Biennial of Contemporary Art, D-0 ARK Underground (Bosnia and Herzegovina). In 2014, she acted as the Art Gallery Chair of (ACM) SIGGRAPH 2014 (Canada) the curator of the Helsinki Photography Biennial 2014 and *the Jerusalem Show VII. Fractures*. She curated the Turkey (2009) and Republic of Macedonia (2015) pavilions at the Venice Biennale. In 2016, she curated the *Lines of Passage (in medias res)* exhibition in Lesvos, Greece. Currently she is working on a long-term research-based art project *Cross Sections* in Vienna, Helsinki, and Stockholm.

Morten Søndergaard is a curator, art critic, theoretician, and associate professor of interactive media art at the School of Communication, Art and Technology at University of Aalborg (based in Copenhagen) as well as a faculty member of the Media Art Histories Master at Blue Danube University in Krems, Austria. As head of research of the research-infrastructure project *Unheard Avant-gardes in Scandinavia* (since 2010), he curated a section of *Sound Art – Sound as Medium of Art* at ZKM, and founded the International Sound Art Curating Conference Series (ISACS), both in 2012. He was curator of the Museum of Contemporary Art in Roskilde (1998–2008), and senior curator at Re-new Digital Arts Festival in Copenhagen (2010–2013); chair of Design Cases at Nordi CHI (2012) and exhibition chair at NIME conference 2017. Søndergaard is member of the curator-board at DIAS artspace in Copenhagen (since 2012) and contributing editor of *Leonardo Electronic Almanac* (2011–2014). In 2017 he co-curated the exhibition *Roads Less Travelled by…* at Ars Electronica which presented projects from students of the Art & Technology and Media

Arts Cultures Erasmus Master courses of Aalborg University.

Alexandra Supper is an assistant professor in the Department of Technology & Society Studies at Maastricht University, Netherlands. In 2013 she held a visiting professorship in "Science-Technology-Society" at the University of Vienna. After being trained as a sociologist at the University of Vienna, she completed her PhD in science and technology studies at Maastricht University in 2012 with a dissertation about the efforts of the auditory display community to achieve scientific legitimacy for sonification. Her research interests include the role of listening and other bodily and sensory practices in the (natural and social) sciences, the emergence of scientific disciplines and interdisciplinary practices, and academic conference practices. Her research has been published in the *Oxford Handbook of Sound Studies* and in various peer-reviewed international journals, including *Social Studies of Science*, *Science as Culture*, and *Information & Culture*.

Atau Tanaka studied electronic music with Ivan Tcherepnin at Harvard University, where he met John Cage during his Norton Lectures (1988–1989). He gained his doctorate in computer music composition with John Chowning at Stanford University's Center for Computer Research in Music and Acoustics (CCRMA). There he began working with electromyogram muscle signals as a musical interface, using the BioMuse. He was in residency in 1993 at the Cité Internationale des Arts in Paris and conducted research at IRCAM, Centre Pompidou. He formed Sensorband with Zbigniew Karkowski and Edwin van der Heide and then worked in Japan, performing with Merzbow, Otomo, and KK Null. He has since been a researcher at Sony Computer Science Laboratory Paris and artistic co-director of STEIM. He is a professor of media computing at Goldsmiths, University of London.

David Toop is a composer/musician, author, and curator based in London who has worked in many fields of music, writing, arts, and theory. He has recorded Yanomami shamanism in Amazonas, appeared on Top of the Pops with the Flying Lizards, exhibited sound installations internationally, and worked with artists including John Zorn, Evan Parker, Bob Cobbing, Ivor Cutler, Akio Suzuki, and Rie Nakajima. His published books include *Rap Attack* (1984), *Ocean of Sound* (1995), *Haunted Weather* (2004), *Sinister Resonance* (Bloomsbury, 2010/2011), and *Into the Maelstrom* (Bloomsbury, 2016). Since 1975 he has released eleven solo albums, including *Screen Ceremonies* (1995), *Black Chamber* (2003), *Sound Body* (2007), and *Entities Inertias Faint Beings* (2016), as a theorist and critic he has written for many publications. Exhibitions he has curated include *Sonic Boom* (2000) at the Hayward Gallery, London, and *Playing John Cage* (2005/2006) at Arnolfini, Bristol, UK. His opera – *Star-shaped Biscuit* – was performed in Aldeburgh, UK, in 2012. He is professor of audio culture and improvisation at London College of Communication.

Peter Weibel, chairman and CEO of ZKM | Karlsruhe and professor of media art (later theory) at the University of Applied Arts Vienna (since 1983/1984), is considered a central figure in European media art through his various activities as artist, theoretician, and curator. He was head of the digital arts laboratory at the Media Department of New York University in Buffalo (1984–1989) and then founding director of the Institute of New Media at the Städelschule in Frankfurt am Main (1989–1994). He was artistic director of Ars Electronica in Linz, Austria (1986–1995), the Seville Biennial, Spain (BIACS3, 2008), and the Moscow Biennale of Contemporary Art (2011). He commissioned the Austrian pavilion at the Venice Biennale (1993–1999) and was chief curator of the Neue Galerie Graz, Austria (1993–2011), since 2018 head of the Peter Weibel research center for digital cultures at the University of Applied Arts, Vienna. He has edited and authored numerous books and catalogs, the most recent of which are three volumes of *Enzyklopädie der Medien: Architektur und Medien* (2015), *Musik und Medien* (2016), *Kunst und Medien* (2019).

Dajuin Yao holds a doctoral degree in art history from the University of California, Berkeley. He is currently a professor in sound art, and the director of Open Media Lab at the China Academy of Art, Hangzhou. In recent years, he has worked on the deconstruction of the Chinese language (sound, script, and meaning), software modulation of traditional music and opera, and emotional sound art. As a pioneer in the Chinese sound art scene, he has organized various important events, including the *Beijing SONART* International Sound Art Festival, the Shanghai eArt Festival, numerous compilation recordings of sound art from China, and *Sound Art China: Revolutions Per Minute,* the first major survey of Chinese sound art, touring across the world since 2013.

Siegfried Zielinski was rector of the Karlsruhe University of Arts and Design from 2016–2018. Until his election in February 2016 he held the chair for media theory, with a focus on archaeology/variantology of the media, at Berlin University of the Arts, where he also was director of the Vilém Flusser Archive. He is the Michel Foucault Professor of Media Archaeology and Techno-Culture at the European Graduate School in Saas-Fee, Switzerland, co-founder of the media consulting course at the Technische Universität Berlin (1972–1982), and founding rector (1994–2000) of the Academy of Media Arts Cologne. He has published numerous books and essays mainly focusing on archaeology of the media and coauthored, edited, or co-edited a number of anthologies. Zielinski is member of the North Rhine-Westphalian Academy of Sciences, Humanities and the Arts; the Academy of Arts, Berlin; the European Film Academy; the Magic Lantern Society of Great Britain; and honorary senator of the Austrian Institute of Media Archaeology.

Bibliography

A

Acquaviva, Frédéric. *Musique pédagogique*. Paris: Editions AcquAvivA, 2012.

Adams, Tyler. *Interstitial*. Los Angeles: Steve Turner Contemporary, 2013.

Adorno, Theodor W. "Art and the Arts." Translated by Rodney Livingstone. In *Can One Live after Auschwitz? A Philosophical Reader*, ed. Rolf Tiedemann. Stanford: Stanford University Press, 2003. 368–88.

Agamben, Giorgio. *Die kommende Gemeinschaft*. Berlin: Merve, 2003.

Akita, Masami. *Noizu Wō: Noizu Myūjikku to sono Tenkai* [Noise War: Noise Music and its Development]. Tokyo: Seikyusha, 1992.

Alertz, Ulrich. "The Horologium of Hārūn al-Rashīd Presented to Charlemagne: An Attempt to Identify and Reconstruct the Clock Using the Instructions Given by al-Jazarī." In *Variantology 4: On Deep Time Relations of Arts, Sciences and Technologies in the Arabic-Islamic World and Beyond*, ed. Siegfried Zielinski and Eckhard Fürlus. Cologne: Walther König, 2010. 19–42.

Altena, Arie, and Sonic Arts, eds. *The Dark Universe*. Amsterdam: Sonic Acts Press, 2013.

Altman, Rick. *Sound Theory, Sound Practice*. New York: Routledge, 1992.

Alves, Isabel Soares. *BES: Bolsa Ernesto de Sousa Arte Experimental Intermedia 15*. Lisbon: Fundação Luso-Americana, 2008.

von Amelunxen, Hubertus, Dieter Appelt, and Peter Weibel, eds. *Notation: Kalkül und Form in den Künsten*. Berlin: Akademie der Küste, 2008.

Amt der Niederösterreichischen Landesregierung, Abteilung Kunst und Kultur, ed. *Bernhard Leitner. Ton – Raum – Skulptur*. Bielefeld: Kerber, 2016.

Anastasi, William. *The Cage Dialogues: A Memoir*. Contemporary Artist Series 6, ed. Aaron Levy. Philadelphia: Slought Books, 2011.

Anfilov, Gleb. "Chapter 10: Composer as a Painter." In *Physics and Music* [1962], trans. Boris Kuznetsov, n. p. Moscow: Mir Publishers, 1966.

Apollonius. *Conics: Books V to VII; The Arabic Translation of the Lost Greek Original in the Version of the Bānū Mūsā*. Edited and translated by Gerald J. Toomer. 2 vols. New York: Springer, 1990.

van Assche, Christine, and Emmanuel Fessy. *Sonic Process: Une nouvelle géographie des sons*. Paris: Centre Pompidou, 2002.

Attali, Jacques. *Noise: The Political Economy of Music*. Vol. 16, *Theory and History of Literature*. Manchester: Manchester University Press, 1985.

Auer, Johannes, ed. *experimentelle literatur und internet: memoscript für reinhard döhl*. Zurich: update, 2004.

Avraamov, Arseny. "Nauchnaja organizatsija khudozestven-nogo materiala" [The Scientific Organization of Creative Material]. *Sovetskoe iskusstvo*, no. 4 (1928): n.p.

Avraamov, Arseny. "Novaja era muzyki" [The New Era of Music]. *Sovetskoe iskusstvo*, no. 3 (1925): n.p.

Avraamov, Arseny. "Smychkovij polikhord" [The Bowed Polychord]. *Muzikalnij sovremennik*, no. 3 (1916): n.p.

B

Baecker, Inge, and Städtische Kunsthalle Düsseldorf, eds. *Sehen um zu hören: Objekte & Konzerte zur visuellen Musik der 60er Jahre*. Exhib. cat. Düsseldorf: Städtische Kunsthalle, 1975.

Bailey, Derek. *Improvisation: Its Nature and Practice in Music*. Ashbourne, UK: Moorland Publishing, 1980.

Barbaud, Pierre. *Schoenberg*. Nice: Editions Main D'Œuvre, 1997.

Barrett, Natasha, and Karen Mair, "Aftershock: A science-art collaboration through sonification." *Organised Sound* 19, no. 1 (2014): 4–16.

Baudelaire, Charles. *Les Paradis artificiels: Opium et Haschisch*. Paris: Libraires – Éditeurs, 1860.

Baudouin de Courtenay, Jan. *A Baudouin de Courtenay Anthology: The Origins of Structural Linguistics*. Edited and translated by Edward Stankiewicz. Bloomington: University of Indiana Press, 1972.

Beil, Ralf, and Peter Kraut, eds. *A House Full of Music: Strategien in Musik und Kunst*. Ostfildern: Hatje Cantz,2012.

Bélis, Annie. "Harmonics." In *The Greek Pursuit of Knowledge*, ed. Jacques Brunschwig and Geoffrey E. R. Lloyd, 161–176. Cambridge, MA: Harvard University Press, 2003.

Benedetti, Lorenzo. "The Shape of Silence." In *MARTa schweigt: Garde le silence, le silence te gardera; Die Kunst der Stille von Duchamp bis heute; Das Mysterium der Etrusker*, edited by MARTa Herford, 90–3. Exhib. cat. Herford: MARTa Herford, 2007.

Benjamin, Walter. *Selected Writings*. Vol. 3, *1935–1938*. Edited by Howard Eiland, and Michael W. Jennings. Cambridge, MA: The Belknap Press of Harvard University Press, 2002.

Bergson, Henri. *Laughter: An Essay on the Meaning of the Comic*. Translated by Cloudesley Brereton, and Fred Rothwell. London: Macmillan, 1911. Originally published as *Le Rire: Essai sur la signification du comique*. Paris: Félix Alcan, 1900.

Beuckers, Klaus Gereon, ed. *Günther Uecker: Die Aktionen*. Petersberg: Michael Imhof, 2004.

Binger, Dietmar. *Backstage: Fotografische Berichte aus der Stadtgalerie Saarbrücken 1994–2000*. Saarbrücken: Stadtgalerie Saarbrücken, 2000.

Birjukov, Sergei. *Zevgma: Russkaja poezija ot man'erizma do postmodernizma* [Zeugma: Russian Poetry from Mannerism to Postmodernism]. Moscow: Nauka, 1994.

Biró, Dániel, Jonathan Goldman, Detlef Heusinger, and Constanze Stratz, eds. *Live Electronics im/in the SWR Experimentalstudio*. Hofheim: Wolke, 2019.

Block, René et al. *Für Augen und Ohren: Von der Spieluhr zum akustischen Environment; Objekte Installationen Performances*. Exhib. cat. Berlin: Akademie der Künste Berlin, 1980.

von Blumröder, Christoph. "Experiment – Experimentelle Musik." *Feedback Papers 29*, (July 1982): 1–24.

Boorstin, Daniel. *The Image: A Guide to Pseudo-Events in America*. New York: Harper & Row, 1964.

Bormann, Hans-Friedrich. *Verschwiegene Stille: John Cages performative Ästhetik*. Munich: Wilhelm Fink, 2005.

Born, Georgina, and Andrew Barry. "Art-Science: From Public Understanding to Public Experiment." *Journal of Cultural Economy* 3, no. 1 (2010): 103–19.

Bosseur, Dominique, and Jean-Yves Bosseur. *Révolutions musicales: La musique contemporaine depuis 1945*. Paris: Minerve, 1999.

Brewster, Michael. "An Interview with Brandon LaBelle." In *See Hear Now: A Sonic Drawing and Five Acoustic Sculptures*, 26–30. Los Angeles: Los Angeles Contemporary Exhibitions, 2002.

Brown, Earle. "The Notation and Performance of New Music." *Musical Quarterly* 72, no. 2 (1986): 180–201.

Bulatov, Dmitry, ed. *Homo sonorus: An International anthology of sound poetry*. Kaliningrad: National Center for Contemporary Arts, 2001.

Bulatov, Dmitry, ed. *Tochka zreniia: vizual'naia poėziia, 90-e gody* [A Point of View. Visual Poetry: the 90s]. Kaliningrad: Simplicij, 1998.

Bull, Michael, and Les Back, eds. *The Auditory Culture Reader*. New York: Berg, 2003.

Busoni, Ferruccio. *Entwurf einer neuen Ästhetik der Tonkunst*. 2nd ed. Leipzig: Insel, 1916.

Byrne, David. *How Music Works*. Edinburgh: Canongate Books, 2012.

C

Cage, John. *Silence: Lectures and Writings*. Middletown: Wesleyan University Press, 1986.

Cage, John, and Morton Feldman. *Radio Happenings I–V: Conversations – Gespräche*. Cologne: MusikTexte, 1993.

Cardiff, Janet, and George Bures Miller. *The Murder of Crows*. Edited by Catherine Crowston. Ostfildern: Hatje Cantz, 2011.

Cardiff, Janet, and George Bures Miller. *Works from the Goetz Collection: Haus der Kunst, 13.04–08.07.2012*. Edited by Haus der Kunst foundation, Munich. Ostfildern: Hatje Cantz, 2012.

Cardiff, Janet and George Bures Miller. "[untitled text]." In *Voices Voces Vox*. Edited by Barbera van Kooij and Christopher Phillips, 89. Exhib. cat. Rotterdam: Witte de With et al., 1998.

Carson, Anne. "The Gender of Sound." In *Glass, Irony and God*, 130. New York: New Directions, 1995.

Cavallaro, Alessio, Shaun Davies, and Annemarie Jonson, eds. *Essays in sound 3: Diffractions*. Sydney: Contemporary Sound Arts, 1996.

Cayley, Arthur. "On the Mathematical Theory of Isomers." *Philosophical Magazine* 4–7, no. 314 (1874): 444–6.

Cayley, Arthur. "On the Theory of the Analytical Forms Called Trees." *Philosophical Magazine* 4–13, no. 85 (1857): 172–6.

Cecchetto, David. *Humanesis: Sound and Technological Posthumanism*. Minneapolis: University of Minnesota Press, 2013.

Celant, Germano, ed. *Art or Sound*. Exhib. cat. Fondazione Prada. Milan: Progretto Prada Arte, 2014.

Celant, Germano. "A Visual Machine: art installation and its modern archetypes." In *Thinking about Exhibitions*, ed. Reesa Greenberg, Bruce W. Ferguson, and Sandy Narine, 371–86. London: Routledge, 2005. Originally published in *documenta 7*. Vol. 2, ed. Rudi Fuchs, XIII–XVIII. Kassel: Paul Dierichs, 1982.

Centre Pompidou, ed. *Sons & Lumières: Une histoire du son dans l'art du XXe siècle*. Exhib. cat. Paris: Éditions du Centre Pompidou, 2004.

Chattopadhyay, Budhaditya. "Auditory Situations: Notes from Nowhere." *Journal of Sonic Studies* 4, no. 1 (May 2013): n.p.

Cheng-Yih, Chen. *Early Chinese Work in Natural Science: A Re-examination of the Physics of Motion, Acoustics, Astronomy and Scientific Thoughts*. Hong Kong: Hong Kong University Press, 1996.

Chion, Michel. *Audio-Vision: Sound on Screen*. New York: Columbia University Press, 1994. First published as *L'audio-vision : son et image au cinéma*, Paris: Armand Colin, 1990.

Chion, Michel. *Guide des objets sonores: Pierre Schaeffer et la recherche musicale*. Paris: Buchet / Chastel, 1983.

Chion, Michel, *Un art sonore, le cinéma: histoire, esthétique, poétique*. Paris: Cahiers du Cinéma, 2003.

Chion, Michel. *The Voice in the Cinema* [1982]. New York: Columbia University Press, 1999. First published as *La voix au cinéma*. Paris: Cahiers du Cinema, 1982.

Chun, Wendy Hui Kyong. *Programmed Visions: Software and Memory*. Cambridge, MA / London: The MIT Press, 2011.

Cianciusi, Walter. "Max Neuhaus: Pioneer of Invisible and (Almost) Inaudible Sound Installations." In *International Computer Music Conference Proceedings*, 248–53. Ann Arbor, MI: Michigan Publishing, 2013.

Clarke, Joseph L. "Xenakis and the Philips Pavilion." *The Journal of Architecture* 17, no. 2 (2012): 213–29.

Classen, Constance. *The Colour of Angels: Cosmology, gender and the aesthetic imagination*. London: Routledge, 1998.
Claus, Jürgen. *Expansion der Kunst: Beiträge zu Theorie und Praxis öffentlicher Kunst*. Reinbek: Rowohlt, 1970.

Cluett, Seth. "Acoustic Projection and the Politics of Sound." *Tacet Revue* 1, no. 3 (2014): 83–119.

Cluett, Seth. "Ephemeral, Immersive, Invasive: Sound as Curatorial Theme 1966–2012." In *The Multisensory Museum: Cross-Disciplinary Perspectives on Touch, Sound, Smell, Memory, and Space*, ed. Nina Levent, and Alvaro Pascual-Leone, 114–5. Lanham: Rowman & Littlefield, 2014.

Coleman, Patrick, ed. *The Art of Music*. Exhib. cat. The San Diego Museum of Art. New Haven: Yale University Press, 2015.

Connor, Steven. *The Matter of Air: Science and Art of the Ethereal*. London: Reaktion Books, 2010.

Conrad, Tony. "LYssophobia: On Four Violins." In *Audio Culture: Readings in Modern Music*, ed. Christoph Cox, and Daniel Warner, 313–8. New York: Continuum, 2004.

Corbin, Alain. *Village Bells: Sound and Meaning in the 19th-Century French Countryside*. Translated by Martin Thom. New York: Columbia University Press, 1998. Originally published as *Les cloches de la terre*. Paris: Albin Michel, 1994.

Cox, Christoph. "Sound Art and the Sonic Unconscious." *Organised Sound* 14, no. 1 (2009): 19–26.

Cox, Christoph, and Daniel Warner, eds. *Audio culture: Readings in Modern Music*. New York / London: Continuum, 2010.

Coyle, Rebecca. "Sound In Space: A Curatorial Perspective." In *Sound in Space: Adventures in Sound*. Exhib. cat. Sidney: Museum of Contemporary Art, 1995.

Crowley, David, and Daniel Muzyczuk. *Sounding the Body Electric: Experiments in Art and Music in Eastern Europe 1957–1984*. Łódź: Muzeum Sztuki, 2012.

Cseres, Jozef, and Georg Weckwerth, eds. *Membra Disjecta for John Cage: Wanting to Say Something about John*. Exhib. cat. Prague: DOX Centre for Contemporary Art, 2012.

Cubitt, Sean. "Video Installation and the Neo-Classical Soundtrack." In *Essays in Sound 3: Diffractions*, ed. Alessio Cavallaro, Shaun Davies, and Annemarie Jonson, 8–20. Sidney: Contemporary Sound Arts, 1996.

Czichon, Joachim, ed. *Klangkörper, Körperklänge*. Karlsruhe: Engelhardt & Bauer, 2004.

D

Daniels, Dieter, and Sandra Naumann, eds. *See This Sound: Audiovisuology; Compendium; An Interdisciplinary Survey of Audiovisual Culture*. Cologne: Walther König, 2010.

Daniels, Dieter, and Sandra Naumann, eds. *See This Sound: Audiovisuology 2; Essays; Histories and Theories of Audiovisual Media and Art*. Cologne: Walther König, 2011.

Davies, Shaun et al., eds. *Essays in sound 2: Technophonia*. Sydney: Contemporary Sound Arts, 1995.

Debord, Guy. *The Society of the Spectacle*, Detroit: Black & Red, 1977.

Dekleva, Dasha. "In Parallel." In *Yasunao Tone: Noise Media Language*, ed. Brandon LaBelle, 39–61. Los Angeles: Errant Bodies, 2007.

Deleuze, Gilles, and Félix Guattari. *Anti-Oedipus: Capitalism and Schizophrenia*. Translated by Mark Seem, Helen Lane, and Robert Hurley. London: Continuum, 2004.

Demosfenova, Galina. "Iz materialov Fonologicheskogo otdela GINKhUKa" [From the materials of the Phonological Department of GINKhUK]. In *Terentev album No.1*, ed. S. Kudrjavtsev, n.p. Moscow: Hylaea, 1996.

Dick, Philip K. *The Man in the High Castle* [1962]. New York: Mariner Books, 2011.

Doris, David T. "Zen Vaudeville: A Medi(t)ation in the Margins of Fluxus." In *The Fluxus Reader*, ed. Ken Friedman, 91–135. London: Academy Editions, 1998.

Drachmann, Aage G. *Ktesibios, Philon and Heron: A Study in Ancient Pneumatics*. Copenhagen: Enjar Munksgaard, 1948.

Dreckmann, Kathrin: *Speichern und übertragen: Mediale Ordnungen des akustischen Diskurses 1900–1945*. Paderborn: Wilhelm Fink, 2018.

Drobnick, Jim, ed. *Aural Cultures*. Toronto: YYZ Books; Banff: Walter Phillips Gallery Editions, 2004.

Duplaix, Sophie, Marcella Lista, and Marion Diez, eds. *Sons & Lumières: Une histoire du son dans l'art du XX. siècle*. Paris: Centre Pompidou, 2004.

Dyson, Frances. *Sounding New Media: Immersion and Embodiment in the Arts and Culture*. Berkeley / Los Angeles / London: University of California Press, 2009.

E

Eco, Umberto. *The Open Work*. Translated by Anna Cancogni. London: Hutchinson Radius, 1989.

Eco, Umberto. "Untitled text." In *Arte programmata: Arte cinetica, opere moltiplicate, opera aperta*. Exhib. brochure. Milan: Galleria Vittorio Emanuele, 1962.

Ensemble musikFabrik, ed. *Saison 2013/2*. Cologne: Ensemble musikFabrik, 2013.

Emmerson, Simon. *Living Electronic Music*. Farnham / Burlington: Ashgate, 2007.

Ernst, Wolfgang. *Das Rumoren der Archive. Ordnung aus Unordnung*. Berlin: Merve, 2002.

Ernst, Wolfgang. *Digital Memory and the Archive*. Edited by Jussi Parikka. Minneapolis: University of Minnesota Press, 2012.

Evers, Elodie, Magdalena Holzhey, Gregor Jansen, and Kunsthalle Düsseldorf, eds. *Henri Chopin. Guy De Cointet. Channa Horwitz*. Leipzig: Spector Books, 2013.

F

Farmer, Henry George. *The Organ of the Ancients: From Eastern Sources (Hebrew, Syriac and Arabic)*. London: William Reeves, 1931.
Felderer, Brigitte, ed. *Phonorama. Eine Kulturgeschichte der Stimme als Medium*. Exhib. cat. ZKM | Center for Art and Media Karlsruhe. Berlin: Matthes & Seitz, 2004.

Feldman, Morton. *Essays*. Edited by Walter Zimmermann. Wasserburg: Beginner Press, 1985.

Fetz, Wolfgang, ed. *4'33": Ein Projekt des Magazin4 – Bregenzer Kunstverein. 14. bis 22. Juli 2007*. Frankfurt am Main: Revolver, 2007.

Fisher, Mark. "Public Space is the Place." *The Wire* 295 (September 2008): 28–33.

Foucault, Michel. *The Archeology of Knowledge*. New York: Pantheon Books, 1972. Originally published as *L'archéologie du savoir*. Paris: Gallimard, 1969.

Foucault, Michel. "Different Spaces." In *Aesthetics, Ethics, and Epistemology*, ed. James D. Faubion. New York: The New Press, 1982. 175–86.

Freke, John. "II. On Electricity." In *A Treatise on the Nature and Property of Fire: In Three Essays*. London: W. Innys, and J. Richardson, 1752.

Freke, John. *A Treatise on Electricity, An Essay to show the Cause of Electricity and why some things are Non-Electricable. In which is also Considered Its Influence in the Blasts on Human Bodies, in the Blights on Trees, in the Damps in Mines; and as it may affect the Sensitive Plant*. 2nd ed. London: W. Innys, 1746.

Fricke, Stefan. "Die Anfänge: 1970–1971." In *Musik im 20. Jahrhundert: 1970–2000; Eine Dokumentation*, ed. Wolfgang Korb, and Friedrich Spangemacher, 11–24. Saarbrücken: Pfau, 2001.

Fritsch, Johannes, Gerhard R. Koch, and Winrich Hopp. *Feedback Studio*. Cologne: Dohr, 2006.

Frye, Northrop. *Divisions on a Ground*. Toronto: House of Anansi Press, 1982.

Fülöp-Miller, René. *Geist und Gesicht des Bolschewismus*. Vienna: Amalthea, 1926.

Fundación Juan March, and Editorial de Arte y Ciencia, eds. *Escuchar con los ojos: Arte sonoro en España, 1961–2016*. Exhib. cat. Madrid: Fundación Juan March, 2016.

Furlong, William. *Audio Arts: Beunruhigende Versuche zur Genauigkeit*. Leipzig: Reclam, 1992.

G

Galeev, Bulat. *Chelovek, iskusstvo, tekhnika (Problemi sinestezii v iskusstve)* [Mankind, Art, Technology (The Issue of Synesthesia in Art)]. Kazan: KGU, 1987.

Galeev, Bulat. *Soviet Faust: Leon Theremin – Pioneer of Electronic Art*. Translated by Sally Brown. Sydney: ETT Imprint, 2010.

Galeev, Bulat. *Svetomuzika: stanovlenije I suschnost' novogo iskusstva* [Light-Music: The Establishment and Essence of the New Art]. Kazan: Tatknigoizdat, 1976.

Galerie Mario Mazzoli, ed. *Douglas Henderson: Poets and Superheroes*. Exhib. cat. Berlin: Galerie Mario Mazzoli, 2012.

Gąsiorowska, Małgorzata. *Rozmowy z Włodzimierzem Kotońskim* [Conversations with Włodzimierz Kotoński]. Warszawa: Warszawska Jesień, 2010.

Gerlach, Julia, ed. *Kirsten Reese: Media Sound Constellations*. Hofheim: Wolke, 2010.

Gerlach, Julia, ed. *Smetak's Inventions: The Interfused Realms of Inventor, Sound Artist, and Musician Walter Smetak (1913–84)*. Hofheim: Wolke, 2018.

Gervais, Raymond. "From sound to image." In *Raymond Gervais 3x1*, ed. Nicole Gingras, 59. Exhib. cat. Montreal: Leonard and Bina Ellen Gallery / VOX, 2006.

Gervais, Raymond. *Via Charles Ives: La Symphonie Universe (via La Question sans réponse)*. Orford: Centre d'arts Orford, 2001.

Gerwin, Thomas, ed. *hören::multisensorial: Jubiläums-Ausgabe 2004–2014*. Berlin: Internationales Klangkunstfest, 2014.

Gesellschaft der Freunde junger Kunst e.V., ed. *Pli selon pli: Porträt Pierre Boulez*. Vienna: Universal Edition Wien, 2004.

Gibson, James J. *The Ecological Approach to Visual Perception*. Boston: Houghton Mifflin, 1979.

Gieryn, Thomas F. "Boundaries of Science." In *Handbook of Science and Technology Studies*, ed. Sheila Jasanoff et al, 393–443. Thousand Oaks, CA: SAGE Publications, 1995.

Gingras, Nicole, ed. *S:ON: Sound in Contemporary Canadian Art / Le son dans l'art contemporain canadien*. Montreal: Éditions Artexte, 2003.

Gingras, Nichole, ed. *Traces*. Montreal: Leonard and Bina Ellen Gallery, 2006.

Glass, Alexie, and Emily Cormack, eds. *21-100-100. One Hundred Sound Works By 100 Artists From The 21st Century*. Melbourne, Hobart: Gertrude Contemporary Art Spaces / MONA. 2008.

Goldsmith, Mike. *Sound: A Very Short Introduction*. Oxford: Oxford University Press, 2015.

Grayson, John, ed. *Sound Sculpture: A Collection of Essays by Artists Surveying the Techniques, Applications and Future Directions of sound Sculpture*. Vancouver: A.R.C., 1975.

Greenberg, Clement. "The Crisis of the Easel Picture." In *Art and Culture: Critical Essays*, 154–157. Boston: Beacon, 1965. Originally published in *Partisan Review* (April 1948): n.p.

Gregory, Richard L. "Knowledge in Perception and Illusion." *Philosophical Transactions of the Royal Society – Biological Sciences* 352, no. 1358 (1997): 1121–8.

Grimshaw, Jeremy. *Draw a Straight Line and Follow It: The Music and Mysticism of La Monte Young*. Oxford: Oxford University Press, 2011.

Groys, Boris. *History Becomes Form: Moscow Conceptualism*. Cambridge, MA / London: The MIT Press, 2010.

Grygar, Milan. *akustické kresby a partitury*. Prague: Galerie Zdeněk Sklenář, 2006.

H

Hall, Donald E. *Musical Acoustics*. Pacific Grove, CA: Brooks/Cole Publishing Company, 1991.

Hamburger Bahnhof – Museum für Gegenwart – Berlin, Staatliche Museen zu Berlin, and Nationalgelerie, ed. *Museum für Gegenwart No 12: Bernhard Leitner*. Cologne: DuMont, 2008.

Hanse-Wissenschaftskolleg, ed. *Transition: knowledge through performance in art and science*. Exhib. cat. Delmenhorst: Hanse-Wissenschaftskolleg, 2012.

von Hantelmann, Dorothea. *How to Do Things with Art*. Edited by J. Gaines and M. Turnbull. Zurich: JRP | Ringier; Dijon: Les presses du réel, 2010.

Harris, Jonathan, ed. *Money, Parties: New Institutions in the Political Economy of Contemporary Art*. Liverpool: Liverpool University Press, 2004.

Hassan, Ahmad Y. al-. *Kitāb al-Hiyal*. Aleppo: University of Aleppo, Institute for the History of Arabic Science, 1981.

Hassan, Ahmad Y. al-, and Donald R. Hill. *Islamic Technology: An Illustrated History*. Cambridge: Cambridge University Press; Paris: UNESCO, 1986.

Hattinger, Gottfried. *Über die Sinne. Geschichten aus einer Wahrnehmungswelt*. Weitra: publication PNº 1, 2006.

Hayles, Katherine N. *How We Think: Digital Media and Contemporary Technogenesis*. Chicago: University of Chicago Press, 2012.

Heister, Hanns-Werner. *Hintergrund Klangkunst: Ein Beitrag zur akustischen Ökologie*. Mainz: Schott Music, 2010.

Heller, Martin, and Nicolas Schärer, eds. *Wolfgang Dorninger: Documentation 2*. Zurich: Sound Development, 2013.

von Helmholtz, Hermann. *On the Sensations of Tone as a Physiological Basis for the Theory of Music*. Translated by Alexander J. Ellis. London: Longmans, Green, Reader, and Dyer, 1875.

Henry Moore Institute, ed. *With hidden noise: Sculpture, video and ventriloquism*. Exhib. cat. Leeds: Henry Moore Institute, 2004.

Hermann, Thomas, Andy Hunt, and John G. Neuhoff, eds. *The Sonification Handbook*. Berlin: Logos Verlag, 2011.

Herzogenrath, Wulf, and Barbara Nierhoff-Wielk, eds. *„John Cage und …": Bildender Künstler – Einflüsse, Anregungen*. Cologne: DuMont, 2012.

Higgins, Dick. *Computers for the Arts*. Somerville, MA: Abyss, 1970.

Hiller, Lejaren A. "Informationstheorie und Computermusik: Zwei Vorträge, gehalten auf den 'Internationalen Ferienkursen für Neue Musik', Darmstadt, 1963." Translated by Peter Jansen. In *Darmstädter Beiträge zur Neuen Musik*. Vol. 8, *Lejaren A. Hiller jr.*, n.p. Mainz: Schott, 1964.

Hiller, Lejaren A., and Robert A. Baker. "Computer Cantata: A Study in Compositional Method." *Perspectives of New Music* 3, no. 1 (Autumn–Winter 1964): 62–90.

Himmelsbach, Sabine, ed. *SOUND// BYTES_ elektronische und digitale Klangwelten*. Heidelberg: Kehrer, 2007.

Hinterwaldner, Inge. "Semiconductor's Landscapes as Sound-Sculptured Time-Based Visualizations." *Technoetic Arts: A Journal of Speculative Research* 12, no. 1 (2014): 15–38.

Hofacker, Ernst. *Von Edison bis Elvis: Wie die Popmusik erfunden wurde*. Stuttgart: Reclam, 2012.

Hollein, Max, Jesper N. Jørgensen, and Schirn Kunsthalle Frankfurt, eds. *Frequenzen [Hz]: Audiovisuelle Räume*. Ostfildern: Hatje Cantz, 2002.

Hopkins, Candice. "Hank Bull: From the Centre to the Periphery." In *Re-Inventing Radio: Aspects of Radio as Art*, ed. Heidi Grundmann et al., 220. Frankfurt am Main: Revolver, 2008.

Huhtamo, Erkki, and Jussi Parikka, eds. *Media Archaeology: Approaches, Applications, and Implications*. Berkeley / London: University of California Press, 2011.

I

Iges, José, ed. *Ars Sonora. Una experiencia de arte sonoro en radio*. Madrid: Fundación SGAE, 2012.

Ikoniadou, Eleni. *The Rhythmic Event: Art, Media, and the Sonic*. Cambridge, MA / London: The MIT Press, 2014.

Izvolov, N. "Iz istorii risovannogo zvuka v SSSR" [On the History of Drawn Sound in the USSR]. *Kinovedcheskie Zapiski*, no. 53 (2001): 293.

Izvolov, N. "Iz istorii risovannogo zvuka v SSSR" [On the History of Drawn Sound in the USSR]. In *Sovietskaia vlast i media* [Soviet Power and Media], ed. Hans Günther, and Sabine Hänsgen, 366. St. Petersburg: Akademichesky proekt, 2005.

J

Jacob, Günther. *Agit-Pop: Schwarze Musik und weiße Hörner*. Berlin: Edition ID-Archiv, 1993.

Jacobsen, Mogens, and Morten Søndergaard, eds. *RE_ACTION: The Digital Archive Experience; Renegotiating the Competences of the Archive and the (Art)museum in the 21st Century*. Aalborg: Aalborg Universitetsforlag, 2009.

Jakobson, Roman. *My Futurist Years*. Edited by Bengt Jangfeldt. Translated by Stephen Rudy. New York: Marsilio, 1997.

Jalving, Camilla. "Inventing reality: On truth and lies in the work of Hayley Newman." In *Performative Realism*, ed. Anne Jerslev and Rune Gade, 145–80. Copenhagen: Museum Tusculanum Press, 2005.

Järvilehto, Timo. "The Theory of the Organism-Environment System: I. Description of the Theory." *Integrative Physiological and Behavioural Science* 33, no. 4 (October 1998): 321–34.

Jostkleigrewe, Anne. "Reaching for the Stars: From The One-All-Alone to Espace." In *Edgard Varèse: Composer, Sound Sculptor, Visionary*, ed. Felix Meyer, and Heidy Zimmermann, 211–9. Woodbridge: Boydell & Brewer, 2006.

K

Kagel, Mauricio. "Komposition – Notation – Interpretation." In *Darmstädter Beiträge zur Neuen Musik*. Vol. 9, *Notation Neuer Musik*, ed. Ernst Thomas, 55–63. Mainz: Schott, 1965.

Kahn, Douglas. "Alvin Lucier, Edmond Dewand und Music for Solo Performer." In *Klangmaschinen zwischen Experiment und Medientechnik*, ed. Daniel Gethmann, 211–29. Bielefeld: transcript Verlag, 2010.

Kahn, Douglas. "Christian Marclay's Early Years: An Interview," *Leonardo Music Journal* 13 (2003): 17–21.

Kahn, Douglas. *Noise, Water, Meat: A History of Sound in the Arts*. Cambridge, MA / London: The MIT Press, 1999.

Kahn, Douglas. "The Lyre's Island: Some Australian Music, Sound Art and Design." *Leonardo Music Journal 6* (1996): 89–93.

Kahn, Douglas, and Gregory Whitehead, eds. *Wireless Imagination: Sound, Radio, and the Avant-Garde*. Cambridge, MA / London: The MIT Press, 1992.

Kahn, Douglas. "Sound Art, Art, Music." *Tacet 3* (2014): 328–347.

Kannenberg, John. *Hours of Infinity: Recording the Imperfect Eternal*. Ann Arbor, MI: Kelsey Museum of Archaeology, 2012.

Kant, Immanuel. *Anthropology from a Pragmatic Point of View*. Translated by Victor Lyle Dowdell. Carbondale: Southern Illinois University Press, 1978.

Kapfer, Herbert, and Joerg Franzbecker, eds. *Eran Schaerf. Frequency-Modulated Scenario*. Berlin: Archive Books, 2015.

Karahasan, Dzevad. "Hārūn al-Rashīd schenkt Karl dem Großen einen Elefanten." *Trajekte – Zeitschrift des Zentrums für Literatur- und Kulturforschung Berlin*, no. 19 (2009): 36–40.

Karkoschka, Erhard. *Notation in New Music: A Critical Guide to Interpretation and Realisation*, Translated by Ruth Koenig. New York: Praeger, 1972.

Karp, Ivan, and Steven D. Lavine, eds. *Exhibiting Cultures*. Washington: Smithsonian Institution Press, 1991.

Katunyan, Margarita, ed. *Sto let russkogo avangarda* [100 Years of Russian Avantgarde]. Moscow: Nauchno-izdatelskij centr "Moskovskaya konservatoria," 2013.

Kelly, Caleb. "Can someone please turn the work up? The Exhibition of Sound in Contemporary Australian Art." *Art Monthly Australasia* 225 (November 2009): 7–10.

Kelly, Caleb. *Cracked Media: The Sound of Malfunction*. Cambridge, MA / London: The MIT Press, 2009.

Kelly, Caleb, ed. *Sound*. Exhib. cat. Whitechapel Gallery, London. Cambridge, MA / London: The MIT Press, 2011.

Kelm, Teresa, and Zygmunt Krauze, *Spatial-Musical Composition*. Warsaw: Galeria Współczesna, 1968.

Khazam, Rahma. "A Brief History of Sound Art" In *Volume: What You See Is What You Hear 1*, ed. Eric Troncy, 6–9. Dijon: Les presses du réel, 2010.

Khlebnikov, Velimir. "On Contemporary Poetry" [1920]. In *The Collected Works of Velimir Khlebnikov*. Vol 1, trans. Paul Schmidt, ed. Charlotte Douglas, 373. Cambridge, MA: Harvard University Press, 1987.

Kiefer, Peter, ed. *Klangräume der Kunst*. Heidelberg: Kehrer, 2010.

Kim-Cohen, Seth. *In the Blink of an Ear: Toward a Non-Cochlear Sonic Art*. New York: The Continuum International Publishing Group, 2009.

Kirkegaard, Jacob. *Earside Out*. Roskilde: Museum of Contemporary Art, 2015.

Kittel, Sören, and Nikolaus Röttger. "Alexander Ljung." *Wired*, February, 2008, 86–91.

Kittler, Friedrich A. *Gramophone, Film, Typewriter*. Translated by Geoffrey Winthrop-Young, and Michael Wutz. Stanford: Stanford University Press, 1999.

Kittler, Friedrich. *Musik und Mathematik*. Vol. 1, *Hellas 1: Aphrodite*. Munich: Wilhelm Fink, 2006.

Kleiner, Marcus S., and Achim Szepanski, eds. *Soundcultures: Über elektronische und digitale Musik*. Frankfurt am Main: Suhrkamp 2003.

Klonk, Charlotte. *Spaces of Experience: Art Gallery Interiors from 1800 to 2000*. New Haven: Yale University Press, 2009.

Knill, David C., and Whitman Richards, eds. *Perception as Bayesian Inference*. Cambridge: Cambridge University Press, 1996.

Koestler, Arthur. *The Act of Creation*. London: Hutchinson, 1964.

Köhler, Michael. "Er lehrt uns Hören mit System." *Frankfurter Allgemeine Zeitung*, no. 265 (November 2010): Z3.

Koller, Gabriele, Edek Bartz, and Gerald Bast, eds. *Secret Passion: Künstler und ihre Musik-Leidenschaften*. Wien, New York: Springer, 2010.

Könches, Barbara, and Peter Weibel, eds. *Stephan von Huene: Grenzgänger, Grenzverschieber*. Exhib. cat. ZKM | Center for Art and Media Karlsruhe. Heidelberg: Kehrer, 2006.

Kostakeva, Maria. *Metamorphose und Eruption: Annäherungen an die Klangwelten Adriana Hölszkys*. Hofheim am Taunus: Wolke, 2013.

Kostelanetz, Richard. *Conversing with Cage*. 2nd ed. New York: Routledge, 2003.

Kostelanetz, Richard. "Glenn Gould: Bach in the Electronic Age." In *Glenn Gould: By Himself and His Friends*, ed. John McGreevy, 134. Toronto: Macmillan, 1983.

Kotoński, Włodzimier. *Muzyka elektroniczna* [Electronic Music]. Kraków: Polskie Wydawnictwo Muzyczne, 2002.

Kovtun, E. "K. S. Malevich: Pis'ma k M. V. Matyushinu" [K. S. Malevich: Letters to M. V. Matyushin]. In *Yezhegodnik rukopis'nogo otdela Pushkinskogo doma na 1974* [Annual of the Manuscript Department of the Pushkin House for 1974]. Leningrad: Nauka, 1976.

Krogh Groth, Sanne. *Politics and Aesthetics in Electronic Music: A Study of EMS – Elektronmusikstudion Stockholm; 1964–1979*, Translated by Juliana Hodkinson, and Isabel Thomson. Heidelberg: Kehrer, 2014.

Krones, Hartmut, ed. *Anestis Logothetis: Klangbild und Bildklang*. Vienna: Lafite, 1998.

Kubiczak, Harald, ed. *Kästrich: Ein unabhängiges Forum für Kunst; Mit einem Report über Independent Art Circuits*. Frankfurt am Main: Circuit-Art, 2009.

Kubisch, Christina, and Bernhard Leitner. *Zeitversetzt / Shifted In Time. Ettersburger Klangräume / Ettersburg Sound Spaces*. Heidelberg: Kehrer, 2004.

Kullberg, Mads, and Jacob Kreutzfeldt, eds. *Simple interactions: Sound Art from Japan*. Roskilde: The Museum of Contemporary Art, 2011.

Kunstverein Göttingen e.V., ed. *Qin Yufen*. Exhib. Cat. Tübingen: Ernst Wasmuth Verlag, 2001.

Kurjaković, Daniel, and Sebastian Lohse, eds. *Other Rooms, Other Voices: Audio works by artists*. Zürich: Memory/Cage Editions, 1999.

L

LaBelle, Brandon, ed. *Manual for the construction of a sound as a device to elaborate social connection*. Berlin: Errant Bodies Press, 2010.

LaBelle, Brandon, and Steve Roden, eds. *Site of Sound: Of Architecture and the Ear*. Los Angeles: Errant Bodies Press, 1999.

Lacan, Jacques. "Le Stade du miroir comme formateur de la fonction du Je: telle qu'elle nous est révélée dans l'expérience psychanalytique." *Revue française de psychanalyse* 13 (1949): 449–55.

Lammert, Angela, and Hubertus von Amelunxen, eds. *iannis xenakis: kontrolle und zufall*. Berlin: Akademie der Künste, 2011.

Lander, Dan, and Daina Augaitis, eds. *Radio Rethink: Art, Sound and Transmission*. Banff: Walter Phillips Gallery, 1994.

Lander, Dan, and Micah Lexier, eds. *Sound by Artists*. Toronto: Art Metropole; Banff: Walter Phillips Gallery, 1990.

Lattner, Heimo. *A voice that once was in one's mouth*. Berlin: Errant Bodies Press, 2014.

Le Corbusier. *My Work*. Translated by James Palmes. London: Architectural Palmes, 1960.

Leitner, Bernhard. *Kopfräume / Headscapes*. Edited by ZKM | Center for Art and Media Karlsruhe. Ostfildern: Hatje Cantz, 2003.

Leitner, Bernhard. *P.U.L.S.E. Räume der Zeit / Spaces in Time*. Edited by Peter Weibel and ZKM | Center for Art and Media Karlsruhe. Ostfildern: Hatje Cantz, 2008.

Leitner, Bernhard. *Sound:Space*. Edited by Christine Traber. Ostfildern: Cantz, 1998.

Lensing, Jörg Udo, ed. *Theater der Klänge: 1987–2012*. Düsseldorf: Theater der Klänge, 2012.

Levent, Nina, and Alvaro Pascual-Leone. *The Multisensory Museum: Cross-Disciplinary Perspectives on Touch, Sound, Smell, Memory, and Space*. Lanham: Rowman & Littlefield, 2014.

Licht, Alan. *Sound Art: Beyond Music, Between Categories*. New York: Rizzoli, 2007.

Ligeti, György. "Neue Notation – Kommunikationsmittel oder Selbstzweck?" In *Darmstädter Beiträge zur Neuen Musik*. Vol. 9, *Notation Neuer Musik*, ed. Ernst Thomas, 35–50. Mainz: Schott, 1965.

Lischka, Gerhard Johann, and Peter Weibel, eds. *Kunstforum: Im Netz der Systeme: Für eine interaktive Kunst; Ars Electronica Linz*. Exhib. cat. Linz: Brucknerhaus, 1989.

Lissa, Zofia. *Estetyka muzyki filmowej* [The Aesthetics of Film Music]. Krakow: Polskie Wydawnictwo Muzyczne, 1964.

Listov, K. "Ekscentricheskij shumovoj orkestr 'Sinej bluzy'" [The Eccentric Noise Orchestra "Blue Blouse'"]. *Sinjaya bluza* 40, no. 16 (1926): n.p.

LIVA, ed. *Der Freie Klang. Ars Electronica 1987*. Exhib. cat. Linz: Ars Electronica, 1987.

Lucier, Alvin. *Klangskulpturen*. Edited by Berliner Künstlerprogramm des DAAD. Berlin: daadgalerie, 1991.

Lucier, Alvin. *Music 109: Notes on Experimental Music*. Middletown, CT: Wesleyan University Press, 2012.

Lucier, Alvin. *Reflections: Interviews, Scores, Writings*. Cologne: MusikTexte, 1995.

Luther Adams, John. *The Place Where You Go to Listen: In Search of an Ecology of Music*. Middletown, CT: Wesleyan University Press, 2009.

M

Magarotto, Luigi, Marzio Marzaduri, and Giovanna Pagani Cesa. *L'avanguardia a Tiflis: Studi, ricerce, cronache, testimonianze, documenti*. Venice, 1982.

Manovich, Lev. *The Language of New Media*. Cambridge, MA / London: The MIT Press, 2001.

Marotti, William. "Sounding the Everyday: The Music Group and Yasunao Tone's Early Work." In *Yasunao Tone: Noise Media Language*, ed. Sheila Jasanoff et al., 13–33. Los Angeles: Errant Bodies, 2007.

Marta Herford, ed. *Booster: Kunst, Sound, Maschine*. Bielefeld / Berlin: Kerber, 2014.

Marx, Ursula et al., eds. *Walter Benjamin's Archive: Images, Texts, Signs*. London: Verso, 2015.

Matejovski, Dirk, ed. *Resonanzräume: Medienkulturen des Akustischen*. Düsseldorf: düsseldorf university press, 2014.

Mathews, Max V., and Lawrence Rosler. "Graphical Language for the Scores of Computer- Generated Sounds." In *Music by Computers*, ed. Heinz von Foerster, and James W. Beauchamp, 84–114. New York: John Wiley and Sons, 1969.

Matossian, Nouritza. *Xenakis*. London: Kahn & Averill, 1986.

Matschoss, Conrad. *Die Entwicklung der Dampfmaschine. Eine Geschichte der ortsfesten Dampfmaschinen und der Lokomobile, derSchiffsmaschine und Lokomotive*. Vol. 1. Berlin: Julius Springer, 1908.

von Maur, Karin, ed. *Vom Klang der Bilder: Die Musik in der Kunst des 20. Jahrhunderts*. Munich: Prestel, 1985.

McLuhan, Eric, and Peter Zingrone, eds. *Essential McLuhan*. Concord, Ontario: House of Anansi Press, 1995.

Mermod, Mélanie. "On the Autoslides: Interview with Jōji Yuasa." In *Jikken kōbō: Atelier experimental/Experimental Workshop*, ed. Mélanie Mermod, 10. Exhib. cat. Paris: Bétonsalon, 2011.

Mertens, Heike Catharina, and Franziska Sollte, eds. *Haroon Mirza: – – {}{}{}*. Berlin: argobooks, 2012.

Metzger, Heinz-Klaus, and Rainer Riehn, eds. *Musik-Konzepte*. Vol. 54/55, *Iannis Xenakis*. Munich: Edition Text+Kritik, 1987.

Metzger, Heinz-Klaus, and Rainer Riehn, eds. *Musik-Konzepte*. Vol. 82, *Jean Barraqué*. Munich: Edition Text+Kritik, 1993.

Meyer, Petra Maria, ed. *acoustic turn*. Munich / Paderborn: Wilhelm Fink, 2008.

Migone, Christof. *Sonic Somatic: Performances of the Unsound Body*. Berlin: Errant Bodies Press, 2012.

Minkowski, Eugène. *Vers une cosmologie: fragments philosophiques*. Paris: F. Aubier, 1936.

Miranda, Eduardo Reck, ed. *Música y nuevas tecnologías: Perspectivas para el siglo XXI*. Barcelona: L'Angelot, 1999.

Mitchell, W. J. T., and Mark B. N. Hansen. *Critical terms for media studies*. Chicago, London: University of Chicago Press, 2010.

de la Motte-Haber, Helga, ed. *Handbuch der Musik im 20. Jahrhundert*. Vol 12, *Klangkunst: Tönende Objekte und klingende Räume*. Laaber: Laaber-Verlag, 1999.

de la Motte-Haber, Helga. *Musik und Bildende Kunst: Von der Tonmalerei zur Klangskulptur*. Laaber: Laaber, 1990.

de la Motte-Haber, Helga, Matthias Osterwold, and Georg Weckwerth, eds. *Sonambiente 2006: Berlin Klang Kunst Sound Art*. Heidelberg: Kehrer, 2006.

MTV Music Television, ed. *MTV's Who's Who in Rock Video*. London: Zomba Books, 1983.

Muneyoshi (Sōetsu), Yanagi. *Cha To Bi* [Beauty and Tea]. Tokyo: Kodansha, 2000.

Museum Tinguely Basel, ed. *Haroon Mirza: hrm 199 Ltd*. Exhib. cat. Cologne: Snoeck, 2015.

N

Nakagawa, Katsushi, and Tomotaro Kaneko. "The Development of Sound Art in Japan: Sound Garden (1987–94) and the Classification of Sound Art by Yoshimura Hiroshi." *Cross Sections* 4 (2011): 56–62.

Nechaev, V. "Vspominaja Kruchenykh" [Remembering Kruchenykh]. *Minuvshee* [The Bygone]: *A Historical Almanac* 12 (1991): 383.

Needham, Joseph. *Science and Civilisation in China*. Vol. 4, *Physics and Physical Technology: Physics*. Cambridge: Cambridge University Press, 1962.

Neisser, Ulric. *Cognition and Reality*. San Francisco: W. H. Freeman, 1976.

Neuhaus, Max. *inscription. sound works volume I*. Ostfildern: Cantz, 1994.

Neuhaus, Max. *drawings. sound works volume II*. Ostfildern: Cantz, 1994.

Neuhaus, Max. *place. sound works volume III*. Ostfildern: Cantz, 1994.

Neuhaus, Max. "Listen." In *Sound by Artists*, ed. Dan Lander and Micah Lexier, 63–67. Toronto: Art Metropole, 1990.

Neuhaus, Max. "Sound Art?" In *Volume: Bed of Sound*. New York: P.S. 1 Contemporary Art Center, 2000.

Nicolai, Carsten. *pionier*. Leipzig: spector books, 2011.

Niebur, Louis. *Special Sound: The Creation and Legacy of the BBC Radiophonic Workshop*. Oxford: Oxford University Press, 2010.

Nikonova-Tarshis, Rea. "Ekologija pauzi" [Ecology of Pause]. In *Experimental' naya poezija: Izbrannie stat'i* [Experimental Poetry: Selected Articles], ed. Dmitry Bulatov, 181. Kaliningrad: Simplicij, 1996.

Nix, Leo M. L., and Wilhelm Schmidt, eds. and trans. *Herons von Alexandria Mechanik und Katoptrik*. Leipzig: Teubner, 1900.

North Whitehead, Alfred. *Process and Reality*. New York: The Free Press, 1978.

Nowak, Nik, and Anna-Catharina Gebbers. *Intermezzo: Nik Nowak*. Berlin: Nik Nowak, 2012.

Noy, Irene. *Emergency Noises: Sound Art and Gender*. Oxford: Peter Lang, 2017.

O

O'Doherty, Brian. *Inside the White Cube: The Ideology of the Gallery Space*. Berkeley, London: University of California Press, 1999.

Ono, Yoko. *Grapefruit: A book of instructions + drawings* [1964]. Lund: Bakhåll, 2000.

Ono, Yoko. *Grapefruit: A Book of Instructions and Drawings by Yoko Ono* [1964]. New York et al.:Simon & Schuster, 2000.

Oswald, John. "Plunderphonics, or Audio Piracy as a Compositional Prerogative." *Musicworks* 34 (Winter 1986): 5–8.

Ott, Claudia, trans. *101 Nacht* [101 Nights]. Zurich: Manesse, 2012.

Ouellette, Fernand. *Edgard Varèse*. Translated by Derek Coltman. New York: Orion, 1968.

P

Packer, Randall. "The Pepsi Pavilion: Laboratory for Social Experimentation." In *Future Cinema: The Cinematic Imaginary after Film*, ed. Jeffrey Shaw and Peter Weibel, 144–9. Exhib. cat. ZKM | Center for Art and Media Karlsruhe. Cambridge, MA / London: The MIT Press, 2003.

Paik, Nam June. "Musik des Zufalls: Darmstädter Ferienkurse für neue Musik." Translated by Kunsu Shim and Gerhard Stäbler. In *Nam June Paik. Fluxus/Video*, ed. Wulf Herzogenrath and Sabine Maria Schmidt, 22–23. Bremen: Kunsthalle Bremen, 1999.

Paik, Nam June. *Niederschriften eines Kulturnomaden: Aphorismen, Briefe, Texte*. Edited by Edith Decker. Cologne: DuMont, 1992.

Panhuysen, Paul, and Johan Goedhart. *Long String Installations: 1982–1985*. Eindhoven: Het Apollohuis, 1986.

Patkowski, Józef et al.,eds. *Res Facta 5*, 1971.

Pereverzev, L. *Iskusstvo i kybernetika* [Art and Cybernetics]. Moscow: Iskusstvo, 1966.

Phelan, Peggy. *Unmarked: The Politics of Performance*. London: Routledge, 1993.

Pichler, Cathrin, ed. *Sinneswerkzeug: Kunst als Instrument. Bernhard Leitner, Max Peintner, Alfons Schilling*. Vienna: Passagen, 1991.

Pisano, Hortense, ed. *words & sounds*. Exhib. cat. suesswasser e.V. Berlin: Revolver Publishing, 2013.

Plasencia, Clara, ed. *I Am Immortal and Alive: Gil J Wolman*. Barcelona: Museu d'Art Contemporani de Barcelona, 2010.

Plato. *The Republic*. Translated by Allan Bloom. 2nd ed. New York: Basic Books, 1991.

Polli, Andrea. "Soundscape, Sonification, and Sound Activism." *AI & Society* 27, no. 2 (2012): 257–68.

Pousseur, Henri. "La nuova sensibilià musicale." *Incontri musicali*, no. 2 (May 1958): 3–37.

Prieberg, Fred K. *Musica Ex Machina: Über das Verhältnis von Musik und Technik*. Frankfurt: Ullstein, 1960.

Priest, Gail. "Preface: Starting Somewhere." In *Experimental Music: Audio Explorations in Australia*, 1–8. Sydney: UNSW Press, 2009.

Puente, Moisés, ed. *CaboSanRoque: The Pataphysical Cobla 2015–2001*. Exhib. cat. Barcelona: Ars Santa Monica, 2015.

Pulkki, Ville. "Virtual Sound Source Positioning Using Vector Base Amplitude Panning." *Journal of the Audio Engineering Society* 45, no. 6 (June 1997): 456–66.

R

Rainer, Cosima et al., eds. *See This Sound: Versprechungen von Bild und Ton*. Cologne: Walther König, 2009.

Rancière, Jacques. *The Politics of Aesthetics*. Edited and translated by Gabriel Rockhill. London: Bloomsbury, 2013.

Razzāz al-Jazarī, Ibn al-. *The Book of Knowledge of Ingenious Mechanical Devices (Kitāb fī ma'rifat al-iyal al-handasiyya)*. Edited and translated by Donald R. Hill. Dordrecht: Reidel, 1974.

Reeder, Roberta, and Claas Cordes. "Light Music in the Soviet Union." In *Ars Electronica 1992: Endo Nano (part 02)*, 225–40. Linz: Ars Electronica, 1992.

Richmond, Cindy, and Scott Watson, eds. *Rebecca Belmore: fountain*. Exhib. cat. Kamloops: Kamloops Art Gallery; Vancouver: The Morris and Helen Belkin Art Gallery, 2005.

Roads, Curtis. "Research in Music and Artificial Intelligence." *Computing Surveys* 17, no. 2 (1985): 163–90.

Robertson, Lisa. *Nilling*. Toronto: BookThug, 2012.

Ross, Alex. *The Rest Is Noise*. Munich, Berlin: Piper, 2009.

Rotta, Christian, and Dirk Katzschmann, eds. *Universitas Orientierung in der Wissenschaft* 61, no. 717 (March 2006).

Rudi, Jøran. *Soundscape i kunsten. Soundscape in the Art*. Oslo: NOTAM, 2011.

Rumyantsev, S. *Ars Novij, ili Dela I prikljuchenija bezustal'nogo kazaka Arsenija Avramova* [Ars Nova, or the Deeds and Adventures of the Indefatigable Cossack Arseny Avraamov]. Moscow: Deka-BC, 2007.

Rumyantsev, S. "Samodejatel'nie orkestry I instrumental'nye ansambli" [Amateur Orchestras and Instrumental Ensembles]. In *Samodejatel' noje khudozestvennoje tvorchestvo v SSSR: Ocherki istorii* [Amateur Art in the USSR: Historical Sketches 1917–1932], 293–313. St. Petersburg: Bulanin, 2000.

Ruschkowski, André. *Elektronische Klänge und musikalische Entdeckungen*. Stuttgart: Reclam, 2010.

Russolo, Luigi. *The Art of Noises*. Translated by Barclay Brown. New York: Pendragon Press, 1986.

Russolo, Luigi. *The Art of Noise (Futurist Manifesto, 1913)*. Translated by Robert Filliou. New York: Something Else, 1967.

S

Sanio, Sabine, and Christoph Metzger, eds. *Musik im Dialog: Jahrbuch der Berliner Gesellschaft für Neue Musik 1997*. Saarbrücken: Pfau 1998.

Schaeffer, Bogusław. "On Audiovisual Music: On Action Music." In: *Atelier '72*, ed. Richard Demarco, n.p. Exhib. cat. Edinburgh: Richard Demarco Gallery, 1972.
Schaeffer, Bogusław. "Z notatnika" [From the Notebook]. In *Forum Musicum 2: Muzyka w studio*, ed. Teresa Chylińska, and Bogusław Schaeffer, 2–11. Krakow: Polskie Wydawnictwo Muzyczne, 1968.

Schaeffer, Pierre. *Treatise on Musical Objects. An Essay across Disciplines*. Translated by Christine North, and John Dack. Oakland, CA: University of California Press, 1966. First published as *Traité des objets musicaux: essai interdisciplines*. Paris: Éditions du Seuil, 1966.

Schaerf, Eran. *Recasting*. Antwerp: Museum van Hedendaagse Kunst Antwerpen, 1997.

Schafer, R. Murray. "Radical Radio." *Ear Magazine* (Winter 1987): 18–20.

Schafer, R. Murray. *The Soundscape: Our Sonic Environment and the Tuning of the World*. Rochester, VT: Destiny Books, 1994.

Schafer, R. Murray. *The Tuning of the World*. New York: Knopf, 1977.

Schäfer, Sabine, and Joachim Krebs, eds. *TopoSonic Arts 1997–2006*. Heidelberg: Kehrer, 2007.

Scherchen, Hermann. "Was müsste ein ideales Tonstudio leisten?" In *Iannis Xenakis: Kontrolle und Zufall*, ed. Angela Lammert and Hubertus von Amelunxen, 34–5. Berlin: Akademie der Künste, 2011. Originally published in *Werke und Briefe*. Vol. 1, *Schriften 1*, ed. Joachim Lucchesi, 147–8. Berlin: Peter Lang, 1991.

Schillinger, Joseph. *The Mathematical Basis of the Arts*. New York: Philosophical Library, 1948.

Schirn Kunsthalle Frankfurt, and Museum Tinguely Basel, eds. *Kunstmaschinen – Maschinenkunst*. Exhib. cat. Heidelberg: Kehrer, 2007.

Schläger, Christof, and Hanno Ehrler. *Urbane Rituale: Der Klangkünstler Christof Schläger*. Bochum: Hanno Ehrler, 2015.

Schmidgen, Henning. *Hirn & Zeit: Die Geschichte eines Experiments 1800–1950*. Berlin: Matthes & Seitz, 2014.

Schoenberg, Arnold. "Composition with Twelve Tones (1)." In *Style and Idea: Selected Writings*, ed. Leonard Stein, trans. Leo Black, 214–45. Berkeley: University of California Press, 2010.

Schöllhammer, Georg, Johannes Porsch, and Channa Horwitz. "Georg Schöllhammer & Johannes Porsch. Artist Talk with Channa Horwitz, 03/18/2012, ZKM | Center for Art and Media Karlsruhe." In *Moments. Eine Geschichte der Performance in 10 Akten*, ed. Sigrid Gareis, Georg Schöllhammer, and Peter Weibel, 426–8. Exhib. cat. ZKM | Center for Art and Media Karlsruhe. Cologne: Walther König, 2013.

Schoon, Andi. *Die Ordnung der Klänge: Das Wechselspiel der Künste vom Bauhaus zum Black Mountain College*. Bielefeld: transcript, 2006.

Schraenen, Guy. *Vinyl: Records and Covers by Artists*. New York: Actar, 2006.

Schuller, Gunther, and Edgard Varèse. "Conversation with Varèse." *Perspectives of New Music* 3, no. 2 (Spring–Summer 1965): 32–7.

Schulz, Bernd, ed., *Christina Kubisch. Zwischenräume*, Saarbrücken: Stadtgalerie Saarbrücken, 1996.

Schulz, Bernd, ed., *Klangräume. Baschet, Bernard et al*. Saarbrücken: Stadtgalerie Saarbrücken, 1988.

Schulz, Bernd, and Irmgard Heesche, eds., *Resonanzen: Aspekte der Klangkunst*. Heidelberg: Kehrer, 2002.

Schulze, Holger, ed., *Gespür – Empfindung -- Kleine Wahrnehmungen: Klanganthropologische Studien*. Bielefeld: transcript, 2012.

Schulze, Holger, ed. *Sound Studies: Traditionen – Methoden – Desiderate: Eine Einführung*. Bielefeld: transcript, 2008.

Schuppli, Madeleine, and Aargauer Kunsthaus, Aarau, eds. *Rhythm in it*. Exhib. cat. Luzern/Poschiavo: Edizioni Periferia, 2013.

Schwartz, Elliott, and Barney Childs. *Contemporary Composers on Contemporary Music*. Boston: DaCapo, 1998.

Schwitters, Kurt. "Merz." Translated by Ralph Manheim. In *Modern Artists on Art*, ed. Robert L. Herbert, 68–75. 2nd ed. Mineola, NY: Dover, 2000.

Seiffarth, Carsten, and Markus Steffens, eds. *singuhr-hœrgalerie in parochial 1996–2006: sound art in berlin*. Heidelberg: Kehrer, 2010.

Seiffarth, Carsten, and Markus Steffens, eds. *singuhr-hoergalerie 2007–2014: sound installation art in berlin*. Heidelberg, Berlin: Kehrer, 2015.

Shanken, Edward E. "From Cyberbetics to Telematics: The Art, Pedagogy, and Theory of Roy Ascott." In Roy Ascott, *Telematic Embrace: Visionary Theories of Art, Technology, and Consciousness*, ed. by Edward E. Shanken, 21. Berkeley: University of California Press, 2007.

Shimamoto, Shozō. "Can a Piece of Cloth Be a Work of Art?." Translated by Reiko Tomii. In *Electrifying Art: Atsuko Tanaka, 1954–1968*, ed. Ming Tiampo, 107–9. Exhib. cat. Vancouver / New York: Morris and Helen Belkin Art Gallery et al., 2004. Originally published as "Ichi-mai no nunokire demo geijutsu sakuhin ka?" *Gutai* 4, (July 1956): 31.

Shirky, Clay. *Here Comes Everybody: How Change Happens When People Come Together*. London: Penguin, 2009.

Sikorski, Tomasz. *Echa II*. Kraków: Polskie Wydawnictwo Muzyczne, 1963.

Skoruppa, Ekkehard, Marie-Louise Goerke, Gaby Hartel, and Hans Sarkowicz, eds. *Choreography of Sound: Between Abstraction and Narration*. Göttingen: Vandenhoeck & Ruprecht, 2015.

Skulpturenmuseum Glaskasten Marl, ed. *Die Stillen: Klangräume – Klanginstallationen – Klangwelten*. Marl: Skulpturenmuseum Glaskasten Marl, 1994.

Slonimsky, Nicolas. *Lexicon of Musical Invective: Critical Assaults on Composers since Beethoven's Time*. New York: Coleman-Ross, 1965.

Small, Christopher. *Musicking: The Meanings of Performing and Listening*. Hanover, NH: University Press of New England, 1998.

Smirnov, Andrey. *Sound in Z: Experiments in Sound and Electronic Music in Early 20th-Century Russia*. Edited by Matt Prize. Cologne: Walther König; London: Sound of Music, 2013.

Šmite, Rasa, and Raitis Šmits, eds. *Acoustic Space / Akustikā Telpa*. Riga: E-Lab, 1999.

Smith, Mark M., ed. *Hearing History*. Athens, GA: University of Georgia Press, 2004.

Smolik, Noemi, and Peter Weibel, eds. *Milan Grygar. Sound on Paper*. Exhib. cat. Karlsruhe: ZKM | Center for Art and Media, 2016.

Snow, Michael. *The Collected Writings of Michael Snow*. Waterloo, ON: Wilfrid Laurier University Press, 1994.

Solev, V. "Risovannij zvuk" [Drawn Sound]. *Radio-front*, no. 3 (1935): n.p.

Sonntag, Jan Peter E.R., ed. *sonArc::project: Der domestizierte Blitz*. Exhib. cat. Berlin: Kadmos, 2008.

Spötter, Walter, ed. *documenta 7*. Kassel: D + V Paul Dierichs, 1982.

Spötter, Walter, ed. *documenta 8: Kassel 1987; 12. Juni–20. Sept.* Kassel: Weber & Weidemeyer, 1987.

Sterne, Jonathan. *The Sound Studies Reader*. New York: Routledge, 2012.

Stevens, John. *Search and Reflect: A Music Workshop Handbook*. London: Rockschool, 1985.

Strawn, J. M. "The Intégrales of Edgard Varèse: Space, Mass, Element, and Form." *Perspectives of New Music* 17, no. 1 (Autumn/Winter 1978): 141.

Sukhanova, Tatiana. *Ljubitel'skoe khudozestvennoe tvorchestvo v Rossii XX veka* [Amateur Art in 20th-Century Russia]. Moscow: Progress-Tradicija, 2010.

Supper, Alexandra. *Lobbying for the Ear: The Public Fascination with and Academic Legitimacy of the Sonification of Scientific Data*. Maastricht: Datawyse / Universitaire Pers Maastricht, 2012.

Supper, Alexandra. "The Search for the 'Killer Application': Drawing the Boundaries around the Sonification of Scientific Data." In *The Oxford Handbook of Sound Studies*, ed. Trevor Pinch, and Karin Bijsterveld, 249–70. New York: Oxford University Press, 2012.

Suzuki, Akio. *"A": Sound Works. Throwing and Following*. Edited by Bernd Schulz. Saarbrücken: Stadtgalerie Saarbrücken, 1998.

Sylvester, James J. "Chemistry and Algebra" *Nature* 17, no. 284 (February 1878): 284.

Szabó, István, ed. *Geschichte der mechanischen Prinzipien und ihrer wichtigsten Anwendungen*. Basel / Stuttgart: Birkhäuser, 1977.

Székely, Katalin, ed. *The Freedom of Sound: John Cage behind the Iron Curtain*. Budapest: Ludwig Museum – Museum of Contemporary Art Budapest, 2013.

T

Taruskin, Richard. *The Oxford History of Western Music*. Vol. 5, *The Late Twentieth Century*. New York: Oxford University Press, 2005.

Tazelaar, Kees. *On the Threshold of Beauty: Philips and the Origins of Electronic Music in the Netherlands 1925–1965*. Rotterdam: V2_Publishing, 2013.

Teltenkötter, Klaus, ed. *resonate: interactive light & sound installation, Luminale 2012*. Mainz: Interior Architecture FH Mainz Design, 2012.

Theremin, Léon. *Fizika I muzikal'noje iskusstvo* [Physics and Musical Art]. Moscow: Znanie, 1966.

Thurmann-Jajes, Anne, Sabine Breitsameter, and Winfried Pauleit, eds. *Sound Art: Zwischen Avantgarde und Popkultur*. Cologne: Salon, 2006.

Tiampo, Ming. *Gutai: Decentering Modernism*. Chicago / London: University of Chicago Press, 2011.

Tiampo, Ming. "Gutai Chain: The Collective Spirit of Individualism." *positions* 21, no. 2 (2013): 383–415.

Tibbits, Skylar. "Design to Self-Assembly." *Architectural Design* 82, no. 2 (2012): 68–73.

Tibbits, Skylar, and Kenny Cheung. "Programmable Materials for Architectural Assembly and Automation." *Assembly Automation* 32, no. 3 (2012): 216–25.

Tomkins, Calvin. *Marcel Duchamp: The Afternoon Interviews*. Cologne: Walther König, 2013.

Toop, David. *New/Rediscovered Musical Instruments*. London: Quartz / Mirliton, 1974.

Toop, David. *Sonic boom: The Art of Sound*. Exhib. cat. London: Hayward Gallery Publishing, 2000.

Transit, ed. *On The Air: Kunst im öffentlichen Datenraum*. Innsbruck: Transit, 1993.

Treib, Marc. *Space Calculated in Seconds: The Philips Pavilion, Le Corbusier, Edgard Varèse*. Princeton: Princeton University Press, 1996.

Tsunoda, Toshiya. "On Portrayals (a reflection)." *Santa: Sound and word review* 10 (2008): 3–4.

Turing, Alan M. "On Computable Numbers, with an Application to the Entscheidungsproblem." *Proceedings of the London Mathematical Society* s2–42, no.1 (1937): 230–65.

U

Ungeheur, Elena, and Oliver Wiener, eds. *Klangforschung in Musik*. Heidelberg: Kehrer, 2012.

Uno-Everett, Yayoi. "Scream Against the Sky: Japanese Avant-Garde Music in the Sixties." In *Sound Commitments: Avant-Garde Music and the Sixties*, ed. Robert Addlington, 187–208. Oxford: Oxford University Press, 2009.

Uthemann, Ernest W., ed. *Stadtgalerie Saarbrücken 1994–2011: Eine Dokumentation*. Saarbrücken: Stadtgalerie Saarbrücken, 2011.

V

Varèse, Edgard. “Rhythm, Form and Content.” In *Contemporary Composers on Contemporary Music*, ed. Elliott Schwartz, and Barney Childs, 201–4. New York: Holt, Rinehart and Winston, 1967.

Varèse, Edgard, and Chou Wen-chung. “The Liberation of Sound.” *Perspectives of New Music* 5, no. 1 (Autumn–Winter 1966): 11–19.

Varèse, Louise. *A Looking-Glass Diary*. Vol. 1, *1883–1928*. New York: Norton, 1972.

Voegelin, Salomé. *Sonic Possible Worlds: Hearing the Continuum of Sound*. New York: Bloomsbury, 2014.

Volmar, Axel, and Jens Schröter, eds. *Auditive Medienkulturen: Techniken des Hörens und Praktiken der Klanggestaltung*. Bielefeld: transcript, 2013.

W

Wagner, Christoph. *Der Klang der Revolte: Die magischen Jahre des westdeutschen Musik-Underground*. Mainz: Schott Music, 2013.

Warner, Daniel. *Live Wires: A History of Electronic Music*. London: Reaction Books, 2017.

Warner, Michael. *Publics and Counterpublics*. Cambridge, MA: The MIT Press; London: Zone Books, 2005.

Weber, Samuel. *Mass Mediauras: Form, Technics, Media*. Stanford: Stanford University Press, 1996.

Weibel, Peter. “Art as Open Practice.” Translated by Julia Scheffer. In *Offene Handlungsfelder / Open Practices / Prassi aperta*, ed. Peter Weibel, 22–35. Exhib. cat. Cologne: DuMont, 1999.

Weibel Peter. *Enzyklopädie der Medien*, Vol. 2, *Musik und Medien*, ed. ZKM | Center for Art and Media Karlsruhe and University of Applied Arts Vienna. Berlin: Hatje Cantz, 2016.

Weibel, Peter. “New Protagonists and Alliances in 21st Century Art.” In *Ars Electronica 2004: Timeshift; The World in Twenty-Five Years*, ed. Gerfried Stocker and Christine Schöpf, 38–40. Exhib. cat. Ostfildern: Hatje Cantz, 2004.

Weinberger, David. *Everything Is Miscellaneous: The Power of the New Digital Disorder*. New York: Henry Holt, 2007.

Wiedemann, Eilhard. “Ueber Musikautomaten bei den Arabern.” In *Centenaro della nascita di Michele Amari*. Vol. 2, *Scritti di filosofia e storia araba [...]*, 164–85. Palermo: Stabilimento Tipografico Virzì, 1910.

Willer, Stefan, Daniel Weidner, and Sigrid Weigel, eds. *Trajekte: Zeitschrift des Zentrums für Literatur- und Kulturforschung Berlin; Zuhören “Listen”* 15, no. 29 (October 2014).

Wilson, Stephen. *Information Arts: Intersections of Art, Science, and Technology*. Cambridge, MA / London: The MIT Press, 2002.

Wittgenstein, Ludwig. *Philosophical Investigations*. Edited by P. M. S. Hacker and Joachim Schulte, translated by G. E. M. Anscombe. Oxford: Wiley-Blackwell, 2009.

X

Xenakis, Iannis. *Formalized Music: Thought and Mathematics in Composition*. Stuyvesant: Pendragon Press, 1992.

Y

Yoshimoto, Midori. “Expo ‘70 and Japanese Art: Dissonant Voices – An Introduction and Commentary.” *Review of Japanese Culture and Society* 23 (2011): 1–12.

Yost, William A. *Fundamentals of Hearing: An Introduction*. San Diego et al.: Academic Press, 2006.

Young, La Monte, and Marian Zazeela. *Selected Writings*. Munich: Heiner Friedrich, 1969.

Yurtsev, Boris. “Orkestr veschej” [The Orchestra of Things]. *Zrelischa*, no. 6 (1922): n.p.

Z

Zakharine, Dmitri, and Nils Meise, eds. *Electrified Voices: Medial, Socio-Historical and Cultural Aspects of Voice Transfer*. Vol. 1. Göttingen: V&R Unipress, 2013.

Zaripov, Rudolf. *Kybernetika y muzyka* [Cybernetics and Music]. Moscow: Znanie, 1963.

Zaripov, Rudolf. “Modelirovanie melodicheskikh variatsii na tsifrovoi vychislitel’noi mashine” [Modeling Melodic Variations on a Digital Computer]. *Problemy Kibernetiki* 40 (1983): 201–54.

Zielinski, Siegfried. “Expanded Animation: A Short Genealogy in Text and Images.” In *Pervasive Animation*, ed. Suzanne Buchan, 25–51. London: Routledge, 2013.

Zielinski, Siegfried. “Im Zustand der Schwingung kann es keine Ruhe geben: Ein kurzes Portrait des Physikochemikers Johann Wilhelm Ritter (1776–1810). Peter Weibel zum Geburtstag.” In *Liebesgrüsse aus Odessa: Für/For/À Peter Weibel*, ed. Ecke Bonk, Peter Gente, and Margit Rosen, 8–12. Berlin: Merve, 2004.

Zielinski, Siegfried. “Theologici electrici: Einige Passagen.” In, *Theologie und Politik: Walter Benjamin und ein Paradigma der Moderne*, ed. Bernd Witte, and Mauro Ponzi, 254–68. Berlin: ESV, 2005.

Zielinski, Siegfried, and Eckhard Fürlus, eds., in cooperation with Nadine Minkwitz. *Variantology 3: On Deep Time Relations of Arts, Sciences and Technologies in China and Elsewhere*. Cologne: Walther König, 2008.

Zielinski, Siegfried, and Peter Weibel, eds. *Allah’s Automata: Artifacts of the Arab-Islamic Renaissance (800–1200)*. Exhib. cat. ZKM | Center for Art and Media Karlsruhe. Ostfildern: Hatje Cantz, 2015.

ZKM | Center for Art and Media Karlsruhe, ed. *Sound Art. Sound as Medium of Art*. Exhib. brochure. Karlsruhe: ZKM | Center for Art and Media Karlsruhe, 2012.

ZKM | Center for Art and Media Karlsruhe, Adam Mickiewicz Institute, Muzeum Sztuki in Łodź, and David Crowley, eds., *Ultra Sounds: The Sonic Art of Polish Radio Experimental Studio*. Heidelberg: Kehrer, 2019.

Index of Names

M

N

O

P

R

Photo Credits

pp. 1–4: © Archive Peter Weibel
p. 13, fig. 1: © Archive Prof. Pietro Verardo, photo © ZKM | Karlsruhe
p. 13, fig. 2: Source: Luigi Russolo: *Die Kunst der Geräusche* (Schott: Mainz, 2000).
p. 14, figs. 3, 4: © Archive Prof. Pietro Verardo
p. 14, fig. 5: Creative Commons
p. 16, fig. 6: © Paul Sacher Foundation Basel, Collection Edgard Varèse
p. 16, fig. 7: © Paul Sacher Foundation Basel, Collection Edgard Varèse
p. 18, fig.10: © Centre Iannis Xenakis, Iannis Xenakis Family
p. 18, fig. 11: © F.L.C./VG Bild-Kunst, Bonn 2019 (Le Corbusier); Iannis Xenakis Family
p. 19, fig. 12: © Centre Iannis Xenakis, Iannis Xenakis Family
p. 21, fig. 14: Source: Dieter Daniels and Sandra Naumann (eds.): *See This Sound: Audiovisuology; Compendium; An Interdisciplinary Survey of Audiovisual Culture* (Cologne: Walther König, 2010) 109. © Ryoji Ito and the former members of the Gutai Art Association, courtesy Ashiya City Museum of Art & History
p. 22, fig. 15: © Liz Philips, photo © CCA Capp Street Archive
p. 22, fig. 16: © Peter Cusack
p. 22, fig. 17: © Thomas Gerwin, photo © ZKM | Karlsruhe
p. 24, top: © Man Ray 2015 Trust/ VG Bild-Kunst, 2019, photo © ZKM | Karlsruhe, photo: Tobias Wootton
p. 24, bottom: © Man Ray 2015 Trust/ VG Bild-Kunst, 2019, photo © courtesy Fondazione Marconi, Milan, photo: Giani Ummarino
p. 25, top: © VG Bild-Kunst, Bonn 2019 (Maurice Henry)
p. 25, bottom: © VG Bild-Kunst, Bonn 2019 (Arman)
p. 26, top, left and right: © Loys Egg, Peter Weibel, photos: Manfred Rahs
p. 26, bottom, left: © Peter Weibel
p. 26, bottom, right: © the artist
p. 27, top: © VG Bild-Kunst, 2019 (Brothers Baschet)
p. 27, bottom: © Galerie Andrea Caratsch, Zurich
p. 28: © VG Bild-Kunst, 2019 (Richard Artschwager)
p. 29: © VG Bild-Kunst, 2019 (Hendrik Neugeboren)
p. 30: © the artist
p. 31: © the artist
p. 32: © VG Bild-Kunst (Jean Tinguely), 2019, photo © courtesy NCAF et Galerie GP & N Vallois, Paris, photo: André Morin
p. 33: © VG Bild-Kunst, 2019 (Jean Tinguely), photo © ZKM | Karlsruhe, photo: ONUK
p. 34: © VG Bild-Kunst, 2019 (Sarkis), photo © Kunst- und Ausstellungshalle der Bundesrepublik Deutschland, Bonn
p. 35: © Archivio Conz Berlin, photo © Supportico Lopez
p. 36, top: Robert Rauschenberg Foundation / © VG Bild-Kunst, Bonn 2019
p. 37, top: Robert Rauschenberg Foundation / © VG Bild-Kunst, Bonn 2019, photo © Rheinisches Bildarchiv, Cologne
p. 37, bottom right: Robert Rauschenberg Foundation / © VG Bild-Kunst, Bonn 2019
p. 38: © VG Bild-Kunst, Bonn 2019 (Susan Hiller), photo © Susan Hiller, courtesy Lisson Gallery. Artwork commissioned by Artangel.
p. 40, bottom: © VG Bild-Kunst, Bonn 2019 (David Jacobs)
p. 41: photo © Hélène Panhuysen, photo: Hélène Panhuysen
pp. 42–43: photo © Hélène Panhuysen, photo: Hélène Panhuysen
p. 44: © the artist
p. 45: installation views Atelier Nord ANX, Oslo, 2013, photo © Atelier Nord
p. 46: © the artist
p. 47: © VG Bild-Kunst, Bonn 2019 (Trimpin), photo © https://soundartarchive.net
p. 48–49, top: © VG Bild-Kunst, Bonn 2019 (Robert Jacobsen)
p. 49, bottom: © VG Bild-Kunst, Bonn 2019 (Rebecca Horn)
p. 50: © VG Bild-Kunst, Bonn 2019 (Rebecca Horn)
p. 51: photo © Bill Fontana
p. 52, top: © Susan Philipsz Turner Prize Installation Photograph © Tate, London 2019. Courtesy the artist, Isabella Bortolozzi Galerie, Berlin and Tanya Bonakdar Gallery, New York / Los Angeles
p. 52, bottom: © bpk / Nationalgalerie im Hamburger Bahnhof, Staatliche Museen zu Berlin / Thomas Bruns. Courtesy the artist, Isabella Bortolozzi Galerie, Berlin, Tanya Bonakdar Gallery, New York / Los Angeles
p. 53, top: © Kunstverein Hannover / Raimund Zakowski
p. 53, bottom: photo © ZKM | Karlsruhe, photo: Steffen Harms
p. 56: photo © Archive of Stockhausen Foundation for Music (www.kalheinzstockhausen.org)
p. 57, fig. 22: © VG Bild-Kunst, Bonn 2019 (Nicolas Schöffer)
p. 58, figs. 23, 24: © VG Bild-Kunst, Bonn 2019 (Nicolas Schöffer)
p. 59, figs. 27, 28, 29, 33: photo © Iannis Xenakis Family
p. 60, figs. 31, 32: photo © Iannis Xenakis Family, photos: Akademie der Künste, Berlin, Hermann-Scherchen archive
p. 61, bottom: © Roy Lichtenstein Foundation, courtesy Experiments in Art & Technology, photo: Shunk-Kender
p. 63: photo © ZKM | Karlsruhe
p. 65: photo © ZKM | Karlsruhe
p. 67, fig. 43: © Paul Sacher Foundation Basel, Collection Earle Brown, photo: Earle Brown
p. 68, fig. 44: The Estate of Yves Klein / © VG Bild-Kunst, Bonn 2019
p. 68, fig. 45: © Anka Ptaszkowska; the negatives and slides are owned by the Museum of Modern Art in Warsaw
p. 68, fig. 46: © VG Bild-Kunst, Bonn 2019 (Jannis Kounellis)
p. 69, fig. 47: photo © Courtesy Galerie Buchholz, Berlin/Cologne/New York and the Estate of Jack Goldstein
p. 69, fig. 48: photo © Christian Marclay. Courtesy Paula Cooper Gallery, New York
p. 71, fig. 49: Association Marcel Duchamp / © VG Bild-Kunst, Bonn 2019
p. 71, fig. 50: © VG Bild-Kunst, Bonn 2019 (Robert Morris)
p. 71, fig. 51: © VG Bild-Kunst, Bonn 2019 (Bruce Nauman)
p. 71, fig. 52: photo © courtesy of the artist
p. 71, fig. 53: Harry Partch Estate, photo: Betty Freeman
p. 72, fig. 55: © Chris Harris, London
p. 73, figs. 57, 58: photo © Atelier Leitner
p. 74: © VG Bild-Kunst, Bonn 2019 (George Maciunas)
p. 75: photo: Keiko Yoshida
p. 76, top: photo © IMAGNO, photo: Franz Hubmann
p. 76, bottom: © VG Bild-Kunst, Bonn 2019 (George Maciunas, Wolf Vostell)
p. 77, top: © VG Bild-Kunst, Bonn 2019 (Günther Uecker), photo © ZKM | Karlsruhe, photo: Steffen Harms

p. 77, bottom: photo © atelier rosalie, photo: Wolf-Dieter Gericke
p. 78, top: © VG Bild-Kunst, Bonn 2019 (George Maciunas, George Brecht)
p. 78, bottom: © VG Bild-Kunst, Bonn 2019 (George Brecht)
p. 79, top: © VG Bild-Kunst, Bonn 2019 (George Brecht, Jim Dine)
p. 79, bottom: screenshots: http://www.openculture.com/2016/11/young-frank-zappa-plays-the-bicycle-on-the-steven-allen-show-1963.html
p. 80: © VG Bild-Kunst, Bonn 2019 (Bruce Nauman, Richard Serra)
p. 81, top: photo © Collection Centre Pompidou, Paris
p. 81, bottom: photo © Loys Egg, photo: Heinz W. Lindinger
p. 82, bottom: photo © CCA Capp Street Archive
p. 83: photo © mumok – Museum moderner Kunst Stiftung Ludwig Wien, formerly Sammlung Hahn
p. 84, fig. 61: © VG Bild-Kunst, Bonn 2019 (Robert Longo)
p. 88, fig. 64: photo © Tate/Audio Arts
p. 89, fig. 66: Association Marcel Duchamp / © VG Bild-Kunst, Bonn 2019, photo © ZKM | Karlsruhe, photo: Tobias Wootton
p. 92: photo: Takahiko Iimura
p. 93: © VG Bild-Kunst, Bonn 2019 (Rebecca Horn)
p. 94, top: photo © ZKM | Karlsruhe, photo: Tobias Wootton
p. 95: © VG Bild-Kunst, Bonn 2019 (Waltraut Cooper)
pp. 96–97: photo © Martin Riches
p. 98: photo © courtesy of the artist
p. 99: photo © Ars Electronica Archive
p. 100–01: © VG Bild-Kunst, Bonn 2019 (Pierre Huyghe), photo © courtesy of the artist and the Marian Goodman Gallery, New York
p. 102, top: © VG Bild-Kunst, Bonn 2019 (Caroline Heider), photo © ZKM | Karlsruhe, photo: Steffen Harms
p. 102, bottom: © VG Bild-Kunst, Bonn 2019 (Joachim Baur), photo © ZKM | Karlsruhe, photo: Tobias Wootton
p. 103: © VG Bild-Kunst, Bonn 2019 (Uwe Bressnik), photos: Uwe Bressnik
p. 105, top: photo © Hirshhorn Museum and Sculpture Garden, Smithsonian Institution, Washington D.C., États-Unis. Gift of Mary and Leigh B. Block, 1988
p. 106: photo © Yale University Library
p. 107, figs. 74–76: © VG Bild-Kunst, Bonn 2019 (Raoul Hausmann)
p. 108: photo © Estate Oskar Fischinger
p. 109: photo © Pfenninger Archive, Munich
p. 110, top right: photo © Lebrecht Music
p. 111, top left, middle left, bottom left: © Light Cone
p. 111, middle right, bottom right: © National Film Board of Canada
p. 113: © Belmont Music Publishers / VG Bild-Kunst, Bonn 2019 (Arnold Schönberg)
p. 114, bottom, right: © Ader, Maison de vente. Courtesy Magne family.
p. 119, top right, left: © VG Bild-Kunst, Bonn 2019 (Anestis Logothetis)
p. 120–121: photo © courtesy of the artist
p. 122: © VG Bild-Kunst, Bonn 2019 (Henning Christiansen)
p. 123, fig. 98: photo © Lisson Gallery, London/New York
p. 124, figs. 99a, b: © VG Bild-Kunst, Bonn 2019 (George Brecht)
p. 124, fig. 99a: photo © ZKM | Karlsruhe, photo: Tobias Wootton
p. 126, fig. 103: © Iannis Xenakis Family
p. 126, fig. 104: photo © ZKM | Karlsruhe, photo: Tobias Wootton
p. 127, fig. 105: photo © ZKM | Kalrsruhe
p. 132, fig. 110: © Iannis Xenakis Family
p. 133, figs. 111, 112: photo © courtesy of the artists, screenshots: © Jens Barth
p. 133, fig. 113: photo © courtesy of the artists
p. 133, figs. 114, 115: photo © courtesy of the artists, screenshots: © ZKM | Karlsruhe
p. 135, figs. 116, 117: photo © ZKM | Karlsruhe, photos: Matthias Gommel
p. 135, fig. 118: photo © Empty Gallery
p. 135, figs. 119, 120: photo © ZKM | Karlsruhe
p. 136, figs. 121, 122: photo © ZKM | Karlsruhe
p. 136, figs. 123, 124: photo © courtesy of the artists
p. 138, fig. 128 : photo: Laszlo Ruszka
p. 139, fig. 129: photo © WDR, Cologne
p. 140, fig. 131: © VG Bild-Kunst, Bonn 2019 (Mary Bauermeister)
p. 140, fig. 132: photo © SWR
p. 140, fig. 133: The Daphne Oram Collection, Goldsmiths College
p. 141, fig. 134: photo © courtesy of Muzeum of Modern Art in Warsaw, photo: Andrzej Zborski
p. 141, figs. 135, 136: photo © EMS Archives, photo: Gunnel Lundholm
p. 142, fig. 137: photo © Raymond Scott
p. 144, fig. 138: photo © ZKM | Karlsruhe, photo: ONUK
p. 146, fig. 139: photo: Richard Alpert
p. 146, fig. 140: © 2019. Digital image, The Museum of Modern Art, New York/ Scala, Florence
p. 147, fig. 141: © 2019. Digital image, The Museum of Modern Art, New York/ Scala, Florence
p. 151: source: Germano Celant, *Record as Artwork. 1959–73* (London: Royal College of Art Gallery, 1973), cover.
p. 152: source: Germano Celant, *Record as Artwork. 1959–73* (London: Royal College of Art Gallery, 1973), pp. 6–7.
p. 155: source: Germano Celant, *Record as Artwork. 1959–73* (London: Royal College of Art Gallery, 1973), pp. 22–23.
p. 163: © VG Bild-Kunst, Bonn 2019 (Benoît Maubrey) photo © ZKM | Karlsruhe, photo: Steffen Harms
pp. 164–67: © Wezel Architektur, photos: Caren-Maria Jörß
pp. 168–81: photo © ZKM | Karlsruhe, photos: Steffen Harms
pp. 176–77: © VG Bild-Kunst, Bonn 2019 (Mogens Jacobsen)
pp. 182–83: photo © ZKM | Karlsruhe, photo: Steffen Harms
pp. 184–85: photo © ZKM | Karlsruhe, photos: Steffen Harms
p. 186, left: photo © Wild Don Lewis Photography / Steve Turner Contemporary
p. 186, right: photo © Wendy McEahern Photography / Dwight Hackett Projects
p. 187: © Tyler Adams
p. 188: © Tyler Adams
p. 189: photo © Wild Don Lewis Photography / Steve Turner Contemporary
p. 191: photos © ZKM | Karlsruhe, photos: Steffen Harms
pp. 192–93: photo © ZKM | Karlsruhe, photo: Steffen Harms
p. 193 right: © Cory Arcangel. Image courtesy Cory Arcangel.
pp. 194–95: © Cory Arcangel. Image courtesy Cory Arcangel.
p. 196: photo © Studio Baghdassarians / Baltschun
p. 197: photo © Studio Baghdassarians / Baltschun
p. 198: photo © Studio Baghdassarians / Baltschun
p. 199: exhibition views: singuhr – hoergalerie, photos: Roman März & singuhr e.v. berlin
p. 200, top: exhibition view: singuhr – hoergalerie, photo: Roman März & singuhr e.v. berlin, Baghdassarians / Baltschun
p. 201: exhibition views: singuhr – hoergalerie, photos: Roman März & singuhr e.v. berlin
pp. 196–201: © VG Bild-Kunst, Bonn 2019 (Boris Baltschun)
p. 202: photo © ZKM | Karlsruhe, screenshots © Jens Barth
p. 203: photo © ZKM | Karlsruhe, photo: Steffen Harms
p. 204: © VG Bild-Kunst, Bonn 2019 (Brothers Baschet), photo © ZKM | Karlsruhe, photo: Steffen Harms
p. 205, top: reproduction with kind permission of Succession Frères Baschet, © VG Bild-Kunst, Bonn 2019 (Brothers Baschet)
p. 205, bottom: source: François and Bernard Baschet, Jacques, Yvonne, and Teddy Lasry, *Les Nouvelles Structures Sonores. Lasry-Baschet*, 1967, 7" EP, Cacophonic, Archive ZKM | Karlsruhe © VG Bild-Kunst, Bonn 2019 (Brothers Baschet)
pp. 206–07: reproductions with kind permission of Succession Frères Baschet, © VG Bild-Kunst, Bonn 2019 (Brothers Baschet)
p. 208: © VG Bild-Kunst, Bonn 2019 (Harry Bertoia), photos © The Harry Bertoia Foundation
p. 209: © VG Bild-Kunst, Bonn 2019 (Harry Bertoia), photo © ZKM | Karlsruhe, photo: Steffen Harms
p. 210: © VG Bild-Kunst, Bonn 2019 (Harry Bertoia), photo William A. Smith
p. 211: © VG Bild-Kunst, Bonn 2019 (Harry Bertoia), photo Tod Greenway (Vancouver Art Gallery)
p. 212: © VG Bild-Kunst, Bonn 2019 (Joseph Beuys), photo © Centre Georges Pompidou, Paris
p. 213: © VG Bild-Kunst, Bonn 2019 (Joseph Beuys)
p. 214: photo © ZKM | Karlsruhe, photos: Steffen Harms
p. 215: photos © Jens Brand
p. 216: photos © Jens Brand
p. 217: photos © Ralf Brunner
pp. 219: photo © ZKM | Karlsruhe, photo: Felix Grünschloss
pp. 220–221: photo © ZKM | Karlsruhe, photos: Steffen Harms
p. 222: photo © Götz Dipper
p. 223: photo © ZKM | Karlsruhe, photo: Steffen Harms
p. 224: photo © Götz Dipper
p. 225: photo © ZKM | Karlsruhe, photo: Steffen Harms
p. 226: photo © ZKM | Karlsruhe, photos: ONUK
p. 228: © John Cage Trust, NY
p. 229: © John Cage Trust, NY / C. F. Peters
p. 230: photo © Janet Cardiff / George Bures Miller, photo: Roman März
p. 231: photo © ZKM | Karlsruhe, photos: Steffen Harms

pp. 232–33.: photo © Janet Cardiff / George Bures Miller, photos: Roman März
pp. 234–35.: photo © Janet Cardiff / George Bures Miller
p. 236: © VG Bild-Kunst, Bonn 2019 (Alexandra Murray-Leslie), photo © Artspace, Sydney, photos: silversalt photography, 2013
p. 237: © VG Bild-Kunst, Bonn 2019 (Alexandra Murray-Leslie), screenshots: © Jens Barth, 2014
pp. 238–39.: photo © ZKM | Karlsruhe, photos: Steffen Harms
pp. 240–43: photos: Stephen Cornford
p. 244: photo © ZKM | Karlsruhe, photo: Steffen Harms
p. 245: video stills: © Chris Cunningham
p. 246: photo © ZKM | Karlsruhe, photo: Steffen Harms
p. 247: photo © ZKM | Karlsruhe, photos: Fidelis Fuchs
p. 248: installation view at the Contemporary Jewish Museum, San Francisco 2015, photo: John Arnold
p. 249: installation views at the DAAD gallery, Berlin, 2010, photos: Paul de Marinis
p. 250: photo © ZKM | Karlsruhe, photo: Steffen Harms
p. 251: photo © Matthias Deumlich
p. 252, left: screenshots © Götz Dipper
p. 252, right: photo © ZKM | Karlsruhe, photo: Steffen Harms
p. 253: screenshots © Götz Dipper
pp. 254–55: photo © ZKM | Karlsruhe, photos: Steffen Harms
pp. 256–57: © VG Bild-Kunst, Bonn 2019 (Ulrich Eller), photos © ZKM | Karlsruhe, photos: Steffen Harms
p. 258: photo © ZKM | Karlsruhe, photos: Christof Hierholzer
p. 259: photo © ZKM | Karlsruhe, photo: Steffen Harms
p. 260, left: photo © Victor Giers
pp. 260, right, 261: photo © Victor Giers, photos: Manfred H. Vogel
pp. 263–65: © VG Bild-Kunst, Bonn 2019 (Sabine Groschup)
pp. 262–63 top: video stills: courtesy of the artist | Bildrecht Wien
p. 263, middle, bottom: video stills: courtesy of the artist | Bildrecht Wien, photos: Barbara Klemm
p. 265: photo © Matthias Bildstein
pp. 266–67: photo © Milan Grygar
p. 268: photo © ZKM | Karlsruhe, photo: Steffen Harms
p. 269: photo © The Artist and Galleria Coninua / Le Moulin, San Gimignano / Beijing / Les Moulins / Habana, photos: Otto Saxinger
pp. 270–71: photo © ZKM | Karlsruhe, photos: Steffen Harms
pp. 272–73: photo © ZKM | Karlsruhe, photos: Steffen Harms
p. 274: photo © ZKM | Karlsruhe, photo: Steffen Harms
p. 275: installation view Galerie Mario Mazzoli, Berlin, 2016, photo: Douglas Henderson
pp. 276–77: installation views Galerie Mario Mazzoli, Berlin, 2012, photos: Douglas Henderson
pp. 278–79.: photo © ZKM | Karlsruhe, photos: Steffen Harms
pp. 280–81.: © VG Bild-Kunst, Bonn 2019 (Gary Hill)
p. 282: photo © ZKM | Karlsruhe, photo: Tobias Wootton
p. 283: photo © ZKM | Karlsruhe, photo: Franz Wamhof
pp. 284–85: photo © ZKM | Karlsruhe, Estate of Stephan von Huene
p. 286: photo © Tod Greenway (Vancouver Art Gallery)
p. 287: photo: Stephan von Huene
p. 288: photo © ZKM | Karlsruhe, photo: Steffen Harms
p. 289: photo © ZKM | Karlsruhe, photo © Fidelis Fuchs
pp. 290–91.: photo © ZKM | Karlsruhe, photos: Martin Wagenhan
p. 292: photo © Yamaguchi Center for Arts and Media (YCAM), photos: Ryuichi Maruo
p. 293: performance at Grec Barcelona Festival and Sonár, Barcelona, 2010, photo © Liz Hingley
pp. 294–95: © VG Bild-Kunst, Bonn 2019 (Mogens Jacobsen), photo © ZKM | Karlsruhe, photos: Steffen Harms
pp. 296–97: © VG Bild-Kunst, Bonn 2019 (Anna Jermolaewa), photo © ZKM | Karlsruhe, photo: Steffen Harms
p. 299: photo © ZKM | Karlsruhe, photo: Steffen Harms
p. 300: © VG Bild-Kunst, Bonn 2019 (Rolf Julius), photo © ZKM | Karlsruhe, photos: Steffen Harms
p. 301: © VG Bild-Kunst, Bonn 2019 (Rolf Julius), photo © estate rolf julius, photo: Jérémie Buchholtz
p. 302: © VG Bild-Kunst, Bonn 2019 (Rolf Julius), photo © estate rolf julius, photos: Kunsthalle Kiel
p. 303: © VG Bild-Kunst, Bonn 2019 (Rolf Julius), photo © ZKM | Karlsruhe, photos: Steffen Harms
pp. 304–05: © VG Bild-Kunst, Bonn 2019 (Timo Kahlen), photo © ZKM | Karlsruhe, photos: Steffen Harms
p. 306: © VG Bild-Kunst, Bonn 2019 (Timo Kahlen), installation view at *Wireless Experience*, ISEA 2004, KIASMA Helsinki, photos: Timo Kahlen
p. 307: © VG Bild-Kunst, Bonn 2019 (Timo Kahlen), installation view at *Sound Art 2007*, nominees of the German Sound Art Prize (Deutscher Klangkunst-Preis), Duisburg 2007, photo: Timo Kahlen
pp. 309–11: © VG Bild-Kunst, Bonn 2019 (Georg Klein)
pp. 312–13: photo © ZKM | Karlsruhe, photos: Steffen Harms
p. 314: photo © Christina Kubisch
p. 315, left: photo © Christina Kubisch
p. 315, right: © VG Bild-Kunst, Bonn 2019 (Rolf Julius), photo: Henry M. Linder
p. 316, top: photo: Jens Ziehe
p. 316, bottom: photo © Christina Kubisch
p. 317: photo © Christina Kubisch
pp. 318–19: photo © ZKM | Karlsruhe, photos: Steffen Harms
pp. 320–24: photo © ZKM | Karlsruhe, photos: Steffen Harms
p. 325: photo © ZKM | Karlsruhe, photo: ONUK
pp. 326–27: © VG Bild-Kunst, Bonn 2019 (Anestis Logothetis, Julia Logothetis), video stills: © Anestis Logothetis / Julia Logothestis
p. 328: photo © ZKM | Karlsruhe, photo: Steffen Harms
p. 329: video stills: © ZKM | Karlsruhe
p. 330: video still: © Christian Marclay
pp. 331–33: photo © Christian Marclay. Courtesy Paula Cooper Gallery, New York
pp. 334–35: photo © ZKM | Karlsruhe, photos: Steffen Harms
pp. 336–37: © VG Bild-Kunst, Bonn 2019 (Benoît Maubrey), photo © ZKM | Karlsruhe, photos: Steffen Harms
pp. 338–39: © VG Bild-Kunst, Bonn 2019 (Benoît Maubrey)
pp. 340–41: photo © ZKM | Karlsruhe, photos: Steffen Harms
pp. 342–43: photo © ZKM | Karlsruhe, photos: Steffen Harms
p. 344: photo © ZKM | Karlsruhe, photo: Steffen Harms
p. 345: photo © Haroon Mirza, photo: Ken Adlard
p. 346: photo © ZKM | Karlsruhe, photos: Steffen Harms
p. 347: exhibition view Fifth Parallel Gallery, University of Regina, Saskatchewan, Canada, 2008, photo: Charlie Fox
pp. 348–51: photos: Gordon Monahan
p. 352: photo © ZKM | Karlsruhe, photo: Felix Grünschloss
p. 353: photo © ZKM | Karlsruhe, photos: Franz J. Wamhof
pp. 354–55.: © VG Bild-Kunst, Bonn 2019 (Bruce Nauman), photo © ZKM | Karlsruhe, photos: Steffen Harms
pp. 356–57: photo © ZKM | Karlsruhe, photos: Steffen Harms
p. 358: photo © ZKM | Karlsruhe, photo: Steffen Harms
p. 359: video stills: © Firefly Pictures 2002, The Estate of Max Neuhaus
p. 360, top left, bottom: photo © The Estate of Max Neuhaus
p. 360, top right: photo © Broadside Television, The Estate of Max Neuhaus
p. 361: photo © ZKM | Karlsruhe, photo: Steffen Harms
p. 362: © VG Bild-Kunst, Bonn 2019 (Carsten Nicolai)
p. 363: © VG Bild-Kunst, Bonn 2019 (Carsten Nicolai), photo © ZKM | Karlsruhe, photo: Steffen Harms
pp. 364–67: scores: © Pauline Oliveros / Smith Publications
pp. 368–69: photo © ZKM | Karlsruhe, photos: Steffen Harms
p. 370: photo © Museum Moderner Kunst Stiftung Ludwig Wien
p. 372: photo © Museum Moderner Kunst Stiftung Ludwig Wien, photos: Manfred Montwé
p. 373, top: image courtesy of Nam June Paik Art Center, photo: Manfred Montwé
p. 373, bottom: photo © Collection Musée d'art contemporain. Division des Affaires culturelles. Ville de Lyon, France, image courtesy of Nam June Paik Art Center
pp. 374–75: © VG Bild-Kunst, Bonn 2019 (Joseph Beuys)
p. 376: photo © ZKM | Karlsruhe, photo: Steffen Harms
p. 367: photo © Marco Peitschopf
pp. 378–79: photo © ZKM | Karlsruhe, photos: Steffen Harms
pp. 380–81: photo © courtesy Studio la Città - Verona, photos: Michele Alberto Sereni
pp. 382–83: photo © ZKM | Karlsruhe, photos: Steffen Harms
p. 384, left: © VG Bild-Kunst, Bonn 2019 (Werner Reiterer), photo © ZKM | Karlsruhe, photo: Steffen Harms
p. 384, right: © VG Bild-Kunst, Bonn 2019 (Werner Reiterer), photo © Werner Reiterer
p. 385: © VG Bild-Kunst, Bonn 2019 (Werner Reiterer), photos © Werner Reiterer
p. 386: photo © ZKM | Karlsruhe, photo: Steffen Harms
p. 387: photo © Steve Roden, photo: Robert Wedemeyer
p. 388: photo © ZKM | Karlsruhe, photos: Bernhard Schmitt, ONUK
p. 389, top: photos: Michael Kneffel

p. 389, middle and bottom: photo © ZKM | Karlsruhe, photos: Bernhard Schmitt, ONUK
pp. 390–91: photos © Tim Otto Roth
pp. 392–93.: installation views Kunsthalle Jesuitenkirche Aschaffenburg, photos © Tim Otto Roth
p. 394: © VG Bild-Kunst, Bonn 2019 (Sabine Schäfer), photo © ZKM | Karlsruhe, photo: Steffen Harms
p. 395, top: © VG Bild-Kunst, Bonn 2019 (Sabine Schäfer), photo © ZKM | Karlsruhe, photo: Bernhard Schmitt, ONUK
p. 395, bottom: © VG Bild-Kunst, Bonn 2019 (Sabine Schäfer), photo © <SA/JO>, photo: Reinhard Vollmer
p. 396: © VG Bild-Kunst, Bonn 2019 (Sabine Schäfer), photo © phaeno Wolfsburg, photo: Clemens Ortmeyer
p. 397: © VG Bild-Kunst, Bonn 2019 (Sabine Schäfer), photo © phaeno Wolfsburg, photo: Johanna Pohlmann
pp. 398–99.: © VG Bild-Kunst, Bonn 2019 (Micheal Saup), photo © ZKM | Karlsruhe, photos: Steffen Harms
p. 400: photos © ZKM | Karlsruhe, photos: Steffen Harms
p. 401: photo © Dawn Scarfe
p. 402: photo © ZKM | Karlsruhe, photo: Steffen Harms
p. 403: photo © Scenocosme: Grégory Lasserre and Anaïs met den Ancxt
p. 404, left: photo © Scenocosme: Grégory Lasserre and Anaïs met den Ancxt
p. 404, right: photo © ZKM | Karlsruhe, photo: Jonas Zilius
p. 405: photo © Scenocosme: Grégory Lasserre and Anaïs met den Ancxt
pp. 406–07: © VG Bild-Kunst, Bonn 2019 (Andreas Brehmer), photo © ZKM | Karlsruhe, photos: Steffen Harms
pp. 408–09: photo © ZKM | Karlsruhe, photo: Steffen Harms
p. 410: photo © ZKM | Karlsruhe, photo: Steffen Harms
p. 411: photos: Jan-Peter E.R. Sonntag
pp. 412–13: photo © ZKM | Karlsruhe, photos: Steffen Harms
pp. 414–15.: video stills: © Joulia Strauss, courtesy of the artist
pp. 416–17: photo © ZKM | Karlsruhe, photos: Steffen Harms
p. 418: photo © ZKM | Karlsruhe, photos: Steffen Harms
p. 419: photos: Junko Wada
p. 420: © VG Bild-Kunst, Bonn 2019 (Takis), photo © ZKM | Karlsruhe, photo: Steffen Harms
p. 421: © VG Bild-Kunst, Bonn 2019 (Takis), with kind permission of Takis Foundation archives, 2019
pp. 422–23: VG Bild-Kunst, Bonn 2019 (Tommi Grönlund, Petteri Nisunen, Carsten Nicolai), photo © ZKM | Karlsruhe, photos: Steffen Harms
pp. 424–25: photo © ZKM | Karlsruhe, photos: Steffen Harms
p. 426: © VG Bild-Kunst, Bonn 2019 (Timm Ulrichs), photo © ZKM | Karlsruhe, photo: Roland Schmidt – Fotografie
p. 427, top: © VG Bild-Kunst, Bonn 2019 (Timm Ulrichs), photo © ZKM | Karlsruhe, photo: Steffen Harms
p. 427, bottom: © VG Bild-Kunst, Bonn 2019 (Timm Ulrichs)
p. 428: © VG Bild-Kunst, Bonn 2019 (Mogens Jacobsen), photo © ZKM | Karlsruhe, photo: Steffen Harms
p. 430: © VG Bild-Kunst, Bonn 2019 (Mogens Jacobsen)
pp. 430–31: photo © ZKM | Karlsruhe, photos: Felix Grünschloss
pp. 432–33: photo © ZKM | Karlsruhe, photos: Steffen Harms
pp. 434–37: photo © ZKM | Karlsruhe, photos: Steffen Harms
pp. 438–39: photo © ZKM | Karlsruhe, photos: Steffen Harms
pp. 440–41: installation views Kunsthalle Düsseldorf, photo © Kunsthalle Düsseldorf
p. 442: photo © ZKM | Karlsruhe, photos: Steffen Harms
p. 443, top: photo © ZKM | Karlsruhe, photos: Steffen Harms
p. 443, bottom: photo: Birgit and Peter Kainz
p. 444, top: photo © Archive Peter Weibel, photo: Belvedere
p. 444, bottom: photo © Archive Peter Weibel
p. 445, top: photo © Archive Peter Weibel, photo: Belvedere
p. 445, bottom: © Archive Peter Weibel
p. 446, top: © Archive Peter Weibel
p. 446, bottom: photo © Archive Peter Weibel
p. 447, top, bottom, left: photo © Archive Peter Weibel, photos: VALIE EXPORT
p. 447, bottom, right: photo © Archive Peter Weibel, photo: Michael Schuster
pp. 448–49.: photo © ZKM | Karlsruhe, photos: Steffen Harms
pp. 450–51: photo © ZKM | Karlsruhe, photos: Steffen Harms
pp. 452–53: photo © ZKM | Karlsruhe, photo: Steffen Harms
p. 455: photo © ZKM | Karlsruhe, photo: Steffen Harms
p. 459, fig. 1: source: René Block et al., *Für Augen und Ohren – Von der Spieluhr zum akustischen Environment. Objekte Installationen Performances*, exhib. cat. (Berlin: Akademie der Künste, 1980), 6.
p. 461, figs. 2, 3: © Creative Commons
p. 463, fig. 4: © Creative Commons
p. 464, fig. 5: source: Annette Beyer, *Faszinierende Welt der Automaten* (Munich: Callwey, 1983), 32.
p. 467, fig. 6: source: Alfred Chapuis and Edmond Droz, *Les automates: Figures artificielles d'hommes et d'animaux. Histoire et technique* (Neuchâtel: Édition du Griffon, 1949), 40.
p. 469, fig. 7: source: Fuat Sezgin, ed., *Compendium on the Theory and Practice of the Mechanical Arts* (Frankfurt am Main: Institute for the History of Arabic-Islamic Science, Goethe University Frankfurt, 2002), 321, 325, 327, and 334.
p. 471, fig. 8: source: A. G. Drachmann, *Ktesibios, Philon and Heron* (Copenhagen: Munksgaard, 1948), 11.
p. 472, fig. 9: source: Michael Markovits, *Die Orgel im Altertum* (Leiden: Brill, 2003), 751.
p. 472, fig. 10: source: Eilhard Wiedemann, *Sitzungsberichte der physikalisch-medizinischen Sozietät zu Erlangen*, no. 46 (1915), 17–26.
p. 473, fig. 11: source: Giovanni Battista della Porta, *Pneumaticorum libri tres* (Naples: Iacobus Carlinus, 1601), 61.
p. 473, fig. 12a: source: Siegfried Zielinski and Peter Weibel, *Allah's Automata: Artifacts of the Arab-Islamic Renaissance (800–1200)*, exhib. cat. (Ostfildern: Hatje Cantz, 2015), 88. Visualization by Clemens Jahn.
p. 474, fig. 12b: photo © ZKM | Karlsruhe, photo: Harald Völkl
p. 474, fig. 13: source: *Guida introduttiva agli strumenti musicali meccanici, Collezione Marino Marini* (Ravenna: Tipografia Musiani, n.d.), 25.
p. 475, fig. 14: source: Athanasius Kircher, *Musurgia universalis* (Rome: 1650), 520.
p. 475, fig. 15: *Guida introduttiva agli strumenti musicali meccanici, Collezione Marino Marini* (Ravenna: Tipografia Musiani, n.d.).
p. 478, figs. 1a–c: © VG Bild-Kunst, Bonn 2019 (Bruce Nauman)
p. 478, fig. 2: © VG Bild-Kunst, Bonn 2019 (Carolee Schneemann)
p. 479, figs. 3a–f: © VG Bild-Kunst, Bonn 2019 (Gary Hill)
p. 479, fig. 4: photo © Bruce Odland
p. 479, fig. 5: photo © Walter Fähndrich, photo: Walter Fähndrich
p. 480, figs. 6a–b: photo © Bill Fontana, photos: Bill Fontana
p. 481, fig. 7: © VG Bild-Kunst, Bonn 2019 (Ulrich Eller), photo © shamrockmedia, photo: Norbert Kesten
p. 482, figs. 8a–b: photos © Tristan Perich
p. 482, figs. 9a–b: photos © Pe Lang + Zimoun
p. 483, fig. 10: photo © courtesy of the artist and Gallery Burster, Berlin
p. 483, fig. 11: © VG Bild-Kunst, Bonn 2019 (Qin Yufen), photo: Zhu Jinsh
p. 484, figs. 12a–c: © 2019 Dennis Oppenheim Estate
p. 485, fig. 13: © 2019 Dennis Oppenheim Estate, photo: Ace Contemporary, Los Angeles
p. 485, fig. 14: photo © Camille Norment Studio
p. 486, figs. 15–16: photo © courtesy gb agency, Paris, photos: Catherine Brossais – CGVO
p. 487, fig. 17: © VG Bild-Kunst, Bonn 2019 (Richard Serra and Nancy Holt)
p. 488, fig. 18: source: Bernhard Leitner, *Kopfräume/Headscapes* (Ostfildern: Hatje Cantz, 2003), cover.
p. 489, figs. 19a–c: © the artist
p. 491, fig. 1: © Alexis Kirke and Martyn Ware
p. 493, fig. 2a, top: image courtesy of TRACE/LMSAL
p. 493, fig. 2b, middle: image courtesy of SST/Royal Swedish Academy of Sciences
p. 493, fig. 2c, bottom: image courtesy of TRACE/LMSAL
p. 494, figs. 3a–d: image courtesy of STFC Rutherford Appleton Laboratory, UK
p. 495, fig. 4: photo © Carsten Goertz
p. 496, figs. 5a–b, top: photos © Natasha Barrett
p. 496, fig. 5c, middle: photo © Henrik der Minassian
p. 496, fig. 6: photo © UAMN, photo: Theresa Bakker
p. 498, fig. 7: photo © Andrea Polli
p. 498, fig. 8: screenshots: © Andrea Polli
p. 505, fig. 1: photo © Thomas Moor, photo: Victor Kolibal
p. 506, figs. 2a–b: photo © Christof Migone, photos: Denis Farley
p. 508, figs 3a–b: photo © courtesy of the artist
p. 512, fig. 1: photo © 1974 Rheinisches Bildarchiv, Cologne
p. 513, fig. 2: photo © 1974 Rheinisches Bildarchiv, Cologne
p. 515, fig. 3: photo © Walter Klein
p. 516, fig. 4: photo © Städtische Kunsthalle Düsseldorf
p. 517, fig. 5: photo © Städtische Kunsthalle Düsseldorf
p. 521, fig. 1: photo © Finnbogi Petursson
p. 522, fig. 2: photo © courtesy of the artist

p. 523, fig. 3: photo © courtesy of the artist, photo: e-Xplo
p. 524, fig. 4: photo © Contour Mechelen
p. 526, figs. 5–6: photo © Learning Site (Rikke Luther and Cecilia Wendt)
p. 528, fig. 1: © F.L.C./VG Bild-Kunst, Bonn 2019 (Le Corbusier); Iannis Xenakis Family, photo: CC BY-SA 3.0
p. 531, figs. 2a–b: © Iannis Xenakis Family
p. 532, fig. 3a: photo © Centre Iannis Xenakis, photo: Bruno Rastoin
p. 532, fig. 3b: photo © Centre Iannis Xenakis, photo: Bruno Rastoin
p. 534, figs. 4a–b: Archive of Stockhausen Foundation for Music (www.kalheinzstockhausen.org)
p. 535, fig. 5: photo © Hanna Haaslahti
p. 536, fig. 6: photo © Luke Jerram Ltd, photo: Richard Deane
p. 537, figs. 7a–b: photos © Studio Thilo Frank
p. 537, fig. 8: photo © Krzysztof Pijarski
p. 539, fig. 9: photo © courtesy Studio Edwin van der Heide
p. 541, fig. 10: photo © courtesy Studio Edwin van der Heide
p. 541, fig. 11a: photo © courtesy Studio Edwin van der Heide
p. 541, fig. 11b: photo © courtesy NOX
p. 542, fig. 12a: photo © Aranda\Lasch
p. 542, fig. 12b: photo © Jakob Polacsek
p. 542, fig. 12c: photo © Hertha Hurnaus | TBA21
p. 542, fig. 12d: photo © ZKM | Karlsruhe, photo: Uli Deck
p. 543, figs. 13a–b: graphics © Aranda\Lasch
p. 545, fig. 14: photo © Patrick Tighe Architecture, photo: Art Gray
p. 546, fig. 15a, top left: photo © Ball-Nogues Studio, photo: Neil Cochran
p. 546, fig. 15b, top right: photo © Ball-Nogues Studio, photo: Benjamin Ball
p. 546, fig. 15c, bottom: photo © Ball-Nogues Studio, photo: Oliver Hess
p. 555, figs. 1–2: photos © ZKM | Karlsruhe, photos: Steffen Harms
p. 556, fig. 3: photo © ZKM | Karlsruhe, photo: Steffen Harms
p. 557, fig. 4: photo © ZKM | Karlsruhe, photo: Steffen Harms
p. 559, fig. 5: photo © private collection
p. 565, figs. 6–8: photo © ZKM | Karlsruhe, photos: Steffen Harms
p. 568, figs. 9–10: © VG Bild-Kunst, Bonn 2019 (Mogens Jacobsen), photo © ZKM | Karlsruhe, photos: Steffen Harms
p. 569, fig. 11: © VG Bild-Kunst, Bonn 2019 (Mogens Jacobsen), photo © ZKM | Karlsruhe, photo: Steffen Harms
p. 570, fig. 12: © VG Bild-Kunst, Bonn 2019 (Åke Hodell), photo © Collection of MFSK, photo: Knud Hvidberg
p. 572, fig. 13: © Morten Søndergaard
p. 573, figs. 14–15: © VG Bild-Kunst, Bonn 2019 (Mogens Jacobsen), photo © ZKM | Karlsruhe, photos: Steffen Harms
p. 578, figs. 16–17: photo © ZKM | Karlsruhe, photos: Steffen Harms
p. 595: photo: Fabio Lugaro
p. 598, fig. 1: source: Ernest W. Uthemann, ed., *Stadtgalerie Saarbrücken 1994–2001: Eine Dokumentation* (Saarbrücken: Stadtgalerie, 2011), cover. Photo: Tom Gundelwein
p. 598, fig. 2: source: Bernd Schulz, ed., *Resonanzen/Resonances. Aspekte der Klangkunst/Aspects of Sound Art.*, exhib. cat. Stadtgalerie Saarbrücken (Heidelberg: Kehrer Verlag, 2002), cover. Photo: Tom Gundelwein
p. 598, fig. 3: source: Bernd Schulz, ed., *Akio Suzuki „A“: Sound Works: Throwing and Following* (Saarbrücken: Herziger, 1998). Publication accompanying the exhibition of Akio Suzuki at Stadtgalerie Saarbrücken, cover. Graphic design: Hans Husel
p. 599, fig. 4: source: Bernd Schulz and Irmgard Heesche, eds., *Christina Kubisch Zwischenräume*, exhib. cat. Stadtgalerie Saarbrücken and HBK Saar (Saarbrücken: Stadtgalerie, 1996), cover. Graphic design: Hans Husel
p. 599, fig. 5: source: Bernd Schulz, ed., *Klangräume* [Soundspaces], exhib. cat. featuring Bernard Baschet, Francois Baschet, Gunter Demnig, Ulrich Eller, Stephan von Huene, Rolf Julius, and Christina Kubisch (Saarbrücken: Stadtgalerie, 1988), cover. Graphic design: Hans Husel
p. 599, fig. 6: photo: Tom Gundelwein
p. 599, fig. 7: © VG Bild-Kunst, Bonn 2019 (Rolf Julius), photo: Tom Gundelwein
p. 599, fig. 8: photo: Tom Gundelwein
p. 600, fig. 9: photo: Tom Gundelwein
p. 600, fig. 10: photo: Tom Gundelwein
p. 601, fig. 11: photo © Sammlung René Block, photo: Werner Maschmann
p. 601, fig. 12: © VG Bild-Kunst, Bonn 2019 (Gunter Demnig)
p. 602, fig. 13: source: Helga de la Motte-Haber, Matthias Osterwold, and Georg Weckwerth, eds., *Sonambiente Berlin 2006: Sound Art* (Heidelberg: Kehrer Verlag, 2006), cover. Graphic design: cyan
p. 602, fig. 14: source: René Block, ed., *Für Augen und Ohren. Von der Spieluhr zum akustischen Environment* [For Eyes and Ears: From the Music-box to the Acoustic Environment] (Berlin: Academy of Arts, 1980), cover. Graphic design: Gruppe 96, photo: Knud Peter Petersen
p. 604, fig. 15: © VG Bild-Kunst, Bonn 2019 (Carlfriedrich Claus), photos: Margit Kreisl
p. 604, figs. 16a, b: photo: Wolf Wachner
p. 606, fig. 17: source: R. Murray Schafer, *The Soundscape: Our Sonic Environment and the Tuning of the World* (Lochester, VT: Destiny Books, 1977), cover.
p. 606, fig. 18: source: R. Murray Schafer, *Sounds Unseen. An exhibition of musical manuscripts and sound sculpture.* Cover image of a flexi disk which was produced 1981 on the occasion of R. Murray Schafer's exhibition at Gallery Stratford in Stratford, Ontario (CA).
p. 608, figs. 19a, b: © VG Bild-Kunst, Bonn 2019 (Raul Keller), photo: Dario Lehner
p. 611, fig. 20: © VG Bild-Kunst, Bonn 2019 (Terry Fox)
p. 613, fig. 1: photo © courtesy of Józef Robakowski and Muzeum Sztuki, Łódź
p. 614, fig. 2: photo © courtesy of Włodzimierz Kotoński and Muzeum Sztuki, Łódź
p. 615, fig. 3: © VG Bild-Kunst, Bonn 2019 (Arnold Haukeland), photo: Ola Nordal
p. 616, fig. 4: photo © courtesy of the Henie Onstand Kunstsenter and Rannveig Getz Nordheim
p. 616, fig. 5: photo © courtesy of Architectural Press Archive / RIBA Collections
p. 617, fig. 6: photo © courtesy of NRK – Norwegian Broadcasting Corporation
p. 617, fig. 7: © VG Bild-Kunst, Bonn 2019 (Anders Kjaer), photo © courtesy of Henie Onstad Kunstsenter
p. 618, figs. 8a–d: photo © courtesy of the artists
p. 618, figs 9–10: photo © courtesy of Muzeum Sztuki, Łódź, photo: P. Tomczyk
p. 619, figs 11, 12: photo © courtesy of Krzysztof Wodiczko and Profile Foundation, Warsaw.
p. 620, fig. 13: photo © courtesy of Igor Hansen
p. 620, fig. 14: photo © collection of the Museum of Academy of Fine Arts in Warsaw, photo: Z. Kapuścik
p. 621, fig. 15: photo © collection of Muzeum Sztuki, Łódź
p. 621, fig. 16: photo © J. Surwiłło
p. 624, fig. 1: photo © courtesy of the Nikolay Nikiforov Collection
p. 625, fig. 2: photo © courtesy of the Multimedia Art Museum, Moscow
p. 627, fig. 3: source: René Fülöp-Miller, *Geist und Gesicht des Bolschewismus* (Vienna: Amalthea, 1926), 247 (Tafel 140).
p. 627, figs. 4–6: photo © courtesy of the Oleg Ivanov Collection
p. 627, fig. 7: photo © courtesy of Andrey Smirnov
p. 627, fig. 8: photo © courtesy of the Bulat Galeev Library, Prometheus Center, Kazan
p. 628, fig. 9: photo © courtesy of Sergey Zorin, photo: A. Glinsky
p. 629, fig. 10: photo © courtesy of the Theremin Centre, Moscow
p. 629, fig. 11: photo © courtesy of Marina Sholpo
p. 630, fig. 12: photo © courtesy of Marina Sholpo
p. 630, fig. 13: photo © courtesy of Andrey Smirnov
p. 631, fig. 14: photo © courtesy of Andrey Smirnov
p. 631, fig. 15: photo © courtesy of the Glinka State Central Museum of Musical Culture, Moscow
p. 632, figs. 16–18: photo © courtesy of the Bulat Galeev Library, Prometheus Center, Kazan
p. 634, fig. 19: photo © courtesy of the Theremin Centre, Moscow, photo: Elena Krysanova
p. 635, fig. 20: photo © courtesy of the artists, photo: Anastasia Blur
p. 635, fig. 21: photo © courtesy of the artist
p. 635, figs. 22 a–b: photo © courtesy of Yury Didevich
p. 637, fig. 1: photo © 2/5bz (Serhat Köksal)
p. 638, fig. 2: source: Fatih Aydoğdu, *HAT (Hybrid Arrested Translation)*, 1998. Magazine cover.
p. 639, fig. 3: photo © NOMAD
p. 639, fig. 4: photo © NOMAD
p. 640, figs. 5–7: photo © Başak Şenova
p. 642, fig. 8, 9: photo © Başak Şenova
p. 645, fig. 1: photo © Stan Godwyn
p. 645, fig. 2: photo © KongFu Wang, Beijing, China, Creative Commons
p. 646, fig. 3: photo © FM3, exploded view by longmo.net
p. 647, figs. 4a–b: © Samson Young, photos © Microwave International Media Art Festival
p. 647, fig. 5: photo © Yan Jun

p. 649, fig. 6: photo © Wenhua Shi, photo: Dajuin Yao
p. 650, fig. 7: photo © Jasper Fung, photo: Jasper Fung
p. 656, fig. 1: photo © Collection of the National Film Center, National Museum of Modern art Tokyo
p. 657, fig. 2: source: Atsuko Tanaka, Drawing for *Work Bell*, 1955. Published in
Gutai journal, July 1956, p. 24.
p. 658, fig. 3: © courtesy Errant Bodies, photo © Gary McCraw
p. 659, fig. 4: photo © Nasaka Senkichiro, Yoshihara Naomi, and the former members of the Gutai Art Association; courtesy Museum of Osaka University
p. 659, fig. 5: © photo courtesy NTT InterCommunication Center [ICC], photo: Shibuya Seiji
p. 661, fig. 6: photo © Nakano Masataka
p. 662, fig. 7: courtesy of the artist, photo © Minoru Sato, photo: Minoru Sato
p. 663, fig. 8: courtesy of the artist, photo © Toshiya Tsunoda
p. 663, fig. 9: courtesy of the artist, photo © Kanta Horio, photo: Kanta Horio
p. 664, fig. 10: photo © Muneteru Ujino, photo: Hiroyuki Matsukage
p. 664, fig. 11: courtesy of the artist and Kobe Art Village Center and MUJIN-TO Production, photo: Kazuo Fukunaga
p. 665, fig. 12: courtesy of the artist, photo © Masahiro Miwa, photo: Masahiro Miwa
p. 665, fig. 13: courtesy of the artist, photo © Atau Tanaka
p. 669, figs. 1a–b: courtesy of the artists and Sarah Cottier Gallery, Sydney, photos: Michael Myers, © VG Bild-Kunst, Bonn 2019 (Joyce Hinterding)
p. 670, fig. 2: © the artist, image courtesy of the artist and Sarah Cottier Gallery, Sydney, photo: Christopher Snee
p. 670, fig. 3: © VG Bild-Kunst, Bonn 2019 (Joyce Hinterding), image courtesy the artist and Sarah Cottier Gallery, Sydney, photo: Ann Shelton,
p. 670, fig. 4: © Ernie Althoff, image courtesy the artist and Museum of Contemporary Art Australia
p. 672, fig. 5: © VG Bild-Kunst, Bonn 2019 (Lucas Abela), photo © Alisha Weng
p. 673, fig. 6: © VG Bild-Kunst, Bonn 2019 (Marco Fusinato), image courtesy of National Gallery of Victoria, Melbourne
p. 674, fig. 7: The development of *Relaxation Circuit* was supported by Underbelly Festival, 2015. © Pia van Gelder, photo: Tom Jones
p. 674, fig. 8: © Vicky Browne, photo courtesy of City Gallery Wellington
p. 675, fig. 9: © Ross Manning, photo © ACCA, photo: Andrew Curtis
p. 675, fig. 10: © photo © Jenn Brewer
p. 675, fig. 11: photo © Marnie Vaughan
p. 677, fig. 1: photo © Brady-Handy Photograph Collection
p. 678, figs. 2a–c: © Firefly Pictures 2002. The Estate of Max Neuhaus
p. 679, fig. 3: © La Monte Young, Marian Zazeela, photo © ZKM | Karlsruhe, photo: Steffen Harms
p. 680, fig. 4: © VG Bild-Kunst, Bonn 2019 (Michael Brewster), © The Estate of Michael Brewster
p. 681, figs. 5a–c: filmed and edited by Viola Rusche, audio recorded live by Juan Parra, camera Martin Zawadzki, Viola Rusche, soundmix Juan Parra, © VG Bild-Kunst, Bonn 2019 (Viola Rusche)
p. 682, fig. 6: © and image courtesy of Alvin Lucier
p. 682, fig. 7: adapted from: *Alvin Lucier. Klangskulpturen*, ed. Berliner Künstlerprogramm des DAAD (Berlin: daadgalerie, 1991).
p. 683, figs. 8a–c: photos © Roman März & singuhr berlin, photos: Roman März
p. 684, fig. 9: © Christian Marclay, photo: Ilmari Kalkkinen – MAMCO, Geneva
p. 684, figs. 10a–b: © Christian Marclay, courtesy of Paula Cooper Gallery, New York
p. 685, fig. 11: private collection, © Steve Roden
p. 685, fig. 12: photo © Heinrich Hermes & singuhr berlin, photo: Heinrich Hermes
p. 686, fig. 13: courtesy of Lisson Gallery, © Allora & Calzadilla
p. 686, fig. 14: commissioned by Creative Time for *Democracy in America: The National Campaign* © Angel Nevarez and Valerie Tevere, photo: Meghan McGinnis
p. 687, fig. 15: photo © Walter Phillips Gallery
p. 689, fig. 1: photo © John Owald
p. 690, fig. 2: source: Michael Snow, *The Last LP*, 1987. LP jacket, cover
p. 690, fig. 3a: photo © Nataliya Petkova, 2011
p. 690, fig. 3b: photo © LPM, 2011
p. 690, fig. 4: photo © Institut du monde arabe, Paris and the artist
p. 690, fig. 5a: source: The Nihilist Spasm Band, *No Record*, 1968. LP jacket, cover [released by Allied Record Corporation, 1968 as LP, later also as CD on other labels]
p. 691, fig. 5b: photo © Ian MacEachern
p. 692, fig. 6: photo © Sarah Ciurysek, courtesy of the Walter Phillips Gallery, Banff Centre
p. 693, fig. 7: photo © Peter Mallet
p. 693, fig. 8: photo © the artists
p. 694, fig. 9a, b: photo © the artist
p. 694, fig. 10: photo © Diane Landry
p. 694, fig. 11: photo © the artists
p. 695, fig. 12: photo © Anita Dambergs, courtesy of the Glenbow Museum
p. 695, figs. 13a–b: photos © Julie Artacho
p. 696, fig. 14: photo © Dwight Siegner
p. 697, fig. 15: photo © Martine Frossard Studio XX Montreal
p. 697, fig. 16: photo © Guy L'Heureux
p. 698, fig. 17a: photo © The Montreal Museum of Fine Arts
p. 699, fig. 17b: photo © The Montreal Museum of Fine Arts
p. 699, fig. 18: photo © Paul Litherland
p. 702, fig. 1: © J. P. Oliveira
p. 703, fig. 2: © Estate of Nam June Paik
p. 704, fig. 3: © Guilherme Vaz
p. 705, fig. 4: © Ass. Amigos Walter Smetak
p. 705, fig. 5: © Paulo Nenflidio
p. 706, fig. 6: © Chelpa Ferro
p. 706, fig. 7: © Tato Taborda
p. 707, fig. 8: purchased jointly by Tate, London (with the assistance of the Latin American Acquisitions Committee) and the D.Daskalopoulos Collection, 2013, as a promised gift to Tate, photo (c) the artist
p. 707, fig. 9: photo © the artist
p. 707, fig. 10: photo © the artist
p. 707, fig. 11: photo © the artist
p. 707, fig. 12: © Daniela Dacorso
p. 709, figs. 1a, b: photos: Albrecht Hotz
p. 711, fig. 2: photo © : Romulo Fialdini
p. 711, figs. 3, 4: photo © Smetak Family
p. 712, fig. 5, 6: photo © Smetak Family
p. 712, fig. 7: photo © Marco Scarassatti
pp. 742–43: © Archive Peter Weibel

Cover image: Installation view daadgalerie, Berlin, photo © Douglas Henderson

If not otherwise noted, the photo copyright belongs to the artists or their rights holders.

Colophon

This book is published on the occasion of the exhibition *Sound Art. Sound as Medium of Art* at ZKM | Center for Art and Media Karlsruhe
March 17, 2012–January 6, 2013

The Exhibition

Curator:
Peter Weibel

Co-curator and project coordination:
Julia Gerlach

Project coordination (July 2012–January 2013):
Till Kniola

Curator Unheard Avant-gardes (in Scandinavia):
Morten Søndergaard

Curators Het Apollohuis:
Hartmut Jörg and Claudia Gehrig

Curators TONSPUR:
Georg Weckwerth and Peter Szely

Curator Broken Music:
Ursula Block

Curators Audio Stations:
Björn Gottstein (soundscape, noise, electronica), Rudolf Frisius (musique concrète), Frank Halbig (radiokunst), Luca Vitiello (radioartemobile), Herbert Kapfer and Katarina Agathos (intermedium)

Concept Maryanne Amacher room:
Micah Silver

Concept Iannis Xenakis room:
Daniel Teige

Artistic direction events:
Ludger Brümmer, Julia Gerlach, Peter Weibel

Visual concept of the exhibition:
2xGoldstein, Peter Weibel

Many thanks to:
Gartenbauamt Karlsruhe; Tiefbauamt Karlsruhe; Regierungspräsidium Karlsruhe; Vermögensverwaltung Baden Württemberg; Zoo Karlsruhe; Fema Farben und Putze GmbH, Ettlingen; Vermögen und Bau Baden-Württemberg; APCOA Autoparking GmbH

The Book

Editor:
Peter Weibel

Editorial staff:
Miriam Stürner, Jens Lutz, Linnea Semmerling (editors)
Martina Hofmann, Caroline Meyer-Jürshof, Anna Straetmans, Patrick Trappendreher, Claudia Voigtländer (assistant editors)

Assistance:
Ida Schneider

Copy editing:
Anne O'Connor, Gloria Custance, ZKM | Publications

Translations:
Gloria Custance, Luca Branchetti, John Laudenberger, Andrea Lerner, Robert Booth

Work descriptions (pp. 182–455):
Martina Seeber (unless otherwise indicated)

Graphic design:
zkm books, Jan Kiesswetter, Leonie Rapp

Lithography:
COMYK, Roland Merz, Karlsruhe

Printed and bound:
NINO Druck GmbH, Neustadt/Weinstraße

Paper:
135 g/m² Luxo Art Samt

This book was set in:
Suisse Works, Styrene

Many thanks to:
Julia Gerlach, Morten Søndergaard, Anett Holzheid, Caroline Jansky, Maraike M. Marxsen, Luise Pilz, Holger Jost, 2xGoldstein, Till Kniola, Jochen Arne Otto, Dörte Zbikowski

© 2019 ZKM | Center for Art and Media Karlsruhe, Germany

Unless otherwise indicated, all texts © by the individual authors and their rights holders.

Despite intensive research and best intentions, it was not possible in every case to establish the copyright holders. We request the holders of such rights who feel they have not been properly acknowledged to contact us.

All rights reserved. No part of this book may be reproduced in any form by any electronic or mechanical means (including photocopying, recording, or information storage and retrieval) without permission in writing from ZKM | Karlsruhe.

Printed and bound in Germany.

ISBN 978-0-262-02966-7
Library of Congress Control Number 2015943532

ZKM | Center for Art and Media Karlsruhe
Lorenzstraße 19
76135 Karlsruhe
Germany

Phone: +49 (0)721/8100-1200
info@zkm.de
www.zkm.de

CEO and Chairman:
Peter Weibel

COO:
Christiane Riedel

Head of administration:
Boris Kirchner

Founders of ZKM:

Baden-Württemberg
MINISTERIUM FÜR WISSENSCHAFT, FORSCHUNG UND KUNST

Karlsruhe

Partners of ZKM:

EnBW

dm

zkm karlsruhe

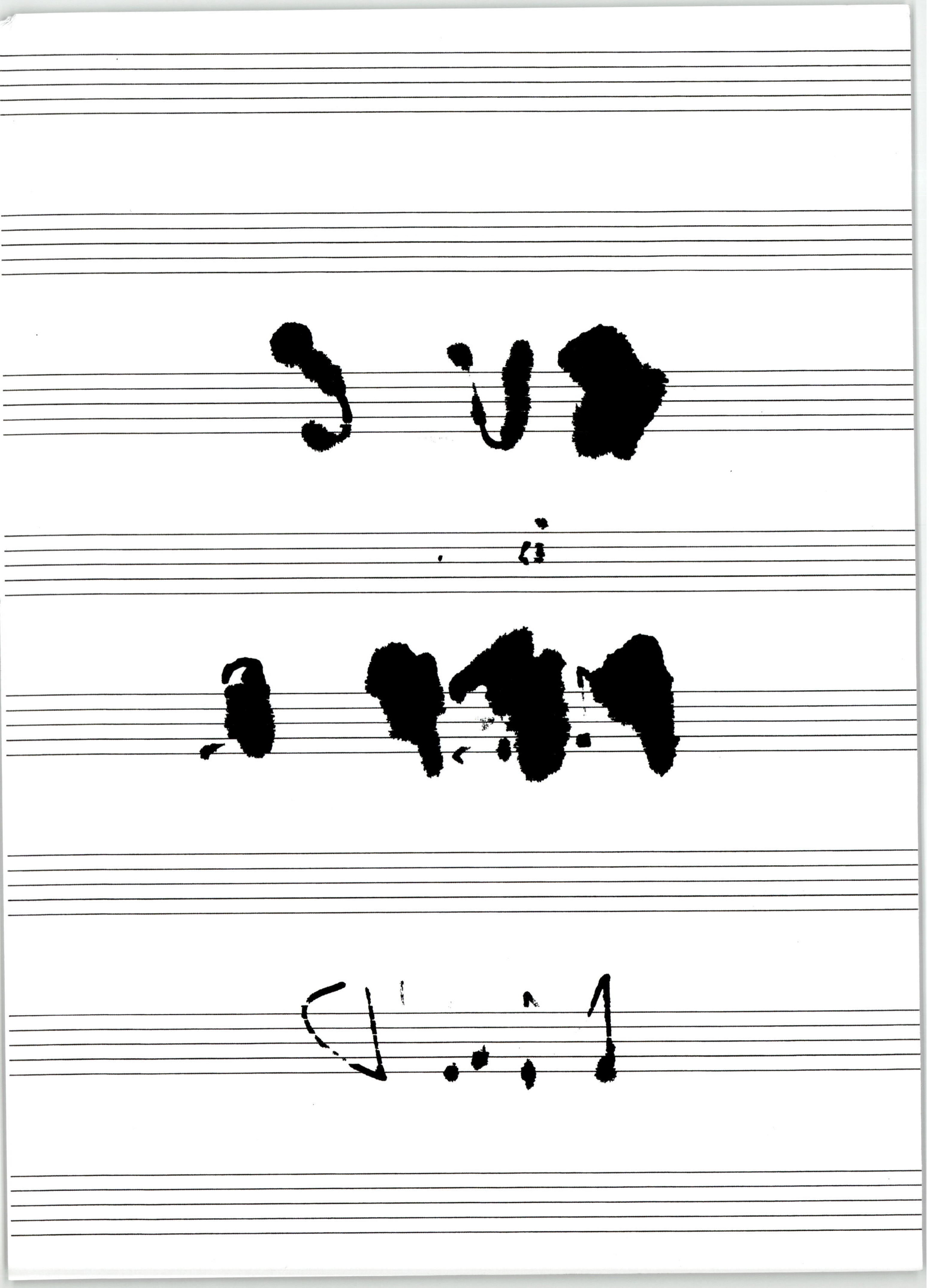

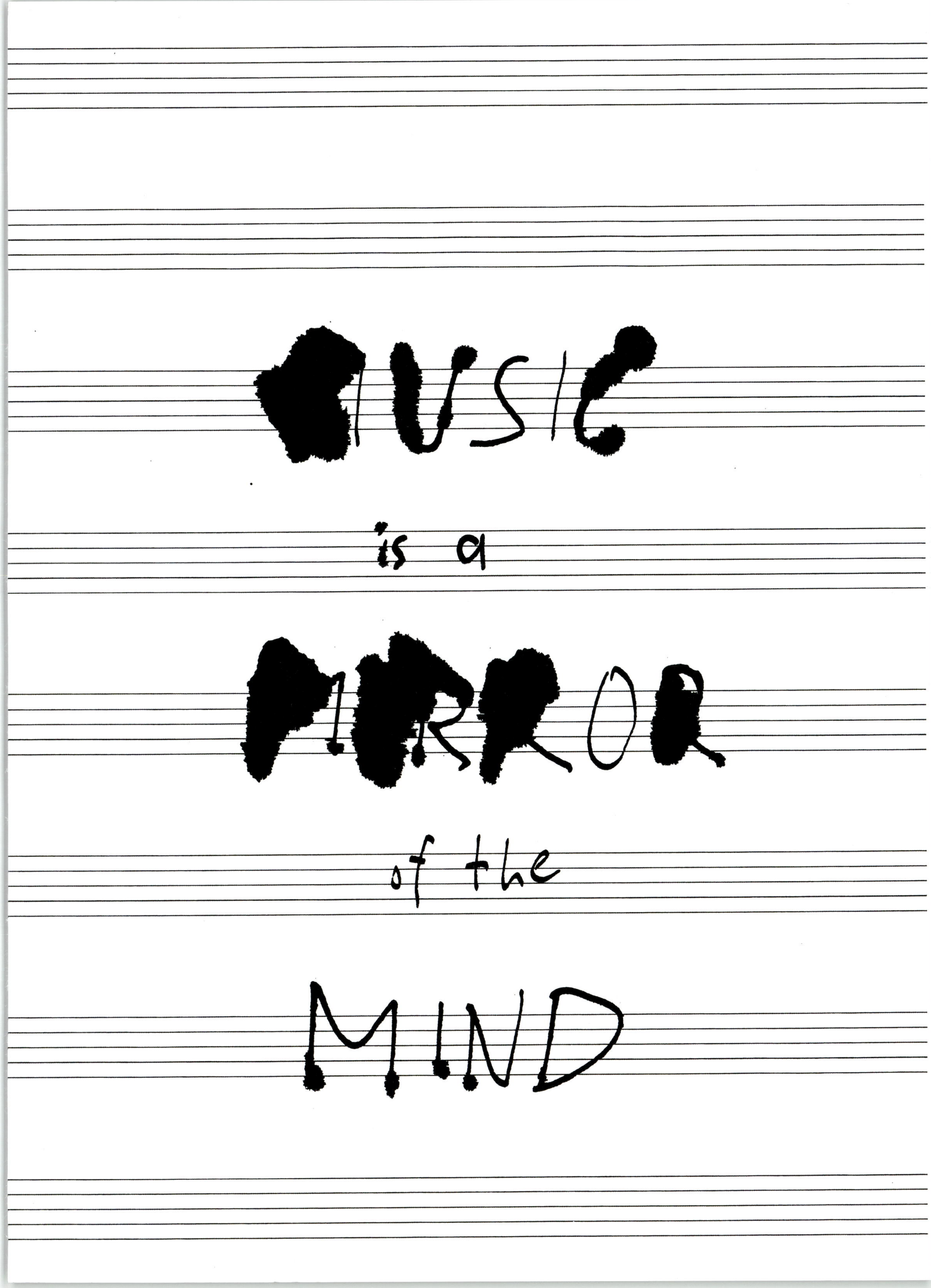
MUSIC
is a
MIRROR
of the
MIND

Cover image:
Douglas Henderson, *stop.*, 2007.
Sound sculpture.

The cover design is based on the exhibition poster by 2xGoldstein.

Endpaper and pp. 1–2:
Peter Weibel, *Transformationelle Musik* [Transformal Music], 1972.
Structure overlay, paper music, visual music.

pp. 742–43:
Peter Weibel, *Music is a Mirror of the Mind*, 1967.
Visual music, graphic notation.

"Music is a hidden arithmetic exercise of the mind, which does not know that it is counting." (Gottfried Wilhelm Leibniz in a letter to Christian Goldbach, April 27, 1712.)